COASTAL ZONE '80
Volume I

Proceedings of the Second Symposium on Coastal and Ocean Management

Billy L. Edge, Editor

Sponsors:

American Society of Civil Engine
(Waterway, Port, Coastal and Ocean Division)
Office of Coastal Zone Management
Edison Electric Institute
Heritage Conservation and Recreation Service
Maritime Administration
National Aeronautics and Space Administration
National Marine Fisheries Service
Office of Sea Grant
Public Works, Canada
U.S. Army Corps of Engineers
U.S. Coast Guard
U.S. Fish and Wildlife Service
U.S. Geological Survey

Affiliates:

American Petroleum Institute
American Shore and Beach Preservation Association
Bureau of Land Management
California Coastal Commission
Environmental Data and Information Service
Federal Emergency Management Agency
Marine Technology Society
National Park Service
State of Florida Department of Environmental Regulation
State of Florida Department of Natural Resources
The Coast Alliance
The Coastal Society
Urban Land Institute
U.S. Department of Energy

Diplomat Resort Hotel
Hollywood, Florida
November 17-20, 1980

Published by the
American Society of Civil Engineers
345 East 47th Street
New York, New York 10017

Library of Congress Catalog Card No. 80-69152
ISBN 0-87262-258-4
Manufactured in the United States of America.

Cover photo by Thomas A. Beckett, New York, NY

FOREWORD

Coastal Zone '80, the major technical event in the Year of the Coast, was designed to bring together planners, engineers, environmentalists, government administrators and interested individuals to discuss the most current coastal and ocean management issues. The purpose of the conference was to provide a forum for discussion of coastal zone management beneficial use, protection and development. This forum has lead to a better understanding of the interrelationships between the environmental, socioeconomic, engineering and regulatory decisions involved. Coastal Zone '80 has fostered more effective and meaningful jurisdiction arrangements, conservation considerations, regulations, enforcement policies, planning activities, and design parameters in the development and implementation of coastal and ocean resource management.

Coastal Zone '80 was sponsored by the following groups:

Sponsors

American Society of Civil Engineers Waterways Division
Office of Coastal Zone Management
Edison Electric Institute
Heritage Conservation and Recreation Service
Maritime Administration
National Aeronautics and Space Administration
National Marine Fisheries Service
Office of Sea Grant
Public Works, Canada
U.S. Army Corps of Engineers
U.S. Coast Guard
U.S. Fish and Wildlife Service
U.S. Geological Survey

Affiliates

American Petroleum Institute
American Shore and Beach Preservation Association
Bureau of Land Management
California Coastal Commission
Environmental Data & Information Service
Federal Emergency Management Agency
Marine Technology Society
National Park Service
State of Florida Department of Environmental Regulation
State of Florida Department of Natural Resources
The Coast Alliance
The Coastal Society
Urban Land Institute
U.S. Department of Energy

The breadth of the sponsorship and the large attendance indicate the far-ranging, multidisciplinary interests in coastal and ocean management. The papers of this conference indicate how far the concept and implementation of coastal zone management has been advanced since Coastal Zone '78 which occurred at the formative states of the National CZM Program. This collection of papers thus represents coastal and ocean management at work.

Billy L. Edge
Chairman

Orville T. Magoon
Co-Chairman

L. Thomas Tobin
Technical Program

Dallas Miner
Co-Chairman

CONTENTS

Volume I

*Papers not available at time of printing

*Papers not available at time of printing

*Papers not available at time of printing

*Papers not available at time of printing

*Papers not available at time of printing

*Papers not available at time of printing

*Papers not available at time of printing

*Papers not available at time of printing

*Papers not available at time of printing

*Papers not available at time of printing

*Papers not available at time of printing

*Papers not available at time of printing

*Papers not available at time of printing

Volume III

*Papers not available at time of printing

*Papers not available at time of printing

*Papers not available at time of printing

*Papers not available at time of printing

*Papers not available at time of printing

PERSPECTIVES ON EROSION: BARRIER ISLAND MANAGEMENT

Ann C. Baker*

Barrier islands are long, thin islands along the eastern and Gulf coasts of our country. Over a third of our states, 18 in all from Maine to Texas, have these islands along their shores. Over a quarter of our country's population lives within 100 miles of these barrier islands.[1] Many, many more people visit them every year.

Erosion and accretion of barrier islands is a process which is essential to the survival of the islands. It is the process by which the islands stabilize themselves in the ocean. Erosion is only viewed as a problem when people build permanent structures on the islands, structures which they then want to protect. Accretion is only viewed as a problem when permanent structures have been placed on the islands as beach front property and ownership of the accreted property becomes an issue.

The response of government to the problem of erosion on barrier islands is the subject of this paper. The paper will begin with an overview of the elements in our environment which contribute to our personal and collective perspectives on erosion and of government's response in the past to erosion on barrier islands. Then the paper proposes alternative policies and responses which government at all levels can apply first to undeveloped barrier islands and undeveloped sections of islands and finally to developed barrier islands.

One of the basic premises of this paper is that in light of present economic scarcity, our limited resources, our ultimate responsibility to the largest number of people, our sheer inability to counter the power of the sea, and our inevitable respect for the natural system, erosion must not always be controlled. We know that whenever it is controlled for some, it

*Planner, South Carolina Coastal Council

may cause or increase erosion for others. We know that sea level is rising. Whenever erosion control structures are placed on the beach at public expense making the beach unsafe and unpleasant for the public to use, we may have protected the private property owner but cheated the public. While we have pumped large amounts of sand onto beaches at public expense in the past, we are being forced to accept the fact that the public may not be willing to continue to defray the increasing cost indefinitely.

Thus, as major new legislation for barrier islands is being considered at the Federal level and as those of us at the state or local level deal with these issues first hand, we must break out of established patterns of thinking about barrier island development and response to beach erosion. We must freshen our perspective and reevaluate the demands and the present realities of 1980. We must consider alternatives which are new or which may not have been reasonable or feasible or necessary in the past. Unquestionably, barrier island management is one of the major issues for coastal management in the 1980's.

Perspectives

The perspective on beach erosion that each of us brings to this place is unique and is influenced by the ways in which the phenomenon of beach erosion has touched us personally - as beachcombers, property owners, policy makers, planners, technicians, neighbors, taxpayers. It is also partially a reflection of our philosophical approach and attitude toward natural processes wherever they occur around us - whether in the water, sky, or our own bodies. It is influenced by our perception of ourselves, our egos, and our sense of place in the universe. Some people strive to live in harmony with natural systems, and others perceive them as a challenge to overcome.

Most of us have been profoundly influenced by the American work ethic and the successes of technological problem solving. As Americans, we have inherited an image of ourselves as an ingenious people who can develop technological solutions to all problems. Our self-confidence feeds our egos until we believe we can always prevail with enough effort and intellect.

However, the world in which we now find ourselves is one of economic instability, scarcity of increasing numbers of vital natural resources, and an increasing number of people using and needing these vital resources. A deeper respect for the natural system of our universe is evolving. Perhaps more than ever before in the history of our country, problem solving must now be designed to function within the dynamics of the natural system rather than to conquer nature's limitations.

Over the past thirty years, the United States has enjoyed an era of prosperity in a technological society unique to the experience of humankind. It has been an era of relative peace and usually adequate supplies of clean water, undeveloped land, energy resources, and construction materials. Prosperity, combined with a rapidly growing population which is highly mobile and has leisure time for recreation, has led to intense waterfront development. While only 3% of the land nationwide is considered urbanized, 14% of the barrier island acreage is urbanized. The rate at which additional barrier island land is becoming urbanized is twice that of the urbanization rate of the nation as a whole.[2]

Because of widespread mobility, we have immense numbers of people living by the sea without an appreciation of the power of the sea, an appreciation usually shared by people who have experienced the ocean for generations. If we walk along some sections of Myrtle Beach, South Carolina, where there is a mixture of old family beach houses and new high-rise development, we can notice a dramatic illustration of this point. New high-rise development is often very close to the waterline with little or no dune field between the structure and the sea. Old, family houses most typically were placed well behind the dune field, which in many places is still fully intact. Families generations ago who spent every summer at the beach from childhood to adulthood typically built their homes in a more secure location having seen and experienced the natural fluctuations of the shoreline over the years. Also, their homes are the ones which are still standing.

In addition to increased mobility, some of the most intense development, especially on the Atlantic coast, has also occurred during a period which experienced fewer hurricanes than usual. Furthermore, the competitive demand for the limited supply of ocean-front property means the beach front property owner usually has a significant financial investment to protect. Our belief in technology, our inexperience with the seas, our sizable investments in shorefront property mean that barrier island development inevitably leads to demand for erosion control.

What has been the government's policy regarding barrier island development and erosion control over the past several decades? During this period, the Federal Flood Insurance Program has taken much of the risk out of waterfront development. Many other Federal programs have encouraged barrier island development in additional ways. Barrier islands are affected by approximately 30 Federal programs administered by 20 different Federal agencies. The programs range from water and sewer projects to acquisition of land to preserve as wildlife refuges. According to Robert Herbst, Assistant Secretary of the Interior for Fish and Wildlife and Parks, half of the programs encourage development of the islands. About a quarter of the programs are protection-oriented.[3] According to the Barrier Island Draft Environmental Statement, half a billion dollars has been spent on development projects over just the last three fiscal years by various Federal agencies through grants and permitting and licensing procedures.[4]

Superficially, it would appear ludicrous for various units of government to be working against each other. In fact, the programs are an outgrowth of a very significant phenomenon - the diversity of the natural functions which the islands serve and the diversity of the ways in which people use and enjoy these islands. Thus, pressures from many directions have been placed on government over the years regarding barrier islands.

We find ourselves, however, at a point where more people are demanding that government cease to respond erratically and inefficiently to development pressures. No longer is there seemingly unlimited waterfront land available to accommodate all the people's needs or to protect the natural balance of the ecosystem. Also, in a period of decreasing economic growth rates, people and government are being forced to reevaluate each and every expenditure of public funds with an eye toward eliminating or decreasing public costs. At the same time, increasing costs for labor and construction materials mean that publicly subsidized development and erosion control projects are more expensive than ever before. Experience has shown that beach nourishment projects cost millions of dollars and often last only a few short years before the beach needs to be renourished again. Seawalls lower the beach face, eventually destroying the public beach and then often fail in storms leaving rubble on the beach and expensive repair costs to be faced. Groins can trap sand and assist in building a beach on the updrift side of the groin but rob the downdrift beach of its longshore sand supply thus having severe impacts on the adjacent shoreline down the beach.

Thus, as policy makers or professionals influencing or implementing decisions, what are our responsibilities and to whom are we responsible? Surely, we have a responsibility to beach front property owners and to the increasing numbers of people who do not own property but seek the beach for recreation. We have a responsibility to taxpayers, to communities, to political officials, to the natural system, and to the nation and its people as a whole. It is not surprising that there are two major pieces of legislation affecting barrier islands presently being considered by Congress in an attempt to come to grips with the complexity of the issues.

Alternatives

Although some overlap is inevitable, my discussion of alternative approaches to beach erosion and to barrier island development will generally be separated into two parts. Initially, approaches to undeveloped sections of the islands will be addressed, followed by a look at the developed islands. Endorsement of specific legislation is not my intent. Rather, suggestions for various responses from multiple sources are being proposed here for consideration. In some cases, these alternatives might only be feasible at the Federal level and require Federal legislative changes. In other cases, the state or local level may be the place to initiate action or policy. My intention overall is to stimulate creative thinking - to approach the complex issues innovatively. I invite you to respond positively or negatively but to respond with an open mind and with dialogue.

Undeveloped Barrier Islands, Undeveloped Sections of Barrier Islands

Before making permanent use decisions, islands need to be evaluated individually. At the same time we need to evaluate the most critical needs of the surrounding areas and their people. Islands which are prime examples of pristine and unique natural areas need to be given serious consideration for permanent preservation. Islands which are located near heavily populated areas with high demands for public access need to be given serious consideration for acquisition and protection as public access areas. After careful evaluation of people and resource needs, alternative approaches to fulfill those needs can be considered.

Government acquisition of barrier islands which are undeveloped is an approach which is receiving an enormous amount of consideration. Response to the damage which occurred in 1979 as a result of Hurricane Frederick crystallized for us a suspicion which has been lurking in the backs of many minds for a long time. Would it, in fact, be less expensive for the Federal government to purchase islands than to pay the flood insurance claims and repeatedly rebuild damaged infrastructure on developed islands after major storms?

Government acquisition represents for many people much more than an equation of pure economics or numbers of dollars. It is symbolic in some people's minds of Big Brother Government infringing on private property rights. Also, if all barrier islands are purchased and placed under one program of management, it could negate our appreciation for the diversity of land form and diversity of use which our heterogeneous population is so masterful at creating. It seems unlikely and politically unfeasible for all or even most barrier islands to be committed to public recreation, or to resort recreation, or to wildlife refuges, or to any other single approach.

Having said that, I hastily add that intense, publicly subsidized development of additional barrier islands is the opposite extreme on the continuum which should not continue. Government acquisition of some islands before they are developed is one approach which can be of real value.

We have seen that even in recessionary times people will take vacations and overwhelming numbers of people will choose to go to the water. During this past summer of recession, the beaches of South Caroline experienced approximately the same numbers of visitors as in recent past years. The people often made shorter visits and drove shorter distances, but the demands for access remained amazingly stable. However, people go to the beach for different experiences and are accustomed to being able to choose from a diverse selection - from the isolated camping experience only accessible by boat to the high density experience with hotels, amusement parks, water slides, etc. It is the high density option on barrier islands which requires bridges, expensive and elaborate water and sewer systems, huge expenditures in insurance claims and publicly financed rebuilding programs after storms, and demands for massive shore stabilization projects. Is the public still able or willing to continue to finance the proliferation and maintenance of that kind of development on barrier islands?

There are access opportunities which can be designed to be compatible, however, with the natural system of islands and can at the same time accommodate large numbers of people. Government acquisition of barrier islands in areas of intense demand for public access has enormous potential. After acquisition, management programs or designs can be developed where inlet areas are maintained as open space available to the public.

Some sections of the islands could be platted in deep, narrow lots along the beach. These deep lots could accommodate the inevitable fluctuations of shorelines without demands for erosion control. The lots could then be leased to individuals with strict lease stipulations for use which would fit into the overall management design for the area. The lease could run for perhaps 5-20 years depending on the geological stability of the area and legal restraints or until a major storm required a reevaluation of the lot lines. A storm requiring reevaluation might be defined as one which resulted in greater than a specified percentage of land lost to erosion or percentage of damage to structures.

Lease stipulations could include deep setback lines for all construction based on the historic erosion/accretion pattern of the area and on the flood elevations of the area. In this way the dune field and natural littoral drift of sand would be protected. Construction could be limited to low density types such as single family structures.

Intensive research into temporary or movable structures needs to be pursued. Kaufman and Pilkey in The Beaches Are Moving describe temporary, nomadic structures which could be designed especially for use on beaches. In highly unstable areas collapsible structures could be used during the spring and summer and removed before hurricane season thereby requiring little or no insurance coverage. In more stable areas or where people are willing to assume more risk for greater comfort, slightly more elaborate modular units could be allowed. Mobile homes or similar structures which prove to be incompatible with the beach could be prohibited.[5]

Another criterion for areas which would be appropriate for the leasing arrangement would be those which could provide necessary utilities without elaborate public systems. For example, islands which have adequate well water and soil consistency for septic tanks might be ideal. Kaufman and Pilkey suggest installing systems such as those used in motor-home parks which can be easily disconnected, and they further suggest never laying lines which connect islands or sections of islands where inlets are likely to form. They go on to say, "Pipelines running the length of an island encourage people to try to prohibit inlet openings and to fortify weak areas. The utility buildings, tanks, and treatment basins should be underground or built with a very low profile and sloping sides, so that they can be overwashed and partially buried without damage."[6]

It can be a challenge to our ingenuity and technological flair to develop ways to live flexibly within natural fluctuations avoiding shore stabilization efforts and discontinuing our present patterns of development which require us to use our talent to try to conquer the natural shore processes. Research into appropriate materials and new engineering design to accommodate nomadic, temporary beach development is needed and offers opportunities to the inventive spirit of our American minds. The technology need not be used only on privately leased, government-owned land but could be adapted to beach development anywhere.

The most obvious advantage to a leasing program such as this would be to fund the acquisition costs. At the same time more new areas would be opened for public recreation. By implementing a lease program rather than a rental program, the administrative costs could be significantly reduced. Leasing stipulations and management designs for the areas should be developed in such a way as to assure that shoreline stabilization would never be required; natural inlet formation, overwash, and erosion/accretion patterns would allow the islands to stabilize themselves; and any areas of special historic or archeological significance or unique natural areas could be left unaltered by development. Significant public expenditures for utilities as well as flood insurance claims could be avoided. Islands which do not already have bridges could, perhaps, be serviced by ferry boats and unpaved roads. State participation in the decisions about the islands to be purchased and management designs of the islands would be possible through Federal consistency in states with approved Coastal Management Programs. Or state control could be assured by purchase of the islands with state money.

An important adjunct to an acquisition program could be institution of a system of tax incentives and disincentives similar to those suggested in the Barrier Island Draft Environmental Statement. These can encourage donations of island areas to public use and discourage intensive development of the islands by decreasing the profit of that type of development. In this way the initial costs of acquisition could be reduced. A well designed disincentive program which applies only to barrier islands could reduce some of the development pressures on islands by making it more profitable to develop in other types of areas.

Another approach could be the adoption of a state policy by individual states or the adoption of a nationwide policy similar to one established in North Carolina. In 1979 the State of North Carolina adopted a regulation which prohibits the building of any erosion control structure to protect development which was not existing on June 1, 1979. This regulation does not prohibit beach nourishment or berm projects. The public was put on notice that construction along the shore after June 1, 1979, carried significantly more risk. The owner is unlikely to be able to acquire permission to construct a seawall, revetment, etc. to protect the property.

Federal expenditures could be prohibited for shore stabilization in front of property which was not developed by the date of the institution of the policy. The intent would be to discourage development by assuring that the property owner assumes more of the risk associated with shoreline development. The approach has genuine potential but the legal obstacles surrounding the taking issue will have to be carefully avoided. The policy could be softened and made more politically feasible perhaps, if necessary, by not prohibiting beach nourishment.

There also could be Federal incentives to states to adopt a state policy of this kind. For example, Federal erosion control money to protect developed property could be tied to a condition that no public money could be spent on shore stabilization for property undeveloped in that state as of the date of the agreement.

In addition, severe restrictions need to be placed at the Federal and state level on publicly financed incentives to development of barrier islands. The most comprehensive approach would be along the lines of the one being considered by the U.S. Senate in a bill introduced by Senator Bumpers. The bill would prohibit Federal money, licenses, or permits for structures, infrastructure, transportation, flood insurance, or shoreline stabilization for barrier islands unless it is related to energy development, navigational safety, public recreation, or protection of the ecosystem of the island.

The State of South Carolina's coastal zone management agency, the South Carolina Coastal Council, addressed the issue of public money encouraging development of barrier islands in its Coastal Management Program which received Federal approval in September, 1979. The program includes policies such as the following:

> "3) New road or bridge projects involving the expenditure of public finds to provide access to previously undeveloped barrier islands will not be approved unless an overwhelming public interest can be demonstrated, for example, provision of access to a public recreation area or other facility. Preference will be given to ferry access in those instances where public funds cannot be expended for road access.
>
> 4) The extension of public services, such as sewer and water facilities, to barrier islands should only be proposed in a comprehensive approach which considers the natural 'carrying capacity' of the island to support development and which integrates these facilities to parallel the level of access which is available to the island."[7]

These policies guide the Council when making decisions on permits which they issue directly, such as road, bridge, or utility pipeline permits; in their certification of other state agency permits such as extension of water supply or wastewater treatment permits; or in their Federal consistency determinations.

The Council is thus on record as being generally opposed to public funds being used to encourage development of previously undeveloped barrier islands.

Developed Barrier Islands

Most of the recommendations for undeveloped islands could be adapted to apply to all barrier islands whether developed or undeveloped. The major incentives to states and localities to require these policies, however, may be through the funds which already developed communities seek.

For example, all Federal Flood Insurance total constructive loss payments, disaster relief and recovery money, and Small Business Administration low interest loans to barrier islands after storms could be tied to major policy changes by the localities. To receive these funds, governmental units could be required to develop:

1. Setback lines for all new construction and all rebuilding of structures. Setback lines should be carefully developed based on traditional erosion/accretion patterns and flood zones which include storm surge and wave height considerations. Construction standards adequate to withstand major storms should be developed for all buildings behind setback lines on barrier islands.

2. Major zoning plans and subdivision regulations applicable to all new development and rebuilding after storms. They should incorporate density restrictions, maximum percentage of vegetation which can be destroyed on each site, maximum amount of impermeable surface allowed for each site, and all platting of lots for new developments to incorporate lot designs which are deep enough to allow for sea level rise and normal erosion/accretion patterns, likely for the duration of the development.

3. Evacuation and hazard mitigation plans. Density requirements should be developed to be compatible with feasible evacuation plans.

4. Plans for the relocation of structures, rather than rebuilding, in rapidly eroding areas. Relocation funds, and not shore stabilization funds, could be made available to communities where erosion is severe and likely to continue. Relocation funds need to include funds to restore the areas for public use such as funds to clean up debris, provide public access walkways, etc.

5. Restrictions on expenditures of public funds which would stimulate development. All public funds for construction of or upgrading of water, sewer, and transportation systems for barrier islands should be limited to serve or accomodate existing densities only or to accommodate growth in very isolated instances. In order to allow public funds to accommodate growth, strict requirements such as those listed above would be a minimum condition to receive public funds.

In the case of sections of or whole barrier islands which are already developed, the primary concern is establishing policies and regulations which apply to expansion of or rebuilding of the existing development after storm or erosion damage. Anticipating the need for changes soon in the face of the intense development pressure in some areas is critical. Anticipating the need for changes before more storms occur would be wise.

Conclusions

In order to reverse previous patterns of development while continuing to preserve the diverse values of barrier islands, major changes must occur both in our perceptions of the environment and in governmental policy and legislation. We need to adapt our personal perceptions and government's policy to the present world with its needs, limitations, and opportunities. Each of us needs to see this as a challenge to stimulate us to create new approaches rooted in wisdom and experience.

Footnotes

[1]"Saving Islands in the Year of the Coast", Robert Herbst, Christian Science Monitor, 29 May 1980, p. 22.

[2]U.S. Department of the Interior, Alternative Policies for Protecting Barrier Islands Along the Atlantic and Gulf Coasts of the United States and Draft Environmental Statement, December 1979, p. xv.

[3]Herbst, p. 22.

[4]U.S. Department of the Interior, p. vii.

[5]Wallace Kaufman and Orrin Pilkey, The Beaches Are Moving (Garden City, New York: Anchor Press/Doubleday, 1979) pp. 276-78.

[6]Ibid, p. 278.

[7]U.S. Department of Commerce, State of South Carolina Coastal Management Program and Final Environment Impact Statement, 1979, p. III-71.

List of References

Brower, David J.; Dreyfoos, William; Epstein, Lawrence; Pannabecker, James; Stroud, Nancy; Owens, David. Access to the Nation's Beaches: Legal and Planning Perspectives. Sea Grant Publication UNC-SG-77-18, February 1978.

Clark, John R.; Gilbert, Stephen; MacLain, Bruce; Turner, Robb; Warner, Langdon. Review of Major Barrier Islands of the United States. The Barrier Islands Workshop with the assistance of the Conservative Foundation, July, 1977.

H. R. 5981, 96th Congress, 1st Session. "A Bill To Establish Barrier Islands National Parks, and for other purposes." Introduced by Representative Burton. 28 November 1979.

"Hands Off These Islands." Christian Science Monitor, 29 May 1980.

Kaufman, Wallace, and Pilkey, Orrin. The Beaches Are Moving. Garden City, New York: Anchor Press/Doubleday, 1979.

Leatherman, Stephen P. Barrier Island Handbook. University of Massachusetts at Amherst: National Park Service, Cooperative Research Unit, The Environmental Institute, 1979.

Leatherman, Stephen P. "Migration of Assateague Island, Maryland, by Inlet and Overwash Processes." Geology, v. 7 (February 1979): 104-107.

Robinson, Alan H. "Planning Considerations for Preservation and Use of the National Seashores." Coastal Zone Management Journal, Vol. 5, No. 1/2 (1979): 5-34.

S. 2686, 96th Congress, 2nd Session. "A Bill To Direct the Secretary of the Interior to provide for the protection of the Barrier Islands, and for Other Purposes." Introduced by Senator Bumpers. 9 May (Legislative day, January 3) 1980.

Sanibel Planning Commission, City of Sanibel, Lee County, Florida. Comprehensive Land Use Plan. City of Sanibel Ordinance No. 76-21.

"Saving Islands in the Year of the Coast." Christian Science Monitor. 29 May 1980.

Shows, E. Warren. "Florida's Coastal Setback Line - An Effort to Regulate Beachfront Development." Coastal Zone Management Journal, Vol. 4, No. 1/2 (1978): 151-164.

Town of Briarcliff Acres, South Carolina. Building Housing, and Urban Development Regulations. General Ordinances, Chapter 4, 1978.

U.S. Department of Commerce. Final Environmental Impact Statement and Proposed Coastal Management Program for the State of South Carolina. National Oceanic and Atmospheric Administration, Office of Coastal Zone Management, 3300 Whitehaven St. NW, Washington, D.C.: August, 1979.

U.S. Department of Commerce, National Oceanic and Atmospheric Administration, Office of Coastal Zone Management. Beach Erosion in South Carolina by Miles O. Hayes, Thomas F. Moslow, and Dennis K. Hubbard. Grant No. 04-7-158-44141, pursuant to Public Law 92-583.

U.S. Department of the Interior. <u>Alternative Policies for Protecting Barrier Islands Along the Atlantic and Gulf Coasts of the United States and Draft Environmental Statement</u>. Director, Heritage Conservation and Recreation Service, Department of the Interior, Pension Building, 440 G Street NW, Washington, D.C.: December, 1979.

A Strategy for Financing Coastal Erosion Control

Waldon R. Kerns, Ph.D.
Resource Economist and Associate Professor

and

Janet Scheid
Former Research Assistant

Water Resources Research Center
Virginia Polytechnic Institute and State University
Blacksburg, Virginia 24060

In the Chesapeake Bay of Virginia, erosion action has caused losses of up to twenty feet of shore depth per year and in several cases structures have actually fallen into the Bay waters. This physical process of erosion generally has not been perceived as a problem until it has induced an economic impact on either an individual or a community. These economic impacts have become more significant and more widespread and consequently the situation has resulted in major emphasis being placed on better management of tidal shoreline erosion. Management is a very expensive operation and a most important component of a management strategy is the cost-sharing formula to be used.

This study provides an economic decision framework for management of the impacts of coastal erosion on shoreland property and structures. The evaluation procedure provided a realistic basis for comparisons of: (1) benefits and costs of selected control measures; (2) various levels of control; (3) distribution of costs between private and public sectors; and (4) ability of localities to support control programs.

Three case study areas in Middlesex County, Virginia, which have experienced substantial erosion action were selected for use in estimating the erosion induced impact on property and structures. Each study area represents a reach. A reach basis was used because a reach is characterized in physical terms as a segment of shoreline wherein the erosion process and responses are mutually interactive. Each reach includes many individually owned parcels of property.

The first study reach (Reach 1) was an open space, undeveloped area of 23 individually owned parcels of property along 2.27 miles of shoreline. The second study area (Reach 2) was a developing area of single family residential and vacation dwellings with 70 parcels of individually owned property along 1.7 miles of shoreline. The third study area (Reach 3) was an established area of single family residential and vacation dwellings with 73 individually owned parcels of property along 1.76 miles of shoreline and formed part of a shoreland point which extended into the Chesapeake Bay. These areas contained a total of 166 individually owned parcels of property. Table 1 provides a summary of lot characteristics in each reach.

Procedure for Application of Economic Decision Framework

An economic assessment methodology was applied to the three reaches. Values were established for costs and benefits for selected management

Table 1. Average Value of Selected Characteristics for Three Erosion Study Areas in Middlesex County, Virginia

Characteristics	Reach 1	Reach 2	Reach 3
Average Value Lot ($)[1]	45,687	20,875	10,377
Average Value Lot/Sq. Ft. ($)[1]	0.32	0.67	0.91
Average Length of Lot (ft)[1]	546	157	73
Average Depth of Lot (ft)[1]	378	328	160
Number Lots with Dwelling[1]	18	52	44
Average Value Dwelling ($)[1]	21,686	18,515	19,643
Average Yearly Erosion Rate 1960-1978 (ft./yr.)[2]	+1.4	-0.5	+0.3
Average Distance Dwelling from Shoreline (ft.)[3]	320	107	28

[1]Obtained from 1977 assessment record in Commissioner of Revenue's Office, Middlesex County, Virginia. Assessment was based on 100 percent market value. A comprehensive study of all property transactions during past 20 years established the valuation procedure.

[2]Erosion rates provided by scientists from the Virginia Institute of Marine Science. Erosion rates (positive for loss of soil and negative for accretion) for each individual parcel were obtained by interpretation of aerial photography for 1960 and 1978.

[3]Distance obtained by interpretation of aerial photography.

strategies in each reach. As part of the technical assessment, the capital costs of implementing proposed structural control measures were estimated on the basis of standard cost guides and expressed in terms of present dollar values.[1] Operation and maintenance costs were discounted to current dollars. Costs were determined for proposed structures on a segment basis and summed for the total reach. Also, the expected effectiveness in percentage terms of control of erosion by the structural controls and time period of effectiveness were provided for each reach.

A second set of costs were derived from estimated decreases in values of property and improvements or losses from restrictions on use of resources because of implementation of selected non-structural control measures. Values were calculated for each reach on the basis of a "without" and "with" approach. That is, values for resources and their uses were estimated for the current situation and compared to their values after implementation of a control measure. Values were estimated for loss of agricultural productivity, loss of soil, loss to structures and other improvements, and loss of tax revenue.

Through use of least-squares multiple regression value of soil loss was estimated. For 49 lots with dwellings and loss of area to erosion, as the erosion rate increased by a foot per year marginal value decreased by $.242 per square foot (or 28 percent of the average assessed value of these 49 lots). Thus, erosion action has a very high cost in terms of value per square foot for each square foot lost to erosion if that lot contained a dwelling. The addition of area or accretion does not add to the value of the lot. For 17 lots without a dwelling, as the erosion rate increased by a foot the marginal value per square foot decreased by $.125 per square foot (or 50 percent of the average assessed value of these 17 lots) a significantly higher rate. As with those lots with a dwelling accretion does not add to the value of the lot.

Decreases in market value of property will be reflected in lost tax revenues. The recent assessment for Middlesex County supports this assumption. Whenever a property owner requested a tax reduction due to decrease in value from soil loss, it was granted. Thus, for purposes of this study the computed soil loss in dollar value was multiplied by the current Middlesex County tax rate of 43 cents per hundred valuation.

Non-structural control measures such as restriction on ownership will impact use of property in a designated area. In order to provide that estimate, present value of property (minus improvements) for each parcel in the study area was used. Value was calculated on basis of square footage in the impact area for use in determining value of loss due to restriction. That is, square footage in the parcel was divided by present value to obtain present value of parcel per square foot. Square feet in each setback or erosion impact area was multiplied by the value per square foot to get a proportionate value per square foot for losses due to restriction.

Present value for dwellings was obtained on each shorefront parcel of property. If the dwelling was located in the erosion area, the total value was recorded. If not, a second cost component for dwellings was based on decrease in distance between dwelling and shoreline as a result of projected erosion action. The historical average annual erosion rate was not a significant explanatory variable for value of dwelling. However, distance of the dwelling in feet from the shoreline is an important explanatory variable. As distance increases value increases, at least in the limit of this analysis. Loss of distance between dwelling and shoreline was interpreted as having a marginal impact of decreasing value of dwelling by $8.64 for each foot lost to erosion.

A third set of costs were those associated with transaction and administration activities involved in the actual implementation and control of a program. Where appropriate, legal costs and the cost of administering compensation programs were included. These costs as with the first two sets were calculated for individual parcels or as an average for a reach.

Onshore benefits of structural control measures were derived by applying a "without" and "with" control analysis. Benefits were derived by calculating future erosion damages which would be prevented by implementing erosion control measures. Evaluation of the impact of erosion control strategies on value of property (including improvements) and uses of that property for each individually owned parcel was based on the value of those resources in a status quo state (that is, let erosion continue without additional control measures) compared to the value with control strategies. Therefore, present value in 1978 dollars for the resources in the identified impact area was established as the basis for calculation of impact costs and benefits.

For purpose of this study, the value of property and improvements of individually owned parcels was determined for both a 100-foot and 200-foot depth frontage as well as erosion rate depth area, based on years of effectiveness of proposed control measures, and then consolidated for each identified reach. These alternative impact areas allow a decision maker to compare the magnitude of costs and benefits of various management strategies. The two frontage area alternatives were selected because the 100-foot and 200-foot depth frontage are commonly suggested management strategies.

Area for each setback or erosion year period was calculated for each property segment within the reach on basis of square feet or portion of acres involved. Areas in each segment to be impacted were determined by multiplying the years in each period by the historical erosion rate in feet per year for each individual parcel transect--the point on the shoreline where a recession rate was calculated. The resultant depth line for the number of feet of erosion for each period was measured from the point of high water mark and noted on aerial photographs for the individual parcels of property.

Consolidation of Values into Reach Basis

Because the management decision framework should be placed on a reach by reach basis, the costs and benefit values to be used for comparative purposes are the consolidated reach values. Individual parcel values were consolidated into a reach summary. The established values for dwellings, other structures, property and loss of building site were expressed in present dollar values. Annual loss of property taxes and annual net return for land-use were also expressed in present value based on the assumption of constant annual stream of benefits with a constant level of net profits and a constant tax rate over the period of analysis. Present value of the stream of benefits was calculated at an 8 percent discount rate.

Costs for implementation of structures for each option are provided in Tables 2A and 2B. These costs could be allocated to either the private or public sector. Costs of technical assistance remains constant for each reach. Benefits for each option were calculated from the potential shoreline erosion loss prevention values based on percentage effectiveness of structures for that option in controlling erosion. These cost and benefit values provide a logical basis for choosing between selected control measures, between various levels of control and distribution of costs between the private and public sectors.

In comparing the costs and benefits of each structural option for Reach 1, none of the options should be implemented. However, other benefits (perceived and non-quantifiable benefits) may dictate selection of an option. If one of these options is selected for implementation only $10,872 under Option I, $18,957 under Option II, $82,540 under Option III, and $251,506 under Option IV should be allocated to private owner costs. Costs were allocated in this manner to equal identified private sector benefits.

These costs and benefits are not simply additive; combinations of costs and benefits must be considered for each proposed management strategy. The maximum monetary costs of restriction on ownership in Reach 1, the open and undeveloped area, was $172,512 for the 100-foot area and $345,904 for the 200-foot area. The cost of an ownership type program with a taking action would be an additional $34,150 (included in $36,519). These options are more expensive than structural Options III and IV.

Other values in the table are not necessary for the benefit and cost evaluation but are necessary for a complete evaluation of available alternatives. Acquisition of property and improvements in the 100-foot or 200-foot impact areas may be a desirable management strategy or may be a requirement under the taking issue. For Reach 1, the maximum cost of acquisition is $139,022 in the 100-foot or $238,394 in the 200-foot impact areas.

A second important factor could be cost of relocation of dwellings. Relocation could decrease cost of acquisition and make an infeasible management strategy feasible. In Reach 1, relocation cost could decrease acquisition cost in the 100-foot area by $23,400 ($116,500 value of dwellings minus $93,100 relocation of dwellings). That difference is $39,000 ($209,150 minus $170,150) in the 200-foot area. Savings from relocation are relatively small for this area as it is characterized by smaller homes where relocation costs are almost as large as the value of the houses. With larger, more expensive homes, relocation costs could be a significant factor in selection of management strategies.

Table 2A. Assessment of Proposed Structural Control Options for Three Reaches

Reach	Option	Effectiveness[1] Years	Effectiveness[1] Percent	PV Structural Costs $	PV Technical Assistance $	PV Benefits Private $	PV Benefits Public $	Transaction Administration Costs $
1	I	10	15	101,000	1,911	10,872	189	2,036
	II	15	20	125,550	1,911	18,957	484	2,036
	III	25	50	462,650	1,911	82,540	2,526	2,036
	IV	40	95	1,372,000	1,911	251,506	8,059	2,036
2	I	15	50	37,000	1,431	96,100	184	4,444
	II	40	95	395,000	1,431	298,959	1,084	4,444
	III	40	95	749,500	1,431	298,959	1,084	4,444
3	I	40	95	558,250	1,482	598,524	1,393	4,425
	II	50	95	1,476,000	1,482	781,365	1,855	4,425

[1] Years indicate the project time period of effectiveness of controls. The percentage indicates effectiveness of structures for that option in controlling erosion.

Table 2B. Assessment of Proposed Non-Structural Control Options for Three Reaches

Reach	Option	Transaction Administration Costs $	Ownership Restriction Costs Private $	Ownership Restriction Costs Public $	Public Acquisition Costs $	Dwelling Relocation Costs $
1	100-foot depth area	36,519	172,512	10,000	319,377	93,100
	200-foot depth area	36,519	345,904	20,000	599,104	170,150
2	100-foot depth area	115,667	256,270	0	964,288	540,200
	200-foot depth area	115,667	512,540	0	1,483,375	731,550
3	100-foot depth area	118,758	470,452	0	1,399,152	634,800
	200-foot depth area	118,758	921,304	0	1,850,004	634,800

In Reach 2, benefits are significantly greater than cost for Option I but significantly lower for Options II and III. Even with added cost of technical assistance and administrative and transaction components Option I appears feasible and total costs should be borne by private owners. Unquantified benefits would have to be extremely large to justify the other two options.

Cost of restriction on ownership is high for this area. The $256,270 plus $115,667 administrative and transaction costs give a total of $371,937. Any added costs of acquisition (minus difference in relocation) due to the taking issue would push this cost past that for structural controls in Option II and possibly Option III. The ownership restriction may have little impact on mitigation of erosion whereas structural Options II and III eliminate 95 percent of the erosion.

Option I is economically feasible. Options II and III appear to be better alternatives than non-structural ownership restrictions in this developing area. If implemented, over half of the cost of Option II should be borne by the public. A significantly larger amount of Option III cost should be borne by the public.

In Reach 3 potential benefits from controlling erosion were relatively high but fairly constant for each erosion rate area in this mature, developed area. The identified benefits for Option I were almost as large as the costs. Only a small number of unidentified benefits would make this option feasible. Almost all of the benefits are private and therefore, most of the cost should be borne by the private sector. Option II appears to be uneconomic. If it were implemented, a very large part of the cost would fall on the public sector.

Cost of ownership restrictions ($470,452 plus $118,758) $589,210 plus any cost of acquisition would make this alternative extremely expensive compared to Option I which also controls 95 percent of the erosion.

The assessment procedure clearly identifies those structural control options for which costs are greatly in excess of the expected benefits; such as all structural options in Reach 1, Option III in Reach 2, and Option II in Reach 3. Benefits other than those identified would have to be extremely large to justify the action. Several of the structural control options do have identified benefits nearly equal to or in excess of costs and would appear to be economically feasible.

The procedure also provides a measure of the relative magnitude of cost for the non-structural measures such as setback requirements. For instance, an ownership restriction is preferable to Options III and IV but would be more expensive than structural Options I and II in Reach 1, the open space and undeveloped area. Of course, acquisition in response to the taking issue would add significantly to that cost but would not constitute a prohibitive expenditure. Non-structural controls become significantly more expensive and rival the magnitude of structural measures in the developing area, Reach 2. Non-structural measures become very expensive for the developed areas and do not control actual erosion as structural controls do.

The division of costs and benefits between the private and public sectors provides a reasonable basis for consideration of allocation of the burden of costs of the program between these two sectors. Successful implementation of a proposed program for any reach will depend on willingness of private property owners to construct or, where necessary, support the expenditure of public funds.

Costs of Total Shoreline Control

Decision makers must have information on the total cost of a program for at least each county. In order to determine an approximate cost of implementing erosion control options on all shoreland of the same predominate type as in each specific reach, information on shorelands use classification for Middlesex County was assembled. The shorelands were divided into two classes depending on use. Class I consisted of open and undeveloped land and included agricultural, recreational, government, preserved, and unmanaged land uses. Class II consisted of developed land and included residential, commercial, and industrial land uses. The miles of fastland in each class were determined, as well as their percentage of the total miles of fastland. An assumption must be made that similar controls would be cost effective for other reaches and that benefits of control would be equivalent for other reaches.

Middlesex County had 213 total miles of shoreland with 152.4 miles in Class I, open and undeveloped, and 60.6 miles in Class II developed. However, based on the scientific projections only 25.3 miles of Class I and 17.0 miles of Class II use were eroding and suitable for management under the proposed options. For illustrative purposes costs associated with structural controls in Option I, the least amount of effort needed, in both the open and undeveloped reach (Reach 1) and the developed reach (Reach 3) were used.

The procedure for the case study area was to allocate projected erosion control costs to the private sector in an amount equal to the identified private benefits. The remaining costs were presumed to be a public responsibility.

As shown in Table 3, the 25.3 miles or 133,584 feet of eroding shoreland classed as I would require a minimum expenditure of $1,124,777 ($156 added cost per person) in total cost with $1,078,023 ($150 added cost per person) of that allocated to the public sector. The 17.0 miles or 89,760 feet of eroding shoreland classed as II would require a minimum expenditure of $5,410,733 ($752 added cost per person) in total cost with $987,360 ($137 added cost per person) of that allocated to the public sector. Based on this analysis, a program for only one county becomes extremely expensive.

Ability of Locality to Support Erosion Program

The additional cost per person for each option is of course a one-time expenditure which could be paid in one year or over several years. Therefore, for comparative purposes these costs should be amortized over a given payment period. A comparison of the ability of different localities to support a public coastal erosion program can be determined

Table 3. Projected Costs of Implementing Selected Control Options for the Total Shoreline in Middlesex County, Virginia

	Class I[1]		Class II[2]	
	Open and Undeveloped Area		Developed Area	
Category	Total Cost	Public Share	Total Cost	Public Share
Feet of shoreland in reach (ft)	12,000	12,000	9,300	9,300
Cost of structures for reach ($)	101,000	96,824	558,250	102,630
Cost per foot ($)	8.42	8.07	60.28	11.00
Total feet shoreland class (ft)	133,584	133,584	89,760	89,760
Cost of structures total county ($)	1,124,777	1,078,023	5,410,733	987,360
Population of county[3]	7,200	7,200	7,200	7,200
Added Cost per capita ($)	156	150	752	137

[1]Figures based on data for Option I in Reach 1.

[2]Figures based on data for Option I in Reach 3.

[3]Assume constant population over period of analysis.

Table 5. Calculation of Additional Fiscal Effort for Erosion Control in Middlesex County, Virginia

	Class I[1]		Class II[2]	
	Open and Undeveloped Area		Developed Area	
Category	Total Cost	Public Share	Total Cost	Public Share
Added Cost per capita ($)	156	150	752	137
Years effective life annualized cost ($)	10	10	40	40
Added % points to fiscal effort (%)	6	5	17	3

[1]Figures based on data for Option I in Reach 1.

[2]Figures based on data for Option I in Reach 3.

by amortizing cost over time and using a measure of current fiscal effort. That measure can also provide guidelines for the distribution of limited erosion control funds.

Measures of fiscal effort have been developed for two primary purposes: 1) to measure the actual financing performance of a government against its estimated financial reach, and 2) to compare differences in relative government financial effort among localities. In dealing with fiscal effort we are seeking to measure governments' use of their potential financing capacity rather than to compare the resulting burdens that fall upon people in various areas.

The formula selected for calculating fiscal effort is revenue from own sources per capita divided by computed revenue capacity per capita. This procedure yields a percentage of fiscal effort.[2]

$$\%\text{ fiscal effort} = \frac{\text{revenue from own sources per capita}}{\text{computed revenue capacity per capita}}$$

This method was selected because it provides a comprehensive picture of local effort and avoids some of the extremes inherent in the use of other methods.

Revenue from Own Sources. Revenue from own sources is calculated for each city and county on an annual basis by the Auditor of Public Accounts.[3] General revenue from own sources includes: property taxes, penalties and interest; local sales taxes; taxes on utility services; motor vehicle licenses; permits, licenses and license fees; service charge on county owned enterprises; other local taxes; revenue from use of money and property; fines and forfeitures; and reimbursement for services to other localities.

Revenue Capacity. Revenue capacity is a measure of the ability of a local government to obtain resources for public purposes. It is especially important to observe that this measure involves the financing capability of governments, rather than the economic well-being of people. The approach used in this study is the result of a 1971 ACIR report, and is the basis for a State of Virginia Revenue Resources and Economic Commission Study (1975).[4] This approach accounts for nontax revenue as well as taxes, and rests on the proposition that, in trying to arrive at a meaningful summary measure of relative revenue capacity for various areas, it is best to weigh various detailed elements of potential capacity according to their relative contributions to the grand total of all revenues raised by state and local governments. Whether applied at the national level or on a with-in-state basis such a set of weights seems more likely than any alternative to give summary capacity estimates with which actual revenue-raising performances can logically be compared. It provides a reflection of the real world, rather than of some other set of assumed circumstances.

The method used in this study for estimating revenue capacity is the "average effort" approach. This method gives a more balanced picture than any single measure. This "average effort" approach is the method the Virginia Revenue Resource and Economic Commission uses to compute local revenue capacity and is based on the Advisory Commission on Inter-governmental Relations (ACIR) approach.[5] For each major tax source ACIR calculated the state's tax base and then multiplied the base by the weighted national average ratio of tax receipts to tax base. For their purposes, the Resource and Economic Commission multiplied each major tax base in a locality by the statewide average effort. For the purposes of this study, the statewide average efforts that the Commission determined are used to calculate the revenue capacity per capita in each of the coastal counties and cities of Virginia. Those calculations are as follows:

The true value of real estate was multiplied by \$.0106, and personal income, a proxy for non-property and non-sales taxes and other revenue, was multiplied by \$.0160. The number of motor vehicles was multiplied by \$27.29, as a proxy for personal property taxes. The resulting products were added to local option sales tax collections to obtain computed revenue which was then standardized by dividing by population. Standardizing by resident population is a common method of making data for different size localities comparable.

Average revenue capacity was calculated for each coastal county and city. The average revenue capacity per capita for coastal counties of Virginia was \$324 with a high in Arlington County of \$485 per capita and a low in Northampton County of \$215 per capita. The average revenue capacity per capita for coastal cities was \$292 with a high in Williamsburg of \$476 per capita and a low in Chesapeake of \$223 per capita. For Middlesex County it was \$367 per capita.

These revenue capacity computations are of little use in an absolute sense. Since they are measurements of the potential reach of local governments in obtaining resources, revenue capacity computations take on meaning only when viewed in terms of the previously discussed fiscal effort.

Application of Fiscal Effort Measure. Measures of fiscal effort must be used in a selective and careful manner with respect to certain limitations. The measure selected for this study is the best available for consideration of state/local grant arrangements but does not provide a consideration of the total fiscal differences. For instance, it does not provide a consideration of federal activities and the many complexities of marked interstate differences. It does not relate to overall service requirements or fiscal needs nor the authority and willingness of fiscal units to provide services. It *is* however a meaningful comparative measure of fiscal capacity and effort for local areas and adequately serves the purposes of this study.

Percentage fiscal effort was calculated for each coastal county and city. For comparative purposes, selected county and city values are shown in Table 4. The average fiscal effort for all coastal counties in Virginia was 75 percent with a high of 179 percent in Prince William and a low of 35 percent in Lancaster. For cities the average was 119 percent with a high of 169 percent in Richmond and a low of 78 percent in Suffolk. It was 39 percent for Middlesex County, our case study area.

Table 4. Fiscal Effort for Selected Coastal Counties and Cities in Virginia

County or City	Revenue from Own Sources Per Capita	Computed Revenue Capacity Per Capita	Fiscal Effort
Counties	$	$	$
Accomack	143	259	55
Gloucester	193	339	57
Lancaster	141	400	35
Mathews	140	325	43
Middlesex	144	367	39
Northampton	126	215	59
Northumberland	165	354	47
Prince William	552	308	179
Westmoreland	145	314	46
York	253	270	94
Cities			
Hampton	288	243	119
Newport News	350	267	131
Norfolk	333	224	149
Portsmouth	303	226	134
Richmond	537	317	169
Virginia Beach	276	304	91

Selected data obtained from report of Auditor of Public Accounts of Commonwealth of Virginia on Comparative Cost of County Government, 1977 and Report of Auditor of Public Accounts of Commonwealth of Virginia on Comparative Cost of City Government, 1977.

These comparative measures of fiscal effort were used: 1) to look at the ability of each locality to support an erosion control program by using its own resources; and 2) to provide a basis for establishing priorities in the allocation of state (or state controlled federal) funds among coastal localities. Grants-in-aid are generally thought of in terms of providing equalization (and defining equalization as support for a level of public service without gross variation) in the financing effort of recipient jurisdiction.

The basis of comparison of cost per person with fiscal effort was to compare the per person increase as an annualized percentage increase in fiscal effort. This procedure provided the important link between fiscal effort and a measured need or service requirement for one service --erosion control--for use in a priority index. Costs were amortized at an 8 percent rate over the years of effective life of the structural option.

The added percentage points to fiscal effort provides a basis for comparison of the cost of an erosion control program for the county. Because of data requirements only Option I in the open and undeveloped area and Option I in the developed area are used to demonstrate the application of this procedure. The use of only Option I represents a minimum necessary effort for each area. This effort would result in significant increases in fiscal effort for both total cost or the public share. As shown in Table 5, added percentage points to fiscal effort was (5 + 3) 8 for only the public share of costs to (6 + 17) 23 for total cost including both the public and private share. These increases in percentage points of fiscal effort do represent a substantial increase in needed effort above the current 39 percent for Middlesex County just to support one public service program of erosion control.

A preliminary study of other local areas is needed to provide the basis for establishing a priority system for grant allocation. Such a procedure would provide the important link between fiscal effort and a measured need or service requirement for the priority index.

In summary, this comparison of the projected public cost of a proposed erosion control program and the locality's current fiscal effort provides a measure of the gross impact on the financial effort of that jurisdiction. Such a comparison should be used to indicate the ability of the locality to pay for the public portion (or total cost) of the proposed erosion control program cost. This comparison can also be used to develop a priority scale for allocation of limited funds among localities.

Notes

1. The technical assessment for each reach was provided by Robert Byrne and Carl Hobbs of the Virginia Institute of Marine Science, Gloucester Point, Virginia.

2. Fiscal Prospects and Alternatives: 1976, A Staff Report to the Revenue Resources and Economic Commission, June 1975, p. 88.

3. Report of Auditor of Public Accounts of Commonwealth of Virginia on Comparative Cost of County Government, 1977, p. 14-15.

4. Fiscal Prospects and Alternatives: 1976, A Staff Report on the Revenue Resources and Economic Commission, June 1975, p. 82.

5. Ibid.

References

Procedures for Determining Ranges of Use-Value...With 1978 Suggested Use Values, State Land Evaluation Advisory Committee, Richmond, Virginia, September 1977.

Armstrong and DenUyl, "An Investment Decision Model for Shoreland Protection and Management," Coastal Zone Mangement Journal, Vol. 3, No. 3, 1977, pp. 237-253.

Brown and Pollakowski, "Economic Evaluation of Shoreline," Discussion Paper #75-14, Institute for Economic Research, University of Washington, November 1975.

Erosion/Insurance Study, by Erosion/Hazard Management Subcommittee of the Great Lakes Basin Commission Standing Committee on Coastal Zone Management, June 1978.

Polinski and Shavell, "Amenities and Property Values in a Model of an Urban Area," Journal of Public Economics, Vol. 5, 1976, pp. 119-129.

Subdivision Ordinance, Middlesex County, Virginia, prepared by the Middle Peninsula Regional Planning Commission and adopted by the Board of Supervisors of Middlesex County, Virginia, effective September 1, 1966.

"Cost Report on Non-Structural Flood Damage Reduction Measures for Residential Buildings Within the Baltimore District," U.S. Army District, Baltimore, MD, IWR Pamphlet No. 4, July 1977.

URBAN WATERFRONT REVITALIZATION EMPLOYING VISUAL PREFERENCE AS A STRATEGY FOR MANAGING VISUAL RESOURCES AND FACILITATING PUBLIC PARTICIPATION

Thomas R. Hudspeth, Assistant Director
Environmental Program, University of Vermont
153 S. Prospect, Burlington, VT. 05405

A significant issue in coastal zone management in the 1980's will be finding ways to manage visual resources of urban waterfront environments and to facilitate public participation in urban waterfront revitalization efforts. It is likely that urban waterfronts will continue to be the site of intense pressures for redevelopment. Strategies to assure that revitalization efforts make use of the public's views and involve citizens in the decision-making process are needed. Use of visual preference as a research methodology offers considerable promise in both these areas.

This paper offers an overview of the topic of urban waterfront revitalization and then demonstrates the applicability of visual preference as a strategy for dealing with two major concerns in urban waterfront revitalization (and coastal zone management in general): managing visual resources and facilitating public participation.

URBAN WATERFRONT REVITALIZATION

"Water is a vital element; not only is it essential to human life, but it provides the aesthetic and recreational needs of our people. It helps create, and is an important part of, our environment" (Beauty for America, 1965). Human settlements have been associated with water throughout history. The uses water affords include drinking, carrying away human wastes, farming, fishing, industrial processes, transporting raw materials and finished products important for trade and commerce, power production, and aesthetic and recreational possibilities. For these reasons, many of our nation's cities are located on rivers, lakes, bays, and oceans.

Role of Waterfronts in History

When Europeans first landed on the edges of North America, they settled at river mouths or where harbors protected them from stormy seas and tides. In time, these settlements became active seaports through which came materials, goods, and people from around the world. The country's commercial well-being was tied to the fisheries, shipbuilding yards, and brisk coastal shipping trade of its Atlantic seaports until the Industrial Age. Systems of canals built to form linkages between larger navigable waterways and to avoid rapids, falls, and erratic river courses helped in America's growth and prosperity by opening up the back country in New England, in the Middle Atlantic states, and at the prairie's edge. During the latter half of the nineteenth century, the development of urban America and the preeminence of certain cities over others depended largely on their combination of transportation systems; those places which combined railroads and port or river access became dominant.

The waterfronts were the center of activity of these urban areas. They had similar characteristics in most of the cities that developed along the water's edge. There were large wooden docks, behind which were warehouses, shops, offices, bars, restaurants, and hotels. Small industrial plants turning out finished goods from the raw materials unloaded at the adjacent docks were usually found nearby. Businesses catered to townspeople, traders, and seamen from far-off lands. Newspaper and telegraph offices were normally close by. Churches and bawdy houses frequently existed side by side. In colonial America, a great percentage of trade was in illegal goods that avoided various taxes and tariffs; as a result, smuggled and pirated merchandise was also to be found. In short, the waterfronts were exciting, bustling places, full of life, turmoil, and legal and illegal activities.

Decline of Waterfronts

With the Industrial Revolution came the first withdrawal of dependence on waterfront trade. Between the Civil War and World War I, over 200,000 miles of railroad track linked towns and cities across America. Railroads became the prime mover of people and goods. Vast railyards exempted all other activities at seaport waterfronts and at ports along the Mississippi, Missouri, and Ohio Rivers. Even greater changes in the use of urban waterfronts have resulted from four changes over the past quarter of a century:

1. Steadily decreasing waterborne passenger travel

"The tourists who once flocked to ocean liners to see their loved ones off now drive to airports, and tourism by boat has become a luxurious anachronism best suited to warmer climes. Even the charm of a ferry ride has given way to the expedience of a quick trip across a bridge" (Cowey in Harney, 1979).

2. Technological changes in marine transportation and cargo handling

The phenomenal growth in size of bulk petroleum carriers has created a need for new and specialized terminals which cannot---and need not---be located at traditional urban waterfront sites. The largest crude oil carriers are so large that they cannot enter any U.S. ports. Increasingly, offshore terminals are being constructed which connect with onshore refineries and tank farms via submarine pipelines. The petrochemical complexes and oil tank farms are located several miles from waterfront areas. The obsolescence of many old fuel terminals, siltation of navigation channels, and fears of oil spills have contributed to this trend.

The movement, storage, and processing of hazardous bulk liquids (e.g., liquified energy gases and toxic chemicals) is also being displaced to concentrated locations beyond urban waterfronts as the public has become more aware of the effects on human health of accidental releases of combustible, radioactive, toxic, carcinogenic, mutagenic, and teratogenic agents, and has questioned the compatibility of such activity in heavily concentrated areas.

The most significant change in cargo handling and shipping is the shift from break-bulk operations to containerization, which allows goods to move more efficiently. With containerization, goods may be shipped directly from a manufacturing plant to a port and then to a point of distribution without breaking bulk, or packing and unpacking the cargo. The container revolution has created major land use changes in waterfront areas. Containership terminals require fewer berths and less frontage on shipping channels than the old ones, but considerably greater acreage for back-up facilities. They are highly land-intensive operations. While a break-bulk cargo vessel needs only a few acres of waterfront property (basically a pier and warehouse), a single containership requires from 12 to 35 acres of land for sorting and stacking purposes. Containerships tend to be much larger than break-bulk cargo vessels, and to carry much greater cargoes. They make fewer port calls per round trip, stopping only at major ports which serve as load centers to funnel cargo between the port and inland hinterland. Consequently, the amount of cargo loaded and unloaded at a single port call is considerably greater with a containership than with a break-bulk cargo vessel. Also, containerships stay docked at the terminal only 18 to 36 hours, whereas break-bulk cargo vessels spend several days at each port. Therefore, the ports for containerships must have enough room to handle all the cargo that will be loaded and unloaded for each arrival.

Few of the older port cities, especially along the East Coast, are able to provide the acreage needed for container terminals. Consequently, newer container facilities are generally developed in areas remote from traditional urban waterfronts where requirements for back-up space, less expensive land, deeper and wider shipping channels, and access to good highway and rail transportation can be met. So traditional finger piers and much of the break-bulk cargo vessel system are being rendered obsolete as a smaller number of specialized container ports are receiving most of the cargo-shipping business.

3. Changes in land transportation modes

Since the late nineteenth century, coastal shipping of goods has yielded considerably to rail and, later, truck transport. With the increasing dependence on truck and rail commerce, many industrial concerns shifted their operations away from the waterfront to rail spur lines and major highway interchanges to move their goods more efficiently. The city began to expand further from the water's edge, and the waterfront was no longer the hub of activity. The decline of our nation's railroads and the rise of the interstate highway system has contributed to the gradual emergence of trucking as the major means of transportation. More and more, freeway corridors began to be located across derelict waterfronts, severely limiting access to the waterfront by pedestrians.

4. Changing economic role of the central city

Many local businesses traditionally located on urban waterfronts have succumbed to centralized mass procuction. Changing space requirements in manufacturing and mass production technologies have led to the shift of this activity to the suburbs. In many places the economic role of the central city has shifted from manufacturing to highly specialized services, thus eroding the traditional link between urban industries and shipping activity. Consequently, the importance of waterfronts in contemporary industrial operations has declined.

A large number of port areas located next to the central business districts of large cities no longer meet the requirements of modern shipping and transportation technology. Cargo and passenger shipping firms, railroads, warehouses, port-related industries, and other traditional users of urban port areas have changed their activities with urban areas to meet changing requirements. As a result, they no longer require the piers, terminals, and land areas of traditional urban ports situated next to central business districts. As these firms departed or reduced the extent of their activities in port areas, the urban waterfronts became occupied by marginal economic enterprises (including some not associated with port functions at all such as storage of impounded vehicles), less desirable functions--e.g., highways, airports, water supply facilities, waste water treatment plants, energy facilities (power plants, tank farms, pipelines, etc.), landfills and incinerators, and low-rent warehouses---or were abandoned. In a vicious cycle, neglect and underutilization of the area further contributed to undesirable social activities and led to its being perceived "as a place with a bad smell and a bad rumor and good to stay away from" (Balchen and Linville, 1971).

These four changes have rendered the traditional uses of urban waterfronts functionally or economically obsolete, and have resulted in a decline of urban waterfronts and in the loss of their positions as the economic, commercial, and industrial centers of cities. They have forced cities to focus their attentions inland and turn their backs on the areas where they were born, their waterfronts. A familiar pattern of events has resulted:

> land values dropped; waterfront enterprises failed, moved to other areas, or just closed their doors; railroads and industries had to abandon and, eventually, release some of their vast waterfront land holdings; income and tax revenues dwindled; once-elegant shore residences became rooming houses; unchanged road and highway patterns and waterfront access routes became inadequate; waterfront amusement and recreation areas became shabby, neglected, and unprofitable (Whitman et al, 1964).

These events have left large sections of urban waterfronts vacant, unused, or underused:

> Paradoxically, the very existence of decay on the waterfront gives Americans a second chance to improve its appearance and amenities. Although there is still competition for land on the water's edge, the existence of decay is evidence that certain older uses are no longer necessary there and that we should be thinking seriously of the kind of uses which should replace them. Some older harbor cities no longer consider the harbor as part of their economy---long-range truck transportation has been a major factor here---and the result is that refuse and objectionable land uses like wrecking yards find their way to the shoreline (Beauty for America, 1965).

Rediscovery of Waterfronts

However, in the late 1960's and 1970's, tremendous interest has been shown in transforming our nation's waterfronts from desolate, underutilized, and nonproductive areas characterized by haphazard and ugly development into attractive, active, and vital areas. Once again, urban waterfronts are becoming the viable, productive, significant portions of the cities that they were when first established:

> Cities across the nation are beginning to realize that their waterfronts, whether on the ocean, a lake or a river, are an asset. They're being eyed as potential sources of new income, as means of bringing life back into ailing downtowns, as recreation or residential areas, all depending on the needs, on the nature of the waterfronts, or on the citizens themselves (U.S. Department of the Interior, 1979).

In 1980, the "Year of the Coast," it appears that this interest in urban waterfront revitalization will continue and grow in the 1980's.

Over thirty percent of the 520 cities in the United States with populations between 25,000 and 50,000 and seventy percent of the 415 cities with populations over 50,000 are located on the edge of a river, lake, bay, or ocean (U.S. Department of the Interior, 1979). As many of these cities have sought to improve their images as good places to live, work, play, and visit, they have re-examined their waterfronts, have recognized that they are special and important places, and have undertaken revitalization efforts:

> The rapid growth and revitalization of American cities has forced us to examine and re-examine every available plot of land for its commercial, residential, and recreational potential. For many years, low income residential areas and central business districts were the focal point of urban revitalization efforts. Today, the urban waterfront is being given a second chance. Suddenly, after years of neglect, America's coastal and riverfront towns and cities are rediscovering their waterfront areas. Cities like Baltimore, with its multibillion dollar Inner Harbor redevelopment effort, or San Antonio's Riverwalk project have demonstrated to other American cities the multiple uses and advantages of redeveloping this long untapped resource (Pittas in Harney, 1979).

They have rediscovered these areas where the natural amenities of water and coast combine with a rich heritage, recreation, and commerce, and they have recognized that these are exciting and unique places with their combination of the bustling activities of fishing boats, structures, sights such as the bright colors or boats and flags, and sounds. As expressed by Daniel Ahern, Executive Director of the Boston Downtown Waterfront Corporation:

> The basic human interest in the amenities and attractions of water bodies, together with the kind of dramatic proposals which waterfront locations stimulate, results in a high degree of public excitement over waterfront development. Where this type of land is available in or close to an urban center, the most obvious re-uses are quality housing and recreation, including marinas, aquariums, restaurants, theaters, promenades, etc. If the location is close enough to the downtown center, it might also be attractive for prestige commercial buildings (Whitman et al, 1964).

His sentiments are echoed by Thomas J. Whalen, mayor of Jersey City: "Here in the cradle of our city may lie its future" (Whitman et al, 1964). And Vaughn P. Call, city planning director for Spokane, sums up the interest in waterfront revitalization along the Spokane River in his city:

> Yesterday it was a forgotten river lost in a maze of industry, railroad yards, deteriorating houses and isolated barren riverbanks. The city turned away from this hidden water cascade. Today the city is turning toward its water treasure. The river is being opened to view, railroads are moving, banks are being cleared, deterioration removed....If Spokane can do it, other cities can also (Whitman et al, 1964).

They have attempted to transform their urban waterfronts into prime assets:

> There was a time when a city's harbor had a gutsy fascination about it and its wharf on lake or river was the natural meeting place to catch up on the latest and enjoy the life. But during the years, with increasing use of land and air transportation, a number of cities have turned their backs to their waterfronts. This is nothing new: the same thing happened in time even to such important harbors as those in Phoenicia, when they were filled with silt due to lack of use. What is different today is that ours are dirty, polluted, rundown. More and more, however, people seem to be waking up to the fact that here are slices of real estate of prime value, too good to go to waste (Balchen and Linville, 1971).

Such efforts have received much attention in popular magazines (Gratz, 1978; Stephens, 1978) and professional journals and reports (American Society of Planning Officials, 1952, 1959; Architectural Forum, 1960; Balchen and Linville, 1971; Barr, 1976; Beauty for America, 1965; Clark et al, 1979; Commission on Marine Science, Engineering, and Resources, 1969; Committee on the Future, 1979; Committee on Urban Waterfront Lands, 1980; Cowey and Rigby, 1979; Davenport, 1978, 1980; Design and Environment, 1974; Doyle, 1979; Environmental Comment, 1978; Forward, 1969; Hanna et al, 1978; Harney, 1979; Kenyon, 1968; Landscape Architecture, 1974; Lehmann, 1966; McCann, 1979; McQuade, 1961; Mann, 1973; Mann and Alling, 1978; Manogue, 1980; Marcus et al, 1980; Morell, 1976; Morell and Singer, 1978; The President's Council on Recreation and Natural Beauty, 1968; Progressive Architecture, 1975; Rick, 1964; Simpson, 1976; U.S. Department of the Interior, 1977, 1978, 1979; Waterfront Memo, 1979; Watson, 1978; Whitman et al, 1964; Wisconsin Department of Resource Development, 1966; Wood, 1965). As reported by Whitman et al (1964):

> In almost all cases, cities reported a more than gratifying degree of citizen, newspaper, business, industry, and official support for proposals to concentrate renewal processes on waterfront areas. Along with the support, many indicate, came a good many suggestions and ideas about just what the best use of valuable and sought-after waterfront land might be. But even in cases where all the ideas, suggestions, and pet projects of the community didn't get incorporated into final waterfront revewal plans, there's been no report of any lessening of community enthusiasm and support for renewal of the waterfront areas.

Presently more than fifty cities have completed waterfront revitalization projects, and another fifty have projects planned or under construction (Lehmann, 1966; U.S. Department of the Interior, 1979). Virtually every issue of the major professional journals of architecture, landscape architecture, and environmental design (e.g., Journal of the American Institute of Architects, Progressive Architecture, Architectural Record, Architectural Forum, Architecture Plus, Design and Environment, Landscape Architecture, etc.) during the 1970's included at least one description of an urban waterfront revitalization effort. These range from small-scale improvements at the end of a street; to linear parks connected by hike-and-bikeways; to large-sclae, multipurpose developments combining recreation, housing, and commercial components.

As a result of these efforts, rotting wharves and piers and pilings, blighted and abandoned buildings, deteriorating warehouses, scrap yards, and vacant lots are being replaced by a mix of structures, uses, and resources:

* Town houses, condominiums, apartments, hotels, motels, and boatels
* Shops, stores, malls, commercial facilities, marketplaces, convention centers, and office buildings
* Fishing piers and wharves, marinas, boat launches, and boat harbors
* Parks and recreation facilities, playfields, open green spaces, picnic areas, children's playgrounds, promenades, walkways and trails for strolling, hike-and-bikeways, observation towers
* Museums, aquariums, maritime exhibits, historic ships, arts and crafts centers, theaters, outdoor amphitheaters, galleries, sculptures
* Tourist attractions and sites for festivals and performances.

A variety of innovative strategies and resourceful and imaginative approaches to urban waterfront revitalization are being tested by many cities.

While most of the projects are multipurpose in nature, some are single purpose. Some are initiated by private groups (e.g., downtown business assocations, professional organizations, corporations, etc.); others by governmental leaders; some by city departments; and some by non-government groups. They are funded by federal, state, city, bonds, and private sources, and usually a combination. Some, such as Boston and San Francisco, involve extensive adaptive reuse of existing structures, while others, such as Baltimore, involve tearing down and clearing away old structures and rebuilding from scratch.

Why are our urban waterfronts being rediscovered now? Increasing numbers of Americans have traveled abroad, where waterfront amenities are more often the rule than the exception. They have seen for themselves what others (Doyle, 1979; Landscape Architecture, 1974; Mann, 1973; Morell and Singer, 1978; and Simpson, 1976) have described:

> In Europe, the river is the heart of the city. Try to think of Paris without the Seine, Florence without the Arno, or London without the Thames. In the United States, the river is usually looked at as a giant sewer, often politely called a drain, and as a convenient route for high speed highways. We Americans, both children and adults, need to educate ourselves about the wonder and beauty of the rivers that flow through our cities (Beauty for America, 1965).

> In Europe, waterfront areas were incorporated into the fabric of city life prior to the advent of industrialization. As a result, citizen access and use of the water's edge as an amenity were woven closely into the European style of urban life (Morell and Singer, 1978).

> Almost every civilized city in Western Europe has a waterfront dedicated to its inhabitants. To list only a few, Copenhagen, Stockholm, London, Paris, Zurich, Florence, and Naples have waterfronts that are a delight to the eye and a relief on hot summer days. American cities have six- or twelve-lane highways instead....In the United States, the highway planner has succeeded in destroying almost every city in which he has been active (Blake in McCann, 1979).

The treatment of urban waterfronts in Geneva, Zurich, Lausanne, Lucerne; Hamburg; Amsterdam, Rotterdam; London; Paris; Stockholm, Gothenburg; Copenhagen; Helsinki; Leningrad, Moscow; Venice, Florence, Naples; and other cities has provided us with models for improving our own. Creative approaches by our neighbors to the north in Toronto, Quebec City, Montreal, Vancouver, Victoria, and Halifax have also made us more aware of possibilities for our waterfronts. The viewing of the Tall Ships under Operation Sail as part of our Bicentennial celebration in 1976 and again in 1980 by millions of Americans also made us more aware of our urban waterfronts.

Factors Contributing to Interest in Urban Waterfront Revitalization

Increased environmental awareness and concern for quality of life over the past decade explain much of the current interest in urban waterfront revitalization. Several interrelated factors have contributed to these concerns:

1. Urban revival

Having experienced a letdown in expectations after moving to outlying areas, many suburbanites are returning to urban settings closer to employment centers and cultural activities. This movement back to center city areas---especially by middle-class professionals, retired people, and young---offers viable alternatives to the high costs of energy and transportation, the lack of social interaction, and the increase of social problems in suburban communities. Also, developers are expressing great interest in making adaptive use in urban commercial and residential structures of considerable age and character as the costs of securing suburban building sites increase. Redevelopment of urban areas makes especially good sense because many public services are already in place. High inflation rates, slow-growth and even no-growth policies and practices, improvement of safety in downtown areas, improvement of mass transportation systems, and increased costs coupled with shortages of gasoline---as well as federal programs to make our cities more livable, to combat the deterioration of urban areas, and to preserve neighborhoods---have all played a role in revitalizing core cities. So has development of programs which show greater sensitivity to human needs than the wholesale clearance, demolition, and razing characteristic of urban renewal programs in the 1950's and early 1960's. As a result of these many factors, development trends have gone full circle and attention is again being focused on the key portions of our city centers which have fallen into decay.

2. Clean-up of waters and restoration of beauty to surrounding lands

The campaign to clean up our nation's waters is gradually resulting in improved water quality in and near many of our coastal and riverfront cities. The Federal Water Pollution Control Act Amendments of 1976 set a clean water goal for 1983 for most of the waters in our country to be "fishable and swimmable." Through 1979, the Environmental Protection Agency has spent $18 billion to construct waste water facilities and to control discharges into waterways.

3. Urban recreation

Demand for recreational opportunities has increased rapidly in response to increases in population, disposable income, leisure time, and mobility. This demand is especailly acute in urban areas where population concentrations are increasing at the same time that the availability of suitable land and water areas to accomodate that demand is decreasing. The Urban Parks and Recreation Recovery Program, passed in 1978, provides funding for urban recreation areas, including decaying urban waterfronts, open space along rivers, and canals fallen into disuse for commercial transportation conduits. Worries about price and availability of gasoline have resulted in even greater use of close-to-home recreation space in urban areas.

4. Coastal zone management

In the Coastal Zone Management Act of 1972, Congress established a policy to preserve and enhance the nation's coasts for future generations by guiding protection and development along the nation's 95,000 miles of shoreline. The Act declared a national policy

> to encourage and assist states to exercise effectively their responsibilities in the coastal zone through the development and implementation of management programs to achieve wise use of the land water resources of the coastal

zone giving full consideration to ecological, cultural, historic, and aesthetic values as well as the needs for economic development (Davenport, 1978).

That legislation has provided 35 coastal states and territories with matching grants to plan and implement their own coastal management programs, which must include guidelines on priority of uses within particular areas of the coastal zone. The funds that have been available through the Act to date have enabled states to address: flooding and erosion problems; fisheries management; energy facility siting; commercial and industrial development on the coasts; public access, private ownership, and recreational development; pollution problems in coastal waters; and waterfront development.

Amendments to the Act in 1976 required state management programs to develop:

"a planning process for the protection of an access to public beaches and other public coastal areas of environmental, recreational, historical, aesthetic, ecological, or cultural value." Although the language does not specifically refer to urban waterfronts, these are the very characteristics found in most in-town waterfront areas. The regulations go on to say that "special attention should be given to recreational needs of urban residents for increased shorefront access" (Cowey in Harney, 1979).

5. Historic preservation and conservation of historic heritage resources

City waterfront areas are old sections, often full of history. In renewing them, cities are finding precious relics of their beginnings and earliest days. In many cases, preservation and restoration of such historic riches are parts of waterfront renewal programs (Whitman et al, 1964).

Restoring historic homes, adapting older buildings to new economically viable uses, and recreating traditional cultural events in these areas of picturesque natural settings, distinctive architecture, colorful heritage, and rich history add to the amenity value of urban waterfronts for residents and to their attraction value for tourists and other visitors. The high costs of new construction make preservation---once considered merely quaint and "in"---economically practical. Tax breaks available to owners of properties listed in the National Register of Historic Places contribute to interest in recycling old buildings.

Adaptive reuse of old structures related to historic seaport and riverport uses---e.g., lighthouses, canneries, ropewalks, warehouses, ship chanleries, custom houses, etc.---to new, economically-viable uses---e.g., restaurants, galleries, shipping pavilions, guest houses, craft shops, museums holding old objects and documents---which do not alter significantly the external appearances creates an excellent setting for reenactment of historic events; music, theater, and fine arts performances; and fairs and festivals (such as sailboat races, blessing of fishing fleets, etc.).

6. Energy concerns

The movement back to the city and reuse of old buildings instead of building new ones are both practices which relate to concern for energy.

7. Economic development

There has been a gradual shift of commercial activity away from urban waterfronts. Rail and truck transport have eliminated much of the small-scale waterborne traffic that long dominated port cities. And much of the shipping that survives (e.g., hauling fuel and containerized cargo) requires such vast facilities that few old harbors can handle it. Consequently, miles of shoreline are ripe for development. Revitalization creates jobs in the construction and service sectors and expands the property tax base. When such revitalization includes historic preservation strategies, there are direct economic benefits and spinoffs: employment gains attributed to rehabilitation's labor-intensive nature; savings in resources and in building and demolition costs; shorter completion time; fewer delays caused by community opposition; strengthened tax base; and improved business climate.

8. Conservation of natural heritage resources

Revitalizing urban areas instead of constructing new developments helps to conserve the natural environment by preventing urban sprawl and the filling of valuable yet fragile wetland ecosystems (especially tidal marsh estuarine ecosystems). Some of the recreation and open space uses incorporated into urban waterfront revitalization projects help to reduce or mitigate the impact of natural hazards and offer non-structural flood control and shoreline protection.

<u>Efforts Related to Urban Waterfront Revitalization</u>

A variety of efforts have been devoted to urban waterfront revitalization. The federal government has played an important role in urban waterfront projects through its programs, studies and reports, and networking efforts. Several conferences and symposia have been held dealing with urban waterfront revitalization. And a graduate-level course on this topic has even been developed.

1. Federal programs and studies and reports

Urban renewal programs initiated by the federal government have stimulated subsequent waterfront development projects in many cities. More than 80 cities have some kind of waterfront project connected with their urban renewal efforts (Harney, 1979). Other influential federal programs and studies and reports include:

A. Panel of Water and Waterfronts at the White House Conference on Natural Beauty in 1965 (Beauty for America, 1965)

B. Recommendation by the President's Council on Recreation and Natural Beauty in 1968 that "federal agencies be authorized to conduct...a coordinated program of urban waterfront restoration that emphasizes recreational, scenic, and aesthetic values, including physical and visual access" (President's Council on Recreation and Natural Beauty, 1968).

C. Conclusion by a Bureau of Outdoor Recreation study conducted in 1967 that water-related activities would constitute the greatest increase in America's recreational interests

D. <u>Our Nation and the Sea</u>, the report of the Stratton Commission in 1969, recommended that increased amounts of shoreline by made available to the public, with priority given to metropolitan areas where the need was the greatest and where integration of recreational uses with other uses would be necessary (Commission on Marine Science, Engineering, and Resources, 1969)

E. Such National Endowment for the Arts architecture programs as

"City Edges" and "Livable Cities" have supported planning studies for waterfront revitalization projects in more than fifty coastal cities since the early 1970's (Cowey in Harney, 1979; Cowey and Rigby, 1979)

F. Acknowledging the need for increased urban recreation opportunities, the National Park Service established national recreation areas in New York City, San Francisco, and Cleveland in 1972

G. _A Preliminary Urban Parks Study_ conducted by the Bureau of Outdoor Recreation in 1977 emphasized the problems of access to urban waterfronts (U.S. Department of the Interior, 1977)

H. In 1978 the Heritage Conservation and Recreation Service completed its _National Urban Recreation Study_, emphasizing the importance of waterfront revitalization in meeting needs for water-oriented recreation in urban areas (U.S. Department of the Interior, 1978)

I. As part of its Urban Waterfront Study, the Heritage Conservation and Recreation service examined factors that facilitated the development of successful waterfront projects, the needs for future urban waterfront projects, and goals for federal urban waterfront policy, and provided case studies of eighteen urban waterfront projects of various scales (U.S. Department of the Interior, 1979)

J. _Reviving the Urban Waterfront_, a joint effort by Partners for Livable Places, National Endowment for the Arts, and Office of Coastal Zone Management, presented case studies of seven imaginative urban waterfront revitalization projects to show what is possible (Harney, 1979)

K. In 1978, the Office of Coastal Zone Management funded special local demonstration projects encouraging the redevelopment of underutilized waterfronts for recreation, open space, and economic development in more than 30 communities in 13 states. It made available over $800,000 in 1978 to study waterfronts and examine public access problems.

L. The Land and Water Conservation Fund administered by Heritage Conservation and Recreation Service supports urban waterfront revitalization through urban land acquisition

M. Housing and Urban Development's community development block grants and urban development action grants have provided funding for many of the urban waterfront revitaliztion projects, as have Commerce's Economic Development Administration grants.

2. Networking efforts

In 1978 Urban Waterfront Action Group (UWAG) was formed under the terms of a Memorandum of Understanding signed by 18 federal agencies and 4 non-federal organizations to:

A. improve coordination of urban waterfront projects and programs and improve delivery of federal assistance on such projects

B. explore cost effective ways for cities to implement these programs

C. heighten awareness of opportunities, benefits, and needs, and address key issues by serving as an information clearinghouse.

It has published a Directory of Participants and has met on a bi-monthly basis in rotating agency headquarters to exchange information and identify areas needing improved cooperation and coordination.

3. Conferences and symposia

Among the conferences and symposia dealing with urban waterfront revitalization are:

A. Coastal Zone, '78: Symposium on Technical, Environmental, Socio-economic, and Regulatory Aspects of Coastal Zone Management. 14-16 March 1978. San Francisco, CA. Sponsored by the American Society of Civil Engineers.

B. National Workshop on Urban Waterfront Redevelopment. 17-19 January 1979. Detroit, MI. Sponsored by the Coastal Zone Laboratory of the University of Michigan.

C. Urban Waterfront Lands Symposium. 28-29 March 1979. Cambridge, MA. Sponsored by the Committee of Urban Waterfront Lands of the Environmental Studies Board of the National Research Council's Commission on Natural Resources.

D. Three workshops on "Urban Development through Waterfront Planning" at City Care: A National Conference on the Urban Environment. 8-11 April 1979. Detroit, MI. Sponsored by the National Urban League, Sierra Club, and Urban Environment Conference and Foundation.

E. Four full-day seminars on "Urban Waterfront Revitalization: The Role of Recreation and Heritage." 18 January 1980, Jacksonville, FL. 1 February 1980, Portland, OR. 15 February 1980, Boston, MA. 14 March 1980, St. Louis, MO. Sponsored by the American Planning Association and the Heritage Conservation and Recreation Services.

F. A session on "Redevelopment of the Small Urban Waterfront: Problems and Challenges" at the APA National Planning Conference. 12-16 April 1980. San Francisco, CA.

G. Waterfront Festivals: Community Impact and Potential. 19-21 May 1980. Virginia Beach, VA. Sponsored by the Coastal Plains Center for Marine Development Services and the VIMS Sea Grant Marine Advisory Services Program of the College of William and Mary.

H. And, of course, the present conference: Coastal Zone, '80: The Second Symposium on Management, Conservation, and Utilization of the Coastal Zone. 17-20 November 1980. Hollywood, FL. Sponsored by the American Society of Civil Engineers and the Office of Coastal Zone Management.

4. Course

In 1980, New York University Graduate School of Public Administration offered a two-week, graduate-level summer workshop on Planning and Management of the Urban Waterfront. It examined the historical role of the urban waterfront; changing patterns of shoreline activity; economic and technological changes in port location and function; the transition from productive to consumptive uses; government involvement in coastal zone management: the role of federal, state, and local agencies; commercial and industrial redevelopment projects; recreational and cultural redevelopment projects; public access and neighborhood uses of the shoreline; and recent initiatives in waterfront revitalization throughout the country.

Problems in Urban Waterfront Revitalization

There are many problems which inhibit or delay the conceptualization, planning, and implementation of successful urban waterfront projects. Among the major needs identified by the efforts previously mentioned are:

1. Increasing public understanding of the opportunities for and environmental, social, and economic benefits of urban waterfront revitalization
2. Improved environmental planning and design through adoption of criteria or guidelines and development of an information clearinghouse related to:

A. Enhancing the special and unique qualities, the sense of place, of each waterfront

B. Providing the full range of water-dependent recreation activities

C. Considering the particular needs of low income people, minorities, the elderly, the handicapped, and the young in providing recreational opportunities and mixed uses

D. Optimizing public access (physical and visual) opportunities

E. Reducing access barriers caused by railroad and highway

3. Technical assistance, such as techniques for gaining the support of citizens and the business sector

4. Encouraging private sector support of public access goals

5. Urban waterfronts constitute a limited resource. There is fierce competition for space along waterfronts among industrial, commercial, residential, recreational, defense, transportation, and other land uses. Some are incompatible and result in conflicts---such as heavy industry, petrochemical or energy storage facilities or marine transportation terminals in the same neighborhood as residences. Others are overlapping and can coexist in mixed use arrangements. It is an arduous task to get all participants involved to agree on what the waterfront should lool like. Because water is common property, the disposition of shoreland is often considered to be a public interest issue. But the public interest often conflicts with private property rights, and the expectations of a developer for an appropriate economic return.

VISUAL PREFERENCE

Use of visual preference offers considerable promise as a strategy for dealing with these problems, especially in managing visual resources of urban waterfront environments and in facilitating public participation in urban waterfront revitalization efforts. While much attention has been devoted to technological and engineering strategies to deal with problems related to urban waterfront revitalization in recent years, relatively little attention has been given to visual resource management strategies. Visual preference, a particular type of visual resource management assessment system, is based on a strong theoretical model.

A significant part of the increased environmental awareness in the past decade has been manifested as a greater appreciation of aesthetics, the visual landscape, and scenic resources. So important is the issue of aesthetics, in fact, that it has begun to be incorporated into legislation and to play a role in various other public contexts. Environmental designers, planners, managers, and decision-makers responsible for implementing federal and state laws and carrying out decisions of the courts dealing with aesthetics are often at a loss as to where to begin. However, in recent years a whole field of endeavor called visual resource management has arisen to assist them by providing technology to address challenges in landscape planning. The assessment systems and technology employed in this field are derived from a variety of academic and professional disciplines (e.g., psychology, architecture, landscape architecture, geography, economics, engineering, natural sciences, natural resource management, and interdisciplinary approaches), and make use of descriptive, computerized and quantitative, and psychometric and social science approaches.

There are two major types of systems employed in visual resource management:

1. Expert of professional judgment: systems using professional planning and design norms to evaluate landscape qualities. In these approaches, variables are selected arbitrarily, and therefore the scoring values are arbitrary.

2. Public preference or empirical: systems evaluating peoples' preferences toward landscape characteristics.

For either type, numbers are assigned to each factor or variable, and then scores are determined. Used to elicit responses are: actual on-site visitation, black and white photographs, color slides, videotapes, line drawings, art pictures, and tachistoscopic projections.

A specific type of empirical assessment system is the visual environmental preference assessment, or visual preference, which asks participants to rate how much they like what black and white photographs show, using a Likert scale. The information-processing approach of S. Kaplan and R. Kaplan (1978) serves as the theoretical model on which this methodology is based.

Information-Processing Approach

The information-processing approach considers the environment in terms of information qualities that make human functioning possible. It provides a basic understanding of what people are like, what their needs are, and how they know and experience the environment:

1. Humans are concerned with perceiving and knowing.

2. Humans care about making sense and being involved.

Humans are profoundly influenced by the conditions of their evolutionary history. In the uncertain, dangerous, vast, and complex environments in which humans evolved, both the ability to build mental models which provided simplified yet workable conceptions of the environment and the desire for clarity through making sense and being involved were essential for survival. Cognitive maps of the environment---connected mental representations of recurring patterns humans experienced which were clear and definite even when all the available information was partial and incomplete---permitted anticipation, prediction, evaluation, and planning, all of which were necessary to deal with large amounts of information quickly. Uncertainty is still with us, and these abilities are still essential for human functioning.

3. Humans prefer environments where they can use their cognitive maps and extend them, and build new maps.

Environments which have fascination value---for which attention comes easily (involuntarily), thus leading to cognitive clarity---due to their contents or processes or a combination of the two are likely to be preferred. Contents relate to coping with the environment and include: wild animals, snakes, wolves, bears, young animals, fellow humans, green things, water, fires, caves, weather, shelters, tools, food, etc. Processes relate to coping with uncertainty, through making sense (by hanging together and being comprehendible) and involving (through richness or variety and mystery).

4. Deviations from preferred environments produce stress. Such stress can cause people to become unpleasant, irritable, aggressive, and destructive; and to withdraw, become uninvolved, and drop out, resulting in a break-down in community.

When humans find themselves in non-preferred environments, they seek to manage such stress by developing coping strategies which either modify conditions or adapt to them. Strategies to keep the information generated by the environment manageable include: (a) choosing to gravitate toward preferred environments, (b) managing such environments so as to enhance their preference values, (c) imposing patterns on environment and experience through shared cultural interpretations, and (d) participating, or taking an active role in what happens in the environment.

The basic understanding of human nature and human requirements provided by this theoretical approach can be used by a wide variety of persons working to solve problems at the human/environment interface or seeking to create more humane environments to live, work, and play in. It helps them to recognize that humans:

1. can be difficult and potentially very dangerous animals
2. are dynamic organisms, actively striving to cope with their environment
3. are highly capable of problem-solving, cooperation, and creativity when effectively informed, and
4. want to be involved, to have a piece of the action, to influence the course of events; they do not want to have things done to them or for them, though.

Such recognitions, in turn, can be used by these persons in determining where and how they can intervene to be most effective.

Application of Visual Preference to Problems in Urban Waterfront Revitalization

In spite of the numerous problems in urban waterfront revitalization mentioned previously, many cities have succeeded in bringing their visions of revitalized waterfronts into reality. While the specific approaches have varied, successfully completed urban waterfront projects described in popular magazines, professional journals, studies and reports, at conferences and symposia, and in UWAG meetings were typically characterized by an ability to build a consensus of community interests. Where such consensus has been achieved among residents, it has typically resulted from a long process of dialogue, which has helped foster community identity:

> Underlying successful...programs is a community's shared vision of a waterfront resource tied into the economic and cultural life of the city . Achieving community consensus on this goal requires dialogue among many separate groups and interests, which must work together to inspire broad community support (Clark et al, 1979).

Visual preference offers a neat tool for getting at peoples' perceptions and preferences related to urban waterfront projects, a necessary prelude to building consensus and mustering support for such projects. Asking people to share their visual images encourages them to participate and to provide feedback to planners and agency members sponsoring redevelopment efforts. Visual preference assessments can provide useful information to such people for all aspects of the planning and decision-making process.

Urban waterfront revitalization may be a valid enterprise for all cities on rivers, lakes, bays and oceans which have rediscovered their long-forgotten waterfronts. But the form the revitalization efforts take should be different---as different as each city's history. The particular combination of tools and techniques used should reflect the needs, interests, resources, and constraints of each city----and visual preference serves as a promising strategy for determining what these are.

Finally, perhaps the most important use of visual preference in urban waterfront revitalization is as an aid for identifying those elements which lend to the sense of place of the waterfront and the city. Revitalization projects should retain and enhance the identity, distinctiveness, uniqueness, and sense of place of the waterfront and the whole city rather than ape successful efforts in other cities. This means retaining the historic character without becoming excessively cute, overpreserved, too prettied up, or too controlled. There is a danger of efforts being copied without regard for the specific site; witness the many urban waterfronts with exposed brick restaruants with names like The Purple Endive or Steake and Steinn that serve quiche and crepes on butcher-block tables; chic boutiques; gourmet kitchen supply shops with names like The Wonderous Whisk; shops with incense or bay-scented candles and hand-carved wooden signs selling asparagus fern and spider plants in pottery containers and macrame plant hangings, hardwood toys, Marimekko fabric, handmake silver jewelry, and T-shirts and funny hats with names embroidered on them and other kitsch. Use of visual preference can overcome this problem of homogenization by identifying peoples' perceptions of the unique qualities of each city's waterfront.

REFERENCES

American Society of Planning Officials (1952) Municipal Waterfronts: Planning for Industrial and Commercial Uses. Planning Advisory Service Information Report 45. Chicago, IL.

------------------------------------- (1959) Waterfronts: Planning for Resort and Residential Uses. Planning Advisory Service Information Report 118. Chicago, IL.

Architectural Forum (1960) "Waterfronts: Bright, breathing edges of a city's life." Gallery. 112(3): 140-145.

Balchen, B. and J Linville (1971) "The waterfront: Let's face it." AIA Journal (April): 17-28.

Barr, C.W. (1976) Waterfront Development: A Partial Bibliography. Exchange Bibliography 1001. Monticello, IL: Council of Planning Librarians.

Beauty for America: Proceedings of the White House Conference on Natural Beauty. (1965) Chapter 7: Water and Waterfronts, pp. 141-145. Washington, D.C.

Clark, J., C. Wilson, and G.L. Binder (1979) Small Seaports: Revitalization through Conserving Heritage Resources. Washington, D.C.: The Conservation Foundation.

Commission on Marine Science, Engineering, and Resources; Subcommittee on Oceanography; Committee on Merchant Marine and Fisheries; House of Representatives, Ninety-First Congress (1969) Our Nation and the Sea. Washington, D.C.: U.S. Government Printing Office.

Committee on the Future; Regional and Economic Development Task Force; The Port Authority of New York and New Jersey (1979) Waterfront Redevelopment Strategy: Phased Redevelopment of the Inner Harbor Waterfront. New York, NY: The Port Authority of New York and New Jersey.

Committee on Urban Waterfront Lands; Environmental Studies Board; Commission on Natural Resources; National Research Council (1980) Urban Waterfront Lands. Washington, D.C.: National Academy of Sciences.

Cowey, A.B. and R. Rigby (1979) "On the waterfront." Planning 45(11):10-13

Davenport, R. (1978) "Urban coastal development: The Chicago experience," pp. 81-87 in Coastal Zone, '78. New York, NY: American Society of Civil Engineers.

------------- (1980) "The use of waterfronts for public and private recreation," in Committee on Urban Waterfront Lands, Urban Waterfront Lands. Washington, D.C.: National Academy of Sciences.

Design and Environment (1974) "Urban edges." 5(4): entire issue.

Doyle, P.J. (1979) "People on the waterfronts of Europe: A preview for Cleveland Lakefront State Park." The Cleveland Foundation Quarterly. (Fall) Cleveland, OH.

Environmental Comment (1978) "Urban Waterfronts." (June):entire issue. Washington, D.C.: Urban Land Institute.

Forward, C.N. (1969) "A comparison of waterfront land use in four Canadian ports: St. John's, Saint John, Halifax, and Victoria." Economic Geography 45(2): 155-169.

Gratz, R.B. (1978) "Hope on the waterfront." New York. October 2, 1978, pp. 58-69.

Hanna, J.W., S.F. Philipp, and G.W. Mullins (1978) "Decision-making in the Development of urban waterfronts for park/business use." Water Resources Bulletin 14(1): 179-188.

Harney, A.L. (ed.) (1979) Reviving the Urban Waterfront. Washington, D.C.: Parnters for Livable Places, National Endowment for the Arts, and Office of Coastal Zone Management.

Kaplan, S. and R. Kaplan (eds.) (1978) Humanscape: Environments for People. North Scituate, MA: Duxbury.

Kenyon, J.B. (1968) "Land use admixture in the built-up urban waterfront: Extent and implications." Economic Geography 44(2): 152-177.

Landscape Architecture (1974) "U.S. runs last along the waterfronts." 64(2): 73-75.

Lehmann, R.A. (1966) "The principles of waterfront renewal: A summary of experiences in fifty American cities." Landscape Architecture 56(4): 286-291.

McCann, T (1979) "Urban Waterscapes," Trends. 16(2):

McQuade, W. (1961) "The suffering shoreline: Waterside bilight is blocking visual escape from the city." Architectural Forum 114(6): 90-97.

Mann, R. (1973) Rivers in the City. New York, NY: Praeger.

Mann, R. and C.E. Alling (1978) "Historic resources and coastal zone management," pp. 57-65 in Coastal Zone, '78. New York, NY: American Society of Civil Engineers.

Manogue, H. (1980) "Citizen groups: New and powerful participants in urban waterfront revitalization," in Committee on Urban Waterfront Lands, Urban Waterfront Lands. Washington, D.C.: National Academy of Sciences.

Marcus, H., N. Taneja, and P. Roberts (1980) "The impact of changes in transportation technology on the use of land in harbor areas," in Committee on Urban Waterfront Lands, Urban Waterfront Lands. Washington, D.C.: National Academy of Sciences.

Morell, D. (1976) "Managing the urban coastal zone: The political realities." Environmental Comment (November): 8-10. Washington, D.C.: Urban Land Institute.

Morell, D.L. and G.L. Singer (1978) "The urban coastal zone: Challenge of redevelopment," pp. 66-80 in Coastal Zone, '78. New York, NY: American Society of Civil Engineers.

The President's Council on Recreation and Natural Beauty (1968) From Sea to Shining Sea: A Report on the American Environment, Our Natural Heritage. Washington, D.C.: U.S. Government Printing Office.

Progressive Architecture (1975) "Urban waterfronts." (June): entire issue.

Rick, W.B. (1964) Planning and Developing Waterfront Property. Urban Land Institute Technical Bulletin 49. Washington, D.C.: Urban Land Institute.

Simpson, A (1976) Waterfront Rrecedents. Toronto, ON: City of Toronto Planning Board, Central Waterfront Planning Committee.

Stephens, S. (1978) "Taking to the waterfront." Horizon (October): 30-37.

U.S. Department of the Interior (1977) A Preliminary Urban Parks Study. Washington, D.C.: U.S. Government Printing Office.

------------------------------ (1978) National Urban Recreation Study. Washington, D.C.: U.S. Government Printing Office.

------------------------------ (1979) Urban Waterfront Revitalization: The Role of Recreation and Heritage. Volume I: Key factors, needs, and goals. Volume II: Eighteen case studies. Washington, D.C.: U.S. Government Printing Office.

Waterfront Memo (1979) "Conclusions and recommendations regarding urban waterfront issues and problems from the National Workshop on Urban Waterfront Redevelopment." 1(2): entire issue. Ann Arbor, MI.: Coastal Zone Laboratory, University of Michigan.

Watson, K.E. (1978) "A public access plan for San Francisco Bay," pp. 175-194 in Coastal Zone, '78. New York, NY: American Society of Civil Engineers.

Whitman, I.L., R.M. David, and S.E. Goldstone (1964) "Renewal of watefront areas." Journal of Housing (June): 236-255.

Wisconsin Department of Resource Development (1966) Waterfront Renewal. Madison, WI.

Wood, D.F. (1965) "Renewing urban waterfronts." Land Economics 41(2): 141-149.

KEEP THE MOSQUITO FLEET IN THE CITY

by

Joanne Peterson*

INTRODUCTION

A continuing demand for use of our nation's coastal area and navigable waterways exists in the United States. When this country was first explored, access was primarily by water, with most of the early colonists settling adjacent to a natural water route or not far from one. Although migration inland has occurred through the centuries, the majority of our country's present population resides within 50 miles of the coast, with particular concentration in urban areas (Ketchum, 1972:10).

Consequently, the U. S. waterways show a marked competition for space. Competition exists among commercial and shipping interests, industrial plants, recreation and tourism, and environment conservation. "Economic growth in the coastal areas has proceeded so rapidly that over 40 per cent of all manufacturing plants in the U. S. are located within the borders of coastal counties" (Duscik, 1974:54). Within the last few decades, power plants, including nuclear power plants, have joined in the competition for space in this coastal resource.

A great percentage of man's food source also comes from the coastal zone. Nearly 5 billion pounds of fish were harvested in 1971 by U. S. fishermen most of it coming from coastal waters.

A study on the changing patterns of activity on the urban waterways, with particular emphasis on the port area

---

*Assistant Professor, Univ. of N. H. at Plymouth State College. Presently on leave of absence as a National Sea Grant Intern working with the House of Representatives - Committee on Merchant Marine and Fisheries.

Any views expressed in this paper are solely the views of the author and do not reflect the views of the Committee on Merchant Marine and Fisheries.

of the city, has shown that a more diversified use can be made of urban port facilities due to the "developments in marine transportation technology and in the economic structure of the central city" (Moss, 1976:223).

The potential for conflict between industrial and non-industrial use of an urban waterway had been a major topic of controversy in Galveston, Texas from 1974 to 1976. This case concerns the shipping industry of the Galveston Port Authority and the briny reapers of the sea, who harvest a large portion of the City's shrimp and fish from Galveston Bay and the Gulf of Mexico.

Pier 19, an area of approximately six and a half acres on the Galveston ship channel has been the home of the City's "Mosquito Fleet,"[1] shrimp boats, for over 130 years. The site is occupied by forty-eight shrimp boat operators, two fish house owners, and three party boat operators with a total of eight party boats (Brown, 1975:3).

The Galveston Wharves shipping operation surrounds the above site on three sides. The Wharves operates 37 piers that import or export commoditiies such as bananas cotton, rice and veneer plywood. It also owns 26 warehouses which occupy 3,500,000 square feet of space. Although the Wharves had seen sharp increases in tonnage from 1972 to 1974, they were operating at close to capacity(De Podwin, 1975:viii).

Jurisdiction over the piers on the ship channel has been given to the Galveston Wharves Board by power of the City Charter. In March, 1976, with the expansion of the shipping industry in mind, the Wharves' Board sent notification of eviction to the tenants of Pier 19 in anticipation of building an open cargo dock at that site. Backed by the local historical society, tourism advocates, and over 10,000 local citizens, the tenants of Pier 19 refused to move (Staff, Dec. 3, 1974).

At the onset, what appeared to be a simple landlord-tenant problem grew to be a matter of what economic and aesthetic directions the City should follow. The question

---

[1]"Mosquito Fleet" - The name is derived from the appearance of a shrimp trawler when it is coming straight on. The trawler's nets resemble the wings of a mosquito while the remainder of the vessel resembles the body of the mosquito.

concerned the source whereby revenue should come into the City, the kind of appearance the City should have, and the job opportunities that could be offered.

Growth of the shipping industry and the effect that this six and a half acre site would have on Galveston's competition as an urban port eventually clashed with the possibility of further development of tourist trade, retention of small-time salty entrepreneurs (a possible dying breed in a city setting), and public access to the waterfront as well as to a historical site (Fugger, Nov. 19, 1976).

Big business versus small did not remain the entire issue. Public interest[2] began to play an important role, leading to the question of the ability of the public to govern the use of cityowned land. Lawsuits, public hearings, televised debates, heated arguments, and strong opinions supporting both sides led to an agreement by negotiators of the involved parties to determine the use of this city-owned pier by public referendum.

RELATED PERSPECTIVES

The constituency of Galveston began to show an interest in the development of the case only by the middle of 1974, but by the close of 1975 they were strongly expressing their opinion. This was inspired by the public's education concerning the elements of the case. The main concern was the fact that the issue was over municipally-owned public property. All parties to the dispute were now agreeing that this property should be governed for the welfare of the citizens of Galveston and it should represent the public interest. As legal action on behalf of the Pier 19 tenants indicates, citizens were beginning to feel that the majority's best interest was not being considered in Galveston Wharves' management of that section of the waterfront. Slightly

---

[2]Held (1970), in a thorough discussion of public interest, points out that no attempts have brought agreement on how "the public interest" should be understood, yet the concept is indispensable. One theory defines it as "those individual interests which all members of a community have in common"(Held, 1970:99). For this paper an operational definition for the term "public interest" has been chosen from Black's Law Dictionary (1968) as: "Something in which the public, the community at large, had some pecuniary interest, or some interest by which their legal rights or liabilities are affected."

over 10,000 Galvestonians, urging the retention of the tenants of Pier 19, signed a petition which was brought before the City Council in December, 1975 (Fugger, Nov. 19, 1976).

Another major thrust given to the case was the historical factor. Pier 19 was designated for a Texas State Historical Marker in February, 1974 (Kirkpatrick, April 22, 1976). It was nominated to the National Register in August, 1974 and was placed on the National Register of Historic Places on April 21, 1975 (Fugger, Oct. 7, 1976).

Galveston's background is heavily dotted with historical facts that were proudly being presented by the media and displayed by local businesses. A Galveston news clipping pointed this out. "In recent years a new enthusiasm has swept, like a wave from the surf, across Galveston Island. Many historic buildings have been reclaimed" (Staff, Dec. 8, 1975). Having the claim of being the oldest City on the Gulf Coast, Galveston is unique among Texas cities. In October of 1975 the Port of Galveston enjoyed its 159th year and boasted its title as the oldest commercial enterprise in Texas (Nesbit, 1976:42). To add to these historic factors, the Strand,[3] which has been called the "Wall Street of the South," is on the National Historic Register and is in the process of restoration. A significant factor in the Pier 19 controversy was the anticipation of linking the waterfront to the Strand. Pier 19 is within walking distance of the Strand and the downtown area of Galveston. The only accessible public route to the waterfront is by a 20th Street entrance to Pier 19. The Lipton Tea Company is located at the end of 19th Street blocking direct access from that street.

Another consideration was that tenants of Pier 19, the shrimpers, fishermen and fish house owners, are at a strategic location for tourism. Development of a Fisherman's Wharf, similar to the famous wharf area in San Francisco, with gift shops, restaurants, and marine-asspciated interests, would afford a desirable tourist attraction and could be linked to the Strand. Because of the public's apparent

---

[3]The Strand is a street adjacent to the wharves that is registered by the state and federal Historical Societies as a Naitonal Historic Place. It is under renovation and construction to regain the aesthetic quality of the once popular "Wall Street of the South." It was named this because of the tremendous commercial influence that it had on the City and all of Texas at one time.

concern for the historic significance of Galveston Island and its desire for tourist trade, these factors seem to play an important role in the development of the case.

BACKGROUND

The waterfront port of Galveston Island that faces the Galveston Ship Channel is public land owned by the City of Galveston. The City Charter of 1940 designated the waterfront area to be controlled, managed, maintained, and operated by the Board of Trustees of the Galveston Wharves. This Board consists of seven members; six members are appointed by the City Council and one member is elected to the Board from the members of the Council to serve as the ex-officio representative of the City's Council. Their jurisdiction extends to 37 piers on the ship channel.

Having such power of authority, the Galveston Wharves investigated the possibility of expanding and modernizing the port area to include Pier 19 through Pier 22 as early as 1960. This modernization would necessitate the relocation of the Mosquito Fleet to provide for the construction of three new ship berths on the Pier 19 to 22 site.

This endeavor was delayed, however, because the Wharves experienced two extremely poor years financially. And the small boat basin, to accomodate the Mosquito Fleet, would not provide substantial amounts of direct revenue to warrant its reconstruction elsewhere. Previously, in trying to determine the best location for a new small craft basin, it was discovered that most of the tenants preferred to stay at the Pier 19 location. In addition, a study conducted by Arthur D. Little suggested that the "Pier 19 area offers the best possibilities for economic justification." That report went on to say that although the income from expansion of the facilities of Pier 19 would probably not justify the enlargement, additional income could be obtained from restaurants, gift shops and the like that would stimulate tourist trade (Staff, 1974:I4).

By June of 1964 the Wharves Board announced the Galveston Fisherman's Wharf concept. Enthusiasm among the Pier 19 tenants for the Wharf concept was poor at this time, however, with only the tenants of Pier 22 accepting a long-term lease. Consequently, nothing developed from this Fisherman's Wharf idea and by 1966 plans came to a standstill. Even though the project ended, some necessary repairs were made on the pier in 1965 and through 1966. Though more repairs at greater costs were made to areas of Pier 19 other than the immediate area used by the small boat fishermen.

Further repairs to the outer side of Pier 19 occurred in 1972 to 1974 to accomodate the increase in banana imports. These repairs again were not in the area of the small fishing boats. The congestion caused by this particular rehabilitation was stated as the reason for closing one-half of 20th Street to vehicular traffic. Twentieth Street was the only public access to Pier 19.

Modernization and expansion were again a major concern of the Galveston Wharves by late 1973. Containerized cargoes were a new method that was becoming very popular in port shipping. The Wharves feared losing their competitiveness as a shipping port if they did not make adequate improvements to accomodate more container cargoes. Early in 1974 attention was again brought to Pier 19 as an area that could be developed to meet the new expansion requirements.

The focus of this study is on the events that occurred from this point on.

FINDINGS

Intent of Eviction

The Mosquito Fleet, has been a tradition on Galveston Island for over 120 years and has been at its present location of Pier 19 for over 70 years (Weinert, July 1, 1974:1&7).

The unanimous decision by the Wharves' Board of Trustees on March 18, 1974 to expand the port's facilities and use the Pier 19 site to accomodate a new general cargo berth could have put an end to the long tradition of that pier's use. And the end of a rugged nautical setting within the city.

The Wharves' Board's decision was concerned with developing new cargo business and revenue. This economic direction was brought out in a report compiled by the Staff of the Galveston Wharves (March 15, 1974) and adopted by the Trustees on the same day of the decision to use the Pier 19 site for shipping facilities. The report estimated a cost of $3.5 to $4 million for the construction of the planned general cargo berth. The Wharves' Board anticipated close to $1.5 million a year, for the next three to four years, in revenue from this new cargo berth, and the possibility of $400,000 a year income for longshoremen and waterfront laborers. The report further estimated a cost of $1.1 million to renovate and repair the area if the slips were kept in the service of the small boats and fish houses. The existing income form the tenants, by way of rents, was only $19,980.60.

It would take eleven times the amount of the rental

revenue with the sale of general revenue bonds to cover the cost of rehabilitation to the site. Although the report mentioned substantial benefits of the new cargo dock, it failed to mention the possible loss of jobs by 100 families should the tenants of Pier 19 be evicted. This loss would approximate an annual take-home pay in excess of $1 million for the owners and crews (Bartholomew, May, 1974:10). This revenue is important to the economy as it supplies approximately 45 boat operator-citizens with their employment, not to mention the crew income (Brown, Nov. 26, 1975:3).

The report was presented to the City Council and then at a public hearing where interested parties could present their point of view.

The public hearing was held on May, 23, 1974, with approximately 300 Galvestonians in attendance. Those speaking in behalf of retaining the Mosquito Fleet at their historic pier out-numbered those in opposition seven to one. The only speakers favoring the eviction of the Mosquito Fleet were representatives from the longshoremen's union. They claimed that the job opportunities declined because of containerization, so more space was needed to provide more jobs.

Shrimpers expressed their sentiments about the loss of jobs also. They feared a possible permanent loss of jobs should they be moved from their present location, especially with no designated place to go as a fleet. The Wharves, however, would not stop shipping expansion if they did not gain the 6.5 acres at Pier 19.

Another viewpoint favoring the retention of the Mosquito Fleet at Pier 19 was brought out by the president of the Galveston Historical Foundation. He stated that the local color and uniqueness of Pier 19 could prove a long-term asset to Galveston's tourist industry. The possibility of a "Fisherman's Wharf" in the Pier 19 area could enhance the historical Strand district and could help to create a more desirable tourist business. This kind of complex would provide a long-term economic diversity for Galveston as opposed to exclusive Port cargo development (Cook, 1977:4).

The public hearing brought out another interest that concerned many Galvestonians during a time of economic expansion. This was the "quality of life" concern, meaning that there are certain intangibles that are irreplaceable and cannot be assigned a dollar value. It was said that Pier 19 reflected an "eccentric bit of the city's maritime history...and a vanishing breed of small-time entrepreneurs" (Cook, 1977:4).

Inspired by this public hearing, the City Council appointed a "Pier 19 Study Committee" made up of sixteen citizens of Galveston which included two members of the Wharves' Board of Trustees. The citizens who participated in the public hearing had high hopes that the City Council and the Wharves would place a strong emphasis on public input when making the final decision concerning Pier 19. The Mayor himself felt that the resulting recommendations of the Study Committee would be followed, even though they were not binding.

Nearly three months of investigation and deliberation by the Pier 19 Study Committee arrived at a unanimous recommendation. The report was presented to the Mayor and the Members of the City Council late in August, 1974, for their consideration. They recommended that the Mosquito Fleet, party boats, and fish processing houses stay at the Pier 19-20 site[4] until an investigation could be made into the construction of a Fisherman's Wharf complex at Pier 9. The Pier 9 location already docked some shrimp boats and could be expanded to accomodate all the tenants of Pier 19.

The Study Committee also considered the need of the Strand having reasonable access to the Wharves area. So it further recommended that:

> ...the city of Galveston arrange for the development of an attractive and adequate observation area...in the Pier 20-25 area which will provide a good view of the waterfront and shipping operations and will be readily accessible to the public (Ware, Aug. 29, 1974).

Even though this recommendation was not followed, it might be well to mention that the direction of the recommendation seemed to be strongly influenced by Wharves' officials.[5]

The year following the recommendation of the Pier 19 Study Committee saw an attempt towards negotiations to

---

[4] Although spoken of as being at Pier 19, the fish houses actually occupied Pier 20.

[5] The sentiment of this statement is discussed further in Peterson, Aug, 19, 1978:24.

acquire the properties east of Pier 9, which would provide enough space for all entities of Pier 19. The purchase of this property, however, was rejected by the owner.

The Wharves' Board continued deliberation of the appropriate site for the location of the small boat basin. One such decision was made in a closed session by the Executive Board in July, 1975. Closed executive sessions were in opposition to the Texas Open Meeting Law, as was brought out during the court hearings mentioned later in this text.

The site decided in this closed meeting was not favorable to the shrimpers for many reasons. The assistant general manager of the Wharves took this site rejection by the shrimpers as a refusal to pay the projected higher rent. In reaction, the Wharves proposed a site on Pelican Island[6] that would be cheaper since it wasowned by the Wharves and already dredged. The proposed site, which was six miles away by road, included 15 water acres and 35 land acres. The facility had the potential to supply electricity and water and to accomodate the fish houses and other shrimp-related businesses, according to the Wharves' Board. But in reality, the Wharves' Board had no intention of developing any facilities other than the boat slips.

The shrimpers learned that the Pelican Island facility would not have fish houses, fuel docks or space for 72 berths. They also objected to the proposed lease "requiring them (shrimpers) to assume all maintenance, all rebuilding in event of natural disaster, and the cancellation of all leases the moment there was a 25 per cent vacancy at the new pier" (Cook, 1977:5).

The shrimpers felt that they were being forced into a

---

[6]Pelican Island, part of the City of Galveston, is an island on the north side of the ship channel directly across from the municipally-owned piers. It is approximately six square miles in area. The question of the Galveston Wharves using their land on this island for the anticipated expansion had been brought out by the Wharves' opposition many times throughout the case. Although much development was needed on the land to make it immediately operable, two months from the time of this hearing the Wharves' Board anticipated receiving a long range plan by Horace P. DePodwin Assoc. Inc. which would include extensive development of Pelican Island.

decision that did not represent their best interest. It appeared that they needed to revise their position and reorganize themselves into a group with more determination towards their ultimate goal...to stay where they were.

Having pulled themselves together and become cognizant of the new lease stipulations, some of the shrimpers and their representatives attended a November 14, 1975 Wharves' Board meeting. The meeting was held to vote on awarding the contract for consideration of the Pelican Island small boat basin. Among the representatives for the shrimpers' interest were the wife of one of the fishermen, the Executive Director of the Galveston Historical Foundation, Chairperson of the Galveston County Historical Commission, Secretary of the Texas Bay and Gulf Fishermen's Association, and a local citizen who had a personal concern in the historical aspect of the case.

It was made quite clear to the Wharves' Board during this meeting that the Mosquito Fleet would not sign the lease for the Pelican Island berths and would not move from Pier 19. It was also pointed out that the recommenation of the Study Committee had not been met.

At this meeting each representative of the shrimp fishermen's interest made a request in the order of not moving the Fleet from their present site. The Executive Director of the Galveston Historical Foundation went on to assure the Board members that his associates, as well as others interested in preserving Pier 19, did not want a confrontation with the Galveston Wharves. He asked them to postpone the awarding of the construction contract. But should they continue, he warned that the Historical Foundation Board would "use all possible means to prevent the move of the shrimp vessels from Pier 19" (Post Galveston Bureau, Nov. 15, 1976).

Notwithstanding the requests, the Wharves' Board voted five to one to award the contract for construction at a cost of $170,500. The necessary permit from the U. S. Corps of Engineers to construct the new small boat basin was awarded on that same day.

Consequently by January, 1976, construction began on Pelican Island and the court action got under way.

Before the lawsuits were filed an attempt was made by the citizens of Galveston to convince the City Council of the public's desires for the use of this city-owned pier. The request to keep the Mosquito Fleet at Pier 19 was in the form of a petition signed by over 10,000 citizens. The petition was submitted to the City Council in December, 1975, in hopes of exercising the power of direct legislation by the citizens of the City through initiative. This right to petition the government is provided for in the United States Constitution as well as the Galveston City Charter. The petition was used as a plea to the City Council to hold a referendum for the citizens to determine the use of the municipally-owned site.

In what manner the petition influenced the City Council and Mayor is not clear. The outcome was a four to three vote by the City Council, ruling that a non-binding referendum would not be held. The Mayor said "he thought the 10,000 signatures stood for themselves," but nonetheless, the construction on Pelican Island began (Cook, 1977:5).

There would appear to be a breakdown of communications here, or possibly the public officials' disinterest in taking a firm stand one way or another. Even though the Mayor thought the petition stood for itself, which would lead people to believe that Pier 19 should remain as it was, construction on Pelican Island began within a few days. Why the construction on Pelican Island and the relocation pursuit continued, even though the constituency of the City and the tenants of Pier 19 showed a pronounced opposition, had not been explained.

It is apparent that there are political strategies involved in the issue that are beyond the scope of this paper. Conflict of interest had been suggested of the members of the Wharves' Board by some critics. Although this topic was not pursued legally, many questionable relationships occurred.[7]

Court Action

The commencement of the Pelican Island construction prompted five Pier 19 tenants to pursue legal measures.

On January 16, 1976 the attorney for the Pier 19 interests, filed for a temporary injuction against the Galveston Wharves' Board of Trustees and the City of Galveston. The reason for the injuction was to halt the small boat basin construction on Pelican Island and to prevent the moving of the Mosquitio Fleet until a vote of the citizens of Galveston could be held.

In their defense, the Wharves filed a petition to the injunction suit to the District Court. Among other points,

---

[7]For further explanation refer to Peterson, Aug. 19, 1978:32.

this petition alleged that the Wharves was not trying to drive the fishermen from the City or put them out of business. The decision to move them was part of the discretionary power granted by the City Charter. They would also assist in relocating them on an economically feasible basis.

On March 3, 1976, after all interested parties had filed their petition, the case for the temporary injunction was heard in court. The suit was dismissed after a motion for such was requested by the attorneys for the Wharves and the City.

The ruling was appealed by the plaintiffs, who felt that the Judge's decision was made due to the landlord-tenant relationship, whereas, the motivation for the injunction was concerned with matters of a far more complex nature.

In the meantime, the tenants of Pier 19 had received letters of eviction. This action, along with the Wharves' application for permit with the Corps of Engineers to fill in the Pier 19 area, resulted in another attempt at a temporary order restraining the Galveston Wharves from evicting the fishermen and fish house owners and from filling in the pier area.

The second suit was filed on behalf of four citizens including a City Councilwoman, who said that she wanted to ensure that the "public's right to be informed and to participate in governmental decisions is protected" (Chronicle Galveston Bureau, April 7, 1976). The suit was filed on the grounds that the City Charter violates the state constitution which forbids such broad delegation of powers as the Wharves' Board exercises; that the Wharves violated the State Open Meeting Law by deciding, in a meeting closed to the public, to move the tenants of Pier 19 and in another closed meeting deciding to request a permit from the Corps of Engineers to fill in the pier area; and that the Wharves' Board violated a state statuted that forbids the destruction of a historical site without a public hearing and agreement of the citizens (Chronicle Galveston Bureau, April 7, 1976). The attorney of the plaintiffs pointed out that "irreparable harm would be caused to the Pier 19 tenants and the historical nature of the area if the restraining order was not granted" (Kirkpatrick, April 7, 1976). The temporary restraining order was granted on April 6, 1976, until the court hearing date on the injunction seven days later.

Testimony on the actual injunction lawsuit began on April 14, 1976. A second petition signed by 10,000 citizens was accepted as evidence since the basis of the entire suit was the question of constitutionality of the City delegating powers to the Wharves' Board.

As the court proceedings continued and evidence was introduced, the Judge extended the restraining order against the Wharves. This, in effect, made it impossible for the Wharves to issue revenue bonds for the construction of facilities. It stopped the rental of berths at the new small boat basin on Pelican Island, which meant they could gain no revenue from them at that time. The cost of stalled projects continued to escalate and there was a loss of staff time. There was also mention of the refusal of other shipping concerns to do business with a port that was involved in litigation.

Anticipating the aboveproblems caused by a lawsuit, the Wharves had withdrawn both the notice of eviction and the request to the Corps of Engineers for filling the area. A plea was made to the court to dismiss the restraining order and discontinue the hearing. The Judge refused the request upon hearing the evidence from the plaintiffs that showed that the Wharves did not plan to give up their efforts of eviction (Kirkpatrick, April 20, 1976).

Seventeen days of testimony, spanning six weeks in court, resulted in the granting of a temporary injunction halting all City or Galveston Wharves activity with the Pier 19 area for an indefinite time. The presiding Judge had hoped that an out-of-court agreement could be reached by the litigants, but an initial attempt at this failed. So, on May 19, 1976, the Judge commended the lawyers on both sides for thoroughly presenting the facts and briefing the law. He summarized the court hearing with the following statement:

> The primary legal question before the court is whether or not, under the constitution and laws of Texas, the Galveston City Council can legally delegate to the Board of Trustees of the Galveston Wharves complete authority and control over the management and operation of the system and the expenditure and application of revenues of the system...
>
> I realize that my decision will be only the first step in the judicial process, and the answers will not be known until the highest court renders its final decision...
>
> The weighing of the equities is the problem before the court. If a temporary injunction is not granted, there is no assurance that the Board of Trustees of Galveston Wharves will hold an open meeting and proceed to evict the tenants from the Pier 19 area and proceed with steps

> necessary to convert the Pier 19 area into an open dock.
>
> By the time the highest appellate court would have had an opportunity to pass upon the legal questions involved, a new open dock might already be in operation in the Pier 19 area.
>
> In order to maintain the status quo until the legal questions are finally decided, I will grant a temporary injunction in this case (Kirkpatrick, "Text..." May 20, 1976).

At this time he ruled that the City Charter giving full power over city-owned land to the Wharves' Board was "probably" unconstitutional, and that the Wharves had no legal authority to evict the tenants of Pier 19 or destroy or alter said historic site. He further ruled that the Galveston Wharves' Board of Trustees violated various provisions of the Texas Open Meeting Law.

The temporary injunction granted at this hearing was a restatement of the original temporary restraining order.[8] The attorney for the Wharves gave a notice of appeal to the ruling. Shortly thereafter the appeal decision of the initial injunction by the plaintiffs in January, 1976 was reversed. The Wharves also appealed that one. Now there were two lawsuits pending.

By this time both parties involved in the lawsuits were well aware that continued litigation would be costly and time-concuming. The private citizens who filed the injunctions were spending their own personal funds, which were limited. The Wharves was unable to sell revenue bonds for construction as long as its constitutional power to issue such bonds was in doubt. This, along with other factors, limited the Wharves' ability to function efficiently and earn much needed revenue. Under such circumstances a settlement between the parties seemed essential.

Settlement

The attorneys involved in the lawsuits spent months discussing the terms of a proposal that would put an end to the court action and bring about a settlement suitable to both parties.

Finally by September 28, 1976 the Wharves' Board of Trustees unanimously agreed on a vote by the citizens of Galveston to determine the use of Pier 19. The actual vote, when approved by the City Council, would be an amendment to

---

[8] Peterson, Aug. 19, 1978:37.

Article XII, Section 4 of the City Charter[9] (Galveston, March, 1977) dedicating Pier 19 to the exclusive use of the shrimp fleet, retail and commerical fish houses, party boats, fuel docks and public parking, with the quarantee that the public shall always have free and full access to the area. The amendment also would provide that the Wharves make such expenditures for maintenance and repair necessary to maintain all of the dedicated uses of the area and would further provide that rental for use of the facilities would not exceed other rents along the Gulf coast for similar facilities. (Staff, Dec. 31, 1976).

Should the amendment carry, the Wharves would also agree to open 17th Street or an adjacent equivalent street so as to provide access to the Pier 19 area from the east as well as from 20th Street on the west. Another provision by the Wharves was the erection of the Pier 19 Historical Marker at the northeast corner of 20th Street and Industrial Boulevard (this is the entrance to Pier 19).

Should the amendment be voted down, the Wharves would have the option to close 17th Street or whatever street may have been opened to the east of Pier 19 at any time. Also, the tenants in the area must vacate in six months from the date of certification of the election results (Kirkpatrick, Sept. 29, 1976).

Other stipulations of this proposed agreement included that the Pier 19 plaintiffs must dismiss both of the lawsuits against the Wharves: the one filed by the shrimpers and the one filed by the citizens. Also, the plaintiffs must retain the petition of the 10,000 Galvestonians in their custody and guarantee that it would not be filed with the City Secretary or otherwise be presented to the City of Galveston for action. In return the Wharves would sign a release of any and all claims for damages that they had planned to file against the plaintiffs in further litigation (Kirkpatrick, Sept. 29, 1976).

The proposed referendum was agreed on by the City Council and the special election was set for Saturday, June 15, 1977. Both parties in the issue proceeded with extensive campaigning for their position.

The Campaign

Although many other points were reiterated in the months of campaigning, the major emphasis seemed to be focused on the economic factor of one side providing more revenue to the City than the other side.

---

[9] This section of the Charter is quoted in Peterson, Aug. 19, 1978:11.

The economic concerns were battled over in many forms. One was the employment concern. If the amendment failed, economic losses would be suffered by at least 100 families of the Pier 19 area. Some shrimpers admitted that they would give up shrimping altogether.

The jobs that would be created by a new cargo dock were claimed to number 45. But as was mentioned by representatives of the longshoremen's union during the public hearing, containerization caused the number of jobs to decline so more space would only mean jobs for longshoremen displaced elsewhere. Overtime opportunities for 45 people was the most that could be expected from the proposed new facility.

Another economic concern was the reference to the almost $20,000 that the tenants of Pier 19 paid annually to the Wharves. The Wharves' Board had mentioned that they should not subsidize the fishing industry, but without all the facts it is not known how many other tenants are subsidized by tax dollars. The Wharves' Board refused to release the names and rents paid by other clients in the Port. These other Port industries are apparently exempt from city taxes, paying only a rental fee to the Wharves. The Wharves in return pays between $119,376 and $160,000 in revenue to the City of Galveston annually in lieu of property taxes. This amount is fixed and not contingent on the economic success or failure of the Wharves in any one year.

Backing the opposing side of the argument, economic views were emphasized by the ship's log from one of the fishing party boats docked at Pier 19. In 1976, 60,000 people from all parts of the world fished from that boat. The estimated income to Galveston generated from these people was close to $3 million (Boggs, Jan. 14, 1977). This revenue plus that generated from the party boats, shrimp boats, and fish houses was a proven income. Futhermore, if the area was turned into a Fisherman's Wharf and linked to the Strand, it would generate much more income to the City.

During debate over the economic factors, it was clear that little was known, with certain, what group contributed what amount to the economics of Galveston.

The months of campaigning brought a mixed group of supporters to the Pier 19 side. There were professors, doctors and nurses from the University of Texas Medical Branch in Galveston, the City and County Historical Foundations, historical buffs or preservationists, and concerned citizens. An amount of nearly $18,000 was raised to support Pier 19. In the same period of time only $9,000 was raised for the support of the Port. Behind the Port were various business groups and organizations in the City.

January 15, 1977 brought 9,568 of the City's 30,030 registered voters to the polls. The charter amendment passed 6,189 to 3,342 to retain Pier 19 for the exclusive use of shrimp and fish-related business. This decision was secure for at least two years, for a charter amendment cannot be reviewed within a two-year period (T.C., March, 1977:67).

SUMMARY

Political and Legal Consideration

The legal right to appeal by the Wharves' Board and the City of Galveston on the District Court decision extended to the questions of 1) constitutionality of the City Charter, 2) excessive delegation of powers to the Galveston Wharves' Board of Trustees, 3) violation of the Texas Open Meeting Law, and 4) destruction of a historical site without due process. As it happened, these charges were not determined because the lawsuits were withdrawn by the plaintiffs. The Wharves' Board of Trustees and the City of Galveston decided that further litigation was detrimental. The immediate eddect of the two lawsuits and the consequent injunction of the Wharves was in itself harming the efficiency of the shipping business. Further litigation would only magnify the situation.

The agreement for an out-of-court settlement was the desire of the Wharves' opponents from the onset of the issue.

Shrimpers and other citizens sat in on City Council meeting and Wharves' Board meetings to press the issue. The petition of 10,000 citizens was the result of the City Council's refusal to act on a previous petition of 3,000 names backing the retention of Pier 19. The voice of the people continued in an undying effort, only to become stronger as time went on.

But the Wharves' Board continued to press for the destruction of the Pier and the City Council continued to remain uninvolved. The Mayor felt the Wharves should not be answerable in any way to the City Council. It would make them subject to political pressure, when they were running a business (Cook, 1977:2).

The Wharves' Board ignored the recommendation of the Pier 19 Study Committee. The City Council ignored the citizens in the form of a petition that was signed by one third of the registered voters of the City.

Recreational Aspects

The Historical Foundation backed the tenants of Pier 19 to the point of jeopardizing their rapport with the City Council. After a brief period of dormancy, they returned to reaffirm their position in a press release of December 28, 1976. Their position was to stand by the fight to retain Pier 19 for its historical value and its need to be linked to the Strand which was only a block away. The development of Pier 19 as a Fisherman's Wharf and its close proximity to the Strand could afford a unique tourist attraction and recreational facility.

Public access to the waterfront was one of the important factors involved in the case. Midway through the conflict, the Wharves had mentioned construction of an observation deck on top of the Lipton Tea building for the public to view the Port activities, but there would be not access directly to the waterfront. The issue being settled as it was, the public now has full access to Pier 19 and the Historic Site from the east and the west.

There is evidence almost everywhere one turns in Galveston of its interest in preservation and history. It is obvious that not only the Pier 19 advocates, but even the Port advocates were interested in preserving the integrity of the Galveston historic atmosphere. The final question then was not a matter of right or wrong, but rather of direction. Some described it as the quality of life.

Many Galvestonians have voiced their desire to keep Galveston unique as a Texas city, favoring its numerous historical aspects and tourism over the growing pains experienced by modern industrialization and increasing populations as Houston was experiencing, only 50 miles away. Being an island, physical aspects limited Galveston's growth. This physical quality was looked upon by Galvestonians as an aesthetic form of security.

Many incidents have and will continue to arise concerning the future of our coastal shores and urban waterfronts.

The process for decision making brought out in this study is an extreme and could possibly have been eliminated by early public involvement.

It is hoped that the initiative taken by the citizens portrayed in this case study will be an example to citizens and government officials elsewhere in the United States of what can be accomplished through the public's education to an issue and participation in the decision making process that governs public land and the public interest.

REFERENCES

Bartholomew, Harland and Associates. May 1974. "Evaluation of Alternatives For Development of Pier 19, Galveston, Texas." Unpublished report on file at office of City Manager, Rosenburg Street, Galveston, Texas.

Black, Henry Campbell. 1968. Black's Law Dictionary. West Publishing Co., St. Paul, Minn.

Boggs, Columbo, Devoy, Hoskins, Moore, and Tramonte. Jan. 14, 1977. Public Forum. Aired on television, Galveston, Texas. Video tape on file with Pat Buffler, 2875 Dominique St., Galveston, Texas.

Brown, Christopher J., and Associates. November 26, 1975. "Economic Impact on Displaced Industries at Pier 19, Galveston, Texas." Unpublished report on file at Galveston Historical Foundation, Galveston, Texas.

Buffler, Pat. January 13, 1977. "Pier 19 Settlement." Galveston, Daily News.

Chronicle Galveston Bureau. August 27, 1975. "Galveston Wharves Will Build New Shrimp Boat Basin on Pelican Isle." Houston Chronicle.

Cook, Alison. 1977. "Open Warfare On The Low Seas." Reprinted from Texas Business magazine. Commerce Publishing Corp., Dallas, Texas.

Committee FOR Pier 19. January 1976. Unpublished campaign document. Available from Pat Buffler, 110 23rd St., Galveston, Texas.

DePodwin, Horace J., Associates, Inc. and Fred Muller Associates. March 1975. "Master Plan For Galveston Wharves 1972-1992." Report on file at the Galveston Wharves Building, Rosenburg St., Galveston, Texas.

Ducsik, Dennis W. 1974. Shoreline for the Public. The MIT Press, Cambridge, MA.

Fugger, Reita. October 7, 1976. PHONE INTERVIEW, 5116 Av. S½, Galveston, Texas.

____________. November 19, 1976. PERSONAL INTERVIEW, 5116 Av. S½, Galveston, Texas.

Galveston City Charter. March 1977. Article XII, Section 4, on file at the City Hall, Rosenburg St., Galveston, Texas.

Held, Virginia. 1970. The Public Interest and Individual Interest. Basic Book, Inc., New York-London.

Ketchum, Bostwick, ed. 1972 The Water's Edge. MIT Press, Cambridge, Ma.

Kirkpatrick, Joel September 29, 1976. "Petitioners Would Withdraw Lawsuit." Galveston Daily News.

____________. May 20, 1976. "Appeal Planned By Isle Wharves." Galveston Daily News.

____________. April 29, 1976. "End to Pier 19 Hearing In Sight, Attorneys Say." Galveston Daily News.

____________. April 20, 1976. "Port Action Fails To Defuse Pier 19 Suit." Galveston Daily News.

____________. April 7, 1976. "Pier 19 Order Is Issued." Galveston Daily News.

____________. November 15, 1975. "'Mosquito Fleet' Move Is Voted." Galveston Daily News.

Moss, Mitchell L. 1976. "The Urban Port: A Hidden Resource For The City And The Coastal Zone." Coastal Zone Management Journal, vol. 2, no. 3.

Nesbitt, Robert A. January 1976. Port of Galveston Bicentennial Appointment Calendar and Compendium. Can be obtained at the Galveston Wharves Building Rosenburg St., Galveston, Texas. Published at Galveston, Texas.

Peterson, Joanne. August 19, 1978. "Pier 19, Galveston, Texas: A Case Study." Unpublished paper on file with Author, 250 S. Van Dorn N319, Alexandria, Va.

Post Galveston Bureau. November 15, 1976. "Isle Boat Basin Contract Approved." The Houston Post.

Staff. December 3, 1974. "10,000 Galvestonians Sign Shrimp Fleet Petition." Brazoport Facts.

Staff. December 8, 1975. "Unique City: The Comeback of Galveston." Dallas Morning News.

Staff. December 31, 1976. "Pier 19 Statements." Galveston Daily News.

Staff of the Galveston Wharves. March 15, 1974. "Report on the Pier 19 Area Port of Galveston." Unpublished report presented to the Members of Galveston City Council. On file at Galveston Wharves Association, Wharves Building, Rosenburg St., Galveston, Texas.

T. C. March 1977. "Shrimp Boats Are A-Staying." Texas Monthly.

Ware, Dale. August 29, 1974. "Study Committee's Recommendations." Report submitted to Galveston City Council. On file at City Hall, Galveston, Texas.

__________. August 29, 1974. LETTER to R. A. Apffel, Mayor of Galveston and Galveston City Council Members. On file at City Hall, Galveston, Texas.

Walter, Paul. January 16, 1977. "They Won't Have To Move." Galveston Daily News.

Weinert, H.Doc., III. July 1, 1974. "The Mosquito Fleet Berth." Unpublished report on file at Office of Texas Historical Commission, Austin, Texas.

MARINE ECOSYSTEM MANAGEMENT

by

George W. Robertson
Resource Management Specialist
Center for Natural Areas
Washington, D.C.

1.0 INTRODUCTION

The importance of improved marine resource planning and management strategies is exemplified by the increasing number of agencies which have been established throughout the world to address the various marine resource problems. In the United States, the U.S. National Park Service (NPS), Bureau of Land Management (BLM), Environmental Protection Agency (EPA), Office of Coastal Zone Management (OCZM), National Marine Fisheries Service (NMFS), and others, are addressing marine resource problems. In addition, many regional and state agencies have been established throughout the U.S. in an attempt to solve the many problems we now face regarding marine resource management.

At the international scale, many countries throughout the world have established specific agencies or groups to improve planning and management of their marine resources. For example, the United Nations Environmental Program (UNEP), Food and Agriculture Organization (FAO), International Union for the Conservation of Nature and Natural Resources (IUCN) and the World Wildlife Fund (WWF) are making major contributions to this effort.

The primary impetus for this global effort to conserve and manage our marine resources was the relatively recent realization that, contrary to previous belief, the oceans of the world are not bottomless sinkholes capable of withstanding all of man's multiple use activity pressures without proper control and management. Conflicts between various uses of marine resources steadily increase and we now realize that we may destroy our fisheries resources vital to our future survival if action is not taken. Because our dependence on the sea for food, energy, minerals, medicine, recreation and various other important needs has increased exponentially over the past decade, the importance of conservation and wise management of these resources has become paramount.

In the past, national and international efforts to conserve and manage the marine environment has primarily been a post facto rescue operation. Although marine parks and preserves have been established to protect critical marine habitats (such as mangrove swamps and coral reefs) and species (whales and sea turtles) it is clear that establishing park boundaries does not insure protection from air, land and water contaminants. As Carlton Ray has noted:

> Even when coral reefs are included in parks, their protection is not assured. For instance, Voss (1973) has reported the death of patch reefs in the John Pennekamp Reef State Park in Florida because of outside, as yet unidentified influences.[1]

So far, man's approach has been to identify rare, threatened or endangered habitats and species and to take efforts to protect them. It is possible that we may save some of these habitats and species, but only as long as the ecological processes are maintained.

While our present strategy of "rescuing" endangered and threatened species and habitats must ultimately evolve into a comprehensive ecosystem management approach, there are many problems which must be addressed before this will occur. These include political, socioeconomic, scientific and administrative, all of which are closely interrelated. However, for the purposes of this paper they will be discussed separately.

The primary objective of this paper is to trace the general evolution of marine conservation and management efforts, to introduce and discuss some of the problems associated with the evolution of marine parks and to recommend initial steps to overcome or at least face these problems. Obviously, these issues are much more complex than can be covered in this paper. It is intended that this brief presentation will stimulate thought, discussion, and future action in the field of marine resource conservation and management.

2.0 GENERAL EVOLUTIONARY TRENDS

In the early 1900's, several national parks and monuments with coastal boundaries were established in the United States by the National Park Service (NPS), although the marine resources of these parks were not the primary rationale for their establishment.[2] The first National Park to include a significant marine area as an integral part of the park occurred in 1934 when the Everglades National Park in southern Florida was established. This was the first time in the world that the marine environment was considered important enough for protection.

Although the primary impetus for the establishment of Fort Jefferson National Monument in the Dry Tortugas, Florida, in 1935, was the preservation of an historic monument, it was one of the world's first truly marine parks; approximately 19,037 hectares of water area was included within its boundary. The NPS has, since that time, established approximately twenty-nine (29) parks with significant island, marine, lakeshore or seashore resources.

The National Park Service was not the only U.S. Government agency addressing marine management and conservation issues. Strong public reaction to the Santa Barbara oil spill of 1969 and several reports of the late 1960s and early 1970s provided the impetus for establishing additional U.S. Federal programs. In 1972, the U.S. Congress passed two acts, the Coastal Zone Management Act of 1972 (CZMA) and the Marine Protection, Research, and Sanctuaries Act of 1972 (MPRSA), which firmly established the U.S. federal government's role in protecting marine environments.

The CZMA provided financial incentives for coastal states to establish their own coastal zone management plans in which protection of Geographical Areas of Particular Concern (GAPCs) is an essential component. These GAPCs can also be marine parks in certain states such as California and Florida.

Title III of the MPRSA established the Federal Marine Sanctuaries Program (MSP) to provide selected marine areas of outstanding natural or cultural value with more comprehensive management. While the program is not an exlusively habitat or species protection initiative, many sanctuaries such as the existing Key Largo Coral Reef Marine Sanctuary in Florida and the proposed Flower Garden Banks in the Gulf of Mexico, are principally designed for that purpose.[3]

Similar to what Ray reported for the John Pennekamp State Park, many of the coral formations in the Key Largo Coral Reef Marine Sanctuary the sanctuary was designed to protect, are still dying. While the precise cause is not presently known, it is possible that the degradation is the result of pollution from sources outside and far removed from the sanctuary's boundaries. The two examples cited for degradation in established marine parks illustrate two fundamental concepts for offshore resources management:

> First, the capacity of the marine environment to support a geographically limited or representative phenomenon of significant value depends not only on the conditions present at a particular site, but also on the prevailing conditions exhibited throughout a much broader marine setting.... Second, the integrity of regional marine quality is dependent on the

> full coordination and capability of other entities, such as the water and air pollution control agency(ies) and marine enforcement agencies, to protect regional marine conditions which in turn sustain significant values displayed in local marine areas.[4]

The U.S. Senate Committee on Commerce, Science and Transportation emphasized that the intention of the MSP is not to establish comprehensive ocean management programs but rather to preserve, protect, and restore specific ecologically important ocean areas.[5] While not presently "systematic" in its approach, the MSP potentially represents a "reserve system" rather than a series of independent parks. It is the only existing mechanism within the United States which has the authority to coordinate the various ocean management strategies as well as continue its role in providing "immediate" protection for specific threatened habitats and species.

While the United States is making progress in defining and refining its marine conservation and management policies, many countries throughout the world are making similar efforts (see Table 1). It goes beyond the scope of this paper to discuss efforts by these countries to manage their marine resources. For the most part, however, these parks have been established to protect specific species and habitats, rather than the marine ecosystems these habitats and species depend on.

Many international groups such as the International Union for the Conservation of Nature and Natural Resources (IUCN), World Wildlife Fund (WWF), United National Environmental Program (UNEP), and the United Nations Food and Agricultural Organization (FAO), have taken a lead role in initiating and assisting efforts by many nations to develop sound conservation and management strategies. The <u>World Conservation Strategy</u> developed by IUCN, represents an important evolution of concept from the non-cohesive independent project orientation to a more cohesive, holistic approach to conservation and utilization of natural resources.

The concept of establishing marine parks and reserves is only the first, and probably easiest, step in the conservation of the marine resources throughout the world. However, unless the concept of managing discrete marine parks evolves into a more holistic approach, linked to conservation and management strategies designed for entire marine ecosystems, rather than "rescue" attempts of critical marine habitats and species, at best we will have protected isolated habitats and/or species within a desolate sea and at worst we will have failed to protect the vitally important marine resources we depend on. This points to the need for a generic, programmatic strategy to address the problems associated with

TABLE 1. Selected list of nations having established or proposed marine parks or reserves (having a major part of the park or reserve occupied by the sea) (adopted from: Bjorklund, 1974; Coral Reef Group Newsletter No. 2, 1979)*

Algeria (proposed ?)
American Samoa
Antarctica
Australia
Bahamas
Canada (proposed ?)
Columbia
Costa Rica
Denmark
Djibouti
Dominican Republic (proposed ?)
Ethiopia (proposed ?)
Fiji
France
French Polynesia
Greece (proposed)
Guam (proposed)
Indonesia
Israel
Italy
Japan
Kenya
Madagascar
Malaysia
Malta (proposed ?)
Meritius (proposed ?)
Mexico
Mozambique
New Caledonia
New Zealand
New South Wales
Northern Marianas (proposed)
Palau
Papau and New Guinea (proposed)
Philippines
Puerto Rico (proposed)
Seychelles
South Africa
Spain (proposed ?)
Sri Lauka (Ceylon)
Sudan (proposed)
Tanzania
Thailand
Tobago
Tonga
Trinidad
Trust Territory of the Pacific Islands (U.S.)
Tunisia (proposed)
Turks and Caicos (proposed)
United States
United Kingdom
Venezuela
Virgin Islands (U.S.) (proposed)
Virgin Islands (U.K.) (proposed)
Western Samoa

* Question marks (?) indicate that information on status is unknown.

managing marine ecosystems. The following section addresses some of the problems to be faced in the evolution from a segmented marine park concept to a more comprehensive, holistic marine ecosystem approach.

3.0 PROBLEMS FACING MARINE ECOSYSTEM MANAGEMENT

The problems associated with marine ecosystem management are extremely complex and span a broad spectrum of interrelated factors including: political, social, economic, scientific and administrative. For the purpose of this paper, however, these issues have been separated.

One of the most basic problems faced in any discussion of marine ecosystem management is a definition of a "marine ecosystem." Ray states that:

> Ecosystems are functional units with properties of homeostasis; that is, they exhibit stability which in turn may relate to species diversity. Ecosystems evolve so as to maximize community productivity through efficient recycling of materials.[6]

He goes on to say, "The question arises as to how to classify evolving systems and to differentiate between evolutionary and successional phenomenon and the perturbations of man...surely we cannot presently define marine ecosystems with great precision."[7] The first problem we face then, is the lack of a universally accepted classification scheme by which we could better conceptually define a marine ecosystem. Preliminary suggestions as to a classification scheme include: biogeographical regions, coastal biotic provinces and habitats.[8] Although these are not necessarily ecosystem classification schemes, they do represent that potential. Thus, the first, and most basic problem faced with marine ecosystem management is a classification scheme defining what a marine ecosystem is.

Political

The topic of political problems is, without doubt, one of the most complex and difficult problems to solve, as well as one of the most important. There are currently many organizations and individuals presently addressing many of these problems, such as the United Nation's Conference on the Law of the Sea (UNCLOS), and the many governments throughout the world. This brief discussion is intended to point out some of the political realities associated with the concept of marine ecosystem management.

We can examine political problems in two very general categories: those being intra-national political issues and international problems. Intra-national problems being those arising between local or state governments and federal government authority while international problems are those arising between nations.

Ecosystem management will require cooperation and coordination between state and federal governments because "ecosystem boundaries," if such a thing exists, do not necessarily coincide with state-federal territorial waters. While this problem exists in many, if not most, of the countries throughout the world, some of the problems faced within the United States best illustrate this.

State governments within the United States are responsible for and have jurisdiction over the coastal seas out to three miles from the coast. However, the Coastal Zone Management Act, a federal program providing incentives for individual states to establish their own coastal zone management action dictates that those states which choose to take part in the program must give adequate consideration to the "national interest." While the states have primary "control" out to three miles from their coast, through the federal planning program, the federal government can still influence the use of the area.

Another area representing conflict is that between the various government agencies themselves. Presently, the Department of Interior (Bureau of Land Management and National Park Service), Department of Commerce (National Oceanic and Atmospheric Administration), Department of Defense (Navy), Department of Transportation (Coast Guard), Department of Energy and Environmental Protection Agency all have interests and responsibilities in ocean management strategies. Any change from the present (non-comprehensive), segmented management strategies would entail a major reorganization of departmental authority.

Presently, state government officials cannot enforce federal regulations within federal waters; and federal authorities cannot enforce state regulations unless specifically asked to do so. For example, while the implementation and administration of the rules and regulations governing the Key Largo Marine Sanctuary is the responsibility of Florida's Department of Natural Resources, the U.S. Coast Guard has responsibility for surveillance and enforcement of the rules and regulations promulgated for the sanctuary.[9]

As we can see then, a major revision in governmental authority would be necessary in order to achieve "ecosystem management," and this is not likely to occur.

On an international scale, the problems are compounded by the fact that, not only are we dealing with inter- and intragovernmental conflicts in one nation, but we are adding all of the jurisdictional problems again, both inter- and intragovernmental of another country. These involve defense considerations, economic considerations, fishing rights, and so forth.

Any restrictions on fishing, diving, or other similar activities on non-coastal states would present serious political ramifications. Related problems which are often dealt with in the political arena relate to the size of the area to be protected, the degree of regulation to be imposed, and even what activities will be subject to regulation. Often, the difficulty of balancing the need for resource exploitation (such as oil and gas) against marine conservation must be determined through the political process. In the U.S., this process has even been institutionalized by the National Environmental Policy Act (NEPA) and the use of Environmental Impact Statements (EISs) for federal actions significantly affecting the environment.[10] Many states have similar environmental assessment procedures.[11]

Socioeconomic

The socioeconomic considerations associated with ecosystem management are closely related to the political problems. The planning and management of marine ecosystems would most probably entail financial and political cooperation from various countries. For example, in the Gulf of Mexico, both Mexico and the United States would have responsibility for planning, managing and enforcing any management regulations deemed necessary by the two countries. The question of how the financial responsibilities would be divided between the nations would be of considerable importance.

Social considerations have often been overlooked in past efforts to manage marine resources. Local populations throughout the world have utilized the ocean's resources since their existence without any type of "management schemes." These local populations often do not understand why any type of "management" is necessary. The creation of "management strategies" is not only difficult for them to understand, but it often adversely affects their means of making a living.

Even the establishment of a relatively small marine park can drastically affect social patterns. For example, subsistance fishermen in an underdeveloped country might be prohibited from fishing in traditional fishing grounds and may have to travel much further to reach suitable areas. With the increasing cost of fuel and necessary equipment, and the marginal level at which many of these people live, this could adversely affect their livelihood.

Another potential problem could be the loss of natural heritage by the establishment of tourist-oriented marine parks. Tourists and the tourist trade could significantly alter their customary way of living.

The concept of "ecosystem management" compounds this problem. Unless strategies can be developed to manage marine systems within "cultural norms" the problems associated with a drastic change in social patterns brought about by tourists could dwarf well-intentional attempts to wisely manage marine systems. While conservation and management strategies are necessary for wise utilization of marine resources, the effects of management strategies on local populations must be taken into full account.

Scientific and Administrative

The lack of adequate scientific data and administrative problems, while not as complex as the previously mentioned concerns, are nonetheless important. Sound policy and management strategies must be based on solid scientific understanding of the biological, geological and chemical factors affecting a particular system.

Presently, we do not have a thorough enough scientific understanding of the marine ecosystems to fully understand the exact causes and effects of all man's activities or of natural causes. For example, what will be the ecological ramifications, if any, of seabed mining, Ocean Thermal Energy Conversion (OTEC), or wave energy conversion? What will be the long-term effects of overfishing certain fishery stocks, harvesting krill, or of transporting icebergs across the oceans? Obviously not all of the necessary scientific data for making the rational decisions is available.

Administrative problems associated with the concept of marine ecosystem management include such problems as defining the ecosystems and enforcing management decisions. In the Philippines, for example, it would be almost impossible to enforce management regulations over a relatively large area, without utilizing a small navy of enforcement officials. Administrative issues point out the need for international exchange of information on management strategies which would be important in no reinventing the wheel, transfer and use of effective management approaches, etc.

4.0 CONCLUSIONS

While efforts are being taken to protect critical marine habitats and species, as of yet, little effort has been made to protect the ecological systems which these habitats and species depend on. Several international organizations, such as IUCN and

WWF, have begun shifting their focus from the purely habitat and species orientation to more of an ecosystem approach. These efforts are exemplified by the development of the World Conservation Strategy[12] and by increasing efforts to establish an acceptable marine classification scheme.[13]

Because of the reactive, rather than prospective, planning, emphasized by present U.S. marine management programs, it is unlikely that the United States will take the necessary action to shift focus to a systematic ecosystem management approach. Such a shift in focus would probably entail a major federal reorganization. Laist and Epting (1980) indicate four major steps that the federal government would have to take in order to achieve a more cohesive management system:

> ...to be responsive to these concerns, the federal government must reestablish a more cohesive marine management system that includes: (1) the authority (be it a reorganization plan, executive order, or Congressional mandate) to unite and simplify the existing marine authorities; (2) a development and conservation plan, with clear and concise policies, to guide resource allocation decisions in a fair and coherent manner; (3) a process for resolving inevitable conflicts between the public, governmental and industrial sectors; and (4) an organizational structure that includes within one house (such as one independent agency, a separate department, or a single departmental branch) all major marine oriented functions and entities.[14]

The National Advisory Committee on Oceans and Atmosphere (NACOA) recommended to the President and Congress in 1979, that a new ocean agency be created to consolidate ocean activities.[15] This same principal, of establishing one agency to handle all marine related activities could also be applied at the international level. The creation of a new ocean agency either on the national or international level, would only be effective if if was, in fact, a consolidation of all ocean related concerns, rather than the creation of another layer of bureaucracy.

Public education and participation is an essential component in developing proper marine management strategies. This is particularly important in underdeveloped countries, where local fishermen may be adversely affected by the establishment of marine parks or reserves. Efforts to inform the general public of marine management policies through public meetings, information centers and brochures are useful, however, the general public is still relatively uninformed of the problems facing the utilization and management of marine resources. More aggressive techniques may be necessary to involve the public to a greater extent in marine management decisions.

In reexamining the issues presented in this paper, several general conclusions can be made concerning the management of offshore waters: (1) the establishment of marine parks and reserves worldwide is useful in protecting critical habitats and endangered or threatened species from localized sources of degradation; (2) these parks and/or reserves are ineffective for protecting the ecological systems these critical habitats and species depend on; (3) political, socioeconomic, and administrative problems, as well as the lack of an adequate scientific understanding of marine ecosystems hinders the evolution of a habitat and species management approach; (4) the segmented marine management policies the U.S. and other nations are currently following retards the establishment of a cohesive marine management system; and (5) increased efforts should be taken to include the public, especially in underdeveloped countries, in marine resource management decisions.

Unless a more cohesive management system is developed on both the national and international levels, critical marine habitats and species will exist only as long as the ecological processes continue to function. As man's utilization of marine resources increases, the need for cohesive marine management strategies and international cooperation in addressing problems associated with marine management, become paramount.

Author's Note: The opinions expressed above are solely those of the author and do not in any way represent the sentiment of the Center for Natural Areas.

NOTES

1. An International Conference on Marine Parks and Reserves. Papers and Proceedings. Tokoyo, Japan. May 1975. IUCN Pub. News Series No. 37. 1976.

2. Robinson, Alan H. 1975. Marine, island and coastal parks in the United States park system: a review and progress report in 1975. Presented at the International Conference on Marine Parks and Reserves.

3. Epting, J.T. and D.W. Laist. Perspectives on an ocean management system. In Ocean Development and International Law Journal. Vol. 7. 1979.

4. Ibid.

5. Ocean Management: Seeking a New Perspective. Office of Ocean Resources and Scientific Policy Coordination. Prepared by the Traverse Group. 1980.

6. Ray, Carlton G. A preliminary classification of coastal and marine environments. IUCN Occasion Paper No. 14. 1975.

7. Ibid.

8. Ibid.

9. An Assessment of the Need for a National Marine Sanctuaries Program. Prepared by the Center for Natural Areas for the Office of Coastal Zone Management, National Oceanic and Atmospheric Administration. 1977.

10. National Environmental Policy Act. 42 USC §4321, et seq.

11. California Environmental Quality Act of 1970. PRC §2100, et seq.

12. World Conservation Strategy. Living Resource Conservation for Sustainable Development 1980. Prepared by the International Union for the Conservation of Nature and Natural Resources (IUCN).

FISH, FLUSHING, AND WATER QUALITY: THEIR ROLES IN MARINA DESIGN

by

Rick D. Cardwell[1], Ronald E. Nece[2], and Eugene P. Richey[2]

INTRODUCTION

This paper examines water quality, flushing, and design attributes of saltwater marinas (small-boat basins) in relation to the biological requirements of saltwater organisms and water quality standards. The work was in response to both a notable local demand for additional marinas (Oceanographic Institute of Washington 1978) in Puget Sound, Washington and a variety of uncertainties concerning adverse ecological ramifications of marinas (Heiser and Finn 1970). Impacts on juveniles of Pacific salmon (*Oncorhynchus* spp.) and baitfish -- e.g., Pacific herring (*Clupea harengus pallasi*), surf smelt (*Hypomesus pretiosus*), and Pacific sand lance (*Ammodytes hexapterus*) -- are of special concern because of their special economic and ecologic value and tendency to frequent marinas (Heiser and Finn 1970; Cardwell et al. 1978, 1980a; Penttila and Aguero 1978).

Saltwater marinas are frequently constructed in bays and estuaries. In the inland marine waters of the Pacific Northwest, these shallow protected embayments are the same ones preferred by salmon and baitfish juveniles (Fresh 1979; Cardwell et al. 1980b). Juvenile fish probably are attracted to and rear in these embayments because they possess lucrative food supplies and may constitute refuges from predators. The food webs of natural embayments typically are the most complex and diverse of any Puget Sound habitat, and the prey are dominated by epibenthic zooplankton and macroinvertebrates (Simenstad et al. 1979). Therefore, proper siting, design, and operation of marinas to minimize their adverse ecological effects were primary considerations.

Over the past several years investigations have been undertaken of the biological, water quality, and hydrological relationships of Puget Sound marinas to identify those features of marina design and flushing that will protect these biological uses (Cardwell et al. 1978, 1980a, 1980b; Nece et al. 1975, 1979a, 1979b, 1980; Richey and Smith 1977). These studies concluded that most adverse biological effects of marinas may be prevented if flushing is sufficient and upland drainage and other pollutants are diverted away from the marina basin.

1/ Formerly, Aquatic Biologist, Washington Department of Fisheries, Olympia. Presently, Principal Scientist, Envirosphere Company, 10800 N.E. 8th Street, Bellevue, Washington 98004

2/ Professor, Department of Civil Engineering, University of Washington, Seattle, Washington 98195

Nevertheless, a number of water quality parameters were identified as requiring special attention, including spills of petroleum, releases of sewage, and leaching of heavy metals from boat bottom anti-fouling paints (Clark et al. 1974; Garreis et al. 1979; Young et al. 1979). However, the most important water quality problems because of their prevalence are water temperature (TEMP) elevations due to solar radiation and dissolved oxygen (DO) dimunitions resulting from respiration, particularly that of phytoplankton.

It has been presumed that changes in TEMP and DO as well as the accumulation of other substances could be controlled by defining minimum flushing levels. Therefore, variations in TEMP, DO, and certain other water quality and hydraulic parameters were investigated in the summer of 1979 in five Puget Sound marinas relative to marina flushing measured with physical hydraulic models (Nece et al. 1980).

MATERIALS AND METHODS

Marinas

The five marinas are situated in the greater Puget Sound basin along a north-south gradient of decreasing flushing with Pacific Ocean water and increasing water quality variability (e.g., increasing TEMP and phytoplankton standing crop) (Fig. 1). Selection of these particular marinas was based upon their exposure to different water quality regimens in the Puget Sound basin, representation of different flushing efficiencies, and previous hydraulic study.

Physical and hydraulic characteristics of the marinas are summarized in Table 1. Three of the marinas (Squalicum, Edmonds, and DesMoines) were constructed primarily seaward of mean higher high water (MHHW) and are enclosed by rubble mound breakwaters. Skyline and Lagoon Point, on the other hand, were constructed mainly inland of MHHW in either existing embayments or saltmarshes. Squalicum is the only marina whose source water is degraded significantly by industrial and municipal pollution and also is most highly influenced by lenses of freshwater from local rivers. Recreational boating constitutes the principal use of the marinas, although commercial fishing vessels use Squalicum marina extensively. Residential developments border much of Skyline and Lagoon Point marinas.

Water Quality Sampling

Water quality sampling sought to describe worst possible case conditions and monitor variability in water quality between marinas, months, water depths, and time of day. Accordingly, sampling was restricted to tidal ranges between neap and mean and to the June-September period. In each of four months, all marinas were sampled on consecutive days. Although most samples were collected at depths of 1 m and 3.1 m during the morning and again in the afternoon, one station was sampled continuously throughout the night at 3.1 m to monitor nocturnal changes in DO (Fig. 1). Water quality was contrasted

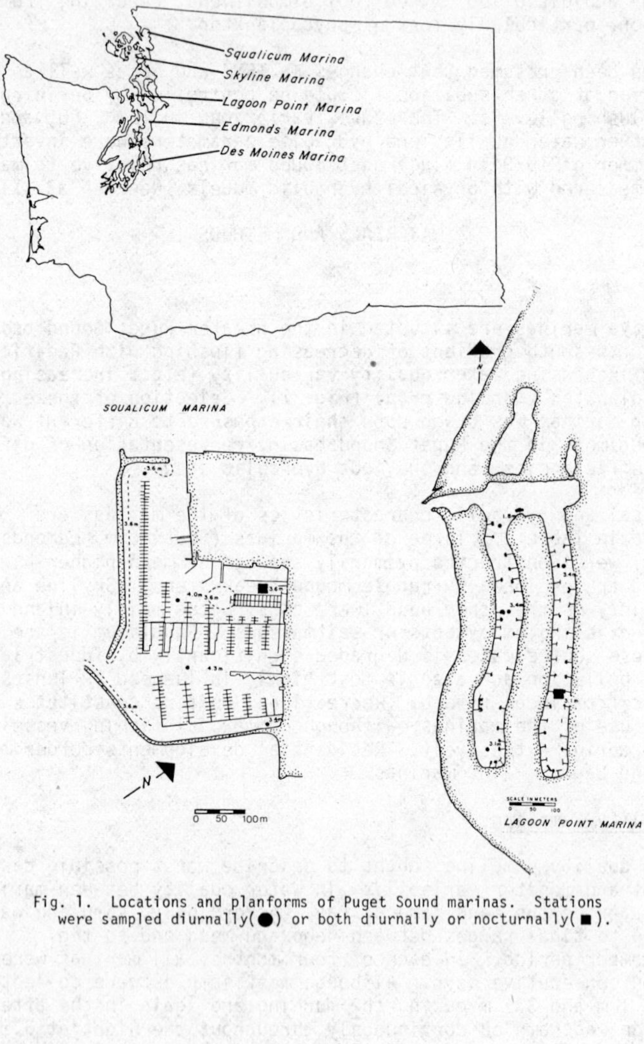

Fig. 1. Locations and planforms of Puget Sound marinas. Stations were sampled diurnally(●) or both diurnally or nocturnally(■).

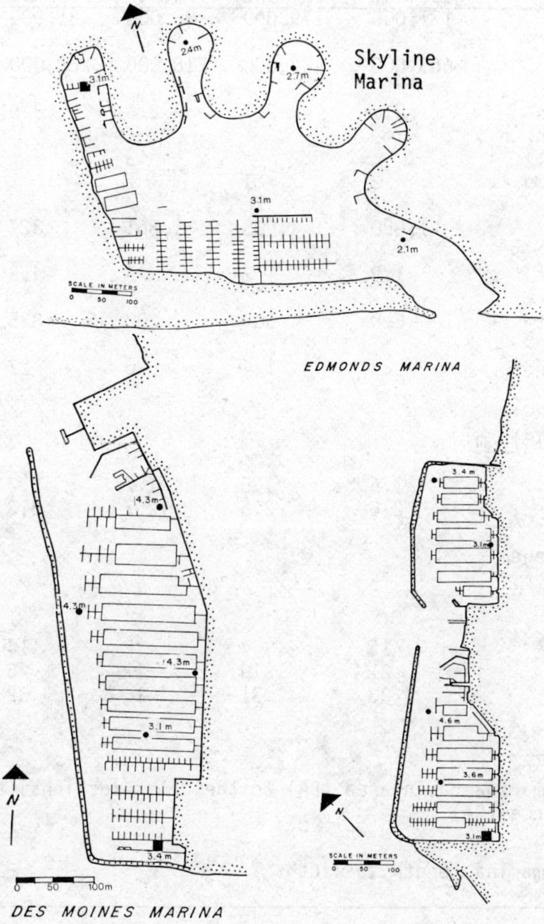

Fig. 1. Locations and planforms of Puget Sound marinas. Stations were sampled diurnally(●) or diurnally and nocturnally(■).

Table 1. Physical and Hydraulic Characteristics of Five Puget Sound Marinas

Characteristic	Squalicum	Skyline	Lagoon Point	Edmonds	Des Moines
Area, m^2	177,000	179,000	83,000	81,000	95,000
Volume, m^3	669,000	480,000	218,000	288,000	365,000
Estimated mean water depth, m	3.8	2.7	2.6	3.5	3.8
No. entrances	2	1	1	1	1
BA/EA[a/]	1,080	1,405	1,861	327	270
Aspect ratio[b/]	1.8	2	9	4.9	3.5
Vessel capacity	620	500	200	825	682
Boat house area, percent	<3	1.4	0	27	22
Tidal range (TR), m					
Mean	1.6	1.5	1.7	2.2	2.4
Diurnal	2.6	2.5	2.7	3.3	3.6
Average exchange coefficients, $\bar{E}$, percent					
0.9 m TR	15	7	7	14	9
1.82 m TR	21	19	11	26	22
2.74 m TR	33	31	16	32	32

a/ Ratio of marina basin area (BA) to the cross-sectional area of the entrance (EA).

b/ Ratio of marina length to width.

between five locations within and three outside (i.e., source water) of each marina. The selection of marina stations sought to bracket the range of local exchange coefficients, an index of flushing. However, due to scheduling constraints, selections were made before the exchange coefficients were measured in the laboratory. Source water stations were located far enough offshore to avoid littoral influence. Because there was not enough time to replicate every sample, we considered all marina and source water samples as separate replicates.

Water quality parameter selection (Table 2) was based upon their potential utility in generating simple models relating marina water quality change to flushing and upon the results of previous studies (Cardwell et al. 1978, 1980a, 1980b). Calculated parameters include marina water temperature change (TCH) and dissolved oxygen change (DOCH), which reflect differences between source and marina water. The mean air temperature for the 7-day period preceding sampling (AIRT7C) was used because it is highly correlated ($r>.9$) with surface source water temperatures (Cardwell et al. 1980b).

Hydraulic Modelling

The hydraulic tests sought to estimate quantitatively the percentage of marina water exchanging with the source water in one tidal cycle, defined in this paper as the time from low water to following low water. Both local exchange coefficients (E), those pertaining to a particular segment of the basin, and average exchange coefficients ($\bar{E}$), those pertaining to the entire basin, were estimated using the following equations:

$$E = 1-R$$

and $$R = (C_n/C_o)^{1/n},$$

where E = exchange coefficient
R = retention coefficient
C_o = initial dye concentration
C_n = dye concentration after n cycle

The apparatus and procedures used in estimating these coefficients (Richey and Smith 1977; Nece et al. 1980) are summarized below.

Tests were performed in a temperature-controlled laboratory using a 2.44 x 3.65 x 0.46 m deep basin in which scale models of the marina were situated. For maximizing model size while maintaining a basin volume sufficient to disperse dye and simulate source water hydraulics, model dimensions were distorted vertically and horizontally to ratios of either 10:1 or 12.5:1 (Squalicum only). Conventional Froude law scaling was employed. Repetitive constant amplitude, constant period tides used were produced by a motor-driven tide generator which was a variable elevation waste weir utilizing gear reducers and a Scotch yoke mechanism to obtain harmonic motion. The tide generator was fed by a constant rate water supply entering the basin through a perforated manifold.

TABLE 2 Times and Methods of Measurement of Water Quality Parameter in Monitoring of Five Puget Sound Marinas (1979)

Water Quality Parameter	Abbreviation	Sampling Times: Morning (0900-1100)	Afternoon (1400-1600)	Night (1600-0800)	Method of Measurememt
Dissolved oxygen, mg/l	DO	X	X	X	Strickland and Parsons (1968) and meter (APHA 1976)
Dissolved oxygen, percent saturation	DOSAT	X	X	X	Truesdale and Gameson (1959)[d]
Water temperature, °C	TEMP	X	X	X	Thermister/ thermometer
Salinity, $^{0}/oo$	SAL	X	X[a]	X	Wet titrimetric method (Stricklnd and Parsons, 1968) and conductivity bridge (Beckman)
Secchi disc transparency, m	SECCHI	X			
Total non-filterable residue, mg/l	TNFR	X[b,c]			APHA (1976)
pH	PH	X[c]			Electrometric
Chlorophylls a, b, c, mg/m^3	CHLA CHLB CHLC	X[c]			APHA (1976)
Phaeophytin a, mg/m^3	PHAA	X[c]			APHA (1976)
Ratio of CHLA/PHAA optical densities at 665 nm	RATIO	X[c]			APHA (1976)
Local exchange coefficient, percent	E				Nece et al. (1980b)
Average (basin-wide) exchange coefficient, percent	$\overline{E}$				Nece et al. (1980b)
Tidal range for day samples collected, m	TR				National Ocean Survey (1978)
Mean air temperature for 7-day period prior to sampling, °C	AIRT7C				. . . [e]

a/ Measured in the afternoon only if morning between station variability exceeded 5 $^{0}/oo$.

b/ Measured only if turbidity was associated only with a non-biological source.

c/ Samples composited separately from all marina stations and from all bay stations.

d/ Equation No. 3 used, i.e., DOSAT = (DO/MAXDO), where $MAXDO = \frac{475 - (283-0.011\ (TEMP)(SAL)}{33.5 + TEMP}$ Temp in °C, and SAL in $^{0}/oo$.

e/ Climatologic data obtained from the National Climatic Center, National Oceanic and Atmospheric Administration, Ashville, North Carolina.

Local and average exchange coefficients were estimated in triplicate for three or more tidal ranges: neap, mean, and spring. Each replicate included measurements made over four tidal cycles. The coefficients listed in Table 1 are conservative because they represent the lowest set of values for each tidal range rather than the average.

Coefficients were estimated with the photodensitometric method of Richey and Smith (1977). In commencing a test, the marina's entrance was obstructed and a dark dye uniformly mixed with the basin. The temporary barrier was removed and the tidal generator started simultaneously. Black and white photographs were taken every one-quarter cycle for four cycles (low slack to low slack). The barrier was then replaced, the basin water thoroughly mixed, and re-photographed. Dye densitites (C_o and C_n) were measured as emulsion density on the negatives with a photodensitometer (Model TBX, Tobias Associates). Calibration was against photographs of the basin when it contained known dye concentrations, from zero to initial concentration used for a flushing test, at the depth corresponding to low water.

Although E and R could be defined using any starting time on the tidal cycle, it is most convenient to use either high water or low water slack. In this study, low water slack was used because residual currents have diminished. In contrast, currents produced during flood tide are very much in evidence at high water slack. The former condition presents two advantages. First, from an experimental standpoint, conditions are more likely to be repeated in more detail from one cycle to the next. Secondly, the relative concentrations of water in the basin should be more indicative of the effective flushing over an entire tidal cycle when based on flushing occurring between two successive low tides. Since effective "flushing times" would be of concern in water quality models, low water measurements are more for applying tidal exchange data.

RESULTS AND DISCUSSION

Minimum Flushing Levels

It was originally assumed, based on theory and a preliminary analysis of the data of Cardwell et al. (1978), that flushing would exert preeminent control over TCH and DOCH, allowing minimum flushing standards to be proposed for Pacific Northwest saltwater marinas. The assumption could not be verified; variability in these non-conservative parameters was much more dynamic and complex than anticipated.

Although several factors discussed later could have obscured their role, local and average exchange coefficients appeared to exert minor control over variation in TCH and DOCH when data for all marinas and sampling depths, times and dates were considered. Consequently, only rough, imprecise estimates of minimum acceptable exchange were derived. These estimates suggested that marinas should possess exchange coefficients averaging at least 20 percent (relative to a 1.82 m tidal range) to minimize water temperature increases which exceed Washington State (DOE 1977) standards. But, because an $\bar{E}$ of at least 30 percent (based on 1.82 m tidal range) appears necessary to prevent excessive noctural D.O. diminutions, the minimum average exchange should be ≈30 percent. These conclusions are rationalized below within the contexts of hydrological characteristics, water quality variation, the biological requirements of marine life, water quality standards, and causes of the weakness in the exchange-water quality change relationship.

Hydrologic Characteristics

The marinas tested have a fairly wide range in planform geometries. Four of the five (Lagoon Point being the exception) fall within an "envelope" of performance with respect to spatial average flushing; their behavior (see Table 1) can be plotted within the envelope formed by the curves $\bar{E} = 0.09$ TR and $\bar{E} = 0.14$ TR, where TR is the tidal range in meters. Higher average flushing efficiences typified the marinas (Squalicum, Edmonds, and DesMoines) where all or nearly all of the basin participates directly in the tidal exchange. This situation occurs when there is a single gyre, or circulation cell, in each basin or sub-basin connecting directly with the entrance. These gyres typically are established during flood tide flowing through an asymetric entrance.

Local exchanges at the stations used for monitoring water quality did not reflect the average exchange for the respective marinas (Fig. 1). These differences are exemplified for stations where noctural water quality was measured. At Edmonds, the station was located in the perimeter of the gyre in the southern basin of the marina. For a 1.82 m tide (prototype), its E's were about 15 percent higher than $\bar{E}$. At Squalicum and DesMoines, for the same TR, the E was about 75 percent of $\bar{E}$. At Skyline the ratio was smaller, about 67 percent; the inner "lobes" of this marina exchange with the central basin of the marina and not with the ambient source water. At Lagoon Point, where the station was located well toward the end of one of the two canals, local E was only about 30 to 40 percent of $\bar{E}$. Almost all of the circulatory tidal flow in this marina occurs in the header channel near the entrance, and only a small portion of the marina basin participates directly in the tidal exchange process (Nece et al. 1980).

Water Quality

Water quality was more perturbed in Skyline and Lagoon Point marinas than in the other marinas. As noted above, tidal exchange in these two marinas does not occur directly with the source water. Although as noted below all marinas except Squalicum appeared to violate Washington State water quality standards (DOE 1977), the most biologically important violations concerned development of hypoxic DO's at night in all marinas in late summer and early autumn. The violations are considered apparent because State standards do not define explicitly the criteria (e.g., means, extremes) for judging the violations. The 95 percent confidence limit for the population of measurements made for a marina was chosen as the criterion because it is statistically reliable and maintains satisfactory water quality in all but 5 percent of the circumstances.

Marina water quality was very dynamic, and was shown by factorial analysis of variance to vary significantly ($p<0.001$) between marinas, months, times of day (morning vs. afternoon) and sampling depths (0.9 vs. 3.1 m). Interactions between these factors usually were significant.

Water Temperature--Water temperature in four of the marinas violated DOE (1977) standards. Squalicum proved the exception only because its source waters are accorded a more lenient water use classification, owing to its location in an industrialized section of Bellingham Bay.

Violations concerned TCH and not absolute temperatures, as summarized in Table 3. TCH deviations were greatest in Skyline and Lagoon Point. Abrupt exposure of organisms to these temperatures probably were not harmful, but if persistent, they might accelerate reproduction rates and food requirements of resident biota, events that could be inimical if dysynchronous with the normal timing of phytoplankton blooms, etc. Temperature increases of at least 5-10°C usually are required to produce functional damage in local biota (Stober et al. 1971), although Andronikov (1975) notes that as little as a 1-3°C increment may harm gametes of some marine invertebrates. Absolute marina water temperatures, which never exceeded 16.2°C, did not seem high enough to cause mortality of several economically important species when interpreted within the context of the species' life history and occurrence (Table 4). Temperatures above 15°C, however, probably would be avoided by and cause diminished growth of juvenile salmon (Brett 1952; 1979).

The assumed preeminence of exchange in predicting TCH was incorrect when the data from all marinas, dates, etc. were considered. The plot of E versus TCH portrayed a broad envelope which suggested a slight overall increase in TCH beginning at E values below about 15 percent (Fig. 2). At higher E's, marina water temperature changes generally were above zero and reached 1.5°C above those of the source water. More of the variation in the TCH generally was explained by E when the data sets were more specific, as exemplified by data for Skyline (Fig. 3). This plot also illustrates the extent to which TCH varies between sampling dates, which suggests the importance of other variables in predicting TCH.

Correlations between all the parameters considered (Table 5) failed to disclose any that had a preeminent influence on TCH variation, suggesting consideration of other parameters and re-examination of the experimental design's efficiency. Within this regard, it is known that temporal changes in marina TCH are very sensitive to local weather conditions, with marinas responding quickly to diel solar heating and

Table 3. Characteristics of Five Puget Sound Marinas in Terms of Mean and Extreme Values of Five Water Quality Attributes, Based Upon Data Collected June - September 1979

Parameter	Squalicum	Skyline	Lagoon Point	Edmonds	DesMoines
Water Temperature, °C					
Mean	14.7	12.0	12.2	12.5	13.5
Upper 95 percent CL[a]	16.2	13.7	14.0	14.1	15.6
Max[a]	16 (15)[b]	15 (1)	15 (1)	14 (9)	16 (3)
Water Temperature Change (TCH), °C					
Mean	-0.4	1.3	0.5	0.1	0.3
Upper 95 percent CL	0.6	3.2	2.1	0.7	1.4
Max	1 (1)	3 (8)	3 (3)	1 (99)	1 (35)
Water quality standard[c]	1.6	1.08	0.9	0.3	0.3
Dissolved Oxygen Concentration, mg/l					
Mean	7.2	7.8	7.6	6.9	9.0
Lower 95 percent CL	4.5	3.9	4.8	4.3	6.9
Min	4 (1)	5 (16)	5 (1)	4 (3)	7 (4)
Water quality standard	5.0	7.0	7.0	7.0	7.0
Dissolved Oxygen Solubility, Percent					
Mean	90	91	92	85	110
Lower 95 percent CL	65	47	59	55	84
Min[a]	55 (1)	57 (3)	68 (3)	55 (3)	87 (3)
Proposed criterion[d]	57	100	100	100	100
Dissolved Oxygen Change (DOCH), mg/l					
Mean	-1.7	1.4	0.6	-0.8	-0.9
Lower 95 percent CL	-3.4	-1.6	-1.6	-1.8	-2.4
Min	-4 (3)	-1 (9)	-1 (6)	-2 (4)	-3 (1)
Water quality standard[c]	-0.2	-0.2	-0.2	-0.2	-0.2

a/ CL = 95 percent confidence limits, max = maximum, and min = minimum.

b/ Value within parentheses represents the percentage of observations the level was attained. The other number is the maximum integer value, which is actually a range (16.0 to 16.9).

c/ DOE (1977).

d/ Davis' (1975) "A" criterion.

Table 4. Lethal Temperatures for Several Marine Organsims indigenous to Puget Sound

Species	L.H.[a]	Acclim. Temp., °C	Incipient lethal temperature, °C	Reference
Pacific Salmon[b]	J	10-11.2	22.5-24.3	Brett 1952; Stober et al. 1971
English sole (*Parophrys vetulus*)	J	11.8	23.0	Stober et al. 1971
Bay mussel (*Mytilus edulis*)	L	18	20-25	Brenko and Calabrese 1969
Geoduck clam (*Panope generosa*)	L	12-14	16	Goodwin 1973
Dungeness crab (*Cancer magister*)	A	10.1-11.3	24.5	Stober et al. 1971
Polychaete annelid				
-*Capitella capitata*	A	...	26.6[c]	Oshida and Reish 1974

[a] L.H. = life history stage; J = juveniles; A = adult.

[b] Includes all five species.

[c] 28-day median lethal temperature for cultured specimens.

cooling. Larger, deeper bodies of water like Puget Sound, on the other hand, are more responsive to seasonal changes. Also, it was not possible to perform the water quality sampling at low tide, the time exchange coefficients were determined, and there was no assurance that the coefficients used in the analysis typified those existing when the stations were sampled. In order to model parameters like TCH, it is apparent that all data be collected at the same times and locations and that additional meteorological variables (e.g., solar radiation) be considered.

Dissolved Oxygen--Concentrations of dissolved oxygen measured during the day had several features, the most important being that they had little bearing on and did not detect the very low DO's that occurred at night. These nocturnal D.O. diminutions not only violated state water quality standards, but also suggested the possibility of kills of immobile, bottom-dwelling fauna. Other features included

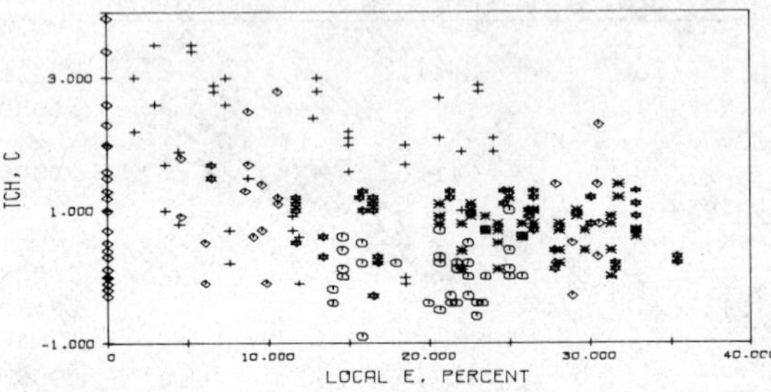

Fig. 2. Variation in TCH in relation to local exchange coefficients for Squalicum(⊙), Skyline(+), Lagoon Point(◇), Edmonds(✳), and DesMoines(★) marinas.

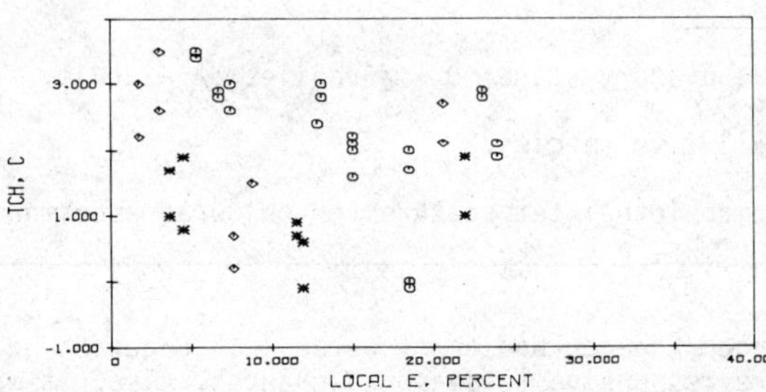

Fig. 3. Variation in TCH in relation to local exchange coefficients for Skyline marina in June(⊙), August (◇), and September(✳) 1979.

significant correlations with phytoplankton pigments, particularly chlorophyll c, and with source water DO's (Table 5.). These correlations had considerable bearing on the significant ($p<0.001$) variation in diurnally-measured D.O.s between months, marinas, depths and time of the day.

During the day, DO concentrations averaged above 6.9 mg/l in all marinas (Table 3). Following the peak in late June, however, they declined to between 3.9 and 6.9 mg/l in August and September in response to source water upwelling (Barnes and Ebbesmeyer 1978) and increasingly senescent phytoplankton populations (Welch 1969). The diurnal declines in marine DO's exceeded the 0.2 mg/l limit allowed by DOE (1977).

A much more serious violation of state DO standards was recorded during monitoring of DO's at night in the innermost sectors of each marina at depths (2.1-3.1m) near the bottom. In late August and September, DO's fell to hypoxic and even anoxic concentrations. In almost every case, the highest concentrations were measured when monitoring commenced, usually 1800, and the lowest levels between midnight and 0600. Although not measured, similar fluctuations in source water DO's would be unexpected due to the strength of tidal circulation. Late August data were used to exemplify the pattern of nocturnal D.O. changes which were greatest during this time (Fig. 4). At the sampling station in the shoreward canal in Lagoon Point marina, where the water motions are more one-dimensional and less complex, the diel DO changes did track the tide curves; i.e., DO levels were highest near high tide and decreased as tide level fell and water which had been in the closed end of the canal moved seaward to the sampling station.

Conclusions concerning the significance of these findings should consider percentage DO saturation (Table 3), the requirements of the indigenous species, and other designated water uses. On the average, diurnal oxygen contents in the marinas generally were satisfactory, although they probably were not high enough to sustain the uncompromised growth and reproduction of sensitive species (Davis 1975). Certainly, monitoring of fish and zooplankton in Puget Sound marinas (Penttila and Aguero 1977; Cardwell et al. 1978, 1980) has not revealed a conspicuously impoverished biota. The nocturnal declines in DO, however, probably killed or severely compromised both benthic infauna and relatively immobile, epibenthic organisms. Reish (1963) derived a similar conclusion following study of benthic organisms in a southern California marina. Though some species, such as certain bivalve molluscs and polychaete worms, can tolerate DO's around 1.0 mg/l before dying, successful reproduction requires higher DO's (Davis 1975; Reish and Ayers 1968).

The principal cause of the nocturnal D.O. declines is presumedly phytoplankton respiration. Skyline marina had both the greatest nocturnal DO declines and the highest concentrations of chlorophyll a, b and c. Of course, the oxygen demand of respiring phytoplankton would

Table 5 Linear correlations for water quality, climatic and tidal parameters measured in five Puget Sound marinas in 1979.

Parameter[a]	TEMP	XCTEMP	TCH	DO	DOSAT	XCDO	DOCH	E	LOGE	$\overline{E}$	SAL	PH	SECCHI	CHLA	CHLB	CHLC	PHAA	RATIO	TNFR	AIRT7	AIRT14	TIDHT
TEMP	1.00																					
XCTEMP	.88	1.00																				
TCH	-.18	-.62	1.00																			
DO	.14	-.14	.53	1.00																		
DOSAT	.09	.05	.04	.15	1.00																	
XCDO	.50	.51	-.26	.54	.14	1.00																
DOCH	-.15	-.68	.83	.52	.03	-.45	1.00															
E	.17	.30	-.34	-.05	.03	.35	-.41	1.00														
LOGE	-.20	.31	-.31	-.05	.04	.34	-.40	.90	1.00													
$\overline{E}$	.44	.55	-.41	.10	.10	.67	-.58	.66	-.63	1.00												
SAL	-.65	-.67	.32	-.27	-.15	-.58	.29	-.20	-.23	-.36	1.00											
PH	.30	.05	.41	.58	.52	.32	.36	-.10	.01	.01	-.40	1.00										
SECCHI	-.30	-.21	-.09	-.18	-.13	-.09	-.13	-.31	.27	.30	.45	-.57	1.00									
CHLA	-.06	-.28	.48	.31	.33	-.23	.59	-.32	-.26	-.42	.14	.49	-.31	1.00								
CHLB	-.10	-.36	.57	.47	.39	.04	.51	-.20	-.11	-.29	-.21	.56	-.48	.50	1.00							
CHLC	.02	-.26	.59	.52	.47	.04	.57	-.18	-.10	-.26	-.15	.60	-.43	.57	.90	1.00						
PHAA	-.17	-.38	.50	.58	.51	.09	.59	-.30	-.27	-.36	.03	.50	-.39	.75	.66	.69	1.00					
RATIO	.40	.26	.14	-.08	.04	-.20	.11	.02	.08	.04	.05	.17	-.17	.55	.03	.12	.0	1.00				
TNFR	-.31	-.22	-.07	-.33	-.28	-.31	-.08	-.23	-.20	-.29	.50	-.21	-.10	.18	-.06	-.17	.05	.03	1.00			
AIRT7	.30	.19	.09	-.02	-.04	.06	-.08	.16	.15	.18	.32	-.11	.28	.13	-.02	.06	.04	.37	-.08	1.00		
AIRT14	.10	.11	-.06	-.19	-.12	-.06	-.14	.14	.10	.13	.40	-.30	.30	.03	-.22	-.16	-.06	.16	.15	.59	1.00	
TIDHT	-.06	.04	-.17	-.20	-.05	-.02	-.19	.18	.12	.25	.41	-.17	.53	-.08	-.39	-.22	-.15	.14	-.42	.17	.32	1.00
TIDRA	.07	.11	-.12	.36	.04	.61	-.24	.47	.37	.74	.04	-.03	.41	-.26	-.19	-.17	-.03	-.18	-.21	.05	.12	.33

[a] Marina water temperature (TEMP), mean source water temperature(XCTEMP), temperature change within marina(TCH), marina dissolved oxygen(DO), marina dissolved oxygen solubility (DOSAT), mean source water dissolved oxygen(XCDO), marina dissolved oxygen change(DOCH), local exchange coefficient(E), $\log_{10}$ local exchange coefficient(LOGE), mean gross exchange coefficient($\overline{E}$), salinity(SAL), chlorophylls *a*, *b*, and *c*(CHLA,CHLB,CHLC), phaeophytin *a*(PHAA), ratio of chlorophyll *a* to phaeophytin *a*(RATIO), total non-filtrable residue(TNFR), mean 7-and 14-day air temperatures(AIRT7, AIRT14), tidal height(TIDHT), and tidal range(TIDRA).

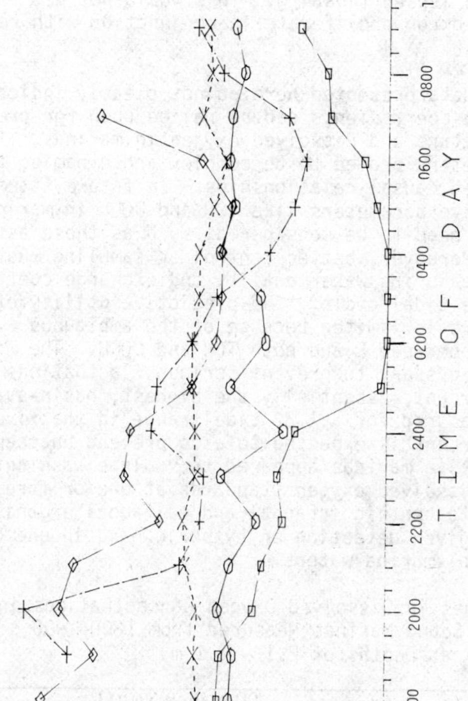

Fig. 4. Diel changes in DO concentrations near the bottom(2.1-3.1 m) in late August 1979 in Squalicum(⊙), Skyline(⊡), Lagoon Point(+), Edmonds(×), and DesMoines(◇) marinas.

be augmented by the other biota. In one study of a Northeast saltwater marina, fouling organisms were judged most responsible for oxygen sags (Nixon et al. 1973).

Changes in marina DO relative to those of the source water (DOCH) were most highly correlated with TCH, phytoplankton pigment concentrations, and exchange coefficients (Table 5). Plots of DOCH vs. local exchange coefficients revealed an ambiguous relationship similar to that observed between E and TCH, although there was some suggestion that DOCH increased with lesser flushing. This would not be unexpected, for phytoplankton proliferate in conjunction with reduced mixing and exchange.

In conclusion, the data presented here do not clearly indicate the extent to which exchange coefficients can be relied upon for predicting changes in water temperature and dissolved oxygen in marinas. The behavior of these parameters proved to be complex and dynamic, two properties which obscured causal relationships. In future attempts at modelling non-conservative parameters like TCH and DOCH in marinas, other control variables need to be considered, such as those associated with solar radiation. Moreover, better control of sampling must be exercised, especially measuring water quality and exchange coefficients at the same times of the tidal cycle. The predictive utility of our data to resource managers is limited because of the ambiguous predictive relationship between E and both TCH and DOCH. The minimum level of exchange ($\bar{E}$) necessary to prevent serious fluctuations in DO seems to be about 30 percent, essentially the highest, basin-averaged exchange coefficient measured for a 1.82 tidal range in the five marinas. An $\bar{E}$ of 30 percent is expected to also prevent unacceptable increases in TCH. All five marinas appeared to violate Washington state temperature and dissolved oxygen standards at one or more times and places, and damage to benthic infaunal and epifaunal organisms appears very possible, given detection of hyxpoxic, and in one case anoxic, DO's in surficial marina waters.

Table 6. Ranges in Dissolved Oxygen Concentrations in Five Puget Sound Marinas Measured from 1800-0900 at Depths of 2.1 - 3.1 m

Marina	DO range, mg/l			
	Mid-June	Late June	Late August	Late September
Squalicum	6.9[a]-9.0[b]	7.2-9.4	3.9-6.3	1.1-7.8
Skyline	6.3-9.5	6.4-10.8	0-5.7	3.4-5.6
Lagoon Point	7.1-10.6	9.4-13.6	2.8-13.2	4.3-7.4
Edmonds	7.2-9.5	...	5.5-6.9	2.8-5.2
Des Moines	7.0-9.0	2.7-8.4	5.2-12.2	4.9-11.5

a/ Minimum
b/ Maximum

LITERATURE CITED

American Public Health Association. 1976. Standard methods for the examination of water and wastewater. 14th Ed., American Public Health Association, Washngton, D.C. 1,193 pp.

Andronikov, B. B. 1975. Heat resistance of gametes of marine invertebrates in relation to temperature conditions under which the species exist. Mar. Biol. (Berlin) 30:1-11.

Barnes, C. A. and C. C. Ebbesmeyer. 1978. Some aspects of Puget Sound's circulation and water properties. pp. 209-227. In: (B. Kjerve, ed.) Estuaries transport properties. Belle W. Baruch Library in Marine Science, No. 7. University of South Carolina Press, Columbia, S.C.

Brenko, M. H. and A. Calabrese. 1969. The combined effects of salinity and temperature on larvae of the mussel *Mytilus edulis*. Mar. Biol. (Berlin). 4:224-226.

Brett, J. R. 1952. Temperature tolerance in young Pacific salmon, genus *Oncorhynchus*. J. Fish Res. Board Can. 9:265-314 + appendices.

Brett, J. R. 1979. Environmental factors and growth, pp. 599-675. In: (W. S. Hoar, D. J. Randall and J. R. Brett, editors) Fish physiology, Vol. III. Bioenergetics and growth. Academic Press, New York.

Cardwell, R. D., M. I. Carr, S. J. Olsen, and E. W. Sanborn. 1978. Water quality and biotic characteristics of Birch Bay Village Marina in 1977 (October 1, 1976 to December 31, 1977). Prog. Rep. 69, Wash. Dep. Fisheries, Olympia, Wash. 90 pp.

Cardwell, R. D., S. J. Olsen, M. I. Carr, and E. W. Sanborn. 1980a. Biotic, water quality, and hydrological characteristics of Skyline Marina, Anacortes, Washington in 1978. Tech. Rep. 54, Wash. Dep. Fish., Olympia, Wash. 104 pp.

Cardwell, R. D, M. I. Carr, and E. W. Sanborn. 1980b. Water quality and flushing of five Puget Sound marinas. Tech. Rep., Wash. Dep. Fish, Olympia, Washington.

Clark, R. C., Jr., J. S. Finley, and G. G. Gibson. 1974. Acute effects of outboard motor effluent on two marine shellfish. Environ. Sci. Technol. 8:1009-1014.

Davis, J. C. 1975. Minimal dissolved oxygen requirements of aquatic life with emphasis on Canadian species: a review. J. Fish. Res. Board Can. 32: 2295-2332.

Department of Ecology. 1977. Washington state water quality standards. Office of Water Programs, Washington Dep. Ecology, Olympia, Wash. 33 pp.

Fresh, K. L. 1979. Distribution and abundance of fishes occurring in the nearshore surface waters of northern Puget Sound, Washington. M.S. Thesis, University of Washington, Seattle. 120 pp.

Garreis, M .J., F. A. Pittman, D. L. Elmore, and R. L. Robison II. 1979. Marina impact on water quality in Kent Island Narrows, Maryland. No. EH:WSC:GS No. 2, Maryland Environmental Health Admin., Baltimore, Maryland. 28 pp.

Goodwin, L. 1973. Effects of salinity and temperature on embryos of the geoduck clam (Panope generosa). Proc. Nat. Shellfish. Assoc. 63:93-95.

Heiser, D. W. and E. L. Finn. 1970. Observations of juvenile chum and pink salmon in marina and bulkheaded areas. Suppl. Prog. Rep., Puget Sound Stream Studies, Wash. Dep. Fish., Olympia, Wash. 28 pp.

National Oceanic and Atmospheric Administration. 1978. Tide tables 1979. High and low water predicions. West coast of North and South America including Hawaiian Islands National Ocean Survey, Rockville, Maryland.

Nece, R. E., E. B. Welch, and J. R. Reed. 1975. Flushing criteria for salt water marinas. Tech. Rep. 42, C.W. Harris Hydraulics Lab., University of Washington, Seattle. 50 pp.

Nece, R. E. and E. P. Richey. 1979a. Tidal flushing in Flounder Bay. Tech. Rep. 61, C.W. Harris Hydraulics Lab., University of Washington, Seattle, 21 pp.

Nece, R. E., E. P. Richey, J. Rhee and H. N. Smith. 1979b. Effects of planform geometry on tidal flushing and mixing in marinas. Tech. Rep. 62, C.W. Harris Hydraulics Lab., University of Washington, Seattle. 74 pp.

Nece, R. E., H. N. Smith, and E. P. Richey. 1980. Tidal circulation and flushing in five western Washington marinas. Tech. Rep. 63, C.W. Harris Hydraulics Lab., University of Washington, Seattle. 51 pp.

Nixon, S. W., C. A. Oviatt, and S. L. Northby. 1973. Ecology of small boat marinas. Marine Tech. Rep. Ser. No. 5, University of Rhode Island, Kingston. 20 pp.

Oceanographic Institute of Washington. 1978. Survey of marine boat launching and moorage facilities in Washington. Ocean. Inst. Wash., Seattle, Wash.

Oshida, P. and D. J. Reish. 1974. The effect of various water temperatures on the survival and reproduction in polychaetous annelids: preliminary report. pp. 63-77. In: D.F. Soule and M. Oguri (eds.), Marine studies of San Pedro Bay, California. Part III. Thermal tolerance and sediment toxicity studies. Allen Hancock Foundation, University of Southern California, Los Angeles, Calif.

Penttila, D. and M. Aguero. 1978. Fish usage of Birch Bay Village Marina, Whatcom County, Washington, in 1976. Wash. Dep. Fish. Prog. Rep. 39. 62 pp.

Reish, D. J. 1963. Further studies on the benthic fauna in a recently constructed boat harbor in southern California. Bull. So. Calif. Acad. Sci. 62(1):23-32.

Reish, D. J. and J. L. Ayers, Jr. 1968. Studies on the *Mytilus edulis* community in Alamitos Bay: - III. The effects of reduced dissolved oxygen and chlorinity concentrations on survival and byssal thread production. Veliger. 10:384-388.

Richey, E. P. and N. H. Smith. 1977. Birch Bay marina: hydraulics of proposed expansion. Tech. Rep. 53, C.W. Harris Hydraulics Lab., University of Washington, Seattle, 38 pp.

Simenstad, C. W., B. S. Miller, C. F. Nyblade, K. Thornburgh, and L. J. Bledsoe. 1979. Food web relationships of northern Puget Sound and the Strait of Juan de Fuca. Environmental Protection Agency, Office Res. Develop., No. EPA-600/7-79-259. 355 pp.

Stober, Q. J., D. L. Mayer, and E. O. Salo. 1971. Thermal effects on survival and predation for some Puget Sound fishes. pp. 537-544. In: D. J. Nelson (ed.), National Symposium on Radioecology, 3rd, U.S. Nat. Lab., Oak Ridge.

Strickland, J. D. H. and T. R. Parsons. 1968. A practical handbook of seawater analysis. Fish. Res. Board Can. Bull. 167. 311 pp.

Truesdale, G. A. and A. L. H. Gameson. 1959. The solubility of oxygen in saline water. Int. Coun. Explor. Sea., J. Conseil. 22:163-166.

Welch, E. B. 1969. Factors affecting phytoplankton blooms and resulting effects on dissolved oxygen in Duwamish River estuary in Seattle, Washington. U.S. Geological Survey, Water-Supply Paper No. 1873-A. 62 pp.

Young, D. R., G. V. Alexander and D. McDermott-Ehrlich. 1979. Vessel-related contamination of southern California harbors by copper and other metals. Mar. Poll. Bull. 10:50-56.

BOATING: CHALLENGED BY COASTAL REGULATIONS

Susan H. Anderson*

Throughout the nation, shorter work weeks and more affluence have increased the demand for recreation, even as population growth slows. Recreational boating has enjoyed a surge in demand perhaps even greater than recreation increases generally. Between 1974 and 1980, the number of people taking part in recreational boating nationally nearly doubled, from 52 million people in 1974 to an estimated 93 million people in 1980.[1]

Recreational boating contributes significantly to the lifestyle and ambiance of the Southern California coast. The climate and the ocean conditions south of Point Conception offer excellent boating opportunities throughout the year. More than 200,000 registered boats in Southern California coastal counties provide approximately 700,000 people with 18,000,000 boating days annually.[2]

Coastal boating is unique with its capability to provide access to the vast open spaces of the ocean. Particularly in this area with its heavy urban concentration along the coast, access to this open space is a major concern. Where else can urban residents get away from the crowding and encroachments of the urban area so quickly? Within an hour's time from launching or casting off the recreational boater can be on his or her own little "island", the boat. Even the small rowboat in a harbor can drift placidly away from urban pressures.

The major concern of those people involved in boating is to continue to enjoy these boating opportunities. They want to be able to relax, cruise and race their boats, just

---

*Ms. Anderson is a Coastal Planner for Moffatt and Nichol, Engineers, Long Beach, California

[1] Matt G. Kaufman, Boat Launching-Reservations Only In: Recreation: Marine Promise, Proceedings of the National Conference on Marine Recreation, S.H. Anderson, ed., USC Sea Grant Program, USC-SG-3-76.

[2] Calif. Marine Affairs & Navigation Conference, Calif. Assoc. of Port Authorities, and Calif. Marine Parks and Harbors Assoc./ California Port and Navigation Caucus October 25, 1979, Port Caucus Issues.

putter on their boats and know that there are harbors of rufuge and destination points available for overnight trips. The occupancy rate of coastal slips in Southern California is 98.6%, indicative of possibly the highest demand for slips in the world.[3] Obviously, to accommodate increasing demands for slips new facilities must be developed. In Southern California, this presents a major problem.

Unlike East or Gulf Coast shorelines, the Southern California coastline has relatively few inlets and natural harbors. Some of the existing harbors have been dredged out of dry lowlands or have been extended out from the coastline, on the leeward side of a headland, such as Dana Point Harbor. The rivers that might be expected to provide protected coastline have been channelized to prevent flood damage, and few, if any, flow year around.

Not only are coastal locations for recreational harbors limited, but also the wave of environmental legislation that was proliferated during the 1970's frustrates and often denies the development of these remaining opportunities. Twelve recreational boating harbors, (including commercial harbors that provide space for recreational boating) are spread along approximately 217 miles of coastline (Figure 1). Already these facilities provide space for enough boats that if all moored and berthed recreational boats were laid bow to stern along the coast they would span the whole distance.

Today, while those people who already have boats are not anxious to increase the crowding of boating facilities or the crowding of offshore island harbors, they are concerned that, if no new facilities are built elsewhere along the coast, their own harbors will reach a level of congestion that is intolerable. For some, even existing facilities are threatened by competition from other coastal demands and several have been removed to make way for some other use.

There is a large concentration of recreational boating industries in the Southern California areas serving both local concerns as well as the interests of boaters nationwide. Statewide the recreational boating industry had an

---

[3] Williams Kuebelbeck & Associates, Inc. Summary of Market feasibility Analysis, Proposed Long Beach Downtown Marina, January 1980.

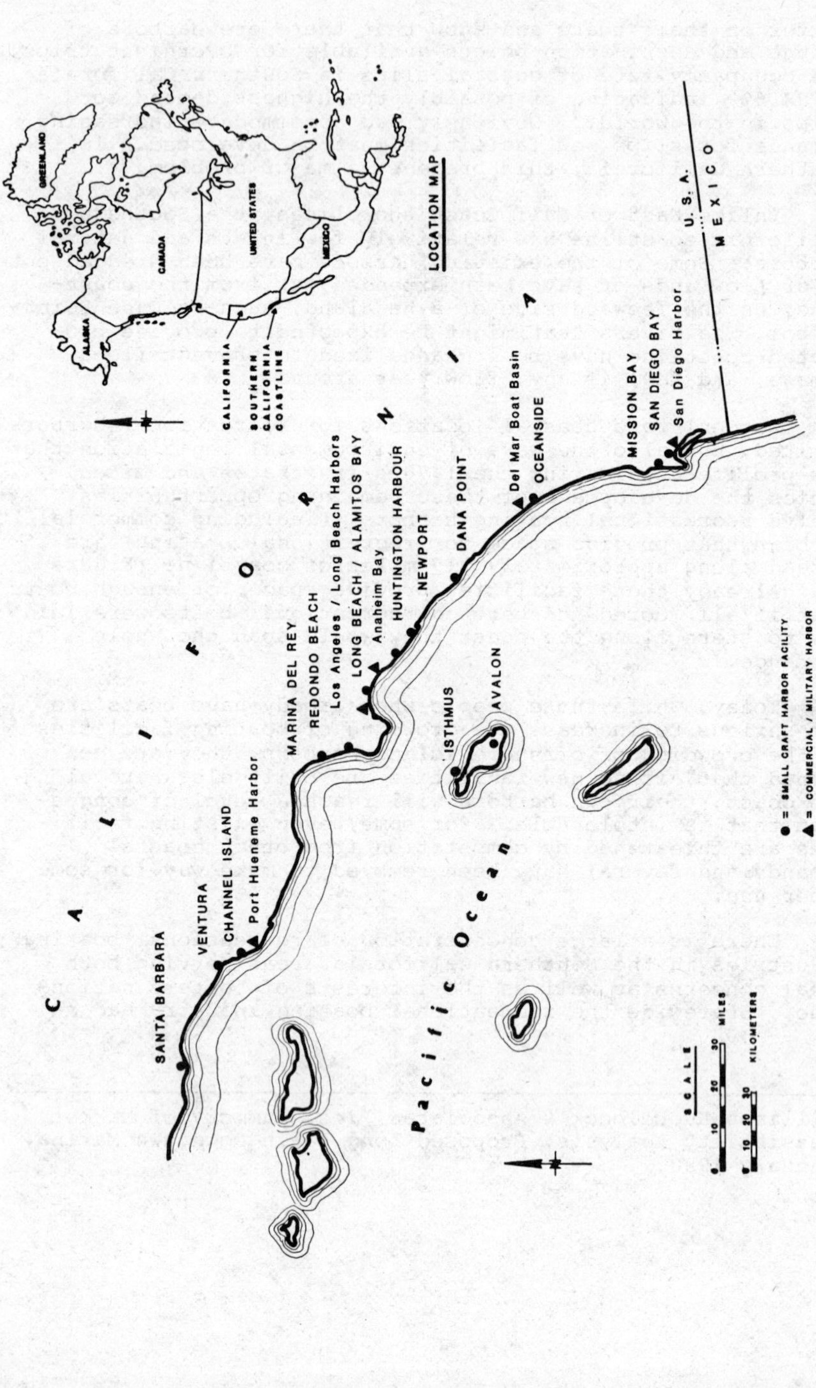

FIGURE 1

SMALL CRAFT HARBORS – SOUTHERN CALIFORNIA COASTLINE

estimated economic impact in 1978 of $2 billion. The number of jobs related to the industry was estimated at 50,000.[4] Annually, boat owners pay approximately $40 million dollars in State and local taxes, in addition to the marine fuel tax which accrues to the Federal Government.[5]

The presence of small craft harbors has frequently enhanced the revenue producing opportunities of restaurants, shops, and apartment complexes that share the same harbor. While the main revenue producers in a harbor such as Marina del Rey are the leases on the restaurants and apartments, without the ambiance of the marina with its 5,600 boats, the revenues from the shoreline restaurants and apartments would undoubtedly decrease, particularly if those establishments had been built adjacent to the mosquito ridden "swamp" (as it was called in 1967 when the project was begun) that characterized the site prior to the development of Marina del Rey.

Thus there is demand for additional recreational boating facilities from both those who already have boats and slips, those who want to own boats and the industries and communities that have financial interests in recreational boating. But recreational boating is only one of many competing interests in a statewide (and nationwide) struggle for access to and use of the coast and coastal waters.

For many decades, it was believed that the ocean and coastline were infinitely able to accommodate new uses and to assimilate our wastes. Protection of the valuable resources provided was not considered and was passive at best. While at one time private ownership encouraged resource protection, the rising taxes and costs of maintaining coastal land as open space and the economic benefits to be gained because of the demands for use of this resource caused the subdivision of much coastal land near urban areas. Resource protection disappeared with these developments. The fencing of private coastal lands in Southern California was greatly accelerated by a court decision (Gion-Dietz c. 1970) which held that vested public easements could occur across unfenced private coastal property, and thus areas traditionally open for casual public recreation were closed.

The tremendous competition for the coastal resource from all concerns has precipitated a complex and extensive

---

[4]Department of Boating and Waterways, Public Information Office, personnel communication, August 8, 1980.

[5]C-MANC, CAPA, and CMPHA op. cit.

set of both national and state regulations. The onslaught of regulations begain in the late sixties, with the formation of the Commission on Marine Science, Engineering and Resources -- the Stratton Commission. The efforts of this commission produced a voluminous report on the condition and uses of our coastal resources and the increasing demands placed on them. The Commission provided recommendations and guidelines for managing and allocating coastal resources in the future. Spurred on by the momentum of the Commission, Congress in 1969 considered over 100 bills related to the protection, management, and allocation of these resources.

Individually, states began to pursue their own more specific regulations. California, much to its credit, was a leader in developing regulations with the intent to allocate for balanced multiple use. These laws were coupled with increasingly strong laws to prevent pollution, protect existing resources, prohibit discharges, preserve species and habitats and to establish sanctuaries where limited activity may be pursued.

The proliferation of these regulations has led to frustration and anger by those who wish to use the coast, whether for an innocent day at the beach or for construction of a multibillion dollar industrial complex. For those seeking access to the coast, whether for development or utilization, even regulations intended to balance use of the coast appear often to be obstructionist and deny access.

These regulations have a direct impact on recreational boating. The regulations designed to allocate can expand the tremendous opportunities boating offers or can effectively force recreational boating to become a sport of the wealthy by eliminating opportunities for new marina development or so delay these opportunities that construction costs and thus use costs skyrocket. Regulations preventing habitat destruction may also cause the costs of new marina construction to escalate to the point that only the wealthy can afford the slips. Regulations preventing discharge may prevent the use of power boats, despite the fact that the Environmental Protection Agency worked with the Boating Industry Association in 1973 to study the impact of recreational boating marine engine use and determined that the impact was negligible. Regulations establishing sanctuaries or preserves may eliminate harbors of refuge for the boater or traditional destination sites so that multiday trips are no longer feasible, except between developed harbors. Recreational boating has lot to loose in this maze of regulations.

By the end of the 1970's the recreational boating community was reeling from the series of regulations that were increasingly affecting their interests. Figure 2 shows the typical process encountered in applying for the necessary permits for marina development or expansion in California. Somehow the politics of the times had out run them, and boating interests realized that they could not ignore the problems and have them go away. The boating community had to get involved in politics to preserve and enhance their recreational pleasure.

Boating interests are not newcomers to the political scene. Because of the lack of natural harbors and estuaries in California, it was recognized early on that development of new boating facilities in California was going to require a cooperative effort between public and private interests. The nature of breakwaters and bulkheading necessary for marina construction would necessarily have to support other related facilities. Thus there is a simbiotic relationship between marinas and restaurants and other shoreline commercial establishments in California -- perhaps more so than elsewhere in the nation.

In 1945, the first statewide boating organization, the California Marine Parks and Harbors Association, was established to help in the development of small craft harbors and the creation of harbors of refuge along the coast. Its interests were primarily administrative and development oriented. Formed by those people involved in harbor administration and seeking the development of expanded facilities, CMPHA worked effectively to gain needed support for the creation of harbors of refuge and the expansion of marina facilities. Under their leadership, a state agency was established to address the concerns of harbors and waterways. Today this agency is called the Department of Boating and Waterways and is housed in the Resources Agency.

More than two decades later, an organization to address the concerns of the individual boater was established in California -- Boat Owners Associated Together, BOAT! Meanwhile nationally, the boating industry had formed two separate industry oriented groups, which today have merged to form the National Marine Manufacturers Association. Before the merger, the Boating Industry Association became aware that their concerns with national legislation could be better pursued with the help of boat users, so under their auspices the National Boating Federation was formed to provide information to the boating community and further their user oriented concerns. BOAT is affiliated with the National Boating Federation.

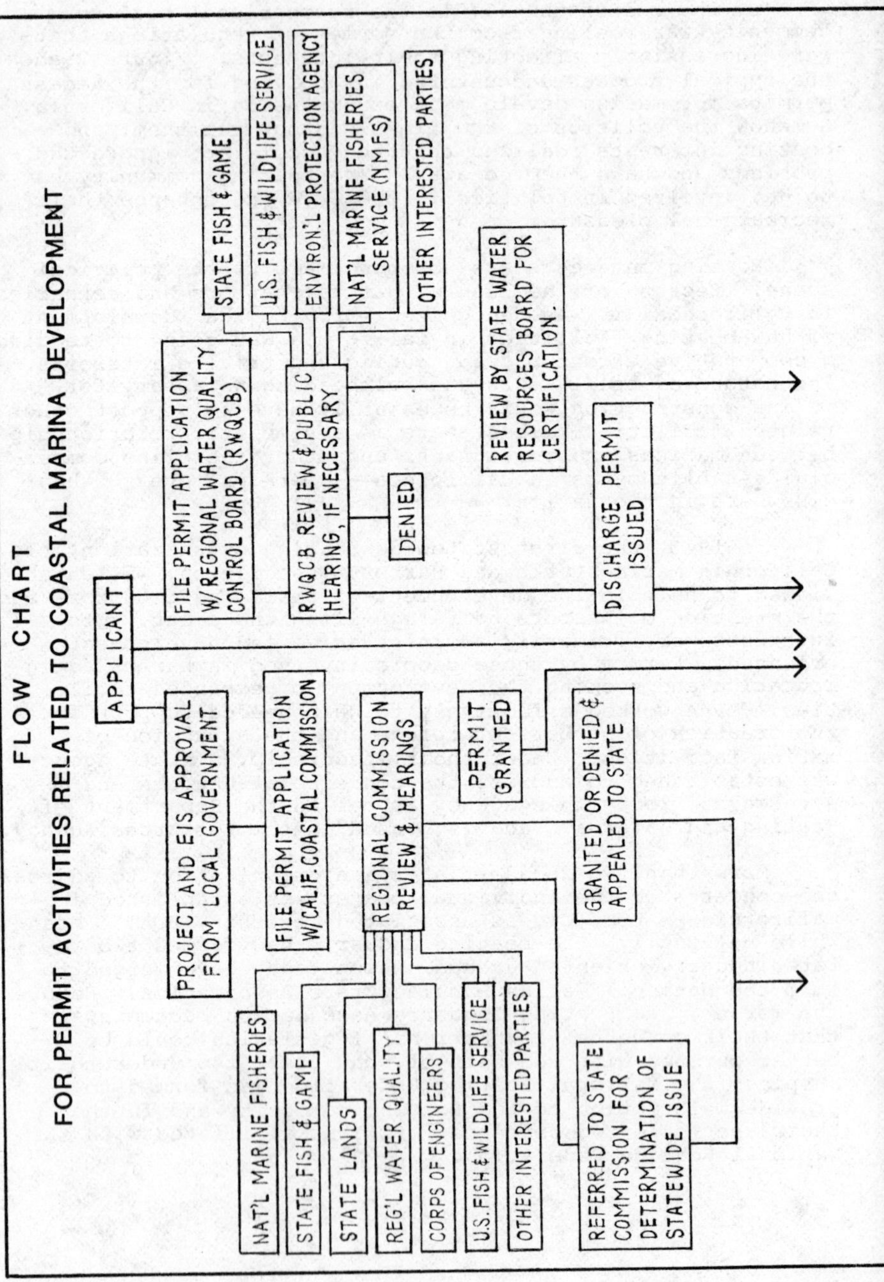
FLOW CHART
FOR PERMIT ACTIVITIES RELATED TO COASTAL MARINA DEVELOPMENT
APPLICANT
FILE PERMIT APPLICATION W/REGIONAL WATER QUALITY CONTROL BOARD (RWQCB)
STATE FISH & GAME
U.S. FISH & WILDLIFE SERVICE
ENVIRON'L PROTECTION AGENCY
NAT'L MARINE FISHERIES SERVICE (NMFS)
OTHER INTERESTED PARTIES
RWQCB REVIEW & PUBLIC HEARING, IF NECESSARY
DENIED
REVIEW BY STATE WATER RESOURCES BOARD FOR CERTIFICATION.
DISCHARGE PERMIT ISSUED
PROJECT AND EIS APPROVAL FROM LOCAL GOVERNMENT
FILE PERMIT APPLICATION W/CALIF. COASTAL COMMISSION
REGIONAL COMMISSION REVIEW & HEARINGS
PERMIT GRANTED
GRANTED OR DENIED & APPEALED TO STATE
NAT'L MARINE FISHERIES
STATE FISH & GAME
STATE LANDS
REG'L WATER QUALITY
CORPS OF ENGINEERS
U.S. FISH & WILDLIFE SERVICE
OTHER INTERESTED PARTIES
REFERRED TO STATE COMMISSION FOR DETERMINATION OF STATEWIDE ISSUE

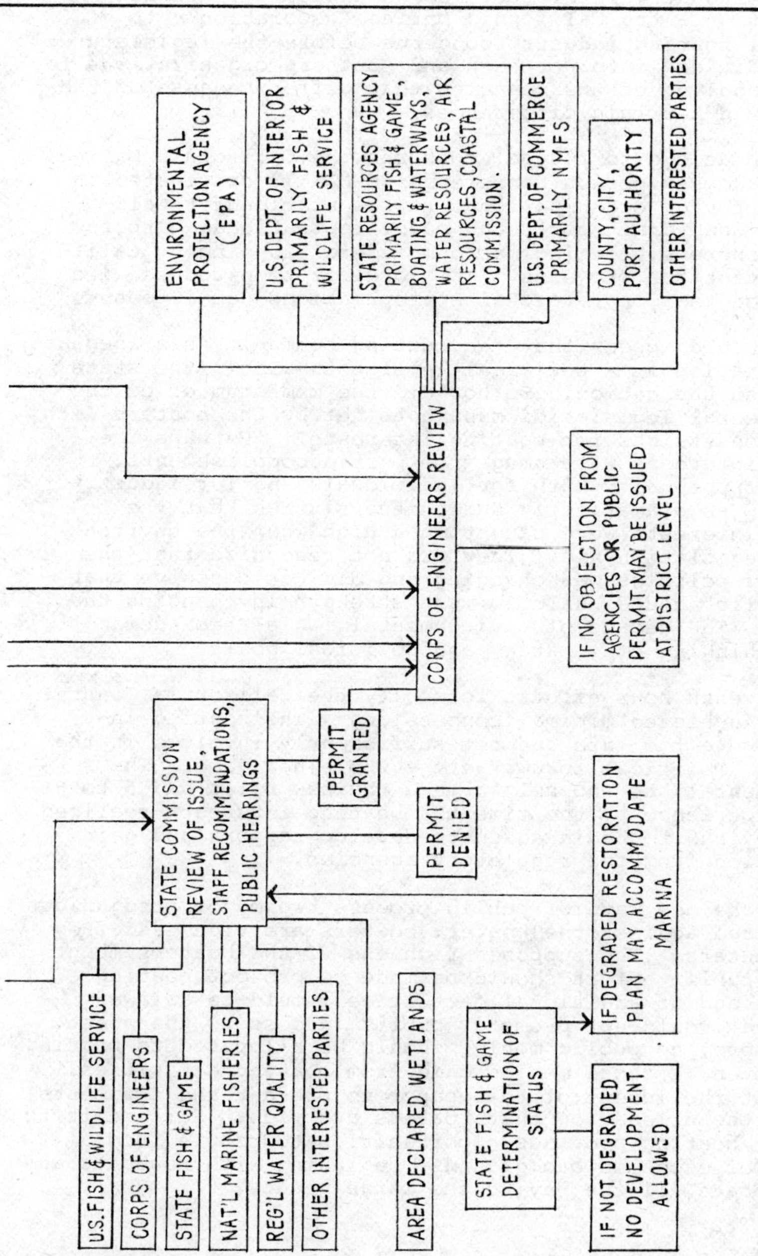

MODIFIED FROM SOURCE: CALIFORNIA MARINE AFFAIRS & NAVIGATION CONGRESS

FIGURE 2

The industry interests also developed a statewide organization - the California Marine Association - to represent boating industry concerns before the legislature. This is divided into northern and southern organizations for better handling of the concerns that differ because of the climate and terrrain differences.

With an eye to federal concerns the California Marine Affairs and Navigation Conference (C-MANC), originated in 1956 to provide organizational representation for California's deep draft ports and major recreational harbors before Congress, seeking maximum federal funding of California navigation projects. This organization focused its efforts on the Congressional Appropriations Committees.

It would appear that the boating community has adequate representation at a political level both within the state and across the nation. So how did the momentum of of the environmental legislation manage to get by the boaters with so few concessions for boating interests?! Perhaps the boating interests were smug from past accomplishments. Boating opportunity both for individuals and for industry picked up so dramatically during the sixties that the boating interests were caught on a high when the environmental legislation hit. They did not recognize that the nature of politics was changing and did not perceive that "the public" could build so much strength in changing the process. The decade of environmental legislation demands some rethinking of strategy and political goals.

Eleventh hour efforts to modify the California Coastal Act only achieved minimal concessions. The boating interests were too late and not sufficiently involved in the process. Political forums were a whole new game. The environmentalists who helped make the new rules had 5 to 6 years experience by the time the boating interests realized that they had to begin again to develop rapport and gain recognition from the regulatory agencies.

In the new wave of public process two major predjudices were voiced against the boater: boaters are rich and they are polluters. The supposed richness of the boaters meant that no public effort should be made to protect boating concerns and no additional facilities should be allowed since that would not promote "public" access to the coast. Furthermore, no public monies should be allocated to boating needs, even if those monies came from boating activities. It is not the intent of this paper to present the arguments against the allegation that boaters are rich. Suffice it to say that boating provides opportunities to people from a variety of economic backgrounds, beginning with rowboats and cartop boats all the way to the large yachts.

With respect to the second allegation, some people contribute to the pollution of the environment more than others -- some of all people, not just boaters. Again, it is not the intent of this paper to develop the arguments concerning boater innocence in matters of pollution. However, it should be noted that some studies have been done that show storm runoff from a single storm to do greater damage to a bay than 5 months of heavy boating activity[6], or that show that marinas can provide a support habitat for juvenile fish species[7].

These predjudices have made it extremely difficult to be heard by the regulatory agencies. Education, through the provision of information, is thus a major component in developing increasing rapport with these agencies.

A major educative effort was mounted by the California Marine Affairs and Navigation Conference and the California Association of Port Authorities in cooperation with the California Marine Parks and Harbors Association. In 1979 they developed a Port Caucus program to provide State Legislators with insight into the serious government related problems facing California's commercial and recreational harbors. Representatives of these organizations found, as is often the case, that legislators were unaware of the way in which certain laws impacted the public benefits derived from California's commercial and recreational harbors. Two basic problems were identified that threaten the benefits provided to the State by the harbors: 1) uncontrolled and uncoordinated regulatory agency activity and 2) lack of a strong State harbor and navigation policy governing the protection, enhancement, development and use of harbors. An educative document was prepared and all the legislators were invited to participate in a day long discussion and tour in San Francisco Bay. Over 40 of the State legislators attended or sent chief aids to explore and discuss the nature of the benefits impacted, the regulatory problems faced by the State's harbors, the perceived causes and actions considered necessary to mitigate or eliminate the problem.

---

[6] Neil Ross, personnel communication, spring 1973.

[7] Scott W. Nixon, Candace A. Oviatt, Sharon L. Northby, Ecology of Small Boat Marinas, University of Rhode Island Marine Technical Report Series No. 5, 1973.

The participation of the boating interests in the decision making process is essential to achieving expanded and upgraded facilities and expanded opportunities for use. The boating organizations play a major role as watchdog, keeping the boating public informed of issues that are being addressed by the legislature and regulatory agencies.

The organizations can also effectively speak with legislators and regulatory agency staff regarding issues. However, representation by organizations at public hearings may not be sufficient to achieve a desired specific goal. There has been some discussion among the boating leaders in the Southern California area of joining forces rather than maintaining the separate identities of each group. Each time the issue is raised it is rejected because, even though some would like to turn over the burden of responsibility to someone else, there is appreciation in today's political scene that diversity of representation is essential. In fact even the organizations alone often find they can not effectively make headway before a regulatory agency without additional representation from individuals, especially when the issue is a specifically local concern.

The individuals are difficult to get out in force, even when the issue is of vital concern to them. It has often been said that pooling any kind of recreational interest group is the most difficult in the world, because people are involved in recreation to get away from the organization and encroachments of the day-to-day routines, rules, restrictions, and politics. They would like to rely on someone else to carry that responsibility. And yet it is the ones who are most immediately affected who can make the biggest impact on regulatory commissions and agencies.

In recent years there have been a number of abomnible failures on the part of the boating interests to achieve their goals and some successes. A brief review of some of these failures and successes should help explain the intricacies of the problem.

Proposition 20, the forerunner of the California Coastal Act of 1976, was a ballot proposition that gained the support of many, including boating interests. Unfortunately, each person who voted for it had a different perception of what it would accomplish and how it would be done. Boaters have a considerable interest in a good ocean environment, for without it their own recreation activities would be spoiled. Thus, to many boaters, Proposition 20 held much promise and there was little perception that the development of a coastal plan resulting from Prop 20 would be something unfavorable to boating interests.

Few if any boaters became involved in the development of the coastal plan as elements were written providing guidelines for development and preservation along the coast. Only when the plan was presented to the legislature did the boaters first begin to realize that boating interests had been only marginally recognized in the plan. The California boating organizations did not become aware of the extent of the problem until it was essentially too late to take organization action. A few boating leaders spent days in Sacramento trying to make a last ditch attempt to have amendments made to the bill before the legislature passed it. A few concessions were reached in the section specifically related to boating, but there were many other sections in the bill that were worded in such a way as to have priority over the recreation section that were left untouched. And there was no statewide uprising of boater interests to prevent the legislature from passing the bill before them.

The ensuing formation of the Coastal Commission and their permitting activities and iron hand over the development of Local Coastal Programs has been a major obstacle since that time.

Unfortunately, antagonism and confrontation, which is often the feeling that boating interests have toward these regulatory bodies, can be their own worst enemy in the political game. The battle for building a new marina in the abandoned seaplane anchorage in Los Angeles Harbor is an example of the effect of antagonism and confrontation on the decision making process.

To begin with, the harbor interests have turned increasingly toward commercial needs and have maintained a long time policy of giving only month-to-month leases to recreational boating facilities. Thus, the harbor is home to some of the most decrepit marinas on the coast. It is in this area that some facilities have been demolished in favor of commercial port related facilities. The long time boating community of the harbor is becoming increasingly discouraged in their attempts to upgrade and expand their facilities.

Over the years, several plans were developed for a new marina complex in the Cabrillo Beach area of the harbor, but each one failed for a variety of reasons. The local boating interests felt that the seaplane anchorage provided a protected harbor suitable for small craft, that could be developed into a small craft harbor much more easily and inexpensively than Cabrillo. Every time the Port developed new plans for harbor expansion, the boating interests would fight them for inclusion of a marina in the seaplane anchorage. The local boating interests gave little if any support for expansion in Cabrillo.

Actually, here the Coastal Commission helped the boating community because the Coastal Act did say that existing harbors were one of the few areas where new recreational boating facilities could be built and should be encouraged. In response to the Coastal Act regulations the Harbor Department began to develop some new plans for a marina in the harbor. Again the plan was for a Cabrillo marina, but the planning looked more carefully at the traditional problems and made an effort to resolve those problems.

The plan was completed with considerable local community support and the concept went before the Harbor Commission for approval. Not only were the local boating interests represented, but also several of the regional and statewide boating groups were present to testify. The local boating interest group came forth with its traditional antagonism and confrontation. The other groups made a major effort to present the issues of concern and to show willingness to work with the Commission and Harbor Department to see this marina accomplished. That had been their intended strategy prior to the meeting and efforts were doubled after the local interest group made their remarks and the Chairman took time out to lambast them saying "if this resolution before us today passes it will be in spite of the local boating interests, not because of them." The resolution did pass because of general community support, but the boating interests remain aloof from the Harbor Commission and have done little to develop an on-going rapport that might ease future participation in the decision making process at this level.

The broadbased statewide and regional boating interests have made some progress in developing better interaction with the Coastal Commission. When the Los Angeles Port Master Plan went before the Commission for approval, boating interests not only provided testimony but met with staff ahead of time to discuss the issues of concern. The West Channel/Cabrillo Recreation Project was included in the plan but with no commitment to completion. Before the Commission, the boating interests took the position that the Coastal Act requires the Commission to actively encourage increased recreational boating facilities within existing harbors -- not just passively accept recreation. The boating interests requested that the Commission condition the Master Plan approval with a commitment to a net increase of new boating slips by 1985. While the Commission staff had suggested that the approval be conditioned with a completion commitment for salt marsh restoration, the commitment to slip completion was a new idea for them and the Commissioners. After some discussion, an additional condition for completion of 1000 net new slips by 1985 was added and the Master Plan was approved. While these additional 1000 slips will only go a short way towards meeting the total projected demand for new slips, this kind of commitment was a positive step for the boating community.

A combination of efforts on the part of individual boaters and the boating interest groups in California has also been successful in changing the wording of the Proposed Marine Sanctuary Designation for the Channel Islands and the related regulations. The Office of Coastal Zone Management, which was responsible for developing the proposal, insisted that they had no intention of restricting boating from the islands. However, the language of the designation document and regulations was at first sufficiently unclear that anyone wishing to interpret a restriction on boating could do so. After the first round of language changes, the regulations very clearly denied boaters the opportunity of anchoring ("disrupting the seabed in any way") or of running their engines within a mile of the islands!

The boating interests finally got press involvement and began to gain momentum on the issue. Twice the Office of Coastal Zone Management arranged to have a representative meet with boating leaders although no public meeting was held in the Los Angeles/Orange County area where most of the affected community resides. A final letter from the Director of the Office indicated wording changes that would satisfy boating concerns and clearly permit traditional boating use of the islands to continue.

The key elements of success have been a combination of 1) competent and dedicated organization representatives, 2) information sharing and education programs, 3) generating and sustaining a vocal local constituency, 4) development of rapport with regulatory agencies, and 5) the ability to adapt boating organization goals to keep up with the times.

There are however some goals that cannot be realistically achieved. The coast is a limited, finite resource and there is only so much that can be done to expand or extend coastline opportunities, whether recreational, commercial or residential. Even at best there will not be enough space along the Southern California coast to meet all of the projected future demand for slips, if in fact the demand continues to increase at present rates. Boat traffic congestion is already enough of a concern in this area that technical information is being refined to measure levels of congestion and administrative measures are being sought to ease congestion. Modern slip configurations may allow for slip increases within existing marinas if sufficient upland is available for parking and other support facilities. Along the Southern California coast there may be opportunity for 2 to 3 new major recreational boating complexes, portions of which would serve to replace existing facilities that are in conflict with surrounding land uses.

Today's regulatory maze will make the achievement of those opportunities tougher, requiring greater determination. In a way, the environmental regulations force all those wishing access to the coast to show how much that access really means to them, how hard they are willing to fight for it.

Some of the demand for recreational boating is certain to fall off in the intensity of the struggle. Those with the greatest determination to achieve expanded opportunities for recreational boating will explore other ways of doing it, for instance, more sophisticated use of dry storage facilities or creation of offshore islands for marinas served by ferry.

In any case dealing with agencies and legislature will require education efforts, sophistication of organizations, generation of vocal local constituencies, and dedication. It is up to the boating interests to determine if their desire and demand is great enough to accept the challenge of regulation and to win in the public forum.

ACCESS IS THE OPIUM OF THE PLANNER

Wallace Kaufman*

Planners, engineers, scientists, government officials and environmentalists all agree that the coastal zone has special problems. The word access names one of these special problems. How and when these people talk about access may be much more important to coastal planning than anyone thinks. Take these examples.

A man in Connecticut frequently loads a station or bus with inner city kids and hauls them off to cozy little beach towns. There by hook or by crook he gets them onto the beach and dares the property owners to deny them their access to the public trust lands.

The California Coastal Commission notifies the exclusive Hollister Ranch Association that its members who paid up to a half million each for ocean front privacy will have no more development permits unless they provide public access. This includes shuttle buses and pedestrian paths along the bluff in front of the picture windows and patio doors.

*Box 54B, Rt. 1, Pittsboro, NC 27312

The biggest coastal engineering project in Florida is the creation of an artificial berm at Miami Beach. This project helps the econnmically depressed tourist industry which ruined the natural beach with its seawalls, bulkheads and groins. In helping the resorts with this $65 nillion temporary beach, the Corps made private interests swallow the public access pill. The industry had spit out this pill repeatedly until conventioneers and vacationers began to seek out places where they could throw a beach ball and watch bikinis on a sandy beach.

All three of these examples have shaped our notions of the "access" problem in the coastal zone. At least they have focused the meaning of the word access. If you examine the literature about "access"in the coastal zone, you will see we are playing a very sad and dangerous verbal trick upon ourselves. The newspapers frequently run articles about private land owners and citizen groups or government fighting over access to the beach. In November the Boston Globe's series on the coastline ran the headline, "Access to Beach A California Battle Cry." The four volume procedings of Coastal Zone 78, a gathering of several thousand planners, engineers and scientists, contain eleven papers about access--all about the beaches.

At this point someone might say, well what do you expect? That's what we are talking about--beach access. Exactly my point. Whenever the word access is mentioned, we are talking about beaches. That was not always the case. And if we look at the traditional and legal well springs of the access debate, we will see the problem is much more important than who gets to jab their beach umbrella in what sand or how far you're allowed to chase your beach ball.

Without certain ancient legal precedents beach access would be as much an issue as access to the gold vaults of Fort Knox or your neighbor's bathroom. As western law and civilization began taking on its modern dress some 2,000 years ago, long before the rise of industrial capitalism and socialism Roman law declared the sea and its shores to be <u>res communes</u>, things of common use by all citizens. The fish of the sea, said the Romans, belonged to all nations. Many kings and feudal lords later tried to erase this public ownership in medieval times. They sank a lot of ships and starved a fair number of people, but they could not hold onto the sand and the water.

When King John signed the Magna Carta in 1215 he conceded any right he had to give away or sell the tidelands. As English common law developed, it became clear that while the king still held title to the tidelands and he could make grants of them to his favorites, his title was burdened with the public trust. Common law wrote on all the grants and deeds made by the king the overriding rights of the general public, rights which were to remain a kind of invisible ink on all conveyances of tidelands, even in the New World. This invisible trust was clearly read and declared by a wise judiciary over several centuries. The accumulated decisions of many courts gave birth to the <u>public trust doctrine</u>. Its existence was not in question when the American colonies sought their independence.

The first important cases under the public trust doctrine did not concern beach access. They were rulings on private claims to submerged lands. In 1869 the Supreme Court declared the Illinois Legislature "incompetent" to give submerged lands under Chicago Harbor to the Illinois Central Railroad. In 1894, in Shively v. Bowlby, the Court applied the same limits to Congress. In that decision the Court repeated the words of Chief Jus tice Taney fifty years earlier. Taney, in Martin v. Waddell, had held that the public trust lands were necessary to the founding of the Republic. Taney allowed that "the men who first formed the English settlements could not have been expected to encounter the many hardships that unavoidably attended their emigration to the New World and to people the banks of its bays and rivers, if the land under the water at their very doors was liable to immediate appropriation by another as private property. Taney allowe also that the king had granted whole colonies to individual proprietors, but he said that the river bottoms, tidelands and seabeds were conveyed only as "a trust for the common use of the new community about to be established."

The reasoning of Romans, medieval Englishmen and Supreme Court justices no doubt reflects the common sense of the times. It was summed up most ably by a Dutch jurist who lived in the seventeenth century when Europeans were rapidly colonizing foreign shores and littering the sea floor with war ravaged ships and sailors. Hugo Grotius, in the treatise Mare Liberum declared:

> . . . that which cannot be occupied, or which has never been occupied, cannot be the property of any one, because all property has arisen from occupation . . . that all that which has been so constituted by nature, that although serving some one person, it still suffices for the common use of all other persons, is today and ought in perpetuity to remain in the same condition as when it was fistt created by nature.

Again in Rights of War and Peace he said:

> There is a natural reason also, which renders the sea, considered in the view already taken, incapable of being made property: because occupancy can never subsist, but in things that can be confined to certain permanent bounds.

Such common sense simply reflected the facts that guman beings cannot breathe underwater and that technology had not adapted them to underwater habitation or work.

In the past thirty years we not only have perfected breathing and diving apparatus, but we have designed huge underwater habitats. We have special motor vehicles for underwater life. We have stationed whole villages of workers on towers in the midst of the most powerful seas on earth. Indeed, in some places we now find it easier and safer to live under the sea than on top of it. I am not entirely joking when I say that some people, especially heart and lung patients, find life more promising in ocean bottom pressures than beneath the Los Angeles or Denver smogs. Not only continental pollution but the exhaustion of many land based resources are turning our eyes seaward. Or are they?

At the very moment in 2,000 years of Western history when we should be thinking about access to the entire range of coastal resources, coastal zone planners and the media have narrowed the concept of access to bathing beaches. Of course the entrepreneurs and multi-national corporations with the technology and finances to exploit the coastal zone have not objected. By nature, private enterprise likes to lead its private life, Privacy means less competition, less scrutiny from nervous and often vicious public and thus less government interference. The only commercial notice we get of activity in the coastal zone are ads emphasizing the extreme risk and human sacrifice of rugged men serving our economic well being. Such work, we are led to believe, is carried out to keep us living in the style to which we are accustomed.

The result of these circumstances is that we are being denied a vital national debate over public access to the whole variety of coastal resources. Fisheries, energy supplies, mineral deposits and the underwater ecosystem in general have a much greater economic and social value than bathing beaches. By tying the concept of access to recreational beaches we may be guilty of selling out our last frontier for beach balls, bikinis and the internal combustion engine.

We the people and our government, of course, do not own any oil drilling towers or any seabed mining ships. As we did with

Lewis and Clark in the Louisiana Purchase, we have equipped and sent out agents to explore and chart and catalog our territory. But should we do with the coastal zone what we did with the American West--give it away piecemeal to the resource developers?

I have been a businessman, a real estate developer and a seller of timberlands for many years. I know what free enterprise has contributed to this country. I know that free enterprise can be inventive and efficient. I don't believe with ardent socialists that the motives of free enterprise are subversive to the public good. And I am enough of a realist to know that whatever coastal resources are utilized will be gathered and delivered to us through a capitalist system. But I know enough of history and my fellow capitalists to know we are about to commit all the mistakes of the old frontier.

Our development of this great continent made us for a while the most secure, the most powerful and the richest nation on earth. Maybe for a brief time we were even the happiest, although I admit the popular consensus is that certain people--ignorant of automobiles, cities and suburbs--are more blissful. Be that as it may, our method of developing the continent has not been an unmixed blessing. As a whole people we were really happy and booming only between 1946 and 1976, out 200th anniversary. A thirty year reward for 200 years of work. And a good part of our people were left out of that celebration anyway or were spending their money and energy trying to get into the party. We are just now discovering and beginning to pay the staggering costs of our development methods. In only five generations, one long lifetime, we have used up most of our high grade ores and most of oil and gas. Almost all our rivers are polluted and it's often dangerous to breathe the air in our cities. We are no long er strong and secure on our island continent, but we dance to the tunes played by tiny dictatorships and racist minority governments.

We now pay hundreds of billions to clean up rivers and lakes. More billions to dispose of toxic wastes. And more to replace timber stands and stop soil erosion. The U.S. Department of Agriculture that replacement of the nitrogen, phosphorous and only 25% of the potassium lost each year in soil erosion would cost $18 billion. It now requires two to five times as much fertilizer for a bushel of Illinois corn than it did a hundred years ago. Who knows how much of our multi-billion dollar health bill is due to uncontrolled carcinogens and to air and water pollution?

Once you have begun to balance the books and update the ledgers in any failing business, it becomes easy to see how and where the deficits began. To risk a broad generalization abo ut our history, it is clear that we forgot to enter many of the costs of development on the ledger. That is to say the government never asked resource developers to pay all the costs of their activity. So now those costs must be entered on our personal ledgers and the tax ledgers. Some of the costs no one knew about. A prudent business person, however, always budgets a contingency factor. In summary, the continent seemed so vast and human beings so small that we couldn't do any real damage. What we lacked most never really seemed necessary--comprehensive planning and detailed accounting.

We cannot make the same mistake in the coastal zone. Some pelple tell us not to worry, we are now overplanning and over-regulating and being too detailed in our consideration of every minnow and microbe. Certainly we spend great amounts of money studying, planning and regulating. We gather together great convocations of scientists, engineers and planners and bring in powerful political officials to kick off the annual debates about coastal zone issues.

All this is promising, but is it an empty promise? So far it seems so. Foodstock harvesting technology increases, creating some temporarily larger catches, but many foodstocks decline in average age and overall numbers. In 1879 Chesapeake Bay provided 117 million pounds of oysters and clams, five times the present harvest and twice the 1979 national consumption. Over 8 million acres of the coastal zone in the Gulf of Mexico alone has been leased for oil drilling. Fishing and lobstering on George's Bank will soon compete with oil. Fully 60% of our undiscovered gas and oil is estimated to be on the Outer Continental Shelf. Despite many vital marine habitats, to date have have only two tiny marine sanctuaries--the 100 acre Key Largo Coral Reef and the Monitor Marine Sanctuary.

The specific facts are a clear alert. The historical and eonomic context is a four alarm fire. Government policy has not changed fundamentally. No it has not changed despite laws with names like Coastal Zone Management Act, Anadromous Fish Conservation Act and the National Ocean Pollution Research and Development and Monitoring Planning Act. The fundamental policy was most clearly spoken by President Gerald Ford during the 1973 oil crisis. He adopted as our destiny energy independence. In fact he simply said something like this: let us find and use up faster our remaining natural resources. This is the same policy Jimmy Carter has followed with more sideshows, with a greater concern for environmental tidiness and some tears for the victims of his "moral equivalent of war." Mr. Carter has put most of his energy money into nuclear power and coal. Last year he directed the Secretary of the Interior to double the rate of Outer Continental Shelf oil and gas leasing, offering 26 million acres between 1980 and 1985. Ronald Reagan would like to do almost the same thing with almost no regulations.

Since rules and regulations are always bent, changed, broken or ignored to the tune of national policy, we are facing a coastal resource development program of awesome proportions and implications. It comes at a time when inflation is making people look toward new supplies of foods, energy and minerals. It comes at a time when world-wide shortages of these basics tempt both profiteers and humanitarians to move fast and throw caution to the waves.

The situation differs radically from our development of the continent. In the last three centuries we had no giant technologies that could destroy whole ecosystems within a decade. The continental U.S. was a much vaster area than the narrow coastal zone. We had to work our way mile by mile into the interior whereas over half of our people already live within fifty miles of the coastal zone, many of them in it.

Business interests and government policy makers will argue--as they do about overcutting in national forests, super tankers and western strip mining--that everybody will benefit from intensified development of coastal resources. They argue everything from increased tax bases from resort building to energy independence.

The picture is clear if we imagine a national timber crisis when private holders, now overcutting their lands 20% faster than regrowth, come to cut the trees on the street in front of your house. It's a public right of way, the crew foreman says. And the town council needs the revenue. Besides,if you go along, you'll be able to buy a set of bookshelves a little cheaper.

Many Americans do not feel as strongly about coastal resources as they do about the view out their front window. 1980, The Year of the Coast, is changing that. And Americans have a right to be concerned. Given where most of us live,the coastal zone is our fro nt yard. Just as we own the trees in our yard and on the public thoroughfares, we own the fish and plants, the oil and minerals beyond the high water mark. Our ownership is as firmly engraven in the law as if great grandmother and great grandfather had given us a recorded deed. It just so happens, as it does in all big families, that a whole bunch of us heirs hold the title, some 220 million of us. We are so many, in fact, that a trustee manages our inheritance. That trustee, for better or worse, is the government. Like most bank trustees it has not been doing a job that really fulfills its fiduciary trust. It invests our wealth with political and economic chums who may not pay the best interest. It assumes the worst about our character, that we are only interested in short term rewards. Worst of all, for its own short term political benefits and to give economic favors to its buddies, our trustee is spending away the irreplaceable capital of our inheritance.

Perhaps we are simply suffering from the committee system. Most trusts are administered by a single trustee. Ours is administered by a committee of several hundred. It is called Congress. Some people say the members of this committee gain their jobs using heavy contributions from big special interests. Thus it is said the committee pays back its supporters with our resources.

This is, perhaps, a cynical view of government. If government does not operate this way, it's certainly producing the same kind of results. To test that conclusion we owners of most of the coastal resources need only ask one simple question--is the value and income from our property rising or falling?

There are pluses and minuses, but so far the answer is clear. Our income is dwindling and our capital is shrinking. Making it easier to dip into our capital--the land, the minerals and the livestock--will only temporarily increase our income and then only if the contractors do not siphon it all off through their own profits and tax writeoffs. Things look even bleaker for those who will inherit our property. More and more people see that beneath the billions spent on laboratories and computers, beneath additional billions for salaries and conferences, our heritage is being sold off too cheaply and quickly.

The professional scientists and ecnomomists who gather the data for our trustees know very well what is happening. Their graphs and charts and tables tell a story as clear as any bank's profit and loss statement. Government officials must make more compromises with politics in their assessment. When the Assistant Administrator for Coastal Zone Management in NOAA writes that North Carolina, my state, has a coastal zone "among the most strictly controlled in the country," I realize the government official requires a little more fantasy to protect his morale. And what happens when we hear Congress debate the amendments to the Coastal Zone Management Act and the Barrier Island Bill? Congress wants only to fine tune its original Coastal Zone bill, an act which at best made things get worse more slowly. The president, who issued a 1977 executive order that no government agency should encourage any more development on barrier islands, now cannot even let us cease our multi-billion dollar subsidies for insurance, utilities and storm relief for people who are about to build billions of dollars of luxury homes and motels in high hazard areas.

Planners have done a little planning in the coastal zone. A few beach communities for the rich preserve many of the coast's more obvious attractions--for the rich. And planners have convinced the Corps of Engineers, local and state governments to provide access to places where citizens can throw beach balls and lie around in bikinis and sometimes without bikinis.

Meanwhile just off shore an over the dunes in the marshes and the bays, our pocket book is being robbed and checks are being written against our future.

The solution is not so hard to see and the tools are at hand. The management of the public's resources in the coastal zone must recognize the realities of nature and law and history. This means comprehensive planning for the region as a unit and it means realistic accounting by developers.

The coastal zone extends from the estuaries to where the continental shelf drops into the ocean's abyss. The only part of this the public does not own and cannot claim access to is that very tiny portion of dry sand and upland vegetation that edges the water. Because this developed area is such a small inholding amidst such a valuable and vast public domain, the public has rights to protect itself and include all coastal areas in its planning.

In the coastal zone no planning is worth much that does not recognize the expanse and integrity of the public domain. The conclusion we must draw is rather simple. We are far behind in extending serious land use planning all the way to the edge of the continental shelf.

Land use planning is often an anathema to citizens who are afraid their rights will be violated. Few states have passed statewide land use plans. But most states can pass coastal zone plans. And most of these plans already cover the more controversial areas from oceanfront landward. It is here that conflict results from a clash between private and public interests. From the high water line seaward it is all public property and yet we have no land use plan for this area. We have only conflicting laws regulating conflicting interests.

We should now extend coastal planning out onto the continental shelf. In many places sensitive claims have already been staked by private enterprise--small fishermen and giant oil firms. Chaos and strife is already brewing. Georges Banks is one of the world's riches fishing grounds. Lobsters from the coasts of Connecticut, Rhode Island and Massachusetts walk as much as 150 miles to winter on Georges Banks. Lobster larvae and many other small forms of marine life are sensitive to oil, yet Goerges Banks may be one of the greatest oil reserves on the continental shelf. We are far behind in planning the future of this submerged land. Since there is no method of making oil companies pay for any damage that doesn't insult the eye and the nose, and since oil brings more profit than lobsters or fish, the oil companies will probably dominate and the lower income people of the world,who us e very little fuel per capita, will eat less protein and pay more for it.

Fortunately in many areas the continental shelf is unclaimed and undeveloped. If there is no comprehensive planning, vested interests will get what amounts to a high class squatters' rights. They will feel, perhaps with some justice, that any exp post facto regulations, are unfair and the costs should be born by the regulators.

We have been lax in our planning because, as on any frontier, there seemed to be a lot of room and very little of interest. Now, we see diverse interests lined up like settlers prepared for the

Oklahoma land rush. They are not poor people with buggies and shovels.

We have only to think of the shelf's obvious resources and uses to realize we are about to see one hell of a fight. Here is a short list:

oil and gas deposits
valuable metal ores
phosphates for fertilizers
sand and gravel for construction and beach restoration
shipping channels
military exercise areas and dumps
garbage dumps replacing big city landfills
sewage sludge depositories
underwater parks
unique wildlife habitats
marine animal migration routes
artificial and natural fishing reefs and nurseries
historic sites
sewage outfalls
deepwater ports
nuclear power plant sites
ocean thermal electricity generating platforms
oil and electrical transmission lines and pipes
geological and biological research sites
commercial fisheries
salvage sites

Some of these things have very high immediate values. Some are difficult to value in any time frame. But they all have value of some kind to their owners--the people of the United States. Because of their location these resources will be developed and handled almost exclusively by large corporations. Large corporations, and small ones for that matter, usually invest by looking at such things as equity yield and the present value of future benefits. Using a discount rate of 15% or even a modest 10% the future value of most things ceases to have any present worth if it cannot be utilized in ten or fifteen years. This is simply hard economics. If we are to develop coastal resources with our free enterprise system, someone has to pay attention to that part of the public inheritance which has a value beyond a single generation. We must especially pay attention to those things which cannot be measured in the same dollars used by business.

It is almost as if we as a nation are being given a second chance to develop a frontier. Isn't this the chance every planner and visionary wants? This time there are no Native Americans to manipulate. There is only ourselves and our fellow citizens. We can accept twenty-four dollars worth of trinkets and turn over our heritage to the same uncontrolled forces and short-sighted policies that have put us in hock for the exploitation of our continent. Or we can begin now to draw up a plan which will protect that heritage. Tha states and the fedederal government must undertake the very difficult political task of acting in harness to protect the income and productive capacity of the entire coastal zone. If they do not, they will probably guarantee economic and social chaos in the near future. If they can plan together and put long term public interest first, they will have guaranteed real public access to the coastal zone.

OPPORTUNITIES AND PROBLEMS IN PLANNING THE NEW JERSEY PINELANDS AND COASTAL ZONE*

John W. Sinton**

Abstract

This paper discusses opportunities and problems in planning for the New Jersey Coastal Zone and its adjacent uplands which are called the Pine Barrens or Pinelands. It is an example of the general opportunities and problems which exist in the Atlantic Coastal Plain from New Jersey to Florida.

Opportunities consist of coordinated efforts to preserve and manage freshwater and estuarine resources and plant and wildlife habitat, all of which are interdependent on the coastal region and adjacent drainage basins. The three major problems are boundary questions, interagency cooperation and policies and decisions about the location and type of development which will occur in the area as a whole.

---

* Presented at Coastal Zone 80, The Second Symposium on Management, Conservation and Utilization of the Coastal Zone, Hollywood, Florida, Nov. 1980.

** Associate Professor of Environmental Studies and Resource Planning, Stockton State College, Pomona, N. J. 08240

Introduction

The coastal area from New Jersey south to Florida is a series of similar physiographic provinces called the coastal plain, a sunken coast with barrier islands, salt marshes and tidal rivers which grades slowly into sandy uplands to the west. With the 1972 passage of the federal Coastal Zone Management Act, the joy of planners, who had been waiting for federal support of their state and local planning efforts, began to be mixed with caution. As planners began to delineate their coastal zone boundaries, it became clear that the success of preservation and management efforts depended as much on adjacent upland areas as on the coast itself. For example, how could coastal managers ensure the health of shellfisheries in estuaries and bays if fresh water entering those bays had been diverted or polluted twenty miles upstream?

The realization that the coast itself and the coastal plain were intimately connected was, of course, not a new concept, but it was a lesson harshly defined by the frustrating exigency to delineate an arbitrary boundary. This article discusses the problems and opportunities of coastal zone and coastal plain planning and management in southern New Jersey.

Description and Boundaries of the Coastal Zone and the Pinelands

There are no distinct hydrologic, pedologic, ecological or cultural criteria which separate the coastal zone from the Pine Barrens. Rather, a series of gradients define the two interdependent regions. The geologic cross section from Route 206 at the edge of the Pine Barrens to the coast clearly shows the interconnectedness of soil formations and aquifers (Figure 1). A vegetation map would suggest some differences in forest types between the interior Pinelands and a narrow intermittent coastal strip. The upland forests contain oaks and pitch pine with some shortleaf pine while Atlantic white cedar or swamp hardwoods (red maple, black gum and sweetbay) dominate the lowlands. Along the coastal strip are persimmon, holly and occasional hickory. The tidal marshes grade from spartina marshes in the saltier sections to cattails, bulrushes and wildrice in the fresher sections.

The Pinelands and coast are also culturally interrelated. Economic activities on the coast, such as fishing, boat building and commerce are dependent on the forests and fields of the Pinelands. One might well find a bayman, who clams on the bays in summer, spending most of his time as a Piney in winter cutting cedar or cordwood well inland.

Given the strong interrelationships between Pinelands and coast, how then is one to define boundaries? The coastal zone boundary was established in the 1973 New Jersey Coastal Area Facility Review Act (CAFRA), and included the ocean and bay areas under state jurisdiction from the Raritan River to the Delaware Memorial Bridge at the head of Delaware Bay. Most of the boundary inland from the coast was dictated more by matters of convenience and compromise than by natural or cultural boundaries which, as explained above, are nearly impossible to determine. The coastal zone's inland boundaries, therefore, generally follow major roads such as the Garden State Parkway on the ocean section, and county roads on the bay section. Boundaries were expanded along the major tidal rivers -- the Mullica, Great Egg Harbor and Maurice.

The definition of the Pinelands boundary, on the other hand, was a more tortuous and politically hazardous process. Early in the twentieth century botanists (Harshberger 1916; Stone 1911) had defined the natural boundaries of the Pine Barrens using soils and vegetation information, and in the mid-1960's the late ecologist, Jack McCormick (1970) defined what has become known as the "central core" of the Pines, an area which includes some of its most noteworthy features, such as the pygmy forests, and which has few inhabitants.

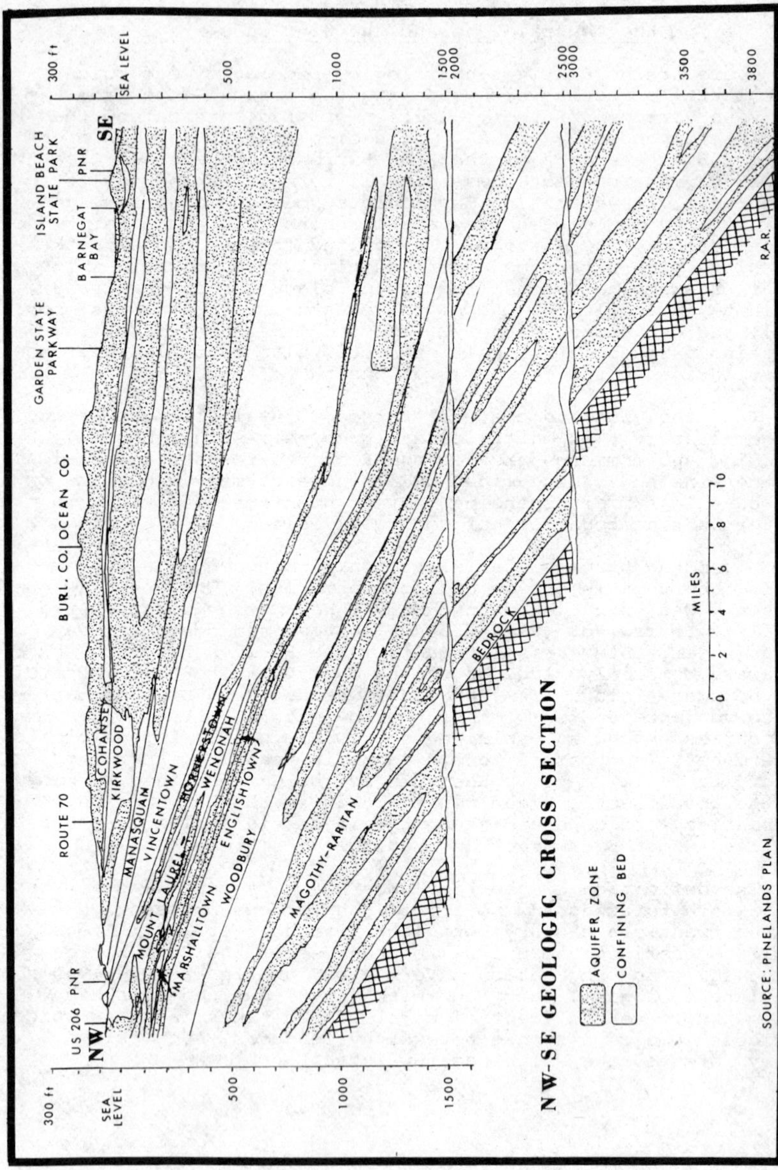

NW-SE GEOLOGIC CROSS SECTION

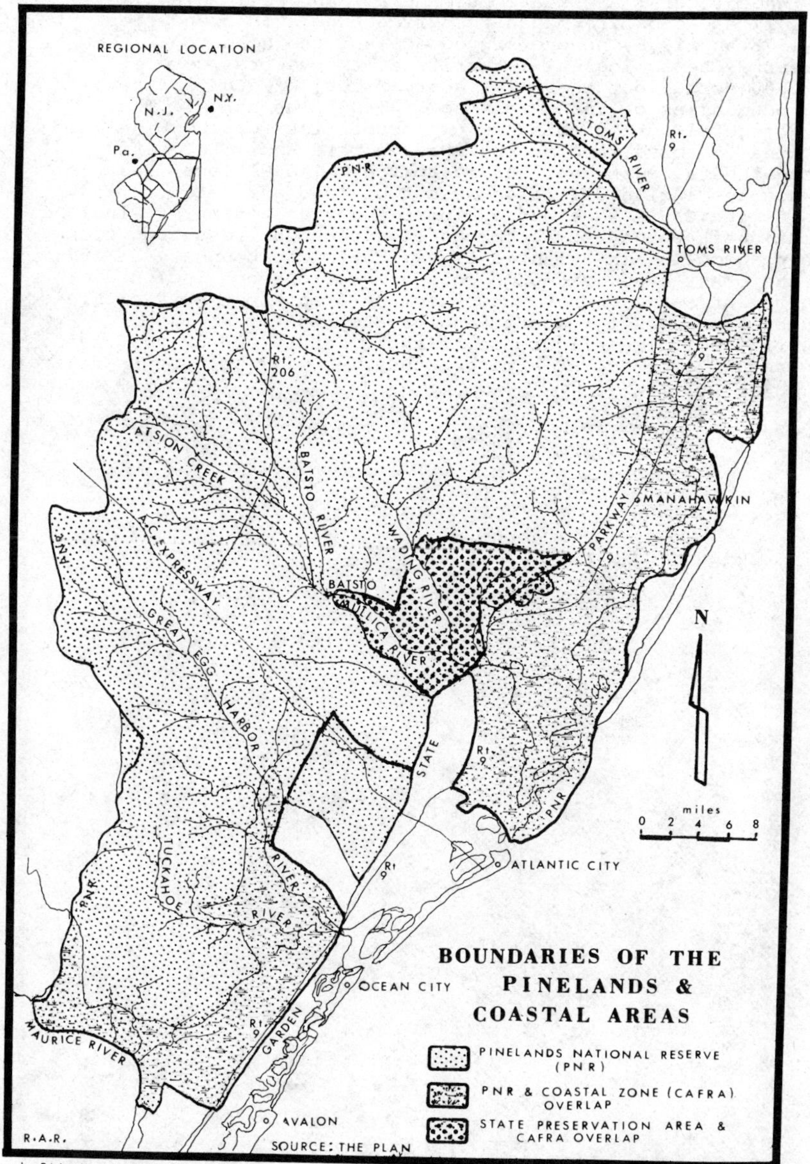

BOUNDARIES OF THE PINELANDS & COASTAL AREAS

n.b. BLANK AREAS FROM TOMS RIVER TO CAPE MAY ARE IN CAFRA ONLY.

When federal and state bureaucrats and politicians, as well as academics, began to define Pinelands boundaries in 1978 and 1979, no one argued against the need to preserve the ecological integrity of the central core of 368,000 acres, but there was a great deal of argument about the boundaries of the remaining 750,000 plus acres. The 1978 federal act established the boundaries for the Pinelands National Reserve, and the boundaries were set at a meeting among federal personnel and academics. The National Reserve boundaries included a large amount of land and water already inside the CAFRA zone and excluded a large section of Pinelands in Atlantic County to accommodate development resulting from casino gambling in Atlantic City. The state boundary of 1979, on the other hand, excluded coastal sections but included central Atlantic County except for an area known as "Perskie's Thumb", so named for State Senator Stephen Perskie, a powerful political figure in Atlantic County and in the state. The result of the overlapping boundaries will be discussed later in this article.

Recent History of Planning Efforts

The Coastal Zone

New Jersey's first major piece of coastal legislation after World War II for the area south of the Raritan River was the 1970 Wetlands Act which prohibited any disturbance of coastal (Spartina) marshes without a permit. Its sister act, the Coastal Area Facility Review Act (CAFRA) passed the state legislature in 1973 and granted what is now the Division of Coastal Resources (DCR) of the Department of Environmental Protection (DEP) authority to regulate developments of 25 or more units and major public and industrial facilities in the coastal zone; the act also gave DCR power to preserve environmentally sensitive areas and to plan for rational development and management of natural resources. In 1977 the New Jersey Office of Coastal Zone Management produced a coastal management plan for the region from Sandy Hook to Cape May and Delaware Bay. In May of 1980, the recently created Division of Coastal Resources, which subsumed the Office of Coastal Zone Management, published the draft of the New Jersey Coastal Management Program, a plan which is expected to be officially adopted in late summer of 1980.

The planning process for the coastal zone began in 1974 and, after more than six years' work, will be completed in 1980. The power to plan and manage coastal resources consists principally in the DCR's power to grant or deny permits for major development, to preserve intact salt marshes and riparian lands and to provide planning funds to counties and municipalities.

The Pinelands

Until the late 1970's planning efforts in the Pinelands, or Pine Barrens, were sporadic and consisted chiefly of a series of state and federal reports on the resources and potential of the region. In 1978, however, Governor Brendan Byrne made it clear that he wanted to make "Pinelands protection" a cornerstone of his two-term legacy to the state. He established the Governor's Pinelands Review Committee, the report from which served as a basis for present federal and state legislation governing the Pinelands.

In 1978 Congress passed the 1978 National Parks and Recreation Act which established the concept of a national reserve and funded the Pinelands project as a model for the legislation. The national reserve concept establishes national interest in a region with unique resources, and grants federal funds for the planning and management of the region on the state level. In other words, it is up to the state to take the lead in planning, and to meet federal requirements for comprehensive planning if the state expects to receive federal funds for both planning and land acquisition.

In 1979 the New Jersey legislature passed the Pinelands Protection Act which, in essence, authorized the state to fulfill the requirements of the 1978 federal act. The state act established a fifteen-member Pinelands Commission -- one member appointed by the Secretary of the Interior, seven by the governor and seven members appointed by the affected counties of Atlantic, Burlington, Camden, Cape May, Cumberland, Gloucester and Ocean. The act's intent was, of course, to insure that major decisions would be made on a broader basis than those based on local interest alone. The state act also provided the Commission with a staff led by the executive director. The Commission is, to all intents and purposes, an independent agency whose decisions can be vetoed by the governor, but cannot be changed by any other state agency nor by the legislature. The act granted the Commission broad powers, including zoning powers if they wished them, in order to preserve, protect and enhance the significant values of land and water resources and to encourage indigenous economic activities and compatible development.

The Pinelands Commission's planning effort began in earnest in late fall of 1979, and the draft plan was completed seven months later, in June 1980. While the plan provides for several resource management programs, its real power consists in the establishment of zones throughout the region which restrict, prohibit or allow all kinds of uses from gravel mining to the construction of a log cabin -- in other words, its zoning powers. Within 18 months of the plan's adoption, which is expected in December 1980, all municipal master plans and zoning ordinances must be certified by the Pinelands Commission as being in conformance with Pinelands standards and land-use requirements.

Opportunities for Planning and Management

Of the many opportunities which exist for coordinated planning of the Pinelands/coastal zone region, none is more obviously important than water resources because the ecological health of bays and estuaries depends on water quantity and quality in the Pinelands. The prognosis is good for continued high quantity and quality water in the area, and both the Division of Coastal Resources and the Pinelands Commission have established policies and procedures to protect and maintain water supply and quality.

In the state-designated Pinelands region all but the sparsest development is prohibited in most recharge zones, and development in other areas must adhere strictly to the Commission's water quality standards for discharge. For example, surface water quality must have the following characteristics:

pH 3.5-5.5
BOD 5.0 mg/l
Nitrate 0.17-2.0 mg/l,
depending on region's designation,
Total Phosphate 0.02-0.07,
depending on region's designation.

In addition, interbasin water transfer is strongly discouraged and export of groundwater from the region is prohibited.

The Division of Coastal Resources lacks the broad powers of the Pinelands Commission and, therefore, cannot enforce as stringent water quality standards. Development in the coastal zone and outside the Pinelands area must adhere to the state's Division of Water Resources standards which, in the coastal zone, are on the high side of Pinelands' requirements (2.0 for nitrates and .1 for phosphates in most areas). Furthermore, most coastal zone streams and lakes are not tied to Pinelands' standards for pH, but rather to general New Jersey standards which require decisions on a case-by-case basis.
Given the nature of the boundaries and bureaucracies, one then might suggest that on the west side of the Garden State Parkway water must have a pH of 3.5-5.5, while on the east side of the road, the pH must be 6.5-8.5. Such an anomaly is unlikely to occur, of course, and the DEP regulations make it impossible to increase pH from major developments on such a grand scale.

The Division of Coastal Resources' power to limit development in the coastal zone is, itself, limited. The DCR has no zoning powers and its permit powers are restricted to residential developments of 25 or more units, energy facilities, large public facilities including wastewater and landfill systems and agriculture, industrial and marine facilities. The DCR has strong policies to protect recharge areas and limit sprawl development, but its powers are not sufficiently strong to prohibit scattered small development. In short, the DCR does not have the power to enforce its policies as has the Pinelands Commission.

Both the Pinelands Commission and the DCR, however, are empowered to prohibit any development or alteration of wetland areas whether salt or freshwater, whether stream, lake, bog, marsh or wetland forest. Many rare and threatened plants and animals inhabit wetlands such as various orchids, the curly grass fern, the Pine Barrens tree frog and the bog turtles. Wetlands also provide habitat for some of the region's most important game such as waterfowl and muskrats.

So the outlook is good for maintenance of healthy water quantity and quality and, hence, good fish and shellfish habitat along the coast. Likewise, the future of wetland habitats of all sorts appears promising. The opportunities to maintain the health of such vital resources have been worth the political battles to enact the state's major pieces of legislation. Still, nagging problems remain and, if left unsettled, may negate a rosy future.

Problems

Attentive readers will have already foreseen two major problems in planning and managing the coastal zone and Pinelands: the question of boundaries, and the two different approaches toward development taken by the Pinelands Commission and Division of Coastal Resources. To these will be added a third problem which is interagency coordination.

Boundaries

There are two major sets of overlapping boundaries: the overlap of state-designated coastal (CAFRA) and Pinelands preservation areas, and the overlap of CAFRA and federally-designated Pinelands National Reserve areas (Figure 2). The first of these poses little jurisdictional difficulty since New Jersey Pinelands legislation states that where CAFRA and Pinelands Preservation areas overlap, the Pinelands Commission's plan takes precedence over the DCR procedures.

The National Reserve/CAFRA overlap, however, presents a more difficult problem. The Pinelands Commission has no state authority to enforce its powers in the National Reserve/CAFRA overlap areas, but the federal act requires that a plan be established for the whole national reserve, not just the Pinelands areas. While it appears clear that CAFRA rules and regulations should apply to this overlap area, the Secretary of Interior may not accept this interpretation. It is possible but unlikely that the federal government would reject the Pinelands plan on the grounds that the federal interest is not fully served because the Commission cannot enforce the plan in the overlap area.

Two solutions to the boundary problem are apparent: First, change CAFRA or Pinelands legislation on the state level, or allow courts to decide on who has jurisdiction. The latter is more likely. The obvious lesson from this experience is that one must take extreme care in establishing the boundaries of a national reserve because sloppiness in the first instance leads to magnified problems in the second.

Approaches Toward Development

Orderly and regulated residential, commercial and industrial development is, of course, the key to planning a healthy environment for human and non-human resources. DCR and Pinelands Commission Policy statements on development are similar: development needs to be clustered and placed in areas which will not degrade water resources or wildlife; development should be contiguous to already developed areas rather than leapfrogging; roads and community services must be available in newly developing areas.

it clearly protects vital water and land resources, spends most of its time describing how to regulate development of any and all major facilities.

The result of the two different approaches is to support the growth trends which have existed in the region since the Second World War -- development in the coastal zone with little growth in the adjacent interior. Even without a federal or state Pinelands act, development would have continued to concentrate heavily in coastal areas, but the Pinelands Draft Plan insures that it will. Coupled with its lack of jurisdiction over the National Reserve/CAFRA overlap area, the approach the Pinelands Commission has taken also insures that the bulk of development will be handled not by itself, but by DCR.

There is an ironic possibility, therefore, that attempts to preserve the watersheds inland from the bays and estuaries may help degrade those very areas. Developments, which would not be permitted west of the Garden State Parkway by the Pinelands Commission, will be built on the east side, especially if the developers are not required to apply for a permit from DCR (e.g., if the development is less than 25 residential units).

Interagency Coordination and Cooperation

During its first four years, the state Office of Coastal Zone Management (now a part of DCR) had difficulty coordinating its efforts not only with municipalities and counties, but with sister agencies in the Department of Environmental Protection. As any experienced planner or resource manager can testify, attempts to implement programs can be severely hampered by lack of coordination on any level of government. More recently DCR has clearly overcome many of its handicaps relating to other agencies, and it is now in a position to implement its programs with the help of other agencies.

To implement its plan, the Pinelands Commission will, of course, have to establish strong ties with other public agencies. The Pinelands Draft Plan includes a section on intergovernmental coordination, parts of which show promise, especially its attempt to tie municipal and county agencies into review and permit processes.

Interagency coordination problems arise where they most obviously should have been first solved -- cooperation between the Pinelands Commission and DCR. From discussion in the previous sections of this paper, it is clear that the Pinelands and coastal regions grade one into the other and that, if at all possible, the two should be treated as a whole. Despite this, and it is no one person's or agency's fault, the region is separated and treated by different agencies with different approaches. Surely the two agencies must exert all their efforts to plan and manage the region in a rational and cooperative manner.

Approaches to development regulation as opposed to policy, however, differ radically, and much of the difference can be traced to the legislative acts which granted DCR permit powers and gave the Pinelands Commission zoning powers. To give an example of the broad powers of the Pinelands Commission: The Pinelands Plan establishes in the central sections of the Pine Barrens a 750,000 acre area composed of the Preservation District in the north and the Forest Districts of the center and south. In the Preservation District, residential development is limited to people in resource-related occupations in the district, and to long-time residents. Recreation, forestry, berry agriculture and some gravel mining are also allowed. In the forest districts, besides recreation, forestry, agriculture and some mining, residential development is restricted to an average of 17 units per square mile with 3.2 acre minimum lot sizes. All municipal master plans and zoning ordinances must reflect the requirements established by the Commission.

The DCR has no power equal to the Pinelands Commission. Instead, it has established in its management program a detailed series of location policies for special areas of ecological or cultural concern, general water areas and general land areas. There are also a series of use policies for particular energy-related, industrial and commercial facilities; and a series of resource policies for air, water, soil, vegetation, wildlife and cultural resources. The power to enforce its policies is limited to DCR permit powers plus legislation which stringently protects water resources and wetlands.

Three other factors affect the different approaches to the two agencies. First, the Pinelands Commission is an autonomous body and, as such, is less prone than the DCR to be influenced by other state divisions of government; for good or ill the Pinelands Commission can develop more free-wheeling approaches than other state agencies. Second, whereas the DCR had six years to produce a final draft of the coastal management program, the Pinelands Commission had only seven months to produce its Pinelands Plan and, therefore, the Pinelands Plan is bound to be less detailed and specific in terms of resources and locational analysis than the Coastal Program. Last, and perhaps most significant, the thrust and general intent of the two plans stem from different philosophies -- preservation for the Pinelands as opposed to regulation of development for the coast.

The tone and content of the Pinelands Plan is geared to preserving the ecology and water resources of the region in their present state, and the message is unmistakable when one reads the draft plan. The plan is a challenge to the limits of power which any regional planning commission in the United States might acquire to restrict land use for the sake of preservation. It almost begs the U.S. Supreme Court to decide on the extent to which a government agency can use environmental criteria to preserve a region; it is at the heart of arguments on the taking issue. The Coastal Management Program, on the other hand, while

Yet there is evidence to the contrary. In DCR's Draft Coastal Management Program there are two sections which suggest how DCR relates to the Pinelands. In one section DCR policy states that "coastal development shall be consistent with the intent, policies and objectives" of federal and state Pinelands legislation. In the other section the program summarizes Pinelands goals and, in referring to DCR's requirement to ensure its policies are consistent with the Pinelands' in overlap areas, the program states, "DEP will report to the [Pinelands] Commission that the Rules on Coastal Resource and Development Policies, which are highly specific and detailed with respect to the soils and vegetation of the Pinelands, are consistent with the purposes of the [Pinelands] Act". On reviewing the Draft Pinelands Plan, one finds not a single reference to the Division of Coastal Resources or its Coastal Management Program - even in the intergovernmental coordination section - although several other divisions of the DEP are mentioned.

Results of the lack of cooperation between the Pinelands Commission and DCR are ominous. It is not this paper's purpose to point fingers of blame, but to suggest that without coordination between the two agencies, preservation and rational development in the region are imperiled.

The lesson here is that enforced cooperation must be built into similar legislation in the future both in New Jersey and other areas of the country. Like New Jersey, other states have no comprehensive regional planning structures and are therefore forced into planning on a case-by-case basis. Lack of interagency cooperation is a frequent, but by no means necessary result of our predicament.

The author looks forward to coordinated planning efforts in other sections of the country which can learn from some of the problems of the New Jersey experience. With the creation of the national reserve concept, coastal planners have yet another tool to protect and enhance coastal resources; but care must be taken to delineate careful boundaries to enforce interagency cooperation and to avoid disparities in the powers and authorities of regional planning agencies which can endanger the health of the environment and our citizens.

References

Forman, Richard T., ed. (1979). Pine Barrens: Ecosystem and Landscape. Academic Press, New York.

Harshberger, J.W. (1916). The Vegetation of the New Jersey Pine Barrens, an Ecologic Investigation. Sower Co., Philadelphia.

McCormick, Jack (1970). The Pine Barrens: A Preliminary Ecological Inventory. Report No. 2, N.J. State Museum, Trenton.

New Jersey Dept. of Environmental Protection, Division of Coastal Resources (1980). Proposed New Jersey Coastal Management Program. Trenton.

New Jersey Pinelands Commission (1980). Draft Comprehensive Management Plan. 2 vols. plus maps. Trenton.

Sinton, John W., ed. (1979). Natural and Cultural Resources of the New Jersey Pine Barrens. Stockton State College, Pomona, N.J.

Stone, W. (1911). "The plants of southern New Jersey with especial reference to the Flora of the Pine Barrens and the geographic distribution of species," Annual Report of the N.J. State Museum, 1910. Trenton.

CORAL REEF PROTECTION DURING BEACH RESTORATION IN FLORIDA

By: Richard Spadoni and Thomas Campbell, P.E.
Arthur V. Strock and Associates, Inc.
Deerfield Beach, Florida

INTRODUCTION

Beach erosion is a significant problem along much of the Florida coastline, threatening structures, property and the tourist industry. The Florida Department of Natural Resources classifies 210 miles of recreational beachfront as in a "critical state of erosion" and 310 miles as near-critical. A majority of the beaches along the southeast Florida coast are badly eroded. Wide beaches provide recreational area and protect coastal development from flood damage in the event of a major storm. Beaches in a critical state of erosion lack the dimensions to provide either ample recreational area or storm protection.

A number of measures are available to control beach erosion and to restore beaches. The dredge and fill operation is presently the most accepted method of accomplishing beach erosion control in Florida. Sand needed for beach fill is obtained from offshore areas where large accumulations of sand occur in depressions in the continental shelf. The fill is removed from these areas, referred to as borrow areas, by dredging and placed on the eroded beaches. The advantages are obvious in that the required recreational space and storm protection are provided with the construction. After the completion of construction an aesthetically pleasing beach remains, similar in appearance to the beach prior to erosion.

Ocean conditions off the southern coast of Florida are condusive to coral growth, much of which is found within a mile of the coastline. The operations which occur during a beach restoration project are potentially destructive to corals and coral reefs.

Water turbidity and sedimentation can have detrimental effects on nearby coral reefs. Water turbidity can block sunlight utilized by small plant-like cells called zooxanthellae which live within the coral animal. The zooxanthellae produces oxygen which is used by the coral.

If sunlight is reduced to the point that the zooxanthellae perish, the effect on coral can be lethal. Likewise, sedimentation over the coral reef can smother coral colonies. An additional potential for coral reef damage exists as a result of the possibility of collision of the cutter head, suction head or anchors with the coral reef. Such destruction is localized but can be quite severe in nature.

The protection of coral reefs in close proximity to a dredge-and-fill operation is an important addition to the administration of beach restoration projects. Protection of coral reefs begins in the initial planning stages of a beach restoration project and continues beyond project completion. With careful planning and supervision, coral reefs can be protected while yet another natural resource, the beaches, are restored.

DELRAY BEACH ENVIRONMENTAL MONITORING PROGRAM

In 1978, a maintenance beach restoration project was completed in Delray Beach, Florida; 570,000 cubic yards of sand was placed over 1.7 miles of shoreline. Beach fill was obtained from an offshore borrow area located approximately one-half mile from the shoreline. The borrow area was situated parallel to the coastline, extending a distance of 6,600 feet. Lying seaward of, and adjacent to, the borrow area is a major coral reef system. The eastern edge of the borrow area, at its closest point, was approximately 400 feet from the coral reef.

As part of the administration of the beach restoration project, the firm of Arthur V. Strock and Associates, Inc. conducted an environmental monitoring program primarily concerned with the preservation of the coral reef. The study began the last quarter of 1977 and extended through 1978.

Conclusions of the monitoring study were that no detectable coral reef damage occurred as a result of turbidity or sedimentation generated from dredging activities. Sedimentation rates and turbidity measurements taken during dredging operations compared favorably with pre-construction data. Through most of the beach restoration project, no damage to corals was observed at established coral reef monitoring stations or any other location on the coral reef. However, a small area of reef damage was discovered soon after the project had been completed and the dredge demobilized and moved northward along the coast. The reef damage was discovered during a diver survey and appeared to have been caused by an anchor and anchor cable from the dredge. The area of coral reef damage occurred directly seaward of the last working position of the dredge and was discovered within a few days of its occurrence. It is most probable that the damage occurred during the dredge demobilization operations.

Subsequent observations of the area of the coral reef damage over a number of months revealed a rapid recovery of most organisms affected by the damage. Except for the single incident of coral reef damage, the beach restoration project did not adversely effect the coral reef system.

The experiences of the Delray Beach environmental monitoring program provided the impetus for this paper.

PRE-CONSTRUCTION CORAL REEF PROTECTION PLANNING

To be successful, protection of coral reef communities must begin in the initial phases of beach restoration planning. Engineers must determine if viable coral reef communities may be threatened by beach restoration activities. Seismic survey information gathered to locate suitable deposits of sediment for beach fill also yields information which can be used to locate coral reefs. Isopach charts generated from seismic information map out hard bottom areas on the seafloor including areas of coral reef. Diver investigations must then be conducted to delineate coral reef areas from rock outcroppings.

Once the coral reef is located, engineers can evaluate the impact of various possible borrow area sites on the coral reef. The ideal situation is to locate borrow areas far enough away from coral reefs to avoid an impact on the reef. Often, however, the location of suitable sediment is near a coral reef and no alternative is available. When the reef is located in close proximity to a proposed borrow area, further evaluation of the coral reef may be necessary to locate reef areas and attempt to avoid these areas by adjusting final borrow area dimensions and boundaries to allow for prevailing currents. Examination of the coral reef by biologists may be needed to delineate the areas of heavy concentrations of benthic invertebrates, including corals, which are particularly susceptible to damage from dredge operations. Borrow areas can be selected to avoid highly productive areas of the coral reef.

Other problems which engineers may encounter in attempts to protect the coral reef include the possibility of the borrow area lying seaward of a coral reef. Floating pipelines over the reef may leak causing excessive sedimentation and turbidity over the reef. Submerged pipelines over the reef can damage delicate benthic invertebrates. Coral reefs which run parallel to the shore often contain breaks where the sandy bottom emerges separating long coral islands. Pipeline lays can be restricted to the reef breaks if adequate diver investigation is undertaken to map these breaks. Another solution to this problem is to restrict the beach restoration project to the use of hopper dredges which are containerized and do not use pipelines to transport sediment. Care should be taken, however, to restrict the dredge load transported across the reef to avoid spillage and subsequent sedimentation over the coral reef.

Prior to finalization of the location of the borrow area, a diver investigation of the borrow area should be conducted to verify that no isolated patches of coral reef lie previously undetected within borrow area boundaries. This procedure will not only limit possible destruction of coral reef, it will also avoid the problems of pumping rock onto the restored beach.

The engineer should provide the contractor a chart which not only illustrates the location of the borrow area, but also delineates the location of the coral reef areas to be avoided, areas of acceptable pipe lays and anchorages. Also, dredge contractors may not be familiar with the environmental sensitivity of coral reef areas and should be instructed as to the importance of avoiding coral reef communities.

REQUIREMENTS OF THE CONTRACTOR

Certain requirements of the contractor should be included in contract specifications for beach restoration projects. Foremost should be the requirement that the dredge and related equipment avoid contact with coral reef areas, including avoiding dropping of anchors on the coral reef. Specific mention of anchors should be made due to the nature of the operation of some dredges. Large dredges sometimes use an anchor-winch system to maneuver the dredge cutter head during dredging activities. Anchors may be positioned many hundreds of feet away from the dredge, well out of the bounds of the borrow area. The potential for reef damage is increased and contractors should be advised that coral reefs are to be avoided.

A number of requirements of the contractor will aid in maintaining the dredge within the borrow area and avoiding reef communities. The following requirements should be included in contract specifications:

1. Electronic positioning equipment should be required for positioning of the dredge. If the dredge is operating near areas of coral reef, electronic positioning equipment will provide the accuracy needed to prevent reef damage due to the dredge straying off position and colliding with the reef.

2. Prior to dredging operations, the contractor should place buoys around the perimeter of the borrow area as a visual aid to maintain the dredge within the borrow area. Electronic navigation equipment should be used to locate the buoy positions. If the contractor plans to dredge at night, the buoys should be lighted.

3. The dredge contractor should notify the engineer of mobilization, demobilization and other major dredge positioning changes prior to the change. If the dredge is maneuvered by use of an anchor-winch system, anchor locations must not extend beyond the borrow area without an approval by the engineer. Additionally, floatation devices should be utilized on cables which extend beyond the borrow area to avoid scraping the reef.

4. If the dredge contractor suspects the dredge has collided with hard bottom, the dredge operation should be shut down immediately until the suspicion has been resolved. This may avoid extensive coral reef damage and the placement of rock on the restored beach.

5. The contractor should use sediment screens during dredging operations when it has been determined by the engineer that sediment and turbidity conditions warrant such usage. A strong current in a direction which carries the dredge plume over a coral reef may be a circumstance where the use of sediment screens are necessary to protect a coral reef system.

REQUIREMENTS OF THE ENGINEER

The engineer has the responsibility of administering the beach restoration project which includes protection of the environment. As such, the engineer should be required in contract specifications, to:

1. Monitor dredge contractor activities to insure compliance with environmental safeguards written into the contract specifications.

2. Utilize electronic positioning equipment to verify dredge and anchor locations and provide approval of proposed dredge or anchor relocation positions. Prime responsibility for positioning will remain with the contractor.

3. If practical, buoy areas of coral reef near the borrow area providing a visual aid as to the location of the areas of coral reef.

4. Provide an environmental monitoring program for coral reef communities in close proximity to the beach restoration project. The purpose of the environmental monitoring program will be to detect any short or long-term changes in the coral reef environment. Short-term changes may require modifications to the ongoing dredging project, while long-term changes may result in changes in administration of future beach restoration projects.

5. If the coral reef lies between a borrow area and shoreline, the engineer should locate "breaks" in the coral reef. If breaks exist and if the break is of sufficient size, the engineer can supervise the placement of submerged pipeline in the break. The engineer should also carefully monitor the line during dredging operations to insure no leaks occur or other problems arise which may be detrimental to the coral reef. The engineer must also supervise disassembly of the submerged line to avoid environmental damage.

6. The engineer should not provide services to, or be employed by, the contractor in any capacity. The duties of the engineer include verification that the dredge contractor fulfills contract specifications to protect the environment. A possible conflict of interest should be avoided.

7. The engineer should be authorized to "stop-work" during a beach restoration project if probable coral reef damage will, or is, occurring. Every effort should be made to correct the problem and resume beach restoration activities.

THE ENVIRONMENTAL MONITORING PROGRAM

Environmental monitoring of a reef system in close proximity to a beach restoration project is part of the engineering inspection and supervision of the project. An environmental monitoring program can provide for early warning of possible sedimentation damage to nearby coral reefs. Inspecting and supervising engineers can then institute directives to modify or cease potentially damaging dredging operations.

Initially, diver surveys utilizing underwater photography conducted along the natural reef and within proximity of the borrow site will determine the general overall makeup of the reef community. Isopach charts can be used in the location of the coral reef. Monitoring stations should be chosen on the basis of two criteria: first, the general location of the stations should be selected to situate stations adjacent to the borrow area spaced at approximately equal distances along the project area. The second criterion is to select the station locations to include a variety of hard coral species.

The stations can be established to provide a specific area of reef surface to monitor. A comprehensive program of photography and photogrammetry conducted at each monitoring station will provide a visual record of the coral reef during the project. Each station should be photographed prior to, during, and after completion of the dredging project. These photographs provided visual records of the monitoring stations from the pre- to post-construction period. Individual coral colonies photographed and rephotographed with a macro-closeup camera will allow both identification and detailed information as to the relative well-being of the organism.

Sediment jars placed at each reef station and periodically collected and analyzed as to content will provide information as to the changes in sediment rates and composition. Background information must be collected prior to the dredging and continued to be collected after the completion of the dredging operations to establish the natural rates of sedimentation. In conjunction with gathering information on sedimentation rates, turbidity data can be measured from water samples taken at each monitoring station. Background turbidity samples should also be collected prior to, during and after dredging to determine the levels of turbidity normally occurring in the vicinity of the coral reef for comparison to turbidity measured during dredging.

A Littoral Environmental Observation Program conducted concurrently with the monitoring study throughout the life of the project will provide information as to the effects of the weather on the reef environment. The ocean variables observed and recorded should include data on wave action, currents, water clarity and weather conditions. This information gathered twice daily, seven days a week and recorded on standarized littoral environmental observation forms, can be used to correlate ocean conditions and sedimentation rates found on the reefs.

CONCLUSION

The protection of coral reef communities adjacent to beach restoration activities requires the joint efforts of the project engineers and dredge contractors to be successful. The development of an environmentally acceptable beach restoration project begins in the initial stages of project planning and concludes with the evaluation of follow-up studies of the coral reef. With proper planning and project coordination, the protection of coral reef communities can be accomplished while valuable beaches are restored.

REFERENCES

Courtenay Jr., Walter R., Hartig, Ben C. and Loisel, Gerard R., "Ecological Evaluation of a Beach Nourishment Project at Hallandale (Broward County), Florida", Volume I Miscellaneous Report No. 80-1, U.S. Army Corps of Engineers, February. 1980.

Courtenay Jr., Walter R., Herrema, David J., Thompson, M. John, Azzinaro, William P., and VanMontfrans, Jacques, "Ecological Monitoring of Beach Erosion Control Projects, Broward County, Florida and Adjacent Areas", Technical Memorandum No. 41, U. S. Army Corps of Engineers, February, 1974.

Marsh, G. Alex, Bowen, Philip R., Deis, Donald R., Turbeville, David B., and Courtenay, Walter R. "Ecological Evaluation of a Beach Nourishment Project at Hallandale (Broward County), Florida", Volume II, Miscellaneous Report No. 80-1, U.S. Army Corps of Engineers, March, 1980.

Raymond, Bill and Antonius, Arnfried "Biological Monitoring Project of the John U. Lloyd Beach Restoration Project", Broward County Erosion Prevention District, July, 1977.

Spadoni, Richard H., "Reef Monitoring of the Delray Beach Erosion Project", Shore and Beach, Volume 47, No. 3, July, 1979.

Issues in Mapping Atlantic Coastal Ecological Data

by

Paul R. Adamus*
Center for Natural Areas
South Gardiner, Me.

Thomas E. Bigford*
National Oceanic
and Atmospheric Administration
Washington, D.C.

James R. Hynson*
Center for Natural Areas
South Gardiner, Me.

INTRODUCTION

Many Atlantic coast planners and decision-makers have long been captivated by the idea of producing preferably one report or atlas which would compile and accurately synthesize all of the best available data on the coast's ecological resources and yet be relevant for most uses, accessible to users, and easy to understand and apply. This grand and attractive idea has been pursued over the years by a wide range of agencies and institutions, with varying purposes, courses of support, and degrees of success. In this paper we selectively review such efforts, especially mapping efforts, describing both their accomplishments and potential limitations to their use. Special consideration is given to characteristics of the Atlantic coast data base which make elusive the goal of a truly holistic, accurate, and practical synthesis of the coast's ecological data.

HISTORICAL PERSPECTIVE

One of the first attempts to comprehensively map important coastal resources was published by the American Geographical Society in 1969. Authored by George P. Spinner, the "Wildlife Wetlands and Shellfish Areas of the Atlantic Coastal Zone" was funded through the University of South Carolina by a grant from the Belle W. Baruch Foundation. Resources were mapped at 1:250,000 scale in a series of 12 maps. Only areas with commercially important shellfish and salt marshes larger than 50 acres were mapped. Ownership was mapped using a flexible series of ownership categories. Color was used to distinguish among these, while pattern symbology was used to differentiate the mapped resources. The importance of each Atlantic coast State was ranked according to its contri-

*The opinions expressed in this paper are those of the authors and do not necessarily represent those of the affiliate organizations. The Center for Natural Areas is a private, non-profit resource management firm.

bution to total acreage of "important open shoal water habitat" and "total estuarine zone habitat." Practically no descriptive narrative information accompanied the maps.

With the onset of outer continental shelf oil and gas lease sale proposals in the early 1970s, the USDI's Bureau of Land Management (BLM) began focusing on its responsibilities for considering the welfare of coastal resources in the context of potential impacts from OCS activities, e.g., facility siting or oil spills. BLM contracted several institutions to compile all existing resource data for three contiguous parts of the coast and shelf north of Cape Canaveral. The Research Institute of the Gulf of Maine (TRIGOM) coordinated the northernmost compilation, published in 1974, for the region from the Bay of Fundy to Nantucket Shoals. The "Coastal and Offshore Environmental Inventory, Cape Hatteras to Nantucket Shoals" was coordinated through the University of Rhode Island (URI) and published in 1973. To the south, the Virginia Institute for Marine Sciences (VIMS) coordinated a compilation, published in 1974, for the Cape Hatteras to Cape Canaveral region.

The intent of these three studies was to provide background data, cite data banks and sources of raw data, list ongoing research, and, in general, provide the BLM staff with as much information as possible to assist them in the preparation of required environmental impact statements. Much of the same information later was published in similar format by the Shell Oil Company (Guscy, 1976) as "The Fish and Wildlife Resources of the Middle Atlantic Bight" and in a companion volume (Gusey, 1977) for the Georges Bank region.

Parts of the data compilations were used by BLM in drafting the visuals accompanying the EISs for Lease Sales 42, 43, and 56. These colored visuals, i.e., resource maps, drafted mostly at a scale of 1:1,000,000, included considerably more resource categories than those of Spinner (1969). Shown, for example, are coastal localities of endangered/threatened animals and wintering ranges and migration routes of several finfish species. The visuals, although referenced in the EIS text, are not specifically related to explanatory narratives.

Beginning in the mid-1970s, another approach to ecological synthesis surfaced within the U.S. Fish and Wildlife Service (FWS). Termed "ecological characterizations," these reports have aimed to emphasize ecological function rather than structure and to synthesize ecological information for each coastal region. Mapping efforts associated with the characterizations have been assigned variable importance, depending on the region. As of 1980, FWS has completed Atlantic coast characterizations in Maine and the sea islands of South Carolina and Georgia; both of those studies were funded by pass-through money from the U.S. Environmental Protection Agency (EPA) Inter-Agency Energy/Environment Research and Development Program. While many scientists would agree that the output has been very professional and useful, some non-scientists and planners have found the characterizations difficult to apply. Moreover, it is fast becoming apparent to some users, unless additional funding is found, the time required to complete the characterization process for the remaining ten Atlantic coast segments is too great to meet immediate needs for decision making on energy siting and oil spill contingency planning issues. Also, the characterizations were not intended to rank specific segments of the coast as more important relative to others. Their greatest strength perhaps will be in providing agency biologists with well-synthesized information of a scope and level of detail not available elsewhere.

In the late 1970s, BLM contracted with the Center for Natural Areas to update the earlier BLM compilations. The products of these contracts were two compilations; one covering the Bay of Fundy to Cape Hatteras (Center for Natural Areas, 1977) and a second from Cape Hatteras to Cape Canaveral (Center for Natural Areas, 1979). BLM program delays and the overall delay of Atlantic lease sales until late 1977 necessitated these updates of previous TRIGOM, URI, and VIMS reports. Moreover, the earlier reports had been criticized for not covering some relevant topics in sufficient detail, and for not adequately reviewing unpublished data pertinent to BLM concerns. For the areas south of Cape Hatteras, the scope of the review was extended seaward from the 200m to the 1500m isobath on the continental slope.

The two BLM data inventories compiled by the Center for Natural Areas included no comprehensive maps of Atlantic coast resources, although some chapters did illustrate the approximate regional distribution of specific resources, generally at scales exceeding 1:2,000,000. The BLM compilations were primarily narrative, extensively referenced, and broad in scope. No effort was made to rank segments of the coast according to their overall resource sensitivity or uniqueness. Topics covered in the BLM reports included nearshore and offshore geology, chemistry, phytoplankton, zooplankton, microbiology, neuston, nekton, benthos, commercial fisheries, sport fisheries, marine benthic flora, marine mammals, marine birds, marine turtles, endangered or threatened species, unique or endangered environments, toxicity and health studies, and meteorology. Each chapter was a treatise of about 50 to 350 pages in length often including discussion of critical or significant resource areas and ongoing research programs. Largely because of their great length and the cost of publishing, the reports were not widely distributed and have yet to contribute greatly to decision making.

RECENT STUDIES

Expanding upon the aforementioned efforts are two recently completed mapping studies, one funded by NOAA and the other by the U.S. Fish and Wildlife Service. (See the two related papers on these studies presented at this symposium; see also Carter, 1980.)

In late 1978, G. Carleton Ray, then of Johns Hopkins University, presented NOAA with the idea of preparing a series of Atlantic coast marine resource maps at a scale of 1:2,000,000. The maps would be conceptually similar to those Ray and his associate, James Dobbin Associates, had prepared for the Caribbean and the Bering Sea. These black and white maps would protray data on physical and living environments, species, economic activity, and jurisdictions. Support for the maps was voiced by the Council on Environmental Quality (CEQ).

One potential user of these maps, the U.S. Fish and Wildlife Service (FWS), expressed concern about several aspects of the approach being contemplated. A key point of concern was the map scale. FWS contended that NOAA's 1:2,000,000 would be of little use for FWS decision-making processes and instead proposed using a 1:250,000 scale, the largest scale of maps available for the entire Atlantic coast at this time.

In 1979, NOAA and FWS initiated complementary projects--NOAA with Carleton Ray and James Dobbin Associates, Ltd. in July and FWS with the firm of Dames and Moore in September. Very similar information would be compiled and mapped by the contractors although in a different manner and scale and with a provision for data sharing. The NOAA study (National Oceanic and Atmospheric Administration, 1980) would contain more information on cultural features, would extend 200 nautical miles seaward (versus 3 nautical miles for FWS study) and would consist of 135 single topic maps capable of being combined with the others via overlays or composites. The FWS study (Fish and Wildlife Service, 1980) would consist of 31 regional maps (U.S.G.S. National Topographic Map Series as base), each map displaying all resources in color on one composite. The NOAA study contemplates a second phase in the near future, which could involve overlaying and synthesizing the mapped information into a series of judgments as to which segments of the entire Atlantic coast are of high, moderate, or low importance for the environment generally. NOAA has not yet clearly specified the process or criteria that will be used in making such highly critical decisions. Both NOAA and FWS anticipate expanding their studies to other coastal regions in the United States.

DISCUSSION

The review which we have just presented should underscore some of the issues involved in the pursuit of the goal we mentioned at the beginning--the preparation of coastal maps which are accurate, comprehensive, accessible, current, relevant for most uses and users, and easy to understand. Additional issues are raised when one examines closely the characteristics of specific Atlantic coast ecological data bases. As examples, four resources (coastal birds, threatened/endangered species, corals, and shellfish) are discussed from the perspective of three key generic issues--format, accuracy/undertainty, and significance judgments.

GENERIC ISSUES

A. Format

1. Choice of scale ultimately hinges on the map's intended uses and the degree to which it will either generalize data presently at a finer scale, or extrapolate data presently at a broader scale. The most precise map, then, will be one which: a) best takes into account the "average" spatial patchiness of the data; and b) portrays the data in a manner best suited for its intended use.

Regarding the patchiness issue, FWS believed that the data in its coastal inventory were sufficiently detailed to allow meaningful and legible mapping at the 1:250,000 scale. They believed that a smaller scale map such as 1:2,000,000 would unavoidably exclude much useful data, such as point-data indicating individual wading bird colonies. NOAA, on the other hand, believed that showing some information, such as wintering grounds of certain finfish, at a large scale such as 1:250,000 could, despite numerous caveats contained in the accompanying narrative, imply a degree of precision in our knowledge which may not exist and potentially mislead decision makers. The varying appropriateness of map scales is portrayed in Table 1, which is the authors' attempt to indicate the average spatial quality of the data for Atlantic coast resources that are typically recognized as significant.

Regarding the issue of intended uses, the FWS believed that only at a scale where 1 inch equalled 4 miles or less would it be possible to adequately show geographic resource "gaps"--unshaded areas where oil pipelines, other rights-of-way, oil spills, and coastal developments--could be planned to minimize major energy/environment conflicts. FWS maintained that if numerous resources mapped at a more general scale, e.g., the 1 inch to 32 miles of the 1:2,000,000 scale, were overlaid, the result would be the "clouding" of virtually the entire coast making impossible even a general determination of where a facility should not be sited. Implicit was the assumption that regional planning for a facility typical of the kind FWS might review under the Endangered Species Act or the Fish and Wildlife Coordination Act would often be mostly confined to an area smaller than 8 miles (the minimum, practically-interpretable unit for the 1:2,000,000 scale), and perhaps to an area as small as 1 mile (the minimum unit for the 1:250,000 scale).

NOAA, on the other hand, was concerned that the 1:250,000 scale might not provide the spatial context essential to their legislative mandates, particularly since the FWS maps would segment the coast into 31 regions rather than portray it on a single map and be restricted to within the 3 mile offshore limit. Important information and relationships, such as the long-distance migratory paths connecting wintering grounds with summer spawning areas, might be missed by the careless user. Moreover, NOAA contended that the densities of shading (determined either visually or by computer) that resulted from overlaying numerous resources could be used to indicate geographic resource gaps where "safe" siting was potentially possible, i.e., the "clouding" of the coast was more apparent than real. Conversely, both the NOAA and FWS maps could be used to indicate resource cluster areas where protection or management might be desirable. The FWS maps, by including all resources on a single map for each region, eliminate the need for manually overlaying the maps to locate resource clusters. This is an important consideration for field personnel. However, by combining data on one map per segment, interpretation may be made more difficult. NOAA correctly believed that in some cases a 1:250,000 scale would imply that knowledge of a resource was sufficient to pinpoint its distribution to within 2 miles, when in fact such knowledge is sometimes not available. Also, both the FWS and NOAA studies leave open the possibility of eventually mapping the Atlantic coast in part at a finer scale.

2. <u>Choice of color and/or pattern symbology</u> depend on legibility considerations and cost. A beginning step is to assess the potential for spatial overlap among the resource categories one intends to map (Figure 1). Next, appropriate patterns and colors are selected. Although color attracts attention to a map, it is usually appropriate only when there is no potential for spatial overlap of categories, as for distinguishing between land and water or between Federal and State-owned lands. This appears to be a serious deficiency of some of the maps published by BLM (for example, Visual 5N in the EIS for OCS Lease Sale 43);

Table 1. "Average"characteristics of important ecological data sets of the Atlantic coast.

1. Data comprehensive and adequate for almost the entire coast; boundaries usually known to within 1 mile; appropriate for 1:250,000.

—Designated protection areas (Refuges and Wildlife Management Areas, National Parks and Seashores, Class I Air Quality Areas, Marine and Estuarine Sanctuaries, areas designated in CZM Plans, etc.)
—State Class I shellfish areas
—Islands
—Beaches and dunes

2. Data nearly comprehensive for the entire coast, and quantified (or quantifiable), subject to some limitations related to its accuracy and interpretation; boundaries usually known to within 1 mile.

—Marshes
—Mangrove areas
—Nesting areas for wading birds and seabirds
—Major shellfish beds
—Reefs
—Marine mammals (seal haul-out areas only)

3. Data available for a geographic cross-section of the coast, and quantified, but confounded because it is normally compiled only as totals for larger and/or ill-defined areas; scales exceeding 1:250,000 may be appropriate, especially if judgments of relative geographic significance are to be made:

—Wintering waterfowl areas
—Wintering areas of shorebirds, wading birds, and raptors (Christmas Bird Count data)
—Migration areas of shorebirds
—Furbearer harvest areas

4. Data available for sporadic portions of the coast, seldom quantified or meaningfully quantifiable; boundaries frequently not known even with a precision of within 8 miles or 1:2,000,000 scale.

—Seagrass beds
—Spawning and nursery areas of finfish and shellfish
—Marine mammals (excepting seal haulouts)
—Endangered/threatened species (varies greatly by species)
—Coral areas

5. Informal data (quality highly variable) available for sporadic portions of the coast, quantified (subject to interpretive caution), but usually from localized but ill-defined areas:

—Migration and concentration areas for most non-game waterbirds.

Figure 1 . Expected potential for overlap among major shading/color categories. ○ unlikely, ◒ possible, ● probable overlap. This material could be used to select shading patterns that will minimize visual conflict among high-affinity categories. This indicates only the <u>probability</u> of overlap, not the <u>amount</u>.

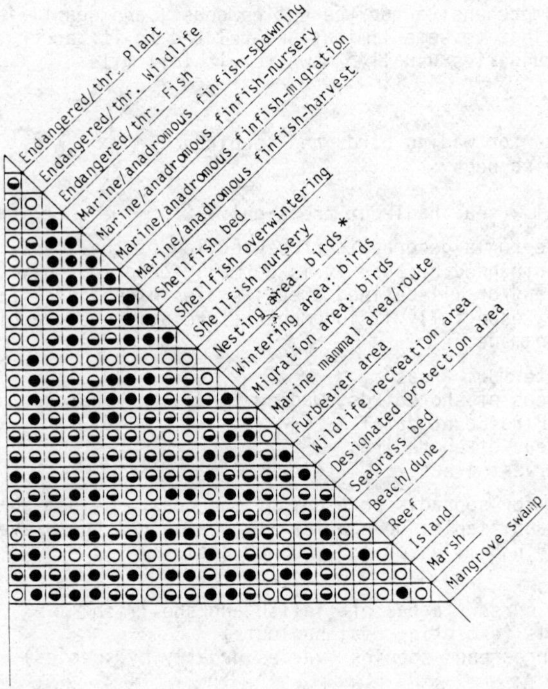

* Can be indicated by a data point (e.g: ◀) rather than by shading/color Subcategories of some of the other topics listed here (e.g., certain endangered plants) could also be shown this way, depending on the specificity of available data.

distribution ranges seemingly are artificially terminated when overlap becomes imminent. The NOAA and FWS maps dealt with potential overlap by employing combinations of shading, numerical, or letter codes to indicate resources present in patterned overlap areas. In some cases in the FWS maps, particularly common resources such as ospreys in Chesapeake Bay are not mapped at all because of problems of scale or because higher priority was assigned to species with higher status. In these cases, a note is added on the map or accompanying text. Although this requires frequent reference to the legend, the representation appears accurate and clear. The FWS maps use five colors to highlight land, water, special land use designations, endangered species habitat, and the coastal zone. The information is being printed on USGS National Topographic Maps, 1:250,000 scale series, and will be published in a pocket-size, folded, roadmap-like format to enhance its availability and use.

B. Accuracy and Uncertainty

The accuracy of a map is as much related to how clearly its assumptions are stated as to how neatly the data are plotted. Beautiful, precise graphics mean little if the data are outdated, incomplete, biased, or unrepresentative. In short, precisely wrong is not as good as vaguely accurate. A map's uncertainties and limitations should be portrayed as plainly as are the data; accurate interpretation should be as much or more the responsibility of the preparer as the duty of the user.

There is a grave danger that mapped areas discussed above as geographic resource gaps could be unwisely developed based on the assumption that they contain little of ecological value. In reality, these geographic resource gaps may merely represent data gaps which upon completion of further surveys may be found to contain resource clusters. Moreover, their phusical support systems may be vital to resources located elsewhere.

In their compilations, FWS and NOAA deal with the danger of misinterpreting resource gaps by providing caveats either on the maps or in the narrative. To enhance awareness of those caveats, any notes to the user should best be noted both on the maps and in the text. Generally, some types of gaps (methodological) are more appropriately noted in text, some (spatial) directly on the map, and others (chronological) on either the map or text.

The FWS narrative discusses these limitations for almost every plotted data point. Other atlas efforts should make similar efforts rather than relying only on introductory disclaimers which address uncertainties generically and not on a very site-specific basis. Each mapped resource should be documented at least according to source of data, its recentness, approximate professional person-months of research represented (is the data a single observation or years of repetitions), frequency of coverage, techniques used, scale of original map, and the relationship of these factors to quality of the data. Such a disclosure system may seem unwieldy, but it should present a more true picture of existing data gaps. To indicate data quality cartographically, codes within each mapped area or resource could have a suffix attached, e.g., "A" for high quality data or "D" for a single data point, or this could be shown on a shaded overlay map.

Although drafting and applying criteria for data quality is subjective and time-consuming, it may diminish the chances of data misuse. Some data compilations (such as the BLM data inventories on marine birds depicting seasonal and taxonomic adequacy of coverage in each coastal interval, NOAA's text accompanying the resource maps, and the FWS maps) evaluate data quality to varying degrees.

C. Significance Judgments

While judging an area's relative significance may seem superficial to some scientists, such decisions are essential to coastal decision making. The NOAA atlas effort hopes to eventually rank segments or areas of the coast as relatively high, moderate, or low sensitivity, taking into account the uniqueness of the resources, their prevalence in the segment, and their sensitivity to perturbations. Each segment would first be ranked for a single resource and then be combined into a cumulative rank for all resources within that segment. Such synthesis procedures are as old as planning, but are being increasingly applied to coastal and ocean uses. This is because after a decade of legislative mandate and program development, many key decisions on such issues as fishery development, energy facility siting, port development, and ocean waste disposal will be made in the next several years. As paraphrased from the May 21, 1980, Doherty lecture by National Oceanic and Atmospheric Administration (NOAA) Administrator, Richard A. Frank, this nation will very soon be faced with making many tough decisions on natural resource issues that have been rapidly escalating since the early seventies.

By their nature, maps of ecological data emphasize areas of special significance. The quantity and size of such areas conveys the relative amount of "critical" resources along the coast. This approach to locate areas with high concentrations of sensitive resources has been used by the Bureau of Land Management (BLM) to develop their matrix of oil sensitive areas (see, for example, Visual 4 and Appendix C of the DEIS for OCS Lease Sale 56). Upon preliminary analysis, regions with more special designations or resource concentrations may be perceived by BLM and other agencies as being potentially more sensitive to perturbations such as oil spills. With that in mind, it is of interest to note that BLM and other agencies sometimes argue against complementary designations such as adjacent wildlife refuge and national park areas or umbrella management efforts such as marine sanctuaries that accentuate a region's sensitivity and may impede development activities. Hence, Federal agencies may work against other agencies and themselves in a numbers game that could ultimately affect research budget allocation, pollutant contingency plans, or resource protection efforts.

Criteria and procedures used to rank significance are not always clearly disclosed, e.g., Rooney-Char and Ayres (1978). Even when criteria are disclosed, there obviously is much subjectivity in the definition of significance. For example, should "resource conflict areas" (as used by G. Carleton Ray in the NOAA atlas to describe where important resources coincide with potential or existing stresses) be considered most significant because the remaining resources may be near or at their coping threshold? Or should remote areas with little apparent stress be assigned highest significance because they may be rare on developed coastlines? One solution is to provide for wider professional participation and peer review not only in reviewing the mapped data, but also in the early drafting of significance criteria. Such input might best be provided during the compilation effort through a questionnaire and/or regional workshop format including as many recognized local authorities on each resource as possible. The alternative is to allow only a few resource managers and well-intentioned professionals or administrators to make crucial decisions on vital resource decisions without sufficient expert input.

It should be noted that data compilations are but one tool needed to understand our resources. Other tools must address the functional dynamic nature of resources including productivity of plants and animals. In fact, productivity assessments may provide one method of assessing more large scale phenomenon occurring between resources along the coast.

RESOURCE EXAMPLES

To apply these perspectives to specific data sets, the following four broad resources are discussed in detail:

A. Coastal Birds

Important areas for coastal birds may include wintering areas, migration areas, and breeding areas of waterfowl (ducks, geese, and swans), shore and wading birds, raptors, and certain songbirds.

Adequacy of Data Base

More is known about the distribution of breeding raptors and wading birds than perhaps any other living Atlantic coast resource. Virtually every colony or nest site has been pinpointed. (See, for example, Custer and Osborn, 1976, Korschgen, 1979, Erwin, 1979). However, relatively little is known about shorebird migration and wintering areas (despite good recent efforts by the International Shorebird Program), pelagic birds (which are not included in the NOAA study, although they frequently occur within the 200 mile seaward limit and have been subject to recent systematic research by Manomet Bird Observatory in Massachusetts), and particularly, about other non-game waterbird migration areas and songbird concentration areas.

Amenability to Mapping

Although nesting and roosting sites are rather discrete, migratory routes and wintering areas may be fairly amorphous, highly variable from year-to-year, and difficult to precisely define and map. Many years of systematic data on wintering waterfowl have been collected by FWS. However, these data have several limitations (Adamus and Drury, 1977, Forsythe and Adamus, 1979): (a) nearly all species totals are tallied by areas no smaller than 30-100 sq. miles--which makes it impossible to use the data in their present form for delineating "major" areas within intervals; (b) although data are collected systematically, considerable variation exists in the sizes of the areas counted, making difficult the calculation and comparison of population densities among areas and between studies; and (c) wintering non-game water birds have never been counted, although research shows that some of these, such as loons, are among the most sensitive species to oil. Although Audubon Christmas Bird counts tally non-game birds and are conducted in standard-sized areas (vs. (b) and (c) above), they too suffer from problem (a), and in addition vary greatly according to level of effort, being ground-level rather than aerial surveys. Another data source is monthly bird counts available from usually well-delimited and measured areas in national wildlife refuges or other special areas. However, this leaves many spatial gaps in coverage of the coast. Considerable untapped bird data, much of it of good quality, is available from dozens of competent amateur ornithologists along the Atlantic coast. This could help plug some geographic data gaps which are especially prevalent for non-game coastal birds and other non-game species.

Significance Judgments

Few attempts have been made to assign relative significance values or levels to coastal bird concentration areas. The Maine coastal characterization assigned regional significance perspectives and the BLM data inventories hypothesized primary and secondary importance areas for nearly all coastal and pelagic species, but in some cases few data were available upon which to base this judgment. Regional compilations have sometimes suffered from poorly-educated definitions of "significance" or overgeneralization of data, e.g., apparently assuming that areas important to wintering waterfowl were inevitably important to other waterbirds or to waterfowl in migration (see, for example, pp. 51 and 54 of Lippson, 1973).

B. Threatened/Endangered Species

Adequacy of the Data Base

Adequacy of this ecological data base varies widely with population levels and behavior patterns of the species. For example, the data base enabling mapping of such relatively visible and fairly abundant (for an endangered species) species as the bald eagle, eastern brown pelican, and Florida manatee is considerably better than that available for the extremely scarce and/or secretive species such as the eastern mountain cougar and the ivory-billed woodpecker. In fact, the Fish and Wildlife Service Office of Endangered Species has for all practical purposes deferred critical habitat designations for a number of species such as the eastern mountain cougar because their elusiveness and extreme rarity makes impossible the determination of what habitat is critical. Data on many rare plants are even more limited and frequently consist of a single herbarium record or inclusions in 19th century botanical manuals.

Amenability to Mapping

The ability to accurately map important endangered/threatened species areas also is somewhat related to the characteristics of the species' range. The smaller the range, the more accurate mapping can be. For example, the Florida Everglade kite occurs exclusively in the Florida

Everglades, and the dusky seaside sparrow is unique to coastal areas in northern Florida. Ranges of these restricted populations can probably be accurately mapped at a 1:250,000 scale. On the other hand, a species like the shortnose sturgeon which has, or may have been, sighted off the coastal waters of Maine, Georgia, Florida, North Carolina, and other states, and whose historical range includes all of the Atlantic seaboard, is very difficult to meaningfully map.

Significance Judgments

Efforts at mapping threatened/endangered species should consider institutional frameworks as well as the biological data base. The definition of what is truly an "endangered species" depends not only upon the biological status of the population, but also upon whether or not the species has been recognized institutionally as endangered or threatened. Several examples have occurred which illustrate the need for thorough understanding of the endangered species "institutional structure" prior to mapping these resources. One example is the case of the proposed Pittston Oil Refinery in eastern Maine, where the endangered status of the northern bald eagle significantly influenced the decision-making process. At the time of initial project planning, the eagle was listed as threatened, later upgraded to endangered. When the eagle's status was changed, additional planning efforts were needed. Therefore, while it is important to consider those species which are endangered or threatened during a mapping process it is also critical to be able to speculate correctly on what species might be listed. This is quite often difficult. For example until January 1980, approximately 1,850 plant species were proposed for listing, but all proposals were withdrawn in February 1980. Federal endangered species regulations provide for both normal and emergency procedures for listing, and State regulations generally provide mechanisms for modifying their lists. Therefore, the resources identified for the mapping process can be subject to significant and sudden change. Moreover, numerous examples of endangered/threatened species lists exist which have no legal basis. The degree of credibility for these lists varies widely.

Finally, most States have the power (through State endangered species legislation) to promulgate lists of their own. Although this is a valuable procedure for conservation at the State level, it may cause confusion when a multi-state region is being mapped. The planner should recognize that there could be several reasons for a species being listed in one State and not in another, including: (a) the range of the species in question does not overlap into other states, or (b) the species is not considered rare by other State's endangered species authority.

C. Coral Resources

Adequacy of Data Base

Coral stocks north of the perennially warm south Florida waters have rarely been the subject of directed field or laboratory research. Most of the data base has been pieced together from incidental observations and collections gathered over the past century. Among the more noteworthy data gaps are (Center for Natural Areas, 1980):

— full life cycle biological and ecological information including population distribution, densities, mortality (natural and human related), reproduction and growth rates (size and weight);
— socio-economic information on the use of coral resources by user groups who do not harvest corals except on on incidental basis such as divers, fishermen, tropical specimen collectors (fish, shells, and plants), and glass bottom boat operators;
— information on the role of coral resources as habitat for associated species, including commercial and sport fish, sea turtles, and hundreds of other species; and
— legal information that could affect use of the resource, for example import trends, the existing domestic jurisdictional hierarchy, and the development of plastic corals for display purposes.

These gaps are crucial to the accurate mapping of the resource. Existing data are too incomplete to allow calculation or even estimation of fishery yields for all but a few of the hundreds of coral species in our Atlantic coastal waters (Center for Natural Areas, 1980). These gaps are magnified by the lack of data on yearly trends and the fact that both natural (hurricanes, temperature, light, etc.) and human (pollution, anchor damage, etc.) impacts play important and confusing roles in coral resource health. The lack of this vital data has affected coral resource management efforts required under the Outer Continental Shelf Lands Act, Fishery Conservation and Management Act (FCMA), Title III (marine sanctuaries) of the Marine Protection, Research, and Sanctuaries Act, and a host of other State and Federal programs.

Amenability to Mapping

The quality fo the coral resource data base places several restrictions on mapping efforts. Available information is point data resulting from indirect collections (trawls, dredge) and direct observations (diver, submersible). Only direct observations allow precise correlation of a species and its habitat to a precise location; non-selective data from indirect collection efforts provide only approximations of resource associations. Ecological parameters also confuse mapping efforts. For instance, maps of larvae (assuming data availability) would probably reveal less about spawning areas than physical factors such as current patterns. On the other hand, maps of the sedentary adults, which again would be based on an inadequate data base, would probably be best represented by separate data points or symbols rather than broad ranges that may apply to finfish. Since hundreds of species of corals inhabit the Atlantic coastal waters of the United States, maps to the species level would have to be very carefully scaled and coded to reduce confusion. It is possible that many dozens of maps or a few synthesis maps would be needed to portray only the most important resources, i.e., those with commercial value, most abundant, or with key ecological roles.

The rigors of coral classification and field surveys of such delicate benthic organisms combine to greatly limit the coral resource data base and mapping efforts. Most coral maps cover only discrete areas of special designation such as national parks or marine sanctuaries in the very southern portion of the Atlantic coast. Many notable coral resource areas, e.g., Gray's Reef (Georgia), Looe Key (Florida), have been inventoried but not mapped. Other areas like deep-water coral banks have not even been inventoried.

Significance Judgments

Several criteria may be used to designate significance to a particular population or other grouping of coral. Examples are value for commercial use, continuation of the species, non-consumptive uses such as photography or research, or habitat of associated species. For example, Key Largo Coral Reef Marine Sanctuary in Florida was designated in 1975 to safeguard that reef's complex and fragile ecosystem, one that is a healthy representative of the coral reefs in the Florida Reef Tract. A second category of critical area, "habitat areas of particular concern" (HAPCs), have been defined by the fishery management plan (FMP) process as areas ". . . which are of particular concern because of a requirement in the life cycle of the stock(s), e.g., spawning grounds, nurseries, migratory routes, etc. . . . (and) . . . those areas which are currently or potentially threatened with destruction or degradation" (Fishery Conservation and Management Act regulations, 50 CFR Part 602.3.b.6ii; July 17, 1977). In the coral FMP drafted for the Gulf of Mexico and South Atlantic Fishery Management Councils, seven areas were rated critical for the coral resource and listed as HAPCs. Since the FCMA and its regulations offer no criteria on selecting HAPCs, the coral FMP proposed a series of standards to which potential areas could be compared. Criteria for ecological, research, exploitation, and recreation values were developed and applied within the context of the definition of HAPCs quoted above.

These different definitions of critical coral resources result in similar but not identical lists of critical areas depending on the program, i.e., marine sanctuaries or fisheries. Still more varieties of "critical" are created by the National Park Service in parks and monuments, by the Bureau of Land Management on the shelf, and by other State or Federal agencies.

Due to the paucity of coral data, maps of this resource often plot more than just critical areas. As one example, Cairns' (1979) mapping effort of deep-water hard corals plotted all collection locations for each of 88 species regardless of density or any ranking criteria. To some, the apparent sparceness of this particular kind of coral implies some special significance. The obvious complexity of mapping all corals in a given area would be further intensified if individual species were plotted rather than community types.

D. Shellfish

Adequacy of Data Base

Shellfish have been studied by two general methods: 1) specific assessments directed at one or a few species, or 2) broad studies directed at the benthic ecology of a larger region. These two approaches have resulted in a detailed but incomplete data base.

The spatial coverage of shellfish data is quite thorough, owing largely to their commercial significance. Fisheries harvests, stock censuses, and baseline studies required for energy facility siting, dredging, or similar actions have contributed to a data base extending the entire length of the Atlantic coast. It is probably safe to say that no large section of the nearshore waters has been neglected, though localized gaps remain. Perhaps the largest gaps are on the continental shelf and slope where some shellfish stocks (lobsters, crabs, clams, scallops) have not been adequately surveyed and where species exist that have recently attained commercial importance, e.g., rock shrimp on the Blake Plateau. In the case of lobsters and some crabs, the largest gaps appear to be in the middle Atlantic region, south of the Georges Bank and southern New England canyon areas.

Despite the broad geographic coverage of the data base, shellfish are beset by numerous ecological gaps which influence their mapping. Three characteristics of shellfish stocks are relevant: 1) all shellfish larvae are planktonic and are therefore subject to the patterns of physical oceanography prior to settling into their benthic lifestyle; 2) juvenile and adult molluscs are sedentary, a fact that renders them susceptible to natural and man-induced impacts; and 3) crustaceans are motile, often exhibiting complex migrations according to environmental variables such as water temperature or day length. These three facts reiterate the difficulties in accurately describing shellfish stocks. For example, annual changes in reproductive viability, water quality, ocean currents, or fishing pressure may mean that areas noted or mapped as significant in one year may no longer be so. These general complications are often compounded by specific gaps in knowledge at the species level, e.g., spawning seasons or predator-prey relationships.

Certain shellfish stocks in New England and perhaps the Florida Keys area are also affected by involvement with our foreign neighbors. In the Gulf of Maine, scallop, crab, and lobster populations most likely transit the boundary of Canada and the United States, the position of which is still being negotiated. Larvae caught in the gyre of the northwest Atlantic coast could easily be transported across national boundaries, as could spiny lobster larvae in Florida and northern Caribbean. This adds an international flavor to the data gaps, studies to fill them, and the possible repercussions of using the data. Adequate studies of species that have international aspects must consider activities in stocks of each involved nation. The recent data compilation coordinated by NOAA does successfully incorporate data from Canada in its mapping efforts.

Amenability to Mapping

As stated above, the shellfish data base is quite complete. Among the molluscs, planktonic larvae and low diversity (few species) beds of benthic juveniles and adults do permit mapping of broad ranges for larvae and specific beds for older individuals. Fishery records and research censuses also offer insight into historical changes of adult population sizes and locations. For example, comparison of soft clam bed locations in Chesapeake Bay as mapped by Spinner (1969) and Rooney-Char and Ayres (1978), among others, and annual catch data compiled by Virginia, Maryland, and NOAA's National Marine Fisheries Service could, with adequate understanding of the assumptions of each study, reveal temporal changes. For crustaceans, planktonic early life stages present the same mapping problems as molluscs; only broad ranges and perhaps some trends in density of larvae have been mapped. Furthermore, most plankton sampling efforts are restricted to certain levels of the water column; thereby limiting conclusions on the distributions of larvae that exhibit vertical diel or seasonal movements each day. Adult crustaceans present a different problem. Many populations migrate in what appear to be annual cycles that oscillate stock loci from nearshore to offshore. Often, these patterns are different for each sex. Rock crab populations in Narragansett Bay, R.I. migrate from the protected inner Bay in the winter to deeper waters near the mouth of the Bay and adjacent Block Island Sound in the summer. Water temperature appears to trigger this movement, with egg-bearing females the first to sense the changing temperatures of early spring. These examples should warn the user of data bases and maps about potential pitfalls. The NOAA maps, for example, note overall range of species without differentiating gross seasonal patterns in distribution.

Other questions should also be asked. Are the data from one season, and if so, which? Are the data partitioned by sex? Do the data represent a composite of several years and thereby help account for migrations and meteorological anomolies that effect animal movements? Are all ages mapped or only mature adults?

However, other problems arise. Information collected from fishermen, commercial or recreational, may be prejudiced by biases in exact catch location, harvest, species type, or other key parameters. Protecting one's favorite fishing grounds or under-reporting catches to avoid catch limitations are well known phenomena that must not be disregarded. The confusion these distortions add to the data base are often unquantified but are suspected to be important. They have not been dealt with in most data compilations. Nevertheless, by dividing the data base into workable segments and by noting any limitations, the shellfish resource can be mapped quite effectively. Spinner (1969) adequately mapped the resource. The recent FWS (1980) maps probably come closest to the objective of mapping while noting limitations.

Significance Judgments

Shellfish resources, their habitat, and their economic values may be ranked or prioritized on the basis of life stage concentrations, unique ecological assemblages, historical catch levels, or harvest potential (as in the case of expanding fisheries such as Gulf of Maine scallops or south Atlantic rock shrimp). Since shell fisheries are so valuable economically, "critical" often denotes the most productive grounds. Occasionally, terms such as "primary" or "secondary" are used to classify harvest areas. An example of specific shellfish habitat protection may be found in the shrimp fishery management plan for the Gulf of Mexico Fishery Management Council (1980). To protect small pink shrimp, the Council established the "Tortugas Shrimp Sanctuary" off southwest Florida. This permanent closure was based on the area's critical role as habitat for juvenile shrimp and also to reduce potential gear conflicts with fixed gear stone crabbers. Another application of critical is for egg-bearing crabs and lobsters; these ovigerous females are protected by most States through landing laws as a type of critical resource. Lastly, on a different perspective, State public health agencies and U.S. Environmental Protection Agency closed area closures to mollusc harvest due to sewage or toxic waste disposal. For example, most upper Narragansett Bay waters are classified SB and portions of the Mid-Atlantic Bight are permanently closed to harvest.

The data base includes many reports that ascribe critical attention to crustacean and molluscan populations. In Maine, Cowger (1975) mapped oyster beds based on commercial potential; Gusey (1976) compiled existing maps of shellfish harvest areas in the middle Atlantic; in Virginia, Rooney-Char and Ayres (1978) mapped oyster settling grounds, blue crab fishing grounds, peeler (recently molted) crab grounds, and blue crab critical habitat (not to be confused with the Endangered Species Act designation of critical habitat for protected species); and Spinner (1969) plotted mollusc grounds along the entire Atlantic coast. Most of these mapping efforts can be more aptly described as pure data presentations rather than determinations of habitat or populations essential to the species.

CONCLUSIONS

This paper does not intend to negate the value of nor attempts to compile or synthesize the Atlantic coast ecological data base. Conversely, our intent is to highlight some issues involved in data base interpretation and to increase its accurate use. Since the decision-making process must and will proceed, the arguments and conclusions presented herein will hopefully enhance our ability to make the best possible resource decisions in the 1980s and beyond.

If there is one conclusion that can be reached by this review, it is that no atlas or mapping effort using a single scale can accurately present all resources, especially those with spatial, temporal, or methodological gaps. However, this does not mean that single-scale mapping efforts of broad geographic coverage are not useful, only that their limitations should be noted and that judgments of relative significance should be based on the widest possible spectrum of professional opinion and input. At the same time, it must be recognized that there are probably as many ways to map a data base as there are potential uses of those maps. Mapping efforts at progressively finer scales are justified to keep pace with advances in our knowledge of the distribution of critical ecological resources. Full coordination among such mapping efforts on issues such as defining mapping terms is essential or the usefulness of the maps will be impaired.

ACKNOWLEDGMENTS

This paper has been reviewed by appropriate offices in the Bureau of Land Management, Fish and Wildlife Service, and the National Oceanic and Atmospheric Administration. Comments from those agencies have been included in the paper. Their input is gratefully appreciated. The views are those of the authors only, not necessarily of any other persons or agencies.

LITERATURE CITED

Adamus, P.R. and W.H. Drury. 1977. Marine birds. In A summary and analysis of environmental information on the continental shelf from the Bay of Fundy to Cape Hatteras. Bureau of Land Management

Contract AA 550-CT6-45 by the Center for Natural Areas. pp. XV-1 to XV-244.

Cairns, S.D. 1979. The deep-water Scleractinia of the Caribbean Sea and adjacent water. Studies Fauna Curacao 57(IBO): 1-331.

Carter, L. 1980. East coast maps to alert industry to ecology conflict. Science 208: 159.

Center for Natural Areas. 1977. A summary and analysis of environmental information on the continental shelf from the Bay of Fundy to Cape Hatteras. Bureau of Land Management Contract AA 550-CT6-45. 3 volumes.

Center for Natural Areas. 1979. A summary and analysis of environmental information on the continental shelf and Blake Plateau from Cape Hatteras to Cape Canaveral. Bureau of Land Management Contract AA 550-CT7-39. 3 volumes.

Center for Natural Areas. 1980. Draft fishery management plan for coral and coral reef resources. Prepared for the Gulf of Mexico and South Atlantic Fishery Management Councils.

Cowger, J. 1975. Oyster beds in Maine: a report prepared for the Maine Critical Areas Program. Maine Department of Environmental Protection, Augusta, Me. 13 pp.

Custer, T.W. and R.G. Osborn. 1976. 1975 survey of Atlantic coast wading bird colonies: atlas. U.S. Fish and Wildlife Service, Patuxent Wildlife Research Center, Laurel, Md.

Erwin, R.M. 1979. Coastal waterbird colonies: Cape Elizabeth, Maine to Virginia. U.S. Fish and Wildlife Service OBS 79/10. 212 pp.

Fish and Wildlife Service. 1980. Atlantic coast ecological inventory. Prepared by Dames and Moore. Maps and text.

Forsythe, D.M. and P.R. Adamus. 1979. Marine birds. In A summary and analysis of environmental information on the continental shelf and Blake Plateau from Cape Hatteras to Cape Canaveral. Bureau of Land Management Contract AA 550-CT7-39 prepared by the Center for Natural Areas. pp. XIII-1 to XIII-286.

Gulf of Mexico Fishery Management Council. 1980. Fishery management plan for the shrimp fishery of the Gulf of Mexico, United States waters. Prepared by the Center for Wetland Resources, Louisiana State University.

Gusey, W.F. 1976. The fish and wildlife resources of the Mid Atlantic region. Shell Oil Co., Houston, Tx. 582 pp.

Gusey, W.F. 1977. The fish and wildlife resources of the Georges Bank region. Shell Oil Co., Houston, Tx. 553 pp.

Korschgen, C.E. 1979. Coastal waterbird colonies: Maine. U.S. Fish and Wildlife Service. OBS-79/09. 83 pp.

Lippson, A.J. (editor). 1973. "The Chesapeake Bay in Maryland": An atlas of natural resources. Contribution No. 500, Natural Resources Institute, University of Maryland.

National Oceanic and Atmospheric Administration. 1980. Eastern United States coastal and ocean zone resource use mapping project. Prepared by G.C. Ray, M.G. McCormick-Ray and J.A. Dobbin. Maps and text.

Rooney-Char, A.H. and R.P. Ayres. 1978. Offshore pipeline corridors and landfalls in coastal Virginia. Virginia Institute of Marine Sciences (VIMS) Report No. 190.

Spinner, G.P. 1969. The wildlife wetlands and shellfish areas of the Atlantic coastal zone. American Geographical Society Portfolio No. 18.

The Research Institute of the Gulf of Maine (TRIGOM). 1974. A socio-economic and environmental inventory of the north Atlantic region including the outer continental shelf and adjacent waters from Sandy Hook, New Jersey, to Bay of Fundy. 3 volumes.

University of Rhode Island (URI). 1973. Coastal and offshore environmental inventory, Cape Hatteras to Nantucket Shoals. Marine Publication Series No. 2. 2 volumes.

Virginia Institute of Marine Science (VIMS). 1974. A socio-economic environmental baseline summary for the south Atlantic region between Cape Hatteras, N.C. and Cape Canaveral, Fla. 5 volumes.

ENDANGERED SPECIES: COMPATIBILITY WITH A UTILITY

J. Ross Wilcox
Ecologist
Licensing & Environmental
Planning Department
Florida Power & Light Company
P. O. Box 529100
Miami, Florida 33152

INTRODUCTION

Florida Power & Light Company (FPL) is the fifth largest utility in the U. S. and serves more than five million people in a service area that covers approximately one-half the geographic area of Florida. Climates range from warm temperate to tropical.

More than 75% of FPL's generation output is used in residential communities in contrast to other areas of the country, where commercial customers dominate. We have summer and winter peaks that supply energy for cooling and heating. Within our service area, which encompasses all or part of 35 counties, FPL operates 10 power plants, having a total capacity of 11,000 MW. Power is distributed via 62,000 km of transmission lines with a maximum voltage of 500 kV. Approximately one-fifth of the generating capacity of the system is provided by nuclear energy with the balance being provided by oil and natural gas. FPL does not burn coal, but has announced plans to build two coal-fired units in the 1980's.

The Company owns and manages large tracts of land in both rural and urban areas located both in the coastal zone and in the interior of the state. Much of this land is undeveloped and is used as a buffer zone to isolate power plants from residential, agricultural, and coastal zone interests. These buffer zones serve as wildlife habitat for many species, both common and endangered, and place FPL in the wildlife business.

To date, FPL has interacted with five federally rare and endangered species that occur in the coastal zone. For the most part, we have experienced no direct conflicts. Four endangered species include the West Indian Manatee, the American Crocodile, the Atlantic Green Turtle, and the Atlantic Leatherback Turtle. The Atlantic Loggerhead Turtle is a threatened species.

In the sense of stewardship and good business practice, the Company is documenting, through research and monitoring, that the production and transmission of electricity and the well-being of endangered species are compatible. Where real or potential conflicts are identified, FPL has taken positive actions to eliminate or minimize adverse consequences.

In this paper, the interaction of FPL with several of these rare and endangered species will be reviewed and our compatibility discussed.

WEST INDIAN MANATEE

The West Indian Manatee, or sea cow, is an air-breathing, aquatic herbivore that attains lengths of 3.5 m and weights of 1,000 kg. The endangered manatee is found in shallow coastal waters, bays, lagoons, estuaries, rivers and inland lakes of tropical and subtropical Central and North America. In the U.S., its present-day distribution is limited to peninsular Florida during the winter, and isolated individuals are found in Texas and the Carolinas during the summer. Manatee population numbers are estimated at 1,000.

The manatee is a slow swimmer, but is capable of quick and swift movements for short distances. The animal is harmless and has no natural predators. In recent years, human activity has been identified as the largest cause of manatee injury and mortality.

One interesting feature of their behavior is that they seek warm-water refuges in Florida during the winter months. There is no such attraction during the spring, summer and fall months. Historically, the animals have congregated around constant-temperature freshwater springs but, with the population build-up of Florida and tapping of aquifers for agricultural purposes, many of these natural springs are not flowing today. What the manatees have done is to seek out industrial warm-water refuges. FPL has five plants that operate in a once-through cooling mode and we have documented these winter congregations by contracting Florida Audubon Society to conduct a year-round aerial census program for us.

During the 1977-1978 winter, the maximum number observed at our Fort Myers Plant located on the west coast was 271. This count was the highest number recorded for any of the five study sites and probably the largest documented congregation of manatees ever recorded. Maximum counts for this plant during 1978-79 and 1979-80 winters were 172 and 142, respectively. At our Port Everglades Plant, located in the busy port of Fort Lauderdale, maximum winter counts varied from 86 to 125 over the three-year study period. Each of these maximum counts was made on days when the minimum air temperature fell below 10°C and occurred during or immediately following each winter's most prolonged cold spell. At a nearby facility, the Lauderdale Plant, the maximum winter counts for the three years varied from 22 to 36. Further up the east coast, the Riviera Plant in West Palm Beach had maximum counts that varied over the study period from 68 to 108. However, another study had documented a larger winter congregation of 146 during the 1975-1976 winter. Again, the Riviera Plant is located in a busy port and the intracoastal waterway is very close to our discharge. The Cape Canaveral Plant, located in the east-central part of the state, had maximum counts that varied from 75 to 83 over the three-year study period.

When the winter sighting data from these plants are correlated with the mean daily air temperature between November and March, an inverse relationship or negative correlation between manatee counts and temperature is found. Thus, when a cold front arrives and the air temperature drops, the number of manatees at the discharges increases. Conversely, when the cold front breaks up and a warming trend occurs, the congregations disperse and the counts decrease.

Summing the maximum number of manatees observed at each plant for one winter produces a total of about 570 manatees. Thus, over the course of a winter, approximately 60% of the population of an endangered species is congregated around these FPL plants, creating a potential for a mass disaster. These congregations are a very dynamic phenomena because some animals only arrive during the brunt of a cold front and quickly disperse to feed when the air temperature warms. Other animals are much more leisurely about leaving the effluent after a cold-front passage.

FPL is working very hard to promote a stewardship with these endangered animals. Through the efforts of the Aerial Census Program, Florida Audubon Society was able to collect scientifically valid data which were presented to the Florida Legislature and were the impetus for the passage of the Florida Manatee Sanctuary Act of 1978. This Act provides for the regulation of boat speed in 13 manatee congregation areas between November 15 and March 31 of each year.

One of the identifiable causes of manatee mortality and injury is collisions with boats. All adult manatees bear some type of scar resulting from contacts with high-speed motor boats. Manatees can easily avoid slow boats, and control of motor-boat speeds in these winter congregation areas, where large numbers of manatees gather, is an effective means of protecting the species. The speed zone signs with the effect of law, as stipulated in the Florida Manatee Sanctuary Act, were in place the fall of 1979. All of the warm-water discharges of the five FPL plants noted have speed zones. Boat compliance to the speed limit zones (e.g., slow speed, idle speed) varied and depended on public awareness, visibility of the signs, and enforcement actions. The Florida Department of Natural Resources, the lead agency responsible for enforcing the provisions of the Manatee Sanctuary Act, is being encouraged to increase compliance.

The Company has also contracted the Florida Audubon Society to conduct a Public Awareness Program for manatees. This effort has been in partnership with federal, state, and local authorities, and is aimed at minimizing human/manatee confrontations. Activities authorized include the production of TV and radio public service announcements, teacher training workshops, public speaking programs, articles for popular magazines, rental of outdoor billboards, participation in radio and TV talk shows, and production of movies and filmstrips. The public speaking program has been very popular, especially during the winter months.

AMERICAN CROCODILE

The endangered American Crocodile ranges in length from about 23 cm at hatching to 4.6 m for the larger males. The crocodile, in contrast to the more common alligator, is a slimmer animal with a longer tapering snout. Present population numbers are estimated at 200-400 individuals with approximately 25 breeding females.

Historically, the animal occurred in the coastal swamps of Southern Florida and Central America. Present-day distribution in Florida is restricted to the extreme southern tip of peninsular Florida, and to the northern portions of the Florida Keys. The Florida population is the only group of crocodiles in the continental U.S. Recent surveys have also documented that, in addition to occupying coastal mangrove swamps, brackish and saltwater bays, crocodiles will preferentially seek out abandoned coastal canals and rock pits for resting and reproduction.

Adults have no known natural predators, but their eggs are taken by racoons, and hatchlings suffer mortality from large wading birds. Adults wander during the non-breeding season, probably as a result of changing environmental conditions and food supply. The primary food of adults and subadults is fish, while juveniles feed on a variety of aquatic animals.

FPL owns approximately 8,900 hectares of land in south Florida, all of which fall within the critical habitat of the crocodile. Approximately half of the area comprises the Turkey Point site, which has an elaborate set of closed-circuit cooling canals for the condensers. The other half of the land is undeveloped.

In 1977, we first discovered that at least one crocodile was living in the southwest section of the cooling canal system at Turkey Point. We began a survey which documented that there is a dynamic but rather extensive population of crocodiles on our property. The adult population segment contains one large male, one other adult male, and at least two breeding females. There are also six to nine subadults that range in size from 1.2 to 2.1 m, as well as several yearlings that are about .7 m in length.

We have also observed recruitment into the population, when our consultants captured a previously unknown male of 1.5 m in length and an unknown female of approximately the same size. We have documented in the 1978 nesting season that at least two nests were laid on berms of the cooling canal system. These nests produced at least 18 hatchlings, which were captured, marked and released. In 1979, at least one nest was documented with 26 hatchlings being captured, marked, and released. Several hatchlings were repeatedly recaptured over a year's period and allowed us to construct growth curves. One 1979 hatchling that was originally measured at 28 cm and 52 g was recaptured in 1980 and had attained a length of 66 cm and 1100 g. Several of the adults are fitted with radiotelemetry collars, which have been very useful in tracking them and in locating nesting sites.

One of the most interesting facets of this program is the confirmation that crocodiles will preferentially seek out and utilize man-made canals and berms for resting and reproduction. The animals are almost in a controlled experiment because adjacent to the Turkey Point site is approximately the same amount of area which is "prime habitat." Our studies have shown that the animals do not utilize this "prime habitat," but instead utilize the disturbed habitat, the cooling canal system.

We have also documented the presence of a crocodile at our Lauderdale Plant. This site is located in an urbanized and developed area of Fort Lauderdale. The plant was built in the 1930's and has approximately 400 hectares of cooling canals and buffer zone around it. The site contains two oil-fired units and 24 gas turbines, and it is located directly on the approach to the Fort Lauderdale International Airport.

During the Manatee Aerial Census Program, an adult crocodile was seen twice over a six-month period in the cooling canal system of the Lauderdale Plant. This crocodile was shot by someone trespassing on our property. Law enforcement of the Fish and Wildlife Service were called in to investigate, but could not make a case. This area is now posted.

SEA TURTLES

Three species of sea turtles -- the endangered Atlantic Green Turtle, the endangered Atlantic Leatherback Turtle, and the threatened Atlantic Loggerhead Turtle -- all nest on the beaches of eastern Florida. During the months of May through August, the female turtle emerges at night from the sea to deposit eggs in a nest excavated on the beach. Each female undertakes from one to seven egg-laying excursions during a nesting season. Approximately 70-130 eggs are laid and require about 60 days for incubation. The hatchlings emerge from the nest at night and return to the water.

FPL has an operational nuclear plant and sister unit under construction on a coastal site of east-central Florida. The plant cools its condensers with a once-through mode, using an ocean intake and an ocean discharge. FPL owns approximately 500 hectares of land on the coast encompassing more than 4 km of pristine beach, which is ideal for turtle nesting.

Environmental monitoring has documented that three species of sea turtles nest on beaches owned by FPL, with the Loggerhead nests being the most numerous followed by the Green and then the Leatherback. Surveys conducted every other year since 1971 have not attributed any permanent decline in nesting frequency to plant effects. There was, however, a localized decrease in nesting frequency due to installation of the ocean intake and discharge pipes. Nests that were laid by the construction zone and in immediate danger were relocated. Once construction activities had been completed, nesting frequency returned to that of control areas.

In cooperation with federal and state officials, our consultants have tagged more than 1,500 sea turtles. This work is forming the data base for population estimates of these endangered species. The consultants are also verifying turtle nesting counts on selected beach transects to compare with aerial counts being obtained by the National Marine Fisheries Service.

When a female emerges from the sea to nest, she is very sensitive to human disturbances, which may deter her from nesting that night. This is called a false-crawl and it is becoming more common in areas where beach development is occurring. Also of importance is the visual cue turtle hatchlings receive when they emerge from the nest at night. If the skyshine over the land area is greater than over the water, the hatchlings will crawl toward land instead of towards water and perish. Thus, it is paramount that human disturbance and night skyshine be minimized if we are to protect these animals.

Because FPL owns 4 km of beach and will use it as a buffer zone for this nuclear plant, sea turtles will be guaranteed an undeveloped stretch of prime nesting beach for at least the next 40 years. To minimize the effects of skyshine on turtle hatchling behavior, FPL has placed reflector shields on security lighting and has planted a light screen immediately behind the beach dune to minimize misorientation of the hatchlings.

CONCLUSIONS

As I have documented with these examples, FPL can co-exist with many endangered species. Large buffer zones owned by FPL in both urban and rural areas attract numerous species of wildlife. Some rare and endangered species inhabit these buffer zones for all or part of their life cycle and readily adapt. Thus, FPL does not need to take any special precautions other than just to be aware of their presence. In other instances, when real or potential conflicts have been identified through research and monitoring, the Company has taken positive actions to eliminate or minimize any adverse consequences. Thus, in the sense of stewardship and good business practice (i.e., to avoid costly delays in permitting and construction), FPL is striving to demonstrate that the production and transmission of electricity and the well-being of rare and endangered species can be compatible.

A Local Endangered Species Program - Florida Manatee, *Trichechus manatus*

Diane D. Barile
Department of Management Science
Florida Institute of Technology
Melbourne, Florida
Environmental Planner, City of Palm Bay, Florida

ABSTRACT

In the preparation of Local Comprehensive plans, many Florida communities have included goals for the preservation of the State official marine mammal, manatees, *Trichechus manatus*. There are, however, no guidelines or descriptions of programs or policies for the implementation of these goals by city or county governments.

This paper describes the development of a manatee preservation program at a local government level using Palm Bay, Brevard County, Florida as a model. The process is described for collection of data, compilation of resource materials and analysis of the impact of local government activities on manatees. Recommendations for program initiation are presented. The process and mechanism is delineated for coordination with State and Federal agencies charged with enforcement of the Florida Manatee Santuary Act, the Federal Endangered Species Act and the Marine Mammal Protection Act. This interdisiplinary approach, coordinated through local government brings citizens, staff of community government, university researchers, conservation groups, and state and federal regulatory agencies together to strengthen the effectiveness of all levels of government in their efforts to protect manatees.

Manatees or sea cows are considered to be the most endangered marine mammal in the United States. Increased human/manatee interaction has reduced the population of these 13 foot long vegetarians to between 800 and 1000 individuals. Often blundering into marinas, waterfront canals, waterways and ports, they are subject to harrassment, vandalism and most critically collisions with motorboats and barges which share their shallow coastal waters.

This study was funded by the Florida Department of Environmental Regulation and the Office of Coastal Management.

Manatee biology and habitat needs are discussed. A matrix of local government activities and possible benefits to manatee populations suggests realms of great positive impacts at a city or county level. These guidelines are then incorporated into the framework for a manatee protection program for education, public awareness, enforcement, public works, recreation and inter-government coordination.

This model of implementation of manatee protection goals for local government can be applied to other communities in Florida who share the desire for maintenance of a viable population of the animals Columbus mistook for "mermaids." A similar process may, in addition, be applicable to community aspirations for preservation of other threatened or endangered species.

INTRODUCTION

In the course of the planning process local government agencies or advisory groups may develop goals for the preservation or maintenance of wildlife or a particular endangered animal. The intent of such aspirations are notable but attainment of the goal is often limited by the lack of community action or implementation policy. There are few guidelines for the preservation of endangered species at a community or local level.

Using the Florida manatee, *Trichechus manatus*, as an example a comprehensive process for endangered species' preservation is developed using the City of Palm Bay and Brevard County, Florida as a model. A similar program can be applied to fulfillment of similar environmental or endangered species protection goals by other communities or for the protection of manatee by other local governments in Florida.

Florida's official marine mammal is threatened with extinction throughout most of its range. The United States population is estimated to be between 800-1000 animals, inhabiting Florida's coastal freshwater streams, springs, estuaries and ocean waters. A natural low birth rate and impacts associated with sharing habitats with people have combined to steadily reduce the population. Coastal development and land use have altered feeding grounds. Targets of harassment from shorelines or divers in the water, struck by power boats and caught in flood control dams, these gentle giants wander through coastal canal systems and loll in congested marinas. Most manatees, 80 - 90 percent, are scarred from collisions with the propellers of power boats. Vulnerable to cool water, manatees have become dependent on the artificially warm effluent water from electrical power plants. Natural curiosity may lead these nearsighted creatures to entanglement in fishing lines or crabtrap ropes.

Manatees, as an endangered species designated by the Federal government, are the subject of active programs of biological research, enforcement of protective legislation and efforts to salvage sick or injured animals. Several permit review agencies consider project impacts on manatee. The State of Florida developed a comprehensive public awareness program. The entire state has been declared a manatee sanctuary by the State Legislature. Protection measures have been undertaken in specific areas where manatee congregate. One city and two counties have local manatee protection ordinances or resolutions. The implementation of these programs for manatee protection are seldom integrated into local plans.

In preparation of Comprehensive Plans, mandated by the Florida Local Government Comprehensive Planning Act, many cities have expressed the desire to protect manatee. Goals and objectives for endangered species protection may be put forth as part of required Coastal Zone Protection or Conservation Elements. Faced with the implementation of more complex elements of the comprehensive plan (i.e., land use, transportation, etc.) strategies for preservation are often delayed or foregone. This paper seeks to present a comprehensive program for protection of the Florida manatees by county and city governmental agencies.

WHAT IS A MANATEE? - DESCRIPTION AND BIOLOGY

The Florida Manatees, also called the West Indian or Caribbean manatee (Trichechus manatus) are mammoth aquatic mammals mistaken by early Spanish explorers for mermaids. With reference to the sirens of Greek mythology, manatees are classified in the mammalian order of Sirenia. Manatees are the only herbivorous mammals adapted to aquatic life. Ancient kinship to the elephant is evident in similarities in teeth, ivory-like bones and toenails on the flippers. As mammals, manatees are air-breathing, bear their young alive and nurse them to young adulthood. Manatees grow to lengths of up to fifteen feet and weigh up to 1500 pounds.

Surprisingly, for their large size and bulk, manatee move inobtrusively in coastal waters or rivers, lakes and streams with access to coastal waters. These gray to brown animals with flat spatulate tails surface quietly to breathe about every four to five minutes. The fusiform body is covered with a thick wrinkled skin, forelimbs are paddle shaped and hindlimbs absent. The cleft upper lip is lobed and surrounded by the bristly hairs of the muzzle. The nostrils close as the animal submerges and open for surface breathing. There are no external ears.

Behavior of manatee is not completely understood. Usually solitary, often lethargic, they tend to be curious and gentle. They have no natural enemies or means of defense. Periods of activity are arhythmical. Sleep occurs during periods of bottom resting or suspension near the surface of the water. Social behavior in manatees includes embracing with the flippers, muzzle contact, bumping, rolling, circling, breathing in unison or swimming in a follow-the-leader fashion.

Manatee may be indiscriminate feeders, grazing for six to eight hours a day on any plant soft enough to be torn by their muscular lips. The 150 pounds of vegetation consumed daily includes bottom growing seagrasses, natant and emergent plants, algae from boat bottoms and pilings as well as overhanging bank vegetation. In Guyana an experiment showed that manatee could clear aquatic weeds from canals. The need for a source of fresh drinking water has not been proven but manatee have been observed drinking from freshwater sources.

Except for large cold spell congregations, manatee are seldom found in large aggregations. More commonly they are found in small groups of three or four individuals. These groups are often short lived with each animal retaining individualism. Some manatee may reside seasonally within a restricted area while others are transients, remaining days, weeks, or months at a time in a suitable habitat. Some individuals return to the same range year after year.

Florida pioneers in the 1870-1900's found an abundance of manatee. Tourist guidebooks often mentioned the thrilling experience of seeing manatee in the state's estuaries, bays and streams. As recently as the 1930's manatee supplied meat for the table of many families. Early in this century, use as a food source exceeded the rate at which manatee could reproduce. A cow matures at about eight years of age, carries a calf for 13-14 months, nurses and cares for the calf up to three years. Manatee were hunted for sport, for oil and for their elephant like bones and even their hides. Concern for the conservation of manatee occured as early at 1893 when Florida Law recognized the danger of manatee population. In 1907 Florida prohibited killing or molesting the animals under threat of a $500 fine or 6-month imprisonment.

As more people were drawn to live near manatee habitat, the number of human-manatee confrontations increased as the animals' natural habitat was modified. Favorite feeding ground on grass beds were destroyed due to sediment loadings or dredging of shallow rivers and estuaries. New habitats, however, were created in drainage canals and waterfront developments.

Finally, the greatest single hazard to the lumbering, blundering, slow moving cows of the deep became popular - motorboats. Manatee can move at speeds up to 16 mph but they dive very slowly; usually just alternating a sinking and floating motion by exchanging up to 50 percent of the air in their lungs. Their dive is sometimes accompanied by powerful downstrokes of the spade shaped tail. Manatee are air breathers. They must surface at least each 10-15 minutes to breathe. No match for the speed and maneuverability of powerboats with exposed propellers and crowded by barges in shallow canals, manatee injuries increased. As deliberate poaching and killing manatee for food or sport decreased there was an increase in manatee mortality due to accidental collisions with boats. In 1979 there were 77 recorded manatee deaths, 23 caused by boat or barge collisions.

Manatee migrate from one area to another during the summer, but tend to congregate in warm water during winter months. They are sensitive to low temperatures. In 1977 several deaths could be attributed to prolonged exposure to water below 21°C or 68°F. Historically manatee moved south or up freshwater streams to warm springs at Crystal River or Blue Springs to survive cool weather. These winter concentrations increase the chances of boat collision and harassment from divers who are also drawn to the warm, clear springs.

Modern development of Florida requires electrical power. This too has impacted manatee; their lives and distribution. Power plants are located on the coast where water to cool equipment is readily available and fuel can be developed inexpensively by barge. The water warmed by power generation is released back into the stream. Manatee have become attracted to warm water effluent. They have become dependent on this heat during extremely cold weather. If during a winter storm the power plant ceases operation, the congregated manatee are denied refuge. Exposed to dangerous cold, they become susceptble to bronchitis, pneumonia or death. Manatee also are injured by becoming ensnarled in flood control gates and dams. Many of these slow moving animals are harassed, wounded or killed by adults and children who shoot, spear, hook or pelt them with stones. During the past twenty years people have been concerned about manatee.

WHERE DO MANATEE LIVE? - MANATEE HABITAT

Manatee are distributed along coastal Florida within the jurisdiction of many cities, towns, and county governments. The environmental, physical and geographical features of these communities provide an adequate place for manatee to live. Manatee habitats have generally been defined but the purpose of this discussion has a slightly different perspective. Based upon an expressed community physical goal to preserve manatee, what features are necessary to insure their continued utilization of the area?

The place where manatee live is determined by those factors which influence the animals' food preferences, need for drinking water, special requirements and behavior. Definition of manatee habitat needs is not exact and research in continuing in this area. However, assumptions can be made on habitat requirements based upon literature review, analysis of distribution and investigation of parameters describing areas of occurrence.

Manatee are found in fresh and salt water in tropical and subtropical climates. They are distributed in estuaries, rivers and streams of coastal Florida waters. Excursion into the saline coastal waters of the Atlantic Ocean and Gulf of Mexico do occur. Manatee do not appear to be influenced by salinity. They may, however, require a fresh water source periodically. They drink fresh water from garden hoses, boat bilges and air to water air conditioner discharge points. There does seem to be a shift in distribution toward freshwater during the winter months. This may be in response to water temperature rather than salinity.

Manatee are voracious feeders on aquatic and marine vegetation consuming between 100-150 pounds per day. Therefore, a food source should be present within their range. In fresh water manatees feed on *Myriophyllum spicatum*, *Hydrilla verticillato*, *Ceratophyllum demersum*, *Vallisneria neotropicolis*, *Elodea densa*, *Panicum molle*, *Eichornia crassipes*, and *Alternanthera*. Brackish and saltwater food sources include the seagrasses, *Ruppia maritima*, *Thallassia sp.*, *Syringodium sp.*, *Halophila sp.*, and *Diplanthera sp.* Seagrass beds in flats in water more than one meter deep adjacent to deeper waters would be considered possible manatee feeding zones.

While turbidity seems to have little or no effect on the manatee directly or their preference for one area versus another, turbidity may affect food supply and perhaps the number of boat/manatee collisions. Submerged vegetation is the preferred food for manatee. Turbidity from dredge and fill projects or sewage waste water disposal may reduce the amount of light penetrating the water. Deprived of light, the grasses can neither function nor grow. Turbidities may also cause sedimentation rates to increase in the estuary. As grass beds are buried in settling silt, manatee must exploit a different food source or move to another area. In the absence of submerged vegetation manatee will consume water hyacinth or other aquatic weeds.

Turbidity and water color may increase the incidence of manatee/boat collisions. Dark water may obscure the visibility of the giant animals to boaters moving through the water. This could account for accidental injury to manatee from propeller blades or boat skegs.

Depth of water in a manatee habitat is generally considered limiting at less than one meter. Manatee do, however, move in shallower water to reach vegetation over hanging shorelines and banks. A female in estrus being pursued by a herd of males, will move into shallow water perhaps to discourage their advances.

Temperature appears to influence manatee distribution and therefore, is a critical feature of the animals habitat. Manatee frequent an area so long as temperatures remain above approximately 20°C or possibly as low as 15°C for short periods of time. These temperatures are probably tolerated as the animals move toward a warm water refuge or during short forages from the refuge in search of food.

Manatee seem to be found often in marinas, waterfront residential canals and protected bays, harbors and the mouth of freshwater creeks. These areas appear to offer sanctuary possibly from wind, current or some other feature as yet undetermined.

It is interesting that these areas offer the features most amiable to boat mooring. It is not understood whether manatee are attracted specifically to marinas or whether marinas are ususally built in locations habitated by manatees. Marinas may be built in sheltered coves historically utilized by manatees. Boaters and manatees then share the cove. Boat bottoms and pilings are convenient back scratching posts. Algae is scraped from boat bottoms and freshwater is available from dock hoses and boat bilges. Manatee eat water hyacinths which block in berths during the summer and provide an interesting entertainment to boaters using the area.

Typically then, optimum habitat for manatee should include in part (1) a source of seagrasses, algae, floating or bank vegetation; (2) water depth of one meter or more graded to shallow sloping banks; (3) possibly a source of freshwater; (4) access to brackish or sea water probably through a deep channel or stream; and (5) water temperatures generally above 20°C and seldom below 15°C.

WHAT CAN A LOCAL GOVERNMENT DO TO PROTECT MANATEE?

The actions of communities, towns, city and county governments can have a signficant impact on the survival of the vanishing Floridian, the manatee. Decisions made at responsible community levels can benefit both the animal directly and its habitat indirectly. A manatee preservation or protection program could be implemented on a community wide scale, by the community's governmental organization or by an appropriate department of local government. The scope would be determined by the degree of interest and motifivation of people to become involved in the project.

If an endangered species protection program is to be initiated within a city or county government how could the greatest benefit be derived? Table I assesses the degree of benefit which could be attributed to manatees through various local governmental functions. Specific programs, policies, and strategies leading to these benefits are discussed in subsequent sections.

The greatest direct benefit to manatee can be accrued through programs of education for all age groups, improved local law enforcement coordinated with state and federal agencies. Community planning and public works projects primarily enhance and protect the manatees' habitat, maintain water quality, and safeguard against potential hazards.

TABLE I
AN ASSESSMENT OF LOCAL GOVERNMENT
MANATEE PROTECTION ACTIVITIES AND PROGRAMS

Activity	Degree of Benefit					
	0	1	2	3	4	5
Education				X		
Grade I-12						X
Higher Education						X
Continuing Education				X		
Public Awareness						X
Youth & Civic Groups				X		
Public Works			X			
Water Treatment	X					
Sewage Treatment				X		
Solid Waste Dispose			X			
Canal Construction & Storm Drainage				X		
Power Plants	X					
Site Selection	X					
Emergency Program				X		
Dams and Flood Control Structures				X		
Aquatic Weed Control				X		
Mechanical				X		
Chemical						
Transportation			X			
Streets, Railroads & Airports	X					
Ports				X		
Law Enforcement						X
Planning						X
Residential Development					X	
Indsutrial & Commercial			X			
Coastal Zone Management				X		
Greenbelt Management		X				
Zoning					X	
Building Permits					X	
Sanctuaries, Preserves & Conservation Areas						X
Recreation					X	
Education Program					X	
Docks, Ramp, Marinas					X	
Budget					X	
Inter-Government Coordination			X			
Fla. Dept. of Environmental Regulation				X		
Fla. Dept. of Natural Resources						X
Fla. Marine Patrol						X
Fla. Dept. of Transportation			X			
US Fish & Wildlife Service						X
US Army Corps of Engineers				X		
US Coast Guard			X			

THE LOCAL PLANNING PROCESS AND ENDANGERED SPECIES - THE MANATEE

The impedance for any action is the desire to attain some result. Planning, then, provides the information, methodology, objectives and means to evaluate results of the process. Utilization of the planning process in establishing a manatee or an endangered species preservation program is described as follows.

A. Establish goals - Does the community want or need a manatee preservation program?

Preservation goals may result from citizen concern, the suggestion of conservation, environmental groups, or may evolve from community goals established in the comprehensive planning process. These ideals may be expressed in recreation, conservation, land use or coastal protection elements of a comprehensive plan. Initial objectives and strategies may be enumerated as a means of further defining the scope of the proposed action. These courses of action can then be evaluated and refined as program development continues.

B. Data Collection and Analysis. What do you need to know about local manatee populations?

Successful management of an endangered species with the jurisdiction of a local government requires that basic information regarding the biology, distribution, habitat, and special needs of the animal be compiled. An assessment should be made as to the number of animals affected locally and their special and seasonal distribution within the community. Special areas of concentration or congregation and potential hazards should be noted.

Data collection and analysis efforts include the review of local ordinances and resolutions for existing measures which could be incorporated into the proposed program. Similarly, state and federal legislation regarding manatee should be made familiar. Sources of information, brochures, literature and handout material for public education programs should be determined. Environmental action groups (Audubon Society, Sierra Club, Wildlife Federation, etc.) or individuals can often be valuable resources.

Manatee are protected by the Marine Mammal Protection Act of 1972, the Endangered Species Act of 1973 and the Florida Manatee Sanctuary Act of 1979. The U.S. Department of the Interior and the Fish and Wildlife Service are engaged in an active program of manatee research and enforcement. The State of Florida Department of Natural Resources promotes an extensive education and public awareness program in addition to the regulatory activities of the Florida Marine Patrol. Universities, particularly the Rosenstiel School of Marine Sciences of the University of Miami and the Florida Institute of Technology, are engaged in on-going studies of manatee.

Sources of information need not however be limited. City staff members may conduct surveys and studies. Citizen volunteers can monitor the daily recurrence of the animals throughout the community. This technique was successful in a recent survey of Brevard County, Florida. Oceaniar, Marineland, Sea World and the Seaaquarium, may also provide assistance.

Data collection and analysis therefore quantifies the extent, scope, and feasibility of an endangered species program proposed for the community. Sources of information and the level of control via legislature are delineated. Additionally resource materials are compiled and the basis for intergovernmental coordination established.

C. Objectives and Programs. What are you going to do to protect manatee?

Goals for sound management of living natural resources are translated into action programs when expressed as measurable objectives and related to a functional program. Basically, objectives in the manatee preservation program intend to reduce adverse impacts, create an awareness throughout the community of the vulnerability of the animals and to protect and maintain aquatic habitats for manatee utilization. The following are objectives which could become part of a local community manatee protection program:

1. Educate citizens to the life history of manatees.
2. Alert the public to the danger of manatee extinction.
3. Support continued research.
4. Improve enforcement of existing legislation.
5. Prepare local manatee protection ordinances or resolutions.
6. Decrease vandalism.
7. Decrease harrassment.
8. Decrease boat/manatee collisions.
9. Maintain and improve water quality, habitats and food sources.
10. Avoid creating new warm water refugia.
11. Protect manatee at existing warm water refugia.

Implementation of these goals can be achieved through a variety of programs undertaken in local government. Education and awareness programs will fulfill the objectives information exchange and increase sensitivity by the public to the dangers facing manatees. Permitting and enforcement programs will allow protection of manatee habitat and the means to discourage harassment and vandalism. Land use and zoning programs can insure maintenance of manatee habitat, water quality, and buffer wildlife habitats from uses inconsistent with their maintenance.

WHO WILL DO WHAT? - IMPLEMENTATION

Achievement of any goal ultimately depends upon the necessity of working to attain some objective; for someone to do something. The following lists the various functioning units of local government and describes activities for each which could be undertaken to protect an endangered species.

A. The city council or county commissioner could by resolution declare manatee aware day, week, or month. This effort perhaps with the sponsorship of a local school, conservation or boating group could prepare a press campaign, obtain available literature, films, and slide shows for use by civic clubs, yachting organizations and other schools. Manatee awareness materials could be distributed at the city hall, through newsletters or chamber of commerce bulletins. Mailings could be made in monthly utility bills.

Youth groups and adult organizations often invite programs featuring issues of local interest. City or county personnel could be made available to present manatee awareness programs. Groups which may express an interest in manatee could include 4-H scounts, church youth groups or YMCA.

State, city and county governing bodies have passed legislation, ordinances or adopted rules for the protection of manatee or their habitat. The city council by resolution could declare the entire county a manatee sanctuary or declare vital portions of the community as an area of critical habitat.

The city budget could include funds necessary for special manatee programs. Professional survey of manatee distribution, occurrence and utilization of local marine habitats could be funded through such agencies as the Florida Audubon Societ, the U.S. Fish and Wildlife Service, local universities, junior colleges or private environmental engineering firms.

Expenditures could be authorized for a patrol boat for use by local law enforcement officers to facilitate monitoring boat speeds and enforce manatee protection measures during periods of manatee concentration or local habitat use. The boat could also be used for surveys and water quality monitoring.

B. Public Works projects also impact manatees. Sewage treatment effluents often increase turbidity and suspended solids in the receiving water body. Turbidity has little or no direct effect on manatee. However, the ability of boaters to avoid collision with manatee decreases in highly colored water. Additionally, increased turbidity decreases light penetration through the water, deduces grassbed vegetation productivity, and a loss of manatee feeding grounds. Manatee will then leave the area in search of food elsewhere. Effluent discharge should be discouraged in areas of manatee congregation or feeding. There was no data reviewed during this project which indicated an impact from water treatment facilities on manatee.

There is no evidence of impact of solid waste disposal on manatee. They are, however, indiscriminate feeders, consuming any vegetable matter in the water. Local government can enforce liter campaigns near waterways to insure that poisonous or potentially hazardous materials are not discarded into lakes, streams, rivers or estuaries.

Canal construction which impacts manatees takes place in the estuaries and freshwater streams as well as in upland dredged canals and channels. Construction of canals adjacent to manatee habitats in shallow estuaries, affects light penetration through the water, limiting food sources preferred by the manatee. Rather than clear relatively shallow water the manatee are forced to maneuver in dark, or turbid deeper water frequently used by boaters.

As canals are constructed in uplands, new avenues for migration and utilization are created. These canals, however, increase the possibility of collisions with boats since manatee tend to use boat channels for cruising from one feeding ground to another rather than use adjacent shallow zones. Residential canal development increase the opportunity for harassment and mortality from people who have easy access to the large animals along relatively narrow or confined corridors.

Food sources for manatee in upland canals is limited. Seawall lined channels have high turbidities, and little or no bottom of shoreline vegetation. Floating vegetation is often eliminated by aquatic weed eradication programs.

Local government should introduce natural shoreline structure design using energy absorbing vegetated slopes for existing channels and shore. This will decrease turbidities and provide supplemental sources of food. Adoption of local dredge and fill ordinances with prohibition of dredge and fill activities in manatee habitat areas would benefit overall water quality.

Manatee congregate near Florida electrical power plants during periods of cool weather when water temperature drops below 20°C. Since manatee are susceptible to bronchitis, pneumonia or death in cold water, they are dependent upon the heated water discharged from the generation of electrical power. These areas are critical habitats for sea cows. Florida Power and Light Co. studies manatee and has awarded grants to the Florida Audubon Society and the U.S. Fish and Wildlife Service to study the utilization of power plant effluent by manatee. The results of these studies could be of use to communities near these facilities.

Initial site selection for power plants should take into account the penchant of manatee to become dependent on warm water effluent. If the plan site is near manatee habitat perhaps rather than create a new artificial refuge, an alternate method of discharge could be incorporated into the design. For instance, discharge water could be recycled rather than allowed "once through cooling." This would avoid the possible responsibility of the power company for manatee deaths during col weather power plant shut-downs.

Local governmental agencies in areas of artificial refuges (either power plants or industrial effluents) could coordinate an emergency contingency plan for the rescue and/or feeding of manatee in case of a shut-down which would deny manatee the protection of heated water. Resources needed for such a program could include high protein food, antibiotics, transportation and manpower. The program could be coordinated with the utility company and research teams through the City Managers Office.

In the southern half of Florida, automatic flood control dams may account for up to 25 percent of manatee mortality. In manatee areas, design of flood gates should allow passage of both adult manatee and calves. Manatee are attracted to flood gates or locks and have been crushed or drowned when trying to swim under the gate as they close. A University of Miami designed flood gate is being tested by the South Florida Water Management District.

The method of eliminating floating and submerged vegetation of canals, streams and channels can impact manatee since these plants can be an important part of a manatee diet. Conversely manatee could be utilized by local government to clear channels. Manatee eat between 100-150 pounds of vegetation per day. In Georgetown, Guyana manatee have been used for weed control for more than 15 years with no expenditure of funds and no maintenance costs for the manatee (National Science Research Council 1974). Since the cost of aquatic weed eradication can range from $10 - $15 per acre, per application, manatee management could represent a considerable savings to local government agencies.

Mechanical aquatic weed control is among the most costly methods of eradication. It requires maintenance of right of way, expensive equipment and a disposal for the wet and decaying materials pulled from the canal. This method may increase turbidities in the canal during operation. There is no evidence of manatee being harmed directly by mechanical cleaning of canals or waterways. Therefore local government should recommend mechanical weed control especially in designated manatee areas.

There is a possibility that chemical control of aquatic weeds can affect manatee. The 2, 4-D often used to spray water hyacinth or products produced as it is diluted or broken down in the water, may be incorporated in the tissues of manatee eating the sprayed vegetation. Manatee may leave an area usually frequented after the weeds have been sprayed. Local government should refrain from chemical weed control in manatee areas.

The noise of streets, railroads and airports has little effect on manatee distribution or occurrence. So long as drainage structures serving these facilities meet criteria listed previously, culverts and adverse water control structures are designed for safe passage of manatee. Adverse impact of transportation systems have not been documented.

The dredging of port existing facilities, channels, turning basins and docking sites in areas routinely used by manatee could increase the likelihood of boat collisions and/or crushing by boats or barges in narrow, shallow channels. If the design of channels were such that the depth and width would allow manatee passage beneath or around the ship perhaps mortalities could be avoided. Additionally, construction could be scheduled for periods when manatee do not utilize the area or during tidal periods when manatee cannot negotiate the shallow shoreline.

C. Recreation Departments.

Recreation programs offering community wildlife or nature courses could hold courses, classes, or lectures on manatee. Manatee awareness programs could become a part of community sponsored boating and sailing classes. Manatee area caution signs should be posted at community owned or operated docks, boat ramps or marines. Manatee awareness literature and information describing manatee protection regulations, enforcement procedures and fines should be readily available at these locations.

Dock masters or marina maintenance crews should be aware of procedures for reporting injured or dead manatees via the manatee hot line. Also procedures should be well established for quickly alerting enforcement officers in the event of manatee harassment or vandalism. These procedures could include summoning local law enforcement officers, district U.S. Fish and Wildlife Officers or the Florida Marine Patrol. Local manatee survey information could also be collected at city operated marine facilities. Data on sightings by boaters could be compiled on data sheets or lists posted prominently at or near docking or ramp facilities.

D. Planning.

The planning, community development and environmental health agencies of local government can play an important role in a manatee preservation program. It may be that the planning department is the most logical division of local government to coordinate the program.

Residential waterfront property development impacts manatee in several ways. Dredging of canals in wetland areas creates new paths for manatee movement while decreasing the total area of manatee habitat. Thus manatee instead of using rather a shallow, dispersed water body with vegetated bottoms are forced to use deeper more confined channels. This removed grassbeds as a food source and increases the possibility of boat collisions in canals. The channel may benefit manatee distribution by allowing the animals to move from one food source to another. Residential development may also decrease the availability of shoreline vegetation if banks are modified by non-vegetated sea walls and revetments. Residential development adjacent to manatee areas can increase incidence of harassment and poaching. Local governments can encourage the use of shore line vegetation for bank stabilization or plan for buffer zone of native negetation along streams and canals.

If commercial or industrial development includes facilities which release heated effluent into streams or canals in waters inhabited by manatee an artificial refuge may be established. Manatee can become dependent on the heat of these refuge areas, just as they are dependent on power plant effluent areas. Local government can plan for non-water dependent commercial and industrial land use in areas other than the coastline. Design for water dependent use can take into account the vulnerabilities of manatee population and methods of mitigating impacts of heat or polluted water release.

Coastal Management Programs, Federal, State and local are consistent in their goals with a manatee management program. Manatee, as a coastal resource could be considered in a Coastal Zone Element of a Florida Local Comprehensive Plan. Implementation of community goals for manatee preservation could be considered in the planning process by the locally designated planning agency. Assistance in the implementation of coastal plans may be available from the Florida Office of Coastal Management in Tallahassee.

Greenbelts planned along creeks, rivers, canals and estuaries frequented by manatee could preserve shoreline vegetation, improve water quality and prevent erosion of banks. Well vegetated banks would decrease accessibility to the animals and may reduce vandalism or manatee harassment.

As a land use implementation tool the zoning process could be used to establish preservation of conservation zones through the community in essential manatee habitats or congregation areas. As a primary tool for implementing local comprehensive plans, zoning could require performance standards related to construction of facilities adjacent to critical manatee habitats.

Building officials reviewing applications for the construction of bridges, boat houses, sea walls or other structures in manatee areas could alert applicants to be presence of manatee. Hand out materials on manatee and construction in manatee areas could be made available. Guidelines describing requirements and legislation for manatee protection could be included. Building officials should be made aware of critical habitat or congregation areas in the community. Building codes for construction of sea walls and revetments could require vegetated, natural shoreline design for bank stabilization. Banks would be protected from erosion while enhancing manatee habitat, reducing sedimentation and turbidity in the stream. Emergent vegetation would trap soil and serve for nutrient uptake from the stream, therefore improving general water quality in the City.

Sanctuary, preservation or conservation areas can be designed through the planning process. Decision makers by resolution can designate the community a manatee santuary and prepare a program of posting signs or conduct a public information program. Conservation zones can be designated as part of the communities' land use planning process.

Special review of development plans in these areas can be required before any development modification can be undertaken.

E. Police

Local law enforcement officers could aid in manatee protection programs by patrolling areas where manatee are known to use habitats within the community. These patrols could be triggered by reports of manatee presence by survey groups or be undertaken routinely as regular surveillance during manatee season. Any local Florida Law Enforcement Officer can enforce Florida Manatee legislation.

Local law enforcement officer could report cases of harassment, poaching or incidents of manatee death to the Florida Marine Patrol or other federally deputized agents for enforcement of the Endangered Species Act and the Marine Mammal Act. Police dispatchers should have access to the manatee hot-line number for reporting manatee information or for distribution to citizens who may make inquiries of the police department concerning manatee.

City or county government may through police or sheriff offices use boats to patrol critical habitats during periods of local manatee concentration. The patrols could warn boaters of "no wake" zones and the utilization of the area by manatee.

While it is understood that Federal laws authorize only Federal agents or other deputized agents to enforce the Federal Endangered Species Act or the Marine Mammal Act, local governments can play a supportive role in enforcement. Local agencies which may become more effectively involved include the Florida Marine Patrol, City Police, County Sheriff's Department, County Commissioners or City Councils and the State Attorney. City enforcement can fill the gap between State and County level enforcement but local law enforcement officials need information related to potential enforcement tools which could be used in efforts to protect manatee. It has been suggested that the Florida Department of Natural Resources, Marine Patrol training officers conduct training sessions for enforcement agents in local government. The DNR should develop a program designed to increase the awareness of enforcement problems, to give training in detection of incidents of animal harassment and outline procedures for notification of appropriate regulatory or recovery teams when problems arise.

F. Schools

Most authors, workshop recommendations and manatee researchers point to education of the public as a prime method of reduction in manatee mortality due to harassment, vandalism and boat collision. While local government other than the school board has little interaction with local school policy, a program could be introduced through resource people available through city or county agencies. Manatee preservation or endangered species education should be encouraged as a part of the Florida school curriculum.

Colleges and universities, both state supported and private, could be a source of information on local manatee. These institutions could also provide professional expertise in monitoring manatee population, habitat needs, water quality and other data necessary for responsible decision making. Sponsorship or partial sponsorship of research on a university or junior college level would encourage the expansion of knowledge about the local population of manatee, present information for effective management and increase the number of professionals involved in manatee research.

Local recreation departments and junior college continuing education programs could incorporate study, field trips or canoe trips to study manatee or use manatee awareness programs in their regular classes on wildlife, fishing or boating.

WHO CAN HELP A COMMUNITY WITH A MANATEE PROTECTION PROJECT? - INTERGOVERNMENTAL COORDINATION

A. Since manatee are oblivious to city, county, and state political boundaries, cooperation among agencies it important in manatee protection and management. Adjacent cities can coordinate their survey, surveillance and enforcement programs. This would maximize time and energy expended as well as produce a more cohesive body of local information and a broader range of enforcement capabilities.

B. State Agencies.

- Florida Department of Environmental Regulation
- Florida Department of Natural Resources
  - Educational materials
  - Manatee hot-line
  - Florida Marine Patrol
- Florida Department of Transportation

C. Federal Agencies.

- U. S. Fish and Wildlife Service
  - Information
  - Enforcement
  - Signs
  - Recovery team

U.S. Army Corps of Engineers
U.S. Coast Guard

REFERENCES

Applied Biology. 1977. Review of the Ecology and Life History of the Florida Manatee for Florida Power and Light Company, Miami. Atlanta, Georgia.

Barile, Diane D. Ed. 1979. Third Annual Brevard Water Resource Conference. Florida Institute of Technology, Melbourne, Florida.

Barile, Diane D., Ky B. Ostergaard, and John A. Tiedemann. 1980. A Local Government Manatee Preservation Program. Florida Institute of Technology, Melbourne, Florida.

Barile, Diane D. and Ky B. Ostergaard. In press. Local Management of Manatee *Trichechus manatus* in the Coastal Zone. Proceedings at *Marine Technology* '80, Marine Technology Society, Washington, D.C.

Brownell, Robert L, Katherine Ralls and Randall Reeves. 1978. Report of the West Indian Manatee Workshop, Orlando, Florida. 27-29. Unpublished manuscript.

Campbell, H. W. and J. Powell. 1976. Endangered Species: The Manatee. Florida Naturalist. 49(2):15-21.

Hartman, D.S. 1971. Behavior and ecology of the Florida manatee, *Trichechus manatus latirostris* (Harlan), at Crystal River, Citrus County. Ph.D. Dissertation. Cornell University. June. 285 pp.

Marine Mammal Commission. 1979. Annual Report of the Marine Mammal Commission, Calendar Year 1978. Marine Mammal Commission, Washington, D.C.

National Science Research Council. 1974. An International Centre for Manatee Research, Report of a Workshop held 7-13 February 1974. Georgetown, Guyana.

Wray, P. 1978. The West Indian manatee in Florida: a summary and analysis of biological, ecological and administrative problems affecting preservation and restoration of the population. U.S. Marine Mammal Commission Report. Washington, D.C. 88 pp.

Atlantic Coast Ecological Inventory, 1980

J. Albert Sherk, Jr.*
and
Larry R. Shanks**

INTRODUCTION

Recently, the U.S. Department of the Interior and the U.S. Fish and Wildlife Service (FWS) played major roles in issues involving the proposed siting of two Atlantic coast refinery projects. The basis for the Department's position on both projects was the National and regional significance of the fish and wildlife resources at each site and the susceptibility of these resources to redundant stress. These recent actions highlighted the need for advance planning for Atlantic coast energy facility siting, since one of the proposed projects (Pittston, Maine) was located in an area of relatively high ecological sensitivity, while the other (Portsmouth, Virginia) was located in an area of significant resources under relatively high ecological stress.

Given the projected needs for oil refineries along the Atlantic coast, and the likelihood for Departmental involvement in similar siting issues in the future, the Under Secretary of the Interior proposed that "...Government should take the initiative in helping industry find locations for refineries, power plants, and other energy stations which would best serve the interests of all involved" (news conference, Norfolk, Virginia, 2 January 1979). To enable the Department to more effectively meet its own responsibilities under the Fish and Wildlife Coordination Act, the Endangered Species Act, and related legislation, the FWS initiated the Atlantic Coast Ecological Inventory during September 1979 which would provide information to help resolve significant environmental and energy issues for that coastal area.

---

* Biological Oceanographer, U.S. Fish and Wildlife Service Office of Biological Services, U.S. Department of Interior, Washington, D.C. 20240

** Assistant Leader, National Coastal Ecosystems Team, U.S. Department of the Interior, U.S. Fish and Wildlife Service, NASA/Slidell Computer Complex, 1010 Gause Boulevard, Slidell, Louisiana 70458

OBJECTIVE AND PURPOSE

The objective was to produce an inventory of those important ecological resources along the Atlantic coast which are potentially vulnerable to coastal siting of oil and gas processing and manufacturing facilities and their transportation systems. The study focused on ecological values subject to the provisions of the Fish and Wildlife Coordination Act, the Endangered Species Act, and related Federal legislation. To enhance the utility of the information compiled by the inventory for Federal, State and local coastal zone planners as well as the energy industry, a set of ecological resource inventory graphics were prepared using National Topographic Maps (1:250,000 scale series) provided by the U.S. Geological Survey as a base.

This inventory of important coastal fish and wildlife species, their habitats, and special land-use designations was initially planned to be the first step to provide an understanding of the relative sensitivities of important living resources of the Atlantic coast to the impacts of energy facility siting technology. Using this information, potential conflicts can be identified early in the planning stages of projects, before a site is even selected. This will lesson the chances for serious disputes later after development has actually begun. Our intent is to provide information which will help initial planning by alerting industries to areas with the greatest potential for conflict with environmental values. Information of this type provides the basis for a much better way of dealing with environmental and economic issues, before all sides are entrenched.

STUDY AREA

The study encompassed the entire Atlantic coastal zone of the Continental United States extending from the Canadian border to the southern-most tip of Florida (excluding the Keys). This zone included all areas:

1. From shoreline oceanward to the Federal-State demarcation (3 miles).

2. From the shoreline inland to the Coastal Zone Management Boundary for the Atlantic States with exceptions (see Narrative Report cited below).

3. From the shoreline to the head of navigable waters under tidal influence which exceed the Coastal Zone Management Boundary.

SCOPE OF THE INVENTORY

Data Collection

Data compiled by the inventory were collected from numerous Federal, State, local and private organizations such as FWS, the National Marine Fisheries Service (NMFS), the U.S. Environmental Protection Agency (EPA), Bureau of Land Management (BLM), National Oceanic and Atmospheric Administration (NOAA), the various state departments of natural resources and environmental protection and conservation, private organizations, such as the Atlantic States Fisheries Commission, regional commissions, environmental research foundations, universities and local experts. Special emphasis was given to collecting that information which had already been assembled by the States for use in their coastal zone planning activities.

Three major categories of information were compiled by the inventory as follows;

- o Important coastal fish and wildlife species, especially those with special designations ("endangered," "migratory," etc.) or high commerical, recreational, and esthetic value;
- o Habitats of these species; and
- o Special land use designations in the coastal area (National Wildlife Refuges, National Parks, and State wildlife management areas, for example).

Inventory Graphics

The base maps on which the inventory information was displayed were prepared from a combination of the black (culture, roads), blue (drainage) and light blue (open water) feature separation plates for the USGS National Topographic Map Series, 1:250,000 scale, the largest scale available in a complete set for the Atlantic coast. The set prepared for the Atlantic coast contains 31 maps.

On each base map the coastal zone (study area) was highlighted in yellow. The type of detail shown on each map varies from resource to resource and map to map, and was dependent on the type and amount of information available. Locations of threatened and endangered species (shown in red) and other species of special interest, as well as special land use designations (shown in green) were the key features presented. Within the biological groups other than those with endangered status, the population size of a particular species generally determined the priorities for display to prevent the map from becoming too cluttered. Large populations had the highest priority and small populations had the lowest priority. For example, on the Washington (D.C., MD., VA.,) Map (See Table 1) it was impractical to display the precise location of individual pairs of dabbling ducks, but known major dabbling duck nesting areas were mapped.

In other cases, for example, the display of meadow voles on the Providence (RI., MA., CT., NY.,) map (See Table 1) where the uniqueness of a particular species to an area was significant regardless of its abundance, these species were mapped also. For all important species the main classifications for the display of resource information were the fish and wildlife resources. For aquatic resources (fish and marine mammals), subdivisions (shown in blue) included spawning areas, nursery areas, shellfish bed locations, and fishing areas and ranges. For wildlife resources, the subdivisions (shown in brown) included nesting areas, wintering areas, migration routes, and habitats for furbearing animals. Because many species were located in a range of areas and not in just one specific area, a notebox on each map contains information on the distribution and habitat uses of organisms which could not be mapped. Areas of special biological concern, such as reefs, seagrass beds, beaches, dunes, offshore islands, marshes, and Class I shellfish waters have been indicated depending on their size, and on their economic, ecological, or scientific importance. Each of these designations, for example, species habitat, range, or status, has been shown on the maps by a combination of symbols keyed to alphanumeric descriptions and color.

In the collection, review, and analysis of the literature for this inventory, certain facts about the adequacy, reliability, and timeliness of the information became apparent. Primarily, the information portrayed on the maps was obtained from a variety of sources through personal communications with Federal and state government agencies and several private institutions; therefore, the data are only as reliable as the source material. In addition, due to the scale (1:250,000) of the base maps, and in some cases, to the large amount of information which could not be portrayed at a single location without cluttering the map, only significant species locations and distributions were presented.

Table 1. List of USGS 1:250,000-scale maps used in the Atlantic Coast Ecological Inventory

1. Eastport, ME, US, NS, NB, CAN (includes portion of Fredericton, NB, CAN; ME, US)
2. Bangor, ME
3. Bath, ME
4. Portland, ME; NH
5. Boston, MA; NH; CT; RI; ME
6. Providence, RI; MA; CT: NY
7. Hartford, CT; NY; NJ; MA (includes portion of Albany, NY; CT; MA; NH; VT)
8. New York, NY; NJ; CT
9. Newark, NJ; PA; NY
10. Wilmington, DE; NJ; PA; MD
11. Salisbury, MD; DE; NJ; VA
12. Baltimore, MD; PA; VA; WV
13. Washington, DC; MD; VA
14. Richmond, VA; MD
15. Norfolk, VA; NC
16. Eastville, VA; NC; MD
17. Manteo, NC
18. Rocky Mount, NC
19. Beaufort, NC
20. Florence, SC; NC
21. Georgetown, SC; NC
22. James Island, SC
23. Augusta, GA; SC
24. Savannah, GA; SC
25. Brunswick, GA
26. Jacksonville, FL; GA
27. Daytona Beach, FL
28. Orlando, FL
29. Fort Pierce, FL
30. West Palm Beach, FL
31. Miami, FL.

To portray as much information as accurately as possible, draft copies of each inventory graphic were reviewed by key personnel with state organizations as well as personnel from regional and field offices of the U.S. Fish and Wildlife Service. During this final review phase, all comments were checked and verified, and then incorporated on the final maps. Therefore, the information presented on the inventory graphics represents the most reliable information available to date for the Atlantic Coastal Zone which can be displayed on maps at this scale.

Narrative Report

The format and organization of the narrative report is structured geographically on a number of discrete intervals located along the Atlantic coast. Each of these intervals covers about 78 to 260 Square kilometers (30 to 100 square miles). The base map series (the USGS National Topographic Map Series, 1:250,000 scale) contains a Universal Transverse Mercator (UTM) grid with 10,000 meter centers which has been employed to provide the basic reference system for describing the biological resources within discrete intervals, or bands, along the Atlantic coast. The flexibility of the UTM grid system is particularly advantageous in this inventory because of the variable orientation of the coastline. Although major sections of the Atlantic coast trend in a north-south direction, other sections are oriented in an east-west direction. As a result, the east-west grid lines are employed to great advantage in describing the biological resourses along a north-south coastal segment, whereas the north-south grid lines are used to describe the biological resources occurring within portions of an east-west trending coast.

This reference system reflects ecological resources and not political boundaries by tying the UTM grid system into Terrell's Level II regional classification scheme for Atlantic coast ecosystems.* The east-west grid lines were used to provide the basic reference system for the major portion of the coast, particularly in the region extending from the southern half of Terrell's New York Bight section (New Jersey shore) southward to Biscayne Bay. The north-south grid was employed to describe the biological resources within selected portions of the study region, specifically the Gulf of Maine coast, the southern New England coast, and the northern third of the New York Bight (South Shore of Long Island).

All resources displayed on the map series were described and located according to this reference system in the narrative report. A complete list of all information sources contacted is included at the end of the report.

Availability of Products

Maps produced for the Atlantic Coast Ecological Inventory Series (Table 1) are for sale by the U.S. Geological Survey, Reston, Virginia 22092. The narrative report "Atlantic Coast Ecological Inventory" is available from Superintendent of Documents, U.S. Government Printing Office, Washington, D.C. 20402.

ACKNOWLEDGEMENTS

Data collection, basic design of inventory graphics, and narrative report preparation for the Atlantic Coast Ecological Inventory were preformed by Dames & Moore, Bethesda, Maryland, under contract to the U.S. Fish and Wildlife Service. Basic design of the map collar for this series as well as final cartographic quality control and map publication were provided by the U.S. Geological Survey, Reston, Virginia.

* Terrell, T.T. 1977. Interim Hierarchical Regional Classification Scheme for Coastal Ecosystems of the United States and Its Territories. FWS/OBS 77/48. U.S. Fish and Wildlife Service.

COASTAL MANAGEMENT POLICY EMPHASIS IN AN URBAN NATIONAL RECREATION AREA

Herbert S. Cables, Jr.
Superintendent
Gateway National Recreation Area

and

John T. Tanacredi*
Natural Resources Management Specialist
Gateway National Recreation Area
Floyd Bennett Field
Professional Services Division
Brooklyn, New York 11234

Abstract

Gateway National Recreation Area, over 26,000 acres of ocean beach, dunes, wetlands, forest and wildlife preserves, is part of a proposed program intended to bring the National Park System closer to the urban areas of the United States. The lands and waters now included in Gateway - although many of them have been degraded, overused and altered - are virtually priceless because they represent the last remnants of the original shoreline and a coastal recreation resource that, when fully developed, may serve more visitors than any other park in the nation.

Recreation, a broad category encompassing a variety of "passive" and "active" activities, is one of the largest and fastest growing uses of our coastal zone. Recreation has in fact become a major economic force in coastal zone management with projected revenues derived from these activities in the billions of dollars.

There are natural constraints on use of Gateway NRA as a coastal area for recreation. Primarily these constraints consist of "limitations of space" and "access to the space" to pursue a particular recreational activity. The particular recreational activity being considered, as well as the public's perception of proper conditions and pleasant surrounding, are also major considerations in the development of this National Park. The proximity to desired recreation facilities is a problem especially for the poor, elderly and handicapped who cannot afford to travel to, or utilize resort areas to enjoy water recreation. Alternative sites must provide for these populations.

The living resources of the coastal zone pose a constraint on recreation since fishing, hunting and wildlife observation are totally dependent upon the natural fauna and flora of the coastal zone. Thus, recreational activities compete with use of the shoreline for wildlife preserves, as well as industrial and residential development. Pollution stemming from

*Forward all requests.

industrial and waste disposal uses of the coastal zone seriously limit or preclude certain recreational activities. Transportation access and population concentrations all contribute short and long term impacts which exert great influence on both the provision for and participation in coastal recreation. If there is to be rational management of the coastal zone and of Gateway as a coastal resource for future generations to enjoy, then the public must be well informed. Wise use, not total protection for its own sake, is the important thrust in protecting our coastal resources. Public involvement in the environmental assessment process and development plans has been an on-going activity of the National Park Service and Gateway NRA.

With 1980 being designated the "Year of the Coast" it will certainly be a year, the beginning of our second environmental decade, where considerably important decisions will be made concerning the use, protection or enhancement of our coastal resources. With the advent of bringing the National Parks to our urban environments, we in the Service begin to wield a double-edged sword. We inherited some of the few remaining natural systems under heavy urban impacts with all the commensurate environmental crises. Yet, we are able to unequivocally provide millions of Americans with park experiences of high quality in attractive environments. This paper will explore the myriad of management activities that will be necessary in the future development of this National Recreation Area. Such specific management scenarios to be discussed are barrier beach and bathing beach stabilization, marsh stabilization, water quality and public health aspects of bathing beaches, wildlife refuge resources, and the coupling of research priorities in the National Parks with management decisions in our coastal zone.

Introduction

The National Park Service was established in 1916 to administer and manage a system of national parks that preserved its natural wonders and scenic masterpieces for future generations. Prior to the 1940's, it had been policy[1] to set aside large areas of land for "public use," so that the traditional national parks were overwhelmingly concentrated in the mountain states and the far west. Our coastal environments were effectively left to the spectre of shoreline development and man's pollutional impacts. Because of the great distances generally involved in getting to the "traditional National Parks," and the need for a National Park experience, Congress was prompted to create in 1972 the first National Urban Recreation Areas in our National Park system. The Congressional authorization establishing Gateway NRA (and its sister park, Golden Gate NRA, in San Francisco) was a bold new step, since it brought the NPS and the national park idea out of the "countryside" and into the environmental crisis of millions of urban Americans. Gateway, therefore, is unique because it is in the coastal zone of one of the most highly developed commercial and industrial regions in the world; it is adjacent to the most densely populated metropolis in the United States; it offers an opportunity to demonstrate that abused coastal resources can be regenerated and, it offers millions of Americans a chance to have a national park adventure and experience, virtually at their own doorsteps. Rene Dubos, in an attempt to describe what he believes one portion of

Gateway will provide our future visitors after development, noted that "a complex landscape design... would help urban people to move progressively and with increasing understanding from the completely humanized world into the wilderness."[2]

Gateway encompasses three basic districts or units; Jamaica Bay/Breezy Point, Staten Island in New York, and Sandy Hook in New Jersey. There is a total of 26,645 acres of land and water which includes a variety of ecosystem types (Figure 1). Recreation activities at Gateway are as varied as those who participate in them. Traditional "active" sports activities make up a good portion of our facilities support; softball, paddleball, bicycling, etc. "Passive" recreational activities, such as tent camping, bird watching, sail boating, and fishing, broaden our visitor use and are the focal point of delicate management decisions. In the long term development plan for Gateway, provisions will be made for "Gateway Villages" which will be cultural, scientific, as well as recreational centers.

Aldo Leopold once defined recreational development as a job "not of building roads into lovely country, but of building receptivity into the still unlovely human mind." "Man and His Environment," the theme of the Gateway Villages, will emphasize the building of this receptivity to man and environment, with particular emphasis on coastal ecosystems.

Several areas have posed problems of controlled access for visitor use at Gateway, and have been handled by special permits, designated trails and limited-access points. For example, in areas of our coastal dune system, ORV's have been restricted to special trails out to the fishing beaches, and surf fishing, a growing recreational activity at Gateway, will require the future establishment of carrying capacities of designated areas.

Russell E. Dickenson, in his acceptance speech as the 11th Director of the National Park Service in May, 1980, noted that, "N. P. Langford, the first superintendent of Yellowstone in the early 1870's, had no salary, no appropriations, no staff, only high hopes for a new concept of natural area protection and public use. It was an inauspicious beginning for world-famed Yellowstone. We have come far. Today, (the NPS is) a... professional organization... serving the American people as stewards of its priceless park lands. One of the strengths of the system is its diversity - history, ecology, recreation, geology, scenery, architecture, reflecting the strength and diversity of America."[3] It is this diversity in coastal management policies which provides for the future development of Gateway National Recreation Area, which will be the topic of this paper.

Gateway: A Plan for Implementation

Recreation is regarded as one of the largest and fastest growing coastal zone uses, with 44% of outdoor recreation participants favoring water-based activities[4,5]. Each American, on the average, spends ten days per year in some form of coastal recreation[6]. During the period from 1940 until 1970, the number of people living within 50 miles of the coastal shorelines has increased by 47.8 million[7]. In 1970, 54% of the United

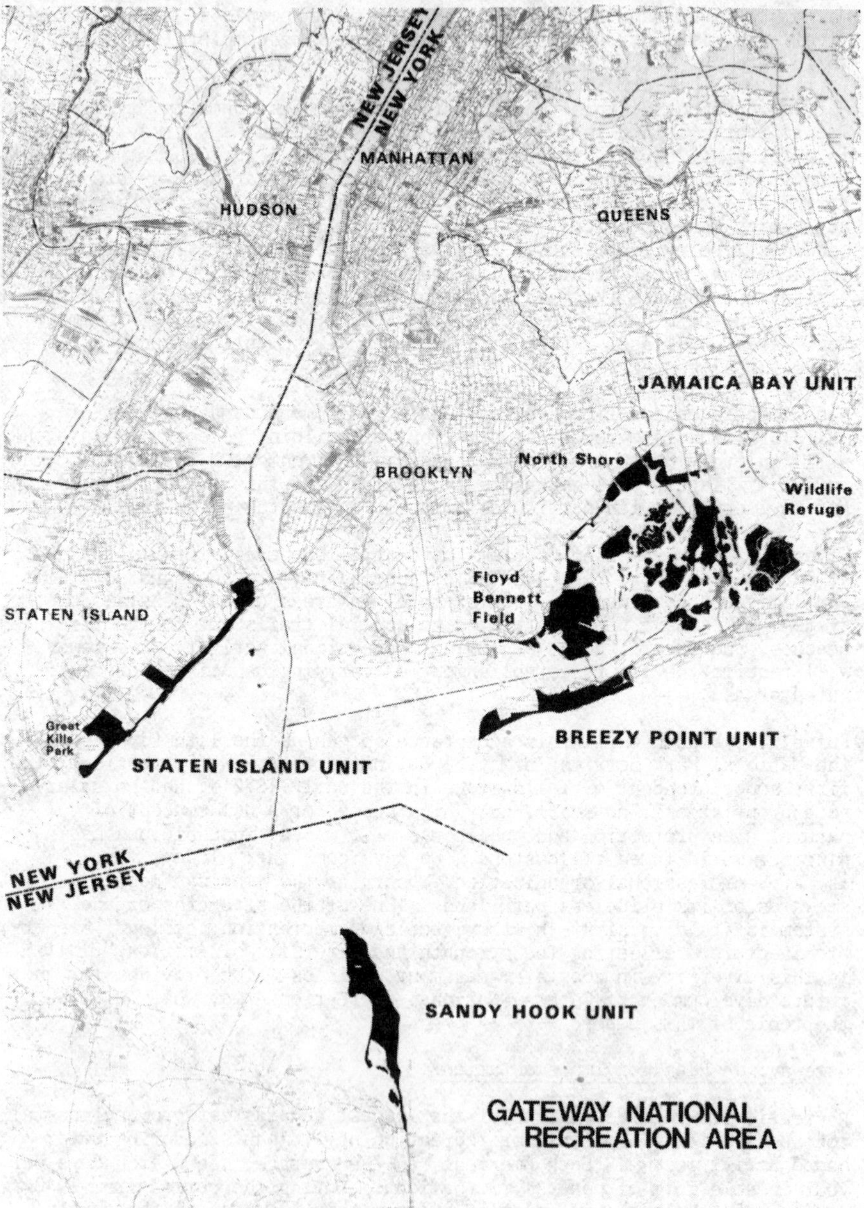

FIGURE 1. Gateway National Recreation Area, New York/New Jersey

States lived within 50 miles of the coast[8]. Not surprisingly, less than 2% of the coastal shoreline of the 48 contiguous states is in public ownership[9], thus running in direct opposition to public access parks such as Gateway.

The paradox of Gateway, and its potential as a recreation resource, is that the advantages of location are largely offset by the severe environmental stresses that the park's lands and waters have undergone because of their proximity to this densely populated and highly urbanized area. The problems are compounded by the fact that heavy demands for shoreline recreation opportunities have outstripped the public access facilities to the shore in support of this demand[10].

However, even with these severe environmental stresses, Gateway's barrier beaches, marshes, upland forests and bay provide an unparalleled experience with a natural coastal system. Thus, although the stated purpose of Gateway is to preserve outstanding natural and recreational features, inherent in this purpose is the need to revitalize the natural landscape so that it can be enjoyed by present and future visitors. It will also encourage new and harmonious expressions of urbanism throughout the greater metropolitan area that will allow the perpetuation and use of Gateway's unique resources.

In order to properly harmonize the varied areas encompassing Gateway, the park has been allocated into six management zones, each of which has specific management strategies and types of allowable use and development. Offshore waters, including those of Jamaica Bay, have not been zoned. The six zones are Protection, Use-by-Reservation, Beach, Unstructured Recreation, Structured Recreation, and Development; the Development zone includes two subzones: Support and Gateway Village.

The Protection Zone includes natural areas so sensitive to human activity that they will be available for use only under special conditions. Areas in this category include high and low salt marshes, primary dunes, freshwater marshes, and any identified critical areas or habitats, such as waterbird-nesting sites. Buffer areas, because of their special protective function, also fall within this management zone.

Within Protection Zone resources, natural processes will be encouraged. Dunes will be allowed to shift with the seasons, marshes to grow or shrink according to normal environmental cycles. Wildlife species, particularly any identified as threatened or endangered, will be protected through appropriate management strategies (i.e., research related structures, such as photo blinds will be allowed). Historic and archaeological resources will receive similar protection. Policies will be established to discourage random access onto protected lands. Where Protection Zones are traversed by stabilized paths, these paths will be designed to minimize possible deviations from prescribed routes. Primary dune systems and dune vegetation will be protected from trampling by means of raised walkways, boardwalks, or other appropriate devices. Protection zones in areas where natural resources have been destroyed or substantially modified will be managed to encourage restoration of a more natural regime; for example, the dunes behind the beach at Sandy Hook and Breezy Point will be encouraged to restore themselves through protection and sand fencing, and some marsh areas on the periphery of Floyd Bennett Field will be replanted. Major new landscaping will not take place except in areas that are being restored or where buffers are established. Thus, uses of lands in the Protection Zones will be strictly regulated and confined to research and maintenance.

Areas of Gateway that are sensitive to overuse will be zoned for use-by-reservation so that management can control access by scheduling groups and issuing permits to individuals. Included in this zone are most forests, some upland marsh areas and secondary dunes, and grasslands with significant natural values. Areas planned for special uses, such as nurseries and horticultural experiment sites, will be placed in this zone, as will certain historic and archaeological sites, structures, and features.

Low-impact uses allowed in Use-by-Reservation Zones will include nature study, environmental education, historical interpretation, and -- where designated -- hiking, jogging, gardening, and planting. Bicycling will be permitted only on designated trails and paths, and camping only at specified sites in supervised groups. Only small unobtrusive new structures (e.g., shelters and field classrooms), wayside exhibits, educational signs, and similar facilities will be permitted in natural areas; and roads and trails will be allowed, but they will be designed and managed to discourage people from striking out on their own. Gardens may include greenhouse structures, maintenance sheds, or similar facilities. Historic structures may incorporate modern appurtenances required for interpretation, security, or visitor safety.

The Beach Zone includes both ocean and bay shores having the capacity to accommodate moderate to substantial recreation use. Management strategies in Beach Zones will, to the extent possible, reflect natural processes affecting specific beach areas. Natural shoreline dynamics will be interrupted only as neccesary to reduce erosion and maintain beach use in major public use areas. Techniques may include maintenance or establishment of dune systems, plantings, and windbreaks, and sand nourishment where necessary on eroding shorelines. Beaches will be maintained or expanded to provide for planned levels of recreation use.

The primary uses of the Beach Zone will be swimming and surfing (where offshore currents and water quality permit), sunbathing, beachcombing, and similar activities. Surf fishing will be an important activity in a few locations identified by park management where such use will not conflict with other activities, primarily during the off-season and in the summer at times when few people are swimming. With the exception of existing seawalls and groins, only mobile structures to support recreational use, such as sun shades, lifeguard stands, ramps for the handicapped, and other minor, temporary structures required for seasonal use, will be allowed on the beach itself. No roads or established trails will be permitted in the Beach Zone.

Areas zoned for unstructured recreation contain predominantly natural areas, other than beach, that can accommodate substantial recreational use. Included are open grassland/thicket areas with no particularly sensitive natural or historic resources. This zone will be managed to preserve natural vegetation and scenery while establishing sites that can support casual outdoor uses. Wildlife species that can coexist in such areas will be protected through appropriate management techniques (for example, Ringed neck pheasants, rabbits, owls, grasshopper sparrows, at Floyd Bennett Field). Land uses may be periodically rotated to permit regrowth of natural vegetation.

All outdoor activities that require little supervision and few support facilities will be allowed in this zone, e.g., picnicking, walking, birding, kite flying, bike riding, and informal athletics.

The Structured Recreation Zone encompasses outdoor areas that can be developed and landscaped to support active organized sports and recreation. They include lands already developed for these purposes and selected open or disturbed uplands where there are no sensitive or significant natural resources and where active use will not conflict with preservation of historic resources.

Structured Recreation lands will accommodate all athletic activities requiring hard surfaces, structured grounds, playing fields, and so forth.

Future visitors will find a Gateway that is substantially different from the park today. At Sandy Hook and Breezy Point, large new beach and picnic areas will be added to those already open to the public. At Floyd Bennett Field, more than a thousand acres of newly landscaped open space and recreational areas will be opened, plus extensive space indoors for community groups, recreation, educational programs, and cultural events.

In response to an often expressed need, the Gateway plan calls for "Gateway Villages" to be established at Floyd Bennett Field, at Fort Hancock in Sandy Hook, and at Fort Wadsworth in Staten Island. In some respects, what is already happening at Fort Hancock is the prototype for the Gateway Villages. The fort is used for educational programs, theatrical productions, research by other Federal, State and local agencies or institutions, community events, day camping, formal and informal athletics, staff housing, historical interpretation, and organized group outings.

The Gateway Village concept derives from the following conclusions:

> A concern for the natural environment must also increasingly be a concern for the impact and implications of urbanization.
>
> It is no longer constructive for those who consider themselves environmentalists to separate their concern for the man-made environment, since the fate of the two are now seen as inseparably related.
>
> Ecological principles have demonstrated the interdependence which binds natural and urban systems together, and we have begun to evaluate the performance of our built environments with concepts and values derived from the study of nature.

It follows from these conclusions that the mission of Gateway NRA, as it matures in line with the growing awareness of these relationships, is to reflect a concern for the environmental impact of urbanization and seek to promote relationships between our works and nature's which are mutually supportive.

Although all three sites will be major year-round activity centers for the park and will focus their programs and facilities around the theme of people and their coastal environments, each village will have a distinct character based on the nature of its site and resources. They will neither serve as nor look like amusement parks or modern theme parks. They will have strong educational and interpretive programs tied to the park's location in a large metropolitan region. All of Gateway will become an object lesson in the relation of man and the environment -- from the holly forest at Sandy Hook and the wildlife refuge at Jamaica Bay to the nurseries at the Gateway Villages devoted to experiments in urban plant ecology, just one example of programs that will illustrate the interdependence of natural and urban systems.

To ensure that Gateway will serve the largest number of people possible, the plan calls for programs to serve those whose ability to get to Gateway is limited: the poor, the handicapped, and the elderly. The plan also seeks to ensure that community participation will be an integral part of the planning process -- and of Gateway's daily operation. Gateway is anticipated to have an important positive impact on the economy of the New York/New Jersey metropolitan region and on the quality of life of its residents.

Recreation Perspectives*

The National Park Service has been administering Gateway National Recreation Area since 1974. At present, it manages some 89 percent of the 8,373 acres of the land that will be open for public use once place under its jurisdiction. Another 1,241 land acres will remain under other jurisdiciton, either as enclave properties or as city rights-of-way. Gateway also includes approximately 17,031 acres in offshore ocean and bay waters and in two inland ponds. Fort Wadsworth and South Beach in Staten Island have recently been donated to Gateway by the United States Army.

Sandy Hook

The Sandy Hook Unit is located on a peninsula at the northern end of New Jersey's Atlantic shoreline just north of the town of Atlantic Highlands. It covers approximately 1,665 land acres, including 7½ miles of ocean beaches and sheltered bayside coves, hundred of acres of ecologically significant barrier-beach vegetation, and at the northern end of the lands, facilities, and fortifications of the Fort Hancock complex. The U.S. Coast Guard maintains properties at the Sandy Hook tip that are

*This section is a synopsis of Gateway NRA based upon a consolidation of the Final Environmental Impact Statement and Decisions Paper, prepared by the NPS Park Planners, Denver Service Center, 1979-80.

excluded from the park; one is the historic Sandy Hook Lighthouse, the oldest lighthouse (still in operation) in the United States.

The previous land uses at Sandy Hook -- a New Jersey state park on the lower portion of the peninsula and the U.S. Army's Fort Hancock covering most of the rest -- have left much of the land area in a natural state. In addition to the beaches, it includes tidal areas, dunes, back-dune areas, "heathlands," and holly forests -- the last being of particular ecological interest. Development is focused at the South Beach developed area near the neck of the peninsula, within the Fort Hancock complex, and in a few other areas, leaving large stretches of undisturbed land in between. These spacious coastal landscapes and natural features are an important part of the Sandy Hook resource and are deserving of protective management and uses that will allow their perpetuation.

Shoreline changes at Sandy Hook during the past century and a quarter indicate a definite trend toward enlargement at the tip and a tendency for the base of the land mass to separate periodically from the mainland. A seawall now protects the narrowest portion of the neck connecting Sandy Hook to the mainland, but a tendency toward breakthrough still exists in the South Beach area north of the wall, and this tendency was particularly pronounced in the winter of 1977-78. Emergency beach nourishment programs have been initiated to preserve the beach and protcct against breakthroughs.

A wide variety of recreational uses currently take place in the Sandy Hook Unit. The South Beach area continues to be used in traditional fashion -- for all types of beach activities -- and the North Beach bathing area is now available for limited public use. Picnicking and fishing are popular on the bay side of the Sandy Hook neck. Environmental education programs and casual nature study occur in the marsh/grassland communities near Spermaceti Cove. Several historic batteries are visited on a reservation basis. The holly forest just north of Spermaceti Cove and the shorebird-nesting sites at Mid and North Beaches are under protected status.

Automobile traffic in and near Sandy Hook has already reached saturation levels on several heavy-use summer weekend days, backups occur at the entrances to the South Beach parking areas, on the single-lane access ramp leading from the Highlands bridge to Sandy Hook, at the bridge itself (a drawbridge for boat access and egress), and along heavily signalized Route 36, which is also a main access road for local residents. Substantial increases over the present number of park visitors (an estimated 30,000 on peak days; 2 million total for FY 1976) will require improved public-transit service and Outreach programming designed to increase use at other than peak times and the introduction of water-borne transit to service the unit.

Staten Island

The Staten Island Unit, extending along the southeastern shore of Staten Island, include (from south to north) Great Kills Park, Oakwood and New Dorp Beaches, Miller Field, Midland and South Beaches, historic Fort Wadsworth, and two small man-made islands, Hoffman and Swinburne, lying just offshore -- for a total of some 1,210 acres of land.

An important influence on planning and management in the Staten Island Unit is the poor water quality in adjacent bay waters, which precludes swimming on the baches north of Great Kills.

Opening and upgrading of water pollution control plants and comprehensive planning for the improvement of water quality are currently being accomplished as part of regional 208 studies. New York City has recently published a 208 study proposal which recommends the implementation of actions that will be directed at providing swimmable waters at all Gateway beach areas.

The shoreline on Staten Island was developed by erosion of the headlands at New Dorp and South Beach/Fort Wadsworth and by subsequent littoral drift that transported these materials along the shoreline to create a beach barrier. More recently, beach nourishment projects have supplemented this natural activity. Most of the tidal marshes behind the barrier have been filled in modern times, and piers, groins, and jetties have slowed rates of littoral drift on the beach. The Crookes Point peninsula at Great Kills Park was created in part by bulkheading.

A small portion of salt marsh at Great Kills Harbor is all that is left of the extensive salt marsh system that once extended all the way north to Fort Wadsworth. Most of the open area of Great Kills Park is dominated by Phragmites sp., with scattered thickets of shrubs and trees, particularly at Crookes Point. There is a small, disturbed forested area in the northwestern uplands near the park boundary. A unique peat deposit occurs just offshore northeast of the Great Kills bathhouse.

Great Kills Park has several major areas that are good places to observe migrating birds and, in recent years, a focal point for study of migration patterns of the Monarch butterfly. Miller Field contains a small swamp white-oak forest that is locally unique and valuable for environmental education.

Most portions of the Staten Island Unit have been previously developed, either as city parklands or as military sites. Great Kills, initially a salt marsh and later a landfill, was developed as a city park over 20 years ago and today offers a wide spectrum of recreational facilities along its beach, interior, and harbor. The beach at Great Kills, particularly the portion directly in front of the bathhouse is quite narrow, largely as a result of the numerous groins and other man-made structures which slow sand movement from the north. No dune line is present in front of the facility. The area is also considered a potential overwash site during severe storms.

The harbor side of Great Kills/Crookes Point contains a small, boat marina with boat-launching and storage facilities. The Crookes Point interior east of this development has been heavily disturbed by random use, and unplanned dirt roads and trails crisscross the area. Rock jetties and groins extend along the inlet to the harbor, and the interior edge is bulkheaded.

Breezy Point/Jamaica Bay

The Breezy Point District of Gateway contains approximately 1,059 acres and 4½ miles of ocean beaches. Included in this district are the popular Jacob Riis Park and most of the lands and facilities within the Fort Tilden military complex. Two communities will remain as enclaves of single-family houses within the Breezy Point district boundary -- the Breezy Point Cooperative and the community of Roxbury in Jamaica Bay.

Making all or most of the Breezy Point peninsula into public parkland is an idea that goes back more than 50 years. The struggle of the tenants at Breezy Point to acquire their homes and of various public and private organizations to preserve the peninsula for public outdoor use is a long and complex story. One souvenir of that struggle -- and its resolution -- was the half-finished complex of high-rise buildings that stood abandoned just to the east of the Breezy Point Cooperative when Gateway was authorized. The complex, acquired by the city and transferred to the park, has been demolished by the Park Service.

Stanley B. Tankel, a leader in regional planning in the New York area, led the movement to halt the construction of the high-rise complex on Breeezy Point and to return the land to designated park use.

The vegetation of the undisturbed areas within the Breezy Point District is typical of barrier beaches in this region. At the western end -- the Tip area -- are some 200 acres of sand dunes, salt marshes, brackish marshes, and grasslands that are in a relatively undisturbed condition. The Fort Tilden area contains beachgrass dunes, grasslands, phragmites, high and low thickets, and coniferous and deciduous forests at higher elevations where the sand dunes have stabilized. Particularly fragile and/or significant natural features include the remaining primary dune systems at the Tip, Tankel Beach, a portion of Breezy Point and in front of Fort Tilden, the productive high and low salt marshes on the bay side of the Tip.

The major park attraction at the present time is Jacob Riis Park. Riis Park was initially established by the City of New York in the 1930's. It features a mile of ocean beach and a highly developed back-beach area. A bathhouse is flanked on both sides by paddle-tennis and hand-ball courts, ballfields, and tot lots; the entry mall area houses a variety of food concessions, recreational open spaces, and shuffleboard facilities. Completing this intensely developed area is an 18-hole pitch-and-putt golf course. The entire Jacob Riis facility is considered a significant historic resource representative of public park development in this country.

The Jamaica Bay District encompasses about 4,450 acres of land and marshland in and near Jamaica Bay. For the purposes of planning, this District has been further subdivided as follows.

Floyd Bennett Field/Plumb Beach/Bergen Beach

This area contains some 1,448 acres of open space and facilities at the former naval air base in Floyd Bennett Field (the present park headquarters site), as well as the nearby shorelands at Dead Horse Bay, Plumb Beach, and Bergen Beach.

Few significant natural features have been identified at Floyd Bennett Field. Most of it was created by landfill, and the open portions that have not been cleared and flattened are dominated by stands of Phragmites sp. Mixed grasslands and low shrub thickets are also present and provide habitat for several kinds of birds. Most of the periphery is bulkheaded, although there is a narrow strip of low salt marsh connecting to the waters of Jamaica Bay. In a few areas, marsh is developing on the outboard side of the bulkhead because of the accretion of sand.

Development at Floyd Bennett Field is extensive. Historically, a civil aviation facility, it is dominated by construction and man-altered landscapes. The extensive fill activities in the past century have connected Barren Island and smaller tidelands in Jamaica Bay to the main portion of the field, and years of development and use have drastically modified its character.

Strategically fronting on Flatbush Avenue midway between the Belt Parkway and the Marine Parkway Bridge, Dead Horse Bay includes two facilities providing recreational services. Developed by private interests under New York City concession permits prior to Gateway establishment, the facilities are a marina, tennis courts, and a golf driving range. Plumb Beach includes a small building which houses maintenance/storage functions and public toilets, and an adjacent parking lot. Bergen Beach has one riding academy built under a concession permit during the city's ownership, including an inside ring, stables, training corral, and parking lot.

North Shore

Gateway lands along the north shore of Jamaica Bay, totalling 702 acres, include the existing developed parks at Canarsie Pier and Frank Charles and Hamilton Beach Parks, the undeveloped lands at Spring Creek Park, and the active landfill sites at Pennsylvania and Fountain Avenues. With the exception of the landfills, whose transfer carried a deed restriction designating their continued exclusive use for fill operations until 1985, all these lands, previously owned by New York City, have been turned over to the National Park Service.

Most lands on the northern border of Jamaica Bay were historically well-developed marsh areas. Seven inlets, which are the remains of this marsh system, have been narrowed by fill activity, and the bay is now bordered by mud flats backed by narrow bands of salt marsh. These salt marshes are considered significant natural resources deserving of protection. Most of the rest of the "natural" portion of the north shore is covered with extensive stands of Phragmites and grasslands with scattered thickets of shrubs and trees.

Wildlife Refuge

This Gateway area includes the lands, marshlands, and other habitats of the Jamaica Bay Wildlife Refuge as well as some of the other islands in Jamaica Bay.

The wildlife refuge, some 2,475 acres, includes uplands and low-lying islands surrounded by saltwater, freshwater, and brackish impoundments (Figure 2). The two artifically created ponds, the brackish East Pond and the freshwater West Pond, increase the variety of habitats available to the over 320 bird species residing in or migrating through the refuge. Because the area is on the Atlantic Flyway, a geographic course birds customarily follow between breeding and wintering sites, its management as a wildlife refuge has proved most beneficial and critical to a great variety of waterfowl and shorebirds.

The coastal shoals, bars, and mud flats in Jamaica Bay provide habitat for a number of invertebrates and shorebirds. Extensive *Spartina* salt marshes surround most of the islands and land masses in and around the bay.

Grassland and woodland/thicket communities are found at higher elevations that are not affected by the tide. The refuge management has selectively introducted a number of upland trees and shrubs to improve food and cover for birds.

Present recreational uses of the wildlife refuge are limited to the visitor center/West Pond vicinity and the larger islands (Canarsie Pol, Ruffle Bar, and Little Egg Marsh), which are visited by fishermen and picnickers in boats. Fishing is also popular along and at the south end of North Channel Bridge. The remainder of the wildlife refuge core is under protected status or is available for nature study on a permit basis. Most of the other Jamaica Bay islands are used for fishing, clamming, and baiting.

Studies indicate that less affluent citizens from the core communities do come to Gateway to swim, to picnic, or to fish along its beaches. The implications of these studies have a definite bearing on immediate planning for Gateway: Transportation and Outreach program must be an integral part of the plan. The Jamaica Bay Wildlife Refuge and Hamilton Beach are the only areas of Gateway served by direct mass transit. There are no direct rail lines or connections to Floyd Bennett Field, Breezy Point, Great Kills, or Sandy Hook. Buses do serve Gateway's units, but -- with the possible exception of local lines on Staten Island -- the routes and schedules are so poor that this service does not begin to provide an alternative to private automobile.

Private autos account for slightly over 90 percent of the trips to Gateway. Although this pattern is common throughout the National Park System, it is not a viable planning solution at Gateway, because fully one-third of the households (over 1.6 million in the region) do not include automobiles.

Much is presently being done to try and alleviate congestion and to facilitate use at Gateway's beach areas on summer weekdays and in the off-season, as well as on heavy-use summer weekdays, by expanding public transit service from nearby mass transit stops. Additional existing-fare bus routes, programmed bus outings from core neighborhoods, and institutional bus transit will also be supported and encouraged wherever possible.

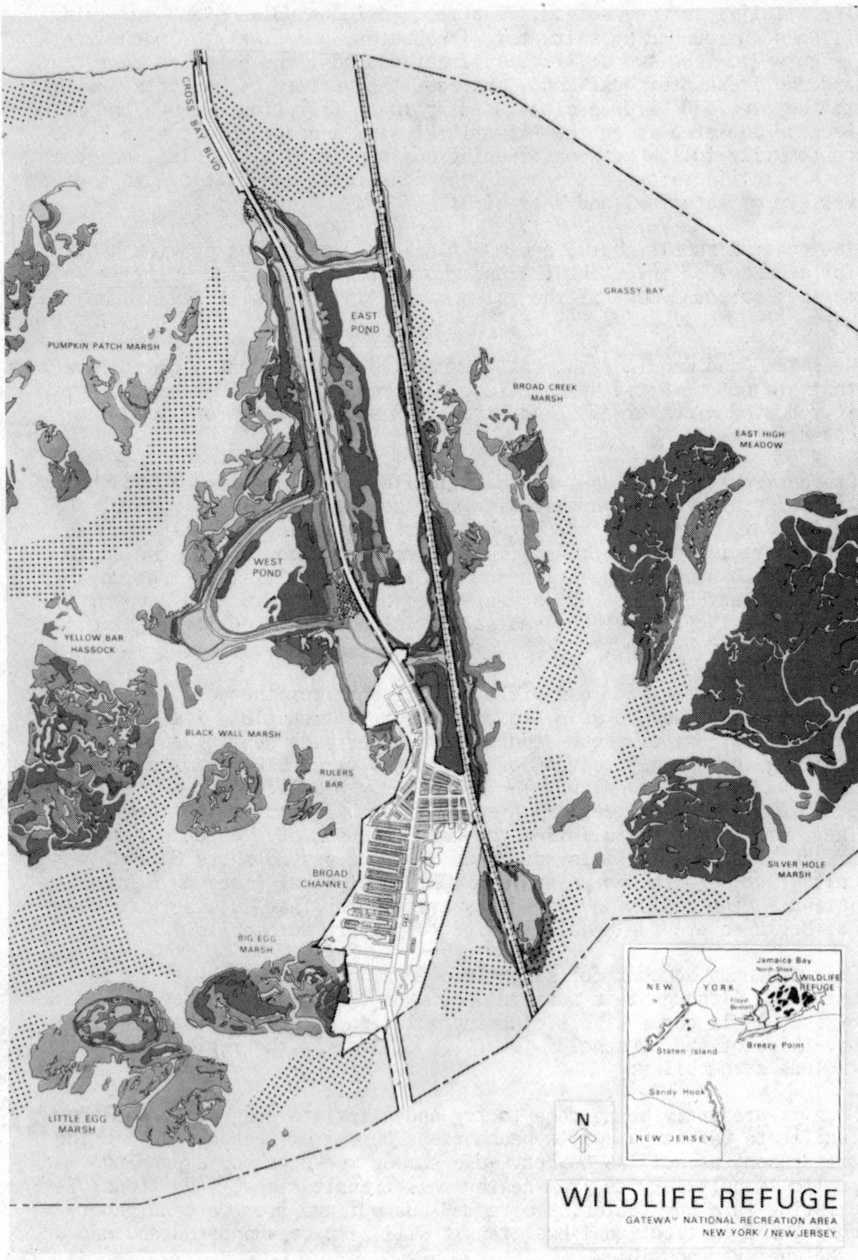

FIGURE 2. Jamaica Bay Wildlife Refuge, Gateway NRA

Protection of the Resource

Coastal zone management is a process designed to achieve a set of stated objectives. In the coastal zone, the stated objective is to maintain and to improve its usefulness for man by ensuring the quality and extent of the natural system upon which he depends.

Gateway's natural environment will be managed to ensure the preservation and improvement of the coastal environment and to provide for compatible recreational activities. Beyond the protection of critical resources and habitats, specific projects are recommended to expand the wetlands, improve water quality, and generally upgrade the park's natural environment. However, it should be noted that many factors involved in the implementation of these proposals, such as improvement of air and water quality, are beyond the direct control of the National Park Service and therefore require cooperative undertakings with appropriate federal, state, and local agencies.

Management policies at the park are designed to reflect specific recreational requirements. Certain beaches are being nourished periodically as part of a coordinated shore improvement program, and landscaping projects have been undertaken to provide natural settings for outdoor activities.

The following are the major natural resource areas to be considered for protection in the overall coastal management policy implementation at Gateway.

Ocean Shorelines

Gateway's 19 miles of beaches will be maintained or expanded, by sand nourishment where necessary, for recreational bathing purposes.

No new permanent structures will be built or allowed on the beach in front of the dune line, and no projects are to be undertaken that would interfere with the active processes occurring in this zone. Natural shoreline dynamics will be interrupted only as necessary to reduce erosion and maintain beach use in major public-use areas.

Primary dune systems will be protected from excessive use by placing them in Protection Zones and by providing pedestrian crossings and boardwalks for access to beaches.

Marshes and Wetlands

Approximately 95 percent of the marshes and wetlands, extremely valuable resources totalling 3,075 acres, will be included in Protection Zones and managed to minimize disturbance. Specific projects for marsh reclamation are to be undertaken in areas of the park with the remainder of the wetlands being placed in the Use-by-Reservation Zone.

Vegetation and Wildlife

Threatened and endangered species are to be indentified and protected through appropriate vegetation and wildlife management programs. Habitat management techniques will be employed within the Jamaica Bay Wildlife Refuge and other selected areas of the park to protect existing wildlife species, including the important migratory bird and butterfly populations.

Plantings to provide food and cover for indigenous species will be incorporated into landscape designs wherever possible. Native species are to have preference over exotics, and exotics will not be introduced except where it can be shown that native species will not meet the needs of the management program.

A study of the role of *Phragmites sp.* in the marsh ecosystem is being undertaken. If feasible and desirable, this monoculture will be removed from selected areas to encourage succession by a broader range of species.

Wherever possible, biological methods will be employed for controlling ticks, mosquitoes, green flies, etc. Well researched chemical methods may be used only to preserve public health and safety and to permit recreational use and enjoyment in major public use areas.

All wildfires will be extinguished, except where prescribed burning is identified through management assessment as an acceptable alternative for achieving the park's resources management objectives.

Pollution Abatement

In accordance with the requirements of Executive Order 1152 relating to prevention, control, and abatement of environmental pollution at federal facilities, the National Park Service is actively seeking assistance from state and federal environmental protection agencies in minimizing air and water pollution and in monitoring air and water quality. Specific projects will be directed toward pollution reduction in Jamaica Bay to reach a water quality that is suitable for contact recreation. The Park Service is also continuing to work with New York City and New Jersey agencies involved in coastal zone management planning and section 208 studies to ensure coordinated implementation of proposals to upgrade water quality in adjacent ocean and bay waters.

Given available scientific evidence as well as recent 208 study proposals, it appears that all waters within the park will be suitable for swimming and fishing sometime in the foreseeable future. Thus, planning for Gateway is proceeding under the assumption that regional water quality will continue to improve rather than decline.

Energy Conservation

As part of a National Park Service commitment to energy conservation and in accordance with the requirements of Executive Orders 11912 and 12003, energy consumption is to be minimized at all Gateway facilities. Environmental technology (wind energy, solar heating and cooling) is to be employed in new buildings to the extent feasible, and retrofitting of existing buildings is presently being undertaken to minimize fossil fuel consumption in these facilities. For example, a gymnasium up-graded to accommodate a special environmental education program at Gateway is having a solar hot water heating system installed.

Groundwater Resources

Continuing checks will be made to detect saltwater intrusion into groundwater resources at the Sandy Hook Unit. Bacteriological analysis is being regularly performed by the Park Service on all surface and groundwater resources.

Visitor Use Protection

Hunting will be prohibited in Gateway under the Secretary's authority to restrict this use should it conflict with other legitimate park purposes. Other appropriate recreational uses will be allowed in specific management zones, with the following stipulations:

Boating, fishing, and baiting will be permitted year-round in all park waters, except for East and West Ponds in the Jamaica Bay Wildlife Refuge and other specifically designated marshes and wetlands where these activities may be controlled or not permitted according to specific management zoning.

In addition, park managers will monitor motorboat use in bay waters, and if results show that such use is causing significant damage to or loss of wetland resources, motorboating will be restricted in designated areas in cooperation with the U.S. Coast Guard and other responsible agencies.

No off-road vehicle use will be allowed, except on special permit by park management.

Boat launching on ocean beaches will be permitted at areas designated by park managers; however, no permanent structures will be built on the beach to accommodate this activity.

Surf fishing will be permitted on all Gateway beaches as long as it does not conflict with use of established swimming areas or with bird nesting sites.

In this regard, for optimum protection of our resource, it will be necessary throughout our development to determine the capacity of our coastal zone in relationship with these desires, and to determine which uses are compatible within a defined area of capacity.

Coastal Resources Policy Implementation: Management Scenarios

Probably the best way to describe the coastal managerial policy approach at Gateway is "flexible." Letting nature take its course is in concert with the conservation/preservation ethic of the National Park Service and, in many cases, is the most economical way to handle a particular ecological situation. However, application at Gateway will depend upon existing environmental conditions and the originally intended purpose of this "recreation area."[11] Management considerations must account for man's manipulation of the landscape. At Gateway, where man has significantly modified the natural patterns of beach dynamics or upland areas, nature would require considerable amounts of time to restore itself. Where and when the natural system is interrupted, it may be necessary to utilize "engineering" approaches on a site specific basis. At the Sandy Hook Unit, for example, the NPS inherited an eroding shoreline with a history of engineered structures. Management approaches for this Unit ranged from "allowing the natural system to run its course, to a beach nourishment program and it ramifications, or to a solid sea wall to stabilize the shoreline."[12] Located downdrift of an extensive system of groins and seawalls, which act to starve sand from this sand spit, inlet truncation of this barrier beach from the mainland seems inevitable due to the much reduced supply of sediment. The significance of this beach as a recreational area precludes the abandonment of the roadway across to the tip. Other considerations involve navigational factors and the construction costs of a causeway or bridge over an inlet of unpredictable character and migration tendencies. The best approach will involve some plan of continued beach nourishment, where the sand naturally accruing at the north tip, is recycled back through the system to the south bathing beach areas[13]. The NPS is presently negotiating with the U.S. Navy and U.S. Army Corps of Engineers to determine the feasibility of having dredged materials associated with Ambrose Ship Channel or Raritan/ Shrewbury River Bay area, to be placed offshore for subsequent beach nourishment activities. Thus, natural tasks are being performed that have previously been rendered unworkable. Another case in point was the original planning concepts for Floyd Bennett Field which envisioned the return of this site to natural processes. Planners failed to recognize that restoration, including the removal of bulkheading stabilizing fill material, would have resulted in the erosion and destructive modification of the entire site by swift tidal currents through the Jamaica Bay inlet system. Flexible coastal management approaches coupled with sound scientific data averted a potentially significant alteration to the Floyd Bennett Field area for the future.

For Gateway, the coastal management policy of letting nature take its course, may be untenable with existing needs. Immediate and localized situations concerning beach stabilization and barrier island protection, may necessitate an engineering solution. It is not unforeseeable to conceive, for example, of offshore or nearshore breakwaters made of recycled tires occurring at Gateway NRA. Several sites at Gateway are ideal for applying breakwater technology. Plumb Beach, an area of only 15 acres, yet under intensive visitor use, has been subject to severe erosion impacts. Several marsh restoration/ revitalization projects[14] have been implemented and are presently being monitored for long term erosion prevention. Once the natural system (replanting of _Spartina alterniflora_) has been able to propagate, with help from man, nature can take its course to develop this marshland. Coupling the breakwater and marsh restoration approaches together will aid these environmentally stressed areas. Ignoring past mistakes which have altered the barrier and allowing the system to oscillate wildly is not appropriate where many factors must be considered, not the least of which is public access and usage of the seashore[15].

As early as the mid-1960's, public health officials studying the problems of sewer sludge disposal in the New York Bight noted that bottom sediment several kilometers from shore had higher populations of coliform bacteria; an indication of continued pollution. Marine biologists had been collecting information on the impact of sewer sludge, dredging spoils and chemical wastes on living resources including plankton, bottom dwelling invertebrates and finfish. Initial results back from sampling stations through the Bight apex indicated that waste materials were not accumulating in a single restricted area but rather were spreading out over much of the NY Bight apex[16]. During the summer of 1976, the low dissolved oxygen (DO) levels (oxygen in solution required for respiration by aquatic organisms) lasted from June through September and extended from Sandy Hook to Cape May, New Jersey. Millions of dollars worth of commerical shellfish were killed. Finfish and shellfish were diseased by pollutants (i.e., fin rot syndrome). Beaches were closed along Long Island and the Jersey Shore. In response to this pollution episode, Gateway has instituted monitoring of bacteriological water quality into its management activities. Presently, weekly samples are collected at some 30 sites within the waters of our jurisdictional boundaries. These samples provide us with water quality data at existing-use beaches, and as of the summer of 1980, now include potential bathing beach areas. Monitoring beach water quality will provide us with trends in improvement or degradation. This information plays an important role in our cooperative efforts with the city and state regarding section 208 and CZM plans. We also monitor our two sewage treatment facilities to make sure they comply with USEPA and USPH discharge requirements. As we move toward increased development, proposals have been developed to provide for tertiary treatment capabilities at these plants. Since bathing and beachcombing appear to be the most active use of Gateway, and one of the most enjoyed activities by the American public[17], water quality plays a critical role in acceptability of its use. Water pollution within the NYC coastal region, and particularly within the harbor area, is well known to the public. Such conditions can and do affect the quality of the beach environment and often result in negative user perceptions of local beaches[18]. Under the terms of the Federal Water Pollution Control Act, the NY harbor and estuary waters gradually are being cleaned up. While much work still has to be done (e.g., end ocean dumping 1981 and reduce waste hydrocarbon contributions[19]), improvements are likely to continue to the extent that "closed areas" or marginal swimming areas presently within Gateway will become sites for aquatic activity.

Considerable amounts of Gateway properties were created by previous landfill activities. Pennsylvania Avenue Landfill, for example, is a man-made peninsula of 110 acres located on the north shore of Jamaica Bay. Refuse was dumped from 1959 until 1962; dried sewer sludge was deposited there from 1962 until 1972, and from 1972 to today, construction debris is being placed so that contour lines of 75-100 feet above MSL exist. The City of New York, under a memorandum of understanding with the NPS, will continue filling and grading at Pennsylvania Avenue Landfill until 1985. At that time, it will be turned over to be managed by Gateway NRA. Part of this cooperative undertaking is the implementation of a plan developed in 1974 by the City of New York. The plan uses various disciplines in coordinating the creation of a recreation facility out of construction fill - from raw fill to finished terraced park[20]. Ordinarily, the time needed for a waste disposal site to be converted to usable land is measured, not in years, but in decades. The building of Pennsylvania Avenue Park is an attempt to echo nature and designers have sought to create an environment that functions both as a recreational facility and as a natural habitat. Vistas, terraced landscapes, public access, energy (methane gas) generation are all advantages assigned this landfill scheme. The problems associated are basically leachates into surrounding waters. The NPS is presently directing research efforts toward determining to what extent these leachates are entering Jamaica Bay.

One of the most unique management scenarios at Gateway today is the park's administration of the Jamaica Bay Wildlife Refuge; 9,155 acres of marsh, wetlands, ponds, and forested areas. The Refuge was created under the direction of New York City Parks and Recreation Department by Herbert Johnson. When the NYC Subway system dredged in Jamaica Bay and created a solid fill right-of-way on which to place a new railroad to the Rockaways, the City agreed to dredge an additional 6 million yards of sand and deposit it as berms to create two impoundments, to be managed by the refuge. This was completed in 1953. NYC Parks Department than began a vigorous planting program of vegetation selected because of their ability to thrive in a coastal environment, and to provide feeding, nesting and habitat for bird species. Today, visitors are required to get a permit to move along approximately 2 miles of trails to enjoy some 320 species of birds that move or reside in the refuge area over the course of the year. Several management problems have become evident since Gateway assumed its administration. Routine grounds keeping duties in vicinity of trails and visitor centers; Phragmites control to prevent its overgrowth dominance of food plant species; avian botulism and water quality; visitor control; fire management and oil pollution.

Due to the vast numbers of bird species and the variety of habitats they occupy in Jamaica Bay, the botulism problem became a focal point for control. Without measures to prevent its outbreak, and plans of action in case of an occurrence it can quickly decimate large numbers of waterfowl and shore birds. Approximately two weeks prior to spring and fall migrations of shorebirds, the ponds water level is reduced to expose mudflats. Signs of botulism are monitored and fresh water is added coupled with removal of any organic material (dead algae) that may accumulate. A limnological study of the ponds is anticipated in order to determine the rate of aging these ponds are undergoing.

Management of wildlife species involves much more than setting aside an area for them, or creating an area for them and letting "nature take its course." Creating habitat has superimposed a new ecosystem still requiring constant care in order to provide maximum efficiency. Natural successional stages will occur to the ponds with brush-covered areas leading finally to climax forest communities. The greater the variety or diversity of vegetative habitat types, the more diverse the species of animals that will occupy these areas. This species diversity has been impacted due to the prolific growth of *Phragmites sp.*, an exotic plant with little wildlife support capabilities. *Phragmites sp.* also invades and out-competes all the other habitat types at the Refuge. Total elimination is not practical since the plants do provide nesting and escape cover for some species of birds and mammals. The management strategy has been to selectively cut back areas and maintain them from re-introduction of Phragmites. No herbicide use will be made to control these plants.

Conclusion

Thus, the emphasis at Gateway NRA regarding coastal management has revealed that national parks should be preserved and, wherever possible, allowed "to permit the evolution of the dominance of natural forces and the resulting landscapes and ecological scenes. This should not be construed as a policy of neglect, but understood as a position of living with nature as opposed to one attempting to control nature."[21] Coupling of Gateway's management "flexibility" with "evolutionary processes" can only be accomplished by our bringing home to our visitors the interrelatedness of all natural events. This can be aided by bridging the gap between scientific research in our parks and traditional interpretive skills. This type of "interpretive connection," that is the taking of complex scientific data and making it palatable to the general public, will reduce the scientific credibility gap that appears to exist in the United States today.

As John Sawhill recently noted[22] "The scientific community as well as teachers and educators must do a better job of conveying to the public the importance of basic research. People do not comprehend the often circuitous paths to scientific discovery. Many people do not understand why men and women should be paid for studying the mating habits of spiders and the sonar sytems of bats when the planet itself seems to be in peril. The growing cynicism about science must be tackled at its roots through better scientific education. People must be taught the limits as well as the possibilities of science and most important, its changing and vastly enlarged role in our national and global destiny."

Similar to Aldo Leopold's "land ethic" concept, Bostwick Ketchum suggests[23] adopting a "sea ethic" responsible to the health of the coastal zone and an ecological conscience in its management. Deliberate changes in man's activities must occur to reverse negative environmental impact trends. More than making information about coastal problems known therefore, is a definition, agreement and implementation of measures that reduce the level of haphazard utilization of valued coastal zone resources. At Gateway, management is integrating this coastal ecosystem consciousness into its daily and long term development.

References

1. Jubenville, A. (1978) "Outdoor Recreation Management" (W.B. Saunders Pub., Co.) p. 290.
2. Mitchell, J.G. (1978) "The Re-greening of Urban America" Audubon, pp. 29-52.
3. Remarks made on May 15, 1980.
4. R.B. Ditton and M. Stephens (1976) "Coastal Recreation: A Handbook for Planners and Managers," USDOC/NOAA; Washington, D.C.
5. Outdoor Recreation Resources Review Committee Report (1962).
6. National Parks & Conservation Magazine, July (1980) Editorial.
7. Taeuber, C. (1970) "Population Trends of the 1960"s," SCIENCE, 4036: 773.
8. U.S. Bureau of the Census, Statistical Abstracts of the U.S., USCPO (1972) Washington, D.C.
9. Ditton, R.B., J.L. Seymour & G.G. Swanson (1977) "Coastal Resources Management" (Lexington Books) 197 pages.
10. In: Power, Pollution & Public Policy (1971) "The Crisis in Shoreline Recreation" by D.W. Ducsik, Cambridge, Mass., MIT Press, pp. 91-186.
11. Leatherman, S.P., Godfrey, P.J., and Buckley, P.A., "Management Strategies for National Seashores" (1978) in: Proceedings Symposium of Technical, Environmental, Socioeconomical & Regulatory Aspects of CZ Planning and Management, San Francisco, Ca., March 14-16, pp. 322337.
12. Psuty, N.P., Nordstrom, K.F. and Allen, J.R. (1976) "Application of Coastal Geomorphology to Management of Beach Resources in Gateway NRA," Rutgers University, Center for Coastal and Environmental Studies Contribution 26-1, p. 4.
13. Niedoroda, A.W., Coch, N., and Godfrey, P.J. (1975) "Preliminary Report on Erosion Problems at Great Kills Park and Sandy Hook, Gateway NRA." U. of Mass., NPS CRU - Report #8, p. 11.
14. "Transplanting the Valuable Marsh Grass Spartina alterniflora in an Environmentally Stresses Area" by B. Gay and J. Enrico (1979) in: Proccedings 2nd Conference on Research National Parks, San Francisco, Ca. Nov. 1979.
15. Godfrey, P.J. (1975) Comments on Present Beach Management Philosophy in Relation to Gateway NRA. U. of Mass., NPS CRU - Report 9, p. 9.

16. Pearce, J.B. (1976) "Our Coastal Waters: an endangered zone" in: The Science Teacher, Vol. 43, No. 9, Dec.

17. Heatwole, C.A. and West, N.C. (1980) "Beach Use and User Constraints in the NYC Coastal Region," NYS Sea Grant Report Series, NYSG-RS-80-01, p. 3.

18. David, E.L. (1971) "Public Perceptions of Water Quality," Water Resources Research 7(3): 453.

19. Tanacredi, J.T. (1977) "Petroluem Hydrocarbons: Contribution to the Marine Environment," Journal Water Pollution Control Fed. Vol. (2) p. 216.

20. "From Landfill To Park" An Experiment in Construction Waste Management at the Pennsylvania Avenue Landfill site; Dec., 1974/ NYC DCP 21-74.

21. Environmental Dynamics and Resource Management in the U.S. National Parks by Dolan, R., Hayden, B., and Soucie, G. (1978) Environmental Management, Vol. 2, No. 3, pp. 249-758.

22. Sawhill, J.C. "The Role of Science in Higher Education," Science, (1979) Vol. 26, No. 4416, Editorial.

23. "The Water's Edge: Critical Problems of the Coastal Zone" (1972) edited by Bostwick H. Ketchum, MIT Press.

URBAN RIVERFRONT EXPANSION

by D. B. Robbins,[1] J. M. Armstrong,[1] C. L. Kureth, Jr.,[1] D. V. Pratt,[2] and T. V. Dobson[3]

Abstract

Development pressures along the City of Detroit's riverfront have produced landuse, space, and access conflicts between residential, commercial, recreational and industrial interests. Of particular significance to the public is the denial of access stemming from development in some riverfront areas. These conflicts are broadly typical of those encountered in other riverfronts located in intensely developed urban areas.

A study of Detroit's riverfront was initiated in 1977 to determine the feasibility of expanding its riverfront areas through land use reorientation and "built-form" techniques. Conceptual expansion techniques were evaluated with respect to planning, economic, legal, navigation, engineering, and environmental considerations. The analysis demonstrated the flexibility of these techniques to resolve development related conflicts at five selected sites along the river. Built-form riverfront expansion techniques (RETs) were shown to be generally feasible within engineering and environmental constraints.

Although expansion has broad, useful applications in riverfront areas, the actual implementation of RETs may not be simple. Development and funding priorities may lead to complex negociations between the various interests in the riverfront. Implementation is further inhibited by the inertia of working within an existing riverfront, where past decisions and commitments have set development and use patterns.

For riverfront expansion to be successful, it should be examined within a planning/management approach that recognizes the unique role, value and importance of the riverfront. A framework of goals, objectives and policies to guide development will help insure that expansion is used to maximize the broad development needs of urban riverfront areas.

Introduction

Urban riverfronts are subject to intense landuse, space and access pressures which are concentrated in a narrow zone adjacent to the river. Open spaces and visual and physical access to the riverfront must be provided while maintaining or allowing for additional residential, commercial and industrial development. One approach to meeting these needs is to create space and access routes by expansion within the existing context of the riverfront. Riverfront areas can be expanded by providing opportunities for mixed landuse, by employing built-form techniques to create new space, by providing space for access or connections between key areas, and by actually creating new land areas.

---

[1]The Traverse Group, Inc., Ann Arbor, MI
[2]Burgoyne and Pratt, Attorneys, Ann Arbor, MI; TGI Associate
[3]formerly with the Coastal Zone Laboratory

In 1978 the Coastal Zone Laboratory, University of Michigan, examined the feasibility of expanding the urban riverfront in Detroit, Michigan. The study benefitted from a cooperative working relationship with the City of Detroit's Interdepartmental Riverfront Task Force and the Michigan Department of Natural Resources, which allowed the analysis to be responsive to current development issues and to concerns of both the City and State.

The purpose of the study was fourfold:

1) collect planning, cost, legal, navigational, engineering and environmental data to be used to examine the feasibility of expanding the riverfront,
2) create conceptual riverfront expansion techniques (RETs) to meet the needs presented by the City,
3) suggest which conceptual RETs would be most appropriate to enable broader use of five riverfront sites selected by the task force, and assess them in terms of the above data,
4) examine the process for implementing riverfront expansion techniques and comment on possible approaches to enhance their implementation.

It is the intent of this paper to summarize the major considerations involved in the expansion of urban riverfronts as learned from the Detroit experience, and to present some diverse conceptual RETs. These considerations and concepts should be viewed as being broadly applicable to other urban riverfront areas.

The Need for Expansion

Expansion of Detroit's riverfront has historically proceeded in response to primarily economic pressures. Since 1873, the entire length of the riverfront has been filled out to the pre-existing channel drop-off. The intense commercial and industrial development which followed utilized all of this "filled" land. In more recent years, developments have replaced some of the older buildings. With the economic stimulus provided by Detroit's Rennaissance Center, a major new commerical development, redevelopment efforts have been increased along the riverfront. Planners, however, have inherited a crowded riverfront with few vacant parcels and access to the public limited along major sections.

Detroit's scenario has elements in common with other urban centers where rivers have played an important role in development and new opportunities are sought to redevelop and improve the riverfront. Under these circumstances, expansion can help meet the need for new development and the desire to provide new opportunities for people to participate in the riverfront. Some of these needs are described below.

ECONOMIC STIMULATION. Where the lack of space is an impediment to development, expansion can provide attractive, usable riverfront to stimulate development. RETs which do not provide expanded areas may also be a positive growth inducing factor, in that they may encourage nearby development because they create a good "atmosphere."

SPACE REQUIREMENTS. Space created by expansion may allow for the growth of existing or the development of new commercial, industrial, or recreational facilities. For example, integrated commercial development such as cafes, shops, or restaurants can be established in space provided by the expansion. Expansion into the river may allow for marina development at key locations along the riverfront. Other commercial and industrial activities which may be suited to riverfront expansion sites include, but are not limited to: power plant development; coal storage; alternative energy development such as solar collector units, microwave relay stations, wind and water generators; waste treatment facilities; port facilities; hydroponics; and aquaculture. Expansion can also provide greater opportunity for recreational use of the riverfront. Detroit's Belle Isle is a major urban recreational park partly created by landfill.

OPPORTUNITIES FOR MULTIPLE LAND USE. Additional space to turn single land use areas into multiple use areas can be established at ground level or through implementation of multi-level designs. Expansion may allow for innovative mixtures of commercial, industrial, and recreational uses with accessible open space at the river edge. RETs can be used to create unobstructed access into industrial areas to develop interpretive areas along riverwalks.

BROADENING WATER-RELATED ACTIVITIES. Reconfiguration of the river's edge by scalloping of the shoreline or creation of canals can provide shallow, quiet zones for water-related recreational activities.

VISUAL AND PHYSICAL ACCESS. Riverfront areas need to provide generous visual and physical access as well as to accomodate development. The provision of adequate vehicular and pedestrian access and circulation throughout the riverfront and along the river's edge can be readily accomplished by RETs.

LINKING RIVERFRONT AREAS. Riverfront parcels having compatible uses may be linked along the river's edge to encourage broader use of the riverfront and to provide better access at the river's edge. Access across canals, influent streams, and harbor mouth entrances and the development of continuous river edge walks can be achieved by RETs. Linkages which provide access along the riverfront can lessen congestion at the river's edge created by activity or development nodes. RETs are also a means to access offshore structures and islands.

Conceptual Riverfront Expansion Techniques

Conceptual RETs were developed in response to site specific problems and issues posed by the Detroit Interdepartmental Riverfront Task Force. They are described here to show the variation of style and flexibility of functions which RETs can achieve. The accompanying descriptions are intended to lead the reader to conceptualize alternatives of applicability to his or her own situation. Costs of materials in 1978 dollars (National Construction Estimator, 1978) are given to show the comparative investment required for implementation of the various RETs. Material prices were generally within 10% of actual contractors prices, although the actual costs including labor will be 2 or 3 times those quoted.

LANDUSE PLANNING AND DEVELOPMENT ORDINANCE. Standard zoning ordinances tend to isolate uses, discouraging the integration of compatible uses which is often desirable in riverfront redevelopment. Flexibility in land use decisions should be insured, while providing a clear and standardized set of guidelines or policies for making land use decisions. A new riverfront ordinance or an amendment to an existing zoning ordinance should go beyond traditional zoning concepts. It should:

1) include a findings and purpose section to help avoid constitutional challenge,
2) define the "riverfront district",
3) create a decision-making body having the dual role of general planning and passing on the proposed land use,
4) specify the body's extent and limits of authority,
5) give authority to make land use approval contingent upon certain improvements, such as public easements, safety or landscaping,
6) define the uses which can be made of land in the riverfront district, and
7) set forth explicit criteria for making land use decisions along the riverfront, including provisions for appeal.

As an alternative to a riverfront ordinance, a planned development ordinance in combination with a well conceived and detailed master plan could provide flexibility within the constraints of an existing zoning ordinance. The Planned Unit Development principle involves density clustering so that the total density of development on a large parcel equals the density permitted by the zoning ordinance on individual lots within the parcel. It has been successfully applied to mixed use complexes, where residential, commercial and park and/or recreational facilities are shared.

LESS THAN FEE SIMPLE PUBLIC ACCESS. Access to the riverfront and particularly at the river's edge is limited if the lands are privately owned. Although the most desirable location for a pedestrian/bikeway would be along the riverfront, this may not be presently feasible. Under present law, the land would have to be purchased or, more probably, condemned as the fee simple owners would not be willing to permanently part with their land. There are riparians, however, who do not actively use the river nor its adjacent land. Under these circumstances, an easement concept which encourages property owners to open up their lands for recreational purposes would be appropriate. Under such a plan, municipalities could obtain easements for recreational pathways and other purposes, while protecting the interests of the grantor. Land owners would have two major objections to granting an easement; first, that they might incur liability for injury to members of the public using the easements and, secondarily, that they might lose title to the land by adverse possession or implied dedication. A proposed statute (Coastal

Zone Laboratory, 1979) has been drafted and submitted to the Michigan State Legislature to confront these two objections and to offer an incentive for the landowner to grant the city an easement. The proposed statute provides options for landowners in both industrial and rural settings. A property owner in a rural area may informally open his premise and be free from responsibility to the public. If the owner chooses, he may formally grant an easement, whereby liability would be to the grantee (if liability were the same as that held for other recreational lands, there would be no increased burden on the grantee), where he could have a property tax reduction and be protected from either adverse possession or implied dedication while the easement exists, a minimum of 20 years. The grantor would be free to sell within the period of easement, but subject to it.

RAILROAD ABANDONMENT. Available land in the riverfront district is at a premium. There are usually, however, abandoned rail lines which crisscross the area and could be used. Spur, industrial, team, switching and side tracks are exempted from the need for an Interstate Commerce Commission (ICC) certificate of abandonment or discontinuance. If the line is entirely within a single state, the parties to the contract may agree to abandon the spur. Abandonment of all other rail lines requires the certification of the ICC according to Section A of the Interstate Commerce Act.[1] The test is that the "present or future public convenience and necessity require(s) or permit(s)" such action. The ICC must consider if the action "will have a serious, adverse impact on rural and community development." The statute outlines a complex and relatively difficult procedure to abandon rail lines. Although the carrier must initiate action, the city can encourage the carrier, recommend approval to the ICC, and participate in the investigation and hearing, if any.

The use of abandoned rail lines might take several forms. The corridors could be used for access, interesting niches, green space or opportunities for industrial interpretation. The estimated materials cost for converting an abandoned rail line to an asphalted promenade is \$320/L.F.[2]

Advantages and disadvantages include:

- use of underutilized riverfront space
- gives access in highly developed areas
- problems associated with land aquisition

PEDESTRIAN/BIKEWAYS. Ped/bikeways can be used to provide access into areas, even when the only available routes are narrow. Further, in situations where there are no existing access routes, ped/bikeways can be adapted to a variety of circumstances to create linear space. Built-form options include: piling supported, cantilevered, floating, covered, building-attached, and multiple tiered forms. The use of these various techniques in international development has demonstrated the feasibility of ped/bikeways as RETs.

Beyond providing access, ped/bikeways in themselves create opportunities for active and passive recreational use of riverfront areas. The experience of the riverfront can be enhanced by creating water con-

[1] 49 USC 10903 (a)
[2] L.F. = linear foot

tact opportunities through the use of ramps. Ped/bikeways may also serve as the locus for shops, restaurants, and other commerical development.

Estimated costs range between $120 to $580/L.F.

Advantages and disadvantages include:

- design flexibility
- simple construction forms
- minimal environmental disruption
- shore edge forms require easement
- offshore forms may limit boaters access to shore

PLATFORMS. In areas where land take is an impediment to certain passive and active uses, elevated platforms can provide the additional space required. Platforms can function as observation decks and rest stops or support small businesses. Used near or over water, platforms can serve as boat touch points for recreational and commercial craft. Tour barges or river "taxi" systems are effective expansion techniques, especially when routes are tied into a series of platform touch points spaced along the riverfront. Street ends represent often underutilized space which can be expanded by platforms extending over the water. Platforms can either function as activity modes at ends of major access routes or can be integrated as a part of the access routes. A platform attached or suspended from an obelisk can facilitate pedestrian and bike circulation where obstructions to access exist, such as in industrial or port areas. The aerial view gained from suspended platforms provides interpretive viewing of river industry. Roof space has platform type benefits which should not be overlooked as an expansion opportunity.

Advantages and disadvantages of platforms include:

- provide space with minimal landtake
- facilitates access
- interpretive area opportunities
- significant ramp length requirements
- potential navigational hazard
- visual impairment
- air and solar rights conflicts

BRIDGES. Bridges have a variety of applications as riverfront expansion techniques. Attached platforms may be employed to provide rest and viewing stops or space for small commercial development. Restriction of navigational passage can be avoided by arch or draw-type bridges. Manual-operative drawbridges are a simple and effective method to provide access across canals or harbor entrances of recreational craft, but may not be efficient in high traffic or congested areas.

The estimated cost of a manual drawbridge with a duraluminum liftspan is $180/L.F.

Advantages and disadvantages include:

- access over obstacles or over congested areas
- avoid interference with recreational craft
- possible visual impairment
- air and solar rights conflicts

SHORELINE MODIFICATION. Modification of the shoreline may be desirable to increase the amount of shoreline available for passive or active recreational use, or to provide visual relief along the shoreline. Modification may be accomplished by scalloping the shore edge or by creating canals or inlets into the existing shore. Alternatively, artificial ponds may be situated near to the shore but separated by a thin strip of land from the river. This configuration would allow users to experience the aesthetic and ambient qualities of the riverfront, while providing a water contact environment free of natural river hazards or pollutants.

When canals or inlets are preferred, gates across the entrance would ensure a measure of protection to active users.

Contamination of the river during dredging activities would be largely avoided in the case of confined artificial ponds as compared to the inevitable short-term contamination which would occur during canal construction. Permit applications for the former would, therefore, have a higher chance for success.

The advantages and disadvantages of shoreline modification include:

- creates "new" shoreline
- provides physical access to the riverine environment
- potential fish and wildlife habitat
- moderate land take
- permit requirements
- short-term environmental impacts
- maintenance costs

ARTIFICIAL ISLANDS AND PENINSULAS. Artifical islands and peninsulas (i.e., fastland) can support a variety of uses. The creation of "finger" island parks for pedestrian/bike pathways is a solution to major restrictions or obstacles which prevent shore parallel access along the river's edge. By driving piles through the island fill to rest upon competant sub-surface strata, fastland can often be used for moderate intensity residential, commercial or industrial development. Proven construction techniques are varied and can utilize a wide range of materials. Some of the more common forms rely upon sheet piling polder-dikes or rock revetments to contain the fill material. Either "clean" fill or dredge materials, or materials requiring placement in a confined disposal facility (CDF) may be employed. These include dredge spoil; upland fill such as sand, gravel, or quarry stone; locally excavated material; lake and river bottom resources; fly ash; and municipal wastes and sludges.

Estimated materials' costs can be expected to range from $34,000/acre to $46,000/acre for sheet piling or stone revetment construction, to $55,000 acre for polder designs. These figures assume a fill depth of 25 feet. Costs of fastland may be reduced through the use of waste materials such as dredge spoil, fly ash, locally excavated material or municipal wastes.

Advantages and disadvantages of fastland include:

- provide recreational or low density mixed-use opportunities
- create shore parellel access
- utilize urban waste materials for construction
- create fish habitat

- may be relatively high cost
- interim disruption of biological communities
- interim water quality degradation
- alteration of river hydraulics

VACANT LAND. Municipalities may purchase vacant land parcels in riverfront areas for future or immediate development endeavors. A commonly overlooked expansion opportunity is afforded by former dredge disposal sites. Many such sites have potential for recreational or small commercial development.

NOOKS AND CRANNIES. Riverfronts contain numerous, small "hidden" places of high appeal to the fisherman, the onlooker, the reader, or the person who seeks privacy along the river. These places are special to the riverfront and add to its character. Opportunities taken to preserve or create nooks and crannies will contribute to the appeal of the riverfront.

INTER/INTRA BUILDING SPACE. Space between or within buildings can be utilized for access routes and a variety of activities. Atriums created between buildings form pleasant niches for arts and crafts viewing, outdoor restaurants, and a variety of other activities. With security precautions, the space between buildings can serve as access for pedestrians and bicyclists. Parking structures can be designed with built-in ped/bikeways to allow access when they may otherwise interfere with or cutoff a desirable riverfront circulation route. Overhead tramways or "people-mover" cars can be routed through certain structures, such as highway overpass viaducts, to achieve optimal layout. Their passage through industrial areas or buildings could provide riders with an interpretive experience.

Contexts of Expansion

Conceptualization, design, and implementation of RETs demands a multidisciplinary approach to the analysis of issues, data collection and evaluation of RETs. Important considerations are grouped by context.

THE PLANNING CONTEXT. A major goal of planning riverfront development should be to provide a framework of goals, objectives, and policies to "steer" development in directions perceived as desirable by the riverfront users and to meet the needs and demands of public and private interests. Certainly not all needs and demands can be met within the limits of riverfront resources, nor are all possible uses mutually compatible. In the case of expansion, as with riverfront development and use in general, decisions must be made regarding what activities or changes are desirable and will be allowed. An anticipatory posture to planning will help reduce ad-hoc decision making or an unprogrammed commitment of resources which is otherwise likely to occur, due to the month to month or day to day dynamics that occur for economic, political or institutional reasons. Before planning can proceed, it will be necessary to address the fundamental question regarding the desirable hierarchy between recreational, residential, commercial, and industrial uses; and what mixtures of these uses are possible and/or desirable.

The evaluation of RETs should include consideration of:

1) conformance with urban patterns,
2) site and environment characteristics,
3) composition of nearby neighborhoods and the social setting,
4) civic and community facilities,
5) availability of special municipal services, and
6) appropriateness and compatibility to site uses.

THE ECONOMIC CONTEXT. Capital and/or fiscal outlays and the commitment of riverfront resources for riverfront expansion will produce changes upon the local and regional economy. These changes should be evaluated in terms of their primary and secondary effects. The results of such an evaluation are important to the decision to proceed with riverfront expansion and whether to emphasize recreational, residential, commercial, or industrial development at particular locations.

Although expansion may cause division or disruption of existing land uses, it generally has potential to stimulate the growth of the economic base. RETs built to provide space for development will have direct and immediate influence. RETs implemented offshore, mainly for access purposes, may encourage future development.

Immediate effects of expansion will be felt by local businesses and neighborhoods. Expansion of riverfront areas to provide for riverfront access and to encourage citizen use of riverfront areas should be highly beneficial to existing businesses and should encourage new commercial enterprise. Although expansion may stimulate new residential development and encourage upgrading of existing neighborhoods, existing neighborhoods may offer some negative effects. Disruption due to increased noise, vehicular and pedestrian traffic, air pollution from vehicular emission, and safety hazards may lead to undesirable deterioration of neighborhoods. As expansion fosters economic growth, neighborhoods may undergo radical change as the population shifts to higher income persons with concomitant cultural and ethnic shifts. The value judgment leading to the possible future displacement of low income groups should be tempered against the potential destructiveness to personal lives caused by relocation. Expansion and development of riverfront areas will provide opportunities to create economic opportunities to local residents, and these should not be overlooked.

THE LEGAL CONTEXT. Expansion of riverfront areas will require adherence to state and federal regulatory statutues and permitting procedures. At the state level, statutes and procedures will be invoked which promote the wise use and management of state coastal resources. At the federal level, the primary federal statutes are the River and Harbor Act of 1899 (RHA), the National Environmental Policy Act of 1969 (NEPA), the Federal Water Pollution Control Act Amendments of 1972 (FWPCA), and the Coastal Zone Management Act of 1972 (CZMA).

Waters subject to federal jurisdiction, as set forth in these acts, have been tested and modified in the courts. Section 33 CFR 329 defines federal navigable waters and lists four categories: 1) coastal and inland waters, lakes, rivers, and streams that are navigable waters as defined in the courts;[1] 2) tributaries to U. S. navigable waters; 3) interstate waters and their tributaries; and 4) all other waters of the U. S. Each of these catagories includes adjacent wetlands.[2] The first three are subject to U. S. Army Corps of Engineers (USCOE) permits under Sections 9, 10 and 11 of RHA, and the fourth catagory gives the USCOE and Coast Guard jurisdiction over further waters for purposes of FWPCA, as detailed below.

RIVER AND HARBOR ACT OF 1899.

U. S. Coast Guard requires a permit under Section 9[3] for bridge or causeway construction in navigable waters.

USCOE requires a permit under Section 10[4] for construction of piers, breakwaters, bulkheads, revetments, power transmission lines and aides to navigation in navigable waters.

USCOE requires a permit under Section 10[4] for dredging, channeling or filling in navigable waters.

USCOE requires a permit under Section 11[5] since May 27, 1970, for all action which would otherwise require a permit, commenced shoreward of the harborline.[6] Existing and future harborlines have since been used to define the offshore limits of structures and fills with respect to navigational interests. The same permit procedure applies to actions taken seaward or landward of the line. Activity beyond the harborline requires modification of the existing line or establishment of a new one.

NATIONAL ENVIRONMENTAL POLICY ACT OF 1969.[7]

If the District Engineer finds "that the decision on the application is not a major federal action significantly affecting the quality of the human environment, no Environmental Impact Statement be filed. In every case an Environmental Assessment must be prepared by the District Engineer.[8]

FEDERAL WATER POLLUTION CONTROL ACT AMENDMENTS OF 1972.[9]

USCOE requires a National Pollution Discharge Elimination System Permit (NPDES) under Section 404, otherwise RHA Section 13[10] prohibits discharge of "refuse matter." FWPCA applies to all waters, not only navigable waters.

---

[1] Daniel Ball v. United States 77 US 557 (1871)
Economy Light and Power Co. v. United States 256 US 113 (1921)
United States v. Appalachian Power Co. 311 US 377 (1940)
United States v. Moretti 478F. 2d 418 (5th Cin. 1975)

[2] United States v. Holland, 373F. Supp. 665 (A. D. Fla., 1974)

[3] RHA 33 USC 401

[4] RHA 33 USC 403

[5] RHA 33 USC 404

[6] 33 CFR 328.5(b)

[7] 42 USC 4321 et seg.

[8] 33 CFR 325.2(a)(6)

[9] 33 USC 1251-1376

[10] 33 USC 407

COASTAL ZONE MANAGEMENT ACT OF 1972.
Section 307(c)[1] requires applicants, in states having an approved Coastal Management Program, to have state certification on projects before a federal permit may be issued.

The general policies of the U. S. Army Corps of Engineers for evaluating permit applications are set forth in 33 CFR 320.4. "A public hearing on a permit application will be held where the application is for a FWPCA 404 or federal project permit and at the discretion of the Department of the Army where a public hearing will assist in decision making."[2]

From the primary emphasis on "public interest" in the permit evaluation procedure, it is clear that a well designed, planned and coordinated RET or system of RETs proposed or backed by a public agency will have an excellent chance of receiving permit approval. Chances of approval will be enhanced if the municipality actively works with USCOE during the planning process.

Two basic riparian rights, the right of wharf out to navigable waters[3] and the right of access to navigable waters[4] are of particular concern in considering the legal ramifications of riverfront expansion techniques. Within these legal constraints, the municipality and the riparian should be able to negotiate a mutual accomodation by contract, which would specifically set forth the rights and obligations of each.

The construction of bridges poses special questions regarding navigability and the rights of riparians. Navigability is protected through federal and state permitting procedures. Interference to riparian boat owners may result where an offshore RET is connected to shore by a bridge, or a canal is bridged. A simple solution would be the construction of a swing, draw or other type of movable bridge. Where the proposed bridge is stationary, passage of recreational craft, particularly fixed mast sailboats, may be blocked, In this event, riparians could assert their right to access which may result in the municipality having to compensate the riparian for damages.

THE NAVIGATIONAL CONTEXT. RETs must avoid interference with recreational and commercial craft. An adequate analysis of potential interference must include consideration of the future demands of navigation upon the river, as well as the existing demands. RETs may interfere with navigation by:

1) creating traffic congestion problems through the constriction of the channel, or near an RET serving as a docking area,

2) usurping docking/loading space for commerical vessels at the river's edge,

---

[1]16 USC 1456(c)
[2]33 CFR 327.4(a)
[3]Obrecht v. National Gypsum Co., 361 Mich 399, 105 NW 2nd 143 (1960); Obrecht v. Director of Conservation, 361 Mich 399, 105 NW 2nd 143 (1960)
[4]Rice v. Naimish, 8 Mich App 698 (1967)

3) posing navigational hazards,

4) changing traffic patterns so that recreational craft interfere with established commercial traffic lanes.

Any one of these interferences would cause a permit application to be denied, and so should be avoided by careful planning with full understanding of future navigational needs.

An analysis of the future needs of navigation provides a basis for determining whether additional marina or port space, expanded through the use of RETS, may be required.

THE ENGINEERING CONTEXT. The feasibility of implementing certain RETs will depend upon local site conditions; but, in most situations, the limiting factor will be cost. Detailed site investigations will be required to determine soil types, their load bearing capacities, characteristics of bed rock, and the slope, condition and mass stability of the river edge.

In northern climates, winter ice conditions pose a potentially serious threat to structures in the river or along shore areas. Structure may be subject to horizontal and/or vertical ice forces. Piling supported structures are particularly vulnerable and it may be necessary to protect these structures by wall-type or clustered-piling ice-deflectors placed upstream. Piling supported fishing piers in Detroit resolved the problem with some success by using decks designed to break-a-way upon ice impact, leaving pilings intact, and thus reducing maintenance costs. Piles having conical sections near the water line are effective in breaking apart ice sheets and deflecting floe ice. The use of concrete filled pipe piles with hinged decking (Chicago Department of Planning, 1979) is recommended for small craft marina docks to compensate for vertical ice forces.

Efforts to collect data concerning the nature and location of ice development should preceed RET design. During the design phase, the U. S. Army Corps of Engineers Cold Regions Research and Engineering Laboratory can provide useful direction.

The placement of artificial islands or peninsulas in a river will alter the hydraulic regime, causing changes in flow and current direction and increased water-surface elevations (backwater) upstream of the constriction. Water-surface and velocity profiles are controlled by four factors: the shape and size of the channel, the volume of water flowing through the channel, the bottom roughness of the channel, and the water-surface elevations of hydraulically connected lakes or streams. The calculation of backwater elevations can be carried out using computer modeling techniques, such as the U. S. Army Corps of Engineers HEC II model.

Applications which request the placement of fill in rivers will require the review and evaluation of U. S. Army Corps of Engineers in cooperation with state agencies to determine the effects on channel hydraulics and water-surface elevations.

THE ENVIRONMENTAL CONTEXT. Potential environmental effects of RETs must be assessed to determine if a particular RET design is acceptably suited to a particular site. The assessment should include evaluation of potential impacts in the following areas:

1) Biology
   a) benthic communiities
   b) fishes
   c) zooplankton
   d) phytoplankton
   e) rooted aquatic plants
   f) terrestrial animals and birds
   g) rare, threatened or endangered species

2) Water Quality
   a) short term contamination during construction
   b) contamination resulting from fill or dredging activities and confinement of those materials
   c) long term contamination from activities associated with RET use

3) Air Quality

4) Noise Pollution

5) Socio-economic Effects

It was found in Detroit that only a few localities in the river were sensitive areas where RET implementation would result in unacceptable adverse impacts. Irretrievable commitments of natural resources would be involved in the creation of fastland, in that the natural river environment would be permanently altered or consumed. Potential irreversibile effects may include river flow modification, discharge modification, creation of navigational hazards, and destruction of aquatic habitat (which may in part be balanced by the creation of new habitat).

To examine the effects of RETs, a list of RET factors which could potentially cause changes in the environment was developed:

1) displaces substantial water volume

2) has high relative cross-sectional area compared to river cross section

3) covers substantial river bottom area

4) has large interactive zone at river bottom/water/RET surface

5) covers substantial riverbank area

6) creates significant underwater, sloped surface

7) produces surface land changes such as contouring, levelling

8) changes existing shore configuration

9) is constructed of material which through discharge, leaching, or corrosion contributes negatively to river water quality

10) covers substantial river surface area

11) creates significant land surface area

12) protrudes significantly into river channel with low

cross sectional area

13) is located and designed to encourage new recreational activities

14) is located and designed to encourage development

These RET attributes lead to a series of environmental impacts. Short-term impacts are those associated with the site development or construction phases and are not enduring after these phases are completed. Long-term impacts for purposes of this discussion include those which endure throughout but not beyond the life of the project, those which extend beyond the life of the project but not permanently, and those which may endure permanently. Cumulative impacts may also be realized from the collective sum of all impacts from individual RETs.

CONSTRUCTION ACTIVITIES. Major construction activities interacting with the water column and the bottom sediment zones may cause disruption to aquatic biological communities by degrading water quality or disrupting the sediment substrate. Construction activities involving dredging and/or discharge of fill and dredge soil represent the most significant potential impacts of any construction phase activities. Other on-site construction phase activities, such as site clearing or site work (excavation, construction and landscaping, etc.), may result in some pollutant materials (e.g., oil and gas) being introduced into the river. Increased runoff from site clearing and site work has potential to create temporary higher turbidity levels near the construction site. For the most part, introduction of pollutants as suspended material during construction is not expected to produce significant impacts which cannot be overcome by proper construction practices as regulated by the Erosion and Sedimentation Control Act (Michigan Public Acts, Section 347, 1972). Noise and dust resulting from construction activities may be temporarily bothersome to nearby residents. Movement of trucks and construction equipment may cause inconvenience to local traffic.

FLOW VELOCITY MODIFICATION. RETS having significant cross sectional area relative to a river channel may increase local flow velocities. Increased velocities may present increased navigational difficulty to certain craft. They may also result in higher erosion of shore areas and increased bottom scouring.

Decreased flows may occur in "shadow zones" at the downstream end of fastlands or other structures emplaced in the river channel. Expansion techniques may be purposively placed to "deflect" flow, to create zones of decreased flow and wave protection near shore.

Increases or decreases in flow may have a direct effect upon macrozoobenthic organisms and other filter feeding organisms, and may produce localized changes in the species composition of these populations. Flow velocity modifications will also affect rooted aquatic plant habitat. If flow velocity increases are significant, areas inhabited by rooted aquatics may be rendered unsuitable to support rooted aquatic plants.

FLOW DIRECTION MODIFICATION. RETs which occupy a significant river cross sectional area may modify the direction of flow or currents. The magnitude of the change in direction will be determined by the size of the river cross sectional area influenced by the structure. A modification of flow or current direction may cause difficulty to small craft maneuvering in "tight" areas such as harbor entrances, and may result in increased shore erosion with a possible shifting of erosional patterns.

DISCHARGE MODIFICATION. RETs having a high cross sectional area in proportion to river cross sectional area have potential to cause significant increases in backwater elevation. Such an increase in water surface levels may lead to an increased hazard for flooding in upstream areas.

CIRCULATION MODIFICATION. Natural circulation of river water may be inhibited through partial blockage of flow or by diverting water into canals and lagoons cut into shore areas. If modification of circulation is significant, lower than normal oxygen levels in river water may result and mixing of essential nutrients to all levels of the water column may be inhibited. This may cause interruption of natural aquatic ecosystems or stress aquatic populations.

DISRUPTION OF THE RIVER BOTTOM. In Detroit, the dredging of river shipping channels has been carried out for many years. The environmental impacts from this activity have generally been viewed as acceptable, particularly, in reference to the large aquatic ecosystems encompassed by the Detroit River and connected bodies of water. Observations and analyses by the U. S. Army Corps of Engineers (1974) showed no immediate or extended effects of dredging upon water quality, as iron, suspended solids, and BOD returned to normal levels within 1/2 mile behind the dredge barge. Disruption of river bottom benthic habitats resulting from dredging has not been documented, however, dredging undoubtably influences some changes in benthic habitat near the dredged channel.

DESTRUCTION OF RIVER BOTTOM HABITATS. Discharge of dredge and fill material, as well as dredging, results in destruction of river bottom environments which may support a variety of benthic organisms and rooted aquatic plants. The construction of fastland may lead to the reestablishment of rooted aquatic plants on underwater slopes, but this depends upon the construction technique used. Regions of lower flow velocity created at the downstream edges of RETs, such as artificial islands, may further encourage reestablishment of aquatic plant communities. The destruction of rooted aquatic plants can be minimized by avoiding discharge of fill and dredge material in zones known to support their growth. Benthic communities will undergo a change in structure or function when covered by discharged dredge and fill. Communities will often reestablish, but are usually of a different ecological structure (Diaz and Boesch, 1977). The mortality of benthic animals is due mainly to direct covering. Adverse impacts will be largely confined to zones immediately adjacent to discharge areas.

RELEASE OF POLLUTANTS FROM SEDIMENTS INTO THE WATER COLUMN BY DREDGING. This effect can be expected to produce short-term degradation of water quality. Certain constituents may become more concentrated in river sediments as they are precipitated out of the water column.

RELEASE OF POLLUTANTS FROM OPEN WATER DISPOSAL OF DREDGE OR FILL. The mixing of fill or dredge material with the water column at the time of the material's discharge may lead to adverse environmental impacts, which will be largely short term. The leaching of contaminants from fill or dredge material may continue for some time after construction.

RELEASE OF POLLUTANTS FROM CDFs. Impacts from the release of pollutants through Confined Disposal Facility effluents will be primarily short-term, although operations may continue over a period of years depending upon the size of the CDF. Effluent quality can be controlled to avoid environmental degradation.

LAND RUNOFF. Runoff from construction activities will be short-term. As sites are developed and landscaped, introduction of nutrients from fertilizers by runoff will become the major source of contamination.

CORROSION OF STRUCTURAL SUPPORT MATERIALS. The introduction of chemicals such as from preservatives used to give longer life to wood pilings or from metals released through corrosive processes (a more important consideration in salt water environments) may have an effect upon water quality or aquatic organisms. A discussion of preventative actions to inhibit corrosion of steel, concrete, and wood piles may be found in Whitaker (1976).

CREATION OF AQUATIC HABITAT. Certain expansion techniques may provide habitats for aquatic organisms, but new biological communities which take over may be significantly different from pre-existing communities. Covering of polluted river bottom substrate with cleaner material may promote establishment of some macrobenthic and rooted aquatic forms if underwater, artificially created slopes are not prohibitively steep. Rubblemounded slopes may attract fishes by providing shelter and creating "collecting" places for phytoplankton, zooplankton, and other small plants and animals.

CONTAMINATION OF MUNICIPAL WATER SUPPLY. Expansion activities should be maintained at safe distances from municipal water intakes to avoid the contamination of water supply during construction or by post construction leaching of RET materials.

MODIFICATION OF NEARSHORE FLOW. RETs should be designed and located so as not to inhibit natural river "flushing" processes. Nearshore flow prevents the impounding of water which may create zones of potentially high-oxygen demand and encourages the efficient removal of sewer effluents which may be discharged into the river.

WATER QUALITY. Development supported by RETS may have impact on water quality and aquatic organisms depending upon the nature of future uses.

AIR QUALITY. Riverfront expansion, in and of itself, will not produce significant air pollution. However, certain future activities arising because of opportunities fostered by expansion may degrade air quality. Increased automobile traffic will be the major contribution. However, industry which takes advantage of fastland may also contribute to lower air quality.

NOISE. Noise resulting from new activities or traffic related to expansion sites may intermittently disrupt local neighborhoods.

It is apparent from the above discussion that a complex array of RET influences and environmental impacts may result from expansion. The interactions between influences and impacts is also complex. Golden (1977) points out the importance of being cognizant of these interrelationships and to analyze environmental impacts individually and in terms of their relationship to the whole environmental picture. These complex relationships can be easily displayed in a matrix to show the more important interrelationships between RET design and potential environmental impacts (Coastal Zone Laboratory, 1979). Environmental impacts can be separated into two levels for convenient handling in a matrix. For example, RETs occupying a large cross sectional area would cause flow velocities to be increased. This would constitute a LEVEL I impact. The increased flow velocity, in turn, might result in a series of second level (LEVEL II) impacts, such as navigational difficulties, alteration of habitats, etc. Such a matrix can serve as a tool to aid in environmental assessment and to anticipate environmental impacts early in the project development phases.

The Planning and Implementation Process

The implementation of RETs can help achieve an attractive riverfront, capable of supporting a variety of desired and/or necessary uses. The process of planning and implementing RETs, however, may not be simple. For example, although there may be a perceived need for expansion, the concept may be incompatible with existing priorities which determine the allocation of funds to recreational, residential, commercial and industrial development. Even in the event that expansion were compatible with existing priorities, the planning and implementation process may be slowed by the inertia of working within the existing riverfront patterns of use and development. The perception of expansion as being an added feature to development, with added costs above and beyond those of a project, further inhibits the expansion process. Expansion should be viewed as an important integral part of a riverfront project, when expansion is needed to ensure the best use of riverfront resources. Negociations which seek compromise resolutions to expansion issues may become long and difficult, further impeding the process of expansion. The "nobody wants to be first" syndrome can stall worthwhile projects in the planning stages. Finally, unfamiliarity with the complexity of permitting procedures may hamper the planning process. Early identification and involvement with permitting agencies will help to clarify environmental concerns, familiarize the developer with permit procedures and restrictions, and speed the planning and implementation process.

The examination and selection of RET's will be most effective when carried out within a planning/management approach that is reflective of the role, values, and importance of the riverfront. An integrated and comprehensive approach to riverfront expansion should form the basis for developing, evaluating, ranking, and finally selecting possible RET concepts. Such an approach would have several advantages to foster the best overall use and development of riverfront resources:

1) It could help minimize the foreclosure of future opportunities by short-term decisions made in the absence of a total view of the riverfront as a complete system.

2) It could encourage an anticipatory approach to development.

3) It might increase the public awareness of the value and potential of the riverfront by communicating the goals, objectives and alternatives for riverfront use.

4) It might increase the efficiency and effectiveness of the use of municipal, state and federal funds and technical assistance.

5) It might enhance the important and required participation of the private sectors by providing clear, well defined and efficient guidelines as to the city's intentions for its riverfront and how it will achieve its objectives.

Two basic activities are involved in the planning and development of the riverfront system. First, the development of riverfront resources and/or their protection must be planned. Secondly, specific projects must be implemented, either within the context of prepared plans or within the scope of riverfront development projects that have arisen outside of previous planning efforts. Without the aid of adopted goals and policies and without a continuing anticipatory process of goal setting, policy formulation and program review, the planning and development of the riverfront will be piecemeal and reactive. What is required is an integrated process of riverfront planning and development that reflects the unique nature of the riverfront's natural and cultural systems. The elements of such a process should include:

1) A continuing process for re-examination and review of development <u>goals and objectives</u>. This should be carried out with the involvement of all of the various interest groups at various levels.

2) The formulation and adoption of <u>policies</u> that are aimed at attaining the goals and objectives.

3) The identification of critical <u>issues</u> that exist and may arise from the stated goals, objectives, and policies.

4) The identification of specific <u>questions</u> that must be answered in order to resolve the identified issues.

5) The identification of <u>data and information</u> required to address the questions and the establishment of <u>projects or efforts</u> to obtain the required answers.

A mechanistic approach towards an integrative riverfront process that blends these elements would be the establishment of a permanent "Riverfront Office." The office would work with the private sector, planning and regulating agencies of the city, state, and federal government and public interest groups and would have the following charges:

1) It would identify and/or develop objectives, goals, and policies for the riverfront.

2) It would address the various elements of the riverfront development process; i.e., economic implications, environmental implications, riverfront planning coordination, public involvement, riverfront status, social implications, and design aspects.

3) It would evaluate and review proposals for development, and could also actively propose particular developments for the riverfront area.

4) It would have decision-making responsibilities to approve or disapprove proposed uses.

5) It would offer desired proposals to the private sector; i.e., it would "broker" development.

6) It would act as an information source on riverfront activities, plans, proposals, studies, etc.

7) It would develop and publish a detailed process for developing concepts and carrying out development of specific sites along the riverfront.

In summary, expansion should be viewed as an activity which is integral with the broader development needs in riverfront areas. The creation of a riverfront which combines the large number of interests, desires, needs, and commitments of the private and public sectors demands a comprehensive, coordinated approach. A strong, coordinated and integrative body that can view the riverfront in a complete and systematic fashion would enhance the achievement of a desired result.

Summary and Conclusions

Riverfront expansion can be accomplished through a variety of land use, legal and built-form techniques. Although there are real limits and practical difficulties involved in pursuing expansion alternatives, expansion provides the planner and developer with powerful flexibility to redevelop and enhance urban riverfront areas. Expansion of the riverfront can best be attained through a comprehensive planning/management approach which seeks the integrated development of riverfront resources.

Acknowledgements

The original study upon which this paper is based was carried out by the Coastal Zone Laboratory, at the University of Michigan, under Contract No. LRP-91 of the Michigan Coastal Program, Department of Natural Resources. Contract funds were provided by the Office of Coastal Zone Management, NOAA, with match from the City of Detroit. The authors wish to express their gratitude for this support. Mr. William C. Williams and Miss Barbara Kauffman contributed to the hydraulic engineering and planning sections, respectively, as part of the Coastal Zone Laboratory study team. Mrs. Harriet Saperstein and Mr. Robert Hoffman of the City of Detroit are thanked for their helpful comments and review of this paper. The views presented are those of the authors and do not necessarily represent those of the City of Detroit or the State of Michigan.

References

Coastal Zone Laboratory, 1979, Riverfront Capability Expansion Analysis, CZL Tech. Rept. No. 116, The University of Michigan, Ann Arbor, Michigan, 532 p.

Diaz, R. and J. Boesch, 1977, Environmental Impacts of Marsh Development with Dredged Material: Acute Impacts on the Macrobenthic Community, Appendix C, Tech. Rept. D-77-23, U. S. Army Engineer Waterways Experiment Station, Vicksburg, Mississippi.

Golden, J., R. P. Ouellette, S. Saari, and P. N. Cheremisinoff, 1977, Environmental Impact Data Book, Ann Arbor Science Publishers, Inc., Ann Arbor, Michigan, 864 p.

U. S. Army Engineer District, Detroit, March 1974, Confined Disposal Facility at Pointe Mouillee for Detroit and Rouge Rivers, Final Environmental Impact Statement, Detroit, Michigan.

Whitaker, T., 1976, The Design of Piled Foundations, Second Edition, Pergamon Press, 218 p.

-------------, 1978, National Construction Estimator, 26th Edition, Craftsman Book Company of America, Solama Beach, California.

WATERFRONT POLICY: RESPONDING TO LOCAL NEEDS
BY

Letitia C. Langord, Ph.D., University of Pittsburgh
Irene L. Murphy, Ph.D., Heritage Conservation and Recreation Service

Introduction

As the interest of cities in revitalizing their waterfronts grew by leaps and bounds over the past decade, so did expectations that Federal urban policies might adapt to changing river and shore fronts. A plethora of programs has been tapped for development in this new and exciting area but far too many communities have experienced long delays in completing projects, or in getting them off the drawing board and into construction. While no local official has expected total support from the Federal government, far too frequently a grant or permit which would have made all the difference in building momentum for a project has been denied or delayed by a Federal source. This paper will describe present Federal policy (or the lack thereof) as it responds to waterfront needs. It will suggest ways in which the government could do more at little extra cost to improve the quality of life <u>and</u> the economies of inner cities by stepping up aid to flagging waterfront projects.

Policy on waterfronts developed as it has in all major issue areas, although its characteristics may be more complex than some. Need for a policy response has been substantiated by community experiences while studies have suggested ways to improve policies. Specific demands for action by decision makers are now being formulated. As in any issue area they roughly divide into four categories: (1) funding in the form of grants or loans for land acquisition and development; (2) expert assistance with planning and implementation of plans; (3) research that will help to keep funding and assistance in line with needs, and (4) regulation of private and public actions which serves the needs of all without unfairly binding any one group or interest. Waterfront policies present some uniquely complex problems in their formulation and implementation.

These policies result from laws and regulations which reflect other problems and earlier times. Developing waterfront projects are regulated by a conglomerate of water resource, housing, floodplain management, environmental, navigational and other policies. Funding for projects is available from forty or more program sources and technical assistance from many more,but all of these programs refer to other needs and must be adapted to waterfront use. This takes time and expert knowledge. Even research needs are met separately and unevenly as they relate to hydrology, biology, and economics among others with no attempt to pull these diverse strands together.

During the past two years there has been a spurt in efforts to bring cohesion to Federal waterfront policy, no doubt occasioned by a substantial increase in demand for policy output by hundreds of local communities. Projects are begun because communities anticipate a variety of benefits from them, from amenities like greenways and parks to jobs and other economic improvements. Some proceed rapidly from concept to plan, but most have required help after experiencing delays in their beginning phases. Federal agencies themselves, in collaboration with state and local agencies and the private sector, are reacting: analyzing needs, establishing goals, suggesting specific program revisions. How well this bootstrap effort to improve policies and programs is succeeding and what more can be done for it is the topic of this paper.

Evidence of Need

Despite the enviable accomplishments of cities to date, information from a number of sources - studies, workshops and training sessions, conferences, meetings with diverse interest groups, and presentations which have been made by city representatives to the Urban Waterfront Action Group (see below) in Washington - clearly identifies needs to which Federal policy can and should be responsive. The recent HCRS study, Urban Waterfront Revitalization: The Role of Recreation and Heritage - describes these as the need "to increase public understanding of the opportunities for and environmental, social and economic benefits of urban waterfront revitalization...to develop sound planning and design guidelines and criteria...to improve and increase technical assistance in key areas...to ensure that the benefits of waterfront revitalization are equally available to everyone...to reduce access barriers caused by highways and railroads...to encourage private sector support of public access goals...and to improve present Federal cooperation and coordination in the delivery of program resources."

Meetings jointly sponsored by HCRS and the American Planning Association early in 1980 in Alexandria, Jacksonville, Portland, St. Louis and Boston were attended by public officials, planners and private developers. At these meetings participants affirmed the urgency of the need for improved Federal policies in the areas cited above. While the level of funding available was always a major concern, requests for information through publications, training sessions, and one to one technical assistance was stressed again and again.

And recognition of the multiple public benefits of urban waterfront revitalization have resulted in requests for policy action by such interest groups as the National League of Cities and the Sierra Club. Interest in urban waterfront revitalization by the National League of Cities, which has been participating in a Federal interagency task force on waterfronts (see below) since its inception, is understandable.

The League's headquarters in Washington, D.C. regularly receives requests for information from cities who want to do something with their waterfronts. and the number of requests has increased dramatically in the last two years. But the Sierra Club? The importance of waterfronts was formally acknowledged by the Sierra Club in April of 1979 when a panel at the City Care Conference (sponsored by the Urban League, Sierra Club, and Urban Environment Conference Foundation) highlighted the potential of waterfronts to help resolve certain issues - water quality, wise land use, and enhanced urban recreation - in which the Sierra Club is deeply interested. And finally, the recognition of urban waterfronts in many of the Year of the Coast activities echo once more the inadequacy of present Federal policy in this important issue area - and the need for action.

The Federal Response to Pleas for Help

Most waterfront projects have received Federal help of some kind: funding, planning assistance, or a combination of the two. For most, Federal funding has been part of, perhaps the nucleus of, financial help from other public and private sources. No one has yet counted the total sum of money presently invested in waterfront revitalization or calculated what percentage represents the Federal slice of the investment pie. A summary of available Federal grant and loan programs serves to show the range and diversity of available Federal aid. It is safe to say that several billions of dollars of money from these programs has aided the revitalization effort.

The Departments of Commerce, Housing and Urban Development and Interior lead the way in the number and size of funding programs which can be used on shore and riverfront areas. Each Department has grant money which follows from select areas of responsibility, although there is some overlapping. A community may look to Commerce's Economic Development Administration to aid a commercial sector of its waterfront; to Interior's Heritage Conservation and Recreation Service to fund a park or the preservation of an historic monument; to HUD for an urban development action grant for housing and related uses. It could seek help from the Corps of Engineers to widen or deepen a channel or to improve flood control measures. A number of other Departments and agencies can contribute. Transportation can help provide access for distant city residents to waterfront amenities, or perhaps help solve the age-old problems associated with the ubiquitous highways and railroad tracks that seal off so many of our cities' waterfronts. The National Endowment for the Arts has funded designs for projects, The Environmental Protection Agency not only sees that city waters are swimmable and fishable within a specific time frame but also funds the clean-up of lakes in the heart of urban areas so that inner-city residents can swim and fish. Altogether, there are at least 18 Federal agencies with funding programs available for waterfront development.

Technical Assistance

Communities need, and customarily get, expert help along with, or perhaps separately from, financial aid. The broad sweep of help available is suggested by the review of the funding programs. Most Federal agencies have aid which, depending on their budgets, can be as simple as a brochure or can involve consultation on a one-to-one basis for long periods of time. Until very recently this assistance did not address specific waterfront problems.

HCRS, for example, because it had extensive experience with recreation planning associated with a variety of water resource projects, has developed a special waterfront expertise. On occasion it has interrelated with other agency programs, particularly the Corps of Engineers urban studies progra, to assist such long-run, large scale greenway projects as that achieved on the Platte River in Denver, Colorado. HCRS and the Office of Coastal Zone Management (Commerce) have developed expertise in the acquisition of waterfront land for public access which has made the difference between success or failure in determining the economic feasibility of a number of waterfront projects.

Waterfront project sponsors, as they seek Federal funding and planning assistance, must chart a knowledgable and painstaking course. They must not only be aware of the existence of 25 or more Federal programs but also know, in detail, about funding application requirements, and about which agencies have what kind of expertise available and on what basis. It is small wonder that the estimated 350 communities which are or would like to improve their waterfronts generated requests for five thousand plus copies of the first waterfront material prepared by the Office of Coastal Management and the National Endowment for the Arts, or that in six months time planners and community leaders snapped up eight thousand copies of the HCRS Urban Waterfront study.

The piecemeal nature of most available technical assistance material reflects the kind of research the Federal government sponsors on waterfront-related issues. The separate interest of Congressional constituencies has supported studies on water pollution with an eye to its control; on water quality so that commercial and sport fishing may be protected; on hydrology so that floods and droughts may be predicted and perhsps controlled. Economic feasibility studies and recreation user surveys receive, in the view of many, insufficient funding support and have not been directly related to other kinds of research which, if intergrated, could be of direct help in the formulation of technical assistance material.

Regulatory Functions Which Apply To Waterfronts

Urban waterfronts present a nightmare labyrinth of overlapping jurisdictions - Federal, State, regional and local - and a parallel maze of regulatory responsibilities by each of these units of government. Amazing variety of permits must be obtained by both public and private entities before any project, large or small, can be undertaken. The process of obtaining all necessary permits is difficult enough for a non-waterfront project. When a waterfront project is involved these difficulties are compunded enormously.

Administration of the 404 permit process, which involves thousands of applications each year, by the U.S. Corps of Engineers, is probably the most critical Federal regulatory role in the urban waterfront. Intended to ensure that potential activities at the water's edge do not adversely impact our nation's waters, the complex review process can be both obstacle and opportunity cities wanting to capture potential waterfront benefits.

Until recently, the review process for a proposed project could easily take 18 months from start to finish. By the end of this time, funding application deadlines may have been missed and political support lost. The state of Maryland has significantly improved these circumstances. Everyone involved in the 404 review now meets once a month to make decisions on all permit applications received during the preceeding 30 days. Most present no problems and for those that are controversial everyone is present with supporting materials.

The advantages of this system have not been overlooked at the Federal level. Agreements recently signed between the Corps and the Departments of Transportation, Commerce, and the Interior, and the Environmental Protection Agency have established a greatly reduced turn around time. The Corps will now make a decision within 90 days after issue of the public notice re an application, thus reducing costly delays.

However, it must not be lost sight of that the 404 process can also be a real opportunity for cities to achieve their waterfront goals. Local governments play a key role in the approval process. The Corps does not normally grant a permit to which a mumunicipality has strong objections. To cite just one example: in Pittsburgh, the City agreed to withdraw its objection to the permit application of a riverfront business, if the company would help to meet the City's public recreation needs in that stretch of shoreline by providing a public access easement on an unused portion of its property adjacent to a planned park. Agreement was reached, with both parties, public and private, being satisfied. The 404 and other Federal regulatory processes, when properly used have the potential to help meet public needs in broader ways than normally perceived.

How Present Policy Fails and Its Effects

The present delivery of Federal services - funding, technical assistance, and research - is now only marginally coordinated through standard review processes, such as A-95, at regional and state levels. In part this reflects the sheer magnitude of the task. But it also reflects the absence of timely, integrated interagency coordinating and decision mechanisms for urban waterfront revitalization projects. (This is of course not a problem restricted to urban waterfronts). The closest thing is the Federal consistency requirement which ensures that all Federal actions in the coastal zone will be consistent with approved Coastal Zone Management Plans. But this provision has no effect on Federal actions in the vast inland waterway system.

While there is great interest at the Federal level in improving the delivery of resources, especially funds, little has been accomplished. Funds are often lost or projects seriously delayed because of differences in grant eligibilities, deadlines and requirements. Yet in many instances the improved coordination necessary to help meet critical needs would require only administrative, not legislative, changes in rules and regulations.

Circumstances are little better when it comes to technical assistance and research. There is endless duplication in too many instances, while little or nothing is done by responsible Federal agencies in other areas. For example, the HCRS report and OCZM studies, identified many specific areas in which research, directly linked with critical technical assistance, was needed. To date, almost one year after their publication, little response has been made. Why?

Federal agency personnel both in Washington and regional offices are often simply not aware of the needs identified earlier in this paper. Or if they are aware - and are concerned - too often their present legislative authority doesn't permit them to respond effectively.

As the HCRS study noted: "Inadvertently, program guidelines and requirements contribute to the lack of concern for a broad consideration of waterfront revitalization needs."

This situation is compounded by the fact that major Federal policy documents - and their counterparts at state and regional levels such as SCORP's, HPP, 208, and Economic Development District Plans, which are prepared in large part as a response to Federal programs - do not discuss shared opportunities, benefits and responsibilities - thus making it very difficult to introduce this dimension into the decision making process. The resultant planning vacuum should be not surprising.

The Urban Waterfront Action Group:, How it Grew and What it Recommended

1979 - A Benchmark year for coordination on the waterfront

There are good reasons why 1979 was a waterfront "watershed" year initiating better coordination of Federal programs to a limited extent. Regional representatives of a number of Federal agencies were reporting increasing requests for information on grant and assistance programs. Few agencies were equipped to deal with the numbers of requests; resources in the form of expert staff people and handouts were clearly inadequate. It is at that point that the "home office" (Washington) must come through with shortcuts. Targeted printed and audio-visual material needs to be developed and workshops sponsored. In this way hundreds of requests for information may be satisfied with the expenditure of only a few workmonths. It was increasingly obvious to Washington and regional offices of Federal agencies alike that an increased, better coordinated effort would be necessary if "output" of funding and services was to keep up with demand.

The HCRS waterfront study was a structured, deliberate attempt to discover why some waterfront revitalization projects succeeded, and to draw from that experience guidelines for others that might be failing and in need of extra help. The study also made certain recommendations (Volume I) for policy and program improvements. Volume II, which gave details of the case studies of 18 waterfront projects, was designed as technical assistance material. It describes funding sources for each project and lists leading participants by name and title. It is being used by project sponsors throughout the country to help improve grantsmanship and acquire ideas for such essential tasks as achieving and maintaining public support throughout the history of their projects.

The study identified three phases for all kinds of waterfront development: conceptualization, when there is recognition by some or several decisionmakers in a town or city that public benefits could be achieved from a developed or revitalized waterfront; planning, when a professional plan for waterfront development is under way and increasing, and, finally, the implementation phase when ground is acquired and construction begun. It is interesting to note that, although no two projects ever duplicate each other, common problems are shared regardless of the purpose (recreation alone or recreation and open space plus other uses) or size of the project.

Two separate efforts to improve coordination were operating during 1978-1979, which should be of interest to policy analysts and project planners alike. While on the one hand local officials were avidly seeking information from any source to help their separate projects and pressuring agencies for help; on the other, need for better coordination among Federal agencies with waterfront responsibilities was becoming more and more apparent. The authors of this article, both members of HCRS staff at the time, and Richard Gardner and Ann Cowey of the Office of Coastal Zone Management, undertook to form an ad hoc task force among these agencies late in 1978. The response on the part of the agencies and by four active and influential private sector groups (the National Trust for Historic Preservation, the National League of Cities, the United States Conference of Mayors and Partners for Livable Places) was purposeful and productive. Wise in the ways of interagency bodies of a more formal nature, which tend to meet once a year to decide when to meet the following year, or which labor long to produce unused and useless reports, representatives of the agencies and the private sector groups set about to be specific about what they would do and when they were going to do. By the simple device of having a kind of "roundrobin" memorandum of understanding signed at the bureau chief/assistant secretary level, actions by the group had official sanction and they could use resources (staff time, typing and printing only) toward two primary goals. One was to be of maximum help to waterfront project sponsors and their constituents by providing coordinated funding information and, whenever possible, coordinated funding, and by pooling technical information and making it separately or jointly available to all who asked. The second was to see over a year's time how coordination among programs might be achieved on a long run basis.

The members of the group, which became known as the Urban Waterfront Action Group (or UWAG), surprised themselves by reaching all of their goals, with only minor slippages past deadlines. Its first timely effort was the production and distribution of a directory of Federal programs which could help waterfront projects. Names and phone numbers had to be updated every few months but this has been an inexpensive, useful handout.

City representatives asked if they could come to UWAG meetings, as word of the group and its cooperative efforts got about. Since cities were willing to come at their own expense and since it seemed very desirable that Federal agency representatives learn more about waterfront development problems from them, several meetings with cities were held. Altogether 17 state and local agencies made presentations to UWAG. All in attendance learned <u>something</u> at each meeting: local planners about grant programs or technical assistance; Washington staffers about successful local strategies which could be applied elsewhere. There were problems, however, which led UWAG members to forego further presentations by cities and suggest other methods of assistance. For one thing, time is of the essence in all waterfront projects; the number of cities which could profit from individual or group sessions with Washington agencies was inevitably small. And then, most agencies are regionalized, with grant giving functions decentralized. It appeared that Washington was being asked to assume regional functions, a move that could delay rather than advance projects.

Conferences were also conducted by the Coastal Zone Laboratory of the University of Michigan and OCZM and by the Maritime Administration (Commerce) on ports and coastal management as they relate to urban waterfronts, among other topics. Technical assistance being developed includes a joint-HCRS-OCZM publication on various phases of waterfront development: management structures, permitting, and land acquisition among them. HCRS, the Corps of Engineers, the Fish and Wildlife Service and the Environmental Protection Agency are co-sponsoring a manual on planning for urban fishing and waterfront recreation. These publications are scheduled for publication in 1980.

UWAG Conclusions and Recommendations

UWAG, in accordance with the requirements of its self-imposed deadline under the informal memorandum of understanding, has produced a brief report, (36 pages) which records its progress and makes succinct recommendations for future policy coordination. "Adequate mechanisms to insure and facilitate coordination of Federal, State, local, and private programs for urban waterfront revitalization do not currently exist," the report states. The cities which made presentations to UWAG were uniformly critical: Federal programs were at cross purposes on occasion, if not all of the time. The availability of and help with the use of Federal funding programs for the express purpose of revitalizing urban waterfronts had never been clearly stated; and generally integrated information and technical assistance were lacking. The lack of coordination required, according to the report, the establishment of a "formal" UWAG. The group would consist of 22

Federal agencies, and work alongside several private sector groups. It would "encourage urban waterfront revitalization; coordinate related programs and services, facilitate an exchange of information, and...develop and deliver technical assistance."

The UWAG report does not provide a mechanism for creating the organization but specifically states that no one agency should have exclusive leadership. It suggested that chairpersonships be rotated among HUD, Commerce, and Interior, and that resources be found to support a permanent "secretariat" which would conduct a number of different functions. These functions would fall into three main categories:

(1) Assistance to projects would be improved and policy and program recommendations made to the Administration for this improvement. All relevant agencies were to evaluate and revise programs so that they would be more responsive to community waterfront revitalization needs.

(2) Joint funding and reviews of specific waterfront projects was to be facilitated.

(3) Information about public and private funding, permitting, planning and project design, public access provisions, economic development opportunities, enlisting and maintaining public support, et al, was to be provided to local communities on an as needed basis.

As noted in the UWAG report, regionalization of UWAG activities was considered to be very desirable. There was talk of the formation of RUWAGs (Regional Urban Waterfront Action Groups) and LUWAGs (Local, etc.) In fact, regional groups have formed in the Southeastern United States (centered in Atlanta, Ga.) and in California. There is no clear way to structure such groups in a uniform fashion, however. Thus, unlike the national group they need to remain ad hoc, responding to special situations. Generally speaking, the regional groups carry out the same goals as the national. The federal agencies and private sector groups get together to mesh programs and provide joint assistance. This can take the form of joint publications or it may include site visits by teams of officials. Committments must be a product of available resources.

A model for a local action group in several instances has included a one-day (or half-day) meeting of local public and private sector officials with Federal and state agencies, experienced project sponsors and consultants from other cities with successful projects, and national experts. The meeting can achieve several goals if properly prepared. It can arouse a local constituency for the waterfront effort. Local planners, in their preparation, are required to become clearer about concepts, their cost, and the possibility of funding them. It means that <u>all</u> interests are involved early on and that local property owners, for example, do not feel that a development is being planned behind their backs. The one day meeting would appear to be an ideal way for a community to move from the conceptualization to the planning stage, or perhaps from planning to development.

How Recommendations Could Be Implemented

The recommendations in the UWAG report represent the most basic steps which must be taken if Federal policy is to become more effective in supporting state and local waterfront goals and objectives. And if it is to ensure that national concerns, which often have a focus and scope beyond the immediate interests of state and local governments, are to be protected in cities waterfront revitalization plans and projects.

The criticism is often made, and often with justification, that Federal bureaucrats are not interested in change and that business as usual is their primary concern. The very existence of UWAG and its report are evidence that this is not always so. Concern, interest, recognition of shared responsibilities, and most importantly, good will - on the part of the directors and staff of the participating agencies have brought things to this point. UWAG has done what was stated it would do in the signed Memorandum of Understanding. However, the agreed upon period of time is now over.

And people, informal policies and a spirit of cooperation can change. While it is possible for UWAG to continue as it has to date, and while some needed Federal policy changes may and indeed probably will take place, there are no guarantees that the report's recommendations will be implemented. This would require action by the President and/or Congress. Each route has advantages and disadvantages. An initiative by either would accomplish a primary goal for all subsequent events; formal Federal recognition of the importance of urban waterfront revitalization and the need for improved Federal policy.

A Presidential initiative, e.g. policy statement or executive order, which formalizes UWAG and directs all Federal agencies to review their programs (1) to identify their present ability to respond and (2) to make those changes in existing rules and regulations which will permit necessary coordination and cooperation has the advantage of speed. Such a directive could be released immediately. The effectiveness of this course of action is well illustrated by Executive Order 11988 re floodplains and wetland.

This approach, however, while it mandates a review process and the allocation of existing resources to this effort, does not create new resources or authorities which may be needed. Only Congress through legislation and the budget process can do this. And this could take valuable time. During this period opportunities may and probably will continue to be lost.

Simultaneous action by both branches of government may be the optimal way to go: Presidential Executive Order based on the premises outlined above and legislation which formally creates UWAG and includes the necessary resources in a major program authorization. With these steps Federal policy will be moving towards meeting identified needs more effectively. And with a Federal mechanism firmly in place, the creation of counterparts at state, regional, and local levels can happen. Without this, experience with other Federal programs has shown that what is pronounced in Washington and written into programs does not always affect local decision making.

Events to date are encouraging and they show promise. But it would be unrealistic to become overly optimistic. Change takes time and the anticipated level of effort is not very great. It will be impossible for these first steps to meet all needs in the near future, let alone meet them in the most effective way. Additional Federal legislation creating new resources and authorities premised on both carrots and sticks may well be needed. And not just at the Federal level.

The role of state and local government in supporting Federal policy is critical. As the HCRS report states: "(Urban waterfront) goals will only be met by a clear commitment on the part of decision makers at all levels, but especially those in urban areas. Without this strong local role the complex problems associated with waterfront revitalization cannot be solved. Success will be determined by the clarity of policy, support provided for it by city leaders, and constant oversight on their part to ensure that programs and projects move to completion as planned."

WATERFRONT MASTER PLAN FOR KENOSHA, WISCONSIN

J. William Allen,*
Thomas W. Wood,* &
Ian J. McAlister,*

INTRODUCTION

The city of Kenosha (population 78,000) is in southeast Wisconsin on the west shore of Lake Michigan, sixty miles north of Chicago, as shown on Figure 1.

This paper presents the development of a master plan for the waterfront area of the city. The study was conducted by the authors in cooperation with officials of the city of Kenosha and the Wisconsin Coastal Zone Management office. A primary goal was identification of opportunities for future public and private development of the central city waterfront. Particular emphasis was placed on provision of additional marina facilities, improvements to existing port facilities, improved public access to the shore, and related water-oriented recreational and commercial activities.

As with most coastal cities, waterfront land is at a premium, with many interests competing for use of the limited resource. Resolution of these interests and integration of various facilities into a cohesive plan was a key task. The work was coordinated with an on-going U.S. Army Corps of Engineers' small boat harbor feasibility study.

An unusual feature of the Kenosha waterfront is the presence of a U.S. Army Corps of Engineers' diked dredged material disposal area adjacent to the main harbor channel. Analysis of potential uses for the water area and/or ultimately reclaimed land formed an integral part of this study.

The organization of this paper follows the major analytical steps undertaken. A brief overview of the regional character and site features is given, followed by an evaluation of the market factors affecting the demand for marina and port facilities. Development and evaluation of alternative plans is outlined, and the recommended plan presented. Financial and institutional aspects related to implementation of the proposed plan are discussed.

---

* Mr. Allen was formerly head of the Water & Waste Systems Planning Department at Stanley Consultants, Inc., and is now with the firm of Williams and Works, Inc. Mr. Wood and Mr. McAlister are principal economist and coastal engineer at Stanley Consultants, Inc.

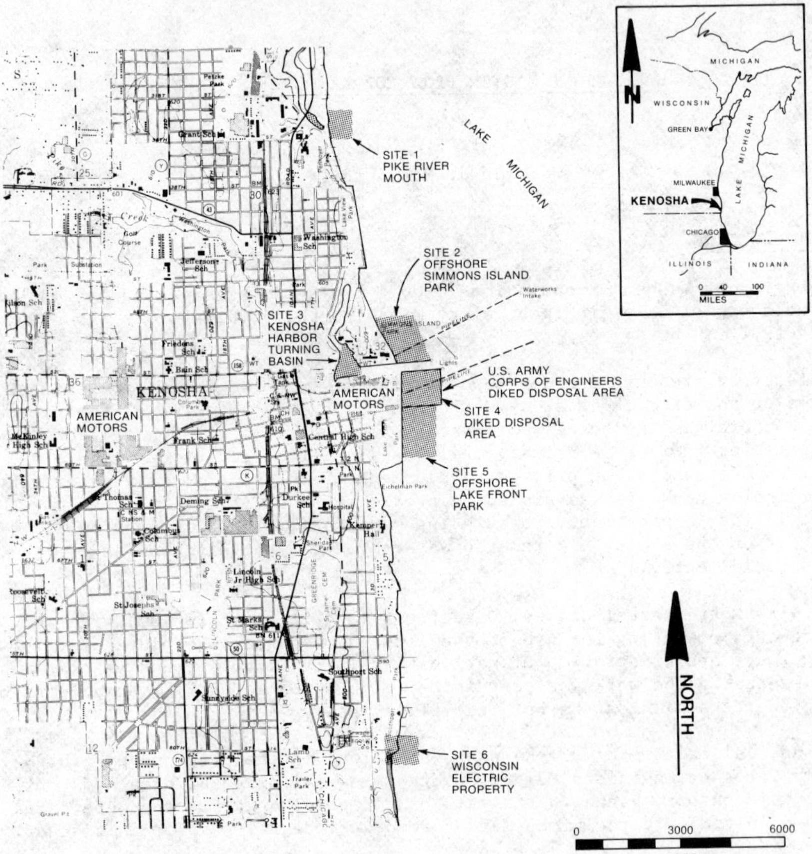

FIGURE 1 • KENOSHA HARBOR AREA LOCATION MAP AND POTENTIAL DEVELOPMENT SITES

SITE EVALUATION

Land Use - The study area for the analysis consisted of approximately 200 acres of waterfront land in the area adjacent to the existing harbor and central business district, as shown on Figure 2. Also shown are the major physical and institutional elements in the project area. It is apparent that the available land is almost completely utilized by a variety of industrial, commercial, residential, and recreational concerns. The generalized land use patterns for Kenosha are indicated on Figure 2. Although the uses are varied, all are well established and have coexisted for an extended period of time. No major change in use of the area is anticipated by local officials.

This development situation dictated that major waterfront improvements would be within the context of the existing use patterns, the lack of available vacant land providing a constraint to development opportunities. The emphasis of the study was therefore placed on expansion of existing uses, including:

1) Development of additional marina facilities and boat launch ramps.

2) Expansion of the port facilities.

3) Consideration of additional marina support facilities.

Kenosha Harbor - The dominant feature of the Kenosha waterfront is the harbor. The harbor has been formed by channelization and dredging of Pike Creek for the lower 1,150 feet of the river, a further 850 feet has been gained by construction of parallel jetties into Lake Michigan. Available navigation widths in the harbor channel vary from 150 to 170 feet. A turning basin with a radius in excess of 250 feet exists to the west of the harbor channel.

The harbor channel and basin are dredged to a depth of 25 feet below LWD. The north channel is maintained at 21 feet below LWD. From the mouth of the harbor to the eastward limit of the offshore breakwater, a 26-foot below LWD transition channel meets the 800-foot wide, 27-foot deep channel which extends 2,800 feet into the lake. An 800-foot long offshore breakwater lies 550 feet northeast of the north harbor jetty.

Diked Disposal Area - The U.S. Army Corps of Engineers built the disposal area in 1975 to contain and consolidate dredge spoils from Kenosha and Racine Harbors. The disposal area lies adjacent to the south pier, occupying approximately 32 acres, of which about 25 acres is available for the disposal of dredge material. The design capacity for dredgings in the structure was 750,000 cubic yards, to be filled over a 10-year design life.

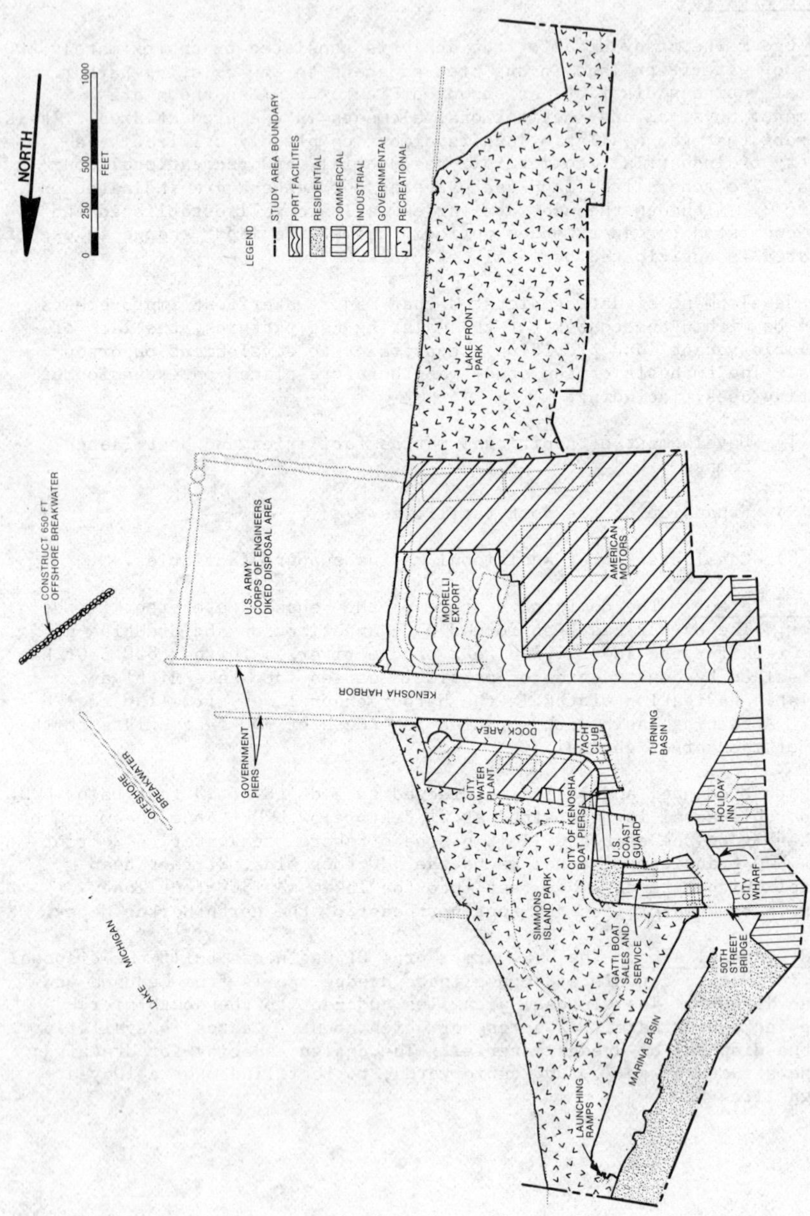

FIGURE 2 • KENOSHA HARBOR AREA EXISTING LAND USE AND MAJOR PROJECT ELEMENTS

Although the structure has reached 50 percent of its design life (five out of ten years in 1980), only 15 percent of the disposal area's capacity, or 110,000 cubic yards, has been filled. The U.S. Army Corps of Engineers has indicated that they may only require 300,000 to 400,000 cubic yards for disposal of dredged material. The reduced estimated dredge quantities result from material never being supplied to the disposal area from Racine Harbor, and reduced quantities being dredged from Kenosha Harbor. At the present fill rate, design holding capacity will not be reached for 30 to 35 years.

Since the diked area is not being used at its design rate, several options for development within the area were investigated.

Existing Marina Facilities - A marina for recreational power boats exists in the body of water north of the 50th Street bridge and west of Simmon's Island Park. The marina area is restricted to power boats because of the low clearance capability of the 50th Street bridge. The area has capacity for 142 boats at the city dock system and about 125 on private moorings. The body of water is approximately 1,500 feet long from north to south, and averages 200 feet wide. Water depths in the basin range from nine to two feet below LWD.

Other existing recreational boating facilities at Kenosha Harbor include the city wharf, a private boat sales and service facility, the city of Kenosha Boat Piers, and the Kenosha Yacht Club. Except for the city wharf, these facilities occupy the southwest end of Simmon's Island between the harbor channel and the 50th Street bridge. The location of these facilities is displayed on Figure 2.

Six boat launching ramps are located at the north end of the basin. Although set in an unfavorable position from a water and land aspect, the ramps experience heavy usage. From the land aspect, the ramps have limited access, poor traffic maneuverability, and inadequate parking capacity. Water-oriented problems include inadequate boat holding capacity, distance from the lake, and traffic movement through the recreational and port mooring areas.

Port Facilities - The port of Kenosha is a general cargo facility located roughly in the center of the harbor study area. This port facility is operated by a private operator without municipal support. Due to its location, facilities, and cargo mix, the port is classified as a "gateway" port by the Wisconsin Department of Business Development. Primary commodities handled through the port of Kenosha include food and kindred products, machinery, and motor vehicles. Current volumes are about 80,000 tons/year.

Facilities at the port include 131,000 feet of closed storage, 185,000 square feet of reefer storage, 19 acres of open storage, 2 container cranes, and 2 forklifts. In addition both truck and rail access are provided and the port has the use of a rail siding on the property. Rail and highway links to the hinterland are excellent. The major constraints to port expansion are market related (proximity of competing ports) and lack of space for additional operations.

Based on analysis of market factors and discussions with the port operators, potential cargo may be as high as 220,000 tons/year.

MARKET ANALYSIS

Analysis of potential markets for both water-oriented recreation and port activities was conducted in conjunction with the development of the Kenosha Harbor Master Plan.

Marina - The importance of reliable and reasonably accurate market or demand information is obvious for the development of either publicly or privately sponsored marina improvement or construction. Typical analytical practice for estimating or forecasting demand for marina activities involves consideration of existing activity levels, socioeconomic factors, and competing facilities. This sort of analysis was conducted in the development of the Kenosha Harbor Master Plan, and was supplemented with an additional measure of potential demand.

The location of Kenosha Harbor on Lake Michigan, near a rapidly growing high income area, makes it ideally suited for the presence and continued development of recreational boating interests. An initial step in marina market analysis was review of waiting lists for the limited number of power and sailboat slips at the city managed docks. This analysis indicated a fairly large market area including most of the northern Chicago suburbs and extending about 75 miles from the harbor.

As noted above, the method of quantifying potential demand for marina or other recreational boating facilities involves forecasting the number of boats owned (or berthed) within a certain area, based on socioeconomic characteristics. Prior to the initiation of the Master Plan Study, this sort of analysis had been undertaken by the U.S. Army Corps of Engineers in their 1974 Lake Michigan Regional Boating Survey & Analysis. A set of regression equations was used to predict the number of boats registered within Kenosha County, based on income, population, and other factors. The table below sets forth the estimates derived by the Corps using these techniques. These forecasts indicate roughly a doubling in the number of registrations within the county over a period of which approximates the normal design lifetime for marina facilities.

BOAT REGISTRATION FORECASTS - KENOSHA COUNTY

Year	Outboards 16'-25'	Outboards 25'+	Inboards	Sailboats 25'+	Total
1970	1,000	16	220	80	1,316
1980	1,512	29	430	581	2,552
1990	2,075	35	459	629	3,198
2020	3,710	50	496	761	5,017

Source: U.S. Army Corps of Engineers

A slightly more sophisticated method of estimation was also used in the same study. This involved allocating so-called "excess" demand among several competing marina facilities located on the western Lake Michigan shoreline. This analysis resulted in a forecast of 265 slips of "excess" demand for Kenosha and Racine Harbors by 1980.

It is noted that the "excess" demand formulation of demand estimation undertaken by the Corps, while more sophisticated (and probably more useful) than a simple boat registration forecast, is also done without explicit regard to the price at which marina services are offered.

In addition to these two fairly traditional measures of potential demand for marina activity, an additional concept was employed during the development of the Kenosha Harbor Master Plan to give an quantitative indication of probable demand for marina services. This method of estimating demand for marina services is a correctly formulated economic analysis in that it includes explicit consideration of price.

The ability to use this method of demand estimation depends on the fairly proximate coexistence of both private and public "markets" for marina services. As noted above, this was the case at Kenosha, with privately owned facilities accounting for about 125 slips in the existing marina basin. Based on discussions with those familiar with the local boating scene, it was reported that the average price per slip among private slip holders, was approximately $1,200 per year, with some slips renting for as high as $1,600 per year. The average annual charge at city docks is about $400 per year, with slips allocated by rationing.

While the quantitative significance of this information for "excess" demand at existing prices is not immediately obvious, this information did present a compelling informal case for the existence of strong demand for additional public marina services. As it turns out, the information can also be exploited in a more quantitative fashion if an independent estimate of demand elasticity is available.

Figure 3 presents an analysis indicating how a market equilibrium in the private sector and an independent estimate of demand elasticity may be compared to the existing (nonmarket) equilibrium at existing prices, to gain insight into the quantity of "excess" demand for slips that may exist at any price. It is noted that the demand elasticity indicated in this example is low, and thus the resulting estimate of "excess" demand at artificially low prices is also low. A more realistic demand elasticity would likely result in higher excess demand estimates.

Unfortunately, no reliable quantitative estimate of demand elasticity was available for use in estimation during the study. The information gained from this exercise, however, does indicate that existing private market prices and independent estimates of elasticity could furnish very accurate and theoretically correct methods of estimating "excess" demand for public marina facilities. These methods could easily be more efficient than traditional "number-of-boat" forecasting techniques, requiring less data and less analytical effort.

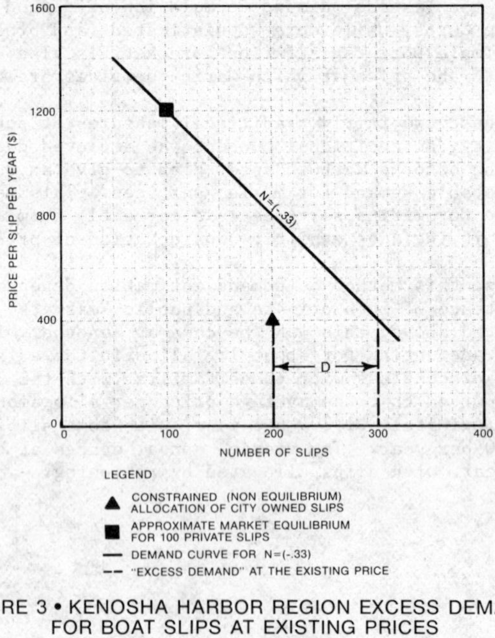

FIGURE 3 • KENOSHA HARBOR REGION EXCESS DEMAND FOR BOAT SLIPS AT EXISTING PRICES

Thus, traditional, informal, and somewhat innovative quantitative techniques all added to the conclusion that demand in the area was strong enough to warrant consideration of new marina facilities. Based on a combination of these analyses, and the judgement of these investigators, it was determined that at minimum, 300 additional slips could be supported at prices significantly higher than those charged at the existing city facility. In order to fully investigate the potential range of marinas possible at the site, several size ranges between 300 and 1,200 slips were chosen for preliminary engineering investigation. Analogous investigations led to similar conclusions regarding the demand for launch lanes, with estimates of from 6 to 12 additional lanes required.

<u>The Market for Port Services</u> - Although market analysis for port services was conducted on a less formal basis than that for marina services, qualitative results were found useful in the master planning process.

On a conceptual level, expansion of cargo volumes at the Kenosha Port could come from three basic sources:

1) Normal growth of businesses and consumer sectors within the port's market area.

2) Switching in transport modes by existing shippers within the market area.

3) Increase in market share due to competitive advantage vis-á-vis other ports.

The first of these potential sources of growth would normally provide for real increases averaging from two to three percent in cargo volumes and fees. Such growth will probably not manifest itself as smooth and steady, but will come in discrete steps. Both the recent historical record of cargo volumes at Wisconsin ports and a generally slow national economy indicated that growth of this type would be minimal in the near term.

The second potential source of market growth appeared to offer more prospect for significant increases in the near future, since water transportation is generally more energy efficient than rail or truck. As energy costs become a larger fraction of total transport costs, the incentive for switching modes increases.

The Kenosha regional economy includes several potential candidates for new port business of this type, and the larger market area includes many potential shippers in the food products and transportation equipment industries.

An increase in market share due to competitive advantage vis-á-vis other ports may be possible with improved facilities. As a general cargo port, Kenosha's competition is primarily at Milwaukee and Chicago. (Racine is a "limited cargo" port oriented to local shippers.) While effective competition with the larger ports of Chicago and Milwaukee for bulk commodity shipments would be difficult, competition for a greater share of the general cargo market could easily result in major increases in Kenosha volumes.

Based on an assessment of these factors, the study team and port operators estimated potential increases for various commodities as set forth in the table below.

CURRENT AND POTENTIAL CARGO VOLUMES,
PORT OF KENOSHA

Shipper	Present	Potential
USDA	12,000	24,000
Steel	0	100,000
Twine	4,000	12,000
Containers	0	3,000
General Cargo	13,000	16,000
Reefer Cargo	40,000	50,000
Vehicles/Tractors	7,000	13,000
TOTAL	76,000	218,000

The basic planning implications of this market analysis were:

1) That no additional technologies were central to a port improvements plan, and

2) That significant additional land would be required to fully meet market potential.

HYDRAULIC CONDITIONS

The most important hydraulic factors are waves, water levels, and ice. Detailed wave and water level analyses were conducted in the study. Only the resulting criteria used in the development of facilities will be presented here.

Design Lake Level	20-year frequency	581.1 feet
Design Wave Height	(for breakwaters)	10.5-11.5 feet from northeast

Ice is potentially an extremely destructive element in Great Lakes marinas. A thick (12 to 24 inches) covering of ice can be expected to form in the harbor and any proposed marina site each winter.

Ice cover itself does not represent a significant problem in the operation of a marina. The proposed dock system would possess sufficient strength to withstand ice forces. Numerous marinas in the Great Lakes region operate their dock system successfully in all year conditions. Sufficient structural-protection for the basin must eliminate exposure of the docks to moving ice.

Another factor in many Lake Michigan marinas is the presence of surge and seiche phenomena. Harbor surge normally takes one of three forms: short-period oscillations, long-period oscillations, or water level responses to climatological conditions. Short-period oscillations result when lake waves reflect off harbor structures and accumulate energy within the basin. Such conditions can be avoided by limiting use of vertical walled structures surrounding the basin perimeter. Seiches are long-period (10 to 60 minute) water level changes resulting from wind or pressure differences on the lake.

Local boaters and city officials have indicated that a surge/reflection problem exists in the turning basin area at the west end of Kenosha Harbor. Severe damage has been experienced by recreational boats moored in this area. Unloading of cargo ships has been difficult or impossible on occasion. The rough water problem in the turning basin is a direct result of waves directly entering the harbor channel from the east to southeast quadrant.

Careful attention to basin geometry and avoidance of vertical wall boundaries will eliminate serious harbor surge in the proposed Kenosha marina. Consideration has been given to measures to reduce surge in the existing harbor. However, the cost of additional breakwaters or wave absorbers does not appear to be justified by damages incurred.

DEVELOPMENT OF THE RECOMMENDED PLAN

Several alternatives were evaluated for additional marina and port operations, and related facilities. The following discussion addresses marina and port options separately, and then presents the overall plan.

The basic criteria for marina development at Kenosha are summarized below:

Item	Criteria at Kenosha
Marina Basin	Dredging; breakwaters, excavation, and landfill; diked disposal area.
Wave Protection	Breakwaters and revetments as required.
Provision of Launch Lanes	Six launch lanes.
Number of Boats	250 boat minimum, 900 boat maximum.
Size Distribution of Boats	Boat lengths, 18-50 feet.
Sailboats	25-40 percent, boat lengths 25-50 feet.
Power Boats	percent, boat lengths feet.
Type of Dock	ating piers-chain, winch, and deadman anchorage.
Utilities on Docks	Electricity and water-possibly telephone.
Fuel Dock, Sewage Pump-Out	To be provided.
Hoist or Crane	Provided near administration area.
Parking Facilities	3-4 acres, launch ramp and berthed car parking required.

In the site selection process, several sites in the Kenosha vicinity were evaluated based on the following criteria.

- Physical and hydraulic factors, as discussed previously
- Water area available, based on a general requirement for one acre of water space for each 20 boats
- Harbor entrance requirements, primarily channel dredging or breakwaters
- Harbor protection required to limit waves to less than one foot in the marina basin
- Economics of each of the above factors
- Environmental conditions, anticipated environmental impacts
- Land area available, one acre for each 100 boats

In the site selection stage, a total of six sites were initially considered potentially appropriate for marina and launch ramp facilities (see Figure 1). Preliminary evaluation dismissed sites 1, 2, 3, and 6 for the following reasons:

Site 1 - Environmentally sensitive area

Site 2 - Economics, negative impact on existing park

Site 3 - Conflict with commercial traffic, limited water area

Site 6 - Site availability, economics

Site 4, in the diked disposal area, and Site 5, immediately south of the diked area, were studied in detail. Three major alternatives were developed at these sites.

ALTERNATIVE 1 - Diked Disposal Area - As described above, the U.S. Army Corps of Engineers diked disposal area is not being filled as rapidly as anticipated. The city of Kenosha has requested the U.S. Army Corps to consider development of a portion of the site as a small boat marina. It was agreed that inclusion of such a plan would be considered in the study. Therefore, a layout was developed that would berth 278 boats in approximately 50 percent of the diked area as shown on the left portion of Figure 4. However, the lack of adequate nearby parking and limited land access required that an artificial island be constructed on the south end of the area. The estimated construction cost for this plan was $5.5 million (1980).

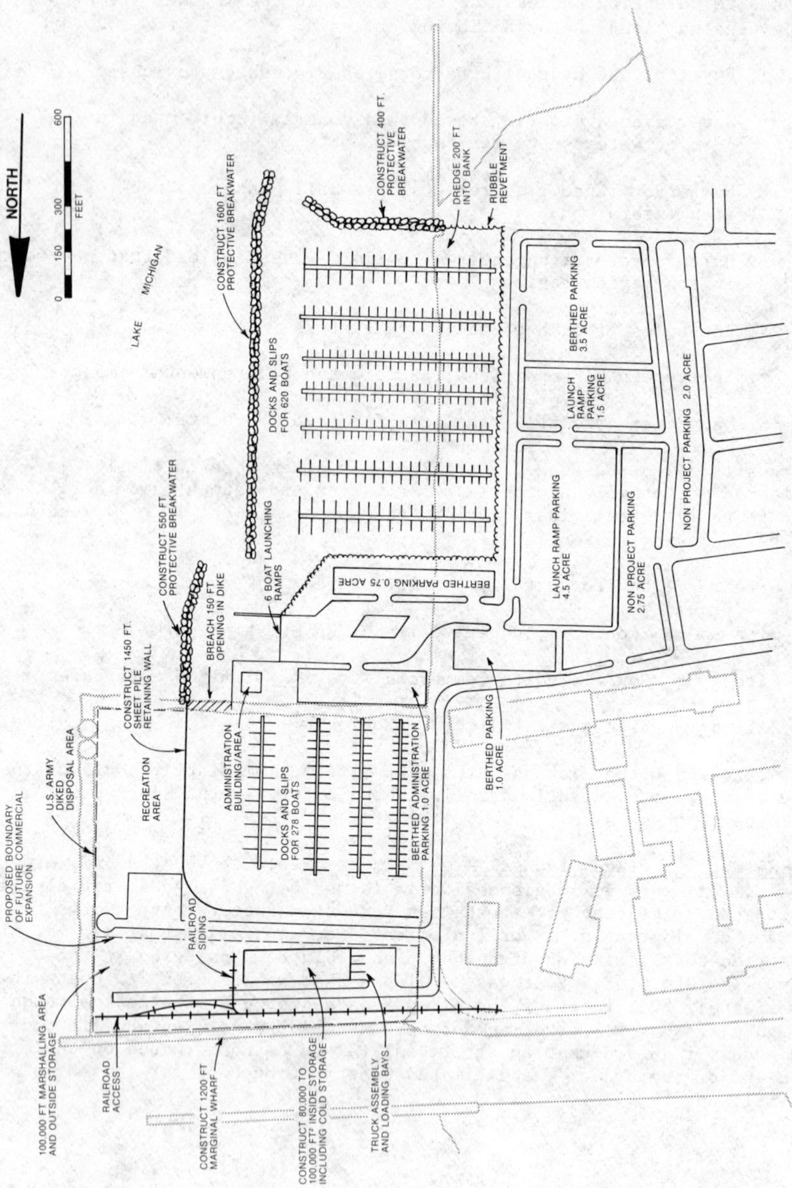

FIGURE 4 • KENOSHA HARBOR AREA MARINA AND PORT DEVELOPMENT PLAN

ALTERNATIVE 2 - Lakefront Park Site - This site is very attractive since the diked area to the north provides excellent wave protection from northerly storm waves. Several possible layouts were considered at this site.

Final selection of the most cost-effective plan involved consideration of breakwater costs, land area available, and marina size. It was determined that 600 boats represented an optimum size for this site. The land facilities required to support a marina of this size were located, and the marina basin was designed, with the aim of minimizing breakwater costs. An optimization analysis between Marina water surface area with the associated boating revenues and breakwater construction costs was conducted. The final layout, shown on the right side of Figure 4, provided slips for 620 boats with rubble mount breakwaters located in 11-13 feet of water, and included excavation 200 feet back from the shore. The estimated construction cost for alternative 2 is $8.4 million.

ALTERNATIVE 3 - Combined Development Plan - Elements of alternatives 1 and 2 were incorporated into an ultimate plan, shown on Figure 4. A total of 898 boats could be moored at this site. The estimated cost for this plan is $12.3 million.

Port Facilities Expansion - The market analysis indicates significant potential for expansion of port activity, primarily along the commodity classes now handled at the facility. Examination of the water and land facilities at the port indicates that the former pose few, if any, operational problems for the port. Channel and turning basin configuration are adequate for freighters of up to 600 feet, which include the majority of Great Lakes general cargo vessels. Larger vessels can be turned with tug assistance in the lake and backed into the shipping channel. Channel width is adequate for simultaneous docking at the north and south wharfs. Although some Great Lakes bulk carriers would have difficulty navigating at the facility, the market for bulk commodities is limited. In summary, major expansions of the channel, or turning basin do not seem warranted based on market potential.

It is apparent from inspection of the port's landside facilities and operating area that a shortage of space poses a serious obstacle to major expansion. The major operating area on the southern side of the channel is between the AMC plant, diked disposal area, and channel. The open and covered storage areas on the northern side are also constrained from expansion by the water plant and yacht club. Covered and cold storage space are also at capacity with existing cargo volumes.

Potential Improvements - Since lack of land is the constraining factor in port volumes (with the existing cargo mix) any major expansion will require creation of land in the existing diked disposal area.

Based on our marina alternatives, and the location of the port facilities, the logical portion of the diked disposal area for port use is a strip along the channel. Based on discussions with the operator, a development scheme was identified which would allow for expansion of dockage, covered storage, cold storage, and open storage areas. This development scheme, as indicated on the far right of Figure 4, would utilize about 9 acres of the approximately 26-acre interior area of the diked spoil disposal area.

This development would be consistent with any of our three marina alternatives, and would be most compatible with the implementation of alternative 1 or alternative 3 (via immediate construction of an interior retaining wall).

A preliminary construction cost estimate for these improvements indicates an anticipated cost of $6.3 million.

ALTERNATIVES EVALUATION

As noted above, the primary issue with regard to development of alternatives for Kenosha waterfront area revolved around the selection of an appropriate marina layout and design. The basic criteria for selection of a recommended alternative from the three developed during the course of the study were environmental, economic, and land use considerations.

With regard to the physical environment, all three of the alternatives developed were highly similar in their impact upon the ecology of the area. In short, none of these alternatives was expected to have a serious or significant negative impact on the biologic communities of the area. The only slight difference with regard to potential impacts on the physical environment is that any alternative which made use of the diked disposal area for marina purposes might have shortened its useful lifetime somewhat, and thus resulted in the ultimate need to construct another such spoil disposal area in the fairly immediate vicinity of the harbor. Given the magnitudes of spoil involved, and the typical lifetimes associated with such spoil disposal sites, this was not a major consideration.

It was expected early during the development of the three alternatives that any alternative utilizing the spoil disposal area for marina purposes would have a cost advantage generated by the fact that a "ready-made" protected harbor was offered by the disposal area containment laws. During the economic evaluation of alternatives, however, it became apparent that this cost advantage did not exist and that all of the three alternatives evaluated were roughly equivalent with regard to cost per boat. The first stage in this economic evaluation was the capital cost estimation for each of the three alternatives. After capital costs had been estimated, assumptions on federal and state participation (through the U.S. Army Corps of Engineers' Small Boat Harbor Program and the Wisconsin Department of Natural Resources) were formulated. Prospective participation in capital costs was quite closely matched among alternatives and varied from 56 to 59 percent.

A basic tenant of the city's feasibility criteria for development of a new marina facility was that any such facility would be completely financially self supporting, e.g., no general tax dollars would be required as a permanent subsidy for the development of facility. Thus, the analysis was geared to estimate revenue requirements on a self-sustaining basis, these revenue requirements to be deferred by user fees at the proposed marina facility. Using these assumptions, development time tables based on legislative and engineering constraints, and typical current costs of capital, the average cost per berthed boat was calculated for each of the three alternatives. Alternative 2 faired slightly better than the other alternatives under this criterion with $844 average annual cost per boat as opposed to $1,180 and $871 for alternatives 1 and 3, respectively.

Even given a slight cost advantage for alternative 2, the major factors leading to its selection as a preferred alternative for near term marina development were not economic but related to institutional constraints and land use compatibility issues. The institutional constraints arose from the significantly higher degree of uncertainty faced by the city if it had opted to pursue a marina alternative using the diked spoil disposal area. Given the fact that the spoil disposal area was constructed as a federal project under strict guidelines with regard to its future uses, any diversion of this facility for use as a marina would have apparently required a lengthy and fairly extensive series of petitions and other legal manueverings. Even the exact procedure for seeking federal permission for such use was uncertain at the time our study was concluded.

This institutional uncertainty was combined with the concern with potential traffic and parking problems on the downtown area. Based on these considerations and on the apparent cost effectiveness of alternative 2, it was selected as the recommended marina development alternative.

Three detailed cash flow and financing schemes were developed. Based on alternative borrowing rates for short- and long-term issues and on policy alternatives for ratemaking in the period prior to completion of the new facility. These analyses indicated some short-term sensitivity to current rate policy, and (as expected) a high degree of sensitivity to interest rates, with high revenue requirements/interest rate "elasticity" of .73.

Port Development - In the light of the results of market and site analyses, all three marina development options were designed to permit the use of the northern half of the diked disposal area for port use. Marina alternatives 1 and 3, which provide for construction of an interior wall in the diked area, would partition off about 10 acres of land for this purpose. Alternative 2 would allow the entire area to fill, and would also allow ultimate use of about 10 acres for port functions.

A long-term improvements plan for this area was developed, including construction of wharf, rail siding and apron, and a building housing additional covered and cold storage as indicated on Figure 4.

Our preliminary estimate of construction costs for the new storage facilities, dockage, and cargo handling systems is over $6 million. Based on a 15-year borrowing at a typical corporate rate of 15 percent, this investment would be required to generate a little over $1,000,000 per year in net operating revenue and tax savings in order to be economically viable.

Thus the recommendation to pursue major new investments in port facilities was not made as a short-term measure. The optimal port development scheme evolved instead as a matter of using created land for additional marshalling and open storage without major new facilities until better and more clearly defined economics can be demonstrated.

SUMMARY

Both methodological and policy-oriented lessons were drawn from the planning process at Kenosha. On the methodological side, our experience suggests the following:

- An efficient marina location and preliminary feasibility study should concentrate on rapid elimination of most sites. In today's environmental and economic climate, most candidate sites will have readily identifiable flaws and can be dropped from further study without impairing the quality of results.

- The traditional method of protecting measures of demand for marina (or other waterfront recreation) services may be supplanted by more theoretically correct and less time-consuming methods if both a private market price and demand elasticity estimates are available.

- Where marinas are to be designed for combined dredge and breakwater construction. A fairly simple technique will optimize the location of breakwaters and the quantity of dredging.

Among the factual observations and policy implications, the most central included:

- Even with high levels of federal and state support, construction of new marinas along unprotected shorelines will require much higher user fees than currently paid at older facilities.

- The apparent economic advantage of converting a diked disposal area into a marina basin can disappear when costs for interior bulkheading and integration with land side facilities are accounted for.

- Even if the economics of conversion of such a facility were convincing, the institutional inertia that must be overcome is significant.

- Due to the large and overlapping market areas for coastal marinas in urban areas and the attendent uncertainties and risks in independent marina development, a continuation of federal and state support is indicated.

- Even though it may be difficult to justify major new investments in smaller port facilities, communities can seek to preserve and enhance the important economic function of port activities by fairly simple measures such as providing storage areas.

- In a multiple use environment where many land uses (that have often been labeled incompatible) existed and functioned within feet of each other, little "conflict" was apparent, and few comments indicated that future recreational development in an industrial/commercial setting would not be desirable.

Regional Planning for the Urban Waterfront

John R. Weingart

New Jersey's Hudson River waterfront may be better known to residents of New York who can see it from across the water than it is to New Jersey's residents who are generally separated from it by roads, railroad tracks and fences. This paper briefly describes this waterfront and the work of the advisory Hudson River Waterfront Study, Planning and Development Commission which has recently completed its final recommendations and report. The second half of the paper presents lessons drawn from the work of the Commission which can apply to future efforts to manage waterfront redevelopment in New Jersey as well as in other urban areas.

The waterfront on the New Jersey side of the Hudson River should be magnificient. It has a direct view of the New York skyline and access to Manhattan by two tunnels, a bridge, and a modern, in expensive subway line. It is adjacent to parts of one of the world's largest ports which, unlike the ports across the river, anticipated the change from break-bulk shipping to containerization, and modernized accordingly. It includes the unusual cliffs of the Palisades, parts of which are preserved in an interstate park and it is within eleven municipalities which need the jobs, housing and parks the waterfront could provide.

Instead of being magnificient, however, New Jersey's Hudson River waterfront is the site of abandoned railroad lines, warehouses and factories left over from the early part of this century when many of the people and much of the goods headed for New York arrived on trains which terminated in New Jersey and then were transferred to boats to cross the river.

The waterfront in New Jersey is at two levels with the Palisades on top, particularly in the northern part of the waterfront, forming a natural barrier between the water's edge and the land further west. The cliff face of the Palisade is largely wooded and undeveloped while the top of the cliffs house a number of high rise apartments buildings, some residential neighborhoods, scattered small parks and vacant land.

Below the Palisades, the waterfront is the scene of underutilized land, with tracks and building abandoned by once successful railroads and industries. Further south, the waterfront is wider with the New Jersey Turnpike and abandoned railroad lines separating residential areas from the water. The Hudson River's southern point in New Jersey is in Bayonne where industrial facilities and oil storage tanks are the dominant waterfront use.

John R. Weingart is Chief of the Bureau of Coastal Planning and Development in the New Jersey Department of Environmental Protection, and served as staff director to the Hudson River Waterfront Study, Planning and Development Commission.

Along this seventeen mile underutilized, theoretically prime piece of real estate, the only major development activity in the last ten years has been the creation of Liberty State Park, a wonderful park opposite the Statue of Liberty and Ellis Island which will, when completed, cover 800 acres. The only other waterfront development in the last decade has been a hospital, an apartment building and a complex which includes a waterfront restaurant, racket ball club and movie theater.

In 1979, Governor Brendan Byrne established the Hudson River Waterfront Study, Planning and Development Commission to "conduct a thorough study and investigation of the various alternatives for the planning and redevelopment of the Hudson River Waterfront south of the George Washington Bridge." (Executive Order No. 69; January 11, 1979) The Governor appointed thirty-nine people to this Commission including the mayors of each municipality potentially affected by its recommendations, four county freeholders, four members of the State Legislature, Commissioners of three state departments, a representative of the Port Authority of New York and New Jersey, a representative of a labor union, and twelve New Jersey residents, five of whom lived or worked near the Hudson River. As chairman, he appointed Kenneth D. McPherson, an attorney from Jersey City, and he designated the New Jersey Department of Environmental Protection to serve as staff to the Commission.

The Commission held its first meeting in September 1979 and submitted its final report to the Governor this past September. The Commission did not have a unique mission. State, county and municipal agencies, as well as both public and private regional agencies, had studied the Hudson River waterfront over the past fifteen years and had written at least fifteen major reports which address the future of the area. The Commission quickly found that most of these studies had reached a similar conclusion; that is, that the waterfront had tremendous unrealized potential for a variety of types of development which all should provide public access to the waterfront. One report in particular, The Lower Hudson prepared in 1966 by the Regional Plan Association, a private group based in New York, articulated a fairly specific plan for the Hudson River waterfront and Palisades which could be applied today with little modification.

The Commission focused its efforts, therefore, on examining the existing decision-making process along the waterfront to determine why the past plans and suggestions had never been adopted. The process it found, which is similar to that found at other waterfronts, was that to develop a piece of land along the waterfront, one must receive approval from the local municipal government, sometimes the county government, several divisions of at least one state agency, and, in many cases, at least one federal agency. In addition, the Commission identified a large number of other agencies at all levels of government and in the private sector which provide often essential technical or financial resources for waterfront redevelopment.

While the system has certain advantages in terms of adhering to New Jersey's respect for home rule and for the separation between its 567 municipalities, it was clearly not leading to the redevelopment of the waterfront. First, the master plans of the individual municipalities are not formally coordinated or linked so that one municipality can now pursue and develop a waterfront project considered undesirable by a neighboring municipality. In addition, several of the Hudson River municipalities either do not have master plans for the waterfront or have plans that are too general or too outdated to be useful.

Second, the waterfronts of the nine municipalities share problems and opportunities which are often distinct from the issues raised in the other sections of the municipalities. The result is unnecessary repetition of some time-consuming activities such as identifying federal funding sources for the waterfront, piecemeal uncoordinated approaches to to other tasks such as seeking to attract potential developers to the Hudson River waterfront, and the virtual neglect of other activities including exploration of the possibility of attracting new industries, such as fish processing, to the area and the feasibility of new or improved forms of transportation both to and within the area. These latter tasks cannot easily be pursued at the municipal level.

Third, the existing system also discourages construction of waterfront parks, since, under New Jersey tax laws, one municipality which builds a park that provides benefit to the entire region, receives nothing in return from the other municipalties. Similarly, municipalities can often suffer the impact of commercial or industrial facilities which produce pollution and traffic congestion for the region while providing direct economic benefit only to the municipality in which it is located.

Fourth, many of the municipalities do not have adequate staff to plan for the waterfront. Of the nine municipalities along the Hudson River, only two have a planner whose responsibilities include waterfront activities, while five hire consultants to do planning related work and two have no planning staff.

Last, and perhaps most importantly the current system presents a complex and often forbidding bureaucratic network to a developer seeking a waterfront site or seeking to predict whether the necessary permits and funding can be secured for a particular project.

While the members of the Commission accepted many of these points and agreed that a regional approach to waterfront redevelopment was needed, they differed over the specific appropriate response. The Commission gave serious consideration to two alternate proposals for permanent regional agencies. Both would have established a regional master plan and provided a full time staff to plan and promote waterfront development, but in one case the plan would have been advisory to the municipalities, while in the other the regional agency would have issued a permit for proposals consistent with the master plan which would have substituted for the presently required municipal approvals and for receipt of a State Waterfront Development Permit issued by the Department of Environmental Protection.

After many months of debate, several proposed compromises emerged and the Commission eventually agreed to recommend a permanent regional waterfront agency with limited powers. The recommended agency would create a master plan for the waterfront, but new development would require approval from both the agency and the affected municipality. This recommended agency, therefore, would have power to block projects inconsistent with the regional plan it developed, but it would not have power to authorize any project opposed by a municipality except for a waterfront park or path proposal. This recommendation must now be considered by the Governor and New Jersey Legislature.

Now that the Hudson River Waterfront Commission has submitted its recommendations and report to Governor Byrne, New Jersey is perhaps one third of the way towards changing the waterfront. The Commission increased the visibility of the waterfront and of its potential, and it has made recommendations which now require serious study, modification and action to prevent its report from simply being added to the regional planning shelves of libraries. Certain general observations and lessons emerged, however, from this Commission's efforts which can enhance the success of future regional efforts to make greater use of urban waterfronts.

<u>First, many groups care about the waterfront</u>. A large number of groups and individuals have a stake in the future of the urban waterfront. This includes nearby and potential residents and businesses, national developers, foundations, and a plethora of government agencies at the municipal, county, regional, state and national levels. New Jersey's Hudson River waterfront and Palisades

area, for example, include eleven municipalities and two counties while the jurisdiction of the San Francisco Bay Conservation and Development Commission covers thirty municipalities in nine counties.

This widespread interest is a resource in that it can provide diverse information, expertise and opinions for consideration in the redevelopment of the waterfront. The agencies and interest groups must have opportunities throughout the regional planning and development processes to have their views heard and seriously considered. It is, however, impossible to have them all included as part of any type of productive regional decision making entity. New Jersey's Hudson River Study Commission tried by having thirty-nine members, with the result that not only was the Commission almost unworkably large, but key interest groups still felt excluded, including local environmental groups, established regional agencies, and current and potential business people and developers.

A regional decision-making commission should be small so that it can work efficiently with its members each taking their responsibilities seriously. It should, however, adopt a process which encourages full participation by all people potentially affected by the waterfront decisions. Such a process can be specified in the legislation or rules defining the agency and it can, in some cases, be promoted through the establishment of advisory committees to provide regular input and feedback from groups with particular areas of knowledge and interest such as developers or neighborhood residents. Meaningful public participation, however, can be assured only by the attitudes of the individual decision makers.

<u>Second, most people have similar visions of what should be located on the urban waterfront</u>. The Hudson River Waterfront Study Commission heard long, sometimes acrimonious, debates both at its public hearings and in its own meetings. Yet if any single member of the Commission or of the public who testified before the Commission were to suddenly be given absolute power and unlimited resources to create the waterfront of their dreams, I am convinced the results would please most of the others. Unfortunately, such super-human powers were not located, so that debate had to focus on how to accomplish the change.

The Hudson River Waterfront Study Commission adopted 23 recommendations which it submitted to Governor Byrne. The Commission devoted approximately 20 percent of its time to agreeing to the 22 policy recommendations which concerned issues including public access and open space, environmental resources, development on the

Palisades, industrial and port development, transportation, energy facilities, wastewater treatment, solid waste, taxes and mixed use development. The twenty-third recommendation, which concerned how the 22 policy recommendations would be carried out, occupied the remaining 80 percent of the Commission's debate. For the local press incidentally, which enthusiastically covered the work of the Commission, the policy recommendations seemed even less controversial and received almost no attention.

The waterfront of the future suggested by these policy recommendation would combine the often sited successes of urban waterfronts in other states with the specific resources available along the Hudson River to produce a variety of residential, recreational, commercial, light industrial and port-related structures which would all be linked by a public waterfront pathway made as continuous as possible for the seventeen miles from the George Washington Bridge to Bayonne. Clearly, a site-specific formulation of this vision would create much room for debate, but the general areas of agreement should provide with great advantage for a urban waterfront planning over the planning for areas where totally different long range interests shape the public debate, such as Alaska's undeveloped areas, barrier islands, or the New Jersey Pinelands.

Third, specific fears raised by proposals for a regional agency should be directly addressed. While some people oppose a regional approach to the urban waterfront for reasons of philosophy or of personal political situations, most opposition along the Hudson River seemed based on specific fears of the effects of such an approach to existing building and activities and of the ability of a regional agency to impose an energy facility on a community against its will.

Proponents of a regional waterfront agency must pinpoint these and other fears which inhibit support for such an approach and address them directly. Virtually all new governmental land use regulations respect existing active buildings and operations by grandfather clauses or by simply having regulatory authority only over proposals for new buildings. Such an approach would also be appropriate along the Hudson River waterfront and proposals for a waterfront agency can gain more local support by recognizing this explicitly.

The fear that energy facilities will be imposed on a municipality is a more difficult issue, and one which should also be addressed directly but separately from decision-making for other types of facilities. It is unreasonable to expect that a regional agency for one stretch of urban waterfront would be able to solve this problem which has continued to confound the United States Congress and a number of State Legislatures, including New Jersey's. By designing a distinct regulatory process for energy facilities as opposed to other land uses, opposition to a regional agency could be diminished. The location of energy facilities could continue to require both the municipal and state approvals they now generally require in New Jersey, for example, while all other activities could be regulated by a new regional agency. This would leave questions regarding the national need to locate certain facilities and questions of state or national override and mobilization boards to be resolved by Congress, without allowing that one issue to thwart the regional planning and redevelopment of the waterfront.

<u>Fourth, the waterfront includes a large amount of developable land.</u> This point applies specifically to New Jersey's side of the Hudson River, but it applies as well as to waterfronts in other large urban areas. Some of the developable waterfront land would require extensive site preparation and some of it is suited to only limited types of activity, but the notion that has sometimes been expressed by a range of organizations from port authorities to environmental groups that the entire waterfront must be preserved for their favorite activity does not appear necessary in New Jersey. This has been clearly demonstrated in Baltimore and other cities where mixed use development has replaced some port facilities which have moved from the inner city to the outer harbor where greater space is available. In fact, the recently completed New Jersey Coastal Management Program revised a proposed policy that would have permitted only water dependent development along the urban waterfront to allow a wide variety of activities that would not pre-empt use of the waterfront for particular water dependent facilities and would benefit from a waterfront location.

This is not to say that the waterfront should not be considered a finite resource, and planned and developed as a region with unique attributes. The range of allowable waterfront activities must be more limited by environmental and social concerns than it has been historically, but it need not be as narrow as it has often been considered in theoretical regional plans.

<u>Fifth, a series of public-private partnerships is essential to redevelop the waterfront in the forseeable future</u>. Government must protect the public interest in use of the waterfront, but it lacks the capital to serve as an independent developer or landlord, while private developers need governmental assistance in the form of regulatory approvals and often financial grants. The goal should be to design governmental rules, policies and procedures that are clearly understood, widely publicized and consistently enforced so that private developers can efficiently apply their skills to implementing public policy on the waterfront.

Conclusion

A regional waterfront agency will not be a success if it contains sufficient checks and balances on its powers to maintain the status quo. Such an agency for a wilderness area might well be desirable, but establishment of such an area for the urban waterfront would probably cause more harm than good by merely creating an additional bureaucratic forum.

A regional waterfront agency should be able to create desirable change happen as well as being able to stop destructive development. It should be established to the maximum extent possible as the single, or at least central, agency responsible for the waterfront. Federal, state and local plans and polcies which apply to the waterfront should be incorporated into the rules, policies and decisions of the agency. The agency should create a regional plan to serve as a basis for future decisions through a process which allows all public views to be heard and explored. It should then, however, be able to make decisions on individual development proposals through a process which again provides for consideration of input from the public, but which is focused on compliance with the previously adopted regional plan. Battles over the powers of a regional agency that are fought and decided when the regional agency is established should not be refought as the agency adopts a regional plan, and similarly, debates over the future general character of the waterfront that are addressed and resolved when the agency adopts or amends a regional plan should not be reopened each time an individual development proposal is reviewed. Instead, these proposals should be reviewed on the basis of their compliance with the previously adopted regional plan.

This is by no means to suggest that a regional plan should be adopted and never respond to new information or attitudes, nor is it to suggest that any individual or group should be given the type of unlimited, unresponsive power that the railroads exerted on New Jersey's Hudson River waterfront or that Robert Moses held on the other side of the river. Instead, desirable public safeguards should be identified and a waterfront planning and regulatory process should be designed which is as simple as possible given the need to incorporate these safeguards, rather then incorporating complicated checks and balances into the approval system.

Such an approach to decision making for the urban waterfront could take many forms. If it was successful, people could always argue that the redevelopment would have occured anyway; Its failure, however, would be obvious for the waterfront would look just the way it does today.

PREPARING FOR OIL SPILLS IN THE DELAWARE RIVER ESTUARY

June Lindstedt-Siva
Senior Science Advisor
Atlantic Richfield Co.
515 South Flower Street
Los Angeles, CA 90071
(213) 486-0741

ABSTRACT

Spill response resources (time, people, equipment) will usually be limited during a spill event. Therefore, response goals and priorities should be developed before a spill occurs so that available resources can be most effectively used. Except where life and limb are threatened, the primary goal of spill response should be to minimize the ecological impacts of oil spills. To accomplish this, a two-part approach is necessary: (1) identify biologically sensitive areas (sites most vulnerable to the ecological impacts of spills) and develop strategies to protect them, and (2) develop guidelines for minimum-impact cleanup for all types of environments within the jurisdiction of the response plan.

Fifty-four biologically sensitive areas have been identified in the Delaware River Estuary. Methods to protect them are being developed by two oil cleanup cooperatives in cooperation with the Coast Guard and other agencies.

Once a spill has occurred cleanup methods should be chosen not primarily to remove visible oil but to minimize the spill's ecological impacts. Low impact, rather than high-impact clean-up methods should be used and the "no-cleanup" option should always be considered.

The participation and guidance of ecologists is important during development of spill response plans as well as during the spill event. Ideally, groups responsible for spill response from government and the private sector should agree on basic spill response goals, priorities and actions before a spill occurs.

INTRODUCTION

In spite of increased efforts to prevent them, oil spills continue to occur worldwide. As long as oil is produced, transported or handled, the risk of spills will never be reduced to zero. Therefore, spill preparedness becomes as important as spill prevention, even though the spill risk at any given location or facility might be very low.

Resources available to respond to oil spills, i.e., trained people, equipment, time, will almost always be limited during a spill emergency. Therefore, response goals and priorities should be developed before a spill occurs so that optimum use can be made of available resources. I have proposed that except where life and limb are threatened, the primary goal of spill response should be to minimize the ecological impacts of oil spills (1, 2, 3). Spill response goals and priorities are as follows:

1. Protect life and limb
2. Minimize ecological impact
3. Minimize socioeconomic impact

These goals and priorities have been supported by the U.S. Coast Guard (4) and by the Council on Environmental Quality in the National Contingency Plan (5).

Past oil spill response has most often emphasized removing visible oil rather than minimizing ecological impact. Consequently, cleanup methods have been commonly used which can increase rather than decrease the ecological impacts of a spill. For example, heavy equipment was used to remove oil from Ile Grande marsh during the Amoco Cadiz spill. Long and Vandermuelen (6) surveyed the marsh and predicted that recovery to prespill conditions will take a century.

To achieve the response goals and priorities stated, a two-part approach is used: (1) identify biologically sensitive areas (sites most vulnerable to the ecological impacts of oil spills) and develop strategies to protect them, and (2) develop guidelines for low-impact oil spill cleanup for all types of environments within the jurisdiction of the response plan.

Protection is the best solution. If oil can be kept from contaminating a sensitive area, both its impacts and subsequent cleanup impacts are prevented. It, therefore, becomes critical to identify sensitive areas before a spill occurs so that protective countermeasures can be planned and practiced.

If oil does contaminate shorelines, decisions regarding cleanup must be made. Cleanup should be directed toward minimizing the impacts of the spill rather than removing visible oil for cosmetic reasons. Many decisions regarding cleanup methods can be made during planning rather than waiting until a spill occurs.

Seven U.S. oil cleanup cooperatives on the east and west coasts and Alaska are implementing this type of goal-directed spill response planning, with input and cooperation from state and federal government agencies. Delaware River Cooperative and Delaware Bay Cooperative are two of these.

This paper discusses spill response planning for the Delaware River Estuary as a case study.

BIOLOGICALLY SENSITIVE AREAS

Because estuaries have major ecological importance, the entire Delaware River Estuary could be termed biologically sensitive, but this is not useful for planning purposes. In an emergency it will never be possible to protect an entire coastline, but with planning and practice, it may be possible to protect some parts of it. Biologically sensitive areas identified have one or more of the following characteristics (7, 8):

High biological productivity. A highly productive habitat is often a source for repopulation of surrounding areas. If such an area is disrupted, a disproportionate part of the area's population could be affected. Marshes are examples of such highly productive habitats.

High ecological significance. If a habitat has particular food-chain importance, acts as a spawning or nursery area, its disruption would affect not only its residents but its users as well; e.g., mud flats are used as feeding areas by birds.

Unique features or uses. If an area has unique features or uses - for example, a habitat of threatened, rare or endanged species - its disruption could significantly affect this uniqueness.

Vulnerability to oil pollution. Some habitats are especially vulnerable to impact from oil spills and/or spill cleanup activities. Low-energy environments like wetlands are vulnerable because oil easily collects in the complex waterways and is very difficult to remove without causing additional ecological impacts.

MINIMUM IMPACT CLEANUP

When oil enters nearshore waters and is deposited on shorelines, decisions must be made about whether to remove the oil and which methods to use. Just as the effects of oil vary with the different environments impacted, so the effects of clean-up will vary as well. Cleanup methods should be chosen for specific environments after considering the ecological effects of the spilled oil combined with those of various cleanup methods. Wherever possible, cleanup methods should be chosen that will minimize the ecological impacts of the oil spill, not add to them. The "no-cleanup" alternative should always be considered.

The Society of Petroleum Industry Biologists formed a task force which reviewed oil spill cleanup technology and recommended low-impact methods for several different habitat types (9).

There may be occasions, such as the cleanup of public amenity beaches, where visible oil must be removed. Every attempt should be made to use low-impact methods if these will remove oil as efficiently as more commonly used high impact methods.

ENVIRONMENTS OF THE DELAWARE RIVER ESTUARY

The Delaware River and its tributaries drain nearly 13,000 square miles of the northeastern U.S., including parts of New York, Pennsylvania, Maryland and New Jersey. The Delaware arises on the western slopes of the Catskill mountains in New York. The lower region of the river basin, below Trenton, NJ, is the estuarine portion of the river. The study covers the entire estuarine portion of the river, including Delaware Bay (Fig. 1). These boundaries were chosen because they represent the maximum up and down river movement of spilled oil within the system. Potential spill sources include refineries in the Philadelphia area, terminals at several locations in the upper estuary and lightering operations in Delaware Bay near Lewes.

Estuaries are complex zones of transition between freshwater and marine environments. Plants and animals living in estuaries must be tolerant of changes in salinity, temperature, turbidity and other physical parameters that change with the tidal cycle. The range of tolerance of a species for these variables will determine its position within the estuary. For example, oysters can live in waters of lower salinity than can their major predator, the oyster drill. Consequently, oysters are most abundant in the estuary at salinities between their lower limits of tolerance and that of the oyster drill.

In spite of previous and continuing environmental stresses, the Delaware River Estuary, particularly the lower estuary, is a productive system. It is an important spawning and nursery area for more than 60 fish species (10) in addition to many invertebrates. Clams, oysters, crabs and shrimp are abundant. These, plus rich planktonic and benthic communities, support large populations of birds. Wetlands provide important stopover points on the Atlantic Flyway for migrating birds and as well as nesting, feeding and overwintering areas.

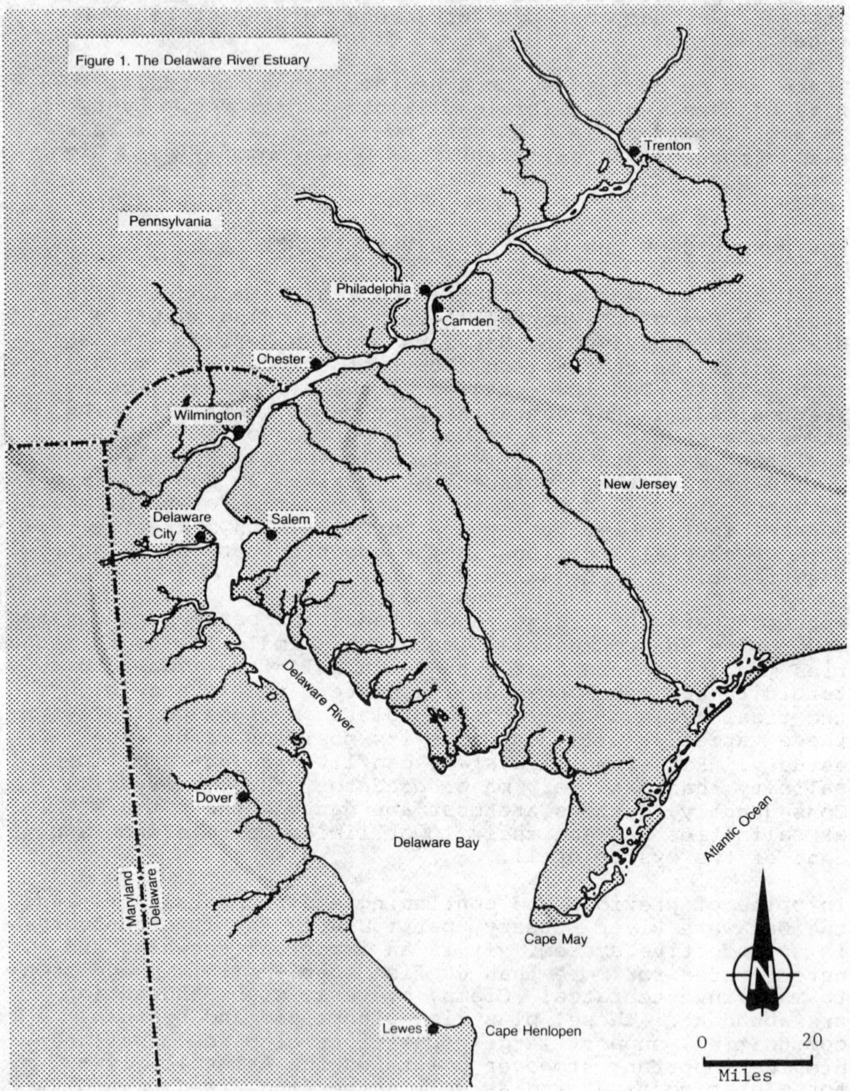

Figure 1. The Delaware River Estuary

Marshes bordering the estuary represent nutrients stored in the form of biomass. It is significant that they release these nutrients to surrounding ecosystems at a fairly steady rate, thereby preventing pulses (11). Estuaries, then, are not only important in themselves but also as they contribute to marine systems.

IDENTIFIED BIOLOGICALLY SENSITIVE AREAS - PROTECTION AND/OR CLEANUP

Fifty four biologically sensitive areas were identified in the Delaware River Estuary (7, 8). For the most part, I have identified the major routes by which spilled oil could enter wetlands systems. Most identified areas are openings of creeks, rivers or other water channels. If these openings could be protected during a spill, oil might be prevented from entering the larger and more complex systems with which they communicate and would be confined instead to the outer fringe of marshes or barrier beaches.

Other biologically sensitive areas include both fresh water and salt marshes, lagoons, tidal flats used as bird feeding, bird nesting or gathering areas.

Identified sites were rated according to their ecological sensitivity and likelihood of impact by spilled oil. Based on these ratings, each was assigned a response priority (7, 8).

Figures 2, 3 and 4 show the locations of identified biologically sensitive areas in the upper and lower Delaware River Estuary, respectively. For each area, protective methods and strategies have been suggested (7,8) and are being developed further by the spill cleanup cooperatives. Delaware River Cooperative, based in Philadelphia, has responsibilities for the upper estuary, and Delaware Bay Cooperative, based in Lewes, Delaware, will respond to spills in the lower estuary. These two cooperatives are planning a merger in the near future.

Marshes

Important freshwater marshes are located just south of Trenton in the upper estuary (No. 2, Fig. 1). They are characterized by high species diversity of vegetation and high use by wildlife.

Salt marshes and mixed marshes are found at various points along the river from Trenton down to the capes. Some of these marshes are heavily influenced by human activity as they are located in the midst of large population centers

Figure 2. The Delaware River: Trenton to Philadelphia*

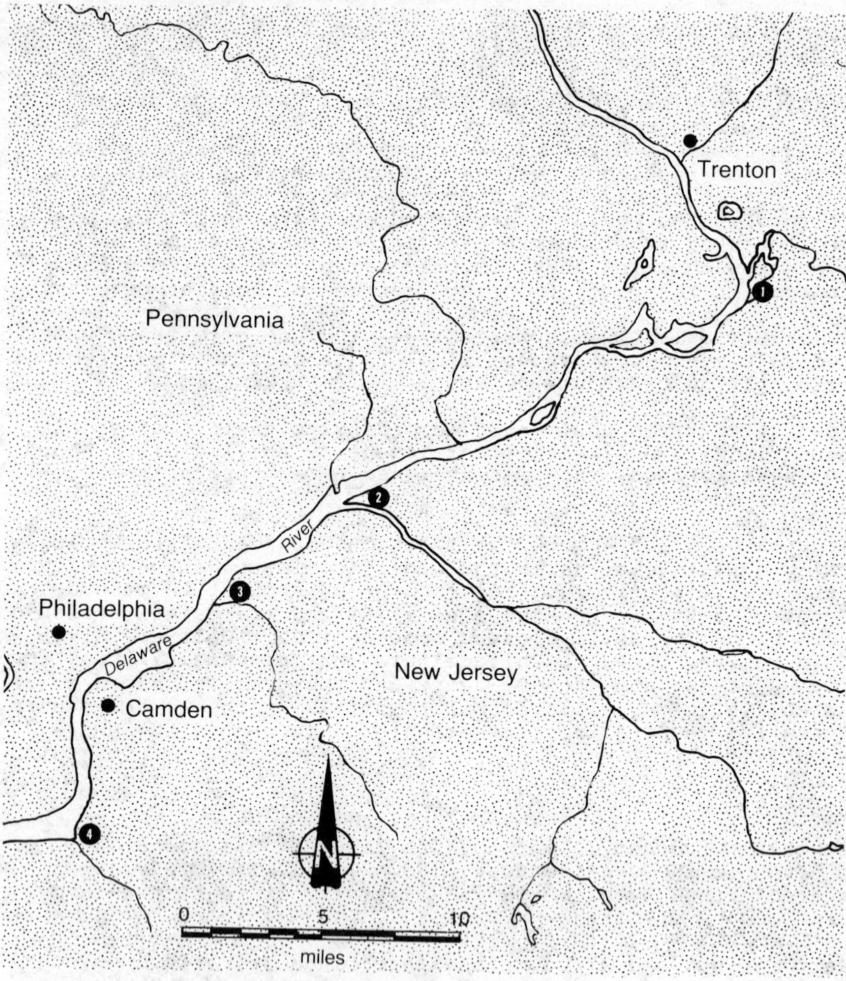

*Identified biologically sensitive areas are numbered.

Figure 3. The Delaware River: Philadelphia to Delaware City*

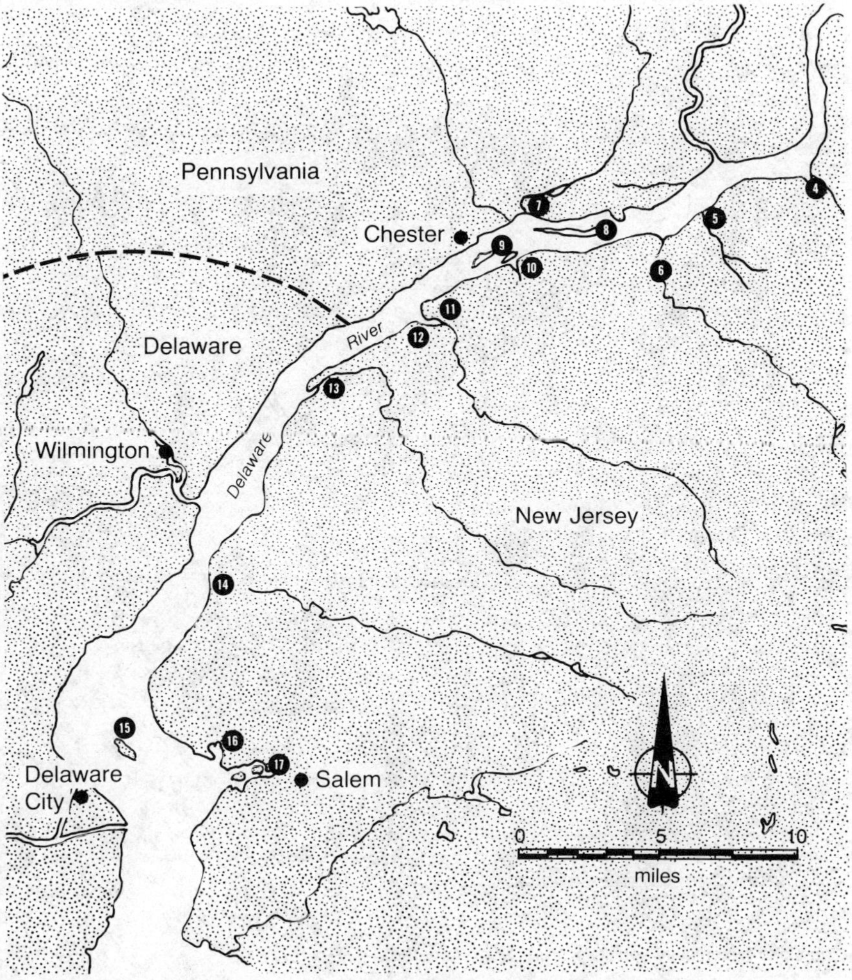

*Identified biologically sensitive areas are numbered.

Figure 4. Lower Delaware River and Delaware Bay*

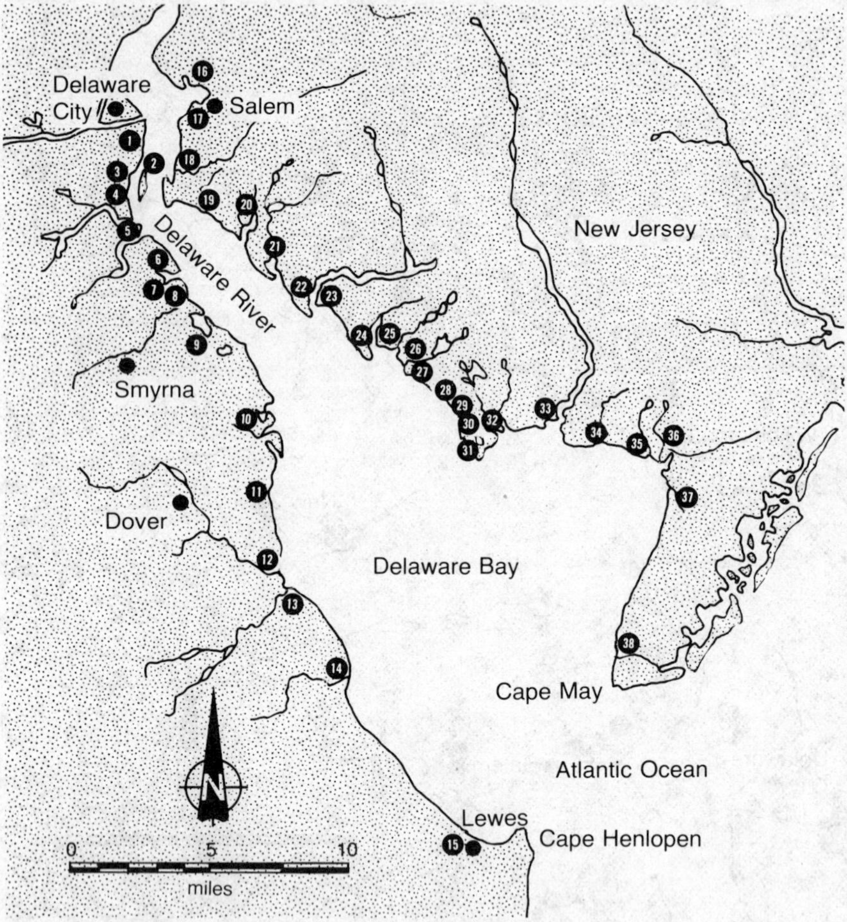

*Identified biologically sensitive areas are numbered.

and industrial development. Others, such as several wildlife refuges on Delaware Bay, are of very large size and are almost undisturbed.

Marshes are low energy environments and very sensitive to oiling. Some of the longest lasting impacts from oil spills have been observed in protected salt marshes where oil was worked into sediments. Less is known about oil effects on freshwater marshes. Further, almost any cleanup method used in a marsh increases the spill's overall impacts (11). Clearly, the preferred option is to keep oil away from marshes whenever possible.

Marshes can often be protected using oil containment booms to divert oil from or block the waterways leading into the marsh. This method has been used successfully in southern California (12) and more recently to protect the lagoons behind the barrier islands during the Ixtoc I spill.

Most marsh ecologists agree that marsh ecosystems are sensitive to all forms of disturbance and difficult, if not impossible to "clean up" without damage (11, 13, 14, 15, 16, 17, 18, 19). Even low level foot traffic can damage plant root systems and disrupt drainage patterns. Vehicles and heavy equipment do more severe, longer term damage. Baker (13, 14, 15) and others (16, 17) have stated that almost anything done to clean up tidal marshes can only add to the ecological impacts already caused by the spilled oil. Therefore, most marsh cleanup methods cannot be justified on ecological grounds (11).

Without further research, use of chemical dispersants in marshes cannot be recommended. Baker (13, 15) found that dispersants penetrated salt marsh plant tissues rapidly and that no amount of subsequent water rinsing could undo the tissue damage which had occurred.

Oil spill containment and pickup devices (booms, skimmers) can be effective in wetlands if water depths are sufficient for their operation. However, they should be deployed from water rather than land, and traffic through the marsh should be avoided. Roads, causeways, seawalls and even sand or gravel beaches should be used for equipment transport rather than traffic through a marsh.

Should oil strand in a marsh, the "no-cleanup" option should be considered first. If some cleanup is necessary, and, for example, contamination of unoiled portion of the marsh is likely, minimum-impact methods should be chosen. Westree (20) and Dibner (21) recommend low-pressure water flushing done from boats at high tide. Free oil is flushed out of the marsh for containment and pickup. This technique was used recently and very successfully in a northern California brackish marsh (22).

Extreme measures such as shoveling, bulldozing, raking, etc., are not recommended. Cutting and removal of marsh vegetation are usually not necessary. They further disturb an already disturbed system and create storage, leaching and disposal problems elsewhere. If cutting must be done to prevent contamination of an unoiled part of the marsh, the preferred method is to use aquatic or mechanical weed cutters from boats. Foot traffic in the marsh should be kept to a minimum and no vehicles should cross the marsh. Even load-spreading vehicles like the "rolligons" used on arctic tundra cannot be recommended (11).

Lagoons

Lagoons can frequently be protected by using containment booms to block or divert oil from the lagoon entrance. If lagoons become oiled, mechanical skimming devices and oil absorbent or adsorbent materials are effective. Bird frightening devices can be used to keep birds away from oiled areas until cleanup is complete.

Tidal Flats

Tidal flats can often be protected by using oil containment booms or booms in combination with fire hoses to divert oil away from them. If tidal flats become oiled, heavy equipment should not be used to clean them. Heavy equipment use results in loss of unoiled as well as oiled sediment in addition to organisms which live in the sediments. Low pressure water flushing can be used to sweep free oil from a tidal flat into an adjacent area for pickup.

Bird Nesting Areas

Bird nesting areas can sometimes be protected using oil containment booms to divert approaching oil. It is important that this diversion be well planned in advance because it should occur some distance (e.g. more than 1/2 mile) away from the nesting area. During the nesting season, many birds are very sensitive to disturbance by human activities (nearby boat traffic, aircraft flyovers, etc.). Efforts should be made to avoid containment or cleanup activities in the nesting area itself.

Oiled Birds

Large numbers of migratory and resident birds use estuarine habitats (marshes, tidal flats, waterways, lagoons) for feeding, resting, over-wintering, nesting, or as stopover points along the Atlantic Flyway. Consequently, a spill in an estuary will nearly always affect significant numbers of birds. Waterfowl are often hardest hit, but other forms, such as waders, may also be vulnerable at certain times of the year.

While cleaning oiled birds is still not highly successful in terms of percentage of birds rehabilitated and released, much has been learned in the last ten years about handling, cleaning and maintaining oiled birds (23). Bird frightening devices have been developed to keep birds away from oil-impacted areas. Improving and refining such devices as well as use of different sounds is now an active field of research (24).

OTHER ENVIRONMENTS-RECOMMENDED CLEANUP METHODS

Low impact cleanup methods can be used not only in biologically sensitive areas, but in all types of environments within the jurisdiction of a response plan.

Rocky Beaches

There are not many rocky shores in the Delaware River Estuary. Where they occur, they are most likely sand beaches to which rocks or cement have been added for beach stabilization. Jetties and seawalls may also be made from rock or concrete. When a solid surface is present, it serves as a substrate to which living plants and animals attach.

Extreme cleanup methods that result not only in removal of oil but also attached organisms are not recommended (e.g., steam cleaning, hydroblasting, scraping). An effective method that removes most of the oil while not harming attached plants and animals is low-pressure water flushing. This operation should be done at low tide and flushed oil diverted to a boomed area for mechanical removal. Trampling and vehicle traffic should be kept to a minimum in these areas.

Sand and Gravel Beaches

The most common oil spill cleanup method used in the United States has been removal of oiled beach sediments using earth moving equipment (25). Methods to remove oil from sand and then replace the "cleaned" sand have also been proposed (26). In the United Kingdom, dispersant spraying and burning have been used as well (27, 28, 29, 30). These methods all may cause ecological damage in addition to that caused by the spilled oil (29).

There has been very little documentation of either the initial ecological impacts or recovery following these several sandy shore cleanup methods or the "no-cleanup" alternative. This research is needed.

Most sand beaches are probably best left to recover by natural means. However, amenity beaches with heavy public use often must be cleaned. This is an instance when the esthetic goal (removing visible oil) may take precedence over the ecological one (minimizing impact). Earth moving equipment is commonly used to remove oiled sand from amenity beaches. Such use is often also common practice to "clean" public beaches of debris (bottles, seaweed, etc.). If earth moving equipment is used, care should be taken to minimize disturbance to surrounding areas, e.g., backshore dunes, marshes, etc., and sediment removal should always be kept to a minimum. Use of hand crews to remove tar lumps or oiled beach sand should also be considered. This usually results in less sediment removal than use of mechanical equipment. The Arctic Marine Oilspill Program (AMOP) in Canada is funding research into low impact methods of sandy beach cleanup (31). These developments should be monitored.

Harbors, Marinas, Terminals

Since the Delaware River is a very busy waterway, harbors and terminals receive heavy commercial use. Smaller marinas, scattered throughout the estuary, are used by recreational boaters and fishermen. Such facilities may support rich benthic and/or fouling communities. They are not generally identified as biologically sensitive areas since their primary function is to serve the needs of their users and not necessarily to maintain viable, healthy ecosystems. Most harbors and marinas in the Delaware River were dredged out of coastal wetlands. Often they must be periodically dredged to maintain depth. Containment booms, skimmers and other mechanical oil recovery devices generally work well in these semi-protected areas. Often, booms can be used to prevent oil from entering (or leaving) a harbor, marina or terminal. Low pressure water flushing is recommended to remove oil from pilings, breakwaters, or seawalls.

IMPLEMENTING THE RESPONSE PLAN

Delaware River and Delaware Bay Cooperatives are implementing the type of spill response planning described here (7, 8). Delaware River Cooperative has purchased trailers which will be equipped with a spill containment boom and other equipment designed to protect the 17 identified biologically sensitive areas in the upper estuary. Delaware Bay Cooperative is in the process of developing strategies to protect the 38 identified biologically sensitive areas in the lower estuary.

The trailer approach has been used successfully by Clean Seas, Inc., in Santa Barbara. Pre-equipped trailers are stored at various locations along the coast, near wetlands sites. Rapid deployment of this gear protected Carpinteria slough from being impacted by a small oil spill which occurred nearby (12).

Another approach is being used by Southern California Petroleum Contingency Organization, based in San Pedro, CA. SC-PCO has two major staging areas at which all of its equipment is located. Equipment is stored in modules that are helicopter-transportable for rapid response.

Effective spill response planning cannot be done by any one group, government or private sector, acting alone. It requires communication and cooperation among the various agencies and groups with spill response duties. For this reason prior agreement on spill response goals and priorites is a necessary first step. Also recommended are periodic meetings or workshops where specific response strategies and tactics are discussed. The decisions that can be made in advance of a spill, should be made so that time is not wasted debating their pros and cons during the emergency. For example, areas to be protected should be identified and priorities and the means to protect them agreed upon. If, as in the case of the marine mammal and bird rookeries of the California Channel Islands, use of chemical dispersants appears to be the only viable means of protection, prior clearance should be obtained from the appropriate government agencies, specifying when and where dispersants can be used. If dispersants are to be used to <u>protect</u> biologically sensitive areas, permission to use them must be granted almost immediately or it will be too late to do any good.

In addition, a very important part of spill response is to have ecological input to response decisions made throughout the course of the spill event. Ecologists should be included on both industry and government primary spill response teams as advisors to on-scene coordinators (OSC) and cleanup managers. The appointment of (NOAA) Science Support Coordinators and Field Response Coordinators (Fish & Wildlife Service) to advise the OSC is a step in the right direction. Likewise, biologists have begun to appear on some industry spill response teams as well (32, 33).

To write a well-conceived response plan is one thing; to put it into practice is another. A spill response plan must be put into practice through regular drills and training exercises. Both planning and practice are necessary components of spill preparedness. Planning sets priorities reduces the number of decisions that must be made after a spill occurs. Practice keeps the plan current and operational rather than only theoretical. Both are essential if one of the objectives of planning is to influence the outcome of a spill situation, i.e., determine where spilled oil will or will not go.

CONCLUSIONS

Except where life and limb are threatened, the primary goal of spill response should be to minimize the ecological impacts of oil spills. Protection of biologically sensitive areas should receive high priority in such response since it prevents ecological impacts from both oil contamination and cleanup activities.

To be effective, protective countermeasures must be implemented rapidly. Yates (34) contends that the most critical period in oil spill response is the first 24 hours. This period may be shorter in a semi-enclosed system like the Delaware River Estuary.

Many decisions regarding spill cleanup can be made before a spill occurs. For the most part, these decisions should be ecologically, rather than esthetically, based. In areas where this type of pre-spill planning has not been done, it is especially critical that biologists be involved in cleanup decision making.

In spite of stepped-up efforts to prevent oil spills, the risk of spills still exists. Preparing in advance for the worst both complements spill prevention efforts and increases our chances that, if the worst ever comes, ecological impacts can be minimized.

REFERENCES AND NOTES

1. Lindstedt-Siva, J. (1977) Oil spill response planning for biologically sensitive areas. Proc. 1977 Oil Spill Conf. API publ. no. 4284: 111-114.

2. Lindstedt-Siva, J. (1979) Why clean up oil spills - another look. Spill Technol. Newsletter 4(1):15-16.

3. Lindstedt-Siva, J. (1980) Minimizing the ecological impacts of oil spills. Environment Intl. Vol. 3 (in press).

4. Valenti, J. (1979) Spill response priorities of an on scene coordinator. Proc. 1979 Fish & Wildlife Service Pollution Response Workshop, C. H. Brown (ed.), U.S. Dept. Interior: 16-19.

5. Council on Environmental Quality (1980) National Oil and hazardous substances pollution contingency plan; Final revision. Fed. Reg. 19 March 1980: 17832-17860.

6. Long, B.F. and J. H. Vandermuelen (1979) Impact of cleanup efforts on an oiled saltmarsh (Ile Grande) in north Brittany, France. Spill Technol. Newsletter 4(4):218-229.

7. Lindstedt-Siva, J. (1979) Oil spill response planning for the Delaware River Estuary I. The upper estuary: Trenton to Delaware City. Atlantic Richfield Co. Rept., April 1979, 104 pp.

8. Lindstedt-Siva, J. (1979) Oil spill response planning for the Delaware River Estuary. II. The lower estuary: Delaware City to the capes. Atlantic Richfield Co. Rept., Sept. 1979, 134 pp.

9. Lindstedt-Siva, J. (1979) Ecological impacts of oil spill cleanup: are they significant? Proc. 1979 Oil Spill Conf. API publ. 4308:521-524.

10. Maurer, D. and H. Wang (1973) Environmental vulnerability of the Delaware Bay area to supertanker accommodation. Vol. II. Biology. Rept. submitted to C.E.Q., Feb. 1973.

11. Cowell, E.B. (1978) Ecological effects and minimizing the impacts of oil pollution cleanup in intertidal marshes and related communities. Working paper presented to Soc. Petrol. Industry Biologists, 1978, 18 pp.

12. Lindstedt-Siva, J. (1979) Resources sensitive to oil spills (Case study of the Santa Barbara Channel). Proc. 1979 U.S. Fish & Wildlife Service Pollution Response Workshop, C. H. Brown (ed.), U.S. Dept. Interior: 79-90.

13. Baker, J. M. (1971). Effects of cleaning. In E.B. Cowell (ed.) _Ecological effects of oil pollution on littoral communities_. Applied Science Publ., Essex:52-57.

14. Baker, J. M. (1971) Oil in salt marsh soil. In E. B. Cowell (ed.) _Ecological effects of oil pollution on littoral communities_. Applied Science Publ., Essex:62-71.

15. Baker, J. M. (1971) The effects of oil on plant physiology. In E. B. Cowell (ed.), _Ecological effects of oil pollution on littoral communities_. Applied Science Publ., Essex:88-98.

16. Baker, J. M. and G. B. Crapp (1971) Predictions and recommendations. In E. B. Cowell (ed.), _Ecological effects of oil pollution on littoral communities_. Applied Science Publ., Essex:217-220.

17. Gundlach, E.R. and M.O. Hayes (1978) Vulnerability of coastal environments to oil spill impacts. Mar. Technol. Soc. Journal _12_(4):18-27.

18. Holt, S., S. Rabalais, N. Rabalais, S. Cornelius and J. S. Holland (1978) Effects of an oil spill on salt marshes at Harbor Island, Texas. I. Biology. Proc. Conf. on Ecological Impacts of Oil Spills, Amer. Inst. of Biol. Sci. publ.: 344-352.

19. Harrel, R.C. and C.A. McCauley (1980) Effects of oil spill cleanup techniques on a salt marsh - Rept. prepared for NOAA, Proj. No. RD0000-87121480-2517, 38 pp.

20. Westree, B. (1977) Biological criteria for the selection of cleanup techniques in salt marshes. Proc. 1977 Oil Spill Conf., API publ. _4284_:231-235.

21. Dibner, P.C. (1977) Response of a salt marsh to oil spill and cleanup. Rept. prepared for USEPA, No. 68-03-2160, 51 pp.

22. Lindstedt-Siva, J. (1980) This method was used successfully to clean up a spill of heavy fuel oil in a brackish marsh in northern California.

23. Berkner, A.B. (1979) Wildlife rehabilitation techniques: Past, present and future. Proc. 1979 U.S. Fish & Wildlife Service Pollution Response Workshop, U.S. Dept. Interior: 127-133.

24. Jehl, J. (1978) Hubbs-Sea World Research Institute, San Diego. Personal Communication.

25. URS (1970) Evaluation of selected earthmoving equipment for the restoration of oil contaminated beaches. Rept. prepared for Fed. Water Qual. Admin., No. 15080 EOS 10/70.

26. Gumtz, G.D. (1972) Restoration of beaches contaminated by oil. Rept. prepared for U.S. E.P.A., No. EPA-R2-72-045, Sept. 1972.

27. Bloom, S.A. (1970) An oil dispersant's effect on the microflora of beach sand. J. Mar. Biol. Assoc. U.K. 50:919-923.

28. Dolan, F.X. and J.P. Bowersox (1973) Development of a mobile system for cleaning oil-contaminated beach sand. Rept. prepared for USEPA, No. EPA-OR2-73-233.

29. Nelson-Smith, A. (1968) The effects of oil pollution and emulsifier cleansing on shore life in southwest Britain. J. Appl. Ecol. 5:97-107.

30. URS (1977) Recommendations for shoreline oil spill cleanup in northern San Francisco Bay. Vol. I. Discussion and Recommendations. Rept. prepared for Clean Bay, Inc. URS-7048-01-01, 41 pp.

31. Russell, L. T., G. D. M. Mackay and W. Carlson (1979) The removal of spilled oil from recreational beaches. Proc. Arctic Marine Oilspill Program Tech. Seminar (March 1979): 24-1 - 24-12.

32. Clean Seas, Inc. (1978) Oil spill cleanup manual. Prepared for Clean Seas by Woodward-Clyde Consultants, San Francisco.

33. SC-PCO (1978) Contingency response plan. Prepared for So.-Calif. Petroleum Contingency Organization by Ultrasystems, Ltd., Newport Beach.

34. Yates, R. A. (1975) The making of a local contingency plan. Proc. 1975 Conf. Prevention & Control of Oil Pollution. API Publ.: 307-313.

IXTOC I OIL SPILL & ATLANTIC RIDLEY TURTLE SURVIVAL

Ens. Donald S. Delikat*

Abstract

The annual summer hatching of Lepidochelys kempii, an endangered species of marine turtle, was threatened by the serious June 3, 1979 offshore oil well blowout in the Gulf of Mexico. The oil from the well traveled in the direction of the Lepidochelys kempii nesting grounds on the eastern Mexican coast. Hatching success rate in 1979 was 70 percent as compared to 58 percent in 1978. Three hundred sixty one nesting females were radio tagged to measure offshore movement and respiration in 1979 as compared to two hundred twenty one females tagged in 1978. Two thousand eggs were tranferred to Padre Island, Texas in early July 1979. Upon hatching, the turtles were immediately recaptured for captive-rearing at the National Marine Fisheries Service Galveston Laboratory. Preliminary hatching success rate for 1980 is 90 percent. Lepidochelys kempii hatchings have suffered no apparent harm from the Mexican oil spill. However, long range studies must be conducted in order to monitor reproduction and ultimate turtle survival.

INTRODUCTION

On June 3, 1979 an offshore rig in the Bay of Campeche (IXTOC I) blew its well head and began delivering an estimated 30,000 barrels of crude oil a day into the Gulf of Mexico. The flow rate estimate fluctuated and exact volumes were impossible to accurately calculate.

For a spill of this size, especially crude oil, some environmental effects on the shore, water, and their inhabitants may be harmful. Some of these environmental effects involve the toxicity of the hydrocarbons, damage to benthic organisms from redistribution of sediments, and effects from the increased turbidity in the water (Brooks et al, 1978).

One of the inhabitants of these waters is the endangered species of marine turtle, Lepidochelys kempii, or the Kemp's Ridley turtle which nests primarily on a single beach in the western hemisphere named Rancho Neuvo, in the state of Tamaulipas, Mexico. The oil from the IXTOC I well traveled in the direction of this annual summer nesting ground on the eastern Mexican coast (see Figure 1). The effects, if any, on Lepidochelys kempii from this oil spill will be examined in this paper.

HISTORY

Kemp's Ridley turtles are the smallest of all sea turtles with neither sex of the Ridley reaching lengths much greater than 28 inches (71.1 cm.) in carapace length (Rebel, 1974). Although this turtle prefers coastal areas, it is sometimes seen in large numbers in the open sea (Rebel, 1974). This turtle was previously thought to be restricted to the Gulf of Mexico, especially around the coasts of Mexico and Florida, but has been seen at other places including Bermuda (Mowbray and Caldwell, 1958). This species feeds primarily on invertebrates, particularly crabs, but also on shrimp, snails, sea urchins, sea stars, medusa, fish, and occasionally marine plants (Ernst and Barbour, 1972; Pritchard and Márquez, 1973; Rebel, 1974).

*U.S. Coast Guard, National Response Center, U.S. Coast Guard Headquarters, Washington, D.C.

The views expressed in this paper are those of the author and are not necessarily those of the U.S. Coast Guard.

The first record of a nesting Kemp's Ridley at Padre Island, Texas was published in 1951 (Werler, 1951). However, the location of a major nesting area was not discovered until an obscure film which was made in 1947 was examined (Carr, 1963; Hildebrand, 1963). This film, which was taken by a Mexican engineer, Andres Herrera, and which was unknown to the scientific community until 1961, showed approximately 40,000 of these turtles present during the filming. The Kemp's Ridley is known to nest in aggregated nesting emergencies, called ARRIBADAS, and it nests primarily during daylight (Rebel, 1974). Subsequent studies on the Ridley population near Rancho Nuevo, Mexico were started in 1966 (Chávez et al, 1968). The turtles were observed to arrive when moderately strong winds blew from the north, east, and southeast and, the beach matrix was of a fine sand (Chávez et al, 1968). The tagging operations in 1966 of 285 females and the subsequent return of nine of these tagged females in 1967, showed the highly developed sense of orientation of this species in the finding of its nesting place (Chávez et al, 1968). In 1970 and 1971, these turtles had an estimated ARRIBADA of 2,000 to 2,500 turtles (Pritchard and Márquez, 1973). There are estimated to be 2,500 to 5,000 native female turtles in the world, making them an endangered species (Pritchard and Márquez, 1973; Federal Register, 2 December 1970).

MEXICAN PLANNING SCHEMES

As early as 1965, the ESTACIÓN DE BIOLOGÍA PESQUERA (Fishery Biology Station) was established in the port of Tampico to undertake the study of Lepidochelys kempii and to develop a plan to protect it (Chávez et al, 1968). A camp site was established on April 15, 1966 in the nesting area of the turtle and staffed with biologists, fish inspectors, and marine personnel (Chavez et al 1968). After the turtle eggs were laid, they were transplanted in plastic bags to a protected area so that coyotes and other predators (including sand crabs and man) would not destroy them (Chávez et al, 1968). The turtle eggs were protected by means of tree trunks and ground wire.

These turtle camps were set up by the Mexican government with marines and infantry soldiers patrolling them. However, in 1972 these patrols were cut down because of budgetary allocations (Zwinenberg, 1977). In that year, a new Federal fisheries law became active (Zwinenberg, 1976). Today, the capturing of Kemp's Ridley turtles is prohibited as well as the taking of its eggs.

UNITED STATES AND INTERNATIONAL CONTINGENCY PLANNING

Drs. Carr and Hildebrand were the two United States scientists who pointed to the extinction danger of the Ridley turtle. The general population decline, human overutilization, and the diurnal nesting on a single beach in Mexico made the turtles susceptible to predation by man and wild animals (Lippincott, 1978). Since Lepidochelys kempii nest on a single beach at Rancho Nuevo, Mexico, any habitat modification could result in the loss of an entire breeding population (Lippincott, 1978). As a result of this extinction notice, a private citizen, Mr. Dearl E. Adams of Brownsville, Texas, tried to transplant this species in 1963 at Padre Island, Texas (Adams, 1966). In the following years, Adams was a bit more successful in his endeavors and noticed that the Mexican government was also working on this species with its survival goal in mind.

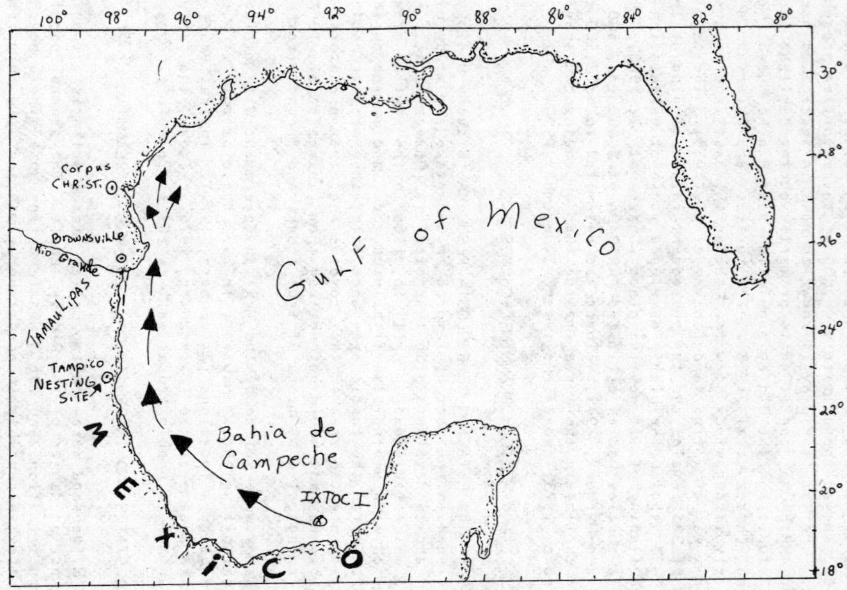

Figure 1. The chart shows approximate route of IXTOC I oil well blowout and the nesting site of Kemp's Ridley turtle at Tampico, Mexico. Drawing by Cathy Gerzel.

In January 1977, state and Federal agencies met in Austin, Texas with Mexican authorities to propose a plan for (1) the enhancement of nesting success and survival at Rancho Nuevo, Mexico (2) the establishment of a second breeding population at Padre Island National Seashore, Texas and (3) an experimental study for headstarting this species (Klima and McVey, 1980). This international contingency plan began in 1978 and had a 58 percent hatching success rate with 221 female turtles tagged (Klima and McVey, 1980).

By July 1979, oil from IXTOC I started to come ashore on these nesting grounds and continued its northward trek to the Texas shoreline. In view of the urgency of the situation and nearness of the oil spill to the nesting grounds, the year old international program was further implemented with special emphasis on the establishment of a second breeding population at Padre Island, Texas. 1,846 eggs were transferred by aircraft to Padre Island, Texas in early July. Upon hatching, the turtles were allowed to enter the surf at Padre Island and were immediately recaptured for captive-rearing at the National Marine Fisheries Galveston Laboratory (see Figure 2). Hatching success rate was approximately 70 percent with 361 nesting females radio tagged to measure offshore movement and respiration (Klima and McVey, 1980).

Meanwhile, a massive effort on the part of the Mexican government, the United States Coast Guard Strike Team, and a host of other agencies was initiated to attempt to stop the flow of oil at the IXTOC I well site and to protect sensitive coastal areas from oil. It was not until late winter 1980 that the oil flow was finally stopped. A combination of a seasonal wind shift in winter, a gradual decrease of oil flow at the well head, massive beach cleanup operations, and use of dispersants in Mexican waters saw an end to oil covered beaches in both Mexico and the United States.

SOME AFTERTHOUGHTS AND SOME UNANSWERED QUESTIONS

The World Conference on Sea Turtle Conservation, held in the fall of 1979 at the State Department in Washington, D.C., considered as one of its topics the survival of Kemp's Ridley turtles. It suggested that like whales, all marine turtles are international; they travel many miles, and know no boundaries. The Conference suggested an international planning scheme with enforcement provisions to assist in controlling and protecting the turtles and also cited the lack of knowledge on the effects of offshore and pelagic pollution (Ehrenfeld, 1979).

Mexican and American cooperation in containing the IXTOC I oil and the protection efforts for newly hatched turtles is commendable. The international headstart program for Kemp's Ridley turtles marked the first time that two countries (Mexico and the United States); four Federal agencies (National Park Service, Fish and Wildlife Service, National Marine Fisheries Service, and Coast Guard); and one state agency (Texas Parks and Wildlife Department) would be working together for a single purpose, namely to save Kemp's Ridley from extinction (Wauer, 1978). An international contingency plan between Mexico and the United States is being fine tuned and hopefully will shortly be finalized through diplomatic channels. International contingency planning has been shown to work with Canada and the United States (Corbett and Beckett, 1977).

Figure 2. A sketch of Atlantic Ridley turtle, Lepidochelys kempii entering the surf at Padre Island, Texas hatching station. Drawing by Anne Lacy. (Based on photo by Fred E. Mang, Jr., National Park Service, U.S. Dept. of Interior.)

As to the ultimate survival of Kemp's Ridley turtles in view of the Mexican oil spill, only time and oil spill studies can tell. At the present time, the oil spill has not seemed to have adversely affected the turtles. The oil which turned to a "mousse" formation or water-in-oil emulsification, was fairly weathered by the time it reached United States beaches. It is still not known if this "mousse" will have any adverse effects on the cues of this species of turtle in its orientation of nesting sites.

The headstarting program is unique in promoting the species' survival, but it is still on an unproven management scheme. Preliminary results for 1980 indicate a 90 percent hatching success rate (Whistler, personal communication). In Bermuda, where some persistent tar balls are being studied (Morris and Butler, 1973; Maynard et al, 1977; Delikat, 1978), there is also a type of headstarting program that has raised green turtles to be "imprinted" to return to their hatching site on Nonsuch Island, Bermuda (Burnett-Herkes, 1968, 1971, 1974, 1976; Wingate, 1967, 1969, 1971, 1972, 1975, 1976, 1977). It would be interesting if this Bermuda experiment could take into account any effect tar has had on these green turtles.

Other questions arise as to the "oil mat" or layers of oil found in the beaches and their effects on turtle egg hatching at Padre Island, Texas and Rancho Nuevo, Mexico. Also, have the dispersants used in Mexican waters had any effect on the plankton and invertebrates upon which Kemp's Ridley depend for survival? It thus seems more questions are left to the answered on Kemp's Ridley survival and how the Mexican oil spill may have affected that survival. If anything good can be said about this or any other oil spill, it is that contingency planning is a must and that something beneficial can always be learned from it in coping with future oil spills and protection of endangered species.

REFERENCES

Adams, D.E. 1966. More about the ridley operation; Padre Island egg transporting. Int. Turt. and Tort. Soc. J. 1(1):18-20, 40-43, 45.

Brooks, J.M., B.B. Bernard, T.C. Sauer, Jr., and H. Abdel-Reheim. 1978. Environmental Aspects of a Well Blowout in the Gulf of Mexico. Env. Sci. and Tech. 12(6):695-703.

Burnett-Herkes, J. 1968. Aquarium and Fisheries, Monthly Bulletin of the Bermuda Department of Agriculture and Fisheries. 38(11):92-93.

____________. 1971. Aquarium and Fisheries, Monthly Bulletin of the Bermuda Department of Agriculture and Fisheries. 41(1):7-8.

____________. 1974. Returns of Green Sea Turtle (*Chelonia mydas*, L.) tagged at Bermuda. Biol. Conserv. 6(4).

__________. 1976. Bermuda Turtle Tagging Programme, Monthly Bulletin of the Bermuda Department of Agriculture and Fisheries. 46(9):69-71.

Carr, A.F. 1963. Pan-specific reproduction convergence in Lepidochelys kempii. Ergebnisse der Biologie. 26:298-303.

Chávez, H., M. Contreras G., and R.P.E. Hernández D. 1968. On the coast of Tamaulipas, Parts I and II. Int. Turt. and Tort. Soc. J. 2(4):20-29, 37 and 2(5):19, 27-34.

Corbett, C.R. and C.J. Beckett. 1977. A dynamic international contingency plan. Proceedings of 1977 Oil Spill Conference. New Orleans, LA. pp79-80.

Delikat, D.S. 1979. Bermuda and the biological station. Underwater Naturalist. 11(2):12-14.

Ehrenfeld, D. 1979. Draft Conservation Strategy. World Conferrence on Sea Turtle Conservation. State Department, Washington, D.C. 10pp.

Ernst, C.H., and R.W. Barbour. 1972. Turtles of the United States. Univ. of Kentucky Press, Lexington. 374pp.

Federal Register, 1970. Vol. 35(233):18319, 2 Dec. 1979.

Hildebrand, H.H. 1963. Hallazgo del área de anidación de la tortuga marina "lora" Lepidochelys kempii (Garmean) en la costa occidental del Gulfo de México. Ciencia. 2(4):105-112.

Klima, E.F., and J.P. McVey. 1980. Headstarting the Kemp's Ridley turtle, Lepidochelys kempii. U.S. Dept. of Commerce, NOAA. Contribution Number 80-08G.

Lippincott, W. 1978. Action Plan, Restoration and Enhancement of Atlantic Ridley Turtle Population. An International Plan. Appendix II. pp21-28.

Maynard, N.G., C.D. Gebelein, and A. Zsolnay. 1977. The effects of pelagic hydrocarbons on the rocky intertidal flora and fauna of Bermuda. Proceedings of 1977 Oil Spill Conference. New Orleans, LA. pp499-503.

Morris, B.F., and J.N. Butler. 1973. Petroleum residues in the Sargasso Sea and Bermuda beaches. Proceedings, Joint Conference on the Prevention and Control of Oil Spills. American Petroleum Institute, Washington, D.C. pp521-529.

Mowbray, L.S., and D.K. Caldwell. 1958. First record of the Ridley turtle from Bermuda, with notes on other sea turtles and turtle fishery in the Islands. Copeia 1958 (2):147-148.

Pritchard, P.C.H., and R. Márquez M. 1973. Kemp's Ridley turtle or Atlantic Ridley Lepidochelys kempii. IUCN Monogr. No. 2 Marine turtle series. Morges, Switzerland.

Rebel, T.P. 1974. Sea turtles and the turtle industry of the West Indies, Florida, and the Gulf of Mexico. Rev. ed. Univ. of Miami Press, Coral Gales. 250pp.

Wauer, R.H. 1978. "Head start" for an Endangered Turtle. National Parks and Conservation Magazine. Nov. 1978. pp16-20.

Werler, J.E. 1951. Miscellaneous notes on the eggs and young of Texan and Mexican reptiles. Zoologica 36(3):37-48.

Whistler, R. 1980. Headstart program at Padre Island, Texas. National Park Service. Personal communication on 11 July 1980.

Wingate, D. 1967. Nature study and Conservation, Monthly Bulletin of the Bermuda Department of Agriculture and Fisheries. 37(8):64-66.

__________. 1969. Nature Study and Conservation, Monthly Bulletin of the Bermuda Department of Agriculture and Fisheries. 39(11):82-86.

__________. 1971. Nature Study and Conservation, Monthly Bulletin of the Bermuda Department of Agriculture and Fisheries. 41(8):58-59.

__________. 1972. Nature Study and Conservation, Monthly Bulletin of the Bermuda Department of Agriculture and Fisheries. 42(9):74-76.

__________. 1975. Nature Study and Conservation, Monthly Bulletin of the Bermuda Department of Agriculture and Fisheries. 45(6):47-58.

__________. 1976. Nature Study and Conservation, Monthly Bulletin of the Bermuda Department of Agriculture and Fisheries. 46(10):82-84.

__________. 1977. Nature Study and Conservation, Monthly Bulletin of the Bermuda Department of Agriculture and Fisheries. 47(1):12-13.

Zwinenberg, A.J. 1976. The olive ridley, *Lepidochelys olivacea* (Eschscholtz, 1892): probably the most numerous marine turtle today. Bull. Md. Herp. Soc. 12(3):75-95.

__________. 1977. Kemp's Ridley, *Lepidochelys kempii* (German, 1880), undoubtedly the most endangered marine turtle today (with notes on the current status of *Lepidochelys olivacea*). Bull. Md. Herp. Soc. 13(3):170-192.

POLLUTION EXTERNALITIES — WHO PAYS?

by Michael W. Christensen *

(A Collection Mechanism For Oil & Hazardous Chemical Pollution Liability & Compensation Legislation)

The thesis of this paper is that parties external to oil and chemical transactions subsidize the production of these commodities in much the same way they would if they were paying a tax to support oil and chemical production; and that it is more efficient, and by one reasonable definition, more equitable for the parties to the transaction (the producers and consumers) to pay these costs.

The purpose of this paper is to describe a manageable and equitable collection system for financing an oil and hazardous chemical pollution liability and compensation fund. This paper assumes that the revenues collected by the fund will be required for the payment of :

(1) cleanup and removal costs and damages resulting from oil pollution incidents;

(2) cleanup and removal costs and damages resulting from the release of hazardous substances to the environment; and

(3) costs for emergency assistance and for the temporary containment of wastes at uncontrolled hazardous waste disposal sites.* (The fund will cover either the temporary containment or the ultimate disposal of the contents of uncontrolled hazardous waste sites, whichever is most cost effective.)

Each of the many proposed oil and hazardous chemical release compensation and cleanup bills has tiptoed around the issue of House of Representatives Ways and Means Committee jurisdiction, by calling the collections a "fee" rather than a "tax". A disadvantage of this "fee" posture is that, though the delegation for collection could fall to the Secretary of the Treasury, the full enforcement

* Uncontrolled hazardous waste disposal sites are defined as those which have no Resource Conservation and Recovery Act disposal permit. It will be impossible to determine who the producers and consumers of some disposed hazardous materials are; therefore, on efficiency grounds, a portion of the cost of abandoned dump site emergency assistance should come from general revenue. This equity arguement is especially evident for dump sites which have been abandoned, and are lying behind the veil of statutes of limitation.

*U.S. Coast Guard, Washington, D.C.

tools which are accorded the Director of the Internal Revenue Service for collection on domestic production and the Director of the U.S. Customs Service for collection on imported products are not employed. Any revenues collected should be considered in their economic effects to be a tax.

The collection mechanism designed must balance maximum equity with a minimum number of collection "gates". Equity is defined as internalizing the environmental damage costs to the producers and consumers of the organic and inorganic elements and compounds which cause the damage. In this paper, the question of equity is distinguished from efficiency, both goals are addressed with equal concern. "It is implicit that many of the equity questions disappear when efficiency is achieved." 1/

The premises on which the collection mechanism are based are:

(1) The burden for financing the fund will be targeted to those persons and industries whose activities or products pose an environmental risk to society; i.e., the burden will be on the consumer and industry rather than on the taxpayer and government.

(2) The collection system must be easy to administer, with a minimum number of collection points (gates).

(3) The burden for payment will be spread over a broad base, so that the costs borne by any one person or company will not be excessive. (The marginal cost of the assessment should not put any firm out of business.)

(4) I recognize that thru the workings of the market, industry may "passed costs through" paid in part by industry and by the ultimate consumer, the energy and chemical consuming public.

GOALS

The goals of the collections proposed in this paper are:

- to have firms more nearly pay the full social cost of their production, to facilitate a more rational use of their resources.
- to finance the emergency cleanup of oil and hazardous chemical spills.
- to finance the payment of claims for economic damages incurred by those who are injured as a result of oil and hazardous chemical pollution.
- to finance emergency cleanup of selected abandoned chemical dump sites.

Determining the appropriate size of the fund is at best a "swag". The information on claims paid for oil and hazardous pollution is sparse, and is scattered among too many industry and insurance sources to discover a "hard dollar cost". Therefore, the size of the fund must be based on estimates of the cost of cleanup, and the availability of personnel and capital resources to perform cleanup. The Fund could equal the consumer surplus for the goods consumed or the cost of cleanup.

"An EPA report estimates that it will cost between $13.1 and $22.1 billion to clean up all hazardous waste that pose a danger to public health and the environment.

- Cleanup costs at the Love Canal have already exceeded $27 million and area residents are seeking more than $2 billion for personal injury and property damage. It is estimated that a properly secured disposal site would have cost only $4 million (in 1979 dollars) in 1952 when the site was closed.

- The state of Michigan has estimated the cost of cleaning up the Montague site at $100 million.

- The state of New Jersey has estimated the cost of cleaning up Chemical Control at $10 million.

- EPA estimates it will cost up to $2.9 million for remedial work to eliminate the danger of phosphate slag in Florida.

- The Colorado Health Department says that cleaning up the radium in Denver could cost up to $25 million.

- Should the containment efforts at the S-area in Niagra Falls and on Long Island fail, the cost of building alternative water supply systems would be astronomical.

Special problems with certain kinds of dump sites:

- Abandoned sites — those where no owner can be found or where the owner cannot afford the cost of cleaning up — have forced State and local governments to bear the costs.

- On-site facilities — those owned and operated by the generators of the waste, either on the plant site or elsewhere — pose serious problems because State regulation of them historically has been minimal. The Subcommittee's survey of the 53 largest domestic chemical manufacturers revealed that 94% of their wastes have been disposed of on the plant site." 2/

SELECTION OF A COLLECTION SYSTEM

A number of alternative collection systems have been considered. The objectives of each have been to balance administrative simplicity (i.e., the fewest number of collection gates as possible) against equity. It is reasonable to expect parties to the transaction, the segment of the population which receives the benefits of the goods, to bear the burden of costs for releases and for abandoned hazardous waste disposal site restoration. It is also reasonable to expect that these costs will be borne by the entire energy and chemical consuming public. A portion of the cost will be paid by a pass-through of the cost to the consumer. A portion will be paid from profits.

Three options for collection were considered:

(1) an assessment on products of the oil, petrochemical, and inorganic chemical industries at the point of sale for consumption,

(2) an assessment on the disposal of hazardous wastes,

(3) The preferred option, is an assessment on petroleum oil and on those feedstocks and raw materials used to produce refined oil, organic and inorganic chemicals.

Option 1 was discounted as administrationally unworkable because the number of collection gates would be massive. Option 2 was discounted because it ignored spills and the disposal of products after their useful life. It did not respond to non-commercial disposal; option 2 also encouraged surreptitous dumping. Option 3, a assessment on oil, petrochemical feedstocks and inorganic raw materials, was selected for detailed analysis because it provides the distribution of the cost without excessive impact on equity and the least number of collection "gates".

In each of the catagories, oil, petrochemicals, and inorganic chemicals, the goal was finding the most efficient collection point. The point where the marginal cost for personnel would be lowest. That is to say, industry would be the least burdened with increases in staff for payment. Coincidentally, this point is where the fewest staff would be required by the government for collection. For oil, the optimum collection point is the refinery. For organic chemicals, the optimum point is the processing facility; for inorganic chemicals, the optimum point is the first processing facility downstream from the mine. The oil, petrochemical, and inorganic chemical collection proposals are described individually in the following sections of this paper.

REVENUE FROM OIL

THE SOURCE AND DISPOSITION OF DOMESTIC AND FOREIGN CRUDE OIL

The following (Figure 1) shows the flow of crude oil through the transportation system, and serves as an introduction to the findings of this report. In the center of figure is a refinery. The left side of the figure depicts the flow of domestic oil to the refinery. The right side of the figure depicts the flow of imported crude and unfinished oils.

Figure 1

Domestic and Foreign Crude Oil Source and Disposition

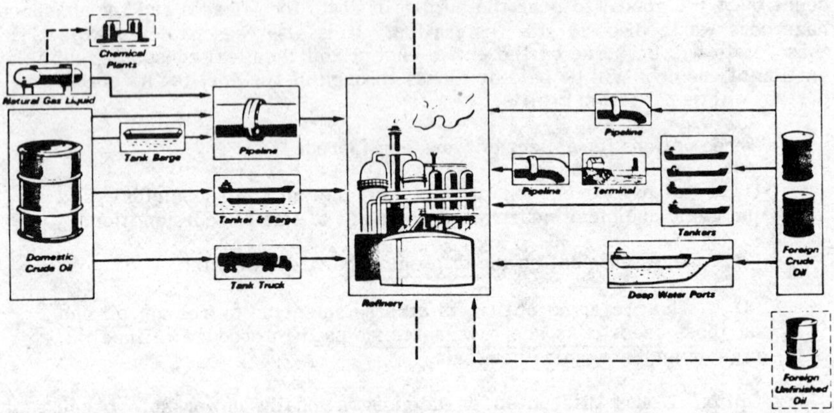

IMPORTED OIL REPORTING MECHANISM

When oil enters the commerce of the United States, all importers of crude oil and oil products are required by the U.S. Bureau of Customs to file Form 7501, Consumption Entry or Form 7505, Warehouse Withdrawl For Consumption, which serve as the U.S. Customs record of entry of goods. The reports record the amount and type of oil imported; the amount of duty due; and the license under which the oil is imported. The report is filed by the importer or his broker and certified by the Customs agent at the port of entry. (Import duties must be paid within 10 business days following entry of the oil into United States commerce.)

I now suggest that all collection of the revenue on crude oil take place at the refinery and that revenue on imported product be collected by the Customs at the point of entry.*

CRUDE OIL REVENUE

I propose that the Superfund collection mechanism for all crude oil parallel the collection of revenues for the Highway Trust Fund. That is to say, the Internal Revenue Service (IRS) mechanism for the reporting and collection of Federal excise taxes can be adapted to accomodate the Superfund collection on crude oil. This proposal will require a modification to the Quarterly Federal Excise Tax Return (IRS Form 720) to allow reporting and collection of the Superfund revenue.

* The import license fee collected by the Department of Energy has been turned on and off like a light switch. In previous papers it was suggested that the administratively efficient way to collect the Superfund revenue for imported oil was as a "rider" on the import license fee. In light of waffling import license fee this approach can no longer be recommended as a rational policy.

I also propose that the oil companies base payments to the Superfund on the number of barrels of crude oil reported as input to refinery on the Department of Energy Monthly Refinery Report (FEA-P-320-M-0).

A detailed description of this process is presented in "A Fee Collection Mechanism for the Oil Pollution Liability and Compensation Legislation".[3]

Table 1 is a summary of the collection mechanism for imported and domestic crude oil, and for imported products.

REVENUE FROM PETROCHEMICALS

Primary synthetic organic chemicals produced in the U.S. are derived largely from petroleum and natural gas hydrocarbons. Almost 100% of the hydrocarbon portion of the primary organic chemicals, used as building blocks for the production of more complex intermediate compounds and end use products, are obtained initially from petroleum and natural gas. The organic chemical industry products, derived from petroleum and natural gas, are referred to as "petrochemicals". The flow from feedstock to finished product is shown in Figure 2. The figure illistrates that organic chemicals are essential building blocks for a large and varied group of industries.

Figure 2

Petrochemical Product Flow

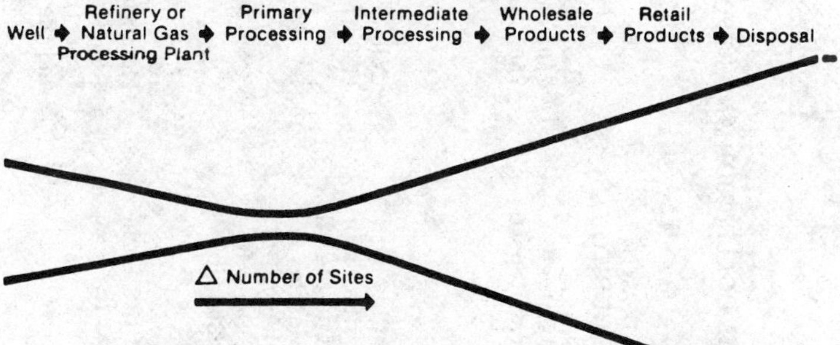

The intricate movement of products within these industries through inter-industry transfers is not indicated on this chart. For instance, ethylene, a building block, when used for polyethylene, flows directly to the plastics materials industry, then to plastic products manufacturers and then possibily to many other manufacturing industries for their products. Methane, a petrochemical feedstock, is used directly as a building block for methanol production and for other organic chemicals such as chlorinated hydrocarbons, and fluorocarbons. Methane also is a source of hydrogen for ammonia, cyclohexane and other products for hydrogenation reactions. Benzene, a building block, is converted into cyclohexane, then to adiponitrile and hexamethylenediamine in the organic chemical sector, and finally is converted to nylon. Nylon as a resin, becomes a plastic product, and as a fiber, a textile product. Such an example of an intricate product flow is only one of the many involving benzene products and chemicals in general.

TABLE 1

OIL COLLECTION AND REPORTING SUMMARY

	IMPORTED AND DOMESTIC CRUDE OIL	IMPORTED OIL PRODUCTS	EXPORTED DOMESTIC CRUDE OIL
REPORTING MECHANISM	Department of Energy Monthly Refinery Report (P320)	U.S. Customs Forms: 7501 Consumption Entry 7505 Warehouse Withdrawl for Consumption	Documentation required by 15 CFR, Part 377.6 (e)(1) Export Administration Regulations, Petroleum and Petroleum Products, Documentation.
COLLECTION MECHANISM	Paid Twice Monthly by Refinery; Deposited with FET Payment Due the 9th & 24th of the Month; Reported Quarterly, IRS FET Form 720, Deposited in US Treasury Superfund Account	Collected by U.S. Customs at Port of Entry	Collected by Department of Commerce via Export License Deposited in US Treasury Superfund Account (minus DOC G&A Costs)
REQUIREMENT	Modification of IRS Form 720 Quarterly Excise Tax Return	Memorandum of Understanding between the Department of Treasury and the EPA	Memorandum of Understanding between the Department of Treasury and the Department of Commerce for collection

The chemical industry is its own best customer and is a supplier of derived products, first to other industries — manufacturing, agriculture and construction — and ultimately to "end use" products.

COLLECTION GATES

The most crucial issue in the development of a revenue collection system for organic chemicals is locating the point to determine the sums due. The first step in organic chemical production is the conversion of petroleum and natural gas into a relatively small number of petrochemical feedstocks. This initial transformation occurs at refineries and natural gas processing plants and produces typical petrochemical feedstocks such as natural gas; methane; the natural gas liquids, ethane, propane, butane, pentane; and petroleum liquids, naptha, gas oil, reformate, raffinate, and carbon black oil. The next step in the process is the production of the primary petrochemicals; such as the olefins, for example ethylene, propylene; and the aromatics, benzene, toluene, xylene, and napthalene. The primary petrochemicals are transformed into thousands of more complex organic compounds (intermediates) and eventually into hundreds of thousands of more complex products. Figure 3 depicts four levels of product classification in the organic chemical manufacturing chain. The cost of cleanup is different for different products, implying that the assessment could vary with the product. The administrative burden of such an assessment scheme would be excessive and costly.

Figure 3

PRODUCT CLASSIFICATION

Feedstocks and Petrochemicals

Typical Petrochemical Feedstocks (Raw Materials) (I)	Petrochemicals: Primary Petrochemicals (II)	Petrochemicals: Petrochemical Intermediates (III)	Petrochemicals: Petrochemical Products (IV)
Natural Gas	Olefins	Ethylene Oxide	Plastic Materials
(Methane)	Ethylene	Ethylene Glycol	● Polystyrene
Natural Gas Liquids	Propylene	Ethylene Dichloride	Synthetic Rubber
Ethane	Butylene	Vinyl Chloride	● Polybutadiene
Propane	Butadiene	Acrylonitrile	Synthetic Fibers
Butanes	Acetylene	Cyclohexane	● Nylon
Pentanes	Aromatics	Ethylbenzene	Surfactants
Petroleum Liquids	Benzene	Styrene Monomer	● Alkarylsulfonate
Naphtha	Toluene	Phenol	Medicinal Pharmaceuticals
Gas Oil	Xylenes (Mixed)	Phthalic Anhydride	● Sulfa Drugs
Reformate	o-Xylene	Terephthalic Acid	Nitrogenous Fertilizers
Raffinate	p-Xylene		● Ammonium Nitrate
Carbon Black Oil	Naphthalene		Phosphatic Fertilizers
	Methanol		● Diammonium Phosphate
	Carbon Black		Pesticides
			● Malathione

To assess thousands of petrochemical products would be administratively impossible. Determination of an equitable assessment would be unduly complex. For these reasons, it was determined that collection of the revenue on organic chemical intermediates and end use products would be untenable.

These considerations argue for the establishment of the collection gate for petrochemicals very early in the organic chemical product and hazardous waste chain. The collection point should be late enough in the chain, however, so that collection on non-hazardous uses of organic chemical feedstocks is avoided, e.g., use of natural gas for home heating fuel.

The system would collect the sums due on two classes of petrochemical feedstocks: liquid hydrocarbon feedstocks and gaseous hydrocarbon feedstocks, received at chemical processing facilities and which are used as either fuel or feedstock to produce primary petrochemicals and organic chemical intermediates. The revenue will be collected at the chemical processing facility

The simplified diagram, Figure 4, depicts the organic chemical manufacturing process and shows the collection gate.

In the event that revenues are collected on liquid hydrocarbon feedstock shipped from refineries, there will be less than 400 collection gates.

Figure 4

COLLECTION ON ORGANIC CHEMICAL FEEDSTOCKS

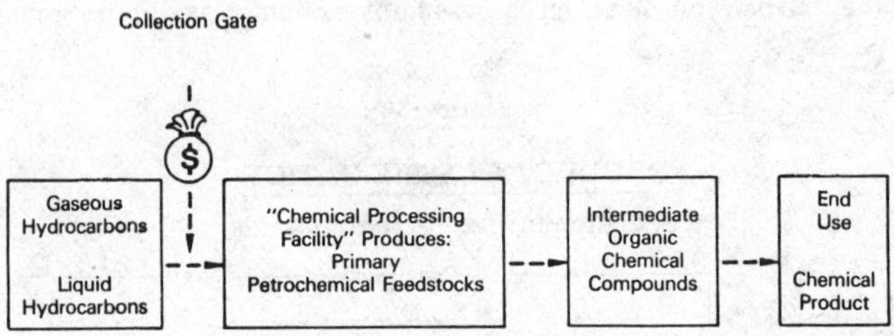

ANNUAL COLLECTIONS

It is suggested by the EPA that the annual collection on organic feedstocks not exceed $200 million. [4/] The revenue from each feedstock should be established by regulation so that the amount contributed to the fund by the petrochemical industry annually will be equitable and not exceed the legislated limit. I recommend that, in the event of the passage of separate oil and chemical funds, the collections should be in addition to the sum collected on liquid hydrocarbons used as fuel or feedstock. There are two separate and distinct risks to the environment, first oil as oil; and second, petrochemicals. The assessment is on fuel and feedstock in order to provide incentive for more economic use of fuels; and to assure that there is no sudden change in factor inputs. In the case of liquid hydrocarbons, the assessment is in addition to assessment paid on oil, there is a new and separate risk.

IMPORTS

The sums due would be assessed on a list of specific organic feedstocks and organic compounds by the U.S. Customs Service. The assessment would be paid by the "importer of record" at the time of entry for consumption.

REVENUE INORGANIC RAW MATERIALS

Inorganic raw materials may be transformed into more complex intermediates, directly into products or combined with organic chemicals into organic-inorganic intermediates or end use products. There are thousands of these more complex compounds. Attempting to assess each of these chemical intermediates and products would be unmanageable and the determination of the sumes due would be unduly complex. The other extreme of the inorganic chemical product chain (the "mine"), was examined to determine if fund revenue could be collected at that point.

Inorganic raw materials are produced from mineral ores, brines, or process wastes, e.g., bromine may be produced from natural lake brines; chlorine can be produced from sea water; mercury is produced from cinnebar ore. Assessing the mineral source of the raw material, however, would result in collection on substances which are not hazardous: e.g., limestone may be used in cement and sulfur in agriculture. These considerations strongly suggest that assessments be on inorganic chemicals at the beginning or the end of the inorganic chemical manufacturing process would not be reasonable or feasible.

The collection point that would provide the minimum number of collection gates, and the point that would capture the chemical substances that are the inorganic chemical building blocks for many hazardous substances, would be the first gate where the ore, brine or process waste is transformed into an inorganic raw material or chemical. Selection of this gate will provide the most efficient collection. This point was defined as the "chemical process facility". The sums due will be assessed when the raw material is consumed on site or shipped from the "chemical processing facility".

The simplified diagram, Figure 5, depicts the inorganic chemical manufacturing process and indicates the collection gate. (Organic feedstocks follow a simular process flow.)

Figure 5

COLLECTION ON INORGANIC RAW MATERIALS

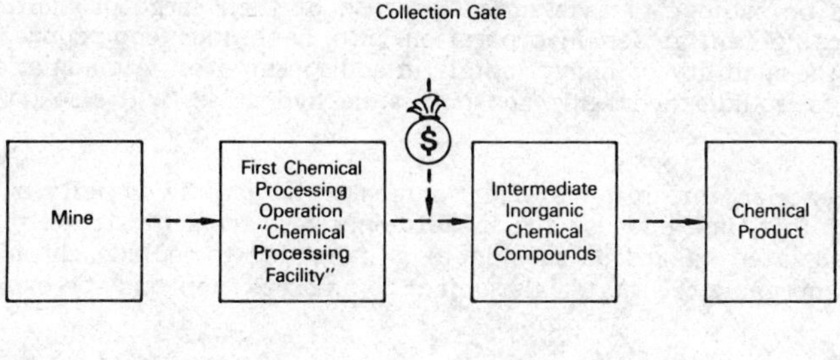

In some instances, the first chemical processing operation may occur at the mine site. For example, in the U.S., cinnebar ore is mined and processed into metallic mercury at the mine.

In certain integrated chemical processing plants, the element assessed will be consumed on site during the production of more complex chemical compounds. For example, chlorine is produced by the electrolysis of brine and is often immediately consumed in the same plant for the manufacture of chlorides.

SELECTION OF INORGANIC RAW MATERIALS

The selection of inorganic raw materials to be subject to an assessment was based upon satisfaction of two or more of the following considerations: the raw material must be hazardous in a number of forms; the raw material in some form must be hazardous if spilled; the raw material must be produced in large quantities, or the raw material must be capable of increasing the hazard potential when combined with other elements. Of the heavy metals, arsenic, cadmium, mercury, lead, and chromium are included in the collection system because they are produced in quantity. Many compounds, incorporating these elements, are hazardous. All of these elements are subject to maximum contaminant levels under the Safe Drinking Water Act.[5]

Lead and chromium are used in chemical manufacturing as compounds, i.e., lead oxide, sodium dichromate, and chromic acid. Antimony, nickel, cobalt, beryllium, zinc, barium and selenium were considered but are not proposed to be subject to assessment at the present time. When additional data demonstrate satisfaction of the criterial listed above, they should be added by regulation.

Of the halogens, chlorine, bromine and fluorine satisfy at least two of the above criteria and are included in the collection system. These elements, or their acids, are: produced in large volumes; hazardous if spilled; can form hazardous materials and wastes; and can increase the mobility of heavy metals, if present in a disposal site in their acidic forms. Elemental fluorine is not produced in sufficient quantities to render it administratively cost-effective to be subject to assessment; however, hydrofluoric acid is. Ethylene dibromide and methyl bromide are produced in large quantities and are quite hazardous. It is proposed that assessments on the element equivalent of the halogen in the following chemicals: chlorine, bromine, hydrochloric acid, hydrofluoric acid, methyl bromide, and ethylene dibromide. (Iodine is not produced in sufficient quantity to justify assessment.)

In addition to the halogenic acids above, sulfuric, phospohoric and nitric acid would be subject to assessment because of their large production volume, spill hazard, potential for incorporation into hazardous compounds, and their effect on the mobility of heavy metals in a disposal site. Ammonia, sodium and the caustics, sodium hydroxide and potassium hydroxide, will also be subject to assessment.

Other inorganic raw materials demonstrate chronic toxicity but were not included at this time because the initial concern is for acute toxicity. When the list of designated hazardous substances is expanded to include chronic toxicity, the list of inorganic raw materials subject to an assessment may be expanded.

NUMBER OF COLLECTION GATES

Table 2 lists the inorganic elements by class and number of companies and plants producing them.

TABLE 2

COLLECTION GATE SUMMARY

Heavy metals:	34 companies,	49 plants
Halogens:	95 companies,	190 plants
Acid Formers:	267 companies,	486 plants
Caustic Formers:	42 companies,	82 plants
	438 companies	807 plants

Assessments are based on element weight or element equivilant weight in compounds. The specific elements assessed are listed in Table 3.

TABLE 3

SPECIFIC INORGANIC ELEMENTS ASSESSED

Heavy Metal	Halogens	Acid Forming Classes	Caustic Forming Classes
Arsenic	Bromine	Sulfur	Potassium
Cadmium	Chlorine	Nitrogen	Sodium
Chromium	Fluorine	Phosphorus	
Lead			
Mercury			

Table 4 shows the number of collection gates for heavy metals, halogens, acid forming classes, and caustic forming classes.

ANNUAL COLLECTION

The liability would be assessed on the equivalent weight of the element contained in raw material at a rate not to exceed an amount specified by the Fund administrator expressed in dollars per ton.

The amount assessed or element equivalent weight on each element compound will be established by regulation so that the amount contributed to the fund by the inorganic chemical industry in one year will not exceed $75 million. A dollars per ton assessment on the element equivalent weight would result in a higher collection for the heavier elements. The density of each element must be considered in the assessment; to assure that the heavier elements do not pay a disproportionate percentage of the fund.

TABLE 4

COLLECTION ON INORGANIC RAW MATERIALS

ESTIMATED NUMBER OF PRODUCERS

RAW MATERIALS	NUMBER OF COMPANIES	NUMBER OF PLANTS
HEAVY METALS		
Arsenic	1	1
Arsenic Trioxide	1	1
Cadmium	6	7
Chromium	2	2
Ferrochrome	5	5
Sodium Dichromate	4	4
Potassium Dichromate	3	3
Chromic Acid	3	3
Lead	4	7
Lead Oxide (Yellow)	4	15
Mercury	1	1
TOTAL	34	49
HALOGENS		
Bromine	6	9
Ethylene Dibromide	5	6
Methyl Bromide	4	4
Chlorine	34	70
Hydrochloric Acid	37	87
Hydrofluoric Acid	9	14
TOTAL	95	190
ACID FORMING CLASSES		
Phosphoric Acid	33	60
Phosphorus	7	10
Sulfuric Acid	71	160
Nitric Acid	50	88
Ammonium Nitrate	43	64
Ammonia	63	104
TOTAL	267	486
CAUSTIC FORMING CLASSES		
Potassium Hydroxide	3	11
Sodium Hydroxide	30	66
Sodium	3	5
TOTAL	42	82

The EPA recommended that an economic incentive for recycling elements and compounds be considered incorporated in the collection system by permitting a waiver of the assessment or a credit on the elements or compounds recycled and recovered from the waste stream and reintroduced into products.[6]

IMPORTS

The assessment would be on a specific list of inorganic elements and compounds by U.S. Customs Service. The assessment paid by the "importer of record" at the time of entry for consumption.

SUMMARY AND CONCLUSION

I have proposed that revenues for Superfund be assessed on:

Oil

- Crude oil at the refinery
- Imported oil products at the point of entry.

Petrochemicals

- Liquid hydrocarbons input to petrochemical plants
- Gaseous hydrocarbons input to petrochemical plants

Inorganic Chemicals

- Specific heavy metals
- Specific halogens
- Specific acid formers
- Specific caustic formers

I believe that the revenue system described in this report best balances equity with ease of administration. I also believe that the system described will help industry and consumer more nearly pay the true cost of the production and consumption of oil chemicals.

I invite your comments and criticism on this paper. Please write:

M. W. CHRISTENSEN

% Commandant (G-WER-4)

US Coast Guard

Washington, DC 20093

REFERENCES

1/ Professor Robert Leone, Harvard School of Business Administration, personal communication, November 3, 1979.

2/ Marinelli and Wentworth, "Pollutors Should Pay", Environmental Action, June 1979.

3/ "A Fee Collection Mechanism For The Oil Pollution Liability And Compensation Legislation", Report No. CG-WEP-78-1, August 1978.

4/ Thomas Jorling, Associate Administrator, EPA, personal communication, April 3, 1979.

5/ "The Safe Drinking Water Act, PL 93-253", December 16, 1979.

6/ Thomas Jorling, Associate Administrator, EPA, personal communication, April 3, 1979.

Fishing and OCS Energy Development--
Charting a Course of Compatibility

Keith G. Hay
Conservation Director
American Petroleum Institute
2101 L Street, NW
Washington, DC 20037

The Coastal Zone is a vital habitat--not only for the creatures that inhabit its vast marshlands, bays, and estuaries, but for the men and women whose livelihood depends upon a clean and healthy coastline. The protection of that habitat for both man and nature is of primary concern to the petroleum industry.

During the past 10 years, I have committed a considerable portion of my time and energy towards understanding the problems between the coastal-dependent industries of commercial fishing and offshore oil and gas. Identifying the real problems from the imagined ones has often been as illusive as the quarry that both industries are pursuing.

It seems that a lot of remarkable achievements and historical compatibility between fishing and oil interests are overshadowed by exaggerated or momentary problems. Please forgive me, therefore, if I devote equal time to some non-problems, some things that do work, and even to some benefits that flow from oil operations on the Outer Continental Shelf (OCS).

Let me begin by paying tribute to the long and colorful history of American commercial fishing. It was America's first industry. It is still a major industry, employing more than a quarter of a million people and generating some $1.5 billion in income. U.S. fisheries provide one-fifth of the world's seafood resources, and our own domestic fleets supply a third of the seafood we consume.

Extension of the U.S. fishing jurisdiction to 200 miles in 1976 has presented an enormous new challenge to marine resource managers and educators, as well as to commercial fishermen. The challenge lies not only in the management and rebuilding of depleted fisheries, but in accommodating national requirements for developing other OCS resources, as well.

Among them, of course, is our need for energy. U.S. energy consumption has risen almost every year. This nation is now dangerously dependent on uncertain foreign supplies. Our imports rose from 3.4 million barrels a day in 1970 to 8.2 million barrels a day in 1979. And the cost from $3 billion in 1970 to an estimated $90 billion in 1980. To effectively reduce this dependence, we must increase our energy conservation efforts and our domestic production of energy.

Fortunately, the U.S. still has some potentially energy-rich frontiers, of which the OCS is one of the most promising. The U.S. Geological Survey estimates that from 12.5 to 38 billion barrels of oil and from 61.5 to 139 trillion cubic feet of natural gas may be locked under the waters surrounding our nation. That's enough to keep lots of homes heated and lots of trawlers fishing for many years to come.

As you know, oil exploration and development are just beginning in the South Atlantic. By contrast, the industry has logged more than a generation of offshore development experience in the Gulf of Mexico and off the California and Alaska coasts. In fact, in 1979 almost 12.5 percent of the oil and about 27 percent of the natural gas produced in the U.S. came from offshore wells.

Highly specialized engineering skills and advanced technology evolved by the oil industry in recent years have enabled successful development of offshore energy resources in such difficult environments as the high Arctic and the North Sea. Within the last ten years, for example, the industry has perfected the state of the art to the point where they can now drill in water depths of more than 5,000 feet. Platforms in the North Sea are built to withstand winds more than 100 miles per hour, waves topping 90 feet and underwater sand waves 30 to 40 feet high.

The record of U.S. offshore oil operations is truly outstanding. More than 24,000 oil wells have been drilled in U.S. waters since 1953, yet there has been only one major spill which reached the shoreline in any significant amount. That spill was in the Santa Barbara Channel in 1969, and its effect on intertidal life was temporary. No permanent damage resulted. In fact, data compiled by the National Academy of Sciences[1] reveal that only 1.3 percent of the petroleum entering the oceans worldwide comes from offshore production, compared to 9.8 percent which comes from natural oil seeps.

An April 1980 report[2], by the Royal Society of

London, on oil pollution in the marine environment, concluded that "The local ecological effects of oil pollution are highly variable, depending upon the type of oil and the environmental situation, but with the possible exception of mechanically-mediated effects on some pelagic bird populations, there is little or no evidence that the biota of the open sea have suffered significantly at the population or resource levels."

The largest and most recent offshore spill was that in the Bay of Campeche in Mexico. Testimony at a U.S. Senate Hearing[3] revealed that the accident in Campeche Bay would have been avoided by U.S. practices and regulations. Although the final environmental impact report is still not available, recent discussions with Mexican authorities indicated that no apparent extensive damage to the environment occurred.

With these facts in mind, let's look at the five major problems--real and potential--which can arise between fishermen and oilmen on the OCS.

First, there is the possibility of lost fishing grounds. Installations like drilling rigs and production platforms actually take up very little sea space; but to protect them, operators are permitted to establish a safety zone around them, generally one-quarter nautical mile in radius. To put it in prospective, though, the size of a platform relative to a lease area is equivalent to a postage stamp in a ten by ten-foot room. Normally, each lease would have no more than two platforms.

With the exception of rather circumscribed areas, therefore, fishing grounds are normally unaffected, and the loss of fishing space is a problem only to the degree that petroleum activities coincide with live bottoms or other key harvesting areas.

Off New England, where fishermen go far out to sea, the Woods Hole Oceanographic Institution[4] has estimated that at full production--not likely until 1985--only between 9.6 and 62 nautical square miles out of 20,000 square miles in the Georges Bank area would be unavailable for fishing. Similarly, a report on relations between the offshore oil industry and the Norwegian fishing industry in the North Sea[5] points out that lost fishing grounds are mostly a political, not a technical problem, and that "offshore installations are not a primary cause of trouble and do not conflict with fishing operations except in isolated incidents."[5]

In fact, fixed oil production platforms can

actually provide some positive benefits offshore. One Coast Guard commander praised the oilmen in the Baltimore Canyon area for providing assistance and communication relays during several marine emergencies. And fishermen have found that the well-lighted rigs serve as valuable navigation points, reliable sources of local weather information and welcomed refuges in a storm.

In terms of the marine ecosystem, offshore oil development has not proven detrimental to either commercial or sport fisheries. In the Gulf of Mexico, for example, where intensive oil development has occurred over several decades, the commercial harvest of fish and shellfish has consistently increased. For example, the average harvest in the Gulf over the last five years has been 3 to 4 times the 1950 catch of 570 million tons.

The oil industry obviously does not take credit for this phenomenal growth, but the resulting sport fishery around the platforms is directly related to the "steel reefs." In both cases, however, it is abundantly clear that platforms, oil, fish and shrimp can coexist in a viable relationship.

Studies now confirm that platforms perform a far more useful purpose than simply attracting fish.[6] They serve as living, productive artificial reefs--sometimes more productive than natural reefs. The underwater steel columns rising from the benthos to the tropogenic region become encrusted with thick layers of fouling animals and plants. The proliferation of mussels and barnacles is often so great that chunks slough off, forming a hard substrate on what otherwise might be a soft bottom. These shelf deposits are rapidly colonized by prolific benthic organisms which form a new, localized community. The benthos, as well as the vertical column organisms, provide food and shelter and a constant rain of detritus and plankton to the surrounding biomass.

Marine biologists have found as many as 200 different species of invertebrates and 45 species of fish on or near a platform; some areas with platforms shelter 20 to 50 times more fish than control areas without platforms.[6]

In short, the physical space which the platform occupies may in the long run be more than compensated for by its continual biotic productivity.

A second, and possibly more serious matter for concern, are bottom obstructions and debris. This is a particularly vital issue in areas where commercial fishing depends to a large extent on bottom trawling. The ocean bottom, of course, is far from pristine. It has been littered by man and nature for centuries. When I visited Charleston's docks last year, I saw one boat bring up a 200-pound loggerhead sponge; a rusted diesel engine block sat nearby from an earlier catch; and I am told that everything from airplane parts to mastadon teeth have been caught in trawls.

Existing regulations forbid deliberate littering. But obviously, there are bound to be some occurrences due to accidents, carelessness and/or rough weather. The frequency can be reduced, however, and the impacts mitigated.

For example, a system for reporting and charting large obstructions can and should be carried out. In this context, it is interesting to note that North Carolina, the Gulf area, and Norway have compiled special "hang guides" listing loran coordinates of major obstructions. However, lack of cooperation from fishermen who consider such information proprietary, has forstalled similar initiatives on the Atlantic Coast.

The OCS Lands Act Amendments call for a survey of all underwater obstructions on the OCS. Obviously, it would be impossible to survey the entire OCS, at least in our lifetimes. But it is probably not unreasonable to survey the major travel corridors between the shore and working platforms.

In addition, there is the question of compensation for lost or damaged gear. The OCS Act Amendments also provide for a special fund financed from oil revenues to compensate commercial fishermen for losses or damage to gear resulting from oil industry operations. Fishermen will have to report precise bearings on the offending object, and the claimant will have to verify its nature and ownership.

The time, paperwork and expense involved in processing such claims will undoubtedly deter some fishermen. Nonetheless, similar approaches in Norway and Scotland are working to varying degrees.

In the context of underwater obstructions, pipelines associated with production phases of petroleum operations deserve mention. Regulations require that a pipeline be trenched to a depth of three feet in water depths of less than 200 feet. With proper backfilling, a pipeline should present no problem for trawlers; however, the dynamics of local bottom sediments and tides could work to expose the pipe at some point. The location of such a marine pipeline would be indicated on all nautical charts. Since trawler captains are well aware of the dangers to their gear in tangling with a pipeline, they soon learn their location and easily avoid them.

A third cause for concern is, unfortunately, what many people think of first--oil spills. This concern is not limited to fishermen. The oil industry, itself, wants nothing less than an oil spill. What it would like, however, is to dispel the widespread misconception, misrepresentation and mystification which plague this whole subject. The fact remains that petroleum exploration has been conducted in the offshore waters of more than 100 countries, over 50 of which are producing petroleum. Since these operations began more than 30 years ago, there have been no known cases of spills causing permanent damage to the environment nor has it "been established that any commercially valuable offshore stock of fish or shellfish has suffered a reduction in recruitment due to damage by floating oil."[2,7]

Oil pollution research, most of which has been initiated since the "Torrey Canyon" disaster of 1967, has generated an enormous amount of literature. The main conclusions of this research were that the effects of oil pollution varied considerably depending on the type of oil, the weather and sea conditions, the clean-up procedure used, the nature of the receiving environment and the amount and duration of the input.

Many biological and chemical investigations are still in progress, but it is fair to say that the main conclusions are that petroleum and its derivatives do not accumulate in the higher levels of marine food webs; or pose a credible threat to the health of the consumer of sea foods and that earlier ideas on the comparatively moderate ecological significance of oil pollution are probably still valid.[2]

The uncoordinated oil pollution research efforts of numerous federal agencies were finally corralled by the National Ocean Pollution Research, Development and Monitoring Act of 1978. The National Oceanic and Atmospheric Administration was given the mandate for such coordination and held three meetings last summer and fall to review the federal marine science program for oil pollution research, development and monitoring activities. The program is divided into two broad categories. Technology for oil and hazardous materials spill prevention, control and clean-up, and environmental impact of oil pollution with respect to generic and site-specific studies. This long-needed management effort will hopefully focus the government's expensive research efforts on pertinent national priority needs and encourage the use of standard methods with recognized levels of precision and accuracy for judging the effects of oil in the environment.

We do have the results of numerous past scientific studies on the effect of spills, which, as I mentioned earlier, have only rarely been related to offshore production. Among the most important are three critical studies conducted in the petroleum producing area of the Gulf of Mexico. The three studies[8,9,10] were separately conducted and funded and came to essentially the same conclusion: the ecosystems around offshore platforms are normal and healthy; they are not adversely impacted by such operations.

Numerous factors account for the apparently minimal adverse impacts. Hydrocarbons have been an integral part of the marine environment over geologic time. Crude oil is a natural, organic, biodegradable compound. It has low solubility in water, and natural forces almost always mitigate its impact. In the case of large spills in confined areas, however, the effects can be severe. But even then, they are usually temporary, as researchers have found to their surprise in the wake of the *Amoco Cadiz* tanker accident in France.

The eggs and developmental stages of many commercial finfish species spend some time in the plankton where they are exposed to oil and a variety of other pollutants which are harmful to them. Fish stock assessments for the North Sea[11] have not revealed any detectable impact of the enormous quantity of wastes reaching this body of water and, certainly, any influence of pollution is insignificant beside the impact of fishing practices.

In short, the potential impact of any spill must be considered on a case-by-case basis, but scientific evidence strongly indicates that offshore production has had no significant ill effects on the health of the marine community.

Nonetheless, the oil companies make preventive measures their number one priority. Through training, supervision and continuous development of new and improved technology and procedures, the industry places heavy emphasis on safety and environmental protection across the board. Its superior safety record is its just reward.

The fourth area of concern are navigational hazards associated with increased vessel traffic. In early stages of development, the major factor is seismic ships, which tow delicately instrumented cables up to 5000 feet long. Because they must move along predetermined paths, they are not as maneuverable as fishing captains may believe, especially if they, too, are towing heavy gear.

Supply boats, which support drilling rigs and production platforms, might also create some awkward moments for commercial fishermen, especially in or near port. In all, however, the amount of traffic is relatively light and solutions relatively simple.

The Norwegians have successfully established a special radio frequency for the sole use of communications between supply boats and fishing boats in heavily trafficked zones. Wider use of radar reflectors and transponders, especially on small wood-hull boats, has often been suggested. And the oil industry has also introduced a few ideas, which I will describe in a minute. But, of course, the success of almost all such efforts requires that the fishermen and oilmen both install the necessary equipment, monitor information sources and participate in periodic meetings.

Finally, there are the potential onshore impacts of offshore petroleum activities. Here the issues are largely social and economic: the competition for labor, the impact on wages, the availability of harbor facilities, adequate ancillary services, etc. Sometimes, the economic infusion from oil operations balance off other local economic effects.

In closing, I want to discuss the many things the oil industry is doing to mitigate problems associated with offshore oil operations.

An outstanding example is the New England Marine Industries Council (NEMIC), established in 1974 by the API and a handful of New England commercial fishermen. Its purpose has been to jointly identify problems and explore solutions to them. Over the past several years, NEMIC has produced several positive results. A major thrust has been to expand and standardize communications between oil operators and fishermen. Channel 16, the emergency frequency, has been identified as the contact frequency between fish and oil captains.

In addition, information placards for display in wheelhouses have been prepared and distributed to fishermen. They show the type of warning signal geophysical ships use to indicate a submerged cable under tow. Another placard for use on oil vessels describes what fixed fishing gear and its accompanying buoys look like. Copies of both, plus a radio placard listing frequencies used by different types of fishermen and procedures for radio contact between vessels, will soon be sent to all operators.

A form has also been approved and distributed to fishermen for reporting lost gear or conflicts at sea.

Most important, perhaps, have been the captains' meetings which NEMIC has sponsored. This kind of one-on-one contact between captains of oil and fishing vessels is vital for establishing a basic understanding and congenial environment in which to head off trouble and deal with issues that arise.

While the successes of NEMIC have been limited, they have nonetheless been real, and the petroleum industry intends to develop similar liaison with fishermen in other areas.

Last February, Governor Garrahy of Rhode Island, in cooperation with API's Conservation Liaison office, hosted an unusual conference between the chief executive officers of eleven companies involved in the exploration of Georges Bank and some forty New England fishermen. The face to face meeting of the fishing and industry captains resulted in a frank exchange of concerns and the pledge by industry to immediately compensate fishermen when their companies are responsible for any loss or damage to fishing equipment.

In the final analysis, we believe that good communications between the two industries go a long way toward mutual respect and cooperation; good faith goes even further. It is with this in mind that the petroleum and fishing industries are charting a course of compatibility.

References

1. National Academy of Sciences, "Petroleum in the Marine Environment," 1975, Washington, DC.

2. Royal Society of London (to the Royal Commission on Environmental Pollution), "Oil Pollution in the Marine Environment: Some Research Needs." April, 1980. pp. 86. London.

3. Kash, D. "Joint Hearing Before the Committee on Commerce, Science and Transportation and Committee on Energy and Natural Resources, U.S. Senate," December 5, 1979. pp. 125-128.

4. Allen, D. W. et al, "Effects on Commercial Fishing of Petroleum Development Off the Northeastern United States," Woods Hole Oceanographic Institution, 1976. pp. 80.

5. Wade, N. "A Report on Relations Between the Offshore Petroleum Industry and the Norwegian Fishing Industry in the North Sea," Scanews, 1976. pp. 55.

6. Bascom, W., Mearns, A. J., and Moore, M. D., "A Biological Survey of Oil Platforms in the Santa Barbara Channel," J. Petroleum Technology, November, 1976. pp. 1280-1284.

7. Hunt, J. M. "Report to National Ocean Policy Study Committee of the U.S. Senate, Committee on Commerce," January 29, 1975.

8. Southwest Research Institute (1979) "Quarterly Conference for Bureau of Land Management Study, "Ecological Investigations of Petroleum Production Platforms in Central Gulf of Mexico," April 18-19, 1979, San Antonio, Texas.

9. "University Study Shows Platform Contribute to Food Chains," Ocean Industry, pp. 86-88, May 1978.

10. Ward, C. H., Bender, M. E., and Reish, D. J., "The Offshore Ecology Investigation--Effects of Oil Drilling and Production in a Coastal Environment," (1980) Rice Universities Studies, Rice University, Houston, Texas.

11. Cole, H. A. and Holden, M. J. "History of the North Sea Fisheries, 1950-1969," In: Goldberg, E. D., (ed.) North Sea Science, MIT Press pp. 337-360.

U.S. - Canadian Agreement on East Coast Fisheries

Sang-Myon Rhee
Harvard Law School
Cambridge, MA

I. Problems

The traditional reciprocal fishing arrangements between the United States and Canada, confirmed in the Treaty of Paris over 200 years ago, were threatened when the United States and Canada eventually adopted the 200-mile fishery zones. It began when the United States Congress enacted on 13 April 1976 the Fisheries Conservation and Management Act(FCMA). This act sought to extend the exclusive fishery zone from 12 to 200 miles, effective 1 March 1977. This dispute between the two gigantic fishing nations escalated when the Canadian government published on 1 November 1976, in the *Official Gazette*, the claimed limits of its 200-mile fishing jurisdiction -- to become effective as of 1 January 1977. As a result, vast areas of high seas within 200-miles, traditionally regulated by the International Commission for the Northwest Atlantic Fisheries (ICNAF), fell under the national jurisdiction of each country. The United States government responded to the Canadian action three days later by publishing its claimed areas adjacent to Canada. These actions thus established as a disputed area the eastern one-third of Georges Bank, an area of approximately 12,000 square miles. The settlement of the dispute in this Gulf of Maine area became more crucial than that of any other areas because of the enormous potential of Georges Bank, historically a famous fishing ground and presently an area of hydrocarbon deposit.

Both governments initiated discussions in late 1976, but did not produce any agreement by 1 January 1977, when Canada began enforcing its 200-mile fishery zone. Fortunately, an agreement on an interim arrangement for 1977 fisheries was reached on 24 February 1977, four days before the enforcement of the corresponding United States' 200-mile fishery zone. The interim agreement attempted to continue traditional patterns of reciprocal fisheries off the coasts of each country subject to coastal state regulation and maintenance of the <u>status quo</u> as to the existing fishing patterns. In the disputed areas, either flag state would enforce interim measures without authorizing third-party participation.

On 1 August 1977, the United States and Canada each appointed a special negotiator — Lloyd N. Cutler for the United States, Marcel Cadieux for Canada — to conduct intensive negotiations to reach a comprehensive settlement, encompassing maritime boundary delimitation, and fishery and hydrocarbon resources. The negotiators turned their attention to the immediate problems of marine resources, and submitted a joint report on 15 October 1977 specifying principles for the division and sharing of marine resources. The report envisaged a general structure of a comprehensive cooperative management of fish stocks of common concern, including guidelines for the basic administrative structure and a mechanism for settling disputes.

Despite the joint report, negotiations stalled several times. The 1977 interim fishery agreement expired on 31 December 1977 without further extension. Based on the negotiators' joint recommendation of 28 March 1978 concerning the terms of interim arrangements to govern reciprocal fishing, both governments agreed on a revised interim fisheries agreement for 1978. However, before its coming into force, the Canadian government unilaterally suspended the provisional implementation of the agreement on 2 June 1978, alleging that "the provisions could not be implemented by the United States under the its law," and that "the agreement did not contain effective mechanisms for cooperation and conservation for the proper management of fish stocks of mutual interest." Thus, the traditional order of reciprocal fishing broke down, and neither side could fish in the undisputed zone of the other. Meanwhile, in the disputed areas, competitive overfishing was caused by fishermen from both countries without coordinated policies, each side applying its own policy.

Furthermore, Canada announced on 15 September 1978 a tentative decision to withdraw its traditional position of equidistance line principle. She proposed as a substitute an "equitable equidistance line," claiming that Cape Cod and Nantucket Island should not be regarded as base-points for the purpose of maritime boundary delimitation. Such a revised Canadian claim would add 2,880 square miles — including 1,500 square miles of Georges Bank — to Canadian waters.

In October 1978, one year after the joint report, the negotiators reached agreement on most of the fisheries problems, shelving the sea boundary delimitation for submission to binding third-party settlement. On 22 November 1978, both governments instructed the negotiators to complete the negotiation of the presently unresolved details and to refer any remaining differences for resolution at the ministrial level by the end of the year. The year-end report and drafts of agreements were generally well received by the governments.

After eighteen months of intensive and protracted negotiations, both governments eventually announced on 14 February 1979 their approval of the recommendations of the special negotiators. They signed the two separate but interlinked agreements on 29 March 1979 in Washington D.C.: one on East Coast fishery resources, the other on the settlement of the maritime boundary delimitation by binding third-party decision.

This agreement created a unique framework of joint and cooperative management of fisheries in the adjacent exclusive fishery zones. It calls in detail for a joint fisheries commission, the sharing of fishery resources, mutual access, to each other's zones, and also a mechanism for settling disputes.

First of all, both parties needed a division of fishing grounds in detail, based on scientific data and distribution patterns, in order to implement their idea of fishery unitization. As a convenient way of identification of stocks, the huge area between 35°N and 55°N and between 42°W and 71°40'W has been divided into 5 subareas, and 25 divisions and subdivisions in the same manner as under ICNAF. As a result, approximately half of the total area falls outside of the 200-mile exclusive fishery zones of the two countries.

For the convenience and efficiency of joint and cooperative management, some twenty-eight fish stocks of mutual interest are divided into three categories. The stocks in category A — those which migrate widely throughout the waters of both countries or in the disputed area — will be jointly managed by both countries. Category B includes the stocks in which one country has the primary interest. The party of primary interest will propose management measures for the consideration and acceptance of the other. Category C covers those stocks which occur primarily in the undisputed fishery zone of one country to which the other has been given entitlement. Management measures are entirely within the discretion of the management responsibility.

The agreement contains detailed sharing and access provisions for stocks of mutual interest, taking account of ecological, historical and jurisdictional factors. For example, percentage entitlement shares in Georges Bank are as follows: for scallops 26.65% (USA), 73.35%(Canada); for cod 83%(USA), 17%(Canada); for haddock 79%(USA), 21%(Canada; and after six years, 66.66%(USA), 33.33%(Canada) for herring. Fishermen of one country may get access to the designated fishing areas off the coast of the other country, where primary management responsibilities are assigned to the coastal State for designated stocks. For example, for at least ten years United States fishermen will be allowed to fish for the agreed 35% share of the total allowable catch of redfish off Nova Scotia; Canadian fishermen will be permitted reciprocally to fish 9% of the annual permissible commercial catch of loligo squid off New England. The agreement also provides for permanent mutual entitlements and access for cod, haddock and certain other ground stocks in each other's designated zones.

The agreement further provides for the establishment of a joint fisheries commission, composed of seven representatives from each country. The commission will make recommendations to each government for management measures for categories A and B stocks, and will serve as a forum for consultation for category C. Each national section will have one vote; decisions of the commission are to be taken by agreement of both sides. Since decisions of the commission require the affirmative vote of both sides, there is a potential threat of occasional disagreement.

In order to overcome such cases, the agreement established a two level dispute settlement procedure in the commission. Two co-charimen, jointly appointed by the parties, will attempt to settle any disputes which may arise in addition to sharing the responsibilities of presiding over the commission meetings. In the event that the co-chairmen are unable to resolve the dispute, it will be referred to an arbitrator who will be jointly appointed by the two parties. Decisions of the arbitrator will be final and binding.

In consideration of the permanent character of the agreement, there are provisions for review and adjustment of the shares of the stocks at regular ten-year intervals. Either side may request such a review. If both parties do not agree on such redetermined entitlement, the problem will call for the arbitrator's binding decision. However, the arbitrator may make adjustment to the original entitlements up to a maximum of 10 percent of the total allowable catch for a stock, if a country has more than 50 percent share of the stock; or up to 5 percent, if a country has less than 50 percent share. Ultimately, any entitlement may not be reduced by more than one-third of its original level. Such adjustments are to be made primarily on the basis of the final delimitation of the maritime boundary by means of a third-party binding settlement procedures. Such, however, may also be redetermined through an account of socio-economic and other relevant factors.

Despite the unique provisions for the sharing of resources, mutual access, and cooperative management rules — along with a dispute settlement mechanism — the agreement has not yet been approved by the United States Senate. This stems from the conflicting views over the fairness of the agreement.

II. Symptoms

In sharp contrast to its reception in Canada, where fishermen, while not overjoyed with the agreement, generally thought it to be a good one, the fisheries agreement came to face strong opposition in New England fishing communities. The fishermen there, particularly those from New Bedford, Cape Cod, and Rhode Island who depend upon scallops, herring and squid, were furious at this intrusion on their livelihood. They formed on 15 May 1979 an emergency group, American Fisheries Defense Committee(AFDC), to block ratification. They have argued that the fisheries agreement is unacceptable "because it tilts awesomely, dangerously in favor of Canadian fishermen," and that it "would lock New England fishermen into an unfavorable situation forever." They even charged that the State Department sought to trade away the valuable fishery resources of Georges Bank in return for Canada's cooperation in submitting a maritime boundary dispute to a third-party settlement in order to get oil and gas development in the area [sic]. AFDC elaborated their opinion on four major issues.

First, they oppose the unfair allocation of stocks, particularly the 26.65% entitlement of scallops in Georges Bank. They would not listen to their Administration's view that the historical share, on which the allocation is detailed, represents the average in the period 1965-77. They still vividly remember the days of the 1950's when their fishermen dominated the scallop fishing in Georges Bank. They say that it was only in the early 1960's that the Canadians moved in, while United States scallopers were moving away to the south for better scalloping or diverting their efforts into other groundfishery where they could make a greater profit. This steady replacement grew until the Canadians rose to dominate the fishery, taking 85% of the scallop catch in Georges Bank in 1976. Dreaming of old days the Americans exclaim: "Had this treaty been signed in 1962, we could have laid claim to 80% of the scallop!" They assert that in 1978, when the competitive fishing was going on, they increased their catch up to 31%, and that in 1979 the estimated share would be as much as 40%.

Second, they assert that the Canadian access to the United States fishing zone is much more favorable than the reciprocal rights granted American fishermen in the Canadian zone. They contend that this is partially owing to the inequality — by an almost 2 to 1 margin — in the relative dollar value of the reciprocal fishing; partially owing to the less meaningful right of access to Canadian waters with the relatively long distances from the United States ports and relatively short fishing period during the winter.

Third, they worry that the agreement would give Canada unwarranted influence in fisheries management decisions by the New England Regional Fisheries Management Council(NERFMC), causing enormous difficulties in implementing FCMA. They argue that the complicated mechanism for settling disputes would eventually deprive the NERFMC of the flexibility to pursue effective and less costly management approaches. Implied in that is a fear that the arbitrator's decision might be beyond their control.

Fourth, they believe that the close interrelationship between the fisheries agreement and the maritime boundary settlement agreement may not reflect the United States interest, since they expect that the United States will probably win in the third-party settlement of the maritime boundary.

These complaints were heard in the Congress. Representative Gerry Studds(D- Mass), who represents the South Shore, Cape Cod and the Islands, took the initiative in blocking ratification. He was joined by some of the powerful senators from New England, particularly Edward M. Kennedy, and Claiborne Pell of the Committee of Foreign Relations.

Members of the House Merchant Marine and Fisheries Subcommittee, who will not have any say on the ratification but will, nevertheless, have to approve technical legislation carrying out its provisions, were highly critical of the agreement at a hearing on 22 June 1979. The Subcommittee found that the settlement of fisheries issues prior to settlement of the boundary dispute was against the United States' interest, and that the permanent nature of the agreement, along with unfair entitlements of stocks, was also undesirable. The Subcommittee recommended that such a complex and unworkable agreement should either "be renegotiated" or "be abandoned in favor of more suitable, and perhaps less formal arrangements."

The agreement had been stalled at the Senate more than a year when a preliminary hearing was held on 15-17 April 1980. The AFDC's struggle to block ratification was highly successful thanks to the support of Senators Kennedy of Massachusetts and Pell of Rhode Island.

The two senators held that the fisheries agreement was not fair and equitable in its present form. Their first objection concerned the permanent nature of the agreement. It would be impossible, they argued, to determine whether shares were equitable until the final settlement of the maritime boundary. Second, they asserted that allocations under the agreement were generally unfair, particularly the 26.65% entitlement of scallop on Georges Bank. Third, they argued that, in light of the FCMA, the United States should reserve exclusive management of scallops in the United States zone. In this regard, Senator Kennedy proposed a three-fold amendment that 1) the agreement should terminate three years from the date of its entry into force; that 2) considering the unfairness of the primary management line of

68°30'W, the United States should not grant Canadian vessels the right to harvest scallops west of a line derived by drawing straight lines connecting the following points: 42°30'N-68°00'W; 41°30'N-68°00'W; 41°30'N-67°00'W; and 40°30'N-67°00'W, and that 3) the United States should have the exclusive right to determine management measures, including scallops located on the United States side of the proposed line. However, the Kennedy amendment was postponed without a schedule being set for future debate. The preliminary hearing in the Committee ended without any conclusion.

Canada has been quite impatient with this lukewarm response on the United States' part. In contrast to the complicated process of ratification in the United States, the Canadian cabinet automatically approved the agreements soon after the signature. On 23 April 1980 the Canadian House of Commons unanimously passed a resolution that urged the Senate of the United States to take early and favorable action for ratification of the agreements since they represented "fair and urgently required cooperative arrangements in the interests of both countries." Furthermore, on 12 June 1980, the Canadian government decided to increase the Canadian quotas — of cod from 5,950 MT to 13,500 MT; that of haddock from 6,800 MT to 10,500 MT; and that of yellowtail flounder from zero to 2,500 MT — in retaliation for the continued delay by the United States Senate. As there was no immediate reaction from the United States, the Canadian government further threatened in early August 1980 to escalate the fisheries dispute into a fish war over Georges Bank scallop stocks by stating that Canada might increase or lift its scallop quota on Georges Bank in an effort to force the United States Senate to ratify the fisheries agreement.

It is desirable that the issues be debated again when the Senate resumes in the fall. Considering the findings and the recommendations of the House Merchant Marine and Fisheries Subcommittee at the hearing of 22 June 1979, and the debates and proposed amendment in the preliminary hearings of the Senate Foreign Relations Committee, it is doubtful that the agreements will be ratified in the present form in the near future.

III. Approaches

First of all, both the United States and Canada should realize that this current situation of non-agreement will bring enormously harmful impacts upon both countries. In the absence of a fisheries agreement, there is no formal mechanism or procedure to coordinate conservation goals and management policies for shared stocks. There can be, and often are, two separate policies to the same stock of fish, resulting in imbalance between potentially divergent interests. As has been proven since 1978, the absence of a fisheries agreement can cause competitive overfishing. Each country may win in a particular fishery in a particular year, but such gains will be offsetting. Such unwise competition will eventually do more harm than good to both countries. Denial of access to the others' fishing zone will cause enormous losses to both countries. The estimated losses of American fishermen related to the denial of access to the undisputed Canadian zone would be as much as $4.3 million in 1980. Canadians will suffer from similar losses as long as current non-agreement continues. Potential long term losses would become even greater because most of the major stocks, such as scallops, cod, haddock, would be depleted through overfishing in the disputed area.

The second controversial issue is whether the current fisheries agreement is fair and equitable. Draft Environmental Impact Statement(DEIS) shows absolute gains will go to the United States by the sharing and mutual access.

> <u>Stocks of Primary Interest</u>
>
> The United States would trade a potential annual catch of 4.1 thousand MT of fish in the United States zone, worth an estimated $4.2 million, for 15.5 thousand MT of Canadian fish potentially worth $7.7 million. After the reciprocal redfish and <u>loligo</u> entitlements expire the potential annual U.S. harvest of Canadian zone stocks is estimated at 7.95 thousand MT, valued at $4.9 million, while

> Canada would be entitled to a potential annual catch of 128 MT in the U.S. zone, worth about $70 thousand.
>
> <u>Transboundary Stocks</u>
>
> U.S. fishermen would be entitled to about 68 percent of the long-term potential transboundary stock harvest, or 424.5 thousand MT annually. The potential value of that share, at 1978 prices is about $200.9 million a year or 61 percent of the value of all transboundary stocks.
> ...
> Canadian fishermen would be entitled to a potential annual harvest of 200.5 thousand MT or 32 percent of the total potential transboundary stock yield. The potential value of that share is $126.6 million a year or about 39 percent.

Opponents of the fisheries agreement, however, argue that the <u>specific</u> entitlements are unfair to United States fishermen. For example, scallopers complain that the 26.65% share is not taking enough account of the recent increase of up to 40% or more in their harvest. They argue that if the United States would reduce its scallop catch by 35% of the current harvest, it will lose $10 million or 1,900 MT scallops. They further argue that such losses would be far greater than the possible gain of $4.3 million by future access to the undisputed Canadian zone. It seems then that while on the national level it is generally fair, there are on the regional level inequalities that have to be resolved.

The third crucial point at issue is the linkage between the fisheries agreement and the maritime boundary delimitation. The reason for the linkage was two-fold: that a "winner-take-all" approach would be so highly risky for communities in both countries that there should be an "insurance" against a possible loss of claims; and that once both sides settle problems of fishery resources, the importance of the ultimate placement of the maritime boundary would be diminished. However, neither government seems to be quite confident in the third-party settlement of the maritime boundary, mainly because applicability of "principles and rules of international law" to the delimitation of this particular case is not certain; and the "equitable principles", which represent the United States position, are competing almost equally with the Canadian position of the "equidistance principle" in the current Third United Nations Law of the Sea Conference. From the beginning Canada insisted upon the linkage between the boundary settlement and the permanent fisheries agreement, in order to insure against any possible disruption of its fishing community as a result of possible defeat in the third-party settlement of the maritime boundary. It is not certain, however, whether Canadians were less confident about the validity of the equidistance principle of the delimitation of the maritime boundary to be upheld in the third-party settlement; nor whether the reason behind their strategy was to lead the United States to conclude a "balanced" agreement in order to buffer their position in the future deliberations of the Court to get a "balanced" division of the Gulf of Maine. The most important factor and immediate concern in the Gulf of Maine dispute is the settlement of fishery resources rather than of hydrocarbon resources. The outcome of the fisheries agreement will be one of the major factors in the third-party settlement of the maritime boundary, which will again have impact upon future adjustment of fisheries allocations. The unity of resource deposit was frequently respected in the delimitation of maritime boundaries. Therefore, both issues of the fisheries agreement and those of the boundary settlement are closely associated and should not be held apart.

Hence comes the controversial issue of the alleged inflexibility and permanency of the fisheries agreement. Inflexibility involves provisions concerning mutual access, share entitlements, and management of certain fish stocks in the other party's zone. Opponents of the fisheries agreement worry that mutual access may not be desirable in implementing sound management of the FCMA. They also believe that the agreements will operate at cross-purposes with the implementation of the FCMA by permanently limiting the total harvest by United States fishermen. They argue further that this situation becomes worse because the United States is expected to win in the maritime boundary settlement. This allegation is not sound, because it ignores the fact that transboundary stocks should be managed through international cooperation to achieve optimum sustainable yield; and because it is risky to predict that the United States will win in the boundary settlement. The provision of 5% to 10% adjustment seems to be relatively inflexible, taking into account the actual fluctuation of fisheries economy in the past. The provision does not warrant full adjustment because it would be difficult to continue to break the agreed entitlement without competitive overfishing unless one side gives up, and furthermore because the arbitrator may just split the difference between the two different claims. Therefore, it is predicted that there would continue to be an almost constant level of entitlements unless one side gives up a certain percentage of shares. As long as the stipulated entitlements would not represent dynamic changes as were seen in the past, it is highly doubtful that this permanent inflexibility should continue even after the final delimitation of the maritime boundary. It is said that Canada rejected a United States proposal of ten year duration for the fisheries agreement, successfully persuading the United States negotiators to establish permanent linkage between the fisheries agreement and the third-party settlement of the maritime boundary. The reason was expressly that in case of defeat there would be disruption of fishing industries. It would be also inequitable that the hiatus between the maritime boundary and the allocation of fishery resources should remain forever without flexible adjustment. In the long run it would be fair and equitable that gradual reallocation should be made without disruption of fishing economy, respecting the final outcome of the boundary, which should be also determined respecting the fair shares of resources.

IV. Proposals

The fisheries agreement should be renegotiated and amended. The fisheries agreement is generally regarded as a fair agreement, excepting a few specific regional complaints, particularly those of New England scallopers on Georges Bank. Their successful lobby brought about overwhelming support from the House Subcommittee on Fisheries and Wildlife, and the Senate approval of ratification is highly likely to be blocked by the probable amendment resolution proposed by Senator Kennedy. Agreements should be fair and equitable, and thereby acceptable to both sides, particularly to the affected communities in both countries. Otherwise, agreements will be violated and disputes will arise. Therefore, it is to be advised that both parties should renegotiate the agreements, particularly respecting the following proposals.

AFDC's arguments are sympathetic and convincing as far as their position about the scallop fishing is concerned. It is not fair to make a nearly permanent allocation of 26.65% from the stock taking account of the limited period between 1965-77. Although the New England scallopers were pushed out of Georges Bank in the 1960's, they still remember their golden days in the 1950's before the Canadians moved in, and their recent harvests show about 31% in 1978, 40% or more in 1979 — well above the provided average. It seems to be desirable that the average should be made at least as long as a span of a generation, or an approximate working period of an average fisherman, e.g., a quarter of a century.

The fisheries agreement provides that Canada shall be the party of primary interest with respect to scallops located east of 68°30'W, whereas the United States shall be the party of primary interest west of 68°30'W. This is absurd and should be amended. It is difficult to imagine how the entire scalloping zone including the undisputed western half of Georges Bank can be permanently under the Canadian domination. Although it is understandable that the line 68°30'W was drawn by the negotiators taking into account the historic title of the Canadian scallopers during the period of 1965-77, it is not reasonable to argue that the historic title allegedly consolidated during that decade should continue forever with only a minor adjustment in each decade. Considering the unsettled fluctuation of scallop fishing in the zone and the relatively premature consolidation of the title, it seems only fair that the share entitlement or zonal allocation should not continue permanently but should diminish gradually in the long run, or be adjusted flexibly respecting the local interest where the zone belongs.

It should be borne in mind that since scallops are basically sedentary species, not belonging to transboundary stocks, they should be primarily managed by the country which owns the continental shelf. As to lobsters, both sides already agreed to terminate joint management plan in category A and to switch to independent management in category C by the end of the fishing year in which the maritime boundary has been delimited. Similarly, as to scallops, the aspects of sedentary species should be respected — taking into account the final determination of the jurisdiction, without ignoring Canadian historical title.

By the same token, it is possible to create an elaborate and sophisticated management mechanism for this particular stock by drawing an acceptable line or by establishing a joint management zone. In this regard, it is useful to know the pattern of distribution of scallops in Georges Bank. Scallops are concentrated in an oval shape in all but the center of Georges Bank. The northern edge of Georges Bank has been the biggest and most consistent producer. The northeast peak and the Great South Channel are consistent producers, but at a lower rate. The northern edge and northeast peak of Georges Bank are located in the disputed area and have been dominated by Canadian scallopers, while Americans prevailed in the Great South Channel. Respecting the historic titles in the areas dominated by each side, it would be desirable to establish a line in the middle or a joint development zone between the two.

In this regard, Senator Kennedy's amendment proposal of scallop management line is noteworthy. He proposed that no Canadian scalloper should be granted the right to harvest scallops in Subdivision 5ze west of a line derived by drawing straight lines connecting the following points: 42°30'N-68°00'W; 41°30'N-68°00'W; 41°30'N-67°00'W; and 40°30'N-67°00'W; and that the United States scallopers should be given the exclusive rights to determine management measures in the above mentioned zone. This calls for the geographical division of Georges Bank by halves, barring against the expanding Canadian scallopers while giving them preferential rights on the important northern and northeast peak, where they have already been flourishing.

Next, the relationship between the fisheries agreement and the maritime boundary settlement agreement should be reconsidered. Under the agreements, both will enter into effect simultaneously, subject to ratification; the former may be terminated by agreement, the latter will remain in force until either party decides to terminate the agreement upon six months written notice. It is understandable that the risk of the third-party boundary settlement should be "insured" and should be remedied in advance. But this does not necessarily mean that the advantage of the "insurance" should continue forever. The expressed purpose of the "insurance" was to safeguard the fishing community in both countries. As long as the possible disruption of the local economy is remedied one way or the other, or the impacts diminished gradually, e.g., over several decades; resources should belong to the party who retains the jurisdiction over the zone. It is expected that factors of natural resources will be given due respect in the third-party settlement. In this regard, it may be desirable for both parties to choose the *ad hoc* Court of Arbitration rather than the Chamber of the International Court of Justice, because there is generally more room for compromise in arbitration then in adjudication. Alternatively, it would be more desirable to settle through mediation or conciliation. If both parties are afraid of the outcome of the third-party decision, they may ask the Court to identify applicable principles and rules instead of "the course of the single maritime boundary."

As both parties have already acknowledged, the importance of the maritime boundary will be substantially diminished once the problems of resources are settled. In fact, fishery resources are much more valuable than hydrocarbon in the Gulf of Maine. According to the DEIS and the Final Supplement to Environmental Statement, proper management of Georges Bank could sustain annual landings of $229 million, whereas hydrocarbon reserves are, by contrast, estimated at 123 million barrels of oil and 870 billion cubic feet of gas, having a net value of $588 million. Once the hydrocarbon has been depleted, the importance of the maritime boundary will be further diminished. Indeed, both parties may arrange a hydrocarbon joint development project as found in the 1974 Korea-Japan Continental Shelf Joint Development Agreement, shelving the issue of the boundary delimitation for future generations.

The crucial question to be answered is which party might gain a possible benefit by concluding the fisheries agreement along with the boundary settlement by a third-party decision. The answer is Canada. In the fisheries agreement, the issue was how to balance and harmonize mutual interest in a fair and equitable manner. This was the very objective that Canada sought. In the maritime boundary claims,

Canada will still pursue "fair, balanced, and 'equitable'" objectives, arguing that the equidistance line would be fair and balanced for Canada; the new Canadian claim will be "equitable" for Canada. The United States will clamor for "more," because she has a three times longer coastline and, allegedly, Georges Bank is a natural prolongation of New England. In this regard, "fair and balanced" fisheries agreement is not "equitable" for the United States. This is especially true where the unfair regional scallop fishery is concerned. That is why Canadian fishermen are quiet and ready to accept the agreement, why AFDC are clamoring. For these reasons, the fisheries agreement and the maritime boundary settlement agreement should be held aloof — even though they are closely associated with each other.

The third important issue is the duration of the fisheries agreement. Since the answer to the relationship between the fisheries agreement and the maritime boundary settlement agreement has been suggested, the answer to this issue is simple. It is desirable that the fisheries agreement should last until the eventual settlement of the maritime boundary. This does not mean however that the fisheries agreement should be wholly discarded. It created a novel framework for peaceful resolution of serious differences between two neighboring countries over fisheries in the exclusive fishery zones, by providing in detail, as proposed, for sharing of resources, mutual access, and cooperative management rules, as well as a mechanism for settling disputes. Both parties may simply renew the agreement or just reserve to some extent by the principle of _pactum de contrahendo_. Agreement should be fair and equitable, but it should be also acceptable by providing flexibility and adjustability following the dynamic course of history.

Once the maritime boundary is settled and subsequently unfair provisions in the fisheries agreement adjusted, the issue of duration of the fisheries agreement will become less important than it is now. Then the remaining major issue would be the flexibility of readjustment at certain intervals with respect to share entitlement and mutual access. More flexible provisions, _e.g._, adjustment of 10% to 20% every 10 years or 5% to 10% every 5 years, would encourage the joint and cooperative management to be more convenient and efficient.

The fisheries agreement opened a new era of fair and equitable solution of fisheries problems between neighboring States by the creation of a unitization mechanism for fish stocks, i.e., division of fishing areas by scientific data, allocation of stocks by preference entitlement, and sharing fishing zones by mutual access and cooperative management. Despite the uniqueness of the agreement, there still remain some unsettled points. Once the agreement is renegotiated and amended as proposed above, it will not only contribute to friendly relations between the United States and Canada, but also to the development of principles and rules for the peaceful solution of fishery resources and maritime boundary disputes between States as a model precedent to be followed by others.

FIGURE 1

U.S. - CANADIAN CLAIM LINES AND FISHERY MANAGEMENT ZONES WITH DISTRIBUTION PATTERNS OF SCALLOPS

Source: Department of State, DRAFT ENVIRONMENTAL IMPACT STATEMENT, p. 69 (April, 1980)

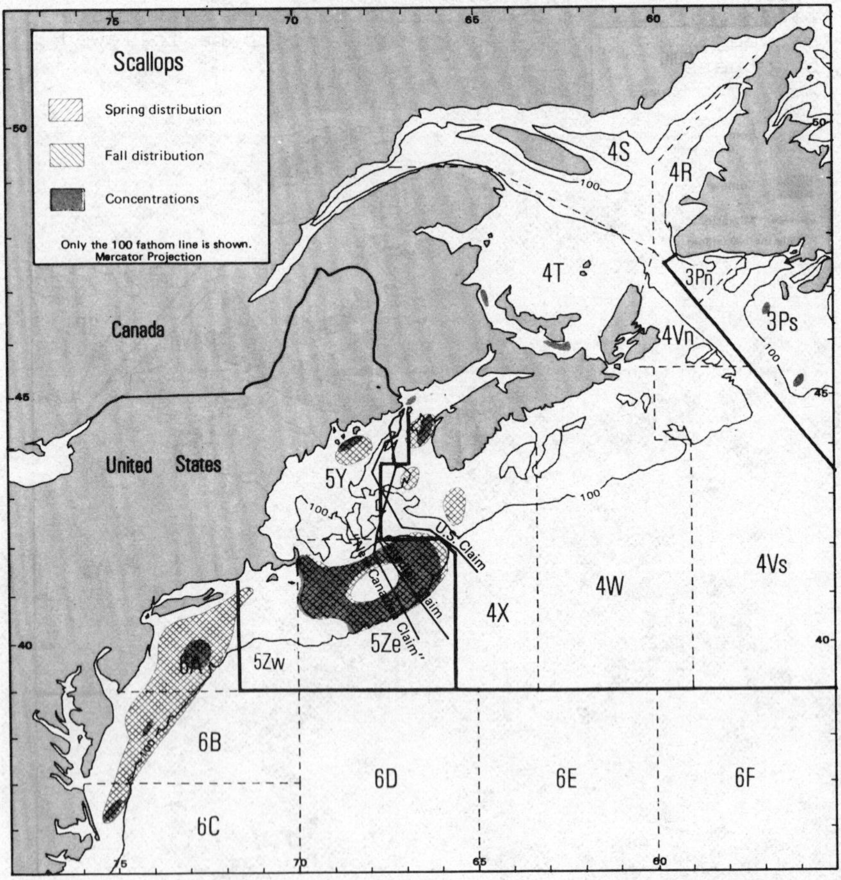

FIGURE 2

FAIRNESS TEST: DOLLAR VALUE OF SHARES

DISTRIBUTION OF THE VALUE OF POTENTIAL ANNUAL LONG-TERM CATCH SHARES*

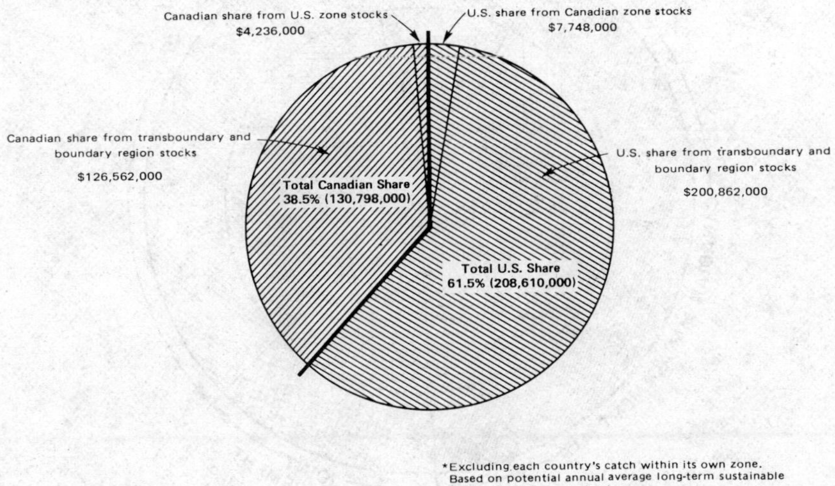

*Excluding each country's catch within its own zone. Based on potential annual average long-term sustainable catch, at 1978 exvessel prices.

3545 12-79 STATE (RGE)

Source: Department of State, DRAFT ENVIRONMENTAL IMPACT STATEMENT, p. 128 (April 1980)

FIGURE 3

FAIRNESS TEST: ALLOCATION OF MANAGEMENT POWERS

ALLOCATIONS OF PRIMARY MANAGEMENT AUTHORITY
BY VALUE OF POTENTIAL HARVEST*

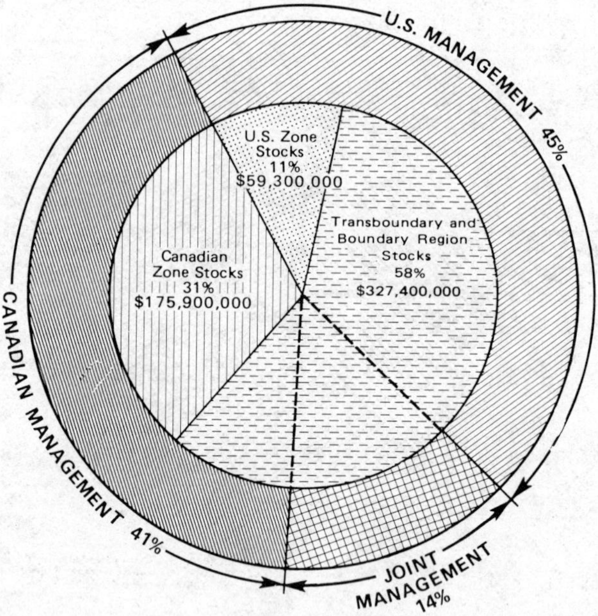

*Potential annual long-term sustainable catch, at 1978 exvessel prices in U.S. dollars.

3544 12-79 STATE (RGE)

Source: Department of State, DRAFT ENVIRONMENTAL IMPACT STATEMENT, p. 125 (April, 1980)

THE PACTEX PROJECT: ANY LESSONS LEARNED?

Richard M. Donaldson, Vice President,
Government and Public Affairs

F. Harlan Flint, Jr., Director,
Policy and Planning,
Government and Public Affairs

The Standard Oil Company (Ohio)

In May, 1979, The Standard Oil Company of Ohio ("Sohio") abandoned its five year, multimillion dollar effort to build a marine terminal and pipeline--the so-called "Pactex project--which would have moved 500,000 barrels a day of Alaskan North Slope crude oil eastward from Long Beach, California to Midland, Texas. The abandonment of this major project that almost everyone believed was in the national interest, sparked an instant coast-to-coast furor. What happened to this project? What lessons can be learned from its failure? Have we learned them?

THE BACKGROUND OF PACTEX

The Prudhoe Bay field was discovered in 1968 on the Arctic Coast of Alaska. It is the largest single oil field ever discovered in the United States. In 1969, Sohio acquired leases from The British Petroleum Company which contained over 50 percent of the oil in the Prudhoe Bay field. As a result, Sohio took on the responsibilities of producing and transporting this oil to consumers in the United States.

Starting in 1970, Sohio participated with seven other oil companies in the long and costly effort to build the trans-Alaska pipeline which was completed in 1977. This pipeline ran 800 miles southward from Prudhoe Bay on Alaska's North Slope to Valdez, an ice-free port on Alaska's southern coast. It was built in one of the most environmentally sensitive areas in the world and posed unusual engineering challenges. The pipeline's construction and operation have been given high marks by virtually everyone in and out of government who has taken the time to study and observe the engineering and environmental aspects of the project.

When the trans-Alaska pipeline was first being planned, forecasts indicated that all of the Alaskan North Slope crude oil could be refined on the West Coast of the United States. Shortly after the Arab oil embargo in October, 1973, it became apparent that modification of those forecasts would be required. The embargo caused an immediate drop in demand for petroleum on the West Coast and various refiners there cancelled plans for additions to refining capacity.

In early 1974, Sohio completed an initial examination of the probable impact of these factors. The key conclusions were: (1) that there would be a lesser growth in the demand for petroleum on the West Coast than had been expected before the embargo, (2) that oil in excess of West Coast requirements would be available on the West Coast beginning in about 1978, just after the start-up of the trans-Alaska pipeline and that a substantial part of the West Coast surplus would be new Alaskan North Slope crude oil, and (3) that since this Alaskan oil could not be legally exported from the United States, the quantity surplus to the West Coast would have to be moved eastward either through the Panama Canal or through a pipeline like Pactex. Sohio informed federal government officials of this developing situation.

During May, 1974, Sohio began feasibility studies to evaluate alternative transportation routes for moving Alaskan oil eastward. Many possible routes were considered, including routes through Canada, across the United States, Mexico, Central America, Panama, and around Cape Horn. Technical feasibility, environmental concerns, cost, time, and other significant matters were carefully studied. This work indicated a clear advantage for a pipeline commencing at San Pedro Bay in California and extending across the southwestern United States to a point near Midland, Texas, connecting with pipelines to refineries east of the Rockies.

The existing ports in San Pedro Bay were Los Angeles and Long Beach. Sohio settled on Long Beach for the marine terminal and began the process of detailed engineering and seeking the necessary governmental permits and approvals for what promised to be a nearly one-billion dollar construction project. Though the investment would be large and one not contemplated when Sohio undertook its Alaskan venture in 1970, the Pactex link appeared to be a less expensive means in the changed circumstances for moving large quantities of Alaskan oil east of the Rocky Mountains where it would be needed during the years of maximum Prudhoe Bay production.

THE PROJECT ITSELF

The Pactex project was basically simple. It contemplated straightforward and technologically conventional construction of a marine terminal and pipeline. It involved no novel engineering or unusual environmental problems.

It would begin with the construction of a new, deepwater tanker terminal at the Port of Long Beach, including a pier for docking tankers at the outer edge of the harbor, a minor amount of dredging, and the prospect of reducing and simplifying existing harbor traffic. The terminal facilities were to include storage tanks of the most modern design minimizing air pollutants and incorporating the usual pumping facilities. Landscaping and arrangements to accommodate recreational activities in the immediate area were integrated into the plans.

The pipeline portion of the Pactex project would carry crude oil eastward from Long Beach 1,030 miles through California, Arizona, New Mexico, and West Texas to Midland. A significant feature from the standpoint of cost, time, and environmental concerns was the fact that the Pactex pipeline would use two existing natural gas lines owned by the El Paso Natural Gas Company and the Southern California Gas Company for about 800 miles of the route. These lines would be modified to carry crude oil from west to east. Only 227 miles of new pipeline would be needed to connect the two existing pipelines and the new terminal with the West Texas terminus of the project.

It also seemed likely that the necessary permits could be secured and all the construction completed to be ready shortly after oil started moving through the trans-Alaska pipeline.

ALMOST EVERYBODY FAVORED PACTEX

Given the rising level of foreign oil dependence, the shock of the Arab oil embargo, and concerns about its impacts on both the country's economy and national security, it was not too surprising that almost everybody favored the Pactex project. As the first west-to-east crude oil pipeline in the United States, the advantage of being able to transport quantities of new Alaskan oil eastward to help balance oil supplies in areas east of the Rockies that were more dependent on foreign oil was apparent. Communities along the pipeline's proposed right-of-way appreciated its potential for supplying oil to their areas as well as the economic stimulus and employment it would bring. Most people recognized that private industry was undertaking this project as the most efficient, safest, best environmentally, and least expensive way to move Alaskan oil eastward, without one cent of government funds being involved.

The President of the United States, the Secretary of Energy and other key federal executives described the Pactex project as being "of critical national importance." Congress enacted legislation to expedite federal decisions on the project.

The Governors of Texas, New Mexico and Arizona strongly favored the construction of Pactex. The Governor of California supported it -- if air quality disputes between the California Air Resources Board ("CARB") and the State's South Coast Air Quality Management District ("SCAQMD") could be resolved. Resolutions of support were also adopted by a variety of governmental bodies in all four states.

And, in the November, 1979 general election, the voters of Long Beach who might be most affected by this project, approved the project by a 61 percent to 39 percent vote when they approved the Lease Agreement between the Port of Long Beach and Sohio.

With all this support, it seems inconceivable that the Pactex project could fail, but it did.

THE PERPETUAL PERMITTING PROCESS

By the time the Pactex project was abandoned, Sohio had identified more than 700 federal, state and local permits required if the project were to be constructed.

The issues were many and they were interrelated. Each aspect of the project required the involvement of a wide variety of federal agencies and every level of state and local government in California, Arizona, New Mexico and Texas. New agencies were created in the midst of the planning process. New federal and state laws were adopted, and numerous new federal and state regulations were put into effect during the period of Sohio's efforts.

In order to comply with new laws and regulations, Sohio was required to do additional work, and sometimes to redo work that was already done under pre-existing laws and regulations. Some of the laws and regulations were mutually inconsistent. Sohio tried to resolve problems which occurred between different government levels, within state and local governments and between various federal agencies. At one point, three different agencies with different viewpoints on an issue, each insisted that it make the "final" decision.

The proliferation of public hearings under various laws and regulations extended the time for decisions on key permits, multiplied the number and complexity of issues concerning many of these key permits, and occasionally made public officials politically gun-shy from making routine decisions.

Any official, agency, or private individual who wished to challenge or delay the project for whatever motive also could take any permit to court, questioning its substance or the procedural details under which the government had issued or proposed to issue it. A number of lawsuits were filed, and others were threatened if the pending suits were not successful in stopping the project.

The key lease of the area for the marine terminal at the Port of Long Beach was politically challenged by a ballot initiative. Though the challenge was soundly defeated by the people of Long Beach at the polls, Sohio did not know how the vote would go, and expended months of work and several hundred thousand dollars trying to communicate the nature of the Pactex project to the people of Long Beach and the benefits they would derive from it.

In retrospect, the cause of the abandonment of the Pactex project does not rest on any single agency, official, law, regulation, hearing or lawsuit. Each challenge to the project was conceived as having merit. The problem was inherent in the overall process, or said another way, the system was the problem and not the solution.

Through five years of work on this project, Sohio's outlook became one of disappearing optimism. Ultimately, time destroyed the underlying economics of Pactex as a project. When Sohio abandoned the project, most of the 700 plus permits had been secured or clearly could be. Sohio knew, however, that construction of a project that large could not be started until all key permits were approved and not tied up in court. By May, 1979, Sohio still didn't know when the key remaining permits would be issued and free from litigation. By the earliest possible time that permits could be issued, litigation ended, and construction completed, most of the Alaskan oil that Pactex should have carried eastward would have already been produced and moved to market by tanker through the Panama Canal, substantially eliminating the principal reason for building Pactex in the first place. Though other oil from Alaska or the West Coast might be discovered in the future, major investments in pipeline systems are not made on the hope that they might have something to transport in the future.

A CLOSER LOOK

Although "perpetual permitting" may summarize the demise of Pactex, a closer look at the successes and failures on some of the major permits gives more understanding of the situation.

By the latter part of 1976, a lot of work had been done. Applications had been filed in mid-1975 with the Bureau of Land Management for the right-of-way, and with the Federal Power Commission to convert the existing El Paso gas line to a crude oil line as part of the Pactex project. The federal draft Environmental Impact Statement was completed, and California's Environmental Impact Report had also been completed and released. Also, in October of 1976, Sohio filed its air quality permit applications for the project with SCAQMD.

Despite the rather advanced stage of the planning and permitting process, there were some rather disturbing and conflicting signals coming from California authorities. They dealt with some very fundamental aspects of the project. In July, 1976, the Chairman of CARB wrote the Federal Energy Administration that:

"Sound public policy would seem to indicate

> that even if the legal obstacles could be overcome, the Sohio facility should not be built in the Los Angeles basin."

And in a U.S. Senate committee hearing, he said:

> "Sohio's Long Beach proposal would create massive environmental problems."

And further that:

> "Other terminal locations--particularly Kitimat, British Columbia, a site at Port Angeles, Washington, and a site on the south-central coast of California are preferable on both environmental and economic ground."

The Chairman of the California State Energy Commission supported CARB's views stating:

> "It appears that the proposed Long Beach/ Sohio terminal project is the least desirable of the three North American pipeline proposals."

California officials were not unanimous in this position, however. In another hearing at about the same time, the California Fish and Game Department testified that if a California site were chosen as the port of entry for Alaskan North Slope crude oil, it considered the Long Beach Harbor to be the best location for it.

The Sierra Club also objected to the CARB suggestion that other California sites be considered. The Energy Coordinator for the Sierra Club testified at the U.S. Senate hearing stating:

> "Evaluating the Sohio proposal from a land-use standpoint, of all the terminal proposals, Long Beach is clearly preferable."

It was very disturbing to Sohio and to others that, after over two years of planning, designing and engineering, and after over two years of permit processing, it was still not even generally agreed that the project should be located in Southern California, or indeed, any other part of California

Everyone concerned, opponents, proponents and Sohio, all agreed that the air quality problems in the Los Angeles Basin would be the most difficult to solve and that the Pactex project should not worsen them. In December, 1976 hearings, Sohio's Executive Vice President said:

> "On the basis of our work over the last three years, we believe that this air quality issue can be resolved at Long Beach through 'trade-offs' and use of appropriate technology, and we have indicated a willingness to employ such measures to implement the project with an acceptable level of emissions."

The underlying problem of air quality impacts and acceptable means for offsetting these impacts, however, would continue to provide the major source of controversy throughout the remaining life of the project.

By early 1979 and after more than four years of work, it was clear that substantial progress had been made:

* The U.S. EPA had issued all federal air quality approvals required by the project.

* All other major federal permits and approvals had been issued or would soon be ready to be issued.

* The Congress of the United States had recently determined that expeditious federal and state decisions on the project were of the utmost priority--and had enacted expediting legislation for all federal permits and approvals.

* Texas, New Mexico and Arizona had issued or were prepared to issue promptly all necessary approvals for the Pactex project in their jurisdictions.

* The Pactex project had been the subject of extensive studies and public hearings before all affecting California state, regional and local agencies. A massive record had been accumulated and many issues had been resolved.

* With the approval of a large majority of the voters of Long Beach, the Port of Long Beach had

approved a lease to Sohio for the marine terminal.

* Agreements had been reached with El Paso Natural Gas and Southern California Gas Company to utilize hundreds of miles of the existing and unneeded pipeline system.

* Sohio had committed itself to a precedent-setting emission trade-off package estimated to cost it in excess of $80 million. The result would have been a net benefit to the Los Angeles Basin air quality.

But much still remained to be done:

* The vital air quality construction permit was still pending before SCAQMD and would later require the approval of CARB.

* A required environmental impact report related to the proposed air quality trade-off package was pending before the California Public Utilities Commission.

* An application to locate two essential oil tanks at the terminal facility was still pending before the California Coastal Commission, and some of its members were concerned about the air quality aspect of the project though that may not have been within the Commission's jurisdiction.

* There were several lawsuits already pending in the courts of California. The most important of these challenged the adequacy of the California Environmental Impact Report on the project, and none of the major permits could be finally approved until that report successfully withstood judicial review. This case had been pending in the California Supreme Court since February of 1978.

* Several other lawsuits were threatened and virtually assured if and when the final permits were issued. Even if the existing and prospective litigation had only a slim chance of success, it still had the potential for delaying final decisions and the start of construction for as much as several years.

THE AIR QUALITY PROBLEM

Symptomatic of the Pactex problems with the whole permitting process was the air quality permit track.

The Los Angeles Basin which includes the Port of Long Beach is described as a "non-attainment" area, having failed to attain the air quality standards set under the Federal Clean Air Act as amended. The 1970 Clean Air Act established that in non-attainment areas, there could be no further industrial growth which would increase emissions. Because a complete suspension of industrial growth was likely to cause severe economic and social hardships, in 1976 the Federal EPA adopted a new policy to permit new growth if emissions from the new source could be offset by reductions from one or more existing installations. This meant that industrial growth even in a non-attainment area like the Los Angeles Basin, need not be halted so long as there could be an overall net improvement in the air quality.

Under the amended rule, any project emissions, including those from tanker operations, had to be controlled to the greatest extent possible at the source. Any remaining emissions had to be offset by an even greater reduction in emissions from other existing air pollution sources in the area. Since Sohio did not have any other operations of its own in the area, it was obliged to seek out sources of pollution by others which could be reduced or eliminated at Sohio's expense. This was the price for allowing the Pactex project to proceed.

To make a long story short, Sohio's Pactex project came along at a time when neither the Federal EPA nor CARB nor SCAQMD had established clear requirements to determine how the so-called "new source offset policy" would be carried out. In the absence of clear federal, state or local requirements, the participants engaged in long and tedious negotiations over such things as how to measure the emissions that would have to be offset, and what would be appropriate offset candidates to meet the requirements of the new laws.

The disagreements were not solely between Sohio and the California agencies. There were also very important differences of opinion between SCAQMD and CARB. Sohio not only had to satisfy the differing government agencies, but had to negotiate the arrangements for the setoffs with third parties who were not altogether willing to be involved.

After many unsuccessful attempts, a plan finally began to evolve. The main elements of the trade-off package were a flue gas scrubber and de-NO_x equipment to be installed on a Southern California Edison Company power plant. Sohio would have to bear the cost of constructing and operating these facilities up to a total cost of about $80 million.

CARB and the Governor of California seemed to believe the scrubber package was essential to the project. SCAQMD, on the other hand, felt that other means were more appropriate, such as requiring incoming tankers to burn low sulfur fuel within a specified distance of the Port of Long Beach and requiring power plants to be fueled with low sulfer oil. Both alternatives were expensive. Sohio was prepared to do either, but not both, and the stalemate on the air quality permit continued. The disagreement between these agencies was still not resolved at the time Sohio finally threw up its hands after almost five years and $57 million of effort and said "We quit!"

HUMPTY-DUMPTY

Although Sohio's announcement of the Pactex abandonment touched off a coast-to-coast furor, the warning signals had been apparent for months as a result of the Company's frustration with the permitting process. Many public officials scurried around to find someone else in government to blame. However, some responsible individuals wanted to know what might be done to reinstate the project.

Sohio met with the Secretary of Energy and other members of the Administration, and with members of the Congressional leadership. It also met with the Governor of California, various state executives, and key members of the California Legislature. Sohio informed them of the seriousness of the problems it still faced and the impact of delay on the project's economics. Hearings were held and legislation was introduced in Congress and in the California Legislature to speed up the permit decision-making process and to try to expedite the then pending and prospective future litigation challenging the project. Both legislative bodies were sympathetic but were understandably reluctant to make the substantive and procedural changes in existing law that would enable the project to start construction promptly.

Ironically, the California Public Utilities Commission approved the impact report on the trade-off package, the

California Coastal Commission approved the two oil tanks for the terminal facility, and CARB decided that Sohio might choose either CARB's or SCAQMD's alternative for the trade-off package. All of this happened after Sohio decided to abandon the project.

After again considering all the alternatives, even under the most optimistic assumptions, Sohio again and finally terminated the Pactex project. Following a resolution by Sohio's Board, Sohio's Chairman stated:

> "The fact of the matter is that neither these government leaders nor Sohio can turn around the results of five years of delay which have substantially eroded the economic attractiveness of this project."

PACTEX REVISITED

Many in the United States are concerned, not only with what happened to the Pactex project, but with the likelihood that the same thing could happen to any other major energy project. Indeed, Pactex is a good case study of the problems that could occur with any major project.

* The project would have employed conventional engineering and design, and would have been constructed and operated in accordance with the highest standards.

* It would have been obliged to incorporate environmental safeguards consistent with the best available control technology, and its direct environmental impacts would have been extremely small.

* It would have involved a one billion dollar private investment, and yet was almost universally deemed to be in the clear national and state interest.

* It fell within the jurisdiction of a large number of federal, state, regional and local agencies and was subject to the review of several states and many local political subdivisions. The objectives and procedures of these governmental bodies sometimes changed in the course of the project, fre-

* quently overlapped, and were sometimes in conflict. All of that cannot be avoided, but perhaps much of it could be in the future.

* There was sometimes a conflict between the project sponsor's need for promptness and finality of decision and the need for full and free participation by all public and private interests in all levels of regulatory and judicial proceedings. Both are important, but a better balance is needed.

WHAT LESSONS CAN BE LEARNED FROM PACTEX?

* There should be some means of imposing reasonable time limits on the total decision-making process at all federal, state and local levels.

* There needs to be some coordinating mechanism for complex projects, particularly those in the national interest and particularly where there is the likelihood of overlapping jurisdictions.

* While the right of full participation in project proceedings by all public and private interests must be preserved, some means must be found to assure that legal challenges are resolved more quickly without reducing full consideration of the substantive aspects of such challenges. It should be recognized that expedited decision-making does not mean weakening environmental standards.

* In the event any federal, state or local government fails to make a timely decision, as distinguished from making a decision which either approves or disapproves a permit, some centralized decision-making body should do so applying the applicable federal, state or local law.

* With regard to the principle of trade-offs, there can be no doubt that the burden of cleaning up historic pollution puts an onerous financial burden on any new project. Legislative guidance is needed here and should be considered by the appropriate legislative bodies in their next sessions.

* The multiplicity of permits required for any sizeable project will inevitably prevent the construc-

tion of some projects which on their own merits and in the national interest should be approved. Since many permits find their ancestry in federal legislation or in state legislation spawned by federal legislation, Congress may from time to time have to exercise oversight approval for a particular project in the national interest. That procedure, however, is a cumbersome and unpredictable track to follow in implementing a major project.

WHAT IS BEING DONE ABOUT PACTEX LESSONS?

Today, not much.

Tomorrow,............?

ENERGY FACILITIES AND THE CALIFORNIA COAST

by William R. Ahern

Introduction. A rancher near Point Arena on California's central Mendocino County coast believes there may be a small oil field under his property.(Figure 1) He applies to the California Coastal Commission for a permit to drill two temporary exploratory oil wells to find out. The wells would be about three miles from the coast on a coastal shelf. But these would be the first oil wells permitted in this rural scenic area. Some residents worry that an approval would signal the U.S. Interior Department that leasing for offshore oil development offshore Mendocino would be acceptable. They strongly oppose such leasing and are in a fight on proposed OCS Lease Sale #53. Lawyers for the Pomo Indians, who have a small reservation near the ranch, worry about the effects of oil drilling on that community and want strict archaeological surveys before any drilling might start. The California Department of Fish and Game biologists oppose the drilling. The two wells would be in the watersheds of the Garcia River and Hunters Lagoon, where wild swans breed. Although the mouth of the River and the Lagoon are about three miles away, the Fish and Game biologists worry that an oil spill during the rainy season might make its way through the gulches to the waterways. On the other hand, local ranchers, many barely sustaining cattle operations on the sparse coastal shelf grasses, see the oil royalties as making them able to keep their ranches.

This proposal for two small exploratory oil wells onshore had all the controversial elements associated with proposed large electric power plants and liquefied natural gas terminals. In some ways the two wells were more controversial. Mendocino County and the regional coastal commission approved them. Opponents, of course, appealed to the State Coastal Commission, which also approved them, but only with a large number of stringent conditions, including a $15 million bond to compensate for any damage that might occur to the wildlife.

This small permit case illustrates our general experiences with energy facilities during the past three and a half years under the California Coastal Act of 1976. The California Coastal Commission has approved all energy project applications made to it, but usually only after much controversy and only with conditions that sometimes make major changes in the proposed project. The Commission has permit authority over all proposed energy developments except power plants and an LNG terminal. With an approved coastal management program, the Commission also has "consistency" review over federal agency energy activities and permits. It is responsible for approving port plans and the coastal plans developed by 68 local coastal governments. And the Commission is the State allocating authority for the Coastal Energy Impact Program (CEIP) grant and loan funds. With permit, planning, "consistency" and CEIP grant authority, the California Coastal Commission has the needed "tools" to affect and control coastal energy developments. But even with these tools, protecting

---

Dr. Ahern is Chief, Energy and Ocean Resources Division, California Coastal Commission, 631 Howard St., San Francisco.

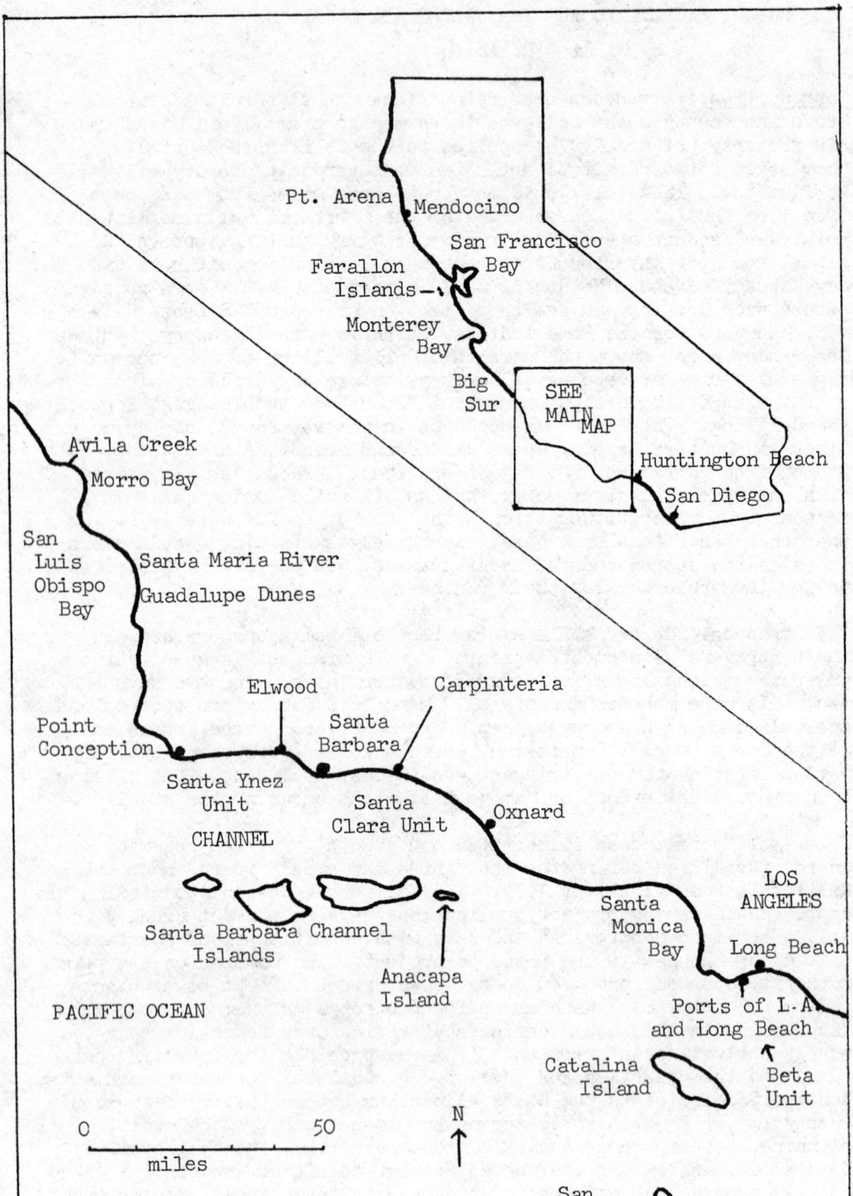

FIGURE 1
THE CALIFORNIA COAST

the coast while providing for needed energy facilities has been a difficult balancing act.

This paper first presents some general reflections on how the general coastal energy facility siting process has been proceeding in California. Then it summarizes a number of cases covering the full range of Commission energy activity. The paper concludes with a few recommendations and reflections on the predictions made in the paper I presented at the Coastal Zone 78 Conference in San Francisco.

GENERAL REFLECTIONS

I have been responsible for the Commission's staff work on energy facilities, ports and ocean resources since the beginning of 1977 when the California Coastal Act of 1976 took effect. These reflections are solely my own.

1. The California Coastal Commission can both protect and enhance the coast and provide for energy facilities, in areas where there are existing energy facilities. On the Southern California coast from the Mexican border to Point Conception there has been oil development since the turn of the century. There are 15 coastal power plants, three industrial ports, and 25 offshore oil production platforms and artificial islands. The proposed new platforms, processing plants, oil wells, marine oil terminals, and other projects for this area have all been approved. But projects have been changed in major ways to minimize risks of oil spills and other adverse effects. Chevron is building a new pipeline connection in order to phase out use of a marine oil terminal next to Carpinteria State Beach. Union is dedicating land at the mouth of the Santa Maria River for wildlife protection as part of its project to drill more wells in the Guadalupe Oil Field. Through project modifications and conditions, the Coastal Commission has been able to not only minimize the adverse effects of energy projects but also to provide net improvements in the coastal environment.

2. In coastal areas where there are no existing energy facilities, new energy project proposals meet stiff opposition. The two oil wells near Point Arena illustrate this point. So does the proposed LNG terminal at Point Conception, which has been in the permit process now for about five years. The California Coastal Act encourages coastal dependent industrial developments to site or grow in already developed areas. The Act then encourages the Commission to protect scenic undeveloped coastal areas. Therefore the Commission strongly opposes any Interior Department offshore oil leases seaward of the rural, scenic and undeveloped areas of the central and northern California coast.

3. When a new project is needed to improve an existing situation, all the approval processes and uncertainties that affect applicants for developments can also affect the coastal management agency. If a company has an existing outmoded marine oil terminal that should be replaced so modern tankers can use it, so new pipelines can be built and so large tankers can avoid transfers of oil at

sea to small tankers to make deliveries, that company faces so many new permits and requirements that the modernization project may die. This can result in net damage to the environment, because there is much inertia to maintain outmoded operations and the status quo. This inertia affects coastal protection programs too. The Office of Coastal Zone Management(OCZM) and the Commission have experienced all the typical difficulties of oil company applicants while we have been trying to establish marine sanctuaries around California's offshore islands. For example, the Western Oil and Gas Association has sued OCZM for an inadequate Environmental Impact Statement on the proposed Channel Islands Marine Sanctuary.

4. It is difficult to plan for energy facilities, and it is easy to plan to prevent coastal energy facilities in specific areas. The Commission can exclude areas for new power plants and has excluded all coastal parks and wetlands and also scenic areas, forests, wildlife habitats and special agricultural areas. The Commission excluded about 75% of the 1700 square miles of the California coastal zone. In areas where an electric utility had once expressed interest in a site, local organizations mobilized to persuade the Commission to find the reasons to exclude those areas. It was easy to exclude, but difficult to provide, ahead of time, areas or sites for new coastal energy facilities. The resistence of local governments and environmental groups is great. When one of the electric utilities did tell us some coastal areas it wanted at least the option of considering in the future, such as the Santa Maria River Valley or Avila Creek area, the Commission and other agencies focused on these and excluded them. Even when coastal planning identifies sites for new oil developments, such as at the existing sites on the Santa Barbara Channel coast, there is no way to determine ahead of a project application if a project there could meet all the future permit requirements of other agencies. The coastal management agency cannot deliver on affirmative planning for coastal energy facilities. Affirmative planning can only be done in cooperation with another agency that may be able to deliver approvals for a project. Thus, the California Energy Commission, which preempts local agencies and which has cooperative agreements with air and water quality agencies, might be able to approve a new site for a plant. So when that agency says a siting area should not be excluded, it has some credibility behind it. A meaningful plan can only be constructed between a protecting agency and an advocating agency. If an advocating agency does the plan, if, for example, Interior planned the OCS, it would keep siting options open everywhere. A protecting agency is inclined to protect everything. It is tricky to legislate a balanced planning process, but our experience with the California Energy Commission may be one. For projects other than power plants, the planning involves a process for considering oil terminals and onshore oil drilling and processing plants. The Coastal Act provides for a plan amendment process that gives special consideration to energy facilities. The local plans are providing policy discussions on this process, listing standards of review for new projects that cannot now be anticipated. Some local plans discuss where to apply for future amendments. But few are providing new sites absent a live proposal for a project.

5. During the last three years the coastal energy facility siting process has increasingly settled down, with companies, agencies and interested parties now mutually aware of requirements and concerns. But there are still a number of irritating glitches in the process. The smoothness of the process very much depends on the energy company involved and the state of agency regulations. In situations where a company has been through a number of permit processes before, the resulting learning benefits everyone, unless the regulations have changed. Unfortunately environmental agency regulations seem to change about every two years. But it takes about two years for agency staff to learn how to apply and implement new complicated regulations. So agencies are still evolving policies and procedures on NSRs, PSDs, NPDESs, 404s, "consistencies", and wildlife and endangered species consultations. The Coastal Commission has just about settled down with "consistency" after two years. We are extremely thankful that Congress did not change the "consistency" provisions of the Coastal Zone Management Act during 1980.

The process glitches that I find irritating are the frequent uselessness of expensive environmental reports, the "late hits" raising issues after the environmental analysis is done, the duplication when the same issue is considered in a number of permit processes, and the lack of a systems perspective given the multiple single purpose permit actions. These irritants will be mentioned next in the case discussions. The limited usefulness of federal agency environmental impact statements(EISs) prepared under NEPA and the state or local agency environmental impact reports(EIRs) prepared under the California Environmental Quality Act continues to be a major disappointment. This is despite the best attempts of agencies and consultants. The focus is still to compile enough material for legal adequacy, not for relevance to permit decisions. The California Permit Simplification Act has tried to remedy this by: defining the lead agency with responsibility for EIR preparation; requiring responsible agencies with permit authority to list all information requirements and to participate in scoping the EIR; and placing a one year deadline on the lead agency for a permit decision and a six month deadline on responsible agencies after the lead agency's decision. If deadlines are missed, approval is presumed. However, all agencies know how to avoid these requirements. A lead agency can delay accepting the application until "enough information" is provided. Responsible agencies can tell an applicant that denial will occur unless added information is provided. Permit agencies simply do not focus on a project until the application is in their own processes. Then late blooming issues arise that were never uncovered in the EIS or EIR. There is no real penalty to an agency for avoiding the EIS or EIR process. Agencies cite the usual problems, lack of staff and time to participate early. But this really reflects poor management. Because if an agency can place its permit analysis requirements into the EIS or EIR, that work is done by a consultant and then paid for by the applicant, and that should save agency staff time and money. If issues arise later, after the EIR, the agency staff must devote substantial research time at the end of the permit process anyway. So total staff time is generally greater the later an agency addresses a project. Ideally, all agencies should have their analytical needs covered in the EIR. Not only is the work then done for them, but the EIR can show how various regulatory requirements interact in affecting the project and its impacts. But there are some problems, as we have learned in trying to do this. First, as a responsible agency we have to persuade the lead agency staff to put our needs in the scope of the EIR. Sometimes there is friction or lack of understanding, and we have to heavily criticize the draft EIR to get our concerns addressed. It is always easier to get information from an applicant in your own permit proceeding. Second, EIR consultants are not trained in analysis needed for permits. Too often the least cost consultant firm cuts and pastes material. It is difficult for an outside firm or another agency's staff to learn the ever changing regulatory requirements. Before discussing a few recommendations to improve these processes, let us see how these observations are reflected in the Coastal Commission's experiences over the last three years.

CASES

Offshore Oil. The Interior Department is trying to accellerate oil and gas production from the federally owned Outer Continental Shelf (OCS) offshore California. Some aspects of offshore oil development have matured and are running smoothly. Other aspects are not.

Southern California Offshore Oil Developments. In my 1978 paper I reported on the approval of the first project to increase offshore oil production in the Santa Barbara Channel after the 1969 oil spill. Atlantic Richfield had proposed a small increase in capacity for its Elwood processing plant. Not only was the project approved by the Coastal Commission and Santa Barbara County, but it was supported by environmental groups because of the stringent conditions placed on the project. Since then agencies have approved Chevron's Santa Clara Unit development and Shell's Beta Unit development. Both projects involve only pipeline transportation of the oil. Chevron has built a new onshore pipeline connection that eliminates oil tanker traffic that had been carrying Channel oil to refineries. Both projects were approved because they met strong conditions, including low air pollutant emissions, pipeline transportation of oil, and updated oil spill contingency plans. The permit processes were not, however, friction free. Despite large joint federal-State environmental reports and many meetings, late blooming issues appeared. Chevron could not start construction on time because Chevron, the regional coastal commission, the City of Carpinteria and the California Department of Parks and Recreation could not agree on a path for a bicycle route that was a condition of the coastal permit. The Coast Guard was involved at the last minute in Shell's project, because it was not clear how deep Shell's offshore pipeline should be buried under the Port of Long Beach anchorage area.

With these experiences, oil companies know what to expect when proposing new oil developments offshore Southern California. The federal, State and local agencies learned how to work together. But these are easy projects, as there are existing industrial facilities all along the coast. There was no real opposition to the projects, and companies such as Chevron and Shell have large staffs that have learned how to work with California's permitting agencies.

Exxon and the Pipeline Working Group. Exxon has now also learned how to deal with the permitting agencies. In 1976 Exxon rejected a Coastal Commission permit condition requiring use of a pipeline, if economically feasible, instead of oil tankers for moving oil to be produced from the Santa Ynez Unit in the western Santa Barbara Channel. The Interior Department then approved an offshore floating marine oil terminal a half mile seaward of the State's jurisdiction. After lengthy wrangling with EPA and State and local agencies, Exxon's floating terminal is still not on station. However, in 1979 Exxon took the initiative to develop a memorandum of agreement with Santa Barbara County and State agencies that would allow their offshore terminal and onshore gas processing to proceed without being challenged or delayed further. Exxon agreed to add air pollution controls on the floating terminal, to make its onshore property available to other companies for oil and gas processing, and to inform agencies of their development plans. The Governor's Environmental Adviser agreed that the floating terminal would meet State air quality protection requirements if the State had such jurisdiction. The Santa Barbara County Board of Supervisors agreed to extend the zoning on Exxon's onshore property and to permit the gas company's project that will buy Exxon's gas. This agreement changed the climate surrounding future offshore oil and gas development in the western Channel. This was fortunate, because the Joint Government/Industry Pipeline Working Group that had been formed to determine if Channel oil could be moved by pipeline was unable to reach agreement.

Offshore Oil Lease Sales. The Coastal Commission has had a fundamental disagreement with the Interior Department on leasing California's continental shelf for oil and gas development. Essentially, the Commission holds that Interior should plan the OCS, determining where oil development should be allowed and where it should not be allowed because other uses, such as vessel traffic lanes or marine life habitat protection, should be the paramount uses. Through previous OCS lease sale proceedings, the Coastal Commission has determined where Interior should not lease offshore in order to be consistent with California's federally approved Coastal Management Program. These areas, shown on Figure 1, include areas within six miles of the Channel Islands and areas offshore the scenic and rural Mendocino, Marin, Sonoma, and San Mateo County coastlines. The Commission is also supporting designation of three marine sanctuaries around the Channel Islands, the Farallon Islands, and Monterey Bay to protect some of the most valuable marine habitat areas from offshore oil development.

The Interior Department, on the other hand, wants to continuously keep the option open for considering all areas on the shelf for oil and gas exploration and development. Interior's National Five Year OCS Leasing Schedule includes four sales offshore California. One for 1981, Sale #53, includes five separate geologic basins extending offshore 700 miles of California's coast. Two sales scheduled for 1983 and 1984 are called "California" sales and may include areas anywhere offshore California's coast. The Commission believes that Interior has violated the OCS Lands Act by including such huge areas as lease sales in the Schedule. This makes it impossible for Interior to balance economic and environmental considerations. The Commission also believes Interior should have deleted OCS areas from the Schedule which have low oil potential and high environmental risks.

Interior has also considered areas for sales that had been set aside by previous Interior Department Secretaries for environmental protection reasons. Santa Monica Bay and the Santa Barbara Ecological Preserve were included in the first areas considered for OCS Sale #68 scheduled for 1982. These had been excluded from leasing by Secretary Kleppe and Secretary Hickle respectively. Secretary Andrus deleted them from Sale #68 just before the California presidential primary election. But the areas can be included again in future lease sales.

To further keep options open, Interior has determined that the decision to lease and delete tracts in an OCS lease sale has "no direct effect" on the California coastal zone. Therefore Interior has avoided a requirement of the Coastal Zone Management Act. That Act requires a federal agency undertaking an activity that directly affects a state's coastal zone to submit to the state a determination that the activity is consistent, to the maximum extent practicable, with the state's federally approved coastal management program. The Coastal Commission believes that a large subdivision of the OCS, creating the right to develop discovered oil and gas resources, does directly affect the coast and is the only stage at which an overall judgment can be made on the lease sale's effects. Because Interior and the Commission continued to disagree on this in formal mediation proceedings, the Secretary of Commerce has directed the National Oceanic and Atmospheric Administration to publish new regulations defining the term "directly affect". Interior appears to have no wish to plan the OCS with affected states and other agencies. The Commission is trying to accomplish such planning through "consistency" reviews and through the marine sanctuary program. But if the OCS landlord, the Interior Department, continues to resist such planning, each OCS lease sale will experience controversy and conflict.

OCS Plans of Exploration. The U.S. Geological Survey approves oil company plans to drill exploratory wells on the OCS. The USGS cannot grant the drilling permits until the Coastal Commission has concurred with the oil companies' certifications that the drilling will be consistent with the California Coastal Management Program. The Commission has now concurred with 15 of these consistency certifications in an average time of about two and a half months. The consistency review process on exploration plans now works smoothly, thanks to a cooperative working relationship with the USGS. But some plans have been difficult and controversial. The Commission's first objection to such a plan will probably be this fall when it acts on Chevron's plan to place a drillship within six miles of

Anacapa Island and in the buffer zone right next to the vessel traffic lane through the Santa Barbara Channel.

MARINE OIL TERMINALS. Permitting marine oil terminals is a complicated process, with a large set of uncoordinated permit agencies. The process is a mess.

The Sohio Project. The Sohio Project has been dead now for more than a year. Sohio owns more than half of Alaska's Prudhoe Bay oil production. The company in 1975 proposed to build a new marine oil terminal in the Port of Long Beach, with four new tanker berths, to receive tankers carrying Alaskan oil. The terminal was to connect to a pipeline which was to carry most of the oil east. Oil would also be delivered to Los Angeles area refineries. In November 1977 the Coastal Commission gave the terminal its first permit. The Commission found that the project was in the national interest and that the Los Angeles ports badly needed new, modern tanker berths able to receive larger tankers.

But the Project was the first major project to experience a New Source Review for air pollutants in the smoggy Los Angeles air basin under the new "tradeoff" policies of the Clean Air Act. The local air pollution control district and the State Air Resources Board had to develop new policies and regulations for the Project, and they often disagreed. Sohio, a Cleveland based corporation, had no operating experience in California. Sohio won a referendum in the City of Long Beach and finally won general endorsements from State officials. But after four years Sohio declared the Project dead in early 1979. Reasons for this are unclear, but the cost of the air emissions "tradeoff" package was to be about $80 million. Sohio's share of Alaskan oil production was likely to decline in the late 1980s, so the longer the delay the less Sohio oil would go through the terminal.

The Alaskan oil being delivered to Los Angeles refineries now still goes through older marine oil terminals, and some is transferred from larger to smaller tankers offshore. Alaskan oil for the Gulf Coast travels by tanker to the Panama Canal. Without the Sohio terminal, transportation of Alaskan oil to Los Angeles and to the Gulf Coast is both more risky to the environment and more costly. All the tankers call at the Ports of Long Beach and Los Angeles either to unload or to take on bunker fuels. But they do this in the water instead of at new fixed tanker berths where oil spills can more easily be controlled. There is no air pollution "tradeoff" package at all. The death of the Project was unfortunate to the coast and the economy.

Moss Landing Marine Oil Terminal. The Pacific Gas and Electric Company(PG&E) has a large electric power plant on Monterey Bay at Moss Landing. In 1953 PG&E built a small marine oil terminal to bring fuel oil to the plant, which ran mostly on natural gas. As gas supplies declined and fuel oil needs rose, PG&E upgraded the terminal in 1974 and applied in 1977 for a larger terminal to receive tankers up to 120,000 deadweight tons in capacity. The project was approved by Monterey County, the State Lands Commission, air and water quality agencies and the Coastal Commission. The Commission found that the project would not change the overall amount of oil carried into the Bay, that the oil would be carried in newer tankers in fewer trips, and that, with the many conditions imposed on the permit, the project would result in a net decrease in the risk of oil spills in the Bay. This was because it was a replacement oil terminal, not a new one. By June 1979 PG&E had all State and local permits.

The Army Corps of Engineers would not consider an application until PG&E had all these other permits. After more than a year, the Corps District Commander denied the application. His reasons were: the Fish and Wildlife Service in its consultation opinion said the project might not assure protection of the endangered sea otter; the local Congressman opposed the project; and there was substantial local opposition in some Bay cities. This local opposition organized after all State and local permits had been granted. This episode showed that Corps permit criteria explicitly include the attitude of the local Congressman. In addition, the Fish and Wildlife Service was not formally involved until well after the environmental analyses had been prepared. The Service's call for more studies and risk assessments came at the end because under the Corps' procedures, the Corps only accepts applications at the very end of the process, and then requests the Service's opinion under the Endangered Species Act.

This episode also showed extensive duplication of jurisdiction over impact areas. One of the worst is oil spills. Under NEPA and CEQA, which are almost entirely duplicatory, the federal and state lead agencies on a project must consider risks of oil spills and provide for prevention and mitigation in their permits. The Regional Water Quality Control Board and the Coastal Commission also have jurisdiction over oil spills and add permit conditions. Then in the Corps permit the same issue is addressed with the Fish and Wildlife Service. These agencies can come to different conclusions. Ironically, the agency responsible for oil spill control has no permit authority, but the Coast Guard must review and approve oil spill contingency plans and terminal operations manuals. The Moss Landing oil terminal proceeding has been a lengthy mess, and coastal management goals to minimize risks of oil spills were jeopardized by the Corps. Unfortunately there is no "consistency" review under the Coastal Zone Management Act of federal permit denials.

Pier E Terminal. In 1976 Shell applied to build a new large oil pipeline to connect an existing ARCO marine oil terminal on Pier E in the Port of Long Beach with inland refineries. Shell and ARCO tankers were then to jointly use the terminal. The existing terminal, however, is on a narrow entrance channel to the inner harbor, and it was built for small tankers after World War II, not for the 150,000 dwt tankers used now. The Coastal Commission, with Coast Guard support, approved the pipeline, but with the major condition that the companies build a new terminal indented back on the channel so other vessels could pass unhindered. Shell and ARCO agreed, since Shell wanted to move its crude oil unloading operation from a shallow, constricted terminal area in the Port of Los Angeles. However it has taken three years to design the new terminal and to obtain the air quality permit and other approvals, even though both companies had existing old terminals available for air pollutant "tradeoffs". Given all the required permits and their conditions, a company must have a very large economic advantage to pursue improving or replacing an existing marine oil terminal.

The Harbor Deepening Project. The Port of Los Angeles had been attempting for more than 20 years to conduct a project with the Corps of Engineers to deepen its main ship channels from 35 to 45 feet. The Coastal Commission approved the project because it would help make more efficient use of existing port terminals and land areas, reducing the need for more fill projects or more ports. The project hit two typical late blooming issues. The air pollution control district applied a new policy that the dredge would be considered a stationary source of air pollution, so "tradeoffs" would be needed under a New Source Review. Then the Fish and Wildlife Service and California Department of Fish and Game decided, the day before the final Coastal Commission vote on the project, that compensation should be required for filling 190 acres of muddy soft bottom water area in the Port. The joint Corps/Port environmental report had indicated there would be no net damage to the Port marine environment. But this "late hit" resulted in Coastal Commission imposed conditions to pay $5,000 an acre in compensation for the fill and the loss of the worm, clam and sea squirt habitat and resulted in putting off the vote for a month.

LOCAL COASTAL PLANNING. The most important part of the California Coastal Commission's responsibilities is the certification of plans. Local governments and ports must develop plans that show where different types of coastal developments, including energy facilities, can and cannot be sited in the future. Excluding San Francisco Bay, the California coast has 68 local governments and four commercial ports. When the Commission certifies the plans, the local jurisdiction takes over coastal development permit authority.

Local Coastal Programs. The land use plans developed by local governments are supposed to designate the permitted uses for coastal land and water areas. Such uses include agriculture, commercial, recreation, residential, habitat and, a new category, coastal dependent industrial uses. Part of the logic of California coastal planning is to preserve areas for coastal dependent industrial uses, giving them priority over private residential and other uses. The Coastal Act requires local governments to consider in their plans all energy facilities that can be "anticipated". This has turned out to be difficult for two reasons. First, although energy companies say they want specific plans that provide sites for their facilities, they really have no idea where they want sites until they have an actual project in the planning and economic feasibility estimation stage. Second, local governments do not want to reserve sites for energy facilities when they do not know what the facility is to be or whether a company would really want the site. Consequently the local coastal plans are generally including a siting process for different kinds of energy facilities, and they do consider the possible need for future facilities and sites. But they do not designate, on a land use map, sites for new facilities, unless there are existing marine oil terminals or oil and gas processing plants or oil fields in their jurisdictions. In those cases, the local government, as the Coastal Act encourages, designates the existing sites as the prime candidates for considering new or expanded facilities. For example, the Santa Barbara County coastal land use plan has a "coastal dependent industrial facility" land use designation with 8 subcategories. These subcategories include marine oil terminals, processing plants, oil and gas drilling, pipelines, electric transmission lines and aquaculture. Processing plants are only permitted at 11 sites of existing processing plants. But pipelines and onshore oil and gas drilling are permitted uses in a number of land use categories, such as in agricultural areas. Each type of energy facility has its own process. After these local coastal programs are certified by the Coastal Commission, an energy company needing a site not provided in the plan must apply for a plan amendment. The amendment must be approved both by the local government and the Coastal Commission.

Where we have been able to use Coastal Energy Impact Program grants for precise planning of areas, such as the Guadalupe Dunes area of north Santa Barbara County, the local government has been able to define where oil drilling can and cannot take place.

Port Plans. Under the Coastal Act, the four Southern California commercial ports must also complete plans. These plans are all completed, and three have been certified. But even in industrial ports, it has been difficult to plan for projects, particularly energy projects involving hazardous materials and projects involving any filling of water areas. The Commission retained permit authority over projects involving hazardous materials and fill, while certifying most of the port areas for container terminals, auto storage and other port activities. The ports had a strong incentive to complete their plans and take over coastal permit authority, because saving a few months of permit processing time can save money on their own and port tenants' projects. But local governments have no such incentive. With CEIP we have been able to fund the development of port risk management plans by the Ports of Long Beach and Los Angeles. With expert consultant help and the participation of the Coast Guard, fire departments and other agencies, these plans are developing siting strategies for hazardous terminals, permit conditions and coordinated responses to accidents. The Commission should be able to certify these risk management plans next year. With respect to fill projects, however, the ports cannot provide the analysis, at the general planning stage, to persuade the Commission the fill will have no major adverse environmental effects or that it is needed. Therefore, without such analysis, generally required in permit proceedings, no parts of coastal plans involving future projects that may have noticeable adverse environmental effects have been approved as parts of port plans. Anticipatory planning for benign projects such as container terminals and storage areas has been easy, but for projects with possible major impacts, impossible. If guidelines could be written into the plans for compensation for wildlife habitat losses, it might be possible. The Port of Long Beach is trying to agree on such guidelines with the Fish and Wildlife Service, Department of Fish and Game, and the National Marine Fisheries Service. These agencies so far have mainly focused on permit proceedings on a case by case basis.

ELECTRIC POWER PLANTS The California siting process for new electric power plants is much different than the process for marine oil terminals. The California Energy Commission is a "one stop shop" that preempts both local and Coastal Commission permit authority. An electric utility proposes three or four sites to the Energy Commission, and that Commission ranks the sites and designates two or more as "acceptable". The utility can then apply to the Energy Commission for one of the "acceptable" sites, and there is no local government or Coastal Commission permit involved.

However, under the Coastal Act the Coastal Commission has designated specific coastal areas where the Energy Commission cannot approve a new plant. These "designations" are extensive and cover wetlands, parks, scenic coastal areas, special coastal agricultural and forested areas, anadromous fish streams, natural plant and wildlife communities and other resources the Coastal Commission is charged to protect. But the Coastal Commission is in a long term cooperative planning relationship with the Energy Commission, which forecasts the need for new power plants in the State. The Coastal Commission is charged with revising these "designations" every two years. In making the revisions, the Commission is to consider the policy recommendations of the Energy Commission. Therefore the Commission has contracted with the Energy Commission, using CEIP funds, to analyze the effect of the "designations" on new power plant siting. Phase I of the project involved the existing 29 coastal power plant sites in California. The Energy Commission study, completed this year, analyzed all the constraints to expansion of the existing sites. It lists the megawatt capacity of gas turbines, coal units, and other types of generating technologies that could be added at each of the sites. The major constraints on expansion are identified as lack of tradeoffs for air pollutant emissions, encroaching residential and industrial development, and water quality problems for sites on enclosed water bodies such as bays and estuaries. The Energy Commission report concludes that the Coastal Commission and Bay Commission "designations" do not preclude the reasonable expansion of existing sites. The second phase of this planning work is an Energy Commission assessment of the feasibility of siting new plants in the "undesignated" coastal areas. When these studies are compared with the forecasted need for new plants, both commissions can obtain an impression of whether the adopted "designations" have left reasonable opportunities for siting new plants that can use ocean water for cooling. This planning effort is successful because both agencies can deliver, that is, the Coastal Commission can guarantee resource protection and the Energy Commission can conduct an overall siting assessment and can approve sites without local government veto. This planning is easier because a number of factors have taken the pressure off coastal power plant siting. The State has an effective moratorium on new nuclear plants that need massive amounts of water for cooling. New coal-fired plants are likely to be sited inland nearer the Utah coal deposits. The California drought of 1977-78 is over. And in populated coastal areas it is extremely difficult to meet air quality protection regulations, especially "tradeoff" requirements. The Energy Commission's most recent forecasts on need to site new plants indicates that, at most, there would be a need for three or four new coastal sites in the next twenty years. By next year the two commissions will have together determined where new electric generating capacity can and cannot be sited on the coast.

THE LNG TERMINAL During 1977 and 1978 the Coastal Commission was involved in a one-time special permitting process for California's first proposed liquefied natural gas(LNG) terminal. Under special legislation, the California Liquefied Natural Gas Terminal Act of 1977, the Coastal Commission was charged with ranking remote onshore sites for such a terminal. With the help of geotechnical, engineering and biological consultants, the Commission completed this site ranking in May 1978 and sent the ranking to the California Public Utilities Commission(PUC). That Commission was required to approve the first ranked site, unless it found a lower ranked site was needed so the terminal could be built fast enough to prevent serious gas supply problems. The PUC made this finding and selected the third ranked site, Point Conception, which also happened to be the site applied for by the gas utilities. The PUC not only made this finding, that Point Conception could be on line at least 2 years faster than the other sites, but it also found the Coastal Commission's other three sites could never be approved because they were not safe, that is, they were too near roads or parks where transients would be exposed to the hazards of the LNG terminal. After rejecting these sites, however, the PUC was unable to approve the Point Conception site because some small earthquake faults were found on the site. Despite a $1.5 million environmental report, the PUC required further studies on these faults. These studies are still going on more than two years later, with added trenching and geologic investigations at the site. Conflicts with Indians about the spiritual value of the site have slowed the seismic evaluation process. Because of the PUC's rejection of the other three ranked sites, there is no other approvable site on the California coast for an LNG terminal. The long delay may, however, be constructive. After gas price deregulation, supplies have been plentiful. The LNG was to come from Indonesia, which has raised gas prices substantially. And the times have been terrible for financing a $650 million LNG terminal or the entire $2 billion project. However, this is a coastal siting process that has been going on now for seven or eight years, since the gas utilities first applied to the City of Oxnard for an LNG terminal. The entire process has cost the utilities about $80 million. This case has shown, however, that special legislation giving sole State permit authority to one agency, preempting local government and the coastal management agency, does not necessarily resolve issues or make for a speedy decision.

RECOMMENDATIONS

We need recommendations, of course, only for problem areas. As this brief tour has shown, the only really serious problem area is in the permitting process for marine oil terminals, where delays and denials of projects can result in harm to the environment compared to what happens without the projects. One possible solution to "delays" and poor coordination is the "one stop shop" permit process. In California that has a mixed record. The California Energy Commission's siting process for new power plants seems understandable and successful. But the special "one stop" LNG terminal approval process has not prevented delay and paralysis from setting in. But in that case the delay is not necessarily bad, as time may show the LNG Project not to be in the State's best interests. Two years ago I predicted there would be moves toward more preemption of state and local agencies on major energy facilities. The Energy Mobilization Board proposal did subsequently arise and die. That proposal would really have only added another complicated step in the permitting process, and the Board would have surely promulgated lengthy regulations and a massive application form for being declared a national interest energy project.

I also predicted that federal agencies would use complex computer models to determine which energy facilities are needed. That too has not progressed very far, and there is a scarcely noticeable federal energy policy presence in the permit processes for most coastal energy facilities. But the Department of Energy has its foot in the door, as it provides offshore oil and gas production goals to the Interior Department. Interior then uses the goals, generated by DOE computer models, to figure out how much OCS acreage should be leased nationally for petroleum development.

One need I do see, however, is for increased federal energy agency participation in coastal energy facility planning and siting. Federal preemption is not needed. But coastal management agencies and local governments naturally tend to avoid planning for energy facilities that are not in the permit process already. Anticipatory planning is difficult, and private industry is naturally reluctant to say they need a site or an area. In general, the immediate reaction to such announcements is negative. There should be dynamic tension in coastal energy planning. But the Coastal Commission's experience is that it is easy to avoid energy planning, and local governments do not want it. It would also be difficult for the Office of Coastal Zone Management to be seen as an advocate for active energy facility planning. Planning for future coastal energy facility needs is fine in principle, but when specific coastal areas are discussed, there is a constant pressure, at least in California, to protect the area, and little pressure to keep energy facility siting options open. Neither the Department of Energy nor the Interior Department have taken steps to provide at least advice and policy positions to the Coastal Commission or local governments on coastal energy planning. Interior is at least trying to plan offshore pipeline corridors under its Intergovernmental Planning Program, but we are having difficulties making the onshore connections. Of course, neither agency would meet a cordial reception if it tried to encourage the Coastal Commission or local governments to plan for offshore oil related facilities in areas where the State of California strongly opposes any offshore oil leasing. But that should not stop them from trying.

On two of the permit process irritants, I have two suggestions. One is that at both the federal and state level there should be penalties for "late hits". A responsible permit agency should make its information and analytical requirements known to the developers of the EISs and EIRs. Each agency should monitor the EIS/EIR preparation, comment on it, and include mitigation measures in it. The EIR/EIS should consider how the project affects all aspects of the environment and how the environmental permit requirements affect the project and the environment. The Council on Environmental Quality or EPA at the federal level and the Governor's Office of Planning and Research or the Resources Agency in California should monitor agency participation in the EIS/EIR processes. They should note when an agency participates early and usefully and when an agency ignores the process and brings up environmental issues well after the EIS/EIR are completed. If an agency displays a regular trend in "late hits", the monitoring agencies should bring such poor performance to the attention of the President and Governor, Congress and the Legislature, and OMB and the California Department of Finance. They in turn should correct the situation. There are legitimate "late hits", particularly when a project is changed in mid-stream. But in general, "late hits" arise because an agency does not focus on a project until the last minute.

The EIS/EIR can only be made useful to permit processes with improved EIS/EIR preparers. Either consultant firms or lead agency in-house staff prepare these environmental reports. Too often such work is regarded as low status backwater work. Instead, EIS/EIR consultant firms and lead agency staff should be certified as trained in permit requirements and analysis needed for environmental protection permits. Agencies should periodically hold training and updating sessions for EIR/EIS preparers and certify EIS/EIR analysts as competent in incorporating their needed information and analysis in the documents. Building a certified set of consultant firms and lead agency staffs qualified in permit requirements would serve to improve the chances for avoiding late blooming issues, cutting delays, and needing added analyses at the end of the permit process. Intervenors could also then focus on the EIS/EIR public hearing process, with less need to participate heavily in many different permit proceedings. This suggestion would make EISs and EIRs more costly at first, but both time and money would be saved in reducing duplicative analyses.

CONCLUSION California has perhaps the most stringent requirements in the world for protection of the State's environment. The California Coastal Commission is regarded as one of the toughest permitting agencies in California, and it operates under strong State legislation to protect the 1100 mile long coast. The Coastal Commission has very straightforwardly indicated where it will not allow new power plants and where it opposes offshore oil development. But the Commission has yet to deny a proposed coastal energy project or to object to a federal permit for an energy project. The first "consistency" objection may be this August - an objection on Chevron's proposal to site a drillship next to the vessel traffic lane in the Santa Barbara Channel within six miles of the breeding area of the endangered brown pelican on Anacapa Island. But where a project will enhance coastal management goals, such as the Sohio Project, the Coastal Commission may give that project its first permit, to assist it with other permit agencies. If the Coastal Commission did not exist, the approved projects probably would have proceeded. But there would have been no pipelines to replace tankers, no bicycle paths, no wetland dedications associated with the projects. The projects might have been conducted during breeding seasons of the nearby harbor seals or brown pelicans. And many residents and users of the coast would still be wondering if a power plant were going to be proposed near them. The Commission has so far been able to deal with energy and port facilities because it has permit, "consistency", planning and CEIP allocation authority all within its jurisdiction. The CEIP funds enable the Commission to encourage local governments and ports to conduct needed planning, while the permit and "consistency" authorities get the attention of industry and federal agencies. My personal impression is that most needed energy facilities have found sites on the California coast, that in siting and approving such projects the Coastal Commission has generally improved the attributes of the coast, and that the Coastal Commission has now clearly indicated the conditions under which projects may be approved and has also clearly indicated those offshore and onshore areas where projects should not be approved. These results are due to the unique provisions of the Coastal Zone Management Act and the California Coastal Act, working together. The Commission and staff, industry, federal agencies, local governments and environmental groups have learned how to implement those laws during the last three years with respect to energy facilities. With this behind us, I can only now pray that the California Legislature not change the Coastal Act next year.

DONALD E. GLASS
SHELL OIL COMPANY
MANAGER, COASTAL ENVIRONMENTAL ACTIVITIES
ANCHORAGE, ALASKA

ENERGY FACILITY PLANNING AND SITING IN THE COASTAL ZONE

INTRODUCTION

This is a report on a study entitled "Energy Facility Planning and Siting in the Coastal Zone." It was prepared for the American Petroleum Institute by Rogers and Golden, with Alan Mallach Associates and Dames and Moore, a group of consultants who have worked together in the fields of environmental and socioeconomic planning. The approaches presented in this study were furnished for use by industry personnel in working with state CZM planners.

The study focused on ways to address two issues related to the Coastal Zone Management Act: namely, energy facility planning; and facility permitting and siting.

DISCUSSION

In the Conference Report on the important 1976 amendments to the Act, Congress notes that "the 1972 Act was enacted before the advent of the current and continuing energy crisis; that is, before attainment of a greater degree of energy self-sufficiency became a recognized national objective of the highest importance and priority."

The Report went on to encourage orderly development of oil and natural gas resources on the Nation's Outer Continental Shelf. As a consequence, the amended Coastal Zone Management Act sets three requirements for states to follow regarding energy facility planning and siting:

The first, from Section 306(c)(8), is that states must adequately consider the national interest in the planning for, and siting of, those energy-related facilities necessary to meet requirements which are other than local in nature. An enforceable mechanism must be in place to assure that regional and national interests receive proper consideration.

The second requirement, from Section 306(e)(2), is that a state program must include a method of assuring that local regulation will not be allowed to unreasonably restrict or exclude uses of regional or national benefit.

The third requirement, from Section 305(b)(8), is that a planning process must be provided for energy facilities which are likely to be located in, or which may significantly affect, the coastal zone.

It is the view of the petroleum industry that Federal regulations and state programs have not addressed these requirements adequately. For that reason, this study was conducted to investigate the possibility of developing a model process for wide use by CZM planners.

Based on the results of this study, however, the industry concluded that it was not possible to develop a model with universal applicability for meeting industry needs and the requirements of the Act. The primary reasons being: wide differences in the law and governmental structures of the various coastal states; variations in their social and political attitudes toward the petroleum industry; and differing economic conditions.

Therefore, this study does not present a complete "model" planning and permitting process endorsed by the petroleum industry. Neither does it offer easy answers to important questions such as: what is adequate consideration of the national interest? What it does, hopefully, is provide insight into classical planning and assessment techniques, and it sets forth one possible approach to considering national interest in a permitting process. The study covered three main elements:

A regional screening process to identify potential areas for energy facilities development.

A discussion of project assessment and cost-benefit analysis.

A permit review procedure for siting energy facilities determined to be in the national interest. Let's take a brief look at each of these elements.

Regional screening is a common part of a planning process such as is required by the Act. The purpose it serves is to identify potential areas for energy facility siting. It is essentially a three-part process: first, a representative example of each type of energy facility is hypothesized. Then, siting and locational criteria are developed for the hypothetical facility; and finally, these criteria are used to examine regional data in order to locate geographic areas meeting the criteria for the hypothetical facility. In simple terms, regional screening is a systematic approach for determining what parts of the coast have those combinations of physical and social characteristics likely to attract certain facilities.

Not all coastal areas are worthy of regional screening. Federal lands or highly developed urban areas, for example would commonly be excluded from the CZM screening process. The study referred to these as Omitted Areas, meaning omitted from screening. Other areas commonly accorded special treatment were called Geographic Areas of Particular Concern or Public Concern Areas - such as wetlands, flood-prone areas, or waterfowl breeding habitat. Those areas which satisfied the minimum requirements of a specific type of energy facility were called Threshold Areas. Superior areas were those areas where screening indicated optimum conditions for facility development. For example, superior areas for locating a marine terminal would have a sheltered, deep-water harbor and would not pose any serious threat to the environment.

The matching of a hypothetical facility's requirements to an area is a flexible one, mediated by geographic, economic and environmental considerations. To think a given type of facility has a fixed set of requirements would be an oversimplification. No single facility requirement (e.g. potable water supply) should be viewed as an absolute limitation on facility development. Therefore, a regional screening process should only be used as a planning tool because, at best, the results are only a first approximation of where applicants are likely to seek facility sites. Screening results, of course, must be updated as new and pertinent information becomes available. It might also be noted that each successive phase of screening and mapping will involve higher levels of planning sophistication, and require additional funding. Maps of regional characteristics are not cheap, and the more comprehensive they become, the more likely they are to become misunderstood as zoning maps.

Screening affords a state the opportunity to examine a range of coastal concerns and goals. At the same time, applicants considering development of energy facilities are provided a reasonable degree of predictability concerning the state's attitudes, plans, specific policies, and procedures. Lines of communication are opened, viewpoints exchanged, interests and goals better understood.

The second element of the study - Project Assessment - includes consideration of environmental impacts, socioeconomics, and cost-benefit analysis. Project assessment should be sensitive to the likely nature of debate about a potential facility and responsive to the scope of conflicts which may arise. Any conflict in facility siting will likely be between environmental and socioeconomic impact considerations, or among local, regional and national interests. Therefore it is important that all factors reasonably related to a conflict be included in the assessment process.

An important part of the assessment process is cost-benefit analysis -- the means by which the environmental and socioeconomic costs, benefits and effects are evaluated and weighed in the decision-making process. The cost-benefit analysis must not only relate the societal costs and benefits associated with the specific project, but must also consider the national interest applicable to the facility, such as the provision of additional energy resources. Cost here, of course, does not mean the economic viability of the private facility itself.

The project assessment submitted by the applicant may be reviewed by many separate agencies with several types of review taking place within each of them -- factual, analytical and substantive review. Where conflict in the project assessment takes place, it is likely to grow out of controversy over whether an identified impact is or is not significant. Or, regardless of the significance, do the benefits provided by the facility outweigh the negative impacts?

A matter of great concern to the industry is that such review is commonly carried out independently by several agencies, each of which has a narrow focus. It is not realistic, for example, to expect an agency responsible for wetlands to value other factors as highly as wetlands protection. In essence, each agency has veto power over the permitting process, and this veto power often takes place without benefit of an integrated overview.

Consideration of how such overview might be provided led to the third and final element of the study - a permit review procedure for the siting of energy facilities. Examples of such facilities would range from those that are complex but few in number, like oil refineries, to those that are less complex but many in number, like oil wells. This permit review procedure did not focus on specific details, but rather on those fundamental attributes of a permit review procedure by which a state might demonstrate that the national interest was to be considered.

The report did not suggest a total recast of state administrative processes. Only the provision of an Energy Review Board - which would assure that state and local agencies reflected the national interest in their decisionmaking - represented a dimension not usually found in standard state administrative processes.

The permit review procedure as discussed in this study envisioned the following steps:

Project Application.

Agency Review.

State Energy Board Review.

Judicial Review.

Applicants usually undertake two kinds of activities in connection and prior to filing a permit application. First, they should consider any prior state planning. Second, they would normally evaluate the many factors associated with siting, such as consideration of candidate areas and the technical, economic, social, and environmental implications of those areas.

Once a company decides to seek approval for development of a particular site, it files an application with the appropriate agencies. In the process discussed in the study, if the application covers a national interest energy facility the company would at the same time file with the Energy Review Board a declaration that the project be entitled to review involving adequate consideration of the national interest -- a procedure the company may subsequently elect to trigger if an adverse siting decision is handed down by a state or local agency, or in the event of undue delays in the issuance of permits.

After an agency receives the permit application, it publishes a notice that public review and comment on the application is invited. After evaluation of public comment, each agency would determine whether a public hearing was to be held. Any hearing would preferrably be of the adjudicatory type, with witnesses being sworn and cross-examination allowed. Upon completion of review, the agency's decision should describe, among other things, how and to what extent, <u>if any</u>, the national interest in energy facility siting was reflected in the decision.

If a company elects to request a national interest review because of a local or state agency's adverse decision or undue delay, the most significant feature of the permit review would be activated -- namely, the Energy Review Board.

The Board as a minimum would be composed of the heads of state agencies, and hopefully would include cabinet-level representatives from the Governor's office. The Energy Review Board as discussed here should <u>not</u> be viewed as a superagency. It exists primarily as a mechanism for ensuring that the national interest is adequately considered in the siting of energy facilities. Other methods of exercising this responsibility may exist. However, any body so charged should have broad representation, and avoid a narrow outlook.

The Board would consider two categories of issues. First, it would make a determination as to whether the national interest is indeed involved. Second, if national interest is affirmed, the Board would undertake a review of facility siting issues on their merits. Specifically, the Board would determine whether the agency's decision did in fact adequately consider the national interest in that particular instance. In its final decision, the Board could uphold, remand, or reverse an agency decision.

Finally, the report briefly discussed judicial review. Final decisions of the Energy Review Board, both on the national interest determination and on its review of other siting issues, would presumably be subject to judicial review of both procedural and substantive issues.

Conclusion

In summary, this study has presented planning techniques and procedures which should be helpful as industry personnel participate in CZM program development. The elements of the study were believed to be consistent with the requirements of the Coastal Zone Management Act and its regulations. They did not, however, form a complete model endorsed by the industry for dealing with energy development in the coastal zone. Instead, the study suggested a number of key elements which may be useful in the overall planning, permitting and siting procedure, specifically:

<u>A regional screening process.</u>

A <u>Project Assessment and cost-benefit analysis methodology</u>,

and a <u>Permit Review Procedure</u> for Siting Energy Facilities.

A state's energy facility siting process should be flexible -- embracing the principle of multiple use -- and not attempting predesignation of sites. It should be a clearly described process which identifies all decision points. This study emphasized the importance of responding to national interest energy needs, and illustrated one way to address this requirement during the development of state CZM Programs.

DEPARTMENT OF ENERGY INTERESTS IN
COASTAL SITING OF ENERGY FACILITIES

Emmett W. Turner
DOE CZM Coordinator

Coastal areas are of major importance for the siting of energy facilities. I plan to review resource production and facility siting data on the extent to which energy resource development is coastal dependent, to examine coastal dependence as a regulatory concept, and to review the national interest in coastal energy facilities.

Siting Requirements

The oil and gas industry is especially dependent upon the coastal zone. Including both crude oil and petroleum products, about 43% of our total petroleum supply is imported from foreign sources. About 9% of domestic production is from Alaska and about 15% from offshore sources. Collectively, this represents a coastal delivery dependence of about 59% of our total petroleum supply. In addition, some domestic production is from onshore areas within designated coastal zones. This is an amount I really cannot pin down, but it certainly would be significant in Gulf Coast areas.

Although internal petroleum distribution is primarily by pipeline, Congressional studies published in 1977 indicate about 22% of domestic transportation by water carrier. A part of this 22%, of course, would be coastal, including tanker and barge movements on coastal waters, the Great Lakes, and the intracoastal waterway system. I really have no way to combine directly all elements of coastal production, delivery, and internal distribution, but it seems probable that perhaps 70 percent of all petroleum supplies are produced, or transported within coastal areas.

Natural gas also has substantial coastal dependence. About 23 percent of domestic natural gas production is from offshore sources. Imports of liquified natural gas are a small element of natural gas supply, about 1.3 percent in 1979 and virtually insignificant for 1980 due to price conflicts and suspension of deliveries from Algeria. LNG supplies are likely to remain small because of preference for bringing Alaskan gas to market, more economical pipeline delivery from Canada and Mexico, and new Federal initiatives for synthetic fuels development.

Electric powerplants are less likely than oil and gas production and processing facilities to be regarded as coastal dependent, but in fact are the most common coastal energy facility. Working from the DOE national powerplant inventory and assuming that all plants in coastal counties are either in coastal locations or have potential to impact coastal resources, there are 2,280 coastal powerplants in the coterminous states. The number of coastal sites is much lower at 557, since many sites have several separate generating units at a single location. The total installed capacity is 181,645 megawatts, or about 39% of total capacity in coastal states.

The near universal impression from the type of information just reviewed is that coastal areas either have been or are about to be overwhelmed by energy facility development. A closer look strongly indicates this is just not the case.

About 60 percent of domestic refinery capacity is in coastal areas, but future expansions of refinery capacity are likely to be limited. Based on the mid-level projection of the 1985-1995 energy demand/supply analysis of the Energy Information Administration, 1978 domestic refinery capacity was about 17.1 million barrels per day, with 1985 estimated at 19.3 mb/day, 1990 estimated at 19.8 mb/day, and 1995 estimated at 20.2 mb/day. The same projections indicate actual crude runs declining from 14.7 mb/day in 1978 to 13.2 mb/day by 1995. New refineries will be built to respond to such factors as new supply sources, shifts in regional markets, and shifts in product mix rather than overall capacity needs.

Current inventory data on electric power plants indicates a total of 115 proposed new units in coastal counties. Several are proposed to be located on sites already in use and others will involve construction of more than one unit at a single site. As a result, only 38 new sites are proposed for siting of the 115 plant units. There is a clear trend to site power plants away from coastal locations. The existing power plants in coastal counties represent 39 percent of total installed capacity in the coastal states. Capacity proposed in coastal counties is only 27 percent of total proposed capacity in the coastal states.

Limited requirements appear to be likely for other types of energy facilities. DOE is currently attempting to encourage exports of coal. New terminal capacity will be needed. Terminals must be large to take advantage of economies of scale. As a result, over any regional area, development is likely to be concentrated in one or two major ports, most likely involving expansion of existing commercial port areas.

As already noted, new LNG facilities are expected to be very few based on comparative costs of alternative supply sources and uncertain reliability of foreign supplies. Good progress continues to assure delivery of Alaskan gas to the "Lower 48" by pipeline rather than LNG carrier.

New OCS support facilities will be needed in frontier offshore oil and gas producing areas. But, experience in supporting mid-Atlantic and New England area operations from a single support base in Davisville, Rhode Island, should be considered instructive. The petroleum industry can operate most efficiently where a variety of service requirements can be co-located and where existing facilities minimize new investment requirements.

I could continue to review other potential energy developments, but hope the point is made. When energy development requirements are examined closely by types of facility and geographic distribution, the real number of facility proposals is invariably much less than general expectations.

Coastal Dependence

Coastal dependence is a commonly applied requirement for approval of energy facility proposals under State coastal program policies. The actual program approval regulations set forth a requirement that coastal programs establish some means of "assessing the suitability of sites". We are concerned that the concept of coastal dependence is subject to overly simplistic application as a measure of site suitability. Coastal dependence may correctly recognize direct physical dependence of navigation or use of coastal waters for cooling or process requirements. But even such seemingly direct physical relationships are in large measure economic in character. Alternative transportation forms and water supplies are generally available if price is no object. Coastal dependence is in truth (and pardon the pun) a concept which just does not hold water. Coastal programs should assess site suitability among alternative coastal sites as well as inland sites. Environmental, economic, and cultural values should be compared. Values which cannot be fully measured in economic terms should at least be described as specifically as is possible. DOE recognizes that flexibility is available in application of differing standards of design, in choice among alternative energy technologies, and in alternative site selections. National objectives to make our energy future less costly, more reliable, less dependent on foreign sources, and more sensitive to environmental and cultural values are all dependent upon use of coastal land and water resources But, this dependence is flexible. Coastal dependence should not be advanced by either the energy industry or environmental interests as a simplistic justificaion for or restraint on regulatory actions.

National Interest

The national interest in energy facilities clearly is a matter of special concern under the Coastal Zone Management Act. But, what is the national interest in energy facilities? How is it to be identified by the States and other public and private interests? There are no easy answers to these questions. The national interest may be better understood by careful attention to three elements. I prefer to think of these elements as national interests (plural), national interest (singular) an national energy policies.

The national interests (plural) follow from requirements of Section 302(a) of the CZM Act and from several closely related parts of the findings and policies of the Act. Section 302(a) states, "There is a national interest in the effective management, beneficial use, protection, and development of the coastal zone." The national interests (plural) element essentially involves balancing the protection and development of coastal resources for various beneficial uses noted in the Act. This element of the national interest has been strongly reinforced by the 1980 amendments of the Act. The national interests (plural) have been set forth in a series of national policy requirements which may be briefly identified as:

A. Protection of natural resources,
B. Management of coastal development,
C. Coastal dependent uses and facility siting,
D. Public access,
E. Coordinated and simplified decision making,
F. Federal agency consultation,
G. Public participation, and
H. Harvesting of aquatic plants and animals.

States may now be required to spend increasing portions of their Federal financial assistance, up to 30 percent, on meeting the various national interest elements.

The national interest (singular) is a specific program requirement of Section 306(c)(8) of the CZM Act. The Act requires coastal programs to provide "for adequate consideration of the national interest involved in planning for, and in the siting of, facilities (including energy facilities...) which are necessary to meet requirements which are other than local in nature." There are three parts of this mandate which require close attention.

First, the CZM Act includes a specific definition of the term "energy facilities". Thus, the term has specific meaning under the Act and mandates that energy planning and siting must be comprehensive to include essentially all types of energy facilities.

Second, the Act imposes a test of scale or significance. Coastal programs must consider the national interest in those facilities "necessary to meet requirements which are other than local in nature". This language picks up one of the major themes of the CZM Act. A major Congressional finding in Section 302(h) of the Act encourages the states to regulate "land and water use decisions of more than local significance." In addition, section 306(e)(2) establishes a condition of program approval that State coastal programs must provide "for a method of assuring that local land and water use regulations within the coastal zone do not unreasonably restrict or exclude land and water uses or regional benefit."

Third, the Act requires an assessment of whether or not the national interest is adequately considered. It is the position of DOE that the test of adequate consideration imposed by 306(c)(8) must be based on the noted statutory requirements that energy planning and siting must: (1) be comprehensive of all types of facilities, (2) apply a test of scale to determine if the facility does meet more than local needs, and (3) be able to prevent arbitrary and unreasonable exclusion based on local land and water use controls.

National interest is a term with specific meaning to the CZM program. I have discussed how this term has a plural sense of balancing many alternative coastal resource uses and a singular sense applying only to the siting of energy facilities. National energy policy is a separate matter, it is not a matter of coastal planning or a statutory mandate of the CZM Act. The review of documents such as the National Energy Plan and of various national energy program statutes is not directly related to the standards for approval of the energy element of a coastal program. Such discussions have been almost universally included in the various State coastal programs and should be encouraged to the extent that a State may thereby be better able to support national energy policies through its coastal program. But, review of national energy policy, however detailed, is not a measure of the adequacy of the planning and siting process which a State must establish pursuant to Section 306(c)(8) of the Act.

Conclusion

Overall, I hope that all coastal interests will accept the fact that America's national energy future is vitally dependent upon continued use of coastal energy sites. At the same time, close examination shows that coastal siting requirements are not so great as oen might expect. Coastal energy facility siting should not be feared as about to overwhelm all coastal areas and resources. Energy needs can be met by careful assessment of site suitability and alternative site or technology availability. If coastal program policies and procedures provide for case-by-case assessment of more than local needs, site suitability, and the best available information on alternative energy technologies and environmental impacts, then programs will meet the test of providing adequate consideration of the national interest in energy facilities.

COASTAL ZONE MANAGEMENT EVALUATION: SOME ALTERNATIVES

John L. Kermond*

ABSTRACT

Section 312 of the Coastal Zone Management Act (CZMA) of 1972, as amended, gave the Office of Coastal Zone Management (OCZM) the mandate to evaluate the coastal zone management (CZM) efforts of each approved Great Lake, Territory and coastal state. OCZM are neophyte evaluators and have undertaken this important task in the absence of formal guidelines or regulations. The CZMA was silent on such matters.

This paper outlines some alternative evaluation strategies (principally Evaluability Assessment) that may be extremely useful for both OCZM and states consideration. Evaluability Assessment may be the optimum evaluation technique available that will adequately handle the superimposition of national priorities over those of the various states. Logic, function and measurement models are presented, alongside of suggested recommendations for improving the OCZM evaluation process.

*Coastal Zone Studies Program, Univ. of West Florida, Pensacola, FL

Section 312 of the Coastal Zone Management Act (CZMA) of 1972, as amended, gave the Office of Coastal Zone Management (OCZM) the mandate to evaluate the coastal zone management (CZM) efforts of each approved Great Lake, Territory and coastal state. OCZM are neophyte evaluators and have undertaken this important task in the absence of formal guidelines or regulations. The CZMA was silent on such matters.

Coastal states have not been as silent with respect to their own evaluations.[1] The Oregon CZM program went as far as to hire consultants to develop an evaluation framework within which their own efforts could be accounted for.[2]

The Coastal Zone Management Advisory Committee (CZMAC) were concerned about section 312 evaluations. At their September 28, 1979 meeting in Charleston, South Carolina, they felt that "......overall, program evaluation was working pretty well; that a detailed record was being compiled".[3] However, they also noted a lack of evaluation criteria, and raised the serious questions of how are the results used, and by whom?

These concerns were not new.[4] OCZM staffer Grant Dehart tried to arouse awareness of this forthcoming, and now manifest, problem.[5] OCZM funded an evaluation study the product of which has not been widely circulated, or apparently appreciated.[6] An evaluation workshop[7] was conducted for OCZM personnel in November of 1979. However, with the CZMA scheduled for re-authorization and subsequent 1980 amendment, section 312 (Review of Performance) is still characterized as being merely procedural, and entirely dependent upon the interpretation of the executive agency.

This paper highlights the results of a study completed in 1980 after a twelve-month fellowship at OCZM (sponsored by the National Sea Grant Program and Texas A & M University). The paper outlines some alternative evaluation strategies (principally Evaluability Assessment) that may be extremely useful for both OCZM and states' consideration.

Both evaluation per-se, and CZM evaluation have been reviewed extensively elsewhere.[8] Evaluation has been aptly described as an emerging art--it is still far from being a science! CZM evaluation literature is relatively scarce. Of interest was that in an 800 page book on CZM readings, only two pages were produced on CZM evaluation.[9]

A most promising evaluation methodology to recently appear has been Evaluability Assessment.(EA.)[10] This methodology grew out of the intensive evaluative work done by the Urban Institute in the early 70's, and has culminated in a sound, common-sense approach to program evaluation. Wholey[11] presents EA in the following manner.

Table 1 (Evaluability Assessment: Steps and Questions Answered)

TABLE 1

EVALUABILITY ASSESSMENT: STEPS AND QUESTIONS ANSWERED

Evaluability Assessment Step	Questions Answered
Step 1. Selecting the Program to be evaluated	• What program activities and objectives are the main focus of the evaluation?
Step 2. Collecting Information on the Intended Program	• What are management's program objectives and expectations? (What resources, activities, objectives, and causal assumptions make up management's intended program?) • What are the objectives and expectations of those at other management and policy levels? • What measures of program performance have been selected by those in charge of the program?
Step 3. Modeling: Synthesizing on the Intended Program	• What is the logic of causal assumptions that link resource inputs, program activities, outcomes, and impacts (from the point of view of the intended users of the evaluation)?
Step 4. Analysis: Identification of Users' Measures of Program Performance	• For which of the anticipated resource inputs, program activities, outcomes, and impacts (including important side-effects) are there agreed-on measures acceptable to the intended users of the evaluation?
Step 5. Collecting Information on Program Reality	• What are the actual program inputs, activities, and outcomes? • What measurements and comparisons are feasible?

TABLE 1 (continued)

Evaluability Assessment Step	Questions Answered
Step 6. Modeling: Synthesizing Information on Program Reality and Analyzing the Plausibility of Program objectives	. What data are obtainable on program performance? (Are there data sources for management's agreed-on measures?) . Is there a program in place that is likely to achieve management's objectives and expectations for the program?
Step 7. Identification of Evaluation. Management Options	. What portion of the intended program is ready for useful evaluation? . How would management use information on program performance? . What evaluation/management options would enhance program performance?
Step 8. Presentation to Management and Management Response	. What are the likely uses of the evaluation information? What range of actions might the intended user take or consider as a result of various possible findings? . What level of confidence would the user require before acting on the information? . To what extent is the intended user inclined or able to change program activities or objectives?

This procedure confirms much of what has been written, but only sometimes practiced, regarding management and evaluation. For example, Weidman[12] offered the following model.

FIGURE 1

EVALUATION/MANAGEMENT LINKAGES

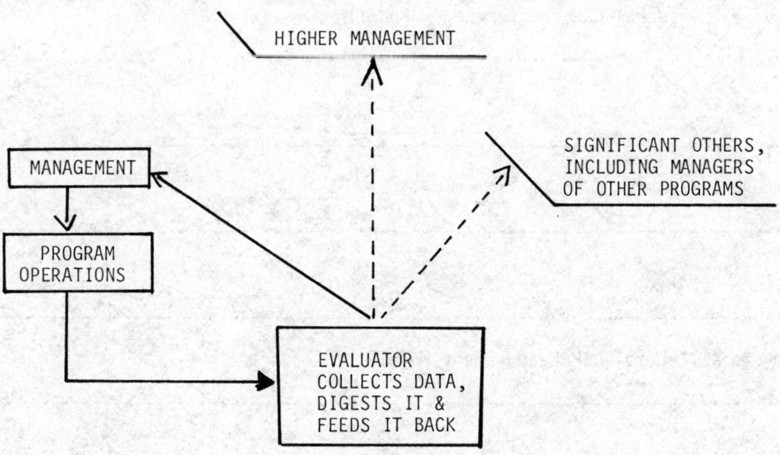

One of the key features of EA is that it demands the production of three basic models: logic, function and measurement. Each model is dependent on the other. In basic terms, the models describe what the program does, how it goes about it, and how all this can be measured. In doing so, program shortcomings can be identified and every effort made to rectify adverse measurement situations. The following models were produced during 1979. They represent this author's attempt to perform an EA on the federal CZM program. Since the time of completion, some internal re-organization has taken place at OCZM, and also the proposed amended (but not yet passed) CZMA (HB 6979) addresses some eight areas of national importance as regards CZM.

Figure 2 (Federal CZM Program Logic Model)

Figure 3 (Federal CZM Function Model)

Figure 4 (Federal CZM Measurement Model)

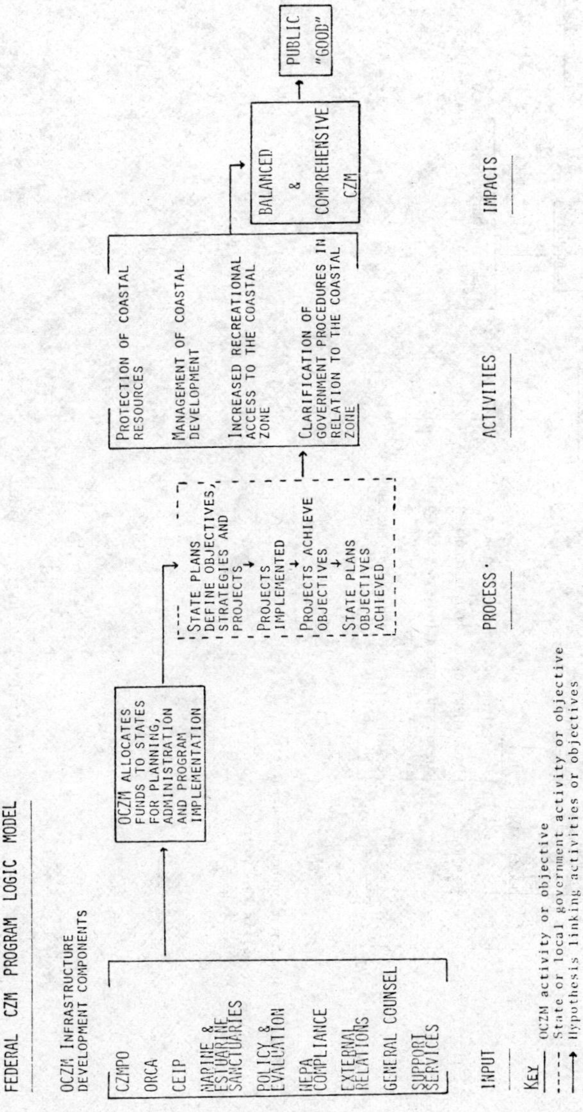

Figure. 2

Federal CZM Program Logic Model

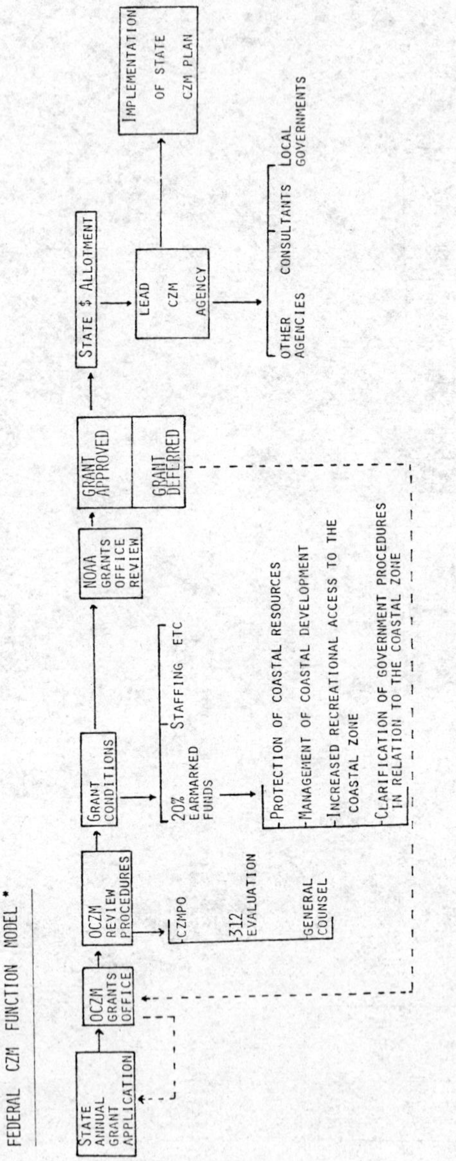

KEY
---> Return loops for document amendment etc
* Tracing the dollar flow process

Figure. 3
Federal CZM Function Model

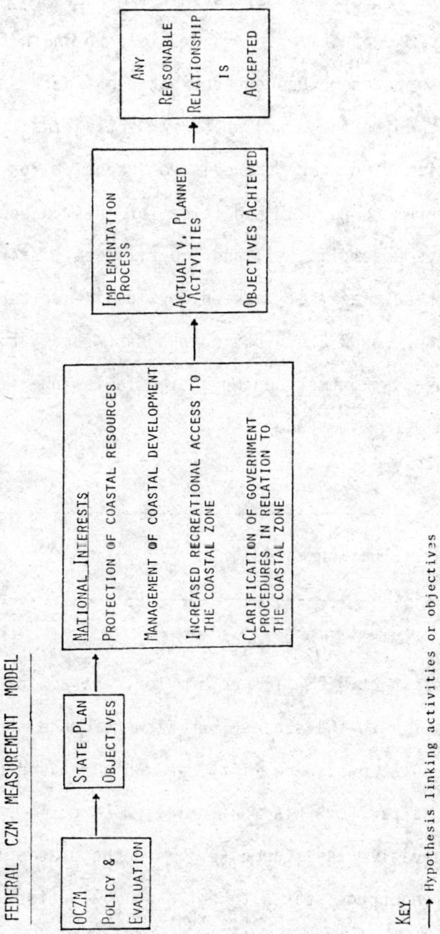

Figure. 4
Federal CZM Measurement Model

Of greatest interest in this discussion is the measurement model. It is here that the federal CZM evaluative effort breaks down in that no strong cause and effect relationship can be brought out between federal objectives (or priorities), state CZM objectives, and state CZM activities (or projects). Essentially, one has a situation where there is no incentive for a state to evaluate what it does, there are no real penalties for this absence (regardless of what a state accomplishes), and superimposed over state objectives are national priorities that lack any evaluative criteria.

OCZM refer to the section 312 evaluation process--rarely has one heard or seen written the 312 evaluation methodology. The following figure suggests an overall framework that OCZM might adopt.

Figure 5 (A Suggested CZM Evaluation Paradigm)

This figure suggests that OCZM (i) perform (or contract out) an EA, (ii) leave program evaluation to the states (but with guidance, and not another set of federal regulations), and (iii) take on a revised posture of performance monitoring. As part of this monitoring, OCZM could provide technical support and evaluative assistance (e.g. aid the states in establishing records and encourage states to build-in evaluation techniques into their management structure).[13]

From the states' CZM program perspective, the paradigm suggests a revision of their objectives (which in the case of Washington State, are now four years old and have never undergone any revision or up-dating despite notable changes in the energy, political and environmental climate).

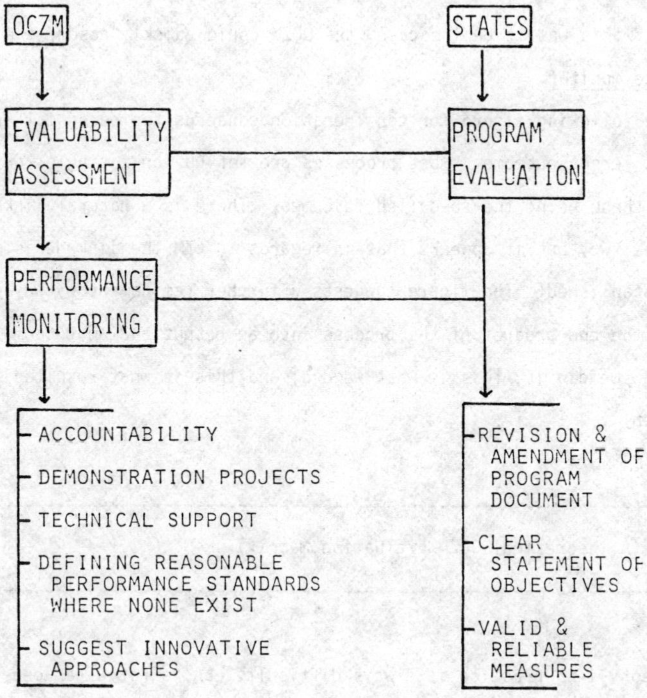

Figure. 5

A Suggested CZM Evaluation Paradigm

State CZM objectives should be drawn up for long term, intermediate and shortrange outcomes. Any state CZM grant document ought specify exactly what it is the state will achieve as a result of the projects for which funding is sought. If this was to be the case, the OCZM could quickly assume the role of performance monitor.

The following figure for consideration concerns the perennial problem of process versus outcome. Most processes are set up for a single, or several purposes--these being the so-called outcomes. There is a natural linkage between the two, but it appears that in regards to CZM the linkage is not yet clearly established. The figure suggests a further refinement by way of filtering the end product of the process into an output and an outcome (viz. this is how we did it, this is what we did, and this is what resulted from that action).

Figure 6 (A Theoretical OCZM Evaluation Matrix)

Theory into practice is always difficult. EA is not a panacea--it is one of several evaluation techniques available. The term "Evaluability Assessment" may belittle its common-sense, management approved approach to program evaluation. If improperly understood, EA may be seen as another high priced, fancy gimmick. It (EA) warrants closer inspection as its real value lies in determining what can and cannot be evaluated. An EA of OCZM's overall and state program operation, and an EA of the various state's programs may be useful at this juncture in the life of CZM in the nation, particularly if the following concluding models were ever pursued, refined and adopted.

	OBJECTIVES	CRITERIA	MEASURES OF EFFECTIVENESS	STRATEGY FOR DATA COLLECTION
PROCESS				
OUTPUT				
OUTCOME				

Figure 6

A Theoretical OCZM Evaluation Matrix

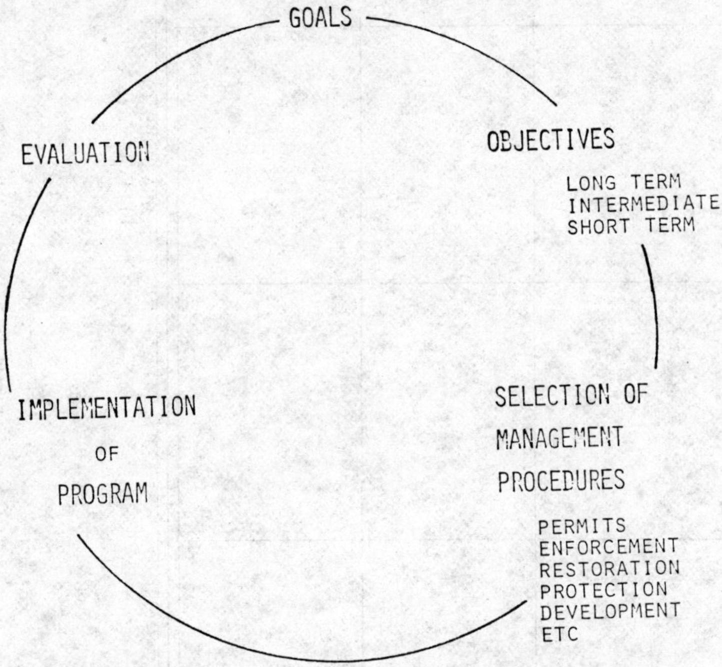

Figure 7

OCZM/States General Working Model

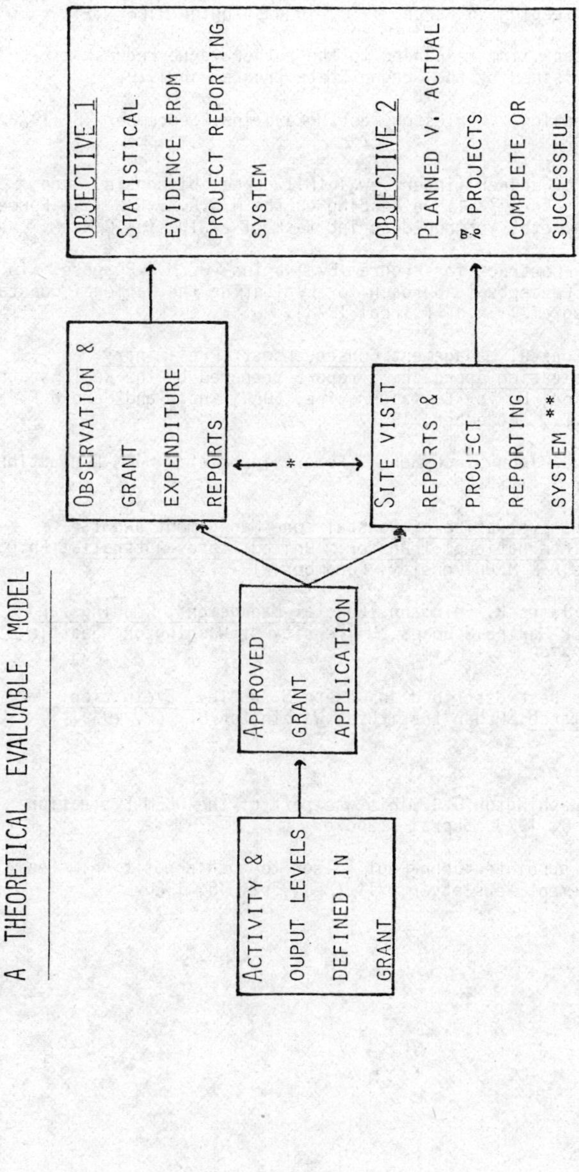

* Not strongly linked at present by OCZM

** Not established clearly by OCZM

Fig. 8 A Theoretical Evaluable Model

Notes and References

1. Both California and Oregon have expressed concern over their 312 evaluations. The most recent expression came out of a Coastal States Organization (CSO) meeting in March, 1980, in Washington D.C.

2. An economic consultant firm responded to the rather vague request-for-proposal (RFP) issued by the Oregon State program in 1979.

3. Minutes of CZMAC meeting, Charleston, South Carolina, September 28, 1979. p.3.

4. According to Richard Gardner (interview in Milwaukee, Wisconsin, June, 1979), Robert Knecht in 1972 (then heading up the N.O.A.A. CZM Task Force) was aware of the mammoth, yet non-descript task of evaluating CZM.

5. See Draft document--Contract for Program Evaluation--OCZM. Purpose: To Develop and Test a Conceptual Approach for Evaluating the National Coastal Zone Management Program". n.d. (circa. 1978).

6. See OCZM and CEQ, Coastal Management Consequences: Preliminary Conclusions and Evaluation Approach, a report prepared by the Real Estate Research, Corporation, De Voy Collaborrative, EDAW, Inc., and Harold F. Wise, Washington D.C., September, 1977.

7. See John L. Kermond & John L. Seymour, "Report in Brief: OCZM Evaluation Workship", November 9, 1979.

8. John L. Kermond, The Evaluation of Coastal Zone Management and Its Ramifications for International Transfer: United States-Australia, Ph.D. dissertation, Texas A & M University, December, 1979.

9. Marc J. Hershman & James H. Feldmann, Coastal Management: Readings and Notes, Institute for Marine Studies, University of Washington, Seattle, January, 1979. pp 227-8.

10. This methodology is best described in Joseph S. Wholey, Evaluation: Promise and Performance, Urban Institute, Washington D. C., 1979.

11. Ibid, pp. 75-76.

12. Don Weidman, OMB, Washington D.C. This was part of the OCZM Evaluation Workshop, November 6, 1979, Supra. 7 above.

13. OCZM have recently appointed John-Paul Tolson to such a position. See Coastal Zone Management Newsletter, VII (28), July 16, 1980.

Evaluation and Inter-Governmental Relations in CZM

G. Kem Lowry, Jr.*
Norman H. Okamura**

I. Introduction

National environmental management programs are based on an apparent contradiction.[1] On the one hand, a fundamental premise of national programs is the failure of state and local units of government to respond adequately to the degradation and depletion of valued natural resources. On the other hand, the planning and administration of national resource management programs must, of necessity, rely heavily on these same state and local agencies which are perceived as having failed.

The Clean Air Act,[2] the Federal Water Pollution Control Act,[3] the Resource Conservation and Recovery Act,[4] the Flood Disaster Protection Act[5] and the Coastal Zone Management Act,[6] among others, all assume some form of state and local participation in program planning and implementation. Yet, each of these programs is either implicitly or explicitly based on the assumption that state and local governments are not performing adequately with respect to these environmental management concerns.

The consequence of the contradiction is that environmental management programs are particularly difficult to develop and implement. Even if it were possible to create a federal force of inspectors and enforcers, management activities would still have to be well coordinated with state and local transportation and land use planning as well as all those other activities undertaken at the local level which impact the environment. Improving the practice of state and local land and resource governance is problematic when effective changes require the understanding, cooperation, and acceptance of those who have not participated in the formulation of national environmental policy but are largely responsible for its local administration.

Congress has sought to resolve this apparent contradiction by requiring some form of federal evaluation of state and/or local planning and management activities undertaken under the mandate of federal environmental programs. There are basically two modes of federal evaluation in these programs: plan review and program evaluation.

Most federal environmental management programs require some form of plan review. Briefly, plan review focuses on three fundamental aspects of a program:

---

* Assistant Professor, Urban and Regional Planning Program, University of Hawaii
** Research Coordinator, Urban and Regional Planning Program, University of Hawaii

state program goals, implementation strategies for achieving goals and the process of plan preparation. The plan review is usually conducted during and immediately after plan formulation, and results in a formal acceptance or approval by the responsible federal authority.

The attention given to the review of state and local programs goals varies considerably among federal environmental programs. In some programs, such as the Clean Air Act, program goals are federal mandates in the form of explicit standards with which implementing agencies are expected to comply.[7] In other programs such as Coastal Zone Management the mandate is broad enough that the goals for any particular state are likely to be the result of federal-state consultation and negotiation.[8] Finally, with regard to some programs, such as the Land and Water Conservation Fund Act, great discretion is granted the states to set their own goals subject to review.[9]

Federal scrutiny of proposed state implementation strategies under federal environmental management programs also emphasize the appropriateness of proposed means for achieving federal and state program objectives. How will land or water uses be managed? What agencies have decision-making authority? What is the legal basis of their authority?

Finally, federal plan review usually involves some sort of evaluation of the processes by which the plan was developed. What is the technical basis for the design of program strategies? Is it adequate? Who was consulted in the process of plan preparation? Were affected parties given ample opportunity to participate in the planning process? What formal and informal mechanisms were used to promote citizen education and participation?

The plan review process may take months--or years--and require a great deal of collaboration and negotiation between state and federal officials. However, the incentives to both state and federal officials to gain plan approval are substantial.

Plan review represents just one type of federal evaluation. A second type of evaluation occurs after a plan has been developed at the state or local level and found acceptable by a federal authority. Congress has mandated, in the case of some federal environmental programs, program evaluations of state performance subsequent to plan review and approval. The purposes of these program evaluations are to assure that state and local governments implement the proposed plan, measure the achievements or failures of the program, and to provide measures for ongoing program and fiscal accountability.

In the case of Coastal Zone Management, for example, Section 312 of the Act outlines the responsibilities of the Office of Coastal Zone Management (OCZM) for ongoing program evaluation. The Act states that:

(a) The Secretary shall conduct a continuing review of:

(1) The management programs of the coastal states and the performance of such states with respect to coastal management;

(2) The coastal energy impact program provided for under Section 308.

(b) The Secretary shall have the authority to terminate any financial assistance extended under Section 306 and to withdraw any unexpended portion of such assistance if: (1) he determines that the state is failing to adhere to and is not justified in deviating from the program approved by the Secretary; and (2) the state has been given notice of the proposed termination and withdrawal and given an opportunity to present evidence of adherence or justification for altering the program.[10]

To implement the above provisions of the CZM Act of 1972, as amended, the federal Office of Coastal Zone Management has utilized the so-called "312 Evaluation" as the principal mechanism to perform program evaluation. The 312 evaluations are intense, yearly reviews of state programs covering the full range of state coastal management activities. However, for a number of reasons, these evaluations have become a source of contention between the OCZM and the states.

Our purpose in this paper is to assess the 312 evaluations conducted by OCZM and, in so doing, make some general inferences and prescriptions both about the 312 evaluation process and, more generally, about post-plan review by federal officials of state environmental programs developed in response to federal mandates. This meta-evaluation focuses on:

1) the 312 evaluation process;

2) the content of 312 evaluations;

3) issues associated with federal review of state programs.[11]

II. The 312 Evaluation Process in Brief

The first four years of the federal CZM Program focused mainly on the development of state coastal zone management programs. Only in the past two years have a large number of states received program approval from the OCZM. Thus, the 312 Evaluation is a relatively new responsibility for both the states and its federal managers. What follows here is a brief summary of the 312 process.

The 312 evaluations are conducted in five phases:

-- Preparation of the "312 Information Request"
-- State preparation of response to the 312 information request
-- Site visit
-- Preparation of draft findings
-- Final 312 evaluation findings

The Section 312 evaluations site visits are usually conducted three or four months before the Section 306 program implementation grants are scheduled for refunding. Approximately six weeks prior to the site visit a detailed list of evaluative questions for the state being reviewed is prepared and forwarded to the state CZM office. These "312 Information Requests" are based on a review of the state's program document, previous evaluations, if any, projects undertaken under the current grant and other relevant material. Although the Office of Policy, Evaluation, and External Relations (PEER) has administrative responsibility for the preparation of these questions, other OCZM offices participate in the formulation of the final set of questions submitted to the state. The "312 Information Request" is forwarded to the state program director about six weeks prior to the scheduled site visit. The state response to these questions provides the substantive basis for the evaluation.

The size and composition of the evaluation team from OCZM varies depending on the state program to be evaluated and the particular problems or issues confronting a state program. A three to four-member team consisting of a mix of personnel from the PEER, the State Programs Office and of the other offices is typical. The exact composition of the team depends on the issues or needs confronted by the state being evaluated.

The site visits are usually conducted over a four or five day period. The first day is spent mostly with the state's CZM staff to review the evaluative questions and the state's formal response to those questions. Subsequent days are spent in meeting with state, local, and federal officials who have administrative responsibilities in coastal resource management, with citizen groups and other who have an interest in the state's coastal zone management program.

Preliminary findings are prepared by the PEER Office, usually within three to four weeks of the site visit. These findings are then circulated within OCZM for review and comment. Following an internal OCZM review, the preliminary findings are sent to the state for review and comment.

When conflict over the evaluative findings and recommendations occurs within OCZM, the formal process of conflict resolution calls for informal discussion, negotiation and, failing those methods of conflict resolution, a decision by the Assistant Administrator.[12] Before the formal evaluative findings and recommendations are issued, the Regional Manager of the area in which the state is located communicates the findings to the state CZM staff for comment.

III. Content of 312 Evaluations

The content of the 312 evaluations varies considerably from state to state. The major agenda-setting device for the content of the 312 evaluations is the 312 Informational Request. OCZM, as noted above, develops a list of evaluative questions which are forwarded to the state program director well in advance of the site visit. A state's answers to these questions provide the substantive bases for making evaluative inferences about the program.

These evaluative questions are based on a number of implicit and explicit evaluative criteria. Some of these criteria are outlined in a September, 1979 memorandum from Robert Knecht, Assistant Administrator for Coastal Zone Management. In the memo, Mr. Knecht notes that the evaluation site visits will focus on:

(1) Whether there are any unjustified deviations from the approved program that warrant initiation of termination proceedings;

(2) What is occurring as a result of a state's implementation of its CMP with respect to achieving the following national purposes of the CZMA:

(a) Protection of significant natural resources and areas such as wetlands, fisheries habitats, beaches, dunes and barrier islands;

(b) Management of coastal development so as to minimize loss of life and property due to inappropriate development in areas subject to coastal hazards; to give priority to coastal dependent development including energy facilities; and to identify environmentally suitable sites for dredge spoil disposal;

(c) Increased access to the coast for recreation purposes, including revitalization of urban waterfronts; and protection and restoration of cultural, historic and aesthetic resources in the coastal zone; and

(d) Improved predictability and efficiency in public decision-making, including increased governmental coordination and cooperation, and improved permitting procedures.

(3) Whether there are areas where the program is weak or not meeting its terms as fully as possible, or where the program could be further enhanced and made more effective by new or refocused program activities; and

(4) What changes have occurred in the CMP during the period covered by the evaluation, and whether these constitute amendments or routine program implementation.

The specific questions developed around these criteria for each state provide the empirical basis for making claims about the "success" of each state's program. Analysis of specific questions to which states are asked to respond reveals that in general there are three types of empirical indicators upon which judgements about program "success" are made: Indicators of program effort; indicators of program compliance; and indicators of program performance.

Many, indeed most, of the evaluative questions posed by OCZM relate to the efforts expended by states to implement their programs. These questions are based on the management activities that the states have described as fundamental to the successful implementation of their programs in the state CZM plans. Did the state officials claim that they would be developing memoranda of agreement with other state and federal agencies to coordinate their coastal management activites? If so, how many such agreements have been signed? Did they propose to hire personnel to enforce compliance with water pollution standards? If so, how many personnel have been hired. How many violations have been reported? What is the disposition of these violations?

Questions related to program compliance constitute a second major category of indicators by which OCZM seeks to establish the success or failure of state programs. Have local plans and ordinances or state agency rules been revised consistent with the state CZM program's policies and objectives? Has the issuance of specific permits for activities in the coastal zone been consistent with the relevant policies and objectives? Are requirements regarding public notification of the hearings, permit review procedures, and public participation being complied with? What are the sanctions for non-compliance?

Finally, a few of the questions relate to actual management outcomes; on the extent to which coastal conditions are actually altered in some significant way as a consequence of the state's management activities. How many acres of wetlands have been preserved for natural ecosystem maintenance? How much shoreline access has been provided? How have land use management activities affected coastal water quality? These performance or outcome questions are based on both state program objectives and national coastal resource management objectives.

IV. Issues Associated with the 312 Evaluation Process

Formal program evaluation has become increasingly sophisticated in the past few years. There are well-established techniques of evaluation, professional journals devoted to the latest developments in the field, university courses on program evaluation, and an increasing commitment at all levels of government to some form of evaluation.

In spite of this formalization, evaluation, in practice, is frequently a chaotic and highly subjective process with profound political implications. Some programs or activities are identified as "successes," others as "failures" with all that such designations imply in terms of allocation of resources, the value of program activities, and worthiness of those involved in such activities.

Given all of this, it is not surprising that Section 312 evaluations are not anticipated with any particular enthusiasm by state program directors and staffs. This resistance to evaluation is both general and specific. Some of the major issues related to the 312 evaluation are:

-- Lack of clarity and agreement about the purposes of evaluation;

-- Misunderstanding and/or disagreement over the respective roles of OCZM and state programs in the evaluation process;

-- Lack of clarity over the methods and outcomes of evaluation; and

-- Inadequate resources within OCZM and some state staffs to conduct the evaluation.

Purposes of 312 Evaluations

There are fundamental differences between OCZM and the states (as well as among the states) about what purposes are served by the 312 evaluation and the priorities among these purposes. Evaluation, as currently practiced by OCZM, appears to serve three basic organizational purposes: 1) providing basic information that can be used to justify both the national and state programs: 2) insuring program accountability; and 3) generating new knowledge about coastal management activities. States, on the other hand, tend to view evaluation in terms of gaining the final clearance for the following year's grant and providing some validation or reinforcement of the state's effort in coastal management.

It is quite common for organizations to use evaluation as a means for generating information that can be used to defend and justify their programs. The coastal zone management program has its share of critics in Congress and among developers, industrialists, organized environmental protection groups, and other special interest groups. Evaluation was seen initially both as a way of identifying program successes in specific states that could be highlighted and as a means of identifying specific issues that were likely to emerge as threats to specific state programs or to the national program. Successes have proved to be difficult to identify, not necessarily because they are not occurring, but because "success" in resource management tends to be contextual, incremental, long-term, difficult to measure and undramatic. "Success" frequently has to be thought of in terms of the bad projects that did not get proposed because of the program, new weightings of criteria to evaluate major projects and shifts in attitudes about appropriate uses of resources. Moreover, documentable success stories tend to be local and specific rather than general and national in scope.

Hence, in 1979 OCZM began to emphasize what it called "national interest objectives" listed above that, presumably, could become the focus of national debate about the value of coastal zone management activities. Three of these objectives--hazard management, managing development, and beach access--lend themselves well to documentation and inter-state comparisons. It is possible to quantify the amount of new beach access that has been developed or acquired since the implementation of coastal zone management programs or the number of states that have sought or are seeking to modify and simplify development review procedures. Success in these areas can be tangible and immediate, although such success does not necessarily imply the existence of "good" programs as judged by other criteria.

Given the existence of organized opposition to coastal zone management, the generation of documentable program successes is certain to be regarded as an organizational imperative by OCZM. Indeed, it would be by most organizations similarly situated. Organizational survival under conditions of scarce resources is seen as dictating that at least some portion of the evaluative efforts focus on the generation of information of potential threats to the program, even though state program officials may not regard this effort as particularly relevant to their activities.

A second basic function of 312 evaluations as currently practiced is that of determining whether state program activities are achieving the goals of the Coastal Zone Management Act, and the state's own program objectives as described in its program plan. This purpose is consistent with the conventional understanding of evaluation as determining "to what extent is the program achieving its goals."[14] It also overlaps somewhat with the purpose of providing information for program justification described above, but it is much broader, because it includes OCZM's efforts to insure accountability.

To OCZM, accountability is a fundamental purpose of the 312 evaluation. Nearly $26 million was allocated by the federal government to underwrite state management activities in 1979[15]. To be credible, OCZM must be able to assure Congressional oversight committees (as well as OMB) that they are both intimately familiar with how these states are using the grants and that those expenditures are consistent with Congressional purposes. In addition, the OCZM must be able to demonstrate the utility of those funds in achieving crude dollar equivalents in gains.

Most state program officials would agree that accountability is important, but they see themselves as primarily accountable to elected state-level officials. They argue that CZM was developed to be a state program, with state-designed goals and objectives. After OCZM approval of the state program, the obligation and focus of accountability should be, they argue, at the state-level.

States are further concerned that OCZM uses the concept of "accountability" as a justification for engaging in "remote management;" a process of conditioning state program grants with detailed requirements indicating both the task to be performed and the way in which it is to be performed. Such efforts at remote management may or may not result in better state programs, but they are a definite source of friction and misunderstanding between OCZM and some state program officials.

Finally, one implicit purpose of the 312 evaluation is to generate new knowledge about the impact of coastal management efforts and to disseminate that knowledge to potential users. To date, this purpose has not been emphasized and it is not currently an issue in inter-governmental relations.

Role Differences

A second basic issue has to do with the respective roles of OCZM and state program managers. Part of the role issue has to do with the authority question noted above: Who is responsible for making fundamental program decisions?

State program staffs, involved in the particulars of program planning and administration and intimately familiar with the local political environments, feel that they understand best what activities are politically feasible and how resources should be allocated to achieve those ends. OCZM evaluators, on the other hand, sometimes feel that their detachment and distance from local management activities makes them better equipped to see the "big picture" as it relates to the management activities of state programs. In addition, they feel that their experience with other states enables them to better judge activities which should and should not be undertaken.

Secondly, as Conners notes, evaluators are basically "doubters" and program managers are "believers."[16] State program managers feel they know what their program strengths and weaknesses are. Evaluators, on the other hand, are skeptics. They want to be convinced. They want data and justifications while state program staffs want congratulations or at least positive reinforcement. State program people see themselves as the "guys in the trenches," having to justify their activities to angry developers, skeptical legislators, and agency heads. They resent what they see as more "Monday morning-quarterbacking" from the feds.

Although it may be observed that these reactions are similar to those of local governments in the period of growing state involvement in a range of land use matters, the issues of authority and jurisdiction have not yet been resolved.

Part of the role problem is augmented by states' concern over the detailed nature of the 312 Evaluation. There is a concern on the part of the states that OCZM often becomes too preoccupied with details of a state's CZM Program. These states argue that OCZM's has too often concerned itself with small projects and activities disproportionate to its overall importance to the program. These states argue that OCZM concern should be with the "big picture," not in "nickel and diming" a state on what the states regard as rather trivial matters.

On the other hand, OCZM seems to believe that some states just don't have the "will to manage." And, for these states, OCZM has sought to strike a balance between what it sees as practical state concerns and the requirements for "good" coastal resource management. Attaching grant conditions and similar efforts at remote management are viewed as a necessary part of making the CZM program a credible management effort.

Indeed, the tying of the grant awards to the 312 evaluation creates natural adversaries of OCZM and the states. States react by trying to protect and defend program actions.

Lack of Clarity Over Methods and Outcomes of Evaluation

Coastal zone management programs are difficult to evaluate. Most programs contain multiple goals, at least some of which are so vaguely worded as to make it difficult to determine what conditions would indicate the achievement of those goals. Many of the goals are long-term. Moreover, the linkages between program efforts and program goals are so elaborate that it would be extremely difficult to establish with a high degree of rigor that program efforts were responsible for the achievement of program goals. This is particularly true of those programs that rely on the "networking" of existing resource management programs to achieve coastal resource management objectives.

Because of the difficulties in measuring the environmental, social and economic impacts of CZM, OCZM has relied on indirect tests of program success. In addition to financial audits, which are an obvious element in determining fiscal accountability, OCZM has relied on two additional key indirect tests of program success: measures of effort by program staff and measures of compliance by state and local agencies in their planning and permitting activities.

Measures of effort involve such variables as number of violations as reported by fish and game enforcement personnel hired with program funds or number of suits filed by the state's lead agency to force compliance with state CZM objectives. They represent attempts to inquire into the details of state program management. OCZM's emphasis on compliance is based on the presumption that compliance by state and local agencies tied into the state CZM program with planning and permitting requirements is a necessary condition of program success.

The specific questions relating to effort and compliance are frequently numerous. Preparing the questions takes a great deal of OCZM staff time just as gathering the required data involves a substantial investment of state staff time. But the pressing issue is that measures of effort and compliance do not provide state officials with program information that is new or necessarily pertinent to the program decisions they face and, for OCZM officials, such data, in the view of state authorities, can provide no more than a rough approximation, a partial and incomplete picture of the program.[17]

Resources

Conducting a 312 evaluation requires a considerable resource commitment on the part of OCZM and the states. OCZM evaluation staff have to prepare for the evaluation, travel to and from the state being evaluated, conduct the evaluation, and prepare findings which must be monitored through both federal and state reviews. The preparation for an evaluation may take each staff anywhere from two to four months, depending on the amount of existing knowledge the staff has of a state CZM program and its grant work tasks. Traveling often involves fairly large distances, especially for those traveling long distances. When in the state, the evaluation team usually goes through a rigorous schedule of meetings with various agency personnel, advisory commmittees, and community groups. Many of these are substantive, but some are largely symbolic. All contribute to a very exhaustive timetable.

After the evaluation has been conducted, the evaluation team must prepare preliminary findings which are reviewed by the OCZM and the state. If there are disagreements over the findings between OCZM and the state, the evalution team must seek resolution. Finally, the final findings are completed and published.

Similarly, states have to prepare for the 312, respond to the information request, host the evaluation, and respond to the draft 312 findings. In preparing for the evaluation, states generally consult the various agencies involved in coastal management, requesting and coordinating the preparation of reports from these agencies for the informational request, and inform the various citizen advisors of the 312. The states must develop and coordinate the site visit schedule for the evaluation team; and, once the evaluation team has arrived, the state program staff time is totally committed to the the evaluation process.

State concern with the amount of effort required for 312 evaluations is reflected both in a draft questionnaire regarding 312 circulated by the Coastal States Organization and in the report by the Evaluation Committee of the Coastal States Organization. On the draft questionnaire twenty percent of the questions dealt with the amount of time and effort required for preparation and visiting a site visit.[18] The committee report noted that "(e)valuations are a time consuming process for both the states and OCZM. The administrative effort needed should be in proportion to the resources available to produce quality reviews."[19]

V. Increasing the Relevance of 312 Evaluations

The issues identified above do not lend themselves to easy resolution. They can be ameliorated somewhat, however, by clarifying purposes and expectations that might be served by the evaluation, developing intergovernmental understanding and agreement on these purposes and amending and developing evaluative strategies appropriate to these purposes.

Providing basic information that can be used to justify and promote coastal resource management efforts at all levels of government and insuring accountability are likely to remain important evaluative purposes served by the 312 evaluation review or some similar process. The challenge is to construct the evaluative enterprise so that it serves both the need for information that could be used for program development at all levels of government. To do that, several questions need to be answered for each state:

-- What, in the view of OCZM, state officials and other interested actors constitutes program success? How would coastal conditions be altered if the program was successful?

-- Whose behavior (agencies, developers, beachgoers) will have to be affected if those conditions of success are to occur?

-- What data would indicate that the program is a success? If that information is not now available how can it be generated? By whom and how frequently?

-- Once generated, how will these data be used and by whom?

A third purpose of the 312 evaluation is to generate new knowledge. The national coastal zone management effort represents multiple experiments in coastal resource management involving a variety of different management approaches to particular coastal management problems such as erosion, maintaining water quality and insuring physical and visual access to the shoreline. Some of these management efforts are different only in nuance, but these differences may be significant. Moreover, management efforts result in untold unintended physical, social and economic impacts on the physical, social and economic fabric of coastal areas. Far too little is known about these impacts.[20]

The 312 process itself, as currently conceived, does not lend itself well to the generation of knowledge about the effects of coastal management efforts, but establishing correlations and even causal linkages among management and coastal phenomena is clearly an important evaluative enterprise.

VI. Future Directions

While there is no painless approach to meaningful evaluation there are some shifts in emphasis and practice which, in our view, might ameliorate some of the current issues between the OCZM and state programs regarding evaluation without undermining the evaluative functions.

1. The OCZM should collaborate with state and local coastal zone management officials to develop a mutual understanding of the policy purposes to be served by evaluation at each level of government.

Some state program officials do not understand the purposes served by federal evaluation of state programs irrespective of the form. Some other state officials understand, but do not support either the purposes or the way in which evaluations are currently conducted. Moreover, few states have tried to identify the policy decisions they confront that might be informed by evaluation research or to develop some sort of evaluation strategy independent of OCZM.

A collaborative effort involving OCZM and state and local officials to identify the purposes of evaluation at each level of government, significant policy decisions that might be informed by evaluation research, and the types of information required for such decisions might not result in universal consensus among all of the actors, but at least there could be greater mutual understanding of the need for evaluative information and the range of costs and benefits associated with evaluation.

2. Explicit agreements should be developed with each state and states, in turn, should work with relevant local officials to identify specific information to support evaluative objectives.[21]

While there are obviously several ways to implement this recommendation, we favor having states develop, as part of their periodic grant applications, specific indicators by which they propose that their "success" in meeting state and national objectives be measured. These indicators would be reviewed by the PEER staff and additional or alternative indicators might be proposed and negotiated.

3. Collection of evaluative information should be systematized.

If there is agreement between OCZM and each state with regard to what information will be required for evaluating a state's program, it should be possible to make the collection of evaluative data a routine part of program management rather than an episodic event associated with the preparation of a response to the "312 Information Request."

The systematic approach to the collection of evaluative information is likely to require the participation of local governmental officials. In most states, a significant portion of the coastal zone management program relies on the planning and permitting activities of local governments. Much of the information required for evaluation will have to be generated and transmitted by these local governments. Hence, their understanding, agreement, and participation in the development of a systematic data collection and retrieval system is essential.

4. The OCZM should assign a higher priority to the generation of new knowledge about the impacts of coastal management efforts

For reasons described above, most of the OCZM's evaluative efforts until now have focused on state program efforts and compliance. Much less is known about the effects or impacts of coastal resource management efforts. And yet knowledge of program impacts is critical if states are to correct program efforts that are ineffective or that create particularly detrimental side-effects.

Impact studies are usually longer-term and involve sophisticated analytical techniques. OCZM has a number of alternatives for engaging these sorts of studies. The OCZM could conduct such studies itself, assist states that choose to undertake such studies by providing the necessary funding, seek to facilitate such studies by Sea Grant and other agencies, and/or serve as a clearinghouse of coastal impact studies and other policy relevant research. This type of research might more productively be utilized by the OCZM and states, and help to lessen the stresses associated with information gathering by the states.

5. The Policy, Evaluation, and External Relations Office should serve as a clearinghouse for coastal impact studies and other policy relevant research.

A recent GAO report on the coastal zone management program recommended that OCZM work "closely with the States to help them to resolve special problems and provide guidance for coordination with other federal agencies."[22] We think that one of the most effective ways to implement GAO's recommendation is for the PEER office to serve as a clearinghouse for policy-relevant technical studies of coastal zone management programs.[23] These studies should be desseminated to personnel in the state.

6. Site evaluations need not be conducted annually.

The OCZM is already moving toward longer grant cycles. Evaluations conducted every eighteen to twenty-four months consistent with longer grant cycles should become the norm. To the extent that states move to more routinized procedures for the collection of evaluative information there is less reason for formal evaluations in the states on an annual basis.

NOTES

1. See, e.g. Richard B. Stewart, "Pyramids of Sacrifice? Problems of Federalism in: Mandating State Implementation of National Environmental Policy, The Yale Law Journal, Vol. 86, pp. 1196-1272.

2. 42 U.S.C. §§ 2857 et seq. and 42 U.S.C. 7401 et seq.

3. 33 U.S.C. §§ 1251 et seq.

4. 42 U.S.C. § 6941 et seq.

5. 42 U.S.C. § 4106 et seq.

6. 16 U.S.C. § 1451 et seq.

7. 40 C.F.R. Part 50, 36 Fed. Reg. 22384 (1971).

8. 16 U.S.C. §§ 1451, 1452, as amended.

9. 16 U.S.C. § 460 et seq.

10. 16 U.S.C. § 1458.

11. The term "meta-evaluation" was coined by Scriven to describe the process of evaluating evaluations. Michael S. Scirven, "An Introduction to Meta-Evaluation," Educational Product Report 2, 1969.

12. Memo from Grant Dehart, Subject: "Procedures for State Performance Evaluation - Section 312 CZMA," April 19, 1978.

13. Memo from Carol Sondheimer to State Program Managers, Subject: "Interim Procedures for 312 Evaluation Site Visits," November 22, 1978.

14. Carol Weiss, Evaluation Research (Englewood Cliffs, New Jersey: Prentice Hall Publishers, 1972), p 24.

15. U.S. General Accounting Office, Problems Continue in the Federal Management of the Coastal Zone Management Program, CED-80-103, June 25, 1980.

16. Ross F. Conner, "The Evaluator-Manager Relationship: An Examination of the Sources of Conflict and a Model for a Successful Union," in Herbert G. Schulberg and Jeanette M. Jerrell (eds.) The Evaluator and Management (Beverly Hills: Sage) 1979, p. 121.

17. This is not to say that such data would not be useful to states. Indeed a more active state management role for programs that rely heavily on local planning and permitting activities would seem to require information on compliance. However, most states have not yet developed a rationale for collecting compliance information or a systematic way of generating and transmitting such information.

18. Undated draft questionnaire prepared by the Evaluation Committee of the Coastal States Organization and circulated in July, 1980.

19. Report of the Evaluation Committee of the Coastal States Organization, April, 1980.

20. See. e.g., Kem Lowry, "Policy-Relevant Assessment of Coastal Zone Management Programs" Coastal Zone Management Journal, Volume 8, Number 2 and Ernie Englander, Jim Feldman and Marc Hershman, "Coastal Zone Problems: A Basis for Evaluation," Coastal Zone Management Journal, Vol. 3, No. 3, pp. 217-236.

21. This recommendation is similar to one made by a panel at a meeting of the Coastal States Organization. Letter from Michael Fisher to Mike Glazer, April 18, 1980.

22. U.S. General Accounting Office, op.cit.

23. Such a role is alluded to in the CZMA requirements for an annual report by OCZM. 16 U.S.C. § 1462.

EFFECTIVENESS OF ASSESSMENTS OF COASTAL MANAGEMENT

by

Robert J. Reimold, Director
Coastal Resources Division
Georgia Department of Natural Resources
1200 Glynn Avenue
Brunswick, Georgia 31523

During the decade of the seventies, the federal government enacted legislation designed to address the concerns of coastal waters and shorelands, the use of which could have a direct and significant impact on coastal waters. Following this action, the national wheels of government began slowly grinding to coerce states to form coastal management programs consistent with the Coastal Zone Management Act of 1972. Proponents of the program indicate that the coastal zone management process is intended to be a coordinate process by which local, state and federal governments, with both public and private input, design and implement a mutually agreeable plan for the use and protection of coastal resources. The program, by design, relies on the state governments for leadership with the input of local and national interest to supplement this leadership role. Coastal management also, according to the draftsman of the Act, is designed to improve the technical and scientific framework on which coastal decision-making is founded.

As we enter the decade of the eighties, many coastal states will be implementing a variety of coastal management programs. The programs are to be specifically tailored to meet the individual needs of the states and, consequently, vary widely in their ability to manage coastal areas. Some of the programs regulate land use, while others are aimed at responsible conservation of renewable and nonrenewable natural resources. Some of the programs are designed solely to meet the requirements of the federal government in order to receive federal funds, while others have a true concern for sensitive areas of their specific coast. In order to compare coastal management programs, effectiveness assessment methodology for coastal zone management is required.

Some approaches for coastal management follow strategies designed to conserve natural resources, while others follow solely a planning approach. Other facets of coastal management involve economic sanctions and new and improved environmental regulations. Finally, perhaps many of the components of coastal management are designed strictly to satisfy requirements of the federal agency providing funding for the program.

An effective coastal management program must be founded on principles which can be repeated in time and space. It must not rely solely on any individual, public official, or conservation leader, but must be founded on a series of goals and objectives established in relation to the wise use of the coast. The federal Office of Coastal Zone Management has pointed out that the program should protect significant natural resources such as barrier islands, dunes, beaches and wetlands. It also has advised that the program should result in a more effective management of the development of coastal shorelands. They also suggest that the program should promote a better use of coastal shorelands by siting water dependent uses adjacent to deep water. The federal government also points to the increasing need for accessability to the coastal areas for recreational uses, while at the same time, assurance of the protection and restoration or enhancement of the cultural and natural resources is needed. They also document the need for increased cooperation in intergovernmental coordination in order to bring about greater predictability in the public decision-making process.

If we assume that these are indeed the goals of the coastal management program as outlined by its previous administrator (Knecht, R. 1979), it appears appropriate to now analyze how effective the program has been in reaching these goals.

Wetlands are a sensitive component of the coastal ecosystem. The wetlands' control resulting from coastal management should not address the ownership of the wetlands, that is whether state or private, but instead, should require permits for activities encroaching on the wetlands, regardless of proprietorship. Standards and criteria adopted through laws and applicable rules and regulations should provide guidelines for permit evaluation and assessment. Guidelines should include review of the public interest, the water dependent nature of the proposed activity, whether the proposed change is essential for safe navigation and/or public health, and also address the public recreational benefits. Activities proposed entirely for private use such as commercial and industrial uses, filling of wetlands for private roadways or parking lots or for use as dump sites for depositing waste materials or spoil, or the building of structures which constitute unreasonable obstruction to the view of adjoining landowners should be considered contrary to the public interest and, therefore, generally not permitted. In addition, an effective coastal management program should provide free technical assistance to the applicant to assist in meeting guidelines of the coastal wetlands regulation program. The state agency involved should play an assertive interactive role with the applicants to help in the completion of the application and acquisition of data requisite for a judicious decision.

Another important component of coastal management is the ability to manage the dunes, beaches and offshore bars. These areas are fragile products of shoreline evolution and are quite easily disturbed. The offshore sandbars and shoals are the shore's first line of defense against destructive energy generated by winds, tides, and storms and consequently protect the onshore portions of the system by serving as a sand reservoir for the beaches. These ocean beaches frequently provide an unparalleled natural recreation resource which has become vitally linked to the economy of the coastal zone and that of the entire nation. Wise conservation of this sand-sharing system is of vital concern to all citizens to maintain their health, safety and welfare. Appropriate conservation regulations for this dynamic sand-sharing system assures that all structures and activities on the offshore sandbars, shoals, beaches and dunes do not adversely affect the values and functions of the sand-sharing system.

Another component of an effective program in coastal management is the utilization of the most recent advances of science and technology as a foundation upon which to base coastal legislation. This involves close networking with scientists of many disciplines in order to assure that recent discoveries of science are intergrated into wise resource management. Too frequently the laws relating to the use of coastal resources are based on legal framework established 30 to 60 years in the past. An effective coastal management program will be one which reviews existing laws and rules and regulations relative to coastal resources and land use and makes recommendations for change relative to recent findings. In this instance, the scientists and government officials cannot work in a vacuum. Care must be taken to assure that the discoveries of science are interpreted for use by the layman and the legislator alike.

Another effectiveness assessment technique of coastal management is consideration of the amount of paperwork as well as the coordinative processes that are implemented as a result of a wise and efficient coastal management program. Presumably, wise use of the coast should involve more expeditious use of necessary forms, not just more paper. For example, a coordinative process between the federal and state agencies should easily result in one-stop permitting. In such an instance, the federal/state agencies should develop a joint permit application for use of renewable resources in the coastal zone. Such a coordinative joint permitting program will benefit the state agencies and the federal agencies implementing laws along the coast. This has already been successfully accomplished in several states and is now being tried in additional areas.

A careful review of various effectiveness criteria documents that a coastal management program needs to be specifically tailored to preserve, protect, enhance, and develop the use of the coastal renewable and nonrenewable resources. The plan must be formulated to prudently manage unique coastal resources, giving equal consideration to the economic and ecological goals. Through the various permitting processes for natural resource uses that have been discussed above, one of the goals of the program should be to resolve conflicts which arise between natural resource conservation and economic development. This resolution needs to be brought about in a timely manner, assuring the greatest long term increase in public benefit for all user groups. For example, renewable resources, such as fisheries and forests, and nonrenewable resources, such as natural gas reserves, must be utilized in ways which best conserve their value for present and future generations. Other resources, including archeological and historical sites, rare plants and animals, and freshwater wetlands, have special values which must be considered in wise planning for the coast. In addition, some resources are suited for development, while others should be managed for maximum protection.

Another important component of assessing the effectiveness of coastal management is to determine whether or not the program encourages the location of private and public developments in the most suitable areas so that adverse impacts upon natural resources are minimized. The program must effectively control the quality and kinds of development which occur in the coastal zone. This can be best accomplished through zoning. In some states the power for zoning is vested in the authority of the state government, while in other states, the power for zoning and land use planning is constitutionally vested with local government.

Another component of wise coastal management is maintenance or improvement of the quality of the waters of the coast. Maintenance and, where possible, improvement of coastal waters is essential to public health as well as improvement of the tourist industry. Great care must be taken to assure that industrial waste and port activities, as well as urban growth, do not adversely degrade the coastal water quality.

Another way of assessing the effectiveness of the coastal management program is to determine whether the program has had impact on public accessibility to coastal waters and beaches. As the coastal area grows and more people use the coast, additional regulated public access is essential. Through the use of state-owned facilities, acquisition of additional public beach access sites, and state and local requirements that beach access be an integral part of any beach nourishment program, several states have assured that the public will continue to have access to the coastal waters.

Historic sites, structures, and relics both man-made and natural are essential cultural resources which must be preserved. Another tool in assessing the effectiveness of a coastal management program is consideration of the programs' ability to identify, preserve, and rehabilitate these important historic sites.

In designing a balanced use of the coast, it is essential that local governments not unreasonably restrict or exclude facilities from being sited in the coastal zone when they can not easily be sited elsewhere. Facilities of regional and national benefit include power plants, transmission lines, public port development, regional airports, sewage treatment plants, regional solid waste disposal facilities, and energy exploration and production facilities.

Another important component of a coastal management program should be public, as well as local government, involvement in decision-making processes regarding coastal resources. It is essential that there be a coordinative process between the state and local governments in all resource planning efforts in the coast. Such a coordinative process recognizes the importance of considering both development and conservation needs in deciding the future use of all coastal resources.

A less effective assessment tool used by some states in coastal management is the determination of whether or not their coastal management process meets the requirements of the federal government. By meeting such requirements, the states may be eligible for financial assistance in better managing their coastal zones.

In analyzing areas of critical concern, such as marshes, and wetlands, beaches, dunes and offshore bars, commercial and recreational fisheries, historical, cultural and archeological sites, and outer continental shelf energy exploration and development, it becomes apparent that the effectiveness of any state's coastal management program is dependent upon the observer. If the program is being implemented solely to acquire federal funds as mentioned above, certain procedures will be required. These procedures frequently are not in the best interest of the taxpayers of the state and local governments.

Perhaps a better way of assessing the importance of the program is to determine whether we are conserving the necessary components of natural ecosystems so that renewable resources can renew themselves. Another way of assessing the program is to determine whether or not the states are prospering in their management of the coastal resources. Yet another method of assessing program effectiveness is to determine whether or not the program is reducing political tension and bringing about a better coordination between state, local and private interests relative to the use of the coast. Frequently political demagoguery results in diverse discussion about coastal resources. Due to the number of qualified staff and their strategic location, a local unit of government, or a state, can more realistically prepare a management plan for its coast than can the federal agency. Due to limitations of staff number, as well as onsite knowledge of the particular coastal zone, the local/state partnership frequently has brought about a better management scheme than any federal expert has been able to draft. Although the federal agencies have in some instances denied the states participation in the coastal management program because of certain inadequacies, as Senator Mark Hatfield said, "the burden of proof should be on the federal agencies to prove that the state's plan is truly inadequate based on scientific and technological facts which could be sustained in a court of law."

Statistical analysts have a variety of tools for assessing effectiveness. Considering the various components of coastal systems, multivariate analyses, and mathematical models must certainly be employed in any coastal management effectiveness consideration. In clinical laboratories, precision and accuracy are the statistical tests frequently used for quality control. These same measurements are used when assessing productivity of assembly lines.

In more recent years, scientists have utilized sensitivity analysis in determining which are the most sensitive components of an ecosystem. Perhaps, sensitivity analysis applied to coastal management would be one cost-effective means of assessing the performance of the program. It is possible to construct conceptual models of all the components in the coastal zone. By making certain approximations of interactions and using the measurements of interactions that are available, one can conceptualize the interworking of coastal management. Application of sensitivity analysis to this conceptual model would provide a means of identifying the most fragile component, or that component most sensitive to slight changes or perturbations in the system. Such a sensitivity analysis can frequently be performed without a more complicated mathematical analysis of the total model.

In other words, if we list the various environmental and economical considerations of the coastal zone, we can make certain predictions as to the interactive relationships of the various components. Once these interactive relationships are identified, they can be quantified based on a certain data base, perhaps on dollars, perhaps on energy. Then mathematical multiplications can be used to determine which interactions are most sensitive to slight changes. Obviously, wise coastal management should focus on the interactions that are most sensitive to bringing about a change on coastal waters and shorelands.

Other ways of analyzing the effectiveness of coastal management relate to the elimination of unnecessary paperwork in government. The single stop permitting outlined earlier is essential. The foundation of coastal laws and rules and regulations based on current science and technology as opposed to historic political dogma is also essential. This coordinated process of permitting should not only involve state/federal agencies, but also state/local agencies as well. Another way of determining the resultant effects of coastal management is to analyze improved harvests of renewable coastal resources, then one should see improved harvests for all user groups resulting from such legislation.

Finally, there should be an improved ability to manage privately-owned as well as state and federal lands. For example, if having a coastal management program approved by the federal government means that the state loses its ability to manage federally owned lands within its coastal zone, the program cannot possibly be effective. It is essential that the state maintain or have an enhanced management control over the use of lands and waters in its coastal zone. It is essential to assure that when designing a coastal management program, a higher authority, for example the federal government, does not usurp the authority normally vested in the state or local government, giving it the power to manage its own coastal area even if the lands are owned by the federal government. It is also essential that the state serve as leader in demonstrating its intent to wisely use state owned lands on the coast. Any type of development of alteration of coastal habitat should be done as a model to show the private sector how the area can be most economically developed with minimal environmental impact.

While some states have proceeded in managing their coast for the sake of carrying out a process to satisfy the federal agency providing limited funding, other states have been managing their coast for the wise use of all resource users. It is essential that future coastal management not be designed or implemented simply to satisfy a bureaucratic process. The effective coastal management must produce a manageable impact on the state's ability to assure that its coast is wisely utilized for a balanced concern between economic and ecological factors. Careful consideration of the above iterated components of effectiveness assessment in coastal management should enable the planner, the manager, the environmental specialist, and the economist to all better understand the effective means of coastal management.

COASTAL PROGRAM EVALUATION FROM A STATE PERSPECTIVE
or
Can an Amoeba Find True Happiness Under a Microscope?

by
William Travis
Assistant Executive Director
California Coastal Commission

Introduction. This is not a scholarly treatise on the theoretical methodologies for evaluating government programs. Nor is it a paper on the evaluation of coastal management program because it is not a "paper" in the professional sense. It doesn't have any footnotes and it didn't require any research. As such, it is more of an essay; a collection of my recollections, observations, and impressions over the three years that the federal Office of Coastal Zone Management (OCZM) has been grappling with the concept of state coastal management program evaluation. As the principal representative of the California Coastal Management Program vis-a-vis OCZM, I have had the opportunity to both formally and informally participate in the evolution of OCZM's evaluation effort and have been in a position to observe the effectiveness of this effort during OCZM's formal evaluations of the California program in 1979 and 1980. Where I have been able to, I have generalized conclusions so the lessons learned can be applied to future OCZM evaluations of state coastal programs and, as is the case in California, state evaluations of local coastal programs. But much of what I have observed does not lend itself to such conclusions. Some of the mistakes that have been made are irreparable. Some of the problems stem from the language of the federal Coastal Zone Management Act (CZMA) itself, and cannot be resolved without junking the substantial investment and years of work that have gone into implementing the CZMA. Other issues have arisen because of political conditions unique to the time and situation. Still others resulted from the will of individual personalities more than from institutional or legal conditions. For all these reasons, on the whole this paper will probably be of greater interest to historians and gossip columnists than to political scientists and coastal managers.

Undoubtedly, my views and conclusions will differ from those of other observers. Perhaps in some cases I have been dead wrong in my speculations as to the thinking behind a decision. Athough there is considerable dialogue and interaction between state program managers and the OCZM staff, the theorizing and planning did take place in Washington, D.C., while I spent my time 3,000 miles away in California which limits my direct knowledge of the details.

And, in a few instances, a conclusion on the evaluation process was presented as a fait accompli to the state managers. If the conclusion was faulty, it was necessary to resort to investigative techniques to understand how the conclusion was put together so we could go about taking it apart. Fortunately, OCZM seems to be staffed by people who have the wisdom to understand that this process is healthy for the program, the courage to invite examination of their thinking, and the integrity to change their minds if they are wrong. Nevertheless, some of my observations are speculative. If I am wrong as to what was behind a decision, imagine what a scientist must look like to an amoeba peering up through the wrong end of a microscope.

What Are You Talking About? To someone uninitiated in the bureaucratic jargon of coastal management, the term "evaluation of coastal management programs" might be confusing. Under the provisions of Section 306 of the CZMA, in order for a state to have its management program approved, the Secretary of Commerce must find that the program contains all of the elements specified in Section 305 of the Act and meets the approval standards set forth in Section 306. The Secretary of Commerce delegated the administration of the CZMA to the National Oceanic and Atmospheric Administration, (NOAA), which, in turn, established OCZM to handle the CZMA. Thus, OCZM makes the determination of whether a state program can be approved under Section 306. Obviously, OCZM must make an evaluation of the state program to make this determination. But this evaluation is not the subject of this paper. Nor is the subject the evaluation called for in Section 315 of the CZMA, which requires:

> ...the Secretary [to] prepare and submit...each year...a report on the administration of this title [which includes]...a summary and evaluation of the research, studies, and training conducted in support of coastal zone management.

This paper is confined to the process required by Section 312 of the CZMA which provides:

> The Secretary shall conduct a continuing review of...the management programs of the coastal states and the performance of such states with respect to coastal zone management.

Because of the number of the CZMA section requiring this procedure, the process is often referred to as the "312 evaluation" or "312 review". It is this evaluation alone that is the subject of this paper; the continuing review and evaluation of state programs that have been approved under Section 306 to determine if the programs are being carried out as approved and if they are meeting the goals of the CZMA.

<u>You Can't Weigh a Frog Without A Scale</u>. Probably the largest single problem that has plagued the evaluation process is the lack of clear standards for measuring the performance of states in implementing their approved programs. Given the context in which the CZMA has been administred, the problem was definitely unavoidable. With the benefit of 20-20 hindsight it is possible to suggest how the severity of the problem might have been reduced. But it could not have been eliminated. Moreover, it will continue to create difficulties in future 312 reviews unless extraordinary (and, I believe, unwarranted) revisions are made to approved state programs.

Unlike the Clean Air and Clean Water Acts, the CZMA is not a federal environmental protection law. With a few exceptions, it does not include precise standards for what must be in an approved state program. Nor does it dictate what sort of uses may or may not be permitted in state coastal zones. It does not establish what sort of administrative and legal mechanisms a state must use to carry out its program. And the CZMA isn't even mandatory; states participate on a voluntary basis. What the law does call for is the establishment of a program--a process for managing coastal resources.

As might be expected (and, in fact, as was desired by the Congress when it enacted the CZMA in 1972), this situation has resulted in as many different types of state programs as there are states participating under the CZMA. Each program is tailored to the unique geographic, historical, political, and social conditions in each state. Because of this, OCZM, has had to use a somewhat different set of criteria for each state to determine program approvability. Quite obviously, subjective judgments have to be made in OCZM's approval review. And, until lately, standards for measuring program performance were not considered in the process of developing state programs or in OCZM's approval of the programs. Given this situation, it is downright silly to expect that a single yardstick for evaluating approved programs can now be developed.

At first blush, it might seem justifiable to criticize OCZM for not considering evaluation earlier in its administration of the CZMA and incorporating performance standards into emerging state programs. But on further analysis, OCZM's record is certainly more understandable, if not inevitable. Although the Congress enacted the CZMA in October, 1972, it was late 1973 before anything was done to implement it because the Nixon Administration chose to not expend the funds appropriated to administer the law. In the early days, much of OCZM's attention was, quite correctly, directed toward setting up its own office, drafting regulations for these sections of the law that had immediate applicability (particularly Section 305), and "selling" the national program in those states where land-use controls and federal intervention were none too popular. Again, it might seem appropriate to suggest that OCZM should have looked ahead and begun forcing performance standards on the states by including them as requirements in the Section 305 or 306 regulations. This ignores the problem that OCZM had with the states at the other end of the spectrum which already had coastal management programs, not because of the CZMA, but almost in spite of it. In particular, Washington, Oregon, and California had in place or were developing comprehensive, innovative, and agressive coastal programs before the CZMA was enacted. At the extreme, California with its San Francisco Bay Plan and San Francisco Bay Conservation and Development Commission, which was established in 1965, had more years of experience in coastal management <u>before</u> the CZMA was enacted than OCZM has <u>since</u> its enactment.

Faced with a law that called for the development of state programs in an impossibly short three year time span, several states that were, at best, unenthusiastic about the national program, and other states that sought and needed OCZM's assistance in developing their state programs, OCZM did the only reasonable thing with the chargers on the West Coast. It crafted enough flexibility into its Section 305 and 306 regulations to accommodate whatever we had done on our own. Having participated in the formulation of the Section 305 regulations, I recall that the regulations always seemed to tell us how to do something we had finished about six months earlier. Our comments, therefore, seemed to run along the same lines: "What you have drafted, OCZM, seems fine and will probably work, but we have already done it a different way, so please revise the draft regulations so that, ultimately, you can approve what we have done." Had OCZM tried to inject substantive new requirements--like the incorporation of performance standards--at this point, some states may have dropped out of the national program. Politically it would have been suicide for OZCM to have created a situation where it could not approve the West Coast programs, and in particular California's because of its notoriety. Wisely, OCZM avoided this pitfall and the first three programs approved were Washington, Oregon, and California.

So far as I can determine, it was 1977 and after the approval of the West Coast programs before OCZM gave any serious thought to the 312 evaluation process. By this time the Section 305 and 306 regulations were in place and offering all the states the same flexibility that allowed OCZM to bring the West Coast states into the national program. In a September 15, 1977 memorandum, OCZM staff member Grant Dehart, in his typically articulate and comprehensive manner, set forth three conceptual approaches for establishing measurable objectives for state programs:

1. Rearticulate the substantive and procedural objectives of the CZMA in measurable terms.

2. Rearticulate the substantive and procedural "issues" and "problems identified in the CZMA in specific terms and define objectives in measurable terms in each state program.

3. Establish only "process" objectives in measurable terms.

OCZM opted for a variation of approach #1. As a result, OCZM is now attempting to evaluate state programs according to standards that were not available during the development of the programs to most of the states whose programs are being evaluated. Moreover, any measures that are truly "standardized" will have to be force-fitted on the assorted state programs which were developed and approved in a climate of flexibility and recognition of individual differences. Finally, states, quite naturally are skeptical about their likelihood of having successful 312 reviews under performance criteria that were not development criteria. It is somewhat like being asked to design and build a vehicle to transport 100 people, building a train under those standards, and then being evaluated on how well the train flies.

If there is any lesson in this, it is that performance standards should be articulated as early as possible and incorporated into the development of coastal programs. Obvious though this may seem, it is a difficult lesson to learn. We began developing local coastal programs in California in 1977 and should complete virtually all of them within the next year. Yet we have done no work on establishing a process for measuring the performance of the local programs. It is human nature to put off facing the inevitable as long as possible. And, with increasing fiscal constraints on public spending at all levels of government, it is prudent management to focus limited resources on the most pressing and immediate problems. But we will surely have to pay for this short-sightedness in the future, just as OCZM is now facing difficulties because it did not raise the issue earlier in its history.

You Can't Weigh a Frog with an Alarm Clock. The point of the facetious heading is that you have to decide what you want to measure before you can choose the proper tool for measuring it. Unfortunately, OCZM seems to be trying to accomplish two diametrically opposing objectives in its 312 evaluations--neither of which has much to do with measuring the performance of the state programs.

OCZM began its 312 evaluations in earnest just about the same time it was gearing up its efforts to justify the reauthorization of the CZMA to the Office of Management and Budget and the Congress. Therefore, focus was placed on having states demonstrate the successes of their programs by emphasizing the positive contributions that had been achieved in addressing goals and policies of the national effort as articulated in Sections 302 and 303 of the CZMA. Through a strange twist of logic, OCZM felt the best way of convincing the powers that be that the CZM program was addressing "national issues" was to concentrate on those issues that were not being adequately addressed by other federal programs or other laws. One could argue, as we did in California, that this sieve approach to prioritizing issues resulted in some of the least important issues being regarded as most important. If an issue wasn't caught by one or another of federal laws, it would eventually shake out the bottom to be handled by coastal management alone. Of course, critical needs, such as increasing access to the theoretically, this national issue could be addressed by other federal land acquisition and recreation programs.

Eventually some OCZM staff members admitted that this approach didn't always make sense. But they contended they were forced to continue focusing on the national issues because OCZM's plans for dealing with OMB called for the use of this strategy. Whether this was true, or whether the concept had appeal at OCZM primarily for other reasons, is something I can't readily determine. The idea obviously found friends at OMB; the Administration's proposal for reauthorizing the CZMA in 1980 included the national issues concept and OMB was a key actor in the development of this proposal.

So, much of the initial 312 evaluation called for states to expound on their glorious successes in saving the coast, with particular emphasis on the newly defined "national issues" to be used in OCZM's public relations campaign. At the same time OCZM was seeking these success stories, it began looking at the programs it had recently wholeheartedly approved with far greater scrutiny so that it could identify every possible weakness in need of improvement. OCZM was saying, in effect, "Tell us of your successes while we look for your shortcomings." This, of course, caused state managers to withdraw somewhat from the total candor that has been characteristic of state-federal relationships in the earlier administration of the CZMA.

How and why did this change come about?

As previously noted, OCZM had wisely, from both a programmatic and political perspective, crafted sufficient flexibility into its Section 305 and 306 regulations to accommodate approval of the Washington, Oregon, and California programs which were developed parallel to, but not as a result of, the CZMA. While this approach was essential to keeping states that wanted coastal managment in the national program, it had the potential of being abused by states that wanted the federal financial assistance provided by the CZMA but were not willing to make the necessary commitments to develop meaningful state management programs. It would have been inequitable and probably futile for OCZM to insist that new and tougher Section 306 approval standards be applied in 1978 and 1979 than had been applied in 1976 and 1977. So it tried to apply new and tougher program evaluation standards to all approved programs.

At the same time, OCZM was undergoing a personality change, both in terms of individuals and in the composite character of the staff. As seems to be the case with most new and innovative government agencies, OCZM has experienced a rapid staff turnover. For people looking for a more typical public service job, the unstructured atmosphere of a new program may seem chaotic. To others, the lack of bureaucratic controls allows their imaginations to soar, until, inevitably, the imaginative initiatives have to rely on bureaucratic contols for implementation. This freewheeling environment also allows young, bright, and energetic staff members to gain power and prestige far beyond their lowly civil service ratings, and, like a Nova, to burn out, or to gain rapid promotion in another agency.

So by 1978, OCZM was well into its second generation of staff members. The newcomers found their abilities to deal with states encumbered by regulations that seemed more directed toward the pragmatic solution of a past political problem, than toward the on-going administration of a permanent federal program. These staff members weren't party to and did not necessarily agree with the implicit deals that had made to keep the front-running states from running away from the national program. Moreover, the new staff members wanted to leave their mark on the national program by forcing all states to make "improvements" in their programs to address the national issues. Times had changed sufficiently so that OCZM no longer found it necessary to display any political deference to individual states. OCZM had approved a sufficient number of programs to have a success record and credibility of its own. It could afford the loss of a state or two from the national program without permanent damage to the overall coastal management effort. If OCZM could take a tough stand with the President's home state of Georgia, why should California, which by now had bogged down into a struggle to implement its program through a lot of controversial permit and local government planning decisions, be treated any differently?

The other aspect of the OCZM personality change involved the awaking or rededication of the first generation staff members. In California, the initial professional staff of the Coastal Commission was made up largely of people who didn't particularly care who they worked for, but rather were more interested in what they did--saving the coast. I suspect much the same condition existed at OCZM. Professionals chose to work there, not because of promotional opportunities, fringe benefits, or working conditions, but because they too wanted to save the coast. And in the early days, they did just that. They conceptualized a whole new federal program, wrote the cookbooks on coastal management in federal regulation form, and participated in decisions on the form and content of emerging state programs. We were the state coastal managers and they were the federal coastal managers. But as the federal program became institutionalized, the regulations finalized, and more and more state programs approved, OCZM should be changing from a coastal management agency to more of a grant administration agency. Its staff members, awakening to this fact, are either moving on to other things or digging in to retain their role as the nation's coastal managers. Unfortunately for them, the nation's coastline comprises all the separate state coastlines. And unfortunately for us, the only way they can manage their coastline is to look over our shoulders. But they are trying. And their only effective means for telling us how to manage our portion of their coastline is to impose conditions on our grants. And the only way of systematically determinig what those conditions should be is the 312 evaluation process. This brings us to the second objective of the current 312 evaluation process--identifying deficiencies in approved state programs and outlining steps for correcting them. It would be difficult for OCZM to identify deficiencies that result from a failure of an approved program to meet the requirements of Section 306 of the CZMA. If a program includes such a deficiency, OCZM should never have approved the program in the first place. Moreover, if a state is carrying out its program in the manner prescribed when the program was approved, (and it is my impression that, for the most part, states are doing just this) theoretically there should be no deficiencies to identify. But this would not fit into OCZM's desire to retain its coastal manager function as its evaluators would be reduced to the role of auditors.

The solution seemed to lie in finding those issues where states, as a whole, were weak and concentrating federal funds on making improvements in those areas, which brings us back to the national issues concept. In late 1978, OCZM proposed that each state program should address a very narrow range of subjects which, through the sieve process described earlier, had emerged as being "national issues". The 312 review would focus on evaluating state achievements in these areas.

Although the concept has been refined, the situation remains that, despite all the good intentions that brought us to this point, the 312 evaluations currently are not designed or intended to determine whether states are carrying out their coastal management programs in the form approved by OCZM. Instead, the evaluations are being used to generate information for a PR camaign (though to a lesser extent now that reauthorization of the CCMP seems assured) and to cajole states into redirecting their efforts from carrying out the activities called for in their approved programs to addressing new issues defined as being nationally important by OCZM. Until this situation is rectified, the 312 process will continue to be plagued by difficulties and the development of adequate tools for measuring program performance will be impossible.

<u>If It's Important in Washington, It Must be Important in Eureka.</u> Because OCZM decided to use the 312 review as a means of moving states toward addressing national issues, it is worth looking at this concept a bit more closely even though "national issues" should not inherently be part of the 312 process.

As explained earlier, one of Grant Dehart's three options for handling the 312 reviews calls for rearticulating the substantive and procedural objectives of the CZMA in measurable terms. In theory this sounds fine. The vague and non-specific issues, problems, goals, and policies found in Sections 302 and 303 of the CZMA could be fashioned into measurable objectives. Unfortunately, this theoretical purety was quickly compromised when OCZM discovered its need to demonstrate the success of the national program and the OCZM staff found that the course of events would take their professional lives out of the direct coastal management business over the next few years.

At this point another human factor came into play. I have found that virtually everyone has a natural tendency to try to find a single solution that can be applied in identical situations. In bureaucrats, who administer government programs, this tendency becomes a compelling drive to reduce all possible approaches for handling a situation to one set of procedures which, ideally, becomes embodied in regulations. And in federal bureaucrats, this tendency reaches a fever-pitched passion to find one solution to all problems, even if the problems aren't very much alike. Perhaps this is in reaction to the great number of variations that one must find in trying to carry out a law that is supposed to apply equally in all situations. Even when a single standard for a single problem can be found, many find exceptions to the rule can also be found. Imagine OCZM's frustration in trying to find solutions and procedures that could be applied equally well to coral reefs off Tinian, the singles apartment complexes in Los Angeles' Marina del Rey, and the bayous of Louisiana. Yet OCZM had to do this if it was to justify its existence as a coastal management agency and perhaps even the need for a federal coastal management program. If there was no common thread, how could the presence of the federal government in state coastal management be justified? And if an overall common solution could not be found, how could OCZM gain its place in the profession of coastal management except as kibitzers, second-guessers, and grant processors?

The process moved on with OCZM trying to distill from Section 302 and 303 those national issues that should be addressed in all state programs. The results were revealed in November 1978 and it became immediately clear that in the distillation process, the fire under the still had been turned up too high. The general language of Sections 302 and 303 had been boiled down to four specific issues which, in the national interest, had to be incorporated into each and every state coastal program. From California's perspective, the most troublesome of these newly identified concerns was fisheries management. It was not that we regarded management of this resource as unimportant. California has a long and successful record in the management of its living ocean resources. Under the leadership of the California Department of Fish and Game, both sport and commercial fishing have prospered in the state. Under the old adage of, "if it ain't busted, don't fix it", the California Coastal Act gives the Coastal Commission little responsibility in this area. Instead, the Commission concentrates on protecting coastal habitat areas for marine species and ensuring that new marina development does not displace commercial fishermen. Both of these problems were beyond the control of the Department of Fish and Game, but are being satisfactorily handled by the

Commission through our regulatory and planning authority. In this context our discussion with OCZM went something like this:

OCZM: Fisheries management is a national issue that must be addresed in every state coastal management program.

California: Fisheries management isn't a major problem in California and if improvements are needed, they should be provided by the Department of Fish and Game.

OCZM: We believe that fisheries management has a place in every state program. If California chooses to administer its program through its Department of Fish and Game, that's OK, just make certain that CZMA funds are used for fisheries management.

California: Fine. Send us some more money for the Department of Fish and Game.

OCZM: Oh, no. You won't get more money. Use some of the CZMA grant you are now receiving to address the national issues.

California: Run that by us once again, slowly.

OCZM's proposal would work something like this. Back in the very early days, OCZM had to figure out how to divide its program development funds among the states. In the classic federal tradition, the problem was reduced to a formula in which 40% of the grant was based on the population of the state's coastal counties, 40% on the length of the state's coastline, and 20% at the discretion of OCZM. Although this obviously has nothing to do with how much each state really needs or the quality of the state program, the formula was reduced to regulations fast enough to keep us states from fighting among ourselves over our share of the pie. This formula was devised for the allocation of Section 305 funds, but it has also been applied in managing the Section 306 grants. Moreover, the 20% discretionary amount quickly got factored into the base amount so that states regarded the full 100% as the amount they would receive each year. California receives $3 million in Section 306 funds each year and our budget for carrying out our approved program anticipates that this full amount will be available. However, what OCZM was suggesting was that henceforth the 20% would have to be used to address national issues and that through its 312 review, OCZM would determine which of the national issues was most in need of attention in a particular state in a particular year. With fisheries management being in vogue in 1978, this meant that in California $600,000 in CZMA funds would have to be spent on fisheries management in the next year. As noted, this money would have gone to the Department of Fish and Game which, like virtually all resource management agencies, most certainly could have put the funds to good use. But where would the funds come from? From a reduction in OCZM's support for carrying out the planning and regulatory activities that OCZM had approved just a year earlier as meeting the program approval requirements of the CZMA. If California decided to continue its coastal management activities at the level needed to implement its approved program, it would have to allocate an additional $600,000 in state funds to make up for the federal funds that would be diverted to the Department of Fish and Game.

This left us in a situation where one state agency, which was doing everything that it said it would, would receive less federal financial support, and another agency, which had had no contact with the federal agency and had not requested any federal funds, would receive a windfall grant. Since there was no assurance that what was the national issues of premier importance in 1978 would be the same in 1979, states were somewhat frustrated but mostly bemused as we prepared our strategy for heading off this wacky idea. Again, we had the good fortune of dealing with a staff at OCZM that was willing to listen and respond to state concerns. By the summer of 1979, at our persistent urging the national issues had been broadened to four major areas identified in Section 923.94(c) of its Coastal Zone Management Program Development and Approval regulations which prescribe what Section 306 funds can be spent on. Although these subject areas were ostensibly drawn from the policies and findings of Sections 302 and 303 of the CZMA, one is hard-pressed to find any correlation between the two. Moreover, in July 1979 former OCZM director Robert Knecht advised state managers that virtually anything we might want to do would be covered by one or another of the broad areas, but that we should go through the exercise of demonstrating that we were addressing the issues nonetheless.

So for the time being, OCZM had found a way for the states to get around the national issues dilemma without ever admitting that the idea was a failure. But we weren't out of the woods yet. Whether the concept was kept live as a face-saving gesture, whether OCZM had been so successful in selling the concept to OMB that it could not drop it without endangering Administration support for the program and reauthorization of the CZMA, or whether the concept has, like so many screwy ideas of mankind, developed a life of its own, I can't tell. For whatever reason, the concept of tying funding to national issues and national issues to the 312 evaluation had found its way into the Administration's proposal for reauthorizating the CZMA. In President Carter's proposal to the Congress, federal support for state coastal programs would be phased down and out over a five year period. At the same time the portion of the assistance used to support national issues would have to be increased each year. Essentially, the Administration was calling for states to handle more and more of the federal government's responsibilities each year with ever decreasing amounts of money. The Congress quickly saw that this approach would be a strong incentive for states to drop out of the voluntary federal program and placed a limit on the percentage of federal funds that would have to be used for program improvements to address national issues and scrubbed the phase down idea entirely. However, broad areas of national concern needing improvement were articulated in both House and Senate versions of the reauthorization legislation and the 312 review is to be used to measure state success in achieving these improvements.

The difficulty with the national issues concept is not that it is fundamentally faulty. Most assuredly the Congress should spell out in law what it wants accomplished in passing the law. If it doesn't do so, the administering federal agency has a responsibility for specifying how funds appropriated under the law should be spent if for no other reason than to keep its employees out of jail for misuse of public funds. If neither of these happens, based on some practical experiences either Congress or the Adminstration should come back and clarify the intent of the program. And finally, any program, no matter how well it is administered can use improvement.

The problem with the national issues idea is the means by which it is going to be administered. No one has suggested to California that there is something basically wrong with its federally approved coastal management program. No one has suggested that we should not continue to do what we said we said we were going to do when the program was approved. What is being suggested is that we should be doing other things too. And that we should be doing more of what we are doing. Some of the federal funds we had counted on to carry out our approved program would be diverted to pay for the other and more. And the overall success of our performance would be judged on how well we are doing the other and more. If we do them well, we will continue to receive funding to carry out the work that is the foundation of our program. Meanwhile, precious little effort has gone into trying to evaluate how well we are doing on protecting California's coastal resources, which is what we were led to believe the whole effort was all about. And that clearly doesn't make much sense.

The Light At the End of the Tunnel. When OCZM suggests that California divert some of its federal funds from program implementation work to improvements that are in the national interest, my typical reply is, "Fine, which work do you want us to stop doing so we will have the funds to divert?" OCZM's usual comeback is something on the order of it is inappropriate for them to get involved in that kind of detail in state program administration. I then volley back with, "How come you can get involved in the details of telling us what to do but not the details of telling us what not to do?" Or sometimes I can manipulate them into suggesting something that they feel could be dropped and then go out and put together an explanation of why what they have targeted for elimination is at the heart of California's program. The obvious goal is to convince OCZM that we know more about what we are doing than they ever will and, thereby, extract some willingness from them to let us call the shots.

Since I am fundamentally a fair, honorable, and reasonable person, I feel it only just that after all my criticism for OCZM's 312 evaluation efforts, I do more than declare it now to be OCZM's responsibility to figure out how to get itself out of the mess I have described. Instead, I will put my jaded reputation on the line and make some recommendations for improving the situation. Undoubtedly, OCZM staff who have been struggling with 312 evaluations on a daily basis will be able to spot the fallacies in my ideas just as I have been able to do with theirs. Hopefully, this will return some equilibrium to our working relationship and may even begin a dialogue where together OCZM and the states can come up with some acceptable and valid means for evaluating approved state coastal management programs.

1. Recognize that Evaluation is Necessary. Despite all the problems I have cited with the evaluation process and despite the difficulty that states have had in putting up with it, we should all recognize that the evaluation process is a healthy idea that should be encouraged. There is enough popular dissatisfaction with government generally and enough accounting myopia within government itself that someone will be asking pretty quickly whether coastal management is doing what it is supposed to be doing. So we might as well be ready for this examination by having our own evaluation process in place, both to identify problems and provide corrections and simply because it is the only sound and responsible way of running a government program.

2. Recognize that a Process for Evaluating State Program Performance Does Not Currently Exist. This may be hard for a lot of people to swallow, but as I have explained in gory detail, there is virtually no effort being made to evaluate the performance of states in carrying out their approved programs. Perhaps because it is so difficult to come up with means for measuring the success and performance of a process, OCZM has succumbed to trying to get states to do something that can be measured. It seems inevitable that the idea of making improvements in state programs to address national goals will continue to be part of the national

coastal management effort. And states should be evaluated on how well they do in making these improvements. But this evaluation process (where OCZM has concentrated its efforts) should be kept separate and apart from evaluating state success in achieving what their programs were designed for in the first place.

3. Treat the Evaluation Effort as a Research Project. Once there is acceptance that the process which has been used to date has not been designed or intended to evaluate program performance, it makes sense to try to put something together that does rather than continue the largely unsuccessful process on a full-scale basis. This means that for the next year or two OCZM will not be able to conduct its attempts at comprehensive annual evaluation of state programs. It might seem that this would allow states to go off and do something crazy with the federal funds. But the current evaluation process is unlikely to find that out. And by stepping up the routine grant monitoring carried out by the State Programs staff and relying on the fiscal audits that are performed on all grants, there are sufficient controls, so there is little to lose in going back to the drawing board.

The research project should look at other governmental (and private industry) evaluation efforts, set up some hypotheses of how one would evaluate this or that aspect of coastal management, conduct experiments in evaluation based on the hypotheses, and evaluate the success of the techniques in measuring what they are supposed to. This approach would require that considerable time be spent in setting up experiments, carrying out the experiments using one or more states as guinea pigs, and then revising and retesting the procedures based on an evaluation of their success. This is quite at odds with the present hit-or-miss approach that seems to have as its goal conducting as many evaluations as possible, with as little time as possible devoted to self-appraisal of the evaluation process.

4. Resist to the Death Adopting Evaluation Regulations. In considering reauthorization of the CZMA, some interests in the Congress expressed dismay that OCZM had not adopted regulations on program evaluation and strongly urged that OCZM move as quickly as possible to remedy this situation. It should be clear by now that about the only good thing about the evaluation process is that regulations have not been adopted. Developing regulations at this time would institutionalize the current process which wasn't designed to measure program performance and prevent the primitive and mostly unsuccessful measuring techniques from being improved. Regulations should come after a workable process has been developed, and that is a long way off.

5. Provide Adequate Staff Support for Program Evaluation. A recent General Accounting Office study illustrated that one branch of the federal establishment can be as bad at evaluating another branch as a federal agency can be in evaluating a state program. Relying on information that is by now quite out-of-date, the GAO concluded the obvious, failed to recognize any of the mundane but nevertheless real difficulties that make it impossible to achieve perfection in running a governmental program, and offered OCZM little in the way of helpful ideas for improving the situation. The one exception is GAO's observation that OCZM's evaluation effort is critically understaffed. With federal hiring restrictions and increasing emphasis on reductions in government spending, there may be little that OCZM can do to increase its overall staff size (which is probably a good thing). And with calls for a federal program review, more states coming into the critical program approval review stage, and the need to develop new regulations to implement the reauthorization provisions, in the immediate future OCZM may be hard-pressed to assign adequate staff to the evaluation effort. But evaluation is handled as a research project as I have suggested, the success of the experiments will be jeapordized if they are not adequately staffed. Understaffing an experimental process is like adding water to gasoline and then concluding that an engine run on the mixture is faulty. Without adequate staff, the evaluation process will never be perfected.

6. Don't Expect the States to Do the Staff Work for Evaluation. Currently, OCZM relies very heavily on states to provide OCZM with the background information needed for evaluation. There are three basic problems with this. First, it imposes a very heavy burden on states to come up with extensive information in a very short period of time, a responsibility that belongs to OCZM. In theory one can justify this approach by contending that since OCZM provides funds to states, it can prescribe that some of the funds be used to support the information gathering effort. But the funds provided to states are for coastal management. OCZM receives a separate appropriation for program administration of which program evaluation is (or should be) a part.

Second, having the states do the staff work puts OCZM in a position of knowing only what the states tell them. OCZM's credibility will suffer enormously should a state manipulate the process to feed in the information and impressions it wants. The process clearly has the potential for being abused.

Finally, because OCZM's evaluation staff gets only a cursory glance at each state program, it is very easy to draw glib and erroneous conclusions. This has happened far too often and causes states to lose respect for OCZM's ideas, results in states having to spend considerable time and effort to dissuade OCZM of dumb conclusions, and adds rifts to an already difficult working relationship between states and OCZM.

On the other hand, if OCZM were willing to abandon the notion that it is the nation's coastal manager, it could provide the states with technical assistance in conducting evaluations and delegate the entire responsibility for program evaluation to the states. Several states (most notably Alaska) have called for this approach, and I believe it has considerable merit. It would require a great departure from the traditional approach of government bureaucracies, which seems to have as its foundation a belief that the particular agency has cornered the world's wisdom and that everyone outside the agency is either a fool or a cheat. Given a chance, I believe the approach would work as well as anything suggested and a lot better than the current evaulation effort. But, human natue being what it is, I doubt that the idea will ever be given a chance to prove itself.

7. Dump the Forced Annual Comprehensive Review. Evaluation should be a continuous process. It is impossible to get anything meaningful out of an evaluation if OCZM continues to try to conduct an annual review of all aspects of every approved program. Coastal management is, by definition, a continuous process that does not lend itself into ready division by months and years. As part of its routine grant monitoring, OCZM should have staff in the field all the time (once it develops a workable evaluation process) doing the research and conducting hearings. Trying to evaluate performance based on a single annual review is like trying to decide how fast a race horse can run by looking at a still photo of him eating hay.

8. Settle on What is to Be Evaluated Before a Grant is Made. Currently OCZM seems to concentrate its evaluation efforts on the subjects Congress is getting the most mail about. In a comprehensive program like coastal management, priorities must be drawn and not everything can be done equally well at all times. It is both demoralizing to the states and misleading to the public for OCZM to focus its review on a state's lowest priorities rather than its highest. If OCZM abandons the idea of a comprehensive annual review, it can come to agreement with each state as to what it expects accomplished on a few important issues over a fixed period of time, and then evaluate the success of this effort. If OCZM disagrees with the state's priorities, it is incumbent on OCZM to come up with additional funds to support the work it believes should be of higher priority.

9. <u>Settle on the Standards You Are Going to Use in the Evaluation</u>. This is the hardest part about the whole process because there aren't any "standards" in the classic sense. I find this to be the most frustrating aspect of government management. In the private sector, everything ultimately boils down to whether or not you are making a profit and how much. In the public sector, it doesn't matter how efficient, honest, fast, or imaginative you are in carrying out a program if the program is out of favor with the public or its legislative representatives. The California Coastal Commission approves about 95% of the permit applications it receives. Does this indicate that we are doing a good job by seeking compromise and conditions that will allow development consistent with the Coastal Act policies to proceed? Or does it mean we are failing to vigorously enforce the Coastal Act? In a program like coastal management that inherently involves many interpretative and subjective decisions, such dilemmas are bound to arise. The real danger in dealing with this problem is that you can fall into one of three classic traps:

a. <u>Trying to Quantify the Non-quantifiable</u>. A major aspect of our work in California is developing good local coastal programs that will reflect the goals and policies of the state coastal act. However, what makes an LCP "good" is largely a matter of judgment. So, rather than rely on judgment, you can try to devise a measuring stick like the number of issues addressed in the LCP or the number of pages devoted to LCP analysis. The result can, of course, be an LCP that does a lousy job in addressing a lot of issues being regarded a better than one that does a superb job in addressing the few that are really critical.

b. <u>Concentrating on the Quantitifiable</u>. I always like to illustrate this predicament with the story of the man who searched for his lost pocket watch under the street light, not because that was where he thought he had lost it, but because there wasn't enough light to search in the area where he had lost it. The number of new boat slips built in a state coastal zone may be easy to measure, but they may have little to do with how well a state is doing in carrying out its program. Yet because the evaluators find it uncomfortable to rely on subjective information, they may concentrate on quantifiable data in their evaluations. And over time, states may move toward work that improves their scorecards even if it doesn't improve their programs.

c. <u>Measuring Things that Didn't Happen</u>. Coastal management programs are supposed to protect wetlands from indiscriminate filling. If a developer proposes to fill 100 acres of wetland for a shopping center that could be built upland and the coastal management agency refuses to issue a permit, have 100 acres of wetland been saved? What if the developer revises his proposal so that 50 acres are filled for a marina development; has 50 acres been saved? Or lost? And what if the developer keeps believing that eventually he can convince the coastal agency to approve the 100 acre fill so he applies three times for the same permit. Are 100 acres saved three times the same as 300 acres being saved? Or what if he talks to a friend who convinces him that he could never get a permit for the shopping center so he never applies for a permit and instead uses the wetland for duck hunting with his friends. How do you save something that nobody officially proposed to destroy?

You get the idea.

Conclusion. After all this, nuclear physics must seem somewhat more elementary than state coastal program evaluation. In fact, I think just the opposite is true. Program evaluation has been made difficult because it has been treated as an exact science instead of a folk art. Virtually all of my criticisms of the effort to date come because someone failed to use common sense. Instead, they accepted that program evaluation, like the emperor's clothes, existed and could be added to. Let's get the emperor out of the streets and state program evaluation out of the states until both of them are decent enough to be seen in public.

IS THERE LIFE BEYOND MITIGATION?

by

Robert M. Snyder,[1] P.E., M. ASCE

"The beginning of wisdom is to call things by their right names." - Chinese Proverb

Prior to the 1960's there were many scientists, naturalists and outdoorsmen who appreciated the value of wetlands. Some in this group were also in positions of influence or had sufficient wealth to cause to have set aside certain wetland areas of significance, not for social reasons but because they thought it worthwhile. Although we can speak with more erudition today about the social significance of wetland interactions, and can agree that the remaining wetlands increase in significance as our population grows, the intrinsic value of wetlands has not changed as a result of public awareness. And an increase in our knowledge does not alter their function. Jon A. Kusler, in summarizing the "Proceedings of the National Wetland Protection Symposium" (1) stated:

> *"A shifted burden is justified for wetland activities. This shifted burden requires that those wishing to alter or develop wetland areas demonstrate that wetland sites are required, that the proposed activities will not be subject to significant natural hazards or individually or cumulatively increase the hazards, and that the activities not damage or destroy wetland functions important to society unless overrriding long-term social benefits can be shown. Where wetland activities are justified, all reasonable measures should be taken to minimize impact."*

This, I believe, is a dangerous oversimplication if our long range goal is to maximize for posterity the real value of our nations wetlands in both quality and quantity. Prohibition, with itself as its goal, has stifled learning since pre-Ptolemaic times, and for periods as long as thirteen centuries. We cannot afford even a short time-out while the politicians, bureaucratic regulators, lawyers and dilenttantes relearn an expensive lesson while our accumulating scientific knowledge simmers on the back burner.

And we have accumulated a great deal of scientific knowledge in the last twenty years based on an already sound foundation of hydrography, aquatic biology, water chemistry and

[1]President, Snyder Oceanography Services, Inc., Jupiter, Florida 33458

hydrology. To ignore this knowledge by acquiescing to the prohibitionist or protectivist philosphy is to place the chickens in charge of the fox.

There are scientific voices in the wilderness, two of which should provide sufficient encouragement to screw our courage to the sticking place. Lord Ashby - a Fellow of the Royal Society, one of the world's leading botanists and educators and for three years a member of the Royal Society on Environmental Pollution - addressed the "protectionist" doctrine of "delicately balanced nature" as follows:

> *"The notion that the 'balance of nature' is delicately poised and easily upset is nonsense. Nature is extraordincarily tough and resistant, interlaced with checks and balances, with an astonishing capacity for recovering from distrubances in its equilibrium."* (2)

Regarding the doctrine of "nature knows best", René Dubos - Professor Emeritus at Rockefeller University - in response to being co-recipient of the first Tyler Ecology Award, remarked as follows:

> *"Nature is like a great river of materials and forces that can be directed in this or that channel by human intervention. Such intervention is justified because the natural channels are not necessarily the most desireable, either for the human species or for other species. It is not true that 'nature knows best'. It often creates ecosystems that are inefficient, wastefull and destructive. By using reason and knowledge, we can manipulate the raw stuff of nature and shape it into ecosystems that have qualities not found in the wilderness."*(3)

This statement certainly fits with our experience. Why do we choose specific areas for National Parks, Wildlife Refuges and Wilderness Areas? Assuming we had had the opportunity prior to colonization, would it have made sense to checkerboard the United States and say arbitrarily that every third square would become public land and then use the Monte Carlo method to determine which would be Parks, which would be Refuges and which would be Wilderness Areas? Certainly not! Nature integrates all the variables extant in a particular location in both time and space and unerringly presents us with the only result possible under the given circumstances. The variation in her results from place to place is obvious and, without trying to define "good" and "bad" ecosystems we must admit that there is reasonable concensus among thinking people that some are "better" and others. By observation and experiment, mankind has learned some of the whys and wherefores of nature's behavior and can intevene in helping to make the results of her integration more "good" or less "bad" or both. Why then have we adopted the posture that scientists, engineers and builders of the '80's, with more than a decade's worth of new knowledge, are incapable of positive intervention while, at the same time, we look upon as hallowed ground the isolated wetland impoundments created by man a quarter century ago who had no concerns other than the cheapest way to control mosquito breeding.

We all know that many human interventions have been destructive. But to take the attitude that all aspects of man's activity are intrinsically evil and destructive to nature is reactionary and counter productive. The one theme that is apparently consistent throughout most of the efforts to resist the 80's builders' attempts to productively intervene, is the belief that man is an outsider on this planet and further, any activity that renders land more useful and acceptable to the private citizen is contrary to the intent of environmental legislation. The 1977 Corps of Engineers Regulations (4) referencesthe Holland Case (United States vs. Holland, 373, F. Supp. 665; M.D. Fla., 1974) as the first case to address whether wetlands beyond the mean high water mark of traditional navigable waters of the United States were subject to the Federal Water Pollution Control Act (FWPCA). The regulations quote the Court's finding:

"The court is of the opinion that the mean high waterline is no limit to Federal authority under the FWPCA. While the line remains a valid demarcation for other purposes, it has no rational connection to the aquatic ecosystems which the FWPCA is intended to protect. Congress has wisely determined that Federal authority over water pollution properly rests on the commerce clause and not on past interpretations of an act designed to protect navigation. And the Commerce clause gives Congress ample authority to reach activities above the mean high water line that pollute the waters of the United States."

The Federal Register goes on to point out that "Other Courts have pursued the same theme, and often use the Holland rational to support their position." What the Federal Register fails to mention is that the same Court in the same case also held:

"By determining that the defendants have violated the provisions of the FWPCA and by enjoining further unlawful activities, the Court has not permanently prohibited defendants from going forward with their plans. All the Court has said is that the activities cannot be continued without a federal permit.

"The Court realizes that the thought of preserving huge stretches of coastline in a natural state and forbidding all commercial development in coastal areas is unrealistic. This is a societal choice which the government must observe. But the government can and should insure that the public interest in protecting all life forms is at least considered in the development plans. Any expense that might be incurred by this evaluative process will be dwarfed by the cost of

neglecting the ecological interests."

This part of the Court's finding has largely been ignored by the regulators with the result that the expense of the evaluative process is no longer "dwarfable" by any measure. The argument that a single human life is infinitely valuable, as is the existence of any and all species that haven't yet been made extinct by unregulated natural forces, is difficult to argue with a fundamentalist whose electricity comes from a socket, his water from a faucet, his food from a supermarket, his pest control from a uniformed robot and, more often than not, his paycheck from the government or his mother's stock certificates.

Since the turbulent '60's it has been popular to accentuate the negative, predict doom and demand sacrifice in proportion to one's potential contribution to the GNP. When the hoped for doom does not materialize, the unspectacular truth is ignored and a new "cause" is taken up with unabated fervor. The deciples of Barry Commoner, *Science and Survival*; Paul Ehrlich, *The Population Bomb*; the Club of Rome, *The Limits of Growth*, etc., seldom look back to observe nature's "astonishing capacity for recovering from disturbances in its equilibrium". Examples of our intervention in helping nature be less "bad" are numerous but unspectacularly reported. They include: the upper Delaware River, the Housatonic in Connecticut, the upper Ohio, Oregon's Willamette, Atlanta's Sope Creek, New Jersey's Hackensack, Lake Erie, Lake Washington, the Thames Estuary, to name a few. These are examples where man reversed his intervention from one of passive neglect to one of active diversion of the upsetting sources. Other examples of nature's "astonishing capacity for recovering from disturbances in its equilibrium" result from accidental spills involving tanker disasters and refinery accidents. These include: the Santa Barbara blowout, the *Torry Canyon* spill, the *Amoco Cadiz*, the *Argo Merchant*, etc. .... Anthony Nelson-Smith, a zoologist from the University College of Swansea concluded, after studying the *Torry Canyon* recovery that: "Serious oil spills cause about as much ecological disturbance as unusually cold winters but, unlike these, are potentially avoidable." (5) The point here is <u>not</u> that we should continue to casually run oil tankers aground just because nature "can handle" it when we stop. We have recognized that we have gone too far and have accomplished a great deal in rectifying our past mistakes. But how do we rectify the widely held belief that Lake Erie and the Hackensack River are "dead"? The question is: "Can mankind recover as quickly from overstated, selfserving, self-flagelating, irresponsible doom predicting, as nature can recover from our physical improprities?"

It might not be worth our time at this moment to revisit the doom predictors if it were not for the *Global 2000 Report* issued by the Carter Administration on 24 July 1980. The modern stage was set in *The Limits to Growth* sponsored by the Club of Rome and published in 1972 in which the bottom line was that without the equality of the rates of birth and death and without the equality of capital investment and capital depreciation, the economic and industrial systems of affluent countries would collapse about the year 2100. Upon examination, it was shown that many of the assumptions fed into the computer were....

> *"not scientifically established and the use of data was often careless and casual."* (6)

For example, one of the inputs fed to the computer was the reserve estimates published by the U.S. Bureau of Mines which warned that 80% of the data had a confidence level of less than 65%. This tremendous error band was withheld from the computer. Regarding such numerical predictions made by extrapolating trends, Lord Ashby commented in 1976:

> *"Computers have made this sort of long-range forecasting not more reliable, but more dangerous; for the computer confers an impression of precision and determinism upon the assumptions fed into it, and this meretricious treatment gives the assumptions an unwarranted air of accuracy."* (2)

Beyond the basic technical weakness of using computer simulations for exponential functions based on unqualitifed assumptions, Lord Ashby continues:

> *"Far more serious is the omission of a parameter essential for any consideration of these grave problems; namely, the mechanism of social and political response to environmental change. The most striking and distinctive feature of all communities of organisms, from Amoeba to Man, is their homeostatic response and capacity for adaptation. To make simulations of environmental change which omit this parameter is to invalidate the whole operation, for they disregard the forces of public opinion and politics."*

Now comes the Carter Administration's *Global 2000 Report* commissioned in the President's first (1977) environmental message. As quoted in the 1 August 1980 *Science* magazine, the report states:

> *"If present trends continue, the world in 2000 will be more crowded, more polluted, less stable ecologically, and more vulnerable to disruption than the world we live in now."*

This report was sponsored by the Council on Environmental Quality (CEQ) and the Department of State but was contributed to by some 11 Government agencies. According to *Science*, the President has asked Gus Speth - the chairman of CEQ to chair an interagency task force on global resources and the environment. The numerous detailed findings were obtained from the

Government's global model and described as being consistent with other global models including the Club of Rome's World Integrated Model. A fact not picked up by the general press, and mentioned but not emphasized by *Science*, is that according to the models the predicted results are likely IF: there are no changes in public policy! IF there are no major advances in technology! and IF there there are no unforeseeable events such as an outbreak of war! Again quoting from *Science*:

> *"The Global 2000 study begins and ends with the confession that there is no single well-integrated government model but rather a patchwork of various agency models that present serious inconsistencies in methods and assumptions. It concludes, however, that the fact the government model is deficient in linkages and feedback actually imparts an "optimistic bias" and makes its projections somewhat less disturbing than they might be otherwise."*

It appears that the words of Lord Ashby have fallen on deaf ears. This newest doom report states that:

> *"Prompt and vigorous changes in public policy around the world are needed to avoid or minimize these problems before the become unmanageable."*

One could hardly agree more, but must we also agree that only politicians and government agencies are capable of diverting disaster? One hopes, but with trepidation and an ear cocked for a clue to the whereabouts of John Galt, that the Orwellian models are no more accurate than those of the doomsayers.

One has to ask the question: "Who gains from promulgating this mythology?" "The delicate balance of nature?" "Nature knows best?" "Lake Erie is dead?" "There are no technological solutions to our problems?" "We must pay heavily for our past abuses?" "Man is not compatible with nature?"

There are some obvious answers: The authors and publishers of best selling books; the National Audubon Society whose annual operating budget increased from about one million dollars to more than ten million in about a decade; the Sierra Club; Friends of the Earth; Nature Conservancy; Defenders of Wildlife; National Parks & Conservation Association; National Wildlife Federation; Environmental Defense Fund, and others; (organizations who, with swollen membership rosters, must keep the fires fed lest they cease to support their inflated overhead resulting from new offices and staffs of highly paid lawyers and "fact finders"); the government bureaucracies who are "needed" more than ever to prevent us from self-destruction; politicians in need of an issue that their opponents cannot ridicule; the news media in need of greater circulation and advertising revenue; and finally the people who have always loved dragons as long as they were in somebody else's garden. But the doomsday dragon has become real indeed to those who would move forward with knowledge and care in meeting the demands of a growing population and, at the same time, acting to enhance the environment.

Enhancing the environment for its own sake is a fairly new idea but there are scientifically proven examples. One that stands out is the establishment of artificial reefs. Using outdated ship hulls or even specially constructed concrete "apartment complexes" we can provide habitat where it once was absent. It works! But even here there is a payoff. The increased fish populations are an asset to skin divers and sports fishermen. This human benefit is socially acceptable. It is true that there is no such thing as a free lunch. Why then do the bureaucratic regulators expect private property owners to give up the use of their property without compensation? The regulators go beyond the law and the regulations in establishing policies which give no credit to privately funded enhancement of private property. For example, the Environmental Protection Agency's Rules and Regulations published in the Federal Register on 5 September 1975, state in section 230.4(b)(8)(ii):

> *"Discharge of fill material in wetlands shall not be permitted unless the applicant clearly demonstrates the following:*
>
> *"(a) the activity associated with the fill must have direct access or proximity to, or be located in, the water resources in order to fulfill its basic purpose, or that other site or construction alternatives are not practicable; and*
>
> *"(b) that the proposed fill and the activity associated with it will not cause a permanent unacceptable disruption to the beneficial water quality uses of the affected aquatic ecosystem, or that the discharge is part of an approved Federal program which will protect or enhance the value of the wetlands to the ecosystem."*

The Corps of Engineers Rules and Regulations published in the Federal Register on 19 July 1977 state in section 320.4(b)(4):

> *"No permit will be granted to work in wetlands identified as important by subparagraph (2), above, unless the District Engineer concludes, on the basis of the analysis required in paragraph (a) above, that the benefits of the proposed alteration outweight the damage to the wetlands resource and the proposed alteration is necessary to realize those benefits. In evaluating whether a particular alteration is necessary, the District Engineer shall consider whether the proposed activity is primarily dependent on being located in, or in close proximity to the aquatic environment and whether feasible alternative sites are available. The applicant must provide sufficient information on the need to locate the proposed activity in the wetland and must provide data on the basis of which the availability of feasible alternative sites can be evaluated."*

As noted, both the Corps and the EPA require that the activity have direct access or close proximity to the aquatic environment. The agencies concede that marinas require access to the aquatic environment. Their policy is, however, that waterfront residences that allow the owner to moor his vessel at his home does not require access to the water. This in spite of the fact that the property is privately owned and that, if properly designed, the activity could considerably enhance the wetlands affected. I would hold that the reasonable intent of the rule would be to discourage paving over large area of marshland for use as an airport or industrial area. The proclivity of agency personnel for denying the possibility that the private sector is capable of environmental enhancement is best understood by reference to their own words. What follows are factual highlights of a permitting effort that is typical of many.

The project property is also typical of much of Florida's east coast bordering the waters through which the Atlantic Intracoastal Waterway was cut during the 1930's. Along Florida's east coast the land and water area comprises some 200 million acres, much of which has been modified. In a five county area (excluding National Wildlife Refuges) some 9211 acres have been diked and isolated by the government from the estuarine system to serve as mosquito impoundments. Considerably more acreage has been ditched by the government for the same reason.

The project property comprises about 75 acres. Of this about 29 acres can be considered to be wetlands and transition areas, if one accepts the most liberal interpretation of dominance. The plan negotiated with the Florida Department of Environmental Regulation (DER), responding federal agencies (U.S. Fish and Wildlife) and the Corps, proposed to fill 5.3 acres of transitional land, fill 6.6 acres of impounded wetlands, retain undisturbed 9.8 acres of wetlands, improve and retain 5.3 acres of wetlands and convert 1.6 acres of wetlands to well-flushed, revegetated wetlands. This amounts to about 12 acres of marginal wetlands filled in exchange for enhancing nearly 17 acres of wetlands, the latter amounting to 23% of the private land involved. The plan was changed many times during the 19 months of negotiation with some of the "environmental" inputs contrary to the accepted literature. State dredge and fill permits and water quality certification were issued by DER on 8 April 1980. During the negotiations the National Marine Fisheries Service (NMFS) commented (after the end of the commenting period) that:

> *"Based on our analysis of the enclosed report, we have concluded that the work would adversely impact fishery production. Moreover, we cannot condone the unnecessary dredging of wetlands for the non-water dependent purpose of filling more wetlands for the trailer park. Therefore, we recommend that the permit be denied."*

The NMFS Field Investigation Report (conducted by a private contrator) stated:

> *"The white mangrove area is not tidally connected to the Indian River."*

Also:

> *"There will be a direct elimination of approximately 22 acres of mangrove habitat that is not presently tidally influenced."*

What the NMFS report failed to mention was brought out in the State Biological Report:

> *"The wetlands east of the uplands have been altered by ditches and dikes (non-mosquito control activities) and by mosquito control impoundment practices. These two areas are divided into north and south segments by an east-west dike/road (the north 1/3 of these wetlands is impounded for mosquito control). The vegetation in both of these areas has been severely affected by the recent severe winters. However most mangroves are regenerating new growth."*

Eight months later (2 months after state permits and water quality certification were issued) NMFS again wrote:

> *"The revised plans still call for filling a considerable area of wetlands that provide water quality maintenance functions and some detrital food sources for the Indian River. Should water quality be reduced by the proposed action, fishery resources would be adversely impacted. Accordingly, we concur with the comments and recommendations of the Environmental Protection Agency on this project."*

The EPA recommendations referred to were written 21 months after submission of the joint application and stated in part:

> *"We have not inspected the site of this proposed activity. The Fish and Wildlife Service and National Marine Fisheries Service recommend denial of this project in their letters dated March 15, 1979; we did not comment on the public notice. The NMFS again recommended denial of revised plans in their letter dated October 17, 1979. It is our understanding that the project has since been further modified, and the amount of wetland to be filled has been reduced from approximately 18 acres to 11.9 acres, and the size of the excavated marina has also been reduced.....*

> *"A narrow, 3-foot high berm exists along the Indian River which impounds a wetland area predominantly vegetated by white mangroves. These mangroves have been cold-killed and damaged by a hurricane. Historically the area has been wetlands, but the vegetative cover has changed from a herbaceous marsh to a mangrove swamp. Tidal exchange has occurred through breaks in*

the berm, and is presently limited to extreme high water events.....

"The proposed project reduces or eliminates a large wetland area for a residential development. Although this wetland has been stressed, it is our opinion that it will recover into a functional mangrove swamp. This recovery could be facilitied by providing increased tidal exchange. The wetland is not unique to this locale, and approval of this project would be precedent setting and result in a large cumulative loss of fringing wetlands. Therefore, we recommend that this permit application be denied."

There is no mention of the fact that the application called for providing increased tidal exchange, water control structures designed to maintain 7.5 acres of low salinity habitat as requested by the state and the Fish and Wildlife Service, the revegetation of the fringe of a 1.6 acre shallow water (-3ft MLW) basin and removal of dead trees and debris, along with careful sculpting of 5.3 acres to provide sheet flow to retain and improve the wetland's water purification processes. Also not mentioned are the extensive modifications of two upland borrow pits (not under jurisdiction) to provide 4.5 acres of improved freshwater habitat. Although the EPA feels that the "recovery could be facilitated by providing increased tidal exchange" they evidently feel that the property owner should accomplish this without compensatory use of his remaining land. Any precedent to be set would be an admission that the private sector has the willingness, the desire and the technical and financial capacity to improve the nations wetlands in circumstances where the wetlands have been highly stressed by man's and government's past activity, if not in more "natural" areas as well. There is no free lunch on the regulator's side either but the private sector is being forced to pay Cordon Bleu prices for the opportunity to contract botulism.

To an optimist familiar with the Corps Rules and Regulations, the above machinations should be no more than a temporary set-back since in section 325.8(b), the regulations state:

"District Engineer's authority. District Engineers are authorized to issue in accordance with this regulation permits and letters of permission..., in all cases in which there are no known substantive objections to the proposed work or activity or in which objections have been resolved to the satisfaction of the District Engineer. Unless otherwise precluded by this regulation, District Engineers may issue permits over an unresolved objection of another Federal agency if that agency indicates to the District Engineer that it does not desire to refer the application to a higher level of authority for review."

Black's Law Dictionary defines "substantive" as "An essential part or constituent or relating to what is essential." While the applicant is required to provide voluminous back-up for all manner of potential influences, the regulators, without having visited the site, state:"..., we have concluded that the work would adversely impact fishery production." This, even though NMFS and EPA tacitly admit that opening up the area to tidal exchange would have a positive effect.

The District Engineer, however, evidently feels that the above quoted statement is "substantive" and his staff initiates a new round of meetings for the purpose of "negotiating" further reductions in property value at increased cost to the land owner. This, in spite of Section 320.4 (a)(1) of the Corps' regulations:

> *"(a) Public interest review. (1) The decision whether to issue a permit will be based on an evaluation of the probable impact of the proposed activity and its intended use on the public interest. Evaluation of the probable impact which the proposed activity may have on the public interest requires a careful weighing of all those factors which become relevant in each particular case. The benefit which reasonably may be expected to accrue from the proposal must be balanced against its reasonably foreseeable detriments. The decision whether to authorize a proposal, and if so, the conditions under which it will be allowed to occur, are therefore determined by the outcome of the general balancing process (e.g. see 33CFR 209.400, Guidelines for Assessment of Economic, Social and Environmental Effects of Civil Works Projects). That decision should reflect the national concern for both protection and utilization of important resources. All factors which may be relevant to the proposal must be considered; among those are conservation, economics, aesthetics, general environmental concerns, historic values, fish and wildlife values, flood damage prevention, land use, navigation, recreation, water supply, water quality, energy needs, safety, food production, and, in general, the needs and welfare of the people. No permit will be granted unless its issuance is found to be in the public interest."*

It was interesting to find that the "Guidelines for Assessment" referred to (33 CFR 209.400) were removed from the Code of Federal Regulations between 1 July 1975 and 30 June 1976, or at least one year before publication of the current rules that reference it. One of the general criteia to "be considered in the evaluation of every application" is (§320.4(a)(2)(iii):

> *"..the extent and permanence of the beneficial and/or detrimental effects which the proposed structure or work may have on the public and private uses to which the area is suited;..."*

I know of no evidence that this is done on the Corps' District level and a great deal of evidence that it is not. It turns out that apparently the Corps' District does not really have to justify denial of a permit but merely refer to "objections of other federal agencies". The applicant has little recourse, at least in the 5th Circuit Court of Appeals, because he is denied an evidentiary hearing on applications for permit under §10 of the Rivers and Harbors Act. In Taylor vs. District Engineer, 567 F2d 1332 (5th Cir. 1978) the court found that:

> *"Procedures used by Army Corps of Engineers in processing application for permit...satisfied due process where, among other things, regulations required public notice, informal hearing if sufficient public interest were shown and provided applicant with opportunity to comment on substantive objections and provided that permit decision was to rest on evaluation of all relevant factors. Rivers and Harbors Appropriation Act of 1899 §10,33 USCA §403, USCA Const. Amend 5"* *

Obviously the Court assumes that the Corps follows its own procedures when, in fact, the applicant is not provided with the results of "the general balancing process" nor any evidence that such process has, indeed, taken place.

If I didn't have a bent toward optimism I wouldn't be here today. I am convinced as a scientist and an engineer that we can live, not only compatibly but symbiotically, with the rest of the creatures (with the possible exception of bureaucrats) with whom we share the planet. But "Optimist" is probably too naive a word when the "windmills" turn out to be "dragons". Ambrose Bierce may have a better word. He defines "cynic" as:

> *"A blackguard whose faulty vision sees things as they are, not as they ought to be. Hence the custom among the Scythians of plucking out a cynic's eyes to improve his vision."*

Even as a cynic I feel that it is important to end on a positive note. Further perusal of the Corps' regulations offers a glimmer of hope if the state in which you live has a governor who is unashamedly concerned with his state's economy and the balancing process as well as the "environmental" vote. Section 325.8(d) states:

> *"Referral to the Chief of Engineers. Division (and District) Engineers will refer to the Chief of Engineers the following cases: .....(2) When the recommended decision is contrary to the stated position of the Governor of the State in which the work is to be performed;...."*

*Elimination of specifics and underlining by the writer.

Section 325.11 then outlines a procedure wherein the District and Division Engineers must assume a responsibility commensurate with their authority in denying applications merely on the basis of: "Should water quality be reduced by the proposed action, fishery resources would be adversely impacted."

I am aware that each state has its own problems which, in many cases, are at least as unsettling as those discussed here. These should be and will be discussed before a more local audience. It would be unfortuate if true environmental concern were to be sidetracked in response to a general lack of bureaucratic responsibility, but there is cause to believe that zero regulation would be better than what we have today. And that can be said for the following reasons. The collective governments in this country already own more than 40% of our nation's real estate. We are adding a million acres yearly to our wilderness recreation areas and wildlife refuges. Another 2.2 million acres of rural land goes out of agricultural production through abandonment, either because of low soil fertility or a terrain unsuited to efficient use of modern machinery thereby being returned by private owners to the nation's inventory of wildlife refuge. To discourage other private landowners from modifying their land to increase wetland productivity merely because it makes their remaining property more valuable to other human beings does not, to me, seem consistent with the intent of our environmental legislation.

The best news is that developers and their customers have begun to appreciate a closeness to nature and a realization that symbiosis adds a precious dimension to our lives, decreases development costs and leaves options open for our children who, if history has any lesson at all, should know a little bit more than we do today. It would be rewarding if we could leave them some positive examples.

REFERENCES

1. PROCEEDINGS OF THE NATIONAL WETLAND PROTECTION SYMPOSIUM Fish and Wildlife Service, FSW/OBS 78-97, June 1977.

2. Ashby, Eric, ENCOUNTER, March 1976, "A Second Look at Doom".

3. DuBos, René, 6 August 1976, SCIENCE, "Symbiosis Between the Earth and Humankind".

4. Department of the Army, Engineers Corps, REGULATORY PROGRAM OF THE CORPS OF ENGINEERS. July 19, 1977.

5. Cairns, J., Jr., K.L. Dickson, and E.E. Herricks, editors, RECOVERY AND RESTORATION OF DAMAGED ECOSYSTEMS, Proceedings of the International Symposium on the Recovery of Damaged Ecosystems held at Virginia Polytechnic Institute, Blacksburg VA, March 23-25, 1975.

6. Meadows, Donella H., Dennis L. Meadows, Jorgen Randers, and William W. Behrens III, A Report for the Club of Rome's Project on the Predicament of Mankind, THE LIMITS TO GROWTH, 1972.

Author's note: The introductory quote was borrowed from Cy A. Adler's book: ECOLOGICAL FANTASIES, Green Eagle Press, New York NY, 1973. Already in 1973 significant advances in clean-up had been made and Adler's book present a factural perspective that is not consistent with the claims still being made by the environmental defeatists.

PERCEIVING THE UNSEEN VALUE OF COASTAL WETLANDS

JAMES D. SEARS
Chief, Environmental Resources Branch
South Pacific Division
Corps of Engineers
630 Sansome Street
San Francisco, CA 94111

Coastal wetlands are considered one of the nation's more valuable natural resources and the benefits to man are many and varied. However the areal extent has been drastically reduced over time. The present system of evaluating coastal wetlands has not prevented the reduction in the number of acres, nationwide. The use of a Benefit-Cost ratio has necessitated the use of paramters which are readily quantified and subject to an economic evaluation. Many private and public entities are continually trying to "rescue" isolated wetland eco-systems from the "developers." It is apparent that these individuals have a feeling that the value of the land is worth more to society in its present state than as a shopping center, water front homes, etc. The increasing scarcity of these wetlands has resulted in an increased awareness of their importance, over and above economics. The public's perceptions are vague and nebulous. If a value is to be used, it must be quantified. Of necessity, the values will be subjective, but will provide a platform from which refinements can be made. This paper will provide one means of arriving at the value of this preceived but unseen value.

As coastal wetlands are filled, abused, and poisoned, they seldom, if ever, recover or return to the biologically productive level previously sustained. Neither is the natural protection provided from storm damage ever regained. Until recently, these losses to the natural coastal systems were ignored or considered inconsequencial. However, as the natural coast disappeared, it became apparent to a few natural scientists, hunters, commerical fisherman, etc., that there had been a dramatic reduction in the productivity of our coastal wetlands in bays, lagoons and estuaries. At about the same time, coastal engineers were wrestling with the problem of protecting coastal developments which had not only encroached upon the barrier islands, but had also been filling the lagoons and bays.

When these adverse cumulative effects were recognized, the concerned professional and lay persons made efforts to introduce reason and judgement into the haphazard proliferation of actions decimating the wetlands, foredunes, lagoons, and bays. Unfortunately, their voices were lost in racket being made by the advocates of "big is better," "Thats progress" and "larger tax base" themes. While their individual voices were not heard, it is even doubtful that if they had been well organized their collective voices would have had an impact. The responses received in these early efforts to bring about change were somewhat derisive as they were told that they were attempting to slow down "progress," take away tax revenues, and eliminate jobs. In basic terms, it was a matter of demonstrating that the economic benefit exceeded the cost of the proposed action. We were told that the marsh or lagoon was worthless in its present condition and was only a breeding ground for mosquitos and rats.

An interesting sidelight on this subject is presented by Justin Brande in an article in the Journal of Soil and

Water Conservation Jan-Feb 1980. "Hippocrates, the ancient Greek man of medicine, warned against locating homes and cities near swamps." In spite of this position, many of the world's great cities while begun on the higher ground near navigation routes, did spread out into wetlands. This encroachment upon the swamps by human habitation resulted in poor drainage and sanitation as well as high rates of serious illness. As a result, the attitude of the "evil" swamps prevailed throughout history and has been prevalent in this country from its founding until recent years.

In order to be properly repsonsive to these very persistent and formitable concepts of the 1728 "Dismal Swamp" syndrome, the proponents of the natural coastal features began to gather basic data. During the past two decades, many major efforts have been initiated and completed which have demonstrated the productivity of the coastal wetlands (Teal 1962, Keefe 1972, Schelske and Odum 1962, Day et al. 1973, Valiela et al. 1975 etc.). Each of the studies provide one or more bricks of facts used to form the foundation of an economic approach to determining the dollar benefit to society from wetland preservation and restoration. Many wetlands advocates have pressed the findings of Eugene Odum, which placed monetary values on south Atlantic and Gulf Coast marshes and estuaries. His data show Georgia salt marshes to be twice as productive as fertile hay field harvests. Other research has followed the energy flow from sun to plants, into the food chain, ending in the total number of calories, BTU's or weight of bio-mass. This research has shown that wetlands provide a far broader spectrum of natural and physical functions than the production of harvestable fish and wildlife resources. A descussion of these functions is found in Research Report 79-R1 WETLANDS VALUES Concepts and Methods for Wetlands Evaluation prepared by the Institute for Water Resources, Corps of Engineers dated February 1979. The defined functions are as follows:

1. Primary Functions

   a. Food Chain Production

   b. General and Specialized Habitat for Land and Aquatic Species

   c. Aquatic Study Areas, Sanctuaries and Refuges

   d. Hydrologic Support Function

   e. Shoreline Protection

   f. Storm and Flood Water Storage

   g. Natural Groundwater Recharge

   h. Water Purification

2. Cultural Values

   a. Commerical Fisheries

   b. Renewable Resources and Agriculture

   c. Recreation

   d. Aesthetics

   e. Other Special Values

It should be noted that 11 of these items can be, and have been, evaluated in terms of dollars. In turn, these monetary values are added to the costs - as a loss - when related to the economic benefit from a proposed development. This then provides a more realistic Benefit to Cost ratio (B/C).

It has been natural to concentrate upon those factors which are readily defined and quantified. The incorporation of benefits or costs from environmental improvements or degradation was described by A. Myrick

Freeman III in The Benefits of Environmental Improvement published by Resources for the Future in 1979. Freeman stated, "Thus amenity differences must be reflected in differences in the price of housing and ultimately, land rents. Even this requires that land is not also an input in the production of traded goods." His efforts are coming closer to the final increment of value.

This seemingly indefinable value has been reflected in recent legislation such as the National Environmental Policy Act, Clean Air Act, Clean Water Act, and Endangered Species Act to name a few. Executive Orders and Regulations promulgated by the President and executive offices have also commented upon the need to protect, enhance or reestablish environmental quality. Federal Agencies proposals must now recognize not only the National Economic Development (NED) plan, but also, the Environmental Quality (EQ) plan.

We find ourself in an untenable position of having to prove that wetlands have value over and above the potential economic return with a vague discussion of improving the human and natural environment. Because of the dedicated work that has been, and is continuing to be done by concerned scientists, we have made great strides beyond "Jobs are more important than a worthless swamp", or "You can't eat a view." Our dilemma is also somewhat alleviated by the fact that the public awareness, and thus the politicians' awareness, of the value of natural systems has become more sharply focused as "progress" gradually decreases the areal extent of such important eco-systems.

It is this awareness which shows that the public perceives a value, even though that value cannot be specifically defined or refined enough to put on a computer. The general public has begun to feel the impact of reduced opportunities for enjoying the vistas and activities of their own childhood or as described by their parents. As in any "supply and demand" discussion, when societal benefits are reduced, the value of the remaining benefits are increased at an ever increaseing rate. In our area of concern, the loss of any additional acres of coastal wetlands results in the remaining acres being more costly in envronmental or social terms as well as economic terms.

Having accepted that the objective evaluations do not quite define the true total value of the wetlands, an effort must now be made to provide a subjective appraisal of the value to society over and above the prescribed economic figure. The procedure that follows is not new or unusual. It is an effort to determine a weighing factor which can be used to show that there is a value over and above the cost of the lands that should be included in anyhdiscussion of a B/C ratio. The base for computation is the determined economic value of the coastal wetland, whether for marinas and water front houses, or for continued biological productivity. The factor to be used is related to scarity of wetlands in the region under consideration. There is a vast difference in the perceived value of an acre of coastal wetlands in Louisiana and in California. Therefore, the method suggested to weigh this difference is:

1. determine the acres impacted directly and indirectly by the proposed action,

2. determine the total acreage of a similar nature in the region of the coast, using professional judgement to delineate similar systems, and

3. divide the regional acreage into the impacted acreage, which results in the factor to be used. This factor, when muliplied by the economic value, gives a figure which is an indication of the unseen value of the coastal wetlands. These computations would be used to indicate the value society places upon these areas over and above the economic value.

The three examples below demonstrate the procedure outlined above. The economic value in each case is assumed to be $100 per acre.

1. Gulf Coast wetlands

   1,000 acres affected

   400,000 acres in region

   .0025 Factor

   $100 x .0025 = $ .25 added value

   $100.25 Total cost per acre

2. West Coast wetlands

   40 acres affected

   4,000 acres in region

   .01 Factor

   $100 x .01 = $1.00 added value

   $101.00 Total cost per acre

3. San Diego Bay coastal wetlands

   40 acres affected

   400 acres in region

   .10 Factor

   $100 x .10 = $10.00 added value

   $110.00 Total cost per acre

In formula form, this is as follows:

$$\frac{AE}{R} + E = T$$

R = Acres of coastal wetlands in the region

A = Acres affected

E = Economic value of land per acre

T = Total cost of land per acre

It is recognized that this is a very subjective determination. However, it is expected that refinements and tighter definitions will be forthcoming. As stated earlier, this approach is not new but a literature search has not shown it used in this context.

Still unanswered is the strong sentimental value placed upon coastal lagoons, msrshes, etc., by landowners, communities, and even states. Another bit of unfinished business is the need to develop a formula which shows the unit value increasing geometrically as the acreage decreases.

BIBLIOGRAPHY

Brande, J.
Worthless, value or what? An appriaisal of wetlands.
Article in Jan-Feb 1980 issue, Journal of Soil and Water Conservation Soil Convervation Society of America.

Clark, John
Coastal Ecosystems.
Conservation Foundation
Washington, D. C.

Cowardin, L. M., Carter V., Golet, F. C. and La Roe E. T., Dec. 1979
Classification of Wetlands and Deepwater Habitats of the United States.
Published by Fish and Wildlife Service, Dept. of Interior.

Finn, J. T. and Leschine, T. M.
Does Salt Marsh Fertilization Enhance Shellfish Production?
Article in Environmental Management magazine Vol. 4, No. 3, May 1980, by Springer-Verlag, New York, Inc.

Freeman, A. M. III 1979
The Benefits of Environmental Improvement Theory and Practice.
Resources for the future by Johns Hopkins University Press.

Galloway, G. E. 1978
Assessing Man's Impact on Wetlands.
UNC Sea Grant publication UNC-SG-78-17

Reppert, R. T. et al Feb. 1979
WETLAND VALUES.
Concepts and Methods for Wetlands Evaluation. Research report 79-R1 U. S. Army Corps of Engineers Institute for Water Resources.

Shepard, F. P. and Wanless, H. R. 1971
Our Changing Coastlines.
Publisher McGraw - Hill Inc.

California's Coastal Wetlands 1979.
Sea Grant Report Series #2
Published by California Sea Grant College Program
La Jolla, California

SCIENTIST'S REPORT National Symposium on Wetlands.
Lake Buena Vista, Florida
National Wetlands Technical Council 1978
Edited by John Clark and Judith Clark

EVALUATING WETLANDS: STATUS AND NEEDS

BY

John J. Balco*

ABSTRACT**

While the value of wetlands has been acknowledged, it has been difficult to develop a methodology by which particular wetlands can be evaluated. There is no widely accepted approach to evaluating wetlands.

The Water Resources Council is in the process of reviewing wetlands evaluation procedures used or under development by Federal and State agencies. A definition of wetlands values and a framework for identifying important characteristics of evaluation procedures has been developed. Twenty-three wetlands evaluation procedures were collected and are currently (August 15, 1980) being analyzed.

Early indications are that detailed evaluation methodologies are in existence for wetlands habitat values. There is a basis for developing a system for evaluating hydrologic values of wetlands. The basis for developing procedures for other values is weak.

While evaluation procedures are important, it should be recognized that they are tools by which wetlands values can be better protected. The evaluation should not become an end unto itself. The level of accuracy expected should not be in excess of that expected from other scientifically based estimating techniques (e.g. estimates of flood heights).

*Policy Officer, Policy Analysis Division, U.S. Water Resources Council, Washington, DC.

**The conclusions and recommendations of this paper are based on initial review of wetlands evaluation procedures. They are not based on the availability of detailed analysis nor do they represent the conclusions of the U.S. Water Resources Council or member agencies.

There is a need to synthesize information on hydrologic values of wetlands, develop a detailed evaluation methodology based on current knowledge and begin an organized effort to fill gaps in knowledge. In addition, there is a need for an easy-to-use, general evaluation methodology which is sensitive to physical differences among regions. Evaluation approaches should consider impacts of wetlands modification in addition to the value of wetlands in their natural state. There is great promise in developing a data base that will relate wetlands values to the "Wetlands Classification System" developed by the U.S. Fish and Wildlife Service.

Introduction

In early 1979, the U.S. Water Resources Council sponsored a series of workshops on "Emerging Issues in Wetlands/Floodplain Management." One of the needs identified during the workshops was the development of a series of wetlands evaluation handbooks applicable to general, regional and site specific uses (Kusler, 1979). In developing a work plan to fulfill those objectives, there was concern that before developing evaluation handbooks, there was need to assess evaluation procedures in use or under development. The Council, via an interagency work group, undertook this task. The detailed analysis is currently (August 15, 1980) in process. This paper will summarize the important factors in the analysis, the approach taken, and preliminary conclusions. The conclusions are based on general review of evaluation procedures and assessment of current information. They are not based on the availability of detailed evaluations nor do they represent the conclusions of the U.S. Water Resources Council or member agencies.

Need for Evaluation Prodedures

During the past decade, there has been a major shift in the public's perception of the role of wetlands. No longer are wetlands viewed as wastelands which should be converted for direct economic benefit, but rather as vital links in our life support system. In response to this view of wetlands, public policies were developed to encourage or in some instances mandate, full consideration of wetlands values in land use decisions. A variety of Federal, State and local programs include permitting (State wetlands protection programs, Federal 404 permit program), permanent or temporary protection (e.g., acquisition, restrictions in use, leasing), management (e.g., habitat preservation and enhancement), development (e.g., site selection and impact analysis) and other administration requirements (e.g. Executive Order 11990 - Protection of Wetlands). Responses to adequately consider wetlands values in decisionmaking requires an evaluation of those values and impacts of actions on values.

While there is strong public support for protecting wetlands values program administrators have found it difficult to link these values to day-to-day decisions. The realities of administering wetlands programs are placing increased demands on providing assessments of particular values for particular wetlands. This in turn, requires the development of approaches in some cases formalized into procedures, for evaluating wetlands values so that decisions and alternatives can be based on scientific realities, rather than on a romantic ideal.

Approach

The assessment of evaluation procedures sought to answer these questions:

What procedures are in use or under development?

What are their characteristics?

What are the program needs for evaluation procedures?

What are the gaps?

What should be done to fill these gaps:

Improved communication

Development of additional or new procedures

Research

Wetlands Values

The starting point for an analysis of wetlands evaluation procedures is the functional values of wetlands. The ending point of such an analysis also must be wetlands values. It is very easy to fall into the trap of loosing sight of what one is trying to accomplish. The purpose of an evaluation should be to assess the values that exist and could be lost due to wetlands modification. The evaluation is not an end unto itself. It is a tool to allow us to make better decisions. The measurement of the effectiveness of a technique should be better decisions, not technical purity.

Wetlands values have been identified and categorized in many ways ranging from site-specific to global perspectives. The "Proceedings of the National Symposium on Wetlands" thoroughly reviews wetlands values. For the purposes of the Council's analysis of procedures, wetlands values were categorized as follows:

A. Habitat:
- Common wetlands and animal species
- Endangered, threatened, or rare plant and animal species
- Game species: aquatic
terrestrial
avian
- Commercial species
- Nongame species

B. Hydrologic:
- Floodwater conveyance and storage
- Wave energy dissipation and shoreline protection
- Ground and surface water supply including recharge and discharge
- Water quality, including waste assimilation and sediment trapping

C. Recreation:
- Water-dependent activities such as canoeing
- Other activities such as photography, bird watching and camping

D. Agriculture:
- Cultivated crops
- Pasture land and hay crops
- Forestry
- Peat

E. Heritage: Landscape: natural and unique areas, open space

Cultural: archaeological sites, historical sites

Scientific: research, education

Procedures Available

Copies of the procedures in use by Federal agencies, non-proprietary approaches developed by private sector sources and others identified in available literature were compiled. Twenty-five States with some form of wetlands management programs were contacted regarding their use of evaluation procedures. Seventeen States responded. However, only four States indicated that they had, or were developing, evaluation procedures per se. In all, twenty-three documents were compiled for the analysis.

Characteristics of Evaluation Procedures

Criteria were defined for the analysis of each functional value which allows the development of a matrix of values and characteristics. Characteristics are technical (Table 1) and administrative (Table 2).

The technical characteristics are:

A. Data needs - What is needed to use the procedures?

B. Biogeographic characterization

What are the biogeographic characteristics which might point to particular values both on a regional and site specific basis - size, location within a watershed, climate, hydrologic connection with open water, origin, topography, elevation, vegetation, vegetation diversity, local land use and others? Are there any key indications of values, such as indicator species? Are the procedures sensitive to certain wetlands types, or to sensitive locations within wetlands?

C. Technical basis

What is the technical or scientific basis for the evaluation? Is it consistent among regions or wetlands types? Should it be?

TABLE 1

Wetlands Evaluation Procedures Technical Criteria

Functional Values	Data Needs	Bio-Geog. Char.	Technical Basis	Reproduce Results	Quantify Values	Sensitive to Type	Assessing Vulnerability
Habitat: Common wetland plant, animal species							
Endangered, threatened, rare plant, animal species							
Game species: aquatic terrestrial avian							
Nongame species							
Hydrologic: Floodwater conveyance, storage							
Wave energy dissipation, shoreline protection							
Ground & surface water supply (recharge & discharge)							
Water quality (waste assimilation, sediment trapping)							
Recreation: Water-required activities							
Other activities (photography, bird watching)							
Agriculture/ Cultivated crops							
Silviculture Pasture land & hay crops							
Forestry							
Peat							
Heritage: Landscape: natural, unique areas, open space							
Cultural: archaeological sites, historical sites							
Scientific: research education							

TABLE 2

Wetlands Evaluation Procedures Administrative Criteria

Functional Values	Purpose of Procedures	Field Implementation	Adaptability	Assumptions	Constraints
Habitat: Common wetland plant, animal species					
Endangered, threatened, rare plant, animal species					
Game species: aquatic terrestrial avian					
Nongame species					
Hydrologic: Floodwater conveyance, storage					
Wave energy dissipation, shoreline protection					
Ground & surface water supply (recharge & discharge)					
Water quality (waste assimilation, sediment trapping)					
Recreation: Water-required activities					
Other activities (photography, bird watching)					
Agriculture/ Cultivated crops					
Silviculture Pasture land & hay crops					
Forestry					
Peat					
Heritage: Landscape: natural, unique areas, open space					
Cultural: archaelogical sites, historical sites					
Scientific: research education					

D. Ability to reproduce results

Can the results of the evaluation yield comparable results by different evaluators and at different times of the year? Is there variability by wetlands types?

E. Quantify values

Do the procedures quantify values? Is the quantification based on an arbitrary scale or some other value system? What are the uses and pitfalls of the quantification techniques?

F. Sensitivity to wetlands types or wetlands values

Do the procedures relate to a wetlands classification system? Do they address selected values? Are the procedures appropriate to a particular geographic region?

G. Assessing vulnerability

Do the procedures assess vulnerability from major or minor changes? Modification to areas within wetlands or adjacent to wetlands? Do they assess irreversible loss or identify critical points at which values are lost?

The administrative criteria are:

A. Purpose of procedures

Was it developed to fill a particular need? What explicit or implicit constraints are contained in the procedure? For example, a wetlands evaluation procedure developed by the Corps of Engineers has as a goal to speed to permit evaluation process i.e. to sort areas for intensive review (Reppert, 1979).

B. Field implementation

How difficult is it to obtain the data needed? Is it generally available? From what sources? Are field tests required? How difficult and expensive is it to obtain data or to conduct tests? What equipment is needed? How much time is needed? What technical skills are needed to implement the procedures? How complicated is the procedure? What training is necessary?

C. Adaptability

Can the procedures be used in a range of situations from preliminary assessment to detailed analysis of large areas or major wetlands alterations?

D. Assumptions

What explicit or implicit assumptions are made? For example, an evaluation technique that assesses flood storage values places a higher value on a wetland with dowstream development versus a similar wetlands with no downstream development (Soil Conservation Service, 1978).

E. Constraints

Has the procedure been field tested? Are there special requirements, such as field inspection or periodic sampling? Can the procedures be applied widely or are they site specific? Do they include consideration of minor alterations of wetlands? What are the legislative, policy or program constraints imposed on the procedures?

General Observations

Ranking of wetlands is a popular approach to wetlands evaluation. Almost all of the procedures reviewed involved wetlands ranking. Ranking here is considered the quanitifcation of the values of a wetland or wetlands, not the quanitification of a particular value of a wetland or wetlands. While ranking systems have appeal because of relative simplicity, comparability and reproducibility, they may result in overemphasis on "the numbers" with little consideration of the purposes of the evaluation itself. Some of the potential difficulties with a ranking system have been identified by Golet (1979).

A ranking system is one that contains de facto policy decisions on the relative importance of various wetlands values, some which may only be implicit. A scale of 1 to 10 for habitat value and a like scale for groundwater automatically establishes relative importance. Habitat is equally as important as groundwater. A decisionmaker may be forced to trade off one unit of habit for one unit of groundwater. Ranking of wetlands is only valid as it relates to a particular value and only then with full consideration of the limitations of ranking schemes.

Approaches to economic evaluation of wetlands were not considered in the analysis. Economic assessments have attempted to place monetary values on wetlands. These efforts have resulted in considerable controversy between the physical and economic sciences (Shabman, 1979). It is unlikely that this controversy will be resolved in a fashion that will be useful for wetlands evaluation purposes in the near term. Even if an acceptable methodology was developed, its usefulness in wetlands evaluation would suffer the same constraints as ranking systems.

There has been some skepticism of wetlands evaluation procedures because of the assumptions and generalities upon which some of the procedures are based. In any assessment, there are assumptions, estimates and generalizations which might result in inaccuracies. There will probably never be a technically perfect solution, but such a solution may not be needed at all. There are few who question engineering calculations of flood heights. Yet, the methodologies used may be non-standard, formulas may be developed for general conditions, assumptions are made, data may be faulty and establishing values for use in formulas may be at best professional estimates. There is an implicit acceptance of "the number" being within a certain range. However, significant land use, property and economic decisions are made based on accuracy to the "two-foot contour." The level of precision implicity suggested for wetlands evaluation procedures should be no more than that of flood elevations estimates. Excessive demand for precision suggests focusing on the tools themselves, not the need to get the job done.

User's needs

An assessment of user's needs from Federal program perspectives is currently under development. When these user's needs are defined, State programs will have an opportunity to review, comment and provide the perspective of State and local needs. While it is not expected that Federal and State needs will be significantly different, there may be differences in priorities needed to fulfill those needs. Initial statements of user's needs are expected to be available in September 1980.

Wetlands Evaluation - Where are we?

The primary thrust of wetlands protection programs has come from the relationships established between habitat, fish and wildlife resources and wetlands. It is not surprising then that detailed evaluation methodologies have been developed to assess habitat values. (Golet, 1973; Larson, 1978; Galloway, 1978; U.S. Army Corps of Engineers, 1979; Fried, undated; State of Maryland, updated). While some of these procedures include nonhabitat values, their primary thrust is habitat.

There has been some work done to define the hydrologic value of wetlands. However, there has been little development of an evaluation methodology for the hydrologic value of wetlands (U.S. Army Corps of Engineers, 1972, 1979; Soil Conservation Service, 1978) One of the major difficulties with technical basis for establishing the hydrologic values of wetlands is an inconsistent and in some cases inconclusive, scientific basis for reaching conclusions (SCS Engineer, 1979).

While there is some literature on wetands values other than habitat and hydrologic values, it is sparce (Smardon, 1972; Galloway, 1978; Foster, 1979). It would appear that there is considerable work needed to begin developing such methodologies.

While the assessment of some functional values has progressed further than others, most are detailed evaluation approaches. Only one general evaluation approach was reviewed (Winchester, 1978). The former schemes tend to require field data collection, to rely on a comprehensive data base, to require an interdisciplinary technical staff, and require some time for implementation.

Wetlands evaluation schemes have concentrated on the values of wetlands as naturally functioning systems. Few evaluation approaches reviewed considered impacts as part of an evaluation approach (Schuldiner, 1979; U.S. Fish and Wildlife Service, 1979). Evaluation systems have generally not considered impacts due to wetlands use incompatible with the characteristics of the wetlands itself. For example, a house located on wet soils may have substantial damage due to settling. An area which is not generally subject to high energy conditions may have abundant wetlands vegetation. Removal of this vegetation may promote serious damages during storms.

While there have been attempts to relate wetlands values to wetlands types (Silberhorn, 1974; California Coastal Commission, 1979), there is no generally available methodology to do so. However, the U.S. Fish and Wildlife Service is developing a data base which will relate scientific literature on wetlands values to the "Wetlands Classification System" (Cowadin, 1978) and a variety of other parameters, such as location and hydrologic unit (Whilen, personal communication).

Wetlands Evaluation - Where sould we be going?

Based on the preliminary analysis of wetlands evaluation procedures in use and under development, the important considerations in future evaluation development should be:

1. A single, standardized evaluation approach is inappropriate because the purposes for wetlands evaluation are variable and a standardized approach will not account for regional physical differences.

2. Various detailed approaches to evaluate habitat value of wetlands are in existence and are being field tested and modified based on field experience. It can reasonably be expected that these approaches will continue to be refined. No new efforts are needed at this time.

3. There are substantial improvements needed in determining the hydrologic values of wetlands and developing evaluation procedures. Curent knowledge should be synthesized and an action program should be developed to produce both general and detailed assessment methodologies and to fill information gaps.

4. There is a need for general evaluation procedures which account for regional variation in wetlands values. The procedures must allow for rapid identification of values of a particular wetlands without extensive field investigation. They should be flexible, require limited data collection and analysis, be capable of being implemented by field staff with limited technical training and must allow for rapid decisionmaking (30 - 60 days). To the extent that the state-of-the-art will allow, key indicators of value should be identified.

5. Evaluation approaches have focused on wetlands as naturally functioning systems. More emphasis is needed on the impact of wetlands modifications to values and the hazards associated with occupying wetlands. For example, in the glaciated Northeast, many wetlands are associated with present stream vallies overlying ancient valley aquifers. However, there appears to be no hydrologic connection between saturated sand and gravels and overlying wetlands. Thus, wetlands tend not be groundwater recharge areas. However, modification to these wetlands, such as by landfill, may penetrate impermeable swamp deposits resulting in pollution of the aquifer. As another example, wetlands by their nature are low lying and may be subject to flooding. Occupancy of wetlands may present serious hazards to life and property. It does not follow that if a wetlands does not perform an erosion control function that converting it into a parking lot would protect property as well as the wetlands (Shabman, 1979).

6. There is great promise in developing an a system to relate functional values of wetlands to the wetlands classification system of the U.S. Fish and Wildlife Service. Such an effort would have characteristics of ease-of-use, standardization and flexible response to varying conditions. It can also be readily related to the wetlands inventory efforts currently underway and would be cost effective. Such an approach also allow a test of the technical basis for conclusions reached on particular wetlands values.

References Cited

California Coastal Commission, 1979, "Statewide Interpretive Guidelines for Wetlands and Other Environmentally Sensitive Habitat Areas (Draft)", San Francsico, CA.

Cowadin, Lewis M., Virginia Carter, Francis C. Golet and Edward T. LaRoe, 1979, "Classification of Wetlands and Deepwater Habitats of the United States", U.S. Department of Interior, Fish and Wildlife Service, Office of Biological Services, Washington, DC FWS/OBS-76-31, 103 pp.

Foster, J.H., 1978, "Measuring the Social Value of Wetland Benefits" in Greeson, Philip P, John R. Clark and Judith E. Clark, ed. "Wetlands Functions and Values: The State of Our Understanding - Proceedings of the National Symposium on Wetlands", American Water Works Association, Minneapolis, MN. pp. 84-92.

Fried, Eric, undated, "Priority Ranking of Wetlands for Acquisition", Department of Environmental Conservation, State of New York Albany, NY.

Galloway, G.E., 1978, "Assessing Man's Impact on Wetlands", University of North Carolina, Sea Grant Publication UNC-SG-78-17 University of North Carolina Water Resources Research Institute Publication UNC-WRRI-78-136, Raleigh, NC.

Golet, F.C., 1973, "Classification and Evaluation of Freshwater Wetlands as Wildlife Habitat in the Glaciated Northeast", Transactions of the Northeast Fish and Wildlife Conference, Vol. 30, pp. 257-279.

Golet, F.C., 1979, "Rating the Wildlife Value of Northeastern Fresh Water Wetlands" in Greeson, Philip P., John R. Clark and Judith E. Clark, ed., "Wetland Functions and Values: The State of Our Understanding - Proceedings of the National Symposium on Wetlands", American Water Resources Association, Minneapolis, MN. pp. 63-73.

Kusler, Dr. Jon A., 1979, "Summary: Emerging Issues in Wetlands/ Floodplain Management Report of a Technical Seminar Series", U.S. Water Resources Council, Washington, DC. 41 pp.

Larson, J.S., ed. 1976, "Models for Assessment of Freshwater Wetlands", No. 32, University of Massachusetts, Amherst, MA.

Reppert, Richard T. and Wayne R. Sigler, 1979, "Concepts and Methods for Wetlands Evaluation Under Development by the U.S. Army Corps of Engineers", in Greeson, Philip P., John R. Clark and Judith E. Clark, ed., "Wetlands Functions and Values: The State of Our Understanding - Proceedings of the National Symposium on Wetlands", American Water Resources Association, Minneapolis, MN. pp. 57-62

SCS Engineers, 1979, "Analysis of Selected Functional Characteristics of Wetlands", prepared for U.S. Army Corps of Engineers, Contract Number DACW-78-R-0017, Reston, VA.

Shabman, Leonard A. and Sandra S. Batie, 1979, "Estimating the Economic Value of Coastal Wetlands: Conceptual Issues and Research Needs", Department of Agricultural Economics, Virginia Polytechnic Institute and State University, Blacksburg, VA.

Schuldiner, P.W., Cope, D.F., and Newton, R.B., 1979, "Ecological Effects of Highway Fills on Wetlands Research Report", and "Ecological Effects of Highway Fills on Wetlands Users Manual", National Cooperative Highway Research Program Report 218A and 218B, Transportation Research Board - National Research Council, Washington, DC.

Silberhorn, G.M., and Barnard, T.A., Jr., 1974, "Coastal Wetlands of Virginia: Guidelines for Activities Affecting Virginia Wetlands", Special Report on Applied Marine Sciences and Ocean Engineering No. 46, Virginia Institute of Marine Science, Gloucester Point, VA.

Smardon, R.C., 1972, "Assessing Visual - Cultural Values on Inland Wetlands in Massachusetts", M.S. Thesis, Univesity of Massachusetts, Amherst, MA.

Soil Conservation Service, 1978, "Water and Related Land Resources of the Coastal Region, Mass.; Wetlands Evaluation Criteria", U.S. Department of Agriculture, Soil Conservation Service, Amherst, MA.

State of Maryland, undated, Department of Natural Resources, "Environmental Evaluation of Coastal Wetlands (Draft) - Tidal Wetlands Study", Annapolis, MD. pp. 181-208.

U.S. Army Engineer Division, New England, 1972, "Charles River; Main Report and Attachments", Waltham, MA.

U.S. Army Engineers Corps of Engineers, Lower Mississippi Valley, 1976, "A Tentative Habitat Evaluation System (HES) for Water Resources Planning", Vicksburg, MS.

U.S. Army Corps of Engineers, 1979, "Wetlands Values - Concepts and Methods for Wetlands Evaluation", Research Report 79-RI, Institute of Water Resources, Fort Belvoir, VA.

U.S. Fish and Wildlife Service, 1979, "Habitat Evaluation Procedures (HEP) Manual", Washington, DC.

Whilen, William, U.S. Fish and Wildlife Service, Office of Biological Services,, Personal Communication.

Winchester, G.H., and L.D. Harris, 1979, "An Approach to Valuation of Florida Freshwater Wetlands", Proceedings of the Sixth Annual Conference on the Restoration and Creation of Wetlands, Tampa, FL. pp. 1-26.

ECOLOGICAL CHARACTERIZATIONS - AN EVALUATION TOOL

Jay F. Watson*

Introduction

The Fish and Wildlife Service, with the assistance of the Environmental Protection Agency and the Bureau of Land Management, has developed ecological characterizations for the Pacific Northwest and the Central and Northern California Coastal Regions. An ecological characterization provides a description, using existing information, of important components and processes comprising an ecosystem and an understanding of their functional relationships. The purpose of these characterizations are varied and include: integration of complex ecological information, identification of data gaps, prediction and assessment of environmental impacts, and assistance in developing mitigation procedures and alternatives for minimizing environmental impacts. An atlas approach is developed for each regional characterization which provides a framework for understanding the important physical characteristics and biological processes occurring in each study area.

The following will describe what an ecological characterization involves, how a characterization is developed, and what interest the Fish and Wildlife Service has in developing, and using these studies. First of all, an ecological characterization is: A study to obtain and synthesize available environmental data and to provide an analysis of the functional relationships between the different components of an ecosystem and the dynamics of that system. The concept of an ecological characterization is not very difficult or complicated. It is simply a structured approach to combining information from physical, chemical, biological and socio-economic sciences into an understandable description of an ecosystem.

Available data may be compiled from diverse sources. Some of the sources utilized in the Pacific Northwest Study include the Bureau of Land Management's _Summary of Knowledge of the Oregon and Washington Coastal Zone and Offshore Areas_ (1), _Oceans_ (3), the _Washington Environmental Atlas_ (4), and others. Data collection does not involve any field studies. It is strictly limited to existing published and unpublished information concerning the coastal ecosystems of the Pacific Northwest and the Central and Northern California Coastal Regions.

---

* Coastal Ecosystems Project, Office of Biological Services, Fish and Wildlife Service, Department of Interior, Portland, Oregon

An ecological characterization's most important function is an analysis of functional relationships. That is, the characterization shows the relationship between major components of the ecosystem and indicates the flow of energy, money, or products between the different components.

For example, one of the major components of the coastal region of the Pacific Northwest and Northern California is forest. This area has the largest standing biomass of trees in the continental United States if not the world. This component supports, in turn, an extensive forest products industry. The functional relationship is that the forest supports a forest products industry. An example of this relationship can be seen along the Oregon coast. The Coos Bay drainage basin has extensive forest lands and also one of the busiest log exporting ports in the world.

These ecological characterizations also provide an analysis of the dynamics of the flow of energy, material, money, or products among the different components of an ecosystem. The dynamics of the flow of logs between the forest component of an ecosystem and the forest products industry can be used as an example. If logs are being taken from the forest at a rate greater than new logs are being produced by growth, then the supporting component, the forest, will cease to exist.

Developing an Ecological Characterization

The final products from an ecological characterization include: (1) a conceptual model, (2) a characterization narrative or text, (3) habitat descriptions, (4) watershed unit descriptions and (5) a data source appendix. That is, a characterization study consists of five parts.

Conceptual Model

The first part of the characterization process is the development of a conceptual model. The conceptual model is an extremely important part of the characterization process. It is a framework for identifying important components of an ecosystem and identifying functional processes.

The conceptual model ties together all the different components of the biosphere-biological processes, physical-chemical processes, and human activities. The inclusion of human activities in the conceptual model is an important step forward in understanding the structure and function of coastal ecosystems.

The conceptual model also serves as a guide to the type of information that needs to be included in the characterization. The conceptual models were completed with the idea that they would lead to a structured collection and synthesis of existing information for the characterization. For example, there is a great deal of information available concerning the distribution of zinc in the lower Columbia River and Willapa Bay, Washington, which is just north of the Columbia River. All of the models to date, for the Pacific Northwest Coastal Region, seem to indicate that zinc distribution is not a key factor in our understanding of the structure and function of coastal ecosystems.

Most of the conceptual models have been constructed using the symbols developed by H. T. Odum (2). A bullet symbol is associated with the producer group, a hexagon with the consumer and decomposer group, a storage tank with the substrate, and a circle for sources of energy or matter. An attempt has been made in the conceptual model to provide enough information in the text so that anyone, given the time and interest, can understand the various conceptual models.

Characterization Narrative or Text

The characterization has a narrative or text providing a comprehensive description of the environmental features as they have been shaped by nature and man. The elements that are described include: (1) physical-chemical environment, (2) biological environment, (3) socioeconomic environment, and (4) regional descriptions of zones and habitats.

For example, the two characterizations point out that the study areas have the highest standing biomass of conifers in the United States, that in the north it is an area of extremely high precipitation, and that it is an area of coastal upwelling. The narratives also contain numerous models and tables describing representative food webs, community composition and ecosystem structure.

Habitat Descriptions

Each characterization provides descriptions of the various habitats found within the study areas. Approximately 140 habitats are described for the Northwest Coastal Region and 65 for the Central and Northern California Coastal Region. The descriptions are cross referenced with other classification systems and include characteristic species, succession models, food webs and descriptions of abiotic and socioeconomic factors. In addition, each habitat is further described by computer generated annotated species lists. The listings include life history information and references.

Watershed Unit Descriptions

The two study areas, Pacific Northwest and Central and Northern California, are divided into 9 and 21 watershed units respectively. The Central and Northern California study area is further divided into offshore basins. These descriptions include information concerning important industries, areas and species of ecological concern, and other site specific information.

Data Source Appendix

The last part of each characterization is a data source appendix which contains all the data from which information was drawn to complete the model, narrative, atlas, and pilot study. It forms a major data base developed during the course of study. This document contains tables and associated narratives. The data are presented in a format suitable for computer storage, retrieval, and manipulation. The Annotated

Bibliographies use the FAMULUS system developed by the U.S. Forest Service at the University of California. The two bibliographies contain about 4000 references.

Responsibilities

The mission of the U.S. Fish and Wildlife Service is to conserve, protect, and enhance fish and wildlife and their habitat for the benefit of the people of the United States. In order to carry out this mission, the Service is authorized or required to: (1) conduct investigations, surveys, and research; (2) acquire lands for refuges; (3) promulgate and enforce regulations to protect fish and wildlife; and (4) provide monetary support and cooperative help through agreements with other Federal and State agencies in the interest of fish and wildlife resources.

As part of its mission to protect and/or enhance fish and wildlife and their habitat, the U.S. Fish and Wildlife Service operates wildlife refuges throughout the study areas. These refuges provide habitat for waterfowl, shorebirds, endangered species, and seabirds. In addition, the Service is active in reviewing and commenting upon proposed activities that could cause adverse impacts upon fish and wildlife and their habitat. We are very concerned about possible impacts of energy development projects upon coastal ecosystems. These projects include foreign oil imports, Alaskan oil transhipment, LNG import, petrochemical industry development, and Outer Continental Shelf activities.

To meet our responsibilities, we need an ability to predict the impacts of these activities upon the estuarine and nearshore ecosystems. Predictive capability is required so that the Service may provide the best possible advice to the Bureau of Land Management, the Geological Survey, construction agencies and companies, and the public. It is necessary for us to predict accurately the impacts of dredging, filling, development, and other related projects upon coastal ecosystems.

Examples

The ecological characterizations completed for the Pacific Coast have been used by the Bureau of Land Management to prepare an environmental impact statement for a proposed Outer Continental Shelf oil and gas lease sale, by a small Oregon coastal community to determine the impact of flooding on a stand of red alder, and by researchers developing study plans to investigate trophic relationships in the Columbia River Estuary. Each of the investigators, planners or others using the characterizations selected the appropriate section(s) to complete their analyses.

Summary

In summary, an ecological characterization is: a study to obtain and synthesize available environmental data and to provide an analysis of the functional relationships between the different components of an ecosystem and the dynamics of that system.

The actual characterization consists of: (1) a conceptual model, (2) characterization narrative or text, (3) habitat description, (4) watershed unit descriptions, and (5) a data source appendix.

The Fish and Wildlife Service's mission to conserve, protect, and enhance fish and wildlife and their habitat for the benefit of the people of the United States can be enhanced by an improved information base. We believe that the characterization process and the products that it generates will provide us with that improved information base.

References

1. Bureau of Land Management (1977). A summary of knowledge of the Oregon and Washington coastal zone and offshore areas. Prepared for the Bureau of Land Management by the Oceanographic Institute of Washington. 3 vol.

2. Odum, H. T. (1971). Environment, Power, and Society. New York, Wiley-Interscience. 331 pp.

3. Sverdrup, H.U., Martin W. Johnson and Richard H. Fleming (1942). The Oceans, Their Physics, Chemistry and General Biology. Prentice-Hall, Inc., Englewood Cliffs, N.J. 1087 pp.

4. U.S. Army Corps of Engineers (1975). Washington Environmental Atlas, Washington, D.C. 114 pp.

Sitka Coastal Habitat Evaluation

Kimbal A. Sundberg*

Introduction and Purpose

The Alaska Coastal Management Program (ACMP) as approved by the Department of Commerce[1] provides for the preparation of District Coastal Management Programs by local governments. District Programs are reviewed by the Alaska Coastal Policy Council for consistency with the State's ACMP Guidelines and Standards[2]. Upon final approval by the State Coastal Policy Council, District Programs become the touchstone for ACMP actions affecting each District. The aggregate body of District Programs is the primary implementation tool of the ACMP. Federal, State, and local activities must be consistent with approved District Programs.

The City and Borough of Sitka embarked upon the development of a District Coastal Management Program in 1979. At that time the Alaska Department of Fish and Game (ADF&G) entered into a contract with Sitka to provide an inventory and analysis of known fish and wildlife resource and habitat information for the area, as well as a summary of the issues and conflicts concerning the conservation of habitats. The ACMP Guidelines[3] require Districts to identify and plan for the "maintenance or enhancement" of coastal habitats when preparing District Programs.

An initial inventory and analysis (Phase I) document[4] was completed and published in October 1979. The Phase I document served to distill important community values and issues and identified data gaps where additional information was needed. As a result of the Phase I effort, the City and Borough of Sitka signed a Memorandum of Understanding (MOU) with the Alaska Department of Fish and Game to furnish additional data on coastal habitats and to present this data in a report that could be utilized for final preparation of the District Coastal Management Program (Phase II). Specifically the MOU called on ADF&G to:

1. Evaluate and map individual streams and lakes along the 15 mile Sitka road system to delineate fish species, spawning areas, other sensitive habitat areas, and adjacent land and water use practices which affect fish habitat; and make recommendations for maintaining or enhancing specific stream and lake habitats in the Sitka "urban" area.

2. Evaluate and map coastal wetlands and tidal flats to delineate important habitat use areas for birds and other wildlife. Document past and present impacts to wetland habitats and make recommendations for the protection of important wetland habitats.

3. Identify and characterize waterfront marine habitats including estuaries, tidal flats, rocky shores, and man-made structures. Describe and document their biological and physical features and develop a management

*Habitat Biologist III, Alaska Dept. of Fish & Game, Anchorage, AK

system for waterfront habitats which is based upon their relative sensitivities to various aspects of waterfront development.

The purpose of this paper is to summarize the results of the Sitka Coastal Habitat Evaluation Study and to suggest that this type of study can be used for future coastal planning in Alaska and in all areas where the primary goal is to balance resource conservation and protection with the orderly development of the coast.

The Setting

The City and Borough of Sitka lies within the Tongass National Forest on two major islands in Southeast Alaska, Baranof Island, and Chichagof Island. Its 4,600 square mile area includes nearly 3,400 square miles of land and about 1,200 square miles of water. It is roughly the size of the State of Connecticut. With the exception of several scattered logging camps and seasonal fishing camps, the permanent population is located at Sitka, a community of 7,200 situated on the eastern shore of Sitka Sound facing the Pacific Ocean's southern Gulf of Alaska. Steep shorelines forested with Sitka spruce *(Picea sitchensis)* and Western hemlock *(Tsuga heterophylla)* rise above the fjords, inlets, and myriad of islands that make up the Alexander Archipelago. The snow capped 3,000 - 5,000 ft peaks of the Baranof Mountains rise sharply from the sea in dramatic contrast to the bluegreen waters of Sitka Sound and its associated insular waterways. The climate of the area is decidedly maritime with mild winters relative to the bulk of Alaska, cool summers relative to the Pacific Northwest, and abundant precipitation. The natural resources of the area are dominated by timber (mostly Sitka spruce and Western hemlock) and seafood (salmon, herring, halibut, groundfish, and shellfish). Sitka has a well-balanced year-round economy based upon pulp processing, logging, fishing, seafood processing, government employment, health and education institutions, and tourism. Sitka's economy, resources, recreation, and overall quality of life are almost entirely dependent upon its coastal environment. Sitka's primary transportation links to the rest of Alaska and "outside" come through regular ferry and barge service as well as daily jet service to points in Alaska and the Pacific Coast. Sitka is not connected by road to the mainland. Its roads serve only the vicinity of Sitka itself.

Study Area

The study area (Figure 1) includes fish and wildlife habitats within the area of urban influence roughly bordered by Nakwasina Passage on the north, Goddard Hot Springs on the south, Silver Bay on the east, and Kruzof Island on the west, including the islands, shorelands, and major drainages up to the 1,000 ft contour. Based upon the Phase I resource inventory and issue

Figure 1 Study Area Map Showing Study Sites

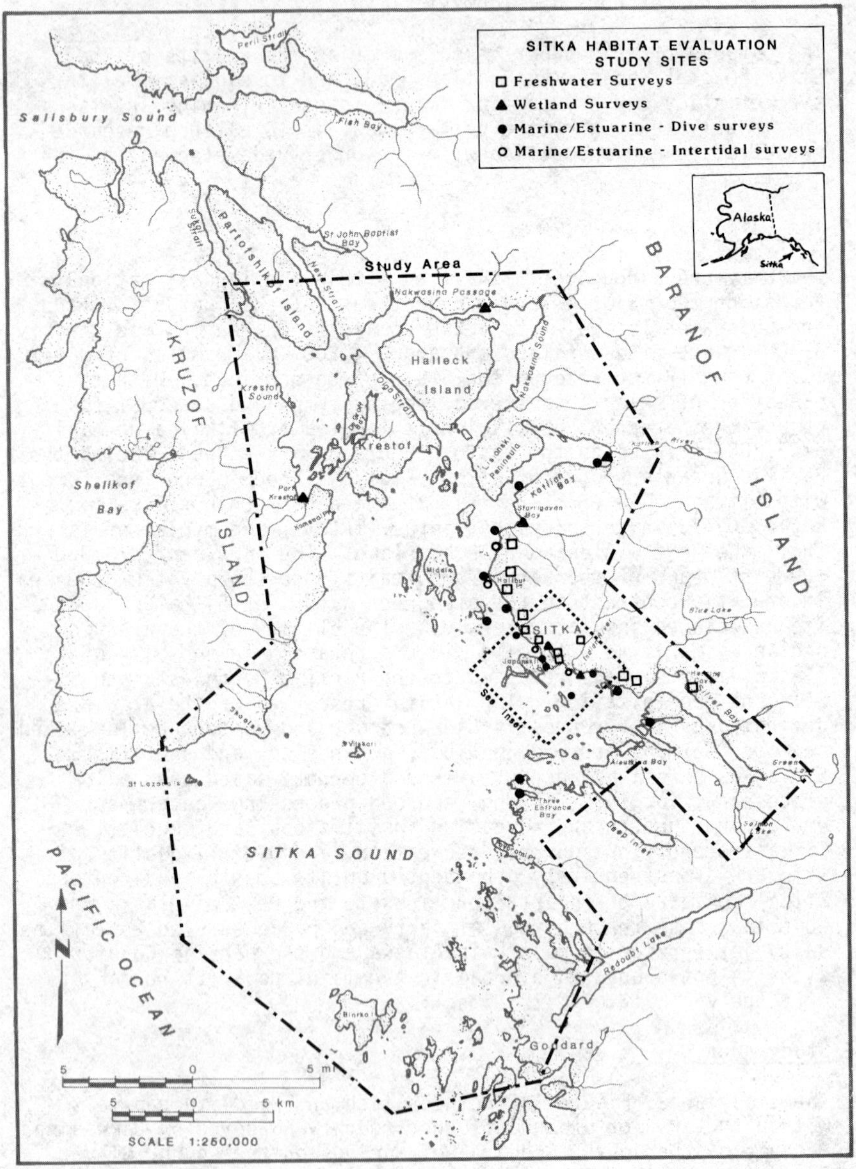

analysis it was determined that most of the habitat impacts occuring from urban growth, apart from logging, would occur within the study area. Habitat impacts from logging on Federal, State, and Alaska native-owned lands are currently handled through ongoing programs with various government agencies including U.S. Forest Service, U.S. Fish and Wildlife Service, National Marine Fisheries Service, Alaska Department of Natural Resources, Alaska Department of Fish and Game, and Alaska Department of Environmental Conservation.

Methods

Standard methods were employed to survey fish and wildlife resources and habitats within the study area. A brief description of each method follows:

1. Freshwater Habitats. Fish species present in freshwater systems were determined using minnow traps baited with salmon eggs. Foot surveys were conducted during peak salmon spawning periods to document spawning areas and estimate numbers of spawners. All streams along the road system were walked their entire length from tidewater to the upper extent of potential fish habitat. General habitat conditions were noted along the way including characterization of spawning gravels, potential rearing areas, obstructions to fish passage (both man-made and natural), streamside vegetation, and adjacent land uses and impacts. Mapping techniques employed the use of low altitude aerial photography taken with 35 mm color infrared Ektachrome, enlarged to a scale of approximately 1 in = 140 ft. Low altitude photography was supplemented with U.S. Forest Service 1 in = 1,320 ft color photos and high altitude color infrared photos (NASA and Ames) at a scale of 1 in = 1 mile. Selected drainages on the high altitude photography were enlarged to approximately 1 in = 600 ft.

2. Wetlands Habitats. Wetland habitats were initially identified using aerial photography and surveys by foot, boat, aircraft, and car. Thirteen areas were initially selected for further study. Of these, six were later selected for intensive ground surveys. Permanent transects were established in six wetlands to characterize plant and soil conditions. One meter wide belt transects were sampled every meter to record plant species composition, percent cover, soil type, and elevation. Elevations were determined with a rod and transit. Bird surveys were conducted during each of the four seasons to document species present, habitat use, and relative numbers. Opportunistic sightings of Sitka black-tailed deer *(Odocoileus hemionus sitkensis)*, brown bear *(Ursus arctos)*, and small mammal use of wetlands were also recorded. Mapping of wetlands was

accomplished using high altitude color infrared photos (NASA and Ames) at a scale of 1 in = 1 mile. Selected wetlands were later enlarged to a scale of approximately 1 in = 750 ft. Areas were determined directly from photos using a polar planimeter. Aerial photos were uncontrolled; however, distortion was assumed to be minimal because all wetlands mapped occur within 30 ft of mean sea level.

3. Marine/Estuarine Habitats. A surface current study was initiated in September 1979 by releasing 3,500 plastic drift bottles at 35 locations within Sitka Sound. Each one inch diameter, 4.5 inch long plastic centrifuge bottle contained an individually numbered international orange card which could be filled out by the finder as to place found, date, time, special circumstances, etc. The cards were self addressed and postage paid. The bottles were weighted to float with approximately 0.3 inch showing above the water surface. One hundred bottles were released per station.

Intertidal and dive surveys were conducted at twenty-one shore sites within Sitka Sound. Sites were selected to characterize a variety of shoreline types, oceanographic conditions, and man-made waterfront development impacts. Intertidal surveys were accomplished by walking along a surveyor's tape stretched from lower low water (LLW) to higher high water (HHW) and recording substrate, epifauna, epiflora, infauna, and elevation within major habitat zones. Elevations were determined with a rod and transit. Flora and fauna were sampled using a bucket and shovel. Infauna samples were washed through a 1 mm seive. Plants and animals which could not be identified in the field were preserved in 10% buffered formalin for later laboratory identification.

Dive surveys were conducted along a 100 meter transect line anchored at mean higher high water (MHHW) and run perpendicular to the shoreline out to 100 meters length, or 80 ft depth, whichever occurred first. Substrate type, depth, epiflora, and epifauna were recorded at 5 meter intervals along the transect. Underwater 35 mm photos were taken along the transect to supplement later identification of marine life. Depth readings were taken using a standard sport diving gauge and corrected to MHHW using the tide tables[5].

Results and Discussion

Freshwater Habitats. Freshwater systems in Sitka are characterized by short, steep drainages with variable flows and relatively low productivity. Nearly all systems produce pink salmon *(Oncorhynchus gorbuscha)*, coho salmon *(O. kisutch)*, and Dolly Varden char *(Salvelinus malma)*. Steelhead/rainbow trout *(Salmo gairdneri)*, cutthroat trout *(Salmo clarki clarki)*, sockeye salmon *(O. nerka)*, and chum salmon *(O. keta)* are also produced in several systems.

Pink salmon spawning populations range from several 10's for small creeks, to over 100,000 for the Katlian River, the largest drainage surveyed. Peak pink salmon spawning generally occurs during odd numbered years with correspondingly low numbers during even years. Low flows combined with periods of sub-freezing weather can severly limit the yearly salmon production of Sitka area streams.

Eleven freshwater systems were surveyed during this study. These systems were selected because of past or anticipated future impacts from urban development which could affect their habitat value. Of these, two are discussed in this paper. Indian River is an important stream system facing impacts from urban development. Swan Lake is a valuable urban lake under impact from shoreline development.

Indian River (Figure 2) flows from a watershed formed at the foothills of the Baranof Mountains and empties into Sitka Sound at the Sitka National Historical Monument. The stream produces pink salmon, coho salmon, steelhead trout, and Dolly Varden char. Pink salmon escapements (numbers of spawners) average approximately 2,000-3,000 with peak years of over 15,000. During August, the lower reaches of the stream are a popular viewing area for pink salmon spawning. Bald eagles gather along the river to feed on the spawned out fish carcasses.

The City derives its major domestic water supply from Indian River. A diversion dam and intake gallery facility located several miles upstream withdraws an average of 3.1 million gallons per day from the river. Peak usage generally occurs during the winter months when residents run water to prevent pipes from freezing. These cold, dry spells are also periods of minimum flow when fish populations may be under maximum stress. We recorded a stream flow of 10 cfs during one particularly dry period. Peak flows have been estimated in excess of 20,000 cfs.

As a result of this study and cooperation from the City, a stream gage will soon be installed to monitor stream flows and enable future quantitative fish habitat studies to be conducted. These studies will provide the minimum in-stream flows necessary to support various critical life events of fish and wildlife populations utilizing Indian River. Armed with this knowledge, the City and State should be able to prevent the eventual dewatering of the lower stream while still producing sufficient domestic water for the growing population.

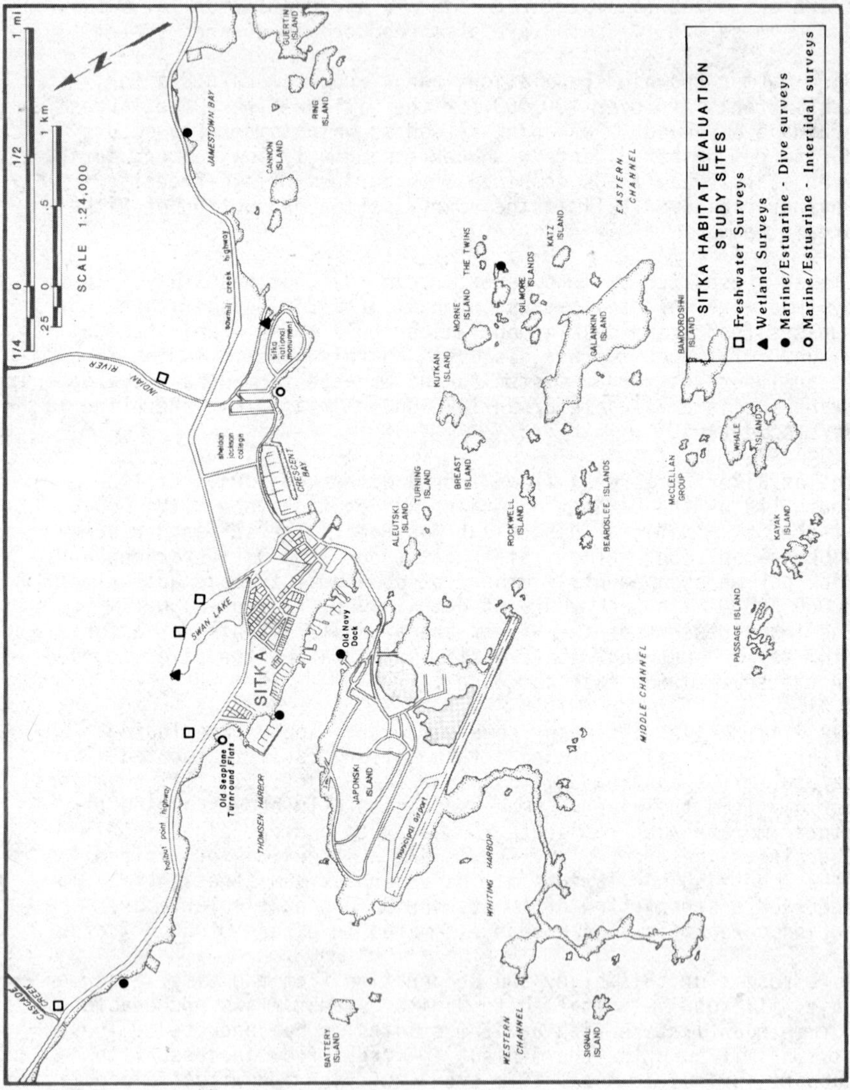

Figure 2 Inset of Study Area Map Showing Waterfront Detail

Swan Lake is a shallow 20 acre lake located within the heart of Sitka. Its two tributaries, Wrinkleneck Creek and Arrowhead Creek, both originate in muskeg bogs and flow through residential areas before entering Swan Lake. The lake outlet flows into a culvert which then discharges into Sitka Sound approximately one-quarter mile away. Previously the lake outlet flowed through an open channel and supported anadromous runs of sockeye salmon, cutthroat trout, and Dolly Varden char. An "urban renewal" project and street improvements in the 1960s enclosed the outlet stream, thus eliminating the anadromous fish. At present the lake supports resident populations of Dolly Varden and rainbow trout. Both species utilize Wrinkleneck and Arrowhead creeks for spawning and rearing.

Swan Lake also attracts waterfowl which utilize the lake for resting and feeding. Some common species include greater scaup *(Aythya marila)*, bufflehead *(Bucephala albeola)*, American coot *(Fulica americana)*, and pied-billed grebe *(Podilymbus podiceps)*. During the spring the lake provides a resting and feeding area for migrating trumpeter swans *(Olor buccinator)*.

Swan Lake is essentially the only freshwater lake within the Sitka urban area. It provides visual interest to the community, supports an annual children's fishing derby, is a recreational resource, and provides an interesting viewing area for waterfowl.

Swan Lake also has important residential land values. Recent Corps of Engineers 404/10 permits for filling marshlands at the north end of Swan Lake to create residential land, were denied because of resource agency concerns over the effects that wetlands filling could have on the biological, hydrological, and water quality values of the lake. These decisions are being appealed by the land owners.

As part of this study and the Sitka District Coastal Management Program, a management plan is being prepared that will designate Swan Lake as an Area Meriting Special Attention (AMSA). AMSAs are special management areas under the provisions of the Alaska Coastal Management Program. The boundary of the Swan Lake AMSA includes the lake, marshlands, tributaries, and a buffer area which varies in width from 15-50 feet depending upon shoreline vegetation, topography, hydrology, land ownership, and existing land use impacts. The AMSA management plan calls for a moratorium on future filling within the lake, creeks, and marshlands associated with Swan Lake, unless it can be shown that such filling will maintain or enhance the habitat, public recreation, aesthetic, and water quality values of the lake. Floating docks and pile supported structures will be allowed, consistent with local building codes. The buffer zone will be managed to maintain or enhance bank vegetation, stream channel morphology, shading and cover, and filtration capacity of the shorelands. Private land owners may be compensated where necessary through fee simple aquisition, property tax relief, land trades, or modified development rights.

The AMSA plan should help to eliminate the current piecemeal approach taken by land owners and permitting agencies in the development of the lakeshore. If approved at the local level and by the State Coastal Policy Council, the AMSA will be managed through coordinated local, State, and Federal authorities.

Wetlands Habitats. Wetlands comprise less than 10 percent of the shorelands of Sitka Sound, yet many were found to be important coastal habitats. The term "wetland" as evaluated in this study, includes four broad categories:

1. Coastal marshes are low gradient shorelands which often occur at the heads of bays and on the alluvial deposits at stream mouths. This type of wetland tends to be relatively small and isolated when compared to the bulk of the shoreline of Sitka Sound which is characteristically steep and rocky. Sizes range from 4 acres at the mouth of Indian River to 200 acres in Katlian Bay. Coastal marshes and their adjacent marine/estuarine waters were found to be high value feeding and resting areas for approximately 70 species of waterfowl and shorebirds. The nesting use of coastal marshes was found to be minimal. Typically five plant communities listed in order from the high intertidal zone landward, to the forest occur in coastal marshes: alkali grass *(Puccinellia nutkaensis)*, bluejoint grass *(Calamagrostis canadensis)*, sedge *(Carex lyngbyaei)*, beach rye-grass *(Elymus arenarius)*, and reed bent grass *(Calamagrostis nutkaensis)*. The upper limit of a coastal marsh is usually sharply delineated by the forest fringe or shrub zone. This has been taken to the upper extent of occasional saltwater influence.

2. Estuarine tidal flats are comprised of deposits of silt, sand, and gravel and often occur in conjunction with coastal marshes. They are generally vegetated with scattered patches of rockweed alga *(Fucus distichus)*, and the red alga *(Gloiopeltis furcata)*. Some areas such as the Old Seaplane Turnaround Flats have extensive low intertidal and shallow subtidal eelgrass meadows *(Zostera marina)*. Invertebrate life is rich on tidal flats and estuaries making them important feeding areas for waterfowl and shorebirds. Eelgrass beds are utilized as spawning areas for Pacific herring *(Clupea harengus pallasi)* and as rearing areas for juvenile pink salmon.

3. Freshwater marshes occur in conjunction with lakes, potholes, and low gradient stream reaches. Vegetation

in freshwater marshes includes the aquatics; pond lily *(Nuphar polysepalum)* and pondweed *(Potamogeton gramineus)* as well as mare's tail *(Hippurus vulgaris)*, horse tail *(Equisetum fluviatile)*, the grasses *(Deschampsia beringensis* and *Calamagrostis canadensis)*, and the sedge *(Carex sitchensis)*. Freshwater marshes provide food and cover for waterfowl and fish. They also export nutrient energy to adjacent rivers, lakes, and streams.

4. Muskegs are localized patches of poorly drained soils found throughout coastal forests and the alpine zone. Typical muskeg plant communities consist of sphagnum mosses *(Sphagnum* spp.*)* and sedges *(Carex* spp.*)* interspersed with lodgepole pine *(Pinus contorta)*, Alaskan yellow cedar *(Chamaecyparis nootkatensis)*, crowberry *(Empetrum nigrum)*, labrador tea *(Ledum palustre)*, skunk cabbage *(Lysichiton americanum)*, and bracken fern *(Pteridium aquilinum)*. Muskegs comprise the largest wetland type within the Sitka area. Bear, deer, and several bird species utilize muskeg areas on a seasonal basis. Their primary importance may be as hydrologic reserves. The peat deposits found in muskeg bogs function as large sponges capable of soaking up peak rainfall to be released into adjacent streams during periods of low flow when fish survival is critical.

Of the four wetland types, we found the coastal marshes and estuarine tidal flats to be of the highest habitat value. They provide feeding and resting areas for a large proportion of the migrant and resident waterfowl and shorebirds of Sitka Sound. The marine invertebrates associated with estuarine tidal flats are an important food source for waterfowl, fish, and shellfish. The plant species present in marshes provide high protein food for Sitka black-tailed deer, brown bear, ducks, and geese. Spawning salmon in streams and estuaries contiguous to marshes provide food for brown bear and bald eagles. Eelgrass beds provide spawning substrate for herring and rearing habitat for juvenile salmon.

Port Krestof is a typical coastal marsh/estuarine tidal flat complex located on Kruzof Island approximately seven miles west of Sitka (Figures 3, 4). The area supports relatively high numbers of waterfowl and shorebirds during spring and fall migrations. Dominant bird species observed include Canada goose *(Branta canadensis)*, mallard *(Anas platyrhynchos)*, bufflehead, least sandpiper *(Calidris minutilla)*, western sandpiper *(Calidris mauri)*, and dunlin *(Calidris minutilla)*. Dungeness crab *(Cancer magister)*, butter clams *(Saxidomus giganteus)*, and littleneck clams *(Prothaca staminea)* are sought after by local residents. The grass flats are utilized for spring feeding by Kruzof Island's large population of Sitka black-tailed deer and brown bear.

Figure 3 Map showing Port Krestof Coastal Marsh/Estuarine Tidal Flat Complex

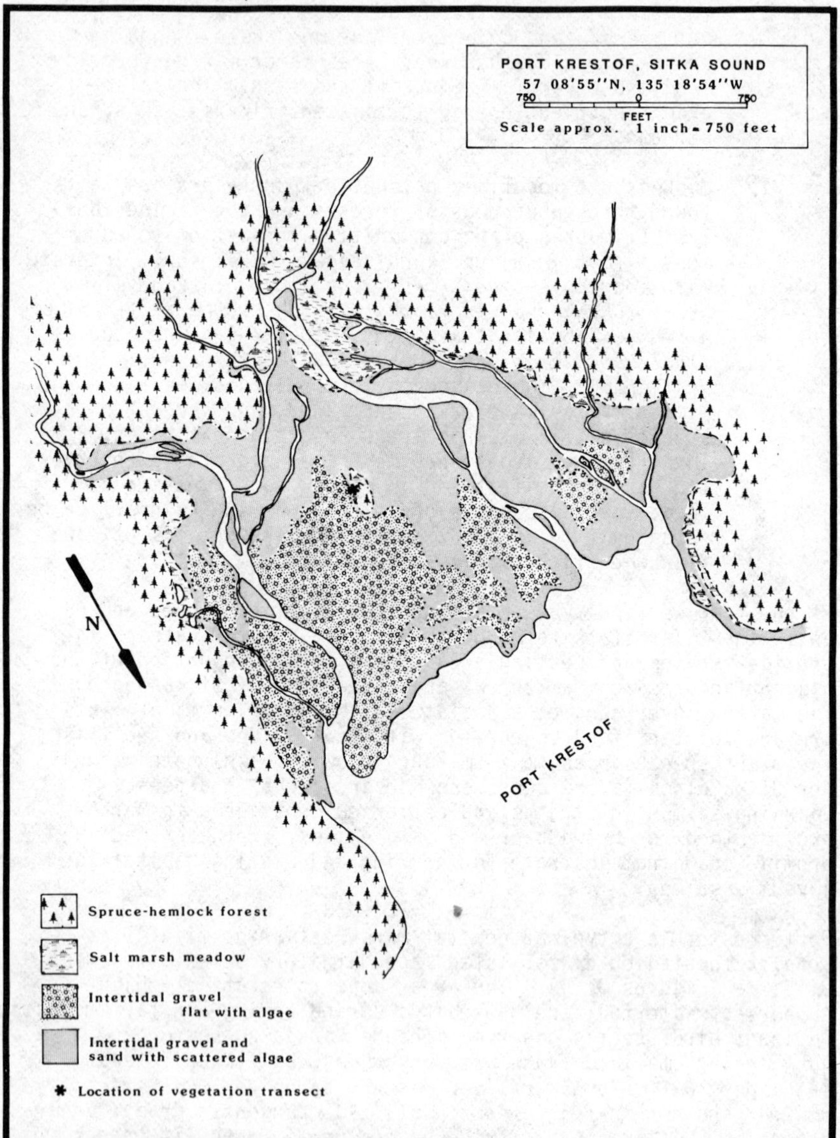

Figure 4 Plant Transect Profile for Port Krestof Coastal Marsh

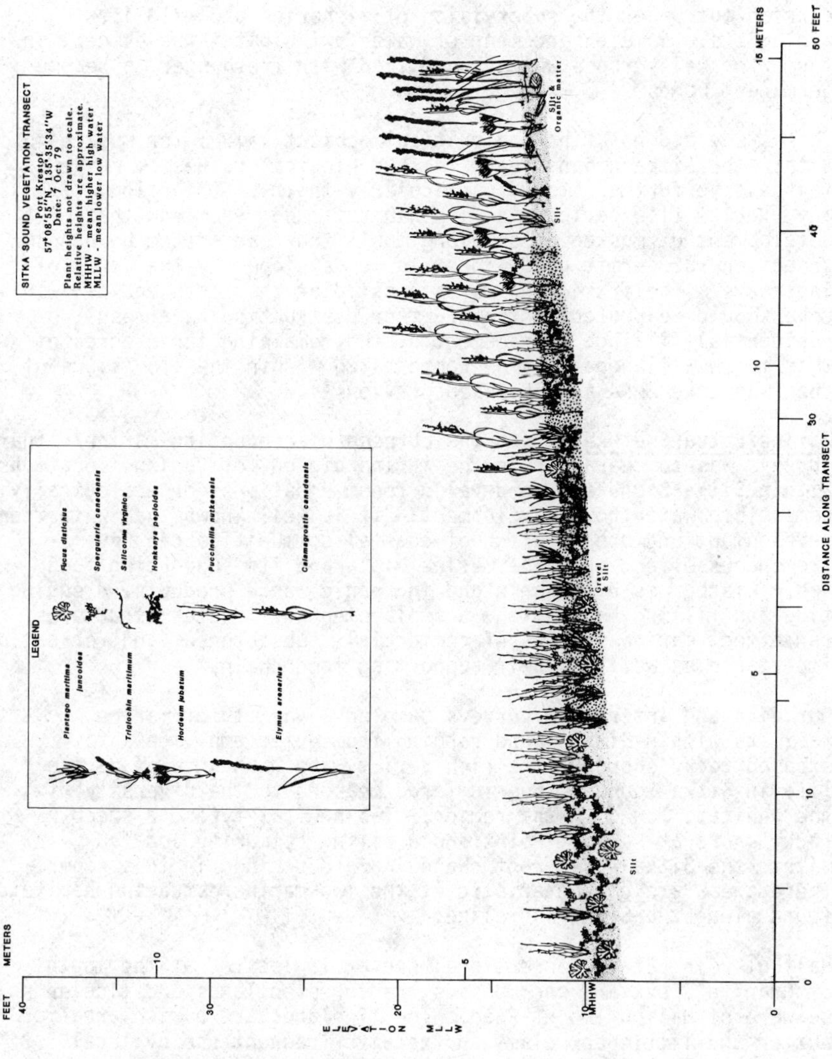

Our management recommendations for coastal marshes and estuarine tidal flats place a high value on protecting these habitats from destruction, pollution, excessive noise, and disturbance. We recommend that these areas be given a low priority for siting industrial operations, log storage, and permanent settlements. We noted the certain activities such as selective gravel removal, particularly in the high marsh areas, may be beneficial when carried out under the supervision of fisheries and wildlife biologists. Some enhancement of waterfowl habitat was evident in former gravel scrapes which had filled with freshwater to become permanent ponds.

This study did not find any critical habitat values for muskegs within the Sitka urban area. The hydrological values of muskegs may deserve further study, particularly their contribution, if any, to low flow periods in fish habitat. We recommend that development of muskeg areas along the Sitka road system be handled under general permit guidelines, to be developed by the Corps of Engineers. The freshwater marsh located at the north end of Swan Lake should be protected from further destruction as a result of residential filling. Management of the remaining three acres of this marsh will hopefully be coordinated within the provisions of the Swan Lake AMSA as discussed previously.

Marine/Estuarine Habitats. The purpose of conducting marine/estuarine surveys was to characterize the marine biology of various locations within Sitka Sound and to develop recommendations for ecologically compatible waterfront development. It is well known that waterfront development and urbanization of coastal communities can have a tremendous effect upon the marine biota and its supporting food web. Discharges of organic and inorganic waste products, dredging, tideland filling, siltation, and placement of inwater structures can effect marine species of commercial, subsistence, and recreational importance as well as their supporting food chain.

Our dive and intertidal surveys sampled a variety of marine habitats within Sitka Sound ranging from quiet embayments to exposed rocky shores. The rich variety and abundance of marine life in Sitka Sound is due in large measure to the diversity of the habitats found in the region. Two habitat types, a steep rocky shore at Halibut Point and a piling community located within the Sitka waterfront channel are described in this paper. These areas are characteristic of the two marine/estuarine habitats found along the Sitka shoreline.

Halibut Point State Recreational Center is located at the mouth of Granite Creek and encompasses the forested lands and tidelands seaward of Halibut Point Road. The tidelands are a rich area for butter and littleneck clams and receive frequent use by local clammers. The tidelands are also an important feeding and resting area for waterfowl. During the fall, harlequin duck *(Histrionicus histrionicus)* were attracted to this area in fairly high numbers at low tide. Surf scoters *(Melanitta perspicillata)*, oldsquaw *(Clangula hyemalis)*, common goldeneye *(Bucephala clangula)*, and bald eagles *(Haliaeetus leucocephalus)* are also commonly seen feeding in this area.

An intertidal survey was completed for the gravel beach and a dive survey was conducted at the small bedrock island off the southwest shore of Halibut Point (Figure 5). Apparently this site is also popular for beginning scuba divers as it affords easy access from the roadway. The flora and fauna at this location are representative of a moderately high energy coast. The intertidal zone is largely composed of barnacles, limpets, and rockweed. The upper subtidal zone is dominated by the five rayed seastar, (*Pisaster ochraceus)* and the green alga, *(Codium fragile)*. Deeper waters at about -25 ft MHHW, have a greater diversity and abundance of marine life. This area is also a transition from steep bedrock to a rubble slope. Schools of dusky rockfish *(Sebastes ciliatus)* frequent this area. Other rockfish include China *(S. nebulosus)*, copper *(S. caurinus)*, quillback *(S. maliger)*, and black *(S. melanops)*. Invertebrate life is chiefly composed of red sea urchin *(Strongylocentrotus franciscanus)*, sea cucumbers, and sea stars with a variety of limpets and snails. A few Pinto abalone *(Haliotis kamtschatkana)* are present. This is a known herring spawning area.

Management recommendations for this area recognize the primary needs to protect the rich intertidal zone for subsistence taking of hardshell clams and the subtidal zone for herring spawning. Activities such as gravel dredging, intertidal filling, waste discharges that degrade water quality within the clambeds or nearshore waters, and vehicle traffic within the intertidal zone, are all considered adverse to this marine habitat. Permissible uses include clam digging, scuba diving , abalone picking, and fishing. All harvest of marine life is subject to restrictions set by the State Board of Fisheries. Pile supported structures are permissible providing that they minimize disturbance of the tidelands and allow for the free circulation of water and movement of marine life.

The Old Navy Dock is a piling supported structure located in the Sitka Waterfront Channel (Figure 6). The dock was built during World War II and has since fallen into disrepair. It is estimated that the piling examined on this dive were approximately forty years old. Marine growth on the piling was remarkably rich and luxuriant reflecting both longevity and good current circulation. The latter can be inferred by the large percentage of suspension feeders including feather duster worms, sea squirts, mussels *(Mytilis edulis)*, bryozoans, and anemones which feed upon suspended food particles such as plankton carried along by water currents. Copper rockfish and kelp greenling *(Hexagrammos decagrammus)* are attracted to the piling as are herring and tubesnouts *(Aulorhynchus flavidus)*. An octopus *(Octopus dofleini)* was found living in a pipe section near the base.

Figure 5 Dive Transect Profile for Halibut Point

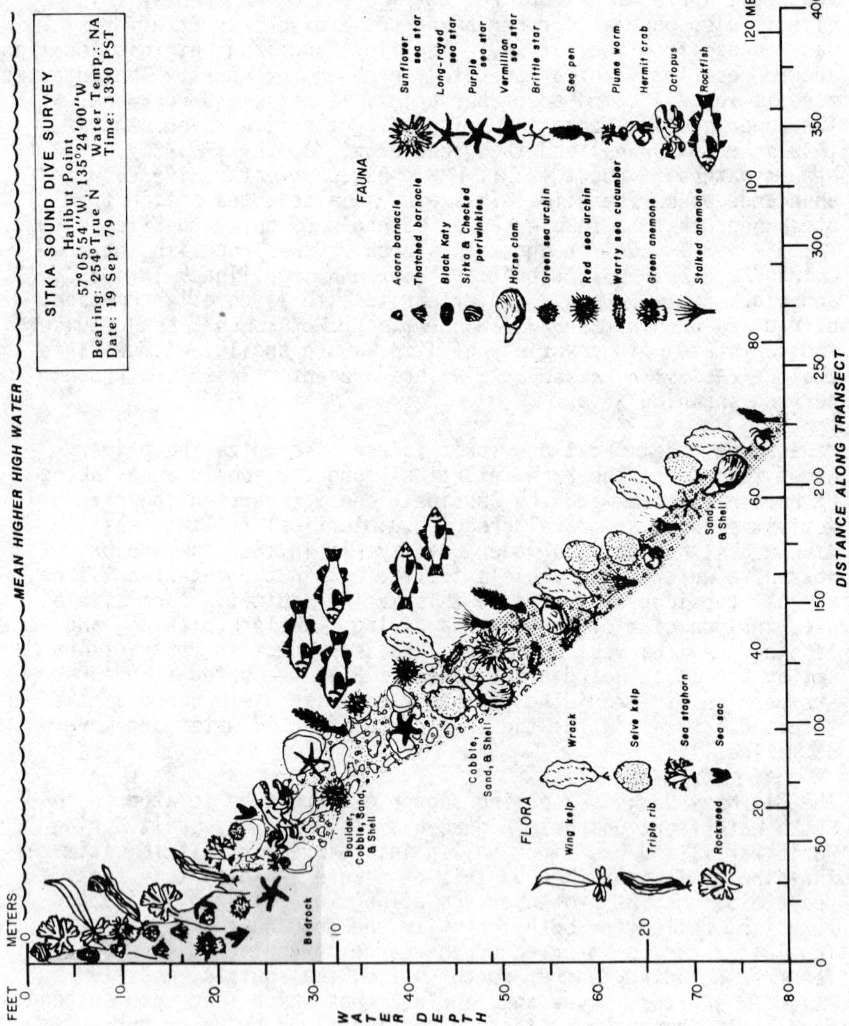

Figure 6 Dive Transect Profile for Old Navy Dock

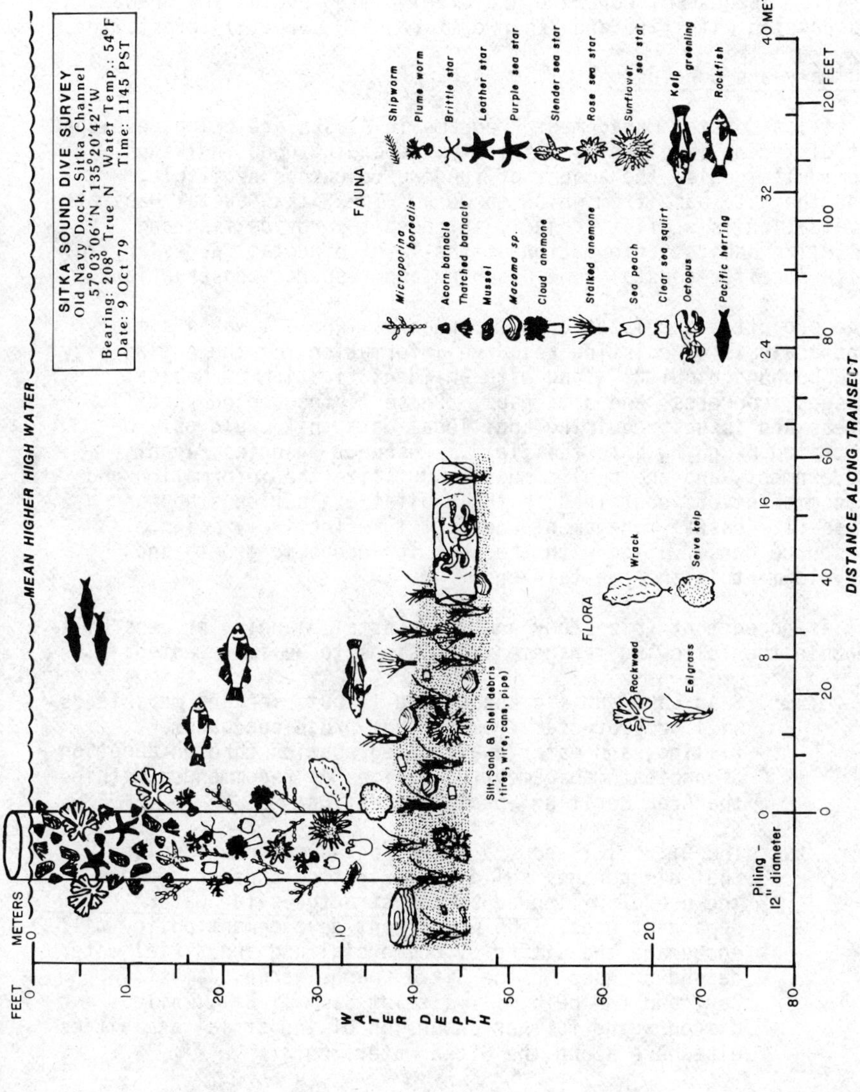

The Old Navy Dock provides a good example of the habitat benefits of piling supported structures over fill in terms of maintaining and perhaps enhancing waterfront marine habitats. Pilings provide increased surface area for attachment of marine life, provide cover for juvenile fish such as salmon fry and herring, attract rockfish for recreational fishing, and have the least impact upon shoreline processes such as nearshore currents and littoral drift. The use of concrete and steel piling avoids the drawbacks associated with fire and "shipworm" *(Bankia setacea)* infestation.

Summary and Conclusions

District Coastal Management Programs in Alaska are being developed at different levels of specificity, depending upon individual community goals, the amount of funding resources available, and the data base from which to work. The Sitka Coastal Habitat Evaluation is a pilot project, intended to provide fish and wildlife habitat information at the level of detail necessary for site specific coastal zone planning and resource conservation.

The project was completed in two phases. Phase I was a summary and analysis of existing resource information for the entire City and Borough of Sitka along with an identification of habitat issues, concerns, and data gaps. Phase II focused on specific areas and issues requiring additional data that could only be obtained by going into the field. Resource agencies, local government, and the public must now utilize the information and recommendations contained in the habitat evaluation study and Phase II Coastal Management Document to effectively balance resource conservation with the need for economic growth and development in the coastal zone.

It is hoped that this study and the coastal planning process will enable the following management decisions to be implemented:

1. Swan Lake and its associated tributaries and marshlands will be protected from further urban encroachment, filling, and water quality degradation through adoption of habitat management guidelines as recommended within the Area Meriting Special Attention;

2. The City will adopt a waterfront development policy that discourages solid fills in tidelands and favors the use of piling supported structures for water dependent uses. The waterfront development policy will encourage the siting of commercial and industrial water dependent uses in the Sitka Channel area, Jamestown Bay, and the pulp mill site at Sawmill Bay, while discouraging further expansion of industrial activities elsewhere along the Sitka waterfront;

3. Future planning, platting, and land use decisions will provide for vegetative buffer zones and public access along all aquatic habitats including streams, lakes, and wetlands. Necessary instream work will be done to maintain or enhance fish habitat through consultation with fisheries biologists; and

4. The City will work with the Corps of Engineers and the resource agencies to develop general permits and guidelines for development activities occurring in wetlands. The City will protect wetlands habitat occurring at Starrigavan Bay, Old Seaplane Turnaround Flats, Swan Lake, and Indian River.

Literature Cited

1. U.S. Department of Commerce, NOAA, Office of Coastal Zone Management. Final environmental impact statement and proposed coastal management program for the State of Alaska. 1979. 578 pp.

2. Alaska Administrative Code, Chapters 80 and 85.

3. Ibid, Chapter 85, Sections 50-60.

4. City and Borough of Sitka Planning Office. Sitka Coastal Management Program - Phase I Report. October 1979. 150 pp. 6 maps.

5. U.S. Department of Commerce, NOAA, National Ocean Survey. Tide tables 1979 and 1980, west coast of North and South America. Issued 1978 and 1979.

CHESAPEAKE BAY DEVELOPMENT PRESSURES: RAMS DATA BASE ANALYSIS

Russell C. Eberhart*
Thomas J. Dolan**

INTRODUCTION

It is now recognized that the environmental and social effects of regulatory actions are important issues which must be considered when decisions are made. Sometimes, the potential effects of a single decision can be relatively large, as is the case when dredging a major navigation channel and disposing of the spoil. At other times, the effects can be relatively minor, as is the case when a private property owner builds a small pier. What is becoming increasingly evident, however, is that these decisions, taken together, result in what has come to be called "cumulative impact".

The Research and Management Shoreline (RAMS) data base is the only single information base from which an assessment of the cumulative effects of Chesapeake Bay development activities can be made. The RAMS data base contains information relative to physical alterations of the shoreline and near-shore areas of Chesapeake Bay for which permits have been granted by the U. S. Army Corps of Engineers Baltimore and Norfolk Districts. The RAMS data base is designed to provide information to persons making recommendations or decisions at the local, state or federal level relative to subjects such as coastal zone management, fish and wildlife management, and land use planning.

Permits applied for in January 1973 are the earliest in the data base; those permits granted in June 1979 are currently the most recent. Over 8,000 permits, representing approximately 38,000 individual structures (such as bulkheads) or alterations (such as dredging projects), are included. Although the computer is physically located at the Applied Physics Laboratory of The Johns Hopkins University (APL/JHU) in Howard County, Maryland, the RAMS Data Base may be queried, modified or updated via data processing terminal and telephone links from any location.

With support from the State of Maryland and the Commonwealth of Virginia, a preliminary analysis of the information in the RAMS data base has been completed. In the process of performing the analysis, significant deficiencies in the data were discovered. This paper discusses the structure of the RAMS data base and the framework within which it operates. Also discussed are the data deficiencies identified and the results of the analysis performed.

* The Johns Hopkins University Applied Physics Laboratory, Laurel, Maryland

** Coastal Resources Division, Maryland Department of Natural Resources, Annapolis, Maryland

DATA BASE STRUCTURE

Information Categories

Categories of information for each granted permit in the RAMS data base include:

Date permit applied for;
Date permit granted;
Application number;
Location of alteration by
    State
    County
    Waterway
    Latitude and longitude;
Type of ownership;
Purpose/use of alteration;
Name of owner;
Type and size of each structure.

In addition, shoreline and marsh types are stored for some permits.

Information Coding

The information for each granted permit is coded into a record structure consisting of three main sections. The first section contains all of the information except that relative to structure type, structure extent, and state permit number. The second section consists of the coding for the structure type and up to three measurements of structure extent. (Examples of structure extent are length in feet, maxiumum extent channelward in feet, and volume in cubic yards.) The third section consists of the state-assigned permit number.

Coding of the information for the RAMS data base has been done so as to require reduced storage capacity while, at the same time, making the information available in useful and logical forms. Following are examples of information coding.

Each county (and independent city) has been assigned a three-digit code. The numerical value of this code corresponds to the number of miles of shoreline in the county or independent city, thereby facilitating calculations of quantities such as structure densities.

The waterway classification scheme for Maryland has been developed in cooperation with the Maryland Department of Natural Resources. The first character of the Maryland waterway code corresponds to the sub-basin level of the "STORET" waterway classification system developed by the U. S. Environmental Protection Agency. The second character corresponds to the sub-sub-basin level of STORET. The last three characters correspond to the unique code assigned to each waterway (down to and including unnamed tributaries in many instances) by the Water Resources Administration of the Maryland Department of Natural Resources. This coding system for Maryland, then, not only identifies each waterway uniquely, but identifies drainage basins as well.

The waterway classification scheme for Virginia has been developed by the Corps of Engineers, Norfolk District Office. The first two or three characters of the code for each Virginia waterway generally correspond to a major tributary, such as the James River. The last two or three characters generally correspond to sub-tributaries, such as creeks, flowing into the major tributaries.

Latitude and longitude are coded from U.S.G.S. 7.5-minute series quadrangles. Each is coded in degrees, minutes, and tenths of minutes. Any fraction of a tenth of a minute is truncated. (The U.S.G.S. 7.5-minute topographic quadrangle which includes the location of the permit is also coded.)

In determining the fastland code, the general configuration of the land is considered. Elevation anomalies are excluded, unless the alteration site is located at one of these anomalies. The type of fastland is determined by the elevation of the shoreline 400 feet inland at mean high water (MHW). In tidewater regions, the MHW level is defined as zero elevation, and the elevation of the surrounding fastland relative to the shoreline can be read directly from the U.S.G.S. topographic quadrangle. In non-tidal waters, which include those areas above the "fall line" of a tidal river, the MHW elevation is greater than zero, and the relative elevations of the water and land are considered. For example, if the MHW elevation of the shoreline is 80 feet above mean sea level (MSL) and the average elevation of the land 400 feet inland is 125 feet above MSL, the difference is 45 feet, resulting in the fastland being characterized as a "moderately high shore", as illustrated in Table I.

Three types of marshes (fringe, embayed and extensive) may be selected as the shore zone type. In the absence of a marsh, the shore zone may be classified as "beach" or "artificial". Selection of the term "beach" implies the existence of a natural shoreline. Selection of the term "artificial" implies a shoreline defined by a man-made structure, such as a bulkhead.

The near-shore category reflects the expanse of water between the MHW level at the alteration and the MLW depth of six feet. The measurement is defined as the shortest distance between the alteration and a six-foot depth, regardless of the angle which the line of measurement makes with the shoreline (as long as it remains on the water). Near shore classifications are listed in Table II.

The structure (or alteration) type and up to three measurements of extent for each structure or alteration granted by a permit are coded. Unique codes are available for new structures, and for those being repaired. Approximately 40 types of structures and alterations are currently coded in the RAMS data base. Table III lists the structure codes and extents for five of the more frequently occurring structures and alterations.

The foregoing examples illustrate the information coding for the RAMS data base. Complete coding instructions are contained in Reference 1.

Table I
Fastland Coding Scheme

Elevation of Shoreline 400 Feet Inland	Fastland Type	Code
<20 feet	Low Shore	1
20 - 39 feet	Moderately Low Shore	2
40 - 59 feet	Moderately High Shore	3
>59 feet	High Shore	4
Dune	Dune	5

Table II
Near Shore Coding Scheme

Distance Between MHW and MLW Depth of Six Feet	Code
<300 feet	1
300 - 600 feet	2
600 - 1200 feet	3
>1200 feet	4

Table III
Structure Type and Extent Coding Scheme

Structure Type	New Structure Code	Repair Code	Extent 1	Extent 2	Extent 3
Bulkhead	BULK	RPBU	Total lenght of new structure (feet)	Maximum extent Channel-ward (feet)	Number (units)
Dredging	DRED	RPDR	Volume (cubic yards)	Length (feet)	Width (feet)
Fill	FILL	RPFI	Volume (cubic yards)	Area (tenths of acres)	-null-
Pier	PIER	RPPR	Total length new structure (feet)	Maximum extent Channel-ward (feet)	Number (units)
Pile	PILE	RPPE	-null-	Maximum extent Channel-ward (feet)	Number (units)

DATA BASE FRAMEWORK

Query Capabilities

The RAMS data base operates within the framework of IBM's Generalized Information System (GIS). Basic query capabilities available within GIS include the ability to obtain subsets of information based upon various search criteria. A wide variety of retrieval problems can be easily structured and transformed into GIS procedures. Search criteria may be compounded, requiring that several conditions be satisfied before further action is taken. A typical query might consist of a request for a listing of the permit number, ownership category, and applicant's name for all permits granted between January 1, 1974, and December 31, 1975, for bulkheads over 150 feet in length on the Severn River.

In addition to the basic query capability, a formal report capability exists within GIS which provides efficient report processing and preparation. This formal report capability is typically used for the retrieval and formatting of subsets of information which are required on a regular basis. Approximately 15 report routines are currently stored on-line for use with the RAMS data base, including routines to summarize permit activity according to waterway, county, and structure type.

Graphics Capabilities

The shoreline of the entire Chesapeake Bay area has been digitized and stored in the APL/JHU computer. This was accomplished by digitizing two charts: Chesapeake Bay, Northern Part (C. & G.S. Chart 77); and Chesapeake Bay Southern Part (C. & G.S. Chart 78). The two charts were divided into 67 15-minute squares for the purpose of shoreline digitization.

Any of the information contained in the data base, such as structure type or purpose/use, can be superimposed on a map printout. A typical plot could consist of a map of the entire Bay watershed with the location of each dredging permit involving more than ten thousand cubic yards depicted by the letter "D".

A wide variety of map scales is available. For instance, the entire Bay watershed can be reduced to fit on an eight-by-ten-inch plot; or, one 15-minute square can be displayed on a plot 30 inches square. Any combination of 15-minute squares can be selected for display.

Hardware Configuration

IBM Model 3033 multiprocessors store the data on Model 3350 disk units and make it available to program investigators and user agencies. The data base may be queried, modified or updated using the Time Sharing Option from within the Applied Physics Laboratory via several types of data processing equipment, including IBM Model 3270 and Hewlett Packard Model 2641 CRT terminals, and IBM Model 2741 communications terminals.

From outside the Laboratory, the data base is generally accessed via a Texas Instruments Silent 700 Model 745 terminal.

Punched cards are generally used for entering information regarding newly granted permits. Answers to queries which result in relatively large volumes of information are usually printed offline on a high-speed line printer. The configuration of the data processing hardware associated with the RAMS system is illustrated in Figure 1.

System Cost

The cost of obtaining an answer to any particular query varies according to the amount of processing required. Typical inquires, such as a request to list all permits for a given structure type on a given waterway, generally cost less than fifteen dollars.

The graphical (map) displays are produced by the APL/JHU computer center with a turnaround time of approximately one day. The cost associated with map printouts is varible, depending upon the area and information required. Typical printouts, such as an illustration of the locations of all permits for structures of a given type within a given area, generally cost less than thirty dollars.

DATA ANALYSIS

Analysis Process

The RAMS data base contains a relatively large amount of information in a variety of categories. Any mandate to carry out an analysis of the information, and summarize and report results and conclusions, must therefore be carried out carefully, and with significant interactions with the program sponsor and the Corps of Engineers District Offices where the data originated.

With support from the Coastal Resources Division (CRD), Tidewater Administration, Maryland Department of Natural Resources, the Applied Physics Laboratory of The Johns Hopkins University (APL/JHU) was asked to perform such an analysis. In the process of deciding how best to analyze the RAMS data base to portray the cumulative patterns, rates and trends of development activities in the Bay area, a preliminary analysis was performed, and the resulting data summary was reviewed with CRD and the Corps of Engineers District Offices in Baltimore and Norfolk.

The preliminary analysis produced interesting and useful results, which are summarized in the "Analysis Results" section of this paper. The interactive process which occurred during and following the preliminary analysis resulted in interesting and unexpected results pertaining to the completeness of the data in the RAMS system.

Prior to the preliminary analysis recently completed, the only time the data in RAMS had been reviewed and examined in aggregate was in

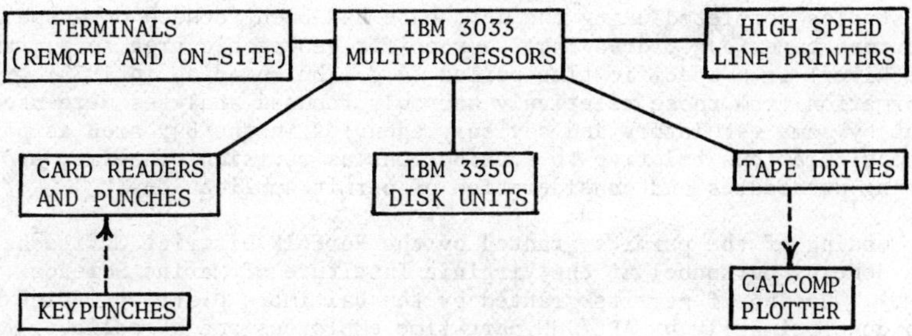

Figure 1. Hardware associated with the Rams data base

1974 as part of the effort which resulted in the publication of "Pressures on the Edges of Chesapeake Bay - 1973" (Ref. 2). Since that time, the studies completed using the data base had been focused on a specific structure type (e.g., dredging), a specific geographic area (e.g., the West River) or a specific time period (e.g., bulkheading in 1975). Information from these relatively narrowly focused analyses were used as input by some regulatory and advisory agencies in the Bay area as part of considerations relative to such actions as revision of permit processing procedures and consideration of permit applications.

Coding of the permits granted by the Norfolk District Office has been done by personnel at the Virginia Institute of Marine Science (VIMS). Coding of permits granted by the Baltimore District Office has been done primarily by APL/JHU part-time employees and by college students employed by CRD. To ascertain that essentially all of the permits granted by the Baltimore Office were being entered into RAMS, the total number of granted permits entered each year was compared with the total number listed as issued (on a monthly basis) by the Baltimore District Office. The discrepancy was typically three to five percent, but this was considered acceptable because of problems such as missing files,etc. It was assumed that a similar comparison was being done for the Norfolk District Office.

At a meeting hosted by the Norfolk District personnel to review the preliminary analysis, it was learned that the assumption was not valid. It appears that a significant number of Norfolk District permits have not been coded or entered into the RAMS system. Although at the time of the writing of this paper the magnitude of the deficiency is not known precisely, it would appear that somewhere on the order of 10-15 percent of the permits were not coded in 1973-1975, and that on the order of 20-30 percent were not coded in 1976-1979.

Upon learning of the apparent data deficiencies for Norfolk District data, it was decided to verify the granted permit totals for the Baltimore District by comparing the number of permits in RAMS for each year with the yearly totals submitted by the Baltimore District to the Office of the Chief of Engineers (OCE). Again, the result was unexpected, in that it appears, from the OCE report figures, that a significant number of Baltimore District permits have not been coded or entered into the RAMS system. Again, at the time of the writing of this paper the nature of the deficiency, or in this case, whether or not a deficiency actually exists, is unknown. If the yearly totals contained in the OCE reports are correct, rather than those obtained from the monthly granted permit listings, up to 30 percent of the granted permits may not have been coded for the entire period (1973-1979).

Personnel at APL/JHU and CRD, together with staff members of both Corps District Offices and the Corps' Institute of Water Resources are currently attempting to determine the magnitude of the problem, and once determined, explore ways to resolve it.

It must be emphasized that the authors are in no way attempting to imply that any person or any organization is at fault with respect to the apparent data deficiency problems. Rather. the situation has been described in an attempt to reflect the context within which the summary of the preliminary analysis which follows must be considered.

Preliminary Analysis Results

Following is a summary of the preliminary analysis of all granted permit information contained in the RAMS data base as of June 1980. The complete results of the preliminary analysis appear in a report entitled "Pressures on the Edges of Chesapeake Bay: 1973-1979" (Ref. 3).

In addition to the probable data deficiencies noted in the previous "Analysis Process" section, it is appropriate to note the inconsistencies which appear in the data. Not everything "adds up", even internally. For example, the total number of permits in RAMS at the time of the analysis was 8187; but if the total number of permits is calculated by summing the totals by state (see Table IV), the figure 8145 results. This is because, for some reason, 42 permit records do not have information in the "state" data field.

Other inconsistencies and errors exist. A limited effort was expended to resolve the more obvious inconsistencies, and a number of errors were corrected in the process of performing the analysis. The limited level of effort available, however, did not permit thorough and complete data verification.

It is hoped, then, that the individual numbers presented in this paper will not be considered in detail. The intent of the analysis is to portray the general patterns, rates and trends of development in the Chesapeake Bay region as reflected by the permits granted by the Baltimore and Norfolk District Offices.

Table IV consists of a tally of permits by state. The vast majority of permit activity occurred in Maryland and Virginia, as would be expected. Table V contains a listing by year, and by the Corps District Office which granted the permit, for permit activity from 1973 through 1978. (Permit information is currently available only through June 1979 and was not included in Table V. Available 1979 data are, however, included in cumulative tallies which follow for such categories as counties and structure types.) It should be noted that permits for the Potomac River shoreline in Virginia are issued by the Baltimore District Office.

Table IV
Tally of Permits by State

State	Permits
Maryland	5,744
Virginia	2,168
Pennsylvania	173
District of Columbia	29
West Virginia	18
Delaware	13

Table V
Tally of Permits by Year and by Corps District

	1973*	1974	1975	1976	1977	1978	Total
Baltimore District	442	1,107	850	1,115	1,223	1,097	5,834
Norfolk District	254	344	302	287	329	201	1,717
Total	696*	1,451	1,152	1,402	1,552	1,298	7,551

* Permits granted in 1973 resulting from applications made after January 1, 1973.

One potentially useful way to examine the pattern of development in the Chesapeake Bay region is to look at granted permit activity by location. The smallest political jurisdiction about which information is stored in the RAMS data base is the county or independent city in which the structure or alteration approved by the permit is located. Tables VI and VII present a tally of granted permits by county or independent city, Table VI for Maryland and Table VII for Virginia.

One way to assess the intensity of the development in an area is to examine the density of new projects approved; one method of examining the density is to calculate the number of permits approved for each mile of shoreline. Tables VI and VII, therefore, also present one measure of development intensity: the number of permits granted for each mile of shoreline in each county and independent city.

The types and numbers of structures permitted in the Chesapeake Bay region from 1973-1979 included in the RAMS data base are listed in Table VIII. It should be noted that each granted permit may involve more than one structure type; furthermore, more than one structure of a given type may be approved by a single permit. The information in Table VIII, then, includes both the number of permits granted for a given structure type, and the total number of structures of a given type authorized. For the purposes of illustration, all repair operations, regardless of structure type, have been combined. All temporary structures have also been combined.

The total length of bulkheading permitted and included in RAMS for Maryland and Virginia during the time period is illustrated in Table IX. The total amount of bulkheading approved is about 83 miles, with about 59 miles of bulkheading in Maryland and 24 miles in Virginia.

The situation with respect to dredging, channelization, fill and spoils disposal is a complex one for a number of reasons. First, the interconnections amoung the categories can be confusing. For example, sometimes the byproduct of a dredging operation is considered to be "fill"; at other times it is "spoils disposal". Second, dredging can occur either as a "new" activity (i.e., the location in question has never been dredged before) or as "maintenance" dredging (i.e., the permitted activity is for the re-dredging of a previously dredged location). Third, the byproduct of channelization can be classified in

Table VI
Permit Activity by County and Independent City:
Rams Data Base Totals for Maryland

County or Independent City	No. of Permits	No. of Permits per Mile of Shoreline
Anne Arundel	1,330	3.67
Baltimore County	577	3.22
Baltimore City	78	2.29
Calvert	350	2.17
Caroline	19	0.12
Cecil	173	0.79
Charles	107	0.66
Dorchoster	353	0.55
Harford	144	0.72
Kent	198	0.74
Prince Georges	51	1.06
Queen Annes	426	1.27
Somerset	180	0.30
St. Marys	535	1.65
Talbot	748	1.42
Wicomico	71	0.45
Worcester	344	0.98

Table VII
Permit Activity by County and Independent City:
RAMS Data Base Totals for Virginia

County or Independent City	No. of Permits	No. of Permits per Mile of Shoreline
Accomack	102	0.41
Alexandria	4	1.33
Arlington	2	0.18
Caroline	5	0.20
Charles City	3	0.03
Chesapeake	50	2.08
Chesterfield	13	0.15
Essex	41	0.55
Fairfax	64	1.94
Gloucester	90	0.24
Hampton (and Elizabeth City)	34	0.51
Henrico	2	0.06
Isle of Wight	12	0.40
James City	29	0.19
King George	8	0.14
King and Queen	3	0.21
King William	5	2.50
Lancaster	217	1.30
Mathews	90	0.42
Middlesex	148	1.10
New Kent	6	0.75
Newport News	41	0.89
Norfolk	168	1.22
Northampton	30	0.11
Northumberland	245	0.82
Portsmouth	11	0.31
Prince George	14	0.22
Prince William	16	0.70
Richmond	30	0.40
Stafford	12	0.24
Suffolk	17	0.39
Surry	9	0.22
Virginia Beach	333	1.63
Westmoreland	153	1.44
York	107	0.56

Table VIII
Permit Activity by Structure/Alteration Type

Structure/Alteration Type	Total Number of Permits Granted	Total Number of Structures/Alterations Approved
Pile	2,988	20,686
Pier	3,933	6,040
Bulkhead	1,869	1,962
Fill	1,848	1,848
Buoy	433	1,810
Jetty or Groin	472	1,698
Riprap	822	1,005
Dredging (New)	888	888
Spoils Disposal	610	610
Dolphin	68	406
Building	236	251
Boat Ramp	206	210
Discharge Pipe	105	169
Maintenance Dredging	115	115
Aerial Crossing	61	109
Bridge	89	89
Pipeline	65	75
Channelization	71	71
Submerged Cable	53	69
Crab Impoundment	38	38
Marine Railroad	25	34
Intake Pipe	20	32
Dam	17	17
Fence	14	14
Tunnel	5	10
Intake Structure	5	5
Artificial Reef	3	3
Wave Gauge	1	1
Duck Blind	1	1
New Structure Subtotals	14,901	38,266
Repair Activities	628	672
Temporary Structures	89	99
TOTALS	15,618	39,037

various ways (e.g., fill or spoils disposal).

Table IX
Permit Activity for Bulkheads in
Maryland and Virginia: 1973-1979

State	Total Number of Permits for Bulkheading Approved	Total Amount of Bulkheading Approved (in feet)
Maryland	1,413	313,577
Virginia	519	124,502

Despite the complexities involved, it is important to understand the status of dredging, channelization, fill and spoils disposal in the Chesapeake Bay area because of the relatively major effects these projects have. These effects range from direct environmental effects to indirect socioeconomic effects which help shape the economy of the entire region.

Some dredging and spoils disposal projects do not require Corps of Engineers permits, but rather are carried out by the Corps for itself or another Federal agency such as the Navy. Because of the significant magnitude of these unpermitted projects, information on them has been coded and incorporated in the RAMS data base. The table which follows, then, includes data on these projects as well as those approved by the standard permit procedure.

Table X summarizes project activity contained in the RAMS data base for new dredging and maintenance dredging in the Chesapeake Bay area during the time period 1973-1979. As can be seen, a total of about 56 million cubic yards of dredging occurred as a result of over 1,100 new and maintenance dredging operations. New dredging accounted for about 45 percent of the activity, maintenance dredging for about 55 percent.

An examination of the information on a project-by-project basis reveals that a vast majority of the material dredged is accounted for by a relatively small percentage of the new and maintenance dredging projects. In fact, of the 1,129 projects reflected in Table X, only 253, or about 22 percent, involve more than 10,000 cubic yards of material; these 253 projects, however, account for about 54.6 million cubic yards of material, or over 97.6 percent of the total volume dredged.

SUMMARY AND CONCLUSIONS

In summary, the preliminary analysis of the over 8,000 permits in the RAMS data base indicated that:

1) Approximately 2,000 of the granted permits involve new bulkheading. Approximately 84 miles of shoreline have been bulkheaded in the Chesapeake Bay as a result.

Table X
Project Activity for New and Maintenance Dredging: 1973-1979

State	New Dredging		Maintenance Dredging		Totals	
	Projects (number)	Volume (10^6 yd^3)	Projects (number)	Volume (10^6 yd^3)	Projects (number)	Volume (10^6 yd^3)
Maryland	575	7.4	81	6.6	656	14.0
Virginia	311	18.1	119	23.6	430	41.7
Other States	37	0.1	6	0.1	43	0.2
TOTALS	923	25.6	206	30.3	1129	55.9

2) About 56 million cubic-yards of dredging has occurred; about one-half represents new dredging projects, the other one-half is maintenance dredging.

3) In the period 1973-1979, an average of one new structure or alteration has been permitted for each 1,100 feet of shoreline over the entire 8,000-mile Chesapeake Bay shoreline.

The inclusion of data for permits which have apparently been granted but not entered into the RAMS data base could significantly alter the figures cited above.

The main conclusion resulting from the preliminary analysis performed is that, despite the increased awareness relative to the environment and the Chesapeake Bay, it appears that the patterns, rates and trends of development in the Bay region have not appreciably changed in the last seven years.

The large volume of new and maintenance dredging being done supports the conclusion that increased attention should be paid to the evaluation of dredging operations. The focus should be on large projects, which account for the vast majority of all material dredged.

The apparent data deficiency problems encountered underscore the need for the constant verification of data quality and quantity. If effective management of coastal resources is to be accomplished by the regulatory and advisory agencies in the Chesapeake Bay region, the establishment and maintenance of critical information sources such as the RAMS data base are essential.

REFERENCES

1. R. C. Eberhart et al., "Research and Management Shoreline (RAMS) Data Bank - Coding Instructions," Publication No. 25, Chesapeake Research Consortium, Annapolis, Maryland, April 1974.

2. R. C. Eberhart et al., "Pressures on the Edges of Chesapeake Bay - 1973," Publication No. 26, Chesapeake Research Consortium, Annapolis, Maryland, April 1974.

3. R. C. Eberhart and D. A. Maurer, "Pressures on the Edges of Chesapeake Bay: 1973-1979," Report No. CPE-7905, The Johns Hopkins University Applied Physics Laboratory, Laurel, Maryland, August 1980.

MAPPING AND SURVEYING FLORIDA'S SUBMERGED LANDS

George M. Cole
Bureau of Survey and Mapping
Florida Department of Natural Resources
Tallahassee, Florida

INTRODUCTION

The State of Florida is blessed with ownership of an extremely large amount of submerged lands in a wide band along its lengthy coast. This band extends from the mean high water line to a line three geographical miles seaward of the mean low water line in the Atlantic and nine geographical miles seaward in the Gulf of Mexico.

The State Government of Florida is likewise blessed with an extremely large problem of managing these submerged lands. Recent emphasis on more active management of this portion of the Coastal Zone has created an obvious need for two essential tools of land management -- a knowledge of the precise location of the boundaries of the submerged lands (or an ability to easily locate the boundaries as needed) and detailed maps or plats of the submerged lands.

Florida was a pioneer in state participation in Coastal Mapping Programs beginning in 1969 with the start of the cooperative Coastal Mapping Program with the National Ocean Survey. This progressive attitude towards this need continued with the passage of the Coastal Mapping Act in 1974 which standardized and required correct and precise on-the-ground location of coastal boundaries by surveyors. Now with a 1979 legislative mandate to inventory and legally describe all state owned submerged lands, the submerged lands are being platted into easily managed parcels similar to the townships and sections of public uplands.

This paper represents an overview of the total program for mapping and surveying Florida's state owned submerged lands. It gives the history of the programs and a description of the techniques used and products produced. It includes discussions on the Florida experience with the location of private/state and state/federal coastal boundaries, production of base maps of submerged lands, and subdivision and platting of submerged lands into easily described and easily managed parcels.

The various states must play an active role in ocean management. Before they can do so, however, they first must have adequate boundary information. This describes the programs used by the State of Florida in developing such information.

DENSIFICATION OF TIDAL DATUMS

One of the most important requirements for surveying and mapping in the coastal zone is a densified network of tidal datums. These need to be relatively dense since there can be considerable variation in the height of a tidal datum, such as mean high water, in even the same general vicinity.

Graphic examples of this undulation occur in the Florida Keys where differences greater than 12 centimetres in the elevation of mean high water have been measured at points less than 200 metres apart on the coast of the same key. Further examples can be seen in tidal marshes or mangrove areas where significant differences can be seen in the elevations of mean high water when determined at the seaward edge of vegetation and at the actual line of mean high water inside the marshes. Thus it may be seen that considerable densification is needed for accurate location of a tidal boundary.

To achieve this in Florida, the State and the National Ocean Survey (NOS) entered into a cooperative Agreement in 1969 for establishing a densified network of tide stations.

The field work for this program was accomplished by combining personnel from the State and the NOS and was completed in 1979.

For the Florida program, a scheme of densification was developed using a series of long-term primary gages as control stations. These gages have operated for a complete tidal epoch of 19 years. Within the control framework, a series of secondary stations was established for operation for at least one year. These sites were selected to define various unique tidal areas. Further densifying the coastline between these stations, a series of tertiary sites for observations of 30 to 90 days was installed. By operating the short-term gages simultaneously with a secondary and primary control station, an equivalent of a 19-year observation was made at all of the almost 800 stations along our coast. Attachments I and II illustrate typical final products for these stations.

At each of these locations, an automatic tide gage was installed. These gages punch the water level reading in binary code onto paper tape ten times an hour. The punched tape output allows for easy processing of the resulting data by computer.

A tide staff, usually consisting of a graduated vitrified glass scale, was installed near the gage. Daily comparisions were made between the gage and the staff to determine the constant difference in reading between the two. The staff was connected by precise leveling to several permanent bench marks in the area. These bench marks preserve the elevation of the tidal datum that was determined even though the gage and staff are removed.

The end result of the tide studies described above is twofold. These stations will serve as control points for mapping of the shoreline of the State and will also serve as control points from which precise, on-the-ground, boundary surveys for resolution of boundary disputes may be based.

It should be pointed out that even though the horizontal position of the tidal boundaries will change with time due to erosion, accretion, etc., the elevations of these boundaries are relatively permanent. Therefore, these control points will serve for repeated surveys and mappings.

The State is now involved in a program to connect all of the bench marks established by this program to the National Geodetic Vertical

Network by precise geodetic levels. This will preserve the datum even if the tidal bench marks are destroyed and will also allow the comparison of the relative elevation of all the datums. In addition, the State is involved in a program to reoccupy all of the tide stations on a 10-year cycle to detect any changes in the datums due to hydrographic changes.

It is noted that even at the density of the Florida tide stations, there are many sections of coastline between these stations where it will become necessary to establish supplemental datums. This can be accomplished on a need basis by making simultaneous observations at the desired point and the nearest tide station with similar characteristics using only two or three days of observations.

COASTAL MAPPING

In addition to tidal datums, the cooperative program with NOS also called for the production of a series of maps depicting the mean high and mean low water line for planning and management purposes in the Coastal Zone.

For this purpose, the mean high and mean low water lines were located by use of tide synchronized black-and-white infrared photography. This photography was controled by tide observers on the ground at previously established tide stations. At the precise instant that the desired stage of tide was reached, the aircraft was notified and the film exposed. The infrared film shows the water/land interface graphically.

The final map product of the Florida Coastal Mapping Program has gone through an evolution process. At the beginning of the program, a new orthophoto map series was begun. Upon analysis of the cost and usage of these maps, however, it was decided that a completely new series of maps was not in order. There already were NOS nautical charts of Florida's submerged lands as well as U.S. Geological Survey (USGS) topographic maps of the uplands. The Coastal Mapping Program was locating a precise dividing line between the two which could be depicted on the existing two series of maps.

Therefore, it was decided to produce only a large-scale informal work sheet showing the precise shoreline, salient natural and man-made features, and control grid lines as the "final" map product. This product would then be available as an informal shoreline map in itself, and could also easily be incorporated into future editions of NOS charts and USGS Topographic Maps bordering the coast. These shoreline sheets are produced on a scale of 1:10,000 in the urban areas and 1:20,000 in the open areas. Rectified photographs are also produced for each sheet at the same scale. The sheets may then be overlayed on the photos if reference to the photo imagery is necessary. Attachment III illustrates a portion of a typical shoreline sheet and Attachment IV illustrates the incorporation of some of the shoreline onto a USGS topo map. Note that NOS bathymetry is also on these maps.

As mentioned earlier, this program began as a cooperative program with NOS in 1969. In 1974, the program was addressed by State legislation which mandates repeat mapping of the entire coast at least every

25 years and of dynamic areas as necessary.

The entire State will be covered by these sheets as a result of the cooperative program by July 1981. Following that, it is hoped to begin an instate program to rephotograph all of the dynamic inlet areas on an annual basis for updating of these sheets. This will allow monitoring of our dynamic boundaries as well as providing working maps for beach erosion control.

A significant omission from these sheets, from the State's point ov view, is in the coastal marsh areas which comprise a large percentage of Florida's coast. Since the land/water interface is obscured from the air by vegetation, the mean high water line is not depicted in these areas. Only the seaward edge of vegetation, labeled as the "apparent shoreline", is shown. If future funding allows, it is hoped to gradually map the real boundary within these areas.

ON-THE-GROUND SURVEYS

In addition to maps of the submerged lands, there is obviously a need, at times, to also locate the boundaries of these submerged lands on the ground. This need has become especially frequent in Florida with the intense development pressure on its coastal lands and the equally intense pressure to protect our submerged lands. Such pressures have also created a demand for precision in such surveys to eliminate chance for dispute.

Reflecting such pressures, Florida's Coastal Mapping Act of 1974 standardized surveys of tidal boundaries. It required that such surveys use the local datums generated by the Coastal Mapping Program as control; that only approved methods be used in locating the boundary; and that a copy of the survey be filed with the State upon completion of the survey.

The Florida Society of Professional Land Surveyors worked closely with the State in developing the approved procedures and has also sponsored frequent continuing education seminars for the surveyors of the State where the latest techniques in tidal boundary surveying are taught. In addition, the State Board of Professional Land Surveyors includes this topic in their registration examinations. As a result, many of the State's professional land surveyors are now proficient in the precise location of tidal boundaries.

SUBDIVISION OF SUBMERGED LANDS

Once the boundaries of the submerged lands of a State are mapped and provision made for precise on-the-ground location when needed; one additional item in the survey and ampping field is needed for easy management. This is subdivision of the submerged lands into easily described, easily located, and easily managed parcels. It is obvious that this is essential for leases, seales or other uses of submerged land just as it is for uplands.

The Federal Government, through the Bureau of Land Management (BLM) has already divided the outer continental shelf (OCS), the area within federal jurisdiction outside of the State's territorial waters, into such parcels. It is a logical extension of the Federal Program for States to do likewise with their portion of the Coastal Zone to prepare for more orderly use and management of our oceans.

The Florida Legislature recognized this in 1979 when it passed legislation mandating that an inventory of all publicly owned lands be prepared which includes "to the nearest extent practical, the legal description or proper reference thereto and the number of acres or square feet for all publicly owned submerged lands".

Therefore, a program developing such a subdivision system is now being planned.

Obviously, it will not be practical to physically occupy every corner of these parcels and to monument as was done in the GLO subdivision of the upland areas of many of the states. Therefore, it is necessary that the subdivision be based on some type of coordinate system which would allow the physical location of the parcel on a "need basis".

Several coordinate systems were considered, and it was decided to use the Universal Transverse Mercator (UTM) grid for two reasons -- the Federal OCS subdivision is based on this system, and all future USGS quad sheets will contain a full line UTM grid system. This would allow for easy interrelation of the two systems and also would allow the use of USGS quads as base maps.

USGS quads will have these grids at 1000 metre intervals. Therefore it was decided to use these 1000 metre by 1000 metre blocks as the basic subdivision unit.

Each block will be identified by the UGM North coordinate of the South line of each block and by the East coordinate of the West line of the block.

As mentioned previously, USGS 7 1/2 minute topo maps will be used as the master maps showing the grids. Future editions of these maps will also be depicted on these maps. The approximate coordinates of Florida's seaward boundary will be generated and plotted on the maps to delineate the outer limits of the subdivided area. (Existing BLM programs for this purpose will be used together with tidal data from the Cooperative Coastal Mapping Program to insure consistency with Federal data.) In Florida, the seaward boundary is three geographical miles seaward of the mean low water line on the Atlantic Coast and nine geographical miles seaward of the Gulf Coast. Where the topo maps do not extend sufficiently seaward, a special printing of maps which will show only a UTM grid and bathymetry will be made to have complete coverage. Attachment V illustrates a portion of a typical base map of the Cape Canaveral area.

A cross reference index of all existing sales and leases of submerged land will then be made from records in the State Archives, and the submerged land parcel in which these are located will be identified.

Finally, all of the parcels will be entered into a computer file along with pertinent information such as acreage (approximate acreage of fractional blocks), outstanding leases, mineral information, etc.

SUMMARY

In summary, Florida's coastal surveying and mapping program has four major elements. These are the establishment of a densified network of stations with known tidal datums; the production, and then periodic revision, of a series of shoreline maps of the coast; the standardization of procedures for the precise, on-the-ground location of tidal boundaries as needed; and the subdivision and inventory of the submerged land within the state's jurisdiction.

Together, these elements will give Florida the basic tools necessary for orderly use and management of our coastal waters and submerged lands.

REFERENCES

1. Hull, Wesley V.,(1969), "Coastal Boundary Mapping", Pan American Institute of Geography.

2. Guth, Jack E., (1974), "Will the Real Mean High Water Line Please Stand Up", American Society of Photogrammetry.

3. Marmer, H.A., (1951), "Tidal Datum Planes", U.S. Coast and Geodetic Survey Special Publication No. 135.

4. Cole, George M., (1976), "Tidal Boundary Surveying", Proceedings, American Congress of Surveying and Mapping.

5. Cole, George M., (1978), "Florida's Coastal Mapping Program", American Society of Photogrammetry (Potomic Region) Coastal Mapping Symposium.

STATE OF FLORIDA
DEPARTMENT OF NATURAL RESOURCES
BUREAU OF SURVEY AND MAPPING

INDEX OF TIDE STATIONS
QUAD 250801
(DATA FROM COOPERATIVE FLORIDA/NATIONAL OCEAN SURVEY COASTAL MAPPING PROGRAM)

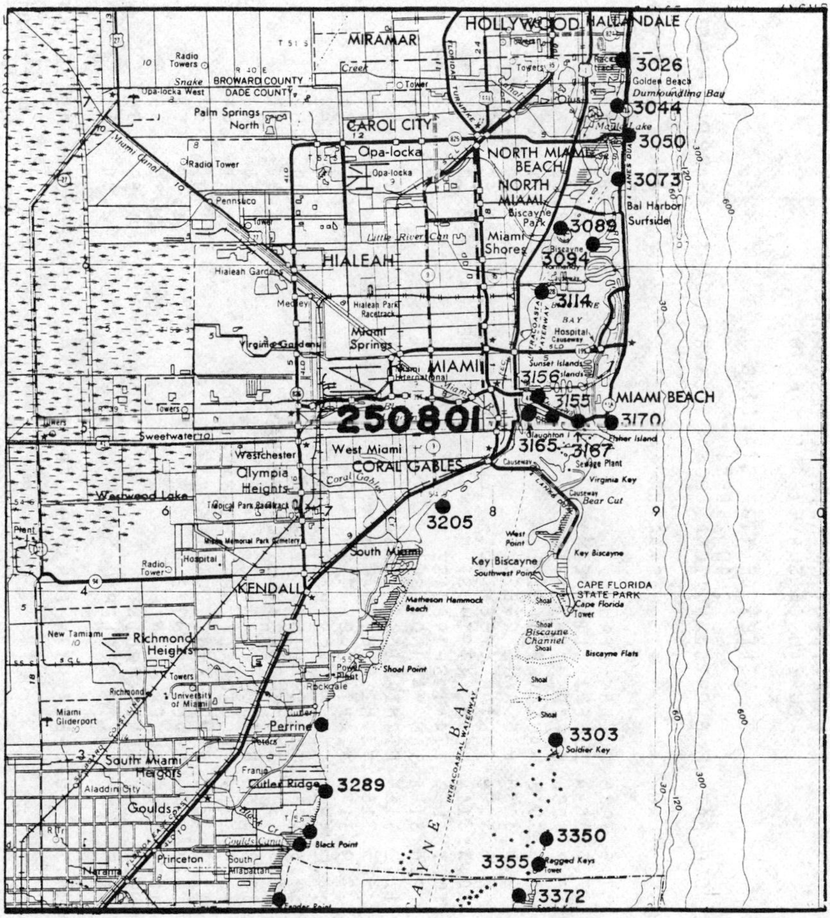

Attachment I
(Page 1 of 2)

STATE OF FLORIDA
DEPARTMENT OF NATURAL RESOURCES
BUREAU OF SURVEY AND MAPPING

INDEX OF TIDE STATIONS
QUAD 250801
(DATA FROM COOPERATIVE FLORIDA/NATIONAL OCEAN SURVEY COASTAL MAPPING PROGRAM)

LAT.		LONG.		NO	STATION NAME	RANGE (FT)	RANGE (M)	MHW W/NGVD (FT)	MHW W/NGVD (M)
25	58.0	80	7.5	3026	GOLDEN BEACH	2.10	.64		
25	56.5	80	7.5	3044	DUMFOUNDLING BAY	2.00	.61		
25	55.8	80	7.2	3050	NORTH MIAMI BEACH	2.50	.76	1.59	.48
25	54.2	80	7.5	3073	HAULOVER INSIDE	2.00	.61		
25	52.8	80	9.7	3089	BISCAYNE CREEK	2.20	.67	1.48	.45
25	52.5	80	8.6	3094	INDIAN CREEK GOLF CLUB BISCAYNE BAY	1.90	.58		
25	50.9	80	10.3	3114	MIAMI (79TH ST CASWY), BISCAYNE BAY	2.00	.61		
25	46.2	80	10.1	3155	MIAMI SHIP BASE FL				
25	47.6	80	9.8	3156	SAN MARINO, FL				
25	46.8	80	11.2	3165	MIAMI BISCAYNE BAY	2.10	.64	1.65	.50
25	46.4	80	8.5	3167	MIAMI BCH, MACARTHUR CAUSEWAY				
25	46.1	80	7.9	3170	MIAMI BEACH CITY PIER	2.50	.76	1.54	.47
25	43.6	80	14.2	3205	DINNER KEY MARINA BISCAYNE BAY	1.90	.58		
25	36.9	80	18.3	3289	CUTLER BISCAYNE BAY	2.00	.61		
25	35.4	80	9.7	3303	SOLDIER KEY, BISCAYNE BAY	1.90	.58		
25	32.0	80	10.3	3350	RAGGED KEY NO3	1.60	.49		
25	31.4	80	10.5	3355	RAGGED KEY NO5, BISCAYNE BAY	1.50	.46		
25	30.4	80	11.3	3372	SANDS KEY ELLIOT KEY	1.40	.43		
25	34.1	80	18.2		BLACK POINT 2	1.90	.58		
25	32.2	80	18.9		BLACK POINT	1.80	.55		
25	31.1	80	19.7		FENDER POINT	1.80	.55		
25	35.5	80	18.5		SOUTH CUTLER	2.00	.61		

USERS ARE ADVISED TO CHECK WITH DNR FOR UPDATED INFO BEFORE USING ABOVE DATA.

Attachment I
(Page 2 of 2)

1/10/76 **3050**

FLORIDA I

U.S. DEPARTMENT OF COMMERCE
NATIONAL OCEANIC AND ATMOSPHERIC ADMINISTRATION
NATIONAL OCEAN SURVEY

TIDAL BENCH MARKS

North Miami Beach
Lat. 25°55'.8; Long. 80°07'.2

BENCH MARK 1 (1973) is a standard disk, stamped "NO 1 1973," set flush in the northwest corner of a concrete manhole cover, 56.2 feet southwest of the southwest corner of the tackle shop and 33.9 feet north of a wire fence (on top of seawall) around the Castaways Hotel. Elevation: 11.25 feet above mean low water (8.75 feet above mean high water).

BENCH MARK 2 (1973) is a standard disk, stamped "NO 2 1973," set flush in a concrete seawall located just north of the Castaways Hotel, 53.2 feet southeast of bench mark "NO 1 1973," and 6.6 feet east of the west end of the seawall. Elevation: 11.67 feet above mean low water (9.17 feet above mean high water).

BENCH MARK 3 (1973) is a standard disk, stamped "NO 3 1973," set flush in a concrete curbing, 50.4 feet west of the southwest corner of the tackle shop and 14 feet west of the center of the exit-entrance gate to the Newport Fishing Pier. Elevation: 12.25 feet above mean low water (9.75 feet above mean high water).

BENCH MARK 4 (1973) is a standard disk, stamped "NO 4 1973," set flush in a concrete curbing, 46 feet north of the centerline of Interama Boulevard leading to the Newport Fishing Pier and 31.5 feet south of the center of the driveway to the parking lot area. Elevation: 12.53 feet above mean low water (10.03 feet above mean high water).

BENCH MARK A-2 is a brass plug, stamped "BM A-2," set on the top of the northwest corner of a catch basin, 43 feet south of the south curb of State Highway 826, 3 feet northwest of the center of a manhole cover, 0.7 foot east of the east curb of State Highway A1A and 0.5 foot above the level of the highway. Elevation: 7.55 feet above mean low water (5.05 feet above mean high water).

(OVER)

Attachment II
(Page 1 of 2)

North Miami Beach

Mean low water at North Miami Beach is based on 16 months of record, May 1973 - May 1975, reduced to mean values. Elevations of other tide planes referred to this datum are as follows:

	Feet
Mean high water	2.50
Mean tide level	1.25
Mean low water	1.00

The estimated highest water level to the nearest half foot is six and one-half feet above mean low water. The estimated lowest water level to the nearest half foot is two feet below mean low water.

Data from State of Florida/National Ocean Survey Cooperative Coastal Mapping Program.

Attachment II
(Page 2 of 2)

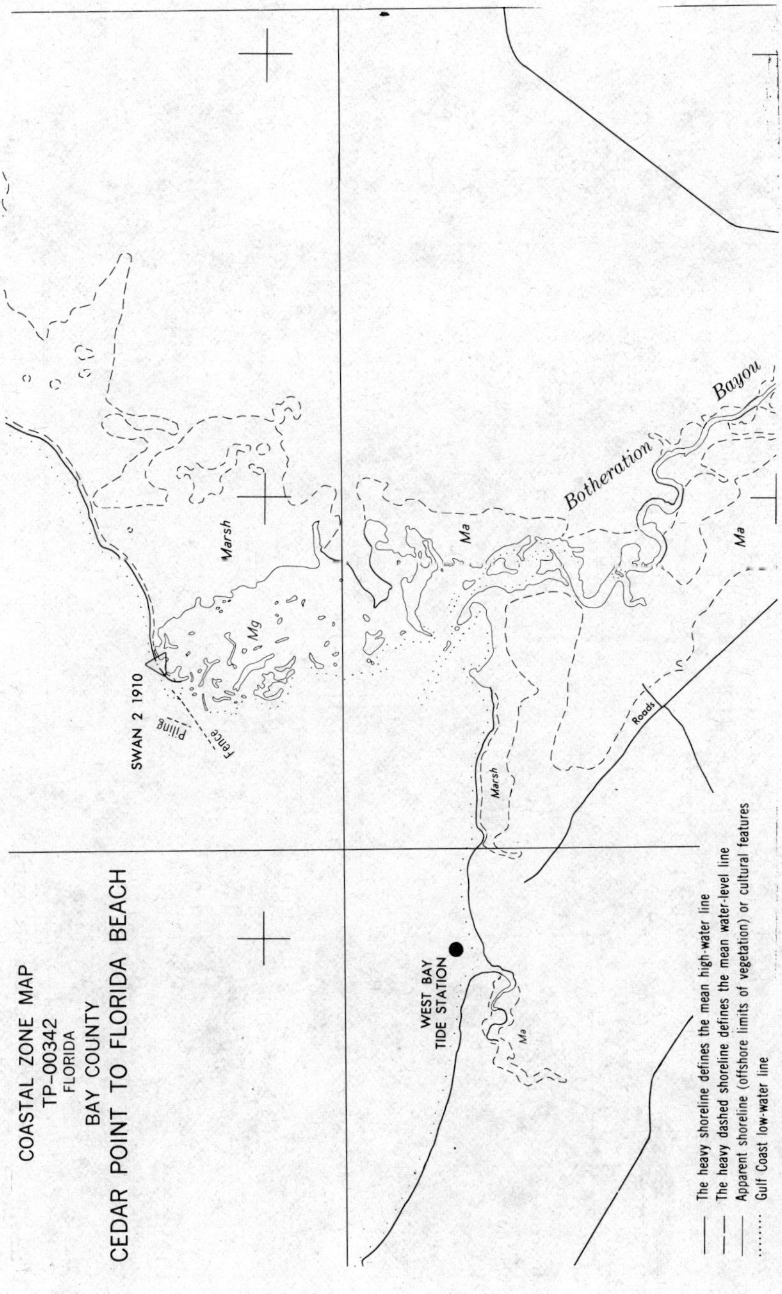

Attachment III

TYPICAL SHORELINE SHEET

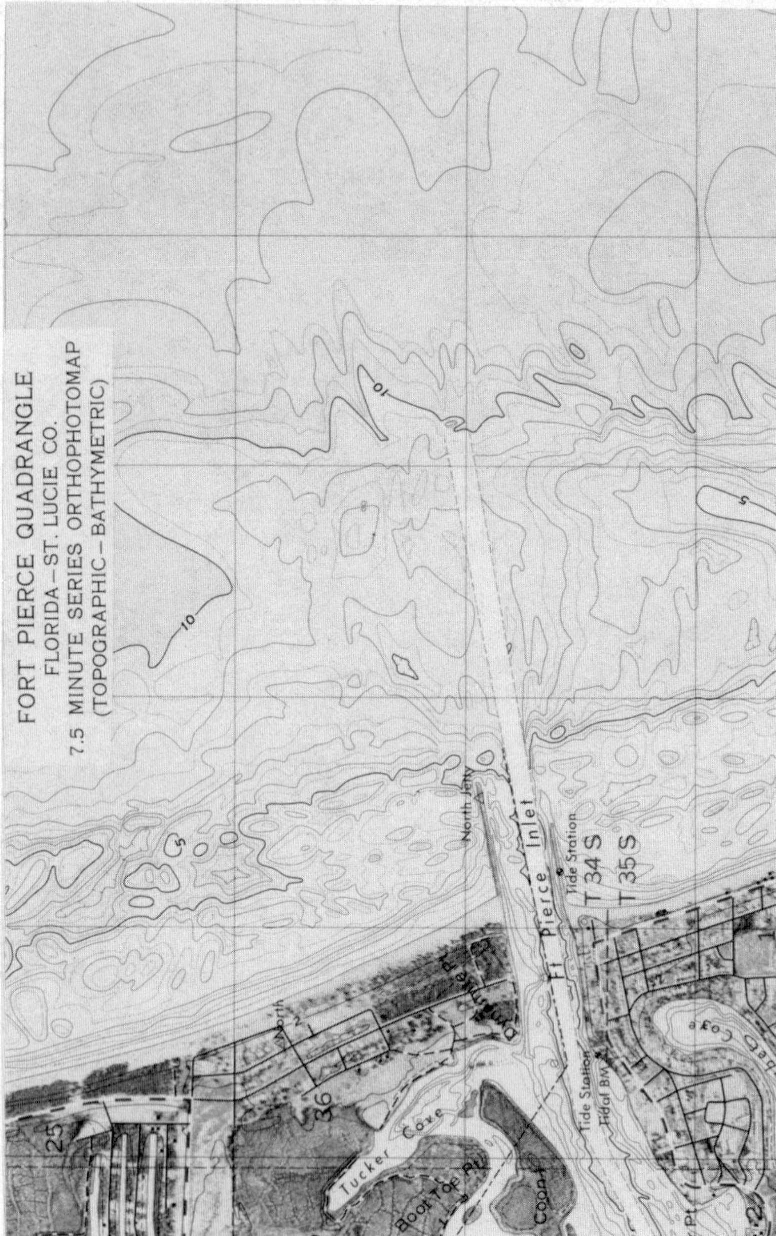

USGS TOPO MAP INCORPORATING SHORELINE AND BATHYMETRY

Attachment IV

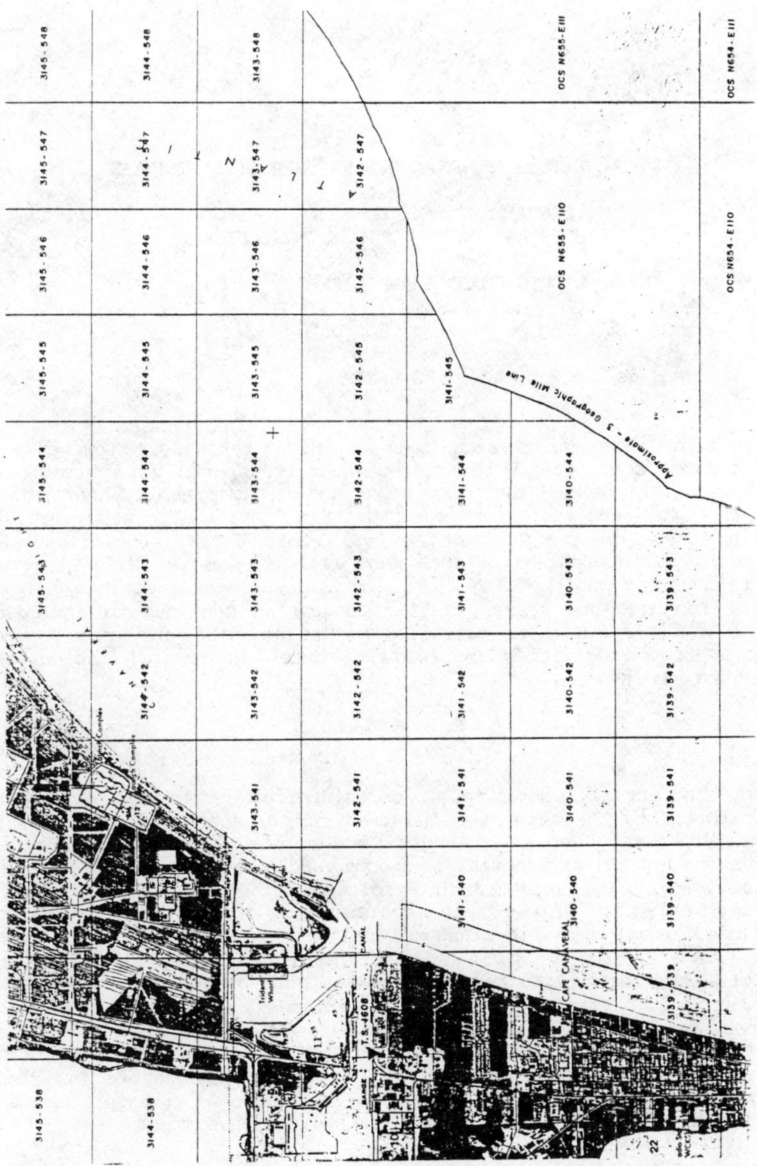

BASE MAP SHOWING SUBDIVISION OF SUBMERGED LANDS

Attachment V

COASTAL SITING OF HAZARDOUS WATERFRONT FACILITIES

LIEUTENANT PAUL C. GOLDEN, USCG
ASSISTANT DISTRICT PLANNING OFFICER

EIGHTH COAST GUARD DISTRICT (dpl)
NEW ORLEANS, LOUISIANA

ABSTRACT

Rapidly expanding petrochemical and chemical refinery, storage, and manufacturing facilities, with related water transportation systems have significantly increased the risk of a major chemical accident in or along an urbanized port or waterway, in particular, the rapidly expanding ports in the Gulf of Mexico. Port safety regulations are engineering and operations oriented and will eventually fail to ensure the safe movement and handling of bulk hazardous substances in urbanized areas as tonnages and vessel traffic congestion continue to increase. The only viable solution is coastal land use planning which can be used to locate these facilities and traffic routes in remote, unpopulated coastal and river areas.

Problem. Industrial, commercial, and residential development represents a major threat to the natural environment of our coastal regions. This development has been challenged under the banner of coastal preservation, with environmental groups and coastal zone management agencies leading the vanguard in a holding action to preserve rapidly dwindling wetlands. However, the overall goal of coastal preservation has blinded us to a reality that directly threatens thousands of lives daily. As a result of environmental opposition, chemical and energy facilities are being blocked from developing in remote, unpopulated coastal and river sites in the name of wetlands and coastal protection. This has resulted in a concentration of water dependent facilities handling bulk hazardous substances in metropolitan industrial corridors. Land use planning agencies have given no serious thought to the risks the local populations are exposed to in the event of a major accident involving these facilities, or their transportation systems.

Background. This problem has paralleled the development and growth of our automative and synthetic based lifestyles since World War II.[1] The problem became acute in the 1960's when cheap energy allowed the rapid expansion of synthetic chemical based industries and created major new demands for petroleum and chemical feedstocks for industrial, agricultural, and energy industries. This has resulted in the bulk transportation annually of hundreds of millions of tons of hazardous substances along our waterways to supply chemical processing and storage facilities located adjacent to our urban rivers, ports, and coastal waterfronts. These facilities located there to take advantage of the cheapest available form of bulk product transportation next to pipelines--shipping. For example, the relative cost (in 1974) of moving a ton of methanol from Beaumont, Texas, to Belle, West Virginia, was \$37.00 by rail, \$62.00 by truck and \$6.88 by barge (table 1).[2]

As transportation fuel costs continue to skyrocket, these differences become increasingly more significant, shifting more and more bulk hazardous substances onto our waterways from other transportation modes.

TABLE 1

COST AND RISK ANALYSIS OF DIFFERENT TRANSPORTATION MODES FOR HAZARDOUS SUBSTANCES

(MILES/TRANSPORTATION COST PER TON/YEARS BETWEEN SPILLS ON THAT ROUTE)*

	BARGE	TRUCK	RAIL	PIPELINE
ACRYLONITRILE (Chocolate Bayou, TX - Decatur, AL)	1400 mi/\$7.50/ 86 yrs	780 mi/\$40.00/ 16 yrs	770 mi/\$11.50/ 6 yrs	-
ANHYDROUS AMMONIA (Donaldson-ville, LA - Palmyra, MO)	960 mi/\$8.73/ 150 yrs	808 mi/\$58.00/ 14 yrs	837 mi/\$13.00/ 22 yrs	700 mi/\$6.26/ 2.6 yrs
BENZENE (Port Arthur, TX-Beaumont, TX)	26 mi/\$2.42/ 1000 yrs	26 mi/\$12.20/ 500 yrs	19 mi/\$7.00/ 110 yrs	-
CAUSTIC SODA (Geismar, LA-St. Marks, FL)	520 mi/\$8.78/ 110 yrs	477 mi/\$32.00/ 40 yrs	536 mi/\$12.15/ 15 yrs	-
CHLORINE (Portland, OR-Eureka, CA)	440 mi/\$6.69/ 9400 yrs	418 mi/-/ 250 yrs	965 mi/\$23.00/ 250 yrs	-
METHANOL (Beaumont, TX-Belle, W.VA)	1700 mi/\$6.88/ 22 yrs	1225 mi/\$62.00/ 2.3 yrs	1121 mi/\$37.00/ .95 yrs	-
STYRENE (Texas City, TX-Addyston, OH)	1500 mi/\$8.90/ 35 yrs	1232 mi/\$56.00/ 4.5 yrs	1035 mi/\$25.00/ 1.1 yrs	-
SULFURIC ACID (Within a major North Atlantic Harbor)	8 mi/\$1.70/ 33,000 yrs	30 mi/\$4.40/ 1700 yrs	30 mi/\$6.80/ 590 yrs	-

Source: A.D. Little

*Based on historical accident statistics, these are predictions of the probability of an accident on that specific route in each mode of transport.

Increasing Shipments of Hazardous Substances. Just how serious the problem is may be inferred from table 2, which shows a doubling and tripling of the bulk hazardous substances transported through the Gulf of Mexico ports and the Lower Mississippi River.[3,4] This growth pattern shows every indication of continuing as each Gulf port competes to announce plans for new multimillion dollar chemical complexes or the enlargement of channels to take ever increasing tonnages through ever increasingly congested waterways.[5]

TABLE 2

HAZARDOUS COMMODITIES* BY TONNAGE SHIPPED THROUGH GULF OF MEXICO PORTS

	MOBILE	LOWER MISS. RIVER BATON ROUGE TO GULF MEX.	HOUSTON SHIP CHANNEL	INTRACOASTAL WATERWAY MISS. RIVER TO SABINE	CORPUS CHRISTI
1966	1,628,000	46,773,000	28,488,000	15,474,000	11,123,000
1977	4,822,000	106,843,000	43,406,000	28,154,000	20,118,000

*Commodities Counted: 2811, 2813, 2816, 2817, 2818, 2819, 2876, 2891, 2911, 2912, 2913, 2914, 2915, 2916, 2917, 2921, 2991

Source: Army Corps of Engineers Waterborne Commerce Reports for 1966 and 1977

Types of Risks. This growth pattern in the water transportation of bulk hazardous substances is centered in the Gulf (figure 1)[6] and raises several questions: (1) How dangerous are these hazardous substances? (2) How probable is an accident in an urbanized area where loss of life would be greatest? (3) What are the alternative approaches to improving the existing problem, and to prevent a worsening situation?

There are four possible hazards associated with an accidental discharge of bulk hazardous substances: fire and explosion, toxic vapor clouds, water toxicity and chemical damage to property. Water toxicity can result in the chemical contamination of commercial and sport fisheries, particularly filter feeders (e.g. oysters). They can bioaccumulate and concentrate low levels of contaminants into toxic concentrations. Another threat is from water contamination that can shut down municipal river water systems. An example of this type of problem is the July 1980 M/V TESTBANK (figure 2) incident in Louisiana, where the vessel sailing under good operations collided with another vessel because of human error. As a result, approximately 12 tons of toxic pentachlorophenol (PCP) with 11-14% secondary chlorinated hydrocarbon contamination (e.g., dioxin), was lost overboard in the Mississippi River Gulf Outlet near New Orleans. The incident is still ongoing, but bioaccumulation in oysters has already raised PCP concentrations in a week from parts per billion to parts per million, which is in the health risk range for humans. Hundreds of square miles of prime shrimp and oyster grounds have been closed indefinitely, with no clear picture of the long term impacts to the regional fisheries, wetlands foodchains or human communities. In terms of risk of life, though, fire and explosion represent the greatest potential for death and destruction in the area immediately surrounding the scene of the accident. However, toxic vapor clouds, especially if they are heavier than air, can represent the greatest risk of life to a metropolitan area. Several detailed gas dispersion models have been developed[7] that can accurately predict the downwind impact area. For example, in March 1972, a fully loaded chlorine barge broke away and wedged in the McAlpine Dam spillway in the middle of Louisville, Kentucky. Fortunately, the highly engineered

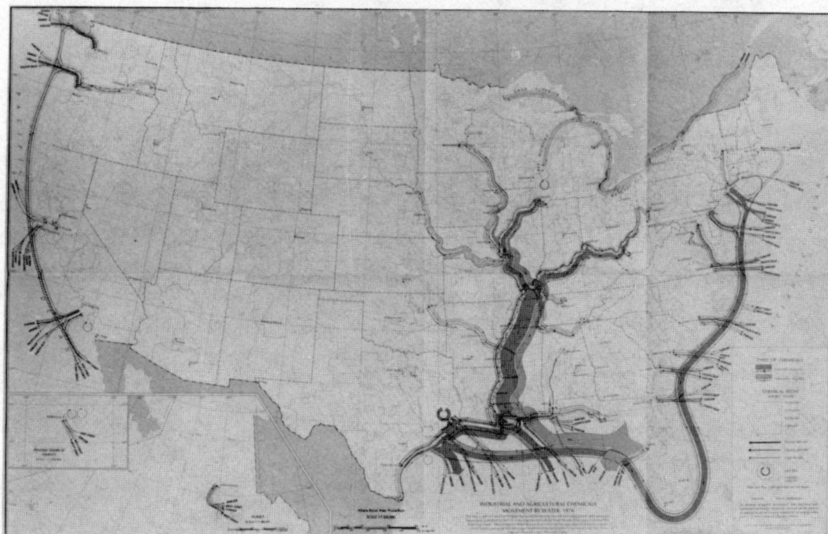

Figure 1: Industrial and Agricultural Chemicals Movement by Water 1976 (Dark Shading - Industrial Chemicals; Light Shading - Agricultural Chemicals)

Source: Army Corps of Engineers National Waterways Study

impact resistant barge and tanks remained intact. However, modeling showed that in the event of a total failure of all the barge tanks a toxic swath of chlorine gas several miles wide and over 100 miles long could have run through highly urbanized regions under the right wind and temperature conditions. The risks depend on the chemical involved. There are thousands of chemicals shipped yearly;[8] many of them can form toxic clouds in the event of an accident.

What are the risks from the bulk shipment of these chemicals through our urban ports? Several studies, in particular, one by A.D. Little for MARAD,[9] show that waterborne transportation by barge is the safest mode of transporting bulk hazardous substances, but that remote but definable risks do exist (table 1). Risk analysis cannot, however, totally predict certain risk elements such as random human error and vessel mechanical malfunctions in very inopportune times and places, such as rush hour in downtown New Orleans or Houston, nor can statistical analysis reflect the greatly increased risks associated with low visibility, or transits and waterfront facility transfers in congested areas associated with urban ports and rivers. What is an acceptable risk along a sparsely populated stretch of waterfront is not necessarily acceptable if it leads into or through a congested, densely populated port or waterway.

Figure 2: M/V TESTBANK following a collision resulting in the loss overboard of approximately 12 tons pentachlorophenol (PCP) into the Mississippi River Gulf Outlet, Louisiana, July 1980

Acceptable Risks. That leads to the question of what is an acceptable risk. The science of risk taking covers a lot of thesis territory from economics to policy making. In terms of economics, it can be the economic benefits of the chemicals transported as compared to an estimate of the (potential) loss of life as a result, and the value of that life (e.g. annual costs per person for life saving services).[10,11] Voluntary risk taking is predominantly an economic choice, such as a chemical plant worker who is aware, to some degree, of the exposure risks and accepts them for monetary reward. Involuntary risk takers who are uninformed (by deliberate choice, or otherwise) as to the risks they are taking do not and cannot make rational risk taking decisions, but rely rather on their political institutions to act on their behalf.[12] That is the core of the problem with the water transport of bulk hazardous substances - the public is totally uninformed as to the risks they are daily exposed to and have unknowingly placed their lives in the hands of political and regulatory institutions that have based their regulatory policy decisions on benefit/cost considerations that historically show high economic benefits as compared to low historic loss of life. This regulatory approach worked exceedingly well when an emphasis on the quality engineering of the vessel and its containers effectively insured a very high probability of safety in the event of an occasional accident.[13]

Figure 3: Naptha laden barge PATCO 202 burning under the Cochrane Bridge, Mobile River, Mobile, Alabama, February 1977

Until the 1960's, this was a very effective approach because of the relatively small quantities of hazardous substances being shipped. However, as discussed above, the picture has radically changed both in terms of tonnages and traffic densities which make engineering questionable as the primary approach to guarantee public safety. Under the right conditions, the best engineered chemical barge or tanker will fail. "Close calls" and accidents involving chemical cargoes are increasing[14] in the Gulf ports and waterways, increasing the possibility, however small, of a major chemical accident in an urban area. One example of a regular occurring event that can lead to an accident is the phenomenon of the "five o'clock rush hour." With stevedores finishing work at 5:00 P.M., many ships will leave their piers at the same time to avoid paying overtime and additional berthing fees. This causes localized vessel congestion (especially on Fridays), often at dusk, when the crew is tired after a full day's work. A recent operations approach to the traffic congestion problem has been the establishment, by the Coast Guard, of Vessel Traffic Services (VTS) in major port areas. These VTS's consist of communication and traffic surveillance systems that allow the forwarding of traffic and safety information to each vessel in the system. Unlike the aircraft control network, the VTS's do not provide positive control over vessel movements, nor do all of the existing VTS's require participation by all of the vessels operating in their surveillance areas. However, even with VTS, traffic congestion problems will continue to be compounded as each port expands to increase their business by adding new facilities for handling larger, less maneuverable ships, which in turn means dredging confined deep draft channels to get them in and out with full loads. A trend throughout the Gulf is the onshore superport[15] justified by the economics of petrochemical, grain and coal supertankers and large bulk carriers.

The combination of the ever increasing water transportation of bulk hazardous substances in larger less maneuverable vessels, in increasingly more congested urbanized ports, will lead to only one inevitable result, the major loss of life in an urban area downwind from an accident involving chemical carrying vessels and/or waterfront facilities that service these vessels.

Existing Safety Regulatory Schemes. What then are the alternatives to this specter of disaster? Ports and waterways safety is a mixed bag of Federal, state and local regulatory schemes. The Coast Guard is the primary Federal agency responsible for regulating maritime, coastal, river, and port safety activities. This authority is mainly derived from the Ports and Waterways Safety Act of 1972; the Natural Gas Pipeline Safety Act (LNG facilities); the Port and Tanker Safety Act of 1978; the Hazardous Materials Transportation Act; and the Federal Water Pollution Control Act of 1972, as amended by the Clean Water Act of 1977. The Coast Guard views its primary role as one of accident prevention through strict safety regulations, inspections, surveillance, licensing and both civil and criminal law enforcement penalties. The Coast Guard presently has regulatory responsibilities for a total of over 2216 waterfront facilities in the country handling hazardous substances. These include over 201 facilities handling cargoes of particular hazard, 15 liquified LNG or LPG facilities, 190 container facilities and 1850 bulk liquid facilities. The reader is directed to the Coast Guard's Environmental Impact Statement on Waterfront Facilities Regulations[16] to obtain an in-depth understanding of Coast Guard safety and regulatory policies for hazardous substances. However, the Coast Guard port and vessel safety programs' emphasis is placed on a cooperative working relationship with the local port authorities and the maritime shipping and industrial communities to ensure that high engineering and operational safety standards are actively implemented and practiced. This safety mechanism has been notably successful in preventing major accidents, the last one taking place in 1947 in Texas City, Texas. Explosions and fires from a vessel carrying fertilizer leveled a 50 acre Monsanto chemical plant, oil and gas tank farms and one-third of the city's buildings. The results were a total of 522 persons dead, with 3000 injured and 200 missing and property damage exceeding $100 million (1947 dollars).

Figure 4: S/S SANSINENA. The empty tanker exploded throwing its midsection ashore in Los Angeles Harbor, December 1976

The states, until recently, have had a limited role in maritime and port safety. Florida and Washington have attempted with limited success to regulate oil tanker activities through tonnage limitations in specified waterways, and by enacting accident liability requirements exceeding those called for by Federal law. Local municipal and port authority controls have focused on facility permits, licenses and safety inspections, the emphasis again being placed on engineering and operations standards. Seldom have these authorities differentiated, and segregated out, waterfront facilities and their water transportation systems which handle bulk hazardous substances. Often, there are no local land use planning and regulatory agencies effectively controlling industrial development outside of a port's municipal borders. This results in the transportation of bulk hazardous substances through the port without any municipal controls on the traffic or bordering facilities. It appears that the general rule is, that if a site is zoned for industrial use, anything can be sited there as long as it generates revenues and meets the prevailing (if any) public safety regulations.

This triad (Federal, state, local) of regulatory schemes has been generally effective in protecting ports and adjacent population centers from catastrophic accidents since 1947. But when virtually every major port and waterway in the Gulf continues to experience major expansions in the shipment of bulk hazardous substances, some waterway carrying capacity limit will sooner or later be reached and exceeded. For example, how much additional growth of petrochemical industries can the Houston Ship Channel absorb? It has had an increase in hazardous cargo shipments from approximately 28.5 million in 1966 to 43.5 million tons in 1977. Indications are that this growth rate has accelerated as major industrial and population migrations from the North have focused on the Gulf central and west coasts. This attraction to petrochemical industrial centers such as Houston-Galveston and the Lower Mississippi River stems from several factors including: high political and public tolerance for "dirty" industries, reliable energy and feedstock sources, and an extensive chemical transportation infrastructure based on existing shipping, rail and pipeline systems. Every industrial growth forecast leads to the inescapable conclusion that petrochemical and chemical manufacturing will continue to sharply expand in the Gulf port and river regions.[17]

This leads back to the earlier point that very few individuals outside of the maritime community understand the dynamic forces that are engulfing them. Without a knowledge of the water transportation risks that the general public is exposed to, there cannot be any rational decision making process to determine what is an acceptable risk in relation to the economic benefits gained from those risks. With the public uninformed and placing blind trust in our political and regulatory institutions to continue to protect them, these policy making institutions are placed in an impossible situation. Assuming they are aware of the potential for some remote risk of an accident in a particular location, should they publicly raise the unsupportable specter of some impending disaster? This could result in the shutdown of major segments of our chemical and energy dependent industrial economy, because of public outcries to severely limit, or prohibit, the transit, or storage, of hazardous bulk chemicals and petrochemicals that are the raw manufacturing stocks of every heavy industry in or adjacent to our urban port and waterways population centers. Obviously, no public official would be willing to make such a "King Solomon" decision based on historical statistics that show a very small risk, and yet we can see similar circumstances, issues and demands with nuclear energy development.

Alternative Approaches to Public Safety. What approaches can be taken to deal with this problem? The status quo implies continued petrochemical growth until a serious water transportation accident brings on demands for change. Perhaps the best that can be accomplished

Figure 5: S/S AMOCO VIRGINIA burning out of control, Houston, Texas, November 1979

is for all levels of regulatory government to recognize that they can no longer completely insure public safety through technical safety regulations and enforcement programs, and instead, attempt to shift a portion of this risk taking responsibility over to public and political forums where risk-benefit land use policies and decisions are more appropriately carried out.

Because these risks are water related, an appropriate mechanism for facility siting decisions would seem to be the state and local Coastal Zone Management agencies. The primary obstacle to this approach is that most state CZM agencies are not adequately staffed, and do not have the political and legal authority under the Coastal Zone Management Act and state implementing laws to make broad, long range land use decisions that would significantly impact on the state's economic well-being. Riverfront industrial development is constantly encouraged in metropolitan areas because of the jobs and tax money it brings into the community. These areas often are already environmentally impacted; this helps generate the philosophy that it is better to allow development there rather than in more remote and environmentally unaltered areas.[18] Although all CZM management plans have provisions for assessing, approving, or objecting to the impacts of riverfront development, this is normally associated with quantifiable local impacts such as the loss of wetlands. There is no pressure to change these procedures because the general public is totally unaware of the activities and risks associated with their local ports and waterways. Without an immediate threat to a park or a popular endangered species, no public pressure can be expected and the economics of industrial development will prevail.

Recommendation. What must be done is to coordinate the planning efforts at the local, state and Federal levels to identify remote, environmentally low quality waterfront sites for hazardous substance facilities.[19] Site criteria should ensure that the shipment of bulk hazardous substances by vessel and rail will not run through metropolitan and urbanized areas. Ideally, the facilities to produce and store chemical feedstocks should be located next to the industries that will use them. Internalizing the use of hazardous substances within remote chemical and manufacturing industrial parks would eliminate many potential risks to the public. These facilities must be downwind of urban areas and strictly zoned to avoid peripheral adjacent commercial and residential development. Finally, a public education program must be initiated. Only if the public is aware of the risks can rational judgments be made as to what are acceptable risks. It is to the credit of the maritime shipping, petrochemical and chemical industries that hundreds of millions of tons of hazardous substances move safety through our ports and waterways each year. However, a policy maker always prefers to maintain a stable status quo. This often means a tendency to ignore changing conditions. The policy status quo on siting waterfront facilities and supporting shipping routes based strictly on economic considerations cannot be maintained. Changes will come either from a major accident and the resulting public outcry for changes in these policies, or change will come from rational coastal zone land use planning with a knowledgeable and participating public making the decisions through responsive elected officials. Admittedly, this has been a simplistic overview of a very complicated, multifaceted problem. However, policy-makers must first learn to understand and appreciate the general parameters of the problem before there is any hope for positive corrective actions, this article was directed towards that goal.

The views and opinions of the author are not necessarily those of the Coast Guard or necessarily reflect its policies.

BIBLIOGRAPHY

1. Commoner, B., The Poverty of Power, Bantam Books, Inc., New York, 1977.

2. Little, A.D., A Modal Economic and Safety Analysis of the Transportation of Hazardous Substances in Bulk, U.S. Dept. of Commerce, Maritime Administration, Office of Domestic Shipping, July 1974.

3. U.S. Army Corps of Engineers, Waterborne Commerce of the United States, Calendar Year 1966, COE New Orleans.

4. U.S. Army Corps of Engineers, Waterborne Commerce of the United States, Calendar Year 1977, COE New Orleans.

5. U.S. Coast Guard, "Gulf Commercial and Governmental Developments (for the 1980's)," Eighth District Long Range Development Plan, Commander (dpl), USCG, New Orleans, 1980.

6. U.S. Army Corps of Engineers, National Waterways Study - Waterways Systems and Commodity Movement Maps, COE, Institute for Water Resources, Fort Belvoir, VA.

7. U.S. Coast Guard (G-WEP-7), Chemical Hazards Response Information System (CHRIS), Coast Guard Publication 446 Vols. 1-4, Washington, D.C., 1980.

8. U.S. Coast Guard (G-WLE-1), Draft Environmental Impact Statement - Waterfront Facilities Regulation, Washington, D.C., 1980.

9. Little, A.D., op cit.

10. Angell, G.R. and J.L. Kalelkar, "The Cost of Transport and Relative Spill Hazards Associated with the Transport of Hazardous Materials by Barge, Rail, Truck, and Pipeline," August 1974, American Institute of Chemical Engineers Conference on the Control of Hazardous Materials Spills, Proceedings, Washington, D.C., 1974.

11. Selman, J. and V. Selman, "How Safe is Safe Enough?" Ibid.

12. Knetsch, J.L. and R.K. Davis, "Comparisons of Methods for Recreation Evaluation," Economics of the Environment, R. Dorfman Edit., Norton & Comp., New York, 1972.

13. National Transportation Safety Board, Risk Concepts in Dangerous Goods Transportation Regulations, NSTB-STS-71, Washington, D.C., 1971.

14. Interviews with Coast Guard Port Safety Personnel in the Gulf of Mexico Region.

15. USCG, Eighth District Long Range Development Plan, Ibid.

16. USCG, DEIS for Waterfront Regulations, Ibid.

17. USCG, Eighth District Long Range Development Plan, Ibid.

18. Morell, D.L. and G.L. Singer, "Energy Siting in Urban Coastal Zone: Compatible or Not?" Coastal Zone Management Journal, Vol. 6, No. 2/3, 1979.

19. USCG (G-DSA-1), Guide for Facility Site Evaluation for Proposed Marine Terminals Handling Cargoes of Particular Hazard, Rpt. No. CG-D-45-79, Washington, D.C., 1979.

MARITIME RISK AND COASTAL ZONE MANAGEMENT

Justus R. Riek, Ph.D.
Mara-Time Marine Services, Inc.
Centerport, NY

INTRODUCTION

Political and economic forces continue to place great pressures on coastal zone managers to permit development of energy resources and energy-related ports/terminals. The absolute question of whether or not to develop is being replaced by questions of how much and under what conditions development is to be permitted. Questions about development raise obvious questions of risk acceptance. Thus, coastal zone management is becoming very much concerned with risk management in energy development.

The Computer Aided Operations Research Facility (CAORF) of the Maritime Administration is a sophisticated ship maneuvering simulator which has become deeply involved in marine transportation risk management relating to energy development. Intensive application of simulation techniques to investigate the human element in shipping has provided answers to questions involving LNG terminal siting, offshore energy development, tanker safety, and the modification of shipping patterns in response to port/terminal development. Such investigations have developed information in support of Environmental Impact Statements and helped define safe operating plans for energy-related terminals and ships.

This paper will briefly discuss the history of several simulator applications and describe how the techniques may be embedded in a comprehensive approach to coastal zone risk management. The specific simulator applications will include the opening of Port Valdez to VLCC traffic, proposed limitation of large oil tankers in Puget Sound, the Point Conception LNG terminal siting, and a risk management program to evaluate the interaction between shipping and oil exploration activities in the Santa Barbara Channel. A demonstration will be made of the utility of marine simulation to decision makers by supplying meaningful answers to questions which permit a more objective analysis of the safety/environmental implications of coastal development.

MARINE TRANSPORTATION IN COASTAL MANAGEMENT

Coastal management programs must resolve the competing demands placed upon coastal resources by a wide variety of uses. In at least three areas, marine transportation plays a substantial role, and if not wisely integrated, can subject the entire coastal zone to significant risks (see Figure 1). The three areas under discussion here are shipping, mining, and power plant development. Of the three, mining in the form of offshore oil and gas development is currently the most sensitive in the public eye. Power plant development, to the extent that it involves the delivery of vast amounts of energy (oil, LNG, coal) by ship, may also be

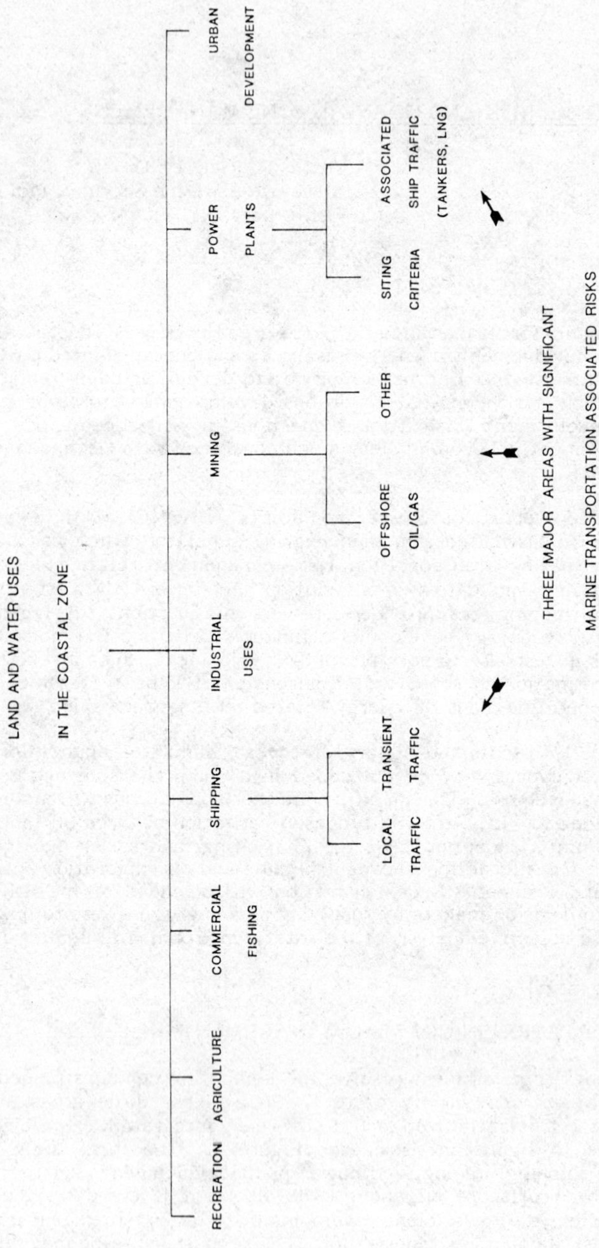

Figure 1 MARINE TRANSPORTATION ASSOCIATED WITH IMPORTANT COASTAL ZONE USES

involved in the marine transportation system. Shipping, as always, forms a major area of concern. As ships become larger and competition for the preferred traffic lanes near the coast becomes more intense, the integration of shipping requirements into an equitable and farsighted coastal development plan becomes all the more important. Clearly, the generation of objective and credible data relating to the various risks and benefits of proposed marine-related projects will be vital in the coordination of such coastal management programs.

It is worth pointing out that the word "risk" is used in a somewhat restricted sense. Risk management here refers to the identification and mitigation of risks of physical damage or destruction to life, property, and the environment which may occur as a result of a marine casualty, or as the cumulative result of normal marine operations (see Figure 2). Because the definition of what is "valued" can vary both as a function of circumstance and point of view, it is the responsibility of various user agencies to translate the risks of such physical damage into dollar values or a degradation of the overall quality of life. Even if such direct translation into objective utility is not possible, the performance of these risk identification and mitigation efforts are useful in a relative sense, to order different situations as to relative risks and relative costs.

CAORF AND HUMAN PERFORMANCE RESEARCH

The Computer Aided Operations Research Facility (CAORF) is the most advanced ship maneuvering simulator of its kind in the world today. It was built for the U.S. Maritime Administration as a research tool to investigate the man-machine interface as it relates to licensed deck officer and piloted control of merchant ships. The type of controlled research performed using this state-of-the-art simulator is referred to as man-in-the-loop simulation.

CAORF is not a part-task simulator. It is a realistic reproduction of the total environment in which the licensed deck officer/pilot operates, comparable to the cockpit simulator of an aircraft pilot. It permits an investigation of the dynamic situation to which the man is exposed. This is crucial, for man always modifies the environment in which he works. Sometimes he improves the design capabilities of the machinery with which he works, and often he degrades those capabilities, but he always modifies.

The phrase man-in-the-loop describes this phenomenon. The officer on the bridge both influences and is influenced by the ship he sails and the environment in which he sails. To understand this total environment, it is not sufficient to study the ship or the waterway or the man in isolation. We must ultimately grapple with the complex total of them all together, including their interactive influences on one another. Ship capabilities, environmental variables, geographical configuration, and human information processing limitations all combine to produce a complex total which must be accurately simulated in order to achieve reliable insight into the dynamic nature of human controllability problems.

Thus far, numerical modeling of this process using digital simulation has proved inadequate. From a mathematical point of view, human performance modeling

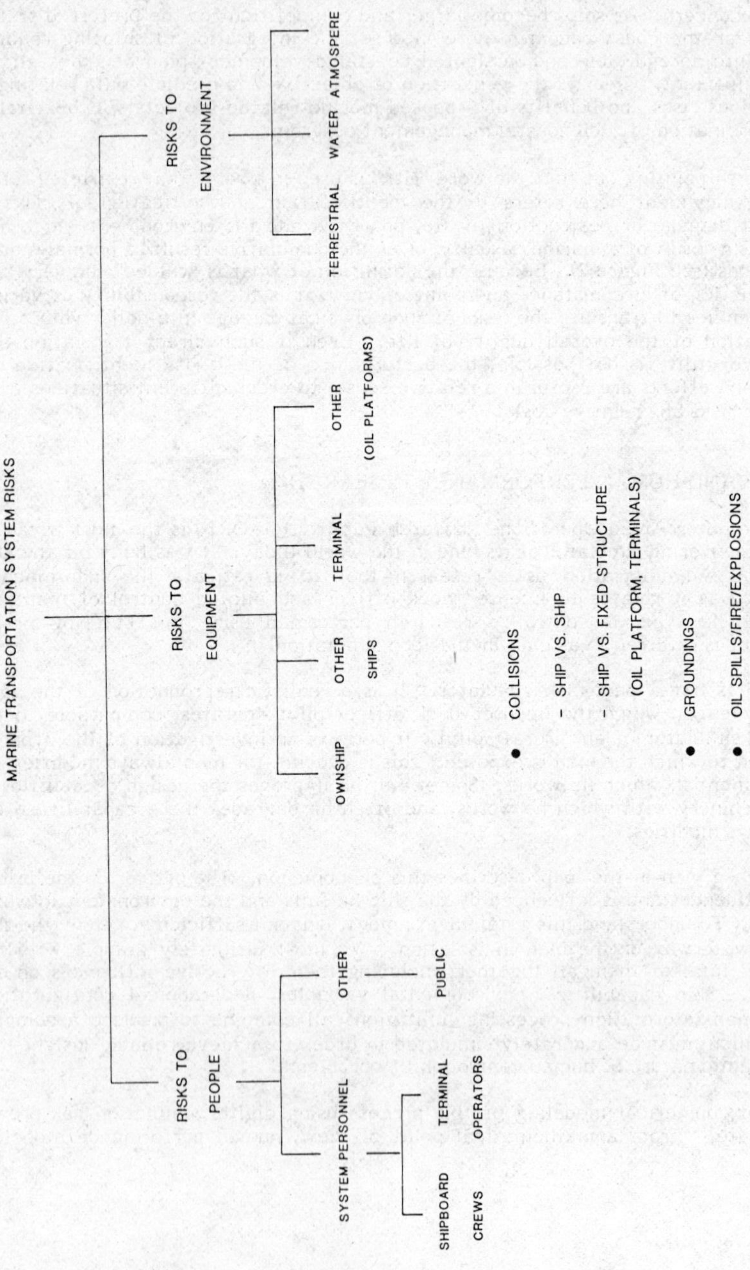

Figure 2 GENERIC TYPES OF MARINE TRANSPORTATION SYSTEM RISKS

with its associated complex feedback control loops has proved to be an intractable problem, producing incomplete and often erroneous results. Even part-task simulation utilizing human subject does not often permit generalization from the part-task situation back to the real environment in which the task must ultimately be embedded (e.g., laboratory results on "pursuit tracking" or "serial learning"). The interaction, between variables present in the real life situation and the sparse part-task environment, produces effects which severely limit the part-task generalizability and permits only incomplete problem solution. It is for these reasons that CAORF was built. CAORF is a realistic and controlled environment in which all important variables can be systematically manipulated to observe their effect upon human performance.

MARITIME RISK MANAGEMENT

Risk management as practiced by the National Maritime Research Center (NMRC) utilizing its CAORF simulator is a comprehensive approach to the identification and mitigation of risks in the marine environment. It involves the three fundamental steps of:

- definition of risk areas
- risk analysis
- formulation of mitigation measures

The identification of risk areas associated with any maritime project is the first step to be taken in the establishment of any risk management program. As such, these areas are generally thought of as associated with people, property, or the environment:

- risks to life
  - shipboard
  - operating facilities (e.g., terminals)
  - public

- risks to property
  - ownship
  - other vessels
  - terminals
  - other facilities

- risks to marine environment
  - atmospheric
  - waterways
  - terrestrial

The definition of risk areas is conceptually simple, yet very difficult in practice. Political decisions must be made concerning what is important and what is not. Budgetary constraints impose limits on how far the coastal zone manager can go in pursuit of the "truth." Some tasks can be attempted, while others cannot. And always there is the "time schedule" looming in the background. Technical feasibility imposes additional constraints. It is difficult to conduct a rational investigation into those areas of risk where the technical tools necessary for a reliable, scientific solution do not exist. The world is full of Environmental Impact Statements in which answers, either because of faulty assumptions or insufficient analytical capabilities, have been provided to difficult questions which are "precisely wrong to the fifth decimal place." In the area of marine simulation, at least, a tool like CAORF is capable of providing reliable answers.

Having identified those areas in which risk management is to be applied, the next step involves the choice of an analytical tool with which to investigate the nature of, or quantify, specific risks. Such tools may include:

- statistical analysis
- digital simulation
- man-in-the-loop simulation

Regardless of the specific method of analysis employed, most approaches have in common that they generally require some form of scenario generation, definition of performance measures, data collection, and the specification of baseline conditions. These elements define the context within which specific problems may be addressed and some definition of what will be accepted as perceived risk and, ultimately, problem resolution. Within the CAORF environment, a destinction is usually made between the "On-line" man-in-the-loop simulation utilizing human test subjects performing in real time, and the other (or "off-line") types of analysis generally performed in support of this primary effort. As such, fast time digital simulation and statistical analysis are usually performed as a prelude to the on-line simulation. Digital simulation utilizing computer-driven algorithms to simulate the behavior of ships under human control is a good example. This activity is used to scope the problem before putting it on the CAORF simulator. By studying the off-line, digital approximation to reality, numerous environmental conditions may be reduced to a more manageable level for real time simulation. Ship speeds, and channel configurations, which are "clearly" beyond the man's ability, are eliminated. Wind and current conditions which, from a hydrodynamic point of view, are not tolerable can be so labeled and more time allocated to a study of the "gray areas" of marginal controllability. It is these areas which are allocated time on the CAORF simulator for further study. These are the areas where there is hope that, by improving training, providing specialized types of information, or defining new/modified procedures that, an otherwise questionable area of human performance may be improved. As the reliability of human performance increases, the level of risk in the marine transportation system decreases.

Historically, the generic types of risk mitigating measures realistically available to the coastal zone manager have fallen into three main (somewhat overlapping) categories:

- modification of the ship, crew, and its operating procedures

- modification of the harbor, facilities, and its operating procedures

- modificaton of the traffic management procedures applied to vessels at sea

A listing of the more important specific mitigating measures which comprise these three major categories is shown in Figure 3. However, a simple listing of risk mitigation measures does not convey the real importance of their application. It does not reveal the impact which their use may have in an actual problem situation. In order to demonstrate the effects which marine simulation may have on real problems, we shall now review four historical applications.

VALDEZ

The first major real world problem addressed by CAORF involved the opening of the Trans-Alaskan pipeline in August, 1977. The pipeline was contructed to carry crude oil from the north slope of Alaska to the Port of Valdez for transport by oil tanker to the lower forty-eight states.

The waters around Valdez (Prince William Sound) constitute one of the world's great salmon fishing areas, and both the fishing industry and environmentalists were opposed to Valdez becoming a major oil port. In addition, environmental conditions appeared to constitute a serious risk to safe navigation. Winds have been recorded for sustained periods at up to 47 knots with gusts to 75 knots or higher. "Williwaws" familiar to this northern region can also occur suddenly with intensities of 100 knots or more. Limited visibility occurs during both winter and summer due to precipitation and fog. And finally, there was the infamous Middle Rock, which lay approximately in the middle of the Valdez Narrows, the only approach to Port Valdez from the sea. These local environmental/geographical conditions, combined with the size and poor low speed handling characteristics of very large crude carriers (VLCC), appeared to constitute a serious marine transportation risk.

The U.S. Coast Guard, concerned with the potential for ship casualties and the resulting pollution, initiated a study of ship operations in the area based upon "worst case" casualties. An experiment was prepared at CAORF in March, 1977, to investigate VLCC operations through the restricted waters of the Valdez Narrows and into the port itself. The objectives of the study were to:

- identify support requirements, such as type of tug escort and configuration, to control a disabled VLCC in the event of a casualty.
- evaluate the effectiveness of a prototype "precise navigation system" and compare to existing radar systems.
- identify safe operating limits for VLCCs navigating in or out of Valdez, either loaded or ballasted with an emphasis on determining an optimal safe speed.

In order to execute such a study, the following preparations were made:

- the mathematical description of the handling characteristics of a 165,000 DWT tanker, taken from actual sea trial data, was developed for the CAORF simulator.

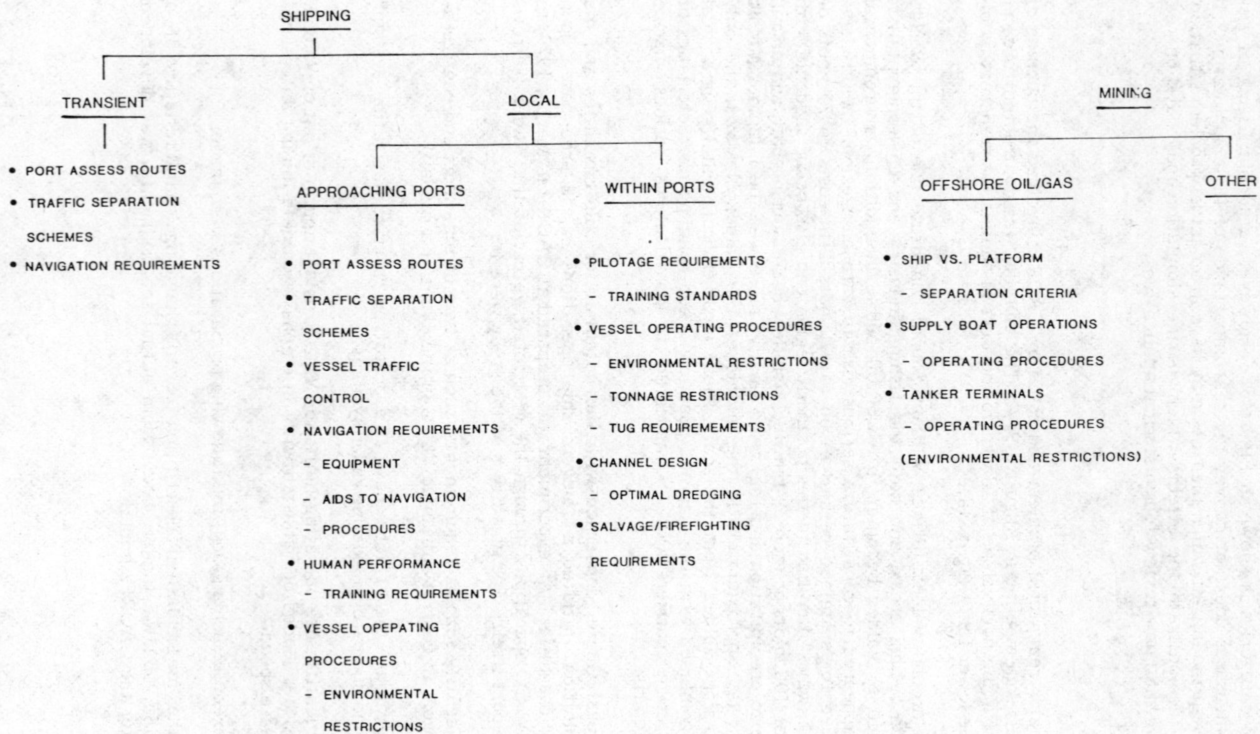

Figure 3 TYPICAL CATEGORIES OF RISK MITIGATING MEASURES

- a visual and radar data base, describing the Prince William Sound and Port Valdez area, was constructed using actual photographs and navigation charts.

- an experimental design was developed to permit an orderly and systematic study of human performance on the above VLCC in the approaches to Valdez as a function of:
  - speed (3, 6, 9 knots)
  - tug configuration
  - power/steering casualties
  - limited/unlimited visibility
  - availability of prototype navigation equipment,

  all in the presence of winds gusting to 40 knots.

Eight Southwest Alaskan pilots took place in the experimental effort. Two of the pilots spent time on the simulator validating the shiphandling characteristics as well as the environmental and visual data bases to insure their accuracy. Four pilots took part in the main statistical experiment, completing a total of 712 runs. An additional 59 runs were made by the remaining two pilots as part of a supplementary experiment.

Directly as a result of this simulation program and other analytical studies carried out by the U.S. Coast Guard, the following regulations have been enacted:

- present VTS rules limit traffic to one-way if the Valdez Narrows are being navigated by a tanker over 20,000 DWT.

- tanker transits of the Narrows are permitted only when wind speeds are below 40 knots, and when gusts do not exceed 55 knots.

- outbound loaded tankers are required to have at least one tug in escort.

- tanker speed limits are as follows:
  - outbound: 6 knots in the Narrows
  - inbound: 12 knots in the Narrows

PUGET SOUND

Early in 1978, the U.S. Supreme Court declared unconstitutional portions of a Washington state law which prohibited the passage of large tankers into Puget Sound. Due to the opening of the Trans-Alaskan pipeline, the state was attempting to stem the flood of oil aboard large tankers that was expected to be transported to several refineries in the Puget Sound area. As a result of this Supreme Court decision, the U.S. Coast Guard was given the responsibility of formulating any new regulations which might be required.

At the request of the Coast Guard, a simulation study was performed at CAORF during the summer of 1978 to provide data which would be helpful in formulating such regulations. It was to be a comprehensive study to assess the safety of passage of large tankers while considering such factors as vessel size, speed, maximum adverse environmental conditions, tug support, and equipment failures.

Six different size tankers, ranging from 40,000 DWT to 400,000 DWT were utilized in the experiment which involved sailing through selected straits in the Puget Sound area. Given the large number of ship types and combinations of wind, current, tug configurations, and failed equipment, a total of 2,400 different ship-environment combinations had to be studied. It was necessary to find a relatively fast and inexpensive method of generating data. The approach taken was to utilize fast-time simulation to investigate the conditions initially and then use CAORF simulation with four practicing pilots as a validation tool.

Two distinctly different types of study were made. One involved an investigation of the ability of the six tankers to navigate the various straits with all equipment functioning normally but in the presence of extreme wind and current conditions (track-keeping runs). The second type involved the effects of extreme wind and current on ships with engine, or engine and rudder failures, when assisted by various combinations of tugs (failed equipment runs). Maximum credible adverse environmental conditions were chosen in accordance with the philosophy that acceptable performance under extreme conditions would automatically guarantee equally acceptable performance under lesser conditions. In addition, during the fast-time simulation runs, steering control was provided by a mathematical model which was essentially a track-keeping autopilot with gains adjusted to simulate human performance.

A summary of results, all conducted under maximum credible adverse environmental conditions, is as follows:

Track-keeping Runs

- Each of the tankers studied was able to successfully transit each of the selected routes if reasonable speeds through the water are maintained.

- At 4 knots, all ships exhibited poor track-keeping ability. At 6 knots, the situation improved, and this may be an acceptable minimum speed for some ships. The lowest acceptable speed, applicable to all ships, appears to be 8 knots.

- At high wind-to-ship speed ratios (about 10 to 1), quartering wind forces can reach levels that overpower the ability of the rudder to turn the ship.

- Vessel size is not a primary variable affecting track-keeping capability. The 80,000 DWT and 400,000 DWT tanker held track about equally well.

Failure Conditions

- The use of tugboats to retard the forward motion of the vessel results in a significant reduction in the transfer. In general, a significant benefit results from the application of the first two tugboats, while the incremental improvement provided by two additional tugs is not so great.

- At speeds of less than 8 knots, reasonable values of transfer can be achieved with retarding tugs. However, these lower speeds may conflict with the requirements for satisfactory track-keeping.

CAORF Validation Runs

- There were definite behavioral differences between the 4 and 6 knot ship speed conditions. All test subjects felt that 4 knots of ship speed was below the threshold for safe navigation.

- Comparison of CAORF and fast-time track-keeping data suggests that 8 knots is a minimum safe navigation speed.

POINT CONCEPTION

In early 1977, Western LNG Terminal Associates applied to the state of California for permission to construct a liquified natural gas (LNG) import facility at Point Conception at the northern end of the Santa Barbara Channel. The facility was to be served by approximately 193 LNG carrier trips per year by ships traveling from Indonesia and Alaska. The need for this LNG facility was based upon projections which indicated a shortfall of energy in California by the middle 1980s.

LNG is an explosive issue in the public eye. There is a great deal of uncertainty and concern on the part of public officials as to the real risk involved with the operation of an LNG facility of the type proposed here. In addition, the LNG carrier used to transport the gas represents a significant departure from the design of previous ships. It has a significantly greater broadside area to be affected by the wind, and the height of the bridge is nearly twice that of most other ships. What would be the effect of this radically different height of eye on the performance of a pilot in a docking situation?

To formally investigate the acceptability of Point Conception as a terminal site, an Environmental Impact Report was generated. As part of this EIR, the CAORF simulator was utilized in a comprehensive study of vessel traffic in the Santa Barbara Channel as well as the environmental conditions under which docking would be permitted.

The CAORF simulator required two major modifications in order to run the simulation. First, the hydrodynamic description of the LNG carrier had to be inserted into the computer to guarantee realistic handling characteristics of the ship. Second, the visual and radar scenes were prepared to represent that portion of the Santa Barbara Channel from Point Conception down to Port Hueneme including all the channel islands.

A total of six masters performed in the scenarios which involved crossing the traffic lanes in the presence of various traffic ships. Subsequently, four pilots performed in the scenarios which involved docking the LNG carrier in the presence of wind and using tugs as is normally the case with actual ships.

The results of these two general types of scenarios indicated that:

- LNG ship masters could be expected, occasionally, to cross between consecutive ships when passing north over the traffic separation scheme, rather than waiting to go behind all traffic ships.
- LNG master collision avoidance behavior was occasionally erratic and unpredictable, indicating a need for some form of control or regulation.
- The docking scenarios raised serious concern about safety of docking under high wind conditions (over 25 knots) especially when there is a component of the wind pushing the ship into the dock.

SANTA BARBARA RISK MANAGEMENT

The Santa Barbara project represents the first systematic attempt by CAORF to formally apply the principles of risk management to a marine transportation problem. We had been performing work similar to this in the past, but without realizing the context in which it was being done. Then in 1979, NMRC formulated its position and methodology with regard to risk management and the result was a publication, Maritime Risk Management Procedures. As is often the case, once the formal statement of the procedure was made, thinking crystalized and it became obvious how to proceed. These principles are now being actively applied in the Santa Barbara Risk Management Project.

The essence of the problem in the Santa Barbara Channel is the risk inherent in the interaction of shipping and oil/gas exploration activities. As such, Santa Barbara represents a generic problem. All along the coast of the United States, the search for oil/gas is accelerating. The potential for conflict between shipping and exploration activities is beginning to intensify. This risk management project represents a comprehensive attempt to investigate oil/gas development within the channel area in so far as it may affect, negatively, the flow of merchant shipping through the established traffic separation scheme. The single most important problem to be solved involves the specification of a separation criteria: how close should fixed structures be allowed to the traffic lanes?

We know that structures on the edge of the lane probably represent an unacceptable risk based upon experience with ship-platform collisions in the Gulf of Mexico. But how far back should they be pushed? Is the 500 m buffer zone criteria proposed by the U.S. Coast Guard sufficient? What criteria should be imposed upon the temporary operations of stationary drillships?

The project began early in 1980 with an intensive data-gathering effort. Forecasts were made of anticipated drilling activities during the next 20 years, and merchant vessel traffic along the California coast was projected. Comprehensive environmental data (wind, currents, visibility) were also gathered to permit realistic simulation of weather conditions.

Next, a Working Group was formed, composed of representatives form the U.S. Coast Guard, California Coastal Commission, and other interested industry representatives. The function of this Group is to advise and, in general, oversee the direction of the work. Its real function is political — to work out difficulties along the way rather than jeopardize the successful implementation of recommendations at the project's conclusion. There should be no surprises in the final report — all will be in general agreement as to what was done, and the implications of the results.

There are two different types of simulation which are being carried out in association with this project, fast-time and CAORF.

The fast-time simulation is designed to investigate the track-keeping capability of merchant ships through the Channel using equipment subject to probabilistic variability (i.e., a Monte Carlo model). The purpose is to estimate how far ships might be expected to "wander" from their intended track lines under various environmental conditions. This information will be used to help determine an acceptable buffer zone for stationary objects along the Santa Barbara Channel.

The CAORF simulation is designed to study the effects of fixed structures on the maneuvering responses of masters while sailing through the Channel. Various configurations involving a platform or drillship at the edge of a traffic lane, and then a pair of such platforms ("gated" one on either side of the lane) will be studied to determine how masters of tankers and containerships behave in their presence.

Work is continuing on all aspects of this risk management program. The results from both fast-time and CAORF simulation are not yet available, since the program is not due for completion until November, 1980. At that time, a final report will be available detailing the results of all simulation together with analysis and a series of conclusions/recommendations.

SUMMARY

We have seen four different applications of simulation to the solution of problems in the marine transportation system. Marine simulation has come of age. It is capable of providing reliable answers to problems about which, ten years ago, one could only speculate.

- In Valdez, marine simulation provided the basis of an operating plan for the most vital energy-related port in this country today.

- In Puget Sound, marine simulation accomplished a comprehensive investigation into the relative operational risks associated with a broad spectrum of oil tankers.

- In Point Conception, simulation permitted a preliminary investigation into the siting adequacy of an energy facility and associated ship transportation system that may ultimately cost $2 billion — all while the project was still in the design stages and capable of modification.

- In the present Santa Barbara Risk Management Project, marine simulation has truly matured. It has finally, and logically, been incorporated into a systematic program to evaluate and mitigate a series of well thought out risks inherent in the marine transportation system as it affects vital coastal zone development questions.

Coastal zone managers can continue to look toward marine simulation with increasing confidence. The tool itself is improving, and users are growing in their understanding of how to apply the tool to the effective solution of problems. Marine simulation has the potential for playing an increasingly important role in the development of our coastal zones.

PUBLIC AND PRIVATE REGULATION OF LNG TRANSPORT

Jonathan David Aronson

and

William Westermeyer

School of International Relations

University of Southern California

Abstract

This paper examines the propositions that (1) the private banking sector is capable of assuring that liquefied natural gas projects are economically viable and (2) the insurance sector is capable of rating the relative risks of LNG projects to assure their operations are as safe as possible. If the markets could assure these outcomes, government regulation of LNG projects could be reduced. Reviews of public and private regulation of safety and economic viability suggest that the private sector cannot serve these functions adequately. Unless private financial and insurance enterprises are pushed by government regulations to smooth their own operating cycles, private self-regulation of major new developments is unlikely to be effective. In the case of LNG, the red tape of governmental bureacracy may have prevented the large-scale development of costly, potentially hazardous white elephants, but for all the wrong reasons.

To surmount the energy crisis the United States must develop multiple new energy sources. To assure that alternative energy sources are available at competitive cost, in appropriate quantities, and without undue delays, it is essential to institute efficient mechanisms to promote cost effective energy supplies which can be provided with relative safety. Any energy source should be subjected to two tests. A cost-benefit analysis should assure that energy from one source is competitive with alternative sources. A risk-benefit analysis should assure that the energy can be delivered with relative safety. These two tests are interrelated. The society may be willing to pay more for clean, safe energy than for more hazardous, cheap energy.

This paper investigates the prospect that private banks and insurance enterprises might assure that the costs and risks of new energy sources stay within acceptable boundaries with greater efficiency than do government controls. If this can be shown to be true, it would suggest that many government regulations can be eliminated. If the private sector does not provide these services, it would be useful to know why not and whether they might be induced to do so. Here, we examine a single case: the costs and risks of importing liquefied natural gas (LNG) into the United States.

I. Background on Liquefied Natural Gas in the United States

In the mid-1970's as the energy crisis deepened and industry and government came to appreciate the dimensions of the energy supply problem facing the industrial nations, the prospect of transporting large quantities of LNG from OPEC to industrial nations brightened. It seemed a natural fit. The United States and Western Europe possessed an extensive gas infrastructure but falling gas production. U.S. gas consumption began outpacing new discoveries in the late 1960's.[1] The falling reserves to production ratio in the U.S. was largely a result of U.S. policy which held gas prices to about 20¢ per million British Thermal Units (BTUs), making it uneconomic to exploit marginal fields. The Netherlands, Europe's largest gas exporter, has more recently begun to move into energy deficit.[2] On the other hand, Algeria, Libya, Abu Dhabi, Saudi Arabia, Brunei, Indonesia, and several other oil producing nations possess an abundance of gas, often found associated with oil and produced as a by-product of oil production. Even after meeting local demands and reinjecting sufficient gas back into the oil deposits to maintain adequate recovery pressure, excess gas remains.[3] Since building pipelines from remote gas producers to the industrial consumers was not feasible or economical, gas was and still is flared in vast quantities. To the producers, this gas had a marginal value approaching zero. All parties were therefore delighted at the prospect of developing LNG projects to tap what was wasted gas. Initially, producers were willing to sell the gas cheaply, making it possible for consumers to invest in the heavy capital costs needed for LNG projects and still bring the gas to market at reasonable prices.

All LNG projects have three integrated parts. In the producing country a liquefacation facility reduces the temperature of the gas to approximately -260 F°, resulting in a liquid which takes up 1/620 of the volume of gas. The LNG is pumped into LNG carriers, the most expensive non-military sea-going vessels in the world, which serve as a floating pipeline, transporting the LNG to the consumer country. There, regasifacation facilities, almost as expensive as the liquefaca-tion plants, restore the LNG to usable gas form as needed.

LNG in the U.S.

The world's first LNG facility was a peak shaving plant built in Cleveland, Ohio during World War II. The idea was to liquefy gas for storage when supplies were large, and to regasify it for use when demand was heavy. Unfortunately, in 1944 the Cleveland plant exploded when one and then a second LNG tank ruptured, causing the spilled LNG to vaporize and ignite. One hundred and twenty-eight people died, at least three hundred were injured, and property damage reached about $7 million. The accident stopped the development of LNG peak shaving and transport projects for two decades.

As memory faded and technology improved, the LNG industry began to revive. The first tests in the transport of LNG came in 1959 and 1960 when the Methane Princess successfully moved seven cargoes of LNG from Lake Charles, Louisiana to Canvey Island near London. On October 7, 1964, two small carriers began transporting LNG from Arzew, Algeria to Canvey Island, the first important import project. France began importing LNG in 1965, and Italy and Spain became regular importers in 1971. Japan began taking Alaskan LNG in October 1969 and subsequently added import contracts from Brunei, Indonesia, and Abu Dhabi.

Without too much ado, the United States began importing small amounts of LNG into Columbia Point, Massachusetts on an experimental basis in 1968. After twenty-one uneventful test runs into the Northeast between 1968 and 1972, Distrigas, a subsidiary of the Cabot Corporation, began small scale, but rather continuous importation of LNG from Skikda, Algeria to Everett, Massachusetts on September 1, 1973, to meet peak energy demands in the Boston area.[4]

However, concern over LNG increased tremendously when on February 10, 1973, forty workers repairing the liner of a 600,000 barrel LNG storage tank in Staten Island, New York were killed when a spark ignited the tank's flammable insulation, setting the little gas remaining in the tank ablaze. The roof was lifted off the structure before collapsing onto the workers below.[5] Questions concerning the safety of the transport and storage of LNG proliferated, prolonging state and federal approval processes considerably.

The El Paso Natural Gas Company's project to import 400 billion cubic feet of LNG yearly from Algeria at a revised price of $1.95 per million BTUs was approved only after a delay of four years. Problems struck at all three corners of the deal. Permission to build the regasification facilities was delayed, thereby raising the price. The construction of the liquefaction facilities in Arzew escalated from $540 million to $2.3 billion when the original contracter had to be replaced by Bechtel. The construction of El Paso's sophisticated 125,000 cubic meter LNG carriers in France proceeded on schedule, were delivered several years before the project was completed, and had to be stored, at great cost, in Norwegian fjords until 1978. Then, in late June 1979 the El Paso Paul Kayser ran aground near Gibralter, sustaining substantial damage but not rupturing her storage tanks.[6] Subsequently, structural flaws were discovered in six of El Paso's nine LNG carriers, mostly relating to weak welds in the containment systems.[7]

By 1979 the United States government was becoming disenchanted with the pricing of LNG and refused to grant permission to Tenneco to import LNG from Algeria to Nova Scotia before pipelining the gas to the United States. This decision began to look economically more sensible when Algeria preempted the LNG pricing debate early in 1980. Rising oil prices brought $9 billion more into the coffers of Algeria in 1979 than in 1978. The transmediterranean pipeline (from the Hassi R'Mel fields, through Tunisia, under the Straight of Sicily at depths of up to 2,000 feet, and across the Straight of Messina to Italy and then as far north as Minerbio near Bologna) is progressing so well that the Algerians decided to lay a second pipeline, parallel to the first, instead of building a third gas liquefaction complex at Arzew. Fortified by their solvency and European prospects, radical Algerian officials, who replaced a more experienced group of technocrats after Houri Boumedienne's death two years ago, demanded early in 1980 that the price of LNG be linked to the price of oil in world markets. Denying that the marginal value of their gas was near zero, in one stroke the Algerians more than tripled their asking price for gas to approximately $6 per million BTUs(f.o.b.), a figure roughly equivalent to $40 per barrel oil. The U.S. government, which had just completed lengthy negotiations with Mexico and Canada which set a price of $4.47 per million BTUs for pipeline gas, was unwilling to allow El Paso to pay $6 (plus shipping costs from Algeria) and expose the North American pricing structure to further leap-frogging. Therefore, on April 1, 1980, Sonatrach, Algeria's national oil company, cut off deliveries of LNG to El Paso thereby idling expensive gas carriers and gas importation facilities at Cove Point, Maryland and Elba Island, Georgia.[8] The smaller Distrigas operation, which relies on Algerian owned ships and a different pricing system, has to date continued to receive small gas shipments.

No other U.S. LNG import projects are currently in operation, but two are at different stages of development. The Panhandle Eastern Pipe Line Company has almost completed construction of a regasification facility at Lake Charles, Louisiana scheduled to import 460 million cubic feet of Algerian LNG daily. It is questionable, however, whether they will be allowed to pay the price the Algerians are now demanding.[9] In addition, a marginally larger project is scheduled to bring Indonesian LNG to Point Conception, California. This joint venture of the Pacific Lighting Corporation and Pacific Gas and Electric was due to open in the late 1970's, but safety and environmental opponents stalled the project. Although almost all permits are now in hand (some seismic testing remains), no Indonesian LNG is likely to arrive before 1985 even if there are no further delays. Although the joint venture partners have already invested over $200 million without ever breaking ground on their reception facility,[10] the economic viability of this project in the long-run remains clouded. No one believes the Indonesians will sell Pacindonesia gas at the contracted price of $2.09 in 1985 when we are now paying Canada and Mexico $4.47 and Japan is presently paying $4.60 for Indonesian gas.[11]

II. Regulating LNG Price and Safety in the United States

There now appears to be a substantial possibility that little LNG will flow into America in the near future despite large investments to promote such flows. This section reviews federal and state regulations that have developed to control the pricing and safe handling and transport of LNG.

The hazardous nature of LNG results from carrying large quantities of gas at cryogenic temperatures. Gerald Greenwald summarizes the potential dangers of handling liquefied natural gas:

> Contact with the extremely low temperature of LNG causes most metals to lose their ductility and become brittle. Thus, brittle fractures may occur when supercold LNG comes in contact with metals that are not suited to low temperature use. If the tanks of the LNG carrier or the LNG terminal were to be breached, the primary safety concerns would be with the effects of LNG spills on land or on water. If spilled outside the specifically insulated tanks which exclude heat, the pool of LNG would naturally boil, evaporate rapidly and generate a cool vapor cloud. Initially heavier than air, the vapor mixed with air becomes buoyant. Once buoyant, the gas would rise and expand, become further diluted by mixing with air, and disperse into the upper atmosphere. If ignition occurred when the gas was within its relatively narrow flammability range, persons and property would be endangered. When the LNG vapor is within its flammability range, the pool of spilled LNG may ignite and burn rapidly or the drifting vapor cloud may reach a source of ignition and burn.[12]

While LNG is indeed a hazardous substance it is not the most hazardous substance being transported today.[13] It is interesting that while the properties of LNG have been systematically investigated, other hazardous cargoes, such as Liquefied Petroleum Gas (LPG) -- which is in some respects more dangerous than LNG[14] -- have received much less attention. For example, the Department of Energy is not involved in the control of LPG facilities or transportation but is very much involved in the regulation of LNG projects[15], and whereas LNG is prohibited from Los Angeles Harbor, LPG has been delivered there on a regular basis. Several reasons for this anomaly exist. The transportation of LNG at cryogenic temperatures requires special materials and techniques. The technology is sophisticated, costly, and relatively new. LPG, in contrast, has been around longer and is shipped in smaller quantities. Through it use as a fuel for such things as camper stoves it has gained widespread public acceptance. The very bigness of the proposed LNG projects -- and indeed of other large scale energy projects -- makes the public wary. The safety record of LNG tankers has been excellent even accounting for the newness of the trade and the high risk cargo involved.[16]

Regulating LNG Safety

Since 1973 questions concerning the safety of the transport and storage of Liquefied Natural Gas have proliferated. Regulatory agencies at all levels of government have become involved in an examination of the potential and problems of an LNG industry in the United States. While LNG is both a non-polluting and efficient source of energy, there has been much concern about both the safety and the economics of importing large quantities of LNG. Vocal opponents of LNG have forced through stringent siting and operating procedures when they were unable to block projects outright.[17] Numerous, but also inconclusive studies, have been conducted on questions of safety.[18]

Many regulatory bodies are concerned with the safety of the LNG industry. Regulations govern such things as LNG tanker design, navigation rules, offloading procedures, design of onshore facilities, and personnel skill requirements. While numerous, these types of regulations have generated little controversy. Of more importance to the pace of development of the industry have been the cumbersome permit and licensing process and the controversy that has developed over the siting of LNG terminal facilities.

Long lead times are required to develop any LNG import project, and the process of permit approval takes up much of this time. The 1977 Office of Technology Assessment (OTA) study on the transportation of LNG points out that the typical LNG import facility will need more than 130 permits from Federal, state, and local agencies before the project can begin operation.[19] In one way or another most of these permits concern either safety or pricing issues. Several studies have described in some detail the responsibilities of the various concerned Federal agencies. The Office of Technology Assessment listed 12 Federal agencies involved in LNG import projects in its 1977 report.[20] The Federal Power Commission has since been replaced by the Department of Energy. Of the twelve agencies listed the Department of Energy, the Department of Transportation and the Army Corps of Engineers are perhaps the most important and all are involved in regulating safety in some way. Different agencies are responsible for varying portions of the safety load. And, each area has seen evolving responsibility as more knowledge became available.

The Siting of LNG Facilities

More knowledge about the potential dangers of LNG and increasing concern about public safety have made an issue of the selection of sites for LNG facilities. New facilities are likely to be sited in remote areas. However, the first LNG liquefaction facilities in the U.S. and abroad were sited in populous areas, near existing markets, pipeline networks, skilled labor forces, and public facilities (such as roads and electricity). The Distrigas facility in Boston, Massachusetts began operation in 1971. At the time little opposition to the project was voiced despite the fact that the facility was located in Boston Harbor in a heavily industrialized area adjacent to a large population center. However, the amount of gas imported into Boston Harbor to date has been quite small. Congestion would increase with the expansion of the facility that has been proposed. This proposed expansion has generated much more controversy since the public is now more aware about safety, social, and environmental issues. Beginning in 1978 the Energy Facilities Siting Council of Massachusetts adopted regulations for the siting of intrastate liquefied natural gas storage facilities. New energy facilities must now also be consistent with the state's Coastal Zone Management program.[21]

Stricter regulations applied to the development of the Cove Point facility. This facility on the Chesapeake Bay in Maryland began operation in January 1978. Unlike the Distrigas facility the Cove Point regasification plant has been remotely sited. From the beginning, environmental and safety concerns were taken into account in the design and operation of the plant. For instance a large buffer zone is maintained around the facility and industrialization is kept to a minimum. However, the site was approved despite the fact that the Calvert Cliffs nuclear plant is located just five miles north of the Cove Point facility.[22]

In California the site selection process for the one terminal allowed by California law to be located within the state has been long and tedious. To date there is still controversy surrounding the Point Conception site. Previously, Los Angeles Harbor and Oxnard had been considered as possbile sites for the terminal. In the wake of the increased safety concerns, however, both the Department of Energy and the State of California established siting criteria which included specific population density standards.[23] L.A. and Oxnard were both eliminated as potential sites, but the preferred site of the utility companies, Point Conception, is still having problems gaining acceptance because it may be subject to unacceptably strong earthquakes and is located in an area which is sacred to the Chumash Indians. The California Coastal Commission has also considered and evaluated potential offshore sites for an LNG terminal.

While siting has been a state regulated matter, minimum safety standards for LNG facilities are formulated by the Materials Transportation Bureau (MTB) of the Department of Transportation. The Federal Energy Regulatory Commission of the Department of Energy for its party, has the authority to impose more stringent rules when necessary and can also require that LNG facilities be located away from densely populated areas.[24]

DOE Concerns

Two agencies within the Department of Energy have regulatory authority over LNG import projects. These are the Economic Regulatory Administration (ERA) and the Federal Energy Regulatory Commission (FERC). Broadly DOE's process of approving LNG projects is that of a licensing authority. A "utility proposes a project and a site, and the Department rejects it, accepts it, or accepts it with conditions, which can be quite substantial."[25] The ERA derives its authority from section 3 of the Natural Gas Act (NGA). It must determine whether LNG import projects are in the public interest. Under sections 4 through 7 of the NGA FERC has statutory functions regarding terminal facility certification as well as the price and other terms under which regasified LNG is sold in interstate commerce.[26] LNG project applications are submitted jointly to ERA and FERC. The approval of an LNG import project is based upon an examination of environmental, economic, and safety issues.

FERC prepares an environmental impact statement in accordance with the National Environmental Policy Act. Potential safety risks are specifically addressed. Furthermore, the application must generally comply with environmental guidelines minimizing damage to fish, wildlife, recreational, scenic, archaeological, and historical values.[27] The ERA, on the other hand, reviews each application in light of such issues as the security of supply, national and regional needs, cost, and the effect on the U.S. balance of payments.[28] Regulations also require applicants to give details about project financing, ownership, pricing, design, engineering, and construction of ships and facilities.

Department of Transportation Concerns

A. Materials Transportation Bureau Responsibilities: The Materials Transportation Bureau (MTB) is more directly concerned with safety than is the Department of Energy. In order for DOE to approve a project, the project must comply with safety regulations promulgated by the MTB. The MTB sets guidelines for the design, construction, and materials used for regasification facilities, storage tanks, and related facilities. The MTB also specifies standards for diking storage facilities and for creating buffer zones around the facilities. The Natural Gas Pipeline Safety Act of 1968 gives the MTB the authority to promulgate regulations designed to guard against risks of fire, radiant heat, explosion, and asphyxiation from leakage or spillage of LNG.[29]

B. U.S. Coast Guard Responsibilities: The U.S. Coast Guard, for its part, is responsible for ensuring the safety of life and property at sea and also at related shoreside facilities. U.S. Coast Guard regulations seek to ensure ship stability and survivability. Criteria for ship hull materials, electrical systems, firefighting methods, ventilation methods, cargo containment systems, temperature and pressure limits, and ship instrumentation are specified by the Coast Guard.[30] Coast Guard and/or MTB regulations also specify the means by which cargo is to be loaded and unloaded, the type of piping materials and insulation to be used, fail-safe valving, instrumentation of transfer operations, and diking around storage facilities.[31] Furthermore, the Coast Guard has established standards for personnel qualifications aboard LNG carriers. Enforcement of ship design standards is carried out in three ways: 1) by inspection of detailed plans and specifications submitted by the vessels' owner; 2) by inspection of documentation that the vessel is accepted by a recognized foreign classification society whose standards are comparable to U.S. standards; and 3) by inspection of the ship as it enters a U.S. port. Certified U.S. vessels receive certificates of inspection while foreign flag vessels receive a Letter of Compliance.

The Coast Guard is also charged with managing traffic control in ports. For example, LNG vessels entering Boston Harbor are controlled by the Coast Guard through a contingency plan developed in cooperation with the maritime industry and state and local officials. The plan requires "prior notification of LNG vessel movement, estimated times for arrival and departure, escort details, vessel inspection and safety checks."[32] As a vessel enters the harbor, a moving safe zone is established. Other ships are required to be at least two nautical miles ahead or one nautical mile astern of an LNG tanker while the tanker is in transit.

Finally, the Coast Guard also works with the Intergovernmental Maritime Consultative Organization in developing uniform worldwide standards for the safe transport of hazardous cargoes.

Department of Defense, Army Corps of Engineers Concerns

The Army Corps of Engineers regulates construction in navigable waters.[33] A permit must be obtained from the Corps for any construction done in U.S. waters. In evaluating an application to construct an LNG terminal the Corps may require an Environmental Impact Statement. The Corps consults with state and local authorities in evaluating a proposal as well as with such agencies as the National Oceanographic and Atmospheric Administration, the Environmental Protection Agency, and the U.S. Fish and Wildlife Service.[34] The safety and environmental concerns of the Corps overlap with those of the MTB, Coast Guard, DOE, and the various state and local agencies.

In addition to these Federal agencies, state and local agencies are also involved in regulating the broad safety concerns of the industry. The result thus far has been a very complex and cumbersome regulatory structure. Inevitably duplications of authority and overlapping responsibilities exist. In some instances "memoranda of understanding" clarify the responsibilities between agenices. The MTB and the U.S. Coast Guard have, for example, agreed to a division of responsibility at waterfront LNG facilities although both have a mandate to regulate these same activities. Likewise, there is some duplication of authority concerning safety issues between the DOE and the MTB. The complex regulatory structure has resulted in a slow pace of development for the industry. While LNG policies in Japan and Europe have encouraged LNG expansion, those in the U.S. have not.[35]

Controlling LNG Pricing

In addition to its licensing concerns the Department of Energy is the agency which sets the price guidelines for natural gas. The pricing issue has often overshadowed other concerns in the application approval process.[36] DOE wants to insure that LNG import projects are advantageous to the Nation as a whole. It does not consider that LNG imported from Algeria at the current asking price of $6.00 per million BTUs will be in the best interest of the Nation. The cost should be competitive with alternative fuels or conservation measures over the life of the project. Imported Liquefied Natural Gas could be a major supplemental source of supply for U.S. energy needs. However, the DOE prefers that our natural gas come first from conventional sources within the U.S.[37] They argue that domestically produced natural gas would be priced lower, would be a secure source of supply, and would not be subject to the same safety concerns that accompany the transport and regasification of LNG.

The relationship between price and amount supplied is complex. Four basic issues complicate the picture.[38] First, the size of the remaining U.S. gas reserves is the natural and absolute limit constraining the amount of domestic gas supplied. As this limit is approached, remaining U.S. natural gas reserves will be more and more difficult to produce, requiring more sophisticated technology and hence ever higher production prices. Second, the lead time must be considered when developing domestic sources, and this time must be compared with the estimated lead time required to develop an LNG import project. The regulatory structure of the natural gas industry has a major impact on both domestically produced and imported natural gas. A third consideration is the extent to which prices in the other energy sectors, particularly oil, could affect the rate of gas exploration and development. Finally, one must consider the extent to which domestic gas is being kept off the market in anticipation of higher gas prices.

The debate about an appropriate price mechanism for LNG in light of the above considerations has centered on whether the price of imported LNG to customers should be rolled-in with the price of cheaper domestic gas or whether it should be priced at its true cost. Arguments against rolled-in pricing are that it masks the true costs of supplemental sources of energy and provides fewer incentives to conserve or to convert to other less costly fuels.[39] Since rolled-in pricing encourages the sale of high-priced LNG, project developers feel it is necessary to secure financing for projects. Without rolled-in pricing the economic viability of LNG projects may be problematic.

Incremental pricing of natural gas is favored by most DOE staff. Under this pricing mechanism, gas from each category is sold at a price that reflects its specific cost. LNG would be sold at a price considerably higher than domestically produced gas or gas imported from Canada or Mexico. The Natural Gas Policy Act (NGPA) of 1978 stipulates that LNG from projects planned after May, 1978 "be priced incrementally and paid for by certain non-exempt large industrial customers."[40] Both the NGPA and the Fuel Use Act of 1978 were established to make more North American gas available to high priority markets. The acts establish incentives for exploration and production and promote long-term conversion of oil and gas-burning facilities to coal. However, imported LNG is considered by the DOE to be a marginal source of supply given its relatively high price, capital intensiveness, and vulnerability to interruption.

III. Private Banks, Insurance Enterprises and LNG Transport

Would America have been better off not worrying about LNG safety and price and allowing the private sector to regulate these developments through the markets? Have we created so much red tape that America will be denied abundant supplies of this non-polluting energy source which is flared in many countries? Utility company executives, unhappy about the 1978 ban on new gas fired facilities warn that coal and other polluting energy sources will be just as expensive as LNG which they will have to replace. Even Barry Commoner sees natural gas (although not LNG) as the key bridging fuel to a solar future.[41]

Or, has the bureaucracy actually saved America from going overboard into a fuel technology which will prove uneconomical and unnecessary? Some critics of LNG point out that forecasts of U.S. energy demands have dropped substantially during the 1970's. The most pessimistic prognosticators in 1977 and 1978 were projecting energy demand in 2000 or 2010 lower than the wildest solar conservationists' predictions in 1972.[42] Perhaps the United States will not need as many new electricity generating facilities in the future as once expected.[43] Already, a strong argument holds that the nuclear industry may ultimately collapse not because of safety questions but because the economics of nuclear energy stripped of governmental subsidies are not cost effective when compared to other energy sources.[44] Is the same argument applicable to LNG? The question remains would the market working on its own have restricted the growth of LNG if it were unsafe or uneconomic and promoted it if it made sense?

The remainder of this paper asks whether the banking system could have determined whether LNG would prove cost effective and whether the insurance market could have maintained high safety standards unaided?

To clarify matters we distinguish between market mechanisms and government regulations in regard to the issue of safety and the issue of price. The following table suggests that within the private market banks should help determine an efficient price at which a project is feasible:

	Market Adjustment	Government Regulations
Price	Banks lend money if they will get it back, assuring appropriate price	Department of Energy sets/negotiates acceptable prices for energy sources
Safety	Insurance enterprises accept risks in order to get premiums. High premiums force greater safety	12 Federal agencies and states check for safe siting, handling and appropriate precautions

Theoretically banks will not lend to projects that they believe are not capable of producing a capital stream sufficient for debt repayment. This is particularly true for LNG projects which have generally been financed on a "project" basis under which banks are only repaid once the project is successfully running.[45] (This is similar but not identical to the all events tariff approach used by utilities under which they can bill their customers for the costs of a project once a thousand cubic feet of gas has flowed through the pipelines, even if the project is subsequently forced to shut down.) For their part, insurance enterprises should "rate" risks to assure that riskier ventures pay higher premiums, causing them, on average, to improve their safety record or go out of business. Whether banks and insurance enterprises are actually able to perform these functions is less apparent.

Can Insurance Enterprises Rate LNG Risks

There is a paradox central to insurance. These enterprises which allow customers to smooth their growth and profits by avoiding the financial burdens of catastrophic losses, have been totally incapable of stablizing their own growth and profitability. Underwriting cycles in each line of insurance have followed set patterns. When profits are high; new underwriters are attracted to business opportunities. Competition pushes rates so low that losses exceed premiums, driving the smaller and less competent underwriters to seek other lines of business. As losses mount, premiums are pushed back up.

In recent years competition has become even more fierce as underwriting capacity exceeded industrial demand for coverage. Many insurance enterprises chose to accept business to generate premium income (which could be invested at a profit) even if they suffered underwriting losses on the business. Investment profits dwarfed underwriting profits for almost every company. Competition from brokers, determined to earn their fees by getting insurance purchasers the best possible terms, and the abundance of available reinsurance capacity pushed insurers towards too much reliance on investment profits. The result has been an increase in the amplitude of underwriting cycles and a simultaneous shortening of the duration of each cycle. Some insurers are resigned to the inevitability of these cycles, arguing that it is impossible to persuade risk takers to behave in an orderly, rational manner. The larger, more expert insurers and reinsurers simply wait for the downturn in underwriting profits and the economy to wipe out the foolish speculators and restore order to the insurance market.[46]

Insurance for the transport of liquefied natural gas is as vulnerable to these cycles as any other lines of insurance. It should be noted at once that there exists no simple package for LNG. Participants in the trade of LNG insure against specific risks individually. Thus the value of LNG carriers are insured by the shipowners under standard marine Hull and Machinery coverage. Insurers accept the risks on entire fleets of vessels for a year at a time, so long as they remain within specified trading areas. LNG carriers are sometimes forced to accept somewhat higher deductables on damages than regular tankers because of the potential cost of repairing cryogenic systems, but when they are insured as part of an oil and gas fleet, the deductables appear to be standard.[47] El Paso, the largest American fleet owner, places their insurance through Sedgwick Forbes Bland Paine (major LNG brokers) who in turn distribute it in the British company market, at Lloyd's, and throughout Europe. The Japanese appear to insure their fleet primarily in Japan. Most LNG carrier insurance is handled within Lloyd's as a standard marine risk so separate from the remainder of the coverages that the lead for LNG carriers in Lloyd's was only vaguely aware of who the chief underwriters for non-marine LNG risks were.[48]

Shipowners rely on their Mutual Protection and Idemnity Club to insure against third party liability arising from the operations of their LNG carriers. They also use the P & I Clubs to cover themselves against such risks as: injury, death, and medical expenses of the crew; strikes; the need for salvage; and any fines, penalties, or legal costs that may arise. A. M. Platt, a leading LNG broker, summarized the situation as follows:

> I think it is fair to say that the Protection and Indemnity Clubs have viewed LNG carriers as being no more hazardous from the explosion liability point of view than that applying on an ordinary crude oil carrier, and it must certainly be true that the pollution risk is considerably less since the LNG vapors evaporate rapidly.[49]

The only exception on unlimited liability has been for American flag ships which P & I Clubs are reluctant to fully cover because American courts have usually assessed higher damages than anyone else and also been stricter in finding and interpreting damage than others. American ships are thus restricted to $100 million in coverage within P & I Clubs and can sometimes purchase additional coverage elsewhere.[50]

Shipowners also carry three other types of insurance protection, none of which are unique to LNG. War risks coverage is traditionally handled separate from regular hull and machinery insurance. Second, shipowners will normally protect themselves against loss of time and earnings brought about when cargoes cannot be delivered because the ship is under repair or idled by breakdowns at the liquefaction or reception facilities. After a waiting period, shipowners begin to collect for delays of up to 180 days (depending on the terms of the policy). Finally, shipowners may insure against business interruption or consequential loss sustained when LNG facilities on either end are shut down for a prolonged period. To date, however, no insurer has been willing to provide coverage protecting tanker owners from delays in the initial start up of LNG facilities. Malaysia, for one, has been caught with finished carriers before their fields and facilities were ready for start up.[52]

Another contract must be taken out by the cargo owner to cover the value of the cargo. Here, LNG is always separated from other cargoes. Until recently, cargo insurers were somewhat concerned that unlike oil tankers, LNG carriers' cargo was worth far less than the vessel that transported it. They feared that in a minor emergency the captian might choose to vent the LNG into the atmosphere. However, now that full cargoes are worth in excess of $30 million, this fear is receding. LNG cargo also presents the special problem in that between 0.13 and 0.25 percent per day of the cargo must be allowed to boil off to prevent pressure from building. The vapor can be used to partially fuel the carrier. Cargo owners can insure against excess boil off caused by bad weather or other conditions. Coverage is available at a cost of about 0.15 to 0.20 percent of the delivered value of the gas.[53]

On land, the owners of LNG facilities can purchase various coverages from either International Ocean Insurers in London or American International in New York. These coverages may handle facilities, pipelines, gas, and through separate coverages liability. However, the values involved are so huge that there does not always seem to be necessary capacity to cover possible losses. It is not clear, for instance, whether Arzew with an estimated replacement value of over $4 billion is fully insured or whether just Algeria's foreign exchange investment in the plant is covered.[54] In some cases it is also not clear what would happen if a major disaster occurred while a LNG carrier was transferring its cargo to the ground facility (or vice versa). It might not be possible to tell where the accident began or easily determine liability and coverage. However, to this point there has been no such disaster and the insurers feel that a major catastrophe is unlikely.[55]

This dispersion of coverage has made it difficult for the insurers to rate the safety of LNG operations as total systems. As of mid-1979 it appeared that various LNG rates were going in opposite directions. Cargo and carrier insurance were becoming cheaper because of the excess competition to write this business even though some accidents had begun to occur at sea and the structural problems of some carriers were beginning to show. In contrast, rates on land based coverage for LNG appeared to be rising.[56] It seems that the lack of capacity when coupled with a major liquefied petroleum gas accident in Qatar persuaded the insurers to raise premiums.[57] The underwriting cycle and the thirst for premiums to invest appears to have had far more to do with LNG insurance rates than the safety of the market. Even after the discovery of problems in the Avondale containment systems and the grounding of <u>El Paso Paul Kayser</u>, rates were too low. There was no incentive for LNG traders to take special precautions.[58] One LNG carrier insurer commented, "We've given them quite a deal to get this trade off the ground and we're still giving it to them."[59]

<u>Do Banks Assure the Economic Viability of LNG Projects?</u>

As with insurance, banks have fallen prey to a difficult dilemma. If a bank loans an individual or a small company $10,000 and the borrower does not repay, the borrower is in trouble. If a bank lends a large company or a country $100,000,000 and the borrower cannot or will not repay, the bank is in trouble. Banks and their largest borrowers are mutually hostages to each other.[60] In addition, banks are vulnerable to a cycle similar, though not as well recognized, as the one which plagues insurance companies. The number of banks willing and able to make loans in large quantity has increased more rapidly than the number of high quality borrowers. As a result, the most desirable borrowers have found themselves able to negotiate ever improving terms with their creditors. The tenor of many loans has increased, and the spread borrowers pay over the cost of money to the banks has narrowed until the best borrowers can take money at less than ½ percent over the banks' cost of funds. Unlike insurance, there is no simple mechanism to reverse the fall in rates. Large companies and countries are not supposed to go bankrupt and default on their debts. In fact, with rare exceptions, strong governments are not willing to allow countries or important private institutions to become insolvent and mount rescue operations which stretch out payments but guarantee eventual recovery of most funds.[61] However, recent events in Iran demonstrating that countries can default have shaken the banking hierarchy and forced them to reexamine their political risk analysis and their exposure to defaults.[62] Still, while there is some indication that banks may extend international credits less recklessly in the future, rates have not widened noticeably in the first two quarters of 1980.

The financing of LNG projects present a special problem for financiers. Liquefaction facilities, LNG carriers, and regasification operations are financed separately, but if any of them falters, the entire integrated system collapses. In addition, the scale of the investment required and the long time frame of projects make it difficult for banks to become involved in all phases of LNG projects. In practice, banks have played a major role in LNG carrier financing, but have been less important to the financing of land-based facilities. Insurance company investment and government guarantees usually have been necessary to assure project viability.

Financial backers require that any integrated LNG project financing meet four criteria. First, the terms of the financing should allow the cost of the entire system to be repaid over a period commensurate with its economic life. Financiers are uneasy that the burden of repaying a LNG project with a twenty year lifetime in only ten years could place unhealthy financial strains on the borrower and the project. They prefer to finance long-assets with long-term debt. Second, adequate credit support must be available. Backers require that all project participants are sound and solvent, not just their own borrowers. Third, the suppliers of funds are unwilling to carry non-financial risks. Government guarantees indemnifying the borrower in cases of _force majeur_ are frequently required. Finally, backers demand minimum cost financing for the project to protect its viability. Since, in any LNG project synergy exists, the financial stability of the whole depends on the separate financing of the parts making economic sense.[63]

The proposed financing package for the Pacindonesia project is indicative. The liquefaction facilities at Arun, Indonesia would be owned and separately financed by Pertamina, Indonesia's national oil company. This financing would in part depend upon the contractual agreements between Pertamina and Pacific Indonesia. Pacific Indonesia would be required to make payments to Pertamina only if Pertamina has available for delivery the contracted volume of gas.[64]

The nine LNG carriers working the Arun - Point Conception run will be owned by several separate corporations. Three vessels were financed and constructed in foreign yards. The six U.S. flag vessels in the fleet will require U.S. government ship financing guarantees under Title XI of the Merchant Marine Act of 1936, as amended. The nine vessels will be chartered by Pacific Indonesia at fixed charter hire rates which are tied to its ability to collect payment for the gas at set rates from the Southern California Gas Company (and its customers). To encourage the U.S. shipbuilding industry, guarantees are available for up to 87½ percent of the cost of the vessel and for terms of up to twenty-five years.[65]

The LNG carrier financing is the trickiest portion of the package. Backers insist that vessels be tied to specific routes before they are constructed. Normally, the carrier owner and the receiving utility sign a "take-or-pay" contract. These come in two forms. "Hell or high water" contracts require that a credit-worthy entity make payments for a product or service under all circumstances for a specifie period of time. More lenient "conditional" contracts commit the taker to pay unless the gas is not made available to the supplier due to its failure to perform. On a "Hell or high water" contract only the solvency of the user matters. On a "conditional" take-or-pay contract the supplier's ability and willing ness to provide the gas must also be considered.[66]

While the Pacific Indonesia project plans to base its financing on guarantees available through the U.S. government, other finacing techniques are available. Lease financing and particularly "leveraged leasing" have grown in popularity because they provide tax advantages to the lender and require less up-front capital for the borrower. These arrangements are closely analagous to "bare-boat charters" in which control of the equity remains with the lenders and the use of the carriers is granted to the operators. Particularly sound borrowers may also borrow from commercial banks for periods of up to five to seven years or tap the commercial paper markets for short periods for inventory financing during some portions of the production cycle.[67]

The California facilities would be constructed by Western LNG and could be financed assuming appropriate financial agreements with the utility assured that in case of failure to complete construction, the California Public Utilities Commission would allow the utility to repay the lenders, by passing the cost on to consumers if necessary. Insurance companies and perhaps pension funds would provide the necessary capital. In fact, one reason the utilities remain eager to import LNG is that only by getting the project moving can they assure that they will be able to recover funds already disbursed in the planning and licensing phases.

Unfortunately, the size of the borrowers involved in LNG projects, the willingness of the government to provide guarantees to stimulate shipbuilding, and the small probability that the government could afford to allow a major utility to declare bankruptcy has tended to get the financiers so involved in the trees of specific LNG projects that they lost sight of the LNG forest. The banks and insurance companies have protected themselves from the likelihood of default, but they have lost their ability to determine whether a specific project is viable without government guarantees. A vicious circle is created. Banks and other financial backers can only assure that a project is economically viable if their funds are at risk, but they are unwilling to risk their funds unless the government guarantees them against most of their exposure. The banks can force the utility and the relevant public utility commission to insure that the gas will be purchased and rolled into the customers' bills at whatever price it is delivered. The project is simply not financable without long-term guarantees that LNG will be used and paid for. Once the infrastructure is in place and the project is operating, the investors cannot allow themselves to be undercut by cheaper fuel or to be priced out of the market by inflated gas prices. The market is biased against making ambiguous decisions on long-term prices and supply factors.

IV. Conclusions and Further Hypotheses

Our review of existing government regulations relevant to the safety and pricing of LNG operations in the U.S. indicates that the regulatory process is expensive, time-consuming, and bothersome. This is as true where new projects are safe and economically viable as when they are potentially hazardous and of unclear economic merit. Distrust and disharmony disrupt corporate-government cooperation to achieve energy breakthroughs except at times of crisis. Mechanisms need to be devised which will allow the public and private sectors to collaborate without losing sight of the public interest.

However, those calling for an immediate return to an unregulated market economy fail to recall that America's regulatory system did not appear miraculously, but was a response to callous market manipulation motivated by greed for wealth and power. The government intervened to insure that the market would not shake itself apart or discriminate against certain segments of the population. Over time, the government became ever more entangled with the market and with private market participants. The government provided a safety net to backstop the largest corporations in critical industries from bankruptcy, allowing them to take larger risks than they otherwise could have accepted. Today, the financial system could not fund huge LNG projects without assurances that the borrower was large enough to absorb losses in other parts of its business or that the government would guarantee the loans. Similarly, the insurance market remains more interested in premium income than in rating risks. Indeed, at the extreme, insurers never want to force unsafe operations out of business since they would then forfeit future potential premiums.[68] Both the financial and the insurance communities need government backstopping to conduct their business when large amounts of funds are involved. It seems probable that government withdrawal from the markets would slow rather than increase the pace of research, innovation, investment, and development in LNG and other expensive industries. As long as financial and underwriting cycles persist, banks are unlikely to monitor economic viability of large projects and insurers will not be able to assure their operations are within acceptable safety and pollution limits.

But, and it is a very big but, the present strained relationship between business and government is also not working. America and much of the rest of the industrial world faces a crisis of falling productivity.[69] Few companies have been willing to invest in new capital facilities or to undertake expensive basic research and development decisions without government backing. High inflation and heavy taxation persuaded many companies that they could not earn an adequate return on their capital pursuing these goals so they started to acquire service sector operations instead.[70] The government is caught in a trap. They cannot efficiently promote industrial innovation in the present phase of capitalist development, but neither can they withdraw from the market and await a Friedmanian miracle to deliver salvation.

Marxists have a simple, one word solution to this dilemma -- revolution. Reformers have a more difficult task before them, to find ways to allow the economic system to adjust to changing conditions and needs before it collapses under its own weight. Before the government is likely to be able to withdraw from its present, over-regulated stance, two major changes in policy process are needed. First, we posit that the government needs to enact policies whichwill encourage industrial corporations to undertake innovative problem solving tasks which will benefit the nation and earn profits for the corporation. The government needs to indicate to the private sector what they want accomplished without telling them exactly how it must be accomplished.[71] There are some encouraging signs that this approach is winning advocates. For instance, the U.S. government's effort to promote synfuel provides the private sector with massive infusions of funds but only minimal regulation of their specific use. On the other hand, the Congress recently killed the proposed Energy Resources Corporation which would have allowed new energy developments to bypass much of the bureacratic licensing and red tape most projects face if it could demonstrate that it was acting in the national interest. In this same vein, there need to be new policies which will allow subcontracting companies to retain patents developed under government sponsored research. Without financial incentive, private researchers are less than forthcoming about the true results of their work, particularly about serendipitous discoveries.[72]

Second, we posit that the financial and insurance communities need to be encouraged to reestablish their economic functions within the market and police would-be industrial sinners. The government needs to distinguish between regulations which curtail the activities of financial intermediaries and those which encourage them to promote economic and environmental sanity. It may make more sense to regulate new developments in LNG and other exotic technology areas indirectly through the banks and insurers than to constrain them directly. By recognizing the catalytic role of service sector industries and encouraging them to act countercyclically and maintain the market, a more efficient, flexible regulatory scheme may eventually emerge.[73]

It is possible that the delays and red-tape surrounding government regulation of new LNG projects in the United States may have saved the operators tremendous sums of money. If this is true, the government made the right decisions in the wrong ways. In order to stimulate sensible economic and risk decisions in the future, new ways need to be devised to smooth the interactive networks betweeen government, financial intermediaries, and industrial firms. Otherwise, the risk-taking that the government is trying to banish from the economy will come back to haunt us in the form of destabilizing catastrophes.

Notes

1. Stuart, Alexander. "El Paso in the Grip of the Frozen Fuel," Fortune, July 14, 1980, p. 71.

2. Zuinlan, Martin. "Moving into Energy Deficit," Petroleum Economist, March, 1980, p. 107-109. See also "Becoming an Importer *in* The Economist, March 17, 1979, p. 88-89.

3. Corkhill, Michael. LNG Carriers: The Ships and Their Markets, Fairplay Publications Limited, 1975, p. 1.

4. Thomas, William Dubarry. "Record of Voyages Completed by the World Fleet of LNG Carriers," El Paso LNG Company, Houston, 1978.

5. Federal Power Commission, Bureau of Natural Gas, "Preliminary Staff Report on Investigation of Disaster at Texas Eastern Transmission Corporation LNG Storage Tank on Staten Island, Borough of Richmond, New York City, New York, February 10, 1973," Docket No. CP73-235, Texas Eastern Transmission Corporation, Washington D.C.: FPC, July 9, 1973.

6. E. F. Shumaker, "Ship-to-ship transfer of LNG, the "El Paso Paul Layser"/"El Paso Sonatrach" gas transfer operation," Proceedings: GASTECH 79 LNG/LPG Conference, Houston, Texas, November 13-16, 1979. (Rickmansworth, Herts, England: gastech Ltd., 1980): 173-175.

7. Stuart, op. cit., p. 74.

8. Stuart, op. cit., p. 70.

9. Forbes, June 23, 1980, p. 62-63.

10. Interview with Albert Hunt, Vice President for Finance, Southern California Gas Company, Los Angeles, July 10, 1980.

11. "LNG Project Hopes Dim in Face of Looming Gas Price Increase," Los Angeles Times, April 20, 1980. Part II, p. 1, 6. Indonesia is already moving towards higher prices - See: "Indonesia-Japan LNG Sale Testing OPEC Price Plan," Asian Wall Street Journal, July 21, 1980, p. 22.

12. Greenwald, Gerald B. "LNG Carrier Safety: A Guide to the System of Federal Regulation," Journal of Maritime Law and Commerce, Jan. 1978, p. 159-160.

13. U.S. Coast Guard. Liquefied Natural Gas: Views and Practices, Policy and Safety. Dept. of Transportation. CG-478, February 1, 1976, p. II-8.

14. Oceanographic Commission of Washington, LNG and LPG Hazards Management in Washington State, submitted by the Oceanographic Commission

of Washington to the 46th Legislature of the State of Washington, December, 1978. It is reported here that "under certain conditions, LPG can be more hazardous than LNG." Unconfined vapor clouds can explode. When stored under pressure there is some possibility of a boiling liquid expanding vapor explosion (BLEVE). And the vapor cloud from an LPG spill, being heavier than air, would take longer to disperse, p. I-8.

15. Ibid, p. IV-2.

16. Corkhill, op. cit., p. 36.

17. The two best known attacks on LNG safety are: 1) Peter van der Linde, Time Bomb: LNG - The Truth About Our Newest and Most Dangerous Energy Source. Doubleday, Garden City, New York, 1978, and 2) the more carefully documented Lee N. Davis, Frozen Fire: Where Will it Happen Next, Friends of the Earth, San Francisco, 1979.

18. The most comprehensive is probably the Comptroller General's Report to the U.S. Congress on "Liquefied Energy Gases Safety." General Accounting office, Washington, D.C., July 31, 1978.

19. Office of Technology Assessment. Transportation of Liquefied Natural Gas. U.S. Government Printing Office, Washington, D.C., September 1977, p. 9.

20. Ibid, p. 96.

21. Oceanographic Commission of Washington, op. cit., p. IV-30.

22. Oceanographic Commission of Washington, op. cit., p. IV-32.

23. Department of Energy/Economic Regulatory Administration. "Opinion and Order on Importation of Liquefied Natural Gas from Indonesia," December 30, 1977, p. 8.

24. Office of Technology Assessment, 1980. Alternative Energy Futures: Part I, The Future of Liquefied Natural Gas Imports, Washington, D.C., March 1980, p. 22.

25. Oceanographic Commission of Washington, op. cit., p. IV-1.

26. Office of Technology Assessment, 1980, op. cit., p. 17.

27. Oceanographic Commission of Washington, op. cit., p. IV-4.

28. Office of Technology Assessment, 1980, op. cit., p. 18.

29. Oceanographic Commission of Washington, op. cit., p. IV-4.

30. Office of Technology Assessment, 1977, op. cit., p. 19.

31. Office of Technology Assessment, 1977, op. cit., p. 19; Greenwald, op. cit., p. 165; Butler, op. cit., p. 801.

32. Office of Technology Assessment, 1980, op. cit., p. 22.

33. Oceanographic Commission of Washington, op. cit., p. IV-3.

34. Ibid, IV-6.

35. Alexander, op. cit., p. 73.

36. Office of Technology Assessment, 1980, op. cit., p. 18.

37. Ibid, p. 17.

38. Faridany, Edward K. and Roger C. Ffooks. LNG Review, 1977. Energy Economics Research Limited, September 1977, p. 8.

39. Office of Technology Assessment, 1980, op. cit., p. 19.

40. Ibid, p. 19.

41. Barry Commoner, The Politics of Energy, Alfred A. Knopf, Inc., 1979, p. 63-64.

42. Eliot Marshall, "Energy Forecasts: Sinking to New Lows," Science 208, June 20, 1980, p. 1353-1356.

43. In this vein, the Lawrence Livermore labs announced in July 1980 that it may not be necessary to build further electricity generating utilities in California until 1995. Los Angeles Times, July.

44. Amory Lovins, L. Hunter Lovins, and Leonard Ross, "Nuclear Power and Nuclear Bombs," Foreign Affairs 58:5, Summer, 1980, p. 1137-1177.

45. See: Stanley Lewand and Glenn Meyers, "Project Financing of LNG Trade," Proceedings of LNG5, Dusseldorf, September 1977 or more narrowly Samir Mankabady, "Financing LNG carriers," Marine Policy 3:3, July 1979, p. 225-231.

46. So claimed the larger actors in interviews in London, Zurich, and Munich in May and June, 1979.

47. However, as with all marine contracts in cases of a total loss or a constructive total cryogenic containment system which do have some potential rating impact. See: "LNG Carrier Analysis (London: The Stedgwick Forbee Group, no date) and John Mellows, "Broker's view of LNG insurance," Lloyd's List, July 5, 1976, p. 4.

48. Interview with Lloyd's lead for LNG carriers, London, May 24, 1979.

49. A. M. Platt, "Insurance of LNG Vessels," Tanker and Bulker International, January/February 1976, p. 29.

50. Ibid.

51. Samir Mankabady, "Insurance and reinsurance requirements of LNG," Marine Policy, October 1978, p. 332.

52. Interviews in London, Spring 1979. On the development of the Malaysian project and the reasons for the delays see: M. B. Hashim, "The Malaysian LNG Project," Proceedings: GASTECH 79 LNG/LPG Conference, Houston, Texas, November 13-16, 1979, (Rickmansworth, Herts, England: Gastech Ltd., 1980), p. 7-13.

53. Interview with Lloyd's lead for LNG cargo insurance, London, March 20, 1979. Also: Samir Mankabady, "Insurance and reinsurance requirements of LNG," Marine Policy, October 1978, p. 332-333.

54. This is unconfirmed but was suggested by a leading Citibank syndicator with prime responsibility for Algeria, London, May 14, 1979.

55. See: "The Evaluation of Estimated Maximum Loss from Fire or Explosion in Oil, Gas and Petrochemical Industries with Reference to Percussive Unconfined Vapor Cloud Explosion," (London: International Oil Insurers, April 1979.

56. Interview with Lloyd's lead for LNG carriers, London, May 24, 1979. Rates on carriers have remained low. We have no information on current non-marine rates.

57. Interviews London and Munich, spring and summer 1979.

58. Note, however, that since these are first line vessels and are not used in the spot market the quality of crews is high and the vessels are relatively new and well maintained. Since companies own and run these vessels insurers are less fearful of carelessness or fraud.

59. Interview, Lloyd's of London LNG specialist, June 25, 1980.

60. See: Jonathan David Aronson, "The Politics of Bank Lending and Debt Renegotiation," in J.D. Aronson (ed.) Debt and the Less Developed Countries (Boulder: Westview Press, 1979), pp. 283-316.

61. For U.S. policy on debt rescheduling for delinquent national borrowers see: Brian G. Crowe, in J.D. Aronson (ed.) Debt and the Less Developed Countries. The most recent cases of government resolve to keep large industrial enterprises and banks from filing for bankruptcy are the cases of Chrysler and First Pennsylvania. See: Statement of G. William Miller before the Senate Committee on Banking, Housing and Urban Affairs, November 14, 1979 for the governments logic on the Chrysler bailout. On the First Pennsylvannia crisis see: Linda Snyder Hayes, "The Bet that Almost Broke a Bank," Fortune 101:11, June 2, 1980, p. 48-50.

62. Jonathan David Aronson. "Iran, the Dollar, and the International Monetary Sys Paper presented at the annual meetings of the International Studies Association, Los Angeles, March 1980.

63. Alan Batkin and Roger Goodspeed, "Financing Aspects of LNG Trade," Proceedings of LNG5, Dusseldorf, September 1977.

64. Testimony of T. Kimball Booker before the Federal Power Commission in the matter of the Pacific Indonesia LNG Company, January 23, 1976. Docket No. CP74-160, et. al. p. 9.

65. Ibid.

66. Batkin and Goodspeed, "Financing Aspects of LNG Trade," op. cit.

67. Ibid.

68. See: Kenneth Arrow, "Risk Allocation and Information: Some Recent Theoretical Developments," The Geneva Papers on Risk and Insurance No. 8 (Geneva: L'Association Internationale pour l'Etude de l'Economie de l'Assurance, June 1978).

69. See: Edward F. Denison, "Explanations of Declining Productivity Growth," Survey of Current Business, August 1979, pp. 1-24.

70. We thank Nancy Peretsman for discussion and insight on this point.

71. Charles Schultze, The Public Use of Private Interest (Washington: the Brookings Institution, 1977).

72. Interviews, Jet Propulsion Lab, Pasadena, July 22, 1980.

73. For instance, until the early 1970's large banks performed the role of equilibrating speculators within the foreign exchange markets. When they withdrew after two dollar devaluations, exchange rate volatility increased. For more analysis of the service sector in the world economy see: Jonathan David Aronson, "Industrial Structure, Surplus Capacity, and the International Monetary System: The Role of the Service Sector in the World Economy," presented at the annual meeting of the American Political Science Association, Washington D.C., August 29, 1980.

THE CEIP: AN OHIO PERSPECTIVE

Michael L. Adams
Programs Coordinator
Ohio Department of Energy
Columbus, OH

Over the past three years, the Coastal Energy Impact Program (CEIP) in Ohio has proven itself to be a useful tool for mitigation of certain impacts resulting from coastal energy facilities and activities. The CEIP, as a sister to the Coastal Zone Management program, has created new interest in the adoption of CZM objectives and priorities at both the State and local levels, and helped to assure orderly development of domestic energy resources.

However, continuing deficiencies in the CEIP's statutory language and regulations inherently limit local enthusiasm for CZM involvement. The program's bias toward labor intensive energy facilities and activities discriminates against states with proportionally larger emphasis on coal. More importantly, recent changes in federal administrative philosophy are threatening the State's prerogatives regarding the administration and distribution of its allotted CEIP funds. The Office of Coastal Zone Management (OCZM) is seeking to regulate the State CEIP by adding additional review procedures, lengthening the federal application review process, and making CEIP subservient to, rather than complementary to, the 306 program.

Ohio believes that these changes in OCZM administration are contrary to the intent of Congress which mandated

> "...that the coastal states and localities, which are closer to and more cognizant of the situation, should make the basic decisions as to the particular needs which result from such new or expanded energy activity; and <u>that the discretion of the Secretary of Commerce and other Federal officials should be correspondingly limited.</u>"[1]

A brief history of this original legislation is given here to illustrate the atmosphere and intents under which the CEIP was created.

Additionally, Ohio's Intrastate Allocation Process and new OCZM initiatives to require additional funding decision documents are explained. Finally, the program's advantages, and the relative importance of coal in funding allocations are examined.

LEGISLATIVE BACKGROUND

With enaction of the Coastal Zone Management Act Amendments of 1976 (P.L. 94-370), Congress took the first step toward integrating general self-sufficient energy production. These two agendas had been causing conflicts in administrative handling of state CZM plans submitted for federal approval. In their review of the Washington and Maine plans, the Federal Energy Administration, the Federal Power Commission, and the Energy Resource and Development Administration (The U.S. Department of Energy's predecessor agencies), objected to the states' treatment of energy development provisions. In the spirit of post-Arab embargo national energy objectives, these agencies protested to a lack of specific designation of "areas of particular concern" which might serve as sites for energy facilities.

The basis of the problem stemmed from the original CZM Act of 1972, which required that state programs provide for "adequate consideration of the national interest involved in the siting of facilities necessary to meet requirements which are other than local in nature."[2] While the Office of Coastal Zone Management agreed with the energy agencies that energy activities and facilities were of a greater than local area nature, it continued to support the view that the states were not obligated to designate specific areas or sites for energy facilities.

This crucial division in the Ford Administration was aggravated when the White House sent a comprehensive energy facilities siting bill for each state to prepare a program for siting of power plants, oil refineries, and other energy facilities which would be acceptable to the FEA Administrator. If the Administrator found any portion of the siting program to be inconsistent with federal energy policy or guidelines, he could override the state's program, and require revisions. Further, the siting program was not to be bound by CZM plans. Rather, the bill required that such a siting program be compatible with CZM plans "to the extent possible."

This final provision prompted Russell W. Peterson, then chairman of the Council on Environmental Quality (CEQ) and a former governor of Delaware, to express his disagreement with the legislation in a letter to Frank Zarb, the FEA administrator. Peterson wrote:

> "I am concerned about the effect such legislation if enacted might have on the existing Coastal Zone Management program. In addition, serious questions

> are raised by provisions in the bill that would allow federal administrative decisions in certain circumstances to override long-standing state and local laws governing the siting of energy facilities and the regulation of land use.
>
> ...Opposition to the measure as proposed can be expected to be strong in the Congress, and to be backed by the state, local, and citizen opposition related to both the need for legislation and the unnecessary federal involvement included in the act. The result cannot fail to be prolonged discussion on the legislation that should be simpler, more palatable, and more in the spirit of intergovernmental cooperation. The answer to our energy siting problems cannot be to try to override other laws and practices with single-purpose wisdom."[3]

The CEQ was joined in its dissent by the Departments of Commerce and the Interior, and the Environmental Protection Agency.

Peterson correctly pointed out that many states had existing energy facilities siting policies which were designed to meld local development priorities with regional and national needs for energy activity expansion. For example, the Ohio Power Siting Commission (OPSC) was created in 1972 to establish a "one stop" state commission for gas and electric utility companies seeking to locate major utility facilities within Ohio. The commission's basic function was to issue certificates of environmental compatibility and public need for the location, design, construction and initial operation of major utility facilities. The utility's formal application must contain two viable sites for each facility and/or transmission line corridor.

The OPSC makes a decision to grant a certificate of environmental compatibility based on the following criteria:

1. The basis of the need for the facility;
2. The nature of the probable environmental impact;
3. That the facility represents the minimum adverse environmental impact, considering the state of the available technology and the nature and economics of various alternatives;
4. In the case of an electric transmission line, that such facility is consistent with regional plans for expansion of the electric power grid of the electric systems serving Ohio and inter-

connected systems; and that such facilities will serve the interests of electric system economy and reliability;

5. That the facility will comply with air and water pollution control and solid waste disposal laws and regulations;

6. And that the facility will serve the public interest, convenience and necessity.[4]

The nature of Ohio's siting process, and similar siting policies of other states, demonstrated three distinct contestable issues in the administration's arguments for comprehensive energy facility siting legislation. First, the administration proponency was based upon the tacit assumption that states were not considering the national priority for energy development in their respective state plans. This was obviously incorrect, as demonstrated by the fact that states began adopting siting legislation which included determination of regional and national need well before the Arab oil embargo of 1973 and the subsequent increase in federal pressure for domestic energy development.

Secondly, the administration's bill required the states to designate potential energy facility sites. This was contrary to many of the operating state siting processes, such as Ohio's, which left the burden (and cost) of proving site acceptability to the utility.

Finally, there were obvious reasons for contesting the bill's authorization of the FEA Administrator's right to preempt or override state siting processes.

Fortunately, Peterson's forecast of Congressional opposition was correct. This bill suffered the same death as previous years' power plant siting bills. It did exhibit, however, the President's position that energy facility siting was an immediate national priority, nothwithstanding CZM plans or policies, and that federal preemption of direct or indirect state energy siting programs was not unconscionable.

The intentions of the Congress were directed along a different line --that of appeasing states who were feeling unprecedented federal pressure in the development of Outer Continental Shelf (OCS) oil and gas. This issue was quantified by Representative John Murphy (D-New York) in his opening remarks to the Subcommittee on Oceanography concerning the 1975 CZM Act Amendments bill:

> "The area on which Congress is focusing its attention is...whether our coastal States will be adequately brought into the entire process of OCS development and energy facility siting and whether they will be given sufficient assistance to research and plan for

> OCS energy-related activities and to ameliorate any adverse environmental, economic, and social impacts which result from those activities."[5]

Up to this time, the Congress had been considering a number of bills containing provisions for energy impact assistance to the states. These legislative proposals generally divided into three categories:

1. As amendments to the Outer Continental Shelf Lands Act of 1954;
2. As amendments to the Coastal Zone Management Act of 1972; or
3. As parts of a comprehensive domestic energy impact assistance program.

None of these proposals were originally supported by the Ford administration. In his testimony before House and Senate Committees during this period, Robert Knecht, Assistant Administrator for OCZM, indicated that the administration was "actively studying proposals on this (assistance) issue, ranging from revenue-sharing plans to direct impact grants...Consequently, we do not support such changes at this time."[5] (The referenced studies were ostensibly being conducted by the Department of Interior, but were postponed indefinitely when Thomas Kleppe was appointed secretary of that agency).

Congressional interest centered on compensating states for OCS activities which were stressing local capability to provide on-shore public facilities and services. Members of the House Committee on Merchant Marine and Fisheries agreed that the final program was to address "net adverse impacts," basing their conclusions upon studies which had shown major economic impacts upon shoreline communities as a direct result of OCS activities.

Although various funding schemes were discussed, including direct sharing of revenue derived from OCS federal leases, the final bill created a new direct grants and loans program to be funded by the U.S. Treasury. The Great Lakes states were included under grants for planning the mitigation of impacts from coastal energy facilities, grants for mitigation of losses in environmental or recreational resource value caused by coastal energy activities, and direct loans.

INTRASTATE ALLOCATION PROCESS

The primary strength of CEIP funds in the past has been the latitude afforded the States in determination of priorities and uses for the funds. A required Intrastate Allocation Process (IAP), developed by the State in consultation with its affected local government, is filed with and approved by the CEIP Washington staff prior to the solicitation of local project applications. The State applies its IAP to project applications, and submits those which most adequately meet the criteria for federal funding.

In Ohio, this IAP has been found to be a simple and highly effective means of portioning our limited CEIP funds. The process has several advantages:

1. All local applicants are aware of the criteria upon which their proposals are to be judged;
2. All State agencies and the Washington CEIP staff are notionally aware of the IAP's inherent incorporation of State energy and CZM objectives;
3. The review of local project applications may be done at the State level in a very short time (often no more than 10 working days); and
4. The allocation and application processes satisfy the State's complicated contractual (or grant) award process.

The last advantage is often overlooked by federal agencies, but may, in fact, be the most vital. In Ohio, all contracts (and grants) of greater than $10,000 must undergo legislative review. This review includes consideration of the process upon which awards are made. The official IAP used for CEIP awards has been an acceptable way of satisfying these requirements, thereby avoiding costly delays in State processing of local applications. Other states have similar legislative or administrative review requirements which may be satisfied by the IAP.

Ohio has exercised two optional policy decisions within its IAP. First, no single type of energy facility or activity has been considered more 'eligible' for basis of CEIP funds awards than any other. For example, projects addressing impacts from electricity generating stations are not considered more desirable than those addressing coal dock impacts.

There are many reasons for this policy. Along Ohio's 140-mile coastline there are eight coal handling ports, nine fossil-fuel electricity generation stations, one operating nuclear power plant, two nuclear units under construction, oil and gas wells and pipelines, an oil refinery, and several oil storage facilities. These energy facilities, and their associated energy activities, are evenly dispersed between Toledo and Conneaut allowing nearly every local government a legitimate claim of attributable impacts. In addition, a proposed U.S. Steel plant in Conneaut will have its own electricity generating plant and fuel handling and storage facilities. With such a wide range of potential impacts from such diverse energy activities and facilities, it has been more rational to prioritize project applications on the basis of impact magnitude, immediacy of need, and demonstrations of local support for the project.

Secondly, Ohio produces substantial quantities of coal, crude oil, natural gas, and biomass products. Therefore, projects which address impacts from any of these fuels benefit the State equally.

Third, Ohio has long supported creation of a separate fund under CEIP for mitigation of coal impacts. The production, transshipment, and use of coal is a national priority equal to the need for OCS fuels, and should be specifically addressed under CEIP.

As of this writing, legislation being considered by the House of Representatives which would reauthorize the Coastal Zone Management Act includes a new 308(c)(3) fund for grants to coal-impacted states. Should such a fund be created, Ohio's IAP would be revised to limit planning and environmental loss grants to non-coal related impacts.

The second policy decision exercised in Ohio's IAP is closely related to the first. Namely, the type of impact to be addressed and its priority relative to other local impacts are determined by the local applicants. The State does not mandate the impacts to be addressed during the funding period, although advice is often sought by local agencies when more than one impact from a single energy activity is apparent. This concept is carried over from the original CEIP regulations of 1977. Those regulations required consideration of the State's CEIP investment priority needs as part of the IAP. Ohio's original program vested this requirement in local government agencies. Although specifically deleted from later federal program regulations, Ohio has maintained an emphasis on local priority determinations.

In July, 1980, OCZM issued an 'action memorandum' requiring State CEIP agencies to submit a "5-10 page energy impact strategy...describing the likely focus of its CEIP program for the next few years." This document was to aid in efforts to establish annual funding for the CEIP.[6] Ohio has no objection to supplying any information which would contribute to this worthwhile objective.

However, in later correspondence from CEIP Washington staff, the States were informed that this 'strategy' was to be also used to "make grant and loan approval decisions on the individual state applications."[7] There are three obvious objections to this latter purpose.

First, a long-term investment 'strategy' cannot replace the IAP as the basis upon which funding decisions are made. This was recognized in 1978 when the priority needs determination required under CEIP/IAP regulations was withdrawn.

Secondly, if this strategy <u>is</u> to have a bearing on funding decisions, then the requirements for its preparation and contents, and the exact effect it will have on those decisions, should be placed in the CEIP regulations. It is inappropriate to have such requirements issued in divergent 'action memoranda' when they affect such a variety of state, local, and regional government agencies. While it is agreed that the regulation process is lengthy, its purpose is to assure all affected parties equal opportunity to examine and respond to major programmatic changes, especially those which affect federal funding decisions.

Finally, it is clear that OCZM has altered its function under CEIP. Under the IAP, the State is responsible <u>to the local applicant</u> to assure that all applications for federal funding meet IAP and other program eligibility criteria. This is the reason for including an appeals process in the IAP. OCZM acts as arbitrator in those appeals involving misuse of the IAP.

The same logic should be applied to any further documents used for funding decisions. The State should be responsible for assuring equitable treatment <u>to its applicants</u> under the investment strategy. Their satisfaction, and that of the 306 agency, should be sufficient documentation for federal approval. Deviation from this procedure, especially the specific inclusion of OCZM review of State adherence to the investment strategy, is contrary to the Congressional directive of federal nonintervention presented above.

<u>CEIP ADVANTAGES</u>

In Ohio, responsibility for the CEIP is designated to the Ohio Department of Energy (ODOE). Administration of 305/306 is assigned to the Ohio Department of Natural Resources (ODNR). While federal authorities feel that this division is inappropriate, State and local officials have been pleased with the arrangement.

Clear advantages lie in the interest of two State Cabinet-level agencies supporting CZM objectives, as opposed to only one. State legislative opposition to some ODNR programs has not affected CEIP, since ODOE does not experience similar legislative difficulties. Additionally, the congressional intent invested in the CEIP includes provision that energy facility siting not be hampered by the program. It is appropriate, then that the CEIP be administered by a pro-development energy agency.

At the project level, CEIP provides a unique opportunity to coordinate with other federal programs which indirectly address energy activity impacts. In several instances, the CEIP assistance offered to a particular project in Ohio has resulted in the final commitment of funds from other local, state, and federal sources. For example, in Ashtabula County,

Ohio, five federal, two state, and three local agencies are contributing to the completion of one CEIP park renovation project. The final project will have used $100,000 in planning grant funds, $33,000 in e-r loss grants, $4.6 million in (d)(1) loan funds, and more than $2 million from other sources.

Similarly, CEIP planning funds are being used to finalize engineering drawings for a flood protection structure in Eastlake, Ohio. These monies are being combined with construction funds from the Army Corps of Engineers, the Department of Housing and Urban Development, and local sources to provide a complete funding package. Such projects demonstrate the most attractive aspects of the CEIP, and permit the most efficient use of limited CEIP funds.

Finally, the availability of CEIP funds has been contributing significantly to increased local interest in formal adoption of a State CZM program. As of this writing, however, this support has not been sufficient to pass the State's needed CZM legislation.

COAL AND THE CEIP

As mentioned briefly above, the specific inclusion of coal transshipment as basis for a CEIP funding category is a prime objective of the Great Lakes States. However, such an inclusion would benefit a number of coastal states, at least eighteen of which ship coal through major ports.

There is ample reason to support a new coal-related category. In 1979, coal shipment in the Great Lakes rose to 45.8 million tons from 37.8 million tons in 1978. In Lake Erie alone, coal shipments increased a dramatic 22% over the previous year.[8] Canadian shipments of coal on Lake Superior have begun, with 4 million tons shipped already this year. Coal port expansions and development are either under study or construction in several Great Lakes cities, including Superior and Kewaunee, Wisconsin, Buffalo, New York, and several Ohio ports.

With the renewed national interest in the use of coal, the production and shipping of domestic coal resources will continue to increase dramatically. Coal-producing states with coastal ports are actively promoting increased exportation to Western European countries, as well as domestic consumption. Deliveries to Canadian power plants continue, and the movement of Western States' coal to the east is increasing.

Current CEIP funding categories fail to meet the needs of coal states in two manners. First, funding under (c)(1) and (d)(4) programs has been inconsistent, including fiscal years with no funds at all in these categories. Administration support for these programs, as evidenced by NOAA/OCZM budget requests, has been similarly lacking, even though individual staff members have emphasized the need for continued funding levels.

Secondly, the formulae by which funds in existing categories are allocated to the states are disproportionately weighted toward employment factors. Coal transshipment facilities do not require large numbers of employees, even though tremendous volumes of coal may be handled. For example, a recent doubling of capacity at the Pittsburgh and Conneaut Coal Company Dock in Ohio produced a net increase in employment of less than thirty individuals, while increasing capacity by 3.5 million tons.

Using the formula currently used by OCZM to award (c)(1) planning grants, this discrepancy becomes quite visible. The need for planning funds is computed for new or expanded energy facilities according to the following formula:

$$PNF = (PCE+POE)(N+S)(PCD)$$

PNF = Planning Need Factor

Where: PCE = peak employment during construction of facility

POE = peak employment required to operate facility

N = environmental factor (1=low, 2=medium, 3=high)

S = safety factor (1=low, 2=medium, 3=high)

PCD = adjustment for regional variation of similar planning costs (1.00 - 1.66).

As an example, assume a fossil fuel power plant with combined operating and construction employment of 1000 individuals and a major coal dock expansion of 3.5 million tons capacity with employment of 50. Both facilities are in the same state. The total environmental and safety factors, according to OCZM documents, are equal. Applying the factors in the formula, the planning need factor for the power plant is 5650, while that for the coal dock is 283. OCZM, in its technical paper on the planning grant formula indicates that:

> "The formula for PNF is designed to quantify the need to plan for social, economic, and environmental consequences likely to occur in a State's coastal zone as a result of constructing or operating a specific new or expanded energy facility. The greater the PNF, the greater the need to plan if adverse effects are to be minimized."[9]

In this case, the formula indicates that the need to plan for the power plant is 20 times greater than the need to plan for the dock. There is no rational justification for this conclusion.

OCZM/CEIP administrative staff members have been unwilling to modify the (c)(1) or the (d)(4) formulae to correct this discrepancy. They suggest that no formula would treat every state and every facility equitably. While it is agreed that no universally satisfying formula is possible, there is certainly a distinction to be made between energy facilities with major impacts resulting from high employment levels, and those whose impacts are more closely associated with other planning factors.

In an effort to more adequately address the needs of coal port states, Congressman David Bonior (D-Michigan) introduced an amendment to 1980 Coastal Zone Management Act reauthorization legislation which would include coal transshipment as a basis for grant allocations. The House Merchant Marine Committee concurred, creating a new (c)(3) category for coal-related projects.

The Senate Committee on Commerce, Science, and Transportation, on the other hand, considered similar amendments, but concluded that existing (c)(1) and (d)(4) categories were sufficient to address coal-related impacts if appropriated annually. In its deliberations, the Senate failed to recognize the above mentioned discrepancy in allocation formulae which biases the funding scheme against coal. Even so, the Senate specifically directed the Secretary to:

> "...work closely with coastal States impacted by coal transshipment activities, and other Federal agencies, to provide the maximum amount of available funds through section 308 (d)(4) and any other appropriate assistance program."[10]

This is a distinct Congressional mandate to OCZM that the unique aspects of coal transshipment be considered in allocation and funding decisions.

Ohio and the other Great Lakes States will continue to support the inclusion of specific coal-related budget categories in the CEIP, and regulations which treat all coal states in an equitable manner. OCS activities were included under the CEIP as their importance in national energy policy increased. The national goal of increased coal production shipment, and use should be similarly adopted.

REFERENCES CITED

1 Committee of Conference on S.586. 1976. "Coastal Zone Management Act Amendments of 1976." HR Rep. 94-1298. p. 24.

2 Coastal Zone Management Act of 1972. (P.L. 92-583)

3 Carter, L. J. "Energy Facility Siting: The White House Plays it Tough." Science 88:4178. February 28, 1975. p. 724

4 Ohio Department of Energy. 1978. Ohio Coastal Energy Facility Planning Process.

5 Committee on Merchant Marine and Fisheries. 1975. "Hearings on Bills to Amend the Coastal Zone Management Act of 1972." HR Ser. 94-11.

6 Correspondence from William Matuszeski, Acting Deputy Assistant Administrator, OCZM to State CEIP and CZM Managers, dated July 7, 1980.

7 Correspondence from Eileen Mulaney, Director, CEIP, OCZM to CEIP State Managers, dated July 31, 1980.

8 Great Lakes Commission. Great Lakes News Letter 23:2. Spring 1980. p.7

9 Office of Coastal Zone Management. 1980. Technical Paper: Planning Assistance - 308(c)(1). p. 4-5.

10 Committee on Commerce, Science, and Transportation. 1980. "Coastal Zone Management Act Amendments of 1980." S. Report 96-783. p. 12-13.

NEW COAST GUARD ICEBREAKERS FOR THE FOURTH COAST

by

ALAN S. GRACEWSKI, LCDR, U.S. Coast Guard

Naval Engineering Division, USCG Headquarters, Washington, D.C.

The Great Lakes truly constitute the Fourth Coast of the United States. Economically and environmentally, this Fourth Coast is just as important to this country as the Atlantic, Pacific and Gulf Coasts. The Coast Guard has recently accepted five new 140 foot icebreaking tugs. These ships will join the existing Coast Guard fleet on the Lakes to fulfill the mission of assisting waterborne commerce during the icecovered winter season.

The Great Lakes' drainage basin comprises a total area of 395,000 square miles. Land area drained by the St. Lawrence River comprises 300,000 square miles. The five Great Lakes and their connecting channels encompass 95,000 square miles. Thirty million people live in the U.S. part of the Lakes' basin. Two of the five largest cities in the U.S. are located there - Chicago and Detroit.

This vast area is of tremendous economic importance to the United States. The eight states which border the Lakes and eleven additional states which provide tributary area contribute 25% of the Nation's general cargo traffic and 16% of the bulk cargo. In 1977 the Great Lakes - St. Lawrence Seaway System carried 186 million tons of cargo.[1] Raw products such as iron ore, limestone, coal, and grain generally are transported from their sources on the upper Lakes to the industrial centers of the lower Lakes. When ice forms on the Lakes, this waterborne commerce requires icebreaking assistance to continue moving.

The Coast Guard is charged with maintaining icebreaking facilities under statute 14 USC 2 and 14 USC 93. Under Executive Order 75211 of 21 December 1936, the Coast Guard was directed by President Franklin D. Roosevelt to "assist in keeping open to navigation by means of icebreaking operations ... channels and harbors in accordance with the reasonable demands of commerce." Under these authorizations, the Coast Guard has developed and maintained a fleet of icebreaking cutters for the Great Lakes.

More recently, Congress, in section 107 of the Harbor Act (PL 91-611), approved 31 December 1970, established the Winter Navigation Demonstration Program. This program provided for studies and actual demonstrations to show the practicability and identify the means of extending the navigation season on the Great Lakes. Typically the shipping season closed in mid-December and opened in early April when the ice conditions again permitted transit without assistance. The heavy ice build-up forced the Great Lakes bulk carriers to shut down operations

Figure 1. U.S. Coast Guard Cutter KATMAI BAY shown breaking ice on Lake Superior

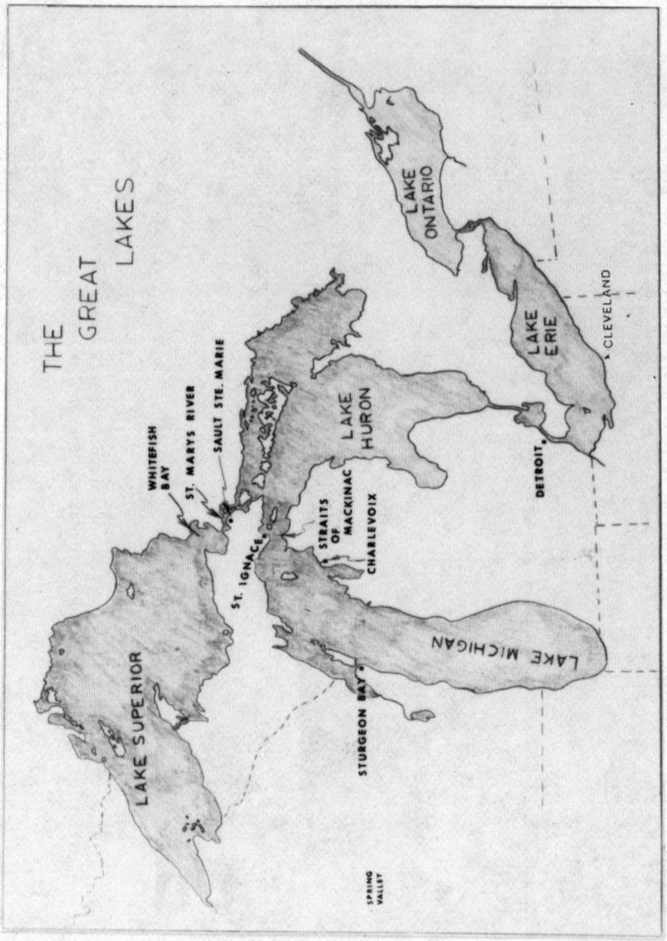

Figure 2. Homeports of the 140' WTGB class cutters on the Great Lakes

for these winter months. An extension of the navigation season would make stockpiling of bulk materials by users unnecessary and allow the carriers to operate profitably for more of the year.

The Coast Guard actively participated in this demonstration program by providing icebreaking support. This included assisting ships which has become beset in heavy ice and maintaining open channels by preventive icebreaking of vessel tracks. In addition, the Coast Guard established an Ice Navigation Center in Cleveland, Ohio which collects and disseminates ice information and keeps track of commercial shipping and Coast Guard icebreakers.

One of the Coast Guard's icebreaking resources for many years has been the 110' WYTM, a medium harbor tug. With many of these tugs over 40 years old and beyond their expected service lives, new icebreaking tugs were required to replace these older cutters. The design finally resulted in the 140' WTGB icebreaking tugs. These cutters were designed as multi-mission ships but their primary mission is domestic icebreaking. Figure 1 is a photograph of the 140' WTGB in the ice.

There are presently five 140' WTGB's on the Great Lakes. Home ports are Sault St. Marie, MI., Detroit, MI, Sturgeon Bay WI, St. Ignace, MI, and Cleveland, OH. These home ports are shown in Figure 2.

The ships were built by Tacoma Boatbuilding Company, Tacoma, WA. The first of the class, CGC KATMAI BAY, was delivered on 16 December 1978. The principal characteristics of the 140 class are shown below.

Principal Characteristics of 140' WTGB

Length Overall	140'
Maximum Beam	37'6"
Mean Draft	12'
Maximum Displacement (Winter)	650 tons
Maximum Speed in open water	14.7 knots
Shaft horsepower	2500

Propulsion - Twin Fairbanks-Morse 38 D 8 1/8 diesel engines driving Westinghouse 1000 KW, 900 volt D.C. generators. These power a single Westinghouse D.C. motor coupled to the propeller shaft. The propeller diameter is 8.5'.

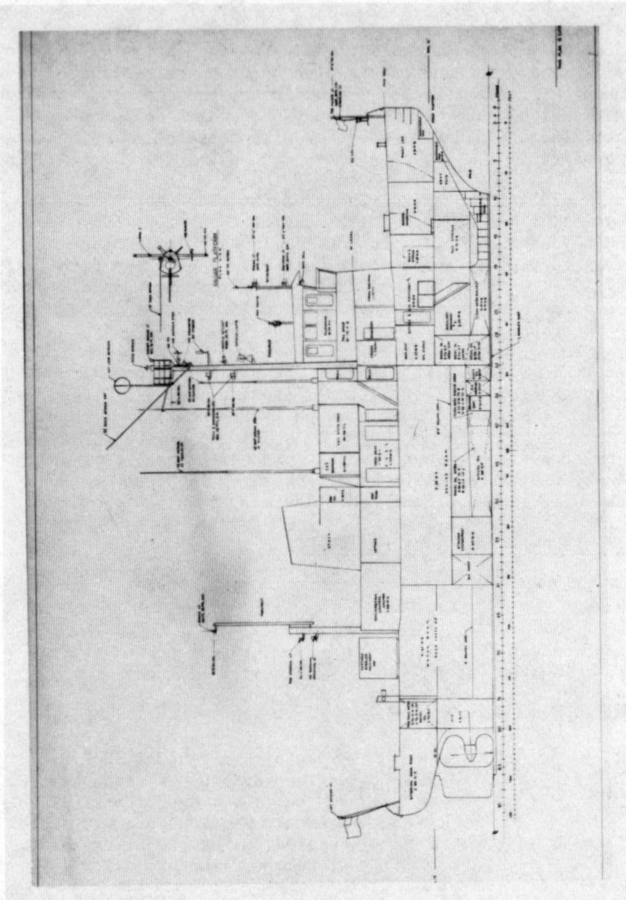

Figure 3. Inboard profile of the icebreaking tug.

Auxiliaries	-	Two Murphy MP24T diesels driving 175 KW Kato A.C. generators.
Crew	-	17 (3 officers, 14 enlisted)
Interesting Features	-	Air Bubbler System for Hull Lubrication while Icebreaking
	-	Vacuum Sewage collection and retention (no discharge) system
	-	10 gallon per minute oily water seperator for pumping bilges
	-	High habitability standard
	-	Pilothouse Control, unmanned enginerooms

In addition to icebreaking, this class of Coast Guard Cutters can perform search and rescue, towing, and law enforcement missions. For navigation and for assistance in performing all missions, the ship has a complete electronics suite. Two radar units on the bridge allow the officer in charge to take range and bearing for navigation and to search the horizon for other ships. A fathometer, Loran C receiver, and two automatic direction finders are also installed. Primary communicaitons are by VHF/FM radio, however, the ship also has HF tranceivers.

Accomodations for the crew are of exceptional quality. All living areas are equipped with an air conditioning and humidity control system for comfort in winter or summer. All berthing and messing spaces are located on or above the main deck. Extensive use of silencing materials and techniques were used to keep the engineroom generated noise as low as possible. Interior decor is colorful and pleasant to the eye due to micarta faced joiner bulkheads, suspended overhead panels, and tile and carpeted decks.

Figure 3 shows an inboard profile of the 140'. One of the unique features of the WTGB is the incorporation of an air bubbler hull lubrication system. A diesel engine and air compressor are located in the bubbler machinery van on the fantail. The van is removable when not required for icebreaking. Air compressed by the van is directed to the air bubbler piping system in the hull. The air rises, carrying water with the bubbles. As this air and water mixture moves toward the surface, hull lubrication takes place and lowers friction against the ice through which the ship is moving.

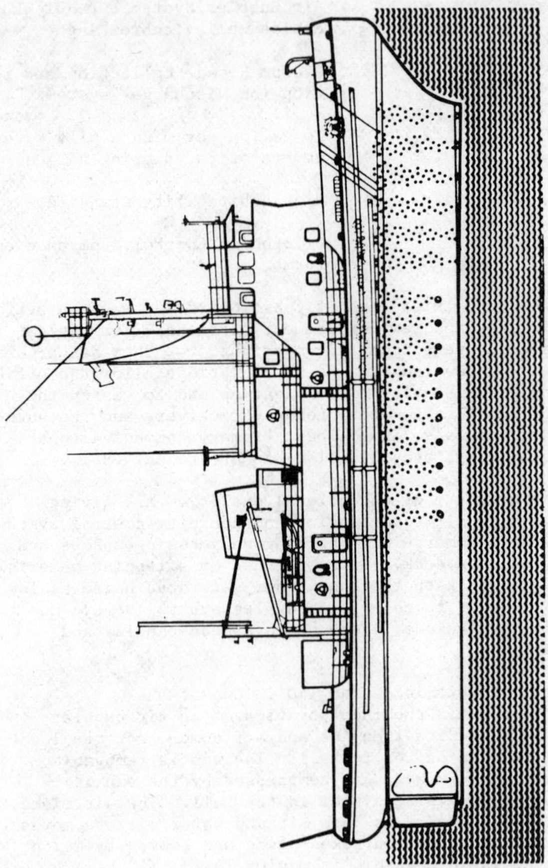

Figure 4. Pictorial showing the air bubbler system orifices and illustrating hull lubrication.

The KATMAI BAY underwent icebreaking trials during the winter of 1978/1979. The ice trials showed the ship was capable of breaking flat, level ice of up to 22 inches thick continuously. Beyond 22 inches of thickness, the ship must back off and ram to continue breaking a track. The ship also demonstrated it could ram up to 5 foot thick ice pressure ridges without sustaining any damage. The bubbler system exhibited effectiveness in level, homogeneous ice and in brash ice by lowering the shaft horsepower required by as much as 25-35% depending on conditions present. Brash ice is generally defined as previously broken ice pieces formed in ship channels by ships passing through. Brash is composed of many different sizes and irregular pieces of ice. These pieces generally float and refreeze in the broken channels only to be rebroken again whenever a ship passes.

It is in the rivers and harbors of the Great Lakes that brash ice usually causes cargo ships to become stuck and require icebreaking assistance from the Coast Guard. Figure 5 shows a Laker in a brash ice channel.

All in all, the WTGB class has met or exceeded all the original design requirements previously established. The crews that sail these icebreaking tugs speak highly of them.

Yet, even with icebreakers assistance, many problems exist and environmental concerns have been voiced. The weather during winter months is hazardous due to many storms. The loss of the Edmund Fitzgerald is a recent example of what can happen to even a modern bulk carrier in heavy weather. Moving ice floes can trap ships and carry navigational aids out of position.

Ice booms have been established by hydroelectric power companies at plant intakes to prevent ice chunks from being sucked into the turbines and clogging the water inlets. The extended navigation season could produce more ice chunks and destroy the stable ice cover, cause flooding, or even change the levels of the different lakes significantly. The Army Corps of Engineers faces problems operating locks during freezing weather. Ice build-up on lock walls, jamming of lock mechanisms, and ice pushed in ahead of the ships entering the locks are some specific problems. In addition, lack of a winter shut down time for lock maintenance causes concern.

Erosion and dock damage may be increased due to additional ice breaking and ship traffic. Vibrations in private homes have been allegedly caused by passing ships in the St. Marys River. These vibrations are said to be caused by the ships' propellers operating at high horsepower and thrust while passing through ice filled channels. These high horsepowers (required to move through the ice blocked waters) also create more propeller wash which can cause bottom scouring. This may disturb the ecology of the benthic community.

Figure 5. Laker Cason J. Callaway proceeding through a brash ice channel on the St. Mary's River.

Disruption of ice fishing areas and removal of stable ice covers also effects recreational activities such as snowmobiling, cross country skiing, and hiking across rivers or bays.

Access to various islands across the ice may also be disrupted. This effects island residents and migration of wildlife such as moose or wolves.

Local weather may actually be altered by icebreaking and additional vessel transits. Ice cover prevents evaporation of water. Exposure of liquid water surface leads to more evaporation and more local snow and ice production.

Vessel traffic control will have to be enhanced since it enables safer handling of traffic in narrow and congested areas. During the ice season, often the channels through difficult ice areas are wide enough for only one ship. This means two way traffic cannot occur. A cost effective way of providing icebreaking assistance is to lead convoys of ships through the difficult channels.

All these concerns could lead one to believe that winter navigation is not worth pursuing because of the damage it might cause to the environment. This dilemma can be resolved by careful study of the problems and their effects on the environment. The Winter Navigation Demonstration Program was successful and showed that the shipping season could be extended to the full 12 months of the year. During this program a significant amount of information was gathered on the effects of an extended season. Careful analysis of the results can help point the way to intelligent decisions regarding this fourth Coast.

The Coast Guard will continue its duty of providing icebreaking services on the Great Lakes. The new additions like the 140' WTGB's will help break the way for an extended navigation season should it prove to be in the Nation's best interest.

References:

1. The Great Lakes and St. Lawrence Seaway Winter Navigation Board Demonstration Program Final Report, Winter Navigation Board, September 1979.

2. Projected United States Coast Guard Icebreaking and Icebreaker Requirements 1975-2000, Plans and Programs Staff, Office of Operations, U.S. Coast Guard Headquarters, February 1975.

COASTAL ENERGY DEVELOPMENT IN SANTA BARBARA COUNTY

By

Kirvil Skinnarland
Project Director, Local Coastal Program
Santa Barbara County Planning Department
Santa Barbara, California

Mel Willis, AICP
Director of Energy Resource Planning
Envicom Corporation
Calabasas Park, California

1.0 INTRODUCTION

Public environmental consciousness in the United States is generally measured from a single event that occurred on January 29, 1969. On that date, a blowout occurred at Union Oil's Platform A in the Santa Barbara Channel. Before the blowout was stopped, television audiences throughout the country were treated to pictures in living color of a massive oil spill that caused the despoilment of scenic Santa Barbara Harbor and the death of thousands of seabirds.

Public response to this event was immediate and angry. Aroused by public demands to protect the environment from the effects of man's destructive actions, Congress passed that same year landmark environmental legislation from which most other federal, state and local environmental laws have their origin--the National Environmental Policy Act of 1969.

A decade later, the County of Santa Barbara is at the forefront of local coastal planning in California and will probably be one of the first counties to have a coastal plan approved by the California Coastal Commission. Since the coastal area onshore and offshore Santa Barbara County continues to be one of the major energy-producing areas in the country, the issue of energy development in the coastal zone has been a major consideration in the development of the Santa Barbara County Coastal Plan.

The purpose of this paper is to examine the issue of energy development in the coastal zone of Santa Barbara County. The paper focuses on the development of an energy facility siting management plan for the Mussel Rock Dunes--an especially unique and sensitive coastal area that is experiencing a significant potential for impacts from onshore and offshore energy development.

2.0 BACKGROUND

2.1 DESCRIPTION OF AREA

The coastal zone of Santa Barbara County, which spans 110 miles, is world-renowned for its beauty and diversity. The south coast area, between Point Conception and the Ventura County line, is characterized by sandy beaches, bluffs, and coastal terraces. Urban development and agriculture are nestled along the narrow coastal shelf in this region against the spectacular backdrop of the Santa Ynez Mountains. Approximately 60 percent of the County's population of 282,000 resides in the south coast area. The coastline from Point Conception north to the Santa Maria River is rugged and rural, consisting of rolling hills, mountains, rocky headlands, steep bluffs, and the extensive sand dunes between Mussel Rock and the Santa Maria River, which are the focus of this paper.

The Santa Barbara County coastal zone has a long history of controversy over the use and development of its resources. Illustrative of this controversy is the example of oil exploration and development. Oil production in the County originates from three sources: onshore, State tidelands, and Federal outer continental shelf (OCS). Total onshore production in Santa Barbara County was 12.65 million barrels of oil in 1979. Most of this production originated from fields outside the coastal zone. Within the coastal zone, the only area that appears likely to experience new or expanded onshore oil activity in the future is the Mussel Rock Dunes. The second source of oil production is the State tideland area which extends three miles offshore. In 1979, total production from the State tidelands in the Santa Barbara Channel was 4.3 million barrels of oil. Since most of the oil and gas fields in the leased tidelands are approaching the end of their productive lifetimes, production from existing leases is expected to decline in the future. Tideland areas north of Point Conception have not as yet been leased by the State. The Federal OCS in the Santa Barbara Channel will be the major source of oil production affecting Santa Barbara County in the future. In 1979, 10.83 million barrels of oil were recovered in the Santa Barbara Channel OCS. This figure is expected to increase through at least the year 1990.

For almost 100 years, Santa Barbara County has tried, with mixed results, to protect its natural and scenic resources from the impacts of oil development in onshore and offshore areas. In the case of offshore oil development, the battle has been particularly frustrating since offshore development has never been subject to local permit jurisdiction. Because agriculture and tourism are two of the major economic sectors, the County has a commitment to preservation of its relatively good air quality and outstanding scenic and natural resources. Therefore, the potential for significant adverse environmental impacts resulting from oil development have in the past, and continue to be, of great concern locally.

2.2 HISTORY OF OIL DEVELOPMENT

The first oil development in the offshore waters of Santa Barbara County occurred in the 1890s when wells were drilled from piers in the coastal waters off Summerland. Offshore exploration and development increased dramatically in the 1920s with the leasing of submerged lands along other portions of the south coast by the State of California. In 1927, the Santa Barbara City Council, reacting to the visual effects of the proliferating wooden oil derricks, acted to prohibit oil drilling anywhere within the city limits. This action marked the beginning of efforts by local government in Santa Barbara County to protect its environment in the face of increasing pressure for oil development. Shortly thereafter, the State Legislature also enacted an emergency moratorium prohibiting new oil drilling in the tidelands, thus temporarily ending oil development in the waters offshore Santa Barbara County.

In the late 1940s, the State of California again began leasing large offshore tracts extending to the three mile limit along the south coast of Santa Barbara County. Excluded from leasing was a sixteen mile strip stretching roughly from the waters offshore the University of California campus, west of the City of Santa Barbara, to the community of Summerland; this area was called the Santa Barbara Oil Sanctuary. By 1966, a total of 39 tracts had been leased in the State tidelands. In 1958, the first platform, Hazel, was erected in the Santa Barbara Channel, a symbol of the County's impotence in controlling activities in its own coastal waters.

In 1967, the Federal government created the Federal Ecological Preserve which prohibited oil development in a two mile buffer zone extending seaward from the State's oil sanctuary. In the same year, the Department of Interior issued a call for nominations of tracts in the Santa Barbara Channel and thus began a leasing program in the outer continental shelf that is expected to continue into the 1980s.

In 1969, the blowout of Union's Platform A caused a major oil spill in the Santa Barbara Channel. This event served to solidify local opposition to offshore oil development and resulted in the formation of several local environmental organizations. As a result of the oil spill, the State of California instituted a ban on drilling in the State's tidelands, a moratorium which remained in effect until 1975. The Federal government added six tracts to the Ecological

Preserve and effected a temporary ban on drilling in the Channel. Soon thereafter, the drilling moratorium in Federal waters was lifted, and leasing and drilling activities resumed in the OCS.

The Santa Barbara Channel is one of the most intensively leased offshore petroleum regions in the United States. To date, the Federal government has sold leases to some 128 tracts in the Channel portion of the OCS. The Department of the Interior recently announced a call for nominations for its proposed OCS Lease Sale 68 to take place in 1982. This sale encompasses the entire OCS of southern California including the Santa Barbara Channel and the once sacred Federal Ecological Preserve. This action has elicited strong negative responses from County and City government and local environmental groups.

Meanwhile, offshore areas north of Point Conception, which have not been subject to oil development in the past, are now being viewed as a new OCS frontier. OCS Lease Sale 53 is scheduled for 1981 and will affect areas north of Point Conception including the waters offshore the Mussel Rock Dunes. The oil industry expressed considerable interest in this particular area during the nomination process. Moreover, the State Lands Commission recently initiated studies to examine the possibility of leasing tideland waters north of Point Conception. Thus, the rural and scenic coastline in the northern portion of the County may soon be faced with the potential for oil spills and pressure to build onshore facilities to support offshore oil development.

Although the Mussel Rock portion of the Nipomo Dunes has not yet been subject to impacts from offshore oil development, the area does have a history of onshore oil activity. A few exploratory wells were drilled in the dunes during the 1940s and 1950s, but were subsequently abandoned. The only existing facilities, which belong to Thriftway Oil, are sited just south of the Santa Maria River mouth and consist of two producing wells and a tank battery. Both Union Oil and Husky Oil hold mineral leases within the Mussel Rock Dunes; Husky plans to submit an application to the County in the near future for exploratory drilling. The area appears particularly promising for oil development because of the existing Guadalupe oil field north of the Santa Maria River in San Luis Obispo County which has been extensively developed. Thus, the pressures for both onshore and offshore oil development are converging on the Mussel Rock Dunes and present a major coastal planning issue for the County to resolve.

2.3 THE COASTAL PLANNING PROCESS

With the passage of Proposition 20 in 1972 and later the California Coastal Act of 1976, there was hope that Santa Barbara County would be able to achieve a greater level of protection for its coastal resources in the face of ever increasing pressure for oil development both onshore and offshore. The coastal legislation set up one State and six Regional Coastal Commissions and granted them permit authority over all new development in the coastal zone, a strip of land ranging from one half to five miles in width running the entire length of California. The guidelines for new development in the coastal zone were established by statewide policies for the conservation and development of coastal resources contained within the Act. The policies address many issues including provision of public access and recreation, protection of environmentally sensitive habitat areas, concentration of urban development, preservation of agricultural lands, and the siting of coastal-dependent and energy related industrial facilities. The Coastal Act requires all coastal governments in California to prepare land use plans and zoning ordinances designed to carry out the statewide policies. These plans must be approved by the Coastal Commission before permit authority can be returned to local government.

The County of Santa Barbara has been a leader in local coastal planning in Californa and will probably be one of the first counties in the State to have its plan approved by the Coastal Commission. Trying to strike a balance among the competing uses for scarce coastal land has not been an easy process for the County and consequently there has been considerable controversy during public hearings on the County's coastal plan. Due to the complexity of issues for the Mussel Rock portion of the Nipomo Dunes, the County was unable to develop a plan that resolved the issue of energy facility siting within the time frame established for completion of the rest of the coastal plan. Instead, the County contracted with Envicom Corporation to conduct a special study of the Mussel Rock portion of the Nipomo Dunes and to develop a plan for its management. Funding for the study was provided by the Federal Coastal Energy Impact Program (CEIP).

3.0 THE MUSSEL ROCK DUNES: A Case Study In Resolving the Conflict Between Energy Development and Coastal Protection

The Nipomo Dunes ecosystem, of which the Mussel Rock Dunes comprise the southern one-third, is a small 18 square mile section of the central California coast, lying in northern Santa Barbara and southern San Luis Obispo counties (see Figure 1), that is considered to be the largest and most beautiful remaining dune-lagoon complex in California (California Department of Parks and Recreation, 1968). Recognizing that the area's ecological and scenic values are of national significance, the U.S. Secretary of the Interior designated all of the coastal area between Point Sal and Pismo Beach, with the exception of a portion of the Guadalupe oil field, a National Natural Landmark and placed

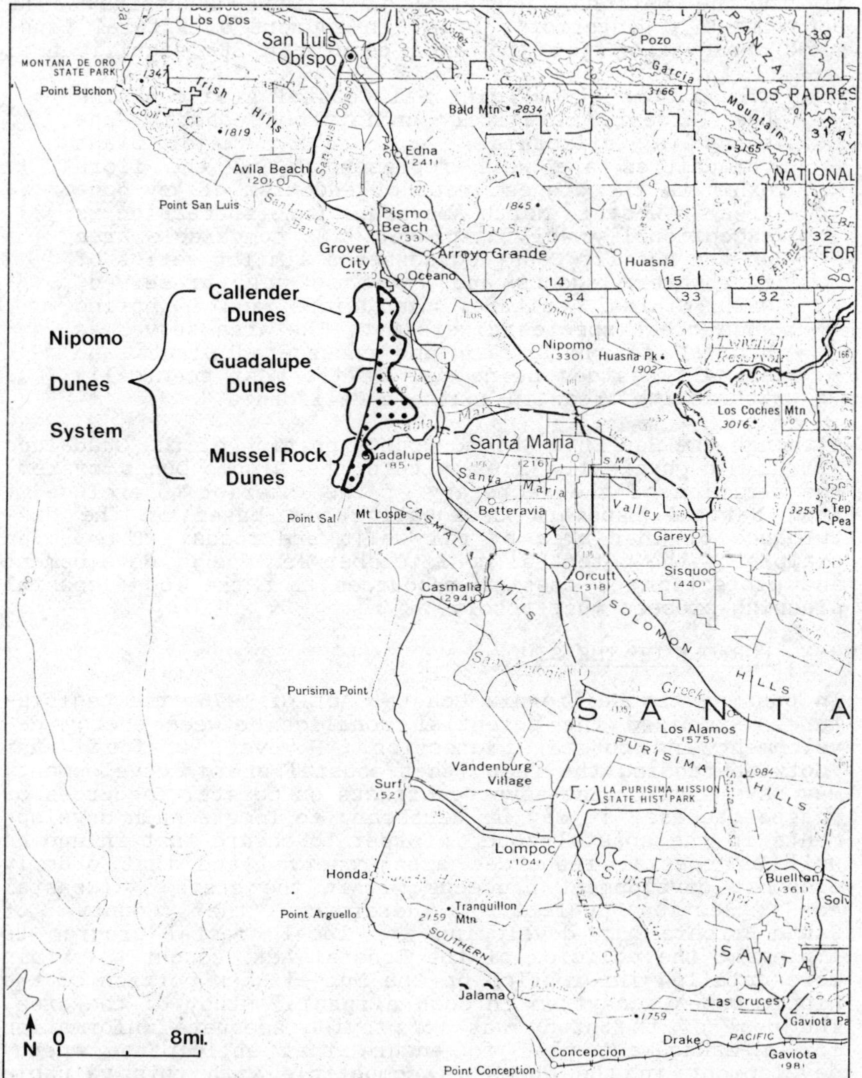

Figure 1. Regional Location of Nipomo Dunes System.

it in the National Registry of Natural Landmarks. The Department of Interior's brief in support of Natural Landmark designation for the area describes its significance:

> "The Nipomo Dunes-Point Sal Coastal Area contains the largest, relatively undisturbed coastal dune tract in California . . . . Five major plant communities are well represented and the flora exhibits the highest rate of endemism of any dune area in western North America. Dune succession is exceptionally well displayed. No comparable area on the Pacific Coast possesses a similar series of freshwater lagoons and lakes so well preserved, with minimal cultural intrusions and harboring such great species diversity. The area serves as habitat for both rare and endangered plants and animals besides being one of the most scenically attractive areas in southern California."

Although the San Luis Obispo County portion of the Guadalupe oil field physically lies within the Nipomo Dunes system, the decision of the Secretary of the Interior to exclude it from Natural Landmark designation was based on the disturbance of that area by oil wells and roads. This is an example of the potential conflict between energy development and protection of coastal resources that the local coastal planning process must resolve.

3.1 PURPOSE OF THE STUDY

In passing the California Coastal Act of 1976, the Legislature recognized the potential conflict between energy development and coastal protection. However, it found that "notwithstanding the fact [that coastal energy development] may have significant adverse effects on coastal resources or coastal access, it may be necessary to locate such developments in the coastal zone in order to ensure that inland as well as coastal resources are preserved and that orderly economic development proceeds within the state." (Coastal Act, Section 30001.2). Therefore, the County of Santa Barbara, in developing its local coastal program to implement the policies of the Coastal Act, sought to obtain more detailed information on the Mussel Rock portion of the Nipomo Dunes ecosystem through a special study of the area. The goal of the study was to provide adequate information for a management plan to ensure that anticipated energy development in the dunes is compatible with this valuable coastal resource and is consistent with the requirements of the Coastal Act. The specific objectives of the project were to:

1. Prepare an inventory of the natural resources of the study area.

2. Prepare scenarios of potential future energy development.

3. Assess the impacts of all potential energy development.

4. Develop mitigation measures and performance standards for anticipated energy facilities including alternative siting strategies.

3.2 METHODOLOGY

This project involved a multidisciplinary team of energy and land use planners, biologists, geologists, archaeologists and cartographers in the preparation of a management plan for anticipated future energy development in the Mussel Rock Dunes. The several steps involved in the preparation of the management plan are summarized below.

In order to develop baseline information for the project, an inventory of the natural resources of the study area was prepared. This information included existing vegetation by plant communities; rare, endangered or protected plant and animal species or habitats; archaeological sites; areas of prime visual quality; and dune areas susceptible to erosion or formation of windblown sand. Because of the absence of suitable base maps for the study area, a separate mapping system was developed. In November, 1979, the study area was flown to obtain coverage of the area with color infrared imagery. Detailed mapping of natural resources was accomplished on transparent overlays of the color infrared photos at a scale of one inch equals 400 feet.

Siting scenarios were prepared to determine the potential for future energy and industrial development in the study area. The types of facilities or activities that were examined included expansion of onshore oil development in the Guadalupe oil field, onshore facilities resulting from OCS Lease Sale 53, and a power plant ocean cooling water pipeline. Based on contact with industry representatives and Federal government officials, scenarios were developed describing the near and long-term plans for the area by industry.

Based on the energy development scenarios, an assessment of the impact potential of projected future energy development in the study area was prepared. This impact assessment did not deal with specific projects, but rather with a broad range of potential facilities that may need to be sited in the study area in order to develop onshore or offshore petroleum resources.

The process of preparing a generic impact assessment of potential energy development in the study area involved several steps. First, the general siting requirements of the facilities that may seek to site within the study area were identified. Second, sites or general locations within the study area that have the potential for energy facility development were identified. Last, by comparing potential facility requirements and sites to resource sensitivity, the potential for impacts and loss of coastal resources was determined.

The management plan for energy facility development in the study area evolved in several steps. Based on an evaluation of public policies related to energy facility siting (e.g., Coastal Act, Santa Barbara County Coastal Plan, County ordinances, etc.), criteria for locating new development within the study area were developed. These criteria were then related to potential impacts of energy development in the study area. From this analysis, basic siting and resource protection strategies, including performance standards, were developed.

3.3 RESULTS

3.3.1 Inventory of Resources

The Mussel Rock dune sheet (see Figure 2) is comprised of an active portion, immediately adjacent to the ocean, and an inland, stabilized portion. The most prominent feature of the active portion of the Mussel Rock Dunes is the well developed complex of transverse dunes that extends from the Santa Maria River southward to Mussel Rock. Because the Mussel Rock dune sheet lies on both the Santa Maria River floodplain and Orcutt Mesa, the height of the dunes ranges between near sea level to 400 feet in elevation. The prominent open sandy nature of these dunes is the basis for their unique visual qualities and their designation as a National Natural Landmark. The variation in elevation and degree of vegetation within the dunes enabled a visual mapping system to be developed with map units ranging between high visibility/high scenic quality area to non-visible/low scenic quality area.

One of the most important characteristics of the Mussel Rock Dunes is its unique biotic resources. The coastal dune habitat is one of the most restricted natural communities on the Pacific coast of North America, occurring only in a few scattered localities. Many of the plants found in the coastal dune habitat have unusual morphological and physiological adaptations that permit them to grow in this relatively harsh environment. Most of the plants of coastal dunes are found only in this habitat and would be considered rare for no other reason than that their habitat is rare.

The Mussel Rock Dunes serve as an important habitat for rare plants and animals due to its location, substrate, and history of limited disturbance. The resource inventory of

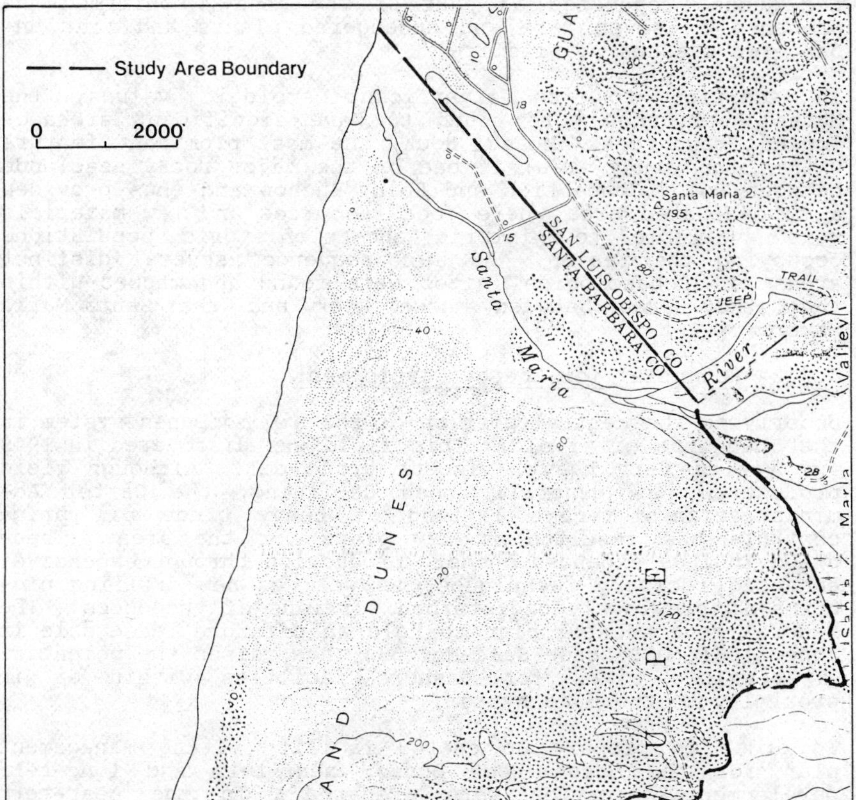

Figure 2. Mussel Rock Dunes Study Area.

the dunes, conducted as part of the study, confirmed the presence of seven rare and endangered plants and four endangered species of wildlife.

In addition to its significant ecologic values, the Mussel Rock Dunes were found to have significant archaeological resources. Mussel Rock, the most prominent feature within the study area, is one of the major rocky headlands between Point Conception and Point Buchon and thus provided a concentrated area where food resources and raw materials were available to historic and prehistoric populations occupying the region. As a consequence, several distinct areas of archaeological sites were found and mapped within the study area between Mussel Rock and the Santa Maria River.

3.3.2 Potential for Energy Development

Underlying the central portion of the Nipomo Dunes system is the Guadalupe oil field. This field was discovered in 1948 and yields very heavy, viscous crude oil. Although field production has generally been declining, the Carter Administration's recent lifting of heavy crude oil price controls has resulted in new plans for the area. These plans include expanded onshore production through expensive, enhanced recovery techniques, as well as new drilling programs in previously undisturbed portions of the dunes. The Federal government's plan to hold an offshore lease sale in June, 1981 (OCS Lease Sale No. 53), has raised the potential of possible sites for onshore facilties within or in proximity to the dune system.

In order to prepare an energy facility siting management plan for the Mussel Rock Dunes, near-term and long-term development scenarios were prepared. In the near-term (1980-85), expanded development drilling programs within the known limits of the Guadalupe oil field are expected to dominate, with additional exploratory drilling outside of the known field boundaries. One oil company has already received approval from the California Coastal Commission for expanded drilling/production operations within its existing lease area in San Luis Obispo County, and another oil company has announced its intention to seek permits for exploratory drilling within the Santa Barbara County portion of the dunes. The net effect during the near-term is that there will be a significant potential for new wells, assorted tankage, access roads, and miscellaneous production equipment to be located within the dunes.

In the long-term (1986-1995), the focus of energy development activity will start to shift offshore. In addition to OCS Lease Sale No. 53 in 1981, two additional lease sales in the waters offshore the study area are scheduled for 1983 and 1984. Although there is currently no offshore production anywhere along the central coast of Caifornia (i.e., north of Point Conception), the Santa Maria Basin offshore the Nipomo Dunes has one of the highest potentials for offshore oil and gas resources in the proposed lease area. Therefore, offshore production scenarios for the area (U.S. Department of the Interior, 1980), assume production to commence in 1986 and peak in 1990 at over 100,000 barrels per day. As this production commences in the Santa Maria Basin, a need to transport the oil to refining centers will be generated. Since the Nipomo Dunes lie across the principal offshore to onshore pipeline corridor, oil and gas pipelines and an onshore processing/treatment facility may need to be located in the study area.

3.4 ENERGY FACILITY SITING MANAGEMENT PLAN

Based on the resource inventory and energy development scenarios for the dune area, an energy facility siting management plan was prepared. The purposes of the management plan were to: 1) enable decision-makers to decide whether new or expanded energy development should be permitted within any portion of this ecosystem; and 2) if so, what types of facilities, what locations, and under what conditions should development be allowed.

3.4.1 Basic Approach

The principal public policy affecting the siting of energy facilities within the Mussel Rock Dunes is the California Coastal Act of 1976. The adopted, but not yet certified, Santa Barbara County Coastal Plan will be the specific vehicle that will carry out the intent of the mandated Coastal Act policies. Both the California Coastal Act and the Santa Barbara County Coastal Plan were used to develop a policy framework within which a specific management plan for the study area was prepared.

In applying the specific Coastal Act policies to the Mussel Rock Dunes, a basic conflict developed between one section of the Act which mandated strong protection for "environmentally sensitive habitat areas" and other policies of the Act which gave the highest priority to coastal-dependent industrial development. However, in passing the Coastal Act, the California Legislature recognized that conflicts between the Act's policies may arise, and directed that "such conflicts be resolved in a manner which on balance is the most protective of significant coastal resources." Therefore, the basic approach that was taken in the development of an energy facility siting management plan for the study area was to permit limited energy development, but only if no alternative locations are feasible outside the dune habitat area, and with performance standards to minimize, to the maximum extent feasible, the loss or degradation of habitat values.

3.4.2 Permitted Facilities

Because of the uniqueness and sensitivity of the area to disturbance, it was determined that the types of permitted energy facilities should be limited to those uses that absolutely require a site within the dune habitat area in order to be able to function at all and for which there are no other feasible locations outside the dunes. Based on this general policy, exploration and production oil and gas wells could be permitted within the dunes. However, accessory equipment and facilities needed for onshore oil and gas production (e.g. separation/treatment facilities, enhanced recovery equipment, storage tanks, etc.) do not actually have to be located at a well site in order to be able to function. Therefore, these facilities are not permitted within the dune habitat area and must locate at sites outside and adjacent to the dunes, unless it can be demonstrated that this configuration is not feasible.

Onshore facilities that support offshore oil and gas development (e.g. separation/treatment plants, marine terminals, pipelines, etc.) do not require a site within the dune habitat area in order to be able to function, and alternative, less environmentally damaging, sites are available outside the dunes. Therefore, these facilities are not permitted within the dune habitat area.

3.4.3 Resource Protection Strategies

Once it was determined what kinds of energy facilities could be permitted within the dune habitat area, it was necessary to develop resource protection strategies and performance standards to ensure the preservation of sensitive coastal resources "to the maximum extent feasible," as required by the California Coastal Act. Furthermore, the Coastal Act requires that all development in environmentally sensitive habitat areas be accomplished in a manner which will not result in "significant disruption of habitat values."

The tool that was developed to minimize disruption of habitat values by permitted energy facilities was a resource sensitivity map which depicted the variation in habitat value and susceptibility to disturbance within the study area. The concept of "habitat value" is not defined by a single resource value, but rather it is the summation of all resource values which serve to make that habitat unique. In defining the scope of work for this study, the resource categories of vegetation, wildlife, visual quality, active dune areas, and archaeological sites were considered to be those factors which make the Mussel Rock Dunes unique. Since each of these resource categories demonstrated some variation in sensitivity and quality within the study area, this provided the opportunity to develop the resource sensitivity map, which is a composite map of all resources inventoried.

The resource sensitivity map was developed by bringing together the staff who conducted the detailed resource inventory of the study area. These specialists then evaluated qualitively the degree of sensitivity of each resource category to impact from energy development. After a number of iterations, the group was able to rank order the resource factors along a continuum which ranged from most sensitive to least sensitive to disturbance. The resource inventory maps provided the basis for determining which portions of the study area had the highest resource value for each of the resource categories. Thus, the final resource sensitivity map was based on an objective inventory of the location of resources and on a subjective ordering of priorities for preservation.

In essence, the resource sensitivity map depicts the variation in "habitat value" and susceptibility to disturbance of all locations within the study area. Its usefulness as a planning tool is that it allows the identification of areas within the sensitive dune habitat which would be least susceptible to disturbance taking into account all factors of importance to its overall quality. Likewise, taking into account all factors, the resource sensitivity map identifies those areas with the highest habitat value and susceptibility to disturbance and which, therefore, should be avoided for siting of energy development.

The resource sensitivity map places all locations within the study area into one of five categories ranging from highest (1) to lowest (5) sensitivity. In the management plan, heavy reliance is placed upon the resource sensitivity map. All facilities are prohibited from being located within those areas with either a rating of 1 or 2 due to their especially high habitat values. Permitted facilities are allowed to locate in areas with a rating of 4 or 5. In areas with a rating of 3, development would only be allowed if it can be demonstrated that alternative locations in areas of lower resource sensitivity are not feasible.

For those energy facilities permitted to locate within the dune habitat area, the management plan includes strict performance standards to ensure the minimum amount of disturbance to the resources. These performance standards include procedures for site abandonment and restoration of the dune habitat, oil spill and clean-up requirements, facility operating requirements, and equipment siting and maintenance standards.

3.4 Summary of Results

The energy facility siting management plan for the Mussel Rock Dunes will ensure 1) that only those facilities are permitted that absolutely require a site within the dunes in order to be able to function; 2) that the permitted facilities are sited so as to avoid those portions of the dunes with the highest resource value and sensitivity to disturbance; and 3) that permitted energy facilities meet strict performance standards to minimize impacts to coastal resources.

The energy facility siting management plan for the Mussel Rock Dunes has been submitted to County decision-makers for public hearings. After action by the County, the plan will be reviewed by the California Coastal Commission. Once adopted, the management plan will become part of the County's Coastal Plan and will provide the framework for review of all new onshore energy development in the Mussel Rock Dunes.

REFERENCES

California, Department of Fish and Game, 1976. The Natural Resources of the Nipomo Dunes and Wetlands. Coastal Wetlands Series #15.

California, Department of Parks and Recreation, 1968. The California State Park System Plan - 1968.

California, Office of Planning and Research, OCS Task Force, 1977. Offshore Oil and Gas Development: Southern California. 2 vols.

Cooper, W.S. 1967. "Coastal Dunes of California." Geological Society of America Memoir No. 104.

Fairchild Collection, Whittier College, California, 1938. Aerial photographs of Santa Barbara and San Luis Obispo County coastal area.

Johnson, G.W. and R.L. Nye (eds.), 1979. Environmental Hazards and Community Response: The Santa Barbara Experience. University of California, Santa Barbara: Public Historical Studies, Monograph No. 2.

Sollen, Robert. "Oil Production Up 40 Percent in State Waters of Channel." Santa Barbara News Press, January 31, 1980.

U.S. Department of the Interior, Bureau of Land Management, 1980. Draft Environmental Statement on Proposed OCS Lease Sale No. 53.

U.S. Department of the Interior, Bureau of Land Management, 1979. Final Environmental Statement on OCS Lease Sale No. 48.

Use of a Porous Dike to Reduce Entrainment and Entrapment in Power Plant Cooling Water Systems

Barry A. Ketschke*

ABSTRACT

The porous dike is a previously untested concept for excluding organisms from power plant cooling systems. The porous dike is a rock or gravel breakwater which completely surrounds the plant intake. All intake water is drawn through the rock dike. New England Power Company (NEP) has been testing the hydraulic characteristics and screening potential of a porous dike on its porous dike test facility located at its Brayton Point Generating Station.

The test facility is a small intake measuring 20 feet wide by 18 feet deep. The test facility is divided into three sections, one with a porous dike made of 8" stone held in wire gabions, another with gabions filled with 3" stones and an open control section. The 3" gabion section is 6 feet thick; the 8" gabion section is 9 feet thick. Water is drawn through the stones by an axial flow, bow thruster, low head pump. Volume is controlled by pump speed and control baffles. The facility has been pumping water continuously since July 1979.

The hydraulic performance, screening ability and fouling potential of the porous dike have been tested and will continue to be tested through 1981.

Hydraulic performance is measured by head loss through the stones, volume flow through the stones and approach velocity. Results to date have shown:

This research is funded by:
U.S. Environmental Protection Agency, U.S. Dept. of Energy
Electric Power Research Institute, New England Power Company

*Senior Fish Biologist, New England Electric System, Westborough, MA

1. Head loss fluctuates between 1.1 feet and 2.8 feet in both test gabions. Head loss is identical in each section. Daily measurements suggest a slight increase in head loss with time.

2. Volume flow measurements were stable at approximately 9 cfs through the first 8 months of operation. Recent data indicates a reduction in volume flow.

3. Approach velocities are uniform at approximately 0.1 fps.

4. Backflushing with a 1-foot head decreased the head loss and increased the flow volume to near initial values.

Screening performance is measured for zooplankton, ichthyoplankton and juvenile and adult fish. Results to date have shown:

1. A significant to highly significant difference in the density of zooplankton between the upstream face and downstream face of the dike. These results suggest entrainment avoidance.

2. A highly significant difference in the density of fish larvae between the upstream and downstream face of the dike suggesting entrainment avoidance.

3. Entrapment avoidance exhibited by most juvenile and adult species tested.

Fouling potential is measured by fouling tubes, a small removable section of the gabions. Results to date have shown:

1. A 7% to 10% increase in the weight of the gabions.

2. A 20% to 28% decrease in the void spaces.

3. The development of a fouling community.

INTRODUCTION

Because of their large cooling water requirements, the coastal zone has historically been a prime site for power plant development. The cooling water intakes for these plants are equipped with screening devices usually designed to protect the pump and cooling water system. Operation of these cooling systems results in entraining - passage through the system of smaller organisms - and impingement - capture on the screen of larger organisms.

As mandated by the Clean Water Act (amended through 1978) which states:

> *"Any standard established pursuant to section 301 or section 306 of this Act and applicable to a point source shall require that the location, design, construction, and capacity of cooling water intake structures reflect the best technology available for minimizing adverse environmental impact."*

continued use of the coastal zone for once-through cooling requires that technologies be developed to minimize the impact of entrainment and entrapment.

Since 1972, a variety of intake technologies designed to minimize environmental impact have been proposed. These devices fall into two distinct categories -- impingement release devices and behavioral barriers.

Impingement release devices are those that capture organisms on a mesh with the organisms being washed or transferred back to the environment. These devices are, in most cases, modifications of the original screening system now designed to optimize survival.

Behavioral barriers are structures that elicit an avoidance response by the organisms. Examples of behavioral barriers include angled screens, louvers, bubble curtains, electric guiding and the porous dike.

A porous dike is a rock or gravel core breakwater surrounding a plant intake through which water is drawn. The advantage of the porous dike is that it allows for a large, relatively maintenance free, intake with a very low approach velocity (less than 0.1 fps).

A limited body of literature (Nece and Bell, 1973; Barker et al, 1977; Fisher et al, 1977 and Bell, undated) suggests that physical exclusion and avoidance response may occur with rock or gravel filter. Both impingements and entrainment could potentially be reduced or eliminated by a porous dike intake.

The porous dike intake design has never been tested, either for intake screening ability or fouling potential. To be considered a viable intake technology, it must be demonstrated that:

1. The system will maintain structural stability and porosity for the life of the plant.
2. That because of either physical exclusion or behavioral avoidance, it will reduce or eliminate impingement and entrainment.

New England Power Company has been testing the screening potential, hydraulic characteristics and fouling characteristic of a small porous dike intake at a test facility since July 1979. This study is designed to demonstrate the value and feasibility of a porous dike intake structure.

Test Facility

The porous dike test facility is located at the Company's Brayton Point Generating Station in Somerset, Massachusetts (Figure 1).

The test facility (Figures 2 and 3) is a reinforced concrete and steel structure measuring 21 feet wide, 60 feet long and 20 feet deep. The chamber is a concrete box, open at the top and front and divided into three cells by wooden timbers. Each cell is six feet wide. A single axial flow pump is located in the back (downstream) end of the box to draw water through the structure. At the front (upstream) end of the box, gabions (wire baskets) are placed in the first and third cell. One cell has gabions filled with 3" stone and the other has gabions filled with 8" stone. Each gabion is 6 feet wide, 3 feet high and 3 feet deep. These are placed on top of each other to form a

FIGURE 1

COASTAL AREAS OF MASSACHUSETTS AND RHODE ISLAND

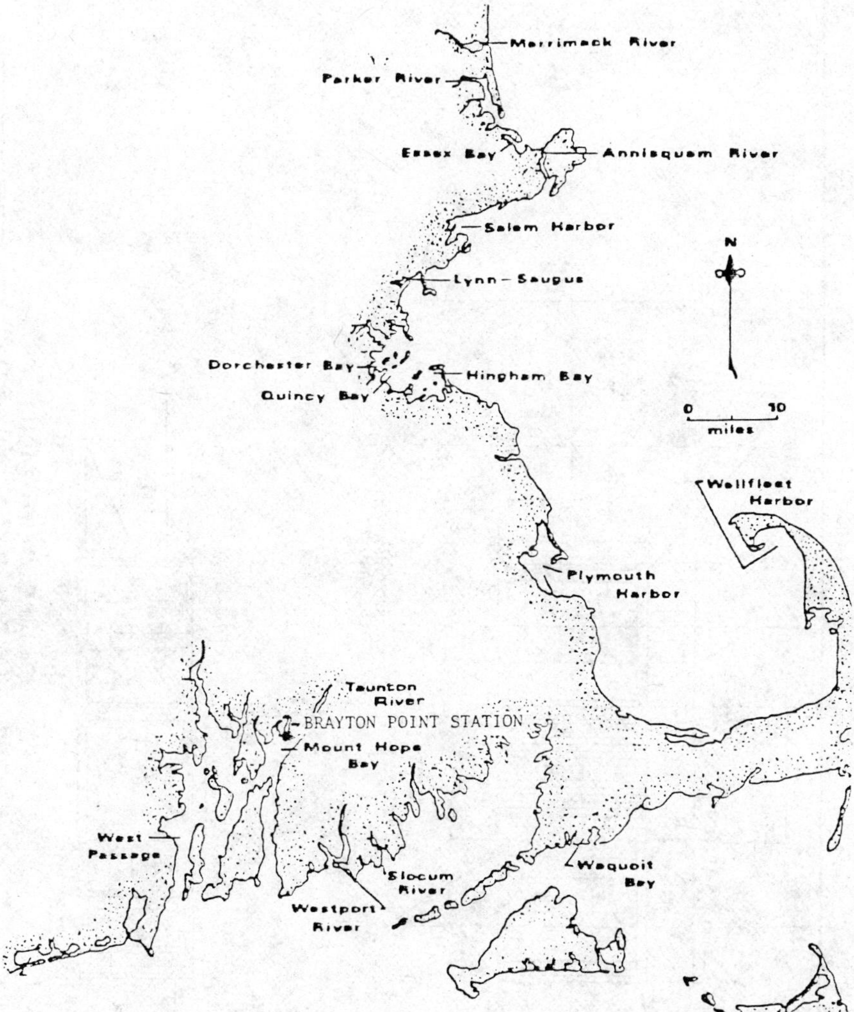

FIGURE 2

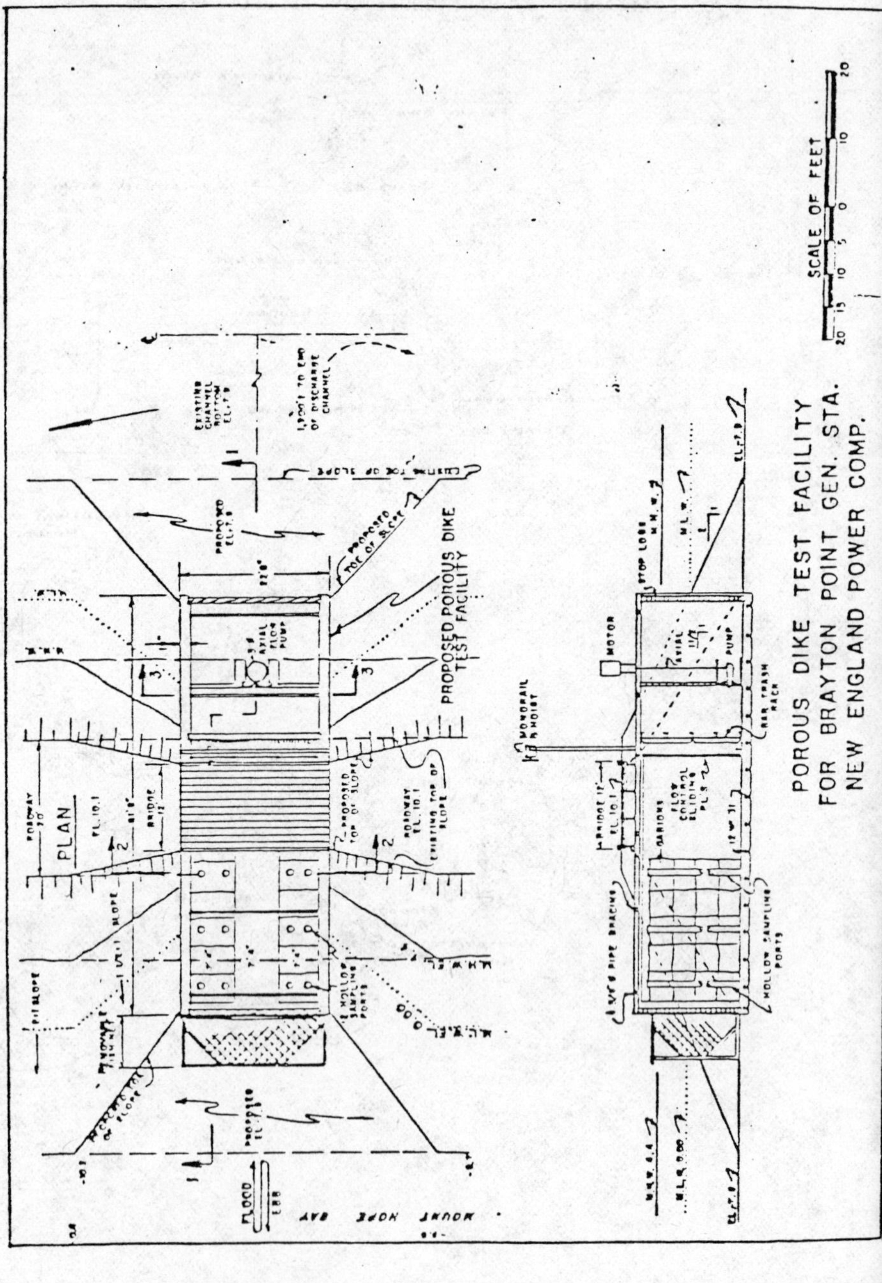
PLAN
PROPOSED POROUS DIKE TEST FACILITY
MOTOR
SCALE OF FEET
POROUS DIKE TEST FACILITY
FOR BRAYTON POINT GEN. STA.
NEW ENGLAND POWER COMP.

FIGURE 3

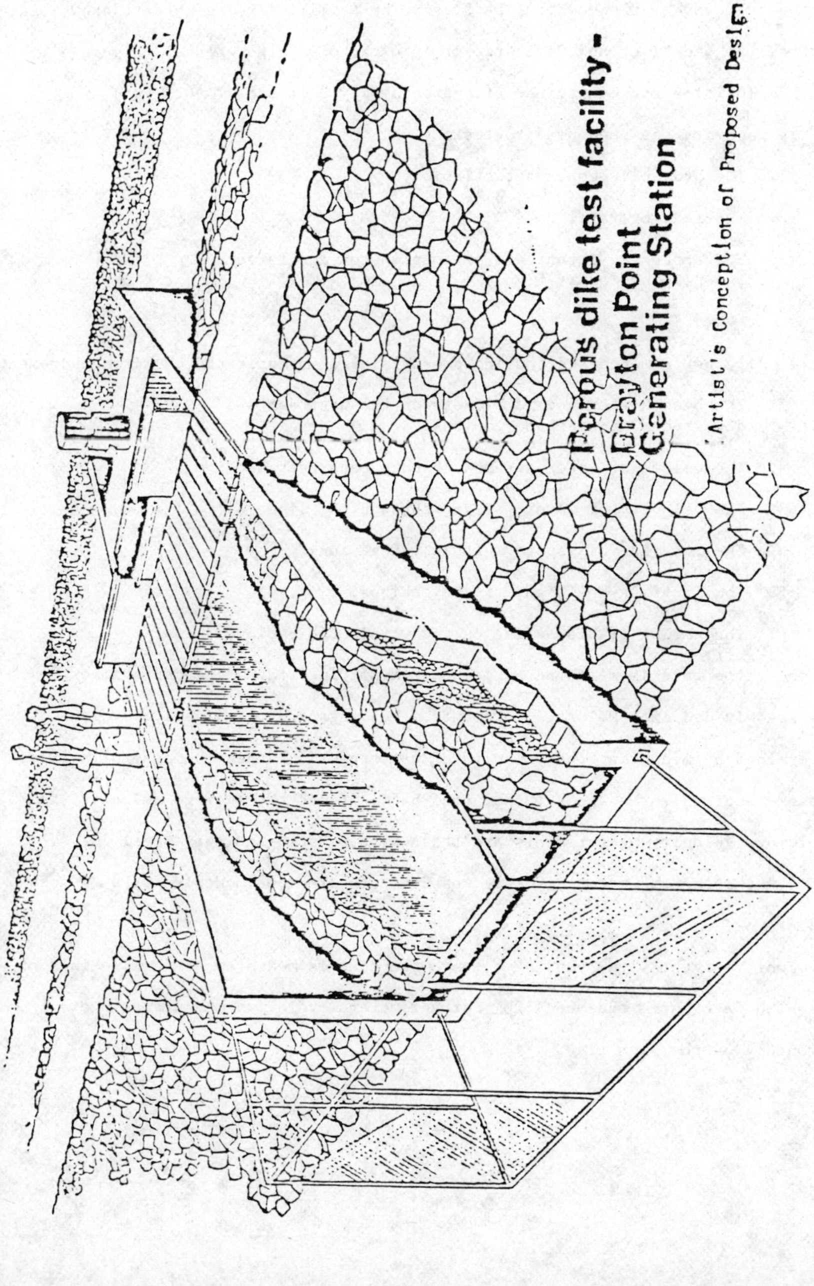

Porous dike test facility - Drayton Point Generating Station

(Artist's Conception of Proposed Design)

wall approximately 13-14 feet high. In the 8" test section, there are three rows of gabions making a 9' thick section. In the 3" test section, there are two rows making a 6' thick section. The center cell is open to act as a control. Flow is regulated by means of flow-control baffles. The facility is designed to draw water from the Lee River and discharge it into the discharge channel (Figure 4). Approach velocities are maintained at approximately 3 cm/sec. ($\approx$0.1 fps).

This facility has been used to test:

1. The hydraulic characteristics.
2. The fouling rate.
3. The screening potential for zooplankton and larval, juvenile and adult fish.

Methods

Hydraulic performance and fouling potential are investigated by six measurements:

1. The head difference (ΔP) between the upstream and downstream face of the gabions.
2. The volume of water passing through each test section.
3. The velocity of water passing through the dike.
4. The loss in void space, within the gabions.
5. The increase in weight of the gabions.
6. The amount of biofouling on and within the gabions.

Head difference is measured by means of measuring staffs in each chamber. Staffs are divided into 0.1 ft. increments. Readings are taken daily. Volume is determined by measuring the velocity of water passing under the control baffles. The cross-sectional area under the control baffle times the average velocity yields the volume. Velocity measurements are taken with a Marsh-McBirney Model 201 portable water current meter. Volume measurements are done monthly for a 24-hour period.

Velocity of the water passing through the dike is measured at 70 locations on each gabion face using the Model 201 current meter. Velocity measurements are done monthly on the high tide.

FIGURE 4

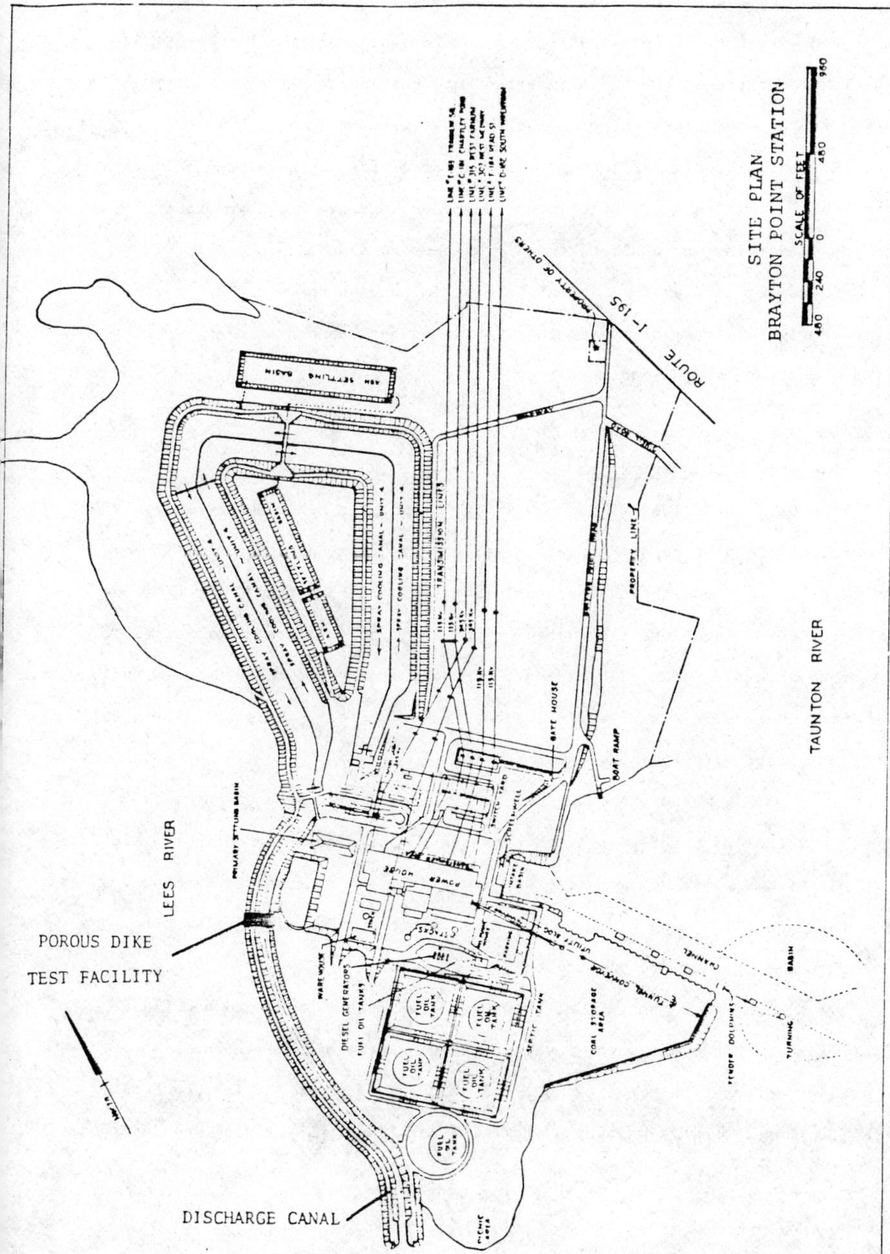
SITE PLAN
BRAYTON POINT STATION
SCALE OF FEET
480 240 0 480 960
ROUTE I-195
PROPERTY OF OTHERS
PROPERTY LINE
ASH SETTLING BASIN
TRANSMISSION LINES
GATE HOUSE
BOAT RAMP
SWITCH YARD
POWER HOUSE
PRIMARY SETTLING BASIN
LEES RIVER
TAUNTON RIVER
POROUS DIKE
TEST FACILITY
DISCHARGE CANAL
WAREHOUSE
DIESEL GENERATORS
FUEL OIL TANKS
FUEL OIL TANK
SEPTIC TANK
COAL STORAGE AREA
CHANNEL
BASIN
TURNING
FENDER DOLPHINS
PICNIC AREA

Loss in void spaces, increase in gabion weight and bioaccumulation are measured with fouling tubes; 9" x 9" x 24" removal sections of the gabions. Thirty-two of these tubes were inserted into the gabions at startup. Two tubes are removed every three months for analysis.

Screening studies of zooplankton and larval fish involve pumping naturally occurring densities of plankton from locations upstream and downstream of the gabions. Samples are analyzed for differences in abundance.

Screening studies for finfish involve impounding known number of fin-clipped fish upstream of the gabions for fixed periods and seining for these fish downstream of the gabions.

Results

A. Hydraulic Characteristics

Daily head loss measurements have been taken since September 7, 1979. Values are plotted on Figure 5. Values have fluctuated between 1.1 and 2.8 feet. Values have been nearly identical in both the 3" and 8" stone sections. Head loss fluctuated around the 1.5 foot level from September through early March. In mid-March, head loss rose to the 2.0 feet level. Since March, levels have fluctuated around 2.0 feet. This pattern is well correlated with the observable fouling.

Monthly measurements of volume have been taken since January 1979. The volume of water passing through the porous dike varies with the tidal height. Minimum volumes occur at low tide. The minimum and maximum volumes for each survey are given in Table 1. Minimum and maximum volumes for the 3" and 8" stones are plotted on Figures 6 and 7.

It should be noted that there have been three unplanned backflush periods. Since the water level on the discharge side of the pump is always greater than the water level at the face of the gabions, each time the pump is shut down, the system will backflush. The system was designed

Figure 5

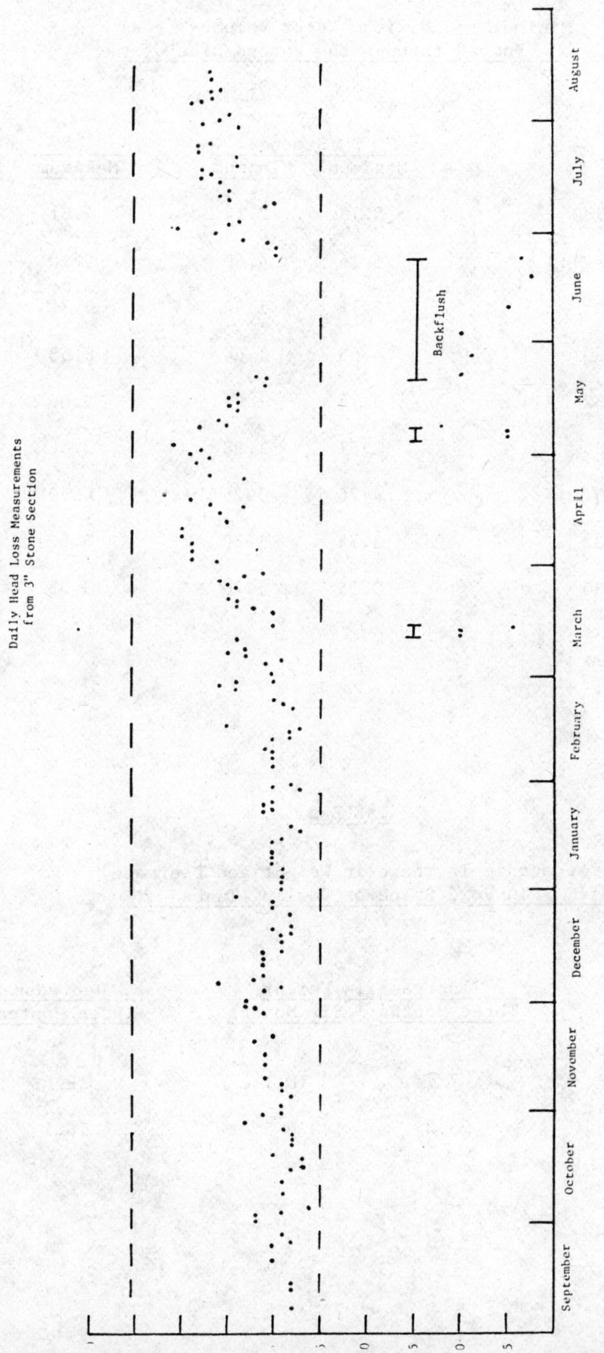
Daily Head Loss Measurements
from 3" Stone Section
Backflush
September
October
November
December
January
February
March
April
May
June
July
August

TABLE 1

Minimum and Maximum Water Volumes (cfs) Pumped through the Porous Dike

Date	3" Stones		8" Stones	
	Maximum	Minimum	Maximum	Minimum
January 8, 1980	5.19	2.14	4.31	2.21
February 12, 1980	6.16	4.62	5.70	2.28
March 6, 1980	2.75	1.44	3.71	1.69
Marcy 17, 1980	4.85	1.64	8.55	2.10
April 10, 1980	3.78	1.94	5.06	1.69
April 28, 1980	2.81	1.88	3.56	1.90
June 26, 1980	4.76	3.25	4.80	3.00
July 10, 1980	3.75	1.30	3.62	1.08
August 1, 1980	3.35	1.70	3.38	1.20

TABLE 2

Percentage Increase in Weight and Decrease in Void Space of Fouling Tubes

	% Increase in Weight		% Decrease in Void	
	Three Months	Six Months	Three Months	Six
3" Stones	7.7	10.1	19.6	2
8" Stones	5.3	6.5	16.1	1

Figure 6

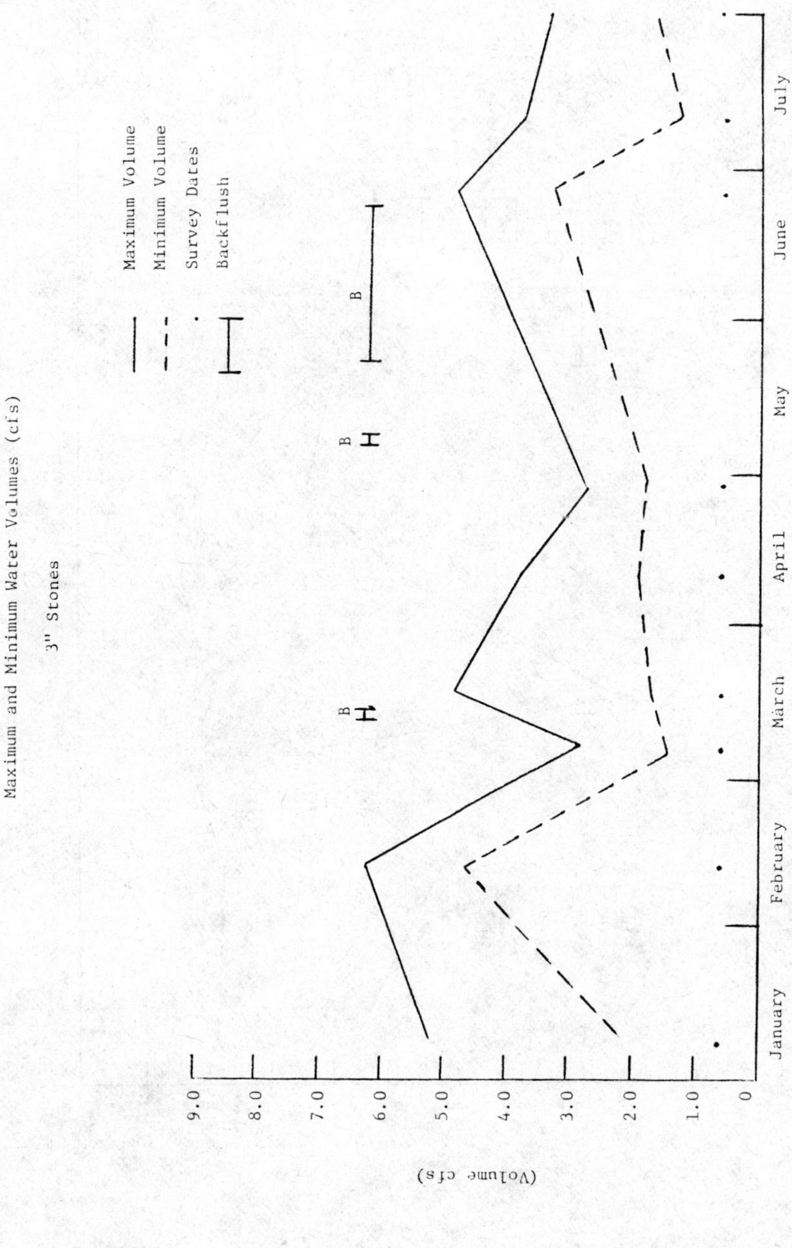
Maximum and Minimum Water Volumes (cfs)
3" Stones
Maximum Volume
Minimum Volume
Survey Dates
Backflush
B
B
B
9.0
8.0
7.0
6.0
5.0
4.0
3.0
2.0
1.0
0
(Volume cfs)
January
February
March
April
May
June
July

Figure 7

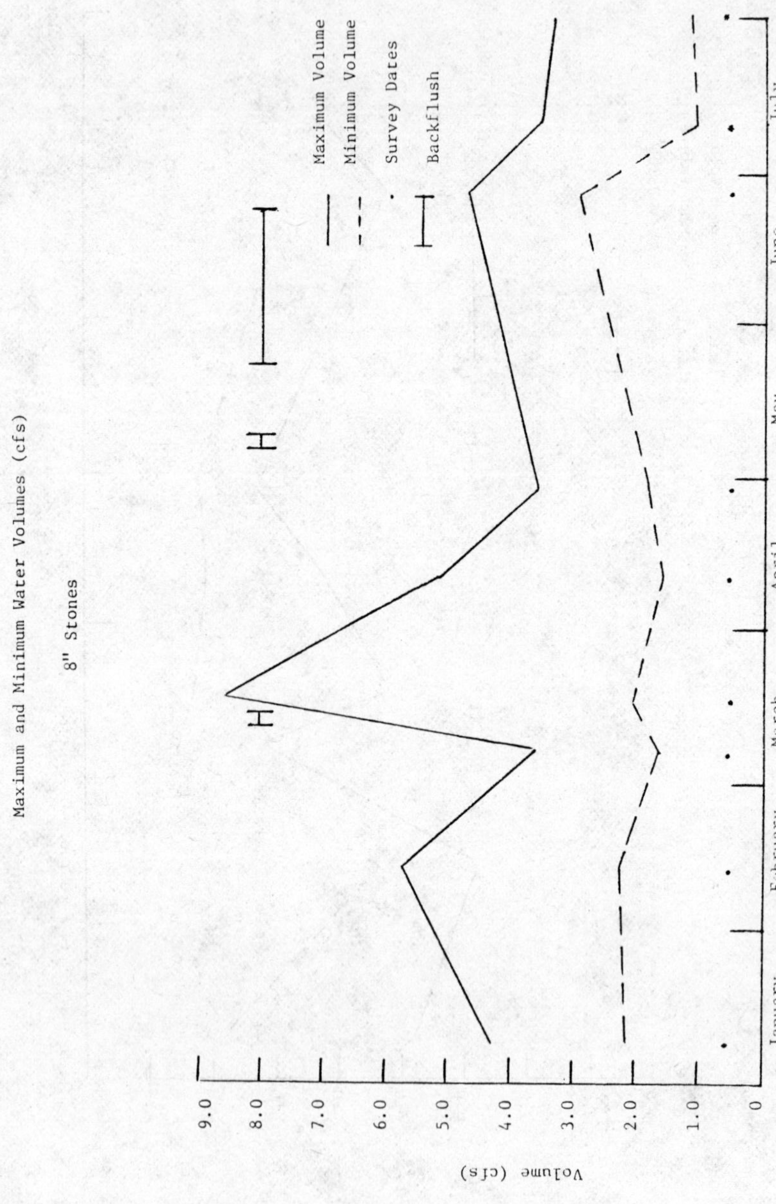
Maximum and Minimum Water Volumes (cfs)
8" Stones
Maximum Volume
Minimum Volume
Survey Dates
Backflush
Volume (cfs)
9.0
8.0
7.0
6.0
5.0
4.0
3.0
2.0
1.0
0
January
February
March
April
May
June
July

to run continuous. Due to mechanical failures, the pump has been out for two three-day periods and one 31-day period.

After each backflush, both the minimum and maximum volumes increased over the previous date. This increase was then usually followed by a slow decline in volume.

The velocity of water entering the face of the porous dike has been both uniform over the face of the structure and stable over time at approximately 0.1 feet per second. Survey data suggests that the gabion section from the low water mark to the bottom is more fouled since more readings lower than 0.1 fps were obtained. Velocity between the low water mark and the high water mark tend to have fewer lower readings suggesting less fouling. These measurements are consistent with the observed fouling pattern.

B. Fouling

To date, 4 fouling tubes, two 8" and two 3" have been removed and analyzed. Changes in weight and void space are given in Table 2. These values represent the change in weight and displacement after three months and six months of continuous operation. After six months of operation, the 3" stone gabions increased in weight by 10.1% and lost 28.4% of the void spaces. The 8" stone gabions increased in weight by 6.5% and lost 19.6% of the void spaces.

Biological analysis of the first fouling tube showed that approximately one-half of the material in the gabions was live organisms. A list of organisms identified from the first set of fouling tubes is given in Table 3. Dominant organisms include the *Neopanope sayi*, a decapod crustacean, *Balanus* sp., a barnacle, and the sea squirt, *Molgula* sp.

C. Screening Potential for Zooplankton and Larval, Juvenile and Adult Fish

Results from zooplankton studies have shown a significant to highly significant difference in the population of zooplankton between the upstream and downstream gabion face. Upstream densities are always higher than downstream

TABLE 3

List of Organisms Identified from the Final Set of Fouling Tubes

3" Stones

- Phylum Cnidaria
  - Class Anthozoa
    - *Metridium senile*
- Phylum Annelida
  - Class Polychaeta
    - *Brania clavata*)
    - *Autolytus* sp.)
    - *Podarke obscura*
    - *Hydroides dianthus*
    - *Polydora ligni*
    - *Sabellaria vulgaris*
    - Ampharetidae
    - *Eumida sanguinea*
    - *Nereis succinea*
- Phylum Mollusca
  - Class Gastropoda
    - *Tergipes tergipes*
  - Class Gastropoda & Bivalvia
    - *Mya arenaria*
    - *Crepidula plana*
    - Mysella planulata
- Phylum Arthropoda
  - Subphylum Pycnogonida
    - *Anaplodactylus lentus*
  - Class Crustacea
    - Order Amphipoda
      - *Aeginina longicornis*
      - *Microdeutopus gryllotalpa*
      - *Stenothoe minuta*
      - *Melita nitida*
      - *Elasmopis levis*
      - *Ampelisca abdita*
  - Subclass Cirrepedia
    - *Balanus* sp.
  - Order Decapoda
    - *Neopanope sayi*
- Phylum Chordata
  - Class Ascidacea
    - *Molgula* sp.
- Algae
  - Mostly *Codium*

8" Stones

- Phylum Cnidaria
  - Class Anthozoa
    - *Metridium senile*
- Phylum Annelida
  - Class Polychaeta
    - *Nereis succinea*
    - Brania clavata)
    - Autolytus sp.)
    - *Eunida sanguinea*
    - *Podarke obscura*
    - Terebellidae
    - *Sabellaria vulgaris*
    - *Polydora ligni*
    - *Hydroides dianthus*
- Phylum Mollusca
  - Class Gastropoda
    - *Tergipes tergipes*
  - Class Bivalvia
    - *Mya arenaria*
    - *Mulinia lateralis*
    - *Mytilus edulis*
    - *Mysella planulata*
- Phylum Arthropoda
  - Subphylum Pycnogonida
    - *Anaplodactylus lentus*
- Class Crustacea
  - Order Amphipoda
    - *Aeginina longicornis*
    - *Microdeutopus gryllotalpa*
    - *Stenothoe minuta*
    - *Melita nitida*
    - *Elasmopis levis*
  - Subclass Cirrepedia
    - *Balanus* sp.
  - Order Decapoda
    - *Neopanope sayi*
- Phylum Chordata
  - Class Ascidacea
    - *Molgula* sp.
- Algae
  - 90% *Codium* sp.

densities. These results suggest either entrainment avoidance or capture within the dike.

Results from icthyoplankton studies conducted during the fall of 1979 have shown far fewer larval fish captured on the downstream side on the gabions. For all samples and species combined, 77.5% of the larval fish collected in the 3" gabion section were taken on the upstream side. These results also suggest entrainment avoidance or capture within the dike.

To date, not a single fin-clipped fish placed upstream of the gabions has been captured on the downstream side. These results strongly suggest that juvenile and adult fish cannot penetrate the porous dike.

Discussion

The porous dike test facility has been operating since July 1979. As anticipated, hydraulic performance has been characterized by an increase in head loss and a reduction in flow volume. The increase in head loss and the reduction in flow is attributed to the accumulation of silt and the biological fouling on and within the gabions. The success of the porous dike is predicated on the theory that this fouling and siltation will stabilize and at that point of stabilization, the dike will remain sufficiently porous. Insufficient data has been collected at this time to confirm stabilization. Hydraulic studies on the porous dike will continue through July 1981.

Biological studies to date have indicated that some larval fish may avoid entrainment through the rock dike and that adult and juvenile finfish avoid entrapment. Additionally, data suggest that zooplankton avoid entrainment.

Both filtration and cropping could produce the same results that are attribuated to plankton avoidance. We are unable at this point to determine if the decreased zooplankton density is due toavoidance or filtration or cropping within the dike. Studies are currently underway to investigate which factor or combination of factors are responsible for the observed difference.

LITERATURE CITED

Barker, A.J., S. Partain, and F.J. Germano, Jr. (1977). The collecting, Culturing and Swimming Ability of Larval American Shad, Alosa sapidissima, New England Aquarium Report, Boston, Massachusetts.

Bell, M.C. (Undated). Handbook of Design Criteria for Fish Facilities, Pre-publication Draft from U.S. Army Corps of Engineers, North Pacific Division.

Fisher, F.W., D.B. Odenweller, and J.E. Skinner (1977). "Recent Progress in Fish Facility Research for California Water Diversion Projects," in Third National Workshop on Entrainment and Impingement. Proceedings of the Workshop held in New York, February 24, 1976 (L.D. Jense, Ed.), Ecological Analysts, Inc., Melville, New York.

Nece, R.E. and M.C. Bell (1973). "Hydraulics of a Gravel Core Fish Screen." Proc. 21st Annual Hydraulics Division Specialty Conference, ASCE, New York.

OFFSHORE PIPELINE ROUTING:
MINIMIZING ENVIRONMENTAL EFFECTS

by

Ann W. Gowen* and Michael J. Goetz*

One of the major sources of concern regarding offshore petroleum development in frontier areas is the potential effects associated with the transportation of oil or natural gas from the platform to the shore. Although pipelines are generally accepted as the safest transportation method, uncertainty is still felt regarding their possible effects on locations chosen as pipeline routes and landfalls.

This report discusses various methods for minimizing and managing environmental effects related to pipelines, both offshore and at the landfall. It is a summary of the results of two reports prepared by the New England River Basins Commission's (NERBC) Offshore Pipeline Project sponsored by the U. S. Environmental Protection Agency (EPA) including: Choosing Offshore Pipeline Routes: Problems and Solutions (NERBC, 1980a); and Pipeline Landfalls: A Handbook of Impact Management Techniques (NERBC, 1980c).

Two basic techniques are available: (1) Choosing a route which has the fewest constraints to pipeline installation; and (2) using construction and restoration techniques which minimize the effects of installation on the chosen route. Offshore, the focus is on the first method, with route selection being made wherever possible by avoidance of areas having the highest risk, either to pipeline safety or the environment affected. At the landfall, on the other hand, experience has illustrated that environmentally acceptable installations can be made in a variety of environmental conditions, if appropriate installation and restoration techniques are employed. Therefore, at the landfall, the primary focus is on the second basic impact minimization technique.

---

*New England River Basins Commission
141 Milk Street
Boston, Massachusetts 02109

MINIMIZING OFFSHORE EFFECTS: SELECTING A ROUTE

The "Straight Line" Route

During the first phase of offshore petroleum development in a frontier area, exploratory drilling is undertaken to determine the existence and extent of petroleum reserves in the lease sale area. During this exploratory phase, industry also begins initial planning for pipelines by selecting very generalized routes to carry potential "finds" to shore. Preliminary routes are delineated based on a few key criteria -- the most important being the most direct path from the lease sale area to the shoreline. This minimum pipeline length is desirable to minimize pipeline construction and operation costs. (In 1979, average U. S. installation costs ranged from $375,000 - $400,000/mile for an 8-inch line to $850,000/mile for a 36-inch offshore line; North Sea offshore installation costs have reached $1.5 million/mile.) From this most direct route, refinements to avoid hazard areas are made based on readily available offshore geological data. Additional modifications may be made to more easily connect the offshore area to existing onshore transmission or processing facilities.

If commercial discoveries are made in the lease sale area, additional onsite data is gathered and the offshore route further refined. At this stage, a number of potential landfall sites may also be identified by industry; the primary criterion remains, however, to minimize the length of the offshore pipeline.

Offshore Siting Criteria

To minimize environmental damage from offshore pipelines, the safest and most stable route from the production area to the shore must be located. To do this, it is first necessary to assess each potential route, identifying those features which are most likely to have an impact on or be impacted by the pipeline's installation. Three major factors should be considered in this assessment: the geologic characteristics of the route; the unique environmental characteristics of the offshore area; and the existing uses of the offshore.

When industry considers a potential pipeline route, the siting criteria are based to a large extent on a geologic analysis of the area. Generally, geologic hazard areas include:

- areas with adverse sediment conditions - This includes soft, unconsolidated fine sediments that provide poor foundation characteristics or could form a fluid backfill in a pipeline trench. Gas charged sediments and sediments prone to liquefaction or other bearing strength failures should also be avoided;

- areas subject to scour - These are areas with swift bottom currents or intense wave activity with sediments that are mobile under those current velocity conditions;

- sand waves - These large, mobile features may present a physical barrier to construction or expose a buried pipeline and subject it to possible spanning;

- eroding shorelines - Shorelines that have a past history of significant erosion should be avoided;

- active faults - Active faults (those that have disturbed recent sediments or have been the location of recent earthquakes) should be avoided in pipeline routing;

- deltas - Deltas have large quantities of rapidly deposited, cohesionless sediments. Their unstable sediment structure makes them susceptible to sediment bearing strength failures. The seaward edges of many deltas are the locations of sediment mass movements. A high organic content within the sediments allows the formation of gases that further weaken the sediment structure.

- heads of submarine canyons - These are sites of sediment mass movements and have an irregular distribution of erosional and depositional areas resulting in considerable bottom instability;

- slump areas - Areas with detachment of masses of sediments near canyon heads and along the edges of continental shelves or deltas should be avoided; and

- buried channels and valleys - Sediments filling buried valleys and channels are generally unconsolidated and in places have slump features that indicate instability. The unconsolidated sediment fill has variable foundation characteristics and may be prone to bearing strength failures.

Additional geologic avoidance areas may include volcanoes and areas of the continental shelf affected by ice gouging from icebergs.

Because of the relatively small amount of disturbance and the short time period involved, it appears that in general, only small-scale, localized environmental damage will result from pipeline installation activities. However, there are particularly sensitive environmental areas which may be easily disrupted by pipeline installation (particularly trenching operations) and which may take long periods of time to recover. Where possible, these types of systems should be avoided in pipeline routing. These "avoidance areas" include:

- spawning and nursery areas - These areas include both offshore and nearshore)particularly estuarine) areas. Construction may be particularly damaging where trenching occurs and in areas used for demersal (bottom dwelling) spawning or larval growth. If routing through these areas is unavoidable, construction should be scheduled to avoid spawning and nursery seasons;

- unique habitats - These include areas of particular ecological significance such as coral reefs, "live bottoms" and "flower gardens;"

- rare or endangered species habitats; and

- small habitats - These include areas where the area disturbed by installation makes up a large percentage of the total habitat, or where high concentrations of organisms exist in small areas. These habitats include: seagrass beds; kelp beds; and (where applicable) shellfish beds.

A potential offshore pipeline route is also carefully examined to identify the existing uses of the environment which could adversely affect pipeline safety. These use conflicts include:

- submarine pipelines;
- submarine cables;
- shipping lanes;
- areas with undetonated explosives;
- dumpsites (dredge spoils, industrial wastes, etc.) and areas with polluted bottom sediments;
- navigational channels;
- archeological sites;
- anchorages;
- commercial bottom fishing grounds.

An Illustration: Preliminary Route Selection, Georges Bank*

For the purposes of the North Atlantic case study, the potential hydrocarbon production area has been determined as being the tracts offered by BLM in Lease Sale #42 on Georges Bank. These tracts lie approximately 80 to 200 miles southeast of Cape Cod, Massachusetts in water as deep as 200 meters (Figure 1).

There is very limited petroleum-related infrastructure in coastal New England. The nearest petroleum refineries are located in New York and New Jersey, approximately 315 miles due west off the potential production area. There is a facility for receiving liquified natural gas (LNG) in Everett, Massachusetts, just north of Boston, also providing coastal access to the natural gas pipeline distribution system throughout New England. A major crude oil pipeline connecting refineries near Montreal, Canada originates at a crude oil receiving terminal at South Portland, Maine.

Based on the infrastructure discussed above, the following straight-line, preliminary pipeline paths have been considered for further analysis:

---

*This analysis is based on a draft report, A Methodology For Deriving Offshore Oil and Gas Pipeline Paths, prepared by the New England River Basins Commission's OCS Programs under agreement with the U. S. Geologic Survey's Resource and Land Investigations Program (RALI).

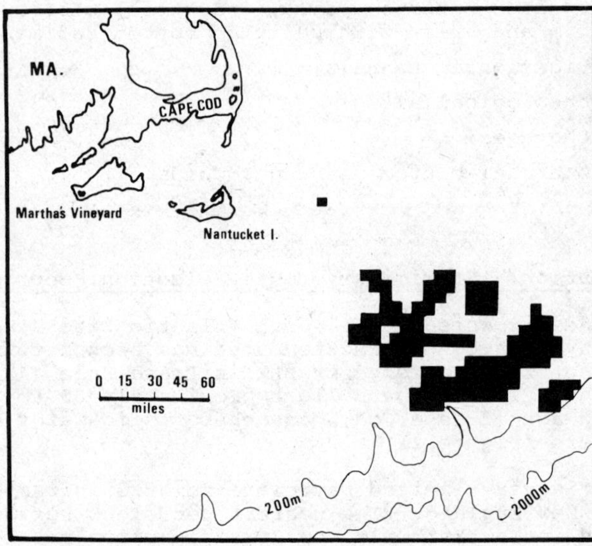

Source: Bureau of Land Management (BLM)

Figure 1: Location of Lease Sale 42, North Atlantic Outer Continental Shelf.

- a route connecting to southeastern Cape Cod (shortest distance);
- a route to the crude oil pipeline in South Portland, Maine;
- a route terminating in the vicinity of Boston, Massachusetts; and
- a route terminating in southern Rhode Island or southeastern Massachusetts (Buzzards Bay vicinity).

Additional straight-line routes could terminate at Montauk Point, New York or the New York City area.

Review of the readily available geologic and environmental information available for the Georges Bank region suggested the following physical and geological constraint areas to be of significance on the North Atlantic OCS:

- areas with adverse sediment conditions (fine sediment);
- slumping/slope instability;
- steep slopes;
- buried channels;
- areas of sand wave development; and
- heads of submarine canyons.

Figure 2 shows the composite physical and geological constraint map produced for the Georges Bank outer continental shelf region. The pipeline routes shown on the map were delineated by avoiding to the greatest extent possible those areas deemed to be undesirable for pipeline routing. The routes as shown have the following general characteristics:

- surface sediments are generally sandy and are likely to have adequate bearing strength and be suitable for trenching if that would be needed;
- a generally flat topography with the slopes between 0 and 2 percent;
- water depths are between 200 and 600 feet -- which is well within present technology for laying, trenching, and burying of pipelines; and
- the width of the path varies from as little as 4 - 6 miles in the vicinity of Welker and Oceanographer Canyons and off the eastern shore of Cape Cod -- to as wide as 50 miles in the vicinity of the Great South Channel.

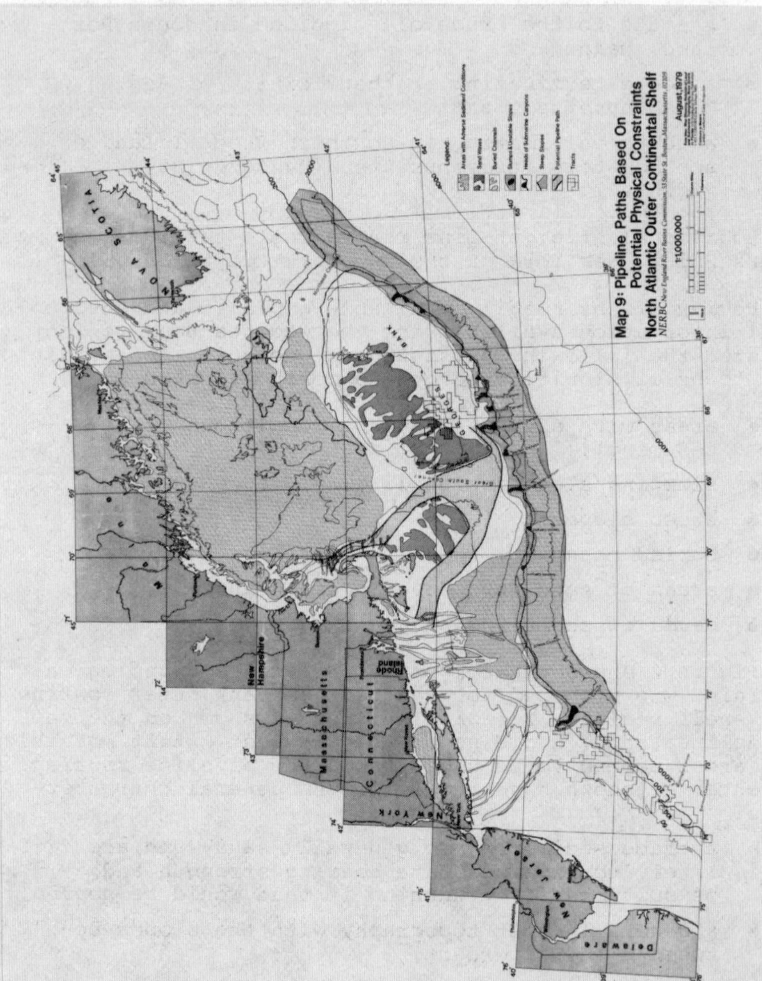

Figure 2: Physical and geological constraints on Georges Bank.

Distances to shore range from approximately 200 miles to Cape Cod to 300 miles to the Boston area and suburban Rhode Island (straight line distances range from 150 to 250 miles).

One constraint area that was difficult to avoid was buried channels. The modified path runs into buried channels in the waters off southern Rhode Island and off eastern Cape Cod. Although not shown on the final composite map, the straight line path to Portland, Maine could still be considered a possibility if crude oil is discovered. In such a case, the sediment conditions in the Gulf of Maine would have to be considered in greater detail to determine the feasibility of that route.

The following environmental and use conflicts were also identified in the Georges Bank region:

- cables;
- undetonated explosives;
- shipping lanes;
- artificial fishing reefs;
- general dumpsites;
- bottom spawning areas;
- canyon habitats (lobster, coral); and
- gray seal breeding areas.

The potentially most significant conflict, however, is the extensive prime bottom fishing grounds located over most of Georges Bank and the adjacent continental shelf. Various estimates* have placed the present economic value of the North Atlantic fisheries at over one billion dollars.

---

*The National Marine Fisheries Service (NMFS), Northeast Fisheries Center estimates a total United States 1978 value for all fish landings (domestic and foreign) from Georges Bank at $167.5 million. Economic multipliers used to assess the economic impact of an industry on the regional economy raise this figure to over one billion dollars.

MINIMIZING EFFECTS AT THE LANDFALL: INSTALLATION AND RESTORATION TECHNIQUES

Landfall Site Selection - The First Step

Although not as critical as in offshore areas, the type of coastal system chosen for a landfall is an important first step toward minimizing pipeline installation effects. Landfall site selection is based on consideration of a number of physical criteria, including the physical/geological structure of the area and adjacent ocean, and the ecological systems at and adjacent to the landfall which may be affected by pipeline installation.

Areas with the following characteristics would be avoided if possible:

- steep slopes - which would make pipeline installation and long-term stable restoration difficult;
- highly eroding shorelines - where pipelines would have to be buried extremely deeply to minimize the risk of future exposure;
- bedrock near the surface - which would require blasting for trenching;
- cohesionless sediments - which are very unstable and thus susceptible to bearing strength failures and sediment mass movements;
- high velocity nearshore currents - which can interfere with pipeline laying barges and equipment and necessitate structural measures (such as jetties) to maintain surf zone pipeline trenches (which would otherwise be quickly refilled);
- large tidal fluctuations - which would also interfere with pipelaying activity and which might necessitate deeper burial depths;
- limited shoreline access - which would limit the usefulness and efficiency of the landfall as a construction staging area;
- unusual and important habitats - such as rare or endangered species habitats, spawning and nursery areas, and areas of particular ecological significance;

- habitats which are difficult to restore - because of their unique sediment or vegetation structure, or aesthetic qualities; and
- habitats with commercial value - such as shellfish flats.

Based on these technical and ecological considerations, certain coastal system types would be more desirable for landfalls than others. For example, those coastal areas which are gently sloping (less than 10% grade), with sandy or firm sediments and no fragile or unusual habitats or species would be most preferable for landfall siting. Unfavorable sites would be those which have characteristics such as eroding shorelines, steep slopes, high velocity nearshore currents, poorly consolidated sediments, or unique and/or rare or endangered species habitats. Based on these criteria, accreting sand beaches with no dunes have the fewest constraints, while marshes have the largest number of undesirable features.

Installation and Restoration Techniques

Based on ten case study installations in a full range of coastal environments, even in circumstances where less than ideal landfall sites must be chosen, environmentally acceptable pipeline installations can be accomplished by using specific installation and restoration techniques chosen to address the site's constraints. Three basic types of techniques were used at these installations:

- general techniques - applicable to most landfall installations;
- specialized technical methods - for use in construction in specific types of coastal features such as dunes and wetlands; and
- planning methods - which stress coordination and interaction between the developer and public officials throughout project development.

General installation techniques include:

- scheduling to avoid key ecological seasons to minimize impacts on local plant and animal species;
- minimizing the size of the area affected and, therefore, the extent of impact;

- restoring the land/water interface area immediately to minimize risk to landward coastal systems (e.g., wetlands or dunes); and
- restoring the entire affected area physically and ecologically to a condition as closely approximating the original as possible.

Certain coastal ecosystems require few additional methods to restore to original condition after installation activities have been completed. These areas include flat or gently sloping beaches with no dunes and industrialized or commercially developed waterfront areas.

The presence of other coastal features, such as dunes, cobble beaches, rock cliffs and wetlands, may require the use of more specialized methods to assure more complete functional and aesthtic recovery. These <u>specialized technical restoration methods</u> include:

- <u>dunes</u> - Artificial stabilization techniques (e.g., snow fences, gabions or Christmas tree tops) may be considered to assure dune recovery, particularly in coastal areas prone to wind or wave erosion. Dunes may also be "naturally" stabilized by replanting with native vegetation.
- <u>cobble beaches</u> - Where blasting is necessary, non-native materials may be required for trench refilling and restoration of the beach form and function. In this case, new fill should approximate the general size, shape and distribution of original surface materials if possible.
- <u>rock cliffs</u> - As with cobble beaches, blasting for trench preparation will require refilling with non-native materials. In this case, however, it is likely that use of concrete would be necessary to restore functional integrity, making total aesthetic restoration impossible.
- <u>wetlands</u> - Because of their very loose sediment structure, wetlands may be the most difficult of all coastal ecosystems to restore to original functional and aesthtic conditions. Restoration programs would need to be designed on a case-by-case basis to address the key impacts expected in the affected wetland system.

Increased government involvement in pipeline landfall siting and regulation will require coordination and interaction between the developer and public officials throughout project development. To encourage this cooperative planning and to facilitate development of acceptable and effective regulations for a particular installation, these planning methods may be considered:

- environmental assessment preparation - This is an important means of identifying major potential impacts and information needs prior to construction.
- single agency planning and permitting - This minimizes the time necessary to obtain permits and variances in original stipulations if necessary. It may also foster better communication between developers and regulators by clearly defining key participants in regulation development.
- ongoing consultations between developers and regulators - This fosters development of installation stipulations appropriate to the landfall's technical and environmental constraints and allows flexibility to change original plans if necessary or desirable as construction proceeds.
- establishing restoration goals - This focuses restoration efforts on minimizing major identified potential effects, rather than on attempting a complete replication of original conditions - which is often impossible to achieve except over a long-term natural restoration cycle.
- long-term monitoring programs - These programs would provide a method for ongoing assessment of restoration success and the opportunity to identify and remedy, if necessary, an initially unsuccessful restoration effort.

AN ILLUSTRATION: THE FLAGS LINE, ST. FERGUS, SCOTLAND

The 1976 installation of the Far North Liquids and Associated Gas (FLAGS) pipeline, a 36-inch diameter line which carries natural gas from Shell/Esso's offshore Brent field to the St. Fergus gas processing terminal on the northeast Scottish coast, is an excellent illustration of the effective use of many of the impact management methods outlined above. The landfall is on a steep, highly dynamic beach backed by high, steep dunes. Landward is a seasonally submerged freshwater wetland - called the Winter Loch - which is considered an area of particular ecological significance both by the Scottish Wildlife Trust and the Royal Society for Protection of Birds. Another ridge of lower more mature dunes back the wetland;

behind the second dune line is the terminal. Because of the landfall's unique ecological character, both Shell/ Esso, the developer, and the Banff and Buchan Council, the lead regulating agency and sole landfall permittor, paid particular attention to the selection and use of landfall restoration methods which would minimize damage to the existing natural system.

All of the general installation techniques suggested above were used at this landfall. Installation was scheduled for June when the wetland area was dry, making installation easier and avoiding the biologically important spring season. The pipeline trench at the land/water interface was immediately refilled after installation and quickly returned to its original condition. Figure 4 shows the landfall shortly after reinstatement.

Special attention was paid to minimizing the extent of the area affected. To do this, the pipeline installation contractor employed an innovative technique - construction of a "causeway" from the first dune line to the terminal with materials excavated in making the pipeline trench. This causeway, which can be seen in Figure 3, served the dual purpose of limiting construction access and activity to a single clearly defined area and providing a storage area for the large amounts of materials to be used later in beach and dune restoration.

Special attention was also given to dune and wetland excavation and restoration. During excavation, the dune's "topsoil" layers were stored separately to be replaced on the dune's surface after topographic restoration. When rebuilding, the dune line was internally stabilized with concrete gabions and the dune face hand planted with Marran grass, the dominant dune grass species. The wetland area was also reseeded with a mix of indigeneous plant species. Follow up examinations of the restored area showed good growth, suggesting that the area was returning to its original condition.

Perhaps the key to this environmentally successful installation, however, was the effective use of many of the planning techniques suggested above. In large part this success can be attributed to the continuing interaction between regulators and developers throughout FLAGS landfall plan development. Prior to seeking permit approval the developer, Shell/Esso, commissioned the preparation of an environmental assessment in which a local,

Figure 3: FLAGS landfall during construction. Note causeway (arrow) across the Winter Loch. (Photograph courtesy of Maire Murphy, NERBC.)

Figure 4: The St. Fergus FLAGS landfall (arrow) shortly after reinstatement. (Photograph courtesy of Charles Fraser, Peterhead, Scotland.)

well respected ecologist detailed possible impacts on the fragile landfall system. Highlighting the foredune and Winter Loch as those likely to suffer the most severe effects, specific recommendations for their restoration - based on physical and functional restoration goals - were made. These recommendations were accepted, in large part, by the Banff and Buchan Council and translated into the stipulations issued with the landfall permit. The Council remained flexible on actual installation methods, however, as shown by their acceptance of an environmentally superior method - the contractor's suggestion for "causeway" construction, which was made after the original permit had been issued. Finally, after restoration, a five year commitment to site monitoring was made by the developer, to assure that restoration had, in fact, been accomplished.

BIBLIOGRAPHY

New England River Basins Commission. 1980a. Choosing Offshore Pipeline Routes: Problems and Solutions. Boston, Massachusetts. 91 pp.

New England River Basins Commission. 1980b. A Methodology for Deriving Offshore Oil and Gas Pipeline Paths. Draft. Boston, Massachusetts. 66 pp.

New England River Basins Commission. 1980c. Pipeline Landfalls: A Handbook of Impact Management Techniques. Draft. Boston, Massachusetts. 253 pp.

MULTIOBJECTIVE MODELING FOR OCS PIPELENE SYSTEMS

by David Engberg, Jared Cohon, and Charles ReVelle[1]

INTRODUCTION

Planning the location of pipeline systems to bring ashore petroleum from the outer continental shelf (OCS) is a difficult problem from both technical and institutional standpoints. In this paper we describe the nature of this location problem and the suitability of a mathematical programming model as a tool for such planning situations. Some aspects of implementing utility corridor models are discussed, and a model for identifying superior onshore pipeline routes is illustrated. The multiobjective solution approach of this model makes it particularly useful for balancing costs and impacts through location planning.

BALANCING OCS DEVELOPMENT IMPACTS

Accelerated development of OCS petroleum resources was an obvious response to the 1973 oil embargo. Estimates have indicated that the potential production from offshore equals the onshore production in magnitude , although in 1972 the OCS accounted for only about 12 percent of total U.S. production (Kash, et al.,1973). Clearly, rapid exploration and development of the OCS could help alleviate the country's reliance on imports.

Such a course of exploration and development (E&D), now underway, is not without its risks and costs. Studies of E&D policy have identified potential environmental and social impacts associated with the rapidly developing offshore technology and its establishment in frontier areas(Kash, et al., 1972; Council on Environmental Quality, 1974; Office of Technology Assessment, 1976). These broad policy studies have helped to clarify that the development of the OCS resource, while necessary, should be undertaken in a manner that balances the need for rapid production at low cost with

---

[1]Department of Geography and Environmental Engineering, The Johns Hopkins University, Baltimore MD 21218

its social and environmental impacts.

A promising approach to avoiding adverse impacts of OCS development is early and anticipatory planning by state and federal agencies for E&D activities. Indeed, this is the objective of the Bureau of Land Management's Intergovernmental Planning Program (BLM, 1979). With regard to onshore impacts, the experience of the Shetland Islands in Scotland has been suggested as a model for such anticipatory planning (Baldwin and Baldwin, 1975; Congressional Research Service,1976). There, the onshore effects of North Sea oil development were controlled, in part, by identifying areas suitable for various shore-based facilities and then taking steps to ensure that development occured in the desired locations. Assessments of OCS development have also emphasized the close relationship between the choice of locations for new facilities and the resulting impacts. Location planning, then, can be an important step in achieving the appropriate balance between development costs and impacts.

General methodologies have been developed for use by public agencies in planning for and assessing impacts of OCS facilities (New England River Basins Commission, 1978; Roy F. Weston, Inc., 1978). They are intended to be comprehensive, outlining steps from initial scenario development to siting analysis and impact assessment. For siting analysis, the general approach has been a process of identifying alternatives, screening these based on industry and public policy criteria, detailed assessment, and ranking of alternatives. While such a methodology is well organized, logical, and practical, it works best when applied to facilities with a few discrete alternative sites.

Route location for pipeline transportation systems presents a more complex planning problem. Several obviously interconnected components make up an offshore-to-onshore pipeline system. For a natural gas system, several platforms may be connected through gathering lines to a manifold platform (where compression may be required) from which the gas is carried ashore in a main trunk line. The onshore portion of the pipeline carries gas to a suitable junction with the interstate gas network; several alternative feasible junctions may exist. Depending on properties of the gas, it may be necessary to locate a facility along the onshore portion of the pipeline to remove valuable liquid hydrocarbons, residual water, and sulfur, and there is usually an economic incentive to locate this processing faclity as close to the landfall as posible (NERBC, 1976). In general, there are few strict, technical siting constraints for pipelines. Rather, the relative costs of construction, operation, and maintenance and the comparative risks to pipeline integrity and useful lifetime dominate the industry's siting considerations (Rooney-Char and Ayres, 1978). Thus, a relatively broad geographic area may be eligible for siting from a technical standpoint.

A system modeling approach has much to recommend itself for this sort of difficult location planning problem. Mathematical programming models are capable of screening huge numbers of alternatives to isolate the most promising solutions to complex problems. Coupled with multiobjective programming techniques, such models can be useful tools for exploring relationships between various objectives and for focusing on alternatives that strike an appropriate balance between conflicting objectives. We are currently applying this multiobjective programming approach to location planning for natural gas pipeline systems in the middle Atlantic OCS. This paper demonstrates multiobjective location planning for the onshore portion of a pipeline corridor, after first reviewing some general aspects of corridor planning models.

UTLITY CORRIDOR PLANNING MODELS

A number of mathematical models have been developed in recent years for the purpose of finding optimal utility corridors, when environmental and socioeconomic impacts are recognized in the evaluation of alternatives.Such models have been applied to the location of pipelines (Stastny, 1975; Wester and Weeter, 1977), power transmission lines (Economides and Sharifi, 1978; Newkirk, 1979), and highway segments (Turner and Miles, 1971), among others. These models have generally relied on a procedure characterized by Church and Clifford (1979) as a four step methodology, summarized below :

1. The study region is divided into a large number of grid cells from which a contiguous string is ultimately selected as the corridor.

2. A number of quantifiable conditions are evaluated by experts who assign an attribute score to each cell for each data category.

3. Each data category is given a weight; weights are used to aggregate the scores for each cell, giving each cell an overall sensitivity score.

4. A shortest path algorithm is used to select the corridor from source to destination with lowest combined (i.e. sum of individual cells') sensitivity score.

Each of these steps raises issues related to the underlying purpose of the model application as well as the details of model implementation, discussed in order below.

Grid Cell Representation. The need for a grid cell system derives from the widely recognized need to use network models for complex location problems. Typically, a uniform square grid has been used, with a grid size as small as the codification and computation burden would allow. A small grid size has been employed to give a more accurate representation of geographic features and a greater location specificity. One argument for modeling a relatively narrow corridor with a small grid is that an accurate estimate of environmental impacts, which tend to be site specific, can't be defined without prescribing a specific corridor route (Newkirk, 1979, p.32).

For the purpose of balancing costs and impacts through location planning, a relatively broad corridor is more suitable for such models. An accurate environmental assessment is actually unattainable with location models, since design, construction practice and scheduling also influence impacts. Optimization models may not be suited for such engineering design applications; rather they are "most useful in the planning of the overall configuration of programs and in the definitions of regions of optimality in which it is profitable to apply the more traditional detailed design" (de Neufville and Chin, 1974, p. 13). This generalization seems to apply well to utility corridor models. They are most useful in defining relatively broad geographic corridors that offer the greatest opportunity for balancing costs and impacts in the final design. While the relatively broad corridor (and its associated cell size) does not allow a precise measure of environmental or economic costs, such precision is neither necessary nor appropriate at the policy planning level where such balancing decisions should be made.

Identifying and Quantifying Objectives. For the process to be modelled, the attributes of corridors must be measurable in terms of the various objectives that apply. Previous attempts at utility corridor models have used the following approach to quantifying objective function attributes. For each cell of the model's study area, relevant physical, biological and cultural conditions are identified and measured, then transformed through some arithmetic operations to give a "suitability value" for each of the objectives intended to reflect the suitability of each cell for siting of the utility line.

This procedure introduces ambiguities to the modeling methodology. It can give arbitrary results in the final objective function values, and the suitability values that emerge do not have a clear meaning and are not expressed in familiar or recognizable units (Hobbs, 1979). The results of the model are difficult to interpret because of this and because of the implicit introduction of preferences. The multiobjective planning and decision analysis literature offers some useful experience that can help clarify this step of the corridor model methodology. Keeney and Raiffa (1976, Chapter 2), for example, discuss some useful guidelines for structuring the relevant objectives and determining appropriate attributes. They stress that a workable set of attributes should be complete, nonredundant, operational, and of minimum size and complexity. Further, they suggest that it is often necessary to use proxy attributes whose values reflect the degree to which an associated objective is met, but do not directly measure the objective.

The use of proxy attributes is consistent with the level at which impacts can be recognized in the utility corridor models. Corridor location models cannot recognize a precise measure of impacts or economic costs, since these are not a function of location alone. They can, however, recognize the presence of conditions that may lead to impacts, when these conditions are used as proxies. As such, the proxy attributes can be measured as the percent of the cell's area that is in such condition. Corridor models then, in essence, are used to minimize the likelihood of encountering, in the corridor, conditions that have associated adverse impacts.

Where a large number of different conditions must be considered, it is useful to organize objectives in a somewhat formal structure, to assure that the objectives reflect a complete and consistent treatment of the various concerns.

A hierarchy also aids a rational aggregation of concerns into appropriate groups that can be treated as objectives. In the utility corridor problem, it may be necessary to treat a number of different conditions as essentially comparable in terms of an objective.

Preferences and Objective Weights. The output of the first two steps of the general corridor methodology is, for each cell of the study area, a vector of attribute scores, one score for each of the objectives. Scalar optimization algorithms have been used by collapsing the separate objectives into a single "superobjective", typically by way of a linear combination. The outcome of this "weighting" approach obviously depends on the values of the weights. In the studies that employed this method, the values for the weights were typically chosen by a panel of experts, sometimes using a Delphi method. When the desired result is a balance of costs and impacts, however, the judgement as to appropriate weights is as much one of preferences between the various objectives as one of tecnical expertise. The weakness of this approach is that the specification of weights for the various objectives is effectively a description of relative preferences, or a utility function. But it is very restrictive in nature, requiring that the marginal rates of substitution (defined by the relative weights) be constant. This is generally not true of observed human behavior. Furthermore, the ability to specify weights representing preferences depends on having either a single decision maker or a group of people who can reach some agreement.

The techniques of multiobjective programming (Cohon, 1978) can improve upon the approach that creates and optimizes a single superobjective. For the utility corridor problem, a method that generates a range of alternatives representing the range of relevent preferences appears to be most appropriate; this generating technique is demonstrated in the last section of this paper. In essence, this generating approach searches out those alternatives that are clearly superior, no matter what relative values are attached to the individual objectives. The methodology then allows a choice to be made among real alternatives rather than allowing only the choice of a set of weights in the absence of implications of the weights.

Network Specification. The application of a shortest path algorithm requires that the discrete cell structure be represented by a network of nodes and connecting arcs. This transformation is rather straightforward; if nodes of the network are taken to represent cells, arcs can represent the connection of contiguous cells, and any path of connected arcs will represent a corridor of continuous cells. Goodchild (1977) and Huber (1980) have illustrated the distortions that are introduced by using such a lattice network to model continuous path problems. They have shown that distortions can be reduced by introducing arcs representing connections between more remote cells besides the adjacent, orthogonal connections. Huber has also shown how arc coefficients can be calculated as an average of the scores of encountered cells, in proportion to the area of the cell that occurs in a corridor represented by the arc.

A MULTIOBJECTIVE ANALYSIS EXAMPLE

Planning for onshore pipeline facilities to transport OCS petroleum resources in a frontier area can benefit from an early multiobjective corridor location analysis. Although such facilities in the United States are constructed and operated by private concerns, they must be approved by an array of state and federal agencies whose responsibilities and concerns introduce multiple objectives to location decisions. Early resolution of conflicts through the identification of generally acceptable corridors can speed the design, approval, and construction of needed facilities. An analysis that utilizes the multiobjective approach and that is suitable for such early corridor location planning is demonstrated in this section.

This demonstration was performed on a study area consisting of the eight southernmost counties of New Jersey. Although the analysis was for demonstration purposes only, an attempt was made wherever possible to incorporate improvements to the traditional corridor location modeling methodology suggested in the previous section.

Model Description. The study area was represented using a uniform square grid of 10 x 10 kilometer square cells[2], as illustrated in Figure 1. Four objectives were identified for this application:

---

[2] In a more recent analysis intended for actual decision making, a grid of 3-5 km dimension was used.

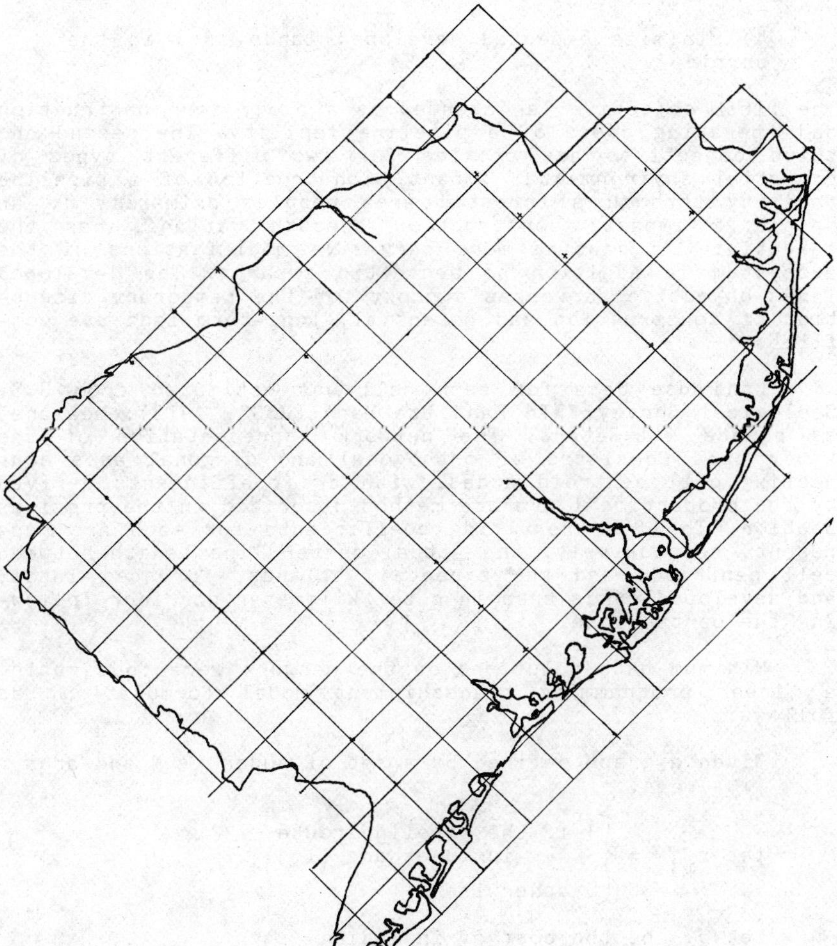

FIGURE 1
CELL STRUCTURE

1) Minimize corridor center line length

2) Minimize expected wetlands area in the corridor

3) Minimize expected forested lands area in the corridor

4) Minimize expected developed lands area in the corridor.

The first objective is intended as a proxy for construction and operating costs of a pipeline facility. The second and third objectives are proxies for two different types of potential environmental impact. Construction of a pipeline facility through a forested area results primarily in an aesthetic impact; construction through wetlands has the potential for causing more pervasive modifications of the ecosystem in addition to aesthetic impacts. The developed lands objective serves as a proxy for the temporary disruption of construction and potential, long-term land use conflicts.

Land use data for each cell was estimated from U.S. Geological Survey ERTS Land Use Maps (USGS, 1972), measured as square kilometers. The network representation of the study area consisted of orthogonal and diagonal arcs connecting cell centroid nodes, with arc coefficients derived by the proportional average method described in the previous section. The four resulting coefficients for each arc represent, respectively, the actual center line length between cell centroids and the expected wetlands, forested lands, and developed lands area in a ten kilometer corridor following the center line.

Minimum path solutions on the network were found using a linear programming transshipment model formulation, as follows:

Given a graph defined by a set of nodes $i \varepsilon N$ and arcs $(i,j) \varepsilon A$,

let $x_{ij} = \begin{cases} 1 & \text{if the pipeline route includes arc } (i,j) \\ 0 & \text{otherwise} \end{cases}$

let c_{ij} be the cost of including pipeline segment (i,j)

let N_i be the set of nodes adjacent to node i; $N_i = \{j:(i,j) \varepsilon A\}$

The minimum cost pipeline route is given by:

$$\text{Min } Z = \sum_{(i,j)\varepsilon A} c_{ij} x_{ij}$$

subject to $$\sum_{j\varepsilon N_o} x_{oj} = 1 , \qquad \sum_{j\varepsilon N_t} x_{jt} = 1$$

$$\sum_{j\varepsilon N_i} x_{ij} - \sum_{j\varepsilon N_i} x_{ji} = 0; \qquad \forall_i \varepsilon N$$

where o is the origin node and t is the terminus node. The meaning and units of "cost" in this model depend, of course, on the particular objective function empoyed.

Noninferior Set Generation. A solution to the sample problem with the objective "minimize distance" (DIST) is summarized as Route 1 (R1) in Table I. Besides the DIST objective function values, the route's performance as measured by the wetlands (WET), forested lands (FOR), and developed lands (DEV) objectives are listed. The significance of this minimum distance route should be interpreted carefully. It is optimum in the sense that no other path in the network between the specified nodes has a lesser length. Since the network for this sample problem is very regular, one could expect several other paths to have the same minimum distance; Route 1 may not be a unique optimum. Moreover, it is optimum only in regard to one objective, although three other objectives are important. If alternate minimum distance routes exist, they certainly deserve examination, as they might be better in terms of the other objectives.

Such an alternative, R2, is also listed in Table I. As can be seen by the data, R2 is also a minimum distance route, but it improves upon R1 in each of the other objectives. Thus, although both routes are optimum in terms of distance, R2 is better when all objectives are considered. This introduces two important concepts in multiobjective analysis, noninferior solutions and the weighting method used to obtain them. R1 is inferior to R2 in the sense that R2 is better than R1 in terms of one or more objectives, without being worse in terms of any other; R2 dominates R1. In a search for "good" routes, R1 can be discarded because R2 is unambiguously better.

TABLE I

OBJECTIVE VALUES OF ALTERNATIVE ROUTES

Route	Optimizing Objective	Single Objective Values			
		DIST (km)	FOR (km^2)	WET (km^2)	DEV (km^2)
R1	DIST	90.5	606.3	52.0	61.9
R2	DIST + 0.01 (WET + FOR + DEV)	90.5	528.3	44.5	48.1
R3	DIST + 0.01 (FOR) + 0.01 (WET + DEV)	90.5	528.3	44.5	48.1
R4	DIST + 0.5 (FOR) + (WET + DEV)	96.4	475.9	45.63	53.9
R5	DIST + FOR + 100 (WET) + DEV	102.3	448.4	12.2	68.38
R6	(.001) DIST + FOR + 0.5 (WET) + .001 (DEV)	122.3	401.1	12.2	380.5
R7	(.001) DIST + FOR + (.001) (WET + DEV)	144.6	387.4	81.7	404.81
R8	WET + .002 (FOR + .001 (DIST + DEV))	130.5	463.1	11.4	379.4
R9	DEV + 0.001 (DIST + WET + FOR)	104.6	603.25	52.2	45.4
R10	WET + .002 (DEV + .001 (DIST + FOR))	110.5	619.3	11.4	54.9

The objective function used fo obtain R2, listed in the table, is a linear combination of all four objectives (the distance objective plus the other three objectives, weighted by very small values). This illustrates the weighting method of multiobjective programming (Zadeh, 1963, discussed by Cohon, 1978, chapter 6.1). Where alternatives have been generated by the weighting method, one can make an additional observation. Not only does R2 dominate R1, no other route in the network dominates R2; R2 is said to be noninferior. In other words, no other route can do better than R2 in terms of any one of the objectives without doing worse in terms of at least one other. The search for good routes for purposes of illustrating the range of choice to decision makers should be confined to alternatives in the noninferior set. The weighting method of multiobjective programming, it should be stressed, <u>is not</u> the attachment of preferences in a linear combination, as discussed earlier. It is a computational device used iteratively to produce the range of choice in alternatives.

In this problem, as in many multiobjective programming problems, the number of noninferior solutions can be very large. It can be very time consuming and may be impossible to find and describe every one of the noninferior solutions. Fortunately, a representative sample of noninferior solutions can give a decision maker much useful information. This is perhaps best illustrated by the tradeoffs between two conflicting objectives illustrated in Figure 2. The points labelled R3, R4, R5, R6, and R7 in this two-dimensional graph of distance versus forested area correspond to routes with the same labels between nodes 21 and 106, shown in Figure 3. Two scales are given for each axis of the graph; one is in absolute units and the other is in percent above the minimum value. Thus, Route 3 or point R3 on the graph is shown to be a minimum distance route between the two end points, with length 90.5 km. Similarly, Route R7 is a minimum forested area route, with an expected 387 km2 forested area in the ten kilometer corridor. Further, R7 has a length of 144.6 km, 60 percent longer than the minimum distance route and R3 has an expected 528 km2 forested area in its corridor, 36 percent more than R7.

Routes R7 and R3 represent the extremes in the tradeoff between forested area and distance. It is also desirable to find points inter mediate to these extremes, such as R4, R5, and R6 in the figure. These points and the dashed lines connecting them are an approximation of the noninferior set of routes between nodes 21 and 106 (when only the two object-

FIGURE 2

DISTANCE VERSUS FORESTED AREA, CELL 21 to CELL 106

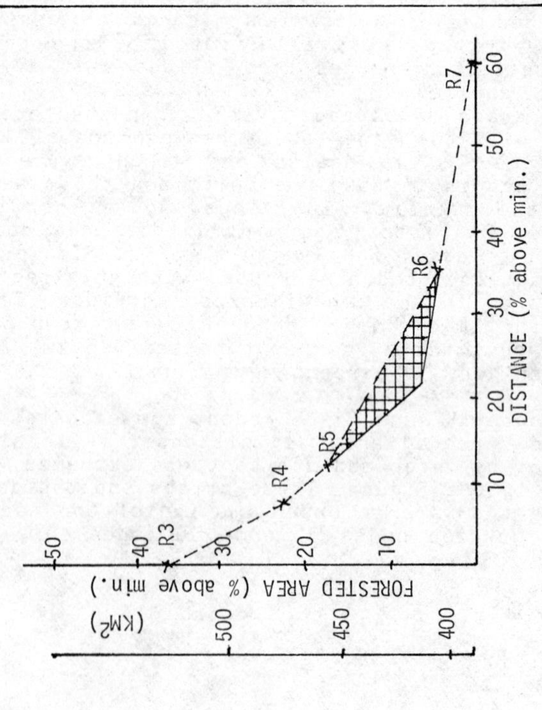

FIGURE 3

DISTANCE VERSUS FORESTED AREA ROUTES, CELL 21 to CELL 106

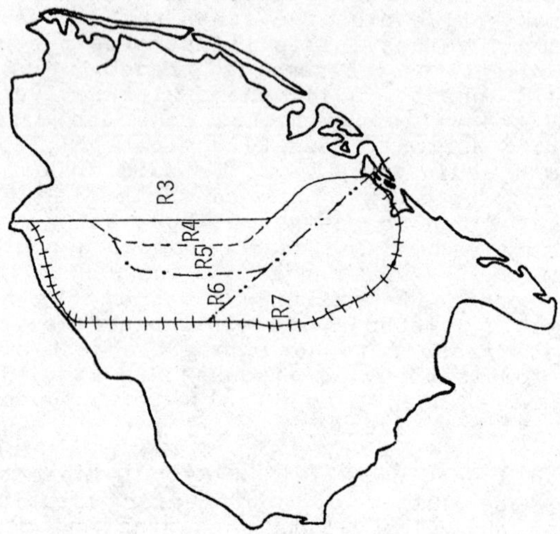

ives are considered). If routes exist with objective values intermediate between, for example, Routes R5 and R6, further solutions can determine these routes and their objective values. Even without further solution, it can be concluded that no route intermediate to R5 and R6 exists with objective values below or to the left of the cross-hatched area shown in Figure 2. Thus, even just a few routes can give a remarkably accurate picture of the outer bounds of the noninferior set.

Unfortunately, the picture is more complex when more than two objectives are considered. One method of illustrating the tradeoffs between objectives in these cases is shown in Figure 4. In this figure, called a value path, instead of a horizontal axis there are columns, one for each objective. The scale of the vertical axis is percent of minimum value. Thus, the minimum distance route (again between nodes 21 and 106) can be identified on this graph by looking at the "distance" column and finding the point with 100 percent on the vertical scale. This point is connected by a solid line to points marked by x in each of the other columns, illustrating that this minimum distance corridor route has 3.9 times the wetlands area as the minimum wetlands route, 1.4 times the minimum forested area, and 1.1 times the minimum developed area. Similarly, the performance of the minimum wetlands route can be seen by reading the values for the line labeled R8, forested lands labeled R7, and developed lands labeled R9. The corresponding routes are shown in Figure 5.

Routes R3, R8, R7, an R9 in this four-objective case are comparable to the extreme points of R3 and R7 of the two objective case. Just as in the two objective case, intermediates between these two extremes can be found if they exist. One such is shown in Figure 4 labeled R10.

This route achieves the minimum wetlands area objective value, and approaches the minima for the distance and developed land objective. It does so with some sacrifice in the forested lands objective. Further exploration of the noninferior set might, for example, look for routes that reduce forested lands by allowing an increase in wetlands or developed lands, in comparison with R10. Or, suppose the relative impacts of R10 are acceptable, but this corridor includes a cell with a special environmental resource that might be adversely impacted by pipeline construction. The model could be solved again wih the same relative values for the objectives, but with the special cell excluded from consider-

FIGURE 4

VALUE PATH DISPLAY OF ROUTES, CELL 21 to CELL 106

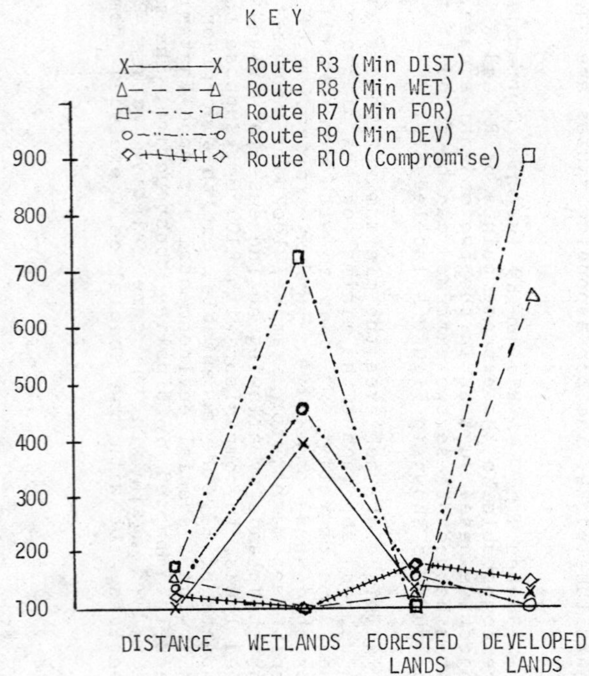

FIGURE 5

ROUTES IN VALUE PATH DISPLAY (FIGURE 10)

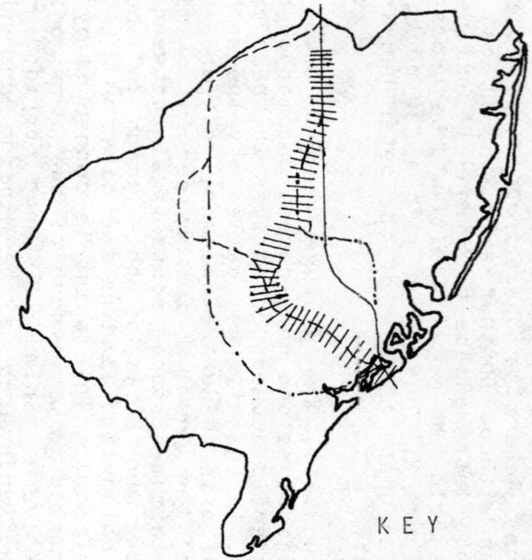

KEY

(See Figure 4)

ation. This new solution would define the sacrifice that would be made in other objectives to avoid the potential impact to this particular area.

In the above examples, it was assumed that landfall and junction nodes were known, and not part of the routing decision. A simple modification to the model allows land fall and/or junction locations to be selected as well as the connecting route. Suppose, for example, any of the nodes 1, 2, 3 or 7 are eligible as landfall sites. The continuity constraint on the origin node

$$\sum_{j \in N_o} x_{oj} = 1$$

can be modified by introducing a dummy node L (for landfall) and variables for dummy arcs

$$x_{Lj}\ , \quad j = 1, 2, 3 \text{ or } 7$$

The new constraint is

$$\sum_{j \in N_L} x_{Lj} = 1 \qquad \text{where } N_L = \{1,2,3,7\}$$

The choice of termination node can be handled in a similar manner. Another simple modification to the model can require any node to be included in the route. Suppose cell 48 is an attractive site for gas processing facilities. The regular continuity constraint on node 48

$$\sum_{k \in N_{48}} x_{48,k} - \sum_{i \in N_{48}} x_{i,48} = 0$$

can be replaced with the following two constraints

$$\sum_{i \in N_{48}} x_{i,48} = 1 \quad \text{and} \quad \sum_{k \in N_{48}} x_{48,k} = 1$$

With these different uses of the model available, it can be a flexible analytical tool. To fulfill this potential, many extensions and refinements to the simple corridor model illustrated above are necessary. A number of other impact considerations can be reflected in such a model's objectives and the geographic scope can be extended to the offshore area, where the location-impact relationship is also important. A pipeline system model should also deal with other components of the system, especially an onshore processing facility, and perhaps an offshore gathering network. We are addressing these topics in ongoing research.

The simple pipeline model described here does, however, illustrate the usefulness of a multiobjective generating approach. The generating approach has some practical advantages when applied to the utility corridor planning problem. It avoids either restrictive assumptions about social preferences or the difficult task of attempting to describe a multiattribute utility function. In a complex institutional setting, where several public agencies exercise some authority and where the public's attention can also come into play, the generating technique focuses debate on tradeoffs between real, feasible alternatives, rather than abstract, hypothetical marginal rates of substitution. Often, an alternative or a small set of alternatives can be readily identified that admit a wide range of different values and thus facilitate effective compromise.

REFERENCES

Baldwin, P. L. and M. F. Baldwin (1975), Onshore Planning for Offshore Oil, Lessons from Scotland, The Conservaton Foundation, Washington, D.C.

Church, R. L. and T.J. Clifford (1979), "Discussion of 'Environmental Optimization of Power Lines', by Economides and Sharifi," Journal of the Environmental Engineering Division, ASCE, vol. 105, no. EE2, pp. 438-439

Cohon, J. L. (1978), Multiobjective Programming and Planning, Academic Press, New York.

Council on Environmental Quality (1974), OCS Oil and Gas, An Environmental Assessment, Washington, D.C.

de Neufville, R. and F. Y. H. Chin (1974), "Role of Systems Analysis in Engineering Design" in Systems Planning and Design, R. de Neufville and D. Marks, ed., Prentice-Hall, Englewood Cliffs, New Jersey.

Economides, S. and M. Sharifi (1978), "Environmental Optimization of Power Lines", Journal of the Environmental Engineering Division, ASCE, vol. 104, no. EE4, pp. 675-684.

Goodchild, M. F. (1977), "An Evaluation of Lattice Solutions to the Problem of Corridor Location," Environment and Planning A, vol. 9, pp. 727-738.

Hobbs, B. F. (1979), "Discussion of 'Environmental Optimization of Power Lines' by Economides and Sharifi," Journal of the Environmental Engineering Division, ASCE, vol. 105, no. EE4, pp. 794-795.

Huber, D. L. (1980), Alternative Methods in Corridor Routing, unpublished M.S. thesis, Department of Civil Engineering, University of Tennessee, Knoxville, Tennessee.

Kash, D. E. and others (1973), Energy Under the Oceans, University of Oklahoma Press, Norman, Oklahoma.

Keeney, R. L. and H. Raiffa (1976), Decisions with Multiple Objectives: Preferences and Value Tradeoffs, John Wiley and Sons, New York.

New England River Basins Commission (1978), Methodologies for OCS - Related Facilities Planning, Boston, Massachusetts.

____ (1976), Onshore Facilities Related to Offshore Oil and Gas Development - Factbook, Boston, Massachusetts.

Newkirk, R. T. (1979), Environmental Planning for Public Utility Corridors, Ann Arbor Science, Ann Arbor, Michigan.

Office of Technology Assessment (1976), Coastal Effects of Offshore Energy Systems, Government Printing Office, Washington, D.C.

Rooney-Char, A. H. and R. P. Ayres (1978), Offshore Pipeline Corridors and Landfalls in Coastal Virginia, Virginia Institute of Marine Science, Gloucester Point, Virginia.

Stastny, F. J. (1975), "Pipeline Corridor Selection Model Concept," Transportation Engineering Journal, ASCE, vol. 101, no. TE2, pp. 337-344.

Turner, K. and R. D. Miles (1971), "The GCARS System : A Computer Assisted Method of Regional Route Location," Highway Research Record, vol. 348, pp. 1-15.

United States Bureau of Land Management (1979), Intergovernmental Planning Program for OCS Oil and Gas Leasing, Transportation and Related Facilities, Government Printing Office, Washington, D.C.

United States Geological Survey (1972), ERTS Land Use - Experimental Edition, 1:250,000, Reston, Virginia.

Wester, T. E. and D. W. Weeter (1977), "Environmental Assessment Utility Route Location," Transportation Engineering Journal, ASCE, vol. 103, no. TE3, pp. 399-407.

Roy F. Weston, Inc. (1978), Methodology for Assessing Onshore Impacts for Outer Continental Oil and Gas Development, West Chester, Pennsylvania.

The work reported here has been supported by the New Jersey Department of Energy, Office of Planning and Policy Analysis.

ENVIRONMENTAL CASE HISTORY OF AN OCEAN OUTFALL

By Noel J. Williams[1]
Richard B. Meighan[2], M.ASCE
Donald D. Treadwell[3], M.ASCE
George J. Murphy[4], F.ASCE

ABSTRACT

The City and County of San Francisco is implementing an extensive Clean Water Management Program to upgrade storage and treatment facilities to significantly reduce raw discharges into the Pacific Ocean and San Francisco Bay from a combined sewer system. When this Program becomes operational during the 1980's, all wastewater will be transported (after treatment) to the southwest section of the City for ultimate disposal through an ocean outfall system. The outfall must accomplish its objective under all oceanic conditions in a manner such that all State and Federal discharge requirements will be achieved.

An understanding of the potential impact of the treated effluent discharges on the ocean environment was of prime importance during the design of the outfall. Oceanographic, water quality, and biological monitoring programs were undertaken to evaluate the impact of the wastefield under varying climatic conditions. Special design features were incorporated in the outfall system to maximize both the utilization and protection of the receiving waters. Regulatory constraints included those of the Environmental Protection Agency, the California State Water Resources Control Board, the California Coastal Commission, and more than 40 other agencies.

---

[1]Senior Environmental Scientist, CH2M-Hill Engineers, San Francisco, CA.

[2]Senior Water Resources Engineer, Parsons Brinckerhoff Quade & Douglas, Inc., San Francisco, CA.

[3]Senior Project Engineer, Woodward-Clyde Consultants, San Francisco, CA.

[4]Project Manager and Assistant Vice President, Parsons Brinckerhoff, Quade & Douglas, Inc., San Francisco, CA.

INTRODUCTION

The precedent established by San Francisco with its earliest sewers in the 1880's has been followed to the present time; San Francisco has a combined (sanitary plus storm) sewage collection system. Three sewage treatment plants presently provide primary treatment for dry weather flows and a small portion of the wet weather flow. When rainfall intensity exceeds approximately 2/100th of an inch per hour, excess combined sewage flows are discharged untreated into San Francisco Bay and the Pacific Ocean at approximately 40 locations. The current level of wastewater treatment is inadequate and does not comply with State and Federal guidelines.

San Francisco is implementing an extensive wastewater management program to upgrade storage and treatment facilities to significantly reduce the number of untreated overflows. This program involves construction of a secondary treatment plant for dry weather flow, one primary treatment plant for wet weather flow, several large storage and transport facilities and an ocean outfall. Storage tunnels and transport facilities will have capacity to store excess combined sewage during storm events. The primary plant will treat combined sewage in excess of that receiving secondary treatment. The combined sewage in storage will be processed as storm flows subside.

All wastewater will ultimately be transported to the southwest area of the City for discharge through the ocean outfall system. Major elements of the wastewater management program are illustrated in Figure 1.

The ocean outfall system is subject to the California Ocean Plan (Amended) which contains discharge regulations set by the State Water Resources Control Board (SWRCB). Major provisions of the Ocean Plan include: 75 percent removal of suspended solids; effluent limits for grease and oil, settleable solids, turbidity, pH and toxicity; and receiving water limits for heavy metals, pesticides, organics, radioactivity and toxicity. The SWRCB regulations recognize assimilative capacity but do not contain effluent criteria for BOD and coliform levels nor specific requirements for initial dilution.

The ocean outfall system is also subject to EPA regulations. The Federal Water Pollution Control Act Amendments of 1977 contain provisions for applications for waivers of secondary treatment requirements for discharges into marine waters. San Francisco has filed an application for such a waiver. Were this waiver granted, a small portion of the sanitary sewage would receive primary treatment and be blended with secondary treated effluent prior to ocean discharge.

The ocean outfall system currently being designed will have an ultimate capacity of 590 mgd. The $2.5 billion Clean Water Program is proposed to be constructed in stages over a ten year period due to the immense size of the project and funding limitations.

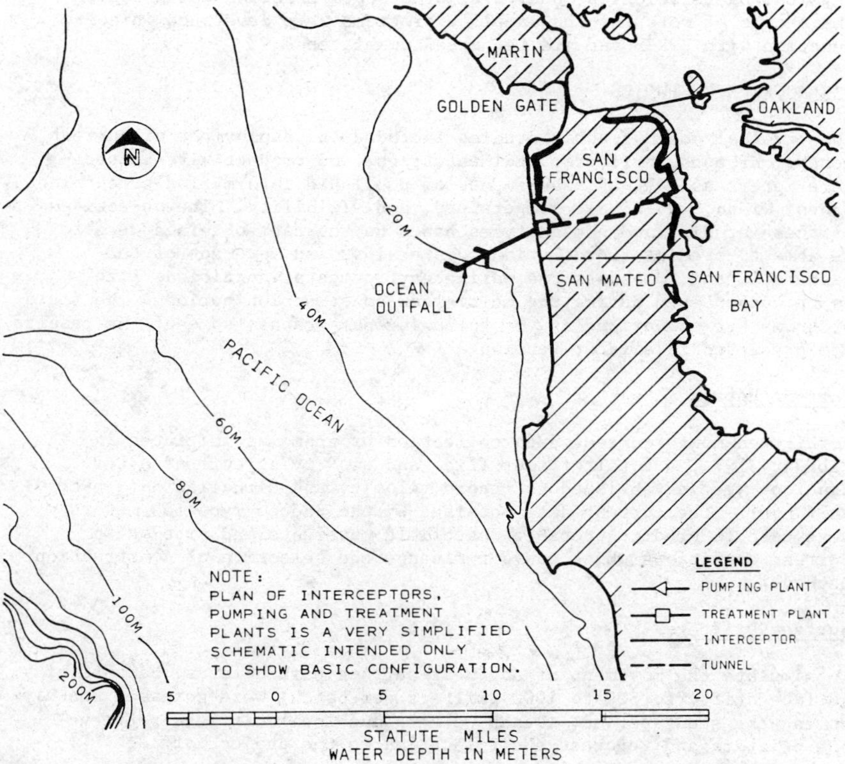

FIGURE 1

SAN FRANCISCO WASTEWATER PROGRAM MASTER PLAN

Extensive predesign studies were undertaken to document the geotechnical, geophysical, oceanographic and water quality characteristics in the study area prior to design of the ocean outfall. Further, an extensive predischarge monitoring program is underway to provide baseline environmental data against which any impacts of wastewater discharge can be measured. This program includes: an infaunal assessment, fish and macroinvertebrate assessment and chemical/physical/bacteriological analyses. The specifications which outline the scopes of work for the sampling programs were developed in conjunction with SWRCB and its technical committees.

OCEANOGRAPHIC STUDIES

The physical oceanographic studies included the deployment of current meters, drogues, drifters, sediment traps, and conductivity-temperature meters as well as assessments of published information concerning tides, winds, waves, air temperature, and visibility. The objectives of these studies and the analyses based on the data obtained were to be able to predict the dilution, dispersion, and movement of the wastefield and to assess the marine environmental conditions likely to be encountered during the multi-year construction period. The field programs are summarized in the following paragraphs; some of the results are presented in a later section.

Current Meters

A thirteen-month current data collection program was initiated in February 1977; a total of four fixed and two roving current meter stations were established to record velocity and direction data every half hour. The current data obtained by the Endeco Type 105 meters were used to predict probable wastefield movements and site ocean current conditions which could influence the selection of construction methods.

Surface Drifters

To simulate the movement of any near-surface particulates, batches of surface drifters (500 to 1000 drifters per batch) were released monthly during the study period. Beach walkers and local citizens retrieved the drifters and reported their location to the project office.

Subsurface Drogues

Window-shade drogues were deployed monthly to check the recorded current meter velocities; they were also deployed in hexagonal arrays on three separate occasions and tracked for extended (12 to 24 hour) periods. The results of the drogue releases were used to determine ocean current speeds and trajectories and to analyze eddy diffusivity.

Density Profiling

In-situ temperature and conductivity measurements were taken monthly at selected locations and depths using an Inter-Ocean Model 513 CTD device. The density stratification information obtained was used to predict wastefield dilution and the rise of the effluent plumes.

Sediment Traps

Site-specific data on the distribution of sediment within the water column was collected by sediment traps installed at the wet weather and dry weather diffuser locations. The data were used to assess the potential for sediment intrusion into the conduits through the diffusers.

Wave Climate

The assessment of significant and extreme wave conditions in the project area consisted of two steps: determination of wave conditions offshore, in deep water; and transfer of these data to the relatively shallow nearshore area. The significant wave height varies with return period and is predicted to range from 10 meters to more than 14 meters for the 10 and 200 year storms, respectively.

WATER QUALITY STUDIES

The purpose of the water quality monitoring program was to establish baseline water quality conditions in the area of the outfall, both as a reference against which any future changes could be measured and to establish water quality conditions for evaluation of permit requirements. Monthly water quality measurements were obtained during the one year predischarge monitoring program. Seasonal water quality measurements were obtained during the two year predischarge monitoring program. Samples were taken at selected locations and depths in the offshore project area. The water quality parameters analyzed are temperature, dissolved oxygen, conductivity, pH, total coliforms, fecal coliforms, total suspended solids, settleable solids, turbidity, ammonia, nitrate, nitrite, total phosphorous, cholorophyll, phaephyton and sulfide. The predischarge program included an analysis of trace metals and organics in the water and sediments. Because of the importance of die-off in the prediction of shore bacteria concentration and the great variability of die-off rates reported in the literature, a bacterial die-off study was conducted in the area of proposed discharge. The study consisted of two parts: 1) measurement of the die-off of bacteria discharged into the ocean by a neighboring outfall and 2) measurement of the die-off of bacteria in the same effluent mixed with ocean water in clear and opaque plastic bags.

Estimates of wastefield transport due to currents and winds were made. A transport model was developed incorporating wind and ocean current data to estimate the resultant surface current for a surfacing wastefield. The same model was used to predict transport of a submerged wastefield by utilizing measured currents. The wastefield transport analysis has been combined with predictions of wastefield dilution, submergence, and bacterial die-off to obtain estimates of shoreline bacterial concentrations.

BIOLOGICAL STUDIES

The purpose of the biological studies was to establish baseline environmental conditions in the area of the proposed outfall. The effects, if any, of wastewater discharge can then be determined by comparison with baseline conditions. The program consists of the sampling and analysis of benthic infauna, fish and macroinvertebrates, and population and life history analysis.

Infaunal Assessment

Because of the relative immobility of benthic infaunal organisms, (e.g. worms, mussels) they are particularly sensitive to changes in their environment. Benthic habitats are, therefore, good indicators of environmental perturbations and have been used as one means of estimating the existing level of biological variability at the proposed outfall region prior to discharge by providing baseline information on infaunal species composition, distribution and seasonal variability.

Twenty benthic infaunal stations were sampled along bathymetric contours in the vicinity of the proposed outfall location in accordance with wastefield transport predictions. All 20 benthic infaunal assessment stations were sampled during the upwelling period. During the subsequent Oceanic and Davidson current seasons, only eight of the original stations were sampled because statistical analysis indicated that the eight stations chosen would adequately represent all major trends shown by analysis of the original stations.

Fish and Macroinvertebrate

The primary objective of this field program is to obtain information on the composition of fish and macroinvertebrate (e.g. crabs) communities inhabiting the project area. Surveys were conducted over a year's time to assess seasonal changes in the benthic fauna. Sampling dates were chosen to coincide with major changes in oceanographic conditions. Twelve stations were sampled by trawling. Station locations were selected on the basis of depth contours, sediment type and distance from the proposed outfall.

Population and Life History Analyses

Life history analyses are being conducted on four species of invertebrates from the study area. Based on the criteria presented in the specifications for the project, two polychaetes and two crustaceans were chosen for analysis. Short term laboratory experiments, and laboratory techniques such as historical sectioning are being used along with field sampling for a more complete understanding of the life histories of these species.

Monthly samples are collected at three stations. Eight replicates are taken at each station to ensure the capture of enough individuals of each species. A minimum of three of the eight replicates are analyzed in order to make meaningful statistical comparison and reduce the effects of small scale patchiness. At least 100 individuals of each species were measured so that all sizes and stages of development were represented.

SUMMARY OF RESULTS

Oceanographic Studies

- measured ocean current speed distributions show that the most frequent speeds vary in the range of 10 to 30 cm/sec. During the entire measurement period, only one reading was exceeded 55 cm/sec. (one knot = 51.5 cm/sec).
- ocean currents in the outfall area are closely related to tidal fluctuations at the nearby Golden Gate entrance to the north. The current directions are predominantly towards and away from the Golden Gate entrance along the coast.
- the magnitude and time of tides at the project site are generally predictable from the published tide tables and correction factors. The extreme tides at the site are approximately +8.5 feet and -2.5 feet (based on MLLW datum).
- wind, wave, and visibility conditions offshore San Francisco are highly variable; it is quite possible for any given year to be 25% more severe than the established long-term average conditions.
- nearshore wave-induced currents may be as high as 10 feet/sec near the seafloor.

Water Quality Studies

- physical and chemical water quality data conform well to the stratification patterns expected for each of the three sampling periods. Simultaneous changes in the oxygen, temperature, salinity, and density values occurred at a depth of 15 m. during the upwelling period. During the oceanic season, the physical profiles showed uniform stratification while during the rainy season (Davidson period) there was a low salinity water mass overlying the sample stations to a depth of about 10 meters. Salinity and density are strongly influenced by freshwater outflow from the Delta.

- nutrient data was quite variable and was influenced by the input of nutrient-rich low density waters from the Bay.

- many of the synthic organic pollutants, including the polychlorinated biphenyls and total petroleum, showed strong concentration gradients from the Golden Gate decreasing southwards towards the Outfall site. San Francisco Bay is therefore a significant input source of these pollutants into the local coastal environment. Operation of the Outfall can be expected to result in higher levels of these compounds in the vicinity of the Outfall with, however, decreased emissions from the Bay.

- data on the levels of individual pollutants in the dissolved and particulate phases of seawater, in mussels, sediments and fish, taken at the same site within a given time interval, have provided information on the partitioning of these pollutants among the components of the coastal environment. This in turn will permit predictions of their behavior in the vicinity of the Outfall site as input increases once operation has begun.

- nearshore bathymetry and the tidal and wave-generated currents of the study area determine the existing conditions and seasonal changes in the sediment characteristics.

- the water column at this outfall site is comprised of two components: 1) the surface bay waters that fluctuate with the tidal cycle and 2) the bottom waters that are more typical of the oceanic environment. Bay waters are characterized by reduced salinities, greater particulate loads and elevated concentrations of some metals. The bottom waters are of high salinity and trace element levels are the same as those observed in pristine oceanic environments.

- metal concentrations in benthic infauna show significant differences in body burdens between mussels from Point Lobos and those collected at Mussel Rock and Point San Pedro. These two southern stations should serve as sensitive indicators to changing metal concentrations which may occur as a result of the development of the outfall structure.

- metal concentrations in the tissues of the English sole showed a high degree of variability but no significant differences between stations. The degree of variability indicates that large numbers of specimens will need to be analyzed in the future if there is to be any hope of detecting changes in the trace metal levels of the population.

- trace metals concentrations in the dissolved and particulate phase of the water column were also defined. Some significant differences were noted in water emanating from the Bay. In evaluating future conditions, it will be important to differentiate between the relative influences of the Bay and outfall structure.

- o the metal levels in the sediments at the outfall site are relatively "clean". There is little variability between sites or between seasons so that any increases in metal concentrations following the onset of the outfall operation will be easily detected.

- o coliform levels cannot be compared directly to SWRCB standards but generally appear quite low. Both total and fecal coliform levels were higher at shore stations but still well within the standards.

- o the sediments in the study area are uniformly sorted fine sands that show decreasing mean particle size with increasing depth sampled. The mean particle size and silt-clay percentages for the sediment stations distinguish three general sediment regions; the northern inshore stations, northern offshore stations and the southern stations.

- o total volatile solids content of the sediments corresponds to the mean sediment particle sizes over the three oceanographic seasons, and there was a progressive decrease in the sediment total volatile solids content in the study area from upwelling to the Davidson seasons.

- o linear regressions of the percent total volatile solids, percent total organic carbon and total Kjeldahl Nitrogen versus mean particle size clearly identifies an inverse relationship between particle size and the organic matter in the sediments. Grease and oil values do not show any correlation to depth, total organic carbon, total volatile solids, or season.

- o total volatile solids, total organic carbon, total Kjeldahl Nitrogen and grease and oil values are all comparable to those found in undisturbed marine sediments.

- o the bacterial die-off study demonstrated that light intensity was the major determinant in die-off rates. Less important were sedimentation and "other" factors. Estimates of all three of these contributions to the total rate were made.

- o currents are dominated by the tidal influence. The first principal component of the currents lies along a direction pointing approximately to the Golden Gate entrance. The vector mean of the currents is toward the northwest. The mean speed is about one mile/day. A submerged dry weather field will virtually never reach shore. A surfacing dry weather field has a very low probability of reaching shore. The wet weather field can reach shore.

- o all coliform discharge standards were predicted to be met except one. On an annual basis, the chance of shoreline concentrations exceeding 10,000 per 100 ml two days in a row is 0.3% for wet weather flow (a recurrence interval of about three years). For dry weather flow, the chance is 0.0%.

Biological Studies

- the Gulf of the Farallones, the biogeographic region of the proposed outfall, is very productive in crab and demersal fisheries. Benthic infauna organisms of the gulf are correspondingly abundant and diverse. Further, these benthic organisms respond to changes in feeding opportunities and the surrounding sediment characteristics by adjustments in the species composition, abundances, biomass and distribution. Changes within the natural variation of benthic infauna assemblages for this region occur both seasonally and yearly in response to environmental conditions and according to the life histories and population dynamics of the benthic infauna species.

- during these surveys, benthic samples from the study area have yielded 310 species from 14 phyla and a total of 128,000 organisms. The data shows the Upwelling and Oceanic Oceanographic season with roughly equivalent total numbers of species and mean total numbers of individuals collected. The individuals in the Davidson Current season was one-half that of the preceding sampling periods and the total number of species declined slightly. Biomass data for the same three seasons did not show any correlation to seasonal changes in benthic infauna numerical abundances.

- life history characteristics of the benthic organisms contribute to the explanation of the seasonal shifts in abundances. The majority of benthic invertebrates in the temperate region reproduce during the spring and early summer, and hence there is a large introduction of newly settled larvae by early summer. By early fall, these new recruits in the benthic community have grown and are of sufficient size to be sampled adequately with the 0.5 mm sieves. Winter sea and swell waves disrupt and shift the sediment characteristics of such a shallow inshore area and benthic infauna assemblages are numerically reduced.

- forty-two species of fish representing 22 families were collected. A total of 17,821 fish were collected in day trawls during this study. The six most commonly collected benthic species comprised 85 percent of the total catch. Another 29 species of benthic fish comprised 4 percent of the total catch. The remaining 11 percent was comprised of species of fish considered pelagic species, longfin smelt, northern anchovy, and Pacific herring.

- the five most commonly collected benthic fish species when ranked in order of numerical abundance were: speckled sanddab, English sole, Pacific tomcod, white croaker, and sand sole.

- three types of abnormalities were observed on five species of fish. Skin tumors were found on one white croaker and 24 English sole. Ambicoloration was observed on three English sole, two curlfin turbot, and one California halibut. One deformed butter sole was collected.

- o approximately 21 species of benthic invertebrates representing five phyla were collected. The four most commonly collected species comprised 97.9 percent of the total catch by number. These species in order of numerical abundance were: Crangon nigromaculata, Cancer gracilis, Pisaster bievispinus, and Cancer magister.

- o Although relatively few specimens of Dungeness crab were collected on this study, the presence of large concentrations of commercial crab traps observed on the March 1979 survey would indicate a considerable stock of this species in the study area.

- o population life history analysis of selected benthic invertebrate species allows study of more sensitive and potentially more important indicators of wastewater discharge than changes in species abundance and composition. Four species (two polychaetes and two crustaceans) are currently being studied in the program. Parameters being studied for these species include growth rates, mortality rates, age structure of the population, fecundity, and reproductive rates.

CLOSURE

The predesign and predischarge oceanographic studies described herein are important chapters in the environmental history of the San Francisco Southwest Ocean Outfall. The information gathered has been utilized in the planning and design of the outfall system and also provides a baseline against which the effects of outfall construction and operation may be measured.

Additional studies and field data collection programs, primarily water quality and biological monitoring, are planned for the periods of construction and operation. It is expected that the results of these continuing studies will be significant interest to scientists and engineers involved with the marine disposal of man-made wastes.

ACKNOWLEDGEMENTS

The prime consultant for the Southwest Ocean Outfall Project is Parsons Brinckerhoff; principal subconsultants are CH2M-Hill, Inc. and Woodward-Clyde Consultants. The efforts of the staffs of these firms are gratefully acknowledged.

The agencies having the primary responsibility for the program are the Clean Water Program of the City and County of San Francisco and the Water Resources Control Board of the State of California. Particular credit is due to Messrs. L.A. Vagadori, A.B. Kenck, W. Tsai, and J.A. Barnickol.

SELECTED REFERENCES

Belvedere, J.A., etal, "Southwest Ocean Outfall: Coastal and Offshore Considerations," Proceedings, Coastal Zone 78, Vol. III, pp. 1845-1861, ASCE, San Francisco, March 1978.

Chin, H., etal, "Comparison of Nearshore Eulerian and LaGrangian Ocean Current Measurements," American Geophysical Union, Annual Spring Meeting, Paper O 141, 29 pp, Washington, D.C., May 1979.

CH2M-Hill, Inc., "Predischarge Oceanographic Study - Final Report - Phase I," report to Parsons Brinckerhoff, February 1980.

Parsons Brinckerhoff, CH2M-Hill, Inc., and Woodward-Clyde Consultants, "Predesign Oceanographic Study," report to the City and County of San Francisco, November 1978.

Treadwell, D.D., etal, "Southwest Ocean Outfall: Geotechnical and Oceanographic Predesign Studies," Proceedings, Coastal Zone 78, Vol. III, pp. 1862-1877, ASCE, San Francisco, March 1978.

MANAGING SHORE DEVELOPMENT IN BRITISH COLUMBIA

W. Bruce Morgan, Masters Candidate
Master of Natural Resources Management Program
Simon Fraser University, Burnaby, B.C., Canada

Jonathan P. Secter, Manager
Evaluation Section, Assessment and Planning Division
Ministry of Environment
Victoria, B.C., Canada

INTRODUCTION

British Columbia possesses approximately 17,000 linear miles of marine coastal shoreline, 7,000 square miles of major lake bed, and 4,600 river bank miles in 5 major river systems. Stream bank miles in other river and stream systems are vast. Likewise, estuarine, intertidal and subtidal acreages are extensive.

However, in spite of this evident abundance of shores in British Columbia, demands and pressures are focused on a relatively small percentage of shore areas. The characteristics of these including configuration, slope, water quality, aesthetic quality and accessibility, are such that they are perceived to possess suitability and economic advantage for a wide array of uses. These same areas are often among the most environmentally fragile, and also may by definition, pose natural hazards. This pattern results in conflicts among competing uses, between use demands and the need to maintain natural shore processes and features, and between the effects of natural hazards and human activity.

In recent decades, shore areas have been subject to rapidly increasing demands for urbanization, industrialization, transportation and recreation. Pollution on some shores has meant widespread shellfish contamination resulting in closure of certain areas to harvesting, the curtailment of a food source and a recreational pursuit and concomitant adverse economic impacts. Poorly planned shore development threatens valuable marshland and wetland wildlife habitat. Excessive demands on estuarine ecosystems have led to adverse effects on the fishery resource with widespread social and economic implications. At the same time, inappropriate development siting has proceeded without accounting either for the risks from natural shore hazards, or for the consequences to society of the project's effects.

Consideration of British Columbia's shores as a natural resource has been overlooked for many years. While various shore resource components have been developed, little overt effort has been made to knit these activities together in a coordinated, planned way. Integrated management of resources in shore areas and of shores as a resource *per se* is essential.

The fragmented nature of responsibilities for the various resources and activities that affect shore areas is such that no one agency, ministry or level of government can act independently to resolve shore problems. Yet the basis for many conflicts stems from the unilateral exercise of specific mandates, either by a single government, a single ministry or by a single agency within any ministry. The very nature of shores and the diversity of activities which affect them necessitate close cooperation within and between all ministries and all levels of government.

With the exceptions of the beds of six federal commission harbours, the beds of the rivers included in the Vancouver Island Esquimalt and Nanaimo land grant, and portions of the Cowichan Estuary alienated in the same grant, all shorelands in British Columbia, including the beds of Georgia Strait and of all coastal embayments are administered in the right of the Province. Similarly the regulation of freshwater and of pollution are provincial responsibilities as are the ownership and regulation of natural resources in general. The Federal Crown regulates all fresh water and marine transportation and navigation, the anadromous fishery, migratory birds and has control over the ocean bed below the mean low tide mark on the open coast.

Juxtaposed upon this scenario is the right of municipalities and regional districts to regulate, plan and zone land use on private lands in the coastal zone and elsewhere, thus affecting use and non-use of shorelands.

Land, water, resource and social management strategies therefore can neither be developed separately nor applied separately. All must be structured to account for the interrelated opportunities and constraints of shore systems. All involved agencies must recognize the potential effects of their respective activities on each other and on shore systems in general. This recognition and an active response to it constitutes the core of proposed provincial shore policy (Ministry of Environment, 1979).

DEVELOPMENT SITING OBJECTIVES

The fundamental objective of a management strategy for locating development in shore areas is the establishment of site-specific selection criteria which integrate the siting of industrial, urban and port development in shore areas. Four major issues are of high priority. These are listed in the sequence in which they should be approached. All are based on recognition of shore area competition, conflicts and problems.

It is necessary to understand the nature and capacity of our shore areas. A sound and broad knowledge of our shorelands is essential, being the basis upon which to make decisions about the future uses of this resource. Obtaining information on the existing natural processes, their importance and sensitivities, as well as the present and potential uses of shore areas, is the main objective of this issue. A broad range of uses requires shore sites. However, numerous uses now located along shores have no real need for direct water access. It is necessary to interpret the demands for shore sites through an understanding of the nature of the uses, their requirements and impacts. This information will supply the basis for development and implementation of "site-specific selection criteria". Sensitivity analyses, site suitability analyses, and capacity analyses are recommended as potential methods for accomplishing this objective.

Development criteria for port, industrial and urban development in shore areas are of major importance in considering development pressures in shore areas. Siting criteria should be developed by taking into account the biophysical, recreational and aesthetic values of the shore areas, and then using this information to evaluate the advantages and disadvantages of specific potential developments. Such an exercise should avoid resource use conflicts.

The critical phase in this process is the allocation of competing uses among available sites. This presents formidable challenges since priorities must be determined among competing uses for an often limited number of suitable sites, and is compounded by further considerations to minimize impacts on various site characteristics. Site selection criteria are extremely important as they provide the basis on which trade-off decisions are made. Legislation and/or regulations can play an important role here.

An appropriate administrative framework is very important. Resolution of jurisdictional problems is paramount. There is a need for both sound design of shoreland uses and effective policing of all uses to assure environmental and social responsibility. Thus the administrative framework necessary for inventorying shore areas, deciding upon their uses, site selection, development design and policy making is of considerable significance to shore management.

The restoration of abused or misused shore areas is the next phase in the process. A related requirement is the need to relocate uses not needing shoreland to non-shoreland sites. Reevaluation of past decisions and activities must be undertaken to measure the success of shore area management and, when faults are noted, to take actions to correct past mistakes and avoid future ones. The overall advantages and disadvantages of development actions, including the design considerations implemented to ameliorate impacts in shore areas, should be considered when locating developments in shore areas (C.C.R.E.M. 1980).

GENERAL FORMS OF SHORE DEVELOPMENT CONTROLS

AREA DESIGNATIONS

The zoning of shore areas for particular types of use such as conservation, development and recreation has been a regularly applied shore management methodology. The focus of this approach has been to determine the capability of a shore area to accommodate various types of activities and then to develop and implement appropriate mechanisms to manage these areas accordingly. This approach not only achieves the objective of siting activities on the most suitable shores but also serves to cluster compatible development.

PERMITTED/NON-PERMITTED ACTIVITIES

Restrictions on the type of activities permitted in shore areas are often used in conjunction with area designation approaches. The most effective screening criteria have been based on an activities physical and/or economic need for a shore location. In general those activities which are dependent upon a shore location and are compatible with the shore system and adjacent uses receive highest priority in allocation decisions, those which are shore oriented receive second priority, and those which are shore dependent receive lowest priority, if they are considered at all. This form of screening permits the allocation of shores to those activities which require shore locations which encourage optimum utilization of a scarce resource.

ENVIRONMENTAL IMPACT ASSESSMENT

In those areas where it is not possible to designate shore areas for specific uses or to restrict the type of activities permitted, site specific environmental impact assessment is often applied as a management technique. British Columbia has applied this approach through use of Orders 908 and 3339 (1977) of the LIEUTENANT-GOVERNOR IN COUNCIL (British Columbia 1977) which respectively apply to the Fraser River Estuary and the Estuary of the Cowichan/Koksilah Rivers. All development proposals within the boundaries set forth in the Orders are subject to mandatory environmental impact assessment in relation to their applications for required statutory development permits. Applications are screened with respect to the magnitude of their probable environmental impact and receive an appropriate level of assessment. The assessment is a staged procedure (Figure 1) whereby an application may be rejected, subject to preliminary environmental assessment, or subject to detailed site specific assessment.

A RECOMMENDED APPROACH TO MANAGING SHORE DEVELOPMENT

A number of approaches to developing shore management plans have been attempted in British Columbia. Due to their perceived suitability for a wide variety of uses, the Province's estuaries have been the focus for numerous shore management conflicts and associated strategies for resolving them. The area designation approach has been the most widely used planning method (including application in the Fraser Estuary, Cowichan Estuary, Ladysmith Harbour and Sooke Harbour and Basin) but for the most part these applications have failed to effectively incorporate developing siting criteria. This exclusion precludes any ability to control the type of activities which will be allowed within an estuary and thus restricts itself to directing <u>all</u> types of development to the most appropriate estuary location.

In order to overcome the shortcoming noted above, an alternative approach to the development of an estuary management plan has been applied in the Squamish Estuary (Squamish Estuary Management Plan Study, 1980). An area designation system has been combined with a permitted activity matrix and specific activity management guidelines to provide a complete management framework for the estuary.

Fig. 1. Flow Chart for the Implementation of O/C 908

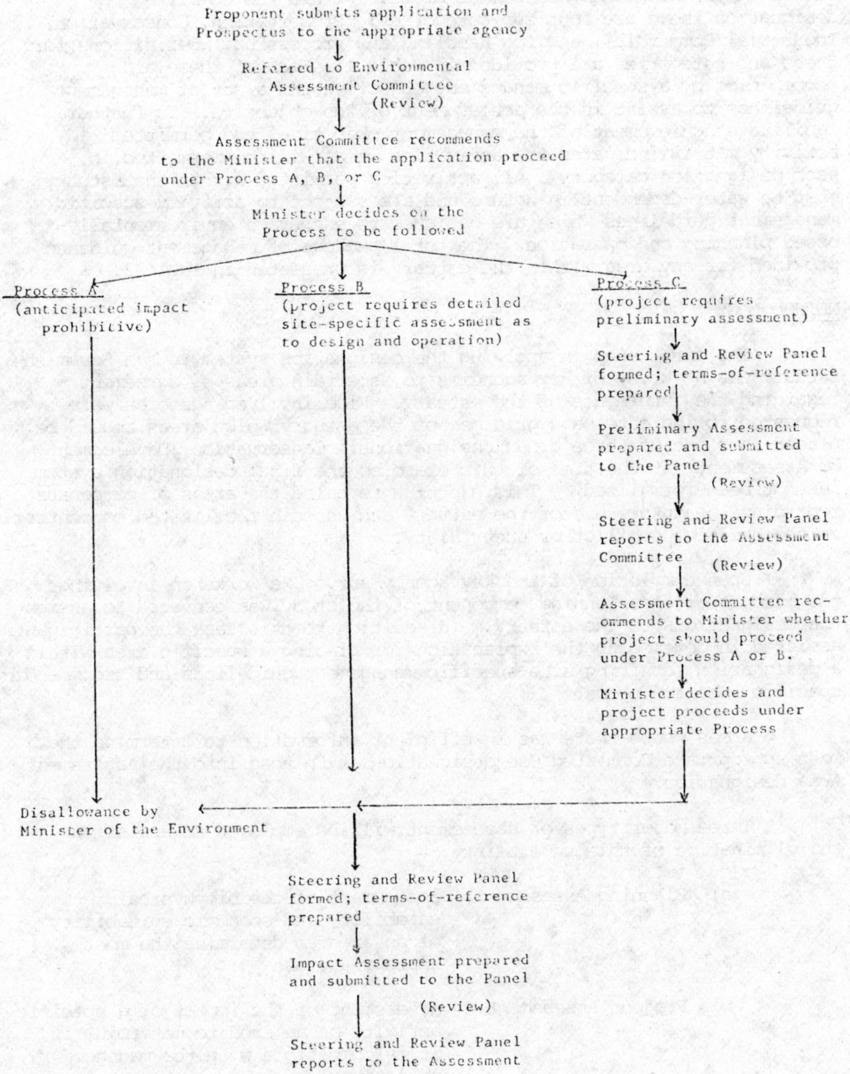

The area designation system consists of two broad categories: Committed Use Areas and Assessment Areas. Within the Committed Use designation there are four sub-categories: Preservation, Conservation, Industrial/Commercial, and Log Handling and Processing. Definitions for these sub-categories are provided in Table 1. Each of these are accompanied by a specific management objective and a set of management guidelines to assist in the achievement of the objective. To further facilitate achievement of the management objective, the permitted activity matrix indicates the activities which will be permitted in each designation category. All activities permitted within the Estuary must be water-dependent in nature and are subject to activity specific management guidelines which are designed to encourage environmentally-based planning and operation. The total package of management guidance provided for any area within the estuary is presented in Table 2.

METHODOLOGY

The first step in applying the designation system to the Squamish Estuary was a survey of key agencies to ascertain areas of agreement regarding the future use of the estuary. Five involved agencies were requested to indicate on a grid map of the estuary which areas should be placed into one of three broad designations: Conservation, Development, or Assessment (at the time of this exercise the final designation system has not been formalized). This exercise revealed the areas of concensus regarding the future use of the estuary and as such facilitated concentration on the areas of conflict or uncertainty.

Upon completion of the key agency survey, a workshop involving the members of the planning component of the study was convened to assign the designations to the estuary. In addition to this task the participants assisted in developing the explanation for placing a specific area within a designation, drafting site-specific management guidelines and recommending specific research needs.

Areas where there was insufficient information to determine the most appropriate Committed Use designation were placed into the Assessment Area designation.

Three major types of assessment will be employed to assist in the elimination of this uncertainty:

a) Planning Assessment: assessment of the biophysical capability and economic suitability of an area to determine the most appropriate use

b) Project Assessment: assessment of the impact of a specific activity on an area to determine if it is compatible with the management objective

Table 1. Definition of Committed Use Sub-Categories

PRESERVATION

The Preservation category pertains to areas characterized by the presence of some unique or sensitive natural feature considered valuable in its undisturbed condition. The primary management objective for these areas will be to ensure their protection from any form of human activity which may alter their natural characteristics.

CONSERVATION

The Conservation category is intended to manage areas of environmental significance in a manner which will ensure the continued operation of the estuarine processes. The existing character of these areas should be maintained, or when appropriate, enhanced to facilitate achievement of this objective.

INDUSTRIAL/COMMERCIAL

This category is intended to identify areas where a variety of water-dependent industrial and commercial activities may locate. The purpose of designating industrial/commercial areas is to promote optimum utilization of shore areas by existing and proposed development. The designation is intended to provide the necessary degree of certainty required to facilitate industrial/commercial development planning.

Source: Squamish Estuary Management Plan Working Documents (Unpublished)

Table 2. Management Guidance Provided in the Plan

Each Committed Use Area Possesses:

1. Overall estuary management guidelines which establish the tenor for decisions throughout the estuary.

2. A specific designation (i.e., Preservation), accompanied by a management objective and guidelines to assist in the achievement of the objective.

3. A list of permitted activities accompanied by specific activity guidelines to encourage environmentally sensitive design and operation.

4. A set of site-specific management guidelines designed to apply the overall and designation to the specific area in question.

5. A list of research requirements to assist in the achievement of the management objective for the area.

Each Assessment Area Possesses:

1. Overall estuary management guidelines which establish the tenor for decisions throughout the estuary.

2. Guidelines which aid in the management of the area until it can be placed in a Committed Use designation.

3. A list of research requirements and/or negotiations required to eliminate the uncertainty regarding the most appropriate Committed Use.

c) Conflict Assessment: assessment of the nature of the resource conflict and recommendation of a solution or an appropriate mechanism for resolution

These studies and negotiations will play an integral role in the ongoing refinement of the Squamish Estuary Management Plan. It is important to note that the Management Plan is a joint Federal/Provincial planning effort which has involved all levels of government. As a result, all agencies with present and/or future decision-making responsibility in the Squamish Estuary are a party to the management policy developed.

CONCLUSION

Shore management in British Columbia has experienced a number of different approaches to planning for the Province's valued shore resource. The majority of these attempts have not taken an expansive enough approach to inclusively manage those activities which desire shore locations. The strategy applied in the Squamish Estuary is intended to achieve this objective. Through the integrative use of area designations, the permitted activity matrix and specific activity guidelines, a holistic approach has been developed for the planning and management of shore activities.

REFERENCES

British Columbia Ministry of Environment 1979. Shore Policy for British Columbia: A White Paper Prepared for the Minister of Environment, Victoria, B.C.

British Columbia 1977. Orders-in-Council 908 and 3339 and Their Respective Guidelines for Implementation, Victoria, B.C.

Canadian Council of Resource and Environment Ministers 1980. Proceedings of the Shore Management Symposium, October 4 and 5, 1978. C.C.R.E.M., Toronto, Ontario.

Squamish Estuary Management Plan Study 1980. Preliminary Plan Outline, Victoria, B.C.

COASTAL DEVELOPMENT AT NILE MEDITERRANEAN OUTLETS

Dr. Ismail Mobarek[1] and Joel M. Goodman[2]

INTRODUCTION

The River Nile is a unique case in modern history with the complete damming of its sediment output by construction and start of operation of the Aswan High Dam in 1964. However, the erosion problem close to the Nile outlets at Damietta and Rosetta was detected at the beginning of the century.

In order to develop the coastal areas of the River Nile Delta, many constraints should be defined and their scope of effects and after effects deligniated. The historical development of the Nile Delta in general, helps to a great extent in understanding the various factors affecting the coastal zone changes, whether by erosion or accretion.

The purpose of this Paper is not to present a scheme for the coastal zone management of the Nile Delta Coast, but to describe the characteristics of the area and problems which must be overcome before such a scheme can be developed.

RIVER NILE, AN OVERVIEW

The River Nile extends from Lake Plateau or the Equatorial Lakes to the Mediterranean Sea. Its main sources of water are the Equatorial Lakes and the Ethiopian Plateau. The flood period starts in July at Aswan, peaks up early in September, and the draw down curve extends to December each year.

The average annual discharge at Aswan reaches 84.5 milliard (109) meter cube per year, with the extremes being 150 and 36 milliard meter cube per year. Before the construction of the Aswan High Dam, the flood water was allowed to pass through the Nile to the Mediterranean, until the peak passes and the recession starts. At this point, the gates of the low Aswan Dam used to be closed collecting the excess water for use during the low discharge periods of the year. During these years the flood water released to the Sea were loaded with sediment.

The Aswan Dam was constructed and started operation in 1964. Since that date the Nile flood water was completely impounded in Lake Nasser, upstream from the Dam,

[1] Pacer Assocs.
[2] PRC Harris Inc.

and Egypt releases water from the Dam according to the needs. Thus the sediment load which previously had been dumped into the Mediterranean stopped.

The sediment constituents of the flood water was determined by the Ministry of Irrigation. Over the period from 1955 to 1963, the flood water sediment was composed of 25% sand, 45% silt and 30% clay. These percentage values vary slightly over the flood months. The Coastal Protection Project estimated an average of 130-140 million tons of sediment would annually reach the Mediterranean through the Rosetta and Damietta River Branches. By assuming that the distribution of sediments between the two river branches below the Delta Barrage were the same as the water discharge distribution (65% and 35% respectively), thus 85 and 50 million tons of sediment are discharged from the Rosetta and Damietta outlets annually.

GLIMPSE OF GEOLOGY AND GEOMORPHOLOGY OF NILE DELTA

Figure 1 shows a geographic definition of the coastal regions. The Nile Valley is composed of fine top sediments that rest upon the coarse sands and gravel, tentatively considered to be of Wurn a/b age. Figure 2 shows the denundation surface of the Wurn gravels.

During historical times the number of Nile Branches were reduced to two from seven or more, and the sources of sediment along the shores have shifted accordingly. The major geomorphic units are shown in Figure 3.

During Roman times, the central headland of the Delta extended much further north than at present and the lake itself is underlain by stiff lagoonal clays sloping seaward under the spit, thus separating the lake from the sea. The top of the clay layer is at or above the present sea level and could only have been deposited well inland from the shore.

CLIMATE

Large-Scale General Features

The Mediterranean weather is highly seasonal in nature and is strongly related to the large-scale pressure systems whose limits overstep the boundaries of the Mediterranean area and extend towards the North Altantic, Eurasia and Africa.

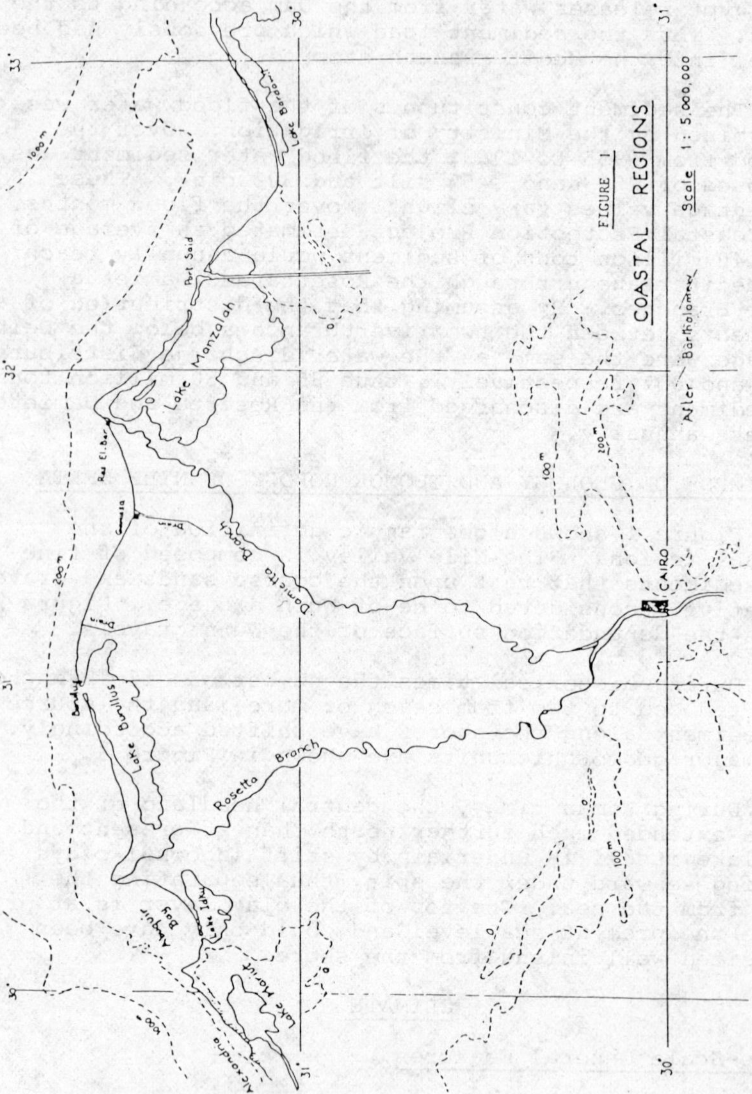
FIGURE 1
COASTAL REGIONS
Scale 1: 5,000,000
After Bartholomew
Lake Bardawil
Port Said
Lake Manzalah
Ras El Bar
Gamasa
Damietta Branch
CAIRO
Burlus
Lake Burullus
Rosetta Branch
Abukir Bay
Lake Idku
Lake Mariut
Alexandria
1000 m
200 m
100 m

Figure 1

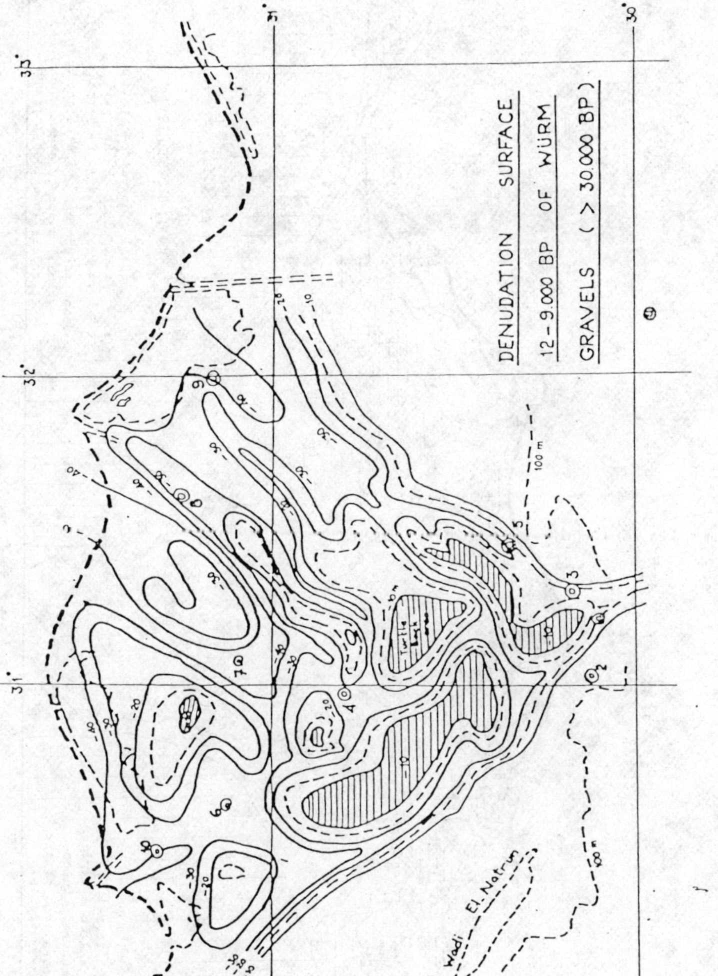
DENUDATION SURFACE
12-9.000 BP OF WÜRM
GRAVELS (> 30.000 BP)
Wadi El Natrun
100 m

Figure 2

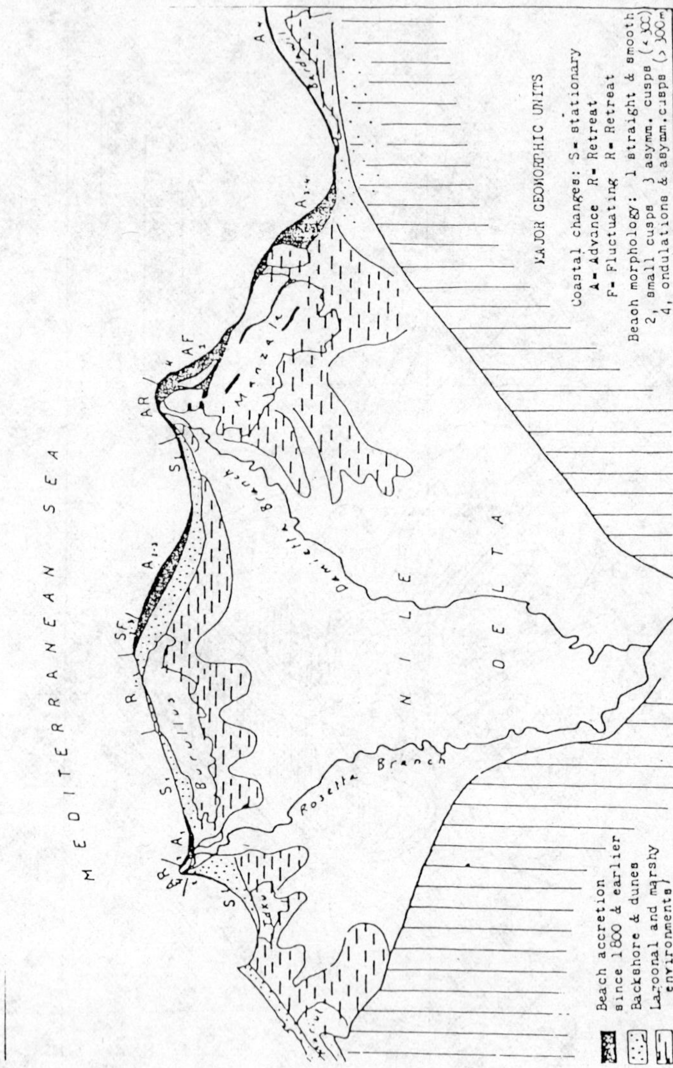
MEDITERRANEAN SEA
NILE DELTA
Rosetta Branch
Damietta Branch
Burullus
Manzala
MAJOR GEOMORPHIC UNITS
Coastal changes: S= stationary
A= Advance R= Retreat
F= Fluctuating R= Retreat
Beach morphology: 1 straight & smooth
2, small cusps 3 asymm. cusps (<300)
4, ondulations & asymm.cusps (>300m)
Beach accretion since 1800 & earlier
Backshore & dunes
Lagoonal and marshy environments

Figure 3

What is known as the cool season covers the period from November to April, while the warm season occurs in June to September. May and October are transitional months. Further details are given below.

Cool Season

The Azores anticyclone often extends over the Libyan Desert, and North Atlantic depressions may enter the Eastern Mediterranean area and bring in masses of cold Arctic air. This cold air, when meeting the warm and moist air of the Mediterranean area, produces vertical instability which often leads to moving atmospheric perturbations associated with meteorological fronts.

Such traveling depressions, when associated with ridges of high pressure over the northwestern Libyan Desert, generate high waves in the eastern Mediterranean and are responsible for most damage on the northern Egyptian Coast.

In Spring, the zone of high pressure tends to shift back westward and air pressure gradients become weaker over the eastern Mediterranean where isobar directions in wind-wave generating areas shift from west-northwest to north-northwest. In May, with the influence of the Arabo-Persian depression, eastward moving depressions become scarce.

Warm Season

During this period a stable, wide Arabo-Persian depression extends towards the eastern shores of the Mediterranean, whereas the Azores anticyclone develops towards western Europe and occasionally the central Mediterranean.

Eastward moving depression systems stop when entering the eastern Mediterranean. In Africa, the Sahara is usually a low pressure zone. The Eurasian winter vanishes in May.

A secondary low pressure cell appears in June over Cyprus and remains stable during summer, whereas the zone between Greece and Tunisia is normally a secondary high pressure area. These features explain the rather constant movement of air masses which is generated between Cyprus and Crete. The corresponding surface (etesian) winds blow mainly from the north-northwest during the period mid-June to mid-September and produce swells which reach the Nile Delta Coast.

From mid-September the Arabo-Persian depression disappears progressively and high pressure belts from southeastern Europe and North Africa extend over central and part of the Eastern Mediterranean, producing a very calm sea there in October.

Climate Characteristics

An average wind rose is presented in Figure 4, based on 50749 ship observations of wind speeds from 1860 to 1970. Meteorological stations are available all along the coastal zone, namely in Alexandria, Abu Qir Rosetta, Damietta and Port Said.

The predominent wind direction is mainly from the west-northwest to north-northwest sector, varying within the sector all along the Delta coastal zone from Alexandria to Port Said.

OCEANOGRAPHY

Waves

The Coastal Protection Institute has been recording the wave characteristics for the last seven years at the specific locations. Combining these measurements with ship observations yielded the probability of exceedance as presented in Figure 5 for the wave heights.

The design wave heights in deep water was computed and its recurrence interval determined and the results are shown in Figure 6.

Water Level Variations

Tide and water level variability gauges are operative continuously in the Port of Alexandria and Port Said. Besides the Coastal Protection Institute measured the water level variability intermittently in between these two locations. Table I summarizes the findings of these measurements all along the coastline from Alexandria to Port Said.

Currents

The predominant current direction in the southeast Mediterranean is due east.

Erosion and Accretion

The coastal changes in the Nile Delta were monitored as far back as the French Expedition in 1805.

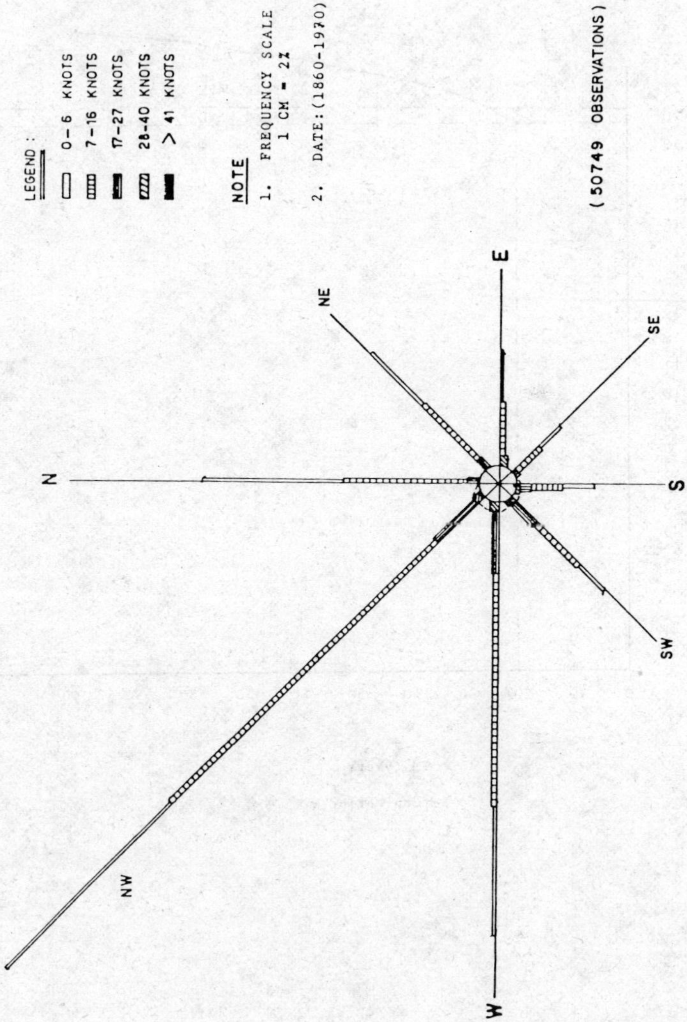
LEGEND:
0–6 KNOTS
7–16 KNOTS
17–27 KNOTS
28–40 KNOTS
> 41 KNOTS
NOTE
1. FREQUENCY SCALE
1 CM = 2%
2. DATE: (1860-1970)
(50749 OBSERVATIONS)
N
NE
E
SE
S
SW
W
NW

Figure 4

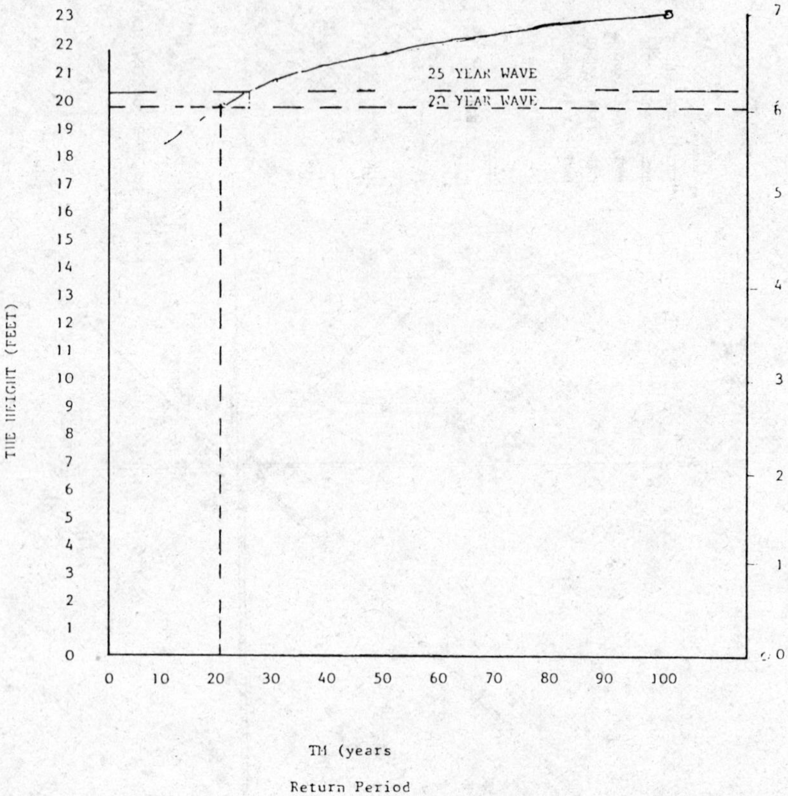
25 YEAR WAVE
20 YEAR WAVE
THE HEIGHT (FEET)
TM (years
Return Period

Figure 5

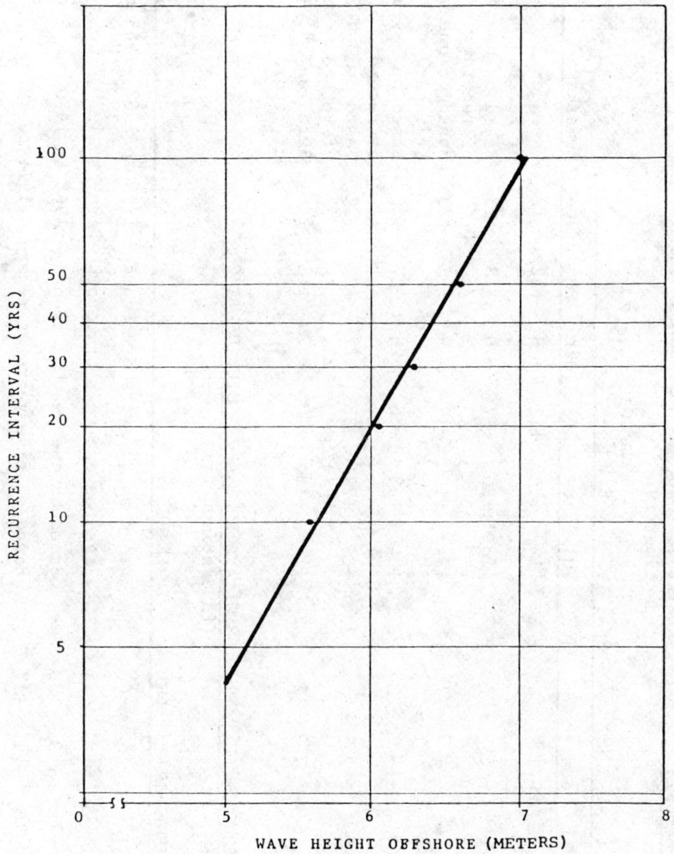
100
50
40
30
20
10
5
RECURRENCE INTERVAL (YRS)
0
5
6
7
8
WAVE HEIGHT OFFSHORE (METERS)

Figure 6

TABLE I

AVERAGE MONTHLY TIDAL VARIATIONS ALONG THE NILE DELTA COAST
(in centimeters)

	MEAN SEA LEVEL												
Locations	Jan	Feb	Mar	Apr	May	Jun	Jul	Aug	Sep	Oct	Nov	Dec	Remarks
Alexandria	+10	+ 6	+ 2	- 1	0	+ 6	+15	+16	+12	+ 8	+12	+12	Based on 23 years of record
Maadia	+22	+10	+14	+ 6	+ 5	+17	+18	+20	+17	+ 6	+ 8	+23	Based on one year of record
Rosetta	+26	+17	+14	+ 8	+ 7	+12	+24	+19	+19	+ 9	+ 7	+23	Based on one year of record
Burullus	+27	+27	+23	+19	+22	+27	+34	+43	+30	+28	+27	+29	Based on 5 years of record
Ras El Bar	+25	+18	+19	+22	+15	+23	+29	+27	+33	+27	+22	+22	Based on 5 years of record
Port Said	+ 1	- 2	- 6	- 7	- 5	+ 2	+10	+13	+10	+ 7	+18	+ 6	Based on 25 years of record

VARIATIONS IN MWL, MHWL, MLWL AND LLWL ALONG NILE DELTA COAST
(in centimeters)

Station	MWL	MHWL	HHWL	MLWL	LLWL	REMARKS
Alexandria	+ 8	+21	+74	- 5	-43	Based on 23 years of record - 1950 to 1962, 1963 to 1967.
Maadia	+18	+20	+44	+ 8	-15	Based on one year of record - 1969 to 1970.
Rosetta	+15	+23	+75	+ 3	-30	Based on one year of record - 1968 to 1970.
Burullus	+28	+35	+76	+21	-17	Based on five years of record - 1971 to 1976.
Ras El Bar	+24	+32	+79	+15	-39	Based on five years of record - 1971 to 1976.
Port Said	+ 3	+26	+68	- 7	-64	Based on 25 years of record - 1923 to 1946.

The Survey Authority also monitored the coastline changes in the 19th Century. Finally in the last ten years, the Coastal Protection Institute kept a bi-yearly record of coastline and nearshore profile changes.

Severe erosion was observed and recorded at several locations, especially at the Nile Delta promontories. Severe erosion was observed at the Rosetta and Damietta promontories.

POTENTIAL DEVELOPMENT AREAS

The following Table shows tentatively the potentialities of the various parts of the coastal zone of the Nile Delta Coast.

	POTENTIAL					
LOCATION	INFRA-STRUCTURE	AGRI-CULTURE	INDUSTRY	FISHING	TOURISM	ENERGY
AbuQir	x	x	x	x	x	x
Rosetta	x	x	x	x	x	
Elburg		x		x	x	
Baltim					x	
Gamasa	x				x	
Ras El-Bar	x			x	x	
Port Said	x		x	x	x	

The infrastructures cover roads, railways, waterways, port sites, either available or need development. The agriculture covers lands already fertile or lands that are viable to reclamation. Industry covers agro-industries, leather, dairy furniture, fish and marine industries, also existing or needs development. The fishing covers both fishing and fish-breeding potentialities. Tourism covers both domestic and foreign tourism potentialities. The energy covers oil, gas extraction and power station for its generation.

THE NILE MEDITERRANEAN OUTLETS

Severe erosion was observed at both the River Nile outlets at Rosetta and Damietta. Figures 6 and 7 show the changes in the promontories at Rosetta and Damietta, respectively. To the best of our knowledge, the erosion of these coastlines started by the end of the last century. This coincided with the start of operating the Aswan Dam in 1902. However, previously Mohamed Ali Barrage, north of Cairo, was built in the 18th century. This might also have had some effect on the coastal changes at these sites.

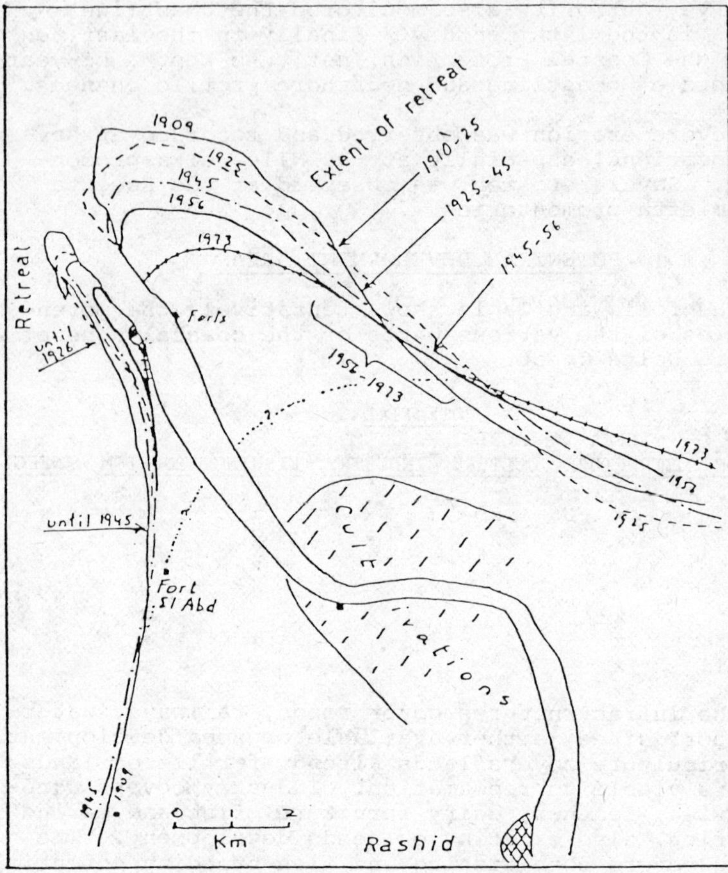

FIGURE 7 - Retreat of the Rosetta mouth promontory, 1909-1973

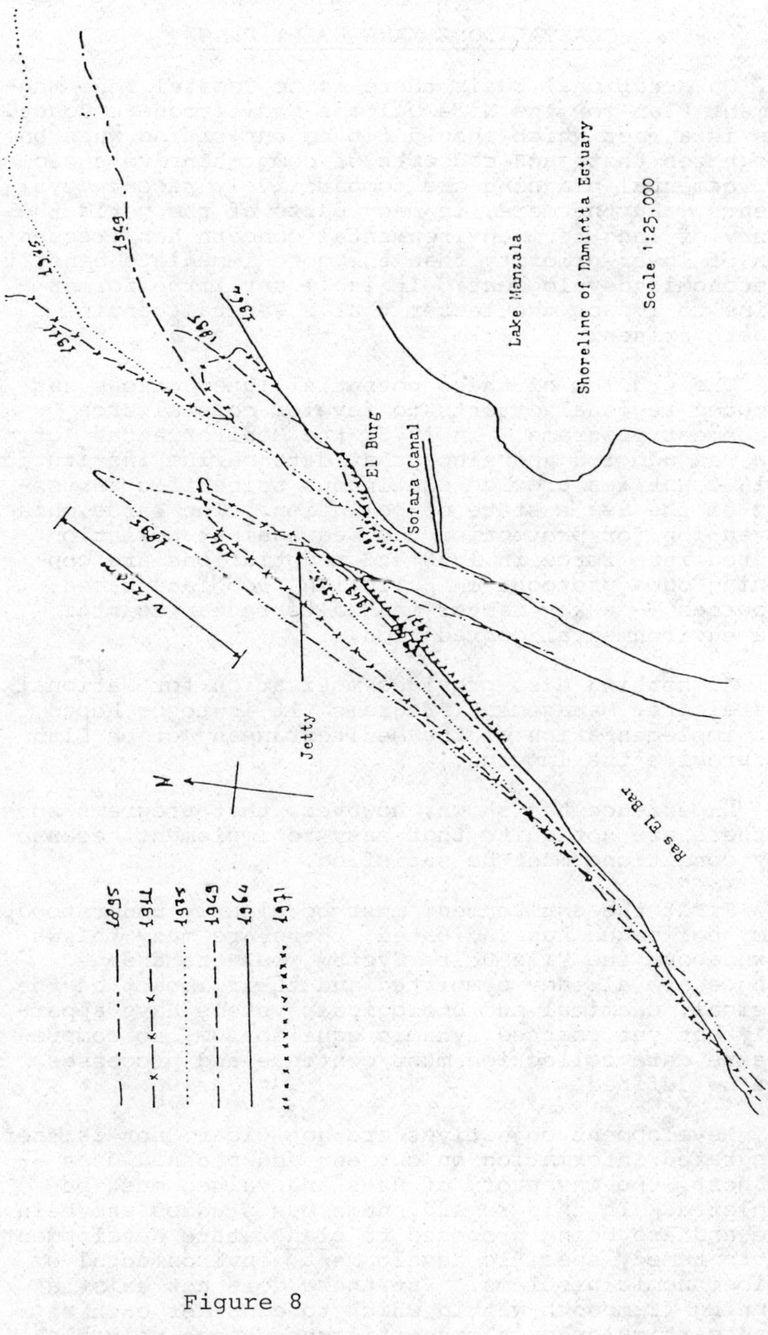
Lake Manzala
Shoreline of Damietta Estuary
Scale 1:25,000
El Burg
Sofara Canal
Jetty
Ras El Bar
N
1895
1911
1925
1949
1964
1971

Figure 8

COASTAL ZONE MANAGEMENT PLAN

On a national basis there is no Coastal Zone Management Plan for the Nile Delta's Mediterranean Coast. This is a fact which should not be surprising when one recognizes that such concepts of comprehensive socio-environmental planning are comparatively recent developments. Furthermore, in many parts of the world the luxury of long-term environmental concern has frequently been of lower priority than the more immediate benefits of economic development. It isn't until the former begins to impact the latter that a general genuine concern arises.

The problem of these potential interactions has prompted regional efforts to develop coastal area management programs. In 1975, the Mediterranean Action Plan was adopted and since that date marine institutions in 16 countries provide continuous scientific assessment of the Sea's state of pollution. The Barcelonia Convention for protecting the Sea against pollution entered into force in 1978 and negotiations are continuing on a protocal on pollution from land based resources -- a key factor in Mediterranean coastal area environmental degradation.

If nothing else provides motivation for National Coastal Area Management Programs, it is to be hoped that implementation of the Mediterranean Action Plan can provide the impetus.

Experience has shown, however, that programs such as these are not quite that easy to implement because many conditions must be satisfied.

First the environment must be clearly understood. As my colleague has indicated, there are many things known about the Nile Delta System, but tremendous changes are already committed and their impact on the physical, chemical and biological systems have apparently not yet reached dynamic equilibrium, so comprehensive data collection must continue and processes must be defined.

Development objectives are not clear, nor is there integrated information on current and planned uses -- in short, the inventory of uses and values must be completed. In this regard, numerous studies are being made and are being proposed to <u>plan</u> future development and to remedy specific development, environmental or socioeconomic problems. But there does not exist a planning framework within which to consider each of these particular problems or issues, hence valuable opportunities are last to acquire information or to format that which is acquired in its most useful form.

This problem has been addressed at several international and regional workshops, conferences and symposia sponsored by the international community and the various individual States. Techniques for developing planning frameworks have been described in literature of the United Nations, but so far as Egypt is concerned, there does not yet exist such a comprehensive plan at either the National level or at the level of the Governorates.

Finally, a suitable management framework must be created that has the authority and responsibility for administering such a program. It should be an Agency or Authority that can survive the rigors of frequent Ministerial change. There is precedence for such Agencies; *e.g.*, Suez Canal Authority, Atomic Power Authority, General Authority for Electricity, and numerous "General Authorities" throughout the many Ministries.

In summary, the same frustrations face the environmentally aware Egyptians today as did their American counterparts about 15 years ago. What, then, should be done?

CONCLUSIONS

It appears particularly timely to develop a comprehensive planning framework for Delta Coastal Area Management; timely because tremendous pressures exist for large scale development supported by very large investments from development agencies throughout the world. The framework would assure that relevant and adequate data is obtained for project development and evaluation in the context of the Deltaic system as a whole. Typically potential major new port developments such as Dekhela and Damietta would be evaluated for impacts on the marine system as well as on land use and the socio-economic system. In a similar way the Alexandria Sewer Project, the additional offshore gas development off Abupir, the hoped for coastal protection program between Alexandria and Port Said and coastal area land reclamation programs around the coastal lakes, could be placed into a comprehensive planning framework that would greatly facilitate the evaluation of the combined effects of all of these projects given adequate knowledge of the environment and its processes. The accumulating experience of other states attests to the value of such a program and its adaptability appears to be simple investiture of cultural or socio-economic differences. For more than 4,000 years, the system has withstood the pressures of civilization. With a little planning, it might even make it through the next onslaught.

TWO APPROACHES TO COASTAL AREA PLANNING IN JAPAN

by
Harvey A. Shapiro[1]

INTRODUCTION

Japan is a 3,000 km (2,000 mi) long archipelago made up of 4 large and some 4,000 smaller islands located on the western edge of the Pacific Basin in the same latitude and climate range as the east coast of the U.S. Its total land area is less than 1/25th that of the U.S., with 3/4 of this being steeply sloping, generally uninhabited mountains. The population of Japan is now over 116 million, most of the people living and working within 13 km (8 mi) of the nation's 33,000 km (roughly 20,000 mi) long coast[1], at an average population density of well over 30 times that of the U.S.[2] The average length of coastline per person in Japan is just 30 cm (about 12 in).[3] In addition to 14 of the nation's 16 cities with over 500,000 population (including 6 of Japan's 8 cities of over a million people) being located on the coast[4], there are some 1,087 commercial ports and 2,831 fishing ports[5] as well as 50 marine parks[6]. This large number of ports is one indication of Japan's great dependence on the sea for both food and trade, while the number of marine parks reflects the superb natural beauty of many parts of the Japanese coastline.

In addition to being the site of over half of the nation's economic activities[7], there are over 14½ million sports fishermen[8] and some 45 million beach users annually[9] in competition with innumerable other uses for a share of Japan's limited, highly vulnerable[10] coastal environment. Coastal area use in Japan is among the most intensive in the world, and yet despite this there is neither a national nor local management-and-planning system designed to deal <u>comprehensively</u> with its coastal environment.[11] This does not mean that management and planning of the nation's coastal area have been neglected. Every portion of the coast is under some kind of management system, with approximately half of its total length being under the jurisdiction of the Ministry of Construction, about 25% being under Ministry of Transportation authority, and most of the remaining being managed by the Ministry of Agriculture, Forestry and Fisheries.[12] These three central government agencies get sectoral authority from some two dozen national laws resulting in the nation's present fragmented coastal management system.

It is within this fragmented legal-political context that each government agency, central and local, and their contracted consultants have developed a wide variety of ad hoc coastal area planning methods, a different one for each coastal area problem.[13] The literature on coastal area planning methods in Japan has been classified as, "basic research": oriented toward developing planning philosophy, planning theory and/or legal theory; and "planning applications": oriented toward policy-making and/or specific physical goals.[14] In this paper, four examples, two by a central government agency, one by a private planning consulting firm and one from academia are present-

---

[1]Instructor, Dept. of Environmental Planning, Kobe National University Kobe City, and Osaka Geijutsu University, Osaka Prefecture, Japan.

ed. The methodology of each is introduced in terms of its goals,and some of the major differences between them are considered both in terms of those who developed the methods,as well as against the background of coastal area management and planning in Japan today.

NATIONAL LEVEL COASTAL AREA PLANNING METHODS

I. A National Coastal Area Development and Protection Planning Method[15]: This coastal area planning methodology was developed by members of the Development Section,Ports and Harbors Bureau of the Ministry of Transportation Headquarters in Tokyo in 1977. It is a "planning applications" method aimed at evaluating the suitability of the coastal areas of the nation for several basic coastal area use categories and developing general guidelines for each area and each use category therein.

The basic methodology is shown in Fig.1. First, mesh data on natural conditions,such as coastline length and types(natural, man-made,estuarine etc.), depth of coastal waters, sea bottom sediments and tides, as well as use conditions,such as percentage of fishing and commercial port areas,natural parks and recreation areas etc. are gleaned from the National Coastal Areas Survey[16]. Data within a 1 km square mesh on conditions within a 2 km wide belt,1 km on each side of the coastline, was used, and additional data was developed or added as considered necessary.

Next, 7 major uses of the nation's coastal area (fishing and commercial ports, heavy and moderately heavy industry , oil storage tanks, coastal recreation and nature study) are analyzed in terms of the conditions which offer the highest "potential" for their location. The concept of "potential" used in this method is defined as a measure of the relative predominance of a use type in one region as compared to other uses in that region.[17] A national government perspective of the total potential (made up of natural, social and surrounding area use sub-potentials),that has been validated according to actual experience,is employed. Each of the 3 sub-potential categories that combine to form total potential are ranked in 10 steps from a minimum of 0 to a maximum of 9. The importance(weight) of each is determined by a panel of experts and from experience. The resulting criteria are applied to the data on coastal areas from Step 1 to give the present potential of each coastal area in the nation for each use type.

In Step 3, a group of oceanographers and other experts is formed to analyze the data from Step 1 in order to first divide the coastal area of Japan into various levels of sub-divisions. Factors such as currents,sea bottom structure and marine living resources among others are considered. This results in the 7 "macro-areas" shown in Fig.2. Next, each macro-area is divided into "meso-areas" (in the order of 100 km in length) according to coastal forms based on the presence of bays and peninsulas etc., as well as into areas of environmental unity based on sea bottom topography and depth, again see Fig.2. Finally, "micro-areas" in the order of 10 km long are determined based on the relationship between adjacent coastal land, i.e. presence of markets,labor force etc. ,and coastal water

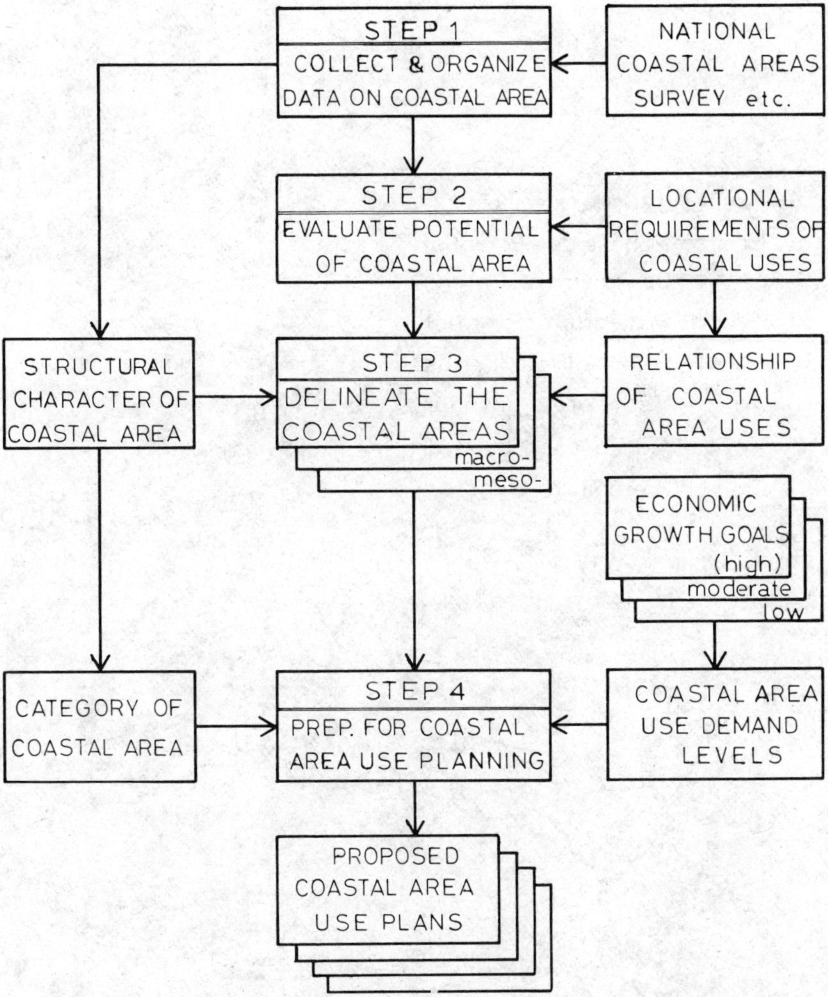

FIGURE 1. A NATIONAL COASTAL AREA PLANNING METHOD

(Source: see note 17)

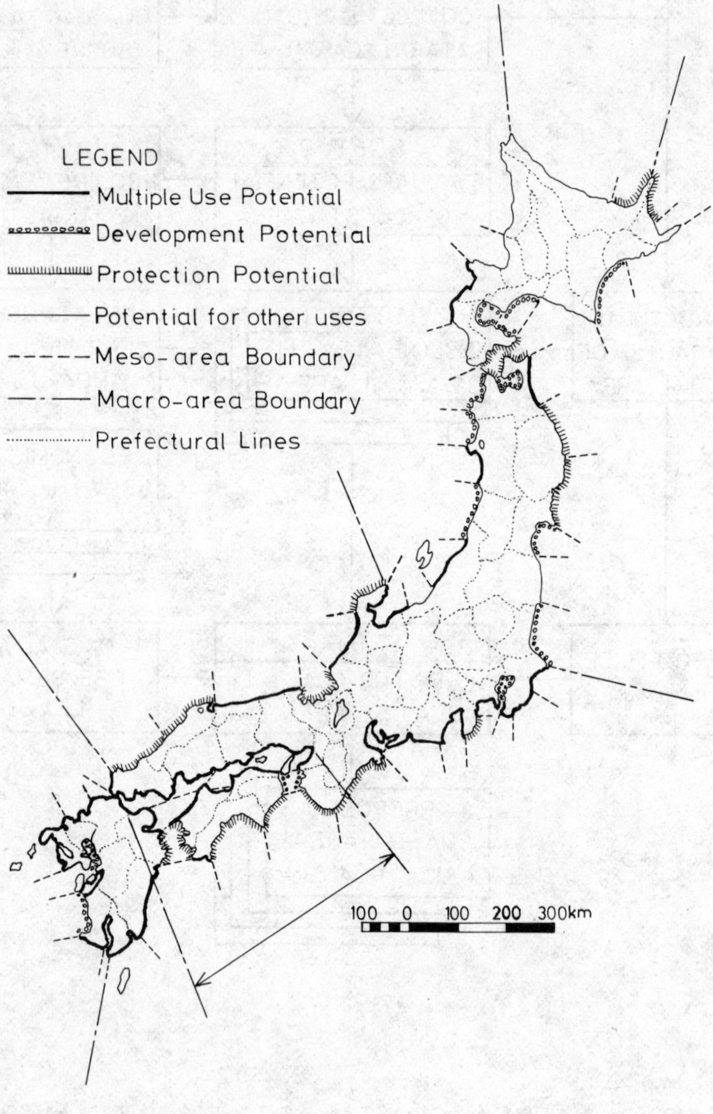

FIGURE 2. COASTAL AREAS AND COASTAL AREA USE ZONES
(Source: see p.20 in Ref.15)

areas. It is at this level that the concrete relationship of coastal areas and use potential is ultimately determined.

In Step 4, preparation for actual coastal area use planning is made. First, several categories of coastal use conditions are developed according to the character of each coastal area, i.e. whether it is "closed" with a strong unity, or "open" with a weak unity. Also the potential of each type of use for protection of the coastal area environment is determined, i.e. compatibility with protection goals is examined. The following four basic categories of uses result: 1) Multiple uses, 2) Development-oriented uses, 3) Protection-oriented uses, and 4) Other uses. Each type varies from "careful overall management" in closed coastal areas to "local coordination" in open coastal environments.[18] Fitting the supply characteristics for each use with its demand as calculated from national and regional economic goals (high, moderate or low economic growth etc.) at varying future times results in a set of alternative comprehensive coastal area plans for the nation's entire coastal area.

II. <u>A National Level Ecological Approach to Regional Planning Based on the Delineation of Coastal Water Units</u>[19]: The method presented briefly here was developed late in 1976 by Japan's only ecological planning firm, The Regional Planning Team Associates, Inc. in Tokyo. It is both a "planning theory" and a "planning applications" method directed not only at the coastal waters but also at regional planning for the whole archipelago as well. The fundamental thought underlying this approach is that in an island ecosystem such as Japan, not only does the surrounding ocean directly influence the climate of the entire environment, but with its many steep mountains, the nation's rivers are, with few exceptions, short and fast-flowing. Thus, they carry most of the influences of land uses directly to the sea with little or no self-purification. This approach advocates that protection, restoration and/or enhancement of coastal area environmental quality could best be achieved by combining coastal area regional planning into a unified process in which the best possible coastal ecosystem function becomes the starting point and controlling factor in overall national and regional land and water use planning.

A schematic diagram of the basic approach is shown in Fig.3. First, coastal water conditions from the coastline to a depth of 200 meters, the depth defined as the outer limit of the "coastal area" in Japan[20], were examined. Emphasis was given to coastal ocean currents and circulation patterns, geomorphology, and coastal ecosystem energetics. These primarily functional factors were found to group naturally into 3 types of coastal water areas according to the degree of openness to the influences of the outer sea, i.e. the degree of potential influence of the land on coastal water quality: (1) Exposed coastal water ecotones, that receive the full influence of the outer sea; (2) Exposed coastal bays and channels, transitional between fully exposed and more sheltered coastal waters; and (3) Sheltered coastal bays and inland seas, i.e. estuarine systems strongly influenced by terrestrial and coastal area uses. The resulting "coastal water units", each with different tolerance to both land and water uses, are shown in Fig.4.

FIGURE 3. AN ECOLOGICAL APPROACH TO COASTAL AREA-BASED PLANNING
(Source: based on Ref. 19)

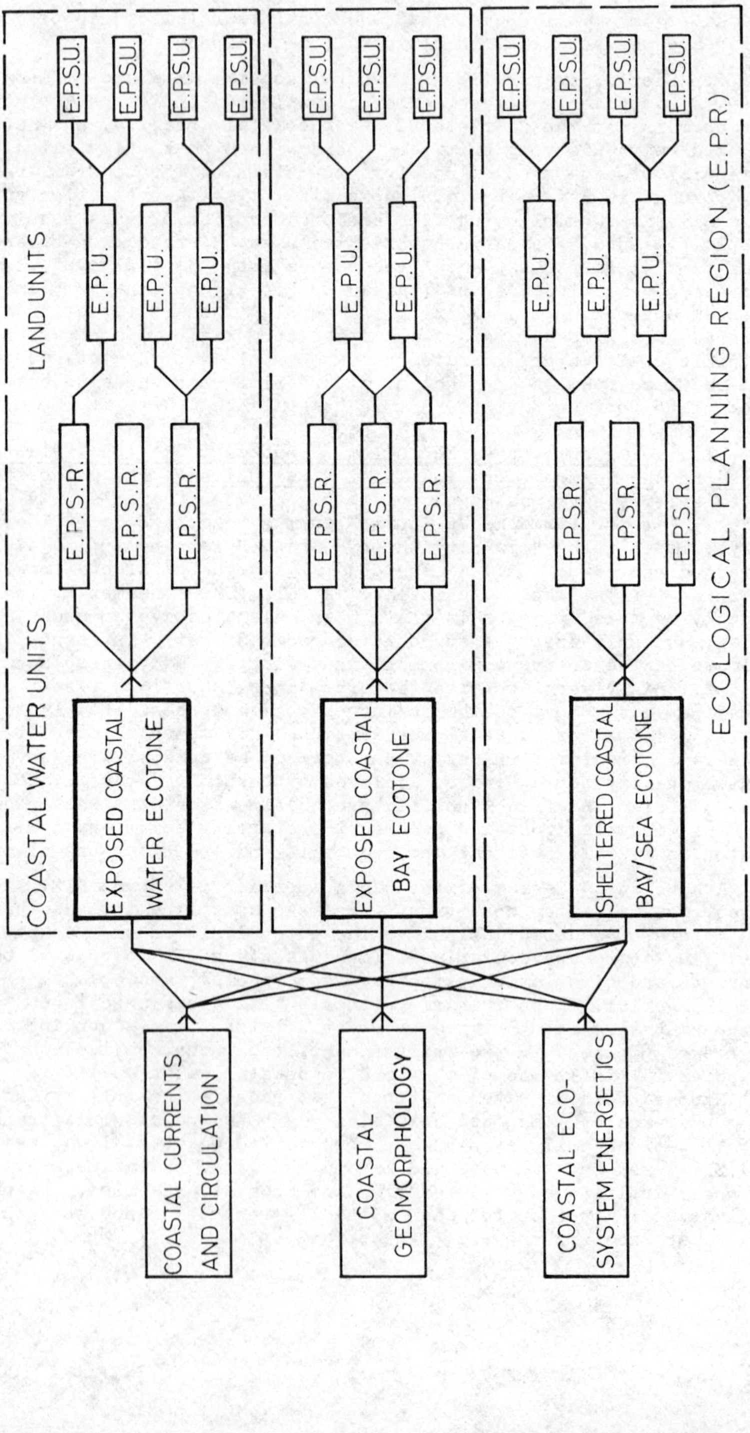

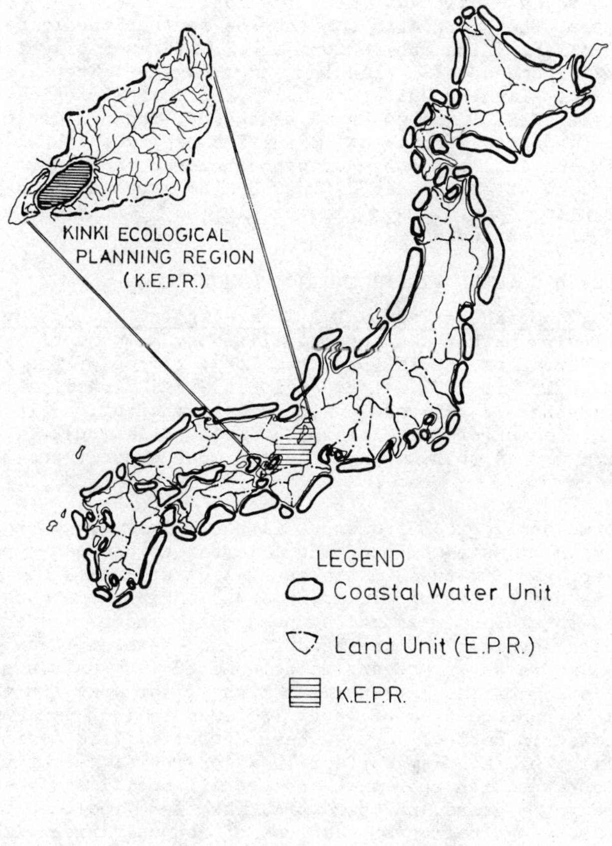

FIGURE 4. COASTAL WATER UNITS & RELATED LAND AREAS

(Source: see p.44 in Ref. 19)

Once these coastal water areas have been defined, the next thing to do is to define the land areas containing all of the water systems that drain into the coastal water units. This is in keeping with the principle that water ,not only the coastal waters but the upland drainageways and land surface , is the primary ecological consideration in coastal area management linking land and sea elements of the coastal ecosystem [21]. First, the largest such land areas, including the coastal water area directly receiving the potential impact of the land and water uses therein, are delineated. These are referred to as "Ecological Planning Regions(E.P.R.)"; See Fig.4 for an example. These E.P.R.'s can then be further subdivided into "Ecological Planning Sub-regions(E.P.S.R.)",many of which are coastal and thus become coastal planning and management areas, smaller urban "Ecological Planning Units(E.P.U.)" , local"Ecological Planning Sub-units(E.P.S.U.)"and so on as necessary. These hydrological units are all parts of geographically unified regions,each with its own identity and inherent range of opportunities and constraints for land,air and water uses as defined by the structure and function of coastal water ecosystems, discoverable and planned for by the use of ecological planning.[22,23]

LOCAL LEVEL COASTAL AREA PLANNING METHODS

I. <u>A Comprehensive Coastal Area Planning Method for Ise Bay</u>[24]: The following method was conceptualized by members of the Ports and Harbors Bureau of the Ise Bay area local office of the Ministry of Transportation in 1977. Although it is aimed directly at furthering port planning and related development in the region (in accordance with the mission of the Ministry of Transportation), it is structured in such a way as to be applicable for most other types of uses as well.[25]

The methodological framework is shown in Fig.5. The approach starts with an estimation of the alternative demands considered likely for various uses in the coastal water area based on several possible regional economic goals,such as high,moderate or low economic growth etc., as measured by per capita income and industrial production levels.[26] Once the difference between these estimated demands and those at present is determined, the next step (Step 2) is to do a study of the "potential" of the water area (in this case,Ise Bay with a surface area of about $1\frac{1}{2}$ times that of San Francisco Bay and located in central Japan). Here "potential" is defined as the suitability of each Bay surface location for different uses in terms of the construction,operation and overall social costs as well as benefits to the users and operators. The items included in the potentiality evaluation of each use of course varies with each use type. The items are arranged in an evaluation tree such as the one shown in Fig.6, the weight of each item being determined by a panel of experienced personnel and after comparison with the importance of each item under existing conditions in the region.

In the third step, a zoning concept is employed in which the surface area of the Bay is divided into 4 basic types of zones:

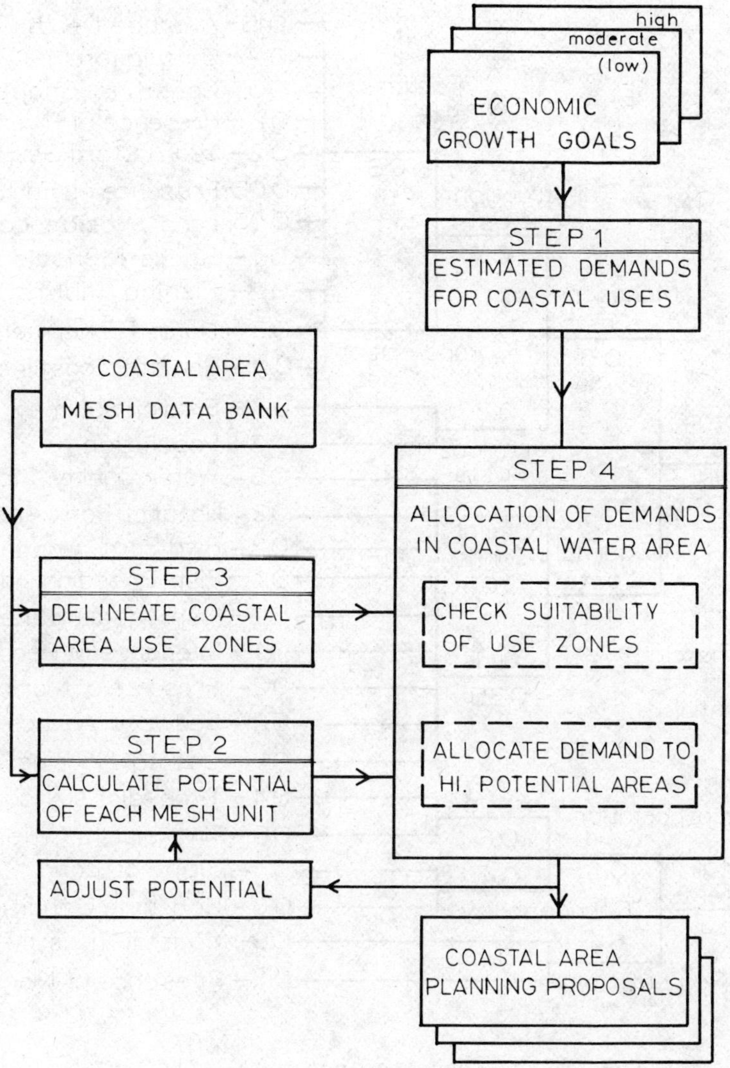

FIGURE 5. A BAY AREA USE PLANNING METHOD

(Source: see p.37 in Ref. 24)

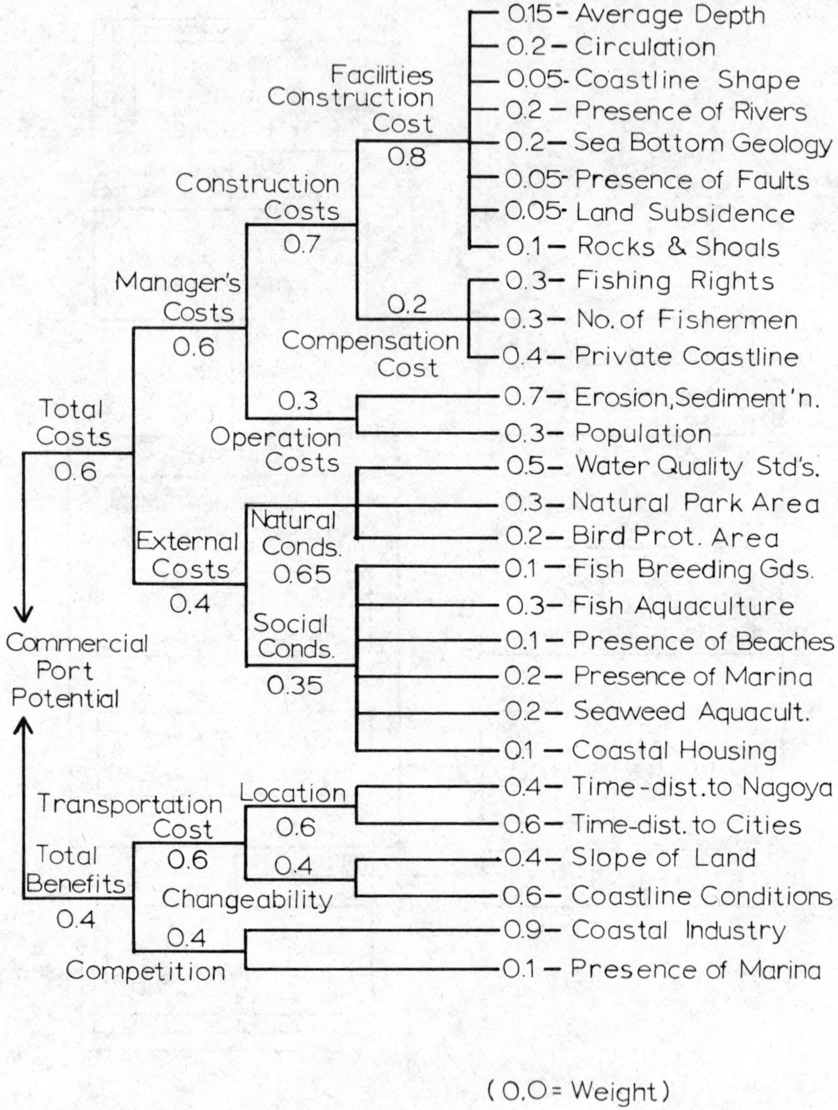

FIGURE 6. SAMPLE EVALUATION TREE FOR PORTS

(Source: see Ref. 27)

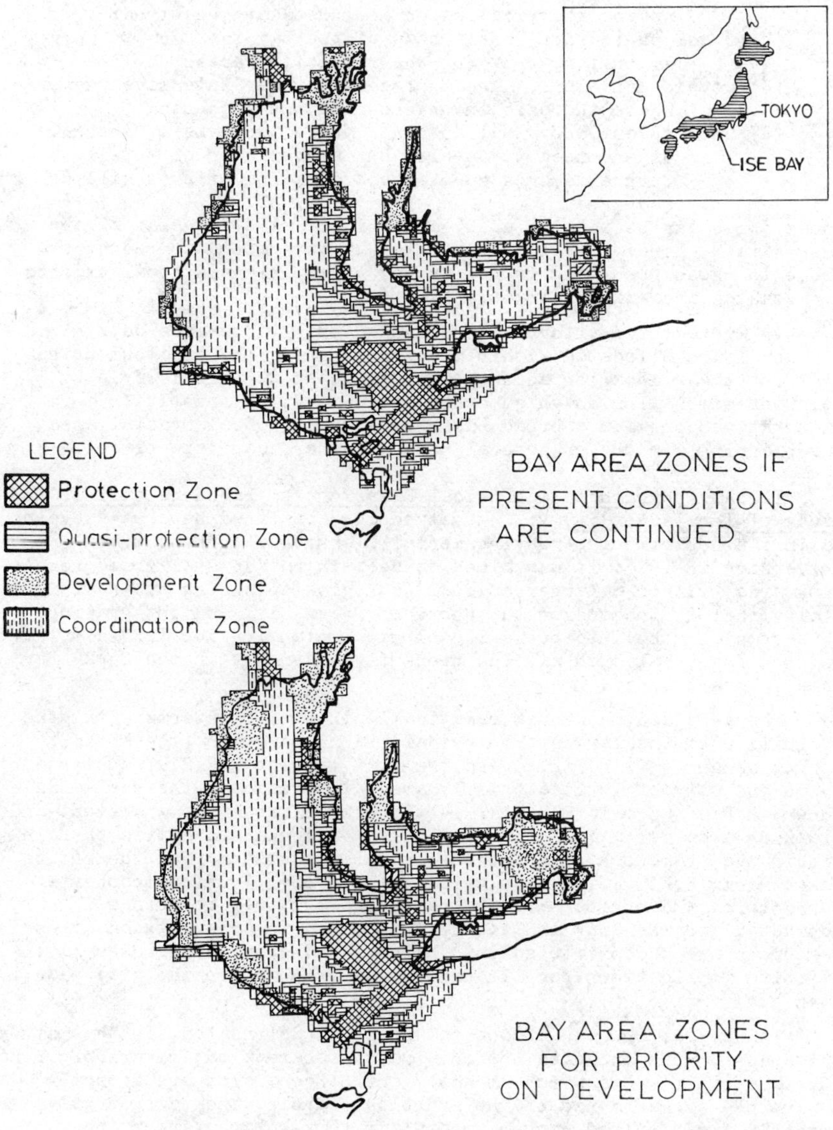

FIGURE 7. USE ALLOCATION ZONES IN THE ISE BAY AREA

(Source: see Ref. 24, pp. 36–37)

(1) Protection Zone: Where the present need for protection and severe restrictions on most uses are required;
(2) Quasi-protection Zone: Areas of the Bay that are to serve as buffer zones for the protection areas;
(3) Development Zone: Areas presently under intensive use or where there is an urgent need for such use;
(4) Development-Coordination Zone: Where there is at present no urgent need to develop but where in order to fill future demands special coordination efforts will be required.

These zones can be identified by analysis of existing coastal area mesh data[28]. Zones identified in this way were found to be very close to those resulting from a survey of experienced local experts and coastal area administrators[29].

In Step 4, the alternative demands for various uses determined in Step 1 are allocated within the zones having the greatest potential for their location as identified in Step 2. This is done by computer simulation in this method, see Fig.7. The resulting alternatives provide a broad basis upon which actual coastal area use planning,for one or several uses together, can take place.

II. An Approach to Ecological Planning for Bayshore Cities:Case Study of Izumisano City[30]: The methodology outlined here was developed in the Department of Environmental Planning at Kobe National University in 1978. It was aimed at defining a "coastal zone"(regulation area) within a larger coastal ecological planning sub-region for a city that is located on the shore of Osaka Bay, and integrating this coastal zone into overall ecological planning for that city. It is one of several such studies presently under way in the Osaka Bay area ,Western Japan.

The approach,shown diagramatically in Fig. 8, starts with the definition of the largest Ecological Planning Region (the Kinki E.P.R. shown in Fig.4) and the next largest E.P.S.R. context in which the City of Izumisano is located, in this case the Osaka Basin shown in Fig. 9 . This E.P.S.R. is the overall coastal management-and-planning area for all of the coastal sub-regions and cities therein[31]. Within the framework of an ecological planning study for the entire Osaka Basin E.P.S.R., an ecological inventory of the atmospheric, terrestrial and marine natural and social-cultural resources is conducted for the City in Step 2. Using a map-overlay method, these resources are reconstituted into natural and social processes having definite implications for all uses to be included in the city plan,in Step 3.[32]

In Step 4, an actual "coastal Zone" is delineated. In this study, the coastal zone includes not only Bayshore areas but rivershore,pond-edge and other areas which directly influence and/or are directly influenced by man's use of the hydrologic cycle, including highly erodable areas, flood danger areas, aquifer recharge areas and others . Furthermore, the resulting coastal zone is divided into types depending on whether it is located on the seashore,river-mouth, riverside or pond-edge etc. Each area contains special regulations and guidelines for all types and intensities of land uses that might

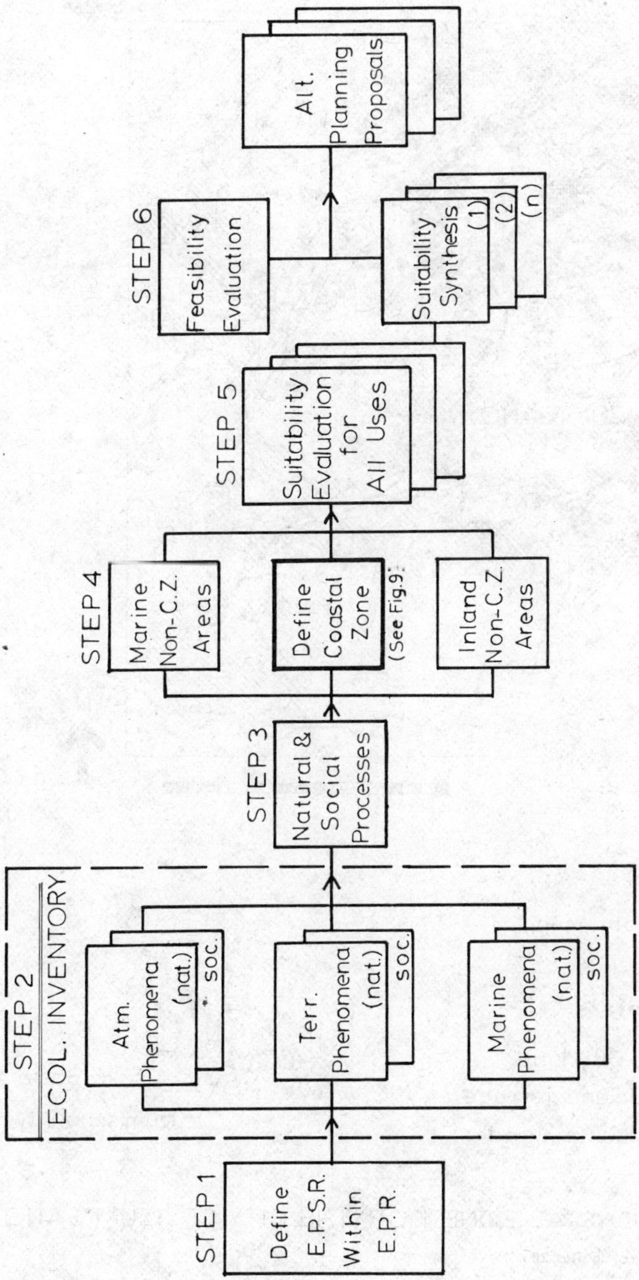

FIGURE 8. AN ECOLOGICAL PLANNING METHOD FOR AN OSAKA BAY COASTAL CITY

(Source: based on Ref. 30)

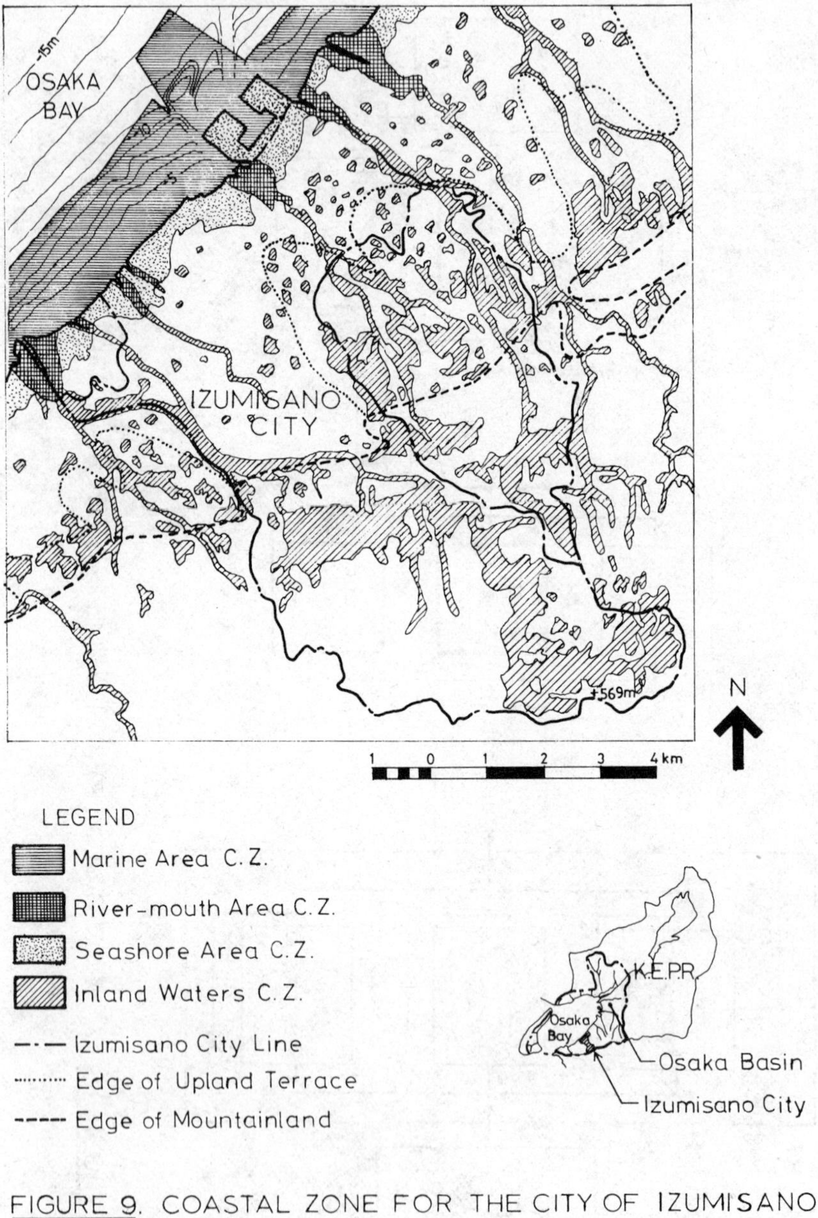

FIGURE 9. COASTAL ZONE FOR THE CITY OF IZUMISANO

(Source: based on Ref. 30)

be located therein. The resulting coastal zone planning-and-regulation area is shown in Fig.9.

In Step 5, a suitability evaluation for each prospective terrestrial, coastal and marine use is conducted using toned evaluation criteria maps developed in Step 3. Each map, accompanied by guidelines for environmental protection, restoration or enhancement etc., may then be overlayed in alternative orders, reflecting the various values and goal-priorities of the people of the area, to develop alternative suitability synthesis maps of mutually compatible uses. These become the basis for feasibility analysis and final urban ecological planning for the City.

DISCUSSION

The preceding four planning methods, still in the developmental stage, were all conceptualized around or since the time Japan declared its 200-mile exclusive fishing zone in mid-1977. They represent some of the earliest, more systematic approaches to comprehensive coastal area planning, still an embryonic field of endeavor in Japan. In addition, they tend to reflect the present state of Japanese land development planning which is divided rather unevenly between engineering-oriented approaches which predominate, and a slowly increasing number of ecologically-oriented methods.[33] This strong engineering-technological inclination can be seen as a reflection of the emphasis of postwar Japan on reconstruction and the building of a thriving, modern industrialized economy. Its success in the achievement of this national goal is now common knowledge, but so too are many of its environmental failures.[34]

The Japanese environmental movement began toward the end of the 1960's, and nearly a decade has passed since the establishment of the government's Environment Agency. Though one of the newest and still relatively weak Agencies of the government, its strength is growing, and with it has come the steadily increasing influence of ecology and environmental factors in the nation's approach to national development planning. For instance, reference to the coastal zone, the basic coastal area classification system and river basin-related land units idea, proposed by the ecological planning firm in the second national level method, was included not only in the 3rd National Comprehensive Development Plan[35] but also the 1978 National Land Use White Paper[36], both by the National Land Agency. In addition, several ecological planning studies have been contracted for by both the National Land and Environment Agencies in recent years.[37,38] However, despite this growing ecological influence, it remains weak, by comparison, in most of the construction-oriented national and local government agencies(Ministries of Construction, Transportation and International Trade and Industry etc.) because environmental considerations are generally considered to be a "brake on development"[39] by these Agencies as well as in industrial and economic circles. The efforts of some of these Agencies and interests have been greatly responsible, for instance, for the lack of the successful passage of the Environment Agency's National Environmental Impact Assessment Bill.[40] The relative weakness of environmental administration in Japan[41] does not provide the kind of atmosphere in which ecological approaches to management and planning of the coastal area, or elsewhere in the nation, are likely to grow to predominance anytime soon.

Another characteristic of the two approaches represented by the four methods is that both at the national and local levels, those of the Ministry of Transportation planners tend to be concerned almost exclusively with the coastal waters and coastline (to 1 km inland), while the ecological methods tend to be more inclusive,including inland areas (to the watershed line), coasts (of the sea,rivers, lakes and ponds etc.) and coastal water ecosystems as the object of their respective planning studies. The exclusiveness of the Transport Ministry's methods may be seen as a direct reflection of departmental sectionalism in planning administration ,mentioned in the introduction. This tends to result in the environment being divided into "easily administered" compartments, such as farmland, coastlines, cities,ports, the sea etc., which are managed and planned for separately.[42] This would appear to be in antithesis to the intrinsic character of the environment which is an integrated unity. This character is recognized, respected and dealt with inclusively by the ecological approaches which cross academic,political and jurisdictional boundaries in an effort to more comprehensively manage and plan for the environment as a whole. The legal-political context of land development planning in Japan does not fit the ecological reality very well reflecting obsolete laws and institutions, "left over" from a time when environmental knowledge,awareness and concern were not as advanced as they are today. This may make "exclusive" approaches appear more "realistic" to administrators,but it tends to make comprehensive approaches,such as those in this paper, and particularly "inclusive" ones difficult to accept as well.

Japan does not, at this time, have a comprehensive management-and-planning program for its coastal environment. The coastal area of the nation is,as in most countries[43],not considered as a separate entity for the purpose of planning or development, nor is there any single agency nor national coastal area law concerned exclusively with it. Furthermore, the central government has been slow in moving toward a strong legal-financial-administrative framework for the nation's coastal area. Though some Ministries have prepared proposals for a relatively comprehensive law[44,45], they tend to be aimed more at consolidating power within existing jurisdictions rather than unifying efforts between them in order to manage and plan for the coastal environment holistically. Nevertheless, as we have seen, in both of the national level methods, the importance of a national system of coastal/coastal area classification was recognized, and preliminary methods for developing such a system as well as possible planning applications of the results were suggested. An I.U.C.N. study states,in part, that "A (coastal) classification scheme is the basis for the establishment of a system... by which marine ecosystems will eventually be conserved,studied and monitored. No matter how preliminary the present effort may appear to be, it is vital that we indulge in the classification exercise...".[46] Furthermore, in both of the local level methods, such concepts as "coastal water area zoning", the "coastal zone", and other ideas which could be important in future coastal area management and planning in Japan, have been introduced and applications attempted. Though it is true that none of these concepts presently have any legal basis in Japan, local areas may be encouraged to include them in their own future coastal area management-and-planning systems. This could set the precedents needed to help move national, coastal area legislation and administration toward realization. To be most effective then, these four methods should probably be part of an overall coastal area management-and-planning concept in which the coastal area is the focal point for management of land and marine, coastal and fresh water resources.[47] However, the extent to which present coastal area-related administrative bodies could be coordinated to work under such a concept, were one to emerge, or even whether the competitive relationship between them could be turned to developing such a concept are questions for which there is as yet no agreement on the answers. However, the answers are likely to offer important clues about the future of coastal area management-and-planning, as such, in Japan; The same may be said for the future of methodologies, such as those presented briefly in this paper, that could be important parts of it too. In any case, it will probably take some time yet for Japan to develop its own unique approach to the management and planning of its highly vulnerable and intensively used coastal environment, a process that has only just begun.

SUMMARY

In this paper, four recently developed Japanese coastal area planning methods, classified as engineering-oriented and ecologically-oriented approaches, were introduced. One of each approach applied at the national level and based on its own coastal area classification system was introduced, after which one of each approach applied at the local level using "coastal water area zoning" and the "coastal zone" concept, respectively, was discussed. There is no national nor local management-and-planning system to deal comprehensively with Japan's coastal environment in that country at this time, and Japanese precedents considered necessary to encourage the establishment of such a system are thus far lacking. Whether present coastal area-related administrative bodies could be coordinated to work under, or be encouraged to work toward developing such an overall system is at present uncertain, but it appears to be an important point and one that could have considerable influence on the future of coastal area management and related planning methods in Japan.

ACKNOWLEDGEMENTS

The author wishes to thank Mr. Tatsuyuki Shishido of the Transport Ministry for permission to use the diagrams from his and his colleagues' papers. Also, thanks go to the R.P.T. Associates Inc. for their cooperation, and to Mr. S. Okamoto of Kobe University for his assistance on the graphics related to the Izumisano City study.

NOTES AND REFERENCES

1. Rivers Bureau,ed. Kaigan tōkei(Coastline Statistics)1978 Edition. (Tokyo: Ministry of Construction,1978), p.20.

2. Nihon tōkei nenkan(Japan Statistical Yearbook). (Tokyo: Statistical Bureau of the Prime Minister's Office,1975).

3. Tanigawa,T. et al,ed. Atarashii kai-iki ryo gijutsu banren (New Coastal Water Areas Use Technology Manual), Fuji-techno-system, Tokyo, Dec. 1978, p.26.

4. Masai,Y. "Distribution and Definition of Bay-coast and Seacoast Cities", in M. Numata,ed. Integrated Ecological Studies in Bay-coast Cities, I . Chiba,Japan, March,1979,pp.3-7.

5. Ibid., Tanigawa,p.27.

6. Marine Parks in Japan, Marine Parks Center of Japan,Tokyo, May 1975.

7. Inoue,S. Toward a Framework for Coastal Zone Management:Comparisons Between Japan and the United States, University of Washington,Seattle, 1977 (unpublished thesis), p.36.

8. This is according to a survey done by the Japan Sports Fishing Association in 1973.

9. Kiuchi,S. et al. Kaigan-sen riyo-no kenkyu hokoku (Report on Coastal Uses in Japan).(Tokyo: Japan Center for Area Development Research,1973),p.107.

10. Aramaki,M. Ikiteru nagisa(The Living Coast),Sanshodo,Tokyo,1972.

11. Ibid., Inoue,p.44.

12. Ibid., Kaigan tōkei(Coastline Statistics), pp.20-21.

13. Kaigan-sen riyo keikaku shiho chosa(Survey of Methods for Coastline Use Planning, Mitsubishi Research Institute,Tokyo, 1977, pp.3-137.

14. Ibid.,pp.229-230.

15. Kuroda,H. & Shingyouchi,H. "Rinkaibu kaihatsu hozen keikaku chosa gaiyo (A Concept for Coastal Area Development and Protection Planning)",in Kowan Journal, Vol.54,No.8,pp. 15-22, 1977.

16. Zenkoku kaigan-iki genkyo chosa (Survey of the Present Conditions of Japan's Coastal Areas). Tokyo: Ministry of Construction,1975.

17. Ibid.,Kuroda & Shingyouchi, p.16.

18. Ibid., p.19.

19. R.P.T. Associates,Inc.,"Chiiki seitai-keikaku no hoho to jissen, II (Ecological Planning in Japan,Its Method and Application, Part II)", in Kenchiku Bunka Journal, Vol.32,No.367, pp.23-152, May 1977.

20. Murakami,Akio. Kaigan no osen (Coastal Pollution), Chikuchi-shokan,Tokyo,1977,p.2.

21. Clark,John. Coastal Ecosystems, The Conservation Foundation, Washington,D.C., 1974,p.4.

22. McHarg,Ian L. Design With Nature, Doubleday & Company,Inc., Garden City,N.Y., 1971.

23. R.P.T. Associates,Inc., "Chiiki seitai-keikaku no hoho to jissen, I (Ecological Planning in Japan,Its Method and Application, Part I)", in Kenchiku Bunka Journal, Vol.30,No.344, pp.43-136, June 1975.

24. Shishido,T. and Arai,T. "Ise-wan kai-iki sogo-riyo keikaku shiki an (Proposed Comprehensive Use Planning Method for Ise Bay:By the Use of Mesh Analysis)", in Kowan Journal, Vol.54,No.8,1977, pp.32-39.

25. Ibid.,p.32.

26. Ibid., pp.35-37.

27. Ibid., p.33.

28. See note 16.

29. Ibid.,Shishido and Arai, p.34.

30. Izumisano-shi oyobi shuhen seitaigaku-teki chiiki keikaku (Ecological Planning Study for the Bayshore City of Izumisano, South Osaka Prefecture),Kobe National University,Dept. of Environmental Planning, 1978-1979 (unpublished research).

31. Shapiro,H.A. "Comprehensive Environmental Management and Planning for the Osaka Bay Coastal Zone and Bayshore Cities, Part I", in M. Numata,ed. Integrated Ecological Studies for Bay-coast Cities,I. Chiba,Japan,March 1979,pp.78-91.

32. See criteria matrix in study,note 30.

33. Shapiro,H.A. "Chiiki seitai keikaku,America to Nihon(Ecological Planning in Japan and the United States)", in Jimbun Chiiri (Human Geography)Journal, Vol.32, No.1,1980,pp.92-94.

34. Shapiro,H.A. "Japan: Torn Between Its Economy and Its Environment", in Planning, Vol.44, No.10, Nov. 1978, pp.15-18.

35. National Land Agency,ed. Dai-sanji Zenkoku Sogo Kaihatsu Keikaku (3rd National Comprehensive Development Plan). Tokyo, Nov. 1977, pp.47-49.

36. ________, Kokudo-riyo Hakusho(National Land Use White Paper),1978 Edition .Tokyo, June 1978, pp.19-22.

37. For instance, RPT Assoc. Inc.,Bosai Toshi Kozo-no-kyoka ni-kansuru Chosa(Ecological Planning for Hazards in Urban Areas), National Land Agency, Tokyo, 1978 and 1979.

38. Also,RPT Assoc. Inc., Kankyo Kanri System-no Kihon Koso ni-kansuru Chosa Kenkyu (Ecological Planning as a Basic System for Environmental Management in Mie and Aichi Prefectures),Environment Agency, 1979, among several other contracted studies.

39. A phrase heard often in discussions between the author and Japanese government officials over the last 10 years(1970-1980).

40. For instance, see Kajiki,T. "Hoseika Yori Hanashi-de (Talking is Better than Laws in Regard to Environmental Assessment)", in Asahi Shimbun,Evening Edition, April 2,1979, among numerous others.

41. For example, see "Kankyo Gyosei-no Kotai Hihan (A Criticism of the Weakness of Japan's Environmental Administration)",in Asahi Shimbun, June 6,1979, among a large number of other articles.

42. Shapiro,H.A. "Coastal Area Management in Japan:Tokyo and Osaka Bays, and Seto Inland Sea", in M. Numata,ed. Integrated Ecological Studies in Bay-coast Cities,II. Chiba,Japan,1980 (in press).

43. Coastal Area Management:Approaches,Problems and Guidelines(Draft), U.N. Ocen Economics and Technology Office,N.Y.,March 1979,pp.3-4.

44. For instance, see Engan Kaiiki-no Kokyoteki Kanri ni-kansuru Hoan (Proposed Law for Coastal Area Management), Ministry of Construction, Sept. 1973.

45. Also, Kaiyo Kanri Hoan-no Seikaku (Characteristics of a Proposed Law for Marine Area Management), Ministry of Transportation.

46. Ray, G.C. "A Preliminary Classification of Coastal and Marine Environments", International Union for the Conservation of Natural Resources. IUCN Occasional Paper,No. 14,1975.

47. Coastal Area Management,Ibid.pp.3-7. This report advocates the need for an"integrated management concept"similar to the type mentioned in the author's paper.

PRINCIPLES & TECHNIQUES FOR SPECIAL AREA MANAGEMENT:
THE GRAYS HARBOR & COOS BAY EXPERIENCE

Gordon E. Davis, A.I.C.P.
Vice President
Wilsey & Ham
Portland, Oregon

INTRODUCTION

Special Area Management, as embodied in the Coastal Zone Management Act Amendments of 1980 is a concept for making collaborative planning decisions. The reason that this concept has evolved is a function of characteristics of the "special areas" to which it has been applied, and the failure of more traditional planning in those areas.

Areas that are "ripe" for such an approach are characterized by complex, interdependent and conflicting environmental, economic and political factors. Even more characteristic, however, is the area's history of conflict that has resulted in extreme positions being taken on most issues. In such situations, the traditional approach of developing a plan by combining technical determinants with planning principles and public input will not be adequate to overcome polarized public positions. Even with the strongest form of advocacy planning by staff and supporters, obtaining a public consensus in these high conflict circumstances is unlikely.

Special Area Management Planning, as a process of building public consensus through issue identification and resolution with techniques of mediation, negotiation and systematic planning, has evolved as much from on-the-ground experience as theoretical ponderings. The need to solve real problems immediately is often the impetus for this new approach. Such is the case with Grays Harbor, Washington and Coos Bay, Oregon.

The Grays Harbor program represents a conceptual model for collaborative planning and decision making. It was conceived with a sense of what was necessary to allow local, state and federal officials to reach agreement on issues. However, the Grays Harbor program did not have the benefit of a full understanding of the systematic steps and techniques necessary to make the process successful. Coming three years after the start of the Grays Harbor program, the Coos Bay program had the advantage of the Grays Harbor experience and, therefore, represents an operational model for collaborative planning and decision making.

This paper will first describe the physical and political context of these two estuaries in the Pacific Northwest. After briefly outlining the planning process and product of each program, the characteristics of the decision making group will be discussed along with the general sequence for preparing for and making decisions. Finally, in three separate sections, specific techniques will be discussed for the use of technical information, citizen involvement and conflict resolution. Throughout the paper, observations about successes and failures in Grays Harbor and Coos Bay will be discussed along with a comparison of the two programs.

GRAYS HARBOR AND THE WASHINGTON COASTAL PROGRAM

Grays Harbor is one of two major estuaries on the coast of Washington State and the only one with an authorized deep-draft navigation channel (35 feet). The estuary encompasses approximately 100 square miles, half of which are tideflats and intertidal land. The harbor watershed drains approximately 2,500 square miles from four rivers in the south-central portion of the state's Pacific coastline.

Biologically, the harbor supports over 50 species of fish and shellfish and 325 species of water, shore, terrestrial and seasonal birds. The estimated 33,000 acres of intertidal lands support extensive eelgrass and saltmarsh communities. In a thirty-five year period from 1940, approximately 3,800 acres of intertidal lands were used for dredged material disposal and are now uplands. Maintenance dredging also occurred on a continuous basis in the preceding thirty-five years although records are incomplete as to the loss of intertidal lands during that time.

Grays Harbor is an industrially developed port which has been centered on the timber industry for the last 100 years. Presently, three major pulp and paper mills operate in the harbor although at one time, as many as 35 mills (of a smaller scale) were located on the harbor. Since the 1960's, almost 400 million board feet of softwoods have been shipped through Grays Harbor annually. Approximately 34,000 people live in the communities of Hoquiam, Aberdeen, Cosmopolis, Westport and Ocean Shores, that surround the harbor. Over three-quarters of the work force are employed in the forest products industry.

Other substantial and growing sectors of the harbor's economy are commercial and sport fishing as well as tourism. Almost one quarter of the state's total annual charter trips originate from Grays Harbor, and Ocean Shores attracts as many as 50,000 visitors on holiday weekends to its long expanse of beaches.

In 1976, under the Washington State Coastal Zone Management Program, the Grays Harbor estuary was designated as "an area of special concern," and under the state's Shoreline Management Act, as a "shoreline of statewide significance." The state cited in its management program the need to "maintain a balance between the filling of intertidal areas and the preservation of wetlands."

Under the Washington management program, each city and county prepared and received state approval for a Shoreline Master Program. The county and city master programs prepared in Grays Harbor were the first approved under the state-wide program, however, they did not resolve conflicts between local governments, the port, and the state and federal resource and regulatory agencies over major development proposals in the harbor. The agencies claimed that the individual plans were done without adequate technical information, that they were too general to provide adequate guidance for permit decisions and that they were done without the agencies direct involvement. Many people at the local level claimed that the plans were too restrictive and did not provide adequate opportunity for growth. The conflict came to a head over a proposal to fill 45 acres of salt marsh for an off-shore drilling platform fabrication facility. Although this proposal was eventually approved, it was contingent on agreement by all parties that no further permit actions could be taken without the completion of a single plan representing agreement by all.

In 1976, an Estuary Planning Task Force was formed by the Grays Harbor Regional Planning Commission (an association of local governments similar to Councils of Government in many states). The Task Force, composed of representatives of each of the local jurisdictions, the port and the key state and federal resource and regulatory agencies, with the help of a consultant, began a four-year, five step process to develop an estuary management plan. The nucleus of the planning process was a series of six, two and three day workshops held over a sixteen-month period. Through a mediation process, the consultant led the Task Force through a systematic effort to:

1) Define an estuary-wide management goal
2) Describe the characteristics and establish guidelines for managing resources and development within eight sub-areas in the estuary, and
3) Define management objectives, allowable uses and activities, and special conditions within 44 specific management units that encompass the entire estuary.

Many issues were cause for debate, however, one issue, a proposal to fill 500 acres of intertidal lands, took two days of negotiating and four additional days of work to revise and refine the final agreements. The initial agreements were reached in the summer of 1977 and a draft plan was published in early 1978. Then began a lengthy review and revision process that will result in a revised public draft in late 1980.

COOS BAY AND THE OREGON COASTAL PROGRAM

There are 22 estuaries on the Oregon coast. Of these, only three are authorized and developed for deep-draft navigation. Coos Bay, the largest estuary on the Oregon coast (excluding Columbia River), has a 35 foot channel and encompasses 13,300 acres. It drains a watershed of approximately 650 square miles with one major river and 30 tributary streams.

Coos Bay is divided into three distinct characteristic areas. Lower Bay, from the jetties to river mile 9 is relatively narrow, ranging in width from approximately 2,000 feet to slightly over a mile. Its hydrology is dynamic and contains fringe area tideflats and intertidal lands. Scattered saltmarsh and eelgrass beds exist throughout this area. Upper Bay, the current and historic center of the developed portions of the bay is relatively wide, ranging up to 2 miles wide in some areas. The majority of the Bay's 6,200 acres of tidelands are in this area. The third portion of the bay is a series of eight sloughs, four of which are river-like in their configuration. The upper portion of South Slough, encompassing an aquatic and watershed area of 4,400 acres, was designated as a National Estuarine Sanctuary in 1977, by the Oregon Legislature.

Like Grays Harbor, Coos Bay is a resting, feeding and wintering ground for migratory birds on the Pacific Flyway. Over 250 species of birds are present at various times during the year. Numerous species of fish and shellfish populate the bay with heavy harvesting of clams and oysters in the lower and mid bays.

The northern configuration of Lower Bay is formed by North Spit, a stabilized dune area, approximately 6 miles long by less than a mile wide. This land mass, rich with a diversity of dune and marsh habitat is also the focus for future development pressures in the bay. Its immediate proximity to the navigation channel and short distance to the mouth of the bay, coupled with the need to shift the development focus from Upper to Lower Bay because of the lack of adequate dredged material disposal sites in Upper Bay, makes the large land expanse of North Spit most attractive for diversifying the area's economy.

Coos Bay is an industrial harbor and like Grays Harbor, is centered on the timber industry. Three major mills along with several smaller operations export in excess of 5,000,000 tons annually of which 70 percent are chips destined for pulp mills in Japan. Oregon, and particularly southern Oregon is approaching the end of its old growth forests resulting in declining timber harvests around Coos Bay. Thus, the community is actively working to diversify its economy.

As a result of the 200 mile limit, Coos Bay anticipates expanding development of the off-shore fishery. Presently, the existing commercial and sport fishery ranks second in size in the state. Half of the commercial fishery is salmon while the sport fishery includes hard and soft shell clams, dungeness crab, shad, striped bass, salmon, steelhead and other popular species. If current Port plans are realized, 200 acres on North Spit will be developed into a major fish processing center, adding substantially to the role of the commercial fishery in the Coos Bay economy.

In 1973, the state legislature established a State-Wide Land Use Planning Program and in 1976 added a coastal component to that program. The principal feature of the program is coordinated comprehensive plans prepared by cities and counties in conformance with prescribed state planning standards called goals. In all, 19 goals direct local governments to deal with issues such as housing, transportation, agricultural and forest lands, urbanization, natural resources, and numerous other topics. Four goals deal directly with coastal resources and establish standards for planning estuarine waters, coastal shorelands, beaches and dunes, and ocean areas.

The Coos Bay area, through an extensive citizen based planning effort, completed and adopted a comprehensive plan for the bay in 1975. Upon evaluation, however, it was found inadequate to comply with the state guidelines. Also, because it lacked direct state and federal agency involvement and subsequent ratification, those agencies refused to accept it as a guide for evaluating state and federal permits, even though a request was consistent with the local plan.

The 1975 plan was scheduled for revision by mid-1979 along with the rest of the county's Comprehensive Plan. By late 1978, it was clear that a consensus was not forthcoming on how to develop an acceptable estuary plan. Based on the Grays Harbor experience, a consultant was hired to manage a process of preparing an Estuary Management Plan.

An Interagency Task Force of local, state and federal officials began work in July, 1979. By July, 1980, the Task Force had reached its agreements and directed the consultant to prepare a first draft of the plan for revision and ultimate adoption by the end of 1980. In all, the Task Force met in 18 two-day workshops to complete its agreements. These agreements are contained in a complex array of approximately 144 aquatic and shoreland management units.

GRAYS HARBOR - THE CONCEPTUAL MODEL FOR SPECIAL AREA MANAGEMENT

Grays Harbor was conceived as a comprehensive planning process moving systematically from broad estuary-wide issues to site specific decisions on use and activity standards. The means to make these decisions was a Planning Task Force consisting of local, state and federal officials, each with some responsibility for the management of resources or the regulation of activities in the estuary. The process began with the preparation of a comprehensive data base of natural and human factors existing throughout the harbor. Those data, reproduced as a series of maps and supporting technical memoranda, were to be the information base for Task Force members to use in arriving at their specific management decisions.

The decision making and plan format were designed to be relatively simple, beginning with a broad, estuary-wide management goal "to manage the estuary for multiple uses." With this goal established, the entire estuary was broken down into eight sub-areas or planning areas, each representing a common set of natural and man-related factors. As each planning area was described in terms of its predominant character, major committed uses, existing or potential conflicts and assets, guidelines were also established to identify how the planning area's natural resources would be managed as well as the economic role it would assume in the future.

With the planning area guidelines established, an overall picture of the harbor emerged to guide the site specific decisions that would be made at the next management level, the management unit. Forty-four management units were established, most representing segments of the shoreline. Each management unit is described by boundaries perpendicular to the shoreline, a boundary parallel to the shoreline in the uplands, and a boundary in the aquatic area at the line of Ordinary High Water. The remainder of the aquatic area is essentially one management unit although there were several special aquatic units designated.

After establishing the boundaries, each management unit was given a management classification. While there were eight classifications to choose from, in general the Task Force had to decide whether the unit should be considered for urban, rural, conservation or natural uses. The assigned management classification included a standard set of allowable uses. Following establishment of a management classification, the Task Force defined a management objective, assigned allowable activities and established any special implementation conditions that it felt were necessary to carry out the unit's management objectives.

By first establishing a direction at the broadest management level, the planning process narrowed the range of choices for each succeeding more detailed management level. It was presumed that it would be easier to gain agreement on more general management policies first than to try to immediately attack site specific management policies, the level at which most of the standing conflicts existed.

The consultant's role was establishing the general framework for decision-making, supplying much of the technical information, and documenting the specific agreements reached. The consultant was to also act as a facilitator to move the Task Force to arrive at consensus decisions. However, because this was the initial effort with the process and the consultant did not have a full understanding of the past conflicts, the process rapidly reached a point of mediating each decision rather than consensus building.

While agreements were ultimately reached and the plan is now in its final stages of completion, the conceptual decision making model that Grays Harbor represented did not work well operationally. A number of factors contributed to the operational problems including: task force members did not understand the role they were being asked to play in this process as a planner/negotiator versus their normal role as regulatory; individuals did not have a clear understanding of the legal and regulatory constraints of other members; technical information was not tailored to the types of decisions that had to be made; agreements reached conceptually by the Task Force were difficult to recall when reviewing specific language documenting that agreement; individuals had different levels of understanding of the issues involved; there was insufficient time for individuals who had developed a general mistrust of each other over years of regulatory conflicts to believe that they could come to an agreement; and leadership roles were not nurtured early enough in the process.

COOS BAY, THE OPERATIONAL MODEL

The Coos Bay Estuary Management Plan represented the opportunity to take the lessons learned from Grays Harbor and build a new approach for preparing special area management plans in high conflict areas. There are numerous differences in the actual content of the Coos Bay plan, some of which are the result of the differences in the two state programs. Others, however, are the result of attempts to correct problems encountered in Grays Harbor with documenting management decisions, boundaries, aquatic versus upland management units, use and activity standards and other items. None of these more technical changes will be discussed in this paper. Rather, this paper will describe the sequences of steps used in Coos Bay to prepare the Task Force for and to better allow them to make decisions. In addition, several key techniques in the Coos Bay process will receive further elaboration.

THE DECISION MAKING MODEL

There are two primary considerations in designing a decision making model such as that used in Coos Bay. First, what is the ability of the group to make decisions, and second, what is the sequential process that the decision makers must go through.

The Decision Making Group

In Coos Bay, the decision making group was an Interagency Task Force made up of representatives of the local, state and federal governments. The local contingent were elected officials from each of three incorporated cities, the county, and the port. In contrast, in Grays Harbor the local representatives were not necessarily elected officials. From the state there were representatives from the fish and wildlife agency, the agency responsible for regulating fill and removal and the agency charged with protecting environmental quality. The federal representatives were the National Marine Fisheries Service, the Fish and Wildlife Service, the EPA and the Corps of Engineers. In addition, two "citizens-at-large" were asked to serve on the Task Force. The state agency responsible for managing the land use and coastal planning program served as an ex-officio member. Each Task Force representative also had an alternate, many of whom attended most or all of the Task Force workshops. Because the management decisions were to be made by consensus, considerable time and effort was required to prepare such a diverse group for decision making. Factors involved in this preparation include:

1. Belief in the Process

   Since there may be substantial outside pressures on individuals and the group as a whole to reach agreements, it is necessary that the participants understand and believe that decisions can be made, otherwise they will not be able to make decisions. In general, individuals must believe that they will be able to achieve what they or their constituency would like to see happen. There are many ingredients in achieving this belief, however, being able to actually see the group make a decision even if it is a small one is the best technique.

2. Knowing the Players

Ultimately for the group to be effective decision makers, they are all going to have to know each other in terms of their:

- Authority - who they represent, how much direct authority they have, how willing they are to go beyond their specific authority to see issues in full perspective, etc.
- Constituency - in addition to the actual authority an individual may possess, some members may be viewed as representing, in fact or by inference, some other external interest. Such outside constituencies may substantially effect the individual's perceived authority in the group.
- Style - everyone has a unique style for dealing with other people; some people come "off-the-wall", others take a hard line to the bitter end, some agree with all positions, some don't say anything until the end, etc.
- Philosophy - everyone has some basic philosophic position, some are very obvious, others are more obtuse.
- Dependability - to what degree can one rely on a person to take a certain position in a situation. Inconsistency will ultimately result in exclusion from the mainstream of decision making.
- Coalition Potential - how willing is the person to develop coalitions to arrive at agreement.
- Knowledge - as people's areas of knowledge become recognized, the group may tap that information regardless of the philosophic position of the individual.

3. Leadership

While consensus decisions are reached through discussion, the discussion process is most effective when there is a coalescing of positions through a spokesman or leader. The natural tendency of a group is to evolve toward identification of a leader. In a group with differing objectives, more than one will evolve. In addition, depending on the personalities involved, leadership may change from issue to issue. Establishing leaders is essential to effective decision making and may require some nuturing by the facilitator. In the absence of leadership, the facilitator can play that role on a limited basis, however, dependence upon this technique will erode the facilitator's effectiveness and may not lead to as strong of a consensus.

The Decision Making Sequence

There are five major steps to the decision making sequence in a Special Area Management Process such as Grays Harbor or Coos Bay. While there is a general order to these steps, in practice there is a great deal of overlap and recycling between steps.

1. Orientation

The stated purpose of this step is to orient the decision makers to the physical setting they will be dealing with, the entire decision making process and other procedural matters. The unstated or secondary objective is to simply get the group together to begin their preparation for decision making.

Observations: An initial orientation does little more than get the group together. Aside from the knowledge they may already have on the issues, following an orientation session, they will have little recall of the specifics of either the physical setting or the program. It will be necessary to reorient the group many times during the process, particularly when they start dealing with specific issues. An effective technique that accomplishes many objectives is to ask members to informally visit specific areas during lunch breaks or before or after sessions with other members that know the area or the issue. Formal field trips are useful but time consuming, and offer less opportunity for one-on-one exchanges.

2. Issue Identification

Occurring early in the process, this step will allow the group to identify, classify and prioritize issues for resolution. In addition, they will be able to identify the types of information that they will need to deal with the issue and how they want that information brought to them (see section on Technical Information for further elaboration). The unstated objective of this step is to provide an early opportunity for the group to agree on something, even if it is only on what they disagree about. In addition, it is an opportunity to get all the issues out on the table so everyone gets a sense of what is coming.

Observations: This may be more difficult than it sounds. Each member knows there are issues, but they are often unclear about the real issue. In addition, there is a reluctance to suggest some issues for fear of "laying your cards on the table." In Coos Bay, no one suggested that development of the North Spit would be an issue, yet at that time the Port was in the middle of a major permit battle for a "T" dock that would allow the development of a major portion of the Spit.

3. Background Information

The knowledge of the decision maker concerning the issues, the setting and the legal and regulatory constraints will be variable and incomplete. This education process will require much repetition and review, particularly when dealing with specific issues.

One of the most important areas of necessary background information is a complete review of the legal and jurisdictional constraints that have bearing on the planning areas.

Observations: Do not assume that state or federal agency representatives know anything about local land use controls. Conversely, do not assume that local officials really understand state and federal regulations and procedures. Even those who should know, rarely do, and often if they do understand the regulations, they do not trust their application. One exposure to the subject of the regulatory process will not be sufficient; many repetitions will be necessary.

The other necessary background information is technical data on the physical, biological and economic setting. This subject will be discussed in a later section in more detail. However, while it may be desirable to develop a comprehensive geographic data base, if that information is not in the format or detail necessary for decision making, it is of little use. Additionally, it is important to remember that if a Special Area Management Program is characterized by a high conflict situation where issues are polarized, technical information will play a less significant role in resolving conflicts between public interests than in a traditional planning process.

4. Decision Making Tools

Prior to embarking on substantive decisions, decision makers must be equipped with a vocabulary of management tools. Management categories, definitions of terms, allowable uses and activities, development standards, are all items that decision makers must agree to and understand.

Observations: The desired level of predictability will be the primary factor that will influence the selection of decision making tools. The tools that are chosen will determine how specific management decisions will be in the plan and what questions will be left to be answered in the permit process. Task Force members will tend to insist on being very specific in making planning decisions and therefore ensure a high level of predictability. However, to do so may require more information than is or can be made available to the planning process. The temptation to deal with project specifics in the planning process is great but will not be possible with tools that are designed to be more universally applicable.

5. Making Decisions

The approach for making management decisions will be shaped, in part, by the specific format chosen for the plan. However, the concept of beginning with broad level decisions and using them as constraints to shape more detailed, and presumably more conflicting decisions is universally applicable to group decision making. The process can be structured as follows:

Sub-Areas

Breaking the planning area into sub-areas organizes a large complex area into smaller, more understandable pieces. Criteria such as physical boundaries, natural features, ownership, conflicts, political jurisdictions, and existing uses can begin to differentiate the sub-areas within the planning area. Having decision makers agree on the areas, describe their characteristics and establish general guidelines for managing the resources and defining the economic role of the areas, creates a general policy framework for the planning area and at the same time progressively increases the level of difficulty of decision making.

Observations: Of all steps in the process, this has been the least successful. Because this is the Task Force's first substantive decision making experience, they have found it very difficult to lay out even general policy guidelines for the future of an area. They immediately realize the implications of the policy guidelines on more specific decisions that will be made at the management unit level. Without knowing the details of management unit decisions, Task Force members are reluctant to foreclose any options, even at this very general level. The experience in Grays Harbor was generally more successful than in Coos Bay although both had similar problems with completing the sub-area process. The facilitator should be prepared with alternative approaches or techniques at this step.

Management Units

The major objective of the process is to make decisions at the management unit level. At this level, the geographic area may be quite small (in Coos Bay one management unit included less than 1,000 feet of shoreline) and the management decision may be very detailed. At this point, the decision makers should be fully prepared with the knowledge of the issues, the information that bears on the issues, and the decision making techniques necessary to make decisions. The format for the specific management decisions will be determined by the management approach and tools selected. However, again a progressive decision making approach appears to be most effective.

Observations: This is the most time consuming step of the process. In Coos Bay this step took nine workshops over a five month period and a total of 19 full days for the Task Force to reach agreements on 144 aquatic and shoreland management units. Of that time, three days were spent on a specific conflict resolution session (see section on Conflict Resolution for elaboration). No matter how much time has been spent preparing decision makers with knowledge about their management tools, the issues, and background information, they do not begin to fully understand the process until they are required to apply it. Therefore, the general format for dealing with each management unit included: 1) a review of the existing physical, biologic and human uses and conditions, 2) a public comment period (10 to 15 minutes) for non-Task Force members, 3) a review of unit boundaries, 4) establishment of a management classification and management objectives, 5) specification of allowable uses and activities, and 6) identification of any special conditions that the Task Force believed were important to achieve the management objectives.

TECHNICAL INFORMATION

In a traditional planning approach a systematic evaluation of technical data plays a primary role in shaping planning decisions. However, as stated previously, "special areas," necessitating the application of the principles of a Special Area Management Program, are areas where traditional planning has failed because of severe conflicts that have polarized public interests. In many cases, technical determinants will not overcome conflicting interests. In fact, getting agreement on the adequacy and significance of data may be as difficult as reaching agreement on the issues themselves.

Planning decisions in a Special Area Management Program are often compromises. Reaching a consensus decision may be dependent upon a realization by Task Force members that it is acceptable and even advisable to make decisions in conflict with the technical data. For example, accepting a resource loss (rejecting a technical determinant) may be acceptable within a Special Area Management process simply because of the guarantees for resource protection gained elsewhere.

In spite of the need to be able, at times, to reject technical determinants, technical data can play a central role in many Special Area Management decisions. The degree of its effectiveness in assisting the Task Force in decision making is dependent on the data being formated in a manner that directly bears on the decision being made.

In the Coos Bay program, the data system was substantially changed from that used in Grays Harbor. Like Grays Harbor, a "bay-wide" resource inventory was prepared at a relatively large scale (1"=3000', 1"=1500') covering a broad range of resource and human use topics. However, in Coos Bay, the Interagency Task Force reviewed the issues early in the process and, therefore, could also review the information requirements that would be necessary to deal with the issues. For some issues, the Task Force believed that the broad resource inventory would be sufficient. Some issues were so complex or controversial, that the level of detail of the information needed had to be substantially increased. Five issues (generally specific areas within the Bay) requiring detailed information were identified and Specific Area Assessments were conducted in each, bringing the level of detail in most cases up to a scale of 1"=400'.

Other issues, while more complex than could be dealt with through the broad resource inventory, did not require a new data collection effort. These issues simply required that knowledgeable people be assembled to deal with the issue through a systematic work program designed to: 1) fully define the issue, 2) describe the alternative courses of action that the Task Force could take to deal with the issue, and 3) describe the implications of choosing one alternative over another. This third form of data collection also served as one of the key elements in "citizen involvement" since the resultant Functional Task Forces were composed of industry representatives, interested citizens, other special interest groups and agency (principally state) technicians.

Perhaps the most consistently useful information to the Task Force was aerial photographs. The state had recently contracted with the Corps of Engineers to fly the coastal zone. Composite color aerials were prepared at a scale of 1"=1000'. These aerials were used throughout the decision making phase and management unit boundaries and other information was placed on them for on-the-ground referencing. In addition, selected enlargements were made to form the base sheets for each of the five Specific Area Assessments. Finally, as specific issues were deferred to Conflict Resolution, additional enlargements were made of the specific areas involved. In the course of discussion on specific issues or management units, it was no uncommon for Task Force members to cluster around an aerial photo to agree on boundaries or other issues.

CITIZEN INVOLVEMENT

Citizen involvement in the planning process is designed to obtain their knowledge, ideas and values (priorities). The Grays Harbor program has been criticized for having insufficient citizen involvement, although numerous techniques were used throughout the process. Perhaps the most severe criticism has been in the inability of citizens to be directly involved with the Task Force in their decision making process. While the Grays Harbor sessions were open and members of the public did attend, direct participation by the public was limited. It should also be pointed out that while there was considerable interest by special interest groups in Grays Harbor, there was very little general public interest expressed through the several processes designed to gain their involvement.

In Coos Bay, the interest of the general community and special interest groups in the planning program was very high. In addition to town halls, community questionnaires, estuary information events, hearings and other typical techniques, several techniques were used to allow direct citizen involvement with the Task Force. First, the Task Force itself included two citizens-at-large appointed by the County Commissioners. While local elected officials theoretically represent the citizen constituency, many are often perceived to have a certain bias. Citizen appointees may also have a bias but their non-elected status provided a balance to the Task Force that was beneficial.

The Functional Task Force process already described in the section on Technical Information used in Coos Bay was a major opportunity for interested citizens to discuss issues and present conclusions directly to the Task Force. The reports from these groups were used extensively by the Task Force in making decisions.

Of particular importance in Coos Bay was allowing public comment at the beginning of the discussion of each management unit. This provided citizens a direct opportunity to tell the Task Force what they felt its decision should be. While it considerably lengthened the time for discussion, it had a significant influence on the decisions that were made.

Finally, the Task Force itself conducted a Town Hall just prior to embarking on the conflict resolution session. In this very spirited meeting, the Task Force was asked to defend many of its decisions and was given opinions on the issues that remained to be resolved. In the Fall of 1980, the Task Force will hold another Town Hall to receive comments on the draft plan before they direct the consultant on the contents of the revised plan.

CONFLICT RESOLUTION

In Grays Harbor, as decisions were incrementally made around the estuary, each issue was carried through to a final agreement. As major issues were confronted, the process moved from a simple consensus discussion into direct negotiation. In doing so, the Task Force began to adopt a negotiating approach for dealing with all issues and as a result, substantially increased the strain of making decisions. Issues became a conflict too quickly and did not allow a more open ended consensus discussion to be fully effective.

In Coos Bay, to ensure that conflict resolution through negotiation would be used only with those issues where there were truly a conflict, it was decided that a specific conflict resolution session would be programmed at the end of the decision making process. Issues that became true conflicts in the course of making incremental decisions around the bay would be deferred to the end of the process. To ensure that only absolute conflicts were deferred rather than every issue on which there was some disagreement, the facilitator made the determination as to when the Task Force had reached such a conflict. In several instances, when Task Force members suggested that an issue be deferred, the facilitator refused to allow it, believing that a consensus could be achieved or was close at hand. Out of the 144 aquatic and shoreland management units, six were deferred to conflict resolution.

To prepare for the conflict resolution session, the deferred management units were categorized by issue of conflict. In this case the issues were moorage (four units) and industrial lands (two units). Special background information was prepared for the Task Force including a review of the existing situation, the decisions the Task Force had made elsewhere in the bay on the issue, and a summary of relevant technical information from the work of the Functional Task Forces, the resource inventory and other available data. During the conflict resolution session, additional technical expertise was brought in to supplement available materials. Over a two day period, conceptual agreements were negotiated; these were then finalized through special sub-committee work and approval of final drafted language.

With the benefit of the perspective of the decisions the Task Force had already made throughout the bay and full confidence in their ability to make decisions as a group, the conflict resolution process was relatively easy in comparison to the very strained sessions in Grays Harbor.

CONCLUSION

The nation's coastal zone contains the most complex array of natural resources, economic values and political and jurisdictional interests of any area in the country. Any planning and decision making process dealing with this area will find difficulty in affecting a balance between such diverse interests. Add to this complexity, a history of conflict and polarized public interests, and the traditional approach to decision making will be ineffective in resolving issues and developing a public consensus. The Special Area Management Program is designed for these areas and circumstances. It is an approach that is not shaped by physical, economic or political determinants, but by the need to reach agreement. The conceptual and operational models developed for Grays Harbor and Coos Bay have established a format on which future Special Area Management Programs can be based. The major task ahead in those programs will be to perfect techniques to make the process more effective in reaching a public consensus.

CONFLICT RESOLUTION: LESSONS FROM GRAYS HARBOR

Nan Evans *
U.S. Department of Energy, Region X, Seattle

Coastal zone management in the 1980's will be increasingly challenged to further coordinate complex government decision making systems in order to eliminate inconsistencies, uncertainties and obstructions. The needs of growing coastal populations; increasing competition between alternate, and sometimes equally necessary, uses; national priorities to accommodate, for example, both exploration for needed energy resources and protection of endangered species populations; and the urgency of protecting scarce resources and vulnerable ecosystems are only a few of the forcing functions which heighten the importance of developing successful management strategies and systems. Meeting the challenges of the 1980's will depend upon the ability of coastal managers, users and protectors to learn from past experiments. Analysis of these experiments will allow the adaptive evolution of new forums for decision making and the requisite adjustments to the traditional relationships between agencies, between levels of government and between competing interests. The purpose of this paper is to examine one such experiment in coastal management for its ability to resolve coastal resource use conflicts.

In 1975 this experiment in coastal zone management began in Grays Harbor, Washington. Grays Harbor, an estuary on the Pacific coast of the state of Washington, is a major port and industrial center for lumber and wood products. Grays Harbor supports major commercial fishing and tourism industries. The biological communities of the estuary and its drainage basin provide the resource base for the primary economic sectors in the Grays Harbor County. Grays Harbor is also one of only five major estuaries on the West Coast of the United States and, therefore, provides crucial biological functions such a nursery areas for marine animals, wintering grounds for migratory birds, and nutrient supply and regeneration for aquatic ecosystems.

Grays Harbor has a long history of resource use. Disputes over resource allocation and use have been persistent and have involved a multitude of issues and actors. Specific and heated conflicts in Grays Harbor are similar to those observed nationally. Each of the conflicts involves disputes over <u>what</u> are the most important and valuable resources, <u>who</u> should determine the character and extent of uses affecting the estuary, and <u>how</u> the resources and their uses are to be managed or controlled. The parties to these resource conflicts include a wide spectrum of actors: Federal resource and regulatory agencies, State resource management agencies, local governments, a port district, private landowners and developers, citizens, and environmental groups.

---

*This paper is based upon a study in which the author participated while a staff researcher at the Institute for Marine Studies of the University of Washington. The major technical report from this study is *The Search for Predictability: Planning and Conflict Resolution in Grays Harbor* which is being published by the Washington Sea Grant Program.

The Grays Harbor Estuary Planning Task Force was formed to resolve these disputes and to avoid such conflicts in the future. Unlike many traditional planning efforts, all agencies and governmental units with decision making responsibilities in the estuary participated in the Task Force. The Task Force undertook to produce a comprehensive management plan for the estuary which would provide a predictable and unobstructive permitting process for shoreline use and development and would provide assurance that the long term viability of the estuary would be maintained.

The Grays Harbor Estuary Management Program did not remain merely an innovative approach to comprehensive planning. The novel composition of the Task Force, the complex decision making environment in the region, and the intensity of specific conflicts moved the dynamics of the effort increasingly toward those of bargaining and conflict resolution by negotiation.

The Grays Harbor effort resides somewhere in the middle of a continuum from conflict resolution by negotiation, such as labor-management arbitration, to problem solving by planning, such as comprehensive planning by an advisory commission. The challenge of coastal managers is to successfully integrate these two critically important, but different, processes in a manner which will attempt to resolve conflicts through comprehensive planning.

Differentiating Comprehensive Planning and Conflict Resolution Through Negotiation and Bargaining

Comprehensive planning as it has evolved in the United States usually involves an advisory body to governments on policies for growth or resource use. "Comprehensive", of course, implies representation of a broad spectrum of public interests and applicability to all aspects of community development and resource use. "Planning" implies the establishment of long-term goals for development and resource use which would allow short-term, specific decisions to fit into a broad, holistic perspective for the future. The plan itself is usually a graphic and written description of uses or activities appropriate for specific regions in the future. General policies about implementing the plan and achieving the desired end-point are also included.

The Grays Harbor Estuary Management Program has all the elements and expectations of comprehensive planning efforts. Memos from the consultants to the Task Force address essential management concepts: a reason to manage, agreement on what is to be managed, decisions on how to manage. The consultants guided the Task Force through a process designed to establish overall estuary management goals, to agree upon the management approach or level, to define criteria for Management Unit boundaries, and to identify use concepts. "The end objective of [the] management plan [was] to provide enough guidance to an elected or agency official to be able to evaluate a specific use or activity and make a decision on that proposal" (Montagne-Bierly, 1977). The end product of the effort is a Management Plan complete with an estuary management goal, general policies, Planning Areas, Management Units, and standard use matrices. (Grays Harbor Estuary Management Plan, 1978).

Modern comprehensive planning, however, often experiences difficulties satisfying the expectations held for it. A major challenge for comprehensive planning is to resolve conflicts between differing community goals. However, such goals (a) may be undeterminable using standard planning methods; (b) probably cannot be completely described regardless of the sophistication or resources available for information gathering and analysis; (c) may have different time scales of concern; (d) may be intricately and inextricably interwoven into the political framework of the region; and (e) may certainly result in intense conflict unless differences are resolved. In such situations the need for additional conflict resolution techniques intensifies.

The group of intergovernmental agencies and decision makers that comprised the Grays Harbor Estuary Management Task Force were faced with a situation having many of these characteristics. Decision making had become highly politicized and the Task Force became an expression or extension of those interests that, as in all political systems, were competing for power, influence and resources. It is little wonder that the expectations held by some Task Force members for a planning activity were overwhelmed by the need for complex negotiation.

The fact that comprehensive planning in Grays Harbor, or elsewhere, is beset with difficulties or inadequacies does not imply that such attempts should be abandoned in favor of a purely incremental decision making system that "muddles through" by making small changes, fragmented choices, and remedial decisions. Instead, strategies for coordinating public and private agendas and for adapting governmental and private decision making to the human environment may be possible through a skillful integration of comprehensive planning and conflict negotiation by bargaining and negotiation.

Conflict resolution by negotiation requires that the parties to the conflict work together in an attempt to reach a mutually agreeable solution, usually involving compromises. Resource management disputes which are resolvable via negotiation procedures are generally those where room for compromise solutions exist and the compromises are capable of being implemented at a local level. In general, negotiable disputes are those in which the primary questions are "where", "when", and "how much" rather than "what should" or "who should". Also, the negotiation process does not guarantee an agreement or set legally binding precedents for future disputes (Cormick and Patton, 1977).

Two questions are important when the Grays Harbor effort is recognized as dynamically situated along a continuum from conflict resolution by negotiation, bargaining and arbitration to comprehensive planning by advisory committees. First, how well was the collaborative comprehensive planning effort designed to resolve persistent and complex conflicts? And secondly, what lessons for future efforts which attempt to integrate these two processes can be derived?

The description of conflict resolution by negotiation used above identifies four major components:

1. The parties to the conflict work together toward a mutually acceptable solution;

2. A solution will usually require compromises, therefore, the parties to the dispute must have room to compromise;
3. The solution must be capable of being implemented; and
4. A solution is not guaranteed.

These four components will serve as the analytical framework for examining the Grays Harbor effort in an attempt to answer the above two questions.

Parties to the Conflict

Relevant parties to resource-use conflicts are those who *define themselves* as significant actors and who *exercise some control* over any of the resources or the decision making process in the dispute, and/or *have the ability to impede* the activities of another party. In any successful conflict resolution (e.g. labor negotiations, international treaties, or environmental disputes), the negotiation team should include all parties to the conflict who have some relative power that they are willing to exercise.

When traditional planning efforts are usually undertaken by an advisory group or a staff of professional planners, plans or policies are recommended to elected or appointed government decision makers. Those doing the planning often do not have any direct control over the resources at stake and they usually have limited or no statutory authority to implement or enforce a plan or its policies. The Grays Harbor effort was innovative in that the Task Force was an inter-agency group with considerable control over the resources and the decision making system.

The restriction of membership to government agencies reflected the perceptions of the conflict situation as one of administrative incompatibilities and inconsistencies. For example, individuals who had experience with the Corps of Engineers' Long-Range Maintenance Dredging Study saw conflicts arising around each large development project in Grays Harbor because:

> the responsibility for making decisions...falls to a variety of local, state, and federal agencies. Each agency has a plan, guidelines, or regulations to decide what may or may not be done; cities and the county have their own comprehensive plans and shoreline master program. Normally, each entity has prepared their plan or guidelines independently, with resulting inconsistencies, conflicts, and complex process of decision making (Weinmann and Malek, 1978).

Membership was also limited to representatives of governmental entities in recognition of the need to keep the Task Force small enough to be able to reach decisions efficiently. Another assumption underlying this limitation was that the range of participating governmental entities could adequately represent all of the public interests.

The parties to the conflict situation in Grays Harbor and the members of the Task Force are not completely overlapping sets. Some important parties are excluded, such as, the Federal Aviation Administration, environmental groups, chambers of commerce, fisherman associations and some of the small towns in the region. Any planning, bargaining, negotiation, and conflict resolution is bound to be extremely difficult (and it was in Grays Harbor), and perhaps impossible, without inclusion in the Task Force of all relevant parties to the conflict. Implementation of the Plan may also be extremely difficult without the commitment that could have come with involvement of the excluded, but relevant, parties.

Disagreements Over Task Force Composition

Actual agency composition of the Task Force was not without its own conflict. Early in the process, local governments questioned the right of Federal agencies to be "at the table" and to make decisions that affected local government behavior or that were counter to local wishes. On the other side, Federal resource agencies questioned the ability of local governments to plan and to make decisions that gave adequate recognition and value to natural resources. Considerable time and effort during the early Task Force meetings were devoted to establishing the right of the separate governmental parties to participate.

One of the remaining disagreements facing the Task Force over the right to participate surrounds the decision to exclude nongovernmental entities. The Task Force believed this exclusion would result in a more efficient process without sacrificing public concerns, since the public would be represented by elected officials and public servants. Also, during the early stages of the planning process, the consultants to the Task Force interviewed approximately sixty-five persons from the community and special interest groups. The consultants provided written summaries of the social, economic and resource use concerns stressed by interviewees (Montagne-Bierly, 1977). Public reactions would also be sought after the plan was drafted. Further, the Task Force believed that the public had other opportunities to affect decisions in the region through the NEPA and SEPA processes (National and State Environmental Policy Acts).

The decision to exclude public and nongovernmental representatives from the Task Force has made the process subject to strong criticism and challenge on legal, ideological and political grounds from environmental groups. They argue that, in order to be implemented, the Grays Harbor Estuary Management Plan is to become part of the Federally-approved State Coastal Zone Management Program (CZMP). Section 306(g) of the Coastal Zone Management Act (CZMA) requires full participation of all interested parties, public and private, in the development of State CZMP's and in the amendment of an approved State program. Environmental groups charge that exclusion of the public from the Task Force violates this criterion for Federal approval. The Office of Coastal Zone Management (OCZM) believes that, while there may be other, or even better, ways to meet the requirements of full public participation, the fact that the Task Force menbership was restricted as it was cannot be construed to be a violation of the CZMA (Kifer, 1980).

As a fundamentally ideological criticism, environmental groups, particularly Friends of the Earth, are objecting that the Task Force refused to recognize the capability of citizens to discuss complex problems in the spirit of cooperation and to identify possible solutions and trade-offs.

The Grays Harbor Plan may be jeopardized by the exclusion of nongovernmental entities from the Task Force because the excluded parties lack the commitment to the Plan shared by the Task Force members and the understanding of why certain trade-offs are included in the Plan. Excluded parties may attempt to block implementation of the Plan by bringing a law suit challenging its legality.

Alternative Task Force Membership

In addition to the need to include all parties to the conflict in the negotiations, two additional improvements to the membership of the Task Force are suggested.

In order to utilize the special expertise of agencies which may not consider themselves as parties to the conflict, a technical or reference team can be formed to provide assistance to the negotiating team. The technical team could be an integral part of the Task Force sessions by continually watching over developments to assure that, for example, the agreements and conditions agreed upon by the Task Force would not threaten the continued approvability of the local Shoreline Master Programs.

In addition to providing a mechanism for recognizing the concerns of neutral agencies, the use of a non-negotiating technical team could provide a mechanism for the continued involvement of resource specialists, legal experts, and special interest groups. Formation of a participating, but non-negotiating, technical team could also provide the means of involving many concerned parties without making the negotiating team too large and cumbersome to be effective.

Another alternative arrangement could be to provide a mechanism whereby the membership of a special area management negotiating team could be dynamic and responsive to changing conditions. For example, by mutual and prearranged agreement, parties to only a specific component or subconflict could become voting members on these issues. Or, for specific types of questions and issues, a member of the technical team could exercise a vote or perhaps veto power. Or, if negotiating a special area management requires a number of years, a flexible membership agreement could provide a mechanism for including parties newly defined by changing laws or by newly perceived challenges to their interests.

To be effective, a dynamic membership arrangement would probably have to be designed and mutually agreed upon during the phase before the negotiation of specific conflicts actually begins. Also, in order to avoid spending extra energy bringing entering members up to date, the entire process would need to be public and well documented. A possible drawback to a dynamic membership arrangement could be the lack of previous opportunity for an entering member to develop the trust of the other members and a commitment to the process.

Room for Compromise

Conflicts which are strongly ideological offer little room for compromise and, therefore, are poor candidates for resolution via negotiation. As a matter of definition, political disputes develop when different interests are deploying political capital and organizational effort to influence an outcome of the decision making or permitting system and the allocation of scarce resources. In Grays Harbor, the political question is "What does or can occur in the estuary?" Ideological disputes arise when beliefs and norms differ about what should be In Grays Harbor, the ideological questions are "What should occur in the estuary?" and "Who should make these decisions?"

In Grays Harbor conflicts, the complex interweaving of political and ideological disputes is well illustrated by the search for predictability. Predictability, although often expressed as a single goal, embraces several fundamentally different political and ideological concepts as shown in Table 1.

The analysis and understanding of the political and ideological components of the conflict situation allow two crucial realizations. First, many of the ideological questions over "what should" and "who should" are and must be decided in forums outside of the Grays Harbor Estuary Management Task Force because they impact much larger populations and regions. For example, agency roles are mandated by Congress or State Legislatures, and resource management systems and environmental protection requirements are often defined by Federal and State law.

Clear understanding of the legal framework for negotiated solutions will avoid futile arguments, tragic misunderstandings and possibly illegal compromises (all of which occured in Grays Harbor at some time). Although the consultants to the Task Force provided descriptive summaries of relevant laws, many of the Task Force members strongly urged as a lesson for the future that the legal constraints be explicitly examined early in the planning process. A legal advisor on a technical, non-negotiating team is a further suggested prerequisite.

The second crucial realization from analysis of the ideological and political components of the conflict situation is that realistic expectations for the results of an effort can be derived. For example, although solutions may be provided to specific political conflicts, future political and ideological conflict situations cannot be avoided. Therefore, a negotiated solution which is to provide predictability for the future must provide a forum for future negotiations.

An OCZM report stated rather unrealistically, "If the Grays Harbor Estuary Management Plan is adopted and implemented as designed, continued conflicts over each individual permit or development activity in the estuary will be avoided" (Dehart, 1978). Quotes from interviews with Task Force members also indicate expectations for final or complete conflict resolution. "No more permit appeals," "avoid issues and controversies," "conflict avoidance mechanism," and "avoid traditional confrontations" were expectations expressed by local, State, and Federal representatives. Yet, resolution of all conflicts is impossible due to the strong ideological nature of many of them.

Also, perhaps as a result of such expectations, the Grays Harbor Estuary Management Plan poorly addresses the need for future deliberations and negotiations and the Plan provides few mechanisms to resolve future conflicts.

TABLE 1

THE GOAL OF PREDICTABILITY

	Political Qualities "What does or can occur?"	Ideological Qualities "What should occur?" "Who should decide?"
Development Interests (esp. Port, local governments and Regional Planning Commission)	Predictability= the assurance that projects will proceed smoothly through the steps from private proposal development, through public administrative review, to approval and completion. (The temporal scale of this predictability is relatively short term.)	Predictability= local determination of estuarine use through the normal operation of the market system. The result would be the most efficient and beneficial set of uses. (Of course, this would be under the umbrella of some government interaction to account for such difficulties as common property and public safety concerns.)
Environmental & Resource Management Interests (esp. NMFS, USFWS, EPA and State resource agencies)	Predictability= the assurance that the system of State and Federal regulations would operate to effectively protect critical resources and ecological systems.	Predictability= protection of the long-term biological viability of the estuary. Ecological concerns would be of major importance in order to protect the resource and the interests of the State-wide and national publics. (The temporal scale of this predictability is long term.)

This categorization is not meant to be rigid or mutually exclusive. For example, resource agencies are committed to the exercise of private property rights as well as to a responsible and effective regulatory system to protect the public interest in a healthy environment. The objective of this categorization is rather to portray the highly complex political and ideological setting of the search for predictability. The table highlights the fact that there may be differing definitions of the single goal of predictability.

Ability to Compromise

At least three factors in addition to political/ideological commitments and legal constraints affect the ability of parties to compromise and to make commitments: (1) the ability of a participant to speak for the agency or interest group, (2) the ability of the agency or interest group to make commitments and compromises, and (3) the skill of the individual participants in negotiations.

In order to move forward toward viable compromises and agreements in a multi-sided dispute resolution process, the participants must not only be committed to the process, but also must have the authority to speak for their constitutents and be able to make some form of commitment on the part of the constituents. Also, whatever the mechanism for obtaining formal commitments (e.g. taking proposed agreements to a board of trustees for a vote, referring a policy decision to a regional director of a agency, etc.) each participant needs to understand the ability of other representatives to make commitments.

The ability of a representative to speak for the agency or entity is partially a function of that individual's position in the agency and the responsibilities delegated to the individual. The organizational distance between the position of the Task Force participants within their own agencies and the locus of agency decision making generally increased along a continuum from local to Federal agencies. The Planning Director of the Port of Grays Harbor, the Port representative on the Task Force, spoke directly for and was responsible to the Port Manager. In fact, the Port Manager attended Task Force meetings with the Port Planner. Although ultimate decisions rest with the Port Commission, the Port representative had the ability to caucus with a significant Port policy setter, to respond rapidly to changing circumstances, and to strategize accordingly. In contrast, representatives of the resource and regulatory agencies were generally quite removed from agency policy setters. In a number of cases, final decision making authority resided in top officials in the National or State headquarters. For example, after a July 1978 memo from EPA regarding the interaction between the provisions of the Clean Water Act (CWA) and approval of State CZMP's, the local office of EPA was effectively removed from the bargaining. Decisions were thereafter handled nationally. One operational result was that communication became more difficult and time consuming. The whole process slowed. As another example of the same basic problem, in the spring of 1979, the U.S. Fish and Wildlife Service (USFWS) representative asked the USFWS Area Manager to request the Director of the USFWS to obtain an opinion from the USFWS Solicitor on the relationship between Federal approval of a State's CZMP and the requirements of section 404 of the CWA (USFWS, 1979). When agency representatives found it necessary to refer to policy makers higher in the organization, the ability to strategize or negotiate was constrained, and representatives could appear weak, indecisive or unreliable to the other side.

Organizational disparities in the ability to speak are often unavoidable. The situation can be alleviated by selecting representatives which have the highest appropriate or feasible delegated authority and by educating all representatives on the limits of the authorities of all participants.

In addition to the varied ability of representatives to speak for their agencies or constituencies, the ability of the separate agencies to speak also varied. For example, the Port has a clearly identifiable, single focus policy of enhancing economic development through encouragement of more industrial growth in the Harbor. Therefore, the Port representative could develop specific strategies to advance these interests. On the other hand, the Federal and State agencies have a more difficult job defining precisely what resources should be protected, to what extent, and in what manner. Also, in the view of at least one resource agency representative, resource agencies cannot "take extreme positions for the sake of bargaining or gaming".

Regardless of the disparities in the ability to speak, bargaining occured. The lessons for the future would seem to be (1) to recognize when these disparities are unavoidable and to adapt procedures and expectations to accommodate them, and (2) to analyze when disparities can be lessened by selection of appropriate individuals to represent a party, by strategizing to minimize confrontations requiring referral to decision makers higher in the bureaucracy, or by designating certain ad hoc powers and responsibilities to agency representatives for the purposes of Task Force negotiations.

Advocates for a particular party to a conflict might argue that a party is in a preferred bargaining position when they are unable to make commitments "at the table" and can use an argument for referral to a higher authority as a means of gaining leverage or bargaining time. While this might be true for a single party in a negotiation, if all parties equally applied this rationale, the efficacy and rationale for the conflict resolution by negotiation would be undermined.

The skills and training of individual representatives are also salient elements in the ability to negotiate, compromise and make commitments. A mismatch in abilities possessed by an individual and those abilities required for negotiations, or substantial inequalities in the distribution of skills among the participants can affect the outcome. In Grays Harbor, the Port representative was especially adept at analyzing the conflict situation and strategizing. These skills allowed the Port planner to be an dynamic and effective leader of development interests. Although not a formal member of the Task Force, the representative for the Regional Planning Commission also seemed to deal proficiently with the group dynamics of conflict resolution and negotiation. Professional training and personal aptitudes combined to make these two individuals strong and effective negotiators. On the resource agency side, able negotiation skills were employed especially by the National Marine Fisheries Service representative, who was also a prime mover, along with the Port planner, in the initiation of the Task Force effort. However, the resource agencies were, for the most part, representated by field biologists or biologists who became resource managers as they ascended the agency bureaucracies. Aside from legal questions of whether the agencies can bargain, negotiations may be foreign to the professional training of biologists and resource managers. One agency representative observed that resource biologists on the Task Force could have greatly benefited from training in the characteristics of group dynamics and methods of conflict resolution.

Especially in the case of resource agency personnel, individuals are not hired for their negotiating skills, but for their technical background and competency. However, based on the Grays Harbor experience, agency administrators are well advised to select individuals who have skills to match the demands of the planning or negotiating activity and to provide training opportunities for agency personnel in the methods and techniques of conflict resolution and planning.

Adopting and Implementing the Plan

A pivotal question for the future of the Grays Harbor effort is whether the Grays Harbor Estuary Management Plan can be successfully adopted and implemented. The answer is not yet clear. The extent to which agencies can participate in intergovernmental planning and conflict resolution and commit themselves to a resultant plan have become issues of national concern as a result of the Grays Harbor experience.

The Task Force began in 1975 with the straightforward assumption that when a Plan was finally agreed upon by the participants, the Plan would be implemented via the existing State Shoreline Management Program and the existing Federal regulatory programs. Local governments would adopt the portion of the Plan relevant to their jurisdictions into the local city or county ShorelineMaster Program (SMP).

The Washington Shoreline Management Act (SMA) has a built-in implementation device. Major proposals require a locally granted substantial development permit. These permit requests are reviewed for compliance with the local SMP at the State level by the Washington Department of Ecology. Other agencies and interest groups participate in this review. Appeals of decisions can be made to the Shorelines Hearing Board.

To implement the Plan at the Federal level, the Washington Department of Ecology would include the Plan as an amendment to the Washington Coastal Zone Management Program and submit the amendment to the Secretary of Commerce for approval and incorporation into the State's Program. Federal agencies would then use the Plan as a specific framework for permit review and approval/denial.

Implementation in theory and implementation in fact have become two entirely different phenomena. The first step toward toward implementation, plan adoption, has become a major difficulty.

In fact, the Department of Ecology would be the only State agency to adopt the Plan because of the Department's role in administering the State SMA. State resource agencies, such as the Departments of Game and Fisheries, are not required to adopt it. The Task Force assumed that State resource agencies would be bound to follow the Plan if it is adopted under the State SMA and by the Federal agencies. If the State resource agencies do not officially adopt the Plan, their commitment to work within its framework may lessen over time as new concerns develop and other laws are employed in the development and review of proposed actions.

Formal adoption of the Plan into the State's Coastal Zone Management Plan was seen by the Task Force members as an indirect way in which Federal agencies would "adopt" the Plan. The Coastal Zone Management Act (CZMA) states under the Federal consistency provisions that Federal agencies must give greater recognition to an approved State CZMP than to non-approved State or local actions. The "carrot" of Federal consistency attracted local commitments to estuarine and coastal planning in the beginning. However, incomplete understanding of the extent of Federal consistency by some of the Task Force participants led to confusion and uncertainty at the outset.

Some Task Force members assumed Federal consistency to have both "positive" and "negative" components. "Positive" consistency would bind or direct Federal activities to follow provisions in a State's CZMP. "Negative" consistency would prevent a Federal agency from undertaking an action counter to a State's CZMP. In fact, only "negative" consistency is provided by the CZMA. Federal consistency does not require agencies to approve projects or activities because the State has approved them. It only precludes Federal agencies from acting where a State has denied that same action under a valid State law. The Federal consistency provisions do not abdicate the legal responsibility of Federal agencies. Federal consistency only limits agency discretion in those cases where an approved State CZMP takes a negative position on an issue.

Some of the confusion over the legal interpretation of the Federal consistency provision of the CZMA could have been clarified early in the process. State and local misconceptions and unrealistic expectations could have been avoided. (Although final regulations on Federal consistency were not issued until March, 1978, long after the Task Force had started its work, agency attorneys knew in 1975 that the CZMA did not provide "positive" consistency).

Even with these limitations on Federal consistency, it appeared that the incorporation of the Plan into local SMP's and into the State CZMP would be a big step toward full adoption and implementation of the Plan.

However, in order to provide complete Federal compliance with the Plan, Federal agencies would have to adopt the Plan and use it in their regular procedures. The Task Force had not addressed the question of how this could or would be accomplished. Individual agencies have offered opinions on how adoption would or would not occur. The Corps of Engineers (COE) representatives indicated that the agency will not formally adopt the Plan because through the normal project review and NEPA process the COE can give considerable weight to the local views as expressed in the Plan--assuming it is adopted locally. Also, the COE has no authority to adopt local plans. For the COE, the Plan will become a general referral document that expresses a preferred alternative, but it will not be a specifically binding agreement on the COE. (This is not the predictability the Task Force initially sought).

From the scanty reference in Task Force background material to potential methods of adoption, one could conclude that the Federal agencies were primarily concerned with Plan content and wished to resolve that problem first before worrying about implementation methods. From their perspective, if no acceptable compromises would result, the debate over implementation methods was academic. As events have turned out, however, acceptable compromises appear to have been generally reached, but substantial adoption and implementation questions remain unanswered.

As a result of the Grays Harbor planning process one of the most important issues facing the management of shoreline resources today has been raised. Can a Plan that prescribes future uses be developed and implemented expeditiously, or will shoreline use decisions be made on a case-by-caes basis with little or no recognition of the Plan?

The central national policy debate is over the integration of the Clean Water Act(CWA) and the Coastal Zone Management Act (CZMA). Can agreements reached in a regional comprehensive planning activity under the CZMA (section 307) substitute for the project-by-project review called for under the CWA (section 404(b)(1))?

In July 1978, the Environmental Protection Agency (EPA) circulated a memorangum that discussed EPA's interpretation of section 404 of the CWA and section 307 of the CZMA. Under their interpretation, any State CZMP, including the Grays Harbor Plan, cannot be approved by the Secretary of Commerce unless the CWA 404(b)(1) guidelines were applied in developing that plan. In 1978 the guidelines did not permit authorization in advance for fill or discharge into aquatic areas, but required a case-by-case determination based on an analysis of the use to which the filled area is to be put. EPA issued proposed revisions of the implementation guidelines for section 404(b)(1) of the CWA in September 1979. The proposed revisions permit CZMP's to provide some of the information needed in project review, but do not allow substitution of CZMP's for that review.

The Grays Harbor Plan is now being revised to comply with the revised section 404(b)(1) guidelines. But, in light of the provisions in the guidelines, the Port and local governments in Grays Harbor have argued that without Federal adherence to an area-wide plan, there is little incentive for coastal planning. Why should local governments bind themselves to a plan when it doesn't guarantee that Federal agencies will go along with it? Such a plan provides no predictability. For others participating in similar planning processes, it is extremely important that the relationship of area-wide plans to site-specific decisions gets early and continuing attention.

Numberable uncertainties remain unresolved regarding Plan implementation. These uncertainties result not only from changing regulations, but also because the Task Force and the Plan do not specifically integrate the laws and regulations under which the agencies could adopt the Plan, do not address the specific level within an agency that the Plan could be adopted, do not identify the timing and form of adoption, or do not examine alternate adoption mechanisms. It is these deficiencies which are now, after five years, major stumbling blocks for the Task Force.

The lesson for the future is clear. Adoption and implementation strategies must be developed as early in the conflict resolution, or comprehensive planning, activity as possible.

Success Is Not Guaranteed

A possibility always exists that mutually acceptable compromises cannot be reached. Early in a conflict resolution process steps can be identified in the process that can act as check points where decisions to proceed or stop can be made by the Task Force without the loss of face and the endangering of future working relationships between parties. Pre-identified check points or a system of staged commitments would allow the negotiating team to move incrementally through the complex and, perhaps, controversial chores at hand. Participants can be given a chance to become comfortable with the progress of the activity before agreeing to move to the next step.

Participants may be able to slowly build a relationship of mutual trust if thay can agree on pieces of a solution. Accepted points of exit from the process would avoid getting to the end of the process and being trapped into an agreement which one does not want to accept but feels one must in order to avoid accusations of bad faith bargaining and the loss of the capability to work together on other issues in the future.

Staged commitment also alters the definition of failure. If incremental agreements can be developed, some success can be measured even if the group eventually concludes that a final plan or compromise agreement is impossible.

Summary

Analysis of the Grays Harbor Estuary Management Program identifies several suggested components for decision making processes designed to integrate conflict resolution with comprehensive planning. The components form the basis of process design and operating procedures when the decision making environment includes a number of diverse competing interests, scarce resources and persistent conflicts. First, the parties in the decision making activity must have realistic expectations and criteria for success. To achieve this, the parties must understand the political and ideological nature of the conflict situation and they must accept that complete accomodation of all interests may not be possible. In addition, the parties to the conflict must expect and be able to compromise. Second, all parties to the conflict who have a stake in the dispute resolution or can affect the implementation of any agreement should be involved in the planning/conflict resolution process. If critical parties are excluded, the political viability of any agreement may be threatened. As a corolary to the inclusion of all parties to the conflict in its resolution, and in order to maintain a task force of a workable size, parties who have no stake in the conflict, who are unable to compromise, or who are relatively powerless are best used as information sources rather than as members of the task force. A non-negotiating technical team is one suggestion for the involvement of such parties. Third, the selected representatives of the parties in a conflict resolution process should have the authority to speak for their agencies and constituencies. Also, the scope of authority of each party and its representative should be clearly understood by all participants. Fourth, representatives should have or be given training in bargaining, negotiation and the dynamics of conflict resolution. Fifth, the mechanisms for adopting and implementing an agreement should be specifically examined and identified early in the process. Finally, a staged commitment process or pre-identified check points for decisions to continue may, in the end, be a highly successful means of reaching mutually agreeable compromises.

The lessons learned from the experience of the Grays Harbor Estuary Management Program should be applicable to efforts elsewhere which attempt to resolve specific resource use conflicts and to provide comprehensive management plans for the future. The Grays Harbor experience has been an invaluable experiment in a form of special area management that is meaningful nationally. Grays Harbor provides a prototype of national coastal zone management issues. The Grays Harbor effort represented and attempted to answer the needs for improved interagency and intergovernmental coordination, for assured and adequate coastal resource protection, for provision of development opportunities to protect the economic health of coastal communities, for the consideration of national interests in State and local coastal management programs, and for the provision of closer connection of management plans and the decision making power and mandates of a multidimensional, multijurisdictional and democratic government.

The succes of the Grays Harbor effort in resolving the many complex and persistent conflicts through an innovative process of comprehensive planning is not total. Integration of the components necessary for successful conflict resolution may have provided additional successes. But the final success of the Grays Harbor effort will be in the learning opportunity it provides coastal managers, users and protectors to meet future challenges.

References

Cormick, Gerald W. and Leah K. Patton. "Environmental Mediation: Defining the Process through Experience." Seattle, Washington: Office of Environmental Mediation. February, 1977.

Dehart, Grant, Office of Coastal Zone Management. "Grays Harbor Management Plan- Case Study Review." Memorandum to Robert W. Knecht, Assistant Administrator for the Office of Coastal Zone Management, National Oceanic and Atmospheric Administration. March 9, 1978 (revised March 21, 1978).

Grays Harbor Estuary Management Plan: Revised Draft. Grays Harbor Regional Planning Commission. (Prepared by Montagne-Bierly Associates and Wilsey & Ham.) November, 1978.

Kifer, Robert R., Office of Coastal Zone Management. Letter to the author. January 10, 1980.

Montagne-Bierly Associates. Grays Harbor Estuary Management Program: Technical Memoranda, Summary of Interviews, Annotated Bibliography; and Phase II, Estuary Planning. 1977.

Weinmann, Fred and John Malek. "Planning for Maintenance Dredging in Grays Harbor Estuary." In Coastal Zone '78: Symposium on Technical, Environmental, Socioeconomic, and Regulatory Aspects of Coastal Zone Management. American Society of Civil Engineers, Vol. II, p. 1346. 1978.

PORT PERMITTING PROBLEMS

Marcia Kenney*

Government regulation of U.S. trade and commerce has been part of our system since Congress first began exercising its powers under the Commerce Clause of the U.S. Constitution. Federal regulation of economic activity has increased substantially, however, over the past decade. In addition, the character of this regulation has undergone significant change. The industry-specific nature of much of our early regulation has been augmented by a new functional type of regulation. Increased concern for the environment, in particular, has generated a considerable amount of legislation and resulted in a proliferation of government regulatory agencies. Today, regulatory agencies affect virtually every sector of U.S. commerce, including port activities. The most common form of Federal regulation is through the permitting process. This paper examines the impact of permit requirements on port development in the United States.

The economic advantages and need for port development are usually readily established. International oceanborne trade has risen steadily since World War II, despite yearly fluctuations in the rate of growth. The United States is the major focal point of this trade. It is a major consumer of oil and raw materials, and the leading distributor of manufactured goods and agricultural products. As international oceanborne trade continues to rise, it generates increased demand for port services. Improving and expanding port facilities to meet the needs of domestic and international trade are necessary.

In the United States, most port planning takes place on the local or port level in response to changes in transportation trends and market conditions. Port planning is an interactive procedure in which requirements for port development and expansion are formulated by planners, designers, engineers, and port users. There is a high degree of uncertainty in port planning, especially long-range planning. Planners must consider rapid changes in shipping technology, loading and unloading techniques, and user needs. Once a plan is approved by the port authority, it must then be submitted for review by government regulatory agencies on the Federal, State, and local levels. The permitting process creates even greater delays and uncertainties for the port planner.

Today, the most important prerequisite to port planning is knowledge of the regulations, permits, and approvals which enable siting and operation Numerous agencies have promulgated a small mountain of overlapping and duplicative regulations. The primary laws governing port development activities and the agencies administering them are:

*Economist, U.S. Dept. of Commerce

Act	Agencies
Rivers and Harbors Act of 1899	U.S. Army Corps of Engineers
Federal Water Pollution Control Act of 1972	Environmental Protection Agency U.S. Army Corps of Engineers State Water Quality Agencies
Fish and Wildlife Coordination Act of 1958	U.S. Fish and Wildlife Service State Fish and Wildlife Agencies
Marine Protection Research and Sanctuaries Act of 1972	Environmental Protection Agency U.S. Army Corps of Engineers
The Port and Waterways Safety Act of 1972	U.S. Coast Guard
Coastal Zone Management Act of 1972	Designated State Coastal Zone Management Agencies through the Federal Office of Coastal Zone Management
National Environmental Policy Act of 1969	All Federal agencies whose actions may significantly affect the human environment
State and local laws and ordinances governing land use, public works, natural resources, health, etc.	State and regional land use planning agencies, natural resources agencies and numerous local government units

Several generalizations may serve to introduce the prevailing problems in discussing this legislation. First, it should be noted that this legislation was not designed to apply only to ports and port development. Thus, the applicability of the legislation is not based on the fact that port activities are conducted, but rather is based on subsidiary factors, the discharge of pollutants, which are not unique to ports. While some industries are subject to specific requirements as a result of the enterprise carried on, the operation of a port is not one of these industries.

It is also important to be sensitive to the Federal-State relationship involved in particular cases. In this regard, consideration was given to the following questions: Have the Federal regulations pre-empted State regulations? Do Federal and State regulations co-exist insofar as they are not contradictory? Is State authority given priority under Federal guidelines? Environmental regulation seems to fall primarily under the latter two categories so that minimum requirements are established nationwide while more stringent requirements may be established locally without the problem of Federal pre-emption. Thus, in any given case, if Federal regulations do not pre-empt State regulations, a particular State's regulations must be checked to determine whether compliance with the State regulations is adequate. Another common provision of Federal law is one requiring appropriate State approval prior to the issuance of Federal approvals.

A further problem is to determine which agency sets the standards, which agency investigates, and which agency prosecutes violation. While it has not always been easy to make this determination as straight forward as possible, it can be said by way of generalization that the Environmental Protection Agency (EPA) sets standards and engages in enforcement, unless another agency has been designated to serve these functions. Where dredging or filling is involved, the principal or controlling agency is usually the U.S. Army Corps of Engineers (COE or Corps). Another Federal agency commonly involved in the port permitting process is the Department of Interior, Fish and Wildlife Service (FWS). However, a total of over 50 Federal agencies may potentially be involved in activities related to the port environment.

The multiplicity of agencies means that there is a considerable amount of substantive jurisdictional overlap among agencies not only at the Federal level but at all levels of government. This jurisdictional overlap causes duplication of procedural steps in the permitting process itself. Similar problems occur in practically all port areas, large and small, throughout the country.

The case studies presented in this paper include three examples of situations where the implementation of Federal, State, and local regulatory programs have led to unnecessary costs and delays due to duplication, inadequate policy guidance, and unnecessarily complex administrative procedures. In addition, they are intended to show that port development, so vital to our National transportation system and totally dependent on coastal land and water use, is actively threatened by unclear and uncoordinated regulatory programs.

The Sohio Oil Terminal Case illustrates the classic conflict between local perceptions and national interests. It examines the implications of the fact that State authority is given priority under Federal guidelines pertaining to environmental quality. Finally, the question of minimizing the adverse impacts of development, rather than prohibiting development, is considered.

The Wando River Terminal Case examines the question of whether the permit process provides the necessary balance between environmental and economic considerations, where port development is concerned. Efforts by the port to mitigate the environmental impact of expansion are also discussed. The Wando River Terminal Case is noteworthy because the permit process caused a three-year delay in construction.

The interrelationship between the Federal agencies involved in permitting is examined in the case of the Port of Tacoma, Small Boat Marina. The case attempts to show how jurisdictional overlaps and lack of common standards result in unpredictable and undependable decisions, create undue delays, and subject ports to bogged-down planning sequences and grossly inflated costs.

Finally, the case of the Grays Harbor Management Program is given in some detail as an example of a situation where initiatives to simplify and coordinate regulatory procedures are being attempted by Federal, State, and local agencies. The result, intergovernmental planning for specific uses and areas, is intended to represent a new approach to resolving conflicts of interest between port development and environmental protection groups.

Sohio Oil Terminal, Long Beach, California

The Proposed System

The Sohio Oil Terminal is intended to be part of a system to transport 700,000 barrels a day of crude oil from Alaska via the West Coast to mid-continent refinery and market points.

The proposed system consists of a marine terminal, 4.92 million barrels of storage/surge tankage within ± 12 miles of the terminal, tie-in to the Southern California refinery system, and 1000+ miles of pipeline from Long Beach, California to Midland, Texas. Almost 800 miles of the pipeline is already in place and presently used to ship some of the natural gas supplied to California from Texas, Oklahoma, and New Mexico. A system already exists to ship crude oil from Midland, Texas to the Midwest and Gulf Coast markets.

The Proposed Terminal

The proposed marine terminal unloading facility would contain three tanker berths to be constructed off the southern portion of the present pier J in the port of Long Beach. Limited dredging would be necessary in this area to accommodate ships up to a maximum of 165,000 dwt. In addition, about 2,600,000 tons of Catalina Island quarry rock would be used to create a breakwater offshore of Pier J in order to provide protection for the marine facilities. A trestle would connect the shore facilities with the berths. Each berth would be connected to the onshore tanks by trestle-supported 48-inch (120 cm) unloading lines.

Unloading rates through mechanical, articulated unloading arms would be approximately 120,000 barrels of crude per hour. The crude oil would be pumped to a 30 acre storage facility on pier J, and then transferred to an inland storage/distribution center via pipeline.

Rationale for the Project

Currently, the West Coast is enjoying a surplus supply of oil, if one assumes some 750,000 barrels per day from the North Slope. By 1982, this oversupply is projected to be about 800,000 barrels per day. Compounding the problem of the surfeit of oil is the Naval Petroleum Reserves Production Act of 1976, requiring the Navy to open up the Elks Hill Naval Petroleum Reserve for nonmilitary production of 350,000 barrels per day.

Proposed Alternatives

Several alternatives have been proposed for moving surplus oil from the West Coast to other markets, such as:

1. Swapping Alaskan crude to Japan in return for Middle East oil that would go to the East Coast of the United States rather than Japan.

2. Use of small tankers to carry oil directly from Valdez through the Panama Canal to Gulf Coast refinery centers.

3. Use of very large crude carriers to carry oil to a point offshore of the Panama Canal for transshipment to smaller vessels for transit through the canal, or through a transisthmus pipeline in Guatemala to Gulf Coast ports.

The Long Beach oil terminal, however, seemed to be the most economically feasible proposal.

Problem

The proposed oil terminal engendered great controversy in the State of California because of its potential impact on air quality in the Los Angeles basin, brought about directly by pollutant emissions from storage tanks and tanker-unloading operations. Another concern was the indirect effect of possible curtailment of natural gas supplies because of the pipeline conversion. These lines are important elements in California's energy strategy. The State of California has helped to subsidize exploration for Mexican natural gas and expected to use the El Paso pipeline to carry any new supplies into the State. The State of California wanted guaranteed access to pipeline gas from Alaska or Canada in return for transporting surplus oil to the Midwest.

Efforts to Mitigate Environmental Impact

In regard to the problem of oil vapors affecting local air quality, Sohio proposed buying out nonessential emitting facilities and installing vapor control equipment as a trade off. The company also proposed scaling down its plan from three berths to two, lowering throughput and reducing pollution potential.

Under the California environmental trade-off policy, a company can build in an area with a high pollutant level if it first eliminates more pollution than its new facility will emit. Trade-offs suggested by the Air Resources Board as conditions for a permit included the following: a) purchase of sulfur scrubbers by Sohio for installation in Southern California Edison power plants (Long Beach, Huntington Beach, and El Segundo); b) switch of tanker operations from 50 percent sulfur to 25 percent; and c) hydrocarbon solvent absorbers at 13 local dry-cleaning plants to be paid for by Sohio.

In late August, 1978, a press conference was held by State Officials and Sohio executives informed the public of antipollution equipment to be installed at the Southern Edison plant. This $80 million device was to be financed by Sohio to offset new pollutants that would emanate from the proposed oil discharge terminal.

Implications

The Sohio/State of California developments illustrate the classic conflict between local perception and the national interest. The State of California has established the most stringent requirements for air and water quality of any State. California has taken a strong position as the interpreter and defender of the public interest despite strong counterpressure from the top administrators of Federal energy policy. To some extent, this strong intervention at the State level tended to downplay public participation at the local level in the case of the Sohio oil terminal.

It is interesting to note the sharp difference of interests between the Port of Long Beach, which stood to realize very substantial public revenue and employment benefits from the proposal, and the State of California, which focused on the adverse emissions impact.

This conflict of varying interests is one of the most interesting facets of this case.

Conclusion

In a precedent setting move, the voters of Long Beach were presented with a municipal referendum in November 1978 to determine whether a terminal lease between Sohio and the Port of Long Beach should be authorized. Long Beach voters overwhelmingly approved the Sohio project. However, terminal construction still had to be approved by State and local air pollution control authorities, which continued opposition to the plan. The final resolution came in March, 1979, when, after seeking approval of the plan for five years, Sohio withdrew its permit applications and abandoned the project.

The Wando River Terminal, Charleston, South Carolina

Background

In the past ten years, general cargo volumes at the Port of Charleston have increased 125 percent. Port facilities at Charleston are now handling tonnages either at or beyond their rated capacities.

Container facilities at Charleston are now virtually saturated. Existing container yards have been expanded to maximum, but they still may be fairly characterized as "wall-to-wall". Substantial operating difficulties are being experienced because of current congestion. The State Ports Authority (SPA) is negotiating with the City of Charleston for the only remaining downtown tract (9 acres) adjacent to the present container facilities. It will add to the port's capacity but will not solve the current congestion problem.

Further expansion of present port facilities is not possible. The downtown terminals are directly impacted by the City of Charleston.

The Search for A New Port Site

Beginning in the late 1960's, the Authority began considering various locations for future port development. A series of separate, comprehensive studies were prepared by professional research agencies, including Kaiser Engineers, Wilbur Smith and Associates, Arthur D. Little, and Palmer and Baker Engineers. In addition, the State Water Resources Commission was consulted. In all, some $500,000 was spent in investigative work to determine the best possible new port location.

From a total consideration of twelve alternate sites for port expansion, one area in the lower part of the Wando River was selected as the most suitable location when environmental impact, costs, and operating requirements were all taken into consideration.

The Proposed Project

To be constructed on a 561-acre tract of prime high ground bordering the river's naturally deep water, the proposed terminal would be adequate to cover the port's anticipated cargo-handling needs through the year 2000. The proposal calls for two 1,000-ft. container berths and two 600-ft. break-bulk berths, with back-up area and warehousing covering approximately 135 acres to be built as soon as possible. The proposal also includes long-range plans for two more container berths and one break-bulk berth with attendant back-up and warehousing.

According to a study conducted by the Bureau of Business and Economic Research of the University of South Carolina, the proposed Wando River port expansion, at full development, will provide: additional transportation cost savings to South Carolina shippers of $4.4 million per year; direct port-related employment of about 9,000 jobs (18,000 statewide); a total economic impact of some $346 million per year; and total "port-induced" state taxes of about $7.8 million per year.

Problem

Despite approval by the Department of Interior, the issuance of a state permit for construction, and the appropriation of $56 million by the State General Assembly for the construction of the facility, some conservationists and local residents strongly objected to the proposal. They attached the proposal as a threat to the Wando River environment, and they feared that the port expansion would attract future industrial development. The opponents of the proposal claimed that the beautiful 14 mile long river which is a tidal, non-flushing body of water surrounded by 100,000 acres of Priority I high-saline marsh, was in danger of eventual industrialization and destruction.

Efforts to Mitigate Environmental Impact

In order to mitigate the potentially harmful impact on the environment caused by the Wando River project, the State Ports Authority agreed to:

- Refrain from building any berths, or turning basins, in Hobcaw Creek. Although in the original plan, this feature has been deleted.
- Set aside a 100-acre buffer zone of natural vegetation to protect the ecology of Hobcaw Creek and to shield the facility from the view of residents to the south.
- Construct a combination storm water drainage and contaminant interceptor system to prevent the run-off of any accidentally spilled contaminants into the Wando River. This will be the only general cargo facility known to exist with such a feature.
- Completely alter the original site grading plan to prevent storm water run-off into Hobcaw Creek. Instead, grading elevations will be changed to provide run-off to the Wando river, where water quality effect will be negligible.
- Provide sanitary sewage hookups for vessels in port, to allow treatment and disposal by the Mount Pleasant sewage system.
- Employ the same techniques commonly applied in sports arenas, airports, parks, and industrial areas, to shield the high-intensity lights from view.
- Limit ships, barge, and other vessel movement under its control during planned boating or sailing regattas so as not to interfere with such events.
- Utilize special construction techniques in bridging Hobcaw Creek to prevent damage to the creek and marsh. No construction machinery will be allowed in the creek or on the marsh area. All construction will be from topside.
- Replace any marsh acreage permanently destroyed by the project on a two-to-one ratio. Site to be selected by either Department of Interior, or South Carolina Marine Resources Division.

Conclusion

After three years of debate, the State Ports Authority won approval from the Corps of Engineers in early 1978 to begin construction on the Wando River Project. Construction is currently underway, with the first container berth scheduled to be operating by early 1981 and the entire terminal open for business by 1985.

Port of Tacoma - Construction Permit Delays

Background

The Port of Tacoma, Wash., proposes to build a small-boat marina of about 1,000 berths in a section of the port's property consisting of 5 acres of uplands and 20 acres of tidelands. The existing small-boat haven and marina in Tacoma are situated close to large ship terminals. This location is undesirable, both because the site is economically better suited for marine terminal use, and because the proximity of the marina causes navigational hazards. Since 1967, the Port of Tacoma has searched for a new small-boat marina site, and has settled on the proposed location.

Construction involves the dredging of part of the tidelands and the filling of another 11 acres for use for boat sheds, parking lots and other facilities.

Under the Washington Shorelines Management Act, the Port of Tacoma was required to get a permit for construction of the marina. The port applied for such a permit on December 1, 1972. The permit was initially issued, but after an appeal from the Washington State Department of Ecology, the port was required to prepare an environmental impact statement. After lengthy hearings and some revisions of the statement, the permit issued by the Tacoma City Council was allowed to stand, and, construction plans for the marina had met all State and local environmental standards.

Problem

A permit for the construction of the marina is also required from the U.S. Army Corps of Engineers. Concurrently with the State permit applications, the Port of Tacoma applied for a Federal permit (December 14, 1972) to place an obstruction in the navigable waters, through the filling process.

Because of a memorandum of understanding between the Departments of the Army and Interior, the Secretary of the Army consults the Secretary of the Interior in such cases when environmental effects are in question. The memorandum states, in part, "If the Secretary of the Interior advises that proposed operations will unreasonably impair natural resources or the related environment, including the fish and wildlife..." the Secretary of the Army "will either deny the permit or include such conditions as he determines to be in the public interest..."

The Department of the Interior objects to the construction of the Tacoma marina and consequently the permit has not been granted by the Secretary of the Army.

Costs Incurred

Because of continuing inflation since 1972, the costs of constructing the marina had risen by $1 million as of October 1976.

Conclusion

On October 21, 1976, the Port of Tacoma filed suit against the United States of America and other persons including the Secretary of the Army and the Secretary of the Interior. The port claimed that the memorandum of understanding represented an unconstitutional attempt to delegate authority to the Secretary of the Interior. The port representatives felt that they had met environmental standards through the State process, and that the Corps had more than sufficient time to consider the granting of the permits. Thus, the port asked for a writ ordering the issuance of the permit and payment of costs incurred because of the excessive delay (over $1 million at present). The case is still pending.

Grays Harbor Estuary Management Plan

Introduction

Grays Harbor Estuary is located on the mouth of the Chehalis River on the southwestern Pacific Ocean coastline of Washington. With an area of about 260 km, Grays Harbor is one of two major estuaries on the Washington coast and is the only coastal estuary in the State with a major commercial navigation channel. The harbor is the focal point of the economy in the region due in large part to the transportation of forest products through the port. The lands surrounding the estuary are heavily forested and provide for the bulk of the region's subsistence through harvest and export. Commercial and recreational fishing, fish processing, and tourism are also important contributors to the existing economy that are tied to the quality of the local environment.

Problem -- The Need for Coordinated Planning

The responsibility for making decisions on the use of land and water resources of the Grays Harbor area falls to a variety of local, State, and Federal agencies. Each agency uses a plan, guidelines, or regulations to decide what may or may not be done; cities and the county have their own comprehensive plans and shoreline management programs. Normally, each entity prepared their plan or guidelines independently, with resulting inconsistancies, conflicts, and complex processes of decision making.

Because regulations are established at each of these levels of government and because most can be applied at some point to local situations, inevitable confusion arises. Arriving at some consistency on the application of these regulations is the main reason for the Grays Harbor Estuary Management Plan. The following is only a synopsis of the major regulations that have a bearing on the estuary.

State Regulations: There are more than of eighteen regulatory and permit processes which apply to shoreline development or alterations in the estuary. Of these, the five most important are:

The Shoreland Management Act
State Environmental Protection Act
State Forest Practices Act
Hydraulics Project Approval
Tideland Ownership and Leasing

The Shoreland Management Act, which is most directly related to the Estuary Management Program, was passed by State Legislature in 1971. The purpose of the Act was to:

- minimize environmental damage and protect the natural shoreline of the State, and

- to encourage public access to shorelines.

The Act requires that local cities and counties prepare local shoreline master plans and establish permit procedures to regulate shoreline activities. Within the Estuary Management Program area, shoreline permits are administered by Grays Harbor County, the cities of Aberdeen and Hoquiam, Montesano, Ocean Shores, Westport and Cosmopolis.

Federal Regulations: The Federal government exerts control over shoreline development through review of state controls, Federal permits and a review of State and local governmental activity for consistency. The most common form of Federal control is through permit approval.

Local Regulations: In addition to the responsibility of local cities and counties to administer permits under the Shoreline Management Act, as noted, most local governments also administer various forms of zoning and land use regulation.

Proposed Port Expansion

The Port of Grays Harbor had for some time proposed to fill approximately 2200 acres of tidal and subtidal lands under its ownership near Bowman Field Airport. The port wanted to fill the land because it is one of the few large tracts adjacent to the deep-water navigation channel in the estuary. The port planned to expand into the area over the next few decades as commerce and industry expanded into the region. Part of the tract had been used for the disposal of dredged material over the years, but disposal has now ceased because of new dredging and disposal regulations. State and Federal fisheries and wildlife management agencies objected to the development because of the potential loss of much tideland now in a semi-natural state.

In response to increasing conflicts, the Grays Harbor Regional Planning Commission formed an Estuary Planning Task Force in late 1975, consisting of representatives from 15 local, State, and Federal governmental agencies and entities with interest or jurisdiction over the estuary. In September 1976, the Task Force, through the Regional Commission, received Federal Coastal Zone Management Section 306 money to develop an Estuary Management Plan.

The Estuary Management Plan

The Estuary Management Plan does not eliminate any local plans or agency regulations, but rather represents a set of guidelines developed by all participants that identifies and attempts to resolve conflicts between the various interests. The intent of the Plan is to allow "predictability", in that an individual property owner, local planning commission, or government agency can consult the plan and be assured that the permitted uses or activities will have the general concurrence of those agencies involved in the plan's formation. While the plan will not change the present procedures for receiving zoning or permit approvals, the speed by which these approvals can be granted should increase, provided the proposals are in conformance with the policies and guidelines of the plan.

Approved Port Expansion

The Estuary Management Plan sets out the principles agreed on for development of the Grays Harbor Estuary area for the next 50 years. The group looked ahead to examine the port's land use needs for the next 25-50 years. The port agreed to limit its expansion during the time period to the eastern-most 500 acres of the 2200 acre tract. The first 250 acres of the tract can be filled immediately without restriction, but permission to fill the remaining 250 acres would hinge on how well the first 250 acres were utilized. The port also agreed to move the Bowman Field Airport from its waterfront location to an inland site in the county. In return for the 500 acre concession to the port, the resource agencies would receive an agreement from the port to leave the remaining 1700 acres in a natural state for at least the duration of the 50-year agreement. During filling of the first 250 acres by the port, the second 250 acre tract would be raised from subtidal elevation to create new intertidal land.

Conclusion

The Grays Harbor Estuary Management Plan provides an example of consideration of long-run development needs and environmental mitigation measures within an estuary. The Grays Harbor Management Plan is composed of representatives from State and Federal resource management agencies, local government, and local business and economic development interests. The action to create this program grew from the frustration on the part of the Port of Grays Harbor and industry groups attempting to obtain State and Federal government permits, and from natural resource protection interests who felt that development was taking a heavy toll on the fish and wildlife resources of the estuary. The plan should provide more predictability for permit applicants because the permitting agencies are expected to use the plan in reviewing subsequent development proposals. The issue of future port development will be moved from a piecemeal approach of filing for permits one project at a time, to a long-term regional perspective.

---

* Fifteen Governmental Entities: Grays Harbor County - Port of Grays Harbor, Grays Harbor County, and the cities of Aberdeen, Hoquiam, Cosmopolis, Westport, Ocean Shores. Washington State Department of - Ecology, Fisheries, Game and Natural Resources. Federal - National Marine Fisheries Service, Fish & Wildlife Service, Corps of Engineers and Environmental Protection Agency.

Several aspects of the Grays Harbor case are noteworthy. If successful, the Grays Harbor Managment Plan will allow for economic development, while providing long-term protection for the tidal salt marsh. Both the port and resource protection interests will have gained a measure of certainty for the future.

Recommendations

As these case studies show, activities and operations related to ports frequently raise issues that involve competing sets of values. Typical of such issues are the effect of economic development and energy requirements on the quality of the environment. The needs of maritime facilities almost by definition, conflict with environmental aspirations in individual cases. Many of the Nation's ports have been created by dredging and filling estuarine and other shallow water areas important to marine life. Adverse environmental impacts may also result from port operations, such as the loading, unloading, and storage of cargoes which may result in cargo spillage or waste discharges.

As concern for the environment has increased in recent years, a considerable amount of legislation has been generated which puts constraints on port development and operation. As noted, most of this legislation was not designed specifically to deal with port development. Rather, port development was affected by this legislation because of the activities and impacts involved. Federal and State involvement in port development has increased substantially. The most common form of involvement is through permit approval. The ability to plan for and conduct port business is frustrated by the uncertainties and delays associated with the permit process as it now stands. In many cases, the permit process has not provided the necessary balance between environmental and economic considerations related to port development. Often inadequate attention has been given to the complex and competitive nature of port operations. The failure to recognize and accommodate diverse interests and concerns evenhandedly and expeditiously in the overall public interest has the potential for serious economic and social disruptions, as the case studies indicate.

There are many ways in which the permit process, as it now stands, can be improved. Basically, the two major problem areas which have been identified and discussed in this paper are: 1) overlapping jurisdictions of Federal, State, and local agencies, and 2) undue delays in processing permit applications. The recommendations reached in this paper include both specific recommendations for permit simplification and general recommendations for greater flexibility in resolving conflicts of interest. The recommendations and their implementation are not directed at completely overhauling the regulatory system. To do so would be a formidable task requiring a large scale legislative effort and the resulting system might turn out to be no more streamlined than the present one. However, definite improvements can be made by reforming present regulatory practices at the Federal, State and local levels and through interagency cooperation.

Recommendations for Permit Simplification

- A "lead agency" should be designated for processing permit applications.

- Multiple public hearings should be consolidated.

- Understandable and enforceable time limits should be established.

- A permit tracking system should be established to provide the applicant with information regarding the status of the application.

- A register of permits and permit requirements should be established.

- Participation of local agencies in the permitting process should be eliminated.

- Standardized forms should be adopted.

- Separate procedures for minor permits should be initiated.

- Greater cooperation between the port industry and government is necessary to resolve land-use conflicts.

Of the recommendations for permit simplification presented in this paper, the first and most important is strengthening of the lead agency principle. Strengthening the lead agency principle is intended as a partial solution to both the problems of overlapping agency jurisdictions, and undue delays in processing permit applications. Because major legislative changes are outside the scope of this paper, no recommendation has been made to alter the number of agencies involved in the port permitting process, or their substantive jurisdictions. Improved agency coordination, however, can help resolve the problem of duplication of agency roles. In many cases, existing legislation and regulations contain mechanisms to consolidate and coordinate regulation of economic activities. One example is the Coastal Zone Management Act.

The "lead agency" principle at the Federal and State levels has in many instances played a constructive role in enlightening and facilitating the development process. With strengthening and elimination of ambiguities and prolonged uncertainties about lead agency designation, the lead agency concept could have a broader and more constructive impact. With a strengthening of the lead agency, a permit application would be submitted to the lead agency, which would then notify the other agencies involved. For most port development activities, the lead agency would be the Corps. The Corps, upon receipt of an application, would then notify and solicit comments from all other Federal and State agencies with jurisdiction over the proposed project.

The idea of joint or consolidated hearings is closely related to the lead agency principle. Public hearings are a procedural step often duplicated by many agencies. While many permit applications can be processed without a public hearing, major projects often require multiple hearings. Each hearing will have its own notice, waiting period, and report on substantially the same information. This represents a significant waste of resources, and an undue delay in the permit process. If the lead agency principle is adopted, the lead agency would be responsible for scheduling consolidated hearings. Representatives from all agencies exercising permit jurisdiction would participate in these hearings.

There is a clear need for thoroughly understandable and enforceable time limits that can be applied to the varying approval and permitting responsibilities of the relevant governmental agencies. Under the present system, agencies have virtually unrestricted processing time because there are no sanctions for delays. All Federal and most State commenting agencies have no time limits. However, the Corps cannot make a decision on whether to issue a permit until it has received comment from the required agencies. Too many agencies can make a decision by making no decision.

Applicants for permits should have the right to a decision, one way or the other. Agencies should be required to respond to permit applications within a reasonable length of time, following completion of hearings. The length of time will vary with area circumstances, the type of project involved, etc. A fixed time limit, written into law may not be flexible enough, but some provision should be made for establishing a deadline. Agencies that, for legitimate reasons, cannot meet a particular deadline can be given the opportunity to justify an extension.

Establishing a permit tracking system to provide applicants with information on the status of their application would be much more feasible under a system with strengthened lead agency participation and established time limits for commenting on and processing applications Under such a system, this type of information could be provided by computer for the applicant's benefit.

Many people do not know that permits are needed for construction, or do not know if a permit is needed for the type of work they propose. This is true of even the larger port authorities which are not always aware of the number of agencies to be contacted under the present system. This problem can be eleviated by the establishment and maintenance of a register of permits at some central location. The register would tell the applicant what permits are required for a particular project and how the permit process works. The register would contain a list of all development related permits required by Federal, State and local agencies and include sample applications. This would abolish the need for registe in each locality. Even if the lead agency principle is adopted, a register would still be useful in informing applicants as to what activit require a permit, how an application for a permit is made, what informati is required, etc.

Besides each agency having its own procedures, many have their own application forms for permits. These usually lengthy forms generally require the same information with minor differences. Often applications for permits are returned to the applicant more than once because of insufficient data or because they were not filled out properly. This causes further delays in the permit process. Too often, applicants do not clearly understand what agencies expect in terms of project requirements. Applicants who file for permits without clear knowledge of agency requirements waste not only their own time, but the agency's time as well.

A single application form (or forms) should be adopted by all agencies. The use of standardized form would greatly facilitate the permitting process by reducing the amount of paper work involved. The format should also increase the clarity of identifying the most significant impacts of the proposed project. This may be done by use of a checklist-matrix system containing the following elements:

a) Impact Checklist

The use of impact checklists is one of the most common means of identifying and standardizing various developmental and operational activities. This is important since it guarantees that all potential environmental issues are considered, regardless of the type of port plan or concerns of the planners involved. Such a checklist would also make explicit what activities are considered significant.

b) Cause/Effect Matrix

An impact checklist, once developed, can then be used as an input to a cause/effect matrix to identify the possible impacts of projected activities.

c) Agency Checklist

The provision of an agency checklist for agencies to indicate that they have reviewed and approve the application would greatly facilitate the permit process. (Additional sheets could be attached for extended comments, when necessary.)

An additional recommendation is to establish separate administrative procedures for processing "major" and "minor" permits. Under the current system, minor projects may be delayed for considerable lengths of time because the agencies involved are investigating major projects. Most of the legislation requiring permits was written with the large developer in mind. While the current permit procedures may prove burdensome to the larger developer, they may be an almost insurmountable burden for the small developer (or private individual) who has limited financial and legal resources.

The issue immediately arises as to what consititutes a minor or major project. While the distinction may at times be difficult, some parameters can be described in advance which will make the decision easier. To some extent this distinction is already made in determining whether or not an EIS and a public hearing are required before a permit is issued.

The Corps, in its Section 404 regulations, is attempting to simplify procedures for small projects, and a few State agencies are beginning to simplify their procedures as well. If simplified procedures exist at only one level of government, however, the entire permitting process will still be only as fast and efficient as its most cumbersome part.

In the permitting process where Federal, State, and local agencies are involved, there is considerable overlap. There are no major gaps in authority in this process except at the local level, where the permit process has low priority. Local agencies often do not participate in decisions on permits because of apparent lack of interest. In addition, they usually lack the expertise and staff time to evaluate each project on environmental, developmental, and economic criteria. Since the Federal and State agencies do have this expertise, local agencies usually rely on their judgments. Finally, the local agencies see a project in terms of a limited jurisdiction. Major port development activities usually affect a larger area--covering several counties or even States. Resolution of conflicts between local agencies are particularly difficult, since, unlike at the Federal and State levels, interagenc agreements and legal guidelines are lacking. One means of resolving these conflicts is for the State to take control. Eliminating the permit issuir functions of local agencies would not, however, prevent local interests from being represented in the permitting process at hearings, etc.

Some of these recommendations are already being implemented. In the Norfolk, Virginia District of the U.S. Army Corps of Engineers, for example, the U.S. Fish & Wildlife Service, National Marine Fisheries Service, EPA, other Federal and State resource agencies and the Corps have established a joint processing program for the Corps section 404 and 10 permits. Each month the Corps convenes a meeting to review 60 to 100 permit applications with each of the agencies with responsibility for the review of these projects. These meetings usually result in agreement on the informational requirements, resources management issues, and, in many cases, concensus of government positions on the less significant projects The larger or more controversial projects may require follow-up action or impact assessment. The joint processing program gives the agencies involved more time to devote to these projects. According to one of the participating agencies, they have reduced the time for permit review considerably, at the same time reducing the number of permits they were taking "no action" on, because of lack of staff. This process also minimizes the expense and disruption to the applicant from duplicative site visits, data gathering and applications from permitting agencies.

An attempt is currently being made to facilitate the permit process in Massachusetts, where two regional coastal offices of the Executive Office of Environmental Affairs provide applicants with all necessary Federal and State forms, and assume concurrent processing by Federal and State agencies. In an effort to establish a permit information center, the State of Alabama Coastal Area Board is currently in the process of identifying all permits required by Federal, State and local agencies for activities within the coastal area, as well as collecting copies of all permit applications, permit regulations, and any other information provided by the permitting agencies.

In a similar program, the State of Oregon has established an Intergovernmental Relations Division which provides applicants with a list of all permits required for a given project.

In addition to the changes recommended here for permit simplification, there is much that the port as a permit applicant can do to speed up the permit process. First, in choosing a project location and design, the port could contact the effected agencies early to determine the acceptability of different locations or design approaches. Many projects are delayed because an environmental impact statement is required. The port can avoid or reduce this delay in two ways:

1) by choosing locations which have less environmental impact, and
2) providing as much information about the project and its effects as possible.

Even with simplification of the permit process, the basic conflicts of interest between port development and environmental protection will continue to exist on a case-by-case basis. For this reason, greater cooperation between the port industry and government to resolve land-use conflicts is recommended.

Most environmental disputes are now fought in court. (See Port of Tacoma and Wando Island Case Studies). Oftentimes, cases are resolved in court on specific points of law and the substantive issues of the case are not always resolved. More recently, however, attempts have been made to find new and more creative ways to resolve these conflicts.

Recently, the California Coastal Commission helped to initiate mitigation/compensation workshops. Such workshops are ongoing, and representatives of State, Federal and local concerned agencies are working to establish a better dialogue between permitting agencies and to develop rational solutions to the conflicts presented by the needs of port areas to expand facilities to accommodate increasing world trade, and the concerns of environmental protection agencies in providing and maintaining habitats and the quality of life in and around port areas.

This idea has been carried even further by the Grays Harbor Management Program, discussed in this paper. As noted, the program is intended to bring together the various Federal, State, and local agencies involved in development conflicts to establish a set of guidelines that identify areas of interest and attempt to resolve conflicts between the various agencies. This type of cooperative land-use planning should result in some degree of "predictability" for permit applicants. A similar approach has been undertaken by the Bay Conservation and Development Commission (BCDC)--an intergovernmental commission interested in the future of the San Francisco Bay. The San Francisco Bay Plan, developed by the BCDC, designated sites that would be appropriate for future industrial or energy facility development. Once a site is so designated, a permit will not be issued for a use incompatible with the area's designation. The existence of an agreed upon development plan in these areas and others should speed up and simplify the permitting process considerably by providing some basic guidelines about where development can be located and where it can not, and an agreement on an organizational network. Participation in such planning provides a forum for ports to express their needs and have them addressed. It also ensures that the plan which results will accommodate port needs as much as possible. The enthusiastic support and involvement of the port industry in the development of such land-use plans is essential to the responsible resolution of use conflicts which will continue to grow in the face of increasing developmental pressures.

BARRIER ISLAND SEDIMENTATION STUDIES PROGRAM

Edward Meisburger, Leita J. Hulmes, Edward B. Hands,
S. Jeffress Williams, Craig H. Everts, Dennis A. Prins,
Richard D. Hobson, Kenneth Finkelstein[1]

I. INTRODUCTION

Coastal Barrier complexes can be divided into two main elements; the barrier itself which may be either a barrier island, barrier spit or baymouth barrier and the bay, lagoon or estuary sheltered by the barrier. These elements can in turn be divided into a number of subenvironments typified in most cases by characteristic morphology, sediment distribution and fauna, and subject to a distinct set of sedimentary processes.

In the past, public concern has been largely focused on specific subenvironments of coastal barrier complexes to the virtual exclusion of others. Thus attempts to preserve or enhance beaches and inlets for public or private use have often had adverse effects on other parts of the barrier environment. In recent years there has been increased awareness of the need to assess human intervention in any part of the barrier subenvironment in the context of its possible effect on the barrier complex as a whole.

The Corps of Engineers through its civil works function is responsible for many engineering projects in the coastal zone. The Coastal Engineering Research Center (CERC) serves this Corps' mission by conducting research to define the characteristics and processes of the coastal environment and by developing design criteria for coastal engineering works. Recognizing a need for engineers and planners to have a more comprehensive view of the barrier complex as a whole, and its response to human intervention, the CERC was directed in 1979 to determine the current state of knowledge concerning the barrier environment, to define the problem areas and to initiate research efforts in areas of inquiry where they seem most needed.

In order to better determine the present state of knowledge concerning barriers, and current research efforts in the US, the CERC conducted two workshops in which the participants were drawn from governmental, academic and private institutions. The first workshop was held at the CERC in September 1979 and was largely devoted to the problems of management, use and preservation of barrier complexes. The participants were persons engaged in these

[1]Engineering Geology Branch, Coastal Engineering Research Center, Kingman Building, Fort Belvoir, VA 22060, (phone 202-325-7144)

aspects of barrier studies. The second workshop, held in November 1979, was concerned mostly with a review of past and current research into barrier problems. The participants presented brief talks illustrating their research efforts followed by a period for questions and general discussion.

From the workshops, our review of literature and individual research efforts over the past year we have endeavored to identify the principal problem areas and deficiencies of knowledge of the behavior of the barrier complex under varying natural or man-made environmental circumstances. Some of these problems are the subject matter of this report.

II. BARRIER ISLAND GENESIS AND EVOLUTION

The question of barrier island genesis has long been a subject of interest to students of coastal landforms and processes. Although there is presently no general concensus on how these unique features came into being speculation has centered around three principal models. The earliest of these was proposed by De Beaumont (1845) who believed that the barriers were initiated by upward growth of an offshore bar which eventually elevated the crest above sea level thus establishing a barrier - lagoon complex. Gilbert (1885) proposed an alternate mechanism in which barrier genesis was initiated by spit elongation from a headland with subsequent segmentation of the spit by breaching and inlet formation. Hoyt (1967) proposed a third model of barrier formation in which the initial step was the development of a mainland beach complex, followed by a rise in sea level which partially drowns the beach complex and floods the hinterland. The crest of the beach remaining above sea level becomes a barrier island which will persist as long as upward accretion can keep pace with rising sea level. Through the years there has been considerable debate as to the primacy of one or more models of barrier genesis. Schwartz (1969) called for the recognition of all existing theories as being potentially valid at certain times and places.

A recently evolved concept of barrier island development which has important implication to the study of barrier genesis is based on the assumption that conditions of sediment supply were far more favorable to barrier development when eustatic sea level was lower. At the present time the partial drowning of streams flowing to the coast and formation of coastal lagoons have greatly limited terrigenous sand supplies to the coast. The initial formation of the barrier occurred, therefore, on the continental shelf during the early stages of the last transgression and retreated to its present position in response to continuing eustatic rise of sea level (see Field and Duane, 1976 for a detailed study of this concept).

The concept of ancestral barriers provides a rational explanation for the presence of barrier islands in places where conditions do not now seem to be adequate for their initiation and subsequent growth in accordance with one of the known genetic models. All that is required is that sediment supply and hydrodynamic conditions favoring barrier development existed at some time during the last transgressions, and that subsequent events allowed the barrier to survive.

There is evidence that most barriers on the North American coast have retreated from their position of formation. For example, in many places Holocene lagoon, marsh and estuarine sediments have been found under sandy barrier island beaches and often cropping out on the lower beach and shoreface. The most probable way to account for this disposition of sediments is to assume that the barrier sands have been driven back over deposits that originally formed in their lee. The mode of origin of the barrier, thus, has little bearing on the present situation since it is no longer in its original location and presumably has responded to a different set of environmental circumstances than those prevailing at the time of formation.

More pertinent questions center around the mechanisms and history of the retreat. When did the retreat begin and how far has the barrier retreated since that time? What caused the barrier to retreat and do the same conditions prevail today? Can it be reasonably expected that human intervention could halt the procedure for a given period of time, or will attempts to halt or slow the retreat rate result in future problems? The answers to these questions have a direct bearing on present day coastal problems and future trends.

The study of barrier islands is one of the most active areas of investigation in marine geology. In our own project we are particularly interested in the history of coastal barrier complexes over the past 500 years. This span would encompass relatively long-term trends that may not be evident from available historic bathymetric and topographic data which, in the United States, does not provide reliable details beyond the latter part of the past century. For this purpose we hope to obtain many peat samples or other organic materials suitable for radiocarbon dating in our extensive sampling and coring programs in barrier environments. We also hope to be able to use pollen and widespread human artifacts such as airborne industrial ash to establish time horizons that can be applied on a regional basis.

III. IMPORTANCE OF SEA LEVEL

The rise in sea level caused by melting continental glaciers at the end of the last ice-age inundated coastal plains worldwide and played a major role in the evolution of barrier islands. Even during the present century, sea level in many areas appears to have risen slowly relative to the shoreline leading scientists to speculate that this continues to be a dominant force reshaping our shoreline. The principal questions regarding relative sea level are: (a) How has it varied in the recent past and what does this suggest about future changes? (b) If future changes were predictable, how could the resulting shore adjustment be estimated?

1. Past Changes in Sea Level

Radiocarbon dating of materials that lived near ancient shorelines and were recovered inplace from presently submerged locations, permits reconstruction of past sea level elevations. About 16,000 years ago, sea level was on the order of 100 m below its present elevation. By 6,000 years ago, the oceans had risen to near their present levels. The principal cause for this world-wide rise was the addition of water derived from melting ice sheets, hence the name glacio-eustatic rise (glacio = glacial in origin, eustatic = world-wide in effect). A contributing factor was the flexure or depression of the coast under the load of the extra water spreading across continental shelves.

With the amelioration of climate 6,000 years ago the glacio-eustatic effect diminished and the relative importance of other factors increased: (a) meteorological (atmospheric pressure, winds, precipitation, evaporation); (b) oceanographic (water density, currents, and momentum flux); (c) river discharge (water and sediment load); (d) geologic (isostasy, tectonism, subsidence, and local compaction); (e) geodetic (frost heave, beach-mark disturbances, deformations of the earth's shape, polar wandering associated changes in rate of the earth's rotation). Though some of these factors are global in effect many are regional or local. Accordingly, sea level curves proposed by various investigators generally agree for the period during which glacial melting predominated (about 16,000 to 6,000 years B.P.) and show increased variations thereafter. Some of the latter discrepancies reflect real regional differences of sea level.

Spatial variations in relative mean sea level during this century are revealed by a system of tide gages. The National Oceanic and Atmospheric Agency's National Ocean Survey operates more than 40 tide gages, widely spread around the United States, that have nearly uninterrupted records extending back to at least 1930. All but five of these gages indicate a trend toward rising relative sea level. Three of the five exceptions are from Alaskan sites where the land appears to be rising. The two others came from short records in the Pacific Northwest; where the apparent decrease in sealevel fails to be statistically significant at the usual 95% confidence level. Thus the total tidal record suggests that relative sea levels are stable or increasing at all gage sites in the contiguous United States. In contrast to the case of the Holocene transgression, causes of this historic rise in sea level have not been established. Fluctuations of the right magnitude could have occurred in response to contemporary changes in ice masses. On the other hand the trend may reflect changing continental elevations. The data are inconclusive. As the cause is unknown future predictions are risky.

2. Effects of Rising Sea Level on Barriers

There are three potential concepts of how a barrier can respond to a rise in relative sea level: *aggrade inplace* (build upward), *retreat*, or be *overstepped*. The first, *aggrading inplace*, is possible only where an external source of sediment supplies a sufficient volume to build the barrier upward at a rate at least equal to the rise in sea level. If the barrier grows vertically, keeping its head above water, it can maintain a roughly fixed geographic location (Figure 1A and B) or even advance seaward. The lagoon will tend to widen as the coastal plain is submerged by increasing sea levels.

Barrier island *retreat* occurs when storm overwash, dune migration and inlet processes carry sand from the seaward side of a barrier to its lagoon side. Seaward erosion and lagoonal accretion move the entire island landward, burying back barrier (marsh and lagoonal) deposits. If retreat continues these back barrier deposits will eventually emerge on the shoreface where nearshore and inlet processes can recycle the sediment. Swift et al (1971) liken this sediment motion to a tank tread moving in a closed elliptical path, and thereby importing a unidirectional movement to the whole system. The width of the lagoon may remain constant, shrink, or expand depending on the rate of inundation at the landward side of the lagoon relative to the rate of barrier retreat at the seaward side (Figure 1C). Barrier island retreat may be considered of more concern than aggradation inplace because retreat threatens to undermine and flood the fixed structures which man builds on barriers.

The third response, *overstepping*, would have the most devastating consequences not only for fixed structures but also for the barrier itself. A barrier that does not grow upward or retreat fast enough becomes increasingly vulnerable to overwash which breaches and flattens the dunes, and sometimes carries sand into the back barrier marsh. The flatter the island becomes, the farther washover deposits penetrate into the marsh and the greater becomes the frequency and intensity of subsequent overwash. Though the ultimate effect of this process has not been observed on a major or developed barrier, it eventually destroys the barrier as an emergent sand ridge. The submerged barrier remnant is reworked by storm waves; back barrier deposits are rapidly transgressed by shoreface processes; and another barrier may be initiated on the higher, coarser grained mainland shore (Figure 1D).

Each of the foregoing responses to higher sea level have been appealed to by various investigators explaining changes in modern barriers, features on the continental shelf, and sequences in the rock column. Retreat seems to be the most frequently discussed response. For our purposes it is especially important because of current economic losses associated with barrier retreat and because it is intermediate between aggradation in place and barrier overstepping.

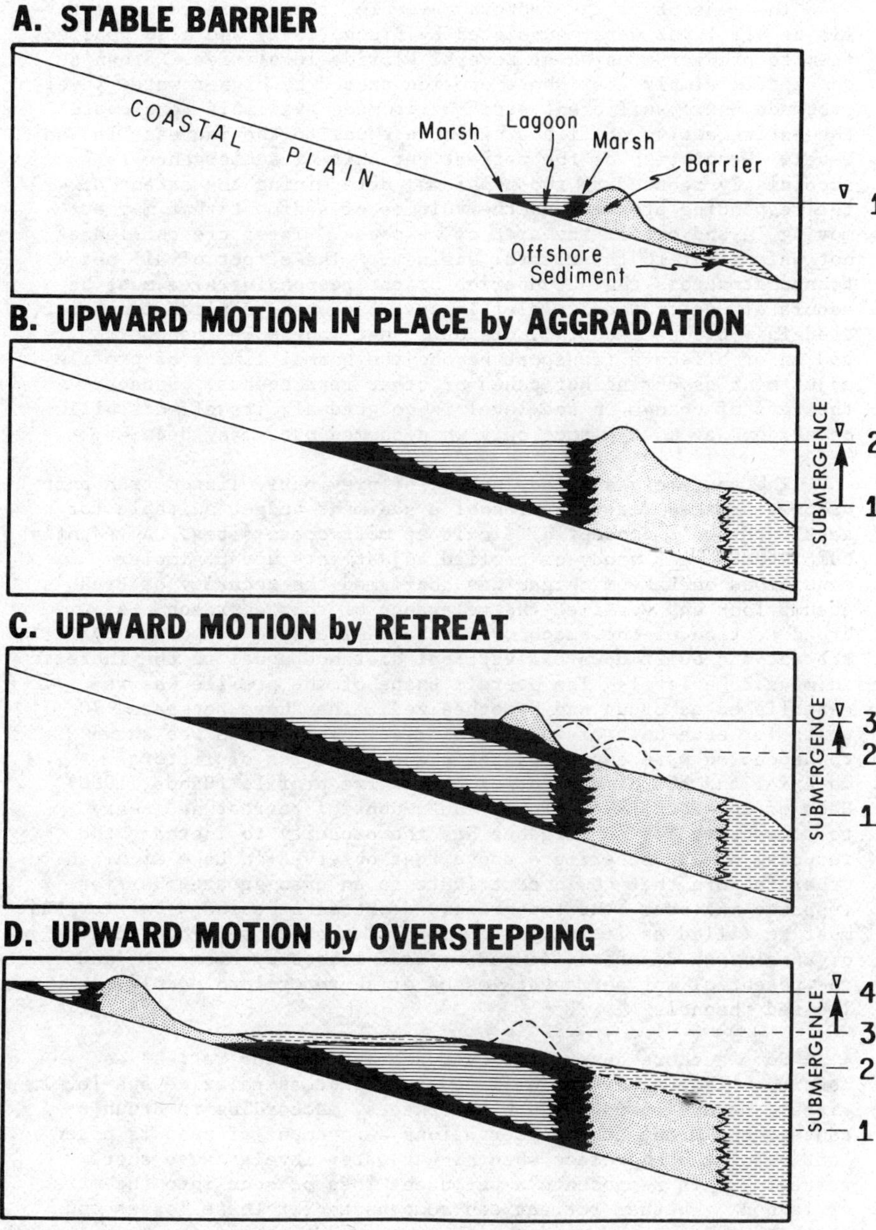

Figure 1. Three responses to rising sea level, each illustrated in the progression from A to D. Ideally the entire sequence is recorded in the geometry of buried strata shown at stage D.

The concept of the factors governing retreat in response to rising sea level were enunciated by Bruun (1962) who also applied them to predict erosion at several Florida localities. Bruun's concept is simply that shore erosion caused by higher water level proceeds until sufficient sediment is made available to elevate the entire active profile a distance equal to the increase in sea level. Prediction of the retreat per unit of submergence is accordingly reduced to two problems; determining the extent of the responding profile and the volumes of sediment that may be moving in and out of the area of response. These are challenges not unfamiliar to the coastal engineer. The effect of all net transport across the boundaries of the responding area must be accurately determined: inlet losses, overwash, aeolian transport, dredging, dredge material, disposal, net longshore transport, and on or offshore transport beyond the normal limits of profile adjustment as during huricanes or other rare events. Because the rate of change in sea level is so gradual, its effects will be evident at most places only when summed over many decades.

The extremely unsteady nature of previously listed transport processes makes determination of a sediment budget suitable for testing Bruun's concept difficult at most ocean sites. A recently completed 9 year study of profile adjustments under simpler conditions on Lake Michigan has confirmed the accuracy of Bruun's assumptions and verified the relevance of this approach. A broad section of the beach over a half mile wide responded by retreat and built upward a vertical distance equal to the increase in mean lake level. The overall shape of the profile was re-established as Bruun had hypothesized. The shore retreated 70 units for each unit of lake level rise. This ratio was shown to depend on wave exposure, the characteristics of littoral material and the dimensions of the active profile (Hands, 1980). Each of the factors governing the amount of retreat necessary to compensate for submergence has the capacity to increase the response of ocean barriers above that observed on Lake Michigan. Other factors that could contribute to an even greater barrier response include: the topographic depression behind a barrier that must be filled as it retreats, compaction of loosely-packed highly-organic marsh deposits, temporary sand losses by overwash, and the potential for permanent losses offshore or into deeply incised channels.

Some authors have cited overwash and barrier retreat as assurance that a barrier will remain intact as relative sea level rises. Nature makes no such guarantees. According to Bruun's concept and Great Lakes observations a pavement of sand is often left below the shoreface when higher water levels cause shore retreat. This represents a permanent loss of sand into the marsh or lagoon. Whether retreat can compensate for these losses and at the same time produce enough extra sand to build the barrier upward with sea level, will depend on the character of the material emerging on the shoreface. Retreat may be the only available mechanism for prolonging barrier life, but it can also be the next-to-last step in destroying it.

Activities underway or planned at the Coastal Engineering Research Center to improve our ability to anticipate barrier response to rising sea level include: coring programs to determine the stratigraphy and properties of sediment under the shoreface, barrier and back barrier environments; studies of long-term changes in both the shoreline position and the shape of the shoreface; investigations into possible onshore-offshore transport; and other elements of sediment budget analysis.

IV. LONGSHORE SEDIMENT TRANSPORT

Transgressing barrier islands are retreating for a variety of reasons. One of these is that some sand is trapped as it moves in an alongshore direction. It is therefore of considerable importance in assessing the behavior of a barrier segment under study to gain some understanding of the regional longshore transport system.

Sediment budget analyses on barrier coasts have been routinely made by coastal investigators for many years. However, techniques for this type study are not perfected in all cases and existing knowledge of how such a longshore transport system functions along a barrier island chain is deficient. One critical aspect of a longshore transport system is the discontinuities which occur at inlets and capes where sediment may be trapped and held for long periods of time, and thus withheld from downdrift areas. Because human activity and infrequent extreme events may create or trigger the formation of inlets and capes, significant changes in the regional longshore transport system may occur abruptly.

Many aspects of the effects of inlets and capes on the longshore transport system and the resultant influence on barrier development are not well understood. We have undertaken a field investigation of the barrier island chain of Virginia's eastern shore (southern Delmarva Peninsula), a relatively undeveloped barrier complex, to achieve a better understanding of this problem. This barrier chain (Figure 2) is rapidly transgressing under conditions of deficient sand supply and is a good example of the coastal compartment shoreline presented by Fisher (1967). Headlands in the north (Rehoboth and Bethany Beaches) are followed to the south by a long thin barrier island (Assateague Island) and finally the short segmented barrier islands of northeastern Virginia. An interesting anomaly is the occurence of the Assateague Island - Wallops Island updrift-offset (Figure 3), the largest updrift offset in the United States and possibly the world. This offset may be considered an accretional cape because of its morphologic setting and magnitude. Accretional capes have developed in many areas despite the general rule of erosion of headlands and smoothing of the coastline (Hoyt and Henry, 1971).

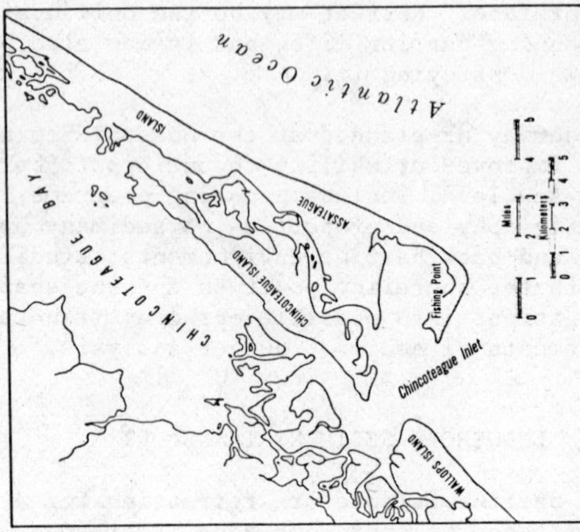

Figure 3.

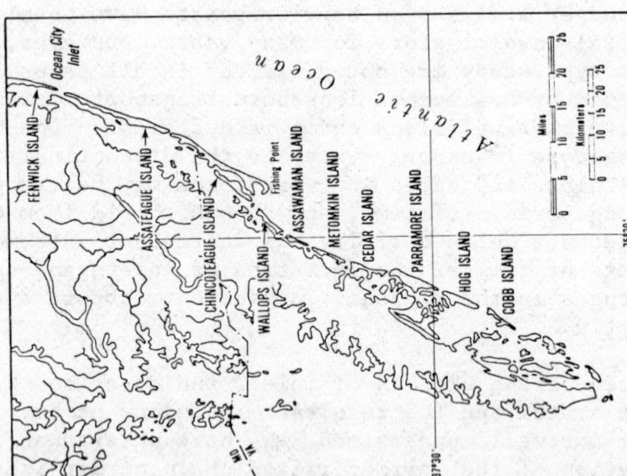

Figure 2.

The historic formation and subsequent growth of southern Assateague Island (Fishing Point) has been explained by Gawne (1966) and Geottle (1978). The implications of such an offset are obvious; large scale trapping of littoral drift and severe erosion downdrift. Why a feature of this scale has developed at all and the possibility of similar accretional capes developing elsewhere are questions that deserve further scrutiny.

1. Updrift-Offsets

The updrift-offset across Chincoteague Inlet at the distal end of Assateague Island is 16,700 feet (Figure 3). This offset is the result of the continual growth of Assateague Island while the sand starved islands to the south have retreated. The Norfolk District, Corps of Engineers (1971) estimated the net average southward longshore sediment transport rate at Fishing Point to be on the average of 380,000 cu yd/yr. It seems reasonable to believe that some portion of this spills off at the end of the spit and is either naturally bypassed to Wallops Island or accumulates on the bar of Chincoteague Inlet, leaving the remainder as the amount available for accumulation on the spit. No information is available on the strength and direction of tidal currents in the inlet; however, visual observations indicate the flood and ebb tide velocity to be in the order of 2 to 4 ft/sec (Norfolk District, Corps of Engineers, 1971).

The development of Fishing Point which began in 1857 (Gawne, 1966) is an extension of the growth of Assateague Island. Littoral drift moving south encounters the tidal currents moving through Chincoteague Inlet. This inlet has a tidal prism of 1.56×10^9 cu ft. (Jarrett, 1976) and drains most of Chincoteague Bay. The advancement of the spit has slowed as deeper water is approached and the bay entrance becomes more constricted, increasing the velocity of the tidal current. Shoreward deflection of wave and flood tide currents tend to curve the spit into the bay where opposing ebb tide currents tend to hold it.

Although updrift-offsets are the exception and not the rule on the Atlantic coast, several are quite evident. Ocrakoke Inlet located northeast of Cape Lookout is an updrift-offset but is not developing into an accretionary cape because of relatively weak tidal currents and the inability of the inlet to maintain its channel as it migrates south. Long Island inlets, for example Rockaway Inlet, East Rockaway Inlet and Fire Island Inlet are all updrift-offsets or overlapping updrift-offset inlets. This complex inlet migration involves the continued growth of the updrift barrier island until it overlaps the downdrift island while the inlet remains open and forces the tidal exchange to flow parallel to the barrier beach before entering the ocean. In this type of inlet the migration is complicated by the size of the tidal prism, volume of littoral drift, and natural bypassing of sediments. This situation outlined for Fire Island and Rockaway Inlets is similar to Chincoteague Inlet except that a combination of tidal and oceanic currents and waves keeps Assateague Island from overlapping Wallops Island.

2. Offset Inlets and Capes

The formation of the Fishing Point offset and other updrift offsets have several important characteristics in common.

1. Large net longshore sediment transport rates.
2. Few widely spaced updrift inlets.
3. Large mostly unvegetated lagoon behind the updrift barrier islands.
4. Small ebb tide delta, therefore no reversal of longshore transport due to refraction around ebb tide delta (Drumstick Model, Hayes et al., 1970).

These characteristics cause a large updrift offset or accretional cape to form partly because of the large tidal prism keeping the inlet open. Water must flow in and out of the inlet to exchange ocean and lagoon water and the strong longshore transport may restrict the inlet but cannot close it. The inlet will most probably be ebb oriented and, if restricted, the tidal currents may be quite swift.

Assateague Island is a prime example. The flood-dominated northerly inlets have closed a long thin barrier island with a large recurved spit. Assateague Island has moved southward in front of Chincoteague Island. Further south the movement of Assateague Island has slowed as the large spit, Fishing Point, has developed in a westward orientation. A possible explanation for this occurrence is large waves from the prevailing southeast direction inhibiting the overlap of Assateague Island past Wallops Island. Because of the orientation of the New Jersey shore, strong southeast winds approaching Long Island have a very short fetch and large waves from this direction may not develop. Thus Fire Island does not turn landward but continues to overlap Jones Island.

Tidal channel currents are important parameters in inlet morphology. High maximum tidal currents (ebb dominated) such as Fire Island (3.88 ft/sec), East Rockaway (3.89 ft/sec), and Rockaway (3.72 ft/sec) are all associated overlap updrift-offset inlet barrier island systems. Ocrakoke Inlet (slightly flood dominated) with a maximum current velocity of 2.36 ft/sec is a small updrift-offset barrier island system.

V. BACK BARRIER MARSH SEDIMENTATION

Marine marshes are some of the most productive lands in the world, providing nutrients for aquatic flora and fauna, but little is known about the effects of man's efforts of shore stabilization on the ecosystems of these marshes. It is first necessary to understand the conditions for the creation and maintenance of marshes and how these conditions are affected by the surrounding environments.

Numerous studies have been conducted that determine the necessary climatic and nutrient conditions required by various marsh plants. Frey and Basan (1978) list several other variables that directly affect marsh development, including

(1) availability, composition, mode of deposition and compaction of sediments;

(2) topography and extent of depositional surface; and

(3) Physical parameters such as currents, waves, tidal range, sea level change and tectonic/eustatic stability.

Several investigators have written on sediment transport in salt marsh tidal creeks, but these studies have not addressed the processes prior to, and instrumental in, the development of the marsh system, or how these processes may change with the migration of associated barrier islands with changing sealevel.

In the spring of 1980, a field study was instituted along the 11 barrier islands of the Eastern Shore of Virginia (Figure 2). To date, samples from 131 gouge auger cores 1 to 11 meters long have been taken along 22 transects in the marshes and tidal flats between the uplands of the Delmarva Peninsula and the barrier islands. These cores are described in the field for grain size, color, macrofauna and flora, and structures as a means of determining the depositional environments through time.

Although the final determinations have not been made, some of the initial field observations have shown that there does not seem to be a preferred substrate for marshes to develop on. In the northern half of the study area, the auger penetrated ½ to 1 meter of mud with up to 50% organics (marsh) and would then penetrate up to 10 meters of organic-free muds. However, in the southern half of the study area, the marsh cores consisted of sandy muds or muddy sands. A preliminary conclusion has been postulated that overwash and floodtidal delta deposits are not necessary for marsh development and growth, but rather marshes will develop on whatever substrate is available. These observations also lead to the supposition that there are multiple sources of sediment for this area.

Samples were also collected for further analyses in the laboratory. Microfauna assemblages, useful in indicating environments of deposition will be determined. They may also prove to be a valuable tool in determining source areas of sediment. Samples containing sufficient amounts of organic material will be C^{14} dated, and clay mineralogies will be compared to additional samples taken along creeks feeding into the bays from the mainland as a possible additional source for sediment infilling behind the barriers. An additional six transects have 34 vibracore drillholes and these cores have been brought back to the lab to be opened and analyzed. Studies by other workers in this same area provide a secondary data base for this study.

The field work has not yet been completed, and the analyses of sediments has just been started. Detailed cross-sections are being constructed and, applying sea level curve information for Delmarva, time lines will be drawn and paleogeographic maps constructed to show the evolution of the various environments. Historical shoreline changes will be used to document recent changes in the morphology of the marshes as the barriers have migrated and ephemeral inlets have opened and closed affecting the amounts and type of sediment infilling the back barrier area.

By looking at the stratigraphy of the marsh, shoreline changes, relative sea level rise and possible sources of sediment for the back barrier area, it is hoped that a model can be developed to predict the origin of marsh development and, therefore, how the marsh will be affected by changing any of the above variables responsible for its development. In this way, future projects planned for any development of our coastlines can be modified to protect the valuable natural resource of back barrier marshlands.

VI. SEDIMENT SOURCES

Each sediment particle in a barrier complex has been derived from a specific source which in many cases can be identified by comparative study of the detrital grains and samples of rock and sediment from possible sources. These sources can be classified into two general categories: immediate sources, and primary sources. The immediate source of a particle is the deposit from which it was eroded immediately prior to its transportation to the deposit where found. The primary source is the parent rock from which the particle was originally detached. Between the time of original detachment and deposition in a study area the particle may have undergone one or more cycles of long-term deposition in secondary deposits.

The immediate source of sediment particle in a barrier complex is of most concern in coastal engineering studies. It is also important to know if the immediate source is still actively contributing sediment or if it contributed only in the past under environmental conditions that no longer exist.

The principal sources of barrier complex sediments are (a) erosion of headlands, (b) fluvial detritus, (c) continental shelf sediments, (d) skeletal remains of organisms and (e) mineral grains precipitated from sea water (authogenic). In addition, redistribution of sediment within a barrier complex by the erosion of sediment in one subenvironment or barrier segment and transfer of this sediment to another subenvironment or segment is an important factor. Unlike the first named sources this transfer does not involve any net increase in the sediment volume of the barrier complex as a whole, but may be important in maintaining certain subelements such as the beach and dunes.

Considering the scope and complexity of the subject it is not surprising that despite considerable study, many aspects of sediment supply to barrier complexes remain obscure. Only a few of the many questions arising from consideration of how sediments are produced and distributed in the barrier environment can be discussed here. We have selected the following because we believe that they are questions of considerable importance and concern areas where present knowledge is deficient: (a) How can techniques for sediment source analysis using natural tracers be improved? (b) What is the importance of relict and recycled sediments cropping out on the shoreface to nourishment of the adjacent barrier beach and dunes? (c) Is the continental shelf a major source of barrier sediments or has it been in the past? (d) Are locally produced biogenic and authogenic sediment particles an important source of sediment in certain locales?

1. Sediment Tracers

Relating a sediment particle to a source area by comparative analysis of minerology or other physical characteristics is a technique widely used in sedimentation studies. Any individual particle in a sediment deposit is potentially useful as a natural tracer, but stable non-opaque minerals are most often employed for this purpose. Opaque minerals, oolites, quartz grains, and organic skeletal debris have also been used successfully.

In coastal plain areas where most barriers are located, the stable heavy minerals, because of their chemical stability and mechanical durability, have often survived many episodes of long-term deposition in coastal plain deposits and have consequently been widely distributed throughout the region. For this reason it is not often possible to locate a localized immediate source for these particles, though the ultimate source can in many cases by established with confidence.

The use of organic skeletal parts as natural tracers has not as yet been widely employed; however, its potential is promising because many marine organisms inhabit a relatively restricted area in which life conditions are suitable. Thus organic skeletal debris eroded from an outcropping fossiliferous deposit may represent a biofacies that is confined to a small area relative to the extent of the parent unit. Similarly, the hard parts of modern organisms with a restricted living range may be transported to areas where it is clearly exotic and potentially an indicator of source.

Other sediment constituents that have been used as natural tracers in coastal or continental shelf deposits are glauconite and phosphorite grains, mica, calcareous oolites, rock fragments, man-made artifacts and quartz grain shape. It is possible that a detailed analysis of some of these particles to reveal differences in chemical composition or some other attributes might be useful to further narrow the number and extent of possible source areas.

In our current barrier island studies we are reviewing existing techniques of using natural tracers for locating the sources of barrier sediment. We are especially interested in sediment constituents other than heavy minerals that might ultimately be more useful for determining immediate source. For example, in work carried out this year we have found relict Cretaceous and Tertiary foraminifera in beach and shoreface sediments on nearly all the beaches between Cape Lookout, North Carolina and Jacksonville Beach, Florida. These may prove to be valuable natural tracers because they may have a less diffuse source than the heavy minerals. Ideally it would be possible to be able to work with two or more natural tracers in any sediment body. Recent research such as the application of detailed shape analysis to quartz grains as a tracer characteristic (Ehrlich, Orzeck, and Weinberg, 1974) is increasing the number of sediment constituents potentially useful for this purpose.

Once a source for a sediment particle is located a formidable problem remains, namely how to determine the net volume of sediments moving from the source to the area under study. In some situations it may be possible to use techniques of sediment budget analysis for this purpose but in many this may not be possible. We hope to make some inroads on this problem during the course of our studies.

2. Relict and Recycling Sediments

On retreating barriers the most likely internal sources of sediment for the beach and dunes are deposits emerging on the shoreface from beneath the barrier. Sediments eroded from these deposits enter the active barrier island sedimentation system and may be distributed to any part of the barrier or transported offshore to the shelf. They are thought to be particularly important in maintaining the beach and dune areas of retreating barriers.

Relict deposits beneath a retreating barrier usually consist of marsh, lagoon, tidal creek and other low energy deposits formed in the lee of the barrier. In some places the pre-Holocene substrate also forms shore face outcrops consisting of a wide variety of sediment and rock lithologies. Recycled sediments have at one time been part of, or have been transported across, the sandy barrier superstructure. Chiefly these sediment bodies consist of overwash deposits, flood tidal shoals, inlet fill and sand from landward migrating dunes. On retreating barriers these deposits are overridden by the barrier superstructure and eventually crop out on the shoreface where they are eroded and again enter the active sediment mass.

Considering the many lithologies of relict and recycled sediments that may lie beneath a barrier it seems reasonable to expect that sediments of different character might be exposed to shoreface erosion at different places along a given segment of coast. It could also be expected that as a barrier retreats it may pass over different sediment bodies which become exposed sequentially on the shoreface. For example a sandy relict flood tidal shoal feeding sand-size sediment into the sedimentation system might be succeeded in turn by a silty lagoon mud which could significantly alter the previous sediment distribution patterns. Significant changes in beaches through time or variations from place to place may be more the product of such variations in shoreface lithology than a result of climatic or hydrodynamic factors.

We expect in our current research effort to obtain continuous drilled or cored sediment samples or barriers to obtain additional data on the nature and distribution of the various internal sediment bodies in a barrier island. We will also consider the possibility of using such three-dimensional data to predict changes in shoreface outcrop lithologies through time on a retreating barrier.

3. Continental Shelf

In our review of the literature concerning sediment sources for Atlantic and Gulf coast barriers we learned that evidence has been found of shelf derivation for at least some of the sand-size material in the majority of studies. This was particularly true of Atlantic barriers south of Cape Cod. These findings were, for the most part, based on natural tracers including oolites, heavy minerals, phosphorite, mica, and quartz grain shape. Two important questions arise as a result of these findings: (a) How much sediment is contributed by shelf source? and (b) Is this sediment being contributed under present conditions or was it contributed in the past under conditions no longer extant?

In regard to the first question a small number of studies provide some pertinent information. These studies were based on sediment budget analysis of barrier coast segments in which it was found that a portion of the sediment accretion in the segment studies could not be accounted for by any other possible source than the offlying continental shelf (see for example Pierce, 1969). The volume of material apparently coming from shelf sources as calculated in these studies was large and indicated that at least in these areas shelf sources may be indeed important.

In regard to the second question the above mentioned studies also indicate that the shelf contribution must be an ongoing process since the accretion was measured from sequential bathymetric and topographic surveys. In addition some of the natural tracers such as the relict foraminiferal tests mentioned previously are delicate and unlikely to long survive in the high energy littoral and beach environments. Consequently their presence suggests frequent replenishment.

Whether or not shelf sources are presently active in only a few places or shelf contributions are widespread is obscure. Some investigators believe that the barrier sediments derived from the shelf were swept off the shelf floor and driven landward at the leading edge of the transgressing Holocene sea; thus this contribution may have been due to special environmental conditions which existed in the past and may not exist today. If this is true the shelf sand contained in the barriers is relict and if lost will not be replaced.

We have available to us over 1200 cores up to 6 meters long from the inner continental shelf of the Atlantic and a small portion of the Gulf coast collected as part of a previous unrelated research program. These cores along with numerous barrier samples obtained for our current program will be used as a data base for investigating the importance of shelf sediments to barrier development and maintenance in selected locales, and whether shelf derived sediments are relict or the product of ongoing processes.

4. Biogenic and Authogenic Sources

Biogenic and authogenic components of a sediment deposit have been formed by precipitation of chemical elements from sea water; one by organic processes and the other through inorganic chemical reactions. Authogenic particles most commonly occurring in sediments are glauconite, phosphorite and calcareous oolites. They are rarely a dominant constituent of sediment deposits and usually if present occur in small quantity. Organic constituents consist mainly of calcium carbonate skeletal material usually dominated by mollusk shell. Locally, especially in areas where sediment influx from the adjacent landmass is low, organic sediments are the dominant constituent in the coastal deposits. In many other places organic components, though not dominant, contribute a significant fraction of the particles.

The importance of shell and authogenic particles in a sediment body is often not apparent if the particles are of the same size as the matrix or tend to be concentrated in localized parts of the deposit. Careful sampling and examination under a hand lens or microscope is often necessary to fully appraise the contribution of these elements to the total sediment mass. However, even where their volumetric contribution is small they may have an important influence on the bulk properties and sedimentological characteristics of the sediment.

We have examined a sizeable number of sediment samples from barrier environments in detail in order to determine the proportion of authogenic and biogenic grains as well as investigate potential natural tracers. We intend to study certain aspects of the non-quartz fraction such as the effective hydraulic diameter and susceptibility of these particles to mechanical degradation in the high energy beach and littoral environments.

VII. EFFECTS OF HUMAN ACTIVITY

A discussion of this problem by Everts appears elsewhere in this volume, consequently it will only be briefly summarized here. The effects of human activity on the coasts since European settlement in North America began have been profound in some areas, while in others there appears to be little change from the pristine condition. In general the impact of human activity began earliest at places where coastal morphology and bathymetry favored development of harbors. Intensive use of the shore area for dwellings and recreational purposes, particularly on barrier islands, is a relatively late development.

Human activities that result in modification of the barrier environment fall into two general categories. First there are those activities that are specifically intended to modify the existing nature or processes of some part of the environment. Most coastal engineering projects would fall into this category. Second there are activities which are primarily intended for some purpose other than modification of the barrier environment, but have incidental effects that may, nonetheless, cause significant changes.

Human activities in both categories have been so pervasive on some barriers that there seems to be little of the original environment remaining. It was suggested in our workshops that on such barriers the modifications in morphology, biology and processes have been so profound that the barrier can no longer be addressed in the same terms that apply to barriers in a pristine state because a highly developed barrier responds in fundamentally different ways to environmental factors than a relatively undeveloped barrier.

We intend during the course of our barrier island project to investigate this question. We will do so by making an intensive study of a highly developed barrier segment to try and understand the behavior of the segment as a whole and the relationship of that behavior to man-made and natural factors. Preferably the barrier segment chosen for this study will be one which has been the object of study prior to development so that pertinent baseline data are available.

VIII. SUMMARY

Coastal barrier complexes front a large part of our Atlantic and Gulf coasts and provide important benefits to the public. In order to better carry out its civil works responsibilities on barrier coasts the Corps of Engineers, through its Coastal Engineering Research Center, has undertaken a study of problems affecting the planning and conduct of engineering works in the barrier environment.

Two workshops attended by specialists in the preservation and management of barrier complexes were held by the CERC in the fall of 1979 in order to obtain information needed to develop a research plan. From these workshops, review of literature and in house discussions with CERC researchers a number of problem areas and research needs were defined.

Some of the principal problems discussed in this presentation are concerned with barrier evolution; effects of sea level changes, sediment supply, development of back barrier environments and effects of human activity.

The CERC barrier sedimentation studies program was actively perused during the past winter with planning and research efforts. During the summer an extensive program of data collection was perused with collection of a large number of cores and sediment samples in the extensive barrier complex on Virginia's eastern shore and on barrier islands of the southeastern Atlantic coast. Analysis of this data is now under way.

ACKNOWLEDGEMENTS

This paper is a product of the Barrier Island Sedimentation Study at the Coastal Engineering Research Center, carried out under the Civil Works Research Program of the U.S. Army Corps of Engineers. Conclusions are not to be construed as officially those of the Department of the Army unless designated by other authorized documents. Permission to publish this paper is appreciated.

LITERATURE CITED

BRUUN, P., "Sea-Level Rise as a Cause of Shore Erosion," Journal of the Waterways, Harbors, and Coastal Engineering Division, American Society of Civil Engineers Proceedings, Vol. 88, No. WW1, 1962, p. 117-130.

DE BEAUMONT, E., "Lecons de Geologie Pratique," in Bertrano, P., ed., Paris, pp. 223-252.

EHRLICH, R., ORZECK, J. and WEINBERG, B., "Detrital Quartz as a Natural Tracer-Fourier Grain Shape Analysis," Journal of Sedimentary Petrology, Vol. 44, No. 1, 1974, pp. 145-150.

FIELD, M.E., and DUANE, D.B., "Post-Pleistocene History of the Inner Continental Shelf: Significance to Origin of Barrier Islands", Bulletin of the Geological Society of America; Vol. 87, 1976, pp. 691-702.

FISHER, J.J., "Origin of Barrier Island Chain Shorelines: Middle Atlantic States," Geological Society of America, Special Paper 115, 1967, pp. 66-67.

FISHER, J.J., "Barrier Island Formation: Discussion," Bulletin of the Geological Society of America, Vol. 79, 1968, pp. 1421-1426.

FREY, R.W., and BASAN, P.B., Coastal salt marshes: in Coastal Sed. Environments, R. A. Davis (ed.), Springer-Verlag, pp. 101-159.

GAWNE, C.E., "Shore Changes on Fenwick and Assateague Islands, Maryland and Virginia," Unpublished B.A. Thesis, University of Illinois, 1966, p. 140.

GILBERT, G.K., "The Topographic Features of Lake Shores," U.S. Geological Survey, 5th Annual Report, 1885, pp. 69-123.

GOETTLE, M.S., "Geological Development of the Southern Portion of Assateague Island, Virginia," Unpublished M.S. Thesis, University of Delaware, 1978, p. 186.

HANDS, EDWARD, "Prediction of Shore Retreat and Nearshore Profile Adjustment to Rising Water Levels on the Great Lakes," U.S. ARMY CORPS OF ENGINEERS, Coastal Engineering Research Center, Fort Belvoir, VA, in prep., 1980.

HAYS, M.O., GOLDSMITH, V., and HOBBS, C.H., "Offset Coastal Inlets," *Proceedings of the 12th Coastal Engineering Conference*, Washington, D.C., 1970, pp. 1187-1200.

HOYT, J.H., "Barrier Island Formation," Bulletin of the Geological Society of America, Vol. 78. pp. 1125-1136.

HOYT, J.H., and HENRY, V.J., "Origin of Capes and Shoals Along the Southeastern Coast of the United States," Bulletin of the Geological Society of America, Vol. 82, 1971, pp. 59-66.

JARRETT, J.T., "Tidal Prism-Inlet Area RElationships," Department of the Army, Corps of Engineers, GITI Report No. 3, 1976, p.32.

PIERCE, J.W., "Sediment Budge along a Barrier Island Chains," Sedimentary Geology, Vol. 3, 1969, pp. 5-16.

SCHWARTZ, M.L., "The Multiple Causality of Barrier Islands," Journal of Geology, Vol. 79, 1971, pp. 91-94.

TANNER, W.F., "Florida Coastal Classification," Transactions of the Gulf Coast Association of Geological Societies, Vol. 10, 1960, pp. 259-266.

GEOLOGY OF DELAWARE BARRIERS: ENVIRONMENTAL PLANNING

by

Kathleen Drew[1] and John C. Kraft[1,2]

[1]College of Marine Studies
University of Delaware
Newark, Delaware 19711

[2]Department of Geology
University of Delaware
Newark, Delaware 19711

ABSTRACT

Land-use planning decisions are frequently based on social variables, political considerations, aesthetics, and recreational opportunities, regardless of existing environmental assets and hazards. The western shore of Delaware Bay from Port Mahon to Broadkill Beach illustrates this problem. Considerable variation exists in geological structure, littoral processes, and land-use planning along this coast, but often there is no relationship between geological factors and land use. This study integrates this information for eleven areas along this coast, to evaluate their present status and make recommendations regarding their management. All except one are estuarine washover barriers. Of these, Bowers (not a washover barrier), South Bowers, Slaughter Beach, and Broadkill Beach were chosen for more detailed study based on data availability, the presence of substantial communities and need for future planning, and demographic and geological variability.

Geological analysis includes structural, meteorological, littoral, and storm data. Planning information includes legal constraints on construction, physical information applicable to planning, and interviews with residents regarding their perceptions of flood hazards and geophysical changes.

Along this shoreline, erosion rates vary from 1-6 meters/year, over a 100-year interval. This is due, in part, to relative sea level rise, which has varied from .17 m/century (over the past 2000 years) to .33 m/century in the present century. Subsidence and compaction also play a role in areas overlying Pleistocene river valleys (e.g. South Bowers). Protective measures that have been used include sheet steel piling and wooden bulkheads, beach nourishment, groin fields, and sediment dikes; all are failing, and thus temporary solutions at best.

To reduce hazards to residents, destruction of coastal ecosystems, loss of lives, and massive expenditures of governmental and individual funds, the importance of considering geological variables in land-use planning of coasts must be realized. Studies like this can help planners achieve the necessary balance between the needs and desires of coastal landholders and visitors and the basically unchangeable geological characteristics of an area, to achieve these goals.

INTRODUCTION

Since the waning of the latest Wisconsin ice mass in North America starting approximately 16,000 years ago, world sea level has risen at various rates and the oceans have transgressed landward. With this coastal transgression, the world's population living in the coastal zone has also been forced to migrate landward and adjust to continually changing coastal morphologies. At present, approximately 50% of the U. S. population lives within 50 miles of an oceanic or Great Lakes coast, and if this trend continues, the need for proper planning of land usage of coastal barriers in these areas will become increasingly critical. Although development is often affected more by political, social and economic factors, many living on the coast are also aware that environmental factors, especially the geology of the area, should also play a major role in determining which areas are suitable for development. This problem is well illustrated by the many towns and villages along the western shore of Delaware Bay (Figure 1). This shoreline is formed of a thin, narrow sandy barrier between Delaware Bay and the coastal marshes, varying from 500 m wide in the south at Broadkill Beach to almost no sand in the transport system at the northern end of Delaware Bay at Port Mahon at the lower end of the tidal Delaware River. Although the villages appear surficially similar, considerable variation exists in the subsurface stratigraphy of these areas, which in turn affects compaction rates, erosion rates, flooding frequency, and, in turn, the advisability of development and/or beach preservation programs (Kraft and others, 1976; Kraft and others 1979).

Information was collected for eleven areas from Port Mahon south to Broadkill Beach, regarding 1) subsurface stratigraphy, 2) erosion rates, 3) longshore transport rates, 4) frequency and causes of flooding, 5) beach preservation measures that have been taken, and 6) legal restraints on building (which are often designed specifically for flood-prone areas). Great geologic variability was found between these areas despite their geographic proximity (the entire length of coastline included is only 50 km). Three villages were studied intensively: Bowers, South Bowers, and Broadkill Beach. In these areas, a series of cross sections were constructed using jet-wash, hand-driven gouge auger, and vibracore techniques. In addition, interviews were conducted to determine demographic characteristics of the town, residents' perceptions of beach changes that have occurred, both natural (e.g. due to storms and seasonal changes) and man-induced (e.g. beach fill, groin construction, etc.). Residents were also asked for their opinions regarding the applicability of different possible beach protection measures to their town, and who should be financially responsible for beach maintenance.

Figure 1. Delaware Bay showing the areas of study. Black area is coastal salt marsh. Rates of erosion are based on Federal coastal surveys over the past 1-1/3 centuries.

Figure 1

WESTERN DELAWARE BAY

LOCATIONS/EROSION RATES

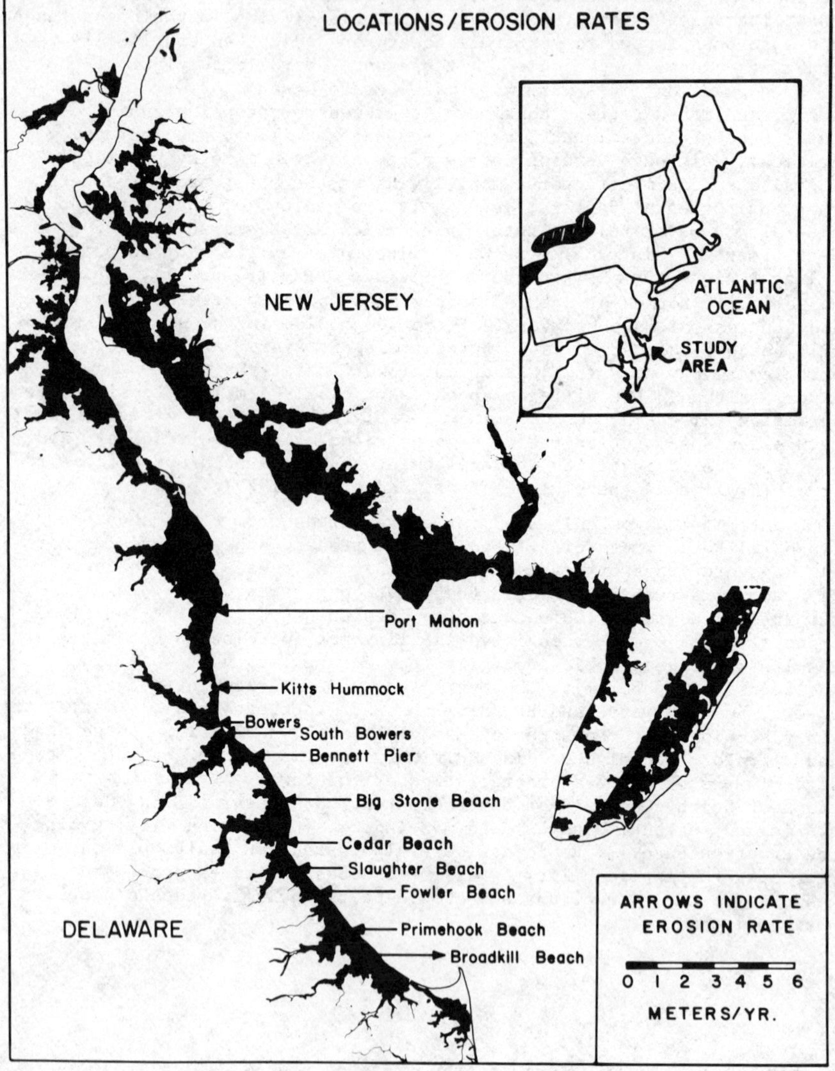

STUDY AREAS

Bowers

Bowers is a fishing village that has been populated since colonial times. It has a relatively high percentage of permanent residents (72% of the people interviewed for this report were permanent residents, and more than 50% had lived there for greater than ten years). The population varied from 324 in 1960 to 268 in 1970, and then increased to 286 by 1977. The town consists of approximately three miles square of land, 3/4 of which is occupied. Bowers is divided into several sections. Most permanent homes, built in the 1900's-1920's, are located in the center of town, north of the Murderkill River, and back from Delaware Bay on higher ground. Along the bayfront, many of the homes in the southernmost area were destroyed in the 1962 Ash Wednesday "northeaster," and were replaced by small, temporary houses and trailers. These are mainly used as summer homes, and many belong to residents of New Jersey, Pennsylvania and Delaware. North of the main town is a section of homes that were built in the 1930's-1940's, belonging to long-term (more than eight years) summer residents, and still farther north lies a section of houses of similar age belonging to short-term seasonal residents. The northernmost section of Bowers is currently wetlands (*Spartina* sp. and *Baccharus* sp.) in which "beachfront lots" are being sold.

Plan view (Figure 2) and cross section A-A' (Figure 3a) show that most of Bowers consists of a thin sand lens directly overlying Pleistocene compact sediments which are an extension of Murderkill Neck. These sediments were deposited during the Sangamon interglacial age, approximately 80,000-100,000 years B. P., and consist of relatively strongly-compacted and oxidized sands and silty muds (Kraft and John, 1976). The wetlands found in the northern section of Bowers are underlain by Holocene marsh muds, but the slope of the late Pleistocene southern valley wall of the St. Jones River is relatively low, so the muds reach a maximum of only about 2 m approximately 200 m south of the river.

Figure 2. A plan view of the Bowers, South Bowers area showing coastal environments, coastal erosion rates (arrows), location of cores and jet wash holes (X), and lines of cross section Figures 3a and b and 4.

Figure 3a. A geologic cross section of the coastal zone at Bowers showing coastal environment sedimentary units encountered by the drill and Pleistocene-Holocene Epoch sediment interrelations.

Figure 3b. A geologic cross section at South Bowers showing coastal environment sedimentary units encountered by the drill. Note the great thickness of soft, compacting Holocene salt marsh muds dating back to 10,000 years before present, overlain by an extremely thin (1.5 m) transgressive barrier.

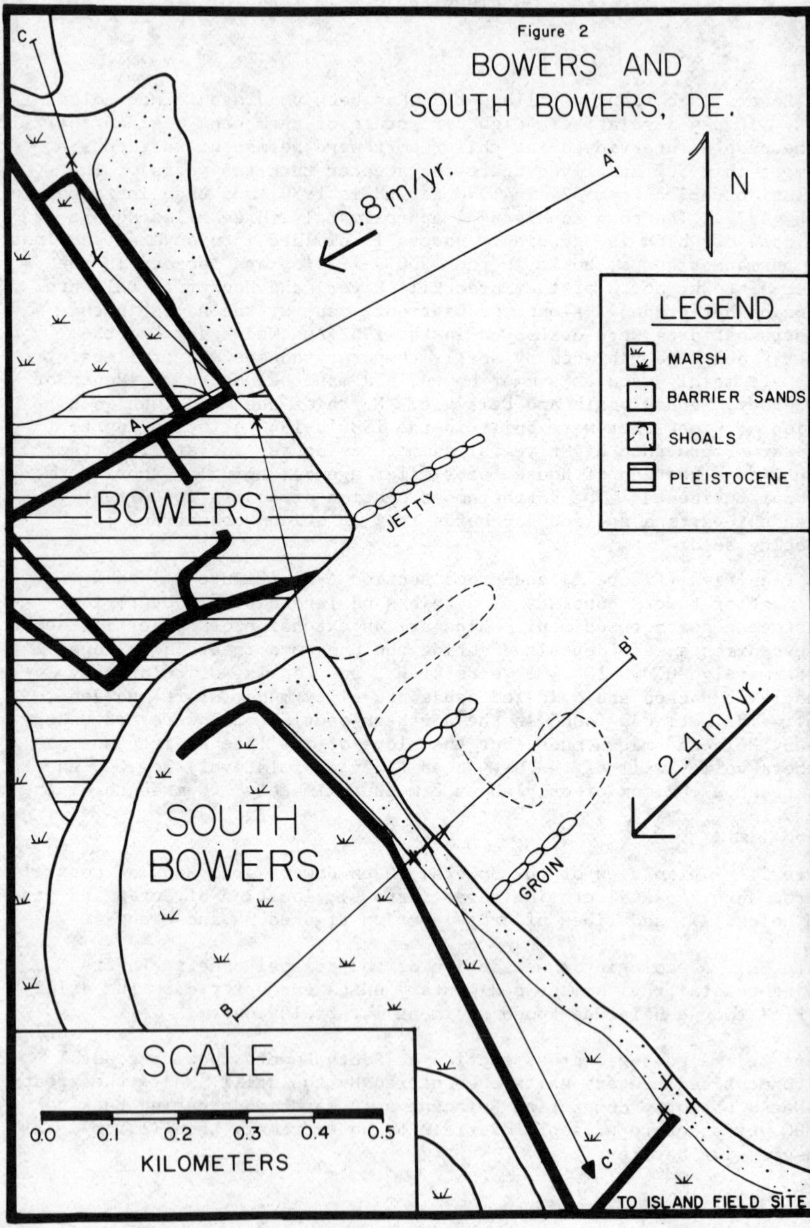
Figure 2
BOWERS AND
SOUTH BOWERS, DE.
N
0.8 m/yr.
2.4 m/yr.
LEGEND
MARSH
BARRIER SANDS
SHOALS
PLEISTOCENE
BOWERS
SOUTH
BOWERS
JETTY
GROIN
A
A'
B
B'
C
C'
SCALE
0.0 0.1 0.2 0.3 0.4 0.5
KILOMETERS
TO ISLAND FIELD SITE

Figure 3

NORTH AND SOUTH BOWERS, DE.

CROSS-SECTIONS

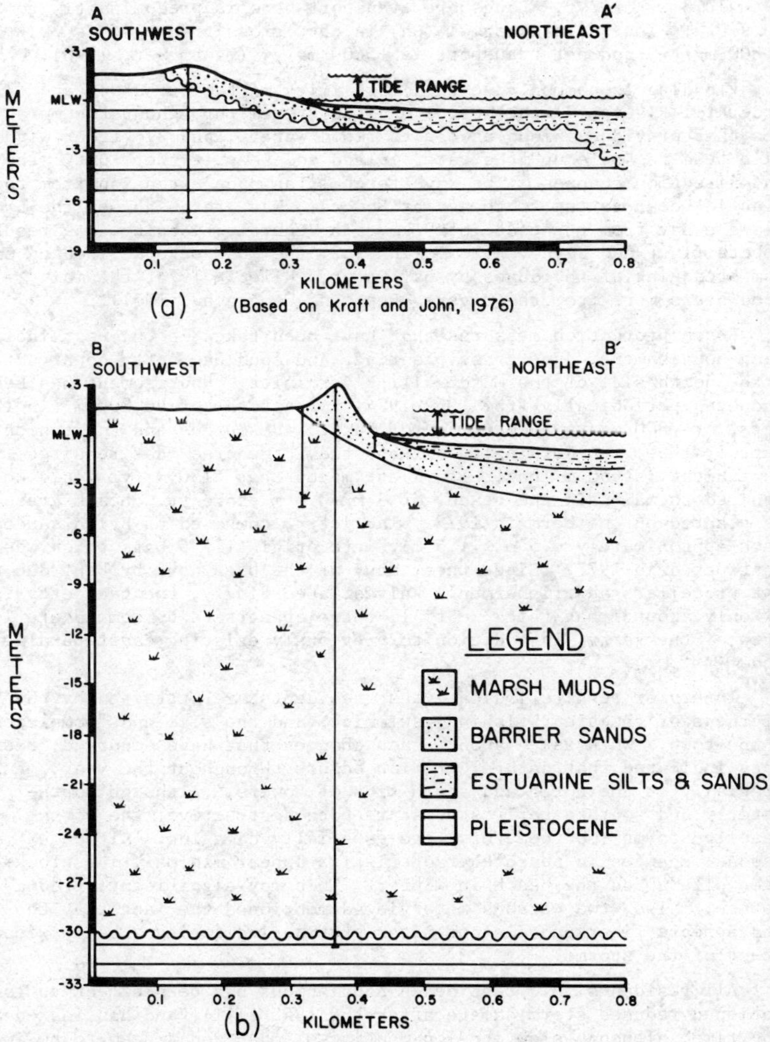

The highlands of Bowers are eroding and supplying sediment by littoral transport to the shores of Delaware Bay, but erosion in Bowers is occurring at a much lower rate than in areas to the north and south. Erosion rates have increased from 1.5 m/yr from 1843-1884 to 2.2 m/yr from 1884-1910; however, from 1910 to 1954 the average erosion rate was only 0.3 m/yr. The erosion rate from 1954-1964 can also be expressed as 0.9 m^3/0.3m shore/yr. Longshore transport here is predominantly toward the south. The outward normal (Θn) is 65°, gross transport is 12,800 m^3/yr, and net transport is $\div$5400 m^3/yr (Maurmeyer, 1978).

Flooding is common, due to 1) proximity of the St. Jones and Murderkill Rivers, 2) the mosquito ditch system throughout the marsh area that provides inland access to flood waters, and 3) strong winds, which tend to "pile up" the water inland and keep it there until the wind direction changes. The wave fetch is an open extension from the Atlantic Ocean to the southeast at Bowers. Wave fetch across the broad Delaware Bay from the east is 29 km. In general, flooding is a result of both river and bay overflow. In 1962, the most severe flooding came from breaching of the dunes by bay water, while in 1974, the main floodwaters were from the rivers (Dept. of Geography, 1977).

Beach protection measures that have been taken in the past include beach nourishment, dune grass planting, and sand bag jetty construction at the north side of the Murderkill River inlet. Nourishment has been conducted periodically from 1962-1974, for a total of 90,800 m^3 in 12 years, or 7500 m^3/yr (Department of Natural Resources and Environmental Control, 1980). According to a 1966 study, the area that requires nourishment is 660 m long. It is estimated that 5150 m^3/yr would be required to maintain the beach, or 3 m^3/0.3 m shore/yr. Beach grass was planted on the berm in 1978. The jetty, composed of nylon sandbags (each approximately 0.5 m x 1.5 m), were originally 5 bags thick when constructed in 1977. Since then, many of the bags have broken, due to boat propellers and fishermen's knives; accordingly, the bag jetty is now only about 1 bag thick. Still, there appears to be some accretion north of the jetty, in addition to previously existing coast-parallel sand bars.

Interview results indicate that residents are attracted by the quietness of the area, fishing, swimming, and the size and "completeness" of the town. When asked about beach changes that have occurred, residents indicated that natural erosion occurs throughout the year, especially at the north and south tips of Bowers. Although northeasterly and southeasterly storms are both destructive, the northeasterly storms (Northeasters) are especially damaging. Although seasonal changes in beach characteristics appear minimal, ice blocks often pile up on the beach in winter. This may significantly reduce erosion. Also, two persons interviewed mentioned the fact that the sand appears finer and cleaner after storms, indicating a beach winnowing effect of the storms.

All residents said that beach nourishment has been effective (one mentioned reduced storm damage after the 1962 fill), and has lasted a long time, although slow erosion continues (most don't seem to be aware of the three times nourishment was performed between 1968 and 1973). There may be some need for further nourishment at present at the north and south ends of Bowers Beach, but not where interviewees live. All residents interviewed except one believe that beach grass has helped reduce erosion, especially erosion due to winds. Bag jetties are generally looked upon favorably, but need constant maintenance. Lengthening the jetties has been considered, and some feel that this may alleviate the erosion problem; the current practice of dredging a channel beyond the jetties and piling the material on the sides has not been effective, since the channel fills in after two or three years. Another respondent said the jetties should not be too long because that could lead to infilling of the channel with mud from the turbid waters of Delaware Bay.

All residents agreed that the State or Federal Government should be financially responsible, and one said that if the beaches weren't signed over to the State for protection as they were circa 1975, it should be left up to the individuals to preserve their own property. Most people would probably not be willing to contribute money since either 1) they only use the area on weekends, 2) they are more fishing-oriented and the beach isn't important to them (this is especially true of the townspeople), or 3) since they signed the beaches over to the State, it should not be the individual's responsibility.

South Bowers

In contrast to Bowers Beach, South Bowers has very few permanent residents. According to one resident, there are now only three permanent households, and a survey in 1965 showed five year-round residents and 120 summer residents; i.e., people who own or rent and reside in the area during the summer, and vacationers who spend more than several days there (Office of the U. S. Corps of Engineers, 1967). There are approximately 45 summer cottages, three-fourths of which are along the bay, and one-fourth along the river. Most of these are on stilts and were built between 1900 and 1930. Although there are several docks and boats along the south bank of the Murderkill River, fishing is less important to South Bowers' residents and visitors than those of Bowers. The town is not subdivided to a great extent, although the few permanent residents tend to live along the riverfront.

The barrier sands of South Bowers, which are found along the bay and river shores (Figure 2), cover a fairly wide area (approximately 70 m) but are quite thin, with a maximum thickness of about 1.5 m beneath the dune crest (Maurmeyer, 1978). These sands directly overly 30 m of Holocene soft marsh muds (Figure 3b) which infilled the late Pleistocene valley of the Murderkill shown by a coast-parallel cross section through Bowers, South Bowers to the Island Field site located to the south of South Bowers (Figure 4). The steepness of the northern

Figure 4. A coast parallel geologic cross section extending from the St. Jones River marsh north of Bowers southward across Bowers highland and the deeply incised, but Holocene sediment-infilled ancestral valley of the Murderkill River at South Bowers, thence southward to the highland at Island Field. Holocene coastal sedimentary environmental units are shown in relation to the town sites and the Pleistocene surface.

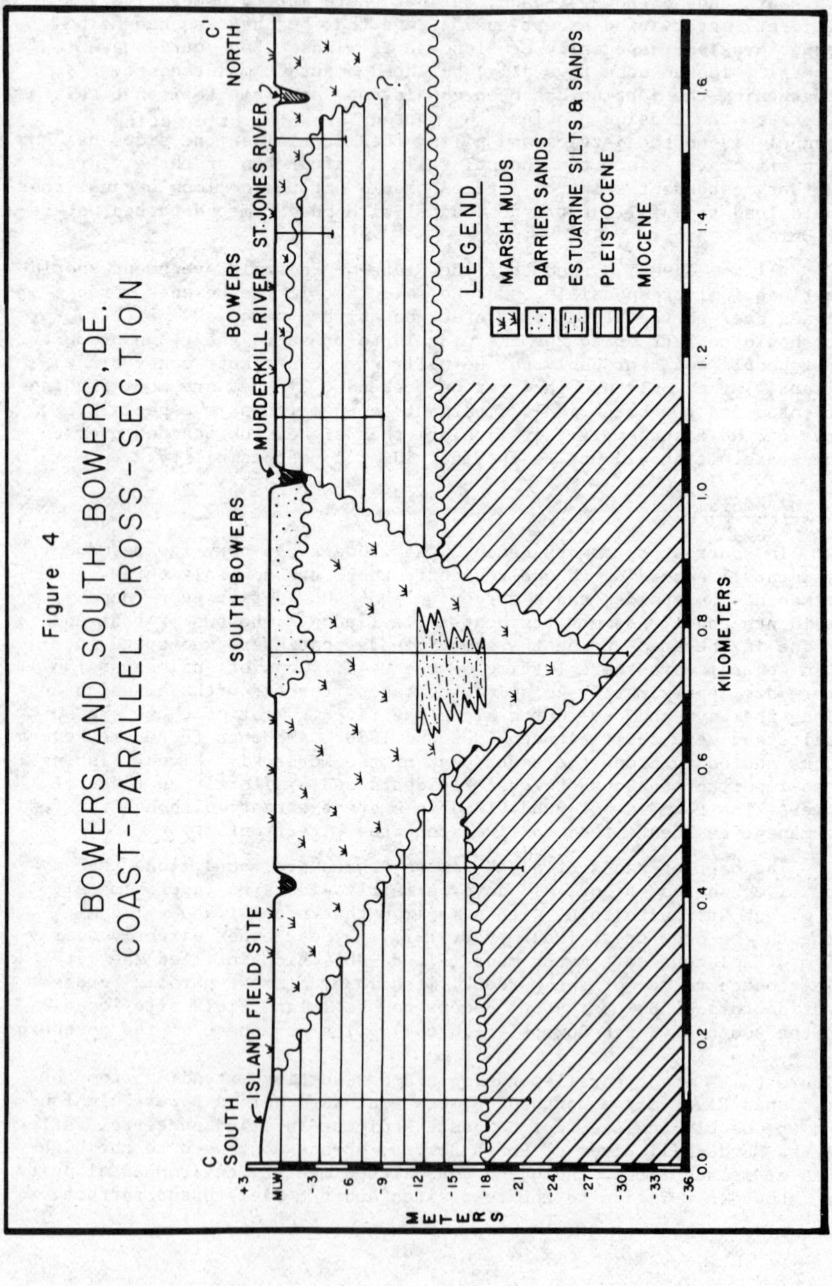

Figure 4

BOWERS AND SOUTH BOWERS, DE.
COAST-PARALLEL CROSS-SECTION

wall of the ancestral valley accounts for the fact that Pleistocene material crops out at the surface of Bowers, while it is found about 30 m below the surface of South Bowers, which is less than 30 horizontal meters away. Except where covered by barrier sands, the marsh muds of South Bowers infill the ancestral Murderkill River valley (15,000 years old), appearing as an extensive *Spartina alterniflora* marsh to the south and west of town.

The gently-sloping beach face (approximately 1:16) and wide, flat washover terrace are characteristic of an overwash-dominated system (Kraft, 1976 and Kraft and John, 1976). In addition, a cross section through the barrier (Figure 3 b) shows internal sedimentary structures of sand, gravel, and shells in landward-dipping, foreset cross-laminations, truncated at the seaward side, suggesting that erosion and washover deposition are dominant processes here. Erosion/accretion rates have varied considerably from 1843 to 1954. From 1843-1884, South Bowers accreted at an average rate of 2.2 m/yr. However, from 1884-1910, there was an average erosion of 0.6 m/yr, which increased to 1.4 m/yr from 1910-1954 (Kraft and others, 1976). Erosion rates were expressed as 2.4 m^3/0.3 m shore/yr from 1954-1964 (Office of the U. S. Corps of Engineers, 1967). However, these figures are somewhat misleading, since the situation at the north end of South Bowers is quite different from that of the south end. The north end has had variable accretion and erosion due to 1) the change in position of the outlet of the Murderkill River, 2) the emplacement and reworking of dredged material (Maurmeyer, 1978), 3) the construction of a coast-parallel dike along the dune line prior to 1966, and 4) the construction of two sand bag groins several years ago. The south end, in contrast, has eroded continuously during the past century at a rate of 2.3 m/yr due to the low elevation of the barrier, permitting frequent overwash and landward transport of beach sands, and the extreme thickness of marsh muds underlying the barrier sands. The combined effects of eustatic sea level rise, subsidence, and wave erosion and overwash led to the high erosion rates found (Maurmeyer, 1978 and Belknap and Kraft, 1977).

Longshore transport rates also vary between the north and south ends of South Bowers. In both areas, dominant transport is to the south. θ_n is 53° and 40° for north and south South Bowers, respectively, gross transport rates are 13,600 m^3/yr and 15,400 m^3/yr and net transport rates are +7200 m^3/yr (Maurmeyer, 1978). These differences are probably caused by the effects of tidal currents at the Murderkill River inlet, and the current-inducing and wave-breaking effects of the sandbag jetty and groins.

South Bowers has been extensively nourished in the past. Between 1961 and 1976, it was nourished five times, for a total of 44,500 m^3/ft in fifteen years, or an average of 3000 m^3/yr (Department of Natural Resources and Environmental Control, 1980). According to a 1966 study, 218 m of coast requires nourishment at an average annual rate of 3100 m^3/yr or 4.4 m^3/0.3 m shore/yr. The two sandbag groins, constructed in 1977, are similar to the jetty erected at Bowers Beach, and are also currently in a state of disrepair. However, they do appear to be holding sediment near shore, as evidenced by the shoaling shown in Figure 2. Although erosion rates for the past several years are not available, the groins have reduced erosion in this area as well.

The thinness of the barrier sands in South Bowers and the depth of the underlying marsh muds have resulted in two major problems facing the summer and permanent residents of South Bowers: 1) high erosion rates, resulting in a need for almost 4.6 m^3/0.3 m shore/yr (as opposed to Bowers, where only 3 m^3/0.3 m shore/yr are required), and 2) subsidence rates of about 0.15 m/century, averaged over the past 3000 years (Belknap and Kraft, 1977). Although this subsidence rate seems fairly low, any construction in this area would greatly increase the subsidence rate, as indicated by conversations with residents, who report having to jack their houses up periodically, as they settle and tilt. According to one source, stable building foundations would require piling at least 100 ft deep (Kraft and others, 1976) at considerable cost to the homeowner.

Broadkill Beach

Broadkill Beach, 18 km to the south-southeast of the Bowers area, occupies a 4.5 km long strip of shore in the southern section of Delaware Bay. Like South Bowers, it is predominantly a summer community (in 1965 the permanent population was 35 and the summer population was 550) (Office of the U. S. Army Corps of Engineers, 1967), but this is changing, since, unlike Bowers and South Bowers, Broadkill is an area of active expansion, with many new lots for sale to the north and south of town. Between 1954 and 1973, the total number of homes increased from 74 to 291 (Department of Geography, 1977). Like Bowers, there is a relatively low turnover rate of residents (excluding those buying new property) as evidenced by the fact that 36% of the respondents for one interview program had been coming to the area for more than 10 years (Department of Geography, 1977). There is a wide range of home types and sizes, from small trailers to fairly large, new houses (circa $100,000). All the houses seem to be located behind the dune line in the center and north of the town but not in the south; most are at least 25 ft back from the dunes. Although most of the houses in the newly developed northern section of town are on pilings, to comply with zoning regulations, most in other parts of town are not. Broadkill Beach can be subdivided into four different sections according to housing characteristics. The northwesternmost end consists of new development, which has occurred mainly since the 1970's. Between this area and the access road from the mainland (see Figure 5, plan view) is a fairly densely populated section of houses that were built in the 1930's-1940's which is partly indicated by the fact that they are situated perpendicular to the beach. Southeast of the access road is another densely populated area of scattered beach houses from 1950 and before, along with houses that infilled the area in the 1950's and 1960's. Southeast of this section is an area of new development, on or

Figure 5. A plan view of the Broadkill Beach area showing coastal environments, coastal erosion rates (arrows), location of cores and jet wash holes (X), lines of geologic cross sections Figures 6a and b, and the approximate position of the coast line in 1900.

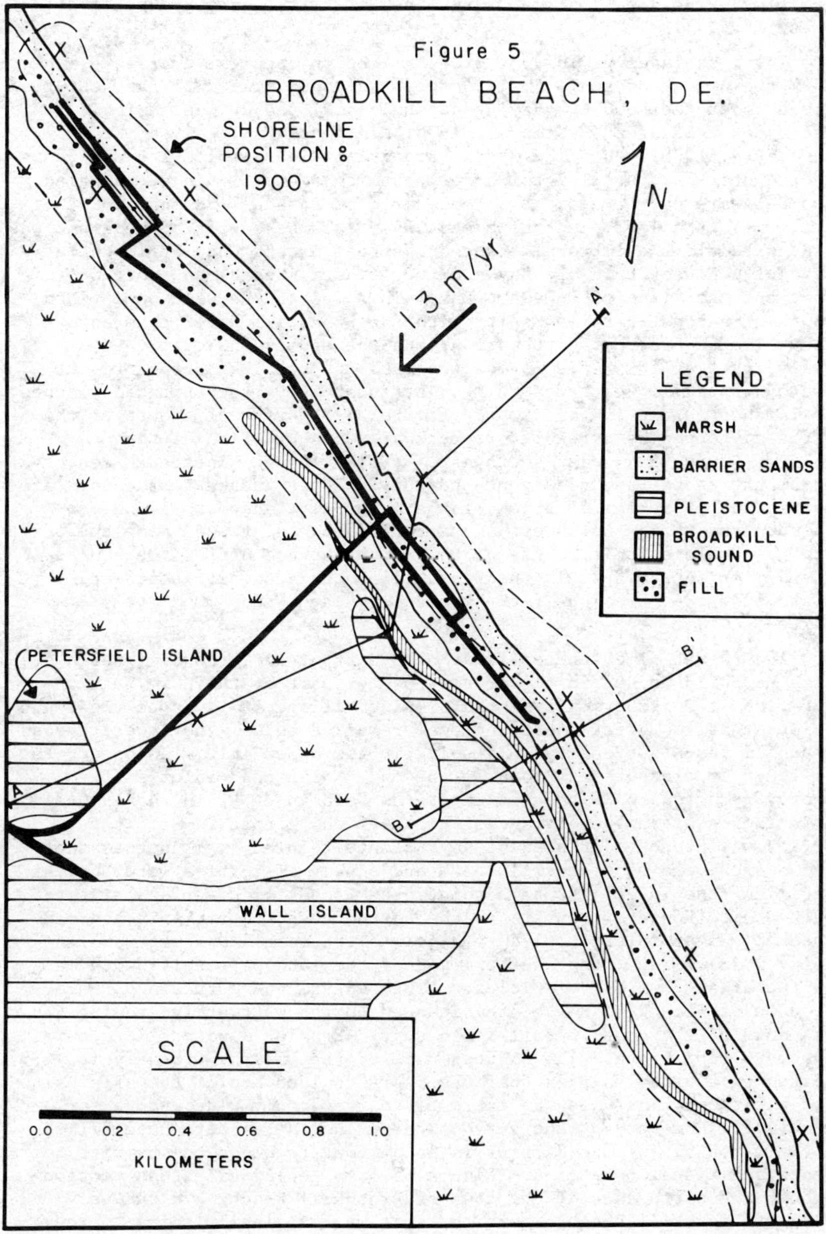
Figure 5
BROADKILL BEACH, DE.
SHORELINE POSITION: 1900
1.3 m/yr
N
A'
B'
A
B
PETERSFIELD ISLAND
WALL ISLAND
LEGEND
MARSH
BARRIER SANDS
PLEISTOCENE
BROADKILL SOUND
FILL
SCALE
0.0 0.2 0.4 0.6 0.8 1.0
KILOMETERS

directly behind a very narrow dune line, of fairly large, expensive homes.

The subsurface geology of Broadkill Beach is highly variable, due to 1) the number of small Pleistocene headlands located to the west that extend under the town at varying depths and 2) the unique geologic history of this area (Maurmeyer, 1974; Kraft and Caulk, 1973; Kraft and others, 1978). Originally the Broadkill River, which has now been diverted to flow parallel to shore, south to Roosevelt Inlet, emptied into Delaware Bay at a point to the north of Broadkill Beach. At that time (up to the late nineteenth century), a large supply of sediment was transported in the littoral transport stream around Cape Henlopen (at the confluence of Delaware Bay and the Atlantic Ocean) and northwest along the shoreline of Delaware Bay. This led to the development and growth of a coast-parallel spit, with the Broadkill River constantly being diverted farther northward around the end of the spit. Eventually, part of this spit formed the landmass upon which the present town of Broadkill Beach was built. By about 1900, the spit had reached its farthest extension northward (Figure 5). Northward growth of the spit had stopped as a result of breakwater construction at Cape Henlopen-Lewes, 8 km east-southeast, earlier in the nineteenth century and at the end of the nineteenth century. These breakwaters caused a restriction and eventual block of the littoral transport around Cape Henlopen and to the northwest and a redirection of the development of the Cape Henlopen spit to the north. With the cut off of the source of sediment to the spit in the Broadkill Beach area, its northward growth stopped. Erosion of the Broadkill Beach spit immediately started.

About that time, the earliest occupants of the Broadkill Beach area began to build houses in the present center part of the town. At the same time, the fishermen from the fishing village of Lewes to the south decided that too great a distance was required to sail northward beyond Broadkill Beach, beyond the old spit tip and a new inlet was opened to the south of present-day Broadkill Beach. This "new" inlet is frequently referred to in the literature as "old inlet." In the 1930's, to further shorten the distance from the fishing port to sea, the barrier was breached at present-day Roosevelt Inlet 4 km southeast of the southern end of Broadkill Beach and a pair of jetties were built to stabilize the inlet. At this point, the Broadkill River was diverted southward to Roosevelt Inlet and the northern end of the spit and old Broadkill River inlet shoaled and joined the mainland (Kraft and Caulk, 1973). Thus, the southern end of the land privately owned is now the area known as "old inlet." The central and northern parts of the town of Broadkill Beach are situated on the westwardly eroding remanants of the coast-parallel old spit. Erosion landward is occurring at a very rapid rate. The southern part of the village probably is located on a sandy barrier that has migrated landward of the original spit position. This area is extremely narrow, only wide enough for one house lot. However, at the very southern end, developers have filled in the marsh behind the barrier in an attempt to widen the barrier to create additional homesites. Figure 6a is a geological cross section through the north part of the town of Broadkill Beach. In the very south of Broadkill Beach, where the spit was originally thinner, the land on which houses are situated may actually be entirely to the southwest of the location of the original spit, on a rapidly migrating portion of the barrier. The narrowness of the barrier in this area is shown by Figure 6b. In general, the marsh muds underlying the barrier sands range in thickness from 3-20 m and are underlain by Pleistocene sediments.

Wave action along Broadkill Beach is often greater than at Bowers and South Bowers, due to its location near the mouth of Delaware Bay, and the resulting greater fetch over which winds can act, especially northeasterly and southeasterly winds, which are often associated with storms. This greater wave attack results in high erosion rates, and a high incidence of flooding and washover (Kraft and others, 1976). Erosion rates since 1843 have been affected by several factors, both natural and man-induced. From 1843-1884, the beach eroded at an average rate of 3.8 m/yr, apparently due to the natural processes of currents and wave attack. However, from 1884-1954, there was an average accretion of 3.5 m/yr (Kraft and others, 1976), probably due to accretion of the coast-parallel spit, and the construction of a system of five timber groins, from 1950-1954. (Beach nourishment was not a factor, since it was not initiated until 1957). In terms of volumetric erosion, an average of 1.15 m^3/0.3 m shore/yr were lost from 1843-1954, and 7.6 m^3/0.3 m shore/yr eroded from 1954-1964 (Office of the U. S. Army Corps of Engineers, 1967). During severe winters, broken ice that accumulates on shore may play a role in affecting erosion rates. Some residents have said that this ice may reduce erosion by protecting the shore from wave action, while other residents imply that the movement of ice against the shore may increase erosion, although the west and northwest winds which prevail during the winter would tend to lessen erosion of the western Delaware shore.

Longshore transport is almost uniform along the length of town. θ_n is approximately 48.5°, the gross transport is 14,200 m^3/yr and net transport is 8100 m^3/yr to the south (Maurmeyer, 1978).

There are three sources of flooding in the Broadkill Beach area--the bay, the marsh to the south and west of town, and Broadkill Sound. Flooding from the bay, caused by high winds and tides, is most common. Flooding from the marsh and Broadkill Sound only occurs when exceptionally high tides and bay waters break through the barrier into the marsh. In 1962, flooding occurred from both the bay and marsh (at the

Figure 6a. A geological cross section of the northern end of the Broadkill Beach barrier showing coastal environment sedimentary units encountered by the drill. This is in the area of the northerly extension of the original coast-parallel spit.

Figure 6b. A geological cross section of the south end of the Broadkill Beach barrier showing coastal environment and sedimentary units encountered by the drill. Note the narrowness of the barrier. Probably the entire barrier is comprised of sands eroded from and washover landward from the old coast-parallel spit.

Figure 6
BROADKILL BEACH, DE.
CROSS-SECTIONS

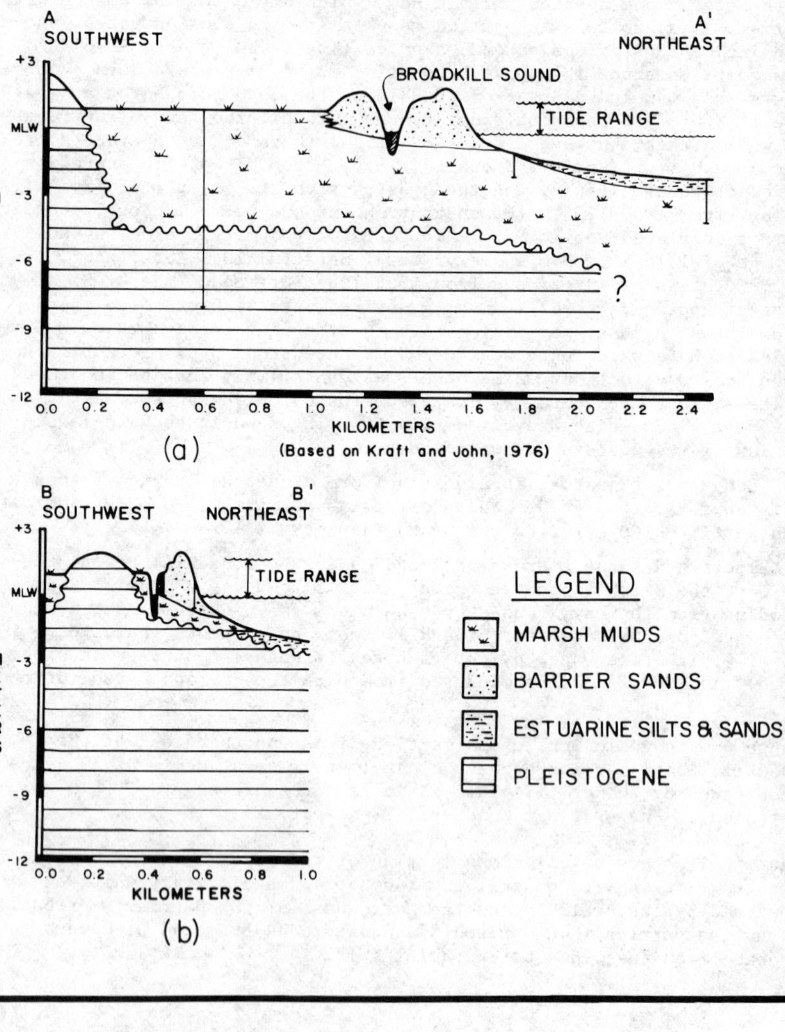

time there was a natural inlet south of Broadkill Beach, which caused water to rise behind the barrier). In the 1974 storm, when the configuration of the area was like it is today, flooding came from the bay alone.

There has been a long history of beach protection in Broadkill Beach to alleviate the erosion that began when sediment supply from Cape Henlopen was greatly reduced. In 1950, three timber groins were constructed, each about 58 m long, near the center of town. These resulted in erosion to the north of the groin field, and two more were constructed in 1954 (Department of the Army, 1972). Nourishment was done six times between 1957 and 1976, for a total of 643,000 m^3 or 22,800 m^3/yr (Department of Natural Resources and Environmental Control, 1980). In addition, in 1969 the State repaired the dune line by dumping broken concrete parallel to the coast, along the dune line, and topping it with sand and gravel. Storm fence was erected in 1975, but this was washed away shortly thereafter. Beach grass was also planted in 1975, and seems to have helped stabilize the dune. As a result of all of these measures, the beach is fairly stable in most areas (with continuing nourishment), and the concrete dike is covered with sand, forming a natural looking dune line. However, erosion has continued north of the northernmost groin, and in this area the concrete rubble is exposed. The necessity of continued nourishment is shown by a section of beach in south Broadkill that is not receiving nourishment, which has eroded at a rate of about 15 m/yr for the past several years. According to a study done by the U. S. Army Corps of Engineers in 1972, the best plan of protection for the beach would involve placing 81,400 m^3 of fill in the central part of town (this was done in 1973), maintaining this with 5400 m^3/yr, along 800 m of coast (approximately 3 m^3/0.3 m shore/yr) and constructing 1400 m of sand fence for berm preservation; however, this apparently was not a sufficient estimation, since 231,000 m^3 of sand have been added to the beach in the three years following the 1973 nourishment.

Interviewees of Broadkill Beach were attracted to the area by the fishing, boating, quietness of the area, and its location on the bay. They report that some erosion occurs continually, but is accelerated by storms, especially those from the northeast. According to two sources, the beach doesn't rebuild itself after storms, but another resident said that while northeast storms result in erosion, the beach can be rebuilt by either southeast storms, or the correct combination of a southeast wind and high tides. According to respondents, the beach sand nourishment placed in 1975 widened the beach by about 100 feet, and one said that if there were no major storms, replenishment would probably be needed approximately every four years. The dumped concrete slabs have generally been an effective measure, but one person said that although beach grass works well to reduce wind erosion, it tends to break off in chunks when hit by waves. He also stated that although the groins were a good idea, they haven't been maintained as they should be. During storms waves top over the groins. If the groins were built higher, there would be less erosion since sand carried by these northeast waves would be trapped north of the groins. When asked for suggestions regarding future beach preservation, interviewees said that nourishment should continue. In addition, two favored breakwater construction, and suggested that reopening the natural inlet that was south of town (old inlet, a man-built inlet) would also help reduce erosion and return the area to its natural state. As in Bowers Beach, most of the interviewees felt that such beach protection measures should be funded by some branch of government. If the people of the towns were to help pay for it (some said that they probably would be willing, others said they wouldn't), the beach should be private.

SUMMARY AND CONCLUSIONS

Although only three villages along Delaware Bay were discussed in this report, the extreme variations found in demographic characteristics, geologic characteristics (both morphologic and coastal processes), the type and effectiveness of beach preservation programs to deal with these environments, and residents' reactions to beach changes and preservation programs are typical of all the barriers along the western Delaware Bay shoreline. The "average" resident is aware of the fact of coastal erosion, although some newer residents are not. Nearly all residents feel that the financial problems of coastal protection are the responsibility of the "government." Almost none of the residents understand (or possibly don't care) that the tremendous costs of coastal protection programs over the longer-term decades or generations of time exceed the lesser present-day value of the shore front property (Figures 7a and 7b). Obviously, an extensive public education program regarding these matters is needed. Probably the reactions to erosion problems are similar elsewhere along the entire American coastal barrier strand. The purpose of this report is not to make conclusions as to how extensively, if at all, these three villages should be developed or condemned. Nor is it to convince developers that the geology of different areas should be their only concern. Its importance lies in the fact that simple but thorough geological and demographic investigations such as this can help prevent the occurrence of houses stranded on pilings over the water, buildings that have structural damage due to tilting and subsiding, homes that have been flooded or carried hundreds of feet by storm surges, and other widespread housing problems and economic losses associated with coastal development.

Figure 7a. Coring in wetlands of northern Bowers. Note the dike in the foreground and background. A developer is preparing and selling lots for homesites. Holocene marsh was only 0.22 m thick, thus wetlands here probably formed as a result of relative sea level rise in the last 100 years. An interesting legal problem is posed as the original (several centuries ago) deeded land was probably dry land.

Figure 7b. Perserverance! This house is relatively new, constructed on the narrow south barrier of Broadkill Beach. Landward migration in less than 5 years has isolated the house over the beach face - Delaware Bay. A permit has been issued to strengthen the piles with intent to reoccupy the house.

Figure 7

(a)

(b)

REFERENCES CITED

Belknap, D. and J. C. Kraft, 1977, Holocene Relative Sea-Level Changes and Coastal Stratigraphic Units on the Northwest Flank of the Baltimore Canyon Trough Geosyncline, Journal of Sedimentary Petrology, v. 47, no. 2, p. 610-629.

Department of the Army, Philadelphia District, Corps of Engineers, 1972, Detailed Project Report - Small Scale Beach Erosion Control Project, Broadkill Beach, Delaware.

Department of Geography, University of Delaware, 1977, Coastal Storm Damage - 1923 to 1974, Technical Report No. 4, Delaware Coastal Management Program, Document No. 1003-78-01-05, Office of Management, Budget, and Planning, State of Delaware, Dover, 442 p.

Department of Natural Resources and Environmental Control, 1980, Beach Nourishment Projects in Delaware, Division of Soil and Water Conservation, Dover, 3 p.

Kraft, J. C., 1976, Geological Reconstruction of Ancient Coastal Environments in the Vicinity of the Island Field Archaeological Site, Kent County, Delaware *in* Kraft, J. C. editor, *Transactions of the Delaware Academy of Science*, v. 5, 1974, p. 83-118.

Kraft, J. C., E. A. Allen, D. F. Belknap, C. J. John and E. M. Maurmeyer, 1976, Delaware's Changing Shoreline, Technical Report Number 1, Delaware Coastal Zone Management Program, Office of Management, Budget, and Planning, State of Delaware, Dover, 319 p.

Kraft, J. C., E. A. Allen, D. F. Belknap, C. J. John and E. M. Maurmeyer, 1979, Processes and Morphologic Evolution of an Estuarine and Coastal Barrier System *in* Leatherman, S. P., editor, *Barrier Islands, from the Gulf of St. Lawrence to the Gulf of Mexico*, Academic Press, New York.

Kraft, J. C., E. A. Allen and E. M. Maurmeyer, 1978, The Geological and Paleogeomorphological Evolution of a Spit System and Its Associated Coastal Environments, Cape Henlopen Spit, Coastal Delaware, Journal of Sedimentary Petrology, v. 48, p. 211-226.

Kraft, J. C. and R. L. Caulk, 1973, The Evolution of Lewes Harbor, *in* Kraft, J. C. and R. L. Salsbury, editors, *Transactions of the Delaware Academy of Science*, v. 2, 1971, p. 79-125, Newark.

Kraft, J. C. and C. J. John, 1976, The Geological Structure of the Shorelines of Delaware. DEL-SG-14-76. College of Marine Studies, University of Delaware, Newark, 106 p.

Maurmeyer, E. M., 1974, Analysis of Short- and Long-Term Elements in a Simple Spit System, Cape Henlopen, Delaware, M. S. Thesis, Department of Geology, University of Delaware, Newark, 150 p.

Maurmeyer, E. M., 1978, Geomorphology and Evolution of Transgressive Washover Barriers Along the Western Shore of Delaware Bay, Ph.D. Dissertation, Department of Geology, University of Delaware, Newark, 274 p.

Office of the U. S. Army Corps of Engineers, 1967, Beach Erosion Control and Hurricane Protection Along the Delaware Coast.

RECENT GEOLOGIC HISTORY OF A BARRIER ISLAND

Richard D. Hobson[1]
Robert K. Schwartz[2]
Frank R. Musialowski[3]

Abstract

Barrier islands form more than sixty percent of our Eastern and Gulf coastlines. In all, there are 280 individual barriers of which 70 are already highly developed for full-time habitation and 100 more are currently being developed. A need clearly exists to understand natural barrier dynamics and rates-of-change in order to protect these limited resources and to manage them effectively.

The present study considers the history of Topsail Island, NC, as determined from the analysis of eighteen cores obtained from the nearshore zone at elevations ranging from +2m to -5.8m MSL and relating the cored stratigraphy to process mechanisms. The cores show the subsurface presence of well-defined beach, shoreface, inlet and back-barrier sedimentary facies. The modern shoreface is composed of fine-to-coarse sand and exists as a prism which thins from about 2m at the present beach-surf zone to about 1m in the core located about 450m offshore. Beneath the shoreface sediments lie the inlet channel facies which contains coarse sand, pebbles and large shell fragments, and which thins from about a minimum of 3m at the surf zone and then "pinches out" approximately 124m offshore. The back-barrier facies occurs seaward of and below the inlet facies and consists of lagoonal muds, oyster reef material, and tidal creek sands. This facies also contains a characteristic lagoonal foraminiferal assemblage and the facies age ranges from 4000 to 6000 years B.P.

The correlation of shoreface profile shapes as surveyed and as reconstructed through stratigraphic analysis indicates a recurring shape that it interpreted to reflect a shoreface condition of "dynamic equilibrium". Shifts through time and space of this surface are brought about to a large degree in response to a shifting middle shoreface zone that appears to play an important role in physically "shaping" the profile during events of shoreline retreat or recession. Although the middle shoreface zone generally lies seaward of the zone of breaking waves, the significant elevation changes observed there suggest that perhaps this zone should be included in calculating sand budgets.

[1]U.S. Army Coastal Engineering Research Center, Fort Belvoir, VA 22060
[2]Department of Geology, Allegheny College, Meadville, PA 16335
[3]U.S. Geological Survey, Woods Hole, MA 02543

Introduction

Understanding of both short and long-term nearshore process-response relationships is a necessary prerequisite to effective coastal planning, protection and development. Of particular interest to the geologist is to understand how waves, currents and tides affect nearshore sediment movement and change the form of the offshore bottom surface and the internal arrangement of the sediments themselves, and, to understand how these fairly quick natural rearrangements interract to produce long-term changes such as shoreline migration in response to changing sea level. Several studies completed recently by the U.S. Army Coastal Engineering Research Center (CERC) provide some unique data that relate to process-response dynamics and to rates-of-change.

Two studies involved the monitoring of dredged disposal sediment placed as "bars" in nearshore dump sites to determine how the processes would redistribute this material, and to determine the value of offshore disposal as a method of beach nourishment. The Phase I study (1976) lasted 5 months and the Phase II study (1978-1979) 12 months and in each, disposal piles were surveyed and sampled frequently to document sediment movement and textural change (see Schwartz and Musialowski, 1980 regarding the Phase I experiment; Phase II data are currently being analyzed). In both phases, coarse sand was dredged from an ebb tidal complex fronting an inlet and placed offshore by the split hull barge Currituck which was being used by the U.S. Army Engineer District, Wilmington, to maintain a boat-access channel into the inlet. Also in both phases, the disposal sediment was placed at specific sites within the study area so that both native and "altered" areas could be monitored. During these monitoring experiments, several shorter experiments were conducted to measure process conditions in order to help explain monitored time-sequence changes in sediment texture and bottom relief. The volume of sediment involved in the Phase I and Phase II experiments were 26,750 m^3 and 40,650 m^3 respectively.

The final experiment involved coring the active profile in order to establish the sub-bottom stratigraphy within the area and to attempt to relate the stratigraphy to both current and longer-term Holocene processes. The term active profile is used here to indicate that portion of the profile located between the beach berm offshore to the fair weather wave base depth. The coring experiment was also conducted in the hope of providing some data relating to the shoreface evolution of a barrier island system, for in many cases, cores are available offshore in fairly deep water and from the exposed barrier surface itself, but few have been drilled within the high energy nearshore zone.

General Setting

The coring was conducted about 2 kilometers south of New River Inlet on Topsail Island, North Carolina, which lies some 60 kilometers northeast of Wilmington, NC (Figure 1). Topsail Island is basically a single dune-ridge type of barrier where a 60 meter-wide beach fronts a vegetated foredune which in turn protects low-lying barrier flats and salt marshes. The nearshore zone is generally characterized by a single, shore-parallel bar and trough system, or by a low tide terrace or ramp. At the time of this study, the area was little developed with a fishing pier, located 1.5 kilometers south of the site, being the only man-made structure on the seaward side of the barrier.

Topsail Island is part of a microtidal (mean tide range = 1.0m), low to moderate wave energy barrier coastline that is oriented normal to prevailing winds from the southeast. During these times, breakers from the south are usually less than 0.6m high, with periods of 6-7 seconds, and wave-driven longshore currents move northeast toward the inlet at speeds of about 30 cm/sec. Significant storm wave heights (H 1/3) range from 0.9 to 1.8m.

Waves and winds from the north and northeast also occur periodically throughout the year with average breaker heights generally greater than 0.6m (6-7 second periods) which produce longshore currents moving southwest away from the inlet at average speeds of 38 cm/sec. Storm waves from the northeast (generally winter storms) have significant heights ranging between 0.9 and 2.7m with breaker periods between 4 and 12 seconds depending upon local sea and swell conditions.

New River Inlet lies about 2 kilometers northeast of the study site. It is maintained by tidal flow from this small coastal plain river estuary and the associated marsh/lagoon system located behind the barrier (Figure 1). Sediment movement appears to be dominantly into the inlet as evidenced by a quite large flood tidal delta (about 40 km^2) versus a much smaller ebb delta/shoal (about 4.5 km^2, see dashed -6m contour, Fig. 1). Although the study area lies outside of the present ebb delta geographic boundaries, it is likely that inlet processes have affected geomorphic development at the site to some degree. Ancient inlet sites are recognizable in the area from aerial photography, but it is unclear if these sites were associated with migration of the present inlet or with short-term storm openings.

Methods

The site for the shoreface coring experiment was where the Coastal Engineering Research Center was completing the second of two long-term experiments studying the behavior and movement of sediments placed in shallow, nearshore disposal sites (Schwartz and Musialowski, 1980). Two aspects of those studies benefiting the coring experiment were that specialized equipment developed for the one was available to the other and that the two years' monitoring of changes in offshore topography and sediment properties provided an adequate data base for evaluating rates-of-change caused by "normal" and storm conditions (Schwartz and Musialowski, 1980). The change rate data were particularly useful for determining a profile shape representing a state of dynamic equilibrium and for seeing how perturbations to the profile caused by the placement of spoil sediments responded to local process conditions.

The specialized equipment developed for the sand placement study included construction by the Wilmington District, Corps of Engineers, of a self-powered, wheeled tripod (Figure 2) that was used for surveying and that could be driven anywhere in the nearshore zone to water depths of 6m, and a minicomputer-controlled positioning system that allowed the tripod vehicle to occupy any desired location. The minicomputer was also used for on-site plotting of survey data during collection and comparison of those data with any previously collected data sets. The vehicle, called the CRAB (Coastal Research Amphibious Buggy), had the further use in this study as an offshore coring platform which could remain on location during vibracoring even when in the surf zone or during times of high energy wave and current conditions.

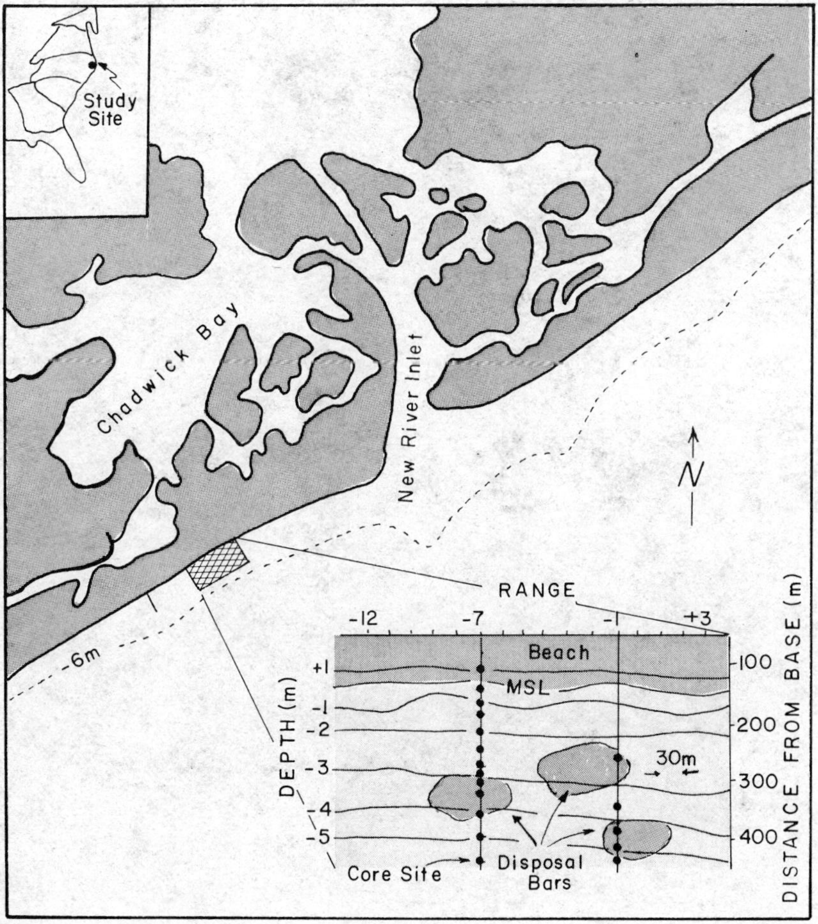

Figure 1: Location Map.

Figure 2: Coastal Research Amphibious Buggy (CRAB) used for Profile Surveys and as Coring Platform.

Coring was carried out over a two-week period (July, 1979) along two range lines (Ranges -7 and -1, Figure 1) using the CRAB, SCUBA divers and a portable piston-vibracoring system (Lanesky, Logan, Brown and Hine, 1979). In all, thirteen cores were taken across Range -7 and five from the offshore portion of Range -1. Additional coring was planned during the two weeks in the field but was not accomplished because of difficulties in penetrating coarse grained and highly compacted sandy layers. These difficulties emphasized the need to further improve the design of these highly useful and portable coring systems.

Stratigraphy

The subsurface sediments (Figures 3 and 4), within coring limits, are comprised of four distinct facies elements: (1) Beach/Sub-Beach; (2) Shoreface; (3) Inlet-Channel; and (4) Back-Barrier. The sedimentary units were assigned to these broad facies elements because the overall characteristics of each facies reflect a unique set of environmental process conditions even though the elements are usually subdividable into several sub-facies components. For example, recognizable lagoonal, tidal flat and tidal creek sediments comprise the back-barrier facies but generally reflect in their composition and organization the lagoonal, washover and tide-dominated processes endemic to a back-barrier environment. Brief descriptions of these facies elements and their components are provided in the following paragraphs. For detailed descriptions, see Schwartz, Hobson and Musialowski (1980). Also, Figure 5 may be useful for identifying processes and geomorphic features associated with the underlying facies elements.

I. The Beach/Sub-Beach facies occurs between the beach profile surface and the underlying inlet-channel facies (Figure 3) and consists of three subfacies containing normal backshore (Ia) and foreshore (Ib) sediments overlying sub-beach storm deposits (Ic). The Beach/Sub-Beach sediments are found in thin sub-horizontal layers and are generally brown-to-gray sands that contain from about 5% (Ia) to 25% (Ic) abraded shell and lithic fragments. Texturally, sediments from the backshore to foreshore to sub-beach become generally coarser and more poorly sorted (well-graded). Normal grading is common within most bedding layers except for a few inversely graded sub-beach layers that are capped by coarse lag concentrate sediments.

II. The Shoreface is subdivided into an upper, middle and lower zone on the basis of position on the profile, profile shape, associated processes, texture and internal sedimentary structures. Upper shoreface sediments (IIa, Figure 3) are found mainly within the surf zone and are comprised of moderately well-sorted, medium-sized sand that contains up to 20% coarse grained shell and lithic fragments in well-defined 0.1 m to 0.8 m thick beds. These beds are contained within the bar/trough couplet or profile ramp and show either normal or reversed grading in the surf zone while normal grading is dominant seaward of the average breakpoint position. Middle shoreface sediments (IIb, Figures 3 and 4) underlie upper shoreface layers and extend disconformably over inlet-channel layers from the toe of the foreshore through the surf zone to a seaward limit where the profile ramp steepens offshore (about 3.5m water depth). Wave shoaling and shore parallel currents are the dominant processes which have deposited these medium-to-fine grained, very well sorted middle shoreface sands. Bedding layers are generally thicker (0.7 to 1.2 m) near the trough zone and thinner (0.02 to 0.30m) offshore. Trough cross bedding (nearshore), occasional biogenic burrows and scour-lag deposits over reworked inlet-channel sediments are typical. The Lower Shoreface (IIc, Figures 3 and 4) extends disconformably over inlet-channel and back-barrier facies to an effective fair weather wave base of about -5.5 m MSL (as determined from the vibracore results). These sediments are thinly bedded, well sorted sands that fine and become increasingly bioturbated seaward. Generally, coarser grained sediments occur as a "basal" layer overlying eroded back-barrier or inlet-channel layers or as filling within storm eroded scours. Active biogenic reworking and finer mud sized material in their matrix causes these sediments to have a mottled or patchy-gray-to-brown (oxidized) appearance.

III. The Inlet Channel (III, Figures 3 and 4) disconformably underlies beach and shoreface facies with back-barrier facies disconformably below. Although contacts are sharp with associated facies, the lower boundary of the shoreward portion of the inlet-channel facies was not penetrated by most cores. The tannish-gray inlet sediments are poorly sorted medium to very coarse grained and contain a considerable amount of coarse abraded shell material (20-40%). Bedding layer thickness ranges between 0.1 and 1.0m and inversely graded layers occur both within the sequence as well as in a reworked contact layer with overlying shoreface layers. In some cores, reworked back-barrier materials are found within lower inlet layers. No sedimentary structures are recognizable in the cores.

IV. The Back-Barrier facies complex contains lagoonal, tidal flat, tidal creek, and tidal creek-associated washover (?) deposits that disconformably underlie inlet-channel and shoreface sediments. Tidal flat sands (IVa) are fine grained and dark gray and contain occasional clay pellets and rip up clasts along with occasionally abundant whole oyster shells. The tidal flat sands are highly bioturbated (mottled) and they grade downward into lagoonal muds through a transition zone of dark sandy mud locally containing a dense (65% by volume) population of oyster shells that are often found in growth position. The lagoonal facies (IVb, Figures 3 and 4) consists of dense gray mud containing occasional thin sand laminations and scattered oyster shell clusters in growth position. Tidal Creek sedimentary sequences (IVc) were penetrated along each cored profile (Figures 3 and 4). These sediments are medium to coarse, poorly sorted and contain several layers of well-rounded quartz pebbles. Mud occurs in thin laminations along with carbonized plant fragments. Relict subhorizontal and trough crosbedded laminations are visible. Two layers thought to be of washover origin occur between the tidal flat and tidal creek subfacies. These washover (?) sands are medium grained, poorly sorted and contain coarse, oxidized oyster shell debris. A coarse grained oxidized layer occurs at the base of one washover horizon.

V. Disposal Sediments (V, Figures 3 and 4) were placed offshore (about 40,650 m^3) on the active shoreface in water depths of 3.0 to 4.3m

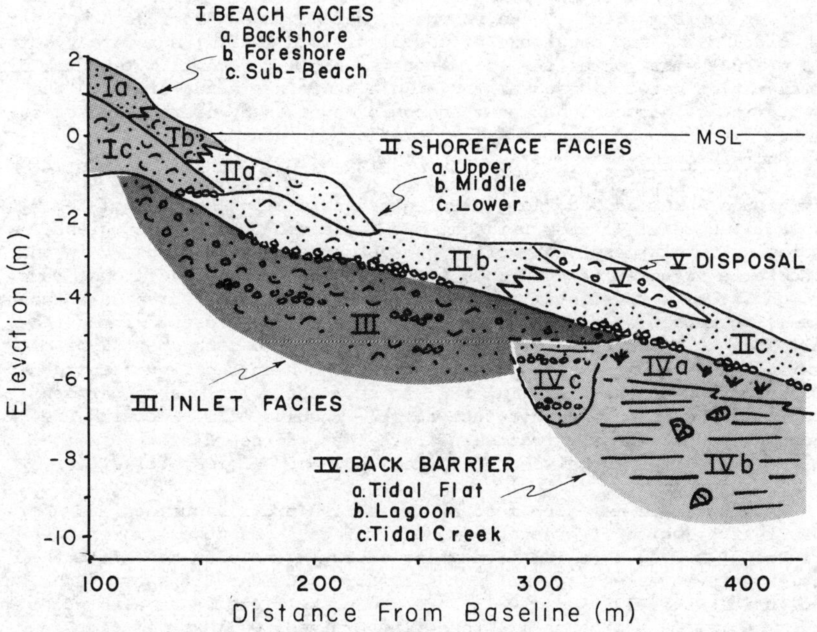

Figure 3: Subbottom Facies Relationships for Range -7.

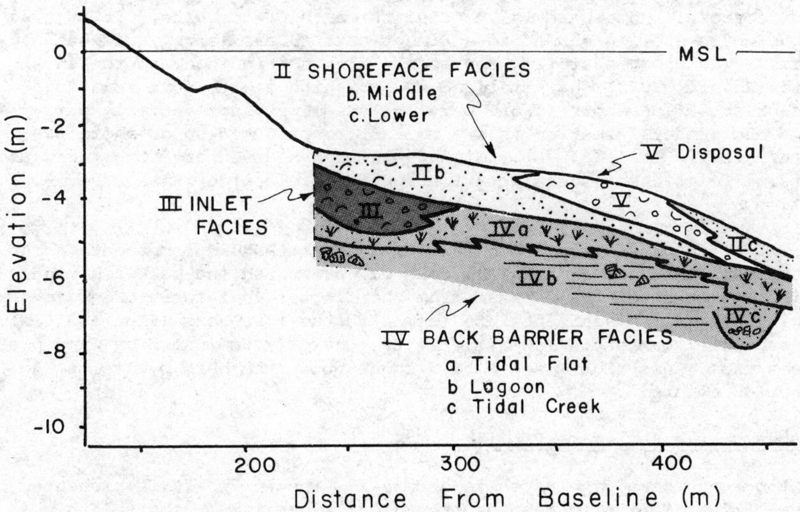

Figure 4: Subbottom Facies Relationships for Range -1.

creating new bar form. These sediments were dredged from the New River ebb tidal delta and consisted of medium grained (.28mm) moderately sorted (Sø = 0.94) sand containing 10-20% coarse shell fragments. Textural dissimilarities between disposal and middle shoreface sands aided in the monitoring of disposal bar as it becomes modified by nearshore processes.

Nearshore Processes

Figure 5 summarized the general process relationship associated with the active profile (after Reinson, 1979) and sub-bottom stratigraphic facies. In this case the profile presented is that of Range -7 in the study area as cored in June 1979. The geomorphic features shown (ramp, trough, etc.) were identifiable on almost all profiles surveyed although a well-formed offshore bar and trough couplet were present within the surf zone as often as the ramp feature shown. The break positions of average waves and maximum storm waves are shown on the figure instead of a generalized breaker zone in order to show the relationship between breakpoint position and shoreface zones. Results of vibracoring and those from earlier experiments indicate that the depth of the fair weather wave base is between -5.5m MSL and -6.5 MSL respectively.

The shoreface makes up most of the active profile surface and therefore is the region of greatest dynamic change. The lower shoreface (Figure 5) serves as a relatively low-energy transition zone from the shelf where waves begin to affect nearshore sediment transport (Schwartz and Musialowski, 1980, p. 10-11) but where shelf processes also occur. Lower shoreface sediments are thus fine grained, well-sorted and usually highly bioturbated (Figure 3, IIc). The middle shoreface is generally a zone of shoaling but where extreme storm waves may also break and where unidirectional shore-parallel coastal currents predominate. It is also a zone where profile shape and sediment texture are largely affected by storms. The upper shoreface represents a transition zone to the foreshore (Howard, 1972) characterized by the high energy surf zone, and wave-produced longshore (shore parallel) and rip (shore normal) currents. Ridge and runnel formation is a common storm-produced upper shoreface feature (Davis et. al 1972). Highly variable sedimentary structures and textural relationships characterize upper shoreface deposits (Fig. 3, IIa).

The beach consists of both foreshore and backshore components. The intertidal foreshore lies in the zone of wave swash and backwash, and exhibits a sharp slope increase from the flatter shoreface. Seaward dipping, laminated sediments are common to the foreshore (Figure 3, Ib). The subaerial backshore (supratidal) is generally dominated by wind processes that redistribute sediment brought above the berm by storms and high tide events.

Changes to the Shoreface Profile

Figure 6 shows five offshore shapes for Range -7. It is presented to document that four of these shapes are quite similar (numbers 1 thru 4, Figure 6) and that this similarity in form reflects that these constitute a definable "equilibrium profile" which shifts position in response to changes in processes or varied sediment inputs. The presentation here is generalized in order to stress a few relationships. For more detailed discussion see Schwartz, Hobson and Musialowski (1980).

The dashed profiles (numbers 3 and 5, Figure 6) show the elevation maxima (upper limit) and minima (lower limit) surveyed at Range -7 for the 12 months preceeding coring. These limits define an "envelope" of erosion and deposition for the year. The lower limit shape is considered to be the "native profile" shape since it corresponds almost exactly with the predisposal profile (not shown) and with profiles surveyed earlier (Schwartz and Musialowski, 1980) except in the shaded region of middle shoreface scour (Figure 6). This scour zone occurred in response to winter storms on all profiles whether or not they contained disposal material. The upper limit profile shows both the effects of disposal sediment placement and natural net accretion through the area during the 12 months preceeding coring.

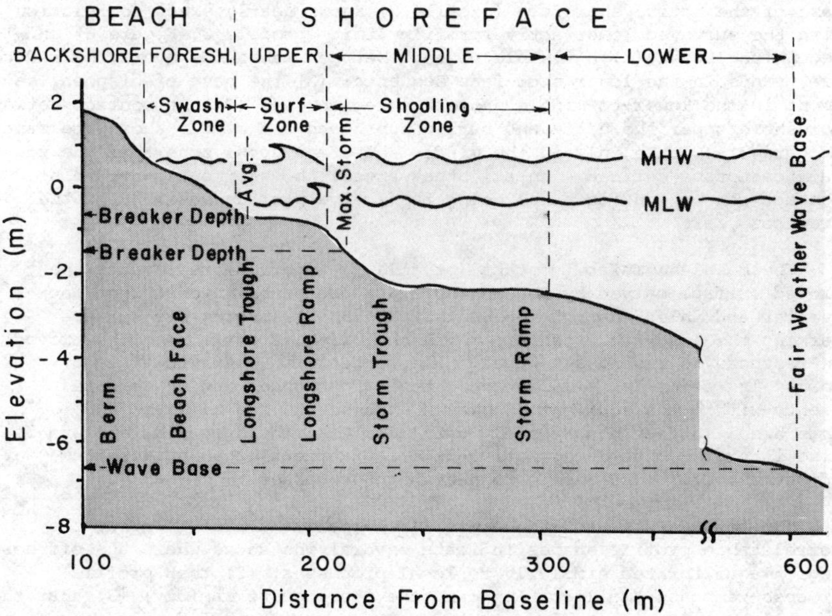

Figure 5: Nearshore and Beach Energy Zones and Associated Geomorphic Features.

The cored profile shape (profile 4, Figure 6) corresponds nearly exactly with the lower limit profile but is shifted upward about 0.6m due to sediment disposition. This shift caused by accretion was observed on most profiles (with or without disposal material) and mainly reflects response to a natural influx of sediment rather than an "artificial" change caused by disposal. There are at least two reasons for minimizing the effects of disposal upon profile shape. First, there simply wasn't enough sediment placed in the area to account for the observed changes in volume and second, natural processes had essentially removed the disposal bars and had recreated "natural profile" shapes about 6 months after placement and much of the accretion observed occurred after that time. The conclusion drawn from these relationships described above is that, like the lower active limit profile, the cored profile also represents a native-like profile in equilibrium with natural processes.

The other native-like profile shapes are identifiable in the subsurface. The first (profile 2, Figure 6) represents the lower limit of active shoreface processes for at least the previous 12-month period. The second (profile 1, Figure 6) is the transgressive surface between the shoreface and the much older back barrier and inlet-channel facies. The base of the active shoreface (profile 2) shows nearly exact correlation with the surveyed lower active profile limit (profile 3, Figure 6) and is recognizable as; (1) a visible contact between tan sands overlying reduced gray sands in the lower shoreface sequence; (2) the base of disposal sediment in the lower-to-middle shoreface region; and (3) the contact between foreshore/upper shoreface and buried sub-beach and middle shoreface sands (Figure 3). It is only in the middle shoreface scour zone that the contact cannot be defined. In all other areas, the sediment above this contact can be considered as being the active profile sediment of the previous year.

The basal shoreface contact (profile 1, Figure 6) represents the eroded surface carved by modern shoreface processes into ancient back-barrier and inlet channel sediments. It is a transgressive surface marking the landward extent to which the offshore environment has moved in response to rising sea level. The contact is clearly defined by coarse lag deposits, significant lithologic changes and by textural trends within the sediment (Schwartz, Hobson and Musialowski, 1980). This basal surface also closely parallels the others on Figure 6 and is also interpreted as being native-like and reflecting an equilibrium adjustment of the shoreface to nearshore processes.

The conclusions from these profile comparisons are: (1) that the correlatable profile shapes indicate several instances where the offshore region equilibrated similarly to local processes; (2) that profile process-response equilibration can take place quite rapidly; (3) that the overall processes themselves in this area have remained fairly consistent in time and space and have repeatedly established the same basic profile shape; (4) and that the greatest and quickest changes in the profile shape seem (at least in this case) to occur within the middle shoreface region suggesting that this region may play a very important role in shoreface migration.

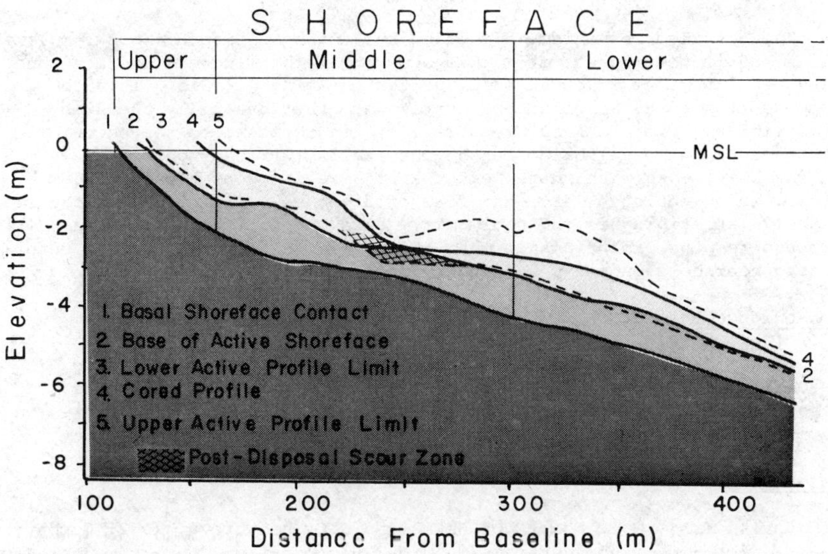

Figure 6: Profile Comparisons. Light Gray region, between Profiles 4 and 2, represents that portion of the Shoreface affected during the 12 months preceeding Coring. Medium Gray region (between profiles 2 and 1) is shoreface region not effected during previous year. Darkest Gray below profile 1 is the region of ancient Inlet-Channel and Back Barrier Facies.

Summary

A coastal site on Topsail Island in North Carolina was used for conducting (1) two 12-month long experiments that monitored the nearshore movement of artifically-placed sediment; (2) the nearshore coring project described in this paper and; (3) numerous short-term process-response experiments in support of long-term studies. The monitoring studies produced a fairly detailed understanding of how processes affect the nearshore and its sediments at this location as well as a large data base for defining normal nearshore conditions. The coring study attempted to document the subsurface stratigraphy and relate it both to active short-term processes and to the longer term process of barrier island migration in response to rising sea level.

The stratigraphic picture that resulted from the coring is basically that of a thin shoreface and modern beach sand sequence overlying older back barrier (4000 to 6000 years BP) and inlet-channel deposits. This facies pattern reflects transgression of the modern nearshore zone.

The detailed mechanisms affecting the transgression are not completely clear but it does appear from this study that the middle shoreface zone plays an important role in "shaping" the changing profile. It also appears from both the cored and survey data that basically the same "equilibrium" profile shape recurrs often in response to nearshore processes and that equilibrium can be reestablished rather quickly (4-7 months here) even with shape perturbations as large as those created by artificially adding a large bar form of disposal sediment to the profile. Finally, stratigraphic and sedimentary data support the often-observed process-response relationship that storms are the dominant events determining nearshore geometry and subsurface relationships.

Acknowledgements

This work was carried out at the U.S. Army Coastal Engineering Research Center, Fort Belvoir, Virginia as part of the Civil Works Research and Development program of the Corps of Engineers. Permission to publish the findings obtained is appreciated.

References

1. Davis, R.A., W.T. Fox, M.O. Hayes and J.C. Boothroyd, 1972: Comparison of ridge-and-runnel systems in tidal and non-tidal environments: Jour. Sed. Pet., v.32, p. 413.

2. Howard, J.D., 1972: Trace fossils as criteria for recognizing shorelines in stratigraphic record: in J.K. Rigby, and W.K. Hamblin, eds, Recognition of Ancient Sedimentary Environments: Soc. Econ. Paleontol. Mineral. Spec. Publ. No. 16, p. 215-225.

3. Lanesky, O.E., B.W. Logan, R.G. Brown, and A.C. Hine, 1979: A new approach to portable vibracoring underwater and on land; Jour. Sed. Pet. v. 49, No. 2, p. 654-657.

4. Musialowski, F.R. and R.K. Schwartz, 197 : The Coastal Research Amphibious Buggy (CRAB) A survey system for the nearshore zone: Joint Proc. of the Am. Soc. Photogrammetry, Am. Cong. Surveying and Mapping 1979, Fall Tech. Meeting, p. 267-273.

5. Reinson, G.E., 1979: Facies models 6; Barrier island systems in facies models: R.G. Walker ed., Geoscience Canada Reprint Series 1, Geol. Assoc. of Canada, p. 57.

6. Schwartz, R.K., R.D. Hobson and F.R. Musialowski, 1980: Stratigraphy of a modern barrier island nearshore zone; (in review) for Northeastern Geology.

7. Schwartz, R.K. and F.R. Musialowski, 1980: Transport of dredged sediment placed in the nearshore zone - <u>Currituck</u> sand-bypass study (Phase I): Tech. Paper No. 80-1, U.S. Army Coastal Engineering Research Center, Ft. Belvoir, VA, 55 pp.

COASTAL ZONE '80

Volume II

Proceedings of the Second Symposium on Coastal and Ocean Management

Billy L. Edge, Editor

Sponsors:

American Society of Civil Engineers
(Waterway, Port, Coastal and Ocean Division)
Office of Coastal Zone Management
Edison Electric Institute
Heritage Conservation and Recreation Service
Maritime Administration
National Aeronautics and Space Administration
National Marine Fisheries Service
Office of Sea Grant
Public Works, Canada
U.S. Army Corps of Engineers
U.S. Coast Guard
U.S. Fish and Wildlife Service
U.S. Geological Survey

Affiliates:

American Petroleum Institute
American Shore and Beach Preservation Association
Bureau of Land Management
California Coastal Commission
Environmental Data and Information Service
Federal Emergency Management Agency
Marine Technology Society
National Park Service
State of Florida Department of Environmental Regulation
State of Florida Department of Natural Resources
The Coast Alliance
The Coastal Society
Urban Land Institute
U.S. Department of Energy

Diplomat Resort Hotel
Hollywood, Florida
November 17-20, 1980

AMERICAN SOCIETY OF CIVIL ENGINEERS FOUNDED 1852

Published by the
American Society of Civil Engineers
345 East 47th Street
New York, New York 10017

Library of Congress Catalog Card No. 80-69152
ISBN 0-87262-258-4
Manufactured in the United States of America.

Cover photo by Thomas A. Beckett, New York, NY

CONTENTS

Volume I

*Papers not available at time of printing

*Papers not available at time of printing

*Papers not available at time of printing

*Papers not available at time of printing

*Papers not available at time of printing

*Papers not available at time of printing

*Papers not available at time of printing

*Papers not available at time of printing

*Papers not available at time of printing

*Papers not available at time of printing

*Papers not available at time of printing

*Papers not available at time of printing

*Papers not available at time of printing

Volume III

*Papers not available at time of printing

*Papers not available at time of printing

*Papers not available at time of printing

*Papers not available at time of printing

***Papers not available at time of printing**

HUMAN INFLUENCE ON THE SEDIMENT BUDGET OF A BARRIER ISLAND

Craig H. Everts, M. ASCE[1]

INTRODUCTION

Natural and man-induced events which may create a need for government involvement on barrier islands include storm flooding; island erosion by waves, currents and wind; and development which results in a loss of ecologically important areas, or which interrupts natural processes that create and maintain such areas. This paper deals with sediment movement on barrier islands, and specifically a sediment budget analysis (SBA) and the importance of various portions of a budget analysis. The results can be used to predict natural shoreline changes. General discussions of the SBA approach are given by the U.S. Coastal Engineering Research Center (1977) and Komar (1976).

The purpose of this paper is to show that the introduction of human processes in the coastal zone can be helpful as well as harmful in slowing or preventing shoreline retreat. This is done in the framework of a barrier island SBA. Descriptions of ways in which human intervention may influence the rate of sediment transport or the paths along which the sediment travels are given in general terms. Their applicability at a specific site will depend on the results of a detailed SBA at that location.

SEDIMENT BUDGET ANALYSIS

Two conditions are necessary to cause a shoreline to retreat. First, the sand at and near the shoreline must be mobilized. Next, a mechanism must exist to transport the mobilized sand away. An SBA addresses the movement of sand and other sediment in the coastal zone. It is, simply, a technique in which all the sediment movements into and out of a fixed coastal region are summed. Shoreline position change is considered the dependent variable in the SBA discussed in this paper. For a change in shoreline position sediment must be added (accretion) or removed (erosion), or the position of the plane of water must change (sea level rise or fall) with respect to land.

Control Volume

Losses or gains of sediment and changes in water surface elevation relative to land must be quantified within a finite, enclosed space in a fixed reference frame (Fig. 1), i.e., a control volume, when the SBA approach is used. An important criterion in defining control volume

[1]Chief, Engineering Geology Branch, U.S. Army Coastal Engineering Research Center, Kingman Building, Fort Belvoir, VA 22060

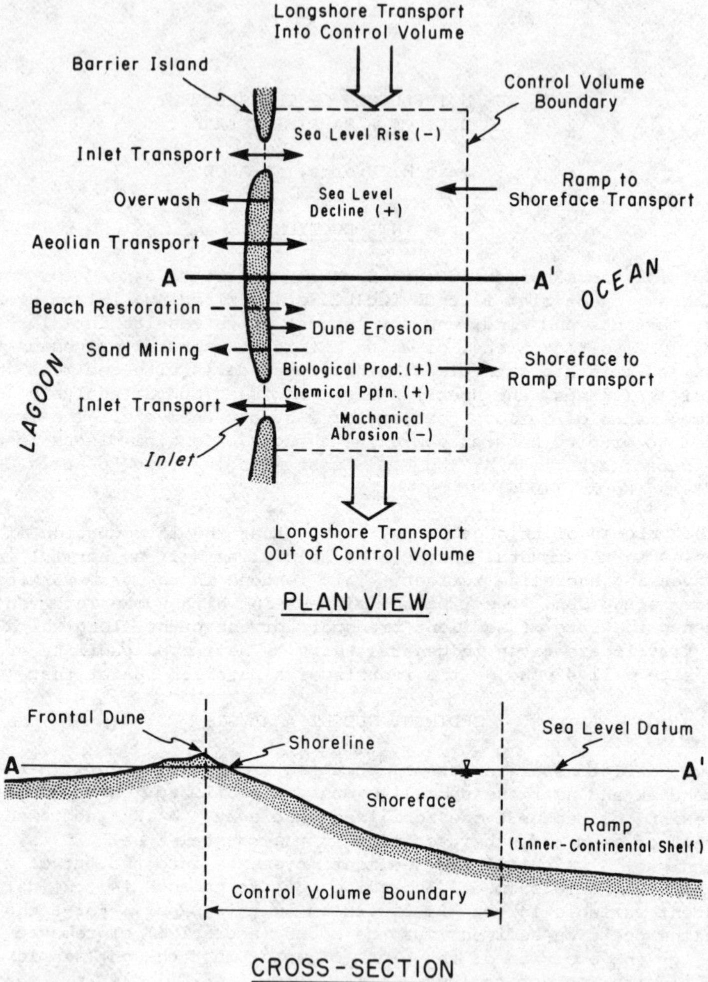

Figure 1. Suggested control volume used in a sediment budget analysis of a barrier island. Transport mechanisms by which sediment enters and leaves the control volume are shown with arrows illustrating the transport path.

boundaries is that they be established so shoreline position changes can be directly related to sediment volume changes. This requires the shape of the shoreface remain constant as it moves horizontally or vertically in space, or if profile shape changes do occur that they are predictable. The shape of the shoreface profile may change when storm waves attack it or as the seasonal wave climate varies. However, over the long term these changes will be assumed reversable and the long-term averaged profile shape will be assumed constant.

Considering long term profile adjustment, this is not an unreasonable assumption. Bruun (1962), for example, considered the shape of the shoreface profile would remain constant (profile of equilibrium) as it moved upward and landward in response to sealevel rise. In a field study on Lake Michigan, Hands (1980) found shoreface profiles maintained their shape as they moved landward from depths of 10-12 m as lake levels rose 0.5 to 1 m in a period of eight years. That rate of lake level rise is much greater than the relative sea level rise along the US marine coasts, and the depth of profile response is similar to that along the marine coasts. This suggests a similar response might be expected for sea level rise along a marine coast. Further, the result of an analysis of historic Natural Ocean Survey charts by Everts (1980, unpublished) shows for the only case where sufficient data are available (Smith Island, VA), the shoreface on that rapidly retreating (eroding) island retained its shape during the retreat.

For the purposes of this paper, the landward boundary of the control volume is defined near the upper limit of wave uprush, i.e., near the base of the frontal dune if one exists (Fig. 1). The seaward boundary is that limiting depth and distance beyond which sediment changes do not affect the barrier island sediment regime. It is here assumed to be at the base of the shoreface, i.e., the location where the near-coast, steep, concave-up profile meets the planar, seaward dipping continental shelf profile (Everts, 1978). This limit is not easy to determine accurately because the juncture of the shoreface profile and the inner-continental shelf profile is assymptotic. Alongshore control volume boundaries are selected according to the needs of the analysis. As they are specified closer together, the resultant mean shoreline change becomes more refined (and the need for accurate data becomes more acute).

Sediment Volume Change Rates

The direction and rate of transport for sediment which enters or leaves the control volume must be quantified. The easiest way to establish sediment volume change is to address each transport mechanism separately. Table 1 summarizes possible transport mechanisms and the direction of sediment movement. It is evident that if more sediment is lost from a shoreface profile than is gained, the profile will be displaced landward. That landward displacement is equal to the volume of sediment lost from the control volume divided by the vertical distance from the base of the shoreface to the top of the frontal dune times the alongshore length of the control volume.

TABLE 1 COASTAL SAND TRANSPORT MECHANISMS

(1) SAND TRANSPORT MECHANISM	(2) TRANSPORT MODE	(3) TRANSPORT PATH AND DIRECTION	(4) CAUSE	(5) MOST SUSCEPTABLE AREAS
Overwash	Wave-driven currents	Beach to island or across island to back-barrier lagoon, deposition on island	Waves and wave-driven currents move water and sand across the island during periods of abnormal high water, usually accompanying storms	Low dunes or absent dunes, on low-lying barrier islands or spits
Aeolian transport	Wind	Usually major component is shorenormal from beach in a landward direction, deposition on island	Onshore winds move available dry sand from beach and backshore to island, often in form of dunes. Dunes may migrate across island to backbarrier lagoon. Sand may be moved back to beach	Low or absent dunes with prevailing onshore winds above the velocity to move beach sand; trapping mechanisms behind backshore, such as vegetation, so offshore-directed winds do not return sand to beach
Inlet transport	Tidal flood and ebb currents	Littoral zone into inlet or less commonly inlet to littoral zone, deposition on flood-tidal shoals (lagoon) and ebb-tidal shoals (ocean)	Tidal currents assisted by waves and longshore currents move littoral sands parallel to shore and into inlets where it is deposited. As a coast retreats, inlet shoals may be eroded thereby supplying the adjacent littoral system	Newer inlets where flood and ebb tidal shoals are not fully developed, where the flood tide has a greater capacity to carry sediment than the ebb tide, and where the longshore transport rate to the inlet throat is large
Continental shelf	Waves and currents	At base of shoreface, major component is shore-normal, transport is to the shoreface from the shelf, or is to the shelf from the shoreface	Infrequent wave and/or current action at bottom usually moves sand and perhaps silt landward and finer material seaward	Onshore movement = quantities of sand available offshore, long periods of long-wave-length waves and offshore winds; offshore movement = fine-grained sediments on shoreface, frequent storms, prevailing strong onshore winds

Shoreface Transport	**Sea level rise or fall relative to shoreface profile**	Rising relative sealevel = sediment usually moves landward on shoreface; falling relative sea level = sediment usually moves seaward on shoreface	To maintain an equilibrium shoreface profile relative to the water surface long term changes in the elevation of the water surface require a sediment readjustment	Rising relative sea level = much compaction of sediments (void ratio is decreased as pore water is driven out of sediments), region of tectonic depression; falling relative sea level = region of tectonic uplift including isotatic rebound
Dune, bluff or cliff retreat	Waves mobilize sediment which is then transported in littoral zone	Movement of sediment into the control volume	High energy waves and currents, usually during storms, erode features just landward of the beach	High dunes, bluffs or cliffs with a toe elevation near high tide and/or storm surge elevation along exposed coasts with a narrow beach
River transport	River currents	River to back barrier lagoon, in some instances river transport is directly to barrier island	Sediment load is carried in fluid flowing down gradient from the mainland	Large river sediment load, high river discharge, narrow lagoons behind barrier islands, perhaps extreme floods which carry river-borne sediments into littoral zone through tidal inlets
Longshore transport	Wave, wind and in some cases tide-driven currents	Transport parallel to shore, most sediment moves in and landward of breaker zone; can be into and out of the control volume at differing rates	Waves approaching the coast at an angle and shore-parallel wind direction create currents which move sediment, especially when the sediment is mobilized by wave action	Interest is in net longshore transport rate which is highest near the apex of a coastal compartment, or along coasts where waves and winds prevail from one direction
Beach replenishment	Dredge, truck haul and others	Sand from outside the Barrier Island system is brought into the system	Man replensihes beaches; result is a net gain in sand volume to the system	Where economic conditions allow
Sand mining	Dredge, endloader, etc.	Sand is removed from Barrier Island system	Man mines littoral sand; result is a net loss in sand volume in the control volume	Where allowed. In most all cases it causes long term harm to the barrier island system

There must be sinks into which the eroded sand is deposited. It may be moved alongshore, but ultimately it must be moved away from the beach. One possibility is a shore-normal loss to deeper water. More likely, on barrier island coasts, it will be carried landward of the shoreline. This could be accomplished by overwash or wind transport, or by alongshore movement to an inlet where it is carried onto flood tidal shoals behind the barrier island. Sand may also be displaced upward (and landward) if sea level rises relative to land.

Changes in Relative Sea Level

A change in water surface elevation relative to the shoreface profile may have a significant effect on the shoreline. A water surface rise, for example, on a sloping beachface will move the still water shoreline landward with no sediment movement involved. A relative water surface rise will also result in a disequilibrium of the shoreface profile. To reestablish the equilibrium will require that the profile move vertically a distance equal to the rise in sea level. Without a large influx of sediment from outside the system, the entire profile will be forced landward as well as upward, liberating sediment from beneath the preexisting shoreface (Fig. 2). Fisher (1979) estimated that the static submergence portion of shoreline retreat on the Rhode Island coast accounted for 15 percent of the total shore retreat. The other 85 percent he attributed to the reestablishment of an equilibrium shoreface profile. In a sediment budget study the static shoreline change must be included. The change that will result as the shoreface profile re-equilibrates is accomplished through the various movements of sediment shown in Figure 1. Thus, sea level change drives those mechanisms which move sediment onto or off of the shoreface profile.

The change in shoreline position resulting from a relative sealevel change may be estimated by graphically overlaying the shoreface profile at the new relative sea level position such that the change in sediment volume is zero (Fig. 2). The horizontal distance the profile must be moved to maintain a net zero sediment volume change is the estimated distance the shoreline will move. This procedure can be used to check the more rigouous SBA technique.

A long-term change in the position of sea level relative to land should be thought of as a site-specific phenomenon. The change may occur because of local sediment compaction, regional tectonic subsidence or uplift, long-term climatic changes, isostatic rebound, or as a world wide response (eustatic) to a changing volume of water in the oceans. The latter is predominately based on the amount of water tied up in glaciers, although tectonic activities can also alter the volume of the ocean basins. Soil compaction, which is the collapse of the soil structure with the concomitant sqeezing of the pore water out of the saturated sediment, may be a locally important phenomenon. Kaye and Barghoorn (1964), for example, found peat deposits which were compacted to one-fifth-their original thickness. Overwash or dune loading of sand on marsh deposits beneath a barrier island may, therefore, produce a relative rise in sea level much greater than recorded historic eustatic rises.

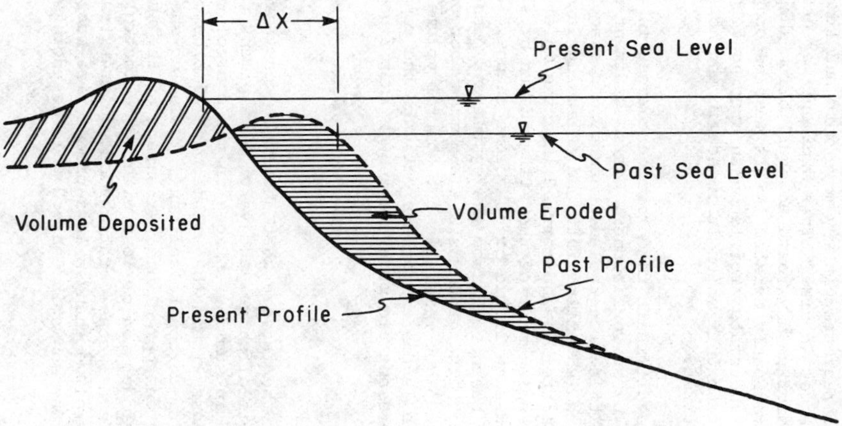

Figure 2. Change in sediment volume and shoreline position when sea level rises relative to the shoreface profile. Without an increase in the net volume of sediment coming into the control volume, the shoreface profile must retreat landward to maintain an equilibrium shape. With a constant volume of sand in the control volume, the shoreline retreat Δx is directly related to sealevel rise for a given shoreface profile shape.

Consideration of Sediment Composition

The constant shape of the shoreface is predicated on the assumption no long-term changes occur in the size of the sediments comprising its surface. Consequently, in calculating sediment volume change rates, the change in volume by sediment size increment must be addressed. Sediments will be eroded and transported off the shoreface and beach where wave, current and wind-produced bed shear stress is greater than that which is necessary to initiate sediment motion. Sediments will be deposited where bed shear stress is less than that critical value.

If the shoreface profile retreats, it may move into and liberate sediment which is smaller or larger in size. This sediment may not remain on the shoreface, or the shoreface may reach an equilibrium shape that is different from that which previously existed. This must be considered. In most cases, however, the shoreface profile on an eroding coast will be moving into material similar to that which it is now in and no sediment weighing factor will be required.

In some cases it will be difficult to determine changes in the volume of sand in a control volume by surveys alone. Coarse sediment may replace finer sediment with little or no volume change. An example is a migrating inlet channel which is cutting into and releasing lagoonal silts and muds, and filling with littoral sand carried into the inlet by tidal currents. The result is a replacement of the silt and mud by sand. Before and after surveys would show little or no volume change. However, the net effect is a loss of beach sand from the system and possibly a consequent shoreline retreat. The eroded silts and muds because of their fine nature generally do not remain in the control volume. They are carried further into the lagoon or perhaps out onto the continental shelf where bottom shear stresses are moderate enough that fine sediments can be deposited.

Sediment Budget

Table 2 is a summary of the factors that should be considered in an SBA. The alongshore, onshore-offshore, and vertical sediment transport elements must be considered in summing changes in the control volume. Natural and artificial changes within the volume must also be considered. Except in areas where calcium carbonate sediment, including shells and shell fragments, constitute an appreciable percentage of the shoreface sediment volume, natural production or loss within the control volume may be considered small. In all considerations of sediment transport into or out of the control volume changes in sediment size must be addressed. The considerations shown in Table 2 will be addressed in the next section.

DISCUSSION

The SBA approach is not a panacea. To work it requires site-specific field data which are costly and time consuming to aquire. Research is still required before the SBA methodology is refined in sufficient detail to accurately quantify all its elements. An SBA is, therefore, an inexact and evolving tool. But, at present, it is one of the most useful approaches to forecast the effect man may have on sedimentation on a barrier island.

Man can modify the sediment transport paths and the rate at which the sediment moves in a number of ways. Some result in a decrease or a termination of shore retreat. Some do not. And some create problems elsewhere, either by causing an accelerated shoreline retreat or by creating ecological problems. Any man-made modification to the overall transport pattern which promotes the movement of sediment onto the shoreface profile or inhibits its movement away can be viewed as a favorable factor in arresting shoreline retreat.

TABLE 2. SEDIMENT BUDGET CONSIDERATIONS

I. ALONGSHORE SEDIMENT TRANSPORT

II. ONSHORE-OFFSHORE SEDIMENT TRANSPORT

1. At Seaward Boundary of Control Volume
2. At Landward Boundary of Control Volume
   a. overwash transport
   b. wind transport
   c. bluff or dune erosion
   d. inlet transport

III. VERTICAL SEDIMENT TRANSPORT (RISE OR FALL OF SEA LEVEL RELATIVE TO LAND)

1. Rise Or Fall On A Static Slope
2. Sediment Transport To Reach A New Profile Equilibrium

IV. NATURAL PRODUCTION OR LOSS OF SEDIMENT (biologic production, chemical precipitation or solution, mechanical abraision)

V. ARTIFICIAL INCREASE OR DECREASE IN SEDIMENT VOLUME

1. Beach Restoration
2. Sand Mining

VI. CHANGES IN SEDIMENT SIZE IN THE CONTROL VOLUME

A number of possible ways in which man can modify the sediment budget are discussed in this section. It does not cover all ways nor does it suggest which method is "best". That decision is often a political one, especially where no clear-cut definition of what is "best" can be agreed upon by all concerned. It is the task of the scientist and engineer to provide the decision maker with the best possible technical evaluation of the consequences of any or no modification to a barrier island sediment budget.

The following sections reference sediment budget considerations listed in Table 2.

Alongshore Sediment Transport

Groins and offshore breakwaters are two structures that impede the longshore transport of sand. Both are usually designed to slow shore erosion and trap sand moving parallel to shore. If successful, sediment will be captured between the groins or landward of the offshore breakwater. Thus, while the volume of sand which enters the control volume is not altered, the volume which leaves is decreased. Downcoast erosion may result because waves and longshore currents responsible for moving sand away from the downdrift beaches are still operating, while the sand supply from updrift is interrupted (Everts, 1979). To offset this situation, it is usually considered good engineering practice to artifically fill the groin compartments with sand from outside the system as a part of a groin construction project. This prevents a disruption of the longshore sediment transport mechanism.

Sediment Transport at the Seaward Boundary of the Control Volume

If sand is moved from the control volume seaward onto the continental shelf, there is little that can be done to restrict that movement. In most cases today data on the long term net volume rate of transport, and direction, is unknown at the seaward control-volume boundary.

In addition to its use in an SBA, a practical reason to determine the direction and rate of sand transport at the base of the shoreface comes in establishing the limit landward from which sand should not be taken on the shoreface. It would be unwise to mine sand, for example, from any location on the shoreface where the excavated pit might fill with sand lost from the beaches during a storm or with sand moving landward during an accretionary period. In both cases sand that goes into the pit is lost from the beach system, i.e., the mining would result in a net sand deficit within the control volume. Hallermeier (1977) has provided guidelines on establishing that limiting depth when wave data are available.

Overwash Transport

Overwash occurs when the island behind the beach is low enough to allow storm waves and storm-driven currents to pass over it. Sand is carried out of the control volume in a landward direction. The most widely used method to lessen the effects of overwash is to construct an artificial dune or to repair the existing dune. The practical reasons for this approach are to prevent flooding of the area behind the beach during periods of storm surge, and to provide a sand reservoir that can be sacrificed during storms.

In some areas overwash is believed to be an important process in moving sand back into the lagoon, thereby forming a substrate upon which marshes will grow. It is not known, however, how prevalent the process is in the creation and maintainence of lagoonal marshes, but in areas such as low-lying Core Banks, NC, it is doubtless a significant factor (Godfrey, 1970). In some cases overwash deposits may be returned to the littoral zone (control volume) by wind. Leatherman (1979), for example, found this to be the case on northern Assateague Island where overwash-deposited sediment usually did not reach the lagoon. In other areas overwash plays only a minor role, if any, in the propagation of back-barrier marshes. In addition to littoral sands, other sources of sediment upon which depth-dependent marsh grasses grow are flood-tidal delta deposits (usually sand), silts and muds transported within the lagoon, and organic debris from the marsh plants themselves. Clearly each barrier island must be studied individually to evaluate the relative merits of using an artificial dune. The trade-off is between the possible effect of the dune on marsh propogation behind the island, and the requirement for flood prevention, protection of structures from wave damage, and the need to inhibit shore retreat. An additional consideration comes on an island where the shoreline is retreating from the sea. Unless the lagoon shore of the island is allowed to prograde landward, perhaps by overwash deposition, the island width will narrow with time. A continuous frontal dune may also prevent new inlets from forming during storms. Inlet creation is often an end member of an overwash event.

Wind Transport

Sand transport by wind is frequently altered by man. Sand fences and planted vegetation are used to slow or halt the movement of sand. Depending on prevailing winds, sand can be moved either to or away from the beach although most frequently transport is away from the beach (out of control volume). Wind-blown sand may move completely across a barrier island, but the sand is usually not deposited far enough out in the lagoon to significantly affect the propagation of marsh grasses. Sometimes the sand, upon reaching the lagoon shore, is moved in alongshore direction. The effect on the ecology of the back-barrier marshes of inhibiting the movement of sand out of the control volume by constructing artificial dunes is less than the effect of inhibiting overwash transport.

Bluff and Dune Erosion

Bluffs form headland sediment sources for some downdrift barrier island systems, i.e., Montauk Point which constitutes some, but by no means all, of the source of sand for the barrier islands on Long Island. Like bluffs, dunes may be eroded by waves, thereby providing sediment to the control volume. While dunes are composed of sand-sized material, bluffs, such as at Montauk Point, may contribute only a small portion of their total volume in sand-sized material.

Revetments and seawalls are shore-parallel structures sometimes used to protect the toe of dunes and bluffs. When these structures are successful, they prevent some or all the sediment volume behind them from reaching the littoral zone. The result is a decrease in the total volume of material entering the control volume. The trade-off then, is between dune or bluff protection at a specific site and the loss of a previous source of littoral sand to the downdrift beaches. Of course if a bluff is composed of silts and muds or other material that will not remain in the littoral zone, the bluff sediment source is inconsequential.

Inlet Sediment Transport

Inlets may have a pronounced effect on adjacent ocean shorelines. They may act as sinks for sand that moves in from the littoral zone, and in some situations, may also act as a source of sand for the littoral zone. Sand moving in the longshore transport system is carried into an inlet from one or both adjacent shores. If the sedimentary deposits in the inlet are in a dynamic equilibrium with the hydraulics of the inlet, that volume of sand which enters the inlet will equal the volume released by the inlet. If, however, the inlet shoals and channels have not reached an equilibrium (stable) volume, the inlet may act as a trap for some of that littoral sand. If the inlet is sand-choked, a sand volume in addition to that volume which enters will be released. In both cases, the volume of sand in the longshore transport conduit (control volume) will be changed.

Natural inlets in equilibrium may migrate in three ways (Fig. 3). They may migrate parallel to shore which is a common occurrance where a large net longshore transport of sand occurs, and it usually results in a loss of littoral sand. As the inlet migrates it leaves flood-tidal shoals composed of littoral material behind. The second way an inlet may migrate is in a shore-normal direction, either landward as the adjacent barrier coast retreats, or seaward as the coast progrades. On a retreating coast ebb and flood-tidal shoal deposits will act as a source of sand for the littoral zone. The opposite will prevail on a prograding coast. The third way an inlet may move is vertically as sea level changes relative to the elevation of the inlet sediment-water interface. If that interface is in equilibrium with the inlet hydraulics, a long-term rise in the water surface, for example, will require a rise in the sediment-water interface. This will probably be at the expense of the littoral zone. Accompanying a relative lowering of sea level one might expect existing channels would be incised, releasing sand to the littoral zone.

Newly opened inlets generally trap much of the littoral sand which reaches them. A good example is Drum Inlet, Core Banks, NC, which was opened artificially in 1972. Shorelines on either side of this inlet have retreated in an inward-flair pattern to provide the large flood-tidal shoal complex and the smaller ebb-tidal shoals (about 18 km^2 in plan area) shown in Figure 4. As an inlet changes toward an equilibrium geometry and sand volume, the rate of sand entrappment declines. The time it takes an inlet to reach a sand-volume equilibrium may vary from a few years to centuries. At Indian River Inlet, DE, for example, the eff-tidal shoal is apparently still accreting at a rate of 100,000 m^3/yr after it began forming in 1938 (Lanan and Dalrymple, 1977).

Jetties are frequently used at inlets to prevent sand from entering navigation channels. A problem encountered with using jetties is that the net longshore sediment transport volume is not passed from one side of the inlet (updrift) to the other side (downdrift). Thus, downdrift beaches do not receive the alongshore supply of sediment that would naturally pass a stable inlet in equilibrium. This problem can be alleviated by an artificial bypass system.

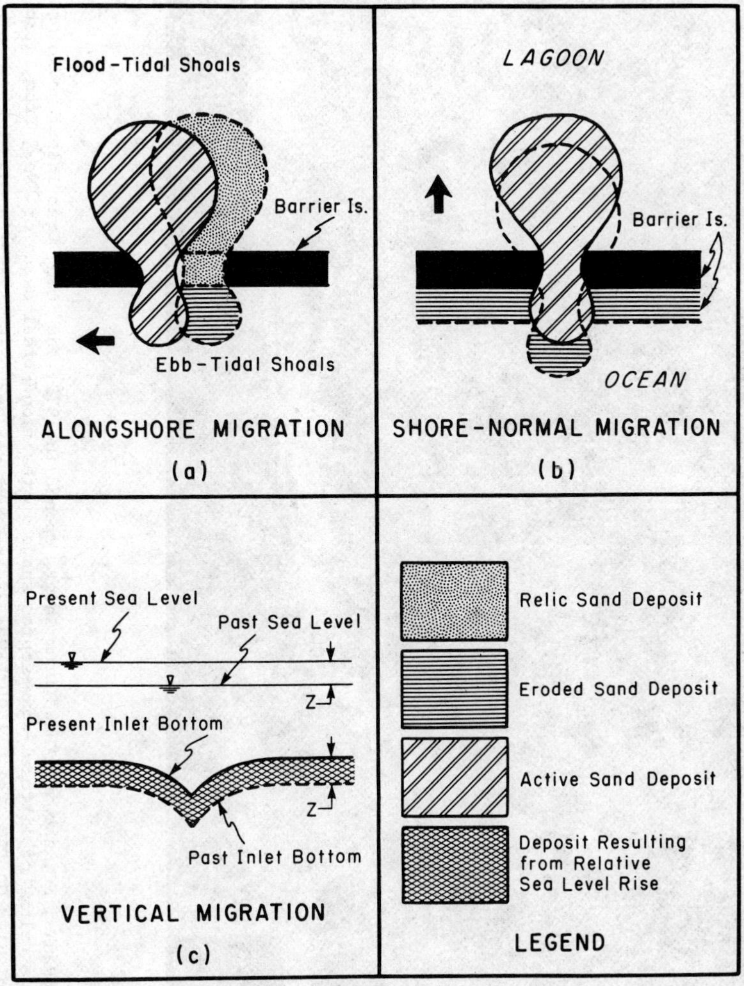

Figure 3. Three ways in which an inlet in hydraulic equilibrium with the inlet sediments may migrate in space.

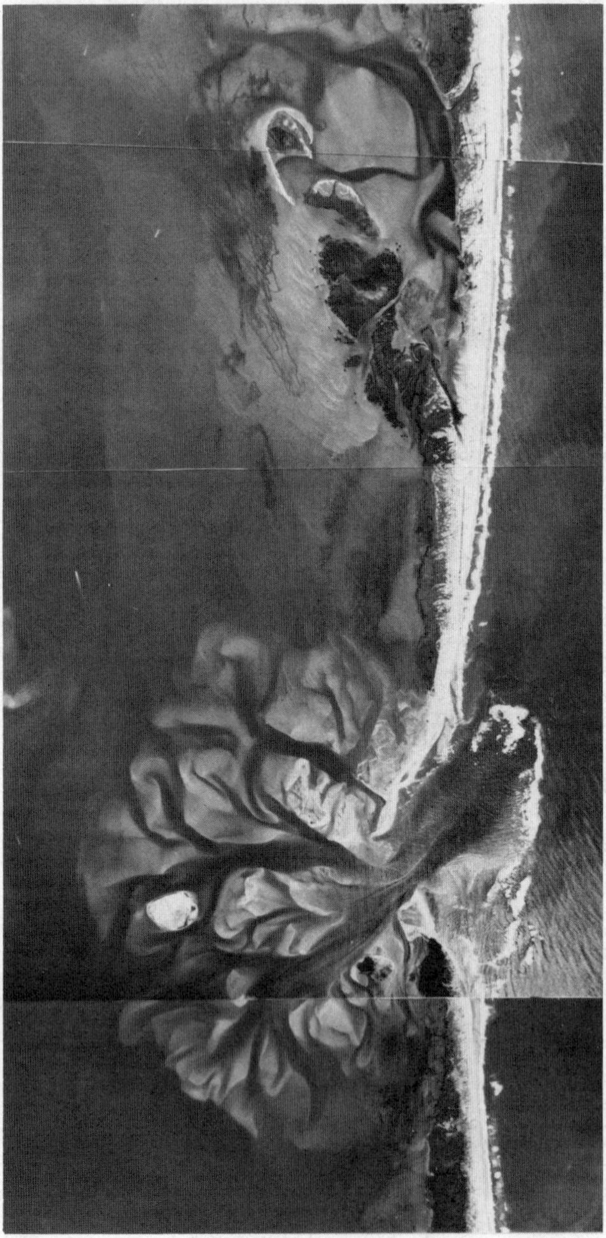

Figure 4. Core Banks, NC, showing "New" Drum Inlet, opened in 1972, and "Old" Drum Inlet, now closed. Note the large volume of littoral-zone-source sands in the flood-tidal delta of "New" Drum Inlet, and the marshes growing on the relic flood-tidal delta sands of "Old" Drum Inlet.

Jetties will have a positive effect on the volume of sand in the control volume if the amount of littoral sand lost to the ebb and flood-tidal deltas of an inlet is reduced. This would occur if the jetties are designed to reduce the volume of sand which passes into the inlet. In turn, the shoreline retreat rate of the downdrift barrier island would be reduced, especially when the inlet is newly opened or is migrating in space. A consideration when impeding the movement of sand into an inlet or in stabilizing it is the effect the reduced sand supply may have on back-barrier marshes. Depth-dependent marsh plants would have a reduced (in aerial extent) substrate on which to grow if the flood-tidal shoal area is reduced. This, of course, assumes that the flood-tidal shoal deposits will form the marsh substrate.

Sea Level Changes Relative to Land

Man does not currently posses the resources to modify the effect of regional and world wide sea level change on the shoreface of a barrier island. However, it is important that he forecast the effect of sea level rise on the sediment budget. Without such knowledge it is impossible to determine the relative importance of the sea level effect on shoreline stability in comparison with other effects (Table 2), and, hence, impossible to evaluate the cost effectiveness or the viability of any human modification.

The task is to predict future relative sea level changes. Geologic evidence (age dating of organic sediments) indicates sea level along the U.S. Atlantic and Gulf coasts has been within one meter or less (lower) than the present sea level for about the past 2000 years (Kraft, 1976; Curray, 1965; and Coleman and Smith, 1964). Evidence compiled by Hicks (1972) indicates an average 0.3 m/100 yr relative rise in sea level near and along the Atlantic and Gulf coast for approximately the period 1920-1970. This rise, for 27 tide stations, varied from 0.15 m/100 yr at Fernandina, FL, to 1.0 m/100 yr at Eugene Is., LA. Although the exact values may be suspect in some cases because of tidal inlet effects, survey control problems, and possibly subsidance of the tide gage support, it appears the relative sea level rise rate in this century is at least a factor of 5 higher than the past 2000 yr. average. The important question is whether the short-term or long-term rise, or another condition, will prevail in the future. It is possible that the short-term rise we experienced in the recent past is one of a series of high frequency sea level rises and falls which Fairbridge (1961) suggested occurred over the past 6000 years superimposed on a near steady-state sea level elevation. Clearly, before truly enlightened predictions of future shoreline change rates can be made, research is needed on the prediction of relative sea level changes, and on the effects of these changes on the position of the shoreline.

Natural Production or Loss of Sediment

Man has little influence on natural production where shell or coral fragments, or chemically-precipitated oolites are an important part of the shoreface sediment volume. From a loss standpoint, however, man might be responsible for a decrease in sand volume caused by abraision if hard quartz grains were placed as beach fill on a beach composed of relatively soft and easily abraided carbonate sand. The abraision of quartz grains, which predominate in most areas, is probably too slight to significantly affect the sand volume within the control volume.

Beach Restoration

Placing sand on an eroding beach and shoreface is an accepted practice to hold a shoreline in place. The success of a beach restoration project hinges on knowing the natural sand loss rate from the control volume which contains the beach. That volume of sand should be replaced with material from outside the control volume. Thus, a key part of any restoration scheme is the selection of the fill source from inland, the continental shelf, or a lagoon if it is environmentally acceptable and not within the hydraulic regime of an existing inlet. Sometimes the cost effectiveness of a beach fill project can be improved by preventing eroded sand from entering sinks such as inlets. A determination of the potential savings is possible when a detailed SBA is made.

The reasons loss rates at beach fill sites sometimes appear greater than the previous natural loss rates may be twofold. In a restoration project sand is usually placed on the upper part of the shoreface profile. After placement, waves and currents distribute the sand over the profile in deeper water, thus giving the appearance of an accelerated loss rate at the shoreline. Secondly, if a large quantity of sand is placed along a short coastal reach, the coast may become out of lateral (plan view) equilibrium. The fill material will spread in an alongshore direction, increasing the sand volume on adjacent beaches. In all cases fill taken from a proper source will result in an addition of sand to the control volume.

A practice which is becoming widespread is that of mechanically moving sand from the beach to rebuild the duneface, i.e., "bulldozing" sand from the upper part of the shoreface to the dune. This can be a stopgap measure to provide a buffer against storm waves, but it clearly does not add sand to the system. There is no increase in the volume of sand in the control volume using this practice.

Sand Mining

If done on the beachface or shoreface, sand mining will decrease the volume of sand in the control volume. The longterm result will be a retreat of the position of the shoreline. Mining sand from the beach, which was once a widespread practice thought to cause no harm because the excavated beach pits were rapidly filled again with sand, is not prevelent today.

SUMMARY

A sediment budget analysis (SBA) is a technique whereby all sediment movements into or out of a fixed coastal region (control volume) are summed. Changes in sea level relative to the shoreface profile and changes in sediment composition must also be considered. Major objectives of an SBA are to determine shoreline change rates and the relative importance of the different mechanisms which cause the changes. In this paper means are discussed in which human processes can be introduced to mitigate a shoreline retreat problem. This is done in the context of an SBA.

Artificial dunes can be created or natural dunes rehabilitated and stabilized to prevent overwash and wind deposition of littoral sands landward of the frontal dune. Jetties can be designed to prevent inlets from migrating parallel to shore, and to prevent the movement of littoral sands into the inlets. If properly done beach sand is kept in the littoral system and bypassed across the inlet throat at a rate euqal to the net longshore transport rate. Offshore breakwaters, sited in regions where the gross volume of littoral sand moving parallel to shore is slight, may prevent littoral sand losses to the offshore region. Properly designed groins may also hold littoral sand along a coastal reach if the compartments between the groins are artificially filled with sand and the longshore transport path is not interrupted. Beach restoration, utilizing sand from outside the shoreface region, will have a positive effect on the shoreline because the volume of sand in the system is increased. Many of the "hard structure" solutions to a shoreline retreat problem present the potential for causing a problem elsewhere. This is especially true in the case where sand is prevented from moving to the area behind a barrier island. The growth and propagation of marshes may thus be limited because the sand substrate on which the marshes grow has not been allowed to develop.

Clearly, site-specific data must be obtained to evaluate the effects of human modification of the sediment transport system of a barrier island.

ACKNOWLEDGEMENTS

This paper is a product of the Barrier Island Sedimentation Study at the Coastal Engineering Research Center, carried out under the Civil Works Research Program of the U.S. Army Corps of Engineers. I thank E. Meisburger, L. Hulmes, K. Finkelstein, S.J. Williams, E. Hands, R. Hobson, D. Prins and N. Parker for many comments they made informally that helped clarify my thinking on this topic. Conclusions are not to be construed as officially those of the Department of the Army unless designated by other authorized documents. Permission to publish is appreciated.

REFERENCES

Bruun, Per, "Sea Level Rise as a Cause of Shore Erosion", Jour. Waterways and Harbors Division, ASCC, WW1, Feb. 1962, pp. 117-130.

Coleman, J.M. and Smith, W.G., "Late Recent Rise of Sea Level", Geological Society of America Bulletin, v. 75, p. 833-840, 1964.

Curray, J.R., "Late Quaternary History, Continental Shelves of the United States", <u>in</u> Wright, H.E., Jr., and Frey, D.G., eds., The Quarternary of the United States, Princeton, NJ, Princeton University Press, p. 723-735, 1965.

Everts, Craig H., "Geometry of Profiles Across Inner Continental Shelves of the Atlantic and Gulf Coasts of the United States", TP 78-4, U.S. Army Coastal Engineering Research Center, Fort Belvoir, VA 22060, 1978, p. 92.

Everts, Craig H., "Beach Behavior in the Vicinity of Groins - Two New Jersey Field Examples", Proceedings of The Specialty Congerence on COASTAL STRUCTURES '79, ASCE, Alexandria, VA, March 1979, pp. 853-867.

Fairbridge, R.W., "Eustatic Changes in Sea Level", Physics and Chemistry of the Earth, v. 4, p. 99-185.

Fisher, John J., "Shoreline Erosion, Rhode Island and North Carolina Coasts - Test of Bruun Rule", in Proceedings of the Per Bruun Symposium, Newport, Rhode Island, Nov. 1979, pp. 32-54.

Godfrey, P.J., "Oceanic Overwash and Its Ecological Implications on the Outer Banks of North Carolina", Office of Natural Science, National Park Service, Washington, DC, 37 p., 1970.

Hallermeier, Robert J., "Calculating a Yearly Limit Depth to the Active Beach Profile", TP 77-9, U.S. Army Coastal Engineering Research Center, Fort Belvoir, VA 22060, 1977, 28 p.

Hands, Edward B., "Bruun's Concept Applied to the Great Lakes", in Proceedings of the Per Bruun Symposium, Newport, Rhode Island, Nov. 1979.

Hicks, S.D., "On the Classification and Trends of Long Period Sea Level Series", Shore and Beach, April 1972, p. 20-23.

Kaye, Clifford A., and Barghoorn, Elso S., "Late Quaternary Sea Level Change and Crustal Rise at Boston, Massachusetts, with notes on the Autocompaction of Peat", Geol. Soc. of America Bulletin, v. 75, p. 63-80, 1964.

Koman, Paul D., "Beach Processes and Sedimentation", chapter 9, Prentice-Hall Inc., Englewood Cliffs, NJ, 429 p., 1976.

Kraft, J.C., "Radiocarbon Dates in the Delaware Coastal Zone (Eastern Atlantic Coast of North America)", Newark, University of Delaware, Delaware Sea Grant Program DEL-SG-1976, 20 p.

Lanan, Glenn A., and Dalrymple, Robert A., "A Coastal Engineering Study of Indian River Inlet, Delaware", Ocean Engineering Technical Report 14, Delaware Sea Grant Program, University of Delaware, Newark, 227 p. 1977.

Leatherman, Stephen P., "Barrier Dune Systems: A Reassessment", Sedimentary Ghology, v. 24, p. 1-16, 1979.

U.S. Coastal Engineering Research Center, "Shore Protection Manual" 3d ed., Fort Belvoir, VA, U.S. Coastal Engineering Research Center, 1977.

NEW TOOLS FOR RESOLVING CONFLICTS
IN PORT PLANNING

Edward J. Linky and Bruce H. Hoff*

Introduction

The Federal Coastal Zone Management Program is coming of age. At least 50% of States still participating in the program have completed the shift from "Section 305" planning mode to the "Section 306" plan implementation mode.(1) All approved state plans contain elements which directly affect ports and energy facility siting. A new federal program involving conversion of oil or natural gas fired electric generating stations to coal combustion will severly test the efficacy of coastal zone management plans. Infrastructure improvements will be needed to accommodate rail-marine barge terminals as well as upgrading docking facilities at coastal power plants. The National Interest sections of the State plans may be tested it port otticials attempt to "move" such projects by cloaking these facilities in that guise. The thrust of this paper is to examine in a case study context how a hypotethical application to improve a now outmoded and low throughput marine rail coal handling facility could be evaluated under administrative procedures currently in place in New Jersey.

New Jersey provides an excellent example because of its bifurcated state procedure for siting energy facilities. More importantly, it has an approved Coastal Management Program and State Energy Master Plan which directly or indirectly impacts all marine facility planning along New Jersey interstate waterways. In the interest of brevity only one case study will be examined, that of Port Reading in Carteret, New Jersey.

Port Reading is located on the Arthur Kill, a waterway that is intensively utilized by ship traffic and has many energy and industrial sites adjacent to it. It is bounded on the east by Staten Island and on the west by the New Jersey counties of Union and Middlesex as shown in Figure 1. Port Reading also is within the planning district of the Port Authority of New York and New Jersey. Finally and most importantly the element of this exercise which accentuates the complexity of the administrative procedure is the requirement of the New Jersey Energy Master Plan that a "NEED DETERMINATION" be required on certain types of new major energy facilities.

---

*Edward J. Linky, Administrator, Office of Planning and Policy Analysis New Jersey Department of Energy, former Staff Attorney, N.J. Office of Coastal Zone Management, Bruce H. Hoff, OCS Coordinator-Office of Planning and Policy Analysis, N.J. Department of Energy. The views expressed herein are the authors and are not necessarily endorsed by the State of New Jersey.

FIGURE 1

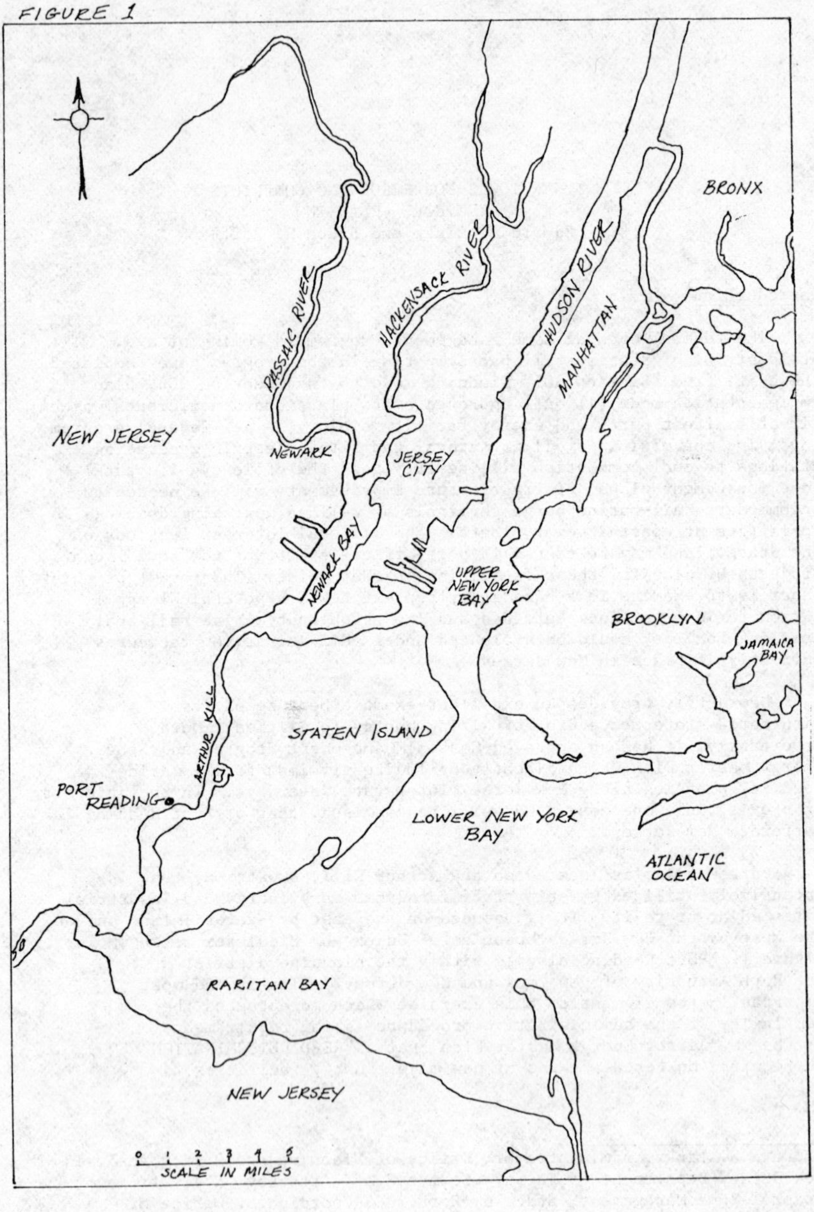
BRONX
PASSAIC RIVER
HACKENSACK RIVER
HUDSON RIVER
MANHATTAN
NEW JERSEY
NEWARK
JERSEY CITY
NEWARK BAY
UPPER NEW YORK BAY
BROOKLYN
JAMAICA BAY
STATEN ISLAND
ARTHUR KILL
PORT READING
LOWER NEW YORK BAY
ATLANTIC OCEAN
RARITAN BAY
NEW JERSEY
0 1 2 3 4 5
SCALE IN MILES

The goal of the analysis is to determine the impact on the siting or refurbishment of a marine rail coal terminal in New Jersey. Can this "Need" process actually expedite permit processing or will it be another complicating factor? Unfortunately in recent years, the planning offices of both the private and public sector are littered with plans for port facilities stalled or defeated by citizen or state agency opposition.(2) Before launching into an analysis of the procedures a background presentation on both the coal conversion program and current port planning practices will be useful to measure against the New Jersey approach.

Coal Conversion Program

The coal conversion program was initiated in 1977 as an administrative program of the U.S. Department of Energy. No power plants in the Northeast converted to burning coal under this program. The new program is based upon the Power Plant and Industrial Fuel Use Act of 1978, P.L. 95-720 which gives the U.S. DOE authority to prohibit power plants from burning oil or natural gas for purposes of generating electricity. Pursuant to this authority a list of 42 power plants has been drawn which have the capability of converting to coal combustion. An Environmental Impact Statement (EIS) must be prepared on the generic program as well as each proposed site specific conversion.(3) The EIS which will be prepared in accordance with Section 102(2)(c) of the National Environmental Policy Act (NEPA) will assess cumulative site specific and regional impacts of the program. One specific impact that must be evaluated is the ability of existing transportation systems to move massive quantities of coal into harbor areas and then transship the coal by barge to specific power plants. Additionally, the transportation system will be required to move the waste products of coal fired plants to approved disposal sites. Thus, the impact of the Conversion Program on near and long term port planning will be substantial. The program could conceivably have negative regional environmental impacts and positive regional economic benefits.

To the extent that port facilities are modernized to accommodate demand for coal it is possible that dredging to accommodate coal handling facilities as well as the facilities themselves will become insnared in a regional need determination for the facility. Evaluating the need for port facilities based on a regional screening process is not popular among port planners and for that reason some description of current attitudes and practices is warranted.

Port Planning and Development Issues

The concept that port planning should be expanded to capture and serve regional and national interests has been gaining momentum within the federal bureaucracy. Port facilities have been planned in the past around immediate hinterland interests. High outlays of capital to create competing and underutilized facilities in nearby ports (i.e. New York and Baltimore Containerization War) have forced the federal government to consider a national port plan as one option of reducing redundancy of service and facilities. The General Accounting Office has noted the potential advantages of a national port plan.

> "The advantages of this choice are the potential for determining the best number and spacing of ports, within the context of an integrated national system...This option would permit an analysis of the costs, benefits, policies with respect to port development and their role in the national transportation network. Airports and seaports face similar problems in keeping pace with technological advancement and coping with environmental and social constraints imposed by Federal, state and local agencies..."(4)

The posture of the port planning community however, is markedly different. The issue of determining port needs was extensively aired at the National Conference on Ports and Coastal Management (July 31-August 1, 1979). Some participants from the environmental community were emphatic that the current port planning process did not sufficiently address environmental issues.(5) They contended that port planners project unrealistic growth rates and do not recognize the scarcity of coastal resources.(6)

Parenthetically, it must be noted that electric demand forecasters for New Jersey public utilities have a similar label for projecting excessive load growth as a means to justify the need for floating nuclear generating stations Atlantic I and II, for example.

The attitude of port planning officials toward regional port planning was stated quite succintly at the National Port Conference.

> "Port agencies in all the workshops were unanimious in opposing regional port planning which led to the allocation of new port facilities. Ports in some workshops like the Gulf strongly opposed regional planning per se because of the fear that the information might lead to the allocation of resources."(7)

Based on the above citation there is every reason to believe that the port planning community will handle the potential of new coal terminals in a business as usual fashion. Indeed Consolidated Rail Corporation and the State of Pennsylvania have announced their intention to develop a coal terminal that would handle 10 million tons annually. Hampton Roads, Virginia is looking to upgrade its facilities for coal export.(8)

Further St. Joe Minerals Corporation has announced its intention to pursue an improvement in coal terminals at Hampton Roads.(9) Finally several groups are investigating the revitalization of Port Reading and other sites to service the coal conversion program or a major coal export center.

The scramble is on and fierce competition between various eastern ports can be expected. Whether this competition is productive and will result in: 1) lower coal transportation costs for the power plant conversion which will be reflected in the cost of electricity, and 2) profits in coal export will be maximized is questionable.

The remaining sections of this paper will examine the existing administrative permit procedure in New Jersey and how this process can address the concerns of competition for regional facilities as well as expedite the permit process. It is important to keep in mind that an alternative analysis on these projects must be performed as part of the NEPA Environmental Impact Statement procedure. It is our thesis that the New Jersey process may allow for a more balanced presentation of the issues than does the NEPA process.

The New Jersey Procedure

The New Jersey Energy Facility Siting process consists of several elements:

(1) A Need Determination by the New Jersey Department of Energy;

(2) Review by the Department of Environmental Protection for consistency with the State Coastal Management Program; and

(3) Review of applications for federal permits for consistency with both the energy Master Plan and Coastal Managment Program.

This two pronged approach to energy facility siting is an outgrowth of the enabling statute of the New Jersey Department of Energy (NJDOE). Specifically New Jersey Statues Annotated Title 52, Section 27F, subsection (c), vests the NJDOE with coextensive siting authority with every other state instrumentality in the siting of energy facilities. An Energy Report as required by statute is submitted to the Department of Environmental Protection, either recommending the facility or giving reasons for denial of the siting application. If the two agencies disagree and differences cannot be mediated, then the matter is referred to an Energy Facility Review Board which makes a binding decision on both agencies. Members of the Board are one from each of the two agencies in disagreement and an appointee of the Governor.

The NJDOE's enabling statute also provides that the Department prepare an Energy Master Plan for a period of ten years on the production, distribution, consumption, and conservation of energy in the state.(10) Further, the statute requires that the instrumentalities of the state government adhere to the Master Plan to the maximum extent practicable and feasible and authorizes the Commissioner of NJDOE to intervene in any state agency proceedings to ensure that the standard is met.(11) The administrative details of cooperative planning and permit processing are set forth in a Memorandum of Understanding negotiated between the two agencies.(12) The Need Determination is the leading edge of the New Jersey process and therefore warrants closer examination. The following sections will show how the Need Determination addresses the national interest sections of the Coastal Management Plan and acts as a facilitator of permit reviews.

The Need Determination

The Energy Master Plan's energy facility siting policy emphasizes:

(a) incorporation of need as the primary criterion which the Department will use to evaluate energy facility proposals;

(b) State initiatives in determining need for future energy facilities; and

(c) Standardization of requirements for energy facility applications to indicate consideration of need.

The actual Need Determination for energy facilities is evaluated against three basic standards:

(1) Whether existing sources of supply will be adequate to meet future levels of demand including careful consideration of the potential impacts of conservation;

(2) Whether better technological alternatives to proposed facility exist; and

(3) Whether better locational alternatives to proposed facility exist.(13)

The need for energy facilities which serve a regional or national market for fossil fuels will be evaluated similarly with certain differences appropriate to private energy industries and to facilities serving a regional market.(14) In such cases, the nation's or region's need for the facility will be considered. The facilities affect on product availability as the New Jersey market will also be considered, as will the effect of a denial on the energy company's competitive market position. In addition, factors such as the State's ecological carrying capacity, levels of acceptable risk, the facility's potential benefit or adverse economic and environmental impacts and the availability of technological or locational alternatives, particularly the availibility of out of state sites, will be incorporated into the determination of need for regional energy facilities. Once the Department has determined that a facility is needed it becomes an advocate with other permitting authoritives to ensure that the facility is constructed.(15)

The Interface of Port Planning and the New Jersey Process

The Coastal Management Plan and Energy Master Plan reinforce one another and intertwine principally on the issue of need. To explain this statement one must examine at least three components of the Coastal Management Program:

(1) port use policies;
(2) national interest section; and
(3) regional benefit decisions section.(16)

Let us explain inturn.

Port Use Policies: New port uses are acceptable only when there is a clear demonstration of need and where sufficient land and wider area is not available in or adjacent to an existing port.(17)

National Interests: The federal Coastal Zone Management Act(18) requires that a State's management program provide for the adequate consideration of the national interest involved in planning for and in the siting of facilities which are necessary to meet requirements which are other than local in nature. The national interest according to the New Jersey Coastal Management Plan need not in every instance be equivalent to an energy or port facility but a strong presumption exists that most port facilities could meet this test.

Regional Benefit Decision: The federal Coastal Management Act also requires that a method of assuming that local land and water use regulations within the coastal zone do not unreasonably restrict or exclude land and water uses of regional benefit. This method must include:

(1) a definition of what constitutes unreasonable restrictions or exclusions, and

(2) an identification of the methods that will be employed to ensure that such unreasonable restrictions or exclusions do not occur.(19)

In each of the instances cited where a port proposal such as Port Reading which could possibly facilitate coal transportation for many of the 42 power plants on the USDOE conversion list, the Need Determination as defined by the Energy Master Plan provides the turn key administrative mechanism whereby:

(a) consistency with the Coastal Management Program is acknowledged with state and federal permits expedited(20);

(b) documentation for eminent domain proceedings is gathered if local communities unreasonably restrict the revitalization of the port loading facility, and

(c) adds to the NEPA EIS process by providing an administrative forum which will more fully weigh the economic and regional advantages of a project than could be accomplished by either NEPA or the Coastal Management Program standing alone.

Conclusions

The basic fear of port planners that a need determination will reduce flexibility in planning and lead to specialization along the lines of a national port plan may be somewhat misplaced. It has been the boon and bane of the planning profession to plan for facilities and not consider whether they are actually buildable given the regulatory environment. The thesis of this analysis is that the fears are largely irrelevant for with more and more states gaining approval of coastal management programs, new or expanded port facilities will enter a regulatory jungle which is principally weighted to environmental considerations.

The New Jersey procedure while not ignoring environmental issues does at least provide a more balanced forum for demonstrating the national or regional need for a facility than either a NEPA, EIS or an evaluation exclusively by a state's coastal management program.

The New Jersey process is as yet untested. However, the port planning community should at least explore the permit expediter role that a need determination could provide to their areas of responsibility. It is our position that this role of the need determination will far outweigh the real or imagined adverse aspects of port specialization.

FOOTNOTES

1. CZM Information Exchange -Status of Coastal Zone Management Program, June 1980 Section 305, 306 P.L. 92-583, 16USC 1451 et seq. Coastal Zone Management Act of 1972 as amended by P.L. 94-370.

2. Sohio at Long Beach, Case II, Seaward Development Case Studies of Citizen Participation, National Academy of Sciences Academy Forum, Washington, D.C. 1980.

3. Federal Register Vol. 45, No. 109, June 4, 1980.

4. American Seaports- Changes Affecting Operations and Development, General Accounting Office November 16, 1979 Washington, D.C., pp. 24-25.

5. National Conference on Ports and Coastal Management, Boston, Massachusetts, 1979, Maritime Administration, National Oceanic and Atmospheric Administration and American Association of Port Authorities, p.6.

6. Ibid.

7. Ibid.

8. Wall Street Journal, June 22, 1980.

9. Wall Street Journal, July 23, 1980.

10. N.J.S.A. 52:27F-11(w).

11. N.J.S.A 52:27F-11(g)(b).

12. See New Jersey Energy Master Plan - Appendix A and New Jersey Coastal Management Program October 1978.

13. New Jersey Energy Master Plan p. 17.

14. New Jersey Energy Master Plan A-15.

15. Ibid, pp. 15-16.

16. Ibid, Draft EIS - p. 213(a), 257, 267 *et seq.*

17. *New Jersey Administrative Code* Title 7, Subchapter 7E, Section 7.9.

18. Subsection 306(c) P.L. 92-583.

19. Ibid, Subsection 306 (e)(2), 15 CER 923.12.

20. *Proposed New Jersey Coastal Management Plan*, Draft EIS, 1980, p. 248.

Endangered Species Coordination for Los Angeles Harbor

By
Daniel Muslin,[1] M. ASCE
and
Charles H. Fisher,[1] M. ASCE

Los Angeles Harbor, which serves the southwestern states, is located on the southern California coast about 25 miles south of the Los Angeles civic center and is one of the largest commercial harbors in the United States. In its natural state, the harbor was comprised of a clustering of islands in an estuarial complex of tidal sloughs, lagoons, and marshlands, which have been obliterated by intermittent dredging and filling operations. These operations have created the existing navigation channels for the harbor as well as the landfill known as Terminal Island.

The Corps of Engineers (Corps) was authorized, through the Water Resources Development Act of 1976, to deepen and widen the existing harbor channels and to create approximately 190 acres of new land within the harbor as expeditiously as possible. As a result of the project, approximately 14.1 million cubic yards of material are to be removed from the existing harbor channels to create deeper channels and the essential landfill at a cost exceeding $60 million. A map of the harbor is shown on Figure 1. A shallower-water, benthic habitat was not originally part of the project, but the explanation of why it is included will be discussed later in this paper. The California least tern, a Federal and State of California endangered species, has historically nested and fed in the vicinity of the proposed landfill. As a consequence of the possible impacts the project would have on the least tern (see Figure 2), the Corps of Engineers was required to coordinate with and receive a biological opinion report from the U.S. Fish and Wildlife Service (Service) on the impacts. This is a requirement of Section 7 of the Endangered Species Act of 1973. In the report, the opinion was stated that the fill project would destroy approximately 40 percent of the feeding area identified as essential to the existence of the Terminal Island least tern colony. In addition, the report stated that no provision had been made to compensate for the direct loss of the aquatic feeding area, nor had any provisions been made for the protection of the Reeves Field or Ferry Street known nesting sites-located on Terminal Island, nor had consideration been given to the future nesting sites which may be established on the new fill. Based on the above, it was the opinion

[1] U.S. Army Corps of Engineers, Los Angeles District

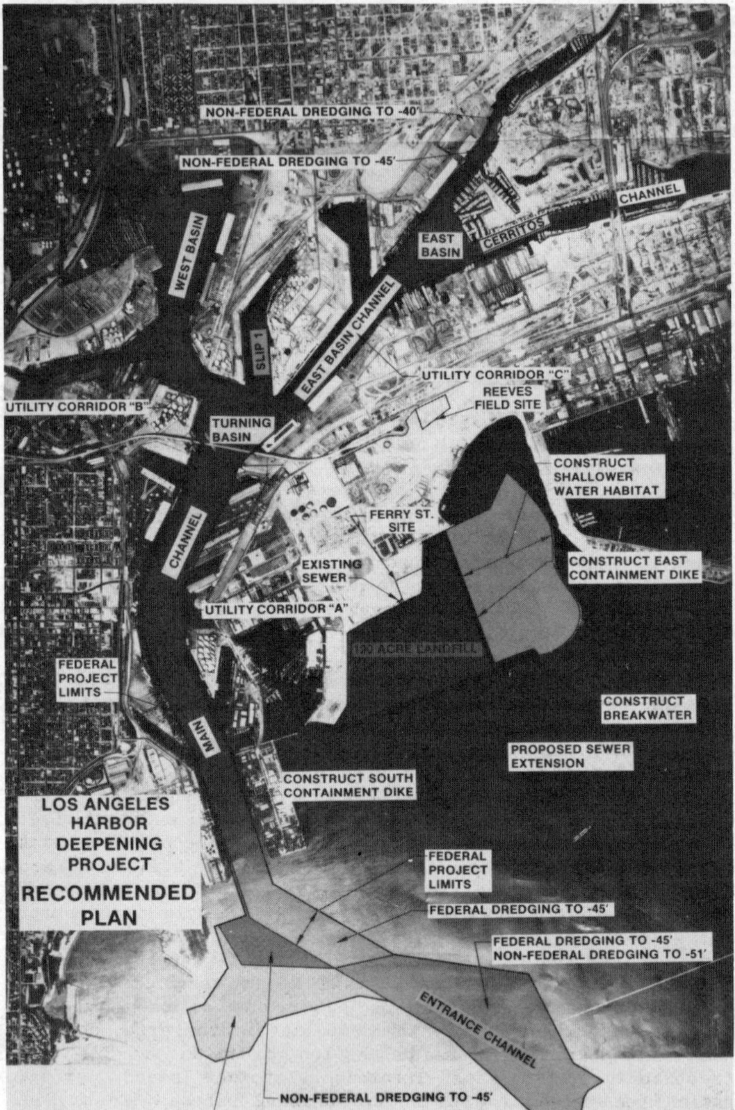
NON-FEDERAL DREDGING TO -40'
NON-FEDERAL DREDGING TO -45'
CHANNEL
CERRITOS
EAST BASIN
WEST BASIN
SLIP 1
EAST BASIN CHANNEL
UTILITY CORRIDOR "C"
REEVES FIELD SITE
UTILITY CORRIDOR "B"
TURNING BASIN
CONSTRUCT SHALLOWER WATER HABITAT
FERRY ST. SITE
CHANNEL
CONSTRUCT EAST CONTAINMENT DIKE
EXISTING SEWER
UTILITY CORRIDOR "A"
FEDERAL PROJECT LIMITS
CONSTRUCT BREAKWATER
MAIN
PROPOSED SEWER EXTENSION
CONSTRUCT SOUTH CONTAINMENT DIKE
LOS ANGELES HARBOR DEEPENING PROJECT
RECOMMENDED PLAN
FEDERAL PROJECT LIMITS
FEDERAL DREDGING TO -45'
FEDERAL DREDGING TO -45'
NON-FEDERAL DREDGING TO -51'
ENTRANCE CHANNEL
NON-FEDERAL DREDGING TO -45'
NON-FEDERAL DREDGING TO -51'

of the Service that the project proposed would jeopardize the continued existence of the California least tern. However, if the project were modified to address these concerns the Service could reinitiate consultation based on the following or similiar recommendations to eliminate the jeopardy:

a. The proposed fill be eliminated and the dredged material be disposed of at sea, or the fill be so located within the harbor such that it would not adversely impact the endangered least tern.

b. The Reeves Field and Ferry Street nesting sites be protected and maintained as nesting sites until such time as they may become permanently abandoned by the terns.

c. Provisions be made to protect any new nesting sites that may become established on the proposed fill area or elsewhere in Los Angeles Harbor area.

d. Compensation, in the form of habitat replacement, be included in the project to make up for the loss of aquatic feeding area that would result from the fill project.

e. Any other similar measures that would result in a net neutral or beneficial impact upon the least tern.

The negative biological opinion report basically vetoed the project as it was originally planned. A meeting was held with representatives from the Service, the Corps and the Los Angeles Harbor department. Based on this and subsequent meetings with National Marine Fisheries Service, the California Department of Fish and Game, and the U.S. Fish and Wildlife Service's Least Tern Recovery Team, a request was made by the Corps to the Service to reinitiate consultation based on the following proposal which was developed among all of the concerned agencies:

a. Maintain an existing fence around the approximately 15-acre Reeves Field nesting site through the 1979 nesting season.

b. Prepare (to the specifications of the U.S. Fish and Wildlife Service) and maintain (until the landfill is suitable for nesting - approximately 4 years) a 15-acre nesting site in the Ferry Street area for the purpose of establishing a new nesting site. Upon successfully establishing the Ferry Street site the Reeves Field nesting site could be abandoned and the land returned for use by the Harbor Department for other port purposes. The proposed Ferry Street site also would be used for an experiment to gather data on methods of establishing nesting habitat.

c. Provide and maintain, until further compensation proposals are implemented as a result of future harbor expansion, a similar 15-acre site on the project landfill as a new nesting site, allowing the harbor to resume using the Ferry Street site for port purposes.

d. Reduce the size of the original landfill to 190 acres.

e. Provide shallower water habitat adjacent to the project landfill.

f. Provide funds for studies of the Terminal Island colony nesting and feeding habitat, to determine the effectiveness of the proposals.

Based upon favorable receptivity by the Corps of the concerned agencies' recommendations the Service responded by reinitiating Section 7 consultation. Numerous staff meetings were then undertaken between all of the concerned agencies, including the Port of Los Angeles. As a result of a good faith effort by the participating parties to resolve the endangered species issue, the Service provided a biological opinion report which stated that the project would not jeopardize the continued existence of the least tern, provided that the following conditions were met:

a. Protect and maintain the existing 15-acre Reeves Field site for at least the 1979 and 1980 nesting seasons (April 1-August 31);

b. After the 1980 nesting season the Reeves Field site could be relocated to an adjacent area with at least a one-third overlap of the 1977 nesting area for a minimum of two nesting seasons. Any subsequent relocations on Reeves Field area will also require one-third overlap of the most recent successful nesting area;

c. The 15-acre Ferry Street site or a 15-acre site mutually acceptable to the Service and the Corps will need to be provided, protected, and maintained beginning with the 1982 nesting season for a minimum of 4 years, or until the new fill site has been utilized for nesting for at least 2 consecutive years, whichever comes first;

d. Provide, protect, and maintain a 15-acre site on the new landfill for at least 4 years from the date the new fill area is mutually agreed to be suitable for nesting. Mutual agreement is to be between the Corps and Service;

e. If no terns have nested at Reeves Field for 4 consecutive years, beginning with 1979 season, the area may be relinquished providing a suitable alternate site is agreed to by the Service and the Corps, and will be protected and maintained by the Corps;

f. If the least terns nest for 2 consecutive years on both the Reeves Field and Ferry Street sites, either site may be relinquished upon mutual agreement by the Service and the Corps and the other site will be protected and maintained for nesting;

g. The existing Reeves Field site must be protected and maintained for a minimum of 2 years without attempting any movement to an adjacent area;

h. Any alternate site could be relinquished upon mutual agreement by the Service and the Corps after 4 consecutive years of non-nesting, or when another alternate site has had 2 consecutive years of successful nesting;

i. Protected sites, suitable for nesting, need only be protected during the nesting season (April 1-August 31) and maintenance will occur prior to least tern arrival for nesting (April 1);

j. Should any site become the only site utilized for nesting, this site will be protected and maintained in perpetuity or until such time as no nesting occurs for 4 consecutive years and it is mutually agreed to by the Corps and the Service that protection and maintenance measures may be abandoned;

k. The size of the landfill shall be a maximum of 190 acres;

l. Shallower-water habitat (less than 20 feet) within the area bounded on the west by the landfill, on the north by the seaplane anchorage breakwater, on the east by a line immediately west of the port boundary, and on the south by a line projected from the south side of the landfill east to the port boundary will be provided. A detailed design for the placement of dredged material to create this shallower-water habitat will be agreed upon by the Corps and the Service, prior to implementation of the project; and

m. The Corps will fund studies of the Terminal Island colony of least terns for a period 7 years. Funding amounts and study plans are to be agreed upon prior to implementation of the project. If during any of the 7 years, no nesting occurs on Terminal Island, the funds allocated for that year's study may be utilized to study other least tern colonies in an effort to gain further information necessary to help conserve the species.

This coordination process has been completed, nearly a year's delay in the project was incurred. This delay, it is estimated, increased the cost of the project approximately $10 million just by inflation along.

The Corps consented to all of the above mutually developed conditions and assured the Service that they would be implemented. In February 1980 the Corps completed their final design report and environmental impact statement, all parties agreeing as to the measures necessary to be undertaken to insure the continued existence of the least tern. The Port of Los Angeles applied to the California Coastal Commission for a permit to contruct certain items of the project which were already included in the Corps' overall project. The Port went before the Commission in March 1980, at which time the Commission questioned the adequacy of the proposed protection of the least tern nesting sites that the "recognized experts" had developed and upon which agreement had been reached. The Commission at first believed that the sites should be protected for 10 years instead of four. The Commission did not take action on the permit until the following month at which time they concurred with the four-year time span and approved the Ports permit application.

Hopefully, other projects and agencies can benefit from the experience that the Los Angeles Harbor deepening project had to endure with regard to endangered species coordination.

References

1. U.S. Army Corps of Engineers, Los Angeles District, Los Angeles Harbor Deepening Project Final Phase II Design Memorandum, January 1980.

2. U.S. Army Corps of Engineers, Los Angeles District and Los Angeles Harbor Department, Final Supplement EIS/EIR, Los Angeles Harbor Deepening Project, February 1980.

RENOVATION OF NEW YORK HARBOR

By Theodore Civetta, Jr.[1] and Philip C. Sears[2]

ABSTRACT

The Port of New York has developed, since the 1600's, into one of the world's greatest ports. However, over the last century, much of the waterfront has become abandoned and dilapidated. Floating drift and debris in the harbor has become a serious threat to navigation and in 1974 the U.S. Army Corps of Engineers was authorized by Congress to stop the floating drift and debris at the source by removing the abandoned piers and derelict vessels.

This paper examines the development of the Corps of Engineer's project to remove sources of drift in New York Harbor, and highlights some of the problems incidental to the project. Additionally, it examines redevelopment of the waterfront in areas where the project is already underway, particularly at Liberty State Park.

INTRODUCTION

The natural harbor of New York was one of the prime reasons for the development of New York City as the leading commercial center of the world. In the early 1600's, purely for trading purposes, the Dutch founded New Amsterdam, which later became New York. Two centuries later, New York Harbor's 650 miles of waterfront had developed into the busiest seaport in the United States. It served as the link between the expanding new nation and Europe. Three centuries after it was first put to use by the Dutch, New York Harbor was the busiest seaport in the world.

During the nineteenth and early twentieth centuries, the shipping and railroad industries controlled most of the port facilities in the harbor. However, by the Second World War most of these facilities were fast becoming obsolete. Automobiles, trucks, and airplanes spelled doom for the railroads and ocean liners. Bridges and tunnels were built across the rivers, which meant that there was no longer a need for ferries. New York Harbor has had a mature infrastructure of piers and other shore structures since the early 1900's. But as the railroad and shipping industries declined, many of the piers fell into disrepair or were abandoned. (See Figures 1, 2 and 3). Graveyards of old barges, tugs, ferryboats, and other abandoned vessels developed throughout the harbor as well (See Figures 3, 4 and 5). Floating debris from old piers and ships, a problem since the early 1900's, worsened, and posed a serious threat to navigation.

---

[1] Project Engineer. New York District. U.S. Army Corps of Engineers.

[2] Project Engineer, Dravo Van Houten. Inc.. New York.

FIGURE 1
PIER NO. 17

FIGURE 2
MANHATTAN, N.Y.

FIGURE 3

FIGURE 4

SHOOTERS ISLAND, N.Y. - N.J.

FIGURE 5

By the River and Harbor Acts, of 1915, 1917 and 1930, the Congress of the United States charged the U.S. Army Corps of Engineers with the Collection and removal of drift from New York harbor and its tributary waters. This authorization was restricted solely to maintenance. That is, only floating material could be collected. In 1915, 32,411 cubic feet of floating debris was collected at a cost of $2,888. By 1974, $1,067,318 was spent to collect 566,057 cubic feet of debris. In that year, floating debris caused $13,102,000 of damage to commercial and public vessels, and $5,749,000 of damage to recreational vessels. It is estimated that 17,800 commercial, public and recreational vessels collide annually with drift and incur damage to propellers, shifts, and hulls. This would indicate that the mere collection of flotsam was not enough to insure a safe harbor. The collection of floating debris continues today, and in 1979, 422,920 cubic feet of drift was collected at a cost of $2,100,000.

Realizing that the collection of drift was a losing battle, on March 7, 1974 Congress authorized the Corps to attack this problem at the source. This new authorization, embodied in the Water Resources Development Act of 1974 (Public Law 93-251), provided for, in addition to the existing activities, the removal and disposal of derelict vessels, deteriorated shore structures and debris along the shorefront of New York Harbor.

DRIFT SOURCE REMOVAL PROJECT

The project was authorized in accordance with the April 1971 Survey Report for the New York Harbor Collection and Removal of Drift Project and is designed to eliminate sources of harbor drift which are the cause of navigation and other difficulties. It is unique in that it involves a combination of federal urban renewal and water resource development objectives. Since the essential nature of the work is the removal and disposal of abandoned waterfront structures and vessels, the established federal formula for cost sharing of urban renewal projects was considered best applicable. For all removal work under this project, the federal government assumes two-thirds of the cost and the local sponsor provides one-third of the cost. Structures which are to remain are repaired by the local sponsor with the local sponsor assuming the entire cost of repair. The local sponsor is additionally subject to the following conditions:

a) Provide without cost to the United States all lands, easments, and rights-of-way for the project.

b) Hold and save the United States free from damage that may result from the considered improvement.

c) Enact and enforce local legislation during and after completion of work to prevent creation of future sources of drift and debris.

The drift source removal project encompasses all of the Port of New York including shores and tidal tributaries of Sandy Hook and Raritan Bays, Arthur Kill, Kill Van Kull, Passaic and Hackensack Rivers, Newark Bay, Upper and Lower New York Bays, Lower Hudson River, Harlem and East Rivers, the Western part of Long Island Sound and Jamaica Bay. It involves the participation of 8 New Jersey and 7 New York counties as well as between 50 to 100 local municipalities. The port covers about 650 miles of shorefront and about 1500 square miles of navigable water. It is currently estimated that there are 23.6 million cubic feet of drift source material in New York Harbor. This material is generated from about 2300 abandoned derelict vessels, over 120 unusable deteriorated piers and wharves, 150 repairable piers and wharves, numerous smaller deteriorated shore structures and debris lying along the shore. Figure 6 shows the project area and Table 1 gives a breakdown of estimated quantities.

PROJECT BENEFITS

A project must be economically justified to warrant federal participation. The benefits associated with the considered improvement must exceed the costs of the improvement. The benefits are both tangible and intangible. Tangible benefits are quantifiable and therefore may be assigned certain monetary values. Intangible benefits are not quantifiable and therefore can only be evaluated qualitatively. The estimated average annual benefits for this improvement is $33,960,000. The estimated annual charges for a 50 year project life at 5 7/8 percent (authorized rate of interest) would be $5,567,000. Therefore the benefit to cost ratio for this considered plan is 6.1 to 1. This means that every dollar invested in this project returns six dollars and ten cents in benefits.

The tangible benefits are as follows:

		Estimated Benefits (Oct 1979)
1.	Reduction of damages to commercial and public vessels.	$19,547,000
2.	Reduction of damage to recreational boats.	12,360,700
3.	Enhancement of real estate values along affected waterfront.	456,800
4.	Reduction of costs for collection of drift under the existing federal maintenance project.	153,700

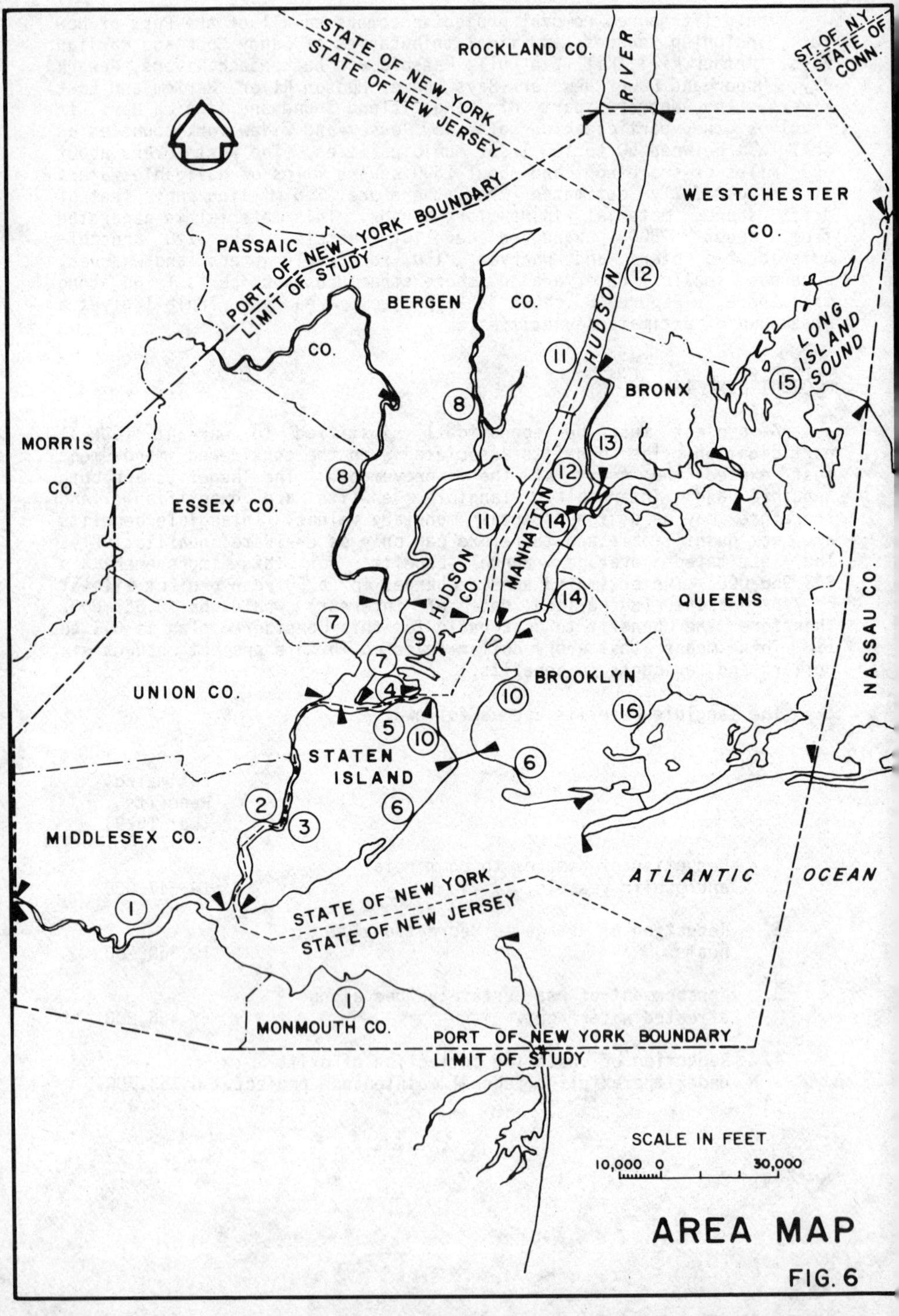
STATE OF NEW YORK
STATE OF NEW JERSEY
ROCKLAND CO.
HUDSON RIVER
ST. OF N.Y.
STATE OF CONN.
WESTCHESTER CO.
PORT OF NEW YORK BOUNDARY
LIMIT OF STUDY
PASSAIC CO.
BERGEN CO.
LONG ISLAND SOUND
BRONX
MORRIS CO.
ESSEX CO.
MANHATTAN
HUDSON CO.
QUEENS
NASSAU CO.
UNION CO.
BROOKLYN
STATEN ISLAND
MIDDLESEX CO.
ATLANTIC OCEAN
STATE OF NEW YORK
STATE OF NEW JERSEY
MONMOUTH CO.
PORT OF NEW YORK BOUNDARY
LIMIT OF STUDY
SCALE IN FEET
10,000 0 30,000
AREA MAP

FIG. 6

TABLE 1
ESTIMATE QUANTITIES TO BE REMOVED

RECOMMENDED IMPROVEMENT

No.	WATERWAY	COUNTRY	REMOVAL VOLUME (CUBIC FEET) STRUCTURE	DERELICTS
1	Sandy Hook, Raritan Bay, N.J.	Mommouth, Middlesex, N.J.	750,700	148,200
2	Arthur Kill, N.J.	Middlesex, Union, N.J.	1,033,300	766,900
3	Arthur Kill, N.Y.	Richmond, N.Y.	154,700	2,864,400
4	Kill Van Hull, N.J.	Hudson, N.J.	481,500	524,000
5	Kill Van Hull, N.Y.	Richmond, N.Y.	121,400	772,800
6	Lower New York Bay, N.Y.	Richmond, Kings, N.Y.	113,800	158,400
7	Newark Bay, N.J.	Union, Essex, Hudson, N.J.	277,700	248,600
8	Passaic atHackensackRiver,N.J.	Essex, Passaic, Bergen, Hudson, N.J.	189,500	244,300
9	Upper New York Bay, N.J.	Hudson, N.J.	516,600	1,953,900
10	Upper New York Bay, N.Y.	Richmond, Kings, N.Y.	2,608,800	73,000
11	Hudson River N.J.	Hudson, Bergen N.J. Rockland, N.Y.	5,646,400	1,978,400
12	Hudson River, N.Y.	Manhattan, Bronx, Westchester, N.Y.	179,200	0
13	Harlem River, N.Y.	Manhattan, Bronx, N.Y.	199,600	113,100
14	East River N.Y.	Manhattan, Bronx, Queens, Kings, N.Y.	865,400	0
15	Long Island Sound, N.Y .	Nassau, Queens, Bronx, Westchester, N.Y.	70,700	244,400
16	Jamaica Bay, N.Y.	Kings, Queens, Nassau, N.Y.	201,300	157,700
	TOTAL		13,410,600	10,248,100

5.	Reduction of costs of removal of drift from recreational beaches.	2,843,200
6.	Negative benefits caused by the detrimental effects from removal of the presently used deteriorated shore structures which would result in loss of use to the owners.	-(1,401,400)
	Total Average Annual Benefits	$33,960,000

Implementation of the considered plan of improvement will result in benefits to commercial and publicly owned vessels by reducing damage losses due to collision with drift. Tug boats, self propelled tank barges, ferries, lighters, deep draft cargo vessels, patrol boats, fire boats and other vessels collide with drift and are subject to direct and indirect losses. Damages normally occur to propellers, shafts and hulls and in some cases to internal gears. In addition to the direct cost for vessel reapir, other costs due to related losses were incurred. These costs include lost revenue, demurrage, crew claims, pilotage, towing, extra stevedoring charges, barging, and boat hire. These losses were estimated as 100 percent of repair costs for commercial vessels and 50 percent of the repair costs for public vessels.

To determine the current value of damage losses to commercial and public vessels, four factors were considered. These factors are:

(1) The annual number of vessel trips,

(2) The number of vessels in the harbor,

(3) The amount of drift material in the harbor,

(4) The escalation of ship repair costs.

The benefits that would accrue by implementation of this project are the difference between projected damage losses without the project and with the project. The benefits are therefore the reduction in projected damage losses.

As with benefits for commercial and public vessels, benefits for recreational vessels are the reduction in projected damage losses. These losses were estimated based upon (1) the number of recreational boats in the Harbor, which is increasing, (2) the amount of drift material in the harbor, (3) the frequency of collision with drift, (4) the cost of repairs to the boat. In 1978 and 1979, a total of five recreational boats were sunk in New York Harbor due to collisions with sources of drift. An intangible benefit associated with this project is the possible saving of lives and prevention of injury to operators and passengers of recreational boats.

Real Estate values will be enhanced as a result of the rehabilitation of the waterfront. Removal of derelict vessels and shore structures opening the waterfront to new and varied land uses. A detailed real estate survey was prepared of twenty sample sites in the harbor. The total enhanced value of real estate for the entire harbor is estimated to be $15,500,000 or $454,800 annual benefit based upon those sites.

Benefits in the form of a reduction in existing floating drift collection will occur. Based upon yearly quantities collected, the overall trend in collection indicates that there is a 1.2 percent compound annual increase in material collected since 1915. The average annual savings for the reduction in collection with the project is $153,700.

The project will reduce the average annual costs of removing drift from recreational beaches in the New York Harbor area. Costs per linear foot for driftwood removal for New York City beaches were used to determine costs of cleaning up beaches in the entire harbor. The average annual savings for the reduction in collection from the beaches with the project is $2,843,200.

The only quantifiable negative benefit associated with this project is from the removal of deteriorated shore structures currently in use. Currently between five and ten percent of the deteriorated shore structures in the harbor which should be removed are being used for mooring of barges and lighters. Although some of these vessels can be relocated at no additional cost, many will have to be moved to reconstructed mooring facilities in the harbor. The current cost estimate for this reconstruction is $22,500,000 or $1,401,000 annually.

As previously stated, when applying these tangible benefits to the costs of the improvement, the project benefit to cost ratio is 6.1 to 1. But for this project, tangible benefits don't tell the whole story. In addition to the benefits that can be evaluated monetarily there are significant benefits that cannot. These intangible benefits include the reduction of fire hazards, air pollution and water pollution, and a lessening of the spread of vermin and marine borers. Improved appearance of the harbor will have a positive effect on the quality of life and morale in the New York area. But most importantly, removing the sources of drift will result in a) the elimination of hazards to life and health from floating drifts colliding with boats and swimmers and b) the elimination of unattractive nuisances in the form of derelict piers and vessels that are dangerous to those who climb on them.

PLAN FORMULATION AND IMPLEMENTATION

In formulating any project, the engineer must try to develop the most engineeringly sound, environmentally sensitive plan for the best price. Often, this is an impossibility and the engineer is forced to make concessions. One of the major problems that has come up in the Drift Source Removal project is how to dispose of the drift material. During the early 1970's, the Corps of Engineers prepared a series of studies on alternative methods of driftwood disposal. A comparison of disposal methods is summarized in Table 2. The Phase I Plan Formulation prepared December 1975, based upon these studies, considered three methods of disposal to be better than the others:

1) Burning at sea.

2) Burying in place

3) Moving to another location for landfill.

Of the three, it was determined that burning at sea was the best method of disposal. The burning must be done 20 miles off the coast of New Jersey and only when prevailing winds can blow the smoke out to sea and away from heavily populated areas. The advantages of burning the material at sea are that 1) there is no material preparation; 2) there is no restriction on material size, and 3) there is little material handling. The disadvantages of burning at sea are that 1) there is destruction of a resource, 2) the operation is hampered by weather, and 3) there is potential adverse environmental impact on air quality. Although the other two methods of disposal tend to reclaim useless land, the use of this land after reclamation is strictly limited, and possible leaching and contamination of the ground water may occur, as well as potential adverse environmental impacts on air and water quality. Additionally land fill areas are scarce today, making this disposal method difficult to implement.

It was further decided that, although it appears that burning at sea is the only reliable alternative for disposal of the timber material, the contractor is permitted to dispose of the material in approved upland disposal sites, as well as recycle the material, if possible. Due to environmental concerns, burning at sea is considered the last alternative. Of the first four contracts of this project, which have been completed the contractors have burned about 65% of the material at sea, 25% of the material has been disposed of in approved landfills, and about 10% of the material has been recycled.

A major environmental concern that has become a formidable problem is the disposal of dredged material. Some minor access dredging is required to remove structures and vessels that are in very shallow water. In order to economically remove the debris sources, access by waterborne equipment through the tidal cycle is necessary. Although the amount of dredged material is small, much of it is polluted, for which environmentally sound disposal methodology is limited. This problem is currently being studied by various groups and its solution will have great impact on all aspects of maritime and coastal affairs.

Table 2

COMPARISON OF DISPOSAL METHODS

Disposal Method	Comment
Destruction	
1 - Burning - on site	Not permitted.
2 - Burning - open pit at city landfill	Not permitted.
3 - Burning - in an incinerator	Problems regarding salt corrosion, emissions and disposal of ash.
4 - Burning - on barges in the harbor	Not permitted.
5 - Burning - on barges in the open sea	Presently being used.
Reuse	
1 - As fuel, primarily for power plants	Would require equipment conversion: Utility companies contacted in 1973 & 1980 were not interested.
2 - As domestic firewood	Problems with salt and creosote; some is being used.
3 - To manufacture charcoal	Possible, but problems with salt and creosote.
4 - To manufacture chips and mulch	Can be sold by local lumber yards, but problems with attached metal.
5 - As scrap lumber	Limited use.
6 - To manufacture hardboard	Problem with preparation of material.
7 - To manufacture pulp paper	Problems with salt, creosote and metal contents.
8 - As telephone poles	Limited salvage, only good piles.
9 - As railroad ties	Limited salvage.
10- For pallet construction	Limited salvage and preparation problems.
Landfills	
1 - Include in existing garbage fills	Space for garbage fill is at a premium.
2 - Bury in place	Good use, particularly for parkland.
3 - Move to another location and bury	Limited by transportation cost.

In the formulation of this project, a number of steps have been taken to deal with the problem of dredged material disposal.

1) Removal operations in shallow areas will be performed by shallow draft barges and channels will be cut to provide just enough water for access and egress.

2) Channels will be laid out in the most efficient manner to provide for the most drift removal with the least amount of channels.

3) Alternative disposal sites are being considered individually as each portion of the project nears construction. Because the amount of dredging for each area is small, an alternative site to ocean disposal may become available.

4) Since this project is not a dredging project, a side-casting alternative is now being considered. In effect, a contractor using this alternative will dig himself into an area and side-cast the material, remove the drift source and then backfill the material. The advantages of this method are that 1) it is cheaper than ocean disposal and 2) contaminated material is not dumped in a new site, but rather stays in an area where the character of the surrounding material is the same.

5) Wherever possible, land-based equipment will be used to remove drift source materials.

6) Ocean disposal of dredged material will only be considered when all other alternatives are exhausted.

PROJECT STATUS

A. New Jersey

One of the recommendations of the plan formulation for the entire harbor is that the project should be broken down into viable "reaches" which are manageable and have local support. The State of New Jersey was the first local agency to show interest in this project and in 1976, the Corps of Engineers prepared a Phase II Project Design for Liberty State Park, in Jersey City, New Jersey. The Phase II was a detailed inventory of drift sources at Liberty State Park. Liberty State Park is a three mile stretch of waterfront on Upper New York Bay, directly west of Liberty and Ellis Islands. This area, once a great railroad terminal complex for the Lehigh Valley and Central New Jersey Railroad Companies, was abandoned during the decline of the railroad industry. In the early 1970's, however, the State of New Jersey realized the potential of this area as an urban park and began developing plans for Liberty State Park. The Corps was enlisted to remove the derelict vessels and deteriorated piers at the park site.

In June 1976, the Corps of Engineers awarded its first drift source removal contract at Liberty State Park. To date, four contracts have been awarded totaling $8.5 million. 430 derelict vessels and 30 deteriorated piers and wharves have been removed at the park. The fifth and final contract is scheduled for award in the fall of 1980. Table 3 gives a breakdown of contract costs. All drift removal work should be completed at Liberty State Park before January 1982.

The State of New Jersey was so pleased with the results at Liberty State Park that, in November 1977, a $10 million waterfront clean up bond issue was proposed by the State. The bond issue passed, giving the State the funds and authority to act as the local sponsor for the entire state on this project. The Corps of Engineers is currently involved in the preparation of Phase II Project Design reports for three more areas in New Jersey: the Jersey City, Elizabeth, and Hoboken to North Bergen waterfronts.

B. New York

The second increment of work on this project is the East River Manhattan Waterfront for which a Phase II Project Design Report was prepared in September 1978. This represents the first work area in New York State. This portion of the project involves the removal of eight piers and several submerged pile fields which are currently hazards to navigation. The work corresponds with city development along the waterfront, including renovation of the South Street Seaport Historic District and the building of a waterfront complex at 21st Street on the East River. The estimated contract cost of this work is $3,300,000. A contract is scheduled for award in September 1980.

WATERFRONT REDEVELOPMENT

Although the primary objective of the Drift Source Removal Project is to improve navigation, this project is a catalyst for the redevelopment of the waterfront. Liberty State Park is the best example of this. The State of New Jersey is converting a wasteland into an 805 acre urban park. It is the first urban park of its kind in the State of New Jersey. The initial phase of state development was the 35 acre Southern Park Overlook, which was completed in time for Operation Sail, part of the Bicentennial Celebration of 1976. In its first year of existance, Liberty State Park attracted over 680,000 people, making it the most popular State park in New Jersey. Future development at Liberty Park includes the full restoration of the Central New Jersey Railroad Terminal built during the nineteenty century, the possible creation of an 18 hole golf course on the park site, building of two marinas on the north and south ends of the park, and the construction of a 2 1/2 mile crescent shaped levee and pedestrian walkway which will form an arc behind the Statue of Liberty and Ellis Island.

Table 3

NEW YORK HARBOR COLLECTION AND REMOVAL OF DRIFT PROJECT
PROJECT STATES

	Contract Award Date	Contract Amount	Removal Items: # Vessels	Removal Items: # Sizeable Shore Structures	Quantity of Materials: Timber (MBM)	Quantity of Materials: Steel (Tons)	Quantity of Materials: Conc (cy)
Liberty State Park							
Contract # 1	6/28/76	$2,498,400	190	4	13,000	1,500	950
Contract # 2	5/15/77	$1,244,500	130	1	14,200	5,000	300
Contract # 3	8/9/77	$2,210,800	40	9	17,000	2,100	18,000
Contract # 4	6/13/78	$2,530,000	70	16	16,000	2,000	5,000
Contract # 5	10/80	$2,000,000	20	9	4,000	80	17,000
East River Manhattan	9/80	$3,300,000	0	8	6,450	379	5,780
Jersey City, N.J.	10/81	$8,000,000	25	20	20,000	2,000	21,000
Elizabeth, N.J.	10/81	$4,000,000	73	14	12,000	470	200
Hoboken to North Bergen	10/82	$10,000,000	500	34	?	?	?

FIGURE 7
LIBERTY STATE PARK PRIOR TO CONSTRUCTION

FIGURE 8
LIBERTY STATE PARK DURING CONSTRUCTION

FIGURE 9
LIBERTY STATE PARK AFTER CONSTRUCTION

FIGURE 10
LIBERTY STATE PARK AFTER CONSTRUCTION

The City of New York in conjunction with the New York State Urban Development Corporation, the South Street Seaport Museum and the Rouse Corporation is planning a major redevelopment of the waterfront in the South Street Seaport Historic District. The "Seaport Marketplace" will be a mixed use, speciality retail center using a combination of historic and new waterfront buildings. Two dilapidated piers in the area are scheduled for demolition under the drift source removal project and two have already been repaired. The proposed development will include expansion of the South Street Seaport Museum, building of waterfront restaurants and hotels, creation of a marina, the incorporation of small retail speciality shops into existing and new buildings, and development commercial office and residential space. Work has already begun, and should be completed by 1984.

The City of New York is additionally redeveloping the waterfront at 21st Street in Manhattan on the East River. Three dilapidated piers are being removed in this area under the drift source removal project. The City plans to build a waterfront marina and restaurant complex on a pile supported platform extending out into the East River. Construction of this platform is already underway.

Waterfront communities can and have been taking advantage of the Drift Source Removal project. After drift sources are removed, a waterfront is ready for redevelopment. In a sense, the Corps of Engineers is offering communities a giant eraser to wipe out unwanted waterfront blight, so that they can go back to the drawing board with a clean piece of paper.

REFERENCES

U.S. Army Corps of Engineers, New York District. Survey Report on Review of Project, New York Harbor Collection and Removal of Drift Project, New York, June 1968, as revised March 1969 and April 1971.

U.S. Army Corps of Engineers, New York District. General Design Memoranda Phase I - Plan Formulation, New York harbor. Collection and Removal of Drift Project, New York, December 1975.

U.S. Army Corps of Engineers, New York District. GDM No. 2 Phase II - Project Design, New York Harbor Collection and Removal of Drift Project, New York, February 1976.

U.S. Army Corps of Engineers, New York District. Final Environmental Impact Statement, New York Harbor Collection and Removal of Drift Project, New York, November, 1975.

Robert Greenhalgh Albion, The Rise of New York Port, New York, 1939, Charles Scribner's Sons - Publisher.

VISUAL IMPACTS OF PORT DEVELOPMENT IN SEATTLE

Iain M. Robertson, A.L.I., Associate
William G.E. Blair, A.I.A., A.S.L.A., Principal

JONES & JONES
Seattle, WA

Introduction

This paper describes the visual assessment of a number of Development Alternatives proposed for a 128 acre waterfront property, Piers 90-91, at the north end of Elliott Bay in Seattle. It was undertaken for the Port of Seattle by Jones and Jones as one of a series of studies of the environmental impacts of redeveloping the property.

It also describes the various steps in the Visual Analysis method used in the study along with the results of the study. Particular emphasis is placed on discussing the advantages which the method offers both the developer, in this case the Port of Seattle, and the public affected by the proposals.

The dynamic context of the study and the significance of this particular site on Elliott Bay are also discussed in terms of recent trends in Port Development and new legislation covering the shoreline and visual considerations.

Context of Study

Along with huge increases in volumes of international trade there have been major changes in the techniques used for shipping and handling cargo over the last twenty years, which have caused correspondingly large changes in the visual character of Ports. An important part of these changes has been a shift from breakbulk to container cargo operations. Among west coast ports Seattle has been in the forefront of the trend towards containerization and the effects of this have caused major changes in the visual character (and the land use patterns) of the City's waterfront.

From its beginning Seattle has been a port city and one of the most vivid images of the city is the series of finger piers and warehouses which line the entire length of the downtown waterfront. With shift towards containers the character of these piers has been changing gradually as port uses are replaced by recreation or commercial uses. Cargo handling operations have moved south into the industrial areas of the city where large, open areas of land have been assembled for modern container terminals by the Port. These new terminals are anonymous and industrial in character partly on account of the character of the handling equipment and the containers themselves and partly because the former diversity of the shore edge has been reduced to long, straight lines by filling because the ratio of length of berth to storage area required for container terminals is much lower than for conventional cargo handling. Thus changes in Cargo Handling Techniques have resulted in distinct benefits for the City of Seattle as public access to the waterfront, particularly along the important central waterfront, has increased dramatically with the consolidation of Port operations to the south of the city center. Not only has public access to the shore increased but the piers themselves have become one of the major tourist destinations in Seattle and a thriving commercial area has grown around them. Redevelopment in the Southeast Harbor has proceeded with little adverse public comment because of its relative isolation from residential areas, and its location in an area which already had a strong industrial character. However, the inauguration of Terminal 37, a new stacked container operation, has resulted in a strong negative public response to its lighting system. This terminal uses taller light poles and higher illumination levels than other terminals and is by far the brightest element in views of the Seattle waterfront at night. Visual problems resulting from this development include disruption of the character of views and concern about safety on adjacent streets as a result of light spillover and glare.

While the majority of Port redevelopment has taken place to the South of Downtown on the former mud flats of the Duwamish River two major projects are located at the north end of Elliott Bay. The visual implications of development in these areas are far more significant for the city as the proximity of residential areas makes this a much more visually sensitive part of the Bay. The first of these projects, the Grain Terminal, has resulted in major visual impacts on adjacent residential areas because of its height and bulk. These impacts include blocking highly valued views of Puget Sound and the Olympic Mountains and disturbance of the adjacent neighborhood by the increased night time illumination levels and glare. This project led to an increase in viewer sensitivity to the effects of Port operations and resulted in the setting up of Neighborhood Advisory Committees which meet with Port officials monthly to discuss mutual concerns. The second major project, the redevelopment of Piers 90-91, the project described here, is still in the early stages of planning. This site is comprised of two finger piers almost half a mile long with a total area of 42 acres, and an additional 86 acres of 'uplands' located beyond an elevated public street, Magnolia Bridge, which bisects the site at the base of the piers. It is flanked on two sides by the residential hills of Queen Anne and Magnolia. Because of the experience with recent Port development described above the Visual concerns of these neighborhoods centered upon the issues of glare and light spillover into residential areas and the impacts on the character and quality of views resulting from introducing a new use, with what is perceived as an industrial character, into these views.

Statutory Visual Issues

At the same time that public awareness of the visual implications of port redevelopment has been growing there has also been an increase in regulations governing the use of the shoreline and the visual impacts of developments. The Washington Shoreline Management Act of 1971 and State Environmental Policy Act (SEPA) of 1971 require that the city develop regulations covering shore edge uses, and that the city assess development on visual grounds respectively. These resulted in 1.) Seattle Shoreline Master Program regulations which are designed to regulate development of the city's Shoreline "in order to (a) preserve, enhance and increase views of the water and access to the water, (b) encourage water-dependent uses, and (c) provide for maximum public use and enjoyment of the shorelines of the city", and 2.) Seattle Guidelines to implement SEPA, Ordinance 107678, which are intended to "reasonably protect" views from public places as well as views of certain designated historic landmarks. This latter ordinance requires that the impacts of light and glare and view blockage be disclosed and assessed by the city but cannot enforce the Port to perform these assessments. Similarly the Port is exempt from the requirements imposed by the Shoreline Master Program. The detailed analysis of the visual impacts of redeveloping the site described here therefore goes far beyond the statutory requirements which would be placed on any private developer proposing to develop the site and represents a strong commitment by the port to making the redeveloped Piers visually acceptable to the public as well as operationally and economically feasible.

The Visual Analysis Study of Piers 90-91

Piers 90-91 were declared surplus Federal Property by the G.S.A. in the early 1970's and were bought by the Port of Seattle with the intention of redeveloping them some time in the "early 1980's". This visual analysis study is one of a series of studies of the environmental implications of redeveloping the presently decrepit underused piers as a modern, intensive port facility. Three reports will lead up to the choice of a use or series of uses for the site, these are: 1.) An Alternatives Document giving factual data on the 10 proposed alternative uses; 2.) A non-use specific E.I.S. discussing all alternative uses; and finally 3.) An E.I.S. on the preferred use(s).

The method used in the study was developed by Jones and Jones initially for assessing generalized visual impacts on a regional basis in rural or wild settings Visual Resource Management. It was subsequently modified and developed to assess specific, project-related impacts first in wild locations (dams, power lines) and more recently in rural and urban settings (highways). Its use here, to assess and compare the impacts of a series of 10 Alternative Uses proposed by the Port, demonstrated the value of a systematic yet flexible approach to visual analysis. The method has been described in detail elsewhere (1) and will, therefore, not be repeated here. The stages of the analysis and the data gathered are illustrated in Figure 1, A Generalized Model for Visual Assessment.

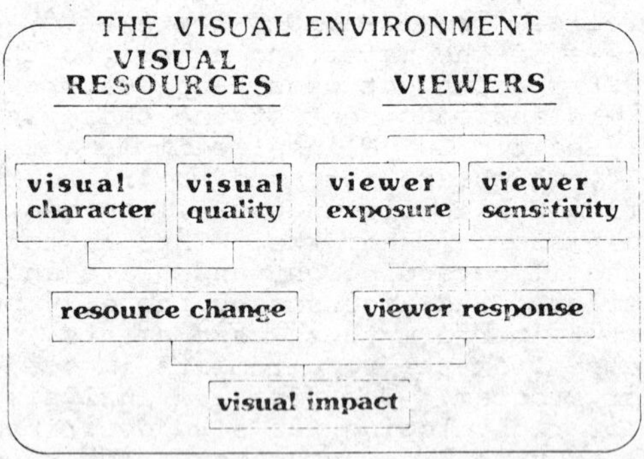

A generalized model for visual assessment (Jones & Jones).

The study may be divided into two stages, first an analysis and description of existing conditions, represented by the top row of boxes in the figure, and second simulation of the changes caused by the ten uses and comparison of these with the existing conditions to determine visual impacts. In this study the simulation of the proposed uses involved retouching 8 x 10 inch black and white photographs of selected 'key views' of the site to give a realistic image of the proposed use from the viewpoint. Because of the central importance of these key views in the analysis method much of the first stage of the analysis was devoted to defining the range of views of the site, the degree of viewer exposure and thus selecting appropriate locations for key views.

The study started with a regional analysis but quickly focused in on a local analysis of the neighborhoods immediately surrounding the site. At the regional scale a Potential Viewshed was mapped, Figure 2, which indicated the maximum area from which the site could potentially be seen, because of intervening structures and vegetation the actual viewshed was considerably smaller than this. This area was then subdivided into 'landscape units' based on physiography and land use patterns, Figure 3. (These same landscape units were later used in the analysis of the character and quality of views.) From this map and a map of City Scenic Routes, Parks and Designated Viewpoints, Figure 4, initial locations from which key views could be chosen were identified.

At the local level a more detailed analysis of Visual Character and Viewer Exposure was conducted. The residential areas on the east side of Magnolia and the west side of Queen Anne Hill were divided up according to the severity of exposure which was defined from slope, elevation, orientation and intervening structures.

The regional and local analysis established the existing Visual Character and quality and the degree of viewer exposure and enabled a series of 24 candidate key views to be selected. Previous experience had made clear the value of public input to the selection of the key views for analysis thus day and night time slides of each viewpoint were presented at meetings and local residents were asked to rate the importance of them. Importance was defined as:

> 'The consideration that should be given to possible changes in this view when examining alternatives for the 90-91 site'.

Based on the public response 8 key views were selected for the analysis (5 day, 3 night). All of these viewpoints were from Scenic Routes and two were from private residences. These key views were assumed to be representative of the types of views of the site and thus suitable for determining the range of impacts resulting from its redevelopment.

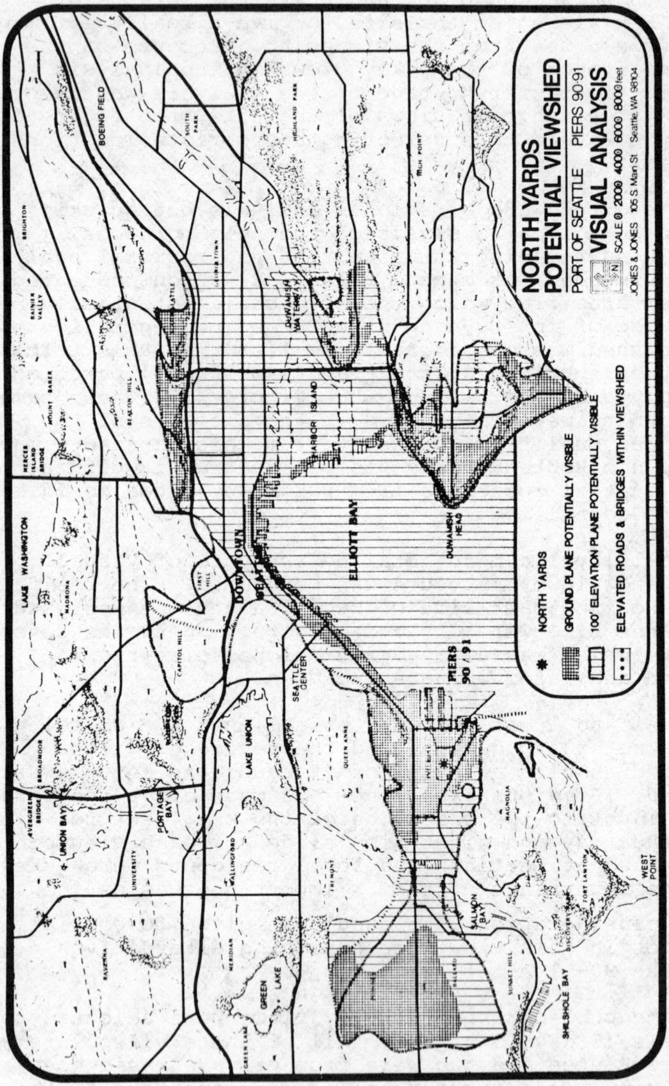

Figure 2. Regional Potential Viewshed

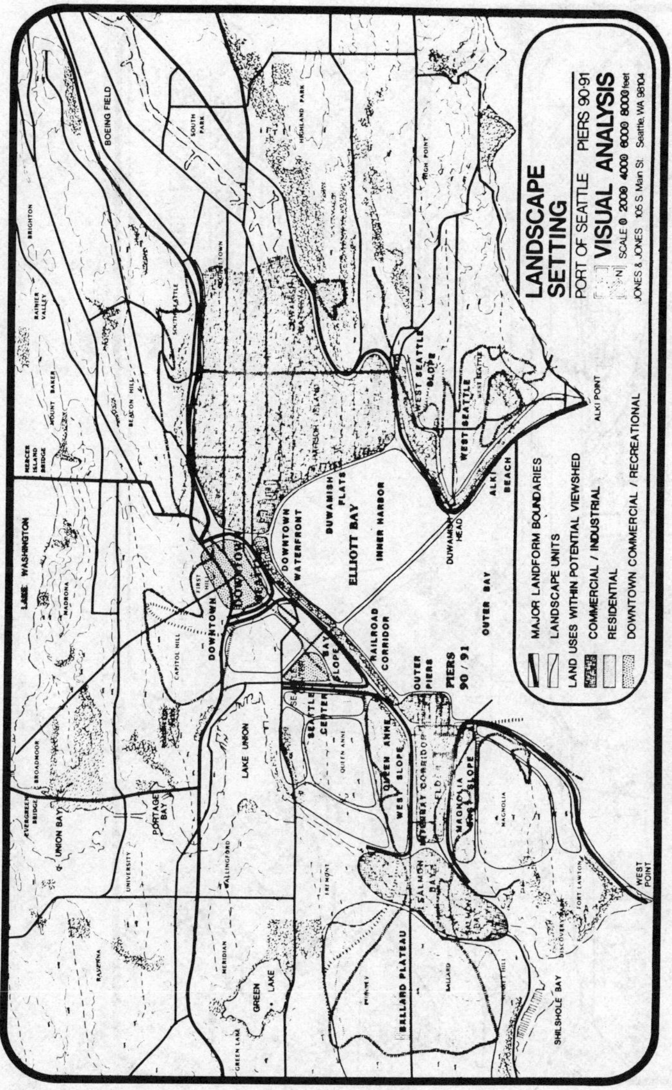

Figure 3. Regional Landscape Setting

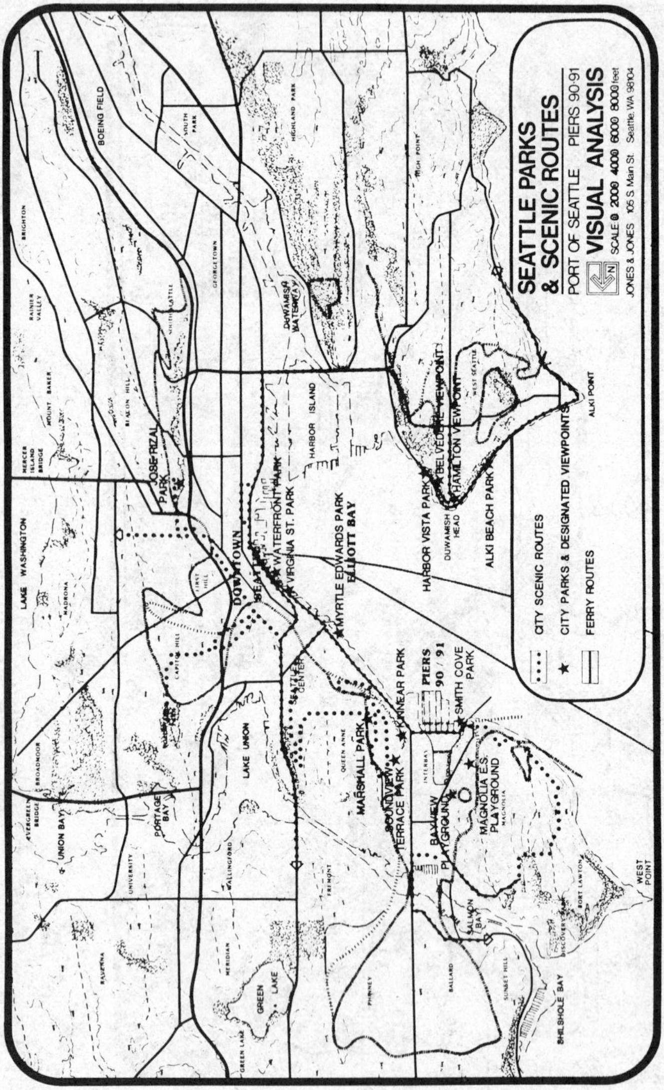

Figure 4. Seattle Parks and Scenic Routes

Although representative of views from an area they were selected to overstate rather than understate the severity of impacts thus making the analysis more conservative in its assessment. Figure 5 illustrates how the characteristics of each key view were summarized.

Importance:
4.1
6th of 24
Location:
Private
Residence
Viewers:
Commuters
Residents
Distance:
Foreground
Elevation:
150-160
Type:
Panoramic - Winter
Wide Angle - Summer
Components:
Territorial

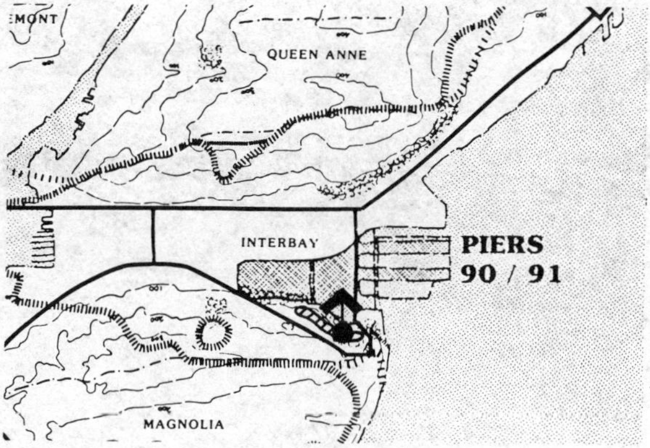

Viewer Exposure:

Approximate Number of Households Potentially Affected

High	Moderate	Slight
40		

PORT OF SEATTLE
KEY VIEW

PIERS 90·91
MAGNOLIA WAY

As not all uses appeared on each key view this enabled the total number of simulations necessary for the analysis to be reduced to 28 from a theoretical total of 80. (10 uses, 8 viewpoints). Examples of the simulations are shown below, the first pair, Figures 6 and 7, show two different uses from the same viewpoint illustrating how comparison of the relative changes and impacts can be made between uses. The second pair of photographs, Figures 8 and 9 show the same use from two viewpoints and illustrate how comparison of the changes and impacts of a single use can be made between views.

Figure 6. Freeze Chill Facility

Figure 7. Breakbulk Facility

Figure 8. Containers Terminal - Yard Crane Operation

Figure 9. Container Terminal - Yard Crane Operation

Each key view simulation was analysed for the change the use produced in the visual character and quality of the view and thus impacts were defined. From this it was possible to propose use-specific measures for mitigating visual impacts.

A similar analysis was performed for the night-time views, Figures 10 and 11, although these presented greater problems in presenting a realistic simulation of the view.

Advantages of the Key View Simulation Method

The method described above has advantages for all parties concerned including the public, the client and the planners. In general the use of a technique which does not require special training to interpret results will lead to clear and immediate comprehension of the proposals by all parties and thus avoid the confusion and mistrust that often occurs if less realistic methods are used e.g. Bird's eye perspectives; Artist's conceptions; Photographs with developed areas cut out; Plans, sections and elevations; or written descriptions. All of these techniques are open to distortions and misunderstandings that do not occur when the actual character of a proposed use is simulated on an actual view familiar to the public. Thus not only does the method clearly present the visual implications and enable direct comparison of different alternatives but it also dispels misunderstandings which are often attached to emotional terms such a 'container cranes' and 'night lighting' by the public. This does not mean that the public will necessarily like an alternative use better but that they will be able to express their dislike objectively in a way in which it will be far more likely to have some effect on the use selection process; e.g. an objection to a use on the basis of a set of objectively defined specific visual impacts is more likely to lead to modifications than a flat statement such as 'I don't like concrete buildings'.

In addition to these general benefits and the public benefits, the technique is valuable for the client also because it presents him with a direct statement of the implications of a course of action. In the case of the Piers 90-91 study the initial photographic simulations led to reappraisal and redesign of several uses. Not only was the Port forced to make the decisions necessary for plans to be prepared and simulations made, thus speeding up the analysis process, but they were immediately presented with the implications of these decisions. This led to reappraisal and change of a number of the uses at an early stage in the planning study before undesirable decisions were irrevocably tied into the proposals. The other major advantage of the method, for both the port and the visual

Figure 10. Container Terminal - Chassis Operation

Figure 11. Container Terminal - Yard Crane Operation

consultants, was that it enable the uses to be ranked in order of their visual acceptability, thus uses were categorised as:

1. Desirable or acceptable as simulated,
2. Acceptable if minor mitigating changes that would not affect their operation were made,
3. Requiring major changes that may affect operational feasibility to make visually acceptable, and
4. Unlikely to be visually acceptable even with major functional changes unless these caused major improvements to the visual character.

Thus the systematic method prevents designer or client from failing to anticipate or comprehend the visual consequences of their proposals from every point of view. In this regard it has been the experience of the authors that there are few simultions which do not hold at least some element of surprise when shown to even the most experienced of planners and there are many that may cause assumptions present throughout the study to be completely reversed.

Conclusion

The visual analysis of the ten alternative uses proposed for Piers 90-91 by the Port did not result in the identification of mitigating measures which would make all uses visually acceptable. However, it did make explicit and comprehensible the reasons why certain uses were undesirable, and ranked the uses according to the type and extent of changes that would have to be made to make them acceptable, thus clarifying the visual implications of the development strategies available to the Port. It also made clear which types of visual impact were the most important, e.g. shoreline changes, thus directing attention to these critical areas.

References

(1) Blair, William G.E.: "Visual Resource Management" in Environmental Comment, pp 4-15, June, 1980.

(2) Jones & Jones, Esthetics and Visual Resource Management for Highways; Seminar Notes 1979-1980 for U.S. Department of Transportation, Washington, D.C., 1979.

MARYLAND'S COASTAL FISHERIES PROGRAM: AN OVERVIEW

by: Mark M. Bundy
Coordinator Coastal Fisheries Program
Coastal Resources Division
Tidewater Administration
Maryland Department of Natural Resources

A. MARYLAND'S COASTAL FISHERIES PROGRAM

In the opening paragraph of his book, Beautiful Swimmers, Watermen, Crabs and the Chesapeake Bay, William Warner describes Maryland's coveted estuary as:

> "The Bay. It is so known through the length and breadth of it's watershed. There is no possible confusion with any other body of water, no need for more precise description. Its waters are rich, the main supply of oysters, crabs, clams and other seafoods for much of the Atlantic seaboard. Its shorelines cradled our first settlements. It is the Chesapeake."

Nearly two hundred miles long and covering just over 4,000 square miles in surface area, the Chesapeake Bay is indeed biologically a very rich and productive estuary. Each year, millions of finfish and shellfish depend on the Bay and Maryland's coastal waters to provide food and shelter. This abundant resource supports a substantial commercial fishing industry and an extremely active recreational fishery in Maryland. The responsibility for managing this considerable resource is that of the State's Tidal Fisheries Division of the Department of Natural Resources.

Major goals of this Division include endeavors to perpetuate and improve equitable opportunities for the various competing user groups to acquire their fair share of the public resource; to assure access to public waters for all fishermen; and to protect the State's living aquatic reosources from environmental degradation and overfishing.

In general, the Division's mission is to conserve and to enhance the State's fisheries, which in this context, is used in the broadest definition of the term. That is, fisheries refers to all of the finfish and shellfish resources as well as the associated commercial and recreational industries that are dependent upon these resources.

Within the Coastal Resources Division of Maryland's Department of Natural Resources (the State's coordinating focal point for its Coastal Zone Program) a Coastal Fisheries Program was established to work in concert with the Tidal Fisheries Division in developing a comprehensive management strategy for Maryland's coastal fishery. Towards this end, the Coastal Fisheries Program initiated a study to define the structure of and interaction within Maryland's Chesapeake Bay commercial fishing industry. Primarily, it consisted of a qualitative analysis of the industry based on extensive interviews with watermen, processors, dealers, and concerned State personnel.

My presentation today, summarizes the findings of this study and presents an overview of Maryland's Coastal Fisheries Program.

Within the confines of the Coastal Fisheries Program, biological considerations serve as the fundamental criteria for developing specific management strategies. For a given species, characteristics of its population dynamics - specifically, the nature of recruitment into its spawning stock - define major population attributes which in turn, dictate the range of possible management options.

Fishery management in reality is the management of the people within the seafood industry and their activity and not the resource itself. This necessitates including nonbiological considerations as we develop fishery management plans. These considerations can be characterized as environmental, economic, sociological, political, jurisdictional and special indexes. As a result purely biological considerations are constrained by these nonbiological considerations to produce a more equitable plan for that fishery.

Maryland's Coastal Fisheries Program seeks to achieve a realistic and practical management strategy through blending these nonbiological considerations with biological ones. To ensure adequate identification and review of all of the pertient considerations, the Coastal Fisheries Program is designed as a long-term, dynamic project that has three phases. They are:

I. Characterization,
II. Plan Development, and
III. Implementation.

Phase I: Characterization

In this Phase, work is focused on developing up-to-date definitions and descriptions of some specific aspects of Maryland's commercial and recreational fishery. Seven specific areas were concentrated on in this phase which are:

1) Life history data on Maryland fishery resources and an assessment of present stocks.

2) General condition and availability of suitable habitat.

3) Institutional framework, i.e., state and federal management activities.

4) Applicability of quantitative fisheries yield and management models to selected fisheries.

5) Infrastructure of the commercial fishing industry, that is, community structure, harvesting activities, dockside sales, handling, processing, and distribution.

6) Economic assessment of the seafood industry.

7) Extent of recreational fishing in Maryland.

Also during this Phase, special emphasis has been given to the use of quantitative fishery models. By serving as a means of organizing, causally connecting and relating various kinds of population, exploitation, and economic data, these models provide the basis for conceptualizing potential problems in management policy. Because the model structures and the data-input from which they are derived are goal-oriented, these models also reveal deficiencies and inadequacies in the data. Detailed quantitative models will be used to develop recommendations for improving the acquisition and processing of biological, commercial and recreational harvest data in the Plan Development Phase.

Another element of Phase I tasks is the collection of data on Maryland's saltwater recreational fishery. The State is carrying out this portion of the study in cooperation with the National Marine Fisheries Service (NMFS). The Coastal Resources Division provided additional funds into the NMFS survey in order to intensify the amount of data collected in Maryland. Over a two-year period, the survey will involve combining an onsite intercept survey with a telephone survey. It should reveal information on frequency, mode (i.e., pier, jetty, boat, etc.), location, gear used, species fished for and species caught, and the economic value of sport fishing in the State.

Still another aspect of the Coastal Fisheries Program's characterization Phase of Maryland's commercial fishery, was documenting the social and institutional framework, i.e., the attitudes, preceptions, and interactions of the various groups associated with the commercial fishing industry. Biological and geographical factors which affect the industry were also documented.

Additionally, the study defined the commercial fishing industry in terms of disaggregated categories. These categories are being used as the foundation for an economic input/output analysis of interindustry transactions of Maryland's seafood industry as a cooperative research effort of the CRD, Agricultural and Resource Economics Department of the University of Maryland, and the Maryland Sea Grant Program. From this project, we hope to learn information about the direct economic value of the seafood industry in the State and the extent to which money generated from it is returned to the local and State economy.

All of the activities in Phase I, are designed to augment existing information or programs and to provide the necessary background data on biological and nonbiological considerations in order to formulate species management strategies - the focus of the next Phase, Plan Development.

Phase II: Plan Development

As an extension of the activities associated with identifying quantitative fishery yield and management models, seven species have been identified to test the applicability of these models to Maryland's coastal fisheries. They are:

1) American eel,
2) Striped bass,
3) Blue crab,
4) Softshell clam,
5) Shad,
6) Oysters, and
7) White perch.

During this second phase of the CFP, the State's efforts will be directed towards obtaining for each species:

A. A detailed assessment of the applicability of selected production or population models.

B. An assessment of data available versus data needs for the application of management models.

C. The development of recommendations concerning data acquisition and processing procedures.

Combining the above with the nonbiological data will contribute to producing a comprehensive management strategy for each of the seven priority species, based upon a defined management objective for each species. As management strategies are developed for each species on the initial priority list, other important commercial and recreational species will be considered.

Phase III: Implementation

Working closely with representatives from legislative, commercial and recreational fishing industries, and others, proposed management strategies will be scrutinized carefully in this Phase of the Coastal Fisheries Program.

Once the strategies have been finalized, the State will promulgate any necessary regulation and recommend modifications to existing laws.

In this first year of the Program, most of the work outlined under the Characterization Phase has been completed. Work is continuing on the sportfishing survey and the economic assessment of the seafood industry. These two aspects of the Program should be completed within the next year. Over the next year and a half, the State anticipates redesigning the collection of biological and commercial harvest data for use with selected quanitative fishery models. Also within the next

18 months, management plans for several priority species should be complete and on their way to being enforced through Phase III efforts.

We firmly believe that we have developed a coordinated State Program that provides the necessary components to formulate comprehensive management strategies for Maryland's coastal fishery resources. It is anticipated that with Maryland's Coastal Fishery Program, we will achieve a balanced partitioning of finfish and shellfish resources among competing users as well as provide the legal-institutional framework for developing a viable commercial fishing industry and maximum recreational activities.

For more details about Maryland's use of quantitative models in fishery management, consult the publication, "A Review and Evaluation of Fisheries Stock Management Models". It is available from the Coastal Resources Division, Department of Natural Resources, Tawes State Office Building, Annapolis, Maryland 21401 or write to the National Technical Information Service (NTIS).

B. CHARACTERIZATION OF THE CHESAPEAKE BAY COMMERCIAL FISHERY

I would now like to briefly share with you what we have learned so far from our characterization of Maryland's Chesapeake Bay commercial fishery industry.

In general, the study revealed that Maryland's commercial fisheries are a complex combination of geographic features, biological fluctuations, harvest regulations and individual preferences. Understanding the Bay's ecology is fundamental to understanding the individual characteristics, social behavior, and product economics of Maryland's Chesapeake Bay Commercial Fishery. Two aspects of the Bay's ecology were found to be major influences: 1) the variations of salinity throughout the Bay and 2) geographic features of the Bay's habitat.

The variation in salinity is caused by fresh water from inland streams and rivers moving out to the ocean mixing with saline ocean water that is moving up into the Bay. As fresh water flows into the Bay, nutrients from the land are washed into the Chesapeake. These nutrients support an enormous population of microscopic organisms, known as phytoplankton, which form the first level of an extensive aquatic food chain, upon which resident and migrating fish can feed.

This nutrient-rich mixture of fresh and salt water provides the basis for the diversity and abundance of fisheries that generates a significant commercial fishing industry, in Maryland.

Great amounts of shallow nearshore waters, intermixed with extensive areas of marsh land and a highly inundated shoreline, which has created numerous protected harbors are characteristics of the Chesapeake Bay (Figure #1). These geographic features have constituted a major in-

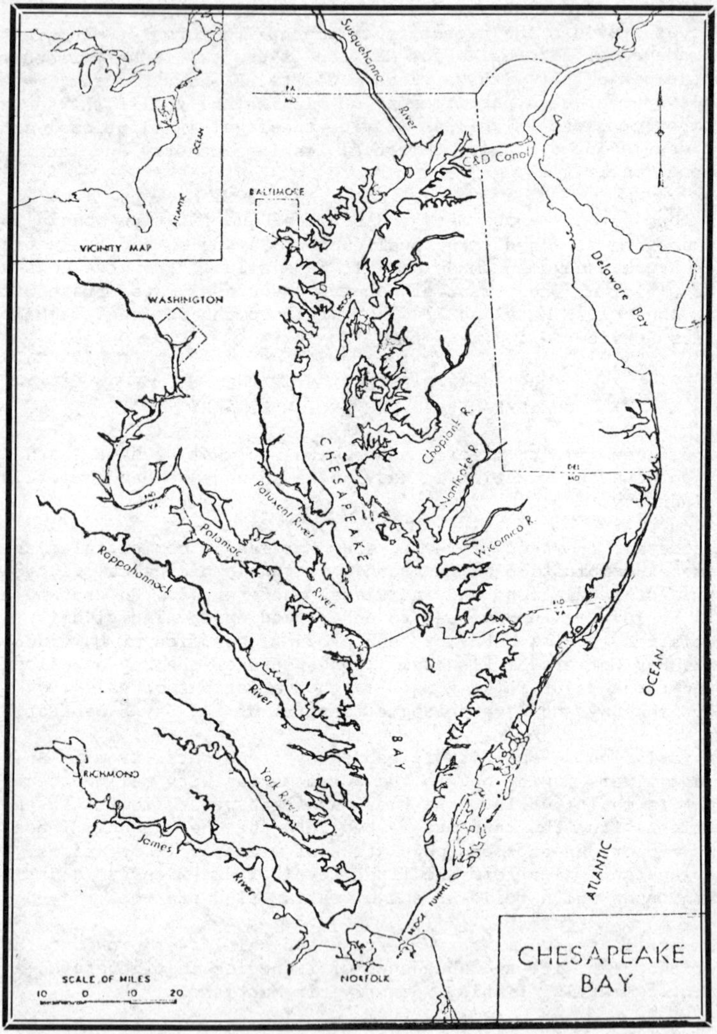
VICINITY MAP
OCEAN
ATLANTIC
BALTIMORE
WASHINGTON
Susquehanna River
C&D Canal
Delaware Bay
CHESAPEAKE
Choptank R.
Nanticoke R.
Wicomico R.
Patuxent River
Potomac River
Rappahannock River
York River
James River
RICHMOND
NORFOLK
BAY
OCEAN
ATLANTIC
SCALE OF MILES
10 0 10 20
CHESAPEAKE BAY

Figure 1

influence on community structure, harvest activities, shoreside support and product distribution.

Before the construction of the Bay Bridge, the Eastern Shore was separated from the major urban areas of the rest of the State. Compounding this was the isolation of remote peninsulas, islands, and secluded small harbors nestled in marshlands. It soon became apparent that the only way for groups of people to survive was to band together and become self-sufficient.

Such geographic isolation has created fishing communities that are based on tightly-knit, extended kin relationships. Bay fishing communities then, developed complete local support systems for their watermen. Watermen and their immediate families carried out much shoreside support work on their own; adapting machinery from auto and farm equipment; performing their own boat maintenance; and making their own gear, as well as buying, processing and transporting the daily catch.

The inherent dynamic nature of a seasonal fishery combined with a close kinship community structure results in services and products exchanged locally on a noncash or barter system.

Turning our attention to harvesting: Collectively, the Bay fishery is a composite of individual fisheries, each one characterized by different harvest methods according to the species sought. With respect to the manner of making a living on the Bay, all Bay fisheries are similar. Except for the Skipjacks, working on the water generally is characterized by a small privately-owned, shoal-draft boat, which often works with rarely more than a crewman aboard. Many watermen are following a family tradition, however, hard physical labor on the water and frequently in the most severe weather conditions is ultimately a personal choice.

Today, more and more young men are engaged in fisheries and operate boats either with their fathers or on their own. Even so, much of the popular literature indicates that very few yound men are entering the profession and because of the high average age of the current watermen, the profession is becoming a doomed one.

The harvest system used by watermen has developed from a need for low-capital investment and low overhead. This reflects their respect for the uncertainty of natural biological cycles and the need to function with a minimum investment in order to survive periods of low species abundance.

Prior to the 1971 Bruce decision, fishing within county jurisdictional boundaries was restricted to county residents only. There was no need for a boat to be equipped for extended cruising and overnight trips. A waterman could not overcapitalize his effort and survive fluctuations in resource abundance occurring in his county. This also contributed to the development of self-sufficient communities and a low investment, highly flexible harvesting system.

Since Bruce, there has been greater mobility of the watermen. It has created a truly "Bay-wide" fishery. This is seen mostly with the oyster patent tongers that will stay away from their home port during the week, working distant bars and unload in any one of a number of nearby communities to trucks sent by a prearranged buyer or to a local buyer, returning home only on the weekends.

The need to respond to unforeseen changes in species abundance requires great personal flexibility and willingness to change gear types and fisheries as conditions demand. Gear types are regulated such that the gear is "inefficient", but inefficiency is compatible with the status of the resource sought. Although, "inefficient" such methods provide a good living for many Chesapeake Bay watermen. Watermen do not speak of their gear as "antiquated", but rather as "in balance". They do not feel they must "overcome" the Bay rather are a part of the Bay, "in synch" with the Bay.

Trends in technological aspects of the Bay fisheries indicate few changes in gear. Surprisingly, the most significant technological trend has been the widespread adoption of the CB radio. CB sets are used extensively for boat-to-boat communication, boat-to-buyer communication, and emergency transmissions.

While many fishermen in the Bay solely fish for clams or oysters, more often true "waterman" will change fisheries several times over the years as well as within the year. A waterman's livelihood is what the Bay provides; if he is unwilling to change as the Bay changes, he is not a waterman for long.

Once harvested, the resource is brought to the dock for sale. Nearly all transactions at the dock are on a cash basis. A waterman unloads his catch, the weight is calculated and he receives cash money. Rarely are "tabs" run up unless by mutual agreement. Watermen with crews function on a "lay" or share system as in most American Fisheries. Boats take one-third of the gross earnings. The boat share covers maintenance, finance costs, and major gear items. Daily expenses are then deducted (fuel, bait, food, and possibly some gear expenses) and the remainder is divided equally between captain and crew.

As for the product once it hits the shore, several buyers handle everything; oysters during the winter, crabs in summers, and clams year-round. Processing is limited to shucking houses for oysters and clams and picking houses for crabs. Sometimes a winter oyster shucking house picks crabs in the summer.

An operation on the western shore of the Bay for example, may handle crabs (steamed and soft crabs sold locally), fish (frozen or sold fresh locally) and, rarely, oysters (shucked, packed, and shipped from a Virginia buyer). Recently, labor has been difficult to find, consequently, between 60 and 80 percent of Maryland oysters leave the state as shellstock to be shucked elsewhere. The majority is shipped to Virginia. Some Eastern shore oyster shucking plants pick crabs during summer at a loss to maintain their labor supply in winter since seasonal open-shut operations generate great problems in finding labor for each season. Many plants have solved their labor shortages by abandoning oyster shucking all together, electing to ship whole oysters to Virginia for processing.

Although all processors interviewed stated a desire to see the development of automated shucking and picking machinery as a means of expanding their business activities, few however, have chosen to invest in such equipment. Processing capabilities, then, within the State of Maryland have steadily declined because many processors are either not willing or not able to make the large financial investment necessary to mechanize their plants. This reflects their desire to maintain low overhead because of the inherent uncertainties of product supplies.

More and more, the seasonal nature of the fisheries has caused many processors to utilize imported products. Blue crabs imported from other states, are picked in Maryland, packaged and distributed to food wholesalers or directly to restaurants. Finfish processing houses exclusively use imported fish rather than Bay fish. Markets for Bay finfish are almost entirely local or sent to freezing establishments handling vast quantities for fast-food chains. Smaller establishments (those without steady supplies of imported fish) have branched into other related areas of the business. For example, some dealers now operate marinas, yacht service facilities, ice-making machines, or transportation-fleets moving a variety of goods.

Unlike finfish, it is normal for shellfish to be processed to some extent before the final product is distributed. Those oysters not marketed in the shell are sold raw, canned, breaded, smoked or frozen. Oysters serve a narrow market based on tradition and a seasonal market demand. The period from Thanksgiving to Christmas is the essence of the business.

Blue crabs are landed and sold live. Those crabs not marketed live are steamed or held in certain instances until they shed and then sold as soft crabs. Many establishments remove the meat from the steamed crabs and sell it either canned or frozen. Soft shell clams are a low-volume, high-priced product primarily as a result of their devastation by Tropical Storm Agnes in 1972. Soft shell clam markets are increasingly filled with surf clams and ocean quahogs, especially for use in chowders. Further processing of the soft shell clam usually involves just the removal of the meat. Many shucked clams go directly to New England. However, for the most part, clams leave Maryland in the shell.

Small buyers and/or processors ship their products to wholesalers, retailer, and distributors. The wholesalers in turn distribute the product widely to many markets located all over the United States.

Most wholesalers own their own trucks (as do the smaller buyers), sending shipments on regular routes as far north as Montreal, south as far as Texas, and west to Chicago and even Los Angeles.

The product distribution system is generally controlled by the telephone and use of trucks. Buyers own small, insulated, one-half to five-ton single chassis units called "buy trucks" which work on short or long-term arrangements with specific watermen. Many coverted pick-ups also serve as buy trucks. These buy trucks will meet watermen at some docking sight prearranged CB radio or telephone in order to make a buy. Many small "gypsie' trucks are sent into the Bay region to fill one or several orders for out-of-state retailers.

There is no central market for product distribution in Maryland. Although, the Baltimore market does distribute fish and shellfish to both the Baltimore and Washington metropolitan areas, it does not appear to be an organizing factor for distribution. Baltimore is considered a market area with a wide variety of retail and wholesale outlets.

Markets tend to be widely scattered and composed of many individual outlets; such as restaurants, retail establishments, chain-stores, and institutions. Markets exist all across the county, each distributor having carefully built-up his or her own clients.

Marketing trends are being influenced by Maryland seafood products fast becoming gourment items. Oyster and crab prices are as high or higher than other shellfish. In fact, today, most shellfish are considered luxury items by many consumers.

Commercial fishing in the Chesapeake is considered by many a way of life. Eventhough technological inovations have been few, the industry is as active now as it always has been, contributing greatly to the local and State economy. Recent declines in the resource along with several legal decisions, changing marketing trends and uncertain processing capability have caused some concern over the future of the Chesapeake Bay commercial fishing industry. Combining the work of the Coastal Fisheries Program with ongoing Federal and State activities to address these concerns, Maryland anticipates that commercial fishing will continue to be an integral part of the State's social and economic structure.

Phillip Alan Cline
Economist
Army Corps of Engineers
211 Main Street
San Francisco, Ca. 94105
(415) 556-5151

SMALL CRAFT HARBOR EVALUATION

Is the Army Corps of Engineers' method for evaluating small boat harbors biased in favor of nonproductive activity? This paper will examine the basic methodologies used in evaluating small craft harbors.

In general, a small craft harbor project can be analyzed by its effects on recreational and commercial activities within the harbor. Recreational benefits are evaluated currently in accordance with EM 1120-3-113, the so-called "small boat formula". The method does not assume that such benefits can be evaluated with mathematical precision, but assumes that reasonable and representative percentages reflecting the net return on the depreciated investment in a for-hire fleet of small boats are reasonable gages of the recreational navigation benefits to the boat users. Commercial fishing benefits, on the other hand, are subjected to a more rigorous methodology which entails a determination of increased fish catch, or more efficient methods of operation as the result of a deeper or more protected harbor.

In both cases, recreational and commercial fishing boats, a purchase of floating stock is required. In most instances, the dollar amount needed to purchase a commercial fishing boat is greater than that required for a recreational boat. That issue aside for the moment, assume two cases in which the purchase price is the same for a recreation or commercial fishing boat. A potential purchaser is faced with the decision of placing his monies into a recreational or commercial boat. If he buys a recreational craft, he can pick up his boat, berth it in XYZ marina and let it remain idle, if he choses. The net effect of his actions is to generate annual benefits for the harbor amounting to 6-12 percent of the depreciated value of the boat.

If, on the other hand, that individual were to place the same amount of money in a commercial fishing vessel, he too could make his purchase, pick up the boat and place it in a berth in XYZ marina. This is where the similarity ends. If the commercial boat owner--like the recreational craft owner--wishes to let his boat remain idle, no benefits will accrue to the harbor. Additional effort is required of the commercial fishing boat owner in order to generate benefits: he must go to work and generate additional net income. The increases in additional fish landings as a result of the boat being placed in the marina can then be taken as a project benefit.

The differences in the methodologies illustrates a basic bias in favor of the recreational craft. Nothing more is required of the recreational boat owner than purchasing a boat and placing it in the harbor. Much more is required of the commercial boat owner: he must become productive in order to generate benefits for the harbor. This basic bias is in favor of nonproductive activity and discriminates against productive endeavor.

The genesis for the methodology used in small craft harbor evaluation of recreational benefits arose out of a Corps of Engineers preliminary examination for harbors of light-draft vessels conducted from 1946 to 1949. The study arose from a desire to provide protection for commercial light draft vessels in the form of harbors of refuge along the California coast. In addition, there was "a great desire for recreational boating and offshore cruising which is hampered by dangerous conditions due to insufficient ports of call and refuge." The study called for the establishment of ports along the coast to provide a federally funded "marine highway", similar in general respects to the Federal-aid land highways furnished for vehicular traffic. Increases in commercial activity was the primary reason cited for the proposed harbor improvements, but added that "Coastal pleasure-boat cruising would be made possible and there would be an increase in local boating with resulting economic, health and recreational benefits to all concerned and a reduction in juvenile deliquency." Thus, recreational facilities were to be included in the benefit analysis of harbor improvements for reasons of health, economic activity as well as for reducing juvenile deliquency.

The 1949 report[1/] established 9 criteria for fishing harbors and the same number for recreational harbors:

Criteria for Fishing Harbors. Sites considered for permanent fishing harbors, whether or not also classed as recreation harbors, should meet the following conditions:

a. Be readily accessible, easy to enter at all times with safety, and provide protection from storms from any quarter.

1/ Preliminary Examination on Harbors for Light-Draft Vessels, Northern California Coast (31 March 1949)

b. Permit increase and more even distribution of the fishing fleet.

c. Provide bases close to fishing banks or grounds.

d. Have adequate communications.

e. Be located where space is also available for cannery or processing-plant location.

f. Provide facilities for drying and storing nets and for sale, receipt, and handling of fish and supplies for boats and crews.

g. Provide facilities for vessel and engine repairs.

h. Have a minimum depth of 12 feet in channels and mooring areas.

i. Have adequate entrance navigation aids.

Criteria for Recreation Harbors. Sites considered for recreation or yachting harbors, whether or not also classed as fishing harbors, should satisfy the following conditions:

a. Be readily accessible, easy to enter, and provide protection from storms from any quarter.

b. Provide facilities permitting increase in boating activities.

c. Provide berths, landings, and storage facilities.

d. Provide for other recreational uses besides boating.

e. Have adequate communications.

f. Provide for sale of supplies of all types needed by pleasure boats.

g. Provide facilities for hull and engine repairs.

h. Have adequate entrance navigation aids.

In response to the District Engineer's request, Congress amended the benefits credited from "commerce" as defined in the ACT[2/] approved 10 February 1932 to "include the use of waterways by seasonal passenger craft, yachts, house boats, fishing boats, motor boats, and other similar water craft whether or not operated for hire."

2/ 33 USC 541.

A variety of procedures were formulated through field studies conducted by various District Offices to deal with the evaluation of benefits for recreational boating and non-commercial fishing. The procedures outlined by these studies provided a "reasonable simple, sound and workable basis for evaluating recreational benefits..."[3/] One of the principals enumerated was "The gross amount individuals or groups are willing to spend for recreation, or the computed return on the amount invested in recreation if invested in commercial enterpises, are not acceptable as a basis for evaluation." The principal considered most logical for the development of a benefit computation is "the net return on the depreciated investment in boats as received by owners of forhire vessels, after expenses have been paid." Studies of boating practices in several parts of the country indicate that the net return for-hire varies between 6 and 15 percent.

The percentage range considered appropriate for each vessel category is as follows:

outboards	10-15	percent
sailboats	8-12	"
inboards	8-12	"
cruisers	6-9	"
auxiliary sailboats	6-9	"

The variation in percentages given for each vessel category allowed judgements to reflect local conditions such as length of season, concentration of population, availability and cost of alternative types of outdoor recreation, cost of access to other small-boat harbors, and income of the using public. It was believed that the predominant craft would be the long range cruiser, but "in relative rare cases, one or two yachts of the larger and deeper draft class may enter into the picture in the consideration of a truly small-boat harbor."

The net-for-hire concept is thus an attempt to measure the nonmonetary "return" to the boat owner in the form of boating enjoyment. The basic idea is that boat owners are expected to receive value or economic welfare gains from boating expenditures equivalent to those they would receive from alternative comparable investment. The recreational boating return is therefore a "shadow" return for foregone alternative investment.

In the evaluation process of public works projects, the purpose of the benefit evaluation process is to attempt to measure the net increases in National Economic Development(NED) as a result of a project. NED benefits are measurements of dollar additions to Gross National Product, and are not measurements of "utility" that has been "translated" into dollar equivalents.

The application of a shadow rate of return should take into consideration the concept of the highest and best use of capital. The idea that an investment yielding a net return of ten percent would yield an equivalent dollar amount of utility to the owner if used to purchase a similar but non-dollar-yielding item is erroneous. While the purchaser might feel "elated" by his purchase, his elation is hardly representative of a net addition to NED.

3/ Evaluating Benefits and Allocating Costs of Small-Boat Harbor and Channel Projects, 3 February 1950.

Why the difference in the methodologies? It would seem that equity would dictate equal methodological treatment for both categories: both should be evaluated solely on the additions to net income generated by their prospective activities, or both should be precluded; on the other hand, both should be allowed to obtain a shadow investment return, or both should be precluded. In either case, they should be evaluated using consistant and equitable criteria. But in no case would equity dictate making one activity easier to evaluate than the other.

If a harbor is open all year, which is typically the case on the West coast, the benefits for recreational boating are on a year round basis. A questionnaire sent to a sample of recreational boaters in Washington State as part of a study conducted for the Sea Grant program indicated that recreational boaters on an average use their boats 85.4 days out of each year. A survey of Oregon State University's Marine Economics Data Sheets indicate the average fishing boat is active 131 days out of each year. Thus, the evidence suggests that commercial fishing boats are active a greater percentage of the year than recreational craft; so that for any given small craft harbor, owners of fishing boats have a higher usage of the harbor than recreation boat owners.

The argument often advanced is that recreational activities such as recreational boating is vital to sustained productivity in other fields of endeavor, eg. the regular place of work. In fact, leisure is often thought of as a reward for labor. Aristotle addressed this issue in this treatise "POLITICA" in which he said "We labor to have leisure". Mental and physical fatigue in the workplace will eventually cause productivity to diminish; consequently, it is necessary to have recreation. That recreation is of value in promoting productivity is not at issue; what is at issue is the discriminatory methodologies employed in benefit analysis of recreational verses commercial boating activities.

An alternate approach to the evaluation of commercial fishing benefits would be to base the rate of return on an average composite of long-term high grade industrial bonds. The rationale behind this approach is that an individual wishing to place monies into floating stock would have foregone investing in high grade industrial bonds. His expected rate of return would therefore at least equal the foregone return. If his apparent rate of return--that is, the actual dollar return on investment--is less than the bond return, the return gap between the rate he actually receives and the bond rate can be a thought of as a measure of various factors such as recreational enjoyment, satisfaction for being "his own boss", the aesthetic improvement in his working environment and a variety of other non-dollar yielding "satisfaction" returns. The sum of the actual dollar return plus the other non-dollar returns would at least equal the average return for a composite of long term high quality industrial bonds.

If such an approach were applied to commercial fishing benefit evaluations, it would tend to place commercial fishing and recreational benefits on a more equitable basis by applying a similar analysis to both benefit categories.

POTENTIAL COASTAL-ZONE IMPACTS OF ALTERNATIVE OCEAN ENERGY SYSTEMS

Volker W. Harms and Nancy Gutshall
Earth Sciences Division
Lawrence Berkeley Laboratory
University of California
Berkeley, CA 94720

INTRODUCTION

The U.S. Department of Energy (DOE) is presently pursuing several new technologies for extracting energy from the oceans. The programs include:

- o Ocean Thermal Energy Conversion (OTEC)

  Configuration:
  shorebased
  bottom resting
  moored (offshore)
  grazing (offshore)

- o Ocean Wave Energy Conversion (OWEC)

  Water motion utilized as is:
  moored floating structure (raft, heaving body, etc.)
  fixed structure (wave-pressure converter, wave paddle, etc.)

  Water motion transformed before conversion:
  wave focusing (converging channel, refraction, etc.)
  cavity resonator (oscillating water column, etc.)
  wave breaking (surge converter, flap, etc.)

- o Ocean Currents Energy Conversion

  ocean turbine system
  drogue-chute device

- o Salinity Gradients Energy Conversion

  osmotic pressure system
  reverse electrodialysis

DOE is concerned not only with environmental impacts in the open ocean but also recognizes the need to anticipate possible effects upon physical and ecological conditions in the coastal zone. The purpose of this paper is to present an outline of the categories of environmental concerns relating to alternative ocean energy systems in the coastal zone. The possibility of adverse coastal-zone impacts must be considered for each of the ocean energy systems listed. For convenience, the coastal-zone-impact categories will be discussed specifically with reference to the OTEC system as it is the most advanced of the alternative ocean energy systems under development, and consideration of its impact upon the coastal zone will reveal many concerns common to the rest.

In view of the many, and sometimes conflicting, definitions that are applied to geographic regions near the coast, some terminology necessary to the subsequent discussions will be defined. Foremost among these is the term "coastal zone". Alexander and Cameron (1978) define the geographic nature and limits of the coastal zone as follows:

> "The coastal zone includes "coastal waters", the shoreline, and the adjacent "shorelands", the uses of which have a direct and significant impact on the coastal waters. Coastal waters are defined by statute as extending to the outer limit of the territorial sea. In the Great Lakes these waters continue to the international boundary with Canada. The shoreline is the point of contact between land and sea, and includes inter-tidal waters. The concept of the "shoreland" is much more difficult to define, since it is a functional unit whose "raison d'etre" is that it is the site of activities which have a direct and significant impact on coastal waters. The Coastal Zone Management Act recognizes no single geographic definition of the landward boundaries of the shoreland, and leaves it up to the coastal states to determine where this jurisdictional limit should be. Some states place this at the landward boundaries of coastal counties, others at the crest of costal mountain ranges, while still others define the landward boundary of their coastal zones in terms of the inland extent of maritime influences."

Since the territorial sea of the United States extends 12 nautical miles seaward from low water mark, in accordance with an almost universal consensus on this matter in recent years (Nanda, 1979), it follows that this defines the seaward limits of the coastal zone as well. Shorelands, the shoreline, and 12-miles-of-coastal-water therefore constitute the coastal zone. For brevity and convenience, the shoreline and shorelands will, for our purposes, also be grouped into an "onshore" region, the 12-mile belt of coastal waters will be referred to as the "nearshore" region, and all waters beyond the 12-mile limit are considered "offshore".

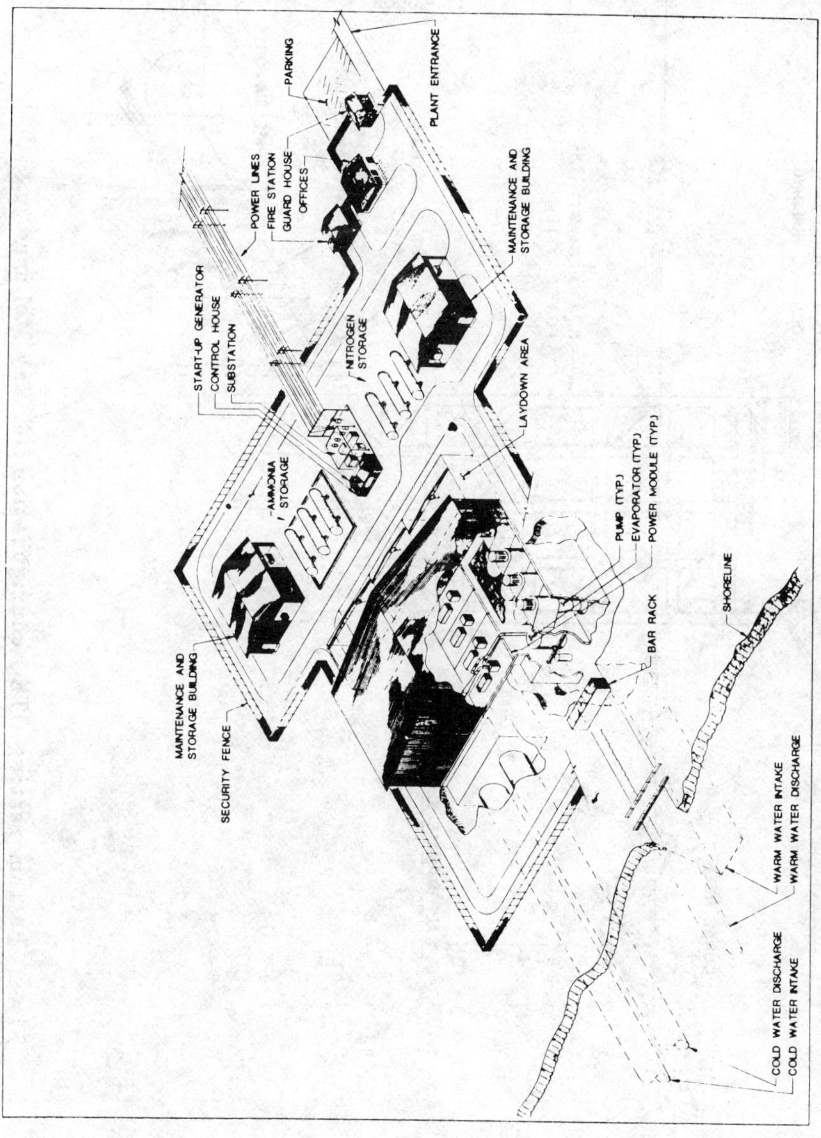

Fig. 1 Shorebased closed-cycle OTEC plant (Deep Oil Technology,Inc.)

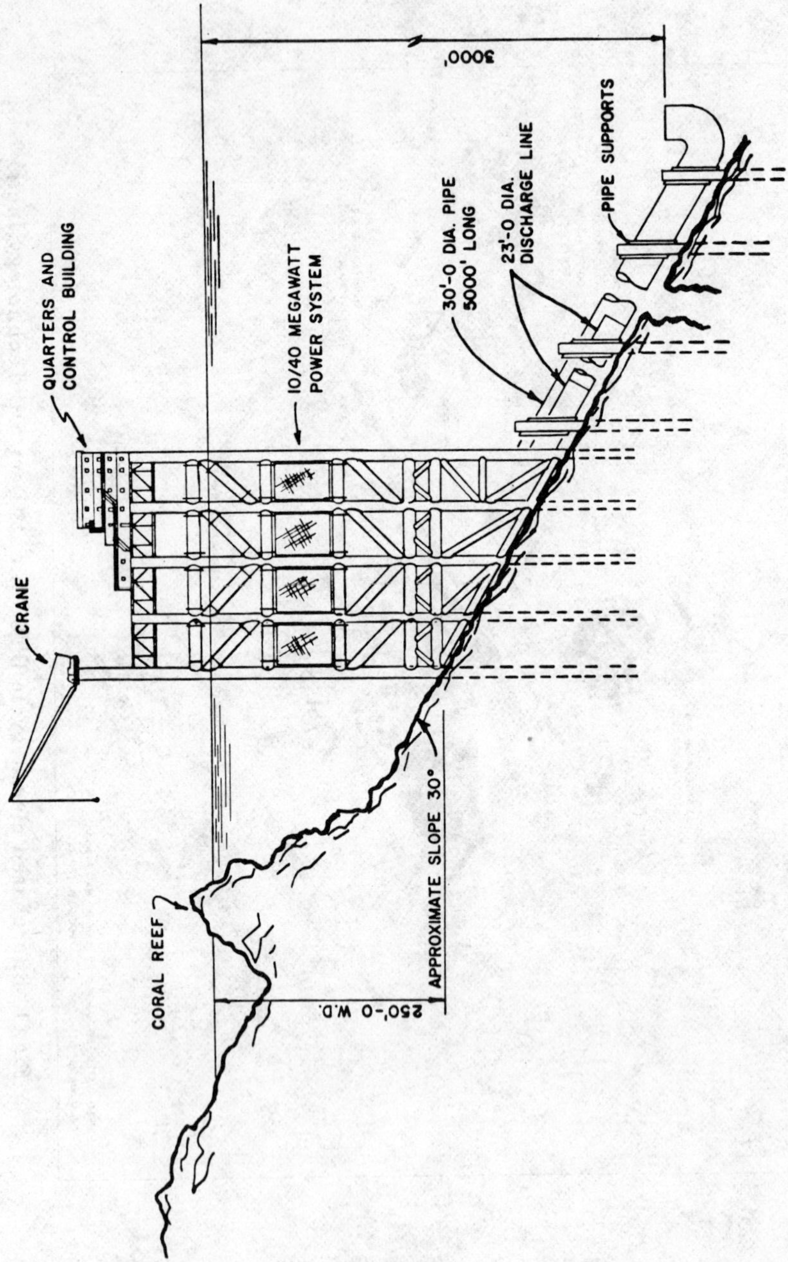

Fig. 2 Bottom-resting OTEC configuration (J. Ray McDermott & Co.)

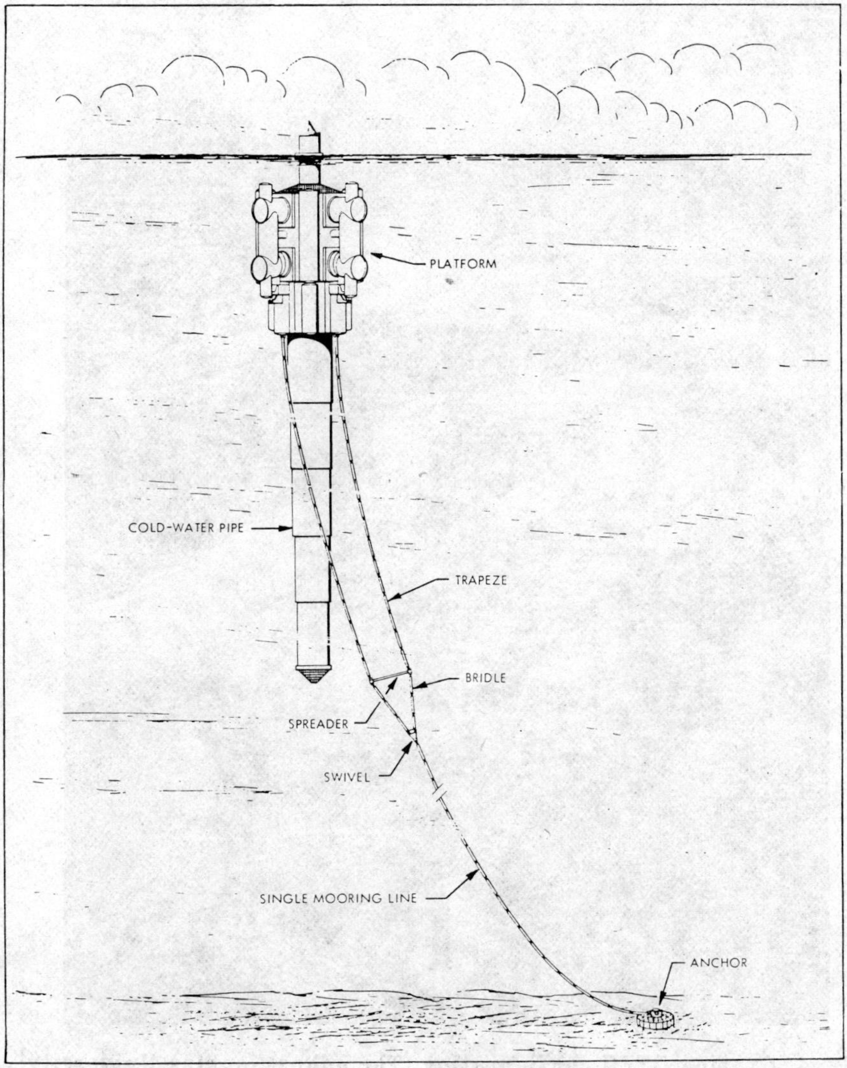

Fig. 3 Moored OTEC plant (Lockheed Ocean Systems)

Fig. 4 Grazing OTEC configuration (The Johns Hopkins University)

OTEC CONFIGURATIONS

Ocean Thermal Energy Conversion is a technology using the temperature difference between warm surface and cold deep water to produce electric power with a gas or steam turbine. An OTEC plant can be operated in either a "closed" or "open" cycle. In the closed cycle configuration, warm surface water heats an evaporator containing an appropriate working fluid (presently ammonia), and the vaporized working fluid drives a gas turbine. After passing through the turbine, the vapor is condensed by cold deep water and then returned to the evaporator for re-use. In the open cycle configuration, sea water is used as the working fluid, warm surface water being transformed into low-pressure steam and expanded across a turbine. Once again, cold deep water is drawn up to condense the steam after passage through the turbine.

A number of OTEC configurations are presently under consideration. They may be conveniently categorized according to level of mobility and ocean exposure. In order of increasing exposure to the ocean environment, as well as mobility, they are:

(1) shorebased OTEC (fixed, fig. 1)

(2) bottom-resting OTEC (fixed, fig. 2)

(3) moored OTEC (floating, fig. 3)

(4) grazing OTEC (cruising, fig. 4)

The grazing configuration (fig. 4) is a self-contained plantship that utilizes the released ocean thermal energy for the production of energy-intensive materials. In this concept, production facilities are moved offshore in order to reduce the onshore demand for electrical energy from existing sources. The OTEC plant is used to generate power which in turn is used to manufacture a product such as hydrogen or ammonia from sea water, or alumina or aluminum from raw material brought to the ship. For the grazing/manufacturing option, the commercial product is not electricity but some goods exported via surface ships to markets. The plant with its own motive power essentially grazes the thermal resource without regard to bathymetry, except that the water depth must be greater than the depth of the cold water pipe. Obviously, this opens up a much larger area which can be used for OTEC. However, it vastly increases the area of potential concerns. With only a minimum depth limitation, the grazing plant could operate either in the economic zone of near-by nations or strictly in international waters. The environmental impacts of an operating OTEC grazing plant could be transferred by ocean currents into neighboring economic zones and coastal zones even though the plant itself was in international waters, thus potentially causing complex jurisdictional problems. Two major categories of potential coastal-zone impacts are associated with the grazing configuration: construction-related impacts, which are generally localized, and those inherent to operation as large offshore industrial facilties that may discharge production-related effluents that are ultimately deposited along extensive regions of shoreline. The main advantage of this OTEC configuration lies in its ability to track the thermal resource and thereby operate more consistently at the largest available temperature difference. Associated with this capability and requirement are several substantial ocean-engineering design problems that will have to be resolved before this concept can proceed to the design stage (e.g. in the presence of waves and currents, hydrodynamic forces on the pipe and gimbal can increase dramatically as the vessel accelerates to the cruise mode). In the phased-development plan for OTEC, it is probable that the data-base necessary to solve these ocean-engineering problems will come from OTEC configurations that require lesser leaps in technology and will therefore be constructed far earlier i.e., the fixed and floating OTEC versions (fig. 1-3). For purposes of this paper, the fixed and floating OTEC plants will therefore be emphasized, with the grazing configuration only treated peripherally.

The shorebased and bottom-resting plants (figs. 1 and 2) are fixed structures with platform (powerhouse) and sea-water conduits resting on foundations. They differ fundamentally with respect to location of the powerhouse: one is constructed in relatively shallow water (typically less than 300m) whereas the other is located onshore. Although the sea-water intakes would probably be at the same location for both options, at a given site as shown in the schematic of fig. 5, the sea-water conduits themselves would not have to be conveyed across the beach and onto shore in the bottom-resting version. The shorebased OTEC plant could generate power for the electric grid or for manufacturing purposes as in the grazing option. This type of plant has the basic attributes of conventional seaside power plants so it would have many of the same environmental concerns.

A moored OTEC plant is shown in fig. 3, This particular version incorporates a tapered cold-water pipe, a mooring trapeze and spar-buoy shape to reduce wave-induced hydrodynamic loads. The moored and bottom-resting OTEC configurations would transmit power onto shore via submarine transmission lines, as shown in fig. 5; depending upon the cost of the submarine cable, it may, instead, turn out to be economically more attractive to consume the generated energy directly on the platform for the production of energy-intensive materials. Since the moored OTEC plant is anchored to the bottom, its operations would come under the jurisdiction of the nearest nation regardless of the boundaries of any economic zone.

It should be recognized that the OTEC versions shown in figs. 1-4 are used only for illustrative purposes and, although under study, do not necessarily represent mature designs receiving special emphasis in the OTEC development program. In their roles as generic OTEC configurations it will be assumed that each utilizes the power-generation system presently assigned the highest probability of achieving early economic viability: the classic steam-power-plant closed cycle (Rankine) with ammonia as a working fluid.

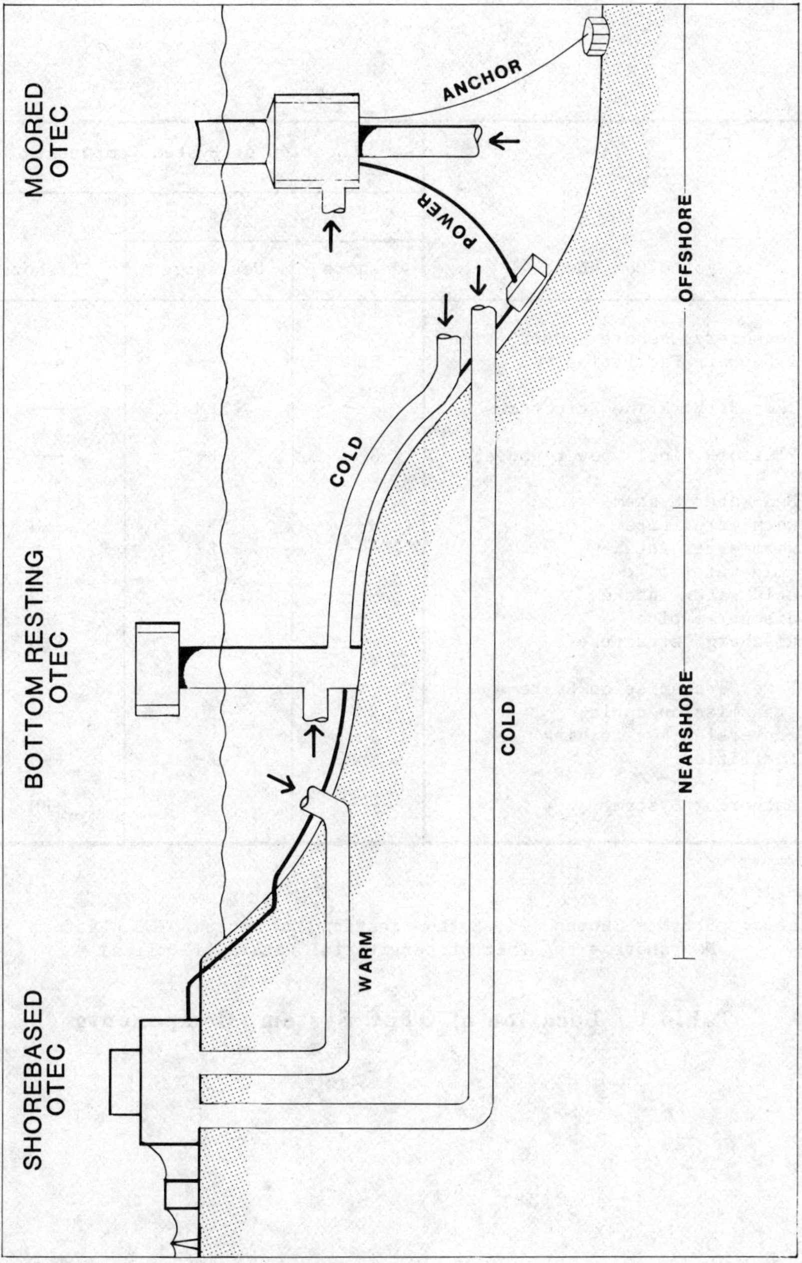

Fig. 5 OTEC configurations, Sea-Water Intakes and Power Cables

System Component	Location of System Component		
	Coastal Zone		
	Onshore	Nearshore	Offshore
o Temporary Onshore Construction & Support Facilities	SBM	---	---
o Temporary Marine Structures	---	SBM	---
o Platform (incl. power house)	S--	-B-	--M
o Sea-Water System			
warm water pipe	S--	SB-	--M
warm water intake	---	SB-	--M
cold water pipe	S--	SB-	SBM
cold-water intake	---	SB-	SBM
discharge pipe	---	SB-	SBM
discharge structure	---	SB-	SBM
o Power Transmission System			
transmission cable	SBM	-BM	--M
conversion & switching facilities	SBM	-B-	--M
o Anchoring System	---	-BM	-BM

Legend: S,B,M = Shorebased, Bottom-resting and Moored OTEC platform
Nearshore = to limit of territorial waters (12 miles)

Table 1 Location of OTEC System Components

POTENTIAL IMPACTS AND IMPACT GENERATING ACTIVITIES

The sources-of-interaction between an OTEC plant and the coastal-zone environment will vary from site to site for a given OTEC configuration (shorebased, bottom-resting and moored); additionally, at a particular site, the distribution of these sources-of-interaction within the coastal zone also will vary as a function of configuration alone. This is apparent in fig. 5, where the sources of interaction with the coastal zone are reduced from power-transmission system, platform and sea-water system (for the shorebased OTEC plant) to only the power-transmission system (for the moored plant) as one moves from onshore to offshore. The geographical location of various OTEC system components has been listed in Table 1 for the three configurations under consideration. It is apparent that the greatest potential for significant coastal-zone impacts is associated with the bottom-resting and shorebased OTEC configurations. The moored configuration interacts with the coastal zone primarily via the power transmission system and temporary construction facilities.

Environmental concerns and key issues associated with OTEC deployment and operation may be conveniently grouped into four categories:

(1) Redistribution of Oceanic Properties (Pumping)

- o current patterns
- o climate alteration
- o receiving water thermal structure
- o recirculation
- o biological entrainment and impingement
- o biostimulation
- o turbidity

(2) Chemical Pollution

- o biocides and toxicants (water quality)
- o working fluid leaks
- o corrosion
- o air quality

(3) Structural Influences

- o biological attachment
- o habitat and migration
- o sediment transport
- o wave climate
- o current patterns

(4) Socio-Legal-Economic Issues

- o worker safety
- o enviro-maritime law
- o aesthetics (noise, visual)

o secondary economic impacts

Concerns for the coastal-zone environment emanate from a variety of OTEC components and activities, and also may be grouped into four major categories:

(1) Construction and Deployment

- o temporary onshore construction and support facilities
- o temporary marine structures
- o dredging and excavation

(2) Structure

- o platform
- o sea-water system
- o anchoring system
- o power-transmission system

(3) Operation

- o platform
- o sea-water system
- o power-transmission system

(4) Inspection and Maintenance

- o platform
- o sea-water system
- o power-transmission system
- o anchoring system
- o permanent support facilities

A matrix relating OTEC activities and environmental concerns raised by these activities is given in Table 2. It should be noted that a substantial number of coastal-zone environmental issues are raised during the construction-and-deployment phase by activities relating to tunnels and trenches (that may require drilling, blasting, excavation and dredging), by onshore pipe-fabrication yards, assembly and lauching areas, backfilling and burial of sea-water pipes, etc. Table 2 also indicates that the physical presence of an OTEC plant in the nearshore zone demands consideration of environmental impacts that may originate from the disturbance of the dynamic equilibrium of bottom sediments and littoral sediments, caused by structure-induced changes in the local wave and current patterns.

DISCRIPTION OF ENVIRONMENTAL CONCERNS

REDISTRIBUTION OF OCEANIC PROPERTIES

The potential changes in the oceanographic properties of sea water due to OTEC pumping operations are a major environmental concern. Because large amounts of cold, deep water and warm shallow water will be

CONCERN	SOURCE OR ACTIVITY						
	Construction and Deployment			Structure			
	Temp. Onshore Construction & Support Facility	Temporary Marine Structures	Dredging & Excavation	Platform	Sea-Water System	Anchoring System	Power-Transmission System
Redistribution of Oceanic Properties (Pumping)							
o current patterns							
o climate alteration							
o receiving water thermal structure							
o recirculation							
o biological entrainment & impingement		x	x				
o biostimulation							
o turbidity		x	x				
Chemical Pollution							
o biocides & toxicants (water quality)	x	x	x				x
o working fluid leaks	x	x	x				x
o corrosion							
o air quality	x	x	x				
Structural Influences							
o biological attachment		x		x	x	x	x
o habitat & migration		x		x	x	x	x
o sediment transport		x		x	x		x
o wave climate				x	x		
o current patterns				x	x		
Socio-Legal-Economic							
o worker safety	x	x	x				
o enviro-maritime law	x	x	x				x
o aesthetics (noise, visual)	x	x	x				
o secondary economic impacts	x	x	x				

Table 2 OTEC Activities and Related Environmental Concerns

CONCERN	SOURCE OR ACTIVITY							
	Operation			Inspection and Maintenance				
	Platform	Sea-Water System	Power-Transmission System	Platform	Sea-Water System	Anchoring System	Power-Transmission System	Permanent Support Facilities
Redistribution of Oceanic Properties (Pumping)								
o current patterns		x						
o climate alteration		x						
o receiving water thermal structure		x						
o recirculation		x						
o biological entrainment & impingement		x						
o biostimulation		x						
o turbidity		x					x	
Chemical Pollution								
o biocides & toxicants (water quality)	x	x		x	x		x	x
o working fluid leaks	x		x	x	x		x	
o corrosion	x	x						
o air quality	x			x				x
Structural Influences								
o biological attachment								
o habitat & migration							x	
o sediment transport								
o wave climate								
o current patterns								
Socio-Legal-Economic								
o worker safety		x	x	x	x	x	x	x
o enviro-maritime law		x	x	x	x	x	x	x
o aesthetics (noise, visual)		x	x	x	x			x
o secondary economic impacts		x	x					x

Table 2 (continued)

pumped to the heat exchangers, likely at some third depth; parameters such as temperature, salinity, density, dissolved oxygen, nutrients, carbonates, particulates, etc., will be modified by mixing with ambient ocean water in the vicinity of the eventual discharge. Discharges in the photic zone may cause biostimulation due to the increased nutrient contribution from the deep waters, with potential changes in either the size, relative abundance or species composition with respect to the resident marine population resulting in secondary effects on the food web. Displacement of sea water also could have toxic effects on ambient species by the introduction of trace chemical substances such as trace metals, organic decay products, etc., from other depths. Certain species, particularly those with low mobility, will be harmed by impingement/entrainment in the pumping system either by contact with the screens and walls of the pipe-heat exchanger system or by the pressure and temperature changes encountered in transit of the system from intake to discharge. Surface discharges may produce climatic effects by alteration of the air/surface water temperature ratio. Such an alteration, at sufficient scale such as in an OTEC park, could affect the microclimate by modification of locally generated winds and currents. Long-term operation of a large number of OTEC plants could result in reduced available heat due to the thermal extraction. Surface discharges also could enhance the release of CO_2 and other gases dissolved in cold deep waters with potential climatic effects for large-scale operations. This is particularly true for open cycle systems where gases, even normally dissolved in the surface ocean, must be separated from vaporized sea water so that gas bubbles do not impede plant efficiency. Such gases will be vented into the atmosphere potentially modifying the local microclimate as discussed above. However, subsurface discharges below the surface mixed layer in the pynocline could mitigate all or most of the potential problems associated with surface discharges.

The recirculation of sea water becomes increasingly more likely as the location of intake and discharge structures approaches the shore and shallower water. This must be prevented because it reduces the available temperature difference and therefore power output (thermal-resource depletion) as well as increasing the concentration of toxicants that may be carried within the discharge plume The combination of large sea-water discharge (typically 7 m^3/sec per MWe), relatively shallow water, and complex nearshore ocean currents demands that oceanographic conditions and intake-discharge hydrodynamics be carefully evaluated at each site.

CHEMICAL POLLUTION

Chemical pollution could result as functions of various OTEC plant operations and maintenance procedures. Of major concern are biocides proposed to keep the system components clean of biological growth. There are alternative ways (various mechanical systems) to clean heat exchanger systems, but these may not clean surfaces to the extent necessary for efficient operation. The major problem with biocides is the levels of concentration needed to impede biological growth in the system which unquestionably will affect organisms in the vicinity of the discharge. Furthermore, if chlorine is used as a biocide and ammonia as a working fluid, accidental combinations of these chemicals can produce compounds even more toxic to ambient organisms than the separate chemicals. Leaks of the working fluid of a closed cycle system also will pollute ambient ocean water. The effects and chemical fate of proposed working fluid leaks into sea water are not well understood. Ammonia, for example, is a nutrient in proper amounts and could stimulate marine growth complicating the biofouling problem. However, the excessive doses associated with a major leak is toxic both to marine and human life. Chemical pollution also will be produced by the corrosive effect of sea water passing through the heat exchanger system. Corrosion would produce metallic ions, and scale particles which could have direct toxic effects or long-term effects through incorporation of corrosion products into the particulate food supply of marine organisms through the process of bioaccumulation.

Some sources of chemical pollution are only important during the construction-deployment phase of an OTEC power plant, as indicated in Table 2. Among these are onshore construction and support facilities with pollution problems similar to those of other large-scale terrestial construction projects, e.g., ground-water pollution from spills of oil and corrosion inhibitors. The installation and operation of marine structures that are to be used only temporarily such as graving docks, breakwaters, wharfs, trestles and caissons, is a source of pollution because of working-fluid leaks from ships and barges, toxicant release from disturbed sediments, smoke, sanitation releases, etc. Dredging operations may release buried toxicants and biocides into the water column, and may cause pollution through the discharge of sewage, antifoulant paints, etc.

STRUCTURAL INFLUENCES

The physical presence in the ocean of a structure the size of an OTEC system itself has an impact on the ocean. The structure of whatever configuration will become an attractive habitat for a wide range of organisms based on experience from artificial reefs. The long-term effects of the structure on the environment will depend on the types, size, and abundance of the organisms attracted to or attached to the structure and this will modify the local population. Regional effects on populations might occur by either interference with, or modification of, nesting habits or migration pathways.

The potential for coastal-zone impacts exists whenever significant alterations in prevailing wave and current patterns are produced by a physical barrier, and this may be the case for OTEC platforms and seawater conduits located in relatively shallow water. Environmental problems may originate from the disturbance of the dynamic equilibrium of bottom sediments and littoral sediments caused by changes in the local wave and current conditions. All OTEC systems include structural components that constitute an obstacle or barrier to normal wave progress and, if not too far from shore, may be responsible for changes in the transport and deposition of sand along the shore. Consequently the interference with the local sediment transport may result in deposition at or near the structure reducing efficiency and reducing its useful life. The environmental impact resulting from the placement of a particular structure at a given distance from shore is generally difficult to assess but must be given careful consideration. This remains true even when the term "environmental impact" is restricted to encompass only structure-induced changes in the wave and current characteristics in addition to modifications in the sediment distribution along the shore, near the structure, and in the intervening region (shore processes).

The ecological effects of oil platforms, artificial islands, and artificial reefs (composed of street cars, old car bodies, concrete blocks and riprap) have been investigated for some time and may serve as indicators of impacts to be anticipated from OTEC ocean structures. Johnson et al., (1978) summarize findings as follows:

> "In general these studies conclude that the habitat features created by the addition of solid substrate are beneficial to the local ecosystem, especially in areas where such substrate is limited. In time, communities or organisms develop which usually support more species than the sedimentary habitat that existed before the addition of hard, high-relief substrate. The biomass of the encrusting flora and fauna is an important food source for species of recreational, commercial, or aesthetic value which would otherwise not populate the area. In addition, physical characteristics of the solid substratum, such as crevices and vertical relief in an otherwise featureless bottom, attract a variety of fishes."

SOCIO-LEGAL-ECONOMIC

The first three major classes of concerns chiefly dealt with impacts on marine life. However, there are human consequences of OTEC operations which are grouped here as socio-legal-economic issues. Worker safety is of prime concern regulated by the Occupational Health and Safety Administration (OSHA) and the Coast Guard for strictly marine occupational concerns. Potential work hazards are the chemicals used or produced by the OTEC system such as ammonia, chlorine, foul weather during marine operations, collision, and systems accidents. Because of the novelty of OTEC operations, standard safety procedures will be augmented by procedures unique to OTEC. The siting of OTEC facilities either in international waters or where the downstream effects of OTEC operations might intrude into international waters, will raise the issues of international rights and responsibilities beyond those treated by conventional maritime law and treaties. At present, the law of the sea is in a state of flux so that the resolution of potential international issues may be complicated and time consuming. Probably multilateral agreements or treaties among concerned parties, as is done for fishing rights, may be the interim solution of potential legal problems which may impede OTEC operations. Finally, the construction and operations of an OTEC facility may affect existing social and institutional structures. New jobs will be created and shore-based "boomtown" growth may occur with its associated impacts on housing, education, sanitation, etc. The electrical energy produced by OTEC plants may be transmitted to consumers either by A.C. or D.C. transmission lines. The cable needed to transmit this power could have impacts on marine ecosystems at the sea bed and at the shoreline. D.C. transmission will require two converter facilities, one at sea and one on shore, causing land use problems. If

OTEC systems are used to produce energy-intensive products, they will produce air/water emissions typical of those produced in similar land-based industries.

Regulation and licensing of OTEC energy facilities is a complicated and unresolved problem. Because of the recent development of the technologies and variety of plant configurations and deployment locations, ocean energy facilities resist catagorization into existing law. They fit somewhere among conventional ships, drilling or oil production platform and power plants. Legal aspects of coastal and nearshore areas are especially complicated because of incomplete Coastal Zone Management (CZM) plans and uncertain state jurisdiction. Major OTEC legislation (PL-96-320), attempting to resolve some of these problems, has recently become law and is discussed in the following text. The remainder of this section is an item by item analysis of legal and socio-economic issues likely to arise in response to OTEC development, including the legislation which will be activiated by the issues. Legislative parallels may be drawn for ocean waves and current technologies in that they are moored structures transporting power to shore via submarine cable similar to the moored OTEC configuration.

On August 3, 1980 the Ocean Thermal Energy Conversion (OTEC) Act of 1980 was signed into law (PL-96-320). This act attempts to resolve overlapping jurisdiction among agencies such as USCG, USGS, EPA and NOAA by establishing NOAA as the lead agency for the licensing of OTEC facilities. NOAA is now charged with designing regulations and programs for OTEC facilities with input from other applicable agencies (e.g., DOE, DOI, EPA, USCG, ACOE and OSHA). The Coast Guard, however, retains jurisdiction over the following: pollution prevention and clean-up, safety and warning devices, vessel transport and maintenance, and personnel equipping and training. In short, the Coast Guard will supervise those areas where their expertise is demanded while NOAA will direct overall management of development of the program, including final approval and monitoring of OTEC licenses and plants.

PL-96-320 also protects the rights and responsibilities of adjacent coastal states by assuring that federal actions are consistent with approved Coastal Zone Management (CZM) plans and other applicable state and local laws. For example, the governor of each adjacent coastal state within an approved CZM plan can stop the licensing of an OTEC plant if he/she determines that the application is inadequate or inconsistent with programs within his/her jurisdiction. Adjacent coastal states without approved CZM plans and all other interested states are to be granted the opportunities to have their views regarding the location, construction and operation of an OTEC facilitiess or plant-ship given full consideration by the administrator of NOAA.

Other provisions of PS-96-320 relevant to the coastal and nearshore areas are:

o The issuance of any license for ownership, construction, and operation of an OTEC facility or plant-ship is deemed a major federal action significantly affecting the environment and thus will require an Environmental Impact Statement (EIS). This EIS will include a detailed documentation and impact analyses of the coastal and nearshore areas affected by construction and operation of OTEC plants in a single application area.

- OTEC facilities and plant-ships licensed under PL-96-320 are not considered to be "vessels or other floating craft" for the purposes of section 502(12)(B) of the Federal Water Pollution Control Act of 1972. Licensees, therefore, are required to obtain National Pollutant Discharge Elimination Systems (NPDES) permits from the EPA for discharge into the ocean of waste material or other effluents. Therefore, the Secretary of Energy in cooperation with other interested federal agencies will establish and enforce standard regulations necessary to assure the safe construction and operation of submarine electric-transmission cables and equipment under United States jurisdiction. These standards and regulations will require the use of the safest and best available technology for submarine electric-transmission cables, shielding and the use of automatic switches to shut off electric current in the event of a break in the electric cable.

Conflicting Use of Nearshore and Coastal Areas

Nearshore and coastal land is in growing demand for oil production, mining, recreational development and wildlife sanctuaries. OTEC facilities will be forced to compete for cable rights-of-way, mooring sites and acreage for onshore facilities. Several of the states likely to be involved in OTEC development have approved CZM plans, i.e., Alabama, Guam, Hawaii, Puerto Rico, and the Virgin Islands. In these plans, as required by the Coastal Zone Management Act of 1972, each state has designated priority uses and possibly prohibited, or strictly controlled, other uses (section 305[§1455]). CZM plans, depending on their individual priorities, could restrict certain plant configurations from being deployed in areas under the plan's jurisdiction (section 307(C)(1)). Also, existing submarine pipeline and communication cables and their federally required rights-of-way may limit OTEC electric transmission cable passage, especially in the Gulf of Mexico where networks are extensive.

Federal and State Jurisdiction

The United States currently claims a territorial sea limit of 12 nautical miles with the adjacent coastal state having jurisdiction over the first 3 miles (except Florida and Texas which claim a 9 mile limit in the Gulf of Mexico). Complications may arise if, for example, the thermal plume of an OTEC plant moored under federal jurisdiction invades an estuary under state jurisdiction. Cables coming to shore and cold water pipes going offshore may also cross jurisdictional boundaries and therefore be subject to different regulations. Similar problems have been presented to the Supreme Court in _California v. Zook_ and _Ray v. The Atlantic Richfield Company_. The Supreme Court decided that if the case falls within an area in commerce thought to demand a uniform national rule, state action is struck down. If the activity is one of predominantly local interest, state action is sustained (Nanda, 1979). Such legal precedents may be useful in simplifying state and federal jurisdiction over OTEC facilities.

Shipyard Requirements

Shipyards with adequate access channel depths to accommodate OTEC construction and fabrication are distributed along the U.S. coast, Hawaii and Puerto Rico. Several shipyards have established drydocks and graving docks (i.e., Norfolk, San Francisco and Puget Sound). If other shipyards are to be used for OTEC construction, facilities will be need to be added or upgraded. Improvement of shipyards for OTEC will open new centers for large scale construction and, in general, enhance U.S. shipbuilding capabilities.

In summary, with the enactment of PL-96-320 it is evident that an effort is being made to clarify interagency jurisdictional complications concerning OTEC development. It is important that state and federal jurisdiction is resolved in the near future so that a comprehensive management system for U.S. coastal areas may be devised. Until such a plan is finalized OTEC energy facilities most likely will be subject to legal action on all governmental levels.

Acknowledgement

The authors would like to thank Dr. Pat Wilde of the Lawrence Berkeley Laboratory, University of California, for his valuable contributions to this paper.

References

Alexander, L.M. and Cameron, F.X., "Jurisdictional Issues in the Coastal Belt", Coastal Zone 80 Conference, Vol. I, American Society of Civil Engineers, San Francisco, March 1978, pp.310-321.

Nanda, Ved P., "Legal and Institutional Aspects of OTEC", Sixth OTEC Conference, Vol. I, U.S. Dept. of Energy, Washington, D.C., June 1979.

Wilde, P., "Environmental Monitoring and Assessment Program at Potential OTEC sites" Environmental Control Symposium, Dept. of Energy. Argonne, Ill. March 1980.

Brewer, J.H. et al., "Feasibility Design Study: Land-Based OTEC Plants" U.S. Dept. of Energy (Deep Oil Technology, Inc.) No. COO-4931-02, Jan. 1979.

Lewis, L.F. "OTEC Environmental and Resource Assessment Program" Sixth Ocean Thermal Energy Conversion Conference, Vol. I, Wash. D.C., June 1979.

Johnson, G.F., et al., "Ecological Effects of an Artificial Island", Coastal Zone 80 Conference, American Society of Engineers, March 1980, Vol. IV, pp. 2837-2856.

MARINE KELP: ENERGY RESOURCE IN THE COASTAL ZONE

Ronald L. Ritschard and Kendall F. Haven

Lawrence Berkeley Laboratory
Energy and Environment Division
University of California
Berkeley, California 94720

INTRODUCTION

The assessment of energy-related impacts in the coastal zone has focused primarily on thermal power plants and OCS petroleum development. However, there is a growing list of energy sources and conversion systems within the coastal environment that are currently being proposed. Marine biomass is one of the most recent entries on this list of coastal energy resources.

A marine biomass farm is one of the few biologically-based systems that has the potential of contributing large quantities of synthetic gaseous fuels to the nation's energy supply. This is especially true from the standpoint of availability of large amounts of space on the sea surface and of plant nutrients in the ocean waters. The California giant kelp (Macrocystis pyrifera), which are well established as a valuable coastal resource and a source of chemical products (algin), is a prime candidate for energy conversion, since it is efficient in converting sunlight into a fixed source of energy. In turn, kelp can be processed by anaerobic digestion or other procedures into methane. Furthermore, other by-products such as food, fertilizer, ethanol, and industrial material can be obtained.

This paper describes an ocean farm system that has been designed and used as an ocean test facility by the Energy from Marine Biomass Program, jointly sponsored by the Gas Research Institute and the Department of Energy and managed by the General Electric Company (Leone, 1979).

The analysis of the ocean farm system includes a description of the types of impacts that might occur if large scale operations become available such as the production of environmental residuals, conflicts with the fishing and shipping industries, and other legal/institutional impacts. Finally, a discussion is given of the relationship of the marine biomass concept and coastal zone management plans.

OCEAN FARM DESCRIPTION

Overall Concept

The basic concept of a marine biomass system is to culture and harvest seaweed plants that are attached to a grid of polyethylene lines suspended fifty to one hundred feet below the ocean surface. These lines are supported by buoyancy-control structures embracing thousands of acres (Wilcox, 1975).

Marine plants require light, carbon dioxide, water and nutrients from the surface layers of the ocean. However, many of the areas along the southern California coast that would support marine algae may be nutrient-limited for as much as six to nine months each year because of a lack of upwelling (North, 1977). Therefore, fertilizing operations are clearly necessary to produce good yields of kelp on ocean farms. The selected process for fertilization is to pump up nutrient-rich waters from depths of a thousand feet or so. While not a general consensus among researchers, it is expected that resultant photosynthetic conversion efficiencies of marine systems will be higher than current terrestial crops.

Design and deployment of the upwelling system has provided the major engineering challenge to the overall farm concept. In order to maintain the adequate nutrient concentration, especially nitrogen, the test farm system uses upwelled water from a depth of 1500 feet through a two foot diameter polyethylene pipe at approximately 9000 gallons/minute. In a prototype or commercial farm, a depth of 300-500 feet would suffice, since nutrient concentrations are relatively constant at depths below 300 feet (Seligman, 1976). The requirement of providing continuously upwelled deep water from the open oceans is similar to the concept required in another energy system, O.T.E.C. (Ocean Thermal Energy Conversion), that is being tested off the coast of Hawaii.

Finally, the seaweed are to be harvested periodically and converted to methane and other by-products (fertilizer, food supplements, etc.) at a processing facility located at an onshore coastal site. Figure 1 shows a generalized diagram of the marine biomass system used in this analysis.

Production System

The California giant kelp (*Macrocystis pyrifera*), which grows along the coasts of California, Mexico and New Zealand, was selected as the biomass source because it is one of the world's fastest growing plants and has been cultivated on an artificial substrate. The reproductive cycle is well understood and in nature it is believed that the plant will reproduce its own weight every six months or so (North, 1971). *Macrocystis* kelp beds have been harvested mechanically for over 60 years along the southern California coast. In the biomass farm system, the plants will be harvested every three months with no replanting expected to be required.

Of primary importance to the basic system is the determination of yield. In general, all aquatic plants have the same basic physical requirements and biological limitations for growth. The physical requirements for growth include light, which is affected by plant density and water temperature, and nutrients, which are controlled by water circulation.

The practical value for aquatic biomass production, on a full year basis, is reported to be eight dry ash-free tons/acre/year, which includes 6.7 tons of organic matter/acre/year (Clendenning, 1971). Klass (1977) reported from laboratory efforts, however, that anchored

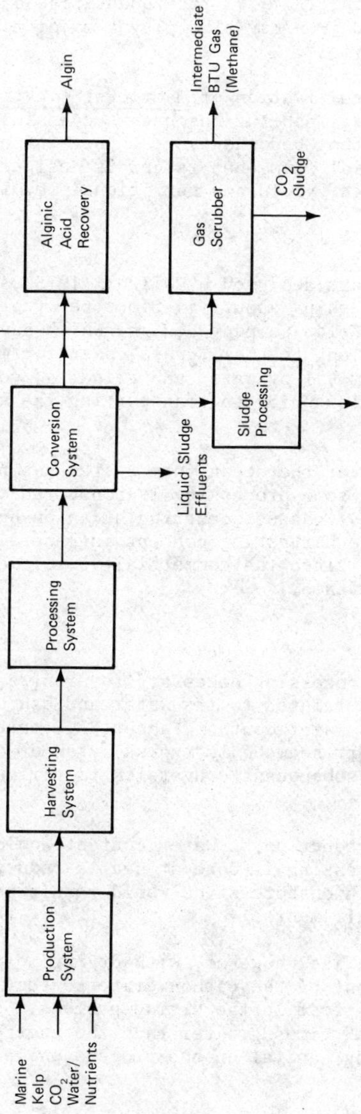

XBL 781-13495

Fig. 1. Complete Marine Biomass System

giant kelp may be expected to yield as high as 50 dry ash-free tons/acre-year. This assumes that nutrients are supplied by the upwelling of deep nutrient-rich water. Nitrogen, an element which most often limits the growth of marine plants, is required at levels exceeding three microgram-atoms per liter (North, 1977). Trace quantities of other micronutrients such as manganese and iron may also play a major role in obtaining maximum kelp growth and yield.

The controlled cultivation of *Macrocystis pyrifera* in water that is deeper than its natural habitat has been successfully accomplished (North, 1979). The test bed, however, is not intended as a miniature version of a commercial farm, but rather is designed to gather data for determining the growth, yield and nutritional requirements of kelp.

Harvesting

In the marine farm designed by Wilcox (1975), the standing crop is harvested by special ships about six times per year. These vessels are patterned after the Kelco Company design, which have been used for commercial harvesting along the California coast for many years. Some preprocessing, e.g., removal of water and grinding, could be accomplished on the harvesting ships prior to transporting the kelp to onshore processing plants.

Since this part of the ocean farm system has not been tested, it will be assumed that some processing is conducted on the harvesting vessels with the final phases occurring at an onshore processing site. One limitation to the harvesting concept in general is the availability of harvesting ships, since the commercial vessels currently used are scheduled long in advance.

Processing

The degree of processing necessary to prepare kelp for conversion into usable fuels is related to its water and ash content. *Macrocystis* typically has a high water content, about 87.5 percent (Wilcox, 1975). The ash content of dry seaweed is typically about 40 percent, which may cause difficulty in subsequent conversion to various products (Leese, 1976).

A substantial reduction in water content would minimize shipping costs. If such processing is done at sea, a reduction in the capital costs associated with onshore sites would result because of the reduction in storage requirements.

A problem may arise, however, with drying kelp for transport. Since about 35 percent of the carbohydrate content is dissolved in water, most of it is lost in the drying process. If the food value of kelp is as much as 15 times greater than the energy resource value, then we may be saving capital on one process while losing it on another (Schneider, 1978).

With regard to ash content, it may be necessary to reduce the level of ash in order to maintain a viable culture for subsequent digestion or fermentation (Hart, et al., 1978). Since kelp ash consists principally of water soluble salts, some market may exist for such products as potassium, calcium and magnesium, since incremental system costs for removing these chemicals from the process stream would not be large.

The degree of mechanical and chemical pre-treatment necessary for increasing kelp separation and digestion has not been fully defined. A process developed for this analysis begins by shredding the wet, harvested kelp by means of hammermill-type grinders. Leone (1979) estimated that about 1.1 kWh/ton of raw kelp is required to produce properly-sized particles from wet kelp. Next, the shredded kelp is treated with a calcium chloride solution heated to 95°C. The material is then pressed (100 pounds pressure) to remove salts and excess water. The resulting mixture acts as feedstock and is fed into the anaerobic digester.

Conversion

The conversion of kelp to methane has been described in previous studies (Wilcox, 1975; Chynoweth, et al., 1978; and Leone, 1979). As mentioned above, there is no standard procedure for pretreating the kelp. Several separation steps are usually used to segregate the electrolytes, carbohydrates, water and volatile solids. The soluble sugars that are pressed out could be fermented to ethanol. The volatile solids (≈60 percent) go into a heated air-tight digester, where methane and carbon dioxide are produced. The feedstock is decomoposed over a period of seven or more days by bacteria in the absence of oxygen. A waste sludge, high in nitrogen, will also result. This material, after further processing, could be used as fertilizer feedstock.

Energy recoveries of methane on the order of 4.9 to 5 standard cubic feet per pound of volatile solids (SCF/lb. VS) have been obtained with 55-60 percent conversion of volatile solids (Chynoweth, et al., 1978). This yield represents approximately 75 percent of that which is theoretically attainable and, on the average, exceeds those of any other known biomass source including feedlot waste or sewage sludge.

In addition to the goal of maximizing methane yields, research is underway in the Marine Biomass Program to increase the digester loading rate and reduce detention times. Digesters have been stabilized at a loading rate of 0.2 lb. VS per cubic foot and 10 days detention time. This rate is compared to earlier values of 0.1 lb VS per cubic foot and 18 days detention time. It is estimated that a loading rate of 0.3 lb VS per cubic foot, a detention time of six days and a methane yield of 5.5 SCF/lb. VS are achievable (Leone, 1979).

Although this study emphasizes the production of methane gas as the primary product, other by-products are possible. In fact, if the marine biomass system is to make a major contribution to the energy sector, it will require that these other products be processed and marketed as well. Potassium chloride, sodium sulfate, sodium carbonate, and other salts are possible moderate value products. Several industrial gum materials such as algin, fucoidan and laminarin are available. Algin has been extracted from kelp for profit for many years and has many uses in the food and chemical industries.

Since all of the carbohydrates in *Macrocystis* are polysaccharides, they contain sugar molecules that may become economically viable food products. Preliminary studies have suggested the potential of the digester solid effluent as an animal feed supplement (Leone, 1979). The data from these investigations indicate that the effluent has a crude protein content of approximately 37 percent and as such, has good potential as an animal feed supplement.

Table 1 lists the baseline design parameters for a marine biomass farm system. Since there is no prototype system in existence, the parameters are given as average values.

POTENTIAL IMPACTS OF OCEAN FARM SYSTEM

Any energy conversion system accrues a variety of potential impacts. This section provides information on the types of impacts that might occur from an ocean biomass system, which converts marine algae to methane gas. Since there are no full scale systems in operation, the data represent a compilation of potential impacts from bench-scale experiments, test farms and the conceptual schemes of production, harvesting and processing components.

Environmental Impacts

The impacts of a massive open ocean farm operation have not been explored. There is a potential for significant climatic modifications. The anticipated climatic changes stem from the massive artificial upwelling that will be required to stimulate kelp growth and maintain it at high rates. When large volumes of cold deep water that are rich in nutrients and supersaturated in carbon dioxide are brought to the ocean surface in order to fertilize the plants, events might occur that could lead to regional and global changes in climate. The culture and harvesting of seaweed over several thousand square miles of ocean surface could result in changes in albedo, air-sea exchanges of materials, and altered ocean surface roughness. The farm structures themselves will reduce or change the patterns of water circulation. These factors, which could change prevailing weather patterns and create additional fog banks, subsequently may have some effect on the productivity of the kelp beds. The possible climatic changes resulting from large scale marine plant culture have been recently reviewed by NOAA (Lehman, 1980).

A potential problem associated with the farm structure itself is the release of numerous chemicals into the ocean from the supports and synthetic lines used to hold the algae. Hruby (1978) noted the possibility of a slow release of toxic metals from the anti-fouling paints and organic chemicals used on the farm structures. The degree of concern regarding these chemicals is unknown at this time. Dissolved organic chemicals, such as phenols, will be released by the marine kelp.

Table 1

Baseline Design Parameters

Marine Biomass Farm System

Parameter	Average Value	
Kelp Composition (Dry Weight)	Ash	38 - 45%
Volatile Solids Composition	Volatile Solids	55 - 62%
	Carbohydrates	(∿28%)
	Algin	13 - 24%
	Cellulose	(3 - 8%)
	Protein	(5 - 7%)
	Fats	(0.5%)
Biomass Yield	50 dry ash-free (DAF) tons/acre-year	
Energy Content	8000 Btu/lb. (DAF)	
Upwelling Depth	300-500 feet	
Surface Water Temperature	20°C (or less)	
Nitrogen (NO_3) at 300 ft. depth	25-30 μg-atoms/liter	
(Biomethanation Parameters)		
Methane Yield	4.9-5 standard cubic feet/pound volatile solids	
Retention Time	10-18 days	
Volatile Solids Converted	65-75%	
Loading Rate	0.1-0.3 lb. VS/ft^2	
Temperature	35°C	
Volume	65% methane - 35% carbon dioxide	
Energy Recovery	∿ 55.5%	

Source: J. Leone, 1979

Sieburth (1969) estimated that these organic compounds could be exuded by the algae at rates as high as 40 percent of the net carbon fixed. Calculations based on available data suggest that the release of extracellular organic compounds by *Macrocystis* will be a problem (Hruby, 1978). The exudations from brown algae have been found to be toxic to some marine organisms.

The upwelling system, which is designed to provide an abundance of nutrient-rich water needed for kelp growth and development, could present several environmental problems. The temperature differences between upwelled waters and those present on the ocean's surface might form large fog banks as warm moist air is blown over the cooler deep water. One consequence of such fog banks is to reduce the amount of sunlight reaching the surface which, in turn, could affect the rate of productivity.

The upwelling of water may alter salinity, temperature, dissolved oxygen, turbidity and nutrient levels. While artificial upwelling may support increased biological production in the kelp beds, it may also increase the growth of less desirable planktonic species that may have long term effects on the residential biological communities.

Furthermore, the upwelling of deep water will entrain marine organisms that cannot resist the vertical inflow velocities. Organisms that are entrained will be subject to mechanical pressure and sheer stresses. The survival rate of mesopelagic organisms in the upwelling streams is species-specific.

A final problem with the upwelling system is related to the use of diesel-powered pumps. There are several possible air pollutants, e.g., particulates, nitrogen oxides and hydrocarbons, that would be emitted to the atmosphere from the pumping operations. The level of emissions corresponds to the type of system selected.

The harvesting aspect of the marine biomass system involves the use of ships with the Kelco Company design. The harvest ships will create some environmental impact due to the emissions during their normal operations. The Kelco ships burn diesel as a fuel source that results in the production of particulates, nitrogen oxides, and hydrocarbons as primary air pollutants. These pollutants, however, will be generated over a more widespread area, since they are released as the ships travel to and from the off-shore kelp farm. In addition to the air emissions, there may be liquid effluents (brine) formed during on-ship processing that will require special handling and disposal. The level of waterborne effluents depends on the degree of processing that is conducted aboard the vessels during harvesting.

Two methods are available for unloading the kelp, depending on the size of the systems involved. For the ocean farm system, the kelp will probably be shredded and pre-processed on the ships prior to reaching the processing plant. In the latter case, a slurry of chopped algae mixture will be piped from the ship to the shore-based facility. Since some pumping will be necessary to move the slurried mixture, a potential exists for the release of certain air pollutants from the diesel-powered engines. The consumption of diesel fuel is low, however, so that this environmental impact should be minimal.

The waste water generated from the shredder, presser and digester during the processing phase will eventually be discharged into the sewer system. The composition of this effluent and the degree of pollution control that is necessary are unknown at this time. It is assumed, however, that the processing plant will conform to EPA discharge permit standards regarding waste water effluents.

As a final step in the marine biomass system, the processed algae is fed into the anaerobic digester. The gas mixture from the digester must be passed through a scrubber to separate carbon dioxide, which is about 40 percent of the gas, from methane. The major environmental residual resulting from this stage of the process is sludge from the scrubber that must be collected for subsequent disposal. The composition of the sludge from the scrubber as well as other phases of processing/conversion is also unknown. However, since marine algae will concentrate various heavy metals, the possibility exists for the sludge to contain considerable levels of heavy metals. If any of this sludge is to be used for fertilizer feedstock, it will require some level of sterilization.

The final concentration of heavy metals and other toxicants, and the biological oxygen demand and organic loading of the aqueous discharge cannot be anticipated without specific measurements from some test, demonstration or prototype facility. Until that information is available, only the potential environmental impacts can be identified.

Other residues from the ocean farm may find their way to onshore locations. For example, the potential exists for an increased amount of kelp debris to break loose (away) from the farm structure and be washed ashore. In turn, there may be a resultant impact on local recreation or on other beneficial uses of the coastal zone.

Legal and Institutional Impacts

Algal farms, depending on their size and location, may have a negative impact on commercial shipping if they disrupt existing shipping lanes. At a projected biomass yield of 50 dry-ash free tons/acre-year, it has been estimated that about 55,000 square miles of ocean surface might be needed to supply the nation's current requirements for natural gas (Leone, 1979). This area (approximately 235 miles by 235 miles), if concentrated offshore of the California coastline, might provide an additional hazard to ocean commerce.

In addition, such large marine biomass farms may adversely affect access to and utilization of coastal fishing locations. The potential exists for impacts on recreational and commercial fishing in the farm area. Since the kelp farms themselves will probably attract certain fish species, there is the further legal issue of trespassing on the marine farm and other liability questions.

Several institutional and legal issues are likely to accompany the research, development and commercial phases of the open ocean system if they are located beyond the 12 mile (territorial seas) or 200 mile (high seas) limit. The current biological test farm is deployed about 4.5 miles offshore from Laguna Beach in southern California. The prototype and commercial farms may be located as far as 20 miles offshore. Not only will the international and domestic legal status have to be analyzed, but a regulatory framework will have to be established to guarantee the various uses of marine resources.

Legal questions include--but are not limited to--liability for: collisions between ships and the substrate or associated fixed structures; blockage of fishing rights and lanes; interference with shipping and navigation; residuals released from the farm structure; and the cold water plume and its impact on coastal areas or fishing grounds.

RELATION TO COASTAL ZONE PLANNING

The federal Coastal Zone Management Act (CZMA) of 1972 established the Coastal Zone Management Program, which gives states federal help to prepare and then administer management programs that "preserve, protect, develop, and where possible, restore coastal resources." Under the 1976 CZMA amendments, state programs are required to include, among other things, a planning process for identifying energy facilities likely to be located in or significantly affecting the coastal zone and for anticipating and managing the impacts from these facilities. With these coastal zone planning tools in mind, we will consider how the introduction of a new energy technology such as marine biomass conversion relates to the planning process.

The ocean farm concept can be divided into its offshore and onshore components. Offshore requirements include the farm site itself that probably will be located outside the legal boundaries of the coastal zone. It is believed that the first commercial farms will be sited about 20 miles offshore of southern California. The harvest ships, however, will follow transit routes through the coastal zone to their onshore terminals. Just how the coastal zone planning process at the local, state or federal levels will affect the offshore activities is undefined at this time.

An immediate question is whether the federal government would establish a permit or licensing system for offshore biomass operations. In the case that a system was established, which federal agency would be given overall lead regulatory authority? There are various possibilities, including the following current responsibilities. Any alteration to the coastline or harbor area, under law, requires an Army Corps of Engineers permit. Vessels including industrial ships such as mobile drilling rigs are subject to close regulatory supervision by the Coast Guard. Fixed structures used in oil/gas exploration and development on the OCS are monitored by the U.S. Geological Survey and the Coast Guard. The U.S. Department of Energy, which supports most biomass research, development and commercialization efforts, must comply with the National Environmental Policy Act and probably would have to prepare one or more environmental impact statements. Finally, discharge of pollutants from the farm structure would probably require a National Pollution Discharge Elimination (NPDE) permit under the provisions of the Federal Water Pollution Control Act. These overlapping roles between various federal agencies will require some degree of clarification.

Another uncertainty with the offshore operations involves the federal consistency provision of the CZMA (Section 307), which has become an important issue with OCS petroleum development. Many states are presently concerned about the fact that a federal agency may be its own judge about whether its action affecting the coastal zone is in compliance. State coastal management programs may not be equipped to adequately address whether the proposed offshore energy activity is in compliance with state or local coastal plans.

An additional concern is how local coastal governments can plan and mitigate for any of the ocean farm impacts previously mentioned. The Coastal Energy Impact Program (CEIP), which is part of the 1976 amendments to the CZMA, is designated to help states minimize the social, economic and environmental disruptions that result from new or expanded coastal energy activities. It is assumed that CEIP assistance can be used for exotic coastal energy resources such as ocean kelp.

Planning for onshore activities, like the offshore operations, may pose a set of different problems for coastal planners. Onshore requirements for a marine biomass system include terminals for the harvest ships, port support facilities, farm fabrication plants, biomass conversion plants, and pipelines for gas supply and distribution. These facilities are not unlike those already included in a natural gas system or in the commercial kelp processing industry.

Although ocean kelp farms and their supporting facilities are not mentioned specifically in most state or local coastal plans, there are usually general policies that favor expansion areas within the coastal zone. In California, for example, ocean-dependent coastal development is encouraged, especially at existing sites. Industrial areas are presently zoned in the ports of San Diego and Port Hueneme for kelp processing plants because of the already existing kelp industry.

An important and unresolved issue with onshore activities is again related to which lead agency is responsible for planning, regulating and mitigating the potential impacts of marine kelp energy conversion. The possibility exists that local jurisdictions would act eventually as the lead agency. Until that time, there may be some confusion with this issue.

With regard to potential air and water quality impacts, it is expected that the air and water quality standards would be satisfied through the issuance of permits by the appropriate federal, state or local agency. The regulation and mitigation of air and water impacts from most onshore energy facilities are adequately considered within existing environmental protection provisions.

In conclusion, the main objective of energy planning in the coastal zone with any energy resource--oil, natural gas, coal, nuclear, or marine kelp--is to promote a timely transfer of pertinent information between the federal, state and local levels of government as well as with the industrial sector where possible. State and local agencies that are responsible for coastal planning (under the provisions of the CZMA), implementing the energy development, and bearing the environmental and socioeconomic impacts inherent to a particular technology must be kept abreast of the plans and potential consequences of the energy technology in question.

SUMMARY

The coastal regions of the United States are relatively rare, biologically important, and vulnerable to human perturbation. The coastal zone has been and will probably continue to be important in the industrial development of the nation. The placement of energy facilities there, however, generates environmental impacts and creates conflicts in the use of our coastal resources.

Marine biomass has been suggested as an energy resource, since it has the potential of contributing significant quantities of gaseous fuels to the nation's energy supply. As part of another project, an ocean farm system, using the California kelp (*Macrocystis pyrifera*) has been designed and used as a test facility offshore southern California. A description of this ocean farm conecpt, which includes a production, harvesting, processing and conversion system is presented.

The potential impacts of the marine kelp system are identified and discussed. Included as environmental impacts are: the potential for climatic modification; the release of numerous chemicals from the farm structure itself; the possible consequences of the upwelling system such as changes in various physical parameters and entrainment of marine organisms; the generation of air emissions from the diesel-powered harvesting vessels; the waste water discharges from the processing and conversion stages; and the waste sludge which may contain considerable levels of heavy metals, from the digester.

Legal and institutional impacts associated with large ocean farm systems are the hazards to ocean commerce; the obstruction of access to and utilization of coastal fishing locations; the questions of liability; and the international and domestic legal status of such an offshore operation.

Finally, consideration is given to how the introduction of a new energy technology such as marine biomass conversion interacts with the coastal zone planning process. Major concerns with the offshore aspects of the ocean farm concept exist including the overall lead regulatory authority; the question of federal consistency; and the planning and mitigation by local coastal governments for any of the potential impacts. Onshore activities will probably pose fewer problems since the proposed facilities are not unlike those already sited in the coastal zone.

The conclusion is that the proponents of a biomass energy system should start early to promote a timely transfer of information between the various institutions (federal, state and local) involved in coastal zone planning. The accurate prediction of environmental impacts and the mitigation of these impacts, required by law, demands a fully-coordinated energy planning and coastal resource management process.

REFERENCES

Chynoweth, D.P., D.L. Klass, and S. Ghosh, 1978, Biomethanation of Giant Brown Kelp, Macrocystis pyrifera, Proceedings of Symposium on Energy from Biomass and Wastes, Washington, D.C., August 1978.

Clendenning, K.A., 1971, Organic Productivity in Kelp Areas, in: Biology of Giant Kelp Beds in California, W.J. North, Editor, Chapter 12, pp. 259-263.

Hart, M.R., D. deFremern, C.K. Lyon and G.O. Kohler, 1978, Dewatering Kelp for Fuel, Feed, and Food Uses: Process Description and Material Balances, Transactions of American Society of Agricultural Engineers, 21 (1): 186-196.

Hruby, T., 1978, Environmental Impacts of Large Scale Aquatic Biomass Systems, in: Cost Analysis of Aquatic Biomass Systems, E. Ashare, et al., Dynatech R/D Company, Cambridge, Massachusetts.

Klass, D.L. and S. Ghosh, 1977, The Anaerobic Digestion of Macrocystis pyrifera under Mesophilic Conditions, Symposium Papers, Clean Fuels from Biomass and Wastes, Orlando, Florida, January 1977.

Leese, T., 1976, The Conversion of Ocean Farm Kelp to Methane and Other Products, Symposium Papers, Clean Fuels from Biomass, Sewage Urban Refuse and Agricultural Wastes, Orlando, Florida, January 1976.

Leone, J., 1979, Marine Biomass Energy Project, Marine Technology Society, New Orleans, Louisiana.

North, W.J., 1971, Growth of Individual Fronds of the Mature Giant Kelp, Macrocystis, in: Biology of Giant Kelp Beds in California, W.J. North, Editor, Chapter 3, pp. 123-168.

North, W.J., 1977, Marine Farm Studies Progress Report, California Institute of Technology, Corona del Mar, California, September 1977.

North, W.J., 1979, Possibilities of Marine Biomass Production as an Alternative Energy Source, American Chemical Society Symposium, Honolulu, Hawaii.

Schneider, Ronald, Integrated Sciences Corporation, Santa Monica, California, 1979, Personal Communication.

Sieburth, J., 1969, Studies on Algal Substances in the Sea, III. The Production of Extracellular Organic Matter by Littoral Marine Algae, Journal Exp. Mar. Biol. Ecol., 3: 290-309.

California Impacts of World Ocean Decisions

by

Robert L. Friedheim
Associate Director
Institute for Marine and Coastal Studies
University of Southern California

and

Robert E. Bowen
Research Assistant
Institute for Marine and Coastal Studies
University of Southern California

This work is a result of research sponsored by the NOAA Office of Sea Grant, Department of Commerce, under grant No. 04-7-158-44113.

Introduction

The international community appears to be concluding its deliberations on the law of the sea. For the past dozen or so years the participants in the Third United Nations Conference on the Law of the Sea (UNCLOS III) have been attempting to design a comprehensive treaty that will define a new international legal regime for the sea. Whether or not the conference actually produces an internationally agreed upon treaty is in some ways not as important as recognizing that UNCLOS is the primary catalyst in a process that has already begun to dramatically change marine legal norms. Unilateral claims over economic activities, such as fishing, is an option being taken by an increasing number of countries. Stronger national control of such factors as vessel source pollution in the coastal zone and transit through straits will likely become accepted international law. In short, we are entering a period in which vast amounts of ocean space are being legally enclosed, primarily by national governments.

It is important to attempt to understand better how this "enclosure movement" will affect our management of marine resources. A moderate amount has been written about the potential impacts of the "enclosure movement" on national governments. It should be recognized, however, that those impacts are not solely national in character and that there will be regional and local permutations as well. This paper will examine some of those more localized effects, specifically a broad-based evaluation of those impacts on the State of California.

California was chosen for this study because its enormous economy is heavily dependent on the sea. International trade is making a significant and growing contribution to the state's gross product. The 1979 Economic Report to the Governor states that the value of trade passing through California customs districts totaled and unprecedented $36.4 billion in 1978.[1] Waterborne commerce accounted for 80 percent of that trade. Fishing has also become an increasingly important factor in the state's economy. By the late 1970s fish landings in California ports exceeded $180 million.[2] Moreover, a number of California-based companies are involved in the development of technologies designed to exploit both marine mineral and energy resources. Most, if not all, of these activities will be influenced by new marine laws. The value of the study discussed in this paper lies in the fact that it is designed to provide information to California decision-makers to aid them in assessing the alternative responses available to them in coping with foreign marine-related decisions.

The study we discuss here involved a three-step process to develop this impact information. Initially, we developed systematic forecasts of the probable ocean policy behavior of those countries that are most likely to directly influence the interests of California.

Given California's dependence on the fisheries stocks, trade routes, and non-living resources of the Pacific Ocean we argue that the countries bordering the Pacific Basin are in a position to most effectively impact California. The probable decisions of the Pacific Basin states were developed by means of a computer-based forecasting model originally developed by Robert Friedheim and Joseph B. Kadane to analyze the UNCLOS III negotiations[3]. The methodology utilized by the model will be outlined in the following section.

The second step was to identify California's marine-related interests. Standard social science research methods were utilized to create a catalog of basic socio-economic data. With such data it was possible to examine the range of and depth of interest in various marine-related economic sectors. The data was collected from various state and federal agencies as well as from private interest groups.

Finally, this data together with the policy forecasts for the Pacific Basin countries, was used to determine the areas in which the activities of those countries could most significantly impact the interests of California.

Research and Methodology

As suggested earlier, the initial step in producing our impact evaluations is to develop forecasts of the probable behavior of the countries bordering the Pacific Basin. The forecasts are based on an analysis of the policy positions identified by those states in international forums. The basic imput data was derived from the statements made by national representatives of those countries participating in UNCLOS III negotiations. We believe that the policy debates at UNCLOS reveal rather well the basic preferences of most states on ocean policy issues. We further believe that if these preferences are basic that these states will, by one means or another, attempt to implement their preferences.

It should be stressed that while our data is drawn from the UNCLOS debates, our interest lies primarily with the stated national preferences of a specific group of countries and not with the output of the conference as a whole. This emphasis is important for a number of reasons. First, there are no guarantees that the conference will produce the comprehensive treaty that many hope for. Moreover, even if the differences over the current draft treaty are resolved, there are many who suggest that ratification of the text by a number of countries, including the United States, is far from certain. If either of these conditions proves to be the case, the degree to which marine legal structures will be defined by individual national decrees will increase dramatically. The importance of understanding the direction that certain key countries will move is evident. Without such information, the ability of decision-makers to act rationally in assessing alternative courses of action is greatly diminished. This remains the case, albeit to a lesser degree, even if the conference produces a treaty that is ratified by all signatories. It is generally assumed that with some revision the current draft treaty, the Informal Composite Negotiating Text (ICNT),[4] will form the core of the treaty. The provisions of that document are broadly stated in many cases, and are open to interpretation. In a number of instances the degree of latitude is sufficiently large to allow state preferences to define policy essentially unencumbered by the articles of the treaty. Therefore, an understanding of nation-state preferences would remain a central factor in rational decision-making.

The technique that we utilized to identify these preferences is derived from a form of quantitative content analysis which allowed us to systematically extract from the proceedings of the UNCLOS negotiations various "themes" that were discussed at the conference. A theme is defined as an expression of national preference for action on a given issue. The conference deliberated well over 1300 different themes during the period 1967-1975.

With this basic data it is possible to create an analytic model of the negotiations. We can create a single expression of a nation's preference on an issue, which is essentially the "average" position put forth by a country over a period of time, weighted by the importance that a country places on the issue. A similar measure can also be calculated for groups of countries. By a different set of calculations utilizing the same basic data, it is also possible to forecast the most likely outcome of the negotiations on a given issue, based on the preferences of the nations involved in the bargaining.[5] With these analytic measures it is possible to produce the forecasts of marine policy behavior for the nation-states of the Pacific Basin.

Pacific Basin States

For the purposes of this study 29 nations states were chosen for analysis. Table I is a list of those countries. These states were chosen because they either border directly on the Pacific or because they are dependent on the waters of the Pacific for access to international markets. A second criteria should also be noted. Only independent nation-states that actively participate in the UNCLOS debates between 1967 and 1975 were included. Therefore, Taiwan, which was not an active participant in all of the UNCLOS process, and non-independent island states were omitted from the study. Fourteen of the major issues debated by the conference were identified as being particularly relevant to the question at hand (Table II).

TABLE I

Pacific Basin States	
Australia	Mexico
Cambodia	New Zealand
Canada	Nicaragua
Chile	North Korea
People's Republic of China	North Vietnam
Colombia	Panama
Costa Rica	Peru
Ecuador	Philippines
El Salvador	Singapore
Guatemala	South Korea
Honduras	South Vietnam
Indonesia	Thailand
Japan	United Soviet Socialist Republic
Laos	Western Samoa
Malaysia	

TABLE II

Major Issues and Areas of U.S./Pacific Basin Preference Conflict.

Issue	Conflict Areas
1. Who shall exploit the international area	
2. Proposed structure of the production control system	
3. Distribution of revenue derived from the area	
4. Decision-making in the council	
5. Delimitation of the territorial sea	X
6. Transit through straits	X
7. Delimitation of the economics	
8. Navigation rights in the economic zone	
9. Delimitation of the fishing zone	X
10. Foreign access to coastal fisheries	X
11. Vessel source pollution standards	
12. Pollution enforcement in the economic zone	
13. Scientific research beyond national jurisdiction	X
14. Scientific research in the economic zone	X

To begin the analysis we first examined the preferences of the group as a whole over our 14 issues. In order to gain some perspective on the relative positions of the countries we compared the group scores for the Pacific Basin states to those of the United States. The U.S. scores were chosen as the point of comparison because the state of California did not independently participate in the conference and because in this type of international bargaining forum the federal government officially represents all 50 U.S. states.

This comparison of policy preferences highlights a number of rather interesting potential conflict areas. Table II shows the specific issues we are examining as well as issues in which there appear to be significant differences between the United States and Pacific Basin countries. Of the 14 issues examined six identify areas in which there is conflict between U.S. and Pacific Basin preferences.

In one of the key issues discussed by the conference--the delimitation of the territorial sea--the Pacific Basin states favor a wider territorial sea zone than does the United States. If one calculates a group preference for the Pacific, the position that is most likely to garner the greatest amount of support would be a 25-30 mile territorial sea. On the other hand, the data indicates that the United States would still prefer a narrower zone of only three-miles.

Another important issue is that of strait transit. A number of key international straits are located in the Pacific. The United States has long favored free, unimpeded, strait transit, stating that such transit not only creates conditions for strategic mobility but also enhances the freeflow of international trade. The Pacific group would find more generally acceptable a position that would place restrictions on strait transit.

A third issue centers around the degree to which these states will allow foreign fishing vessels access to their coastal stocks. This concept is rather central to United States domestic fishing interests as more than 90% of all fish taken worldwide are taken within 200-miles of a coast. The question for California's fishing industry is focused primarily upon the various highly migratory species of the eastern and western tropical Pacific. The Pacific group tends to favor less foreign access to coastal living resources than does the United States. While the U.S. position has admittedly changed drastically in the past five years, its current position remains less restrictive than a position defined by the preferences of the countries of the Pacific Basin would be.

The final issue we will consider here is marine scientific research. The question revolves around the amount of restrictions a country can place on the conduct of marine science within the economic zone of a coastal state. The U. S. has strongly favored a position that would retain worldwide freedom for marine scientific research. Once again, the states of the Pacific Basin for the most part favor a more restrictive environment for marine science research than the United States would prefer.

It is important to note here that although the group preferences of the Pacific Basin states are dealt with, this group is not a political coalition in the traditional sense. They do not meet in any organized fashion nor is there an attempt on their part to produce a common position paper on these issues. Therefore, we should stress that in our study we are not attempting to forecast the actions of the Pacific Basin states as a political unit, rather, we are examining a geographic region that manifests some degree of socioeconomic similarity across political boundaries and has the potential to significantly impact upon the state of California.

Given these group preferences it would be important to be able to offer some measure potential ability of the group to achieve those preferences. The notion of being able to define an absolute measure of political power is something that political scientists have tried to address for some time. We do not propose to create such a measure for this study. We are more interested in the degree to which these countries define a common position. In other words, we are interested in how cohesive the group is. The greater the degree to which the members of the group define a similar position, the greater the probability that it will be able to draw together to exercise political influence. We attempted to identify the level of group cohesion by measuring the degree to which the expressions of national preference derived in this study deviated across the group as a whole. The less those scores deviated the greater the amount of cohesion the group manifested. In order to give some meaning to those measures of cohesion we compared them to those for other United Nations regional and caucusing groups. The results of that analysis are not particularly impressive. Placing particular emphasis on the issues previously discussed reveals the following: Regarding the delimitation of the territorial sea, we find that virtually every United Nations caucusing group defines a greater amount of commonality on this issue than do the nation-states of the Pacific. This is also the case with questions of strait transit, foreign access to coastal fisheries, and freedom of scientific research. Nevertheless, if the comparison is expanded to the full range of countries participating in the Conference, the Pacific Basin states show a greater amount of cohesion on all 15 of the issues chosen for analysis. One can interpret this to mean that the nation-states of the Pacific do manifest some regional preferences that could be significant.

However, it should be recognized that the Pacific is not an environmentally homogenous body. Resources, both living and non-living, are not evenly distributed throughout the area. Thus, certain economic interests could be significantly impacted by the actions of small groups of states, or, indeed, by the activities of individual states. This can be illustrated by considering the questions surrounding the utilization of highly migratory fisheries stocks. While the countries of the Pacific Basin do favor a greater amount of coastal state control over foreign access to these stocks, the group is not particularly cohesive on the issue. On the other hand, if one limits the analysis to a subregional group within the Pacific, in this case the Latin American group, not only does the preferred position point toward even more coastal state control but the group manifests a remarkable amount of cohesion. This is an important consideration because the vast majority of the tuna stocks of the eastern tropical Pacific migrate along the western coast of Latin America. There are a number of other examples were subregional groups or individual states could independently impact California's interests in the Pacific. These will be examined in detail in the later sections of this paper. The following section will examine a number of specific issue areas in evaluating those impacts.

Fishing

The fishing industry in the state harvests a wide range of stocks from both the near shore areas of North America and from the waters of the tropical Pacific. By the late 1970's the total value of fish landings in California ports was approaching $200 million annually. Our concern here is not with the impacts of fishing on the state's economy as a whole, as the contribution of the industry is not particularly impressive in relation to other economic sectors. Rather, we are primarily interested in the effects that changes in current practices could have on localities that are more directly dependent on fishing for their economic well-being.

Table III lists the totals for both amount and value of fish landed in California during the period 1973-1977. Most species identified in the table are coastal stocks. They are primarily harvested in waters over which the United States claims jurisdiction. Therefore, the degree to which

TABLE III: California fish landings and value received by fishermen, 1973-1977.

	Landings					Value of Landings				
	1973	1974	1975	1976	1977	1973	1974	1975	1976	1977
	(1000 pounds)					($1000)				
Wetfish										
Northern anchovy[1]	265,272	165,433	318,586	257,063	231,892	6,646	3,432	4,817	5,675	4,900
Jack mackerel	20,616	25,458	29,664	38,519	110,246	992	1,496	1,256	1,925	5,512
Pacific bonito[2]	30,788	18,818	31,874	8,897	22,654	3,202	2,512	3,747	1,232	3,624
Pacific mackerel	57	133	440	367	10,246	4	10	19	18	512
Squid[3]	12,062	28,905	17,240	20,356	19,410	435	1,437	646	504	679
Sub-total	328,795	238,747	397,704	325,202	394,448	11,279	8,887	10,485	9,354	15,227
Market Fish										
Salmon	9,669	8,749	7,066	7,817	4,531	8,992	8,013	7,555	9,801	8,944
Rockfishes	25,775	21,499	21,525	26,196	22,711	2,310	2,814	4,305	3,930	5,048
Soles & Flounders	32,550	27,985	29,671	33,507	29,750	3,913	3,971	4,154	6,366	6,443
California halibut	274	306	522	578	608	160	204	391	426	516
Sablefish	8,550	12,039	14,132	12,755	20,000	661	1,202	n/a	2,288	3,000
White sea bass	809	752	1,123	1,099	964	475	456	730	644	820
Sharks	417	497	492	740	812	82	105	88	136	146
Swordfish	613	650	877	58	486	773	887	1,403	145	1,336
Sub-total	78,667	72,477	75,408	82,750	79,662	17,366	17,652	18,625	23,735	26,253
Tuna	292,336	329,788	334,250	448,185	309,220	71,138	94,654	86,967	134,415	116,439
Other fish	7,130	16,301	11,338	13,782	32,188	864	2,039	3,348	1,451	5,171
Shellfish	11,087	15,031	20,510	29,936	58,858	3,863	4,960	6,909	16,163	18,973
Grand total	718,005	672,344	839,210	895,668	874,376	104,510	128,192	126,335	185,124	182,053

[1] Not including live-bait catch.

[2] Includes substantial quantities of bonito caught by tuna boats.

[3] Includes squid caught by dip nets in southern California and by lampara vessels out of Monterey, California, as well as that caught by San Pedro wetfish vessels.

Source: Daniel Huppert and Jane McMillan, "California's Wetfish and Market Fish Fisheries - Economic Values". Mimeographed. National Oceanic and Atmospheric Administration, National Marine Fisheries Service, Southwest Fisheries Center, La Jolla, CA, 1978.

their utilization will be directly effected by decisions made by Pacific Basin countries will not be particularly significant. This may not be true, however, for tuna.

Pacific tuna stocks migrate over vast areas of the ocean moving through the coastal areas of a number of different nation-states. This stock would, indeed, be significantly impacted if key Pacific Basin states act in accordance with their stated national preferences regarding foreign access to coastal fisheries.

The figures from Table III clearly illustrate the dominant position of tuna in California's fishing industry. In 1977, tuna accounted for approximately two-thirds of the total value of landings in the state. If the utilization of tropical Pacific tuna stocks by California based vessels are significantly restricted the economic repercussions could be extreme. Most directly effected would be the purse-seiners and bait boats operating out of San Diego and San Pedro in southern California. These vessels fish primarily off the coasts of Mexico and the west coast countries of Latin America with the bait boats and smaller purse-seiners limited almost exclusively to Mexican waters. It is doubtful that these vessels could continue to operate out of California if restrictions for access to these stocks became too severe.

Threats by Latin American governments to place restrictions on United States vessels are not new. Indeed, the first sign of difficulty arose in 1952 when Chile, Ecuador, and Peru, issued a proclamation laying claim to a territorial sea extending 200-miles from their coasts. The purpose of this move was primarily to extend national control over tuna stocks migrating through those waters. Under the provisions set up by these countries, access to the stocks required that foreign vessels purchase permits before fishing. For example, fees for fishing in Ecuadorian waters had reached $77,000 by 1974.[6] The United States has refused to recognize such claims. Instead, it has set up a fund that compensates the owners for fines incurred when vessels are seized for fishing without the required permits. This conflict reached somewhat of a climax early in 1975 when seven tuna vessels were seized and detained by the government of Ecuador. Fines associated with those detentions, which lasted from 26 to 47 days, exceeded $4 million.[7]

The most important development in the evolution of this situation is the degree of international legal legitimacy now afforded the Latin American claims.

One of the major aspects of the UNCLOS debates is the degree to which the international community has accepted an exclusive economic zone (EEZ) extending 200-miles from the coast. Within this zone the coastal state is granted jurisdiction over economic activities, such as fishing. The basic differences between this type of economic zone and a territorial area is that within the EEZ the coastal state is granted a much more limited set of legal rights. Many countries have already begun to unilaterally claimed such zones. The United States did so in 1976. The major policy difference, in this regard, between the United States and the Latin Americans is the exemption by the United States from the EEZ of highly migratory species. The U.S. claims that because these stocks migrate over vast areas of the sea they should be managed regionally by a competent international organization. If management of these species is carried out by individual countries there is fear that conflicting and perhaps arbitrary regulations will arise.

These reservations are manifest to a limited degree in the draft treaty currently being discussed at the UNCLOS negotiations. The relevant section in the ICNT is Article 64,[8] which states, in part,

> The coastal state . . . shall cooperate directly or through appropriate international organizations with a view to ensuring conservation and promoting the objection of optimum utilization of such (highly migratory) species, both within and beyond the exclusive economic zone.

However, the wording here does not obligate the coastal state in any significant fashion. The guidelines set forth in this article, in essence, to grant maximum flexibility to the coastal state.

If key coastal countries take advantage of the ambiguity in Aricle 64 and define stringent restrictions for foreign access, even the the short to medium term, the impact on California's tuna fleet could be devastating. Vessel owners would have few options other than to sell their vessels to buyers in other countries or to switch the registry of the vessels to other countries where they could gain sufficient access to stocks. In either of these circumstances the economic loss to southern California ports would be substantial. Data from 1975 indicates that in San Diego alone, loss of the tuna industry would mean losses of more than $75 million in tuna catch receipts, over $120 million in projected cannery production, and an estimated $30 million in new tuna vessel construction.[9] California government would also face substantial losses in tax revenue from the projjected 4,000 to 5,000 jobs in fishing, canning, and shipbuilding that would be a lost. San Pedro could face similar losses. Although fewer vessels operate out of San Pedro the canning facilities are more extensive than those in San Diego.[10] These figures may not precisely represent the economic loss to ports, as the future of processing and shipbuilding is somewhat uncertain. It may well be that canning plants will not lose as much as has been anticipated if, for instance, fish are landed elsewhere and then transported for pro-cessing. Nevertheless, the total estimated loss to the ports of San Diego and San Pedro would clearly be in excess of hundreds of millions of dollars.

The degree to which these impacts will actually be felt is unclear, but the future does not look particularly promising. United States negotiators have been attempting to reach agreements on tuna with Mexico and other countries but the bargaining has not been fruitful. The emphasis in these negotiations has been the attempt to reach an agreement with the government of Mexico. It is felt that if such an agreement could be reached it would serve as a model for negotiations other for other Latin American countries. Those talks have not been progressing well and in July of this year the Mexican government actually seized a number of United States vessels fishing in waters over which the Mexicans claim jurisdiction.

This section has thus far addressed the question of the potential impact on California of the national decisions made by foreign governments. It would seem appropriate to offer a similar evaluation of the decisions made by the United States government. The interests of the state of California will not always directly coincide with those of the federal government. Actions taken by the United States are the result of a long, often arduous process of political compromise. The concerns of a wide variety of special-interest and regional groups influence the final outcome of virtually any debate. The interests of the state of California are often considered only in terms of their contribution to the broader set of national concerns. Furthermore, the structure of the fishing industry often complicates this process. Interests and concerns vary with the stocks fished. Representatives of distant-water, coastal and anadromous stocks quite often disagree over the direction they feel should be taken in the definition of fisheries policy. The considerations of these differences is quite important because decisions designed to increase productivity in one sector of the industry may well adversely affect another sector.

The choice of the United States government in legislating the Fisheries Conservation and Management Act (FCMA) in 1976 was to emphasize the protection of coastal fisheries. The overwhelming contribution of distant-water stocks to California" fishing industry might suggest that that choice may not have been an optimal one for the state. However, the losses that may be felt by the tuna industry may be at least partially mitigated by increases in productivity from near shore areas. The general impacts of the FCMA on California's coastal fisheries, indeed, appear to be positive. It is difficult, however, to carry out too detailed an analysis in this regard because of the lack of recent data. The last year for which this data was publicly available is 1977, barely a year after the effective date of the Act. In that data, which is included in Table III, there is at least a suggestion of positive impact. This is particularly true of jack and Pacific mackerel, squid and to a lesser degree of rockfish, sablefish and white sea bass. It should be stressed that while this data does indicate an upward movement in some areas it is certainly insufficient to identify a trend.

There are data, collected at the national level that do suggest that if data were available for California it would show continued increases in productivity for coastal stocks. The National Oceanic and Atmospheric Administration (NOAA) has released figures indicating that the number of foreign vessels fishing off the United States coast has dropped significantly. Prior to 1977, as many as 2,500 vessels fished within 200-miles of the coast. Since the FCMA went into effect, fewer than 950 foreign vessel permits have been issued each year. This has resulted in a 30% drop in the catch taken by foreign vessels in United States waters since 1977.[11] While this drop in foreign catch does not necessarily mean there will be an immediate attendant increase in the United States catch it does increase the potential for United States fishing interests within United States coastal waters. Also because fish are not being taken by foreign interests, assuming constant demand in those markets, the potential export market for U. S. fish products increases measurably.

The United States Department of Commerce recently produced a report that attempted to outline the fisheries stocks presently underfished and therefore candidates for increased commercial untilization.[12] According to that report there are a number of such stocks presently available to California's fishing interests. The major species named included Pacific hake, squid, Pacific mackerel and jack mackerel.

Both Pacific mackerel and jack mackerel are currently being harvested by members of the San Pedro fleet. However, vessels are currently being held to quotes for these species by canners who restrict production because of their primary involvement in canning tuna and because of the lack of strong market demand for mackerel. A change in market considerations or in the availability of tuna could lead to a significant increase in the amount of mackerel harvested.

Squid are also harvested by the San Pedro fleet as well as by lampara seines in the Monterey Bay area. The world market for frozen squid is large and the potential for increased productivity from California is substantial.

But perhaps the largest increases are possible through intensifying the effort for Pacific hake. There are currently limited attempts at utilizing these stocks along the northern California and Oregon coasts, however, the stock is remarkably underutilized. More complete utilization would require a rather major investment in nearby processing facilities as hake spoils rapidly. If these investments were made productivity could increase sharply as worldwide demand for the stock is quite large.

The results of this study, if taken overall, suggest the California's fishing industry will suffer a net negative impact over the next several years. It faces the potential, through the loss of the tuna industry, of an 80% drop in the value of fish landings in the state. It is probable that those losses will be mitigated to at least some degree by an intensification of effort in other areas. However, even while the impacts of those new efforts are unknown it is doubtful that they will match the losses incurred through the loss of the state's distant-water tuna fleet.

Trade and Transportation

The importance of international trade to the economy of the state of California cannot be overstated. Trade figures for the state are currently approaching $40 billion annually. The overwhelming majority of that trade is transported by ship. Therefore, if the nation-states of the Pacific Basin could substantially increase transportation costs by altering current maritime practices the impacts on California's economy could be very significant.

The area in which there is the greatest potential for impact is strait transit. A substantial number of trade routes in the Pacific pass through a surprisingly small number of straits. At issue here is the ability of certain countries to halt or otherwise restrict the free, unimpeded transit of traffic through key straits. The straits that are of most concern to California are located in the southwest Pacific and are listed in Table IV. The majority of California's Asian and Middle Eastern trade traffic pass through these straits.

The concern here is evident if one examines the location of the straits and the countries bordering them. With the general international acceptance of the 12-mile territorial sea, control of these areas falls essentially on three countries; Indonesia, Malaysia and the Philippines. Our forecasts for these countries on the issue of strait transit indicate a significantly more restrictive environment for transit than either now exists or the United States would prefer. The concerns of these states center primarily around the setting of standards for the prevention of pollution from vessels, particularly tankers. These countries contend that they should or do have the right to regulate shipping passing through their waters due to the delicate environment of these narrow straits. Current international law grants the setting of these standards almost entirely to the country in which the vessel is registered.

The degree to which transit will be restricted in these areas is uncertain. For analytic purposes we will assume the most extreme, and, admittedly, the most unlikely option open to these countries; complete closure of the straits. We chose this option because even under these circumstances the impact on California's trade figures would not be significant. A 1972 report by Robert R. Nathan Associates attempted to estimate the total cost to the United States of the closure of three important straits. The straits chosen for analysis were the English Channel, Gibraltar, and Malacca. The results of that study indicated that the total cost to the United States from those closures would be slightly in excess of $50 million.[13] While some have critized a number of the assumptions built into that model the limited number of other reports attempting to produce similar evaluations have also produced similar results. Considering that California's share of those losses would obviously be proportionately lower, the economic impact of the closures would not appear to be great.

One analyst has offered reasons why these conclusions are, for the most part reasonable.[14]

First, closing off a certain number of straits does not mean closing off the entire ocean. Alternative routes can always be found even if it means in the extreme case, circumnavigation. Second, the per-mile operating costs of shipping are relatively low as compared to other costs. Therefore, the added costs of alternative routes should not be extreme. Third, since the shipping industry is, in most parts of the world, a heavily subsidized enterprise one would assume that these new costs would be subsumed under such programs.

It is, therefore, fairly certain that the direct impacts on trade and marine transportation in California will be minimal.

Table IV: Major Pacific Straits Carrying California Import/Export Trade

Strait	Countries Bordering
Lombok	Indonesia. Access Indian Ocean to Java Sea
Makassar	Indonesia. Connects Celebes Sea and Java Sea
Malacca	Malaysia, Indonesia, Thialand. Connects the Indian Ocean and South China Sea.
Mindoro	Philippines.
Sunda	Indonesia. Connects Indian Ocean and Java Sea.
Surigao Strait	Philippines.

Non-living Resources

The nations of the world have only recently begun to recognize the economic potential for nonfuel minerals of the sea. The range of these resources is quite wide and their potential value impressively large. Once regarded merely scientific curiosities, manganese nodules are acknowledged as being one of the world's most significant mineral deposits. Phosphorite nodules, concentrated bottom sediments, offshore placer deposits, and other marine mineral deposits are becoming more widely recognized as important future reserves. California is in a unique position to gain positive economic benefit from the development of a number of these resources.

Manganese nodules offer the state a particularly interesting set of opportunities. A number of the companies most actively involved in the development of the technology for mining the nodules from the deep seabed are headquartered in California. Although the specific economic parameters can only be guessed at, when, and as some would agree if, the commercial development of these nodules begins the economic benefits to these companies will be significant. Estimates of world nodules reserves vary. There may be as much as one trillion tons of nodules in the Pacific along. However, various market considerations, particularly those centering around the nickel, copper and cobalt contained in the nodules will determine which of those reserves will be economic and which will not. Suffice to say that there appear to be sufficient economic reserves for the industry to prosper.

Guaranteed access to the reserves is the central issue to this young industry. That access is dependent upon the decisions made on three different sets of decision-makers. The primary body currently considering these issues is, of course, UNCLOS. Indeed, questions surrounding the development of these resources served as the major catalyst in initially convening the conference. Over the last dozen years the Conference has considered an almost endless set of alternative development schemes, and appears finally to be near consensus. If the current draft treaty emerges without major revisions manganese nodule development will be carried out jointly between an international authority and individual companies. Although there are still a number of specific provisions to be worked out, and, again there are clearly no guarantees. The negotiations appear to be closer to resolution than at any time in the past decade. There are those who suggest that, indeed, there will be a treaty by the end of this year.

The second major influence in the development of this resource is the government of the United States. With the uncertainty that has surrounded UNCLOS the industry has requested that the federal government define a system by which their investments could be protected. By 1977 the industry had already invested over $150 million in private funds on research and development.[15] The technological advances were impressive but future development it was argued, could not be carried out in the uncertain legal environment created by the apparently stalled UNCLOS negotiations. The United States finally acted, in June of 1980, to define a domestic regulatory system for nodule development. The law which has an effective date of January 1, 1988, would be superseded by an UNCLOS treaty, but, importantly, guarantees that investments made by U. S. corporations will be secure.

Beyond the influences exerted by the UNCLOS debates or the federal government there is a third set of decision-makers who could register an impact on manganese nodule development. The decisions made by the national governments of Pacific Basin countries could take on increasingly important dimensions in the future. It should be recognized that while the majority of the manganese nodules of the Pacific are located in the open ocean a significant proportion of them do lie on the continental shelves of a number of coastal states. Therefore, their development would be regulated, not by an international authority, but by the coastal state. This distinction is quite an important one. First, if trouble does develop at UNCLOS in these late stages or if the treaty is not ratified there are nodules that could be mined within areas of national jurisdiction. Second, even if the treaty does go into effect mining companies may well be able to conclude more profitable arrangements with certain national governments than with the international authority. However, our data is insufficient to be able to reliably forecast which countries of the Pacific Basin are most likely to agree to more positive arrangements. Perhaps it is sufficient to say that the investment environment for manganese nodules created by these developments is certainly more positive than has existed in recent years.

In addition to the economic benefits California will gain from large-scale mining operations one must also consider the possible impacts if nodule processing plants were to be located in the state. It has been estimated that fully two-thirds of the total value of manganese nodule mining activity would be value added by onshore processing. The Deep Seabed Mining Act recently passed into law stipulates that the processing of manganese nodules must, at least in part, be carried out in the United States. Hawaii appears to be a likely candidate, but, it is clear that California must also be considered. The transportation and construction infrastructure are well developed in the state, and port facilities are among the most advanced in the world. However, California also has one of the most advanced state environmental regulatory systems in the country. Although processing systems for manganese nodules vary widely in design, and there is little to suggest which direction the industry will move, there are a number of similarities that suggest that locating a processing plant in California would be difficult. First, specialized port facilities must be developed to unload the nodules. Second, transportation systems to remove the nodules from the port and deliver them to the processing plants must be designed. Third, the actual processing plants must be constructed. Fourth, some system must be developed to deal with whatever wastes are left over when the mineral extraction process is complete. From this it is fairly easy to recognize that the range of regulatory agencies, both state and federal, that would have jurisdiction of this activity would be remarkably broad. However, if requirements can be met and the necessary permits acquired the benefit to the economy of the state could be substantial.

Conclusions

It is clear that whatever formally emerges from the UNCLOS debates, the future law of the sea will be dramatically different then it was a decade ago. Freedom of the high seas has given way to a system of national enclosure. Marine resources are now, or will shortly be, predominantly under the jurisdiction of individual nation-states. These changes in legal structures have and will continue to have impact on the marine-related interests of the State of California.

The discussion of this study has outlined the following potential impacts. The fishing industry will most likely suffer a net economic loss. It is questionable as to whether or not the loss of the southern California tuna fleet can be offset by increased productivity of coastal stocks. Marine transportation and trade will most likely not suffer significantly. The impacts on non-fuel marine mineral development remain somewhat inclear. Major decisions in this area have only recently been made and, at this time, it is uncertain how the industry will respond.

FOOTNOTES

1. State of California, Economic Report of the Governor, 1979 (Sacramento: California Office of State Pointing, 1979), p. 31.

2. Daniel Huppert and Jane McMillan, "California's Wetfish and Market Fish Fisheries Economic Values." Mimeographed, National Oceanic and Atmospheric Administration, National Marine Fisheries Service, Southwest Fisheries Center, La Jolla, CA, 1978 (unnumbered).

3. For an example of previous work utilizing this model, see: Robert L. Friedheim, Joseph B. Kadane with the assistance of J. K. Gamble, Jr., "Quantitative Content Analysis of United Nations Seabed Debates: Methodology and a Continental Shelf Case Study," International Organization, XXIV:3 (Summer 1970), pp. 479 502.

4. U.N. document A/conf. 62/WP.10/Rev.2, 11 April 1980.

5. A complete description of the methodology can be obtained by contacting the author's at: Institute for Marine and Coastal Studies, University of Southern California, University Park, Los Angeles, CA 90007.

6. Ross D. Eckert, The Enclosure of Ocean Resources (Sanford: Hoover Institution Press, 1979). pg. 133.

7. August Felendo, "Problems Confronting Distant-Water Fisheries in a New Law of the Sea Environment" in Don Walsh, Law of the Sea: Issues in Ocean Resource Management (New York: Praeger Publishers, 1977). pg. 141.

8. U. N. document A/Conf. 62/WP.10/Rev. 2/Part V/Exclusive Economic Zone, Art. 64.

9. F. David Froman, "The 200-Mile Exclusive Economic Zone: Death Knoll for the American Tuna Industry," San Diego Law Review 13 (1976), pg. 732.

10. Ibid., p. 732.

11. Gerald D. Hill, Jr., "Fishermen Enter 4th year of Conservation," NOAA, vol. 10, no. 3 (May/June, 1980), PP. 14 - 17.

12. U. S., Department of Commerce, National Oceanic and Atmospheric Administration, National Marine Fisheries Service, A Study to Determine the Export and Domestic Markets for Currently Underutilized Fish and Shellfish (Washington, D.C.: Government Printing Office, 1978).

13. Eckert, Enclosure, p. 72.

14. Ibid., pp. 71 - 72.

15. U. S., Senate, Committee on Energy and Natural Resources and Committee on Commerce Science, and Transportation, Mining of the Deep Seabed, 95th Cong., First session, 1977, pg. 70.

Shuttle Diplomacy and the Other Side of Coastal Management

by
Louis C. Burney*

The National Coastal Zone Management Act enbodies a basic dichotomy. On the one hand, it contains the widely recognized belief that states should exercise their full authority to regulate activities harmful to the environment. On the other hand, the Act also espouses the belief that states should foster affirmative problem solving endeavors as well as take advantage of opportunities for compatible development and greater beneficial use of coastal resources.

A state coastal management program cannot be judged successful until it moves beyond the reactionary processes inherent in purely regulatory programs, to the ideals contained in the second part of the coastal management formula. If such movement is to occur, different, less quantifiable criteria for judging success will also have to be applied. For instance, does a coastal management program facilitate mutually supportive actions between the planning and regulatory programs? Does it stress the avoidance and resolution of conflicts? Does it improve the efficiency of governmental decision making? Does it provide timely and consistent guidance to the private sector? Does it result in increased beneficial use of coastal resources? Does the private sector view it as part of the problem or part of the solution?

In addressing these questions, program managers must fully understand the needed relationship between the two sides of coastal management and be prepared to openly discuss and promote direct supporting linkages between the key regulatory and non-regulatory programs. This includes providing mechanisms for the non-regulatory components to influence the regulatory programs in meaningful ways. To successfully make this connection, state programs must recognize and come to grips with the subtle but ever-present interagency rivalries, tunnel vision, public apathy, ignorance and suspicions that seem to pervade the governmental and private sectors. They should also recognize that there usually exists a latent sense of public responsibility, patriotism, or even self interest which, if properly stimulated and nourished, can result in "success stories" far beyond the horizons of traditional regulatory endeavors.

In this regard, I would like to share with you a recent experience in Florida which illustrates how, through affirmative actions and diligent use of "shuttle diplomacy", a complex and precedent setting resource allocation issue was transformed from a bitter confrontation between competing interests to a positive working relationship which attained objectives much broader than originally contemplated by the regulatory agencies initially faced with the problem.

---

* Louis C. Burney, Environmental Specialist, Bureau of Water Management
Florida Department of Environmental Regulation

The Setting

The example involves the City of St. Augustine and St. Johns County in northeast Florida, a semi-rural, politically conservative area where personal independence and distrust of state agency motives are noticable traits. The very productive adjacent estuarine waters contain significant oyster and clam beds which support a modest but growing shellfish industry. Faced with the new experience of rapid urbanization, city and county governments were trying to cope with water, sewer and transportation problems, were confused by multiple state and federal programs affecting the area, and were not committed to long term protection of water quality, shellfish and other natural resources. Similar to many areas, these local units of government simply were not prepared to manage rapid growth.

Several state planning and regulatory programs were actively involved in the area, with the primary regulatory programs falling under the Department of Environmental Regulation (DER) and the Department of Natural Resources (DNR). As part of their responsibilities, these two agencies have shellfish management related programs which, theoretically at least, are mutually supportive. In this regard, DER classifies those waters whose most beneficial present and future use is determined to be shellfishing, and, through several regulatory programs, strives to insure water quality at a level which will permit shellfishing in areas so classified. Essentially, this protective function represents a type of resource allocation process. DNR, on the other hand, classifies those shellfish waters which are actually safe to harvest and regulates the actual harvesting, processing, and marketing of shellfish. Coordination between the two programs, while improving of late, has traditionally been inadequate, and has contributed to local confusion over responsibilities of the two agencies.

The Issue

Within this setting, several competing interests were bitterly complaining of state over-regulation of their activities, but from opposite points of view and the stage was set for a classic debate over allocation and use of coastal resources. An important factor in this debate was a precautionary public health measure: DNR had prohibited shellfish harvesting in certain large areas on the basis that scientific data was not available to confirm acceptable water quality. As might be expected, the local shellfish industry contended that such prohibition by DNR was unwarrented and resulted in under-utilization of shellfish resources. Additionally, the industry was demanding increased state and local regulation of new development, which was viewed as a critical threat to the industry. Development interests, on the other hand, contended that those areas designated by DER as Class II shellfish waters were too extensive, particularly in the urban area. Since shellfish harvesting had never been permitted in this area, application of the existing strict water quality regulatory standards to development activities in the urban area was considered unjustified and economically damaging. The issue, then, centered upon determining the most beneficial present and future use designation for waters adjacent to the City of St. Augustine.

The triggering mechanism for setting off formal state level debate on the issue was provided by local government. In the course of conducting a comprehensive sewage treatment facilities study (under the EPA 201 program), the City of St. Augustine determined that the cost of complying with existing shellfish water quality standards in the urban area would be prohibitively expensive. Subsequently, the City petitioned the Florida Environmental Regulation Commission (which sets policy for DER) to reclassify the urban waters as Class III recreational and fishing waters. Bacteria standards for Class III waters are lower than for Class II shellfish waters and are less costly to attain. In support of their request, the City contended that (1) the existing shellfish classification was an administrative mistake and should have originally been Class III, (2) that existing standards were technically unattainable due to irretrievable man-induced conditions, and (3) that if existing standards were applied, unreasonable economic burdens would result to the City. Rather convincing evidence was provided to support the City's contentions.

The shellfish industry, naturally, felt further threatened by the City's action and strongly argued against the reclassification. They were joined by state and local conservation interests and formally organized to fight the petition.

Initial State Response

Subsequent to traditional public workshops and a bitter public hearing on the matter, the Florida Environmental Regulation Commission reluctantly agreed to the City's request "in concept", with final approval to be contingent upon a number of conditions having been met. Inspired by consciencious regulatory staff members, who feared the precedent-setting repercussions of granting a "downgrading", these conditions required completion of additional state level studies, special water quality protective ordinances by the County as well as other actions that the City felt were beyond its control.

Frustrated by the two-year effort and faced with further indefinite delays in getting the matter resolved, the City viewed the state conditions as unnecessary bureaucratic red tape and felt compelled to suspend work on its sewage treatment facilities plan until the reclassification question was decided. Considering the rapid population growth in the area, the proliferation of package sewage treatment plants, the continued installation of septic tanks under less than optimum conditions, as well as recent closure of several traditional shellfishing areas due to unexplained high levels of bacteria, such a delay had potentially dire consequences for all interests involved.

From a broader coastal management perspective, the situation could hardly have been worse: credibility of the state regulatory interests was in serious question; the state regulatory interests had become frustrated in their attempt to force local regulatory actions; local planning efforts were thrust into a state of confusion; millions of dollars of EPA sewage treatment grant assistance were in jeopardy; the shellfish industry felt threatened and alienated; and all parties were preparing to resolve the conflict through legal action. In short, the limited response by the state regulatory agencies had intensified rather than resolved the conflict, and helped precipitate actions counter productive to all of the competing interests.

The Positive Approach

Realizing that the worsening situation had very serious ramifications, not only for the St. Augustine area, but for the entire state water classification system, the State began to soften its position and initiated several actions designed to aid in resolving the conflict. As part of this initiative, an experienced coordinator from one of the non-regulatory DER programs was assigned as sort of a roving ambassador to deal directly with the various interests. While there later evolved a much broader set of tasks, his initial objectives were simply to:

* Obtain necessary data to address the technical aspects of the reclassification issue, and

* Do whatever was possible within existing departmental capabilities to facilitate local cooperation and participation in state efforts to protect water quality in the area if reclassification was granted.

Limited available funding for data gathering and the existing hostile atmosphere made prospects for achieving even these limited objectives appear bleak.

In view of the volatile nature of the matter, the project coordinator chose to abandon, at least initially, the common approach of forming committees and holding workshops in favor of a more personalized "shuttle diplomacy" between the various interest groups. This action resulted in multiple informal person to person discussions with shellfish industry representatives, city and county officials, state agency staff members, legislative staff members as well as affected business interests. The basic thrust of the meetings was to identify perceived issues, problems and needs, and to solicit positive suggestions for achieving the desired objectives.

These meetings, although time consuming, tedious, and sometimes even intimidating to the state coordinator, provided a different atmosphere for discussion of the issue. It became evident in these meetings that, even though the competing interests had difficulty in communicating with one another, there were many common denominators from which to establish a dialogue. In addition, it was found that many of the perceived problems and objections were the result of misinformation, misinterpretation of state and local actions, as well as ignorance of matters assumed by the state to be common knowledge. The personalized approach proved invaluable for establishing the sincerity and credibility of the state's efforts to resolve the issue in an equitable manner.

The dialogue thus created at both the state and local levels also greatly expanded the perspectives of the competing interests and resulted in a number of tangible products. Among these were the following:

* DNR under the state shellfish management program, conducted a sanitary survey of the waters in question, and subsquently opened several new shellfish areas for harvesting;

* DER, utilizing funds of the EPA 208 program, conducted a supplemental bacteriological survey to identify existing pollution sources in the area;

* Utilizing the resources of the solid waste program, DER provided legal and technical assistance to the County regarding solid waste management;

* The County, through the state-supported public health program, increased local surveillance and enforcement;

* DER selected the area as a study area for development of the "Septic Tank" and "Urban Stormwater Management" elements of the State 208 Water Quality Management Plan; and

* DER obtained $24,000 from O.C.Z.M. for a demonstration project to explore ways that local governments can systematically take shellfish management considerations into account in their decision making.

While each of these items represented positive initiatives by the State, the demonstration project was especially important in that it defined active roles for each of the competing interests involved in the initial conflict. In addition, it sought to establish a state and local partnership for managing water quality and shellfish resources, as well as to establish linkages between the regulatory and non-regulatory programs affecting shellfish resources in the area. Specific tasks of the O.C.Z.M. demonstration project included:

* Identification of significant shellfish resource areas;

* Assessment of present and future impacts of urbanization on these resources;

* Identification of geographic areas of particular concern;

* Assessment of existing city and county management capabilities;

* Development of recommendations for specific corrective and preventive actions by the City and County; and

* Where possible, incorporation of the study results into city and county comprehensive plans and ultimately into zoning, subdivision regulations and building codes.

A particular value of the demonstration project was that it provided a needed stimulus and mechanism for the competing interests to become actively involved in the State's efforts, if for no other reason than to protect their own self interests. As a result of this involvement, the State was able to thoroughly reexamine, refine and improve the workability of its water classifications in the area. When the Florida Environmental Regulation Commission took final action on the City's reclassification request, it included a combination of selective reclassifications which actually provided more extensive protection to shellfish resources while at the same time satisfying the concerns of the City. This action was enthusiastically supported by all of the competing interests.

Subsequent to the reclassification issue being resolved, the City and County have made substantial progress in their planning for expanded sewage treatment facilities; several significant pollution sources have been eliminated; the shellfish industry has taken an active role in providing field surveillance regarding new pollution sources; the State is moving forward with its 208 efforts; the County has initiated special development review procedures and is actively developing ordinances and other measures to protect shellfish resources; and there is a much greater degree of coordination between federal, state, and local programs. In all probability, none of these accomplishments would have occurred in the regulatory arena.

The momentum from the St. Augustine initiatives has spilled over into other areas, with similar state and local shellfish protection efforts now being undertaken in two other Florida Counties. In addition, DER, DNR, the Department of Health and Rehabilitative Services, as well as several federal agencies are now working through a formal shellfish sanitation task force to maximize the mutually supportive aspects of their shellfish related programs.

The St. Augustine experience, I think, clearly illustrates the value of shuttle diplomacy to coastal management in that it facilitated establishment of a dialogue between regulatory and non-regulatory programs at both the state and local levels and resolved the basic resource allocation issue in a manner consistent with sound resource management practices and equitable to each of the competing interests.

Conclusions

The most important values of the St. Augustine experience transcend the matter of specific conflict resolution. Linkages have now been established between several of the regulatory and non-regulatory programs affecting the area, thus increasing the efficiency and effectiveness of the state regulatory effort. Increased beneficial use of coastal resources has occurred. Local government is now in a much better posture to effectively address its sewage treatment and other support services problems. And spinoff activities resulting from the experience promise to yield substantial benefits in other areas.

Perhaps the St. Augustine situation reminds you of similar experiences. Allocation issues strike at the heart of coastal management and are quite common. The experience does contain several lessons, not only for state programs, but for federal programs as well. In view of the nation's competing needs for energy, economic development, resource protection, etc., the players in the regulatory arena and the tactics employed are increasingly being criticized by legislators, congressmen, and special interests. For coastal managers, the answer to this criticism should not, in my opinion, be a hardening of positions by the regulatory programs to "cut the losses" from future attacks. Nor should it be a wholesale slackening of standards. Instead, increased effort should be made to demonstrate that the regulatory programs can be adapted to accommodate the other side of coastal management without sacrificing long term resource protection objectives.

Coastal management, by necessity, must be closely allied to regulatory programs. However, the influence, or for that matter, the very survival of state coastal management programs may well depend upon their ability to focus attention on their problem solving abilities and avoid being "simply one more regulatory program applied to the coastal zone". Shuttle diplomacy may well provide the key ingredient in accomplishing this task.

In closing, I must emphasize that success of state efforts to balance the coastal management formula will also depend in large part upon positive actions at the federal level. The Office of Coastal Zone Management must strongly encourage and support state problem solving initiatives with both funding and active participation. It is absolutely necessary that O.C.Z.M. take stronger initiatives with federal agencies to help assure that less comprehensive federal programs such as those arising under the National Clean Water Act do not unnecessarily obstruct positive state accomplishments. In view of recent events in the St. Augustine experience, federal and state cooperation takes on special importance. Much to the consternation of all of the state and local participants in the matter, EPA has indicated that it will not approve the state's reclassification of the urban waters at St. Augustine. Citing certain authorities under the National Clean Water Act, and in apparent disregard of the facts and circumstances leading to the state's action, this federal agency has, in effect, substituted its more narrow judgement for that of the state, and jeopardized the multiple benefits attained by the state. As of this writing, the state is attempting to convince EPA to reverse its action and is prepared to contest the matter in court if necessary. It appears that the City and County will join the state in such a suit if necessary.

It would indeed be unfortunate and ironic if Florida and the local governments involved were forced into court, not against each other, but simply to prevent unnecessary federal disruption of their cooperative ventures into the other side of coastal management.

THE COASTAL LAND DEVELOPMENT PROCESS: DEVELOPERS' PLIGHTS

Valinda Parker[1]

and

Kent S. Butler[2]

ABSTRACT

The private land development process takes place in a dynamic and uncertain world. Factors of economic risk can significantly affect development in coastal areas and at large: impending adjacent developments, availability of suitable land and reasonable financing terms, reliability of continued market demand, and other factors. Land and water regulatory procedures operant in coastal areas further complicate the development process in the form of greater uncertainties, increased costs, and particularly increased time in the design-to-construction period. A study was conducted in 1979 to identify and analyze the problems and opportunities facing developers as they plan, design, and build out recreational-residential coastal developments.

We have formulated a generalized model of the private development process applicable to large scale recreational-residential developments in coastal areas. The model contains seven interactive stages and indicates the points at which permitting procedures enter the development process. Major risk stages and methods used by developers to minimize these risks are examined. Delays and costs incurred by developers in their attempts to obtain project approval are analyzed. We recommend a number of modifications to the permitting systems used on the Texas coast and elsewhere to effect a more timely and expedient, yet environmentally protective review process.

---

[1] Land Planner/Economist, Public Service Company of New Mexico, Albuquerque, NM 87103.

[2] Assistant Professor, School of Architecture and Planning, University of Texas at Austin 78712.

INTRODUCTION

The consumer demand for second home and recreational development along the Nation's shorelines has increased markedly during the past two decades. Developers and investors in Texas and other coastal states have attempted to fulfill this demand through construction of high-rise hotels, condominiums, residential canal developments, and the like. As scenic coastal vistas and access sites became developed, however, and as wetland acreages were dredged and filled to provide additional building sites, public concern arose regarding the depletion and degradation of coastal resources.

As a result of this public concern, the Federal government initiated a series of legislative and administrative actions in the late 1960s and early 1970s to preserve and protect these resources.[1] Consistent with these Federal initiatives, states have enacted a multitude of laws, some comprehensive and others single-purpose in nature, to manage and protect wetlands, estuaries, dunes and beaches, and other critical natural resources. Local governmental regulations have also been expanded to consider various on-site impacts and protect public amenities such as visual and pedestrian access to beaches.

Collectively, this "regulatory maze" of environmental protection has made coastal land development a more risky venture. Developers actively investing in the coastal areas over the last decade have indicated that the complexity of the development process has increased substantially as a result of increased and changing regulations. While aiding in the maintenance of environmental quality, the expanding regulatory process has added as much as two to three years to the project planning period for many large developments. Because the degree of risk involved in real estate investment rises sharply as the time span of the design-to-construction period increases, delays of this nature can pose serious threats to the continued economic viability of a project.

Equally significant is the volatile economic climate and associated factors of risk. Unexpected changes in the money market, a sharp rise in the rate of inflation, changing consumer preferences in the housing market, and escalating costs for construction materials can render the original project design infeasible or obsolete. Because of these risks, developers and investors have devised a complex system of checks and double-checks on critical financial decisions to try to minimize their potential losses.

We have made surveys of land developers along the Texas coast and have characterized the major activity stages in the land development process.* Several stages were identified as points where economic risk and environmental regulations have tended to impinge upon or jeopardize the profitability of housing development projects. In the following section, we outline the major steps in a model coastal land development process, emphasizing the stages in which risk and uncertainty are prominent. We then evaluate the manner in which the development

---

*Research support for this project, entitled Implementation of Workable Land Use Guidance Techniques for the Promotion of Environmentally Sound Development on the Texas Coast, was provided by the Bureau of Business Research, University of Texas at Austin.

process is affected by environmental regulatory procedures, and seek workable solutions to the delays and added costs attributable to regulation.

MODEL DEVELOPMENT PROCESS

Figure 1 indicates major steps of the land development process and delineates a likely sequence.[2] The first three stages of the process are devoted to planning, design, and testing the economic feasibility of the project. If, at any time during this period, the project appears infeasible, the design can be substantially altered or the entire project postponed indefinitely.

Idea Inception Stage

During this initial phase of project planning, the developer may use market analysis techniques to establish the strength of market demand and the size, location, and type of development that will satisfy consumer preferences.

A variety of market study techniques are used by developers as preliminary indicators in deciding whether to go with a project. Formal market analysis approaches are often employed by firms undertaking a new type of project or entering an unfamiliar geographic area. The formal approach usually involves an analysis of general market conditions, a determination of the demand for certain types of housing or other developments, and the generation of a buyer profile using conventional survey techniques.

Coastal developers in Texas tend to take a more informal approach, beginning with the assumption that market demand for vacation housing is strong. Because their operations are often located in or adjacent to small cities and towns, they tend to be familiar with the absorption schedules of similar projects, their price range, and other specific features. They can thereby rely to a large extent on past experience, intuition, and the success of their predevelopment sales in gaging market preferences. The completed analysis, whether performed through formal or informal means, can minimize risk by preventing the developer from designing a product inappropriate for the market.

The market study must be correlated with an investment analysis. A rough *pro forma*, delineating maximum project cost and minimum necessary income, is usually drafted as an initial test of project feasibility.

If the project appears profitable, a development "team", composed of lawyers, environmental consultants, engineers, and architects, may be assembled. By coordinating activities and performing a number of functions simultaneously, the team has the potential to reduce the design-to-construction time period and avoid some unnecessary permitting delays, thereby reducing carrying costs and generating early cash flow receipts.

Idea Refinement Stage

During this second stage of the process, the developer locates a suitable site and acquires or secures the land through some form of

Figure 1

MODEL OF THE PRIVATE LAND DEVELOPMENT PROCESS

	TIMING SEQUENCE (0 →)
IDEA INCEPTION STAGE	
a. Market analysis	a
b. Assess market share	b
c. Initial *pro forma*	c
d. Set up development team	d
IDEA INCEPTION STAGE	
e. Site location & selection	e
f. Gather & analyze pertinent legal & environmental data	f
g. Land acquisition	g
h. Initial discussions with lenders & equity interests	h
PRECOMMITMENT STAGE	
i. Feasibility analysis	i
j. Preliminary plans & design	j
k. Initial construction cost estimates	k
l. Prodevelopment sales	l
m. Begin permit review process	m
n. Discussion on permanent & construction loans; submission of financial package	n
COMMITMENT STAGE	
o. Finalize plans & design	o
p. Complete permitting process	p
q. Arrange financing	q
r. Choose contractor & type of construction management	r
INTITIATION OF CONSTRUCTION STAGE	
s. Receive permanent loan commitment	s
t. Close construction loan	t
u. Sign construction contract	u
v. Close & purchase/pay off land	v
x. Switch to formal accounting system	x
CONSTRUCTION STAGE	
y. Install improvements	y
z. Construct buildings	z
a_1. Sell/lease space	a_1
b_1. Landscaping	b_1
c_1. Overruns funded	c_1
FINAL STAGE	
d_1. Select outside management firm	d_1
e_1. Connect utilities	e_1

purchase option or lease. As ideal sites are rarely found, the final decision in choosing a parcel involves a number of tradeoffs. The site must be physically amenable to development and the original owner must be willing to relinquish all or part of his controlling interest in the property. There is, of course, no guarantee at this stage that the project will meet the necessary permit requirements. The presence of wetlands, dunes, critical wildlife habitats, or other natural features can raise legal and administrative concerns among regulatory agencies and private interest groups, drastically affecting the site's development suitability.

In site acquisition, as in every other major activity stage, the developer seeks to minimize his carrying costs. He may, therefore, attempt to gain control over the site through a purchase option rather than fee simple ownership. In large, multi-phase projects, expansive tracts may be purchased outright and inventoried for later development. Carrying costs on such projects can be disproportionately large; the proceeds from initial lot sales and early phases of development must usually be reinvested to meet the continuing costs. Becasuse of the tight cash flow and the level of risk in making large early commitments, unforseen regulatory changes or delays may have significant impacts on the completion schedule and total project cost.

While negotiating site acquisition, the developer begins his initial discussions with lenders and equity interests, sounding out possible investors and determining the level of interest rates to be expected. At this point, the developer is ready to begin his feasibility analysis.

Precommitment Stage

The planning and design phases of the development process are concluded in the precommitment stage. The feasibility analysis is also conducted at this time for two reasons: it provides an assessment of the final value (profitability) of the undertaking, and it helps in ascertaining whether early project assumptions support a developer's intuition about the project.

The feasibility study consists of four steps. First, total project cost is estimated. The sales or rental revenue is then projected. The schedule of payments and revenue is then converted to a year-by-year, pre-tax cash flow analysis. Finally, the yearly after-tax cash flow is established. The resultant rate of return is the salient figure in the analysis because it will attract or dispel lender and equity interests in the project. Estimations of time and front-end costs (engineering and legal fees) to be expected in permit application and processing can have pivotal significance in the feasibility study. Since the economic and physical phases of project design must be closely coordinated, the feasibility analysis, preliminary plans, and construction cost estimates usually are drafted, modified, and redrafted simultaneously.

Upon completion of these activities, the developer may initiate predevelopment sales. (Some developers prefer to postpone this activity until after completion of the permitting process as the preliminary design may be changed to such a degree that the unit price must go up.) The relative success or failure of the predevelopment sales is a useful indicator of market demand and product-consumer match, providing the developer with feedback that will enable him to revise his plans before making any further commitments.

Permit applications are usually submitted to the appropriate government agencies upon completion of the preliminary design, oftentimes during the design stage itself. The permitting review process can be lengthy--delays on the order of months are common when difficulties arise in obtaining approval from all the agencies involved. During this time, a number of plan revisions that mitigate environmental impacts may be proposed and evaluated. The developer can negotiate with contractors and loan officers, as well as with the staffs in dissenting agencies, but he cannot finalize any agreements until all permits are approved.

Commitment Stage

All arrangements that were previously negotiated are finalized during this stage. Permit applications are granted (or denied) and final site plans are completed. The construction contract is drawn up and financing arranged.

One of the most critical problems faced by the developer of vacation properties is the difficulty that may be encountered in obtaining project financing. Two forms of debt financing, the construction loan and permanent loan, are typically involved.

Three areas of consideration are generally involved in negotiating a construction loan--the real estate, the developer, and terms for permanent financing. In considering the real estate, it is important for the lender to be certain that the property will actually sell, and to know the price and rate at which it will be sold. Knowledge of these factors provides an indicator of the total sellout value of the project, which is the important figure on which the loan amount is based. The developer's "track record" of past successes and failures and his current financial status are also important factors in the loan decision. As a rule, construction loans are more readily available to developers with whom the bank has previously worked or whose past experience with similar projects in the area has been successful. If the project appears particularly risky or the money market highly uncertain, the construction lender may require that the developer obtain a commitment for permanent financing before negotiating the construction loan.

While construction financing is being arranged, the developer also negotiates the terms of the construction contract. The contract usually specifies a series of progress payments to be made subsequent to the installation of improvements. The developer pays interest on the note throughout the construction process. This obligation can carry a significant risk if construction delays are encountered; interest on these notes can run in the thousands of dollars per day. Proper planning and design, construction management and tight project scheduling thus became critically important.

Initiation of Construction Stage

In this stage, the agreed terms of the commitment stage are finalized and construction agreements are initiated. The permanent loan commitment is received, the construction loan closed, the land payed off, and the construction contract is signed. Upon closing the construction loan, the developer switches to a formal accounting system which is based on the provisions of the construction contract, the project budget, and the system of bank draws agreed upon in the loan contract. The developer is now ready to initiate construction.

Construction Stage

During the construction stage, roads and utilities are installed and buildings built. Often, developers who specialize in single family subdivision development will sell the improved lots to individual purchasers or builders. Involvement of the developer would normally phase out at this point. If condominiums are being built, the developer usually remains involved through a rental program.

Throughout the construction process, two other major activities are in progress--marketing and sales. Because the construction loan is often procured without a permanent loan commitment, it is usually critical that sales proceed at or near the projected absorption rate. Some projects may "sell themselves". For many projects, though, an agressive marketing campaign is in order. The initial market search may specify the target population, project features to be emphasized, and the areas in which advertising should be concentrated.

If the development is to be built in phases, the planning and design process begins again with an update of the market analysis, and so forth.

Final Stage

When construction has been completed, the project is ready for occupancy. Although the developer's role is largely completed by this time in projects involving individual home sales, he may aid in establishing a property owner's association. In the case of condominiums or single family rental residences, the developer may hire a manager to run the initial rental program.

ECONOMIC CONSTRAINTS AND ENVIRONMENTAL REGULATION HURDLES

Real estate development schemes such as the model described in the foregoing section carry a high degree of implicit risk and uncertainty which can seriously hinder or delay the completion schedule. In fact, a number of development projects never reach the final stage. Two major considerations--economic uncertainties and governmental regulations--have placed significant constraints on the timing, cost, and general feasibility of housing developments along the Texas coast.

Economic Uncertainty

A number of closely related economic conditions may act to constrain developers of coastal recreational housing projects. The volatile money market and the rapid rate of inflation in housing prices in the 1970s presented problems for developers at every stage of their investment and development process. It became increasingly difficult, for example, for entrepreneurs developing vacation properties to obtain financing when money supplies were tight. During such times, lenders tended to reserve what mortgage money they had for the purchase of primary homes, as second home paper was more difficult for the lending institutions to sell in the secondary mortgage market.

The recent period of price inflation in housing markets has been a particular problem. Changes in regulatory procedures or snags and backlogs in the review of permit proposals, as well as a host of other problems, have caused substantial delays that significantly increased final costs for the developer and consumer. The developer can sometimes pass on to customers the added costs of housing due to delay, but not during periods of excessive supply or unusually tight cash flow. The reality of the development industry is that occasionally the developer will sell at little or no profit rather than hold on.

Growth in Regulatory Complexity (or, Burgeoning Regulatory Bureaucracies)

While the "regulatory maze" faced by coastal developers and homebuilders in Texas is neither so extensive nor complex as those in a number of other coastal states, it has grown significantly larger and more burdensome in the past few years (Table 1). In the context of coastal resource protection, these controls represent an enlightened public awareness of the ecological, aesthetic and economic value of natural coastal areas. From the developers' perspective, conformance and compliance with environmental regulations at all levels of government is a lengthy and troublesome process, potentially involving dozens of local, state and Federal permits.

Local Government. Depending on the site location, developers may have to comply with a multitude of municipal and county land use and environmental controls. Coastal municipalities in Texas have enacted zoning ordinances, building codes, subdivision ordinances, various special standards (relating to floodplains, water and sewer provisions, drainage, subsidence, beach and dune protection, and the like), as well as explicit policies on growth within their extraterritorial jurisdictions (such as on annexation or creation of special purpose districts).[3] Galveston, Nueces, and Matagorda Counties, in addition to their subdivision plan review and sanitary regulations, have adopted dune permit programs to limit the degradation of beach and dune areas by vacation housing developments. Developers that were interviewed along the Texas coast felt that municipal and county ordinances, with the possible exception of dune permits, were reasonably specific concerning the rights to use or modify their property.

State and Federal Government. Environmental regulations administered by State and Federal agencies are intended to protect air, water, submerged land resources, and critical coastal resources of national or statewide significance.[4] The most significant growth in environmental regulatory programs in the past decade has occurred among the Federal agencies. The Army Corps of Engineers is the lead agency for permitting related to navigation in coastal waters, and dredge and fill activities in or adjacent to almost all waters.[9] Developers on the Texas coast have repeatedly criticized the administration of State and Federal

TABLE 1. REGULATORY AND PERMITTING PROCEDURES THAT AFFECT TEXAS COASTAL DEVELOPMENT

GOVERNMENT ENTITY	GOVERNMENTAL FUNCTION	CIRCUMSTANCES OF INVOLVEMENT	ACTIVITIES REGULATED
Municipal			
Planning Department	Review and Comment	In City; in ETJ (Limited)	Subdivision; Construction
Building Inspector	Permit Issuance; Fee	In City	Construction
Zoning Board of Adjustment	Zoning Appeals	In City	Land Use; Density
Planning Commision	Review and Comment	In City; in ETJ (Limited)	Subdivision; Construction
City Council	Permit Issuance	In City; in ETJ (Limited)	Subdivision; Construction
County			
Commissioners Court	Plat Review and Approval	In Unincorporated Areas (Limited)	Subdivision; Roads; Sanitary Drainage Systems
	Dune Permit Issuance	In Foreisland Dunes of Participating Counties	Subdivision; Construction
State			
Parks and Wildlife Department	Permit Issuance; Fee	On Submerged Lands; Habitat Areas	Dredge and Fill
Department of Water Resources	Permit Issuance, Certification	In All Navigable Waters	Dredge and Fill; Spoil Disposal; Waste Discharges
General Land Office/School Land Board	Review and Comment	In Foreisland Dunes (Limited)	Alteration of Dunes
	Easement Issuance; Fee	On State-owned Lands	Drilling; Pipelines; Spoil Disposal; Dredge and Fill
Antiquities Commission	Review and Approval	Wherever Historic Sites are Present	Alteration of Historic Sites
Air Control Board	Review and Approval	Anywhere	Potential Sources of Air Quality Degradation
Governor's Office of Budget and Planning	Review and Coordination of State Permits	If Above State Agencies are Involved	If Above Activities are Involved
Federal			
Environmental Protection Agency	Review and Approval	Anywhere	Pollution Discharges
Fish and Wildlife Service	Review and Approval	In Game and Nongame Habitat Areas	Construction; Dredge and Fill; Spoil Disposal; Pollution Discharges
National Marine Fisheries Service	Review and Approval	In Commercial Fisheries Habitat Areas	Dredge and Fill; Spoil Disposal
Army Corps of Engineers	Permit Issuance	In Navigable Waters; Wetlands	Dredge and Fill; Spoil Disposal Wetland Alteration
Flood Insurance Administration	Permit Issuance	In Floodplains	Subdivision; Construction
Federal Housing Administration	Review and Approval (EIS)	Anywhere	Subdivision; Construction of Large Housing Projects

regulations for being overly ambiguous and subjective, and yet more restrictive than local regulations. Some have claimed that the lack of specific guidelines to aid in project planning has too frequently resulted in extensive and costly revisions of project plans during the permit review period.

Multijurisdictional Coordination. One of the most perplexing challenges of coastal environmental regulation is the networking of overlapping regulatory and review responsibilities. Agency staff have an obligation to process developers' proposals as expeditiously as possible (while still upholding the respective administrative rules). Yet, the sheer magnitude of agencies and private sector interests that request an opportunity to review permit proposals has created unavoidable delays and added costs for developers who must apply for permits.

The salient example of a complex, time-consumptive interagency review process affecting coastal developments is the Army Corps of Engineers permit program for dredge and fill activities or wetland alteration. Upon receiving a permit application for a coastal housing development, the Galveston District of the Corps forwards copies of the proposal to more than 200 agencies and interested parties. The U.S. Fish and Wildlife Service, National Marine Fisheries Service, Environmental Protection Agency, and five State agencies respond regularly upon notification by the Corps, often with highly detailed, site specific observations and recommended stipulations on permit approval. If serious objections to the permit are submitted by one or more of the principal review agencies, the applicant then has to meet with the respective staffs to negotiate a solution or provide additional information regarding the impacts of the project. Each agency is theoretically allowed a 30-day review period, but extensions are routinely granted upon request if large or particularly objectionable projects are being reviewed.

Under the National Environmental Policy Act of 1969, some of the larger project proposals also require an Environmental Impact Statement, in which case developers must submit an Environmental Impact Report for consideration by the Corps before the agency writes its Statement. Public hearings may be requested by agencies and affected citizens. Permits granted by the Corps must be activated within three years, or the application has to be resubmitted and the entire process repeated, possibly including a revised Environmental Impact Statement.

In the attempt to preserve and protect fragile coastal ecosystems, the increasing complexity of regulations and review procedures has rendered the development process much less efficient. While many of the additions to the regulatory process have effectively served to protect the coastal resources for which they were directed, the inefficiencies and resulting increases in permit processing time are clearly reflected in increased costs to the consumer. Hopefully, the Texas Coastal Management Program, which is currently under development, will make significant inroads into the problems of multijurisdictional coordination.

Delays Caused by Permitting Review

A number of factors have been offered as reasons for permit delays: significant increases in the number and complexity of regulations in recent years, inefficient coordination of review and approval by the many agencies involved, poor or insufficient preparation of application materials by developers and their consultants, and understaffed agency offices that encounter backlogs in permit processing. These and many other causes for delay in project completion (weather, money availability, investor indecisiveness, etc.) are still not resolved or fully understood.

The Urban Land Institute sponsored three case studies to determine the effects of regulation on housing costs in San Jose, California; Jacksonville, Florida; and Houston, Texas.[6,7] The Houston study, conducted by the Rice Center for Community Design and Research, evaluated several factors of regulatory delay and estimated their effects on final housing costs. The cost estimations in the Houston study did not fully consider the regulations affecting coastal development, but they are among the most conclusive figures available. A marked increase was observed in the average total time required to develop a housing project over the period from 1966 to 1977. While it took about 21.6 months to complete the average project in the 1966-1970 period, the average length had increased to 27.4 months by the 1974-1977 period. Almost 20% of the projects surveyed took 40 months or longer to complete. Increased regulation was identified as the principal cause of increased delay over this period.

Inefficiencies and delays in receiving approval from regulatory agencies was also cited in our survey of coastal land developers and investors as the most critical problem related to government regulation. One development on Mustang Island that was initially proposed in 1974 is still held up as of 1980, largely on account of environmental regulations. Another development firm has been negotiating a permit for a residential canal development on Galveston Island for seven years, and was obliged to redesign the master plan two times after receiving preliminary approval from the Corps.

Lengthy delays can substantially increase land acquisition carrying costs and construction costs as a result of inflation in the intervening period. Market preferences can change over a period of several years, necessitating an update of the market study and a revamping of engineering and architectural plans. Money market conditions may fluctuate wherein a shift to higher interest rates or a lack of mortgage funds can seriously challenge the feasibility of the project. Not all projects, however, have encountered serious delays in complying with regulations, especially relatively small, innocuous projects that were sited away from shorelines and critical natural areas.

The Cost of Environmental Regulatory Delays

The cost of land development can be increased through the application of environmental regulation in two ways: directly, through compliance with administrative regulations;or indirectly, as a result of encountering protracted, costly delays in the permitting process. The costs of complying with administrative regulations can be substantial for developers and consumers alike. In recent years, rules on floodplains, wetlands, water quality, and other regulations discussed above have so increased in number and specificity that their effects on cost have become evident even in the earliest stages of the development process--in the purchase price of raw land. Restrictions on the manner in which a developer may build on or alter portions of a coastal site has made the supply of remaining suitable land more scarce. The cost of prime developable land is thereby driven up, and, ultimately, the price of marginal land becomes more expensive through required improvements. The net effects are increased front-end carrying costs and final consumer costs, and a heightened risk of foreclosure. Ultimately, there are no guarantees that permit approval will be granted by regulatory agencies upon purchasing or securing an option on land or before paying architects, engineers, and legal consultants to draft a complete set of plans.

The costs of protracted delays during the permitting review process are usually a greater problem than compliance related costs. In the worst cases, negative reviews by environmental regulatory agencies can lead to an absolute halt in the development schedule until (and if) the conflicts are able to be resolved. In almost every case, the number and complexity of regulations and multi-agency review procedures elevates the front-end carrying costs.

The Rice Center study of housing developments indicated an average delay increase of 5.8 months over the 10-year period from 1967 to 1977, attributable primarily to regulation. On the basis of a formula designed to calculate monthly costs attributable to this delay (involving land interest, property tax, overhead, construction loan interest, inflation, and capital tie-up), it was determined that the increase in delay costs for the developer ranged from $390 to $590 per subdivision lot, or alternatively, $78,000 to $118,000 for a 200-unit subdivision.

One developer on the central Texas coast estimated that he incurred $60,000 in carrying costs and additional engineering fees, solely for the purpose of complying with the reviews of one agency concerned with the shoreline configuration at a canal entrance. On a nationwide basis, regulation-oriented costs per lot or per house have ranged up to $7,500.[8] Such magnitudes are only approximations of expected delay costs--they do not necessarily reflect variations in site location and value, project complexity, or developer-agency cooperation.

MECHANISMS FOR MINIMIZING REGULATORY CONSTRAINTS

In this section, we propose some techniques that can be employed readily in Texas and that should effect a more expeditious, less costly review process while maintaining high standards of environmental quality.

Improved Dissemination of Regulatory Procedures

In an attempt to specify agency policies and regulatory requirements without limiting their flexibility, a more intensive program of information dissemination could be established. Published, updated information regarding Federal, State, and local agency policies, permitting review criteria, elements to be included in environmental impact reports, and information specifying which permits are necessary for various types of development could be made available to developers in a consolidated format. In our conversations with developers and agency staff alike, no one professed a working knowledge of all control devices bearing on development activity. Local governments would be appropriate clearinghouses for such information and for distributing new or revised regulations.

By making regulatory procedures and related information readily available to the developer, some of the uncertainty regarding regulatory requirements would be reduced. By designing projects to satisfy the policies and standards found in this information package, original project designs should be more closely aligned with agency objectives and criteria. The developer may thus be able to reduce the duration of the review period as agency objections may be fewer and the requested design changes less substantial. The states of Washington and Oregon have initiated such public information programs and have found them to be valuable in expediting the permitting process.

Developer's Preapplication Conference

A permit preapplication conference attended by the developer and representatives of relevant regulatory agencies would also seem a viable method of reducing uncertainty and delay in the development process. New Jersey has instituted such a procedure for interested developers at the "sketch plan" stage of project development. Initiating the review process at an early juncture allows the developer to avoid many "missed actions" and discover and correct problems as early as possible, perhaps avoiding the need for more expensive changes in the completed master plan. The procedure for improved information dissemination outlined in the preceding section could be administered at such a conference.

Preapplication conferences may also serve to lessen the duration of agency reviews. Agencies with overlapping review or comment authorities could use such meetings to their advantage by developing an early dialogue about their intentions for the forthcoming formal review.

Time Limits on Permit Review

In a further effort to reduce the time frame of the permitting process, an absolute time limit for agency review could be established. As a result of ever-increasing permit application requirements, as well as frequent backlogs of projects to be reviewed in areas of rapid growth (such as the Houston-Galveston area), agency review schedules are often constrained by inadequate staffing. Agencies have the responsibility, however, to give development proposals the most rapid possible response.

Such a time constraint would require a careful study of the most cost-effective means for implementing regulations. It may also dictate that agencies be staffed at fuller levels, that more staff members be trained to carry out multiple responsibilities, and that agencies contract with environmental and engineering consulting firms for review assistance on an *ad hoc* basis. Some agencies already contract with private firms on a retainer basis to aid in the technical aspects of project reviews. Wider adoption of this latter approach would probably necessitate a "cost plus" form of budgeting, which may be rather difficult to negotiate with legislative budget committees.

Coastwide Mapping of Development Suitability

In an effort to facilitate development while still protecting the environment, the State of Texas could undertake a synthesis of coastal resource mapping and planning efforts for the purpose of delineating zones of varying development priority and constraint. The excellent environmental geologic maps and accompanying reports of the Texas coastal area, prepared by the University of Texas Bureau of Economic Geology,[9] and the ecological characterization mapping program of the U.S. Fish and Wildlife Service, currently being prepared by Texas A&M University,[10] would provide an appropriate initial information base. The Texas Coastal Management Program, administered by the Texas Energy and Natural Resources Advisory Council, would be a logical coordinating body to undertake the project.

Maps delineating coastal ecological systems and their intrinsic suitability or hazard for various types of development could be overlayed with derivative maps of prime and marginally suitable development zones. Various functions such as biologic productivity, flood susceptibility, and the like would be assigned to each ecological map unit. Suitability functions relating to existing or potential development factors would be assigned to the development zones. The most updated map information on ecological systems, existing development, floodplain boundaries, and other features would be used in conjunction with the base maps and information. A number of technical considerations such as conflicting information or inconsistent classifications among different map sources would have to be resolved.

Areas could thus be classified as high or low development priority through a consideration of their development value and intrinsic environmental value. The mapping effort would provide regulatory agencies with more clearly established general review criteria as well as a geographic basis upon which to make their permit decisions. North Carolina has successfully used such an approach in implementing the state Coastal Area Management Act.

The maps and associated development priorities would not substitute for on-site inspections by agency staff, or detailed local land use plans, or site specific data prepared by developers. Rather, they would provide general guidance for areas likely to encounter rapid development, especially for areas that are as yet unplanned and sparsely developed.

CONCLUSIONS

The model land development process outlined herein is not intended to be a precise representation of even the majority of coastal development strategies. Each project concept, its site and economic conditions, and the reasoning behavior of the developer creates a unique development plan. The model process does, however, reveal the activity stages that are especially sensitive to governmental review for most, if not all, projects. It also indicates the vulnerability of developers and investors to protracted delays and added carrying costs throughout all stages of the development process.

This analysis emphasizes the plight of the developer. There are equally numerous observations to be made on the "pro" side on behalf of agencies and their mandated environmental regulations. For example, while the Corps' dredge and fill review process is seemingly extended beyond the review agencies' capabilities to effect a timely review and response, such a system of checks and balances on the decisions of the lead agency is indeed a valid process for assuring that environmental protection is given fair consideration.

Our suggestions for mitigating regulatory constraints are limited to initiatives on the part of government. One might also develop many remedial suggestions for developers, such as to improve the integration of their legal, engineering, and financial planning activities, to provide more complete data and information upon applying for permits, and to heighten their awareness of hazardous or sensitive coastal environments and the tradeoffs involved in developing these areas. The challenge is not to cull the most troublesome regulations and erase them from the rulebooks, but first to seek more subtle procedural remedies that will make them operate more efficiently.

REFERENCES

[1] See, for example, Aspen System Corporation. Land and Natural Resources Management: An Analysis of Selected Federal Policies, Programs, and Planning Mechanisms. Prepared for U.S. Council on Environmental Quality. National Technical Information Service, Springfield, VA. 1979.

[2] The framework of the model is derived from Miles, M. and Wurtzebach, C. Risk Analysis in the Real Property Development Process: A Conceptual Framework and Computer Simulation Model. Journal of Business Research, December 1977, pp. 325-357.

[3] Butler, K.S. and Nichols, K.N., The Role of Local Governments in Promoting Environmentally Sound Development on the Texas Coast. <u>Texas Business Review</u> (draft, in press). Austin, TX. 1980.

[4] Texas General Land Office. Current Permitting Processes in State and Federal Natural Resource Agencies. Vol. II. Texas Coastal Management Program. Austin, TX. 1976.

[5] Section 10, River and Harbor Act of 1899 (as amended), 33 U.S.C. § 404 <u>et seq</u>.; and Section 404, Federal Water Polution Control Act Amendments of 1972, 33 U.S.C. § 1344. See also Federal Register, Vol. 42, No. 138, July 19, 1977.

[6] Urban Land Institute and Rice Center for Cummunity Design and Research. The Cost of Delay Due to Government Regulation in the Houston Housing Market. ULI Research Report No. 20. The Urban Land Institute. Washington, D.C. 1979.

[7]Urban Land Institute. Effects of Regulation on Housing Costs: Two Case Studies. ULI Research Report No. 27. Prepared by the Research Division of the Urban Land Institute and Gruen, Gruen, and Associates. Washington, D.C. 1977.

[8]Ibid.

[9]Brown, L.F. Jr., et al. Environmental Geologic Atlas of the Texas Coastal Zone. 7 vols. Bureau of Economic Geology, University of Texas at Austin. Various dates, 1972-1980.

[10]U.S. Fish and Wildlife Service, Texas Barrier Islands Region Ecological Characterization Study. Mapping contract conducted at: Texas A&M University, Remote Sensing Center, College Station, TX. Contracting agency: National Coastal Ecosystems Team, Office of Biological Services, U.S. Fish and Wildlife Service, Slidell, LA. 1980.

GENERAL PERMITS: METHOD OF CRITERIA DEVELOPMENT

James W. Holton, Jr., P.E., L.S., M.ASCE[1]

INTRODUCTION

The purpose of this paper is to present a methodology to establish criteria for general permitting of new coastal structures (which was developed by the author and the Norfolk District, Corps of Engineers). The manifold benefits of consistent criteria accrue to both potential developers and regulatory personnel. The criteria has been structured around permitting specified activities in small-craft harbor systems. One such harbor system has been completed and five others are under study. It is important to note that the procedures and methodologies discussed are applicable to areas other than those for which the criteria was developed since the step-by-step approach outlined allows flexibility to adapt the concepts to local conditions and regulatory restraints.

The reader is probably aware of the Department of the Army (Corps of Engineers) requirements for permitting construction activities in wetlands and navigable waters; therefore, the applicable laws and legislation will not be addressed. However, it might be helpful to summarize the concept of issuing a General Permit.

The concept of a General Permit is explained in Department of the Army Regulations published in the Federal Register on 19 July 1977. In brief, these regulations give a District Engineer the authority to issue General Permits for any activities which are similar in nature, which cause minimal adverse environmental impact when performed separately, and which cause a minimal adverse cumulative effect on the environment. The decision to issue a permit is based on evaluation of the probable impact of the proposed activity on the public interest. The decision reflects the national concern for both protection and utilization of important resources. The benefit which reasonably may be expected to accrue from the proposal must be balanced against its reasonably foreseeable detriments. All factors which may be relevant to the proposal are considered, including conservation, economics, aesthetics, general environmental concerns, historic values, fish and wildlife values, flood damage prevention, land use classification, navigation, recreation, food production, water supply, water quality, energy needs, safety and, in general, the needs and welfare of the people.

The General Permit is put out on public notice with an appropriate comment period. Once effective, it relieves the requirement for further public notice of individual projects which meet the criteria stated in the General Permit. While some activities have been given nationwide

---

[1] President, Waterway Surveys & Engineering, Ltd.
5596 Greenwich Road, Virginia Beach, Virginia 23462

permit status, the General Permit as it applies to this discussion is local and site specific - considering the problems and public needs within an individual Corps of Engineers District.

After issuance of a General Permit, a prospective developer need only apply for an individual permit under the General Permit and, if his or her project qualifies, the work can be approved in a matter of days. An accelerated permit procedure is one of the most important outcomes of issuing a General Permit as it not only creates goodwill with the public but relieves congestion at the regualtory level, and, if properly structured, promotes well planned growth. It is this last benefit that gives the most support to procedures described in this paper for developing specific general permitting criteria and to which most of the following discussion will be aimed.

OBJECTIVES OF THE GENERAL PERMIT STUDIES

Discussions held with Corps regulatory personnel have indicated the need to make more extensive use of their general permitting authority, while retaining control over development that would adversely impact navigation and the marine environment. One general permit which appears to have worked well allows the replacement of existing bulkheads with new structures of like kind. In looking toward issuing further general permits, it was recognized that a rationale that identified general criteria but that would not require an extensive and costly "full-blown" engineering and environmental study would be much more desirable than an arbitrary approach. It is the author's understanding that some regulatory agencies have issued broad general permits covering new construction of certain structures within their entire jurisdiction while others have looked to in-depth site specific studies to solve their permitting problems.

With these divergent approaches to solving permit issues in mind, overall objectives were established that would

1. Result in the issuance of a General Permit;

2. Be economical when compared to future savings in manpower utilized for reviewing individual permits;

3. Protect the waterways from adverse impacts due to general permitted development;

4. Be accepted and generally understood and used by the general public.

Looking at the history of problems with regulatory aspects of development in the Tidewater region of Virginia, it became clear that activities could be grouped according to their impact. The matrix shown in Table 1 below summarizes a typical grouping of activities which were considered for general permitting.

TABLE 1

PLANNED CONSTRUCTION	PRIMARY IMPACT ON NAVIGATION	PRIMARY IMPACT ON ENVIRONMENT	SIGNIFICANT IMPACT ON BOTH NAVIGATION AND ENVIRONMENT
Dredging/Disposal			X
Bulkheading			X
Shore Protection		X	
Vessel Moorings	X		
Boathouses	X		
Piers/Docks	X		
Miscellaneous Open-Pile Structures	X		

Based on experience and judgment, the Corps decided not to pursue a general permit at this time for structures which potentially impact on the marine environment. It is generally agreed that open-pile structures have minimal impact on wetlands in the Tidewater region and, therefore, could be general permitted. However, as their impact on navigation can be most substantial, the final objective of the work was to apply the four general objectives to open-pile structures and formulate criteria to insure against misuse of the program. Extension of the general permit work to include bulkheading and shore protection structures is contemplated and the author will address this later in the paper.

THE SELECTION OF STUDY AREAS

In the Tidewater area of Virginia there are specific areas that can be readily identified as having an expanding potential for waterway development. With some exceptions, most of these areas are rural in nature and primarily situated on or around small, protected inlets, bays or creeks leading from a much larger water body such as the Chesapeake Bay. The character of the waterway development is typified by small and moderately sized docks and piers, small private boathouses, large multi-craft covered commercial boathouses, small seafood industries, repair yards, small and large power and sail marinas and hybrid structures of unique design. As will be discussed more specifically, development in some of these harbors has been at a faster pace than in others. Some are already overcrowded from the standpoint of navigation.

Waterborne and water related activities in and around these small-craft harbors include commercial fishing, crabbing, oystering, and water sports such as skiing and swimming. Transient traffic consists primarily of large, private, pleasure motor and sail craft. Navigation charts of these areas are of such a scale as to render them almost useless.

The size of these harbors cannot be easily typified. Some very active waterways vary from just a mile or two of shoreline to many miles along a popular estuary. Likewise, there are very active areas that do not fit the concept of a small-craft harbor but are most exposed to wave action and storms on our large rivers and bays. For the most part, the throngs of boating and commercial users are looking for protected areas.

One of the most important points to observe and to consider when selecting areas for a general permit study and subsequent issuance is the existing character of development and usage of a harbor system. In assisting the Corps in their selection of potential candidate harbors, we noted that categories could be used to identify harbors where a general permit for certain structures could yield a high return on their investment. Briefly, these classes can be described as follows:

Class 1: A densely populated harbor, having residential development interspersed with commercial and/or industrial activities where navigation has been substantially impacted by channelward extension of structures.

Class 2: A moderately developed harbor where navigation fairways have remained adequate although the potential exists for encroachments on navigation and conflicts between commercial and residential usage.

Class 3: Virtually undeveloped waterways in areas of high waterway activity having the potential for rapid and significant development.

The first small-craft harbor system selected for study fit the Class 2 description. It is not situated in a generally rural area but rather at the southernmost extent of a massive, urban development, Virginia Beach, Virginia. An inlet to the Atlantic Ocean was formed during the 1929 hurricane which inundated low lying areas behind the dune line creating a shallow estuary. Subsequently, the City of Virginia Beach has sponsored dredging activities in the inlet and the new estuary to obtain sand for beach replenishment. The waterways are now deep due to this dredging and the inlet is currently maintained by the City. Presently, two rapidly developing estuaries are prominant and are inappropriately named Lake Wesley and Lake Rudee. The study work and recommendations for a general permitting project has been completed for Lakes Rudee and Wesley. The General Permit is out on Public Notice at this writing and the author hopes to report on comments received and the status of the General Permit during the presentation of this paper.

Five small-craft harbors are currently under study for the Corps and General Permit guidelines are being developed for each. To give the reader a feeling for the types of situations we are addressing in these other harbors, the following brief description of each is offered:

Carter Creek is located partially in the Town of Irvington, Virginia and in Lancaster County, Virginia. The creek currently has wide fairways but development is accelerating and fairways are becoming restricted in upper reaches. Several commercial seafood businesses are located near the mouth while large marina facilities exist well within the creek. Land adjacent to the waterway is largely held by upper income native individuals and much of the permitted development is for private docking and housing facilities for large motor and sail craft. The marinas serve transient traffic and provide storage for private boat owners who do not have land bordering on the creek. Cater is considered to be a Class 2 harbor.

Taylor Creek is located entirely within Lancaster County, only a couple of miles north of Carter. It joins the Corrotoman River, a large deepwater estuary off the Rappahannock River. It is virtually undeveloped at this time with only scattered small seafood operators having docks and piers. Development of Taylor Creek is almost certain in the very near future and the General Permit guidelines developed will insure that fairways and channels are preserved for navigation. Taylor Creek is the classic example of a Class 3 harbor.

Broad Creek is also a tributary of the Rappahannock River and is situated in Deltaville, Virginia. The early development of Broad centered around commercial seafood. The land is still held primarily by native business and residents. Recently, the creek has become a crowded haven for large private motorboats and sailboats. Large covered boathouses provide shelter and facilities for boat owners who live onboard their vessels. Both small commercial seafood and marina operators are complaining about congestion and shoaling and one must use extreme caution while navigating in this harbor. Broad Creek is a classic example of a Class 1 harbor.

Jackson Creek is adjacent to Broad Creek, but is a tributary of the Piankatank River. Development trends in Jackson have been substantially different than in Broad as there appears to be a good mixture of commercial seafood and residential recreational activity. It has a high potential for becoming overcrowded. In fact, one developer is now considering a condominium and marina complex while another is planning a large industrial railway yard. Jackson Creek is considered to be a Class 2 harbor.

Urbanna Creek is a tributary of the Rappahannock River and is located in the Town of Urbanna, Virginia and in Middlesex County, Virginia. Urbanna Creek is bisected by a low, fixed span concrete highway bridge which effectively restricts the size and type of vessel which can pass.

> The assumption for the study is that the structure will remain, but as it has not deterred development, the upper reaches should be included in a general permit. The lower half of Urbanna Creek is considered a Class 2 harbor while the upper half would best fit a Class 3. It is known that a large new marina complex being planned could substantially impact harbor navigation. Also, some structures have been built very close to the Federal Project Channel and it is suspected at this writing that some actually encroach.

As can be seen, all three classes of harbor systems are being looked at in the light of establishing permitting guidelines and issuing a general permit. This is not to say that all small-craft harbors that can be so classed require the work to promulgate such guidelines. Indeed, the selection of harbors or waterways must be made by those who best know the scope of problems associated with the permitting program. Very likely, the individuals who can best make that selection are those who process, review, and inspect permit applications. They must also examine the "tools" they already have for administering their program mainly with respect to reviewing permit drawings, analyzing navigation, protecting wetlands, inspection, detection of un-permitted works, and prosecution of violators. These individuals may also be in a good position to predict future trends in undeveloped systems and to recommend certain of these areas for study.

For a harbor system to be a candidate for a general permit as discussed here, there must be either significant problems or voids in the "tools" needed or high development potential. Selection then becomes a matter of simply classifying the candidate harbors and determining those harbors whose problems are most urgent and will yield immediate improvements in the permitting process.

DISCUSSION OF THE STUDY PHILOSOPHY

It is recognized that the results of the small-craft harbor studies will get considerable use well beyond the issuance of a general permit. In fact, the additional side benefits to the regulatory agencies may prove to far outweigh the original intent - that of establishing a sound general permit. We are cognizant of the potential uses of the data and are striving to be accurate and consistent with our recommendations and conclusions.

For instance, a Class 1 harbor system obviously yields little potential for a good general permit program. To adopt a liberal permitting philosophy in a Class 1 system would be to invite, and in fact encourage, the disasterous situation we are trying to prevent. The philosophy to adopt in a Class 1 system must be a conservative one, so that the General Permit when implemented, would prevent further encroachment on congested fairways. Herein lies a fundamental result of the study which is a good example of side benefits to the regulatory agency. Once the general permit guidelines are established in, say a Class 1 harbor, regulators have a good technical basis from which to deny permits that request development exceeding the recommended limits.

In general, then, the philosophy toward conducting these studies must not only take into account the prime purpose of issuing a general permit but also must consider the other uses to which the data and information will be put in the future.

DISCUSSION OF STUDY OBJECTIVITY

A general outline has been developed which acts as a guide when scoping the work required for an adequate small-craft harbor study. Many aspects must be looked at even if only briefly. It is very important to approach the project without any pre-conceived ideas or bias. In fact, it may be advantageous to procure professional assistance from firms outside the immediate geographical region where the work is to be conducted.

Most harbor studies the author has been exposed to have justification of a particular project or a position on a specific issue as their objective. For instance, a typical study might be undertaken to justify and plan a marina complex while another might be made to prove the adverse impacts of such a project. Both studies would likely reach some very positive conclusions supporting the position desired.

Unlike the work described above, the small-craft harbor study this paper addresses is unbiased and tends to compromise on issues which oppose one another. For example, the wishes and needs of commercial fishermen and those of pleasure sailboat operators may conflict in a particular situation. Recommendations must then be made which allow due consideration for each group, while still making use of the general permit.

With the above in mind, we can proceed with discussing our study outline and briefly comment on each element. It must be strongly emphasized that the outline offered here is general in nature and would certainly be modified to fit other harbor situations.

OUTLINE OF SMALL-CRAFT HARBOR STUDY

Gathering Available Information: Sources of existing "hard" data should be carefully identified at the outset. The amount and quality of the information received will most likely dictate the requirements for additional work. The sources shown in Table 2 would probably apply to any harbor in the U.S. and should be added to as the local situation requires.

TABLE 2

SOURCE	TYPE OF DATA AVAILABLE
National Ocean Survey National Geodetic Survey	Small-craft navigation charts, tide and current data, aerial photography, tidal datum benchmarks. Geodetic monumentation information.
U.S. Geological Survey	7½' and 15' Quadrangle map series (frequently the best maps available). Horizontal and vertical control monumentation.
U.S. Army Corps of Engineers	Large scale maps, hydrographic data, tide and current data, aerial photography, horizontal and vertical control monumentation, water quality data, and existing and future channel plans.
U.S. Coast Guard	Navigation aids and accident reports.
State Fisheries Agencies	Oyster ground information, shellfish restrictions, general and specific information on commercial seafood industry.
City and County Governments, Local Courthouses, Chamber of Commerces, etc.	Planning documents, zoning information, historical data, tract ownership data, public relations and promotional data, and data on tourism and industry.

Table 2 is by no means a complete source list for conducting a site-specific study. In fact, there are many other sources currently being investigated for the on-going work but are not considered to be general enough for inclusion in Table 2.

Interviews: A great amount of useful information about small harbors can be gained via the interview. People whose lives are dependent on or associated with these harbors are observant of changes and what produces them. Fortunately, these people are usually also willing to discuss quite freely their observations and opinions. The author has found that this work is best left to a non-technical person or persons who have been properly briefed and who have had previous experience in public contact. It is very important to gain the respect of those to be interviewed as some of the issues at hand may be sensitive. Table 3 summarizes the type of information gained through conducting interviews. It is very interesting to note that when asked the right questions, even those closest to the project are prompted to give additional thought to information they had taken for granted. That a great deal of interview information is repetitive serves as a cross-check of those issues and concerns which are most important to local residents and businessmen.

TABLE 3

SOURCE	TYPE OF INFORMATION GAINED FROM INTERVIEWS
Marina Operators	Navigation problems - expecially those involving depth of channel, inadequacies in channel marking, expansion plans affecting waterways, expansion plans affecting land development adjacent to waterways, local situations affecting development, general attitudes toward creek development, seasonal traffic pattern, daily traffic patterns, recreational/commercial conflicts, pollution and drainage problems, erosion problems, population composition - summer vs. yearround, available public facilities, general attitudes and recommendations toward permitting.
Commercial Seafood	Seasonal and daily periods of operation, peak recreational traffic times, traffic conflicts between recreational and commercial operators, commercial expansion plans affecting waterways, probable future land use adjacent to waterway, types of commerical seafood operations, pollution problems, attitudes and recommendations concerning permitting.
Shipyards	Dredging requirements for operation and expansion, expansion plans affecting waterways, local opposition to expansion, condition of channel markings, tidal variations in depth of water, seasonal and daily peak activity periods, traffic problems with recreational craft and commercial seafood craft, shoaling due to prop wash of recreational craft.
Residents	Local land sales and consequent planned waterfront development, impact of seasonal population, attitudes toward types of development desired, rate of waterway development, channel depth and marking problems, traffic congestion problems, conflicts between recreational and commercial craft, attitudes of environmental groups, effects of wave action.
Local Officials	Zoning restrictions, pollution problems especially concerning waste disposal, erosion problems, rate of development, attitude toward industrial development along waterways, population composition, seasonal population impact, existing public recreational facilities and planned future facilities, citizen complaints about fishing industry, citizen complaints about permitting process.

Field Investigations and Mapping: This is an initial phase of the small-craft harbor study in that it establishes the technical base upon which the final analysis will depend. Existing mapping and survey data and the adequacy of such work must be considered for any harbor under study. For the purpose of this paper, it will be assumed that no data of this nature exists; therefore, the outline offered can be appropriately modified to reflect actual conditions for a specific study.

1. Horizontal Geodetic Control

The purpose for establishing a sound horizontal control system in the harbor system is manifold and the importance of this cannot be overstated. This type of work is typically a Professional Surveying discipline and the reader is strongly urged to consult a surveyor when planning a study. A visual inspection of the harbor under study will indicate the best locations for concrete control monuments which will have a high degree of intervisibility and will be located close to the waterway. After securing permission from the property owners, monuments such as those detailed in Figure 1 will be placed and stamped with an appropriate designation and referenced as in Figure 2.

Traverse or triangulation procedures are used to establish a grid position on each monument. It is highly recommended that a tie be accomplished to the state plane coordinate grid if practical.

With today's modern electronic distance measuring instruments (EDM's), traverse procedures are more commonly utilized for coordinating the control. The author recommends that Third-Order, Class 1 specifications apply for this work as outlined in CLASSIFICATION, STANDARDS OF ACCURACY, and GENERAL SPECIFICATIONS of GEODETIC CONTROL SURVEYS prepared by the Federal Geodetic Control Committee, U.S. Department of Commerce, National Oceanic and Atmospheric Administration (NOAA), Rockville, Maryland.

2. Project Mapping

Photogrammetric mapping is by far the best procedure for producing up to date, accurate base maps of the harbor. Existing photography must be carefully analyzed and caution in its use is recommended. The photography must have been acquired at the proper altitude for the desired mapping and have stereo coverage. In addition, if the imagery is several years old in an area of high permit activity, then many new structures and developments will have to be field located on the maps at considerable expense.

The author recommends that final project maps be at a scale of 1" = 100'. Larger drawings tend to become unweildly and smaller scale ones loose much of the smaller detail. However, the final decision with respect to scale must be made by the

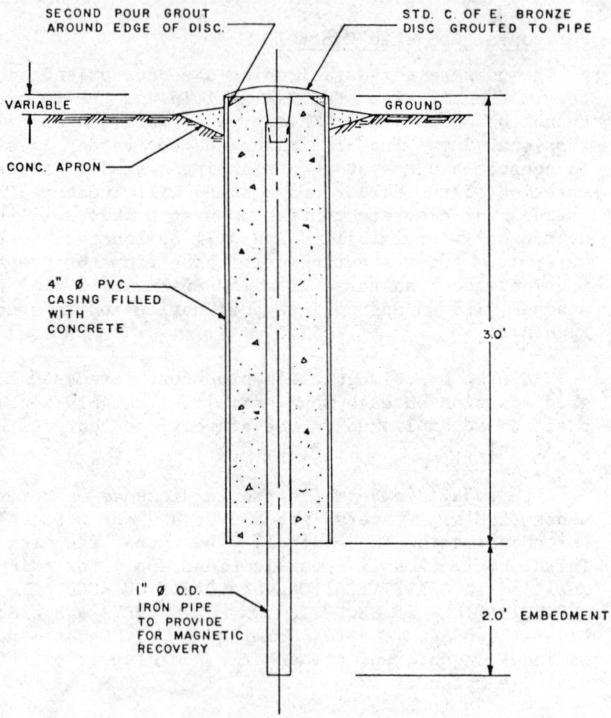

FIGURE 1

CONTROL MONUMENT DESIGN

NOT TO SCALE

Waterway
Surveys & Engineering, Ltd.
Virginia Beach, Virginia

MONUMENT DESCRIPTION

NAME OF STATION: GRF LR-11	PROJECT: Navigation Study/Lake Rudee
LOCALITY: Virginia Beach, Virginia	CITY/COUNTY STATE: Virginia Beach, Virginia
COORDINATES: N 188,387.30 E 2,738,892.03	DATE: June 1979 QUADRANGLE: Virginia Beach
PROJECTION: Virginia State Grid	CHARACTER OF MARK: Bronze Disk in Conc. Mon.
ZONE: South	STAMPING: GRF LR-11 1979
ORDER OF SURVEY: 3rd	ELEVATION: 13.69
ESTABLISHED BY: W S & E, Ltd.	DATUM: Mean Low Water
PARTY CHIEF: W.C.H.	ADJUSTMENT: 1972

VICINITY MAP

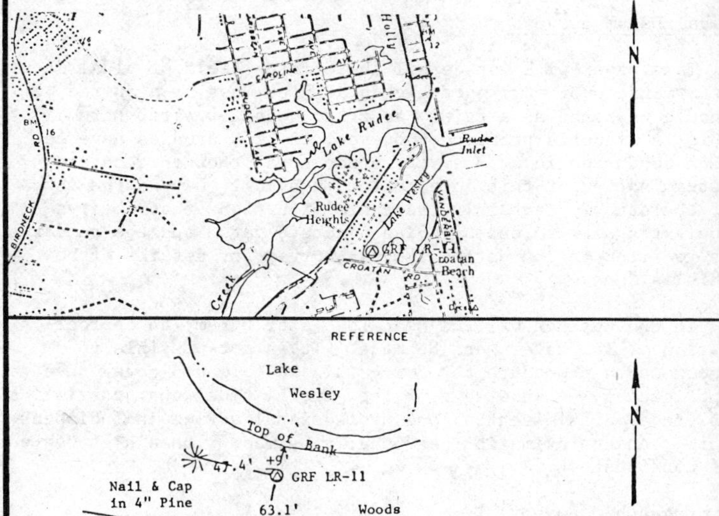

REFERENCE

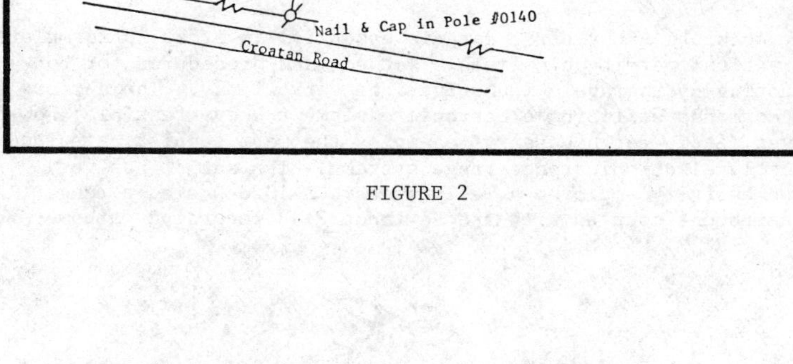

FIGURE 2

project manager after consideration of the factors involved. Only planimetric maps are required, thereby eliminating the costly work of developing contours and elevations.

Photo panels are placed at the control monuments previously established prior to acquisition of new imagery. This will afford the necessary horizontal control for the photogrammetric compilation. If it is inconvenient to panel prior to the flight, the photography can still be controlled by coordinating physical features visible on the imagery.

This paper will not address in detail the actual process of map compilation from aerial photography. Suffice it to say that a professional mapping company must be employed if in-house ability is not available. Rough map manuscripts are thoroughly field checked and annotated by surveyors before final drafting. Standard mapping procedures are recommended and the final base project drawings will be indexed and appropriately matched one to the other.

3. Tidal Datum Surveys

This paper will not detail the complex field of tidal datum determinations. Hydrographic data which must be acquired should be based on a fairly accurate Mean Low Water Datum. This is standard practice; however, if tide studies have not been conducted in or near the study area, then an accurate determination of this local datum is usually beyond the scope of the study. Certainly there are situations that deserve the extra work to establish an accurate datum and the reader is referred to the National Ocean Survey for details of how this is done.

In the absence of true Mean Low Water Datum, an approximation of low water must be made. It is not possible to recommend a procedure for doing this which would have universal application. A knowledge of the general tidal characteristics in the area is essential and professional advice is indispensable. An approximation can usually be made with a high degree of confidence.

4. Hydrographic Surveys

Accurate bathymetric data is essential in order to formulate practical permit guidelines. Methods and procedures for conducting hydrographic investigations vary widely. The author recommends utilizing electronic equipment where feasible; however, it is not necessary to employ the more sophisticated and costly electronic range-range systems. The author has very successfully employed a Tellurometer CA 1000 D single range instrument coupled with the Raytheon 719B recording fathometer.

Location of hydrographic cross-sections and profiles are selected utilizing aerial photography and the new maps. Experience has shown that with careful attention to the waterway features, the scope of the hydrographic work can be held to a minimum while still accurately defining the bathymetry. This is an important point in that a large volume of data improperly taken is not nearly as good as less data which properly reflects the bathymetry.

5. Other Field Investigations

Aside from the technical field work, other investigations are useful in developing data on a harbor system. When there is a moderate to high density of vessels using the system, a vessel inventory is useful in predicting trends in usage and in documenting the current status of vessels in the system. In addition, a vessel traffic survey conducted at strategic locations along the waterway allows a relative comparison of traffic patterns and points up problem areas.

When conducting an inventory of moored vessels it is best to select a time when most boats are sure to be in the system. This work would typically be done during the summer months on a weekday and when the weather is such as to discourage vessels leaving the system. The author has found that developing sample categories into which each vessel would fall based on estimates of vessel size is more practical than physical measurement. For instance, one category might be 15' to 25' length, inboard - outboard. The categories chosen should be tailored, of course, to suit the types of vessels using the system.

When conducting a vessel traffic survey, the objective is to establish a pattern for a typical day primarily to find the mixture of vessels navigating the waterways. This survey would likely be conducted under fair weather conditions. To investigate the maximum vessel traffic pattern, a weekend or holiday should be selected when fair weather conditions exist. Typically, this would give data primarily on recreational traffic since commercial seafood vessels are at a minimum.

Another investigation which should be considered is an inventory of wetlands in the system. This can be as extensive or minimal as required by the project objectives. In the work conducted thus far, only the location of the wetlands has been mapped without cataloging individual plant species. The existance of wetlands and fringe marshes is thereby established and this data will serve investigators in many ways not connected with the General Permit. Information as to the quality and extent of the wetlands in a harbor system gives a good indication as to the direction of future development. For instance, in Tidewater, Virginia, the trend is to discourage construction of bulkheads, revetments, and shore protection structures which would impact on wetlands. Quite obviously, where this situation exists, then the demand for open-pile structures will be high.

Depending on the nature of the harbor under study, it may be useful to investigate in some detail the local commercial seafood industry. Answers to questions such as the following may help to support decisions and recommendations resulting from the study.

What is the distribution system?

Where is the source of the raw product?

What is the schedule of operations?

What are the seasonal variations?

What are current and future docking, loading, and unloading needs?

This information is useful for considering other permit issues outside the General Permit.

ANALYSIS OF HARBOR AS A SYSTEM

The foregoing has been a basic discussion of the approach to be taken in developing a data base for a general permit study. It is by far the most involved data in connection with supporting a general permit and the most costly.

Once the project engineer has the pertinent data base at hand, then an assessment can be made of the various elements which combine to form a system analysis. For example, the following topics can be examined:

<u>Current level of Development</u>:

Questions: Is the system primarily recreational?

Is the system primarily commercial - seafood, recreational, etc.?

What is the impact of industrial activities?

What are the operational aspects of seafood industries, eg. product distribution, source of raw product, scheduling of operations, seasonal variations, etc.?

What is the character of existing waterway construction, eg. private boathouses, piers, docks, offshore mooring facilities (buoys, etc), bulkheads, revetments, etc.?

Where are the marshlands located and what is the general character of the wetlands?

Why has the harbor system adopted the observed growth pattern?

Are the existing facilities adequate to accomodate harbor uses?

Future Level of Development

Is the system expected to continue developing at the same rate?

What forces are at work to change development patterns?

Are there landside activities such as planned residential or commercial projects which might impact the system?

What is the trend in the economic importance of recreation and industry (particularly seafood) with respect to predicting future development?

Is the general geographic economy expanding, static, or declining?

Existing and Future Trends in Harbor Usage

What are the characteristics of vessel traffic?

How significant are problem navigation situations and where are they located?

What are the conflicts, if any, among recreational, industrial and commercial users?

Are sailboats creating or will they be expected to create navigation problems? (Note: There is an upward trend in the use of sailboats for recreation due to the increased cost of fuel,)

Should areas be designated as anchorages? If so, where should they be?

What are the draft and fairway requirements for current and future navigation?

ADMINISTERING THE GENERAL PERMIT

Once the study has been completed, then it supports the decisions which must be made with regard to issuing a general permit. As discussed earlier, only certain types of structures may be selected for a General Permit.

The author and the Corps have adopted the following approach to structuring, issuing, and administering the General Permit resulting from the foregoing described work:

1. The Project Maps are finalized to show

   a. Sufficient bathymetric contours to indicate channels, shoals, and general underwater features;

   b. All existing waterway and landside development including buildings, piers, docks, revetments, boathouses, etc.;

   c. Marshes and fringe wetlands;

   d. Navigation aids;

   e. Permit limit lines which are designed to limit the channelward extent of construction allowed under the General Permit and which outline the reserved fairways. The rationale for designing these lines is site specific and wholly depends on the results obtained via the study. Generally, they should allow construction where possible and allow boat docks and piers to extend channelward to waters deep enough to moor vessels without impacting navigation. (Note: this last benefit may be impossible to achieve, but if it can, it obviates the need for dredging);

   f. Accurate location of recommended anchorages or other reserved areas;

   g. Notes, legends, grid lines, location maps, tables, etc. necessary to complete the drawings.

2. A basic report is prepared to document the study procedures and results and to describe the harbor features and conditions. This report is non-technical in nature but it does address the overall rationale for structuring the General Permit. This report is reproduced in sufficient quantities to distribute on specific request to major respondents to the Public Notice of the General Permit. It would be expected to be distributed to other agencies and local governments but is <u>not</u> intended for general public distribution.

3. A brief supplemental report is prepared for in-house use by the Corps. It contains the technical back-up for decisions and recommendations with regard to criteria, Public Notice, and administration of permit requests. It also contains the geometric data on all permit limit lines and anchorages - usually in the form of plane coordinates.

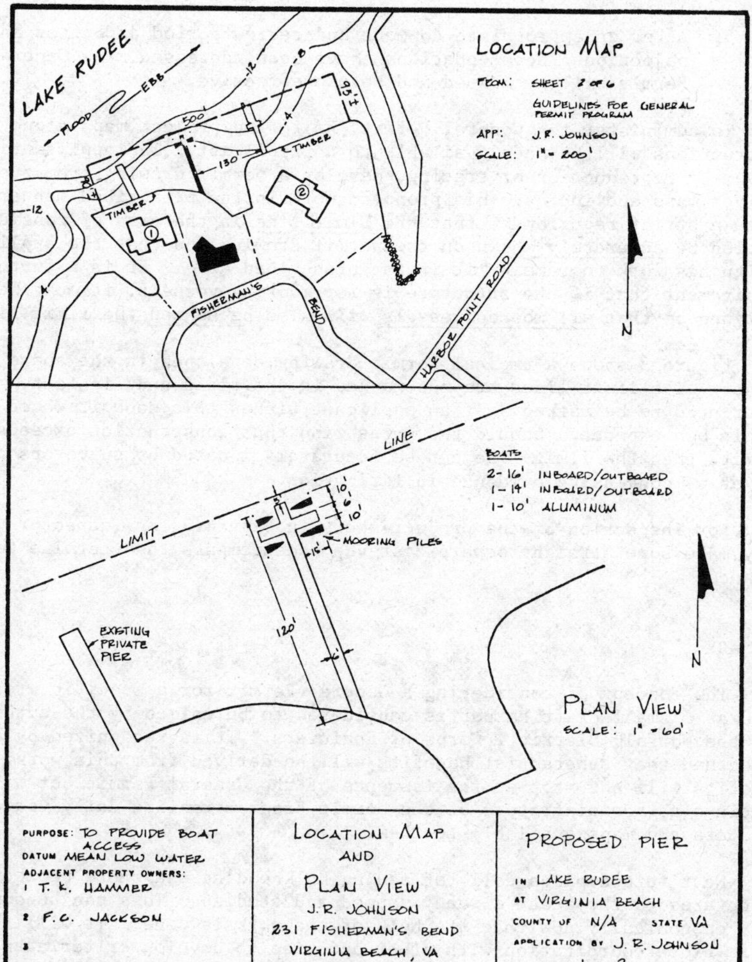
LAKE RUDEE
FLOOD
EBB
LOCATION MAP
FROM: SHEET 2 OF 6
GUIDELINES FOR GENERAL PERMIT PROGRAM
APP: J.R. JOHNSON
SCALE: 1" = 200'
500
130
TIMBER
FISHERMAN'S BEND
HARBOR POINT ROAD
N
BOATS
2- 16' INBOARD/OUTBOARD
1- 14' INBOARD/OUTBOARD
1- 10' ALUMINUM
LIMIT LINE
MOORING PILES
10'
6'
10'
15'
120'
EXISTING PRIVATE PIER
PLAN VIEW
SCALE: 1" = 60'
PURPOSE: TO PROVIDE BOAT ACCESS
DATUM MEAN LOW WATER
ADJACENT PROPERTY OWNERS:
1 T. K. HAMMER
2 F.C. JACKSON
3
LOCATION MAP AND PLAN VIEW
J.R. JOHNSON
231 FISHERMAN'S BEND
VIRGINIA BEACH, VA
PROPOSED PIER
IN LAKE RUDEE
AT VIRGINIA BEACH
COUNTY OF N/A STATE VA
APPLICATION BY J. R. JOHNSON
SHEET 1 OF 2 DATE

FIGURE 3

4. A Public Notice is issued by the District which, among other things, outlines what is covered by the General Permit and the geographical area it covers is described. It refers specifically to the project maps which explain the permit in more detail.

5. After an appropriate comment and review period and after any objections/recommendations have been addressed, the General Permit will be issued and become effective.

To administer the General Permit, half-size project maps along with instructions will be made available to an applicant. The applicant will, either by reproduction or tracing, develop a permit drawing from the project maps and indicate his proposed work in the prescribed manner. It is a strict requirement that the Limit Line in the area of consideration be accurately shown on the permit drawing and that the applicant design his work to remain within the prescribed area. It is a further requirement that if the structure is for mooring vessels, it must be designed so that all moored vessels will also be behind the limit line.

Figure 3 shows a typical Permit drawing developed in the above manner. The limit lines are not marked in the field and, in fact, may never need to be marked. If an applicant wishes this done it would be at his own expense. Should the Corps find that construction exceeds the permit, then the limit line can be accurately located by surveyors and serves as conclusive evidence in litigation.

For inspection of the project shown in Figure 3, an inspector need only make some straightforward distance measurements to determine compliance.

SUMMARY

The concept of considering a General Permit for a specific area such as a small-craft harbor is considered to be unique by the author and the Norfolk District, Corps of Engineers. It is recognized by all concerned that substantial benefits will be derived from this work. The benefits will not stop at the issuance of the General Permit but will continue as administrators of the permit program realize daily usage of the data and maps available to them.

Near future plans call for adding bulkheading and shore protection structures to the areas already mapped and studied. This can be done very economically now that the bulk of the work is done. It will require close coordination with other agencies to develop criteria and establish guidelines.

In the author's opinion, the approach to the General Permit concept outlined in this paper has wide application within the Corps' and other regulatory programs. It is a definite and positive approach to assisting regulatory agencies, reducing processing time, insuring orderly development, improving respect and awareness of the permit program, and protecting our COASTAL ZONE.

N. C. OIL REFINERIES - WHO MAKES THE DECISIONS?

Anne Taylor, Director
Office of Regulatory Relations
North Carolina Department of Natural Resources
and Community Development
Raleigh, North Carolina

Two refineries are proposed for the coast of North Carolina. The Brunswick Energy Company (BECO) is proposing a 150,000 bpd oil refinery on 1,900 acres of land one and a half (1½) miles west of the Brunswick River in Brunswick County, North Carolina.

Carolina Refining and Distributing (CR&D) Company is proposing a 30,000 bpd refinery in Carteret County North Carolina, on the Newport River.

Both proposed refineries fall within the requirements of the National Environmental Policy Act. The Wilmington District of the Corps of Engineers conducted Scoping meetings for each of these two refineries within a month of each other giving one a perspective on where they stand in the decision-making process.

The present status, however, places the CR&D at least a year behind the BECO project although a form of "jockeying" for position is continually taking place.

As of the date of this draft of this paper (August 1980) we anticipate receiving the BECO Environmental Report, which as you know is the company's draft of the Corps' eventual Draft Environmental Impact Statement, in September of 1980.

We do not anticipate receiving the Corps' Draft Environmental Impact Statement on the CR&D project until perhaps the Fall of 1981.

The purpose of this paper is not by any means to promote our approach to public involvement in government decision-making as the answer to the debate and sometimes the debacle on the Art of Public Participation.

Having personally participated in government decision-making as a volunteer for eight years prior to taking a position "on the inside", I believe we could meet the letter of every local, state and federal requirement for public participation and still not have effectively involved the public in the decision-making at hand. The whole issue of this new name for democracy rests fully in the <u>attitude</u> of those government officials charged with reaching the decision. If you truly believe in the right of those people affected by your decisions to be a meaningful part of that decision, then you will find a way, and you will have the courage to make the decision that will serve the most people for the longest time.

So the purpose of this paper is simply to share with you our approach and our experiences to date in attempting to involve the North Carolina public in the decision-making processes underway for these two refineries.

The attitude of the BECO offficials has been one of total openness and cooperation with government officials and all interested publics. The attitude of CR&D has been that they will do only what is legally required of them. My Department's official reaction to this latter attitude is that it makes it harder for us to do our job, to plan our work schedules, or to be helpful to the public and the company.

Briefly, let me set the tone for our general approach to the decision-making on these two refineries.

Public Participation - What is it?: To be Jeffersonian, I would simply say that good public participation is an informed populace making sound judgments. Or even more simply stated, public participation is democracy.

Who is responsible for it?: Every case will be different, but the responsibility remains with each government official making any decision which affects the lives of one or more citizens of this country. I would like to say that private industry officials are equally responsible to the public, but I accept that a lawyer is responsible to his or her client - guilty or not - and I accept that an industry is responsible to its stockholders - hopefully good business interests will not conflict with good public interest. Equally important, and not yet a reality in North Carolina, is the public's responsibility to take the time to find out and participate in what is happening around them, instead of reacting to efforts well underway.

How does it work?: Federal, State and local governments have requirements for public participation. Meet those requirements, analyze who will be affected, and what they need to know to make sound judgments on the issue. Above all, listen so that you hear. If you have already made the decision, then public participation is fruitless, because then you are selling a decision.

What are the problems?: Public confidence. It is not apathy - if the decision will affect them or their families, they care to be responsive to individual cares and concerns. The problem we face is the low credibility of government bureaucracy.

Virtually all of the local, State and Federal environmental laws were enacted during the decade of the 70's. Each was enacted at different times, by different congresses or legislators with varying levels of funding.

During these ten years the task of implementation has been delegated to various levels of government and departments, divisions and sections within government.

Each subdivision of government may be competently implementing its delegated piece of one or more of the many environmental laws. The public is feeling the brunt of, not overregulation, but fragmented management of regulation.

The basis of our approach to regulatory relations in North Carolina is that effective management of the full array of environmental laws will reveal whether the popular charge of overregulation is valid. Our experience to date has been that the laws and their regulations <u>are</u> in the best public interest, but that governments have not quite gotten their show together to implement the laws in a well managed fashion.

Fourteen states have some form of permit coordination, centralized permitting or consolidation of permits, so North Carolina is not pioneering into a newly identified problem area.

One obstacle or hurdle to comprehensive environmental management is government itself where professionals are reluctant to allow changes in well run programs. An invaluable educational exercise for these professionals is to ask them to play the role of a major industry, an interested citizen, or a local government and try to figure out what is expected of them in bringing an industry on line, developing a new water supply, or finding out who is protecting the environment of their community. Tunnel vision may not reveal any problems where a bird's eye view is a drastically different one.

The point is that while each subunit of government is doing its job well, they are not often doing it together.

The user, the affected public, and the involved regulators need to resolve conflicts through mediation, communication and negotiation - leaving the courts to a last resort. The analogy could be drawn with good labor management attitudes where the extra mile, beyond what is required by law, makes the difference.

As simple as a management approach should be, it is an unlikely event unless strong leadership is asserted and I believe this management leadership to be the challenge of the eighties.

Before I share with you the approach we used on the proposed BECO project, I will have to say that a major element was and still is missing.

Who made the first decision? There are hundreds of decisions the public, the user and the governments can and will make, but the very first domino is the major dilemma.

The citizens committee I will talk about later, and especially those members opposed to the refinery, have heard our industry recruitment arm of government say that it is initially up to the local government whether any indistry considers locating in their area; the company involved has indicated that the "state" was receptive to inquiries, others say a study funded by a federal grant identified sites in the state. It is an interesting question - hard to track down. The point is the general public was not involved in that first conscious decision. I would speculate that this is where meaningful land use planning becomes dramatically important. When a major industry decides to locate in your community, it is too late to start asking, "What do you want your community, your state, to be."

The North Carolina Action Plan for the BECO refinery was developed as follows:

I. INTRODUCTION

A. Natural Resources and Community Development is the primary agency of state government for environmental protection.

B. Natural Resources and Community Development is looked to as the primary agency for implementation of the state's environmental policies.

C. The proposed Oil Refinery represents a major action affecting not only the economic and environmental well being of the local community but will have an effect on the region and the state.

II. FUNCTION OF NATURAL RESOURCES AND COMMUNITY DEVELOPMENT WITH REGARD TO PROPOSED OIL REGINERY:

A. To ensure adequate analysis of environmental/natural resource factors for decision-making purposes.

B. To serve as the company's contact for coordination of state permits in cooperation with federal and local permitting agencies.

C. To facilitate public involvement in the total decision-making process.

To carry out the functions of Natural Resources and Community Development with regard to the proposed Oil Refinery, Natural Resources and Community Development proposes to implement a plan of action with the assistance of three committees:

1. TECHNICAL STEERING COMMITTEE

A. A Technical Steering Committee has been formed within the Department of Natural Resources and Community Development which will call upon scientific expertise on all issues of environmental concern to ensure adequate analysis of issues for decision-making purposes. The Technical Steering Committee is made up of:

TSC, Chair, Bob Jamieson, Regional Manager, Wilmington
Natural Resources & Community Development

Dr. Neil Grigg, Assistant Secretary
Natural Resources

B. J. Copeland, Sea Grant
N. C. State University

Bill Ross, Legal Affairs
Natural Resources & Community Development

Frank Yelverton, Wilmington Office
Corps of Engineers

Charles W. Hollis, Wilmington Office
Corps of Engineers

Ken Stewart, Coastal Management
Natural Resources & Community Development

Everette Knight, Consultant
Department of Commerce

Steve Conrad, Director, Land Resources
Natural Resources & Community Development

Page Benton, Environmental Management
Natural Resources & Community Development

Connell Purvis, Marine Fisheries
Natural Resources & Community Development

Bill Raney, Attorney General's Office
Department of Justice

ROLE OF TECHNICAL STEERING COMMITTEE

1. To establish a primary contact person in state government for the purpose of permit coordination (Bob Jamieson).

2. To establish an understanding and close focal point of communication between the company and the state on the legal (permitting) requirements during the decision-making.

3. To relate local and federal permitting requirements to the state's permitting requirements to ensure total permit coordination.

4. To serve as liaison with the NRCD Citizens Advisory Committee and the State Government Interagency Task Force:

    a. to translate the permit coordination process into lay language for public understanding and to facilitate public awareness of opportunities for citizen involvement in the decision-making process.

    b. to respond to technical issues raised by the Citizens Advisory Committee and other state agencies.

5. To serve as a technical resource for the state's primary contact person (Bob Jamieson) and for the local field office technical contacts.

6. To identify and involve experts in the scientific community on all issues of environmental concern.

2. STATE GOVERNMENT INTERAGENCY TASK FORCE (reviewers)

A. A state government interagency liaison capability has been established to further ensure adequate consideration of the concerns and interests of other departments of state government in the decision-making process.

B. This Task Force is to encourage early and continued involvement of every department of state government to ensure that all areas of interest are adequately considered.

Members of the Task Force include:

Department of Human Resources Division of Health Services	NRCD Division of Environmental Management
Department of Education Division of Community Colleges	NRCD Parks and Recreation
Department of Cultural Resources	NRCD Economic Opportunity
Department of Administration Office of Marine Affairs	NRCD Housing Finance
Department of Insurance Division of Engineering	NRCD Community Employment
	NRCD Forest Resources
Crime Control and Public Safety	NRCD Wildlife
Department of Agriculture	NRCD Soil and Water
Department of Corrections	NRCD Water Resources
Department of Labor	NRCD Office of Coastal Management
Department of Revenue	NRCD Land Resources
Department of Transportation	NRCD Division of Marine Fisheries

3. CITIZENS ADVISORY COMMITTEE (public interest)

An 18 member Citizens Advisory Committee has been appointed by the Secretary of Natural Resources and Community Development made up of citizens from across the state representing a broad range of interests.

A. To advise the Secretary of the Department of Natural Resources and Community Development of interests, issues and concerns relating to the location of the oil refinery in Brunswick County.

B. To facilitate two-way communication between the Citizens Advisory Committee and their respective constituents.

C. To participate fully in each point of permit and other legal decision-making.

D. To advise the Secretary of the Department of Natural Resources and Community Development of issues/decisions to be made which are not addressed in the legal requirements of the decision-making process.

E. To serve as liaison with the Technical Steering Committee.

F. To advise the Secretary on ways to more fully involve the general public in the decision-making process.

G. To facilitate full public understanding of the complex regulatory process.

Membership of the Citizens Advisory Committee includes:

Tom B. Rabon, Jr.
Paul Foster
Dr. Jay Langfelder
Grace Bonner
Dr. Orrin Pilkey
Denny Shaffer
Wayland Vereen
Gilliam Wood
Graham A. Phillips, Jr.
Sue Bulluck
Henry S. Goodwin
Dr. John Calfee
Ellen Williams
Ron Newman
Kenneth Bellamy
Steve Varnum
Billy Russ
Anne Taylor

Briefly, activities undertaken to meet these purposes have been as follows:

CAC APPOINTED BY SECRETARY February 1979
TSC SELECTED February 1979
INTERAGENCY COMMITTEE CONTACTED April 1979

BECO LETTER RECEIVED REQUESTING PERMIT INFORMATION February 28, 1979

PERMIT COORDINATION MEETING HELD April 11

CAC MEETINGS HELD May 15, June 14, Sept. 1979; June 1980

CORPS OF ENGINEERS SCOPING MEETINGS May 22-23

CAC TOUR OF REFINERIES IN TEXAS AND MISSISSIPPI June 27-29, 1979

CAC EMB RESOLUTION October, 1979

TWO CONSULTANTS HIRED January 1980 (Holt) and March 1980 (Muga)

SEA GRANT STUDY PUBLISHED February 1980 (Requested by TSC) "A Study of Federal and State Legislation Concerning the Construction of Proposed Oil Refineries"

BECO PROPOSED OIL SPILL SCENARIOS-consultants comments

PSD PERMIT APPLICATION RECEIVED BY NRCD May 30, 1980

PERMIT COORDINATION AGREEMENT CONFIRMED, PUBLIC HEARINGS CONSOLIDATED
June 3 Permit Meeting

SHIFT FROM TSC TO PERMITTING AGENTS

BECO ENVIRONMENTAL REPORT DUE July 15; now anticipated September 15. 64 copies CAC, TSC, Clearinghouse, Interagency

SPECIAL REPORTS REVIEW

DRAFT EIS

During the time all of this was taking place, rules and regulations as required by North Carolina's Administrative Procedures Act were being formulated for oil refinery facility permits.

SUMMARY OF PUBLIC INVOLVEMENT IN THE DEVELOPMENT OF THE OIL REFINERY PERMIT REGULATIONS

1. In March of 1979 approximately 700 notices of public hearing were mailed to citizens throughout the State. Legal notices of public hearing were published in three newspapers in the State. Formal public hearings were held on the first draft of proposed regulations on April 11, 1979 in Raleigh and April 13, 1979 in Wilmington.

2. The hearing record was open for public comment on the first draft of proposed regulations from March 26, 1979 (the date the hearing notice was first published) to July 16, 1979. A period of approximately five (5) months.

3. Approximately 75 people attended these two hearings and an additional 30 letters commenting on the regulations were received while the hearing record remained open.

4. The regulations were revised based on public comments received, and while not required by law, additional public hearings were scheduled for further public comment on the revised regulations. Governor's request.

5. Notice of public hearings were published in three newspapers on the State and copies of the regulations were mailed to approximately 950 individuals and organizations in North Carolina. Twenty-one press releases were mailed to newspapers across the state.

6. Formal public hearings on the second draft of the proposed regulations were held on March 4, 1980 in Raleigh and on March 5, 1980 in Wilmington.

7. Approximately twenty people appeared at the Raleigh hearing and approximately eighty people at the Wilmington hearing.

8. The hearing record is required by law to remain open for ten days after public hearing, but upon Governor Hunt's request, the time limit for public comment was extended.

9. Prior to the closing of the public hearing record on April 8, 1980, an additional thirty five letters commenting on the second draft of the regulations were received.

10. In addition to the formal public participation outlined above, briefings and discussions on the regulations were held at regularly scheduled meetings of the Brunswick Energy Company Citizens Advisory Committee, and the Brunswick Energy Company Technical Steering Committee established by Natural Resources and Community Development and with all individuals or group of individuals requesting a meeting.

After which one of the three major newspapers of the State published the following editorial:

"REFINERY SKIDS GREASED"

> The Hunt Administration is rushing state regulations on oil refineries with disturbing speed. Little publicized hearings were scheduled March 4 and 5 in Raleigh and Wilmington respectively. March 14 was set as the deadline for subsequent written comment. Nine days is hardly adequate time to make the hearing transcripts available and to get comment on the issues raised.

Moreover, the state's Environmental Management Commission is scheduled to meet on March 14, the day of the comment deadline. The EMC, which will have the duty of deciding whether to issue certain permits for refinery operation, can hardly be in a position on that day to discuss and comment on the regulations and analyses of them intelligently.

The EMC's April meeting would be a more appropriate moment for review of the regulations. Delaying consideration of this issue would allow extending the comment period by at least 30 days.

The regulations finally adopted, of course, will have much to do with whether North Carolina approves the plan of the Crown Petroleum Company and its partners to build a 150,000 barrel-a-day refinery on the Cape Fear River in Brunswick County. Another, smaller refinery in Carteret County may also be at stake, but the Hunt administration's more immediate concern is that the Brunswick refinery pass environmental muster.

A key question attached to the project is whether North Carolinians will get enough information on the refinery's effects on the state and its coastal region and the opportunity to debate the pros and cons. The process now being followed by the N.C. Department of Natural Resources and Community Development suggests that the administration wants as little public participation and comment on the basic regulations as possible.

The impression, in other words, is that the skids are being greased for the approval of coastal refineries. If that is not the case, the Department of Natural Resources and Community Development should extend the hearings comment period and do more to publicize the details of its proposed regulations.

As we anticipate the Environmental Report on this project, the public confidence in the role of the department of state government charged with the wise use of the state's natural resources becomes critical. If we are truly into a decision making process, then who makes the final decision? If after full mitigation, mediation, and negotiation among all the affected publics; the user, and the various levels of government come together in a non answer or a hundred pieces of answers, who will make the decision?

Is it realistic to think that the general public really has a choice? If all requirements of the various environmental laws are met by the company and permits are issued, then North Carolina will have a refinery. The economy may make the choice in spite of whether the project is needed and wanted or not. It would not be impossible, prior to North Carolina's succession vote, that three or even four governors of the state would be credited or blamed for the crossroad decision to have or not to have a refinery. But it is not factually correct to say one governor made the decision. The Corps of Engineers issues the "biggest" permit, but our department issues more permits. But it is still not the Corps or Natural Resources and Community Development who makes the decision.

Right now the Draft Environmental Impact Statement is looked to by the public as the accountability or disclosure document to determine whether the project goes or not.

But a birdseye view of the entire process goes back to that first domino. When a community and a state consciously decide what they want their future to be, and remain aware and involved in any change in that decision, the decisions that follow the first domino are directed toward how to get the very best of what we want. I believe the laws are sufficient to ensure the very best. But the laws will not decide for a community or a state what it is they want.

When asked by a friend recently, "Who really decides whether this refinery is best for us in the long run?", I told her the people decide. But just as government can do a much better job of carrying out its responsibilities in implementing the laws of the land in the best public interest, the people must do a better job of deciding what it is they want and demanding a government responsive to their chosen quality of life.

North Carolina has local land use plans in all twenty coastal counties and the county and Coastal Resources Commission are now well into a scheduled update of the Brunswick County plan. The BECO project is not inconsistent with the existing plan. We are assured extensive public participation in the update partially because of the proposed refinery. Here again, however, meeting the minimal requirements of the State Coastal Area Management Act is a major public participation effort, but the extra mile need not and probably cannot be legislated. The local community can and should use existing systems of democracy - local elections, attitudes of public employees, daily involvement in all aspects of decision-making to plan and implement their chosen future - not just the opportunities provided under the Coastal Area Managment Plan.

RESOLUTION OF SHORELINE DISPUTES

By
Peter L. Buck,[1] J.D.
and
Jeffrey A. Layton,[2] M., ASCE

1. NEED FOR DISPUTE RESOLUTION

1.1 Risks Involved in Disputes with Agencies. Disputes between project proponents and agencies can spell the death of a project. If an agency's concerns cannot be eliminated, then necessary approvals may not be granted. A problem of equal importance is the years of delay which can ensue due to a dispute with an agency.

1.2 Risks Involved in Disputes with Citizen Groups. Although citizen groups do not have formal approval power, they can exert significant influence on the process. Serious citizen opposition can result in greatly lengthened agency processing and improve the chance of denial. Additionally, such opposition greatly increases the chance that the agency will require an environmental impact statement which can lead to years of additional delay.

Even after the approval process is completed, citizen opposition can seriously impede a project through litigation. The state of environmental law changes frequently enough that there is always some significant danger of approvals being overturned by courts. Also, citizen-initiated litigation can delay utilization of permits for several years.

1.3 Effects of delay on Proponents. The effects of delay can be disastrous for the private sector. Although a project may be feasible when starting the approval process, delay can work a significant change in a variety of manners. Market conditions may change making the project unviable. Financing which was available at the start of the process may no longer be available. Other agencies may adopt more restrictive regulations or requirements which make a formerly feasible project infeasible. Meanwhile, various carrying costs associated with the land and with the proposed project will continue to add up.

1.4 Effects of Delay on Citizen Groups. Although conventional wisdom suggests that citizen groups benefit by delay, that is often not the case. Citizen groups rarely have anything to gain financially from opposing a project. Nevertheless, citizen groups normally will have to pay a great deal in attorney and expert fees to effectively mount opposition. If the process is delayed through a number of different procedures or through protracted litigation, the costs for

---

[1] Attorney, Law Offices of Peter L. Buck, P.S., 1111 Third Avenue, Seattle, Washington 98101.

[2] Consulting Coastal Engineer, Layton & Associates, 1111 Third Avenue, Seattle, Washington 98101.

citizens can mount up tremendously. Additionally, citizen involvement, both to participate in agency proceedings and to raise money, requires a tremendous amount of time and energy which is rarely productive and often frustrating. The longer a process goes on, the greater the chance that significant segments of the opposition group will be lost through attrition.

2. TYPICAL MYTHS BARRING DISPUTE RESOLUTION

2.1 Myth No. 1: Issues Involved are Non-negotiable. It is often suggested that environmental issues are non-negotiable. Most project proponents believe that they are fighting for their economic survival and cannot give up any portion of their project. Many citizens believe that they are protecting the future quality of their life and that of future generations and that there is no room for compromise.

As a practical matter, however, neither side can predict with certainty what will happen through protracted disputes before agencies or in the courts. There is, therefore, great danger that the proponent will lose its entire project or that the citizen group will gain no significant environmental protection if an extended battle ensues. Consideration of these realities has led to resolution of a number of hard-fought, emotional, environmental disputes which were originally considered non-negotiable. Settlement of these cases rebuts the belief that environmental disputes cannot be settled.

2.2 Myth No. 2: Resolution with Some Citizens Will Not Bar Other Citizens from Challenging the Project. Most proponents are concerned that if they settle with some identifiable citizens, then others will materialize to challenge a project. There unfortunately is no way to tie down the entire universe of potential litigants.

This problem has been resolved through careful drafting of non-opposition clauses. (*See* Appendix A) Even the most careful drafting, however, cannot eliminate the possibility of other opposition. One solution is to wait, if possible, until applicable statutes of limitations have expired. Unfortunately, for Corps of Engineers approvals there is no helpful statute of limitations.

As a practical matter, it is the authors' uniform experience that almost everyone who is concerned enough about the project to effectively litigate against it will appear during the first round of litigation. Any further opposition would be by a small, relatively weak segment of the population. The success many proponents have had in settling with the known opposition rebuts the belief that this fear should bar settlement.

2.3 Myth No. 3: Settlement Provisions Cannot Be Enforced. Inevitably, lay people are concerned that various agreements to settle a dispute will not be enforceable. This is particularly a concern in the pre-approval and pre-litigation stages.

Fortunately, almost any agreement can be reduced to writing. There is valid consideration running from both sides which makes such an agreement a binding contract. If litigation is pending, the agreement can be in the form of a stipulated order and judgment. Additionally, any agreement can be recorded and bind successor owners of the affected properties.

2.4 Myth No. 4: It Is Too Early to Settle a Dispute Prior to Approvals Being Issued. Most people feel that settlement of disputes prior to agency approval are premature. This is particularly persuasive since it is more difficult to identify all potential objectors and since the agency will still be involved in the final detailing of the project to be approved.

Actually, it is often easier to resolve a matter prior to the agency's final stamp of approval or disapproval. Both sides are still left with uncertainty as to the actual outcome. Additionally, the agency can be more flexible in its decision-making role prior to approval. The main drafting concern is to provide for a contingency if an agency will not subsequently approve what was agreed upon between a project proponent and a citizen group.

2.5 Myth No. 5: It Is Too Late to Settle a Case After Agency Approval. Many people feel that after an agency has actually approved a project it is too late for citizens and project proponents to settle a dispute whether or not litigation has ensued. Often it is believed that an agency will be reluctant to modify its approvals.

Almost any dispute between citizens and proponents, however, embroils an agency in controversy which it would like to avoid. For that reason an agency is usually quite desirous of avoiding potential litigation or terminating actual litigation to which it is probably a party. Agencies are also willing to assist settlements since the settlement typically involves greater environmental protection than was contained in original approvals.

If the dispute is between an agency which has denied approval and a proponent, settlement obviously can result from reconsideration by the agency. One caveat is that the agency will usually have to comply with required procedures such as hearing and notice.

3. SETTING THE STAGE FOR NEGOTIATION

3.1 Convincing Citizens They Should Consider Settlement. Usually settlement discussion must be initiated by the proponent of a project. One disadvantage for whatever party initiates settlement discussion is that it signals a weak position which reduces bargaining power. For that reason, a proponent should consider asking a third party such as an agency or an agency's attorney to initiate settlement discussion.

Unfortunately, many citizen groups do not feel the need to settle until well into the process. In that instance, it is necessary to cause such groups to encounter the negative effects of these disputes as early as possible. This can occur through extensive motions or discovery in litigation or through a countersuit if appropriate grounds exist. Those tactics can backfire if abused, however.

Another means of getting citizens interested in settling is to proceed with construction. Even if the entire project cannot be constructed pending resolution of litigation, often visible aspects can proceed without great financial commitment. It is also sometimes possible to proceed with certain elements of a project which effectively reduce or eliminate the citizen group's major concern. For example, if a citizen group is primarily concerned about preserving a particular greenbelt but does not want to discuss settlement, threatened or actual removal of the greenbelt will either encourage settlement or effectively weaken the citizen group's ability to retain citizen support.

Another technique for making citizen groups more amenable to settlement discussion is to demonstrate a willingness and ability to eliminate errors which are potentially their best issues in litigation. For instance, if citizens allege that some required procedural step was not conducted properly, the proponent and agency should redo the process while litigation is pending to demonstrate the futility of continued litigation.

3.2 Creating a Framework for Citizen Settlement. It is always important to know the identity of the group a proponent believes it is settling with. If a citizen group is involved, the extent of the group's membership and the authority of its spokesperson to settle on behalf of that membership should be ascertained.

At the outset it is important to establish an express understanding that both sides are negotiating in good faith to resolve a dispute through settlement. Such an attempt is to be distinguished from meetings to simply communicate information or exchange views.

An express understanding that both parties are negotiating in good faith can be effectively accomplished through a letter outlining the proposed settlement procedure. If both sides will agree upon a proposed set of meetings between spokespersons, review by principals, and further meetings, there will not be a misunderstanding about the seriousness or intent of the discussion.

It is important to create time pressure in which a settlement should occur. Hopefully, this can be done through utilization of some real time pressure such as a motion deadline. Usually it is possible for a proponent to create time pressure by some tactic such as filing notice of depositions and then continuing those to a date certain so long as settlement discussions are progressing.

3.3 Creating a Framework for Settlement with Agency. Most agency personnel will be interested in finding ways to resolve problems. If not, the problems which are in dispute should be identified as ones which will lead to protracted conflict involving higher levels of agency personnel and undesirable publicity. Occasionally, there are also questions of agency jurisdiction or authority which lend themselves to mild or implied threats of litigation.

A proponent should constantly demonstrate a good faith willingness to take any reasonable steps to solve potential problems. This attitude must be communicated clearly to agency personnel. Additionally, it is important to meet as early as possible with the agency and involve agency personnel in design decisions.

4. DISPUTE RESOLUTION METHODOLOGY

4.1 Overview. In any dispute resolution, the first task is to catalog all important concerns which must be mitigated to satisfy the agency or citizen objector. Then an agreed upon mitigation should be developed for each concern. Mitigations will take the form of (1) project plan modifications, (2) operating conditions or (3) off-site tradeoffs.

4.2 Identifying Concerns. Ideally each side in a dispute should separately make a detailed list of problems that they believe are present. The objector should prioritize those concerns.

These concerns should be reduced to writing and lists exchanged. Such an exchange will often help clarify whether or not there is an understanding of the objector's position. The proponent can utilize its list to demonstrate a good faith understanding of problems.

4.3 Conceptualize Solutions. Each side should also be asked to independently develop concepts for solving problems. At this stage each side should be asked to identify at least one project modification, operating condition or tradeoff that could potentially solve the problem. It is critical at this stage for all parties to use their imagination.

Lists of potential solutions should then be exchanged. At that point it is also appropriate to have a meeting and begin discussion of reducing the concepts to precise measures.

4.3 Memorializing Solutions. As potential solutions become more obvious, it is important to develop precise language that protects the objector but at the same time allows the proponent to have a viable project. Plan modification should be made part of definite site plans which are attached to the settlement agreement. Operating conditions should also be drafted with a high degree of specificity and incorporated in an agreement. Both sides should be prepared to develop numerous drafts of such a document.

5. PROPONENT EVALUATION OF A SETTLEMENT

Typically, the proponent of a project will give up a great deal more than originally contemplated. Unfortunately, in such a settlement almost all of the "giving" which is easily identifiable is from the proponent's side. For that reason it is important for the proponent to carefully evaluate any settlement.

A proponent must first evaluate his chances of ultimately prevailing in a dispute with either the agency or a citizen objector. This involves evaluation of chances of obtaining the necessary approval and of defending it successfully in litigation.

The proponent must also evaluate the cost of a full fight both for itself and for citizen or agency objectors. Typically, there will not be substantial costs for an agency in such a dispute, but the costs for both the proponent and a citizen group may be significant. Additionally, the proponent must evaluate the cost of delays which can be realistically expected in a full fight.

The proponent must also consider the precedential effect of a loss either at the approval stage or in litigation. Most private proponents will not suffer serious precedential effects except for the particular project involved. Agencies, on the other hand, have greater reason to be concerned about precedent.

The final feature to evaluate is what extent of harm the proponent will encounter if it ultimately loses either the approval or litigation. At the minimum, a great deal of delay will be encountered requiring reprocessing of approvals. At the maximum, all of the development costs and the value of any real property acquired may be effectively wiped out.

6. DEVELOPING SETTLEMENT AGREEMENTS

6.1 Citizen or Agency Concerns.

6.1.1 Binding future property. All agreements should provide that they bind successors and run with the real property involved to bind all future owners. They should be filed of record to become an encumbrance on the property. There should also be an agreement to provide a copy of the document to any future owners.

6.1.2 Specific plans. It is important to attach to the settlement document a copy of specific plans and diagrams which specify what will be done. As much detail as possible should be included.

6.1.3 Time for performance. The settlement agreement should specify the time within which certain things shall be accomplished. If mitigating measures or other types of amenities are included, it should be specified that those shall be completed prior to or simultaneous with use or occupation of the particular project.

6.1.4 Continuing obligations. Continuing obligations should be specified as such. For example, landscape buffers should be developed with a performance standard as to maintenance in a healthy condition.

6.1.5 Monitoring. Monitoring for both one time and continuing obligations should be provided by disinterested professionals. Such monitoring should be paid for by the proponent with original copies of reports to the objecting citizen or agency.

6.1.6 Enforcement. The agreement should specify that it may be enforced by the objector and perhaps by any other affected government agencies or citizens. It should specify that there is no "remedy at law" and provide for the appropriateness of mandatory and prohibitive injunctions. There should be an agreement as to practical means of accepting service and shortening times for court hearings. Attorney and expert witness fees should be provided to a successful party in litigation. Costs provisions should be enforceable if it is necessary to institute litigation but the proponent complies prior to resolution of the litigation.

6.1.7 Materiality. The settlement agreement should further provide that all provisions are material and substantial and may be enforced by injunction.

6.2 Proponent Concerns.

6.2.1 Delineation of parties. The objecting parties who are settling should be specified with clarity. It should be made evident who in a group is actually being bound. Additionally, the position of signators and their authority should be warranted.

6.2.2 Schematic plans. The proponent should try to keep any attached site plan conceptual or schematic in nature, with the right to modify. Obviously, there will have to be some constraints on the right to modify.

6.2.3 Amendment provisions. It is important that the objector be required to designate an authorized group to improve revisions in the plans or in the agreement. This may be certain designated officers of an organization or a specified percentage of individuals involved.

6.2.4 Support of non-opposition if possible. The objector should be asked to support the revised project. If that is not feasible, a non-opposition clause for both the objectors and their attorney should be included.

6.2.5 Time limit. If additional permits and approvals are required, there should be some time limit specified within which the proponent can reasonably expect to receive those approvals. There should be a right of extension by the proponent as long as good faith applications for permits are pending.

6.2.6 Return of parties to original position. If the settlement cannot be implemented due to other parties bringing litigation or due to governmental agencies which will not approve plans, then the parties should be returned to their original positions as much as possible. Alternatively, the settlement agreement should be held in escrow until it can be determined whether or not agencies will approve plans or until statutes of limitations barring other litigation have run.

7. ARBITRATION OR MEDIATION

Serious consideration should be given to the use of mediators to assist the parties in reaching an agreement. If all parties believe that they are operating in good faith, it may be appropriate to utilize a mutually agreed upon arbitrator. Assistance in this regard can be achieved from the various groups such as the American Arbitration Association.

Additionally, certain organizations have been developed to resolve environmental conflicts through mediation. For instance, the Office of Environmental Mediation, Institute for Environmental Studies, University of Washington, has been funded by grants from the Ford Foundation and Rockefeller Foundation to assist in resolving environmental disputes. It has been used to resolved disputes involving flood control, wetlands protection, port development, transportation system development and a variety of other issues. The office has suggested five critical determinants of whether or not mediation is appropriate in a given situation:

1) Mediation involves the use of a "third-party" intervenor who operates from an impartial base and whose primary role is to promote agreement among the conflicting parties.

2) The mediation process is most appropriate at the point in a dispute where the issues have been defined, the parties are visible and highly involved, and there is some sense of urgency for resolving the conflict. Such circumstances typically occur near the point of impasse.

3) In order that the parties to a dispute will be willing to enter with "good faith" into the negotiation-mediation process, there must be some relative balance of power: Each of the parties must have the ability to exercise effective sanctions against the other(s).

4) Mediation is a decision-making process and, if the parties themselves cannot directly implement an agreement, it is appropriate in situations where there are reasonable assurances that the responsible authority or authorities will implement an agreement reached by the disputing parties.

5) Mediation will result in compromises being made

and certain issues of principle may not, therefore, be appropriate for mediation. (See Reference 5)

8. CASE HISTORY

The Point Roberts Marina located in Washington State south of Vancouver, British Columbia, is a 1,000-berth basin dredged from former agricultural land. Through the course of the planning, design, and construction of the marina, many environmental and social issues were encountered and addressed. Resolution of the public access issue is representative of the techniques used to finish the approval process in only 12 months.

The public's interest in and opportunity to enjoy the physical and aesthetic qualities of natural shorelines of the state are well documented in the Washington State Shoreline Management Act of 1971. The objective of increasing public access to publicly owned areas of the shorelines are explicitly stated in the act; however, guidelines for managing public access on privately owned areas are not explicitly defined.

Since the marina project was a private venture, local and state agencies attempted to use the "intent" of the shoreline act to insist on total public access to all of the project's marine shoreline (approximately 8,000 feet). As an example, the developer proposed a 60-lot single-family subdivision fronting on approximately 1,300 feet of natural shoreline adjacent to the Strait of Georgia. This proposal was consistent with existing land uses and zoning on the Point Roberts Peninsula, which is primarily single-family residential with privately owned shorelines and tidelands. A dispute arose between the developer and regulatory agencies over public access to the subdivision's shoreline. The agencies wanted this beach area opened completely to the public. The developer felt that allowing unrestricted public access would significantly reduce the market value of the waterfront lots. Without the projected maximum income from the sales of these lots, the developer felt the entire project's financial success would be in jeopardy. Thus, an impasse was reached between the agencies and developer over this issue. The issue was finally resolved by a process of negotiation and compromise between the developer and regulatory agencies, with the developer's consultants acting as mediator. The results of the negotiation process are described below.

To break the deadlock, the consultants negotiated a three-part solution with the agencies. First, the agencies agreed not to require open public access to the subdivision's shoreline as a condition of project approval. However, the developer would be required to provide access to the shoreline by all residents of the project, even though they might not live in the subdivision. This meant that at maximum development approximately 1,200 residents could have access to the shoreline. Although this solution was not as attractive as limiting access to the shorelines to only residents of the subdivision, the developer agreed to this condition.

The second element of the compromise concerned the loss of natural shoreline dredged for the marina entrance channel (approximately 200 feet wide). To compensate for the loss of this shoreline, the local agency requested that 350 feet of shoreline adjacent to the east side of the entrance channel remain as open, undeveloped shoreline dedicated for public access. The developer was of the opinion that at a market value of $1,500 per linear foot, dedicating over one half million dollars worth of private shoreline for public access was unreasonable. However, the developer concluded that the chances of favorable approval for the marina project were greatly reduced if public beach access were not provided. Therefore, the developer agreed to dedicate the 350 feet of shoreline for public use.

The final element of the compromise concerned the creation of the new boat basin shoreline. Approximately 6,000 feet of new shoreline was created with construction of the boat basin and entrance channel. Local agencies viewed the shoreline as an extension of the existing shoreline and requested open public access adjacent to its entire length. The developer was concerned with the additional loss of land area, use conflicts between the public and boat customers, subdivision residents and condominium living unit owners, security for the development, the need for additional public services, and maintenance-related problems. These issues were again resolved by negotiation and compromise with the developer agreeing to provide an unobstructed 10-foot-wide walkway around approximately two-thirds of the perimeter of the boat basin complemented with rest areas and support facilities.

In view of the developer's not wanting to give away large amounts of valuable marine shoreline and in view of local and state agency directives to increase public access to shorelines, irrespective of explicit guidelines and regulations, the final recreational plan for the marina project was a result of tradeoffs of private and public interests. Both the developer and the agencies are responsible for the recreational amenities and extensive public access currently provided by the marina project. The developer maintained his overall project objectives by receiving early project approval and retaining control of much of the project's shoreline; the regulatory agencies maintained their overall perspective of maximizing public benefits by gaining a considerable amount of shoreline for public access. (See Reference 6)

9. ENVIRONMENTAL DESIGN PROCESS

To help insure a project proponent's chances of successfully resolving a shoreline dispute, it is very important that the project be carefully planned from its beginning. A weak design concept coupled with poor execution of the concept during the permit approval process will greatly weaken the proponent's negotiating position if disputes should occur. A successful planning method available to developers to expedite the shoreline approval process is the environmental design process. This process is a cyclic planning process which allows a project to remain flexible in its design during the permit approval process so that disputes with agencies or citizen groups can be resolved without significant detriment to the proponent's project. A detailed discussion of the environmental design process is provided in Appendix B. Also, refer to "Expediting the Coastal Approval Process," Coastal Zone 80 Proceedings, and Reference 7 for additional discussion.

REFERENCES

1. Buck, P.L. Environmental Litigation. "Proceedings of the Practising Law Institute's Course on Environmental Law and Practice," Litigation and Administrative Practice Course Handbook Series, Number 117, San Francisco, July 1980.

2. Buck, P.L. Challenges to EIS Adequacy. "Proceedings of Washington State Bar Association Course on Land Use and Environmental Law Practice," Seattle, June 1980.

3. Buck, P.L. A Practical Approach to Land Use Litigation and Land Use Dispute Resolution. "Proceedings of Practising Law Institute Conference on Modern Control of Land Development," Real Estate Law and Practice Course Handbook Series, Number 175, New York City, 1980.

4. Buck, P.L. Citizen Litigation. "Proceedings of Litigation Workshop," sponsored by League of Women Voters, Seattle, 1979.

5. Cormick, G.W. Resolving Environmental Conflicts Through Mediation: Experience, Process and Potentials. Institute for Environmental Studies, University of Washington, Seattle, 1978.

6. Layton, J.A., J.D. Hansen and P.L. Buck. Recreational Planning Features of a Large Marina. "Proceedings of ASCE's Recreational Planning and Development Conference," Snowbird, Utah, 18-21 April 1979.

7. Layton, J.A. and J.D. Hansen. Environmental Design Process: Small-Craft Harbor. "Proceedings of Ports 77 Conference," ASCE, Long Beach, California, 9-11 March 1977.

APPENDIX A

IN THE SUPERIOR COURT OF THE STATE OF WASHINGTON FOR PACIFIC COUNTY

SAVE OUR SHORELINES ASSOCIATION, a Washington non-profit corporation, Plaintiff, vs. TOWN OF LONG BEACH, a municipal corporation of Washington, SHORESIDE APARTMENT ASSOCIATES, Defendants.	NO. 89591 STIPULATION AND AGREEMENT

WHEREAS, Save Our Shorelines, a Washington non-profit corporation (hereinafter "S.O.S.") is the plaintiff and the Town of Long Beach (hereinafter "Town") and Shoreside Apartment Associates (hereinafter "Applicant") are defendants in the above-captioned cause which involves the challenge of a decision by the Town not to prepare an environmental impact statement in connection with development by the Applicant of property in Long Beach; and

WHEREAS, Applicant has proposed to construct a 36-unit low-income apartment complex on certain waterfront property in the Town and to fill the remainder of that property and have received certain governmental approvals authorizing such construction; and

WHEREAS, Applicant has obtained a decision from the Town that an environmental impact statement is not required in connection with the proposed project; and

WHEREAS, all parties believe that they are entitled to a decision on the merits, but at the same time are aware of the unavoidable delays, costs and uncertainty of results if this case proceeds through the normal litigation process and recognize the desirability of terminating this action without further cost to anyone; NOW, THEREFORE,

IT IS STIPULATED by and between S.O.S., the Town, and the Applicant, through their attorneys who are authorized to execute this Agreement as follows:

The Applicant agrees that at its own cost it shall forthwith cause preparation of an environmental impact statement prepared pursuant to the Washington State Environmental Policy Act, RCW Ch. 43.21C with respect to the proposed apartment complex and filling of the subject property.

LAW OFFICES OF
PETER L. BUCK, P.S
13TH FLOOR SEATTLE TOWER
1218 THIRD AVENUE
SEATTLE, WASHINGTON 98101
(206) 382-9540

The Applicant further agrees that it shall not apply for governmental approvals or permits or commence construction prior to the preparation and final acceptance of such final environmental impact statement, except for filling of the subject property.

The Town agrees that it shall act as lead agency with regard to preparation of the impact statement and that should it, upon review of the final impact statement, determine that significant adverse environmental impact will occur as a result of the proposed project and that reasonable measures can be effectuated to minimize such adverse environmental impact, then the Town shall require the Applicant to effect such modifications or perform such conditions as may be deemed lawful and reasonably appropriate by the Town in connection with pending or future applications for permits.

S.O.S. agrees on behalf of itself and its successors that it shall not oppose any existing or future governmental approvals, including but not limited to permits or licenses, or any other governmental actions (hereinafter "governmental approvals") which the Applicant has sought or in the future may seek for construction or operation of the uses on the subject property including filling of the property or construction of a 36-unit multi-family building, including any mitigating modifications of said activity. Such agreement not to oppose governmental approval includes non-opposition through administrative appeal or litigation, but shall not preclude testimony before an approval-granting authority. S.O.S. further agrees not to take any other actions which might have the effect of stopping or delaying construction of the project or increasing the cost of the project. S.O.S. further agrees that it shall not in any way assist, aid, encourage or cooperate with other persons who oppose development of the subject property in any manner described in this paragraph.

The parties further agree that this court shall retain jurisdiction of this matter until an adequate impact statement has been prepared and accepted. At that time this litigation shall immediately be dismissed without cost to either party. The parties further agree that the dismissal shall be conclusive and binding and that none of the parties shall seek judicial or other review of any question concerning compliance with the State Environmental Policy Act, RCW Ch. 43.21C, in connection with the subject property and subject activity nor assist, aid, encourage or cooperate with other persons who might seek any such review.

The parties further agree that, notwithstanding anything herein to the contrary, S.O.S. may take the following steps at

LAW OFFICES OF
PETER L. BUCK, P.S
13TH FLOOR SEATTLE TOWER
1218 THIRD AVENUE
SEATTLE, WASHINGTON 98101
(206) 382-9540

its own cost to ensure that the final impact statement is adequate. Upon publication of the final environmental impact statement, if S.O.S. believes it is inadequate, then it shall give notice, stating the alleged inadequacies with particularity, to other parties through their counsel of record not later than seven (7) days subsequent to publication of the final impact statement. In that event, the parties shall immediately select an arbitrator to review the question of whether or not the impact statement needs correcting or supplementation to remedy any inadequacies. Such arbitrator will be selected jointly by the parties from a list of names supplied by the City of Seattle Department of Buildings and King County Department of Buildings taken from their approved list of consultants. The parties shall agree on the designation of such arbitrator within fourteen (14) days of receipt of notice from S.O.S. that the adequacy is to be challenged. Within fourteen (14) days after selection, the arbitrator shall hold a hearing at a location to be determined by the arbitrator. The arbitrator shall be authorized, if necessary, to retain legal counsel for the purpose of assisting him or her in conducting the arbitration. The arbitrator shall be authorized to declare that the impact statement is adequate or to remand the matter to the Town with specific direction for correction or supplementation of the impact statement. The arbitrator shall render his or her decision within fourteen (14) days after the hearing. The cost of the arbitrator shall be borne by S.O.S. Decisions of the arbitrator shall be binding, conclusive and subject to review solely by the superior court and only reviewable by superior court if a motion for review is filed within five (5) days of such decision. Such review shall be based upon the records of the arbitrator which shall be transmitted to the court immediately by the arbitrator. The parties agree, pursuant to CR 52(a) 5(A), that the decision of the superior court judge will not be subject to appeal or review. The parties further agree that any disputes concerning arbitration shall be immediately presented to the presiding judge in the above-entitled court for binding resolution.

Time is of the essence in this Agreement and the parties agree that all actions taken hereunder shall be accomplished as expeditiously as possible and all parties shall cooperate in that regard.

DATED this 13th day of March, 1979.

/s/
Jeremy Randolph
Attorney for Plaintiff

/s/
August Hahn, Attorney for
Town of Long Beach

LAW OFFICES OF PETER L. BUCK, P.S.

By /s/
Peter L. Buck
Attorneys for Defendants Shoreside Apartment Associates

STIPULATION AND AGREEMENT - 3

LAW OFFICES OF
PETER L. BUCK, P.S
13TH FLOOR SEATTLE TOWER
1218 THIRD AVENUE
SEATTLE, WASHINGTON 98101
(206) 382-9540

APPENDIX B

DETAILED DISCUSSION OF ENVIRONMENTAL DESIGN PROCESS

A. INTRODUCTION

An environmental design process has been developed that systematically directs study efforts immediately to problem areas with the result that a project receives continuous change, modification, and refinement directed to meet stringent environmental quality standards. By following this process a project increases its chances of receiving the required permit approvals and reduces the total processing time.

B. CYCLIC PLANNING PROCESS

The environmental design process is a cyclic planning process rather than a conventional rigid one-shot approach characteristic of many development proposals. The cyclic nature of the design process allows both the proponent and permit agency to cooperate in a mutual learning, exploring and probting activity that can result in a project that meets stringent environmental standards but still accomplishes the goals of the proponent. To accomplish this, however, the proponent must keep the following criteria in mind at all times.

C. PLANNING CRITERIA FOR ENVIRONMENTAL DESIGN PROCESS

1. Feasibility. Assuming project constraints change over time, technological and fiscal resources continually need to be kept in focus. The political, social and economic feasibility of the project will significantly alter the planning process.

2. Flexibility. Mitigations of adverse project impacts frequently require design solutions. The planning process should provide flexibility to resond or adapt to changing conditions throughout the project. The design process will, without exception, require tradeoffs between different flexible elements.

3. Credibility. The basis for favorable project decisions and joint actions by permit agencies requires an open process of interaction and cooperation. Design solutions satisfying social/political concerns may not be the most technically efficient alternatives but must be given full consideration. Credibility is established by the development of professional working relationships directed to designing an environmentally acceptable project.

4. Accountability. An obligation exists on behalf of both the proponent and the permit agencies to implement approved projects to an agreed set of environmental standards. Mitigations of adverse project impacts identified during the environmental reviews will likely become permit conditions by reviewing agencies. Final project design and construction will require contingency plans for unexpected problems to achieve project goals.

D. ENVIRONMENTAL DESIGN TEAM

In addition to the above criteria, a key element of the environmental design process is the early formation of the project design team. A multiple-disciplinary team consisting of engineers, scientists, planners and lawyers is now required to design an environmentally acceptable project. Further, it is the collective interaction among key team members that produces highly effective design solutions to environmental problem areas. Emphasis is placed on the ability to communicate facts about technical fields to specialists in other fields and to community groups and non-specialists.

E. ENVIRONMENTAL DESIGN PHASES

The environmental design process has four distinct design phases: (1) feasibility studies, (2) preliminary studies, (3) environmental reviews, and (4) final design.

1. Feasibility studies. The first stage of the environmental design process is project feasibility. The purpose of this stage is to (1) develop project scope, (2) analyze economic feasibility, and (3) estimate project's environmental compatibility.

2. Preliminary studies. Preliminary studies are required to provide baseline information necessary for project design and agency review.

Typical studies include preliminary engineering design, architectural rendering, and ecological investigations as well as formal preparation of permit applications and supporting documentation.

Prior to any technical research, lines of communication should be established with all responsible and interested governmental agencies and groups. Early in the project, letters should be sent to all applicable permit agencies and interest groups briefly describing the project location and concept and requesting the following:

a. Permit applications and information requests.

b. Assignment of a staff person responsible for agency coordination and project review throughout the project.

c. An opportunity to schedule a meeting with the agency or group prior to preliminary studies being initiated.

The key objective of this project coordination is to identify and place priorities on key project issues. Research efforts should be proportional to the importance of agency and group concerns. All important communications throughout the project should be in writing.

3. Environmental review. This phase of the environmental design process is most critical. All of the information provided in the engineering and environmental studies is now used for preparing the draft environmental impact report as may be required by NEPA or state environmental laws. This environmental review phase is an open process where both citizen and government agencies evaluate the merits and environmental impacts of the project. If the preliminary studies have not been thorough, substantial delays can be expected during the review phase. If there are major impacts associated with the project that cannot be mitigated, lengthy delays or project termination can be expected.

It is critical that legal counsel be involved in this phase. Each of the permit agencies has both procedural and substantive requirements which may be ignored or overlooked by non-lawyers. Missing a step in the process can result in costly delays and litigation.

Project flexibility is also critical during this phase. There will be numerous differences of opinion, conflicts, and established positions among the project sponsor, reviewing agencies and involved citizens. These differences arise from interpretation of public policies, project impacts, proposed impact mitigations, and project/agency time delays and extended costs.

Without exception, the project sponsor should be prepared to modify the proposed project to eliminate significant environmental impacts. It is also important for the project sponsor to establish a base of credibility during the environmental review phase. Reluctance on behalf of the project sponsor to address agency and citizen concerns sets the stage for distinct and polarized positions.

The public hearing process should be an integral part of the environmental process. Proper use of the feasibility and preliminary studies phases will avoid surprises arising from the public hearing process. Citizen and agency concerns should be identified during these early stages while there is both time and budget to research and address significant concerns. During the preliminary stages, all concerns should be treated as significant. If significant impacts are not identified until the public hearing stage, they will result in additional reviews and project delays.

Emphasis should be placed on the importance of mitigating impacts. All projects of any size will have environmental impacts. It is important to provide sufficient time, budget, and design talent to identify feasible mitigations. Not all impacts can be mitigated by design solutions; some will require construction and operation actions. It is also important to note that impact mitigations identified during the environmental review phase will become permanent conditions if the project is approved.

Successfully completing the environmental review phase of the design process is a milestone for any project. At this point, it may appear unwarranted to halt the project. However, a thorough financial re-evaluation of the project should be conducted to totally assess the aggregate costs of permit conditions, and to once again determine the project feasibility. The combined cost of land acquisition, preliminary studies, project mitigating conditions, and current financing rates frequently alters the project feasibility, as do changed market conditions.

4. Final design construction. Upon receiving project approval and deciding to continue the project, the final design and construction phase of environmental design process is commenced. Conditions applied to the project by regulatory agencies and information developed from the previous design phases must now be synthesized into a final design that can be economically constructed.

Final design consists of preparing engineering and architectural plans and specifications that are developed to a sufficient level of detail to permit qualified contractors to build the project.

Final design should be followed up with construction inspection services to ensure that the contractor conforms to the plans and specifications and special permit conditions imposed on construction activities. Construction management services (CMS) may also be retained by the owner to coordinate the myriad of efforts taking place during construction. CMS relieves the owner of the construction management burden and helps to minimize construction costs.

It should be noted that certain restrictions placed on various permits during the environmental review process may significantly alter the optimum design and construction schedule. Thus, the developer must continue to remain flexible in his planning efforts. Further, other changes in construction schedules will undoubtedly occur as the project proceeds, but constant monitoring through the final design and construction phase of the environmental design process helps to ensure project quality and compliance with permit restrictions.

A STRATEGY FOR STATE PARTICIPATION IN OUTER CONTINENTAL SHELF EXPLORATION DECISIONS*

Marie Murphy*

For almost a decade the states have been arguing in the courts, in the Congress and during public hearings, that they should be directly involved in OCS decisionmaking. The enactment of the OCS Lands Act Amendments (PL 95-372) offers states significantly greater participation in the entire OCS oil and gas decisionmaking process from prelease through field development and production, transportation and onshore processing. Among the more important provisions directly affecting states are:

- Increased funding to states to ameliorate the adverse effects of OCS activities;
- Provision for increased involvement by states and local governments in the OCS decisionmaking process;
- A requirement that exploration and development plans be formulated by the lessees and approved by the Secretary of Interior, after consultation with affected states, to ensure orderly development of the nation's oil and gas resources;
- Provision for an information program to affected states so that they can base their decisions and recommendations on up-to-date estimates of offshore resources;
- Provision of an offshore oil spill pollution fund to ensure that damages from OCS activities are speedily compensated.

The increased participation which is mandated by the Act imposes correspondingly greater demands on states' staffs if the full benefits are to be derived from it. The effectiveness of any state's participation will depend to a large degree on the amount of time devoted to it, but also, and more important, on the extent of its preparedness. There are points in the process where maximum influence can be exerted through just the right type and quantity of state input to achieve state objectives. It is now up to the

---

*New England River Basins Commission

states to organize and do the work necessary to achieve effective participation. The framework established for dealing with exploration and the attendant policy decisions made during this period will serve as the underpinning for dealing with all subsequent stages of offshore oil and gas development.

States tend to overlook the exploration stage because of its short timespan, its limited job opportunities and the relatively few onshore facilities associated with it. They underestimate the speed of industry's transition from exploration to development related decisions. In the North Sea the interval between the discovery of a commercial field and the development drilling has been 3 - 5 years. The Department of Interior suggests that there will be a 40 - 47 month interval between a commercial discovery and the start of development drilling in U. S. frontier regions. In order to obtain the equipment and facilities on time industry must make quick decisions about field development, transportation, onshore processing and marketing the petroleum. Engineering will be at an advanced stage and contracts for field development will be let within a year, or a year and a half, of a marketable discovery. Participation in the exploration decision making is, therefore, perhaps the most significant opportunity states will have to shape the overall circumstances under which OCS related activities will occur in their respective onshore areas. Because of its crucial importance and in view of the fact that the formal review period is relatively short (as little as 15 - 20 days for the Exploration Plan in states without an approved CZM Program) states must take the initiative and set up contacts with industry and with federal and local officials immediately after the lease sale so that the possibility of mutually damaging and time-consuming confrontations when industry submits its proposals can be avoided.

- State OCS exploration decisionmaking responsibilities cover a large number of diverse activities and include (1) the review and comment on plans and permits, (2) consistency certificate determination for plans covered in approved state CZM programs, (3) consultations with federal, local and industry officials, and (4) the preparation of economic and facility siting plans. They play a pivotal but complex role in OCS exploration decisions. They are called upon to review, comment or are otherwise given information about industry's plans for OCS exploration activities. For each Exploration Plan they will receive five different documents for review and comment. These are:

Exploration Plan - from the lessee via USGS,

Environmental Report - from the lessee via USGS,

NPDES Permit - from EPA,

Section 10 Permit - from the Corps of Engineers,

Consistency Certificate - from the lessee via USGS.

Other documents, (the lessee's Application for Permit to Drill, the Summary Report, the Index of Lease Sale Information, etc.) will be submitted periodically for information purposes or as required by significant events. States will also be involved in the preparation of a Regional Transportation Studies Plan - Phase II of the Bureau of Land Management's Intergovernmental Planning Program - at this stage. A large number of diverse review and planning activities will require state involvement during the course of exploration for oil and gas on the OCS. As depicted in Figure 1, the review, comment, or information exchange activities cover topics from air and water discharges to determining consistency with CZM plans. Figure 2 puts the exploration stage in the context of the entire OCS decisionmaking process and illustrates the ongoing transportation planning and environmental studies elements which will concern states throughout the process. Planning for new OCS-related economic development and serving as liaison between local governments and the Secretary of Interior will place an additional load on state staffs.

- The consultation process can be used to make clear statements regarding the states' attitude to certain aspects of OCS activity. Economic policies and environmental safeguards articulated at the pre-lease and leasing stages

FIGURE 1

SCHEDULE[1] OF DOCUMENTS WHICH STATES WILL RECEIVE IN THE COURSE OF THE

OCS EXPLORATION PLAN PROCESS

Document	Period
Consistency Certificate[2]	3 months * 3 months * (if extension required)
Exploration Plan and ER[3]	30 days *
NPDES Report[4]	180 days *
Corps of Engineers Permit[4]	2 months
Environmental Assessment[4]	30 days*
Permit To Drill[5]	6 weeks
Summary Report and Index of Lease Sale Information	possibly at six month intervals or following significant events

MONTHS 1 2 3 4 5 6 7

USGS sends EP & ER to states (month 1) — Drilling begins (month 7)

*Indicate periods for review or decision which are fixed by regulation; all other time are discretionary and approximate.

Notes

1) Assumptions

(a) Exploration EIS deemed unnecessary

(b) State concurs with Consistency Certificate without seeking further information

(c) EPA processes NPDES Permit without delay and no major objections raised.

(d) Corps of Engineers processes Section 10 Permit without reference to H.Q.

2) Direct state action - concurrence/objection

3) Major opportunity for effective comment - imposition of conditions etc.

4) Routine consultation by the regulating agencies - state imput can be effective but probably less so than in 3)

5) Notification only

FIGURE 2

DECISIONMAKING PROCESS

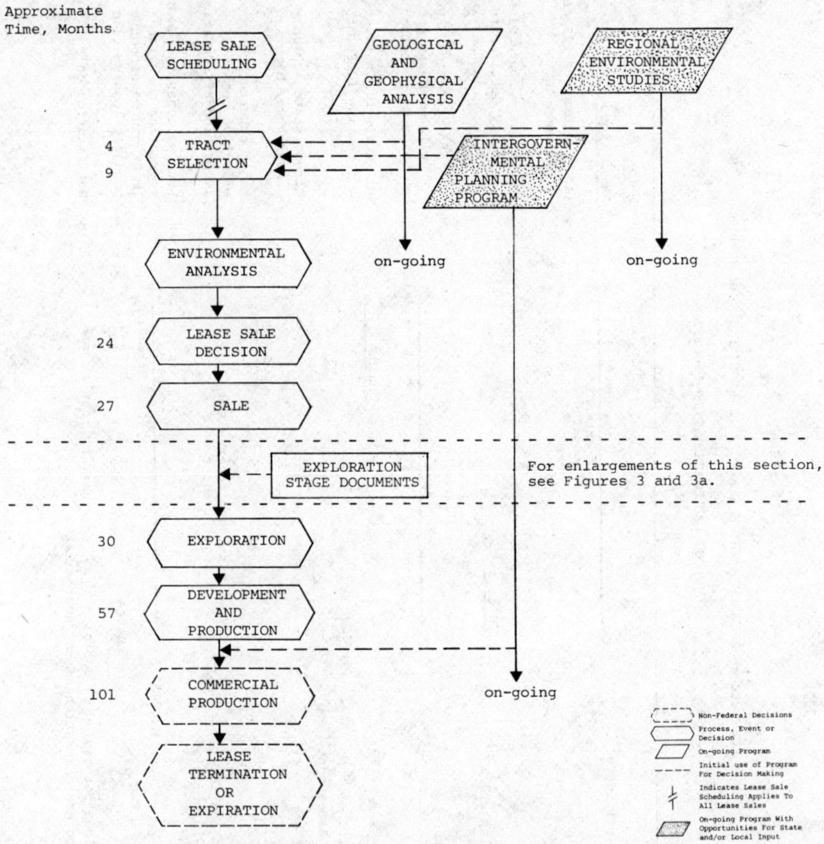

SOURCE: Atlantic Index, January 1975-April 1979. USGS and BLM. p.7.

can now be reiterated and emphasized, clarified or changed.

- States, in cooperation with industry and local officials, need to develop plans for OCS exploration-related activities in order to take advantage of the business opportunities that exist and reduce the environmental damage that might occur. In order to gain the benefits of exploration activity, plans for the siting of temporary service bases, district offices and ancillary industries are required. Training programs will help provide jobs - both onshore and offshore - to the locally unemployed. Education programs will assist local industries obtain new oil related business.

- As shown in Figure 2, the exploration phase of OCS development is a small part of the entire OCS process. However, the handling of this stage will set the pattern for states' involvement in development, transportation, and onshore processing. If these subsequent stages are kept in mind when the exploration strategy is being developed it can serve as an early warning of the potential need for legislation, coordination and/or siting decisions.

- A consistent response to OCS exploration initiatives from industry, the federal government, and local officials is essential. State responses to the environmental protection measures, natural resources management issues, and development aspects of each OCS exploration activity must be consistent. The other government agencies and, perhaps more important, industry, will give greater attention to states' concerns if they can be confident that state comments on one aspect of the exploration package will not be contradicted, in substance of effect, by the response of another agency to another part. Having a single point of contact with a consistent policy orientation will also help to persuade industry and the other government agencies that early, informal discussions are useful and necessary.

- There is a need for a formal mechanism to provide the policy context and carry out OCS ex-

FIGURE 3

TIMETABLE DURING WHICH STATES ARE INVOLVED IN OCS EXPLORATION STAGE DECISION-MAKING PROCESS

(THREE MONTH CONSISTENCY DETERMINATION)

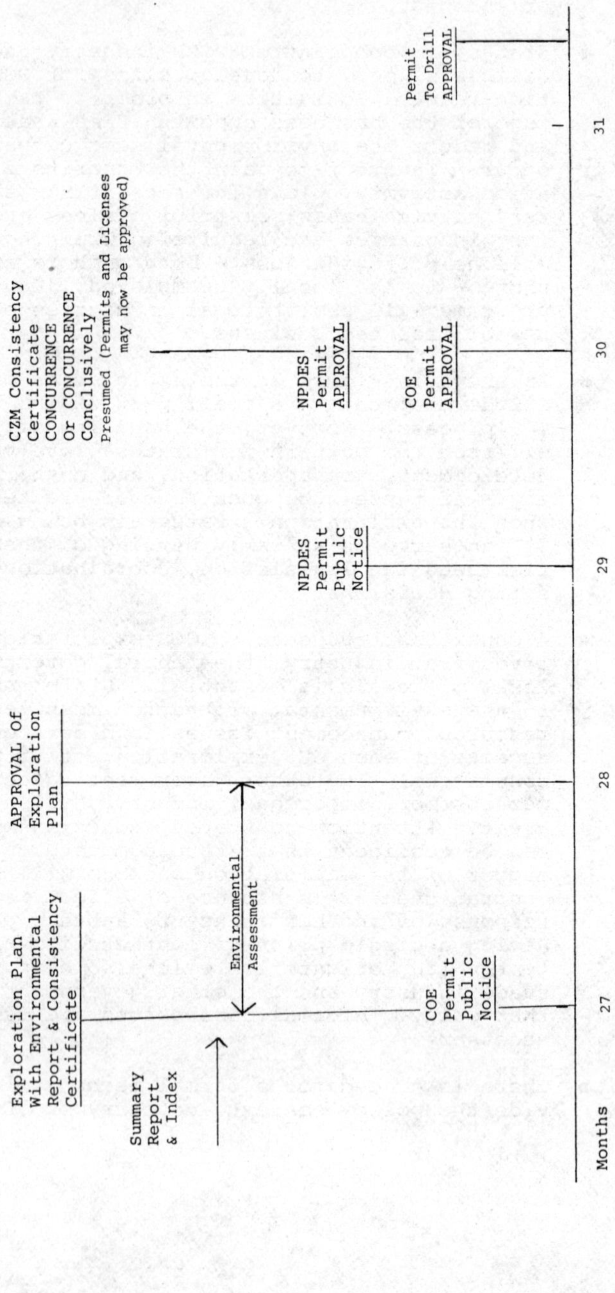

FIGURE 3a

TIMETABLE DURING WHICH STATES ARE INVOLVED IN OCS EXPLORATION STAGE DECISION-MAKING PROCESS

(SIX MONTH CONSISTENCY DETERMINATION)

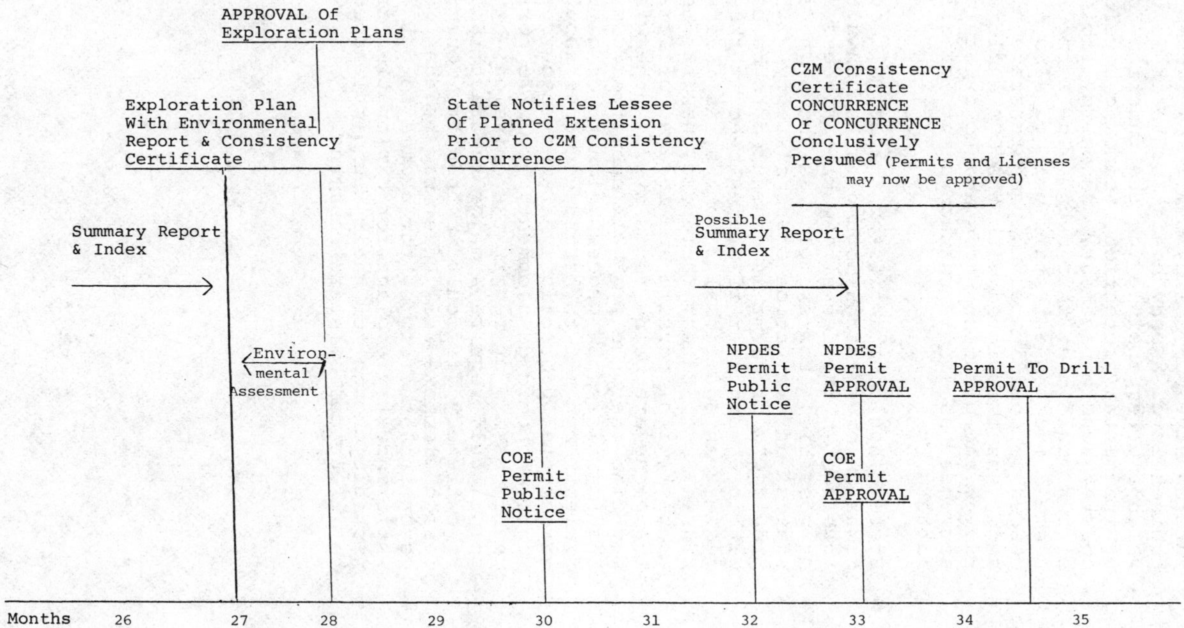

ploration decisionmaking responsibilities. An interagency "OCS Policy Committee", comprised of cabinet level agency heads, or their deputies, which are responsible for the functions of environmental protection, economic development, coastal zone management, and energy facility siting would meet this need. The formation of a small independent staff unit is also recommended to coordinate and prepare state comments and plans associated with the OCS exploration review and planning needs and to otherwise serve as the single contact point for all OCS matters.

- States should not wait for the formal consultations on exploration documents but should get involved in the decision making process as early as possible. For example, as shown in Figure 3, states' determination of consistency with approved CZM programs is scheduled to be completed within 3 months or if states need it they can request an additional 3 months, as indicated in Figure 3a. States, therefore, need to be in contact with industry and federal officials while the plans are being formulated. Otherwise they may find that the proposals submitted, as part of the formal review process, ignore or give insufficient emphasis to, state concerns about environmental protection, natural resources management, economic development, etc. To correct the emphasis, or repair the omission, at that point would require a major re-draft, and would be extremely difficult as industry would have design elements well under way by then. It is essential, therefore, for states to get involved while industry's plans for exploration are still flexible enough for their technical aspects to be influenced by state concerns.

THE #53 WORKING GROUP: LOCAL PARTICIPATION IN THE FEDERAL OCS LEASING PROCESS

by

CELIA LAWREN

INTRODUCTION

At a time when local jurisdiction over major energy facilities intended to serve regional or national markets is being preempted, an experiment in California for local government participation in the federal Outer Continental Shelf (OCS) leasing process is proving quite successful. California is using OCS State Participation Grants under the Coastal Energy Impact Program (CEIP) to educate coastal counties and communities affected by upcoming OCS Lease Sale #53, a frontier OCS development area which contains little or no onshore industrial facilities. Before this experiment, citizens in most of the nine affected counties had never seen or heard of an oil rig, much less imagined viewing one from their living room windows. Now there are CEIP planners in each county, organizing public workshops on the lease sale and its effects on local coastal policies and resources, enlisting local experts to contribute information on sensitive resources, and incorporating OCS planning and energy facility siting into their Local Coastal Plans. At the heart of this network is the #53 Working Group, organized by the California Coastal Commission and comprised of Commission staff and the local CEIP planners, to collectively deal with planning and siting problems and comments to the federal government.

Midway through the leasing process for OCS Lease Sale #53, the #53 Working Group has already reaped benefits for local government participation in this process.

1. There is now a cadre of trained OCS planners at the local government level.

2. There is an overriding consensus among the nine counties and the State in response to OCS Lease Sale #53.

3. Coordination among the affected counties and the State is excellent.

Ms. Lawren is on the Energy and Ocean Resources Division staff, California Coastal Commission, 631 Howard St., San Francisco.

4. There is record local participation in the OCS leasing process on this lease sale.

5. Local governments, though reluctantly, are planning for onshore impacts by considering OCS energy facility siting options in their Local Coastal Plans.

6. Not least of all, communication has opened between the federal and local governments, and the U.S. Interior Department is more attuned to local concerns.

BACKGROUND

The OCS Lands Act Amendments of 1978 significantly modifies the decisionmaking process for OCS lease sale activities. While responsibility for implementing OCS leasing procedures rests with the federal Bureau of Land Management (BLM) under the U.S. Interior Department, opportunities for State and local government participation exist in several steps of the process, including the environmental study program, the development of OCS orders, the call for resources reports, the call for nominations and comments, tract selection, the preparation of and public hearings on the draft environmental statement (DES), the proposed notice of sale, and the review of development plans. (see Figure 1)

Local governments can supplement the federal OCS leasing process and assist in State OCS review by providing information on local coastal resources, policies and potential land uses and impacts that should be considered in developing environmental impact statements (EIS), and in reviewing those statements. While they are not excluded from communicating directly with the U.S. Interior Department or BLM, local governments, however, must submit their comments via the governor of their state for these to be officially considered in the OCS decisionmaking process.

MANAGEMENT PROBLEM: How to Deal with OCS Development in a Frontier Area?

When the U.S. Interior Department proposed leasing areas offshore central and northern California for oil and gas exploration and development, the California Coastal Commission faced the management dilemma of how to deal with OCS development in a frontier coastal area. The two previous lease sales, #35 and #48, had occurred off the heavily urbanized coast of southern California, where residents were somewhat familiar with oil rigs on the seascape in State tidelands. But even in these two lease sales, the federal OCS leasing process did not run smoothly. Key affected local governments did not participate and the environmental impact statements were deficient in such major impact areas as air quality, onshore development and transportation alterna-

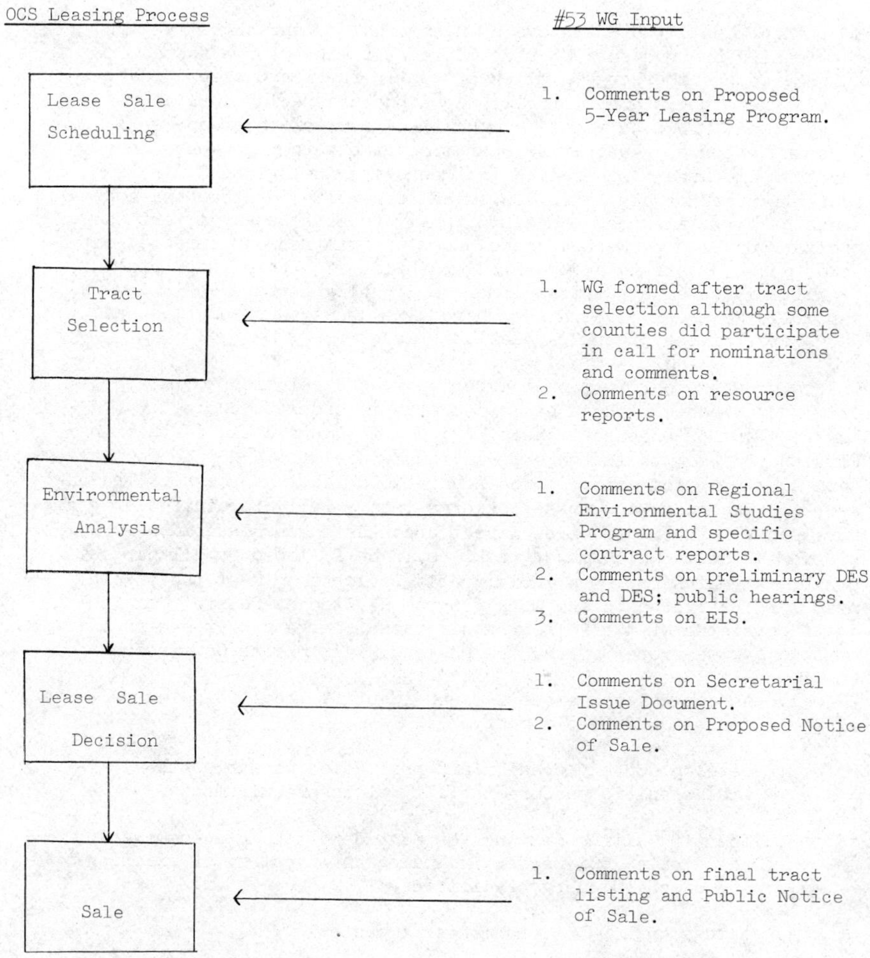

FIGURE 1

#53 WG INPUT INTO THE FEDERAL OCS LEASING PROCESS

tives.

How, then, could the State avoid these problems and actually improve the leasing process for OCS Lease Sale #53, where the affected onshore area encompasses 700 miles of mostly rural and scenic coastline? Where there are virtually thousands of special coastal resource areas, such as scenic communities, recreation areas, wetlands, bays and esteros, aquaculture operations and harbors? Where, for example, one area alone called the Mendocino Ecological Staircase system represents a peak in ecological diversity for the near-coastal areas of North America, containing pygmy forests, rocky intertidal habitat and bird rookeries, sand dunes, virgin redwood forests, wetlands, coastal strand and coastal brushfields? Where industrial facilities are equated with commercial fishing harbors and a few paper mills by local residents? Where nine county jurisdictions and 24 coastal cities govern? Need I say more?

One thing was clear from past experience; the solution to this problem had to address three basic concerns in order to succeed: (1) how to plan for onshore impacts, (2) how to improve the forthcoming EIS, and (3) how to increase public awareness of OCS development and gain local participation in the federal OCS leasing process. The answer for incorporating these concerns into a well-coordinated program for such a large and diverse impact area was through a common work program---one in which major tasks could be followed by all participants, yet which could be individually tailored to meet the special needs of local areas. The work program devised by Coastal Commission staff consisted of the following six tasks, largely borrowed from methodology developed by the New England River Basins Commission.

1. Assemble local expertise and identify local policies relevant to OCS development.

2. Develop OCS exploration and production scenarios to determine specific onshore facility siting requirements.

3. Identify siting options for consideration in accommodating anticipated OCS needs, including an inventory of existing and proposed energy facilities.

4. Assess onshore environmental impacts.

5. Develop policies and mitigation strategies for incorporation into local zoning and regulatory plans.

6. Participate in BLM environmental assessment process.

The remaining concern, of course, was money. Thanks to the far-sightedness of Congress, these State concerns closely matched the objectives of the OCS State Participation Grants made available through the CEIP under the 1978 OCS Lands Act Amendments. Being manager of the State CEIP allocations enabled the Coastal Commission to more easily develop a comprehensive program for local participation in OCS Lease Sale #53. CEIP grants were limited to the county government level with requirements to involve cities, towns and citizen groups interested in OCS issues.

Humboldt, San Mateo, San Luis Obispo and Santa Barbara Counties were first to complete grant applications and receive funding. The Association of Monterey Bay Area Governments, a regional council of government, was next to receive a CEIP grant for #53 planning, with pass-through funds to Monterey and Santa Cruz Counties. Santa Cruz would later receive its own grant and Monterey would drop out during the second year. The last three counties, Mendocino, Sonoma and Marin, were more recalcitrant about actually planning for onshore impacts because of their strong opposition to the lease sale. Thus, compromises had to be reached before their grants were awarded by the Commission. (see Figure 2) But by fall of 1979, at the beginning of the environmental assessment stage in the OCS leasing process, all nine counties and one council of government were participating through the funding of individual #53 planners in each jurisdiction.

SOLUTION: The #53 Working Group

While having #53 planners in each affected county allowed flexibility in dealing with local OCS issues, a centralized forum was needed to bring the network together. The formation of the #53 Working Group (WG), comprised of the local #53 planners and appropriate State and Regional Commission staffers, provided such a forum. Consistent with the overall work program, the objectives of the WG were to gather and discuss information on the onshore impacts and to provide the BLM Pacific OCS Office with local information for the DES. Furthermore, the WG could supplement the training and knowledge of the local planners and prevent isolation from one another. Not to mention the fact that the WG also could serve as an effective management tool for the Coastal Commission in keeping track of all the grants' diverse activities.

Organization and Activities

Membership in the #53 WG has been purposefully limited to enhance group communication and the exchange of ideas and experiences. Since it serves as an in-house technical planning project of the Coastal Commission, the WG sessions are not open to the public. Information, scenarios, and impact assessments and mitigation strategies are made available, however, to local governments, state agencies, the federal

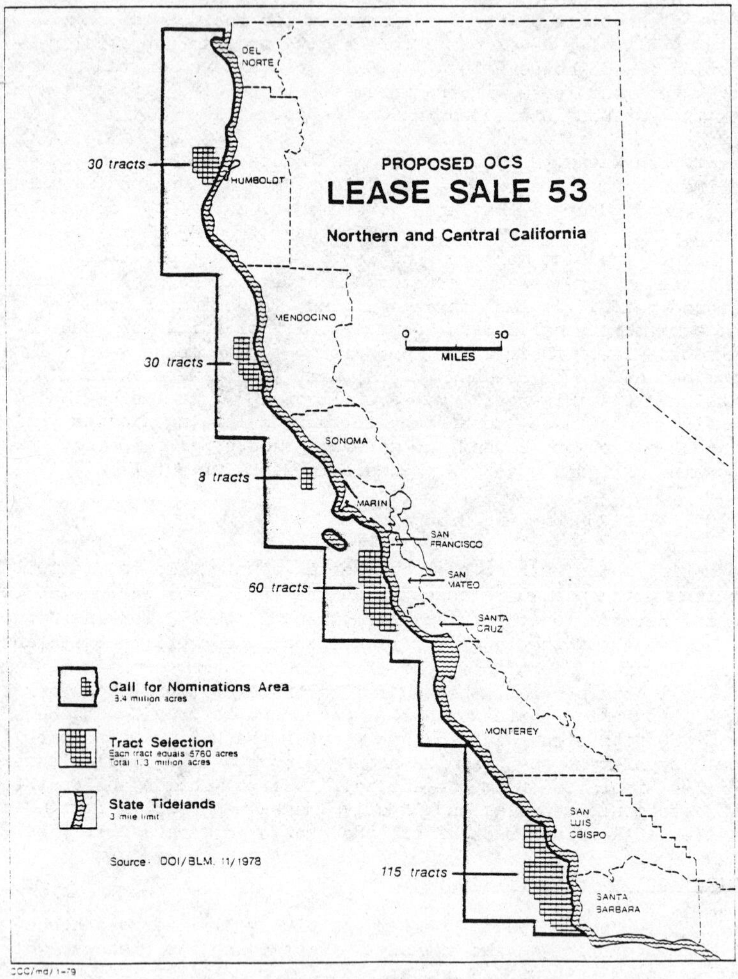

FIGURE 2

THE OCS #53 LEASE SALE AREA

government, environmental groups and any other interested persons. This material can be used by these groups as a basis for recommendations to the State and the U.S. Interior Department on OCS Lease Sale #53, but it is not the function of the WG to prepare or coordinate these recommendations.

Guided by the overall work program, organization and activities of the WG have remained flexible, depending on changing needs in response to the leasing process. Likewise, meetings are scheduled when necessary, but at least once every two months. The State OCS Coordinator on the Coastal Commission staff oversees the group and tentatively plans the meeting agendas with the help of group members and other Commission staff.

The first few sessions of the WG concentrated on familiarizing the local #53 planners with the activities and facilities associated with OCS exploration and development, the alternatives of transporting OCS oil and gas, and the OCS leasing process in general. We invited BLM staff from the Pacific OCS Office to describe how leasing takes place, how petroleum resource estimates were determined for each of the five basins in the lease sale and the kind of information that they needed from the local governments to draft the EIS. We reviewed various development scenarios together to prepare the planners for developing scenarios for their own areas so that they would know the kinds and intensity of facilities to expect. We then linked this OCS planning process with the ongoing Local Coastal Plans (LCPs) in each county since several of the planners also were responsible for completing the energy components of their LCPs. Finally, a veteran OCS participant from the Santa Barbara County Department of Environmental Regulation and a local planner from San Diego County who was deeply involved in OCS lease sale #48 shared their experiences with the group.

As the WG members delved into their individual programs, their knowledge and questions came back to enrich the entire group. The Humboldt County planner, who completed the first development scenario, lead the group through his particular exercise, then received a critique of his assumptions and methodology from the other members. Similarly, the planner from AMBAG, which had produced a slide show on the lease sale and its potential impacts on Monterey Bay, showed it to the group for comments on how to improve it. An oil industry representative attended one session to help answer the many technical and planning questions of the WG, including those on facility siting requirements and preferred locations for particular support facilities. These are examples of the types of "sharing" that has taken place both within the WG and between the WG and outside parties. The liveliest of these sessions, of course, have centered on the DES, with BLM staff on hand to discuss WG comments and "share" a lot of "heat!"

Other sessions have focused on specific topics such as oil spill contingency planning and the effectiveness of containment equipment and the development of lease stipulations. Furthermore, we have reserved several meetings to discuss energy facility siting and planning in the counties since this task seems to fall behind the other, more glamorous ones. Since the planning element also involves other planners and familiarity with onsite locations appropriate for OCS-related industry, we conducted "on-the-road" planning sessions with each WG member on his or her turf, including visits to existing harbors and industry sites where OCS facilities might possibly be consolidated.

EVALUATION

Sharing this description of the #53 WG and its activities hopefully has given you ideas on how such a mechanism could be useful to your situation. It also may help you to know the effect the #53 WG has had thus far on the OCS leasing process for OCS Lease Sale #53. To begin this evaluation, I return to the OCS Lands Act Amendments and its objectives. The most obvious is, of course, to increase State and local government participation within an established decisionmaking framework. Underlying the need for increased participation, however, is the presumption that this will improve the quality of decisionmaking through a more thorough environmental assessment of the impacts. Concurrently, it will avoid stalemates among disparate points of view by allowing for their input at various decisionmaking points throughout the process rather than as one loud cry at the end. Finally, increased participation can benefit local communities by preparing them for OCS energy development, by either reinforcing opposition or acceptance.

The #53 WG, together with the work of the individual grants, has certainly met these objectives. More specifically, six benefits have resulted from this experiment which would not have happened in its absence.

<u>There is now a cadre of trained OCS planners at the local government level.</u>

With the fiscal cutbacks in California resulting from Proposition 13, local governments are hard pressed for funds to support supplemental programs, such as planning for OCS development. CEIP monies have made it possible for cities and counties to become involved on a steady basis in OCS planning and decisionmaking. The #53 WG members have received specialized training and gained a 'greater-than-local' perspective under Coastal Commission guidance that would not have been available to them at the local level. We anticipate that this core of trained individuals will have a "trickle" effect in educating other citizens in the coastal counties of central and northern California.

<u>There is an overriding consensus among the nine counties and the State in response to OCS Lease Sale #53.</u>

While some would not interpret this as a benefit since the State officially opposes OCS Lease Sale #53, developing a consensus on an issue which covers a 700-mile area certainly is an accomplishment of some sort! The fact that the area, except for San Francisco Bay, is rural agriculture with a fishing industry base largely contributes to strong local public sentiment against the lease sale. State agencies are mostly concerned with the valuable offshore resources. Thus, a perfect match! In addition, all affected County Boards of Supervisors, the Governor and the entire California Congressional delegation oppose this lease sale at this time.

The #53 WG has served as a vehicle whereby local and State representatives can informally discuss the issues and reach consensus on certain points which are ultimately taken to the U.S. Interior Department. While the #53 WG has functioned in a technical capacity, its members interface with the political process when they discuss the issues with their Boards of Supervisors.

Presenting such a strong, unified front before the federal government has captured its attention to our concerns about the lease sale. Furthermore, the WG has helped the local planners to focus on the issues, rather than the emotionalism, involved in such a lease sale.

<u>Coordination among the affected counties and the State is excellent.</u>

Correlating with the development of an overriding consensus is the fact that coordination among the local governments and State agencies has greatly improved. At the heart of the participation network is the #53 WG, where comments and deadlines for input on the DES and related working papers are thoroughly discussed. Concurrently, the local planners brief their governing Boards of Supervisors, as mentioned above, on the issues and develop official local comments and positions. The State subsequently incorporates these comments into its official position to the Department of Interior, thereby truly carrying out the intent of the OCS Lands Act Amendments. This has resulted in a closer working relationship among all decisionmakers.

<u>There is record local participation in the OCS leasing process on this lease sale.</u>

Approximately 750 California citizens testified at the six public hearings which were held on the DES for OCS Lease Sale #53 last June. Because the number of people scheduled to testify in a few areas exceeded the twelve-hour time period allotted for each hearing, some hearings extended well into the next day. In the words of one BLM staffer, "Nobody has ever sandbagged the hearings before." In the case of Santa Cruz County, which originally was not targeted for a public hearing, leverage exerted by local officials and citizen groups resulted in a local hearing site. This is a great improvement over OCS Lease Sale #48 in southern California on which one public hearing was held.

Local governments, though reluctantly, are planning for onshore impacts by considering OCS energy facility siting options in their Local Coastal Plans.

Considering the resistance to plan for onshore impacts associated with OCS development in the Lease Sale #53 area, we assume that little or no planning would have taken place without the CEIP grant incentives and the attention this subject received in the #53 WG sessions. Nevertheless, progress has been slow, justified in part because of the difficulty in formulating policies for facilities whose need and location cannot be determined until after the exploration phase. This is especially true in a frontier OCS area where virtually no coastal industry exists.

Timing also has been a problem in that the leasing process for OCS Lease Sale #53 does not correspond with the legislative deadline for completion of the LCPs. Furthermore, the CEIP work programs for the WG were developed separately from the LCP schedules. We have made an effort to correct this error on our part by accelerating OCS planning tasks, so that this work will become part of the LCPs submitted for Commission approval. If not, OCS-related facilities will have to proceed through an elaborate amendment process in order to be considered under the LCP.

So far, three of the counties have stated outright that there will be no OCS-related development along their coastlines; two counties have provided siting criteria and environmental assessment processes related to such development; and the remaining four counties are still developing their planning strategies.

Not least of all, communication has opened between the federal and local governments, and the U.S. Interior Department is more attuned to local concerns.

While tension still remains between the federal agencies responsible for maintaining the leasing schedule and State and local governments, there are several improvements over previous lease sales. First, the BLM Pacific OCS Office has scheduled more public hearings than ever before. And, secondly, it released preliminary draft copies of the DES to the WG to allow more time for review and comments. Unfortunately, this has not meant that these comments would be incorporated into the subsequent draft, but it has been a step in the right direction.

One ear the State and counties have managed to capture is that of Secretary of Interior Cecil Andrus. On a trip to California last spring, he commented that he was sympathetic to our concerns and alluded to dropping the middle three basins from OCS Lease Sale #53. This kind of response and action on the part of the Interior Department would not have occurred without the hard work of the #53 WG and its members.

BALANCING FISH AND FUEL ON GEORGIA'S OCS

Jeannette H. Phillips
Georgia Department of Natural Resources

James O. Barber
Georgia Office of Planning and Budget

The State of Georgia has taken a positive stance with regard to possible hydrocarbon develoment off its coast. The Governor's policy is that exploration and development activities in the marine environment can be compatible with the State's infant fishing industry if enough planning and research is done to anticipate and avoid potential problems. Although the South Atlantic States were involved extensively in planning for Lease Sale 43, held in March 1978, they were unsuccessful in focusing the federal environmental studies program on the needs they perceived as most critical. Nor did the studies produce sufficient information for sound management decisions prior to Lease Sale 43.

Georgia is anxious that this not be the case for Lease Sale 56, scheduled for late summer, 1981. It views the federal environmental studies program as an opportunity to explore the effects of oil and gas extraction on the marine fisheries, and at the same time enhance the State's knowledge of offshore fishery resources. This attitude has brought about the involvement of Georgia Department of Natural Resources in the OCS process.

Georgia's commercial fishing industry is expanding, although it still depends on shrimp for the major part of its revenue. Resource managers need to learn more about the habits and abundance of finfish and other potential commercial species before they can effectively aid commercial fishermen in diversifying their harvests. Potential for a broader marine recreational fishing industry exists also, but needs more study to support effective management.

Department of Natural Resources has alerted the Department of the Interior regarding the critical lack of information about marine habitats and species abundance. Without more comprehensive studies, there can still be no responsible tract selection for Lease Sale 56. These efforts and those of the other South Atlantic States have sharpened studies priorities in the Bureau of Land Management OCS process and resulted in an improved information exchange process between the States and the federal agencies.

Jeannette H. Phillips is acting Chief of the Coastal Management Services Section, Coastal Resources Division, Georgia DNR, 1200 Glynn Ave., Brunswick, Georgia 31523 (912/264-7289)

James O. Barber is a Planning Coordinator for the Physical and Economic Development Division of the Office of Planning and Budget, 270 Washington St., SW, Atlanta, Georgia 30334

Georgia Department of Natural Resources and South Carolina Wildlife and Marine Resources Department are directly involved in the environmental studies program, as contractors of the Bureau of Land Management. Teams of biologists from both states are looking at living marine resources in three bathymetric zones in the Georgia Embayment. Results of the study will help the federal government and resource managers from both states to assess the value of live bottom communities in the South Atlantic, working toward harmonious co-existence of renewable and non-renewable marine resource harvests.

At present, Georgia and the other South Atlantic States can recommended tighter lease stipulations for energy operations on the Outer Continental Shelf. Through an expanded and responsive environmental studies program, however, the States could proceed with OCS planning in a more informed manner, benefitting all offshore resource users and meeting the public needs.

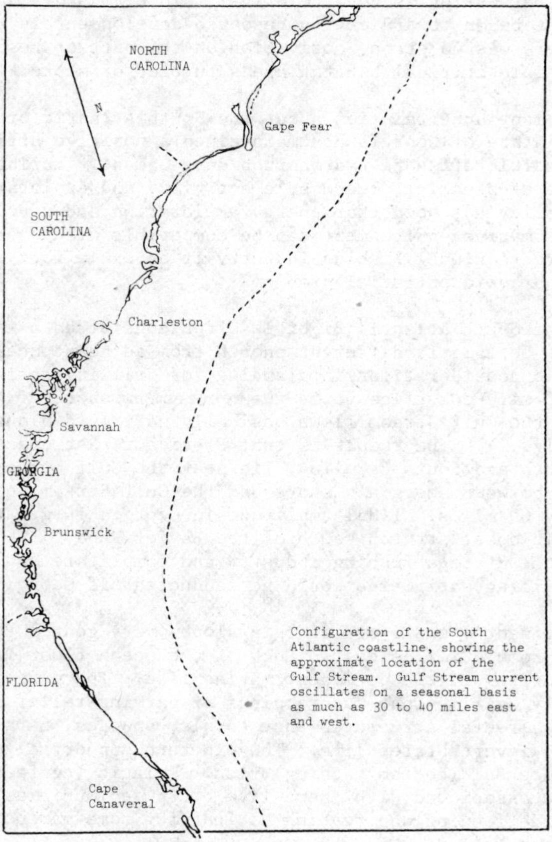

FIG. 1

BACKGROUND

Georgia's Coastal Setting and OCS Policy

The South Atlantic Bight is a frontier area for development both of hydrocarbon and fisheries resources. While portions of the Outer Continental Shelf and and shelf edge off the South Atlantic coast may cover reserves of oil and gas, much data is lacking about how best to extract those resources. There are oceanographic factors influencing the execution of drilling in the South Atlantic that are incompletely understood, or present major impediments to exploration and production activities, and might complicate oil spill recovery efforts.

Large gaps exist also in current knowledge about the living marine resources off the coasts of North and South Carolina, Georgia, and northeast Florida. Although each state has an active fishing industry, most seafood harvesting is done relatively close to shore and, until a serious effort began toward energy resource development in the South Atlantic, there was no strong motivation on the part of resource managers to study possible critical habitat areas further offshore.

Despite many uncertainties about the South Atlantic offshore environment, the State of Georgia has maintained a positive attitude with regard to possible offshore hydrocarbon development. During the past several years of planning, lease sale activity, and exploration, the Governor's policy has been that energy exploration and development operations in the marine environment can be compatible with Georgia's commercial and recreational fishing industry if enough effort is made to anticipate and avoid potential problems.

While the South Atlantic Bight is often considered as a whole, the area offshore Georgia is different enough from adjacent ocean areas to present unique considerations in planning for drilling activities there. Georgia's 120-mile coastline forms the westernmost shore of the Atlantic Ocean, while the Gulf Stream flows basically northward along the Atlantic Seaboard. (Fig. 1). The result is that Florida's east coast, and the coasts of North and South Carolina, lie near the Gulf Stream channel. The distance between Georgia's shore and the Gulf Stream, however, is approximately 60 miles. Tidal amplitude in Georgia is greater than any other Atlantic coast stretch south of the Bay of Fundy. These factors affect both the State's fishing industry and complicate conditions under which any drilling activities would be conducted off Georgia's shores.

The continental shelf off Georgia slopes very gently toward the shelf break, some 40 miles east. Much of the ocean floor is sandy and barren, supporting little permanent marine life. The expanse is broken intermittently by limestone outcroppings of varying relief and extent. It is on these reefal structures that corals, sponges, barnacles and other benthic invertebrates live. They in turn support sizeable populations of demersal (bottom feeding) fish. Pelagic (surface swimming) fish also are associated with these live bottom communities. Live bottoms provide feeding and resting habitat for some marine mammals and sea turtles, as well.

These reef environments do not occur close to shore; the live bottom nearest Georgia's coast is Gray's Reef, 17.5 miles east of Sapelo Island in 65-70 feet of water. Some recreational and commercial fishing occurs near and on these live bottoms.

For the past century the economy of Georgia's coastal area has depended heavily on seafood harvesting, but the commercial fishing fleet traditionally has stayed close to shore and focused on a very few species. Even today, when Georgia's fleet numbers over 1200 boats, most of which are larger and have more sophisticated equipment than their early prototypes, most fishing activity is concentrated within 12 miles of shore. Primary harvests consist of shrimp and blue crabs. By-catches of bottom fish such as flounder, whiting, and sea trout are sold at the dock; the returns provide nothing more than grocery money for the vessel crew.

These fishing traditions, coupled with the remoteness and infrequency of the ocean live bottoms, have provided little impetus for scientists to study limestone reefs and the marine communities associated with them. The situation is changing, however.

Until a few years ago, crab and shrimp fishermen could support themselves -- at least marginally -- by working six to eight months a year. Those days are gone. The economics of fishing in these times of rapidly rising fuel, equipment and supplies costs demand an almost year-round operation of what used to be a seasonal industry. Some shrimp fishermen attempt to meet their expenses by moving to seasonal shrimp grounds: late winter and spring in western Florida; summer in Louisiana and Texas; late summer, fall and early winter in Georgia and South Carolina. But unless a vessel is in good shape and the captain is knowledgeable, this transient mode can be financially disastrous.

The other option is to diversify: using the boats year-round in waters closer to home, fishing for other marketable seafood. Recently some Georgia shrimpers have ventured further offshore to catch snapper, grouper, calico scallops and rock shrimp. Although these fisheries are far from developed in Georgia, the possibility for exploiting them has awakened much interest in Georgia's offshore waters, especially around the live bottoms and in the Gulf Stream area.

During the past decade, there has been increased focus on the waters offshore Georgia by the oil industry and the federal agencies involved in Outer Continental Shelf (OCS) energy exploration and development. There continues to be a lack of information concerning the renewable marine resources in these waters which would equip State and Federal decision-makers to evaluate the possible adverse consequences of drilling near or on a live bottom area. Further, the value(s) of the reefs to Georgia's commercial and recreational fisheries have not been established, so managers are unable to make a well-informed decision should they be forced to choose between siting a well in a promising location or protecting a live bottom habitat.

Georgia policy-makers and resource managers have engaged in planning for OCS development in the face of inadequate information and a rather tight federal lease sale schedule. This paper describes the efforts made to increase needed data and resolve some potential resource conflicts.

OCS Planning Structure in Georgia

The Governor of each state is the source of ultimate state input into the OCS planning process and policy development. In Georgia, the Governor designated staff within the Office of Planning and Budget (OPB) to represent him on the U.S. Department of Interior's national OCS Advisory Board and the South Atlantic Regional Technical Working Group.

The Georgia Department of Natural Resources (DNR) cooperates with OPB to provide scientific and technical resource assistance with regard to OCS-related planning. The unit within DNR best equipped to perform this function is the Coastal Resources Division (see Fig. 2). The Division, headquartered in the coastal community of Brunswick, is concerned with marine fisheries management, enforcement of marshlands and shoreline protection legislation, and coastal management planning and coordination. Coastal Resources Division works with OPB to review and comment on OCS development planning documents, reports, and permit applications. Cooperative efforts have resulted in identification of environmental studies needs and priorities for the South Atlantic region, as well as modifications to OCS lease stipulations. It is the combined expertise of Coastal Resources' fisheries biologists and resource planners that supports OPB's policy decisions and official comments that are transmitted to the Department of Interior.

Georgia's Coastal Management Program

Georgia's coastal zone is not typical of other coastal areas in the country. While over 50 percent of the populations in most coastal states reside within 50 miles of the coast, Georgia's coastal population is six percent of the State's total.[1] The State's beaches are located on its 18 barrier islands. Of these islands, four allow unlimited public access, six have controlled public access, and three are virtually off-limits to the public. As a result, the coast of Georgia maintains a high environmental quality with limited tourism influx, and thus little of the cheek-to-jowl beachfront development that pervades other stretches of the Atlantic coast.

Cooastal Georgia's industry is largely pulp and paper mills, chemical plants, and associated shipping. The port of Savannah, 22 miles inland on the Savannah River, is the most active port in the State. The port of Brunswick is a natural harbor which, if developed, has the potential of serving a wide range of cargo vessels. Brunswick became a base of operations for three oil companies that have drilled exploratory wells off the South Atlantic coast, while a fourth chose a Savannah site. In addition, oil spill contingency equipment and response teams were located in Brunswick.

Georgia, like most coastal states, began in 1974 working toward the development of a federally-approved Coastal Zone Management Program in

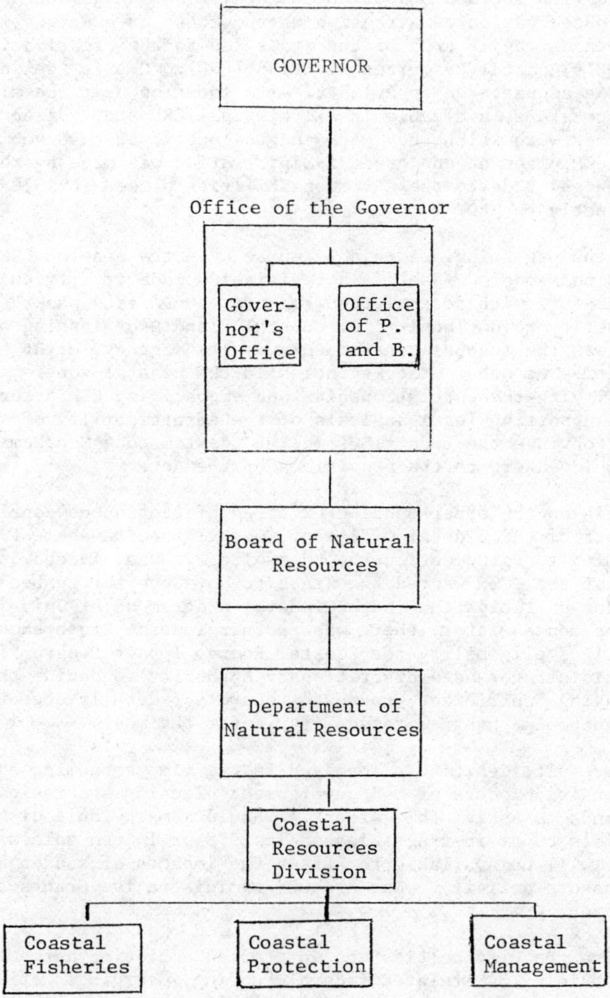

A simplified organization charg showing the relationships between Georgia State agencies involved in OCS development planning. Only principal functional units of Coastal Resources Division are shown.

FIGURE 2

conformance with Section 305 of the Coastal Zone Management Act of 1972. This new concentration of attention on the coast as a unique, essential, attractive and fragile area of the state led to the creation in June 1978 of DNR's Coastal Resources Division (CRD). CRD is the only division of DNR not headquartered in Atlanta. Its location near the midpoint of Georgia's coast is invaluable in allowing all CRD staff direct access to the resources with which they work and the public they serve. Although the early CZM planning and grant administration was done by the Office of Planning and Budget, these responsibilities passed in 1978 to DNR and thence directly to CRD.

From the beginning, Georgia's course upon the seas of CZM participation was not smooth. Georgia's legislation made it difficult for the State to conform with federal requirements. One issue was the lack of State authority to do local (i.e. coastal) land use planning and zoning; the other was the tenuous control DNR, as lead agency, might have over the activities of other State agencies in the coastal zone. An Executive Order clarifying these relationships and emphasizing CZM priority, as well as a supporting legal analysis of the enforceability of the Order, failed to convince the federal Office of Coastal Zone Management that Georgia could adhere to the provisions of the Act.

Georgia on the other hand, grew tired of continuous conflicting directives from the Federal Office of Coastal Zone Management. Despairing of coming to agreement on the unresolved issues; indeed, uncertain exactly what the OCZM wanted the State to do, Governor Busbee withdrew Georgia from participation in the federal program as of July 31, 1979. Nearly nine months later, there was another federal-State showdown over Georgia's eligibility for Coastal Energy Impact Program funds. OCZM's Administrator used discretionary authority to decide that the State was not making "sufficient progress" toward a federally-acceptable coastal management program to qualify it for the money.

As a result, Georgia is involved in the time-consuming and sometimes expensive process of OCS development planning and review with no federal funds to assist this effort. Should a marketable discovery occur off Georgia's coast in tracts leased in 1978 or in new sale areas, no CEIP money will be available to offset the impacts of sudden burst of related onshore activity. Nor are OCS Administrative Grants available to the State.

Despite the difficulties encountered in achieving agreement with OCZM or obtaining OCS-related federal funding, Georgia is still committed to working toward compatible harvests of offshore renewable and non-renewable resources. This philosophy is an integral part of the State's ongoing Coastal Management Program. Efforts to comply with federal requests as well as protect State priorities and marine resources are described in the following section.

DEVELOPMENT OF STATE AND REGIONAL ADVOCACY

Planning for Lease Sale 43

Lease Sale 43, held in March 1978, was the first such sale in the South Atlantic. The Federal Bureau of Land Management (BLM) offered 225 nine-square-mile tracts for lease by the oil industry. Most of these tracts (or blocks) were 40 to 70 miles east of Jekyll Island, Georgia. A total of 43 tracts were leased, and industry obtained permits to drill exploratory wells on 11 of them. To date, six dry holes have been drilled an(further activity on these leases seems unlikely in the near future.

The State of Georgia's endeavors to prepare for offshore energy exploration and development began several years before Lease Sale 43. The Governor's position was one of "support and encouragement",[2] but the State was anxious to safeguard the natural and cultural resources of its coastal environment. The Governor thus called for an "effective planning program" to anticipate needs for both fisheries and fuel resources on the Outer Continental Shelf.

The federal OCS development process originally involved little participation by the affected states. During the mid 1970's, states intensified their efforts to modify OCS regulations so that Department of Interior would consult with them prior to making decisions on OCS development matters. As a result, CFR 250.34 of the OCS Orders was revised to require federal consultation with governors of the affected states, and include states in the review process for exploration and development plans.

Another step toward fuller State participation was the formation of the Intergovernmental Planning Program (IPP). States represented on the OCS Advisory Board urged expansion of cooperative decision-making through regional focus. The Regional Technical Working Groups were thus conceived. A Scientific Advisory Committee balanced the OCS Policy Committee on the national level. This IPP structure is still in its shakedown stages.

Prior to Lease Sale 43, Georgia and other South Atlantic states received the proposed lease stipulations. They urged U.S. Geologic Survey to put more teeth into the stipulations, especially those regarding transportation on the OCS and biological resources. Although the states had only modest success in obtaining tighter language, the stipulations did allow South Atlantic states to insist that one oil company monitor exploratory drilling on a site near potential live-bottom resources.

Despite the process changes made by the states prior to Lease Sale 43, South Atlantic representatives were disappointed in their attempts to focus federal attention on critical environmental studies needs. The orientation of BLM has been toward physical oceanography, computer modelling, and satellite photography. Studies of this sort were in place, and continued as high priority items.

State representative urged BLM to examine the nature and interrelationships of South Atlantic marine organisms and their habitats. There were questions about what kinds of habitat areas were essential to fish species in the South Atlantic Bight. Offshore food webs were also poorly understood. The State of Georgia[2] called for the BLM environmental studies program to be an "integral part" of OCS development planning. It appeared that, instead of predicting potential problems from study data, BLM was gathering data for use after development occured. Georgia felt that BLM was interested only in evaluating impacts after the fact; a "post mortem examination". However, BLM did not alter its environmental studies plan. As a last line of defense, state representatives then pressed for more stringent lease stipulations and notices to lessees to provide a means of protecting marine renewable resources potentially threatened by oil and gas exploration until a more complete understanding was achieved of the resources' vulnerability to human disturbance.

It was the belief of the South Atlantic states that environmental studies performed prior to Lease Sale 43 did not produce sufficient information to enable state or federal resource managers to make sound management decisions regarding the tract selections for Lease Sale 43. In Georgia's opinion at least, this same lack of information hampered OPB and DNR staff in their attempts to thoroughly review and comment on industry exploration plans, environmental reports, and applications to drill, discharge drilling materials, and transport crew and equipment on the OCS.

Developing a coalition

Within a few months after Lease Sale 43, BLM began asking the affected states to respond to new environmental studies plans and priorities for the next South Atlantic Lease Sale, Sale 56, scheduled for August 1981. There was increasing coordination among the four South Atlantic states, and between OPB and DNR in Georgia, to evaluate studies needs and establish priorities for use of the limited studies funds. It was reasoned that a list of ranked studies developed and approved by all four states would reflect more accurately a regional viewpoint and would carry the weight of four governors behind it. The ad hoc coalition of state representatives formed a solid core for the fledgling South Atlantic Regional Technical Working Group, one part of the newly-implemented OCS Intergovernmental Planning Program (IPP). The Working Group membership includes BLM, USGS, U.S. Fish and Wildlife Service and the U.S. Coast Guard as well as representatives of the oil industry and environmental groups. However, it is probably fair to say that the South Atlantic states representatives have taken the lead during this group's infant stages. This was largely due to the States' accumulation experience with the OCS planning process and acute desire to preserve their own ocean resources.

As mentioned earlier, Georgia is represented on this committee by a member of OPB staff. However, the alternate representative (who usually attends all meetings) is a DNR staff person. Any state coastal management or fisheries concerns thus can be addressed immediately, and a team approach toward OCS planning in Georgia maintained.

As the process continues and State representatives have more contact with each other and with federal representatives, more effective ties have developed between the States and to some extent with federal agencies. These formal and informal relationships have resulted in a more effective information transfer, a better understanding of each agency's concerns and constraints, and ultimately, an increasingly efficient use of time and energy because State representatives have a better grasp of where and when their input really counts.

DNR's expanding role in Georgia's OCS planning and a mutually supportive attitude between the two state agencies has allowed Georgia to become more assertive in alerting the U.S. Department of the Interior about the critical lack of information regarding the biota of the South Atlantic Bight. This position was echoed by South Carolina, Florida and North Carolina. All four states agreed that intelligent resource decisions simply could not be made because almost nothing was known about live bottom habitats and their importance to offshore fisheries of the South Atlantic. The States stressed the need for more data about the relationships of the live bottom to the species that use it. This is the first step toward placing a value on these ocean floor "oases". With that assessment, scientists and resource managers can begin to weigh the relative merits of hydrocarbon exploration and production against possible or certain damage to a live bottom area when rigs or platforms are proposed near or on such habitat.

Due to the high priority placed on study of offshore biota by the four governors, BLM allocated funding for a study to examine living marine resources offshore South Carolina, Georgia, and Florida. South Carolina Wildlife and Marine Resource Department (SCWMRD) successfully bid on the contract, subcontracting a portion of the site sampling and taxonomic work to Georgia DNR's Coastal Resources Division. Work on the project began in January 1980, and the first phase of this three-part study will conclude in the spring of 1981.

As originally conceived, the first phase involved doing both remote and diver sampling of three sites in three bathymetric zones (a total of nine sites) to photograph and collect benthic organisms. While most of the stations are off Georgia, there are two off South Carolina and one off northeast Florida. Selected fish species were sampled and the contents of their stomachs analyzed to establish food chain relationships. Juvenile fish of many species were collected also to allow further study of their sometimes baffling appearances and habits. The benthic organisms sampled were taken either to the CRD or SCWMRD laboratory and keyed to the most specific taxonomic category possible.

After the joint South Carolina-Georgia study was underway, the state of North Carolina became increasingly alarmed at the number of promising tracts off its shores being nominated for inclusion in Lease Sale 56. An oil company consortium proposal to drill a Continental Offshore Stratigraphic Test (COST) well offshore Morehead City indicated serious industry interest in the North Carolina OCS. Because the North Carolina Governor's office made a strong case to BLM regarding insufficient biological information on these tracts, the agency expanded its Living Marine Resources study to include participation by North Carolina. A team of scientists from Duke University is conducting sampling activities off North Carolina using similar techniques to those followed by South Carolina and Georgia. SCWMRD retains its role as lead agency, maintaining data banks and coordinating the work of all three parties.

What are the benefits from this effort so far? One of the most evident is the better organization, and more effective communication among marine scientists and resource managers in the three states. Sharing of information, equipment, and expertise has helped the researchers understand offshore resources on a regional basis. It has reduced duplication of effort and largely allowed each entity to perform portions of the study for which it is best suited.

Another benefit of having three of the affected states doing the research is that study information will be available to decision makers as soon as it is obtained. Historically, BLM has had trouble getting final reports on the environmental studies it funds in time to be useful for tract selection and other important decision points. More critical, however, is the difficulty experienced by state resource managers in securing copies of these reports when they are completed. A federal regulation governing printing has forced BLM to severely limit the number of copies of each study it can produce.

Because of the cooperative relationship between Georgia's OPB and DNR, preliminary study findings and hypotheses can be relayed quickly to the State's policy-makers. Because of the team approach to research demonstrated by North Carolina, South Carolina and Georgia, those same data are shared very early among those three states and with Florida. And because of a few strong state-federal agency ties, findings and potential causes for concern can be transmitted rapidly by the contractors to DOI. This kind of immediate data sharing is not always possible when a study contractor is a private company, concerned with competition and the proprietary nature of its work.

Because the South Atlantic States representatives had direct access to the personnel engaged in Phase I of the LMR study, they were able to get the researchers' recommendations for changes in the scope and nature of the work proposed for Phase II. As a result, some sampling tasks may be expanded, some reduced or eliminated - thus ensuring maximum possible efficiency of study procedures and expenditures.

These are a few of the benefits to regional OCS planning of contracting directly with agencies within the affected states to conduct the Living Marine Resources study. Additionally, DNR's involvement in the project has helped Georgia's coastal resources programs by furnishing new information that could otherwise not be obtained, or completing partial data gathered through state fisheries research. For instance, findings from the Living Marine Resources study might aid Georgia fisheries biologists formulate life histories of fish species with commercial or recreational potential. The life history information would then be used to help shrimp fishermen diversify during the off season, or inform sports fishermen about possible catch.

With increasing concern about the future of threatened and endangered species such as the sea turtle, divers with the Living Marine Resources study are paying particular attention to any turtles sighted on or near live bottoms. It is essential that any habitat type used by these turtles be protected, so that the animals are not further threatened.

Georgia's decision-makers are encouraged by the progress made so far in planning cooperatively with other states and with federal agencies for an informed, responsible development of the South Atlantic OCS. However, with Lease Sale 56 just months away, there are a few unresolved issues and some points on which the South Atlantic States disagree with BLM or U.S.G.S.

At a recent meeting of the Interagency Committee on Ocean Pollution, Research, Development, and Monitoring[3], new information was revealed concerning possible hazards for energy development in the South Atlantic. The major concerns were as follows:

- Gulf Stream currents in deep water off Georgia's coast (about 60 miles from shore) may create major problems for drilling operations. With present drilling technology, the combined factors of swift currents and deep water will make exploration difficult and production nearly impossible. Eddies spinning off the western edge of the Gulf Stream send large volumes of water in a southwesterly direction toward the Atlantic shore. The force of these eddies could send oil from a major Gul Stream spill onto South Atlantic beaches, severely damaging the nearshore environment and affecting South Atlantic coastal fisheries, threatened and endangered species, and supporting habitat - - as well as the coastal tourist industry.

- Recent geological studies in the Blake Plateau, a deepwater area of the South Atlantic, show structures that were probably river beds when the sea level was lower, but are now filled with soft material. A drilling platform located on the edge of such a geologic hazard could easily topple.

- The Blake Plateau also has large faults. If a drill pipe entered this subsurface cavity, the apparatus could lose pressure and result in a blow-out.

- Limestone caverns in the southern Blake Plateau, off Georgia's coast, are apparently closely related to the coastal freshwater supply. Any lessening of environmental quality in those areas due to oil and gas activity could affect the water quality onshore.

Such recent information underscores the South Atlantic States' concerns about the status of knowledge regarding the South Atlantic Bight environment. The federal government seems considerably more sanguine, apparently confident that past and present studies fill the data gaps to give a complete picture of the South Atlantic OCS.

One point of disagreement is the current model developed with BLM funding to predict the fates and effects of possible oil spills in the South Atlantic. While BLM is pleased with the effort and seems convinced that the computer-produced predictions accurately reflect real world conditions, South Atlantic oceanographers and State representatives have grave reservations about the real world applicability of the results.

In the absence of much study information for the South Atlantic OCS, and the need to make decisions concerning Lease Sale 56, the South At= lantic states have urged more stringent lease stipulations that could be applied to exploration activities during the next round of drilling. Enforcement of stipulations, although not an ideal solution to the matter of resource protection, at least offers a short-term means of controlling activity that could potentially harm a biological, geological, or cultural resource. The South Atlantic states worked diligently to develop mutually acceptable lease stipulation revisions, and submitted these to BLM as part of the states' responses to the Lease Sale 56 Preliminary Draft Environmental Impact Statement. While the suggested revisions did appear in the DEIS, they were listed as alternative wordings not recommended by BLM.

The South Atlantic Regional Technical Working Group evaluated environmental studies proposed for FY82 at the request of BLM. The ranking procedure set by the agency was followed by the Working Group in an arduous two-day session, and a prioritized list of studies resulted. However, the subsequent FY82 Environmental Studies Plan developed by BLM did not reflect accurately the outcome of that meeting; numerical rankings were changed to allow the upgrading of one study that BLM favored and the Working Group felt had low priority.

However, the commitment within the State of Georgia to plan for responsible OCS energy development still exists. With properly-focused studies and supporting lease stipulations, harvest of both fisheries and fuel resources can probably proceed compatibly. Results of studies on offshore biota can benefit both the states and the OCS development effort.

There are promising indications that the federal government is looking at all ocean resources programs and problems in a more comprehensive, cohesive fashion. One sign is the formation of the Interagency Committee on Ocean Pollution, Research, Development, and Monitoring. This body is comprised of representatives from all federal agencies concerned with ocean resources. It will attempt to avoid duplication of programs, funding, and look broadly at threats to the marine environment.

How effectively Georgia and the other South Atlantic states can work toward wise OCS development will depend largely on their efforts to build stronger relationships with accountable federal agencies, continue to work with each other, and their willingness to allocate state resources toward this goal. Georgia is handicapped by the loss of Coastal Zone Management and Coastal Energy Impact Program monies. However, if a balanced approach to fish and fuel resource harvests is to be achieved in the waters offshore Georgia, it is up to the Office of Planning and Budget and Department of Natural Resources to translate that commitment from philosophy to practice.

CITATIONS

1. Coastal Area Planning and Development Commission, Brunswick, Georgia. Demographic data. 1970.

2. Testimony given on behalf of George Busbee, Governor of Georgia, at the March 1977 public hearing on the Lease Sale 43. Savannah, Georgia.

3. Meeting of the Interagency Committee on Ocean Pollution, Research, Development and Monitoring. Boston, Massachusetts, July 1980.

PROBLEMS FACING ALASKA OUTER CONTINENTAL SHELF RESOURCE DEVELOPMENT

by
Robert J. Brock, Assistant Manager
Dr. Raymond R. Emerson, Environmental Scientist
Gordon M. Euler, Environmental Specialist
Bureau of Land Management, Alaska OCS Office

I. Introduction. The Outer Continental Shelf Lands Act of 1953, as amended, charges the Secretary of the Interior with administering mineral exploration and development of the Outer Continental Shelf (OCS), as well as conserving natural resources of the shelf. The law requires that the Secretary of the Interior make oil and gas resources available to meet the Nation's energy needs as rapidly as possible, and to balance orderly energy resource development with protection of the human, marine, and coastal environments.

In view of the growing imbalance between domestic oil and gas production and use, especially the expected decline in production from prime areas such as the Gulf of Mexico and the North Slope, there is a greater need to develop resources from the OCS frontier areas.

The following facts demonstrate the continuing decline in crude oil production in the contiguous United States, the growing importance of Alaskan oil, and the dramatic rise in petroleum imports.

- By 1985, petroleum imports are projected to be 80 percent higher than in 1975. (The administration's latest energy program proposals, based in large part on conservation and increased domestic energy production would reduce our dependence on foreign oil by 4.5 million barrels per day by 1990, according to the President's message on the environment of August 2, 1979.)

- Through 1990, the shift in the pattern of supply is projected to continue; 36 percent of petroleum liquids production in 1990 would come from Alaska or from the U.S. OCS. In 1977, offshore areas and Alaska accounted for only 19 percent of domestic production.

In contrast to development in other oil importing countries, the volume of U.S. oil imports rose more because domestic production was lower, not because domestic use was higher. Between 1973 and 1977, U.S. oil production declined by 1.2 million barrels per day, while use increased by 1.0 million barrels per day. In addition, the higher price of oil has been the most important factor underlying the large U.S. trade deficit.

Action to expand U.S. energy production is necessary now if we are to ensure that sufficient energy sources will come on-stream when needed. Let us look, however, at the problems that confront oil and gas development in Alaska.

II. State Characteristics.

A. Geography. Alaska is a huge state, with limited transportation, and few, but a wide variety of people. The state's land area, at 586,000 square miles, is well over twice that of Texas, and almost six times the size of California.

There is everything from high-rise buildings in a city of 200,000 to villages of less than 50 people who live off the land and have no running water. There are few major newspapers. Many communities have but one telephone. Television is just reaching some communities. Much communication is still by two-way radio. It may take days to reach somebody in the "bush," as rural Alaska is referred to.

The OCS off Alaska is 830,000 square miles, which is 74 percent of the nation's total. In comparison to the continental U.S., this area would extend from Atlanta to San Francisco, and from North Dakota to New Orleans.

Most of the Alaskan OCS lies in the Bering Sea region, and in the Arctic, which includes the Chukchi and Beaufort Seas. Most of the estimated offshore oil and gas reserves are thought to be in these frontier areas.

B. Population. Alaska has about 400,000 people, half of which live in the Anchorage area. Seventy-five percent of the total population lives within 10 miles of the shoreline. Not counting the population centers of Anchorage and Southeast Alaska, there are about 21,000 people scattered along the 34,000 miles of shoreline. Approximately 17,000 people live in the regional centers of Bethel, Nome, Kotzebue, and Barrow, leaving less than 4,000 people along the coastlines of the Bering Sea and Arctic Ocean. If Manhattan Island were as densely populated as Alaska, it would have 24 inhabitants!

C. Transportation. Alaska has fewer miles of paved road than Washington, D.C., or about 1,500 miles. There are roads connecting Anchorage with Fairbanks and the Kenai Peninsula. The Alaska Highway stretches through Canada to Fairbanks, but there are no roads over the majority of the state. One cannot even drive to Juneau, the state capital. The Alaska Railroad operates from Seward to Fairbanks. The White Pass and Yukon Railroad operates from Skagway in Southeast Alaska to Whitehorse in the Yukon, with combined total track of about 620 miles.

There is water access to all coastal communities (weather permitting), but only a few ports exist that can handle ship traffic (e.g., Kodiak and Dutch Harbor). Lighter barges have to be used in most other areas. Perhaps the most important transportation network is air travel. Every town, village, and hunting camp has an airstrip to land on. However, only about 20 airports exist that can accommodate jet traffic. With the exception of Prudhoe Bay, all of the Arctic and Bering Sea coastal communities depend on air and water travel to move equipment, supplies, and people in or out.

D. Government. The governmental structure in Alaska is unique. There are no counties. In some areas of the state, boroughs have been formed to provide necessary services. Only 15 percent of the coastline is controlled by local government. The other 85 percent is in what is called the unorganized borough, where the state has the planning and zoning authority.

This presented a unique challenge to the legislature when they were developing Alaska's coastal management legislation. The legislature believed that the initiative for coastal planning should come from the local government. Even if all local governments planned, a large area would still not be covered. The legislature solved the problem by allowing, under certain circumstances, creation of special purpose districts in the unorganized borough solely for the purpose of coastal planning. To date, three of these districts have been organized, and others may soon follow. It is important that these areas be created and coastal zone management plans developed, as these areas will be adjacent to frontier OCS lease sales.

III. Regions of Alaska.

A. Gulf of Alaska. The Gulf of Alaska is primarily ice-free in the winter except for Cook Inlet. However, some of the worst storms on record have been recorded here. Seas of 120 feet are not uncommon. This is the area where commercial crab and bottomfishing activity is concentrated. Large salmon fisheries are located in Kodiak also.

Modern technology has influenced this region the most. Although some subsistence fishing takes place here, this is the least use made of the marine resources. There are seven species of whales that summer or migrate through this area.

B. Bering Sea. This area is the most diversified. It is by far the largest OCS area, and has the most potential for major oil and gas reserves. But it is also the area that the least is known about. This region is the transition area between the Gulf of Alaska and the Arctic for a number of reasons: (1) It is a region of commercial fishing activity in the south which decreases northward and is replaced by subsistence hunting and fishing. (2) Primarily ice-free conditions occur in the south to pack-ice conditions in the north. (3) This is the transition zone ranging from modern man's technological world to the practice of Native customs of generations ago.

The commercial salmon fishing industry in the Bering Sea is the largest in the world. The salmon come through the passes on the Aleutian Chain and into Bristol Bay. A bottomfishery is just starting here, but has a potential of importance equal to the salmon fishery. Fifteen percent of the world's marine harvest is taken into the docks of Kodiak and Dutch Harbor.

During the summer and winter this area is a major breeding and calving ground for many marine mammals and the Natives use a variety of these mammals (whale, walrus, and seal) for subsistence.

C. Arctic. The arctic region is a very harsh environment. Known for its ice pack, harsh winters, isolation, and total darkness for two months of the year, this area is very different from the world as we know it. There is usually only one annual barge trip that supplies the arctic villages with fuel and other essential items. There is no commercial harvesting of any marine resource. Whales, seals, and a few species of arctic fish make up the major diet of the coastal Natives. Additional food sources include caribou, moose, and reindeer, and what food may be purchased in the local village store.

This region is where the Native tradition and the fear of the unknown are the most prevalent. Oil and gas exploration in the ice pack is one of the biggest unknowns to the Natives. The concern the Natives have about the effects of oil and gas development with particular reference to an oilspill is also of concern to BLM.

IV. Socioeconomic and Environmental Studies. In an attempt to develop potential oil and gas reserves on the OCS without causing irreparable harm to the social, cultural, and biological resources of Alaska, a massive studies effort has been ongoing for the past 5 years. These studies attempt to focus on potential impacts with respect to socioeconomic and environmental factors. The socioeconomic studies program is funded by the Bureau of Land Management and administered through its Anchorage Outer Continental Shelf Office. These studies are specific to each of the ten leasing areas and are concerned with the effects of oil and gas development on the infrastructure of a village or small community. Projected impacts are potentially greater for the smaller villages than they are for a larger community. Available power and transportation systems are evaluated, and in some cases these systems would need to be upgraded and vastly improved to accommodate the added needs.

Other socioeconomic studies are concerned with the subsistence lifestyle and the effects of oil and gas development on hunting and fishing activities. What compensation is there for lost resources in the event of a major oilspill? This is a basic question to consider. Similar concerns are also addressed for commercial fishing activities.

In conflict with the concern of potential oilspill impacts on subsistence and commercial activities, is the fact that these activities also require oil and gas resources. The Alaskan dog sled, a major mode of transportation for years, has been replaced with the snow machine. Also, the sealskin boat of the Eskimo hunter is now an aluminum or fiberglass craft powered by an outboard motor.

Environmental studies, which include oceanographic, geological, and biological aspects are also funded by the Bureau of Land Management. These studies will be funded in the neighborhood of $18 million for the 1981 fiscal year. The funds are administered primarily through the National Oceanic and Atmospheric Administration (NOAA), under BLM direction, with the exception of the marine mammals program which is administered directly by BLM.

The initial focus of these studies was to conduct reconnaissance surveys. These studies have provided initial information to define the water circulation systems, contaminant trajectories, ice hazards, seafloor faults, and seismic activity. They also have provided information on present levels of hydrocarbon and trace metals in seawater along with population and ecosystem studies. The more recent efforts have concentrated on developing a predictive model system for pollutant transport along with vulnerability indices of resources to a potential oilspill. Additional emphasis is presently being given to cleanup technology and contingency planning.

The information generated from the studies program is readily available to interested parties. In addition, several trips are made to the local villages to inform the people about the leasing program and to present information from the various study efforts. During this process, which is referred to as "scoping," the villagers are encouraged to evaluate the information base and contribute their first-hand experience about the biological resources, wind, ice, and circulation patterns.

The scoping process also provides an opportunity for the villagers and interested groups to consider modifications of the selected tracts for leasing. These modifications may include tract deletions, mitigating measures (special rules), and other considerations which make the area more acceptable for leasing.

V. Case Study - Yakutat. Prior to the first offshore oil and gas lease sale in Alaska in April 1976, there was considerable concern in the city of Yakutat of what could happen there if a discovery were made. One person was quoted as saying "there would be hundreds of rigs in the Gulf drilling for oil." Others felt there could be a doubling or tripling of the population in Yakutat.

The major concern of the mayor and city council was the possible large influx of people during exploration and a major decline in population if nothing was found. The city reached an agreement with the oil companies, whereby the companies would not purchase anything in local stores; nor would anyone arriving by jet destined for an offshore platform be allowed into town. It was hoped this would eliminate any potential problems between oil workers and the local community.

The city zoned the south side of Monti Bay, the town's waterfront, for industrial purposes, which allowed the oil companies to construct docks and warehouse space. A total of 10 acres was used. A small crew of shoreside people were stationed in Yakutat to run the support base. Exploratory drilling activities, which resulted in 11 dry holes, were easily handled out of Yakutat, although much support came from Seward.

There were a few incidents where service boats were alleged to conflict with fishing boats and gear. One oil company worker did leave the airport once on an unauthorized visit to see the city. Aside from these incidents, there was very little impact that resulted from the experience, and Yakutat is now much better prepared to handle activities associated with future oil and gas development.

VI. Summary. The Federal Government is required by law to manage the exploration and development of oil and gas resources on the OCS. To help meet the energy needs of the nation in an environmentally safe manner, these resources must be developed as rapidly and yet as carefully as possible. While overseeing this development, the Federal Government must balance orderly resource development with protection of the human, marine, and coastal environments, and ensure that the public receives a fair return for these resources.

The Alaskan OCS oil and gas deposits are potentially the largest national source of petroleum during a time of critical need. In each OCS area for which development is proposed, extensive environmental studies must be conducted before such development is allowed. If these studies show that development of specific areas will result in unacceptable environmental risk, those areas will not be leased.

Basic management questions further define information needs that the studies must address. At each step of the OCS management process a variety of resource use conflicts may be encountered. The decision to proceed with one resource use, such as OCS oil and gas development, in the presence of other uses of the OCS, will implicitly result in trade-offs between potential conflicting activities. The purpose of pursuing basic management questions concerning major multiple-use conflicts is to make these trade-offs explicit prior to the time the decision is made.

THE EVOLUTION OF MITIGATION IN A HIGHWAY SYSTEM

Rodney Edwin Emmer, Partner
Coastal Environments, Inc.
1260 Main
Baton Rouge, Louisiana 70802

Louisiana, located on the northern margin of the Gulf of Mexico, is the product of the great rivers which drained the interior of the United States. The alluvium forming the state drapes the older Mesozoic formations and plunges off the continental shelf to gather as a curtain on the Gulf floor. The subaerial topography can be separated into two geologic units: the Recent, or that material which has been deposited since the last rise and still stand of sea level, and the older surfaces which pre-date the last low stand of sea level. The Recent surface is composed of three major entities: the Mississippi River deltaic plain of southeast Louisiana; the Chenier Plain of southwest Louisiana; and the alluvial valleys of the Mississippi River, the Red River, and numerous small water courses.

When combined, the deltaic plain and chenier plain form the natural coastal zone of Louisiana (McIntire et al. 1975). The coastal zone is a very complex mixture of physical and biological elements and processes which interact among themselves, thus providing a difficult base upon which to establish the cultural elements of society. For example, the deltaic plain consists of six deltaic lobes which are made up of meanderbelts, interdistributary basins, and marine and fluvial-marine environments (Kolb and van Lopik 1958). Meanderbelts are the higher and drier natural levees, point bars, and abandoned channel courses (Figure 1). The meanderbelts trend generally north-to-south across the deltaic plain and decrease in elevation (from 35 feet to near sea level) in a downstream direction and away from the channel (from 35 feet to below 5 feet). Soils are mainly prime farmland and are the most suitable for residential, commercial, or industrial development because of better physical characteristics for supporting structures.

Between the meanderbelts are interdistributary basins; the cypress-tupelo gum swamps; fresh, brackish, and saline marshes; and lakes and water courses which people perceive when visualizing Louisiana bayou country. The basins are the broad, paludal environments of low elevation and low relief, usually less than 5 feet in the inland ends of the basins to sea level nearer the coast. The soils of clays and organics are not prime farmland and are not suitable for development because of flooding and poor foundation strength.

The interdistributary basins merge into the marine and fluvial-marine environments closest to the coast. Reefs, beaches, and bays constitute the marine system, while prodelta and intradelta deposits form the fluvial-marine system.

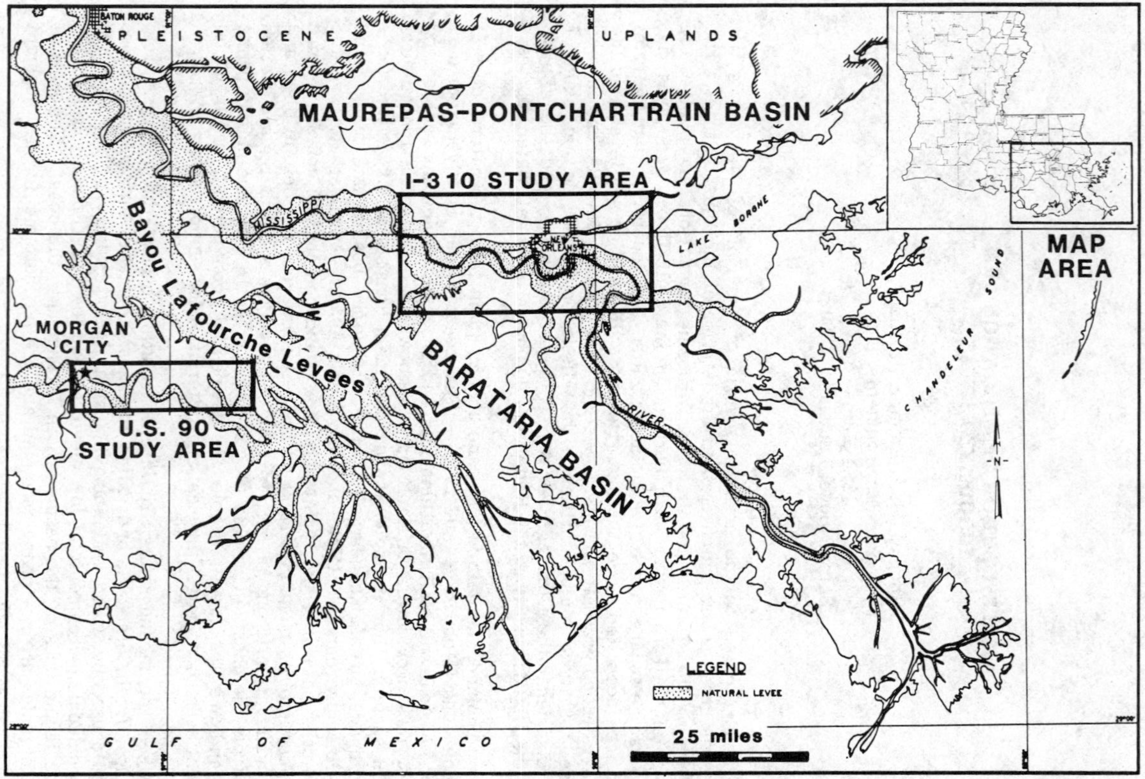

Figure 1. Meanderbelts trend north-to-south across the delatic plain, while commerce needed to move east-to-west. I-310 study area (project area) and U.S. 90 study area (project area) are shown (modified from Kolb and van Lopik 1958).

Both zones are of low elevation (less than 5 feet) and low relief. Development is restricted to selected barrier islands and abandoned beach ridges (cheniers).

Man's use of the coastal zone of Louisiana began before the last still stand of sea level (7300 B.P.). Europeans explored the state in the sixteenth century but did not establish a permanent settlement until the eighteenth century when, in 1718, New Orleans was founded on the Mississippi River. Early settlements in coastal Louisiana concentrated on the crest of the natural levees where the land was the highest and the best protected from flooding, the soil characteristics afforded suitable foundation conditions for structures, and communication systems of roads and water courses were more accessible for commerce and personal travel. However, the natural communication network was oriented north-to-south along the natural levees and water courses, while the commerce network was oriented east-to-west along the coast and between areas of settlements. To get from one community to another by land transportation meant traveling an extended journey on circuitous routes of interconnecting levees, or crossing interdistributary basins by boat along meandering channels. Canals were dredged for waterborne commerce in the coastal zone (Davis 1973), but road travel was indeed a problem as late as the 1920s. For example, the recommended route from Baton Rouge to New Orleans was 139 miles long, a trip which crossed the Mississippi River twice even though both cities are east of the River (Prophit 1980). Such an indirect route was necessary because it followed the high, natural levees and crossed the interdistributary basins where no alternative path existed. The air distance from Baton Rouge to New Orleans is only 70 miles.

To compound the transportation issue, Louisiana draws a tremendous wealth from the renewable and nonrenewable resources of the coastal zone. The renewable resources include commercial catches of shrimp, menhaden, oysters, catfish, crabs, and crayfish, and agricultural production from sugarcane and soybeans. Nonrenewable resources are led by the petroleum industry. The majority of people in Louisiana live in the coastal zone, where they work in the petrochemical and aquatic resources industry. The always-present difficulty of traversing the physical landscape by this large segment of the population resulted in the state legislature creating the Louisiana Highway Commission in 1921, the predecessor of the present Louisiana Department of Transportation and Development (LDOTD). The Commission was charged with the construction, maintenance, and managment of highways in the State and strove to develop and implement programs for safe and efficient transportation systems in the most economic manner. The State shifted roadways from the crest of the natural levees which had become congested with industrial and residential development to the open fields of the backslopes of the natural levees. Interdistributary basins were no longer avoided, but crossed on straight embankments, the most economical installation technique. Structures bridged water courses, and causeways were placed where unstable foundation conditions precluded embankments. The design of the project drainage protected the road from flooding or washout. Canals were dredged and spoil banks were deposited where engineering plans directed their location.

Some consideration was given to protection of the environmental setting, but this was limited in scope and extent (Louisiana Department of Highways 1971). Sediment control included slab sodding with native vegetation on barren ground at the earliest feasible time. Highway embankments would not result in the filling of streams or waterways. Destruction of vegetation during construction was limited to as small an area as feasible. Streams were to be disturbed as little as possible and then returned as near as possible to their original condition. Collection and disposal of solid waste, including oils and chemicals, would be at designated landfills approved for waste disposal.

In 1969, the procedures for planning and implementing public works projects were drastically modified with the passage of the National Environmental Policy Act (NEPA)(42 USC 4321 et seq.) The decision-making process was radically changed nationally, although the full effect of the law in Louisiana was not to be felt for several years after the passage of NEPA. Traditionally, planning for highways was based on the need for an improved transportation network, a system that was economically feasible and could be constructed within the engineering capabilities of the day. With the introduction of NEPA, the planning and implementation process expanded to include environmental systems as part of the evaluation procedure. These systems were to be treated as equals with the established engineering, economic, and social parameters, a process not readily accepted by all decision-makers. Established policies and techniques for estimating economic effects, selecting social policies, and evaluating engineering feasibility were available. However, environmental considerations based on theories of holistic systems and interrelationships between and among physical and biological elements were new and analysis techniques for projecting environmental impacts were unknown, in the developmental phase, or discarded or overlooked as not definitive.

The NEPA requires documentation of environmental impacts and consideration of alternatives. In addition, the Council on Environmental Quality (CEQ), in its guidelines for the implementation of NEPA (Federal Register 1978), instructs that some forms of mitigation be considered in the preparation of the project plans. Mitigation measures are those actions which can reduce damages resulting from the proposed action and partially compensate affected entities for project-related losses. The mitigation measures to be investigated include:

1) the consequences of eliminating selected activities or combinations of activities associated with the proposed action;

2) the limitation of the extent, degree, or magnitude of the associated activities;

3) the restoration, repair, or rehabilitation of the affected physical, biological, or cultural environments by specific structural and non-structural techniques;

4) the reduction of adverse impacts by selecting feasible actions; and

5) the compensation of affected parties by replacement of lost resources or substitution of lost resources.

Before the mitigtion measures result in the modification of any plans, they are evaluated on their advantages and disadvantages for achieving the objectives of the proposed action while at the same time corresponding to public policy for resource allocation. For example, resource values of wildlife habitat and nursery grounds or of prime farmland for food and fiber can be compared to the social worth of the proposed action. The comparisions are scientifically quantified estimates of impacts of the physical, biological, and cultural costs for implementing an action.

The objective of this paper is to demonstrate the influence of NEPA on the planning and implementation of two highway projects in the coastal wetlands of south Louisiana. An evolving planning process at LDOTD in response to NEPA requirements and CEQ guidelines led to a drastically changed philosophy and procedure for locating projects and for incorporating mitigation measures into the project design. Highway I-310 (née I-410) is discussed first because the decisions made during the planning for this project influenced subsequent projects, such as U.S. 90, which were modified to reflect the new philosophy of highway construction.

I-310 (née I-410)

The Dixie Freeway was proposed in 1964 as a southern bypass of New Orleans by the West Bank Council of the Chamber of Commerce. Initially, the route was only 18 miles long through Jefferson Parish (County), but as time progressed the length of the project increased to 54 miles, from I-10 in St. Charles Parish west of New Orleans to I-10 in eastern New Orleans (Figure 2). Correspondingly, costs escalated from $31,067,000 in 1964, to $450,000,000 in 1972. In 1968, the loop became part of the interstate system, with the Federal government paying 90% of the cost. The proposed I-410 alignment was primarily designed to minimize costs by passing through undeveloped wetlands in as near a straight line as possible, and secondarily to serve as a catalyst for economic development of the region. Development was expected to occur in the vicinity of strategically located interchanges that would afford access to vacant wetlands. The highway was on-grade except where it crossed water courses or interchanged with existing or proposed arteries.

Such a project design would have irreversible adverse impacts on the coastal zone. The highway would cross parts of the Maurepas-Pontchartrain and Barataria interdistributary basins, some of the state's most productive wetland systems. The highway would pass through a 300-foot right-of-way plus additional area for interchanges. This acreage would be lost as nursery ground for estuarine-related organisms. At the same time, the on-grade embankment would traverse the Barataria interdistributary basin at right angles to the general north-to-south sheet flow and, as such, would interrupt water movement and associated nutrients from reaching the nursery grounds to the south. Secondary development, one of the principal reasons for the project, would result in destruction of many thousands of additional acres of swamps and marshes which would have to be leveed and drained by pumping before residential, commercial, or industrial development could occur.

Once so modified, the resource no longer functions as habitat or nursery because it is physically separated and modified from the rest of the estuarine system. Water exchange ceases; all water is pumped from the reclaimed lands into the estuary. Juvenile, spawning, and nesting species no longer use the drained areas. Pollution from street runoff, industrial wastes, and sewage from the development is placed directly into the system and does not filter through wetlands which would remove some of the pollutants. Finally, land developers and house builders receive immediate monetary benefits, however, the long-term expense reverts to the home owner who lives on soils notoriously poor for foundations of all types of structures (Earle 1975, Snowden et al. 1979, U.S. Soil Conservation Service [SCS] 1970).

Louisiana, along with the rest off the nation, was moving into the environmental era of the 1970s after the passage of NEPA. At the public hearings (1971-1973) held on I-410, individuals and groups began to question the need for the project in light of the adverse impacts. In response to environmental pressures (Hurrey n.d.), LDOTD contracted for an independent study to propose ways to reduce the adverse impacts of I-410. The study (Coastal Environments, Inc. [CEI] 1973) made several suggestions: 1) relocate the highway to the north in selected segments; 2) eliminate interchanges in the wetlands which stimulate secondary development in unleveed or undrained areas; 3) design interchanges which are in leveed areas to exit to the north only, preventing pressures to extend development to the south; and 4) elevate segments of roadway where they cross unmodified wetlands. All of these suggestions were eventually adopted except for the half interchanges in the protected areas. These were left as full interchanges.

However, even these project changes, which were to this time unheard of modifications in deference to environmental impacts, were not sufficient to satisfy the demands of concerned citizens. In March 1974, the Ecology Center of Louisiana, Inc., the Sierra Club, the Louisiana Shrimp Association, and the Orleans Audubon Society sued the Federal Highway Administration (FHWA) and the State of Louisiana for a permanent and/or temporary injunction preventing construction of I-410. The suit claimed implementation of the NEPA requirements was deficient on several important points. More than one Environmental Impact Statement (EIS) was to be issued on the project, thus piecemealing the proposed action to segments which could be justified environmentally. Because the project was addressed in sections, insufficient information was generated on the impacts from water pollution, biological productivity of the estuarine system, and the cost of secondary development. All feasible alternatives were not thoroughly evaluated, including no-build and mass transit. In addition, air quality standards might be exceeded by the development resulting from I-410. Finally, the highway was not of interstate status because it did not serve the nation, the national defense, or connect metropolitan areas.

In 1974, the U.S. District Court hearing the case dismissed that portion of the suit from I-10 to U.S. 90, citing the Doctrine of Laches (FHWA 1977). In 1976, all parties to the suit except Jefferson Parish entered a stipulation and consent decree for the preparation of a Supplement to the Final Environmental Impact Statement (FEIS) on the western segment of I-410 between I-10 and U.S. 90, St. Charles Parish. Only the Luling Bridge location was not to be changed and served as a common point for all alternatives. The state withdrew approximately 35 miles of interstate between U.S. 90 and the Mississippi River-Gulf Outlet (MRGO) Bridge in Orleans Parish. The eastern segment from the MRGO Bridge to I-10 in Orleans Parish is now designated I-510 Spur, and the final environmental documents are being prepared. The western segment under consideration here is now known as I-310 Spur (Figure 2).

A special committee was organized to advise and oversee the preparation of the Supplement to the FEIS. The committee was composed of personnel of the FHWA, the LDOTD, three technical experts appointed by the plaintiffs (Dr. John W. Day, Jr., Dr. Paul R. Templet, and John R. Hammond), and the consultants to LDOTD, Coastal Environments, Inc. Regular monthly meetings convened and

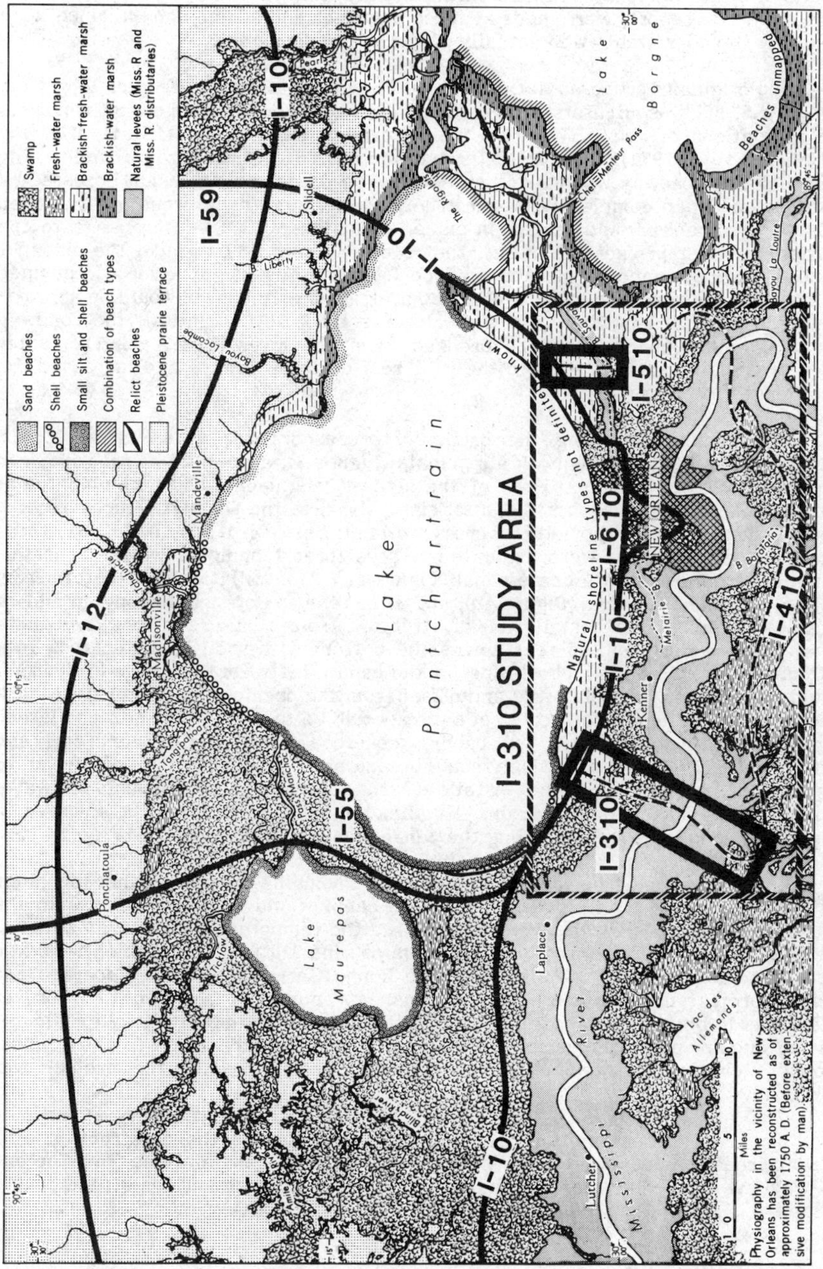

Figure 2. The relation of the interstate highway to the natural setting of the project area (modified from CEI 1973).

several working sessions were held during the preparation of the Supplement to the FEIS. Progress reports were made at these meetings and advice and guidance were sought by the consultants who actually prepared the document.

The originally proposed action (Figure 3), approximately 10.4 miles long from I-10 to U.S. 90, was an interstate highway which would conform to design standards and regulations of the FHWA and the American Association of State Highway Officials (FHWA 1971). From I-10 through the U.S. 61 Interchange (3.7 miles), the twin-elevated roadway was to be installed from a dredged barge canal bordered by spoil banks. Upon completion of construction, the barge canal would remain open and the spoil banks would remain in place. From the U.S. 61 Interchange to the Mississippi River bridge (1.7 miles), the roadway would be on grade; the embankment was to be composed of sand dredged from the Mississippi River and pumped to the right-of-way. The high-level, fixed bridge across the river would be approximatley 2 miles in length. From the bridge to U.S. 90 (3.3 miles), the roadway would again be on embankment dredged from the river (FHWA 1971). Later calculations in the Supplement to the FEIS revised these figures and increased the total length to 11.9 miles.

The committee, using the originally proposed route as a starting point, developed ten alternate highway alignments (Figure 3), a no-build alternative, and a mass-transit alternative. Each of the highway alignments had a common point with the Luling Bridge across the Mississippi River, as the District Court directed. For the installation of structure across wetlands between I-10 and U.S. 61, three construction techniques were evaluated. These three techniques were: 1) installation of structure from a barge canal (Figure 4a); 2) installation of structure from a barge canal and fill road (Figure 4b); and 3) installation of structure by an end-on technique (Figure 4c). Installation of structure from a barge requires dredging a canal ten-feet deep and 65-feet across the bottom with spoil banks up to 6 feet high and 320 feet across on both sides of the canal. Fifty-foot openings in the spoil would be located every 2000 feet and offset from the openings on the opposite bank to avoid erosion. Fifty-to-sixty-foot barges would be trucked to the site and lashed together to support cranes for driving concrete piles, mixing concrete, and installing the 70-foot spans which would be poured in place. At the completion of the project, a weir would be installed across the canal to prevent saltwater intrusion from Lake Pontchartrain. Finally, the spoil areas would be lowered to original surface level by backfilling the canal as much as possible.

The second construction possibility was to combine the barge canal technique just described with a temporary road. The canal would extend from I-10 to the better soil conditons within the swamp where the temporary road would extend to the U.S. 61 interchange. The road would be of sand 2-3 feet thick which would be trucked to the construction site from the Bonne Carre Spillway, a Federal flood control project needing constant maintenance removal of sediment. Wherever piling was to be installed, sand tongues would be built to support all equipment for driving and for pouring of slabs. The sand fill would be removed to the original ground elevation to prevent interruption of sheetflow when the project was complete.

The end-on construction technique does not involve the construction of a canal or road as described above. Instead, smaller cranes install piles from the end of previous slabs. Then, prefabricated caps are set in place and pre-stressed slabs

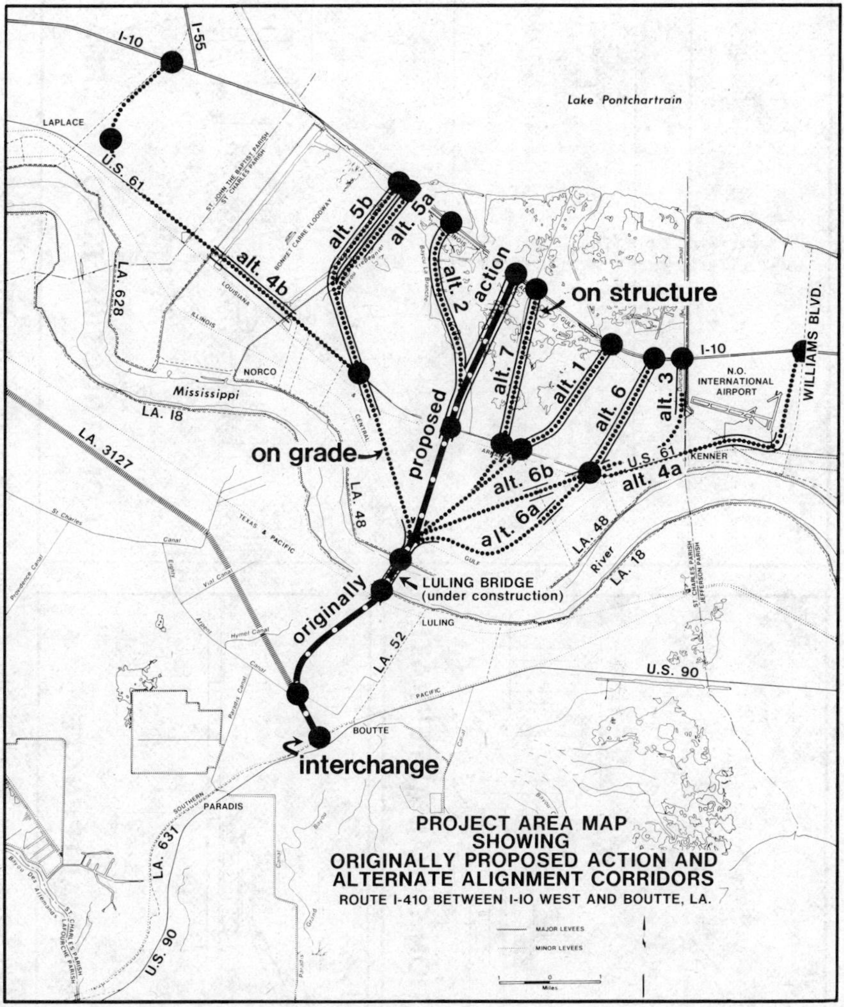

Figure 3. Project area map showing originally proposed action and alternate alignment corridors (after FHWA 1977).

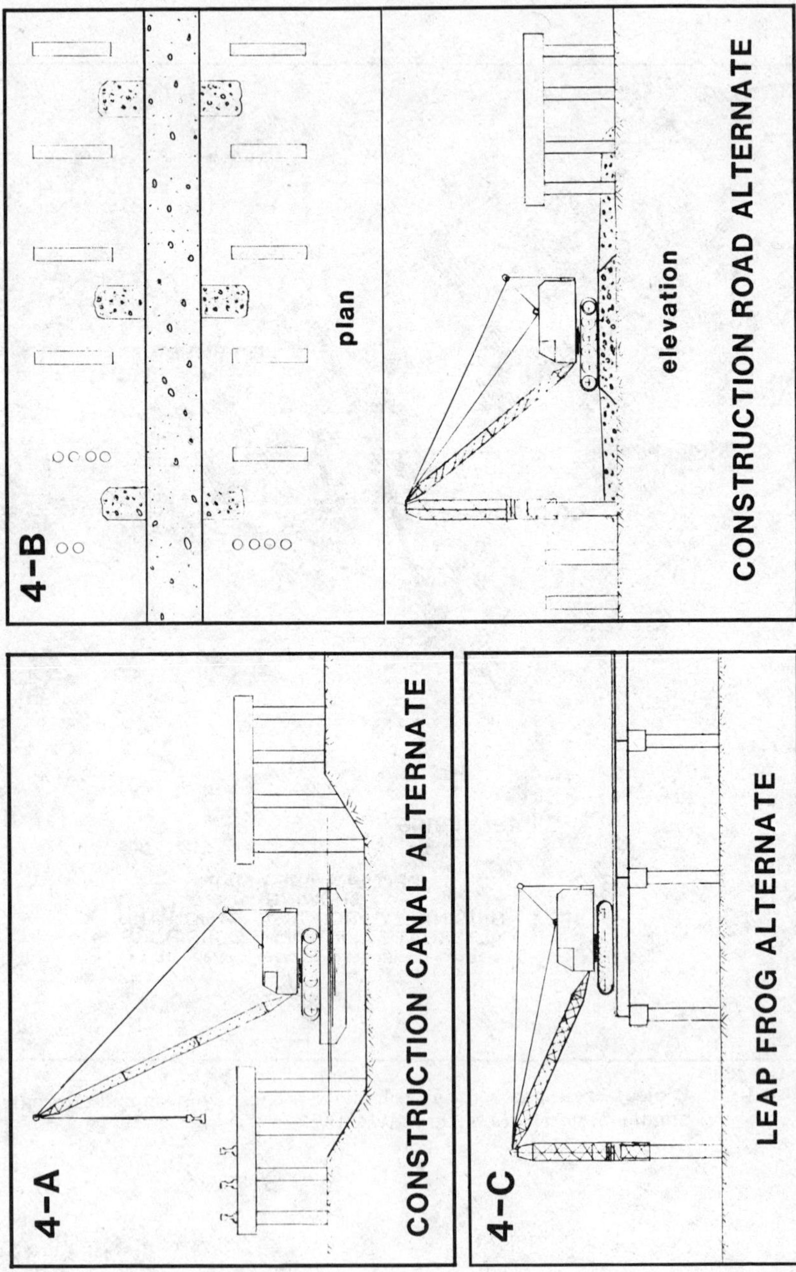

Figure 4. Various installation techniques considered during the I-310 studies (after FHWA 1977).

35 feet long are laid between piers. Once the contractor installs all of the shorter slabs, a concrete deck is poured as the finished road surface.

The environmental consultant, using an interdisciplinary team of natural and social scientists as well as members of the environmental design professions took the alternatives proposed by the committee and generated the Supplement to the FEIS. The analytical document prepared (3 volumes) considered the primary and secondary adverse and beneficial impacts of the proposed action on the environmental setting of the study area (FHWA 1977).

The environmental setting was a controlling factor in the alternative selection. The project area is on both sides of the Mississippi River. On the west bank, the route is on the natural levee of the river and only one alignment was studied because of environmental, engineering, and economic opportunities. Across the river on the east bank, the project area is within a large meander of the Mississippi River. A typical environmental section (Figure 5) from the River to I-10, the zone to be traversed by the highway, shows a natural levee cleared for residential and industrial development, a backswamp which is gradually being leveed and drained for development, and a transportation corridor formed by a mainline railroad embankment and a regional highway, U.S 61, also on embankment. These two embankments define the backswamp unit to be reclaimed for urban development, and it is understood by planners and officials that this zone will inevitably be drained. North of U.S 61 is a relatively undisturbed swamp-marsh wetland. The marsh is the last large marsh system in the Lake Pontchartrain basin and serves as a nursery ground for aquatic species. It became important, therefore, to protect this area from primary, and especially secondary, development encouraged by large public works projects such as this highway. A great deal of pressure from land fills and industrial development already exists without adding to a more rapid loss of wetlands. The objective of the planning process was to devise a project that would protect the wetlands north of U.S. 61 to the greatest extent possible while at the same time providing a safe and efficient transportation system.

In working meetings, and in individual consultation with representatives of the FHWA and LDOTD, the alternatives were reduced from 12 to 4: the originally proposed alignment, Alternate 1, and Alternates 6a and 6b. After another series of evaluations, Alternte 6b was selected as the most desirable considering the social, environmental, and economic factors. In summary, 6b had the least adverse effect on the wetlands while satisfying the traffic needs of the region. (For a detailed discussion of rationale for route selection, see FHWA 1980.)

A number of mitigation measures were introduced into the planning and implementation of this project which to this time had not been significant issues with LDOTD. The LDOTD will eliminate activities or combinations of activities associated with the proposed action which would result in adverse impacts by not using a dredged canal and spoil banks through the wetlands. The LDOTD will limit the extent, degree, and magnitude of the associated activities by not having a 320-foot wide spoil servitude and a dredge canal and by locating the project within the tree line of the swamp, visually screened from recreationists in the marsh area. Use of the end-on construction technique will reduce the area cleared of vegetation and habitat because of the lesser width needed for the project. The LDOTD will eliminate the adverse impact on the most biologically productive ecosystem in the study area (the marsh) by avoiding it completely. By using end-on

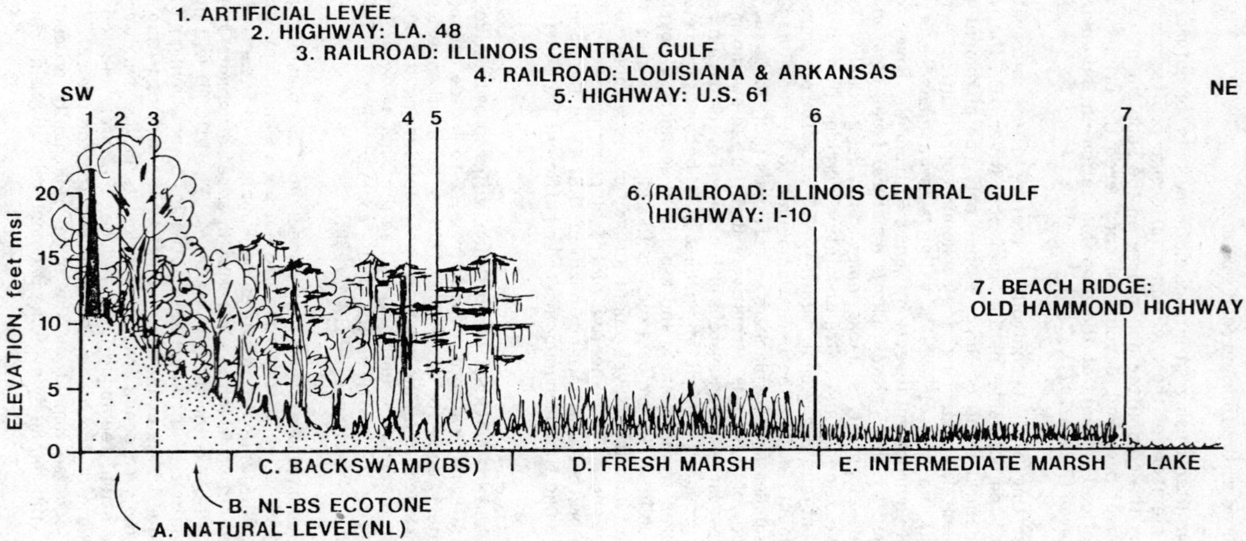

	MORPHOLOGICAL ENVIRONMENTS	VEGETATION ASSOCIATIONS
A	NATURAL LEVEE	MIXED HARDWOOD FOREST, AGRICULTURE
B	NATURAL LEVEE-BACKSWAMP ECOTONE	BOTTOMLAND HARDWOODS, CYPRESS, TUPELOGUM
C	BACKSWAMP	CYPRESS, TUPELOGUM, FLOATING AQUATICS
D	FRESH MARSH	FRESHWATER GRASSES, HERBS, SEDGES, SHRUBS, SCATTERED CYPRESS, FLOATING AQUATICS
E	INTERMEDIATE MARSH	GRASSES, SEDGES, HERBS, SHRUBS (WITH SLIGHT SALINITY TOLERANCE)

Figure 5. Schematic diagram relating vegetation associations to morphological environments in the project area (after Wicker, personal communication 1980).

construction (the first time it will be used for a highway in Louisiana) the sheet flow and general surface hydrology will not be disrupted, and as a result there will be no flooding or drainage of wetlands. No canals will be dredged to allow saltwater to penetrate directy into freshwater wetlands. Finally, secondary impacts will be avoided in the wetlands because the section of I-310 from I-10 to U.S. 61 is all on structure with no ground level access provided.

U.S. 90

At the same time I-410 was being challenged by the environmental community, LDOTD was moving forward with another controlled access facility in the south Louisiana coastal zone. The project, the relocation and upgrading of U.S. 90 between Morgan City and LA 311 (Figures 1, 6), involves the construction of a four-lane, limited-access, divided highway. In 1956, the Louisiana legislature initiated a plan to upgrade U.S. 90 between New Orleans and Lafayette, Louisiana, to meet the traffic needs of the area. U.S. 90 is the only east-west highway across the coastal zone and is therefore the primary transportation corridor of the region. By 1972, only two segments of the Legislature's plan had yet to be started, one of these was from Morgan City to LA 311. The LDOTD desired to complete the authorized facility as quickly as possible. In order to avoid the complicated environmental issues and possible injunctions which were threatening to and eventually did delay I-410, LDOTD contracted with Coastal Environments, Inc., to assist in planning for the proposed upgrading and relocation.

The environmental consultant group began in December 1973, in cooperation and coordination with LDOTD, to prepare a document which would incorporate the environmental opportunities and constraints into the location and design of the proposed action. The interdisciplinary team of natural and social scientists and environmental design specialists systematically collected baseline data on the physical, biological, and cultural systems. The data were synthesized and mapped to reduce the information to a form which would be more easily handled and of more value to decision-makers.

Once the baseline data were in hand, the environmental team projected general impacts which might be expected from any highway project through a wetlands system. In meetings with the engineers from LDOTD, the environmental consultant group developed a series of alternative highway alignments which would theoretically satisfy the objective of moving people and freight in a safe and efficient manner across the project area (Figure 7). Equal consideration was given to economic engineering and environmental opportunities and constraints when selecting the proposed action. In this case, for the first time in Louisiana the preparation of the EIS became an integral part of the planning for a major highway (CEI 1974).

In order to understand the integration of mitigation measures into the planning for the proposed action, a description of the environmental setting is essential. The project area is in the transition zone between two deltaic lobes. The topography is a series of meanderbelts separated by interdistributary basins. The meanderbelts have been cleared for residential, industrial, and agricultural development, while the interdistributary basins are cypress-tupelo gum swamps which are primarily used as wildlife habitat and for recreation. Runoff on the meanderbelts is controlled for drainage purposes by canals, ditches, and pumps,

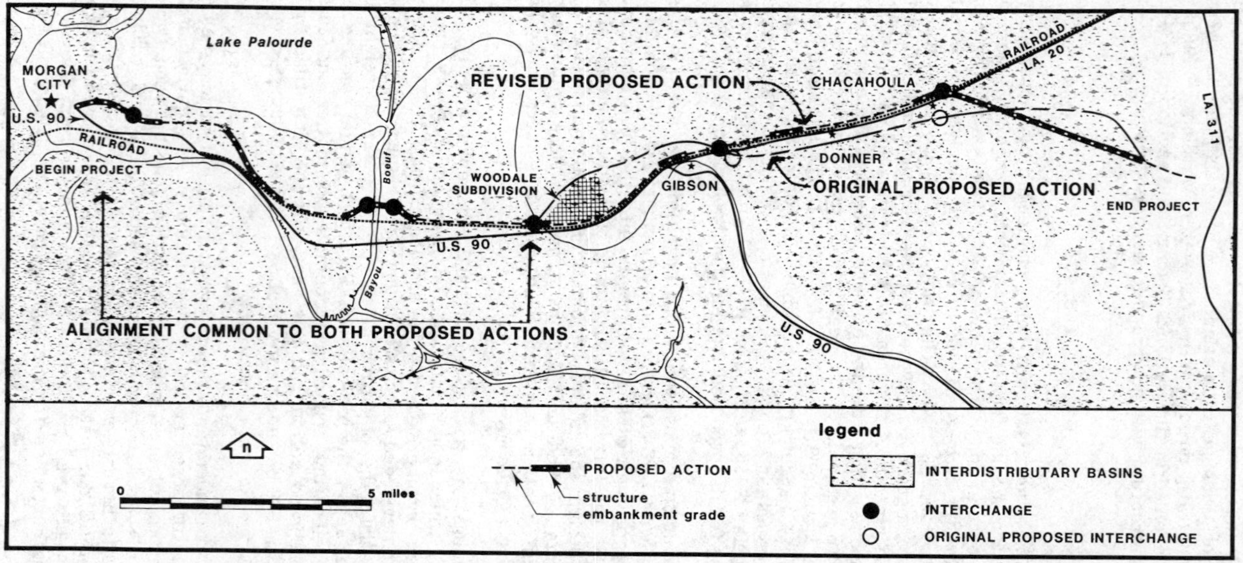

Figure 6. The upgraded and relocated U.S. 90 in relation to the physical setting of the project area.

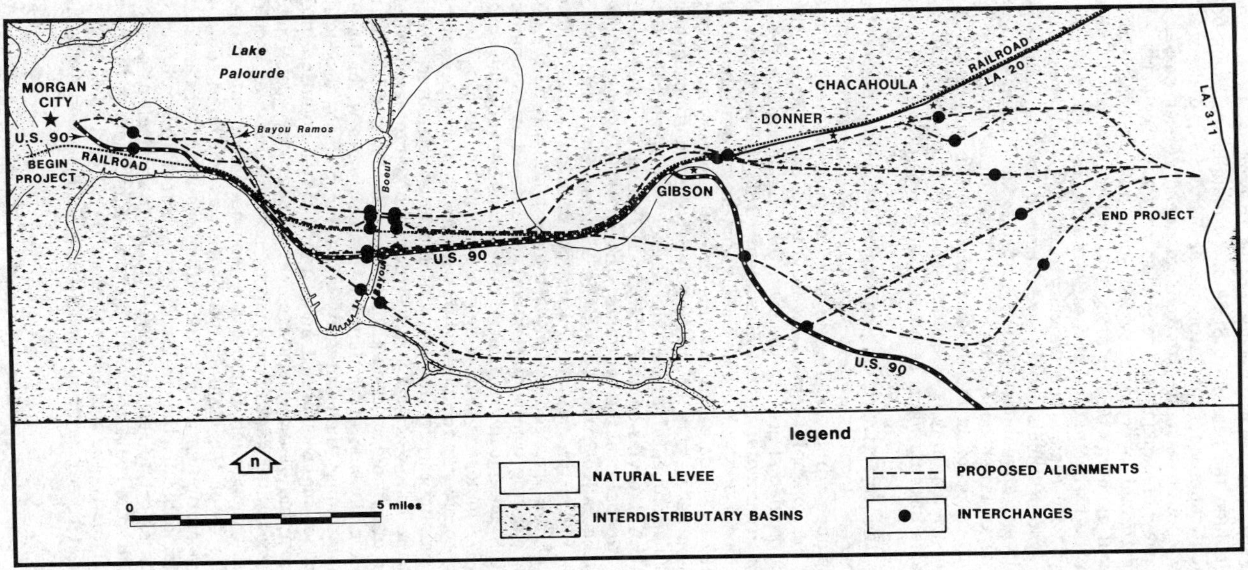

Figure 7. The highway alternative alignment corridors investigated during the planning phase of relocating and upgrading U.S. 90.

while runoff in the swamp is natural sheetflow except where existing transportation facilities cross the basins at right angles to regional flow. The transportation facilities include the present U.S. 90 and a mainline railroad, both of which are on embankment, and the Gulf Intracoastal Waterway, a dredged navigation canal flanked by spoil banks. A limited amount of water passes through the railroad and highway embankments, occurring only at irregularly spaced water courses. This artificial hydrologic system does not approximate the natural sheetflow. Cultural elements cluster along and radiate from U.S. 90 and for the most part are on the higher natural levees, although some pump drainage projects are in the Morgan City area. The majority of the project area is a productive coastal wetland system which contains endangered species habitat and nursery ground for aquatic and amphibious species. At the same time, it is experiencing loss of habitat due to urban encroachment into the wetlands in the Morgan City area and industrial expansion in the Bayou Boeuf and Gibson areas. The objective of the planning for the highway is to encourage wise use of the existing systems by reducing the adverse impacts which result from this type of project.

To achieve the objective of minimizing adverse impacts of the proposed highway, a number of location and design options were instituted. First, the highway will be placed within existing leveed and pumped areas to the maximum extent possible. For example, from Morgan City to Bayou Ramos the project is completely within drained areas. To the east, the Original Proposed Action (Figure 6) in the Draft Environmental Impact Statement (DEIS) (Department of Transportation 1978) was north of the Woodale Subdivision in the wetlands because during the planning in 1974, Woodale was a viable project. However, except on paper, Woodale Subdivision never grew beyond a levee system and one pumping station. Thus, when the Environmental Protection Agency (EPA) and the U.S. Fish and Wildlife Service (FWS) reviewed the DEIS in 1978 and objected to the highway passing north of Woodale in the wetlands, the LDOTD agreed to relocate the highway next to the railroad and within the Woodale Subdivision levee (Revised Proposed Action, Figure 6). Another location decision which will reduce the adverse impacts of the proposed action on the wetlands will be the placement of the highway in the existing transportation corridor formed by the mainline railroad and the existing U.S. 90 and LA 20. Initial discussions between the consultants and the engineers in 1973 indicated that as near a straight line as possible between the termini was being considered for the highway segment. This placed the highway on embankment and passing through the heart of the interdistributary basins. Such a location required the following: 1) the destruction of large acreages of wetlands; 2) the interruption of hydrologic circulation; 3) the alteration of habitat by flooding and drainage; 4) the division of wetlands into small units which have a reduced probability of maintaining their integrity as a functioning, productive system; 5) the introduction of highway noises and air and water pollution into relatively undisturbed swamps; and 6) the encouragement of secondary development around interchanges in areas where development does not now exist or should not occur because of poor soil foundation conditions or encroachment onto prime farmland. To reduce the effects of the highway, the route was relocated and placed within the existing transportation corridor to the extent possible. Presently, the route is adjacent to and north of the railroad embankment from Bayou Ramos to Chacahoula. In an effort to reduce adverse impacts as much as possible, LDOTD again agreed to shift the route proposed in the DEIS in response to objections from EPA and FWS. Originally, the route was to cross LA 20 east of Gibson and parallel the drainage district levees south of Donner (Original Proposed Action). The route was moved north of the railroad to reduce the amount of wetlands needed for the project (Revised Proposed Action).

It will also be possible to reduce adverse impacts by incorporating selected design features into the highway. Where the route parallels the railroad enbankment, cross-corridor waterflow will be maintained by bridges or culverts equal to or greater than those which now occur in the existing embankments. Where the road is on embankment, shell will be used as the road base to the maximum amount to which it is available. The use of shell as foundation eliminates the need to muck and dispose of spoil before backfilling with sand. Where sand is used as foundation, it will come from the Atchafalaya River, a water course which must be dredged for flood control and navigation anyway. Thus, there will be no borrow pits and accompanying adverse impacts. As much sand foundation as possible will be placed without mucking, eliminating spoil disposal problems. The areas where this is possible depend on the soil types and their suitability for this technique. Finally, the roadway will be on structure from Chacahoula to the LA 311 interchange. At first, it was thought that the integrity of the basin could be maintained by embankment adjacent to a logging canal and spoil bank. However, when the route was moved north of the railroad between Gibson and Chacahoula, it became apparent that an embankment was not the solution to crossing the basin. The basin will now be crossed on twin-elevated structures which will be either cast in place or precast and installed from a temporary construction road. Hydrologic flow will be preserved during construction by maintaining a series of culverts or bridges through the temporary road. Once the structure is complete, the temporary road will be removed to the elevation of the original swamp surface, allowing sheetflow to return.

In summary, the planning and eventual implementation of the upgrading and relocation of U.S. 90 has incorporated mitigation measures to reduce adverse impacts. The LDOTD eliminated selected activities associated with the proposed action when the Department decided not to use barge canals to install structure. Borrow pits will be eliminated by dredging sand from the Atchafalaya River. The LDOTD limited the extent, degree, and magnitude of associated activities. Shell will be used when it is available, limiting the extent and magnitude of spoil areas. Use of the existing transportation corridor will reduce the magnitude of secondary development in undisturbed areas. The LDOTD will restore the affected environments to approximate their original condition when they remove the temporary construction road used to install elevated structures. Adverse impacts will be reduced by LDOTD relocating the route from the center of the wetlands to the transportation corridor of the railroad and highway. Hydrologic circulation will be maintained along this alignment by the strategic placement of culverts and bridges, and where the route crosses the center of an interdistributary basin the roadway will be on structure. Finally, the LDOTD will purchase compensation lands of equal habitat value for the wetlands destroyed by the project. These lands will be given to the Louisiana Department of Wildlife and Fisheries to be used as part of the wildlife refuge management system.

Summary and Conclusions

The LDOTD over the past ten years evolved into a public agency which has given more than superficial recognition to the environmental problems and issues associated with a public works project. The LDOTD, as early as 1973, began considering the environmental consequences of its projects in the design and location of the facilities. Department procedures started with the relocation and redesign of I-410 and carried through to U.S. 90, the first planned highway which fully incorporated the environmental opportunity and constraints from the initiation of the project. Mitigation measures, such as eliminating or limiting the extent, degree, or magnitude of associated activities, restoration of environments, selection of feasible alternatives, and compensation for lost resources, are now part of LDOTD projects in Louisiana, especially in the coastal zone. Environmental planning is now an important element in the planning and implementation of projects. Such planning, when properly executed, does not delay highway projects and results in the protection of one of Louisiana's and the nation's most sensitive and biologically productive areas, the coastal wetlands.

REFERENCES CITED

Coastal Environments, Inc.
1973 Environmental considerations, Interstate I-410. Prepared for Louisiana Department of Highways, Baton Rouge.

____
1974 Environmental evaluation, Morgan City - Gibson Highway. Prepared for Louisiana Department of Highways, Baton Rouge. Contract No. S.P. 700-08-61. 2 vols. May.

Davis, D. W.
1973 Louisiana canals and their influence on wetland development. Unpublished Ph.D. dissertation. Louisiana State University Department of Geography and Anthropology, Baton Rouge. 199 p.

Department of Transportation.
1978 Draft environmental impact statement, relocated and upgraded U.S. 90, Morgan City - La 311, St. Mary Assumption and Terrebonne Parishes, Louisiana. New Orleans, Louisiana. May. U.S. Coast Guard, Lead Agency.

Earle, D. W., Jr.
1975 Land subsidence problems and maintenance costs to homeowners in east New Orleans, La. Ph.D. dissertation, Louisiana State University Department of Marine Sciences, Baton Rouge.

Federal Highway Administration
1971 Final environmental statement, administrative action for State project No. 700-07-72, Federal aid project no. I-410-4(28)217 Interstate Route 410, Junction Interstate 10-Junction U.S. 90, St. Charles Parish. FHWA-La-EIS-71-10-F. Revised Dec. 29, 1971.

Federal Register
1978 Council on environmental quality regulations for implementing the procedural provisions of the National Environmental Policy Act, 43 FR 55978-56007. November 29. 40CFR Parts 1500-1508.

Hurrey, R. C.
n.d. An unpublished document prepared under the supervision of Associate Professor J. A. Barach, Graduate School of Business Administration, Tulane University, New Orleans, Louisiana. 30 p.

Kolb, C. R. and J. R. van Lopik
1958 Geology of the Mississippi Deltaic Plain, southeastern Louisiana. U.S. Army Corps of Engineers, Waterways Experiment Station, Technical Report 3-482. 2 vols.

Louisiana Department of Highways
1971 Standard specifications for roads and bridges. Baton Rouge.

McIntire, W. G., M. J. Hershman, R. D. Adams, K. D. Midboe, and B. B. Barrett
1975 A rationale for determining Louisiana's Coastal Zone. CZM Series, Center for Wetland Resources, Louisiana State University, Baton Rouge. Sea Grant Publication LSU-T-75-006.

Prophit, W.
1980 Louisiana...from byways to highways. Louisiana Contractor, p. 8-33.

Snowden, J. O., W. B. Simmons, and E. B. Traughber
1979 Subsidence of marshland peat in the greater New Orleans Area, Louisiana. In: J. W. Day, Jr., D. D. Culley, Jr., R. E. Turner, and A. J. Mumphrey, Jr. (eds) Third Coastal Marsh and Estuary Managment Symposium, Louisiana State University, Division of Continuing Education, Baton Rouge. pp. 273-289.

U.S. Soil Conservation Service
1970 Soil survey of portions of Jefferson, Orleans, St. Bernard Parishes, Louisiana. U.S. Department of Agriculture, Alexandria.

______
1977 Draft supplement to the final environmental impact statement, State Project no. 700-07-72, Federal aid project no. I-410-4(67)217, Interstate Route I-410, Junction Interstate 10-Junction U.S. 90, St. Charles Parish. FHWA-La-EIS-71-10-F(Draft Supplement) Region 6, Fort Worth, Texas. 3 vols.

______
1980 Final supplement to the final environmental impact statement, State project no. 700-07-72, Federal aid project no. I-410-4-(67)217, Interstate Route 310 (Formerly I-410), Junction Interstate 10-Junction U.S. 90, St. Charles Parish. FHWA-La-EIS-71-10-F(Final Supplement) Region 6, Fort Worth, Texas. 3 vols.

PROTECTING SUISUN MARSH -- 85,000 ACRES OF HABITAT

WILLIAM J. KOCKELMAN
U.S. Geological Survey
345 Middlefield Road, Menlo Park, CA 94025

JEFFRY S. BLANCHFIELD
San Francisco Bay Conservation and Development Commission
30 Van Ness, San Francisco, CA 94102

Basic and interpretive scientific information was used in the preparation and implementation of a plan for protecting the San Francisco Bay marine estuary in 1970 and the Suisun Marsh, an irreplaceable aquatic and wildlife habitat, in 1977. In the Suisun Marsh Protection Plan, a combined use of regulations, acquisition, local protection, fish and wildlife management, water management, and preferential tax assessment has resulted in the preservation of a unique estuarine environment. One of the purposes of this paper is to illustrate the use of scientific information in protecting a coastal area.

The San Francisco Bay is the largest estuary along the California coastline. It is a resting place, feeding area, and wintering ground for millions of birds, and nearly a hundred species of fish live in the marshlands, mudflats, saltponds, managed wetlands, and open water of the estuary. For man, the bay serves as a source of food, a recreation area, a scientific research laboratory, and an unusual living environment.

Salt marshes are extraordinarily fertile. Large numbers of shore and water birds, including ducks and geese, come to the marshes to feed on the lush vegetation or on the animals that thrive in the brackish water. Most marine life in the bay either depends directly on the marshes and mudflats for sustenance or indirectly upon the other marine life nourished there. Shore birds depend upon the marshes and mudflats for both food and shelter (San Francisco Bay Conservation and Development Commission, 1979, p. 9).

More than 50,000 acres of managed wetland, adjacent to the bay but diked off from it, are maintained as duck-hunting preserves, game refuges, and occasionally as farming areas. In most of these areas, tide gates permit occasional intakes of bay water. The diked wetlands, particularly when adequately managed, are as important to wildlife as the tidal marshes and mudflats (San Francisco Bay Conservation and Development Commission, 1979, p. 25).

The Suisun Marsh, located where the salt water of the Pacific Ocean and fresh water from the Sacramento and San Joaquin Rivers meet and mix (fig. 1), is the largest wetland area in California. The marsh contains approximately 85,000 acres of tidal marsh, managed wetland, and waterways -- more than 10 percent of California's remaining

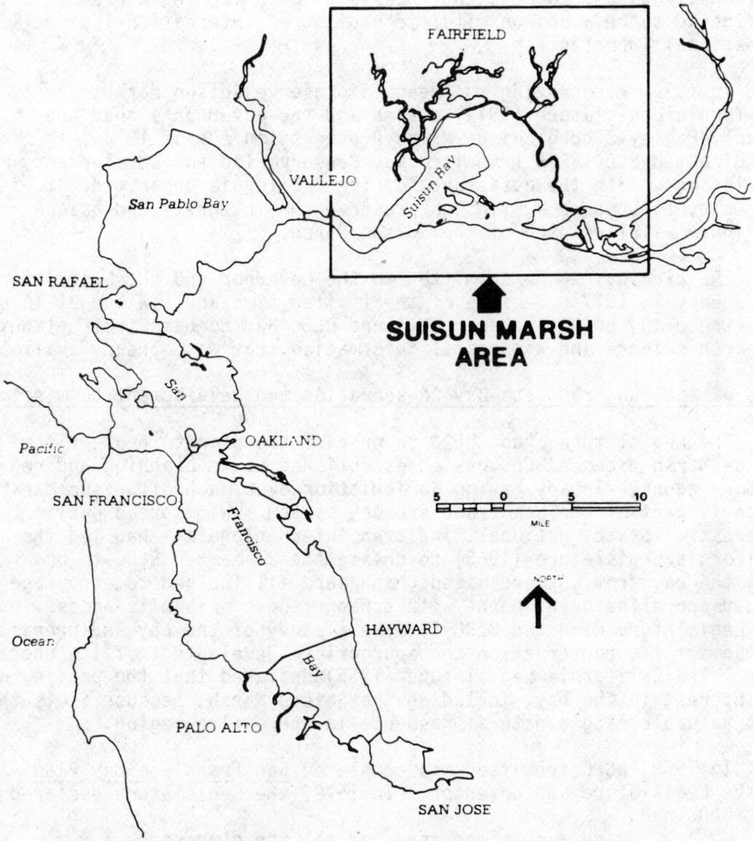

Figure 1.--The Suisun Marsh Area. The San Francisco Bay Conservation and Development Commission has jurisdiction over the San Francisco Bay system including San Pablo Bay and Suisun Bay; all sloughs, marshlands, tidelands, and submerged lands; a 100-foot shoreline strip inland from the bay system; diked saltponds; managed wetlands; and certain tributary waterways.

wetland area. The marsh supports a great diversity of plants, fish, and wildlife, all of which are essential to the life chain. It also provides an important resting place for waterfowl on the Pacific Flyway from Canada to Mexico. Because of Suisun Marsh's close proximity to other wetlands and to adjacent grasslands with varying terrain, it is considered to be a unique wildlife habitat of international as well as national importance.

In 1974, recognizing the need to preserve Suisun Marsh, the California Legislature (1974) passed and the Governor signed the Nejedly-Bagley-Z'berg Suisun Marsh Preservation Act of 1974. The Act directed BCDC (San Francisco Bay Conservation and Development Commission), with the assistance of the California Department of Fish and Game, to prepare a plan to "preserve the integrity and assure continued wildlife use" of the Suisun Marsh.

The plan was to be submitted to the Governor and the Legislature by January 1, 1977. Because of the limited time and low budget to prepare the plan, both the Department and BCDC had to base their planning on earth-science and biological information that was already available.

Role of the San Francisco Bay Conservation and Development Commission

The Legislature chose BCDC to prepare the plan to protect the Suisun Marsh because BCDC was an established State planning and regulatory agency already having jurisdiction over much of Suisun Marsh which is part of the San Francisco Bay system. Widespread public interest -- State, national, and even international -- had led the California Legislature (1965) to create the 27-member BCDC to protect the bay from uncoordinated, haphazard filling and to encourage proper shoreline development with arrangements for public access. The Legislature directed BCDC to make a study of the bay and prepare a plan for its conservation and appropriate development of its shoreline. The California Legislature (1965) declared that the public has an interest in the bay, including the Suisun Marsh, because it is the most valuable single natural resource in the entire region.

In 1969, BCDC submitted its completed San Francisco Bay Plan to the Legislature and Governor. In 1970, the Legislature declared that BCDC had

> ... made a detailed study of all the characteristics of the bay, including: the quality, quantity, and movement of bay waters, [and] ecological balance of the bay ... and that on the basis of the study ... prepared a comprehensive and enforceable plan for the conservation of the water of the bay and the development of its shoreline ....

and, adopting most of the recommendations contained in the Bay Plan, made BCDC a permanent agency with powers to implement the Bay Plan.

In carrying out the Bay Plan, BCDC is authorized to grant, deny, or grant subject to conditions, permits to place fill, extract materials, or make any substantial change in the use of any water, land, or structure within its area of jurisdiction (California Legislature, 1965, Sec. 66632). Pursuant to this authority, BCDC adopted an application and permit system and prescribed the content and procedures required for obtaining permits. In February 1977, BCDC management program for the bay system -- its enabling legislation and the Bay Plan -- were approved by the Secretary of Commerce as the Coastal Zone Management Program for the San Francisco Bay system.

Use of Available Information for Coastal Planning

Information regarding the location and character of the marshlands, geologic hazards, existing land uses, and natural resources is necessary to secure public and legislative support before preparing a plan. Furthermore, accurate, clear, and large-scale maps showing the boundaries of the lands to be acquired or regulated are prerequisite to the effective and legal implementation of the plan.

Most of the earth-science and biological information necessary for preparing and implementing plans to protect marsh and coastal areas in the San Francisco Bay region is available, much of it at relatively large scales (1 inch = 2,000 feet) from the U.S. Geological Survey, the California Division of Mines and Geology, and others.

The geologic and seismic history, physical and engineering properties, development problems, and general land-use capabilities of bay mud are discussed in reports on flatland deposits by Helley and others (1979), flood-prone areas by Waananen and others (1977), and relative slope stability by Nilsen and others (1979).

Wildlife resource requirements are discussed by Rollins (1973), and the production of waterfowl food plants by Mall (1969). Bird counts, wildlife inventories, and marsh and wildlife management techniques are documented in materials published by the California Department of Fish and Game, as well as in environmental impact statements and reports prepared for projects within the marsh planning area.

Information on landslide deposits, historic marshlands, landslide susceptibility, land use, flood-prone areas, steep slopes, and active faults in the marsh planning area was obtained from published and unpublished sources, transferred to regional topographic work maps, and then summarized on a Natural Factors Map (fig. 2). From this map, the marsh protection plan map (fig. 3) was prepared. Recommendations for land-use practices in the Benicia Hills are also based upon earth-science information, as were some of the legal boundaries and areas of jurisdiction for BCDC's marsh planning and regulatory functions.

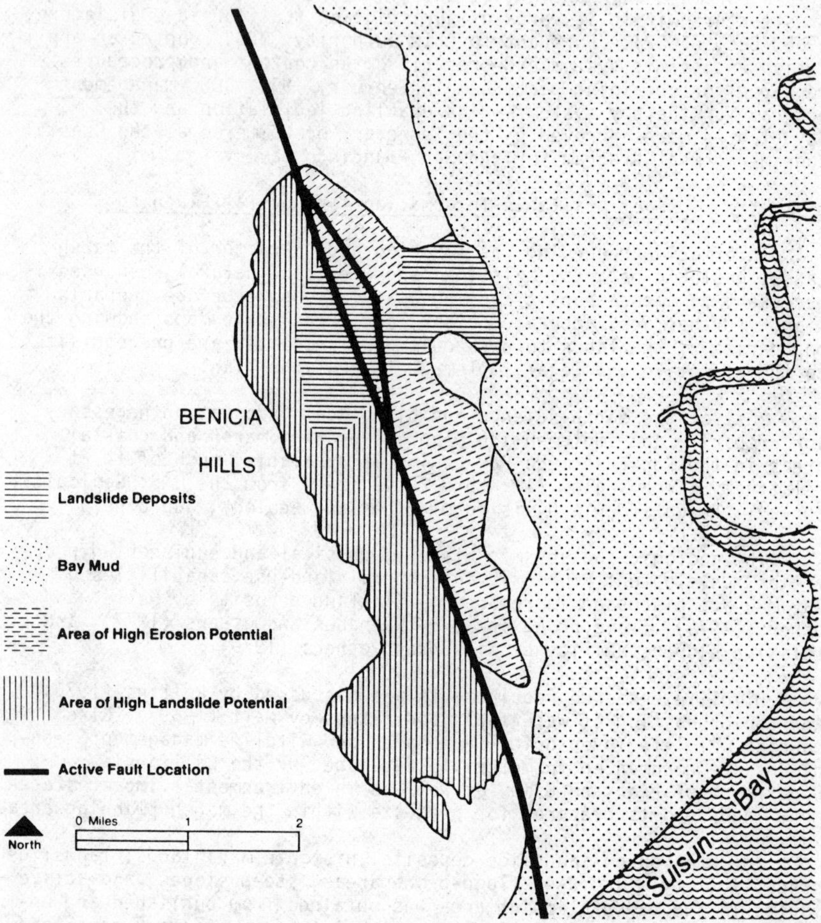

Figure 2.--Part of the Natural Factors Map used in preparing the Suisun Marsh Protection Plan showing some of the information used -- landslides, bay mud, and active faults. The San Francisco Bay Conservation and Development Commission (1976) derived the location of the primary and secondary management areas shown on figure 3 from the natural factors shown here.

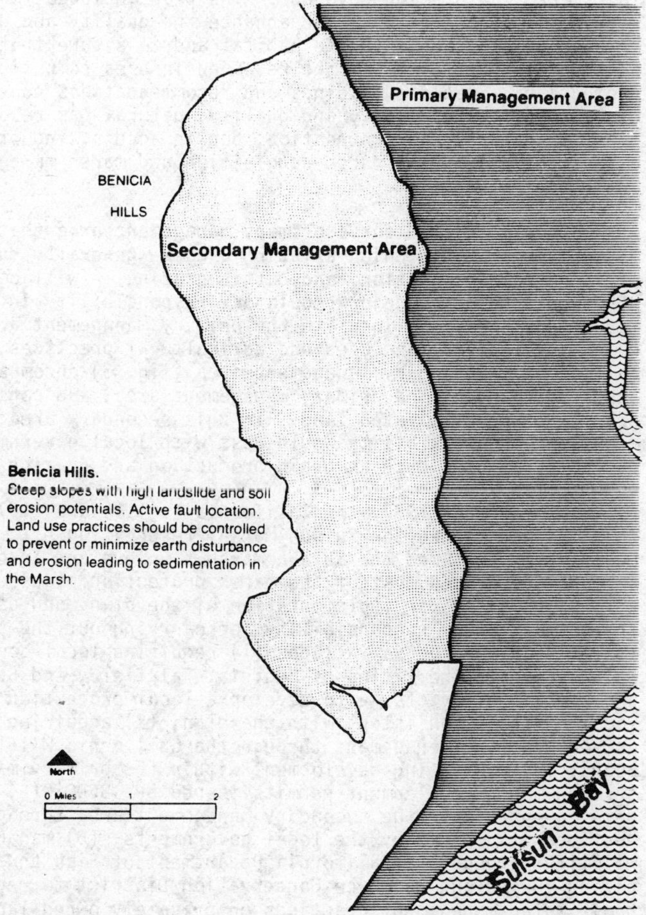

Figure 3.--Part of the Suisun Marsh Protection Plan showing a part of the primary management area that encompasses 89,000 acres, including Suisun Marsh, and 5,000 acres of adjacent lowlying grassland, and a part of the secondary management area that encompasses 22,500 acres.

Implementation of the Suisun Marsh Protection Plan

In December 1976, BCDC submitted the Suisun Marsh Protection Plan to the Legislature and Governor. The plan included findings and policies designed to preserve and enhance the quality and diversity of the marsh's aquatic and wildlife habitat and to assure that upland areas adjacent to the marsh will be retained in uses compatible with its protection. Specific findings and recommendations concerning marsh environment, water supply and quality, natural-gas resources, utilities, transportation, recreation, public access, industry, land use, assessment practices, marsh acquisition,and marsh management were also included in the plan.

The plan called for a primary management area that encompasses 89,000 acres within which existing uses -- generally duck hunting, limited grain production, and cattle grazing -- will continue. BCDC would have the major regulatory responsibility in the primary management area. To insulate the primary management area from incompatible upland land uses and agricultural practices, the plan also called for a secondary management area (fig. 3) encompassing 22,500 acres surrounding the primary management area, and consisting of upland grassland and cultivated land. In this secondary area, where the major regulatory responsibility would rest with local governments, existing agricultural uses such as grain production and grazing will continue.

In September 1977, the California Legislature (1977) passed and the Governor signed the Suisun Marsh Preservation Act. In it the Legislature declared the Suisun Marsh a unique and irreplaceable resource, approved most of the marsh protection plan recommendations, provided for implementation of the plan, and assigned to BCDC the primary State responsibility for carrying out the plan. The implementation process includes (1) requiring local governments -- Solano County, the cities of Benicia, Fairfield, and Suisun City, and special districts -- to develop a local protection program and local controls consistent with the plan; (2) acquiring lands for public use or resource management through the California Wildlife Conservation Board; (3) regulating development within the primary management area through marsh development permits issued by BCDC; (4) regulating development within the secondary management area through marsh development permits issued by the local governments; (5) managing fish and wildlife through the California Department of Fish and Game; (6) requiring the Suisun Resource Conservation District to regulate and improve water-management practices on privately owned lands; (7) requiring preferential tax-assessment practices; (8) providing for cease-and-desist orders; and (9) appealing the issuance of marsh development permits by local governments to BCDC.

The types of development that are regulated in the primary and secondary management areas include placement of any material or erection of any structure, discharge or disposal of any material, grading or removal of any material, change in density or intensity of land use, land division, alteration or demolition of any structure, and the removal or harvesting of vegetation other than for agricultural purposes.

BCDC has integrated a marsh development permit system into the existing bay development permit system. The system requires that any proposed development be consistent with the policies of the marsh protection plan, the provisions of the marsh preservation act, and the local protection program. Using its enforcement authority, BCDC monitors development in its area of jurisdiction and has issued cease-and-desist orders for filling and other work in the marsh area. In addition, BCDC reviews and comments on environmental impact reports for actions affecting the marsh and has already filed a lawsuit to halt a proposed development that is based on a report felt to be inadequate.

BCDC has prepared a detailed map for the marsh at a scale of 1 inch = 2,000 feet (1:24,000) showing both the primary and secondary management areas. BCDC has also prepared guidelines for preparation of the local protection program, and the county, cities, and special districts are preparing the program for BCDC review and certification.

About $4 million is available to the California Wildlife Conservation Board for acquiring and improving lands in the marsh. Almost 1,400 acres have been acquired, and additional acreage is being considered for acquisition.

Conclusion

This example shows how basic and interpretive scientific information was used in preparing and implementing a plan to protect an irreplaceable aquatic and wildlife habitat of an area of more than 110,000 acres.

The combined use of regulations, acquisition, local protection, fish and wildlife management, water management, and preferential tax assessments has resulted in the preservation of specific resources.

Scientific information can be applied similarly to resource protection planning and plan implementation in other estuarine environments, including those in the coastal areas of the United States affected by the Federal Coastal Zone Management Act (U.S. Congress, 1972).

References Cited

California Legislature, 1965, McAteer-Petris Act, as amended: California Government Code, section 66600 and following, West's Annot. Codes.

_____1974, Nejedly-Bagley-Z'berg Suisun Marsh Preservation Act of 1974: California Fish and Game Code, section 1850 and following, West's Annot. Codes (repealed January 1, 1978).

_____1977, Suisun Marsh Preservation Act of 1977: California Public Resources Code, section 29000 and following, West's Annot. Codes.

Helley, E. J., and others, 1979, Flatland deposits of the San Francisco Bay region, California -- their geology and engineering properties, and their importance to comprehensive planning; selected examples from the San Francisco Bay region, California: U.S. Geological Survey Professional Paper 943, 88 p.

Mall, R. E., 1969, Soil-water-salt relationships of waterfowl food plants in the Suisun Marsh of California: Department of Fish and Game Wildlife Bulletin No. 1, Sacramento, California, 59 p.

Nilsen, T. H., Wright, R. H., Vlasic, T. C., and Spangle, W. E., 1979, Relative slope stability and land-use planning in the San Francisco Bay region, California: U.S. Geological Survey Professional Paper 944, 96 p.

Rollins, G. L., 1973, Relationships between soil salinity and the salinity of applied water in the Suisun Marsh of California: Department of Fish and Game, California Fish and Game quarterly, v. 59, no. 1, p. 5-35.

San Francisco Bay Conservation and Development Commission, 1976, Suisun Marsh Protection Plan: San Francisco, California, Plan 48 p., Supplement 496 p.

_____1979, San Francisco Bay Plan: San Francisco, California, 41 p.

U.S. Congress, 1972, Coastal Zone Management Act of 1972, as amended: Public Law 92-583, 86 Stat. 1280, 16 U.S. Code 1451 and following.

Waananen, J. T., and others, 1977, Flood-prone areas and land-use planning; selected examples from the San Francisco Bay region, California: U.S. Geological Survey Professional Paper 942, 75 p.

THE ISSUE OF FRESHWATER INFLOW TO ESTUARIES

Norman G. Benson[1]
U.S. Fish and Wildlife Service
Slidell, Louisiana

INTRODUCTION

It is becoming increasingly difficult to provide a suitable quantity and quality of freshwater to sustain estuarine ecosystems on many sections of the U.S. coast. By definition, estuaries are coastal areas where freshwater mixes with saltwater from the ocean. The freshwater needs of estuaries must be planned for along with freshwater needs of humans, industry, and agriculture, if estuaries are to continue to furnish food and recreation at reasonably high levels. Estuaries are our most productive coastal areas and they require freshwater inflow for their regular ecological processes. Real progress towards allocating freshwater inflow for estuaries in water resource planning has only been made within the last few years, although scientists and natural resource managers have recognized the fact that estuaries require freshwater. Coastal zone management programs cannot preserve or protect estuarine ecosystems unless freshwater inflow is provided.

The importance of freshwater inflow and the efforts to forecast the ecological effects of various amounts of freshwater inflow are reviewed in this paper. Freshwater inflow management programs are also discussed.

THE IMPORTANCE OF FRESHWATER INFLOW TO ESTUARIES

Gunter (1963) identified the importance of freshwater inflow when he called the section of the Gulf of Mexico at the mouth of the Mississippi River the "Fertile Fisheries Crescent." He found it to be one of the most productive fishery areas in the world and attributed the high fish production to the flow of nutrients and sediments from the Mississippi River and the creation of the northern Gulf of Mexico coastal marshes. Other studies on the Atlantic and Pacific coasts verified the importance of low salinities and marshes in estuaries to finfish, shellfish, birds and mammals.

Estuarine habitats develop where river inflow contributes various amounts of freshwater, sediments, and nutrients to the coastal zone. Some of the effects of freshwater inflow on the production processes of estuaries are listed below.

- River flow provides nitrogen and other nutrients that are commonly deficient in coastal waters (Copeland 1966; Gosselink et al. 1979);

[1]National Coastal Ecosystems Team, Slidell Computer Center, 1010 Gause Boulevard, Slidell, Louisiana 70458

- River flow creates a salt wedge which results in the creation of both a nutrient and a sediment trap environment that increases nutrient retention time and promotes the production of organic matter (Correll 1978) see Figure 1;

- Inflow provides sediment and nutrients for the creation and nourishment of saline and brackish marshes and provides a medium for the flushing and exchange of materials between marshes and downstream coastal waters (Gosselink et al. 1979) see Figure 1;

- The fine sediments provided by freshwater inflow are more effective for storing and recycling nutrients than the coarser sediments present along much of the seashore (Correll 1978);

- Low and variable salinities in estuaries cannot be tolerated by many predator species and this fact enhances the survival of important young finfish and shellfish particularly oysters (as well as menhaden and shrimp that use estuaries as nurseries) (Gosselink et al. 1979; Copeland 1966);

- Freshwater inflow contributes micronutrients in the form of vitamin complexes and trace elements which enhance the production of organic matter in estuaries (Copeland and Bechtel 1970).

Freshwater inflows have been changed by watershed management, development of river channels for navigation, construction of reservoirs, consumption of water, changes in seasonal flow patterns, reduction of sediment and nutrients, introduction of pollutants or contaminants, or by direct physical modification of estuaries through leveeing or construction of navigation channels. Some of the effects of modifying freshwater inflows on estuarine habitats, production and organisms are summarized in Table 1. This summary indicates that many ecological processes and many of our major commercial finfish and shellfish groups are affected by modification of freshwater inflow.

Estuaries that have been most adversely affected by changes in freshwater inflow include San Francisco Bay (Smith and Kato 1979), Florida Bay (Snedaker 1977), Chesapeake Bay, the Louisiana coastal marshes, and the Texas Gulf Coast. The ecological consequences of altering freshwater inflow have been most evident in large estuarine ecosystems such as San Francisco Bay and the Louisiana and Texas estuaries. In Chesapeake Bay freshwater inflow quality rather than quantity has been reduced. Influxes of nutrients from the watershed have resulted in noxious algal blooms that were rare in the past, and contaminents from industry and urban areas have accumulated in the sediments (U.S. Army Corps of Engineers, 1974). The estuaries along the west coast of Florida have been subjected to reduced freshwater inflow, but the ecological effects have not been quantitatively described. The adverse effects of reducing freshwater inflow have been most evident in such water deficient areas as the Texas Gulf coast and in central California.

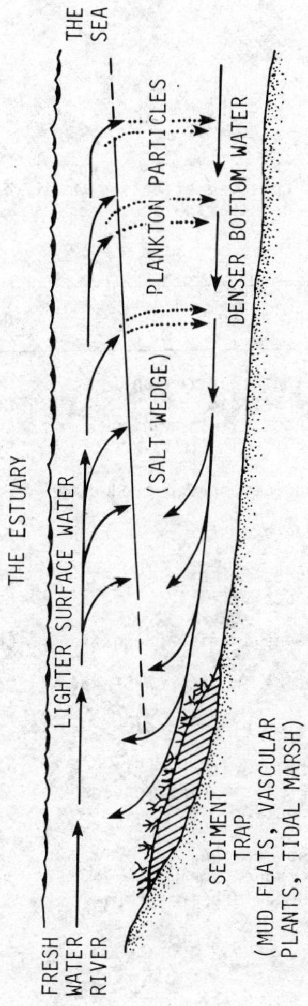

Figure 1. Schematic diagram of recycling mechanism in estuaries, including the salt wedge, sediment trap, marshes, current patterns, and recycling through the sinking of plankton particles (modified from Correll 1978).

Table 1. Some potential and observed effects of modifying freshwater inflow on estuarine ecosystems.

Type of modification	Major effect	Reference
Reduction of inflow volume	Saltwater intrusion	Craig et al. 1979
	Marsh erosion	Craig et al. 1979
	Reduces fish spawning	Smith and Kato 1979 Stevens and Chadwick 1979
	Reduces menhaden production	Gosselink et al. 1979; Schaff 1975
	Reduces oyster production	Copeland 1966
Reduction of nutrients in freshwater inflow	Reduces primary productivity	Copeland 1966; Correll 1978
Changing natural seasonal variations in flow by dams	Decline in fish production	Copeland 1966; Gosselink 1979
	Decline in blue crab harvest	Copeland 1966; Gosselink 1979
	Erosion of wetlands	Craig 1979
Reduction of sediment inflow	Marsh erosion	Craig 1979
	Lower shrimp and menhaden production	Gosselink 1979
Prevent normal distrubution of freshwater inflow to estuaries through leveeing and navigation channel development	Marsh erosion	Craig et al. 1979
	May reduce shrimp harvest	Turner 1977
	Reduced furbearer harvest	Gosselink 1979

The effects of diverting river flow directly to the Gulf of Mexico rather than allowing overbank flooding into estuarine marshes has resulted in significant land loss in the Louisiana marshes (Craig et al. 1979). The effects of modifying the distribution of freshwater inflow from the Mississippi River on the estuaries were considered following several Congressional Resolutions in 1967 to study (1) the effects of diverting 12 million acre-feet of water to the high plains of Texas and (2) the advisability of improvements in the coastal areas of Louisiana in the interest of hurricane protection, prevention of saltwater intrusion, preservation of fish and wildlife, prevention of erosion and related water resource purposes. A draft study (U.S. Army Corps of Engineers 1976) gave preliminary estimates of the riverflow needs -- freshwater and sediment -- necessary to maintain fish and wildlife resources within the Louisiana coastal marsh and associated embayments.

The estimated 27 million acres (10.9 million hectares) of estuaries along the U.S. coast have high recreational and commercial fishing value in addition to their use for transportation and industry. They supported an estimated 190 million days of sport fishing in 1975 (U.S. Department of the Interior 1977). From 60% to 80% of our commercial finfishes and shellfishes are estuarine dependent, i.e., they require estuaries for breeding, nursery, or feeding purposes. The commercial catches of the major estuarine-dependent finfishes and shellfishes in 1977 and 1978 had dockside values of 1.7 billion dollars and 1.3 billion dollars, respectively (National Marine Fisheries Service 1979).

Finfishes and shellfishes that utilize estuarine habitats include (1) permanent residents or those that complete their life cycle in estuaries, (2) those that use estuaries for nursery areas, and (3) transients that move through estuaries.

Permanent residents are usually either sessile (oysters) or weakly mobile (plankton). From the standpoint of commercial fish production, estuaries are most important because they furnish nursery habitat for shrimp, menhaden, crabs, and several sciaenid fish species. All of our important anadromous species -- salmon, striped bass, shad -- are estuary transients and recent studies have found that estuarine habitat is important for the survival of some species of salmon young.

EFFECTS OF MODIFYING FRESHWATER INFLOW

Losses of coastal marsh and changes in fish species abundance and harvest have been caused by modification of freshwater inflow to estuaries. Probably the most widely documented change is the annual loss of 16.5 square miles (42,7 square kilometers) of coastal marsh in Louisiana that has been related to the leveeing of the Mississippi River (Craig et al. 1979). The area of coastal marsh is directly correlated with penaeid shrimp production in Louisiana and throughout the world (Turner 1977) and the reduction of coastal marsh will presumably reduce shrimp production.

Chapman (1966) estimated that if the Texas Water Plan were constructed as originally planned, it would reduce the average annual flow of freshwater to estuaries to one-half its historic amount; this would result in an annual loss of 118 million pounds (53.1 million kg) of commercial fishery products and three million days of sport fishing. The estuaries associated with San Francisco Bay have been drastically changed by water diversion and development, pollution, land reclamation, and dredging. Freshwater inflow has been reduced to about one half of the historic natural volume (Stevens and Chadwick 1979). Only specific ecological changes in the San Francisco Bay complex can be directly attributed to alteration of freshwater inflow (Smith and Kato 1979) because land reclamation (filling), navigation channel development, hydraulic mining and pollution have occurred simultaneously with reservoir and water diversion projects that reduced and altered freshwater inflow. Dam construction, water diversions, and reduced flows have eliminated about 80% of the spawning area for chinook salmon. Water diversion in the Sacramento-San Joaquin Delta has been credited with causing reduced reproduction of striped bass (Stevens and Chadwick 1979). All the major finfish and shellfish stocks have decreased and much of the change has been attributed to reduced freshwater inflow.

The roles of freshwater inflow on Florida west and north coast estuaries were reviewed, but the effects of reductions in freshwater inflow on estuarine production have not been evaluated quantitatively (Snedaker et al. 1977). Lugo and Snedaker (1974) calculated that a 50% reduction of freshwater inflow would reduce mangrove production by about 50%; many valuable sport and commercial fish species utilize mangrove stands for feeding and breeding (Odum and Heald 1975). Livingston (1976) showed that the number and species of fishes and invertebrates in Appalachicola Bay, Florida, fluctuated with physical-chemical changes induced by river-flow characteristics.

An excess of freshwater in estuaries, as caused by floods, kills those organisms that cannot move to areas where salinities or temperatures are suitable for survival (Copeland 1966; Gunter 1953). Oysters are particularly vulnerable to such conditions because they are sessile. Estuarine production increases, however, with additional freshwater inflow as shown from studies describing the effects of three releases from the Bonnet Carre Spillway, Louisiana (Gunter 1953). The greatest value of freshwater inflow is the simple fact that the volume and area of estuarine habitat -- the most productive habitat in the coastal zone -- are directly related to the amount of freshwater inflow.

Although it would be desirable to be able to attribute definite quantities or values of fish products, biological production, or marsh for each acre-foot of freshwater inflow or ton of sediment, it has been rarely possible to develop these numbers. Craig et al. (1979) estimated that the annual loss of 16.5 square miles (42.7 square kilometers) of Louisiana coastal marsh results in a minimum annual loss of 8-17 million dollars of fisheries products and services.

PROJECTING THE EFFECTS OF ALTERING FRESHWATER INFLOW

The management and planning of freshwater inflow to estuaries requires the capability to forecast the effects of various amounts of freshwater on the production and ecology of estuaries. What will be lost or gained by modifying the quantity of freshwater, its sediment and nutrient content, or the timing of its entry into estuaries? Definite progress is being made but methodologies and/or data for prediction are weak.

On the national level, the U.S. Water Resources Council (1978) has provided "approximations" of the monthly river inflow needs of estuaries for the major rivers of the U.S. These "approximations" were developed through (1) calculation of historic median flows by month, (2) technical committees, and (3) technical studies. These recommendations alert planning and development agencies to the estimated instream flow needs of estuaries but were not intended to be used for detailed river and estuary planning.

The Texas Department of Water Resources (1979) has estimated the freshwater inflow requirements for Texas estuaries. The associated studies were conducted following the enactment in 1975 of Senate Bill 139 by the Texas legislature to establish the freshwater needs of six estuaries for Texas water planning. The Texas freshwater inflow model, a comprehensive quantitative system for estimating freshwater inflow needs, was developed. Three alternative goals of freshwater inflow were established including the following: (1) subsistence - includes marsh inundation, nutrient transport, and salinity control; (2) fisheries harvest maintenance - supports average commercial fisheries harvest using 1962 through 1976 catches as a base; and (3) fisheries harvest enhancement - increases harvest of a single speices of finfish or shellfish. Seasonal and monthly flow needs were calculated for each alternative goal in each estuary. It was recognized that Texas estuarine systems have historically received a wide range of freshwater inflows due to drought ane wet or hurricane years. Estuarine ecosystems have adapted to these annual and seasonal cycles and these variations are considered necessary for ecosystem functioning. Therefore, the estimated freshwater inflow needs of different estuaries by month or year were called "statistical long-term control tendencies". Presumably, it will be possible to evaluate the effects of river development programs on freshwater inflow to specific estuaries by evaluating how each development plan would influence the "statistical long-term central tendencies." Although the study revealed that there was inadequate long-term data on many estuaries to calculate freshwater inflow needs, the recommended amounts of freshwater needed for each estuary were considered adquate to fulfill the three alternatives goals. Sediment was not modelled as a component of freshwater inflow but with that exception the model was based on the best available data and the important ecological products of estuarine ecosystems were included.

The U.S. Army Corps of Engineers has been studying and evaluating the effects of levees, navigation developments, and flood control programs in estuaries for over 50 years (Hudson et al. 1979). Most of the early studies involved the construction of physical models that provided information on the effects of various development alternatives on currents, salinities, and depth. Sediment deposition was difficult to evaluate with physical models. Recently, physical models have been combined with numerical computer models (McAnally et al. 1980). These hybrid models make it possible to predict sediment transport and deposition with much more precision. Significant progress is being made to project the effects of river inflow on the salinities and the morphology of estuaries.

It will be difficult to model ecological systems because of their complexity with the same degree of precision as the primary physical attributes -- depth, substrate, currents, salinities -- of estuaries. Ecological models of the complexity needed for treating species in the higher trophic levels of the food chain in estuaries are yet in a formative stage (Armstrong 1980). Also, there are few data bases that cover the range of freshwater inflows. It has been possible to relate variation in freshwater inflows to commercial fish catch, shrimp abundance, and striped bass reproduction (Armstrong 1980; Stevens and Chadwick 1979). Although many ecologists resent using commercial or sport finfish or shellfish speices for evaluation of ecological effects, it is doubtful if long-term data bases will ever be developed for most other species. Remote sensing will probably be used more for monitoring habitat change in estuaries in the future. Increased knowledge of estuarine habitats, innovative methods for using small bits of information, and advances in data processing and analysis will probably guide most near-future estuary modelling.

MANAGEMENT OF FRESHWATER INFLOW

Three management programs to provide freshwater inflow to estuaries are in various stages of development on the Texas Gulf coast, the Mississippi River delta of Louisiana, and in the San Francisco Bay complex of California. The Texas Freshwater Inflow Model mentioned earlier was prepared to assist water planners in providing adequate freshwater to sustain and flush estuarine marshes, salinity levels, and finfish and shellfish production. Application of the model to water management is presently under legislative study.

The U.S. Congress passed a resolution on 3 September 1976 to study the possibility of reintroducing freshwater and sediment into Louisiana marshes and estuaries because of the loss of coastal marshes and the inevitable loss of finfish and shellfish due to saltwater intrusion at the mouth of the Mississippi River. This study effort is underway but an appraisal of progress is not available. The State of Louisiana has also initiated a program to reintroduce freshwater to estuarine habitats.

The California State Water Resources Board (1978) has provided freshwater inflow to the Sacramento-San Joaquin Delta and Suisun Marsh by Water Right Decision 1485. The Board has mandated authority to protect vested water rights and to protect water in the public interest. Decision 1485 established and revised terms and conditions for salinity control and protection of fish and wildlife. Striped bass reproduction and Suisun Marsh preservation were the target fish and wildlife objectives. Water quality and quantity objectives were established for various sampling points in the estuarine system and a specific monitoring program was established. The Decision has a 10-year life; then it may be modified to consider updated information and water demands.

DISCUSSION

It will be difficult to bring the freshwater needs of estuaries into water management planning effectively until we improve our capability to project the consequences of varied amounts and qualities of freshwater inflow on estuarine ecosystems. Although we know that freshwater inflow is needed and know its essential roles in estuarine production, we have not always been successful in projecting potential effects for several reasons.

First, most man-caused modifications of freshwater inflow to estuaries -- dredging, filling, leveeing -- have taken place concurrently with other environmental perturbations and it has been difficult to isolate the direct effects of altered freshwater inflow. It has often been impossible to develop data that quantitatively relate freshwater inflow to various levels of biological production and to quantify the dollar benefits that could be expected by providing additional freshwater inflow.

Second, estuarine ecosystems are extremely dynamic and have evolved with great natural variation in tides, salinities, freshwater inflow and sediment inflow. Estuary ecosystems are durable and have developed resiliency to these natural variations. Some animals, e.g., the rangia clam, living in estuaries require fluctuations in salinity for their reproduction and survival. Estuaries also have short-term (1-2 years) rather than long-term successional communities. Rapid change in species composition and abundance is expected and is the rule. Estuaries contain more opportunistic species than permanent residents. This variability has made it difficult to set quantitative ecological goals.

Third, the setting of goals for management of freshwater inflow has varied greatly. Many ecological models use commercial fishery catch statistics mainly because they are the only long-term ecological data available for many areas of our coast. Other models are designed to maintain selected salinity regimes, to inundate marshes, or to provide suitable habitat for one species of finfish or shellfish. Estuaries serve many publics and some of the goals established for management of freshwater inflow are not necessarily acceptable to a broad clientele.

The most logical way to establish goals for management of freshwater inflow to estuaries is to develop plans that describe the products and services we expect estuaries to provide us in the future. If we expect estuaries to produce such products as shrimp, oysters, crabs, or menhaden, or to provide such services as hurricane protection through the buffering benefits of coastal marshes, we can prepare plans that will provide the necessary freshwater, sediment and nutrients to create the environments necessary to produce these products and services.

It may never be possible to estimate explicitly the monthly and annual freshwater needs of estuaries, as desired in water management planning, because natural conditions vary greatly. The effects of major changes in river inflow characteristics on estuarine ecosystems can be evaluated with our present state of knowledge. Only a larger data base on ecological effects, and additional experience, will permit us to project effects more quantitatively. If estuarine ecosystems are to exist in the future, we must bring freshwater inflow into the planning process.

REFERENCES

Armstrong, N. E. 1980. Effects of altered freshwater inflow on estuarine systems. Pages 17-31 in Proceedings of the Gulf of Mexico coastal ecosystems workshop. U.S. Fish and Wildlife Service, Albuquerque, NM. FWS/OBS-80/30. 214 pp.

California State Water Resources Board. 1978. Water Right Decision 1485. Sacramento-San Joaquin Delta and Suisun Marsh. 44 pp.

Chapman, R. 1966. The Texas Barrier Project. Pages 83-92 in R. Smith, A. Schwartz, and W. H. Massman, eds. A symposium on estuarine fisheries. Am. Fish. Soc., Spec. Publ. 3.

Copeland, B. J. 1966. Effects of decreased overflow in estuarine ecology. J. Water Pollut. Control Fed. 38(1):1831-1839.

Copeland, B. J. and T. S. Bechtel. 1970. Impact of water development on ecology of bays and estuaries. Water for Texas, 15th Annual Conference, p. 55-90.

Correll, D. L. 1978. Estuarine productivity. Bioscience 28(10):646-650.

Craig, N. J., R. E. Turner, and J. W. Day, Jr. 1979. Land loss in coastal Louisiana. Environ. Manage. 3(2):133-144.

Gosselink, J. G., C. L. Cordes and J. W. Parsons. 1979. An ecological characterization study of the Chenier Plain coastal ecosystem of Louisiana and Texas. Vol. 1. U.S. Fish and Wildlife Service, Office of Biological Services. FWS/OBS-78/9. 302 pp.

Gunter, G. 1953. The relationship of the Bonnet Carre Spillway to oyster beds in Mississippi Sound and the "Louisiana marsh" with a report on the 1950 opening. Publ. Inst. Mar. Sci., Univ. of Texas 3(1):17-71.

Gunter, G. 1963. The fertile fisheries crescent. J. Miss. Acad. Sci. 9:286-290.

Hudson, R. J., F. A. Hermann, Jr., R. A. Sager, et al. 1979. Coastal hydraulic models. U.S. Army Corps of Engineers, Coastal Eng. Res. Center, Spec. Rept. 5, Fort Belvoir, VA.

Livingston, R. J. 1976. Diurnal and seasonal fluctuations of organisms in a north Florida estuary. Estuarine Coastal Mar. Sci. 4:373-400.

Lugo, A. E., and S. C. Snedaker. 1974. The ecology of mangroves. Annu. Rev. Ecol. Syst. 5:39-64.

McAnally, W. H., Jr., W. A. Thomas, and J. V. Vetter, Jr. 1980. Physical and numerical modelling of estuarine sedimentation. Presented at the International Symposium on River Sedimentation, Beijina (Peking) China, March 1980. 29 pp. MS.

National Marine Fisheries Service. 1979. Fisheries of the United States. 1978. Curr. Fish. Stat. 7880. 120 pp.

Odum, W. E., and E. Heald. 1975. The detritus-based food web of an estuarine mangrove community. Pages 265-286 in E. Cronin, ed. Estuarine research, Vol. 1. Chemistry, biology and the estuarine system, Academic Press, N. Y.

Schaff, William. 1975. Status of the Gulf and Atlantic menhaden fisheries and implication for resource management. Mar. Fish. Rev. 37(9):1-9.

Smith, S., and S. Kato. 1979. The fisheries of San Francisco Bay: past, present and future. Pages 445-468 in T. J. Conomos, ed. San Francisco Bay: the urbanized estuary. Pacific Div., Am. Assoc. Acad. Sci., San Francisco, CA.

Snedaker, S., D. de Sylva, and D. Cottrell. 1977. A review of the role of freshwater in estuarine ecosystems. Univ. Miami, FL. 126 pp.

Stevens, D. E., and H. K. Chadwick. 1979. Sacramento - San Joaquin estuary - biology and hydrology. Fisheries 4(4):2-6.

Texas Department of Water Resources. 1979. The influence of freshwater inflows upon the major bays and tributaries of the Texas Gulf coast. Executive Summary LP-115. 51 pp.

Turner, R. Eugene. 1977. Intertidal vegetation and commercial yields of penaeid shrimp. Trans. Am. Fish. Soc. 106(5):411-416.

U.S. Army Corps of Engineers. 1974. Chesapeake Bay, existing conditions report. Baltimore District, 7 vols.

U.S. Army Corps of Engineers. 1976. Fish and wildlife study of the Louisiana coastal area and the Atchafalaya Basin Floodway. Unpublished draft.

U.S. Department of Interior. 1977. 1975 national survey of hunting, fishing and wildlife-associated recreation. Fish and Wildlife Service, Washington, D.C. 91 pp.

U.S. Water Resources Council. 1978. The Nation's water resources, 1975-2000 - the second national water assessment. Vol. 1: Summary. Washington, D.C. 86 pp.

The Columbia River Estuary Data Development Program

by
Gary Lawley*
Robert J. Moulton**
David H. Kent***

ABSTRACT: The Columbia River is the second largest river in the U.S., yet less scientific information is known about the Columbia River estuary than any other U.S. estuary of comparable size or importance. Major land and water use conflicts now exist; land and water management decisions are being made with limited information and even less understanding of the Columbia River estuarine system. If the estuary is to continue to provide a wide range of benefits, more knowledge of its physical and biological elements and their interactions is essential. Toward that end, the Pacific Northwest River Basins Commission set up the Columbia River Estuary Data Development Program to manage and guide research/data development efforts on the Columbia River estuary.

Introduction

The Columbia River Estuary is an artifact of the Missoula floods, which occurred 18,000 to 20,000 years ago, and of the rise in sea levels throughout the world during the last 2000 years.(1) The first evidence of human habitation in the Columbia River area, about 15,000 years ago, indicates that nomadic peoples followed the Pacific coastline and major tributaries in their quest for food and shelter. By the time Sebastian Viscaino saw the Oregon coast in 1602, Indian cultures were well developed with trade links as far north as Alaska.(2) In 1792, the British Captain, Robert Gray, discovered and mapped the entrance to the Columbia River. Based on these maps and Lewis and Clark's reports about the New World in 1804, John Jacob Astor and the Hudson's Bay Company made the first major land use decisions regarding the estuary. The town of Astoria was founded in 1811 by 30 British subjects and 3 Hawaiians.

From that time until 1968, little planning for development of the Columbia River estuary occurred. Decisions were based on the

---

* Project Manager, CREDDP, Envirosphere Corporation, Bellevue, Washington.
** Program Manager, CREDDP, Pacific Northwest River Basins Commission, P.O. Box 908, Vancouver, Washington 98666.
*** Assistant Program Manager, CREDDP, Warrenton, Oregon.
( ) Refers to bibliographic citations listed at end of text.

needs of the user, rather than the limits of the resources of the estuary. That minor planning was designed to control the river; little thought was given to the resources supported by the river with the exception of the fisheries and power development.

During the early Twentieth Century, substantial declines in fish resources, particularly salmonids, occurred. These declines have been attributed primarily to the construction of numerous dams which have undoubtedly had a major effect. However, substantial reductions in fish stocks occurred prior to dam construction due to intensive fishing pressure.[3] The need to allocate the use of the Columbia River's limited resources has recently forced government to plan, make judgments, and enforce these judgments through a variety of mechanisms. Most decisions have been based on limited information about discrete parts of the estuarine system.

The Estuary. The Columbia River estuary is the ninth largest in the United States, yet little scientific information exists on this estuary in light of its relative importance (Figure 1) . Depths are generally less than 30 feet except in the lower estuary. For purposes of the Columbia River Data Development Program, the estuary has been defined as the area of high tide from the Columbia River bar to River Mile 46, even though salinity seldom encroaches beyond River Mile 23. However, tidal effects may be detected over 100 miles upstream

Planning. During the late 1960's, concern by the Atomic Energy Commission[4] regarding the fate of radioactive materials originating at the Hanford Research Facility near Richland, Washington led to the first organized attempt to study the Columbia River estuary. Except for the AEC's limited investigations in 1965, no intensive study of the estuary's resources was undertaken until recently. Fisheries concerns continued to increase as a result of increased fishing pressure, the effects of upstream dams, substantially enlarged fisheries programs and facilities and the U.S. Army Corps of Engineers maintenance of the navigation channel. Heightened concern for the Columbia's resources was brought about by increased awareness during the 1960's of the limits of natural systems. Passage of the National Environmental Policy Act (NEPA: P.L. 91-190) of 1969 led to renewed efforts to coordinate investigations of man's effect on the Columbia River and its estuary. Local efforts to bring about a state or regional integrated research/planning effort were initiated in 1970.

At the same time, concern for the coasts of the United States was increasing nationally. In response to the President's Commission on Marine Resources and Engineering Development,[5] the Congress passed the Coastal Zone Management Act of 1972 (P.L. 92-583). This Act led to formation of coastal planning organizations in 28 States, including Oregon and Washington. Washington state furthered the intent of the Coastal Zone Management Act with the passage of its own Shorelines

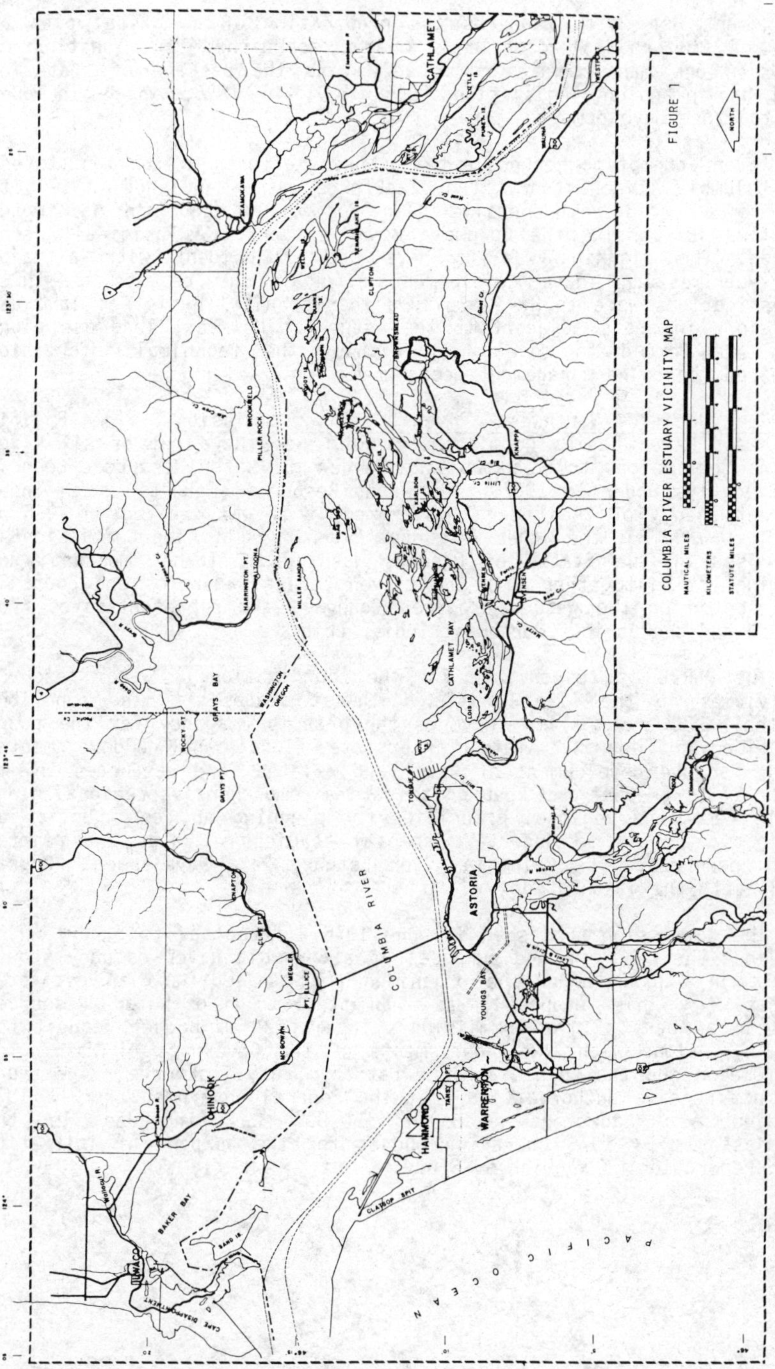

map furnished courtesy of the Columbia River Estuary Study Task Force

Management Act which requires the preparation of a coastal plan by each of the coastal counties. Oregon approached the question of coastal zone management more broadly with the passage of statewide land use planning legislation (S.B. 100) in 1975, which includes coastal zone development.

In an attempt to reconcile the planning efforts of the two states, the Columbia River Estuary Study Taskforce was established in 1975 to prepare a local comprehensive plan for the Oregon and Washington counties surrounding the Columbia River estuary. As originally conceived, the planning was to have been coincident with a major five-year estuary data development effort. The results of these investigations were to have provided the technical basis for land use decisions in areas adjacent to the estuary. However, land use planning was forced to move ahead without the technical information required for sound management decision-making.

The Pacific Northwest River Basins Commission. The Pacific Northwest River Basins Commission (PNRBC) was the first of six River Basins Commissions to be established under the Water Resources Council by Executive Order in 1967. The Water Resources Council is an independent agency of the Federal government. It was created in 1965 by Public Law 89-80, the Water Resources Planning Act. The Commission is comprised of the States of Oregon, Washington, Idaho, Montana, and Wyoming acting together with the eleven federal agencies and councils represented on the Water Resources Council, and representatives from Canada and the various Northwest Indian tribes.

The PNRBC is responsible for the coordination of water resource activities in the five Pacific Northwest states.(7) The functions of the PNRBC are: 1) to serve as the principal agency for the coordination of Federal, state, interstate, local and nongovernmental plans for the development of water and related land resources in its area; 2) to prepare and keep up to date a comprehensive regional plan; 3) to recommend regional priorities for planning and construction of projects; and 4) to undertake special studies of water and related land resources. The Columbia River Estuary Data Development Program falls into the last category.

The Commission published in June 1976 a "Proposal to Study" which outlined the need for and purposes of a Columbia River estuary study, how such a study would be organized and managed, and an estimated budget.(6) This proposal was submitted to the Water Resources Council as part of the Commission's total program budget request for FY 1978, along with the endorsements of the Governors of Washington and Oregon and the Columbia River Estuary Study Taskforce. The study did not receive authorization from the Federal government for FY 1978 funding as requested, although $50,000 was included in the Commission's FY 1978 operating budget for the purpose of initiating the preparation of a "Plan of Study."

The Commission resubmitted its request for authorization and full federal funding for the program as part of its total budget request for FY 1979. Subsequent Congressional support and action resulted in authorization of the program and funding was provided under the appropriation authorities of the October 1978 Public Works Bill, P.L. 95-482. The Plan of Study, detailing the scope, organization, work elements, schedules and budgets for the program, was completed and approved by the PNRBC on March 15, 1979.

Purpose. The purpose of the Columbia River Estuary Data Development Program is twofold: to increase understanding of the Columbia River Estuary as a system, and to acquire information needed for making specific land/water management decisions. When completed, the program will provide much of the information necessary to support future management decisions concerning use of the Columbia River estuary through greater insight into how the Columbia estuarine system functions.

P.L. 95-482 provided $828,900 for FY 1979 to complete program planning and initiate FY 1979 program activities; the FY 1980 budget provided an additional $991,000 to initiate field investigations; the FY 1981 budget provides $1,365,000 to continue field data acquisition and begin analysis. Program participants include agencies of state and federal governments, universities, and private organizations having demonstrated expertise in particular technical subjects. The program's elements are divided into 26 work units that are being accomplished by this mix of participants. Requests for Proposals were solicited and after review of qualifications, approved organizations were asked to submit formal proposals for CREDDP work. A total of 28 program contractors were selected to perform the 26 work units and four special reports. The total cost of the program over a six-year period is estimated at $6,204,000 with 100 percent federal funding.[8]

The Data Development Program is based on the following assumptions:

- The Columbia River estuary is an interdependent system (one in which physical and biological elements relate)
- The Columbia River Estuary Data Development Program will use and expand present data and programs, not replace them
- Data will be developed to facilitate management of the Columbia estuary
- The program will focus equally on the biological and physical processes; little emphasis will be placed on land resources, social, or economic studies
- The program will report the data collected in a manner immediately usable by managers and interested local parties

Process. The program is being accomplished in seven phases over a six-year period. These include program planning, contracting, implementation, field data acquistion, and a summation or report phase. Each contractor's effort is reviewed annually and individual contracts are modified on the basis of their previous performance and the availability of funds. The results of the program include a variety of technical reports for each work element. These include "summaries" for decision-makers, an annual report which outlines program progress and results, and a series of conferences and special reports for technical and lay audiences, newsletters, and reports on key management issues. A list of CREDDP publications currently available can be obtained from the PNRBC.

Organization and Management. A schematic diagram of the program organization and management is shown in Figure 2. Policy guidance is provided by the PNRBC, primarily through its Program Committee and the Commission Chairman. Technical guidance for the program comes from a Technical Advisory Committee (TAC) working through a Program Manager who is a member of the Commission staff. Public input through a Citizen Advisory Committee (CAC) is used throughout the study to identify local problems that require information or application of program results.

Figure 2
Organizational Chart

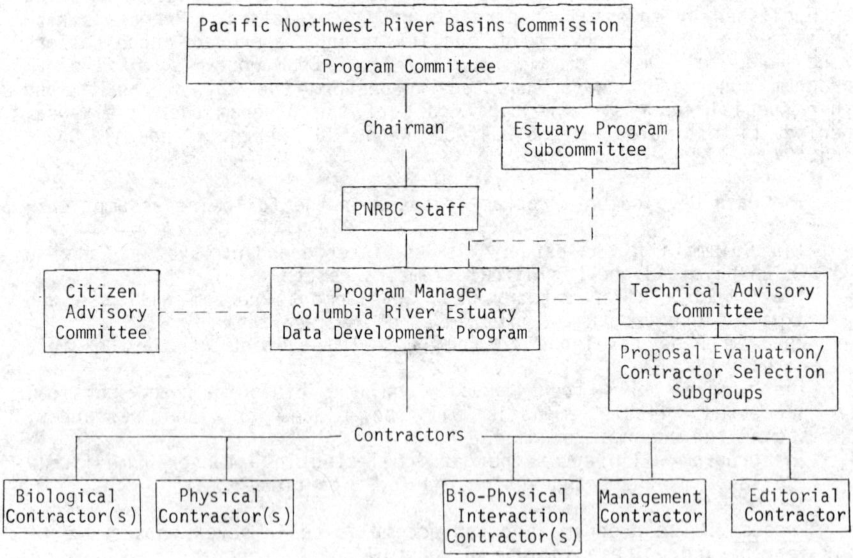

Policy Guidance

Pacific Northwest River Basins Commission. Policy guidance comes from PNRBC. The policies of the Water Resources Council are provided through the Commission and the PNRBC Chairman. The functions of the Commission as they relate to the Columbia River Estuary Data Development Program are as follows:

1. Review and approval of annual budget submissions prior to transmittal to the Water Resouces Council.
2. Provision of general policy guidance and coordination of the related policies or actions of member agencies.
3. Resolution of major policy issues or disputes arising from contractor selection or performance if referred to PNRBC by the Chairman or Program Committee.

Program Commmittee. The Program Committee consists of all 11 PNRBC member agencies. This committee is concerned with all phases of the PNRBC planning program. Its responsibilities to CREDDP include:

1. Review and recommend to the Commission appropriate action on the initial and all revised Plans of Study.
2. Review program progress and budgets and recommend Commission action.
3. Review contractual agreements and resolve conflicts on contracts or memoranda of agreement referred to it by the Commission Chairman or the Estuary Program Subcommittee.
4. Perform other activities assigned by the Commission.

The Estuary Program Subcommittee of the Program Committee consists of Commission representatives or alternates from Oregon and Washington and from the Departments of Defense, Commerce, Interior, Agriculture, and the Environmental Protection Agency. The committee has the following responsibilities related to the data development program:

1. Provide, on behalf of the Program Committee, guidance to the Program Manager, including review of the Plan of Study, overall work priorities and study objectives; make recommendations to the Program Committee.
2. Consider recommendations from the TAC and Program Manager concerning contractor activities.
3. Resolve policy or contractor selection disputes when referred by the Chairman. Major policy issues may be referred to the Committee or Commission as appropriate.
4. Monitor progress of the CREDDP work program. Provide a forum for discussion of agency concerns and disputes regarding the CREDDP work program.
5. Review the annual report and budget request prior to submission to the Program Committee.

6. Perform other program related responsibilities as determined by the Program Committee.

Chairman. The Chairman of the PNRBC and his staff are responsible for implementing CREDDP. The Chairman, working with the approval of the Commission, performs the following functions:

1. Provides detailed policy and program guidance and review of CREDDP progress.
2. Coordinates program activities with member agencies as needed, and informs the Commission of CREDDP progress.
3. Approves contractor selections, contract negotiations, contracts, and Memoranda of Agreement.
4. Resolves policy or contractual disputes referred to him by the Program Manager.
5. Reviews and approves an annual report and budget request prior to submission to the full Commission.
6. Approves limited use of consultants upon recommendation of the Program Manager.

Technical Guidance

It is the responsibility of the Technical Advisory Committee (TAC) and Program Manager to give the program direction and continuity. Implicit to the program is the assumption that all data acquisition is to be undertaken by contract on a competitive basis. The roles of the Program Manager and TAC are identified as follows:

Program Manager. The Program Manager is a member of the PNRBC staff and responsible to it. His functions are as follows:

1. With TAC assistance, develops "Requests for Proposals" based on the approved Plan of Study.
2. Establishes a proposal review and contractor selection process for the program work components.
3. Serves with Commission's Planning Director and TAC member as contractor negotiating panel chairman.
4. Manages all contracts and Memoranda of Agreement entered into by the Commission under this program including fiscal management, monitoring of field work, and coordination of related work components.
5. Provides for development of a system for storage and retrieval of data generated by program work components so that data are quickly available in a usable format.
6. Prepares for the Commission quarterly progress reports and an annual report to the Water Resources Council.
7. Testifies before the Water Resources Council or Congress as requested by the Chairman.
8 Provides liaison and identifies appropriate Commission participation in the CREST program.

9. Keeps Commission member agencies and others informed of program progress.

Technical Advisory Committee. The TAC is an advisory rather than a decision-making group and plays a major technical role in guiding the program.

The designation of specific TAC members is based on the following assumptions:

1. All major disciplines that are a part of the program are represented where possible by people who perform or manage research in each of the disciplines.
2. Important management interests are represented, but no agency has more than one representative.
3. A balance between disciplines and between science and management is sought.
4. Membership is voluntary and members are reimbursed for services rendered, either by the program or by their agency.
5. TAC members may also be contractors, but will not be allowed to evaluate those specific tasks for which they are competing.

Most important is the assumption that the TAC functions principally as a disciplinary forum; policy considerations and agency concerns are primarily the responsibility of the PNRBC and conflicts are resolved there.

The functions of the TAC are as follows:

1. Review and comment on the program. During the first year, the TAC played a major advisory role to the Program Manager in the formulation of the Plan of Study and the work program which derived from it. Subgroups to the TAC reviewed the Plan of Study, the Requests for Proposals, and suggested hypotheses, methods, instrumentation, and/or techniques needed to implement the Data Development Program. Annually, the TAC reviews the draft of the annual report and develops a list of appropriate tasks for use in preparing subsequent contract modifications.
2. Advise on the selection of contractors. The TAC evaluated conceptual proposals and statements of contractor qualifications provided in response to Requests for Proposals issued by PNRBC following approval of the Plan of Study. Working with the Program Manager, it designated disciplinary work groups to evaluate and recommend those contractors from whom detailed conceptual and cost proposals were or will be requested. Based on the review of the detailed proposals received, the TAC recommends proposal modifications and contractor rankings for consideration by the Program Manager and approval by the Chairman and the Commission prior to negotiation of a contract.

3. The TAC evaluates the program, identifies new research directions, data needs, or program emphasis. It reviews contractor reports, evaluates contractor performance, and keeps abreast of the information developed for the estuary. In order to do so, the TAC chairs or participates in an annual research conference and quarterly workshops. It annually reviews the program and makes recommendations to the Program Manager concerning program emphasis or reduction. In formulating recommendations, the TAC is cognizant of the research programs of federal and state agencies and the needs of local, regional, and governmental managers.

Contractors. The work of the Data Development Program is implemented by contract. A basic program assumption is that a mix of federal/state, institutional, and private contractors or individuals is to be used, based on the merits of their proposed work. Contractors meet quarterly or at the request of the Program Manager to discuss and coordinate the work program.

Three of the contractors are of particular management interest. The first is the "management contractor." This contractor has the difficult function of insisting that the TAC and program contractors think in holistic, systems terms and of helping them to do so. The management contractor does not collect field data. Many of the key management and key research questions cannot be answered within the confines of any one discipline. Thus, some of the most important results that will emerge from CREDDP will come from the management contractor or the management contractor working with other program contractors. The management contractor must also ask the difficult and important questions as to the meaning of the information already obtained, so that the Program Manager can redirect the program.

The management contractor has a number of special requirements. He must have access to the data of all contractors before the publication stage, since publication will generally occur too late for the management contractor to provide meaningful feedback to redirect activities. Interchange of data and ideas between contractors will also be necessary on a continuing basis and the Program Manager will have to assure this in contracts and see that it takes place in a timely manner. Workshops must be organized, literature reviews compiled, research questions formulated, and technical conflicts resolved by this contractor. The function of the management contractor in organizing data collection and providing an overview is greatest at the beginning and toward the end of the program; and thus, this contract may continue after other contracts are completed. The functions of the management contractor are as follows:

1. Organize and coordinate an annual research conference.
2. Prepare an annual program report to the TAC following the annual meeting. This report will identify technical

problems, report on performance by individual contractors, identify new areas of research, and perform other appropriate activities.
3. Obtain technical procedures from each contractor.
4. Prepare a quarterly, impartial program evaluation for the Program Manager to be used in the Annual Report and with the annual budget request.
5. Perform other duties as necessary and appropriate to technical program coordination.
6. Review technical aspects of proposals and make recommendations to the program manager.
7. Coordinate technical portions of quarterly meetings.
8. Meet with contractors regularly basis and at the end of each fiscal year.
9. Set up and coordinate meetings between contractors, consultants or technical committees to resolve conflicts between work units.
10. Give technical advise, as required, to the program manager concerning proposals and contract negotiations.

The second integrative contractor is called the "Editorial Contractor." Individual program contractors are required to provide executive summaries as well as technical reports but it may be too much to expect that the contractors will provide reports in language that can be understood by the Citizens Advisory Committee, the informed layman, and decision-makers on the local, state, and federal level. The translation step is viewed on the local level as extremely important; a separate contractor is used to perform these functions.

The function of the Biological-Physical Interactions contractor, is to integrate the various program elements conceptually. This integrative and analytical role requires that substantial time be spent creating the necessary conceptual models to describe estuarine processes. This contractor is working with the individual program contractors to synthesize and evaluate the outputs of individual program elements. The integrative function generally does not require the acquisition of new data; the data of other program researchers is used or modified to meet the needs of the contractor. The activities of the biophysical contractor are supplemented by the integrative efforts of all the other program contractors. The end result of this contract will be to present an understanding of the current level of estuary interrelationships and creation of a systems model of the estuary.

PNRBC Services. The PNRBC provides the following services to the Columbia River Estuary Data Development Program:

1. Provides management and accounting services for program funds.
2. Prints research reports as required.
3. Oversees contractor payments and billing procedures.

4. Provides program office space and clerical staff.
5. Performs other administrative services as needed.

Citizen Involvement. To assure that management needs are addressed, the data development program has an independent link to the local community. The Citizen Advisory Committee's functions include:

1. Advise the Program Manager on the overall focus of the program and local problems that should be addressed.
2. Review products for usefulness to and understandability by the public and local decision-makers.

Other opportunities for citizen involvement include:

1. Quarterly and annual meetings that provide the public an opportunity to comment.
2. Concerns which may be sent directly to the Program Manager.
3. Comments about the program made through the local Citizen Advisory Committee.
4. Citizen views may be provided through the Commission.
5. Annual contract modifications that are responsive to changing public needs and citizen concerns.

Progress. As of August 1980, 28 contracts have been awarded. The program has leased a 39' vessel and purchased a 22' vessel as well as several smaller boats and equipment. A local office and field station has been established in Warrenton, Oregon. Field work was initiated in October 1979, with full scale field effort continuing since March 1980. When Mount St. Helens erupted in May, 1980, the CREDDP research effort was well underway. During the weeks prior to this event, detailed information had been gathered in the estuary on the distribution and abundance of infaunal and epifaunal organisms, zooplankton, ichthyofauna, avifauna, marine and terrestrial mammals, benthic and planktonic primary production, and the distribution of sediment types. Within a week after the eruption, the impact of the ash and flooding on all of these system components were being assessed as part of the regularly scheduled research. Additional, and more intensive, research was also funded to study the changes in the water chemistry, avifauna, marine mammals, benthic infauna, zooplankton, and ichthyofauna.

The fisheries research, under contract to the Hammond, Oregon laboratory of the NMFS, detected significant reductions in both the absolute numbers and the diversity of ichthyofaunal assemblages in the estuary.[9] Subsequent research suggests a general recovery of these communities although the absolute numbers still appear low for this time of year based on previous collections (Durkin, personal communication). There has been no indication that any of the other biological components suffered significant or permanent alteration as a result of the volcano.

A number of reports have been completed by the CREDDP. These include:

- A Plan of Study
- Wetlands Maps. (Department of Interior, NWI)
- Bibliography of Columbia River Estuarine Research (Oregon State University)
- Review of Historic Literature and Research activities (all contractors)
- Guide to Organisms of the Columbia River estuary (Department of Commerce/Corps of Engineers)
- History of the Columbia River Estuary (Oregon Historical Society)
- Quarterly newsletters (Oceanographic Institute of Washington)
- A Style Guide for Report Preparation (Oceanographic Institute of Washington)
- Basemaps at a variety of scales (Oceanographic Institute of Washington)
- A Procedures Manual and Data Users Guide (Envirosphere)
- Historic bathymetric maps (Northwest Cartography)

Others planned for release in 1980 include quarterly newsletters, Data Quality Assurance Manual (December), Key Species Life Histories (December), and a variety of short technical reports.

Status of the program as of August, 1980, is as follows: All work unit contractors have been selected. Several contractors have completed their tasks and reports are available; complete literature reviews have been completed and are available; a style guide and a procedures manual have been prepared; field data is being collected; effects of the first major volcanic eruption in the U.S. in this century have been monitored; coordination activities such as standard taxacodes, data sheets, reporting units, and meetings between various disciplines take place continually.

Summary. The Columbia River Estuary Data Development Program is a product of the many resource concerns and investigations which have preceeded it. The historic limitations of earlier Columbia River estuarine efforts have been evaluated and management techniques for dealing with these shortcomings have been included within CREDDP. Recognition of past deficiencies does not assure future success; it does provide a base from which to plan for future contingencies, to integrate the activities of the many participants, and to understand the function and resources of the Columbia River estuary more fully. Early progress and success of CREDDP and its organizational structure has been substantial. It is the Commission's hope that the longer term results are equally worthwhile.

BIBLIOGRAPHY

1. Norwood, G., "History of the Pacific Northwest Power". 1979. (unpublished manuscript)

2. Oregon Historical Society, "A History of the Columbia River Estuary". Portland, Oregon. 1979. Published by the Pacific Northwest River Basins Commission, June 1980.

3. Pacific Northwest River Basins Commission, "Second Status Report on Anadromous Fish Runs of the Columbia River", prepared by the Fish and Wildlife Committee. Vancouver, Washington. 1974.

4. Pruter, A. T. and D. L. Alverson, Columbia River and Adjacent Ocean Waters. University of Washington Press. Seattle, Washington. 1972.

5. National Commission on Marine Resources, Engineering and Development, Our Nation and the Sea, Government Printing Office. Washington, D.C. 1970.

6. Pacific Northwest River Basins Commission, "Columbia River Estuary Special Study, Proposal to Study". Vancouver, Washington. June 1976.

7. Pacific Northwest River Basins Commission, "Annual Report for 1978". Vancouver, Washington. June 1979.

8. Pacific Northwest River Basins Commission, "Plan of Study for the Columbia River Estuary Data Development Program". Vancouver, Washington. March 1979.

9. Durkin, J.T., "Downstream Fisheries", The Effects of the Mount St. Helens Eruption on Water Resources, PNRBC, Vancouver, Washington, June 12, 1980 p. 49-54.

INFLUENCES ON PLANS FOR
THE SANTA ANA RIVER CHANNEL MOUTH
BY

FALA POWERS[1]

INTRODUCTION

The Santa Ana River Basin covers 3,200 square miles and includes three counties. It is the largest river system in southern California. The Santa Ana River originates in the San Bernardino Mountains, runs onto an alluvial flood plain and through a densely populated coastal plain before emptying in the Pacific Ocean, a total distance of 75 miles. A standard project flood[2] from the Santa Ana River Basin would inundate over 160 square miles of heavily developed land primarily in Orange County. Losses in the range of $9 billion would be likely, and the lives and safety of approximately 2 million people would be jeopardized. Existing flood control improvements along the river reduce damages from small floods but provide an insufficient level of protection for the highly urbanized flood plain in Orange County. Responding to this problem, the Los Angeles District of the Army Corps of Engineers has investigated a number of possible solutions and prepared a recommendation for Congressional review. The plan for the mouth of the river is a key element in the overall flood control package.

The overall plan requires that the 1.5 mile channel at the mouth be improved so its capacity is increased from current capacity of about 20,000 ft^3/s to 47,000 ft^3/s. The design for such improvements is complicated by the fact that the area available for improvement is severely constrained. Two drainage channels run parallel to the main channel, levees contain sewage outfall pipes 10-feet in diameter, and houses, oil wells, State beach and a sewage treatment plant abut the current channel near the mouth of the river (see Existing Conditions Map on the following page). The "most engineering-efficient" design would be to construct a new channel, with vertical concrete walls, capable of carrying the required 47,000 ft^3/s.

[1]U.S. Army Corps of Engineers, Los Angeles District.

[2]This is a flood that would result from the most severe combination of weather and hydrologic conditions considered reasonably characteristic of the region. For the Santa Ana River Basin, the standard project flood has a 1 in 200 chance of happening in any given year.

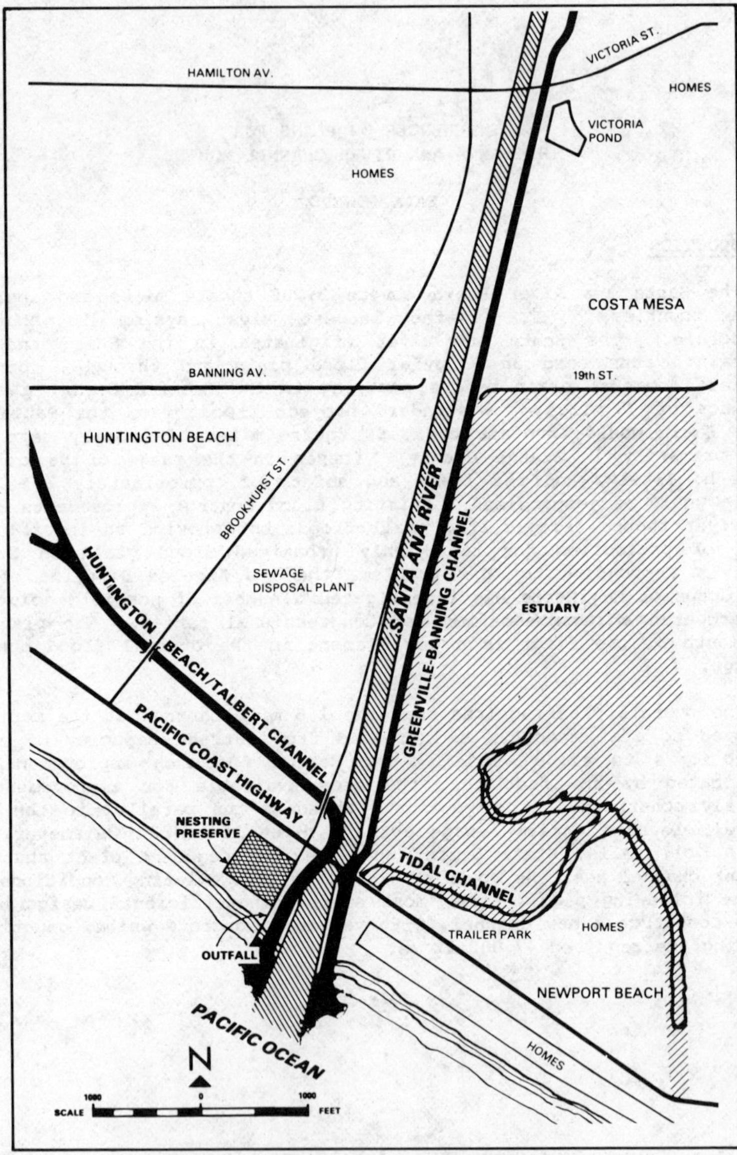

SANTA ANA RIVER MOUTH: EXISTING CONDITIONS

This design for a rectangular concrete-lined channel was determined to be infeasible due to a number of social and environmental constraints. The concerns of a number of interest groups made it necessary to adjust the design of the channel significantly. There were seven major social and environmental issues to be dealt with in design of the final river-mouth channel: (1) the need for an earth-bottom channel; (2) the need to preserve habitat for the California least tern nesting colony; (3) the need to preserve and restore the marshlands in the vicinity of the river mouth; (4) interest in the area in using the mouth of the river for a marina; (5) the effect of the channel on Victoria Pond, a 13-acre pond of significant environmental value; (6) the need for recreation in the area; and (7) the interest in replenishing sand on local beaches. This paper addresses the changes which were made in the river-mouth plan in reponse to these issues. It shows that the final design of the river-mouth channel represents a balance of engineering with social and environmental concerns. The channel was designed to carry the necessary capacity of floodwaters with a minimum of damage to the existing ecosystem. The river mouth's plan also includes a number of social and environmental benefits to the area. Although the Santa Ana River mouth is unique, similar issues are likely to be raised during design of other flood control measures near environmentally sensitive river delta-estuarine areas and in this respect this paper would have significance in addressing problems common to other flood control projects.

ISSUES

Earth-Bottom Channel

Biological studies conducted during the planning process by the U.S. Fish and Wildlife Service revealed that the existing earth-bottom Santa Ana River, the Greenville-Banning, and the Talbert channels provide valuable habitat for intertidal and benthic organisms. These organisms feed fish which are in turn fed upon by birds including the California least tern, a Federal and State endangered species. Consequently, any structure designed for the river mouth must have an earth-bottom.

Engineering studies to deal with these concerns were conducted. The results of these studies revealed that the Santa Ana River and Greenville-Banning channels can remain earth-bottomed about one mile upstream of the Hamilton Avenue-Victoria Street bridge to the ocean. The earth-bottom channel, however, will cost considerably more than the concrete rectangular channel since it will require the toe of the channel walls to be heavily protected by a 15-foot deep cutoff wall and a layer of riprap from the wall, on a one to four slope, to a depth of 25 feet below the channel invert. The Talbert channel can remain earth-bottomed and must be relocated 200 feet to the west. The Greenville-Banning channel must be relocated 200 feet to the east, since the sewage outfall pipes preclude widening the Santa Ana River and Greenville-Banning channels to the west. This would eliminate about 16 acres of the salt marsh on the east side of the channel, causing an environmental concern that will be discussed in the next section.

Consequently, the feasibility of taking less right-of-way by combining the Santa Ana River and Greenville-Banning channels was studied. These studies disclosed that the channels could be combined at about 2,000 feet below the Hamilton Avenue-Victoria Street bridge to the ocean. Combining the channels further upstream was not feasible since the flow depths of the Santa Ana River would begin to interfere with drainage of the low-lying area serviced by the Greenville-Banning channel. This wider combined channel will have the added benefit of an increased area of tidal wetland. The combined channel will only require moving the channel 130 feet to the west and will eliminate only 8 acres of marshland (see Proposed Plan Map). Prior to construction, additional studies will determine the effects of freshwater flows and velocities on the channel organisms.

California Least Tern Nesting Colony

The proposed widening of the Santa Ana River will require realignment of the Talbert channel, a local storm drain channel. This in turn, will displace one-third of the 4.5-acre least tern nesting preserve on the west side of the river mouth at the beach. As stated earlier, the California least tern is a Federally protected endangered species. Therefore, the Board of Engineers for Rivers and Harbors, a review agency within the Corps requested that we investigate adequate means and measures for preserving this existing nesting colony. Consequently, our planning also concentrated on alleviating impacts to the least tern preserve.

A series of meetings for the Corps of Engineers, hosted by the Coastal Commission to coordinate plans for the Santa Ana River mouth with all concerned agencies, uncovered that the Talbert channel was their most important concern. These meetings included representatives from the U.S. Fish and Wildlife Service, California Department of Fish and Game, California Department of Parks and Recreation, California Department of Transportation, Orange County Environmental Management Agency, Orange County Sanitation District, and the Cities of Costa Mesa, Huntington Beach, and Newport Beach. During these meetings five possible relocations for the Talbert channel were suggested. Two suggestions were deleted. One,--to cover the Talbert channel under the nesting area and to top it with soil, marsh, vegetation and sand--was too expensive (cost: over $2 million). The other,--to empty the Talbert channel into the Santa Ana River--was infeasible. This was because the Talbert channel is at a lower elevation than the Santa Ana River channel; combining these channels would result in backflow up the Talbert channel during floods.

This left three possible alternative locations for the realignment of the Talbert channel. These relocation sites, along with public and various agencies reactions are displayed in Table 1.

Although it appears that moving the houses on the east side of the channel rather than moving the nesting preserve is more acceptable to the agencies, it is socially unacceptable as well as more costly. Moving the Talbert channel upcoast received criticism since it would isolate a large portion of the public beach. According to biological studies, the tern colony will probably not be adversely affected by taking a third of the nesting area from the east side and replacing it with an equivalent area to the west side. However, the nesting area must be moved gradually, over a five-year period, and its success must be monitored annually. The relocation must be initiated before channel construction, and in the non-nesting season (August 15 to April 15).

Consequently, we determined that the original "engineering solution" for the location of the Talbert channel, with modifications, is the most feasible and will remain as part of the plan.

TABLE 1

ALTERNATIVE SITES FOR RELOCATING THE CALIFORNIA LEAST TERN NESTING PRESERVE

Comment	Raised by (x) General Public	Coastal Commission	U.S. Fish and Wildlife Service	State Dept. of Fish and Game	State Dept. of Parks and Recreation
1. Shift SAR channel to the east, taking homes as necessary to avoid any impacts to the least tern preserve.	-	+	+	0	+
2. Shift the Talbert channel upcoast from the existing nesting site.	0	-	-	-	-
3. Move the Talbert channel 200 feet to the west eliminating 1.5 acres of the 4.5 nesting site on the east side and replace this loss with an equivalent area on the west side.	+	+	+	+	-

Legend: + Favors
0 No opinion expressed
- Opposes

Marsh Restoration

Concerns expressed by environmental groups including U.S. Fish and Wildlife Service (USFWS), California State Department of Fish and Game (CDFG), Sierra Club, and Audubon Society, disclosed that a marsh area to the east of the river may be of some environmental significance. Follow-up studies confirmed that this is a viable salt marsh habitat, although poor tidal flushing and localized dumping of fill have reduced the marsh in size and quality. This 400 acre salt marsh and uplands, however, remains a functioning ecosystem: it provides essential feeding, breeding and nursery grounds for birds, reptiles and mammals and in the winter the marsh serves as a natural resting ground for migratory birds of the Pacific flyway. Since the number of coastal marshes in southern California is rapidly diminishing, populations of most birds have been reduced. Among the birds which rest at this marsh are three endangered species, the light-footed clapper rail (Federal and State endangered), the Belding's savannah sparrow (State endangered and Federal Candidate species) and the California least tern (Federal and State endangered). This site is particularly significant as a feeding area for newly hatched California least terns.

Since the combined Santa Ana River and Greenville-Banning channel would be moved 130 feet to the west, the proposed project would add to the marsh's deterioration by destroying 8 acres of marshland. The planning response was not only to propose acquiring 8 acres of marsh

Tidal gate between the tidal channel adjacent to the salt marsh and Greenville-Banning channel. Proposed tidal gate will improve tidal flushing between these water bodies.

to mitigate for habitat loss but also to acquire 84 acres under the authority of the Endangered Species Act of 1973 for preservation and enhancement of endangered species habitat. Without this acquisition, chances are that the marsh will undergo continued poor tidal flushing, increased siltation and localized dumping which will result in its further deterioration. In addition, as pressure to develop coastal areas in southern California increases, there may be development of this area, although, this may be limited by the Coastal Commission, or by the Corps of Engineers which has jurisdiction over wetlands.

The U.S. Fish and Wildlife Service, California Department of Fish and Game, and Corps of Engineers biologists felt that the most important marsh restoration goal should be to enhance the California least tern's feeding habitat, particularly by providing these birds with fish to feed newly-born terns (the marsh's current primary function). A secondary goal should be to provide habitat for the light-footed clapper rail and Belding's savannah sparrow. To accomplish these goals, tidal flushing in the existing marsh area has to be increased; and the marsh recontoured to allow for an increasing number of tidal channels and elevations suitable for an array of vegetation zones, ranging from subtidal to upland areas.

After extensive biological studies, a tentative marsh plan was developed. This plan called for (1) replacing the existing tidal gate between the marsh and the Greenville-Banning with a larger gate and adding 4 to 6-foot-diameter culverts within the proposed marsh restoration area; (2) maintaining 10 to 15 percent of the surface area above high tides and 10 to 15 percent subtidal, and leaving the remaining area subject to tidal influence; and (3) filling and planting the marsh in stages while monitoring animal populations and water quality.

Marina

At the public workshops on issues concerning the Santa Ana River mouth, some participants opposed marsh restoration because they wanted a marina developed in that area. In fact, some thought that the Corps of Engineers was going to study marina feasibility in this area. However, strong sentiment against a marina existed, opponenets saying that it would eliminate wildlife and open spaces. This public workshop served as a forum for explaining to the public that a marina could not be a project purpose of the Santa Ana River flood control project. Although the Corps of Engineers has authority to study a marina, the study was not funded.

The Corps of Engineers' response to the public outcry for marina development, was to make every possible effort to assure that marina development was not precluded on adjacent lands. When the U.S. Fish and Wildlife Service suggested that the project include a wildlife corridor between Victoria Pond and the proposed 92-acre marsh acquisition area, the Corps refused, because acquiring the marsh restoration area alone will meet a better balance between environmental and marina interests.

The proposed marsh restoration will adequately meet the environmental objective of preserving the California least tern while not precluding marina development which would occur if the corridor were added.

Victoria Pond

Biological studies of the area impacted by widening the Santa Ana River revealed that one-third of this 13-acre freshwater pond habitat just southeast of Hamilton Avenue-Victoria Street bridge will be destroyed. This pond provides some diversity of habitat for mammals and waterfowl species within the coastal region. The pond attracts large numbers of water insects; these in turn provide an essential food supply for birds. Waterfowl and shore birds primarily use the pond for nesting and foraging purposes, but such use diminishes considerably in the spring. At that time the pond becomes an important foraging site for the California least tern. Victoria Pond is included as essential habitat for the California least tern by the U.S. Fish and Wildlife Service (USFWS) and they requested that we replace this pond.

The USFWS, as well as all other agencies concerned and the general public agreed that extending the pond to the south would be an acceptable solution. This solution was agreed upon after USFWS and Corps of Engineers biological studies showed that extending the pond adjacent to the existing pond will cause the least damage. Physical barriers on the north and west excluded its extension in these directions so the pond had to be extended to the south or east. Since the high quality vegetation to the east should be preserved, an extension to the south of the pond became an acceptable solution. The pond's restoration plan will include a vegetative buffer zone and fencing to provide feeding and habitat areas for species where foraging areas will be eliminated during construction. The pond's 4.5-acre extension will be added prior to project construction.

Recreation

Construction of the Santa Ana River channel will impact the existing recreational trails along the west side of the channel at the river mouth. The trails, some of the most heavily used in the state, are part of a twenty-mile trail system that runs through the highly urbanized area of Orange County.

Participants at a Corps of Engineers' public workshop, including numerous recreational special interest groups, requested that the proposed project should preserve these trails, to the greatest extent possible. Bicycle enthusiasts wanted safer crossings at Pacific Coast Highway locations. Currently, bicyclists must cross this high volume highway in the traffic lane. Coordination with California State Department of Highways (CALTRANS), revealed that CALTRANS plans to replace Pacific Coast Highway, subsequent to the flood control project, and the new bridge will include a separate bicycle lane on the bridges ocean side and will include a trail underpass under Pacific Coast Highway's west side.

As a result of these recreational concerns, the proposed project includes replacement of the existing trail. The bicycle and hiking trail will be on the proposed paved service road and the equestrian trail will be adjacent to the service road on a graded surface. Both trails will be to the west of the channel (see Proposed Plan Map). Although the current trails turn to the west upstream of the Talbert channel, the proposed project will include extending the bicycle and hiking trail to the Pacific Coast Highway. Thus, these trails can tie into the described trail links and the proposed CALTRANS bridge underpass and crossing.

Beach Replenishment

Deepening and widening of 10 miles of the Santa Ana River will require the excavation of about 7.5 million cubic yards of material. Since Orange County loses about 300,000 cubic yards of beach material yearly to littoral drift, the State of California requested that the Corps of Engineers study the feasibiliy of using some of the excavated material for beach replenishment.

Determining feasibility of using this material for beach replenishment is a two-fold process. We first determined how much of this material would meet the environmental standards for beach replenishment. Only materials meeting environmental standards can be used for this purpose. Once we knew how much sand was suitable for beach replenishment (up to 3.7 million cubic yards), the next step was to determine where the sand was needed most. With the assistance of U.S. Army Corps of Engineers Waterways Experimental Station (WES) hydraulic laboratory, we studied where to put the sand and what the temporal distribution of the sand would be, once placed. The WES study determined that the sand's predominant littoral drift at the Santa Ana River mouth appears to be downcoast. The beaches upcoast of the Santa Ana River are nourished by the feeder beach at Surfside-Sunset Beach, 10 miles upcoast and west of the Santa Ana River mouth. Every five years the Corps of Engineers places 1.5 million cubic yards of sand on this beach. Some of this sand is being transferred downcoast past the river mouth (Newport Beach), and the City of Newport Beach has indicated that it is satisfied with the present beach configuration. Significant capacity for placement of this material, however, still exists downcoast of the river mouth.

At the Corps of Engineers' public workshop concerning Santa Ana River mouth issues, we asked the participants their reaction to placing the sand downcoast of the river mouth (Newport Beach). Some workshop participants requested that the sand be placed in the winter months so as not to interfere with summer recreation. Others questioned why we were only placing the sand at Newport Beach. Corps of Engineers personnel explained that most of the sand will drift downcoast. In general, the workshop participates favored placing the sand on the beaches downcoast of the river mouth.

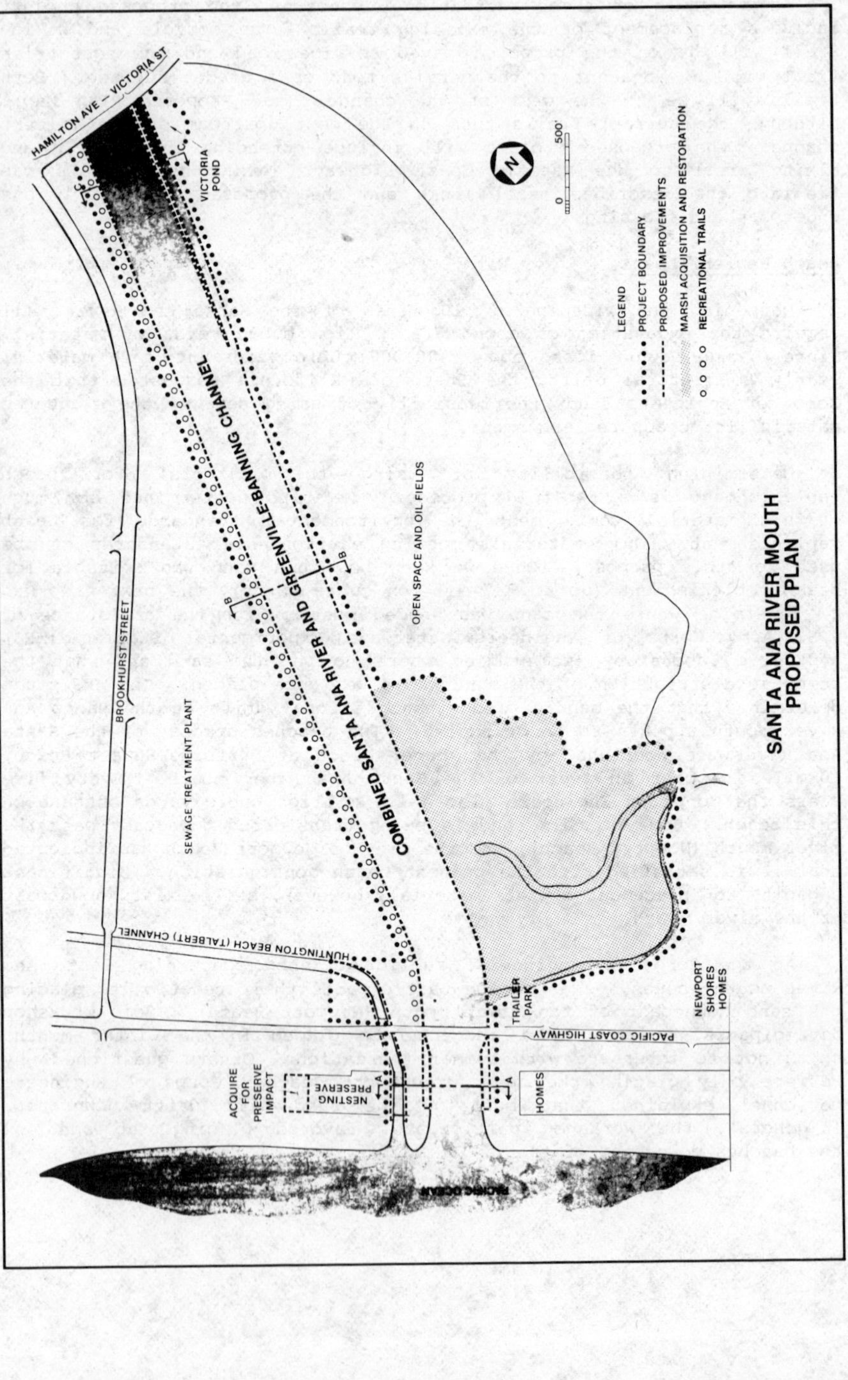

SANTA ANA RIVER MOUTH
PROPOSED PLAN

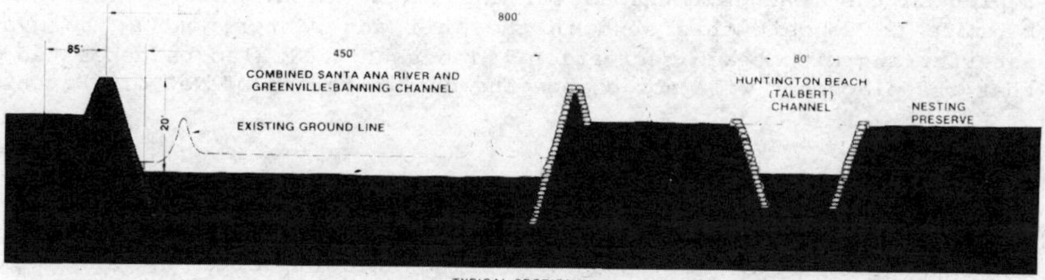

TYPICAL SECTION A
(River Mouth)

NOT TO SCALE

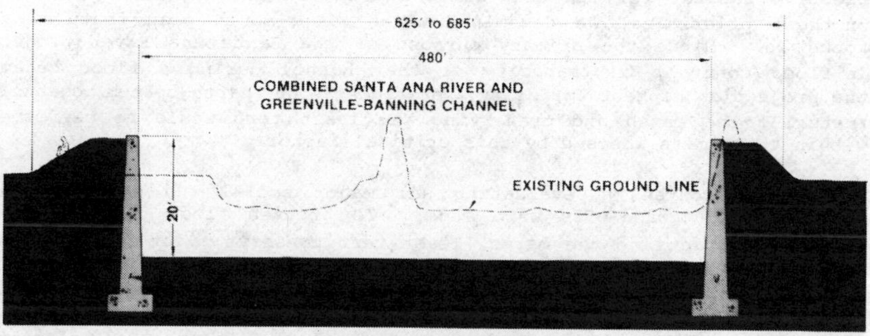

TYPICAL SECTION B
(Below Hamilton Ave. to River Mouth)

NOT TO SCALE

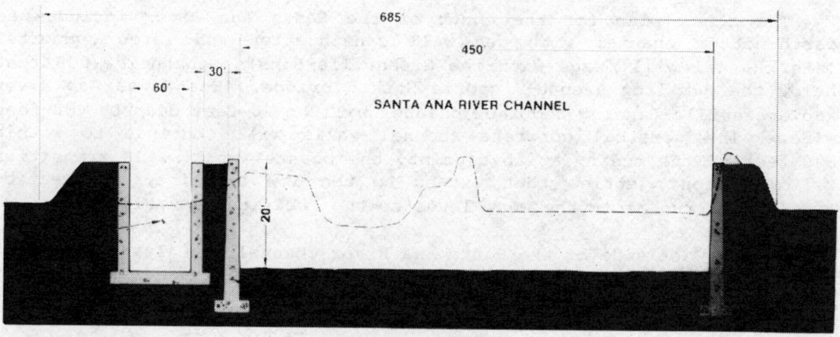

TYPICAL SECTION C
(Below Hamilton Ave.)

NOT TO SCALE

SANTA ANA RIVER MOUTH - PROPOSED PLAN
TYPICAL SECTIONS

Consequently, up to 3.7 million cubic yards of sand will be used to replenish the beaches downcoast of the river mouth. Every effort will be made to deposit this sand in the fall and winter months, to avoid interferring with beach recreation. The sand will also be deposited so that its placement will not change the configuration of Newport Beach.

CONCLUSION

Although the primary purpose of the Santa Ana River project is flood control, the river-mouth plan also included some features related to environmental and social issues. During the course of the study each issue was given a priority. The final priorities were (1) the need for an earth-bottom channel, (2) the need to preserve and restore the marshlands in the vicinity of the river mouth, and (3) the need to preserve habitat for the California least tern nesting colony. Based on these priorities, we detemined which river-mouth plan features to focus on. Since the primary purpose of the Santa Ana River project is flood control, the capacity of the channel during a flood became the project's highest priority, providing the earth-bottom channel, restoring the marsh and preserving the least tern would be performed within the limits imposed by this critical factor.

In the course of evaluating all major social and environmental concerns some trade-offs were made. To provide flood control, some temporary impacts to the marsh, least tern preserve, Victoria Pond and existing recreation trails were inevitable. However, every attempt to avoid and compensate for these impacts was made. Although the provision of an earth-bottom channel will increase the channel's area of tidal wetlands, temporary adverse impacts on other environmental elements such as the marsh, Victoria Pond, and the least tern nesting preserve would occur. In order to restore the marsh, some lands which otherwise might be available for marina development are proposed for acquisition.

The final plan for the mouth of the Santa Ana River includes an earth-bottom channel with tee wall construction and levee supports. This channel will merge with the Greenville-Banning channel 2,000 feet below the Hamilton Avenue-Victoria Street bridge. The Santa Ana River and Greenville-Banning combined channel will be 20 feet deep by 480 feet wide. The vertical concrete channel walls will continue to within 500 feet of the ocean; at that point, the channel walls will transition to rock stone jetties that extend to the low tideline. Recreation trails will run along the west levee to the Pacific Coast Highway.

On the west side of the Santa Ana River channel, the last 2,100 feet of the Talbert channel before the ocean will be realigned. Realignment of the lower portion of the Talbert channel will require moving one-third of the California least tern sanctuary, about 1-1/2 acres. Relocation of the least tern nesting site will commence before flood control construction and be moved gradually over a five-year period. The success of the relocation will be monitored annually.

Adjacent to the Santa Ana River's east side, 4.5 acres of the 13-acre Victoria Pond will be lost and a like amount of pond will be added to the preserve's east end before flood control construction begins. The restored pond will include a vegetative buffer zone and fencing. Further downstream, to mitigate for the loss of 8 acres of marsh from widening the Santa Ana River channel, 8 acres of salt marsh will be acquired. An additional 84 acres of marsh including adjacent waterways and uplands will be acquired for enhancement and preservation of an endangered species habitat. The marsh will be restored by increasing tidal flushing within the site; and by recontouring the site so that it provides more tidal channels and elevations suitable for a diverse range of target species vegetation. Like the proposed least tern nesting site extension and Victoria Pond, the marsh restoration will begin prior to initiation of flood control construction. It will be moved, in stages, over a five-year period. Monitoring the success of the restoration will continue through the entire Santa Ana River project's construction period (eleven years). Deepening and widening of about 10 miles the river channel will provide up to 3.7 million cubic yards of which could be used to replenish the beach downcoast of the river mouth. Table 2 (on the following page), Construction Sequence, presents a schematic display, in five stages, of how construction of all the project elements at the mouth will fit together. Every effort is made in this schedule for project construction to avoid adverse impacts to the plan's environmental and social provisions.

This proposed plan will provide the flood control critically needed in this area while incorporating some significant environmental and social concerns. It does not preclude marina development. Provisions in the plan offset adverse environmental impacts by enhancing the salt marsh habitat, and by replacing habitat loss at the California least tern nesting site and Victoria Pond. The wider earth-bottom channel (combined Santa Ana River and Greenville-Banning channel) will increase the area of tidal wetlands. Although project construction will cause temporary loss of the existing Santa Ana River recreation trails, these trails will be replaced and improved. Project construction will also result in providing up to 3.7 million cubic yards of sand that will help replenish the constant sand lost by the beaches downcoast of the river mouth.

The final design of the river channel represents a balance of engineering concerns with social and environmental concerns. The channel will carry the required 47,000 ft^3/s, will do so with a minimum of damage to the existing ecosystem, and will provide a number of social and environmental benefits to the area. Similar efforts to adjust engineering to meet the needs of river delta-estuarine regions will be necessary whenever such areas are targeted for major construction projects. The experience of this project shows that designs can be modified to minimize adverse impact and to provide long-term benefits--without sacrificing the engineering integrity of the basic design.

TABLE 2
CONSTRUCTION SEQUENCE

	STAGE 1	STAGE 2	STAGE 3	STAGE 4	STAGE 5
CHANNEL AT THE RIVER MOUTH	ACQUIRE LAND		CONSTRUCT		
LEAST TERN NESTING COLONY	ACQUIRE LAND	MOVE AND RESTORE	MOVE AND RESTORE	MONITOR	MONITOR
MARSH RESTORATION	ACQUIRE LAND	MOVE AND RESTORE	MOVE AND RESTORE	MONITOR	MONITOR
VICTORIA POND	ACQUIRE LAND	MOVE AND RESTORE	MONITOR	MONITOR	
RECREATION			TEMPORARY TRAIL	REPLACE TRAIL & LANDSCAPE	
BEACH REPLENISHMENT				NOURISH BEACHES	
OTHER PROJECT ELEMENTS		ACQUIRE LAND	ACQUIRE LAND & CONSTRUCT	ACQUIRE LAND & CONSTRUCT	CONSTRUCT

REFERENCES

1. Collins T. C., Bender K. E., and Dorothy D. R. "Report on the Feeding and Nesting Habits of the California Least Tern, Orange County, California", Contract No. DACW09-78-C-008, Work Order No. 3, U.S. Army Corps of Engineers, Los Angeles District, June 1979.

2. Hales L. Z. "Littoral Processes Study, Vicinity of Santa Ana River Mouth from Anaheim Bay to Newport Bay, California", Miscellaneous Paper HL-80-9, U.S. Army Engineer Waterways Experiment Station, CE, Vicksburg, Miss., June 1980.

3. Massey, B. "Avifauna, Lower Santa Ana River", Contract No. DACW09-78-C-008, Work Order No. 11, U.S. Army Corps of Engineers, Los Angeles District, June 1980.

4. Newling, C. J. "Feasibility Report on the Santa Ana River Marsh Restoration and Habitat Development Project", (unnumbered report), U.S. Army Engineer Waterways Experiment Station, CE, Vicksburg, Miss., September 1979.

5. U.S. Army Corps of Engineers. "Draft Phase I GDM on the Santa Ana River Main Stem including Santiago Creek and Oak Street Drain", Los Angeles District, July 1980.

FLOOD EMERGENCY EVACUATION PLANS

By Ronald E. Hilton,[1] M. ASCE

INTRODUCTION

The Corps of Engineers Flood Plain Management Services Program in the Jacksonville District includes conduct of Flood Emergency Evacuation Studies. In May 1979, the Lee County, Florida, Flood Emergency Evacuation Plan was completed by the Jacksonville District. In September 1979, a Flood Emergency Evacuation Plan for the Tampa Bay Region (Hillsborough, Manatee, Pasco, and Pinellas Counties, Florida) was initiated by the Jacksonville District and is scheduled for completion prior to the 1981 hurricane season. A Flood Emergency Evacuation Plan provides a comprehensive detailed plan for responding to flood threats from hurricanes on both county and municipal levels. The "plan" extends through the evacuation and shelter phase and can be used as a base for the development of post flood recovery plans by county and municipal officials.

Flood Emergency Evacuation Plans in Florida are prepared at the request of the Boards of County Commissioners with the endorsement of the State of Florida Department of Community Affairs, Bureau of Disaster Preparedness. The studies are funded under continuing authority provided by Section 206 of the 1960 Flood Control Act, as amended.

GENERAL

Coastal counties in Florida are composed almost entirely of low-lying and heavily populated areas. Many of these areas have only limited access routes to higher inland areas. Those routes that do exist are characterized by points of dangerously low elevations subject to early inundation from rising storm waters. In developing a decision making framework for the evacuation of such areas having these characteristics, the most important factor is time. The response to any major storm situation in Florida will require the safe evacuation and sheltering of thousands of persons, all within a constrained availability of time. Some parameters used to formulate flood emergency evacuation plans in Florida by the Jacksonville District, Corps of Engineers, are described in this paper.

[1] Chief, Flood Plain Management Services Branch, Engineering Division, Jacksonville District, Corps of Engineers.

HAZARD ANALYSIS

It is impossible to formulate evacuation plans for every type of situation caused by tropical storms. A plan must be based on probable storm-caused situations. It also must be geared to cope with "worst-case" hazard situations so that lesser hazard situation needs will be met by the recommended plan actions. For this reason, the locations and directions for the storms selected for simulation should be those which maximize the hazard created by the storm of given intensity. The hazard analysis studies in Florida for evacuation plans are formulated based on five distinct contingency levels of response and on three types of storm movement.

The five levels of response are correlated to the five categories of storm intensity on the Saffir/Simpson Hurricane Scale. The Saffir/Simpson Hurricane Scale is used by the National Weather Service to give public safety officials a continuing assessment of the potential for wind and storm-surge damage from a hurricane in progress. Scale numbers are made available to public safety officials when a hurricane is within 72 hours of landfall. Scale numbers range from 1 to 5. The Saffir/Simpson Hurricane Scale ranges for atmospheric pressure, winds, and surge heights at the coastline are listed in table 1. The Weather Service emphasizes that the scale numbers are not forecasts but are based on observed conditions at a given time in a hurricane's lifespan. They represent an estimate of what the storm would do to a coastal area if it were to strike without a change in size or strength.

TABLE 1. SAFFIR/SIMPSON HURRICANE SCALE RANGES

SCALE NUMBER (CATEGORY)	CENTRAL PRESSURE		WINDS (MPH)	SURGE (FEET)	DAMAGE
	MILLIBARS	INCHES			
1	>980	>28.94	74-95	4-5	Minimal
2	965-979	28.50-28.93	96-110	6-8	Moderate
3	945-964	27.89-28.49	111-130	9-12	Extensive
4	920-944	27.18-27.88	131-155	13-18	Extreme
5	<920	<27.17	>155	>18	Catastrophic

The three types of storm movement are landfalling, paralleling, and exiting. The storm surge modeling using the five levels of response and three types of storm movement is conducted by the National Oceanic and Atmospheric Administration (NOAA), National Hurricane Center (NHC), in Coral Gables, Florida. The NHC utilizes two computer programs to conduct the storm surge modeling: (1) SPLASH II (Special Program to List Amplitudes of Surges from Hurricanes) and (2) SPLOSH (Sea, Lake, and Overland Surges from

Hurricanes). The storm surge modeling results provide: (1) hourly values of storm positions, speed and direction of motion, pressure drop, and size of the storm, (2) height field for various hours of the storm, (3) time histories of surges at selected locations, (4) computed wind speeds at selected locations, (5) computed wind directions at selected locations, and (6) a surface envelope of maximum surge values. The storm surge modeling is conducted for various storm tracks for the five levels of response and three types of storm movement. The storm surge results are used to determine the evacuation zones, the time needed for evacuation, the location of shelters, and the determination of evacuation rates.

EVACUATION ZONES

In order to specify particular coastal and inland areas that may require evacuation under various situations, the study area must be delineated into evacuation zones pertinent to local conditions. Two alternatives are available for forming evacuation zones. One alternative is to base the formation of the zones on some previous partitioning of the study area, such as: jurisdictional boundaries, census tract boundaries, traffic analysis zones, and natural barriers (rivers, bridges, railroads, etc.). This alternative provides the advantage of ease in obtaining information. A disadvantage is that such zones may not reflect the areas needing evacuation due to flooding. The second alternative is to form the zones to correspond to the flood contours produced by the hazard analysis. The main disadvantage for this alternative is that evacuation zones may not coincide with existing data bases relating to population, traffic analysis zones, etc., and would require reformulating or directly preparing data for the new geographical units created. Therefore, the creation of the evacuation zones for each particular study area needs to be looked at very carefully.

POPULATION DATA

The methodology used to gather population and vehicle data for an evacuation plan is highly dependent on the existing data base available to the agency formulating the plan. U.S. Census or other estimates of only the permanent population are not appropriate for evacuation planning studies in Florida due to the lack of census data regarding transient population. Two important population data requirements for evacuation planning studies are: (1) the estimated number of residents/tourists within each evacuation zone, and (2) the estimated number of vehicles within each evacuation zone. Both of these requirements can be derived from an inventory of dwelling units and can be developed for five types of dwelling units within each zone: (1) single-family units, (2) mobile home units, (3) multi-family units, (4) hotel/motel units, and (5) travel trailer spaces.

The number of single-family units in each zone can be determined through aerial photograph "rooftop counts," local planning department surveys, local special censuses, and local water company data. The number of mobile home units by zone can be determined through aerial photograph "rooftop counts," local planning department surveys, and local special censuses. The number of multi-family units (apartments and condominiums) by zone can be determined through data furnished by the Division of Hotels/Restaurants - Bureau of Licensing, local planning department surveys, and local special censuses. The number of hotel/motel units by zone can be determined through data furnished by the Division of Hotels/Restaurants - Bureau of Licensing, local planning department surveys, and local special censuses. The number of travel trailer spaces by zone can be determined through aerial photograph "rooftop counts."

Because of the seasonal cycle in resident and tourist occupancy of dwelling units in Florida, it is imperative to select an appropriate functional population for accurate planning of evacuation procedures. The planning for evacuation should consider the highest level of population within the hurricane season. The generally accepted Atlantic hurricane season is June 1 through November 30. The population level should include permanent and seasonal residents and tourists. The number of occupied dwelling units for each zone can be obtained from an occuancy sampling survey, if the data is not readily available. The total population of each zone (total potential evacuees) can then be estimated by multiplying the most current persons/dwelling unit rates by the total occupied dwelling units in each zone.

The number of vehicles egressing from each evacuation zone is also needed for planning evacuation. This is best accomplished through reliable estimates of the number of vehicles per type of dwelling unit (i.e., 1.4 vehicles/single-family dwelling unit) in each evacuation zone. The information on vehicle unit rates in Florida is available from the Florida Department of Transportation. The rates can then be applied to the number of occupied dwelling units for each evacuation zone and results in a figure for the total number of vehicles in each evacuation zone.

BEHAVIORAL DATA SURVEY

A statistically significant behavioral data survey should be conducted in order to estimate the number of vehicles to be used for evacuation, the destination of the evacuees (shelter, friends or relatives house, travel out of study area, etc.), the evacuation route preferred by evacuees, and other related evacuation indicators in formulating an evacuation plan.

The survey should include all the census tracts in the study area. Households within the census tracts should then be randomly selected. Names, telephone numbers, and addresses of potential respondents can be obtained from location sections of local cross-reference directories. A method that has been successfully used to conduct the behavioral survey was a telephone interview. Before the survey, each potential respondent received a post card alerting them that a representative would be calling to ask a few questions on evacuation.

Following are sample questions that can be asked during a behavioral survey:

(1) Assuming everyone is home and were to be ordered by a governmental authority to evacuate, how soon could you be ready and would you leave?
- a. Immediately
- b. Certain number of hours
- c. Never

(2) Do you live in a:
- a. Condominium building
- b. Apartment
- c. Mobile home
- d. Single-family home
- e. Other

(3) How many vehicles are there in your household?

(4) How many vehicles would you use during an evacuation?

(5) Would you need transportation such as a bus or taxi?

(6) How many people live in your home including yourself?

(7) Is there anybody who could not be evacuated without help from outside your home?

(8) After leaving would you:
- a. Go to a Red Cross shelter?
- b. Go to the home of a friend or relative?
- c. Look for a hotel or motel room?
- d. Don't know where you would go?

(9) What route would you take to get to the shelter, friend, or relatives, or hotel/motel?

(10) How long have you lived in the study area?
- a. 0 to 10 years
- b. 11 to 30 years
- c. 31 to 60 years
- d. Over 60 years

(11) Have you ever lived in an area that experienced the direct hit of a major hurricane?

Results of the behavioral data survey can be used to determine the estimate of the total number of vehicles that would possibly be egressing from each evacuation zone. The survey can also be used to determine the total number of persons in each evacuation zone requiring shelter space.

SHELTER DATA

The responsibility for operation of shelters in Florida is assumed by the American Red Cross in coordination with county and municipal governments. Three categories of shelters are identified:

(1) Primary. Provides the highest degree of structural safety and resistance to flooding;

(2) Secondary. Not as structurally sound or flood resistant as primary shelters, but may be used if the intensity of the storm requires additional shelter space; and

(3) Refuge. Not recommended for shelter, but designated as a refuge in the event portions of the population cannot or do not evacuate in time to reach the primary or secondary shelters.

The number of shelters needed for evacuation is dependent on the shelter capacities. It is recommended that capacities range from 10 to 40 square feet per person. The 10 square feet per person leads to extremely cramped conditions and shelter operations at this capacity should only be used for the length of the storm and the initial recovery period. After persons are released to go back to their homes, the 40-square-foot capacity is recommended for caring for families whose homes have been damaged beyond occupancy. It is also recommended that first floor elevations of primary shelters be obtained to insure they are not located within a flood-prone area and that a structural analysis of the primary shelters be accomplished.

EVACUATION ROUTES

Evacuation routes are established based on their availability and proximity to shelters and the number of vehicles that would use the routes. In coastal areas of Florida the luxury of being able to pick and choose evacuation routes does not exist since there is generally only one route leaving a barrier island. The elevation of the roadway is an important factor in estimating when storm caused hazards will interrupt evacuation routes. Many roads connecting islands to the mainland and other coastal roadways have very low elevations. Experience from past storm occurrence in Florida has shown that low-lying areas and evacuation routes will become inundated before actual landfall or closest approach of the storm. The critical question in the evacuation of residents on coastal islands is when will the evacuation routes become inundated and therefore impassable. For this reason, knowledge of the exact elevation of critical evacuation routes is necessary in determining those zones whose access areas would be cut off first. Evacuation from the zones having very low points along their evacuation routes must be completed before those low points are inundated. The Coastal Hazard Mitigation Office, NOAA, National Ocean Survey, has completed storm evacuation map coverage for many coastal areas of the United States. Road elevations on those maps were obtained by spirit leveling with particular attention being given to low points along roads. The storm evacuation maps are available from NOAA.

Another factor in estimating when storm caused hazards will interrupt evacuation routes is the presence of gale force winds (winds from 38 to 55 miles per hour). Especially hazardous on evacuation routes during the presence of gale force winds are travel trailers, campers, and other high-standing vehicles. The high winds might overturn such vehicles.

EVACUATION RATES

The largest determinant of how long evacuation from critical areas would take is the evacuation rate attainable on an evacuation route. This rate is dependent upon the characteristics of the roadway in question. Determining the number of vehicles expected to use specific evacuation routes is basic to estimating the amount of lead time necessary for an evacuation order before the arrival of the storm. The evacuation rate is measured as the number of vehicles per lane per hour that can pass over a given section or roadway under prevailing roadway, traffic, and ambient conditions. Roadway conditions include the number of available lanes, width of lanes, roadway alinement, influence of intersections, and lateral clearance. Traffic conditions include the presence of trailers, trucks, and buses in the traffic flow and the overall ratio of service volume to capacity. Ambient conditions include rain, wet pavements, wind, darkness, and fallen tree or powerline interruptions. The evacuation rates for the Lee County study were determined utilizing the 1965 Highway Capacity Manual, Special Report 87, Highway Research Board, National Academy of Sciences. Following are the assumptions made for determining evacuation rates for the Lee County study:

(1) At least one traffic lane of each evacuation route would be restricted for <u>emergency</u> <u>vehicle</u> movement only.

(2) All evacuation traffic would be one-way.

(3) At some point in time or location, all evacuation traffic lanes would be functioning at maximum traffic-carrying capacity under prevailing conditions (Service Level F - unstable flow).

(4) On the average, adequate passing site distance (greater than 1,500 feet) is restricted to approximately 20 percent of the time vehicles are evacuating. Wind, rain, amount of daylight, traffic density in a single lane, fallen trees, and powerlines all contribute to traffic flow interruptions and minimal visibility.

(5) Vehicle operating speeds would range between 25 and 45 miles per hour resulting in average operating conditions approximating Service Level D for most of the time. Approximate operating speed = 35 mph.

(6) Intense storm conditions eliminate the effect of posted highway speed limits in computing traffic volume. Therefore, the working value for restricted average highway speed may approximate 40 mph.

(7) Lateral clearance will, at various times, approximate 0 feet due to roadside shoulder saturation and flooding swales.

(8) Local government should adopt an emergency evacuation ordinance restricting the use of trucks, buses, trailers, and other high-profile, wind-resistant vehicles on areawide roadway facilities during storm evacuation periods. As a result, traffic-carrying capacity calculations will incorporate a minimal 3 percent truck correction factor (3 percent of all evacuation vehicles will be composed of trucks, buses, boats, campers, etc.).

(9) Intersections of evacuation routes would be highly controlled to minimize traffic flow interruptions.

Traffic movements or evacuation rates were calculated for the identified evacuation routes based on the nine assumptions identified, as well as the information obtained from the roadway, traffic, and ambient conditions. The evacuation rates for the Lee County study varied from 284 to 810 vehicles/lane/hour.

The Tampa Bay Region study involves a highly complex roadway system and will require the use of a computerized transportation model to determine the evacuation rates. The U.S. Department of Transportation has developed a computer model package, "Urban Transportation Planning (UTPS)." UTPS is a package of computer programs, attendant documentation, users guides, and manuals providing state-of-the-art methods for multimodel urban transportation planning. Several models within the overall UTPS package could be utilized to simulate evacuation vehicle movment; however, preliminary investigations point to the "UROAD" model as being capable of determining the evacuation rates and time needed for evacuation tasks in an evacuation study. The UTPS computer model package, or a similar transportation analysis and planning computer model program, will be used in the Tampa Bay Region study.

The objective of the transportation modeling task is to produce the following hurricane evacuation plan elements: (1) the assignment of vehicle volumes from evacuation zones to "minimum path" evacuation routes and shelter destinations, and (2) the calculation of time (in hours) necessary for the evacuating vehicles from each evacuation zone to reach the assigned shelter destinations. The assumptions made for determining evacuation rates in Lee County will also be used in the Tampa Bay Region study. Also, reductions in highway capacities will be based on roadway, traffic, and ambient conditions.

TIME NEEDED FOR EVACUATION

The time needed for evacuation is a product of the vehicles/hour that can be moved over each segment of the evacuation route and the distance persons must be evacuated. The time needed for evacuation can be conceptualized into two separate portions of time termed the "time to pass" and the "time to travel." These two portions of time are calculated and then added together to form the time needed for evacuation. To determine the "time to pass," the evacuation rate assigned to each segment of roadway must be divided into the estimated number of vehicles that must pass through that segment on route to the assigned shelter. This gives a time in hours and minutes that it takes all vehicles to pass through the segment being analyzed. The most restricted segment (that segment with the lowest evacuation rate) that the vehicles must pass through on route to the shelter defines the time it takes for all vehicles to pass. The "time to pass" generally constitutes the greatest portion of the overall time needed to evacuate to the shelter. This time, however, must be supplemented by the time it takes a vehicle to travel to the shelter ("time to travel") from a designated starting point. The "time to travel" estimate should be based upon the point (by roadway) farthest from the shelter in a set of evacuation zones. Using this farthest point, "time to travel" is then added to the "time to pass" to give the resultant total time needed to evacuate. The Lee County study determined that the barrier islands would require total evacuation time ranging from 15 hours for a Category 1 hurricane to 20 hours for a Category 5 hurricane. These times are excessive because the road elevations are low and the evacuation routes can be cut off with a minimal surge of 3 to 4 feet.

TRAFFIC CONTROL/SECURITY

Heavy traffic flow and the adverse road conditions of a storm situation will require unusual traffic control throughout much of the study area. In addition to traffic control, actions may be necessary to order high-standing vehicles removed from traffic and to re-route traffic due to adverse conditions. Also, one lane of all evacuation routes is left open and reserved for emergency vehicles. Traffic control and security (anti-looting, etc.) are the responsibility of the local law enforcement agencies. Critical traffic control, security, or blockade points must be identified in the evacuation study and coordinated with the local law enforcement agencies. During an evacuation, those traffic control, security, and blockade points must be fully manned for the proper movement of vehicles and for the security of the evacuated area.

SPECIAL EVACUATION PROVISIONS FOR ELDERLY/HANDICAPPED

Generally, several public and private institutions within a study area require special evacuation procedures and assistance in transporting patients or residents out of endangered areas. The institutions should be identified in the evacuation study and provisions made to provide safe shelter during a storm situation. The movement of persons in these institutions may require the use of buses or other high-standing vehicles. These persons should be evacuated when winds are less than 40 mph, if possible, prior to the evacuation of the general population.

EMERGENCY OPERATIONS CENTER

In Florida, The Emergency Operations Center (EOC) is the center for disaster preparedness in a county. The EOC is designed for full living accommodations for executive operations, communications, and support personnel over extended periods of time during disaster situations. The EOC should possess independent power and water supplies and should withstand the forces of any foreseen natural disaster. All decisions regarding the response to a countywide natural disaster situation will be made and disseminated from this operations center.

PUBLIC INFORMATION PROGRAM

The overall response to potential hazards, including a safe and orderly evacuation, requires that the public be informed of what they would need to do at all times, not just during the approach of a storm. All available channels must be utilized for dissemination of general educational and training information. The most effective approach will probably be through the use of printed material supplemented by public meetings, seminars, and presentations to various groups. Local newspapers can also provide an effective means of disseminating information through news articles, feature stories, and public notices. Local officials should encourage newspapers to carry these articles prior to the beginning of each hurricane season. The area's television and radio media should also be active in informing the public of the general precautions that are necessary in the event of storms. This is a valuable service of providing the background information necessary for an informed public.

The Lee County study included pamphlets that were distributed to residents and tourists to inform them of needed actions in the event of a hurricane. The pamphlets included general information on how to prepare for a hurricane, courses of action to take in the event evacuation was ordered, and addresses of shelter locations. The pamphlets also included a hurricane tracking map, a map showing evacuation routes and shelter locations, and zone specific evacuation instructions describing the evacuation route to the shelters or leaving the county.

HURRICANE RECOGNITION

The National Hurricane Center (NHC) in Miami, Florida, is the official "hurricane warning" agency. The NHC provides timely information on the progress, direction, and intensity of storms. This information is in the form of advisories every 6 hours during the approach of a storm and supplemented by more frequent bulletins as a hurricane watch is issued. The "watch" period addresses a wide range of coast that an existing hurricane some distance offshore may hit. As particular areas for a hurricane making landfall become more firmly identified, a hurricane "warning" may be issued. The "warning" period is expected to begin a minimum of 12 hours before hurricane landfall. This is the time period which is recommended as a minimum to begin mass evacuation of areas to be flooded. Evacuation of some low-lying areas may have to begin during the "watch" period.

In Florida, the Division of Public Safety Planning and Assistance, Bureau of Disaster Preparedness, is responsible for coordinating warning and response efforts of, and among, local (county and municipal) disaster preparedness organizations. This includes the recognition and communication of impending regional and local hazards.

COORDINATION

No flood emergency evacuation plan, however well devised, can be successful unless the participants understand the reason for the plan and the role they must play in its proper execution. The participants include local public officials, business leaders, and private citizens.

During the course of an evacuation study, coordination must be maintained with various Federal, State, and local governmental agencies. The coordination can be in the form of letters and meetings.

ACKNOWLEDGMENTS

The writer wishes to note that the Southwest Florida Regional Planning Council, Ft. Myers, Florida, compiled data and prepared the Technical Data Report for the Lee County, Florida, Flood Emergency Evacuation Plan under contract and guidance of the Jacksonville District, Corps of Engineers. Much of the material contained herein has been drawn from that report. The Tampa Bay Regional Planning Council, St. Petersburg, Florida, is currently under contract with the Jacksonville District, Corps of Engineers, to compile data and prepare a Technical Data Report for the Tampa Bay Region, Florida, Flood Emergency Evacuation Plan. Some of the material contained herein has been drawn from their work to date on the contract.

REFERENCES

1. U.S. Army Corps of Engineers, Jacksonville District, Lee County, Florida Flood Emergency Evacuation Plan, May 1979.

2. U.S. Army Corps of Engineers, Jacksonville District, Tampa Bay Region, Florida (Hillsborough, Manatee, Pasco, and Pinellas Counties, Flood Emergency Evacuation, Plan of Study, September 1979.

3. Jelesnanski, Chester P., SPLASH (Special Program to List Amplitudes of Surges from Hurricanes), Part 2, General Track and Varient Storm Conditions, NOAA Technical Memorandum NWS TDL-52, Silver Springs, Maryland, March 1974.

4. Jelesnanski, Chester P. and Chen Jye, SLOSH (Sea, Lake, and Overland Surges from Hurricanes, National Oceanic and Atmospheric Administration, National Weather Service, Silver Springs, Maryland, unpublished.

5. Southwest Florida Regional Planning Council, Lee County dwelling unit occupancy survey, Ft. Myers, Florida, June 1978.

6. Highway Capacity Manual, 1965. National Academy of Sciences, Highway Research Board, Special Report 87, Washington, D.C., 1966.

7. U.S. Department of Transportation, Urban Mass Transportation Administration, Planning Methodology and Technical Support Division and the Federal Highway Administration, Urban Planning Division, Washington, D.C., Urban Transportation Planning System (UTPS), January, 1976.

FLOODPLAIN MANAGEMENT: NEW FEDERAL GROUNDRULES

by

Timothy E. Maywalt*

Introduction

Last summer, Mr. Frank Powledge, a free lance writer and longtime friend of the southern seacoast, published an article titled "Gone With the Wind" in the Washington Post Magazine. His thesis, summarized in the article's subtitle, was that when the next big hurricane came along, we could kiss the southern seacoast goodbye. Within a few weeks of the article's publication, hurricanes David and Frederick were taking their best shots at fulfilling his prediction.

His article very concisely stated the nature and origins of the expanding flood risk and environmental degradation within coastal areas and especially on barrier islands. He was right on point as well in identifying the Federal contribution to these problems. Perhaps the only significant oversight in his very timely and comprehensive article was the omission of reference to a major tool for getting control of these problems--President Carter's 1977 Executive Order 11988, entitled Floodplain Management.

This Executive Order and its companion document, Executive Order 11990, entitled Protection of Wetlands, require a thorough evaluation of any Federal actions proposed in or near floodplains and wetlands. Within the next few months--three and a half years after the Orders were signed--the last of the specific implementation procedures required of all Federal agencies and programs are expected to become effective.

What is the significance of these Executive Orders to those interested in ensuring sound coastal management? It means that we already have in place the evaluation process necessary to control the Federal contribution to increasing flood losses and environmental disruption resulting from coastal and inland floodplain and wetland development.

The Evaluation Process

The evaluation process required by the Executive Orders is described in detail in the Water Resources Council's "Floodplain Management Guidelines" (43 FR 6030). It calls for Federal agency staff to determine first if a proposed project (either a totally Federal project or one supported by Federal grants, loans, licenses or permits) would be located in a floodplain or wetlands area; or if it is to be located near one, whether the project could potentially affect either kind of

*Acting Director, Policy Analysis Division, U.S. Water Resources Council, Washington, DC.

area. Next, alternatives to floodplain and wetlands locations must be evaluated; if no alternative sites are found, the potential effects of the project on the floodplain or wetlands must be evaluated and steps identified to mitigate them. The Orders also require that the public be involved early in the evaluation process, and throughout its duration.

Until the 1960's, federally-supported unsound floodplain and wetland development was carried on without a systematic evaluation of its effects. The result of these unsound actions was the building of flood control works; flood disaster assistance outlays became the primary response to ensuing losses. More recently, State efforts and local action under the National Flood Insurance Program are based on the understanding that if floodplain development is to occur, some constraints are necessary to reduce potential losses. Protection of the environmental values of floodplains and wetlands has just begun to be integrated into floodplain management efforts.

The Executive Orders build upon these recent efforts. Neither Order allows a Federal agency to assume that floodplain or wetland development is acceptable in all cases as long as steps are taken to reduce losses. The Orders go beyond simple mitigation by mandating an evaluation of non-floodplain and non-wetland alternatives to any proposed development. The agencies are required to evaluate what the Orders term "practicable" alternatives. These are alternatives that are useable in light of such existing constraints as those posed by the natural environment (topography, habitat, hazards), social concerns (existing development patterns, aesthetics), economic aspects (costs of space, services, construction), and legal constraints (deeds, leases). It should be realized that on its own, no single constraint should be viewed as determinative in deciding whether a given alternative is practicable. For example, non-floodplain and non-wetland alternatives are not to be dismissed as impracticable solely on the basis of increased costs or the need to develop a unique design to adjust to non-floodplain or non-wetland topography. The agencies' evaluation must consider three options: use of alternative sites outside the floodplain or wetland; alternative actions that serve the same purpose but have less potential for adverse impacts; and the option of no action at all.

This "test or practicability" permits balancing the need to locate a proposed action within a floodplain or wetland against the need to minimize flood risk and environmental values disruption. What this test asks, essentially, is what there is in the nature of any proposal that requires a floodplain or wetland location. Although a case can often be made for locating water-dependent uses (docks, loading and unloading facilities within ports, highway or utility crossings) within a floodplain, the need to locate a residence, hospital, post office or port-related warehouses there is considerably less clear. Even in the case of water-dependent uses, for which non-floodplain sites may never be "practicable" alternatives, evaluation of the alternative action and no-action options is essential. The potential loss of the facility itself, or its potential adverse effects on the flood risk and natural values, may outweigh the need to construct the facility.

In situations where the agency cannot identify a practicable alternative to locating in a floodplain or wetland, the Orders' provisions for mitigating impacts come into play.

Impacts are to be mitigated by:

- designing or modifying the project to minimize harm to or within the floodplain;
- minimizing the destruction, loss or degradation of wetlands;
- restoring and preserving natural and beneficial floodplain values; and
- preserving and enhancing natural and beneficial wetland values.

The requirement to minimize harm is more comprehensive and rigorous than the provisions of the National Flood Insurance Program; it requires reducing flood loss exposure and the potential for environmental disruption to the smallest possible degree. Restoration of disrupted floodplain values involves returning the site of development to a condition in which the natural values or functions of the floodplain can again operate. Preservation involves preventing changes in natural conditions and maintaining the values and functions of floodplains and wetlands in their natural state. Enhancement involves increasing, heightening or improving the natural and beneficial values associated with wetlands.

The "natural values" of floodplains and wetlands are often critically important in protecting life and property against flooding. Naturally vegetated floodplains can provide a broad area to spread and slow flood waters, thereby reducing the velocity of flood waters and flood peaks. Stream meander, dune formation in coastal areas and other natural processes can reduce the force of flood waters. In coastal floodplains, sand dunes, mangrove stands, and other natural barriers can reduce the impact of waves. Development in or adjacent to floodplains and wetlands can reduce or eliminate these functions or values with a corresponding increase in the loss potential.

Implementation Roles

Each of the conference participants has one or more important roles in ensuring adequate and consistent implementation of these Orders. Most importantly, we each have an educational role to perform. In some respects, progress in implementing the Orders has mirrored that of the National Environmental Policy Act. Progress has been slow. Approaching four years after their signing, some floodplain and wetland actions are still being performed without going through the required evaluation. Renewed effort is necessary on the part of implementing staff and the affected public to make the Orders' provisions known and understood.

The representatives of consulting firms among you may find yourselves called upon by Federal agencies complying with the Orders or State and local communities affected by them. You may be asked to assist in determining whether a proposed site for development is in a floodplain, and if so, to define the degree and characteristics of the risk. You may be asked to aid in identifying alternative sites for the development that pose less risk. You may be faced with a similar analysis for wetlands, or you may need to develop a design that can withstand flooding while not aggravating the existing risk or disrupting the natural values of the floodplain or wetland.

Those of you representing the academic community may be asked to contribute in similar site specific applications. More likely, however, you can expect a role in meeting the research needs that will grow out of the Orders' implementation. In almost all respects, we know more about protection against flooding than we do about protecting the natural values of floodplains. The identification of what constitutes a wetland, the determination of the effects of development upon them, and the mitigation of these impacts all need further investigation. Practical solutions to these and related problems need to be identified for use by Federal agency field staff responsible for implementing the Orders.

Federal agency representatives have the responsibility to ensure that effective implementation of the required evaluation process is performed for all affected actions. We need to be more than simply aware of what's required and prepared to respond, however. We need to be able to evaluate the effectiveness of implementation, to identify potential impediments and to adapt successful applications to our own activities.

State and local government staff should recognize that part of their battle is over. A system is now in place at the Federal level that has significant potential to support State and local floodplain management and wetlands protection efforts. On the other hand, some State and locally supported actions that have previously been assisted by the Federal agencies will no longer be supportable. State and local development proposals involving Federal funding or approval will need careful analysis at their inception to ensure consistency with agency implementation of the Orders.

Conclusion

In summary, the point needing greatest emphasis is that new Federal groundrules to ensure improved floodplain, wetlands and coastal management exist. The focus of your attention should be twofold. First, the rules must be understood and used. Second, their use must be integrated, along with a range of related tools, into a combined coastal management approach. It's either that; or, as Mr. Powledge warns: "kiss the seacoast goodbye."

CURRENT PRACTICE IN MANAGING COASTAL FLOOD HAZARDS*

Steven P. French
Senior Staff Associate
Rice Center for Community Design and Research
Houston, Texas 77046

and

Raymond J. Burby
Assistant Director for Research
Center for Urban and Regional Studies
University of North Carolina at Chapel Hill
Chapel Hill, North Carolina 27514

Coastal flooding is of major concern along the Atlantic and Gulf coasts of the United States. The damage potential from wind and wave action associated with hurricanes in these areas has been magnified in recent years by the continuing migration to the nation's coasts and by rapid construction in coastal flood-hazard areas. In order to decrease the potential flood risk, many governments have adopted some form of land use management as a means of guiding development away from hazard areas or insuring that development is constructed so as to withstand some specified level of flooding.

The National Flood Insurance Program and various state coastal management programs have played key roles in encouraging coastal jurisdictions to manage their flood-hazard areas. Although severe sanctions were required to secure widespread participation in the flood insurance program by riverine communities, in coastal areas most communities had joined the program prior to the initiation of sanctions in 1974. Apparently the threat of property damage from hurricanes and coastal storms alone was great enough to induce communities to enact the land use management measures required to qualify for federal flood insurance. In addition, participation in the NFIP has given coastal communities access to Flood Insurance Rate Maps (FIRMs) and other information needed to undertake an effective land use management effort.

Although coastal communities have adopted the minimum land use management measures needed for participation in the National Flood Insurance Program, several authors have raised serious questions about the effectiveness of this approach to the coastal flood hazard. For example, Sheaffer and Roland (1980) have noted that with the current federally mandated land use management measures in

---

* The research reported in this paper was supported, in part, by funds from the National Science Foundation under Grant No. DAR78-07603. Of course, any opinions, findings, and conclusions or recommendations expressed here are those of the authors and do not necessarily reflect the views of the National Science Foundation.

place, national flood losses will continue to rise, in part, because the measures do not deal with existing development in hazardous locations. Baker and McPhee (1975) have suggested that the intense economic pressure to develop on the seashore often precludes the possibility of effective land use management. In fact, an empirical study by Cheatham (1979) found that total annual construction along the Mississippi coast had not decreased significantly as a result of the land use management requirements of the NFIP.

In response to these questions, this paper seeks to (1) identify land use management methods being used in coastal flood hazard management, (2) isolate key factors associated with the pursuit of broader, more stringent management programs, (3) assess the effectiveness of the management programs that are currently in place, and (4) identify community characteristics and other factors which are associated with more effective coastal land use management programs. These analyses provide a basis for suggestions to improve federal, state, and local flood plain land use management policies.

The Data

The data reported here are based on information obtained from 178 local governments with jurisdiction over coastal hazard areas. These communities are a subset of respondents to a national survey of 1,203 communities participating in the National Flood Insurance Program. The sample included 100 percent of the NFIP "regular" program communities and a 20 percent random sample of "emergency" program communities with a 1970 population of 5,000 or more and which processed at least one hazard area building permit in 1978. Data from the sample communities were obtained from a mail survey conducted during the spring of 1979. All letters and questionnaires were sent to the person listed in the files of the Federal Insurance Administration as the local flood plain coordinator, or if none was listed, the jurisdiction's planning director or chief executive officer. The survey was conducted using techniques developed by Dillman (1978). The overall response rate was 85 percent. Data from the mail survey were supplemented by additional information from secondary sources, including the Federal Insurance Administration. (Detailed results of the entire survey are reported in Burby, French, Kaiser, and Moreau, 1980, and French, 1980.)

The average population of the sample of coastal cities, towns and counties is 56,949. The median population is 20,364. A majority (73%) of the communities considered flooding a problem, but not very serious. As a result, most (88%) placed low priority on the solution of flood problems among the whole array of community needs. Although coastal communities do not often rate flooding a serious problem, the potential for flood damages continues to mount. As shown by the data summarized in Table 1, coastal invasion of hazardous areas is strikingly larger than the corresponding invasion of riverine flood plains in inland areas.

TABLE 1. COMPARISON OF HAZARD-AREA CONSTRUCTION IN COASTAL AND NON-COASTAL COMMUNITIES

	(Average Values)	
	Number of Flood Plain Permits	Percent of New Development in Flood Plain
Coastal Communities	181.5 N = 150	22.4 N = 139
Non-coastal Communities	24.0 N = 891	4.9 N = 804

The State of Practice

Local land use management programs were measured in terms of three component parts: goals, methods and implementation. The current status of coastal flood plain land use management in terms of these three components of local programs is summarized in Table 2.

Although communities reported an average of four goals for their flood-hazard areas, only two specific goals--reducing property loss from flooding and protecting the safety of the population--had been adopted by a majority. The goals communities have adopted provide a reading on the results of a number of federal programs that are designed, in part, to affect local land use policies. For example, the data in Table 2 show that the National Flood Insurance Program is having a clearly discernible effect at the local level. The property loss mitigation goals of the NFIP are evident in the goals most frequently adopted and in the sharp differences between regular and emergency phase communities in the adoption of property loss mitigation and safety goals. Other federal programs have had less effect. Environmental goals were adopted by a strikingly lower proportion of the communities. It is also surprising that there were no significant differences in goals being pursued by coastal and riverine communities.

Communities can employ a number of different types of methods to achieve their objectives for flood-hazard areas. These methods include police power regulations, such as zoning and floodproofing requirements, planning, public land acquisition and relocation in the hazard zone to remove existing development and prevent further private development, public facilities location designed to attract new development to locate outside of flood-hazard areas (or at least not induce new development in hazard zones), incentives to maintain low-density uses in the hazard zone, and the provision of public information about the hazard. The use of each of these methods is summarized in Table 2.

Regulations were clearly the most frequently used land use management methods--particularly elevation requirements, subdivision regulations, zoning, and floodproofing requirements--all of

TABLE 2. ADOPTION AND IMPLEMENTATION OF FLOOD PLAIN LAND USE MANAGEMENT (percent of jurisdictions)

	National Flood Insurance Program Status	
	Regular Program (N = 154)	Emergency Program (N = 24)
<u>Community Goals for Hazard Area</u>		
Reduce property loss from flooding	75%	56%
Protect safety of population	74	52
Reduce flood damage to public property	46	39
Preserve natural areas	42	35
Minimize the fiscal impacts of floods	41	35
Preserve open space and recreational areas	38	39
<u>Management Methods Adopted</u>		
Police Power Regulations:		
Elevation requirements	92	87
Subdivision regulations	75	83
Zoning	67	75
Floodproofing requirements	68	54
Septic tank permits	45	58
Wetlands protection regulations	43	37
Sand dune regulations	31	13
Critical areas designation	27	17
Sedimentation and erosion control	25	33
Planning:		
Comprehensive planning	62	58
Land Acquisition:		
Land acquisition for parks, open sapce	30	33
Relocation of existing uses	4	13
Public Facilities Placement:		
Location of public facilities outside of hazard areas	18	17
Incentives/Disincentives:		
Preferential taxation	5	8
Density transfer	4	13
Public Information:		
Public information about hazards	36	31
<u>Implementation</u>		
Expenditures in Last Fiscal Year:		
Less than $1,000	42	54
$1,000 - $4,999	30	33
$5,000 or more	28	13
Staff Time Devoted to Flood Plain Land Use Management per Week:		
Less than 1 person hour	38	61
1 - 7 person hours	45	26
8 person hours or more	17	13

which were being used by a majority of the communities in both the NFIP regular and emergency phases. These regulations, aimed primarily at reducing potential property damage, are promoted by the National Flood Insurance Program. Regulations not promoted by the NFIP, such as those designed to reduce environmental damage (wetlands protection, sand dune regulations, critical areas designation, sedimentation and erosion control), were being used by a much smaller proportion of these coastal communities.

A majority of the communities in both the regular and emergency phases of the NFIP were using comprehensive planning in their land use management programs. Other land use management methods, however, were being used by far fewer communities. Only about a third were using land acquisition and public information and less than one in five coastal communities was employing public facilities placement and locational incentives in its program.

Although coastal communities were likely to have adopted goals for the future use of their hazard areas and to be using an array of management methods, in most cases they devoted few resources to program implementation. Overall, 42% of the NFIP regular program communities spent less than $1,000 during the last fiscal year on hazard-area management; only 28% spent $5,000 or more. Similarly, less than one in five communities was devoting as much as one person-day a week to hazard-zone management.

Program Scope and Stringency

Analyses were conducted to explain why communities varied in their interest in and commitment to hazard-area management. These analyses focused on two measures of local commitment: program scope and program stringency. Program scope measures the range of planning and management instruments employed by a community. The assumption is that the more methods a community uses to achieve its hazard area objectives the more likely its program will be effective. Program stringency measures the use of succeedingly more difficult to employ methods that are, in turn, succeedingly more likely to result in the mitigation of flood losses and achievement of other objectives for hazard areas. Variation in program scope and stringency is illustrated in Table 3.

Although the average coastal community has adopted six land use management measures, a sizable proportion of the communities had much more robust programs consisting of nine or more measures, while some communities were barely complying with the minimum requirements of the National Flood Insurance Program and had adopted as few as one or two different land use management measures. Similar variation is apparent in the stringency of the program adopted. Less than a quarter of the coastal communities had very stringent programs in place that included subdivision or zoning regulations, elevation requirements, and land acquisition or relocation. Half of the NFIP regular program communities had gone as far as enacting special elevation requirements, while about a quarter of the regu-

TABLE 3. HAZARD AREA MANAGEMENT PROGRAM SCOPE AND STRINGENCY (percent of jurisdiction)

	National Flood Insurance Program Status	
	Regular Program (N = 154)	Emergency Program (N = 24)
Program Scope		
Number of Measures Adopted:[a]		
0 - 2	7%	4%
3 - 5	32	42
6 - 8	35	21
9 or more	26	33
Total	100	100
Program Stringency		
Type of Goals and Measures Adopted:[b]		
0 - Minimal program[c]	5%	4%
1 - Subdivision or zoning regulations	24	42
2 - Elevation requirements	51	37
3 - Land acquisition	19	17
4 - Relocation of development	1	0
Total	100	100

[a]This variable consists of the number of measures listed in Table 2 the community has adopted.

[b]This variable is a Gutman scale (coefficient of reproducibility = .95; coefficient of scalability = .80) formed by combining community goals and management measures adopted. The scale is cumulative so that communities with more stringent program elements also adopted the less stringent items.

[c]Community has not adopted subdivision or zoning regulations, elevation requirements, land acquisition, or relocation of development in its hazard area management program.

lar program communities had adopted programs that were limited to the use of traditional subdivision or zoning regulations. Very few coastal communities had adopted only "minimal" programs.

Factors Affecting Program Scope and Stringency

Based on the work of Hutton and Mileti (1979), community adoption of hazard-area management measures was viewed as a function of four sets of factors: (1) objective characteristics of the hazard zone that contribute to flood risk; (2) community perception of flood risk and other hazard-area problems; (3) community concern for solving hazard-area problems; and (4) objective characteristics of a community that affect its ability to respond to problems in an effective manner. Each of these sets of factors is discussed in turn as it has influenced the scope and stringency of coastal communities' hazard-area management programs.

As shown in Table 4, flood risk has a direct, positive relationship with coastal communities' pursuit of land use management. The broadest, most stringent land use management programs in coastal communities are found in those communities with the largest proportion of the community in the hazard zone, with the greatest intensity of development in the hazard zone, and with the greatest absolute amount of hazard zone development. In contrast, in riverine communities interest in land use management tailed off with high levels of flood plain development. One reason for coastal communities continued interest in land use management as a hazard-mitigation adjustment, even after high levels of hazard-area development, is that structural solutions are not as viable an alternative. While riverine communities can turn to channel improvements, dams, levees, flood walls and similar measures to mitigate the hazard, structural measures, such as sea walls, have not proven to be particularly effective in preventing losses from hurricanes and major coastal storms.

Coastal communities that were experiencing various problems in their hazard areas were more likely to have adopted broader, more stringent programs. Also, communities that were more concerned with flood problems, that had adopted goals for their hazard areas and that assigned a high priority to to the solution of hazard-area problems, were more likely than others to have adopted a broader range of management measures and to have moved as far as the adoption of elevation requirements in terms of program stringency.

Coastal communities with more resources to devote to land use management (higher governmental revenues and higher property values) and with greater previous land use management experience tended to have the most developed programs. Also like riverine communities, urbanization indicators (population size and recent, 1975-1977, growth) were positively associated with the strength of the flood plain land use management program.

One final difference between factors affecting riverine and coastal communities should be noted. In riverine communities, the availability of sites for development outside of the hazard area was associated with the use of land use management measures in mitigating the flood hazard. When alternative sites were limited or not available, riverine communities looked for other types of solutions to their flood problems. In coastal communities, there was only a slight tendency in this direction. The most likely reason for this difference is again the viability of alternatives to land use management. As noted above, riverine communities can turn to structural solutions when nonhazard sites for development are limited. Coastal communities have little choice but to use land use management measures to deal with their hazard-area problems.

TABLE 4. FACTORS ASSOCIATED WITH PROGRAM SCOPE AND STRINGENCY

	Gamma: A Measure of Association[a]	
	Program Scope[b]	Program Stringency[b]
Hazard-zone Characteristics		
Percent of community in hazard zone	.23*	.17
Number of dwelling units located in hazard zone	.18	.20
Percent of hazard zone developed in urban use	.15	.14
Hazard-zone land uses:		
Multi-family	.43**	.39*
Stores or offices	.37**	.35
Industry or warehouses	.18	.18
Public ownership of hazard-zone property	.35***	.17
Perceived Hazard-zone Problems		
Encroachment on natural areas	.46**	.41*
Erosion	.43**	.15
Buildings in disrepair	.34	.18
Flood damage to public facilities	.26*	.21
Disruption of economic activity due to flooding	.24	.35*
Property loss from flooding	.21	.42**
Concern for Problems		
Adoption of goals for hazard zone	.42*	.90***
Priority assigned to solution of hazard-zone problems	.37**	.34
Community Characteristics		
Type of jurisdiction	-.19	-.10
Fiscal resources:		
Total revenue	.35*	.29
Property tax revenue	.33	.37
Extent of land use management experience prior to joining NFIP	.49***	.37***
Affluence:		
Median family income, 1969	-.11	-.14
Median dwelling value, 1970	.18	-.02
Urbanization:		
Population	.30	.29
Number of building permits issued in 1978	.30*	.27
Percent increase in permits issued, 1975-1977	.19	.21

*Statistical significance of Chi-Square test:

*$p \leq .05$ **$p \leq .01$ ***$p \leq .001$

-TABLE 4. CONTINUED-

[a]Gamma is a nonparametric statistical measure of the correlation between the variables, varying from -1 (+1) indicating a perfect negative (positive) monotonic association between two variables and zero indicating the lack of any association. It requires an assumption of ordinality in the measurement of two variables, but takes advantage of that assumption to provide more consistent and readable results than chi-square, which varies with sample size and size of table.

[b]See Table 3 for definition of these measures.

Program Effectiveness

If avoiding urban encroachment on hazard areas is viewed as a key objective of hazard-area management, then land use management in the coastal zone must be viewed as a failure. As shown earlier in Table 1, coastal communities are continuing to invade hazard areas and at rates far in excess of the invasion of inland flood plains. On the other hand, programs may still be viewed as effective, if they protect the population and reduce flood losses through building design (elevation and floodproofing) adjustments. In addition, if steps are taken to protect significant natural areas, various local environmental goals may also be achieved through the land use management program while allowing considerable invasion of the coastal flood hazard area.

Perceptions of local officials who have worked closely with flood plain land use management were used to provide measures of these aspects of local program effectiveness. As shown in Table 5, nearly half of the officials rated their land use management programs very effective in terms of exposure of future development. Slightly over one-fourth of the entire sample of programs were rated very effective in terms of encroachment on natural areas, and regular program communities were somewhat more likely to be effective. Land use management was considered least effective in dealing with exposure of existing development. Roughly a third of the respondents rated their land use management programs not effective in this regard. It is somewhat surprising that enrollment in the regular phase of the National Flood Insurance Program did not show a stronger influence on these effectiveness ratings, particularly in regard to existing hazard zone development.

TABLE 5. LOCAL RATINGS OF PROGRAM EFFECTIVENESS (percent of jurisdictions)

	National Flood Insurance Program Status	
	Regular Program (N = 154)	Emergency Program (N = 24)
Effectiveness in Terms of Exposure of Existing Development to Flooding[a]		
Very effective	11	14
Moderately effective	57	50
Not effective	32	36
Effectiveness in Terms of Exposure of Future Development to Flooding[a]		
Very effective	49	43
Moderately effective	49	52
Not effective	2	5
Effectiveness in Terms of Encroachment on Natural Areas[a]		
Very effective	29	16
Moderately effective	51	58
Not effective	20	26

[a]Includes ratings of only those officials who considered exposure of existing development, exposure of future development, or encroachment on natural areas as a problem in their community.

Table 6 shows that a number of community and program characteristics are weakly associated with program effectiveness in terms of the exposure of future development and encroachment on natural areas. However, only a few of these relationships are statistically significant at the .05 level. The scope of the local land use management program (i.e. the number of methods used) was the only factor significantly related to effectiveness in terms of future development. Comprehensive planning and the adoption of goals related to the preservation of open space and natural areas were significantly related to effectiveness in terms of encroachment on natural areas. No community or program characteristics were significantly related to effectiveness in terms of exposure of existing development. This analysis suggests that to be effective in dealing with future development, communities should adopt a program which uses multiple management methods. Comprehensive planning coupled with the adoption of specific preservation goals seems most likely to produce a local program which is effective in dealing with encroachment on natural areas.

TABLE 6. FACTORS ASSOCIATED WITH PROGRAM EFFECTIVENESS

	Gamma: A Measure of Association[a]	
	Effectiveness in Terms of Exposure of Future Develement	Effectiveness in Terms of Encroachment on Natural Areas
Hazard Zone Characteristics		
Percent of community in hazard zone	.10	.27
Percent of hazard zone developed	.23	.20
Concern for Problems		
Interest by public officials	.27	.35
Priority assigned to solution of hazard zone problems	.25	.25
Program Characteristics		
Goals		
Reduce property loss	.40	.21
Preserve open space	.30	.62***
Preserve natural areas	.27	.67***
Management Methods		
Scope	.41*	.43
Stringency	.31	.18
Comprehensive planning	.31	.37**
Enrollment in regular program	.13	.24
Direct state involvement	.30	.45

*Statistical significance of Chi-Square test:

*p ≤ .05 **p ≤ .01 ***p ≤ .001

[a]See Table 4 for definition of this statistic.

Policy Conclusions

The results of this study have a number of implications for the management of coastal flood hazard areas.

First, the invasion of hazard areas in coastal communities, as measured by the number of construction permits and the percentage of total new construction located in the hazard zone, is markedly higher than in non-coastal communities. This higher rate of hazard area invasion indicates that potential flood damages are growing much more rapidly in coastal communities than in inland communities. This suggests that more attention must be given to coastal flooding by federal, state and local agencies.

Second, enrollment in the regular phase of the National Flood Insurance Program is associated with the adoption of several goals and methods directly related to flooding. Regular program communities were more likely to have adopted the goals of reducing property losses and protecting the population; they were also more likely to have adopted elevation and floodproofing regulations. Regular program communities also tended to devote somewhat more funding and staff time to implementing their land use management programs. Other community characteristics, such as the size of and land uses in the hazard zone, perception of the hazard, and prior experience with land use management, were related to the scope and stringency of the management program adopted by the community.

Third, local officials rated land use management most effective in dealing with exposure of future development to flooding. Land use management was considered less effective in dealing with exposure of existing development. This suggests that, while land use management is relatively effective in reducing the exposure of future development and preventing encroachment on natural areas, existing development which is exposed to coastal flooding will remain a risk dispite programs which rely on land use management. Thus, coastal flooding will continue to be a serious natural hazard despite the current efforts of federal, state and local agencies.

References Cited

Baker, Earl J. and Joe Gordon McPhee. 1975, Land Use Management and Regulation of Hazardous Areas: A Research Assessment. Boulder: Institute of Behavioral Science, University of Colorado.

Burby, Raymond J., Steven P. French, Edward J. Kaiser, and David H. Moreau. 1980. Managing Flood Hazard Areas: A National Assessment. Chapel Hill: Center for Urban and Regional Studies, University of North Carolina at Chapel Hill.

Cheatham, Leo R. 1979. An Assessment of Some Economic Impacts of FIA Land Use Requirements on Urban Coastal Zone Development. Mississippi State, Miss.: Water Resources Research Institute, Mississippi State University.

Dillman, Don A. 1978. Mail and Telephone Surveys: The Total Design Method, New York: John Wiley and Son.

French, Steven P. 1980. Flood Plain Land Use Management: An Analysis of Context, Programs and Impacts. Ph.D. Dissertation. Chapel Hill: Center for Urban and Regional Studies, University of North Carolina at Chapel Hill.

Sheaffer and Roland, Inc. 1980. Evaluation of the Economic, Social and Environmental Effects of Flood Plain Regulation. Washington: Department of Housing and Urban Development, Office of Policy Development and Research.

Regional Coastal Zone Management Coordination: Dealing With Coastal Flooding in New England

by: Steven L. Colman*

Introduction

This paper does not deal with the so-called "hard" aspects of flood plain management such as calculating storm surge or erosion rates. Rather, the emphasis of this paper is on improving the intergovernmental coordination necessary to achieve more effective flood plain management. A regional perspective -- that of New England and New York will be found throughout the paper. The regional perspective, however, stresses the institutional arrangements that have evolved to foster coordination rather than the traditional regional differences like climate and geography.

The paper describes the process whereby steps to improve coastal flood plain management are given credibility and are institutionalized. The immediate results of the process, the formal acceptance of a set of recommendations by the New England River Basins Commission may seem very prosaic. The recommendations themselves may seem rather mundane. In reality, though, the process leading up to the formal acceptance of the recommendations is important in its own right. Further, the fact that the coastal flood plain management policy recommendations underwent close scrutiny by state and federal officials has served to focus increased attention on the need to revise existing programs and policies. Finally the formal acceptance of the coastal flood plain management policy recommendations has helped to stimulate action by states, federal agencies and by the President and Congress to improve the region's ability to respond to the problem of coastal flooding.

* Coastal Zone Liaison
New England River Basins Commission
141 Milk Street, Boston, Massachusetts 02109

Background

Among the amendments to the Coastal Zone Management Act passed by Congress in 1976 was the addition of Section 309 - Interstate and Regional Coordination. As stated in the CZM Act the major purposes of Section 309 are to encourage states to give high priority to:

- coordinating state coastal zone planning, policies and programs with respect to contiguous areas of such states and
- studying, planning and implementing unified coastal zone policies with respect to such areas.

It is likely that the 1980 amendments and reauthorization of the Coastal Zone Management Act will retain Section 309. Despite the worthwhile policy statements and language of Section 309, however, no funds have ever been appropriated to carry out this Section.

At the same time a growing realization was developing in many parts of the country that a regional approach towards solving coastal problems made more and more sense. Indeed as the understanding of coastal planners-scientists as well as political leaders grew and became more sophisticated, increasingly coastal problems became perceived as being regional or interstate in nature. In fact, Jacques Cousteau at the Coastal States Organization meeting in September 1979 urged American coastal planners to accept an international approach towards coastal and ocean management. As the regional nature and dynamics of coastal ecosystems became better understood the scope and complexity of the systems became clearer.

In 1976 the Landmark Fisheries Conservation and Management Act was passed by Congress. The key feature of this Act is the establishment of Regional Fishery Management Councils that set quotas on the take of selected commercial species by both American and foreign fishing vessels. This Act, which created 10 distinct geographic ocean management regions, is the clearest legislative statement of the regional management principal to date. The key difference that distinguishes the FCMA from other possible regional schemes is the fact that the management area is under the exclusive domain of the federal government. In contrast the coastal zone is typically a mix of federal, state and local jurisdiction.

For this reason the regional coastal planning efforts that have been undertaken thus far are characterized by an emphasis on coordination with management in the traditional sense, principally the responsibility of state governments.

In diverse sections of the country groups of states have organized together to form regional entities. Among the most prominent of these are the:

New England-New York Coastal Zone Task Force - coastal states from Maine to New York - affiliated with the New England River Basins Commission;
The Great Lakes Standing Committee on Coastal Zone - affiliated with the Great Lakes Commission;
The Coastal Plains Organization - North Carlina, South Carolina, Virginia, Georgia and Florida;
The Sea Use Council - Alaska, Oregon, Washington, Hawaii, British Columbia, NOAA, Navy, Coast Guard and MarAd.

New England River Basins Commission

The New England River Basins Commission (NERBC) was established in 1967, by executive order under the authority of Title II of the Water Resources Planning Act of 1965, P.L. 89-80. The Commission is comprised of representatives of the governors of the six New England states and New York, ten federal agencies, six interstate agencies, and a Presidentially appointed Chairman. The primary responsibilities of the Commission are

- to be the principal agency for coordinating water and related land resource plans for New England
- to prepare a comprehensive coordinated joint plan for managing the region's water and related land resources
- to recommend long-range priorities for meeting the region's natural resource information and planning needs, and
- to undertake studies needed to carry out these missions adequately.

Early in the Commission's history a decision was made that the coasts of New England constituted one of the region's most vital resources under its jurisdiction. Two of the Commission's early major studies, "The Long Island Sound Study" and "The South Eastern New England Study" were heavily oriented towards the coast. Currently the Commission devotes a significant portion of its resources towards dealing with regional coastal problems. Among the on-going coastal projects are: A Long-Range Dredging Management Program A Regional Ports and Harbor Study, An Offshore Oil and Gas Information and Planning Program and an Offshore Pipeline Study.

New England-New York Coastal Zone Task Force

In 1974 the newly designated state officials charged with developing coastal zone management programs decided to use the Commission as the vehicle for future cooperation on coastal issues of common concern. Early in 1974 the Coastal Zone Task Force was formally established. The purpose and objectives of the Task Force are to:

1. serve as a means for interstate coordination among coastal zone management programs in New England and New York;
2. serve as the forum for discussion and where appropriate resolution of coastal issues of common, interstate, federal-state or regional concern;
3. identify interstate or regional issues and the measures for addressing them;
4. serve as the means for disseminating information of common concern on coastal issues/problems; and
5. facilitate federal agency participation and review, and obtain federal-state cooperation in the development and implementation of state CZM programs.

The state coastal zone program managers from Maine, Massachusetts, New Hampshire, Connecticut, New York, and Rhode Island comprise the membership of the Task Force. Professional staff support to the Task Force is provided by the Coastal Zone Liaison located at the New England River Basins Commission.

NERBC Regional Policy Statement on Flood Plain Management

Flooding, both riverine and coastal, has been and continues to be a significant problem in New England. NERBC has devoted considerable attention towards dealing with the flooding issue in recent years. In 1978 the Commission adopted a "Regional Policy Statement on Flood Plain Management in New England." The adoption of the regional policy statement was the culmination of four years of intensive effort by the NERBC Task Force on Flood Plain Management. It should be noted that according to River Basin Commission regulations plan and policy, "adoption" is a highly significant process. "Adoption" requires formal review and approval by the Governors of the six New England states and New York and by the heads of member federal and interstate agencies. Thus after such a review the Regional Policy Statement became an adopted element of the Comprehensive Coordinated Joint Plan for managing New England's water and related land resources in June of 1978.

The Regional Policy Statement addresses both riverine and coastal flooding. With respect to an overall flood plain management strategy three basic policies are adopted:

1. The natural and beneficial values of flood plains should be preserved and enhanced.
2. The wise use of flood plains should be promoted to the maximum extent practicable.
3. The cost of flood plain use should be shifted to those who create the risk by occupying the flood plains.

The Policy Statement, while reorganizing the need for increased attention in coastal flood plain management, did put forth specific policies for the coastal flood plain. The following policies were adopted for managing the coastal high hazard area (v-zones as mapped by FEMA)

- prohibit non-essential uses
- prevent encroachment in wetlands
- acquire, where needed, unbuilt land in fee for public management and use
- acquire, where needed, built lands when damaged
- publicly display maps delineating flood hazard areas, and
- require that prospective buyers of flood prone areas by officially informed of the hazard.

Winter Blizzard of '78

At the same time the NERBC Regional Policy Statement was out for review by the region's Governors the coasts of Massachusetts, New Hampshire and Maine were struck by the "Great Blizzard of 1978." Described as the most damaging weather event to strike the northeast since 1962, the Great Blizzard hit the New England coast on February 6 and 7, 1978. The storms statistics are awesome: more than 27 inches of snow was recorded in Boston, wind velocities peaked at 79 miles per hour in Boston and 92 miles per hour on Cape Cod, storm surges averaged 4 feet above mean spring tide, high tides in Boston were recorded at 10.4 feet above mean sea level, and wave heights on Cape Cod were measured at 9 feet. In Massachusetts 29 lives were lost and over 11,000 homes were damaged or destroyed. Total storm damage has been estimated at $300 million.

NERBC Coastal Hazards Report

In January of 1979 NERBC published a staff draft of a report entitled: "Dealing With Coastal Hazards: Implementing the Regional Policy Statement on Flood Plain Management." The Report evaluated the region's response to the coastal flooding associated with the Blizzard of 1978 and proposed a series of specific recommendations for reducing the vulnerability to future coastal storms. A discussion of the Report's ten major recommendations follows:

Recommendation #1: Implement Programs to Acquire Previously Developed High Hazard Properties in Lieu of Redevelopment

This recommendation is directed at those properties located in the "Velocity-Zones" (V-Zones) as mapped by the Federal Emergency Management Agency. It recognizes that, in many ways, the purchase of storm damaged property represents the most cost-effective flood plain management strategy. It was found that after the Blizzard of 1978 there were no Federal or State programs available for purchasing storm damaged property. Immediately after the storm a bill to establish a $10 million bond issue to purchase damaged property was defeated in the Massachusetts legislature. To assist in the implementation of Recommendation #1 the following steps are advocated:

- Section 1362 of the National Flood Insurance Act should be funded and implemented. This Section of the Flood Insurance Act authorizes the acquisition of property which is substantially damaged in lieu of payments of a claim on the property's flood insurance.

  Although recognized as important when passed in 1968, Section 1362 did not receive appropriated funds until Fiscal Year 1980. The $5.5 million which was appropriated for F.Y. 1980, although minimal on the national scale, did provide some opportunity to purchase properties damaged by Hurricane David and Frederick.
- State standby bond authority and state damaged property acquisition programs should be established.
- Tax relief programs for communities depended on tax revenues from properties located in the V-Zones should be established.

Recommendation #2: Strengthen the Requirement That Heavily Stormdamaged Homes and Businesses Be Rebuilt According to Flood Proofing Standards

It was found that immediately after the 1978 Blizzard many, the exact number is still undetermined, homes were rebuilt without meeting the flood proofing standards of the Federal Flood

Insurance Program (FIA). It was noted that there is a loophole in existing F.I.A. regulations that require rebuilding to flood proofing standards only when the amount of repair costs exceeds 50% of the structure's value. Some homeowners utilized 50% or less of the home's pre-storm value in repair costs and thus were permitted to rebuild, even though the actual damage was more than 50% of the structure's value.

An additional factor cited in the inadequate enforcement of flood proofing regulations is that responsibility for their implementation rests with local building inspectors. Apparently there is both post-storm sympathy and small town pressure that make it difficult for local building inspectors to declare a property to be substantially damaged. Substantially damaged properties, however, in towns such as Hull, Massachusetts were required, if rebuilt, to have a minimum first floor elevation of 21 feet above mean sea level in certain V-Zones. Thus, if the ground floor is elevated sufficiently and construction meets F.I.A. standards, it can be argued that damage to such structures from future storms will be reduced significantly. At the same time requiring strict flood proofing standards considerably raises the cost of rebuilding and clearly illustrates the danger of inhabiting a given coastal V-Zone. This, in turn, discourages rebuilding and makes selling the storm damaged property to the state more attractive. To implement Recommendation #2 the following actions were proposed:

- Amend F.I.A. regulations to close the 50% repair cost loophole.
- Provide better training and technical assistance to building inspectors.
- Amend state building codes to include flood proofing requirements.
- Identify flood hazard areas more accurately. F.I.A. should accelerate its program to incorporate wave heights and storm surge into its calculation of the coastal 100 year flood plain. Existing maps should be revised based on the wave height-storm surge data.

Recommendation #3: Keep Development Away From High Hazard Areas That Are Still Undeveloped

Although very basic to any flood plain management strategy this recommendation is difficult to implement. Nevertheless since the '78 Blizzard a number of positive steps have been achieved. Four states in the region now have approved Coastal Zone Management Programs: Connecticut, Maine, Massachusetts and Rhode Island. Thus the federal consistency provisions of the CZM Act should help assure implementation of state coastal Flood Plain Management policies. Two states, Massachusetts and Maine, have passed new laws or regulations that protect barrier beaches and dunes. In August of 1980 the Governor of Massachusetts issued an Executive Order designed to protect undeveloped barrier beaches and discourage rebuilding of storm damaged property on developed barrier beaches and islands. On the federal level, where it has been claimed many of the incentives for promoting development of the coastal flood plain originated, a number of important events are occurring.

President Carter, in his 1979 Environmental Message, directed the National Atmospheric and Oceanic Administration to undertake a systematic review of all federal programs that effect the coastal zone. The results of the NOAA review are expected to be released in late fall, 1980. In January of 1979 the Heritage Conservation and Recreation Service of the Department of Interior released a Draft Environmental Impact Statement recommending Alternate Policies for Protecting Barrier Islands Along the Atlantic and Gulf Coasts of the United States. The Department of Interior's effort, when issued in final form, is expected to make a significant change in federal policy towards barrier islands which constitute the most hazard prone of all coastal areas. In addition barrier island protection legislation has been introduced in both houses of Congress. Finally the establishment in April of 1979 of the Federal Emergency Management Agency promises to bring improved coordination to programs such as Flood Insurance and Disaster Relief.

Recommendation #4: Improve Construction Standards for Floodproofing

Recommendation #5: Take Account of Natural Coastal Processes in Clean-Up, Construction and Repair of Protection Measures

Recommendation #6: Re-Focus Data Collection and Analysis Activities

Two years after the Blizzard of '78 it is still not possible to provide accurate statistics on storm damage. The Corps of Engineers, in its 1979 storm damage assessment "Coastal Storm Damage Study" reported that their data on storm damage was collected from 46 separate communities and over 20 state and federal agencies. Some data, the report states "was based upon the judgment of a community official."

The Corps, in September of 1978, procured color aerial photography of most of the storm damaged area from Cape Cod to the New Hampshire-Maine border. The Corps' aerial photography meets photogrammetric specifications and will provide an excellent reference to establish baseline conditions prior to a future storm. No such comprehensive aerial photography existed prior to the '78 Blizzard. Thus, the lack of comparable "before and after" aerial photography hampered the collection of accurate storm damage statistics. The lack of baseline data and the lack of a centralized data collection system were serious obstacles encountered after the '78 Blizzard. To implement Recommendation #6 the following steps are suggested:

- Data collection activities should be centralized and coordinated by F.E.M.A.
- The necessary authority to collect storm-related information in a timely fashion should be obtained.
- Baseline information against which to measure storm-related data should be established.

Recommendation #7: Improve Coordination in Storm and Disaster Planning and Recovery Operations, as Well as Flood Plain Management

The establishment of the Federal Emergency Management Agency in April 1979 should assist in implementing this recommendation. F.E.M.A., because it is a comprehensive agency that deals with both disaster mitigation and relief, has indicated a capability to coordinate storm and disaster planning and recovery. Both the Office of Disaster Response and Recovery (formerly the Federal Disaster Assistance Administration) and the Office of Insurance and Mitigation are under the direction of the regional administrator of F.E.M.A.

In July of 1980 the New England River Basins Commission reactivated and expanded its Flood Plain Management Task Force. The objectives of the Task Force are: (1) to implement the NERBC Regional Policy Statement on Flood Plain Management; (2) to exchange information and encourage consistency on current, local, state and federal flood plain management activities; (3) to discuss implementation of policies and work towards solutions to common problems and to provide a forum to discuss assistance for state program development, including data needs in the region. Membership in the Task Force, in addition to federal and state officials, has been expanded to include: representatives from business, local governments, the New England Congressional Caucus, and private conservation organizations. To strengthen the coastal flood plain management focus of the Task Force, representation from the region's CZM programs has also been added. Both actions, the establishment of F.E.M.A. and the reactivation of the NERBC Flood Plain Management Task Force, are consistent with the actions suggested to implement Recommendation #7.

Recommendation #8: Improve and Utilize Educational Programs to Support the Programs, Plans, and Policies of Flood Plain Management Agencies

Recommendation #9: Shift the Cost of Protection to Those Who Create the Risk and Enjoy the Benefits of Locating in Coastal Hazard Areas

Recent studies have shown that a variety of federally supported programs sink millions of dollars into directly or indirectly encouraging development in the coastal flood plain. The Department of Interior's Barrier Island E.I.S. reports that some $500 million was utilized by federal agencies to support development projects on barrier islands in the last three fiscal years. Agencies such as E.P.A., E.D.A., F.H.A., and the Corps all invested heavily in barrier island projects. In 1979 the Federal Flood Insurance Program also utilized $426 million in federal funds to remain solvent and has requested an additional $600 million this year. While these figures are nationwide, it has been documented that the ratio of payouts to premiums collected in coastal hazard areas is even higher than the national average.

To eliminate the federal subsidies that tend to encourage development of coastal hazard areas and to implement Recommendation #9 the following actions are suggested:

- Assure that FIA actuarial rates fully reflect the true cost of insurance coverage.
- Consider raising flood insurance rates on properties which suffer repeated damage.
- Eliminate or reduce subsidized interest rates in federal flood disaster loans.
- Re-draw maps for more accurate determination of the 100-year flood plain and incorporate wave heights associated with the 100-year storm.

Recommendation #10: Utilize Executive Order 11988 (Flood Plain Management) and Executive Order 11990 (Wetlands Protection), as well as Federal Consistency Provisions for Approved Coastal Zone Management Plans, to Strengthen Disaster and Flood Plain Management Planning

NERBC Coastal Activities Committee - Review of the Coastal Hazards Report

Subsequent to the Great Blizzard of '78 and the adoption of the Regional Policy Statement on Flood Plain Management an organizational change occurred at NERBC. A "Coastal Activities Committee" (CAC) was established to guide and review the coastal studies of the Commission.

State CZM Program Managers were appointed to represent their respective states on the CAC. Membership of the CAC includes representatives of federal agencies such as EPA and the Corps as well as the six coastal state members of NERBC: Connecticut, Maine, Massachusetts, New Hampshire, New York and Rhode Island.

The Coastal Hazards Report originally was published as a staff report and thus had no official status as a NERBC policy document. One of the first assignments given to the CAC was to review the Coastal Hazards Report and its recommendations and recommend an appropriate action to the New England River Basins Commission. Options available to the CAC to recommend include: (1) adoption of the Report as part of the region's Comprehensive Coordinated Joint Plan; (2) acceptance of the Report as a policy and planning guidance document; (3) rejection of the Report altogethter; and (4) acceptance or adoption of selected parts of the Report.

At its March 1979 meeting the CAC debated the various options but failed to reach a consensus on any one option. The New England/New York Coastal Zone Task Force, acting as the state's subcommittee of the CAC, agreed to undertake a review of the Report's recommendations and recommend an option to the full CAC.

As part of the review process the Coastal Zone Task Force decided to concentrate its effort on examining the ten major policy recommendations contained in the Report. The recommendations were reviewed for consistency with state Coastal Zone Management Programs and with the NERBC Regional Policy Statement on Flood Plain Management. The review revealed that there were no major conflicts between the Report's recommendations and CZM Programs or the Regional Policy Statement. A second phase of the review consisted of a meeting between the Coastal Zone Task Force and the Regional Director of the Federal Flood Insurance Program. The meeting had two main purposes. The first was to improve coordination and communication between the region's C.Z.M. Programs and the Federal Flood Insurance Program. The second was to discuss the Report's recommenations and receive the reaction of the Regional Director of the Flood Insurance Program and the Federal Emergency Management Agency. The Regional Director of the Flood Insurance Program expressed support for the Report's recommendations and indicated that they were consistent with reforms already underway within F.E.M.A.

Based on the findings that the recommendations were consistent with existing state and NERBC flood plain management policy and that there existed a high probability of their implementation by F.E.M.A., the Coastal Zone Task Force recommended that the Coastal Activities Committee should make the following recommendations to the New England River Basins Commission:

The Coastal Activities Committee recommends that the Commission formally accept the report, "Dealing With Coastal Hazards: Implementing the Regional Policy Statement on Flood Plain Management" for publication and distribution and for use as a basis for the coordination of state and federal coastal flood plain management and as a guide for future planning and program development.

On October 25, 1979 the Coastal Activities Committee concurred with the Coastal Zone Task Force and agreed to submit the recommendation for acceptance of the Report to the New England River Basins Commission for its consideration. At NERBC's December 1979 quarterly meeting a representative of the CAC addressed the Commission and concluded with the following: "Acceptance of the Report would formally recognize the need to take action in the region to improve coastal flood plain management and would, through distribution, help educate the public on coastal hazards. Perhaps most importantly, acceptance of the Report implies a commitment on the part of the Commission to actively support and effectively implement the appropriate recommendations so that New England will be better prepared to face a major coastal storm." Agreeing to make such a commitment, the New England River Basins Commission agreed to accept the Coastal Hazards Report and its policy recommendations.

RESPONSE TO A CHANGING COASTAL ENVIRONMENT

by

Kenneth C. Caldwell, A.S.L.A.
Environmental Planner
Jones & Jones, Seattle

Here in the 1980's we all know that the ostensibly stable planet Earth is a dynamic and changeable place. The continents beneath our feet float and move, drifting around the globe. In Washington State, we have a volcano that has recently demonstrated the dynamics of our planet: today's scenic mountain lake is tomorrow's steaming mud pool. And of all our commonly used environments, the shoreline is perhaps the most dynamic and changeable: there is a constant interplay between land and water.

Unfortunately, our behavior and our laws do not allow for an unstable world. In the law of ownership the earth is divided into Euclydian parcels that are theoretically fixed for all time. Your acre of land, 208 x 208 feet, is expected to eternally remain an acre on the tax rolls. Thus our disregard of shoreline dynamics stems first from our system of land law. The geometry of ownership is fixed, yet the earth or water that is owned is not fixed: it shifts, accretes, erodes, rises and falls.

And there are other reasons for this disregard of the landscape's ephemeral nature. In the United States and Canada our patterns of land use are fairly new. Going back to Mt. Saint Helens in Washington State, widespread agriculture in Eastern Washington has been in place for less than 100 years, and in many areas less than 50 years. Such a short period of time is just not sufficient to develop a cultural awareness of volcanic activity. Right here in Florida, there are vast newly-populated areas that will be severely threatened by an extreme hurricane tide, something that recurs with a regular frequency.

The changeability of the earth is nothing new, but it always seems new when it happens. For instance, let's look at climate for a moment. From 1877 to 1886 there were 10 years of abnormally bountiful rainfall on the Great Plains. The ranchers and farmers, many recently arrived, felt that the plains were the Garden of Eden; land costs skyrocketed and settlers ran up huge investment debts. At the height of the boom, in 1887, there was a terrible blizzard, followed by a hot, dry summer, followed by a new 10-year cycle: little rainfall, and scorchingly hot summers. Parts of Western Kansas and Nebraska were almost depopulated. So, when we hear of the scorchingly hot summer the southern plains have had in 1980, we must remember that the Great Plains change faces, like the rest of the earth, and that heatwaves and droughts are part of the prairie scheme of things.

The essential question is, "How should society respond to the on-going changes inherent in our environment?"; and more to the point, "How should we adjust to dynamic shoreline environments?" This is a *public* question because individuals have become accustomed to public relief when "natural disasters" occur. We are a cohesive society, and individuals expect social help when struck by "Environmental Hard Times," just as they expect social help during times of economic depression.

This subject has large-scale and small-scale aspects. The large-scale side involves issues such as urbanization in known flood hazard areas. The small-scale side involves the construction of residences, roads, etc., in areas subject to land slip, river meander advance, wave erosion, storm tides, and so on. The source of public concern stems from the likelihood that "natural disasters" (of either the large- or small-scale type) will incur a public cost, in terms of rescue, loss of economic productivity, publicly funded remedial actions (e.g., shore armoring, levees), or other social involvements. So the question is one of public/social costs, and the actions governments should take to either promote or avoid those costs.

Let us look at two simple examples.

In the Puget Sound region of Washington, a large percentage of the marine shore consists of high, steep erosion bluffs--"feeder bluffs" that steadily recede. The tops of these bluffs are typically highly accessible, and the bluff crests are highly valued as residential "view properties." In several instances chains of landowners have found themselves victims of an impending "natural disaster": their homes are being undermined by natural bluff recession. In one instance, the Army Corps of Engineers has studied the problem (at great public expense), and proposed heavy armoring at the base of the bluff, with the intention that this armoring would be heavily subsidized by public funds.

Now: could this impending public cost (stemming from a "natural disaster" scenario) have been avoided? The answer is clearly "Yes," and the detailed solution is presented below:

1. The local or state government could analyze (or require the analysis of) the historical and projected rate of bluff recession, based on historical mapping and aerial photographs, and geological studies.

2. The local or state government could prohibit the placement of habitable structures (and other structures with a value in excess of X dollars) within the boundary of the 50, 100 or 200 year projected bluff recession zone.

Another example is the "river meander advance zone." Lowland rivers shift and translocate over time. At Jones & Jones we have had experience (through the expertise of Dr. Thomas Dunne, an eminent fluvial geomorphologist with the University of Washington Department of Geolgoical Sciences)* in discovering the historical locations of a river, and projecting its likely future behavior. This has been done by researching historic mapping, historic aerial photographs (e.g., through the U.S.D.A. Soil Conservation Service and the Army Corps of Engineers), and performing an analysis of erosion/accretion rates, boundary shear force, and other factors. Similar to the bluff recession example, this can lead to a map showing likely 50, 100, or 200-year meander advance lines, which can be translated into laws prohibiting substantial development within the endangered zone (the "river meander advance zone"). (See Figure 1.)

I think these two examples display the conceptual framework, which can be encapsulated as follows:

Proposition 1: Shorelines are dynamic interfaces that are subject to natural changes, and extremes of environmental condition.

Proposition 2: These natural changes are increasingly predictable and quantifiable.

---

* He is the co-author of a highly recommended text; see Dunne and Leopold, Water in Environmental Planning, W. H. Freeman & Co., San Francisco, 1978.

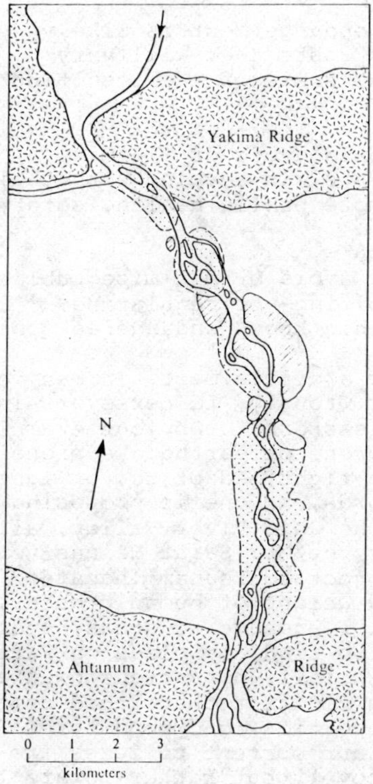

Map of relative stability of sites along the Yakima River. The lined areas are relatively stable, and the dotted areas are very unstable. The map was constructed after measuring rates and directions of channel shifting and diversion from sequences of aerial photographs and after field examination of bank conditions and the microtopography of the valley floor.

Figure 1.

EXAMPLE OF A RIVER ADVANCE HAZARD ZONE

Yakima River, Washington

This figure, based on a project performed by Jones & Jones ("The Yakima River Regional Greenway") is taken from Dunne and Leopold (op.cit). It shows areas having a high likelihood of river channel invasion, and could be used as a rationale for regulating development within the hazard zone.

Proposition 3: These natural changes are potentially endangering to human life or property.

Proposition 4: This potential endangerment is likely to imply public costs that will vary according to the scale of the resulting disruption.

Proposition 5: Governments should avoid or minimize these public costs, in keeping with their mandate to promote public health, safety, and welfare.

Proposition 6: Governments may avoid or minimize public costs by prohibiting--or regulating--development within known endangered zones.

After all that, let me register a disclaimer. In many cases society will decide, rightly or wrongly, to persevere in the face of predictable "natural disasters." Obvious examples include the occupance of floodprone or earthquake-prone lands, where the decision is to fight, adapt to, or ignore the known dangers. In other words, no one is proposing the abandonment of San Francisco, the Ohio River Valley, all areas potentially endangered by the Mt. Saint Helens volcano, or all Hawaiian areas subject to tsunami hazard. In many instances the overwhelming weight of human pressure argues against the abandonment of endangered areas; and in fact, we would be hard-pressed to find any place on Earth that is entirely safe from potentially-disastrous natural changes. Nevertheless, it is a sad fact that flood damages, for instance, have risen in proportion to expenditures for flood control, indicating that our current policies are not geared toward Proposition 6 (above), but rather toward the public subsidy of private development in hazardous areas.

DUNE EROSION AND RECOVERY ON A NORTHERN BARRIER

John W. Armon
Department of Geology
Memphis State University
Memphis, TN 38152

ABSTRACT

Dune destruction and recovery characteristics of a Canadian Maritime barrier system are investigated, and the results indicate ways in which dune damage can be minimized and dune recovery best accomplished. Dune blowout formation is currently restricted to exposed dune ridges and seaward locations. The risk of more pronounced blowout damage is only likely with breakdown of the vegetation cover or the formation of additional high dunes. Results indicate that high dunes accentuate wind effects and blowouts commonly form from the wave-trimmed seaward dune face. Dune destruction by storm waves has been severe in some areas, but elsewhere a partial or complete recovery has occurred between episodes of storm damage. The frequency of widespread dune erosion is low - once in 2-4 years. Patterns of dune recovery on washover surfaces have varied with shoreline exposure, orientation, and local effects. Washovers are relatively low lying (1.5-1.7 m above low low water) and overwashing occurs more frequently than episodes of dune erosion. Natural dune recovery patterns in the west of the barrier system suggest that dune restoration attempts on exposed barriers of the southern Gulf of St. Lawrence should focus on establishing dunes in patches across washover surfaces, 100-200 m landward of the berm. This would permit more rapid dune development, continued overwash during the initial years, and dune growth landward of the accreting dune nuclei.

INTRODUCTION

Barrier island and spit shorelines form the most extensive and economically significant coastal sections of the southern Gulf of St. Lawrence. In spite of the region's northerly location and the relatively low number of coastal users, barrier shorelines remain of considerable importance to the governments and peoples in New Brunswick and Prince Edward Island. Small harbors and local fishing economies in those provinces are generally centered on tidal inlets. Two National Parks in coastal locations play a role in attracting tourists to the region, and the expanses of barrier-lagoon coast provide perhaps the main summer recreational focus for many of the region's inhabitants. Significantly, there remains unrestricted access to many of the islands and spits for people possessing boats.

The barrier systems fronting the southern Gulf present similar problems in coastal management to the open ocean barrier shorelines further south. However, the perception of problems is often different and the related costs are lower because of the lack of roads, installations or buildings on the barrier islands and spits. The normal actions of erosion by wind and waves, and of inlet formation, migration or closure are of little consequence in many areas. At Cavendish National Park on the other hand, coastal actions assume more importance. The sand dunes are fragmented, dunes encroach on the road and parking facilities, and beach sand volumes are occasionally inadequate. At the more recently established Kouchibouguac National Park in New Brunswick, initial attempts were made to assess beach and dune stability (McCann et al., 1973) and to minimize the adverse effects of large numbers of park visitors on the dune system. Outside the National Parks the perception of coastal problems may change in the next few decades if landownership patterns alter as they have done elsewhere, or if changes occur in the recreation pressures upon barrier coasts.

This paper is concerned with an investigation of dune destruction and recovery in the Malpeque barrier system, situated along the north coast of Prince Edward Island. The dune system is focused upon because it is potentially the most fragile component of the subaerial barrier system there. The Malpeque barrier system is suitable for this study because of its exposure to wind directions and wave approaches associated with the major storms crossing the Gulf of St. Lawrence (Fig.1). It is located 25 km west of Cavendish National Park in a comparable situation. However, this barrier system exhibits considerable variety in barrier and dune character, and a substantial data base enables some assessment to be made of the degree and rate of dune stability and recovery. Results also allow comments to be made on methods of dune restoration in southern Gulf barriers.

Description of the field area has been given elsewhere (McCann, 1972; Armon, 1975; Armon and McCann, 1977). There are three islands and a spit present along a curving shoreline arc. Different sections of the shore encounter the full force of storm effects from easterly and northerly winds. The tides are mixed semi-diurnal tides with a range of 1.1 m in springs. Wave dimensions are limited by fetch distances and water depths in the southern Gulf, as well as by the presence of sea-and fast-ice in the winter.

METHODS

The basis for this study is provided by extensive field surveys between 1972 and 1978, by aerial photography flown in 1935, 1958 and 1968, and by assorted imagery recorded at other times. Profiles were surveyed across the dune zone at 1.6 km intervals along the coast in 1972. Additional surveys were carried out in areas that underwent dune restoration following 1935. These surveys enabled the calculation

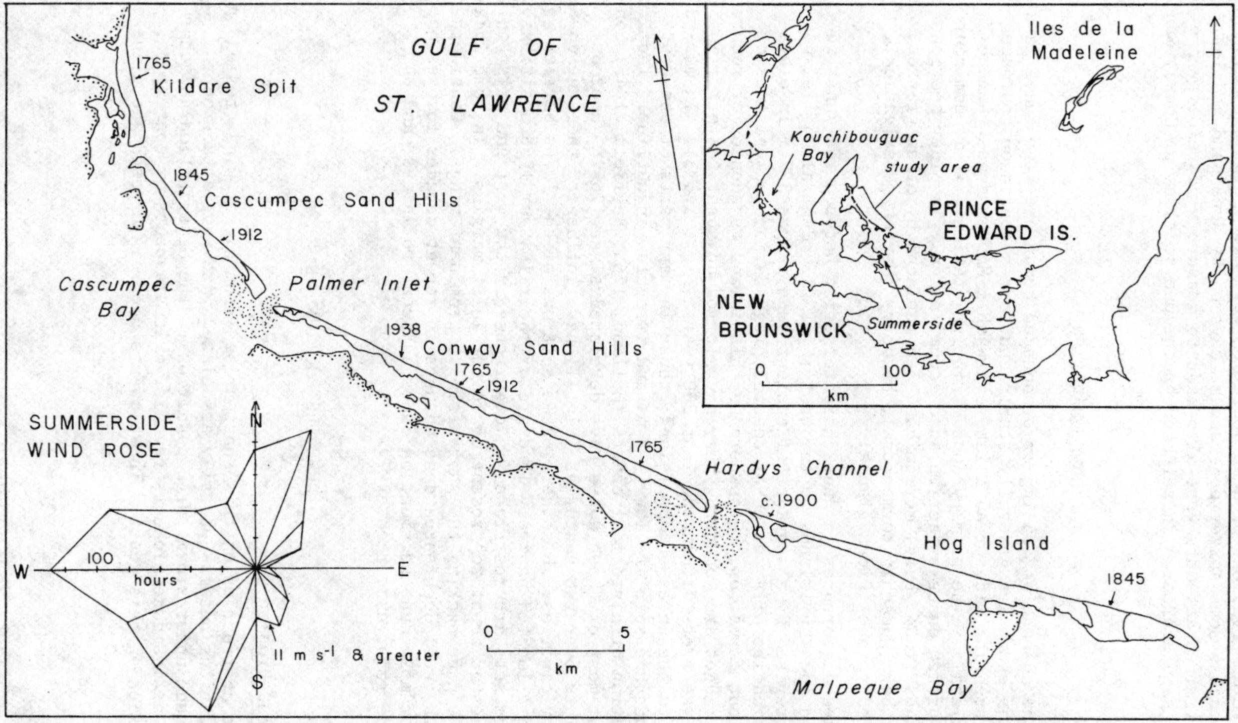

Figure 1: Location of Malpeque barrier system and wind rose for Summerside, Prince Edward Island, (9 month period, April-December). The years indicate the dates when inlets were last recorded at particular locations. Source: Wind data - Hourly Data Summaries #45, Summerside Airport, 1968, Atmospheric Environment Service, Toronto, Ontario.

of dune accretion rates for the barrier chain (Armon and McCann, 1977). Qualitative observations of dune accretion were also recorded during surveys because some survey pegs became covered by dune sand and lost. In 1977, limited attempts were also made to measure wind velocity profiles and sand accretion in selected beach-dune sections of Conway Sand Hills.

DUNE CHARACTER

In the absence of grazing or pronounced human use, the dunes landward of the shore are generally stable and form a continuous surface cover, preventing widespread island overwash. Marram grass (Ammophila breviligulata) is the primary dune fixing vegetation. It is dominant in areas of new dunes or active sand accumulation. In older sections of the dunes (100+ years) marram has been replaced by other grasses, false heather (Hudsonia tormentosa), bayberry (Myrica pensylvanica), and stunted spruce (Picea glauca).

Marked differences in dune character exist between the dunes on Hog Island and the islands to the west (Fig. 2). Central Hog Island is characterised by a sequence of ancient dune ridges 5-7 m in elevation, indicative of a period of coastal progradation and episodes of dune blowout formation. Elsewhere on this island, sequences of recurving dune ridges are younger. They have formed adjacent to former inlets and washovers within the last 200 years (Figs. 1 & 2). In marked contrast to the dunes on Hog Island, those in many parts of the western islands and spit are low (1-2 m) and irregular in outline. The investigation of historical charts (Holland, 1765; Bayfield, 1845) confirms that most of these dunes have developed since 1765. Numerous tidal inlets over two centuries (Fig. 1) have influenced the dune cover in their vicinities. Elsewhere, some sections of the 1970s barrier islands are located entirely landward of their 1845 or 1765 positions. Dune ridge sets are restricted in occurrence in the west and exist only as recurves adjacent to present or former inlet margins. In some locations, the dune backing the shore forms a prominent ridge. Consideration of the well documented dune changes in recent decades focuses on the seawardmost dune section, on dunes adjacent to inlets, and on zones undergoing dune recovery.

DUNE DESTRUCTION

Direct Wind Erosion

Dune destruction on this coast arises from direct wind erosion in a few locations but is generally associated with wave attack on the seaward dune under storm surge conditions. The distribution and character of active dune blowouts reveals various aspects of contemporary wind erosion on this coast. The blowouts occur in exposed shoreline

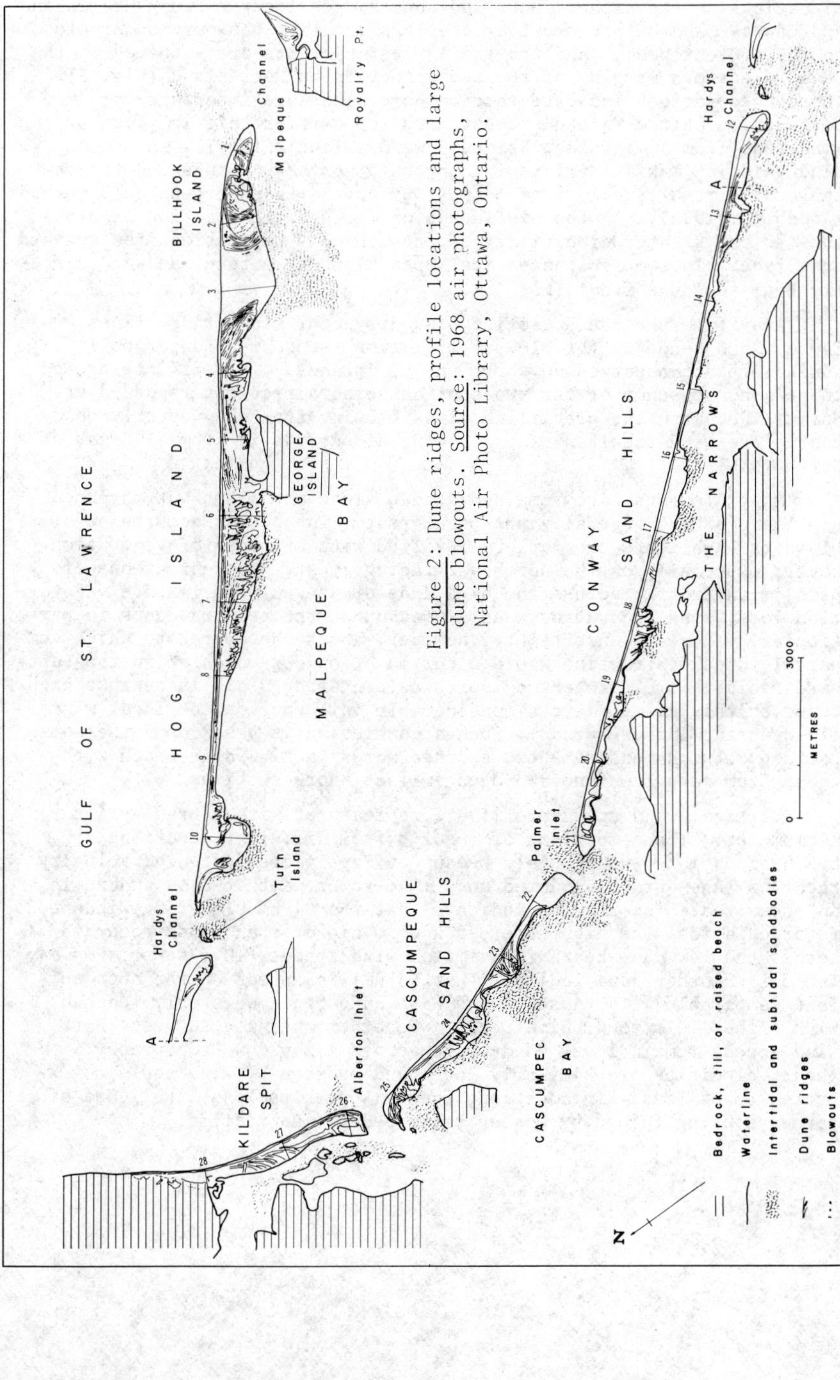

Figure 2: Dune ridges, profile locations and large dune blowouts. Source: 1968 air photographs, National Air Photo Library, Ottawa, Ontario.

situations-on the higher dunes and dune ridges located along Hog Island and Conway Sand Hills. Most of the larger active blowouts occur close to the present beach, and the two largest blowouts are situated on the seaward eastern margins of the two washovers on Hog Island (Fig. 3). These observations indicate that onshore winds are important for dune destruction on the Malpeque coast, and support for this is given by the wind rose from Summerside, Prince Edward Island (Fig. 1). A secondary wind velocity maximum exists for north to northeast winds and the onshore wind resultant is from 6° west of north (Landsberg-Bagnold method; Landsberg, 1956). Another influence on the seaward location of blowouts is the greater erodibility of the commonly vegetation-free seaward dune face. In some instances scalloped blowouts extend landward directly from the dune front (Fig. 3).

Away from the beach, active blowouts occur along ridge lines. There is no tendency for blowout migration to occur, and parabolic dunes were not observed anywhere on the islands. This is interpreted to be a consequence of the two dominant wind directions recorded at Summerside, with the prevailing winds blowing from the westerly quadrant. The wind resultant for all wind directions is from 73° west of north (WNW).

Many blowouts have remained active since the 1930s but morphologic changes have generally not been great. The common occurrence of blowouts highlights the dangers involved with any loss of vegetation cover, especially on the northeast facing coast. Concentrations of pedestrian traffic across the high dune areas would facilitate vegetation loss, sand excavation, and a breakup of the seaward dunes in particular. There is uncertainty, however, about the degree to which comparable dune destruction would occur in a low dune area or in the eastward facing, more "protected" northwestern area. This is because high dunes on this coast exert a considerable influence on the local wind field during onshore winds. Such a conclusion is based upon the comparison of wind measurements recorded across a high dune beach with those from both low dune and open surface shore-sections.

Vertical wind speed profiles were measured with a Bradley 6 cup anemometer. The comparison of results from different locations or beach situations was possible because of an independent wind velocity recorder (anemograph) mounted on a 10 m tower, situated on a dune in an open area. The results indicate that the 54 cm elevation windspeeds recorded across the high dune beach were higher than those recorded across the low dune beach for similar wind speeds (10 m tower; Fig. 4, Table 1). Other data indicate that 54 cm windspeeds on the open surface comparable with those on the high dune beach were only reached with higher 10 m tower windspeeds. Together the data suggest that wind speeds remain lower on average across a low dune topography. Shear velocities are similarly lower and erosion by wind could be expected to be less. In addition, there is perhaps less likelihood of pathways being formed by concentrated pedestrian traffic.

Figure 3A: Washover on western Hog Island showing blowout locations and dune ridge sequences. The blowouts in the foreground extend from the seaward dune. The blowout on the left side is pictured below. 1974. View south.

Figure 3B: View southwards of large blowout adjacent to the washover on western Hog Island. 1974.

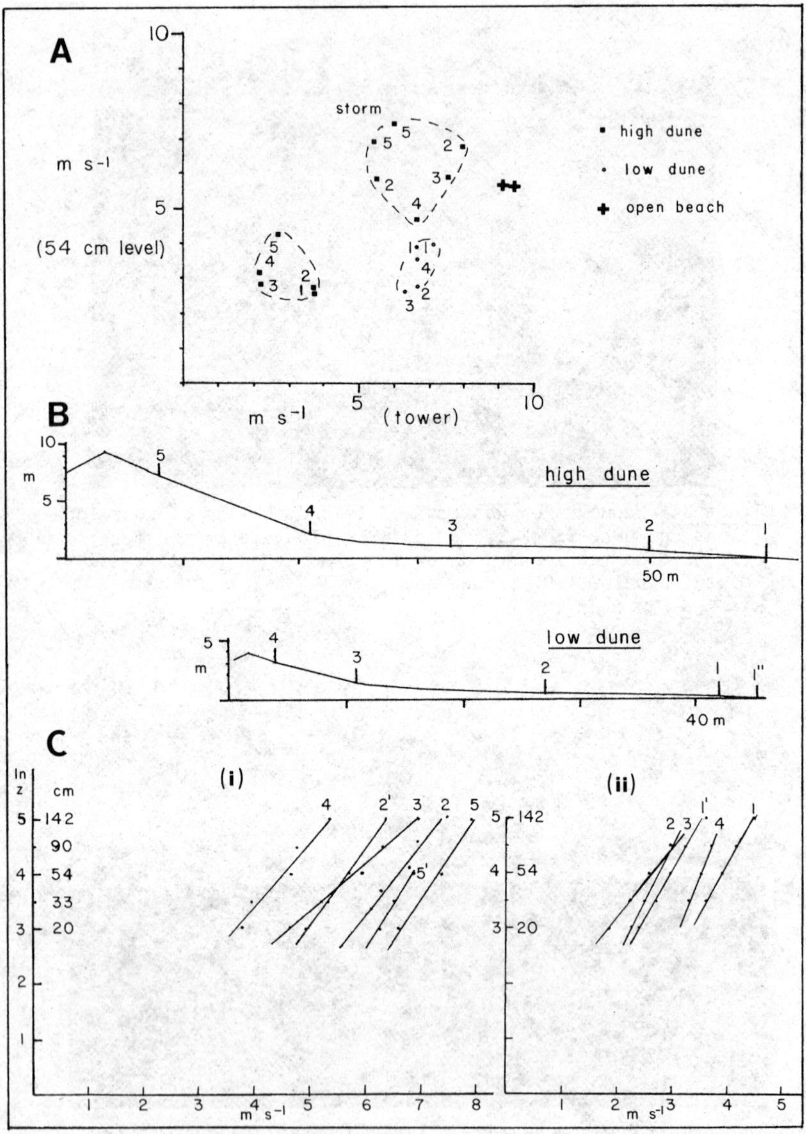

Figure 4: Windspeed data from Conway Sand Hills. A: 10 m windspeed vs anemometer windspeed, different situations. B: High and low dune profiles with anemometer locations. C: Windspeed profiles for (i) high dune storm and (ii) low dune (7/27/77) conditions.

TABLE 1: WIND MEASUREMENTS, JULY 1977

Loc.	Situation *	Windspeed 10 m tower ($m\ s^{-1}$)	Windspeed 54 cm level ($m\ s^{-1}$)	Shear Velocity ** ($cm\ s^{-1}$)
High dune beach, 7/23/77				
#2	lower foreshore	8.0	6.8	32.63
#3	mid foreshore	7.6	5.95	47.06
#4	base of dune	6.7	4.7	33.70
#5	upper dune slope	6.0	7.4	26.49
#5	upper dune slope	5.4	6.95	27.03
#2	lower foreshore	5.4	5.8	29.63
Low dune beach, 7/27/77				
#1	lower foreshore	6.7	3.9	22.77
#2	mid foreshore	6.7	2.8	19.55
#3	base of dune	6.3	2.65	32.00
#4	upper dune slope	6.7	3.55	16.81
#1	lower foreshore	7.2	4.0	23.00
Open beach (flat with 30 m rippled, 30 m in low roughness, 40 m smooth sand, upwind). 7/26/77				
A		9.4	5.7	23.87
A		9.2	5.7	22.91
A		?	6.2	26.49

* See Figure 4 for locations.

** Shear velocity,

$$u_* = \frac{\kappa (u_1 - u_2)}{\ln y_1 - \ln y_2}$$

where κ is von Karman's constant (approx. equal to 0.4)
u is velocity
y is elevation
and 1, 2 are different levels.

Temperature data at the various anemometer levels indicated that profiles were close to neutral.

Sand movement (0.2-0.25 mm diameter modal sizes) begins with u_* of 21-23 cm s^{-1} (fluid threshold curve, Figure 3, Horikawa, K. and Shen, H.W., 1960. Sand movement by wind action--on the characteristics of sand. B.E.B. Tech. Memo. 119, 51p.)

The absence of dune blowouts in the northern area provides some support for these conclusions. Dunes are generally lower there with elevations of 3-5 m above LLW. However, that part of the shoreline is protected from north to northeast onshore winds, and east to southeast winds are less significant. This area appears to be one with a lower potential for dune destruction by wind erosion.

Wave Erosion

Destruction of the seaward dune by storm wave action reflects the long term erosional tendency for the Malpeque barrier system. This is evident from the truncated appearance of the seaward dune margin. The longshore variations in beach and dune erosion have been discussed elsewhere (Armon and McCann, 1979) and will be only briefly reviewed here. Average 1935-1968 erosion rates measured from air photographs were generally 0.5 m year^{-1} and higher along the western half of the shoreline, in contrast to 0.0-0.3 m year^{-1} along much of the eastern half. This general variability reflects inlet effects, together with downcoast changes in both the southeastward longshore sediment transport rate and the subtidal extent of the barrier sediment body. On this coast the shallow subtidal zone is an important sand source affecting the volume of sand in the beach system and hence, dune erosion. Fathometer profiles seaward of the islands plus other data (Fig. 5) suggest that sediment thicknesses over bedrock are generally 2-3 m but that the sediment cover extends seaward for much greater distances along Hog Island. The resulting larger sediment volumes in the southeast make for a greater potential sand input there for the same amount of shoreline retreat.

Two important aspects of dune erosion by storm waves have become evident during the fieldwork. First, widespread dune erosion occurs once in 2-4 years on the average, permitting some recovery of the dune front between episodes of storm damage. In the 1972-1978 period storm erosion was only observed along the entire coast from storms that occurred in June 1976 and December 1977. Dune erosion during other storms was restricted to short sections in several locations along the barrier system. Figure 6 summarizes the changes on profiles surveyed across the dune front and indicates the low degree of erosion in the 1972-1974 period. The December 1977 storm produced the most severe erosion, with measured recession generally in the range 1-4 m. Average wind speeds reached 120 km hour^{-1} from the north and strong winds blew through three high tides. Signs of erosion were present along the entire coast, including sections where no erosion had been observed in the 1972-1977 period.

A question arises as to the degree that the 1972-1978 period is representative of longer time intervals. It appears to be normal for recent decades, based on the examination of storms using 15 years of hourly wind data from Grindstone Island, Iles de la Madeleine (Fig. 1).

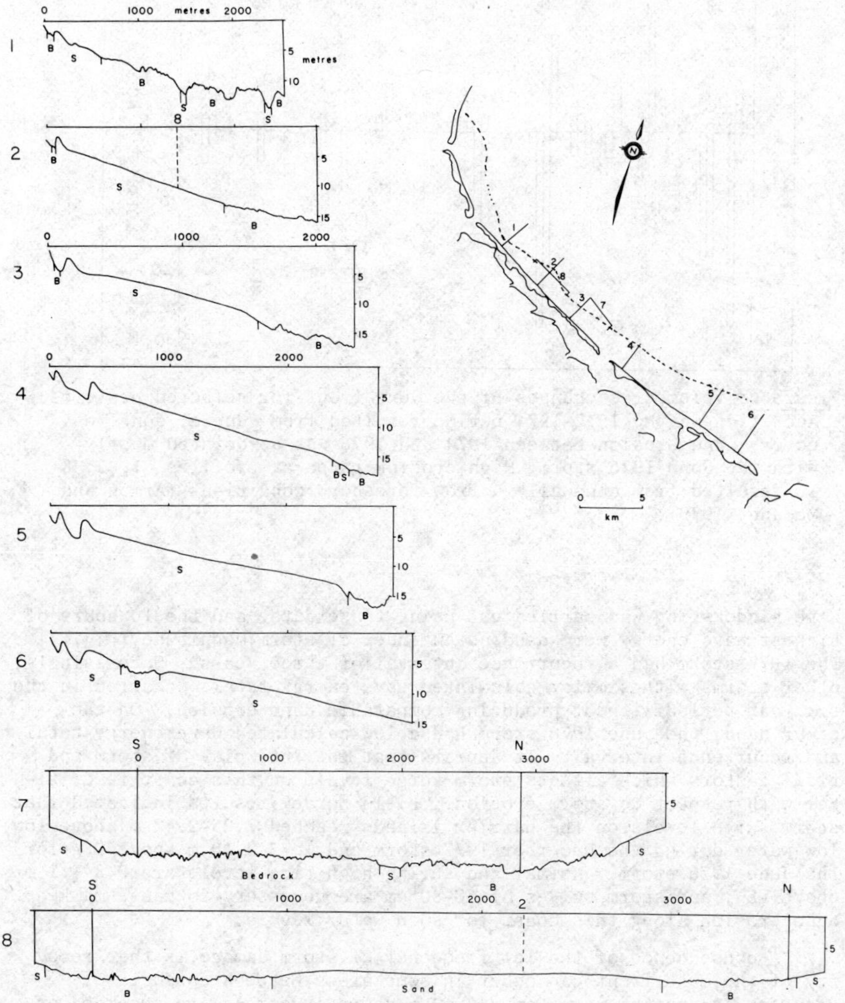

Figure 5: Selected fathometer profiles seaward of Conway Sand Hills and Hog Island. The dashed line on the location map indicates the approximate location of the sand/bedrock boundary. Source: Armon,1975.

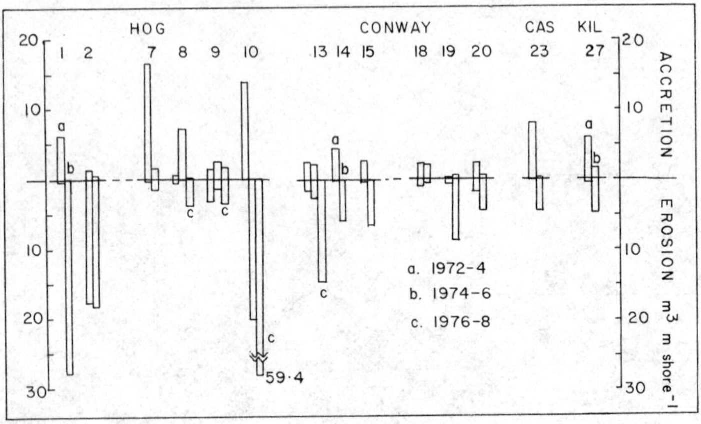

Figure 6: Volumetric changes of the dune front for selected profiles. Accretion in the 1972-1974 period resulted from general dune recovery, and erosion between 1974 and 1976 was associated mostly with the June 1976 storm. High erosion rates at profiles' 1, 2, & 10 resulted from unusually narrow foreshore conditions (Armon and McCann, 1979).

Wave hindcasting was carried out using these data, and the 10 hours of highest wave energy were used as an index of storm magnitude (Fig. 7). The 1977 storm had a recurrence interval of three years. Surprisingly, other storms with similar calculated wave energy totals occurred in the six year period without producing comparable dune erosion. On the other hand, the June 1976 storm had a low calculated wave energy total and recurrence interval. It appears that the interplay of storm and tidal factors which affects storm surge levels on this coast is critical with respect to storm erosion. Field survey results indicated that storm surge levels on the barrier islands reached 2.25-2.42 m above low low water during the December 1977 storm and 1.75-2.10 m above LLW for the June 1976 storm. Normal and spring high tide levels are 0.8-1.1 m above LLW, and storm surges of 60-80 cm are necessary to cause marked dune erosion along this coast for such water levels.

A consequence of the low frequency of storm damage is that recovery of the dune front can occur in some areas between episodes of erosion. Accretion rates of 2-4 m^3 m $shore^{-1}$ $year^{-1}$ were recorded on the seaward dune faces of some profiles undergoing recovery (Fig. 6).

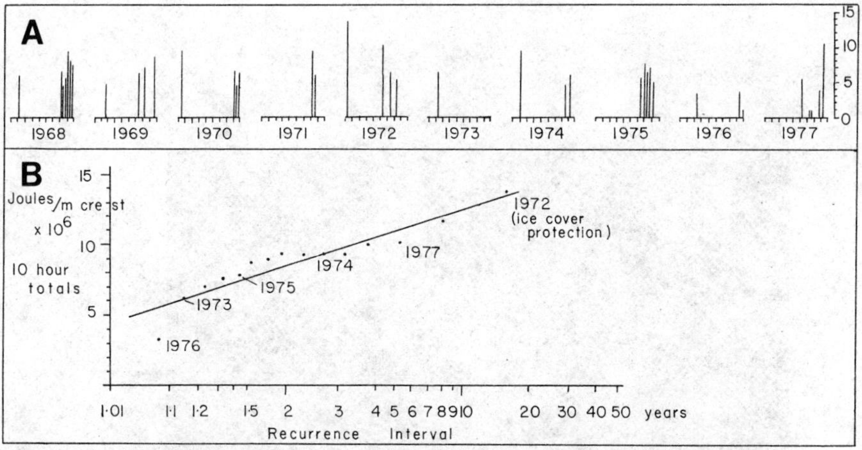

Figure 7: Storm magnitudes established by wave hindcasting - 10 hour totals of highest wave energy. Only the April-December period is considered. A: Storm magnitudes by month, 1968-1977. B: Plot of annual storm series for 1963-1977 data.

Marram grass was also observed growing on the seaward slopes in several locations. These results combine with an absence of net erosion between 1935 and 1968 along the eastern half of the coast to suggest that dune recovery can be complete. Sand transport onto the landward side of the dune front is also significant because of the generally bare seaward slope. Minimum individual storm sand transport rates of 0.035 m^3 m $shore^{-1}$ were measured on the dune top, and sand accretion rates on the landward dune slope were commonly 2-20 cm $year^{-1}$. Locally, accretion rates up to 40 cm $year^{-1}$ were established by changes at survey pegs. In the absence of long term erosion, such wind action has commonly led to the growth of a seaward dune 5-10 m high (Fig. 8). Elsewhere though, the beach profile has been translated landward such that dune recovery and dune dimensions remain limited.

A second aspect of storm actions on the dune zone concerns the dune heights sufficient to prevent storm overwash. The minimum dune height is about 2.0 m above the berm (or 3.5 m above LLW). Areas with dunes of this elevation show no evidence of being overwashed during the 1970s or in the years preceding the 1935, 1958 and 1968 photography. The 3.5 m height is approximately 1.0 m above 1977 surge water levels which may well have been the highest levels since the 1968 photography. Seaward dunes with vegetated beach slopes appeared the same in 1972 and were substantially eroded only in 1977. Surge levels on this coast appear to be relatively low compared with open coast situations.

Figure 8: A high seaward dune along a narrow section of Conway Sand Hills, 1972. View southeast. This photograph was taken close to the high dune beach section (Fig. 4).

DUNE RECOVERY

Dune recovery occurs in two situations along the Malpeque coast. Consideration has already been made of the dune restoration and accretion observed on the seaward dune margin along relatively stable sections of shore. The other, volumetrically larger dune developments have taken place at present inlet margins and past inlet locations (washovers). The patterns of dune recovery at washovers show considerable variability and reveal the impact of shoreline orientation and wind exposure on dune building in the southern Gulf barriers.

The wind rose for Summerside, Prince Edward Island, together with wave data suggest that there is a progressive increase in exposure to north winds and waves in a southeastward direction along the barrier coast. This is particularly evident from an examination of the nearshore bar system. The bar crest-to-trough depth ranges increased to the southeast while the depth of the outer bar crest became greater. This increase in bar crest depth, from about 2 m to 3.5 m by Hog Island, reflects larger wave and breaker heights in the southeast and the greater exposure to north winds. In this more exposed section of coast the two large washovers have remained free of dunes for at least 45 years. The recovery has been restricted to washover margins, where dune ridge sets are developing (Fig. 9). At both washovers the dune ridges are higher by 1.5-2.0 m along the southeastern margins in response to the predominant westerly winds (Fig. 1). The low washover surfaces are close to the water table, most of the sand surface is moist, and sand movement by wind is generally inhibited. This makes for unfavorable conditions for dune development but other factors may also contribute. The strong westerly and southwesterly winds can readily remove dry sands from the open washover surfaces to the margins where they are trapped by vegetation. Dune accretion at the margins of 4-6 m^3 $year^{-1}$ per meter of margin or shore were calculated from field surveys and mapped 1935-1968 changes. Also, the marram grass which does occasionally grow in isolated favorable situations on the open surfaces can be readily removed by storm overwash. The surfaces are low (1.5-1.7 m above LLW) so that low to moderate storm surges can cause overwash.

West of Hog Island, the patterns of dune recovery on open areas have been more variable. This is considered to be a result of the generally less exposed coastal position, changes in shoreline orientation, and smaller open area dimensions in several instances. Some locations might also be sheltered from the full force of westerly winds. The widespread low dune cover on southern Kildare spit shows no sign of overwash influence during dune development. Further south, the pattern recorded for two former inlet sites since 1935 was instead one of isolated dune formation across the lagoonward side of open areas, followed by progressive confinement of storm overwash to channels between the accreting dunes (Fig. 9). This occurred as the dunes extended seaward. Eventually a stage was reached where dunes formed across the seaward ends of the washover channels. Such accretion can be rapid on this coast, given favorable conditions for several years. At the former inlet site on Cascumpec Sand Hills, 1-2 m high dunes formed between 1974 and 1976, fully closing off most of the washover channels there.

This latter pattern of dune recovery possibly reflects less frequent overwashing or less damaging overwash effects on average. The change in shoreline orientation and location toward the northwest suggests that storm effects might be less damaging there for the more common, northerly storm winds. Longshore changes in wave height have been recorded (Armon, 1975) but there is no information to evaluate longshore changes in storm surge elevations. Local factors have had an influence on dune recovery in some localities too. The site on central Cascumpec Sand Hills is partly sheltered by the ebb tidal delta of the inlet to the north. Also, the rapid and widespread dune development updrift of Hardys Channel on the more exposed coast has probably been a response to partial protection by the ebb tidal delta there. That inlet has been migrating steadily to the southeast since 1935.

Local factors might also contribute to the contrasting response on the updrift margin of Palmer Inlet in the western half of the barrier system. An open surface has existedfor 600 m north of the inlet since the 1950s. The small ebb tidal delta offers no protection there and the surface is fully exposed to the predominant westerly winds

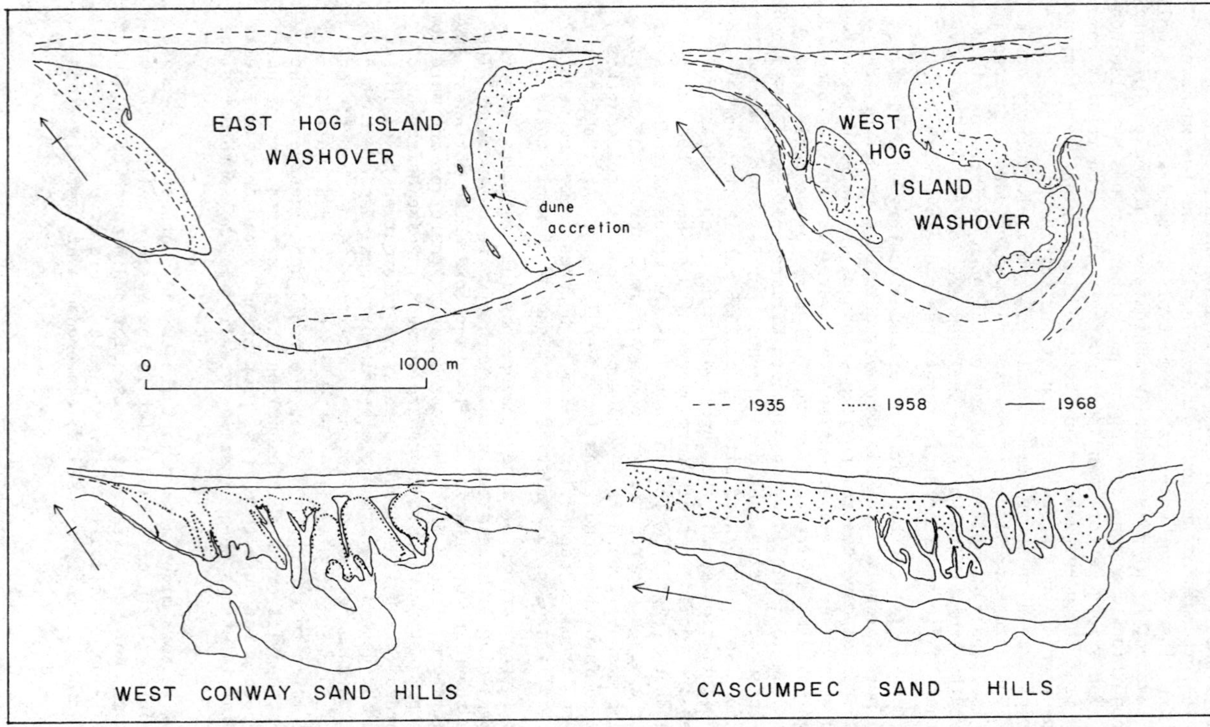

Figure 9: Dune recovery patterns on present or former open surfaces. An inlet was open along west Conway Sand Hills in 1935; consequently, only the 1935 duneline is shown there. See Figure 2 for locations.

that blow across Cascumpec Bay. Northerly winds blow parallel to the beach and westerly winds blow seaward.

DISCUSSION AND CONCLUSIONS

The Malpeque barrier system exists in a relatively windy region but dune erosion directly by wind is limited to dunes in exposed situations. In the absence of agricultural use or concentrated pedestrian traffic the dune vegetation maintains a generally continuous protective cover. Significantly, many active dune blowouts have shown no indication of growing substantially since the 1930s. The major amount of dune destruction along this coast has been associated with wave damage under storm surge conditions. Along eroding sections dune sand loss into the nearshore system is permanent, but in non-eroding (1935-1968) areas dune recovery is often accomplished between storms. Recovery of the seaward dune is possible because damaging storms are relatively infrequent (2-4 year recurrence interval) and onshore winds are common. This occurs without any significant post-storm recovery of the foreshore zone by constructive wave processes. Long period swell waves are not encountered along northern Prince Edward Island, and foreshore profiles show only minor storm changes.

Dune recovery in open areas has been most successful along the western half of the Malpeque coast. In the east recovery has occurred along washover margins whereas in the west the more usual pattern has been one of scattered dune formation across washover surfaces. The latter action involves a progressive confinement of storm overwash to narrow passages and leads to a more rapid dune recovery. The washover surfaces on this barrier are low (1.5-1.7 m above LLW) and are affected by lower storm surges than those producing the widespread dune erosion. Varying dune recovery responses are tentatively related to differences in (1) shoreline exposure, (2) local protection from storm overwash damage, and (3) orientation and exposure to prevailling winds.

Variations in the patterns of dune destruction and recovery at Malpeque point to several aspects of importance for those attempting to maintain or restore dune areas on the southern Gulf barrier. First low dunes 1.5-3.0 m above the backshore or berm appear to be more suitable than high dunes. At Malpeque a high dune backing the beach does not appear to provide a reservoir of sand for the beach system during storms in the way described for open ocean coasts (U.S. Army Corps of Engineers, 1973). Eroded dune sands move into the inter- and subtidal zone during storms but the lack of foreshore recovery prevents their return. The sands that accumulate on the seaward dune front are blown from the upper foreshore. A high dune backing the beach also affects wind conditions in its vicinity and, with a vegetation-free dune front, enhances blowout formation. Low dunes might be subject to greater potential destruction during severe storms because of their smaller volume, but they are less likely to undergo wind erosion.

Second, attempts to restore dunes along open parts of this coast or the barriers to the east should center on establishing dunes in patches, rather than building a continuous duneline. Developing dune nuclei should stand a better chance of survival during storm overwash than a continuous dune barrier. Also, the creation of small patches would facilitate rapid dune accretion, which is so important for dune formation on these low surfaces. The measurement of accretion in 30 m square plantings of marram grass on Core Banks, North Carolina, revealed very high accumulation rates over the first 15 months (Savage and Woodhouse, 1969, in U.S. Army Corps of Engineers, 1973, p. 6-51). Sands can be moved into the dune nuclei from all sides. The presence of dunes in patches would also mean that sands washed between the dunes would enable continued dune accretion in the more landward areas. The formation of a continuous dune could follow successful accretion of the dune nuclei.

The natural establishment of dune nuclei across open barrier surfaces at Malpeque has involved dune initiation 100-200 m landward of the washover crest (Fig. 9). The low surface elevations and limited dry sand in this area might make for lower initial accretion rates than adjacent to the berm crest. However, the landward situation favors deposition rather than erosion during storm overwash, and dunes have a greater chance of survival there. Dune restoration attempts at such distances from the crest would call for the use of slat-type snow fences first, to establish a base for marram planting. The recommended spacing of 5 m between fences 1.2 m (4 feet) high would allow coverage of a 30 m deep area with 6 or 7 fence lines placed parallel to the shore. Adjacent dune nuclei could be separated by 80-100 m, the natural spacing along the western coast. Conservative dune accretion rates at the margins of the open washovers are 4-6 m^3 $year^{-1}$ m $shore^{-1}$ (1935-1974 changes) and initial accretion rates could be significantly higher. Consequently, the planting of marram could follow fence installation by perhaps one season.

Coastal actions along many parts of the souther Gulf barriers favor the formation of washovers, either from the closure of temporary tidal inlets or from general dune destruction. The redevelopment of dunes ensures that both the sand volumes and the general character of the islands are maintained. Results of this study indicate that natural dune restoration occurs over a few decades in more protected shoreline sections. In other locations, however, it may take longer than a century. Attempts to restore dunes to such areas should focus on establishing dunes in patches rather than building a continuous duneline.

ACKNOWLEDGMENTS

The early phases of work were undertaken for a doctoral dissertation at McMaster University, supervised by Dr. S. B. McCann. Thanks are expressed to him for suggestions, discussions and financial support during various phases of the work. The field program was supported by funds from National Research Council, Canada, Grant Numbers A5082 and A4227, and an award from the Science and Engineering Research Board, McMaster University, Hamilton, Ontario. Acknowledgment is also made of the use of the Sigma 9 computer, Memphis State University Computer Center. Conversations with Mr. E. Peck, formerly of Freelands, Prince Edward Island, were valuable.

REFERENCES

Armon, J.W., 1975. Dynamics of a Barrier Island Chain, Prince Edward Island, Canada. Thesis, McMaster University, Hamilton, Ontario, 546 pp. (unpublished).

Armon, J.W. and McCann, S.B., 1977. Longshore sediment transport and a sediment budget for the Malpeque barrier system, southern Gulf of St. Lawrence. Can.J.Earth Sci., 14(11):2429-2439.

Armon, J.W. and McCann, S.B., 1979. Longshore variations in shoreline erosion, Malpeque barrier system, Prince Edward Island. Can.Geographer, 23(1):18-31.

Bayfield, Captain, 1845. Richmond Bay. National Map Collection, Public Archives of Canada, Ottawa, Ontario.

Bayfield, Captain, 1845. Cascumpeque Harbour. Number 2694, Hydrographic Office, Washington DC, 1912.

Holland, S., 1765. L'Isle de St. Jean. National Map Collection, Public Archives of Canada, Ottawa, Ontario.

Landsberg, S.Y., 1956. The orientation of dunes in Britain and Denmark in relation to wind. Geog.J., 122:176-189.

McCann, S.B., 1972. Reconnaissance survey of Hog Island, Prince Edward Island. Maritime Sediments, 8(3):107-113.

McCann, S.B., Bryant, E.A., and Seeley, R.S., 1973. Barrier Island, Shoreline and Dune Survey, Kouchibouguac National Park. Final Report, National and Historic Parks Branch, No. 72-36, Ottawa, (unpublished).

U.S. Army Corps of Engineers, Coastal Engineering Research Center, 1973. Shore Protection Manual. U.S. Government Printing Office, Washington DC.

MANAGING COASTAL DUNES IN SOUTH AUSTRALIA

by

Peter W. Cullen[1] and *Eric C. F. Bird*[2]

[1]Principal Lecturer in Natural Resources,
Canberra College of Advanced Education.
Australia.

[2]Reader in Geography,
University of Melbourne,
Australia.

INTRODUCTION

Sand dunes are common features behind most of the 1900 km of sandy beaches in the State of South Australia. They cover an extensive area, and in many places their stability is threatened, especially by the impacts of man's activities. This paper will consider the sand dune systems of South Australia, the land use pressures they are subject to, and the planning controls available to the State to ensure their stability.

THE BEACH AND DUNE SYSTEMS

The Sand

On the ocean coasts of South Australia beach sands are generally calcareous (usually at least 50%, and locally more than 90% calcium carbonate), dominated by shells and shell fragments, and the remains of other marine organisms, especially algae, foraminifera and bryozoa, carried in from the sea floor. The proportion of non-calcareous material, mainly quartz sand and disintegrated rock particles, is higher on the East Coast of Eyre Peninsula and on the Adelaide Coast.

Coastal dunes derived from these various South Australian beaches are typically well-sorted very fine, fine or medium sands, with particle diameters ranging from 0.06 to 0.5 mm, and some proportions of coarser material incorporated during occasional episodes of strong wind action. In old, stable dunes the calcareous sands become lithified by precipitation of carbonates from percolating ground water to form a relatively coherent dune sandstone, an aeolian calcarenite (American term - eolianite) which impedes subsequent wind erosion, but has been exposed as rocky outcrops in coastal cliffs and shore platforms shaped by marine erosion. In some cases, a thin crust of cemented calcareous sand (calcrete) forms, and tends to preserve the dune topography.

Sand Supply

Sand is supplied to the beaches mainly by onshore movement, since there are no rivers that yield sand to the South Australian coastline.

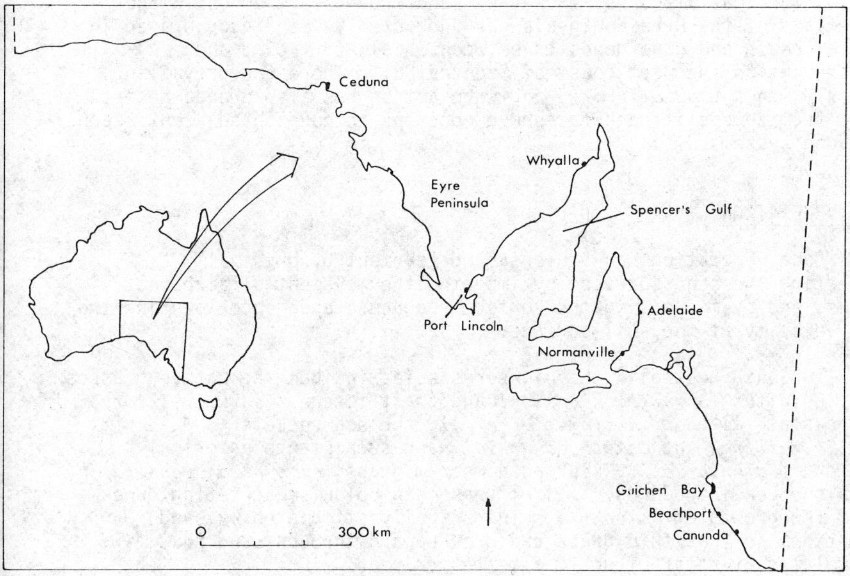

FIGURE I : Location Map

Short and Hesp (1980) estimated that over the past 7000 years rates of onshore sand supply to beaches in the south-east of the State ranged up to 4.7 m^3 per metre of beach per year, with an average of 1.5 - 2.5 $m^3m^{-1}yr^{-1}$, but as a sequel to this large-scale deposition these sandy shorelines are now generally receding. Beach depletion and shoreline erosion have become widespread phenomena elsewhere in South Australia, as on many other coasts (Bird, 1980). It does seem that in general offshore sources of sand have diminished, and sandy shorelines that were previously prograding are now either stationary or receding. As the shoreline retreats, sand lost from the beach and eroded from backshore dune systems is either withdrawn to the sea floor, carried away by longshore drifting, or swept inland by wind action. Where backshore dunes are moving inland, shoreline erosion accelerates because there is less volume of sand to be consumed per metre of shoreline retreat. In this situation, now prevalent on much of the South Australian coast, conservation of backshore dune areas is particularly necessary to reduce the rate of shoreline recession.

As prevailing winds are usually onshore along the South Australian coast, hinterland sand deposits are of little importance as sources for coastal dunes. However, sand from desert dunes in the hinterland of the Eyre Peninsula has undoubtedly been incorporated in coastal beach and dune deposits between Shoalwater Point and Plank Point on the west coast of Spencer Gulf, where the prevailing westerly winds blow *offshore*, and wave action has also eroded desert sands from some cliff and foreshore outcrops to nourish adjacent beach and dune systems.

DUNE VEGETATION

Dune vegetation can prevent sand movement by physically protecting the sand surface, by impeding the movement of airborne grains, and by increasing the surface roughness and hence lowering the wind velocity at the soil surface.

In South Australia the pioneer species include the native grasses, hairy spinifex *(Spinifex hirsutus)* and coast fescue *(Festuca littoralis)*, the coastal saltbush *(Atriplex cinerea)*, the sea rockets *(Cakile maritima* and *Cakile eduntula)*, and such plants as prickly saltwort *(Salsola kali)*. European marram grass, *Ammophila arenaria*, has also been introduced. All of these can colonise accreting dune sand, the growth of *Ammophila* being actually stimulated by sand accretion, so that this grass can grow upward through sand that accumulates at rates of up to a metre per year.

These pioneer dune plants bind sand by their root rhizome networks, their stems and leaves impeding wind action and sheltering the sand surface from the effects of wind scour. In addition, they add organic humus and conserve soil moisture, thereby producing a ground surface that is less readily winnowed than bare incoherent sands.

Dense tussocks of marram grass are particulary effective in stabilising dunes, any sand blown into them being quickly trapped and built up into ridges or hummocks by vertical accretion. *Spinifex* is also an effective sand-trapping agent on the South Australian coast; spreading from the backshore it builds grassy terraces behind the sandy beach. When storm waves attack the beach, the seaward edge of these terraces is cut back, and sand washed or blown up on to the eroded margin may then initiate a new foredune, defined as a ridge of sand built up parallel to the shoreline back of the beach.

The role of vegetation in colonising and stabilising dunes is influenced by ecological factors, especially the availability of water supply. Where coastal rainfall is relatively high, as in the south-east of South Australia (typically 500-700 mm/yr), the growth of dune grasses, shrubs, scrub and woodland is rapid and luxuriant, but on the drier west coast of the Eyre Peninsula (typically less than 500 mm/yr) the vegetation colonising coastal dunes is sparser and poorer, and much less effective in trapping and retaining blown sand. Growth of dune vegetation is also influenced by such factors as fire, salinity and grazing, each of which can weaken the plant cover, reducing its sand-trapping capacity and perhaps allowing previously stabilised sand to become mobile. Grazing and trampling of dune vegetation by native fauna, and by introduced animals such as sheep and rabbits, can initiate blowouts, as can damage where man deliberately clears dune vegetation or damages it by driving vehicles through. Blowouts are wind-excavated hollows through a foredune or other vegetated dune terrain, with sand spilling out to leeward.

The available pioneer vegetation in a coastal dune area is initially restricted to the native species, but has often been supplemented by the introduction of alien species from similar coastal dune environments in other parts of the world. In South Australia the chief introduction has been European marram grass, which grows well on humid temperate coasts but it is less vigorous and less effective in semi-arid environments.

The rate of sand supply to a dune is also a factor influencing the effectiveness of pioneer vegetation. If sand accretion is faster than the capacity for vertical growth of dune plants the vegetation will be overwhelmed and killed, and the dune will become mobile. As plant growth varies seasonally, and is related to such factors as rainfall incidence, it is possible for a dune to be overwhelmed by sand accretion in a drought period, and recolonised and restabilised by vegetation in a succeeding period more favourable to the germination, growth, or regeneration of dune plants. The rate of dune accretion is largely determined by sand supply to the beach and the frequency and strength of onshore wind action.

DUNE TYPES

The South Australian coastline is 3264 km long and some 60% of this is sandy. The Commonwealth Scientific Industrial Research Organisation (CSIRO), has recently completed a National Inventory of the Australian Coastal Zone. The entire coast was divided into 10 km segments and relevant data assessed from random points in each segment, located on recent aerial photographs.

The CSIRO Study defined some 12,680 km^2 as coastal lands in South Australia. If the land within 0.5 km of the coastline is considered then the following land uses can be identified for the main dune types:

TABLE 1 : Vegetation (sq km) on Coastal Sand Dunes within 0.5 km of the Coastline

(Galloway, pers.comm.)

VEGETATION	DUNE TYPES					TOTALS
	Transverse Dunes	Parabolic Dunes	Irregular Dunes	Beach Ridges	Sand[1] Mantle	
Bare	141	186	135	9	78	549
Agriculture (crops)	0	0	0	24	0	24
Pasture	3	3	9	63	18	96
Grass	9	0	9	63	6	87
Shrub	129	273	252	147	165	966
TOTAL	282	462	405	306	267	1722

[1] Sand Mantle refers to a thin layer of sand whose surface is determined by the bedrock form.

WIND EROSION OF COASTAL DUNES

The stability of dunes, whatever their form and origin, is a function of the completeness of their vegetation cover, the degree of coherence of sand bound by organic soil material or cementing carbonates, and the frequency and vigour of the wind action to which they are exposed. Vegetation is clearly the critical factor, determined largely by the available (including introduced) flora, the moisture and nutrient content of the dune soils, the gradient of the dune surface, the aspect of the dune site in relation to wind action, rainfall and continuing accretion (if any), and the impacts of grazing, burning, trampling and traffic. The achievement and maintenance of stability is primarily a matter of the establishment and persistence of a vegetation cover.

(i) Topographic Factors

Lateral variations in exposure to onshore winds are responsible for lateral changes in dune topography, the more stable dunes of sheltered sectors in the lee of a headland giving place to actually or potentially unstable dunes in adjacent sectors beyond that protection.

(ii) Marine Processes

Marine processes are obviously critical in determining the amount of sand deposited on a beach that may subsequently be moved inland by wind action.

They are also critical in removing sand from the dune during periods of wave erosion. During storm wave episodes, substantial volumes of sand may be moved offshore in a few hours. The dune also serves to protect hinterland areas from wave attack.

Wave attack on foredunes cuts a vertical scarp on the seaward margin of the dune. Subsequent sand deposition may rebuilt the seaward margin but during the period when the foredune is scarped it is especially likely to form blowouts.

(iii) Climatic Factors

Not only does much of the South Australian coastline receive little rain, but the rain it does receive is variable. Sequences of dry years occur and dune vegetation is damaged unless the plant roots are able to penetrate to ground water.

Patterns of dunes, transgressive in form but now stabilised beneath vegetation indicate that there have been major periods of erosion in the past with massive sand movements by wind. There are a number of possible causes of such instability, but the weakening or destruction of retaining vegetation during periods of drought is the most likely.

(iv) Impact of Man

Coastal dunes can be modified both directly and indirectly by the impacts of man's activities. Direct modifications include the effects of mining dune sand (as at Normanville) or constructing roads and buildings on dune topography (as near Middleton). Where the coastal dunes were held in place by a vegetation cover, these activities involve at least partial clearance or destruction of the vegetation, which can result in mobilisation of dunes. Indirect modifications also result from damage to dune-retaining vegetation cover by grazing animals, especially sheep, cattle and rabbits, and the driving of vehicles over vegetated dunes.

Rates of Dune Movement

The rate of advance of coastal dune is related to the size and shape of the dune, its orientation in relation to the wind, and the strength and duration of the wind. Measurements from dated sequences of air photographs show that in coastal regions a dune front 10 metres high may advance inland at up to 20 metres per year. Armstrong (1977) has shown that the dune spilling into the north-west of Lake George, near Beachport, has advanced intermittently, particularly during periods of strong onshore winds, by up to 15 metres per year over the period 1945 to 1975. Such information can be used to predict the pattern of dunes and the extent of sand inundation to leeward at future dates. In making such predictions it should be borne in mind that dunes developing behind beaches that are still receiving a sand supply from offshore or longshore will be augmented by onshore movements of wind-blown sand, whereas cliff-top dunes, and those behind beaches that are stable, or losing sand, represent a finite volume of sand that can be redistributed to leeward by wind action. The extent of such redistribution will depend partly on hinterland topography: dunes will move rapidly across low-lying terrain, such as salt flats, but will be impeded where they spill into deeper lakes, or ascend into hilly country.

INSTITUTIONAL SETTING

A number of government agencies have responsibilities for managing sand dunes along the South Australian coast. South Australia was the first Australian State to introduce comprehensive coastal planning for the entire State when it established the Coast Protection Board in 1972.

The Coast Protection Board

In 1960 there was extensive storm damage to the Adelaide metropolitan coast where dunes had been built upon, and local governments pressed the State of South Australia to assist with coast protection and repair. Various Committees investigated the position and a major study of the littoral system around Adelaide was carried out by the University of Adelaide. In 1971 further major storms caused considerable damage, and in 1972 the Coast Protection Act was passed. This established the Coast Protection Board, comprising the Director of Planning, the Director of Marine and Harbours, and the Director of the Tourist Bureau, together with experts in local government, coast protection and the biological sciences. The Board is serviced by the Coast Protection Division of the State Planning Department and is responsible to the Minister of Planning. The duties of the Coast Protection Board are to protect the coast from erosion, deterioration, pollution and misuse, to restore any part of the coast after damage has occurred, to undertake coastal improvements, and to carry out research, cause research to be carried out, or contribute towards research into matters relating to the protection, restoration or development of the coast.

The coast is defined as extending 3 nautical miles seaward and 100 metres landward of High Water Mark, except where declared otherwise by regulation. This interim inland boundary presumably indicated that the greatest concern in South Australia was with erosion occurring in built-up urban areas around Adelaide,and it is likely that regulations will be used to extend this area where dune management is necessary.

The Coast Protection Act also empowers the Governor, on the advice of the Board, to declare Coast Protection Districts. Declarations must be discussed with local governments, and for each district a Consultative Committee made up of nominees from local governments in the district must be established. Districts have been proclaimed in the South-East, Fleurieu, Metropolitan and Yorke Peninsula Areas. Once a District is declared the Coast Protection Board is required to investigate and develop a management plan, which must also be discussed with local governments and the general public prior to its approval by the Minister and the State Governor. Management plans have been developed for the Metropolitan and Fleurieu Coast Protection Districts. These include general coastal policies as well as specific policies for each district.

Once an area is declared as part of a Coast Protection District the Coast Protection Board is authorised to execute all works required to implement an approved plan. Local councils wishing to carry out works on the coast must seek the approval of the Board, and once this is obtained the Board may grant funds to assist with the works. The amount of the grant is determined by the Board subject to the following provisions:

- where the works consist of storm repairs, the grant may cover the whole or any portion of the cost to be incurred by the Council;
- where the works consist of the provision or repair of a coast facility, the grant may cover up to one-half of the cost to be incurred by the Council; and

in any other case, the grant may cover up to four-fifths of the costs to be incurred by the Council.

Where the Board itself carries out works in an area it may recover the appropriate contribution from the Council.

The Board also has the power to control land use in a Coast Protection District. No works of a prescribed nature may be carried out in a Coast Protection District without the approval in writing of the Board. To date no regulations prescribing the works that require a permit have been proclaimed. The development of these regulations has been delayed awaiting the aproval of management plans for the various districts.

The early activities of the Board were dominated by the restoration of erosion damage and the construction of coastal protection works but since 1975 the emphasis has switched to avoiding problems rather than repairing them. With this change of emphasis has come an appreciation that sand dune management is integral to effective management of the South Australian Coastal Zone and that incompatable uses need to be defined and restricted from sand dune areas.

State Planning Authority and Subdivision Controls

Subdivision of coastal sand dunes has occurred in many areas of South Australia. In old established areas, such as Metropolitan Adelaide, engineering works have been used to protect structures on the dunes. In other places erosion has damaged private and public property. The South Australian government has been active in purchasing house blocks in hazardous areas but this is obviously a costly solution for any community to adopt.

Preventing unwise subdivision of coastal dunes is clearly an essential part of any dune management programme. In South Australia, subdivision of lands into blocks of 40 ha or less requires the approval of both the Local Council and the Director of Planning under the Planning and Development Act 1966-1973.

Section 49 and 52 of this Act identify the grounds on which the Director or the Council may refuse approval for a plan of subdivision. The grounds relevant to subdivision of coastal sand dunes are as follows.

49 (a) Land liable to inundation by tidal waters;

(g) Natural slope of the whole of the land in any allotment is steeper than 1 in 4;

(h) The proposed subdivision would destroy any site of exceptional natural beauty (but this section required the government purchase the land).

52 (a) Should the Engineering and Water Supply Department advise that water and sewerage services to each allotment have not or can not be met;

52 (d) The development is considered by the Director to be premature with regard to provision of services, community facilites and nearby unbuilt allotments;

(e) If the natural features (including trees and vegetation) would be materially impaired.

(a, e) If the development does not form a compact and continuous extension of a township or developed urban areas;

(f) If the land or adjacent lands are liable to undue erosion from surface waters or any other cause.

Section 53 of this Act establishes a coastal set-back line of 30 metres in areas that an authorised development plan shows as desirable for preservation or protection.

Controls over Rural Lands

The Soil Conservation Act 1939-1975 provides two mechanisms to control soil erosion. Under Section 6j of the Act is is an offence to cultivate, burn or graze land if that creates erosion, and under Section 13b a Soil Conservation Order can be made to require certain land management practices.

Section 6j(l) states,

> 'A person shall not by cultivation, burning off or stock grazing on any land create on that land conditions as a result of which sand drifts from that land to any other land and detriment or damage to such other land or loss to any other person is caused.'

Any affected or interested person can request the Police to take action under this section, and if convicted of the Offence a penalty of one thousand dollars is incurred.

This Section of the Act has not beenused by the Department of Agriculture who have preferred to use Soil Conservation Orders under Section 13 to direct how the land is to be managed.

Soil Conservation orders can control land management practices and require eroding areas to be treated. There are strong powers to enforce compliance. These powers have been used in coastal areas but not directed towards protecting coastal sand dunes.

There are also restrictions to agricultural uses of coastal dunes that can be applied under the Crown Lands Act 1929-1978. Public lands abutting the coast may be made available for grazing under yearly licences, and under Section 244a of the Crown Lands Act, the Minister of Lands could require the licencee to fence such lands which would assist with controlling grazing. Other lands may be made available to grazing or cultivation for 21 year leases under Section 77 of the Act. Conditions regarding fencing, grazing pressure and possibly rabbit control could be applied with such leases.

While much of the dune vegetation of South Australia has been affected by grazing, there are some sectors where the dune vegetation is still intact. It is urged that these remnants be conserved, both to prevent further instability developing, and more importantly to retain some reference areas in undamaged coastal dune ecosystems.

DEFINITION OF SAND DUNE MANAGEMENT AREAS

While it may be convenient to make general policy statements 'there will be no further development on sand dunes' this is unrealistic since very substantial areas in the coastal regions of South Australia are made up of sand dunes. Even if one considers sand dunes within the land declared as constituting part of the coast, such a blanket prescription is unrealistic since the capability of the dunes to support various uses differs between the wetter South-East and the drier western part of the State.

Clearly, if regulations or planning guidelines are to be developed it is necessary to be able to define the coastal areas to which such guidelines should apply. Such definition should preferably be obvious to non-specialist observers on site, and so some landform or vegetation criteria might be appropriate.

Nordstrom and Psuty (1980) have suggested the research requirements of this problem of delineating dune management areas and identify empirical analysis of dune forms, determination of erosion rates storm frequency analysis, distance of storm penetration related to dune dimensions and effective return distances of overwash to dunes. In the absence of such analytical data for South Australia a number of more empirical approaches have been attempted.

For practical purposes, coastal dunes may be defined as accumulations of unconsolidated sand in the coastal fringe immediately above High Water Mark. In some places they are held in position by a cover of vegetation; in others, where vegetation is sparse or absent, this unconsolidated sand is likely to be mobilised by wind action and carried inland as drifting dunes.

(i) Landform Criteria

It is attractive to consider defining the dune management area on the basis of dune landforms. In locations where the hinterland consists of different landform than the coast this approach can be effective. In the U.S.A., Delaware for example uses the 10 ft contour to define the landward limits of a coastal planning area, and Oregon uses the crest of the first mountain range.

A number of different dune landforms were identified in Table 1, and it is apparent that one landform, 'beach ridges', supports a substantial amount of agriculture in comparison to the other dune landforms. This, and the relatively small proportion of bare sand on this landform unit certainly suggests it is more stable than the other types. If one examines the other landform types in Table 1 it is apparent that each of them have a substantial area of bare sand (up to 50%) and this indicates that these dune types are all unstable.

Furthermore, in substantial areas of South Australia the coastal dunes are backed by stable dune landforms that may still have natural vegetation, or may be in use for farming. This means that any boundary defined simply by landforms will not be satisfactory because substantial stable areas will be included.

(ii) Vegetation Criteria

Any areas of bare sand, and of sand colonised by primary stabilisers (species of *Ammophila, Spinifex, Festuca* or *Cakile)* should certainly be included in a dune management area.

This vegetative criterion does enable a dune management area to be readily identified. Unfortunately it does not encompass all the lands that need specific management. It is necessary to include some of the dunes colonised by the secondary stabiliser vegetation - the heaths and low shrublands. These communities may extend inland many kilometres and it is often not necessary to include all of them in a dune management area.

This secondary stabiliser vegetation is less tolerant to over-use by people and vehicles than the dune grasses (Godfrey, 1977) since it takes much longer to recover from any damage.

(iii) Linear Criteria

Another approach to this problem is to proclaim all lands within a set distance from High Water Mark (HWM) as dune management area. Florida for example, has included all lands within 50 feet of the High Water Mark, and has physically marked this boundary with survey points in some areas. The 1972 California Coastal Act established a permit area of 1000 yards inland from High Water Mark.

There are a number of problems with this approach. A line drawn to include the dune areas that need to be managed is likely to contain considerable areas that do not need such management. The owners of these lands will either be required to observe unnecessary land-use controls, or will need to be specially exempted from the controls following field investigation.

Another difficulty is that the dune area will increase with time if erosion occurs and sand is blown farther inland, so that the advancing landward face of the dune moves outside the dune management area.

Defining a Dune Management Area

It is therefore not possible to define coastal sand dune management areas in South Australia on such simple and obvious criteria as those above.

To define a dune management area where specific controls can be applied requires an expert judgement based on field surveys of each locality rather than any blanket prescription.

The dune management should include the following types of unconsolidated coastal sands:

(i) All area of bare and mobile sand;

(ii) All areas with primary stabilising vegetation (species of *Ammophila*, *Spinifex*, *Festuca* or *Cakile*);

(iii) Any areas with existing breaches of the dune;

(iv) Areas covered with a thin calcrete crust (easily broken by wheels);

(v) Dunes of unconsolidated sand that are not well covered with secondary stabilisers;

(vi) Any dunes that are exposed to wind (rear dunes may be protected by a stable backshore dune and will not be so susceptible);

(vii) Areas in front of advancing sand mass in an eroding area.

MANAGEMENT OF SAND DUNE AREAS

The objective of a sand dune management area is to ensure that sand dunes and their vegetation are protected from damage by man and his activities. Dunes are maintained to act as protective buffers between sea and inland areas during storms.

Once dune management areas are identified then it is necessary for managers to do the following:

(i) Decide what uses are permissible in the dunes;

(ii) Establish effective legislative controls;

(iii) Implement these controls by some system of surveillance and policing;

(iv) Educate local communities about the values of dune systems;

(v) Monitor usage to ensure controls themselves are appropriate, as well as that they are being implemented.

Permissible Uses

As these areas are actually or potentially unstable, they are not capable of supporting uses that will increase the number of people in the area or damage vegetation in any way.

In such areas there should be no further subdivision of land for housing, or further residential development if the area is already subdivided. It may be desirable to purchase properties and return them to public ownership.

Agricultural uses may be tolerated in some dune management areas, but it is important to control the grazing and burning of dune areas. Unfortunately it is during drought periods that heavy grazing pressure is likely, and it is at such times that the tolerance of dune vegetation to grazing and trampling is probably lowest.

Recreation pressures from people wanting to drive vehicles in dune management areas need careful control. In some areas pressures may exist from Off-Road Vehicle Clubs for areas of bare sand. In other areas less structured vehicle use can cause problems. This may be from off-road vehicle recreation, or from fishermen wanting access to the sea for themselves and boats. Pressures will exist for coast roads, car parks, boat ramps, camping areas and other facilities. Decisions can only be made on a case by case basis after a consideration of the dune system itself, the pressures on it and the management skills and resources available. Environmental Impacts of such activities in dunes has been recently reviewed by Vogt (1979).

Methods of Control

Subdivision throughout the State is controlled jointly by local government and the Director of Planning. Further controls could be introduced by regulation, under the Coast Protection Act, but before such over-lapping controls are introduced an attempt should be made to have the Director of Planning accept the validity of dune management areas and to have him refuse subdivision proposals in such areas. Such an agreement seems possible at either the agency or the Ministerial level and seem preferable to duplicating these controls.

Agricultural uses can also be controlled under existing legislation covering Soil Conservation and Crown Lands. Again, there seems little point in duplicating these controls unless it can be shown they are ineffective.

Other uses that need to be controlled such as off-road vehicles, sand removal, roads, car parks, caravan parks, boat ramps and other buildings appear not to be adequately controlled under existing legislative arrangements. It is necessary to control these uses if the sand dunes of the State are to be protected, and it has been proposed that these uses should be controlled under the Coast Protection Act. This requires the promulgation of regulations that define any works associated with such uses as 'prescribed works'.

Policing of Controls

Having legislative control is a necessary but not sufficient condition to ensure protection of dune management areas. Ideally the Coast Protection Board should have field officers who could ensure compliance with permit conditions. It may however be necessary to develop a collaborative arrangement with local government or other State agencies such as National Parks, to ensure that controls are enforced.

Public Education

The long-term efficiency of any attempts to control coastal land-uses is dependent on how well the community understands the controls and the need for them.

It has been proposed to the Coast Protection Board that a public awareness programme be designed to reach land holders and the general public and tell them about the controls and the reasons for such controls. It is necessary to gain the support of the community if the Coast Protection Board's objectives are to be achieved.

Monitoring of Controls

As well as policing of controls to ensure they are implemented, it is necessary to monitor the condition of vegetation in a dune management area, as well as the rates of dune movement, to see if the controls applied are being effective. This can be done by periodic aerial photography of dune management areas.

CONCLUSION

It is necessary to protect sand dunes from the impacts of man and his activities especially in arid regions where dunes can be destabilized readily and may be impossible to re-stabilize. To achieve this protection it is necessary to identify dune management areas over which certain controls can be implemented.

In the South Australian situation simple criteria such as landform, vegetation or a linear distance were not found to be appropriate for identifying dune management areas. A number of considerations have been identified and it is proposed these be applied in each area by an expert panel using aerial photographs with some ground inspection.

Once dune management areas are identified in a way that can withstand legal challenge then the planning criteria for dunes need to be further developed. Existing controls over subdivision and agricultural uses seem adequate, but additional controls over recreation uses will need to be imposed. The difficulty of policing controls as well as monitoring their effectiveness has been identified, and the need for a programme of public education has been stressed if community acceptance of such controls is to be obtained.

REFERENCES

Armstrong, D. (1977). *Drift sand migration, north-west corner of Lake George*. Department of Mines, Sth. Aust. Tech. Serv. Report 77/133.

Bird, E. C. F. (1980). *Recent changes on the world's sandy shorelines.* I.G.U. Commission on the Coastal Environment, Report No. 1. (In press).

Cullen, P.W. & E.C.F. Bird (1980). *The Management of Coastal Sand Dunes in South Australia.* Report to Coast Protection Board, South Australia.

Godfrey, P. J. (1977). *Recreational Impact on Shorelines Research and Management.* In Proc. 'Coastal Recreation Resources in an Urbanising Environment'. Univ. of Massachusetts, pp.113-118.

Nordstrom, K. F. and N. P. Psuty (1980). *Dune District Management : A Framework for Shore-front Protection and Land Use Control.* Coast.Z.Mgmt J. 7, pp.1-23.

Short, A. D. and P. A. Hesp (1980). *Coastal Engineering and Morphodynamic Assessment of the Coast within the South East Protection District.* Report to Coast Protection Board.

Vogt, G. (1979). *Adverse Effects of Recreation on Sand Dunes : A problem for Coastal Zone Management.* Coast.Z.Mgmt J., 6, pp.37-68.

DELINEATION AND IMPLEMENTATION OF A DUNE MANAGEMENT DISTRICT

by

Paul A. Gares[1]
Karl F. Nordstrom[2]
Norbert P. Psuty[3]

INTRODUCTION

Like their counterparts in other areas of the United States, inhabitants of New Jersey's coastal areas are faced with shoreline erosion and its concomitant effects. Substantial efforts have been made in New Jersey to reduce the problem through construction of groins and bulkheads. However, people are discovering that these traditional methods of shoreline protection are expensive and they are only short term solutions. The underlying causes of shoreline erosion are a rising sea level and decreased sediment sources. It is desirable to develop protection strategies which are compatible with these constraints of the natural system. One protection strategy directed toward this objective is the implementation of a Dune Management District. The proposed management area is located in the dune zone because of certain protective properties associated with the dunes. Dunes provide a natural barrier to flooding

[1]Research Associate, Center for Coastal and Environmental Studies, Rutgers--The State University of New Jersey, New Brunswick, N.J. 08903

[2]Associate Research Professor, Center for Coastal and Environmental Studies, Rutgers--The State University of New Jersey, New Brunswick, N.J. 08903

[3]Director, Center for Coastal and Environmental Studies, Rutgers--The State University of New Jersey, New Brunswick, N.J. 08903

during coastal storms, and they represent storage of sediment which can be released during storms and which can be used to build beaches. Dunes also offer other advantages in terms of coastal aesthetics and wildlife habitat. Dune districting is both a method of creating shoreline protection and a framework for a comprehensive shoreline management strategy. The method offers a form of protection which is essentially a compromise between natural and developed systems (Nordstrom and Psuty, 1980: pg.8). The full rationale for the establishment of Dune Management Districts is discussed by Nordstrom, Psuty and Fisher (1978) and by Nordstrom and Psuty (1980).

This paper presents the results of a study conducted for the State of New Jersey whose objectives were to identify the functions of coastal dunes and to identify and map dune dimensions. Further, we developed a methodology for identifying the degree of protection afforded by existing dunes and we also established a process for delineating a Dune Management District along the New Jersey shoreline. This methodology is described in greater detail in Garès, Nordstrom and Psuty (1979). A dune management model is developed here which includes both a method for establishing the boundaries of the district and the identification of practices which should be encouraged within the district. This methodology can be employed in other areas by inserting the appropriate data.

DUNE MANAGEMENT PRINCIPLES

The establishment of a Dune Management District is a component of the total concept of managing the beach/dune system. The application of the use of a Dune Management District derives from a theoretical base and is developed from guidelines for its implementation. As the Dune Management District concept becomes accepted by shoreline managers it is important to remember certain basic principles which need to be observed:

1. The beach and dunes together form the beach profile in which sediment is continually being exchanged between the two sub-units, and to treat one without recognition of the other is shortsighted.

2. Dunes provide for life safety and property protection during the elevation of storm water and they buffer rates of shoreline erosion.

3. The beach/dune system is migrating and thus any Dune Management District boundaries are assumed to be dynamic.

4. Because of the dynamic nature of the dune zone, Dune Management District boundaries should be reviewed at predetermined time intervals, or following a catastrophic storm which significantly modifies the beach/dune system.

5. In order to be defensible in a court of law, a Dune Management District must be given absolute dimensions which are obtained through the use of scientific theory and empirical data.

6. Planning for development in or adjacent to the Dune Management District should accomodate natural shoreline migration by providing for

development that is compatible with the natural processes.

COMPONENTS OF THE DISTRICT DELINEATION METHODOLOGY

An important feature of the Dune Management District is the natural protection which is provided against flooding by raised water levels associated with coastal storms. It is, therefore, desirable to be able to predict the elevation to which storm waters will rise for different intensities. This information may then be used to determine the dune dimensions necessary to provide protection against those storms. In addition, because the dune district concept allows for the migration of the dune, this factor must be incorporated into the Dune Management District methodology.

Dune Height

The height component of the dune determines the extent to which inland areas may be flooded by ocean water as a result of waves overtopping the dune. The overtopping is a function of wave runup which is calculated using a procedure described in The Shore Protection Manual (CERC, 1977). Although the CERC procedure applies primarily to the prediction of runup on engineering structures, it has been successfully used in the prediction of runup on beach/dune systems (Maurmeyer, 1980).

Principal components of the wave runup calculation procedure are the foreshore and nearshore slopes, dune foreslope, beach width, deepwater wave height, wave period, and storm surge. Recognizing the geographical variation of these components along the shore, the coastline is divided into segments. This segmentation should, optimally, result in the identification of shoreline stretches of only a few miles in length. The degree of segmentation is, however, dependent on the quantity of the beach and ocean data that are available. If the required data are available in abundance, it may be possible to establish many short segments. If, however, the data are available for only a few locations, segmentation may have to be done at a more gross scale. Once the shoreline has been divided into segments, design beach profiles are then constructed by averaging profiles gathered in each segment by field survey. The empirical profile data represent storm conditions because the absence of a berm and a narrower beach result in the highest wave runup, and, therefore, give a conservative estimate of the protection requirements for each segment. In this paper, the Dune Management District delineation procedure will be illustrated by focusing on two design profiles which represent Long Beach Island and Ludlam Island (Figure 1).

In New Jersey, ocean conditions required for the runup calculations were gathered for a number of years at Atlantic City (U.S. Army Corps of Engineers, 1976; Thompson, 1977). This is the only location in New Jersey where wave data have been collected over the long term. The network of tidal gauges is more dense, but the data do not always cover a long period of time and they are not always analysed in terms of frequency. Although site specific data should be employed where possible, the Atlantic City data are used in the calculations presented here because they already exist in the desired form (Figure 2).

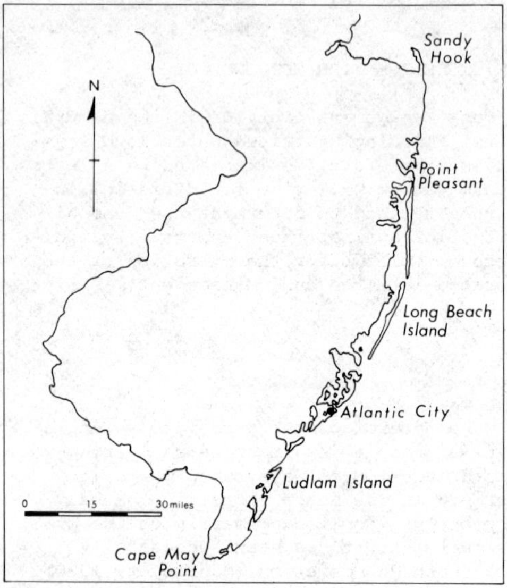

Figure 1: Dune Management District sample locations

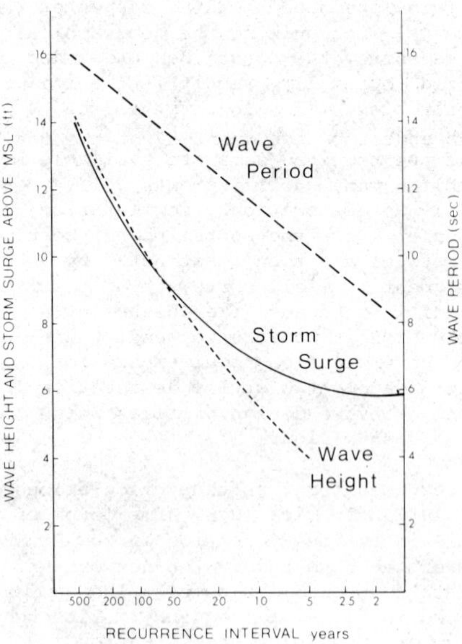

Figure 2: Wave and storm surge data used in wave runup calculations (Source: U.S. Army Corps of Engineers, 1976 and Thompson, 1977)

The height of wave runup associated with a storm of any frequency magnitude is calculated using the wave and tide frequency data (Figure 2) and the design profiles. The objective is to compute the elevations reached by the combination of storm surge and breaking waves during storms of different intensities. To accomplish this, wave height, wave period, and storm surge data are obtained from the frequency curve (Figure 2) for sample storm frequencies and a data set is, thus, formed (Table 1). The CERC runup calculation procedure for a composite slope is then followed to obtain a wave runup value for each of the sample storm frequencies. These wave runup values are plotted on frequency graph paper to produce a wave runup prediction curve for each segment (Figure 3).

Once the wave runup values are obtained, it is necessary to transform them into equivalent dune heights. In this case, dune height is defined as the difference between the highest point on the dune portion of the profile and the maximum beach elevation, located usually on the backbeach at the foot of the dune foreslope. To obtain the dune height required to prevent overtopping during a storm of a given frequency magnitude, it is necessary to subtract the backbeach elevation from the predicted wave runup for that storm. In this fashion, dune height required to protect against overtopping may be represented on the same graph as the wave runup (Figure 3). In the case of the two examples, the backbeach elevations on the design profiles for Long Beach Island and for Ludlam Island are 10 feet and 8 feet respectively (Figure 4). These elevations are subtracted from the wave runup values on the prediction curves for each location and the dune height may also be predicted from Figure 3 by reading from the outside y-axes.

The height of wave runup varies from shoreline segment to shoreline segment as a result of different characteristics of the beach profiles for each segment. For Long Beach Island (Figure 4) a storm with a 20 year recurrence interval produces a runup of 12.9 feet, whereas a storm of a 77 year recurrence interval overtops the dune. At Ludlam Island (Figure 4) the same 20 year storm has a runup of 10.8 feet. The dune on this profile will be overtopped by a 40 year storm. These comparisons are made on the basis of one set of wave data from Atlantic City. Where additional data are available, the differences may be accentuated due to differing ocean conditions at each shoreline segment.

The depiction of the wave runup differences between Long Beach Island and Ludlam Island not only shows how the profile modifies the wave runup, but it also indicates how this methodology may be used to determine the protective capabilities of the existing dunes. This information is useful to shoreline managers because it may be used to point out areas along the coast where hazard potential is great. An analysis of the protective capabilities of the New Jersey dunes has been accomplished by using actual dune heights obtained from forty-four profiles taken between Sandy Hook and Cape May Point (Figure 5). Dune height for each profile was obtained from the survey data and each value was assigned to an appropriate classification category. Because the dune height at each survey line is categorized, the dune line within each segment is represented not in terms of a specific value, but rather in terms of a range of values.

Table 1. Design Specifications for Runup Calculations (from Figure 2).

Storm Frequency (years)	4	5	10	20	50	100	200	500
Surge (ds)	6.1	6.3	6.9	7.8	8.9	10.1	11.6	14
Wave Height (H) (USACE)	-	4	5.6	7	9.1	10.6	12.1	14.1
Wave Period (T)	9.7	10.2	11.2	12.4	13.6	14.2	14.9	15.6
Wave Runup (R) above ds	2.6	2.8	4.9	6.7	9.3	12	13.7	17.6
Wave Runup Above MSL (R+ds)	8.7	9.1	11.8	14.5	18.2	22.1	25.3	31.6

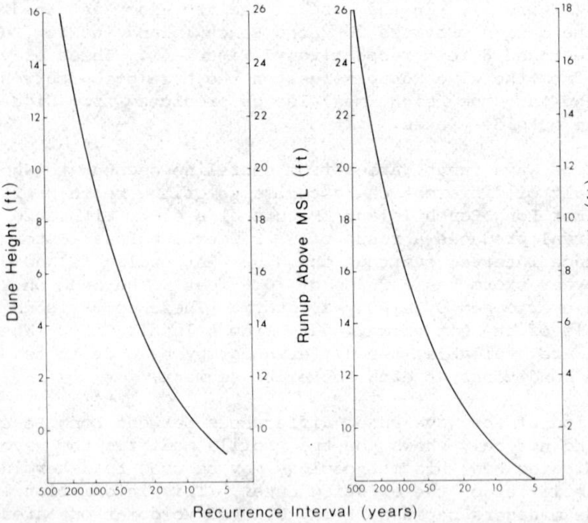

Figure 3: Runup predicton curves for Long Beach Island (left) and for Ludlam Island (right).

Once dune height is mapped, it is necessary to determine the frequency magnitude storm against which the range of dune heights for each segment provides protection. This is accomplished by entering the wave runup frequency curve from the dune height axis with the appropriate dune height, and the storm frequency value is read off the x-axis. Because the dune height for each segment is represented as a range of values, so too are the protective capabilities of the dunes. The data shown in Figure 5 are rather crude due to the scale of the illustration, but it is quite apparent that this procedure may be applied to very small stretches of the shoreline. If a site specific analysis were required, a wave runup curve could be constructed for an actual rather than average beach/dune profile.

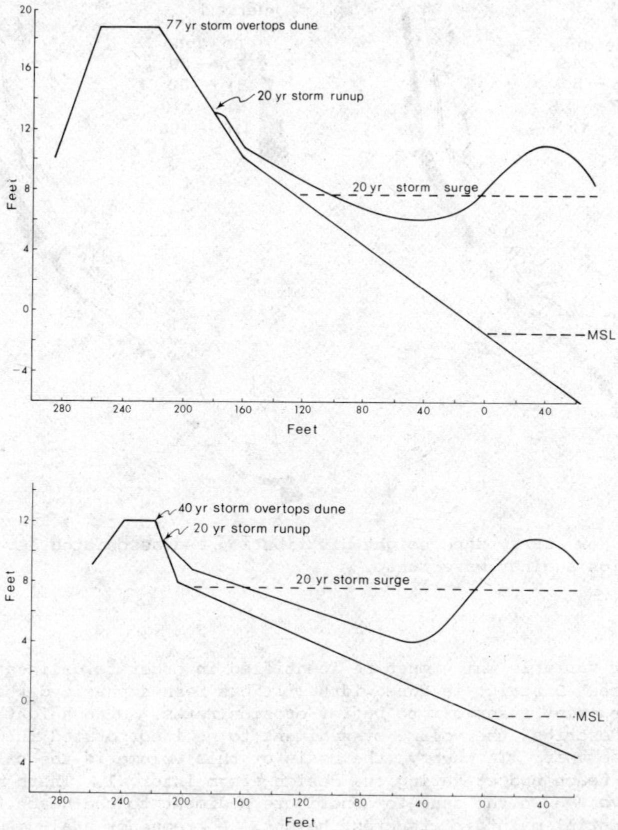

Figure 4: Predicted wave runup on design profiles for Long Beach Island (top) and for Ludlam Island (bottom).

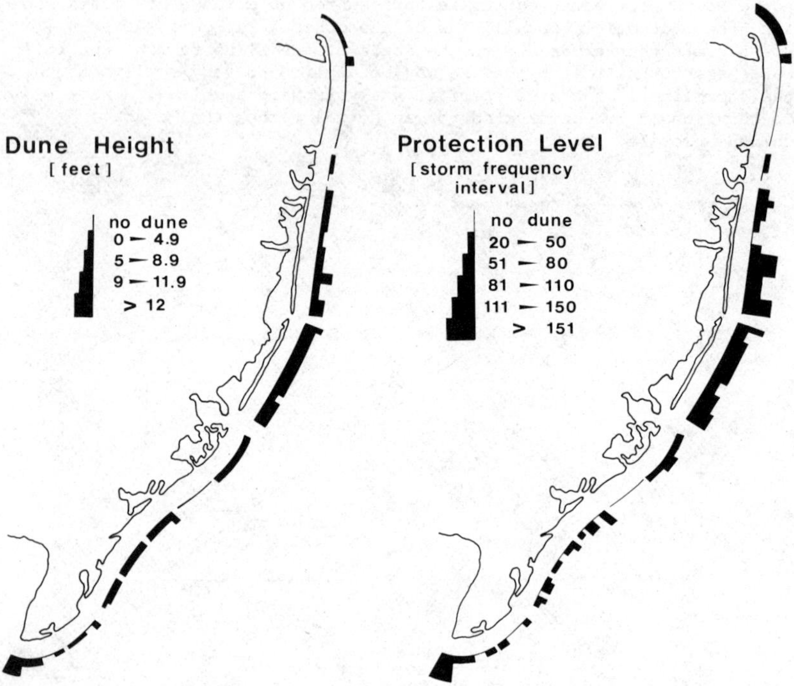

Figure 5: New Jersey dune height distribution and associated level of protection against wave runup.

Dune Width

Another variable which must be identified in order to delineate the Dune Management District is dune width. It has been emphasized that dunes provide a sediment reservoir to buffer erosion rates. Dune height plus dune width describes the volume of sediment to be incorporated into the sediment reservoir. In theory, the basis of this volume is the calculation of the beach budget during the design storm interval. There must be sufficient volume in the dune to contribute sediment to the beach to buffer erosion and still maintain its crest height. A frequency analysis in which it would be possible to predict dune erosion rates associated with individual storms would constitute the best data source for this part of the methodology. A dune erosion frequency prediction curve would indicate how much erosion would be expected for a storm of a given frequency magnitude. This amount of lateral dune retreat would, then, constitute the minimum dune width for that particular storm. However, the available models (Edelman, 1969; U.S.Army Corps of Engineers, 1973) which perform this predictive function are still too crude to yield reliable dune erosion estimates. Another thechnique is, therefore, required.

Using the concept of dynamic equilibrium, it seems reasonable that dunes of given heights must also have an equilibrium width. Ideally, this represents an appropriate approach to determining dune width. However, there is a lack of information about the equilibrium dimensions of dunes. One approach to solving the problem of establishing a dune height/width relationship, would be to examine empirical dune height and width data and to correlate the variables using regression. This correlation analysis was performed for New Jersey dunes (Figure 6) using the data from the beach/dune profiles and from profiles published by Birkemeier (1979). The data set purposely includes both natural and artificial dunes. The north-easterly storm of March, 1962 destroyed many of the dunes which existed along the New Jersey shoreline. Following the storm, dunes were created by the U.S. Army Corps of Engineers along the developed portions of the shore and these remain today. Both natural and artificial dunes were included in the regression analysis because it was felt that to include only natural dunes in the analysis would not be representative of the developed nature of the New Jersey shoreline.

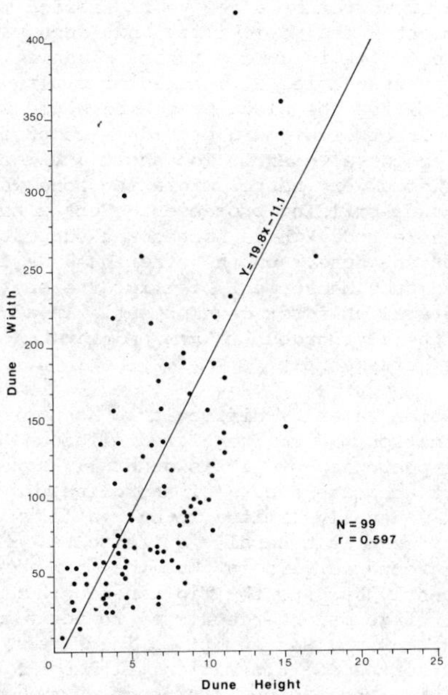

Figure 6: Dune height/width regression line.

For the methodology presented here, dune width can only be computed once the dune height value is known. The dune height value for a given storm is obtained from the wave runup curve. The height/width regression line is, then, entered on the x-axis using that dune height value. The corresponding dune width is read off the y-axis. Although this approach does not take the beach budget into account, it is felt that this constitutes an acceptable method of establishing dune width for storms of given recurrence intervals. Once dune erosion prediction techniques have been improved, they should be substituted for the method employed here.

Erosion Trends

One basic principle of Dune Management District delineation is that the district will migrate along with the barrier island and that its boundaries, therefore, must be readjusted after a predetermined time interval. The Dune Management District is designed to be broad enough to allow for the design dune whose dimensions are obtained as described in the preceeding paragraph, as well as for the migration of that dune. This migration may be accomodated by providing a migration zone inland of the dune that is related to the rate of shoreline displacement. The landward boundary of the Dune Management District is determined by adding the design dune width to the amount of erosion predicted to occur before readjustment takes place. It is assumed that in the future erosion will proceed at the same rate as in the past. Erosion rates used here are obtained by examining the historical record of shoreline change. It is recognized that short-term fluctuations will occur where, for a few years, erosion may cease, or accretion may even take place. The possibility that such variations might occur should be accounted for in dune district planning by implementing a review of the district boundaries at the end of a planning period. An appropriate length of time for the planning period would be on the order of 10 years. This amount of time would provide a compromise between resource conservation and land development. Too short a time period would result in the creation of a very narrow migration zone which would limit the effects of natural dune building processes. Such a zone would replicate conditions at the shore today where houses and support infrastructure leave little room for the dune. On the other hand, a long planning period would create a broad Dune Management District thereby severely limiting the amount of space available for development. In addition, a 10 year planning period has the advantage of reducing administrative problems by requiring review fairly infrequently.

The determination of past erosion rates is difficult to accomplish. Three sources of this type of information may be identified: 1) ground survey; 2) maps and charts; 3) aerial photographs. It is generally agreed that ground surveys provide the most accurate record of shoreline migration. However, this type of survey is usually limited to only a few shoreline locations, and the historical record is generally quite short. Maps and charts have the advantage over ground surveys in that they can provide long-term information, dating frequently back to the mid-nineteenth century. Unfortunately, consistent coverage may be restricted to certain specific locations and the early cartography was sometimes inaccurate. Aerial photographs represent a good source of information, although coverage exists only for the last 40 years. Photographs represent a compromise between the other two potential sources of information in that they cover a longer time period than surveys, although they are less accurate. They can also be more accurate than maps and charts, though they cover a shorter time span. There is, however, measurement error associated with lens distortion, variations in flying height, quality of older photographs and human fallibility (Nordstrom and Psuty, 1977; Dolan et al., 1978).

In spite of these inherent problems associated with aerial photograph analysis, it is suggested that they constitute an important source of erosion data. These data should be complemented by using any other sources of information that are available. A comprehensive approach to the problem of determining erosion rates, therefore, is required.

In the case of New Jersey, erosion rates were readily available from Nordstrom *et al*. (1977). This data set included both long-term erosion data computed from U.S. Army Corps of Engineers shoreline change maps dating from the 1830's and more recent erosion information derived from the comparison of shorelines shown on aerial photographs taken in 1952 and in 1971. For the Dune Management District delineation procedure, the erosion data pertaining to the 1952-1971 period were used. It was felt that the shoreline had been subjected to intensive emplacement of engineering structures which modified the erosion rate available from historical maps and charts. The more recent data were, therefore, considered to be more representative of current erosion trends along the New Jersey shoreline.

DESIGNATING THE DEGREE OF PROTECTION

Before calculating the width of the district, management decisions must be made as to the degree of protection that the dune will be required to provide. This is not necessarily an easy decision to reach. In the Dune Management District concept a conflict arises between the amount of overwash that is required to maintain the dune and the amount of protection desired against flooding. Dunes may be built to protect against the extraordinary storm and to maintain the shoreline in a static position. However, it has been suggested that dune migration is accomplished by a combination of overwash processes and of aeolian reworking of washover sediment (Godfrey, Leatherman, and Zaremba, 1979). A critical question which comes out of this discussion concerns the degree to which the dune should function as a barrier to overwash. If dune districts are created in which no overwash is allowed, there will be no sediment delivered by this process to build up the landward edge of the dune. Unless there is net onshore migration by aeolian processes, the dune crestline and the landward limit of the dune will remain in a relatively stable location. The result could be the gradual elimination of the beach from in front of the dune and eventual dune destruction with elimination of its protective function. The Dune Management District concept must provide for the maintenance of the dune system as well as for the protection of areas landward of the dune.

Leatherman (1979) suggests that storms with a 50 to 100 year recurrence interval are responsible for the majority of the landward transfers of sediments on the northeast coast. In terms of natural processes, there appears to be no reason why a natural dune can not afford protection against a storm with a recurrence interval of 100 years. However, the width of the dune zone required to maintain a dune of sufficient dimensions would be considerable. A dune which protects against a 100 year storm will reduce the amount of overwash and will thus limit dune migration.

A major factor to be considered in terms of protection level is the condition of the existing dunes. In New Jersey, for instance, there exist very few locations where the beaches and dunes have the height/width characteristics necessary to protect against a storm with a 100 year frequency magnitude (see Figure 5). If the 100 year storm is selected as the design storm, it will be necessary to build the beaches and dunes to the required dimensions using sand nourishment and stabilization techniques. The cost associated with beach and dune building projects would become a factor in a decision regarding the desired level of protection.

In New Jersey, a 50 year storm is suggested for use in the Dune Management District delineation procedure for several reasons. New Jersey dunes are such that little modification is necessary to conform to the dimensions of a 50 year dune. In addition, a 50 year dune allows for natural washover processes to occur, and dune migration in response to a rising sea level will proceed. The design dune still has the advantage of providing protection against low intensity storms which occur relatively frequently. The dimensions of the design dune will also be such that a minimum amount of land will have to be set aside for the Dune Management District. Thus owners of property within the district will be able to draw some benefit from their investment. It appears, therfore, that a 50 year storm represents a logical design interval. However, the selection of the design storm interval should be modified if new information concerning dune processes becomes available.

DELINEATING THE DUNE MANAGEMENT DISTRICT

The determination of the width of the Dune Management District is accomplished through an analysis of wave runup, dune height, dune width and erosion rate. Examples of the delineation procedure are provided here for the two sample locations (Figure 7) for a storm with a 50 year recurrence interval. The first step is to obtain from the appropriate wave runup curve the dune height which is necessary to protect against the 50 year storm. Next, a dune width value is obtained from the dune height/width regression line. Finally, the erosion rate for the shoreline segment in question is obtained. This erosion value, given in feet per year, is then multiplied by the number of years in the planning period. In the examples presented here, the width of the dune migration zone is computed on the assumption that a 10 year planning period is in use. The erosion rates shown in Figure 7 represent the average annual erosion multiplied by 10. The width of the Dune Management District is, thus, computed by adding the width of the design dune to the amount of erosion predicted to occur during the time span of the given planning period.

Looking at the two examples, one notices variations in the width of the district. These result from two causes. First, the different beach and dune slopes cause the runup values to differ for each of the shoreline segments. This difference would probably be accentuated if site specific wave and tide data were utilized. The dune heights of each sample differ because of the variations in wave runup, and as a result the dune widths are also different because of the relationship between dune height and dune width. Secondly, the erosion rates vary along the shoreline. These differences in erosion rates are accounted for in the width of the dune migration zone.

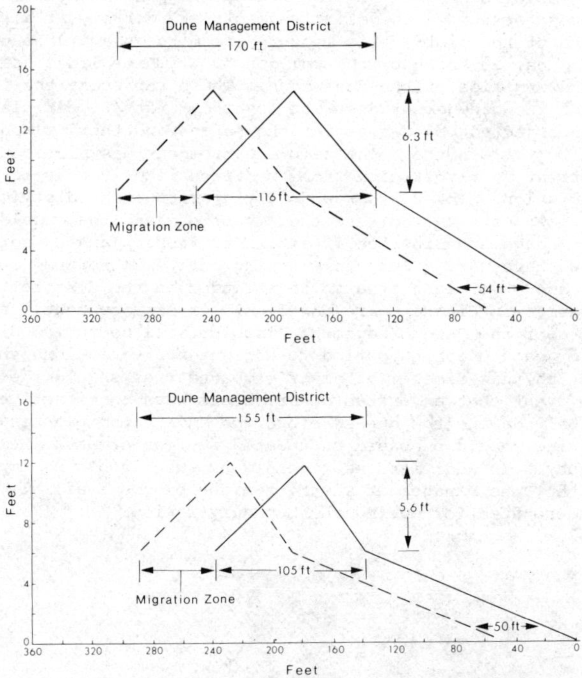

Figure 7. Dune Management District delineation procedure for Long Beach Island (top) and Ludlam Island (bottom).

DUNE MANAGEMENT DISTRICT MAPPING PROCEDURES

The Dune Management District must be mapped in order for it to be implemented by the responsible state and local government agencies. There is a need for precision in the mapping of the district because it must be carefully identified in relation to existing property lines. Therefore, the Dune Management District was delineated on maps whose scale was 1:2,400. A stereo zoom transfer scope was used to facilitate the transferral of information from aerial photographs to these large scale maps. Mapping for this project was done using overlays designed to fit over New Jersey Department of Environmental Protection wetlands photomaps (Figure 8).

Two sets of information are represented on the maps. A polygon identified as "D" indicates the dune in its most recent form. This provides a perspective on the difference between the existing dune and the proposed Dune Management District. A second area, identified as "DMD", depicts the proposed Dune Management District. The width of this area, calculated according to the procedure presented herein, must be recomputed for each of the shoreline segments. The present computation procedure used erosion rates for the 44 shoreline segments (Nordstrom, et al., 1977). The width of the district is measured from the seaward boundary of the existing dune, as shown by the polygon "D". Three lines are located to the landward side of the dune zone. These represent the initial landward limit of the management district for the first 10 year planning period, and the projected boundaries for the second and third planning periods, 20 and 30 years hence. The sample management district boundaries are based on current yearly ersoion rates modified into 10 year intervals of shoreline recession. The 20 year and 30 year management district boundaries are shown here to indicate the amount of land that would fall within the Dune Management District if either of those planning periods were adopted. We have used a 10 year planning period to reduce as much as possible the amount of land area to be included in the district. The projected boundaries in Figure 8 are based only on data currently available. The limits of the Dune Management District will be reviewed at the conclusion of each planning period to account for variations in the erosion rate during that time. The district boundaries should also be reviewed in the event that an extraordinary storm were to effectively destroy or severely damage the dunes before the end of the planning period. All boundary review should be based on updated sources of data including aerial photography. In addition, new techniques which could be employed in delineating the Dune Management District might become available, and they should be accounted for during the boundary review.

Figure 8. Example of Dune Management District maps showing the predicted locations of the landward boundary of the district for selected 10 year intervals.

DUNE MANAGEMENT DISTRICT MODEL REFINEMENT

The primary objective of the study is to identify a method by which a Dune Management District can be delineated. The procedure for calculating the dimensions of the district is applicable immediately. However, an important feature of future research should be concerned with refining the model. There is much to be learned about dune processes both on undeveloped and developed shorelines. As information about these processes becomes available, it should be incorporated into the procedure.

The foremost requirement for the refinement of the model is the accumulation of additional data. It has been shown that many factors influence wave runup. Yet, for this project it has been necessary to assume that information about storm surge, wave height and wave period gathered at one location is representative of the entire coast of New Jersey. Whereas the tidal range on many coastlines is sufficiently consistent to allow this assumption to be acceptable, wave characteristics show greater variability from place to place. In order to improve wave runup prediction, it would be desirable to obtain wave and storm surge data for several locations along the shore. In the case of wave data, this may involve the installation of new wave gauges. Tide gauges may already exist though the data collected may have to be reduced. Once more detailed wave and storm surge information become available, the wave runup prediction curves should be recalculated to increase their accuracy.

A second means of refining the Dune Management District delineation procedure involves calculating the wave runup for smaller segments of the coastline. Beach and dune characteristics vary from place to place, but these variations may not be accounted for if beach profile information is lacking. It is, therefore, necessary to carry out intensive beach surveys in order to account for the variability of the profile over both time and space. Shoreline segmentation could then be carried out in such a fashion as to assure a more site specific approach to delineating the district.

A third means of refining the model involves the development of beach/dune interaction models for the particular area in question. Information about these relationships has come from site specific studies of the North Carolina (see Dolan, 1973; Godfrey and Godfrey, 1973) and of the Massachussetts (see Godfrey, 1976; Godfrey, Leatherman, and Zaremba, 1979) coastlines. While the application of that information to other areas may help to understand the processes that can be expected to be operative, the responses specific to those other areas are not as predictable. Research into these relationships would help to determine how dune formation could be enhanced and how dune migration could be accomodated in the Dune Management District.

There is also opportunity for research to improve the wave runup calculations. The current calculation technique applies to impermeable engineering structures. While this is not an ideal approach, it is felt to be acceptable because it provides a conservative wave runup value. Wave runup on a permeable beach can infiltrate through the sand. To increase the accuracy of the runup prediction technique, runup studies on beaches need to be conducted. Such a refinement would help to increase the precision of the wave runup predictions and would, thus, strengthen the model.

Although the refinement of the Dune Management District model is considered a necessity, it is suggested that these districts delineated according to the methodology described herein, be established without delay. Modifications indicated by future research can be made at the time of the review. At that time the performance of the district should also be assessed. This assessment should, for the first planning period, be done very deliberately. Factors to be considered include dune formation and migration, overwash frequency, shoreline erosion, storm frequency protection afforded to existing buildings, and condemnation procedures. In addition, policies which apply to areas of the barrier islands outside of the Dune Management District should be reviewed by officials to assess their compatibility with the regulations existing within the district.

The methodology described here applies to barrier islands and spits. Headland areas present a different situation. The higher elevations which characterize the headlands are not subject to flooding, but erosion remains an important concern. Consequently, shoreline management in the headland zone need not focus on dune protection. The erosion problem may be dealt with through the use of construction setback lines. It is important to recognize that these setbacks are related to the erosion rate. The establishment of setback lines, thus, requires a different methodology.

COASTAL DUNE DISTRICT POLICIES

Land use controls are necessary to ensure the maintenance of the protective dune and the viability of the Dune Management District. A second objective of this study is to provide policy guidelines for legislation designed to control development in the Dune Management District and to direct development on the barrier islands. These policies are presented in such a manner that they may be adopted at the state level, or they may be modified so that they can be implemented locally through the use of a municipal ordinance. A model Dune Management District ordinance, based on the policies listed below, is presented in Gares, Nordstrom and Psuty (1979).

The policies presented in this section apply to two areas: 1) the Dune Management District; and 2) the areas landward of the Dune Management District, referred to hereafter as the Barrier Island Zone. It is necessary to treat these two areas in relation to each other because the Dune Management District is migrating into the area to its rear, and that area will eventually be subject to the policies which apply presently to the Dune Management District. The policies which govern the entire barrier island, therefore, should not conflict with, but rather should support the Dune Management District policies. It is recognized that controlling land use on the barrier islands may be difficult to implement. Policies applicable to the barrier island are presented here, however, to demonstrate how land use planning in that zone could be supportive of the Dune Management District concept.

Policies for the Dune Management District

a) The boundaries of the Dune Management District are considered to be dynamic and migrate inland in response to rising sea level and a negative sediment budget.

b) The boundaries of the Dune Management District should be reviewed after a predetermined period of time to determine if boundary relocation is necessary. In addition, boundary review should take place following a major storm if extensive erosion has resulted.

c) Dunes should be allowed to form and to migrate in a natural fashion. Dune building through artificial means, such as planting dune grass or erecting sand fences, may be encouraged. Where present dune lines are breached by beach access gaps, these gaps should be closed.

d) Natural processes should not be interfered with. Overwash fans should not be bulldozed to relocate the sediments. The growth of natural vegetation should be encouraged. Bulkheads or other similar engineering structures should not be permitted.

e) All new construction in the Dune Management District should be prohibited. This includes construction of buildings or support structures and expansion of existing structures.

f) Any existing structure in the Dune Management District which sustains damage as a result of storm action should be evaluated to determine if it should be condemned and removed.

g) Access to the beach through the dune area should be via designated walkways which are maintained in such a way so as not to lower the dune.

h) When overwash occurs, any clearing of access roads, driveways and parking areas to allow continued usage should not be permitted. When these areas are no longer functional, access to the housing should be accomplished via common walkways.

i) Any vacant land should be converted to public land through direct acquisition, transfer of development rights, donation, or any other such mechanisms.

Policies for the Barrier Island Zone

a) New or expanded residential development in the Barrier Island Zone is acceptable provided that the following performance standards are complied with:
   i. All buildings should be elevated on adequately anchored piles or columns.
   ii. A minimum elevation for the first habitable floor and an optimum ratio of pile spacing to pile diameter should be established by each locality. These requirements shall be determined by considering Federal Flood Administration construction recommendations and the requirements to maintain the integrity of the dune.

b) Natural processes should be subjected to minimal interference. Overwash fans may be bulldozed, but under no circumstances should the sediment be removed from the system.

c) Any construction project should be reviewed by a review board on a unit by unit basis to insure that it complies with performance standards.

d) Day users should be encouraged to park off the islands, and appropriate public transportation should be made available to connect off-island parking lots with the beaches.

e) All public and private resort/recreation development adjacent to coastal waters must provide for reasonable public access to the beach.

f) Hotel-motel developments are acceptable in the Barrier Island Zone, but they should be encouraged to locate on the mainland.

g) New or expanded public facility development is acceptable in the Barrier Island Zone.

h) New or expanded industrial-commercial development should be discouraged in the Barrier Island Zone.

CONCLUSIONS

The Dune Management District concept offers an alternative to traditional methods of coping with shoreline erosion and with flooding of land areas during coastal storms. This management approach promotes the formation of coastal dunes by reducing interference with natural processes. This is accomplished by implementing land use controls to eliminate development in the dune zone. The use of the dunes as a means of shoreline protection is based on their inherent properties. Dunes provide a barrier to flooding and they buffer shoreline erosion by constituting a sediment reservoir. However, these properties will serve no long term function if it is not recognized that the beach/dune system is migrating in response to a rising sea level and decreased sediment budgets. In order to take these facts into account, the boundaries of the Dune Management District must be assumed to be dynamic. Becuase of these dynamic characteristics of the beach/dune system, Dune Management District boundaries must be reviewed at predetermined time intervals, or following an extraordinary storm which significantly modified the beaches and dunes. The methodology for establishing the district boundaries cannot be arbitrary. Rather, it must be based on scientific theory and it must employ empirical data. This strict scientific approach is necessary if the concept is to resist challenges in the courts.

The Dune Management District concept falls into the framework of comprehensive shoreline management. Consequently, planning for development in or adjacent to the Dune Management District should accomodate as much as possible natural shoreline migration by providing for development that is compatible with natural processes. The total management framework should not be seen as an attempt to remove all development from the barrier islands. Rather, it is designed to reduce expenditures of public monies on erosion protection structures and on post-disaster relief by taking advantage of a feature of the shoreline environment which naturally provides the protection desired by all. In addition, the approach would result in the enhancement of the beach/dune system by encouraging the growth of dunes and the widening of the beaches. These outgrowths of the use of Dune Management Districts will in turn increase the attractiveness of the recreation resources of the municipalities and the revenues derived from the recreation industry.

ACKNOWLEDGEMENTS

Research was supported by a contract from the New Jersey Department of Environmental Protection, Division of Coastal Resources, with financial assistance from the U.S. Department of Commerce, NOAA, Office of Coastal Zone Management.

REFERENCES

Birkemeier, W.A., 1979. "The effects of the 19 December, 1977 coastal storm on beaches in North Carolina and New Jersey." Shore and Beach. 47(1): 7-15.

Dolan, R., 1973. "Barrier islands: natural and controlled." in Coastal Geomorphology, D.R. Coates, ed. Binghampton, N.Y.: State University of New York. pp. 263-278.

Dolan, R., B. Hayden, and J. Heywood, 1978. "A new photogrammetric method for determining shoreline erosion." Coastal Engineering. 2:41-51.

Edelman, T., 1969. "Dune erosion during storm conditions." in Proceedings of the 11th Conference on Coastal Engineering. New York: American Society of Civil Engineers. pp. 719-722.

Gares, P.A., K.F. Nordstrom, and N.P. Psuty, 1979. Coastal Dunes: Their Function, Delineation and Management. Tech. report NJ/RU-DEP-14-12-79. New Brunswick, NJ: Center for Coastal and Environmental Studies, Rutgers--The State University of New Jersey. 112 pp.

Godfrey, P.J., 1976. "Barrier beaches of the east coast." Oceanus. 19(5):27-40.

Godfrey, P.J., and M.M. Godfrey, 1973. "Comparison of ecological and geomorphic interactions between altered and unaltered barrier islands and systems in North Carolina." in Coastal Geomorphology. D.R. Coates, ed. Binghampton, N.Y.: State University of New York. pp. 239-258.

Godfrey, P.J., S.P. Leatherman and R. Zaremba, 1979. "A geobotanical approach to classification of barrier island systems." in Barrier Islands. S.P. Leatherman, ed. New York: Academic Press, pp. 99-126.

Leatherman, S.P., 1979. "Barrier dune systems--a reassessment." Sedimentary Geology. 24:1-16.

Maurmeyer, E.M., 1980. "Quantification of overwash threshold conditions, Delaware Bay Shoreline." Paper presented at the 15th annual meeting of the Northeastern section of the Geological Society of America, Philadelphia, PA.

Nordstrom, K.F. and N.P. Psuty, 1977. "Comments on study methods and regulatory policies presented at the National Conference on Coastal Erosion." in Proceedings of the National Conference on Coastal Erosion. Washington, D.C.: Federal Flood Administration. pp. 105-127.

Nordstrom, K.F. and N.P. Psuty, 1980. "Dune district management: a framework for shorefront protection and land-use control." Coastal Zone Management Journal. 7(1):1-23.

Nordstrom, K.F., N.P. Psuty, and S.F. Fisher, 1978. "Empirical models of dune formation as the basis for dune district zoning." in Proceedings of Coastal Zone '78. New York: American Society of Civil Engineers. pp. 1489-1507.

Nordstrom, K.F., S.F. Fisher, M.A. Burr, E.L. Frankel, T.C. Buckalew, and G.A. Kucma, 1977. Coastal Geomorphology of New Jersey, Vols. I and II. New Brunswick, N.J.: Center for Coastal and Environmental Studies, Rutgers--The State University of New Jersey.

Thompson, E.F., 1977. Wave Climate at Selected Locations along the United States Coasts. Technical report no. 77-1. Ft. Belvoir, Va.: U.S. Army Corps of Engineers Coastal Engineering Center. pp. 364.

U.S. Army Corps of Engineers, 1973. General design memorandum: hurricane wave prediction--beach erosion control, Brunswick North Carolina beach projects, Yaupon Beach and Long Beach segments. Wilmington, N.C.: Office of the District Engineer.

U.S. Army Corps of Engineers, 1976. New Jersey Coastal Inlets and Beaches. Hereford Inlet to Delaware Bay Entrance to Cape May Canal. House Document no. 94-641. Washington, D.C.: U.S. Government Printing Office. pp. 396.

U.S. Army Corps of Engineers Coastal Engineering Research Center, 1977. Shore Protection Manual, Vols. I, II, and III. Washington, D.C.: U.S. Government Printing Office.

THE ROLE OF A STATE GEOLOGICAL SURVEY IN BARRIER ISLAND MANAGEMENT: A CRITICAL LINK BETWEEN RESEARCH AND POLICY

Randall T. Kerhin, Kenneth N. Weaver, and Jeffrey P. Halka
Maryland Geological Survey
The Johns Hopkins University
Baltimore, Maryland 21218

Abstract

State Geological Surveys can be a critical line between on-going geological research and the formulation of public policy: communicating with the scientific communities, acting as technical advisors to public officials, and conducting their own research functions.

In a recent survey of 27 coastal states geological surveys, a majority of the surveys view their roles as absolutely essential to good coastal zone management practices. The most important contributions are the applied research programs directed to a family of problems and the intergration of these results by direct communication and dialogue with state environmental managers.

The Maryland Geological Survey has maintained a close cooperative relationship with the State Park Service since 1965. The Maryland Geological Survey has regularly surveyed the dune system at Assateague Island State Park and has communicated the results to the Park Service.

Since 1965, the dune system has shown a steady, vertical and seaward growth, indicating aeolian transport from the back barrier island. By 1975, the growth rate of the dune has stabilized as the dunes reached an equilibrium configuration with the natural processes. Only one section of the dune has deteriotated since the initial growth period, the section seaward of the day-use parking lot.

Management policy and strategy depends on cooperation between the various governing agencies on Assateague Island. The policy of non-intervention on northern Assateague Island by the Federal Government will influence the shoreline stability of Assateague Island State Park.

INTRODUCTION:

In our concepts of barrier island management the legal authority to manage is delegated to government at the federal, state or local levels. The question of "who" is secondary to the question of "how" the government agencies are going to manage barrier islands. In any adequate management program, the first priority is a clear understanding and definition of the issues. Secondly, policies and strategies must be selected that are both realistic and workable in terms of human and environmental needs. Finally, these policies and strategies must be executed using the appropriate statutory controls and managerial procedures.

Barrier islands are affected by dynamic forces requiring that the policies and strategies selected are adaptable to this changing environment. A flexible strategy of management, based on the regional geomorphology of the barrier island system has been advocated for the Assateague National Seashore (Leatherman, et, al, 1978). Yet the manager must not only demonstrate flexibility in the selection of the policy or strategy, but also in the implementation of these policies. Good management requires a constant stream of information and a line of communication between the scientific community and the environmental manager as new information and research data become available.

The state geological surveys can play a major role in management decisions regarding coastal zone resources and issues. Most of the state geological surveys are part of a larger departmental organization. The Maryland Geological Survey, for example, operates as a line agency in a Department of Natural Resources and is mandated by law to investigate the geology and mineral resources of the State of Maryland. By this mandate, the Maryland Geological Survey develops and implements long-term research efforts through its programmatic budget cycle and at the same time acts as a technical advisor in communicating this information to the various state officials. By being within the Department of Natural Resources, a direct line of communication is open at all levels of government, e.g., between the Director of the Survey to the Secretary of Natural Resources and the elected public officials and at the lower levels of government between the survey program geologists and project manager of Natural Resources management agencies. In essence, the state geological surveys are a scientific organization within state government that have direct access to the state officials in charge of environmental management.

This paper makes the assertion that state geological surveys can be a critical link between geological research and the formulation of public policy and illustrates the statement through a discussion of the research and advisory role of the Maryland Geological Survey in barrier island management at Assateague Island State Park.

OVERVIEW OF STATE GEOLOGICAL SURVEYS:

In order to determine the involvement of state geological surveys in coastal zone activities, we devised a questionnaire which was circulated to state surveys in 27 coastal states (including states bordering on the Great Lakes). We received returns from 25 of the 27 states. It should be noted at the outset that the questionnaire did not identify specific areas of geological research, e.g., coastal erosion, sedimentation, and stratigraphy. The activities that might be considered in the questionnaire could range from groundwater studies to geochemistry of estuarine sediments and from mineral resources investigations to subbottom profiling. The results of the questionnaire are summarized in Table 1.

TABLE 1

Summary of questionnaire sent to 27 state geological surveys.

Within your state, who conducts geological research in the coastal zone?	
State geological agency	23 states
Federal government	13
State government	9
Academic institutions	20
Consulting firms	8
How is the coastal zone research you conduct generated?	
Internal priorities	21
Requests from other state agencies	17
Mandated legislation	7
Orders from the Executive Department	3
How would you classify the type of research you conduct in the coastal zone?	
Applied research pertaining to a specific problem	11
Applied research that can be used for a family of problems	19
Basic research	8
Do you have regulatory/management functions in the coastal zone?	
No	17
Yes	7
How do the results of your coastal zone research activities get into the regulatory/management policy-making process?	
Publications of results	19
Communication with planners/managers of regulatory agencies	21
Serve as technical advisors on committees	21
Testify as expert witness in a court of law or public hearing	9
How do you view the role of the geological agency in your states's coastal zone management program?	
Absolutely essential to good management practices	18
Significant, but not essential	3
Not very significant	2

All of the state geological surveys polled either conduct or are embarking on a program to conduct research on the coastal zone. Academic institutions carry out research on the coastal zone in 20 states and research by the federal government was indicated by 13 states. It should be noted that answers to the survey questions are not mutually exclusive. The type of research conducted by state surveys in the coastal zone is classified by most as applied research that can be used for a family of problems (19 states) rather than pertaining to a specific problem (11 states). This is consistent with the primary missions of most state geological surveys; geological data gathering, interpretation, and mapping to provide the basic geological framework of the state in addressing a variety of problems.

Thus the majority of state geological surveys conduct applied research in the coastal zone which is potentially applicable to a wide variety of management problems. Furthermore, they view this work as absolutely essential to good management practices (18 states). It is not clear, however, that a direct mechanism exists for insuring the incorporation of the results of these research efforts into management plans. The majority of surveys (21) establish their own internal priorities for coastal zone research programs while seventeen state surveys consider requests from other state agencies in setting their research goals. Only seven are legislatively mandated to conduct coastal zone research. In only one instance is a survey mandated to provide written portions of state generated management plans.

The results of the state survey's coastal zone research programs are communicated to regulatory/management agencies through a number of routes including publication of results, communication with planners and managers of regulatory agencies and serving as technical advisors on management committees. It is primarily through the active role of communicating and advising that geologic information is input to the process of managing the coastal zone. It is assumed (although not actually asked in the survey) that the future research activities of the state geological surveys are based in part upon the managers needs which become apparent through this direct contact.

Although the coastal zone studies carried out by state geological surveys encompass a wider range of topics than barrier islands, such as energy resources, mineral and aggregate resources, and water supply, the problems inherent in transferring the results to the management process are similar. In some respects the problems associated with barrier island management are exacerbated by the dynamic nature of the processes at the land-sea interface. An ongoing dialogue should be maintained between the scientist who has a better understanding of the changing natural system and the manager who is entrusted with the responsibility of guiding the human demands placed on the system.

Most scientific studies in the coastal zone which are designed to provide input to management decisions involve either 1) defining base-line conditions at a particular point in time such as the Environmental

Geologic Atlas of the Texas Coastal Zone Survey or 2) specific applied projects designed to respond to immediate questions and needs such as utilities siting, sewage outfalls, or severe erosion problems. There appears to be an absence of long-term monitoring programs. Such programs would allow the scientist to continually update and expand the data base giving a better understanding of the system, and also simultaneously providing insight into the response of the system to particular management policies (for both the scientist and manager).

The state geological surveys are well suited to fulfilling the role of maintaining long-term monitoring research programs. They are stable state agencies, are oriented to conducting applied research suitable for a variety of problems, and have a direct or advisory role in developing management programs for coastal areas. This is in contrast to academic institutions which have a continual turnover of students and are therefore more oriented toward short-term studies of a specific nature with a defined beginning and end point.

A conceptual model for developing a monitoring program could proceed as follows: a baseline conditions report would be developed which included studies of many aspects of the natural systems. From this report necessary and useful data types could be defined and a set of representative locations identified where these data would be collected on a periodic long-term basis. The data collection technique should be designed such that a minimal amount of time and personnel are required to obtain the maximum amount of information. If suitable the surveys could be conducted by local groups with an interest in coastal zone problems and management. In Maryland, park personnel have supplemented our data collection with direct observations of the system under study. In another state park, the park service personnel are actively involved in data collection (for a monitoring program) established and overseen by the Maryland Geological Survey staff. It is this type of long-term program that is in effect for Assateague Island State Park (Figure 1).

INCEPTION OF ASSATEAGUE STATE PARK:

The initial idea for establishing a state park along the Atlantic Ocean shoreline was proposed in a 1940 report by the Maryland State Planning Commission. A second report by this same commission in 1952 again recommended the development of a state park on Assateague Island. These reports provided the basis for legislative action by the Maryland General Assembly in the 1959-1962 sessions: approval of a master development plan for a state park and appropriation of $750,000 for initiation of the plan and purchase of suitable land. By 1964, implementation of the master development plan had begun.

Assateague Island State Park is bounded on the north and south by Assateague Island National Seashore and is located approximately six(6) miles south of Ocean City Inlet. The State Park provides both day-use and overnight camping with the necessary facilities and services. Protection of these facilities from storm surge penetration is accomplished by means of a primary dune system which has been maintained with vegeta-

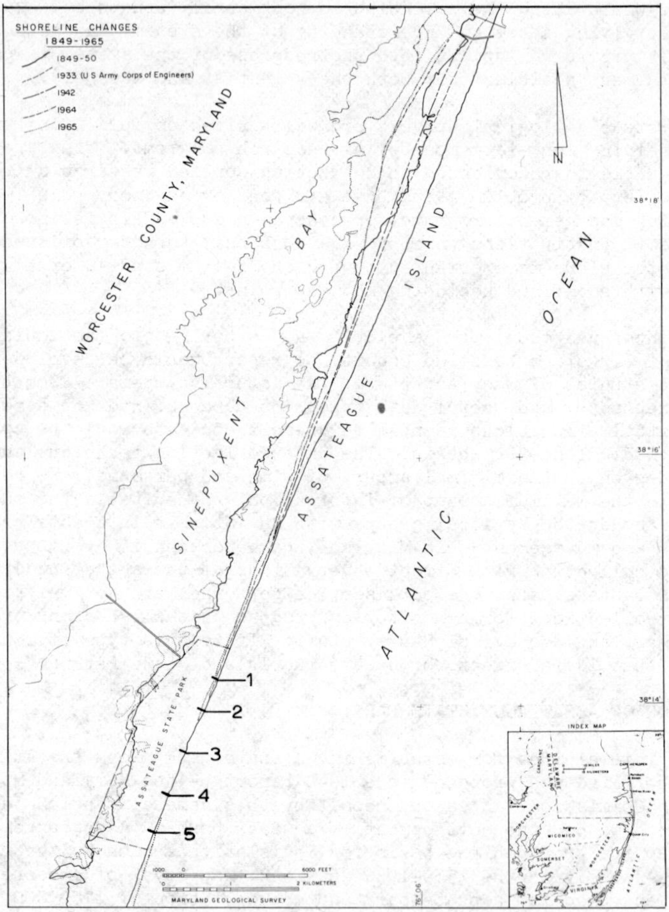

FIGURE 1: Location of Assateague Island State Park on a base map exhibiting shoreline changes since 1849. Beach profiling stations are located at the north and south boundaries of the Park and at the five numbered locations.

tion, dune fencing, and designated traffic paths.

In the initial planning reports for the General Assembly, a preliminary survey of the island determined that the maximum dune elevation for the tract of land under consideration was +19 to 20 feet, mean sea level. By the time the implementation of the master plan had begun natural processes had reduced the elevation of some of the dunes to +10-11 feet msl, with numerous washover channels as a result of the March 1962 (Ash Wednesday) Storm. Continuency plans were drawn up and amended to the master park plan to start a dune development and stabilization program to provide protection for the new park facilities.

INITIATION OF DUNE DEVELOPMENT:

The Maryland Geological Survey is cooperation with the former Department of Forest and Parks devised a program to develop an artificially induced dune system and to periodically measure the growth of this system. Initial development of the dune system was accomplished by installing dune fencing. The fencing was emplaced along the approximate location of the dune line that existed prior to the 1962 northeaster. The fencing ran the entire length of the state park and 600 feet into the National Seashore on the south. The plan for dune development called for a series of fences, the exact number depending on the growth rate, followed by vegetation stabilization.

The first series of dune fencing was erected approximately 200 feet landward (westward) of the mean high tide line in the fall of 1964. A year later (1965) the second series was constructed 30 feet seaward (eastward) of the first series and at the same approximate elevation. In the spring of 1966, the third and last series of fencing was emplaced 15 feet landward of the second series but seaward of the first series, marking the projected dune crest (Figure 2). Concurrent with the construction of this last series, American Beach Grass was mechanically and hand-planted on the landward slope of the incipient dune followed,in the spring of 1967,with an additional planting on seaward slope.

To document the growth rate of the incipient dune after each succeeding installation of fencing, a beach and dune profile was collected at seven different locations. Initially, these profiles provided the necessary information to judge the performance of the dune fencing. Once a dune system had been established, updates on the profile configuration were provided to the Park Personnel on a semi-annual basis. These measurements are continuing today. They provide the basis for updating the park personnel on the yearly performance of the dune system, and modification,if needed, of park management policies.

FIGURE 2: Incipient dune formation following emplacement of the three series of dune fences (looking south; east is to the left). The initial fence (1) was erected 200 feet landward of the mean high tide line in 1964, followed by the second fence (2) in 1965 and the third fence (3) in 1966. Note wooden walkover in the distance.

DUNE DEVELOPMENT AND GROWTH PATTERNS:

A plot of the volume changes at five of the profile locations since the second baseline was established illustrates a rapid growth for the period between 1969 and mid-1973 (Figure 3). By 1974 the dunes at four of the five locations had reached a near equilibrium condition with respect to volume. Since 1974 the characteristic pattern has been one of fluctuations in the volume of sediments in the dunes except at profile 2 which has experienced a dramatic loss in total dune sediment volume. The factors which may have contributed to the cessation of rapid dune growth and the development of an equilibrium configuration are: a complete coverage of the dune system with vegetation, increased frequency and intensity of coastal storms between 1973 and 1975, and an equilibrium between the wind regime and the supply to and removal of sand from the dunes.

The volume increases exhibited by most of the dune system until mid-1973 was the result not only of a vertical growth but also of an overall seaward progradation. Figure 4 is a compilation of survey data at Profile 4, which exhibits these trends rather clearly. Between 1965 and 1966 the dune at this location expanded its base and increased in elevation following emplacement of the three sets of dune fences. From 1966 to 1971 the dune elevation increased relatively little when compared to the seaward progradation of the foredune slope and a concomitant elevation increase of the backbeach area. During the next period, 1972 to 1975, which was marked by a rapid increase in dune volume at Profile 4 (Figure 3) the elevation increased to +17 feet msl, while by 1973 the crest had migrated to a position approximately 60 feet seaward of its 1966 location (Figure 4).

Between 1974 and 1975 there appears to have been a landward shift of the entire dune system with erosion along the foredune slope matched by accretion on the back-dune. The dune crest remained at near +17 feet but was now located only 30 feet seaward of the 1966 position. Since 1976 the dune has shown only relatively minor variations in elevation, configuration, and volume as portrayed by the cross sectional profiles in Figure 4. The only notable change during this time is a gradual coalescence of the foredune slope with the backbeach producing a gently sloping configuration from the dune crest to the waterline.

The recession of the dune line between 1974 and 1975 was a cause of concern to personnel and managers who feared for the integrity of the dune protection system. It was pointed out to them, however, that the foredune erosion, which was most evident to them, was matched by accretion of the backdune resulting in a landward migration of the dune system with no dramatic loss in sand volume. As surveying has continued since that time the relative stability of the major part of the dune system in this new position has been verified. It is now felt that a critical limit of seaward progradation of the dune had been exceeded by 1973-1974 and that the system made a necessary readjustment to a more stable configuration relative to the prevailing coastal processes with-

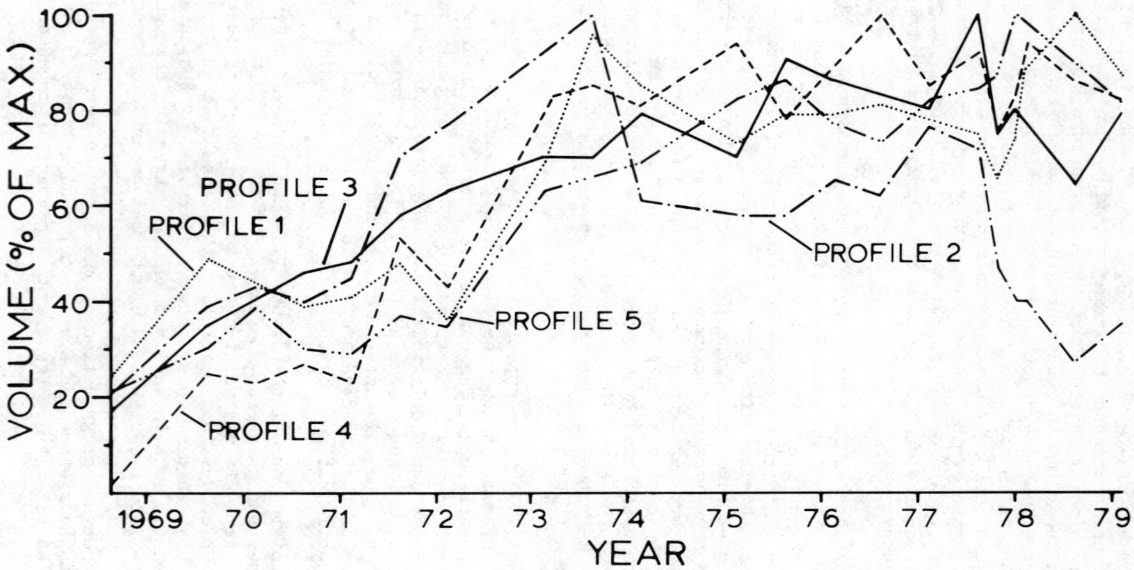

FIGURE 3: The volume of the dune at each profile location for a ten year period as calculated from the survey data. Values expressed as the percent of the maximum volume for each location. Note the rapid growth rate until mid-1973 followed by a slower rate at all profiles except Profile 2.

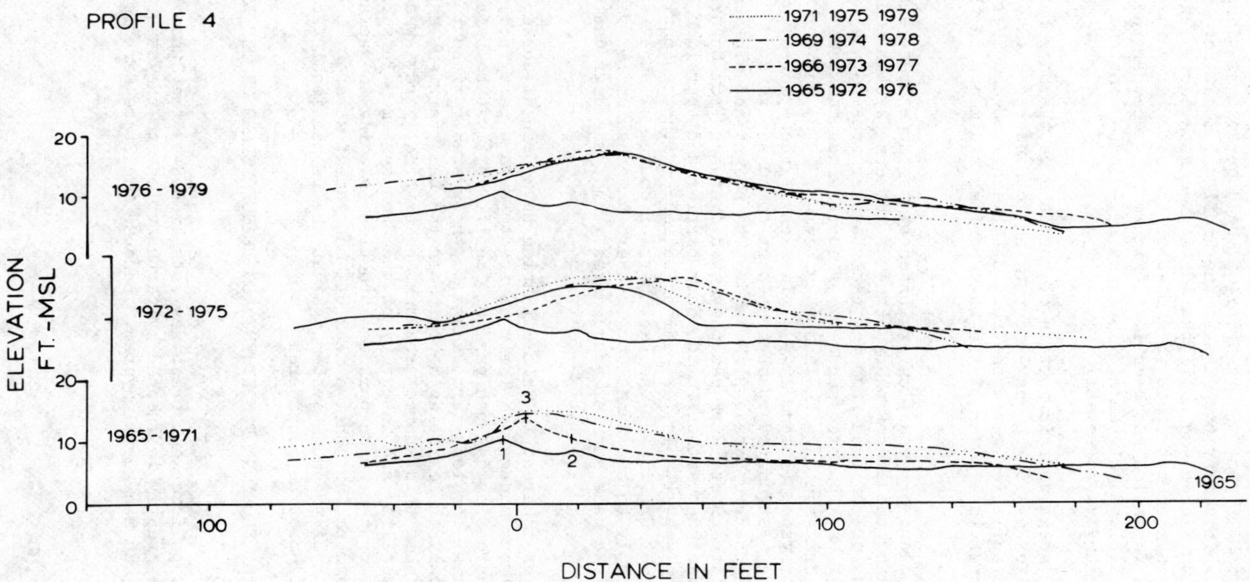

FIGURE 4: A comparison of dune and beach profiles from 1965 to 1979 at Profile 4 (see Figure 1 for location). Note the upward and seaward growth of the dune until 1973. Since 1975, the dune has maintained a stable configuration.

out a significant change in volume.

The relative stability of the majority of the dune systems since 1974-1975 is not exhibited by the dune at the Profile 2 location. In this area dune volume increased until that date at a rate equivalent to the rest of the system but since then most of the material gained during the initial formation has been lost (Figure 2). Whether the volume increase exhibited between mid-1978 and 1979 is to continue will remain for future surveys to reveal.

Immediately adjacent to this portion of the dune on the back barrier section of the State Park is the day-use parking lot and recreational pavilions. These facilities were constructed between 1968 and 1971, and produced an impervious cover of the back-barrier sediments in this area.

The general seaward progradation exhibited by the dune crest and foredune slope during the accretionary phase of the dune system indicates aeolian transport of sand from the back barrier area to the dune. A redistribution of back barrier sands by offshore winds has been reported for northern Assateague Island (Leatherman, 1976) and for the Tabusintac barrier system (Rosen, 1979). In each of these investigations there existed a strong overwash transport component which resupplied the back barrier areas with sand. The height and width of the dune system at the State Park has prevented overwash from occurring, thus denying sediment deposition in the back barrier and therefore decreasing the supply available for redistribution. With the conditions currently existing at Profile 2, the back barrier source of sand is not available for resupply to the dunes nor is there a mechanism for reintroducing sand to the back barrier source area, thus dune degradation has occurred.

The difficulties inherent in this situation have been discussed with the park personnel and managers. Their desire to expand park facilities, particularly by increasing day-use parking facilities, must be balanced against the potential for long-term damage to the system. Without a mechanism to resupply the existing deposits in the back barrier, any physical alteration of this source area must be critically reviewed relative to the stability and integrity of the dune system. This is not to say that a moratorium on park construction is going to guarantee an adequate dune system, but that a flexible management strategy that balances the health of the dune system against the recreational facilities for the public must be considered.

THE INFLUENCE OF THE EROSION OF NORTHERN ASSATEAGUE ISLAND ON THE STATE PARK:

The management policies and practices established by the State Park administrators are greatly influenced by perturbations in the system, both the physical and managerial system. Because the State Park is bounded by the National Seashore, the policies selected by the National Park Service will affect or possibly alter the State Park management plans. A case in point is the influence of the rapidly erosing northern Assateague Island, controlled by the National Park Service. The Park Services' policy of non-intervention in this section has a potential and predictable impact on the State Park policy of dune maintenance.

Ocean City Inlet, located 6 miles north of Assateague Island State Park, was formed as a result of a breach in the barrier island which occurred during the 1933 hurricane. The breach was stabilized in 1935 with a pair of stone jetties. The impact of the stabilized inlet on northern Assateague Island has been to accelerate the erosion rate from 12 feet per year to 30-40 feet per year. Figure 1 displays the the shoreline recession which has occurred at the north end of Assateague Island since the island was mapped in 1849-1850. The accelerated erosion of the north end of the island can be traced to two causes: 1) entrapment of sand updrift of the northern jetty (Singewald and Slaughter, 1949; Gawne,1966; Leatherman, 1979) and 2) passage of sand over and through the north jetty into the inlet through which it is cycled into the ebb and flood shoals (Dean and Perlin, 1977).

With the rapid erosion occurring at the northern end of the island the State Park Managers questioned whether or not the sand-starved conditions in this area would have any influence on the stability of the Park shoreline. In answer to this concern, Slaughter (1975) superimposed a series of aerial photography (1938, 1952, 1956) and the USGS quadrangles (1942, 1972) and located the intersection of the various shorelines with a baseline established as the southward extension of the Ocean City broadwalk. The boardwalk approximates the position of the Assateague Island shoreline prior to 1933. He concluded that there has been a progressive southward expansion through time of the erosional conditions which occur at the upper end of the barrier island. Since 1938, the erosional conditions and rapid landward migration have extended an additional two miles further south. It is possible to project that the erosional processes occurring on northern Assateague Island will incorporate a greater length of shoreline and eventually affect the stability of the State Park. There is evidence that the north boundary of the State Park is undergoing a transistion from relative stability to increasing erosion. Volume changes calculated from 10 years of beach profile data show an overall erosional trend which has greatly increased in the past few years (Figure 5). In the early to mid-1970's erosion of the shoreline after major storms was followed by accretion and rebuilding. With the southward movement of the sand-starved conditions of northern Assateague Island, the response of the State Park shoreline has become totally erosional.

Given that the erosional processes prevailing on northern Assateague Island will eventually jeoparize the stability of the beach-dune and recreational facilities at the State Park, it is critical that a dialogue be established between the State and Federal officials in charge of Assateague Island. This dialogue needs to be conducted with an

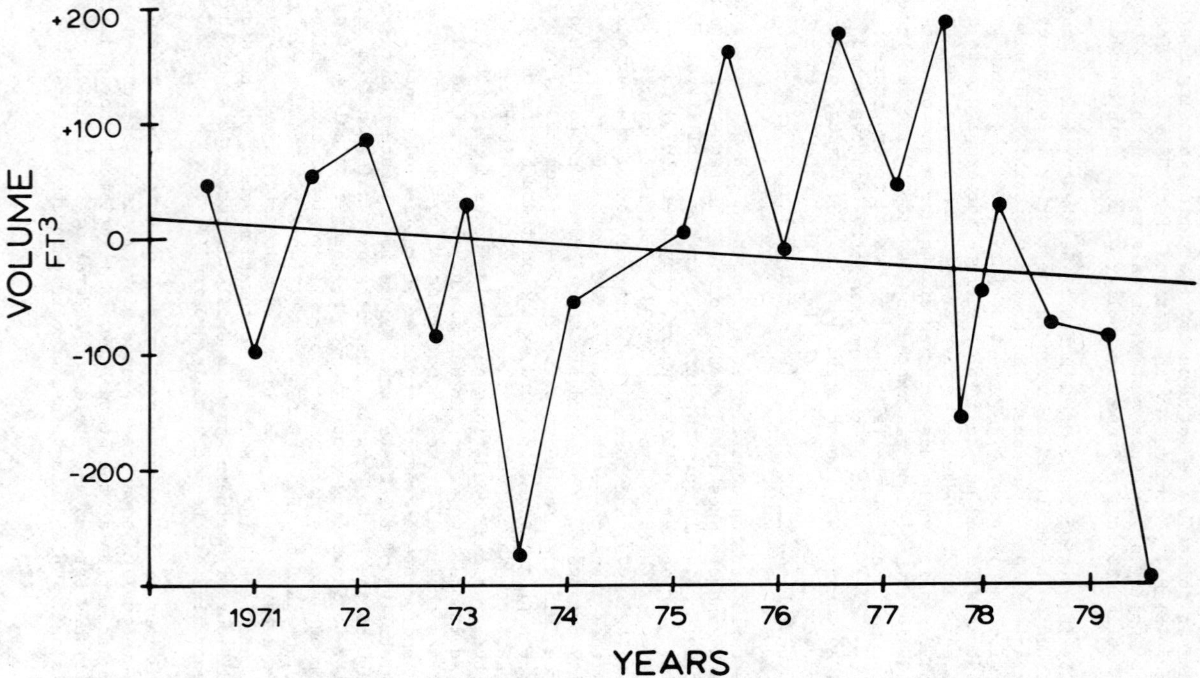

FIGURE 5: The volume increases and decreases of the beach-dune system at the north boundary of the State Park relative to the 1970 volume. The regression line fitted to this data exhibits a long-term erosional trend within the more widely fluctuating data values.

appreciation for the dynamic nature of barrier island systems, and with consideration given to this data which has been collected on Assateague Island. Now is the time to consider a variety of management options in terms of maintaining the recreation potential of the State Park.

ACKNOWLEDGEMENTS:

We would like to express our sincere appreciation to the various State Surveys for taking the time to respond to our Questionnaire. Without their valuable input, the formulation of ideas regarding state surveys' involvement in coastal zone management may not have been accomplished. We would also like to acknowledge the State Park personnel for the many years of dutiful service in our beach and dune surveillance program. Their cooperation is greatly valued.

REFERENCES:

Dean, Robert G., and Perlin, M., 1977, Coastal engineering study of Ocean City Inlet, Maryland: Proceedings, ASCE Conference, Coastal Sediments '77, Charleston, South Carolina, P. 520-542.

Gawne, C. E., 1966, Shore changes on Fenwick and Assateague Islands, Maryland and Virginia (B.S. Thesis): Chicago, University of Illinois, 140 p.

Leatherman, Stephen P., 1966, Barrier island dynamics: overwash processes and eolian transport: Proceedings, 15th Coastal Engineering Conference, Honolulu, Hawaii, p. 1958-1974.

________, 1979, Migration of Assateague Island, Maryland, by inlet and overwash processes: Geology, v. 7, p.104-107

________, Godfrey, Paul J., and Buckley, P. A., 1978, Management strategies for National Seashores: Proceedings, Symposium on Technical, Environmental, Socioeconomic and Regulatory Aspects of Coastal Zone Planning and Management, San Francisco, California p. 322-337.

Rosen, Peter S., 1979, Aeolian dynamics of a barrier island system: in, Leatherman, S.P., ed., Barrier Islands from the Gulf of Saint Lawrence to the Gulf of Mexico: Academic Press, New York, p. 81-98.

Singewald, Joseph T., Jr., and Slaughter, T. H., 1949, Shore erosion in tidewater Maryland: Maryland Department of Geology, Mines and Water Resources, Bulletin No. 6, 141 p.

Slaughter, T. H., 1975, Coastal landforms of northern Assateague Island, Maryland: in, Weaver, K. N., ed., Investigations of the Geology, Mineral Resources, and Water Resources of the State of Maryland: Final Report, NASA-SKYLAB, 10 p.

GRAVEL BARRIER GENESIS AND MANAGEMENT : A CONTRAST

R.W.G. Carter[1] and J.D. Orford[2]

Abstract

Fringing barrier beaches in Co. Wexford, SE Ireland are dominated by gravel-size material (-4 Ø to -1 Ø). While these barriers show many sedimentological and geomorphological affinities with their sand counterparts, there are major differences which do not fit easily into existing geodynamic or facies models for transgressive barrier complexes. Within a macro-tidal environment the steep gravel barriers display a tendency for vertical and horizontal accretion, via a continuum of overspill, overtop, and overwash mechanisms, followed by little or no seaward sediment transfer. Many of these processes are controlled by precursory beach changes in both processes and bedforms, with most of the latter having only a poor preservation potential.

The longer term development of the Wexford barriers appears to be related not only to variations in sea level and sediment economy, but also to the underlying topographic and geological basement, so that evolution of equilibrium planforms may occur around stable hinge-points. However many of the more transient features (outlets, overwashes, etc.) can be related to longshore variations in the inshore wave field.

Lack of understanding of the barrier dynamics has resulted in poor resource management. Uncontrolled gravel extraction has caused non-uniform shore erosion and repeated efforts to cut new lake outlets have led to rapid morphological adjustment of the barrier. Furthermore, attempts to artificially build up the ridge crest including construction of "dunes", appears in places to be inducing a further type of instability involving percolation and seepage erosion, possible heralding a significant new phase in barrier mobility.

Introduction

The literature available on sandy barrier coasts far outweighs that on gravel barriers, despite the fact that the latter form an important shoreline type in high latitudes, for example in NW Europe, Alaska, NE Asia, New Zealand and Antarctica (Zenkovich 1967, Hayes and Kana 1976, Randall 1977, Kirk 1980). In all these places heterogeneous-sized glacial deposits have been reworked into depositional coastal features. Randall (1977) has calculated that on the British coast alone 900 km (approximately 10%) comprises

---

1 The New University of Ulster, Coleraine, Co. Londonderry, BT52 1SA, Northern Ireland

2 The Queen's University, Belfast, BT7 1NN, Northern Ireland

gravel features, with sizes ranging from boulders to fine grit/coarse sand.

There are a number of distinctive differences between sand-dominated and gravel-dominated barriers (Table 1) and it would be unfortunate if the recent profusion of geodynamic and sediment facies models relating to sand barriers (qv. Hayes and Kana 1976, Elliott 1978, Hayes 1979, Reinson 1979) were to be simply extended without question to include their gravel counterparts. Table 1 illustrates some of the extreme differences between sand dominated and gravel dominated barriers, which manifest themselves in both external and internal barrier morphology and influence facies variation and preservation.

TABLE 1. Comparison between sand and gravel barriers

SAND	GRAVEL
Beach and seashore geometry	**Beach and seashore geometry**
Flat	Steep
Dissipative	Reflective
Rapid adjustment rate to process variations	Slow adjustment rate to process variations
Nearshore bedforms common	No nearshore bedforms
Barrier morphology	**Barrier morphology**
Extensive dunes	Dunes absent or rare
Inlets/stream exit common	Inlets/stream exit uncommon
Tidal lagoons common	Tidal lagoons rare
Freshwater lagoons rare	Freshwater lagoons common
Large overwash fans	Generally small scale overwash fans, though broad shorewards crestal disslocation by sluicing overwash can occur under extreme events
Crestal overtopping deposits rare	Backbeach seepage features common
No backbeach seepage features	
Sediments	**Sediments**
Fine	Coarse
Non-porous	Porous to non-porous due to high internal size variation from stratification
Low roughness	High roughness
Low entrainment threshold	High entrainment threshold
Normal sediment grading units	Coarsening upwards units common
	Clast form may be as important as clast size in conditioning sediment mobility and zonation

The most obvious and fundamental difference between sand and gravel barriers is that of sediment size. In gravel barriers the larger maximum clast size creates a steeper beach face slope, leads to an absence of large-scale nearshore bedforms, and produces a variable porous ridge structure. Also, because of the relatively high entrainment threshold of the gravel population, many morphological forms are associated with low frequency, high energy events, and possess a longevity which exercises a major control over subsequent shorter and/or periodic developments.

Sand barriers are known to be highly unstable forms, changing in shape and size in response to both short and long term process variations. However many sand barriers exhibit a degree of kinematic equilibrium through a balance of landward and seaward transport vectors, although these may operate on several different scales. Gravel barriers, on the other hand are dominated by net shoreward transport with the crestal landward mobility of large clasts occurring only under extreme events. Whether crestal accretion by swash overtopping or crest displacement by sluicing overwash occurs, depends on the magnitude of the storm. Both of these sedimentological responses are generally irreversible, in that seaward vectors of sediment transport on upper reaches of gravel barriers are negligible (Hardy 1964, Orford 1977, Orford and Carter in prep.). It would appear almost axiomatic that gravel barriers are strongly shorewise transgressive relative to their sand equivalents. The gravel barrier constantly attempts to "roll over" itself (Fig. 1a) while the sand barrier retains its position (Fig. 1b). This implies that in any transgressive phase associated with an economy of sediment scarcity, then the preservation potential of a coherent gravel barrier is high due to its ability to 'roll over' and keep station with mean sea level. The coherence potential of a sand barrier under the same conditions is by no means as assured. The porous nature of the gravel precludes in many instances the development of inlets, as entrapped water, be it derived from terrestrial run off or marine overwash, simply soaks towards the tidal base-level. Finally the lack of surface sand-sized material reduces the importance of eolian mechanisms both to construct dunes and even to return marine sand to the beach face. Thus gravel barriers have few mechanisms by which they can remain stationary, even when sea level remains constant.

Between the ideals of pure gravel and pure sand barriers, occurs a continuum of mixed gravel-sand forms with varying sediment size modal domination and multi-modal association. In practise, even the pure gravel barriers include a subsidiary sand component which is usually located subsurface due

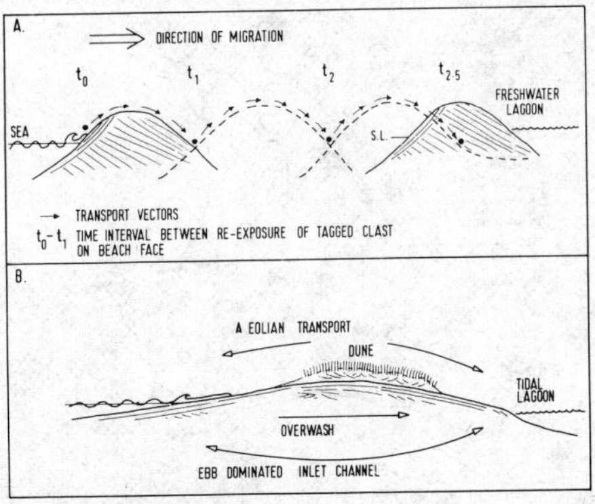

Fig. 1 Shore-normal comparison of gravel and sand barrier dynamics. In 1A the gravel barrier rolls onshore in the absence of effective seaward transport processes. The tagged gravel clast moves over the ridge, is buried and is ultimately reexposed on the seaward face. In 1B, over the same time period, a particle may pass through the barrier cross-section several times without contributing to a landward shift.

to the "seiving" affect of the larger particles under swash action. Kirk (1980) has made a strong plea that such mixed gravel-sand barriers should be considered as separate entities from their unimodal extremes. The S.E. Ireland barrier examples discussed in this paper are principally of this gravel-sand type, with gravel/coarse grit domination. Although little information exists for this barrier type, it appears to exhibit some characteristics of both pure gravel and pure sand barriers.

Gravel dominated barriers of SE Ireland

The extreme SE coast of Ireland comprises a series of swash aligned, fringing gravel-dominated barriers (Fig. 2). These barriers are composed of sediments derived from local glacial deposits, reworked and incorporated into the coastal system during the Holocene. Present sediment supply to the coast is relatively limited, and the major headlands (Carnsore Point, Forlorn Point and Hook Head) act as permanent groynes to the widespread longshore transport of material. The present sediment economy is therefore one of scarcity. The sediments normally exhibit a strong size bimodality with a dominant gravel

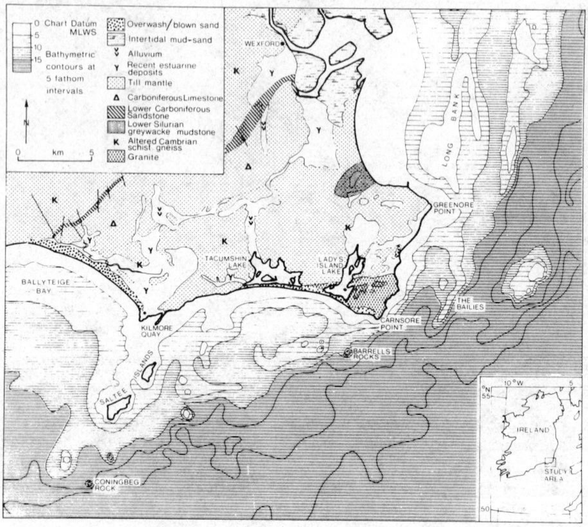

Fig. 2 Geology and bathymetry of SE Wexford. Note the complexed offshore bathymetry.

mode between -1.0 Ø and -4.0 Ø and a subsidiary sand mode near +1.0 Ø to +2.0 Ø. Variation in the proportions of the dominant size mode occurs west-east along the barrier. Carnsore Point has the coarsest mode (gravel/coarse grit) while at Tacumshin Barrier the dominant mode moves into the fine grit/ coarse sand range. Throughout most of its 40 km length the barrier ridge rests up against underlying and flanking glacial till, Carboniferous granite or Cambrian schists, although, in two places the ridge cuts-off coastal lagoons (Fig. 3). In general, cross-sections of the ridge display a transgressive structure, with only a shallow active beach prism, which on occasions is storm-stripped to reveal landward dipping (5°), back ridge and back beach coarsening upwards sedimentation units.

The Wexford coast is affected by both local wind waves and Atlantic based swell waves, resulting from the frequent SW to NE passage of low pressure cyclones. Although SW and W winds and waves have a statistical dominance (Hogben and Lumb 1967), all directions are represented for more than 6% time during a year while extreme storms are S and SE generated. Significant deepwater wave heights may exceed 4m for approx. 10% of the time (Martin 1971, Carter and Orford, in press) and wave periods are usually in the range 4-5s,

Fig. 3 Oblique air photography looking west across the Lady's Island section of the barrier. Note both the distinctive pseudo flood structures from overwash incursion into the man made lake outlet or drainage cut across the barrier, and the vegetation disturbance by gravel extraction equipment.

and rarely above 10s. Spring tidal range is just over 3m. Extensive and complex refraction of all incident waves across the variable bathymetry (Fig. 2) results in a transient but characteristic pattern of non-linear shoreline wave processes. Lowry (in prep) has used a number of criteria to identify process "cells" based on the abc ... model advocated by May and Tanner (1974). A long-term integration of longshore power and cell structure is helping unravel the evolution of the barriers. At a more local scale extreme focussing of inshore waves within caustic zones may account for high-level barrier sedimentation (Carter and Orford in press).

Barrier morphology and processes

(a) Cross-sectional structure

The cross-sectional structure of the barrier is well displayed at Lady's Island Lake (Fig. 4). However some caution should be applied before accepting this as a valid representative section, because of human activities in the area; the cross-section is marginal to an artifically excavated channel or "cut" which is used to regulate water level in the lake. At present the cut is opened once a year, remaining operational for two or three weeks before it is closed by upper beach accretion. In historical times the cut was opened less regularly (probably only once every 7 to 10 years) but local informants suggest the practice dates back to at least the eighteenth century. Not withstanding the human element, the section does exemplify many of the structures which are known to occur on more pristine parts of the barrier and as such is a useful guide to barrier processes and sedimentation.

Figure 4 suggests the barrier may be divided into two distinct zones.

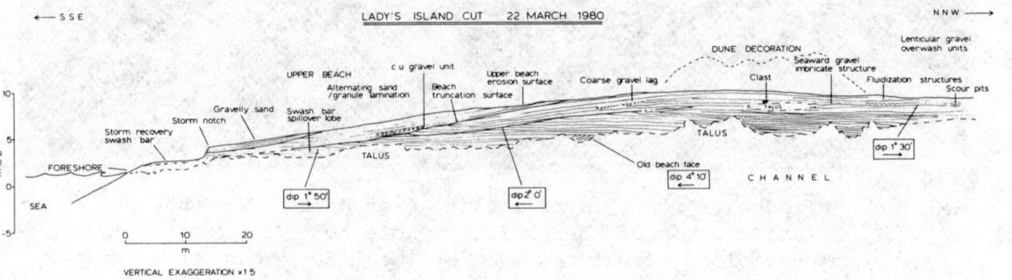

Fig. 4 Cross-section of the seaward barrier at Lady's Island Lake. Stratified overwash gravels persist for a further 200m to the righthand (lakeward) side of the section.

First on the seaward side there is a beach zone composed of a series of seaward dipping (4-7°) concave, gravel and gravelly-sand units, which show several truncations and general inpersistances most likely related to spring/neap and storm/fairweather process sequences. Some evidence of swash bar/berm development is visible in the middle foreshore. The upper beach units pass laterally into the second, more extensive, zone of overwash deposits. At lower levels there is a sharp, concave contact between zones one and two, probably indicating the lower limit of the beach scour or sweep zone. The overwash units dip gently (1-2°) landward and comprise a series of persistent, lenticular, coarsening-upward, seaward imbricated gravel units, 5-20cm thick. Individual units may be traced for over 350m, with little variation in thickness or gravel size (although the sand sub-population tends to increase from around 5% to nearer 50% towards the lake, thus decreasing the overall mean size). In places large clasts, up to -10.0 Ø, outcrop within the gravel units, and appear to have been deposited contemporaneously, with only marginal disturbance of the gravel units. Also occasional fluidization structures are present associated with either saturated bed stress perturbation by an overpassing surge, or are due to some dewatering process.

Where fan splay deposition has occurred in the lake, small tabular delta foresets may exist, often defined only by the laminated sands. Subsequent reworking of the fan surface by lake processes tends to obscure many of the original forms.

Small eolian dunes have developed as decoration on the highest part of the barrier fed from a limited supply of suitable sand exposed between the tide marks. The dunes show low-angle, concave-up cross-bedded stratification with numerous root intersections. There appears to be no consistent pattern of dune growth merely slow, spatially irregular accretion around stabilizing vegetation.

(b) Short-term morphological changes

The barrier ridge, including the dune decoration, has a maximum height of around 10m OD, or 6m above HWMOST. On the foreshore there is an active seasonal sweep zone of 1-2m depth. Relaxation and reaction times of gravel-dominated beaches are longer than for sand equivalents so that inherited controls affect many subsequent events. For example a mid-beach supra-tidal terrace constructed under moderate energy conditions is able to survive into storm periods and help determine upper beach changes. High level, storm processes including overwashing and overtopping, are instrumental in causing ridge migration. In order for storm waves to generate run-up capable of significantly altering the dimensions of the barrier, a certain amount of supra-elevation of water levels will be required. On this rather open coast it is unlikely if meteorological or wave-induced surges would exceed the 0.5m/100 years recorded on the Welsh side of St. Georges Channel by Pugh and Vassie (1978), so that it is necessary to look for mechanisms capable of producing either high inshore breaking waves or high run-up. Both depend on the initial shore profile, which varies between concave or convex (Fig. 5). The profile shape introduces something of a paradox; because while the concave slope would allow higher inshore breakers, run-up would be muted on what is essentially a reflective form. Conversely the convex, terrace profile would encourage dissipation and high volume swash, but the final breaker height of waves capable of reaching and breaking on top of these terraces would, from the available wave data, be no greater than 0.75m in order to accomplish overtopping of crestal dunes, (Orford and Carter, in prep.). The importance of the structure and organisation of the inshore field in upper storm beach sedimentation has been noted by Orford (1977) who observed that spilling mode breakers are essential to provide the constructional thrust for generating gravel ridge crestal deposits.

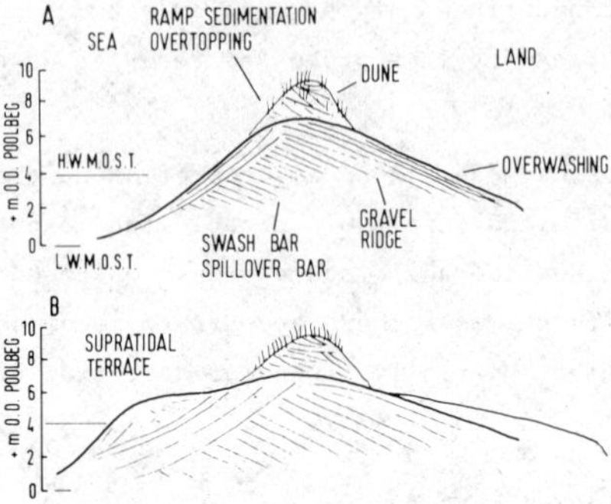

Fig. 5 Barrier profiles, showing the two distinctive types of beach cross-section without (A) and with (B) the supra-tidal terrace.

The nature of a dune crestline is important in SE Wexford in that it provides a control over ridge migration. If there is any semblance of breach points along the ridge (within a height range of 5-7m) then overwashing is likely, but if no such access points occur then the mixed sand and gravel ridge acts more as a barrier and induces vertical development by swash ramp accretion and finally overtopping. Breach points tend to direct sediment-charged swash flows through the crestline forming characteristic overwash fan splays. (The presence of a coarse gravel element appears to act in greatly reducing the size of these fans when compared to their sand equivalents.) The origin of individual breaches is often contentious, but they appear to fall into two classes; overwashes which show no dominant longshore periodicity occur at Carnsore (Carter and Orford, in press), related to vehicle tracks, footpaths, eolian blowouts, and possibly, local wave focussing. Periodic spacing exists further west at Tacumshin where a series of 53 overwash fans have a basic throat spacing of about 55m. Such periodicities may be attributable to resonant inshore wave activity under storm conditions. While the reflective nature of the steep (6-8°) upper beach slope at Tacumshin is probably conducive to the generation of low horizontal mode edge waves, it is probably the only beach position occupied for long enough during surges to form rhythmic shoreline features. At other times, tidal stage elevation is insufficient to cover these reflective parts of the ridge, and a dissipative wave mode prevails on the lower beach face.

Excavation of the coarser overwash deposits at Carnsore has revealed a number of distinctive structures and facies arrangements. (Carter and Orford in press, Orford and Carter, in prep). Often at the seaward edge of the throat a coarse plug conglomerate with vertical stonework has been emplaced into an elliptical scour envelope, and acts as a barrier, to control subsequent overwashing of coarse material.

Where the barrier remains as a coherent boundary to horizontal sedimentation, the barrier builds vertically. Eroded dune faces are subjected to a protective armouring of coarse material, followed by the construction of swash ramps which may reach 9-10m OD. (Carter and Orford in press). Ramps consist of highly stratified gravel and coarse sand units, built up against the seaward dune cliff. Where ramps reach the dune cliff top extensive overtopping of wave-lain material may occur onto wind-blown sediments.

In places where the barrier is not now being overtopped or overwashed, the back slope is being extensively modified by tidal seepage streams from ponded-up beach face water (during extreme events), eroding headward into the ridge crest, and forming deep incising channels with fan splays at their distal ends. Carr and Blackley (1974) have described similar features, called "cans" on the back of the gravel-based Chesil Bank, in S. England.

(c) Long-term morphological changes

At this juncture it is only possible to speculate on longer-term barrier changes, pending more thorough investigation. However most field evidence indicates that the barrier is strongly transgressive with little or no ability to remain stationary via compensatory seaward sediment transfers. Eroded balls of indurate, but fissile, blue-black clay with lacustrine floral remnants can be found on the beach face, particularly after onshore storms. Following Carr and Blackley (1974), this circumstantial evidence suggests that the barrier has been displaced landwards over the old lagoonal bottom beds, which are then eroded by storm action, when they are exhumed on the seaward barrier face.

The barrier has presumably evolved since the last glacial maximum (about 20-25000 years BP), and has been supplied by a variety of both local and non-local terrestrial and marine sources. Over the last 10-15000 years it would be reasonable to suggest that material has been added at a varying but generally decreasing rate, within the overall context of an exponentially decreasing rate of post-glacial eustatic sea-level rise. If one extrapolates from the work of Kidson and Heyworth (1978) on the adjacent coast of Wales (about 65 km distant) then local sea-level has been within 5m of the present sea level for approximately 6000 years.

The planform of the barrier appears to be determined by two factors working in concert. First the underlying basement rocks appear to provide a series of "hinge-points" of differing erosional potential around which the barrier evolves. These hinge-points are of two types, firstly those composed of hard ancient rocks - granites and schists - and unlikely to suffer substantial erosion within the life-span of the barrier, and secondly, those made up of semi-consolidated Pleistocene deposits - mostly tills - which if exposed not only erode quickly but also act as source of material.

The second factor is that of wave refraction across a complex offshore bathymetry (Fig. 2), which when integrated through time provides a swash-aligned quasi-equilibrium plan form, and may in part control the position of tidal inlets and sediment transport along and across the barrier. In the case of the most resistant rocks, headlands emerge creating drift divides, permanent groynes and irreversible changes in plan form. However where hinge-points are formed of less durable material they act not only as shape controls but also as suppliers of sediment to the adjacent barrier. In spite of this primary control is retained by the resistant headlands. Further evidence for the domination of swash alignment is afforded by the Tacumshin barrier, which despite many recent modification shows no tendency to become offset.

Barrier management

Basically there is no formal management of the barrier. It is not used for recreation or agriculture and most people would regard it as a somewhat sterile stretch of coast. However the barriers must be viewed as providing substantial protection to the SE coast of Ireland and in particular, to the proposed first nuclear power plant in Ireland, although long-term environmental benefits are often disregarded for short-term commercial profit. There are three important human activities which have potential major effects on the future of the barrier system. These are: (i) gravel aggregate extraction, (ii) cutting of drainage outlets and (iii) the encouragement of coastal "dunes".

(i) Gravel extraction - there has been particularly severe extraction between the east end of Tacumshin Lake and Carnsore Point, while the rate of removal has increased markedly since 1970 under commercial auspices. Gravel is usually taken from the upper foreshore (Fig. 6A) destroying both the supra-tidal terrace feature (especially near Carnsore Point) and any swash bars that may be present. Heavy vehicles move along the ridge crest, and excavators have created trackways onto the foreshore. A conservative estimate would be that 300,000 tonnes have been removed since 1970. Extraction leads directly and indirectly to geomorphological changes. First, finite removal of beach material, probably orders of magnitude greater than supply leads to inevitable shoreline adjustment and erosion. At Carnsore, the site of the greatest extraction pressure, the shoreline has retreated by up to 8m in the last two years, over a 1.5 km stretch. Ironically some of the gravel is being supplied to a shore protection/harbour improvement scheme at Rosslare Harbour, 8 km away. Second, the vehicles are destroying fragile sand dune and gravel ridge based habitats. Much of the affected area supports only a degenerate vegetation cover (Fig. 3), and where it has been completely removed vehicles have simply moved onto a new adjacent track in order to attain traction, so that a disproportionate area becomes afflicted. Even the designation of a portion of the Lady's Island barrier as a National Nature Reserve for gravel substrate flora has done little to halt this damaging process. Third, trackways that open onto the beach offer many suitable breach points for overwashing. On 14 March 1980 a SE storm caused overwashing at a breach W of Lady's Island which had been made for gravel excavator access to the beach, and where no recent activity of this type had been recorded. A "bird's foot" shaped fan, several hundred square metres in extent has now formed (Fig.6B) based on this man-made breach.

There are clear signs from the juxta-positioning of recent and fossil overwash sites that the increasing pressure on the barrier, by top beach gravel extraction is ensuring that a new phase of barrier transgressive behaviour is being felt along several sections of the Wexford barrier due to oversteepened beaches and diminishing longshore supply.

(ii) Inlet construction - the cutting of inlets through the barrier has been attempted at Lady's Island and Tacumshin. This practice has been extant since, at least, the eighteenth century, aimed both at regulating lake water levels and for the possible reclamation of

Fig. 6 Human activities affecting the barrier

A Gravel aggregate extraction from the top of the now-eroded supra-tidal terrace near Carnsore Point. Terrace failure is indirectly attributable to gravel extraction. Note coarseness of included material on the beach face

B The 'birds foot' overwash at the western end of Lady Island occupying an excavator's exit breach in the dune line

C Artificial "dune" ridge constructed on western Tacumshin Barrier to prevent further overwashing.

marginal land. Most inlets experience rapid closure, usually with two to three weeks due to a combination of low discharge and active beach transport. Only a recent (1974) and fairly elaborate sluiceway keeps the small Tacumshin outlet open throughout the year. Despite the ephemeral nature of the outlets they appear to have a significant if somewhat local affect on the barrier. At Lady's Island repeated reopening of the exit channel has resulted in a local lowering of the barrier ridge height allowing overtopping and overwashing. A number of mostly fossil and ill-defined flood tide-delta structures exist at the lake/barrier margin. At Tacumshin, at least three outlets have been cut in the last hundred years (Murphy 1974). In the case of two attempts (1879, 1920) the outlet rapidly closed, with the outlet position switching back to the traditional swash aligned outlet position at the western end of the barrier. As yet it is not known if the outlets have migrated down the barrier or quickly sealed up under longshore drift, while the traditional westerly exit re-activated at the point of lowest barrier elevation. The extreme morphological change apparent on the western end of Tacumshin barrier relative to the stabilised vegetated eastern end of the barrier act as a witness to human interference with lake drainage.

(iii) "Dune" construction - many local landowners have tried to encourage "dunes" by bulldozing up the fine grit/coarse sand beach face sediments to form a continuous crestal ridge which blocks off active overwash throats (Fig. 6C), and to reduce the possibility of overwashing. The ridges have been planted with marram grass (*Ammophila arenaria*) but obviously show few signs of natural accretion, due to the paucity of suitable blown beach material. The building of this artificial crest renders the barrier morphology more susceptible to the effects of even lower frequency, high magnitude events. In addition the sealing of overwash throats often acts to increase back beach seepage and it is speculated that this may lead to the reactivation of a more dynamic barrier phase, via renewed overwash activity.

It may appear paradoxical that on one scale longshore elements of the barrier show increasing transgressive activity, while on a larger scale, the whole barrier (in particular from Kilmore Quay to Carnsore Point) shows a uniformity of swash aligned control. Without any human interference, the rate of barrier transgression would be partially dictated by the rate of overall bay swash-aligned adjustment as a function of refraction patterns set-up by hinge-point attitude. The very recent (on a geomorphological scale) activity of man on the barrier in promoting differential transgression may eventually break the swash aligned control. In principle, new patterns of longshore sediment supply must be developing to account for continuing swash alignment of 'risk' areas (i.e. Carnsore Point and west Tacumshin Barrier). How long these adjustments can continue is not yet known.

Conclusions

Pure gravel and gravel-dominated barriers are in many respects different to their sand counterparts, not least because of their inherent tendency to move onshore, even under relatively stable sea-levels. Transgressive barrier migration is largely accomplished by low frequency, high magnitude storms, initiating a range of upper and back beach sedimentation processes which include overtop veneers, swash ramps, overwash fans and landward crestal dislocation. In many instances on the barriers of SE Ireland, these sedimentation events are controlled by precursory lower and mid-beach changes.

Long-term evolution of the Wexford barrier is believed to be related to the overall Holocene sediment economy and the emergence of headlands of varying resistance to erosion.

Human activities tend to interfere with the natural barrier processes, mainly by introducing new phases of instability, at irregular intervals.

Acknowledgements

We would like to thank Philip Lowry for field assistance and allowing us access to his as yet unpublished wave refraction work. Thanks also to John Snow and Edna Doherty for technical assistance, and to Sharon Wright for typing the paper.

References

Carr, A.P. and Blackley, M. (1974) Ideas on the origin and development of Chesil Beach, Dorset. Proceedings of the Dorset Natural History and Archaeology Society, v.95, p.9-17.

Carter, R.W.G. and Orford, J.D. (in press) Overwash processes along a gravel beach in south-east Ireland. Earth Surface Processes.

Elliott, T. (1978) Clastic shorelines in Reading, H.G. (ed.) Sedimentary Environments and Facies, Blackwell, Oxford, ch.7.

Hardy, J.R. (1964) Sources of some beach shingles in England. 20th Int. Geogr. Congr. London, Abstracts Paper 1651.

Hayes, M.O. (1976) Transitional-coastal depositional environments. In Hayes, M.O. and Kana, T.W. (eds) Terrigenous clastic depositional environments. CDR-Tech. Report 11, University of South Carolina, I32 - I111.

Hayes, M.O. (1979) Barrier island morphology as a function of tidal and wave regime. In Leatherman, S.P. (ed) Barrier Islands, Academic Press, N.Y., p.1-27.

Hogben, N. and Lumb, F.E. (1967) Ocean Wave Statistics. H.M.S.O., London.

Kidson, C. and Heyworth, A. (1978) Holocene eustatic sea level change. Nature v.273, p.748-750.

Kirk, R.M. (1980) Mixed sand and gravel beaches : morphology, processes and sediments. Progress in Phys. Geogr., v.4, p.189-210.

Lowry, P. (in prep) Computer simulation of the synoptic wave climate and delimitation of sediment cells on the barrier coast SE Wexford.

Martin, A.D.H. (1971) Wave data for the Kish Bank Lightvessel, Dublin Bay, Barrells Lightvessel, Carnsore Point and Daint Rock Lightvessel, Co. Cork. Irish Lights Office, Dublin, 48p.

May, J.P. and Tanner, W.F. (1974) The littoral power gradient and shoreline changes. In Coates, D.R. (ed), Coastal Geomorphology, Binghampton, p.43-60.

Murphy, H. (1974) The Light at the end of tunnel? Kilmore Parish Journal.

Orford, J.D. (1977) A proposed mechanism for storm beach sedimentation. Earth Surface Processes, v.3, p.381-400.

Orford, J.D. and Carter, R.W.G. (in prep) The structure and origin of recent sandy gravel overwashing and overtopping deposits at Carnsore, SE Co. Wexford.

Pugh, D. and Vassie (1978) Extreme sea levels from tide and surge probability. Proc. of the 16th Conf. Coastal Engin., Amer. Soc. Civil Engin. New York, v.1, p.911-930.

Randall, R.E. (1977) Shingle foreshores. In Barnes, R.S.K. (ed) The Coastline, Wiley, London, p.47-61.

Reinson, G.E. (1979) Barrier Island Systems. In Walker, R.G. (ed) Facies Models, Geoscience Canada, Reprint series 1, p.57-74.

Zenkovich, V.P. (1967) Processes of coastal development. Oliver and Boyd, Edinburgh.

CUMULATIVE IMPACT ASSESSMENT-CALIFORNIA COASTAL LAW

William Milan Boyd
Chief Staff Counsel
California Coastal Commission

The passage of the California Coastal Act of 1976 provided the vehicle for the transition from the permit-by-permit form of analysis required by the terms of the California Coastal Zone Conservation Act of 1972 (Proposition 20) to a statutory scheme premised in large part on a formalized system of comprehensive environmental impact assessment often transcending individual local government jurisdictions. Not only were the four basic policies of the 1972 Coastal Act (e.g. "avoidance of irreversible and irretrievable commitments of coastal zone resources," former Public Resources Code Section 27302(d) replaced by an elaborate policy basis for resource management but a system of local coastal planning was created in order to apply Coastal Act policies at the general plan and zoning level. These local plans are referred to as local coastal programs. In contrast with the limited substantive requirements of California's General Plan law (Government Code 65302-65303), the policy requirements of the California Coastal Act constitute a set of substantive performance standards to which all local coastal programs prepared by local governments must adhere (see Public Resources Code Sections 30200-30264, 30512 and 30513). While the Coastal Act of 1976 contains important policy groups dealing with specific, resource-oriented policies (e.g. wetland policies contained in Public Resources Code Section 30233), use priorities for shoreline and upland support areas, the provision of public access to the coast, provisions for siting coastal dependent development and special policies applicable to the major ports, this paper will focus solely on the manner in which the cumulative impact policies of the Act relate to the local coastal program process.

Cumulative impact assessment is often regarded as an esoteric exercise that is generally relegated to an obscure subsection of a massive environmental impact report. Public perception of a bulldozer filling a marsh is quite direct so that the objective of a policy regulating the filling of wetlands can be readily perceived. In contrast, even though the siltation resulting from poorly regulated grading in the watershed of a wetland may have the same net effect as the bulldozer, both public and legislative understanding of policies dealing with cumulative impact are often dim vague with overtones of planners' abstractions at best and environmentalist quackery at worst. While the many actions of the California Coastal Commission based on cumulative impact assessment have not yielded a magic formula for escaping controversy (e.g. Sea Ranch) or for conveying graphically the message of cumulative impact assessment to the public at large, the Commission does have seven years of experience in applying cumulative impact analysis; in a number of instances the regulatory actions deriving from this analysis have yielded concrete, measurable results with a positive impact on the preservation of coastal resources and on assuring public access to the coast. This paper will focus on the legal basis for the methodology used in cumulative impact assessment and will present several examples of instances where cumulative impact assessment has been translated into formal permit guidelines, general plan policies and local plan implementation programs. (see figure I on the following page for a graphic example of cumulative impact analysis).

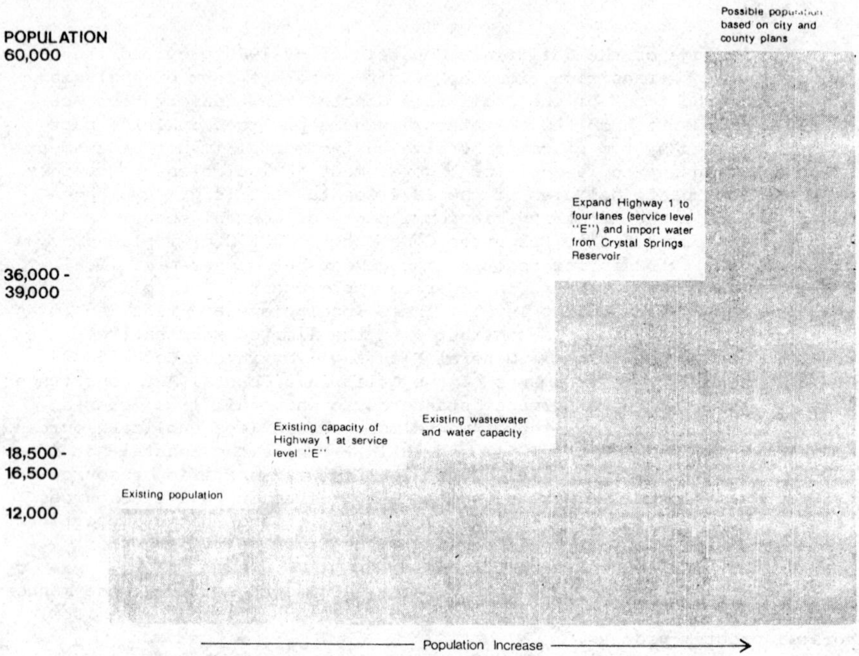

EXISTING AND MODIFIED LEVELS OF PUBLIC SERVICE CAPACITIES
— HALF MOON BAY CASE STUDY —

a. Legal Basis for Cumulative Impact Assessment

The legal basis for cumulative impact assessment can, and must, be traced to the nuisance law underpinnings of resource protection legislation. In upholding the constitutionality of the California Coastal Zone Conservation Act of 1972, the California Court of Appeals stressed that:

> The power of the state to declare acts injurious to the state's natrual resources to constitute a public nuisance has lone been recognized in this state...Contemporary environmental legislation represents an exercise by government of this traditional power to regulate activities in the nature of nuisances. (CEEED v. California Coastal Zone Conservation Commission 43 Cal. App. 3d 306, 318)

The court noted that, due to the nuisance law premises of environmental regulation, "the nature of the governmental interest differs substantially from that involved in conventional zoning." (id., 312). The court then went on to relate this nuisance law premise to the issue of state regulation of matters traditionally within the purview of local government:

> ". . . where the activity, whether municipal or private, is one which can affect persons outside the city, the state is empowered to 'prohibit or regulate the externalities.' In this connection courts have recognized that the impact of an activity which in times past has been purely local, may under changed conditions transcend municipal boundaries." (emphasis added id., 321)

In support of its position the court cited the example of the Tahoe Regional Planning Compact as an example of a situation where "'only an agency transcending local boundaries can devise, adopt and put into operation solutions for the problems besetting the region as a whole.'" (id., 327)

The major test of the Coastal Commission's legal authority to regulate the externalities of private development located within a single local jurisdiction came in the case of Coastal Southwest Development Corp. v. California Coastal Zone Conservation Commission, (55 Cal. Appl 3d 525). In that case, the Commission had denied a motel project, not on the basis of the project itself, but rather on the basis of the impacts on wetland and scenic resources that would occur if the development trend represented by the project were to continue. The court held that cumulative impact considerations "apply not only to the separate phases of an overall project but to any single project in relation to the conditions then existing and to conditions that would inevitably or probably result from accelerating or setting in motion a trend productive of adverse impact upon environment and ecology." (emphasis added, id.,537). In a similar appellate case decided under the California Coastal Act of 1976, the permit applicant contended that the coastal commission "improperly considered the cumulative environmental impact of future restaurants which might be built" in the area (Stanson v. San Diego Coast Regional Commission, 101 Cal. App. 3d 38, 47-48). The court cited the precedents dealing with cumulative impact assessment and concluded that the applicant's construction of the Coastal Act "would reduce the Regional Commission's planning function to a shambles resulting in a piecemeal approach which would guarantee the destruction of coastal resources". (id. 48).

b. Cumulative Impact Provisions of the California Coastal Act of 1976

The Coastal Act of 1976 contains two types of cumulative impact policies, namely those which are applicable to development in general terms (e.g. location of development) and those which are to be applied to the protection of specific types of resources. The more general type of provision is contained in Public Resources Code Section 30250 and 30254. Under Public Resources Code Section 30250, new development must be located in areas where there are adequate public services and "where it will not have significant adverse effects, either individully or cumulatively, on coastal resources." However, recognizing that in some areas of the coast it may not be possible to expand public services consistent with resource protection policies, the Act in Public Resources Code Section 30254 requires that, where the amount of public services is in fact limited, the capacity of existing public services must be budgeted to give priority to coastal dependent development, essential public or private facilities and recreational uses over other types of development.

Other sections of the Act deal with cumulative impact issues relating to specific resource protection and recreational use concerns. Section 30230 of the Coastal Act (all references to sections of the Coastal Act are to the Public Resources Code) provides standards for protecting the marine environment that can be related directly to cumulative impact concerns while Section 30231 deals specifically with controlling runoff into wetlands and preventing depletion of groundwater supplies, both of which involve cumulative impact assessment. Section 30253 of the Act also relates to this section with its focus on erosion control. Section 30241 of the Act contains an elaborate set of policies dealing with prime agricultural lands aimed at protecting the agricultural economy of the area as a whole, thereby mandating an assessment of the cumulative impact of the conversion of prime lands in terms of regional agricultural viability. Section 30252 requires the review of the cumulative impact of development on coastal access and, among other policy mandates, specifically requires assurances that "the recreational needs of new residents will will not overload nearby coastal recreation areas by correlating the amount of development with local park acquisition and development plans with the provision of onsite recreational facilities to serve the new development."

c. Specific Applications of the Coastal Act Cumulative Impact Policies

Under the provisions of the Coastal Act of 1976, the Coastal Commission undertakes the review of specific development projects at the same time that local governments are preparing local coastal programs designed to provide the basis for the future transfer of coastal permit authority from the Coastal Commission to local governments. The Coastal Act recognizes the functional and legal relationship between the Commission's permit and planning functions by requiring the Commission to find, in approving a development application, not only that the proposed development is consistent with the policies of the Coastal Act but also that the development will not prejudice the ability of the local government to prepare a local coastal program (PRC 30604). The Act also provides that decisions of the Commission, including permit decisions, are to guide local governments in their future actions under the provisions of the Coastal Act. Thus, the Commission's interim permit function is related directly to the ongoing coastal planning program.

Because the Commission's decisions constitute precedents for future development, the Commission has taken considerable effort to assess the potential cumulative impacts of its permit decisions. As a consequence, many of the issues that are being addressed in local coastal programs were either originally identified and fleshed out in Commission permit decisions or have been defined by analogy to permit decisions through the Commission's interpretation of the Act's policy requirements (i.e. Interpretive Guidelines) as they apply to specific types of development impacts (e.g. standards for reviewing grading impacts).

One substantive policy area where this permit/planning relationship can be readily perceived is in the assessment of the cumulative impact of development on groundwater supplies. While the major urban areas of California are served by water supplies imported from areas remote from the urban centers, many of California's smaller coastal communities (e.g. Santa Barbara, Morro Bay, Monterey) obtain their water supplies from local groundwater basins or nearby storage facilities with limited capacities. Quite often adjoining local jurisdictions (e.g. adjacent cities, or a city and a county area) will draw water from the same source but, short of a water moritorium, no effective system for relating the amount and location of development to available water supply exists. The severe drought that California experienced in the late seventies demonstrated that many communities were perilously close to exhausting their local water supplies. The drought also demonstrated that natural resources will be subordinated to human needs whenever water shortages force such a choice. Even during normal rainfall periods, the dry summertime conditions of most of the California coast leave many areas extremely close to the limits of locat water supplies. However, where development approaches or exceeds water supply limits, state law requires that the interests of natural resources such as fisheries must be subordinated to human needs, as was pointed out in the recent past by a decision of the California State Water Resources Control Board dealing with the impact of Sea Ranch water usage on the water levels of the Gualala River.

The relationship of the cumulative impact of groundwater usage to coastal resources was vividly demonstrated in a series of permits in the Morro Bay area of the central California coast. The City of Morro Bay is a small coastal community adjoining one of the most important bays in California. The city is surrounded by unincorporated areas composed partly of agricultural lands and partly of lower density subdivisions providing retirement housing for many individuals. Both the incorporated and unincorporated areas draw their water supplies from the same groundwater basins. In a series of agonizing permit decisions (nothing the Commission does is easy), the Coastal Commission determined that overdrafting of water supplies resulting from the continuation of the current level of buildout into the future would cause "significant adverse effects on anadromous fishery resources, riparian and upland habitats and coastal agriculture." (City of Morro Bay, Appeal No. 12-78) The results of the Commission's actions were a permit rationing system, a groundwater monitoring study and more extensive analysis of water conservation and water supply alternatives such as wastewater reclamation. A change in the composition in the local city council may also have been something more than a coincidence. Because many of the resource impacts would have occurred outside the city limits, the Commission was better able than the local governments, both in planning and in legal terms, to determine the potential cumulative impacts of current development trends and to take steps to assure that the choice between serving human water needs and maintaining riparian habitat and a fisheries resource would not have to be made.

In the case of the Carpinteria Valley area of Santa Barbara County, the resource conflict was between coastal agricultural uses primarily in unincorporated areas and further urban development in the city of Carpinteria. The Carpinteria Valley is characterized by a mild coastal climate that is very conducive to orchard (lemons, avocados) and greenhouse use. The same mild climate has begun to attract additional residents. Due to regular shortages of water occurring in many areas of Santa Barbara County, the presence of the Coastal Act and the Commission were not necessary to create an understanding of cumulative impact concerns. However, prior to the advent of the Commission's permit authority and the provisions of the local coastal program there did not exist an effective mechanism for allocating water usage between the city and county areas. Not only were there conflicts between urban residents of the city and agricultural users in the county area but additional demands for water usage were being created in the county area by large lot residential development and the expansion of the greenhouse industry. After a series of hotly contested permit hearings, a public workshop was held in the area in the fall of 1978. One of the outcomes of the workshop was a State Commission decision on water allocations for additional urban development for the coming year. But the Commission decision was intended to serve more as a catalyst for a system of allocation of water supplies to be developed through the local coastal programs for the area rather than as an annual water allocation event.

With the permit proceedings focussing attention on the need for a comprehensive water allocation system, the local coastal programs for the city and county areas began to address the water supply issue. After calculating the potential buildout allowed under existing zoning, both the city and the county proposed to down-zone buildout potential to numbers closer to the limits of water supply. (see Santa Barbara Coastal Plan, January 1980, p. 163) However, because even the reduced buildout potential would exceed existing water supplies, it was necessary to allocate the existing water surplus to city and county areas. The ratios for city/county usage also included a ratio of city to county residential development (the county was allowed a small percentage of water for non-agricultural development with the remainder going to agricultural use). Future buildout would be contingent on demonstrating long-term water availability resulting from the implementation of a water management plan. Allocations by type of use for additional development would be based on Coastal Act use priorities.

The Carpinteria Valley proposals are not as significant for their specific content as they are for what they reveal about cumulative impact assessment under the Coastal Act. Despite historic problems with water supply in Santa Barbara County, both the city and county area of Carpinteria Valley were zoned for development far in excess of water supply capacity. A likely consequence would have been either the elimination of agricultural users by urban users and/or the effective exclusion of priority urban uses under the Coastal Act (such as commercial recreation uses,); based on past experience, proposals for recreational developments might not be prepared until some point in the future, well after local water supplies had been exhausted. By applying an analysis of "the conditions then existing and" the "conditions that would inevitably result from accelerating or setting in motion a trend" (Coastal Southwest, supra) toward the depletion of available water supplies, the local coastal programs identified a need for allocating scarce water resources. Coastal Act priorities for protecting agriculture not only created an awareness of the need for reserving sufficient water supplies for future needs but also created the basis for re-assessing proposed urban expansion into agricultural areas. As a result, an urban limit line was drawn resulting in only a minimal conversion of agricultural lands compared with prior plans and trends. The policy directives of 30231, 30241, 30250, and 30254 of the Coastal Act provided a firm basis for incorporating the results of the cumulative impact analysis into local general plan and zoning regulations.

Another substantive area of cumulative impact analysis involved in many of the Commission's actions is the area of watershed protection, particularly as watershed management relates to grading and urban pollutant impacts. Much of the initial analysis of watershed protection issues arose from permit actions, particularly involving developments in the highly erosive areas of southern California. In Appeal No. 510-77 (Pardee), the Commission found the following impacts of development in the watersheds of coastal wetlands:

> Urbanization has three different but related effects, which are of concern to the productivity of Los Penasquitos Lagoon, quite apart from the problems of road capacity discussed above. First, the development process increases the production of sediment from land, at least during the construction process. Second, urbanization increases the rate of runoff by increasing the amount of impermeable surfaces. The USGS publication "Effects of Urbanization on Sedimentation and Floodflows in Colma Creek Basin, California" indicated that between 1964 and 1970, an increase in urban areas from 34 percent of the watershed to 46 percent of the watershed resulted in an increase in runoff from 25 percent to 45 percent. Similar increases are predicted in "Hydrology for Urban Land Planning--A Guidebook on the Hydrologic Effects of Urban Land Use." The increased runoff in turn increases the rate of sediment deposition into the lagoon and decreases the purifying effects of overland flow. The third major effect of urbanization is an increase in the concentration of pollutants contained in runoff.

> All of these three factors are of concern in this permit. The project site was graded and foundations and driveways were installed approximately 5½ years ago; no replanting was done on either of the two lots graded. Due to the installation of the foundations and partial drainage and the failure to revegetate, considerable erosion has occured on the site with visible deposition of sediment into the lagoon. Coverage with impervious surfaces is also a major concern. The proposed development on Lot #2 would result in over 73 percent of the site being covered with impervious serfaces. Information "Hydrology for Urban Land Planning--A Guidebook on the Hydrologic Effects of Urban Land Use" (Leopold) indicates that development in a watershed that results in 80 percent impervious surfaces will increase the peak discharge of the annual storm by between 3.6 and 6.0 times. The cumulative effects of development involving such large amounts of impervious surfaces is thus of very major concern. Such development would also increase the concentration of pollutants in runoff, and the decrease in pervious surfaces and increased runoff would increase the significance of this impact as less filtration and infiltration of urban runoff occurs.

The Commission also noted that the cumulative impacts of development in this area would overload the nearby access road, thereby creating pressure for the expansion of the road into the adjacent wetland in a manner parallel to the fact situation described in Coastal Southwest.

Subsequent to the initial permit actions, the Commission staff began to develop a series of local plan requirements intended to elaborate on the statutory language as applied in the permit decisions. The first elaboration of the statutory requirements came in the form of the staff recommendation on the Irvine local coastal program. The Irvine coastal area of Orange County, comprised of 10,000 acres, is the last major open space in that portion of the southern California coast and is characterized by a series of ridges interspersed with small canyons containing extremely important habitat areas. The coastal zone extends several miles inland in order to embrace the small watersheds of these important canyons. Offshore is an extremely significant marine habitat area with kelp beds that are highly susceptible to sediment impacts from on-shore development (past hillside development had already generated erosion problems). The development plan contained ridgetop areas designated for several thousand residential units. The county plan required the creation of a watershed program but did not define the substantive elements of such a program. Because of the potential impacts of accelerated runoff and landform modification, the staff recommendation contained the outlines of a watershed management program (sentences underlined below were proposed additions to the County's watershed policy):

> "To implement a watershed program that is phased with development to prevent excessive erosion and sedimentation and emphasizes the prevention of excessive siltation in marine

waters. Erosion and Sedimentation. Prior to division or development of any parcel in the watersheds of Los Trancos Canyon, Muddy Canyon, Moro Canyon, or Emerald Canyon, a watershed hydrology study analyzing the effects of development shall be completed, and drainage facilities meeting the policy requirements of the plan shall be master-planned. The drainage facilities necessary to meet the policy requirements of the land use plan shall be constructed in a phased manner with development of subdivisions, and shall ensure that the rate of runoff does not increase over natural runoff rates.

Subdivision Design. A grading and erosion control plan shall be developed as an integral part of subdivision design. The grading and erosion control plan shall serve as the implementation mechanism for the erosion control policies of the plan.

(a) The grading plan shall include the following types of information:

(1) topographic map showing existing contours, drainage areas, and rock outcroppings; (2) subsurface soil and geology for areas to be graded; (3) proposed contours; (4) hazard areas, unstable slopes; (5) limits of construction; (6) existing vegetation to be protected; (7) existing topsoil to be protected; (8) waste disposal areas; (9) schedule of areas, phasing of operations;

(b) The erosion control plan shall include the following types of information:

(1) site map of soil type, depth, and erosion potential; (2) soil description, existing coefficient of runoff; (3) soil compaction specification for cut and fill slopes; (4) drainage structures, retention devices, dissipation structures; (5) grading season; (6) temporary stabilization of disturbed areas; (7) permanent stabilization of disturbed areas; (8) coefficients of runoff for natural slopes and for finished slopes; and (9) professional certification."

Recently, a plan for another large paracel of land in Orange County (6,000 acres) has been submitted with a runoff management plan that is intended to meet the basic substantive requirement of not increasing runoff over the natural rate. While details of this plan are still being worked out, the major features of the Aliso Viejo runoff control plan and the focus on cumulative impact of grading are agreed upon. (see Appendix I) Attesting to the fact that cumulative impact concerns have been acknowledged to a significant degree by the development community is the participation of several major landowners in San Diego County in discussions to settle on erosion control programs for substantial developments proposed in coastal lagoon watersheds.

The third major area of cumulative impact analysis and regulation has been in the area of reviewing the impact of traffic generated by coastal development on coastal access roads. The origins of regulatory concern with traffic impacts can be traced to the permit review of second home subdivisions in rural areas (e.g. the Sea Ranch development in Sonoma County at one time proposed to comprise 5,000 units in a rural area of the coast) and the impacts of primary residential and commercial development on beach access in the urban areas of southern California (the latter had been one of the prime factors in the passage of Proposition 20). Topography and the desire to maintain the rural quality of the central and northern California coast have limited the size of the main coastal road to two lanes along the coast (codified in Section 30254). In many of these areas (e.g. Big Sur, San Mateo County, Sonoma County) the presence of a major coastal ridge and rugged inland topography have limited the number of connecting roads. The two remaining undeveloped areas in close proximity to metropolitan Los Angeles (the Santa Monica Mountains and the Orange County coast) are also typified by rugged terrain, limited road systems and established communities limiting new road options. Several studies (one by Sea Grant in San Mateo County, one by Commission staff and a consultant in Orange County, one by Commission staff in Big Sur and one by private consultants for Marin and Sonoma counties) have demonstrated that buildout at existing zoning could effectively utilize all existing road capacity, thereby effectively precluding inland residents from access to the coast. In a state where 80% of the population lives within 30 miles of the coast, such an impact would have a significant adverse impact on recreational access. While the debate continues in some quarters over the precise methodology to be used for calculating road capacities and traffic impacts from development, the appropriateness and basic approach to be used in capacity budgeting for coastal access roads have been accepted in a number of local jurisdictions.

An example of the cumulative impact assessment approach is the case of the expansion proposal for the Ventana Inn in Big Sur. While the type of development proposed was a priority use under the Coastal Act (i.e. commercial recreation as opposed to residential development), the Commission's subregional cumulative impact analysis had shown that the existing zoning would allow development far in excess of the road capacity required to served projected visitor use. In its findings, the Commission noted that the Big Sur subregional analysis had found:

> (a) that existing zoning would allow an additional 9,371 to 14,727 residential units; (b) that recreational use could be expected to double in the next 20 years to approximately 2,800,000 visitors per year and that a high proportion of these visits would continue to be for pleasure driving; (c) that Highway 1 between Malpaso Creek and Big Sur Village was already at level of service D, during summer weedends occasionally reaching level of service E during peak periods, and was thus an immediate development constraint; and (d) since the capacity of the highway cannot support the future recreational travel demand, no significant capacity appears to remain for future residential development, and future

recreational use must be regulated as well.

Accordingly, the Commission determined that it could only approve a portion of the requested units based on the following analysis:

> The Regional Commission staff noted that a recent survey conducted by the California Department of Transportation established average peak traffic in the vicinity of the Ventana Inn to be 6,800 vehicle trips per day (vtpd), thus leaving a remainder of 3,200 additional vtpd until level of Service D, estimated at 10,000 vtpd in the subregional analysis, is exceeded. The Regional Commission found using standard estimating techniques that the proposed expansion could use from 3 to 19 percent of the remaining highway capacity. State Commission staff analysis of vehicle use surveys submitted by the applicant indicate that guests in Ventana Inn's existing 24 units contribute an average of 105 vtpd or ± 1.5 percent of the existing 6,300 vtpd. The Commission recognizes that the Ventana Inn is both an existing development and a "visitor-serving" land use that is given priority under Section 30254 of the Coastal Act. However, the Commission also recognizes the mandate of the Coastal Act requiring the careful allocation of scarce and valuable coastal resources among competing uses. In the subject appeal one of the scarce and valuable coastal resource is the remaining capacity of Highway 1, which must be allocated among the residents of Big Sur and visitors seeking either public or private visitor-serving accommodations in the area. The Commission finds that an interim method of allocating the remaining capacity of Highway 1 prior to a more comprehensive allocation to be undertaken in the Local Coastal Program would be to allow the Ventana Inn a "fair share" of the remaining capacity of 3,200 vtpd, based on the Ventana Inn's current percentage of highway use (Exhibit 4). Such an allocation would allow a 12 to 16-unit expansion of the existing facility while at the same time allowing the final apportioning of highway capacity to occur in the Local Coastal Program. Therefore, the Commission finds the proposed development, as conditioned, to be consistent with the provisions of Section 30254 of the Act.

Cumulative impact concerns relating to coastal access have in fact become central issues in the local coastal program effort. In both San Mateo County and Big Sur, the calculation of the level of buildout that would be allowed under current zoning has led to proposals to limit rural land divisions commensurate with the carrying capacity of the coastal access road system. Calculations of recreational traffic impacts resulting from projected buildout in southern Orange County have provided impetus for a transit-only corridor through a valley to the beaches to facilitate coastal access from inland parking areas. Local government concerns voiced by the cities adjoining the Irvine coastal area in Orange County were a factor in the Irvine Company's decision to reduce drastically the buildout projected for their 10,000 acre coastal property. Cumulative traffic impacts implicit in older subdivisions in the Santa Monica Mountains have been a major factor in the creation of a transfer of development credits system (to be discussed later).

The preceding discussion has focussed on cumulative impact assessment as it applies to water supply, watershed management and coastal access issues in order to convey an understanding of current legal and planning techniques as they apply to particular types of resource and access concerns. The combination of the legacy of subdivision practices in effect prior to the enactment of the Coastal Act and the inherent complexity of coastal resource management have necessitated the fashioning of an array of plan implementation techniques in order to be responsive to the wide range of cumulative impact issues. In the following section, we will examine a number of the techniques currently being employed in California.

d. Implementation Techniques for Resolving Cumulative Impact Issues

The legal techniques employed for resolving cumulative impact concerns range from capacity budgeting of limited public services to phased development to transfers of development credits. While each of these techniques differs in significant measure, what they all have in common is a basic planning methodology that provides the factual groundwork for the legal implementation tools.

In most states, the scope of judicial review applying to quasi-legislative actions (e.g. the adoption of a general plan) is more limited than the scope of review applicable to a quasi-adjudicative action (e.g. the approval or denial of a permit action). While the differences in scope of judicial review may in fact be very real, the complexity of cumulative impact review and the often involved and abstract manner of analysis may not enlist the degree of judicial and political sympathy that may be inherent in forms of regulation aimed at at more readily perceived direct impacts such as marsh fill. Notwithstanding the casebook distinctions in scope of judicial review, I would suggest that any regulatory technique keyed to cumulative impact analysis follow the California Supreme Court's admonition in Topanga regarding the nature of findings, that any chosen implementation technique must explicitly set forth its factual basis (i.e. findings) for bridging "the analytical gap between the raw evidence and ultimate" conclusions of law (Topanga Assn for a Scenic Community v. County of Los Angeles, 11 C 3d 506, 515). In the case of cumulative impact review, this judicial standard not only requires exhaustive documentation of the cumulative impacts but also a careful analytic statement relating the impacts to the particular implementation techniques. For, as the California Supreme Court noted:

> Among other functions, a findings requirement serves to conduce the administrative body to draw legally relevant sub-conclusions supportive of its ultimate decision; the intended effect is to facilite orderly analysis and minimize the likelihood that the agency will randomly leap from evidence to conclusions, (emphasis added, Topanga, id. 516)

While the above holding was applied to a quasi-adjudicative function (zoning variances), the Court's statement is a clear admonition for any agency action involving factual findings and ultimate conclusions of law.

The basic methodology for cumulative impact assessment is set forth in Collaborative Land-Use Planning for the Coastal Zone: Volume II - Half Moon Bay Case Study (University of California, Sea Grant Publication No. 53 - by T. Dickert, J. Sorensen, R. Hyman, J. Burke, 1976). Among the techniques described are: (1) an exhaustive inventorying of existing parcels in the defined geographic region (2) a comprehensive assessment of buildout allowed under existing zoning (3) a thorough assessment of the capacities of major service systems including water supply, sewage treatment and road system and (4) the delineation of critical decision points in the buildout process (See Appendices II , III , IV). The techniques will obviously vary according to the substantive cumulative impact issues of concern in any particular area. For instance, watershed management concerns may involve relating the inventory of existing parcels to cumulative impacts of development in areas with steep slopes in particular geographic problem areas; in turn, watershed erosion issues may interact with water supply and road capacity concerns to form the basis for defining overall cumulative impact issues. (see Appendix III).

Once the precise nature and scope of cumulative impact concerns have been identified, the choice of implementation options can be narrowed. (see Appendix IV). Where existing public service capacities have definable limits and/or resource protection concerns (e.g. steep slopes in a significant watershed area) create absolute limits on buildout, it is necessary to allocate any remaining public service capacities to future development. In California, the use priorities in situations where public service capacities are finite are prescribed in Section 30254 of the Coastal Act. The term generally applied to this technique is capacity budgeting. Examples of capacity budgeting are the water supply-agricultural/residential/industrial use allocations described above with reference to the Carpinteria Valley (supra, p. 6) and the Ventana Inn permit (supra, p.11).

Where ultimate buildout will exceed existing public service capacities but where public service capacities can be expanded consistent with the Coastal Act, the phased development technique can be employed. Future population levels are related to existing or projected public service capacities and critical decision points for facility expansion are identified. This technique is not particularly novel but it tends to be talked about in planning literature more than it is actually practiced. Under this approach actual development approvals are keyed to the timing of public facility expansions (additional water supply, added capacity in a sewage treatment plant, road or transit expansion). One twist that has been added in California coastal planning is the notion of reserving capacity for priority uses. Because non-priority uses such as residential development may actually be constructed at a faster pace than priority uses such as commercial recreation, it may be necessary to reserve capacity in a public service system for priority uses at such time as approval is given for the particular facility.

Perhaps the most difficult application of cumulative impact regulation is where the existing pattern and/or amount of subdivision is in excess of buildout consistent with a coastal management statute. To our continuing chagrin, the issue of existing, single family lots may well be the most vexing question facing California coastal regulation. Land division practices prior to the advent of coastal regulation led to the creation of thousands of lots in areas where public service and resource carrying capacities will not be able to cope with full buildout. The human equities inherent in the ownership of a single lot make such textbook techniques as merger and enlarged minimum lot sizes of limited applicability.

One technique currently being employed in the Santa Monica Mountains, the transfer of development credits, holds promise of implementing a private market mechanism for resubdivision that is responsive to the equities inherent in the ownership of lots severely limited by the terrain. The Santa Monica Mountains are characterized by almost every known geologic hazard on the California coast combined with a propensity for terrifying fire storms such that most individuals would be deterred from considering the area for human habitat. However, the Santa Monicas are the only rural, dramatic landscape in close proximity both to the employment centers of the northern portion of Los Angeles and to the ocean. In combination with the mystique of Malibu, lot owners are drawn to an area where they actually present plans for structures that can be burned over without any damage.

After several years of attempting to deal with individual permit applications, Roy Gorman of the Coastal Commission's legal staff devised a system in which the private market is encouraged to effectuate the transfer of development credits from areas with severe development constraints to areas that are appropriate for development under the Coastal Act. At the heart of the concept is the notion of the donor area (areas where existing lots should be extinguished) and the notion of the receiver area (areas where development can be encouraged so long as overall bouildout in the sub-region is not increased). The system is dependent on factual findings that no further commitment to total buildout can be made on a cumulative impact basis and that the buildout of particular geographic areas will have direct or cumulative impacts on coastal resources or coastal access roads within that particular area. A summary outline of the transfer of development credits system is set forth in Appendix V.

e. Legal Prospects for Cumulative Impact Regulatory Mechanisms.

A recent California appellate case has caused concern because the court overturned a Coastal Commission denial of a proposed land division in San Mateo County that had been based on cumulative impacts. The Commission's findings had relied heavily on the precedential impact that approval of the land division would have on future permit applications. However, even while reversing the Commission's decision, the court concluded that "the Commission could not base its refusal of the permit on such a speculative future contingency" precisely because "the

Commission clearly has the authority to prohibit any future development whose cumulative effect is both significant and adverse"(Billings v. California Coastal Commission, 103 Cal. App. 3d 729, 741). The outcome of the case clearly reiterates the importance of thoroughly substantiated findings. Its long-term significance may will be its implicit affirmation of overall cumulative impact assessment rather than permit by permit review, a function that is entrusted to the quasi-legislative function of the Commission and local governments in the local coastal program process. The latter view is buttressed by the fact that the State Supreme Court allowed to stand an appellate decision affirming the Commission's cumulative impact review authority (Stanson, supra) and a case reiterating the importance of cumulative impact assessment under the California Environmental Quality Act (Whitman v. Board of Supervisors of Ventura County, 88 Cal. App. 3d 397).

Cumulative impact assessment may be one area where the quasi-adjudicative versus quasi-legislative distinction is significant, not so much in terms of the scope of judicial review as for the focus on the governmental function inherent in the review. The local coastal program process is intended to define and resolve coastal regulatory concerns through local planning legislation subject to the review of the Coastal Commission. If the Commission's findings are precise and articulate in identifying "the impact of an activity which in times past) has been purely local, may under changed conditions transcend municipal boundaries (CEEED, supra p. 3), it is likely that the courts will defer to a quasi-legislative determination. Where cumulative impacts cross jurisdictional lines and are regional in scope, the courts are even more likely to defer to a well-documented quasi-legislative determination. In a setting analogous to the Coastal Commission's regional perspective, the California Supreme Court determined that a regional agency (in that case the agency charged with reviewing city annexations) was "better qualified" than local government to prepare and evaluate the regional aspects of an environmental impact report with "special emphasis on environmental resources peculiar to the region" and with "reference to projects: existent and planned, in the region so that the cumulative impact of all projects in the region can be assessed" (Bozung v. Local Agency Formation Commission Ventura County, 13 Cal 3d. 263, 283). The reality of the litigation process is that the relevance of cumulative impact review to coastal management is much more tangible and comprehensible in the quasi-legislative function of reviewing a plan for 30 or 40 miles of the California coast than it is to relate a proposal for a land division to past and future permit applications. In light of the volume of Coastal Act litigation over the past seven years in California, one suspects that we will not have to wait long for a case in controversy to test the application of cumulative impact analysis to local coastal programs.

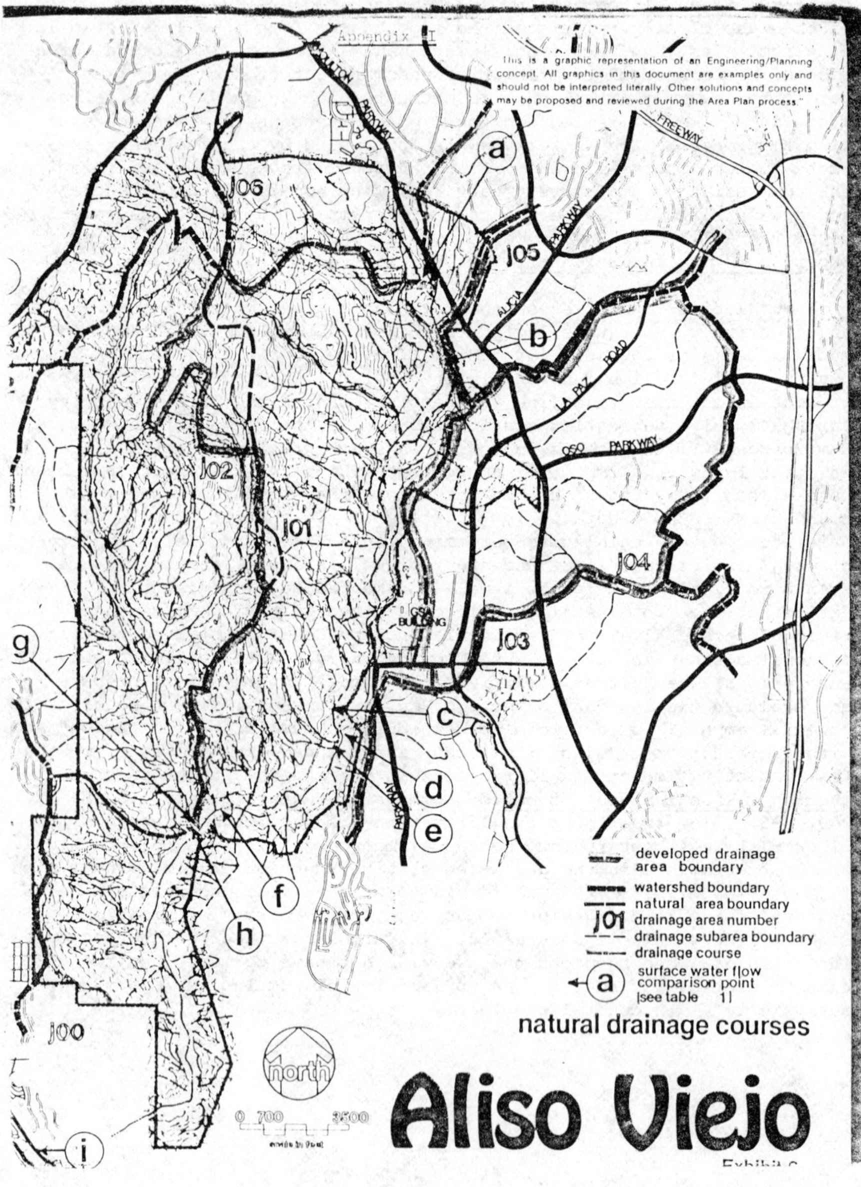
Appendix I
"This is a graphic representation of an Engineering/Planning concept. All graphics in this document are examples only and should not be interpreted literally. Other solutions and concepts may be proposed and reviewed during the Area Plan process."
MOULTON PARKWAY
FREEWAY
ALICIA PARKWAY
LA PAZ ROAD
OSO PARKWAY
GSA BUILDING
PARKWAY
J06
J05
J02
J01
J04
J03
J00
a
b
c
d
e
f
g
h
i
developed drainage area boundary
watershed boundary
natural area boundary
J01 drainage area number
drainage subarea boundary
drainage course
a surface water flow comparison point (see table 1)
natural drainage courses
north
0 700 3500
Aliso Viejo

APPENDIX II

CUMULATIVE IMPACT BOUNDARIES

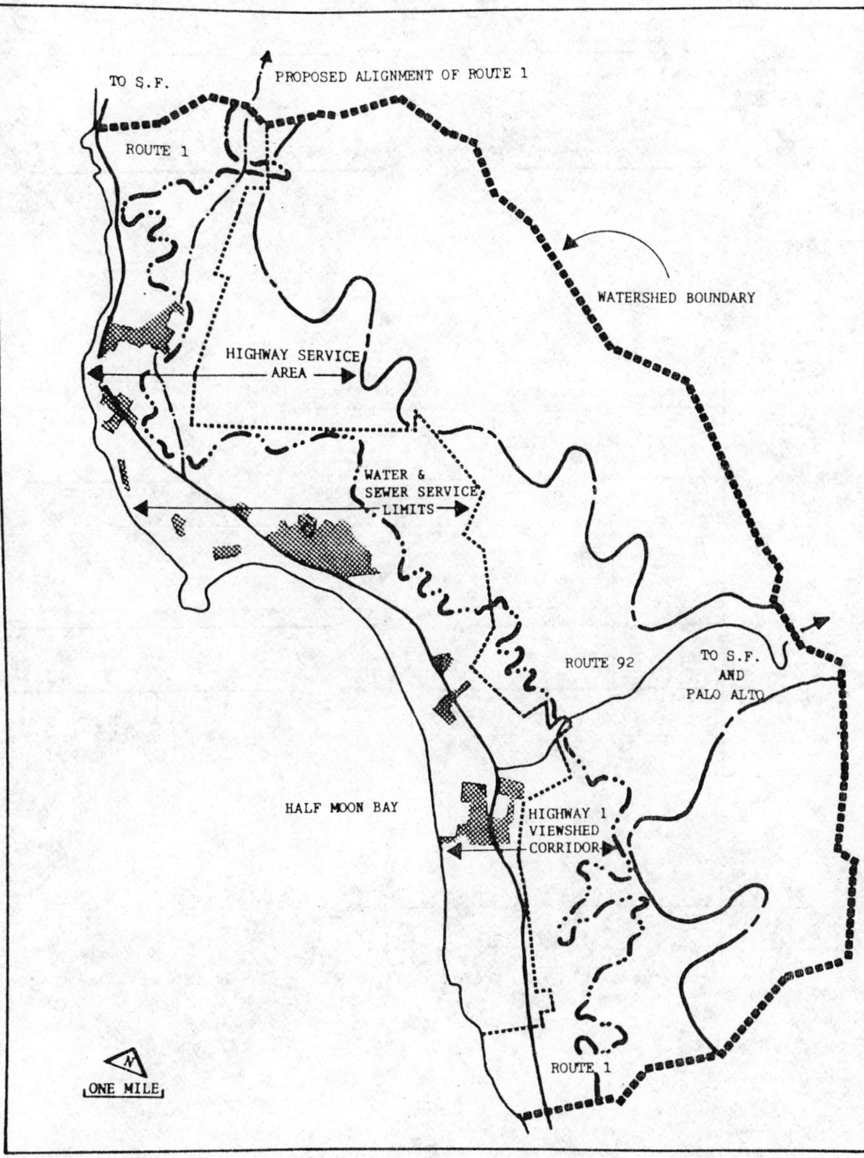

APPENDIX III

Existing and Modified Levels of Public Service
Capacity Comparison with Alternatives

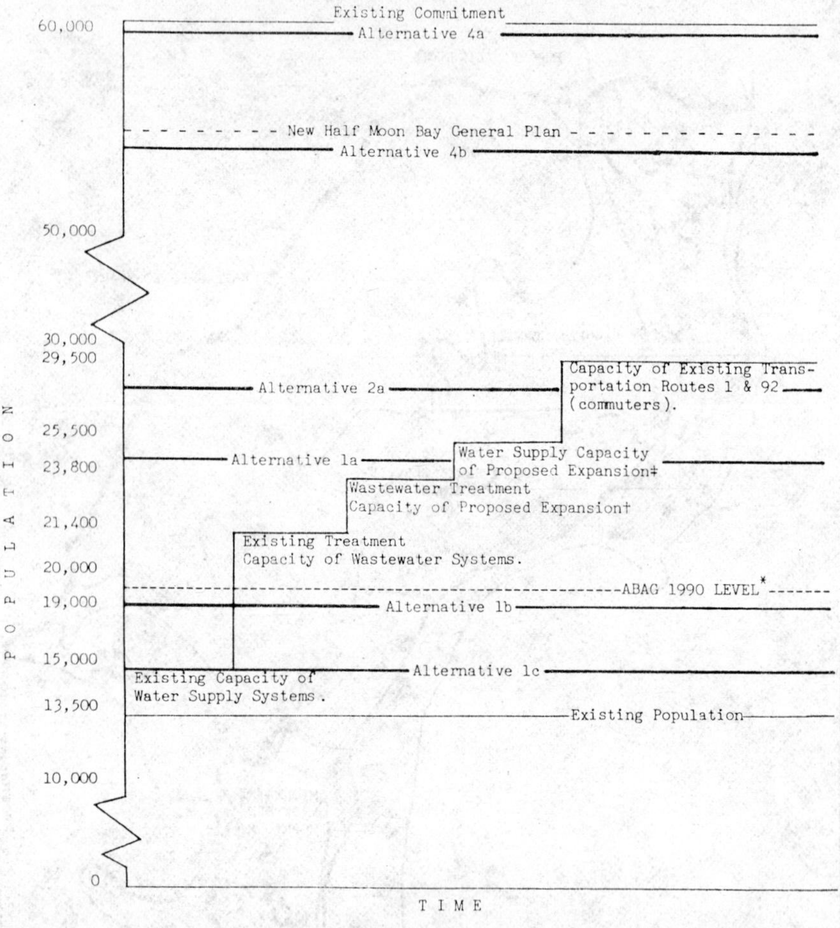

*ABAG will recommend approval of water or wastewater increases commensurate with this 1990 population level (ABAG/MTC, 1975)

†Midpoint of high and low projections in "Wastewater Treatment Allocation and Development"

‡Midpoint of high and low projections in "Water Supply Allocation and Development"

APPENDIX IV

A Generalized Process for Applying the Coastal Commission Policies to Water Supply Allocation and Water Development Projects

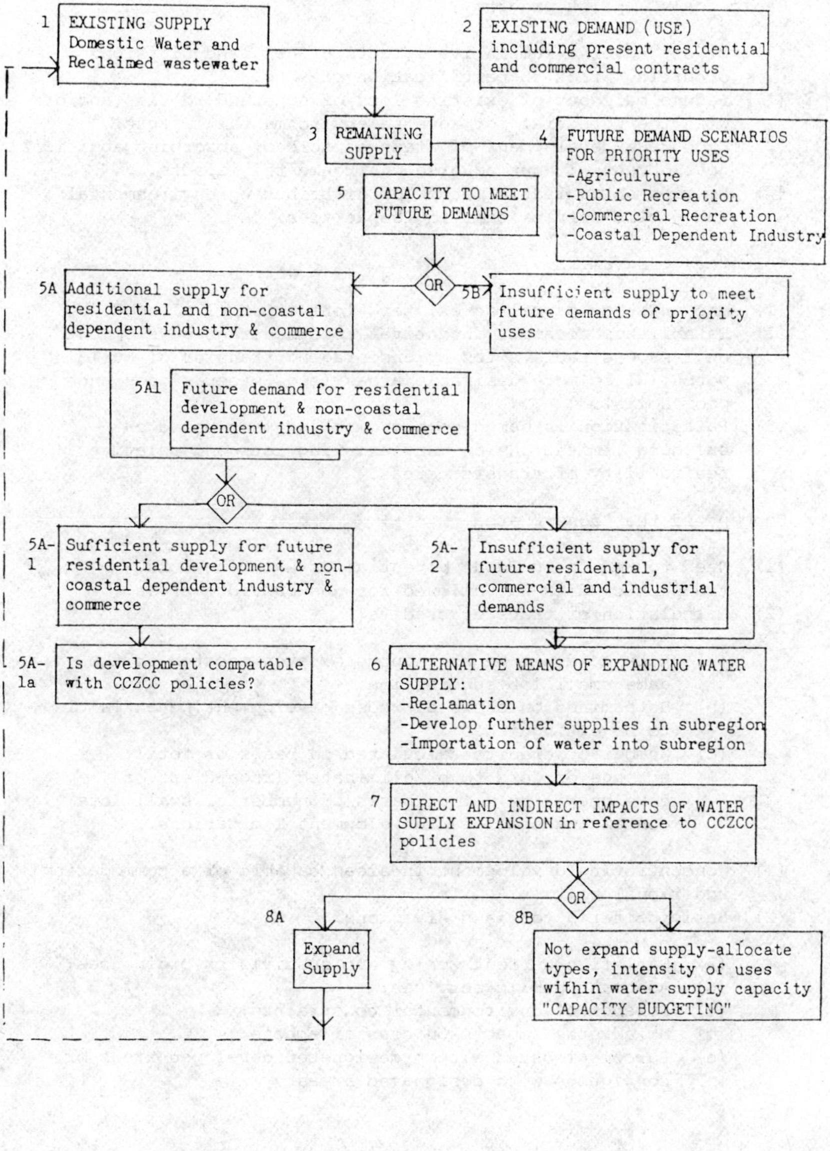

APPENDIX V

Transfer of Development Credits.

a. Basic goals - pilot program.

(1) Offset adverse cumulative effects of new land divisions occurring prior to certification of LCPs.
(2) Reduce buildout of existing lots of substandard size and/or with the potential for adverse environmental impacts.
(3) Encourage development of areas capable of absorbing additional development without adverse environmental impacts.
(4) Provide an equitable method of distributing environmental and public service carrying capacities.

b. Voluntary program.

(1) One unit per parcel on existing lots is still allowed.
(2) Establishes economic incentives to discourage development on less desirable sites by encouraging transfer of building potential to more desirable areas (e.g. steep slopes and poor soils).
(3) Participation is based on economic judgment based on building limitations on many existing lots and relative desirability of transfer areas.

c. Concept of the "donor" area.

(1) Need to reduce buildout potential of existing lots - transfer credits not allowed for unsubdivided areas.
(2) Calculation of transfer credits.

  (a) Lots need not be contiguous or even located in the same small lot subdivision.
  (b) Determined on basis of whose development potential is to be extinguished.
  (c) Number of credits calculated on basis of total acreage of lots to be eliminated (regardless of developability) or on basis of number of small lots containing specified development limitations.

d. (1) Concentrating development in areas capable of accommodating additional units.
(2) Basic criteria for land divisions.

  (a) Satisfies current zoning and subdivision ordinances;
  (b) Adequate public services;
  (c) Absence of environmental constraints;
  (d) No adverse impacts on coastal access;
  (e) Parcels located within designated developed areas or contiguous with designated areas.

Legal Standards for Coastal Development Permission

Jonathan M. Davidson
Department of Urban and Regional Planning
Florida State University

Concerns for predictability and fairness underly the enactment of policy guidelines and minimum procedures in coastal management legislation. The setting of substantive objectives such as environmental protection, economic development, and public access to the shoreline may be supplemented by more specific criteria for permit issuance and denial. Procedurally, safeguards can include notice to concerned persons, and assuring the applicant a hearing with the right to judicial review. Though these legal standards lack the precision of construction or engineering codes, they reflect a balancing of interests in prescriptive standards to guide decisions, reliance on scientific and administrative expertise, and ensuring fairness in the permitting process.

This paper examines the impact of procedural and substantive standards on coastal regulatory decisions. After outlining established approaches in four development control programs, it reviews efforts to adapt legal doctrines to the coastal management context. It concludes with recommendations for integrating procedural fairness, efficiency, and policy objectives with coastal regulation.

I. Overview of Regulatory Approaches

The legislatively created regulatory programs of Maine, Washington, Florida, and California illustrate a range of delegated authority to state and local authorities. The Maine Site Location of Development Law[1] establishes a state-administered permit program which reviews certain larger-scale development activity for potential environmental impact. In Washington, the Shoreline Management Act[2] vests initial permitting authority for development with local coastal governments, which must establish master programs to guide these decisions. The Florida regulatory program is statewide in scope, though its dual focus on critical environmental areas and developments of regional impact have already generated controversies in the state's coastal areas. Finally, the California coastal program offers the most comprehensive approach to development controls. That program is shifting interim state permit controls (1973-76) to local administration, reserving certain critical area controls to a permanent state Coastal Zone Conservation Commission.

As the name of the Maine statute implies, its regulation is concerned primarily with the location of development in the coastal zone.[3] The Site Location Law authorizes the state Environmental Improvement Commission to review proposed developments of greater than 20 acres, all drilling or excavating of natural resources, and structures over 60,000 square feet[4] for their potential environmental impact. The statute sets out criteria for issuing permits, including findings of: sufficient financial capacity in the developer to meet state air and water pollution standards, provision for solid waste and odor control, traffic movements, and for "fitting the development harmoniously into the existing natural environment."[5] A site location permit does not exempt the applicant from compliance with other state and local land use and environmental controls.[6]

In Washington, the concern with overlapping permit requirements has encouraged integrating coastal development permits with other governmental regulations. The basic coastal development control is the "substantial development permit"[7], which the locality administers in accordance with a state-approved master program for regulating shoreline uses.[8] A state shorelines hearings board is authorized to hear appeals from local permit decisions.[9] Permit applicants may also choose a single development review procedure outlined in the Environmental Coordination Procedures Act.[10] Under this law, a person proposing a project may submit a master application to the state Department of Ecology requesting the issuance of all necessary permits.[11] The local government may certify that the proposal complies with planning and zoning requirements, but must continue to exercise its delegated discretion under the Shoreline Management Act.[12] A developer following these coordinated procedures may consolidate his claims for all necessary permits before the Shoreline Hearings Board and pursue a right to judicial review of the entire record.[13]

Coastal regulation in Florida has elements similar to both the Maine and Washington programs. The statewide Environmental Land and Water Management Act[14] includes provisions for local regulation of large-scale developments, and for state-designated "areas of critical state concern."[15] The statute establishes three such areas: Big Cypress,[16] the Green Swamp,[17] and the Florida Keys.[18] In addition, developments of regional impact must also be locally regulated, subject to state guidelines but not review.[19] The Act provides separately for a coastal program centered in the state Department of Environmental Regulation.[20] However, the program must be compiled from existing rules and regulations, and is prohibited from amending existing laws for reasons other than ensuring federal consistency.[21]

By comparison, the California coastal program offers the most complete and controversial[22] regulatory approach. An initial state-administered interim permit program (1973-76)[23] shifts to local coastal programs under terms of the Coastal Act of 1976.[24] Under interim authority, a state coastal commission and six subordinate regional commissions reviewed over 25,000 permit applications for consistency with coastal planning objectives, potential environmental impact, and effect on public access to the shoreline.[25] The 1976 law establishes the state commission as a permanent authority which retains powers to designate critical areas, and require amendments in local policies to conform to state coastal objectives.[26] It also provides for a transition from state regulation to local administration of coastal development permits. However, localities must establish coastal programs, subject to state commission certification, which include a land use plan and zoning regulations.[27] The local coastal program can then issue development permits consistently with the planning and management policies of the current coastal act.[28]

II. Legislative and Judicial Guidance for Permitting Decisions

Once a regulatory scheme is established, the coastal program must show legally sufficient standards to guide its permit process. The legal issues in this context concern first, the validity of the legislature's delegation of discretionary authority to the agency; second, whether the application and review procedures have sufficient safeguards for applicants and concerned parties to present evidence and appeal adverse decisions to the courts.

The initial issue, nondelegation, is based on the principle that legislative policy decisions must not be entrusted to administrators who could exercise judgment beyond the law's initial intent. The traditional means for addressing this concern is to require ascertainable statutory standards to guide the permitting agency.[29] A reviewing court will look first to whether such standards exist, then consider whether the agency has exceeded whatever range of authority has been granted. Strictly applied, the nondelegation doctrine permits the court to invalidate any administrative program which makes independent policy judgments. In fact, the legislature may purposefully delegate complex matters requiring scientific and technical analysis with explicit recognition that precise standards are impracticable. Particularly in environmental and energy regulation, statutory prescriptions to limit the scope of discretion might inhibit administrative response to changing conditions and needs.

The difficulty in applying nondelegation doctrine to coastal regulation is illustrated in the Florida Supreme Court's decision in Askew v. Cross Keys Waterways.[30] That controversy centered on designations by the Florida Administration Commission of the Green Swamp and Florida Keys as critical areas under the Environmental Land and Water Management Act.[31] The statute authorized the commission, upon recommendation by the Division of State Planning, to designate boundaries and land development regulations to protect economic and ecologic values of the area.[32] The reviewing court was not unduly concerned with the legislature's entrusting the commission with determining "significance" of developmental or environmental activity, or with recognizing statewide importance of environmental and natural resources.[33] Rather, it invalidated the relevant sections of the Act because the legislature failed to delineate priorities among competing areas and resources.[34] The court concluded that:

> ...the standard by which land development regulations are to be measured is not a standard articulated by the legislature but one determined by the Administration Commission through formulation of principles for guiding development. In short the primary policy decision of the critical state concern to be designated as well as the principles for guiding development in that area are the sole province of the administrative body.[35]

Following the Cross Keys decision, the critical areas provisions in the statute were amended, requiring the commission to obtain direct legislative approval for critical areas designations.[36]

To avert the harsh and confining effects of prior specification of legislative standards, many courts have adopted a view of discretionary justice emphasizing procedural fairness.[37] This theory treats the basic thrust of the delegation doctrine as assuring that administrative discretion be exercised reasonably, without arbitrary or capricious conduct by officials. While it recognizes the desirability of legislative standards, its primary focus is on procedural safeguards including the applicant's right to a hearing before the decision-making board, and judicial review of agency action for fairness, rationality based on available evidence, and adequate prior notice to persons affected by the decision.[38]

The California, New Jersey, and North Carolina courts have relied in varying degrees on the presence of procedural safeguards to uphold coastal regulations. The California decision in CEEED v. Coastal Commission[39] found sufficient guidance for permitting decisions in the standard requiring a finding of "no substantial adverse environmental or ecological effect"[40] and consistency with other findings and objectives in the 1972 Coastal Act. The court also affirmed the need for deferring to the expertise of the coastal commissions:

> ...The fact that the Commission is required to weigh complex factors in determining whether a development will have a substantial...effect does not...

> mean that unbridled discretion has been conferred on it. A statute empowering an administrative agency to exercise a judgment of high order in implementing legislative policy does not confer unrestricted powers....[41]

The New Jersey court also found adequate standards for permit issuance in the state's Coastal Area Facilities Review Act.[42] These statutory criteria include: conformance with other environmental regulations (air, water, radiation); adequate waste disposal; minimum feasible impairment of water supplies and natural processes; safety factors; and minimal degradation of unique or irreplaceable land types, or those with historic, archeologic, scenic, or aesthetic attributes.[43] Though some of these standards are arguably no more precise than those invalidated in Florida, the New Jersey decision found that the coastal law defined an adequate checklist to guide regulation in light of changing conditions and evolving scientific knowledge of environmental protection.[44] In addition, the court noted that the act provides procedural safeguards against unreasonable or unwarranted administrative action through its hearing and judicial review requirements.[45]

The North Carolina Supreme Court supported even greater administrative flexibility under the Coastal Area Management Act.[46] The legislation establishes general policies and goals to guide the creation of a coastal management system, including: sound ecological management, orderly and balanced use and preservation of resources, economic development, recreation, historic and cultural preservation, and other pruposes deemed necessary.[47] The court upheld this broad delegation in light of "the obvious need for expertise in the achievement of legislative goals" and the infeasability of laying out a detailed statutory agenda covering every conceivable problem which might arise.[48] It also tempered this support with an inquiry to determine the adequacy of the act's procedural framework - to insure that the Coastal Resources Commission "will perform its duties fairly and in a manner consistent with legislative intent."[49]

III. Ensuring Procedural Fairness in Coastal Regulation

Beyond broad-scale standards affecting the scope of agency discretion, administrative law and due process may impose additional procedures on permitting agencies. The relevant references in this context are hearing requirements specified in the program's authorizing statute or presence of a constitutionally-protected "property" interest[50] entitling an applicant or affected party to an agency hearing. If a right to a hearing exists, the form and basic procedures are prescribed by applicable provisions in the state's administrative procedure act. Depending on the nature of the discretionary process and facts considered, the permit hearing will resemble either legislative or judicial proceedings. If no hearing is specified, only general standards for unreasonable, arbitrary, or patently unfair actions could reverse an agency's grant or denial of a permit.

Once the state's administrative procedure act(SAPA)[51] is invoked by a hearing requirement,[52] the permitting agency determines whether quasi-legislative (rulemaking) or adjudicatory (contested case) procedures should be followed. If the decision is characterized as a generalized policy-type matter, SAPA rulemaking provisions would apply.[53] For example, setting criteria for what constitutes a "development of regional impact" involves reviewing data and testimony from a variety of public and private sources (e.g. commercial and industrial developers, soil scientists, land use and transportation planners). SAPA rulemaking provisions typically require the agency to provide general notice to interested persons at least two weeks in advance of its scheduled hearing describing the intended action or agenda.[54] At the actual hearing, any person or group may present evidence, though the agency may limit oral testimony and cross-examination in the interests of efficiency.[55]

Thus, the primary purpose of a rulemaking proceeding is to provide information and views to agency decision-makers in a manner comparable to legislative hearings. The agency may base its final rule on information presented at the hearings or through other reasonable means; however, the rule must generally be accompanied by a concise statement specifying the rationale and data underlying the decision.[56] If appealed, the rule will be upheld by the court if it appears supported by substantial evidence.[57]

However, if the facts presented to the agency relate uniquely to legal rights, duties, or privileges of individuals, contested case procedures will likely control the agency hearing.[58] Under these circumstances, many of the evidentiary elements of a judicial trial are present. For example, a developer seeking a coastal permit would be entitled to prior notice of the facts and issues which may be in dispute. Other parties, including nearby landowners, may also be granted participation.[59] A contested case hearing includes the right to present and rebut evidence and cross-examine witnesses[60]- though the agency need not follow rules of court, so long as the basic adversary nature of the hearing is maintained.[61] The agency's decision must then be based solely on the record of the proceedings, and a reviewing court may overturn decisions which are clearly erroneous in light of the evidence presented.[62]

The Model State Administrative Procedure Act and most state court decisions characterize the permitting function as quasi-judicial, and apply contested case standards to determine the adequacy of agency hearing procedures. For example, the Wisconsin Supreme Court required such procedures for the State Public Service Commission to determine whether construction of a breakwater met prescribed statutory standards.[63] The Washington Shoreline Management Act specifies that contested case procedures apply to the Shorelines Hearing Board's review of a locality's grant or denial of a substantial development permit.[64] In reviewing the adequacy of the Board's record of decisions, Washington courts accord deference to the agency's specialized knowledge and expertise in applying shoreline management policies to individual cases.[65]

California courts characterize comparable permit proceedings under its interim coastal management program as quasi-judicial.[66] In distinguishing the permitting process from broader policy and planning functions of the coastal commissions, the supreme court called it "a delicate balancing of the effect of each development upon the environment of the coast."[67] In Patterson vs. Coastal Commission,[68] an intermediate appellate court rejected a contention that the interim permit function was so inextricably interwoven with the concurrent preparation of a comprehensive coastal plan that the permit function was quasi-legislative.[69]

In contrast, a New Jersey appellate decision concluded that the hearing provision in the Coastal Area Facility Review Act is quasi-legislative.[70] The provision gives parties the right to inform the Commissioner of Environmental Protection of their position on a proposed application, and provide data regarding the environmental effects of the proposed facility.[71] The court held that rule-making adminstrative procedures were sufficient to meet this statutory hearing requirement, and affirmed the agency's denial of rebuttal and cross-examination rights to a public interest group challenging a permit application for a nuclear facility.[72] By including this regulatory decision as part of a broader policy and planning function,[73] the New Jersey approach runs counter to the trend toward judicializing proceedings designed to resolve disputed facts.[74]

In reviewing hearing procedures under its site location law, the Maine Supreme Court avoided making a distinction based on legislative or adjudicative facts. Rather, the court examined the Environmental Improvement Commission's procedural rules for principles of due process and fair play.[75] It found no inherent illegality in agency restrictions on oral cross-examination, so long as parties are not unfairly prejudiced by these limitations.[76] Since the board did provide written methods for challenging and rebutting evidence, the court decided not to invalidate the permit hearing as denying due process on merely procedural grounds.[77] Unless the applicant could show that unfair treatment would result, the board could consider the interests of orderliness, efficiency, and reliance on its own expertise,[78] in fashioning its rules for permit hearings.

Among these states, standards imposed by administrative and

constitutional law create an outline for coastal programs to establish a fair policy-making and adjudicatory process. Within this framework, agencies can promulgate rules and procedures that account for reliance on expertise, and demands for efficiency and equity. Judicial interference will likely be limited to clear excesses or omissions by authorities which affect substantive rights of permit applicants or other parties with a stake in the proceeding. The agency can then proceed to translate legislative objectives into a consistent program of coastal development regulation.

IV. The Role of Substantive Standards

Once a coastal program exceeds the requisites of administrative law and procedure, its regulatory task focuses on the "delicate balancing" process to determine when, and with what conditions, it will allow development to proceed. Even the fairest procedures offer little assistance to resolving multiple objectives and demands for coastal resources. Furthermore, the legislature may create concurrent planning and policy roles for the program, with few guidelines on how individual permit decisions relate to larger and longer-range considerations. This final section considers the administrative role of translating policy objectives and criteria in coastal legislation into principles to guide individual permitting decisions.

From a substantive perspective, legislative commands to "promote and enhance the public interest,"[79] or to foster all reasonable and appropriate uses in shoreline management[80] provide little indication of priorities for regulation. Often, policy statements affirm the values of natural and economic coastal resources without recognizing potential conflicts among competing management objectives. Directing the coastal program to consider conservation and developmental uses merely passes the crucial trade-off decisions to the administrative process:[81] for example, determining whether environmental interests should prevail in locating energy facilities; whether public access or ecological principles should control in considering the intensity of shoreline use; or whether unique or fragile coastal areas can be protected from development by regulation alone. These crucial decisions fall to the coastal boards and commissions, which must arbit the competing demands and value preferences forwarded by public and private actors.[82]

One legislative approach that aids permitting decisions is the setting of developmental priorities. The Washington Shoreline Act provides general use priorities along shorelines of statewide significance. In descending order, these include preserving the natural character, protecting the resources and ecology, and increasing public recreational opportunities.[83] Development permission for such areas is also prioritized, with preference to residential, shoreline recreational, coastal-dependent industrial and commercial, and other developments which enable substantial numbers of people to enjoy the shoreline.[84] The California Coastal Act of 1976 also establishes a priority for coastal dependent developments,[85] with industry taking preference over recreational use; but coastal-dependent commercial recreation retains priority over general industrial, commercial, or private residential development.[86] The act's planning and management policies favor concentrating new activity within or adjacent to existing developed areas,[87] establishing clearly defined buffer zones between urban and rural areas,[88] and encouraging agricultural lands preservation.[89] Compared to the four criteria set out in the Maine Statute,[90] or the seven-item checklist for New Jersey coastal permits,[91] California offers permit issuers an extensive range of land and water policies for its coastal management framework. Ironically, the task of translating these myriad objectives into principles for granting or denying a single development may lead to less consistency with broader policies than a more limited approach.[92]

V. Conclusion

The establishment of a legal framework for coastal policies and regulation can best serve as an enabling device for administering program fairly, efficiently, and with due deference to expertise and balancing competing interests. Toward these ends, both legislative and adjudicatory decision models can ensure an informed management process. Substantive standards are most appropriately set out in the authorizing legislation. As the agency refines and adds detail to statutory criteria, rule-making procedures allowing the broadest possible range of interests to be heard should precede these decisions. For individual permitting decisions, contested case procedures provide an optimal format for focusing the issues in whether to approve a developer's proposal. Finally, the coastal program should compose a method for integrating policy and planning functions with the regulatory system.

FOOTNOTES:

1. Maine Revised Statutes, title 38, §481 et seq.
2. Washington Revised Code §90.58. See generally Crooks, "The Washington Shoreline Management Act of 1971", 49 Washington Law Review 423 (1974).
3. See In Re Belgrade Shores, Inc., 371A.2d413, 415 (Me. 1977).
4. See Maine Rev. Stat. §482(2)
5. Ibid §484(3).
6. See generally F. Bosselman, D. Fuerer, and C. Siemon, The Permit Explosion 19-20 (1976).
7. See Wash. Rev. Code §90.58.140.
8. Ibid §§90.58.080 - .100. See generally D. Mandelker, Environmental and Land Controls Legislation 256-259 (1976).
9. Ibid §90.58.170 - .190.
10. Wash. Rev. Code §90.62. See generally Roe, "Simplifying Regulatory Procedures: A Commendable Washington State Approach," in A. Kolis, ed., Thirteen Perspectives on Regulatory Simplification 43-51 (1979).
11. Ibid §90.62.040.
12. Ibid §90.62.100(4)
13. Ibid §90.62.080.
14. Florida Statutes Annotated §380. See generally T.G. Pelham, State Land-Use Planning and Regulation 29-63, 100-110 (1979); R.G. Healy and J.S. Rosenberg, Land Use and the States 126-169 (2nd edition, 1979).
15. Fla. Stat. Ann. §380.05. See generally Pelham, supra, pp. 75-93; Mandelker, "Critical Area Controls: A New Dimension in American Land Development Regulation," 41 Journal of the American Institute of Planners 21, 21-25 (1975).
16. See Fla. Stat. Ann. §380.055.
17. Ibid §380.0551.
18. Ibid §380.0552.
19. Ibid §380.06.
20. Ibid §380.19.
21. Ibid §380.23.
22. See Duddleson, "How the Citizens of California Secured Their Coastal Management Program," in R. Healy, ed., Protecting the Golden Shore 3-15 (1978); Adams, "Proposition 20 - A Citizen's Campaign," 24 Syracuse Law Review 1019-46 (1973).
23. See California Public Resources Code §§27400-27405, superceded by California Public Resources Code §§30600-30626. See generally Healy, The Role of the Permit System in the California Coastal Strategy," in Healy, ed., Protecting the Golden Shore 67-95.
24. Cal. Pub. Res. Code §§30500-30522.
25. See generally Note, Coastal Zone Management and Planning in California: Strategies for Balancing Conservation and Development,15 Urban Law Annual 253, 260-74 (1978).
26. See Cal. Pub. Res. Code §30502.
27. Ibid §30512.
28. Ibid §§30200 to 30264.
29. See K. Davis, Administrative Law Text §2.06 (3rd edition 1972).

30. 372 So.2d913 (Fla. 1978). _See generally_ Pelham, _supra_ note 14, at 118-128.
31. 372 So.2d at 916-918.
32. _Fla. Stat. Ann._ §380.05(1)(a).
33. _372 So.2d at 919._
34. _Ibid_ at 919.
35. _Ibid_ at 924.
36. _See Fla. Stat. Ann._ §380.05(1)(c).
37. _See generally_ K. Davis, _Administrative Law Text_ §2.06; I.F. Cooper, _State Administrative Law_ 61-70.
38. _See_ K. Davis, _Administrative Law Text_ §2.08.
39. 118 Cal. Rptr. 315 (Ct. App. 1974).
40. _See Cal. Pub. Res. Code_ §27402.
41. 118 Cal. Rptr. at 329-30.
42. _New Jersey Statutes Annotated_ §§13:19-1 to 15. _See_ Toms River Affiliates v. Department of Environmental Protection, 355 A.2d679, 684-85 (N.J. Super. Ct. App. Div. 1976).
43. _N.J. Stat. Ann._ §13-19-10.
44. 355 A.2d at 684.
45. _Ibid_ at 684-85.
46. _North Carolina General Statutes_ §§113A-100 at 128 (1975). _See_ Adams v. North Carolina Department of Natural and Economic Resources, 249 S.E.2d 402 (N.C. 1978). _See generally_ Glenn, The Coastal Management Act in the Courts, 53 _North Carolina Law Review_ 303.315 (1974).
47. _N.C. Gen. Stat._ §113A-102(b). See 249 S.E.2d at 412.
48. _Ibid_ at 411.
49. _Ibid_ at 411.
50. _See, e.g._, Goldberg v Kelly, 397 U.S. 254 (1970). _See generally_ Reich, Law of the Planned Society, 75 _Yale Law Journal_ 1227 (1966). _See also_ L. Tribe, _American Constitutional Law_ 10-7 (1978). The fifth and fourteenth amendments to the federal consitiution provide that "life, liberty, or property" are not to be deprived without "due process of law." Professor Tribe states that:

> These procedural safeguards have their historical origins in the notion that personal freedom can be preserved only when there is some institutional check on arbitrary government action....
>
> The element of due process analysis characterized as "procedural due process" delineates the constitutional limits on judicial, executive and administrative _enforcement_ of the legislative dictates. This has traditionally involved the elaboration of procedural safaguards designed to accord the individual "the right to be heard before being condemned to suffer grievous loss of any kind" as a result of governmental choices which can take the form of acts or, less commonly, of omissions.

Ibid.

51. Each state has enacted legislation prescribing minimum administrative procedures for agencies. There are currently 27 states which have a form of the Model State Administrative Procedure Act first adopted by the Commissioners on Uniform State Laws in 1944. The references in this section are to the current Revised Model State Administrative Procedure Act §§1-19 (U.L.A. 1980 Supp.)
52. Administrative procedure acts are not directly concerned with substantive determinations of regulatory agencies. Rather, they exist as overlays which require uniform procedures whenever agencies adopt rules, or adjudicate the interests of parties entitled to a contested case hearing. APA procedural standards will apply only when there is a hearing requirement in the statute authorizing the development permit program, or when a constitutional or otherwise protected interest "triggers" the APA rule or contested case procedures. See J. Davidson, Administrative Procedures Required for Development Permits in Hawaii, Technical Supplement No. 18, Hawaii Coastal Zone Management Program 2 (1980).
53. See Revised Model State Administrative Procedure Act §§2-6.
54. Ibid §3(a)(1).
55. Ibid §3(a)(2). See, e.g., Public Interest Research Group v. Department of Environmental Protection, 377 A.2d 915, 922 (N.J. Super. Ct. App. Div. 1977), cert. denied 384 A.2d 517 (N.J. 1977).
56. See Revised Model State A.P.A. §3(a)(2).
57. See Ibid §7.
58. See Ibid §§9-11.
59. See, e.g., Wash. Rev. Code §90.58.140(b)(i); Daly v. Natural Resources Board, 208 N.W.2d 839 (Wis. 1973), cert. denied, 414 U.S. 1137 (1973).
60. Revised Model State A.P.A. §10.
61. Ibid §10(1)
62. Ibid §15(g)(5).
63. See Hixon v. Public Service Commission, 146N.W. 577, 583-84 (Wis. 1966).
64. See Wash. Rev. Stat. §90.58.180(3).
65. See Hayes v. Yount, 552P.2d 1038, 1044 (Wash. 1976); Department of Ecology v. Ballard Elks Lodge, 527P.2d 1121, 1124 (Wash. 1974).
66. See City of Coronado v. California Coastal Zone Conservation Commission, 131 Cal. Rptr. 241, 243-44 (Cal. App. 1977); Patterson v. Central Coast Regional Commission, 130 Cal. Rptr. 169, 173-74 (Ct. App. 1976).
67. State v. Superior Court, 524 P.2d 1281, 1288 (Cal. 1974).
68. 130 Cal. Rptr. 169 (Ct. App. 1976).
69. Ibid at 173.
70. See N.J. Stat. Ann. §13-19-9(a); Public Interest Research Group v. Department of Environmental Protection, 377 A.2d at 922.
71. N.J. Stat. Ann. §13-19-9(a).

72. 377 A.2d at 923-24.
73. Ibid at 923.
74. See K. Davis, Administrative Law Text §§7.02-7.04.
75. In Re Maine Clean Fuels, 310 A.2d 736, 744 (Me. 1973).
76. Ibid at 744-76.
77. Ibid at 745.
78. Ibid at 748.
79. See Wash. Rev. Code Ann. 90.58.020. But see also Klosterman "A Public Interest Criterion", 46 Journal of the American Planning Association, 323 (1980).
80. Ibid.
81. See J. Hite and F. Laurent, Environmental Planning: An Economic Analysis 3(1972). Hite and Laurent refer to this conflict between preservation and exploitation of resources as the "environmental dichotomy", since in many cases (particularly in the coastal zone) it involves either/or decisions. Ibid.
82. See Hufschmidt, Knox, and Parker, "A Policy Analysis Approach", in J. Hite and J. Stepp, eds., Coastal Zone Resource Management 107(1971); Slonin, "Coastal Zone Resource Allocation: Some Legal and Economic Considerations, 1 Sea Grant Law Journal 369, 374-75 (1976).
83. Wash. Rev. Code Ann. § 90.58.020 (1) to (7).
84. Ibid.
85. Cal. Pub. Res. Code §§ 30255, 30001.5(d).
86. Ibid § 30222.
87. Ibid §§ 30250, 3007.5.
88. Ibid § 30241(a).
89. Ibid § 3024(b) to (e).
90. See note 5 supra and accompanying text.
91. See note 43 supra and accompanying text.
92. See e.g., Billings v. California Coastal Commission, 163 Cal. Rptr. 288 (Ct. App. 1980). The appellate court reversed the commission's denial of a permit based on findings of inconsistencies with agricultural preservation, concentration of development, and coastal dependent use policies. The courts reversal appeared to rely on the conclustion that the proposed development was too small to significantly affect these broader policy concerns. Ibid at 295.

THE COAST IN WASHINGTON STATE

Charles H. Odegaard*

INTRODUCTION

In his environmental message to Congress, President Carter endorsed the Year of the Coast. By so doing, he brought to bear the prestige and emphasis of his office on a subject that has long needed our attention, yet seldom received it to the degree needed.

1980 - the year of the coast was declared by the Coast Alliance, a coalition of 18 U.S. conservation groups who promote knowledge of our coastlines.

1805 - On November 10, Lewis and Clark cited the mighty Pacific Ocean fulfilling President Jefferson's directive to find a way to our western shore. Shortly thereafter, Captain Lewis and five men strolled north on that beach reaching what is presently the City of Long Beach.

Since that eventful November, 175 years ago, the residents of what became and now is the State of Washington have been enjoying active and passive recreation on the bountiful sand - sun - and water.

1961 - The Ninth Circuit Court of Appeals held that the western boundary of the upland property owner was mean high tide.

1966 - The Washington Supreme Court held that the western boundary of the upland property owner was the line of 1889 ordinary high tide.

1967 - The Supreme Court of the United States overturned the State ruling by holding that the western boundary of the upland property owner was, in fact, mean high tide.

In that same year the Washington State Legislature assigned administrative responsibility for the Pacific Ocean beaches to the Washington State Parks and Recreation Commission.

This paper will address itself as to how that agency tried to preserve and conserve their coast and what were the results.

*Deputy Regional Director, National Park Service, Pacific Northwest Region, Seattle, Washington

THE AREA

INTRODUCTION

The Pacific Ocean Beach of the state of Washington is somewhat analogous to the other ocean beaches in the United States. The entire marine shoreline in Washington is 2,337 miles in length, subdivided into areas of 1,784 miles bordering Puget Sound and the Straits of Georgia, 34 miles of the Columbia River, 129 miles in Willapa Bay, 89 miles in Grays Harbor, and 157 miles along the Pacific Ocean. It is to the 157 miles that this paper is addressed.

TOPOGRAPHY

The northern approximately 2/3 of the coast is a rocky, mountainous terrain consisting of narrow, steeply sloping rocky beaches backed by high forested bluffs. Numerous rocky outcrops exist just off shore. The winters are wet and mild.

The southern portion is characterized by wide sandy beaches with extensive dunes in front of grasslands and forests. Two major estuaries, Grays Harbor and Willapa Bay, occurr in this portion. The winters are mild here also and not nearly so wet as their neighbors to the north.

During the winter, Pacific currents set toward the north, while during the summer months, they set to the south. Associated with the summer currents, is a general offshore movement of the surface water, resulting in upwelling of waters from lower depths. The upwelling water is cold, high in salinity, low in oxygen content, and rich in nutrients. Since the northern flow contains more energy than the southern flow, the Columbia River is a key factor in the amount of sand deposited on the southern beaches.

POPULATIONS

The northern beach is sparsely populated and remains largely unaltered. There are two Indian Reservations, one to the extreme north, and one where the rocky northern portion blends into the southern. The remainder of the area is primarily national park.

The southern beach is more populated but would still be characterized as very rural or low density. Since a considerable amount of the low bank upland is private, there are many year round and seasonal homes as well as transit overnight accommodations.

For this paper, it is important to know that though the beach population is less than 2 percent of the total state population, over one-half of the state's population resides in the overall marine area as described above and is within close commuting distance to the beach.

ECONOMICS

The overall marine area is also the major industrial and business area of the state. These activities include the development of a nuclear submarine base, proposed nuclear power plants, large port facilities, refineries, timber cutting and processing and--Boeing.

The area under consideration is almost pastorial in comparison. Fishing, commercial and sport, is the main industry with an estimation that over one-half of the state's 4 million people and thousands of visitors fish. Commercial fishermen annually harvest over 6 million salmon, 20 million pounds of bottom fish, and 6 million pounds of shellfish. The total value of sportfishing is even greater. Directly on the beach itself, razor clam digging is the name of the game, and with the tasty morsels come tourism. The same tourists also come to walk the sandy beaches and watch the ever-changing interface between the shore and the water.

The taking of commercial logs from the storm-swept beaches is another commercial enterprise directly related to the beach, as is the taking of sand for both construction and for the production of cranberries. Mining is also a factor as "black sands" (found on the southern beach) contain gold and titanium among other metals. There is the possibility of petroleum production.

THE SITUATION

LEGAL STATUS

Coastal Zone Management (CZM)

On October 27, 1972, the President signed into law, P.L. 920583:86, stat. 1280, the Coastal Zone Management Act. The Act affirms a national interest in the effective management, beneficial use, protection, and development of the coastal zone and provides assistance and encouragement to the coastal states to develop and implement rational programs for managing their coastal zones. Three financial assistance grant programs are authorized by the Act:

1. Program Development Grants
2. Program Administrative Grants
3. Program of Estuarine Sanctuaries

As an additional incentive for state participation, the Act stipulates that federal actions within the coastal zone shall be, to the maximum extent feasible, consistent with approved state management programs. Washington State's coastal program was approved by the National Oceanic and Atmospheric Administration in June, 1976, making it the first such approved plan in the nation. Basic to this program were the already existing Seashore Conservation Act and Shoreline Management Act.

Seashore Conservation Act

The state of Washington became increasingly concerned about the Pacific Ocean Coast in late 1965, approximately 4 years before Congress started to consider the Coastal Zone Management Act. At the request of Washington State Parks and Governor Daniel J. Evans, the 1967 legislature considered and passed RCW 43.51.650-685, the Seashore Conservation Act.

The Act gave tremendous authority to State Parks except within the boundaries of the Indian Reservations. The authority was very specific within the broad declaration that "nonrecreational use of the beach must be strictly limited. Even recreational use of the beach must be regulated in order that Washington's unrivaled seashore may be saved for our children in such the same form as we know it today."

The principals as set forth in the Act stated, "Where feasible, the area shall be preserved in its present state: Everywhere it shall be maintained in the best possible condition for public use. All forms of public outdoor recreation shall be permitted and encouraged in the area, unless specifically excluded or limited by the Commission. While the primary purpose in the establishment of the area is to preserve the coastal beaches for public recreation, other uses shall be allowed as provided in RCW 43.51.650-43.51.685, or when found not inconsistent with public recreation use."

For reference later in this paper, the following specific items should be noted.

43.51.670---"Shall seek cooperation and assistance of federal agencies, other state agencies, and local political subdivisions."

---"All state agencies, and the governing officials of each local subdivision shall cooperate."

43.51.675---State Department of Fisheries shall regulate the taking of food fish and shellfish.

---State Department of Game shall regulate the harvesting of wildlife except no hunting shall be permitted in any state park.

43.51.680---State Parks, after agreement with State Highways, shall establish reasonable regulations for use and control of vehicular traffic on and along the ocean beach highway---provided that automobile driving shall be permitted on beaches subject to authority of State Fisheries to prohibit driving over clam beds.

43.51.685---Lands may be leased for exploration and production of oil and gas, provided that oil drilling rigs and equipment will not be placed on the seashore conservation area.

---State Parks shall sell sand to cranberry growers if State Parks finds said sale not harmful to the character of the land.

Shoreline Management Act

Washington Shoreline Management Act, passed in 1971, was born in the Wilbour v. Gallagher case before the State Supreme Court in 1969, subsequently appealed to the U.S. Supreme Court in 1970.

The SMA, termed by many as the most comprehensive tool for guiding shoreline uses to be found in the nation, established as policy "management of the shorelines of the state by planning for and fostering all reasonable and appropriate uses." The Act mandated a use permit system and a solid environmental planning program.

The basic concerns which prompted the legislature to adopt the SMA are very similar to those adopted a year later in the CZMA.

Court Decisions

There have been many court decisions directly or indirectly affecting the coast. There are four such decisions affecting the Pacific Ocean.

1. Ghione v. State of Washington, 26 Wash 2d 635, 175 P2d 955, December 12, 1946.

Where a navigable river, over a period of years, imperceptibly shifted its bed toward the west and subsequently dried up, the owners of the uplands on the east were entitled to the accretions to the eastern bank, and the state was entitled only to the bed and shores of the river as they were situated when it ceased to be navigable.

2. Hughes v. State of Washington, 67 Wash 2d 799, January 20, 1966.

The Washington Supreme Court held that all accretion to land bordering waters where the tide ebbs and flows, which has occurred since November 11, 1889, constitutes an addition to state-owned property, and not to upland property, since when the state acquired ownership of tidelands (to the line of ordinary high tide) at the time of its admission as a state, the littoral rights of upland owners was terminated; hence, the dividing line between state-owned tidelands and upland property is a fixed line, which is the line of ordinary high tide (and which is the same as the line of mean high tide) as it existed on November 11, 1889.

Appeal from a judgment of the Superior Court for Pacific County, No. 14546, Warner Poyhonen, J., entered February 11, 1964. REVERSED.

The State appealed to the U.S. Supreme Court.

3. Hughes v. State of Washington 389 US 290. (Also 19 LEd 2d 530, 88 SCt 438) Argued November 6, 1967. Decided December 11, 1967.

On certiorari, the Supreme Court of the United States reversed. In an opinion by Black, J., expressing the view of seven members of the court, it was held that the question was governed by federal, not state, law, and that under federal law the property owner who traced her title to a federal grant prior to statehood, was the owner of accretions.

In a separate but concurring opinion one Justice argued that the State Supreme Court decision represented an abrupt change in law and resulted in the taking of property without due process. In Ghione v. State of Washington, the Washington Supreme Court established the legal precedent that accreted land belongs to the upland owner.

4. Wilbour v. Gallagher 77 Wash 2d 306 December 4, 1969

a. Navigable Waters - Rights of Public-Private Land Periodically Submerged.

By analogy with changing rights of riparian owners and the public in the case of navigable lakes whose waters are periodically raised or lowered--Owners of upland, not riparian, which are regularly and periodically submerged by an artificial raising of the water level of a navigable lake, cannot fill or otherwise use their land so as to interfere with the public's expanded right of navigation and rights corollary thereto during the periods that the water level is raised. This is particularly true where the title of such non-riparian upland owners is deraigned through deeds which make the property subject to be "perpetually unundated and overflowed."

b. Adjoining Landowners - View of Water - Periodic Separation of Intervening Property.

In the absence of zoning or other restrictions, a landowner did not have a year-round right of an unobstructed view of a navigable lake, where another's land, which was submerged at times and positioned between the objecting landowner and the water, was exposed periodically and during those times was subject to the other's unrestricted and exclusive use, including the possible erection of temporary structures.

This case was appealed to the U.S. Supreme Court; however, on October 26, 1970, certiorari was denied (400 US 878).

THE PROBLEMS--AND THEIR SOLUTIONS

Prior to the 1960s, the problems on the Pacific Ocean Beach were basically twofold in nature: (1) Owners of upland property who wanted to sell were unable to secure title insurance until they were able to affix their western boundary, i.e. the ocean. To do this many many of them sued the state of Washington and fixed the western boundary as the line of 1889 i.e. statehood; (2) The other major problem was that of county governments' growing expenses. For some time, the two southern counties had requested state assistance as more and more visitors from out of the county came to enjoy the sandy beaches--a use requiring more police, more litter pickup, and more concern for adequate sanitary facilities.

Circa 1960, the citizens of the state became aware that the beaches were not an infinite resource and that the beaches not only needed to be preserved but conserved. From that point to the "Year of the Coast", the problem in trying to preserve and conserve the beaches greatly accentuated.

After the Hughes case decision, the problems associated with the beaches were the result of three factors:

1. That land which Washington citizens had been using since before Washington became a state, suddenly became primarily private. The southern portion of approximately 53 miles was now 43 miles private and 10 miles public.

2. That the persons who successfully fought to have their land legally declared their own were, at least in part, not desirous of having it become public.

3. The Legislature had just passed the Seashore Conservation Act.

These problems and their solutions are best divided into five categories--maintenance, administration, ownership, machine movement, and resource protection.

MAINTENANCE

The first problem was an acceleration of one that had continued to exist--that of maintenance. It might seem unusual that maintenance would serve as a catalyst for all that followed, but such was the case.

When a new administration took over State Parks in late 1963, the local county officials made contact requesting that their purpetual pleas now be heard--pleas that the state would assume responsibility for the ocean beach maintenance. The need included such items as litter pickup, providing for and collection of waste disposal receptacles, meeting sanitation needs, and providing for beach access road grading and replacement.

These items were successfully addressed prior to and as a part of the administrative change and implemented thereafter. To bring about the administrative change, State Parks needed the local governmental support. Consequently, the locals could, for the first time, get seriously needed state funding when and if they gave that support.

The offering of that type of support is exemplified by the quote from the local two county health district, "The entire beach is one large park. It should be managed as such. One agency should be overall responsible for the management of the beach. The scope of the problem is beyond the local village or county."

The agreements, to later be modified over the years, were:

1. Parks would start a program of replacing pit toilets with flush comfort stations as monies became available. Further, on extremely heavy use days, parks would contract for additional portable sanitary facilities.

2. Litter--A similar agreement was executed to have litter picked up daily plus placing sufficient receptacles at all access roads and to contract the pickup with local firms as needed.

3. Beach Access Maintenance--The beach access roads are hard-packed sand that frequently need regrading and sometimes replacement. Rather than doing its own grading, the State agreed (and subsequently consummated) to contract with the county for their crews to do the grading thus providing the county with a broader financial base and also providing for the county to be answerable for the condition of the roads. It should be noted that prior to 1967 all beach access road maintenance was borne by the counties.

ADMINISTRATION

As previously noted, the Seashore Conservation proposal was drafted in 1966. This concept was deemed necessary to preserve and conserve the ocean beaches. It seemed necessary to have administration of the beaches transferred from an agency charged with generating income from its lands (Department of Natural Resources) to one charged with preservation and conservation (Washington State Parks and Recreation Commission).

It also seemed advisable at that time to address the beaches in the Indian Reservations as one reservation included considerable low bank beach area which was highly sought for use by non-Indians. It must be kept in mind that this was about the time of considerable court decisions concerning Indian rights. Therefore, from both a legal as well as a practical aspect, the Indian Reservations were deleted from the provisions of the proposed act.

In Washington State, the climate was right for a series of conservation bills. Therefore, it came as no surprise that the 1967 Executive Request Bills included one for the seashore. However, though the climate was right, the political party composition was not. The Legislature and the elected head of the DNR were of one party--not that of the Governor. Since DNR wanted to retain control of the beaches, it was easily understood that the focus of the legislative hearings would be on party issues and personalities. Therefore, it was important to have at least one sponsor of the bill be a beach legislator and one of the governing party. This was accomplished as a result of considerable concentrated effort.

After the bill was introduced, the hearings commenced and, as predicted, turned into a party/personality conflict. To alleviate the situation, if not totally eliminate it, the Parks Department was able to successfully have the legislature not address personalities/parties but, rather, express what they, the legislature, believed to be the prime purpose of the beaches and then assign the administration by virtue of that purpose as compared with the purposes of the respective agencies. The legislature did so and transferred the administration to the Parks and Recreation Commission--the state's preservation and conservation agency.

It was also necessary to address many other items in the legislation.

COOPERATION

Though the Parks Department had organic authority to cooperate, it seemed prudent to be area/activity specific in this instance. Therefore, not only was the general authority to cooperate conveyed to the Department but the specific reference also. These references included: federal agencies, as it would be needed with both the U.S. Fish and Wildlife Service and the National Park Service; the State Departments of Fisheries, Patrol, and Game; and local government.

The Act also provided for sand needed by cranberry growers, for vehicular control, and mining. One more vital aspect of administration was the basic authority of the Park Department to appoint and utilize the services of advisory committees.

OWNERSHIPS

Ownerships on the ocean beaches were public and private, with that division not only being latitudinal but also longitudinal. That is, the public and private are not only interspersed up and down the coastline, but also east and west between the ocean and the upland.

Within the marine area, are 38 cities and 15 of the state's 39 counties, each of which applies a variety of building codes, land-use regulations, and health codes. The Pacific Ocean beaches are in 4 counties with no ocean beach municipality having over 1,500 residents.

As previously noted, there also is a large federal ownership and influence on the ocean beach in terms of the National Park Service, the U.S. Fish and Wildlife Service, the Bureau of Indian Affairs, and to a lesser degree, other bureaus.

The State of Washington also plays a large governmental role, primarily through the Washington State Parks and Recreation Commission, both in operation and control as well as ownership. In addition, though they do not own land, the Department of Ecology plays an important role in the administration of the state's Shoreline Management Act.

As previously noted, there would need to be special arrangements made with two federal agencies. One was with the U.S. Fish and Wildlife Service for a landlocked tip of a peninsula which will be discussed under resource as well as another agreement with the National Park Service for administration of the state parks beach in front of Olympic National Park.

There are two special ownerships which need more detail as they relate to preserving the beach.

Quinault Indian Reservation

As previously noted, the two Indian Reservations were excluded from the provisions of the bill. For some time, the Quinaults had allowed non-Indians unlimited use of their beach. Then, in the 1960s, the Tribe closed the beach to non-Indians on the basis of too much litter and improper use. There were some individuals who thought this closure might also be part of a negotiation for other considerations sought by the Tribe. Regardless of the reasons, State Parks met with the Tribe in an attempt to have the beach re-opened. The proposal was made on the basis of the needs and desires of the Tribe. These included: (1) hiring Tribal children to clean the beach; (2) hiring Tribal refuse collectors to remove the debris; and (3) adopting Tribal beach rules for use and hours of use. State Parks concurred with all of these requests. However, for whatever reasons, the Tribe decided not to enter into such an agreement and their beaches remain closed to the non-Tribal persons. Though the resource is not able to be used by non-Indians, it is preserved and is an acceptable situation.

Accreted Lands

As previously noted, one of the most important far-reaching cases affecting the future of Washington's Pacific Ocean was the Hughes Case.

Basically, the court ruled that the upland owner also owned to ordinary high tide. Effectively, this meant that tremendous acreage of dunes and sand beach could now be built on, subject to local zoning and codes. It should also be noted that this ruling took place in December of 1967, three and one-half years prior to the state's Shoreline Management Act, and almost five years before the Coastal Zone Management Act. It should also be noted that the court ruling came only months after the administrative responsibilities were transferred to State Parks. In some cases the land shift was a mile in depth (EW). In anticipation of this possible court ruling, the gubernatorial request for the Seashore Conservation Act contained language providing for the establishment of a seashore conservation line with all lands west of that line to be governed under the Act. Though this Act was passed, it was modified to exclude private lands. Therefore, miles of beach were removed from the applicability of the Act just six months after its passage. To offset this, State Parks implemented three planned approaches--local zoning, local advisory committee, and land acquisition.

1. Local Committee--It was deemed advisable to establish a local committee from which to seek advice as well as to provide local support to State Parks. Further, it seemed advisable to have the committee to be composed of:

a. Official representatives of the local governmental units to be appointed by the governmental authority.

b. At-large local citizens to give a balance of concerns, interests, and acceptability. This committee to be selected by the Governor.

c. Representatives of those state agencies legislatively involved in the beach program.

This committee proved of some value although not as much as anticipated as the committee's recommendations were not acceptable to a vociferous minority who cast doubt on the committee's validity by noting the heavy state influence, i.e. state agencies and Governor selected representatives.

2. Local Zoning

It was crucial that State Parks work closely with local officials and local citizens so as to secure some sort of zoning which would help stave off any possible building on the beach lands.

The two counties at the southern part of the ocean, as well as the largest town, responded admirably and placed various distant and height regulations into effect so that only a few structures were built.

3. Land Acquisition

There were two major aspects to the acquisition program--purchase and donation.

a. Purchase

The state of Washington operates on a biennial basis. Therefore, the first budget request time period after the Seashore Organic Act was for the 1969-71 biennium. From that biennium forward, State Parks requested and received considerable funds for acquisition and development with approximately 635 acres being purchased in 37 separate transactions to date for $3,500,000. These acquisitions ranged from .1 acre to 205 acres. Developments necessary to handle the millions of visitor days were also forthcoming in the amount of $4,388,000.

b. Donations

The cooperation between the citizens and State Parks was most gratifying and also most financially rewarding. Many of the citizens who owned the beaches did not want any more building and were anxious to take any steps to help prevent such building. Others had hopes of building on the dunes but did not want to pay taxes on the land in front of such building. Still others were concerned about maintaining such land, especially if a whale were to be beached or some other such calamity were to occur.

For any donation program to be effective, it was important that the potential donors be assured that all were being treated equally. Therefore, a master deed was prepared in cooperation with some of the owners. The basic thrust of the master deed concerned: (1) Setback distance; (2) no construction allowed; and (3) maintenance.

With this master deed in hand, State Parks went about securing such donations. From June 3, 1968, (the first such deed) until 1980, 127 deeds were tendered for over 25,000 front feet of ocean water front ranging from 20 to 1,800 front feet. Further, 291 more acres were acquired through special cooperative agreements with other agencies. There were another 14 deeds for over 3 acres of beach east of the seashore conservation line.

MACHINES

Once the process for preservation was set into motion, it became important to accentuate the implementation of sound conservation practices. One of the major aspects to that thrust was the proper regulation of vehicles, i.e. automobile, dune buggies, ORV, and airplanes.

Airplanes were the least controversial, although more complex than one might assume. For years some pilots of small planes would land on the beach, dig an ample supply of razor clams, and fly out again. There were rumors that this method provided an almost "catch proof" method of avoiding being caught with several limits. Nevertheless, with cooperation by State Aeronautics and the existence of two nearby inland small airstrips, the beaches were placed off-limits to airplanes except for emergency landing.

Of much greater concern and more difficult to regulate were the various off-road vehicles. All vehicles on the beaches were of concern but the most resource damaging were the dune buggies. These wide, soft-tire travelers could and usually did leave the hard-packed sand and went <u>over the dunes</u>. This latter action resulted in: (1) breakage of the dune thus opening the way for serious water errosion; (2) damage and even destruction of the fragile dune vegetation; and (3) causing injury to sunbathers lying behind the dunes. After a series of work sessions and public hearings, all non-hard-packed sand driving was declared illegal.

Left to be resolved was driving on the beach per se. It will be recalled that the legislation provided for regulating driving on the beach but precluded removal of vehicles. There were those who believed that, in the interest of safety, all vehicles should be banned from the beach. Others expressed that cars have always used the beach and that, in fact, the area had been legislatively established as a public highway.

Under the administrative code, various approaches were taken to address the situation. First though, why the controls?

--Saftey. People walking the beach are often unable to hear an approaching vehicle due to the pounding of the surf, and are also not able to see the vehicle due to their viewing the ocean and the horizon.

--Parking. The beach was also used for parking, i.e. those same people who walked the beach for pleasure, collected driftwood, or dug clams, also drove to the beach, with a majority parking on the beach. To provide "off-beach" parking for all of them would literally have required building thousands of parking stalls on the precious upland.

Therefore, to address the need for safe areas in which to walk while yet avoiding the need to build the off-beach parking, the first legal approach tried was to allow and limit driving to one-fourth mile north and south of each access road. By so doing, there was provided a non-driving zone on all beaches where adjacent access roads were more than one-half mile apart. This and all subsequent approaches were challenged and still have not been resolved in the courts.

However, though the basic aspect of where driving is or is not legal has not yet been settled, all other aspects of beach rules have not only been adopted, but also are accepted by the driving public, i.e. speed limits, equipment, and safety rules such as no "squirreling or "cutting donuts."

It should be noted that horses were included in the rules governing vehicles, since the agency did not want horses on the dunes nor on the areas where the people were walking. This was changed in 1979 by placing horses in a category somewhat between humans and automobiles.

RESOURCES

Lastly, the Pacific Ocean Beach is first and foremost a resource, a resource needing to be addressed.

Sand--As noted under Administration, sand removed for cranberry growing was to be allowed if its removal did not adversely affect the beach. This removal proved to be no problem and continues at a low level.

Of much greater concern and more difficult to administer, was the removal of sand for construction purposes. To ascertain whether such sand removal would cause a problem, a test plot was done which showed that removal of sand in large quantities from a small area would be a problem as the sand did not replenish itself. Therefore, rules were initiated to preclude the taking of sand from state-administered land. These proved to be fairly effective.

Logs--The taking and/or cutting of logs was far more difficult to address. For years the local residents considered the logs their own both for personal home heating as well as for commercial sale. What further complicated the situation was the lack of scientific basis for any claims by the state as to whether or not the removal of logs would have an adverse effect on beach stability. Over several controversial years, rules were adopted which basically gave the right for people to cut wood in designated woodpiles, during specified days, and during specified times. These latter factors were to control the audio disturbance to visitors.

Clams--Since the State Department of Fisheries controlled clams elsewhere in the state of Washington, State Parks agreed to follow their rules and regulations and to aid in enforcing those rules. This has worked very well.

Minerals--Mining was a permissible activity if the mining did not interfere with the recreational aspects or beach preservation. An application for mining was filed with the State, who subsequently granted a temporary use permit. This was for the mining of black sands, subject to certain provisions calling for measurements. After a sufficient period of time, State Parks concluded that mining would adversely affect the beach, and, therefore, the operation was closed and all equipment was removed.

Animals--As noted under Administration, hunting was allowed in the Seashore Conservation Act, but not in a state park, since hunting of any type is illegal in any state park. This was one reason why the Act was titled The Seashore Conservation Act, only the second such area so classified in the state. Within the Seashore Conservation Area, are 6 state parks and 16 ocean beach access areas, in none of which is hunting allowed. Therefore, the only area of any concern is that of the landlocked area administered by the Fish and Wildlife Service.

At one time this area was to be sold to State Parks by the Bureau of Land Management. For reasons not germane to this paper, BLM withdrew that sale and transferred the property to the Fish and Wildlife Service. The ensuing problem was that the Fish and Wildlife Service has no resident manager located at the area with the closest manpower being many, many, miles distant. Therefore, an unofficial agreement was developed whereby State Parks assisted Fish and Wildlife Service in preventing vehicles from driving on the fragile land. However, hunting is allowed on the Fish and Wildlife Service land.

Patrolling the Resource

Though all of the previously noted statutes and rules are of value, their effectiveness would leave much to be desired if they were not able to be enforced. This enforcement has continued to be a workable example of excellent cooperation.

In addition to the already mentioned cooperation with the State Department of Fisheries, the following four patrolling aspects should be given special note:

1. The Washington State Patrol, an agency with a very fine image, frequently has their patrol cruise the beach which not only has a direct immediate effect but also a lasting one for those visitors who see the specially-marked cars.

2. National Park Service--In front of the large Olympic National Park is the state-administered beach. For efficiency and economy in government, State Parks and National Park Service have entered into a legal agreement providing for the National Park Service to police that area.

3. County Sheriffs--State Parks provided vehicles to the sheriffs of each of the two southern counties and subsequently entered into contracts with them. These contracts provided for yearly state financing of patrols by county sheriffs on all the beaches thus aiding the financial base of the counties and enhancing county/state relationships.

4. State Parks personnel have police powers of the state, and, as such, the authority to cite violators. These personnel are daily on

the beaches and are most effective with their administration of the statutes and rules.

CONCLUSION

The preceding is a factual representation of one state's efforts to preserve and conserve the Pacific Ocean portion of its coast. Additional efforts have been made and are continuing in a desire to preserve and conserve other portions of Washington's marine land. One of these exciting efforts is that of Padilla Bay where the Straits and Puget Sound merge.

The two decade story has no ending. The accomplishments brought about to date could easily be wiped out unless all of us continue our vigilance.

Whatever the successes, they are due to a willingness of dedicated people to cooperate, to reason, and to act for the common good. The accomplishments are a living credit to all three branches at all levels of government, to a citizen involvement and dedication to be sought by all, to an understanding and supportive media, and to a plan -- a plan to preserve our priceless heritage -- our coast.

ACKNOWLEDGMENTS

The National Park Service was pleased to contribute to Coastal Zone '78, held in San Francisco, and has encouraged its personnel to participate in CZ '80.

I would certainly be remiss if I did not publicly express the gratitude I feel for the support given to me by National Park Service Director Dickenson and the staff and Regional Director of the Pacific Northwest Region, National Park Service.

Appreciation is also expressed to the staff of the Washington State Parks and Recreation Commission, who so graciously gave of their time and effort to refresh my memory of the events that made up my nearly 17 years as Director of that fine agency.

To the Chairman and the staff of Coastal Zone '80, I extend my warmest expression of thanks for presenting me the opportunity to offer this paper in hopes that it might in some small way be of benefit to others.

LEGAL ASPECTS OF COASTAL NATURAL HAZARDS MANAGEMENT

by Richard G. Hildreth*

INTRODUCTION

Development in the coastal zone is exposed to a variety of natural hazards. Development activities increase the risks of personal injury and property damage in several ways. This article explores the legal aspects of public management of privately-owned coastal land threatened with damage from geologic hazards such as earthquakes, erosion, landslides, and land subsidence, with emphasis on erosion management.

The most common natural hazard in the coastal zone is erosion, the removal and transport of rock or soil by moving water. Erosion is a vital process in the coastal zone because it produces a source of sand for the beach. A beach acts as a shock absorber; it provides a buffer zone which dissipates wave energy before it reaches the shoreline.[1] When the sand supply is reduced, both the size of the beach and its buffering abilities are diminished. More wave energy reaches the shore, thereby increasing seacliff erosion and the risk of floods or landslides.

In areas where development is located close to the beach, this presents a major problem. Where coastal properties are threatened by an encroaching shoreline or cliff erosion, owners of improvements may want to construct protective devices such as groins, rip rap, seawalls, and jetties. This often accelerates the problems by further reducing the supply of beach sand. Many coastal erosion problems are the direct result of unwise development in the past. Erosion damage is likely to increase in the future.

Two basic situations confront coastal land use managers: 1) land on which existing residential or commercial development is threatened with hazard damage; and 2) land in a hazardous area for which development permission is sought. Examples of both situations can be drawn from recent experience with eroding bluffs on the Pacific Coast. Frequently the blufftops are desired locations for residential development, and development has in fact been permitted along the bluff edge in various locations. Occupied structures, some of them less than five years old, are now being undermined by erosion in several locations. The landowners are responding to the problem by requesting permission from regulatory agencies to construct protective works on the bluffs and on the beaches below.

The following common law materials further introduce coastal natural hazards management. Most of the cases involve developed rather than undeveloped property. Several of the cases illustrate the familiar sequence of haphazard development leading to construction of shoreline protective works which cause damage to third parties.

---

*Professor & Co-Director, Ocean and Coastal Law Center, University of Oregon School of Law.

A clear understanding of common law principles promotes better natural hazards management. A coastal manager should know the legal implications of decisions involving permits for construction in the coastal zone. Most importantly, the common law framework is the starting point for legislative alteration of the system. The strengths of the common law can be built upon while its weaknesses are corrected.

COMMON LAW PRINCIPLES

Coastal Property Boundaries

Each coastal state has a special interest in determining coastal property boundaries due to the public trust doctrine. Under the doctrine, although the state owns the submerged soil and the foreshore of all navigable bodies of water, the state's title is a "trust devolving upon the state for the public and which can only be discharged by the management and control of property in which the public has an interest [and] cannot be relinquished by a transfer of property."[2] The majority of coastal states use the mean high tide line as the dividing line between public and private ownership.[3] This is defined as the "line of high water as determined by the course of the tides," meaning the intersection of the tidal plane with the shore.[4] Since the shoreline is constantly changing, an ambulatory boundary results.

The general rule is that where the shoreline is gradually and imperceptibly changed by natural accretion or erosion the boundary line is extended or restricted in the same manner. The owner of the upland property acquires title to additions to his land from accretion and loses title to lands lost through erosion.[5] With respect to accretion caused by the upland owner, for example by dredging and filling, or construction of groins, rip rap, seawalls, or other protective devices, the general rule appears to be that the upland owner does not acquire ownership of the accreted soil.[6] Awarding ownership would allow the upland owner to encroach upon the state's public trust lands at will. However, if construction activities of third parties cause accretion, the accreted soil belongs to the upland owner.[7]

In some coastal boundary disputes a boundary line which is to remain fixed despite future changes in the shoreline has been established by judicial decree, administrative decision, or special legislation. This solution is especially attractive to developers because of the relative certainty it provides in comparison to the traditional ambulatory boundary approach. With a fixed ownership boundary established, the next development step is to try to stabilize the shoreline with such protective devices as groins, rip rap, and seawalls.

To the extent that accretion and erosion rules and undertainty about them are hindering effective coastal natural hazards management, the recent decision of the United States Supreme Court in Oregon v. Corvallis Sand and Gravel[8] appears to encourage state legislatures and courts to clarify them. The Court concluded that once the equal footing doctrine vested title to a riverbed in the state, state law thereafter applies to ambulatory boundary and ownership questions. However, some uncertainty remains as to whether Corvallis applies to disputes involving coastal tidelands as well as inland riverbeds.[9] But two principal legal constraints on state modification of the ambulatory boundary rules remain: 1) constitutional protection of private property from governmental takings without just compensation; and 2) public trust doctrine limitations on transferring public tidelands into private ownership.

The Common Enemy Doctrine

Coastal property owners quite naturally seek to protect their land and improvements from damage by constructing various shoreline protective devices. The crucial issue is the extent to which a landowner may take protective measures without incurring liability for injury to his neighbor's land. Three theories have developed; although most of the litigation has involved inland rivers rather than ocean coasts, the doctrines are said to be applicable to both areas.[10]

The common enemy doctrine states that a landowner may do anything to repulse surface floodwaters from his land without liability for damage to others caused by his actions.[11] Neighboring landowners have a corresponding right to protect their own property. Any injury to neighboring land is said to be *damnum absque injuria*.[12] The doctrine applies only in the face of imminent harm. Many courts have modified this doctrine by adding requirements of reasonableness to prevent negligence causing unnecessary injury to neighboring land.

The common enemy doctrine obviously operates in favor of those able and willing to improve their property. Unfortunately, coastal property owners who construct seawalls, rip rap, and groins to protect valuable improvements may increase the damage suffered by unprotected neighbors and interfere with public access to and use of the shoreline.

Under the civil law rule a servitude is imposed on downstream property requiring it to receive all surface water flowing through its natural course. But the upper owner may not do anything to increase this burden. The justification for the rule is that those acquiring land should accept it subject to the burden of natural drainage. The advantage of the civil law rule is that rights are predictable. However, it tends to inhibit the development and improvement of land. Because of this tendency, some courts have modified the civil law rule to permit a reasonable use of land.

Under the reasonable use rule, the rights of the parties are determined by assessing all of the relevant factors. These include: 1) surrounding circumstances of the particular case; 2) nature of the improvements to be made; 3) extent of interference with surface water; 4) value of the improvements compared with the injury to neighbors; and 5) foreseeability of the harm. Thus, the three theories converge on the concept of reasonable use, thereby paralleling the law of nuisance.

The coastal erosion cases indicate that a littoral owner has more rights against another private party than against the state. A beachfront property owner may collect damages and enjoin unreasonable conduct by his neighbor which causes erosion. However, he has no cause of action where the erosion is caused by public navigation improvements.

Caveat Emptor

When coastal property erodes or suffers other natural hazards damage, the present owner may attempt to hold the person who sold him the property responsible. At common law the owner had no recourse; the doctrine of caveat emptor prevailed. The buyer had the responsibility of inspecting land and improvements prior to purchase. Absent fraud, courts were reluctant to impose liability on sellers after they had parted with ownership and control of the property. Acceptance of a deed by the buyer ended the seller's liability except as to any express warranties contained in the deed.

However, the law in this area is changing. Courts first found an implied warranty that houses under construction be completed in a workmanlike manner.[13] Courts have held that an implied warranty exists on the sale of a new home by a builder regardless of the stage of construction; only used homes have escaped the implied warranty. Liability includes both personal injury and property damage.

A recent Oregon case, Beri v. Salishan Properties, Inc.,[14] gives the buyer added protection. The court imposed a duty on the land developer-seller to exercise reasonable care to determine whether homesites offered for long-term lease or sale are fit for residential use. In Beri the plaintiffs leased prestigious beachfront lots with condominiums which were being destroyed by erosion. The court based liability not on implied warranty but on a theory of negligence. By holding themselves out as highly skilled land developers, the defendants were liable for failure to exercise reasonable care in the project.

Liability has also been extended to a commercial construction loan lender, again based on negligence.[15] In this California decision the court cautioned that liability would be imposed only where there is substantial involvement by the lender in the development and sales of the homes above and beyond customary loan procedures.

With increasing pressures for development in hazardous coastal areas will come more litigation over liability for damages suffered, and further evolution of the liability rules. However, a legal response which merely allocates responsibility for damages after loss is suffered may not be satisfactory here. Preventing losses from occurring is the ultimate goal. If the common law liability rules do not operate as a sufficient disincentive against development in hazardous coastal areas, then stronger regulatory measures become necessary.

REGULATORY APPROACHES

Programs regulating coastal natural hazards have been implemented at the federal, state, and local levels with varying degrees of success. Most of these programs are directed at undeveloped land. Development is either prohibited or else required to be sited and designed to withstand the hazard. Relatively little attention has been paid to the problems of existing development threatened with hazard damage, to disclosure of hazard information to prospective purchasers, and to treating affected landowners as fairly as possible. The main impetus for identifying and mapping coastal natural hazards has come from the federal flood insurance and coastal zone management programs.

Federal Flood Insurance

The 1973 Flood Disaster Protection Act amended the National Flood Insurance Act to cover erosion and the undermining of shorelines by waves or current.[16] To be covered, the erosion must be both unexpected and a result of flooding.

The two primary goals of the National Flood Insurance Program are: 1) to provide flood insurance based on the actual flood hazard at a particular location; and 2) to encourage wise management of flood-prone land to minimize flood-related damage. Federally subsidized flood insurance is available to property owners in participating communities. To qualify for participation in the program, communities in flood-related erosion-prone areas must, in part: 1) require the issuance of a permit for all proposed development in the area of flood-related erosion hazards; 2) review each permit application to determine whether the proposed development will be reasonably safe from such hazard; and 3) if the proposed development is found to be in the path of flood-related erosion or to increase the erosion hazard, require the improvement to be relocated or adequate additional protective measures to be taken.[17] Within six months after special flood-related erosion hazard areas are delineated, the community must require a setback for all new development to create a safety buffer of natural vegetation. Additionally, the Act encourages adoption of an overall comprehensive management plan. A Flood Insurance Rate Map is published outlining the community's flood hazard areas, with appropriate insurance rates for each area.

The benefits to communities and their citizens are so great that most coastal communities do participate in the program. However the program has not been successful in discouraging development in hazardous coastal areas.[18] The fear that restricting coastal development would induce many communities to drop out of the program prompted the Federal Insurance Administrator to reject a ban on development in the "velocity zone," the highest hazard area.

Instead of deterring development, the program has had the effect of promoting it. The availability of federally subsidized insurance has eliminated much of the economic risk involved in coastal development.

Coastal Zone Management and Coastal Erosion

The 1976 amendments to the federal Coastal Zone Management Act (CZMA) signalled increasing federal interest in coastal erosion problems.[19] After October 1, 1978, federally funded state coastal zone management programs must contain a planning process for shoreline erosion problems. This process must include:

> (1) A method for assessing the effects of shoreline erosion;
> (2) Articulation of State policies pertaining to erosion, including policies regarding preferences for non-structural, structural and/or no controls;
> (3) A method for designating areas for erosion

> control, mitigation and/or restoration as areas of particular concern or areas for preservation and restoration, if appropriate;
>
> (4) Procedures for managing the effects of erosion, including non-structural procedures; and
>
> (5) An identification of legal authorities, funding programs and other techniques that can be used to meet management needs.[20]

States also should consider erosion losses along shorelines and in estuarine areas from both man-made and natural causes, and the effects of erosion, mitigation, and restoration on natural shore processes as well as adjacent land and water uses. The critical areas concept is one approach the states could use to meet the CZMA's erosion planning requirements.

State Critical Areas Management

Greater attention to development impacts on sensitive environments such as wetlands has led to the critical areas concept in land use control.[21] To control development in critical areas state government assumes a greater role vis-a-vis local government. The notion is that local government's perspective on development may be too narrow given the important environmental resources at stake.

Several coastal states included hazardous areas within a state-level critical areas process, stimulated in part by the federal Coastal Zone Management Act, which requires federally-funded state coastal management programs to include "an inventory and designation of areas of particular concern within the coastal zone."[22] With respect to natural hazards, federal regulatons then required states to consider as areas of particular concern:

> Areas where, if development were permitted, it might be subject to significant hazards, due to storms, slides, floods, erosion, settlement, and salt-water intrusion;
>
> Areas needed to protect, maintain or replenish coastal lands or resources, including coastal flood plains, aquifers and their recharge areas, estuaries, sand dunes, coral and other reefs, beaches, offshore sand deposits, and mangrove stands.[23]

For land designated as an area of particular concern, the state was required to establish priorities for use of the area.

North Carolina has made extensive use of the critical areas concept as a natural hazards management tool. Under the North Carolina Coastal Area Management Act, the Coastal Resources Commission may designate "areas of critical environmental concern"; legislative approval of critical areas designations is not required.[24] Along the North Carolina coast groups of critical areas with erosion problems have been identified and, in accordance with the federal regulations described above, use priorities have been established.

Major developments within critical areas must be approved by the Coastal Resources Commission; other developments approved by local governments may be appealed to the commission. In designated critical natural hazard areas development permits must be denied where development would unreasonably endanger life or property. In all erosion-prone areas concern is expressed for avoiding construction of commercial and public facilities that would encourage further growth and development. Research on specific regulatory and management techniques for implementing these use priorities has been carried out. The next step in using the critical areas process as a coastal natural hazard management tool in North Carolina is the selection and application of specific development controls in the designated critical areas.

A full evaluation of the critical areas process as a natural hazards management tool at this time is premature. It is a state level approach to undeveloped hazardous lands which does not attempt to deal with the problems of existing development in hazardous areas. And from the legal perspective, difficulties remain to be overcome. For example, delegation to state commissions of the power to designate critical areas without subsequent legislative approval has been successfully attacked as unconstitutional in Florida. The Florida Supreme Court found the Florida critical area process defective because the legislature had not specified priorities for 1) state designations of critical areas, and 2) state determinations of local government inadequacy justifying such state intervention.[25] Elsewhere the courts have been more sympathetic to legislation delegating significant environmental and land use decisionmaking power to state agencies and commissions.

While coastal state use of a critical areas process is encouraged by the federal Coastal Zone Management Act's "areas of particular concern" language, it is not clear whether coastal state critical areas designations must be followed by the imposition of state development controls in the designated areas. If specific development controls are not imposed, then the critical areas process becomes mostly meaningless, because one of its primary purposes is to focus the application of more stringent development controls on areas of critical need. The more stringent development controls applied in critical areas inevitably will be challenged as governmental takings of private property for a public use without just compensation. Whether such challenges succeed will depend largely on the scientific accuracy of the information relied upon by the state in designating the area as critical. Where the information on potential developments hazards in a designated critical area is inconclusive, the regulatory techniques providing full or partial compensation to those regulated described later in this article may be utilized. Even where development hazards are relatively well established, compensatory regulatory techniques may be used for reasons of fairness and to reduce the potential for litigation.

Other State and Local Approaches

During the 1970's, coastal states were becoming more concerned about shoreline erosion and other coastal geologic hazards. For example, Oregon included specific rules with respect to shoreline erosion management in its federally-approved Coastal Management Program: generally non-structural solutions are preferred to structural responses; permits for beach front protective structures may be issued only for development existing on January 1, 1977; building is prohibited on active foredunes, conditionally stable foredunes that are subject to ocean undercutting or wave overtopping, and on deflation plains subject to ocean flooding; and foredunes may be breached only to replenish sand supply in interdune areas or temporarily for emergencies.[26]

Oregon local governments have adopted various types of planning guidelines to meet the Oregon Land Conservation and Development Commission (LCDC) statewide land use goals and guidelines, several of which relate to coastal natural hazards management. Local plans to implement these goals include site specific studies in areas of potential hazards, zoning ordinances pertaining to uses on active dunes, on-site inspection prior to issuing development permits, designation of natural resource zones to protect those resources, and preservation of scenic areas.

In California the 1976 California Coastal Act requires local governments to prepare local coastal programs to implement the act, including restrictions on bluff-top development and shore protective works. A variety of techniques are being proposed and utilized by coastal local governments to control development in hazardous areas. For example, bluff-top development has been approached as part of an open-space hazard district, and through coastal overlay zoning. However, these approaches do not differ greatly from the techniques used by local governments prior to the advent of statewide coastal zone legislation in California.

Perhaps local governments are not being innovative in responding to coastal natural hazards because they lack both data on the hazards involved and information on innovative techniques in land use control. Perhaps they also fear being held liable for having "taken" property if they impose new and severe restrictions on the use of land in hazardous areas. But they may also be held liable if they do approve development in hazardous areas and injuries to persons and private and public property result.[27] In attempting to resolve local government's dilemma, it is useful to consider first the full range of land use control techniques available.

Setbacks and Related Techniques

Setbacks have been the classic regulatory response to shoreline erosion problems. A fixed setback requires all structures to be set back a designated distance from a specified point, often the mean high water line or the bluff edge. Fixed setbacks are probably more useful for bluff tops than ocean shorelines where beach and dune configurations are likely to shift drastically. In some cases fixed dune setbacks have led to the elimination of protective dunes. Even setbacks based on adequate scientific data can be rendered ineffective if variances from the setback requirement are liberally granted by administrative agencies.

While courts have tended to accept building setbacks imposed as part of urban zoning laws, the judicial response to fixed shoreline setacks has been mixed. A successful setback ordinance should be based on scientific evidence of its necessity and should be imposed only after providing notice and an opportunity for all affected property owners to be heard. A provision for periodic review should be included since the conditions of the beach or bluff may change, warranting a change in the setback requirements. The stronger the evidence supporting a setback, the more likely it is to be upheld in the courts.

Dedication of lateral public access along the shoreline as a condition of development approval is becoming an accepted coastal land use control technique. Additional width for structural safety can then be added as a hazard setback. Imposition of a public access condition is not appropriate if it will unreasonably expose the public to injury or significantly increase the natural hazard. For example, public access along the faces and edges of severely eroding coastal bluffs must be carefully controlled or disallowed altogether.

Another alternative to fixed building setbacks is the planned unit development (PUD), which can be used to cluster development on part of a parcel, with the remainder devoted to open-space uses. PUD's can be used for development along hazardous shorelines and bluffs only if the parcel is large and extends far enough inland to permit clustering of structures on the safe inland portions, with the undeveloped seaward portion serving as a buffer or safety zone which reduces the risk of a hazard loss. Administration of a PUD ordinance does require adequately trained staff to review and improve PUD development proposals.

Density credits can be used in combination with a fixed setback line. Permission for higher than normal development density can be awarded to the portion of the parcel behind the setback line, based on the development disallowed in the setback area.

Related to the density credit idea are compensable setbacks, which are especially useful when the imposition of fixed setbacks or access dedication and clustering requirements could render a narrow shorefront parcel undevelopable. The more generous schemes compensate landowners for market value reductions caused by imposing setback requirements; others only compensate for interference with ongoing uses. It should be remembered that where adequate data indicates a hazard to life or property, reasonable setbacks may be imposed constitutionally without compensation; however, for reasons of fairness to property owners affected by the availability of better hazard information and greater public willingness to utilize it, authorities should consider compensable setbacks.

There are additional land use control techniques not heavily emphasized in past efforts at natural hazards management which could be useful in future efforts. These techniques include the traditional planning, zoning, and subdivision review processes; various types of beach, dune, sand, and shore protection laws; erosion and sediment control laws; grading and slope ordinances; special building codes for hazardous areas; substandard and unsafe building abatement; statewide and regional land use controls based on soils capability such as the Vermont Land Use and Development Law and the Tahoe Regional Planning Agency; and development performance standards based on sophisticated scientific investigation of local natural systems.

A more market-oriented approach would be to impose special fees or assessments on development in hazardous areas. The fees collected could be used for community hazard warning and protection activities. Special assessment mechanisms to finance and construct shoreline protective devices necessitated by bluff-top development could be required as a condition of development approval in order to internalize some of the development's costs, which might otherwise fall on the public. But more generally, this type of regulation seems ill-suited to natural hazards involving danger to life or property such as bluff erosion because development is merely made more expensive rather than prohibited or strictly regulated. If the true costs to society of development in hazardous areas could be ascertained and quantified, which is doubtful, then a development fee system based on those calculations used in conjunction with other management techniques would have more merit.

Hazards Disclosure

Requirements such as setbacks are attractive to regulatory agencies because the concepts are simple and the publicly-funded research required is not extensive. But, however combined, setbacks and related techniques to not constitute a comprehensive shoreline natural hazards management program. At a minimum, they should be supplemented with effective hazards disclosure so that potential investors in hazardous lands can react accordingly. Hazards disclosure can occur through environmental impact reports, hazard warnings in deeds, mortgages, subdivision plats and covenants and other recorded documents affecting land titles, laws requiring hazards disclosure in real estate sales, and general public awareness efforts ranging from governmental-published handbooks for coastal developers and property owners to hazard information enclosed with utility bills.

There are two components to hazards disclosure: disclosure of hazard information and disclosure of development restrictions, if any, imposed because of the hazard. Site-specific geologic reports and environmental impact reports can be used to disclosure hazard information. The land title recording system should be used to disclose development restrictions.

Site-Specific Geologic Reports

Setbacks and the alternatives just discussed ordinarily to not operate as absolute development prohibitions; instead, they serve to place structures on the safer portions of parcels located in hazardous areas. However, in some hazardous areas, site-specific investigation may be required in order to determine whether development anywhere on a given parcel, particularly small frontage lots, would be reasonably safe. In such situations, the developer may be required to provide the regulatory agency with a professionally prepared site-specific geologic report which indicates that the proposed development would be safe. Otherwise, the agency may reject the development proposal. A federal district court so ruled in Kopetzke v. County of San Mateo.[28]

In the Kopetzke case, the United States Geological Survey had discovered significant soil instability in the Moss Beach-Seal Cove area of San Mateo County, California. The county commissioned further geologic investigation by an engineering consultant. Based on the consultant's report, the county required applicants for building permits in the area to provide a professional geologic soils report showing that the specific building site was safe or could be made safe.

The plaintiff landowners presented considerable evidence that the effect of the county's action was to render their properties unmarketable in a practical sense. The costs of obtaining the geologic report were estimated to equal the value of the land; moreover, investing in the required soils report did not guarantee issuance of a building permit. Therefore, the plaintiffs claimed that the county in effect had taken their property without just compensation.

The trial court upheld the soils report requirement as a valid exercise of the state's police power. Based upon information that the soil instability in the area presented a danger to both life and property, the court ruled that the county's actions were reasonably necessary and not unduly oppressive. The opinion is probably the strongest judicial statement in support of geologic hazards regulation to date. Regulatory actions based on credible scientific evidence of a hazard to life or property will be upheld without compensation to affected property owners. The court rejected any notion that it is unfair to saddle individual landowners with the costs of determining the geologic stability of their property.

Environmental Impact Reports

State environmental impact assessment laws can aid coastal natural hazards management efforts significantly. The California Environmental Quality Act (CEQA), for example, requires state and local agencies to consider the environmental impacts of proposed developments prior to project approval.[29] If a proposed project may have a significant adverse environmental effect, and Environmental Impact Report (EIR) must be prepared. CEQA does not prohibit development in geologically hazardous areas. However, a well-prepared EIR disclosing natural hazards may motivate a public agency to deny approval. If denial of approval is based on such an EIR, the courts are not likely to overturn the agency's decision.[30]

However, local governments should not use the EIR process as their primary coastal natural hazards management tool because it is not a comprehensive hazards management scheme. Instead the EIR process uses relatively expensive case-by-case scientific investigation and data assembly to produce general assessments of a particular proposed development's environmental impacts, with no special emphasis given to hazard problems.

Recorded Hazards Disclosure

Some may question the effectiveness of using the land title recording system for hazards disclosure. But one of the merits of recorded hazards disclosure is that it does not rely solely on better informed investment decisions by developers and lot purchasers, but also involves real estate lenders and title companies in natural hazards management. To minimize conflict with recording system practices, hazard information should be included in traditionally recorded documents rather than in separate hazard instruments. For subdivisions in hazardous areas, the ideal document for hazard information is the recorded subdivision map which is in the chain of title to all deeds to individual lots or condominiums within the subdivision.

COMPENSATION TO AFFECTED LANDOWNERS

When private property is taken for public use under the power of eminent domain, the owner is entitled to receive just compensation from the government taker.[31] But if a governmental regulation restricting use of the property is held to be a valid exercise of the police power, no compensation is required. Although the demarcation line between valid exercises of the police power and regulation constituting a taking remains unclear, coastal resource managers have been accorded a fair degree of latitude by the courts in many states. However, fairness to property owners adversely affected by new and better hazard information and greater public willingness to use it suggests awarding compensation even where it is not constitutionally required.

Several land use control techniques which could be utilized in coastal natural hazards management provide full or partial compensation to affected landowners. They include acquisition voluntarily and by eminent domain of complete ownership, open-space easements, development rights, or similar controlling interests in land located in hazardous areas; preferential tax assessment to reduce development incentives in hazardous areas; and the awarding of transferable development rights to landowners in hazardous areas where development is prohibited. Public acquisition of significant interests in land does reduce the local government tax base; in addition, government is exposed to possible liability for injuries upon or caused by publicly-owned hazard lands. These factors may lead local governments to prefer techniques such as transferable development rights in which hazard lands remain in private ownership.

Transferable Development Rights

Transferable development rights should be considered as a potentially fair, effective approach to several coastal natural hazard problems. For example, where there is pressure to develop eroding coastal bluffs for residential purposes, a coastal building height limit can be used to create a market for transferable development rights awarded to landowners prevented from developing because of severe erosion problems discovered upon detailed, site-specific investigation of their property. If the building height limit is two stories, a very simple scheme would award landowners prevented from developing the right to sell two stories of additional height to landowners whose tracts are found upon site-specific investigation to be safe for the structure's projected useful life. The amounts paid by landowners with safe tracts for the right to build above the general height limit would partially compensate those with hazardous tracts for the limitations on or additional expenses of development. In Oregon and California the general coastal height limit and a transferable development rights ordinance could be adopted as part of the local government's coastal plan required by state legislation.

If all or most of the undeveloped tracts subject to the coastal height limit are too hazardous to develop, this simple development rights scheme would fail and alternative hazards management approaches would be necessary. The above example assumes sufficient residential development pressure to create a market for the right to build above the general coastal height limit. It also assumes that such a height limit imposed by a state or local government to promote public safety, coastal aesthetics, and related coastal and general welfare policies is a valid exercise of the regulatory police power not requiring compensation to affected landowners. Older zoning cases have upheld building height limitations imposed for reasons of health, safety, and aesthetics; the factual and policy bases for imposing coastal building height limitations seem analogous.

Transferable development rights are basically an untested concept. In New York the judicial response to such schemes has been mixed. However, these were cases concerning open-space and historic building preservation, where the landowners' claims for compensation for development restrictions are strong.[32] When credible scientific evidence indicates that development would be hazardous to life or property, the claim to compensation for development restrictions is not nearly so strong, as discussed previously in connection with site-specific geologic reports. Thus, in hazardous situations the awarding of transferable development rights to landowners prevented from developing is more a matter of fairness than of constitutional necessity.

California Coastal Conservancy

Application of several compensatory land use control techniques to California coastal management problems is the primary responsibility of the California Coastal Conservancy.[33] The Conservancy was created by the legislature in 1976 to supervise and fund state and local efforts to preserve coastal agricultural land, to preserve and increase public access to the coast, and to restore, enhance, and protect significant coastal resources. It is authorized to acquire voluntarily and through eminent domain fee simple ownership, development rights, easements, and other interests in land; to sell, lease, and exchange such interests; and to award grants and interest-free loans to state and local agencies for coastal access, restoration, enhancement, and protection projects.

Until the Conservancy establishes its priorities for coastal restoration and protection, the potential impact of its programs on California coastal natural hazards management is difficult to gage. While the Conservancy's programs are not expressly linked to hazards management, they are closely tied to the local coastal plans that are the primary vehicle for coastal natural hazards management in California. Several of the programs could be applied to natural hazards problems, particularly the provisions for restoration, enhancement, and resource protection.

CONCLUSION

All too frequently, responses to natural hazards begin after a disaster. Damaged parties sometimes turn to the courts. But historically the courts were hesitant to impose liability for coastal hazard damage.

Homebuyers suffering damage due to coastal erosion had no recourse against the seller. Neighbors suffering damage from protective actions taken by other property owners often could not recover. In recent years, courts have taken a more active role in determining the legal responsibilities of parties in suits resulting from coastal disasters. Property owners now have more protection than in the past. Although a clear understanding of these common law developments promotes better coastal hazards management, regulatory schemes must be developed further to eliminate or reduce damage.

The enactment of the 1976 amendments to the federal Coastal Zone Management Act reflected increasing federal interest in coastal natural hazards. However, most regulatory responses to coastal natural hazards are implemented at the state and local level.

A comprehensive regulatory response should: (1) generate and disseminate information on coastal natural hazards; (2) avoid shoreline protective devices for new development and review protective devices proposed for existing development for adverse effects on other private and public lands; (3) prohibit development on land that is too hazardous to develop; and (4) treat landowners prohibited from developing their property fairly.

Several land use control techniques can be combined to achieve those purposes. Hazard information can be generated and disseminated through site-specific geologic reports required as a condition of development permission in potentially hazardous ares. Shoreline protective devices can be avoided by requiring new development to be set back a safe distance from the shore. Land that is too hazardous to develop can be zoned accordingly and the landowners awarded transferable development rights as compensation for the development prohibition.

FOOTNOTES

1. F. Shepard & H. Wanless, Our Changing Coastline 9 (1971).

2. Illinois Central R.R. v. Illinois, 146 U.S. 387, 453 (1892).

3. Fifteen states use the mean high tide line; nine use the low water mark. Hawaii, Illinois, and Michigan have boundaries peculiar to their jurisdictions. MacGrady, Florida's Sovereignty Submerged Lands: What Are They, Who Runs Them and Where is the Boundary? 1 Fla. St. U.L. Rev. 596 (1973).

4. Borax Consolidated Ltd. v. Los Angeles, 296 U.S. 10, 22 (1935).

5. Board of Trustees of the International Improvement Trust Fund v. Medeira Beach Nominee, Inc., 272 So. 2d 209, 212-13 (1973).

6. Burns v. Forbes, 412 F.2d 995 (3d Cir. 1969); State v. Sause, 217 Or. 52, 342 P.2d 803 (1959); but see Roberts v. Brooks, 78 F. 411 (2d Cir. 1897).

7. County of St. Clair v. Lovingston, 90 U.S. (23 Wall.) 46 (1874).

8. 429 U.S. 363 (1977).

9. See Hughes v. Washington, 389 U.S. 290 (1967).

10. Katenkamp v. Union Realty, 11 Cal. App. 2d 63, 53 P.2d 390, rev'd 6 Cal. 2d 765, 59 P.2d 473 (1936), reh. denied, 36 Cal. App. 2d 602, 98 P.2d 239 (1940).

11. Rex v. Pasham Commissioners, 89 B. & C. 355 (1828).

12. A loss which does not give rise to an action for damages against the person causing it.

13. See, e.g., Vanderschrier v. Aaron, 103 Ohio App. 340, 140 N.E.2d 819 (1957).

14. 282 Or. 509, 580 P.2d 173 (1978).

15. Connor v. Great Western Savings and Loan A., 69 Cal. 2d 850, 447 P.2d 609, 73 Cal. Rptr. 369 (1968).

16. Pub. L. No. 93-234 (1973) (amending 42 U.S.C. § 4001 et seq. 1976).

17. 40 C.F.R. part 60 (1979).

18. COASTAL ZONE MANAGEMENT, Nov. 7, 1979 (Nautilus Press).

19. 16 U.S.C. §§ 1451-63. See generally Hildreth, The Federal Coastal Zone Management Act as Amended, 10 Nat. Resources Law 211 (1977).

20. 15 C.F.R. § 923.26 (1979).

21. See generally D. Mandelker, Environmental and Land Controls Legislation 66-86 (1976); American Law Institute, Model Land Development Code §§ 7-201 through 7-204 (1975).

22. 16 U.S.C. § 1451 et seq. (1976); 16 U.S.C. § 1454(b).

23. 15 C.F.R. § 923.21(h)(i) (1979).

24. N.C. Gen. State. § 113A-113 through 113A-121 (1978).

25. Askew v. Cross Key Waterways, 372 So. 2d 913 (1978), reh. denied (1979).

26. Oregon Land Conservation and Development Commission, Oregon Coastal Management Program 199-206 (1976).

27. See Somerville, Government Tort Liability, 10 The Urban Law. 376 (1978).

28. 396 F. Supp. 1004 (N.D. Cal. 1975).

29. Cal. Pub. Res. Code §§ 21000-21176 (West 1977).

30. See, e.g., Carmel Valley View, Ltd. v. County of Monterey, 58 Cal. App. 3d 817, 130 Cal. Rptr. 249 (1976).

31. U.S. Const. Amend. V.

32. Fred R. French Investing Co. v. City of New York, 39 N.Y.2d 587, 350 N.E.2d 381 (1976), appeal dismissed 429 U.S. 990 (city plan to preserve private parks invalidated); Penn Central Transportation Co. v. City of New York, 42 N.Y.2d 914, 366 N.E.2d 1271 (1977), aff'd., 438 U.S. 104 (1978) (scheme to preserve Grand Central Station utilizing transferable development rights approved).

33. See Cal. Pub. Res. Code §§ 3100-31406 (West 1977). For further discussion of the issues considered in this article, see Hildreth, Coastal Natural Hazards Management in Volume 59 of the Oregon Law Review.

Designation of All Public Rights of Way to the Tidal Areas of Rhode Island

Joseph F. Turco*

The first white settlers to Rhode Island found that the Narragansett Tribe, the dominant native group, practiced a communal approach to natural resources. Their practice was to utilize the shoreline and the sea in common, with private ownership strictly forbidden. They extended their generosity to the colonists as well.

The settlers were quick to see the commercial advantages of easy access to the sea. They wanted to avoid the hardships that private riparian rights had caused the non-owners back in England. Toward this end, they petitioned the Parliament and the King for extensive rights of fishery and associated shoreline uses. The Charter of 1663 granted by Charles II was a model of liberal rights granted to the Rhode Island colonists, their heirs and the neighboring colonies. The R. I. Constitution of 1843 which superseded the Royal Charter preserved these rights.

By the latter part of the 19th century many homes, often intended for summer use, had been built along the coastline. The free and easy access of earlier times had begun to disappear.

In this century, the Depression and World War II had the effect of nullifying efforts to reopen the public rights of way to the shore which dated back to colonial days and had become obscured by the ravages of time and the periodic violence of the sea. But the greatest deterrent to public use of the shoreline was found to be private ownership. Since most of the state's shoreline is privately owned, legal access to the coast is now impossible in many locations.

By the 1950's, private shoreline development was accelerating and private interests exerted greater efforts to limit access to the shore by the general public. As of today, more than 90% of the coastline is under private ownership. The State recognized, however, the historic right of the people to have access to their common resource, the sea and its shoreline. In 1958 the General Assembly established a permanent Public Rights of Way Commission whose mandate was to carry on a continuing discovery of the public rights of way to the water areas of the state, to legally define, and to mark and cause to be opened for public use all discovered rights of way. The duties of this Commission were transferred to the Coastal Resources Management Council in 1977 by legislative action.

*Chief of Test & Evaluation, Construction Engineering, Electric Boat Division, General Dynamics, Groton, CT and Chairman Subcommittee on Rights of Way, R.I. Coastal Resources Management Council

The Rhode Island Legislature in 1977 amended the general laws of Chapter 46-23 entitled "Coastal Resources Management Council" to include Section 46-23-6E "Rights of Way," which read as follows:

> The Council shall be responsible for the designation of all public rights of way to the tidal areas of the state, and shall carry on a continuing discovery of the appropriate public rights of way to the tidal water areas of the state.

In the Spring of 1978, the Coastal Resources Management Council's subcommittee on Rights of Way was established to carry out the mandate of the legislature.

The subcommittee drafted rules and regulations for the designation of rights of way. In accordance with the Administrative Procedures Act, a public hearing was held on the proposed regulations after which the subcommittee made its recommendation to the full Council. With the unanimous approval of the full Council, the policies, regulations and procedures for the designation of rights of way were incorporated into the R. I. CRMP in July of 1978. See enclosure (1).

The subcommittee decided that it would begin its task by reviewing for accuracy the existing rights of way data compiled by the former legislative Commission on Public Rights of Way for possible designations. Following this action, a systematic search for unidentified rights of way would be initiated.

The procedure, which the subcommittee on Rights of Way has been using to designate a right of way is as follows:

a. Assignment of Title Attorneys to Research Prospective Rights of Way

   Attorneys are asked to research and review the land records in the various cities and towns for any information that could be used as a legal basis for designating a right of way. Initially eleven attorneys from the coastal communities involved were employed to perform this task.

b. Review of the Title Attorney's Findings

   Upon completion of their title searches, the data are reviewed for accuracy and relevance by the subcommittee's counsel.

c. Legal Counsel's Recommendations to the Subcommittee. Recommendations include:

1) Additional information is needed
2) Designate it as Public Right of Way - State Jurisdiction
3) Designate it as a Public Right of Way - City or Town Jurisdiction
4) Designate it as a Public Right of Way - Private Jurisdiction
5) It is not a Right of Way

d. Public Hearing Scheduling

A public hearing on the prospective right of way is held in the community in which it is located. The governing body of that community, other public officials and abutting landowners within 200 feet are notified of the hearing. Public notice of the hearing is also made in an appropriate newspaper.

e. Engineering Survey

With the descriptive information supplied by the title attorneys, an engineering survey of the prospective right of way is conducted, and an engineering drawing of the site is made by a registered land surveyor.

f. Appraisal

An independent appraiser makes a determination as to the value of the prospective site.

g. Decision and Report of the Subcommittee

Based upon the information obtained from the title search, the public hearing, the engineer's survey and appraisal, the subcommittee makes its recommendation to the entire Coastal Resources Management Council.

h. CRMC Hearing/Decision

Based upon the subcommittee's report and any new evidence made available to the Council, the CRMC makes a final decision as to whether the right of way is to be designated as:

1) Public - State jurisdiction
2) Public - City or Town jurisdiction
3) Public - Private jurisdiction
4) Not a public right of way

An affirmative decision by CRMC designates the right of way as a Public Right of Way.

The designated public right of way is recorded and placed on file in the Town or City Hall in which it is located, and in the office of the Secretary of State.

The Department of Environmental Management is notified of the CRMC's designation and is responsible for the maintenance of all public rights of way under the state's jurisdiction.

The Division of Coastal Resources of the DEM is responsible for marking the designated rights of way as requested by CRMC.

The current status of the work being done by the CRMC subcommittee on Rights of Way is summarized on the Status Report Chart, enclosure (2).

The subcommittee has found that the task of designating all of the public rights of way to the tidal water areas of the state is a long, tedious, sometimes frustrating, but worthwhile and necessary process that should be beneficial to all of the people of the state.

The designation of these rights of way relate to the utilization of recreational areas in accordance with the recreation element of the State Guide Plan in order that there be better utilization of the natural resources of the State.

In conjunction with this ongoing program, the Rights of Way Subcommittee is also addressing the following relate areas of interest:

a. Lateral access along the tidal waters.

b. Research, identification and recommendations for acquisition of potential rights of way by DEM and the State Properties Committee.

c. Search for unidentified existing rights of way.

d. Encourage and promote the dedication of new rights of way by private citizens, developers, corporations or other groups.

Enclosure (1)

RIGHTS OF WAY, POLICIES, REGULATIONS AND PROCEDURES

1. Findings:

A. Rhode Island is richly endowed with public shorefront access opportunities via public parks and beaches, conservation areas, scenic view points and rights of way. However, efforts to enhance opportunities through more effective utilization of existing areas and identification of additional sites are needed.

-The distribution of coastal rights of way is uneven.
-Public rights of way along substantial lengths of shoreline remain unverified.
-Public rights in many presumed rights of way are contested.
-Rights of way must be posted to encourage use and prevent trespass on private property.
-Parking space at popular rights of way is frequently inadequate.
-Vandalism, littering and trespass on adjacent private holdings are issues of growing concern.

B. The scenic qualities of the Rhode Island coastal region and shoreline are one of the state's greatest natural assets.

-Much of the state's shoreline is highly scenic.
-Many scenic areas are readily accessible to the public from designated overlooks and scenic highway routes.
-Scenic viewing is a major recreational and tourist activity.
-Additional scenic opportunities exist at local, state and municipal highways. These opportunities require careful evaluation if they are to be utilized to their full potential.
-Scenic view sites are highly sensitive to the aesthetic impacts of adjoining development.

C. The State has numerous public and private boat launch ramps.

-Ramps are unevenly distributed with deficiencies in upper Narragansett Bay where demand is concentrated.
-Additional ramps should be sited in or near the major recreational boating harbors of the Bay. Acquisition costs can be extremely high in such areas, however.

-Ramp siting can have major environmental impacts where it contributes to increased boat activity in sensitive marine areas.
-Ramps can have major social and aesthetic impacts where service roads are insufficient to handle increased traffic and/or where inadequate parking is provided.

D. Development of additional rights of way, scenic viewpoints and launch ramps will place additional management responsibilities on state personnel.

2. Policies:

A. The Council finds that the public has a right to the use and enjoyment of the shoreline guaranteed under Article 17 of the State Constitution. Enhancement of public access to coastal waters is consequently a high priority use of the coastal region.

B. Applicants for Council permits for any development or activity contiguous to a posted public right of way, boat launch ramp or public view site shall be called upon to demonstrate by a fair preponderance of evidence that such development or activity or any development or activity related thereto will not interfere with or prevent public use and enjoyment. Where such interference or prevention is found the Council shall prohibit or require appropriate modification of the proposal in question.

C. The Council, working in coordination with the Department of Environmental Management, shall prepare periodic updates of the state rights of way, scenic view sites, and public launch ramp map commencing in 1978. It shall ensure maximum publicity and distribution of these maps to the general public.

D. The Council shall, commencing in 1978, designate all verified public rights of way, scenic view sites and launch ramps.

3. Legislative Authority:

Chapter 270 of the Public Laws of 1977 granted to the Coastal Resources Management Council the following authority:

"Rights of Way.--The Council shall be responsible for the designation of all public rights of way to the tidal water areas of the state, and shall carry on a continuing discovery of appropriate public rights of way to the tidal water areas of the state.

"The Council shall maintain a complete file of all official documents relating to the legal status of all public rights of way to the tidal water areas of the state.

The Council shall, subject to the provisions of chapter 6 of title 37 of the general laws, as amended, have the power to designate for acquisition and development by the Department of Environmental Management land for tidal rights of way, parking facilities and other Council related purposes.

In conjunction therewith every state department controlling state owned land close to or adjacent to discovered rights of way are authorized to set out such land, or so much thereof as may be deemed necessary for public parking.

No such use of land for public parking shall conflict with existing or intended use of such land, and no improvement shall be undertaken by any state agency until detailed plans have been submitted to and approved by the governing body of the local municipality."

4. <u>Regulations and Procedures</u>:

A. There shall be a duly formed rights of way subcommittee appointed by the Chairman of the Council consisting of not fewer than three (3) members of the Council. The Chairman of the Council shall appoint a Chairman of the Rights of Way Subcommittee. The rights of way subcommittee shall elect its officers thereafter.

B. The rights of way subcommittee shall meet at the call of the rights of way subcommittee chairman at a place so designated by said chairman.

C. The rights of way subcommittee shall be responsible for the designation of all public rights of way to the tidal water areas of the state and shall carry on a continuing discovery of appropriate public rights of way to the tidal water areas of the state in accordance with the regulations and procedures established herein. The rights of way subcommittee shall implement the responsibilities designated under Section 46-23-6E of the General Laws relative to rights of way.

D. In considering designation of a public right of way, the subcommittee shall evaluate all existing data including attorney's title reports, municipality land records, court cases and any other pertinent information available. The subcommittee may enlist additional professional help should it deem necessary. After compilation of the information it deems necessary to designate a public right of way, the subcommittee shall hold a public hearing at which time it shall allow all interested parties to be heard in accordance with the rules and regulations of the Coastal Resources Management Council and the Administrative Procedures Act. Prior to the hearing the subcommittee shall duly advertise said hearing and shall notify all abutters within a two hundred foot (200') radius of the proposed right of way. This public hearing shall require a quorum of at least three members of the rights of way subcommittee.

E. Where on the basis of the evidence and/or demonstrations presented the subcommittee finds a reasonable probability that a public right of way does exist, it shall so declare the public right of way as existing by a vote of the subcommittee. A quorum of three members of said subcommittee shall be required for this vote and a majority of those present and voting shall be required for passage. All proceedings shall be in accordance with the rules and regulations of the Coastal Resources Management Council and the Administrative Procedures Act.

F. Upon designation by the rights of way subcommittee of a public right of way the subcommittee shall forward its designation to the full Coastal Resources Management Council and the Coastal Resources Management Council shall as a whole act thereon in accordance with its procedures and the Administrative Procedures Act a public right of way shall be deemed to exist. The Coastal Resources Management Council shall then forward to the appropriate state agencies its designation thereof.

G. The rights of way subcommittee may from time to time review its prior actions and nothing contained herein shall be deemed to preclude the rights of way subcommittee or the full Coastal Resources Management Council from reconsidering its prior actions.

H. The rights of way subcommittee shall, subject to the provisions of Chapter 6, Title 37 of the General Laws, as amended, have the power to designate for acquisition and development by the Department of Environmental Management land for title rights of way, parking facilities and other council related purposes. The procedure for such designation shall be the same procedure as stated herein for designation of public rights of way.

I. In those instances where the rights of way subcommittee designates the acquisition and development by the Department of Environmental Management, land for rights of way, parking facilities, or other council related purposes it shall be the duty of the rights of way subcommittee to submit for approval to the governing body of the local municipality its plans and the plans of the Department of Environmental Management therefore prior to proceeding with final designation.

AMENDMENT TO RIGHTS OF WAY REGULATIONS

Add

Regulation 4-(J)

The following Prefixes shall be the designated prefix for Rights of Way in each coastal community:

Town	Designated Prefix
Westerly	A
Charlestown	B
South Kingstown	C
Narragansett	D
New Shoreham	E
North Kingstown	F
Jamestown	G
East Greenwich	H
Warwick	J
Cranston	K
Providence	L
Pawtucket	M
East Providence	N
Barrington	P
Warren	R
Bristol	S
Tiverton	T
Little Compton	W
Portsmouth	X
Middletown	Y
Newport	Z

Upon official adoption of a Right of Way, the Rights of Way Subcommittee shall utilize the designated Prefix for the community in which the right of way is located and then assign in numerical order the next number to the Right of Way (e.g. the first Westerly right of way shall be A-1; the second Westerly right of way shall be A-2, etc.)

Enclosure (2)

STATUS REPORT - PUBLIC RIGHTS OF WAY

CITY/TOWN	No. of Titles Searched	No. of Titles Reviewed	Date of Public Hearing	Engr'g Survey Completed	Evaluation Completed	Sub-Comm. Recommen-dation	CRMC Public Hearing	ROW Desig-nation Recorded	Recommend-dation to DEM	ROW Marked
Westerly	19	19	11/30/78 1/4/79 9/79	17	12/14/79	12/11/79	1/30/80			
Charlestown	3									
South Kingstown	12	12								
Narragansett	10	10	11/29/79							
North Kingstown	16	26	4/17/79	16	5/80	5/27/80				
East Greenwich										
Warwick	24	28	10/11/79 2/27/80	14						
Cranston										
Providence										
East Providence	13	13	1/24/79	8/79 (13)	9/1/79	9/11/79 (13)	10/79	2/29/80	3/20/80	
Barrington	4									
Warren	20	20	7/31/79	2/14/80 (20)	3/24/80	5/13/80	7/30/80			
Bristol	18	18	4/17/80	8/80						
Tiverton	1	4	6/3/80	4						
Little Compton	1									
Portsmouth	16	17	7/31/80	17						
Middletown		1	8/28/80	1						
Newport	2	2	8/28/80	2						
Jamestown										
New Shoreham										
TOTALS	159	170		104						

GUARANTEEING PUBLIC ACCESS IN PUBLICLY-FUNDED SHORELINE PROJECTS

By

Kenneth C. Caldwell A.S.L.A.
Environmental Planner
Senior Associate, Jones & Jones, Seattle

What I will be discussing is whether or not the general public should have a right of access to what they pay for. This is a very simple and straightforward subject.

The discussion is divided into four parts:

1. Background

2. The present situation

3. Examples (benefit areas vs. assessed population, and extent of benefits)

4. Possible solutions

1. BACKGROUND

Large complex societies evolve public institutions to provide what are considered to be essential public services. Whether they are local, state, or national, our governments operate--constitutionally--to promote the public health, the public safety, and the public well-being. In my practice I have noticed many situations where government agencies have played their individual parts in promoting the health, safety, or well-being of the community, yet have produced no widespread positive results.

In any community, we are faced with a multitude of rigid-purpose public and quasi-public entities. We find road departments that provide roads, water departments that supply water, sanitation departments that dispose of solid waste, sewer utilities that dispose of liquid waste, utilities that distribute energy, parks departments that provide parks, social service agencies that provide human services, building departments that uphold building codes, diking districts and hydraulic departments that control coastal shifts and provide flood protection. Even that listing is a gross simplification, for like a four-dimensional chessboard we have municipal, county, state, and federal agencies acting on the same stage.

Each of these agencies provides a service that society considers to be essential, and this paper will not address the timely issues of over-regulation or misguided public services. Rather, this paper will neatly focus on one class of "socially-beneficial activities": publicly funded (or assisted) shoreline works; and, this paper will further refine the topic by addressing the provision (or nonprovision) of public benefits (especially public access) in these publicly-financed shoreline works.

2. THE PRESENT SITUATION

An anonymous U.S. author once proposed that the principal function of public works in the United States is to benefit the private economy. This statement touches the issue at hand: if public shoreline works are to be justified by their public benefits, who is the "benefited public"?

Shoreline works are often financed by the public-at-large, through general tax revenue. On the other hand, when a locale builds a water system, a sewer system, or a solid-disposal system, those who are directly benefited are expected to bear the full financial burden. If you receive public water you pay your fair share of the cost of providing that water. If you do not use the water system, you provide nothing.

We do not typically view flood protection or shoreline erosion-control projects as items that must be funded by their "consuming publics" (those that receive direct benefits). Instead, many shoreline projects are financed from general tax revenue, and in those cases the "benefited public" must be defined as the public-at-large, be it municipal, county, state, or national in scope.

This paper proposes that the "benefited public" is not synonymous with the "funding public"; specifically, general tax revenue is being used for shoreline projects that do <u>not</u> benefit most of the public that finances them.

3. EXAMPLES

A. The Green River, King County, Washington

The Green River Valley is a classic case history of public expenditure and private benefit, and a poignant example of our contemporary use of a river system adjacent to a growing urban area. This river system has received an enormous amount of public funds. From the 1940's to the early 1960's, much money was spent to promote, plan and build an Army flood control dam in the river's Upper Basin. The county hydraulics department has spent large sums assisting diking and bank-armoring. The Port of Seattle, a quasi-

Photos of diking along the Green River, King County, Washington. Top: Landowner has set out home-made "No Trespassing" sign. Dike was publicly assisted and is publicly maintained, but is privately owned. Bottom: County "No Trespassing" sign. The county has a dike maintenance easement, but no public access easement.

public body, has transfigured the lower reaches of the river for shipping. Public expenditures are planned to continue into the future: the U.S.D.A. Soil Conservation Service will (with local sponsors) pay for a system of massive canals to drain the valley, and the Army is studying a multi-million dollar dike-raising project that will further guarantee flood protection.

At this point, this enormous investment has not yielded any benefit to the public-at-large. No public access has resulted, and the prime tangible benefit is an economic boon to the landowners in the floodplain. The Army flood control dam was built within a closed municipal watershed, and neither it nor its reservoir is publicly accessible. Existing drainage canals and pump stations are generally not usable by the public, and nearly all publicly-assisted diking is still privately owned. As far as projects planned for the future (the dike-raising and drainage canals), either public access is not anticipated or it will be the responsibility of local (i.e., municipal and county) sponsors.

The Green River case history displays a record of general public expenditure that has not resulted in general public benefits. The public benefits are primarily spillover benefits resulting from private economic gains: the increase in employment and capital flow produced by the public realization of a flood-free and well-drained valley adjacent to an expanding rail-served port city. In terms of the needs of the funding public, these vast expenditures of public funds have not resulted in increased aesthetic amenities, expanded public amenities, or expanded recreational enjoyment of the river system. It simply was not the policy of the sponsors to provide "full-spectrum" projects.

Unlike most other public works projects, flood control works "function" for only a small percentage of the time: in the case of levees, floodwaters are actually held back perhaps as long as 20 days each year, or as short as a few days every 20 or 30 years. This is a theoretical "range of functionality" from 5% to 0.005% of the average year. Compare this to public roads, water supplies, or waste disposal systems, which provide a functional benefit close to 100% of the time. During most of its life, a typical flood control project is nonfunctional: it is just so much rock, concrete, etc., piled along (or across) the shore. As such, it not only does not provide benefits to the public at large, but rather it often produces negative effects.

B. Onondaga Creek, Syracuse, New York

In the City of Syracuse, New York, there was--in the late 1960's when I lived there--a creek that ran through the heart of the city. The creek periodically flooded, and the local public works authorities--probably with federal assistance--had solved the problem by reconstructing the creekbed as a flat-bottomed "V" channel with concrete walls at a slope of about 45°. This was a very efficient solution, both hydraulically and economically.

However, this creek flowed through an old residential neighborhood where the people were rather poor, and there were lots of children. Each summer the heatwave came. There were no nearby public swimming pools, and the city's lake (Lake Onondaga) was far too polluted for swimming. So each summer, the local children tried to splash and swim in this creek. Each summer several drowned, for one misstep would send a child tumbling into the swift-flowing concrete ditch that was once a natural creek, and there was no escape up the steep sides.

The city responded to the drownings by building security fencing along the creek, but the children climbed over or under the fencing and continued to drown. So as a final measure, watchtowers were built along the creek at the hub of strategic sight lines. The towers were equipped with monstrous loudspeakers, and each summer they were manned by guards. If a child was seen climbing the fence or approaching the ditch, the loudspeaker would bark out and scare them away.

4. POSSIBLE SOLUTIONS

Now, that story of Onondaga Creek may be a bit archaic. We don't do things like that anymore, do we? If the creek had originally been designed not just as a sluice for floodwaters, but as a public feature that was safe and holistically beneficial to the community, would the life cycle cost of the project have been more, or less, in terms of human life, human needs, neighborhood impacts, or ongoing cost to the municipality? Are we still building Onondaga Creeks, or have we learned that shoreline works cannot merely meet functional requirements: they must meet the fullest range of needs.

We have seen that the problem of narrowly-conceived projects exists at all levels of government--municipal, county, state, and national. However, the problem can increase in seriousness as the scale of government increases, and it may be that large, federally-sponsored projects are of greatest concern.

In the following material, I will therefore begin by outlining federally-related solutions...

I. The "Block Grant" Approach
II. Qualifications for Federal Funding

Then a third locally-related solution is outlined...

III. Revision of Criteria for Local Funding Assistance.

As you will quickly note, I have serious misgivings about several federal agencies that supply shoreline projects, especially the U.S. Army Corps of Engineers. It is only through a quirk of fate (or rather, a quirk of our heritage) that an engineering branch of our military has for so many years been called upon to provide so many of our major hydraulic works, and their mandates have remained narrow in scope and largely utilitarian in purpose. This is anathema to the need for holistic full-spectrum projects, and it may be that a civilian "water resources agency" (of some sort) will one day replace the U.S. Army as our national overseer of water resource management, with actual projects being designed by the individual states.

I. <u>The "Block Grant" Approach</u>

Local bodies (or states) could directly receive federal funds for all shoreline projects, and they would have discretion in expenditures (just as they are given discretion in expending certain H.U.D. and H.C.R.S. funds). Under the existing situation, a federal entity--such as the Army--is called onto the scene to deliver an hydraulic project. The project will be planned by the federal government, and often there are rigid parameters (e.g., 100% funding for "structural solutions," with no federal provision for other project elements, such as public use of the completed work).

What are the implications of a state-oriented (for example) "block grant" approach to federally-assisted shoreline works?

A. States would be required to produce long-range plans for hydraulic projects.

B. States would be given federal funds in proportion to their documented historical flood losses, other demonstrable shoreline needs, population level, and so forth.

C. States would be asked to engage in effective land use planning to reduce future flood hazards and other adverse shoreline conditions.

D. States would be given great latitude in expending appropriated funds for a broad range of projects.

E. The federal government (e.g., through a water resources agency) would act in a technical review and advisory capacity. Projects would be planned and designed by the states.

F. Where interstate waters are involved, states would be required to form interstate authorities (or compacts) to coordinate shoreline works.

These sketchy implications display the point that major shoreline works and hydraulic projects could be planned and designed by the states, and that states could have reasonable freedom to creatively use appropriated funds.

The following item addresses possible provisos and qualification criteria attached to such funding.

II. Qualifications for Federal Funding

In recent times the federal government has engaged in backdoor legislation by attaching elaborate riders to federal funds disbursed to local governments. A recent example is a proposal to withhold federal highway funds from states that do not legislate a mandatory jail sentence for drivers that fail a breath test.

Still, federal funding sources are justified in attaching some kinds of provisos. For example, it is reasonable to require that local governments have an adopted plan for the use of funds. It is reasonable to ask that local governments act broadly to meet public health, safety, and welfare needs; that is, to require that they act in the general public interest.

If federal shoreline funds were disbursed as grants-in-aid to state and local governments--similar to the way H.C.R.S., H.U.D., or F.H.W.A. funds are disbursed--what might this imply regarding publicly-funded shoreline works?

A. Projects must demonstrate broad public benefits above and beyond the reduction of damages.

B. Projects must produce a net enhancement of visual, recreational, and biological resources.

C. Project planning must directly involve a wide spectrum of agencies and interests.

D. Projects must be part of an adopted community plan that addresses long-term public issues, including visual resources, recreational opportunities, biological effects, and neighborhood health.

E. Public access must be provided, unless it is demonstrated that this is infeasible or undesirable, with other benefits in attendance. No public funds could be spent on private property unless public access (if feasible) was part of the project.

The above implications are a preliminary attempt at expressing an underlying goal: that federally-funded projects will meet the fullest spectrum of public needs.

III. Revision of Criteria for Local Funding Assistance

This primarily applies to shoreline works such as diking or shore armoring that are subsidized by local public funding. For instance, in King County, Washington, landowners can often get a 50% matching grant for dike construction or bank armoring along rivershores. In return, the general public receives nothing.

The solution is obvious:

A. The public body would offer a matching grant only in those cases where the landowner offered either a land dedication (fee simple or a perpetual public easement) or comparable public benefits.

B. Local or state governments can pass laws requiring that no general public funds be spent on privately-owned shoreline projects unless a perpetual right of public access (or "other tangible public benefits," if public access is infeasible) accrued.

C. Through park and recreation plans and coastal zone planning, the local or state body would develop realistic long-range plans for shoreline works, and for public access.

CONCLUSION

I admit that the foregoing material is a rather superficial look at the issue of providing public access in publicly-funded shoreline projects, but I feel it at least frames the problem, and begins to explore some solutions. The major points are as follows:

1. The essential issue involves narrow-purpose projects vs. general-purpose projects; i.e., shoreline projects that are designed to fulfill narrow utilitarian needs (such as bare flood control or shoreline armoring projects), compared to projects that are designed as well-integrated community features.

2. Agencies that provide narrow-purpose shoreline projects are found at all levels of government.

3. Many shoreline projects--especially flood control works--are funded by a general public that receives no direct utilitarian benefit (e.g., where the funding public is largely outside a flood hazard area, and thus receives no utilitarian benefit from single-purpose flood abatement projects).

4. Funding publics have a reasonable right to demand direct benefits in return for public expenditures.

5. All levels of government should adopt laws that prohibit the expenditure of public funds for shoreline projects on private property unless perpetual public access is provided. (In cases where public access is infeasible or undesirable, other tangible public benefits would have to be demonstrated.)

6. States should have the principal responsibility for planning and designing federally-assisted shoreline projects. States would receive federal grants-in-aid (with appropriate provisos and qualification criteria attached to such funding), with federal agencies acting solely in an advisory and review capacity.

7. Many major shoreline works are federally assisted or federally planned; thus effective reform must involve revisions in federal law and policy.

PROVIDING PUBLIC WATERFRONT ACCESS IN MARYLAND

by:

Kathy H. Fitzpatrick
Marine Recreation Specialist
Coastal Resources Division
Tidewater Administration
Maryland Department of Natural Resources

Trisha A. Bednarz
Marine Recreation Specialist
Coastal Resources Division
Tidewater Administration
Maryland Department of Natural Resources

Earl H. Bradley, Jr.
Chief, (Local) Technical Assistance Program
Coastal Resources Division
Tidewater Administration
Maryland Department of Natural Resources

Public demand for waterfront access is a subject about which every coastal state is concerned. When the need for public access and its corresponding management alternatives are discussed, several variables must be taken into consideration: 1) the geographic area being studied; 2) the associated demographic characteristics; 3) the corresponding legal ramifications; 4) the environmental impacts; 5) the present level of opportunity for public access within the area, and 6) pertinent public and private, recreation and open space programs. This paper will present Maryland's approach to providing public waterfront access.

The State of Maryland is unique in that its coastal area includes both bay and ocean shoreline frontage. However, with 30 of its 31 total ocean shoreline miles presently available for public use, the State's concern for providing public access is more pronounced in the Chesapeake Bay region.

The Chesapeake Bay and its tributaries have over 4,000 miles of shoreline. The Bay shoreline can be divided into two very distinctive regions, both physiographically and culturally - the Eastern and Western shores.

The Eastern Shore of Maryland is predominantly rural, with its economy heavily dependent upon agricultural and seafood harvesting. The shoreline is characterized by wetlands, wooded swamps and farmland.

The Western Shore contains two major metropolitan areas, Baltimore and Washington, D.C., as well as the rural southern Maryland region. The topography of the western shore is that of gently rolling uplands bounded by deep cut stream valleys. Much of the shoreline is cliff or bluff with narrow (15'-20') sand beach at the base.

The area seaward of mean high tide is in public ownership, however, this does not guarantee public access for recreational purposes. As stated in the Shorefront Access and Protection Planning Element of Maryland's Coastal Zone Management Program,

> The small tidal range, the lack of broad sandy beaches, the need to protect extensive marsh lands adjacent to shorelands, and the small percentage of adjacent shoreland in public ownership combine to limit public access. Moreover, because Chesapeake Bay beaches are typically isolated, non-extensive, and narrow, there are few places where wet-sand areas alone are suitable for intensive recreation. Therefore, the provision of public access to beaches generally must include the acquisition of upland areas to provide adequate recreational space.

Demand for water-oriented recreation opportunities on the Chesapeake Bay is intense. Recreational demand can be divided into two major categories - recreational boating and other forms of water-related recreational activities (swimming, fishing, viewing, etc.).

The major source of information about recreational demand throughout the state is the Maryland Outdoor Recreation and Open Space Plan (SCORP), which was updated in 1978. The SCORP is prepared by the Department of State Planning in cooperation with the Department of Natural Resources. State law requires that "every acquisition and development project funded by the State in whole or in part shall meet needs identified in the SCORP" (N.R. Article, Section 5-906).

The SCORP discusses virtually all recreational activities, but it addresses water-related recreation specifically:

> The demand for water oriented recreation opportunities is increasing significantly. The high demand for water recreation opportunities in a natural setting is not limited to a particular geographic region, but occurs in all other areas of the State. However, the major water resources of the State are located in only a few areas of the State. Therefore, the provision of and access to water oriented recreation opportunities pose serious problems to people who do not live adjacent to the State's water bodies...

Maryland's Coastal Zone Management Program addresses the demand for water-oriented recreation in several of its objectives:

a. to promote increased recreational opportunities in shoreline areas,

b. to promote increased public access to tidal waters, and

c. to assure that these occur in a manner which protects the quality of coastal resources and which maintains public health and safety.

In complying with these objectives the Coastal Resources Division (CRD) of the Tidewater Administration is the coordinator for the Maryland Coastal Zone Management Program (MCZMP) and is presently involved in an effort to achieve a balanced relationship between the State's currently existing access areas and the increasing public demand for additional sites. Several methods are being utilized in this effort, foremost among these is the Bay Access Park Study.

The Bay Access Park Study

The basic function of the Bay Access Park Study (BAPS) is to consolidate the various inventories of open space areas which have potential for use as waterfront access sites.

All existing inventories were reviewed and sites which are appropriate for shorefront access were identified.

Data is being collected to bring all sites up to the same level of information available. Once this data is collected, it will be computerized. The program will be such that it can be updated periodically to reduce the possibility of the information becoming obsolete. The program will also be designed in such a manner as to allow a list of sites to be called out based upon criteria or needs of individual counties, state agencies or other interested private organizations.

An inventory of existing public and private waterfront recreation facilities has also been conducted. The locations of both the existing sites and the potential sites identified in BAPS will be compared so as to minimize any duplication of services. It will also avoid any possibility of competition between public and private facilities and the failure to meet the needs of the public. Publication of the results of this inventory will also make the public aware of existing public access sites that they were unaware of, thus in effect increasing public recreational opportunities.

A user guide will be developed to describe how local jurisdictions and other users can benefit from this study, how to request and use the computer information and how to interpret the information. The users will play a role in updating the study periodically by notifying CRD of any sites which need to be deleted or sites which need to be added.

A Statewide priority ranking will be done based upon Maryland's Outdoor Recreation and Open Space Plan ranking system which will be modified to be applicable to shorefront access needs. Once these needs are identified, the sites will be evaluated using the guidelines and criteria which have been developed by relevant State acquisition agencies (Land Planning Service, Capital Program Administration, Maryland Park Service, etc.). A priority listing of sites recommended for acquisition at the state level will be the result.

The county priority ranking will be conducted on an individual basis with consideration for each jurisdiction's special needs and guidelines. This will be done in a manner compatible with Maryland's Outdoor Recreation and Open Space Plan.

Intergovernmental Coordination

While the Bay Access Park Study is a major effort of Maryland's Coastal Zone Management Program it is not the Program's only effort aimed at improving shorefront access. The Coastal Resources Division staff has been working with existing recreation and open space programs, and the state and local governmental agencies that administer them, to promote increased shorefront public access. Such coordination is important since the Coastal Resources Division is not an acquisition agency and does not have the principal authority for the provision of recreation and open space opportunities in the State.

At the local level, the Coastal Zone Management Program is funding technical staff in the planning office of the counties located in Maryland's coastal zone. Such funding allows the local governments to give increased attention to coastal zone management concerns including the provision of shorefront access.

The Coastal Resources Division itself is involved in a cooperative effort with the Maryland Department of State Planning to administer the State's Urban Waterfront Program.

<u>Urban Waterfront Planning Program</u>

Another method for providing public shorefront access in Maryland is through the Urban Waterfront Planning Program. This recently established program encourages redevelopment and reuse of, and increased access to, waterfront areas and facilities which have fallen into disrepair or disuse.

The funding is on a 95%-5% (State-municipality) matching basis, with CRD providing the state portion.

The program was initiated in April, 1979, and a Memorandum of Understanding was signed with the Department of State Planning on February 27, 1980. The State program involves a call for applications, review of the received proposals and selection of proposals that will receive financial assistance. Each proposal is evaluated on the basis of the following criteria:

a. Extent of Distress in Project Area

    1) Structural deterioration
    2) Vacant buildings
    3) Deterioration of public facilities and services
    4) Environmental blight
    5) Economic stress

b. Public Need for Shoreline Access

    1) Responds to identified shortage
    2) Serves a variety of public access needs

c. Potential for Implementation

    1) Local staff and/or financial resources
    2) Public ownership of land in project area
    3) Private sector interest
    4) Coordinated with other local projects and programs
    5) Consistent with comprehensive plan

The first year of implementation has seen four projects get underway: three on the Eastern Shore and one on the Western Shore. This work has helped to stimulate interest and response from other jurisdictions interested in providing access to their revitalized waterfront areas.

Coastal Energy Impact Program

CRD is also responsible for administering the State's Coastal Energy Impact Program (CEIP). This program has provided funding for cooperative efforts with local governments to provide shorefront recreational areas to mitigate the adverse impacts of coastal energy facilities on coastal recreational opportunities.

Through the Coastal Energy Impact Program the Coastal Resources Division has assisted Calvert County, Baltimore City, and Kent County in the selection of shorefront sites and is providing financial assistance for the acquisition of the sites. Description of these projects follow:

Flag Ponds - Calvert County

The Calvert County Board of Commissioners have initiated efforts to provide the residents of Calvert County with access to the Chesapeake Bay and the Patuxent River via public beaches and landings.

On the Chesapeake Bay a 327-acre site known as the Flag Ponds has been selected as the most desirable location. This site is an extremely unique natural area which will provide a rich diversity of possible uses; including swimming, fishing, nature trails and associated Environmental Interpretive Programs.

This is a new project which will be completed in the phases detailed below:

Phase I - Acquisition

Phase II - Development of a Master Plan for the area which will establish appropriate activities, and level of public participation that may occur without resultant environmental damage to the unique area.

It is assumed that any and all activities in the area will be of low intensity use due to the sensitive nature of the site.

Phase III - Construction - To include at least a road improved with gravel base and a stabilized parking area near the beach. Possible Construction includes:

nature trails,
a group camping area (primitive only), and
a swimming area and beach house.

Acquisition of the Flag Ponds Site is being funded by approximately 75% CEIP funding and 25% local funding.

Reedbird Park - Baltimore City

The Reedbird Park is a 118-acre component of the Middle Branch Park System located within the City of Baltimore at the confluence of the Patapsco River and the Middle Branch of the Baltimore Harbor. The site is divided by the river with 28 acres to the east and the largest segment, 90 acres, to the west. The larger, western parcel is the site of a completed landfill and former incinerator. The eastern parcel was also used as a landfill for a number of years. Both landfill operations have been phased out and the incinerator demolished.

The proposed project would involve three primary segments. The first would be landscaping along the railroad right-of-way to reduce the visual impact on residential property and passive areas of the park. The second, involves construction of pedestrian pathways under the rail bridge to insure a continuous trail system. This will require special design solutions to isolate pedestrians from rail movement. The third requires the construction of recreational facilities, adequately removed from the rail right-of-way, which can respond to the needs of the residents and park users.

Planned facilities include a small fishing and boating pier on the Patapsco River shoreline slightly north of the rail bridge. This will insure park users access to the shoreline in an area which is not negatively impacted by movement of rail cars.

This project will be largely funded through CEIP funds with Waterway Improvements funds used for the construction of boating-related facilities.

Betterton Beach - Kent County

Plans for upgrading the public beach of the town of Betterton on the Chesapeake Bay and Sassafras River involve the acquisition of land and structures within the beach area, the demolition of some deteriorated structures and the upgrading of existing facilities. This effort, when completed, will provide a high quality water access facility for use by residents and visitors.

Betterton is located within 80 miles of the Baltimore Metropolitan Area by car. These two areas are also linked by the Port Welcome, a touring cruise ship operated by the Maryland Port Authority. The Port Welcome makes two trips per week bringing approximately 180 visitors from Baltimore City to Betterton Beach. When the acquisition and improvements are made, the number of visitors brought by the Port Welcome is projected to increase by more than 100. Port Welcome authorities are anxious for improvements to be made and it is probable that a third weekly trip will be added when the project is completed. They have also indicated that failure to improve the facilities could result in the discontinuation of the Port Welcome-Betterton Baltimore City residents and its visitors.

There are no areas on the Upper Eastern Shore which provides the recreational opportunities available at Betterton. This Beach is the only public beach in the Upper Eastern Shore region, and the State Comprehensive Outdoor Recreation Plan (SCORP) has as its first recommendation for that region, the acquisition and development of the Betterton Beach site. Further, no area is so well linked to the Western Shore with direct transportation available.

Acquisition of Betterton Beach was initiated with POS funding. CEIP funding was used this past winter to acquire adjacent property which otherwise might not have been available by the time other funds could be found. A special appropriation by the State legislature this past legislative session will allow Kent County to complete acquisition this fall.

In each of these projects CEIP funding was used to supplement other sources of funds to allow the provision of recreational opportunities that otherwise would likely not have been possible.

CRD utilizes two major State funding programs in Maryland that provide public recreational opportunities throughout the State:

1) The State's Program Open Space (POS) is the major source of funding for the acquisition of recreation and open space areas; and

2) the Waterways Improvement (WWI) program is the major source of funding for developing and maintaining public recreational boating facilities.

Program Open Space

Program Open Space was established in 1969 to make funds available to State agencies, counties and Baltimore City in order: "(1) to expedite the acquisition of outdoor recreation and open space areas before potential areas are devoted to some other use; and (2) to accelerate the development of needed outdoor recreation facilities."

POS is unique in that it provides its own self-generating funding source; namely, the State Property Transfer Tax. The idea behind the transfer tax is that the individual who buys a home or other property for private use had hastened the decline in available open space land. By paying a tax at the rate of one-half of one percent (0.5%) of the property purchase price, that same individual would help to support the buying of land which could ultimately be used and enjoyed by the general public. In addition to the transfer tax, funding for Maryland's POS is derived from the issuance of general obligation bonds and from federal grants from the Land and Water Conservation Fund.

Program Open Space has essentially become a "pay-as-you-go" program. Only $20 million in bonds have actually been sold since 1970 because the transfer tax revenues have been sufficient to meet the program's yearly cash outlay. During the period from 1970 to 1978 the annual transfer to collection has risen from $7,349,868 to $22,049,192.

These revenues are divided on a 50-50 matching basis between the State and the local governments. Distribution of POS funds among local jurisdictions is determined by a formula which utilizes the following factors: 1) percentage of the State population; 2) percentage of the State population in the next 10 years; 3) percent of change; and 4) percent of transfer tax collections. POS funds are allocated to the subdivisions on a 50%-50% basis; 50% for acquisition purposes and 50% for either acquisition and/or development. The main prohibition of use of POS monies is that they not be designated for maintenance and operation costs.

With regard to the funds received from the federal government, the State of Maryland negotiated an agreement in 1973 with the Federal government and the subdivisions. Through this arrangement, the Department of Natural Resources (DNR) would submit a limited number of projects for federal reimbursement rather than an onslaught of separate State and local projects. Once the State's reimbursed with Land and Water Conservation Fund monies, the amount is split 50%/50% between the State and the local jurisdictions and can be utilized for either acquisition and/or development. This arrangement greatly simplifies the processing of recreational acquisition projects in Maryland by allowing POS to act as coordinator of all the pertinent recreational and open space acquisition programs.

The ultimate goal of POS is the acquisition of 397,000 acres for the State. So far, Maryland has acquired 296,000 acres, leaving 101,000 acres to be purchased by 1980. At the local level, 93,585 acres is presently in public use.

Through cooperation with POS, the CRD reviews proposed acquisitions in coastal areas and provides recommendations regarding potential acquisition sites that are located along the Chesapeake Bay and its tributaries. Once completed, the Bay Access Park Study will provide a more informative basis for such recommendations. Through this cooperative effort POS is placing more emphasis on the acquisition of lands bordering tidal waters capable of providing various types of coastal recreational opportunities.

POS has required that in order to receive funding each county must submit a comprehensive recreation and open space plan. This plan must include waterfront recreation facilities planning and determination of the potential for additional public coastal recreation opportunities. POS has recently required local governments to submit by June 1, 1981 a new complete recreational and open space plan or an update which describes the local governments plan for the next ten years regarding recreational and open space acquisition and development. The utilization of the data provided by the Bay Access Park Study will play a vital part in the counties meeting this request. The CRD has also offered its technical assistance to local governments in meeting this requirement and the local governmental staff funded by the Coastal Zone Management Program will assist local recreational planners in addressing this requirement. Once developed, CRD will review the plans with particular attention to those sections that deal with shorefront access, waterfront facilities planning and the potential for providing additional public coastal recreational opportunities.

Waterway Improvements Program

The Waterway Improvements Program which administers the State's Waterway Improvements (WWI) Fund was established to meet the demand for improved recreational boating safety, services, and facilities. The WWI Fund finances the following activities:

a. marking channels and harbors and establishing navigation aids, in cooperation with the United States Coast Guard;

b. clearing obstructions, such as vegetation and debris, from waterways;

c. constructing marine facilities beneficial to the boating public;

d. dredging channels and harbors and constructing jetties and breakwaters in cooperation with the U.S. Army Corps of Engineers;

e. altering or removing bridges that interfere with navigation;

f. evaluating water-oriented recreation needs in Maryland; and

g. developing plans for waterway improvement.

The Waterway Improvements Program of the Maryland Department of Natural Resources annually receives $125,000. The Fund is financed by a 5% boat sales tax, boat title fees, and boat manufacturer or dealers license fees. These funds may be used without obtaining legislative approval on small projects not exceeding $5,000. For projects greater than $5,000, legislative approval is required. Projects up to $25,000 can be funded entirely by the WWI Fund, but for projects requiring larger sums of money, funds are available only on a 50% matching basis. Exceptions are made for projects involving dredging and marking channels and harbors, constructing jetties and breakwaters, and clearing debris and vegetation from navigable waters. These projects will be funded entirely by the WWI Fund.

The State Boat Act which established the WWI Program has also given authority to local governments to establish waterway improvement districts. In established districts, projects may be undertaken to improve boating facilities, for which WWI will provide technical and financial assistance for engineering design and project supervision. Interest free loans of up to $75,000, amortized over a 25 year period, may be made to the governing body for the district. All other costs must be met by the governing body. A special assessment or tax may be levied against each property owner whose property lies within the district.

Through the Fund many boat launching ramps and similar facilities have been constructed in Maryland's coastal areas. In recent years, considerable attention has been given to using WWI funds to construct boating facilities as part of state or local park development programs. It is also anticipated that WWI funds will be used to provide boating facilities and waterway improvements associated with specific CEIP projects.

Public Participation and Awareness

Public awareness of available facilities is an important part of improving the utilization of existing public and private waterfront facilities. Presently, Maryland's efforts include an in-depth inventory of boat launching ramps and their corresponding facilities and services as well as an inventory of private and public waterfront recreation facilities. A set of brochures identifying these sites, their locations and fees is being designed. This effort is a communicative attempt to make both Maryland residents and tourists aware of existing access opportunities on the Chesapeake Bay and its tributaries.

Public participation is also a very important element in the overall process of providing shorefront access in Maryland. Maryland's CZMD has a unique group, the Coastal Resources Advisory Committee (CRAC), which consists of more than 90 private individuals, government representatives and private company personnel. This group has a very prominent role in the coastal zone management process within the State. They provide input into the budget process, review proposals for new projects and aid in the annual Federal evaluations of the CZM Program.

Having a citizen-oriented group allows for input, questioning and comment on a broader, less technical level. This process assists the CRD staff in maintaining a comprehensive, yet understandable, communication network with Maryland's general public.

With regards to shorefront access, CRAC has established a Recreational Boating Task Force which deals specifically with boating issues. Thus far, the Task Force has been mainly involved with a boating utilization study, conducted cooperatively with the University of Maryland, which should result in a comprehensive picture of boating activities and patterns on Maryland's tidal waters. The Task Force was involved in the Scope of Work development and will ultimately review the draft report and make recommendations for use of the study's results.

The Task Force is also involved in the development of a comprehensive management plan for one of the heavily utilized rivers on the Eastern Shore, the Sassafras River. This is the initial stage in a full-scale attempt to define and enforce recreational boating management policies for the State.

In summary, Maryland's efforts to improve shorefront access include in large part the coordination of several state, local and federal programs. The problem of public waterfront access is only one of many faced by coastal states. In the case of Maryland, it has begun to receive increased attention in an effort to ensure continued access to the Chesapeake Bay for both the present and future water-oriented public.

This paper is thus a description of how one state proposes, through the above measures, to implement the Shorefront Access Element it developed to meet the requirements of the 1976 amendments of the Coastal Zone Management Act and the possible application of its approach in other states.

Preserving Public Access: The Bay Experience

by

Nancy Wakeman Twiss

Coastal Analyst

San Francisco Bay Conservation and Development Commission

The coastal management agency for San Francisco Bay is the San Francisco Bay Conservation and Development Commission (BCDC). Over the last 15 years BCDC has required that about 50 linear miles of shoreline be set aside for public access purposes as part of its permitting process. Attractive, permanent public access to the shoreline of San Francisco Bay has been provided by the Commission through a strong legislative mandate, design criteria for public access that are flexible, but well defined, and a consistent administrative review process.

This paper will discuss how BCDC has utilized its powers, design criteria and administrative structure to provide public access. The Commission's legislation and plan policies are discussed to illustrate the legal basis for requiring access areas. The paper also covers the way project design, designers and the staff work to achieve a quality product.

The Importance of a Clear Legislative Mandate

BCDC was the first coastal management agency in the nation, created by the California Legislature in 1969. The Commission's enabling legislation, the McAteer-Petris Act, states that the Commission shall regulate filling, dredging and construction within the Bay and within 100 feet of the shoreline of San Francisco Bay. BCDC may not approve any development which does not provide maximum feasible public access consistent with the project. Thus, the Commission clearly has the police power to require public access as part of any development.

By comparison the California Coastal Commision is an agency which also has a legislative mandate to require access, but has less definitive authority and has difficulty getting public access areas constructed. One difficulty is that one California Coastal Plan policy states that access areas shall not actually be required to be opened to the public until a public agency or private association agrees to accept responsibility for maintanance and liability. Few jurisdiction's have been willing to accept dedication of access areas and thus few have been constructed.

The Importance of Reasonable Access Requirements

Another important aspect of the Commission's legislation is that the law defines what constitutes a reasonable access requirement. By requiring that "maximum feasible" public access be consistent with the proposed project the Legislature provided a test of reasonableness for the administrative agency to use. In total, approximately 500 permits for major projects have been issued by the Commission. These have included office, commercial, large residential development, and industrial development. Generally about 15 percent of each site has been reserved for shoreline access. The width of access on each site varies greatly, generally ranging from 20 to 200 feet. The width and design of the area is dependent mainly on the height and bulk of adjacent structures.

In one instance the Commission's public access requirements were not upheld. In 1966, the Commission required complete access along the shoreline of an existing marina for the reconstruction of a restaurant that predated the McAteer-Petris Act and that had been destroyed by fire. In subsequent litigation the Court struck the public access requirement from the permit. Although the Court did not so state, in this instance the Commission may have overreached its authority by requiring a very large new public access area, without adequate demonstration that the new public access was consistent with the reconstruction proposed.

Public Access Criteria

Guidance regarding amount and type of access is found in the San Francisco Bay Plan and in the Public Access Guidelines. Maximum feasible public access must be provided in all new major developments except where it is clearly inconsistent with public safety. In such rare cases, access must be provided at an adjacent site. Access areas are permanently guaranteed. This means that a permanent restriction in favor of the public must be placed on the area proposed for access. The Commission has had guidelines for development of public access areas since its inception. These include suggestions regarding clustering of development, placement of structures relative to the shoreline, taking advantage of a Bayside location, and providing visual access. In addition, access must be connected to the nearest public thoroughfare. The Bay Plan states that access area improvements should be consistent with the project, be designed to encourage Bay-related activities, and should include an ongoing maintenance program.

Administrative Framework for Achieving Public Access

A competent staff is critical to achieving reasonable and well designed public access. The staff must be thorougly familiar with the legislation and plan and act solely in response to the written materials. These, not personal opinion or good intentions, define the limits of the police power. Let me site an example of well-intentioned but improper staff work. Most projects submittted to BCDC are designed to maximize economic return. Every square foot of land is covered by parking or structures. New staff members usually "feel" these projects are too dense. BCDC does not have the authority to control density, only public access.

The phrase "maximum feasible public access" has been interpreted by the staff to mean that improvements and maintenance mentioned in the Bay Plan policies should be installed as part of the project. Improvements are routinely required as a condition of approval of a permit. The staff interpretation has not been tested in the court. Thus, administrative procedures and the staff's ability to reach creative design solutions with the cooperation of the developer is a critical part of getting public access areas constructed. Developers need to be shown how the provision of public access can enhance development in the case of commercial projects, and that access need not interfere with development in the case of residential or industrial developments. Staff requests for access must be reasonable and consistent with past projects of like type and size.

Assuring continuity as well as quality design is a job both for the staff and the Design Review Board (DRB). Our staff includes a full time design analyst, trained in architecture or landscape architecture, to work with developers and to review working drawings before construction. He reviews projects to determine how much access is feasible for a particular shoreline and what type of improvements are most appropriate.

The Design Review Board is made up of prominent Bay Area architects, planners, landscape architects, and engineers. They meet once a month to review proposed projects. Many members of the Board have taught in universities around the country and thus have an effect that was not anticipated when the law was passed. They have a subtle positive influence on former students, who now represent developers that have projects within the Commission's jurisdiction. The DRB also advises the developer's designers and the BCDC staff regarding maximum feasible public access and the quality of improvements to be placed in public access areas. The DRB routinely encourages developers to provide the following: salt tolerant plant species, signs to indicate public access areas, all-weather pathways, trash containers, and permanent maintenance programs. While designs vary from location to location they are in most instances quite simple. Improvements consist mainly of paths, landscaping, signs, and benches. Parking is usually located away from the shoreline. Widths of areas depend on the height and bulk of adjacent structures.

Examples of Problems in Implementation

Even though the law, design guidelines and administrative structure exists, the Commission has not always been able to provide attractive public access areas. From experience we have learned what the common problems have been and these are discussed in detail below. Problems generally occur when: (1) public access areas appear to be solely a private extension of building landscapes rather than for public use, (2) inappropriate designs and uses are proposed in the shoreline, (3) there is no separation between public and private uses, (4) areas are difficult or expensive to maintain, and (5) there is distinct competition between land use proposed and public access.

1. Public Areas Not Designed for Public Use

A theme restaurant was built in a city south of San Francisco. The Japanese garden which was designed as part of the shoreline access appeared to be a part of the restaurant rather than for public use. As time passed the walkways through to the shoreline were blocked by trash containers, utility structures, and parking since even the employees were not aware that the areas were anything other than landscaping for the enjoyment of diners looking out the restaurant windows toward the Bay. The Commission is presently negotiating with the owner of the restaurant to remove unauthorized structures.

2. Inappropriate Design

Some design solutions are clearly inappropriate for the shoreline. Some years ago a park was proposed for the shoreline in Oakland which lies between a residential and an industrial area. The site chosen for the park had traditionally been used for fishing. The architect proposed a large space frame structure over a concrete area containing ten picnic tables. No facilities for fishermen were included in the plan. The structure provides shade in an area which is somewhat windy and where the sun would be most welcome. The area is now seldom used.

3. No Separation Between Public and Private Use

A restaurant which was built over the water in Tiburon was required to reserve deck areas on the exterior of the restaurant for the public's use to view the water. The restaurant deck and the deck for public use were at the same level and no separation between the two was provided. The Commission's enforcement staff continually receives complaints that the restaurant's tables are being placed in the public area. The problem will not be resolved unless a physical barrier can be constructed which will discourage this encroachment.

Encroachment by adjacent uses is common where no physical barrier is constructed. This often occurs where yacht sales uses are placed in close proximity to public access ways, especially where no grade separation or barrier is constructed. Trailers with yachts are rolled into the access areas which simply become expanded sales lots. In new projects the Commission requires physical barriers as separators.

4. Maintenance

Quite often developers propose landscaping that is too fragile or too intricate for shoreline sites. Landscape plans also often include exotic plants rather than native and salt tolerant plant materials. These are difficult to maintain and generally require the installation of an irrigation system.

Other developers often propose the most inexpensive materials. Public access constructed as part of an industrial development provides an example. Plant materials were not self maintaining and pathways were constructed with crushed rock. The area is now overgrown and unkempt. Enforcement problems often result when this type of landscaping plan is approved. The staff now has more rigorous landscape review procedures especially regarding landscape and construction materials.

5. Land Use Competition

The largest and most talked about commercial development on the San Francisco waterfront is Pier 39. A large deteriorated finger pier which was formerly used for maritime purposes was demolished and a new pier was constructed for commercial use. As part of the project a public accessway around the exterior of the pier and a four-acre park was constructed along The Embarcadero. The park is the entry to the project and being the first and most visible part of the project is naturally the place where the developer feels most inclinded to place advertising, signs, and ticket booths. As a result, the area which was originally designed as public access has slowly evolved so as to be a part of the commercial development.

Future Maintenance Problems

The major problem the Commission will face in the future is the lack of resources in local jurisdictions to maintain public access areas. Tax initiatives in California have had an immediate effect on recreation and parks department funds. Shoreline access areas have been among the first to suffer. These access areas are spread out and thus costly for a public agency to maintain. As revenues are decreased, these areas become more and more degraded from lack of maintenance. The Commission has no source of funding for maintenance and very few other sources for maintenance exist.

Maintenance by homeowners' associations is one alternative. The Commission's experience with this, however, has been that homeowners have very little interest in the public using shoreline areas within a residential project. As a result they often neglect to maintain these areas or they erect barricades to preclude public access.

In some cases it is possible to establish a special district, one of whose responsibilies is the maintenance of public access areas. This has been done for areas where shoreline erosion problems have occurred and maintenance could easily extend to public access improvements. It is essential in this case, however, for the regulatory agency to exercise some legal control since a special district can be voted out of existence.

Enforcement Action

One solution which the Commission has investigated in order to lessen future difficulties has been to use its enforcement program to generate funds for maintenance. The Commission established an enforcement program in 1978 and is presently reviewing all previous permits for compliance. In cases where violations occur it is possible that fees could be collected in lieu of providing additional access if the Commission finds no other administrative remedy. The Commission has the ability under its legislation to give grants. The monies collected as a result of enforcement actions could be placed in a fund which could be used to provide maintainance grants to local jurisdictions.

Acquisition by Conservancies

The California Legislature has established a conservancy for the California coast that is enpowered to acquire or to grant funds to local agencies to acquire public access to significant coastal resources. The same type of conservancy is proposed for San Francisco Bay. Grants to local government from the conservancy would include funds for construction and maintenance of public access areas. Enactment of legislation for San Francisco Bay would enable the Commission to identify areas of significant public access value and recommend their acquisition or maintenance by a conservancy.

Conclusion

Coastal agencies throughout the nation will experience similar difficulties in preserving public access. For states that are drafting new legislation it is important to include specific language requiring public access. The legislation should also define what reasonable public access would be for a project. The language need not set standards but as is the BCDC case, merely state access must be consistent with the project. Language requiring improvements and maintenance would be an improvement on BCDC's legislation.

Agencies preparing plans should ensure that there is consistency between enabling legislation and plan policies. Policies should provide the direction for implementation, not to constrain it.

Another important aspect of sucessful implementation of coastal legislation is a competent permit staff. The staff must be thoroughly familiar with the legislation and plan and act solely in response to the written material. In order to be credible in the eyes of a developer, the Commission and the court, if it comes to that, the staff must learn to confine its comments to the interpretation of the agency's legally defined authority. New agencies should strive to employ an interdisciplinary staff, consisting of planners, lawyers, and designers who have had some pervious regulatory experience.

For agencies already involved in regulation, permit authorization and conditions should not only describe the project in detail, but also describe what areas are required for access, how and when it must be constructed and who is responsible for maintenance. To that end BCDC permit conditions routinely require that public access areas be legally restricted and improvements be in place prior to use of the authorized facilities. Maintenance by the permittee is routinely required.

The problems highlighted in this paper are not unique to San Francisco Bay, nor are they insurmountable. Public access to the coast can be preserved with adequate legislation, design review and staffing.

HURRICANE IMPACT AT DAUPHIN ISLAND, ALABAMA

by

Shea Penland, Dag Nummedal, and William E. Schramm
Department of Geology
Louisiana State University
Baton Rouge, Louisiana 70803

ABSTRACT

Hurricane Frederic made landfall near Pascagoula, Mississippi at midnight, September 13, 1979. At the time of landfall, the central pressure had dropped to 946 mb; onshore winds in excess of 200 km/hr were lashing the Alabama coastline, and the open coast storm tide peaked at 365 cm at Gulf Shores, Alabama.

Vertical aerial photography obtained in 1976 and again 9 days after Frederic made landfall, combined with multiple reconnaissance overflights and ground surveys by the authors, provided the data base for determination of shoreline erosion and the distribution of hurricane scour and sedimentary deposits.

Erosion of the Gulf beach at Dauphin Island proved to follow a predictable pattern, controlled by nearshore bathymetry, whereas retreat of the shoreline of the Mississippi Sound margin was an unexpected occurence. Apparently this retreat was due to a hydraulic jump as washover currents entered the deep water of Mississippi Sound. Large-scale sediment redistribution on Dauphin Island proper was a consequence of the storm surge flood. The ebb surge, however, was responsible for the reopening of three inlets across Little Dauphin Island.

The wave-induced property destruction on Dauphin Island was most intense immediately west of the area of high dunes. This segment of the island, the easternmost portion of the Holocene spit, has been breached twice in this century. Wave refraction analysis demonstrates that this is an area of wave energy focusing. Therefore, during future storm events, breaching, or at the very least severe property destruction, in this area seems inevitable. A sensible land use plan for Dauphin Island should include a search for alternative, and potentially safer, areas for development.

INTRODUCTION

During the 1979 hurricane season, six tropical cyclones were spawned in the North Atlantic; four of these made landfall in the United States, three of which were of hurricane strength (fig. 1). Hurricane Frederic

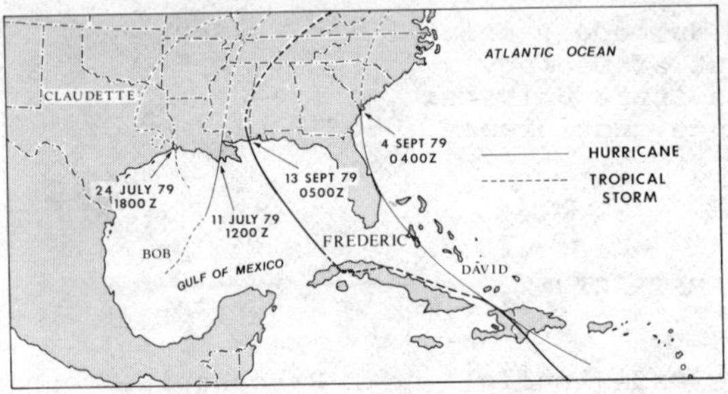

Figure 1. Tracks of the four tropical cyclones which made landfall in the United States during 1979.

made landfall near Pascagoula, Mississippi after crossing over Dauphin Island, Alabama at 00:00 CDT, September 13, 1979 (fig. 2). Total property damage in the coastal counties of Mississippi, Alabama, and northwest Florida was estimated at near 2 billion dollars (Wall Street Journal, Nov. 23, 1979), making Frederic the costliest hurricane disaster ever to hit the United States.

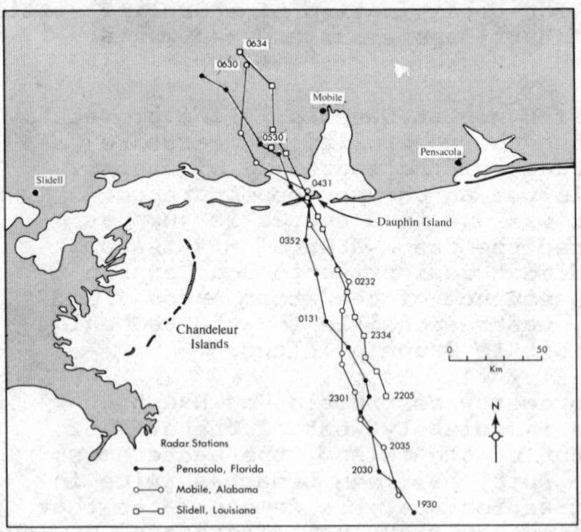

Figure 2. U. S. Weather Service radar positions of hurricane Frederic, between the evening of September 12 and the morning of September 13, 1979, Greenwich Time. (Data from: New Orleans Area Weather Service Forecast Office).

The objectives of this paper are: (1) to describe the changes in the physical environment of Dauphin Island caused by Frederic, (2) to assess the factors responsible for variations in shoreline erosion along Dauphin Island's beaches, (3) to review pertinent aspects of historical hurricane impacts on the upper Gulf Coast, and (4) to use this integrated information on hurricane response to evaluate the presently established pattern of island land use. It will be shown that the residential development which has occurred on Dauphin Island over the last 25 years is environmentally unsound, and very vulnerable to storms. An analysis of all of Dauphin Island suggests alternative and safer sites of development.

EVOLUTION OF UPPER GULF COAST BARRIERS

Otvos (1970, 1979) suggests that the Mississippi/Alabama barriers originated through vertical shoal-bar aggradation (the deBeaumont theory) some 3 to 4 thousand years ago. Non-barred, open, marine nearshore sediments underlie the islands. The location of the barriers is related to the presence of the Pleistocene core at the east end of Dauphin Island, which may have controlled the formation of the incipient shoals further west. Once the shoals became subaerial barriers, they migrated westward by erosion of their updrift end and spit growth on the downdrift end.

Two distinctly different barrier island morphologies have developed: low-profile and high-profile islands. Generally, low-profile barriers are transgressive sand bodies, whereas high-profile barriers are regressive. Low-profile barriers are characterized by (1) narrow widths, (2) low, irregular dunes, and (3) high washover density. High-profile barriers, in contrast, are (1) wide, (2) high, with well defined fore-dunes and multiple parallel accretionary ridges, and (3) have few washover features (Morton and McGowen, 1979).

Dauphin Island is a combination low- and high-profile barrier (fig. 3). The eastern end of the island consists of a Pleistocene core 5 km long and 2.6 km wide, with large landward-migrating sand dunes up to 14 m high. This higher portion of the island is vegetated by a dense stand of pine. Prior to *Frederic*, the 19 km long Holocene spit west of this core was characterized by an almost continuous washover terrace with small discontinuous dunes. The barrier width ranged from 300 to 600 m. The other barrier islands on the upper Gulf Coast, Petit Bois, Horn, and Ship Islands, are all high-profile regressive barriers.

The upper Gulf Coast barriers are frequently and extensively modified by hurricanes. A major event was the separation of Petit Bois and Dauphin Islands during the 1740 hurricane (Otvos, 1979). U.S. Coast and Geodetic

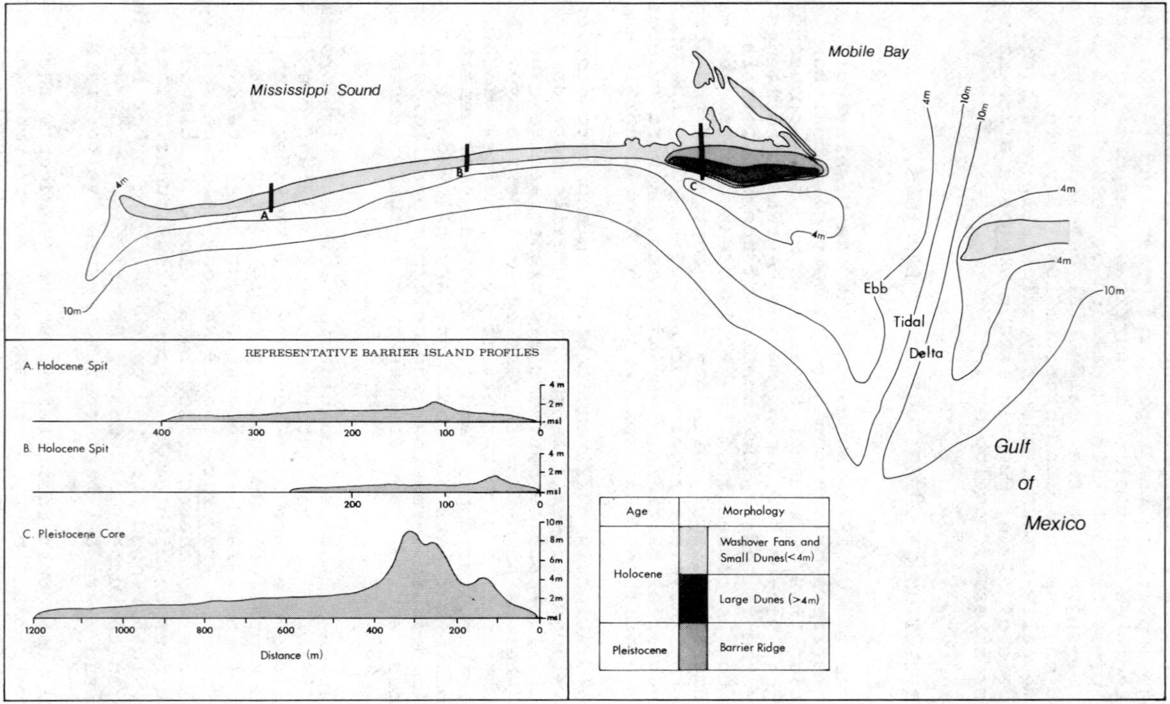

Figure 3. Morphological map of Dauphin Island, Alabama, illustrating the high-profile Pleistocene core and low-profile Holocene spit.

Survey charts from the 1850's indicate that Dauphin Island was breached by the 1852 hurricane. In this century, Dauphin Island has been breached twice. An 8.5 km breach was initially opened west of the Pleistocene core (fig. 4) by the July, 1916 hurricane (Hardin, et al.1976). Hurricanes in October of 1916, 1917, and 1923 helped to prolong the existence of this breach. The last recorded breach took place in 1947. This was a fairly narrow breach (0.5 km wide) which closed quickly. The 1947 breach occurred within the segment originally breached in 1916.

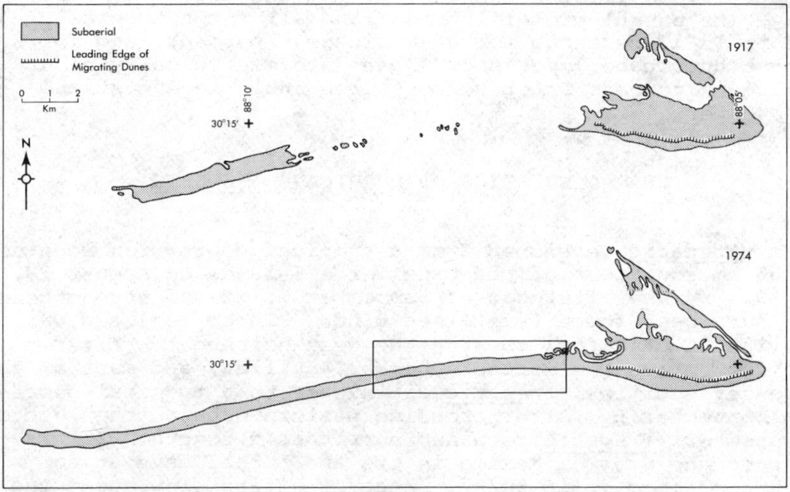

Figure 4. Maps of Dauphin Island in 1917 and 1974. The breach in the island was created by the July 5, 1916 hurricane. The black box encloses the area of extensive residential development on the Holocene spit and is also the location of the area covered by the vertical mosaic in figure 13. (Maps from: Hardin et al. 1976).

According to Waller and Malbrough (1976), Ship Island has been divided by hurricane breaches at least four times in the past 130 years (1852, 1893, 1947, and 1965). Hurricane Camille in 1969 created "Camille Cut," permanently separating the east and west portions of Ship Island. A tidal channel 3.5 m deep developed in this new breach, with an associated ebb-tidal delta. Numerous other examples of hurricane breaching have been documented by Otvos (1979) for Petit Bois (in the 1940's), Ship Island (1916-17), and Horn Island (in the 1700's and 1800's).

HURRICANES AFFECTING COASTAL ALABAMA

After the settlement of Mobile in 1702, the record of hurricanes affecting Alabama is fairly complete. In a 269-year period between 1711 and 1980, 56 tropical cyclones of hurricane intensity have crossed the Mississippi/Alabama coast, or near enough to have affected Dauphin Island (U.S. Army Corps of Engineers, 1967a and 1967b). Table 1 lists the physical characteristics of most storms affecting the region since 1772. Simpson and Lawrence (1971) predict that between Biloxi, Mississippi and the mouth of Mobile Bay, the annual probability of landfall for a tropical storm is 13 percent, for a hurricane 6 percent, and for a great hurricane 1 percent. These storms typically occur between June and October, most frequently in August and September.

CHARACTERISTICS OF HURRICANE *FREDERIC*

Frederic developed from a tropical depression located 1050 km southwest of the Cape Verde Islands on August 29, 1979. By the afternoon of September 1, it had strengthened to hurricane force (sustained winds 120 km/hr) and was moving at 32 km/hr towards the west-northwest. After moving across the northern Lesser Antilles, and much of the Greater Antilles, *Frederic* diminished to a tropical depression on September 6. After crossing western Cuba on the 10th of September, *Frederic* regained hurricane strength. It traversed the Gulf of Mexico in two and a half days at an average speed of 16 km/hr. Because of its slow speed and huge lateral extent, *Frederic* developed a large storm surge. The peak open coast storm tide during landfall on September 13 was recorded at Gulf Shores, Alabama, where the water reached 365 cm above MSL (fig. 5). This peak storm tide occurred 30 km to the east (right) of the point of landfall.

Storm tides decreased rapidly to the west of the point of landfall. Pascagoula recorded only 185 cm and Bay St. Louis a mere 100 cm above MSL. Figure 5 demonstrates the asymmetry with regard to the observed storm tide distribution at the point of hurricane landfall.

Figure 6 presents synoptic charts of hurricane *Frederic's* circulation between midnight (CST) on September 12 and 0600 (CST) on September 13, 1979. The maps demonstrate the rapid shift in winds over Dauphin Island from northeast via north to west and southwest during hurricane passage. The sequence of geological events recorded at Dauphin Island reflects quite distinctly this sequence of

Table 1

Hurricanes Affecting Coastal Alabama (1772-1979)

Date of Landfall	Landfall Location	Name of Storm	Saffir/Simpson Scale[1]	Lowest Recorded Pressure[2]	Maximum Recorded Wind[3]	Elevation of Storm Tide above Mean Sea Level[4]: Bay St. Louis	Pass Christian	Gulfport	Biloxi	Pascagoula	Bayou La Batre	Coden	Dauphin Island	Mobile	Gulf Shores	Source
4 Sept. 1772	*	*	*	*	*	*	*	*	*	*	*	*	*	249	*	U.S. Army Corps of Engineers, 1967a
23 Aug. 1852	*	*	*	*	*	*	*	*	*	*	*	*	*	243	*	U.S. Army Corps of Engineers, 1967a
11 Aug. 1860	*	*	*	*	*	*	*	*	*	*	*	*	*	195	*	U.S. Army Corps of Engineers, 1967a
15 Sept. 1860	*	*	*	*	*	*	*	*	*	*	*	*	*	213	*	U.S. Army Corps of Engineers, 1967a
30 July 1870	*	*	*	*	*	*	*	*	*	*	*	*	*	213	*	U.S. Army Corps of Engineers, 1967a
19 Aug. 1888	Lake Charles, La.	*	*	*	*	*	*	*	*	*	*	*	*	220	*	U.S. Army Corps of Engineers, 1967a
2 Oct. 1893	Pascagoula, Miss.	*	*	974.1	116	329	*	*	*	*	*	*	*	256	149	U.S. Army Corps of Engineers, 1967b
15 Aug. 1901	Grand Isle, La.	*	2	994.5	145		Mississippi Sound 180-240 cm above normal					*	*	226	*	U.S. Army Corps of Engineers, 1967a
27 Sept. 1906	Mobile, Ala.	*	3	955.4	145	*	*	*	190	*	*	330	*	277	360	U.S. Army Corps of Engineers, 1967a
20 Sept. 1909	Grand Isle, La.	*	4	993.8	105		Mississippi Sound 240-365 cm above normal					*	*	213	*	U.S. Army Corps of Engineers, 1967a
14 Sept. 1912	Mobile, Ala.	*	1	998.6	96		Mississippi Sound 120-150 cm above normal					*	*	135	*	U.S. Army Corps of Engineers, 1967a
29 Sept. 1915	Grand Isle, La.	*	4	952.3	200	360	390	275	275	*	*	304	*	195	*	U.S. Army Corps of Engineers, 1967b
5 July 1916	Gulfport, Miss.	*	3	954.7	172	*	*	131	*	*	*	330	235	330	345	U.S. Army Corps of Engineers, 1967b
18 Oct. 1916	Pensacola, Fla.	*	2	977.8	208	*	*	*	*	*	*	*	*	98	*	U.S. Army Corps of Engineers, 1967a
28 Sept. 1917	Pensacola, Fla.	*	3	969.3	200	*	*	*	*	*	*	*	*	38	*	U.S. Army Corps of Engineers, 1967a
21 Sept. 1920	Houma, La.	*	2	985.6	*		Mississippi Sound 150-180 cm above normal					*	*	*	*	U.S. Army Corps of Engineers, 1967b
15 Oct. 1923	Houma, La.	*	1	995.9	98		Mississippi Sound 150-180 cm above normal					*	*	*	*	U.S. Army Corps of Engineers, 1967b
26 Aug. 1926	Houma, La.	*	3	962.5	112		Mississippi Sound 120-150 cm above normal					*	*	*	*	U.S. Army Corps of Engineers, 1967b
20 Sept. 1926	Pensacola, Fla.	*	4	958.8	245	70	182	182	165	*	*	*	*	137	*	U.S. Army Corps of Engineers, 1967a
1 Sept. 1932	Bayou La Batre, Ala.	*	1	987.0	145		Mississippi Sound 120-150 cm above normal					*	*	137	*	U.S. Army Corps of Engineers, 1967a
6 Aug. 1940	Port Arthur, Tx.	*	2	999.6	82		Mississippi Sound 150-180 cm above normal					*	*	*	*	U.S. Army Corps of Engineers, 1967b
10 Sept. 1944	Mobile, Ala.	*	3	*	*		Mississippi Sound 180-240 cm above normal					*	*	*	*	U.S. Army Corps of Engineers, 1967a
19 Sept. 1947	New Orleans, La.	*	4	969.3	177	463	408	426	338	2.74	250	186	*	143	241	U.S. Army Corps of Engineers, 1967b
4 Sept. 1948	Grand Isle, La.	*	1	993.1	126	180	*	152	170	120	180	180	*	118	*	U.S. Army Corps of Engineers, 1967b
4 Sept. 1949	Grand Isle, La.	*	*	*	*	152	*	*	131	*	*	*	*	119	*	U.S. Army Corps of Engineers, 1967b
30 Aug. 1950	Mobile, Ala	"Baker"	1	983.2	120		Mississippi Sound 120-180 cm above normal					*	*	*	*	U.S. Army Corps of Engineers, 1967a
24 Sept. 1956	Fort Walton Beach, Fla.	"Flossie"	2	983.6	192	82	90	121	100	97	192	192	100	66	174	U.S. Army Corps of Engineers, 1967a
15 Sept. 1960	Pascagoula, Miss.	"Ethel"	1	975.8	145	*	*	150	153	137	*	*	150	118	*	U.S. Army Corps of Engineers, 1967a
3 Oct. 1964	Franklin, La.	"Hilda"	3	965.6	217		Mississippi Sound 120-180 cm above normal					*	107	128	100	U.S. Army Corps of Engineers, 1967a
29 Aug. 1965	Grand Isle, La.	"Betsy"	3	952.1	245		Mississippi Sound 120-180 cm above normal					*	*	*	*	U.S. Army Corps of Engineers, 1966
17 Aug. 1969	Bay St. Louis, Miss.	"Camille"	5	905.0	320	*	689	673	563	387	*	*	*	225	*	Roberts, 1969
13 Sept. 1979	Pascagoula, Miss.	"Frederic"	3	946.0	234	100	90	125	150	185	275	275	310	365	365	U.S. Weather Service, 1979

[1](Neuman et al, 1978) [2]millibars [3]kilometers per hour [4]centimeters * no data

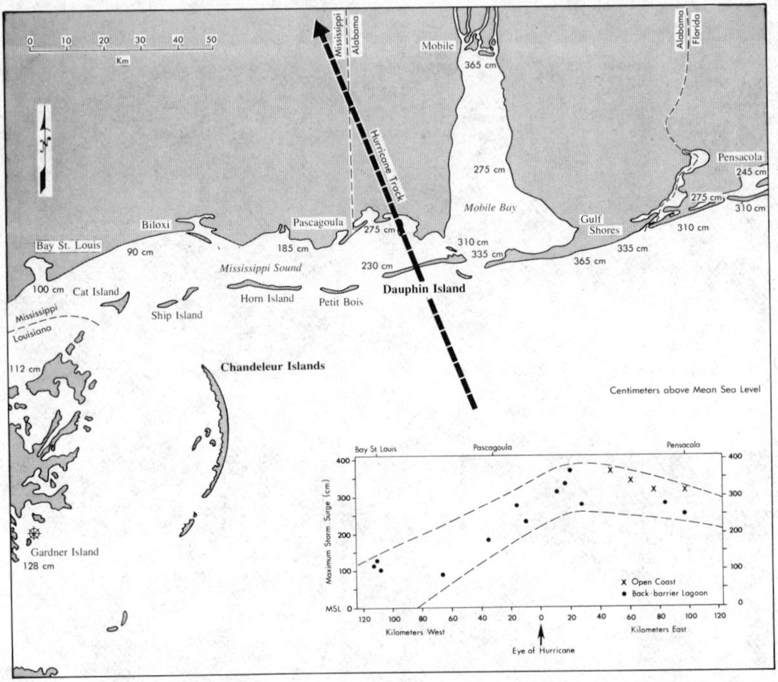

Figure 5: Distribution of maximum storm tide elevations during landfall of Frederic. The inset shows the asymmetry of the storm tide at the point of landfall.

wind change.

NOAA's data buoy EB 42003 located at 26° N, 86° W was directly in the path of Frederic and recorded its deep water wave and meteorological characteristics (fig. 7). This time series records a maximum wave height of 9.1 m coinciding with a maximum wind speed of 118 km/hr. At landfall the strongest winds reached 205 km/hr , associated with a minimum central pressure of 946 mb (fig. 8). The highest winds at the Dauphin Island causeway were recorded at 234 km/hr.

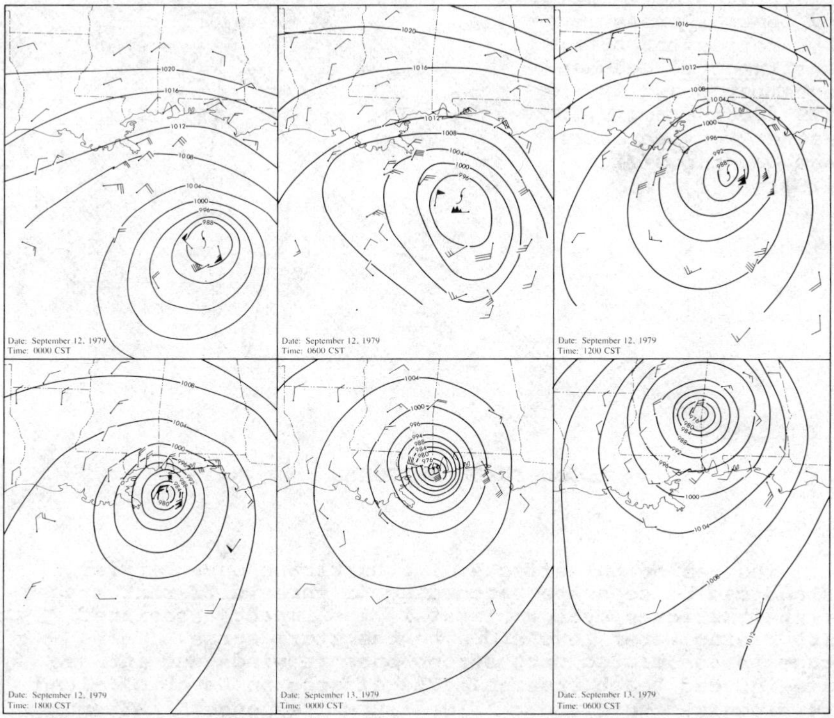

Figure 6. Synoptic surface weather maps of Frederic. (Data from: New Orleans Area Weather Service Forecast Office.)

Figure 7. Deep water time series of Frederic's physical parameters recorded at NOAA's data buoy EB 42003 (from Diez, 1980).

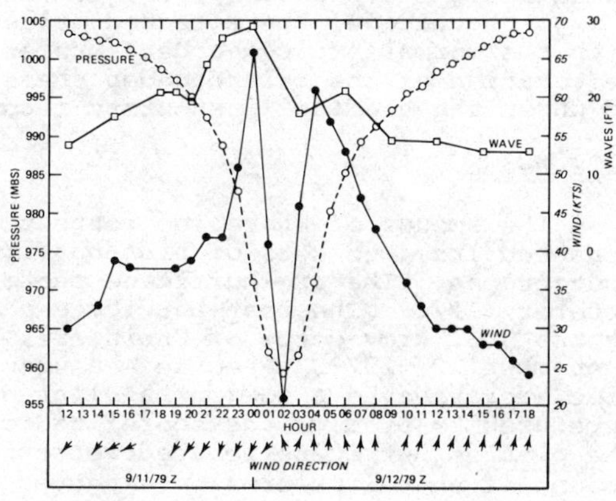

Figure 8. Wind velocity and central pressure time series for hurricane Frederic between September 8 and 19, 1979 (Data from: U.S. Weather Service Hurricane Warning Office, 1979).

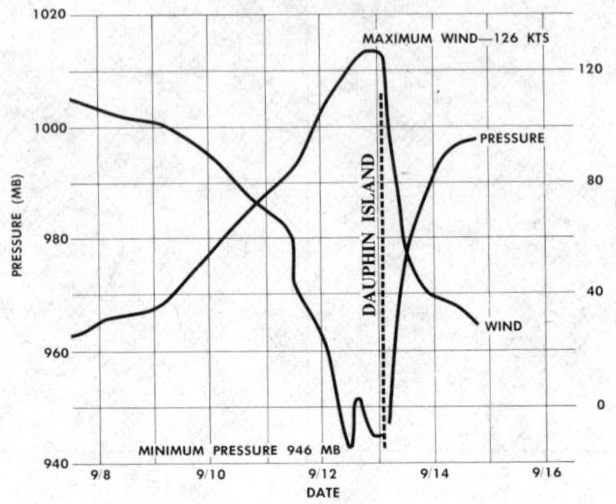

FLOOD-SURGE RESPONSE

General

The geological effects of a hurricane on a barrier island can be separated according to three different stages of the hurricane cycle. First is the impact associated with rising water levels due to the storm surge. This stage is associated with strong onshore winds and intense wave-induced beach erosion. The effects on Dauphin Island are discussed here under "flood-surge response." After landfall, the winds generally blow offshore or alongshore; the water level drops rapidly, and another set of processes, the deepening or cutting of tidal passes, dominates. The specific effects on Dauphin Island are discussed under "ebb-surge response." Finally, for months or perhaps years after a major hurricane, the affected island is out of equilibrium with the normal "fair-weather" processes. The gradual restoration of the island under these conditions is discussed under the heading "post-storm recovery."

Gulf Beach

The amount of shoreline retreat at Dauphin Island was measured from two sets of high-quality vertical aerial photographs. The pre-hurricane photographs are dated October, 1976. The post-hurricane photographs were obtained by the U.S. Army Corps of Engineers, Mobile District, on September 22, 1979. Precise measurements to the high water line, relative to a common baseline on both sets of photos, were used to produce the shoreline retreat map in figure 9. The plotted hurricane-related retreat is corrected for the amount of shoreline erosion expected to have taken place under "normal" processes between 1976 and 1979. Hardin et al., 1976 estimate this value to be 3 m/year, or a total of 9 m.

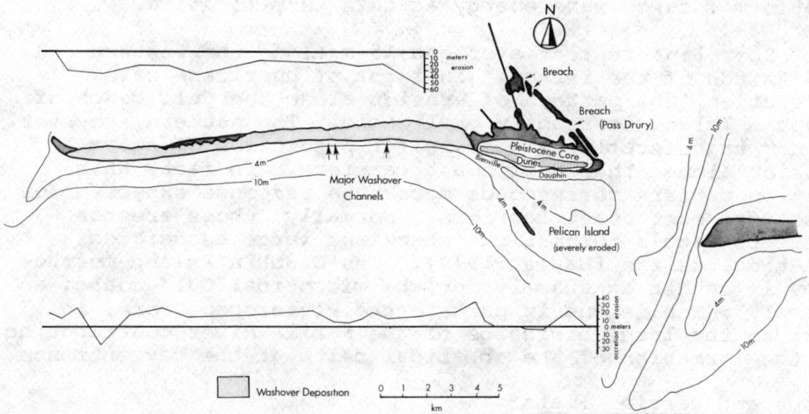

Figure 9. Shoreline retreat values for the Gulf and Sound shores of Dauphin Island.

The amount of shoreline retreat varied considerably along the Gulf beach of Dauphin Island (fig. 9). The least amount of erosion took place near the center of the Pleistocene core at Dauphin and Bienville Beach areas. This small amount of erosion is related to the sheltering effect of the supra-tidal shoals on the west flank of the ebb-tidal delta. These shoals, Sand and Pelican Islands, were nearly leveled to subtidal shoals by the hurricane. These sand bodies greatly reduced the amount of wave energy reaching the shoreline. It appears that hurricane wave refraction around the ebb-tidal delta platform and these shoals may have created a zone of longshore transport convergence at Dauphin and Bienville Beaches. East and west of these beaches, the amount of shoreline retreat was much greater. The increase in erosion along the east end of Dauphin Island reflects an increase in distance and water depth between the shallow ebb-tidal delta margin and the island shore. Based on the track of Frederic, surface weather charts, and post-storm barrier morphology, the peak wave energy that struck the island appears to have arrived from the southeast. The east end of Dauphin Island was openly exposed to hurricane waves coming from that direction. The east end shore retreated about 20 m.

A maximum shoreline retreat of about 40 m occurred

immediately west of the high dunes fronting the Pleistocene core of the island (fig. 3 and 9). At this location, the downdrift margin of the ebb-tidal delta is attached to the shoreface. Refraction analysis for hurricane waves arriving from the SE or SSE suggests that the ebb-tidal delta swash platform focused wave energy at this very location.

Shoreline retreat averaged 15 m along the western two-thirds of the island. In light of hurricane wave refraction, the pattern of erosion along the Gulf beach of Dauphin Island is readily explicable. The pattern, however, is not transferable as a basis for prediction of storm erosion along other Gulf coast barriers. In fact, this erosion pattern corresponds more to a response expected for mesotidal east coast barriers. Normally, those are the barriers with a "drumstick" shape and large associated ebb-tidal deltas (Hayes, 1979). The Dauphin Island morphology is rather an anomaly for the microtidal Gulf coast; an anomaly which is due in part to the Pleistocene core, in part to the large discharge of the Mobile River contributing to the formation of the ebb-tidal delta at the bay entrance.

<u>Dunes and Barrier Flats</u>

The dune complex on the Pleistocene part of Dauphin Island was eroded along most of its front but only locally breached by minor hurricane channels (fig. 10).

Figure 10. Two breaches, now closed by landward bar migration, through low areas in the dune complex at the east (Pleistocene) end of Dauphin Island. (Courtesy of Irving Mendelssohn. Photo taken September 28, 1979.)

The low-profile Holocene spit of Dauphin Island had a typical elevation of 1.5 to 2 meters above MSL (fig. 3). Prior to <u>Frederic</u>, only small discontinuous dunes resting on a nearly continuous washover terrace existed. The response of this spit to hurricane overwash includes the formation of: (a) a continuous washover terrace (fig. 11), (b) large scour holes (fig. 12), and (c) major and minor washover channels (fig. 13).

The continuous washover terrace on Dauphin Island extends across the island to the Mississippi Sound shoreline (fig. 11). The thickness of the sand sheets varies

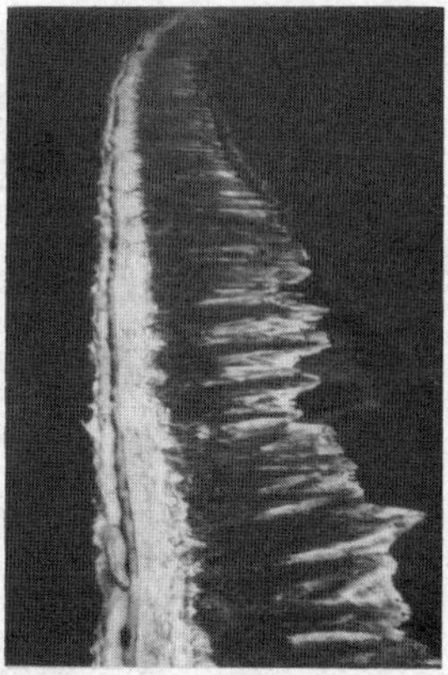

Figure 11. View of washover terraces toward the west end of Dauphin Island, immediately after the passage of Frederic. (Photo taken Sept. 15, 1979.)

Figure 12 (A) (above). View towards the north of scour holes on the barrier flat.
Figure 12 (B) (right). Ground photo of a single scour hole in the washover terrace. (Photo taken June 21, 1980).

from 110 cm to 20 cm and tends to be thicker towards the back-barrier side of the island. From east to west, the washover thickness generally decreases as the relative elevation of the barrier increase , and the height of the storm tide decreased.

Large scour holes, some many meters in diameter and up to a meter deep, had formed in what appeared to be a random distribution across the barrier flat (fig. 12). Pre-Frederic air photos revealed an abundance of similar depression, most likely dating back to earlier hurricanes.

Minor hurricane channels[1] were abundant along the entire island, particularly within the housing development (fig. 4 and 13). Typically, these minor channels are scoured deepest into the old storm berm and then shoal landward. They are oriented at an oblique angle to the beach toward the northwest, reflecting the dominant wave approach during the flood surge (fig. 14). Only three

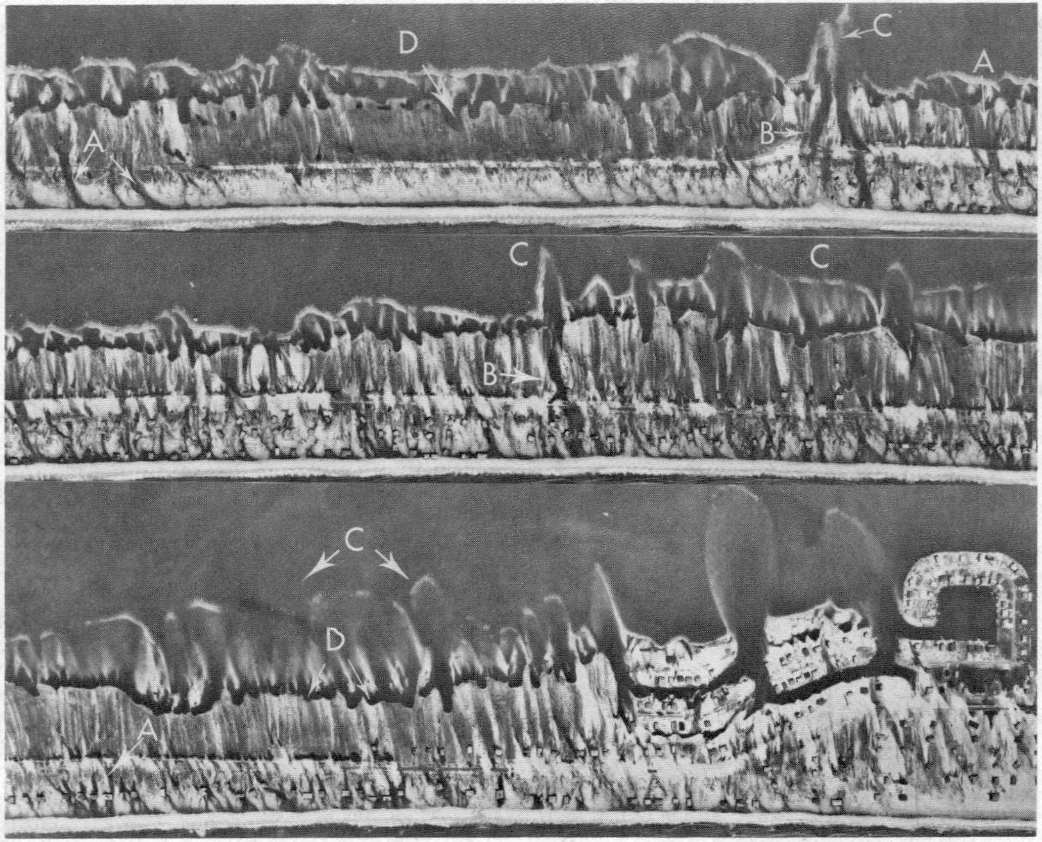

Figure 13. A sequence of vertical air photos, west to east, along the central portion of Dauphin Island after _Frederic_. Note minor (A) and major (B) washover channels in addition to flame-shaped washover fans (C) and back-barrier scour (D). Gulf is at the bottom of each photo, Miss. Sound at top. (Photo taken Sept. 22, 1979) For location of this mosaic, see figure 4.

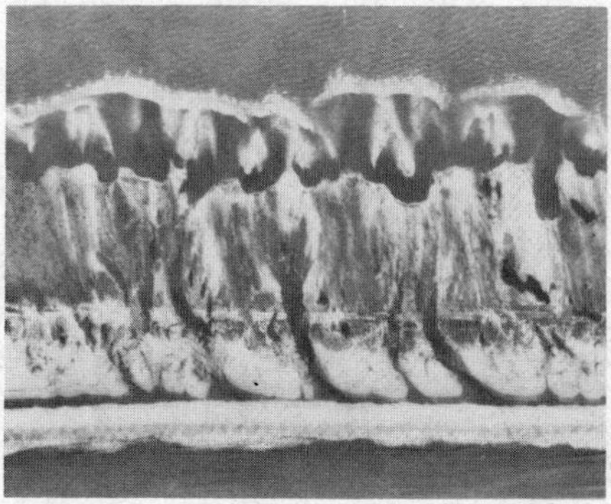

Figure 14. Minor washover channels scoured into the beach crest oriented towards the northwest. (Photo taken on Sept. 22, 1979.)

major hurricane channels[2] developed. These, however, are not natural occurences; they are associated with two pre-existing drainage canals and a dirt road (see figure 13, B').

Back-Barrier Margin

To our surprise, the Mississippi Sound shoreline of Dauphin Island retreated more than the Gulf shore; the average amount of retreat was about 25 m (fig. 9). Back-barrier erosion was caused by an intense hydraulic jump (Chow, 1959) that developed where the shallow sheet of water flowing across the barrier flat in an upper flow regime entered the deeper water of Mississippi Sound (Schramm *et al*. 1980). The scouring action of overwash is responsible for shoreline retreat and for the formation of a nearly continuous deep trough along the back-barrier margin with large flame-shaped washover fans extending into Mississippi Sound (fig. 13, 14, and 15). Deep ramps lead from this scoured trough onto the washover fans. The fan size is generally in proportion to the size of the ramp channel. As overwash entered Mississippi Sound, scouring the back-barrier margin, the entrained sediment was deposited in flame-shaped fans, due to mixing of overwash and Sound water and consequent reduction in sediment transport capacity. The higher the velocity of the overwash current, the more elongated the fan, because mixing is most effective along the lateral margins of the jet.

---

[1]Minor hurricane channels are those cut through a beach berm or foredune ridge above MSL.

[2]Major channels are those which are cut below MSL and remain active after subsidence of the storm surge (Hayes, 1967).

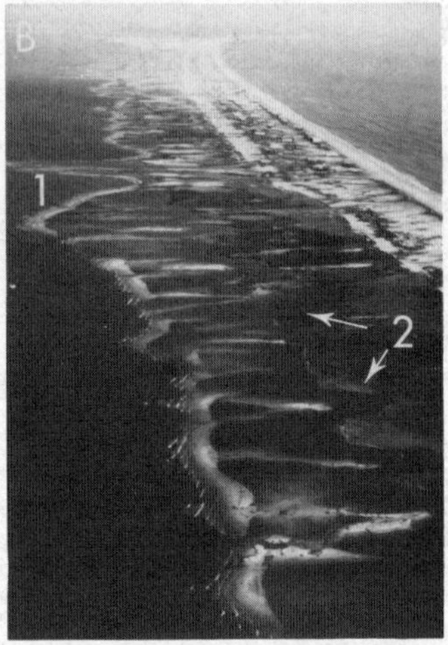

Figure 15 (A) (above). Pre-*Frederic* photo of the back-barrier margin of Dauphin Island. (Photo taken Dec. 17.)

Figure 15 (B) (above right). Post-*Frederic* photo of same area. Note flame-shaped washovers (1) and scour trough (2). (Photo taken Sept. 22, 1979.)

EBB-SURGE RESPONSE

The ebb-surge primarily affected Little Dauphin Island by breaching or reopening three inlets and forming a deep scour trough along the Mobile Bay margin with flame-shaped washover fans extending into the bay (fig. 16).

The reopening of Pass Drury by an ebb-surge flowing from Mississippi Sound to Mobile Bay via Little Dauphin Island Bay is documented by the following data (fig. 17): (1) Immediately after opening, there was a large sand body on the Mobile Bay side of the inlet, i.e. the ebb-tidal delta of Pass Drury. (2) Several channel markers in Dauphin Island Bay, west of the inlet, were bent to the east in the direction of ebb flow. (3) Post-hurricane hydrography in Pass Drury immediately after opening indicated that the inlet is flood dominated, in spite of the existence of the "ebb-tidal delta." However, on Dauphin Island proper, the morphologic evidence indicates that hurricane channels were cut by flood flow. The most common sequence of inlet-cutting through barrier islands is probably one in which the storm-surge flood initiates the breach, and the ebb currents deepen it into a major channel (Hayes, 1967; Greenwood and Keay, 1979).

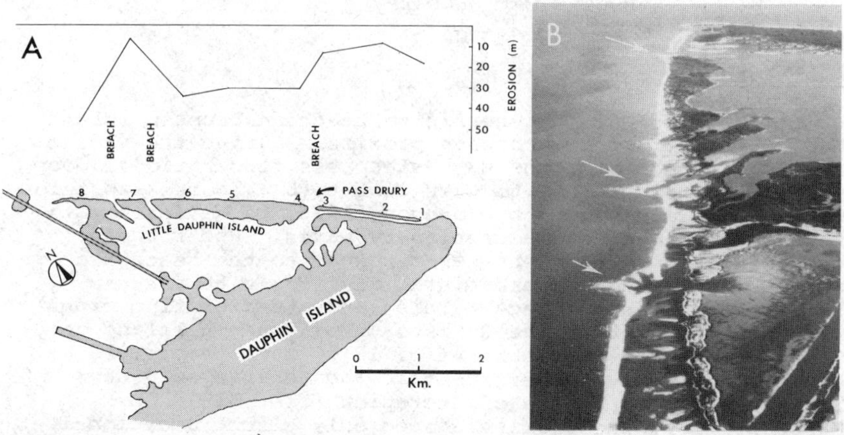

Figure 16(A). Shoreline erosion and breaching at Little Dauphin Island.
Figure 16(B). An aerial view along Little Dauphin Island towards the southeast. Note the flame-shaped sand bodies associated with each newly breached inlet extending into Mobile Bay. (Photo taken March 12, 1980).

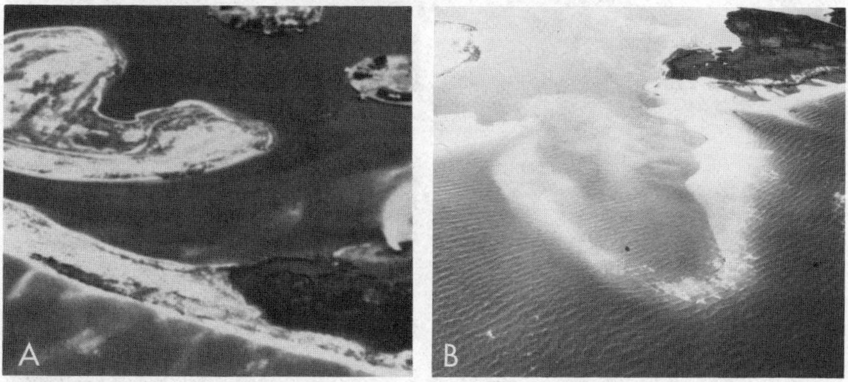

Figure 17(A). Pre-Frederic air photo of the location of Pass Drury before breaching. (Photo taken Dec. 17, 1978.)
Figure 17(B). Post-storm air photo of Pass Drury with ebb-surge oriented sand body extending into Mobile Bay. (Photo taken October 25, 1979.)

POST-STORM RECOVERY

Gulf Beach

The immediate post-hurricane beach on Dauphin Island had a smooth, upward - concave profile. During the waning stages of the storm, the wave steepness, the ratio of deep water wave height (H_o) to wave length (L_o), decreased below 0.025, the value found by Johnson (1956) to differentiate between erosional and accretionary waves. Under these wave conditions, sediment was returned to the beach face in the form of a landward-migrating ridge with a steep landward-dipping slipface. This landward-migrating ridge was quite evident on the Gulf beach of Dauphin Island only 9 days after the hurricane (fig. 18). A beach profile established 3 days after Frederic and resurveyed 8 days later measured 9 meters of accretion (fig. 19). By March 8, 1980, the profile showed only slight additional accretion.

Figure 18. Oblique air photo of a major hurricane channel (A) closed by a large accretionary ridge only 9 days after Frederic. (Photo taken Sept. 22, 1979.)

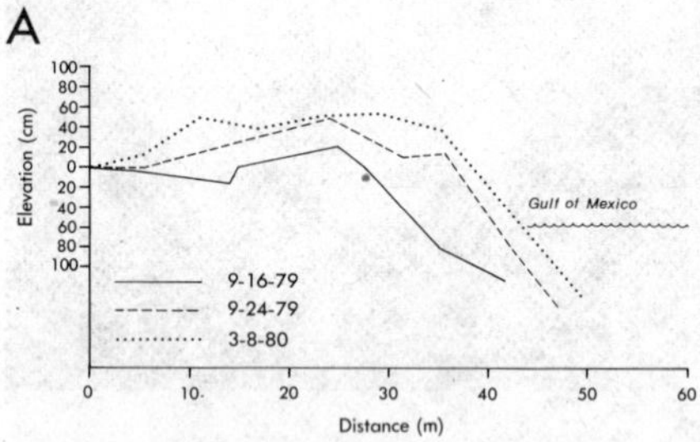

Figure 19 (A). Beach profile illustrating beach recovery after Frederic.
Figure 19 (B). Ground photo of accretionary ridge (light) overlapping the smooth post-storm profile. (Photo taken Sept. 24, 1979.)

Mississippi Sound and Mobile Bay Shoreline

The post-storm recovery along the back-barrier margin of Dauphin Island took place at a slower pace because of the lower wave energy of Mississippi Sound. The termini of the large washover fans were slowly reworked and smoothed. A beach profile established 3 days after Frederic made landfall was resurveyed 8 days later showing only minor change (fig. 20). By March, 1980, considerable change had

Figure 20. Back-barrier beach profile illustrating the reworking of the washover fan margin. (Photo taken June, 1980.)

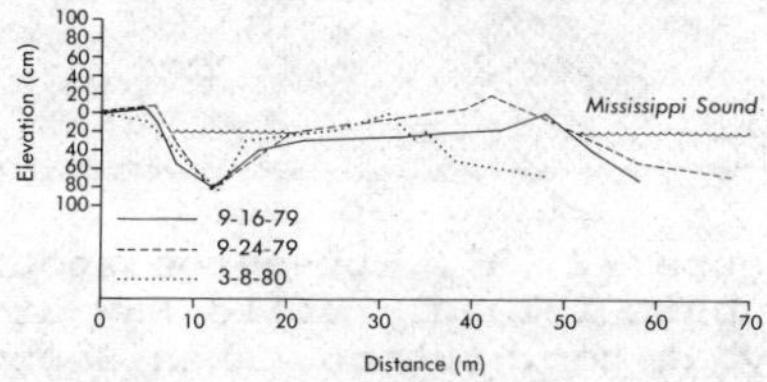

taken place. The flame-shaped washover fans formed by Frederic had been smoothed and reworked landward, forming a nearly continuous ridge along the entire sound side of Dauphin Island, enclosing a series of inter-connected ponds (fig. 21). These ponds fill with fine sediment and provide the framework for the development of back-barrier marshes along Dauphin Island. A similiar pattern of recovery was observed along the Mobile Bay shoreline of Little Dauphin Island (fig. 22).

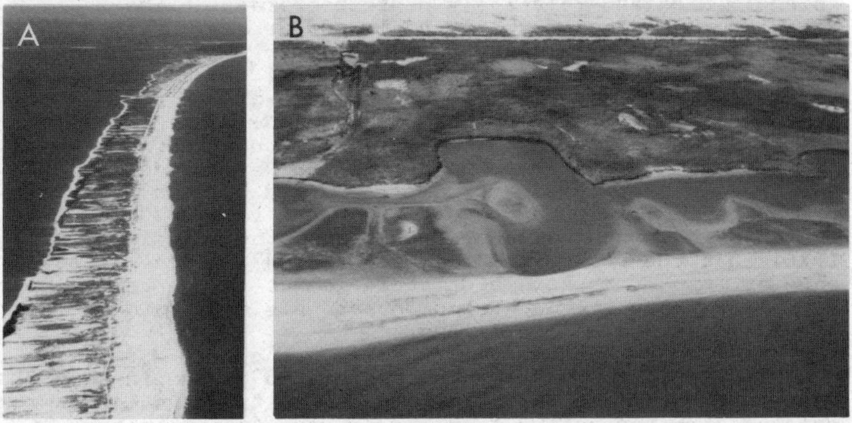

Figure 21 (A). An oblique air photo of the smoothed, reworked washover fan margins forming a nearly continuous ridge.(Mar. 12, 1980).(B) Air photo of a small pond. Note the straight ridge (June 21, 1980).Compare w/ fig. 13 and 15.

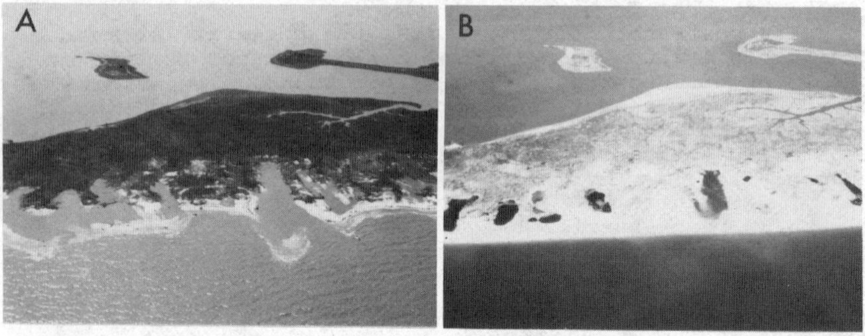

Figure 22 (A). Ebb-surge scour and deposition at Little Dauphin Island. Mobile Bay at bottom and Dauphin Island Bay at top. (Photo taken Sept. 15, 1979).
Figure 22 (B). Oblique air photo of the barrier beach which had welded onto Little Dauphin Island by June 21, 1980. (Photo taken June 21, 1980).

EFFECTS OF MAN-MADE STRUCTURES

Dauphin Island was connected to the Alabama mainland by a causeway across Mississippi Sound in 1955. The bridge connection initiated rapid residential development on the island. High and relatively protected areas on the Pleistocene eastern core of the island were subdivided and made ready for the housing boom. However, the low Holocene spit west of the dune ridge (fig. 4) caught most of the attention of land developers because of the more aesthetic view; it offered views of the Gulf of Mexico unobstructed by any foredune ridges. It was known to the original founders of the Dauphin Island Development Corporation that this particular area had been an open breach between 1916 and some time in the late 1940's. It was also known and documented by air photos (Hardin et al. 1976) that the breach had reopened in 1947. These facts, however, appear to have had little impact on the decision to center the western development at this most vulnerable of all locations on the island.

The intensity of storm-surge related damage in the western development was primarily due to two first-order effects: the refractive focusing of wave energy by the western flank of the Mobile Bay ebb-tidal delta and the low profile inherited from earlier breachings. There were, however, a wide range of second-order scour effects, many of which were clearly directly caused by man-made structures and features.

Shore-normal low features, for example driveways, drainage and boat canals, and marina entrances, acted as conduits for the rising flood waters, became preferentially scoured, and caused the formation of all the major, and most minor, hurricane channels. The presence of strong currents through the marina in the western development is quite evident in the existence of large Mississippi Sound flame-shaped fans immediately behind the navigation channels (fig. 23). The two major hurricane channels at the western extremity of the development (fig. 13) occupy the previous locations of a dirt road and a drainage canal.

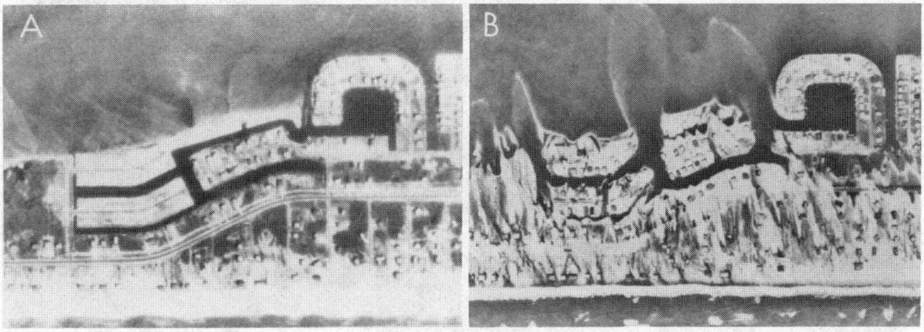

Figure 23 (A). Pre-*Frederic* vertical air photo of marina-style subdivision. (Photo taken October, 1976.)
Figure 23 (B). Post-*Frederic* vertical air photo of same subdivision. Note the beach erosion and breaches through the marina into Mississippi Sound. (Photo taken Sept. 22, 1979.)

House-support pilings induced turbulence in the overwash currents. This, in turn, accelerated scour and produced linear channels (fig. 13, 24) or crescentic scour zones downcurrent of many homes (fig. 25). Areas between many crescentic scours became zones of sediment deposition, locally causing sand accumulations as much as one meter thick (fig. 24).

It should be pointed our that the 1916 and 1947 hurricanes, both of which had smaller storm tide elevations than Hurricane *Frederic* (Table 1), breached Dauphin Island; *Frederic* did not. The reasons for this are unclear, because data from the former hurricanes are scanty. The possibility should not be ruled out, however, that there might be some element of strengthening of the island with the construction of multiple homes, paved roads, and lawns. The location of the western development is certainly unwise from the point of view of property safety. A development within the island's Pleistocene core, which was

Figure 24. Scour channels and zones of sediment deposition downcurrent of some houses within the western development.

Figure 25. Crescentic scour zones related to "obstacles" on the back-barrier side of the western development.

hardly damaged by Frederic, would have made much more economic sense. However, it does not follow that the sediment budget for the developed portion of the Holocene spit has been changed in an adverse way.

DISCUSSION AND CONCLUSIONS

The many instances of hurricanes breaching Gulf coast barriers, the significant alteration of Dauphin Island documented in this paper, and observations of the multitude of geological alterations on the south Texas coast caused by Hurricane Allen a week before this writing, have convinced the authors that hurricanes play a dominant role in the evolution of barrier island morphology and stratigraphy along the U.S. Gulf Coast. The significant geological effects of hurricanes include the following:

(1) Island breaching and the formation of multiple major and minor hurricane channels.

(2) On low-profile barriers hurricanes normally deposit continuous washover terraces as the barrier flats

and deep erosional channels and fans in the back-barrier lagoon.

(3) On high-profile barriers one generally experiences extensive dune scarp retreat and localized foredune breaching with the deposition of attendant interdune fans (Scott and McGowen, 1975).

The frequency of hurricanes along the Gulf Coast is high enough to cause repeated impacts during the lifetime of any given man-made development project. Therefore, major efforts should be undertaken to try to predict the distribution of different hazard zones on any given island prior to its development. The observations presented in this paper of Hurricane _Frederic_'s impact on Dauphin Island should help identify some of the factors which must be considered in such an assessment. These factors include:

(1) The island topography. On Dauphin Island the eastern high-dune region was cut by only a few minor hurricane channels, whereas the low western Holocene spit was completely overwashed.

(2) The nearshore bathymetry. Maximum shoreline erosion and property damage on Dauphin Island occurred immediately downdrift of the ebb-tidal delta flank. This is a zone of wave energy focusing due to refraction around the tidal delta shoals.

(3) The geometry of man-made features. Roads, canals, and other shore-normal structures acted to localize erosion and created both minor and major hurricane channels. Property damage was especially severe adjacent to these channels.

Hurricane Frederic was a severe hurricane of the magnitude which should be considered in proper planning of Gulf Coast barrier development. Many of its destructive effects could have been predicted, and property damage could have been greatly reduced if an effort had been made to do so prior to the construction of the western development.

ACKNOWLEDGEMENTS

Support for parts of this study was received from the Office of Naval Research (Contract no. N00014-78-6-0612 to LSU), the LSU Foundation, and the LSU Department of Geology. Support for ongoing investigation of Pass Drury has been received from the U.S. Army Corps of Engineers, Mobile District, (Contract DACWOI-80-M-9487). Logistics support from the University of Alabama Sea Lab, Dauphin Island (George Crozier, director) is greatly appreciated.

Rosemary Manty, Robert Gerdes, and Perry Howard lent support both in the field and in data analysis. Robert Muller, LSU, kindly provided meteorological data on Hurricane Frederic. The drafting is by the Cartography Section, School of Geoscience. Typing and editing is by Mary Penland.

BIBLIOGRAPHY

Chow, V. T., 1959. Open-Channel Hydraulics. MacMillian Publishing Co., New York.

Diaz, H. F., 1980. Fall Coastal Climate Review. Coastal Oceanography and Climatology News, 2(2):14-15.

Greenwood, Brian and P. A. Keay, 1979. Morphology and Dynamics of a Barrier Breach: A Study in Stability. Canadian Journal of Earth Sciences, 16(8): 1533-1546.

Hardin, J. D., C. D. Sapp, J. L. Emplaincourt, and K. E. Richter, 1976. Shoreline and Bathymetric Changes in the Coastal Area of Alabama. Geological Survey of Alabama, Information Series No. 51, p. 177.

Hayes, M. O., 1967. Hurricanes as Geological Agents: Case studies of Hurricane Carla, 1961, and Cindy, 1963. Texas Bureau of Economic Geology Report of Investigations 61, 56 p.

Hayes, M. O., 1979. Barrier Island Morphology as a Function of Tidal and Wave Regime. Barrier Islands, S. P. Leatherman, ed., p. 1-27.

Johnson, J. W., 1956. Dynamics of Nearshore Sediment Movement. American Assoc. of Petroleum Geology Bulletin, 40(9):2211-2232.

McGowen, J. H. and A. J. Scott, 1975. Hurricanes as Geologic Agents on the Texas Coast. Estuarine Research, L. E. Cronin, ed., p. 23-46.

Nummedal, Dag, Shea Penland, Robert G. Gerdes, William E. Schramm, Jacob Kahn, and Harry H. Roberts, 1980. Geologic Response to Hurricane Impact on Low-Profile Gulf Coast Barriers. Gulf Coast Association of Geol. Society, v. 30 (in press).

Otvos, Jr., E. G., 1970. Development and Migration of Barrier Islands, Northern Gulf of Mexico: Reply. Geological Society of America, Bull. v. 81:3783-

Otvos, Jr., E. G., 1979. Barrier Island Evolution and History of Migration, North Central Gulf Coast. Barrier Islands, S. P. Leatherman, ed., p. 291-320.

Schramm, W. E., Shea Penland, R. G. Gerdes, and Dag Nummedal, 1980. Effect of Hurricane Frederic on Dauphin Island, Alabama. Shore and Beach, v. 48(3): 20-25.

Simpson, R. H. and M. B. Lawrence, 1971. Atlantic Hurricane Frequencies Along the U.S. Coastline. NOAA Technical Memorandum NWS SR-58. U.S. Dept. of Commerce, National Weather Service, 14 p.

U.S. Army Corps of Engineers, 1967a. Hurricane Survey of the Alabama Coast. U.S. 90th Cong., 1st Session, House Doc. 108, 74 p.

U.S. Army Corps of Engineers, 1967b. Hurricane Survey of the Mississippi Coast. U.S. 90th Cong., 1st Session, House Doc. 99, 92 p.

Waller, T. H. and L. P. Malbrough, 1976. Temporal Changes in the Offshore Islands of Mississippi. Water Resources Institute, Mississippi State University, 109 p.

HAZARD MITIGATION ON BARRIER ISLANDS AND BEACHES

Dinesh C. Sharma

ABSTRACT

From Maine to Texas, the Atlantic and Gulf coast of the United States is laced with over 300 barrier islands and barrier spits. These low-lying unconsolidated land forms are highly dynamic and vulnerable, constantly adjusting to the sea level changes, ocean waves, tides, currents and winds. They protect the bays, estuaries and mainland behind them from direct ocean waves and storm attacks, provide habitats for diverse species of finfish, shellfish, and wildlife, and offer diverse recreational opportunities to the public. During the last two decades, urban development on the islands and beaches has proceeded as though they were the interior mainland. Lack of appreciation and understanding of the dynamic nature of island resources and processes, inadequate building location and construction codes, and failure to recognize the carrying capacity of the barrier islands' natural and man-made resources has caused increased loss of natural resources, human lives, property damages and tax burden on communities.

Coastal hazard mitigation planning has not been addressed by local, state and federal governments in a comprehensive manner. In this paper, recent trends in federal and state programs and policies toward coastal hazard mitigation are identified. Innovative hazard mitigation programs from 18 coastal states are summarized. Comprehensive resource planning and improved coastal building codes developed, adopted and implemented in several local coastal communities are highlighted. Ten specific preventive (non-structural) and corrective (structural) solutions for improving coastal hazard planning and mitigation for local, state, and federal governments are presented. If these modest recommendations are not implemented, a major hurricane or nor'easter striking a developed coast will give new meaning to the term 'natural disaster' in the 1980's. The burden on the nation's taxpayers and individuals to rebuild the devastated areas will be astronomical and disaster relief subsidies unsupportable. The decade of the 1980's offers unique opportunities to minimize loss of resources, lives and property by prudent island planning and management.

KEY WORDS: Barrier islands and beaches, coastal zone management, hurricane disaster mitigation, hurricane and wind resistant building codes, coastal construction setback lines, erosion mitigation and controls, island preservation.

Environmental Resources Consultant, 2750 Rhode Island Avenue, Fort Myers, FL 33901. Tele: (813) 334-0883.

INTRODUCTION

Islands conjure up certain images in our minds -- sandy beaches and vertical cliffs, ocean waves and surf, exotic flora and wildlife, fishing camps, pirate camps or historic forts, desolation and solitude, condo rows and crowded beaches, hurricane storms and destruction and so on. Of all the island types, barrier islands are the most unique, dynamic, fragile, vulnerable, unstable landforms on earth. The Atlantic and Gulf Coast of the United States coast is laced with over 300 barrier islands and barrier spits covering an area of 1.6 million acres with 2,650 miles of sandy beaches (Table -1). They may range in length from several hundred yards to many miles, and in width from a stone's throw to several miles across. Barrier islands can be relatively pristine such as Cape Hatteras, Cumberland, or Chandeleur to highly urbanized Miami Beach, Ocean City, or Galveston Islands.

Composed of unconsolidated sand and shell, barrier islands are constantly evolving, changing and migrating in shape, size and location. They are generally 4-5,000 years old although some islands could be as old as 30,000 years. These islands are fronted on one side by the ocean and on the other by a bay or an estuary. Since they protect the mainlands and bays behind them from constant ocean waves and storm attacks and act as a barrier, they are called barrier islands. Each barrier island is somewhat unique. Hurricanes, storms and a slow but steady rise in sea level at a rate of 1 foot per century are the primary means by which the islands are created and reformed (2,14). Natural erosion rates on the islands range from 2-5 feet per year and may be as high as 10-15 feet per year (8). Beaches and dunes migrate as the shoreline currents, winds and waves undermine their stability. Storms can wipe out whole sections of an island, creating an inlet, and can just as easily close that inlet as sediments, sand and shells are redeposited when the islands are overwashed by storms. The islands literally roll over on themselves through interaction of the rising sea level and periodic overwash. It remains an ecological paradox that change provides an island's only stability. They are not a fixed piece of real estate as are most other landforms.

The fundamental human urge to vacation and live near the ocean has led to a dramatic increase in new ocean-front communities along the Atlantic and Gulf Coast barrier islands. During the last three decades, the population within the narrow coastal zone has been growing three to four times faster than the rest of the national average. More than 53% of the U.S. population lives in the counties and cities within 50 miles of the ocean (32). Between 1950-73, forty barrier islands that had no urban acreage in 1950 had urban acreage by 1973, and over 137,000 acres were urbanized on all barrier islands. The urban acreage on barrier islands increased to 14% of the total area of the islands in comparison to only 3% urban area for the rest of the United States (19). Table -1 shows the status of barrier islands of the United States. Although barrier islands are more prone to flooding, erosion and hurricane damage than the safer interior mainland, local zoning laws generally allow grid-pattern subdivisions with 10-30 dwelling units per acre (25).

While the positive attributes of the coast, such as sun, surf, beach, blue waters, fishing and carefree living have been oversold, the hazards have been underemphasized. The coastal areas are afflicted with distinct arrays of hazards. At one place or another along the coast, the problems of hurricanes, floods, erosion, landslides, tsunamis, earthquakes, snow avalanche or land subsidance are severe and may appear in different combination (38). Since 1940, the average annual fatality figures from extreme events in nature has more or less stabilized due to improved forecasting and warning systems. However, damage to property has increased tremendously (7, 38). The average loss per hurricane has been estimated at $500 million but could range from $4 million to 4 billion (30). The average annual loss due to erosion has been estimated at $300 million (28). Compounding the problems of natural coastal hazards is the fact that about 80% of the people moving to and living on the coast have never experienced a major hurricane or northeaster storm (11). Most people know little about the mechanics and behavior of a hurricane, tend to underestimate its devastating affects, and view the odds of one striking "their" coast as remote and hypothetical (18).

The basic premise of this paper is that barrier islands, beaches and coastal floodplains are highly critical, vulnerable and fragile resources that should be treated with the utmost care if we are to minimize losses of resources, lives and property from incompatible and hazardous development. The management of barrier islands and beaches must have a fundamental goal of conservation, i.e., wise use of the island ecosystems at the highest achievable carrying capacity for human enjoyment and other uses (3). The first priority in coastal hazard mitigation should be public acquisition of yet undeveloped barrier islands to meet our recreation and conservation needs and to minimize future losses from hurricanes and erosion in these areas. Some barrier islands can be maintained at a high level of health and productivity while limited urbanization may be permitted, provided there is effective planning. With effective planning, trade-offs can be accomplished without serious penalties by utilizing innovative growth management programs (4). However, planning and goal-setting are essential for proper island use and mangement and island and beach communities should not wait for Federal Big Brothers or State bureaucracies to take the action or force the initiative. Federal, state and local governments need to re-evaluate their programs which increase hazards to lives and property and are inconsistent with the sound fiscal and environmental policies and long-term public interest and welfare. The next section presents an overview of major Federal, state and local programs for hazard mitigation on barrier islands.

FEDERAL ROLE

A comprehensive strategy to address coastal hazard management in the United States is yet to evolve. Early federal programs were dominated by engineering construction solutions for navigation, flood control, shore erosion and hurricane protection projects under the aegis of the U.S. Army Corps of Engineers. Federally funded or subsidized capital improvement projects treat barrier islands and coastal flood hazard areas the same as if they were the interior mainland. Presently, there are twenty federal agencies with thirty different programs affecting the barrier islands. Most programs encourage and subsidize development and redevelopment in the high hazard zones. In the last three fiscal years, the federal agencies committed nearly $500 million to barrier island development projects, in addition to hundreds of millions in disaster relief and insurance payments (33). Federal programs also encourage public acquisition of barrier islands through the Department of the Interior's conservation programs.

TABLE - 1

STATUS OF THE BARRIER ISLANDS AND BEACHES OF THE ATLANTIC AND GULF COAST

	Island Acreage-Base Year 1973-74 (1)				Sandy Beach Length (miles)(2)				
State	Total	Developed	Protected	Undeveloped Unprotected	Total	I	II	III	IIID
Alabama	28,200	5,200	1,300	21,700	48.3	29.3	3.4	14.9	0.7
Connecticut	2,362	670	1,398	286	18.3	9.7	2.2	6.4	---
Delaware	10,100	2,900	3,100	4,100	24.2	8.8	14.0	1.4	---
Florida	467,710	103,405	233,735	130,570	715.2	325.6	118.3	167.6	103.7
Georgia	165,600	37,000	71,000	85,900	99.3	17.6	54.6	26.2	0.9
Louisiana	41,120	3,980	9,650	27,490	181.5	18.2	55.2	108.1	---
Maine	2,640	1,640	760	240	24.5	15.9	5.2	3.4	---
Maryland	14,300	2,400	10,500	1,400	30.5	8.8	21.7	---	---
Massachusetts	37,600	8,390	19,580	9,630	119.9	53.6	14.5	47.7	4.1
Mississippi	9,500	200	6,100	3,200	36.0	---	26.8	9.2	---
New Hampshire	1,100	700	100	300	8.1	8.1	---	---	---
New Jersey	48,000	22,700	9,600	15,700	106.8	74.4	24.9	5.4	2.1
New York	30,310	11,700	12,260	6,350	155.5	66.6	70.4	18.5	---
North Carolina	146,400	23,700	57,100	65,600	310.0	91.0	123.8	65.3	29.9
Rhode Island	3,660	1,360	850	1,450	26.0	8.3	5.4	12.3	0.9
South Carolina	144,150	20,525	38,725	84,900	156.3	48.1	51.1	45.8	11.3
Texas	383,500	22,850	120,200	240,450	372.9	58.2	134.3	150.2	29.7
Virginia	68,900	1,350	28,660	38,890	108.6	18.6	71.5	18.5	---
TOTALS	1,605,152	242,370	624,610	738,156	2,623.0	861.0	877.8	700.9	183.3

Sources: (1) BIWG (1979), A Plan for Protecting the Barrier Islands, Heritage Conservation & Recreation Service, Washington, D. C.
(2) Robert Peoples, <u>Personal Communications, October 10, 1979</u>, U. S. Fish and Wildlife Service, Washington, D. C.

Categories of barrier islands and beaches thereof defined by current land use and protective status:

Acreage: Developed-more than 1000 acres or 75% urban; Protected-under Federal, State, Local and private conservation use. Undeveloped-no development or less than 10% urban or more than 5000 undeveloped acres.

Miles: Category I-Developed; Category II-Unspoiled, Protected; Category III-Unspoiled Unprotected; Category IIID-Unspoiled, Developing.

It should be noted that the number of islands are different under acreage (295) and beach length (302) category because several islands which are not true barrier islands and barrier spits are included in the later category.

1. PUBLIC ACQUISITION AND HAZARD MITIGATION

About one-third of the Atlantic and Gulf coast beaches are protected from urban development under Federal, State and local government ownership and management (Table -1). Since most barrier islands and beaches, by their nature, are subject to constant erosion, migration and flooding, these hazards can be minimized by public acquisition of all suitable and available barrier islands. During 1934-78, at least twelve comprehensive studies of the U.S. shoreline have been conducted by the Federal government, recommending public acquisition of several barrier islands and beaches (15). During this period, 717,383 acres of barrier islands and beaches have been brought into eleven National Seashore Systems. An additional thirty-one National Wildlife Refuges comprising 388,582 acres have been established along the Atlantic and Gulf Coast (13). The Barrier Islands National Parks Bill (HR-5981 and S-2686) now pending before the U.S. Congress would, if passed, minimize loss of lives and property, reduce the tax burden, and provide recreational areas to our growing leisure-oriented population. Preliminary data show that fee simple acquisition of undeveloped barrier islands by the government for public recreation, conservation and hazard mitigation is five times more economical than federally subsidized development (27).

2. FLOODPLAIN MANAGEMENT AND HURRICANE PROTECTION

The federal strategies for floodplain management have been applied mainly to riverine flood hazard areas. The passage of the Flood Control Act of 1936 was the first major federal involvement in flood control which emphasized structural solutions. Presidential Executive Order No. 11296 in 1966 was an effort in the federal administrative policy to move away from strictly engineering solutions, but it had little impact. Subsequently, the National Flood Insurance Act of 1968 and the Flood Disaster Protection Act of 1973 were passed, with the intent of providing alternatives to structural flood control measures and to encourage non-structural solutions to floodplain management.

The National Flood Insurance Program is presently administered by the Federal Insurance Administration (FIA). The program's stated goals are "to...encourage state and local governments to make appropriate land use adjustments and to constrict the development of land which is exposed to flood damage and minimize damage caused by flood losses," and "to...guide the development of proposed future construction, where practicable, away from locations which are threatened by flood hazards." There is ample evidence that neither of these goals are being met. FIA's minimum structural requirements have somewhat encouraged improved building construction, however, these requirements do not reflect a total awareness of the hazards of building on barrier islands. FIA has just begun to incorporate wave height and runup in the velocity zones of the coastal high hazard areas to revise minimum flood elevations (15). The National Flood Insurance Program covers over $60 billion worth of policies and has become the second largest liability of the federal government after Social Security (16). For each dollar collected in insurance policy premiums on barrier islands, over three dollars are paid out in losses (15). There are no minimum building codes in effect to withstand hurricane winds or wave uplift and horizontal pressures. Further, none of the existing requirements are concerned with the location but only with the design -

mainly elevation. Thus, buildings may meet the minimum FIA requirements but at the same time can be located in highly exposed and hazardous areas and destroyed in hurricanes, only to have the same structure re-erected in the same place, rebuilt with federal subsidies. This is happening on Dauphin Island and Gulf Shores in Alabama after Hurricane Frederic (24).

In spite of the existing laws and Presidential Executive Order No. 11988 and Executive Order 11990 on Floodplain Management and Wetland Protection, barrier islands and floodplains continue to be urbanized in a haphazard manner. The most important reforms that are urgently needed include faithful and vigorous implementation of the existing laws and Executive Orders to minimize short-sighted and costly development on the high hazard areas such as barrier islands (37). The FIA needs to include location as well as structural building codes. Although the existing FIA regulations stipulate protection of "dunes and vegetation" in the velocity zone, implementation of this requirement by local communities has been weak. FIA structural codes need to be strengthened to incorporate hurricane winds, wave uplift and horizontal pressures, and relocation after storms to safer places. It has been shown that the added benefits more than offset the small cost increases (15, 17).

3. DISASTER PLANNING AND HURRICANE HAZARD

The Disaster Relief Act of 1974 authorized FEMA to establish disaster preparedness plans in cooperation with the states. Hurricane evacuation is a critical problem on barrier islands and coastal floodplains. Few quantitative studies have been conducted to date to assess the magnitude of the problem. The Texas Hurricane Evacuation Study and the Lee County Flood Emergency Evacuation Plan are the only quantitative studies on hurricane-flood evacuation. The findings and conclusions of these studies of these studies are alarming. In both studies it was found that due to heavy concentration of population in the narrow coastal zone, in combination with low topography, narrow roads, vulnerable bridges, insufficient hurricane warning systems, and lack of coordination between various agencies, it would not be possibel to evacuate the existing population to safer areas even if they were willing to evacuate (36, 29). In spite of these problems, post-disaster relief programs do not recognize these hazards and continue to subsidize more redevelopment after hurricane and storm damage on the barrier islands. Following the Blizzard of '78 on the New England coast and Hurricane Frederic on Alabama in 1979, intensive redevelopment efforts are underway (24).

During the last twenty-five years, losses to property from major hurricanes have been increasing (Table -2). The federal disaster relief programs exceed $3.36 billion annually. The annual flood disaster relief payments now exceed $2 billion, the Small Business Administration loans $400 million, and the Federal Insurance Administration payments to policyholders exceed $400 million. All other disaster relief payments amount to $600 million (39). Hurricane Frederic was the biggest hurricane "blow" in the U.S. history, and inflicted over $2.2 billion in property damage. It provided a unique opportunity to the federal, state and local governments to implement innovative hazard mitigation programs on barrier islands but subsidized rebuilding efforts are well underway (24).

TABLE -2

DAMAGES AND DEATHS FROM MAJOR HURRICANES IN THE UNITED STATES: 1954 - 1979

NAME OF HURRICANE	TIME MONTH - YEAR	DAMAGES (MILLIONS)	DEATHS (NUMBER)	REGION OF LOSSES
CAROL	SEP 54	$456	60	NORTHEAST
HAZEL	OCT 54	232	94	SOUTHEAST
DIANE	AUG 55	832	180	NORTHEAST
AUDREY	JUN 56	150	556	SOUTHEAST
DONNA	SEP 60	360	13	SOUTHEAST
CARLA	SEP 61	431	46	SOUTHCENTRAL
THE GREAT ATLANTIC STORM	MAR 62	245	33	NORTHEAST
BETSY	SEP 65	1,420	81	SOUTHEAST
BEULAH	SEP 67	169	44	SOUTHCENTRAL
CAMILLE	AUG 69	1,421	256	SOUTHCENTRAL
CELIA	JUL 70	440	13	SOUTHCENTRAL
AGNES	JUN 72	3,000 (EST)	122	NORTHEAST
CARMEN	SEP 74	107	5	SOUTHCENTRAL
ELOISE	SEP 75	229	0	SOUTHEAST
BLIZZARD OF 78	FEB 78	1,087	29	NORTHEAST
FREDERIC	SEP 79	2,200	17	SOUTHCENTRAL

SOURCES:
1. DACY AND KUNREUTHER (1969), THE ECONOMICS OF NATURAL DISASTERS, **YEARS 1954 - 1965**
2. U.S. ARMY CORPS OF ENGINEERS, HURRICANE REPORTS, 1966 - 1978
3. PLATTE, R. AND McMILLAN, G. (1978), THE COASTAL FLOODING IN MASSACHUSETTS DURING FEBRUARY 1978

4. SHORELINE PROTECTION AND NAVIGATION

The U.S. Army Corps of Engineers administers three programs which greatly influence barrier islands. These are: navigational dredging, shoreline protection, and dredge and fill permits. Authority for these programs is found in the Rivers and Harbours Act of 1899, the Flood Control Act of 1936 and the Federal Water Pollution Control Act Ammendments of 1972.

Erosion has been recognized as a major problem in the coastal zone, especially along the urbanized shoreline. It has been estimated that over 15,400 miles of the 37,000 miles of shoreline in the continental U.S. is experiencing significant erosion. Various engineering solutions to "halt" erosion for 2,700 miles of "critically eroding" shoreline would cost over $1.8 BILLION in 1970 prices (31). In the fiscal year of 1977, the Corps of Engineers had $272.8 million in 19 authorized erosion control, hurricane protection and flood control projects on barrier islands and beaches (23). Seawalls, revetments, groins, jetties and breakwaters are costly as well as ineffective in "halting" erosion. Beach nourishment at the Dade County-Miami Beach project will cost $63 million for 10.5 miles of beach (1). Beach nourishment costs over $1 million per mile initially and requires frequent renourishment. Such erosion control projects give a false sense of security and encourage high density development after the projects because the Corps does not require any population or density ceiling as a pre-condition for funding these projects. The cost of beach nourishment would become prohibitive in the future as labor, materials and energy costs soar. During the 1970's, some recognition was given to low-cost structural, non-structural and land use management techniques, but none have been implemented on a significant scale. These include sand fencing, native vegetation planting, building setbacks, dune and vegetation protection, relocation to safer areas and public acquisition of undeveloped barrier islands. In the 1980's the national policies and programs would have to move away from costly structural solutions to preventive non-structural solutions for erosion hazard mitigation.

The Corps of Engineers navigation projects are costly and cause many adverse environmental impacts on the barrier islands in two ways. First, inlet stabilization, dredging and deepening causes severe erosion on the down-drift barrier islands. Examples: stabilization of Moriches and Shinnecock Inlets in New York have exacerbated erosion problems on the south shore of Long Island; jetties at Cape May, New Jersey, have robbed the old resort of its beach sand and forced it into poverty; channel jetties at Ocean City, Maryland, have accelerated erosion of the north end of Assateague Island; Charleston jetties in South Carolina have almost obliterated Morris Island and are blamed for washing away two rows of houses on Folly Beach; the jetty at St. Lucie Inlet, Florida, is causing rapid erosion of the wildlife refuge on Jupiter Island. Second, the disposal of huge quantities of spoil materials from the dredging operation on the barrier islands and tidal marshes causes resource destruction. In the last three fiscal years, the Corps has spent over $88 million to maintain and operate the Atlantic and Gulf Coast Intercoastal Waterways. In some instances, navigation projects are essential, but excessive dredging of the waterways for the benefit of a few boats or vessels is not in the public interest. The dispoal of dredge spoil material needs improvement and careful planning to minimize adverse impacts. In Florida, the Corps encourages placement of dredge spoil from inlet maintenance on the public beaches whenever feasible.

5. PUBLIC INFRA-STRUCTURES AND PROJECTS

Federally subsidized growth-inducing projects on barrier islands include roads, bridges, water supply, waste water treatment facilities, small business loans and economic development grants. The bridge and highway construction programs are administered by the Department of Transportation. DOT administrative procedures do not treat projects on barrier islands any differently than those on the interior mainland area. Bridges are essential to barrier island development because without the easy access, development costs would be prohibitively expensive. In the past five years, the Coast Guard has granted 24 bridge permits for barrier island-related projects and six applications were pending in February 1978. The Federal Highway Administration funds roads if they are part of a planned primary or secondary highway system. During the fiscal year 1976-78, over $37 million in 70% Federal and 30% state or local matching grant monies were provided for development of roads and highways on barrier islands (13).

Wastewater treatment and water supply facilities are funded by the Environmental Protection Agency, the Department of Housing and Urban Development, the Farmers Home Administration and the Economic Development Administration. However, the agency regulations do not designate barrier islands as geographic areas of environmental concern or hazard zones. In most instances, these projects encourage more intensive development after completion because of excess capacity providing for 20-years population growth. During 1972-78, communities with barrier islands received $395 million in grants from EPA. In 1975, HUD granted $54 million to coastal communities with barrier islands. FmHA grants and loans to barrier island communities during 1976-78 exceeded $26 million (13). Implications for hazardous growth on barrier islands are clear. Federal agencies need to reassess their programs affecting barrier islands to minimize hazardous and dangerous development in high hazard coastal zone areas.

6. COASTAL ZONE MANAGEMENT

The National Commission on Marine Science Engineering and Resources identified the coastal zone of the United States as an "area of particular importance, requiring special planning and management" (21). In 1972, Congress enacted the Coastal Zone Management Act (CZMA). The Act and its 1976 Ammendments provide for coordination of all Federal activities in the coastal zone and the appropriation of funds to aid states in coastal planning and research through the office of Coastal Zone Management in the Department of Commerce. Three guidelines apply to hazardous zones: First, the management programs must establish permissible activities within the zone; second, the management program must designate critical environmental areas; and third, the management of the program requires promulgation of guidelines on priority uses in the zone. Under Section 315(1) of the Act, funds are available to aid the states in acquiring, developing, and operating estuarine sanctuaries. Through 1979, five sanctuaries have been established at a cost of $5.9 million. Section 315(2) of the Act provides for grants to states to acquire land to provide public access to the beaches and for the preservation of the barrier islands. As of 1980, no funds had been appropriate for Sec. 315(2) of the program. However, the CZMA and its administrative regulations still treat coastal hazard management incidentally and coastal programs which tie the goals of coastal resource protection and coastal hazard management have yet to be developed. In light of the original objectives of the Act and in the context of the Nation's need for coastal zone management today, the CZM program is falling short of its original goals, and the vital natural systems on the nation's coastline are more in need of protection today than they were in 1972 (20).

STATE ROLE

Thirty-five coastal states and territories have recognized the special attributes and problems of coastal resources and are participating in developing coastal zone management programs (22). As of July, 1980, 14 states had adopted coastal zone management programs approved under the CZMA. Coastal hazards from hurricane, erosion, flooding, land subsidance, salt-water intrusion, and hurricane evacuation is a relatively new arena for the states. Some state CMP's reflect substantial awareness and concern with coastal high hazards and recommend structural and non-structural solutions to mitigate losses. In other states, weak or merely symbolic efforts require significant upgrading and revisions to reflect current scientific, technical and fiscal knowledge about the coastal hazard management.

Relevant features of coastal hazard mitigation in thirty-five states are summarized in Table -3. Further details can be found in individual coastal programs. It should be qualified that this review and assessment is only preliminary and identifies only the existing laws and regulations. However, the effectiveness in implementing these programs to prevent loss of critical resources or to minimize hazardous development on barrier islands has not been systematically analyzed. The groundwork for formulating comprehensive coastal hazard mitigation has been laid, but comprehensive strategies and programs are lacking.

Protection of barrier islands and beaches is of primary concern to states along the Atlantic and Gulf Coast. However, comprehensive programs which address whole barrier island ecosystems have yet to evolve. Most states address barrier island through beach and dune protection or setback ordinances, wetland protection laws, or public facilities location policies. Some states have declared barrier islands as areas of particular concern or high hazard areas. Undeveloped barrier islands are protected from future development by either prohibiting development, discouraging subsidized public projects, or by public acquisition. On partially developed islands, development is controlled through minimum building setbacks, storm and flood-proof structural codes or by discouraging location of public infrastructures. Density limitations and hurricane evacuation have not been addressed by any state in a comprehensive manner.

Twenty-six states are addressing protection of their dunes and beaches by establishing construction setback lines, requiring permits before construction, prohibiting removal of sand and vegetation, requiring public access to the beaches before issuing permits, designating

TABLE - 3

COASTAL HAZARD MITIGATION AND CRITICAL RESOURCES PROTECTION IN STATE CZM PROGRAMS

	Comprehensive CZM Law(L) or Special Legislation(S)	New Management Agencies(A) and/or Commission(C)	Barrier Islands	Open Beach Laws or Dedication of Access	Beaches and Dunes	Erosion-prone Areas	Floodplain Management	Wetlands and Marshes	Offshore Mining	Subsidance and/or Saltwater Intrusion
Alabama	L	A+C		P	P	P	P	N		E
Alaska	L	A+C	N	E		N	N	N	N	N
California	L	C		N	E	E		E	E	E
Connecticut		A						E		
Delaware	S	A		N	E	E		E		
Florida	S	A			E	E		E		
Georgia	S	C			N	E		E	E	E
Hawaii	L			EE	E	N	N		E	N
Illinois					E		E	E		
Indiana							E	E	E	
Louisiana	L	A+C	P		P	P	P	N	P	P
Maine	S	A	E			E	N	E	E	
Maryland		C		E	E	E	E	E		
Massachusetts		A			E	N	E	N	E	
Michigan	S	A			E	N	E	E	E	
Minnesota	S						E	E	E	
Mississippi			E		E			E		
New Hampshire		A	P	P	P		N		P	P
New Jersey	L	A		E	P	N		E	E	
New York		A				P	P	E	E	
North Carolina	L	C	E		N	N	N	E		E
Ohio		A			P	P	P	E	E	
Oregon	L	C		E	N	N	N	E	N	
Pennsylvania		A		P		N	N	E	E	
Rhode Island	L	A+C	N	N	N	N	N	E	N	
South Carolina	L	C	N	N	N	N		N	N	E
Texas	S				N		E	N	P	E
Virginia		A+C			E			E		
Washington	S				E	E	E	E	E	
Wisconsin	S	A		E	E	E	E	E	E	
American Samoa										
Guam				E	E		N	N	E	
Puerto Rico	L	A		E	E	N	N	E		
Virgin Island			N	EE	N	N	N	N		

Source: Office of Coastal Zone Management, March 1979

Explanations: E - Pre-existing law or program incorporated into CZM programs.
N - New or expanded law or program attributable to CZM Act.
P - Proposed law or program to be part of State Coastal Management Plans.
Blank - No law/program or information not available.

areas of particular concern, or by acquisition. Twenty states have incorporated some non-structural erosion mitigation programs by adopting construction setbacks or beach and dune preservation laws. Thirty-one states have new wetland statutes and regulations or have improved implementation of the existing laws. Providing adequate vegetation buffers between the water's edge and structures minimizes estuarine shoreline erosion as well as provides fish and wildlife habitat. Concentrated withdrawal of groundwater, oil, and gas in coastal areas could cause salt-water intrusion and/or land subsidance. This could cause lowering of ground elevation and accentuate flooding such as in the Galveston-Houston area of Texas. In the future, as the population in the coastal area grows, rigorous enforcement and improvements of groundwater laws to protect freshwater supplies and to mitigate land subsidance and flood losses would be essential. The effectiveness of the state CMP would depend upon the willingness to enforce the maximum protection under the law and to deny variance requests by private interests.

State acquisition of barrier islands for public recreation and conservation provides additional benefits for hurricane protection and hazard mitigation. Sixteen of the 18 states have set aside selected barrier islands, or sections thereof, as State Parks, Wildlife refuges, or natural areas. The management programs vary greatly within and among the states. A good example is the Delaware Seashore State Park where all dunes, vegetation and wetlands are carefully protected while providing public access for recreation. In contrast, on Jekyll Island State Park, Georgia, 200 acres of dunes and 175 acres of wetlands were destroyed for development and convention facilities were built in the hazardous area. Most state programs fall somewhere inbetween and as awareness of hazards increases, intensive structural programs are avoided.

LOCAL ROLE

The most important and lasting land use decisions in the United States are made by local government. Until recently, comprehensive planning at the local level was non-existent or was primarily a mandatory ritual for receiving Federal and State grants. Lack of understanding and appreciation of the dynamics of coastal resources and processes, improper building location, inadequate setback and construction codes, and failure to recognize the carrying capacity of natural resources and man-made structures has caused increased loss of resources, lives and property, and pollution of water and fishing areas. The adverse impacts of the "any growth is good growth" philosophy are most visible at local levels affecting the health of local citizens, resources, and the economy.

In the late 1960's and early 1970's, awareness of the unfortunate consequences of uncontrolled and unsafe urban growth began to seep into the American consciousness. Few local communities have taken the initiative of developing and formulating comprehensive growth management plans which take into account the carrying capacity of natural resources and processes, man-made structures and fiscal capabilities. Places like Sanibel, Petaluma, Ramapo, Fairfax, Boulder, Gasparilla and Boca Raton are showing the way for other communities. Their plans appear impressive, however, coastal hazard mitigation has not been addressed in local plans in any comprehensive manner. Even with these comprehensive plans,

5-30 dwelling units per acre continue, with individual water and sewer systems, inadequate setbacks, elevations and structural building codes, and traditional grid-pattern subdivisions which are unsuited to the barrier islands and beaches (25). Life cycle fiscal impact analysis of the comprehensive plans is a rarity. Natural disaster mitigation and evacuation planning is in its infancy. If hurricanes like Donna, Carla, Camille or Frederic strike an intensively developed coast, loss of lives and property could run in the tens of thousands and to billions of dollars due to high densities, inadequate building codes and low-lying limited capacity evacuation routes and bridges. If we are to minimize losses in urbanizing barrier islands, local comprehensive planning needs to be accelerated, strengthened and expanded in its scope, level and types of analysis.

Recreation and tourism is a major economic industry in the coastal zone. Local public and private acquisition of undeveloped and remote barrier islands and beaches could virtually eliminate hazards to lives and property and at the same time help the local recreational economy. It should be noted that more than 79% of our population lives in the coastal or Great Lakes states and more than 53% of the national population lives in coastal counties. Of all recreational resources of the coast, none has greater appeal than the beaches. However, the National Urban Recreation Study (1978) indicated that in virtually every one of the 17 urban SMSA's examined, all areas had serious problems in meeting the recreational demands. Glynn County, Georgia, Brevard and Collier Counties and the City of Sanibel is Florida are examples of innovative and aggressive beach protection and acquisition programs. In many instances, the economic welfare of a community will depend upon adequate, clean and aesthetically pleasing barrier islands and beaches for recreational enjoyment.

Some natable examples of local hazard mitigation planning on the coast are listed in Table -4. The Lee County Coastal Construction Code (26) and the Lee County Hurricane Evacuation Study (29) are the two most innovative location-specific quantitative studies on the subject. However, it should be noted that Lee County Codes do not adequately protect against hurricane winds and storm surges. Texas codes provide better wind protection (17). It should be emphasized that although the basic concepts and methodology of the Lee County and Texas Codes are applicable to other coastal communities, specific numbers and factors must be developed for each site to incorporate local natural features. The City of Galveston Island developed erosion and hurricane mitigation plans. As a result, automobiles and ORV's have been prohibited from the Galveston eaches and dunes -- the first such effort in Texas (26). Perdido Key and Longboat Key in Florida have developed draft plans which incorporate the carrying capacity, life cycle fiscal impact, and hurricane and erosion hazard mitigation analysis. The State of Georgia has developed coastal construction codes for all of its populated barrier islands (26). Sanibel and Gasparilla Islands located in Lee County, Florida, provide an insight and methodology for comprehensive barrier island planning for other island communities (4). The Sanibel Plan is the first plan to incorporate hurricane and erosion hazard mitigation by adopting beach and dune vegetation protection, coastal setbacks, minimum building elevations, structural codes and hurricane evacuation plans. It includes capital improvement and fiscal analysis and defines specific policies

TABLE - 4

COASTAL HAZARD MITIGATION IN COUNTY AND LOCAL BARRIER ISLAND PLANS

	COMPREHENSIVE PLAN	PROTECTION OF CRITICAL RESOURCES	CARRYING CAPACITY ANALYSIS	FISCAL IMPACT ANALYSIS	BEACH & DUNE EROSION MITIGATION	HURRICANE - PROOF BUILDING CODES	HURRICANE EVACUATION
FLORIDA							
Brevard County	Y	P	-	-	Y	P	-
*Collier County	Y	Y	-	P	Y	-	-
*Gasparilla Island	Y	Y	Y	Y	Y	Y	Y
*Sanibel Island	Y	Y	Y	Y	Y	Y	Y
Lee County	-	-	P	-	Y	Y	Y
*Longboat Key	Y	P	P	P	Y	Y	P
Perdido Key	Y	Y	P	P	Y	P	P
GEORGIA							
St. Simon & Sea Islands	P	Y	-	-	Y	-	-
*Jykll Island (State)	P	P	-	P	-	-	-
Little Cumberland Island (Private)	-	Y	Y	-	Y	Y	-
TEXAS							
Galveston Island	P	P	P	-	Y	P	P
SOUTH CAROLINA							
St. Phillips Island	Y	Y	Y	Y	Y	-	-
*Kiawah Island (Private)	Y	Y	Y	Y	Y	Y	-

Explanations: Y - Yes
P - Partial
* - Adopted and Implemented

and standards for the conservation of environmentally sensitive areas. It delineates lands that can be developed (6,372 acres) and those to be protected and preserved (4,409 acres). On developable land, densities were assigned and vegetation removal and paving criteria were formed (4). The Sanibel Plan has withstood four court challenges on fairness, reasonableness, taking and scientific issues. In 1978, Sanibel adopted an annual growth limitation of 180 dwelling units per year. In 1980, the City embarked on a comprehensive hurricane hazard mitigation study to further strengthen its plan. Gasparilla Island took the lead from Sanibel, formulated a Comprehensive Plan, and went to the State Legislature for help to enforce the island plan because the County Government had failed. The State Legislature passed a Bill authorizing a public referendum and creating an Island Development Authority.

Sanibel, Gasparilla and other plans are no panacea but provide examples and hope to other island communities. Some plans are better than others but good progress is being made by the communities. When political skills, active citizen concern, and technical and scientific knowledge are working together for sound community goals, the natural and cultural heritage of the barrier islands can be preserved and conserved. During the 1980's, increased citizen awareness and understanding of the dynamic, fragile, and hazardous nature of barrier islands would bring about improved hazard mitigation programs.

PROSPECTS FOR THE 1980'S

Federal recognition of the development pressure and hazards of barrier islands has heightened recently due to hurricane activities and increasing fiscal burden on the nation's taxpayers. Recognizing the problems of federal inconsistencies and loss of fragile resources, President Carter expressed his concerns in his first environmental message to Congress in 1977. The President ordered the Department of the Interior to conduct a study of all Federal programs which might be causing loss of valuable resources and encouraging Federally subsidized hazardous development, and asked the Secretary to recommend a plan to protect undeveloped barrier islands. The Department of the Interior released a Draft Environmental Statement on barrier islands in January 1980 (33).

In his second environmental message to the Congress on August 2, 1979, President Carter again expressed his concerns for loss of coastal zone resources and the need for protection of barrier islands and coastal hazard mitigation. President Carter directed the Secretary of Commerce to conduct a detailed study of federal programs and their consistencies for resource protection and development of the coastal zone resources. The Office of Coastal Zone Management within NOAA is conducting this study and the findings and recommendations of the study are to be submitted in the near future.

In addition to Executive initiative, Congressman Phillip Burton (D-CA) and Senator Dale Bumplers (D-AR) have introduced Barrier Islands Protection Bills (HR-5981 and S-2686) in the current session of Congress to protect and preserve yet undeveloped barrier islands for public recreation, conservation, and hurricane protection, as well as fiscal savings. The basic provision of these Bills are to prohibit incompatible, hazardous and subsidized development on barrier islands through denial of federal grants, loans, permits and licenses, and acquisition of undeveloped barrier islands for public use and hazard mitigation. However, these Bills do not affect already urbanized or developing barrier islands and beaches in any way and additional efforts are needed.

Possible adjustments to coastal hazards may range from wholly non-structural to totally structural solutions; from strict preservation of natural resources to uncontrolled and unsafe development; from complete withdrawal of funds for publicly funded projects to total government apathy and a status quo. Such extremes are unncessary and undesireable. Strategies which combine public acquisition of undeveloped and available barrier islands and beaches and land-use planning, zoning and improved building-structural codes on developed or developing areas which permit wise use and safe development would be more acceptable, as well as desireable. The initial approach to mitigate the coastal high hazards on barrier islands and beaches may include, but should not be limited to:

(1) Inventory and analysis of the natural and cultural resources and delineation of critically unique, sensitive, hazardous and available areas for earliest possible public acquisition for recreation, preservation and hazard mitigation. A ten-year cooperative Federal, State and local acquisition program would be an economical and sensible solution for island protection and hazard mitigation (27).

(2) Assessment of risks to life and property for 100-year and 500-year natural events at various population and growth levels. Carrying capacity of natural and man-made resources and life cycle fiscal impact analysis of growth in all coastal high hazard areas must be determined to assist citizens, policy-makers, and tax-payers to improve land use decisions.

(3) Adoption of appropriate land uses, densities and building setback lines landward of the full primary dunes, berms, mature pioneer vegetation and all freshwater and salt-water wetlands on urbanizing islands and beaches. They must also take into account the natural sea level rise and erosion rates. Multiple setback zones to provide hurricane, flood, erosion, pollution hazard mitigation, and adequate public beach access are preferred over single arbitrary defined setback lines. Once these setback lines and regulations have been adopted, they must be rigorously enforced and variances denied.

(4) Dredging and filling of all wetlands and estuaries should be absolutely prohibited, because this nation has already lost more than 40% of its freshwater wetlands and more than 73% of the estarine areas to destructive human activities (6,35).

(5) Establishment of hurricane resistant building codes which incorporate horizontal and vertical impacts of hurricane surge, wave uprush, scour and winds. Since we cannot control or modify hurricanes, we must build safer

structures, specifically suited to the coastal high hazard areas. Neither the Southern Building Codes nor the South Florida Building Codes are adequate for protection against hurricane winds and storm surges. Improved building codes add only 1-3% to the construction costs, but minimize damages by 60 to 70%.

(6) Public capital improvement projects such as roads, bridges, water and sewers should recognize the high risk on the barrier islands and federal and state governments should effectively prohibit and discourage direct or indirect expenditures of public funds on undeveloped islands. Where necessary, all major roads and bridges should be built above 100-year flood levels and the provision of sewerline to control water pollution should be conditional so that intensive urbanization as a result of new sewerline is prohibited.

(7) Development of comprehensive hurricane evacuation plans which identify evacuation routes, shelters and coordinating agencies. Lee County's evacuation plan could serve as a model for all barrier island and coastal communities.

(8) Non-structural solutions for erosion control must be used to maximum possible extent. Relocation of public facilities and buildings to safe areas under post-disaster programs and relocation and beach nourishment for erosion mitigation are preferred alternatives. However, relocation has not been evaluated by any federal or state agency in a systematic manner, but offers a unique opportunity in the wake of Hurricane Frederic.

(9) Provide additional warning to prospective buyers of property located on barrier islands and hurricane hazard zones that "this property is likely to be flooded or destroyed from hurricane winds, storm surge and scouring." Such a warning can be incorporated in the National Flood Insurance Program's existing floodplain warning program.

(10) Public information and education programs to improve the general understanding of the nature of barrier islands and coastal high hazard zones and the risks and benefits associated with alternative and wise use of these resources. Public education programs of the Texas Coastal and Marine Council and the Barrier Islands Coalition provide the prototype for other states and local communities.

REFERENCES

1. Adams, James W.R., 1978: "Rebuilding the Beaches of Florida," Florida Sea Grant, Technical Report No. 7, Beach Seminar '78, Gainesville, FL, pp 102-121.

2. Bloom, Arthur L., 1971: "Glacial-Eustatic and Isotatic Controls of Sea Level" in The Late Cenozoic Glacial Ages, Karl K. Turkian, Editor, Yale University Press, New Haven, CT, pp 357-379.

3. Clark, J.R., 1977: "Coastal Ecosystem Management: A Technical Manual for the Conservation of Coastal Zone Resources," John Wiley & Sons, New York.

4. Clark, John, 1976: "The Sanibel Report," The Conservation Foundation, Washington, D.C.

5. Congressional Research Service, 1978: "Past Federal Survey of Shoreline Recreational Potential," prepared for Senate Committee on Energy and Natural Resources, U.S. Govt. Printing Office, Washington, D.C.

6. Council on Environmental Quality, 1978: "Our Nation's Wetlands: An Inter-Agency Task Force Report," Washington, D.C.

7. Dacy, D.C. and Kunreuther, H., 1969: "The Economics of Natural Disaster: Implications for Federal Policy," The Free Press, New York.

8. Dolan, R., B. Hayden and H. Lins, 1980: "Barrier Islands," <u>American Scientist</u>, Vol. 68 (1), pp 16-25.

9. Federal Emergency Management Agency, 1979: "Proposed Federal Emergency Management Agency Disaster Relief Policy for Barrier Islands, Beaches and Spits," October 1979, Disaster Relief and Recovery, FEMA, Washington, D.C.

10. Federal Insurance Administration (no date): "The National Flood Insurance Program's Treatment of Coastal Wave Heights," FIA, Federal Emergency Management Agency, Washington, D.C.

11. Frank, Neil, 1978: "Personal Communications and letter dated October 4, 1978," National Hurricane Center, Coral Gables, FL.

12. Godfrey, Paul J., 1976: "Barrier Beaches of the East Coast," <u>Oceanus</u>, Vol. 19 (5), pp 27-40.

13. Heritage Conservation and Recreation Service, 1979: "Report of the Barrier Island Work Group (BIWG)," U. S. Department of the Interior, Washington, D.C.

14. Hicks, S.D., 1973: "Trends and Variability of Yearly Mean Sea Level 1893-1971," NOAA Tech. Memo No. 12, Rockville, MD.

15. Jiminez, Gloria, 1980: "Testimoney Before the Subcommittee on National Parks and Insular Affairs on HR-5981," Federal Insurance Administration, Federal Emergency Management Agency, March 24, 1980, Washington, D.C.

16. Jiminez, Gloria, 1979: "Coastal Storms, Development and National Flood Insurance Program," A paper presented at the 'Development of the Coast: Facing the Tough Issues' Conference, September 27-28, 1979, Charleston S.C., Coastal States Organization.

17. Lesso, William G., 1979: "Economic Incentives to Build Improved Hurricane-Resistant Building Standards," Texas Coastal and Marine Council, Austin, TX.

18. Levy, C.C. and Smith, J.K., 1978: "Apathy and Inexperience: A Study of Hurricane Hazard Perception in Sarasota County," Environmental Studies Program, New College of South Florida, Sarasota, FL.

19. Lins, H., 1980: "Patterns and Trends of Land Use and Land Cover on Atlantic and Gulf Coast Barrier Islands," U.S. Geological Survey Professional Paper 1156, Department of the Interior, Washington, D.C. (In Press).

20. National Advisory Committee on Oceans and Atmosphere, 1979: "Coastal Zone Management 1978: Finding and Recommendations," A Special Report to the President and the Congress, Washington, D.C.

21. National Commission on Marine Science, Engineering and Resources, 1969: "Our Nation and the Sea: A Plan for National Action," A Report to the President and the Congress, U.S. Government Printing Office, Washington, D.C.

22. Office of Coastal Zone Management (OCZM), 1979: "The First Five Years of Coastal Zone Management: An Initial Assessment," U. S. Department of Commerce, Washington, D.C.

23. Pollack, K., 1980: "Letter to Congressman Phillip Burton, Subcommittee on National Parks and Insular Affairs on HR-5981," April 7, 1980, Natural Resources Defense Council, Washington, D.C.

24. Sharma, Dinesh C., 1980: "Dauphin Island and Hurricane Frederic: Implications for Hazard Mitigation on Barrier Islands," Testimony before House Subcommittee on National Parks and Insular Affairs on HR-5981, March 27, 1980, Washington, D.C.

25. Sharma, Dinesh C., 1979: "Barrier Islands and Beaches of Florida: Unique Resources, Problems and Prospects," in the Proceedings of Florida Shore and Beach Preservation Association, 23rd Annual Meeting, October 3-5, 1979, Bal Harbour, FL.

26. Sharma, Dinesh C., 1979: "Development in Hurricane and Erosion Hazard Zones," Proceedings of the III World Congress on Water Resources, Mexico City, International Water Resources Association, Washington, D.C., pp 2235-2247.

27. Sheaffer, J.R. and L.Rozaklis, 1980: "Barrier Islands Purchase: A Cost-Effective Approach to Management," Statement Before the Committee on Interior and Insular Affairs concerning Barrier Islands National Parks Bill HR 5981, March 27, 1980, Washington, D.C.

28. Sorensen, John H. et. al., 1975: "Coastal ERosion Hazard in the United States: A Research Assessment," Institute of Behavioral Science, Monograph #NSF-RA-E-75-014, University of Colorado, Boulder, CO.

29. Southwest Florida Regional Planning Council (SWFRPC), 1979: "Lee County Flood Emergency Evacuation Plan Report," prepared for the U.S. Army Corps of Engineers, Jacksonville District, FL.

30. Sugg, A.L., 1967: "Economic Aspects of Hurricanes," Monthly Weather Review, Vol. 95, pp 143-146.

31. U.S. Army Corps of Engineers, 1971: "National Shoreline Study," U.S. Govt. Printing Office, Washington, D.C.

32. U.S. Bureau of the Census, 1976: "Statistical Abstract of the United States," U.S. Govt. Printing Office, Washington, D.C.

33. U.S. Department of the Interior, 1980: "Alternative Policies for Protecting Barrier Islands Along the Atlantic and Gulf Coasts of the United States and Draft Environmental Assessment," HCRS, Department of the Interior, Washington, D.C.

34. U.S. Department of the Interior, 1978: "National Urban Recreation Study," U.S. Govt. Printing Office, Washington, D.C.

35. U.S. Fish and Wildlife Service, 1970: "National Estuary Study, Main Report, Vol. 1," U.S. Department of the Interior, Washington, D.C.

36. Urbanik II, T., 1978: "Texas Hurricane Evacuation Study, A Technical Report for Texas Coastal and Marine Council," Texas Transportation Institute, Texas A & M University, College Station, TX.

37. Water Resources Council, 1978: "Floodplain Management Guidelines for Implementing Executive Order 11988," Federal Register, Vol. 43 (29), February 10, 1978.

38. White, G.F. et. al., 1976: "Natural Hazard Management in Coastal Areas," a Report prepared for the Office of Coastal Zone Management, U.S. Department of Commerce, Washington, D.C.

39. Wilcox, W.H., 1979: "Closing Remarks - National Conference on Hurricane and Coastal Storms," in the Proceedings of the Conference, May 29-31, 1979, Orlando, FL, Earl J. Baker (Editor), Sea Grant Technical Report, University of Florida, Gainesville, FL (in press).

BARRIER ISLAND MANAGEMENT

Stephen P. Leatherman[1]

INTRODUCTION

Barrier islands tend to be dynamic in nature, creating problems for human management and development of these mobile landscapes. Most barriers are retreating landward due to the worldwide rise in sea level; this migration is manifested in the form of seaside beach erosion which further complicates the situation for human developments. Therefore, primary interest has been placed on the ocean beach to the exclusion of other barrier environments. Shoreline engineering projects have been designed and undertaken along mainland and barrier coastlines with no apparent distinction in application. Whereas a mainland shore, such as the cliffs along the Great Lakes or Outer Cape Cod, must continue to erode in-place, barrier islands have the distinct ability to migrate landward with sea level rise.

A holistic management approach is needed, where the interrelationships among natural processes (principally inlet dynamics and overwash) and responses (primarily beaches, dunes, and salt marshes) are clearly defined. This type of analysis would delineate the relative contribution of each process within the time frame for landward barrier migration.

It is clear that most barrier islands have been migrating landward during Holocene time (Kraft 1971; Swift 1975). Within the context of a geologic time frame (last several thousand years), islands have moved landward on the order of several kilometers (Sanders and Kumar 1975; Moslow and Heron 1979). In considering shoreline engineering projects which are designed to hold the barrier in a static position, the present natural rate of island migration must be addressed. Engineering projects such as seawalls and dune stabilization tend to minimize or preclude overwash and inlet processes, thus affecting barrier migration. As sea level rises, the island may be placed in a disequilibrium position. Most Army Corps of Engineers projects have a stated lifetime of 50 to 100 years; during this time interval the natural tendency for barrier retreat may be miniscule to highly significant. In addition to the high costs and sometimes futility of these efforts, shoreline stabilization has had the effect of encouraging continued development.

---

[1]Director, National Park Service Cooperative Research Unit, University of Massachusetts, Amherst, MA 01003

This article was written to serve as a generalized review paper on present management practices and alternatives. The discussion is basically focused on the various approaches to barrier management -- static vs. dynamic, particularly along developed shores. Although an exhaustive treatment of these management techniques is not attempted, key issues are conceptualized. New legislative initiatives are clearly required for presently undeveloped barrier islands, and quantitative geomorphic research is needed to precisely define the dynamics of already developed barriers. This scientific information should serve as a primary input to the economic analysis when shoreline engineering projects are contemplated.

APPROACHES TO BARRIER STABILIZATION

At the onset it is necessary to separate the islands into two distinct categories: natural barriers with little manipulation of the environments or interference with natural processes (Fig. 1), and developed barriers with large-scale alterations of the environments and various shoreline engineering projects in effect (Fig. 2). In the case of natural areas, management guidelines can be quite simple since "nature is allowed to take its course." Only where there have been previous misuses of the land or present "upstream" or external interferences with the system should restorative or protective alternatives be considered or implemented. Such is the case at Sandy Hook, NJ, a unit of Gateway National Recreation Area, where updrift seawalls and groins have sediment-starved this recurved spit (Leatherman et al. 1978). Severe erosion just north of the last major groin threatens to cause barrier truncation by inlet breaching, thus creating Sandy Hook Island.

Typical engineering projects to stabilize barriers have been directed toward the ocean beach. Such devices as groins and seawalls have long been in use, and their inherent problems are now being recognized. Seawalls only protect what is landward or behind them, and these massive structures often accelerate the rate of beach erosion (Silvester 1977). Groins have limited successful application in some environments, but have been found to be ineffective or enhance present problems in other areas (Everts 1979). In addition to environmental problems, there are the aesthetic considerations where massive concrete structures have been emplaced along an otherwise natural beach. The expressed purpose of these engineering structures is to interfere with beach dynamics, but their long-term application, particularly in the case of long-standing seawalls, may have pronounced adverse effects on barrier dynamics and equilibrium.

Barrier dune building and stabilization offer a viable alternative to these previous "concrete" solutions. Barrier dunes can be quickly built (Knutson 1977a; Woodhouse et al. 1977), but their stabilization is

Figure 1. Along natural (undeveloped and unaltered) islands, barrier dynamics are not of principal concern (Assateague Island National Seashore, MD, 1974).

Figure 2. Barrier development is often accompanied or followed by engineering works in an attempt to stabilize the shoreline (Ocean Beach, Fire Island, NY, 1978).

actually a relative term. Where shoreline erosion is rampant, the dune structure will eventually be cut away by wave and surge attack during storms. A solution to the dune lifetime dilemna involves beach nourishment, where large quantities of new sand are introduced to the beach-nearshore system.

During times of rising sea level, most barriers can be expected to migrate landward, except where large quantities of sand are locally available, such as near river deltas. Rigid engineering structures, as previously mentioned, only interfere with sand transport processes for a limited amount of time; their failure ultimately occurs with shoreline recession. Barrier dunes can be utilized as a "limited" protection system within the framework of a dynamic shoreline and landward migrating barrier system (Gares et al. 1979).

DUNE DYNAMICS AND STABILIZATION

Where there is a slow rate of shoreline recession, landward barrier dune migration (with the dune moving in toto as a geomorphic unit) may be possible. Along many eroding shorelines where the beach backshore is quite narrow, there is a limited source area for windblown sand. Also, barrier dunes, which provide a measure of protection against overtopping, would be reasonably high and probably scarped on their seaward face in this situation. Therefore, the amount of sand that can be blown or washed over a dune of this description would be insignificant compared to the amount of seaside erosion, and the dune would continually diminish in size. Eventually, overwash surges would flatten the dune, carrying much of this sand landward to form backbarrier fan deposits. New dunes would naturally form on these washover fans in a more landward position than the previous dune line. Sand fencing and beachgrass planting can be used to accelerate the dune building process.

In order to utilize a dune as a protective structure, this morphological feature must have space to evolve and migrate. Where development has already proceeded in the backdune or even in the dune area itself, a conflict results. The dune is artificially restricted on its landward (advancing) side, and its eventual demise is insured by a retreating shoreline. As suggested by Gares et al. (1979), plans must be implemented that allow for landward dune migration with a dynamic shoreline situation.

There are two approaches to barrier dune management: static vs. dynamic. Where the shoreline is stable or accreting, a static approach can be easily employed. However, in most cases the shoreline is eroding, so that the choice is much more difficult. Quantitative, site-specific data on the rate of beach recession and island migration must be obtained. The two principal data sources for this information are historical aerial photographs and NOS shoreline manuscripts along with stratigraphic cross-sections derived from geologic cores. Some islands may be found to be so mobile that stabilization would be virtually impossible without unprecedented expenditures of money. The decision to allow the shoreline to migrate or to attempt stabilization should be based on an economic analysis for developed barriers. The cost of a beach fill project and continued maintenance may be affordable and justifiable in some cases, e.g., where the barrier is highly urbanized and the natural rate of retreat is relatively slow. It should be noted that most engineering projects, including beach nourishment, have proceeded without a basic knowledge of or quantitative data on the rate of barrier migration. Therefore, the true costs associated with such a project cannot be objectively assessed, and large-scale nourishment projects, intended to last for many years, may be quite ineffective, requiring continued expenditures of large proportions.

Assuming that there is truly an economic justification for beach nourishment and dune stabilization, the next issue concerns the environmental consequences of such an approach. For most heavily-developed barriers, the island ecosystems have already been greatly altered. Much of the land surface has been covered with pavement and buildings, and the remaining land has been revegetated with lawn grass and ornamental trees and shrubs. Consideration of natural terrestrial ecosystems on developed barriers is largely academic, as virtually all the land has been altered in the urbanization process.

Perhaps more significantly, bayside salt marshes have been dredged and filled, and some bayshores have been bulkheaded. A major objection to dune stabilization stems from consideration of marsh creation and productivity. Godfrey and Godfrey (1975) stated that salt marshes at Core Banks, N.C. are rejuvenated by a new influx of sand by overwash or inlet activity every 20-30 years, providing for maximum estuarine productivity. Therefore, long-term barrier stabilization may significantly lower the productivity of this system without compensating measures.

Fortunately, it is possible and quite simple to build new salt marshes, which exhibit ecologic characteristics and biological productivity levels similar to natural marshes (Knutson 1977b; Woodhouse et al. 1974). In order for the natural processes (i.e., overwash or inlets) to build new marshes, sand must be transported from the beach to the bayshore. On developed barriers, the surging water would destroy or severely damage all "improved" property in its path. To mitigate the adverse affects of human development and shoreline engineering projects, salt marshes can be built with dredge spoil. Sediment from the island system should not be dumped offshore, since this type of activity represents a net loss to the barrier sediment budget. Where downdrift beach erosion is a chronic problem, artificial sediment by-passing at jettied inlets should be undertaken. However, a portion of the material dredged from the inlets should be placed near the bayshore to form spoil islands or along the bayshore where possible. In this fashion, new and highly productive salt marshes can be continually created on a short time interval (most inlets are dredged at least every five years), and there will be no net loss of sediment to the barrier. Although it could be argued that the barrier island is being "rolled over" artificially, the quantity of sediment actually transferred landward through these projects may be insignificant compared to the amount required over the long term for true island migration.

LEGISLATIVE INITIATIVES

Presently there are two bills pending in Congress (H.R. 5981 and S. 2686) concerning barrier islands. These legislative actions would serve to cut off federal programs and grants that facilitate and even encourage development on presently undeveloped barrier islands and spits. If either of these bills become law, future development on undeveloped barriers would probably be minimal and restricted to low-cost summer cottages.

For already developed areas, the problem has not yet been fully addressed. An Environmental Impact Statement, released by the Department of the Interior in 1979, lists three levels of alternatives for protection of barrier islands which range from maintenance of the status quo to redirection of federal programs. Although there are over twenty federal programs that contribute to barrier island development, the two most important federally-sponsored programs are flood insurance and erosion control. It would be virtually impossible to make major modifications to the flood insurance program with so many homes currently under this coverage. However, following disaster relief and insurance payoffs to damaged structures, the federal government, with state support, should buy out the victims and acquire title to this highly vulnerable real estate. In this manner, the federal flood insurance program under FEMA would not have to continually recompensate people after each major storm for damage to structures that have been rebuilt in the same exact location. The acquisition clause has always been a part of the flood insurance program, but lack of funding has precluded its usage.

Shore erosion abatement and hurricane protection have been undertaken by local and state groups, but the major projects have been directed by the U.S. Army Corps of Engineers. An economic analysis incorporating a cost-benefit ratio has been utilized to justify these projects, but the cost side of the formula may have been well understated in many instances. It may be possible to compensate for sediment loss due to storm-induced changes with engineering projects (such as beach fill) over the short term, but if sea level rise is proven to be the driving force in barrier migration, then the fight is certainly futile. Assuming that the present trend of sea level rise continues (Hicks 1978) and large-scale beach erosion is associated with subtle increases in mean water level, then barrier stabilization will be an expensive and fruitless experiment. Of course, each pulse of new sand introduced into the beach system by artificial nourishment will help to forestall the inevitable, but taxpayers' money would be best spent in adopting some approach of land-use control and compensation (with federal acquisition of storm-damaged property). In effect, island residents would be forced to live with a dynamic shoreline management approach, as their property will eventually be lost with landward barrier migration.

States, particuarly New Jersey, are presently attempting to enact legislation that would restrict development in the critical buffer zone between the beach and backbarrier environments. Although alteration and development of coastal wetlands has been restricted by many states, dunes and particularly adjacent backdune areas remain prime real estate. Development restrictions placed on this zone without compensation is considered by island residents as a "taking" of their land. This type of approach raises constitutional and legal issus which will not be discussed in this paper.

An alternative to these restrictions and controls would be the institution of a "home rule" policy, which is being advocated by some island residents. In this case, federal flood insurance programs would be revoked and Corps erosion programs terminated. Development practices and shoreline stabilization would be left to the local citizens without "outside interference." Some island residents are convinced that they can solve their own beach erosion problems, but this view ignores the fundamental dynamics of barrier islands and the present rise in sea level. After a major storm has ravaged the coast, it is certain that these same people would call upon the federal government for disaster relief and assistance. At a time of great destruction of property and loss of life, it would be virtually impossible to cite geologic principles as the reason to ignore this dire situation and "abandon" our fellow Americans; it would be politically expedient to declare the area a federal disaster due to natural causes and fully compensate the people for their own development follies.

SUMMARY AND CONCLUSIONS

Where barriers are presently undeveloped and geologic investigations show that islands are inherently unstable, development of a substantial nature should be prohibited by federal legislation. Along rapidly transgressing (landward migrating) barriers, only low-density, low-cost "summer" cottages with an expected short lifespan should be allowed. No federal funds should be authorized or spent to aid development, and residents should not be eligible for any disaster relief or erosion assistance.

Unfortunately, major development has already proceeded on many barrier islands. Construction of high rise condominiums at Ocean City, MD, may be the worst case of uncontrolled and unwise barrier island development in recent times (Fig. 3). Although there has not been a comprehensive assessment of this barrier's dynamics, shoreline erosion problems are indicative of the critical nature of the present situation. Along short shoreline segments, such as Atlantic City, NJ and Miami Beach, FL, there does not seem to be any alternative to shoreline stabilization due to the high capital investments.

For the single-dwelling type development that characterizes many urbanized barriers, the economic analysis becomes more critical. Here the cost of massive shoreline engineering projects over relatively long stretches of coast may make a static shoreline approach economically prohibitive. In this case a flexible posture of dynamic barrier management should be instituted. This solution will be unpopular to affected residents, as an underlying assumption of this strategy is the erosional transformation of the environment with the landward retreating barrier system. In the process, buildings adjacent to the shoreline would be destroyed by storm activity, based on their distance from the advancing water edge and severity of each individual storm. Redevelopment in storm-damaged areas would be disallowed, and new development on accreting bayside areas would be a subject of much debate and probably prohibited due to wetlands restrictions. In essence, all development would be removed with complete barrier rollover (landward island migration equal to its own width). The time frame for this eventuality with continued sea level rise is not known and deserves detailed investigation on a per island basis. Dune stabilization will help to minimize overwash and inlet dynamics and hence slow down the rate of landward retreat, but this type of strategy cannot stop barrier migration. In fact, there are no engineering projects that can ultimately prevent this natural evolution if sea level rise is the forcing function. Rigid engineering structures will generally serve only to exacerbate an already critical situation.

Clearly there is a need for a detailed understanding of barrier island dynamics, in principle as well as on an island by island basis. Although many coastal scientists and engineers recognize the geomorphic logic of barrier dynamics, the principal unresolved issues involve the relative contribution of various processes in relation to the geologic time frame. Since most coastal engineering projects are based on a useful life of 50 to 100 years, it is argued that in such a short time period, geologic processes such as sea level rise play a minor role in shaping barrier islands. Questions of time frame and mechanism for barrier rollover can only be addressed by an extensive coring program and geomorphic analysis. Without this type of basic information, speculation abounds and neither viewpoint can be validated or totally rejected.

Figure 3. High-rise condominum development along an eroding shoreline invites future disaster (Ocean City, MD, 1974).

ACKNOWLEDGMENTS

Continued support by the National Park Service of our barrier island research is gratefully acknowledged. The concepts espoused in this paper are strictly those of the author and do not represent official NPS policy. This manuscript was critically reviewed by Ms. Sheila Pelczarski and typed by Mary Ellen Miller.

REFERENCES

Everts, C.H. 1979. Beach behavior in the vicinity of groins -- two New Jersey field examples. Proceedings of Coastal Structures 79. pp. 853-867.

Gares, P.A., K.F. Nordstrom and N.P. Psuty. 1979. Coastal dunes: their function, delineation and management. Center for Coastal and Environmental Studies. Rutgers University. 112 pp.

Godfrey, P. and M. Godfrey. 1975. Some estuarine consequences of barrier island stabilization. Estuarine Research, Vol. II. New York, NY: Academic Press, Inc. pp. 485-516.

Hicks, S.D. 1978. An average geopotential sea level series for the United States. J. Geophysical Research 83:1377-1379.

Knutson, P.L. 1977a. Planting guidelines for dune creation and stabilization. CERC TA77-4. 26 pp.

Knutson, P.L. 1977b. Planting guidelines for marsh development and bank stabilization. CERC TA77-3. 21 pp.

Kraft, J.C. 1971. Sedimentary facies patterns and geologic history of a Holocene marine transgression. Geol. Soc. of Am. Bull. 82:2131-2158.

Leatherman, S.P. 1979. Barrier Island Handbook. Nat. Park Serv. Coop. Res. Unit. UMass, Amherst, MA. 101 pp.

Leatherman, S.P., P.J. Godfrey, and P. Buckley. 1978. Management strategies for National Seashores. In Proceedings of the symposium on technical, environmental socioeconomic and regulatory aspects of coastal zone planning and management. pp. 322-337.

Moslow, Thomas F. and S.D. Heron. 1979. Quaternary evolution of Core Banks, N.C., Cape Lookout to New Drum Inlet. In Barrier Islands (S.P. Leatherman, ed.). New York, NY: Academic Press, Inc. pp. 211-237.

Sanders, J.E. and N. Kumar. 1975. Evidence of shoreface retreat and in place "drowning" during Holocene submergence of barriers, shelf off Fire Island, N.Y. Geol. Soc. of Am. Bull. 86:65-76.

Silvester, R. 1977. The role of wave reflection in coastal processes. Proceedings of Coastal Sediments 77. pp. 639-654.

Swift, D.J.P. 1975. Barrier island genesis: evidence from the Central Atlantic Shelf, Eastern U.S.A. Sedimentary Geology 14:1-43.

Woodhouse, W.W., Jr., E.D. Seneca, and S.W. Broome. 1974. Propagation of *Spartina alterniflora* for substrate stabilization and salt marsh development. CERC Tech Memo. 46. 155 pp.

Woodhouse, W.W., Jr., E.D. Seneca, and S.W. Broome. 1977. Ten years of development of man-initiated coastal barrier dunes in North Carolina. UNC Sea Grant Pub. 77-01. 53 pp.

Management of Coastal Resources in Arctic Alaska

Gretchen Keiser[1] and Wayne Pichon[2]

With 33,904 miles of coastline, Alaska has greater than one-third of the coastal zone under United States jurisdiction (Figure 1). Unlike other regions in the United States, where restoration and enhancement of coastal productivity may be a primary concern, Alaska has a coastal zone which is largely intact and remains relatively productive. As national attention focuses on Alaska's offshore oil and gas potential, the management of coastal resources and activities has become an important issue in Alaska.

The establishment and implementation of a system where competing demands upon coastal lands and waters are balanced has presented a difficult challenge. Nowhere is this challenge more evident than along the Arctic coast. Conservation of suitable habitat for migratory animals and preservation of local hunting and fishing practices ("subsistence") while industrial oil and gas production activities accelerate, are key management concerns. The Alaska Coastal Management Program (ACMP) is seen as the mechanism for this process.

In this paper we discuss briefly the history and structure of the ACMP, highlighting elements of particular importance to the Arctic coastal zone. After presenting a short background on the local government and coastal resources in this area, we discuss the evolution, public review, and current status of local coastal management planning. A prospectus on coastal management and energy-related activities in adjacent areas of rural Alaska also is offered.

The Alaska Coastal Management Program

Alaska's coastal zone management planning was initiated in 1974 when the state first received federal Section 305 funding (Coastal Zone Management Act of 1972, as amended in 1976). An initial legislative proposal, emphasizing direct state planning and implementation with limited public participation and local government involvement, was defeated in 1975. Renewed efforts resulted in enactment of the Alaska Coastal Management Act in June 1977.

---

1 Northern Alaska Ecological Services, U.S. Fish and Wildlife Service, 101 12th Avenue, Fairbanks, Alaska 99701

2 Alaska Area Ecological Services, U.S. Fish and Wildlife Service, 1011 E. Tudor Road, Anchorage, Alaska 99503

NORTH SLOPE BOROUGH COASTAL MANAGEMENT PROGRAM PHASES

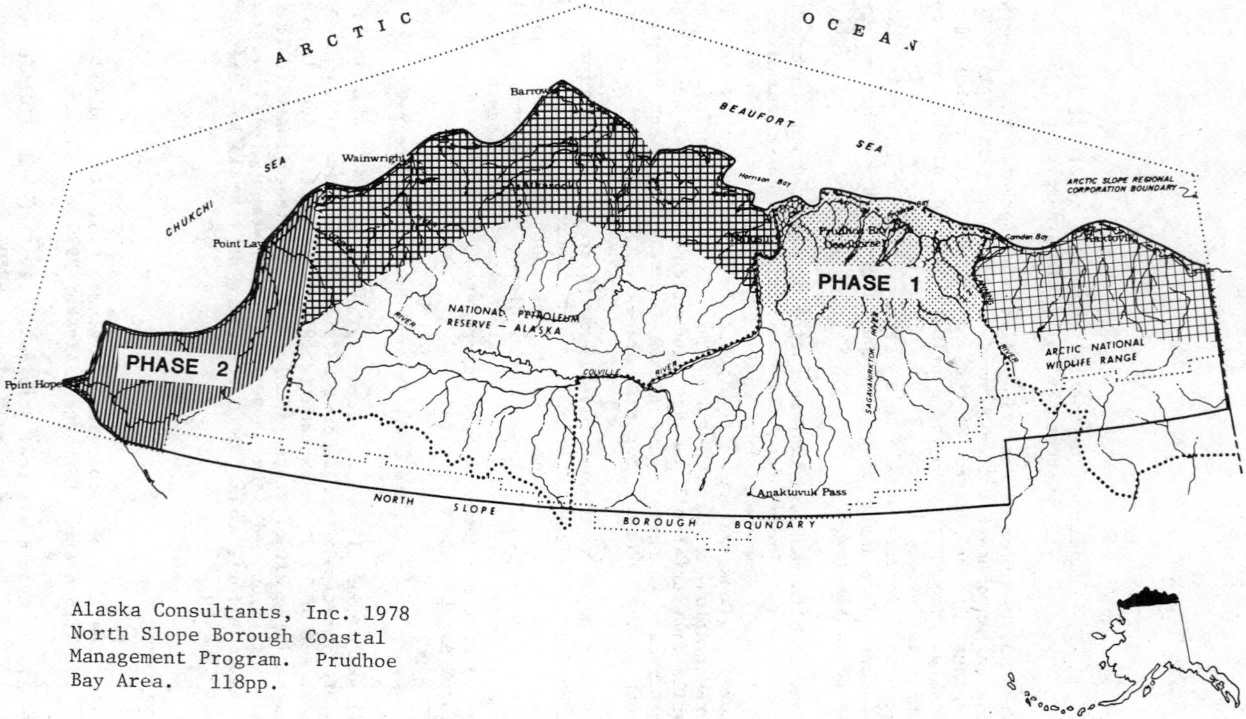

Alaska Consultants, Inc. 1978
North Slope Borough Coastal
Management Program. Prudhoe
Bay Area. 118pp.

Figure 1

The Act created the Alaska Coastal Policy Council, a regulatory and review body, charged with the development of a federally-approvable program. The Act also created the Office of Coastal Management (OCM) as the lead state agency within the Division of Policy Development and Planning and as staff to the Council. The legislation required the 16-member Council (composed of public members and government officials) to establish, by regulation and subject to legislative approval, state guidelines and standards for the management of coastal land and water uses and to develop program procedures and policies. Local governments were to develop specific Coastal Resource District programs utilizing their existing planning and zoning powers. Local planning efforts were to be guided by ACMP guidelines and standards, funded by federal Section 305 or Section 306 funds, and subject to Council review and approval prior to local implementation. Coastal Resource District programs in areas where no zoning authorities existed (a significant portion of western and southwestern Alaska) were to be implemented by appropriate state agencies following Council and legislative approval[1]. The legislation also included a provision for public and governmental participation through the public hearing process.

Initial ACMP efforts began in early 1978 with the development of draft guidelines, standards, and program document. The U.S. Fish and Wildlife Service (FWS) became involved in the review process at this stage. Our concerns repeatedly emphasized the need for (1) incorporation of specific procedures assuring federal agency involvement in the ACMP planning process, (2) greater specificity within the standards and guidelines through definition of terms, (3) strengthening of wetland and floodplain habitat standards, and (4) unambigious procedures for conflict resolution. Comments from rural communities highlighted the Native concern that subsistence hunting and fishing be given priority among all proposed coastal uses and activities.

The ACMP was approved by the federal Office of Coastal Zone Management on July 6, 1979 (1). The ACMP provides the skeletal framework including criteria and procedures to be utilized by local governments in their specific planning process. The ACMP, reinforced by Administrative Order No. 54 (enacted by Governor Jay Hammond on April 23, 1979) requires state agencies to conduct their activities in a manner consistent with ACMP standards and ultimately approved local programs.

---

[1]An April 29, 1980 Attorney General Opinion (State of Alaska Department of Law) eliminated the requirement for legislative approval of amendments to the ACMP, including approval of local district programs.

Standards for nine major categories of coastal uses and activities, ranging from coastal development and energy facilities to recreation and subsistence, are specified. The Coastal development standard prioritizes permissible uses and activities as follows: (1) water-dependent, (2) water-related, and (3) neither water-dependent nor water-related. Local governments may designate subsistence as the dominant use and activity of coastal areas in their district programs. With this one exception, the ACMP does not specify a hierarchy among the proposed uses and activities outlined in the nine standards. Local programs must, however, describe proper and improper uses and activities within their coastal area.

Coastal uses of greater than local concern, such as state and federal interests in energy activities, are assured adequate consideration in the ACMP. A balancing of interests is incorporated into the state program whereby documentation of (1) consultation with appropriate state and federal agencies, (2) the availability of reasonable alternative sites, and (3) the incompatability of a proposed use of state concern with a proposed site must be provided by the coastal district. Mediation procedures were outlined in ACMP and further amendments are pending which specify the services of a third-party professional arbitrator when a conflict arises between the Council and a coastal resource district during the local program approval process. The federal Office of Coastal Zone Management (OCZM) also reviews all district programs and may require a formal amendment process including an environmental impact statement prior to federal approval.

In addition, the ACMP outlines criteria for the maintenance of eight coastal habitats. Given the vast wetland resources of coastal Alaska, many looked to the ACMP for the implementation of a coordinated coastal wetlands management system. Initial Section 306 funding from OCZM included monies earmarked for such a management task to be undertaken by OCM in 1979 and 1980.

The ACMP defines the following two zones which must be included within all local coastal management programs and consistency determinations: (1) the zone of direct interaction and (2) the zone of direct influence. The zone of direct interaction is that portion of the coastal area where physical and biological processes are a function of direct contact between land and sea. That portion of the coastal zone extending landward and seaward from the zone of direct interaction and closely affected and influenced by the close proximity between land and sea is defined as the zone of direct influence. A series of biophysical boundary maps of the coast have been produced and incorporated into the ACMP. Along the low-relief Beaufort Sea coast, for example, the coastal zone as legally required in the ACMP extends 40 to 48 kilometers inland to include optimum waterfowl and shorebird nesting and polar bear denning habitats (2). Local program boundaries may diverge from those in the ACMP maps provided they (1) include areas to the extent necessary to manage uses that have or are likely to have direct and significant impact on coastal waters, (2) include all transitional and intertidal areas, and (3) are compatible with boundaries of adjoining coastal districts.

Vast portions of western Alaska, including the Aleutian Islands and lands adjoining the southern Chukchi Sea and most of the Bering Sea, are classified as Unorganized Borough without local planning and zoning powers. The state has services in these areas through Rural Educational Attendance Areas established along culturally distinct geographic units in 1975. The ACMP responds to these boundaries in coastal management planning by establishing a procedure for the organization of Coastal Resource Service Areas (CRSA). Following educational efforts and a petition and election process sustained by grants and matching state funds, local leaders can organize a CRSA for the purpose of local coastal management planning. In practice, the CRSA's have drawn heavily on the organization, expertise and leadership within the existing Native Regional Corporations. Regional centers, such as Kotzebue, Nome, Bethel and Dillingham serve as the headquarters for CRSA planning activities.

As of late April 1980, three district programs have been approved by the Council, while another 29 were in varying stages of development. One of these 29, the North Slope Borough Coastal Management Program (CMP) for the mid-Beaufort region, was submitted to the Council for approval in November 1979, and subsequently was withdrawn by the local resource district with great controversy in January 1980. Second-year Section 306 funds in the amount of $400,000 have been requested in addition to Borough funds to conduct further planning efforts. The OCM anticipates the resubmission of the mid-Beaufort region CMP for Council consideration in early 1981 (3).

The North Slope Borough

Before presenting a case history of the mid-Beaufort CMP, a brief description of the North Slope Borough is appropriate. Following a long cultural history of "local government" within nomadic and isolated groups and the strong stance given to local governing bodies in the Alaska Constitution, the Inupiat Eskimos filed for home rule municipality incorporation as the North Slope Borough in 1971. The desire for local government formation was an immediate response to the impacts and revenue potential of the expanding oil and gas development in the Prudhoe Bay area. Native lands claims activities in Alaska in the 1960's and early 1970's stimulated local leaders and provided an indirect, yet significant influence on desires for local control on Alaska's North Slope.

Following stormy legislative and judicial sessions, the Borough was incorporated in 1972. The Borough has full municipal planning and zoning authorities and somewhat limited taxation powers. The Borough is a powerful form of local government with regulatory authority over coastal land and water uses not specifically prohibited by statute or charter. Mayor Eben Hopson provided strong, independent leadership in the Borough government from its inception until his untimely death in June 1980. Also, the Borough is a wealthy local government, prohibitive costs notwithstanding, drawing roughly 62 percent of its projected FY 1980-81 operating budget of $95.1 million from the Prudhoe Bay oilfields (4).

The Borough extends from the Brooks mountain range (68° N) north to the Arctic coastline (up to 230 miles in distance) and roughly 600 miles west-to-east to the Alaska-Canada border (Figure 2). Approximately 1,200 miles of marine coastline fall within the Borough and six of the eight permanent communities are coastal. Roughly 60 percent of the 4,000 permanent residents (1977 population estimate) live in the City of Barrow, the seat of Borough government. The Borough is immense: 55.6 million acres, encompassing approximately 15 percent of Alaska's land and approximately one-third again as large as the combined acreage of the New England states. Several entities administer lands within the Borough's boundaries, including the (1) United States government, (2) State of Alaska, (3) Borough municipality, (4) Native village corporations, (5) Arctic Slope (Native) Regional Corporation; and minimally, (6) private individuals. Pending onshore and offshore boundary and jurisdictional litigations, the uncertain future management of 23.4 million acres in the National Petroleum Reserve-Alaska and differing management philosophies all combine to create a rather confused land management picture on the North Slope.

The coastal environment of Arctic Alaska is dominated by sea ice during 8-9 months each year. Marine fishes and motile invertebrates migrate offshore to deeper waters or concentrate in shrinking nearshore refuges of water as the sea ice thickens. Late fall storms can literally toss large ice floes onshore, gouging as they plough up the beaches. Offshore, in 13-18 meter water depths, the juxtaposition of Arctic pack ice and nearshore annual ice creates an active zone of colliding pressures, often resulting in the formation of ice ridges greater than 10 meters in height. This sea ice environment is the domain of the polar bear and ringed seal, both amply adapted to the rigorous environment. Onshore, seasonal freezing of the soil deepens to the underlying permafrost. Anadromous and freshwater fish winter in limited water pockets of river deltas, upstream springs, and deep lakes.

During the spring, the deltas and coastal floodplains of major rivers provide the first open water for resting and feeding birds during their lengthy migration to breeding grounds (5). Once the warming

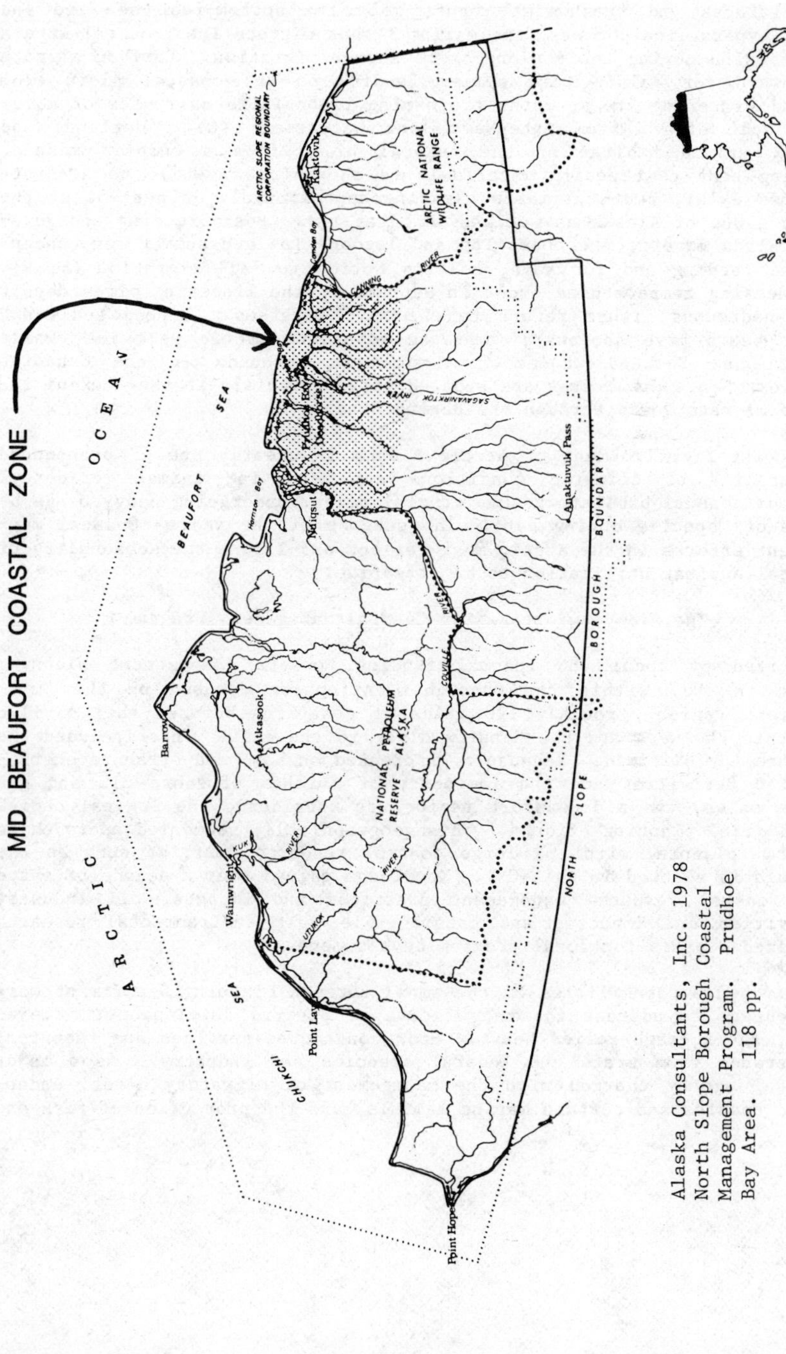

Alaska Consultants, Inc. 1978
North Slope Borough Coastal
Management Program. Prudhoe
Bay Area. 118 pp.

Figure 2

temperatures and freshwater runoff melt the bottom-fast sea ice and flush hypersaline pockets, the marine fishes migrate into coastal waters for feeding during their along-shore summer migration. Caribou migrate northward for calving and summer feeding on the coastal plain from inland wintering areas, often preferring floodplains as routes of movement and refuges from intense insect harassment (6). Wetlands and drier, upland habitat on the coastal plain form a complex mosaic, playing host to breeding waterfowl and shorebirds. While no accurate figures exist, best estimates classify approximately 65 percent of the north slope of Alaska as wetlands habitat (7). Post-breeding and juvenile birds move to the coastline and lagoons in late summer for concentrated feeding and for some, molting, prior to fall migration (8, 9). As freezing temperatures occur in September, the lingering birds depart and anadromous fishes return to the river systems. Ringed seals and polar bears move shoreward with the advancing Arctic pack ice, while Beluga and Bowhead whales from summering grounds in the Canadian Beaufort Sea and walruses and spotted seals primarily in the Chukchi Sea coast migrate south through the Bering Straits.

The coastal environment of Arctic Alaska is clearly one of pronounced seasonality of climatic conditions and migrating animal residency. Various coastal habitats assume crucial roles, undergoing heavy usage by migratory species during particular seasons of the year. Coastal management efforts in the Arctic must reflect and insure the seasonality of coastal habitat utilization to be effective.

The Mid-Beaufort Region Coastal Management Program

Supported by local and federal funding, coastal management planning began in 1977 within the Borough. Attention focused on the "mid-Beaufort region", roughly 120 miles of coastline between the Colville River to the west and the Canning River to the east. This included the Prudhoe Bay oilfield. Because of increased onshore and offshore energy-related activities and the imminency of further offshore oil and gas lease sales, the mid-Beaufort region was considered the logical choice for initial planning efforts. This geographically segmented approach to program planning within a large coastal resource district such as the Borough is allowed in the ACMP. Knowledge gained over a decade of state and federal resource management practices and impacts, oil industry activities and impacts, and considerable past environmental research provided a basis for local program development.

The in-house capabilities of the small Borough Planning Department were inadequate to address the many facets of program development. Therefore, the Borough relied heavily upon contracted services and technical assistance from state and federal agencies and industry. As a major federal agency charged with the management of migratory birds, endangered species and certain marine mammals, and the protection of fish and wildlife habitats, FWS became involved in the mid-Beaufort coastal program early in 1977. Our participation included (1) a synthesis of current research information for coastal resource inventories and special conservation area designations, (2) an analysis of current resource management practices, and (3) assistance in the development of environmental standards and specific habitat definitions within a coastal management ordinance (10). Several other parties also participated in the extensive review process.

In November 1979, the Borough's CMP for the mid-Beaufort region was submitted to the OCM for review and approval by the Council. A sense of urgency prevailed during the latter half of 1979 as the Borough strived to have a coastal program approved, at least in concept if not in a final implementable form, prior to the Joint Federal/State Beaufort Sea Oil and Gas Lease Sale which was scheduled for December 1979. The mid-Beaufort coastal program contained the following:

A. Framework Plan - a conceptual document outlining the intent (objectives and policies), area (boundaries), implementation (zoning regulations through a local permitting system), priority of uses, and administration;

B. Resource inventory map - a map delineating habitats important to migratory fish, birds and mammals;

C. Coastal Zone District map - a zoning map designating six land/water use districts; and

D. Ordinance (No. 33-79, North Slope Borough Assembly) - the specific legal mechanism whereby uses and activities were to be regulated.

In addition, the Borough submitted numerous reports to the OCM as supplemental information to the coastal resource program. These supplements document several crucial aspects of the program including: (1) public participation; (2) an analysis of alternative energy development scenarios; (3) an analysis of the sensitivities and incompatibilities of selected habitats with proposed uses for energy activities; and (4) traditional and current subsistence activities along the Beaufort Sea coast.

The ensuing review process rapidly positioned the mid-Beaufort region Coastal Management Program (CMP) in the middle of an extremely polarized arena. With state and national interests in oil and gas resources and local interest in subsistence resources as the primary coastal management issue, the Borough CMP and fledging ACMP were subject to widely divergent interpretations. From the plethora of lengthy dissertations on the approvability/disapprovability of the program, a short list of major issues are evident:

1. The legal authority vested in the Borough which allows local government regulation of oil and gas development;

2. The unreasonable restriction or exclusion of a use of state and federal interest, namely energy-related activities, through;

   a. the extent of zoning and priority designated for subsistence uses and activities; and

   b. the extent of use limitations applicable to specific districts established within the coastal zone;

3. The divergence of the Borough's CMP from the ACMP guidelines; and

4. The incompleteness of the CMP ordinance.

Does the North Slope Borough have the legal authority to regulate, through a local permitting process, the development of oil and gas resources on state and federal lands? A number of state, federal and private reviewers requested that a legal determination of this question be made by the state. Usurpation and diminishment of state agency authorities contrary to ACMP guidance was highlighted by some, whereas others (including OCM) argued that the apparent regulatory duplication in the form of local oversight may be necessary to assure effective and comprehensive coastal management. A subsequent opinion issued by the Attorney General, State of Alaska Department of Law, in May 1980 has helped clarify these questions (11).

Home rule municipalities, such as the North Slope Borough, have the authority to "exercise all legislative powers not prohibited by law or charter" (AS 29.08.010). Following review of all appropriate Alaskan statutes, the Attorney General's Opinion concluded that the only prohibition against Borough regulation of energy resources extraction is that outlined in the Alaska Coastal Management Program itself. The Council is prohibited from approving a local program which "arbitrarily or unreasonably restricts or excludes uses of state concern" (AS 46.40.060(a)). The Council, however, must approve the restriction or exclusion if the coastal district has (1) consulted with and considered the views of appropriate state and federal agencies, (2) based the restriction on the availability of reasonable alternative sites, and (3) based the restriction on an analysis which demonstrates the incompatability of the proposed use with the proposed site (AS 46.40.070(c)). Thus, the test of reasonableness regarding the restriction of a use of state concern rests upon the thorough and adequate documentation of the above criteria by the coastal district. Although the Borough provided OCM with numerous reports in support of this documentation, these reports did not receive wide public review by interested parties. Many commentors felt that this was a serious omission in the review process and requested the materials. Consequently, the OCM has requested that the Borough produce a detailed synthesis of these materials as part of an overall Program Document to be included with the Borough's CMP at the time of program resubmission to OCM.

In answer to the basic legal question, then, the Borough is within its rights on state lands, with the above-mentioned qualifications. According to the Attorney General's Opinion, the permitting system as outlined in the Borough's ordinance is also a legitimate mechanism for the execution of these rights, provided that basic constitutional safeguards of due process and equal protection are assured. The Opinion, however, further cautions that the exercising of local authority through a permitting process does not imply control over basic state policy formulation. The development of policies reflecting "the utilization, development, and conservation of all natural resources belonging to the state . . . for the maximum benefits of its people" (Alaska Constitution, Arcticle VIII, Section 2) remains in the hands of state government. The local ordinance and permitting requirements, then, guide the specific management process reflecting these state policies by establishing coastal management standards. The Borough ordinance, for example, contains one highly contested requirement for long-range exploration plans from industry. The OCM concluded that the Borough was within its rights in establishing this condition on future energy-related activities.

Borough oversight of activities on federal offshore lands is treated differently than that on state lands. The Borough's permitting system would apply only to activities in support of Outer Continenal Shelf (OCS) development which were located onshore or within the state's 3-mile offshore jurisdiction.

Another aspect of the mid-Beaufort region CMP subjected to serious disagreement and contention was the designation of the entire coastal zone as a subsistence zone, with the exception of the Petroleum Service Base and Production District (namely, the existing Prudhoe Bay oilfield and service infrastructure and the adjacent Kuparuk Development area to the west). The Borough's CMP gives priority and protection to subsistence activities and areas, and allows competing uses only where they do not conflict with the maintenance of subsistence resources. Several reviewers argued that the designation of such a large area as a subsistence zone (1) exceeded the intent of the ACMP guidelines, (2) would unreasonably restrict or exclude energy-related activities, and (3) was not justified based on current subsistence activities.

Clarification is needed to explain the complexities of the subsistence designation and the resultant controversy. The ACMP specifically allows for the designation of areas where subsistence uses have priority over all other uses in the coastal zone. There is no specified limitation on the areal extent of these subsistence areas, but there is a prerequisite for consultation with interested parties. The Borough submitted several reports to the OCM in support of this designation, but again these were not readily available for public scrutiny during the review process. The ACMP subsistence guidelines do not exclude competing uses, rather they require a study of possible adverse impacts and the generation of adequate safeguards before a potentially conflicting activity will be allowed in a subsistence area. The Borough's CMP takes this stipulation further by requiring that competing uses be allowed only where they do not threaten subsistence resources and their habitat (Framework Plan, Objective No. 3). The intent of these coastal management guidelines is to require prior consideration of the fish and wildlife subsistence resources and not to exclude other uses and activities within the coastal zone.

The concept, management, and practice of subsistence activities constitute a complex and evolving issue within the State of Alaska. The full range of differing viewpoints were highlighted forcibly during the review of the Borough's CMP. We suggested that the local government subsistence concept incorporate areas utilized in both current and traditional fishing and hunting activities. Several reviewers, however, focused only on the location and intensity of current subsistence activities. The difference, particularly in the mid-Beaufort coastal region, is that current subsistence activities encompass only a portion of that geographic area utilized over a much longer historical period. The areal extent of traditional subsistence activities reflects the inherent long term variability in location and population size of the various fish and wildlife resources along the Arctic coast. Annual variations in environmental conditions dictate the abundance and location of any particular protein resource available to the resident human population. Hunting and fishing activities for annually variable migratory species must be flexible and wide-ranging. Likewise, the management of habitat significant to the maintenance of these fish and wildlife resources must be broad in areal scope.

An additional element of the Borough's CMP subjected to considerable debate was the restriction placed on uses and activities within three specific districts: (1) Conservation, (2) Geophysical Hazard, and (3) Deferred Development. Conservation districts (C) as defined, are those coastal areas vital to the perpetuation of wildlife or of great cultural significance. They include the major river floodplains and deltas, two offshore barrier islands and an unusual offshore kelp community. Energy-related activities are allowed only when there is no feasible alternative site and the proposed use is compatible with the particular site. There is no total prohibition of industrial activities, as earlier program drafts stipulated; rather, special caution is implied so that inappropriate activities which could reasonably occur elsewhere would be scrutinized. A serious argument arose in the C's designated within the Sagavanirktok and Kuparuk River floodplains, as these areas coincide with sizable portions of identified oil and gas reservoirs. While acknowledging the values of the various C's, the OCM requested that the Borough coordinate additional planning with industry to specify procedures which would allow necessary industrial activities in these areas.

Although the ACMP requires local planners to identify known geophysical hazard areas where coastal development is likely to occur, many reviewers disagreed with the prohibition against any drilling and production activities within the designated Geophysical Hazard District (GH). Few could deny the hazards associated with the GH as it delineates the pressure-intense transition zone between shorefast and pack ice beyond the 12 meter isobath. Nevertheless, the OCM requested that the prohibition be eliminated and that the CMP incorporate language which facilitates cooperation among interested parties in the deployment of experimental drilling platforms in order to obtain necessary technological data.

The third area under serious contention was the Deferred Development District (DD), or that area between the 8 and 12 meter depth interval where petroleum exploration is prohibited for five years until experience and technological advances warrant such activities. Several reviewers argued that a 5-year moratorium constituted an unreasonable restriction. Others questioned the logic behind the DD designation and instead urged the Borough to encourage technological experimentation. The OCM called for the reconsideration of this DD, and recommended that the Borough include modifying terms and mediation language.

The final issues considered major points of contention in the Borough's CMP reflect disparate interpretations of the fledgling ACMP. Many criticized the definitions included in the CMP Ordinance. Reviewers believed the definitions to be inconsistent with or different than those incorporated in the ACMP guidelines. Others championed their specificity. The ACMP review process reveals that a major justification for the general nature of ACMP definitions was so that local district programs would provide the necessary specificity. We suggest the explicit terminology, while often difficult to attain, is a commendable goal, since the end product clarifies the management process.

While the ACMP allows a geographically segmented approach to program development, it does not condone incompleteness of program elements within one coastal area. The Borough's CMP Ordinance contained several reserved sections including those covering permit standards for petroleum development and production, pipelines, and roads. We contend that the hurried time frame contributed to the incompleteness of the Ordinance. At this relatively early stage of the ACMP system, however, the incomplete CMP also may stem from a misinterpretation of what constitutes an approvable local program. Piecemeal program development frustrates the intended goal of comprehensive coastal management and leads to confusion and inconsistencies within a program.

Current Status and Prospectus on Arctic Coastal Zone Management

The Borough's CMP for the mid-Beaufort region is currently being revised and expanded. Completion of the Ordinance and the synthesis of a comprehensive Program Document explaining the rationale behind various CMP elements and outlining ACMP requirements are two primary tasks still facing the Borough. Best estimates suggest resubmission for program approval early in 1981. The OCZM has indicated that they will require the formal amendment approval process at the federal level with an environmental impact statement likely. With an approved local program unlikely before 1982 at the earliest, management of coastal uses and activities in the mid-Beaufort region falls back on the general ACMP standards. Those familiar with the ACMP guidelines and consistency procedures are in a minority, but repeated reference to the ACMP will conceivably raise its effectiveness as a management tool.

What are the substantive products of the Borough's planning process to date? Awareness of the values of and conflict of uses in Alaska's Arctic coastal zone has increased within Alaska. It is doubtful, however, that sustained interest in coastal management in Alaska has occurred elsewhere. However, a recent resolution adopted by the Inuit Circumpolar Conference indicates international interest among the Inuit community in this topic (12). A valuable lesson for those who felt constrained or threatened by the Borough's CMP is the necessity for early and constructive involvement in the coastal resource district planning process. After 1 ½ years of participation, we felt that many of FWS environmental management concerns had been addressed adequately in the Borough's CMP. Another outcome is a much-diluted Borough Interim Zoning Ordinance which facilitates local government tracking of industrial activities without providing substantive new directions in coastal management.

Coastal planners and government leaders of other coastal regions of rural Alaska are observing the North Slope Borough's program evolution and approval process. Likewise, Borough personnel are sharing their knowledge and experience with coastal communities along the southern Chukchi and Bering Seas. With less elaborate governmental structure and reduced revenue-generating potential, these Boroughs and Coastal Resource Service Areas (CRSA's) will benefit from North Slope Borough experiences. While the specific coastal environments in western Alaska differ from that to the north, subsistence activities and future oil and gas exploration in these regions are similar. Given an accelerated OCS lease schedule and the rudimentary stage of coastal planning, it is questionable whether local programs will be in effect in western Alaska prior to lease sales. This underscores the importance of the ACMP.

The controversy over the North Slope Borough's CMP highlighted some inadequacies within the ACMP and precipitated the development of proposed regulatory amendments to the state program. The ACMP can perhaps best be described as in the turmoil of adolescence - subject to not only wide interpretation but also clarifying refinements as the program matures. It is paramount that this evolution of the ACMP maintain its integrity and increase its clarity on key management issues since the ACMP will be the only tool which approximates comprehensive management of coastal regions in western and Arctic Alaska in the near future.

References

1. State of Alaska Office of Coastal Management and the U.S. Department of Commerce Office of Coastal Zone Management. 1979. State of Alaska Coastal Management Program and Final Environmental Impact Statement. 578 pp.

2. Alaska Department of Fish and Game. 1976. Coastal Biophysical Boundaries Beaufort Sea: Canning River - Colville River. Anchorage, Alaska.

3. Lawson, T. 1980. District program status report. Office of Coastal Management, Juneau, AK. April 25, 1980.

4. North Slope Borough Budget Document ordinance 80-3. FY 1980-81. North Slope Borough, P.O. Box 69, Barrow, AK 99723. 341 pp.

5. Schamel, D. 1978. Chapter 5: Birds. Pages 152-173. In Environmental Assessment of the Alaskan Continental Shelf Interim Synthesis: Beaufort/Chukchi. National Oceanic and Atmospheric Administration. Boulder Colorado. 362 pp.

6. Cameron, R. and K. R. Whitten. 1979. Distribution and movement of caribou in relation to the Kuparuk Development Area. First Interim Report. Alaska Department of Fish and Game. Fairbanks, AK. 32 pp.

7. Selkregg, L.L. 1975. Alaska regional profiles, Arctic region. State and Joint Federal-State Land Use Planning Commission for Alaska. Anchorage, AK. p. 16.

8. Connors, P. 1979. Shorebird dependence on Arctic littoral habitats. Outer Continental Shelf Environmental Assessment Program, Research Unit No. 172. Annual Report. 60 pp.

9. Bergman, R., R. Howard, K. Abraham, and M. Weller. 1977. Waterbirds and their wetland resources in relation to oil development at Storkersen Point, Alaska. U.S. Department of Interior, Fish and Wildlife Service. Resource Publ. 129. 38 pp.

10. Keiser, G. 1979. Coastal wetland habitat along the Beaufort Sea coast. U.S. Fish and Wildlife Service, Fairbanks, AK. 10 pp.

11. Alaska Department of Law. 1980. An Attorney General's opinion on local zoning authority over State oil and gas development (J-66-536-80), May 12, 1980. 16 pp.

12. Crampton, B. 1980. Culture, land claims top ICC resolutions. Fairbanks Daily News Miner. Fairbanks, AK. July 7, 1980.

The New Jersey Coast: Going, Going, Gone?

Sheldon Abrams, President
D.W. Bennett, Executive Director

American Littoral Society
Highlands, N.J. 07732

The New Jersey coast has it all...or soon will. It has 127 miles of sea beach, heavily indistrialized coastline at its ends (Camden on the Delaware River and Newark/Jersey City on the Hudson River). With Philadelphia and New York on its borders, it is neighbor to two of the nation's great seaports. New Jersey has Atlantic City, the only place in the United States where a person can gamble (legally) within the sound of the surf, yet just a few miles northwest of Atlantic City is one of only two breeding pairs of peregrine falcons east of the Mississippi.

New Jersey's leading industry is petrochemicals, and it, Pennsylvania, and Delaware have the only important refining capacity in the northeast. The state's second leading industry is tourism. Boosters of Atlantic City are proud to point out that one-sixth of the U.S. population can drive there on a tankful of gas, which means they can drive to any of the state's beachfront communities also, so seashore recreation is the major part of New Jersey tourism.

New Jersey also has the Hackensack Meadowlands, about 10,000 open acres of land within 10 minutes of Manhattan. It is about the most valuable piece of undeveloped real estate of that size in the world.

Everything mentioned above lies in the coastal zone: refineries, the Hackensack Meadows, the peregrines, the petrochemicals, and the seashore recreation, both the wild beaches and Atlantic City. It is this coastal zone that state officials are busy trying to either regulate, preserve, or develop, depending on one's point of view. It is our thesis that despite good, tough state laws, and despite some eight years of planning and five years of experience, New Jersey's coastal zone plan, one of the nation's first and one of its most comprehensive, is failing to protect the coastal environment. What follows is an investigation of the state plan, some specific cases to illustrate the point of this paper, and then a discussion of whether the planning here -- or in any state -- can do much more than slow the process of coastal development and guide it ever so slightly into areas of less but still significant environmental sensitivity. First some recent history:

SOME RECENT HISTORY

Beginning in the mid-fifties, New Jersey underwent 15 years of massive dredge and fill development of its wetlands. One such development alone, on the west shore of Barnegat Bay, called Beach Haven West, consumed 2000 acres of wetlands in the process of digging 21 miles of lagoons and producing upland (though barely upland) a small city of one story vacation houses. In those days, such a house sold for about $8000. From then through the sixties and early seventies, New Jersey lost about 24 percent of its salt wetlands -- 62,000 acres. This development was encouraged by the state. Indeed, it was in the active real estate market, selling off tidelands as quickly as an applicant could come to the window. Tidelands were being filled and dredged without permits. Sometimes, when such a dredger or filler was caught in the act, he was told gently to apply for a permit which he was assured of getting.

A number of events came together around 1970 to halt much of this sort of development and environmental damage.

First, a developer busy erecting a bulkhead below the tideline and filling behind it with unpermitted dredge material, was reported to the state which replied to the reporter that it was going to go ahead and grant a permit, after the fact because, in the state's own words, "the damage had already been done, and, besides, someone up river had done much the same thing and they couldn't apply tougher standards downstream, could they?" As it turned out, they could indeed. On the investigation, it turned out that the citizen council dealing with state-owned tideland was (1) holding its "public" meetings privately, (2) rushing through controversial items, and (3) neglecting to consider the value of the tideland in its natural state as part of its deliberations.

Second, the state went through a gubernatorial election, and one candidate, William T. Cahill, pledged that he would break the Department of Conservation and Economic Development into two departments, one of which, the Department of Environmental Protection, would have that one function to perform, rather than having to trade environment for development.

Third, the early seventies marked a rebirth of environmental interest and a surge of legislators eager to be a part of a new, vocal political movement. It was a time when a piece of environmental legislation could fly through legislatures.

The direct results of all this in New Jersey was the passage of two strong laws, the Wetlands Act, which regulated land use on about half a million acres of tidal wetland in the state, and two years later, the Coastal Area Facility Review Act (CAFRA), which extended state land use jurisdiction upland of wetlands to include another 900,000 acres of lands close to the coast. Already on the books (actually from the time of the English kings) were regulations dealing with state lands from the high tide line out to the three-mile limit. These lands belong to the people and are held in trust by the state, which is allowed to sell or lease them, if such action is in the public interest.

Put together, the three laws mean a decision has been reached that coastal lands are so valuable to all the people that the state now decides what can be done with them. Starting from the ocean and proceeding inland, here is the general effect of the laws:

1. Tidelands, those either always underwater or washed by normal tides can be owned (privately or publicly) but it is extremely difficult now to get permits to fill those lands, to make dry upland out of them. Licenses are given to do things on them: place nets, drive piles, or put in docks, but long term occupancy is not easy to get.

2. Wetlands, defined by vegetation and water level, but generally wet and supporting productive salt marsh grasses, can be modified only if in the public interest, if the damage to the environment is minimal, and there is no upland site to use instead. Wetland filling, diking, ditching, and otherwise modifying has slowed in New Jersey but it still happens.

3. Coastlands -- the coastal zone as defined by CAFRA -- has lesser land use controls. The law, simply stated, allows almost anything as long as the developer goes through an environmental impact statement process. The state has published a long set of guidelines, most of them couched in enough loose language to give administrators slack during decision making. An important loophole in the CAFRA legislation is the exemption of housing developments of fewer than 25 units, a compromise put into the bill to help get it passed, and a loophole that the state is now trying to plug (more about that below).

With that general outline of how the state's coastal zone management laws are supposed to work, let us list a set of case studies to show how the laws do actually work, cases to illustrate the administration of tidelands, wetlands, and coastal lands, and some comments on the state's ideas for more or different legislation.

TIDELANDS

Decisions on the sale or rent of tidelands are made by a non-paid governor-appointed Natural Resources Council, made up of citizens who gather in the state capital once or twice a month and go through an agenda of possible land deals, proposed by applicants who wish to buy, fill, build on, and occupy land that is now underwater. Ten years ago, most of this land was dealt away casually, with little effort to charge a fair market price for it and with almost no regard for the environmental question involved. In fact, the state was really in the land game, a sort of tidelands hustler. Money from such sales went and still does go into a school fund, as if accumulating money in such a non-political simon pure fund somehow cleanses the sale. Most tideland sales today are for small parcels, sometimes as little as 200 square feet where an upland owner along a creek needs that much land to gain a building lot big enough for local zoning. Some large parcels in urbanized areas -- Camden, Jersey City, and in the Hackensack Meadows can be considerably larger, up to 40 acres, and the money is also considerable -- millions of dollars.

In general, the state is tough on such applicants, often sending them away with a suggestion that they scale down the development proposal for which they want the extra land. Exceptions prove in most cases to be where applicants can prove an absolute need for waterfront access, where adjacent areas have already been sold off and developed, and where the resultant development can be shown to provide lots of jobs. (Economic considerations are always important.) Often, the applicant trades similar land (and tideland) in a nearby area in order to be able to develop the section in question.

(The Corps of Engineers, under Section 404 of the Clean Water Act Amendments and under the Rivers and Harbors Act of 1899, has jurisdiction over these lands also. More about the Corps later.)

It is the feeling among many environmental groups that it is and should be almost impossible for an applicant to be able to convince the state that <u>any</u> filling of tidelands is necessary, given that New Jersey spent the first 180 years of its life peddling tidelands as fast and as aggressively as it could, resulting in mile upon mile of bulkheaded, riprapped, filled, polluted, dredged waterway, from the George Washington Bridge to the backside of Staten Island, and in the Delaware River from Trenton south past Camden.

"Enough," much of the public is saying. Better the state encourage the redevelopment and reuse of abandoned already bulkheaded and stressed waterfront before new parts of the coast are sold off for development.

WETLANDS

New Jersey's wetlands act is broad-ranging biologically based state zoning; it is patterned after a similar Connecticut law, that is in turn very close to the Massachusetts tidal wetlands law.

The wetlands law is short and clear. It says that wetlands are those lands along the coast that are affected by tidal action (fresh and salt). They are defined by vegetation. Such lands are subject to state permits. As far reaching as the law is (and in some towns it has ruled out virtually all development), it has withstood attacks through the courts, partly because it is based on sound biology and partly because uses by the owner, while limited, are still enough to have convinced judges that the law is not a "taking." Also, an upland owner with wetlands fringes can still develop his upland leaving the wetland section alone. Thus, judges have ruled that the owner has not been deprived of all of his land's uses. And indeed, people are still willing to buy wetlands, albeit for a sharply depressed price. (A duck hunting club recently bought 200 acres of prime waterfowl and muskrat habitat -- all of it wetland -- in south Jersey for $57 an acre. If that land could have been bulkheaded and filled, it would have been worth 100 times that.)

Since the passage of the Wetlands Act, the amount of such land that the state and its citizens are losing to development has dropped, but exceptions are made, some of them disturbing: wetland in Cape May County was turned into a softball field and an Atlantic City casino was granted state permits to fill wet- and tideland for a freshwater reflecting pool. The plan for development in the Hackensack Meadows calls for filling about 4000 acres of wetlands for such non-waterfront-dependent uses as residential housing and football and basketball stadiums.

But the wetlands law is in place and wetland destruction has slowed.

COASTAL ZONE MANAGEMENT

This third (in order of passage) law, the Coastal Area Facility Review Act, passed through the New Jersey legislature at about the same time as the passage of the federal coastal zone management act. It was followed by several years of rule making and the filing of environmental impact statements so it would be approved by the federal Office of Coastal Zone Management. At each level of writing and rewriting the regulations, the state has been exposed to pressures from developer interests wanting as much leaway as they can get, and public interest groups like the American Littoral Society wanting the tightest, most environmentally protective language possible. The result is a book of rules surrounded by a collection of tensions: an owner of coastal land proposes a development and begins the process of getting his state permits. Public notice of the application results in comments and suggestions from environmental groups. The state's division of coastal resources moderates. Most applicants get their permits -- about 90% of all CAFRA applications are approved (though some applicants are turned away early in the process and don't show up as rejections).

Here are some sketchy examples of how the CAFRA process works:

<u>A developer</u> seeks approval for 400 units of residental housing that will encroach on two small tidal streams, impinge wetlands, and cause large non-point water pollution problems. The state directs him to cut the density of development, avoid both streams, and design better drainage systems to handle runoff. The permit is granted.

<u>A municipality</u> seeks to sell 30 building lots on a barrier island spit. The state says yes, but only if the town submits conservation plans for the remaining undeveloped spit (about one mile of beach, dune, upland and marsh). The town submits such a plan which the state accepts. The permit is granted (and now under appeal by environmental groups).

<u>The Hackensack Meadowlands Development Commission</u> plans a massive riverfront residential development -- town houses, high rises, shopping malls, and movie theaters. The development is exempt from CAFRA regulation because it falls outside the law's boundary.

<u>A 50-unit motel</u> is planned for a resort community in southern New Jersey. Preliminary talks indicate that the proposal will be denied. The developer changes his plans and builds two 24-unit motels on the site. Because each motel has less than 25 units of housing, it is exempt from CAFRA.

From these few examples, it can be seen that the state's coastal zone planning laws are not strong enough to protect the coastal environment. Major defects are that (1) the state has not defined and set aside critical areas, claiming that to do so would open it to suits claiming a taking of the land without compensation, (2) the plan fails to protect the coast against the cumulative impacts of many small (less than 25-unit) developments, and (3) barrier islands are bearing much of the brunt of these weaknesses, most pointedly in the failure of barrier island communities to control construction on sand dunes and too close to the ocean.

One avenue of environmental protection we plan to pursue is through more stringent exercise of the provisions of Section 404 of the Clean Water Act, which we read as giving the Corps of Engineers greatly expanded jurisdiction over coastal land use.

This latter weakness was recently addressed by the introduction of another state law, called the Dune and Shorefront Protection Act, which, in short, would control barrier island development by closing some areas to any development and banning the reconstruction of homes damaged to more than 50% of their value by coastal storms. The bill ran into a storm of protest, led for the most part by segments of the construction industry -- builders, real estate agents, banks, and mayors (sometimes one and the same).

THE FUTURE

Lack of development pressure on New Jersey's coastal lands and waters in the late 1970s has been as much a product of economic difficulties as it has been of coastal zone management laws, but with Atlantic City now developing very fast, a new surge of land interest along the coast emerges. The coast is in for another bout of environmental damage, tempered slightly by the laws we have been discussing. But money can be made along the coast, and economic interests have a way of exerting great pressures on the state agencies whose role it is to protect the environment.

Balanced against those pressures are the less well financed and staffed interests of the environmental community which often finds itself at odds with the Department of Environmental Protection, the agency it expects to do what its title says.

The future of the New Jersey coast is thus: pressures to develop, pressures to preserve, and a state agency in the middle. The result: a coast looking much the way we have learned to expect, developed the way development interests want it with slight design improvements to lessen, but not drastically cut, the environmental impacts.

The Networking Approach to Coastal Zone Management

By
Earl H. Bradley Jr.
Chief, (Local) Technical Assistance Program
Coastal Resources Division
Tidewater Administration
Maryland Department of Natural Resources

Introduction

Maryland's Coastal Zone Management Program was one of the first programs to be approved which uses the networking approach of management of its coastal resources. Rather than passing comprehensive new legislation, Maryland established its program largely by linking together existing programs through an Executive Order, Memoranda of Understanding with the other state agencies, and contractual agreements with local governments in its coastal areas.

The program basically entails (1) providing a unified framework to guide the activities of state and local governments agencies concerned with coastal resources and activities; (2) providing technical and financial assistance to such agencies so that they can undertake their activities in a more effective manner; and, (3) promoting coordination among such agencies to ensure that they carry out their activities in a manner consistent with the program's policies. The extent of the program's coverage includes the tidal waters of the State and Baltimore City and the sixteen Maryland counties bordering the Atlantic Ocean, Chesapeake Bay, and Potomac River below Washington, D.C.

The purpose of this paper is to discuss the strengths and weaknesses of the networking approach to Coastal Zone Management based on two years experience of program implementation in Maryland.

Development of Maryland's Coastal Zone Management Program

The basic reason for adopting the networking approach to coastal zone management in Maryland was that by the summer of 1974 when the State first received funding under the provisions of the Federal Coastal Zone Management Act, the State had already enacted substantial legislation concerning the management of coastal resources and there was a reluctance in the state legislature and elsewhere to pass coastal zone management legislation which could be viewed as adding another layer of bureaucracy to the management of coastal resources. Prior to 1974, the State had already established a wetlands permit program, a waterways obstruction/construction permit program which covered the 100 year riverine floodplain, a sediment control program, a power plant siting program, a program for the acquisition of open space and recreation areas at the State and local levels, and a State water quality program. The 1974 legislative session enacted two additional pertinent pieces of legislation, a State Land Use Act which established a State Critical Areas program and gave the State the right to intervene in any local land use decision of greater than local concern, and a Beach Erosion Control District Act which protects the State's Atlantic Coast beaches from encroachment. In addition, all local governments had developed comprehensive plans, zoning ordinaces, and water and sewer plans pursuant to State law.

There was felt a need to pass legislation to fill two gaps in the existing legislative framework. This recognition led to the passage of a Coastal Facilities Review Act in 1975 which provided a mechanism for regulating the siting of oil and natural gas and associated support facilities in Maryland's coastal counties, and a Flood Control-Watershed Management Act in 1976 which provided a mechanism for coordinating State floodplain management efforts, the Federal Flood Insurance Program and related efforts.

Program Development

The main thrust of the program development efforts in Maryland was focused in three areas: first, on developing a set of goals and objectives relating to coastal resources and activities to provide a unified policy framework for implementing State and local governmental programs dealing with coastal resources and activities; second, on strengthening the information base on which such programs depend for their effective implementation; and third, on promoting coordination among such programs to ensure that all relevant coastal resources issues are addressed in a manner consistent with the common set of goals and objectives.

Thus, much of the Coastal Zone Program development effort in Maryland was spent on identifying and obtaining agreement on a set of goals, objectives and policies pertaining to each coastal activity of concern based on existing laws and and regulations, and on developing an implementation mechanism to ensure that coastal resource-related programs are carried out in a manner consistent with the program's goals, objectives, and policies. An Executive Order was issued mandating that State agencies carry out their activities in a manner consistent with the goals, objectives and policies of the program. Memoranda of Understanding between the Department of Natural Resources and other State agencies were agreed upon specifying the mechanisms that would be utilized by each agency to carry out the program, and a Secretarial Order was issued within the Department of Natural Resources to ensure consistency of Departmental actions. One unit with the Department of Natural Resources, the Coastal Zone Unit of the Energy and Coastal Zone Administration (since reorganized into the Coastal Resources Division of a new Tidewater Administration) was given responsibility for acting as a catalyst to ensure that that program was effectively implemented. In addition, contractual agreements were established with local governments to provide technical and financial assistance to them so that sufficient attention would be directed to the management of coastal resources at the local level.

An advisory body, the Coastal Resources Advisory Committee, was also established to obtain the input of the interested "public" in the development and administration of the program. This committee consists of over 90 members with local governments, citizen groups, industrial, and commercial organizations, environmental groups, and other special interests represented on a voting basis. State and federal agencies and academic interests are represented on a non-voting basis.

To provide a sound technical basis for implementation of Maryland's Coastal Zone Management Program, a number of technical studies were undertaken including the Tidal Wetlands Study, the Upland Natural Areas Study, the Historic Shoreline and Erosion Rates and the Shoreline Structures in Maryland: Their Location and Extent map portfolios, the Major Facilities Study, support of the Chesapeake Bay Earth Science Study, and flood-plain delineation efforts in conjunction with the State's flood-plain management program. The results of these studies are being used to improved the effectiveness of the management programs which provide the regulatory basis for the program. Such programs include the State's Wetlands Permit Program, Water Quality Programs, Critical Areas Program, and Flood Control-Watershed Management Program.

Two foci of the program were established: project evaluation and program review. The purposed of the project evaluation process is to insure adequate review of projects in coastal areas by identifying and comprehensively evaluating those projects needing special attention because of their potential impact on coastal resources. Particular attention is focused in cases where existing information is inadequate upon which to base management decsions and additional study is needed to insure that all relevant issues are adequately addressed.

The purpose of the program review process is to provide a means for reviewing existing programs and procedures dealing with coastal resources to ensure that adequate consideration is given to Coastal Zone Management concerns and to ensure their consistency with the CZM Program. It is used to review the implications of proposed legislation affecting coastal resources, issuance of new or amended regulations, development or revision of state and regional plans and local comprehensive plans and zoning ordinances, and similar actions. The Coastal Zone Unit of DNR was given the responsiblity for conducting these processes in conjunction with other State agencies and local governments.

Actual Implementation of the Program

The remainder of this paper will be devoted to an analysis of how well the networking approach to program implementation has worked in Maryland. The program's first year of implementation had some difficulties but a framework was laid for more effective implementation of the program is succeeding years. To a certain extent, it was found that many of the features of the program that give it much potential for success have pitfalls that needed to be avoided if the program is to be truly successful.

Since the unit that has direct reponsibility for implementing the program is not in a dictatorial position, considerable effort had to be expended to define and evaluate the relationship of its activities to ongoing programs and permit actions. Within the permitting activities of the Department of Natural Resources, inhouse working relationships were generally established and several positive contributions were made both with regard to providing a better information base on which management decisions can be made and on resolving agency conflicts. For example, on a proposal regarding the siting of a marina financed by the State Waterways Improvement Program, a revised plan developed by the Coastal Zone Unit reduced the adverse impact of the proposed marina on tidal wetlands while providing approximately as many recreational boating docking facilities. The Unit also provided needed expertise on recreational boating facility proposals to the State Wetlands Permit Program, developed State mooring buoy regulations, and has assisted one county in revising its regulations dealing with the siting of marinas and other recreational boating facilities.

However, in some of its project evaluation efforts, the Coastal Zone Unit tried to address too many issues in a comprehensive fashion and as a result much of the information developed was not developed in a form or a timeframe useful to agencies making regulatory decsions on projects in the areas studied. As a result, during the past year more attention has been focused on narrowing the issues to be evaluated in project evaluations to those which can be analyzed in a timely manner to improve relevant regulatory decisions. In addition, some difficulty was encountered in establishing mechanisms by defining the areas in which Coastal Zone Unit could make positive contributions in plan formulation and project development over which other state agencies have the principal review and coordination responsibility such as State transporation facility and sewage treatment plant planning and construction.

The lack of legislatively mandated regulatory responsibility on the part of the Coastal Zone Unit for routine regulatory activities allowed considerable flexibility on which aspects of coastal zone management would be given priority. Much emphasis during the first year was placed on undertaking research efforts on subjects of coastal zone management interest, with result that program review activities and intergovernmental coordination efforts were somewhat shortchanged. During the past year, greater attention has been paid to these areas. During the past legislative session pertinent legislation was reviewed under the purview of the program's advisory committee with recommendations made both to the pertinent state agency and the legislature. More attention has also been given to intergovernmental activities both at the state and local level. At the state level, considerable effort is being focused on establishing detailed working arrangements with on-going programs with the emphasis on bolstering existing efforts. For example, financial assistance has been given to the States floodplain management program so that 100-year floodplain delineation information produced by Federal Flood Insurance Studies at a scale of 1:24000 can be mapped at the scale of the tax maps (1" = 600') used by local governments thus putting the information in a format that is more usable by local governments. Also considerable staff effort has been expended in working with the Maryland Department of State Planning in refining the management approach to areas designated as State Critical Areas.

At the local level, progress has been made in incorporating coastal zone management concerns into local governmental planning and regulatory activities. The reliance on previously legislated programs has allowed the program to develop a more cooperative working arrangement to be developed with local governments than if the program had been initiated as a major new regulatory program. With pass-through funds provided by the program, the counties have worked hard to integrate coastal zone management concerns into their planning and regulatory activities. In the Baltimore Metropolitan area, a major effort has been undertaken by the Coastal Resources Division, other State agencies, the Regional Planning Council, and the local governments in the area to identify areas in their planning and regulatory activities needing attention. The purpose of this effort has been to insure that coastal zone management concerns are adequately addressed at the local level by identifying issues of major concern and to develop implementation mechanisms to address such issues. Similar action on a less formal basis has been undertaken by other coastal counties.

One deficiency that was noted in the first year of program implementation was a lack of visibility of the program's actions both in documenting the positive results of its actions and in establishing its proper relationship to other state agencies. This problem was alleviated by a reorganization in the summer of 1979 which transferred the Coastal Zone Unit into a new Tidewater Administration with the State Waterways Improvement Program and Tidal Fisheries Program plus State Boat Act Administration. These programs have objectives more closely allied with program than was the case in the Energy and Coastal Administration. This reorganization also promoted closer involvement by the program in development of a comprehensive approach to fisheries management for Chesapeake Bay since it promoted day-to-day interaction between the Coastal Zone Unit (renamed the Coastal Resources Division) and the Tidal Fisheries Program.

Conclusion

Based on the experience provided by the first two years of implementation of Maryland's Coastal Zone Management Program, it can be concluded that the networking approach provides the opportunity to concentrate on pressing issues previously not adequately addressed by State and local governmental programs. It also promotes a more cooperative approach to coastal zone management, encouraging greater willingness of State and local agencies to participate in the program since their present authorities are not threatened. However, if a State is to make a networking approach work, a strong base of relevant coastal resource management laws needs to exist and major efforts need to be undertaken in four areas. These are in: 1) establishing an agreed upon set of relevant goal, objectives and policies for the program; 2) establishing effective coordinative mechanisms;3) establishing support for the program by other state agencies, local governments, citizen organizations and special interests; and 4) defining the areas in which the program can effectively supplement existing activities to accomplish more effective management of coastal resources and activities. Particular attention needs to be paid in identifying the positive contribution the program can make to the more effective implementation of the programs on which is based.

LOCAL ASSISTANCE: A PROBLEM IN TRANSLATION

STEPHEN OLSEN
DIRECTOR, COASTAL RESOURCES CENTER
GRADUATE SCHOOL OF OCEANOGRAPHY
UNIVERSITY OF RHODE ISLAND
KINGSTON, RI 02979

Coastal Zone Management and planning in Rhode Island

Rhode Island's 419 miles of shoreline comes under the jurisdiction of 20 cities and towns. In 1971, at the height of the environmental movement, the state passed sweeping legislation that created the Coastal Resources Management Council (CRMC) and granted it broad powers to plan for, and to manage, coastal and marine resources. The Council's eighteen members represent coastal communities of various sizes and are appointed by the legislature, the Governor, and the Lieutenant Governor. The CRMC has exclusive state permitting authority for activities proposed in tidal waters (excluding fishing). Council permits must also be obtained for activities proposed within 200 feet of mean high water or on coastal physiographic features such as saltmarshes, barrier beaches, and sea cliffs. In addition, the CRMC may require a permit for selected activities proposed elsewhere in the state if it can be demonstrated that there is a reasonable probability of conflict with the CRMC plan or damage to the coastal environment. In situations where other permits, such as local permits, are also required, the CRMC will not act until those permits have been obtained. The CRMC, therefore, reserves the right of the ultimate veto.

The CRMC views its policies and regulations over terrestial activities and uses as minimum standards. Local governments, however, have limited authority for controlling environmental alterations. Their principal authorities stem from the Zoning Enabling Legislation passed in 1921, the purpose of which was to protect single-family housing in built-up areas from "nuisances and adverse influences." The zoning legislation is oblivious to air and water pollution or the concept of land as a limited natural resource with widely-ranging characteristics and suitability for various uses. It does not provide municipalities with authorities to recognize natural hazards or natural constraints on developments.

In addition to zoning, many coastal communities have prepared, or have had prepared for them, comprehensive development plans and recreation and open-space plans. Many of these are excellent documents but too often they have not been approved by the town council and are ignored and forgotten. All of Rhode Island's coastal communities enjoy the full benefits of the National Flood Insurance Program. Unfortunately, this has been a mixed blessing since federally-subsidized insurance has encouraged building on such highly vulnerable and fragile features as the barrier spits that line the state's ocean shore. The barriers were platted in the 1930's, but after the almost complete destruction of buildings by both the 1938 and 1954 hurricanes, local insurance firms would not provide insurance for renewed construction. Flood insurance removed this constraint and neither the towns nor the CRMC have been able to halt building on many barriers. The CRMC has, however, held the line on barriers that could be classified as undeveloped in 1973. The Flood Insurance Program has brought problems in areas other than barrier beaches since it makes no provisions against building in erosion-prone areas and since enforcement of flood proofing and elevation standards has been uneven.

For the past three years, there have been unsuccessful attempts to pass comprehensive land management legislation that would provide local governments with greater authorities to control and direct development. The bill also provides for state level review of selected major proposals. The local governments, although eager for authorities that would permit them to adopt such resources as cluster zoning and to recognize lands unsuitable for development are very jealous of their perogatives. The state-level review provision has been a major stumbling block to passage of the land management bill for this reason. The bill has been extensively rewritten and will hopefully pass eventually. If and when this happens, many of the severe limitations to present zoning can be addressed by local governments.

Development of the Local Assistance Program

One phase of the preparation of the comprehensive Rhode Island Coastal Resources Management Program, or plan, that was formally adopted by the CRMC in 1977 and approved under the 306 requirements by the federal office of coastal zone management (OCZM) in 1978, was a review of the conformance of local zoning and planning with the state's CZM program. It was quickly reaffirmed that local zoning made little or no distinction between lands suitable for development and such features as saltmarshes and dunes and that environmentally enlightened development plans had often never been adopted or were ignored. The initial response to this, on the part of OCZM and concerned groups such as the National Resources Defense Council (NRDC), was to press the coastal program to work with local governments to change their zoning and ordinances as quickly as possible. The state's Statewide Planning Program, therefore, undertook a detailed review of the zoning adopted plans of all coastal communities which revealed that almost without exception the problems were the same in all communities. For example, virtually all saltmarshes and barrier spits are zoned for intensive residential development.

In 1979, the Coastal Resources Center at the University of Rhode Island which had prepared the state's CZM program with the CRMC began to use 306 funds to assist local governments in meeting the spirit and substance of the adopted CZM plan. It became increasingly clear why the state had chosen in 1971 to create a state agency that would oversee development in the coastal zone rather than decentralize planning and authority to the 20 cities and towns. One reason is the state's small size, Rhode Island is the size of the average American country, and the other is the enormous difficulty of developing and adopting local comprehensive development plans. The Town of South Kingstown, for example, which had obtained broader authorities from the state legislature than most other communities, undertook such an effort in 1973. After enormous effort over a period of three years, a new town plan was adopted. It still leaves much to be desired. Even without such efforts, however, the CRMC, through its permitting function, can and does protect fragile and important coastal features. At the same time, many important prerogatives still reside with the local governments. They initiate projects (subject to CRMC permits) on town-owned shorefront property, and most importantly, they exert considerable control and influence over developers at the initial stages of planning activities in the coastal zone. The cooperation of local building inspectors and other local officials is essential if the CRMC's permitting process is to proceed in an efficient manner.

The first step taken by the Center in the local assistance program was to visit, with the chairman of the CRMC, the top officials in each coastal community. The goals of the state's CZM Program were reviewed, the purpose of the assistance program set forth and local problems discussed. Those exploratory meetings were extremely useful. It became clear that there was little to be gained by pushing local governments to change their zoning until the land management bill passed. We found few disagreements on the goals and even the priorities for improved coastal management in each specific city or town. The major problems were also similar. Local governments are shorthanded, most towns do not have a planning staff and decisions are made by boards of local citizens who receive no monitory compensation for their work. Most communities had specific problem areas where they needed technical assistance and many asked for help in reviewing major proposals for developments in their coastal zone. The requests for assistance resulting from these and subsequent meetings with local officials have produced the following numbers of requests. These data reflect the responses as of July, 1980, of sixteen cities and towns of which only two have made no specific request. Four municipalities have expressed no interest in the program.

Topic	Number of requests
Dredging and spoils disposal	5
Coastal erosion and flooding	20
Access to the shore and tidal waters	10
Review of development plans, preparation of recommendations for management of specific areas	9
Environmental degradation (water pollution, filling)	10
Other	7
Total	60

The Translation Problem

After completing our initial formal contacts with local communities, we reassessed our goals and priorities. Given that the state's CZM Program does not have significant amounts of money to give to communities to hire their own staffs or consultants and that 306 funds cannot be used for "bricks and mortar," it was clear that we at the Center would have to make the most of our own limited resources and time. The project was funded in FY80 and the current fiscal year at a level equivalent to somewhat less than one coastal resource specialist with very limited operating funds. It appeared to us that even more than money, local governments are in need of ideas and encouragement. Throughout the state, we saw problems and opportunities where the municipality should take action but nothing was happening. The perception on the part of many municipal officials is that any initiative, be it to dredge, to nourish an eroding beach, or to expand a municipal park or conservation area, will incur the wrath of some groups of citizens and that the tangle of red tape in obtaining the necessary permits and/or funding will be horrendous. The environmental pros and cons as well as the permitting problems are often viewed as too complex and abstruse for an overworked official to grapple with. It was clear, therefore, that we could play a useful role by trying to translate and transfer existing information and ideas into a form that was readily understandable and geared specifically to a specific problem in a specific locality. For example, the Center has over the years produced and distributed technical reports on a range of coastal management topics. Although we had distributed copies to local officials, few were even aware of their existence. In many cases much of the answer to questions on such matters as the more efficient use of space in marinas or the feasibility of building saltmarshes with dredge spoils are in these reports. Through the local assistance program, we can cull the requested information from existing sources and carry it a step further by having the appropriate person at the Center visit the site, discuss the problem and help the local officials proceed to the next stage which often involves application for one or more permits.

The Center is particularly suited to a role as a translator or mediator for information by being a part of the University. We find that most University "experts" are eager to help a municipalty solve a problem that falls within their discipline. It is important, however, for the Center to work at establishing a dialogue between the two parties rather than a mere exchange. It is easy, for example, for a University expert on coastal erosion to visit an area, make his analysis and write a letter. The municipal officials, however, may not really understand why the expert advised one course of action rather than another. The expert on the other hand is probably unaware of the town's financial constraints or the political ramifications of adopting a given strategy. Our role through the local assistance project is to fill in these blanks and work to assure that a meaningful dialogue takes place. One problem here is to protect the expert from putting in too much time and becoming needlessly embroiled in these ultimately important, but for the expert, peripheral issues.

As well as mediating between muncipalities and expertise at the Center and the University, the local assistance project needs to help local governments establish personnel contacts within regulatory agencies and explain what lies behind their policies and regulations. A case in point is upper bay muncipalities with dredging problems. The spoils that need to be moved are not great in volume but they are highly organic and polluted with high levels of various metals and petroleum hydrocarbons. Virtually, no dredging has taken place in the state for ten years because of the seemingly insurmountable problem in establishing an open water disposal site for the spoils. Meanwhile, in the towns in question, marinas are losing business and anything larger than a sailfish touches bottom at low tide in the marinas and boat basins. Our role is to explain that using the spoils to build a saltmarsh in the immediate area is a viable option and explain in real terms the issues that lie behind the lists of criteria and requirements provided to an applicant by the regulatory agencies. We must alert the officials to the major issues of pollutant resuspension, of the unavoidable objections that will be raised by residents apprehensive about bad odors, mosquitoes, and an eyesore. We need to explain how these problems can be surmounted, which problems will remain unanswered, and how such a project can be most effectively presented to the public. Finally, the town officials must understand that the new state regulations on hazardous waste disposal will be applicable if high enough levels of some pollutants are found in the spoils and elucidate the procedures that must be followed to obtain CRMC and Army Corps permits. Then there is the matter of findings. All of this takes time and patience. At present, one town with a promising and well-advanced saltmarch building project is stalling until after the November election because the project has become heavily politicized. Another town is eagerly moving through the preliminary steps of selecting a site and finding the funding. It will be well over a year before any dredging takes place in either area.

Some projects, but unfortunately few of them,are pleasantly cut and dried. The Town of Narragansett, for example, asked for assistance in generating ideas for making better utilization of a former salt pond behind the town's extensive and popular barrier beach. The pond is now fresh and it is known as Canonchet Lake. It is separated from bathing pavilions and the large town beach by a highway and behind it lies an extensive tract of farmland recently acquired by the town for recreational and conservation uses. We were asked to develop ideas for appropriate use of Canonchet Lake that would enhance the recreational values of the area. The Center made a quick but thorough survey of the pond as a fish and wildlife habitat, evaluated the needs of the surrounding areas and made specific recommendations on how recreational fisheries could be enhanced and how the aesthetic qualities of the beachfront could be greatly improved by using the lake and its shore as an important feature rather than a convenient place to dump rubble. A brief report detailing our findings and recommendations was warmly received by the town. When action is taken, the aesthetic appeal of the area will be greatly enhanced and the problems caused by trying to pursue some of the inappropriate uses first suggested to us by the town will have been avoided.

Priorities for the Local Assistance Program

Our experience with working with local communities has influenced the planning and research undertaken at the Center for the state's continuing coastal management program. It is also providing an important feedback loop to assist us in helping the CRMC refine its permitting process and sharpen the program focus on important coastal issues. The present priorities for the local assistance program fit into the broader context of the Rhode Island CZM Program and specifically the projects undertaken at the Center as described below.

Coastal Erosion and Hazards

The Northeast is due for another major hurricane. In Rhode Island, evacuation procedures are in place, but houses have been built in areas that saw great destruction during major hurricanes in 1938 and 1954. Rebuilding in these areas was well underway before the state developed a CZM Program. Using CZM funding, a URI coastal geologist has worked with the Center to map the state's shoreline in terms of its susceptibility to erosion and storm flooding, and to identify particularly hazardous areas. Among the results of this work is the startling statistic that 25% of the Narragansett Bay shoreline is in engineered structures, most of which have been built at great expense by local property owners. We have also learned that most erosion takes places during major storms and that intra-storm erosion, with few exceptions, is minor. Unfortunately, a great many of the bulkheads, riprap walls, and other "erosion-control" structures are far more elaborate than what would be necessary to control erosion between major storms, but not substantial enough to withstand the periodic "high-energy events" that, therefore, will continue to shape our coastline. One priority of the local assistance program is to circulate this information and to advise local governments on appropriate measures to cope with erosion. This is a very active aspect of the program. Every municipality is concerned about erosion, and in many cases it is not easy to convince them that the do-nothing option or a low-technology measure is the most appropriate. Citizens and local officials alike are, unfortunately, too often unaware that hurricanes are a reoccurring phenomena and not historical disasters that lie behind us.

Management of Salt Ponds

Two years ago, the Center began to coordinate a major interdisciplinary study funded by various sources directed toward answering the many complex problems posed by the salt ponds that line the state's ocean shore. These problems include siltation, burgeoning residential development, deteriorating water quality, overfishing, and multiple-use conflicts of every kind. The research being undertaken by six principal investigators is aimed specifically at providing information of pratical usefulness when attempting to address these management problems. Although we are yet to complete the research phase of the project, we can already provide useful insights into some of the problems brought to us by local governments. The results of this work are also transferable to other small estuaries. We are working with various communities to evaluate the results of small-scale dredging proposals, possible measures to increase the flushing of embayments cut off from the main pond by causeways, and the advisability of adopting greater set-back requirements for leaching fields for individual houses. Working with the towns on such matters provides us with important insights into the goals for management that will ultimately shape the direction taken in the future management of these areas.

Upper Bay Revitilization

Upper Narragansett Bay is a heavily-urbanized area with large expanses of derelict or low-quality waterfront. The waters are highly polluted, and for the past twenty years, the state's most productive quahog grounds have been off limits to fishermen. Providence, at the head of the Bay, is slowly expanding as a general cargo port as petroleum imports decline, but port management and expansion presents a tangle of administrative and financial problems. The Center has been working for a year to characterize the many problems and opportunities in the upper bay and to formulate a strategy for the revitilization of the area. Here again, the local assistance program provides an important vehicle to interact with local governments as we work with them on various mini-projects. These projects enable us to better appreciate how such initiatives are received in the area, the desires of local residents for an improved coastal environment, and a laboratory to try out some of our ideas for revitilization. The upper bay project is not as far along as the effort in the salt ponds, but we hope to see exciting initiatives in the areas of port management and increased recreational opportunities during the present fiscal year. One exciting possibility is to utilize an old railroad right of way as the basis for a linear park along segments of the most derelict and presently inaccessible shoreline of the East Bay. Initiatives involving local dredging projects have already been discussed.

Review of Major Development Plans

The final priority for this year's local assistance program is to advise town authorities that are reviewing major development projects within their coastal zone. We find that this is a productive exercise since developers are frequently more willing to adjust their plans at this early stage than later in the permitting process. In towns which have no planning office, the citizens that serve on various review boards are usually open to our suggestions of ways in which they can work toward achieving the goals of sound coastal management. Our suggestions focus on the identification of important natural features, measures to minimize nonpoint pollution impacts and aesthetic considerations.

We view, therefore, the local assistance program as an important and integral part of the state's CZM Program. As presently structured, the Center is able to provide assistance on a broad range of topics at a very low cost. We view the program as an excellent measure toward improving the transfer of information between university community, and local government, and as an important feedback mechanism on how well the overall CZM Program is functioning in its efforts to properly manage the coastal resources of the state. Ultimately, a successful coastal management program must have the support of local government and citizens. The local assistance program, if it is successful, can go far to assure that the state program is on target and that the message is getting through.

TRINIDAD: A SOLUTION FOR COASTAL LAND-USE CONFLICT

by

Steele Wotkyns

Architect, Design and Planning Consultant

I have titled this address essentially with the single word "Trinidad" because it is the name of our city and also because of the evocative symbolism of the word as it pertains to our story. It is a story about the ongoing resolution of coastal land-use conflicts which involve the interest of the basic triad in American affairs: the interests of the private citizen, the interests of the general public expressed in the institutions of central government, and the responsibilities and interests of the local community. It is a report from a city on the north California coast which covers in microcosm the range of coastal problems and suggests means for the resolution of conflict. As the story unfolds I will tell you about the establishment of an alternative mechanism for the implementation of land-use regulation, a local non-profit land trust. This local alternative organization has helped to restore a large measure of understanding and cooperation between the parties in this often adversary triad. First, I must set the stage.

Trinidad is the smallest and second oldest incorporated city in California. It is a semi-rural community on the fringes of Eureka, the only moderately large population center which is 300 miles north of San Francisco. For many in California and most who do not know the state, the vast north coast is a hidden and unknown land. Roads, rail, and air service are minimal. It is a haven for adventurous travellers and exurbanites. It is home to a significant and remarkably stabile native population whose ancestors came for gold and trees. Trinidad has been a minor port for the region's products in conjunction with Humboldt Bay and is an important local fishing center. It retains in a large measure the virtues and vices of a traditional, independent, and self-sufficient small American town. These virtues and vices are inseparable from the physical environment of the city. The several miles of coast which it and its hinterland command are incomparably beautiful. Rare on the western coasts of the Americas, the presqu'isle of Trinidad Head offers protection in the summer from the prevailing northwest winds. It was here in the 18th Century that Spanish explorers sought refuge, named the town for the Feast of the Trinity, and gave the first accounts of the native American population, the Tsurai of Yurok stock. The prehistoric village site is within the city limits, and the descendants of its inhabitants are among our influential citizens. Huge, rocky sea-stacks dot the ocean, while the coast consists of a variety of pocket beaches nestled under the brooding palisades which are so characteristic of the California coast. The steep coastal mountain range rises behind, still covered in part with the redwood forest for which our region is famous. One third of the incorporated land within the city limits is a long established state park. Within ten minutes of the city are two flanking state parks of generous proportion. Other parks, including the new Redwood National Park, are less than an hour's drive from the town. Nearly all of the other privately owned beaches and coves within the town limits and the hinterland have always been available to the public by a network of informal trails leading from the nearest coastal road. Development along the coast is relatively unobtrusive. Moderate private surveillance in the immediate past has managed to preserve the near pristine quality of the coastal experience for visitors and the local population. In the dry summers there are morning fogs with brilliant afternoon sunshine; in the winter it rains and rains. A price for paradise!

Crab and salmon are the principal fisheries, while the coast is alive with birds and a rich tidal life. Sea lions and a pod of harbor seals cover the off shore rocks within sight and sound of the city. Fishing boats anchored in the lee of the Head, with a fringe of white houses on the palisades above, rival the beauty of the most famous fishing villages of the Mediterranean coast. Those of us who live in this setting have never been unaware of the value of our incomparable resource. We have always been resentful of the undisciplined elements of the larger general public, but we have never been ungenerous in sharing our riches with them.

As a result of the Coastal Act of 1976 in California, our community was suddenly overwhelmed by rigid demands for coastal preservation, conservation, and planning. We were also threatened with state purchases within the city limits and the adjoining hinterland. These purchases would have further constricted the naturally limited land-base of the town.

You can imagine the consternation of a community of 300 souls which is suddenly beset by the two pronged descent upon it of powerful and remote state agencies: one, the California Coastal Commission, demanding rigid adherence to an elaborate and inflexible planning standard geared to the infrastructure of a large city; the other, the State Department of Parks and Recreation, planning to remove yet another third of the city's land by purchase. Our part-time city clerk and then unsophisticated city council were overwhelmed. Threatened land owners, both within the city and nearby, faced the situation with a mixture of disbelief, resignation, and mounting anger. The perspectives of the parties involved were sharply polarized. This resulted in an adversary and conflict laden confrontation.

In the early Seventies the conservation community of California, in despair with the inactivity of the State Legislature, promoted and won a public proposition which provided for general planning, conservation and public access to coastal resources. In 1976 the California Legislature superseded Coastal Proposition 20 with a comprehensive planning act mandated by the original proposition. Many of us in the Trinidad area had actively worked for the passage of the original proposition, anticipating the need for more rational planning standards in general and with good will toward the legitimate needs of the larger public. The job seemed well done. Those of us for the proposal and the many against it relaxed. It seemed clearly the policy of the State to guide local communities into a path of planning procedure which once accomplished would return local autonomy on a new and sounder footing for the conservation of coastal resources and provision for public access.

Remote from the pressing coastal problems of the populous south where rights of access and unlimited development had reached crisis proportions, we on the north coast were unaware and even indifferent to the Draconian measures and ultra-legal regulations which an active and dedicated staff began to invent in order to implement the provisions of the Act. As residents of a remote and unspoiled area, we did not see ourselves as an enemy to be controlled at all cost. Distance and modest financial capability prevented our participation in the hearings which were set up to discuss and formulate policy. The pro forma clauses in the Act which required public participation simply did not work in reaching us or in galvanizing us to take action on our own behalf. As a result, an organized minority of single purpose conservation interests had a major influence on the early formulation of policy and the identification of "crisis" priorities. All of this happened while we were sleeping.

In contrast to southern California where coastal resources had been seriously abused, the beauty of our pristine environment attracted attention and was forthwith designated for special treatment. An understandable policy, based at least subconsciously on the principle of triage, was adopted. Trinidad was a <u>saveable</u> resource of high priority. This made for an unnatural focus on the "problems of the north coast." For the planners and policy makers of the State to acquire and restore already spoiled lands was an obvious financial impossibility. Subsequent study by the Commission and its advisors then proceeded to identify and designate special resources, under the provisions of the Act, for immediate public acquisition. The community of Trinidad slept serenely on as funds were quietly appropriated for purchases in our area, purchases which logically were assigned to the Department of Parks and Recreation, a huge bureaucracy with immense internal resources.

In the meantime the Coastal Commission, in order to get the Coastal Program underway, selected several communities on the California coast as pilot projects. Trinidad was one of them. In due course a federal/state administered grant was made for initial studies. Planning consultants were hired, ostensibly to help the city with the development of a plan to conform with the requirements of the Act.

It all began in a low key while the Commission and the city reviewed their respective parts in the process. Citizens were surveyed concerning their desires for the community. This seemed to all of us participating, benign and hopeful. We had been consulted, had done our part, and were satisfied. And so, the technical part of producing the planning document (known in California as the Local Coastal Program (general plan)) began. Subsequently, the desultory public hearings, which probably satisfied the letter of the law in receiving public input, failed entirely to reach the understanding of the community. The ultra-legal language of the Commission's regulations clouded the issues and, because they were without precedent, could not be adequately interpreted by either the Commission's Regional Staff or the people for their ultimate effect on local autonomy. Finally a document, foreign in many ways to the expressed desires of the citizens, was produced, and the certification hearings were upon us. Some of us began to suspect that the return of reasonable local jurisdiction, upon certification, was a chimera. Exhausted and intimidated by endless wrangling with the Commission's staff, the consultants and city officials urged the adoption of the document from a somewhat uncritical point of view. Fortunately, by vigorous citizen action, certification was delayed. And then the axe fell!

Thirty-three coastal landowners received notification from an agency of the Department of Parks and Recreation that their lands were being considered for purchase by act of eminent domain in order to establish yet another park in our town and immediate vicinity. The shock to the landowners was paralleled by the shock registered in the community at large. Already surrounded and partly invaded by public parks, we were able to see immediately that the future economic and social structure of our community would be subservient to the whims of a distant, impersonal, and unapproachable bureaucracy. The rush to restudy the largely imposed plan was on. Sure enough, subtily embedded in policy statements and boldly tacked on as an addenda was the stated acquiescence of the city to large scale public purchases for the establishment of park lands. These provisions had been represented by the Commission staff as unimportant and remote contingencies which would assure certification. They were appended to the document without adequate public hearings.

California planning law, of which the Coastal Act is a part, makes the certified general plan the principal document in law which guides subsequent ordinances and requires them to conform in order to be legal and effective. Needless to say, the exact formulation of the Local Coastal Program (general plan) was therefore of key importance to the community's future.

Today, a couple of years after the event, I am amused by the naive name which the threatened landowners gave to our informal association. We called ourselves "The Concerned Citizens of Trinidad," knowing little of the many others in similar circumstances who had found the same words to express their distress. We soon found ourselves at sea, without voice, without connections, and without access, confronted by remote, unresponsive, and unaccountable bureaucracy. This appeared to be an unimpeachable power which seemed to hold all of the favorable cards.

In the early stages of our association an expensive consultation by one of us with leading attorneys in San Francisco produced the intelligence that our only recourse was to hold out for a maximum purchase price by forcing the State to go through the complex procedure for purchase by right of eminent domain. Our unified stand, forcing the State to this cumbersome procedure, bought our association some time. However, we soon recognized that there was no real recourse in legal procedure and that our only alternative was to organize for a political solution. Our collective very moderate means precluded the endless cost of legal services which could only gain for us a maximum invasion of the public purse. As tax payers it was not an attractive course for action.

Because the production of a sense of crisis seems key in producing an active citizen involvement, and perhaps necessary in general, it is worthwhile for me to trace for you the evolution of our initial local organization, in that my conclusion suggests a possible alternative to the acceptance of traumatic crisis as a means of getting direct citizen involvement. It is undoubtedly true that a sense of impending and profound personal loss, unconnected with the collective good, drew us together. We were all preselected and deeply involved with our physical environment by the very fact of our choice of land on which to build houses and make our homes. A substantial number of us had actually constructed these houses with our own hands. A number of them approach an independent artistic value as architecture. Our commitment in terms of generally limited financial ability and individual emotional attachment runs very deep. It is this circumstance which made us all take the next step in the face of apparently insuperable odds. Those of us within the city of Trinidad questioned the proposed general plan sharply and called to the attention of the City Council the implications of admitting within the corpus of the plan the validity of further public purchase in our community. Others organized themselves to attract wide public support which was done by a circulated petition in opposition to the unnecessary and costly public purchase of local land. Others arranged for area-wide publicity which included television talk shows, newspaper articles, and even television publicity as far away as San Francisco. The community was soon thoroughly aroused. It was a surprise to the landowners that their seemingly private dilemma became a local cause célèbre. The community perceived collusion (true or false, the question has never been resolved) between the Department of Parks and Recreation and the planning staff of the Coastal Commission. This perception, injected into the atmosphere of protest generated by the tax revolt of the now famous Proposition 13 (which in my opinion was also an inchoate protest against the growing invasion of bureaucratic direction over people's personal lives), coupled by the visible and maddening arrogance of the agencies involved, grew into a highly charged town meeting, attended by virtually the entire population of Trinidad. The feature of this meeting was an emissary from the Department of Parks and Recreation who blandly announced that our general plan called for public land acquisition. Moreover, she added that it was "too bad for us that we happened to live in such a beautiful place" and that the dwellings purchased by her department would simply be allowed to fall to pieces in the name of the "restoration" of the landscape. This was the first public news of the almost secret plans which the central government had developed for our community. All direct and indirect inquiries had been ignored and unanswered, even when demanded under the Freedom of Information Act.[1]

---

[1]A quirk of law (in California and perhaps elsewhere) makes public agency planning for acquisition cautious and secretive because of the possible damage which threatened owners could claim. The result of this quirk is to seed suspicion, enforce bureaucratic secrecy, and produce the appearance of a conspiracy to defraud private owners of timely recourse.

These were dreary and dispiriting times, but they did awaken many of us to our wider responsibilities as citizens and the duties which we all have to make sure that the basic tenets of our democratic philosophy be made to work. The enemy in the end is not the arrogance of bureaucracy, but a selfish preoccupation with exclusively personal affairs which does not include responsible and sustained action as citizens.

As the group of Concerned Citizens began working together, it soon became apparent to us that the issues raised by Coastal Commission procedure with its tendency to withdraw responsibility and jurisdiction and by public acquisition of our land base amounted to a greater loss than we could sustain singly.

The lay of the land precluded efficient park management and would result in substantially more actual development along the coast: development in terms of parking spaces, traffic increases beyond the ability of roads to expand, and the inevitable "tables-toilets-and-trash" which while necessary in some places was not in ours. In short, the aim of conservation and access for the public to a rich coastal experience could not, in all probability, be produced by public ownership and management. Somewhat grimly, we all began to understand that even though access to our beaches had always been easy for the public, the general public did in fact have rights which were not recognized de jure.

The Coastal Initiative and the Act which superceded it withdrew by public consensus certain property rights where public interest and public welfare superceded the traditional bundle of private property rights. The writing of things to come was on the wall. At the same time it was clear to us all that the preservation and pristine condition of our coastal resources was due to two causes: first, the lack of pressure on them by the local population, and second, and not least important, the passionate responsibility each of us as private owners had exercised for their maintenance in the past. In fact I, who was by chance near the State Capitol in Sacramento and therefore the "self-elected political spokesman," in talks with an high official on the Coastal Commission staff, was told, gratuitously, by him that "the only beaches in California in pristine condition were those in private hands."[2]

So here was a dilemma. On the one hand the public consensus had expressed a new interpretation of private property rights in places of high resource value, while on the other it could unequivocally be shown that unlimited public use of those resources results in degradation of them. Thus under certain specific circumstances there is a kind of built-in defeat in the very objectives of the Coastal Act. We felt that those certain specific circumstances could be proved to exist in our area on objective grounds.

We also began to realize that there was something new in the air that needed to be accounted for by the conservation community, a community which in its single purpose zeal (border to fanatic disregard for other social values) had brought us to this crisis. We felt that inasmuch as the "conservation ethic" is a human idea, it cannot be singled away from the rest of civilized and human response to the environment. In short, it could not be considered separately from the other social values which made our community viable. Among the most important of these values are the social and economic factors which assure the continued health of our city and region. In the end, real conservation depends upon people with and in the environment. To remove them destroys the structure of community and transfers personal responsibility from each single individual to a faceless central mechanism which cannot generate independent intelligence and is incapable of specific judgments. This idea is put more directly by Jane Jacobs in her discussion of safe neighborhoods in her splendid book, <u>The Life and Death of American Cities</u>. She suggests that the intelligence behind the "eyes" that continually survey a loved and diverse neighborhood accounts for its safety and cohesion. In a like way the conservation of valuable resources could depend on vigilant personal responsibility keenly felt. This equates with people in the landscape.

---

[2]There are so many actors in my report to you that I have chosen not to name names so as not to encumber the paper with either thanks or damnation. On personal request I would be glad to supply relevant information.

Thus the question before us was redefined. It turned out to go something like this: What kind of an alternative would balance the abstract general welfare against the specifics of the environment and the community and would at least be attractive to private interest and partly palatable to more conservative private sensibilities? The idea of looking for an alternative to outright public ownership of land and the dictation of land-use criteria was born among us. We went to work with a will, and believe it or not these ideas flew!

We rediscovered Professor Abraham Maslow's law that there are only five steps to personal communication with anyone, whomever, wherever. Someone always knows someone who knows someone who . . . is the person who makes the decisions. We confined ourselves rigorously to finding just and only those decision makers. With the bomb of our fate already fused and lit and the clock ticking, it was a waste of time to do otherwise. We had found within our own group a latent access to power which is the key to effective citizenship. As an aside, it makes me feel good to tell you how generously we were received by those in power and of the intelligent and careful advice they gave, even though they might not have agreed wholly with us in principle. Breaking the mutual isolation between "the grass-roots" and "the powers-that-be" is a sub-theme of my report to you.

Now back to Lady Luck. A prominent talk-show MC was interested in our cause because of the malaise which I have alluded to in the atmosphere of Proposition 13 and a personal connection with our region. A long distance telephone interview on his program with one of us was heard by the director of the Land Trust Program of the Trust for Public Land which is based in San Francisco. The Trust for Public Land is a private non-profit organization which as its principal objective acquires private land for subsequent resale to public agencies unable to move fast enough to secure land in the public interest. There are many other inventive facets to the Trust for Public Land's operation, among which is the encouragement of the formation of local non-profit land trusts. (Of interest in our particular situation is that the Trust for Public Land is the brainchild of the present Secretary for Resources in the State of California.) The Trust for Public Land official who heard our complaint offered her services. A coincident conference sponsored by her was being held in San Francisco to which a member of our organization was invited. This proved to provide a viable alternative for our dilemma.

Events moved rapidly. The extraordinary help of our assemblyman, now senator, Barry Keene, was crucial. He devised a special bill which withdrew appropriations from the Department of Parks and Recreation for the contended purchases, recognized the scarcely formed Humboldt North Coast Land Trust as an appropriate entity to serve the public interest, and granted it $100,000.00 as seed money for the purchase of lands where owners had been adversely affected by coastal regulation. The bill then attracted wide support in the California Assembly and Senate, chiefly for its potential merit in resolving growing conflicts over the adversary climate of coastal regulation in the State. By this act of essential faith in local initiative our California Legislature has gone miles to restore mutual understanding between government and the citizens of our area.

The Coastal Commission and the city stumbled through to completion of the Local Coastal Program (general plan) for Trinidad. On the one hand, procedure was modified to some extent, and on the other, apprehensive acquiescence was achieved. Apprehensive acceptance by the city of Trinidad and its active and very aware citizenry is not too strong a term. Proposition 20 and the Coastal Act which replaced it in 1976 are both documents essentially requiring long range coastal planning in a designated zone. However, the requirements of start-up gave to both these resultant organizations a permit orientation which in my opinion has plagued the procedure and the intention of both coastal measures. The appointed, uncompensated Commissioners and the scale of coastal problems in California make the Commissioners heavily dependent on staff opinion and findings. The Commissioners simply lack time to cope with the overwhelming detail. And it is often detail which determines the ultimate function of overall planning. Pro forma and unimaginative public commentary is provided for by the Act in ways which are often procedurally out of sequence. When these built-in circumstances are combined with the structurally unapproachable and unaccountable influence exerted on the Commissioners by a staff trained in a "school" which is permit oriented and leans heavily on ultra-legalistic regulatory models, the requirements of long range planning often go by the board. Coastal citizens, because of the atmosphere engendered by the situation in general, are therefore viewed by the staff as "the enemy." The natural tendency is for the staff to be cautious (even to be negative as a primary reaction, for reasons well known to those familiar with the study of bureaucratic decision-making) in accepting the least public criticism or suggestion which affects the operation of the planning process and its practical outcome. The result is to place the citizenry in an adversary role without real and tangible procedural recourse within the system itself. The best that the Commissioners can do is to acquiesce in unimportant matters suggested by the citizenry. By this means they give the illusion of public participation. I wish to emphasize that I believe the problem to be circumstantial and structural and not remotely connected with willful malevolence. Unfortunately, the latter is the public's perception of the process.

Again, the Commissioners simply do not have time to reconsider real policy issues which come up with specific circumstances and in the course of planning. The staff, on the other hand, is committed to previous policy by pride of authorship. Like Marxist theory which postulates the withering of the state, the California Coastal Act provides for a "withering" of its directed powers. Those of us with an ironic twist of mind indulge in cynical smiles. Indeed, complicated and quite likely onerous post certification procedures, included almost as an afterthought in the Act, will assure the continuance of permit orientation beclouding broader planning issues. It is for this reason that many of us in Trinidad have acquiesced both in exhaustion with the process and with apprehension to final certification of our Local Coastal Program (general plan). For example, the very first permit (for the addition of a modest porch) under a supposedly returned jurisdiction was questioned by the post permit process for reasons which had nothing to do with the addition itself.[3] I would add a plea to the framers of similar acts to retain a necessary and restrained "Sword of Damocles" but to allow for greater

---

[3]The Act empowers the Commission with a kind of legal blackmail in which permits are granted in trade for other objectives, for example, public access. The invasive powers which this procedure grants the staff are, in my opinion, ultimately damaging to the encouragement of local voluntary responsibility without which all measures are futile.

flexibility which promotes a healthy diversity, more effective mechanisms for real public participation, and, of most importance, to ensure that the "atmosphere" of the law does not preclude trust and cooperation with the citizenry in its efforts to be completely ordered and effective.

The dedicated agitation and publicity of the conservation community, even though its methods to achieve effectiveness in the past have sometimes resulted in unanticipated social difficulties, have raised the awareness of coastal citizens to a very high level whatever their philosophical stripe. Regulation should recognize this phenomenon. There is a pervasive "last-chance-gone-forever" tone to regulatory legislation which is simply unrealistic, can be undemocratic, and is certainly anti-human. The perceived crises of every generation are exaggerated and their physical projects are made of sand, like the warning in Shelley's poem to Ozimandias, King of Kings. Relax a trifle, for the angel of the Lord is delayed! Don't sacrifice dealing with the big things for trifles.

Here, I must make it clear that my criticism of the Act and the Commission is not intended to be destructive of very real accomplishment. The citizens of Trinidad have been exasperated and often ignored, but their thrust has always been to improve the system and to offer constructive criticism. It may be pretentious for so small a town to feel a real responsibility for precedent which could affect fellow citizens elsewhere. That, however, has been the tone of our objections and continued criticism. Real local effort to achieve imaginative alternatives to the dangerous tendency towards centralization and the decay of local responsibility and judgment is vouched for by the establishment of the North Coast Land Trust and its success in helping to resolve conflict.

So Trinidad, the smallest city in California, became the first to have a genuine certified coastal plan under the California Coastal Act of 1976, and the area, a genuine, locally administered alternative to some aspects of land-use on the coast.

Land trusts are more common on the east coast than they have been in the west. The best recent example among several which have existed for many years, aside from the extremely effective Nature Conservancy, is a land trust organization of the Lincoln Township in Massachusetts. Mr. Robert Lemire's recent book on the experience in Lincoln, Creative Development, is the best general reference on a synthesis of cooperation between local government, private land trusts, state and federal land-use programs, and the host of other programs offered by peripherally interested agencies at many levels. The title of the book is misleading to some extent because activities in Lincoln have been oriented to the conservation of land in conjunction with needed and balanced development. In the case of the North Coast Land Trust, our organization, our objectives are not focused as much on development pressures and their resolution, but more on public access, open-space conservation, viewshed protection, balanced ecological emphasis on carrying capacity for man and nature, and considered build-out where it is appropriate and harmonious. Our land trust, like others, has the enormous advantage of promoting the enlightened self-interest of our patrons who by recognizing and providing for genuine public needs can achieve certain advantages for themselves. These are protection of their residual rights after suitable gifts are made in the public interest, tax and estate planning advantages, and, above all, the ultimate satisfaction of real participation in the community's interest. That negotiations between the Trust and its patrons can remain low keyed and unpressured is an enormous advantage in obtaining grants of easements or gifts of land from the conservative members of the community who most often control critical lands. Aside from the financial advantages that are often possible for Land Trust patrons, negotiatiations are conducted with friends and neighbors, and emotional attachments to land are released gently. Patrons retain not only the sense of directing the destiny of their erstwhile land, but a very real participation in the formulation of policy and active participation in its continued management. Along with this active participation through the Trust, their involvement provides a very real continuity of responsibility. In short, the implicit demand to continue caring promotes the best kind of social responsibility and community solidarity.

Our Trust is very young. In the two years of our existence we have been given land of nearly a half million dollars in appraised value. These gifts include easements and below-market-value purchases. Considering that the original appropriation for purchase in our area was

$1,200,000.00, representing a minor portion of the completion of the Parks and Recreation project, and that to date the Trust has drawn only $5,400.00 on the funds granted to it by the State Legislature, the difference on the drain from the public purse is already very substantial. For example, one below-market-value purchase tripled the effective size of an adjacent county park. Another case, financed wholly by the Trust, proposes the sale of two acres by the Trust of already developed land within the eighteen plus acres of the purchase. This will assure permanent open space of more than sixteen acres and will preclude further build-out costly to the environment. This land is particularly interesting because it had been repeatedly offered to the State by its conservation-minded owner and repeatedly refused in spite of its obvious value for conservation purposes. The moderate profit which the Trust expects from this partial sale will finance other more critical purchases and assure the financial viability of the Trust. Built-in care and surveillance for the land assured by conditions placed on the sale, augmenting a county tax base which is chronically in need. Humboldt County is not rich. Within the year other promised gifts, purchases, and granted easements will much exceed in public benefit the original plans of the California Department of Parks and Recreation. All of this is being accomplished with relatively little pain in an atmosphere of cooperation as opposed to confrontation and in a way which strengthens the social fabric of the community rather than threatening its ultimate decline. The Trust has also exciting plans in cooperation with the city of Trinidad and the county of Humboldt for the establishment and management of trails to beaches, the establishment of a recycling septic tank management district, and various aspects of our harbor's development. It is true that all of this came about by an imposed crisis, which brings me to my conclusion and to some suggestions for a creative approach to coastal land-use management.

The possibilities for the functional evolution of the Trust are large within the city and within the region. These many possible opportunities will require conscious restraint, but will certainly, because of the voluntary nature of the Trust, be self-limiting. In any event the organization itself has provided a forum for local initiative, garnered significant political support, and has become a rallying point for dealing with future crises.

Aside from its concrete objective which is to protect and make available coastal resources, the Trust is having considerable success in areas of social influence not contemplated on its formation. Points of view, political and organizational orientations which were polarized by the onslaught of regulation and the threat of acquisition, have been resolved by membership within the organization itself, making possible the cooperative rapprochement between distant state agencies and local people. For example, the previously unapproachable director of the State Department of Parks and Recreation came to Trinidad to address a workshop of the trustees and generously offered the cooperation and expertise of his staff for the planning of coastal access trails which the Trust owns and operates. Our views and philosophy do not coincide, but civilized discourse has been reestablished between citizens and their serving agency. This is not a single incident.

Over a hundred and fifty years ago Alexis De Tocqueville, especially in the second volume of Democracy in America, stressed over and over again the crucial function of free citizen associations in the dance of the balance of power in democracies. He feared the tyranny of the majority; today we face tyrannies from the majority and single purpose minorities. However if there is no place to talk about these twin tyrannies at home, we will certainly be prey to both. Expressed in a fading popularism, modern civilization is "into" management and more management. In the opinion of many it is unavoidable and necessary. However, total management is not inevitable or desireable. I believe firmly that institutions like local non-profit land trusts can leave the element of necessary centralization unencumbered by the detail which paralyzes overall policy. Unquestionably the resultant division of responsibility will "leak" a bit in the ecologist's sense of the word, but the gains in variety and diversity which are prerequisite for future adaptation and evolution outweigh minor losses. Percy Shelley's warning to King Ozimandias leaps once again to mind.

I hope that I have sufficiently emphasized the roles that luck and the conjury of circumstance played in Trinidad for winning through to at least partial solutions in its complex of problems with regard to the protection of coastal resources. I have focused only on two major issues;

we have a long way to go with others that in the end may be crucial. It is true that once over the initial shock we found astonishing internal human resources. Our tiny community contains expert land planners, geologists, a resource manager, ecologists, an economist, a physicist, skillful entrepreneurs, attorneys, architects, an executive secretary, artists, a state officer with bureaucratic experience, and, most importantly, a generally cohesive and caring citizenry. It is true that crisis found us to each other and that this is a rich harvest. However, I expect that almost any community has a richness of this kind which is unperceived. Why, then, is crisis a seeming necessity for the organization of talent and effective citizen participation? Is there a gentler way? Is there a way to fill gaps in local resources when they are absent? I think there is the possibility for reasonable experiment.

If there had been a provision in the California Coastal Act for a citizen's advisory, an idea which incidentally has been in the wind in California, and this advisory provided for on a long time basis in order to make person-to-person contact in communities slated for coastal regulation, I believe that the atmosphere of conflict and surprise could be much reduced. The discipline of Community Development is underused by state and federal agencies for reasons which have much to do with the structure of the agencies themselves. It's not much fun, nor is it quite natural, to play one's own devil's advocate. However the advisory service might be organized, its objective for activating citizens before the event, finding and organizing the kind of local human resources (accidental in the case of Trinidad), and, finally, offering a variety of functionally equivalent alternatives would certainly take no more time than the confrontations and adversary haggling which have taken place in California. I think of such a service as being a kind of artificial crisis, but one which has the elements of human direction rather than of chance. It is hardly arguable that without some measure of coercion there is a tendency to resist change in social structures and community organization. The question is whether the coersion should be a de facto stick or a proffered carrot. If, for example, a land trust does offer a functionally equivalent alternative to direct regulation and has the advantages which ours certainly seems to have, why not make way for it as an alternative? Why not find the local people and encourage the local gumption which would make this mechanism a reality? As time and experience go on I am sure that it would be possible to tailor a variety of local mechanisms which would be functionally equivalent and meet the objectives of regulatory acts. The social advantages far outweigh whatever extra effort might be required. I am also convinced that this sort of advance preparation would reduce the time frame of emplacing needed regulation.

Again, it can't be argued that the California Legislature is unaware of the ponderous, slow-moving creatures of its work, nor even that the idea of public advisory has not been considered. In Trinidad's experience a new creature of the Legislature has been of great help in reaching the many accommodations which have been necessary to resolve conflict. The California Coastal Conservancy is an in-house expediting and holding agency which is modeled to some extent on the operation of the Trust for Public Land mentioned above. In its youth it is still flexible and perhaps because of its fluidity, (part of the original concept for the agency) has acted in the role of the citizen advisory which I have suggested. Its still small and sympathetic staff and director have been important ombundsmen for us. They even seem sympathetic to the idea that in order to be effective our alternative local organization, the Humboldt North Coast Land Trust, must stay clear of cooptation by any state agency. It seems possible to me that citizen advisory functions granted to and administered by the California Coastal Conservancy could begin the process of experimenting with that concept. I have stressed the importance of the atmosphere and tradition of cooperation rather than confrontation. The California Coastal Conservancy is young enough to be molded in that tradition.

It is a temptation for me to expand upon this particular theme, but that is clearly beyond the scope of this paper and the information which it brings you on a specific experience in the west. It is meant as a suggestion which is worth exploration and offers a possible way out of the systemic dilemmas which are explicit or implicit in my account.

To summarize: I am convinced that a means should be found to curb the limited viewpoint of necessary advisory staffs, to humanize the system by making use of an inteligent citizenry capable of directing, for the most part, their own affairs, even for the public interest at large, to return a sense of responsibility and judgement to local communities, and, finally, to relieve the makers of high level policy from the crushing and stultifying burden of specific detail. This is, I believe, a central issue in the effective regulation of coastal resources; I shall not belabor its wider application.

I have told this story from an individual viewpoint and take the customary responsibility for its objectivity. I have told you, for a change, an up-beat success story. It is possible for active intelligent citizens to make a difference. There is, where you find it, responsiveness in government. Creative cooperation is an option. Institutions can be human.

Acknowledgements: I want to thank the legion of actors, necessarily unnamed, in this story; and personally to thank Professor Bonnie Mesinger for her help with the production of this paper.

Inquiries to: Steele Wotkyns
c/o Humboldt North Coast Land Trust
Box 457
Trinidad, California 95570

Submersed Aquatic Vegetation in the Tidal Potomac

Virginia Carter, U.S. Geological Survey
James E. Paschal, Jr., U.S. Geological Survey
G. Michael Haramis, U.S. Fish and Wildlife Service

ABSTRACT

The results of a continuing survey, begun in 1978, show that submersed aquatic vegetation is distributed unevenly in the tidal Potomac River. The largest and most diverse aquatic plant populations are found in the transition zone where freshwater and saltwater mix; there is almost no submersed vegetation in the freshwater tidal river north of Quantico, Virginia, and very sparse vegetation in the lower (mesohaline) estuary. This distribution is very different from that reported near the turn of the century when the tidal flats just south of Washington, D.C., were covered with a luxuriant growth of plants.

The factors that might cause this unusual distribution, and the decline of submersed aquatic vegetation in general, are multiple and interrelated. They include nutrient loading, increased sedimentation rates, substrate composition, and various types of chemical pollution associated with human population growth. The decline of plants in the freshwater tidal river and in other freshwater ecosystems, such as the Illinois River, indicates environmental imbalance. It has implications for the decline in waterfowl and fish populations in previously productive areas, and for replacement of sensitive species by more pollution-tolerant species.

INTRODUCTION

Submersed aquatic vegetation (SAV) is one important component of the aquatic environment. In the Chesapeake Bay and the Potomac River, as in other coastal ecosystems, aquatic plant communities have many functional values. Whether growing in beds along the littoral zone or floating in masses moved by tidal currents, SAV provides a habitat for large numbers of small invertebrate species which are eaten by waterfowl, fish, and larger invertebrates. The plants are at the base of detrital and macrophyte food chains. Seeds, leaves, stems, tubers, and epiphyte populations on leaves and stems are consumed by such diverse species as ducks, muskrats, snails, and turtles (Martin and Uhler, 1939; Lippson and others, 1979). SAV provides shelter for juvenile fish, turtles, and macroinvertebrates such as shrimp and crabs (Orth and others, 1979). The stems and leaves retard current flow and permit roots to stabilize the bottom. Thus slowing erosion and encouraging the accumulation of sediment (Allen, 1979). SAV has also been cited for its role in oxygenating the water (Cumming and others, 1916; Haslam, 1978) and in cycling nutrients from the sediments to the overlying water (McRay and McMillan, 1977).

The results of vegetation surveys since the 1950's, as well as reviews of historical documents, herbarium records and personal observations, show that populations of SAV in the Chesapeake Bay and its tributaries, including the Potomac River, have declined in recent years (Kerwin and others, 1976; Munro, 1976; Southwick and Pine, 1975; Elser, 1969, Orth and others, 1979; Stevenson and others, 1979; Bayley and others, 1968, 1978; Stevenson and Confer, 1978). Bartsch (1954) and later, Stewart (1962) reported that the upper Potomac (tidal section above Chicamuxen Creek), which once had extensive beds of wildcelery (*Vallisneria americana*), naid (*Najas*), and sago pondweed (*Potamogeton pectinatus*) was devoid of SAV. Stewart found an abundance of SAV in the central Potomac (Nanjemoy Creek-Port Tobacco River area), but reported only narrow zones of SAV in the lower estuary. In 1972, 1973, 1977, and 1978, personnel of the Migratory Bird and Habitat Research Laboratory (MBHRL) of the U.S. Fish and Wildlife Service (FWS) found no SAV in the tidal river; only 4 percent of 150 sampling stations on the Potomac Tidal River and Estuary were found vegetated (unpublished data, MBHRL: Haramis, 1977).

In 1977, the U.S. Geological Survey began a comprehensive interdisciplinary Potomac Estuary Study which includes extensive water quality sampling and studies of hydrodynamics, geochemistry of bottom sediments, transport, low-flow conditions and ecology of the tidal river and estuary. The wetland ecology group in the Potomac Estuary Study has as its objectives:

1. To document the present distribution and abundance of SAV in the tidal Potomac River.

2. To compare the present distribution with the historic distribution.

3. To attempt identification of the factors responsible for the present distribution of SAV.

4. To consider the implications of the present distribution and abundance of SAV.

In 1978, 1979, and 1980, USGS and MBHRL carried out a cooperative survey of the present distribution and abundance of SAV in the riverine and estuarine environments of the tidal Potomac River. The survey will be repeated in 1981. This report summarizes the findings of this survey through June 1980, and addresses the remaining three objectives listed above.

DESCRIPTION OF STUDY SITES

The Potomac River is the second largest tributary entering the Chesapeake Bay in terms of drainage area and discharge, contributing about 18 percent of the total freshwater inflow (Pritchard, 1952). The Potomac River enters the western side of the Chesapeake Bay between Point Lookout and Smith Point (Fig. 1) and the tidal section extends 183 km from the mouth to Chain Bridge in Washington, D.C. The tidal Potomac River can be divided into three zones: (1) the tidal river above Quantico, Virginia (upper Potomac), where the water is fresh and the net flow is directed seaward at all depths, (2) the transition zone (middle Potomac) where salinity is low and extensive saltwater-freshwater mixing occurs, and (3) the estuary proper (lower Potomac) which exhibits an internal circulation with reverse bottom flow, strong tidal currents, moderate vertical stratification and considerable longitudinal variation in salinity (Elliott, 1976: Wilson, 1977).

During 1978, only the Maryland side of the river was studied. Four regions including six tributaries were selected for intensive sampling (Fig. 1): (1) the Piscataway-Mattawoman Creeks region in the tidal river, (2) the Nanjemoy Creek-Port Tobacco River region in the transition zone, (3) the Wicomico River region in the estuary and (4) the St. Marys River region in the estuary. The Wicomico River, which receives substantial freshwater runoff, retains a freshwater to saltwater gradient and thus was also subdivided into tidal river, transition zone, and estuary. In 1979, sampling was repeated on the Maryland side and also extended to the Virginia side of the river. Five regions including 6 tributaries were selected on the Virginia side (Fig. 1): (1) the Gunston Cove region in the tidal river; (2) the Aquia-Potomac Creek region in the transition zone; and the (3) Machodoc Creek, (4) Nomini Bay, and (5) Yeocomico River regions in the estuary. Because the Virginia side of the river generally has smaller tributaries and vegetation patterns are similar to those on the Maryland side, sampling in 1980 and 1981 is being conducted only on the Maryland side.

DISTRIBUTION AND ABUNDANCE

Sampling methods for the SAV survey are discussed in Carter and Haramis (1980). Sampling was done with oyster tongs on transects oriented perpendicular to shore in both tributaries and along the main river on either side of the tributary mouth. Sampling was conducted seasonally, in the spring, summer, and fall of 1978 and in the spring and fall of 1979. Sampling has also been completed for the spring of 1980. Figure 1 shows the pattern of abundance of submersed aquatic species in the study area for 1978 and 1979. A total of 13 species of submersed aquatic macrophytes was recorded for the survey (Table 1). These included 11 species of vascular plants and 2 species of algae in the genus *Chara*. The highest species diversity (Table 2) occurred in the Nanjemoy Creek-Port Tobacco River region, specifically in the Port Tobacco River, the mouth of Nanjemoy Creek, and the adjacent Potomac

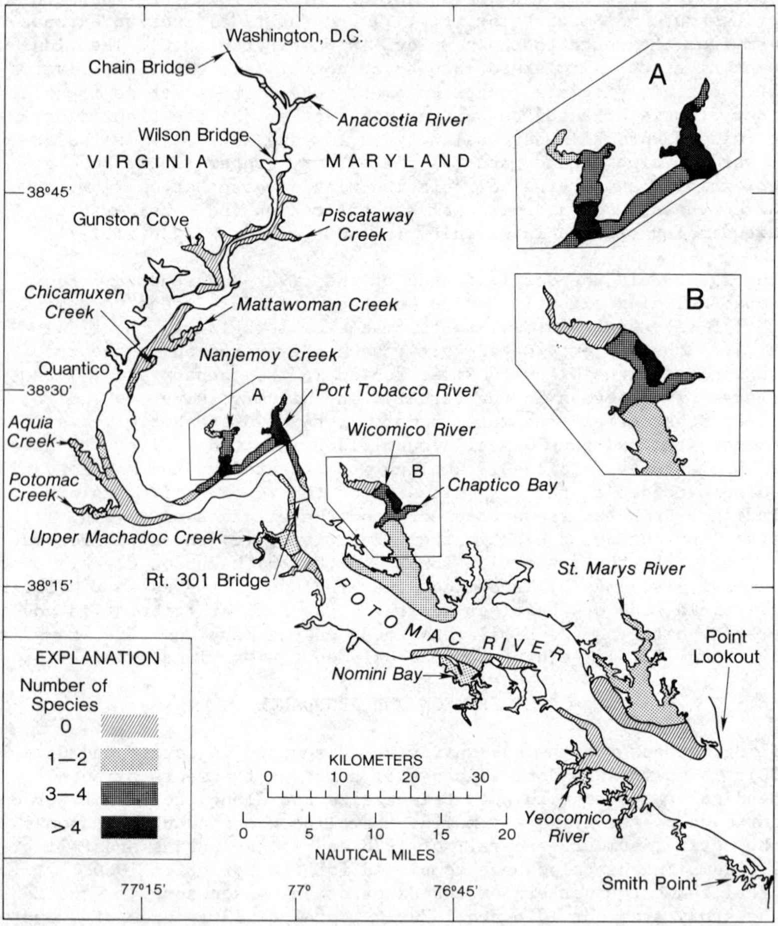

Figure 1: Map of study area showing patterns of abundance of submersed aquatic vegetation, 1978 and 1979.

Table 1. Species list of submersed aquatic macrophytes found during the tidal Potomac River survey. Taxonomy follows Hotchkiss (1950, 1967) unless otherwise noted.

Family	Species	Common Name
Characeae[a] (muskgrass family)	*Chara braunii* Gm.	Muskgrass
	Chara zeylanica Km. ex Wild.	
Najadaceae (pondweed family)	*Potamogeton perfoliatus* L.	Redhead-grass
	Potamogeton pectinatus L.	Sago pondweed
	Potamogeton crispus L.	Curly pondweed
	Potamogeton pusillus L.	Slender pondweed
	Ruppia maritima L.	Widgeongrass
	Zannichellia palustris L.	Horned pondweed
	Najas guadalupensis (Spreng.) Morong	Southern naiad
Hydrocharitaceae (frogbit family)	*Vallisneria americana* Michx.	Wildcelery
	Elodea canadensis (Michx.) Planch.	Common elodea
Ceratophyllaceae (coontail family)	*Ceratophyllum demersum* L.	Coontail
Haloragidaceae (watermilfoil family)	*Myriophyllum spicatum* L.	Eurasian watermilfoil

[a] Keyed from Wood (1967).

Table 2. Percent frequency of occurrence of submersed aquatic macrophytes in vegetated grabs for each of 4 Maryland sampling regions of the tidal Potomac River, 1978, 1979, and 1980.

Region	Species	Sampling Period 1978 Spring	1978 Summer	1978 Fall	1979 Spring	1979 Fall	1980 Spring
Piscataway-Mattawoman Creeks (freshwater river)	*Vallisneria americana*	0	100	100			a
	Total vegetated grabs	0	3	2	0	0	a
Nanjemoy Creek-Port Tobacco River (transition zone)	*Vallisneria americana*	51.6	62.4	86.6	56.7	84.8	64.8
	Potamogeton perfoliatus	21.4	36.9	46.2	26.3	14.3	17.9
	Potamogeton crispus	17.5	12.7	3.0	17.5	0	6.2
	Potamogeton pectinatus	3.2	7.6	3.7	2.1	0	6.2
	Zannichellia palustris	10.3	0.6	0	14.9	0	9.0
	Myriophyllum spicatum	0	5.1	4.5	6.7	24.1	3.4
	Ruppia maritima	3.3	3.2	8.2	1.0	8.0	5.5
	Ceratophyllum demersum	0	1.9	0.7	7.2	11.6	7.6
	Potamogeton pusillus	2.4	1.3	0	4.6	0	0.7
	Chara spp.	2.4	1.9	0	1.5	0.9	1.4
	Elodea canadensis	0	0	0	1.0	0.9	0
	Total vegetated grabs	126	157	134	194	112	145
Wicomico River (transition zone and estuary)	*Zannichellia palustris*	76.6	11.1	0	82.8	0	77.2
	Ruppia maritima	34.0	88.9	100.0	15.7	89.7	14.6
	Najas guadalupensis	9.7	38.9	19.2	5.2	20.5	9.6
	Chara spp.	27.2	0	0	16.4	0	19.2
	Potamogeton perfoliatus	1.0	13.9	30.8	2.2	12.8	1.8
	Myrophyllum spicatum	0	0	0	1.5	0	1.4
	Potamogeton pusillus	0	0	0	0	0	0.9
	Vallisneria americana	0	0	0	0	0	0.5
	Potamogeton pectinatus	0	0	0	0	0	0.9
	Total vegetated grabs	103	36	26	134	39	219
St. Marys River	*Zannichellia palustris*	84.0	0	0	95.8	0	100.0
	Ruppia maritima	16.0	100	100	8.3	0	0
	Total vegetated grabs	25	7	1	24	0	19

[a] Sampling not repeated in 1980

littoral. Wildcelery (Vallisneria americana) and redhead-grass (Potamogeton perfoliatus) were the most frequently encountered species in the region.

High diversity was also observed in the upper Wicomico estuary or Wicomico transition zone. This was the only location where muskgrass (Chara braunii) and southern naiad (Najas guadalupensis) were found in abundance. Additionally, this was the location where widgeongrass (Ruppia maritima) achieved its most extensive and luxuriant growth and seed production in 1978.

No SAV was found in the Piscataway-Mattawoman Creeks region or the Gunston Cove region in 1978 or 1979, except for the farthest downstream Maryland transect across from Quantico, Virginia, where wildcelery was collected in the summer and fall of 1978. During the summer of 1978, 6 extra transects were sampled in the tidal river in the District of Columbia. None of these were vegetated. The Wicomico River below Chaptico Bay, and the St. Marys River region were characterized by sparse growths of widgeongrass and horned pondweed (Zannichellia palustris). Eelgrass (Zostera marina), which at one time vegetated shoal zones up to Cobb Island at the mouth of the Wicomico River, was not found during the survey. Vast areas of the lower estuary were almost devoid of aquatic plant growth during summer and fall.

Table 2 summarizes the vegetation occurrence in the Maryland study area in 1978, 1979, and 1980. Submersed aquatic macrophytes were most widely distributed in the spring sampling periods; 28 percent of all transects had at least one vegetated sample in both 1978 and 1979. Horned pondweed and widgeongrass, which were more widely distributed in the Wicomico and St. Marys River regions during spring than during the summer or fall sampling periods accounted for these results. In spring of 1980, the Mattawoman-Piscataway Creek region was not resurveyed using the same transects because there was little chance that vegetation would be found on any except the most downriver station. Instead, new transects were established on the Maryland side of the tidal river to eliminate the possibility that some small vegetated areas might have been missed by the original survey. One small pocket of wildcelery (Vallisneria americana) and slender pondweed (Potamogeton pusillus) was located in Mattawoman Creek about 5 km upstream of the tributary mouth.

The 1978-1979 survey, supplemented by frequent observations during the growing season, showed a paucity of SAV in the tidal river except for tiny pockets within marshes or small populations in small side tributaries. It also showed the general lack of vegetation in the expanses of the lower estuary and identified the transition zones of both main river and adjoining tributaries as the remaining stronghold for SAV. These findings are in general agreement with those of Bartsch (1954) and Stewart (1962).

HISTORIC DISTRIBUTION

Historically, the Potomac estuary and tidal river contained numerous species of aquatic macrophytes in abundance. Seaman (1875) reported wildcelery {(Vallisneria spiralis (V. americana Michx.)}, coontail (Ceratophyllum demersum), naiad (Najas flexilis), and elodea (Anacharis alcinastrum) (old world elodea) in the vicinity of Washington, D.C. A map of the river at low tide from Cumming and others (1916) (Figure 2) shows a narrow channel and wide shallow vegetated margins and flats containing beds of cury pondweed (Potamogeton crispus), coontail and wildcelery. Hitchcock and Standley (1919) reported additional pondweeds in the mouths of tributaries below Washington including Potamogeton amplifolius, P. perfoliatus, P. zosteraefolius (P. zosteriformis Fern.), P. pusillus, P. pectinatus, which was reported as cosmopolitan, and P. robbinsii. These historical records suggest that the highest diversity of aquatic macrophytes occurred in the tidal river and transition zone and the number of species declined to fewer, more salt-tolerant species, in the lower estuary. Similar trends are seen in other tidal riverine-estuarine systems (Alexander and others, 1935:75; Conover, 1958).

FACTORS AFFECTING DISTRIBUTION

The loss of SAV's from the tidal river probably occurred in the 1920's and 1930's because by 1939, Martin and Uhler (1939) noted the loss of aquatic plants. They suggested the decline was due to increased turbidity caused by suspended silt and clays. The decline of SAV populations in the tidal freshwater riverine environment is not limited to the Potomac River, but appears to be the case throughout the Chesapeake Bay area (Orth and others, 1979; Stevenson and others, 1979). Submerged macrophyte populations have also declined sharply in other freshwater environments such as lakes (Phillips and others, 1978) and rivers (Moll, 1980; Mills and others, 1966).

Many factors, both natural and man-related, have been implicated in the decline of SAV populations in the Chesapeake Bay and in other aquatic ecosystems (Stevenson and Confer, 1978; Stevenson and others, 1979; Mills and others, 1966; Phillips and others, 1978; Haslam, 1978). Haslam (1978) groups rivers into three arbitrary categories: (1) stream with continuous, severe pollution with either no plants at all or very sparse population of pollution-resistant plants, (2) streams with accidental, temporary, very poisonous spillage, where plants will recover, although effects may be longterm if the plants are already under stress from another cause, and (3) streams with continuous mild pollution where the vegetation differs in species presence or abundance from those characteristic of a clean site. The first results of pollution are general reduction in plant diversity and sometimes lush overgrowth of the more pollution-tolerant species which replace more sensitive species.

Figure 2: Upper part of freshwater tidal Potomac River at low water showing distribution of aquatic vegetation (reproduced from Cumming and others, 1916).

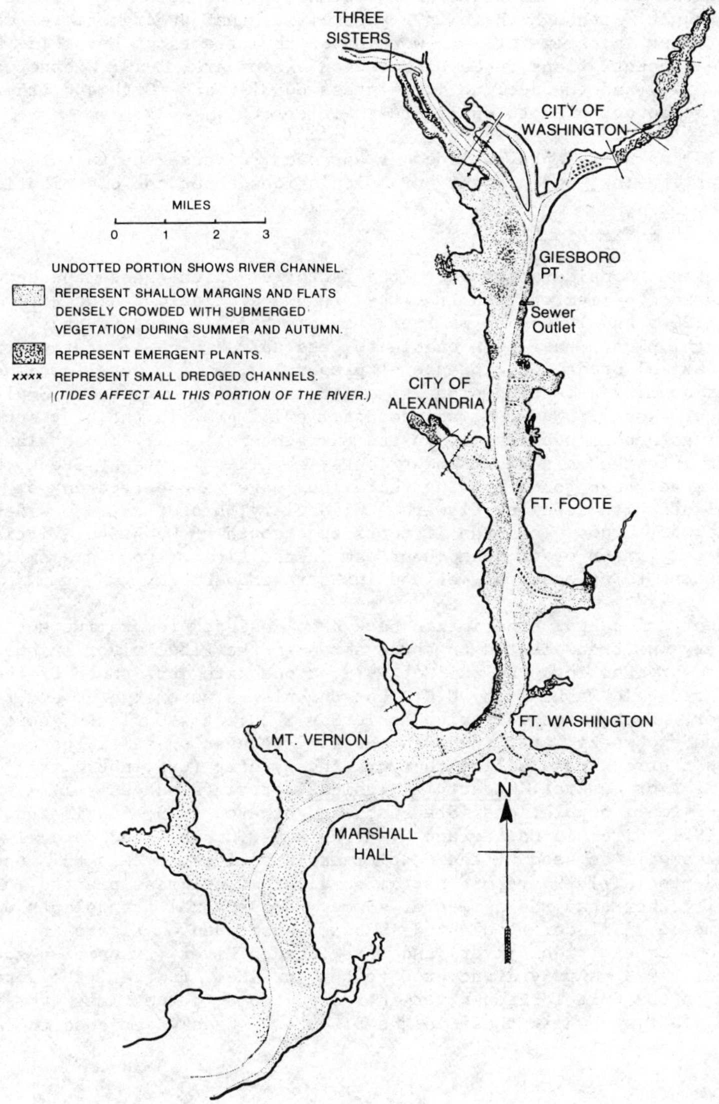

Isolation of the factors responsible for the decline of plants in order to attempt corrective measures is generally difficult because of the presence and interaction of multiple factors. We have chosen to approach the problem of the present distribution with multiple hypotheses which can be used to design laboratory and field experiments and to isolate single parameters or closely related factors. In this way, perhaps some factors can be eliminated from further consideration or reduced in importance, although there is a possibility that synergistic effects exist. Other investigators are working on the effects of herbicide and chlorine concentrations on SAV (Stevenson and others, 1979; Webster and Rawles, 1976) and the decline in eelgrass populations (Orth and others, 1979). We do not plan to duplicate their efforts.

The following paragraphs outline our approach, discuss the factors we are investigating, and present our speculations about the distribution.

Survival in the tidal river

One important consideration about the factor(s) which caused the abrupt disappearance of macrophytes from the freshwater tidal Potomac River in the 1920's and 1930's is whether they are still operating today. Perhaps the plants have been unable to reestablish because of factors such as animal predation, absence of propagules, or a general decline in water quality affecting plant vitality and, thus, survival. Small pockets of elodea, pondweed, and coontail still grow in the tiny subtributaries and in pockets within the emergent wetlands. These plants wash out into the main river and tributaries like Piscataway Creek, but they have not been found growing there, nor have we observed any of these plants with seeds or flowers. Wildcelery, historically a widespread plant, grows along the littoral zone south of Quantico, Virginia, but the only trace of it found upstream in the tidal river is a small bed adjacent to an emergent wetland in Mattawoman Creek.

Transplant and caging experiments have been designed to examine survival. Wildcelery was transplanted to three sites in the tidal river in spring 1979 and survived into the fall of 1979 at one site just south of the Wilson Bridge in Washington, D.C., but the plants were stunted and few in number. Four sites were planted in May 1980; of 400 plants, none was found two weeks later, nor was the 1979 vegetation still there. There is a strong possibility that grazing by carp (*Cyprinus carpio*), other fish, or muskrat (*Ondatra zibethicus*) prevents successful revegetation of small populations (Stevenson and others, 1979; Slavik and Uhler, 1951; Virginia Carter and G. M. Haramis, unpublished data, 1979). Exclosures will be used at the next transplant sites to test this theory. Orth and others (1979) report better survival of eelgrass planted in the fall rather than the spring or summer, so seasonal transplants will be attempted. Wildcelery planted in boxes suspended on floats in Piscataway Creek (tidal river) and Port Tobacco River (transition zone) in summer, 1979, survived and grew to the surface. However, the populations in the Port Tobacco River increased two to three times over the initial planting, while those in the Piscataway Creek decreased to less than a third of the original number of plants. This preliminary experiment was terminated when fish moved into the Piscataway Creek boxes to nest and eliminated the remaining plants.

Coontail, a nonrooted submersed aquatic plant, was suspended in mesh cages in the tidal river during the summer and fall of 1979. In this preliminary experiment, plants survived in both the tidal river and the transition zone, however, those in the tidal river became discolored, lost many leaves, and were coated with slime. In 1980, measured amounts of coontail and elodea are being systematically placed in mesh cages in the tidal river and transition zone to measure growth and survival.

Salinity

Salinity is a major factor affecting plant distribution in any estuary and tidal river system. As far as can be determined, no prolonged or unusual alterations in the salinity regime have occurred in the tidal Potomac River since the turn of the century. Those plants common in freshwater (pondweeds, wildcelery, coontail) do not presently grow in the tidal river, nor does eelgrass, a salt-tolerant species, grow in the lower estuary. Salinity dynamics in the transition zone, combined with a high nutrient load, may however, be responsible for SAV survival in that reach (see Nutrient Loading).

Substrate

During the 1978-1979 survey, we recorded substrate composition on all sampling sites, vegetated and unvegetated. The majority of plants occupied a narrow band of water depth near shore and in 1978, 90 percent of 620 vegetated samples were on sand-based sedimenta; 42 percent were on silty sand. Only trace amounts of vegetation were found on softer silts or silt-clay sediments in deeper water. Fine mud sediments and marsh detritus did support plant growth in shallow protected areas; such sites were uncommon and generally dominated by Eurasian watermilfoil (*Myriophyllum spicatum*), curly pondweed, and coontail. Extremely fluid sediments characteristic of high sedimentation rates were generally devoid of aquatic plants. During 1979 and 1980, we found wildcelery growing on a silty-clay site in Upper Machodoc Creek and in the Port Tobacco River, but in general, such sites are not vegetated. Box transplant experiments conducted in 1979 showed no evidence of reduced growth in sand substrates from the tidal river as contrasted with sandy substrates from the transition zone.

The literature suggests that aquatic plants thrive best in mud substrates followed by sand-based substrates and gravel (Denny, 1972; Spence, 1964). The distribution of plants on substrates in the Potomac River led us to consider the possibility of some growth-reducing substance adsorbed on the finer sediments - herbicides, trace metals, or some other chemical. We are presently conducting two experiments, one related to plant growth in a variety of substrates and one looking at the effects of trace metals on growth (see Trace metals).

Nutrient loading

Macrophyte decline may be associated with the long term increase in nutrient enrichment from the Washington Metropolitan area. In 1971, Jaworski and others reported that about 325 mgd (million gallons per day) of sewage discharge enter the tidal river, contributing 450,000 lb/d (pounds per day) of ultimate oxygen demand, 2,400 lb/d of phosphorous (P), and 60,000 lb/d of nitrogen (N). This is in contrast to loadings of about 5,000 lb/d P and 7,500 lb/d N during the decade between 1910 and 1920. The introduction of sewage waste and nutrients from non-point sources may have changed the balance between phytoplankton and macrophyte growth and favored a major ecological shift in primary productivity. The literature contains much information which supports the competitive relationship between phytoplankton and macrophytes in enriched waters (Spence, 1976; Jupp and Spence, 1977; Moss, 1976; Mulligan and Baranowski, 1969; Morgan, 1970; Goulder, 1969; and others). Recently, Phillips and others (1978) have reviewed this relationship and have presented evidence for a complex interaction between macrophytes and phytoplankton. They acknowledge the ability of certain macrophytes to secrete phytoplankton suppressants, and identify heavy epiphyte and filamentous algal growth as the causative agents of initial macrophyte decline under enriched conditions. Following this initial decline, shading from increased phytoplankton populations can eliminate the remaining vegetation. Experiments with fertilization of fish ponds also document the elimination of macrophytes in this way (Smith and Swingle, 1941). Although we can neither corroborate nor reject the existence of such complex macrophyte-phytoplankton interactions in the tidal Potomac, our observations show that macrophyte growth is sparse or absent in the freshwater region and the mesohaline estuary where abundant plankton and periphyton burdens do exist. The salinity variations and related osmotic changes, as well as perhaps flushing rate, may limit the existence of, or exclude, certain freshwater-adapted and true mesohaline plankton and periphyton from the transition zone and permit macrophytes to survive. Major shifts in the salinity gradients caused by floods or droughts would permit these organisms to enter the vegetated zone, but because several years may be needed to eliminate SAV populations, the effects may be noticeable, but temporary.

We have designed and set out on the tidal river, transition zone, and lower estuary, racks of plastic dishes to look at epiphyte populations, to test the reduction of light by coatings of epiphytes, and to test the resistance of epiphyte populations from the freshwater and mesohaline environments to fluctuations in salinity.

Light

Reduced light penetration has often been cited as a major cause for SAV decline in the Chesapeake Bay area and the Potomac River. Reduced light penetration is generally due to increased turbidity from suspended sediments or large algal concentrations. Decreased light penetration confines plant growth to shallower water, decreasing the total amount of vegetation present and selectively eliminating the most light sensitive species. Martin and Uhler (1939:120) and Slavik and Uhler (1951) cite the loss of aquatic plants in the Potomac as a result of increased turbidities due to water laden with suspended silts and clays. Excessive siltation is also implicated in the loss of SAV in the Illinois River (Mills and others, 1966), and in coal streams in Britain (Haslam, 1978). Sedimentation associated with high turbidity may cause unstable substrates and extra sediment moving during storm flows; plants may be smothered by excessive deposits or washed away. Recently, Steenis (1971:26) stated that sewage effluent was the main cause for increased blooms of green and blue-green algae which have increased turbidity and promoted progressive loss of seed-bearing plants for some 64 km below Washington since 1930. Nutrient enrichment has prompted blooms of the nuisance blue-green algae *Anacystis* which has been persistent in the tidal river since the early 1960's, often forming green mats of cells at the water's surface (Jaworski and others, 1971; Pheiffer and others, 1972). Such blooms of blue-green algae are particularly severe during summer low flow periods which high temperatures, low flushing rates and high nutrient availability combine to promote growth. Extensive blooms of brown and red tides of dinoflagellates which appear in the lower estuary have also been suspected to be a result of nutrient enrichment and have increased turbidity in that region. Density of these dinoflagellates may reach 500 mg/liter Chlorophyll *a* and persist for several months (R. Cohen, unpublished data, 1980).

Cumming and others (1916) emphasised the effects of the vegetated flats in the freshwater tidal Potomac on reducing turbidity in the river. They found high turbidities on the flats during the winter and spring when vegetation was absent, but in spite of this turbidity, the vegetation grew some 1.2 to 1.5 meters to the surface and the water on the flats was generally clear in summer and fall. Today vegetation is found mostly in water less than a meter deep at low tide and seldom does it extend more than about 100 to 150 meters from shore, even in shallow areas. Secchi disk transparencies measured in spring and fall in the tidal river, transition zone, and lower estuary show that the water is much clearer in the lower estuary, but that very little if any difference exists between transparencies in the transition zone where there is abundant vegetation and in the tidal river, where there is none.

Heavy metals

Heavy metal accumulation in benthic soils or heavy metal concentrations in water could inhibit plant growth, although Haslam (1978) notes that a survey of Britain showed no detectable correlation of plant pattern with the variability shown by any or all of ten heavy metals, even in polluted sites. He does point out that the effects could be masked by those of other pollutants. A recent literature review (Leland and others, 1977) shows relatively little information on the bioaccumulation of heavy metals by aquatic plants or the effects, if any, on aquatic plant growth. However, recent work by Marchyulenene and others (1978) and Dusanskene-Duzh and Pohkarpov (1978) indicate that SAV accumulate lead (^{210}Pb) and that it can have a toxic effect on some species. Peter and others (1979) and McIntosh and others (1978) also discuss macrophyte cycling of heavy metals.

Pheiffer and others (1972) detected concentrations of lead, cobalt, copper, nickel, barium and lithium in substrates in the vicinity of Woodrow Wilson Bridge. Concentrations generally declined down river, but increased again near the U.S. Route 301 bridge. We are presently concentrating on laboratory experiments to determine the effects on growth of dissolved concentrations of zinc (Zn), manganese (Mn), and lead (Pb).

ENVIRONMENTAL IMPLICATIONS

The decline in SAV reflects a change in historical ecosystem conditions. In an unpolluted riverine or estuarine environment, macrophytes and phytoplankton both utilize a limited nutrient supply and provide a vital link in the food chain. Populations fluctuate in response to natural stresses such as increased salinity caused by drought or heavy flows from unusual storms. When subjected to gross or even subtle man-made environmental perturbations, the ecosystem may change. The decline or absence of SAV in freshwater environments has implications beyond a disturbance of ecosystem components. It is generally accompanied by a decline in macroinvertebrate, fish, turtle, and waterfowl populations and (or) replacement of sensitive species by more pollution-tolerant species. Although this paper primarily addresses the Potomac River, examples of similar environmental disturbances and their effect on the Illinois River and Loch Leven, Scotland, are discussed below to demonstrate that this is a widespread problem.

Historically, the tidal waters of the Potomac, which typify the diversity of aquatic habitats found throughout the Chesapeake Bay system, were the prime wintering areas for thousands of ducks, geese, and swans. So important was the Chesapeake Bay, that large portions of the continental population of some waterfowl species could be found wintering there. Similarly, the Illinois River was a popular stopping place for diving ducks on their fall and spring migrations (Miller and others, 1966), and Loch Leven, Scotland, was used during the summer by mute swans (*Cygnus olor*), coot (*Fulica atra*), and pochards (*Aythya ferina*) (Jupp and Spence, 1977).

In the 1930's, Martin and Uhler (1939) documented the importance of SAV in the diets of waterfowl. They found pondweeds (*Potamogeton* spp.), wildcelery, and widgeongrass to be the most abundantly consumed aquatic plants in a general sampling of both dabbling and diving ducks. Of the total consumed foods analyzed, plant foods accounted for 70 percent of the volume and animal material accounted for the remainder. Among the dabbling duck species, the widgeon (*Mareca americana*) and the gadwall (*Anas strepera*) are dependent upon water plants for food (Miller and others, 1966). Cottam (1939), who specifically examined the diets of diving waterfowl, found pondweeds, particularly sago pondweed and redhead-grass, widgeongrass, northern and southern naiads (*Najas flexilis* and *N. guadalupensis*), wildcelery and muskgrasses to be of major importance to the inland diving ducks (tribe Aythyini), commonly called pochards or bay ducks, and also to a distant relative, the ruddy duck (*Oxyura jamaicensis*). He found plant food accounted for the following percentages of total food volume: 90 percent in the redhead (*Aythya americana*), 81 percent in the ring-neck (*A. collaris*), 81 percent in the canvasback (*A. valisinera*), 60 percent in the lesser scaup (*A. affinis*), and 72 percent in the ruddy duck. The remaining species of genus *Aythya*, the greater scaup (*A. marila*), consumed predominantly animal foods.

Some of the most important species of SAV for waterfowl are found in the Potomac River. Sago pondweed, with its wide salinity tolerance, has almost a continental distribution and thus was recognized by Cottam as possibly the single most important submersed aquatic waterfowl food plant known. Sago pondweed was abundant in the Potomac River in the early part of the 1900's and as late as the 1950's (R. E. Stewart, oral communication, 1979), but has declined sharply and presently occurs in only scattered patches in the transition zone and estuary. Widgeongrass was also formerly well distributed in the Potomac Estuary, but presently remains only locally abundant. Wildcelery and redhead grass are abundant in the transition zone from Aquia Creek to Upper Machodoc Creek. This area, including Nanjemoy Creek and the Port Tobacco River, and the upper Wicomico River provides most of the remaining submersed plant resource for waterfowl in the tidal Potomac River.

The vanishing SAV's in the Chesapeake Bay area do not represent an isolated or unusual phenomena. Aquatic plant beds along the Illinois River were luxuriant in 1930; coontail, elodea, pondweeds, naiads, and wildcelery filled the river and flood-plain lakes. Following a series of severe population declines and temporary recoveries, the aquatic plants were largely eradicated between the mid-1950's and 1966 (Mills and others, 1966). During the same period, most of the bottom fauna, including mollusca such as the fingernail clam (Sphaeriidae) disappeared. Submerged macrophytes declined in Loch Leven between 1950 and 1972 following eutrophication (Jupp and Spence, 1977).

Loss of the valuable aquatic plant resource, and the concurrent loss in some cases of benthic invertebrates, has had adverse effects on the quality of wintering habitats for many waterfowl, and particularly for the pochards. Continental populations of canvasbacks and redhead ducks, two prized game species, have declined sharply during this century, and although losses of prairie breeding habitat, vulnerability to hunting, and other factors are often implicated in this decline, it is clear that loss of quality wintering or migratory habitat is an additional important factor. Waterfowl must modify their food habits to adjust for the declines in vegetation or shift their distributions to more favorable habitats. A good example is that of the canvasback and redhead. Recent work by Perry (1974, 1975) and Perry and Uhler (unpublished data, 1976) indicates that wintering canvasbacks in the Chesapeake Bay area now consume over 90 percent animal food, predominantly the pollution-resistant clams Rangia and Macoma, both of which are abundant in the Potomac Estuary and much of Chesapeake Bay. Unlike the canvasback, the redhead apparently has been unable to adjust to animal foods and is no longer generally abundant in Bay waters. The loss of aquatic food resources has drastically affected the numbers of migratory diving ducks using the Illinois and Potomac Rivers. Redheads were forced to seek food elsewhere and populations of ring-necked ducks, lesser scaup and canvasbacks dropped far below those in 1955 and previous years (Mills and others, 1966; Linduska, 1959). The long term changes in the incidence of some herbivorous waterfowl species, coots, swans, and pochards, has been correlated with decline of submerged macrophytes (Jupp and Spence, 1977).

Loss of aquatic plants has a variable effect upon wildlife species depending upon their adaptability. For example, although the waterfowl populations using the Illinois River have been impacted by the loss of aquatic vegetation and bottom fauna, the turtle populations are doing well, except for one or two species. Moll (1980) explains that turtle survival amidst drastically changing environmental conditions is due to the turtle longevity and its capacity for behavioral and physiological adaptation. Although turtles once used rooted aquatic plants for food and shelter, they now thrive without benthic invertebrate fauna or SAV, eating drowned insects, detritus and carrion, thus switching from a herbivourous to carnivorous diet. Pollution-resistant species, such as the Asiatic clam, Corbicula fluminea, which has recently invaded the Illinois River (Mills and others, 1966) and the Potomac River (Dresler and Cory, 1980), may serve as new food source for turtles and for waterfowl. Similarly, metropolitan debris and fish kills may provide substitute shelter or food for remaining wildlife species.

CONCLUSIONS

This paper has discussed the decline of SAV in the tidal Potomac River, some potentially contributing factors, and the environmental implications of such a decline. To demonstrate that this decline is not a single isolated event, but is duplicated in many freshwater aquatic systems, we have also used as examples the Illinois River and Loch Leven in Scotland. The loss of a significant environmental component is a signal that the system itself is under stress. Chronic pollution changes and damages habitat at the expense of sensitive and often desirable species. It reduces the carrying capacity of an ecosystem for economically important wildlife, such as crabs, fish, and waterfowl, by depleting the food supply, eliminating shelter and nursery areas and increasing competition by pollution-resistant populations. This, in turn, affects the hunting and fishing industries, as well as impacting recreational use of the waters. Perhaps, the ultimate question posed by all this is whether we are willing to settle for a depauperate flora and fauna made up of pollution-tolerant, sometimes less desirable species, or whether we can isolate the factors responsible for habitat degredation and make the appropriate changes necessary to restore a more diverse, productive, and attractive environment.

LITERATURE CITED

Allen, H. H., 1979, Role of wetland plants in erosion control of riparian shorelines; in Greeson, P. E., and others, eds., Wetland Functions and Values: the State of Our Understanding: American Water Resources Association, Minneapolis, Minn., p. 403-414.

Alexander, W. B., Southgate, B. A. and Bassindale, R., 1935, Survey of the River Tees, II. The estuary - chemical and biological: Water Pollution Resources W.S.I.R. Technical Paper no. 5, 171 p.

Bartsch, A. F., 1954, Bottom and plankton conditions in the Potomac River in the Washington Metropolitan area-Appendix A in A report on water pollution in the Washington Metropolitan area: Interstate Commission on the Potomac River Basin, 57 p.

Bayley, Sandra, Rabin, H., and Southwick, C. H., 1968, Recent decline in the distribution and abundance of Eurasion milfoil in Chesapeake Bay: Chesapeake Science, v. 9, no. 3, p. 173-181.

Bayley, Sandra, Stotts, V. D., Springer, P. F., and Stennis, John, 1978, Changes in submerged aquatic macrophyte populations at the head of Cheaspeake Bay, 1958-1975: Estuaries, v. 1, no. 3, p. 73-84.

Carter, Virginia, and Haramis, G. M., 1980, Distribution and abundance of submersed aquatic vegetation in the tidal Potomac River - implications for waterfowl: in Proceedings of Audubon Naturalist Society 1st Mid-Atlantic Natural History Symposium on "Bird Populations - A Litmus Test of the Environment" (in press.).

Cottam, Clarence, 1939, Food habitats of North American diving ducks: U.S. Department of Agriculture Technical Bulletin, no. 643, 140 p.

Conover, J. T., 1958, Seasonal growth of benthic marine plants as related to environmental factors in an estuary: Publication of Institute of Marine Science, University of Texas, v. 5, p. 97-147.

Cumming, H. S., Purdy, W. C., and Ritter, H. P., 1916, Investigations of the pollution and sanitary conditions of the Potomac watershed: Treasury Department, U.S. Public Health Service Hygienic Laboratory Bulletin, n. 104, 231 p.

Denny, Patrick, 1972, Sites of nutrient absorbtion in aquatic macrophytes: Journal of Ecology, v. 60, no. 3, p. 819-829.

Dresler, P. V., and Cory, R. L., 1980, The Asiatic clam, Corbicula fluminea (Muller) in the tidal Potomac River, Maryland: Estuaries, v. 3, no. 2, p. 150-151.

Dusanskene-Duzh, R. F., and Pohkarpov, G. G., 1978, Extraction and accumulation of ^{210}Pb by aquatic organisms in lakes: Hydrological Journal, v. 14, no. 3, p. 62-70.

Elliott, A. J., 1976, a mixed-dimension kinematic estuarine model: Chesapeake Science, v. 17, no. 3, p. 135-140.

Elser, H. J., 1969, Observations on the decline of the watermilfoil and other aquatic plants, Maryland, 1962-1967: Hyacinth Control Journal, v. 8, no. 52, p. 52-60.

Goulder, R., 1969, Interactions between the rates of production of a freshwater macrophyte and phytoplankton in a pond: Oikos, v. 20, p. 300-309.

Haramis, G. M., 1977, Vegetation survey in cooperation with Maryland Department of Natural Resources: U.S. Fish and Wildlife Service, Memorandum, Migratory Bird and Habitat Research Laboratory, Laurel, Maryland, 4 p.

Haslam, S. M., 1978, River plants: New York, Cambridge University Press, 396 p.

Hitchcock, A. S., and Standley, P. S., 1919, Flora of the District of Columbia and vicinity: Contributions from the United States National Herbarium, v. 21. Smithsonian Institution, 329 p. + 42 pl.

Hotchkiss, Norman, 1950, Checklist of marsh and aquatic plants of the United States: U.S. Fish and Wildlife Service, Wildlife Leaflet 210, 34 p.

Hotchkiss, Norman, 1967, Underwater and floating-leaved plants of the United States and Canada: U.S. Fish and Wildlife Service Resource Publication 44, Washington, D.C., 124 p.

Jaworski, N. A., Lear, D. W., Jr., and Villa, O., Jr., 1971, Nutrient management in the Potomac estuary, in G. E. Likens, ed., Nutrients and eutrophication: the limiting nutrient controversy: American Society of Limnology and Oceanography, Inc., Lawrence, Kans., p. 246-273.

Jupp, B. P., and Spence, D. H. N., 1977, Limitations of macrophytes in a eutrophic lake, Loch Leven. I. Effects of phytoplankton: Journal Ecology, v. 65, p. 175-186.

Jupp, B. P., and Spence, D. H. N., 1977, Limitations of macrophytes in a eutrophic lake, Loch Leven. II. Wave action, sediments and waterfowl grazing: Journal Ecology, v. 65, p. 431-446.

Kerwin, J. A., Munro, R. E., and Peterson, W. W. A., 1976, Distribution and abundance of aquatic vegetation in the upper Chesapeake Bay, 1971-1974, in J. Davis, ed., The effects of tropical storm Agnes on the Chesapeake Bay estuarine system: Chesapeake Research Consortium, Inc., p. 393-400.

Leland, H. V., Luoma, S. N., and Wilkes, D. J., 1977, Heavy metals and related trace elements: Journal of the Water Pollution Control Federation, p. 1340-1365.

Linduska, J. P., 1959, Waterfowl and the Potomac River: Maryland Conservationist, v. 36, no. 6, p. 4-8.

Lippson, A. J., and others, 1979, Environmental Atlas of the Potomac Estuary: Martin Marietta Corp., Baltimore, Md., 279 p.

Marchyulenene, D. P., Dushauskene-Duzh, R. F., Moteyunene, E. B., Nyanishkene, V. B., Polikarpov, G. G., and Vorob'ev, L. N., 1978, Accumulation of ^{210}Pb in freshwater plants: Hydrobiological Journal, v. 14, no. 6, p. 73-77.

Martin, A. C., and Uhler, F. M., 1939, Food of game ducks in the United States and Canada: U.S. Department of Agriculture Technical Bulletin 634, Washington, D.C., 308 p.

McRoy, C. P., and McMillan, C., 1977, Production ecology and physiology of seagrasses, in C. P. McRoy and C. Helfferich, eds., Seagrass Ecosystems: A Scientific Perspective: New York, Marcel Dekker, Inc., p. 53-87.

McIntosh, A. W., Shephard, B. N., Mayes, R. A., Otchison, G. J., and Nelson, D. W., 1978, Some aspects of sediment distribution and macrophyte cycling of heavy metals in a contaminated lake: Journal of Environmental Quality, v. 7, no. 3, p. 301-305.

Mills, H. B., Starrett, W. C., and Bellrose, F. C., 1966, Man's effect on the fish and wildlife of the Illinois River: Illinois Natural History Survey Biological Notes, no. 57, Urbana, Illinois, 24 p.

Moll, Don, 1980, Dirty river turtles: Natural History, v. 89, no. 5, p. 42-29.

Morgan, N. C., 1970, Changes in the fauna and flora of a nutrient enriched lake: Hydrobiologia, v. 35, p. 545-553.

Moss, B., 1976, The effects of fertilization and fish on community structure and biomass of aquatic macrophytes and epiphytic algal populations: and ecosystem experiment: Journal Ecology, v. 64, p. 313-342.

Mulligan, H. F., and Baronowski, A., 1969, Growth of phytoplankton and vascular aquatic plants at different nutrient levels: Verhandlungen International Vereinigung Limnology, v. 17, p. 802-810.

Munro, R. E., 1976, Distribution and abundance of submerged aquatic vegetation in the upper Chesapeake Bay - 1976 compared with 1971-1975: U.S. Fish and Wildlife Service, Migratory Bird and Habitat Research Laboratory, Laurel, Md., Mimeograph, 7 p.

Orth, R. J., Moore, K. A., and Gordon, H. H., 1979, Distribution and abundance of submerged aquatic vegetation in the lower Chesapeake Bay, Virginia: Environmental Protection Agency Report Number 600/8-79-029/SAV 1, Office of Research and Development, U.S. Environmental Protection Agency, Washington, D.C., 198 p.

Peter, R., Welsh, H., and Denny, Patrick, 1979, The translocation of lead and copper in two submerged aquatic angiosperm species: Journal of Experimental Botany, v. 30, no. 115, p. 339-345.

Perry, M. C., 1974, Looking out for the canvasback, Part IV: Ducks Unlimited, v. 38., p. 21-22, 25-27.

Perry, M. C., 1975, Looking out for the canvasback, Part V: Ducks Unlimited, v. 39, p. 14-16, 23.

Pheiffer, T. H., Donnelly, D. R., and Possehl, D. A., 1972, Water quality conditions in the Chesapeake Bay system: Environmental Protection Agency Technical Report 55, p. I-1 to V-9.

Phillips, G. L., Eminson, D., and Moss, B., 1978, A mechanism to account for macrophyte decline in progressively eutrophicated freshwaters: Aquatic Botany, v. 4, p. 103-126.

Pritchard, D. W., 1952, Salinity distribution and circulation in the Chesapeake Bay estuarine system: Journal of Marine Resources, v. 11, no. 2, p. 106-123.

Seaman, W. H., 1875, Remarks on the flora of the Potomac: Field and Forest: Bulletin of the Potomac-side Naturalists' Club, v. 1, p. 21-25.

Slavik, W., and Uhler, F. M., 1951. The Potomac basin. 3. Where are the ducks?: Atlantic Naturalist, v. 5, no. 4, p. 143-150.

Smith, E. V. and Swingle, H. S., 1941, The use of fertilizer for controlling the pondweed, Najas guadalupensis: Transactions of the Sixth North American Wildlife Conference, p. 245-251.

Southwick, C. H., and Pine, F. W., 1975, Abundance of submerged vascular vegetation in the Rhode River from 1966 to 1973: Chesapeake Science, v. 16, no. 2, p. 147-151.

Spence, D. H. N., 1976, Light and plant response in freshwater, in Light as an ecological factor: Second Proceedings, Sixteenth Symposium of the British Ecological Society, March 26-28, 1974, Blackwell Science Publications, London, P. 93-133.

Steenis, J. H., 1971, Aquatic weed problems in estuarine areas: Hyacinth Control Journal, v. 9, no. 1, p. 25-26.

Stevenson, J. C., and Confer, N. M., 1978, Summary of available information on Chesapeake Bay submerged vegetation: Fish and Wildlife Service, Office of Biological Services FWS/OBS-78/66. 335 p.

Stevenson, J. C., Confer, Nedra, and Pieper, C. B., 1979, The decline in submerged aquatic plants in Chesapeake Bay: Office of Biological Services, U.S. Fish and Wildlife Service, FWS/OBS-79/24, 12 p.

Stewart, R. E., 1962, Waterfowl populations in the upper Chesapeake region: U.S. Fish and Wildlife Service Special Scientific Report on Wildlife, no. 65, 208 p.

Wester, H. V., and Rawles, S. D., 1976, Impact of chlorine pollution in the upper Potomac and Anacostia estuaries (abs.), in First Conference on Scientific Research in the National Parks: November 9-13, 1976, New Orleans, La.

Wilson, R. E., 1977, A model of dynamics in the lower Potomac River estuary: Chesapeake Science, v. 18, no. 2, p. 177-187.

Wood, R. D., 1967, Charophytes of North America: Stella's Printing, West Kingston, R. E., 72. p.

SEASONAL WATER CHEMISTRY IN THE TIDAL POTOMAC RIVER

By

David H. Peterson[1/]

ABSTRACT

During high winter discharge the upper Potomac River is a major source of carbon, silica, and nitrogen (CSiN) to the downstream tidal regime. This is indicated in two ways. First, time variations in the concentrations of these substances upstream (e.g., due to a storm event) can be observed later downstream; second, estimates of the magnitudes of these river inputs (from CSiN concentrations and river flows) appear large in comparison to those from other sources.

Concentrations of dissolved nitrate and silica during winter, in the river upstream of Washington, D. C., for example, are both approximately 100 μg at ℓ^{-1}. Such a concentration represents approximately 260 td^{-1} (using a typical winter flow of 500 $m^3\ s^{-1}$). Similarly, for an area of $1.8 \times 10^8\ m^2$ (approximately the downstream region of river and estuary from Chain Bridge to Maryland Point) this source is equivalent to 24 mg at $m^{-2}\ d^{-1}$.

In contrast to the situation in winter, identifying major CSiN sources and sinks in summer is complex. Summer river flow is on the order of 50 $m^3\ s^{-1}$. Concentrations of nitrate and silica are typically less than 25 μg at ℓ^{-1}. Making the same calculation for summer as above for winter, the river supply is only 0.6 mg at $m^{-2}\ d^{-1}$. To put this into perspective, using nitrogen as a measure of waste source, the Blue Plains sewage plant accounts for 3 mg at $m^{-2}\ d^{-1}$; and summer phytoplankton nitrogen assimilation averaged 24 hours over the photic depth accounts for about 5 mg at $m^{-2}\ d^{-1}$. Preliminary estimates by others of benthic and atmospheric exchange rates (of CSiN where appropriate) further indicate that these sources and sinks may equal or exceed the contribution of river sources and sinks during summer.

The typical condition of low river flow during late summer, when water residence or replacement time in the tidal regime is very long (months), magnifies the importance of other sources and sinks. Preliminary results suggest another, more subtle but potentially important, phenomena related to low summer river flows. During the dry late-summer of 1977, concentrations of dissolved silica were not maintained above the concentrations that limit diatom growth. Diatoms were found to be relatively scarce (as is apparently typical). However, during the wet late summer of 1978 silica concentrations were considerably higher and diatoms appeared to be more abundant. Such a covariation is also suggested by the seasonality in dissolved silica and, presumably, diatom abundance. This may be critical to the extent that diatoms are favorable as a primary food source.

[1/] U.S. Geological Survey, 345 Middlefield Road, Menlo Park, California 94025

INTRODUCTION

Riverine-estuarine systems (RES) are dynamic and complex and are not yet well understood as complete systems. This paper discusses some of the factors influencing oxygen, carbon, silicon and nitrogen (OCSiN) distributions in the tidal Potomac river and upper estuary. A variety of data sources are used including preliminary results of water-chemistry studies made in 1977 and 1978 (Smith and Herndon 1979, 1980a, b, c, Cole and Alpine 1979, Cole and Harmon 1980). The purpose of this paper is to present a simplified view of the Potomac system on a seasonal time scale. A full discussion of the relevant processes and a complete review of previous work is beyond the scope of this paper.

SOME FEATURES OF THE POTOMAC SYSTEM

It is often useful to first attempt to put the major sources and sinks into some simplified quantitative perspective. The following discussion is given in an approximate order of how much we know about these elements of the system and not necessarily in their order of importance.

River Inflow

Potomac river flow peaks in March and declines through September (Fig. 1) and the September river flows can be extremely weak for such a large estuary especially in a dry year. Seasonal river flow is, of course, a fundamental parameter influencing seasonal water circulation and the associated salt field (Elliott 1976). Note, however, that there is a lag in response of the salt field to the river flow as indicated by the minimum landward salt encroachment in May and maximum in October (see Lippson et al., 1979, Fig. 3-10).

River inflow, then, directly affects chemical distributions by altering rates of supply, dilution and removal and typically the lowest rates (of river supply, dilution and flushing) are in the late fall. Effects of wet versus dry summer flows can also be observed. As a well-known example, the region of lowest dissolved oxygen in the Thames estuary is less severely depleted in oxygen in summers of higher river flows and is shifted further downstream relative to its position during lower flows (Department of Scientific and Industrial Research 1964). A similar response to increased river flow has also been observed for the dissolved oxygen minimum near Washington, D. C. (Durum and Langbein 1966, Fig. 6, p. 8). These apparently simple down-stream distributions are often further complicated by upstream seasonal and short-term variations in concentrations of river-borne substances, especially for the elements CSiN. Thus, many observations are generally needed to define the annual river source of CSiN. For our purposes, however, the following estimates are considered adequate although simplistic.

The annual supply of oxygen is estimated by assuming the annual flow of 300 m^3 s^{-1} has an average concentration of 700 μg at of oxygen per liter. This concentration is probably low (about 30%) during the high discharges and low water temperatures of winter and early spring and therefore this estimate is perhaps conservative (Table 1).

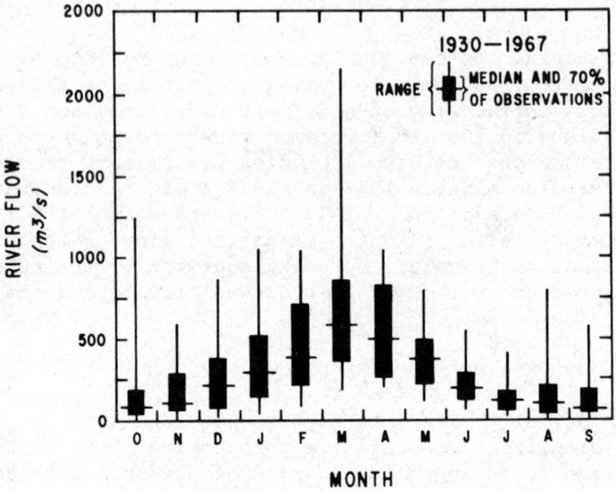

Figure 1. Mean-monthly river flow Potomac River. Data from U.S. Geological Survey surface water supply of the U.S. publications for Point of Rocks.

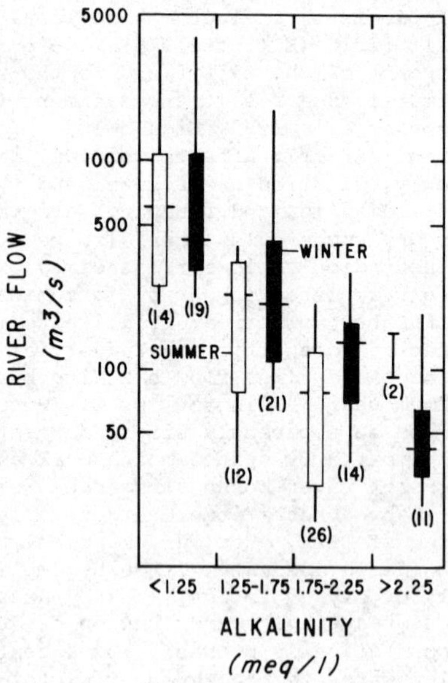

Figure 2. Potomac River alkalinity in relation to river flow. Analyses for water samples taken at Point of Rocks and published in U.S. Geological Survey Water Resources Water Quality Data Reports.

TABLE 1. PRELIMINARY ESTIMATES OF SOME OCSiN SOURCES AND SINKS FOR THE UPPER TIDAL POTOMAC[a/]

SOURCES/SINKS[b/]	OXYGEN	CARBON (kilo tons/year)	SILICON	NITROGEN
RIVER				
DISSOLVED	100	180	30	10
PARTICULATE	-	60	-	6
WASTE[c/]	-20	10	?	9
PHYTOPLANKTON				
NET PHOTIC DEPTH	210	-80	-20	-9
APHOTIC DEPTH	-140	50	?	7
BENTHIC BOUNDARY[d/]	-40	20	10	5

[a/] See text for assumptions.

[b/] Sources (positive values) mean the element is introduced to the water-column as a dissolved species (with the exception of river-borne particles); sinks (negative values) mean the dissolved species is converted to a particulate phase or leaves the system.

[c/] Adapted from Bennett, unpublished manuscript, see text.

[d/] Adapted from Hammond and Callender unpublished manuscript, see text.

The alkalinity of Potomac water is mostly determined by the concentration of the bicarbonate ion, the major dissolved inorganic carbon species. Dissolved inorganic carbon is estimated as 1.5 mg at per liter for a river flow of 300 $m^3\ s^{-1}$ (Fig. 2, alkalinity 1.25 to 1.75 meq. per liter) and dissolved silicon is estimated to be 100 μg at per liter (Fig. 3). Alkalinity, hence, bicarbonate, tends to decrease in value with increasing river flow (probably because of dilution effects). Note that a time-weighted series of observations may show a slightly higher value than 1.5 meq. per liter. Similarly, in our simple analysis dissolved silica tends to increase with increasing river flow (except possibly for extreme winter flows) and a time-weighted estimate may show a slightly lower value than 100 μg atoms per liter.

Estimates of river-borne nitrogen sources probably include larger errors than for OCSi as given above. In winter, nitrate often closely follows dissolved silica, increasing with increasing river flow, although these large sources of nitrate may be largely anthropogenic and silica sources are natural. I have used an observed winter concentration of nitrate of 100 μg at per liter to estimate the river sources of inorganic nitrogen. This estimate may be somewhat high, but contributions of other relatively minor dissolved inorganic nitrogen forms such

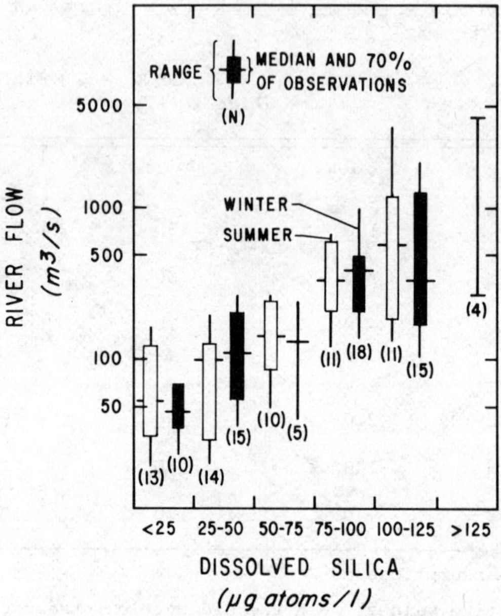

Figure 3. Potomac River dissolved silica in relation to river flow. Analyses for water samples taken at Point of Rocks and published in U.S. Geological Survey Water Resources Water Quality Data Reports.

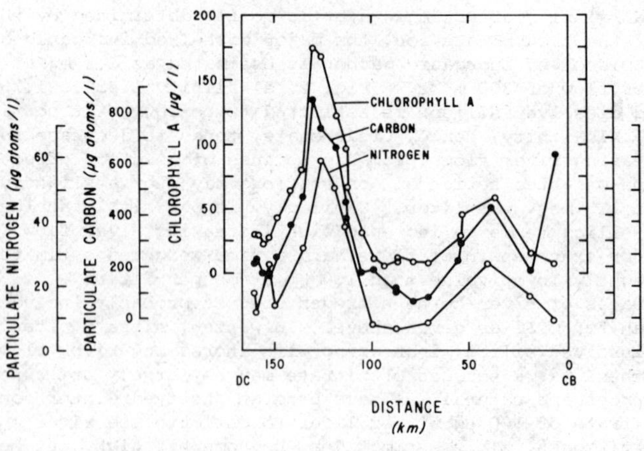

Figure 4. Distribution of chlorophyll a, particulate organic carbon and nitrogen from Washington, D. C. to Chesapeake Bay, August 1977 (Smith and Herndon 1979).

as ammonium and nitrite have been ignored. Also, this estimate does not include dissolved organic nitrogen.The importance of this latter simplification is uncertain because the role of total dissolved organic nitrogen in biochemical circulation processes is poorly known.

Particulate carbon is estimated as 3% of an approximate annual sediment yield of 2 X 10^6 tons per year (Wark and Keller 1963). Erosion of upland soils are probably the only major source of river-borne particulate organic carbon except in summer and soil organic carbon in the Potomac drainage basin ranges from 1 to 3% (Parker *et al*. 1907).

Similarly, particulate nitrogen is estimated from a commonly observed ratio of soil organic C:N = 10:1 by weight.

Waste

Estimates of Potomac waste-related sources are continually being refined and improved and the estimates reported here are not necessarily intended as the most accurate available but are considered realistic for our purposes. The Blue Plains sewage treatment plant serving the metropolitan Washington, D. C. area is used as the measure of waste sources. It is appreciated, however, that other sources are also important. For example, nonpoint releases (biochemical oxygen demand and nitrogen in the region downstream from Washington, D. C. and almost to Maryland Point (U.S. Geological Survey, unpublished manuscript) are estimated to be nearly equivalent to the source reported here for metropolitan Washington, D. C. (Table 1).

Oxygen consumed by waste is an estimate from biochemical oxygen demand (5-day) of waste from Clark and Roesch (1978, cited in Bennett, unpublished manuscript). Carbon and nitrogen sources are 52% and 87% respectively of the earlier values reported in Jaworski *et al*. (1972), presumably due to more recently improved waste-treatment practices at the Blue Plains plant.

Production and Loss of Phytoplankton

Phytoplankton, macroscopic algae and rooted aquatic plants are major elements of primary production in the Potomac, however, only the production of phytoplankton is considered here. Their direct effects on water chemistry (dissolved and particulate matter), as observed along the main channel, are clearly seen (Figs. 4 and 5).

In spring, when insolation is relatively high, phytoplankton biomass is generally not yet at the maximum concentrations observed in late summer probably because of high rates of river dilution and other factors. Maximum phytoplankton development is more typically in fall, when river flow is relatively ineffective and insolation is still relatively high. This suggests but does not prove that nutrients play a secondary role in the seasonal timing of peak phytoplankton biomass development. Phytoplankton are, of course, maintained by many factors including sunlight and nutrients and peak development is generally downstream of the large Washington, D. C. waste source (Carpenter *et al*. 1969). Furthermore, it is generally accepted that the present and historically high ambient phytoplankton concentrations in the upper Potomac are related to relatively large anthropogenic nutrient sources.

A few observations of *in situ* daylight and 24-hour phytoplankton productivity have been made in the Potomac RES (Cole and Harmon 1980.

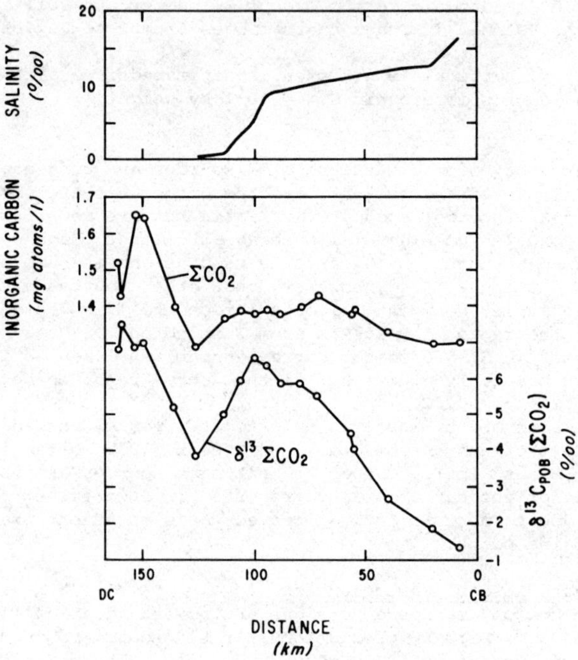

Figure 5. Distribution of salinity, total dissolved inorganic carbon and its isotopic (δ^{13}) values from Washington, D. C. to Chesapeake Bay, August 1977 (unpublished data of Elliott Spiker, U.S. Geological Survey).

These observations provide information on the relation between light and phytoplankton productivity in the water column and will be discussed later. Because sufficiently large numbers of observations appear not to have been made for making an average annual estimate of phytoplankton productivity in the Potomac, perhaps a better approach is to utilize results from another area. Studies of the New York Bight showed 1 g Carbon m^{-2} d^{-1} to be the average annual net phytoplankton productivity over the photic depth (e.g., approximately the depth to 1% surface light penetration Malone 1975). Intuitively, this would seem to be a low estimate of phytoplankton productivity in the Potomac RES when compared to the New York Bight. Thus a value of 1.2 g C m^{-2} d^{-1} was assumed in estimating carbon assimilation for the Potomac (Table 1) and the statistically average O:C:Si:N proportions 212:106:16:16 respectively (by atoms) was used in calculating the OSiN values (cf., Redfield _et al_. 1963). But what might be a realistic upper value of this annual average is difficult for me to evaluate. Is it higher than even 1.5 g C m^{-2} d^{-1}? It would seem unlikely to be higher than 2 g C m^{-2} d^{-1} because of the cold winter season and, possibly, because of high spring river flows.

Even less is known about phytoplankton loss rates than the little known about phytoplankton production! A survey using 12- and 24-hour dark bottle oxygen consumption in the fall 1978 indicated rates of about 1 to 6 μg at oxygen per liter per hour from main-channel samples collected between Washington, D. C. and the Chain Bridge (beyond Maryland Point). Most values were about 3 to 4 μg at per liter per hour (Cole and Harmon 1980) and a value of 3 μg at per liter per day was used in estimating aphotic (dark) oxygen consumption. The aphotic depth was assumed to contain one-half the water volume from Washington, D. C. to Maryland Point (approximately one-half of 6.5 X $10^8 m^3$). Again, the stoichiometric phytoplankton assimilation O:C:N of 212:106:16 was used to estimate carbon and nitrogen mineralization. Thus these values, which do not include observations obtained during the winter season, are most likely a high estimate of annual aphotic phytoplankton consumption in the water column. Also, these observations must include oxygen consumption rates of all organic matter in the sample such as waste-derived detritus, bacteria and zooplankton as well as phytoplankton respiration. Obviously, then, we know very little about phytoplankton loss rates.

Benthic and Atmospheric Exchange

There is considerable interest in understanding seasonal and spatial variability in benthic (and atmospheric) exchange in the Potomac as well as other systems and the methodology is an active area of research. Typical values from ongoing studies by D. Hammond (USC) and T. Callender (USGS) in the Potomac of about -40 mmol O_2 $m^{-2}d^{-1}$, 20 mmol CO_2 $m^{-2}d^{-1}$, 5 mmol SiO_2 $m^{-2}d^{-1}$, 7 mmol N_4^+ $m^{-2}d^{-1}$ and -2 mmol NO_3^- $m^{-2}d^{-1}$ (U.S. Geological Survey, unpublished manuscript) were used to estimate exchange rates with the bottom (approximately 1.8 X 10^8 m^2 from Washington, D.C. to Maryland Point).

Atmospheric exchange rates appropriate for the dissolved gases O_2, CO_2 and N_2O are being investigated by D. Hammond and others but no attempt was made here to estimate the importance of the atmosphere as a source or sink for O_2 or CO_2 on an annual basis, or, as a sink for N_2O (see also McElroy _et al_. 1978).

Embayments and Marshes

Their significance as OCSiN sources and sinks is expected to be important but is difficult to evaluate and was not attempted.

Exchange with Chesapeake Bay

Chesapeake Bay is a potential source of substances to the upper Potomac estuary as well as a sink but an estimate of exchange rates at the seaward boundary of the Potomac was not attempted. This problem is complicated but numerical modeling methods may be very useful for studying net exchange of dissolved OCSiN between the Potomac and Chesapeake Bay (cf. Rattray and Officer, in press).

SEASONAL RIVERINE-ESTUARINE WATER CHEMISTRY

Annual estimates of some of the sources and sinks need more detail and refinement to illustrate how the Potomac RES works on a seasonal basis with regard to its water chemistry. However, the importance of some factors can be inferred from annual estimates.

During winter river sources must increase over annual average values because of the higher river flows. Similarly, phytoplankton production is considerably lower because of high river flows, low insolation and cold water temperatures. Also, benthic exchange rates may be lower in winter with regard to oxygen (Nixon *et al*. 1976). Seasonal variations in point waste sources are assumed to be negligible.

During the winter high river flows, high silica, nitrate and suspended sediment loads are typical. Then river sources dominate over other sources and sinks in influencing their downstream distributions. Thus, as seems apparent in the winter downstream CSiN distributions, much of the winter dissolved bicarbonate, silica and nitrate is probably transported through the upper Potomac to the lower estuary and in some instances to Chesapeake Bay.

Spring is a transition period with the diminishing importance of river sources and the emerging importance of waste and phytoplankton. Fall, with weak river flows commonly 50 to 100 $m^3\ s^{-1}$, is a period apparently little influenced by river flows but strongly influenced by phytoplankton production, waste loads and benthic exchange processes. The enormous importance of phytoplankton to OCSiN distributions, then, is seen both in field observations (cf. Figs. 4 and 5) and in annual estimates of net photic production relative to other factors. Therefore, for brevity, illustrative calculations of summer sources and sinks such as given in the abstract are not repeated.

THE POTOMAC AND SAN FRANCISCO BAY

Why are phytoplankton such an impressive element of the Potomac RES? There appears to be no simple quantitative answer partly because many factors maintain phytoplankton in RES. Budget calculations, as discussed above, serve as an approximate measure of the importance of phytoplankton production in comparison to other sources and sinks. Another interesting perspective is provided by comparing the same factors and their effects in different systems. For example, the biochemical response of a wide variety of complex systems to waste sources has been simplified by considering the relative rates of waste supply to the flushing time (Officer and Ryther, 1977).

Northern San Francisco Bay (NSFB) and the Potomac RES receive comparable levels of waste (Tables 1 and 2). Under present conditions phytoplankton development in NSFB is minimal in contrast to the Potomac, except locally as along some Bay margins with restricted circulation. One of the reasons for this remarkable difference is most likely the opposite locations of waste disposal in each system (certainly not the possible differences in levels of waste treatment). Most of the point sources of waste in NSFB are located near the mouth of the estuary (San Francisco and Oakland) whereas in the Potomac the source is located far inland (Washington, D. C.). The accompanying differences in dilution capacity with the landward or seaward location of the waste outfall might be considered as a simple extension of the discussion by Officer and Ryther (1977). Is is appreciated that other differences in dilution capacity in addition to location are expected between NSFB and the Potomac because the tidal currents are stronger in San Francisco Bay, the mean Sacramento-San Joaquin river flow is about two times the Potomac river, the volume of the NSFB is about 6 times that considered for the Potomac and the area is about 3 times larger (Conomos 1979). In this simplified analysis one might conclude that the "dilution capacity" of the Potomac has been exceeded but perhaps not so for NSFB. What OCSiN distributions would result if their locations were reversed (e.g., San Francisco was located farther inland and Washington, D. C. farther seaward)?

TABLE 2. PRELIMINARY ESTIMATES OF SOME OCSiN SOURCES AND SINKS FOR NORTHERN SAN FRANCISCO BAY ESTUARY[a/]

SOURCES/SINKS[b/]	OXYGEN	CARBON	SILICON	NITROGEN
		(kilo tons/year)		
RIVER				
DISSOLVED	210	280	110	9
PARTICULATE	-	100	-	10
WASTE	-30	10	?	8
PHYTOPLANKTON				
NET PHOTIC DEPTH	160	-70	-60	-10
APHOTIC DEPTH	-120	40	?	3
BENTHIC BOUNDARY	-230	62	30	10

[a/] Adapted from Peterson, 1979, see text.

[b/] Sources (positive values) mean the element is introduced to the water-column as a dissolved species (with the exception of river-borne particles); sinks (negative values) mean the dissolved species is converted to a particulate phase or leaves the system.

SPATIAL VARIATIONS IN WATER CHEMISTRY

To the extent that phytoplankton activity represents an important element of Potomac water chemistry, what limits the maximum phytoplankton biomass in the Potomac and why is it generally located where it is? Obviously, a most basic part of this problem is our knowledge of the light field because of the fundamental relation between phytoplankton photosynthetic production and light intensity. For example, *in situ* incubation studies (Cole and Alpine 1980) demonstrate a quantitative relation between light intensity and phytoplankton productivity:

$$P = A_o - A_1 e^{-\beta I} \qquad (1)$$

where P is net productivity in mmol oxygen photosynthetically produced phytoplankton or of carbon, silicon or nitrogen consumed by phytoplankton per unit water-sample volume per unit time; the derived coefficients, $A_o - A_1$, provide an estimate of dark respiration; β is a derived coefficient apparently including such effects as shade adaptation and temperature; and I is *in situ* light intensity (e.g., photons cm^{-2} s^{-1}).

Light intensity attenuation with depth in the water column is largely dependent on the concentration of suspended particulate matter. Thus an additional distinction between San Francisco Bay and the Potomac systems might be that average concentrations of summer suspended particulate matter are relatively low in the Potomac (commonly less than 50 mg per liter, Smith and Herndon 1979 and 1980a, c) relative to NSFB (greater than 50 mg per liter, Conomos *et al.* 1979). Availability of light is a major parameter in seasonal mathematical models of estuarine and coastal phytoplankton dynamics (cf. Kramer and Nixon 1978). But the problem is further complicated when vertical mixing processes are considered because water depth as well as suspended particulate matter and phytoplankton concentrations influence the average water-column light intensity as indicated in mathematical models where phytoplankton development begins in the shallow margins (Di Toro *et al.* 1977, Cloern and Cheng in press).

A bimodal chlorophyll distribution (Fig. 4) is a common feature in the Potomac (Lear and Smith 1976). The chlorophyll minimum approximity between Maryland Point and Chain Bridge (about 90 to 60 km, Fig. 4) is probably partly related to the increased average cross-sectional water depth and increased suspended particle concentrations (turbidity maximum). Incubation experiments suggest phytoplankton development is less favorable there because observed compensation water depths are shallower than the total water depth which is probably also commonly the mixing depth in this portion of the estuary (cf. Sverdrup, 1953 and Fig. 6). This partly, but not completely, explains why high dissolved inorganic phosphate and silica (Fig. 7) and to a lesser extent nitrate concentrations persist there during the fall (Smith and Herndon 1979 and 1980b, c). An additional complication is that this is the low salinity segment of the estuary where both fresh-water and marine phytoplankton have difficulty in maintaining their population (Lear and Smith 1976). That this salinity effect is the sole explanation is not clear, for example, because similar low-salinity segments in other estuaries (with relatively

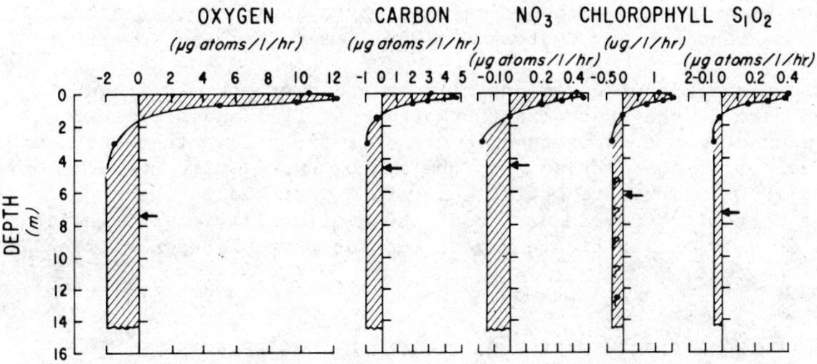

Figure 6. Depth variation in photic (phytoplankton) production/assimilation and aphotic mineralization rates for oxygen, carbon, nitrate, chlorophyll a and silica (Cole and Harmon 1980). Arrow denotes compensation depth, where net production equals zero as indicated by 24-hour incubation experiments (near Maryland Point, August 1977), but not necessarily in the real system. Also note that the aphotic mineralization rates for dissolved silica are possibly erroneous.

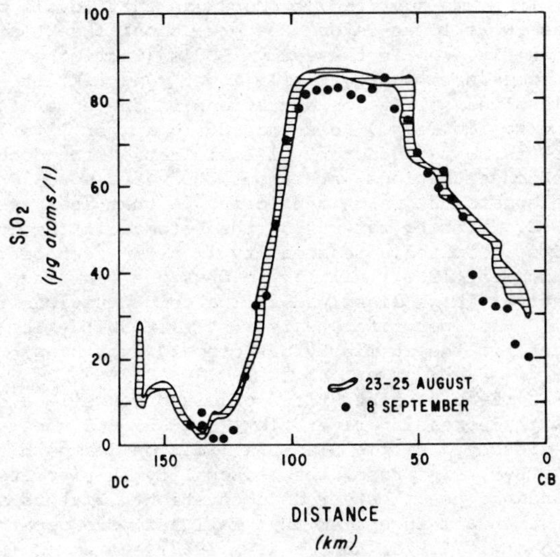

Figure 7. Distribution of dissolved silica from Washington, D.C. to Chesapeake Bay, fall 1977 (Smith and Herndon 1979).

shallower water depths) support maximum phytoplankton populations (e.g., San Francisco Bay, Peterson, 1979; Tamar, England Mommaerts 1971).

Concentrations of nutrients silicon and nitrogen may frequently approach zero in the phytoplankton maximum in fall and dissolved carbon dioxide may be depressed to concentrations less than 30% saturation values in equilibrium with the atomosphere (Smith and Herndon 1979 and 1980c) Thus, to better understand what places the upper limit on phytoplankton biomass, both the availability of nutrients as well as effects on light attenuation such as phytoplankton self-shading appear important.

SOME PRELIMINARY IMPLICATIONS OF THE EFFECTS OF ANTHROPOGENIC NUTRIENT SOURCES

Man has increased the dissolved concentrations of major elements of many rivers of eastern United States by more than 30% in the last 100 years and possibly by more than 100% for nutrient elements (Meade in press). It is also widely accepted that man has substantially increased the nutrient (nitrogen and phosphate) concentrations in the upper Potomac RES (Carpenter et al. 1969). It is reasonable to assume that rates of dissolved silica production by natural weathering processes have not changed over the past 100 years or longer (Zobrist and Stumm in press). However, rates of phytoplankton silica removal have probably increased in some lakes (Schelske and Stoermer 1972) and rivers (Wollast 1978) along with anthropogenic increases in phosphate and nitrogen. It seems justified to conclude that levels of phytoplankton biomass were lower before the pre-colonial settlement of the Potomac and possibly even in the early 1900's (Cumming et al. 1916) when nutrient loads were substantially lower (Jaworski et al. 1972). Summer-fall dissolved silica concentrations in the upper tidal Potomac are presently very low (e.g., less than 25 μg at per liter) but to my knowledge there is no satisfactory historic data which documents what typical summer concentrations were about 100 years ago. One clue, considered inadequate but worth mentioning is that analyses of Washington, D. C. drinking water from the Potomac in the late 1800's and early 1900's indicated a seasonality in nitrate concentrations (U.S. Senate Document 1900), similar to what one would expect naturally, and relatively high dissolved silica concentrations through summer (about 70 μgat per liter, Parker et al. 1907). The question remains, however, if the chemistry of Potomac River "tap" water was a valid sample.

Because the river concentrations of dissolved silica can decrease dramatically with decreasing river flow (Fig. 3) and because the concentration of dissolved silica places a limit on diatom biomass (Officer and Ryther, in press) the seasonality in river-borne silica and diatom abundance needs better documentation. Diatoms are typically more abundant in spring than fall and were apparently more abundant during a wet (1978) than a dry (1977) summer (Cole and Alpine 1980). It seems significant, as noted by Officer and Ryther (in press), that a similar seasonality is not obvious for NSFB (Cloern 1979) where summer river dissolved silica concentrations remain high and maximum summer-fall phytoplankton development is commonly diatoms (Cloern, personal communication). To conclude, then, I would like to learn from my biologist colleagues: would the role of phytoplankton in the tidal Potomac River be different if diatoms were the predominant species throughout the year?

ACKNOWLEDMENTS

The cooperation of many colleagues is too numerous to mention. I must, however, thank the San Francisco Bay group: A. Alpine, B. Cole, J. Conomos, L. Dedini, S. Hager, D. Harmon, R. Herndon, A. Hutchinson, L. Schemel and R. Smith for their excellent field work in the Potomac. This paper presents only a small part of their results.

REFERENCES

Bennett, J. P., 1979, Interdisciplinary Potomac estuary study work plan: U.S. Geological Survey, unpublished manuscript.

Carpenter, J. H., Pritchard, D. W., and Whaley, R. C., 1969, Observations of eutrophication and nutrient cycles in some coastal plain estuaries. In Eutrophication: Causes, Consequences, Correctives, National Academy of Sciences, Washington, D. C., p. 210-221.

Clark, L. J. and Roesch, S. E., 1978, Assessment of 1977 water quality conditions in the upper Potomac Estuary: U.S. EPA Annapolis Field Office, Annapolis, Maryland, EPA 90319-78-008, 75 p.

Cloern, J. E., 1979, Phytoplankton ecology of the San Francisco Bay system: the status of our current understanding, *in* Conomos, T. J., ed., San Francisco Bay, the Urbanized Estuary: American Association for the Advancement of Science, Allen Press, Lawrence, Kansas, p. 247-264.

Cloern, J. E. and Cheng, R. T., in press, Simulation model of *Skeletonema costutum* population dynamics in Northern San Francisco Bay, California: Estuarine and Coastal Marine Science.

Cole, B. E. and Alpine, A. A., 1980, Biological water properties of the Potomac River estuary: August 1977-August 1978: U.S. Geological Survey Open-file Report 80-001.

Cole, B. E. and Harmon, D. D., 1980, Potomac River phytoplankton productivity, nutrient regeneration, and respiration, August 1977 - August 1978: U.S. Geological Survey Open-file Report 80-000.

Conomos, T. J., 1979, Properties and circulation of San Francisco Bay waters, *in* Conomos, T. J., ed., San Francisco Bay, the Urbanized Estuary: American Association For the Advancement of Science, Allen Press, Lawrence, Kansas, p. 47-84.

Conomos, T. J., Smith, R. E., Peterson, D. H., Hager, S. W., Schemel, L. E. 1979, Processes affecting seasonal distributions of water properties in the San Francisco Bay estuarine system, *in* Conomos, T. J., ed, San Francisco Bay, the Urbanized Estuary: American Association for the Advancement of Science, Allen Press, Lawrence, Kansas, p. 115-142.

Cumming, H. S., Purdy, W. C., and Ritter, H. P., 1916, Investigation of the pollution and sanitary conditions of the Potomac watershed: Treasury Department, U.S. Public Health Service, Hygienic Laboratory - Bulletin No. 104, 239 p.

Department of Scientific and Industrial Research, 1964, Effects of polluting discharges on the Thames estuary: Water Pollution Research Technical Paper No. 11.

Di Toro, D. M., Thomann, R. V., O'Connor, D. J., and Mancini, J. L., 1977, Estuarine phytoplankton biomass models - verification analyses and preliminary applications, in Goldberg, E. D. McCave, J. M., O'Brien, J. J., and Steele, J. H., The Sea Volume 6: Marine Modeling, p. 969-1020.

Durum, W. H., and Langbein, W. B., 1966, Water quality of the Potomac River estuary at Washington, D. C.: U.S. Geological Survey Circ. 529-A, 9 p.

Elliott, A. J., 1976, The circulation and salinity distribution of the upper Potomac estuary: Chesapeake Science 17, p. 141-147.

Jaworski, N. A., Lear, D. W., Jr., and Villa, O., Jr., 1972, Nutrient management in the Potomac estuary, in Likens, G. E., ed., Nutrients and Eutrophication: The Limiting-Nutrient Controversy: Special Symposia Solume I, p. 246-272.

Kremer, J. N. and Nixon, S. W., 1978, A Coastal Marine Ecosystem, Simulation and Analysis, Springer Verlag, New York, P. 217.

Lear, D. W., and Smith, S. L., 1976, Phytoplankton of the Potomac estuary in, The Potomac Estuary Biological Resources Trends and Options, Interstate Commission on the Potomac River Basin Technical Publication 76-2, p. 70-74.

Lippson, A. J., Haire, M. S., Holland, A. F., Jacobs, F., Jensen, J., Moran-Johnson, R. L., Polgav, T. T., and Richkus, W. A., 1979, Environmental Atlas of the Potomac Estuary, Maryland Department of Natural Resources, Annapolis, Maryland, 279 p.

Malone, T. C., 1975, Phytoplankton productivity in the apex of the New York Bight: Environmental regulation of productivity/chlorophyll a, in Gross, M. G., ed., The Middle Atlantic Shelf and New York Bight Limnology and Oceanography Special Symposium 2, p. 260-272.

McElroy, M. D., Elkins, J. W., Wotsy, S. C., Kolb, C. E., Duran, A. P., and Kaplan, W. A., 1978, Production and release of N_2O from the Potomac estuary: Limnology and Oceanography 23: p. 1168-1182.

Meade, R. H., in press, Man's influence on the discharge of fresh water, dissolved material, and sediment by rivers to the Atlantic coastal zone of the United States, in Burton, J. D., ed., River Inputs to Ocean Systems:UNESCO.

Mommaerts, J. P., 1971, Note sur le mannoplancton de l'estuaire du Tamar (Plymouth, Grande Bretague) Societe Royal de Botanique Belgique Bulletin 104: p. 173-180.

Nixon, S. W., Oriatt, C. A., and Hale, S. S., 1976, Nitrogen regeneration and the metabolism of coastal marine bottom communities, in Anderson, J. M., and Macfadyen, A., ed., The Role of Terrestrial and Aquatic Organisms in Decomposition Processes: Blackwell Scientific Publications, Oxford, p. 269-283.

Officer, C. B., and Ryther, J. H., 1977, Secondary sewage treatment versus ocean outfalls: an assessment: Science 197, p. 1056-1060.

Parker, H. N., Willis, B., Bolster, R. H., Ashe, W. W., and March, M. C., 1907, The Potomac River Basin: geographic history - rainfall and stream flow - pollution, typhoid fever, and character of water - relation of soils and forest cover to quality and quantity of surface water - effect of industrial wastes on fishes: U. S. Geological Survey Water-Supply and Irrigation Paper No. 192, 364 p.

Peterson, D. H., 1979, Sources and sinks of biologically reactive oxygen, carbon, nitrogen and silica in Northern San Francisco Bay, in Conomos, T. J., San Francisco Bay, the Urbanized Estuary, American Association for the Advancement of Science, Allen Press, Lawrence, Kansas, p. 175-194.

Rattray, M., and Officer, C. B., in press, Discussion of trace metals in the waters of a partially mixed estuary: Estuarine and Coastal Marine Science.

Redfield, A. C., Ketchum, B. H., and Richards, F. A., 1963, The influence of organisms on the composition of sea water, in Hill, M. D., ed., The Sea v. 2: Interscience Pub., New York, p. 26-77.

Schelske, C. L., and Stoermer, E. F., 1972, Phosphorus, silica and eutrophication of Lake Michigan, in Likens, G. E., ed., Nutrients and Eutrophication: The Limiting-Nutrient Controversy, Special Symposia Volume I, p. 157-171.

Smith, R. E., and Herndon, R. E., 1979, Physical and chemical properties of Potomac river and environs, August-September, 1977: U.S. Geological Survey OpenOfile Report 79-1635.

Smith, R. E., and Herndon, R. E., 1980a, Physical and chemical properties of Potomac river and environs, January 1978: U. S. Geological Survey Open-file Report 80-742.

Smith, R. E., and Herndon, R. E., 1980b, Physical and chemical properties of Potomac river and environs, January 1978: U.S. Geological Survey Open-file Report 80-745.

Smith, R. E., and Herndon, R. E., 1980c, Physical and chemical properties of Potomac river and environs, August 1978: U.S. Geological Survey Open-File Report 80-746.

U.S. Geological Survey, 1979, Potomac estuary study 1979 fiscal year annual report, unpublished manuscript.

United States Senate 1900 Document No. 259, 56th Congress, 1st session.

Wark, J. W., and Keller, F. J., 1963, Preliminary study of sediment scources and transport in the Potomac River basin: U.S. Department of Interior Geological Survey and Interstate Commission on the Potomac River Basin, Technical Bulletin 1963-11, 28 p.

Wollast, R., 1978, Modelling of biological and chemical processes in the Scheldt estuary, in Nihoul, J. C. D., ed., Hydrodynamics of Estuaries and Fjords, Elsevier, New York, p. 63-77.

Zobrist, J., and Stumm, W., in press, Chemical dynamics of the Rhine catchment area in Switzerland; consideration on the "pristine" Rhine river input into the ocean, in Burton, J., ed., Review and Workshop in River Inputs to Ocean Systems (RIOS); UNESCO.

A PLANNING PROCEDURE FOR THE EAST EVERGLADES

Narendra Juneja
Senior Associate Partner, Wallace Roberts and Todd, Philadelphia, PA

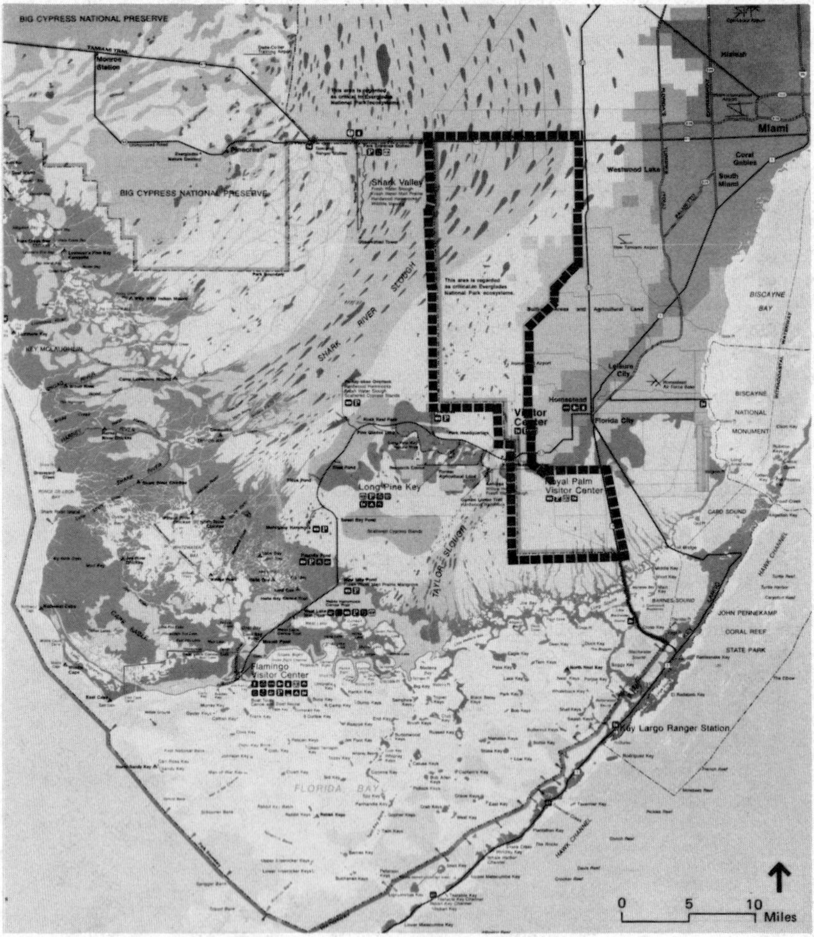

EAST EVERGLADES RESOURCE PLANNING PROJECT AREA LOCATION

INTRODUCTION

This paper presents a summary of the planning procedure used in developing of a Management Plan for an extensive part of South Florida's wetlands. The East Everglades Resource Planning Project area covers 242 square miles of land in western Dade County abutting the eastern boundary of the Everglades National Park, and is a remnant of the once vast Everglades Basin. With the exception of a small State Park, all of the land is privately owned and approximately 12% of it is currently in active residential and agricultural use. The burgeoning growth of metropolitan Miami is expanding over land elsewhere in Dade County which had previously been farmed, thus increasing the pressure for utilization of these Everglades lands for agricultural and other development uses. Accompanying alterations of the water regimen and the irretrievable loss of native biota are matters of great concern not only to the government and people of Dade County, but also to the Everglades National Park, the Atlantic and Gulf of Mexico fisheries, and the water supplies of Dade County and the Florida Keys.

South Florida's natural environment, comprising a primeval and vast wilderness of bays, marshes, swamps, hammocks, forests and flatlands, was a relatively stable and immensely productive ecosystem. Warm weather and abundant freshwater and sunlight produced a diverse proliferation of tropical and subtropical vegetation and an abundance of fish and other aquatic animals, which in turn supported large numbers of reptiles, amphibians, birds and mammals. Human beings, in availing themselves of this abundance of resources, have modified the environment to an extent which in some instances has stressed the natural systems' capacity to survive or recover. The result has often been an irrevocable loss of valuable resources. Drainage of wetlands for agricultural use has been a pioneering mission since the middle of 19th century, reaching its height in the 1930's when an extensive network of canals was built to provide rapid removal of storm water to the ocean. The folly of this uncontrolled modification became apparent in the wake of 1943-45 drought. As freshwater levels were lowered, sea water intruded into both the canals and the aquifer of the area. Fresh water supplies were contaminated by salt, causing a problem which persists to the present-day and requires continuous and skillful water management. Corrective measures entailing flood gates, pumping stations and impoundment areas have been employed for the past 30 years at enormous economic as well as environmental costs.

The conversion of wetlands to agricultural use has modified the natural systems in the region and led to the irretrievable loss of native plants and animals. The region's subtropical climate, and unique hydrology and substrate are only suitable for the growing of specialized crops. Historically, this form of agriculture is a high-risk venture and notable for its impermanence. Experience in the region demonstrates that abandoned land will never be revegetated by the flora which existed prior to farming.

The current planning effort, the East Everglades Planning Project, was conducted under the guidance of an inter-agency Steering Committee representing Everglades National Park, the Florida Department of Environmental Regulation, the South Florida Water Management District, the Dade County Department of Environmental Resources Management, and as the coordinating agency, the Metropolitan Dade County Planning Department. The study was funded by a special matching grant from the U.S. Environmental Protection Agency under Section 208 of the Federal Water Pollution Control Act Amendments of 1972 for studies of areas adjacent to National Parks, such as the East Everglades, where activities having adverse impacts on Park environments might occur. The cooperating agencies developed a Plan of Study and provided matching funds for the Project. The Plan of Study stated that the objective of the Project was to:

> obtain the scientific data that will provide the basis for understanding of the East Everglades ecosystems; analyze, synthesize, and project the effects of various land and water uses upon these ecosystems, such that land and water use management regulations can be prepared and implemented. The purpose of these resource use regulations will be to allow land owners maximum use of their property without jeopardizing the long term economic, social, and environmental health of the community and the region.

The Project's initial task was to obtain the scientific data which would provide the basis for understanding the East Everglades ecosystem. A multi-disciplinary team of highly qualified scientific consultants (see listing at the end of the paper) was engaged to review existing information, collect field data and present their findings to the Steering Committee on a regular basis. This continuous monitoring of data and analyses enabled appropriate mid-course changes in the study design. Unproductive investigations were terminated and unforeseen or important phenomena discovered during early investigations were persued in greater detail.

The second interrelated task was undertaken by Wallace, Roberts and Todd (WRT), environmental planning consultants to the Metropolitan Dade County Planning Department. It was WRT's responsibility to organize the data provided by the scientific and technical consultants into a series of ecological relationships and to establish the Maintenance Requirements for sustaining these relationships in response to the Project's objectives. Best Management Practices (BMPs) could then be developed for each activity performed in the region by public agencies, such as the operation of canals, or by private agents of change, such as farmers, home builders or recreational users. It was also possible, through reference to site data, to identify the geographical regions over which these management practices needed to be applied. The step by step procedure which was employed permitted a rigorous accounting of all prospective management recommendations with reference to the ecological relationships and the Project's objectives. Continuous revisions and updating were made throughout the course of study through regular reviews by the Steering Committee and the scientific technical consultants.

The final task of the Project was to prepare a Management Plan for the East Everglades. WRT assisted the Dade County Planning Department in synthesizing the findings from the previous tasks to enable identification of management regions for which future land use and regulatory policies could be formulated. The boundaries of each management region were determined by the presence within that region of discrete ecological relationships which entail an associated set of management practicies for a prospective land use. Where the implementation of the recommended set of management practices for a land use was determined to be economically, socially or legally infeasible, the Plan recommends prohibition of that use. Otherwise, the use is permitted subject to employment of the recommended Best Management Practices (BMPs). The preparation of the Management Plan proceeded through regular review by the Steering Committee and the scientific technical consultants. A draft Plan was presented and reviewed by an independent technical committee composed of scholars and professionals knowledgable of the natural environment, land use and resource management of South Florida. Based on their comments and those of the Steering Committee, a final draft Plan was presented for public review. At the time of writing this paper, the review procedure and consequent revisions have been completed and public hearings are in process. After adoption of the Plan by the County Commission, modifications to land use regulations and permit procedures will be enacted.

THE PLANNING PROCEDURE

Before describing the interpretive procedure, it is necessary to highlight the major factors of concern explicated by the wealth of data provided by numerous technical, economic and social planning studies as part of this planning exercise. Water is abundant in the East Everglades. Most of the rainfall occurs during the wet season and moves as sheet flow over the land surface. Much of this water courses through the Shark River Slough in the northwest and Taylor Slough in the central and southern parts of the study area. Both of these channels discharge into the adjacent Everglades National Park. Excessively flat topography and a general absence of soil cover over much of the region enables infilteration directly into the groundwater system, making it susceptible to pollution and disruption from activities at the land surface. The underlying limestone Biscayne Aquifer is one of the most porous and productive aquifers in the world. Its porosity makes it vulnerable to overdrainage through uncontrolled canal operation. In the past, this has indeed occurred, leading to lowered freshwater levels and thus reducing the freshwater's opposition to the inland movement of salt water. Under these conditions, the canals often became conduits for the movements of sea water inland. Recent water management policies are remedying some of these deleterious effects. Costs of remedial measures are high; directly, in terms of costs of structures and energy use; and indirectly, by committing extensive natural areas into "conservation" impoundments. Rapid releases of water from these areas have on occasion drowned established alligator holes and bird nests and otherwise disrupted the breeding and feeding patterns of downstream wildlife.

Most of the natural vegetative communities which form the rich mosaic of plant life in the area are supported by an extended period of flooding and respond to very slight differences in elevation (see figure 1). A number of rare, endemic, threatened and endangered species occur within these natural communities. Naturally occurring fires have been historically essential to the maintenance of these fire climax communities, and to the ecosystem for their role in recycling nutrients, rejuvenating dying vegetation and stimulating plant growth. Approximately 10% of the study area is covered by disturbed vegetation communities, mostly as a result of farming and other human uses, and occasionally due to hot dry season

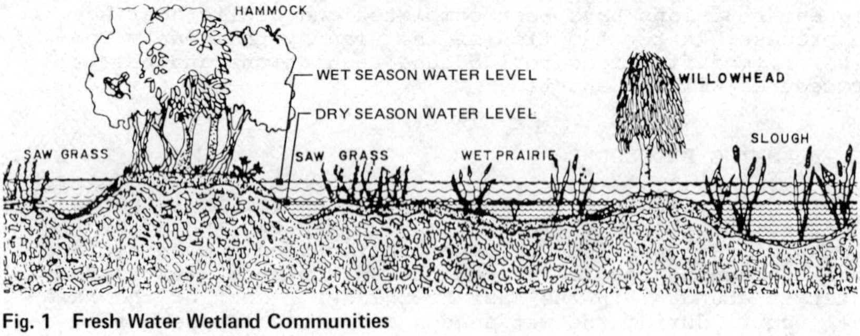

Fig. 1 Fresh Water Wetland Communities

fires resulting from arson and carelessness. Disturbed areas are often colonized by exotic species which pose a constant invasion threat to the natural vegetation. Once established these exotic species are nearly impossible to remove and tend to out-compete and replace indigenous species.

Wildlife is abundant and probably the most visible and unique feature of the study area. Field studies indicate 354 species of fish, amphibians, reptiles, mammals and birds to be present, of which 30 species appear on the various protected lists compiled by federal or state governments and the Florida Committee. These include several of the nation's most rare and unique animals which depend upon a diversity of habitats through which they move frequently and freely. Changes in the extent and duration of standing water have had direct repercussions on wildlife population size, distribution and diversity.

Existing land uses are essentially restricted to the eastern edge of the study area. Agriculture is the predominant use. Crops commonly grown include winter vegetables, citrus and tropical fruits, tropical vegetables and plant nursery stock. Decreasing availability of land elsewhere in Dade County is placing greater emphasis on utilizing the more flood-prone land of the East Everglades. Residential development in the study area is scattered and rural in character. Existing public services for the current population of 400-450 persons are poor and County land use policies explicitly discourage their improvement. While Chekika State Recreation Area is the only designated public recreation facility, recreation activities actually extend over the whole area. With or without the landowner's permission, off-road vehicles (ORVs), commonly airboats and glade buggies, but also jeeps and two/three wheeled motorcycles, are driven throughout the East Everglades with consequent adverse impacts on the environment.

Interpretation of the environmental and planning data to formulate future plan and management strategies proceeded through a step by step procedure shown in Fig. 2. A summary description of each step with illustrative examples is described below:

Step A(1): ECOLOGICAL RELATIONSHIPS

The technical and scientific papers describing the East Everglades environment were carefully studied for the purpose of identifying all known interactions among the elements of the natural system and the effect of man on them. These interactions were expressed in the form of Ecological Relationship statements and encoded as shown in Fig. 3. The encoding device allows for updating and revisions as the relationships become more fully understood or better data becomes available.

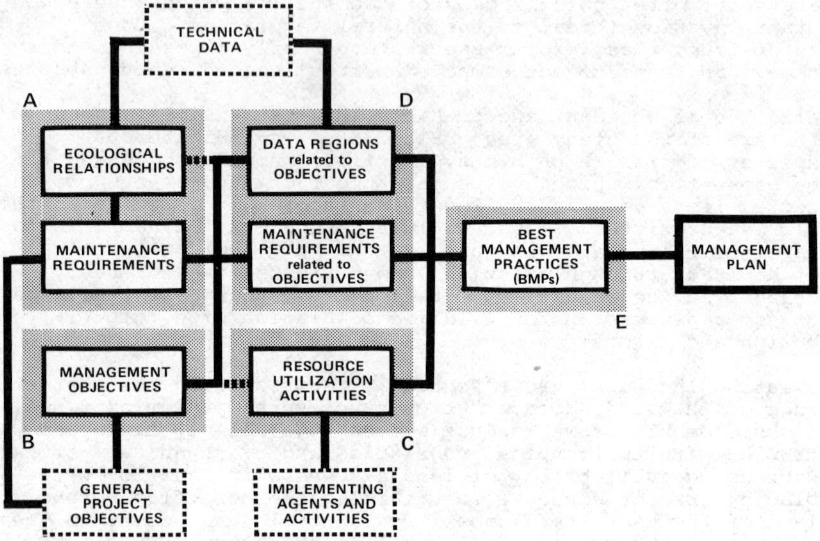

Fig. 2 East Everglades Resources Planning Procedure

ELEMENT	INTERFACING with								
	Climate	Topo-graphy	Surface Water	Ground-water	Soils	Fire	Vege-tation	Wildlife	Land Use
SURFACE WATER	SW/C	SW/T		SW/GW	SW/S	SW/F	SW/V	SW/WL	SW/LU
GROUND-WATER	GW/C	GW/T	GW/SW	GW/GW	GW/S	GW/F	GW/V		GW/LU
SOILS	S/C	S/T	S/SW	S/GW		S/F	S/V		S/LU
VEGETATION	V/C	V/T	V/SW		V/S	V/F	V/V	V/WL	V/LU
WILDLIFE	WL/C		WL/SW	WL/GW	WL/S	WL/F	WL/V	WL/WL	WL/LU
LAND USE	LU/C	LU/T	LU/SW	LU/GW	LU/S			LU/WL	LU/LU

Fig. 3 Code Designations of Ecological Relationships

Unfilled spaces indicate no ecological relationships were recorded.

Step A(2): ECOLOGICAL MAINTENANCE REQUIREMENTS

The Steering Committee had identified the goals and objectives which the management plan should attempt to achieve. These objectives were based upon existing laws, regulations and policies of all levels of government which are applicable to the study area in particular, and freshwater wetlands in general. These objectives, encoded 1 thru 4, are listed in Fig. 4.

This planning interpretation step (A2) attempted to relate each ecological relationship statement from step A1 to the Project Objectives; and then outline "what" must be done to maintain the relationship in consistency with the applicable objective. Selected examples of ecological relationships, related objectives and maintenance requirements, appear below:

ECOLOGICAL RELATIONSHIP Code; Source	**PROJECT OBJECTIVE (Ref. No.)**	**MAINTENANCE REQUIREMENT**
SW/GW-2; Schneider & Waller, U.S.G.S.: Surface water flow is largely determined by groundwater levels.	Preserve surface water movement.(2A, 2B)	Regulate groundwater withdrawals and permit no reduction in groundwater levels.
SW/LU-5; Schneider & Waller, U.S.G.S.: Canals alter the basin drainage pattern and tend to change the annual discharge per unit of rainfall in the study area.	Maintain flood level which reinforces natural conditions. (2A, 2B)	Operate the SDCS to avoid major changes in the basin drainage pattern and discharge.
GW/SW-1; Parker: High surface water levels and high southerly flows resulting from the normal wet season rain and storm events prevents intrusion of saline water into the aquifer in the southern regions of the study area (C-111 area).	Preserve groundwater quality above established water quality criteria. (1A, 1B)	Permit no diminishment of surface water flows and levels to maintain groundwater levels and inhibit saline intrusion.
GW/S-4; Parker: Peat soils have a high moisture content (saturated) and act to hold water in the hydrologic system.	Preserve groundwater quality. (2A, 2B)	Regulate land uses and activities which alter peat soils, and maintain saturated conditions during appropriate times of the year.
V/T-3; Hofstetter & Hilsenbeck: Narrow beardgrass-three-awn-grass prairie occupies slightly elevated "mesas" of oolitic rock with or without thin marl soil layer.	Preserve natural prairie vegetation. (3)	Retain elevated areas and small-scale, elevated topographic features.
V/SW-2; Hofstetter & Hilsenbeck: East Everglades' spike rush-beak rush flats typically occupy areas with 6 to 10 month hydroperiod.	Preserve natural wetland vegetation; maintain hydroperiod for biological health. (2B, 3)	Retain 6 to 10 month hydroperiod which approximates natural conditions.
V/F-19; Hofstetter & Hilsenbeck: Australian pine forests have often developed in areas which once supported bayhead or tropical hammock communities which were removed by fire.	Ensure regeneration of natural vegetation association following disturbance. (3)	Prevent establishment of this alien species by protecting selected areas against fire, especially when seeds of the alien species are abundant.
WL/SW-9; E.N.P.: Extended hydroperiod benefits Pomacea paludosa, the apple snail, which is the sole food source of the endangered Everglade kite; drying of the marsh area decreases this food supply.	Preserve endangered native species, and maintain water quantity and movement for biological health. (2B, 3)	Retain water regimen to protect food sources of endangered species.
LU/LU-1; Parker: Pumping of wells located near canals may result in replacement of high quality aquifer water by poorer quality canal water.	Preserve groundwater quality above established water criteria.(1A, 1B)	Regulate pumping of wells located near canals to prevent accelerated recharge.

Fig. 4

RESOURCE SPECIFIC **MANAGEMENT OBJECTIVES** / PROJECT GOALS & OBJECTIVES	**1. Water Quality** A. **Protect the water quality characteristics of surface and ground-waters in the East Everglades** to insure optimal biological health for the native flora and fauna in the study area and in Everglades National Park.	B. **Protect the water quality characteristics of surface and ground-waters** to insure the future availability of clean water for public and private water supplies in Dade County.	**2. Water Quantity** A. **Maintain a pattern of water periodicity and a quantity of water in the East Everglades** which will insure a sufficient water supply for users in south Dade County and prevent further salt water intrusion.	B. **Maintain a pattern of water periodicity and quantity of water in the East Everglades** which will optimize the biological health of the natural wetland environments in the study area and Everglades National Park.	**3. Natural Features** A. **Protect the existing natural wetland function and characteristics of the East Everglades and restore functions and characteristics where practicable** in order to preserve the biological productivity, animal and plant habitat, and natural biotic diversity.	B. **Protect the aesthetic value of the natural environment in the East Everglades**	C. **Protect unique natural features of the East Everglades study area**	**4. Land Use** **Provide for the use of land by private property owners consistent with the protection of Dade County and Everglades National Park water supplies and the public health, safety, and welfare.**
HA Maintenance of Water Quality through regulation of:								
1. Surface water levels, stage, and duration			●	●			●	
2. Groundwater levels and fluctuation			●	●				
B Ensure Water Movement through regulation of:								
1. Surface water flows		●	●	●				
2. Groundwater movement	●	●		●				
C Maintenance of Water Quality through:								
1. Maintenance of substrate function			●	●				
2. Restriction of chemical application	●	●						
3. Prevention of saline intrusion	●	●	●					
VA Ensure Occurrence through protection of:								
1. Rare and endemic species					●		●	
2. Habitat value for wildlife					●			
3. Value for water quality maintenance	●	●			●			
4. Scarce tree cover					●	●		
B Ensuring Continued Growth and Vigor through:								
1. Maintenance of substrate function					●			
2. Maintenance of water regimen	●			●	●			
3. Maintenance of wildlife role					●			
4. Management of fire occurrence					●			
5. Prevention of invasion of alien species					●			
WA Ensuring Occurrence through protection of:								
1. Native wildlife species					●		●	
B Ensuring Continued Presence through:								
1. Protection of critical wildlife habitat					●			
LA Maintenance of Life and Property Values associated with existing:								
1. Residential development		●	●					●
2. Agriculture		●	●					●
3. Recreation uses		●						●
B Ensuring Sustained Availability of Resources to meet demands for:								
1. Unique recreation					●	●	●	●
2. Agriculture		●	●					●
3. Residential development		●	●					●

Step B: MANAGEMENT OBJECTIVES

The Project objectives referred to in Step A2 were general in nature and, as mentioned, derived from existing laws, regulations and policies. In order to proceed with specifying management practices for the East Everglades, it was necessary to define resource specific objectives. This was accomplished through identification of management objectives in the resource categories of water (H), vegetation (V), wildlife (W), and land use (L). Fig. 4 shows the relationship of these resource specific objectives to the general Project objectives.

Fig. 5

LAND USE: Agriculture	LAND USE: Residential	LAND USE: Recreation	ACTIVITIES / MANAGEMENT OBJECTIVES	HA1 Surface Water Levels, Stage + Duration Regulation	HA2 Groundwater Levels + Fluctuation Regulation	HB1 Surface Water Flows Regulation	HB2 Groundwater Movement Regulation	HC1 Maintenance of Substrate Function (recharge)	HC2 Restriction of Chemical Applications	HC3 Prevention of Saline Intrusion	VA1 Rare and Endemic Species Protection	VA2 Habitat Value for Wildlife Protection	VA3 Value for Water Quality Maintenance Protection	VA4 Scarce Tree Cover Protection	VA5 Community Diversity Protection	VB1 Maintenance of Substrate Function (growth medium)	VB2 Maintenance of Water Regimen (refer to HA1,A2,B2,C2,C3)	VB3 Maintenance of Wildlife Role (refer to WA1,WB1)	VB4 Management of Fire Occurrence	VB5 Prevention of Alien Invasions	WA1 Native Wildlife Species Protection	WB1 Critical Wildlife Habitat Protection	LA1 Residential Development Protection	LA2 Agriculture Protection	LA3 Recreation Areas Protection	LB1 Providing for Unique Recreation	LB2 Providing for Agriculture	LB3 Providing for Residential Development
■	●	○	P1 Use Allocation								●	○	○	●	●				○		■	■	■	■	■	●	■	■
			P2 Canal Operation	■	■	○	■		○	■		○			●					○		●	○	○	○	○	○	○
			P3 Wellfield Operation	●	■		■	○		○					○								○		○			○
	●		P4 Road Building	○		■		○			○	○	○	○	○	●			○	○		●	●	●	●	○	●	●
			P5 Fire Management					○			●	●	●	●	■	○			■	●	■	■	■	■	■	●	●	●
■			F1 Rockplowing			○		●			○	○	○	○	●	■			○	●		●		○	○	●		
●			F2 Berm Formation	○		■		○			○	○	○	○	○	●				○		●	○	○	○	○		
●	●		F3 Ditching & Dredging	○	○	●	○				○	○	○	○	○	●				○		○	○	○	○	○		
	■		F4 Filling	○		●		○			○	○	○	○	●	●				■		○	○	○	○	○		
●	●		F5 Water Withdrawal		●		●			○	○	○	○	○	○					○		○	○	○	○	○		
	■		F6 Effluent Disposal		○				●														○		○			
■	●		F7 Chemical Application						■			○				○					●	○	○	○	○	○		
■	●		F8 Clearing & Planting		○	○		○			■	○			■	○			○	■		■	○	○	○	●		
		●	M1 ORV Use					●			●	●	●	●	●	●			○	○	●	●	●	●	○	●		
		○	M2 Hunting & Fishing								○	○	○						○		■		●	○	○	●		

IMPORTANCE OF THE ACTIVITY TO ACCOMPLISH THE MANAGEMENT OBJECTIVE

Primary ■
Major ●
Minor ○

Fig. 6

AGGREGATED MAINTENANCE REQUIREMENTS / MANAGEMENT OBJECTIVES	HA1 Surface Water Levels, Stage + Duration Regulation	HA2 Groundwater Levels + Fluctuation Regulation	HB1 Surface Water Flows Regulation	HB2 Groundwater Movement Regulation	HC1 Maintenance of Substrate Function (recharge)	HC2 Restriction of Chemical Applications	HC3 Prevention of Saline Intrusion	VA1 Rare and Endemic Species Protection	VA2 Habitat Value for Wildlife Protection	VA3 Value for Water Quality Maintenance Protection	VA4 Scarce Tree Cover Protection	VA5 Community Diversity Protection	VB1 Maintenance of Substrate Function (growth medium)	VB2 Maintenance of Water Regimen (refer to HA1,A2,B2,C2,C3)	VB3 Maintenance of Wildlife Role (refer to WA1,WB1)	VB4 Management of Fire Occurrence	VB5 Prevention of Alien Invasions	WA1 Native Wildlife Species Protection	WB1 Critical Wildlife Habitat Protection	LA1 Residential Development Protection	LA2 Agriculture Protection	LA3 Recreation Areas Protection	LB1 Providing for Unique Recreation	LB2 Providing for Agriculture	LB3 Providing for Residential Development
Maintain groundwater levels and movement		●		●	●		●						●						●						
Maintain groundwater recharge		●		●	●	●															●	●			
Prevent saline intrusion								●	●	●	●														
Maintain surface and groundwater quality				●	●	●	●	●	●	●	●	●	●					●	●						
Maintain surface water flows	●		●				●												●						
Maintain flood duration & frequency	●				●			●	●		●	●	●	●	●	●		●	●		●	●			
Regulate use and operation of canals & ponds	●	●	●	●															●		●	●	●	●	
Maintain topography	●		●		●			●	●		●	●	●	●						●	●	●	●	●	●
Maintain substrate	●			●	●			●	●			●	●	●					●		●			●	●
Maintain natural vegetation	●	●	●		●	●	●		●			●	●		●				●						
Ensure periphyton growth	●		●		●		●		●	●		●	●												
Protect rare and endangered species								●				●	●	●		●	●								
Permit natural succession								●	●		●	●	●			●									
Restrict colonization by exotic species								●	●		●	●	●	●		●		●		●	●		●	●	
Regulate and permit fire management								●	●	●	●	●	●		●	●	●	●	●						
Protect wildlife									●						●			●				●		●	
Ensure movement of wildlife																		●							
Ensure opportunities for reproduction																		●	●						
Regulate use of critical habitats																			●	●	●	●	●	●	●
Regulate human use and activities								●	●		●	●	●		●	●		●	●						
Regulate placement of roads and levees			●																						
Regulate placement of fill and drainfields	●							●	●		●	●	●			●									
Regulate drainage												●	●			●	●								
Regulate rockplowing and agriculture				●	●	●		●	●			●	●								●			●	
Regulate chemical applications						●														●	●			●	●
Regulate wellfields and water demand	●		●				●													●	●			●	●
Restrict extent of disturbed areas																			●						
Regulate soil fires					●								●												

Step C: RESOURCE UTILIZATION ACTIVITIES

Resource management in the East Everglades is the responsibility of several public agencies at the federal, state, regional, county and local levels. Through their respective authorizations and the means available to them, they perform actions, some of which directly affect the environment, while others of which are indirectly affectuated through their regulatory functions. Regulations need to be developed for a variety of private users who, through the activities related to their individual interests, are affecting the regions resources. For responsible management, it is important that recommended management practices be developed for each resource specific activity in the region. Fig. 5 identifies these 15 resource specific activities, their relationship to general land uses, and their importance in the realization of management objectives.

Step D(1): MAINTENANCE REQUIREMENTS RELATED TO OBJECTIVES

The "what" statements from Step A2 outlined the maintenance requirement for each identified ecological relationship. Inevitably, many of these were mutually reinforcing and repetitive. It was now possible to aggregate these and express them in direct relationship to the resource specific management objectives identified in Step B. Fig. 6 shows these aggregated maintenance requirements for which management practices need to be developed. As can be seen, the requirement for one element, such as maintenance of flood duration and frequency, relates to a host of management objectives.

Step D(2): DATA REGIONS RELATED TO OBJECTIVES

The data, as it was originally gathered and mapped, was comprehensive and detailed in order to fully explain the regional phenomena and the operative processes. With the identification of resource specific management objectives (Step B) and the aggregated maintenance requirements (Step D1), it became necessary to interpret and aggregate data into directly relatable categories. For example, except for some water management practices which are specifically applicable to canals (HC3), and others (HC1, HC2) which are areawide, the remainder are responsive to water Stage Duration in the region (Fig. 7). Reference to ecological relationships and other management objectives (in this case land use related) determined the particular data aggregation and mapping categories shown on the map. Similar aggregations of data regions were prepared for vegetation, wildlife and land use resource categories.

Step E: BEST MANAGEMENT PRACTICES

The previous steps had ensured that the region under study was well understood. The ecological relationships were recorded (Step A1) and "why" one must be concerned about these was established through enunciation of resource specific objectives (Step B). The relationship between the two led to identification of "what" must be done (Steps A2 and D1), and the objectives' relationship to data directly led to

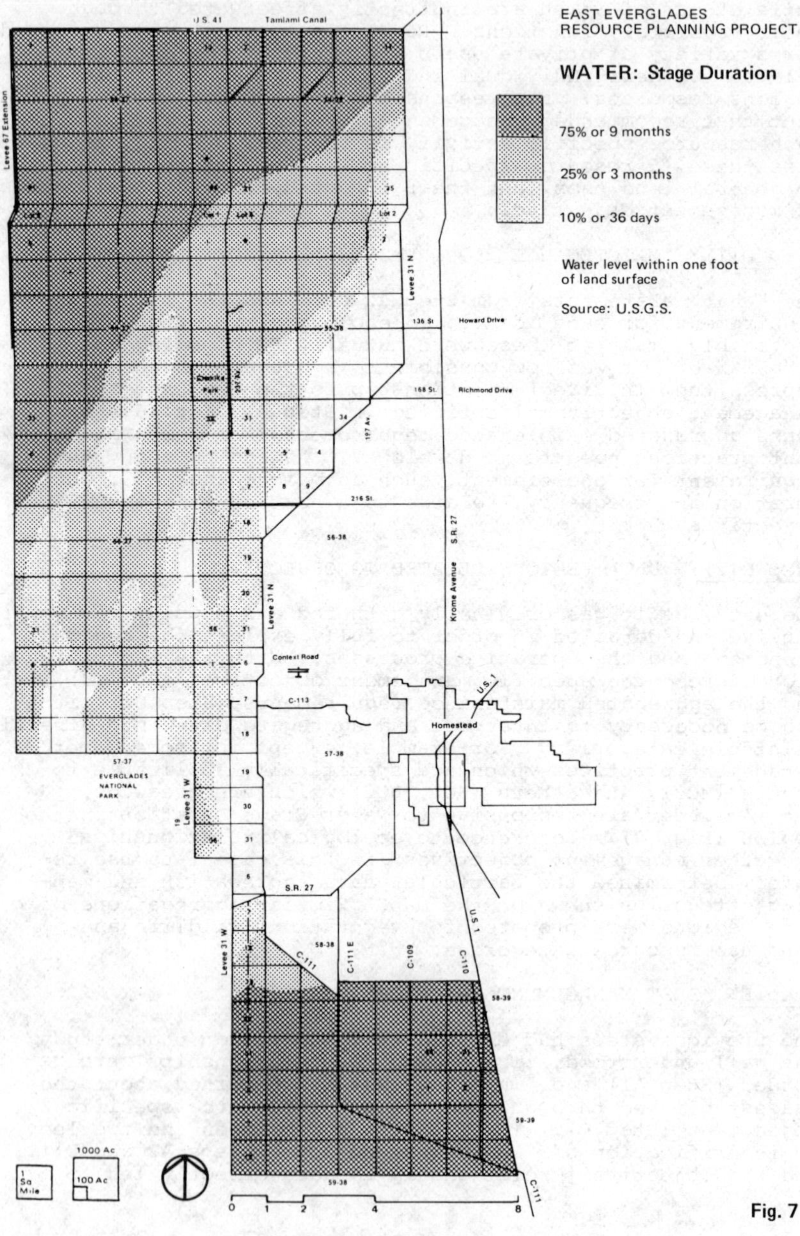
EAST EVERGLADES
RESOURCE PLANNING PROJECT
WATER: Stage Duration
75% or 9 months
25% or 3 months
10% or 36 days
Water level within one foot
of land surface
Source: U.S.G.S.
U.S. 41
Tamiami Canal
Levee 67 Extension
Levee 31 N
136 St.
Howard Drive
168 St.
Richmond Drive
216 St.
S.R. 27
Krome Avenue
Context Road
C-113
U.S. 1
Homestead
EVERGLADES
NATIONAL
PARK
Levee 31 W
S.R. 27
US 1
C-111
C-111 E
C-109
C-110
1000 Ac
100 Ac
1 Sq Mile
0
1
2
4
8

Fig. 7

the identification of "where" it must be practiced (Step D2). Also, it had been established "who" the agents of change are to whom the management recommendations need to be addressed and the nature of their activities (Step C). It was now necessary to indicate "how" the recommended management must be realized. Below are examples of the best management practices (BMPs) expressed as outline specifications for each activity within each management region. The encoded symbols under the ecological relationship column permit tracing back to all other preceding steps:

BEST MANAGEMENT PRACTICES (BMPs)	ACTIVITY	MAPPED REGION	MANAGEMENT OBJECTIVE	ECOLOGICAL RELATIONSHIPS
o Canal stage must be maintained to meet low flow demands coincident with 9-month flood level (low flow). o Canals must not be relied upon for removal of flood water.	P2: Canal Operation	Flood Period: Low Flow (75% stage duration)	HA1: Regulate surface water levels, stage, and duration	SW/LU-2, 7 SW/C-2 LU/GW-5 SW/LU-1, 4, 5, 17 GW/LU-4, 5, 22
o An average freshwater head of at least 2.5 above MSL must be maintained between well and salt water contact zone.	P3: Wellfield Operation	Areawide	HC3: Prevent saline intrusion	GW/LU-24 LU/SW-2 GW/C-2-11 GW/SW-2
o To the extent possible, orient all new and existing roads parallel to the historic flow pattern. o Roads aligned across the flow pattern must provide adequate perforation (culverts) to allow free passage of flood water resulting from 100-year storm.	P4: Road Building	Areawide	HB1: Regulate surface water flows	SW/LU-6, 11
o Any reduction of infiltration rate must be compensated for by withholding artifact (such as a ditch or a depression) which will retain the surface runoff within the site. The design of this withholding artifact must ensure that the collected water will infiltrate into the ground.	F1: Rockplowing, Cultivation	Rockland, Marl Soils	HC1: Maintain substrate function	S/LU-2 LU/S-4 SW/GW-4 SW/S-4 GW/C-6-8 GW/S-1, 2, 4 GW/LU-8 GW/LU-10, 12, 14
o The amount, type, and timing of chemical applications must ensure that potentially polluting residues are assimilated by local vegetation and soils and not transported off-site (residential lot or agricultural field). At no time must the levels of these chemicals down-gradient, and at the edge of the site exceed those specified for Class III Surface Waters and Class IB Groundwater, and those established for designated National Resource Waters (except where natural background levels exceed standards).	F7: Chemical Application	Areawide	HC2: Restrict chemical application	SW/LU-12 S/LU-3 LU/S-3 SW/V-2, 3 GW/GW-1 GW/S-3
o Following abandonment of agriculture, the owner must revegetate the site according to approved practices or carry out a control program using fire and/or mowing. Time of release must be chosen to minimize invasion by alien species and artificial microtopographic variation must be eliminated (i.e., regrade beds and furrows).	F8: Planting and Clearing	Areawide	LB2: Provide for Agriculture	V/LU-1-13
o Wheeled ORV's must only use designated trails or use areas at specified times of year.	M1: ORV Use	Peat, Peaty Marl Soils	HC1; VB1: Maintain substrate function	LU/T-3, 4 LU/SW-7

THE MANAGEMENT PLAN

The step by step procedure described above is rigorous in interpreting all data in response to all objectives. Inevitably, the conflicts which exist between different land uses and management objectives became not only apparent but were in fact highlighted. In certain instances, a recommended best management practice can be theoretically met by innovative technology. Raising structures on stilts, providing access over trestled roadways, and utilizing packaged chemical waste treatment systems could indeed enable the stringent requirements dictated by wetlands protection to be met. But in other instances, such as the conflict between consumptive agricultural land use and natural vegetation protection, an exclusionary decision must be made and justified.

The East Everglades Management Plan, the final draft of which is under review and revision at the time of the writing of this paper, has persued the following method for conflict resolution:

Firstly, the social need for accommodating the future use demand was evaluated in the context of existing and proposed public policies for the East Everglades. Existing County land use policies explicitly discourage improvements of public services. Furthermore the South Florida Water Management District has recommended that additional structural drainage methods are not likely to be implemented within the present planning horizon. These considerations, in combination with previous experience which has indicated that positive drainage needs to be inevitably provided even for rural residential development, led the Management Plan to recommend that residential use in the East Everglades should be discouraged. Agriculture was acknowledged to contribute to the County's economic diversity despite the fact that it entails irretrievable committment of native plants and animals. Past experience has shown that the frequent impermanence of agriculture has resulted from either its displacement by unrestricted urban growth or from wasteful abandonment when farming operations have ventured into areas which are simply too wet to be profitably farmed. Consequently, the Steering Committee recommended that agriculture should be given permanence in the East Everglades by designating suitable areas as exclusive agricultural use zones where new residential development would be prohibited.

Secondly, the Management Plan affirmed the necessity for ensuring that the final recommended Best Management Practices (BMPs) be strictly enforced. The numerous BMPs developed in Step E of the Planning Procedure were carefully edited and aggregated into discrete packages which are relevant to specific users in specific parts of the East Everglades. Areawide BMPs were essentially directed to the various public agencies operative in the region. BMPs applicable to residential and agricultural uses were made specific to the anticipated types of each use and to locations within specified suitable areas for these uses.

Finally, the Management Plan relied upon the division of the East Everglades into five management areas, each of which could be defined using explicit criteria for its spatial delineation, the recommended land uses within it and the related BMPs which must be followed. Fig. 8 diagrams the defining environmental elements and the related use recommendations. Fig. 9 shows the spatial distribution of these management areas. Brief descriptions of each management area follow:

<u>Management Area 1</u>: MODIFIED ENVIRONMENT

Includes all land which has been permanently altered by human activity. Currently 10,600 acres are in agricultural use, 600 acres in residential use and 3,500 acres are vacant. The well drained rockland is relatively high, and due to its proximity to canal L-31 is subject to a lesser flood risk.

The Plan recommends that future residential development in this area should be discouraged. Existing agricultural and vacant land should preferrably be reserved for agriculture, thereby reducing the pressure for land clearance and alteration in other parts of the Everglades which are still natural wilderness.

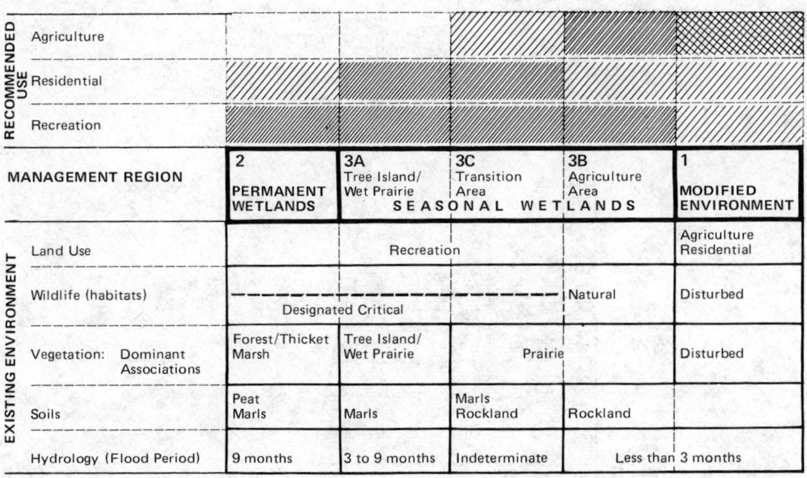

Fig. 8 DIAGRAM: EAST EVERGLADES MANAGEMENT AREAS

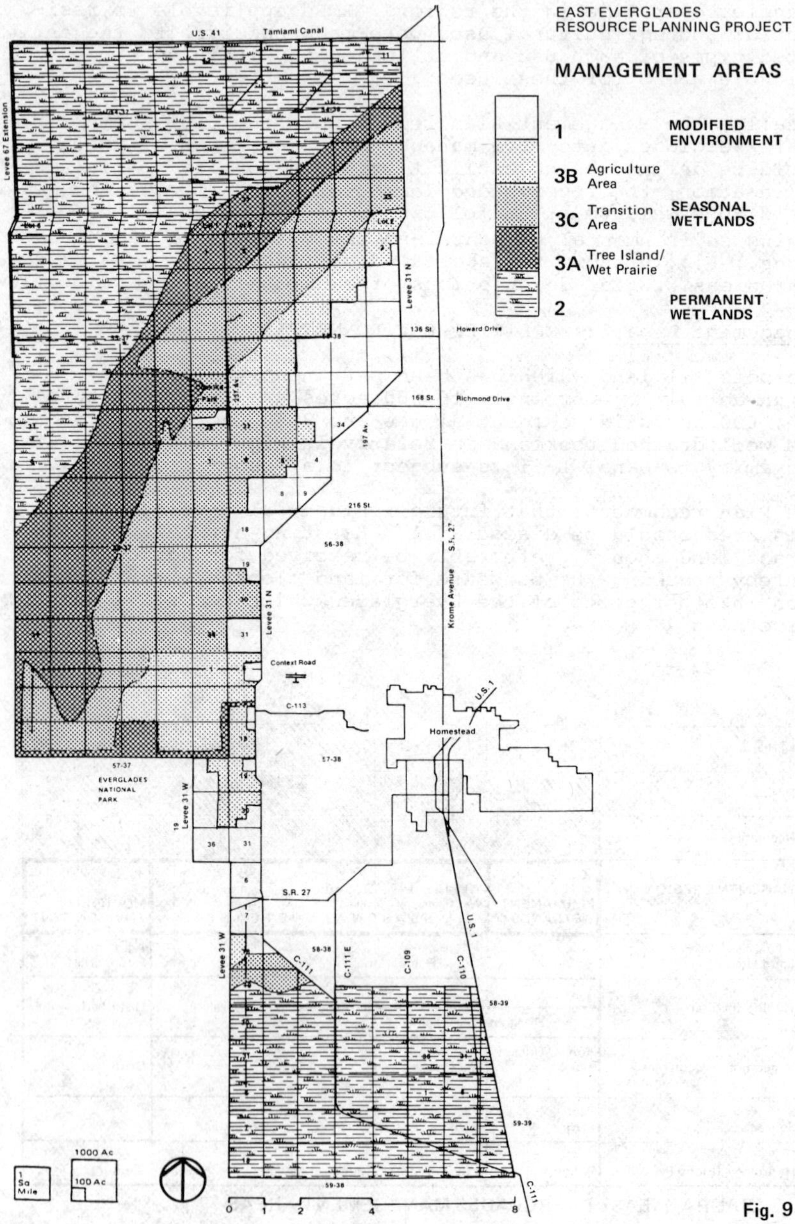
EAST EVERGLADES
RESOURCE PLANNING PROJECT
MANAGEMENT AREAS
1 MODIFIED ENVIRONMENT
3B Agriculture Area
3C Transition Area
3A Tree Island/ Wet Prairie
SEASONAL WETLANDS
2 PERMANENT WETLANDS
U.S. 41
Tamiami Canal
Levee 67 Extension
Levee 31 N
136 St.
Howard Drive
168 St.
Richmond Drive
216 St.
S.R. 27
Krome Avenue
Context Road
C-113
Homestead
U.S. 1
57-37
57-38
56-38
EVERGLADES NATIONAL PARK
Levee 31 W
S.R. 27
C-111
C-111 E
C-109
C-110
58-38
58-39
59-39
59-38
1000 Ac
100 Ac
1 Sq Mile
0 1 2 4 8

Fig. 9

Applicable BMPs relate to the control of undesirable exotic species; the maintenance of surface water flow pattern through regulation of access roads, raised planting beds, ditching and ancillary structures; the maintenance of water infilteration through compensatory on-site retention and recharge; and the maintenance of water quality through regulation of all chemical applications.

Management Area 2: PERMANENT WETLANDS

Includes all land flooded for a least 9 months of a normal water year. Most of the valuable vegetation communities of Forest/Thicket and Marsh, along with almost half of Prairie communities, lie within this area. Also included within its boundaries are the federally designated critical habitat of the American crocodile and substantial portions of valuable feeding grounds and habitats for wading birds and other wildlife species. Most of peat and marl soils also occur within this area.

The Plan recommends exclusion of agriculture from this area because of its infeasibility due to high water levels. Residential development of a density and character which ensures that water quality, water movement, fire management, natural vegetation and wildlife distribution are not altered can be permitted. Regulated recreational use is permitted. No new roads are permitted to be built, thus restricting access by foot or by the use of off-road vehicles.

Applicable BMPs relate to the restriction of dredge and fill activities to the minimum required for an on-site wastewater disposal system; strict water quality maintenance; raising of habitable structures on stilts to a height above the 100 year flood level; and restriction of ornamental plantings to native species.

Management Area 3: SEASONAL WETLANDS

The remainder of the East Everglades, in recognition of its range of hydrologic characteristics, wildlife habitats and locational attributes, is divided into the following three subareas:

Subarea 3A: TREE ISLAND/WET PRAIRIE

Includes all land flooded from 3 to 9 months during an average year. Valuable vegetation communities of Tree Island and Wet Prairie are naturally related to this flood period and are included within this area. Scattered "islands" which may flood for less than 3 months, and a 300 foot wide buffer strip adjacent to the Everglades National Park boundary and the feeder canal (L 31 W), are also included.

The Plan recommends exclusion of agriculture from this area because of the presence of inundation risk during the growing season. Residential development is permitted with the same restrictions as specified for Management Area 2. Roads are permitted to be built as long as water flow patterns are not altered.

Applicable BMPs are essentially the same as described under Management Area 2, except for a relaxation of fill restrictions: fill pads (not to exceed ½ acre) to support a house, as well as on-site disposal systems, are permitted.

Subarea 3B: AGRICULTURE AREA

Includes all land, except federally designated critical habitat of the Cape Sable sparrow, flooded for less than 3 months in an average year and contiguous to Management Area 1.

The Plan recommends discouragement of residential development from this area, because this is the only additional land in the region with acceptable flood risk for agriculture. It is recommended that future agricultural expansion should proceed in an orderly fashion. Except where unusual circumstances prevail, new agricultural uses should be adjacent to lands already in agricultural use.

BMPs described under Management Areas 1 and 3A apply.

Subarea 3C: TRANSITION AREA

Includes the remaining land. A substantial part of this area is the federally designated critical habitat of the Cape Sable sparrow.

Due to incomplete hydrologic information for this area, the Plan recommends permitting agriculture in this area only if (1) there are no alternative locations available, (2) the proposed farming site is adjacent to land currently in agriculture, and (3) the site is known to be hydrologically capable of supporting farming. Additionally, federal permits within the designated critical habitat must be obtained. Residential development is permitted as described under Subarea 3A.

Applicable BMPs for agriculture are the same as Subarea 3B, and for residential development as described under Subarea 3A.

CONCLUSION

The East Everglades Resources Planning Project demonstrates a responsible planning procedure. The area of concern is unique, its environment highly sensitive, and its future use or abuse is of interest well beyond its geographical limits. Yet any regulation of its use must "allow land owners maximum use of their property" as stated in the Plan of Study for the Project. Even though the above quoted statement is qualified by the addition of "..without jeopardizing the long term economic, social and environmental health of the community and the region", the task is not easy.

The Project from its inception has been a cooperative venture of practically all the agencies which have jurisdiction and/or a particular interest in the East Everglades. Background studies conducted by the Project and the project maps and reports represent the most complete and most accurate body of knowledge about the East Everglades that exists today. Planning interpretations are fully explicit both in reference to scientific technical data as well as social goals and objectives. The final recommedations, currently under review, still face the challenge of public acceptance. A gamut of implementation strategies have been proposed and are being evaluated. Hopefully, the Plan recommendations will meet approval of the East Everglades constituents, and the thoroughness and explicitness of the procedure will benefit resource managers and planners elsewhere.

PROJECT CONSULTANTS:

VEGETATION:
- Dr. Ron Hofstetter, Associate Professor of Biology, University of Miami
- Dr. Taylor Alexander
- Chuck Hilsenbeck, Research Associate, University of Miami
- Dr. Joan Browder, University of Miami

WILDLIFE:
- Dr. James Kushlan, E.N.P.
- Oron Bass, E.N.P.

CLIMATE:
- Dr. Paul Orth, Agricultural Extension Service

WATER RESOURCES:
- Bradley Waller, U.S. Geological Survey
- Howard Klein, U.S. Geological Survey
- James Schneider, U.S. Geological Survey
- Abe Kreitman, U.S. Geological Survey
- Peter Rhoades, South Florida Water Management District
- Ron Mireau, South Florida Water Management District
- James Lane, South Florida Water Management District
- James Hartwell
- Dr. Jerry Steinber, Water and Air Research, Inc.
- Garald Parker, Certified Professional Geologist

GEOLOGY:
- Ellis Donsky, U.S. Geological Survey

TOPOGRAPHIC MAPPING:
- Tony Jurado, U.S. Geological Survey

AGRICULTURAL ECONOMICS:
- Dr. Cary Mathis, University of Florida

PLANNING:
- William Roberts, Wallace Roberts and Todd
- Narendra Juneja, Wallace Roberts and Todd
- Richard Collier, Wallace Roberts and Todd
- Peter Diana, Wallace Roberts and Todd

Port Reception Facilities and The Coastal Zone

by

Henry S. Marcus, Associate Professor of Marine Systems

Massachusetts Institute of Technology

Port reception facilities can receive basically three types of wastes from a vessel: (1) sanitary wastes, (2) oily bilge water, and (3) tank washings. The most difficult category to handle is tank washings when they come from a chemical carrier. Recent international agreements and pending federal legislation will result in an increase in port reception facilities for chemical carrier wastes. This paper will dicuss the impact of these events on the U.S. coastal zone.

BACKGROUND

In 1973 seventy nations met under the sponsorship of the Intergovernmental Maritime Consultative Organization (IMCO) to negotiate and adopt the International Convention for the Prevention of Pollution from Ships, 1973 (MARPOL '73). The subject of MARPOL is pollution of the marine environment with particular emphasis on the major source, oil, but also including chemicals, sewage and garbage. For each of these classes of pollutants an annex was added to the Convention; these annexes state general regulations governing discharge of their respective pollutants. MARPOL '73 has subsequently been modified by the 1978 Protocol.

Annex II constitutes a set of general regulations whose aim is the preservation of the marine environment. For the purposes of this paper it has three parts:

a. restrictions on the amount of substance which may be discharged into the ocean;

b. a list of substances to which the above restrictions apply; and

c. a requirement that ports provide reception facilities for volumes of restricted substances above permissible discharge limits.

The requirement for ports to provide reception facilities has significant implications. It imposes economic costs on port owners/operators, the magnitude of which is not explicitly determined or limited. In order to determine the magnitude of these costs it is necessary to know the size and characteristics of those reception facilities.

The substances covered by Annex II are mostly chemicals (as the term is commonly understood); the volumes requiring discharge to reception facilities arise from tank cleaning. Thus, to determine the required size and characteristics of reception facilities the chemical trades must be analyzed.

U.S. seaborne movement of chemicals can be roughly divided into two segments: a domestic segment which is relatively stable in terms of cargo movements; and an international segment, which fluctuates greatly in response to changing supply/demand balances for different chemicals around the world. Geographical extent, cargos carried, direction of carriage and volumes lifted may all change rapidly and drastically. Because of constant changes in cargos, tank cleaning operations are relatively more frequent in the international segment. The international segment will, therefore, drive the demand for shoreside reception facilities.

<u>Annex II Substances</u>

The chemicals or "noxious liquid substances" of most interest are classified as categories A, B and C, as identified in Tables 1 through 3, respectively. Category A chemicals represent "a major hazard" and justify "stringent anti-pollution measures". Category B substances represent "a hazard" and justify "special anti-pollution measures". Category C substances represent "a minor hazard" and require "special operational conditions".

The discharge regulations specify how much of a substance in a particular category may be discharged and under what conditions. Category A substances must be discharged to a reception facility until the concentration of the effluent from tank washing is below a prescribed residual level (usually 0.1% to 0.01% by weight). When the concentration is below this level, remaining residue may be discharged at sea under restricted operational conditions. Further dilution is required, and minimum vessel speed (7 knots), distance from land (12 nautical miles) and water depth (25 meters) are specified. Discharge must be below the waterline.

TABLE 1: LIST OF ANNEX II CATEGORY A SUBSTANCES

Acetone Cyanohydrin	Naphthalene (molten)
Acrolein	Naphthenic Acids
Carbon Disulphide	Phosphorus (elemental)
Creosote	Sodium Pentachlorophenate (solution)
Cresols	Tetraethyl lead
Cresylic Acid	Tetramethyl lead
Dichlorobenzenes	

TABLE 2: LIST OF ANNEX II CATEGORY B SUBSTANCES

Acrylonitrile	Ethylene Dibromide
Alkylbenzene Sulfonate (branched chain)	Ethylene Dichloride
	2-Ethyl, 3-Propyl-Acrolein
Allyl Alcohol	Hydrofluoric Acid (40% aqueous)
Ammonia (28% aqueous)	Methylene Chloride
Benzyl Chloride	2-Methyl, 5-Ethyl-Pyridine
Butyl Butyrate	Monochlorobenzene
n-Butyraldehyde	Pentachloroethane
Butyric Acid	Perchloroethylene
Camphor Oil	Phenol
Carbon Tetrachloride	beta-Propiolactone
Chloroform	Pyridine
para-Chlorotoluene	Toluene Diisocyanate
Crotonaldehyde	Trichloroethylene
Dichloroethyl Ether	Tritolyl Phosphate
D.D. Soil Fumigant	Turpentine (wood)
Epichlorohydrin	Vinylidene Chloride

TABLE 3: LIST OF ANNEX II CATEGORY C SUBSTANCES

Acetaldehyde
Acetic Acid
Acetic Anhydride
Acetyl Chloride
Acrylic Acid
Alkylbenzene Sulfonate (straight chain)
Allyl Chloride
iso-Amyl Acetate
n-Amyl Acetate
Aniline
Benzene
Caustic Potash
Chloroacetic Acid
Chloroprene
Chlorosulphonic Acid
Cumene
Cyclohexane
Dibenzyl Ether
Diethylamine
Diethylbenzene
Diethylenetriamine
Diethylene Glycol
Monoethyl Ether
Diisopropanolamine
Diisopropylamine
Dimethylamine (40% aqueous)
Dimethylethanolamine
1,4-Dioxane
Dodecylbenzene
Ethyl Amyl Ketone
Ethylbenzene
Ethylenediamine
2-Ethylhexyl Alcohol
Formaldehyde (37-50% soln)
Furfuryl Alcohol
Hexamethylenediamine
Hydrogen Peroxide (greater than 60%)
Isobutyraldehyde
Isopropylamine
Mesityl Oxide
Methyl Acrylate
Monoisopropanolamine
Monomethyl ethanolamine
Mononitrobenzene
Monoisopropylamine
Morpholine
Nitric Acid (90%)
ortho-Nitrotoluene
Nonyl Alcohol
Nonylphenol
n-Octanol
Oleum
n-Pentane
Phthalic Anhydride (molten)
n-Propyl Acetate
n-Propylamine
Sodium Bichromate (solution)
Sodium Hydroxide
Styrene Monomer
Sulphuric Acid
Tetrahydronaphthalene
Toluene
Trichloroethane
Triethylamine
Trimethylbenzene
Vinyl Acetate
Xylenes (mixed isomers)

For Category B substances, discharge at sea of an amount up to 1 cubic meter or 1/3,000 of tank capacity (whichever is greater) is permitted. This volume must be diluted and discharged such that the concentration in the wake of the ship is at most 1 part per million. In addition, the same speed, distance and depth restrictions as for Category A discharge operations apply.

The regulations for Category C substances are similar to those for B. The discharge limitation is raised by a factor of 3 to the greater of 3 cubic meters or 1/1000 of tank capacity. The concentration limit in the wake is 10 parts per million. The same speed, distance and depth restrictions apply.

Adequacy

The regulation dealing with reception facilities requires that ports have facilities "adequate for reception without undue delay to ships" of those residues which will remain onboard as a consequence of the application of the Annex. In addition, ship repair ports are required to have facilities for receiving residues from chemical tankers undergoing repairs.

There is some uncertainty concerning what constitutes an "adequate" facility. The USCG appears to be in favor of accepting mobile facilities (e.g., tank trucks, barges, etc.) within their definition of adequate facilities; this will enable many ports to avoid spending funds on fixed facilities.

Another area of uncertainty is the size of facilities. The USCG will apparently only require that ports receive those wastes which must be disposed of ashore. Therefore, the port will typically not be forced to handle Category C substances. Nevertheless, there are still uncertainties with handling just Category A and B substances. First, must the port handle wastes from tank washings which did not take place within the port? Second, even when tank washing occurs within the port, how much water will be used; thorough washing versus less through washing? The regulations the USCG will eventually issue will probably provide guidelines rather than set minimum capacity requirements for reception facilities.

Under proposed federal legislation the USCG, as a representative of the Secretary of Transportation, will develop regulations to compensate chemical carrier operators when "undue delay" takes place in port because of Annex II regulations. Uncertainty arises as to the extent that ports causing "undue delay" will not be considered "adequate".

Disposal

Annex II merely mandates reception facilities. Yet the facilities themselves are only one aspect of the problem which port operators face; they must also find some means of disposing of the wastes which they receive. This brings the issue within the scope of domestic pollution regulations.

A major piece of U.S. environmental law governing the disposal side of the chemical waste problem is the Resources Conservation and Recovery Act (RCRA). This legislation provides authority for regulating all aspects of handling hazardous wastes (which includes tank washings from chemical carriers).

In principle, a port operator can contract with a disposal firm to have his wastes removed and treated off-site. Of course, such services must be paid for, and RCRA regulations require accurate recordkeeping and control of the wastes; but providing these conditions are met no further regulatory barriers are encountered. However, such off-site disposal may be difficult and expensive in some geographic areas. RCRA regulations governing final disposal sites for hazardous wastes are so strict and existing sites are generally so inadequate that relatively few will be eligible for permits when the RCRA regulations are implemented. Even allowing for temporary permits, the availability of disposal sites will be drastically reduced in the short run. In addition, because of local restrictions on the siting of hazardous waste disposal facilities, the development phase for new sites may be lengthy.

The alternative to offsite disposal is the construction of on-site disposal facilities. The cost of this alternative nationwide would be astronomical, particularly if each port developed its own facilities. The total volume of wastes arising from the application of Annex II is, by comparison with all other U.S. hazardous wastes, trivially small. When this volume is divided among many ports and when it is considered that economies of scale in the waste disposal/treatment business are significant, the potential economic waste is enormous. Even if such economic waste were acceptable, the ability of ports to finance expenditures or find land for disposal facilities of such magnitude is doubtful.

To summarize, the interactions of Annex II, the nature of the chemical trades, the economics of waste disposal, and RCRA generate a number of potential problems for participants in the chemical trades - particularly port operators.

Impact of Annex II To Date

Over the last two years federal legislation has been introduced twice giving the United States Coast Guard (USCG) the authority to implement Annex II. Such legislation has not yet passed. The main results of Annex II to date appear to be: (1) a large degree of uncertainty as to the impact of Annex II and related federal legislation as it applies to an individual port; and (2) a general lack of communication among the parties involved.

The second factor may be related to the low priority placed on these problems by many parties. This low priority may be based on the knowledge that the new regulations concerning Annex II will not go into effect before 1984. However, while 1984 may seem to be a long way off, it is not too early to start planning.

NATIONAL PORT PERSPECTIVE

In the long run, the national port system, considered as a whole, should not be greatly affected by either Annex II or related domestic environmental legislation. As noted above, the ports will be supplying only a small amount of hazardous waste when compared to all other U.S. sources. Eventually, the U.S. will meet its environmental objectives - or the objectives will be relaxed. In either case, facilities capable of handling shore-generated hazardous wastes will also be suitable for treating Annex II wastes.

Further, new chemical carriers will be constructed to meet Category B and C pumping requirements by having efficient stripping of tanks and carrying high melting point products in properly insulated tanks. Since total movements of Category A chemicals are smaller than either the B or C categories, the eventual replacement of existing vessels with inadequate pumping systems by modern carriers will help to reduce the aggregate impact of Annex II on the ports.

INDIVIDUAL PORT PERSPECTIVE

The individual port has the problem of responding to new environmental regulations wisely during this transition period when the U.S. is determining how to solve its national problem of hazardous waste disposal. While the long run impact on the overall national port system may be small, the effect on an individual port in the short run may be major.

IMPACT ON THE COASTAL ZONE

The impact of Annex II on a particular coastal zone region will depend on answers to the following questions:

.. Will the ports in the region require additional port reception facilities? If so, how much more capacity is needed?

.. Should a regional approach be taken for providing port reception facilities (e.g., regional tank farm, a barge that travels to each port)?

.. Where will the chemical carrier wastes received in the port be treated and disposed of ? Are there RCRA - approved hazardous waste disposal facilities in the area?

.. If more hazardous waste treatment and disposal facilities are needed in the region, should any be placed in the coastal zone?

.. Will the combination of Annex II and RCRA place chemical-related industries in this region at an advantage or disadvantage versus competitors in the area?

.. Will actions planned in this region encourage new chemical-related industries to come to the area or will it discourage such industry? What is the desired outcome relative to the growth of local industries?

FACTORS IN A REGIONAL COASTAL ZONE ANALYSIS

An analysis for a particular coastal zone region should consider: local ports, cargo flows, demand for port reception facilities, existing port reception facilities, choice of technology, treatment and disposal facilities, potential government funding, and the overall coastal zone management plan.

Types of Ports

Ports can be categorized by the three following types of operators:

1. public port authority

2. public terminal operator; or

3. private terminal operator.

Public port authority refers to a quasi-governmental body which has a legal mandate to promote terminal operations in a given geographic area. Public port authorities typically concentrate their efforts on general cargo movements rather than bulk cargo flows. Often the public port authority acts only as a landlord and does not actually operate the terminals.

Public terminal operators are privately-owned companies which offer their services to the public. This paper is concerned with those public terminal operators which provide docking and storage services for owners of chemical cargos. The public terminal operators typically do not own either vessels or cargos.

The private terminal operator is a privately-owned firm that typically handles only cargos which it owns. Examples are oil or chemical companies.

Cargo Flows

A port will find little published data to help it determine the amounts of Category A and B substances moving through its boundaries. Tables 4 and 5 show estimates of cargo flows of Category A and Categories A and B combined, respectively. These values were derived from Bureau of Census data and apply only to international movements. However, published data has two key shortcomings: it is often too aggregated to provide an exact match with Category A and B substances; and the geographic boundaries included in the data are very broad, making it difficult for an individual port to pick out its cargo movements.

Demand For Port Reception Facilities

Even if the port knew the exact cargo flows of Category A and B substances, it would not be able to exactly determine the amount or even the number of tank washings to be received. The number would depend on whether the vessels involved can meet the pumping requirements for disposal at sea. Even when the ships cannot, they may choose to dispose of tank

TABLE 4 : CATEGORY A CHEMICAL MOVEMENTS

Customs District	Exports('77)	Imports('78)
Portland, Me.	*	*
Bridgeport, Conn.	*	*
New York, N.Y.	1800	6100
Philadelphia, Pa	13600	4200
Baltimore, Md.	1500	*
Norfolk, Va.	*	7900
Wilmington, N.C.	*	3400
Charleston, S.C.	*	*
Savannah, Ga.	*	*
Miami, Fla.	*	*
Tampa, Fla.	*	10400
Mobile, Alab.	*	*
New Orleans, La.	8000	20500
Galveston, Tx.	*	*
Houston, Tx.	39900	14800
Los Angeles, Ca.	2500	*
San Francisco, Ca.	5200	*
Portland, Ore.	*	4200
Seattle, Wa.	*	9800

* = less than 1000 tons

TABLE 5 CATEGORIES A+B CHEMICAL MOVEMENTS

Customs District	Exports('77)	Imports('78)
Portland, Me	*	*
Boston, Ma.	*	*
Bridgeport, Conn.	*	*
New York, N.Y.	7900	44100
Philadelphia, Pa.	13900	5200
Baltimore, Md.	2800	*
Norfolk, Va.	*	7900
Wilmington, N.C.	*	3400
Charleston, S.C.	*	*
Savannah, Ga.	1200	*
Miami, Fla.	1600	*
Tampa, Fla.	*	10400
Mobile, Ala.	78300	*
New Orleans, La.	156100	20500
Port Arthur, Tx.	75700	*
Galveston, Tx.	258000	4900
Houston, Tx.	85700	22500
San Diego, Ca.	26000	*
Los Angeles, Ca.	3700	3500
San Francisco, Ca.	6200	*
Portland, Ore.	22400	4200
Seattle, Wa.	1300	9800
Anchorage, Alaska	79000	*

* = less than 1000 tons

washing at another port.

Existing Port Reception Facilities

Surveys to date have had limited success in determing the number and type of existing port reception facilities. Future general legislation giving the USCG authority to make such surveys should provide a more accurate data base. Until that time many ports will have an unclear picture on existing facilities in their areas.

Choice of Technology

Given the uncertainties described above, it is difficult, if not impossible, for a port to determine in detail the optimal choice of technology concerning port reception facilities. However, it is obvious that many ports will favor the use of mobile facilities (i.e., tank trucks and barges) rather than constructing fixed tanks.

Funding

Most regions have limited funding for planning in the area of port reception facilities, much less acquisition of equipment. It is unclear to what extent the Federal Government will aid in the process. (A demonstration project, funded partially or totally by the Federal Government, would prove extremely helpful). The Coastal Energy Impact Program within the Office of Coastal Zone Management may be a potential source of funds, if the analysis is part of an overall program looking at the impact of coastal energy on the region.

Treatment and Waste Disposal

Under the new RCRA regulations, there will probably be a lack of facilities in the area to treat and dispose of hazardous wastes. Typically the amount of hazardous wastes coming ashore as chemical carrier tank washings will be trivially small compared to land-based industries.

CZM Plan

If the region is covered under a state coastal zone management plan, this will also provide guidance.

CONCLUDING COMMENTS

Within a region it appears that (privately-owned) public terminal operators will be most affected by Annex II. The private terminals of oil and chemical companies will have full knowledge of the chemical wastes to be received and will also generally have the technological expertise and equipment to handle these wastes as part of their shoreside petrochemical complex. On the other hand, the public port authorities typically do not handle bulk chemicals at their facilities so that most publicly-owned facilities will not be affected.

The (privately-owned) public terminals now handling large amounts of bulk chemicals may be able to adapt easily to Annex II requirements. Typically they have dealt with the problems of chemical wastes in the port. In addition, they are normally located in areas of the country (e.g., Gulf of Mexico) where the existing petrochemical industry has already brought about the necessary infrastructure, in terms of transportation, treatment and disposal, to handle chemcial tanker wastes. In contrast, public terminals handling small or medium amounts of bulk chemicals may be more dramatically affected.

If the public terminal only receives chemicals affected by Annex II on an infrequent basis, the operator may feel that it is easier to refuse to handle such chemicals rather than comply with new regulations. Even if this operator wished to handle such chemicals, he would tend to want any related services provided by outside contractors.

The situation of the public terminal operator handling a medium amount of chemicals affected by Annex II shows the potential opportunities and dangers that exist, to some extent, for all terminal operators. Where Annex II chemicals make up a significant proportion of the terminal operator's revenues, he will be reluctant to refuse such cargos. However, he is also reluctant to handle such cargos if his costs due to regulations might exceed his revenues from these chemicals. Consequently, the danger is that he will lose money through trying to comply with new regulations. On the other hand, the terminal operator or region has the opportunity to set up an efficient system for handling chemical carrier wastes and thereby divert cargos from ports not as well equipped.

While Annex II and related federal legislation will not have a major impact of the nation's coastal zone, taken a whole, it may have a significant impact in specific areas.

Consequently, it is suggested that each region perform the following analysis:

.. check published data on Category A and B substances and reception facilities in the area.

.. conduct a mini-survey among terminal operators, chemical carrier operators, shippers, etc. to better define the supply and demand for port reception facilities.

.. locate disposal centers in the region which will meet RCRA requirements.

.. determine the range of alternative technologies which are potentially suitable.

.. determine whether the region should apply for federal funds for planning or hardware.

ACKNOWLEDGEMENTS

This paper is based on the report, "Port Planning Strategies and Policy Alternatives Governing Waste Reception Facilities for Chemcial Tankers" by Henry S. Marcus and Moray P. Dewhurst at the Massachusetts Institute of Technology, July 1980. This study was funded by the U.S. Department of Commerce, Maritime Administration under contract number DO-A01-78-00-3134.

MITIGATION OF PORT DEVELOPMENT AT TAMPA, FLORIDA

William K. Fehring
Director of Environmental Affairs
Tampa Port Authority

Abstract

The Port of Tampa is currently undergoing rapid development including several large dredge and fill projects. These include channel deepening, shipyard expansion, relocation of existing industries, and maintenance dredging. These activities have been coordinated to minimize necessary filling and to provide long-term disposal sites for maintenance dredging. Several mitigation projects to offset the impacts of these activities have been undertaken, including a variety of habitat-creation programs. These efforts have been coordinated through the use of a port environmental advisory committee to produce maximum ecological benefits from very limited financial resources.

INTRODUCTION

The Port of Tampa has experienced dramatic growth over the past three decades. Cargo tonnage through the port has grown from five million tons in 1949 to 50 million tons in 1979. This growth was accompanied by considerable dredging and filling of submerged lands to create new port facilities. Most of this work occurred during the period of 1963-1971 when cargo tonnage increased from 14 million to 40 million tons per year. As a result, nearly 2,000 acres of new port lands was created. Overall, port development, along with other commercial development and the development of waterfront residential property, has caused the loss of approximately 83 percent of the historical wetland vegetation around the Hillsborough Bay portion of Tampa Bay. For the Tampa Bay estuary as a whole, this figure is 44 percent (Lewis 1976).

This rapid development of filled land ended quite suddenly in the early 1970's due to economic factors and the passage of new environmental regulations. Development of the port has continued, however, involving primarily development of terminal facilities on the previously-filled land. Even with the availability of land for terminal development, the need for dredging and filling has not been entirely eliminated. In 1976 work began on the deepening of over 45 miles of the main shipping channel from a depth of 34 feet to 43 feet. This project will allow more efficient use of large bulk carriers carrying petroleum and fertilizer products in and out of the port. Other necessary projects such as the renovation of an old drydock, construction of new shrimp facilities, and maintenance dredging of existing berths, have also required dredging and spoil disposal. The purpose of this paper is to explore methods used to coordinate these activities,reduce the environmental impact of this work, and to maximize financially-limited efforts to mitigate unavoidable impacts.

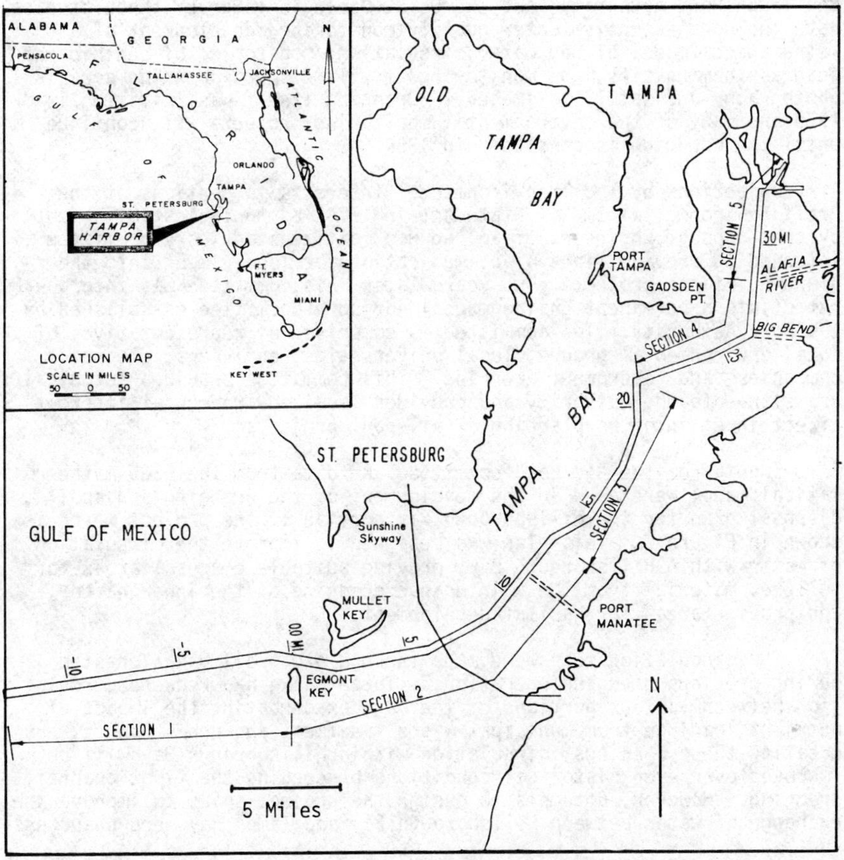

Figure 1. Map of Tampa Bay showing main ship channel.

TAMPA HARBOR PROJECT

Congressional authorization for the Tampa Harbor Project in 1970 contained a requirement for extensive environmental studies associated with development of the project. This requirement was one of the first of its type. Beginning with a 1971 contract between the local sponsor, the Tampa Port Authority, and the U. S. Geological Survey these studies have included extensive water quality monitoring, development of a mathematical model of bay water circulation, monitoring of benthic and seagrass communities, and monitoring of silt dispersion using aerial photography and satellite imagery. Dredging itself was initiated in 1976 and many of the environmental monitoring projects will continue until the dredging is concluded in 1984.

Objections by local environmental interests to portions of the Draft Environmental Impact Statement in 1974 led to the establishment by the Corps of Engineers of an "Ad Hoc" citizens advisory committee intended to provide lines of communication for local imput into the planning of the project. Two years later this committee was incorporated into a permanent Environmental Advisory Committee established by the Port Authority. The committee is comprised of representatives of local environmental groups, local universities, engineers, electric companies, and government agencies. This committee provides technical assistance to the Authority and provides local environmental interests direct input into the planning of all port projects.

Input from the "Ad Hoc" committee and data from the USGS mathematical model were used in the development of the dredging and spoil disposal plan for the Hillsborough Bay portion of the project which is shown in Figure 2. This plan was designed to improve the circulation of water within Hillsborough Bay, provide suitable disposal areas for polluted material from the maintenance dredging of the inner harbor, and provide areas for habitat development.

The circulation cuts were made through old spoil banks created during previous dredging operations. These banks had reduced circulation between various portions of the bay, exacerbating the impact of nutrient loadings from municipal waste treatment and urban runoff. By creating these openings, circulation within Hillsborough Bay will be improved over even historical conditions preceeding the first channel dredging. However, attempts to design the project so as to improve the exchange of water between Hillsborough Bay and Tampa Bay were unsuccessful.

The original plan for the deepening project called for the placement of dredged material from the channel to form elongate horseshoe-shaped islands parallel to the channel. Future maintenance dredging material was to create shallow berms within the horse-shoe shapes which were to be used to create marshes. Unfortunately, this plan left the Port Authority and the Corps with the task of repeatedly acquiring permits for the open-water disposal of polluted silty material from the inner harbor. With ever more stringent regulations on the horizon, this was deemed inadvisable and a new plan was developed.

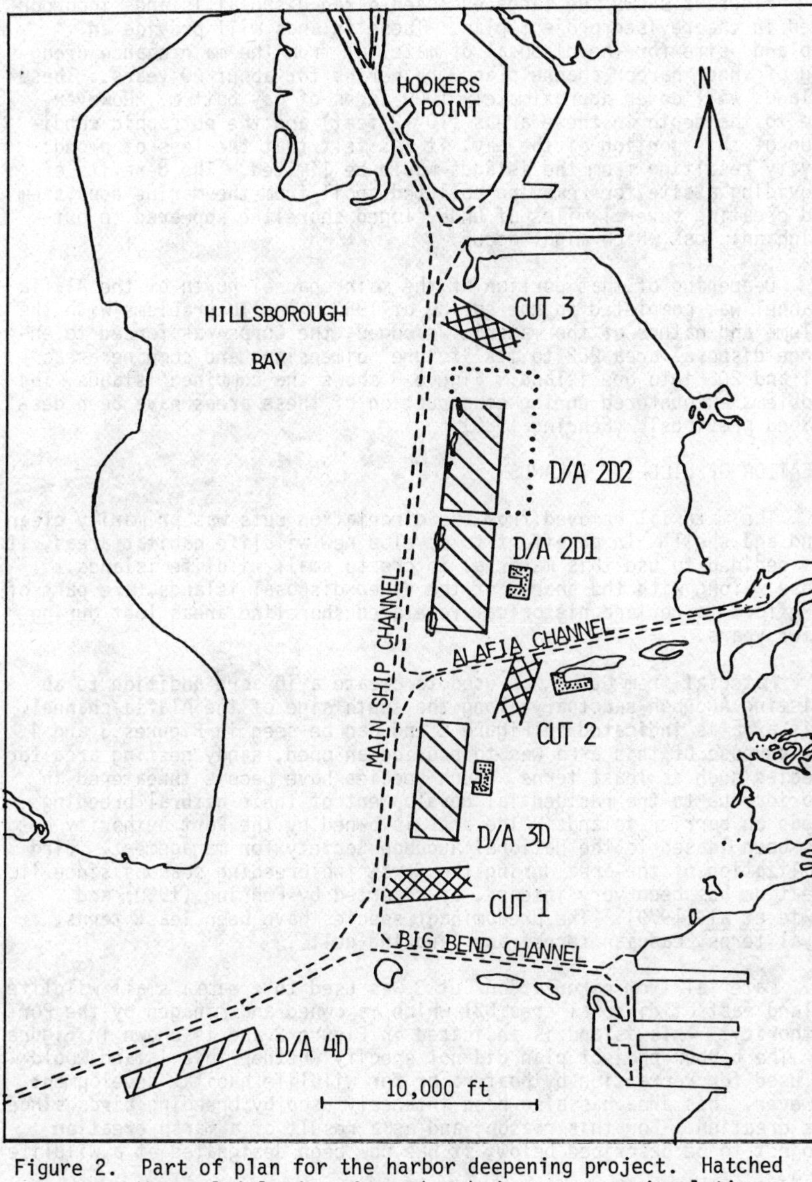

Figure 2. Part of plan for the harbor deepening project. Hatched areas are disposal islands. Cross-hatched areas are circulation cuts. Stippled areas are wildlife islands.

Figure 2 shows the large enclosed diked disposal islands incorporated in the revised project plan. These islands will provide an "upland" site for the disposal of material from the maintenance dredging of inner harbor channels and the berths for about 30 years. These islands will cover approximately 1000 acres of bay bottom. However, due to the depth in these areas (10-12 feet) and the eutrophic condition of this portion of the bay, it was felt that the loss of productivity resulting from the islands would be limited. The benefits of providing a site for removing polluted spoil from the marine ecosystem and creating several miles of undeveloped shoreline appeared to outweigh any loss which might occur.

Deepening of that portion of the main channel north of the Alafia channel was completed in the spring of 1980. Due to problems with the volume and nature of the material dredged, the Corps was forced to enlarge disposal area 2D2 to its "future" dimensions and combine areas 2D1 and 2D2 into one island. Figure 3 shows the combined islands. The problems encountered during construction of these areas have been described previously (Fehring 1980).

CREATION OF WILDLIFE ISLANDS

The material removed from the circulation cuts was primarily clean sand and shell. In an effort to develop new wildlife habitat areas, it was decided to use this material to create small wildlife islands. These, along with the shores of the diked disposal islands, are part of an effort to replace historical vegetated shoreline areas lost during prior years.

Material from Cut 2 was used to create a 10-acre addition to an existing Audubon sanctuary along the south side of the Alafia channel. This area is indicated in Figure 2 and can be seen in Figures 3 and 4. The purpose of this area was to provide an open, sandy nesting area for species such as least terns. Such species have become threatened in Florida due to the residential development of their natural breeding areas on barrier islands. The area is owned by the Port Authority but has been leased to the National Audubon Society for management. Bird utilization of the area during the first two breeding seasons since its creation has been very intense, as reported by Fehring (1980) and Lewis *et al*.(1979). The predominant species have been least terns, royal terns, caspian terns, and laughing gulls.

Material from circulation Cut 3 was used to create a small wildlife island east of disposal area D21 which is owned and managed by the Port Authority. This island is indicated in Figure 2 and is shown in Figure 3. The harbor project plan did not specify whether this island would be used for recreation by boaters or for wildlife habitat development. However, this area has also been intensely used by breeding birds since its creation. For this reason, and as a result of a marsh creation project to be described below, it has now been designated at a wildlife area.

Figure 3. Aerial view of completed disposal areas 2D1 and 2D2, the wildlife island, and the addition to the National Audubon Society sanctuary.

Figure 4. Aerial view of the Audubon sanctuary addition.

The plan for the deepening of the main ship channel includes a third wildlife island to be located east of disposal area 3D which is currently under construction. However, due to a shortage of good material for the construction of the dikes of the disposal island, this wildlife area will not be built at this time. Rather, it will be considered for possible inclusion in future projects involving the deepening of side channels.

ENVIRONMENTAL RESEARCH

For almost a decade the Tampa Port Authority has supported a variety of environmental studies. Since 1976, the port's Environmental Advisory Committee has developed ideas for such studies and has reviewed research proposals received by the Authority. The results of these studies have been reported elsewhere (Detweiler et al. 1975, Lewis & Dunstan 1974, Lewis & Dunstan 1975, Lewis & Dunstan 1976, Lewis & Lewis 1977, Mosiera & Estevez 1977, Lewis 1977). While these studies have each been small, ranging from $5,000 to $12,000 in cost, their benefits to local habitat development and port development have been significant. Several of these studies have concerned techniques for creating marshes on local spoil areas. The information developed has contributed to the design of the harbor deepening project and to habitat creation and mitigation projects described below.

A current research concern of both the Audubon Society and the Port Authority is controlling natural plant encroachment on the clear breeding areas of the new islands. Natural plant succession will soon make these areas unsuitable for the more threatened tern species. Finding an acceptable, inexpensive, and effective way to remove these plants has presented a significant challenge to efforts to maintain the productivity of these areas.

For the past five years, the Port Authority has also cosponsored an annual conference on coastal revegetation held at a local community college. This program has grown significantly over these years and now attracts attendees from across the nation and several foreign countries. The cross-pollination of ideas at the conference has contributed significantly to local mitigation programs.

AUDUBON MARSH CREATION

The National Audubon Society has undertaken a project to create a marsh adjacent to the dredged material addition to their sanctuary which is described above. The purpose of this project is to provide a feeding area for the young of wading species nesting on the adjacent sanctuary islands. Using techniques developed in earlier studies supported by the Authority, over four acres of new marsh has been created by transplanting plugs of *Spartina alterniflora* from natural marshes around the bay. The marsh is shown in Figure 5 and after 18 months can no longer be distinguished from older natural marshes. The creation of these marshes has been described by Lewis et al.(1979) and by Hoffman and Rodgers (1980). The Audubon Society is currently working on an additional project to protect the more exposed shorelines of the

Figure 5. Planted marsh on the Audubon sanctuary.

Figure 6. Sea oat plantings on the disposal island.

sanctuary through the construction of automobile tires.

Much of the labor for these projects came from the government-supported Young Adults Conservation Corps program of the Tampa Marine Institute. Because of this, the cost for the marsh was only about $2,000 per acre. Partially, as a result of the coordination provided through the port's Environmental Advisory Committee, the YACC program has also become involved in numerous other small marsh-creation projects around Tampa Bay. Many of these projects have evolved as a result of the dredge and fill regulatory programs of the Port Authority, the State of Florida, and federal agencies.

DIKED ISLAND PLANTING

In an effort to reduce erosion of the dikes of the new disposal island, the Corps of Engineers has undertaken a planting program. In addition to planting grass on the upper portion of the dikes, 150,000 plants of native species have been planted along the toe of the western, southern, and eastern dikes of D/A 2D2. These have included sea oats, seagrapes, red mangroves, and two species of *Spartina*. A portion of the sea oats planted is shown in Figure 6. The plan for this planting was developed by members of the port Environmental Advisory Committee. While the plantings of red mangrove seedlings and *Spartina alterniflora* plugs have not survived erosion by waves, the plantings of sea oats, seagrapes, and *Spartina patens* appear to be doing well. Since these plantings extend for almost 1.5 miles, they will represent an additional, significant habitat when mature. Further, during the 1980 breeding season, numerous pairs of terns were found nesting on the bare interior of the dikes. Since vegetation on these areas may be greatly reduced by the periodic flooding resulting from maintenance dredging operations, the interior of the island may become a significant breeding habitat in future years.

SHRIMP FACILITY

The development of new docking facilities for the Tampa shrimp industry probably represents most clearly the benefits of the long-range planning, environmental research, and environmental coordination to port development. In 1978 the Authority found itself faced with a dilemma. A local ship repair firm which leased its facilities from the Authority was experiencing rapidly declining profits because of its old graving dock was too small to adequately serve the ever-larger ships calling on the port. In order to keep this important tenant, a decision was made to convert three old abandoned graving docks adjacent to the shipyard into a new, large graving dock and a new wet slip. This work involved little, if any, environmental impact and the new facilities would result in over $40 million in annual impact on the local economy.

The problem with the plan was that the old slips served as a port for the local shrimp fleet. In order to upgrade the shipyard, a new shrimp facility would be needed. The only economically viable site for this facility was a causeway adjacent to one of the new turning basins.

Figure 7. Site of the new shrimp facility.

Figure 8. Transplanted mangroves on the wildlife island.

This area had been created through dredging many years ago and had been enlarged by dredged spoil from a flood-control project in the early 1960's. The area is shown in Figure 7.

Even though the area bordered both a major roadway and a deep-water port area, it contained numerous black mangrove trees and a productive littoral zone. In acquiring the necessary permits for the new facilities, the regulatory agencies demanded that the Port Authority replace the five acres of wetland which would be disturbed on an acre-for-acre basis.

Fortunately, the newly-created island adjacent to D/A 2D1 provided a site for the necessary mitigation. First, 1500 young black mangroves were transplanted from the proposed shrimp facility site to the new island. The techniques for doing this had been developed during previous port-funded studies and other studies presented at the restoration conference. This work created 1.3 acres of new marsh, a portion of which is shown in Figure 8. The details of this portion of the project have been presented previously (Fehring *et al*. 1979).

The remainder of the required acreage will be created in the next year by transplanting plugs of *Spartina alterniflora* to both the new wildlife island and protected shorelines of the diked disposal island. The same techniques refined during the creation of the Audubon marsh will be employed. Previous studies have shown that *Spartina alterniflora* provides both an excellent trap and protection for mangrove seeds. With the previously-planted trees nearby, it is anticipated that eventually this entire new marsh area will be dominated by mangroves which are the climax species of local marine marshes.

The plan for this mitigation program was developed in conjunction with the Environmental Advisory Committee. Arrangements were also made through the committee for the YACC program to provide the labor for at least the initial phase of the project. Conversely, the committee also served as a vehicle for the Authority to lend its support to the efforts of the Tampa Marine Institute to obtain continued appropriations for the YACC program. This program offers job training in biological sciences to disadvantaged youths and is funded under the Comprehensive Education and Training Act. Thus, the program which has been established has simultaneously provided environmental, social, and economic benefits without significant expenditures.

CONCLUSIONS

As with any growing urban area,there are many additional projects either underway or under consideration which will affect the marine environment in the Tampa area. Many of these projects are currently under discussion by the Environmental Advisory Committee, which serves as a forum for informal discussion of projects while they are still in the conceptual stage and not mired in design costs and regulatory red tape. Discussions regarding reducing the impacts of necessary projects, combining projects, and reducing conflicts between projects serve a vital role in implementing local coastal zone management plans and goals.

One concept which has evolved through our discussions and which bears particular attention is that of "mitigation banking." Heretofore, mitigation for project impacts has been a last-minute requirement of regulatory agencies, the cost of which could not be included in projected project costs until the necessary permits had been issued. Several major projects in the Port of Tampa have been abandoned because these costs could not be pinned down. In addition, regulatory agencies have approached mitigation for each project separately under the excuse that this close tie-in is necessary to "justify" the individual permit. As a result, numerous small projects of questionable biological value have been undertaken at an unnecessary cost. This approach has led to an unfortunate and unnecessary animosity and distrust of the concept of mitigation by many port interests who view regulatory agencies as making mitigation as costly as possible as a method of discouraging all port development.

It appears that all interests would be served by a more eclectic view of mitigation by regulatory agencies. Far greater biological benefits might be achieved if the mitigation efforts for several small projects were combined into a single well-designed, biologically important project. Further, if agencies such as port authorities were able to undertake long-term mitigation projects, such as marsh creation programs, with the knowledge that these efforts would "count" as offsetting future environmental impacts in some specified way, they would probably be far more agreeable to mitigation as a standard part of port development. Such long-term projects would have the advantage of allowing biological reproduction to reduce costs, would spread the costs out over a longer period, and would make them more predictable. This approach would have the added benefit of providing the biological productivity of both the mitigation project and the ecological amenity it would eventually replace until the project involved was actually undertaken. It is a concept which bears serious consideration by regulatory agencies and coastal planners.

REFERENCES

Detweiler, T., F. M. Dunstan, R. R. Lewis, W. K. Fehring, "Patterns of Secondary Succession in a Mangrove Community, Tampa Bay, Florida," in R. R. Lewis, ed., Proceedings of the Second Annual Conference on Restoration of Coastal Vegetation in Florida, Hillsborough Community College, Tampa, Florida, 1975, pp. 52-80.

Fehring, W. K., Artificial Islands in Tampa Bay, Florida, in Proceedings of the Ninth World Dredging Conference, 1980 (in press).

Fehring, W. K., C. Giovenco, and W. E. Hoffman, "An Analysis of Three Marsh-Creation Projects in Tampa Bay Resulting from Regulatory Requirements for Mitigation," in D. P. Cole, ed., Proceedings of the Sixth Annual Conference on Wetlands Restoration and Creation, Hillsborough Community College, Tampa, Florida, 1979, pp. 191-206.

Hoffman, W. E. and J. A. Rodgers, Jr., "A Cost/Benefit Analysis of Two Large Coastal Plantings in Tampa, Florida," in D. P. Cole, ed., Proceedings of the Seventh Annual Conference on Wetlands Restoration and Creation, Hillsborough Community College, Tampa, Florida, 1980 (in press).

Lewis, Roy R., III, "Impacts of Dredging in the Tampa Bay Estuary, 1876-1976, in Proceedings of the Second Annual Conference of the Coastal Society, 1976, pp. 31-55.

Lewis, R. R., Potential for Establishing a Tidal Marsh and Seagrass Beds on Dredged Material in Tampa Bay, Tampa Port Authority, Tampa, FL 1977

Lewis, R. R. and F. M. Dunstan, The Spoil Islands of Tampa Bay, Tampa Port Authority, Tampa, FL 1974.

Lewis, R. R. and F. M. Dunstan, "The Possible Role of *Spartina alterniflora* Loisel. in Establishment of Mangroves in Florida," in Lewis, R. R., ed., Proceedings of the Second Annual Conference on the Restoration of Coastal Vegetation in Florida, Hillsborough Community College, Tampa, Florida, 1975, pp. 82-100

Lewis, R. R. and F. M. Dunstan, "Use of spoil islands in reestablishing mangrove communities in Tampa Bay, Florida," in G. Walsh et al., eds., Proceedings of the International Symposium on Biology and Management of Mangroves. Vol. II. Inst. of Food and Agricultural Sci., University of Florida, Gainesville, Florida, 1975, pp. 766-775.

Lewis, R. R. and F. M. Dunstan, "Results of Mangrove Planting Experiments on a Spoil Island in Tampa Bay," Tampa Port Authority, Tampa, FL 1976.

Lewis, R. R. and C. S. Lewis, "Tidal Marsh Creation on Dredged Material in Tampa Bay, Florida," in R. R. Lewis and D. P. Cole, eds., Proceedings of the Fourth Annual Conference on the Restoration of Coastal Vegetation in Florida, Hillsborough Community College, Tampa, Florida, 1977, pp. 45-61.

Lewis, R. R., C. S. Lewis, W. K. Fehring, J. A. Rodgers, Jr., "Coastal Habitat Mitigation in Tampa Bay, Florida," in G. A. Swanson, tech. coord., The Mitigation Symposium: A National Workshop on Mitigating Losses of Fish and Wildlife Habitats. July 16-20, 1979, Fort Collins, Colo. General Technical Report RM-65, 696 p. Rocky Mountain Forest and Range Experiment Station, Fort Collins, Colo., 1979, pp. 136-140.

Mosura, E. L. and E. D. Estevez, "*Rhizophora* Litter Production and Freeze Effects in Tampa Bay," in Lewis, R. R. and D. P. Cole, eds., Proceedings of the Fourth Annual Conference on the Restoration of Coastal Vegetation in Florida, Hillsborough Community College, Tampa, Florida, 1977, pp. 107-113.

EUROPEAN DREDGING - A STATE OF THE ART

John F. Hoffman, Ph.D., P.E.[1]

INTRODUCTION

The U. S. Army Corps of Engineers is tasked by the Rivers and Harbors Act of 1899 to maintain the depth of navigable waterways with funding appropriated annually by the Congress. The U.S. Navy has the responsibility of maintaining the depths of pier slips, turning basins and approach channels to their facilities. Dredging of Navy pier slips, etc. is either done under the auspices of the Corps of Engineers, by private contractors hired by the U.S. Navy or, in some instances, by the U.S. Navy itself.

Coping with shoaling problems in the maintenance of water depths in pier berths by dredging requires a different approach than that used by the Corps of Engineers in the maintenance of water depths in channels. The problems are different.

In order to be informed of advances in dredging equipment and techniques in Europe, the Naval Facilities Engineering Command and the Office of Naval Research (London) sponsored the author for a three-month tour of pertinent facilities in six European countries.

The agencies or activities visited may be categorized as follows: dredging firms (2), dredging equipment manufacturers (1), port and river authorities (8), governmental departments (other than those included elsewhere in this list)(3), laboratories concerned with dredging and/or sedimentation in harbors (6), and universities (3). During the time spent in Europe conferences were held with more than 40 people in Belgium, England, France, Germany, Holland, and Scotland.

Answers were sought to the following questions:

a. What methods are being used presently to dredge a subject waterway?

b. What methods, equipment, or techniques are in use or are contemplated as being put into use?

---

[1]Professor and Associate Chairman, Oceanography Dept., U.S. Naval Academy

c. What slip maintenance problems, from a dredging aspect, have been or are being encountered?

d. What research is being carried on in the dredging field?

Discussed below are the results of the effort. This discussion is divided into three categories: dredging equipment, dredged spoil disposal and research concerning shoaling in piers and harbors and in the design of dredging equipment.

EUROPEAN DREDGING EQUIPMENT

The following types of dredges are in use on the European scene:

Trailing Suction Hopper Dredge. Shown in Figure 1 is one of Europe's largest trailing suction hopper dredges, "Geopotes IX.", owned by the Royal Adriaan Volker Group headquartered in Rotterdam, Holland. This type of dredge is essentially a ship built around a series of holding tanks - hence the name hopper dredge.

The trailing suction hopper dredge consists basically of two 10-ton drag heads, one on either side of the ship, with projections to scarify the bottom. The heads are dragged along the bottom to stir up the sediment. A long pipe connects each head to a large capacity centrifugal pump which pumps the dredged spoil to the hoppers. When the hoppers have been filled, the dredge steams to a designated dump site, the bottom doors opened and the material stored in the hoppers released through the bottom.

Dimensionally "Geopotes IX" is 126 meters (412 feet) long and 21 m (69 feet) wide. It has a hopper capacity of 6392 cubic meters (8360 cubic yards) and an installed power plant of 11000 hp. The dredge can be used to dredge soft sediments in a depth of water of 34 m (110 feet).

Positioning of the dredge in a Rotterdam channel, when it is being dredged, is accomplished within one meter of the desired position by use of a Motorola system with a series of transponders located along the shipping channel. Positioning the same dredge in the North Sea, at the dump site, is by means of a Decca system with two transponders located on the coast. A government inspector accompanies the ship to insure the proper disposition of spoil.

Sophisticated equipment on board the "Geopotes IX" includes indicators of the slurry level in the hopper and for the positions of the trailing heads in relation to the bottom, an automatic compensator for swell that keeps the trailing heads on the bottom at all times, a system of lights to indicate whether the hopper doors are open or closed and to indicate the position of the various valves, and an automatic pilot to steer the ship on a straight course toward the dump site.

Figure 1: Trailer Suction Hopper Dredge
(Royal Adriaan Volker Group, Holland).

Figure 2: Split Hull Trailing Suction
Hopper Dredge (IHC, Holland).

Some of the newer trailing suction hopper dredges have water jets at the drag heads to fluidize the sediment thereby increasing the dredging rate.

Perhaps the newest design in this type of dredge is the Split-Hull Trailing Suction Hopper Dredge, shown in Figure 2. By opening the entire hull the hopper is emptied of its contained spoil very rapidly. However, the basic concept of dredging has not changed from that of a trailer suction hopper dredge. The main feature of the design is that the entire hopper splits open at the dump site for a more rapid and a more complete dumping of spoil. The main part of the ship with controls, etc., remains in an upright position. Such a design eliminates much of the complicated off-loading mechanism.

The Self-Propelled Cutterhead Suction Dredge is perhaps one of the latest developments in the dredging field. The commonly encountered cutterhead dredge depends upon spuds driven into the bottom to provide the resistance against which the dredge can push during the cutting operation. Dredged spoil is usually pumped to shore through a pipeline supported on pontoons.

The self-propelled cutterhead suction dredge shown in Figure 3, "Vlaanderen XIX", Zeebrugge, Belgium uses a spud for anchoring purposes, however, a steel piston pushing against the spud advances the dredge 6 meters (19.7 ft) before repositioning of the spud is required. Supported by a nylon and rubber casement, the pipeline floats thus making the operation of the dredge less dependent on the sea conditions than if pontoons were used enabling the dredge to work in a relatively heavy sea. Its transoceanic crossing capability without disassembly makes this dredge a competitor in the foreign dredging market.

The Bucket Dredge, shown in Figure 4, utilizes an endless chain of large capacity buckets moving between two ladders or guides that extend into the water at an angle to the deck. The bottom material scooped up is chuted to an adjacent barge. A bucket dredge can dredge in depths of water up to about 40 m (131 ft) and at rates up to about 800 m^3/hr (1046 cy/hour). There are no bucket dredges in use in the United States.

The Multi-Grab Hopper Dredge is used in the United Kingdom. It is also used supplementally in maintaining the harbor at Antwerp, Belgium. Good for close-in work such as dredging at bulkheads, alongside piers, and in pier slips, this type of dredge has been used in water depths up to 18 meters (60 ft). In construction the multi-grab hopper dredge is a boat built around a hopper with one to five cranes mounted on the deck. Clam-shell buckets, similar to the one shown in Figure 5, attached to the cranes are used to fill the hopper. When filled the dredge moves to the disposal area where the bottom doors are opened and the contained sediments released.

Figure 3: Self-propelled Cutterhead Suction Dredge "Vlaanderen XIX" (Decloedt, Belgium).

Figure 4: Ladder-bucket Dredge (Royal Adriaan Volker Group, Holland).

Figure 5: Clam-shell Bucket used for Multigrabber Dredging.

The Dipper Dredge, shown in Figure 6, is useful when rock or tight clay constitutes the bottom of pier slips or waterways. It has been used more widely in Europe than in the United States. Equipment manufactured by shipyard De Donge for this purpose was exhibited at the Europort '78 Exhibition. Basically, the dipper dredge consists of a backhoe mounted on a barge that has three or four spuds penetrating the bottom to secure it. In one type the bucket cuts downward in the same manner as a backhoe; in another type the bucket has been rotated 180° so that it cuts upward. The dredge swings laterally to enable the bucket to be emptied into an adjacent barge.

The Vertical Bucket Wheel, shown in Figure 7, was in the process of development by the IHC Company (Netherlands) in the early fall of 1978 and was operational by late Fall 1978. The cutterhead is a round hollow cylinder with diameters up to 3 m (10 ft), mounted vertically in guides. Affixed to the peripheral wall of the cylinder are small scoop-like buckets. The "wheel" is powered by an axially-mounted hydraulic motor. A centrifugal pump, connected to the hub of the cylinder, pumps the spoil either to an adjacent barge or to the shore.

The Amphidredge, shown in Figure 8, is a patented dredge manufactured by IHC Holland. This dredge has the capability of traveling in areas where the load bearing capacity of the soil is low. The maximum reach of the excavator is 6.70 meters (22 ft) and the maximum dredging depth is 4.00 meters (13 ft).

This device has also been used effectively to break the crust formed on the surface of diked disposal areas. Such a practice accelerates the evaporation of the subsurface water and accelerates soil stabilization.

SPOIL DISPOSAL

Spoil disposal is a vital aspect of the dredging operation. Places where dredging is most necessary usually offer very few sites closeby for spoil disposal. Accordingly, in the dredging operation, the price of spoil transport becomes an important economic fact. Europe is aware of the impact of dredge spoil on the environment but appears to be less conservative in its concerns about these matters than the United States.

Listed below are spoil disposal practices in six European countries.

BELGIUM

The Port Authority of Antwerp, the largest port in Belgium, utilizes both bucket and multi-grabber hopper dredges to maintain pier slips and main channels. The latter dredge is used to excavate close to bulkheads and piers.

Spoil, amounting to about 2.5 million m^3 (3.3 million cy) annually, is pumped to industrial sites to increase the land area.

Figure 6: Dipper Dredge (De Donge, Holland).

Figure 7: Bucket-wheel Dredge (IHC, Holland).

Figure 8: The Amphodredge (IHC, Holland).

Land disposal is carried on as follows: a barge with a pump and pipeline attached is moored at the shore. Hopper barges and multi-grabber hopper dredges offload to the moored barge by pumping from which the spoil is then repumped inland.

Such a procedure is followed with non-toxic spoil. The dredge spoil in only one harbor in Belgium, the harbor of Ghent, contains toxicants. Amounting to about one million m^3 (1.3 million cy) annually, the spoil from dredging this harbor is pumped to an onshore disposal site 10 hectares (25 acres) in area. In order to prevent the leachate from contaminating the ground water a water-tight membrane seals the bottom of the disposal area.

ENGLAND

The amount of spoil dredged annually from the Thames River is about 600,000 m^3 (460,000 cubic yards) along a rather extensive reach. The unit cost for dredging and offloading spoil is about $2/$m^3$ ($1.53/cubic yard).

The Port Authority of London land spoil disposal program started in 1969. The spoil disposal site is a diked area with overflow weirs spilling to a peripheral drain which empties into the Thames River. The site, underlain by relict riverine sediments, is in the vicinity of a range used for artillery practice. Evaporation from soil surfaces and transiration from plant growth dries the spoil fairly rapidly causing cracks to develop on the surface. No use has been, as yet, planned for the area.

Another port of interest from a spoil disposal standpoint is Garston-on-the-Mersey. Dredge spoil in previous years has been dumped 48 kilometers (30 miles) downstream from the harbor. Studies have shown that dumping spoil only three miles downstream resulted in a minimal return of the spoil to the dredge site. Presently, spoil from channel dredging is now dumped in the Mersey River three miles downstream. Spoil from dredging pier slips, however, is still dumped 48 km (30 miles) downstream.

FRANCE

Spoil is disposed of in two ways. Offshore dump sites and onshore spoil areas for creating additional land near expanding ports. Offshore disposal, by law, requires an analysis of the spoil and an environmental impact study. However, environmental control is less rigorous than in the United States.

GERMANY

In Germany, the Federal Government maintains shipping channels only. The Federal States maintain their own harbor channels and slips. The German Navy maintains its own channels within the naval base.

Dredging in the Elbe River is accomplished by means of two trailing suction hopper dredges; one has a hopper capacity of 2500 m^3 (3270 cubic yards), the other 4500 m^3 (5886 cubic yards). Each has a draft of 6.7 m (22 ft) and the capability of bottom dumping. The smaller hopper-capacity dredge dumps the spoil in the North Sea; the larger one has the spoil pumped off to form small artificial islands. Any dumping on marshland is resisted by the environmentalists.

Artificial islands are formed by outlining a shoal area in the Elbe River with a single layer of copper slag rock. Sand pumped into the middle of this area is spread by a bulldozer when the sand surface is above the water surface. One island, so formed, is 5 km (3 miles) long and 1 km (0.6 miles) wide. To date (1978), these islands have not been utilized for any purpose.

About one-half of the eight million m^3 (10.5 million cubic yards) dredged annually from the Elbe River in the 110 km (68 miles) reach from Hamburg to the North Sea is used to build artificial islands. The other half is dumped in the North Sea within 20 km (12.5 miles) of the shore.

The material dredged originates from the river channel bed and river banks and is basically one-third very fine sand, one-third medium sand, and one-third coarse to very coarse sand. There is a 30-km (48 miles) reach of river, however, where flocculation has caused silt to be separated out and to be deposited.

The disposal of dredge spoil in the Port of Hamburg differs from the method of disposal of spoil from the Elbe River. Sand spoil is used to build up land adjacent to the harbor. Inasmuch as there are no ground water wells in the vicinity, the leaching of toxicants to the water table is considered not to be a problem. Silt spoil, on the other hand, is deposited on farmland rented for a period of about five years. If the land is sandy and ground water wells are close by, a one-meter (3.3 ft) layer of clay is initially placed. Where the soil is silt and has a poor bearing-capacity, a one-meter (3.3 ft) layer of sand is placed over the layer of silt to facilitate drainage.

About 100 hectares (247 acres) of land are presently used for the disposal of sand and silt. Optimum operation requires that an estimated area of between 100-200 hectares (247-494 acres) is required. Annually, this area should be increased by 20 hectares (49 acres) to cope with the removal of land that has been filled to capacity.

THE NETHERLANDS

About 20 million m^3 (26 million cubic yards) are dredged from the Netherlands' waterways annually. Fifteen million m^3 (19.6 million cubic yards) of the spoil is dredged from the Port of Rotterdam alone. Dredging in the port is accomplished by means of trailing suction hopper dredges, bucket dredges, and bucket and scow dredges. Spoil dredged from the various inner harbors of the port is discharged at five permanent off-loading pump stations onto polders. Pump capacities vary from 1000 to 2000 m^3/hr (4400 to 8800 gallons per minute).

Prior to 1960, dredged spoil was pumped onto farmland to increase its fertility and hence crop production. Sediment in the middle of the Port of Rotterdam, however, was found to be contaminated by heavy metals brought into the harbor by inflow from the Rhine River. Investigation by the Institute of Soil Fertility determined that the crops were absorbing the heavy metals contained in the spoil posing a potential threat to the consumer. Now the receiving land is devoted for use as parks, houses, golf courses, etc. Inasmuch as the high sodium content in the spoil is deleterious to vegetative growth, the water runoff from spoil placement is collected carefully and drained to the river. Leachate from seepage of water through the placed spoil is skimmed from the water table by shallow wells and returned to the river to avoid contamination of the deep ground water supply.

Sediment dredged from Europort, seaward of the Port of Rotterdam, is dumped in the North Sea under the supervision of Rijkswaterstaat, Nordzee Division. This Division has three functions: supervising the disposal of spoil in the North Sea, evaluating the movement of the spoil after dumping, and studying the concentration and effect of any contaminants that create a problem on the Dutch portion of the Continental shelf. They issue permits for spoil disposal in the North Sea and maintain a staff of inspectors. The function of the inspectors is to travel with the trailing suction hopper dredges to insure that the spoil is deposited in the proper place offshore.

Based on the investigation of the offshore movement of spoil in dumping areas, it is estimated that 50% of the silt present in coastal areas is derived from spoil dumped at the disposal site.

An interesting approach to the dumping of spoil from the waterways of Europort is its disposal in Lake Oostvoorne for recreational purposes. The lake, 40 m (131 ft) deep, is too deep for recreation. The disposal plan is to fill the lake to an acceptable level of 5 to 10 m (16.4 to 32.8 ft) and cover the spoil with a plastic sheet. The plastic sheet weighted down with sand will provide stability as well as converting the lake to a more conventional recreation area. Involved is a computer model study of the seasonal turnover of lake water and monitoring for toxicants in the sediments to be dredged, including analyzing the interstitial water.

SCOTLAND

The River Clyde Authority is responsible for 1036 square kilometers (400 square miles) of water in the Port of Glasgow. Dredging of both main channels and slips is carried out by the Authority using the following equipment: one bucket dredge, four self-propelled hopper barges, and one multi-grabber hopper dredge. Four hopper barges are necessary to keep the bucket dredge working continuously. The frequency of dredging slips varies from 10 to 15 months depending upon the siltation rate.

Some idea can be obtained as to the time consumed in the maintenance dredging of pier slips from the following. Dredging a berth 30.5 m x 198 m (100 ft x 650 ft) removing 0.61 m (2 ft) of material in a water depth of 7.6 m (25 ft), takes 3 days. Two days are spent dredging with the bucket dredge, and one day is spent dredging with the multi-grabber dredge. The bucket dredge requires an additional day for mobilization and demobilization. Before and after each berth is dredged, the bottom is mapped by sonar in order to determine the volume of material removed.

Three siltation basins in the bottom of the harbor are used in the Port of Glasgow for trapping sediments. The dimensions are 91.5 m (300 ft) wide, 1006 m (3300 ft) long, and 2.5 m (8 ft) deep below the channel bottom. These basins trap the silt and allow compaction before dredging and leaves little doubt in the Conservancy Engineer's mind as to their efficiency in preventing siltation in berths.

LABORATORIES CONCERNED WITH SEDIMENTATION and/or DREDGING

Seven major European laboratories concerned with the development of dredging equipment and/or sedimentation studies are described below. One each in England, France, and Germany and three in the Netherlands.

ENGLAND

The Universities of Strathclyde and Manchester have noteworthy hydraulic and coastal engineering laboratories, but their orientation is more academic then commercial. Understandably these laboratories do not compare in size or activity with the laboratories subsequently discussed.

Hydraulics Research Station, Wallingford. The Hydraulics Research Station has a staff of 260 and an annual budget of about $4 million. Problems are investigated in the fields of civil engineering and hydraulics for organizations in the United Kingdom and overseas. Research in support of this work is also carried on. A charge is normally made for carrying out investigations, but advice is given free when a substantial amount of staff time is not involved. Investigations are carried out for engineering consultants and contractors, industrial concerns, multi-national companies, port trusts, water authorities, and British and foreign government agencies.

Numerical and physical models are often used for examining the effects of proposed schemes. The method of approach depends on the type of examination required and sometimes both are used. Additionally, these may be preceded by a site visit or a field survey.

FRANCE

The Laboratoire Central d'Hydraulique de France (LCHF) - Geohydraulique, Maisons-Alford -- This is a private group of consultants acting on behalf of public or private organizations who deal with hydraulic engineering or water management. The Laboratoire Central d'Hydraulique de France (LCHF) was founded in 1939, and Geohydraulique was founded in 1967, at which time they were combined. The staff, consisting of more than 100 persons, half of whom are engineers gives the laboratory expertise in mathematics, hydraulics, hydrography, oceanography, hydrology, geology, hydrogeology, and sedimentology. Its services cover all stages from planning and preliminary studies to the supervision of site work and the training of qualified personnel.

This group provides research and technical skill in the fields of: harbor engineering, maritime structures, offshore structures, coastal protection and planning, river and estuary planning.

Scale-model testing laboratories cover about 15,000 m^2 (161,500 square feet) and include wave tanks, fixed or variable-slope glass-sided waves flumes, and a large flume for offshore structures towing or hull studies. Laboratory facilities also include the following: electronics, water and rock analysis, and sedimentation.

LCHF-GEOHYDRAULIQUE has its own computer center that includes: a PDP 11 computer (128K) for real time treatment of physical data recorded on laboratory scale-models, and an INTERDATA 8/32 computer (256K) for scientific calculations, together with the appropriate peripheral equipment; printer, display terminal, plotter, etc. Also, LCHF-GEOHYDRAULIQUE has access to centers equipped with IBM 360/125 and CDC 6600 and 7600 computers.

GERMANY

Federal Institute of Waterways Engineering, Hamburg This Institute was founded in 1948 as a centralized institute of waterways engineering with the purpose of conducting scientific and practical hydraulic research. Governmental funds (Ministry of Transport) and consulting activities by contract (public and private orders) amounted to about $10.5 million in 1977. The staff numbers more than 300.

Studies include but are not restricted to river training and canalization, interactions between ships and waterways, bank protection, morphology of rivers and estuaries, transport mechanisms of suspended sediment, effect of inlets and outlets on navigation, and harbor and maritime structures.

Franzius-Institut for Hydraulic Research and Coastal Engineering, Hanover This institute is part of the University of Hanover. The Director simultaneously holds the Chair of Inland Waterways and Coastal Engineering. The Institute, with roughly 60 staff members employed by the State of Lower Saxony, is a non-profit organization under the supervision and control of the Ministry of Arts and Science of the State of Lower Saxony.

In addition to the teaching, the tasks of the Institute cover basic scientific research as well as applied research. In addition, the Institute advises Federal, State and Municipal Institutions and private firms such as industrial and engineering firms, and supports them in the planning and constructions of hydraulic projects and coastal works in Germany and abroad.

The Institute's methods are basically concentrated on investigations of hydraulic models, mathematical and theoretical studies, and field investigations.

THE NETHERLANDS

The Delft Hydraulics Laboratory (DHL), Delft Funds to support this facility, amounting to about $33 million annually, come from the proceeds of consulting activities and from revenues for special services.

The DHL has the following facilities: wind-wave flumes, current and wave basins, testing setups for pumps and for sand-water mixture (dredge pumps), calibration stands for flow meters (free surface and closed conduit), density current provisions with warm water and climate control for cold water problems, a saline water supply for salt water intrusion and waste water problems, morphology basins, sediment transportation flumes, sediment mixing plant, remote-control ship maneuvering, hydrographic instrumentation, and computer programs.

The scope of the activities include: theoretical studies; mathematical model investigations; hydraulic model investigations in the fields of hydraulics, hydrodynamics, and hydrology; providing data for the design of hydraulic structures, river, coastal, and ocean engineering works; carrying out surveys in situ; processing and interpreting observations. Fundamental research is also carried on in cooperation with educational and other scientific research institutes, and as support to consulting work.

De Voorst Hydraulic Laboratory, Kampen The "De Voorst" laboratory began operation in 1951. The area of this site, located between the Vollenhover Kanaal, Zwolse Vaart, and Voorsterweg, amounts to about 121 hectares (300 acres). About 225 staff are employed at this laboratory.

The 35 open-air models are supplied with water originating from the Vollenhover Kanaal. A storage basin holds this water for distribution throughout the laboratory by means of a network of open channels. The distribution of water through the models is controlled by weirs, racks and gates to be ultimately discharged to the Zwolse Vaart. Some large models are covered by a shed. The largest building has an area of 6 acres and accommodates the tidal model of the Oosterscheldte region, the main part of the Delta plan.

There are three types of models available depending on the phenomenon studies: flow in waterways, tidal and wave motion and navigation. Often these are used in combination.

Model studies on flow problems in rivers, coasts, estuaries have their supply of water controlled by Romijn-sluices, and water levels controlled by adjustable flap gates. Wave models for harbors and coasts are equipped with wave generators of the flap-type. In the navigation models, ship models are applied to study the nautical problems encountered in hydraulic engineering projects such as harbor entrances, canals and river crossings. These ships can be steered in different ways, depending on model scales. They can be steered electrically (by cables interconnecting ship and shore), by a real helmsman aboard the shipmodel, or by remote control radio.

Detailed studies on the effect of wind or wind waves utilize a wind-wave flume (width 4 m (13 ft) height 2 m (6.6 ft), effective length 100 m (328 ft) which a giant blower can generate high velocity winds.

Studies requiring high velocity flow, a discharge flume (cross section 3 x 3 m (10 ft x 10 ft), length 100 m (328 ft) is available. This flume has a flow capacity of about 10 m^3/s (350 cubic feet/second).

Besides model studies commissioned by governmental authorities, contractors and consultants, the Laboratory carries out fundamental research concentrating on the theoretical and mathematical approach to hydraulic problems.

Delft Hydraulics Laboratory and Delft Soil Mechanics Laboratory, Delft These laboratories are the liaison between the practical aspects of dredging and research. The latter laboratory is one of the most complete soils laboratories this writer has ever seen. Furthermore, it is equipped for the analysis of large numbers of sediment and rock samples despite the fact that detailed tests are required.

Among the various activities inspected was a dynamic resistance test on various sediments using a cone penetrometer moving at the vertical rate of one meter (3.3 ft) per second. This is the estimated speed at which the bucket of a bucket dredge moves through bottom sediments while dredging. The purpose of the test is to enable better time estimates for a specific dredging job by in situ testing of bottom sediments.

IHC Manufacturing Co., Holland maintains a complete soil mechanics laboratory for the design of its equipment. Dredging equipment is designed for a specific type of bottom material. A complete analysis of the properties of the submitted bottom samples is made.

Properties such as sediment shear strength, density, and grain size distribution are examined. Once determined, the results are typed into a computer and the required horsepower and other information needed for equipment design is displayed as a readout on a cathode ray tube (CRT).

INTERAGENCY PERMIT REVIEW
LESSONS OF EXPERIENCE

Major Joseph S. LeGath, USA (Ret.)
Dames & Moore
Washington, D.C.

Robert L. Lippson, Ph.D.
National Marine Fisheries Service/NOAA
Oxford, Maryland

INTRODUCTION

A major factor in current coastal development is the U.S. Army Corps of Engineers permit program for construction and fill activities in the waters of the United States. Under this program the Corps is responsible for evaluating the impact of these activities on the public interest. A key factor in this process is the interagency review of permit applications. This review involves federal and state agencies who must comment on the desirability of the proposed activity in terms of their particular area of responsibility, most of which pertains to ecological and physical impacts on the human environment.

Obviously, this interagency review process has significant value in terms of comprehensive consideration of potential adverse effects. There are, however, potential disadvantages to this process due to the time it takes each agency to review and comment on a particular application. Among the problems with the interagency review process is the impact on the applicant and public in terms of cost incurred while waiting for a decision. This can be in terms of both construction and utilization costs. This does not imply permits should be approved automatically, but, rather that a timely decision, even for denial, may be more in the public interest including the applicants, than a lengthy review resulting in approval. Another factor is that most applications are for activities that do not require major environmental reviews, i.e., EIS's. If the agencies, however, expend unnecessary time in reviewing these applications they will not have the time or resources available for the major activities requiring intensive reviews.

As the permit responsibilities of the Corps and the other federal agencies grew after the passage of the National Environmental Protection Act (1969) and the amendments to the Clean Water Act (1972 and 1977) the problem of review coordination became a more vexing one for all concerned. Efforts were made at various levels to alleviate these problems. This paper describes some of those efforts and the initial results obtained. It is based on the operational experiences of a former Corps of Engineers officer and a National Marine Fisheries Service official during a period from 1975 through 1979. The primary geographical area was the Northeast and Middle Atlantic Coastal Regions of the United States.

BACKGROUND - THE CORPS OF ENGINEERS PERMIT PROCESS

Under the provisions of Section 10 of the Rivers and Harbors Act of 1899 and Section 404 of the Clean Water Act, as originally included in the 1972 amendment and 1977 update, the Corps is responsible for the regulation of construction and fill activity in the waters of the United States, which includes the navigable waters of the United States. Functionally, without getting into the legal swamps, this means that anyone doing work below mean high water line along the coast, in rivers and streams, lakes and adjacent wetlands may require a Corps of Engineers permit prior to beginning the work. In addition, it is likely that other federal, state, and local permits may be required. The Section 10 permit is one of the oldest permits required, dating from 1899. Originally, it was only intended to insure movement of interstate commerce. The principal review being whether or not proposed construction would interfere with movement of shipping on the navigable waters of the United States. It should be mentioned that Section 103 of the Ocean Dumping Act also falls under the Corps permit program. However, the experiences described here relate only to Sections 10 and 404 permits.

The Corps of Engineers permit program is managed through the regulatory functions branches located in 35 District offices, in the continental United States, Hawaii and Alaska. Each of these offices, acting in the name of the District Engineer, is authorized to make decisions involving permits. It should be noted that these regulatory functions offices also have an enforcement responsibility. This involves inspection of waterways for illegal activity and monitoring for compliance of permitted activities.

Permits issued by the Corps fall into two general categories, individual and general. The individual permits require the most review and involve application approval for specific activity at a specific time and place. It may be either a major or minor activity and does not usually require an environmental impact statement. The general permit, of which there are two categories, nationwide and regional, relates to minor activities. The regional type permit is for a simple activity within a large geographic area, such as for residential piers along a given waterway. It was created primarily to assist in expediting the permit process and is still in the process of being reviewed. The general permit process incidentally is an example of interagency coordination.

The procedures for permit processing are relatively straightforward. The initial action is the filing of a standard permit application with the District Regulatory Functions Branch. A public notice is then issued describing the project and inviting comments over a 30-day period. This includes those by other federal, as well as, state and local agencies. After all the comments are received, a public interest review is conducted by the Corps. A decision, called Findings of Fact, is then made whether to: (a) issue the permit as applied for; (b) issue with conditions; or (c) deny the application.

The current Corps regulations, as published in the Federal Register, establish certain criteria that must be considered during the public interest review process. The four basic criteria are: need for the project; alternatives; direct impact, and cumulative impact. The review factors are comprehensive and cover both biological, physical, and socio-economic impacts on the human environment.

CASE STUDIES

1. Joint Processing

The interagency review of permits has an immediate impact on management strategies and social perception in the coastal zone. Permit applications are reviewed by the National Marine Fisheries Service, the Environmental Protection Agency, and the U.S. Fish and Wildlife Service for ecological suitability; however, it is the responsibility solely of the District Engineer, to render a decision on permit issuance. When the permit action is challenged, the final decision may ultimately rest with the Secretary of the Army.

As mentioned previously, the interagency review process, although useful and required under law, was becoming exceedingly cumbersome and inefficient. It became apparent to us, as well as other state and federal representatives responsible for reviewing permit applications, that there was need to expedite the process while maintaining adequate protection for the environment.

In order to streamline the permit review process, the joint processing procedure was initiated in the area served by the Baltimore District Corps of Engineers in 1975, and shortly thereafter in the Norfolk District.

The joint processing procedure has been structured to efficiently utilize the technical resources of state and federal agencies concerned with the review of applications for construction or other work in waters of the United States.

The keyword of this process is coordination. To eliminate unnecessary delays in reviewing applications, a method was developed which brings together agency representatives in a monthly meeting to review and discuss applications for Corps permits. State and federal concerns are freely aired at these meetings, and generally, if the proposed project is not a major or controversial one, agreement on the recommendations for disposition of the project is reached in an expeditious manner. Each agency retains their integrity and the right to respond separately to any and all permit applications regardless of the majority sentiment within the joint processing group. This avoids the "smoke-filled room" connotation of the review process.

Another advantage which cannot be overstated is the savings in time to the applicant and agencies through the elimination of unnecessary paperwork. Agency comments are recorded at the joint processing meeting, thus precluding the necessity for individual agency letters to be submitted at a later time. Implicit in the joint processing meeting is the added advantage of exchanging current topical data and the requirement of being prepared to comment on the applications under review.

What has joint processing actually accomplished? To answer this question we will describe a typical application for a Corps permit and the process that could ensue without joint processing and then with joint processing.

For example, in a case not reviewed under the joint processing procedure, the applicant, on June 1, proposes to bulkhead his slightly eroding shoreline 20 feet seaward of the mean high water line and to construct a 100-foot pier. The state upon receipt of this application assigns a biologist to make a site visit and write a report including recommendations for approval, denial or modification of the project. On June 27, the biologist submits his report in which he recommends that a state permit be issued with the condition that the bulkhead be constructed no greater than 10 feet seaward of the mean high water line.

The District Engineer receives a similar application in the requisite Corps format also on June 1. A Public Notice dated June 15, detailing the applicant's proposed project, is published and circulated. Copies are mailed to the National Marine Fisheries Service, the Environmental Protection Agency, and the U.S. Fish and Wildlife Service, as well as state agencies as required under the Fish and Wildlife Coordination Act.

The applicant's site is then visited by representatives of the concerned federal agencies. Two of the federal agencies subsequently submit reports to the Corps; the remaining federal agency requests an extension of the 30-day Public Notice because they were unable to submit a report during the time period. An extension of 14 days is granted to the agency by the Corps.

On July 1, one federal agency writes the Corps that the proposed pier is acceptable, but the subtidal bottom in front of the applicant's property is a highly productive habitat and should not be destroyed. Therefore, this agency recommends that the bulkhead be constructed at the mean high water line. The second federal agency is in substantial agreement with the first.

On August 17, the third federal agency recommends that the bulkhead be constructed no further than 6 feet seaward of the mean high water line. The seaward extension of the bulkhead is recommended because of recognized difficulties due to the height of the bank.

The District Engineer now must resolve the differences or he can decline to issue the permit, or issue over the objections of the consulting agency or agencies. The course he chooses is to request that the agencies meet at the site on September 10 to discuss alternatives with the applicant. It should be noted that this is the first occasion during which all agency representatives came together to discuss the application.

Resolution of the project is reached during the meeting and agency letters are subsequently submitted to the Corps. A permit for construction of the bulkhead and the pier is issued on October 29; one hundred and fifty-one (151) days after the application was submitted.

It should be emphasized that the lengthy time required for final issuance of the permit was engendered, in part, by the applicant's insistence that he be allowed to acquire additional land to the detriment of the marine habitat and the public.

Now, consider the same case handled under joint processing; the applicant submits his application to the state and the Corps on June 1. A state resource agency representative visits the site, takes photographs and notes and prepares a written report which is utilized for the state's purposes on deciding whether to issue, deny, or modify the permit.

In addition, a Corps representative will also review the application, visit the site, and submit a report to his agency.

On June 27 at the monthly joint processing meeting, the Corps representative and quite often the state representative will present the results of their investigations to the federal and state agency professionals. During this review process, the individual agency concerns are made known to the others. Because all the agencies are now dealing with the same data base and clearly understand the engineering and biological problems, they are able to reach a consensus on the applicant's project.

Subsequent to the joint processing meeting, the Corps advised the applicant that the consulting agencies recommended, with the Corps concurrence, that the bulkhead be constructed no further seaward than 8 feet from the mean high water line. The applicant accepted the condition and the permit was issued on July 27; fifty-seven (57) days after the application was submitted.

This is only one of several types of applications which are reviewed under the joint processing procedure. There is no question that a once cumbersome process has been fashioned into an efficient, timesaving procedure. The average time for review and issuance of a permit in the Mid-Atlantic and northeast section of the United States is approximately two months contrasted with about four to six months prior to joint processing.

The process has been streamlined but not at the expense of the environment. Minor cases, such as the one discussed above, no longer take an inordinate amount of time to resolve; which allows the agencies the necessary time to devote to the amelioration and mitigation of substantial impacts arising from a major project.

2. Portsmouth, Virginia Refinery

An example of a type of interagency review related to major activity is that concerning permit application for a refinery at Portsmouth, Virginia, in the Hampton Roads area. In 1975 an application was filed with the Corps to build a 250,000 barrel-a-day facility near the mouth of the Elizabeth River in Portsmouth, Virginia. Since this was a major controversial case, an EIS was prepared, published, and public hearings held. This was a lengthy and time-consuming process. Finally, in the spring of 1978, the application reached the Office of the Chief of Engineers in Washington, D.C., for a final review and decision.

During the internal review it was decided that the EIS did not contain adequate discussion of comparison of alternative sites. As a result, the Chief of Engineers asked the heads of the other federal agencies to appoint representatives to a Federal Task Force to accomplish the task. Both authors of this paper participated on that Task Force. Agencies represented were: the Corps of Engineers, National Marine Fisheries Service, Department of Energy, Environmental Protection Agency, U.S. Coast Guard, and the Fish and Wildlife Service.

It is the purpose of this paper to discuss the procedures for coordinating this review effort, not the specifics germane to the refinery application. The Task Force first met in late May of 1978. It was agreed at that meeting that the purpose of the Task Force would be to insure that the Hampton Roads site would be compared to other east coast sites and adequate information collected so that the Chief of Engineers and the public could be provided all relevant information available.

Equally important was the fact that it was agreed by all agencies, baring discovery of unforeseen or unanticipated information that the Task Force would complete its report by September, i.e., within four months.

The Task Force met on the average of twice a month during this period and made three separate flights to different sites along the east coast from Maine to Florida. A total of nine boxes of information was collected to support the review.

Meetings of the Task Force were never dull. Agency opinions were pungent, cogent, and often disparate. They were working sessions with an agenda developed prior to each meeting. There were differences of opinion, but due to the fact that a methodology and format had been agreed upon at the beginning of the summer, consensus could be reached. The final report was prepared and delivered on time.

In evaluating this activity in terms of interagency coordination, several points might be noted. Among them: the Task Force defined its plan of action early on and adhered to it throughout. The members agreed they would insure enough information was available on which to base a decision. They did not resort to delaying administrative or legal barriers. The participants were qualified, professional engineers, scientists, and planners. Finally, there was a great deal of trust and respect among the members. This did not mean agreement on what the final decision should be. That controversy still rages on. It is the opinion of the authors that had a Task Force been formed early in the application process, the time to decision would have been significantly shortened.

SUMMARY

Development of an expeditious, yet, effective permit program by the Corps of Engineers is dependent to a large degree upon the manner in which the interagency review is conducted. By 1975, due to increasing scope and volume of applications, the program was in danger of becoming a bureaucratic nightmare. Through methods such as described above, many of the obstacles to efficient processing have been reduced. There are still many things that can be done but at least efforts have been implemented to improve the system.

Obviously, what has worked in the Northeastern and Middle Atlantic may not be appropriate for other sections of the country. What is universally appropriate, however, is that a deliberate, willingness be attempted on the part of representatives of all agencies. There are other techniques available which have not been discussed in this paper.

In light of the recent Council on Environmental Quality's regulations for conducting more efficient environmental reviews through scoping, timing, etc., obviously the concept of interagency review coordination is a concept whose time has arrived.

* * *

The Authors

Major Joseph S. LeGath, USA (Ret.) recently retired from the U.S. Army, Office of the Chief of Engineers, where he served as the federal coordinator for regulatory functions. He also has been the Chief of the Regulatory Functions Branch in the Baltimore Engineer District. Presently he is employed as a consultant by Dames & Moore, Washington, D.C., and is manager of their International Port Development Planning Group. He holds a Master of Landscape Architecture degree and a Master of Technology for International Development degree from North Carolina State University.

Dr. Robert L. Lippson is a Marine Ecologist concerned with a broad range of environmental problems. He is a research coordinator for Environmental Assessment of the National Marine Fisheries Service/NOAA, Oxford, Maryland. Dr. Lippson formerly conducted research on blue crab propulation dynamics at the Chesapeake Biological Laboratory, University of Maryland. He holds a M.S. and Ph.D. from Michigan State University in Marine Biology.

SOUTH PADRE ISLAND COASTAL ZONE MANAGEMENT STUDY

W.G. Roeseler*

PURPOSE AND OBJECTIVES

The Coastal Zone Management Act of 1972 (16 USC. 1451) authorizes a wide variety of exploration and conservation measures concerning the vast coastal regions of the United States. The act and its numerous programs are administered by the Office of Coastal Zone Management (OCZM), National Oceanic and Atmospheric Administration of the U.S. Department of Commerce.

In Texas, for valid and compelling reasons, the thrust of coastal management activities has been the exploration for oil and gas resources on the continental shelf. There exists, however, another area of public concern covered by the federal statute which is the provision of beach access for recreational tourism purposes in general and for residential water front development in particular. Geologic and man-made constraints provide formidable obstacles to the orderly and rational development of beach areas along the coast line of the Gulf of Mexico in the states of Texas, Louisiana, Mississippi, and Alabama.

OCZM considers beach access in this context to be the key issue in a very broad sense, extending beyond physical participation in recreation activities and encompassing "visual, legal, social, and economic access, the barriers that inhibit them, and the tools that are available to enhance them."

The Department of Urban and Regional Planning at Texas A&M University initiated a series of pilot studies with assistance from the Department of Geology in coastal areas in order to develop general guidelines and criteria for a broader based, more comprehensive research effort to be undertaken later. The first of these studies was undertaken in an urban setting and addressed issues of Corpus Christi Beach. The present study emphasizes problems in a non-urban setting.

The South Padre Island Beach access study consisted of a comprehensive assessment of the physical-geological environment, an analysis of the man-made development; and several conceptual solutions reflecting varying conditions.

Within the scope of this study, an attempt was made to gain better insight into the interactions of natural and man-made conditions and circumstances. Also to develop in concept, clearly discernible methods of approaching each major component and the total issue that underlies any effort to improve waterfront accessibility.

Initial orientation was accomplished through a series of briefings by officials of the cities of South Padre Island and Port Isabel. A review of previous pertinent studies and of available records was included.

Research consisted of compilation of data from sources in the town of

*Head, Dept. of Urban & Regional Planning, Texas A&M Univ., College Station, TX

South Padre Island, Chamber of Commerce, Coastal Management Office of Texas, U.S. Geological Survey, records of Texas A&M University, and others. In addition, geologic and land use surveys were made to complement the data and information base.

THE SETTING

Regional Location. The community of South Padre Island is located on the Gulf of Mexico at the southernmost tip of Texas, 20 miles north of the mouth of the Rio Grande. The island is about two miles offshore from the Texas mainland. It is joined by a causeway to Port Isabel on the Mainland with Laguna Heights and Laguna Vista just north. The Laguna Madre, a shallow lagoon, lies between the mainland and the island.

South Padre Island forms part of the smooth, continuous barrier island chain that stretches from Tampico, Mexico, to the Texas-Louisiana border. The smooth continuous shore of the barrier island chain is marked in contrast to the mainland coast which is locally cut by erosion and deeply embayed coastal waters.

Climate. The climate in the area is semi-arid subtropical with an average temperature of 74°, summer average temperature of 82°, and winter average temperature of 65°. Maximum rainfall occurs in May, June, and September with an average rainfall of 26 inches.

Government. South Padre Island was incorporated in April, 1973, and operates under the aldermanic system with one mayor and five aldermen elected for two-year terms.

Economy. Tourism is the major industry of South Padre Island's economy. Almost all of the labor force is seasonal and employed in the service sector of the economy. Motels, restaurants, quick service food stores, tourist shops, fishing, boating service shops, and some shrimping comprise the economic base of the island.

In 1960 beach front property values were $100 a front foot; today beach front footage values are over $1500. Rapidly increasing property values are an indication of the desirability of owning beach front property on South Padre Island. This trend will likely continue as prospective vacationers and seasonal home buyers discover the island as an attractive tourist and recreation center.

Population. The population of South Padre Island is seasonal. Peak winter resident population occurs during the months of December, January and February. Winter resident population is more than 40 times greater than the stable population of the island.

Population estimates by the Lower Rio Grande Valley Development Council show that the island population has more than doubled between 1970 (249 inhabitants) and 1977 (596 inhabitants). Barring any major disasters, this trend is likely to continue with the prospect of more employment opportunities in the tourist industry of the island. Based on counts of causeway crossings in 1978, the total daytime population often reaches 26,000. Total island population is expected to increase as the island continues to grow as a tourist center.

A significant portion of the winter resident population comes from the midwestern United States and the harsh regions of Texas. A large number of winter vacationers from these areas are finding the warm subtropical climate and the beautiful shores of South Padre Island attractive. Hence, it is probable that winter population will increase at a considerable rate.

Moreover, there is an international market of Mexican tourists seeking U.S. coastal resort establishments as vacation spots. As a growing market and with the strategic closeness of South Padre Island to Mexico, it is conceivable that this may become a major factor in the development of the island as an international resort.

Activity Nodes. Activity nodes on South Padre Island are centers that generate commercial and/or recreational activities. They are the primary drawing points of the island's population. The island's activity nodes are two parks, two marinas, numerous hotels and motels, some commercial centers, and others. Main activity nodes consist of the following centers: Isla Blanca Park, Andy Bowie Park, and several large and luxury motels. Isla Blanca Park has a marine biology lab, picnic facilities, amphitheater, trailer hook-ups, and cabanas. Andy Bowie Park offers beach access, fishing, boating and trailer hook-up facilities.

Utilities and Streets. Sewer connections are provided by the Cameron County Fresh Water District #1. There are three contact stabilization plants with maximum capacity of one million gallons each, and maximum daily use of an average of 500,000 gallons each (Cameron County Fresh Water District #1, 1978). The sizes of the sewer lines are 6", 8", 14" and 15" in diameter. Average daily use is presently 50% of design capacity of 1.5 million gallons per day. Plans are being made by the Cameron County Fresh Water District to increase plant capacity.

Water is furnished by the Cameron County Fresh Water District #1. The water system has a maximum daily capacity of 40,000 gallons and a maximum daily use of 2 million gallons. During peak days, water demand exceeds the supply(Cameron County Fresh Water District #1, 1978). It is required that the Water District will make plans in the near future to increase capacity to meet peak demands. The sizes of the water lines are 2", 4", 6", 12" and 16" in diameter indicating an inadequate network.

The main street on the island is Padre Boulevard, a paved thoroughfare running north-south from which the east-west secondary streets branch out. Of the latter, approximately 15 are completely or partially paved, some are surfaced with shell and the remainder are sand. This last group requires constant maintenance due to the inherent instability of loose sand.

HURRICANES

Hurricanes have been a major threat to life and property along the coasts of the Gulf of Mexico. Accordingly, the hurricanes and their potential impact must be clearly understood in connection with any coastal development effort. Fortunately, catastrophic loss of life due to storms has not occurred since the 1900 hurricane which killed 6,000 persons in Galveston, Texas. Conditions are now developing, however, which lead officials to fear another major disaster. The contributing factors include the following:

- -Rapidly increasing development in low-lying coastal areas, many of which are accessible only over long stretches of exposed low highways.
- -Destruction of natural protection systems such as dunes and vegetation.
- -Massive influx of persons from non-coastal areas who fail to appreciate the potential devastation powers of a major hurricane. This problem is compounded by the fact that many persons have experienced a near miss or only a minor storm, and thus, have a casual attitude toward these storms.
- -Warning systems have improved much in recent years, contributing to complacency among both officials and the public. However, a technological plateau has been reached and additional significant improvements are not anticipated" (McGowen and others, 1975).

These potentially dangerous conditions are particularly evident on South Padre Island. Twenty years ago sand dunes and vegetation stretched unbroken along the beach. Today they have been bulldozed and in their place stand hotel, condominium, and apartment complexes. Other areas behind the remaining fordunes are relatively dense beach house developments whose structural designs are vulnerable to hurricane force winds (exceeding 75 miles per hour) and an on-rushing tide of several feet.

Hurricane Classification. During the past 70 years, most coastal areas in Texas have experienced severe weather resulting from direct impact or nearby passage of a hurricane. Not all areas, however, have experienced varying types of hurricanes and their respective physical impacts.

Using meteorological and hurricane data accumulated over the past several decades, it is possible to recognize at least four general kinds of hurricanes and to predict their impact on different parts of the Texas coast. Table 1 illustrates a system of hurricane classification (Dunn and Miller, 1964) which measures hurricane intensity through the measurement of wind speed in miles per hour, and minimum central pressure in inches of mercury. A minor, minimal, major or extreme hurricane is determined by the numerical ranges of each parameter as indicated in Table 1.

Combining knowledge of hurricane classification and the parameters listed below, it is possible to predict the physical impacts and

TABLE 1: HURRICANE CLASSIFICATION

CLASSIFICATION	MAXIMUM WINDS (MPH)	MINIMUM CENTRAL PRESSURE (INCHES OF HG)
MINOR	LESS THAN 74	MORE THAN 29.40
MINIMAL	74 TO 100	29.03 TO 29.40
MAJOR	101 TO 135	28.01 TO 29.00
EXTREME	MORE THAN 136	28.00 OR LESS

SOURCE: NATIONAL OCEANIC AND ATMOSPHERIC ADMINISTRATION, NATIONAL HURRICANE CENTER (1900-1974).

TABLE 2: HURRICANES - CARLA, BEULAH AND CELIA

YEAR HURRICANE	DATE & LOCATION OF LANDFALL	CLASSI-FICATION	APPROACH SPEED (MPH)	WIND VEL. MAXIMUM (MPH)	PRESSURE AT EYE	STORM SURGE HEIGHT	DOLLAR VALUE OF DAMAGE (MILLIONS)	DEATHS
1961 CARLA	SEPT. 11 PORT O'CONNOR	EXTREME	6	150	27.49"	22' PORT LAVACA	408	46
1967 BEULAH	SEPT. 20 E. OF BROWNSVILLE	EXTREME	8	140	27.98"	12' PORT ISABEL	200	115
1970 CELIA	AUG. 3 CORPUS CHRISTI	EXTREME TO MAJOR		130	27.80"	9.2' PORT ARANSAS	453	111

SOURCE: NATIONAL OCEANIC AND ATMOSPHERIC ADMINISTRATION. NATIONAL HURRICANE CENTER (1900-1974).

potential damage of a hurricane to different parts of the Texas coastline:

- -estuary shape
- -shoreline configuration
- -track of hurricane relative to the coastline
- -nature and distribution of physical and biological environments
- -population density

Table 2 shows the impacts of three hurricanes, Carla, Beulah and Celia, that made landfall of the Texas coastline. The table indicates when the three hurricanes hit mainland, their classification and the damages and loss of life that resulted when each hurricane made landfall. Approach speed in miles per hour and storm surge height in feet indicate the force of each hurricane.

Brief History of Hurricanes Carla, Beulah, and Celia. The following discussion is a historical account of three hurricanes, Carla, Beulah, and Celia, that have made landfall of the Texas coast over the past two decades.

Hurricane Carla was spawned in the Western Caribbean on or about 3 September 1961. She became a hurricane on 5 September and moved into the Gulf of Mexico between Cuba and the Yucatan Peninsula on 7 September (Hayes, 1967). Carla moved toward the Texas coast at 9 mph, making landfall near Port O'Connor on 11 September. Her travel time over the warm waters of the Caribbean and Gulf of Mexico was about nine days. Maximum sustained wind at landfall was about 175 mph, and pressure in the eye was about 931 millibars (mb) of mercury. The Galveston weather station was under effects of gale-factor winds for 49 hours (Colon, 1966). Corpus Christi, only 50 to 60 miles from the storm center, experienced peak gusts of 85 mph, and pressure of 977 mb. Hurricane wind diameter was approximately 300 miles. Carla was probably the largest Atlantic hurricane for which there is reliable data (Colon, 1966).

Carla was characterized by extensive storm surge flooding and severe shoreline erosion. Surge height in the Port of O'Connor area was in excess of 10 feet above mean sea level (MSL), and at Port Lavaca, the surge reached a maximum of 22 feet above MSL (U.S. Army Corps of Engineers, 1962). Parts of Matagorda Peninsula were breached by storm channels, and shorelines were eroded as much as 800 feet (Shepard, 1973; McGowen and Brewton, 1975). Dunes on Mustang Island were eroded landward as much as 150 feet (Hayes, 1967).

Carla's track across the Gulf of Mexico was northwestward. After landfall, her course curved to the northeast, and she crossed the United States and entered Canada in the Great Lakes area.

Hurricane Beulah was spawned in the Atlantic, becoming a hurricane on 7 September 1967 (Scott and others, 1967). She moved west-northwestward into the Caribbean, lost considerable energy in the mountains of Haiti, reformed and assumed a more westerly course crossing the Yucatan Peninsula on 17 September. After becoming a hurricane, her travel time over the Caribbean Sea and Gulf of Mexico was 13 days. Maximum wind velocity at landfall was 125 to 16o mph. In Texas, winds of hurricane force extended from the Rio Grande northward approximately 250 miles. Storm surge was about 10 feet above MSL at Brazos Santiago and tides were 6 to 7 feet between Port Mansfield and Port Aransas and 5 feet near Cedar Bayou (Behrens, 1969; Scott and others, 1969). After making landfall, Beulah traveled north-northwestward inland into Duval County, changed her course to the southwest, and moved back into Mexico. The long path overland slowed the storm, resulting in heavy rainfalls, in some areas rainfall was in excess of 30 inches during the four or five days of aftermath storms.

Hurricane Celia was spawned in the Caribbean Sea near Cuba. A tropical squall struck the western part of Cuba on 31 July 1970. On the morning of 1 August, the disturbance became a tropical storm, and on the afternoon of 1 August, Celia became a hurricane (McGowen and others, 1970). Celia's course was west-northwest toward the Texas coast, and her rate of forward movement was 10 to 15 mph. She made landfall at Port Aransas on 3 August; her travel time on the Gulf of Mexico was only three days. At about the time she made landfall, the eye decreased in size by about 40 percent. The wind velocity increased from 90 to 130 mph with gusts of 160 to 180 mph. The wind diameter of Celia's destructive path was about 150 miles, and her hurricane winds had a diameter of 80 miles. Celia's inland path was west-northwest to Del Rio where her progress became irregular. The storm expired in the mountains near Chihuahua, Mexico. Celia was accompanied by high velocity winds and a few tornadoes. Rainfall was minimal and storm surge was restricted to a very narrow zone. Maximum surge (determined from debris lines and, therefore, not indicative of stillwater level) was about 9 feet along the Gulf shore near the Aransas Pass jetties, 12 to 14 feet along the bay shore at Aransas Pass, and up to 9 feet at Corpus Christi Pass. Hurricane Celia was characterized by her destructive winds; storm surge, flooding and rainfall were relatively insignificant.

<u>Classification of Hurricane Hazard Zones on South Padre Island</u>. Classification of hurricane hazard zones specifically identifies different levels of exposure to the physical forces of hurricanes. Once these zones are delineated, appropriate hurricane resistant building standards can be applied that will significantly reduce hurricane damage and loss of life on a wide scale.

A first step in classifying hurricane hazard zones is to describe the potential coastal impacts of the physical forces of a hurricane in a quantitative manner. A relatively simple procedure was developed to be used in conjunction with developing and applying minimum hurricane resistant standards.

The result is four zones reflecting four different levels of exposure (McGowen and others, 1975). They are:

Zone A: -exposure
-battering with debris
-flooding

-wind (+140 mph)
Zone B: -battering with debris
-flooding
-wind (+140 mph)
Zone C: -flooding
-wind (+140 mph)
Zone D: -wind (+140 mph)

South Padre Island is entirely within Zones A and B. A brief discussion of what could happen to building structures in Zone A and B on South Padre Island during a hurricane provides insight about the magnitude of potential disaster.

Zone A: A hypothetical situation, for example, may occur on a barrier island and near the beach where the full fury of a storm strikes. Much of the area is likely to be 15 feet above sea level, and flooding is very possible. The waters could be moving at the velocity of several knots and be topped with violent waves of 5-10 feet. The water mass itself can exert a significant force, but this is compounded greatly with floating objects such as boats, vehicles, parts of other structures, etc., which in essence become dangerous missles. Few, if any, residential structures could be expected to survive the impact of a one-ton object moving at ten feet per second (seven miles per hour). This violent water action can also be a very effective "ditch digger" and cause scour around foundations, walls, etc., and undermine structures that otherwise survive. While this condition will usually occur on a barrier island, or on a Gulf front area of the mainland where there is no barrier island, it could also occur on the shores of the major bays.

It is important to note that many building structures on South Padre Island have been precariously located in Zone A, several of which are high investment hotels and beach front homes. Hence, it is evident that many of these structures will suffer heavy damage when the next major hurricane hits in the vicinity of the island.

Zone B: A hypothetical example of Zone B site may occur at the middle part of a barrier island or just inland from a bayshore on the mainland, where the site is not at an exceptionally high elevation, and away from a washover channel. Such a place would be subject to the same destructive forces as Zone A.

Standards for Construction. Adopting standards for building construction as part of a city's building code is a necessary course of action to minimize damage and loss of life in identified hurricane hazard zones. For the most part, South Padre Island has regulated development under a county zoning ordinance adopted by the city. The ordinance standards are in need of re-evaluation, however, in order to meet the needs of the future, to foster conservation and minimize property damage. Otherwise, building structures will continue to be erected which are considered to be "blowaways" (Burka, 1978).

A "Model Minimum Hurricane Resistant Building Standards for the Texas

Gulf Coast" has been developed by the General Land Office of Texas (GLO, 1976). These standards will vary in accordance to the type of classification (A, B, C, or D) of a particular hazard zone.

GEOLOGY OF SOUTH PADRE ISLAND

While not as obvious and dramatic as hurricanes, other natural processes severely impact coastal zones. To ensure prudent and safe land use, environmental problems arising from natural systems and human modification of these systems must be identified. An awareness of coastal processes and the geological response to man's activity is crucial in guiding responsible development. On South Padre Island, daily shoreline erosion and the island's susceptibility to destructive wind and water from major storms are the two most important problems that must be dealt with.

The shoreline of Texas and northeast Mexico forms a broad, arc-shaped boundary between the Gulf of Mexico and the Gulf Coastal Plain. Five hundred miles of the 650-mile long coastline between Tampico, Mexico, and the Texas-Louisiana border consists of barrier islands. The smooth, continuous shore of the barrier island chain is in marked contrast to the mainland coast which is locally cut by erosion and deeply embayed by coastal waters.

Recent events and processes, relative to geologic time, have controlled the development of the shoreline. A drop in sea level due to glaciation 30,000 years ago brought about entrenchment of streams and erosion of their valley systems of the old coastal plain. As the glaciers melted and sea level began to rise, the coastal lowlands were flooded, thus producing the present bays and estuaries (LeBlanc and Hodgson, 1959).

The gulf coastal plain of Texas is mainly drained by nine rivers. There is a direct relationship between the size and sediment load of each river and the shape of the shoreline at the mouth of the river. The larger rivers (Brazos, Colorado, and Rio Grande) have developed deltaic plains which reach into the Gulf. The smaller rivers have yet to fill their corresponding inlets and are in the process of constructing small deltas at the head of shallow bays.

The Rio Grande distributary system transported sufficient material to allow progradation of the shoreline seaward of its present position. Subsidence of the delta, lateral shifting of the rivers, and flooding of the abandoned sub-deltas have resulted in the present configuration of the coastline. Wind activity and the dry sub-humid to semi-acrid climate of the region are important in the movement of sand and the development of the barrier island chain. Prevailing winds from the southeast are the driving force for the longshore currents carrying sediment to the barrier from the south. The sediment of Padre Island and Brazos Island is composed of fine sand which contains abundant volcanic rock fragments and volcanic quartz. This sand was derived from the volcanic areas of West Texas and Mexico 600 miles upstream from the mouth of the Rio Grande (Bullard, 1972).

The barrier island can be divided into several geomorphic provinces. These are the beach, the foredune ridge, the vegetated flats and the wind-tidal flats. On South Padre Island, the beach is narrow (about 150 feet wide) and steep in areas with a foredune ridge. The beach profile is controlled by wave action. Although daily changes in the beach appearance reflect changing conditions (wind direction and velocity, wave height), the beach conditions may be similar over long periods of time except during following major storms (Morton and Pieper, 1975). The foredune ridge is 15 to 20 feet high and is stabilized by vegetative cover. The low-lying areas behind the foredune ridge are the vegetated flats. These grade lagoonward into the wind-tidal flats which have sparce vegetation. The wind-tidal flats are covered periodically by the hypersaline lagoonal waters.

In areas where the foredune ridge has been breached by storm action, the beach is wide (about 200 feet) and gently sloping. These channels are easily flooded by successive storms and usually have little vegetative cover due to their frequent inundation. Back island dunes, or shifting sand dunes, are highly mobile. These dunes are formed after a storm, when an opening is cut in the foredune ridge. Beach sand is blown through the chute where it accumulates into dunes. Once the chute has been closed by natural processes, the dunes begin to migrate across the island in the direction of the prevailing winds (i.e., towards the northwest) (Mathewson, and others, 1975). Maintenance crews are constantly removing this sand off Park Road 100 several miles north of Andy Bowie Park.

Barrier islands bear the brunt of storm surge floods during hurricanes. The most destructive storms on record to strike South Padre Island and its immediate vicinity were the storm of 1933 and Hurricane Beulah in 1967. All of the geomorphic provinces mentioned above are affected by major storms. Beaches may be severely eroded and regraded to gentle, almost planar, slopes. The foredune ridge is eroded by breaking waves. Storm surge floods breach the foredune ridge in some areas, creating washover channel (Hayes, 1967). During storms, sand and water are transported through these channels into the back-island areas. Although they may be inundated, vegetated flats do not normally experience erosion. Vegetative cover is an important factor in stablization of back-island areas. The 1933 storm caused extensive erosion on South Padre Island due to the combined effects of a winter freeze and a summer grass fire which obliterated all vegetation on the island before the hurricane struck. Both the 1933 storm and Hurricane Beulah were accompanied by surges of 12 to 15 feet (Morton and Pieper, 1975). Low elevations and occurence of many breaches in the foredune ridge on South Padre Island greatly increase the chances for back-island flooding even with the occurrence of moderate storms. In addition to the high water, canals and channels may be filled in with sand and debris transported by the storm surge. The best protection for the back-island areas is a continuous foredune ridge and maximum vegetative cover on dunes and flats. It should be pointed out that the entire island will be inundated when a hurricane strikes. With proper planning, however, the destructive force of the storm surge may be diminished by a stable and continuous foredune ridge.

Shorelines are very dynamic environments. Shoreline changes result from natural phenomena such as tides, storms, sediment supply, and sea-level changes. Wind direction controls the angle at which waves attack the beach as well as the direction of the longshore currents. Waves both erode and deposit sand along the beach and most coastlines undergo short-term erosion and accretion. Erosion produces a net loss in land and accretion produces a net gain in land. It is the long-term effect of wave action which governs whether a beach is eroding or accreting. Longshore currents (or littoral drift) are currents that are aligned parallel to the coast. ERTS satellite photography of South Texas indicates that the longshore current carries sediment from the mouth of the Rio Grande northward along Padre Island. As long as sediment is available offshore for waves to deposit on the shoreline, the beach will undergo accretion. However, if for some reason the flow of longshore currents carrying sediment parallel to the coast is inhibited or interrupted, then erosion will occur. Historical air photography from the National Oceanic and Atmospheric Administration and the U.S. Department of Agriculture indicate that South Padre Island is undergoing long-term erosion.

Sea-level rise is one reason for the present trend of erosion along South Texas beaches, however, a deficit in sediment supply is the most important cause of accelerated erosion along South Padre Island (Morton and Pieper, 1975). The Brazos Santiago Pass jetties constructed in 1935 to maintain a navigable channel, have entrapped much of the sediment flowing north from the mouth of the Rio Grande. In addition to the jetties, the completion of Falcon Dam on the Rio Grande in 1953 seriously decreased the discharge of suspended sediment into the Gulf. Although most of the island is eroding, the southern one to two miles of South Padre Island is accreting due to an eddy effect produced by currents flowing around the north jetty and the northeasterly attack of winter waves.

In summary, prevention of shoreline erosion requires cautious planning. "The trouble with man-made structures is that they are rigid and assume a static environment, whereas coastal processes are not static, but highly dynamic", (Mathewson, 1974). Seawalls do not prevent erosion of the beach; they merely protect the back-island areas. In many cases, a newly constructed seawall has accelerated erosion of the beach as well as property bordering it. The construction of groins will stabilize a beach assuming enough sediment is available in the littoral drift system. On South Padre Island, however, the jetties along Brazos Santiago Pass have entrapped most of the sediment transported by the northward moving longshore current.

As long as nothing is done to retard the erosion on South Padre Island, the problem will become increasingly difficult to solve. The Bureau of Economic Geology in Austin estimates that the island may be eroding at rates of 10 to 13 feet per year. This is a long-term trend, of course, and the figures may or may not be a valid estimate for the next 10 to 20 years. The U.S. Army Corps of Engineers in Brownsville has plans to extend the jetties another 900 feet into the Gulf. This would further restrict sediment flow to the island. Without renourishment of the sand, the beach will simply disappear.

It is also important to note that dozing down the foredune ridge for hotel construction is another example of what simply can be termed total disregard for coastal processes. Recently, tropical storm Amelia caused extensive washouts and backed up sewer lines with its heavy rains, although it was only a moderate storm. Hurricanes are the greatest storms on earth. A continuous foredune ridge and a wide beach are essential if the back-island areas are to escape the devastation caused by the tremendous surge of storm waves. Even with this protection, the island will still become inundated. Disasters can be avoided and the costs of destruction can be kept to a minimum if man can recognize the danger and is willing to act accordingly.

Stabilization of Shoreline Erosion and Prevention of Foredune Destruction. In light of this geologic analysis, consideration should be given to stabilize shoreline erosion and prevent foredune destruction as follows:

1. Renourish the depleted sediment supply by distributing sand along the beach and foredune ridge. The sand should be placed in low spots in the foredunes and then stabilized by vegetation to strengthen the foredune ridge. Sand should also be placed along the beach and allowed to be reworked by the waves to make up for the sediment being taken out of the littoral drift system by the jetties. On South Padre Island, there are several sources from which the sand could be taken:

   a) The shifting dunes blowing across Park Road 100. Presently, maintenance crews are removing this and using it to fill low areas on the mainland.
   b) Maintenance dredging of the back-island canals and channels. It may be feasible to pump sand directly onto the beach.
   c) There are many construction pits, due to the rapid development of the island, that are sources of sand.
   d) Maintenance dredging of Brazos Santiago Pass.

The following ordinances, as related to Texas Coastal Legislation, are recommended for adoption by South Padre Island:

2. Regulations prohibiting development in the beach area extending from the line of mean low tide of the Gulf of Mexico to the line of vegetation bordering on the Gulf of Mexico, or to a line 200 feet inland from the mean low tide, whichever is nearer the mean low tide, or some other appropriate demarcation which would accomplish the desired objectives.

Monitoring of beach erosion or accretion would be critical to the implementation of these legal requirements.

3. Establishment of a dune protection area for purposes of preserving remaining sand dunes which offer a defense against storm waters and erosion of the shoreline. These areas should be vegetative dunes.

4. Regulation of the use of motor vehicle traffic in the beach and vegetative dune areas. Beach access points should be designated in

areas susceptible to flooding where there are no foredunes.

5. Regulation of excavation, taking, removal and carrying away sand, marl, and shell from the island. Substantial damage to critical foredune areas can occur where there are no strict guidelines for excavation.

A PROGRAM FOR DEVELOPMENT

The foregoing exploratory investigations suggest that the South Padre Island area must be carefully planned if financial disappointment and potential loss of life and property are to be avoided or at least minimized. To this end, further analysis of geologic processes are needed and particularized site studies required to produce effective standards of guidance for growth. These studies will be complex for the reason that no absolute certainty exists with respect to the future physical form which this section of the barrier island is likely to assume. The study team outlined three scenarios, each based upon a different geologic framework. These notions do not represent formal plans or recommendations. They merely illustrate the issues of the area which are largely beyond man's control, but which nevertheless must be faced in any realistic approach to the stabilization of the area.

Scenario 1 - Island Remains Intact. Scenario 1 assumes that there will be an indefinite continuation of existing haphazard development and a continued disregard of geological processes. No major alterations to existing design concepts are considered. Recommendations of improved vehicular and pedestrian movement, more public access points to the beach, (i.e., parking facilities and parks), landscaping along major boulevards, and containing development along the existing street pattern are proposed.

Scenario 1 is based on a continuation of the following existing conditions: a) unpaved streets, b) major hotels and condominiums are built without much protection, c) marginal development along South Padre Boulevard, d) beach houses continue to proliferate without storm resistant structural designs, e) foredunes and vegetative cover are destroyed, and h) services remain low. Moreover, it was assumed that the existing causeway which has severe settling problems can be kept in acceptable condition.

In this context, the following improvements might be explored:

1. Isla Blanca Park, a major public research area, is located at the southernmost tip of the island. It is recommended that the construction of a large saltwater pool in the park would allieviate the problems inherent to swimming along the Texas coast. The murky water, treacherous under currents, dangerous jellyfish, natural and man-caused oil slicks, and seaweeds are not inducive to swimming. The saltwater pool would be enclosed by the jetty extending to the east and the sand beach which runs along the park. An arc-shaped seawall would provide a safe, clean and attractive swimming environment. The saltwater pool would be a major attraction

to the area.

An advantage of the saltwater pool is in the shoreline erosion protection that it would provide the beach and amphitheater immediately behind it. The enclosing break-water would isolate the enclosed beach from wave attack and currents. However, further engineering studies are required to determine design details of such a pool.

In addition to the existing cabanas, trailer hook-ups, picnic facilities, marine biology lab, and amphitheater, such improvements as pedestrian walkways, bicycle pathways, parking and tennis courts would make the park an attractive public amenity.

2. Andy Bowie Park, located on the northern edge of the city, has public beaches located on the south and east side of the park. Increasingly the number of parking facilities in the park will provide the community with more opportunities to use these beaches. With additional pedestrian walkways, bicycle pathways, and landscaping improvements, in addition to existing camping, fishing and boating facilities, Andy Bowie Park would provide a wide variety of beach related recreation activities as well as being a beautiful public facility.

3. Landscaping of Padre Boulevard, Gulf Boulevard, Laguna Drive, and other secondary streets will improve the general appearance of the island's urban setting. Trees and plants contribute environmental beauty and positive solutions to some pollution, soil erosion, temperature variation, headlight and sun glare, and other problems.

4. All secondary streets with a dead-end on the Laguna Madre side should be redesigned as cul-de-sac streets with a 50 foot radius. This will improve vehicular movement.

5. A 100 foot right-of-way for Padre Boulevard and an 80 foot right-of-way for Gulf Boulevard should be designated. Generous rights-of-way will protect vehicular turning movements, improve pedestrian and bicycle movement and safety, provide additional space for landscaping, and would give the city the option of on-street parking during off-peak traffic hours. Moreover, public transportation can be accommodated here more readily than an undersized right-of-way.

6. A proposed pedestrian and bike trail with preferential bike and pedestrian crossings approximately every 3 to 5 blocks would be desirable. This would be a shared pedestrian-bicycle right-of-way.

<u>Scenario 2 - Washover Areas</u>. Scenario 2 is based on the physical capacity of the island in addition to the open space and circulation ideas of Scenario 1. The processes of shoreline erosion, wind and wave destruction, and inundation are intregrated into this scenario. It was developed under the following assumptions: a) destruction of buildings will occur in washover channels; b) areas of severe erosion and accretion continue to exist; c) complete inundation of the island is experienced when a severe hurricane strikes; d) temporary

interruption of island traffic falters; and e) sewer back up and water shortages occur.

To minimize these impacts the following improvements might be considered:

Washover areas susceptible to flooding should be designated as open space areas that will provide opportunities for beach access and serve as passive transitional zones between developed areas. The improvements considered under Scenario 1 apply to Scenario 2 as well.

The low areas act as relief valves for water passage during intensive storms by reducing water pressure from waves and storm surge against the island. Any development in these areas creates backwash, increasing flooding and wave damage to other parts of the island. Extensive damage to structures located in the washover areas will occur as well. There is substantial development already located in these areas. Hence, it is desirable that future development be curtailed.

Residential development should be located on the southern part of the island where beach accretion safeguards long-term development. Medium and low residential density development is located in areas of the mid to southern sections of the island where beach erosion is occuring and the possibility of flooding and accompanying wind and wave destruction will increase over time.

Renourishing the depleted sediment supply is required by distributing sand along beach and foredune ridge. Sources of sand are as follows:

a) Shifting dunes blowing across Park Road 100.
b) Maintenance dredging of back-island canals and channels. It may be feasible to pump the sand onto the beach.
c) Due to rapid development of the island, there are many construction pits that are possible sources of sand.
d) Maintenance dredging of Brazos Santiago Pass.

A realistic engineering feasibility study should be made to determine projected needs of the island's future sewer and water requirements. During peak days, water demand substantially exceeds the supply.

Beach monitoring program would be desirable that would identify beach accretion and erosion zones. Knowledge of the location of these zones is critical to the implementation of beach set-back development requirements. Moreover, the city should adopt higher performance standards for building construction as part of the building code in order to assure minimum damage and loss of life in identified hurricane hazard zones. It should also consider evacuation systems.

Scenario 3 - Island Cuts. Scenario 3 includes conditions of Scenario 2, but in addition considers the occurence of major flooding and erosion in washover areas. Potentially, this condition might divide South Padre Island as it exists today into four small, partially separated islands. The consequences of such disasters are difficult to anticipate. In any event, state and local authorities should be keenly aware of the issues at South Padre as crucial development decisions have lead to public subsidy requirements at the expense of taxpayers and property insurance policy holders.

REFERENCES

Bullard, F.M., (1942), "Source of Beach and River Sands on Gulf Coast of Texas," Geological Society of America Bulletin, V. 53.

Burka, Paul, (1978), "Trouble in Paradise," Texas Monthly, Vol. 6.

Cameron County Water District Number 1, Port Isabel, Work Papers and Utility Data.

LeBlanc, Rufus J., and W.G. Hodgson, (1959), "Origin and Development of the Texas Shoreline," GCAGS Trans., V. 9.

Lower Rio Grande Valley Development Council, (1978), McAllen: Population Projection.

Mathewson, C.C., (1974), "The Physical Environment and Coastal Development" or "Beware the Planned Disaster" in Recreational Land Use and Coastal Zone Management - Issues and Perspectives in Texas, Proceedings of the Conference on Recreational Land Use on the Texas Coast, Compiled by R.B. Ditton and J.L. Seymour, Department of Recreation and Parks, Texas A&M University.

Mathewson, C.C., J.H. Clary, and J.E. Stinson, II, (1975), "Dynamic Physical Processes on a South Texas Barrier Island - Impact on Construction and Maintenance," IEEE Ocean 75.

McGowen, J.H., C.G. Groat, L.F. Brown, Jr., W.L. Fisher and A.J. Scott, (1970), "Effects of Hurricane Celia: A Focus on Environmental Geologic Problems of the Texas Coastal Zone," University of Texas at Austin, Bureau of Economic Geology Circular.

Morton, Robert A and Mary J. Pieper, (1975), "Shoreline Changes on Brazos Island and South Padre Island (Mansfield Channel to Mouth of the Rio Grande) - An Analysis of Historical Changes of the Texas Gulf Shoreline," University of Texas at Austin, Bureau of Economic Geology Circular.

Port Isabel - South Padre Island Chamber of Commerce, (1977), Population and Economic Statistics.

Roeseler, W.G., (1976), General and Principles for Prototype Zoning Ordinances and Related Measures, EMR Publications.

_________. (1978), "Corpus Christi, Texas, An Urban Beach Access Study," Proceedings of the Symposium on Technical, Environmental, Socioeconomic and Regulatory Aspects of Coastal Zone Management ASCE/San Francisco, California/March 14-16, 1978.

The Texas Coastal and Marine Council, (1976), Model Minimum Hurricane Resistant Building Standards for the Texas Gulf Coast, General Land Office of Texas.

Texas Coastal and Marine Council, Texas Coastal Legislation, (1977).

THE SAN DIEGO REGIONAL COASTAL ACCESS STUDY

Dennis A. Prescott

The California Coastal Act of 1976 contains a number of policies which require that current levels of recreational accessibility to and along the Coastal Zone be maintained or enhanced without implementing new transportation facilities which might cause environmental disruption. The purpose of the San Diego Regional Coastal Access Study was to develop a new element of the Regional Transportation Plan which would assure compliance with this new policy directive.

The study, completed in the fall of 1978, was conducted in three phases. The first phase was designed to obtain information on the conditions and factors affecting regional access to coastal areas. The second phase of the study involved the detailed examination of four case study sites. Each case study site was selected because it offered unique characteristics. The case study approach was selected for the San Diego Regional Coastal Access Study as a means of identifying access problems common to many areas in the Coastal Zone. It also serves as a mechanism for recommending solutions to prototypical coastal transportation problems.

Subsequent to the investigations at each of the case study areas, an analysis was conducted in order to determine the types and magnitudes of recreational access problems for each site. A set of recommendations keyed specifically to those access problems were developed.

PHASE I: DATA BASE DEVELOPMENT

The San Diego region's beaches and climate have been a major influence in its economic and social development. Tourism, drawing on regional, national and even international markets, is the third most important source of income and employment to the region. Furthermore, many of the area's permanent residents have been attracted to the region because of its excellent natural recreational facilities—a factor that contributes to the above-average participation in recreational activities.

Coastal Resource Areas

For this study, the region was divided into nine zones as shown in Figure 1. Five of these are non-coastal areas which were used to find out where people are coming from when they visit one of the four coastal zones on recreational trips. The number of resource areas and the total amount of annual recreational visitation to each of the coastal zones is as follows:

1. The North County Coast Zone, stretching from the Orange County border through Carlsbad, contains three state beaches and six

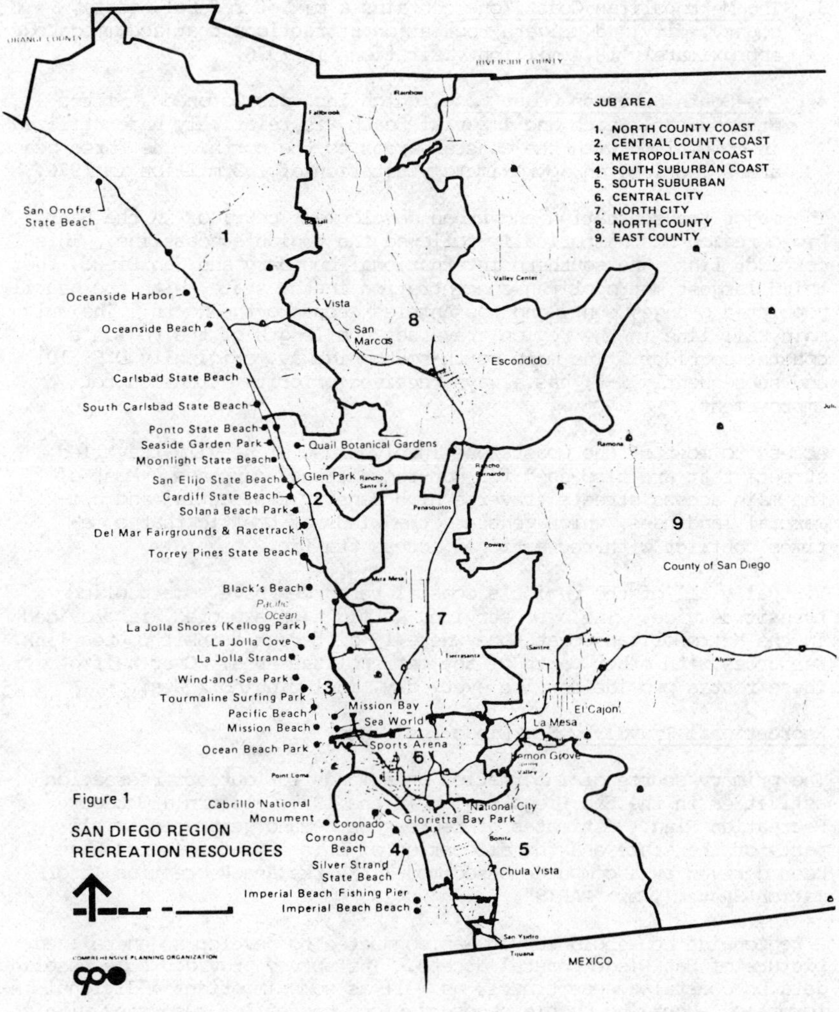

Figure 1
SAN DIEGO REGION
RECREATION RESOURCES

county or local resources that accommodated a total of about 4.3 million visitations in 1976.

2. The Central County Coast Zone contains six state and six county recreation facilities that accommodated a total of about 4.5 million visitations in 1976.

3. The Metropolitan Coast Zone contains a mix of federal, state, county, city and private recreation attractions that accommodated approximately 18.4 million visitations in 1976.

4. The South Suburban Coast Zone, which includes Coronado, Silver Strand State Beach and Imperial Beach, is relatively underutilized in comparison with the coastal areas to the north. The three beach areas received an approximate visitation of 1.3 million in 1976.

The major transportation and urban development corridor in the San Diego region has historically followed the region's coastline. This corridor links the southern international boundary and San Diego, the third largest metropolitan concentration in the state, with the heavily populated Orange County and Los Angeles areas to the north. The only main rail line in the region traverses the length of the region's coastal corridor. The main north-south highway, originally U.S. 101 and subsequently I-5, has always received priority attention for improvement.

Access to most of the Coast Zone involves use of heavily travelled streets that are also used for non-recreational purposes. Most of the main access streets traverse high density residential and commercial land uses, which generate the kinds of traffic that sometimes conflict with recreational access traffic.

Virtually all of the region's coastal resources have some public transit service. The best service, as might be expected, is provided in the Metropolitan Coast Zone where 13 different transit routes link resources with other parts of the metropolitan area. Over half of these routes provide service every day, including holidays.

Recreational Travel by San Diego Residents

The primary source of information on demands for outdoor recreation activities in the San Diego region is the 1975 California Outdoor Recreation Plan. Estimates of recreation demand generated in all parts of the state and the allocation of this demand to zones have been derived by a computer-based model, "Parks and Recreation Information System", or "PARIS".

A regionwide household survey was conducted to develop a generalized picture of San Diego coastal access. The survey provided some baseline data on coastal access travel, as well as some important additional information about why people choose various recreation resources within the region.

Thus, both the recreationists' origins and resource locations were allocated into one of nine zones in the region. Table 1 is a trip table for September, 1977, recreational trips to the Coastal Zone developed out of the data from the household survey. It is probably no surprise that the zone with the largest population, Central City, generated the largest aggregate number of recreational trips. It normally would be expected that the most populous zones would generate the greatest number of trips to recreational resources. However, it would be expected also that zones closer to the coastal area would generate considerably higher coastal trips per capita than inland zones.

TABLE 1

PRESENT MONTHLY RECREATIONAL TRIPS TO COASTAL AREAS
(Thousands of Trips by San Diego Residents)

	Destination Zone				
Zone of Origin	North County Coast	Central County Coast	Metro-politan Coast	South Suburban Coast	Total
1 - North County Coast	20	11	4	0	35
2 - Central County Coast	4	40	4	0	48
3 - Metropolitan Coast	4	22	137	2	165
4 - South Suburban Coast	2	13	16	36	67
5 - South Suburban	0	0	31	24	55
6 - Central City	1	8	199	16	224
7 - North City	3	12	145	9	169
8 - North County	33	20	22	0	75
9 - East County	6	20	137	4	167
TOTAL	73	146	695	91	1,005

Source: Household Survey, September 1977.

Private automobiles, trucks or other recreational vehicles were employed as the mode of access for approximately 80 percent of the total recreational trips to coastal zones. Table 2 shows a breakdown of transportation mode for each of the four recreation zones. The four-zone total column has been normalized so that it adds up to 100 percent. The next highest frequency of modal use is walking, at about 11 percent, followed generally corroborated by the on-site case study survey results.

People who use public transportation for access to coastal recreation areas constitute a minority of recreational participants in the San Diego region. According to the household survey, about four percent of the respondents indicated some use of bus systems for recreational access; a figure of this general magnitude is also confirmed by the on-site case study surveys.

TABLE 2

MODE OF ACCESS

	Percent of Respondents Employing Mode				
Travel	North County Coast	Central County Coast	Metro-politan Coast	South Suburban Coast	Normalized Total
Auto, etc.: Drive	60	67	67	74	57
Auto, etc.: Passenger	39	32	33	26	22
Walk	18	13	12	8	11
Bicycle	10	9	8	8	6
Bus	5	2	6	12	4

Source: Household Survey, September 1977.

Table 3 shows selected characteristics of transit riders who used the bus for recreational trips. Compared with the population at large, people who use transit for access to recreational resources are typically very young or very old. The age breakdowns are substantiated by occupational responses, in which almost half of the recreational bus users listed "student" as an occupation, and nine percent were retired. Tourists, at 18 percent, also account for a significant share of bus ridership.

TABLE 3

CHARACTERISTICS OF TRANSIT RIDERS
ON RECREATIONAL TRIPS TO COASTAL AREAS

Characteristics		Percent of Respondents Recreational Trips	All Trips
Age:	24 and Under	71	51
	60 and Over	13	12
Occupation:	Student	45	33
	Retired	9	9
	Armed Forces	6	4
Status:	Tourist/Visitor	18	8
Ownership of Vehicle in Household		78	60

Source: Transit Survey, 1977.

Resource Choice Factors

The way in which various people select which coastal recreation resource they will visit is important to any examination of recreational travel behavior. Accordingly, respondents to the household survey were asked to place a judgmental rating of very important, important, not so important or unimportant to a series of factors that influence the resource choice decisions. The responses to these questions are summarized in Table 4. As might be expected, the overall quality of the beach resource, however judged, is the most important factor in choice of recreation resources visited. What might not have been so obvious is that the next three ranking factors all relate more directly to access and transportation than they do to the resource itself. Accessibility questions generally ranked ahead of resource-related factors that normally would be considered to be quite important—availability of recreational facilities, water quality and availability of lifeguard service.

TABLE 4

RECREATIONAL RESOURCE CHOICE FACTORS
(San Diego Residents)

Factor	Importance Ranking (Most Important = 100)
Beach Quality	100
Parking Availability	83
Easy to Get to	79
Lack of Traffic Congestion	71
Availability of Public Facilities	66
Availability of Lifeguard Service	60
Water Quality	59
People	53
Familiarity with Beach	40
Availability of Shops and Restaurants	14
Pets Allowed	12

Source: Household Survey, September 1977.

One interpretation of these data is that recreational demand patterns can be altered as much or more by alterations in transportation supply than by alterations in resource supply. These findings also indicate that, to a degree, recreational access problems can be self-correcting, or at least self-limiting.

If accessibility and parking availability are as important as indicated by these responses, substantial numbers of recreationists will search for less crowded and more easily accessible resources automatically when their traditional resources are perceived as becoming too crowded or inaccessible.

Future Recreational Travel

According to the demand forecasts provided in the PARIS Model, recreational participation in coastal areas should increase at a compound annual rate of approximately 3.2 percent per year through 1990. Of this total growth, about one third is accounted for by an increase in per capita recreational participation, and the remaining two thirds are accounted for by inccreased local and regional populations. Under these conditions, total trips to coastal recreation areas during a summer month should increase from 1.00 million in 1977 to 1.55 million in 1995.

Table 5 shows that the highest growth increments and highest growth percentages in recreational trips generally correspond with those zones where total population growth and growth rate are expected to be greatest. In terms of recreational visitation, the Metropolitan Coast remains the biggest recipient of coastal recreation trips in the

County; it also shows the greatest absolute gain, at 315,000 trips. The North County Coast will show the highest rate of growth—1995 recreational visitation trips are expected to be more than double the figure for 1977. The Central County Coast will show a relatively high growth rate (80%) and a relatively high absolute growth increment of 116,000. The South Suburban Coast will show a relatively low total increment and a growth rate only slightly above that in Metropolitan Coast.

TABLE 5

MONTHLY RECREATIONAL TRIPS TO COASTAL AREAS IN 1995
(Thousands of Trips by San Diego Residents)

	Destination Zone				
Zone of Origin	North County Coast	Central County Coast	Metro-politan Coast	South Suburban Coast	Total
1 - North County Coast	39	22	7	0	68
2 - Central County Coast	8	88	9	0	105
3 - Metropolitan Coast	5	30	188	2	225
4 - South Suburban Coast	3	19	23	51	96
5 - South Suburban	0	0	54	40	94
6 - Central City	1	9	241	20	271
7 - North City	5	20	246	15	286
8 - North County	78	46	51	1	176
9 - East County	8	28	187	6	229
TOTAL	147	262	1,006	135	1,550

Coastal Access Problems

As a result of the Phase I Data Collection activities, a generalized assessment was made of the kinds of transportation problems encountered when one attempts to access coastal resource areas.

The following discussion summarizes the problems that were identified. These problems are grouped into four broad categories: (1) parking, (2) resource area congestion, (3) line-haul congestion, and (4) public transportation. They are discussed below and presented in order of severity for each resource area in Figure 4.

Parking

Inadequate parking is the most frequently observed recreation access deficiency. The result is that visitors to recreation resources park elsewhere. This parking pattern can cause a number of problems:

o In some cases, recreationists are forced to park in residential or commercial areas. This inflow of recreational parking creates conflicts and competition for limited parking spaces.

  In less developed areas, overflow parking from recreation resources can take the form of unauthorized parking along the side of highway access routes. In some instances, this type of parking can restrict the capacity of these highways. Also, parking on major access routes can lead to serious safety problems as recreationists attempt to walk along or cross high-speed highways that have no pedestrian facilities. Highway parking also often conflicts with bicycle lanes.

o Parking that is remote from the resource results in severe inconvenience to recreationists, particularly those who wish to bring recreation equipment as picnic baskets, ice chests and surfboards.

Inability of on-site parking facilities to accommodate peak loads was observed at all of the major recreational resources except for the Silver Strand, Imperial Beach and Solana Beach. Severe peak-period conflicts between recreational parking and commercial/residential parking were observed at Oceanside, Carlsbad, Black's Beach, La Jolla, Pacific Beach and Mission Beach. Cardiff was also cited as a severe problem, primarily because of extensive use of the Pacific Coast Highway for parking. Occasional or moderate parking problems were observed at Leucadia, Encinitas, Del Mar Beach, Torrey Pines, Mission Bay, Ocean Beach, Point Loma and Coronado.

Resource Area Congestion

Vehicular traffic caused by people who are coming to or from recreation areas, or searching for off-site parking spaces, can often result in serious congestion of streets used for internal circulation within recreational zones. This problem is particularly severe when the same street network is used to accommodate high volumes of recreational traffic as well as traffic generated by local residents and local commercial/retail activities.

Severe resource area congestion of this type was noted at Oceanside, La Jolla, Pacific Beach, Mission Beach and Mission Bay. In all of

FIGURE 4

COASTAL ACCESS STUDY
RECREATION ACCESS PROBLEMS: 1977

	Severity of Access Problem			
Resource	**Inadequate Parking**	**Resource-Area Congestion**	**Line-Haul Congestion**	**Inadequate Public Transportation**
Oceanside	●	●	◉	○
Carlsbad	●	◉	○	○
Leucadia	◉	◉	○	◉
Encinitas	◉	◉	○	◉
Cardiff	●	○	○	◉
Solana Beach	○	○	○	◉
Del Mar Beach	◉	○	◉	◉
Del Mar Fairgrounds	◉	●	●	◉
Torrey Pines	◉	○	○	●
Black's Beach	●	◉	◉	●
La Jolla	●	●	●	○
Pacific Beach	●	●	●	○
Mission Beach	●	●	●	○
Mission Bay	◉	●	◉	○
Ocean Beach	◉	◉	◉	◉
Point Loma	◉	◉	◉	○
Coronado	◉	○	○	◉
Silver Strand	○	○	○	●
Imperial Beach	○	○	○	●

● **Severe or Frequent Problems**
◉ **Moderate or Occasional Problems**
○ **No Problems or Minor Problems**

these cases, streets used for recreation circulation are also heavily used for other types of local and/or through traffic. The problem is particularly acute in La Jolla, Pacific Beach, Mission Beach and Mission Bay areas, where the roads used for internal recreation traffic circulation are also heavily used for urban through traffic and for local shopping trips.

Resource area traffic congestion was also noted near Del Mar Fairgounds. Here, the problem is aggravated by the severe peaking associated with the opening and closing times of the special events held at the fairgrounds. Modest or infrequent resource area congestion was noted in Carlsbad, Leucadia, Encinitas, Ocean Beach, Point Loma and Blacks Beach.

External Access Congestion

Major arterial highways that connect recreational zones with residential zones are congested by recreation traffic less frequently than local roads within recreation areas. Nevertheless, recreational traffic occasionally can become sufficiently great enough to congest even high capacity arterial highways.

The most notable example of external access or line-haul congestion was near the Del Mar Fairgrounds where significant traffic on Interstate 5 was observed at the major off-ramp. Similar congestion was noted in La Jolla, Pacific Beach and Mission Beach because, in these areas, the major external access routes are also used for internal circulation traffic.

In general, with its excellent high-capacity interstate highway system, the San Diego region is less susceptible to line-haul congestion than many other parts of the country. Still, modest or occasional congestion problems were observed in Oceanside, Del Mar Beach, Mission Bay, Ocean Beach, Point Loma and Blacks Beach.

Public Transportation

Public transportation was considered to be inadequate when the following conditions were encountered:

- Service frequencies were less than two buses per hour.
- One or more transfers were required for a trip from residential areas to recreation resources.
- Public transportation was overcrowded and unpleasant for recreational trips.

Torrey Pines, Silver Strand, Imperial Beach and Blacks Beach were found to be the most inadequately served by public transportation at the present time. In the case of Torrey Pines, there is no provision for a stop to serve recreational traffic, even though service passes near the beach. Except for residents in the immediate vicinity, recreationists would have to transfer at least once to reach Silver or Imperial Beach by public transportation.

In general, the coastal area from Leucadia to Del Mar was found to have only moderately useful public transportation for recreational purposes. The bus routes serving this area run north-south, essentially interconnecting the various resources rather than connecting resources with residential areas.

The best public transportation services were observed in the heavily urbanized La Jolla, Pacific Beach, Mission Beach, Mission Bay and Point Loma areas. Transportation was also good in Oceanside and Carlsbad. In all of these areas, adequacy is based on a route structure that tends to link residential areas with coastal recreation areas. Even in these cases, however, the reduction of public transit service on weekends resulted in many instances of buses being uncomfortably crowded by recreational traffic during peak weekend periods.

PHASE II: THE CASE STUDY PROCESS

The case study approach was selected for the San Diego Regional Coastal Access Study as a means of identifying access problems common to representative areas in the Coastal Zone and as a mechanism for recommeding solutions to prototypical coastal transportation problems.

The case study portion of the project was structured in several work elements. The first task included the selection of specific resources within the San Diego Coastal Zone to be utilized as case study areas. This process included participation by the Coastal Access Study Committee. As an outcome of the case study selection process, the following resource areas within the Coastal Zone were selected for detailed, on-site study:

- Mission Bay-Mission Beach
- Del Mar Fairgounds
- Cardiff
- Imperial Beach-Silver Strand

Once the case study areas were determined, a study program was established for each specific resource, designed to investigate current use, physical attributes and other characteristics of the case study area. On-site surveys were used to collect data on travel characteristics and attitudes of visitors.

Problem Identification

The primary methodology for identifying problems followed the same basic format at each of the case study locations. The first step was the identification of existing conditions and patterns of use. This phase included a series of on-site research activities at each case study resource aimed at fully documenting the existing situation in terms of physical characteristics and patterns of use. Questionnaires were administered on peak-use weekends during the summer of 1977 to determine visitors' attitudes, visitation patterns, reactions to transportation conditions, party composition, activities pursued, primary and secondary origins and associated demographic data. Concurrently, license plate surveys were conducted to further verify the origin distribution of visitors to case study resources.

Data collected at the case study resources were analyzed and evaluated to develop a rational listing and prioritization of problems. Observed problems at each site were classified into the same four major problem areas used during the first phase of the study. These are as follows:

- Parking
- Internal Circulation
- Line-Haul (External) Access
- Public Transit

These four problem areas were used to structure the potential improvement measures for each of the case study sites. They also allow for the development of prototypical regional improvement strategies.

CARDIFF CASE STUDY SITE

Cardiff was the first case study site. It stretches three miles along the Central County Coast. Four major coastal recreational resources are located in the case study site. Close to one million visits per year come to these sites. There is a small commercial-tourist center in the central part of the town.

Usage patterns were determined through license plate and on-site interview surveys. The largest number of visitors to Cardiff came from immediately adjacent Central County Coast. A substantial number of visitors came from the North County and North City areas. Out-of-county use accounted for slightly less than one fourth of the total. This is consistent with the findings at Del Mar and Mission Beach.

Approximately half of the people arrived between 11:00 a.m. and 1:00 p.m. The automobile was used by 80 percent of the respondents. Walking proved to be the only other major mode of access, with 5.9 percent of those surveyed walking to the beach. The average vehicle occupancy was 3.1 persons per vehicle. This compares to an overall weekday vehicle occupancy of 1.5 persons per vehicle. Approximately 90 percent of the respondents parked less than 1,000 feet from the beach, and over half of them parked within 100 feet. Numerous persons noted that as parking facilities filled, they would continue to drive until they found a beach with parking available.

Observed Problems

The major recreation access problems observed at the Cardiff case study site involved parking. Several specific parking and access problems were encountered:

- o There is extensive parking on the shoulder of the major coastal highway. This results in some potentially serious safety problems for pedestrians and bicyclists travelling along and crossing an unprotected, high-speed highway.
- o There is limited on-site parking at the coastal beaches.
- o The configuration of the parking facilities at some beaches does not lend itself to the maximum, most efficient use of the available facilities.
- o The major internal circulation problem at the Cardiff relates to the lack of traffic regulation to and from the beach parking lots.
- o There is no Sunday or holiday public transportation on the east-west transit routes in the vicinity.

DEL MAR CASE STUDY SITE

The Del Mar case study site was selected as representative of a special event area with high levels of visitation. The Southern California Exposition and the Del Mar Racetrack are the most popular of the special events. The fairgounds is situated on a 200-acre triangular piece of land bounded by a five-point intersection north-east of the fairgrounds. Interstate 5, the major north-south interstate freeway, is included in the intersection.

There are 8,000 on-site parking spaces and 3,200 unpaved overflow parking spaces at the resource. Trams are used to transport people between the parking areas and the fairgounds entrance.

Del Mar is one of the three scheduled AMTRAK stops in the San Diego region. There are six trains in each direction daily to San Diego and Los Angeles from Del Mar.

Usage Patterns

Surveys were performed during the Southern California Exposition to determine usage patterns. Approximately 83 percent of the people surveyed were residents of the San Diego region. East County generated approximately 25 percent of the people surveyed.

More than 95 percent of those surveyed responded that they used a car, truck, or camper to access the site. Only two percent of the respondents used public transit. As in Cardiff, the use of private vehicles was reasonably efficient; average vehicle occupancy was 3.6 persons per vehicle.

Observed Problems

The types of problems found at Del Mar consist mainly of traffic congestion on arterial highways leading to and from the event. Occasionally, there is inadequate parking and traffic congestion on the interchanges and roads linking the main access highways with the parking lots. The main transit difficulty arises because San Diego Transit's authority terminates almost two miles south of the fairgrounds complex.

MISSION BEACH CASE STUDY SITE

Mission Beach and Mission Bay Park are centered in a heavily urbanized environment. As a planned recreation and tourism center, Mission Bay Park includes extensive commercial hotel and tourist attractions. The areas immediately adjacent to Mission Beach are also major visitor attractions.

The street network in Mission Beach is comprised of only one major north-south arterial. The rest of the system is basically a grid network of alleyways, which are also used heavily for parking. As a comparatively new and large-scale master plan development, Mission Bay Park incorporates numerous on-site parking spaces. The case study site is well-served by public transit routes. This is not surprising in view of the fact that these areas are within a highly developed metropolitan complex. There is an extensive bicycle route system in the area.

Usage Patterns

Slightly over one quarter of the visitors were from the Metropolitan Coast Area, in which Mission Beach and Mission Bay are located. More than 12 percent of the visitors were from out-of-state and an additional two percent were foreign visitors. Of the people who drove to Mission Beach, 24.4 percent are not San Diego County residents.

As in Cardiff, approximately half of the people arrived between 11:00 in the morning and 1:00 in the afternoon. Fewer than 70 percent of the visitors access the site by car, truck or camper. Almost one quarter of the respondents walked or skated to reach the site. Twice as many people rode bicycles than arrived by bus.

Almost 90 percent of those people who walked to the beach indicated a distant residence: East County and out-of-county. Higher percentages of visitors had to park at greater distances than at other case study sites. Almost 20 percent of the respondents parked more than 500 feet away. Among those who used an automobile, the occupancy was 3.0 persons per vehicle. The reason that people did not use the bus was due to the convenience of the auto.

Observed Problems

Extensive and severe parking problems were found at the Mission Beach-Mission Bay case study area. Limited parking is available on the ocean side of Mission Beach, and there is a great deal of competition among recreationists, residents and patrons of local businesses for the available on-street parking spaces.

Peak-period resource-area traffic congestion was observed. One reason for this congestion is the simple fact that these highways are required to perform double duty as both line-haul and internal transportation links for both local and recreational traffic. The congestion problems are also compounded by parking. A significant portion of the local traffic consists of automobiles driving around in search of parking places. Cars will often block one or two available lanes in either direction for considerable periods of time while a driver waits for another party to enter and load a parked car and then vacate a space.

The major problem with public transit is that, even though buses account for a small percentage of total access volume, the buses that operate during peak recreational periods are frequently more overcrowded than the buses that operate during commute peak hours.

The Mission Beach/Mission Bay area represents the most severe coastal access problems. With the anticipated minimum growth of 45 percent and a possible doubling of visitation by 1995, these problems will become worse.

SILVER STRAND-IMPERIAL BEACH CASE STUDY SITE

The Silver Strand and Imperial Beach case study site is situated in the southwest corner of the San Diego region. Located in this study site are some of the most underutilized beach resources in a major urban area. The Silver Strand-Imperial Beach case study area extends from the south boundary of the Naval Amphibious Training Center to the south end of Imperial Beach.

Transit service to the case study area is limited. Only recently was weekend transit service along the Silver Strand provided. There is transit service to Imperial Beach from downtown San Diego and Otay Mesa. The San Diego Bay Route Bikeway is approximately 23 miles and produces a complete circuit around San Diego Bay. San Diego Transit buses serving Route 9 on the San Diego-Coronado Bridge are equipped with bicycle racks, and enhances the use of bicycles as an alternative mode of travel.

The Silver Strand beach facilities are immediately adjacent to Silver Strand Boulevard. Imperial Beach and the fishing pier are at the west end of Palm Avenue. Both major resource concentrations are somewhat isolated from the rest of the County and its major population center. The Coronado Bridge consitutes a financial barrier to the Silver Strand. The Coronado central business district constitutes a traffic barrier. Alternate access to Coronado through Imperial Beach involves considerable driving distance from all zones except South Suburban Coast and South Suburban.

Usage Patterns

Use of Silver Strand-Imperial Beach resources is heavily concentrated among residents of the South County. Well over half of the visitors lived in the South Suburban and South Burban Coast Zones. Use of these resources by out-of-county visitors is less than the other three case study areas.

Observed Problems

The significant access problem observed was the relative isolation of Silver Strand due to topography and the resultant requirement that traffic cross the Coronado Bridge and then pass through the Coronado central business district to reach the State beach. There are also financial barriers that create access problems. The overall transit service to this case study site is much less than the service offered to other coastal areas. The highway system appears to have ample capacity to accommodate the rather modest 49 percent growth forecast through 1995. The resources would be capable of accommodating greater numbers of visitors.

PHASE III: IDENTIFICATION OF PROTOTYPICAL SOLUTIONS

This section utilizes the work from the case studies to identify prototypical solutions to the types of coastal access problems that were found. Urbanized areas are discussed first using the Mission Beach-Mission Bay case study site as the basis for developing the prototypical solutions. Both the Cardiff and Imperial Beach-Silver Strand case study sites are applied to a discussion of non-urbanized areas. Finally, special events are discussed based upon the Del Mar case study.

Urbanized Areas

The Mission Beach-Mission Bay case study site was selected to represent the region's coastal areas that are located within urbanized areas. Many of the region's urban areas are located along the coast.

The problems found in the urbanized areas include congestion of roads, the use of the same highway system for both internal circulation and line-haul access, and insufficient parking. While urbanized recreational areas generally are accessible to transit service, transit is primarily optimized for work trips rather than for recreational trips. Moreover, buses have to use the same street network as private vehicles and are severely impacted by traffic congestion. Transit use is also complicated for beach users bringing bulky equipment.

Improved traffic control measures appear to be the most useful technique for making improvements in urban areas. Traffic congestion can be decreased by left-turn limitations and selected regulation or removal of parking spaces from heavily travelled roads. The provision of preferential treatment for transit vehicles, the development of special pick-up and drop-off zones and the establishment of special regulations for large recreational vehicles can also relieve congestion. These traffic control measures may require highway improvements such as the construction of left-turn pockets, minor shifts in traffic lanes, and improved signing. This type of solution can only be marginally useful in improving the worst local traffic problems. The basic problems are caused by the presence of more private motor vehicles than can be adequately handled by the highway system and parking facilities. The provision of additional parking spaces would relieve the conflict between local traffic and recreationists competing for available street parking. This would have a positive effect on reducing traffic congestion.

For these reasons, long-term improvements in traffic problems in the urbanized coastal areas will require some shifting of traffic from private vehicles to transit. Park-and-ride services can be developed from available parking areas. Transit services to park-and-ride terminals can be operated using some existing routes. It will probably also be necessary to reroute and reschedule some weekend transit services to optimize service for recreationists. In addition, some form of vehicle modification to accommodate recreational equipment may be necessary.

Land use control is also a critical element of any long-term strategy for improving coastal access problems in urbanized areas. Land use controls should require adequate provisions for off-street parking for residential and commercial developments. Careful consideration will have to be given to total permissible residential and commercial densities in some of the coastal areas.

The capacity of heavily-used coastal areas is often limited by the ability of the area to accommodate automobiles rather than by the ability to accommodate people. This would suggest that growth in recreational, residential and commercial activities could be accommodated in coastal areas if they were tied to non-automotive transportation systems. However, it is difficult to implement this type of development.

Non-Urbanized Areas

The Cardiff case study area was chosen to reflect the residential access problems associated with a series of North County coastal resources that are comparatively undeveloped and non-urbanized. It is also an area of the county expected to show the highest growth in population and recreational demand over the next 20 years.

The Imperial Beach-Silver Strand case study site was selected as a coastal area currently experiencing low recreational demand. This case study represents a non-urbanized area which can accommodate future recreational demand.

The problems in these areas are normally related to relatively inefficient use of resources. Parking is a problem due to poor organization of existing parking areas, lack of adequate on-site facilities and inconsistent parking fees. Traffic problems involve safety hazards, inconveniences and intermodal conflicts. Transit services are minimal, even when provided.

Relatively minor traffic engineering improvements can provide most of the improvements needed for the problems observed. Traffic signals can be installed to eliminate pedestrian safety hazards and to smooth traffic flows in and out of major recreational areas and parking lots. Left-turn pockets or lanes on the main access roads can be used to eliminate hazards and traffic inconveniences. Parking capacities can be increased by paving and striping existing parking areas. Intermodal conflicts can be eased by provision of separate pedestrian, vehicle and bicycle lanes. Improved signing can improve the flow of all modes of traffic. These relatively inexpensive improvements could solve many of the problems at comparable coastal resources in the North County.

North County coastal areas such as Cardiff will show the highest growth in recreational use. Use will more than double at some of them within the next 20 years. Under these circumstances, traffic engineering solutions alone may be insufficient to accommodate growth in the long term. Accordingly, some considerations will have to be given to somewhat more extensive projects. For example, too much space immediately adjacent to the beach is probably dedicated to parking today in the Cardiff area and in comparable recreation areas. As the demand increases, it would be highly desirable to remove parking from areas immediately adjacent to the beach and to construct parking facilities on the eastern side of the major north and south highway.

Some consideration for increased use of public transit is also warranted in these areas. Transit operators should start considering potential future recreational traffic in its route planning and equipment acquisition plans.

Appropriate land use control strategies for commercial developments should be implemented to minimize future coastal access problems before they develop. New commercial developments on or near the beach should be reviewed carefully in terms of trip generation and attraction. Provision of adequate off-street parking should be assured.

Visitors often drive great distances to reach the Imperial Beach-Silver Strand case study site. These visitors have the option of approaching the site via the Coronado-San Diego Bridge or by travelling around the south end of San Diego Bay. Those using the bridge must pay a toll. At the site there are additional barriers: Imperial Beach parking lots have parking meters; Silver Strand State Beach charges an entry fee. Transit service is minimal. There is no weekend or holiday service to the Silver Strand and only limited service to Imperial Beach.

Improved transit service directly to these underutilized sites in non-urbanized areas from major population centers in the San Diego region would be beneficial. A strategy to allow admittance to Silver Strand State Beach and other resources that charge user fees by showing a transfer slip would increase facility and transit usage. Transit service should be coordinated with bicycle paths and routes in the area. Recreational equipment storage and bicycle storage facilities should be installed.

Marketing and communications should be implemented to make beach users aware of the recreational resources available at low demand areas. Public service messages, beach reports and improved signing should be used to increase public awareness of these areas.

Special Events Sites

The Del Mar Fairgrounds and Racetrack was chosen as a case study to represent a number of sites within the region at which special recreational or tourist events take place. Although the Del Mar Fair is undoubtedly the largest traffic generator of special event traffic within the region, a large number of other public and private attractions exhibit the same types of characteristics.

The types of problems found at Del Mar and at other special event sites consist mainly of traffic congestion on arterial highway leading to and from the event, occasionally inadequate parking and traffic congestion on the interchanges and roads linking the main access highways with the special event parking lots.

Access to special events lends itself to the use of public transit better than any other category of recreational participation. Participants tend not to be encumbered by bulky recreational equipment. From a transit standpoint, the concentration of total demand into a fairly short time period and to a single destination site makes public transportation considerably more efficient and easy to schedule than for the more dispersed beach areas. The high efficiency of transit vehicles can be used to minimize traffic congestion both on traffic arterials and in the immediate vicinity of the recreational site. Specific improvements to transit services would include additional direct service to special event sites, including possibly express service on existing routes and routes linking special event sites with available remote parking lots. Service could be improved through preferential treatment for transit vehicles at or near the site of major special events. Finally, parking and admission charges might be adjusted to provide more economic incentives for the use of transit.

Enhanced use of transit services might require some minor traffic engineering improvements. Special access lanes for transit vehicles might be required. These could be developed either by segregation of existing traffic lanes or by widening certain highway links. Traffic congestion is normally due to relatively short, concentrated bottlenecks. Thus, providing a second lane on off-ramps, widening of short stretches of access highways, improved intersection design and giving priority to transit vehicles might alleviate some of the more severe special event traffic congestion.

Traffic control measures may also be used to ease highly congested areas. Coordinated manual traffic control is probably more effective at special event sites than in any other traffic situation.

Special event parking fees are frequently collected from the motorists as they enter the parking lot, which slows down the flow of traffic.

Problems resulting from this practice could be solved by collecting the parking fees either as part of the total admission charge or by having people pay the fees on their way out of the special event.

Long-term improvements in access to special event sites may require some more extensive traffic engineering projects. Special off-ramps, new highway interchanges, or widening of access lanes may be required to handle increased traffic volumes. These solutions are most useful when applied to locations that are used for a series of special events throughout the year.

BIBLIOGRAPHY

California Coastal Act of 1976, SB 1277, as amended by AB 2948.

California Coastal Commission, California Coastal Plan, Sacramento, California, December 1975.

California Coastal Zone Conservation Act, California Public Resources Code, Secs. 27000-27650.

California Department of Parks and Recreation, California Outdoor Recreation Resources Plan, Sacramento, California, February 1974.

California Department of Parks and Recreation, "PARIS-Park and Recreation Information System", Sacramento, California, November 1966.

City of Carlsbad, General Plan, Carlsbad, California, August 1975.

City of Chula Vista Planning Department, Revision of Chula Vista General Plan: 1990, Chula Vista, California, March 1970.

City of Coronado, Citizens' General Plan Advisory Committee and Gruen Associates, Coronado - 1990 The General Plan, San Diego, California, (undated).

City of Del Mar, General Plan 1990, Del Mar, California, May 1975.

City of Imperial Beach, General Plan 1990, Imperial Beach, California, 1968.

City of National City, General Plan, National City, California, September 1974.

City of Oceanside, Beach Parking Study, Oceanside, California, September 1972.

City of San Diego, Mission Beach Precise Plan, San Diego, California, April 1974.

Coastal Zone Management Act of 1972 (P.C. 92-583), amended Coastal Zone Management Act Amendments of 1976 (P.L. 94-370).

Comprehensive Planning Organization of the San Diego Region, San Diego Regional Coastal Access Study, Phase I - Working Paper, San Diego, California, January 1978.

Comprehensive Planning Organization of the San Diego Region, San Diego Regional Coastal Access Study, Phase II - Case Studies, San Diego, California, April 1978.

San Diego County Planning Commission, San Diego County General Plan, San Diego, California, December 15, 1967.

RACE, INCOME AND ATTITUDE TOWARD BEACH CLEANLINESS[1]

Charles A. Heatwole[2] and Niels C. West[3]

Because of the limited spatial mobility possessed by some socioeconomic groups, a kind of de facto segregation often characterizes the clientele who use different beaches in an urban area. Typically, beaches which are located near low income neighborhoods or which are easily accessible by mass transit draw large numbers of urban poor. By contrast beaches which are relatively far from poor neighborhoods and which are accessible more readily by automobile than by mass transit tend to draw a more affluent crowd. Efforts to make the latter beaches more accessible to the less affluent are often resisted by the "rich" beach's clientele and fringing property owners. These peoples' exclusionary philosophy is often based on a belief that the outsiders will ruin the beach--that the minorities and the riff-raff will soon have their garbage strewn all over the place, thus undermining beach cleanliness, general environmental quality, and, perhaps, the value of adjacent property as well.

Consider a case in point. Jacob Riis Park is located near the tip of the Rockaway Peninsula in Queens, New York City (Fig. 1). Its beach is regarded as the cleanest of the ten public beaches in the City (Heatwole and West, 1980). The user population is overwhelmingly White and middle or upper-middle class. Relative absence of minorities and lower income groups may be attributed to two things. First, few people in these categories live within an immediate radius of Riis Park. Second, Riis Park is easy to get to for people who own an automobile but difficult to get to for people

---

[1] This research was sponsored by New York Sea Grant Institute under a grant from the Office of Sea Grant, National Oceanic and Atmospheric Administration (NOAA), US Department of Commerce.

[2] Associate Professor, Department of Geology and Geography, Hunter College of the City University of New York, New York, NY 10021.

[3] Associate Professor, Department of Geography and Marine Affairs, The University of Rhode Island, Kingston, RI 02881.

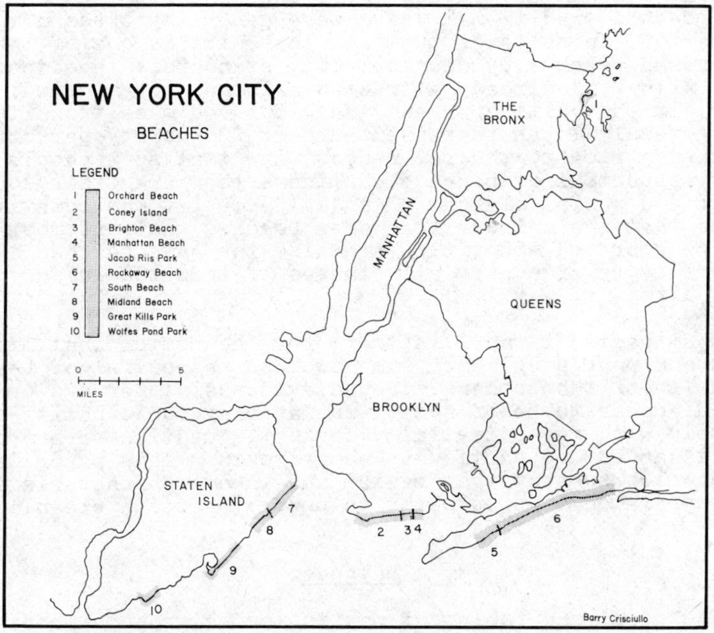
NEW YORK CITY
BEACHES
LEGEND
1 Orchard Beach
2 Coney Island
3 Brighton Beach
4 Manhattan Beach
5 Jacob Riis Park
6 Rockaway Beach
7 South Beach
8 Midland Beach
9 Great Kills Park
10 Wolfes Pond Park
0 5
MILES
THE BRONX
MANHATTAN
QUEENS
BROOKLYN
STATEN ISLAND
Barry Crisciullo

Figure 1

relegated to mass transit. Indeed, due to the relative remoteness of the Rockaway Peninsula, double or even triple mass transit fares may be required. Thus, the beach's location guarantees not only a time-consuming trip, but a costly one as well. As a result, there are strong inducements for users of mass transit to go elsewhere.

Some time ago it was proposed to make Riis Park more accessible to a greater variety and number of people than were using that facility. This was to be accomplished primarily by improving mass transit linkages by means of extension of several public bus routes. Riis's neighboring residents and the beach's traditional users were generally opposed to this scenario. In spite of the fact that Riis is one of the more underused beaches by New York City standards, these people felt Riis Park already attracted a large enough crowd. Moreover, it was quite apparent that improved mass transit access would result in an increased number of minority personnel. This gave rise to the referenced view that such people are less respectful of beach cleanliness than are other socioeconomic and racial classes. As "evidence" it was sometimes cited that the metropolitan area beaches that are most used by the poor and minorities are also the same beaches that are generally regarded (by all classes of individuals) as the most dirty.

What basis, if any, is there for the notion that minorities and poor people are somehow less respectful of beach cleanliness than others? Sound evidence either pro or con would appear to be of some value as many municipalities grapple with the potentially delicate question of increasing public access to particular shorefronts. This paper presents data which suggest that racial and income characteristics are unrelated to propensity to undermine beach cleanliness.

Methods

The ideal and foolproof way to test a group's propensity to make a beach dirty would be to quantify every bit of trash they generate and then compare this total with that of another group. Suffice it to say this sampling procedure is frought with many difficulties of cost, observation and measurement. Thus we opted for a more readily obtainable data set: attitudes toward beach environments. Several hundred people were interviewed on two New York City beaches, Coney Island and Orchard Beach (Fig. 1). These sites were chosen because they atttract some of the most multi-racial user populations among the various New York City beaches, as well as people who typify a full range of income levels.

Subjects were chosen by a team of student interviewers who had been given instruction as to the importance of random selection of subjects and given procedures designed to guarantee a random stratified sample. These procedures involved establishment of a series of paths either zig-zag, parallel, or perpendicular to the shoreline, and along which the interviewer attempted to question every fifth individual or member of every fifth group of individuals. Only one exception to this sampling procedure was allowed: individuals who appeared to be younger than fourteen years old should not be interviewed. Because Coney Island is much larger than Orchard Beach and attracts much larger crowds, more interviewing was done on Coney Island to help guarantee survey of a representative population. Accordingly, it will be seen that the Coney Island sample is substantially larger than the Orchard Beach sample. On both beaches the interview range extended from the water to the landward border of the beach as delimited by either the presence of vegetation, a boardwalk, or similar barrier. Interviewees were asked what qualities they looked for in an ideal beach, and what qualities they liked the most and the least about the beach on which they were being interviewed.

Results

More than ninety specifically different likes and dislikes were reported in response to these questions. In order to facilitate the tabular presentation of these responses, and to make the responses more amenable to tests of statistical significance, it proved convenient to collapse the gathered responses into three positive and three negative attitudinal categories. The positive attitudinal categories included a clean physical environment, a pleasant social environment, and good access and amenities. The negative categories included a dirty physical environment, an unpleasant social environment, and poor access and amenities.

Attitudes and Racial Characteristics

With regard to attitudes toward ideal beach qualities, tabulations based on race revealed remarkable similarities. For each group on both beaches a clean physical environment was regarded overwhelmingly as the most ideal beach quality (Table 1). Whites tended to rate this quality somewhat higher than did the Blacks and Hispanics who, by contrast, placed greater value than the Whites on the social aspects of the recreational experience. (In the case of the Coney Island sample the difference is statistically significant. However, the contingency coefficient is small, and with a

TABLE 1

WHAT QUALITIES DO YOU LOOK FOR IN AN IDEAL BEACH?

	Whites		Blacks		Hispanics	
	N	%	N	%	N	%
Coney Island Respondents						
Clean Physical Environment	519	80.7	43	66.1	133	72.3
Pleasant Social Environment	91	14.1	21	32.3	43	23.4
Good Access and Amenities	33	5.1	1	1.6	8	4.3
Totals	643	100.0	65	100.0	184	100.0
Orchard Beach Respondents						
Clean Physical Environment	186	73.8	30	65.2	24	68.6
Pleasant Social Environment	52	20.6	13	28.3	10.	28.6
Good Access and Amenities	14	5.6	3	6.5	1	2.8
Totals	252	100.0	46	100.0	35	100.0

large data set it is not difficult to generate a statistically significant relationship. In reality the raw data suggest these differences are relatively minor.) Good access and amenities were regarded by all as being of decidedly lesser importance than the physical and social environments on the hypothetical ideal beach.

With regard to attitudes toward the interview sites, respondents again were fairly uniform about their likes and dislikes (Tables 2 and 3). Generally speaking, Coney Island and Orchard Beach were praised for their accessibility and amenities--somewhat more so in the case of Coney Island, which just happens to possess a large and famous amusement area as well as the best mass transit linkages of any beach in the City. On the other hand both beaches were looked down upon primarily because of the perceived uncleanliness of their physical environments. Again, Coney Island was given somewhat higher (worse) marks on filthiness than was Orchard Beach. Social qualities assume intermediate importance on both tables, being neither as highly regarded as access and amenities on the listing of positive qualities (Table 2) nor as lowly regarded as the lack of cleanliness on the listing of negative qualities (Table 3). Indeed, it is the cleanliness (or lack thereof) and access/amenities issues that really stand out on these tables, particularly the emergence of a dirty physical environment as the least liked characteristic of the two interview sites. As in the case of Table 1, the emphasis placed by the three racial groups on the attitudinal categories vary somewhat. For the Orchard Beach respondents in Table 3 the differences are statistically significant, though all three groups are very much in accord in their condemnation of the perceived uncleanliness of that beach. Essentially what these tables tell us is that the White, Black and Hispanic samples all place highest value on the cleanliness of beach environments; and that although they detest the perceived uncleanliness of the interview sites, they continue to patronize Coney Island and Orchard Beach because these beaches are easy to get to.

Attitudes and Income Levels

Attitudes toward beach cleanliness were also analyzed with respect to level of family income. Income data were collected by showing each respondent a card on which nine income categories were listed. Each interviewee was asked to indicate the income category which corresponded to his or her family situation. For purposes of the present analysis the nine income categories have been collapsed into three: less than \$9,000, \$9,001 to \$18,000, and more than \$18,000.

TABLE 2

WHAT QUALITY DO YOU LIKE THE MOST ABOUT THIS BEACH?

	Whites		Blacks		Hispanics	
	N	%	N	%	N	%
Coney Island Respondents						
Clean Physical Environment	86	15.7	9	16.7	27	17.5
Pleasant Social Environment	101	18.5	14	25.9	33	21.4
Good Access and Amenities	360	65.8	31	57.4	94	61.1
Totals	547	100.0	54	100.0	154	100.0
Orchard Beach Respondents						
Clean Physical Environment	62	29.4	4	11.4	9	30.0
Pleasant Social Environment	60	28.4	9	25.7	8	26.7
Good Access and Amenities	89	42.2	22	68.9	13	43.3
Totals	211	100.0	35	100.0	30	100.0

TABLE 3

WHAT QUALITY DO YOU LIKE THE LEAST ABOUT THIS BEACH?

	Whites		Blacks		Hispanics	
	N	%	N	%	N	%
Coney Island Respondents						
Dirty Physical Environment	445	85.4	40	93.0	111	85.4
Unpleasant Social Environment	39	7.5	2	4.7	13	10.0
Poor Access and Amenities	37	7.1	1	2.3	6	4.6
Totals	521	100.0	43	100.0	130	100.0
Orchard Beach Respondents						
Dirty Physical Environment	96	70.0	11	64.7	14	82.3
Unpleasant Social Environment	32	23.4	0	0.0	2	11.8
Poor Access and Amenities	9	6.6	6	35.3	1	5.9
Totals	137	100.0	17	100.0	17	100.0

Attitudes toward ideal beach qualities were remarkably similar (Table 4). A clean physical environment accounted for at least three-quarters of the preferences of each income class sampled on Coney Island, and at least two-thirds of the preferences of each income class sampled on Orchard Beach. In every case a pleasant social environment is a very distant second place in terms of perceived ideal qualities, followed by good access and amenities. Naturally the precise figures vary from beach to beach. It is interesting, for example, that the Orchard Beach percentages on clean physical environment are uniformly lower than those for Coney Island while the percentages for a pleasant social environment are uniformly higher. In any event Table 4 does not present statistically significant differences between income level and perceived ideal beach qualities.

Table 5 shows what the different income classes liked the most about the interview sites. Again there is general agreement as to the favored characteristics. Indeed there is not a statistically significant difference between income level and attitudes regarding the most positive aspects of the interview sites. As was the case with the ethnic data, good access and amenities were the most favored characteristics, and overwhelmingly so in most cases. Importance placed upon the clean physical and pleasant social environments tended to be fairly well split while lingering well behind the issue of good access and amenities.

General unanimity again obtains with respect to the least liked interview site qualities (Table 6). A dirty physical environment was the most cited characteristic by each sample with the Coney Island group being the more negative of the two. An unpleasant social environment tended to be the second most objectionable quality with the Orchard Beach group placing greater emphasis on this negative aspect. Poor access and amenities tended to receive the least amount of disfavor. Although the samples placed differing emphases on these negative qualities, analysis once again revealed no statistically significant differences between income levels and expressed attitudes. Essentially these last three tables that reflect level of income lead us to the same conclusions as did the tables reflecting racial composition; i.e., all respondent categories tended to place highest value on clean beaches, dislike the interview sites because of their perceived dirty physical environments, but patronize them anyway because they are easy to get to.

TABLE 4

WHAT QUALITIES DO YOU LOOK FOR IN AN IDEAL BEACH?

	Less Than $9,000		$9,001 to $18,000		More Than $18,000	
	N	%	N	%	N	%
	Coney Island Respondents					
Clean Physical Environment	363	77.7	257	79.1	85	75.9
Pleasant Social Environment	85	18.2	52	16.0	19	17.0
Good Access and Amenities	19	4.1	16	4.9	8	7.1
Totals	467	100.0	325	100.0	112	100.0
	Orchard Beach Respondents					
Clean Physical Environment	68	70.8	120	75.5	52	66.7
Pleasant Social Environment	22	22.9	35	22.0	18	23.1
Good Access and Amenities	6	6.3	4	2.5	8	10.2
Totals	96	100.0	159	100.0	78	100.0

TABLE 5

WHAT QUALITY DO YOU LIKE THE MOST ABOUT THIS BEACH?

	Less Than $9,000		$9,000 to $18,000		More Than $18,000	
	N	%	N	%	N	%
Coney Island Respondents						
Clean Physical Environment	67	17.1	40	14.0	15	16.7
Pleasant Social Environment	90	23.0	45	15.7	15	16.7
Good Access and Amenities	234	59.9	201	70.3	60	66.6
Totals	391	100.0	286	100.0	90	100.0
Orchard Beach Respondents						
Clean Physical Environment	14	18.4	45	32.1	16	26.7
Pleasant Social Environment	22	29.0	40	28.6	15	25.0
Good Access and Amenities	40	52.6	55	39.3	29	48.3
Totals	76	100.0	140	100.0	60	100.0

TABLE 6

WHAT QUALITY DO YOU LIKE THE LEAST ABOUT THIS BEACH?

	Less Than $9,000		$9,001 to $18,000		More Than $18,000	
	N	%	N	%	N	%
Coney Island Respondents						
Dirty Physical Environment	305	86.4	222	85.1	79	87.8
Unpleasant Social Environment	23	6.5	24	9.2	7	7.8
Poor Access and Amenities	25	7.1	15	5.7	4	4.4
Totals	353	100.0	261	100.0	90	100.0
Orchard Beach Respondents						
Dirty Physical Environment	32	65.3	62	75.6	27	67.5
Unpleasant Social Environment	9	18.4	17	20.7	8	20.0
Poor Access and Amenities	8	16.3	3	3.7	5	12.5
Totals	49	100.0	82	100.0	40	100.0

Implications

Analysis of the data reveals very little difference in attitudes toward beach cleanliness along either racial or income lines. For each sample and class within, the most desirable beach characteristic was cleanliness (usually by a wide margin). Conversely, dirty beaches were universally condemned. Of course a person's actions may not always reflect stated attitudes. Anyone can preach cleanliness and still generate filth. But at least in the case of our data, attitudes toward cleanliness were too strong and too similar from group to group to merit cynical qualifications.

If the data fail to support the contention that minorities and poor people tend to make beaches dirty, then whence comes the accusation? In New York City it appears related to the fact that these people largely are relegated to the few beaches that are served adequately by the city's mass transit system. These same beaches, in turn, tend to be the most crowded and thus the most difficult to manage effectively from a sanitation point of view. By contrast beaches removed from the mass transit network attract relatively few minorities and poor people and tend not to be very crowded by comparison. Accordingly, beach cleanliness appears most related to crowd size and not to the kinds of people who make up the crowd. This suggests that efforts to improve regional beach quality will profit more by efforts to spread users more evenly over existing facilities than by strategies which have the effect of consigning particular populations to particular beaches. At the same time, projected shifts in users should be reflected in the distribution of funds, manpower, tools, and trash containers used to help maintain beach cleanliness.

Reference Cited

Heatwole, Charles A., and West, Niels C. Beach Use and User Constraints in the New York City Coastal Region. Albany: New York Sea Grant Report Series, NYSG-RS-80-01, February, 1980.

Beach Access Problems, Programs and Progress

David J. Brower
Associate Director
The Center for Urban and Regional Studies
The University of North Carolina at Chapel Hill
Chapel Hill, North Carolina

Introduction

Man throughout history has been drawn to the coast. The coast has been and continues to be a source of replenishment for mind, body and spirit in all cultures. In this country where the amount of disposable income and time is rapidly increasing, water related recreational activity continues to be the most popular form of outdoor recreation. The notion of simply "going to the beach" is still one of the most popular outdoor pastimes. But going to the beach is not what it once was. It used to involve quietly lying in the sun, peaceful picnics, beach grass blowing in the breeze, the cry of gulls, long, isolated walks. But now "going to the beach" often involves long lines of hot station wagons filled with grouchy parents, sullen children, and soggy sandwiches trying to get to a crowded beach filled with bodies, beer cans, styrofoam and stereos. In the process the beach-goers run the danger of conflict with hostile residents, defensive land owners and local governments who regard the beach as "theirs." For many, going to the beach has become unpleasant; for others, impossible.

As the popularity of coastal areas of this country increases, development in the coastal areas also increases to take advantage of that popularity. As development increases to meet increased demand, the supply decreases, and quite according to economic theory the price increases. The result is that the coast is becoming increasingly the haven for only upper income people. It is only the upper income groups who can afford the facilities, such as a motel room, a cottage, or a condominium, to assure that going to the beach is a real possibility.

The argument for a strong public policy encouraging public access to the nation's beaches is a simple one. There has always been a strong demand for the use of beaches, and that demand is increasing and will continue to increase for some time to come. The public now owns and/or has the right to use vast stretches of the coast. The problem is that often there is no way for the public to get to those portions of the coast that they have the right to use because it is separated from the nearest public road by privately owned land. The result is that many stretches of public beach are accessible only by long hikes, by boat or by trespassing on private land. This problem is becoming increasingly intense because private land that has been traditionally used by the public is being converted to more profitable uses by the owners, and in some cases accessways that have been dedicated to the public are being withdrawn. Unless there is a strong public policy to obtain new accessways and to protect existing ones, the amount of beach available to the public use will continue to decrease at the same time the demand for it is increasing.

There are two separate questions involved. The first, does the public have a right to use the beach? Then, if the public does in fact have the right to use the beach, how can the public get to the beach? It is the latter question that is the real "beach access" question.

The paper will first explore the legal questions of beach ownership and of accessways to the beach. Then it will discuss the beach access requirements of the federal Coastal Zone Management Program and will conclude by describing eight state beach access programs and how they have responded to the federal program.

The Legal Right Of The Public To Use The Nation's Beaches

Can the owners of a coastal motel fence off the beach in front of the motel and reserve it for the exclusive use of motel guests? Can the public fish from the surf anywhere they want? Or use any beach for swimming, sunbathing, or picnicking? Supposing that they do have the right to use the beach, how can they get to it? To what extent can owners of water front cottages prohibit anyone from sitting on "their" beach or walking across their property to get to the beach?

These questions raise the issue of who owns the beach, and to what extent the public has a legal right to use the beach, and to gain access to it. The answers to these questions are based on sometimes complex legal issues and doctrines.

For the purposes of this discussion, the "beach" consists of four parts: the sea, the wet-sand, the dry-sand, and the upland. First, that area seaward of the mean low tide line is termed the sea, or sea bed (lake or lake bed in non-oceanic situations). Second, the area between the mean low tide and mean high tide lines, which is covered by the usual flow of tides, is termed the wet-sand. "Foreshore" and "tideland" are generally synonymous with this term. Third, the area between the mean high tide line and the line of vegetation, or dune line, an area inundated only during severe storms, is termed the dry-sand. Fourth, that area landward of the vegetation line is termed the upland. The following diagram illustrates this division.

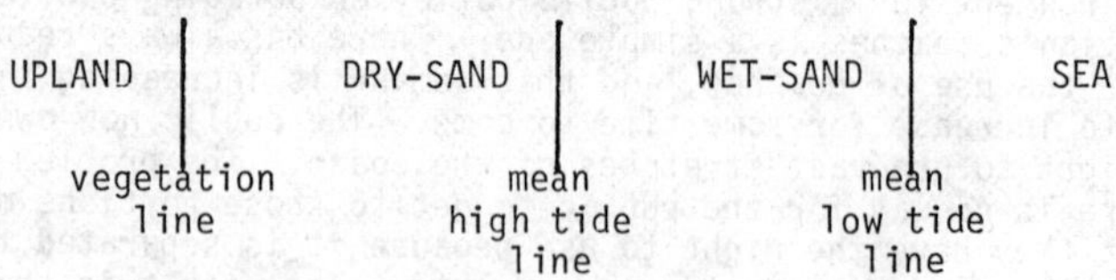

The question of the extent to which the public has the right to use coastal beaches has come before the courts on several occasions. Major cases have been decided in New Jersey, California, New York, Oregon, Florida, and Texas. Each of these cases involved different factual situations and different aspects of the beach use and access questions. Each decision was also based on different legal theories.

The Sea

American courts have held consistently that navigable waters and associated tidelands are held in trust by the government for the public. In the leading case on this point, the U. S. Supreme Court in 1892 said that these lands are held in trust for the people so that public rights of navigation, fishing, and commerce could be preserved. The court further said that the state could not abdicate its control over these areas in any fashion that would diminish the public's right in the trust lands.

The Wet-Sand

As the following chart indicates, approximately three-fourths of the oceanic states have held that wet-sand beaches are publicly owned:

Publicly Owned		Privately Owned
Alabama	New Jersey	Delaware
Alaska	New York	Maine
California	North Carolina	Massachusetts
Florida	Oregon	New Hampshire
Georgia	Rhode Island	Pennsylvania
Hawaii	South Carolina	Virginia
Louisiana	Texas	
Maryland	Washington	
Mississippi		

In most of these states, the public ownership of the wet-sand beaches is subject to a "public trust."

The Dry-Sand

In almost all states the dry-sand is privately owned but in a number of states the courts have used a variety of theories or concepts to find public ownership or a public right to use the dry-sand portion of the beach:

Public Trust: Avon is a small resort community in New Jersey bordering the Atlantic. The city owned a dry-sand beach that for many years was free to all comers. In the 1950s, the New Jersey legislature allowed cities to start collecting user fees for the beaches to defray maintenance costs. Avon established such a fee, which was the same for residents and non-residents. In 1970, Avon changed their beach fee ordinance so that it would charge non-residents about twice as much as residents to use the beach. Neptune City, a neighboring inland city, brought suit to challenge this change in the ordinance.

The court invalidated the ordinance. They said that where a municipality owns a dry-sand beach and dedicates it to public beach purposes, the public trust doctrine requires the beach to be open to all on equal terms. So while New Jersey cities could charge reasonable user fees, they may not discriminate between residents and non-residents.

Implied Dedication: In 1970 the California Supreme Court found that the public had a right to use two privately owned beaches.

The court said that when the public has used a beach for a long period of time, paying no attention to the fact that the beach is privately owned, the public acquires a legal right to use that beach. Unlike the common notion of "express dedication," where the private owner overtly conveys land to the public and the public body formally accepts the dedication, there are no formalities required for "implied dedications." The owner's intent to give the land to the public is implied from his conduct of not preventing public use of the beach. And the public's acceptance of the dedication is implied from public use of the beach.

Customary Rights: In 1969, the Oregon Supreme Court in Thornton v. Hay, employed the legal doctrine of "customary rights" to confirm the right of the public to use Oregon's dry-sand beaches.

This case involved the efforts of Hay, a motel owner, to fence in a portion of the dry-sand beach adjacent to his motel for the exclusive use of motel guests. The publicity generated by this action helped trigger legislation recognizing the public's right to use the dry-sand beaches of the state. However, since legislation of this sort cannot change or determine property rights, the matter of Hay's fence was sure to wind up in court.

And wind up in court it did when, after a winter storm destroyed his fence, Hay replaced it without obtaining a permit required by the legislation. The state went to court seeking an order to force Hay to remove the fence.

In reviewing Hay's claim of a right to exclude the public from this dry-sand area, the Oregon court noted that the public had been freely using the state's dry-sand beaches as if they belonged to the public for over sixty years. At no time had anyone sought or obtained the upland owner's permission to use the dry-sand area. Virtually everyone, including the upland owners, had always assumed that the public had a right to use the dry-sand area. For these reasons, the court held that the public had obtained the right to continue to use the dry-sand beach.

Prescriptive Easements: If it is found that the public has acquired a prescriptive easement for the use of a specific beach area, then the public has a right to continue to use the area in the same fashion as before. What had started out as a trespass becomes, when enough time passes without the owner evicting the trespassers, a legal right to use the beach. The owner still "owns" the property, but cannot do anything on it that would reduce the public's established rights to use it.

The Upland

In all states, the upland, that area above the vegetation or dune line, is privately owned with the exception of land acquired by the public. Many of the "adverse possession" doctrines applied to the dry-sand portion of beach have been applied and are available to be applied to the upland portions of the beach as well. Indeed many of the cases referred to in the Dry-Sand section dealt with public rights in the upland as well as in the dry-sand areas.

Typically, however, public rights in the upland portion of the beach are obtained through some sort of acquisition program. The National Seashores and the many state and local parks that are found on the coast are the product of these programs. A few states (Oregon, Rhode Island and California) have been especially active while others have done very little.

From this description it can be seen that vast portions of the nation's beaches are owned by the public and/or the public has the right to use them. The pattern of public rights varies from state to state. In some states, the public owns only the sea, in some they own from the vegetation line seaward, but in all states, the public does have the right to use some portion of the beach even if it is only the land below the water which can be used only by wading.

The Great Lakes

While the Great Lakes are technically "inland lakes" rather than oceans, their shorelands share many of the attributes and problems of oceanic coasts. Some of the areas have a number of the physical characteristics of ocean beaches; many areas are also increasingly being faced with the same public versus private use conflicts that led to the litigation discussed earlier.

Over the years, state and federal courts have developed a strong position that the states must manage the Great Lakes in a fashion to benefit all residents of the state; that the Great Lakes are a public resource that cannot be appropriated for private gain or use. In fact, the U. S. Supreme Court's landmark decision on the public trust doctrine, Illinois Central R.R. v. Illinois, set aside an attempt of the

Illinois legislature to convey part of the lake bed of Lake Michigan. However, as is the case in many oceanic states, this doctrine has generally been interpreted only to protect public rights lakeward of the "ordinary high water mark".

What Can Be Done About Access?

Concern about the ability of the public to have access to the coast and to make use of the coastal beaches is by no means a new issue. The public trust doctrine was developed by the Romans in part to ensure that the public could make use of the shore. The public trust doctrine was incorporated into English law, brought to the colonies and later adopted by the United States. In Hawaii, a system of trails leading to and along the coast, that under tribal law could be used by all of the people, led to what is now known as the "Ancient Trails Doctrine." In 1959 the Texas legislature adopted the Texas Open Beaches Act which was passed after private landowners attempted to close some Galveston Island beaches to the public. Shortly before the turn of the century, a part of the Oregon beach was declared to be a public highway in an effort to keep it open to the public. In 1913 that policy was extended to all of the Oregon wet-sand beaches by the Oregon legislature.

From 1913 until the early '60s, most Oregonians were secure in their belief that they owned the entirety of the Oregon beach. Certainly they had been free to use the entire beach without interference until the owner of a motel erected a fence on the dry-sand portion of the beach and posted a sign that read: "Surfsand Guests Only Please." This led to the passage of the Oregon Beach Law in 1967 which was similar to the Texas Open Beaches Act. This Act codified into law already existing public rights to dry-sand beaches and gave the State Highway Commission the authority to police, protect, and maintain them. The Act said that the legislature:

> . . . recognizes that over the years the public has made frequent and uninterrupted use of the ocean shore . . . sufficient to create easements in the public through dedication, prescription, grant or other use . . . the Legislative Assembly hereby declares that all public rights . . . are vested exclusively in the State of Oregon.

Both of these acts codified the common law of beach ownership and gave the Attorney General legislative authority to protect that ownership. Both acts have been challenged several times but have been consistently upheld by the courts.

Other states have dealt with access through park programs, but few acted with the force and direction of Oregon and Texas.

Coastal Zone Management Act

In 1976, in order to give added direction and focus to the coastal zone management program, Congress added Section 305(b)(7) the Shorefront Access Amendment to the Coastal Zone Management Act. The amendment requires that "the management program for each coastal state include . . . . a definition of the term 'beach' and a planning process for the protection of, and access to, public beaches . . . ." The Report of the House of Representatives Committee on Merchant Marine and Fisheries commented on Section 305(b)(7):

> The Committee wants, by this requirement, for state Coastal Zone Management Programs to identify their publicly held coastal areas and to devise policies which will either provide for their protection, where that is appropriate as with ecologically significant wildlife areas, and for their ready access as is appropriate with a public beach. Whereas the present management programs must include an inventory and designation of 'areas of particular concern,' this new requirement focuses particular attention on publicly held properties and directs that plans for their best management be included in the state program.
>
> Access to public beaches and other attractions in public ownership in the coasts has come to be identified as one of the critical problems facing local and state governments . . . . The Committee position is that action is needed now to help provide the needed access, especially, in urban areas, and that to wait will only mean additional expense to the taxpayers. The key gain is that the purchase of such access, as is provided in the addition to Section 315 (2) . . . be tied to a comprehensive plan. That is the intent of this new requirement under 305 program development--that all such purchases fit into an overall program for each state.

Regulations implementing 305(b)(7) were hurriedly drawn, and the state managers were advised to "expedite" their shorefront plans. This was interpreted by them, apparently correctly, to mean "don't spend much time on this, it isn't very important."

The result of this less than enthusiastic performance of OCZM in implementing the shorefront access planning amendment resulted in a wide range of uneven responses from the states. Thus far twenty-one states have prepared shorefront access planning elements. Of these, seventeen have been approved by OCZM. They vary from being almost totally devoid of substantive content to being very imaginative and creative. Generally the states that had been concerned with beach access prior to passage of 305(b)(7) prepared the more significant programs, and the states that were uninterested before were still uninterested.

The substance of the programs fell into several areas:

Access to access

Several states concluded that the chief problem was not the lack of accessways but the inability of people to get to the accessways. From this observation a number of programs were developed which involved some sort of mass transit. New Jersey and North Carolina experimented with shuttle buses running from remote parking lots to accessways. Hawaii attached surfboard racks to some of the buses serving the coast of Oahu so that surfers could use mass transit to get to the beach. In the San Francisco metropolitan area, special buses go to the Golden Gate Recreation Area and Point Reyes National Seashore. Time sharing programs for industrial and commercial parking lots have been suggested whereby the public would lease private parking lots or spaces that are not used on weekends. On the negative side, the lack of parking and inability to provide it has been used as a rationale for denying access and as a means of limiting it.

Management of existing access

Perhaps the most interesting access programs have been developed in this area. States have developed programs to systematically review their coastal areas to discover accessways that are already in the public ownership and areas of beach that the public has the right to use. Sometimes known as "protection of public property rights programs" public ownership and use rights are documented and programs to protect them in the future are developed. These range from the ancient trails of Hawaii and early rights of way contained in Rhode Island's King James Charter to deeds "lost" in an eastern North Carolina courthouse and "unowned" islands in Maine.

Acquisition

There are any number of variations on this theme that are used by various public bodies to obtain accessways.

Purchase: A number of states have active programs to purchase accessways. Oregon and California, for example, have been systematically buying relatively small accessways along the coast with the intent, in the case of Oregon, of providing an accessway every 1.5 to 3 miles. Up until now the U. S. Land and Water Conservation Program has been a reliable source of federal money for this sort of purchase program. Some states (Massachusetts, Michigan, Oregon and New Jersey) have or are studying the possibility of creating special state funding sources to acquire accessways. One of the interesting variations is the rights that are suggested for purchase. For example, an easement acquired by the public to use an accessway restricted to the use of lot owners in a subdivision or the acquisition of an access easement on top of a utility right of way, etc.

As a condition of a permit: Several states and many communities require that accessways be dedicated to the public as a condition of obtaining a permit to develop or subdivide a parcel of shorefront land. In some states, the developer is simply precluded from closing existing accessways, but in others they are required to provide new accessways, and in others they must provide and develop (pave and landscape) the accessway.

As a condition of public expenditure: Some programs such as the Corps of Engineers beach nourishment program and a number of state programs require that before public funds can be spent on a beach project some arrangement to ensure public access to that beach must be established.

Acquisition as a part of a coastal regulatory program to mitigate loss in market value: It has been suggested that oceanfront lots should be acquired for access in situations in which the regulations imposed by the coastal zone program come close to leaving the owner with no reasonable use of his land and thus tempting a taking challenge. A related suggestion is the acquisition of properties destroyed or seriously damaged by a natural hazard, thus providing access and preventing the redevelopment of a hazardous area. An obvious advantage to the latter program is that Federal Insurance Administration funds may pay a part of the purchase price.

Acquisition through usage: The acquisition of an easement by the public through continuous usage under some form of implied dedication or customary use doctrine has already been discussed and is often a part of management programs.

Federal land: The federal government and the Department of Defense in particular own a great deal of the coast, and much of it is located close to urban centers. It has been suggested that at least a part of the coast owned by the military could be opened for civilian recreation without jeopardizing the national security. In some few cases military bases have been declared surplus and made available to state and local governments for public use.

Case Studies

The eight state beach access programs that are described in this section were prepared in response to the federal Coastal Zone Management Act although as will be seen several of the states were active in beach access long before Section 305(b)(7) was adopted. All of the programs have been approved by the Office of Coastal Zone Management except for Texas which, as of this writing, is not participating in the federal program. As will be seen, these are not presented as models but rather as examples of how states are dealing with beach access.

California

California public policy has long recognized the recreational value of its coastal resources but access to the beaches decreased as coastal development increased. The state constitution and federal Coastal Zone Management Act provided the incentive and authority for an aggressive state program which among other things requires new development to provide public access to the shore.

While the definition of the public portion of the California beach is the area from mean high tide seaward, the California Constitution guarantees access across private property to all coastal tidelands. Over the years this right has been defined by various statutes and court decisions. California courts have recognized an implied dedication if the public has used a privately owned area for at least five years.

California's coastal management program addressed the access issue extensively prior to the adoption of the 1976 amendments to the federal Coastal Zone Management Act. The California basic access policy is to "maximize public access to and along the coast and maximize public recreation opportunities in the coastal zone consistent with sound resource conservation principles and the constitutionally protected rights of private property owners." The policies of the management program expand on this theme asserting that development shall not interfere with the public's right of access to the shorefront where it has been acquired through use or legislative authority, including use of the dry-sand portion of the beaches and rocky coastal beaches to the first line of vegetation. The policies also call for public access from the nearest public roadway to the shoreline and along the coast wherever possible and in addition for the provision of adequate public facilities (such as parking lots and bathrooms) in order to distribute use throughout the area and prevent overcrowding.

At the heart of the California coastal management plan's access element is the consistent exercise of development review and permit conditioning. All proposed coastal developments must be reviewed for their consistency with state policies among which is the provision of adequate access from the nearest public road. This access provision usually results in the dedication of an easement, but can also include a requirement that the developer provide a stairway or ramp to the beach, a public parking lot, or a bikeway. The dedication requirement has been unpopular with developers, but continues to be upheld in court. The major problem with the dedicated rights-of-way is finding a public body willing to accept and maintain them. The California legislature requires that a public body must agree to maintain a right-of-way before it may be opened to the public. The financial cutbacks imposed by Proposition 13 have discouraged local governments from accepting the burden of maintenance, security and liability that accessways carry with them. The California Coastal Commission and California Coastal Conservancy are currently working together to find agencies and homeowner groups which are willing to accept these responsibilities.

The coastal management program also coordinates the efforts of many other agencies to acquire coastal land for public recreational purposes. The State Urban Coastal Bond Act which was passed by California voters in 1976 provided $145 million for the acquisition of coastal areas and $10 million for the California Conservancy to begin its acquisition program. Approximately $250 million of state and local bonds have been approved for acquisition purposes during the last six years.

The California Coastal Commission has recognized that the lack of public transportation prevents many citizens from getting to the coast. Thus, although coastal management program funds do not directly subsidize any transportation programs, local governments are being encouraged by the Coastal Commission to arrange for transportation to the beaches through their transit authorities. Transportation programs also work to reduce the demand for parking at the point of access.

Coastal management officials continue to seek solutions to four major shorefront access problems in California. In southern California there is inadequate transportation to the coast and inadequate parking to serve coastal recreational demands. In many areas of northern California agricultural land must be crossed to get to the shorefront. In rocky coastal areas, topographic conditions such as steep bluffs pose hazards for those seeking to use the beach below. Finally, as is the case in most coastal areas, the delicate balance between the right of public access and the protection of natural resources must be struck.

Maine

Shorefront access is a delicate political issue in the state of Maine, which has a long tradition of private property rights along the coast. As a result, state coastal management officials have been cautious and conservative about developing new public access opportunities and/or attitudes.

Except in a very few cases, private ownership extends to mean low tide or 100 rods (503 meters) from mean high tide (whichever is less). The Colony Ordinance of 1641 authorized private land ownership in the intertidal zone, subject to public rights for fishing, fowling and navigation. Recreational rights are notably absent from this list. Since that time there has been no litigation or legislation redefining shoreline ownership and the attitude toward public access to the shoreline remains basically closed.

Only a small percentage of Maine's predominantly rocky coast is sandy beach. An inventory by the Maine State Planning Office found that only 75 miles or 27% of the state's coast was sandy. Of this, approximately .5% belong to the federal government, 8.5% belong to the state government, 12.5% belong to local governments and 78.5% are in private ownership. Thus, publicly owned sandy beaches are extremely limited in Maine. Boating access is perhaps a more important issue in the state, not only for recreational boaters and fishermen, but also for commercial fishermen whose livelihood depends on their access to the water.

Traditionally, Maine's coastal communities have provided access to the shorefront for local residents only. Where a town beach or a town harbor has been established, its use is generally restricted to local residents. State coastal program officials had intended to work with the Attorney General to set forth a broader state position for public shorefront access and use rights, but the case upon which this process was to start was dismissed and the program has not been pursued.

There has been little or no change in Maine's policy toward shorefront access since the passage of the 1976 amendments to the federal Coastal Zone Management Act. State officials are wary of introducing shorefront access legislation for fear that it will be defeated and that a "legislative intent" argument could then be used against any future attempts to provide public access. They also assert that there is little demand for increased access. Thus, there has been little activity at the state level to develop new access policies or to create new access opportunities. Provision of shorefront access remains at the discretion of the property owners.

Given this political environment, the shorefront access policy statement of the Maine coastal management program is, not surprisingly, weak. It contains some definitions, ideas for planning and inventory and a description of existing Maine legislation which protects existing shorefront access. The two major pieces of legislation cited are the Coastal Wetlands Alteration Act and the Mandatory Shoreland Zoning and Subdivision Control Act. The former requires permits for dredging, draining, filling, and the construction of permanent structures on or over any tidal or subtidal land and conditions these permits on the protection of existing navigational and recreational uses. The latter enables municipalities to plan, zone, and control the subdivision of land to "conserve shore cover, visual as well as actual points of access to inland and coastal waters, and natural beauty." Both of these existed prior to 1976 and neither takes an aggressive stance toward access (i.e. both protect "existing" access only). The policy statement does not propose any stronger methods for expanding access opportunities.

Some coastal management program projects have had an affect on shorefront access. Coastal Zone Management funds have been granted to approximately a dozen municipalities for feasibility and design studies for the development of urban waterfront parks and fish ports and have also sponsored a State Bureau of Parks and Recreation study to identify potential sites for acquisition as state shorefront parks. Approximately fifteen communities have taken advantage of the Title Search Assistance Fund for legal research to identify existing public access points. Finally, CZM funds were used for a study which identified potential sites for fish piers. The state is now using Economic Development Administration and bonds funds to construct six of these facilities.

The legal definition of shorefront ownership and the conservative political attitude toward public access are the major obstacles which state coastal program officials face.

Michigan

The rights of private owners of lakefront property are so extensive in Michigan that state officials have decided that providing access across private property to the shore is an inadequate solution to meet recreational needs. Instead, they have focused on acquisition and development of whole shorefront parcels to be managed as public recreation areas.

According to Michigan law, the high water mark is the legal boundary between public and private ownership. The state holds the "bottomlands" located on the lakeward side of the ordinary high water mark in public trust under the provisions of the Submerged Lands Act. Riparian private property owners hold the land above the ordinary high water mark and have exclusive use of the state-controlled bottomlands when they become exposed by declining water levels. Thus, the public has a right to use the water, but access to the shore is limited by the "one foot in the water at all times" rule.

Historically, the state of Michigan has sought to increase public access opportunities by actively encouraging state, county, township and municipal acquisition of shorefront property. Three major sources of funds have been used to finance and acquisition. The Michigan Waterways Funds, which has been administered by the Michigan Waterways Commission since 1945, matches a 90% federal contribution with a 10% state or local contribution to develop facilities for public access to Michigan waters. These funds have been used for the construction of boat launching ramps to docks. Nearly 1,000 public access sites have been built and equipped for boaters since this program began. Since 1965 the Michigan Department of Natural Resources has allocated federal Land and Water Conservation Funds for recreation-related acquisition and development programs throughout the state. Shorefront properties have been a major priority of the Michigan LAWCON program. In 1976 the Michigan Recreational Land Trust Fund Program was established. The fund raises revenues from leases and royalities from oil, gas, and mineral extraction activities on state lands and uses this money to acquire property as public recreation sites.

Michigan's response to the 1976 access amendments to the Coastal Zone Management Act has not been particularly innovative. Basically, the coastal management program has been used to reinforce existing state access policies and programs. The Department of Natural Resources, which has the responsibility for the state's environmental and recreational programs, is the lead agency for coastal management. Policy is set at the state level and technical assistance and grants are made available to the local level to implement access programs.

The Michigan shorefront access policy statement follows the OCZM guidelines and refers extensively to the policies defined in the 1974 Michigan Outdoor Recreation Plan (SCORP) and those expressed by various divisions of the Department of Natural Resources. The emphasis on public acquisition of land for recreational purposes is reinforced, but few other management techniques for increasing shorefront access are presented.

Coastal management officials have defined their responsibilities toward access in terms of providing technical assistance and "seed money" to local communities to help leverage additional funding for the acquisition and development of access sites. Approximately half of the Michigan coastal management program access budget goes toward grants to local communities. The majority of these grants fund projects which create, improve, or preserve shorefront access. At the time of application for state coastal management funds for project feasibility review and design, local communities are asked to identify potential funding sources for land acquisition and/or project development. After the design phase is complete, state coastal management officials provide technical assistance to help procure this outside funding.

Urban waterfront parks have been popular local projects which increase shorefront access opportunities. In Detroit three small, existing parks will be tied together by a riverfront walkway, part of which extends through an industrial section of the city. Similarly, several small public parcels in Mackinaw have been tied together with a walkway which necessitated the granting of easements across private property. In St. Ignace, a 3/4 mile long promenade will provide visual access to the shore, but no physical access. Bike path designs have been funded in Detroit, Marquette, Grand Haven and Sault Ste. Marie.

Michigan coastal management funds have also been granted to seven state parks for the construction of boardwalks which provide access across dunes and wetlands while preserving these environmentally fragile areas.

The major access problems which Michigan coastal management officials cite as needing ongoing attention are a lack of shorefront access in urban areas (especially Detroit) and protection of sensitive environmental areas from the pressures exerted by those seeking access to the shore.

New Jersey

Although the legal definition of the "public" portion of the beach allows for liberal use of New Jersey beaches, a history of shorefront development and exclusionary local regulations has made access to the beach itself difficult. The New Jersey Department of Environmental Protection is currently implementing regulatory, financial incentive and transportation programs designed to open up more shorefront access opportunities for the public.

The public portion of the New Jersey beach is from the vegetation line seaward. Thus, the public has the right to use both the wet-sand and dry-sand portions of the beach. However, most shorefront developments have blocked visual and/or physical access to the shore. Some municipalities which own land adjacent to the vegetation line have enacted laws or regulations which make shorefront access inconvenient, expensive or impossible for non-residents. Inland residents have been deprived of adequate access to shorefront recreation areas.

The state of New Jersey placed the responsibility for administering its coastal management program within the Department of Environmental Protection. The Department has begun to take a more aggressive stance on providing public shorefront access. Actions of the Department are based on New Jersey's simply stated shorefront access policy:

> "Coastal development adjacent to coastal waters shall provide maximum practicable public access to the shorefront, including both beach and built-up waterfront areas and both visual and physical access. Shorefront development that limits public access and the diversity of shorefront experiences is discouraged."

The Department of Environmental Protection is currently using four methods to implement its shorefront access program. First access regulations are imposed on new shorefront developments. The Department of Regulation requires that each project provide "maximum practicable access" to the shorefront as a condition for the issuance of a development permit. Projects can provide access through such means as public walkways, boat launch ramps, or public parking lots. Second, shorefront access conditions are also placed on Shorefront Protection Funds, which finance dune restoration, erosion control and beach nourishment projects. Municipalities which accept these funds must pledge to establish and maintain public access to the area being funded. Third, provision of waterfront access is an eligible activity encouraged by CZM grants to local communities. Grants of this type are being used for such projects as developing urban waterfront parks. Fourth, the Department of Protection sponsors shuttle bus transportation each summer weekend from a parking lot near the Garden State Parkway to a state beach 12 miles away. Approximately 8,000 people each summer take advantage of this service which provides shorefront access to inland residents of New Jersey.

New Jersey coastal management officials continue to search for solutions to the state's four major shorefront access problems: (1) the best beaches are difficult to get to without a private auto; (2) almost none of the beaches are free any more (most charge an admission price); (3) the supply of bath-houses and parking lots is inadequate to meet demands; (4) the large amount of development on the shorefront is causing major erosion problems which could necessitate the closing of beaches to use by the public.

North Carolina

Unlike most coastal management programs, the North Carolina Coastal Area Management Act places major responsibility at the local level. As a recent addition to the coastal management mandate, the issue of shorefront access is being treated in a similar manner.

Lack of adequate public access to beaches had been a concern in North Carolina prior to the passage of the 1976 amendments to the federal Coastal Zone Management Act. Each year coastal communities were experiencing an increase in the volume of day visitors to their beaches, yet few of these municipalities had the incentive or tools to meet demands for access. According to North Carolina law, the public has a right to use the wet-sand portions of the beach, but in order to get to these areas, privately owned upland areas must be crossed. Coastal communities perceived their financial resources to be inadequate to acquire, police, maintain and service accessways to the public portion of the beach on a large scale. Private property owners were generally unwilling to allow public access across their land. Some communities had passed ordinances which required subdivison developers to dedicate the ends of streets running perpendicular to the beach and other accessways to the public. However, these accessways have traditionally been unmarked and tend to have little or no parking capacity.

In keeping with the coastal management framework which had already been established, general shorefront access policies were established in 1978 at the state level, but the primary responsibility of insuring adequate public access to the public portions of the beach was delegated to the local level. The shorefront access policy statement in the North Carolina Coastal Management Program declares that state policies are designed to "supplement and strengthen any local efforts."

The state's shorefront access policies assert that:

-development shall not interfere with the public's existing rights of shorefront access (i.e. those acquired through public acquisition, dedication, or customary use established through the courts);
-public beach area projects will not receive state or federal funding unless provisions are made for adequate public access

(including rights, identification and parking);
-coastal dunes shall be protected;
-the state shall continue to support and improve highway, bridges and ferry access to coastal areas;
-the state shall continue to support the development of fishing pier facilities;
-access shall be made available to people of all socio-economic groups.

The North Carolina Coastal Resources Commission (CRC) had mandated coastal communities to include a shorefront access element in their updated land-use plans, which are due to be submitted to the CRC for review and approval during the fall of 1980 and 1981. Specific issues to be addressed in these plans include the definition of "access," identification of existing and potential access points within the jurisdiction, identification of methods for improving access and provision of adequate parking to meet access needs. The CRC hopes that the state's shorefront access program will be adequately defined by these local plans; the Commission does not anticipate designing a program at the state level unless the local plans prove inadequate.

Approximately $100,000 of 306 funds have been granted to local communities to assist in shorefront access planning. Twelve municipalities have been granted funds to either develop beach access plans or design waterfront parks which will increase public access opportunities. Edenton a peninsula which juts into Queene Anne Bay will be developed as a park; a bulkhead will be installed to stem erosion, an icehouse will be restored for use by the public, and a parking lot will be developed. Wilmington is planning a six block long riverfront park. Cape Carteret, a non-beach community, is studying the feasibility of acquiring land for a parking lot at Emerald Isle to allow Cape Carteret residents access to beach areas. Holden Beach, Washington Park, Kill Devil Hills, Nags Head, Surf City and Carteret County are among the localities which have conducted studies to identify potential and existing access points and tools for developing more.

The North Carolina Office of Coastal Management has imposed provision of shorefront access as a condition on some major Coastal Area Management Act (CAMA) permits. Large oceanfront projects such as hotels and condominiums have been encouraged to provide public walkways and parking lots, but the permit review officials are unsure of their legal authority and have therefore been somewhat inconsistent in exercising this option.

During the summer of 1979 the North Carolina Department of Natural Resources and Community Development's Office of Coastal Management and Wrightsville Beach jointly funded a shuttle bus to bring people from a feeder parking lot in Wilmington to Wrightsville Beach. By all accounts, the experiment was a failure; ridership was extremely low and the project lost money. Some reasons cited for the failure include inadequate advertising and locating the feeder lot too near the beach (i.e. people who had driven as far as the feeder lot decided that they might as well drive the remaining distance to the beach). There are no plans to fund a revised attempt at providing mass transportation for beach access in North Carolina.

Oregon

The state of Oregon has been a forerunner in guaranteeing public access to the shore. Sanctions for public use of the beach date back to the nineteenth century. Oregon had passed a Beach Law which opened all beaches to public use long before the federal Coastal Zone Management Act was passed and thus had no problem complying with the 1976 amendments. State and local officials are now using the coastal management program to extend access rights into estuarine areas.

In 1893 the beaches of Clatsop County, Oregon were declared to be a public highway. In 1913 the Oswald West Act extended this policy by declaring that all wet-sand beaches were public highways. The "public" portion of Oregon beaches had been generously defined as the area from the non-aquatic vegetation line seaward. Thus, Oregon residents have historically had good use of the shorefront.

By the mid-'60s the state legislature was becoming increasingly concerned about the potential loss of public use of and access to Oregon's beaches as development pressures increased. In 1967 the Oregon Beach Bill was passed; it declared all beaches to be open to public use. A concerted effort has also been made to keep as much of Oregon's coastal land as possible in public ownership. Thirty state parks are located in coastal areas. The goal of providing public accessways every 1.5 to 3 miles along the coast has nearly been met.

The Oregon Department of Land Conservation and Development (DLCD) has been designated the lead agency for coastal management. Prior to the adoption of the 1976 amendments to the federal Coastal Zone Management Act the DLCD had defined the state's policies for shorefront access in the Statewide Planning Goals. Oregon DLCD officials consider shorefront access provision an important element in the state coordinated land-use planning process. The DLCD also coordinates the many access-related programs which various state agencies administer. Among these are the Oregon Beach Access and Parks Acquisition Program, the Department of Fish and Wildlife Angler Access Program. Each update of Oregon's Statewide Comprehensive Outdoor Recreation Plan (SCORP) provides supply and demand information and policies on shorefront access which are consistent with Statewide Planning Goals.

The Department of Land Conservation and Development is in the process of extending access program development and enforcement to the local level. Cities and counties are presently preparing comprehensive land use plans which are to comply with the Statewide Planning Goals.

Among the shorefront access issues to be addressed in these plans are the identification of further public access needs and the development and adoption of tools and controls to meet them. The local land use plans are due for LDCD review in July 1980. Based on the information gathered from these plans, state level coastal management officials will design a program of technical assistance to help strengthen and/or supplement the local plans and programs.

Oregon coastal management officials are presently working on the next step toward increasing public access to the shore. The issue of balancing access to estuarine areas with the protection of their fragile ecosystems is currently being studied for future policy and program development.

Rhode Island

The state of Rhode Island has a plentiful supply of shorefront access opportunities via public parks and beaches, conservation areas, scenic view points, and rights-of-ways. The physical configuration of the state and its shoreline allow many people to live within walking and biking distance of the coast. Coastal management program officials have given public recreational use and enjoyment of the coastal region a high priority. Thus, the shorefront access element of the Rhode Island coastal management program stresses the ongoing verification of public rights-of-way, publication of their location, and continued evaluation of development proposals to insure that they do not jeopardize the public's right to use and enjoy the shore.

Legal rights to shorefront access have a long history in Rhode Island. During the seventeenth century the King James Charter designated rights-of-way to the shoreline for fishermen and seaweed collectors. Since that time, some of these rights-of-way have been built upon, some forgotten, and others kept in continuous use as accessways. In more recent times, the state and many municipalities have been active in setting aside coastal land for development as public recreational areas. Eighty-five parks of various sizes currently provide public access to the shoreline; 66 of these are locally managed, 18 are state parks, and 1 is managed by the federal government.

Thus, even before the Coastal Zone Management Act was passed and a state CZM program was set up in Rhode Island, state and local officials had developed management capacities for providing public access to the shore. The Rhode Island Department of Environmental Management is the lead agency for most forms of recreational access; it implements various recreational programs and policies and has the authority to acquire and manage parks, beaches, launch ramps and conservation or management areas as set forth under Title 42, Chapter 17 of the Rhode Island General Laws. The amendment to the Coastal Resources Management Council's enabling legislation has assigned the Council the responsibility for identifying and opening public rights-of-way to the shore. Local municipal recreation authorities also establish and maintain recreation areas which provide shorefront access.

The shorefront access policy statement in Rhode Island's coastal management program is one of the best written to date. It recognizes that the state is already richly endowed with public shorefront access opportunities, but also points out that the distribution of coastal rights-of-way is geographically uneven, that public rights-of-way along substantial lengths of shoreline remain unverified, that public rights in many presumed rights-of-way are contested, that parking at popular rights-of-way is frequently inadequate, and that vandalism, littering and trespassing on neighboring private areas are issues of growing concern. Four access policies are set forth:

(1) Public recreational use and enjoyment of parks and beaches is a high priority use of the coastal region.

(2) No permit shall be issued for development if it interferes with public use or enjoyment of the shore.

(3) Legally protected rights-of-way are to be verified and their existence publicized.

(4) The Coastal Resources Management Council recognizes the management responsibility of the Department of Environmental Protection and municipal recreation authorities as they relate to public beaches and parks, and shall endeavor to support these responsibilities to the maximum extent, consistent with other Council policies and requirements.

The coastal management program in Rhode Island is funding efforts to solve two of the major shorefront access problems in the state: lack of lateral access at beaches (i.e. access to both the "wet-sand" and "dry-sand" portions of the beach) and ill-defined rights-of-way to the beaches. Coastal management officials are also exercising the permit review power to place conditions on new coastal development in order to protect shorefront access opportunities for the public.

The "public" area of Rhode Island beaches is generally defined as the area from the mean high water mark seaward. 306 funds are currently underwriting a study to investigate the potential of broadening this definition of public beach. The constitutional, legislative, and judicial principles of shorefront access are being examined in anticipation of the drafting of future legislation to open up more of the beach to the public.

306 funds are also being used to complete the process of verifying the King James Charter rights-of-ways. This effort was begun with mostly

volunteer assistance in the mid-1960's. The earlier study identified as many rights-of-way as possible without doing a full scale search of legal documents. The rights-of-way which were identified by that study are presently being maintained through a partnership between the Coastal Resources Management Council (which uses CZM money to fund two maintenance workers) and the Rhode Island Marine and Fisheries Division (which uses state funds for four to six maintenance workers). Maintenance includes grass cutting, trash collection, rock clearing and stairway construction and maintenance.

Using CZM funding, the state has hired twelve attorneys to document every possible right-of-way. Public meetings are being held on a town-by-town basis to identify public perceptions of rights-of way, land is surveyed, and title searches are conducted. Program administrators intend to file new deeds which will document existing public accessways. All practical rights-of-way will be opened; those in environmentally sensitive areas and those located very close together will not necessarily be kept open and maintained.

The third major element of Rhode Island's shorefront access program is the stringent use of the development permit review power. Permits for development within the coastal zone are always conditioned on protecting the existing public access to the shorefront.

Thus, the Rhode Island coastal management program is currently working to solve two of the major shorefront access problems (beach use, ownership and identification of access), but other access issues remain to be dealt with. Among concerns voiced by coastal management officials are the lack of parking near access and the disappearance of natural coastal areas as development pressures grow.

Texas

The state of Texas does not yet have a coastal management program approved by the federal Office of Coastal Zone Management, but public access to the shorefront is protected under the provisions of the Texas Open Beaches Act of 1959. Access planning and program management falls under the jurisdiction of the Texas Parks and Wildlife Department.

The Texas Open Beaches Act was passed in reaction to attempts by private landowners to close some Galveston Island beaches to the public, thereby challenging a right which the public had acquired through the continuous use of the beaches, bays and gulf waters of Texas since the time of Spanish ownership. The Texas Open Beaches Act guarantees the public unrestricted access to and the use of (1) the state owned portion of beaches below the mean high tide line (wet-sand), and (2) the larger area extending from the mean high tide line to the first line of vegetation (dry-sand) in all areas where the public has acquired a right of use of easement by presecription, dedication, or continuous use. The

The provisions of the Act do not apply to any beaches on barrier islands or peninsulas that are not accessible by public road or common carrier ferry. This includes San Tose Island, Matagorda Island, and the Matagorda Peninsula west of the Colorado River channel. A coastal area is considered accessible if it meets any of the following three criteria:

(1) if it is presumptively legally open

(2) if it is in public ownership

(3) if for any reason the public has a legally protected right of access to and use of the area.

The Texas Open Beaches Act is enforced by the Texas Attorney General, county attorneys, and criminal district attorneys. Early judicial tests of the Open Beach Act by littoral landowners failed to either invalidate the Act or find any beach covered by the Act to be closed to the public. These early failures, combined with a policy of rigorous enforcement by the State Attorney General and local legal officers, have discouraged any subsequent attempts to close a beach which is covered by the Act.

The Texas Parks and Wildlife Department (TPWD) has employed two methods to obtain public access to the shorefront. First, a program has been established to provide boat ramps in coastal areas. Second, a land acquisition program has been instituted to provide public recreation areas along the coast.

The TPWD Special Boat Fund is supported by revenues collected from boat registration fees and a portion of the unclaimed water boat fuel taxes. The state provides 70% of a boat ramp project's cost if the locality where the access facility is to be located provides the remaining 30%. A completed project includes a boat ramp, parking area and access road. Once the facility is constructed, it is maintained by the TPWD. Recently, approximately three boat ramp projects have been constructed in coastal areas each year.

The TPWD has the authority to acquire new public shorefront recreation areas as the public need arises. Determination of this need is based on the Texas Outdoor Recreation Plan (TORP). Acquisition is funded by the TPWD's Fund 31, which is supported by a portion of the state cigarette tax revenues. Approximately $11 million are available each year for acquisition and development. Local acquisition projects are assisted by federal Land and Water Conservation Funds which are administered by TPWD.

The provision of increased public access to Texas beaches has caused some management problems. Among the major problems which must be addressed are congestion (even though other accessible beaches are not crowded, Galveston and Mustang Beaches and the North Padre Islands sustain heavy use), vehicular traffic on the beaches (which can endanger non-vehicular beach users), and litter.

COASTAL ZONE '80
Volume III

Proceedings of the Second Symposium on Coastal and Ocean Management

Billy L. Edge, Editor

Sponsors:

American Society of Civil Engineers
(Waterway, Port, Coastal and Ocean Division)
Office of Coastal Zone Management
Edison Electric Institute
Heritage Conservation and Recreation Service
Maritime Administration
National Aeronautics and Space Administration
National Marine Fisheries Service
Office of Sea Grant
Public Works, Canada
U.S. Army Corps of Engineers
U.S. Coast Guard
U.S. Fish and Wildlife Service
U.S. Geological Survey

Affiliates:

American Petroleum Institute
American Shore and Beach Preservation Association
Bureau of Land Management
California Coastal Commission
Environmental Data and Information Service
Federal Emergency Management Agency
Marine Technology Society
National Park Service
State of Florida Department of Environmental Regulation
State of Florida Department of Natural Resources
The Coast Alliance
The Coastal Society
Urban Land Institute
U.S. Department of Energy

Diplomat Resort Hotel
Hollywood, Florida
November 17-20, 1980

AMERICAN SOCIETY OF CIVIL ENGINEERS FOUNDED 1852

Published by the
American Society of Civil Engineers
345 East 47th Street
New York, New York 10017

Library of Congress Catalog Card No. 80-69152
ISBN 0-87262-258-4
Manufactured in the United States of America.

Cover photo by Thomas A. Beckett, New York, NY

CONTENTS

Volume I

*Papers not available at time of printing

*Papers not available at time of printing

*Papers not available at time of printing

*Papers not available at time of printing

*Papers not available at time of printing

*Papers not available at time of printing

***Papers not available at time of printing**

*Papers not available at time of printing

*Papers not available at time of printing

*Papers not available at time of printing

*Papers not available at time of printing

*Papers not available at time of printing

*Papers not available at time of printing

Volume III

*Papers not available at time of printing

*Papers not available at time of printing

*Papers not available at time of printing

*Papers not available at time of printing

***Papers not available at time of printing**

ACCESS TO COASTAL RESOURCES:

The Access Concept Comes of Age

William W. Dreyfoos
Environmental Planning Coordinator
Berkeley-Charleston-Dorchester
Council of Governments
Charleston, South Carolina

In the context of coastal resources, access has been an evolving concept. Initially, access meant beach access, and referred to physical ingress and egress in sandy beach areas. Over the past few years, the access concept has expanded: first to include parking, transportation, maintenance and similar accoutrements of access; then to expand the areas served by accessways to include non-beach waterfront areas; and finally to broaden the scope of access considerations to include recreation opportunities and the quality of recreation experiences. The literature fleshed out the access concept, providing guidance on the mechanics of acquisition and suggestions concerning the nature of governmental involvement in access matters.

As the concept has matured, governmental concern over access has proliferated. Spurred by the 1976 amendments, OCZM has promoted access considerations in the state coastal management plans. Access programs are highly visible in Oregon, Rhode Island, and in California's local coastal plans. In some waterfront communities, local governments have taken the initiative in addressing access concerns, and there are stellar examples of waterfront redevelopment and local programs to increase public access to coastal resources. There now seems to be widespread recognition that access is important. Writing in the July-August, 1980 issue of South Carolina Wildlife, Dr. John Armstrong notes that both governments and the public "are beginning to become aware that access is a key word in coastal use. We must have access to the many coastal resources of the state, the beaches, the fish, the water surface, and so on. All of this is what we now call coastal management."[1]

While most national attention has been lavished on the access programs that have proved successful, the great majority of both state and local access efforts have been poorly organized and largely ineffectual. Moreover, most governmental attempts to deal with access issues have lacked a clear conception of access problems, program goals, and the techniques available for addressing access issues. The state efforts in particular have displayed an almost myopic preoccupation with acquisition. California, for example, measures the success of its local access plans in terms of miles of accessways acquired, as opposed to more people-oriented measures such as increased recreational opportunities. Acquisition is undoubtedly an important aspect of access to coastal resources, but it is only one among a number of components that combine to produce effective public access. The literature has advanced the access concept to include a wide variety of concerns. Governmental programs have not kept pace.

The 1976 amendments and OCZM's call for access elements in the state coastal management plans provide the vehicle for addressing coastal access concerns. But to date, the opportunities afforded by these mechanisms have been poorly used, and at best OCZM's input has made only incremental differences in the state access programs. The problem has been a lack of direction, and the failure to enunciate clear guidance concerning the purpose and content of state access elements. Despite good intentions, most states simply haven't known what to do in developing their access programs. As a result, their attempts to piece programs together have failed to take advantage of the insights and approaches developed in the literature. The literature does suggest ways of resolving the current malaise in access programs. The starting place is a reexamination of what the state access programs can and should be doing to facilitate the use and enjoyment of coastal resources.

Defining the Scope of Access Programs

Most governmental programs treat access as physical ingress and egress in waterfront areas, perhaps with an adjacent place to park cars. Recreational use of the waterfront is generally presumed. This approach usually embodies a "more is better" bias, which assumes that acquisition is feasible and that public use of coastal resources is best promoted by acquiring accessways. Under this view, access becomes a thing that can be acquired, an end in itself.

Simply put, this traditional approach is the wrong way

to define access. And not surprisingly, the root of many states' problems in developing access programs lies in this restrictive definition.

Access is not, or should not be, an end in itself. Access is a tool. In a coastal context, it is the means of getting people from where they are to where they can enjoy the use of coastal resources. So defined, access consists of three components:

1. the support services, such as transportation and parking, that enable people to reach the point where an accessway begins

2. an accessway - i.e., a strip of land that ends in the coastal resource to be enjoyed

3. the coastal resource itself, and the characteristics that affect its viability and its use.

All three components must be addressed if meaningful access is to be secured.

However, most access programs deal primarily with accessways, component #2. They talk briefly if at all about component #1, and ignore component #3. The resulting access programs lack both perspective and a direct answer to the question "Why do we need an access program?"

In contrast, this question <u>can</u> be answered with reference to the more comprehensive scope of access, as outlined above, that accentuates the resources to which access is provided.

The Resource Approach

The 1976 amendments (Section 305(b)(7) of the CZMA) call for the development of planning processes "for the protection of, and access to, public beaches and other public coastal areas of environmental, recreational, historical, esthetic, ecological, or cultural value." The framers of this provision knew what they were doing. Access is placed in the context of coastal resources and their protection - first you have the resources, then you provide access to them.

Most state access programs have ignored the resource

approach outlined in the amendments. The reasons for this vary - many programs try to take the "easier" approach of just focusing on ingress/egress/parking. But examine how the contents of an access program seem to fall into place when using the resource approach to access problems.

Step 1 - Identifying the Resources

What are the resources to which access is desired? Depending upon the characteristics of the coast, these resources are likely to include the beach, the water surface for boating and fishing, shallow bottom areas for shellfishing, scenic vistas for more passive uses, and other similar segments of the coastal zone. The guiding principal is that access is a prerequisite to the enjoyment of any coastal resource - or in other words, you have to get there before you can use the resource. This raises the point that access is a part of resource use. And this means that many concerns should be raised in the initial portions of a state's access program:

> What quality of the resource, and resource use, does the state want to promote? How will its access policies, or lack of them, affect this quality?
>
> Are there existing policies (e.g., in the CZM plan) regarding the resources to which access is desired? To what extent are these policies inconsistent with increased public use of the resource?
>
> What is the demand for resource use? Is current access adequate?
>
> To what extent are access needs accommodated by the private sector? How much of the access shortfall should be borne by government?

In the resource approach, these questions must be answered for each resource the access program will address. These are difficult questions, despite their superficial simplicity. For instance, several involve the problem of crowding - what level of resource use begins to destroy the resource, what level reduces user enjoyment because just too many other people are there, what levels of use require accommodation because of demand? There are no easy answers to these questions.

However, these questions are not too complex to

preclude the development of access policies. Most state coastal agencies have worked in coastal matters long enough to know what their resources and resource problems are. Given this knowledge, the first step in developing access programs involves

1. picking the resources - i.e., identifying which ones the public uses, or wants to use

2. deciding which resources are underutilized or poorly enjoyed

3. evaluating whether increasing the access to these resources is consistent with state coastal policies

4. selecting resources, resource uses, and geographical areas in which to focus state access efforts.

By focusing on resources, the identification step produces an agenda that spans a state's coastline. Specific resources, in specific locations, are targeted for state attention. The goal is to enhance resource use and enjoyment. "Access" is defined to be the methods that facilitate this goal. Specific governmental actions will vary depending upon the resource and the problems currently encountered in its use and enjoyment.

As described above, this step is similar to the planning process used in developing the state plans. Some of the issues may be complex, but the orderly development of policy positions regarding resource use is eminently feasible. Most states have already addressed this issue in their coastal plans, so the process outlined here will be a refinement of their existing work. The resource focus provides the key that gives direction to a state's access element. Given the selection of resources, resource uses and geographical areas in which to concentrate the state's access efforts, plan development can proceed fairly easily.

<u>Step 2 - Evaluating the Access Problems</u>

Where the selected resources are poorly used, or underutilized, or where the quality of experiences involving these resources is low, the access program should develop strategies to enhance the use or enjoyment of these resources. What this amounts to is a planning process to upgrade the utilization of selected resources in coastal areas: the problems interfering with better resource utilization should be identified; methods for resolving the problems evaluated; alternative strategies developed and prioritized; and specific programs of action selected and organized.

Resource utilization problems are likely to fall within the three components of access previously identified: support services, accessways, and resource characteristics. The access program should contain an assessment of each component for each resource selected in Step 1.

Support Services. Most resources are likely to be lacking in some support services. In a beach context, for instance, roads and places to park are generally indespensible if beach resources are to be used for recreation purposes. Other support services such as restroom facilities, accessway maintenance, picnic areas and lifeguards may not be indespensible, but can contribute to the quality of beach recreation experiences. Other coastal resources have similar service concommitants: recreational boating is facilitated by boat ramps; passive recreation is facilitated by parks with benches and tables; the use of public shellfish areas is aided by signs and markers; and so forth. Some support services affect the quality of the public's use and enjoyment of coastal resources; other services determine whether the public is able to use the resources at all. The access program needs to identify whether deficiencies in support services are serious deterrents to resource utilization. Making better use of existing services should be the first priority. Increasing support services to selected resources should follow.

The access plan should address several specific questions in evaluating the adequacy of support services:

Are there transportation problems in reaching these resources? If so, could the problems be solved by additional roads? road improvements? better road maintenance? better traffic management? more parking area?

Do available transportation services have the effect of excluding certain groups from using the resources - e.g., does the lack of transit effectively exclude poor or elderly people without automobiles?

Which support services are necessary for use of the resources? Which increase public enjoyment of resource use? What capital and operating costs are involved?

These questions are designed to show which support services would actually increase the availability and and quality of experiences involving the resources. Additional support services will always be desirable. The job of the access program is to identify which service deficiencies are seriously constraining the public's use and enjoyment of the resources under study.

Accessways. Sometimes, the resource problem will be too few accessways. And in these instances, more walkways, more boat ramps, more parks may indeed be needed to promote better utilization of coastal resources. The literature has stressed access techniques and methodologies. A variety of guides are available to provide guidance in acquiring new accessways and preserving existing ones.[2] But under what circumstances should acquisition be used?

The resource approach to access creates a context in which accessway acquisition should be placed. Ideally, acquisition is appropriate

- when the new accessway will serve groups previously lacking access to the resource
- when the new accessway opens up a previously unused (or underused) resource to the public
- when the new accessway opens up an important new location of the resource to public use
- when the new accessway increases the variety or quality of resource uses available to the public
- when the accessway facilitates the use or enjoyment of multiple resources (e.g., a beachfront park that promotes beach use, passive recreation, boating and fishing).

The program should develop criteria for the selection of accessway acquisition as a strategy in addressing resource utilization problems. Because accessway acquisition is usually expensive, either in monetary outlay or administrative costs, programs should steer away from acquisitions that fail to accomplish one of the goals listed above - other and probably less expensive strategies are likely to be more effective in enhancing resource utilization.

Resource Characteristics. The assumption that always underlies access programs and presentations is that a resource exists at the other end of the accessway. This resource is presumed to be viable, usable (e.g., for recreational purposes) and stable enough to weather the increased use occasioned by the accessway.

The oversimplicity of this assumption is illustrated in three brief examples. Hilton Head, South Carolina, was developed as a resort island in the late 60's and 70's. Hilton Head boasts a beautiful man-made harbor and many waterfront homes. The marina and private docks provide primarily private access for recreational boating. Unfortunately, the development of the island and near-shore dredging have changed the current patterns at the south end of the island. As a result, the harbor is silting up, and many residential docks - boat accessways - are in danger of being landlocked even at high tide. The accessways are still there - the docks can still function like docks. But the resource at the end of the docks, the water, has changed. The lesson: effective access requires management of the resource to which access is provided.

Another Hilton Head example has been repeated in other areas. There are public shellfish areas off Hilton Head, as in other parts of coastal South Carolina. These areas provide the public with access to oysters, an important coastal resource. However, Hilton Head has developed in the absence of a sewer system, and the combination of malfunctioning septic systems and stormwater runoff have degraded the water. The shellfish areas have been closed.

The third example concerns beach erosion, best typified by Miami Beach, although many other communities illustrate the problem. In Miami Beach, there is no beach at high tide - only water at the edge of the seawall. Beach use is possible only at low tide, when the sand reappears.

These illustrations demonstrate that resource integrity is intrinsic to effective access. Access, as a tool, makes it possible for people to use a resource. It can be effective only so long as the amenities of that resource are preserved. In the example of the Hilton Head docks, the docks are still there - there is ingress and egress associated with a site - but a dock that ends in a highland marsh is of questionable value as an access instrument for boating.

The resource approach argues that access cannot be divorced from resource management. Accordingly, access programs must take cognizance of the status of coastal resources, and must tailor access activities to the dynamics and characteristics of these resources. In addressing the use of coastal resources, access programs must consider a number of questions concerning the resources themselves:

> How fragile is the resource? Is it stable enough to accommodate the proposed increase in use?
>
> Is the resource currently managed by state or local programs? If so, how is access/public use managed?
>
> What activities, at what levels, will adversely affect the viability of the resource? What techniques are available to avoid or mitigate these adverse effects?

The final question is based upon those listed above:

> Is increased public use/access appropriate for the resource, and if so, under what limitations.

This final question makes clear the point that access may not be considered appropriate for all coastal resources, or in all areas. For many resources, there will be a trade-off between public utilization and the preservation of the integrity of the resource. Most of Georgia's barrier islands, for example, are inaccessible except by private boat - the policy gives ecosystem integrity precedence over public use.

The resource approach argues that resource vitality must be considered in the process of providing access - so that, at a minimum, the state will understand the trade-offs involved in pursuing various access alternatives. Each state program must develop its own policies concerning these trade-offs, and concerning the scope of management activities subsumed within its coastal program. It should be noted that the resource depletion evidenced on Hilton Head was preventable only by wide-ranging controls over both land and water use. Depending upon the scope of resource management, the states may wish to scale down their access programs to contain fewer, but more manageable, resources.

Step 3 - Administration

The 1976 amendments place responsibility for access programs at the state level. But frequently, the responsibility for coastal resource management is shared by a number of state agencies, only one of which may be the agency that administers the CZM program. And many of the techniques for accessway acquisition, maintenance and enforcement are appropriate for local, rather than state, governments. Much as in the coastal management plans themselves, the access programs must determine how access activities will be administered, and which agencies will do the work. Networking may be possible, and a model based on California's local coastal plans seems plausible, with local efforts guided by firm state policies (though different ones from those used in California). As with the coastal management plans, the organizational frameworks for the access programs will differ from state to state, depending upon a variety of administrative and jurisdictional characteristics.

In determining the administrative structure for its access program, each state should pay particular attention to the agencies charged with implementing and enforcing the program. Again, a Hilton Head example shows the necessity for program enforcement. Hilton Head has some public beach accessways, which are platted and marked with signs. One such accessway is adjacent to a beachfront home. There, the adjoining homeowner paved over the accessway, connecting it with his driveway, removed the sign, and erected a basketball goal. No governmental enforcement action followed. This particular accessway is still platted as public property, but its presence has not increased public utilization of the beach. The lesson here is that programmatic activities must include implementation measures, and must be tailored to those activities that can actually be carried out.

Put It All Together And ...

The premise underlying the resource approach is that the state access programs are ineffectual because they have lacked a focus. Resource utilization can serve as that needed focus. This approach may have the effect of making the access programs a capstone to the state coastal management plans, and may be criticized as elevating access beyond its proper role in the totality of coastal management. However, this criticism seems to fade in light of the opportunity that the access programs provide.

The coastal management plans (or at least most of them) attempted some rough allocation of coastal resources between developmental and governmental (deemed preservation) interests. Although the governmental interests were viewed as representing the public against private attempts to use up coastal resources, people seemed to get left out of many of the plans.

In the context of coastal resources, the access programs now provide a way of integrating people back into the resources and the activities that the state programs manage. Their goal, if the states choose to adopt it, can be to maximize public use and enjoyment of coastal resources, while first and foremost preserving the integrity of these resources and of coastal processes. To many, this goal is the promise of CZM - managed use. Now that most states have recovered from the trauma of the planning process, there is time and knowledge available for bringing people and their resource-related activities into the CZM framework. For states that are serious about managing their coastal resources, the access programs provide an opportunity for them to do so.

Notes

[1]Armstrong, "Coastal Development and Planning for the Future," South Carolina Wildlife and Marine Resources Department, South Carolina Wildlife, July-August 1980, p. 9.

[2]See, for example

Brower et al., Access to the Nation's Beaches: Legal and Planning Perspectives, Sea Grant Publication UNC-SG-77-18, 1978.

Brower & Dreyfoos, "Public Access to Ocean Beaches: If You Find a Parking Space, How Do You Get to the Beach?" in Coastal Zone Management Journal, Vol. 5, No. 1/2, 1979.

Dreyfoos, Planning for Beach Access: A Manual for Florida Local Governments, FAU-FIU Joint Center for Environmental and Urban Problems, Ft. Lauderdale, 1979.

Office of Coastal Zone Management, Shorefront Access and Island Preservation Study, 1978.

An Environmental Study to Assess the Effect of Drilling Fluids on Water Quality Parameters During High Rate, High Volume Discharges to the Ocean.[1]

R. C. Ayers, Jr., T. C. Sauer, Jr., and D. O. Stuebner
Exxon Production Research Company
P. O. Box 2189
Houston, TX 77001

R. P. Meek
Ecomar Inc.
158 Santa Felicia Drive
Goleta, CA 93017

ABSTRACT

This paper presents results from a drilling fluid dispersion study which was conducted to determine mud component concentrations in the water column during high volume, high rate mud discharges representative of the type that normally occur when the drilling of the well is completed. The test was conducted in the summer of 1978 in the Gulf of Mexico at a water depth of 23 meters.

In two discharge tests, 250 barrels of mud were discharged at a rate of 275 bbl/hr, and 389 barrels of mud were discharged at a rate of 1000 bbl/hr. The mud was a chrome lignosulfonate-clay type having a bulk density of 2.09 g/cc (17.4 ppg). Water samples were taken at the bottom, top and most dense portion of the discharge plume using a rosette sampling array suspended from a helicopter. This novel sampling technique allowed rapid and almost continuous sampling of the plume. Transmittance, dissolved oxygen, temperature and salinity were monitored as a function of depth and distance from the source using instruments suspended from the helicopter. Divers took samples and underwater photographs of the discharge plume close to the discharge source. Water samples were analyzed for solids content, and barium, aluminum and chromium concentrations.

During both discharges the mud formed two plumes, a lower plume and an upper plume. The lower plume contained the bulk of the discharged material and descended quickly to the sea floor. As the lower plume descended an upper, near surface plume was generated by turbulent mixing of the lower plume with seawater. The upper plume was several meters thick, existed in the water column for a much longer time than the lower plume, and drifted away from the source with the current. All sampling and hydrographic measurements were made on the upper plume. Suspended solids and metal tracer concentrations in the plume reached background levels about 500 meters from the discharge source during the 275 bbl/hr test and about 1000 meters from the discharge source during the 1000 bbl/hr test. Transmittance was the only hydrographic variable affected by the discharge. All other hydrographic variables remained unchanged

[1] This work was funded by the Offshore Operators Committee and Exxon Production Research Company.

from ambient conditions within the monitored interval (40-1500 meters from the discharge point).

A comparison of these results with previous work at lower discharge rates indicates that transport time (distance/current velocity) required for suspended solids values to reach background levels increases with discharge rate; however, even at high discharge rates background levels are reached at times of the order of 100 minutes. It is shown that most of the drilling mud settles rapidly and the material remaining in the water column (estimated to be 5-7% of the discharged material) has a minimal effect on ocean water quality.

INTRODUCTION

Upon completion of an offshore well it is often necessary to discharge excess drilling mud. In exploratory wells this is done rapidly to avoid costly delays in moving the rig off location. This paper presents results from a drilling fluid dispersion study which was conducted to determine mud component concentrations in the water column during high volume, high rate mud discharges representative of the type that normally occur when the drilling of the well is completed. Previous investigators[1,2,3] have studied the effects of drilling discharges on water quality. However, most of this work was directed toward determining effects during low rate cuttings discharges or was primarily designed to answer specific questions about a particular drill site. These studies did not directly address the effect of high volume, high rate discharges. In addition the development of improved sampling equipment designed specifically for this type of test allowed a more complete description of plume properties in the water column.

The field test was conducted in the summer of 1978 from the jackup drilling rig, Penrod 63, in the Gulf of Mexico. The rig, under lease to Exxon USA, was located in East Cameron Block 121 approximately 96 km south of Cameron, Louisiana in a water depth of 23 meters. The test was carried out using a high density mud discharged under calm sea conditions in order to demonstrate a "worst case" condition for dispersion of drilling fluids offshore.

METHODS

During each discharge test the mud was pumped from the mud pits through a system of pipes across the mud ditch and down a discharge pipe immersed three meters below the surface. The rig's mud pumps were used to pump the mud at a constant rate. Two discharge tests were performed on consecutive days. On the first day 250 barrels of mud were discharged for 54.5 minutes at a constant rate of 275 bbl/hr and on the second day 389 barrels of mud were discharged for 23.3 minutes at a constant rate of 1000 bbl/hr.

Current data was obtained with an array of three self contained Endeco Model 105 ducted impeller current meters. The meters were positioned approximately 15 meters from the platform's bow support leg at depths of 7, 14, and 21 meters. A Weather Measure Corporation Model W103 anomometer and vane were used for wind velocity measurements. The wind speed and direction system was installed on the rig 35 meters above the water surface. Wave height data was recorded with a Senso-Metrics Model SP-91 pressure transducer. Measurements of wave-induced water motions (speed and direction) were collected with a Marsh-McBirney Model 511 electromagnetic current meter (EMCM). Both of these systems were placed at 3-meter depths.

Divers observed and photographed the plume from its source at the downpipe through the water column to a distance of 20 meters downcurrent. A 12-minute 16-mm color underwater film was made during each discharge to document the behavior of the mud as it entered the receiving water.

Divers collected water samples within 15 meters of the discharge source using a meter tape attached to the discharge pipe and a depth gauge for positioning. Ten-liter Van Doren bottles were used to collect water samples in the plume near the discharge source. Plume sampling at distances greater than 40 meters from the discharge source was accomplished using hydrographic and water sampling equipment suspended from a helicopter.[4]

During each test the helicopter visually located the lateral center of the plume at various distances downcurrent from the discharge source. When over the plume, the helicopter lowered the water sampler and sensors through the plume. Transmission records were continuously monitored to describe the density and vertical extent of the plume. Profiling continued until the lower depth limits of the plume were observed on the transmissometer records. The sampler was then slowly raised back through the plume to points of maximum vertical density as indicated by the transmission records, where water samples were taken.

After collecting a full compliment of water samples (12), the helicopter returned to the heliport where the water sampler was exchanged for an empty one. Within five minutes hydrographic sensors and transmissometer were removed from the first sampler and placed into the second. This exchange permitted almost continuous sampling of the plume by the helicopter. After the exchange water samples were drained into acid washed, distilled water rinsed, 20-liter plastic containers and stored for analysis. The sampler was then prepared for the next exchange. This profiling and sampling sequence was continued as the visible discharge progressed downcurrent until no plume could be found by profiling through the water column. Control hydrographic profiles and water samples were taken at various depths upcurrent from the platform before and after each discharge test.

Water column samples were filtered through pre-weighed 0.45 micron Nuclepore filters. Solids for aluminum, barium, and chromium from water were analyzed by instrumental neutron activation analysis.[5] Filtering and analysis details are given in Ecomar.[4]

RESULTS

The composition and metals analysis for the mud used in both tests is shown in Table 1. The mud was a high density chrome lignosulfonate type, typical of most high weight mud used offshore.

TABLE 1

MUD COMPOSITION AND METALS ANALYSIS

Component	%
Barite	62.0
Low Gravity Solids	5.9
Chrome Lignosulfonate	0.9
Lignite	0.9
Inorganic Salts	0.5
Water	29.8
Density	2.09 g/cc
pH	12.4

METALS ANALYSIS OF MUD AND MUD SOLIDS

Component	Mud, mg/l	Solids, mg/kg
Solids	1,430,000	
Barium	642,000	449,000
Aluminum	15,360	10,800
Chromium	540	378

Particle size distribution for the mud solids were determined by the hydrometer method[6] to estimate settling rates (Figure 1). Distributions were determined in both deionized water and synthetic seawater.

Environmental conditions were extremely calm on both test days. Wave heights were 1-3 feet and winds were out of the east at approximately three knots. During the 275 bbl/hr discharge test, the current speeds and directions were 16.8 cm/sec and 216°, 22.0 cm/sec and 324°, and 1.0 cm/sec and 154° at 7, 14 and 21 meter depths, respectively. For the 1000 bbl/hr test, the currents speeds and directions at 7, 14 and 21 meters were 15.8 cm/sec and 210°, 15.8 cm/sec and 298°, and near 0 cm/sec and 48°, respectively.

Visual Observations

The discharged mud formed two plumes as it entered the water column, a lower plume and an upper plume. The lower plume descended rapidly and was relatively small in diameter, roughly cylindrical in shape, and extremely turbid. As the plume descended turbulent mixing occurred and water entrainment was rapid causing a billowing and swelling effect. The divers noted that the expanding column of discharged mud continued downward striking the ocean bottom just a few meters downcurrent from the discharge source. Upon impact the lower plume formed a fluffy cloud that spread slowly along the bottom. During both tests bottom currents were essentially nil, so little drift with current was noted. During the 1000 bbl/hr discharge test the flexible pipe broke free of its securing lines and jetted mud in several directions causing the lower plume to be more dispersed. However, diver observations and underwater photography confirmed that the bulk of the material quickly settled to the bottom in the vicinity of the rig.

In the immediate vicinity of the discharge the upper plume was being generated by turbulent mixing of the lower plume with seawater. The upper plume, composed of fine particles and soluble components, separated from the lower plume during its descent. The upper plume was much more diffuse and puffy than the lower plume, was several meters thick, and drifted away from the source with the current. In both tests the discharge plume moved away from the rig with the same velocity and direction as the current measured at seven meters. During the 275 bbl/hr discharge the plume thickness varied from four to 13 meters and appeared to move along the 18-20σ_t pycnocline (~14 meters deep). During the 1000 bbl/hr test the plume was thicker (12-16 meters) and more dense, and moved along the 22σ_t pycnocline (~17 meters deep). All of the sampling operations were directed toward determining properties of the upper plume.

Temperature, Salinity, and Dissolved Oxygen

During both tests salinity and temperature measurements in the discharge plume were equal to background at the distance the closest helicopter samples were taken (40 meters from the discharge source). This was generally true for dissolved oxygen values also. However, during the 1000 bbl/hr test two readings of reduced oxygen concentration were obtained during the first sampling cycle. These occurred at distances

FIGURE 1

PARTICLE SIZE DISTRIBUTION OF TEST MUD SOLIDS

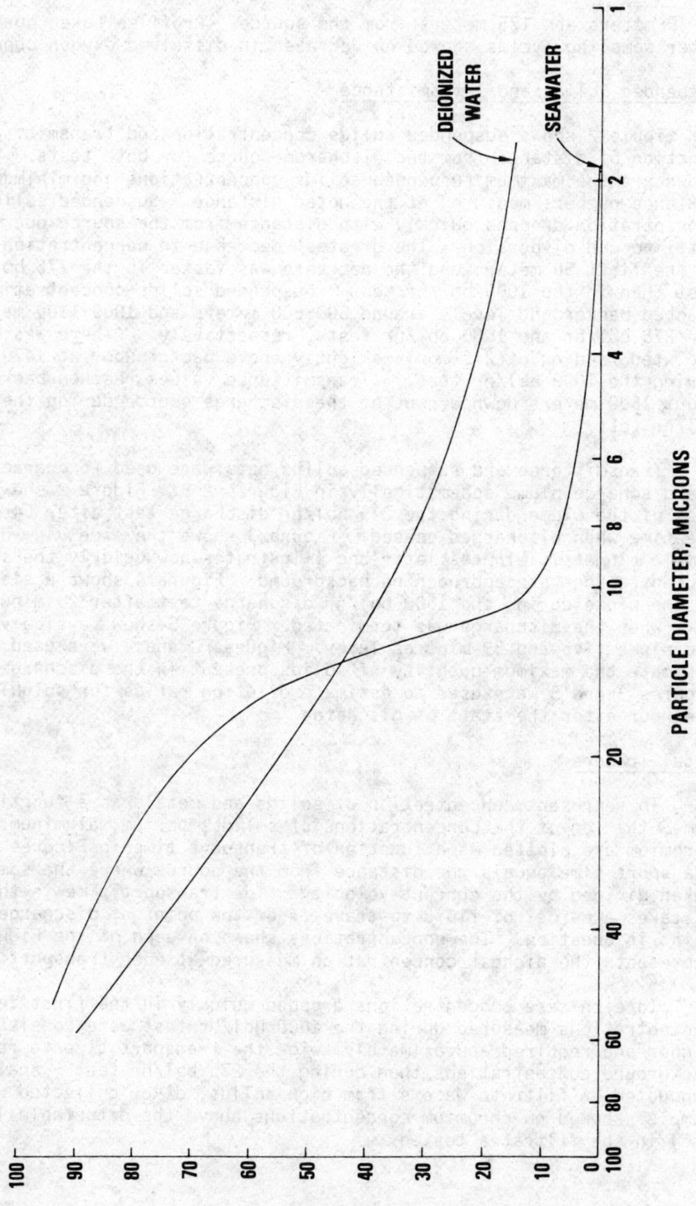

of 40 meters and 175 meters from the source. Profiles taken during later sampling cycles showed no decrease in dissolved oxygen concentration.

Suspended Solids and Transmittance

Table 2 shows suspended solids concentration and transmittance as a function of distance from the discharge source for both tests. Values shown are the maximum suspended solids concentrations and minimum transmittance numbers measured at the noted distance. Suspended solids concentration dropped quickly with distance from the source due to settling and dispersion. The greatest decrease in concentration occurred in the first 50 meters and the decrease was faster in the 275 bbl/hr test than in the 1000 bbl/hr test. Suspended solids concentrations reached background levels around 500-600 meters and 1000-1100 meters for the 275 bbl/hr and 1000 bbl/hr tests, respectively. (There was one isolated reading of 2.2 mg/l--slightly above background, at 1470 meters during the 1000 bbl/hr test.) Transmittance values reached background about 1500 meters downcurrent of the discharge source during the 1000 bbl/hr test.

Transmittance and suspended solids data were used to characterize the discharge plume schematically in Figures 2-5. Figure 2 shows a side view of the plume during the 275 bbl/hr discharge test after 54 minutes, the time when discharged ceased. Figure 3 shows the side view of the sample plume six minutes later and illustrates how quickly the solids concentrations are approaching background. Figure 4 shows a side view of the plume during the 1000 bbl/hr discharge test after 23 minutes, the time when the discharge was terminated. Figure 5 shows a side view of the plume five and 39 minutes later. Figures 2 and 4 were used to estimate the maximum quantity of solids present in the discharge plume. Figures 3 and 5 were used to estimate dilution ratios for soluble components one hour after the start of discharge.

Metal Tracers

To represent concentration of solids and metals as a function of time, the log of the concentrations of solids, barium, aluminum, and chromium are plotted as a function of transport time in Figures 6 and 7. Transport time equals the distance from the source where the sample was taken divided by the current velocity. The transport time is the time it takes a parcel of fluid to travel from the point of discharge to the point in question. The concentrations shown on each of the Figures represents the highest concentration measured at each transport time.

In each case concentrations dropped quickly in the first few minutes. Concentrations measured during the 1000 bbl/hr test were consistently higher and required approximately twice the transport time to reach background concentrations than during the 275 bbl/hr test. Analyses conducted on filtrate waters from high solids, diver collected water samples showed no chromium concentrations above the detectable limit (20 ppb) in the filtrates tested.

TABLE 2

SUSPENDED SOLIDS CONCENTRATION AND TRANSMITTANCE VS DISTANCE

DURING HIGH RATE DISCHARGES

275 Barrel/Hour - 250 Barrels Discharged

Distance From Source, Meters	Depth, Meters	Solids Concentration, mg/l*	Transmittance, %*
0 (Whole Mud)	-	1,430,000	-
6	8	14,800	-
45	11	34	2
138	9	8.5	56
250	9	7.0	48
364	9	1.2	37
625	9	0.9	71
Background		0.3-1.9	76-85

1000 Barrel/Hour - 389 Barrels Discharged

Distance From Source, Meters	Depth, Meters	Solids Concentration, mg/l*	Transmittance, %*
0 (Whole Mud)	-	1,430,000	-
45	11	855	0
51	12	727	0
152	11	50.5	2
375	16	24.1	4
498	14	8.6	23
777	13	4.1	21
878	2	1.2	71
957	12	0.83	76
1470	11	2.2	82
1550	9	1.1	82
Background		0.4-1.1	80-87

*Maximum concentration and minimum transmittance measured at noted distances downcurrent of the source.

FIGURE 2

SIDE VIEW OF PLUME DURING 275 BBL/HR TEST AFTER 54 MINUTES

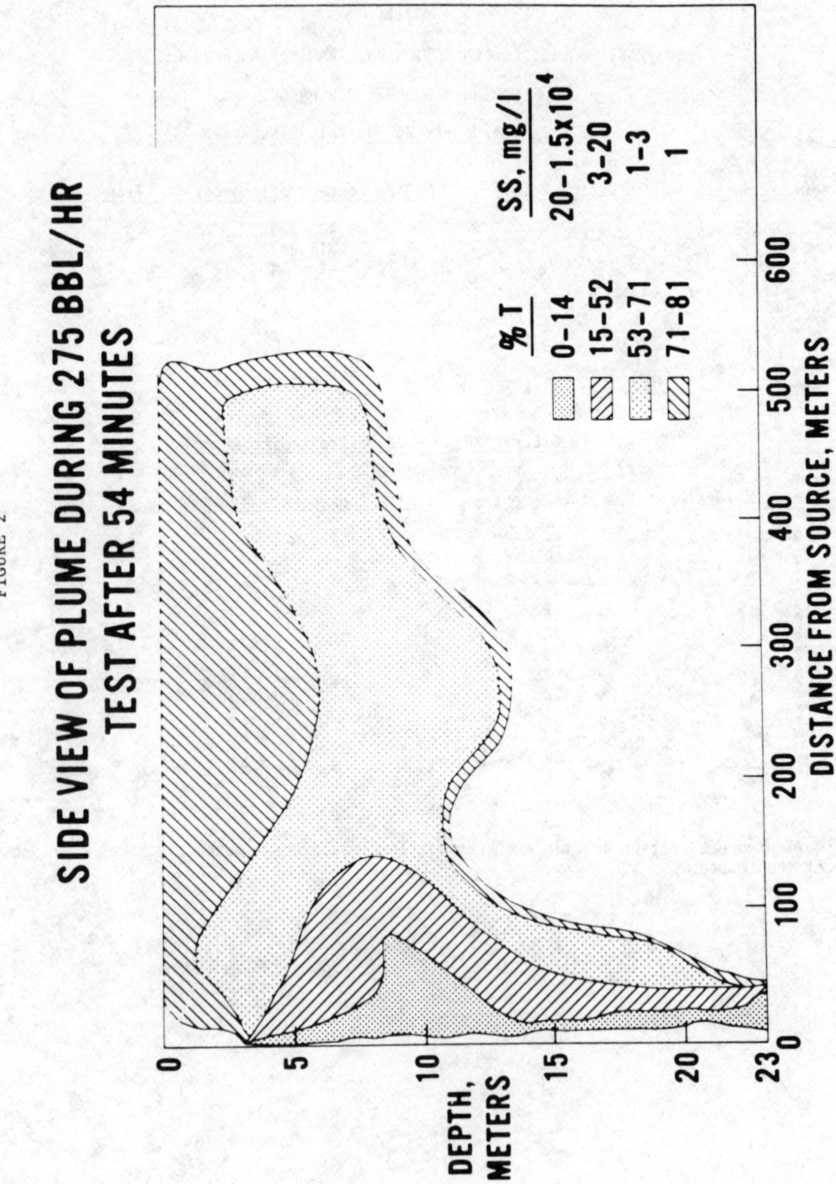

FIGURE 3

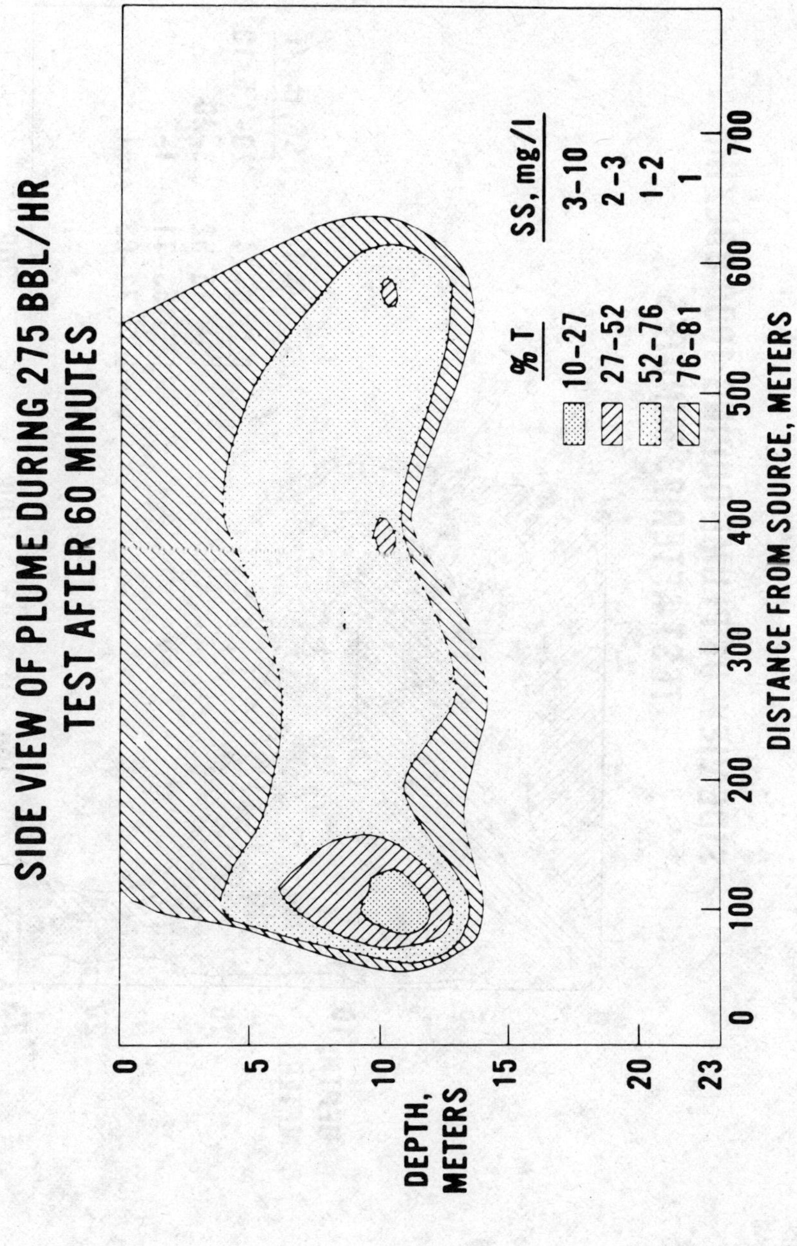
SIDE VIEW OF PLUME DURING 275 BBL/HR
TEST AFTER 60 MINUTES
DEPTH, METERS
0
5
10
15
20
23
0
100
200
300
400
500
600
700
DISTANCE FROM SOURCE, METERS
% T
10–27
27–52
52–76
76–81
SS, mg/l
3–10
2–3
1–2
1

FIGURE 4

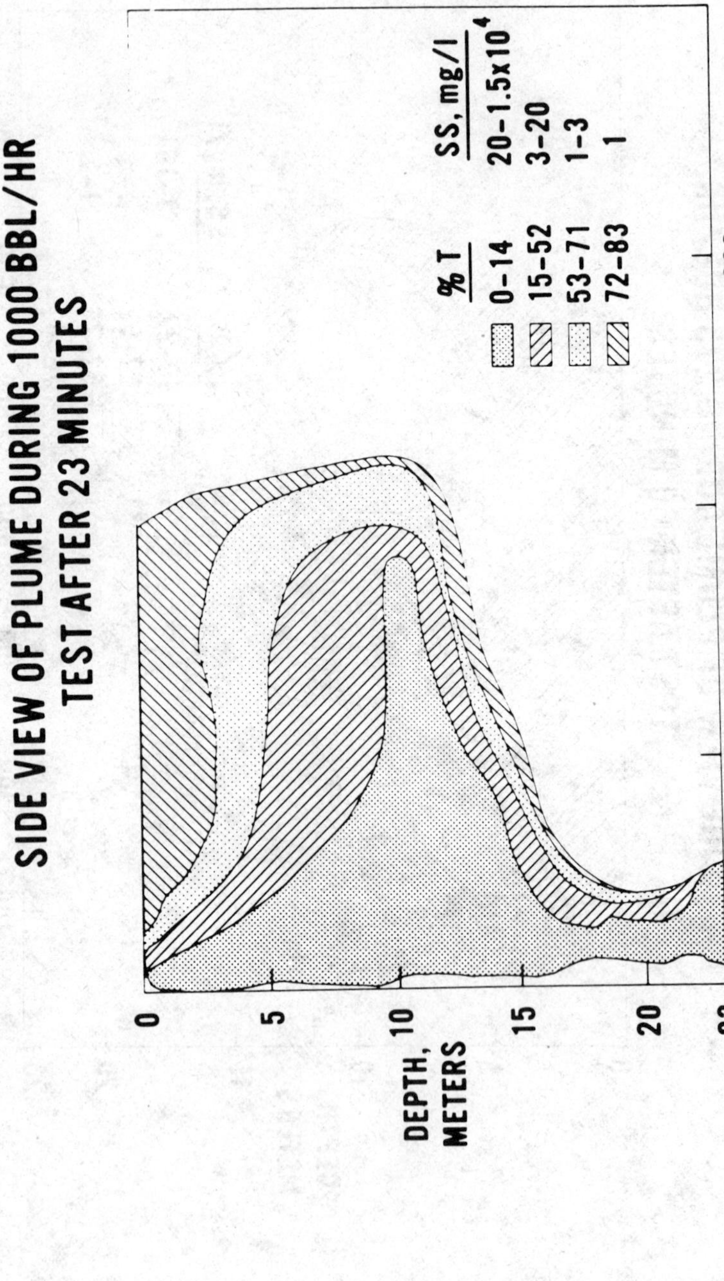
SIDE VIEW OF PLUME DURING 1000 BBL/HR
TEST AFTER 23 MINUTES
DEPTH, METERS
0
5
10
15
20
23
0
100
200
300
DISTANCE FROM SOURCE, METERS
% T
0–14
15–52
53–71
72–83
SS, mg/l
20–1.5x10^4
3–20
1–3
1

FIGURE 5

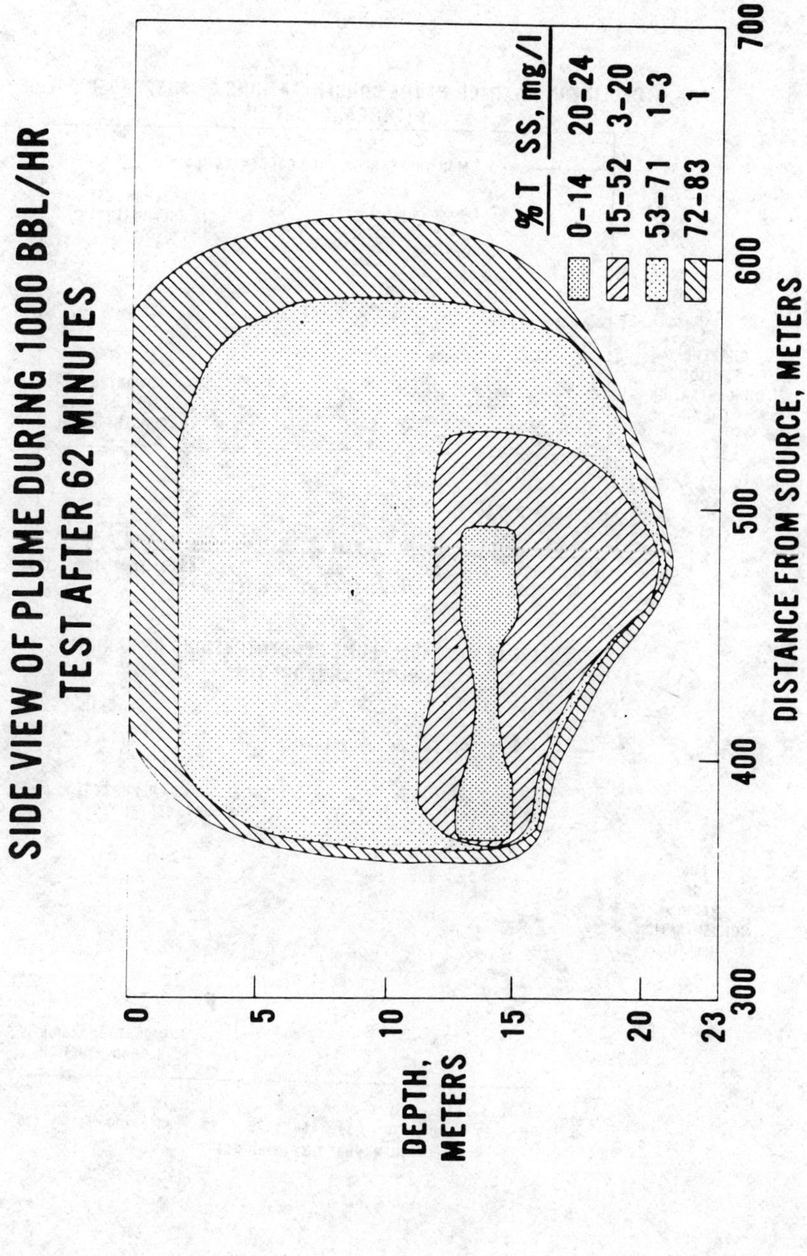
SIDE VIEW OF PLUME DURING 1000 BBL/HR
TEST AFTER 62 MINUTES
DEPTH, METERS
0
5
10
15
20
23
300
400
500
600
700
DISTANCE FROM SOURCE, METERS
% T
SS, mg/l
0–14
20–24
15–52
3–20
53–71
1–3
72–83
1

FIGURE 6

DECLINE IN MAXIMUM PLUME CONCENTRATIONS OF SUSPENDED SOLIDS WITH TRANSPORT TIME

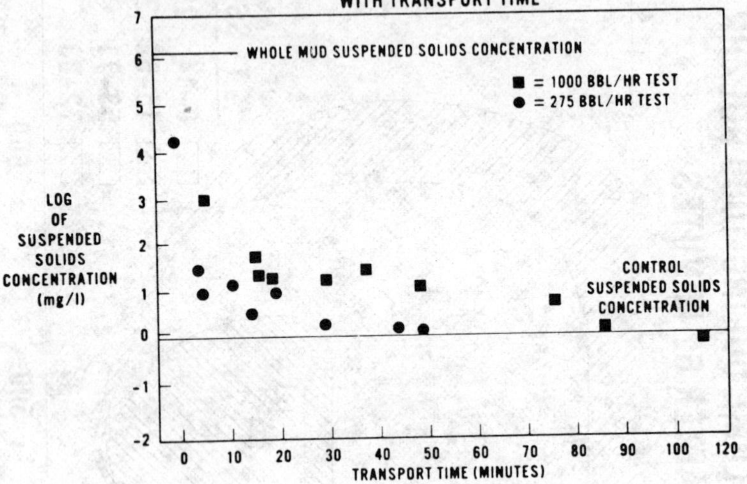

DECLINE IN MAXIMUM PLUME CONCENTRATIONS OF BARIUM WITH TRANSPORT TIME

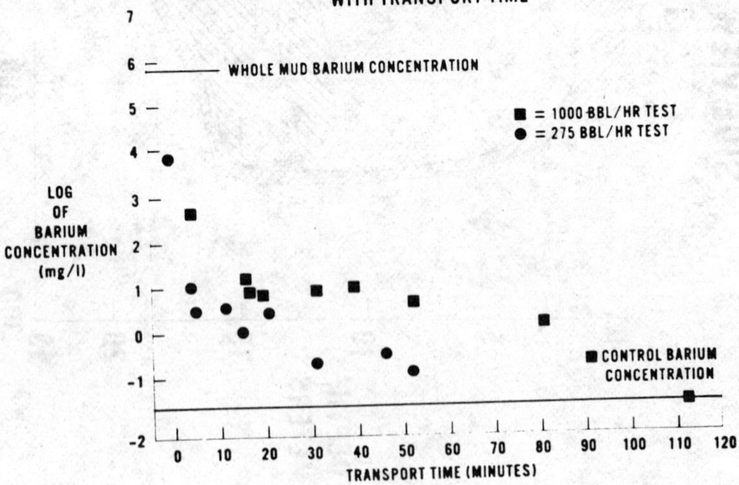

FIGURE 7

DECLINE IN MAXIMUM PLUME CONCENTRATIONS OF ALUMINUM WITH TRANSPORT TIME

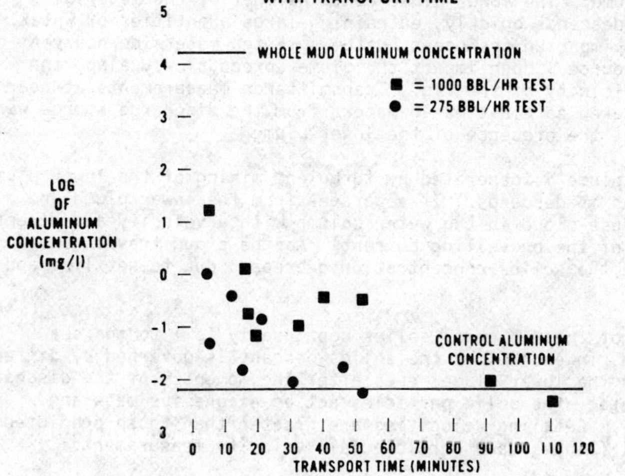

DECLINE IN MAXIMUM PLUME CONCENTRATIONS OF CHROMIUM WITH TRANSPORT TIME

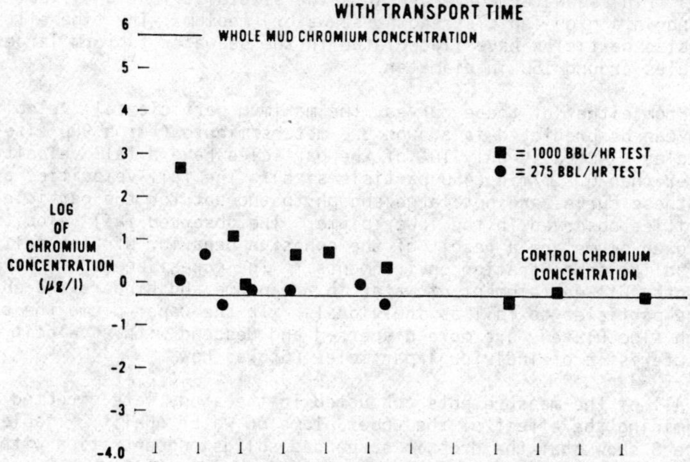

DISCUSSION

It is appropriate to describe the short term fate of drilling mud discharged at a high rate in terms of a two plume model; a lower plume and an upper plume. The lower plume contains most of the material discharged and descends quickly, entraining large quantities of water. In our study it impacted the ocean bottom some ten meters downcurrent of the discharge source. Upon impact the plume spread slowly along the bottom in the vicinity of the rig. Transmittance measurements at near bottom depths taken as close as 40 meters from the discharge source were unable to detect the presence of the lower plume.

The upper plume is generated by turbulent mixing of the lower plume with seawater as it descends. It separates from the lower plume and moves as a diffuse cloud in the water column with a velocity and direction characteristic of the prevailing current. As the cloud travels away from the source the solids concentration decreases due to settling and dispersion.

From both of these plumes, solids continually fall to the sea floor. However, in each plume the solids descent is governed by different mechanisms. In the lower plume, even after the momentum of the discharge jet has dissipated, the solid particles act as a cohesive mass and descend rapidly. Settling velocities are greater than those predicted by conventional (Stokes Law) particle fall velocity measurements.

Figure 1 shows particle size (fall velocity) distributions of mud solids used in the two discharge tests. The freshwater (deionized) curve illustrates the size distribution of independent free-falling solids. The seawater curve presents the size distribution of solids falling in a high ionic strength (seawater) medium. The fine silt and clay size particles have flocculated in the seawater to form larger particles around 15μ in diameter.

From either of these curves, the maximum particle fall velocity which can be predicted is around 0.8 meters/minute (for a 90μ size particle). Actually only 10% of the particles have a fall velocity greater than 0.2 m/min (45μ particle size). The fall velocities predicted from these curves are not large enough to account for the particle velocities observed in the lower plume. The observed fall velocities in the lower plume are a result of the cohesive behavior of fine solids in high solids concentration environments.[7] The cohesiveness of the particles prevents full entrainment of water in the plume and hinders the ability of the particles to fall as individuals. In the upper plume the solids though flocculated, are more dispersed and descend with a velocity more characteristic of individual particles (Stokes Law).

All of the measurements conducted in the study were directed toward determining the effect of the upper plume on water quality. Table 2 and Figure 6 show that the drop in suspended solids concentration with distance (or transport time) due to dilution and settling was rapid. Concentrations measured near the discharge source were higher than those measured in other tests conducted at lower rates[2] however, the rate of decline in concentration with distance agrees well with previous results. Solids concentrations in the water column were three to four orders of magnitude less than concentrations in the whole mud at distance of 50 meters downcurrent from the discharge source.

Barium, aluminum, and chromium concentrations (Figs. 6, 7) also showed this decline and tracked the suspended solids data closely. Due to the higher density of barite it was anticipated that barium levels would fall off faster than aluminum levels (tracking the low gravity solids) however this did not occur to a significant extent. Consistently, the 1000 bbl/hr rate showed concentrations at four to five minutes transport time one to two orders of magnitude greater than those found in the 275 bbl/hr discharge. Also, the time to reach background levels was twice as long for the 1000 bbl/hr discharge test. This effect may not be due completely to rate since in the 1000 bbl/hr discharge test the flexible pipe broke free of its securing lines and jetted mud in several directions thus creating an uncharacteristically high concentration of solids in the upper plume.

The discharge rate versus transport time required for plume solids to reach background concentrations for this study and for the Tanner Bank Study[2] is plotted in Figure 8. The equation for the increase of transport time with rate is $\text{Time} = 5.95\ (\text{Rate})^{0.42}$, with a correlation coefficient = 0.99. The good data fit is an interesting result considering that the tests were done in different offshore areas with different mud compositions, discharge volumes, oceanographic conditions, and discharge configurations. The above equation or Figure 8 combined with discharge rate and ocean current data provide an estimate of the distance from the source that bulk drilling discharges affect water quality.

Previous work[2] using lower discharge quantities and rates found no effects (except for light transmission) from discharged drilling fluids on receiving water hydrography. This was true here as well except for the two isolated readings of lower dissolved oxygen concentrations that occurred during the first sampling cycle of the 1000 bbl/hr discharge.

During this study no pH measurements were taken due to equipment failure. However, no effects of pH were measured in other work.[2] Actually, pH should reach background levels even faster than the other hydrographic variables. Hydroxide ions in the mud react with magnesium ions in seawater to quickly drop the pH of the discharged mud to ten. After which less than two orders of magnitude of dilution is required to drop the pH to that of the receiving water.

Percent light transmittance was the most sensitive measure of the plume within the water column. Table 2 shows that transmittance values remain slightly below background even after suspended solids levels become indistinguishable from background. This is due to the large number of colloidal particles present in drilling fluid. These particles continue to scatter light effectively even when present in concentrations too low to significantly contribute to the weight of the suspended solids. This explains why in some cases mud plumes remain visible to the eye several hundred meters downcurrent from the discharge source.

Some effects of background hydrography appeared to control the general character of the plume within the water column. These effects

FIGURE 8

TRANSPORT TIME REQUIRED FOR SOLIDS TO REACH BACKGROUND VS DISCHARGE RATE

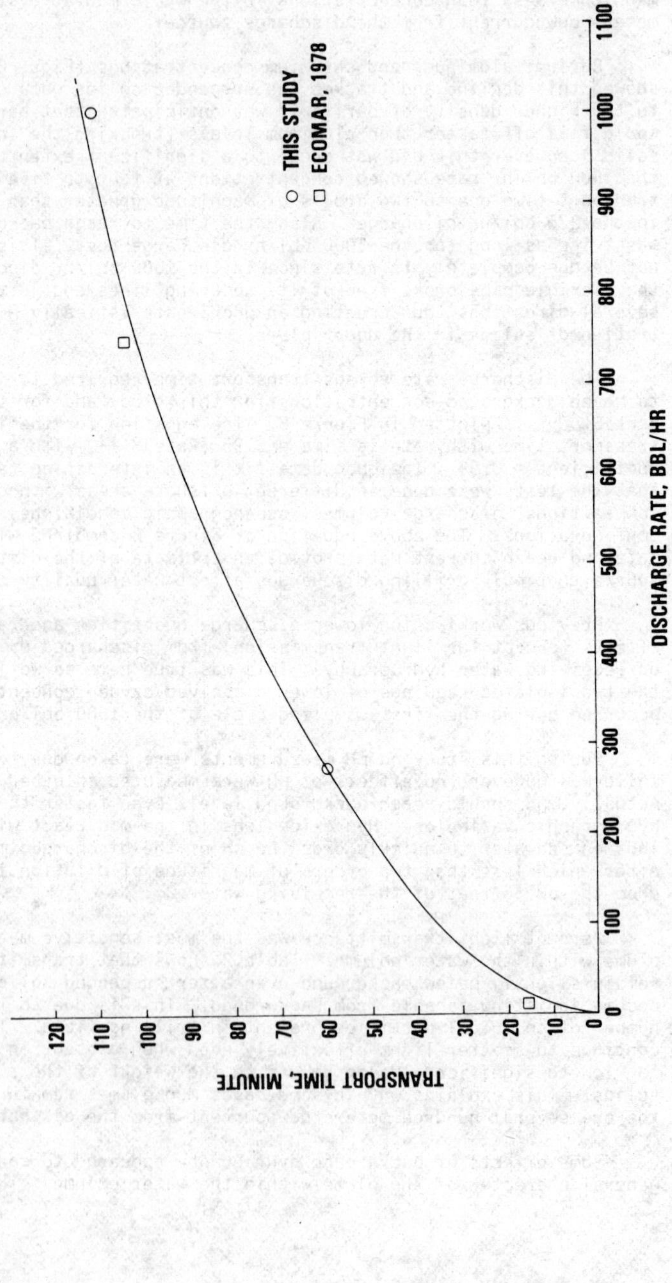

were primarily related to density stratifications. During both discharges the lower boundary of the upper plume (as defined by transmittance measurements) followed one of the major pycnoclines. These effects are illustrated in Figures 2 through 5 which show the behavior of the plume schematically.

It is important to note, however, that this effect of density stratification on the lower boundary only occurred where the plume was very dispersed. In both tests clouds of high solids concentration near the source were unsupported by the pycnoclines. It should also be noted that the pycnoclines did not concentrate, nor did they prevent solids within the upper plume from settling. However, the density clines did support the fine colloidal clay particles that cause the light transmittance reduction.

The maximum quantity of solids in the upper plume, based on the schematic diagrams presented in Figure 2 through 5 and measured suspended solids concentrations, was estimated to be 2.5 and 6.0 metric tons for the 275 bbl/hr and 1000 bbl/hr tests, respectively. The plume width was calculated using a simple model estimate.[8] Approximately 5% of the discharged solids was present in the upper plume during the 275 bbl/hr test and 7% during the 1000 bbl/hr test. The quantity of solids present in the upper plume decreases rapidly after discharge ceases as illustrated in Figure 3 for the 275 bbl/hr test and in Figure 5 for the 1000 bbl/hr test.

In this study we were unable to detect the presence of soluble chromium in the filtrate of the diver collected samples. However, Figures 3 and 5 allow us to estimate dilution ratios for possible soluble components we did not analyze for. This is done by assuming that all of the soluble material is contained in the upper plume, estimating the volume of the upper plume from Figures 3 and 5 (using Pritchard's model to estimate plume width) and then dividing the estimated volume by the volume of the liquid phase present in the discharged mud. In the case of the 275 bbl/hr test (Figure 3), the estimated plume volume at 60 minutes was 796,000 m^3 and the volume of liquid phase discharged was 24.7 m^3. This yields an average diluation ratio of 32,000 after 60 minutes. During the 1000 bbl/hr test (Figure 5) the estimated plume volume at 62 minutes was 562,000 m^3 and the volume of liquid phase discharged was 38.5 m^3. This yields an average dilution ratio of 14,500 after 62 minutes. An example of the method used in calculating these plume volumes is given in Ecomar.[4] In both cases the dilution ratio at the leading edge of the plume would necessarily be greater than these average values. The dilution ratios for suspended solids at these distances are some two orders of magnitude greater than the dilution ratios for the soluble components due to settling of the solids.

The quantity of solids trapped during the 1000 bbl/hr discharge was significantly greater than in the 275 bbl/hr test. This probably reflects the increased concentration of solids in the plume during the 1000 bbl/hr test. Note that the ratios of barium to suspended solids, barium to aluminum, and barium to chromium were closer to the mud solids value in the 1000 bbl/hr test. This may also be a reflection of the increased near source mixing that occurred when the discharge pipe broke free.

CONCLUSIONS

- The short term fate of drilling fluids discharged at a high rate to the ocean can be described in terms of a two plume model, a lower plume and an upper plume. The lower plume contains most of the solids and descends faster than the rate predicted by Stokes' Law for individual particles. The upper plume remains in the water column and drifts away with the current. In this study the upper plume was estimated to contain 5-7% of the solids discharged.

- Due to settling and dilution, suspended solids levels and metal concentrations drop quickly with distance near the discharge source. Concentration decreases of three to four orders of magnitude occur within 100 meters downcurrent of the discharge source. Concentrations continue to fall but at a slower rate as the plume moves further away. Concentration decreases of five to six orders of magnitude occur at distances of 500-1000 meters. Soluble component concentration decreases were estimated to be some two orders of magnitude less than those observed for the suspended solids.

- The tranport time required for suspended solids concentrations to reach background levels was found to be an increasing function of rate. The relationship appears to be a simple power law function. This correlation does not seem to be strongly influenced by other discharge conditions and may be used with ocean current data to estimate distances where discharged solids will reach background levels.

- The only hydrographic variable affected was light transmittance. Percent transmittance values approached but remained slightly below background levels even after suspended solids levels were indistinguishable from background. In the 1000 bbl/hr test transmittance values reached background about 500 meters past the point where suspended solids levels reached background.

- Because of the rapid decrease in suspended solids and metal tracer concentrations with distance and the temporary effect that the discharge have on seawater hydrography (transmittance only), it is concluded that drilling fluids have a negligible effect on open ocean water quality during high rate, high volume discharges.

REFERENCES

1. Ray, J. P. and E. A. Shinn. 1975. Environmental effects of drilling mud and cuttings, p. 533-545. In Conference proceedings on environmental aspects of chemical use in well-drilling operations, Houston. U. S. EPR-560/1-75-004.

2. Ecomar, Inc. 1978. Tanner Bank mud and cuttings study. Conducted for Shell Oil Co.

3. Dames and Moore. 1978. Drilling fluid dispersion and biological effects study for Lower Cook Inlet C.O.S.T. Well. Conducted for Atlantic Richfield Co.

4. Ecomar, Inc. 1980. Maximum mud discharge study. Conducted for Offshore Operators Committee/Exxon Production Research Co.

5. Ellis, K. N. and A. Chattotadahyay. 1979. Multi-element determination in estuarine suspended particulate matter by Instrumental Neutron Activation Analysis. Anal. Chem. 51:942.

6. ASTM. 1972. Standard method for particle-size analysis for soils, p. 69-79. In Annual book of ASTM standards. Amer. Soc. Testing Mater., Philadelphia.

7. Bowers, G. W. and M. K. Goldenblatt. 1978. Calibration of a predictive model for instantaneously discharged dredged material. Corvallis Environ. Res. Lab. U. S. EPA-600/3-78-689.

8. Pritchard, D. W. 1971. Design and sitting criteria for once-through cooling system, p. 224. In Proceedings of AIChE 68th annual meeting, Houston.

Coastal Management Planning
In Anchorage, Alaska
by

Anthony Burns
Municipality of Anchorage

INTRODUCTION

Alaska has in excess of 47,000 miles of coastline; however, 1/5 of 1 percent of the State's total coastal zone supports 60 percent of the total population. The location is Anchorage, Alaska with a coastline of approximately 100 miles in length. In less than twenty years, this area is projected to increase in population from 200,000 to just over 500,000 people.

The Cook Inlet area of southcentral Alaska is undergoing the most intense and rapid development in the State. The Municipality of Anchorage, the State's largest and most populated city, is located at the head of Cook Inlet on a roughly triangular piece of land between the Knik and Turnagain Arms of Cook Inlet (Figure 1).

Anchorage is located at the head of navigation for oceangoing vessels and is the terminus of four major land transportation routes as well as the hub of the State's air transportation network. Anchorage thus serves as a natural center for commerce and government.

The Municipality of Anchorage contains approximately 1950 square miles of land, but only 15 percent may be considered suitable for urban development and habitation. When one considers that the Chugach Mountains, the Chugach State Park and the Chugach National Forest prevent growth to the east and that Cook Inlet effectively blocks continuous development to the north, west and south (in addition to the two military bases immediately adjacent to metropolitan Anchorage), it becomes readily apparent that population growth is restricted and that densities throughout the Municipality are bound to rise as population growth continues.

PHYSICAL SETTING

A brief description of the physical environment will provide some understanding of the geographic situation in which coastal management planning has taken place.

The majority of the turbid water in Cook Inlet originates from Knik and Turnagain Arms, adjacent to Anchorage. Offshore oil activities are also largely confined to Upper Cook Inlet. Cook Inlet supports a major salmon fishery, and serves as a major migration route for spawning salmon.

Suspended sediment is currently the most detrimental and serious pollutant found in Upper Cook Inlet. The mean suspended sediment concentration in water near Anchorage was measured at 1,280 ppm. During the summer months the Knik and Matanuska Rivers (two major rivers flowing into the Knik Arm), discharge up to 150,000 tons of silt per day into the inlet.

Figure 1

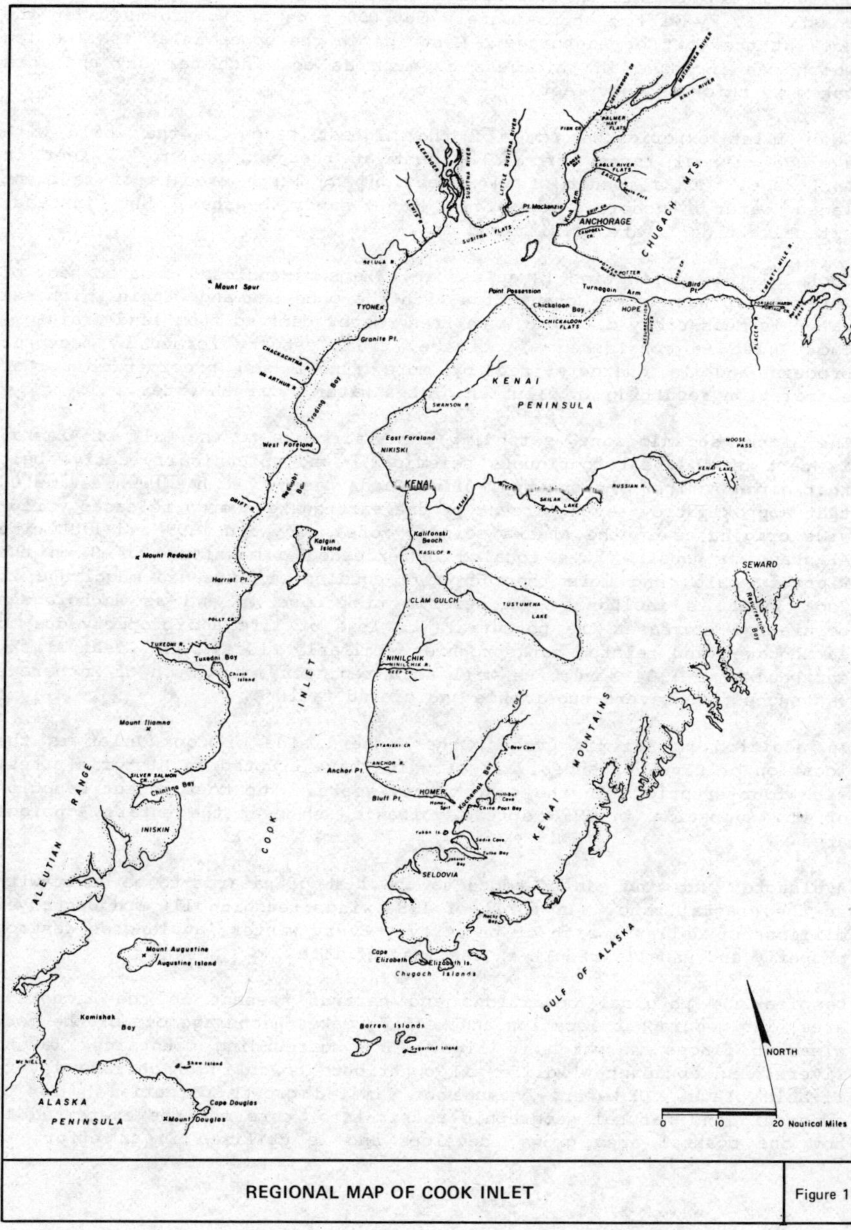

REGIONAL MAP OF COOK INLET

Figure 1

Upper Cook Inlet is subject to severe ice problems in winter. For approximately four months of the year, the inlet is covered with ice, acting as impediments to ship navigation and as scouring agents along the coast. Every winter approximately 500,000 tons of ice forms under the dock at the Port of Anchorage. Most ice in the upper inlet is floe ice which can increase in thickness as much as one inch per day and form cakes as thick as four feet.

Cook Inlet experiences some of the highest tides in the world with average diurnal ranges from 13.7 feet at the entrance to 33 feet in Anchorage. As a result of tidal action and large amounts of sediment laden water, Anchorage does not have sandy beaches, but instead, extensive tidal flats of silt and mud.

Cook Inlet by definition is an estuary: a semi-enclosed coastal body of water which has a free connection with the open sea and within which sea water is measurably diluated with fresh water derived from land drainage. Cook Inlet is considered a positive, tidal estuary formed by tectonic process and is characterized by more runoff and precipitation than evaporation resulting in dilution of sea water by fresh water.

The Alaska seismic zone, extending from Fairbanks to the Gulf of Alaska, is part of the near-continuous seismically and volcanically active belt that circumscribes the entire Pacific Ocean Basin. It has been estimated that approximatley seven percent of the earthquake energy released world-wide originates in the Alaska seismic zone. Between 1899 and 1980 nine Alaskan earthquakes have equaled or exceeded a magnitude of 8 on the Richter scale, and more than 65 have equaled or exceeded magnitude 7. Cook Inlet is included in a seismic risk zone 4, and as Anchorage's population increases the potential for loss of life and property due to earthquakes and related catastrophies similarly rises. The Great Alaska earthquake of 1964 serves us well as a reminder, when much of Anchorage was subject to severe subsidence and ground failure.

In addition to seismic events, the western side of Cook Inlet is the location of five volcanoes, all of which have erupted in historic times, with four eruptions in the past twenty years. The most recent eruption of Mt. Augustine in 1976 spread volcanic ash over the entire Achorage area.

Avalanches and wind can be added to the list of hazards to be dealt with in the coastal zone. In April of 1980 winds reaching 130 mph destroyed millions of dollars worth of property. Every winter, avalanches destroy property and usually result in the loss of life.

Despite the physical conditions and hazards present in the Anchorage area, its geographic location and setting makes Anchorage one of the more pleasant places in which to live. The surrounding mountains, lakes, rivers, and abundant wildlife all contribute to making Anchorage a very livable place. However, because of limited growth potentials (due to ownership and natural geographic constraints) care must be exercised in how the coastal area grows, develops and is utilized in an effort to balance preservation needs with development needs and maintain the resource values making Anchorage a good place to live.

Biological Resources

The five North American species of Pacific salmon (chinook, sockeye, pink, coho, and chum) are found in Cook Inlet. However, the waters of the Knik and Turnagain Arms serve primarily as a migration route for salmon migrating to freshwater spawning streams during the period May through September. Limited set net fishing is conducted in the upper Inlet, however.

The greatest potential for impact to fishery resources in Cook Inlet is development activity. The fishery resources are important not only because of significant commercial and sport fishing purposes, but secondary economic and social benefits associated with tourism and sport fishing. Fishery resources are sensitive to numerous factors related to human activities, and effective planning in the coastal zone is required to ensure maintenance of these important renewable resources. The intertidal areas at the mouths of streams need special management to ensure that damage caused by siltation, or disturbance of the stream bed, or toxic pollutants does not occur. Other fish species, such as smelt, are also common in the streams of upper Cook Inlet.

Numerous bird species are found in the Upper Cook Inlet region, particularly in the coastal wetlands and marshes.

Ducks and migratory waterfowl are common to several coastal wetlands in the Anchorage area, particularly in the Palmer Hay Flats (geese, swans, and ducks), and in Potter, Eagle River and Portage marsh lands.

These wetlands areas, used as stopover resting areas and as nesting areas, are subject to severe damage from human activity. Pollution from urban runoff and encroachment of urban development pose serious threats either by elimination of important habitat, food and cover or by fouling of the habitat.

Marine mammals are occasionally found in the waters surrounding Anchorage but are most common in the lower Inlet; however, increased water pollution in the upper inlet could cause an impact. No data is available to determine what types of pollutants or levels of pollutants would cause an impact.

Moose and bear are common in the Anchorage area and contribute to the natural resource values found in the coastal zone.

The Need For Coastal Planning and Mangement

Quite clearly then Anchorage is situated in an exceptionally severe but beautiful estuarine environment, however, competing uses and user groups in this coastal environment are increasing. Anchorage is dependent upon coastal water transportation for a majority of its supplies. The competing uses in the coastal area combined with limited available or suitable land, and an area subject to a variety of natural hazards makes coastal management planning a considerable challenge. Increased development within Anchorage and its various drainage basins will most likely be accompanied by a lowering of water quality and result in direct and significant impacts on coastal waters given a situation of little or no controls. The severity of impact is almost totally related to the intensity of development.

The livability and economic productivity of the coastal area makes it important, but why should there be concern about it? The answer is that the natural resource base which makes possible the livability and the economic activities of the coastal area is being changed by these very activities. The abundant natural resoruces of the Anchorage coastal area contribute to both economic development and attractiveness of the coastal areas as a place to live. The economic activities based on these natural resources affect coastal waters and shorelands in many ways: by diverting fresh water supplies; by encroaching onto wetlands; by producing goods that require shipping lanes and thus dredging; disposal of dredge material; and creating a public demand for roads and other public facilities. For example, a marsh is partially excavated and filled for residential development, marsh productivity is diminished as is the water recharge function that it serves. To assure a lasting and desirable mix of benefits from coastal resources in the face of growing demands upon then coastal management is being undertaken in Anchorage.

Coastal management has become a major concern because of a need to control and alleviate the environmental distrubances that are and will be associated with this rapid growth in the coastal plain while accommodating new urban growth and services. Although continued coastal growth and development is both desirable and probable, it is not assured, for the coastal area's ability to supply the resources needed for continued increases in productivity is uncertain. Man's actions do and will continue to cause direct and significant impacts on coastal waters at an accelerated rate unless sound management of our coastal resources is encourage and carried out. Man shapes his communites and structures his habitat through the use of land. Land use activities determine the nature and vitality of his environment. The entire spectrum of our human society depends on development practices which alter the natural state. Land is thus the base upon which must exist a man-made environment as well as the natural environment which support life. The need for effective planning to accommodate, in a balance manner, the growing demands in the Anchorage coastal area is needed now. The scope of this paper is on the methodology used to develop and implement the Anchorage coastal management plan.

Requirement of the Alaska Coastal Mangement Act

The Alaska Coastal Management Program provides that the State Office of Coastal Management establish guidelines and standards for local implementation, subject to administrative review and enforcement of compliance. The State Coastal Management Act created the Alaska Coastal Policy Council, a 16 member body composed of 7 state commissioner's and 9 local government officials. The Policy Council has responsibility for making sure coastal boroughs, Municipalities and cities look at all the opportunities for development, preservation and conservation consistant with the state guidelines and standards, and to make sure that all federal and state agencies respect the local plans once they are completed and approved. The State standards applicable to Anchorage include the following:

1. Coastal Development
2. Recreation
3. Historic, Prehistoric and Archaeological Resources
4. Engery Facilities and Facility Siting
5. Transportation and Utilites
6. Fish and Seafood Processing
7. Timber Harvesting and Processing
8. Mining and Mineral Processing
9. Subsistence
10. Geophysical Hazard
11. Air, Land and Water Quality
12. Habitats
13. Shoreline Access
14. Coastal Erosion

For each of the fourteen uses or activities the state has promulgated a standard, which have the force and effect of regulations.

All coastal resource districts, such as Anchoraqe, are required by State law to develop and adopt coastal management plans in accordance with the regulations adopted by the Alaska Coastal Policy Council. In addition to the above listed state standards, ten specific program elements must be followed:

1. Statement of needs, objectives and goals.
2. Municipal program organization necessary to implement and carry out a management plan.
3. Identify and delineate a coastal management boundary.
4. Conduct a resource inventory sufficient for program development and implementation.
5. Conduct a resource analysis.
6. Include a description of the land and water uses subject to the plan.
7. Identify proper and improper uses and activities.
8. Develop policies that will be applied to land and water uses and activities subject to the district program.
9. Develop implementation methods and authorities which will be used to implement the plan.
10. Develop effective public participation opportunities.

Methodology

The key to the methodology resided in three specific tasks: (1) a detailed resource inventory, (2) a resource analysis, and (3) a thorough legal review and analysis. Figure 2 illustrates the planning methodology.

Figure 2

COASTAL MANAGEMENT RESOURCE INVENTORY AND ENVIRONMENTAL ANALYSIS FLOW CHART

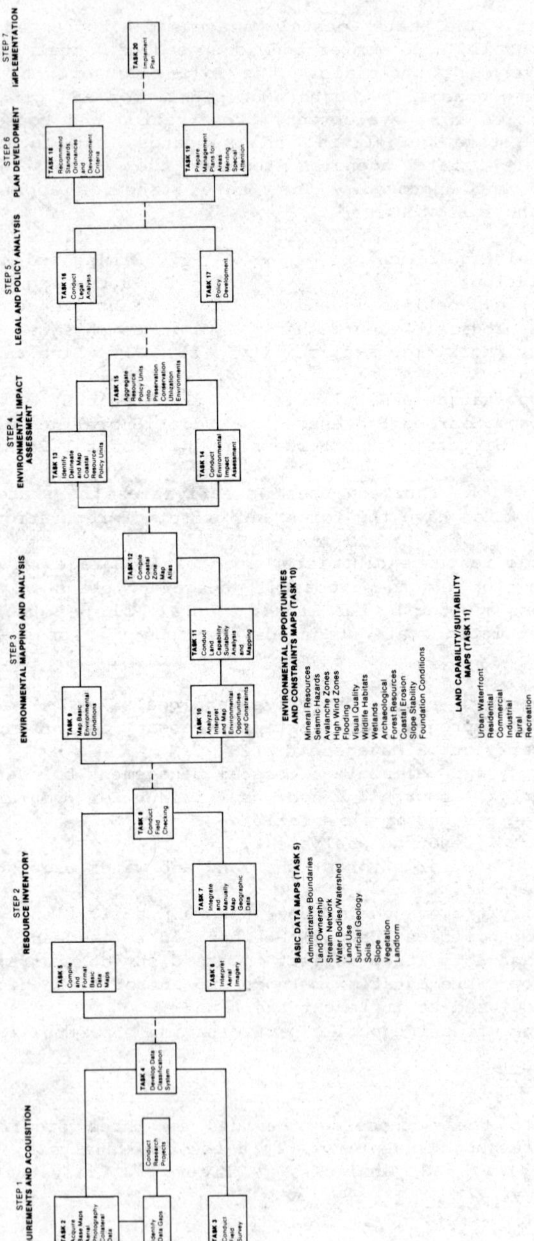

Figure 3
RESOURCE POLICY UNITS

	Preservation Environment		Conservation Environment		Utilization Environment
1	Potable Surface Waters	1	Anadramous Fish Propogation Streams	1	Ocean Waters for Commerce, Trans. & Industry
2	Selected Coastal & Upland Freshwater Wetlands	2	Recreational Waters	2	Urban Residential
3	Tidal Flats	3	Scenic Corridors, Areas and Vistas	3	Urban Development
4	Salt Water Marshes	4	Park and Recreation Areas	4	Urban Waterfront
5	Coastal Habitats	5	Marginal Lands	5	Rural
6	Coastal Bluff/Cliffs	6	River Floodplains		
7	Hazardous Lands	7	Open Space		
8	Historical, Archaeological Sites and Natural Areas	8	Forestry Management Areas		
9	Coastal Flood Zone				

Figure 4

QUANTITATIVE INFORMATION FOR DETERMINING COASTAL ZONE BOUNDARIES

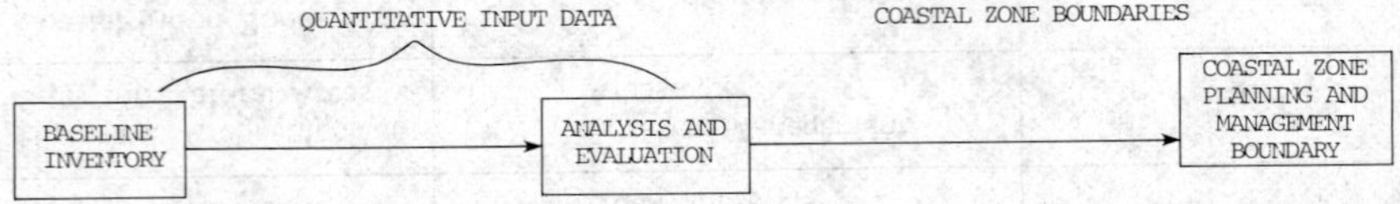

- o SOILS
- o GEOLOGY
- o HYDROLOGY
- o CLIMATE
- o OCEANOGRAPHIC
- o PHYSIOGRAPHY
- o VEGETATION
- o WILDLIFE
- o NAT. HAZARDS
- o TENURE
- o LAND USE
- o POPULATION
- o HISTORICAL SITES
- o RESOURCES
- o RECREATION
- o CULTURAL
- o CIRCULATION
- o (MAPABLE DATA) (GEOGRAPHICALLY LOCATED)

- o EVALUATE NATURAL HABITATS
- o EVALUATE PHYSICAL FEATURES/LAND FORMS
- o EVALUATE AREAS OF GEOLOGIC AND TOPOGRAPHIC IMPORTANCE
- o EVALUATE VEGETATION OF AREA
- o EVALUATE WILDLIFE HABITATS
- o EVALUATE AREAS WHERE DEVELOPMENTS AND FACILITIES ARE DEPENDENT ON ACCESS TO COASTAL WATERS
- o EVALUATE AREAS OF POTENTIAL HAZARDS BY TYPE
- o EVALUATE OWNERSHIP OF AREA
- o EVALUATE CURRENT LAND USE/OPEN SPACE/NOT IN USE
- o EVALUATE POPULATION DENSITIES WITH RESPECT TO PROXIMITY TO COASTAL ZONE
- o EVALUATE HISTORICAL SITES/CULTURAL AND SCENIC AREAS
- o EVALUATE RECREATIONAL AREAS
- o EVALUATE AREAS OF URBAN CONCENTRATION WHICH MAY BE IN COMPETITION WITH COASTAL USES
- o EVALUATE PATTERNS AND ACCESSIBILITY
- o EVALUATE AREAS NEEDING PROTECTION, MAINTENANCE OR REPLENISHMENT OF COASTAL LANDS OR RESOURCES
- o EVALUATE AREAS WHICH ARE UNIQUE, SCARCE OR FRAGILE
- o EVALUATE AREAS WHICH HAVE THE GREATEST AMOUNT OF UNIQUE, SCARCE AND FRAGILE CONCERNS
- o EVALUATE EACH AREA & SELECT A KEY CONCERN WHICH MIGHT REMEDY OTHER CONCERNS IN THE AREA

- o DETERMINE ENVIRONMENTAL SYSTEMS OF COASTAL ENVIRONMENT
- o DETERMINE COASTAL RESOURCE CATEGORIES
- o DETERMINE JURISDICTIONAL UNITS
- o DETERMINE NATURAL/CULTURAL SYSTEMS
- o DETERMINE SOCIAL COMMUNITIES
- o DETERMINE NATURAL/MAN-RELATED INTERRELATIONSHIPS
- o DETERMINE NATURAL COMPONENTS OF COASTAL ZONE AND ONSHORE WATERSHED
- o IDENTIFY VEGETATION OF AREA
- o IDENTIFY WILDLIFE OF AREA
- o IDENTIFY AREAS OF EXCLUSION DUE TO HAZARDS
- o IDENTIFY LAND USE OF AREA
- o IDENTIFY LAND OWNERSHIP OF AREA
- o IDENTIFY ACCESSIBILITY
- o IDENTIFY COMPATIBLE USES
- o IDENTIFY UNIQUE, SCARCE AND FRAGILE AREAS WITH REGARD TO:
  - WILDLIFE
  - HISTORICAL SITES
  - VEGETATION
  - GEOLOGY AND TOPOGRAPHY
  - CULTURAL & SCENIC SITES
  - DEVELOPMENT SITES
  - RECREATIONAL SITES
  - POPULATION
  - HAZARDS
  - RESOURCES

SCENARIO: DETERMINING COASTAL ZONE BOUNDARIES

During the resource inventory and analysis phase of the planning program it was found that certain geographic areas of the Municipality possessed varied landscape characteristics, both physical and cultural. In essence, the resource inventory and analysis permitted the identification of the subunits of the total landscape. These geographic subunits (called coastal resource policy units) provided the framework upon which the Anchorage coastal management plan was built. How these areas were identified is explained in the following discussion.

A Concept for Coastal Planning

The emphasis of the Anchorage coastal management plan is primarily centered around the special patterns and segregations of the landscape, the uses occurring on them, and the capability/suitability of an area to accommodate a given land use. This approach is based on the following assumptions: Not all areas of coastal waters or shoreland are alike. Not all areas of the coastal zone are suited to the same uses or intensity of uses.

The coastal zone, while a continuous system, can be viewed as a composite of numerous and distinct coastal landscapes; each with its own particular character. While the coastal zone should be thought of as being a continuous system, effective land use and natural resource planning and management are based on plans which take into account the natural diversity of the coastal zone. The concept of geographic segmentation of the coastal zone into units of a similar nature allows planners with a means of examining the coastal area in light of its natural diversity and to plan for and manage it accordingly. In this way, land use controls can be tailored to fit the needs of specific sections of the coastal zone. The purpose of applying this concept is to permit a differentation between areas whose geographical features imply differing objectives, capabilities and suitabilities regarding their use and future development.

To understand each of these areas in the context of the coastal system as a whole, the coastal system may be broken into a manageable number of subunits and composite environments. These composite environments may be referred to as coastal environments since they group together into functional units various physical, biophysical and cultural features which are typically associated in nature. Subunits of these coastal environments are referred to in the Management Plan as coastal resource policy units, whether natural or manmade, and are mapable entities defined by local characteristics of processes, land form, soils, biota, and other factors that naturally support certain levels of human activities.

These coastal resource policy units must be taken into account if public and private decision makers are to harmonize the intensive use and development of coastal resource systems with the continued economic productivity and livability of the Anchorage area. These coastal environments and their subunits Coastal Resource Policy Units - are the basic framework and the units upon which the Anchorage Coastal Management

Plan is designed. Figure 3 illustrates the grouping of coastal resource policy units into their respective environment classification.

The attempt then of the Anchorage coastal management plan has been to identify those geographic areas of the Municipality's coastal zone that are best suited for future growth, expansion and economic activity while recognizing that other coastal areas of the Municipality are less tolerant and less suited for such uses or intensity of uses. These may be areas best suited for preservation or conservation classifications.

Resource Inventory

The resource inventory was conducted utilizing field investigations, interpretation of color infrared aerial photography, digital analysis of landsat imagery, and integration of all collateral information. All geographic data capable of spatial expression was manually integrated and mapped as a series of overlays on U.S. Geological Survey basemap at 1:25,000 scale. For purposes of analysis, weighting and ranking of variables, and data manipulation, a uniform geographic grid was superimposed atop all the mapped data. This permitted overlaying and analyzing different combinations of geographic characteristics. The various characteristics were then weighted according to their importance, the opportunities or constraints for a given use, capacity for change, mitigation and impact, etc. A five acre grid size was used for this task.

For purposes of defining data requirements and data evaluation needs, scenarios were developed for each major program element. Figure 4 illustrates this process.

Impact assessment was made possible by overlaying selected environmental characteristics such as floodplains, noise zones, steep coastal bluffs (slopes), and seismic hazard classes on specified boundaries of a geographic area being proposed or anticipated for a development. This approach permitted extracting and summarizing the types of possible environmental impacts that might result if a specific type of use or activity occurred. It should be noted that this approach cannot go much farther than flag those areas that represent a possible conflict with the use or activity being proposed.

The resource inventory and analysis resulted in the production of a map atlas comprised of three types of maps, basic data maps, environmental opportunity/contraint maps; and land capability/suitability maps. The atlas shows the distribution and illustrates the location of all basic environmental phenomena which bear on the description and analysis of the municipality. The basic data maps illustrate the basic environmental characteristics of the municipality, and comprise such fundamental phenomena as geologic formation, slope, and vegetation association. These maps provided a standardized framework and reference for planning and assessment activities in Anchorage. The environmental opportunity and constraints maps illustrate interpretations of environmental patterns and processes, and encompass interpretations related to such conditions as geologic/geotechnical hazards, ecological importance, visual quality, viewing opportunities, flood hazards and drainage conditions. These maps served as a foundation and reference in relation to the final land capability/suitability maps and plans against which plan elements could be compared in the process of environmental impact assessment.

The data generated through this interpretive and analytical process were subsequently applied in the assessment of land capability/suitability. The land capability/suitability maps were based on further analysis which drew upon both basic and interpretive data. These maps were designed to provide interpretations in relation to such activity types and land use functions as urbanization, recreation, preservation, conservation, waterfront development, mining and mineral processing. They provided a graphic guide to the allocation of land resources and the location of land use activities within the municipality. In this particular case this land capability/suitability analysis was utilized to identify broad general areas suitable for residential, industrial, commercial, waterfront development, recreation and forestry areas.

The map Atlas was then used to identify and delineate coastal resource policy units; mapable geographic areas for which policies could be prepared, potential environmental impacts identified and mitigation measures prepared. A concept which is central to the Anchorage coastal management plan is the division of the coast into different coastal resource units, each representing a particular type of environment. These resource policy units are not defined solely on a biophysical basis, but rather, may represent an environment which takes on a special character due to man's activities there. The coastal resource policy units were identified and delineated as a result of the methodology and analysis described above. While the coastal zone should be thought of as being a continuous system, effective land use management is based on plans which take into account the natural and cultural diversity of the coastal area. The geographic segmentation of the coastal area into units of a similar nature permits the preparation of plans which address the particular characteristics of a given area. In this way land use controls can be tailored to fit the needs of specific landscape units of the coastal zone. The act of segmentation reveals a recognition that the coast is not homogenous. Land uses which exist in harmony in one coastal environment may be entirely inappropriate for another. The process of geographic segmentation provides a means of examining the coastal zone in light of its natural diversity and to plan for and manage it accordingly.

Coastal resource policy units were then aggregated to correspond with three broad land use suitability classifications. Such a concept of land designations into three broad use environments was needed in order to plan for the management of coastal areas according to sensitivities and hence, suitabilities and capabilities for specified land and water uses or preservation needs. The purpose of these three broad designations is to differentiate between areas whose geophysical, biological and cultural features imply differing objectives regarding their use and future development. Each environment represents a particualr emphasis in the types of uses and the extent of development which should occur within it.

The generic terms used for these environment classifications are Preservation, Conservation and Utilization (PCU). Each of the three broad suitability environments is a composite of numerous coastal resource policy units. This concept allows for grouping into three broad environmental classes general policy statements for coastal landscape units requiring preservation status; conservation status or suitable for future growth and development. The coastal resource policy units contained within each broad environment class permits more site specific policies and development of management plans as necessary for each coastal resource policy unit.

Planning Framework

Following the resource inventory and analysis tasks, the coastal resource policy units could be identified and delineated and grouped into their respective environment classifications: Preservation, Conservation or Utilization. This process provided a basic framework upon which environmental impact analysis, legal analysis, policy development, mitigation planning and management plans could be developed. The planning framework which was developed utilizing the environmental data and analysis was subsequently evaluated in relation to its projected environmental impact; that is, each coastal resource policy unit was evaluated and weighted against a series of land uses to determine possible impacts and as a method to prioritize uses for each resource policy unit.

This task represented the culmination of an overall process in which environmental considerations were accounted to and integreated into the planning process. The process was carried out in relation to a set of potential impact concerns and issues.

An environmental impact matrix was constructed to provide for the systematic ranking of areas within each coastal resource policy unit in relation to select classes of human activity. The development of this land capability/suitability matrix required a thorough understanding of the environmental requirements and demands of each activity type. The two part matrix, illustrated in figures five and six identifies a possible stress, induced change and environmental impact that could result from various land and water uses. This table was in turn used to rank probable impact and capability/suitabilities for each coastal resource policy unit. The application of this process resulted in a ranking of land uses against coastal resource policy units by three levels of probable impacts for each particular activity: high, moderate, low. For each land use a cross correlation could be made indicating the type of impact that could result if the use/or uses were conducted in the resource policy unit. Each pertinent environmental variable was assigned a numerical value in relation to the degree of constraint which it represents. For example, steep coastal slopes received a higher respective constraint value than gentle slopes. Those areas with the lowest constraint values were interpreted as the most capable for development.

Figure 5A

POSSIBLE ENVIRONMENTAL IMPACTS

STRESS	INDUCED CHANGE	ENVIRONMENTAL EFFECT
1. Ground Water Withdrawal	1. a. Salt water intrusion b. Subsidence c. Flooding	1. a. Degradation of potable water supply b. Property damage c. Loss of fish nursery areas because of altered water chemistry at depth
2. Dredging and Spoil Disposal	2. a. Increased water turbidity b. Loss of photosynthetic process c. Loss of marine resource d. Destruction of habitat e. Alteration of physical & chemical character of estuaries f. Erosion of coastal areas g. Loss of wetlands	2. a. Loss of marine resources b. Possible increase in productivity of estuary c. Destruction of fish and shellfish nurseries d. Destruction of valuable wetlands e. Destruction of upland habitats
3. Devegetation	3. a. Increased runoff b. Removal of wildlife habitat c. Increased erosion	3. a. Loss of soil resources b. Loss of wildlife habitat c. Loss of shorelines d. Flooding e. Loss of water quality
4. Offshore Construction	4. a. Increase in sedimentation b. Potential introduction of harmful chemicals into air and water c. Potential for ice damage	4. a. Damage to wildlife resources b. View obstruction c. Increases need for dredging and spoil disposal d. Loss or damage of structures
5. Shoreline Construction	5. a. Alteration of physical character of estuary b. Alteration of circulation c. Alteration of erosion & deposition d. Loss of coastal access for public use e. Seismic hazards, slope stability	5. a. Degradation of coastal lands b. Production of new shallow water habitats c. Loss of public use of coastal area d. Loss or damage to shoreline structures
6. Inland Construction	6. a. Increased runoff other pollutants b. Decrease in soil fertility c. Devegetation d. Loss of habitat/wetlands e. Soil foundation conditions altered f. Seismic hazards, slope stability g. Subsidence	6. a. Loss of estuary productivity by siltation of shellfish beds — changes in photosynthetic process b. Loss of recreational waters c. Flooding d. Destruction of bird and mammal habitats e. Damage or loss of structures
7. Solid Waste Disposal	7. a. Surface water pollution b. Ground water pollution c. Air pollution	7. a. Threat to public health disease — aquifer pollution b. Loss of aquatic habitat c. Loss of water recreation d. Pollution of downstream water bodies

Figure 5B
POSSIBLE ENVIRONMENTAL IMPACTS

STRESS	INDUCED CHANGE	ENVIRONMENTAL EFFECT
8. Liquid Waste Disposal	8. a. Degradation of surfce water b. Degradation of ground water	8. a. Threat to public health disease — aquifer pollution b. Loss of aquatic habitat & life c. Loss of water recreation d. Pollution of downstream estuaries
9. Gaseous Waste Disposal	9. a. Degradation of surface water	9. a. Threat to public health respiratory diseases, etc. b. Loss of agricultural productivity c. Property damage
10. Excavation	10. a. Potential leaks and spills b. Potential ground and surface water pollution c. Potential for land subsidence d. Destruction of wetlands e. Destruction of natural vegetation	10. a. Loss of marine resources & nursery areas b. Loss of agricultural productivity c. Loss of natural amenity resources c. Threat to public health aquifer pollution e. Erosion and sedimentation
11. Filling	11. a. Destruction of natural vegetation b. Alteration of ground water flow and recharge	11. a. Loss of wildlife habitat b. Loss of natural amenity resources c. Threat to public health aquifer pollution
12. Draining	12. a. Potential destruction of vegetation b. Increased sediment load in runoff water	12. a. Loss of wildlife habitat b. Loss of water recreation
13. Increased Vehicular Traffic	13. a. Destruction of vegetation b. Decreased soil stability c. Loss of shore protection d. Increase in noise levels	13. a. Loss of property b. Loss of natural resource c. Loss of biotic community
14. Subsurface Waste Disposal	14. a. Pollution of ground water b. Pollution of surface water	14. a. Threat to public health disease — aquifer pollution b. Loss of aquatic habitat & life c. Loss of water recreation d. Pollution of downstream waterbodies

Figure 6

PRESERVATION-CONSERVATION-UTILIZATION MATRIX

COASTAL ZONE LAND & WATER USES
SUITABILITY CLASSIFICATION

Weighting Values
3 – High Impact
2 – Moderate Impact
1 – Low Impact

ENVIRONMENTS → PRESERVATION ENVIRONMENT / CONSERVATION ENVIRONMENT / UTILIZATION ENVIRONMENT

COASTAL RESOURCE POLICY UNITS →

LAND & WATER USES	PRESERVATION ENVIRONMENT									CONSERVATION ENVIRONMENT								UTILIZATION ENVIRONMENT				
	Potable Surface Waters	Selected Coastal & Upland Freshwater Wetlands	Tidal Flats	Salt Water Marshes	Coastal Habitats	Coastal Bluff/Cliffs	Hazardous Lands	Historical, Archeological Sites and Natural Areas	Coastal Flood Zone	Anadramous Fish Propogation Streams	Recreational Waters	Scenic Corridors, Areas and Vistas	Parks and Recreation Areas	Marginal Lands	River Floodplains	Open Space	Forestry Management Areas	Ocean Waters For Commerce, Transportation & Industry	Urban Residential	Urban Development	Urban Waterfront	Rural
Navigation & Transportation																						
Roads/Highways	3	3	3	3	3	2	3	3	3	3	2	2	2	2	2	2	2	2	1	1	2	2
Railroads	3	3	3	3	3	2	3	3	3	3	2	2	3	2	2	2	2	2	2	2	2	2
Airports/Airstrips/Float Base	3	3	3	3	3	3	3	3	2	3	2	3	2	2	3	3	2	2	2	2	2	2
Ports	3	3	2	3	3		3	3	2	3	2	3	2	2	2	3	2	2	3		1	3
Spoil Disposal	3	3	3	3	3	3	3	3	2	3	3	3	3	2	3	2	3	2	3	3	3	3
Pipelines	3	3	2	3	3	2	3	3	2	2	2	2	2	2	2	2	2	2	2	2	2	2
Navigation Improvements/Aids	3	3	2	3	3	2	2	2	2	2	2	2	2	2	2	2	2	2			1	
Urban																						
Low Density Residential	3	3	3	3	3	3	3	3	3	3	2	2	3	2	2	3	3		1	3	3	1
Medium Density Residential	3	3	3	3	3	3	3	3	3	3	2	3	3	2	3	3	3		1	3	3	2
High Density Residential	3	3	3	3	3	3	3	3	3	3	2	3	3	2	3	3	3		1	2	3	3
Commercial	3	3	3	3	3	3	3	3	3	3	2	2	3	2	2	3	3	2	3	1	1	2
Industrial	3	3	3	3	3	3	3	3	3	3	3	3	3	2	3	3	3	3	3	3	1	3
Public Lands & Institutions	2	2	2	2	2	2	2	2	2	2	2	1	1	2	2	1	2	1	1	3	3	1
National Defense	2	3	2	2	3	1	3	2	2	2	2	2	2	1	2	3	2	2	2	3	2	2
Mining & Mineral Extraction	3	3	3	3	3	1	3	3	2	3	3	3	3	1	2	3	3	3	3	3	2	3
Fish & Wildlife Habitat	1	1	1	1	1	1	1	1	1	1	1	1	1	1	1	1	1	1	2	3	3	2
Agriculture	3	3	3	3	3	3	3	3	3	3	2	2	2	1	1	2	2		2			1
Forestry	3	3	3	3	3	3	3	3	3	3	2	3	2	1	1	2	1		2			2
Recreation																						
Active	2	2	1	2	3	1	2	2	1	1	1	2	1	1	1	1	1	1	1	3	3	1
Passive	1	1	1	1	2	1	2	1	2	1	1	1	1	1	1	1	1	1	1	3	3	1
Gas & Electric Utilities	3	3	2	3	3	2	3	2	2	2	2	2	2	2	2	2	2		1	1	2	1
Water Supply & Wastewater Trmt.	2	3	2	3	3	3	2	2	2	2	2	2	2	2	2	2	2	2	2	3	2	2
Solid Waste Treatment Facilities	3	3	3	3	3	3	2	3	2	3	3	3	3	2	3	3	2	3	2	3	2	2
Parks	1	1	1	1	2	1	1	1	2	1	1	1	1	1	1	1	1		1	3	3	1
Cultural/Entertainment Facilities	3	3	3	3	3	2	3	2	3	2	1	2	2	2	3	2	3	2	2	2	3	2
Open Space/Undeveloped	1	1	1	1	1	1	1	1	1	1	1	1	1	1	1	1	1	1	1	3	3	1

Collectively, the completed matrix and resulting maps provided a coherent and consistant environmental direction and framework for the formulation of specific policies, mitigation needs, and the legal analysis.

Legal Analysis

In an attempt to reduce the need for additional levels of regulation, and to avoid duplication of existing regulation, a legal analysis was conducted. All major existing federal, state and local statutes, regulations and ordinances were identified that were applicable to each individual coastal resource policy unit. In many instances it was found that existing regulations were sufficient to comply with state CZM standards and provided an existing tool to mitigate most possible environmental impacts. In those cases where existing regulatory mechanisms did not exist or were insufficient to mitigate impacts or comply with state coastal management standards, new regulations were and are still being developed. Two notable examples are wetlands and hazardous lands. In both cases additional studies were authorized and are currently being carried out.

Policy Development

Completion of the legal analysis permitted the policy development task to begin. Policies were written to address the general needs of the Preservation, Conservation and Utilization environments, e.g. policies to protect resource values, public values and associated functions in the Preservation environment; policies that recognize the need for special precautions in use and development in the Conservation environment; and, policies directing new growth to appropriate areas, encouraging water-related/water-dependent uses in waterfront zones in the Utilization environment.

Additional policies were written for each coastal resource policy unit. This involved the development of policies for minimizing significant adverse impacts and for increasing positive impacts.

A series of public workshops were held to solicit public input in the policy formulation phase. Staff from a variety of federal, state and local agencies, having expertise in a variety of land use planning, resource management, environmental science, and engineering backgrounds contributed much of their time in assisting the Municipality in developing policy statements.

Implementation

The final step in the planning process involved the identification and recommendation of measures which were practical, useful and effective in mitigating potentially unavoidable adverse impacts. In general, the impacts which invite mitigation are those which are adverse and significant. Positive impacts, however, can often be enhanced and enriched and moderately adverse ones can often be neutralized by mitigation measures.

In addition to mitigation measures, recommendations were made that would:

1. Adopt the coastal management plan by ordinance as an element of the comprehensive plan.

2. Develop site specific land use and management plans for "area meriting special attention" which include:
   a. historical, archaeological and prehistoric sites,
   b. critical habitat areas,
   c. significant scenic and visual qualities, viewing and/or recreational opportunities,
   d. unique biologic and/or geologic significance, and
   e. lands available for future waterfront and port development.
3. Provide the general public with information about the purposes and values of coastal planning.
4. Provide for a continued and on-going process of acquiring new data and incorporating it into the coastal management plan.
5. Provide for on-going studies for better identifying the nature and extent of specific geophysical hazards, and establish guidelines and criteria for uses in these areas.
6. Map and classify wetlands, identify their values and functions, and develop criteria for their use and /or preservation, and
7. Incorporate 208 water quality planning into the coastal management planning process.

In summary, the Anchorage Coastal Management Plan is designed to illustrate the way in which environmental considerations can be intigrated into the planning process and to indicate the extent to which this systematic process capitalizes upon environmental opportunities and circumvent environmental constraints. In effect, the plan format reflects the comprehensive and systematized flow of environmental data through the entire planning process and fosters a broad view of the process of conflict identification and response. The plan is responsive in meeting the requirements of the Coastal Management Act which established a balanced-use philosophy in its requirements to establish a definition of land and water uses which have or could have direct and significant impacts on coastal waters. The Anchorage Coastal Management Plan was one of the first in the state to be approved by the Coastal Policy Council and State Legislature. Local adoption is now pending.

A Procedure for Evaluating Program Alternatives

By Robert B. McDonald, Forester[1]

A newly developed procedure for evaluating program alternatives is presented in this paper. Its main application is in implementing the National Environmental Policy Act. The advantage of this analytical tool is that it makes visible what factors were considered, how much weight is given to each factor, by whom, and why. In addition, the visuals prepared document the entire process. Available alternatives for decisions are compared against three classes of criteria: Priceable, quantifiable, and ill-quantifiable. Each class is described and its use is explained for reaching a preferred alternative. The decisionmaker's own values are explicitly considered together with trade-offs among the options, to avoid masking important considerations. The system is designed to avoid double-counting of criteria as well as to count only the important factors.

A series of about five tables is presented to lead decision-makers through the process. First, the factors that fall within the three classes of evaluation criteria are listed. A second table summarizes priceable and quantifiable criteria by amount (dollars, acres or other measurements). Another table displays, in order of preference, the ill-quantifiable factors (e. g., opportunities for the disadvantaged).

Each factor used in evaluating the alternatives is ranked in the fourth table. This table readily shows which (if any) factor is dominant, or dominated by other factors. Further evalation may end at this point if one factor is dominant. If further consideration is necessary, the lowest-ranking factor(s) may be eliminated at this point.

A weighting system is used to compare the remaining alternatives. Each alternative is weighed against the other alternatives. Visual aids (figures 1 and 2) are then constructed to display the most advantageous choice(s). The example presented pertains to alternative uses of forest land for wilderness areas, timber production, cattle grazing, various recreation options, wildlife, et al. Other applications can be made with the same process. Decision-makers and others can readily assess the trade-offs among the alternatives and decide which values are most important.

[1]USDA Forest Service, Southeastern Area, 1720 Peachtree Road N.W., Suite 711, Atlanta, Ga. 30309

A TRADE-OFF EVALUATION PROCEDURE

The following evaluation system is designed to facilitate the comparison and trade-offs among alternatives so that a preferred alternative can be identified and the decision rationale documented.

This evaluation methodology requires a number of assumptions.

- All decisions are value-laden. Decisions are made in light of an individual's own set of values and preferences; those values and viewpoints that affect the decisionmaking process should be explicitly displayed.

- Decisions should be made on the basis of differences (trade-offs) among alternatives. Techniques which mask these differences (such as index weighing techniques) lose information and should be avoided.

- Techniques should be used which avoid "double counting" in comparing alternatives and evaluating trade-offs. For example, significant trade-offs might occur between two alternatives in terms of water quality and recreation; water quality should not be given undue importance by comparing a number of water quality parameters (such as BOD, suspended solids, salinity, nutrients, etc.) unless those parameters are each important individually and changes in them are caused by different management mechanisms.

- Techniques should address only the important factors or criteria. The inclusion of relatively unimportant factors can create an information overload problem.

Measurement criteria include those important outputs and effects for each alternative that could be used to compare differences and trade-offs among alternatives. Criteria previously selected for this example include economic, social, physical, and biological factors. 1/ These all fall within one of three types or categories of criteria called "priceable," "quantifiable," and "ill-quantifiable."

Priceable Factors. Those factors that are directly priceable in the market are the "priceable" factors, i.e., things that one would be able to purchase in the market place.

1/ These criteria are defined in Appendix I.

Quantifiable Factors. Factors not directly priceable in the market, but which are measurable are called "quantifiable" if they can pass the test of indifference. For example, two different alternatives might each provide 100,000 acres of roadless areas. In this case, the factor "acres of roadless area" is measurable and if one is indifferent between the 100,000 acres in the first alternative and the 100,000 in the second, the factor is "quantifiable.

Ill-Quantifiable Factors. If the 100,000 acres of roadless area for the first alternative consisted of mountainous land with little timber, and the 100,000 acres of roadless area for the second alternative were fully stocked longleaf and slash pine stands, then the factor "acres of roadless area" would be judged to be "ill-quantifiable." Even though the amount of roadless area is measurable for both alternatives, if the actual acres are of a different type of land, the quantities are not directly comparable.

The evalution combines the priceable, quantifiable, and ill-quantifiable factors with preferences and opinions to arrive at a preferred alternative and a rank order of all alternatives. Judgments used by the decisionmakers are documented in the comparison/evaluation process.

The major steps are: (1) organize the data about the various alternatives, (2) compare the alternatives and document these comparisons, and (3) summarize the comparisons in a graphic display called a profile.

In the example that follows, four alternatives have been evaluated. These alternatives are characterized and displayed in Figures 1 and 2. These figures are useful in organizing the data and provide initial insight to the relative values and implicit trade-offs.

Figure 1

COMPARISONS OF DIFFERENCES

These charts compare differences among alternatives. For example: gains in wildlife habitat improvement and timber opportunities realized through implementing Alternative B can be compared to increased dispersed recreation and wilderness use realized through implementing Alternative C.

These comparisons help decision makers to assess the trade offs between alternatives, thus most decide which values are most important.

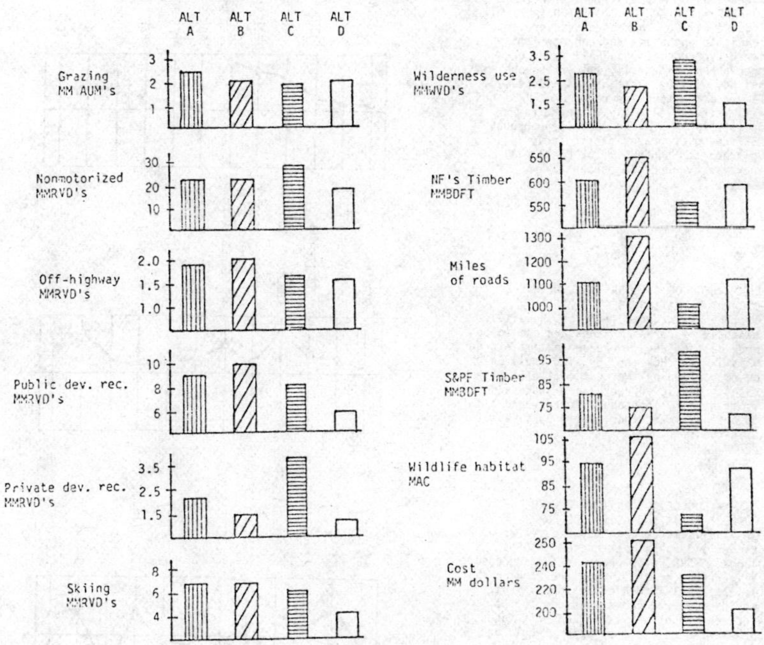

MM = 100,000
AUM = Animal Unit Month
RVD = Recreation Visitor Day
BDFT = Board Feet
S&PF = State and Private Forestry

FIGURE 2 - COMPARISONS OF ALTERNATIVES

1. CONTINUATION OF CURRENT MANAGEMENT
The projected output levels for continuation of the current program has previously been displayed in Table 3 as Alternative D. The profiles in Charts 2, 3, and 4 show the changes from the current management approach: therefore, all the factors are zero; i.e., no change for Alternative D. In most cases, the "current" or zero point in Chart 1 represents the current average output or yield projected for a particular factor.

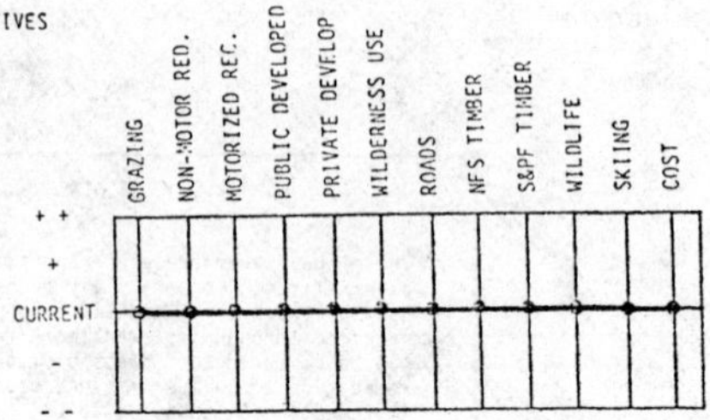

2. COMPARISON OF ALTERNATIVE A WITH THE CURRENT SITUATION AS PROJECTED (ALT. D)
Grazing will be increased over present levels to encourage and enhance family ranch-type operations. All types of recreation, especially wilderness use and skiing, will be substantially increased. Timber production and roads on NF's lands will increase slightly. Timber from private lands will be increased through S&PF co-op programs, particularly in Colorado and South Dakota. Costs will increase by about 30 percent.

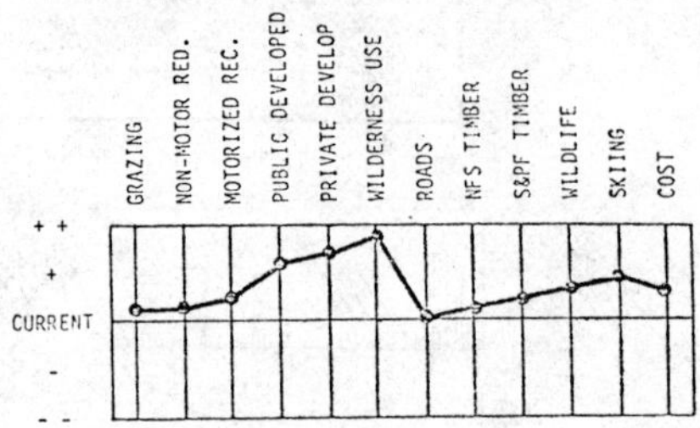

3. COMPARISON OF ALTERNATIVE B WITH THE CURRENT SITUATION AS PROJECTED
Grazing use capacity will be increased slightly as forested lands are opened to provide the large increase in timber from NF's lands. However, grazing would only be emphasized on the most cost effective sites. Developed recreation would also be substantially increased. There would also be a significant increase in cost, about 50 percent.

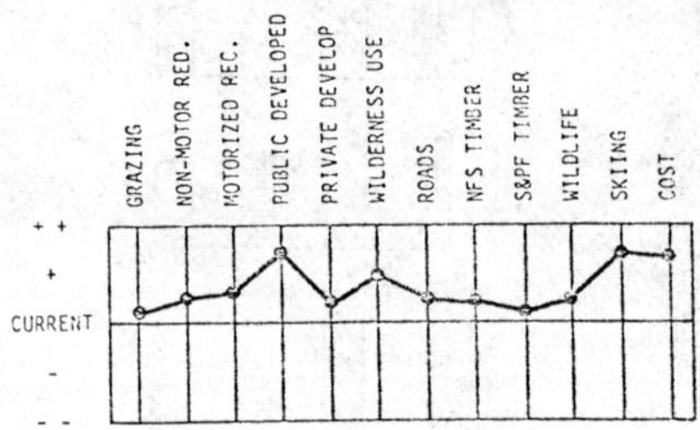

4. COMPARISON OF ALTERNATIVE C WITH THE CURRENT SITUATION AS PROJECTED
Grazing would be slightly reduced as would timber and roads constructed on NF's lands. S&PF co-op programs for increasing timber from private lands are at high levels. Wildlife hibitat improvement would be at reduced levels due to the reduced timber harvesting on NF's lands. However, big game populations would be maintained. Dispersed recreation, including wilderness use, would significantly increase. Costs would be increased by less than 30 percent.

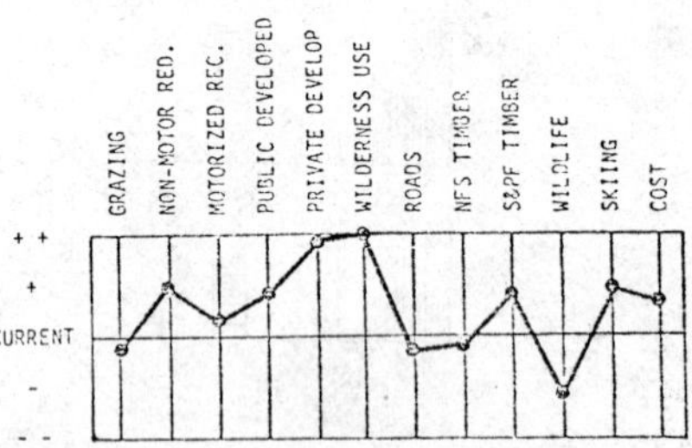

The four diagrams are also intended to gain insight into trade offs among alternatives.

The information concerning the four alternatives is further organized by classifying the previously selected evaluation criteria by the three categories and displayed in Tables 1 and 2.

TABLE 1 - Classification of Evaluation Criteria

Priceable	Quantifiable	Ill-Quantifiable
cost efficiency (net benefits)	soil productivity timber riparian protection soil loss grazing wildlife habitat improvement	energy conservation opportunities for the disadvantaged issues, concerns and goals employment

A tablular summary of the priceable and quantifiable factors for all alternatives is presented in Table 2.

TABLE 2 Summary of Priceable and Quantifiable Factors

ALT	Net Benefits	Timber Sales	Grazing Use	Wildlife Habitat	Soil Productivity	Riparian Protection	Soil Loss
	MM$	MM Bd Ft	MM AUM	M Acres	lbs/acre	M Acres	in/acre
A	298	604	2.40	92.6	585	518	.9
B	270	657	2.25	107.2	595	528	1.0
C	319	550	2.01	69.8	600	498	.04
D	218	582	2.18	90.8	530	508	.06

The ill-quantifiable factors are also used in the evaluation process. Comparative descriptions are shown in a table that precisely and thoroughly describes each ill-quantifiable factor. The comparative descriptions for the alternatives are close to each other so that it is easy to compare alternatives. If the factor is the same for two or more alternatives, the description is written only once, so the reader can see instantly that the alternatives are alike. If the description must be written twice, exactly the same words are used, or it is indicated that, for example, "Alternative C is the same as "A" for that factor. In this manner, all relevant information about the ill-quantifiable factors get into the trade-off evaluation procedure. Table 3 displays the comparative descriptions of the ill-quantifiable factors for the alternatives.

TABLE 3 Comparative Descriptions of Ill-Quanitifiable Factors

<table>
<tr><th rowspan="2">FACTORS</th><th colspan="4">ALTERNATIVE</th></tr>
<tr><th>A</th><th>B</th><th>C</th><th>D</th></tr>
<tr><td>Opportunities for the Disadvantaged</td><td>Emphasize social benefits in the Region, especially in the Southern Rocky Mtn. area where most social problems exist. Provide most opportunities on Western slope and San Luis valley.</td><td>Only slightly lower opportunities are provided than those in Alt. A. Emphasis is on development and economic gains rather than social.</td><td colspan="2">Fewest opportunities for disadvantaged under these alternatives. (Program thrust is toward environmental enhancement (C) or continuation of current (D).</td></tr>
<tr><td>Preference Order</td><td>(1)</td><td>(2)</td><td>(3)</td><td>(3)</td></tr>
<tr><td>Energy Conservation</td><td>Moderate levels of outputs require large amounts of energy consumption.</td><td>Highest energy consumer because of high level of output for timber, roads and recreation.</td><td>Lowest user of energy.</td><td>Same as A</td></tr>
<tr><td>Preference Order</td><td>(2)</td><td>(3)</td><td>(1)</td><td>(2)</td></tr>
<tr><td>Private Sector Employment</td><td>132 person yrs. of new employment primarily in areas of high unemployment.</td><td colspan="2">132 person and 129 person years of new employment with no special attempt to focus on providing new jobs in high areas of unemployment.</td><td>92 person years of new employment</td></tr>
<tr><td>Preference Order</td><td>(1)</td><td>(2)</td><td>(2)</td><td>(3)</td></tr>
<tr><td>Responding to Issues, Concerns and Goals</td><td>Addresses all issues and most responsive to the major issues. Exceeds R-2 targets for National recreation goals.</td><td>Most responsive to issues and concerns. Has the potential to achieve most National and Regional goals.</td><td>Same as A</td><td>Least responsive to issues and goals.</td></tr>
<tr><td>Preference Order</td><td>(2)</td><td>(1)</td><td>(2)</td><td>(3)</td></tr>
</table>

To faciliate pair-wise evaluations of the alternatives, the factors from Tables 2 and 3 have been ranked in Table 4. The alternative having the preferred output level for each given factor is entered in the top box along with the output level itself. The other alternatives are entered vertically according to their preference. All factors--priceable, quantifiable, and ill-quantifiable--are ranked in this manner

TABLE 4 Rank Order of Alternatives for Each Evaluation Factor

Net Benefits	Timber Sales	Grazing Use	Wildlife Habitat	Soil Product.	Soil Loss	Riparian Protect.	Opport. for Disadv.	Energy Conserv.	Employ.	Issues Concerns Goals
MM$	MM Bd Ft	MM AUMs	M Acres	lbs/Acre	in/Acre	M Acres	Rank	Rank	Rank	Rank
C 319	B 657	A 2.40	B 107.2	C 600	C .04	C 498	A	C	A	B
A 298	A 604	B 2.25	A 92.6	B 595	D .06	D 508	B	A,D	B,C	A,C
B 270	D 582	D 2.18	D 90.8	A 585	A .90	A 518	C,D	B	D	D
D 218	C 550	C 2.01	C 69.8	D 530	B 1.0	B 528				

A question to be asked at this point is, "Is any alternative clearly dominant or dominated?" If an alternative were dominant over all others, it would occupy all of the boxes of the top row in Table 4. If one alternative is clearly dominant, it would be the preferred alternative and the evaluation process could stop. Similarly, a clearly dominated alternative could be eliminated from further consideration.

Since Table 4 indicates that no alternative is dominant or dominated, four additional steps are required to compare and evaluate the alternatives.

The first step: Select two alternatives for comparison and, using the information from Table 4, fill out the comparison table (as done for Alternatives A and B in Table 5). The table is completed by indicating, in the bottom row, the differences in the factors between the two alternatives.

TABLE 5 Comparison of Alternatives A and B

ALTERNATIVE	FACTORS										
	Net Benefits	Timber Sales	Grazing	Wildlife Habitat IMP	Soil Produc-tivity	Soil Loss	Riparian Area	Opport. for Dis-advant.	Energy Conser-vation	Employ-ment	Issues Concerns Goals
	MM$	MM Bd Ft	MM AUMs	M Acres	lbs/ac	in/year	M Acres	Rank	Rank	per yrs	Rank
A vs. B	A 298	B 657	A 2.40	B 107	B 595	A .9	A 518	A Max.	A Mod.	A 132	B Max.
	B 270	A 604	B 2.25	A 93	A 585	B 1.0	B 528	B High	B Max.	B 132	A High
DIFFERENCE	A 28	B 53	A .15	B 14	B 10	A .1	A 10	A	A	A	B

The second step: Examine the difference in the factors between the two alternatives and judge which difference (not factor) is most important.

The third step: Place the factor difference judged to be the most important at the "100" position on a scale as illustrated in Figure 3. The column labeled "Factor Difference" indicates the relative importance of each factor difference. The column entitled "Explanation of Weights" is used to document why a particular scale value was chosen.

In Figure 3, for example, the evaluation continues by deciding how important the remaining factor differences are relative to the difference in satisfying "issues, concerns, and goals." To do this, any other factor difference in the comparison table (Table 5) is selected and placed on the scale according to how important this factor difference is judged in relation to the difference in "issues, concerns, and goals." The process is repeated until all the factor differences have been placed on the scale in Figure 3.

The final step: Check to be sure that each factor difference is positioned in relation to all other differences according to the decisionmaker's actual preference.

Figure 3 - Indifference Scaling

SCALE	FACTOR DIFFERENCE	EXPLANATION OF WEIGHTS
100 -	Issues, Concerns and Goals	most significant difference since responding to these items is why we are planning. Contributes to an effective National program while meeting local needs.
90 -	Net Benefits	difference of $28 million important because it represents a very high marginal cost for the production of any other factor difference.
80 -		
70 -		
60 -	Opportunity for Disadvantaged	difference important since benefits represents important social impacts in depressed rural areas and contributes to those communities' social well being.
50 -	Timber Sales	difference judged to be about half as important as the $28 million it would take to produce it.
45 -	Wildlife Habitat Improvement	difference of 14,000 acres worth only about half as much as net benefits difference and less important than timber difference.
40 -	Riparian Protection	an increase of 10,000 ac. provides greater wildlife habitat and scenic beauty benefits. Those unique and often fragile ecosystems provide an abundance and diversity of plant and animal populations.
30 -		
25 -	Grazing	difference important because of current use in near capacity.
20 -	Soil Productivity	difference represents less than 2% of current production.
15 -	Energy	difference important because of potential energy shortage but difference is slight and greatly estimated.
10 -	Employment	difference is that A concentrates jobs in areas of chronic high unemployment, but difference in <u>amount</u> of jobs is insignificant.
5 -	Soil Loss	1/10 inch--not very significant--may be not even measurable difference.

The judgments made about the trade-offs between two alternatives are illustrated by a weighted factor profile, or histogram. The histogram for Table 6 was constructed by using Table 5 and Figure 3 and recording factor differences and their scaled importance values in the tabular portion of Table 6. Going from top to bottom on the scale in Figure 3, if Alternative A is superior to Alternative B for a given factor, then the factor difference and its scaled value are entered in the <u>left-most</u> column of the table. If Alternative B is superior for this factor, then the information is entered in the <u>right-most</u> column. Continue down the scale in Figure 3 until the columns in Table 6 are filled. Then, the profile is drawn by plotting the scale of values as illustrated by the graphic portion of Table 6.

The alternative with the largest histogram area is theoretically the preferred one. If the decisionmaker is not satisfied with the implication of the histogram profile, he/she should go back and analyze his judgments. If the profile is accurate, continue the evaluation by comparing the superior alternative with another alternative that has not yet been examined. Comparisons and evaluations of the remaining alternatives and resulting profiles are displayed in Tables 7 through 10.

TABLE 6 Weighted Factor Profile Comparing Alternatives A & B

	← A is preferred			FACTORS			→	←		B is Preferred →	
	Net Benefits	Opport. for Disadv.	Riparian Protect.	Grazing	Energy	Employ-ment	Soil Loss	Soil Prod.	Wildlife Habitat	Timber Sales	Issues Concerns Goals
	MM$	Rank	M Acres	MM AUMs	Rank	Years	in/yr	lbs/ac	M Acres	MM Bd Ft	Rank
	A 298	A Max.	A 518	A 2.40	A Mod.	A 132	A .9	B 595	B 107	B 657	B Max.
	B 270	B Mod.	B 528	B 2.25	B Max.	B 132	B 1.0	A 585	A 93	A 604	A High
DIFFERENCE	28	A>B	10	.15	A>B	A>B	.1	10	14	53	B>A
SCALED VALUES	90	60	40	25	15	10	5	20	45	50	100

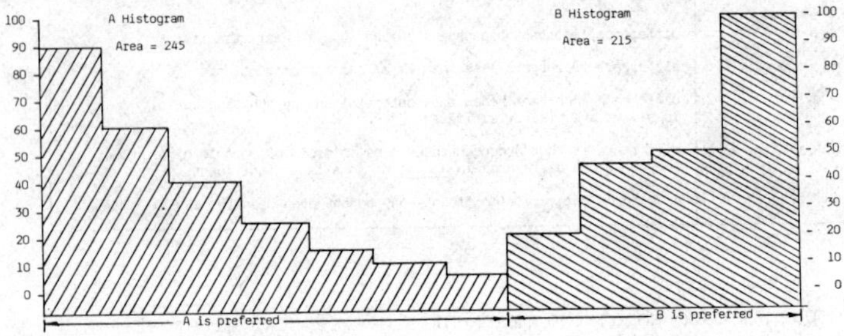

Note that this trade-off evaluation procedure as described was initially done by an interdisciplinary planning team, however, it is the decisionmaker's judgments that are presented here. These drafts were reviewed by the Executive Group: Regional Forester and Deputies of Resource, Administration, and State and Private Forestry, where difference in judgment, criteria, etc. arose, the Executive Group's position prevailed.

TABLE 7 Evaluation of Atlernatives A and C

ALTERNATIVE	FACTORS										
	Net Benefits	Timber Sales	Grazing	Wildlife Habitat IMP	Soil Produc-tivity	Soil Loss	Riparian Area	Opport. for Disadv.	Energy Conser-vation	Employ-ment	Issues Concerns Goals
	MM$	MM Bd Ft	MM AUMs	M Acres	lbs/ac	in/year	M Acres	Rank	Rank	per yrs	Rank
A vs. C	C 319	A 604	A 2.40	A 93	C 600	C .04	C 498	A Max.	C Max.	A 132	A=C
	A 298	C 550	C 2.01	C 70	A 585	A .90	A 518	C Min.	A Mod.	C 129	
DIFFERENCE	C 21	A 54	A .39	A 23	C 15	C .86	C 20	A>C	C>A	A>C	0

SCALE	FACTOR	EXPLANATION OF WEIGHTS
100	Net benefits	most important difference - reflects efficient use of finanical resources.
90		
85	Soil loss	difference of almost 1 inch important in redeeming stewardship and USDA mission.
80	Grazing	difference of 390,000 AUM cause economic hardships in local ranches, loss of income and local dependent ranches may go out.
75	Timber sales	difference significant because of impact on dependent timber industry and rural economics.
70		
60	Opportunity for Disadvantaged	difference not too great but important because of the opportunities afforded communities and disadvantaged.
50	Wildlife Habitat	difference of 23,000 acres of significance because of greater opportunities to enhance wildlife numbers and diversity.
45	Riparian Protect.	difference of 20,000 acres contributes substantially to environmental quality and provides greater plant and animal diversity.
40		
30	Employment	difference of 3 person years important because of focus on areas of high unemployment by Alternative A.
20	Soil Productivity	important because of demand for additional livestock grazing on NFS lands.
15	Energy	difference not too important compared to 21 million net benefits.
10		
0	Issues, concerns and goals	no difference.

TABLE 8 Weighted Factor Profile Comparing Alternatives A & C

	A			FACTORS						C	
	Grazing	Timber Sales	Opport. for Disadv.	Wildlife Habitat	Employ-ment	Issue Concerns Goals	Energy	Soil Product	Riparian Protect.	Soil Loss	Net Benefits
	MM AUMs	MM BF	Rank	M Acres	Years	Rank	Rank	lbs/ac	M Acres	in/yr	MM$
	A 2.40	A 604	A	A 93	A 132	C=A	C	C 600	C 498	C .04	C 319
	C 2.01	C 550	C	C 70	C 129		A	A 585	A 518	A .90	A 298
DIFFERENCE	.39	54	A>C	23	3		C>A	15	20	.86	21
SCALED VALUES	80	75	60	50	30	0	15	20	45	85	100

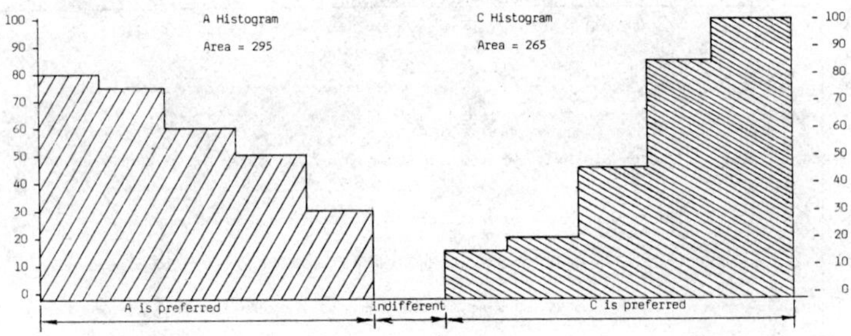

TABLE 9 Evaluation of Atlernatives A and D

ALTERNATIVE	FACTORS										
	Net Benefits	Timber Sales	Grazing	Wildlife Habitat IMP	Soil Productivity	Soil Loss	Riparian Area	Opport. for Disadv.	Energy Conservation	Employment	Issues Concerns Goals
	MM$	MM Bd Ft	MM AUMs	M Acres	lbs/ac	in/year	M Acres	Rank	Rank	per yrs	Rank
A vs. D	A 298	A 604	A 2.40	A 93	A 585	D .06	D 508	A Max.	A=D	A 132	A
	D 218	D 582	D 2.18	D 91	D 530	A .90	A 518	D Min.		D 92	D
DIFFERENCE	A 80	A 22	A .22	A 2	A 55	D .84	D 10	A>D	0	A 40	A>D

SCALE	FACTOR	EXPLANATION OF WEIGHTS
100 -	Net benefits	most important difference as it reflects a much more efficient program.
90 -		
80 -		
70 -		
60 -	Issues, concerns	difference significant because of contributions to National Program.
50 -	Grazing	difference of 220,000 AUM important because of positive economic and social benefit to local ranches and community.
45 -	Soil loss	almost 1" difference because of basic stewardship responsibilities.
40 -	Opport. Disadv.	difference important because of direct benefits to disadvantaged.
30 -	Soil productivity	difference important because of opportunities to meet demand for more forage production.
25 -	Timber sales	difference of 22 million bd. ft. significant because of the greater opportunities afforded the timber industry for expansion.
20 -	Riparian Protect.	difference of 10,000 acres important because of plant and animal diversity and scenic beauty.
19 -	Employment	important because of focus on reducing unemployment in areas where unemployment and underemployment is high.
10 -		
0 -	Energy Wildlife	no significant difference

TABLE 10 Weighted Factor Profile Comparing Alternatives A & D

	A				FACTORS					D	
	Net Bene- fits	Issues Concerns Goals	Grazing	Opport. for Disadv.	Soil Product.	Timber Sale Off	Employ- ment	Wildlife Habitat	Energy	Riparian Protect.	Soil Loss
	MM$	Rank	MM AUMs	Rank	lbs/ac	MM BF	Years	M Acres	Rank	M Acres	in/yrs
	A 298	A	A 2.40	A	A 585	A 604	A 132			D 508	D .06
	D 218	D	D 2.18	D	D 530	D 582	D 92			A 518	A .90
DIFFERENCE	A 80	A D	A .22	A>D	A 55	A 22	A 40			D 10	D .84
SCALED VALUES	100	60	50	40	30	25	19	0	0	20	45

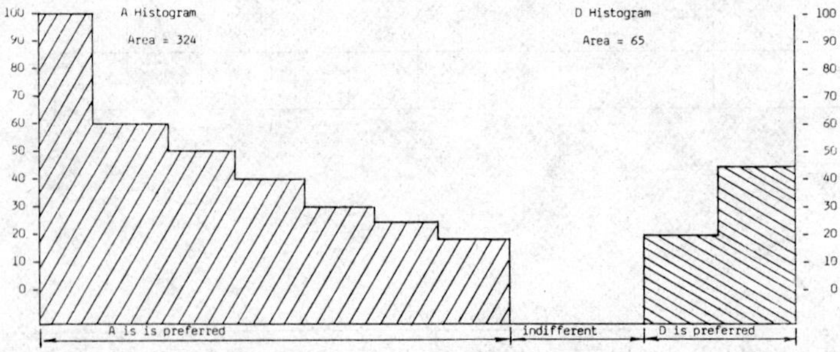

APPENDIX I

EVALUATION CRITERIA

Following are the criteria to be used in the evaluation of the four alternatives:

Economic - Cost efficiency

Social - Employment and opportunities for the disadvantaged in carrying out Forest Service programs. Responsiveness to public issues, management concerns, National Program Goals (1975) and Regional Goals.

Biological - Soil productivity, riparian area protection and recovery, timber sales, livestock grazing and wildlife habitat improvement.

Physical - Soil loss and energy dependency and conservation.

The primary sources for the determination and formulation of the criteria are:

- RPA, A Recommended Renewable Resources Program (1977 to 2000).
- Chief's Memo 8100 dated January 31, 1977: Regional, Area and Station istribution of the 1975 Renewable Resources Program.
- FSM 1951.31, Evaluation Criteria.
- Interviews with Region's Executive Group, State Foresters, other appropriate State officials and other Federal agencies' officials.
- Oral and written opinions resulting from 32 public workshops that led to formulation of the public issues.
- Inputs from Regional Forest's Staff Directors which determined management concerns.
- Recommendations made in a report1/ prepared for the Forest Service by a consultant firm, INTASA, entitled "Investigation of Public Response to Preliminary Area Guides: What Can The Forest Service Learn From It."

1/ A copy of this report is on file in the Regional Office, Lakewood, CO.

DESCRIPTION OF EVALUATION CRITERIA

Cost efficiency, the matter of cost measurements, is sometimes a source of considerable confusion. Many people seem to strive for degree of accuracy which is often not attainable, due to the nature of the problem.

In this analysis concentration on using cost and benefit estimating methods which (1) stress consistency in treatment of alternatives, and (2) emphasize measurement of cost sensitivity stemming from differences in market versus non-market outputs were used.

Two independent estimates of value were used to make the analysis. The first estimate is a willingness to pay estimate and the second an income generated estimate.

The willingness to pay (WTP) values are the values developed for Region 2 in the draft 1980 Renewable Resource Program required by the Renewable Resource Planning Act of 1974. They are based on a series of studies commissioned by the Forest Service. These reports attempt to place comparable values on major resource outputs of National Forests by Region. The standards were estimates of willingness to pay for outputs at the point outputs leave the forest site.

The income generated estimate of value is based on the amount of income generated in a Social Resource Unit.[1/] In this case income is the compensation paid the household sector (private taxpayer) for inputs provided to the purchasing industry. Includes wages, salaries, dividends, interest, rent, royalties, etc. Income was calculated using Social Resource Unit Input-Output (Leontief type) models developed during the Rare II process. Changes in sales to local consumers and exports were determined for each resource output or activity from the National Forests. This user expenditure data was independently estimated to be consistent with the area being modeled. When the actual product is sold, the expenditure is the sale price of the product. For those activities where a product is not actually sold, the expenditues accounted are the expenditures the user incurs to participate in the activity. Camping, for example, requires expenditures for gas, food, sleeping equipment, etc.

Once user expenditures are determined, the input-output model will calulate the income generated by subtracting all other costs of

1/ Multicounty areas having unique similar social characteristics

production. Using an I-O model with constant coefficients provides a consistent estimate between alternatives.

The following display shows the estimated costs, benefits, and benefit-cost ratio by alternatives, together with the marginal costs, benefits and ratios between alternatives and the current situation as projected (Alt. D).

Based on this analysis, the net benefits using the income generated estimate of value will be used in evaluating the cost efficiency of the four alternatives.

ECONOMIC ANALYSIS - WILLINGNESS TO PAY VALUES
(1955 output, 1978 prices)
Table 1

ALTERNATIVE	TOTAL				MARGINAL			
	COST MM$	BENEFIT MM$	NET BENEFITS	B/C	COST MM$	BENEFIT MM$	NET BENEFITS	B/C
D1/	126.86	131.46	4.60	1.036				
A	166.50	170.02	3.52	1.021	39.64	38.56	-1.08	.973
B	194.39	167.29	-27.10	.861	67.53	35.83	-31.70	.531
C	162.92	190.47	28.55	1.176	35.06	59.01	23.95	1.683

ECONOMIC ANALYSIS -INCOME GENERATED ESTIMATE OF VALUE
(1955 output, 1978 prices)
Table 2

ALTERNATIVE	TOTAL				MARGINAL			
	COST MM$	BENEFIT MM$	NET BENEFITS	B/C	COST MM$	BENEFIT MM$	NET BENEFITS	B/C
D	126.86	345.43	218.57	2.722				
A	166.50	464.51	298.01	2.790	39.64	119.08	79.44	3.004
B	194.39	464.78	270.39	2.391	67.53	119.35	51.82	1.767
C	161.92	481.41	319.49	2.973	35.06	135.98	100.92	3.878

1/ Alternative D is the current situation as projected.

ENERGY DEPENDENCY AND CONSERVATION

The use, development and management of National Forest resources requires energy. Most of the energy will be consumed in the form of petroleum products used for one of the three following groups of activities:

- -- Timber harvest
- -- Construction, reconstruction and maintenance of roads.
- -- Recreation travel (to both developed and dispersed sites).

These three activities groups will be used as indicators to predict the relative petroleum consumptions for the various alternatives. It is assumed that petroleum consumption varies directly with the level of timber harvest and road construction. Gasoline and oil consumption by recreational users, while proportional to the amount of use, is also a function of the modes, party size and distances involved while getting to and from and engaging in such uses. The outputs estimates for recreation use provide little insight to the origin-destination relationship of the users. This weakness is compounded by the uncertainty of shifts in user preference for modes and party size in reaction to gasoline prices. Hence, the "energy-effectiveness" of each alternative program can only be subjectively evaluated based on differences in the total outputs in the above activity groups.

RIPARIAN AREA PROTECTION AND RECOVERY

In this region the water system corridor, which is commonly referred to as the riparian area, contains about three and a half million acres. Of this about 14 percent is estimated to be in need of some treatment for improvement or restoration. This is the area classified as being in fair to poor condition.

Since the treatments and uses proposed by the alternatives will have various effects on the riparian areas, the number of acres in the fair to poor condition will be evaluated with the alternative having the least number of acres in this condition being most desirable.

Recreation use, timber harvest, and livestock grazing are the activities having greatest potential for adverse impact on riparian areas. Because of the natural attraction that streams and valleys have for recreation visitors and the relative ease of construction in flat alluvial areas, recreation is considered to have the greatest potential for adverse impact.

TIMBER SALES OFFERINGS (NFS)

The volume of timber sales (live and dead) scheduled to be sold in 1995. The unit of measure is the timber volume in million board feet.

GRAZING

The expected sustained annual production of forage that is suitable and available for livestock and can be grazed without inducing damage to vegetation, soils and other related resources. Measured in Animal Unit Months (AUM's)

WILDLIFE HABITAT IMPROVEMENT (NFS)

The increase that would result from changes or additions in management and habitat improvement activities. Includes all structural and non-structural habitat improvements that benefit big game, small game and non-game wildlife species. Measured in acres of improvement.

SOIL PRODUCTIVITY

This criterion will be used in the evaluation according to the Glossary of Soil Science Terms published by the Society of Soil Scientists of Amercia (SSSA) which defines soil productivity as "the capacity of a soil in its normal environment, for producing a specified plant, or sequence definition emphasizes the capacity of the soil to produce and is expressed in terms of yield.

Because of the importance of livestock grazing on NF's lands in this region, forage production in pounds per acre is selected as the measure of soil productivity.

Since the "...specified system of management..." is a major part in the definition, it is felt that productivity varies among alternatives as the intensity and area under management varies. Also, the management activity affects productivity as grazing of livestock has a direct relationship to compaction, organic content, water holding capacity, soil depth and structure and other soil characteristics.

SOIL LOSS

That is the movement of soil from one place to another by wind or water and measured in inches per acre per year loss. The three most significant soil characteristics influencing soil loss are:

- -- infiltration capacity
- -- structural stability
- -- removal of vegetation

The infiltration capacity is influenced greatly by structural stability, especially in the upper soil horizons. In addition, soil texture, the kind and amount of swelling clays, soil depth, and the presence of impervious soil layers all influence the infiltration capaity.

The stability of soil aggregate affects the extent of erosion damage by the resistance of surface granules to the beating action of rain. A good vegetation cover will help improve the infiltration rate and the soil structure, thereby reducing soil loss.

The major activities that affect soil loss in the Rocky Mountain Region are timber harvest, road construction, cattle grazing and uncontrolled motorized vehicle use. Each alternative will be evaluated in terms of the maximum expected soil loss.

EMPLOYMENT

Is measured in person years of employment. One person year of employment is approximately 2000 hours. This may be filled by one person working year long or by several people filling seasonal employment. An input/output model was used to determine the number of jobs created by Forest Service program and activities. Since individual models were developed for each Social Resource Unit (SRU), the employment factor was available by these units allowing the determination of the impact on areas of high unemployment.

OPPORTUNITIES FOR THE DISADVANTAGED

Insight into how well the Forest Service programs and activities contribute to improving the lot of the disadvantaged was aided by the I/O model mentioned above. Basically, it was determined by looking at the profiles or characterization of each SRU, and estimating the contribution of Forest Service programs to the soical well-being of the disadvantaged people and the communuities. The major indicators used in making these judgments were jobs, income generated, gross regional product and opportunities for the disadvantaged to participate in activities provided through Forest Service programs.

RESPONDING TO ISSUES, CONCERNS AND GOALS

Because of the process for formulating the alternatives, each of the alternatives would achieve this established evaluation criteria, but to different degrees. Therefore, each alternative will be evaluated as to its potential for satisfying the criteria. Since the interdisciplinary team has more complete and intimate knowledge concerning the intent and associated interactions of each alternative, a core team of 3-5 members will be formed to make the initial assessment and rank how each alternative responded to the issues, concerns and goals, The results will be a consensus of the core team's subjective evaluation. Where possible, specific match up of output to issue, concern or goal will be established and made visible.

Wetlands Regulations and Public Perceptions In Massachusetts

Thomas M. Leschine and Stephen R. Cassella

Marine Policy and Ocean Management Program
Woods Hole Oceanographic Institution
Woods Hole, Massachusetts 02543

Purpose

This paper reports on the results of a 1979 survey of wetlands property owners in two Massachusetts coastal towns. The towns in which the survey was conducted were chosen to represent two trends perceived to be affecting the present course of shoreline development outside of large urban areas in the state. Shifting patterns of regional development along the Massachusetts coast over the past two decades are reviewed as a prelude to the report on the survey. The socioeconomic characteristics of the populations interviewed separate quite clearly into categories which mirror the general development trends outlined.

In addition to socioeconomic information, the survey gathered information on the characteristics of wetlands property and its ownership, on property owners' perceptions of wetlands and wetlands values, and their perceptions of and experience with wetlands regulations at the state and local level. The state and local mechanisms which regulate wetlands use and development in Massachusetts are described as contextual background for the latter portions of the survey. The survey results themselves are presented mostly in descriptive terms. Future work will focus on explaining apparent differences in property use and development and in attitudes and perceptions of wetlands value and wetlands regulations in terms of other characteristics described in the survey. The consequences of this information for wetlands regulation in Massachusetts are discussed briefly.

Regional Growth in the Massachusetts Coastal Zone

The coastal zone management problem is often framed in terms of the effect rapid population growth in coastal areas has on fragile coastal resources. Frequently cited statistics point, for example, to the fact that 42% of the United States' population lives in the 400 counties that the federal Office of Coastal Zone Management considers to be coastal and to the fact that the population continues to grow faster in coastal areas than in the rest of the U.S. (U.S. Department of Commerce 1978).

The population trends of the past two decades in the Massachusetts coastal zone illustrate several of the state's coastal zone management problems and pinpoint some of the factors which motivate

Table 1. Population and Percent Change by Coastal Region in Massachusetts, 1970-1977 (Source: U.S. Bureau of Census, Current Population Reports 1979).

	1970 Population	1977 Population	%Change
Plymouth Bay	32,241	48,941	51.8
Nantucket	3,774	5,469	44.9
Cape Cod	96,656	136,919	41.7
Martha's Vineyard	6,177	7,981	30.5
South Shore	58,139	68,502	17.8
Mount Hope Bay	156,218	166,643	6.7
Buzzards Bay	173,925	185,361	6.6
Upper North Shore	78,978	82,689	4.7
Boston Harbor	984,107	958,825	-2.6
Lower North Shore	259,451	240,881	-7.2

this research. The state Office of Coastal Zone Management has grouped for planning purposes the state's 74 coastal cities and towns into ten regions, each region comprised of towns of similar geographic and demographic characteristics (Figure 1). Table 1 above lists these regions in decreasing order according to percent population increase over the eight-year period from 1970 to 1977.

While Massachusetts as a whole experienced a 1.6% population increase during this eight-year period, its coastal region grew by 2.8%. Although such figures mirror the national trends, substantial differential growth patterns evident in Table 1 (ranging from a 51.8% regional increase to a 7.2% regional decrease) serve to define more clearly the nature of the pressure currently being brought to bear on the Massachusetts coast by demographic changes. The state's southeast shore, encompassing the Plymouth Bay, Cape Cod, Martha's Vineyard and Nantucket regions leads the rest of the state in growth

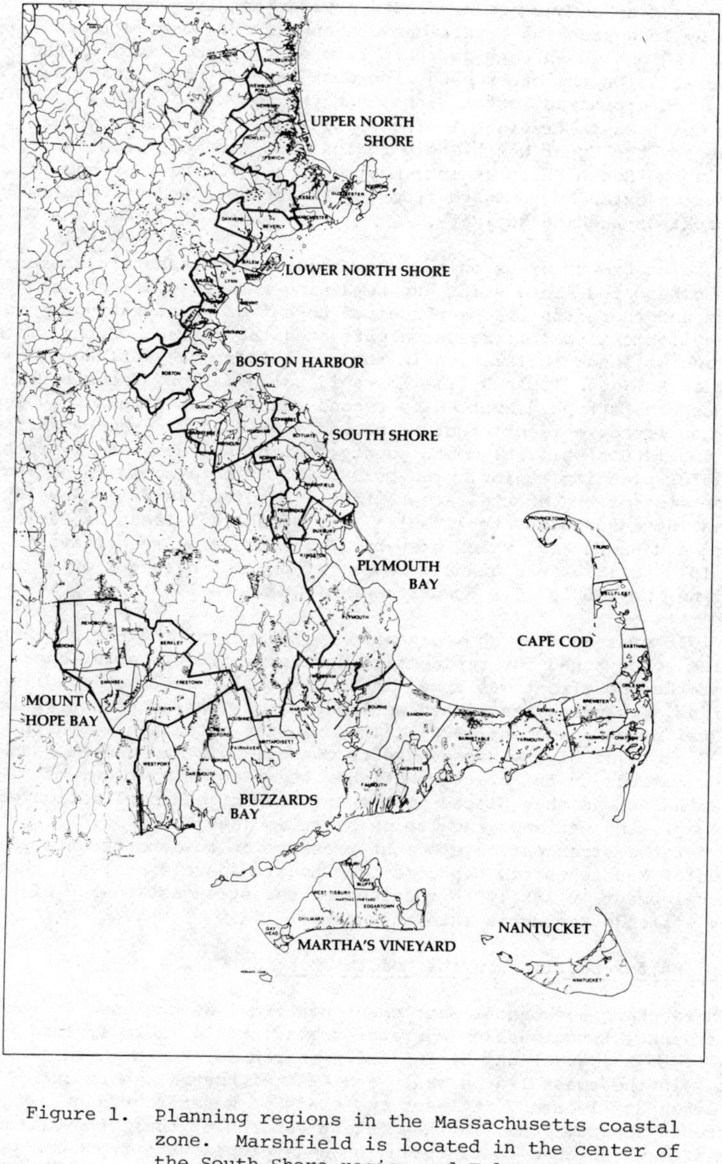

Figure 1. Planning regions in the Massachusetts coastal zone. Marshfield is located in the center of the South Shore region and Falmouth is in southwestern Cape Cod. Figure adapted from an illustration in the Massachusetts Coastal Zone Management Plan.

by a substantial margin. These four regions are characterized by abundant open space and ocean-related environmental amenities as well as by long-standing traditions as centers for tourism (Each of these regions is an easy day trip from most of the metropolitan Boston area.) On the other hand, the most heavily populated coastal regions, the urbanized Boston Harbor and the Lower North Shore regions, are the only coastal regions losing population. The modest growth exhibited by the Upper North Shore region, and the more substantial growth in the South Shore region, may reflect a mix both of suburbanization extending outward from the Boston area and of tourism and recreation-seeking activity.

The high growth areas on the Massachusetts southeastern shore also encompass the state's largest remaining reservoir of undeveloped wetlands and shorefront. The perceived threat of accelerating shorefront development to this resource motivated the passage of the state's pioneering wetlands legislation in the mid-1960's (the Wetlands Protection Act; see Table 3). The rapid colonization of the southeastern shore, strongly abetted by second and retirement home development, was already evident then: Cape Cod's population grew by 37.5% during the 1960's while Plymouth County, which includes the Plymouth Bay coastal planning region, grew by 39.4%. It is also worth noting that the size of the coastal area actively involved in rapid development has increased since the 1960's: the offshore islands of Martha's Vineyard and Nantucket, which grew by 30.5% and 44.9% respectively during 1970 to 1977, had annual growth rates which averaged only about .5% during the 1960's (U.S. Department of Commerce 1978).

A 1976 state survey showed nearly two-thirds of the Massachusetts coastline to be zoned for residential development: only about one-sixth of the shorefront was zoned in specifically conservation-oriented categories, including restricted wetlands zones (Massachusetts Office of Coastal Zone Management 1976). This was the situation in which the 1978 approval of the Massachusetts Coastal Zone Management Plan gave new impetus to the state's wetlands use regulation programs. Renewed emphasis was then placed on the enforcement and application of several existing wetlands regulation statutes (see Table 3). It is clear that the strongest development pressure on the shorefronts and wetlands of Massachusetts has come from housing development and that such development is likely to continue as the strongest source of pressure on wetlands resources into the future.

Motivations for Settling in the Coastal Zone

These changing coastal settlement patterns, as expressed through the differential regional growth rates summarized in Table 1, lead naturally to the questions behind the research described in this report. In the coastal area south and east of Boston, where our study sites are located, at least two kinds of waterfront settlement trends appear. One, in the distant suburbs of the metropolitan Boston area where open space including undeveloped waterfront and wetlands property is still plentiful, seems oriented toward "bedroom community" development which has an emphasis on waterside living.

The second more dramatic trend, in the more distant areas where ocean-oriented recreational opportunities are greatest, seems oriented toward second and retirement home development as part of a general regional emphasis on coastal tourism. This phenomenon was well documented in a Cape Cod study (Booz, Allen & Hamilton Inc. 1972). The study found that at least 90% of Cape Cod's population increase from 1965 to 1970 was due to in-migration. Furthermore, the populations in the 45-to-64 and 65-and-over age groups were found to have grown at rates substantially greater than that of Cape Cod as a whole between 1960 and 1970, with the latter category showing a much larger increase than the former. The suggested pattern is one of retired persons seeking second or retirement homes at a distance from present or former jobs. Such regional differences in motivation for occupying the shorefront may also entail differences in attitudes and concerns related to shorefront property ownership. Such factors may partly explain regional variations in the acceptability of and participation in coastal planning initiatives which affect wetlands development.

The factors which influence attitudes related to coastal programs are difficult to determine - they are certainly not readily apparent in shorefront settlement patterns themselves. It is not surprising that in general little socioeconomic information has been incorporated into coastal planning programs for social equity and other considerations (Dickert and Sorensen 1974). Where wetlands management is concerned, such information can nevertheless help to define more precisely the environment to be managed, the concerns which can redirect the emphasis in wetlands management programs, and the attitudes and perceptions which may limit the ability of program managers to succeed in implementing some forms of regulation. The wetlands environment thus becomes an area defined by social and economic parameters as well as by physical ones.

The research described here was conducted to assess several kinds of information related to the twin themes outlined above. First, this report attempts to build a socioeconomic profile of wetlands property owners in two coastal areas which appear to be developing in response to different kinds of motivation for settlement. Second, it aims to delimit the extent to which socioeconomic and other factors might explain differences in attitudes towards wetlands and their management.

In order to obtain such information, a survey of wetlands property owners was conducted during the summer and fall of 1979. The socioeconomic data collected consisted of personal data on the property owners themselves (such as age, income and occupation) and of details of property ownership (such as length of ownership and pattern of use of property). Other information collected falls broadly into the following categories: 1) history of modifications made in wetlands areas, including future plans for such modifications; 2) knowledge of and attitudes toward state and local wetlands use regulation programs; 3) personal experience with the regulatory apparatus and the ways wetlands regulation influence future plans for property disposition and development; 4) concerns related to wetlands property ownership; and 5) perceptions of wetlands values and the effects of regulations on those values.

Study Areas

The survey information reported here was gathered in the town of Marshfield, located on the Massachusetts South Shore about 30 miles south of Boston, and in the town of Falmouth, located on western Cape Cod (Figure 1). (Table 2 summarizes population and land use characteristics of the two towns.) Although Falmouth is more than 50% larger than Marshfield in land area, the two are otherwise remarkably similar in terms of land use patterns: the two towns have roughly the same amount of wetlands area, their populations are nearly the same, and they have grown at roughly the same rate over the last decade. Also, in both towns, the developed shoreline is almost entirely given to housing. There is some evidence that growth has slowed in Marshfield over the last few years.

The principal difference between Marshfield and Falmouth may well lie in the characteristics of the in-migrants who have contributed to their respective rapid increases in population. The two towns appear to represent the poles of the differential trends in coastal settlement discussed above rather well: Falmouth is widely thought of in the region as a resort and retirement community with an abundance of ocean-oriented recreational opportunities, while Marshfield has long been viewed as a working community of Boston area exurbanites who place a high value on pleasant natural surroundings. These characteristics emerged to clearly differentiate the interview populations of wetlands property owners in the two towns in the survey.

The wetlands of Falmouth and Marshfield are regulated at the state and local levels through similar but not entirely equivalent mechanisms. (Table 3 summarizes information on the most important state and local regulatory programs.) The principal vehicle for wetlands protection in each town is the permit system for wetlands alterations run by the local conservation commission under the state Wetlands Protection Program. Both towns also participate in the federal Flood Insurance Program and both have local bylaws to implement the required floodplain building regulations.

The state-run Coastal Wetlands Restriction Program, which places absolute restrictions on certain kinds of activities in designated wetlands, was applied to some wetlands in Marshfield early in the Program's life. The more comprehensive Restriction Program, which has recently evolved from the Massachusetts Coastal Zone Management Program, is currently being applied to wetlands in Falmouth; the existing restrictions for Marshfield wetlands are to be updated considerably in the near future.

A unique natural resource of Marshfield is the scenic and surprisingly undeveloped North River corridor, which forms the town's northern and western boundary and which passes through five other towns. In the spring of 1979, this river became the first in the state to be designated a Scenic River under the state Scenic Rivers Program. Such a designation carries with it a restriction on certain kinds of building activity within a corridor of 300 feet on each side of the river. Our Marshfield sample includes several owners of property

Table 2. Land Use and Population Data for Marshfield and Falmouth.

Land Use[1]	Marshfield	Falmouth
Forest Land	9,498 acres	16,657 acres
Agr. or Open Land	1,144	3,069
Wetlands	3,469	4,053
Mining & Waste Disposal	280	364
Outdoor Recreation	359	816
Residential	3,524	5,658
Commercial & Industrial	226	656
Open & Public	178	231
GRAND TOTAL	18,678 acres	31,504 acres
Tidal Shoreline	18.5 miles	55.0 miles
Population, 1977	21,052	21,832
% Change, 1970-1977	38.3	36.9
Summer Peak Population, 1977	31,000[2]	54,100[3]

[1] MacConnell (1974).

[2] Town of Marshfield estimate.

[3] Based on Cape Cod Planning and Economic Development Commission estimates.

whose land is newly restricted under this program.

The tradition of home rule is a strong one in New England towns, and its application is evident in wetlands management programs. The considerable power vested in local conservation commissions composed of citizen volunteers constitutes a most important acknowledgement of local decision-making power. Marshfield and Falmouth have both joined a growing list of Massachusetts towns which have passed local wetlands bylaws that mirror the language and intent of the state Wetlands Protection Program. While these bylaws generally extend the application of the Program by adding interests to those specified for protection, they also have the effect of localizing the decision-making process on wetlands permits. This happens because appeals of conservation commission decisions are directed under home rule charters to local courts rather than to the state administrative agency with oversight authority for the Program. This issue is attracting increasing attention in Massachusetts at present.

Table 3. Relevant State and Local Wetlands Protection Measures In Massachusetts.

REGULATION	PURPOSE	ADMINISTRATION
Wetlands Protection Program (MGLA Ch. 131, Sec. 40)	Protects seven interests: Public and private water supply, groundwater supply, flood control, storm damage prevention, prevention of pollution, protection of land containing shellfish, and protection of fisheries.	Local conservation commissions of citizen volunteers issue permits for projects in wetlands areas. Permission for proposed construction can be denied if a commission finds that any of the seven interests will be significantly degraded.
Wetlands Restriction Programs (MGLA Ch. 131 Sec. 40A, Ch. 130, Sec. 105)	To protect coastal and inland wetlands against environmentally harmful development. Designed to supplement the regulatory approach of the Wetlands Protection Program.	The state Department of Environmental Management is authorized to restrict development in designated coastal and inland wetlands areas.
Scenic Rivers Program (MGLA Ch. 21, Sec. 17b.)	To protect irreplaceable wild, scenic and recreational river resources, wildlife, fresh water fisheries, and public and private property.	The state Department of Environmental Management is authorized to designate, regulate and restrict the alteration of selected rivers and adjacent land for scenic and recreational purposes.

Table 3.

AREA OF APPLICABILITY	PROCEDURE	APPEAL
Regulated area extends to 100 ft. beyond either the landward edge of a wetlands or the 100 year flood, whichever is greater. Eleven resource areas are protected: land under the ocean, designated port areas, coastal beaches, coastal dunes, barrier beaches, coastal banks, rocky intertidal shores, salt marshes, land under salt ponds, land containing shellfish, and fish runs.	Any person or agency proposing to alter a wetlands must file a Notice of Intent with the local conservation commission. After a public hearing, the Commission issues an Order of Conditions that approves, conditions or prohibits the activity.	An applicant, an abuttor, an aggrieved person, any ten residents, or the overseeing state agency can appeal a local Order of Conditions to the agency. Other appeals are judicial.
Coastal and inland areas include all freshwater marshes, meadows, swamps, salt marshes, shellfish beds, salt ponds, flats or other lowland subject to tidal action or coastal storm flowage, and contiguous land that is deemed necessary to affect a protective order.	Proposed restricted wetlands are mapped from aerial photographs. After a public hearing the restrictive order, a copy of the map, and a list of permitted and prohibited uses is attached to the deed of each property containing a restricted wetland.	Any owner of land affected by a restrictive order can appeal the restriction to the Commissioner of Environmental Management within 60 days after receiving the order. A two-thirds vote of the Massachusetts Legislature can also repeal a restrictive order.
Includes all Department designated rivers and streams of the Commonwealth or portions thereof and such contiguous land not to exceed 100 yards on either side of the natural bank that is reasonably deemed necessary to be protected.	Same as for Wetlands Restriction Program.	Any property owner of land affected by a restrictive order can appeal to Superior Court to determine whether such order is so restrictive that it constitutes an unlawful taking of property without compensation.

Table 3.

REGULATION	PURPOSE	ADMINISTRATION
Local flood-plain bylaws	Congress created the National Flood Insurance Program to minimize annual flood losses through planning and low cost insurance.	Administered in the same manner as building codes and other zoning ordinances; enforcement is a local community's responsibility. New buildings must be flood proofed or elevated above required levels.
Local wetlands bylaws	To protect wetlands by controlling activities deemed to have a significant effect on wetlands values including but not limited to the following: flood control, erosion control, storm damage pervention, wildlife, recreation, and aesthetics.	Administered locally by town conservation commissions.
Other local zoning bylaws	Various purposes related to health, environmental protection, aesthetics, and other zoning related concerns. Subdivision control laws often dictate that 80%-90% of subdivided lots be upland or that wetlands not be included in satisfying minimum lot size requirements. Some towns have specific wetlands overlay zones. (See text.)	Purely local, under the State Zoning Enabling Act (MGLA Ch. 40A, Sections 1-22). Various town boards and offices participate, such as planning boards, health officers, and conservation officers.

Table 3.

AREA OF APPLICABILITY	PROCEDURE	APPEAL
U.S. Department of Housing and Urban Development provides flood prone communities with Flood Hazard Boundary Maps that outline flood prone areas within the community. In addition HUD provides Flood Insurance Rate Maps which show flood elevations and risks zones for insurance purposes.	After HUD provides a community with Flood Insurance Rate Maps and completes a detailed on-site survey that establishes flood levels, a community qualifies for the program by adopting comprehensive floodplain management measures (bylaws).	Appeals are generally directed to the town zoning board of appeals, and to district courts if necessary.
May differ from town to town. Generally the boundary extends to 100 ft. from any wetland or to 100 ft. from the 100 year storm line.	The interests protected by wetlands bylaws overlap the interests of the Wetlands Protection Program. The Notice of Intent filed with the local conservation commission satisfies the requirements of both regulations.	An applicant may appeal the commission's decision to Superior Court.
All town wetlands areas as defined by any other state or local regulation, or to specially defined districts.	Various procedures.	Variances can be granted by the town zoning board of appeals. Judicial remedies go through district courts.

The Marshfield zoning ordinances contain one feature not found in those of the town of Falmouth, but one shared by a number of other Massachusetts communities with substantial wetlands areas, namely that wetlands zones are explicitly superimposed over the other zoning districts on the town zoning maps. Though the associated zoning ordinance in principle does nothing more than apply existing state and local regulations to land in wetlands zones, in practice the ordinance adds a potentially important local administrative function to the wetlands regulation system. Proposed construction projects near wetlands in a town with such a regulation are often referred by the town building inspector or by the builders themselves to the town conservation officer for a "delineation" of the portion of the wetlands zone which lies on the property. Such a delineation process occasionally accompanies a proposed property sale, at the request of either seller or potential buyer, and the service is provided without cost. This process in effect constitutes an informal negotiation between potential builders and the town, often with the result that building plans are prescribed in a way that no dealing with the town conservation commission is then necessary. Table 4 below summarizes information on construction activity and associated regulatory activity in the two towns.

Table 4. Residential Construction Activity and Wetlands Regulatory Activity, Marshfield and Falmouth.

	Marshfield	Falmouth
Total housing units, 1979[1]	7646	14823
Per cent increase since 1970	42.7	44.3
Total units authorized since joining Flood Insurance Program[2]	111	1200
Total units authorized in floodplain since joining F.I.P.	3	62
Conservation Commission cases, 1972-1979[3]	64	164
Marshfield Wetlands Delineations, 1976-1979[4]	77	--
Acres restricted under Coastal Wetlands Restrictions Program	2167	965
Number of parcels restricted under Scenic Rivers Program	221	--

[1] 1970 U.S. Census figures plus units authorized by building permit, 1970-1979. 59% of Falmouth units are estimated to be year-round.

[2] Marshfield joined F.I.P. in 1978, Falmouth in 1977.

[3] Includes both housing and non-dwelling structures.

[4] Private surveyors did an unknown number of additional delineations not recorded in town records.

As might be expected, the boundaries which define regulated wetlands under the various federal, state, and local programs just discussed are far from coincident, leading to much confusion in administration and public perception of wetlands programs. While the floodplain probably constitutes the greatest regulated wetland area in either town, the definition of wetlands used under the state Wetlands Protection Program is also a comprehensive one which includes such "dry" features as dunes, beaches, and coastal banks, as well as some small isolated ponds and intermittent wetlands. These wetlands are defined in the legislation, but are not mapped. Because much of the local topography, particularly on Cape Cod, is determined by rolling glacial till, many waterfront "wetlands" are high enough to be out of the floodplain. The wetlands eligible for restriction under the state Wetlands Restriction Program have more restricted definitions, requiring that they be "subject to tidal action or coastal storm flowage" or be contiguous land whose protection is necessary to such wetlands (see Table 3). This has made the mapping and the designation of these wetlands a slow and difficult task for state officials.

Survey Methods

State and local records were surveyed to give a detailed picture of wetlands protection activity at the local level in both Falmouth and Marshfield. Special attention was paid to records of permit applications and reviews under the Wetlands Protection Program, building inspector's reports, assessor's maps and records, the Marshfield wetlands zone delineations and state determinations of critical wetlands habitats in private hands. State and local personnel connected with all phases of the governmental process affecting private property development near wetlands were interviewed or consulted, and public hearings related to wetlands protection activities attended.

This survey produced information about how wetlands protection functions at the local level. It also helped identify wetlands areas with varying degrees of development around them. Property owners were then selected for in-person interviews in "clusters" from areas in which modification activity either had or should have produced encounters with the regulatory system. The selection of interview subjects within clusters was essentially arbitrary, though all individual property owners in the area who had filed for wetlands alteration permits from the local conservation commission were included. One hundred and one Falmouth property owners were interviewed in person, usually in their homes, and 76 Marshfield property owners were interviewed either in person or by telephone. The survey was not limited to those actually known to have encountered wetlands regulations in some official way to ensure that a full range of opinions would be surveyed. Specific areas for interviews within each town were identified on assessor's maps to represent the diversity of shoreline development uncovered in the town hall surveys. Places where particular schemes of regulation of interest, such as the Massachusetts Scenic Rivers Program, were in effect were also selected. Local knowledge helped in Falmouth, while state and local officials assisted in Marshfield.

Beyond that, efforts were made to sample property owners in a given area in rough proportion to the area's total number of wetlands property owners. This notion of 'area' or neighborhood of interest seldom coincided with any convenient local boundary lines. In the one area selected which did coincide with the area represented on a local assessor's map, the sample interviewed was about 15% of the wetlands property owners on the map.

Efforts were made to control sources of bias; interviews were conducted on weekends and in the evening when necessary, during summer and fall, to avoid such problems as overrepresenting retired property owners or housewives, or underrepresenting weekend or seasonal residents. Owners of undeveloped property were sought out in their businesses or elsewhere, though these persons probably were underrepresented. All persons selected for interviews were sent a letter of introduction asking them to return an enclosed card to indicate their willingness to be interviewed. Some property owners who did not return cards consented to be interviewed when they were telephoned. Cooperation was generally very good. Cards returned comprised more than 40% of those mailed to Falmouth property owners, and more than 30% of those mailed to Marshfield property owners.

The term 'wetlands property owner,' as applied to the interview subjects selected for this survey, refers specifically to town residents whose property satisfied the following criterion with respect to the Wetlands Protection Program: at least a portion of their property had to fall within 100 feet of a wetlands resource area as defined by the legislation. In most cases the associated resource area was a salt or fresh water marsh, or open water.

Results of the Survey

Personal characteristics of the interview population.

The Falmouth and Marshfield property owner populations interviewed differed substantially in several ways. Sixty-seven percent of Falmouth interviewees were over age 55, while ages of Marshfield property owners were fairly evenly distributed. In Marshfield 33% were between the ages of 25 and 40, while only 9% of Falmouth interviewees were in this age category. The proportion of retired people in the Falmouth sample was more than twice that in Marshfield (49% vs. 20%).

Income distributions also differed substantially in the two towns. Fifty-eight percent of Marshfield property owners reported gross annual family incomes between $20,000 and $60,000, while 39% had incomes under $20,000. Three percent reported incomes over $60,000. In Falmouth nearly 40% reported incomes in the middle range, while 18% reported incomes greater than $60,000, and 42% under $20,000. In addition, many of the retired people in the Falmouth sample reported current retirement incomes which we judged to be much less than their most recent full employment incomes.

Of the interviewees for whom we determined level of educational attainment, comparable portions were high school graduates in both towns. While 48% of these individuals were also college graduates in Falmouth, 25% of those in Marshfield were college graduates. Likewise, for those for whom we determined present or former occupation 54% of Falmouth families had a member who was a present or former professional worker, while 44% of Marshfield families did. We interviewed a much higher proportion of males than females in Falmouth, but the situation was reversed in Marshfield. This suggests that a male member of the household was much more likely to have the discretionary time to be home when we appeared for interviews, usually during the day and during the week, in Falmouth than in Marshfield.

While 49% of Falmouth interviewees identified their developed wetlands property as their principal residence, 74% of Marshfield interviewees did so. We asked all interviewees who owned developed property to tell us what price they would like to obtain for it were they to sell it at the time of the interview (i.e. 1979). The mean of these property valuations was $113,000 in Falmouth, while it was $75,000 in Marshfield.

Thus Falmouth interviewees were older, more likely to be retired, better educated, had higher incomes and somewhat better jobs, had more money invested in their properties and were more likely to consider some other place (usually an inland location) as their principal residence than their counterparts in Marshfield. Though we did not ask respondents their place of employment, the available evidence suggests that Marshfield residents are more locally oriented than their counterparts in Falmouth in general.

Characteristics of wetlands property and its ownership.

We pointed out above that the Falmouth residents surveyed were 25% less likely to consider their in-town wetlands properties to be their principal residences than their counterparts in Marshfield. On the basis of our survey, developed waterfront and wetlands area properties in Falmouth not only have a stronger likelihood of having been acquired as second homes, but the dwellings on them now are more likely to have been built by their present owners. While 8% of those we interviewed in Falmouth presently own undeveloped wetlands property, only 36% of the sample had originally acquired developed property. The remaining 56% had built homes on wetlands properties they acquired as undeveloped land. In Marshfield on the other hand, 74% of those interviewed had property which was already developed when they obtained it.

In both towns, the primary reason given for choosing a wetlands property was related to the enjoyment of environmental amenities. Property owners seemed well satisfied with their investments in both towns; the owners of 68% of the developed properties in Falmouth planned to keep their property into the forseeable future, as did the owners of 80% of the properties in Marshfield.

The fact that housing is generally older on the properties whose owners we interviewed in Marshfield partly explains the lower average asking price reported above. The age of housing has other consequences more directly relevant to wetlands management as well. A higher proportion of residential septic systems, a potential source of coastal pollution, were located in close proximity to wetlands in Marshfield, where 51% of surveyed owners of developed property reported septic systems within 100 feet of wetlands, than in Falmouth (38%).

Despite the fact that the Marshfield wetlands area housing in the survey was generally older than that in Falmouth, Falmouth property owners were more than twice as likely to have built a shoreline modification structure of some kind, most likely a dock, boat slip, rip rap or sea wall. Fifty seven percent of properties in the Falmouth survey had some kind of modification of wetlands areas on their properties, versus 24% of surveyed properties in Marshfield. This difference may be accounted for by several factors, such as a greater emphasis on recreational activity, more disposable income, or more severe shoreline erosion problems, at least in the case of that portion of the Falmouth activity which was for shoreline stabilization. It is also possible that newer housing construction in a developing recreation-oriented market is more likely to include a general reworking of the whole property than was true in the past. This point is of interest because a significant number of enforcement problems under Massachusetts wetlands programs stem from this kind of small scale modification activity to wetlands.

We asked property owners to choose from a list of potential problems associated with owning wetlands property those which concerned them the most, in order of their concern. The choices presented to them were erosion, flooding, storm damage, sewage backup, insects, odors, invasion of privacy, drinking water supply, 'other', or 'none'. In Falmouth, erosion appeared among the first three concerns selected far more often than any other, while problems with insects led the choices by about the same margin in Marshfield. Falmouth's relatively more exposed tidal shoreline, typically with a fringing marsh at the base of a shallow bank, may indeed present a different set of management problems than that in Marshfield with its more sheltered expanses of open marshland, often forming a significant barrier between upland and open water.

Property owners and wetlands protection.

We asked property owners a series of questions about the value of wetlands, their perception of the need for laws to protect them, and their knowledge of and opinions about several types of wetlands protection measures in effect in their towns. In Marshfield many of these questions were asked only of the 26-person subsample which was interviewed in person. Thus care must be taken in comparing results from the two towns pending the outcome of tests of statistical significance which have not yet been done. Some comparisons of the two samples are nevertheless interesting and will be discussed here with this cautionary note.

Respondents were asked to rate on a numerical scale the degree to which the presence of wetland on or near their property enhanced or detracted from its value. Similar large majorities (77% Falmouth, 80% Marshfield) believed that wetlands enhanced or strongly enhanced the value of their property. Only 7% in Falmouth and 5% in Marshfield thought that wetlands detracted or strongly detracted from property value.

We also asked respondents what effect they thought the presence of restrictions on wetlands use and development had on their value. A majority in each town (58% Falmouth, 56% Marshfield) thought such regulations added or substantially added to wetlands value, while minorities (16% Falmouth, 14% Marshfield) thought regulations detracted or substantially detracted from wetlands value. Though the exact numbers are difficult to assess accurately, small minorities (3% Falmouth, 5% Marshfield) volunteered their belief that the effect on value depended on whether the property was already developed or not when regulations were imposed, or whether development options were foreclosed by the imposition of such regulations.

Some possible differences between the populations surveyed emerged over other issues related to wetlands protection. We tried to ascertain whether property owners felt the same about the need for protective laws and regulations to control use of the kinds of wetlands property they owned versus other types of wetlands resources generally thought of as "public" (i.e. beaches and dune areas). In Falmouth about the same percentage (about 90%) of property owners thought protective laws were necessary to protect the resource type they owned (typically waterfront with a fringing marsh) as they did beaches and dunes. In Marshfield, 73% of the subsample of those interviewed in person thought the kind of wetlands they owned (typically marsh) needed legal protection, while 85% of the subsample thought beaches and dunes needed such protection. We asked respondents to tell us in addition how they felt about the filling of marshes. Seventy two percent of respondents in Falmouth and 88% of the subsample in Marshfield thought marsh filling should be generally prohibited. In each case, however, about 65% of those who said marsh filling should be prohibited were willing to grant exceptions to this prohibition if a "public benefit" were involved.

It is not yet clear what if any significance can be attached to these differences, nor what factors in the survey best explain them. One possible interpretation is that fringing marsh is regarded as a nuisance by many property owners, while those who own or live near larger marshes both see more value in them and believe that they and their neighbors are adequate stewards for such land without additional protective regulations. In a more direct sense, the substantial majorities involved in all responses above indicate a strong belief in the need for environmental protection, albeit somewhat tempered by a sense that the public welfare must also be served.

We asked a number of more specific questions about state wetlands programs and alternative modes of wetlands protection. We tested respondents' knowledge of the state Wetlands Protection Program, the chief wetlands program in the state, and judged the public level of information about its purpose and how its permit system works to be very low. Sixty-three percent in Falmouth and 75% in Marshfield were judged to have little or no knowledge of the Program. Thirty-three percent of our sample in Falmouth and 17% of the sample in Marshfield had been directly involved in local conservation commission cases under the Program, either as applicants for wetlands alteration permits or as abuttors appealing commission decisions. Substantial majorities of those with some degree of experience with the system judged it a fair system, despite complaints they had about it. Direct involvement in the permit process was no guarantee that an individual had much knowledge of how the system worked or what its intent was.

Of the other state and local wetland programs in effect, even awareness of their existence was minimal. About 90% of respondents in each town had never heard of the Wetlands Restriction Program. Sixty-five percent of the Marshfield subsample had not heard of the state Scenic Rivers Program, despite considerable publicity a few months earlier when the North River was officially so designated. The only wetland program which had general recognition was the federal Flood Insurance Program, about which half of all repondents in both samples had some knowledge. More than 70% of all respondents said that the presence of wetlands regulations did not materially affect their plans for use or disposition of their property. A comparable percentage of the 32 people we interviewed in the North River corridor in Marshfield also expressed this sentiment despite the relatively great controversy that attended its designation as a scenic river.

A final pair of questions attempted to deal with the issue of how the "ideal" system of wetlands protection should be run. When asked what level or combination of levels of government was "best" suited to run a wetlands protection program, straight local control led the other choices by substantial margins in both towns. State control was the second choice in Falmouth, while combined state - local control, essentially the present system, was second with the property owners in the Marshfield subsample. The federal government was not a popular program manager in either town. Interestingly, the reasons most often given for answers were negative ones, such as the state government being the best choice because local governments were controlled by special interests and the federal government too remote.

We were somewhat surprised by the response to our question as to who would make the best wetlands owners in the ideal world: private individuals, conservation trusts, local, state or federal governments. While 37% in Falmouth and 46% of the subsample in Marshfield chose private individuals, essentially the present system of ownership in Massachusetts, 31% in each town chose conservation trusts, the second most popular choice by far. This response may indicate a general sense of frustration with traditional modes of stewardship for environmental resources, namely government control of the use of resources held in private hands.

Summary

Falmouth and Marshfield have both experienced periods of rapid population growth in recent decades which mirror larger scale demographic changes in their surrounding coastal regions. Both have seen substantial housing development along their shorelines and wetlands as a result. Though a view of the developed portions of the shorelines in the two towns does not reveal it, substantial differences in motivation for occupying the shore separate the wetlands property owner populations in these two towns, reflective of the trends which have driven the development of their respective regions.

These differences are reflected not only in the socioeconomic character of the property owners themselves, but also in some characteristics of the property they own. The tendancy to modify the shoreline, for example, seems in part coupled to several characteristics of those who own the property. The concerns people have for their environment and their relationship to it as property owners depend in obvious ways on the nature of the environment itself, and in less obvious ways on other factors. Their support for environmental protection of wetlands is strong, but not unqualified. Those wetlands resources which are viewed as public may evoke different sympathies than those which are viewed as private.

The public sense that wetlands resources are valuable and worth protecting is stronger than the public awareness of what the goals and procedures of existing state and local regulatory programs designed to protect wetlands are. The power to regulate matters such as wetlands use which are perceived to be of local concern is not easily relinquished to the state government in Massachusetts. Wetlands property owners in the state seem to support local initiatives to retain (and regain) some measure of control over wetlands.

The work remaining in the project will be aimed at determining which factors, among all the information uncovered in this survey of wetlands property owners, are most and least strongly correlated. Such information should give a clear picture not only of what wetlands property owners want, but how their wants can be successfully accommodated within a regulatory framework which guarantees effective protection of the state's wetlands.

Acknowledgements

We wish to acknowledge the assistance of Nancy Pribish in conducting interviews, Kyle Hallsteen in providing background research, C.E. Carlson for editorial assistance, Ann Goodwin for the typing, and of the many state and local officials and office personnel in Boston, Marshfield and Falmouth without whose support and assistance this project could not have been conducted. Prepared with funds from the Pew Memorial Trust and the Department of Commerce, NOAA Office of Sea Grant under Grant #NA 79AA-D-00102, and the Institution's Marine Policy and Ocean Management Program.

References

Booz, Allen & Hamilton, Inc., 1972. Cape Cod economic base study: recommendations for beneficial changes. Prepared for the Cape Cod Planning and Economic Development Commission, Barnstable, Massachusetts.

Dickert, T. and J. Sorensen, 1974. Social Equity in Coastal Zone Planning. Coastal Zone Management Journal, Vol. 1, pp. 141-150.

MacConnell, W.P., and N.A. Pywell, D. Robertson, W. Niedzwiedz. 1974. Remote sensing 20 years of change in Barnstable, Dukes, and Nantucket counties, and in Plymouth County. Publication Nos. 95, 623, Massachusetts Agricultural Experiment Station, University of Massachusetts, Amherst.

Massachusetts Office of Coastal Zone Management, 1976. Zoning Ordinances of the Massachusetts Coastal Communities (Draft). Publication no. 9979-108-25-10-77-CR, Boston, Massachusetts.

U.S. Department of Commerce, Bureau of Census. 1979. Current population reports. Population estimates and projections. Series P-25, No. 834. Government Printing Office.

U.S. Department of Commerce. 1978. U.S. Ocean Policy in the 1970's. Status and Issues. Government Printing Office.

BARRIER BEACH PROTECTION IN MASSACHUSETTS

Stanley M. Humphries[(1)]
and
Jeffrey R. Benoit[(2)]

ABSTRACT

Resource variability and complexity along 1200 miles of coastline in Massachusetts pose a unique management challenge for the protection of barrier beach environments. The North Shore, South Shore and Cape Cod Bay shoreline of Cape Cod comprise an area consisting of 64 mesotidal barriers exposed to a predominant storm influence of extratropical cyclones (nor'easters). In contrast, the Buzzards Bay shoreline, Martha's Vineyard, Nantucket and Nantucket Sound shoreline of Cape Cod comprise an area consisting of 93 microtidal barriers that are primarily influenced by tropical cyclones (hurricanes).

Despite the existence of state wetlands protection legislation since 1963, it was not until 1978 that the Massachusetts Coastal Zone Management Program was approved and there was a concerted effort to protect barrier beach resources. Currently, the Department of Environmental Management and the Department of Environmental Quality Engineering administer deed restriction and regulatory programs respectively. On a town-by-town basis, implementation of the Coastal Wetlands Restriction Act is completed when a list of permissible and prohibited uses is recorded on the property owner's deed. Additional responsibility includes production and distribution of orthophoto base maps having all wetlands greater than 1/2 acre in size typed and numbered. Implementation of the Wetlands Protection Act occurs on a permit-by-permit basis for all dredging, filling, altering, or removing within 100 feet of any coastal wetland. The permits, if approved, include performance standards which either minimize or create no adverse effects on the adjacent wetland.

Technical staff consisting of coastal geologists and marine biologists have the responsibility to interpret and apply research findings for the purpose of determining ecological significance of a barrier beach resource. Based upon an accepted understanding of the major resource functions and the physical and biological characteristics critical to their function, the interaction between barrier dynamics and land use is evaluated. Information required for substantive evaluation includes existing data sets and associated research findings, storm history and vulnerability, long term shoreline changes, overwash dominance and extent of structural development. For the purposes of

(1) Coastal Geologist, Wetland Restriction Program, D.E.M., Boston, MA.

(2) Coastal Geologist, Wetlands Protection Program, D.E.Q.E., Lakeville, MA.

the restriction program, large-scale application of data and information is adequate to identify the importance of a system. However, in the regulatory program, there is a greater need for site-specific application of this data. Additional scientific research related to assessing the impact of structural development on barrier beaches and dunes is recommended.

INTRODUCTION

Massachusetts became the first state on the East Coast to have a Coastal Zone Management Program (MCZMP) approved, in April, 1978. Coastal property effected by the plan includes national wildlife sanctuaries on Plum Isalnd and Monomoy Island, the Cape Cod National Seashore, several state parks, and town-owned conservation land, all of which comprise only 22% of the state's 1200 mile coastline (U.S. Army Corps, 1978). The remaining majority of ocean-front property in Massachusetts is under private ownership, the rights of which extend to mean low water. Physiographically, the coastal resources along the Massachusetts shoreline are quite variable and complex due to glacial modification. Barrier beach resources, in particular, exemplify a state-wide range in variability and complexity. Consequently, only barrier beaches will be addressed for purposes of discussing some of the substantive aspects of implementing resource management through guidelines set forth by the MCZMP. The term "barrier beach" shall be used to collectively represent barrier spits, bay barriers and barrier islands.

Concurrent with the establishment of coastal zone management has been the increased understanding of barrier beach systems through progressive research in the last ten years (Leatherman, 1979). In one publication alone, over 230 references have been cited for research work done on beach and barrier environments (Hayes and Kana, 1977). In the past eight years, a number of East and Gulf Coast states have had resource inventories compiled that serve as useful baseline information. These include geological atlases in Texas (Fisher _et al._, 1972) and Delaware (Kraft _et al._, 1975), a beach erosion inventory in South Carolina (Brown, 1976 and Hubbard _et al._, 1977), a dune management guide in New Jersey (Gares, 1979), a Rhode Island barrier inventory (Olsen and Grant, 1973) and papers on sediment budgets in North Carolina (Jarrett, 1977) and barrier formation in Georgia (Oertel, 1979). Such detailed information only has direct application for the specific state management plans. The effort to acknowledge generic similarities between barrier beach morphology and coastal dynamics, (Clark, 1976 and Leatherman, in press), has provided the understanding of basic functional values from which all states can benefit. In general, inlet dynamics, dune migration and overwash processes comprise primary mechanisms that maintain barriers as a buffer against floods and storm waves. If site-specific research findings are not available for use in determining the influence of these mechanisms, use of qualitative observation and knowledge of basic principles in geomorphology must provide a reasonable basis for exercizing professional judgement in management decision making.

The purpose of this paper is to discuss some of the land-use management policies on barrier beaches in Massachusetts and their implementation with the use of two state wetland protection laws. A generic approach to the evaluation of coastal wetland importance and the use of a qualified technical staff comprise two fundamental elements in the current MCZM plan. In addition, a review of the state's glacial history, a regional classification of existing barriers and a summary of the dominant coastal processes are presented to summarize the physical setting.

PHYSICAL SETTING

Late Pleistocene glaciation is responsible for creating the surficial geology in Massachusetts. The last of four ice advances across North America was the Wisconsin Stage which ended approximately 10,000 years B.P.(U.S. Army Corps,1979).Distinct lobes of glacial ice advanced across the area we recognize today as the Massachusetts shoreline. The southernmost advance and standstill of several lobes formed the terminal moraine resulting in the islands of Martha's Vineyard and Nantucket. Glacial retreat continued, but during a second standstill, Cape Cod and the Elizabeth Islands were formed. Continued retreat of the ice northward resulted in the glacial meltwater topography of the North Shore.

Corresponding to the retreat of glacial ice was the relative rise in sea level, inundating and exposing the glacial formations and deposits to wave action. Once subjected to wave action, the various formations (i.e., drumlins, outwash plains, etc.) were eroded and reworked, becoming sources of sediment for the littoral system. Dumlin-related tombolos, rivermouth bars, pocket beaches and shingle barriers are some of the complex features which have formed by various coastal processes.

Coastal ecosystems along the Massachusetts shoreline reflect a regional differentiation and are classified into two distinct biogeographical zones (CZM Act P.L. 92-583). One zone is the Arcadian shoreline which extends along the North Atlantic coast south to Cape Cod and is characterized by a glaciated shoreline subject to winter icing and boreal biota. The other zone is the Virginian shoreline composing the Middle Atlantic coast from Cape Cod to Cape Hatteras and characterized by lowland streams, coastal marshes, muddy substrates and primarily temperate biota with some boreal representatives.

The type and distribution of barrier beaches along these two shorelines reflect extensive variation. With the use of topographic maps, nautical charts, orthophotos and oblique aerial photos, a tentative list of 157 open-coast barriers has been compiled for Massachusetts and classified into three major barrier types (Leatherman, in press) (Tables 1&2). A total of 64 barriers comprise approximately 85 miles (137 km.) of the Arcadian shoreline, which include the North Shore, South Shore, north side of Cape Cod and the outer Cape, and 93 barriers comprise approximately 110 miles (177 km.) of shoreline on the south side of Cape Cod, easterly and westerly shores of Buzzards

NORTH SHORE

Salisbury Beach—S
Plum Island—I
Crane Beach East—S
Crane Beach West—S
Coffins Beach—S
Wingersheek Beach—S
Pebbly Beach—B
Cape Hedge—S
Long Beach—S
Good Harbor Beach—S
Gray Beach—B
Black Beach—B
West Beach—S
Devereux Beach—B
Phillips Beach—S
Nahant Beach—S
Little Nahant Beach—S
Pond Beach—B
Revere Beach—S
Short Beach—S
Winthrop Beach—S
Yirrell Beach—S

SOUTH SHORE

Stony Beach—S
Crescent Beach—S
Minot Beach—S
Mann Hill—B
Egypt—B
1st Cliff Beach—S
Pegotty Beach—S
Rivermoor Beach—S
Humarock Beach—S

SOUTH SHORE, Cont.

Sunrise Beach—S
Duxbury Beach—S
Saquish Beach—S
Long Beach—S
White Horse Beach—B
Ship Pond—B
Center Hill Pond—B
Ellisville Harbor—S

CAPE COD—NORTH SHORE

Sagamore Beach—S
Town Neck/Spinghill Beach—S
East Sandwich/Scorton Shores—S
Sandy Neck—S
Bass Creek—I
Chapin Memorial—S
Sesuit Neck—B
Quivett Creek—S
Freemans Pond—S
Namskaket—S
Little Namskaket—S
Boat Meadow River—I
First Encounter Beach—S
Sunken Meadow—S
Jeremy Pt.—S
Great Is. Beach—S
The Gut—S
Duck Harbor—B
N. Bound Brook Is.—S
Pamet Harbor—S
Pilgrim Beach—B

OUTER CAPE COD

Province Lands—S
Coast Guard/North Beach—S
Nauset Beach—S

Table 1. A tentative list of 64 barrier beaches located along the Arcadian shoreline of Massachusetts. The types include bay barriers (B), barrier spits (S) and barrier islands (I).

CAPE COD—SOUTH SHORE

Monomoy Island—I
Morris Island/Harding Beach—S
Cockle Cove—S
Forest Beach—S
Red River—S
Sea Gull Beach—S
Great Island—S
Lewis Bay—S
Squaw Island (East)—S
Squaw Island (West)—S
Long/Dowes Beach—S
Dead Neck—I
Popponesset Beach—S
South Cape Beach/
Washburn Is. (East)—S
Eel Pond—S
Bournes Pond—B
Green Pond—B
Great Pond—B
Little Pond—B
Falmouth/Quissett Beach—B
Nobska—B

BUZZARDS BAY (EAST)

Flume Pond—B
Gunning Point—B
Little Sippewisset Beach—S
Great Sippewisset Beach—S
Chappaquoit Beach—S
Squeteague Harbor—S
Bassets Island—I
Toby's Island—I
The Neck—S
Copicut Neck (East)—S
Copicut Neck (West)—S
West End Pond—S

BUZZARDS BAY (WEST)

Little Harbor Beach—S
Long Beach Point—S
Sedge Cave—B
Planting Is.—S
Aucoot Cove—S
Pine Island Pond (East)—S
Pine Island Pond (West)—S
Crescent Beach—B
Eel Pond Beach—S
Brant Island Cove (East)—S
Brant Island Cove (West)—B
Round Island Beach—S
North Point Beach—I
West Island Beach—S

BUZZARDS BAY (WEST), Cont.

Sconticut Neck Beach—S
Pope Beach—S
Apponagansett—S
Barekneed Rocks—S
Round Hill—S
Salters Point—B
Cow Yard—S
Deepwater Point
(Slocums River)—S
Slocums Neck/Little Neck Beach—S
Horseneck Beach—I
Westport Light—S
Cockeast Pond—B
Richmond Pond—B

MARTHA'S VINEYARD

Quicks Hole Pond—S
Cuttyhunk Pond—S
French Watering Place—B
Tarpaulin Cove—B

Cedar Tree Neck—S
Lamberts Cove—B
Lake Tashmoo—S
Mink Meadows—S
East Chop Beach—S
Farm Pond—S
Sengekontacket Pond—S
Little Beach—I
Cape Poge—I
South Beach—B
Stonewall Beach—B
Mashaquitsa—S
Long/Squibnocket Beach—S
Lobsterville/Menemsha Pond—

NANTUCKET

Muskeget Island—I
North Pond—S
East Pond—S
Esther Is.—I
Eel Point—S
Gapaum Pond—B
Reed Pond—B
Coatue—S
Great Point—S
Head of the Harbor—B
Sesuchacha Pond—B
Cisco Beach—B
Madaket—S

Table 2. A tentative list of 93 barrier beaches located along the Virginian shoreline of Massachusetts. The types include bay barriers (B), barrier spits (S) and barrier islands (I).

Bay, Martha's Vineyard and Nantucket representing the Virginian shoreline (Figure 1). Massachusetts may very will have the largest number of barriers for any state along the East and Gulf Coasts. This speculation is supported by evidence provided in a preliminary inventory completed by the federal government which shows that 30% of the undeveloped and unprotected barriers are found in Massachusetts (USNPS, 1980). In addition, these types of barriers represent 20% of the total shoreline mileage along the East Coast, but only 6% of the total acreage. These statistics are indicative of the average barrier configuration in the state which, in general, are long and narrow. Typical barrier beaches along the Arcadian and Virginian shorelines are shown in Figures 2a and 2b.

Active processes of tides, waves and storms can be summarized to reflect a notable contrast between the two regional shoreline types. It has been shown (Hayes, 1979) that wave climate and tidal changes are dominant influences on barrier beach morphology. The Arcadian shoreline is a mesotidal coast, as classified by Davies (1964), dominated by tidal ranges of 6-10 feet (1.8-3.0 meters) along the open coast from the Merrimack River Entrance in the North Shore south to Chatham (outer coast) on Cape Cod. The wave climate is characterized by deep water, long period, swell generated waves. In contrast, the Virginian shoreline is a microtidal coast dominated by tidal ranges of 1.5-4.5 feet (.5-1.4 meters) from Monomoy Point west to the Westport River near Rhode Island. The wave climate varies considerably due to numerous shoreline orientations along Buzzards Bay and distinctly separate basins of Buzzards Bay, Vineyard Sound and Nantucket Sound. However, shallow water, short period, locally generated waves characterize the general wave climate over a majority of the shoreline south of Cape Cod. The two storm types effecting the coast of Massachusetts are hurricanes and "northeasters" and have been well documented (U.S. Army Corps, 1978).

Environmentally, barrier beaches protect against coastal storm hazards while providing for numerous public values. Wave and flood damage from hurricanes and northeasters has been reported 5 to 25 feet (1.5-7.6 meters) above still water levels and tide range levels have been recorded to rise 6 to 10 feet (1.8-3.0 meters) (U.S. Army Corps, 1978). As a first line of defense against wave and flood hazards, beaches and frontal dunes on the barrier provide protection for landward lying areas. The unconsolidated sediment composing these resources allows them to respond to excess hydraulic forces by shifting and conforming in a manner which maintains the integrity of the barrier. As a result, salt marshes, salt ponds and tidal flats behind them are able to continue providing food and habitat for finfish and shellfish. In terms of their direct value to man, barriers protect upland development and other wetland areas from storm damage and flooding. It is a public misconception that coastal engineering structures will protect barrier beaches and their associated public values. However, as the expense to construct and maintain these structures rises man will realize his structural barriers are not as effective as nature's in providing for storm damage prevention. Left alone, barrier beaches will protect themselves. Laws that restrict and/or regulate appropriate building on barrier beaches are a more effective way of

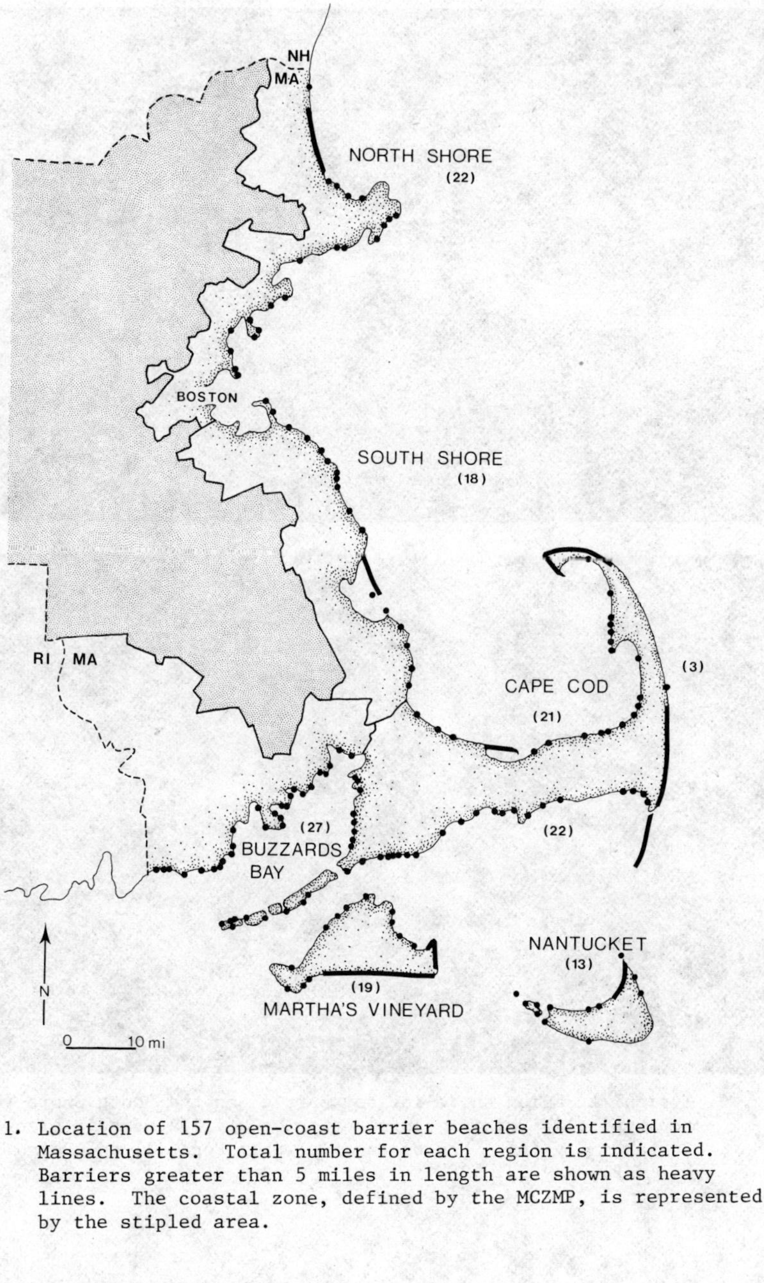

Fig. 1. Location of 157 open-coast barrier beaches identified in Massachusetts. Total number for each region is indicated. Barriers greater than 5 miles in length are shown as heavy lines. The coastal zone, defined by the MCZMP, is represented by the stipled area.

Fig. 2. Typical Arcadian shoreline bay barrier on the South Shore (A) and typical Virginian shoreline inlet-affiliated spit along Buzzards Bay (B). Photos taken of Ship Pond (A) and Round Hill (B), May 23, 1980.

protecting the public values.

LEGISLATION AND ADMINISTRATIVE AUTHORITY

The earliest wetland protection legislation in the United States was established in Massachusetts (Dawson, 1978). As knowledge of the public values related to the values associated with salt marshes and shellfish beds, and an awareness of the indiscriminate alteration of these wetlands developed, the effort to protect wetlands was initiated by the passing of a regulatory law in 1963. The statute was referred to as the Jones Act. Through this law, a permit system became effective in curtailing many of the destructive activities in productive wetlands; however, a form of longer term protection was desired. In 1965, the first form of the Coastal Wetlands Restriction Act, c.131, s.105 (Restriction Act), was passed, allowing the state to exercize its police power through a deed restriction process. The Jones Act was incorporated into the more comprehensive and newly enacted Wetland Protection Act, c.131, s.40 (Protection Act) in 1972.

The concerns of the two wetland protection laws are essentially identical. The Restriction Act states: "The commissioner of environmental management. . .may. . ., for the purpose of promoting the public safety, health and welfare, and protecting public and private property, wildlife and marine fisheries, adopt, amend, modify, or repeal orders regulating, restricting or prohibiting dredging, filling, removing or otherwise altering, or polluting, coastal wetlands." The Protection Act identifies seven "interests" or public values to which the characteristics of inland and coastal wetlands may contribute. These interests are: public or private water supply, groundwater supply, flood control, storm damage prevention, prevention of pollution, land containing shellfish, and fisheries. The existing Restriction and Protection Acts contributed to the early approval of the MCZMP. Each law had an administrative program established through which new policies and regulations could be readily adopted. Among 26 MCZMP policies (Office of Coastal Zone Management, 1978), the "protection of ecologically significant resource areas" is the first to be addressed. The specific wetland types of primary importance are salt marshes, shellfish beds, salt ponds, coastal beaches, coastal dunes and barrier beaches. The language in each Act is general enough to allow for the use of these more specific terms and is clarified in the form of Rules and Regulations. In addition to the adoption of policies, the MCZMP encouraged the wetland programs to add scientists to their staffs. The most desired fields of background included coastal geology/geomorphology, marine botany, marine fishery biology and ecology. As a result, teams of two scientists became an integral part to the wetland programs by accepting responsibilities that range from interpretation of basic and applied research, translation of technical information and wetland evaluation to courtroom testimony, public education presentations and on-site visits with landowners, developers, and government officials. The two wetland programs, formerly under the Department of Natural Resources, now exist under separate administrative departments.

The Restriction Program is currently administered by the Department of Environmental Management (D.E.M.). On a town-by-town basis, the program is mapping all wetlands greater than 1/2 acre in size on a 1:5000 scale orthophoto base map. These maps serve as a work product for ground truth verification of wetlands as well as the final plan which is filed in the Registry of Deeds showing those wetlands that have been restricted. In addition, the maps may be used by state and local officials as an aid used in determining jurisdictional boundaries needed to implement the Protection Act.

Once the wetlands have been mapped and verified, an evaluation and selection process of determining the wetlands to be restricted occurs. Then, owners of wetland property proposed for restriction are identified, given notice of a formal public hearing, and granted the opportunity for an on-site inspection. The on-sites are conducted by the science staff for the purpose of: 1) locating the wetland boundary on their property and 2) answering any questions about the law, program, selection process and practical meaning of the restriction order. The restriction process is completed when the plan and an order listing the permissible and prohibited uses are filed with a County Registry of Deeds.

The prohibited activities listed in the restriction order consist of permanent alterations that would destroy the natural vegetation, substantially alter existing patterns of tidal flow, obstruct the movement of sediment or alter the natural contour of the coastal wetland. These activities include filling, dumping, dredging, excavating, discharge of hazardous substances, and construction of habitable dwellings. The list of permissible activities includes more temporary alterations such as piers, wharfs, docks, floats, duck blinds, beach nourishment and maintenance dredging. Permits for the permissible activities, however, are not issued under the Restriction Act, rather, they are granted through the Protection Act.

The Protection Program is currently administered by the Department of Environmental Quality Engineering (D.E.Q.E.) in cooperation with individual town boards of Conservation Commissions. Jurisdiction is asserted on a permit-by-permit basis over any person who intends to remove, fill, dredge or alter a) within 100 feet of a vegetated fresh water wetland or a coastal wetland, either of which borders on the ocean or any estuary, creek, river, stream, pond or lake, b) any land under "said waters" or c) land subject to tidal action, coastal storm flowage, or flooding. Notice of such intention is filed with the Town Conservation Commission and D.E.Q.E.. The Conservation Commission initially responds to an applicant's proposed activity by conducting a public hearing to solicit comments and evaluate data or information submitted with the applicant's proposal.

The Commission's responsibility is to determine if the wetlands to be affected by the activity are significant to one or more of the seven interests of the Act. If significance is determined, an evaluation of the possible impacts on the wetland characteristics is conducted. The Commission then issues an "Order", or permit, specifying the "conditions" which dictate the manner in which the work shall be performed. The Conditions may, for example, limit the scope of the project, establish siltation controls, or require compensation for flood storage.

If the Commission finds that no Conditions could be imposed to adequately protect the interests of the Act, the proposed activity may be denied.

An appeal process exists for those applicants receiving either an Order of Denial or an Order of Conditions which they believe is excessively restrictive. The first level of appeal resides with the regional office of D.E.Q.E.. Following a site-specific evaluation of project impact on wetland character, this office issues a Superseding Order which may agree or disagree with the Order issued by the Town or may be more or less restrictive. The Superseding Order may be appealed to the Commissioner of D.E.Q.E. and adjudicatory hearings are held to resolve the outstanding differences. The final avenue of appeal has the Commissioners Final Order directed to the Superior Court of Massachusetts.

Whatever appeal level is reached or if no appeal is filed, the accepted Order must be recorded against the property deed of the applicant before any work begins. This insures that until the work is completed as per the Order there remains an outstanding encumbrance on the property. The encumbrance is only cleared after the work satisfies the specified conditions and is certified by the Order issuing authority.

In summary, the two wetland programs interrelate with and complement each other to a large extent. Some of the fundamental aspects with which the programs are consistent include 1) use and application of the same wetland terms, 2) the same mapping product, 3) a team of scientists and 4) the same basic environmental policy described above. However, the complementary aspects need to be explored in further detail. The primary aspect to consider is the differing scope of review. A regional, non-regulatory approach used by D.E.M. and a site-specific regulatory approach in D.E.Q.E. constitute a comprehensive wetland protection program for the state. The regional approach is a limited scope of review requiring the placement of land use restrictions only on certain wetlands or portions therein. Consideration of the total barrier beach system is exercized and conservative restrictions based on consistent policies are applied. In contrast, the site-specific approach is a broad scope of review requiring the regulation of land use within 100 feet from all wetlands. Consideration of all proposed activities within specific resource types is exercized and performance standards are applied.

It is extremely important to understand that a wetland is no less important because it hasn't been restricted. Simply stated, resource value is interpreted by D.E.M. in terms of black and white, whereas, D.E.Q.E. interprets value in terms of the grey tones. In general, the major goal of barrier beach protection in Massachusetts is to prohibit improper activities in those areas which have overwhelming public interest values, yet, allow proper activities to occur elsewhere. A substantive description of each wetland program is presented below and is designed to further clarify this statewide goal, demonstrate how the use of wetland legislation serves as a technique in barrier beach management and provide a means for discussing its success in Massachusetts.

BARRIER BEACH RESTRICTION

Utilization of the Coastal Wetlands Restriction Act as a land-use management technique provides for long term protection of barrier beach resources as long as 1) scientific theory or empirical data is applied; 2) it is procedurally sound and legally defensible; 3) the effected landowners can understand the necessary purposes. The primary interests considered in the Act are marine fisheries and public safety, health and welfare. Other interests such as wildlife and property protection are ancillary objectives. Marine fishery interests are protected when landward lying low-energy environments can allow marine grasses to grow, the base of the marine food chain to form, spawning and nursery grounds to exist, and shellfish habitats to expand. Public safety, health and welfare interests are protected when the high-energy barrier environments are able to respond and conform to storm processes, control flooding and prevent storm damage to upland development. It must remain clear that the use of police power by the state government is in the public interest and not for the purposes of preserving open-space land.

Evaluation of the naturally functioning barrier resources is a fundamental aspect in the restriction process that satisfies legal requirements of identification, verification and documentation (Humphries, 1980). Barrier beach resources are identified on an orthophoto base map (at a scale of 1:5000) which is produced from delineated vertical aerial photos (at a scale of 1:25000) and which meets National Map Accuracy Standards. Therefore, all restriction boundaries must be repeatable on the ground either by the identification of a vegetative boundary or a surveyable line connecting points of a predetermined position, as later discussed. The resource characterization is instrumental for defining the wetland type. Verification and documentation forms have been devised for each of the 6 primary wetland types with use of <u>A Guide to the Coastal Wetlands Regulations</u>(1978). Each form is used as a check list to 1) record the type of geomorphic features, tidal influence, shoreline orientation, and/or dominant vegetation and wildlife species observed, 2) note the manner in which the wetland boundary will be defined, and 3) record the primary wetland function as it relates to one or more of the statutory interests. A quantitative method of documenting geomorphology and vegetation is also used by taking beach profiles (after Emery, 1961). Beach width and slope, dune height and width and barrier width can be computed from the profile data. Also, data recovered from the same profile site at a later date will enable a comparative analysis used for determining long-term shoreline changes.

Data collected during the ground truth verification and any other existing information related to storm history and vulnerability, shoreline changes and overwash dominance is utilized in the process of selecting wetlands for restriction. The manner in which this information is obtained and used parallels a methodology used for identifying coastal high hazard zones in the southeastern and Gulf states (U.S. Army Corps, 1975). Sources include detailed engineering reports, newspapers, photographs, knowledge from local residents, and flood-plain maps.

Decisions concerning the wetlands to be restricted and the extent of the area covered concur with a number of policies developed from the input of scientists, planners and lawyers.

A strong emphasis is placed on the evaluation and restriction of the primary foredune area of a barrier beach because of its ability to dissipate storm waves, supply sand to the beach and control flooding up to its crest elevation. This area has direct consequences to man (Riggs, 1976). The contribution of naturally functioning dunes to the statutory interest of protecting the public safety, health and welfare is becoming as demonstrable as the contribution salt marshes have to marine fisheries. Over the years, the value of salt marshes has been well documented in the literature, has become readily acceptable by the public and was responsible for inspiring many early wetland protection laws. Current research demonstrates that most all salt marshes should be considered ecologically significant and that use of quantitative rating systems to recognize different levels of significance is inappropriate for management purposes (Oviatt _et al_., 1978). With the concept of a dune management district (Nordstrom, 1978), an analogous presumption of significance was demonstrated. As relationships between beaches and dunes (Leatherman, 1979), vegetation and overwash sediments (Godfrey, 1976), inlets and beach morphology (Hayes _et al_, 1978) and dune scarps and crests (Rosen, 1979) are recognized, the justification for restricting barrier beaches and their components has been strengthened. In short, the land-use policies of the Restriction Program, as described below, rely heavily on coastal zone research for support and reinforcement. However, basic research is not required for the purpose of developing or implementing a policy as it has been recently proposed in New Jersey (Gares _et al_., 1979).

Formulation of practical and reasonable policies for the current Restriction Program was initiated with an evaluation of the guidelines set forth in Policy #1 of the MCZMP (Office of Coastal Zone Management, 1978). Basically, the Plan recommends that 1) unaltered barriers be restricted in total and 2) altered barriers be restricted in such a manner that beaches and dunes are included at a minimum.

D.E.M. has adopted more specific policies regarding the restriction of unaltered barrier beaches (Fig. 3a) and altered barrier beaches (Fig. 3b) ranging from the inclusion of total unaltered barriers to the exclusion of total altered barriers. The policies are as follows:

<u>Restriction of Unaltered Barriers</u>

--Evaluate and consider the total barrier to be included.

Explanation: This is consistent with Policy #1 of the coastal plan for Massachusetts and is reasonable in cases where low, narrow barriers are dominated by overwash, tidal inlet shifting, and dune migration.

--Exercize professional judgement in determining the extent of the partial area to be included.

Explanation: In cases where high, wide barriers are relatively stable and not dominated by overwash or migrating inlets and dunes, the area seaward of the primary foredune will be considered as the partial area.

Fig. 3. Unaltered barrier spit (A) and altered barrier spit (B). Photos taken of Springhill Beach (A) and East Sandwich Beach (B) on November 3, 1978.

--At a minimum, the area seaward of the approximate primary foredune crest to mean low water will be included.

Explanation: Maximum protection against flooding is represented by the highest point of the dune or its crest. This area is dominated by on and offshore movement of aeolian sands, is vulnerable to direct wave attack resulting in dune scarp erosion, undercutting and slumping, and is the primary storage area for sand utilized for beach dynamics.

Restriction of Altered Barriers

--Respect existing land-use by excluding the altered areas.

Explanation: In order to avoid unduly encumbering private property, areas dominated by residential, industrial and commercial development are excluded as a contiguous unit which may or may not include undeveloped parcels.

--Exercise professional judgement in determining the extent of the partial area to be included.

Explanation: This is consistent with Policy #1 of the coastal plan of Massachusetts by including those areas exhibiting natural functioning characteristics. The area seaward of the altered area will be considered as the partial area.

--The total barrier will be excluded if the extent of alteration is the predominating characteristic.

Explanation: In cases where low, narrow barriers are dominated by the presence of groins, jetties, seawalls, roads, parking lots and underground utilities the barrier will not be considered. All activities will be regulated through the Wetland Protection Act.

The restriction maps are used to identify the general location of the wetland on the ground. A specific boundary which best represents the transition from wetland to upland vegetation can then be determined. The seaward edge of existing man-made structures may replace the vegetative boundary. Since barrier beaches may be partially restricted and the boundary would be located within the wetland, a surveyable line establishes a non-vegetated boundary.

BARRIER BEACH REGULATION

Massachusetts promulgated Coastal Wetland Regulations in August, 1978 for the purpose of amplifying and clarifying general regulations existing under its Protection Act. Their intent is to "ensure that development along the coastline is located, designed, built and maintained in a manner that protects the public interests in the coastal resources listed in the Act."

The regulations apply to eleven coastal resource areas. They are: Land under the Ocean, Coastal Beaches, Coastal Dunes, Barrier Beaches, Salt Marshes, Coastal Banks, Rocky Intertidal Shores, Land under Salt Ponds and Anadromous/Catadromous fish runs. The physical and/or biological characteristics of each resource area are explained in a Preamble to the appropriate resource area Section. Performance standards establish the level of protection to be imposed on each resource area.

Each of the eleven resource areas is presumed to be significant to one or more of the interests. The presumption may only be overcome upon a clear showing that the resource area does not play the presumed role in the protection of the interest(s). The burden of overcoming this presumption is placed on the proponent of the project.

The following discussion deals with the regulation of proposed activities on barrier beaches. As defined by the Coastal Regulation, "Barrier Beach means a narrow low-lying strip of land generally consisting of coastal beaches and coastal dunes extending roughly parallel to the trend of the coast. It is separated from the mainland by a narrow body of fresh, brackish or saline water or a marsh system. A barrier beach may be joined to the mainland at one or both ends."

This definition reflects the fact that a barrier beach is actually a complex resource area composed of coastal beaches and dunes. For this reason it is necessary to briefly review the regulations on coastal beaches and dunes.

Coastal beaches, which include tidal flats, are presumed to be significant to the interests of storm damage prevention, flood control, and marine fisheries. The characteristics which are critical to the protection of those interests are, in part:

a. volume (quantity of sediments) and form;
b. the ability to respond to wave action;
c. distribution of sediment grain size;
d. water circulation;
e. water quality, and;
f. relief and elevation.

Similarly, coastal dunes are presumed to be significant to the interests of storm damage and flood control. Those characteristics critical to the protection of these two interests are:

a. the ability of the dune to erode in response to coastal beach conditions;
b. dune volume;
c. dune form which must be allowed to be changed by wind and natural water;
d. vegetative cover, and;
e. the ability of the dune to move landward or laterally.

The performance standards of the regulations prohibit a project from having an adverse effect on any of the characteristics listed for either coastal beaches or coastal dunes.

The resource section in the Regulations which pertains to barrier beaches combines, at a minimum, those presumptions for coastal beaches and coastal dunes. Barrier beaches, including all of its coastal dunes, are presumed to be significant to the interests of storm damage prevention, flood control and marine fisheries. The performance standards clearly state that the characteristics of coastal beaches, tidal flats and coastal dunes, previously listed herein, and their ability to respond to wave action, including storm overwash and sediment transport, are critical to protect said interests.

As an aid to help both prospective project proponents properly design their activity within coastal resource areas and Conservation Commissions to properly and consistently review such activities, a guidebook (D.E.Q.E., 1978) was developed. The Guide was designed as a reference document and working tool, and outlines a six step review process for each proposed project.

Step 1. Identify The Resource Area(s): This paper shall limit the resource to a barrier beach.

Step 2. Identify Each Proposed Activity: A proposed project may involve more than one activity; for example, construction of a house on a barrier beach may propose removing, filling, grading, bulkheads, boardwalks and groins.

Step 3. Identify Project Constraints: The guide contains a table which cross-references common activities against barrier beaches and indicates whether or not an activity is likely to be permitted.

Step 4. Identify Adverse Impacts: Each activity should be evaluated for the expected impact to the barrier beach.

Step 5. Identify Measures to Meet Performance Standards: The measures necessary to meet the performance standards for a barrier beach, as established in the Regulations, should be identified.

Step 6. Design the Project/Prepare an Order of Conditions: The Conservation Commission of the D.E.Q.E. must determine if the project is designed to meet the performance standards for a barrier beach. If the proper measures, as identified in Step 5, can be complied with,"conditions" are imposed through the Protection Act permit. If no such conditions can be imposed to meet the standards, the project may be denied.

In summary, the level of authority reviewing a proposed project on a barrier beach must evaluate proposed activity vs. barrier beach impact. The characteristics of a barrier beach, including the individual resource areas of coastal beaches and dunes, are very explicitly protected by the Protection Act. No adverse effect may be incurred to any of those characteristics listed in the Regulations, and a project may be denied if such effect is to be expected.

It should be noted that the Protection Act is meant to insure that a project is undertaken in a manner that will protect the seven public interests listed in the Act. However, if the proponent of a project has little concern about propoerly designing his project or if a project cannot be designed properly, the ultimate decision may be a denial.

CONCLUSION

Through the implementation of the Coastal Wetlands Restriction Act and the Wetlands Protection Act, Massachusetts is taking an aggressive land-use management role in protecting its barrier beaches. The identification of 157 open-coast barriers has been completed in an effort to understand the regional differentiation, variability and complexity of these resources. Interpretation and application of scientific information used in decision making is a current responsibility of qualified professionals in deed restriction and regulation programs.

Policies used to determine areas that should be limited to temporary activities have been formulated with a generic knowledge of resource functions by the Restriction Program. Determination of the potential impact of an activity on resource components and issuance of permits for all activities by the Protection Program also reflect a broad understanding of barrier dynamics. The two programs complement one another to comprise a comprehensive wetland protection element of the Massachusetts Coastal Zone Management Program.

Special addendum: On Friday, August 8, 1980, Governor Edward J. King issued Executive Order No. 181--Barrier Beaches. In effect, a state policy has been established which discourages further government expenditures directed towards encouraging continued and additional development of barrier beaches.

ACKNOWLEGMENTS

The authors would like to thank the staffs of the Restriction and Protection Programs for their cooperation and encouragement. In particular, gratitude is extended to Jeffrey Mills for manuscript review and clerical preparation, Delvine Okereke for administrative assistance, and Sheldon Shapiro for director support. Financial support was contributed from the Office of Coastal Zone Management through contract with the Massachusetts Department of Environmental Management.

REFERENCES

1) Brown, P.J. (1976). Variations in South Carolina coastal morphology. In "Terrigenous Clastic Depositional Environments" (M.O. Hayes and T.W. Kana, eds.), Tech. Report No. 11-CRD, Dept. Geol., U.S.C., p. II-2-II-15.
2) Clark, J.R., ed. (1976). Barrier islands and beaches. Tech. Pro. of the 1976 Barrier Island Workshop. The Conservation Foundation, Wash., D.C., 149 p.
3) Clark, J.R. (1977). Coastal ecosystem management, a technical manual for the conservation of coastal zone resources. The Conservation Foundation. John Wiley & Sons, N.Y., 928 p.
4) Davies, J.L. (1964). A morphogenic approach to world shorelines. Zeit fur Geomorph. Bd. 8, s. 27-42.
5) Dawson, A.D. (1978). Protecting Massachusetts wetlands. In Suffolk University Law Review. Vol. 12, no. 4, p. 755-795.
6) Departemnt of Environmental Quality Engineering. (1978). A guide to the coastal wetlands regulations of the Massachusetts Wetlands Protection Act (G.L. 131, S. 40), Comm. of Mass., E.O.E.A.
7) Emery, K.O. (1961). A simple method of measuring beach profiles. In Limnology and Oceanography. Vol. 6, p. 90-93.
8) Fisher et al. (1972). Environmental geologic atlas of the Texas coastal zone: Galveston-Houston area. Bur. of Econ. Geol., Univ. of Texas, 91 p.
9) Gares, P.A., Nordstrom, K.F. and Psuty, N.P. (1979). Coastal dunes: their function, delineation, and management. Center

for Coastal and Environmental Studies, Ritgers, New Brunswick, N.J., 112 p.
10) Godfrey, P.J. (1976). Comparative ecology of East Coast barrier islands. In "Barrier Islands and Beaches." Tech. Pro. of the 1976 Barrier Island Workshop. (J.R. Clark, ed.) The Conservation Foundation. Wash., D.C., p. 5-34.
11) Hayes, M.O. and Kana, T.W., eds. (1977). Terrigenous clastic depositional environments. Some modern examples. Tech. Report No. 11-CRD, Coastal Research Division, Dept. Geol., Univ. of S.C., 315 p.
12) Hayes, M.O., FitzGerald, D.M. and Hubbard, D.K. (1978). Tidal inlet control of beach erosion-deposition cycles. In Pro. of Amer. Shore and Beach Pres. Assoc. Annual Meeting, Tallahassee, Fl., Oct., 1978, p. 121-140.
13) Hayes, M.O. (1979). Barrier island morphology as a function of tidal and wave regime. In Barrier Islands from the Gulf of St. Lawrence to the Gulf of Mexico. (S.P. Leatherman) Academic Press, Inc., p. 1-27.
14) Hubbard, D.K., Hayes, M.O., and Brown, P.J. (1977). Beach erosion trends along South Carolina coast. In "Coastal Sediments '77", WPCO Div. of ASCE, Charleston, S.C., p. 797-814.
15) Humphries, S.M. (1980). Characterization of beach and dune resources used in the implementation of the Massachusetts Coastal Wetlands Restriction Act. In Tech. Pro. to the Fifth Annual Conf. of the Coastal Society.
16) Jarrett, J.T. (1977). Sediment budget analysis, Wrightsville Beach to Kure Beach, North Carolina. In "Coastal Sediments '77", WPCO Div. of ASCE, Charleston, S.C., p. 986-1005.
17) Kraft, J.C., Allen, E.A., Belknap, D.F., John, C.J. and Maurmeyer, E.M. (1976). Delaware's changing shoreline. Tech. Report No. 1. DCZMP. Delaware State Planning Office, Exec. Dept., Dover, 319 p.
18) Leatherman, S.P. (1976). Barrier dune systems: a re-assessment. Sedimentary Geology. 24: 1-16.
19) Leatherman, S.P., ed. (1979). Barrier islands from the Gulf of St. Lawrence to the Gulf of Mexico. Academic Press, Inc., 325 p.
20) Leatherman, S.P. (in press). Barrier island handbook. Nat. Park Ser., Coop. Res. Unit, Envir. Inst., U. Mass., 101 p.
21) Nordstrom, K.F., psuty, N.P. and Fisher, S.F. (1978). Empirical models of dune formation as the basis for dune district zoning. In "Coastal Zone '78". ASCE, N.Y., p. 1489-1507.
22) Oertel, G.F. (1979). Barrier island development during the holocene recession, Southeastern United States. In Barrier Islands from the Gulf of St. Lawrence to the Gulf of Mexico. (S.P. Leatherman, ed.) Academic Press, Inc., p. 273-290.
23) Office of Coastal Zone Management. (1978). Massachusetts coastal zone management program and final environmental

impact statement. U.S. Department of Commerce, Wash., D.C., 439 p.

24) Olsen, S.B. and Grant, M.J. (1973). Rhode Island's barrier beaches. Marine Tech. report No. 4, The Coastal Resources Center, Univ. of R.I., Vol. 1 and 2.

25) Oviatt, C.A., Nixon, S.W. and Garger, J. (1977). Variation and evaluation of coastal salt marshes. In Environmental Management, Vol. 1, No. 3, p. 201-211.

26) Riggs, S.R. (1976). Barrier islands as natural storm dependent systems. In "Barrier Islands and Beaches". Tech. Pro. of the 1976 Barrier Islands Workshop. (J.R. CLark, ed.) The Conservation Foundation. Wash., D.C., p. 58-75.

27) Rosen, P.S. (1979). Aeolian dynamics of a barrier island system. In Barrier islands from the Gulf of St. Lawrence to the Gulf of Mexico (S.P. Leatherman, ed.), Academic Press, Inc., p. 81-98.

28) U.S. Army Corps of Engineers. (1975). Guidelines for identifying coastal high hazard zones. Galveston Dist., C.O.E., p. 23-26.

29) U.S. Army Corps of Engineers. (1978). Massachusetts coastal study. N.E. Div., C.O.E. 105 p.

30) U.S. Army Corps of Engineers. (1979). Cape Cod easterly shore beach erosion study. N.E. Div., C.O.E. Vol. II, 1-A-1--14.

31) U.S. National Park Service. (1980). Undeveloped and unprotected barrier islands of the United States: a working copy. U.S. Dept. of Interior, Wash., D.C.

WETLANDS REVIEWS - A CALIFORNIA EXPERIENCE

Hugh Converse and Jill Shapiro [1]

INTRODUCTION

Humboldt Bay is a large estuary on the Northern California Coast about 230 miles north of San Francisco. The bay contains a harbor serving an adjacent urban and industrial area and also provides important habitat for fish and wildlife. As is presently the case throughout coastal California, a comprehensive coastal management program for Humboldt Bay is being developed at the local and State level. Federal involvement in this program is mostly indirect. The Corps of Engineers, however, has regulatory and specific development authorities which overlap the State/local program. Under certain of these authorities, the Corps has prepared the Humboldt Bay Wetlands Review and Baylands Analysis (HBWR) for Humboldt Bay, which assembled and analyzed data for use in permit and planning decisions by the Corps and other agencies. Planning for the HBWR and its execution were carried out in close coordination with the groups involved in the Local Coastal Program of the State's coastal management effort to insure that the data gathered would be of both specific and general value.

The review consists of detailed resource profiles which identify, describe, and map natural and cultural values, generally using existing data. A key new study was an identification of biologic habitat by vegetation associations. In addition, governmental policies and existing and historic land and water uses and trends were reviewed. With this data base, sensitive areas in and around the bay were identified, evaluated, and designated either as Areas of Importance - critical areas generally to be maintained in their present state; Areas of Environmental Concern - where careful controls are required, or as General Use Areas - where only normal controls are needed. A discussion of the types of activities that could be proposed for the areas and their impacts was also prepared.

---

1 Hugh Converse is a Civil Engineer, Coastal Engineering Branch, South Pacific Division Corps of Engineers, San Francisco. Jill Shapiro, Ph. D., is President, Shapiro & Associates, Seattle, Washington.

Acknowledgement is gratefully made to the Corps of Engineers, U. S. Army, for access and permission to use this study material. The views of the authors do not purport to reflect the position of the Corps of Engineers or the Department of Defense. The assistance of personnel of the San Francisco District of the Corps of Engineers is deeply appreciated.

The structure of the findings is intended to assist in planning and permit decisions by the Corps and other agencies, notably those carrying out the Local Coastal Program, so that they can more readily determine the suitability, or non-suitability, of a proposed use or activity for an area, and prepare criteria for such use, as appropriate. The HBWR habitat mapping was also specifically designed to fulfill the habitat mapping needs (wetlands and related adjacent areas) of the Local Coastal Program.

The HBWR is a part of a larger Congressionally-authorized investigation by the Corps of Engineers to develop economic, environmental, and social information about the Humboldt Bay area to improve planning at all governmental levels. The review also identified information gaps that are likely candidates for additional studies within the framework of the larger investigation. In future years, it is hoped that such related studies can be carried out to expand the data base developed by the HBWR, and thereby assist further planning in the Humboldt Bay area.

HUMBOLDT BAY AREA

Humboldt Bay is a shallow tidal estuary located about 230 miles north of San Francisco in a relatively remote portion of coastal California (Fig 1.). A jettied entrance and interior channels are maintained by the Federal Government for deep-draft vessels and commercial fishing boats. The bay, which is separated from the ocean by two long, narrow sandspits, is elongated, and dumbell-shaped, widening at its southern and northern extremities. It has an area of about 25 square miles at high tide.

The area, with its moist and moderate oceanic climate, is well suited for continued timber production, which has always been the economic mainstay of the area; pastoral agriculture, fisheries, recreation, tourism, and associated service industries, comprise the balance of the economy.

The immediate bay margins have been cleared for urban and agricultural uses, and a high proportion of the original wetlands has been converted to pastureland by levee construction and draining. Two cities, Eureka and Arcata, border the bay. Eureka, with a population of about 26,000 is an industrial and commercial center, serving a large hinterland. Several miles of its urban waterfront abut the bay. Arcata is the home of Humboldt State University and has a population of about 12,000. The city is a residential and commercial center; its direct use of the bay is comparatively limited.

The remoteness of the area from major growth centers, and its relatively small population, have fostered continued dependence on natural-resources-based industries and precluded the explosive economic growth and diversification characteristic of so much of California. In fact, the timber industry is now in a long-term decline and the county is considered to be a depressed area. Efforts are being made to attract new industry and to expand use of existing

LOCATION MAP

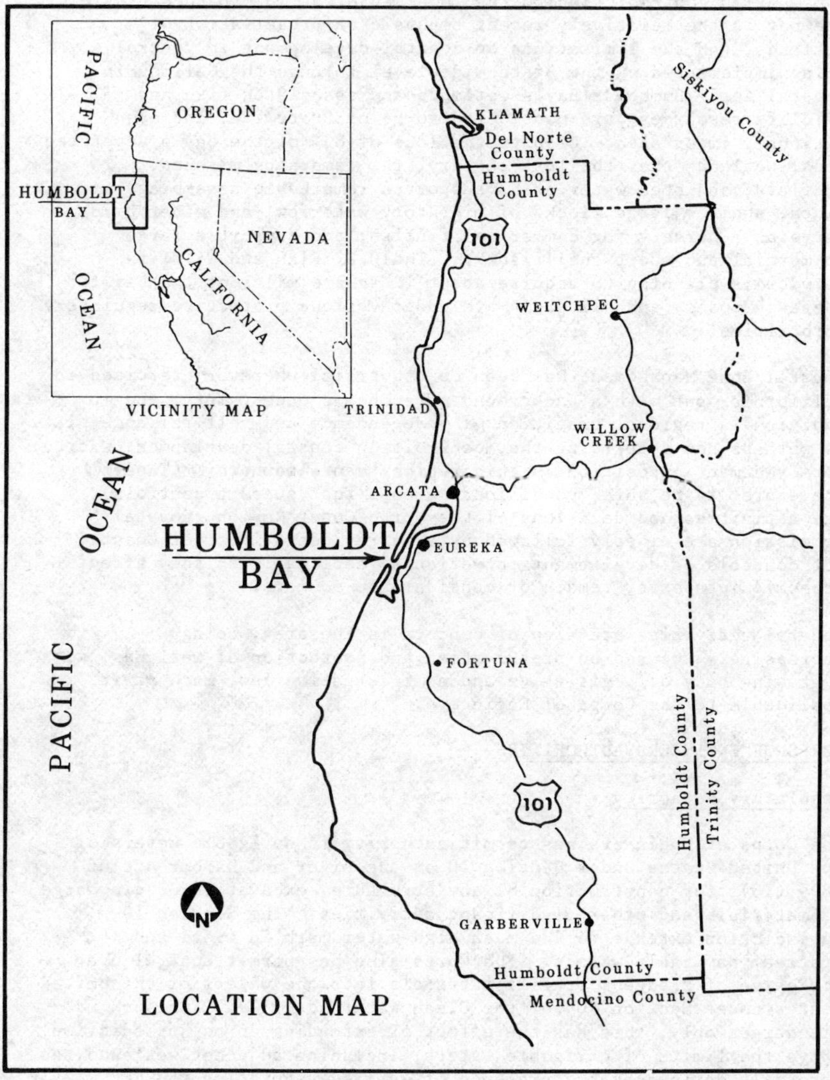

FIGURE 1

resources, notably fisheries, which may have considerable potential. Expanded use of the bay is believed to be important to these efforts.

Somewhat in contradiction to the local interest in expanded uses of the bay is the relatively recent emphasis on preservation of wetlands, and the limitations on coastal development in general, being implemented on the State-wide level through the California Coastal Act. Humboldt Bay's wetlands and associated fish and wildlife resources, are considered to be of Statewide, and even national, importance. Despite the loss of 85% of the bay's vegetated tidal wetlands over the last century, the remaining marshes, mudflats, and open water areas support a remarkable assemblage of biota; shelter large flocks of migratory waterfowl and other birds; serve as a nursery for commercial finfish; and support a large commercial and sport shellfishery. The U.S. Fish and Wildlife Service is planning to acquire about 12 square miles of the bay for these purposes, and all areas are under various protective regulatory authorities.

Coastal Zone Management has been controversial wherever exercised in California, but with a background of economic uncertainty, and in addition, a regional tradition of independence and self reliance, it is perhaps not surprising that controls on coastal development excite more vehement opposition in this region than elsewhere. (Though there are, to be sure, vocal local groups that support controls). The activities and decisions of the North Coast Region, Coastal Commission are closely followed by area residents. Strict Coastal Act controls on development in wetlands, especially as they affect economic interests, remain of considerable concern.

Federal activities are also of concern in the area, being increasingly focused on preservation and protection of wetlands, with a growing body of legislative and administrative law, much of it applicable to the Corps of Engineers.

CORPS OF ENGINEERS AUTHORITIES

Regulatory Authority

The Corps of Engineers has permit authority in navigable waters of the United States under Section 10 of the River and Harbor Act of 1899 (16), for construction of any structure, excavation or discharge of material, and other significant activities. The Section 10 jurisdiction extends to the mean high water mark in tidal and upstream navigable waters. The Corps also has permit control over discharge of dredged, or fill materials into the waters of the United States under Section 404 of the Clean Water Act (17). For such discharges only, this has the effect of extending Corps jurisdiction above the limits of navigable waters, including adjacent wetlands, as generally defined by the presence of wetland vegetation (33 CFR 323.2(c) (12).

Over the years, several hundred requests (350 between 1913 and 1979) for Corps permits have been made in Humboldt Bay, generally relating to fixed and floating structures, dredging, filling, pipelines, and cables. At present, a permit can only be granted if issuance is in the public interest. In evaluation of a permit request, relevant factors must be evaluated, including effects on navigation, fish and wildlife, ecology, pollution, aesthetics, and wetlands. (Corps permits are separate from permit requirement of other agencies and are normally handled on a case-by-case basis. However, permit decisions generally must be consistent with the Coastal Zone Management plan of the State of California, under the provisions of the Coastal Zone Management Act of 1976, as amended). Also, because of the especially important physical and biological functions of the wetlands, the Corps is authorized to undertake wetlands reviews, such as the HBWR, to identify resource values and to assess cumulative impacts of proposed activities (33 CFR 320.4).

PROJECT AND INVESTIGATION AUTHORITY

Besides regulatory authorities, the Corps has, as authorized by Congress, long-established responsibility for numerous commercial ports, small-craft harbors, waterways and beach development projects, and for studies of new project proposals. Within Humboldt Bay, the Corps has constructed the entrance jetties and several deep-draft channels. Work began in the 1880's with regular maintenance and incremental improvement as needed. The latest deepening was completed in 1979, and added deepening of a South Bay channel is now under study.

As noted, there is also a Congressionally-authorized investigation underway to develop planning information about the Humboldt Bay Area. The Congressional resolution (House Committee on Public Works, 11 April 1974) requested the Corps to develop inventories of land use, survey government jurisdictions, evaluate lands for suitability of use, and obtain other data as needed for coordinated planning at all governmental levels. This resolution, along with the above-mentioned regulatory authority for wetlands review, provides dual authority for the HBWR.

PLANNING FOR THE HBWR

The California Coastal Act, as discussed subsequently in this paper, provides for the development of coastal zone plans called Local Coastal Programs (LCP's) by local entities, Initially, a detailed scope of work of the HBWR was circulated in draft to interested agencies, including those responsible for the LCP's. As planned, the wetlands review was to follow the model of Clark (4), and the experience of the Portland and Seattle Districts of the Corps in its application (6, 9, 10, 11). The goal was to identify and evaluate coastal areas, including wetlands, with important environmental resources and functions; and to recommend standards for preservation and use of these areas, as appropriate (as is discussed below, this latter goal was not fully achieved). As it developed, the specific portions of most immediate interest to local and State coastal planners are the habitat identification and mapping, as included in a resources profile.

Agencies who reviewed the draft scope, both formally and informally, included the Humboldt Bay Harbor, Recreation and Conservation District; Humboldt County; the California Coastal Commission, North Coast Region (and its broadly-based Humboldt Bay Advisory Committee); the city of Arcata; and the U.S. Fish and Wildlife Service. Coordination was also maintained with these agencies and others during the course of the review, and interim drafts were circulated to the agencies and the public for comments.

The tangible product of the HBWR is a 3-volume, 800-page study including: Findings regarding resource values; background resource profiles and various technical and legal appendices, (notably discussing wetlands definition and mapping); a summary of public participation; and a bibliography. The resource profiles describe physical and biological characteristics, land use and governmental policy, cultural resources, and economics. The twenty-six specific profiles were: geography, climatology, geology, geologic hazards, mineral resources, soils, tidal characteristics, hydrology, currents and circulation, bottom sediments, water quality, habitat classification and mapping, mammals, birds, fish, invertebrates, reptiles and amphibians, productivity, food webs, nutrient cycling, land and water use, land and tideland ownership, governments, culture, aesthetics, and economics. The material includes 1:24,000 scale maps of bay resources, and numerous figures and tables. As discussed below, the key profile was that of habitat classification and mapping which, along with portions of the aesthetic profile, consisted of original work.

STUDY FINDINGS

Habitat Classification and Mapping; Wetlands Types

As indicated, identification of wetlands and closely related uplands was the first major finding of the HBWR. Approximately 48,000 acres of aquatic, wetland, and upland habitat types were identified, classified, and mapped, primarily by vegetation associations, but also by land use. The classification system is modified from the USGS Land Use and Land Cover Classification System; that portion describing the wetlands vegetation associations can be equated to the National Wetlands Inventory classification (5,7) of the U.S. Fish and Wildlife Service and various Corps of Engineers classifications (as shown in a chart in the report). Habitat identification required color-infrared aerial photography at a scale of 1:6000, extensive field examination, and use of existing county 1:6000 topographic base maps (1:24,000 scale mapping was utilized within open bay waters). The product consisted of twenty-four 1:6000 habitat maps covering the complete study area. A wetland-upland boundary was also defined for consideration by the Corps as a Section 404 jurisdiction limit. Nine major habitat types were identified: urban, agricultural, grassland, shrubland, forest, water, wetland, dunes, and jetties, with further divisions into subtypes. (Fig. 2)

Habitat mapping also included locating levees. This requirement provided an opportunity to locate potential compensation sites: that

NORTH SECTION OF HUMBOLDT BAY

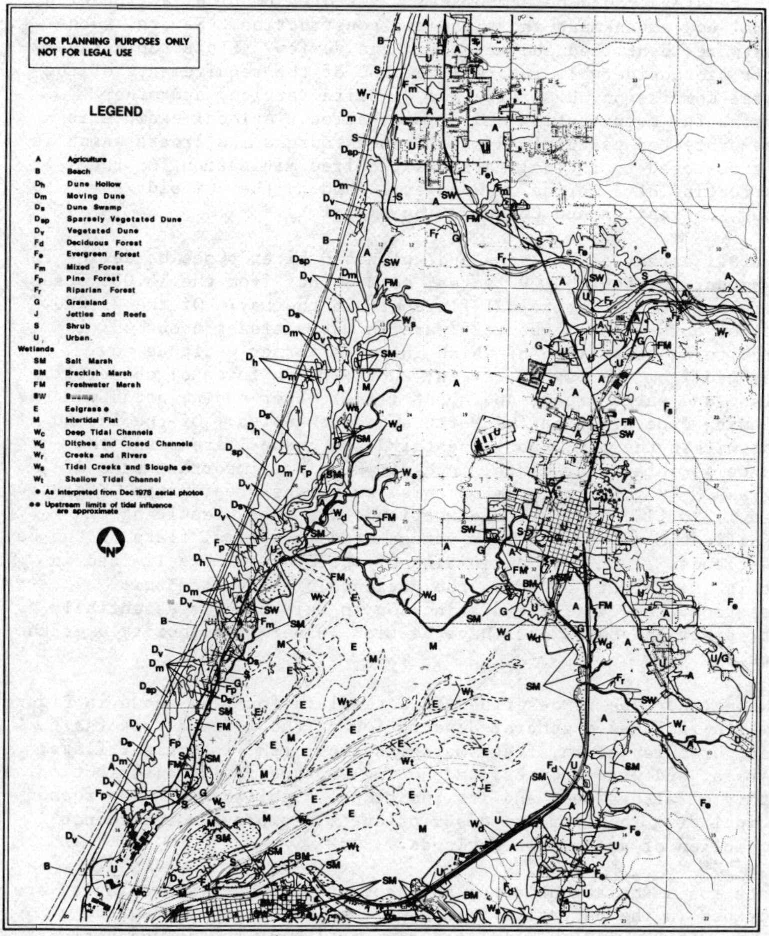

PLATE NO 10 NORTH
1978

FIGURE 2

is, areas where historic wetlands sites might be economically restored, (especially as compensation and mitigation sites for future development which may impact on existing bay wetlands). The mapping of levees included both those presently protecting pastureland from inundation, and older levees or railroad embankments inside the presently-used levees. Older levees present a situation where discrete areas could be economically restored to tidal action with minimal new embankment and tide gate construction. Several such areas were identified and mapped in the review. (This topic is a matter of considerable interest because of the requirements of the Coastal Commission, U.S. Fish and Wildlife Service, and other agencies for compensation and/or mitigation. A local example is a levee-protected pastureland site between Eureka and Arcata which is being restored to a tidal marsh as required mitigation for the construction of a marina on Woodley Island by the Humboldt Bay Harbor, Recreation and Conservation District).

The areal extent of wetlands and uplands habitat types bordering the bay amount to about 33,000 acres, as distinct from the 15,000 acres of open water and intertidal flats within the bay. Of the 33,000 acres of borderlands, about 25% are in urban areas; about 43% in agricultural uses, much of which includes former wetlands now separated from the bay by levees; about 10% in forests; about 11% wetland and water areas; and about 11% of other types, notably including dune areas on the North and South Spits. Of the 15,000 acres within the bay, about one-third are channel areas and the balance is intertidal flats, much of which has important beds of eelgrass. (For the purpose of this study, the Presidential Executive Order 11990 (14) definition of wetlands was used, resulting in classification of the largely unvegetated intertidal flats of the bay as wetlands. The Corps definition of wetlands is more limited in that it is strictly based on the presence of saturated-soil vegetation. However, the distinction in this case is essentially moot, since the Corps does have Section 10 permit authority over the flats.)

Each wetland type is described in a template form, as shown in Figure 3, which includes a general description of the wetland type (salt marsh, brackish marsh, fresh marsh, swamp marsh, intertidal flats, eelgrass, and water bodies, and noting subtypes), its distribution, a history of alterations and its functional significance. The reason for including the template descriptions is to provide a succinct description of study area wetlands.

Vegetation associations of upland habitats, as well as wetlands, are described in the narrative. Information in the other resource profiles (for example, wildlife use) was linked to the vegetation associations and/or locations so that the significance of any area to the bay ecosystem could be identified.

Areas of Importance and Areas of Environmental Concern

The second major finding of the HBWR was identification of valuable and important areas in the Humboldt Bay coastal zone. As noted, the

WETLAND TYPE (Example)

Wetland: Eelgrass

Description

Eelgrass occupies intertidal flats from about -3 feet (MLLW) to about +1 foot. This rooted plant often forms large dense beds; the roots and rhizomes form thick mats which resist erosion and aid in stabilizing the flats.

Distribution

Eelgrass is widely distributed in both North and South Bays, occupying almost every area of appropriate depth. Dredged channels and the North Bay oyster beds, where disturbance is frequent, are locations where eelgrass is not found. Eelgrass is also not found in Entrance Bay, where strong currents and wave activity create a highly energetic environment, unsuitable for eelgrass.

History

The major impacts on eelgrass in the bay have been the result of dredge activities. Navigation channels and turning basins occasionally require dredging beyond the confines of the natural channel. An even more significant impact, however, has been that caused by oyster dredging in North Bay. The dredging operation removes all the eelgrass, including roots and rhizomes, from the substrate. Since the operation is repeated every few years, the eelgrass never does become firmly reestablished.

Significant Relationships

Eelgrass provides food, shelter, and substrate for a diverse population of organisms. Numerous algae, bacteria, and invertebrates reside on eelgrass blades. Many of these in turn are food for the fish and crabs which inhabit the beds. The dense vegetation ahd thick root mass provide shelter for fish, crabs, and benthic organisms, and food for fish and waterfowl, especially brant. All of these in turn are food for a diversity of larger fish, waterfowl, and occasional raptors. Shellfish, including cockles and some clams, are found in eelgrass beds. Detrital material, the remains of dead plants such as eelgrass, are known to be food for a variety of filter feeding organisms, including many which are residents of unvegetated flats. Eelgrass beds are also important spawning and rearing grounds for herring and smelt.

FIGURE 3

model of Clark for identifying and evaluating coastal resources was generally followed beginning with identification of those areas which have important natural physical or biological functions (also referenced in Corps regulations 33 CFR 320-329) (12), or which met ancillary criteria. Important natural functions encompass: biological values, including primary and secondary productivity, vegetation density, plant and animal diversity, and threatened species habitats; ecosystem support, including hydrologic periodicity, location or elevation, and areal extent and ecological importance; physical protection, including shielding from wave and storm damage; storm and floodwater storage; groundwater recharge; and water filtration and purification. Ancillary criteria consist of general and specific preservation policies of local and state agencies; archaelogic, historic or aesthetic values; values for education, scientific study values or as refuges or reserves; recreation and public access values; and economic benefits, such as an area with harvestable renewable resources.

As indicated above, identification of the functional significance of an area and its ancillary values was made by evaluation of the resource profile information. Those areas which met all or most of the criteria were generally designated as Areas of Importance (analogous to "vital areas" in Clark's concept which are those of critical importance and high value that should be preserved and protected intact. (Ancillary criteria alone were insufficient to designate an area as an Area of Importance; fulfillment of some important natural function was necessary.)

Areas which met fewer criteria were designated as Areas of Environmental Concern, which include those adjacent to, and often shielding, Areas of Importance and those where development and use must be controlled to protect the environment. In addition, General Use Areas, in which only normal caution in development controls need be exercised, were designated.

In all, 17 Areas of Importance and 20 Areas of Environmental Concern were identified, with total areas of about 16,000 and 18,100 acres respectively as compared with the total study area of nearly 48,000 acres. (Figure 4)

Each Area of Importance or Area of Concern is also described in a template format (Fig. 5). The location and total acreage of each area is shown and its habitat types are discussed, as well as its history in terms of levee construction or other alterations, and current development pressures. The rationale for designation as an Area of Importance or an Area of Environmental Concern is given by explaining its natural functional and ancillary importance. (Fig 6) General Use Areas were summarized in terms of development pressure and a discussion of future development possibilities.

Activities and their Impacts

As indicated previously, it was intended that the review would recommend standards and criteria for suitable uses and activities in

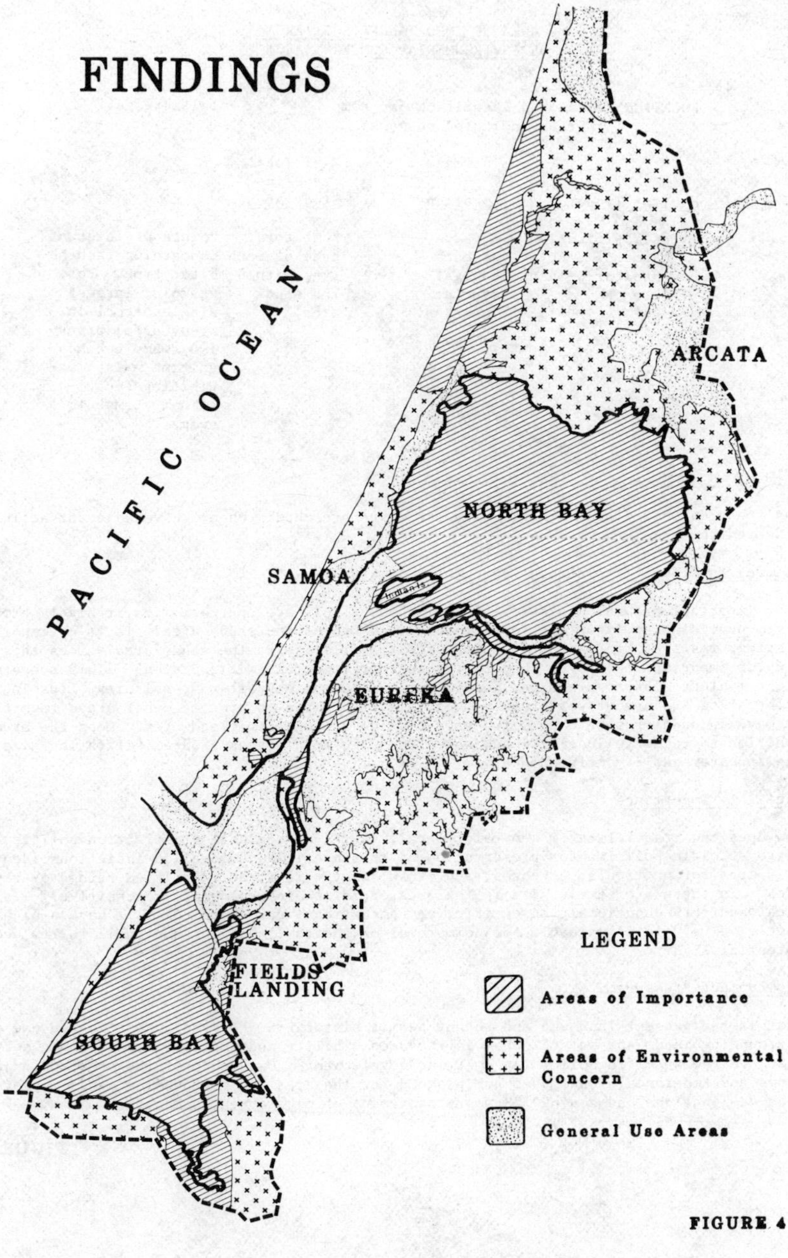
FINDINGS
PACIFIC OCEAN
ARCATA
NORTH BAY
SAMOA
Indian Is.
EUREKA
FIELDS LANDING
SOUTH BAY
LEGEND
Areas of Importance
Areas of Environmental Concern
General Use Areas

FIGURE 4

Template 2 (Example format)

AREA OF IMPORTANCE (EOI)/AREA OF ENVIRONMENTAL CONCERN (AEC)

LOCATION: Name and identification number as indicated on maps. Total Acreage:

Habitat Types	Acreage at Location	% of Total type in Study Area	
1. Classification 2. name of each habitat within 3. area. 4.	Acreage of each habitat type at location.	% of total area of each type within the study area.	Points of Interest concerning each habitat type. For example, largest single parcel in study area; brackish swamp behind saltine marsh; exhibiting successional change to swamp.

History of Area:

Has the area been diked? When? Were the dikes breached? When? Have adjacent activities affected the area? How? When?

Natural Functional Importance:

What important natural biological functions does this area perform? Is it highly productive, providing an important input to the food web of the study area? Is it an important nesting, resting, breeding or feeding site for wildlife in the study area? Does this area provide important ecosystem support for another important wetland area? (This support might include aids to drainage, control of sedimentation patterns, salinity, flushing characteristics or current patterns). Does this area serve to protect other areas from storm or wave induced erosion? Does the area provide storage for floodwater? Does the area function to recharge local groundwater? Does the area act as a water filtration site, aiding in water quality maintenance?

Ancillary Importance:

How does the area relate to general federal, state, or regional preservation policies? Are there specific policies for preservation of this area? Have local jurisdictions identified this area for particular purposes, and if so, what? Is it an identified refuge or reserve area? Is there a historic site in the area or is it archaeologically sensitive? Is the area used for educational, scientific, or recreational purposes? Does it have a high aesthetic value? Is it valuable for commercial or sports fishing or hunting? Is its economic potential high?

Development Pressure:

What is the present land use and recent permit history in the area? What is the use of surrounding lands and water? What is the accessibility and availability of urban services? What are the specific policies permitting development? Ownership of the area and adjacent lands and tidelands? What uses are planned for the area and surrounding lands; general plan designations and zoning? Summary statement of development pressure.

FIGURE 5

AREA OF IMPORTANCE (Example)

LOCATION: (11) SOUTH BROADWAY WETLANDS Total Acreage: 21.5

Habitat Type	Acreage at Location	% of Total type in Estuary	Remarks
Swamp	9.0	4.9	
Brackish sedge marsh	6.5	23.2	
Fresh cattail marsh	0.8	2.7	
Fresh oenanthe marsh	0.1	0.2	
Fresh rush marsh	1.6	3.5	
Ponds (closed ditch)	0.3	0.2	
Shrub	3.0	0.9	
Grassland	0.3	<0.1	

History of Area

Essentially the entire shoreline from Elk River to Eureka was wetland (probably salt marsh) in 1850. By 1870 most of the area south of Vigo Street was diked as agricultural land. The Northwest Pacific Railroad was completed along the shore of this area in 1901. Encroachment, resulting in increased runoff, inadequate drainage, and accessation of agricultural activities, has converted this area to a mixture of brackish and freshwater wetlands.

Natural Functional Importance

Both cattail and sedge marshes exhibit high primary productivity, which may be exported to the estuary during high water conditions. The marshes provide habitat for small birds and rodents which in turn are prey to marsh hawks, kites and other raptors. Herons and egrets use the swamp as a resting area during the winter. The small pond is probably used by some waterfowl for at least portions of the year.

Ancillary Importance

As productive wetland/swamp habitat, the area falls under EO11990, the Resource Agency's wetland policy, and general preservation policies of FWS, NMFS, EPA, SCS, CCNCR, CEC (south portion only), and OPR. It is an area designed unsuitable for power plants. The Eureka General Plan shows the southern part of the area as Agriculture/Open Space. The area is archaeologically sensitive and is identified as a potential public access point.

Development Pressure

The area is unused open space with industrial development on the north and major transportation corridors through and west of it. It is generally accessible and has available urban services. The Eureka General Plan and the Harbor District's Ordinance 7 both designate the area for industry. The area is privately owned. The southern part is not zoned; the northern part is zoned partly Agriculture and partly Commercial. Development pressure is considered high because of local policy and lack of alternative industrial sites.

FIGURE 6

each particular Area of Importance or of Concern, in part through compilation of existing land and water use regulations, and in fact, a draft edition did recommend activity criteria and performance standards. However, following study coordination, it was decided that the report would not recommend criteria, either to the Corps or to other agencies within their respective jurisdictions. As completed, the review does, however, contain discussion of each type of activities likely to be proposed in the study area, including associated uses; possible activity conflicts; physical, chemical, biological, and socioeconomic effects; and listing of likely regulatory processes. Also, a general approach is described for the assessment of the impacts of a proposed permit activity. This material is designed to assist the Corps and other agencies in their analysis of future permit requests on a case-by-case basis and in selection of appropriate conditions and guidelines for permitted activities.

Other Findings

In addition to the above results, the HBWR also identified areas potentially useful as compensation sites for development which impacts study-area wetlands. These compensation sites are generally areas in which wetlands could be restored by selective levee breaching with a minimum of new construction, as described above. Finally, the HBWR identified information gaps and procedural needs and studies needed to fill gaps in the data base: these are discussed below.

THE LOCAL COASTAL PROGRAM EFFORT

In discussing the use of the HBWR in local and State Coastal Planning, some general explanation of pertinent government jurisdiction and of the California Coastal Act is necessary. In all, at least fifteen Federal and 20 State agencies have a significant direct interest in the Humboldt Bay Coastal Area. Well over a dozen local agencies were identified; those most involved with the bay are the Harbor District, (Humboldt Bay Harbor, Recreation and Conservation District), Humboldt County, and the cities of Eureka and Arcata. The Harbor District is the chief trustee of State tidelands in the Bay, with jurisdiction over privately owned tidelands, but has no significant holdings on uplands. Eureka and Arcata both control adjacent State-entrusted tidelands. Humboldt County has jurisdiction over bay borderlands outside of the incorporated areas (Eureka and Arcata, only).

All of the foregoing local organizations are involved in the current development of a Coastal Plan for the area, which is part of a total Statewide action, set in motion in 1972 following approval of a controversial Coastal Zone initiative by the California electorate. This initiative set up a State Commission and six regional commissions empowered to strictly control Coastal Zone activities and construction during a 3-year period, while the staff was to prepare a comprehensive conservation and development plan for stated subject areas or elements. This plan was eventually adopted by the State legislature in a considerably streamlined form as the California Coastal Act of 1976 (1). The Act provides general policies for fourteen elements, specifically: shoreline access; recreation and visitor-serving facilities; housing; water and marine resources; dredging, filling and shoreline structures; commercial fishing and recreational boating; environmentally-sensitive habitat areas; agriculture; hazards; forestry and soils resources; locating and planning new developments; visual resources and special communities; public works; and, industrial and energy development.

A major feature inserted by the legislature in the Coastal Act was delegation of much of the planning and permit responsibility to local agencies. By mid-1981, regional commissions will be phased out, with local agencies holding a major share of planning and permitting powers, which must be in accordance with the State Coastal Plan (Tidal, submerged and public trust lands, including coastal wetlands and lagoons, will remain under State control). To accomplish this, the Local Coastal Program (LCP) process was devised, where during the transition period, local governments are required to make plans and take implementing actions to conform with the Coastal Act. The State Coastal Plan will ultimately consist of all the Local Coastal Plans developed by the local agencies.

LCP preparation (3) is a three-phase process: a) issue identification, wherein the adequacy and possible conflicts of existing local plans and zoning are evaluated with respect to the Coastal Act policy elements; b) preparation of an appropriate land-use plan and Environmental Impact Report; and c) zoning and implementation. (As described below, the HBWR was designed to be useful in both Phase II and III of the LCP.) A successful work program for the LCP requires careful study of local conditions and development of a consensus of local and State interests to resolve conflicts and create mutually acceptable plans and ordinances. Most involved agencies have marshalled forces to complete the program on schedule. Arcata and Humboldt County are preparing LCP's for their jurisdictions and the County is preparing the LCP for the Harbor District's area. The North Coast Regional Coastal Commission is preparing the LCP for the City of Eureka.

As of mid-1980, the Arcata LCP Land Use Plan has been completed and is nearing approval by the State; the County's plan should be completed late this year. The Eureka LCP is to be completed in 1981. Notable common issues being dealt with include preservation of agriculture, water quality, wetlands habitat protection, public access and development, and resource improvement. To develop and maintain public understanding, extensive hearings were conducted and a public participation process was used. The County, for instance, prepared a widely - distributed coastal planning newsletter, discussing progress of the LCP, for public distribution. Its LCP was also guided by a County-wide advisory committee. The North Coast Regional Coastal Commission (CCNCR) has a comprehensive hearing and public participation procedure. The CCNCR set up a Humboldt Bay Advisory Committee for LCP advice; the Committee includes representatives of local private groups, industry and governments, and State and Federal agencies. (The HBWR has been publicized appropriately and presented to interested parties through these forums, and has also been presented at workshops and hearings specifically required by the Corps of Engineers for public input).

CURRENT AND EXPECTED USE OF THE HBWR IN THE STUDY AREA

The content of the HBWR has been well received by local and State agencies and the report is considered valuable both for the current LCP process and for future permitting actions. The HBWR largely fulfills requirements for mapping of "environmentally sensitive" areas and adjacent habitats (areas whose biota are valuable and easily damaged including threatened-species habitat, all coastal wetlands and lagoons, dunesplant habitats, and biologic protection and research reserves). The designation of Areas of Importance and Areas of Environmental Concern, when combined with the information on the impacts of likely activities is expected to be of lasting importance to local and State agencies and the Corps of Engineers. Also, mapping of suitable marsh restoration sites for compensation, as noted previously, should prove helpful in determining appropriate compensation or mitigation for future permit actions.

As mentioned above, the HBWR also identified procedural needs and information gaps which should be addressed by additional Corps-sponsored studies. Procedural recommendations included a suggestion that the Corps set up an archaelogical/historic resources advisory committee for its projects and permit applications; and various proposals regarding improved compensation procedures, including the development of a compensation "land bank" program. Recommendations for further studies included determination of tidal datums and the limits of historic wetlands; a numerical (mathematical) circulation model of the bay for water quality and sediment management purposes; and various environmental and socio-economic studies. If carried out, these studies will provide a continuing flow of information to update and expand the data base assembled in the HBWR.

SUMMARY CONCLUSION

The Humboldt Bay Wetlands Review, and Baylands Analysis prepared under the regulatory and general investigation authorities of the Corps of Engineers and by continuous coordination with State and local interests appears to be a data base of genuine value to those responsible for management of this complex estuary. The report is viewed as very useful to the Corps in its Section 10 and Section 404 jurisdiction, to other Federal agencies in their permit and coordination responsibilities, and to involved local and State units of government in their exercise of primary decision-making power for the area.

Specifically, the key item of original work was identification and large-scale mapping of wetland and upland habitats which, when coupled with existing information about resource values of the area, allowed determination of those areas of the high natural and cultural value. Summary information about the likely impacts of permit activities and a methodology for environmental assessment of a proposed permit or project are provided for the use of regulatory and planning agencies, and proponents. The mapping accomplished in the review is of current value in development of California Local Coastal Programs, and, in combination with impact assessment data, will be of service in the direct implementation of the Coastal Plan.

The results of the wetlands review technique, as demonstrated by the HBWR and the predecessor prototype studies in Oregon and Washington, indicate that it could be applied generally throughout coastal California. The data assembled would help fulfill the expanded need for information on wetlands and adjacent habitat as the State coastal effort moves beyond its recent roles of enforcement and planning into active coastal zone management (2).

References

1) California Coastal Act of 1976.

2) California Coastal Commission, Coastal News, California Coastal Commission, San Francisco, Ca., August 1979.

3) California Coastal Commission, Local Coastal Program Manual, California Coastal Commission, San Franciso, Ca., 1977.

4) Clark, John, Coastal Ecosystems: Ecological Considerations for Mangement of the Coastal Zone, Conservation Foundation, Washington, DC, March 1974.

5) Cowardin, L.W. et al Classification of Wetlands and Deepwater Habitats of the United States, Government Printing Office, Dec. 1979.

6) Moulton, R.J. Wetlands Reviews; New Approach in the Corps Permit Program, American Society of Civil Engineers Proceedings, Coastal Zone 78, 1165-1181.

7) Nyc, Rudolf, In Search of Wetlands, Water Spectrum Corps of Engineers, Water Spectrum, Corps of Engineers, Water Resources Support Center, Fort Belvoir, Va., Spring 1980.

8) Reppert, R.T. et al, Wetlands Values; Concepts and Methods for Wetlands Evaluation, Corps of Engineers, Institute for Water Resources, Fort Belvoir, Va., February, 1979.

9) U.S. Army, Portland Engineer District, Alsea Wetlands Review, Corps of Engineers, Portland, Oregon, March 1976.

10) U.S. Army, Portland Engineer District, Nehalem Wetlands Review, Corps of Engineers, Portland, Oregon, June 1977.

11) U.S. Army, Portland Engineer District, Siletz Wetlands Review, Corps of Engineers, Portland, Oregon, March 1976.

12) U.S. Army Corps of Engineers, Regulatory Programs of the Corps of Engineers, Federal Register, Vol. 42, No. 138 - Tuesday, July 19, 1977.

13) U.S. Army, Seattle Engineer District, Snohomish Estuary Wetlands Study, Shapiro & Associates/A. L. Driscoll, Seattle, Wash, August 1978.

14) Presidential Executive Order 11990, Protection of Wetlands, May 24, 1977.

15) Coastal Zone Management Act of 1976.

16) Section 10, River and Harbor Act of 1899.

17) Section 404, Clean Water Act of 1972, as amended.

NATIONAL WETLANDS INVENTORY PROJECT

BILL O. WILEN
PROJECT LEADER, NATIONAL WETLANDS INVENTORY
UNITED STATES FISH AND WILDLIFE SERVICE

RUDOLF NYC
REPRESENTATIVE TO NATIONAL WETLAND INVENTORY
FROM THE UNITED STATES CORPS OF ENGINEERS

BACKGROUND

The United States Fish and Wildlife Service has major responsibility for the protection and proper management of fish and wildlife and their habitat. The Fish and Wildlife Coordination Act authorizes the Secretary of the Interior "to make surveys and investigations of the wildlife of the public domain." This Act expands the concept of wildlife and wildlife resources to include not only animal life but also all types of aquatic and land vegetation upon which wildlife is dependent.

In 1974 the U.S. Fish and Wildlife Service directed its Office of Biological Services to design and conduct an inventory of the Nation's wetlands. The mandate was to develop and disseminate a technically sound comprehensive data base, concerning the characteristics and extent of the Nation's wetlands. The purpose of this data base is to foster wise use of the Nation's wetlands and to expedite decisions that may affect this important resource. To accomplish this, state-of-the-art principles and methodologies pertaining to all aspects of wetland inventory were assimilated and developed by the newly formed project. The following is a summary of how this project, formally known as the National Wetlands Inventory Project, is organized, how it operates and how it is accomplishing its goal.

NWI ORGANIZATION STRUCTURE

The National Wetlands Inventory (NWI) operational office is the focal point for coordinating all activities concerning the inventory. Located in St. Petersburg, Florida, the staff consists of Fish and Wildlife Service (FWS) personnel and a representative from the U.S. Army Corps of Engineers, the Soil Conservation Service, and the U.S. Geological Survey (USGS). Other agencies not represented at the central office have established formal points of contact who are responsible for coordination with the NWI. A service support contractor is also located in close proximity to the central office. In addition to the central office staff there are seven Regional Wetland Coordinators, one in each of the six FWS Regional Offices, plus Alaska. The project leader, involved principally in budgetary and planning efforts, is located in Washington, D.C.

WETLANDS CLASSIFICATION SYSTEM

The first National wetlands inventory completed in 1954, utilized a classification system that compartmentalized wetlands into 20 broad types (*Circular 39*, Martin *et al*., 1953). This was satisfactory at that time, since the purpose of that inventory was principally to assess waterfowl habitat. The Martin *et al*. system, however, could not be used to effectively achieve the broader goals of the new NWI.

In recent years, the characteristics and natural values of wetlands have become better understood, creating the need for a broader set of definitions with which to classify wetlands. Such factors as dominant animal and plant life, soil type, and water chemistry are all used in the Fish and Wildlife Service's new system to define and group wetlands.

The new classification system (Cowardin *et al*., 1979) is designed to meet four main objectives:

- To describe areas that have similar natural attributes;
- To arrange these areas in a system that will enable better resource management decisions to be made;
- To convert these areas into a format for inventory and mapping; and
- To provide a uniform classification system and terminology.

The system is hierarchical, progressing from five systems (marine, estuarine, lacustrine, riverine and palustrine) at the most general level to dominance types based on plant or animal communities at the most specific level. This hierarchical structure permits the aggregation and disaggregation of data at different levels of detail. The system is purposely open-ended at the dominance level. It is assumed that users will supply dominance types as wetland habitats are described in the future. To supplement the hierarchical portion of the system, the classification system allows for additional information, referred to as modifiers, to be added at any level in the classification. These modifiers are considered essential in determining wetland values by almost everyone who is working with the wetland resource. The modifying terms used are: (1) water regime, (2) water chemistry, (3) soils, (4) and man-made disturbance. There has been a great increase in our understanding of wetland ecology in the last 10 years. This increase is expected to continue at an accelerated pace. It is expected that this will result in the identification of new data needs. This was anticipated and can be accommodated through the addition of new modifying terms.

Each element of the classification system is identified by either a letter or a number. The composite code identifying a particular wetland consists of a series of letters and numbers and is referred to as an alpha-numeric. The following is a typical alpha-numeric illustrating how a wetland is classified to the water regime, with special modifier level of detail.

E	System	Estuarine
2	Subsystem:	Intertidal
EM	Class:	Emergent
5	Subclass:	Narrow-leaved persistent
N	Water Regime:	Regular
d	Special Modifier:	Partially drained/ditched

E 2 EM 5 N d

Wetland classes and subclasses can be mixed as is shown in the following example:

E2F03/EM5N = Estuarine intertidal, forested, broad-leaved evergreen (E2FO3), mixed with Estuarine intertidal, emergent, narrow-leaved persistent (E2EM5) with a regular water regime (N).

In Florida, this would indicate a wetland plant community consisting principally of mangroves (*Rhizophora mangle*) and salt marsh cordgrass (*Spartina alterniflora*).

SELECTING THE REMOTE SENSING SYSTEM

Remote sensing was the obvious answer to the problem of inventorying the Nation's wetlands in a relatively short period of time. What was not so obvious was which tool to use in the ever increasing technological arsenal of remote sensing systems. Because of the immensity of the task, the number of choices was quickly narrowed to LANDSAT and high altitude aerial photography. After comparing LANDSAT's capabilities with the expressed needs of FWS regions and local agencies for wetland inventory information, it became evident that LANDSAT could not provide the desired level of detail without what appeared to be an excessive amount of collateral data such as aerial photographs and field work. When it also became apparent that the new wetland classification system that was being developed for the NWI could not be used to conduct an inventory with LANDSAT, the decision was made to use high altitude aerial photographs. Realizing that the state-of-the-art would continue to evolve, the NWI has been participating in cooperative demonstration studies for mapping wetlands with LANDSAT. The most promising application of LANDSAT data appears to be for change detection and updating of wetland maps.

ACQUISITION OF HIGH ALTITUDE AERIAL PHOTOGRAPHS

The scales of the aerial photographs used by the NWI range from 1:60,000 to 1:130,000 with 1:80,000 quad-centered color infrared being the preferred scale and type of imagery for conducting the wetland inventory. When taking into consideration the combined cost of air photo acquisition, plotting work areas on the photos, photointerpretation and transferring the annotated information to overlays/maps, the cost of inventorying the Nation with large-scale photography is prohibitive.

Priorities for wetland mapping are based principally on needs of FWS regions and those of other Federal and State agencies. As soon as an area has been designated as a priority and the necessary funding has been allocated, a complete search for suitable high quality aerial photographs is conducted. Usually, available aerial photographs are obtained for the conterminous United States through the EROS Data Center and for Alaska through ASCS aerial photography field office in Nevada, though other sources are used on occasion. At the same time as the aerial photographs are ordered, additional pertinent information, including soil surveys, regional wetlands inventories, topographic maps, nautical charts and reports, are assimilated at the central office.

When suitable imagery is not available in priority areas, contracts for aerial photographic acquisitions are prepared by the central office. This is generally accomplished in conjunction with other agencies, such as the U.S. Geological Survey and the Corps of Engineers. Since 1978 the NWI has obtained all its new imagery for Alaska through the State and Federal Interagency Agreement for High Altitude Airphoto Acquisition. The imagery being obtained is 1:60,000 CIR. Beginning in 1980, the NWI will obtain most of the imagery it needs for the conterminous U.S. from the newly formed Federal High Altitude Aerial Photography Data Base Program. To meet the different needs of this interagency group, 1:80,000 black and white and 1:60,000 CIR aerial photographs will be imaged simultaneously. All of the conterminous United States will be imaged by this program on a 3-5 year cycle. The U.S. Geological Survey is the lead agency in this program.

AERIAL PHOTOINTERPRETATION

Upon receipt of the aerial photographs at the central office, the imagery is sorted and filed by USGS 1:100,000-scale quadrangle map areas. Work areas are drawn on the photos to aid the photointerpreter and to ensure that all adjacent photographs overlap so that wetlands on the edge of work areas "tie" together. The aerial photographs and all collateral data are forwarded to the Regional Wetland Coordinator.

The aerial photointerpretation is performed by contractors which are generally either States, Universities or private organizations. The Regional Wetland Coordinators have the principal responsibility for the successful completion of the photointerpretation phase of the wetland inventory.

After the aerial photointerpreter (API) has been trained in the use of the new wetland classification system and is familiar with all available collateral data, he proceeds with a preliminary classification of wetlands. Generally, this consists of about 10% of the work area, with emphasis placed on identifying potential problem areas. After a sufficiently large number of photographs have been interpreted in this manner, the API is accompanied in the field by the Regional Wetland Coordinator to resolve any problems. Usually the most difficult part is delineating the upland from wetland boundary and identifying the correct water regime. As the photointerpretations are completed for each 1:100,000 scale quadrangle, the API again proceeds to go to the field to verify the photointerpretations. If the Regional Wetland Coordinator is satisfied with the work, he forwards the annotated photographs with all collateral data to the central office for quality control review. In addition to accuracy of wetland boundary delineations and classification, adherence to mapping conventions is also reviewed during this quality control step to ensure National consistency. If the photointerpretation passes this quality control review, it is forwarded to the cartographic laboratory for production of large scale overlays and small scale maps.

CARTOGRAPHIC PRODUCTION

The service support contractor, under the guidance of the NWI staff, is responsible for the cartography involved in preparing large scale overlays and small scale maps.

LARGE SCALE WETLAND OVERLAYS

The wetland interpretation data on the aerial photographs are transferred to an overlay on a large scale USGS base map using a zoom transfer scope. During this process, most distortions, are eliminated. Quality control review following the zoom transfer process also may reveal errors due to photointerpretation. Draft overlays are then reviewed in the field by the Regional Wetland Coordinator for accuracy of photointerpretation and zoom transfer. Draft wetland overlays are also reviewed by other Federal and State agencies, including the Corps of Engineers and the Soil Conservation Service. After all corrections have been made, the final overlays are produced.

SMALL SCALE MAPS

The final large scale overlays are used to prepare the NWI 1:100,000 scale topical wetland map series. Each overlay is photographically reduced on a copy camera, composited, and paneled to a USGS base map. (There are 32 7.5-minute quads and 8 15-minute quads that comprise one 1:100,000 scale map quadrangle.) Where 1:100,000-scale USGS base maps have not yet been prepared, the NWI produces only a composite overlay.

MAPS AND OVERLAYS COMPLETED

As of September 30, 1980, the NWI has completed over 7,000 draft or final large scale overlays encompassing an area of over 350,000 square miles. Also, 130 1:100,000 scale maps or composite overlays, have been completed.

DISSEMINATION OF CARTOGRAPHIC PRODUCTS

The NWI has a limited reproduction and dissemination capability that was implemented in order to make maps and overlays available to potential users. To date, 70,000 copies of draft and final overlays have been distributed under this interim dissemination program. The long-term dissemination strategy is to utilize, through a cooperative agreement, the U.S. Geological Survey's reproduction and distribution facilities for dissemination of NWI products. For information concerning the availability of NWI products, contact the appropriate Regional Wetland Coordinator shown in Appendix A.

WETLAND INFORMATION AND REPORTS

In addition to cartographic products, the NWI is developing technical data and providing reports in order to fully comply with the mandate with which it has been charged.

WETLAND AREA MEASUREMENT DATA

Area measurement data is not being produced in an operational mode. When area measurement becomes operational the large scale overlays will be used to obtain area measurement data that, after computer processing, will provide statistical information concerning the number of acres of a specific wetland classification type that is located in a particular state, county, Bailey Ecoregion, Hammond's Land Form, USGS hydrologic unit and USGS large and small scale maps.

LIST OF HYDRIC SOILS

The Soil Conservation Service representative assigned to the NWI is preparing a list of hydric soils for the entire Nation, including Alaska and Hawaii. A preliminary list for selected portions of the country will be available by 1 January 1981. In addition to helping identify wetlands in the field, this list will make existing products, such as SCS's soil survey maps, more useful as a data base for wetland evaluation.

LIST OF HYDROPHYTES

The Fish and Wildlife Service is developing a list of the plants found in the wetlands of the United States. This will support our definition of wetland and will form the appendix of hydrophytic plants cited on page 3 in the Fish and Wildlife Service's new Classification of Wetlands and Deep-Water Habitats of the United States (Cowardin *et al*., 1979). A tentative alphabetical listing of wetland plant species was developed by Dr. Lewis Cowardin concurrently with his writing of the Service's new wetland classification system. This tentative list contained many of the wetland plants of the United States, but was recognized by Dr. Cowardin to be fraught with taxonomic errors, synonym duplications, and was thought to be lacking up to 2,000 additional species. The initial tasks now being undertaken to correct the list are (a) to edit the list for taxonomic problems and cross reference synonyms, (b) add and delete species based upon habitat citations in botanical manuals and ecological papers.

REGIONAL AND STATE REPORTS

For each 1:100,000 scale quadrangle inventoried, the NWI is preparing "Notes to Users" which provide information about the imagery used and documents the Regional Wetland Coordinators' observations concerning wetlands in that area. This report also includes correlations between the common alphanumerics found on the map overlays with the common plant species found on the ground.

As states are completed, the "Notes to Users" along with pertinent collateral data will be consolidated into "State Reports."

TREND ANALYSIS

In order to document the changes that have occurred to the Nation's wetlands over the past 20 years, the NWI is conducting a wetland trend analysis that will utilize 1950 and 1970 era aerial photographs, the new wetland classification system, and sophisticated statistical procedures. The statistical work is expected to be completed by the end of March 1982.

USE OF NEW WETLAND MAPS AND OVERLAYS

When using NWI maps and overlays, it is important to remember that the NWI is inventorying all wetlands without emphasis on any particular type - such as in the first national wetlands inventory - and not restricting the mapping to only those wetlands regulated by Federal, State, or local regulatory agencies. Also, since the wetland definition in the new classification system involves consideration of several parameters: vegetation, animals, soils and hydrology, it may appear broader in concept than a definition based on fewer parameters. For example, if a definition is based principally on vegetation, then technically, an unvegetated mud flat would not be considered a wetland.

In anticipation of possible technical differences in concept and the requirement to consider the additional statutory criteria for determinations of jurisdiction, wetlands regulated by Federal, State and local agencies may be a subset of the wetlands depicted on NWI maps/overlays. A special note is contained on all NWI products, stating that there was no attempt to define the limits of proprietary jurisdiction of regulatory agencies and alerting the user to seek the advice of appropriate Federal, State or local agencies concerning possible regulatory jurisdiction if any work is contemplated within or adjacent to a wetland area depicted on the NWI map or overlay. Section 404 of the Clean Water Act of 1977 (CWA of 77), as amended, identifies the categories of wetlands subject to the federal regulatory responsibility for certain of these wetland categories to States that comply with pertinent regulations. By knowing where all the wetlands are located, it should be easier to identify those wetland categories addressed in the CWA of 77.

INTERMEDIATE SCALE MAPS (1:100,000)

- o Basin-wide planning
- o General overview of wetland information
- o Display of additional information pertinent to wetlands such as
  - Alternate project sites
  - Navigable waters of the U.S.
  - Section 404 wetlands
  - Wetlands covered by nation-wide permits
  - Wetlands covered by general permits
  - Critical habitat
- o Display of cumulative impacts on wetlands.

LARGE SCALE OVERLAYS (1:24,000 to 1:62,500)

- To assist in design and project alternatives that avoid or minimize impact on wetlands
- To communicate by referencing NWI code, coordinates, and name of map
- As data base with file referencing NWI code, coordinates, and name of map with information concerning field visit and rationale for determination of jurisdiction, information on vegetation, etc.
- To depict jurisdiction, on a case by case basis, following field review
- To show where nation-wide permits have been issued and for issuing general permits
- For mitigation with similar wetland types
- For habitat characterization
- For periodic reinventorying to determine cumulative impacts and trends
- To assess general or range of values for wetlands (functional curves)
- Display of cumulative impacts on wetlands
- To facilitate field review and setting of priorities for inspection

CONCLUSION

The need for a comprehensive, technically sound data base concerning the characteristics and extent of the nation's wetlands has long been recognized as a precursor to fostering wiser use of the nation's wetlands and for expediting decisions that could affect this important resource. In recognition of this need, other agencies have been providing technical and financial support to the NWI to achieve this goal.

With this continued support and with high quality imagery of the desired scale and type virtually assured by the new Federal High Altitude Aerial Photography Data Base Group, the NWI is more assured that the goals of the National Wetlands Inventory will be achieved in an expedient and cost-effective manner.

APPENDIX A

Region	Geographical Area	Regional Wetland Coordinator
1	*Alaska, California, Nevada, Hawaii, Oregon, Washington	Dennis Peters, US Fish & Wildlife Service, Lloyd 500 Bldg., Suite 1692, 500 NE Multnomah St., Portland, OR 97232 COM: 503/231-6116 FTS 429-6154
2	Arizona, New Mexico, Texas, Oklahoma	Warren Hagenbuck, U.S. Fish and Wildlife Serv., PO Box 1306, Albuquerque, NM 87103 COM: 505/766-2914 FTS: 474-2914
3	Minnesota, Wisconsin, Illinois, Indiana	Ron Erickson, US Fish & Wildlife Service, Federal Bldg., Ft. Snelling (AS/BSP), Twin Cities, NM 55111 COM: 612/725-3593 FTS: 725-3593
4	Arkansas, Louisiana, Florida, Mississippi, Alabama, Georgia, S.Carolina, N. Carolina, Tennessee, Kentucky, Puerto Rico	John Hefner, U.S. Fish & Wildlife Serv., R.B. Russell Federal Bldg., 75 Spring St. SW Atlanta, GA 30303 COM: 404/221-6343 FTS 242-6343
5	Maine, Vermont, New Hampshire, New York Massachusetts, Connecticut, Rhode Island Pennsylvania, New Jersey, Delaware, Maryland, West Virginia, Virginia	Ralph Tiner, US Fish and Wildlife Serv., One Gateway Center, Suite 700, Newton Corner, MA 02158 COM: 617/965-5100 FTS: 829-9217
6	Montana, Wyoming, North Dakota, South Dakota, Nebraska, Utah, Colorado, Kansas	Charles Elliott, U.S. Fish and Wildlife Service, PO Box 2548, Denver Federal Center, Denver, CO 80225 COM: 303/234-5586 FTS: 234-5586
*1	Alaska	Art Laperriere, US Fish & Wildlife Service, Alaska Area Office, 1011 East Tudor Road Anchorage, AK 99503 COM: 907/2976-3800 x412 FTS: 399-0150

SCIENCE AND PLANS: CALIFORNIA'S COASTAL WETLANDS

by

Christopher P. Onuf*

INTRODUCTION

The traditional role of scientists in the management of natural resources has been to provide technical information. The traditional explanations (by scientists anyway) of the shortcomings of most attempts at management have been that the necessary technical information was not available, or if available, that it was not applied, or if applied, that it was applied ineptly. The remedy to bad management followed obviously: enlist the participation of high caliber scientists by supporting the necessary research and pay heed to their findings.

This is clearly a caricature of the relationship between science and management. Nevertheless, something of this attitude has motivated the involvement of the University of California Sea Grant College Program in coastal wetlands research. The *California Coastal Plan* (1975) starkly stated the need for the management of remaining coastal wetlands by enumerating what has already been lost:

> Of the original 197,000 acres of marshes, mudflats bays, lagoons, sloughs, and estuaries in California (excluding San Francisco Bay), the natural productivity and open space values of 102,000 acres (52 percent) have been destroyed by dredging or filling. Of California's remaining estuaries and wetlands, 62 percent have been subjected to severe damage and 19 percent have received moderate damage. In southern California, 75 percent of the coastal estuaries and wetlands have been destroyed or severely altered by man since 1900. Two-thirds of 28 sizable estuaries existing in southern California at the turn of the century have been dredged or filled.

Coupled with the recognition of the need for management was the realization that firsthand scientific knowledge about these systems and understanding of their function was limited. Furthermore, there was no reason to believe that the much greater body of knowledge already assembled for Atlantic and Gulf coastal wetlands would be directly transferable to the rarer, smaller, and more sporadically distributed wetlands along the Pacific coast of the southwestern U.S., because of the vast differences in climate, topography and geology.

---

*Assistant Research Biologist, Marine Science Institute, University of California, Santa Barbara, California, 93106.

Funded by University of California Sea Grant Program Grants USDC R/NP-1-8E to C. P. Onuf and R/CZ-52 to C. P. Onuf, R. W. Holmes and C. H. Peterson.

The outgrowth of this perception of the problem and how to cope with it was the initiation, in 1976, of a series of on-going research projects on coastal wetlands and an escalating effort by the Sea Grant Program and the staff of the California Coastal Commission to involve scientists in the planning process in a substantive way. I have been a participant in this collaborative effort as a scientist. Here I shall report my observations on what has been achieved; how the plans that are now being formulated are deficient in their protection of natural resources; and what changes in attitudes and perspectives might facilitate the implementation of more effective measures.

My principal conclusion is that the plans are not deficient for lack of relevant scientific information, nor for failure to incorporate such information into the planning process. It is there in the preliminary stages. Unfortunately, in many cases no vestige of that input actually emerges in the final statement of policies of the management plans. I suggest that this regrettable outcome results from the esoteric aura which attaches to scientists and their work (and which, incidentally, we seem disinclined to dispel, for the most part) and the attitude that only scientists are competent to use the information that they produce. In this situation, scientists must actively promote the use of the relevant technical information throughout the planning process if adequate protection for natural resources is to be achieved.

I shall use a few examples from my own experience to illustrate the kind of deficiencies that are occurring in plans and how they developed; what alternatives might more appropriately have resulted; how common sense, not scientific expertise, was necessary to arrive at those more appropriate alternatives, and what procedural changes might expedite the achievement of those acceptable ends (i.e., consistency with the legal mandate, in the proximate sense, and protection of a rare and vulnerable natural resource, in the ultimate sense).

ANALYSIS OF EXISTING PLANS

As set forth in the Coastal Act of 1976, California's coastal zone management program has two phases in time and two levels of organization. The initial phase is the preparation of land use plans and implementation measures by all local governments which include parts of the coastal zone within their jurisdictions and the certification of those plans (Local Coastal Programs) for consistency with the provisions of the Coastal Act by the California Coastal Commission. During the initial phase, six regional commissions are also in operation to administer the Act until certification of the Local Coastal Programs by the state Commission and to assist local agencies in the preparation of their Local Coastal Programs. The second phase commences after the certification of the Local Coastal Programs and the disbanding of the regional commissions (scheduled for January 1981), with local governments assuming full responsibility for most aspects of coastal zone management. The state Commission will retain authority in such matters as siting of energy facilities; serve in an advisory capacity to the local programs; and hear appeals on local actions. Eleven other states have organized their coastal zone management programs along similar lines (Sorensen, 1979).

This system will only be effective if local plans contain adequate policies and implementing measures to manage coastal wetlands. After the certification of a plan, there will be little opportunity for revision to repair mistakes or to tune the planning instruments. Whatever guarantees for the protection of natural resources must therefore be incorporated before certification, if protection is to be provided.

The intent of the Coastal Act with respect to natural resources is explicitly stated in its first sentence (after naming itself).

> Section 30001. The Legislature hereby finds and declares:
> (a) That the California coastal zone is a distinct and valuable natural resource of vital and enduring interest to all the people and exists as a delicately balanced ecosystem.
> (b) That the permanent protection of the state's natural and scenic resources is a paramount concern to present and future residents of the state and nation.
> (c) That to promote the public safety, health, and welfare, and to protect public and private property, wildlife, marine fisheries, and other ocean resources, and the natural environment, it is necessary to protect the ecological balance of the coastal zone and prevent its deterioration and destruction.

Since the California Coastal Commission has consistently regarded coastal wetlands as environmentally sensitive habitats, the primacy of consideration for natural resource protection is even more emphatically stated:

> Section 30240. (a) Environmentally sensitive habitat areas shall be protected against any significant disruption of habitat values, and only uses dependent on such resources shall be allowed within such areas.
> (b) Development in areas adjacent to environmentally sensitive habitat areas and parks and recreation areas shall be sited and designed to prevent impacts which would significantly degrade such areas, and shall be compatible with the continuance of such habitat areas.

Realizing that local governments needed assistance in preparing plans that would meet the special requirements of the "environmentally sensitive habitat" designation, the Commission provided grants, essentially to assemble Environmental Impact Reports that delineated the relevant resources (for instance, presence of endangered species, resting or feeding areas for resident or migratory birds, nursery areas for fishes, etc.); assessed the additional constraints to development imposed by the presence and sensitivity of those resources; and explored the consequences of alternative uses. The assumption was that the final specific plans for wetlands would be based upon these preliminary assessments of the resources. Members of the technical staff of the Commission then asked me to review some of the early plans and preliminary reports on coastal wetlands to evaluate their adequacy for the protection of natural resources.

As is apparent from Table 1 below, some of the planning documents examined were woefully deficient in protecting these resources. (Implicit in this analysis is the assumption that human activities are generally destructive of natural resource values* of coastal wetlands, notwithstanding that planned manipulations can be used to enhance specific values.)

Table 1. Solutions to planning problems extracted from existing planning documents.

Proposed Use	Mitigation Measure
1. Put high intensity human uses (4-20 residential units per acre and roads) adjacent to a wildlife preserve	Erect artificial barriers to keep uses separate
2. Put 4-20 residential units per acre on the edge of a lagoon where the environmental constraints are moderate to severe while maintaining some other areas away from the lagoon at 0-4 residential units per acre where environmental constraints are minimal	Protect the lagoon by engineering the roads and drainage systems to contain the greater stresses that would otherwise be transmitted to the lagoon
3. Build a park far from the center of city (despite the _Plan's_ claim of its "centralized location,") develop it for intense use either in the existing public utility transmission corridor or on the slopes immediately above the wildlife preserve	Move the transmission towers and lines? (not mentioned in _Plan_) Convert agricultural land. Or grade the slopes with provision for arresting the concomitant acceleration of erosion. Erect artificial barriers to prohibit uncontrolled incursions into the wildlife area
4. Limit planning for environmental impacts to ∿700 acres immediately around the lagoon	Except as provided by catchment basins built as conditions for development within the _Plan_ area, deal with sedimentation of the lagoon by dredging
5. Develop portions of the floodplain to 10-20 units per acre	Require more expensive construction practices to assure safety in a potentially hazardous area

*By natural resources values I mean the utility of an area of land and/or water for the support of wild, especially native or migratory animals and plants. Commonly the natural resource value of an area is regarded higher when the area is essential to sustain exploitable, renewable resources, such as fish and game; supports rare organisms; or is exceptional for its aesthetic appeal in its natural state.

Table 1 (cont'd)

Proposed Use	Mitigation Measure
6. Place a facility for the testing of armed missiles in the middle of a mudflat where shorebirds are more heavily concentrated than elsewhere in the lagoon. Deploy the fill in an elongate, lobed configuration	Compensate for the lost area by replacing a culvert with a bridge to allow regular tidal inundation of an added area predicted to be larger than the area lost
7. Zone for medium density residential development an area that includes 20-30 acres currently used as a breeding colony by endangered least terns.	Set aside 8 acres for the endangered species in a different location from that presently in use, half of which is under the jurisdiction of another agency (an agency that has already expressed official opposition to a preserve.) Protect the preserve with a 100 foot wide buffer strip

1-5. Extracted for the City of Carlsbad's Aqua Hedionda Specific Plan.
6. Extracted from the Final Environmental Impact Statement for Military Construction Project P-875 "Ready Missile Test Facility".
7. Extracted from the City of Chula Vista's Bayfront Plan.

The point is not that the measures cannot accomplish the stated objective of protecting the natural resource values; rather, it is that they consist of artificial structural modifications that will often require strict monitoring and diligent maintenance in order to be effective. Such programs are usually very expensive. In addition, the possibility cannot be ignored that the need for active law enforcement will be increased by placing unoccupied areas of restricted access immediately adjacent to areas of high intensity human use. The system could work, but any lapse in the active and conscientious execution of the program would compromise its effectiveness.

In contrast, the following measures would have the distinct advantage of being less vulnerable to the pitfalls mentioned above and would in addition be cheaper to institute and maintain: surround critical habitat areas with buffer strips; ensure that the highest intensities of human use are arranged to be distant from a preserve; exploit natural barriers to discourage human access to restricted areas; encourage development in areas with the fewest environmental constraints and discourage it in areas with the most severe constraints; design fills and developments to be compact in shape and restrict them to the edge of wetlands to minimize their perimeter of contact with the wetland; and, in general, make efforts to control potential adverse impacts at their sources.

I convened a workshop of a small group of planners, regulatory agency representatives and scientists, all who were deeply involved in coastal wetlands work, to address the evident failure of the local coastal planning process for coastal wetlands. It was a simple matter to achieve consensus on a set of recommendations that, if followed, would almost certainly result in sounder plans than those now being proposed. They are presented below in outline form and are available from the California Coastal Commission in their entirety under the title "Guidelines for the Protection of the Natural Resources of California's Coastal Wetlands: Proceedings of a Workshop Held at University of California, Santa Barbara, 24-26 May 1979."

RECOMMENDATIONS FOR THE PREPARATION OF MANAGEMENT PLANS FOR COASTAL WETLANDS

I. PREREQUISITES FOR PLANNING THAT WILL ADEQUATELY PROTECT THE NATURAL RESOURCES OF COASTAL WETLANDS

A. Specify the relevant resources. (Maps and transparent overlays to the same scale of ground [and water] cover, habitat utilization by animals, outstanding views, etc.)

B. Set the environmental context. (Maps and transparent overlays to the same scale of topography, slope, soil, floodplain, earthquake hazard, and existing land use; and the full series of available aerial photography.)

C. Identify present and potential adverse impacts on the coastal wetland. (Mainly by comparisons between A & B above.)

II. RECOMMENDED MEASURES AND POLICIES TO PROTECT THE NATURAL RESOURCES OF COASTAL WETLANDS

A. Design the land use plan to minimize adverse impacts. (Based on the assumption that human activities are generally destructive of natural resource values.)

1. Zone low intensity uses close to areas of high natural resource value.

2. Zone the intensity of human use to be least in areas of severest constraints on development (for instance, steep slopes, floodplains).

3. Where development is permitted adjacent to coastal wetlands, require a buffer zone.

4. Where environmental constraints to development exist, zone in large enough units that parcels include some portion that has few constraints. Cluster development in that portion.

B. Exploit natural barriers to separate incompatible uses

1. Set the size of the buffer area necessary for a particular habitat according to the sensitivity of the most readily disturbed species that is known to be present on that site.

a. Intertidal sand and mud flats \
b. Vegetated intertidal areas / (requires widest buffer)
c. Transition zone (intermediate buffer)
d. Permanent open water bounded by steep shore (narrowest buffer)

C. Regulate construction activities and site planning to minimize adverse impacts on coastal wetlands.

1. Phase construction activity to minimize the expanse of bare soil exposed during the rainy season.

2. Minimize the area of impervious surfaces.

3. Discourage "technological fixes" as alternatives to planning that accommodates to environmental constraints.

4. If development is allowed with a condition for mitigation, require bonding, inspection, and follow-up monitoring to assure that the mitigation is successful.

D. When any filling or extensive disturbance is permitted in a wetland consistent with the Coastal Act, require the acquisition of an area of greater natural resource value in compensation for the area of coastal wetland lost.

1. Plan developments within coastal wetlands to minimize the perimeter of contact between human uses and natural areas by:

a. restricting development to the extremities of the wetland and
b. requiring the developed area to be compact in shape, other factors being equal.

III. IMPLEMENTING THE MANAGEMENT PROGRAM

A. Insure that there is accountability in the administration of the management program by clearly defining and delegating responsibility.

B. Provide for the on-going review of the management program by organizing an advisory committee composed of wetlands scientists, agency representatives conservation groups, and adjacent land users.

It is disconcerting that, regardless of the special qualifications of those who actually proposed the recommendations, virtually anyone with sufficient interest to think about the effects on wild plants and animals of the situations posed in Table 1 would have arrived at similar suggestions. On the one hand, it is reassuring that better plans are attainable by straightforward application of existing information, rather than requiring additional inputs from experts. On the other hand, it is perplexing and discouraging that, given the relative simplicity of arriving at sound plans for natural resource protection, they have not been forthcoming. In order to provide guidance to local governments that are still in the process of preparing their plans for wetlands, it is essential to ascertain at what stage(s) the plans diverge from the stated goals of natural resource protection. In the case of the City of Carlsbad's Agua Hedionda Specific Plan, it is possible to make a comparison with its precursor Environmental Impact Report, to determine whether the shortcomings of the Plan are attributable to an inadequate compilation of relevant technical information or to some other cause.

The comparison revealed that the necessary input was accessible in the Plan's own EIR. The problem evidently lay in failing to actually utilize the EIR in the preparation of important parts of the Plan. This is most obvious in the apparent insensitivity of the Land Use Plan section of the Agua Hedionda Specific Plan in its consideration of impacts related to slopes, hazardous geological conditions, and floodplain hazards, all delineated (described and mapped) in the EIR.

If it is true that the failings of the Plan were not due to inadequate technical input, then clearly, supplying more and better information in the preliminary stages, as would be the scientist's wont, will not be a solution. It would seem that merely contributing the technical foundation for a plan is insufficient. That information must be promoted throughout the planning process.

Given that we now know where in the planning process the technical input got lost and natural resource protection got side-tracked, it is still necessary to explain how it happened, before redress will be possible. Many factors undoubtedly conspire; however, there are a few important factors about which something can be done. First, the Coastal Act attempts to do all things for everyone, or so it seems. In addition to the protection of natural resources the Act includes policies which are intended to preserve or promote public access (including "lower cost visitor and recreational facilities and housing opportunities for persons of low and moderate income"); to maintain coastal dependent industry (including agriculture, forestry, commercial fishing, aquaculture, small boat harbors and major ports); and provide for the siting of essential energy facilities (even when their presence will have adverse effects on most of the other resources, natural and man-made, of the coastal zone).

Second, although planning is supposed to be an integrated process, in fact, it is not. Because the task is so large and so all encompassing, it must be subdivided both by subject matter and in time. Thus, the assembling of background material for a segment of the plan will often be financed by a special grant and the work will often or usually be contracted out. At a later date the related segment of the plan will be formulated, probably by others than those who gathered the relevant background information.

Third, planning that includes an ecological perspective is a relatively recent activity. It is not yet a well defined academic discipline. Older, more experienced planners probably have received no formal training with that perspective. Younger planners will more likely have been introduced to that perspective but cannot have had much experience in applying it.

My interpretation of how these factors interact to the detriment of natural resources in coastal wetlands is as follows. The Coastal Act serves so many masters that is must be very hard indeed to know which master takes precedence in each of the possible situations that could confront a small planning staff in the preparation of its Local Coastal Plan. I with my one axe to grind can deduce (correctly, I think) from the environmentally sensitive status of coastal wetlands that the protection of natural resources in this instance is the highest priority of the management program. It is not surprising that those who have to consider and respond to the whole welter of conflicting demands (and answer for it to established constituencies) might often or even usually arrive at some other hierarchy of objectives.

Obviously, the purpose of requiring an inventory of the resources of an "environmentally sensitive habitat area" and an assessment of their environmental context is to give those factors the fullest possible consideration in the formulation of the local management plan. Unfortunately, singling out this phase of the process as a special task that must be completed before proceeding to the construction of the plan seems to have the effect of divorcing it from the planning process. It is as if having fulfilled that requirement (cleared that obstacle) it could then be ignored when the real job was tackled. Since the background of most planners is not strong in natural sciences, and since others, perhaps a local consulting company, compiled the information on natural resources, the material is likely to be foreign to the planning staff that formulates the final plan. The commonly held attitude that science is for scientists and that no one else is equipped to handle such material will insure the segregation of the technical input, even though common sense by itself is entirely adequate to reach an acceptable conclusion (i.e., given that the resources of note are concentrated in such locations and are subject to adverse impacts from such source transmitted in such ways - presumably the form of the technical input - the relative merits of various measures and activities ought to be obvious).

The final factor that works toward the diversion of proper attention from the protection of natural resources in the local planning process is that many of the other goals of the Coastal Act have deeply entrenched constituencies, whose vested interests are well understood and whose systems of exerting influence on public policy are well established. Thus, the planning agency will know the interests of land developers and how their interests affect the local community. It is predictable how this will focus the attention and perhaps channel most of the effort of the planning departments into reconciling the mandates of the Coastal Act for public access, low cost housing and public and private visitor-serving and recreational developments with the proposed projects of land developers. Considerations for natural resources are not so represented, at least in the case of the local stewardship of coastal wetlands.

REMEDIES

If this reconstruction of the local coastal planning process as it applies to coastal wetlands is essentially correct, it seems that some actions, entirely within the realm of the possible, could yield substantial improvements. First, it is critical that local agencies be directed, presumably by the California Coastal Commission, that protection of natural resources is the uncontestable highest priority in the management of coastal wetlands and will take precedence over all other considerations except public safety in planning. This by itself could counteract the factors listed above that militate against the use of the input of natural science information. Instead of having to interject this one unfamiliar element into the familiar framework that encompasses all the other competing demands (and in which it would understandably get short shrift because of its unfamiliarity) its relation would be clearly set as preeminent to those other demands. Dominance-subordinance being established, the necessity of coming to grips with the constraints imposed by a specific environment will also be established. When inescapable, natural sciences become knowable (sufficiently to put together sensible plans, anyway).

It is important for scientists to participate in this phase. The main reason for this is there is as yet no generally recognized and effective constituency at the local level for exerting influence on behalf of natural resource protection in coastal wetlands. Until that vacuum is filled, planners will have to resort to the authority of the so-called expert opinion of scientists to confer legitimacy on their own common sense findings. Such outside consultation and review can only be beneficial to the planning process. On the other hand, the appeal to scientists as the repositories of the special knowledge that will invest a policy with the necessary weight of authority can only perpetuate the aura of othernessness about scientists and their work, which I regard as counterproductive, however unavoidable it may be in this situation. Concluding this line of reasoning, the second action that will help sustain the consideration of natural resource protection throughout the planning process will be to involve scientists as a political instrument. Although in the abstract this is not necessary and is a corruption of the traditional conception of the scientist as an investigator of natural processes and a provider of knowledge about such processes, in practice it appears to be necessary.

Unless this paper is read or heard by a lot more people than I have any reason to expect, there is no chance that either scientists or local planners will realize how their interaction might yield better plans. The initiative must again be taken by the Coastal Commission, in this case most appropriately channeled through the regional commissions. It should take the form of advising all members of the Local Coastal Program staffs as to how the organization of the coastal planning process seems to have obscured the explicit intent of the Coastal Act in regard to coastal wetlands, and why and how scientists should be enlisted to provide the balance of perspective that will most easily achieve consistency with the statutory requirements. Perhaps the same communication would serve to inform interested scientists about the state of the program and how they might be asked to contribute to its furtherance. The regional commissions are the appropriate match-makers, because they will have identified the appropriate scientists from previous consultations in relation to permit hearings and the like.

The third action is already underway: to provide a systematic set of recommendations for dealing with different situations, (i.e., the set of guidelines outlined previously). I am now convinced that they can serve no useful function until the more basic problems of attitude and perspective are addressed by the other two actions.

In this paper I have dealt with the role of scientists in providing technical information in the planning process of wetland resource management. I have belittled the contribution of their on-going research and recommended turning them into political pawns. It would be very misleading to leave them there. A sound plan may be an essential prerequisite of an adequate management program, but it is not sufficient by itself. In the case of California's coastal wetlands, for instance, restoration and mitigation will play very prominent roles. This is because much of our surviving coastal wetland is currently in a degraded condition. Also, some conversion to other uses will be unavoidable, though law requires compensation by replacement with equal or greater value of wetland (in terms of natural resources) elsewhere. Research has barely begun and is desperately needed on all these issues, if our management of coastal wetlands is going to be ultimately successful.

REFERENCES

1. Sorensen, J. 1979. State-local relations in coastal zone management: implications for change. Coastal Zone Management Journal, 6: 295-302.

2. California Coastal Zone Conservation Commissions, 1975. California Coastal Plan, Sacramento, California.

UNPREDICTED RAPID EROSION, TYBEE ISLAND, GEORGIA

BY
Frank H. Posey, Jr. PE[1]
F. Wade Seyle, Jr.[2]

Introduction.

Tybee Island, Georgia, is directly south of the confluence of the Savannah River and the Atlantic Ocean and is about 17 miles east of the city of Savannah, Georgia. It is the northernmost barrier island along the Georgia coast which provides a buffer zone for the mainland against storm-generated winds and waves. The island is about 3.5 miles long, and has an average width of about 1/2 mile. The north end is bounded by the entrance to Lazaretto Creek and the Savannah River bar channel lies approximately 1 mile from the shore. The south end is bounded by the entrance to Tybee Creek. Tybee has been developed as a residential, commercial, and recreational area. Approximately 2.5 miles of beachfront are available for recreational activities. The principal concentration of commercial activities on the island is near the south end of the beach. The permanent year-round population of Tybee is about 1,950 and the summer population is about 12,000. On special holidays and weekends, visitors increase the population to 25,000 with some regularity. The beach serves the recreational needs of a tributary area extending some 300 miles inland and having a maximum width of about 200 miles. In addition to regular users from the tributary area, tourists from all along the eastern seaboard frequent the beach. Suitable access to approximately 75 percent of the beach is available by means of street ends and public lands. Access to 2,200 feet of beach on the north end is somewhat limited because of the location of historic homesites along Officers Row in the Fort Screven area. The State of Georgia is currently conducting a study to formally determine the eligibility of the Fort Screven historic district for the National Register of Historic Places.

Historical Changes.

Historically, Tybee Island has experienced severe beach erosion and lowering of the beach profile in areas protected by seawalls, and erosion and recession of the shoreline in unprotected areas. Erosion has placed the seawall and other structures in a position vulnerable to severe damage, especially during storms and hurricanes when the island is subjected to high winds, waves, and tidal surges.

Both the north and the south ends of the island have experienced almost continuous shifts in the location of the mean high water and mean low water lines. In 1867, the mean high water line on the north end was

[1]Civil Engineer, U.S. Army Corps of Engineers, Savannah District, Savannah, Georgia.

[2]Civil Engineer, U.S. Army Corps of Engineers, Savannah District, Savannah, Georgia.

approximately 2,800 feet seaward of the Tybee Island lighthouse. However, by 1930 this line had receded 1,500 feet. The shoreline along the middle of the island has remained relatively stable during this period fluctuating seaward and back approximately 900 feet.

The south end has experienced the most extreme and frequent changes of the location of the mean high water line. In 1867, the mean high water line had eroded to a point 1,200 feet behind the location of the existing seawall. By 1900, the distance between the location of the seawall and the mean high water line was 3,500 feet. However, by 1933, the large beach at the south end had again eroded and plans were developed for shoreline protection measures.

Aerial photographs, taken periodically since the 1930's, show the changing hydrographic patterns that have existed just off the south end of the island. Sediment movement by waves and inlet currents in these shallow areas dictate the size and shape of the bars and shoals. As these waves and currents deposit sand, large spits that extend 2,000 to 3,000 feet from the seawall form. These spits are generally visible at low tide. The change in flow patterns and velocities associated with storms cause channels to breach these spits resulting in erosion of the shoals and in some cases loss of beach material.

Previous Construction.

As shown on page 3, existing protective structures along the recreation beachfront include concrete and rock seawalls, timber bulkheads, and a system of groins. The seawalls and bulkheads are in good condition although the groins are in a poor state of repair contributing little to erosion control.

The construction of protective works to control erosion of the Tybee Island beach was undertaken as early as 1882 with placement of three rock groins at the north end of the island. Since this initial effort, numerous attempts have been made to control erosion by structural measures. However, many of these structures have been damaged or destroyed by the elements. In 1933, a 1,000-foot steel sheet pile bulkhead and groin system was completed on the north end of the island to protect the adjacent property. These structures are now in poor condition, as a result of exposure to a marine environment, and offer little protection from waves and currents. A 2,000-foot treated timber pile bulkhead and five groins of similar construction were completed on the north end in 1938 as a Works Progress Administration project. This bulkhead was constructed with a top elevation of 12.0 feet above mean low water and is generally in good condition. A reinforced concrete sheet pile seawall built in 1941 extends along the beach from about First Street to the southerly tip of the island at Tybee Creek, a distance of 11,000 feet. The top elevation of the seawall varies from +12.0 m.l.w. at the south end to +15.2 m.l.w. at the north end. Construction also included the addition of 19 treated timber groins perpendicular to the seawall. The timber groins were built to a top elevation of +10.0 m.l.w. adjacent to the seawall and sloped seaward to elevation +2.5 m.l.w. Hurricane Dora made landfall at Tybee Island in September 1964 and damaged a section of the seawall between First and Center Streets. Approximately

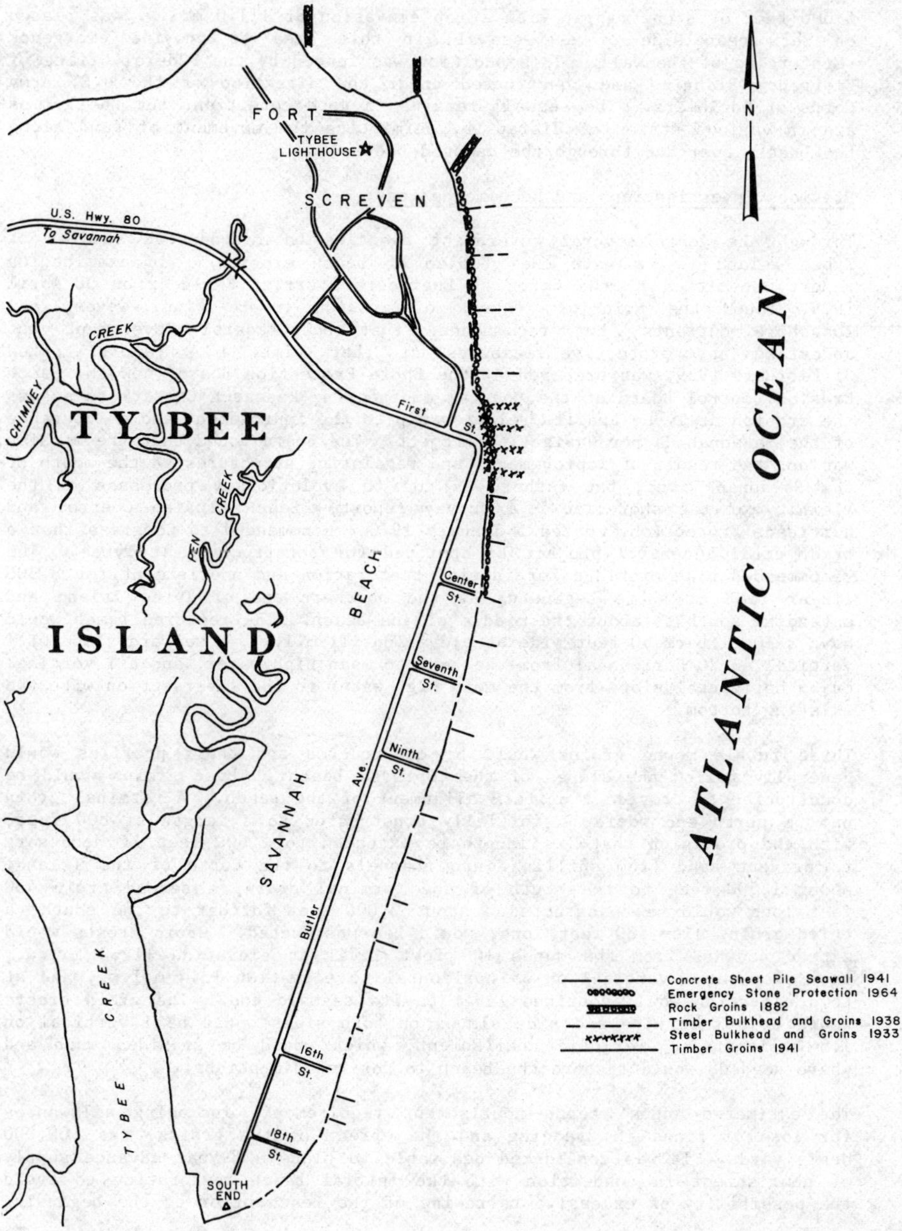
FORT
TYBEE LIGHTHOUSE
SCREVEN
U.S. Hwy. 80
To Savannah
CHIMNEY
CREEK
TYBEE
ISLAND
PEN CREEK
First St.
Center St.
Seventh St.
Ninth St.
16th St.
18th St.
SAVANNAH BEACH
Butler Ave.
TYBEE CREEK
SOUTH END
N
ATLANTIC OCEAN
Concrete Sheet Pile Seawall 1941
Emergency Stone Protection 1964
Rock Groins 1882
Timber Bulkhead and Groins 1938
Steel Bulkhead and Groins 1933
Timber Groins 1941

5,000 feet of stone riprap with a top elevation of +11.0 m.l.w. was placed on the ocean side of the seawall in this area to provide emergency reinforcing of the wall. This addition was funded by the Federal Office of Emergency Planning and constructed under the direction of the U.S. Army Corps of Engineers. The seawall remains in good condition, but the groins are in various states of disrepair, permitting the movement of sand along the beach, over and through the damaged sections.

Previous Investigations and Recommendations.

Through the years several government agencies have conducted studies of Tybee Island to evaluate the problem of beach erosion. An examination report, submitted by the Corps of Engineers District Engineer on 30 April 1929, found the principle causes of erosion to be wind, waves, and longshore currents, but recommended that the Federal Government not undertake any protective measures at that time. A report, dated 25 February 1939, was prepared by the Shore Protection Board (now the Beach Erosion Control Board at the Coastal Engineering Research Center) to assess the erosion at Tybee and its relationship to the improvement and maintenance of the Savannah Harbor Navigation Project. The Board found that the erosion was not the result of improvements and regulating structures at the mouth of the Savannah River, but rather was due to evolutionary processes in the vicinity of the shoreline. A survey report, "Beach Erosion Control and Hurricane Protection," dated 5 January 1970, recommended to Congress that a beach erosion control project be approved for construction at Tybee. The recommended plan provided for initial restoration and nourishment for 8,300 linear feet of shore beginning at the northern end of Tybee Island and extending south to about the middle of the beach. The restored beach would have a level berm 60 feet wide at elevation +11 m.l.w., a seaward slope of 1 vertical on 20 horizontal from the berm to mean high water, and a 1 vertical on 35 horizontal slope from the mean high water to the intersection with the existing bottom.

Three rubble mound groins would be constructed and their profiles would generally follow the slopes of the improved beach. These groins would be constructed to provide a uniform alignment of the beach. A terminal groin on the north end would be initially constructed to a length of 800 feet, with the provision that its length be extended to 2,000 feet if necessary to prevent sand from spilling into channels to the north of the island. About 1,500 feet to the south of the terminal groin, a second groin 480 feet long would be constructed. About 3,000 feet further to the south, a third groin, also 480 feet long, would be constructed. Groin crests would extend seaward from the dunes 100 feet level at elevation +12.0 m.l.w., then slope at 1 vertical on 35 horizontal to elevation +3.4 m.l.w., and at this elevation would continue level to its seaward end. The groin crests would be 6 feet wide and side slopes on both sides would be 1 vertical on 1.5 horizontal. Periodic nourishment, which would be provided when and where needed, would restore the beach to desired dimensions.

The estimated annual beach nourishment requirement, including allowances for loss of fines in dredging and the effect of the groins, was 100,000 cubic yards. It was considered desirable to place a 3-year advance supply of nourishment in connection with the initial beach restoration to avoid the possibility of excessive narrowing of the beach prior to the beginning of subsequent scheduled nourishment operations. The top elevation of the 3-year advanced nourishment would not exceed the design grade of the groin. Material for initial restoration and future renourishments would be obtained from the shoal at the south end of the island. Investigations indicated that sufficient amounts of suitable sand were available from this shoal for these planned purposes.

Conditions Prior to Initial Nourishment.

Prior to initial beach nourishment in 1975, the north end of Tybee had suffered the more extensive erosion. Between 1920 and 1972, the northernmost 5,800 feet of beach eroded at a rate of 6.7 horizontal feet a year reducing the beach width by approximately 350 feet. The adjacent 3,200 feet southward eroded at an annual rate of 2.9 feet horizontally landward. The beach south of this area, approximately 4,800 feet, intermittently eroded during severe storms and gradually built up when the sea subsided. Accretion occurred generally in the offshore area along the southernmost 1,700 feet of the island, and a relatively small volume of accretion was moved off the northeast tip of the island. In August 1975, the average dry beach width along the total project was 87 feet and the average beach width at mean low water was 284 feet.

Initial Nourishment Project.

As shown on page 6, the beach nourishment project deviated slightly from that originally proposed in the January 1970 report. On the recommendation of the Board of Engineers for Rivers and Harbors, the project was extended about 5,000 feet southward to provide protection for the entire ocean shoreline of Tybee Island. The berm width was reduced from 60 feet to 40 feet thereby helping distribute the recreation area evenly along the beach. The Board also recommended the two intermediate, short groins be omitted from initial construction. The effectiveness of the 800-foot-long terminal groin at the north end would be monitored after construction to determine the need for its extension, or alternatively constructing the two additional intermediate groins.

Construction of the north end terminal groin began in September 1974. The rubble mound groin extended 800 feet seaward and 225 feet landward of the existing sheet pile seawall and sloped from elevation +11.0 m.l.w. to +3.4 m.l.w. The purpose of the groin was to anchor the beach fill against wind, waves, and currents. Beach restoration began on 30 July 1975 and consisted of placement of 2,262,100 cubic yards of sand fill. The beach fill, excavated by hydraulic pipeline dredge from the shoals approximately 2,000 feet off the south end of Tybee Island in the inlet to Tybee Creek, extended approximately 13,615 feet from 18th street on the south end to the terminal groin at the north end. A berm, which averaged 100 feet wide for the length of the project, was pumped in at elevation +11.0 m.l.w. The fill then sloped at 1 vertical on 20 horizontal to elevation +6.8 m.l.w. and then 1 vertical on 35 horizontal to the intersection of the existing profile at or below mean low water. The average width of the dry beach (that above mean high water) after nourishment was 357 feet and the average width of the beach at mean low water was 524 feet. Construction was completed on 30 March 1976.

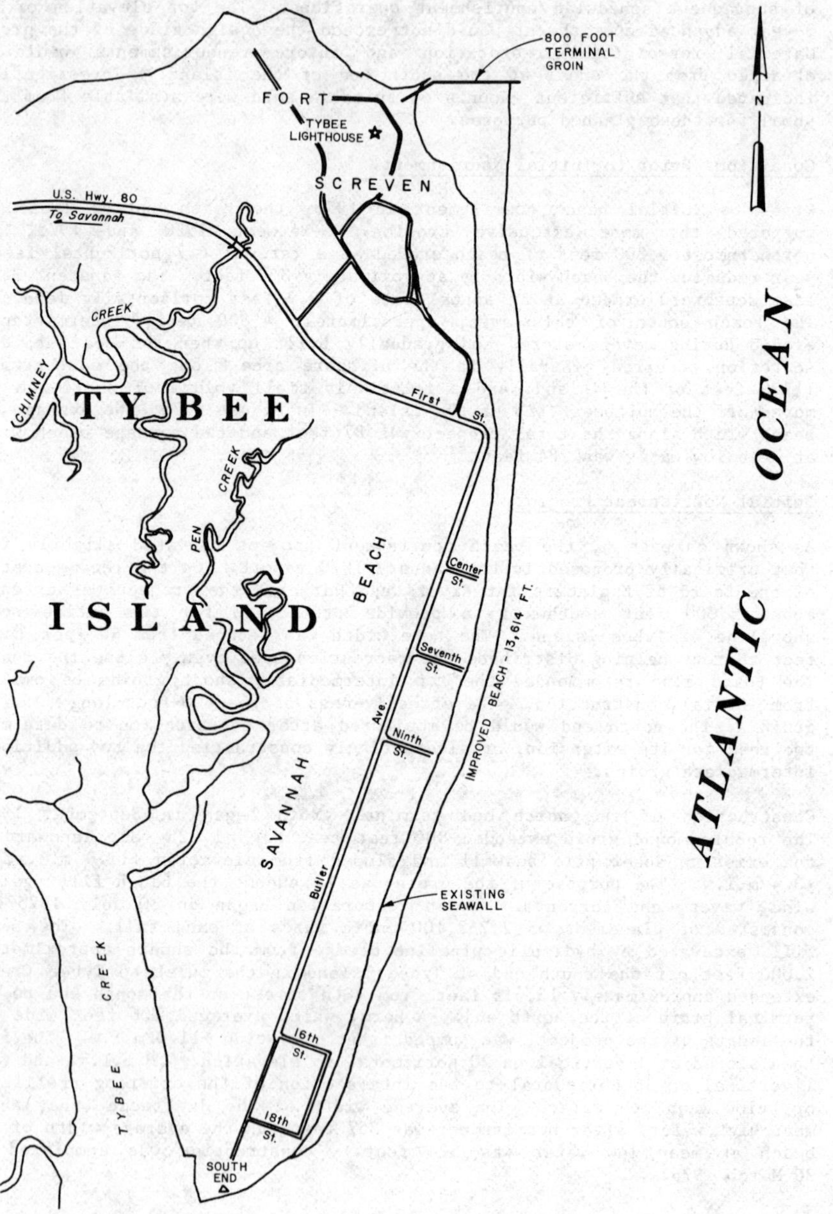
800 FOOT
TERMINAL
GROIN
N
FORT
TYBEE
LIGHTHOUSE
SCREVEN
U.S. Hwy. 80
To Savannah
CREEK
CHIMNEY
TYBEE
First
St.
PEN
CREEK
Center
St.
BEACH
IMPROVED BEACH - 13,614 FT.
ISLAND
Seventh
St.
Ave.
Ninth
St.
SAVANNAH
Butler
EXISTING
SEAWALL
CREEK
TYBEE
16th
St.
18th
St.
SOUTH
END
ATLANTIC
OCEAN

Project Performance.

Prior to the beach nourishment, a monitoring program was established to document the results of beach restoration and erosion. The program consisted of land and hydrographic surveys taken along established profile lines within and beyond the southern and northern limits of the project. The profiles extend landward a sufficient distance to identify the restored beach and seaward from 0.5 to 1.5 miles to define the offshore contours. Since the inception of the monitoring program, additional profile lines have been added to obtain a more accurate description of the hydrography in the offshore zone. Monitoring surveys are taken at least four times a year and after any major storm and are used as a basis for estimating volumetric erosion and accretion quantities and rates. Also, the Littoral Environmental Observation Program is conducted on a daily basis by volunteer observers at the beach. The type, magnitude, frequency, and direction of the forces that act upon the beach can be determined from the information collected as part of this program. Data collected includes wave breaker period, height, type, and direction; foreshore slope, surf zone width, and littoral drift current and direction.

During construction, the beach experienced some erosion due to loss of fine grain material, sorting and settling activities, and tropical storm "Dottie" which passed near the Georgia coastline in August 1976. The total net loss of material over the 43-month period between May 1976 and March 1979 was 460,140 cubic yards or 20.3 percent of the material placed. Of this volume, 339,400 cubic yards passed over, around or through the terminal groin and can now be observed on the beach north of the project limits. The remainder of the material has either settled in the shoal area on the south end of Tybee, has shifted to the accreting area in the middle of the beach, or has been carried off by the littoral currents. Pockets of erosion and accretion which have developed on the beach are shown on page 8.

The middle 5,200 feet of the beach has remained relatively stable since construction. However, the beach north and south of this stable area has experienced erosion. Approximately 44 percent of the material placed on the northern portion of the beach has eroded away. This is not evident from a visual observation because this area was overfilled during construction. The beach south of this stable area has lost 62 percent of the material placed there. There is no dry beach available at the extreme south end for recreational activities at mean high water when the tide is against the seawall. The loss of beach on the south end is of great concern due to the concentration of commercial and recreational activities in this area. The maximum erosion rate occurred during the period May to August 1976. This was partially a result of tropical storm "Dottie" which passed offshore in August 1976. Contrary to general trends observed at nearby islands, the periods of highest erosion rates have occurred in the spring and summer months, while only minor erosion has taken place during the fall and winter months which are usually accompanied by high energy

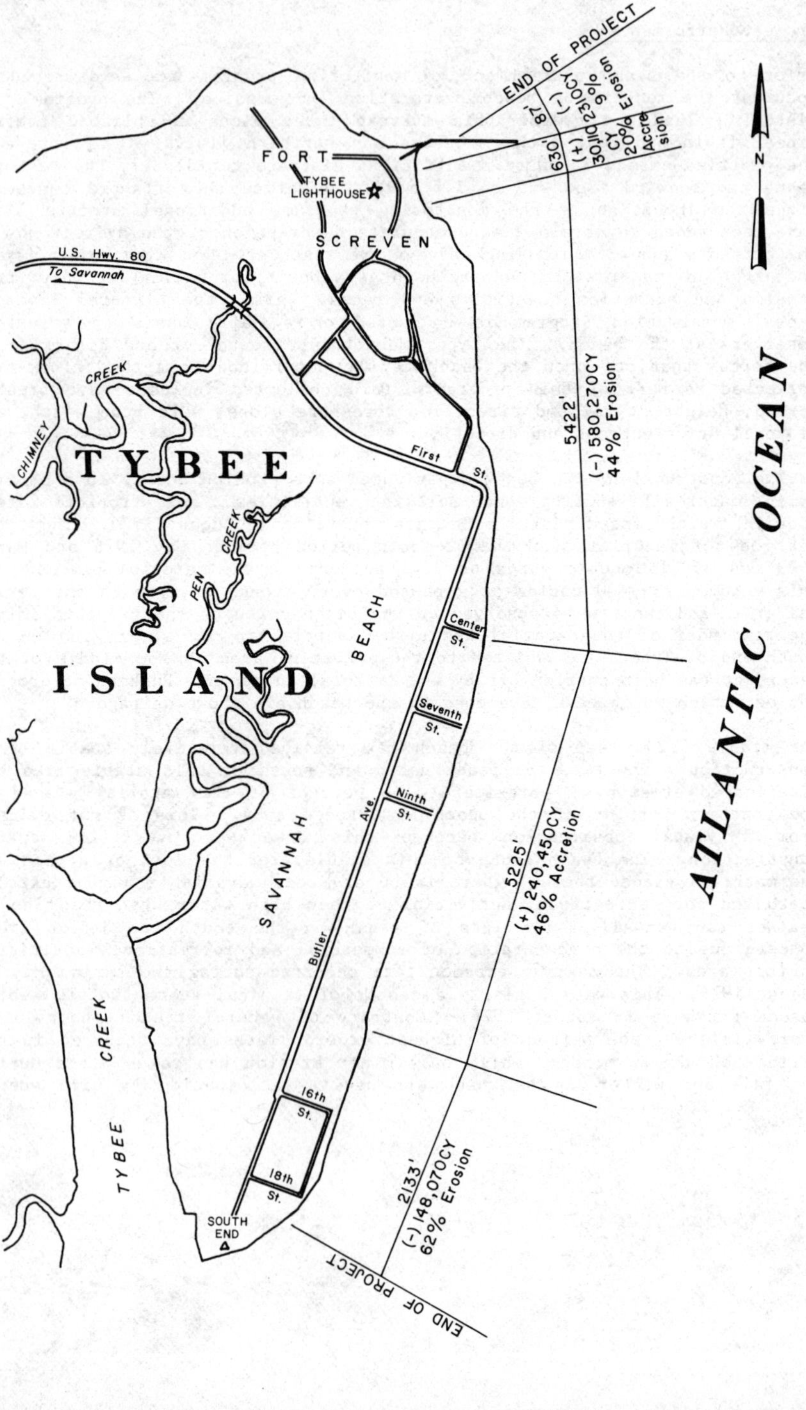
END OF PROJECT
87'
(-) 2310CY 9% Erosion
630'
(+) 30,110 CY 20% Accretion
N
FORT SCREVEN
TYBEE LIGHTHOUSE
U.S. Hwy. 80
To Savannah
CHIMNEY CREEK
TYBEE ISLAND
PEN CREEK
First St.
5422'
(-) 580,270CY 44% Erosion
ATLANTIC OCEAN
BEACH
Center St.
Seventh St.
Ninth St.
SAVANNAH
Butler Ave.
5225'
(+) 240,450CY 46% Accretion
TYBEE CREEK
16th St.
18th St.
SOUTH END
2133'
(-) 148,070CY 62% Erosion
END OF PROJECT

waves. During the period December 1977 to April 1978 the volume of sand within the project area actually exceeded the volume of sand placed during nourishment. This seasonal anomaly is a result of the island's orientation, approximately 30° to a north-south line, to prevailing winds and currents.

The total beach surface area above mean low water after nourishment in July 1975 was 4.01 million square feet. Twenty-six percent of the beach area was above mean high water and 74 percent was between mean high and mean low water. By March 1979, the total area had increased to 6.85 million square feet of which approximately 51 percent of this area was above mean high water and 49 percent was between mean high and mean low water. This indicates an increase in the total surface area of the beach above mean low water, and a steeper beach slope between high and low water; a condition which is more vulnerable to wave attack and erosion during storms.

Study of the South End.

At the time the nourishment project was initiated, a large beach existed at the southern tip of Tybee and this area was not included in the project. Six months after completion of the beach nourishment, the south end began to experience a steady increase in the erosion rate. By the summer of 1978, the southern 2,100 feet of the project had lost 62 percent of the sand that was initially placed. The mean high water beach width had been reduced approximately 100 feet, and in some areas the water was against the seawall during high tides. The large sand dunes that existed on the ocean side of the seawall just south of the project had eroded away along with approximately 1,200 linear feet of beach. During periods of spring high tides, water splashing over the seawall on the south end eroded the face of the dunes behind the seawall. On 16-17 October 1978, the U.S. Army Corps of Engineers, Savannah District, a contract consultant, and personnel from the Coastal Engineering Research Center and the Georgia State Department of Natural Resources conducted a field study at the south end of Tybee to determine the causes of this erosion. Detailed surveys were made to define the hydrography in the shoal area. Page 10 shows the location and relationship of three well-defined channels within the inlet to Tybee Creek. A velocity meter was placed in each channel and three across Tybee Creek, and velocity measurements were made at approximately 30-minute intervals. Measurements were taken near the surface, at mid-depth and near the bottom when there was sufficient water depth. Otherwise velocity readings were taken near the bottom and surface. Measurements were taken over one complete tidal cycle beginning at 1440 on 16 October and ending at 0330 on 17 October. Velocities and cross-sectional areas of the channels that flow around the sandbars were used to compute the volume of water that flows into and out of Tybee Creek. During maximum flood tide, float and dye release studies were conducted. Dye packets were released at the same locations as the velocity measurements were taken and aerial photographs of the dye trails were made. This helped determine surface current directions and define the channels. Float studies involved plotting the path of jugs suspended just below the water surface at a point immediately seaward of the breaker zone.

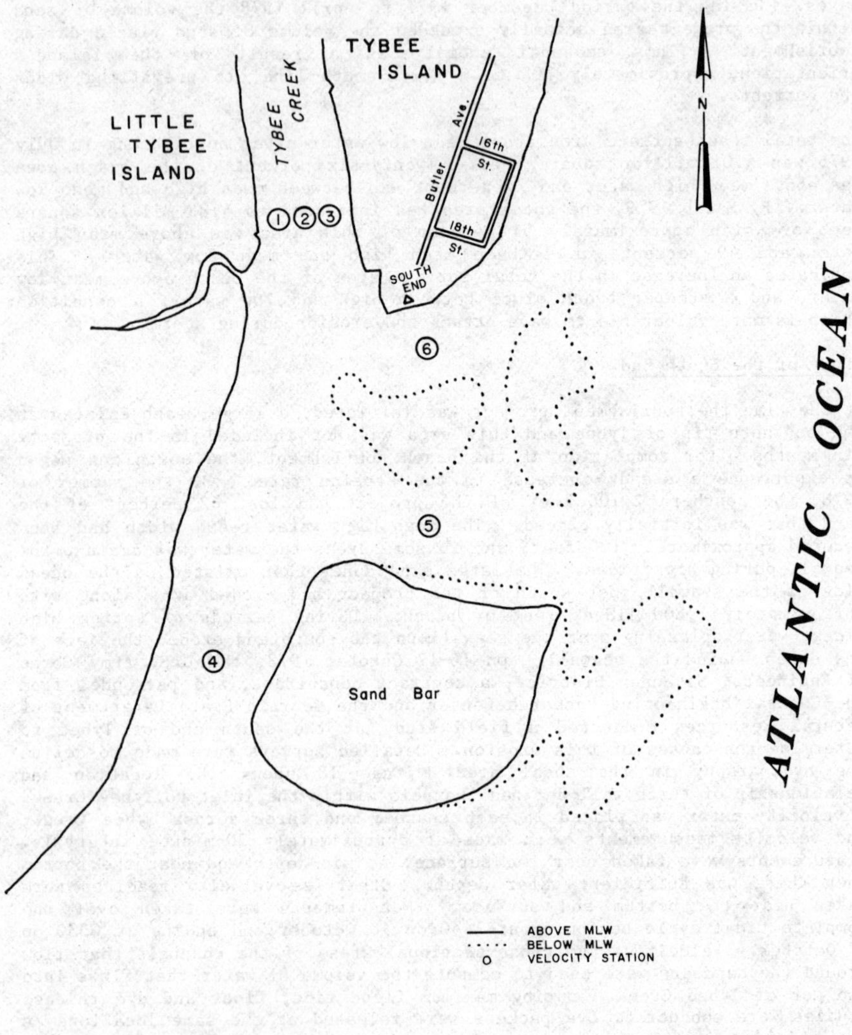
TYBEE ISLAND
TYBEE CREEK
LITTLE TYBEE ISLAND
Butler Ave.
16th St.
18th St.
SOUTH END
N
ATLANTIC OCEAN
Sand Bar
ABOVE MLW
BELOW MLW
VELOCITY STATION

Results of the Study.

Results from the study indicate that three channels flow into Tybee Creek with some water passing over the sandbars during the higher stages of the tide. As shown on page 12, one channel is within 600 feet of the seawall on the south end, another is between Little Tybee Island and the large sandbar 3,400 feet offshore, and the third channel is between this bar and the submerged shoals off the southern end of Tybee. As the tide begins to flood, the flows are basically confined to the three channels flowing into the inlet. During the first 4 hours of the flood tide, most of the water flows from the ocean through the middle channel and into Tybee Creek. At this point, approximately equal volumes of water flow through each channel and a small percentage flows over the shoals. During the last hour of the flood tide, 59 percent of the water is flowing through the channel which is parallel to Little Tybee Island. As indicated on page 13, during the maximum ebbtide, which occurs 2 hours after high tide, approximately 7 percent of the flood volume flows out the channel nearest to the beach, and about 46 percent of the volume flows through each of the other two channels.

As the water flows along the beachfront in the nearshore zone, it carries sand in suspension. On the flood tide, this sand is carried in the channel nearest the southern tip of the island and into Tybee Creek. As the tide begins to ebb, the suspended sand then flows out of the other two channels and bypasses the channel flowing along the beach. Some of the sand settles out in the inlet area resulting in the sandbars, and some is placed back into the littoral drift system and flows around Little Tybee Island. The floats released beyond the breaker zone verified that a flood-dominated channel exists relatively close to the beach. Historical photographs indicate that this channel will migrate into and away from the shoreline. This in turn will determine the size and shape of the beach of the south end of Tybee.

Conclusions and Recommended Renourishment.

The initial beach nourishment and groin construction on Tybee Island functioned well over the last 4 years. Although the erosion rate was underpredicted, the beach has remained relatively stable even though the southern end of the project has virtually no recreational beach at high tide. As of March 1979, 150,000 cubic yards of sand had eroded from the southernmost end of the project plus 1,200 linear feet of beachfront south of the project end which averaged 300 feet wide. This half mile of beach, which historically was most used, is no longer available for recreational activities at high tide. The flood channel nearest the beach is continually eroding sand at a rate faster than it can be naturally replaced. To reduce this erosion, the next scheduled renourishment plan will include the construction of a weir groin perpendicular to the seawall at the south end of the project. The groin will close off the flood channel forcing and redirecting the currents to flow over the shoals southeast of the beach. This should permanently reduce the erosion in this area. The precast concrete sheet pile groin would be 850 feet long and slope from elevation +10.0 m.l.w. to +1.0 m.l.w. A weir, 75 feet long at elevation +3.4 m.l.w., would permit passage of some sand beyond the groin and prevent starvation of the beach on the down drift side. A sand fillet

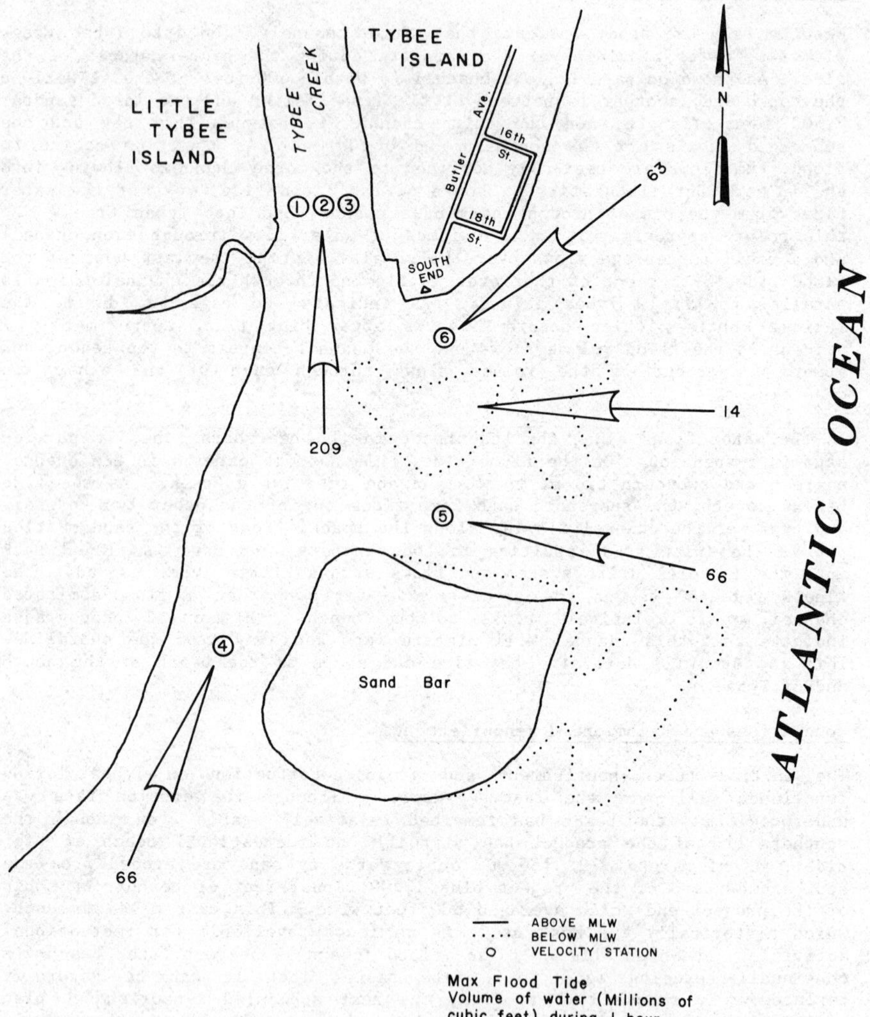

Max Flood Tide
Volume of water (Millions of cubic feet) during 1 hour period occurring 4 hours after low tide.

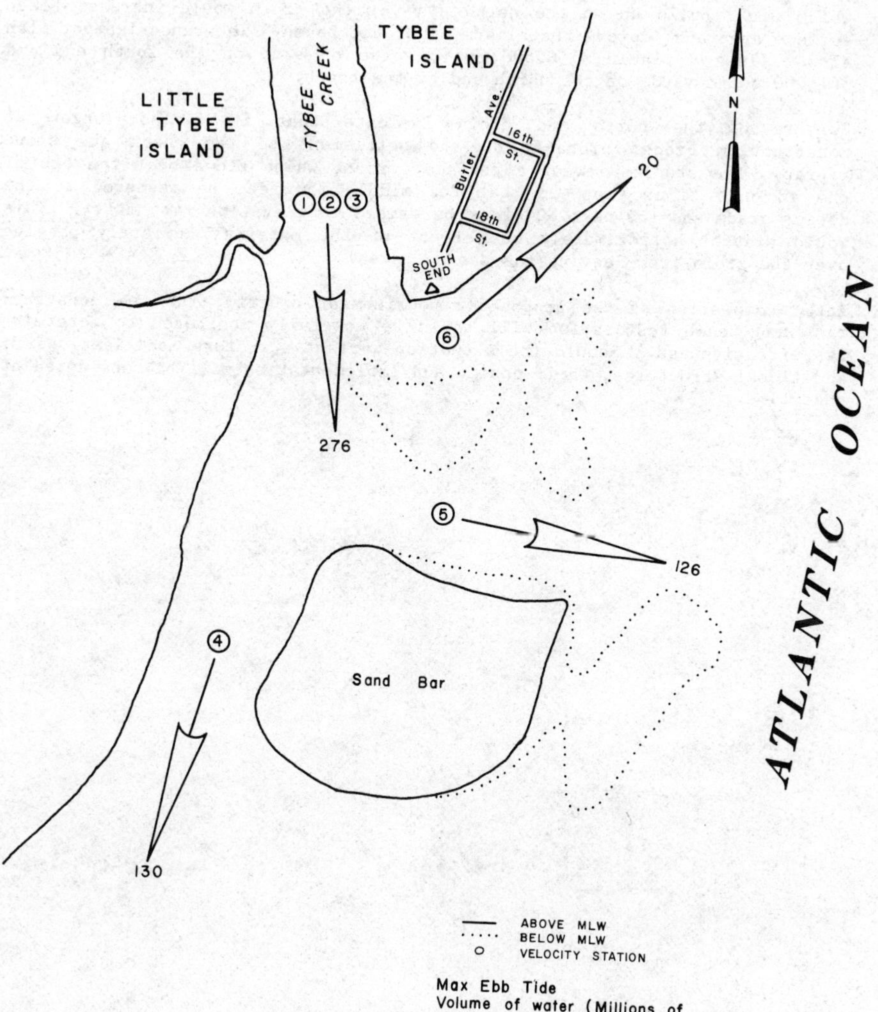
TYBEE ISLAND
LITTLE TYBEE ISLAND
TYBEE CREEK
Butler Ave.
16th St.
18th St.
SOUTH END
N
20
1 2 3
6
276
5
126
4
Sand Bar
130
ATLANTIC OCEAN
ABOVE MLW
BELOW MLW
VELOCITY STATION
Max Ebb Tide
Volume of water (Millions of cubic feet) during 1 hour period occurring 2 hours after high tide.

which would build up on the north side of the groin would increase beach surface area and improve the alignment of the beach. The renourishment plan also calls for placing 180,500 cubic yards of sand on the south end and 361,000 cubic yards on the north end of the beach.

Surveys of the north end groin indicate that it is no longer at construction grade; probably due to settlement of the beach and stone riprap. Dye studies reveal a surface current which flows over the top of the groin. It is proposed that the middle 150 feet be restored to the design grade and the next 300 feet be raised to elevation +4.5 m.l.w. This would prevent approximately 60 percent of the material currently passing over the groin from leaving the project area.

After completion of the proposed renourishment program, the new construction and beach restoration will again be carefully monitered to determine its effectiveness. Should the situation warrant structure modifications or additional structures, their design and implementation will be considered at that time.

MANAGING LOCALIZED EROSION

OF COASTAL BLUFFS

H. Bokuniewicz and J. Tanski
Marine Sciences Research Center
State University of New York
Stony Brook, New York 11794

ABSTRACT

High bluffs of unconsolidated sand form large sections of the shore in formerly glaciated areas, such as New York, Washington, and around the Great Lakes. The erosion of this valuable coastal property is episodic and localized, and presents difficult management problems. The characteristics of bluff erosion were studied along an 80-kilometer length of the eastern north-shore of Long Island, New York. The line of bluffs is nearly unbroken along this section. The bluffs are composed of unconsolidated sand and they are 50 meters high in some places. About 75% of the bluff face is vegetated (covered) and not actively eroding at this time; 25% is stripped of its vegetation (exposed) and receding at rates as high as 2 meters/year. The transitions between the exposed areas and the covered sections are sharp. The exposed sections are typically only 100 meters wide although the widest continuously exposed area is 1,000 meters wide. The actively eroding areas most likely began during storms. Since 1930, at least six newly exposed areas have appeared in formerly a continuous section of covered bluff face. Most of the exposed areas have persisted at least as long as several decades and, during this time, they receded rapidly. One eroding area that is 33 meters wide and 50 meters high has lost more than 4,800 cubic meters of sediment within six years of its inception. The exposed areas also spread laterally and may coalesce before they are revegetated and become stable. Long-term management of such a shoreline should include a method for quickly identifying new erosional features. A good set of oblique or horizontal photographs of the bluff-face is needed and aerial inspections should be made regularly especially after major storms. Areas of incipient erosion are likely to be small and, therefore, more easily stabilized by bulkheading, terracing, and replanting. (This research was supported by the New York Sea Grant Institute.)

INTRODUCTION

High bluffs of unconsolidated sand form large sections of the shore in formerly glaciated areas such as in New York, Washington, and around the Great Lakes. The collapse and erosion of bluffs present serious managment problems. Every year the erosion causes millions of dollars of damage to property, structures, roads, and other facilities. On the north shore of Long Island, New York, erosional losses are estimated at over five million dollars annually (Koppelman and others, 1976).

The causes of erosion are complicated and extremely variable (Gray and Wilkinson, 1979). The contributing factors include undercutting of the bluff by waves or tidal currents, ground-water seepage, overland runoff, too much or too little vegetation, as well as the composition and geometry of the bluff itself. As a result, specific problems in bluff erosion cannot be adequately answered in general terms and require special studies at the site.

Where detailed studies are lacking, managers are not able to predict future erosion and must instead rely on direct measurements of the long-term recession rates. The long-term recession rates may be calculated from comparison of sequential sets of aerial photographs and charts. Once these values have been determined, management decisions are made on the assumption that the erosion rate in the near future will be the same as the average rate in the past. The average recession rates can be misleading, however, because severe erosion does not occur gradually or uniformly. Most of the erosion is the result of local catastrophic events that are often associated with storms. The purpose of our paper is to describe the size, distribution, and lifetime of erosional features along a bluffed coast and to discuss how the presence of small, erosional features might affect management decisions.

THE STUDY AREA

The study area was an 80-kilometer section of the eastern portion of the north shore of Long Island, New York (Fig. 1). This stretch of coastline is particularly suitable for our study because the beach is backed by an almost continuous line of bluffs; long stretches of gently curved beaches separated by subdued headlands front the bluffs. Beach widths range from less than 3 meters to almost 100 meters, but typically the beaches are near the narrower rather than wider end of this range. There is rarely a dune line behind the beach and in most cases the beach ends at the toe of the bluff. The toe of the bluff is typically 1.8 meters above mean high water but it may be as much as 2.9 meters or as little as 0.3 meters in elevation (U.S. Army Corps of

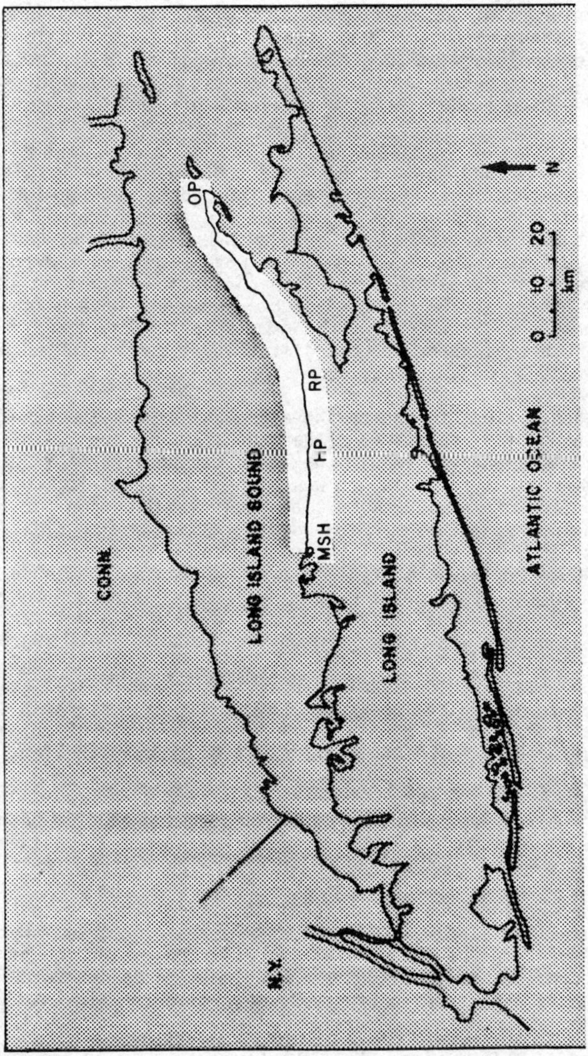

Fig. 1 Index map of Long Island area. Unshaded portion is the study area. MSH - Mount Sinai Harbor, HP - Herod Point, RP - Roanoke Point, OP - Orient Point.

Engineers, 1969). The bluffs are higher along the western end of the section and decrease in height to the east. The bluffs are 30 to 50 meters high between Mount Sinai Harbor and Herod Point (Fig. 1) and decline to less than 3 meters near Orient Point. They are composed of a heterogeneous mixture of unconsolidated glacial drift. The predominant component is sand which comprises about 70% of the bluff material. The composition of the bluff becomes slightly more silty to the east. A terminal moraine lies on top of the bluff in most places and large boulders that have been eroded from the moraine are found in many places along the shore.

Long Island Sound borders the study area to the north. The narrow eastern passage of the Sound protects the shoreline from large ocean waves. Waves that impinge on the study area are generated by local winds and the limited fetch in the Sound prevents the buildup of large waves. Under normal conditions, the waves striking the coast are generally steep, short-period waves with an average height of 0.5 meters (Wise and others, 1978). During storms, however, wave heights may be about one meter, although the U.S. Army Corps of Engineers reported waves 10 meters high on the Connecticut coast during the hurricane of September 1938 (Ellis, 1962).

The beaches along the north shore are not always adequate protection for the bluffs against even the usual, small waves. This is because changes in the water level in the Sound may be several meters. The tidal range increases from east to west; the mean range is 1.04 meters near Orient Point and 1.95 meters at Mount Sinai Harbor (U.S. Army Corps of Engineers, 1969). The range at times of spring tides may be 15% to 25% greater. In addition, storms passing over southern New England may cause water levels to rise by over a meter at the shore (Miller, 1958). In the winter, the water level can be expected to rise 0.5 meters above normal several times a month and to rise 1.0 meter above nromal about once a year (Bokuniewicz and Gordon, 1980). During the hurricane of 1938, tides ran 3 meters above normal (Ellis, 1962). As a result of the tidal and nontidal changes in the water level, waves might be expected to attach the toe of the bluff directly at least once or twice a year. Undercutting of the bluff and the subsequent collapse of the bluff face appears to be the major cause of erosion (Fuller, 1914; Wise and others, 1978).

There are few previous studies of the erosion problem on the north shore of Long Island. In 1969, the U.S. Army Corps of Engineers conducted a beach erosion control and interim hurricane study for this area. The report gives a detailed description of shoreline conditions and describes the erosion problems at specific areas. In an extension of this work Davies and others (1972) not only gives a good summary of the physical features and of the general processes affecting the north shore, but also provides estimates of the long-term recession rates. Comparison of charts and aerial photographs were used to calculate the average recession rate over an 80-year period at 158 locations along the north shore. As you might expect the rates were variable; they ranged from an erosion rate of 1.8 meters/year to an accretion rate (for the beach) of about 0.6 meters/year. The shoreline was found to be generally receding at a rate of about 0.5 meters/year. Most of the erosion is associated with storms (Wise and others, 1978) and the storm-induced erosion is catastrophic and localized, although it may be widespread. A single storm frequently cuts a meter or more into the base of the bluffs (Fuller, 1914). As an example, the hurricane of 14 September 1944 cut back the bluffs at one location a horizontal distance of over 12 meters while creating a vertical cliff 3.3 meters high (from the report of the New York Joint Legislative Committee 1947 as reported in Davies and others, 1972).

The bluff face is divided into two types of sections. Long stretches of the face are thickly vegetated and appear dark from a distance (Fig. 2). The vegetated sections are separated by sections that have been stripped of vegetation. These exposed sections appear light and the transition between the vegetated and exposed areas of the face are very sharp. In many places you can cross from a thickly vegetated area to an exposed face in one step. The smaller, exposed sections of the face are typically parabolic in outline with the closed end of the parabola pointing up. At the base of some of the light areas, the remnants of a slump block can be seen (Fig. 3). The hypothesis is that these features are formed by undercutting and collapse of a vegetated bluff face. The sediment and vegetation that reach the beach are quickly removed by waves. The denuded bluff face is then subject to runoff erosion and ablation. Where large sections of the bluff face are eroding the bluff appears to be receding uniformly. At the base of some of these features, sand can be seen cascading to the beach and the top edge of the receding bluff face is often notched with hanging valleys. These were first described by Fuller (1914) in his classic description of Long Island's geology. He found that "examples of hanging valleys are usually very numerous, especially within the first year or two after a severe northerly storm ..." Where the erosion is local, deep notches or amphitheaters are cut back into the bluff face; there are no comparable features inland. These were also noted by Fuller (1914). He describes the formation of amphitheaters with the undercutting of the toe of a bluff by rapid, ground-water seepage over local, impervious strata. Notches, he claims, are produced by "an agency working from above instead of from beneath ..." By this he probably means erosion by overland runoff and, with our hypothesis, a notch would be started after collapse of a section of the bluff face had removed the protective vegetation. Some of these features are unvegetated, and deltas of freshly removed sediment may be found at their base. Others have been revegetated and apparently stabilized. As we shall discuss later, it appears that the exposed sections of the bluff are receding at many times the rate of the vegetated sections. The fact that the shoreline remains relatively straight in the face of such intense, localized erosion means that the features have a relatively short lifetime.

Fig. 2 Typical section of the bluff face showing both exposed and vegetated areas.

Fig. 3 Section of bluff face showing a slump block (in the center of the photograph).

Eventually areas that are now actively eroding will be cut back, vegetated, and stabilized while formerly vegetated sections are undercut, denuded, and start as new sites of erosion.

Because of the sharp contrast between the vegetated and exposed faces of the bluff, these features are easily mapped. To study the size, distribution, and lifetime of erosional features, we examined and compared sets of aerial photographs and oblique photographs. In addition, we mapped several small features in the field. Aerial photographs were available from 1930 and 1976 at a scale of 1:7,200 and 1:10,000 respectively and the oblique photographs were taken in 1978 at a scale of 1:3,000. Examination of the oblique photographs shows that about 22% of the bluff face between Mount Sinai Harbor and Orient Point is exposed. This percentage is the fraction of a line along the bluff at mid-height that passes through exposed sections. The same results were obtained from the aerial photographs from 1976. In 1976, 24% of the bluff face was exposed. The difference between this value and the percentage measured from the oblique photographs could easily be due to distortion on the prints. From the 1930 photographs about 26% of the bluff face was found to be exposed.

The exposure is not uniform. If we look at 5-kilometer sections of the shoreline, some sections are found to have as much as 55% of their face exposed while others have less than 5% (Fig. 4). The percentage of the bluff exposed in these sections shows a distribution similar to that of the 80-year average recession rate as determined by Davies and others (1972). In general, those areas that have been measured as having higher recession rates also have a greater fraction of the bluff face exposed. This seems reasonable given our hypothesis for the recession. In sections where more of the bluff face is exposed, a greater percentage of the shoreline is rapidly receding so that on the average the shoreline is rapidly receding faster than in areas where most of the shoreline is stable.

In an attempt to quantify this relationship we performed a linear correlation between the percentage of bluff face exposed per 5-kilometer section and the long-term recession rate. The recession rates were recalculated for the same intervals as those used to describe the distribution of exposed surfaces. Based on the exposure measured from the 1976 aerial photographs the relation is approximately given by

$$R = 0.7\, L_e + 0.1$$

where R is the recession rate in meters/year and L_e is the linear fraction of the bluff face exposed. The correlation coefficient is 0.6. The correlation between average recession rate and fraction of bluff face exposed must be viewed cautiously, however, due to the fact that the recession rates were averaged over an 80-year time span while the percentage of the bluff face exposed was measured from aerial photographs which represent a single point in time rather than an average

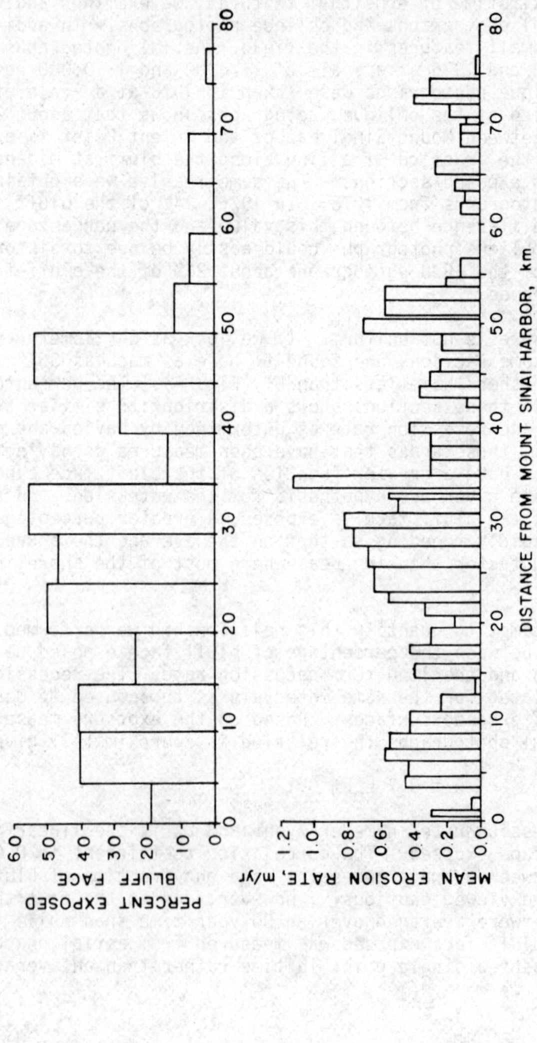

Fig. 4 Distribution of the percentage of exposed bluff face by 5-kilometer sections of the shoreline and recession rates (from Davies et al.) for the same area.

over time. An area that was relatively stable for most of the 80-year period of record but that experienced increased recession (and, in turn, increased bluff erosion) in the latter few years of the interval could have a relatively low average annual recession rate for the 80 years while exhibiting a high percentage of exposed bluff face. Even a single catastrophic event such as a storm could increase the amount of exposure while having little effect on the long-term average recession rate. The reverse could also occur. A long period of intense erosion followed by stable conditions that allowed the bluff face to revegetate could result in high average recession rates with a low percentage of exposed bluff face. Disparities in Figure 4 could be due to these differences.

Most of the exposed features are small (Fig. 5). About 21% of the total exposed length is composed of discrete features that are less than 100 meters wide and 60% of the length is composed of features less than 500 meters wide. The erosion rate of small features is rapid. One small amphitheater found near Roanoke Point was measured in the field to have a volume of 4,800 cubic meters. By searching for this feature in earlier sets of aerial photographs we found that the erosion began between 1970 and 1974, that is to say, the feature was present in 1974, but not in 1970. Residents of the area recall that the feature appeared in 1972. In six years 4,800 cubic meters have been removed from this feature. At the time the measurements were made, sand could be seen flowing down the axis of the amphitheater in a thick slurry feed by ground-water seepage through the walls. A delta of sand several meters high had formed at its base during the summer of 1979. Within a period of a week, however, most of the delta was removed by winter storm waves. The amphitheater was 33 meters wide and 50 meters high; so the average recession rate has been about 0.5 meters/year.

As the birth of this newly eroding feature shows, the distribution of exposed bluff face is not unchanging. A comparison of the 1976 aerial photographs with the 1930 aerial photographs showed, however, that the total percentage of exposed bluff face did not change significantly between the two years. In 1930, 26% of the bluff was exposed while 24% was exposed in 1976. Lateral variations in bluff erosion were also very similar for the two years (Fig. 6). The percentage of the bluff face exposed was relatively low in the eastern portion of the study areas and higher in the central and western section in 1930 as well as 1976.

Although the total percentage of the bluff face exposed was relatively constant for the two years, there were some definite changes in the distribution of features (Fig. 6). The largest changes were in the section between 30 kilometers and 35 kilometers from Orient Point and the section between 45 kilometers and 50 kilometers from the Point. In both of these areas the portion of the bluff face that was exposed increased by approximately 40% during the 46-year period. Almost half of the exposed surfaces in these areas are less than 46 years old. Most of the differences between the distribution of features in 1976

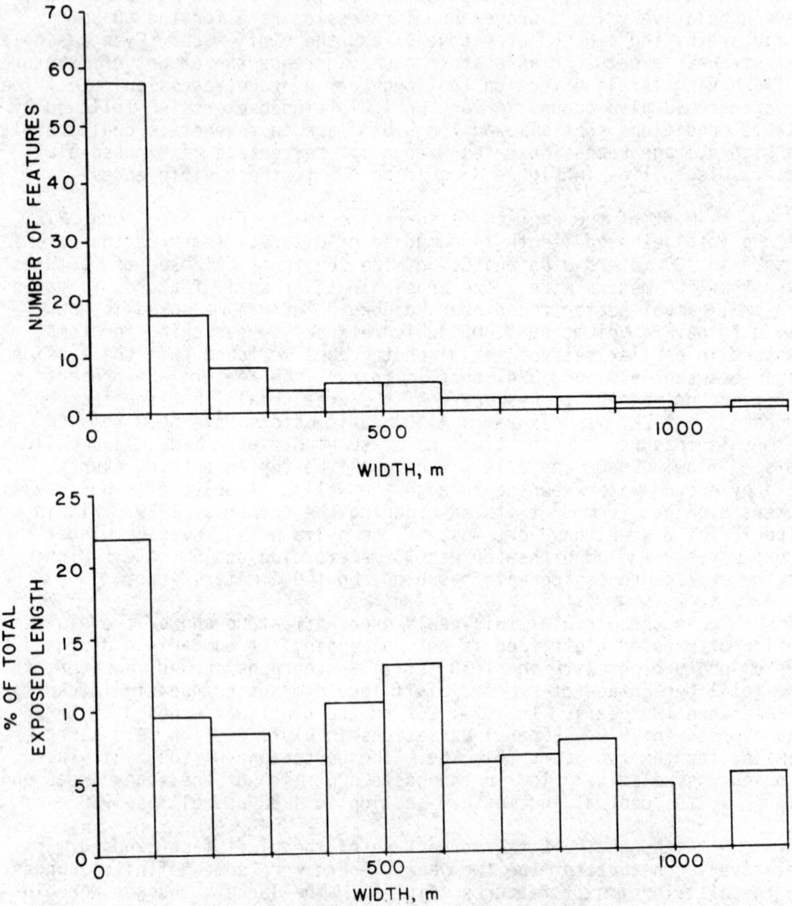

Fig. 5 Size distribution of erosional features and contribution of each size class to the total percentage of exposed bluff face.

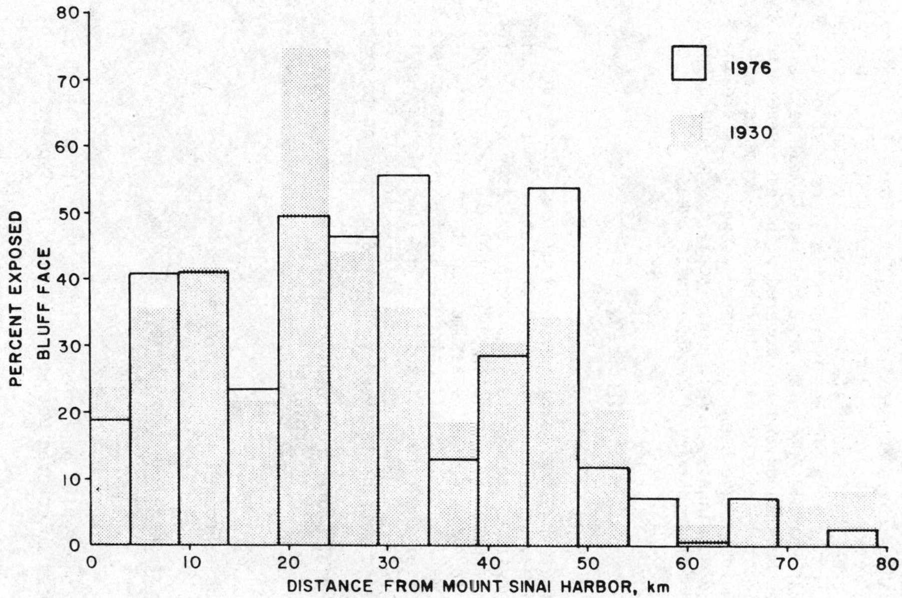

Fig. 6 Variations in the distribution of the percentage of exposed bluff face between 1930 and 1976.

and that in 1930 were due to a widening or narrowing of existing exposed areas and the coalescence of several small features into one larger one. Some of the differences however, are attributable to the appearance of newly exposed areas in sections of the bluff that had previously been heavily vegetated. At least six new features, such as the Roanoke Point site described above, have appeared since 1930. These features range in size from 30 meters to over 200 meters. During this same period, however, other areas had become stable and were revegetated, either naturally or with the help of man. Three kilometers west of the erosional feature at Roanoke Point, an area that was completely exposed in 1930 had been bulkheaded and was covered with dense vegetation by 1976. Since almost 15 kilometers of bulkheads, revetments, and other protective structures have been constructed along this shoreline (Bartholomew and McGuinness, 1972), man's activity has undoubtedly had an effect on the distribution of these exposed areas.

DISCUSSION

The fact that the total length of exposed bluff face has remained relatively constant while the distribution of the eroding areas has changed suggests a steady-state condition exists where the rate of transition from exposed to vegetated must equal the rate of transition from vegetated to exposed. If we assume that this steady state does exist and that man's activity affecting the bluffs remains constant, a crude estimate of the rate of transition and the lifetime of these exposed features can be calculated. The rate of transition for a change in either direction (exposed to covered or covered to exposed) is:

$$\Delta = \frac{\text{fraction of shoreline trasnformed from exposed to covered}}{\text{time}}$$

This can also be expressed as:

$$\Delta = \frac{\text{meters of shoreline transformed}}{\text{kilometer of shoreline x time}}$$

Examination of the aerial photographs showed that 11% of the bluff face went from exposed to covered between 1930 and 1976, so the rate of transition is:

$$\Delta = .11/46 \text{ years,}$$

or,

$$\Delta = \frac{110 \text{ meters}}{\text{kilometer x 46 years}} = 2.4 \text{ meters/kilometer/year.}$$

Because approximately 25% of the shoreline is exposed, we can assume that a single kilometer of the shore in likely to have a feature 250 meters wide. The expected lifetime of that feature T is then:

$$T = \frac{250 \text{ meters/kilometer}}{\Delta},$$

or,

$$T = \frac{250 \text{ meters/kilometer}}{2.4 \text{ meters/kilometer/year}} = 104 \text{ years.}$$

Furthermore, we might expect that the rate of recession of the exposed sections would be about four times higher than the long-term average recession rate for the whole shoreline, because only about one-quarter of the bluff face is exposed at any one time. The data are not adequate, however, to test these conclusions.

MANAGEMENT IMPLICATIONS

In a long-term, regional management strategy the importance of small features should be recognized. As we have shown, most areas of rapid erosion are small features and, when new erosional features appear, they are likely to appear with widths less than a few hundred meters wide. If bluff erosion is to be stopped, effort could most effectively be directed at bulkheading, grading, and revegetating small features. Smaller projects have a greater likelihood of being successful than larger projects and, as we have seen, these small features comprise a substantial portion of the bluff erosion.

A management plan might begin by cataloging the exposed sections on a linear coordinate system. This survey should be updated periodically, especially after winter storms, in order to locate and identify new sites of erosion. Annual, aerial photographic surveys could be utilized for this purpose.

Because the erosion of the bluff supplies sand and gravel to maintain the shore's beaches, management strategies should not be aimed at stabilizing the entire bluff face. Ziegler and others (1964) determined that since Cape Cod beaches receive roughly 50% of their sediment from bluff erosion, complete stabilization of the bluffs would make them disappear in 86 years. A similar situation exists on Long Island. On the north shore the fraction of the bluff face that is exposed is thought to be contributing almost all of the supply of sand and gravel to the beach. In other words, to maintain the status quo about 25% of the shoreline should be allowed to erode, so that on a regional scale controlled erosion should be considered. Erosion may be retarded in some areas but allowed to continue, or even start, in other suitable areas.

In order to define the erosional risk of a particular stretch of the shoreline we would like to know the probability that an erosional feature will appear there, the rate at which it should be expected to erode, and its expected lifetime. These statistics are akin to the information concerning river discharge which are used to calculate a recurrence interval for floods of different magnitudes. If these statistics were available for a coastline, the management of bluff erosion could be approached by techniques similar to those used in flood control. Unfortunately, because the lifetime of the features is longer than the period of recorded changes, data are not on hand to calculate the necessary parameters in more than the general way we have done here. A new meaning might be assigned to the long-term average recession rate, however, because of the correspondence between it and the fraction of bluff face exposed. In terms of transient, erosional features, the recession rate gives a relative measure of the probability of occurrence of exposed sections of the shoreline. The long-term recession rate is still one of the most valuable parameters in managing coastal erosion.

ACKNOWLEDGEMENTS

We are indebted to the many residents of the Long Island shore who contacted us to discuss their erosion problems, guided us on surveys of their property, and supplied us with photographs, news clippings, and other valuable information during our work. We would also like to thank DeWitt Davies of the Long Island Regional Planning Board for helping us obtain access to the aerial photographs. We are also grateful for the cooperation of Captain Chris Stuebe of the R/V ONRUST and the help of Marjorie Sumner and Mary Jane Hamilton, who typed the manuscript. This work was supported by the New York Sea Grant Institute. Contribution 281 of the Marine Sciences Research Center of the State University of New York at Stony Brook, New York.

REFERENCES

Bartholomew, F.L. and W.V. McGuinness, Jr. 1972. Coast stabilization and protection on Long Island. Center for the Environment and Man, Inc., Hartford, CT. 47 p.

Bokuniewicz, H.J. and R.B. Gordon. 1980. Storm and tidal energy in Long Island Sound. Adv. in Geophys. 22: 41-67.

Davies, D.S., E.W. Axelrod, and J.S. O'Connor. 1872. Erosion of the north shore of Long Island. Tech. Rept. 18, Marine Sciences Research Center. State Univ. of New York at Stony Brook. 101 p.

Ellis, C.W. 1962. Marine sedimentary environments in the vicinity of the Norwalk Islands, Connecticut. Conn. State Geol. Nat. Hist. Sur. Bull. 94.

Fuller, M.L. 1914. Geology of Long Island, New York. U.S. Geol. Survey Prof. Paper 82. 231 p.

Gray, H.G. and B.H. Wilkinson. 1979. Influence of nearshore till lithology on lateral variations in coastline recession rates along southeastern Lake Michigan. Internat. Assoc. Great Lakes Res. 5(1): 78-83.

Koppelman, L.E., P.K. Weyl, M.G. Gross, and D.S. Davies. 1976. The Urban Sea: Long Island Sound. Praeger Publishers. New York. 223 p.

Miller, A.R. 1958. The effects of winds on water levels on the New England coast. Limnol. Oceanogr. 3: 1-14.

U.S. Army Corps of Engineers. 1969. North shore of Long Island, Suffolk County, New York, beach erosion control and interim hurricane study. New York District, New York. 271 p.

Wise, W.M., P. Sanko, and J.R. Schubel. 1978. Erosion and erosion control on the north shore of Long Island. Workshop on Erosion in Bays, Sounds, and Estuaries. Sea Grant Marine Advisory Services. 22 p.

Ziegler, J.M., S. Tuttle, G. Giese, and H. Tasha. 1964. Residence time of sand composing the beaches and bars of outer Cape Cod. Amer. Soc. Civil Engineers, Proc. 9th Conf. on Coastal Engineering: 403-416.

COASTAL EROSION IN SAN DIEGO COUNTY, CALIFORNIA

Gerald G. Kuhn* and Francis P. Shepard**

Introduction

During the past two decades there has been a tremendous boom of land development along the cliffs at the seaward margins of the coastal terraces of San Diego, particularly to the north of La Jolla. Until enactment of the California Coastal Zone Conservation initiative in late 1972, local agencies frequently permitted construction of single-family residences and multiple-unit residential structures within only a few feet of bluffs and on the beaches. Developers justified this practice by stating that none of these sea cliffs were retreating at an appreciable rate. What was overlooked was that the cited low retreat rates were usually based on only the experience of the last 25 or 30 years, an unusually benign and quiescent time, characterized by low rainfall and few local storms capable of producing heavy surf. Earlier studies of coastal erosion in the area which provided a far less optimistic picture of the stability of sea cliffs were ignored or discounted.

In 1889, during a wet, stormy period which lasted from 1884 to approximately 1893, the United States Coast and Geodetic Survey (U.S.C.G.S.) was conducting topographic and bathymetric surveys along the coast of San Diego County. The U.S.C.G.S. (1889) topographic notes indicate the bluffs showed ". . . new erosion during each winter storm as the characteristic feature of this coast." Additionally, Hanna (1926, p. 233) noted that the cliffs north of La Jolla were actively retreating during the wet and stormy years of the 1920's. Vaughan (1932) also reported that alluvial bluffs near Scripps Institution of Oceanography in La Jolla retreated 10 to 20 feet between 1923 and 1930. Shepard and Grant (1947) found erosion of the same area occurring at a rate of about 1 ft/yr during the stormy periods that preceded 1947.

From various sources summarized in Shepard and Wanless (1971, p. 265), there are indications that cliffs of Eocene and Cretaceous slightly dipping strata have retreated episodically due to large rock falls, although the rock cliffs in many places showed no indication of having had appreciable retreat since photographic records began some half century earlier.

The stormy period with unusually high rainfall that began in early 1978 has considerably changed the picture, especially on the beaches

---

*Geologist - Research Associate, Geological Research Division, A-015, Scripps Institution of Oceanography, Univ. of California, San Diego, La Jolla, California 92093

**Professor Emeritus of Submarine Geology, Geological Research Division, A-015, Scripps Institution of Oceanography, Univ. of California, San Diego, La Jolla, California 92093

and along the bluffs north of Del Mar (Figure 1), extending to Oceanside where much building has been done on the low terrace close to the bluff edge in recent years. Also, the alluvial cliffs near Scripps Institution in La Jolla, which had appeared to stabilize after 1947, became active areas of erosion once again during this same period.

Beginning in 1973 Kuhn (1977) has made extensive and continuing studies of the coastal erosion in the area from La Jolla to San Onofre State Park, including repeated measurements and photographic coverage of critical points. Kuhn has also investigated the erosion history through a search of old maps and land ownership records, and interviews with long-time residents whose families have inhabited these coastal areas for long periods of time. Some of these personal stories of erosion have been verified by examination of old newspaper files, historical maps and old weather bureau reports of storms reported by the residents.

Erosion of the sea cliffs, bluffs and canyon heads is episodic, site specific, directly related to prevailing meteorological conditions and, in recent years, to man's altering of natural drainage patterns.

Historical Aspect

In evaluating the danger of building on the beaches and near the bluff margins, we should not overlook evidence of erosion during the entire century or more that the coast has been occupied by California residents. Examination of old maps, road, railroad and topographical surveys dating back to 1876 indicates that the beaches and the bluffs were markedly different than those of the present day. Perusal of these maps has shown that in the area north of San Diego city limits there are tax assessor records of property that has since ceased to exist. In 1883 Encinitas was the only surveyed town in San Diego County between San Juan Capistrano in the north and the city of San Diego in the south. An extremely stormy wet period began in 1884, following the eruption in August 1883 of the volcano Krakatoa; this wet period lasted until 1893 (Ganus, 1976). During this period entire surveyed blocks of real estate were annually devalued and disappeared at Encinitas. As the seaward block property value decreased, the land parcels directly inland increased. The seaward blocks decreased in value until they were removed from the tax rolls by 1895. Streets were then legally closed by the County Board of Supervisors after the turn of the century.

Since 1884, the Santa Fe Railroad maps and records indicate that tracks have had to be relocated on repeated occasions at several sites due to collapse of the bluffs along the coastline as well as along the river valleys. In particular, the Sorrento valley between Del Mar and Torrey Pines Mesa was hard-hit during February and March of 1884. The San Diego Union of March 2, 1884 states that at that time all telegraph lines were down, the railroad through that area was completely impassable, and that two trains were ". . . hemmed in on one side by landslides in the big Soledad cut, and on the other by the unsafe condition of the track, which is mostly underwater through the whole of Soledad flats." This same valley presently contains not only the same railroad

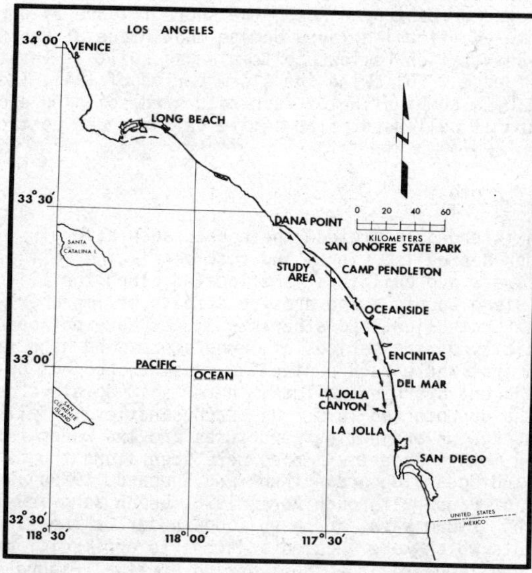

Figure 1. Location map. Arrows indicate predominant longshore sand transport current direction.

line, but portions of major freeways and an extensive industrial development, all in the flood-plain area.

A particularly bad place is in the southern portion of Del Mar where trains have been derailed on various occasions and even fallen over the cliffs. During an extremely stormy period in the winter of 1940-41, the tracks were undermined and a freight train went off the cliff on New Year's Eve, 1940; the cause was a series of heavy rains, large waves and unusually high tides (Kuhn and Shepard, 1979a). Beach cottages were damaged or destroyed along the shore (Figure 2) during the storm, and some additional delayed damage took place at Encinitas. In 1938 the Self-Realization Fellowship temple was built at least 30 feet from the sea-bluff edge. Following the storm period of 1941, the bluff had been weakened by a combination of rain-saturated soil and erosion by wave action, and it collapsed. The temple building was destroyed (Figure 3).

Artificial Coastal Protection

Construction intended to protect the coast, such as groins, jetties, concrete and wooden sea walls, riprap and rock debris, as well as concrete lifeguard towers and buildings, are located along the cliffed shoreline of San Diego County. The erosive effects of improperly located artificial protection, and other structures have not been fully understood, especially during periods of large storms and high tides. It has been observed that where such protective measures project or extend seaward beyond adjacent unprotected lots, there is immediate erosion and notching of the unprotected lots. As beach sand levels fall, storm waves tend to converge on projecting structures and the waves refract toward unprotected lots. Recent examples were seen along the La Jolla Shores, Carlsbad and Oceanside areas (Kuhn and Shepard, 1979b). During the storms of January through April 1978, beach sand levels dipped to a very low point, storm waves and high tides undercut the slope toe, and unprotected lots which were located adjacent to artificial projections were eroded landward in a short period of time (Figure 4). Since 1978 the sandy beach continues to decrease in width and at many locations has been replaced by extensive cobble beaches. The cobbles in turn have greatly accelerated erosion by moving over, and abrading behind existing seawalls, riprap and other structures located along the shore, even during summer months (Figures 5, 6 and 7).

Beach Condition Prior to 1978

Prior to 1900 the sandy beach served as the most heavily traveled "highway" along the California coast. During the 1880's the coastal sand beaches varied in width from 100 to 300 feet (Figure 8). The U. S. Coast and Geodetic Survey (1889) noted that " . . . from Mussel Rocks [Bathtub Rock at Torrey Pines] there is an unbroken sand beach for forty or fifty miles or as far north as the valley of San Juan Capistrano and in former times this stretch of beach was used whenever specially fast time [was] to be made on the route Los Angeles, via Capistrano, to San Diego." Wide, sandy beaches existed along this shore until the early 1940's.

Figure 2. Beach cottage was damaged at Carlsbad during a severe sea storm in 1941. Note: In recent years apartments and homes have been constructed at this site. Photo courtesy of Shore Processes Lab, Scripps Institution of Oceanography; January 5, 1941.

Figure 3. Collapse of Self Realization Fellowship temple along bluff at Encinitas following stormy period in the winter of 1940-41. Photo courtesy of Self Realization Fellowship; 1941.

Figure 4. March 1978 view of the eroded beach at Del Mar, California. The wooden seawall collapsed and the adjacent unprotected lots were eroded. Photo courtesy of Sea Grant; March 5, 1978.

Figure 5. 1980 view of a site in Carlsbad eroding even during the summer. Beach cobbles have eroded behind and around seawalls and riprap and adjacent to building along the shore. Photo courtesy of Sea Grant; July 25, 1980.

Figure 6. (Before.) Beach cobble abrasion along the shore at South Oceanside. The cobbles moved up the beach 18 to 20 ft and covered riprap seawall in places, and eroded the lawns directly behind. Photo courtesy of Sea Grant; January 1978.

Figure 7. (After.) Same view at South Oceanside in June 1978. Only remnant pockets of cobbles remain. Photo courtesy of Sea Grant; June 1978.

The sandy beaches which still exist along the cliffs north of Oceanside at San Onofre State Park and Camp Pendleton, are the result of both man-induced and natural causes (Kuhn, Baker and Campen, 1980). During the recent wet years of 1978-80, erosion has accelerated at canyon heads. Placement of storm drains and culverts beneath the rail-road bed and highways have increased erosional rates by concentrating surface water runoff along pre-existing drainage avenues. Also, land-slides and blockfalls have occurred during this period, contributing sediment to the beaches.

Today the formerly wide sandy beaches south of Oceanside have, for the most part, been replaced by beach cobbles resting on exposed bed-rock outcrops. This is the result of numerous factors, both natural and man-induced. 1) Agricultural storage dams have been constructed on the principal rivers in San Diego County, thus cutting off the sand transport to the beaches (Inman, 1976). 2) No flood of any significance has occurred on the undammed segments of the river basins during the previous forty years, and consequently very little sediment has reached the coast. 3) In 1942 the Camp Pendleton Boat Basin was constructed directly north of Oceanside and began silting up immediately following construction (Figure 9).

Erosion from January 1978 through August 1980

Beach Cobble Abrasion:

Beach cobbles found along the shore in San Diego County are derived from both the rivers in flood and also directly from the bluffs at some localities (Emery, 1955, 1960, p. 184).

During the 1880's lenses of these rocks were so extensive that at one location across the mouth of Penasquitas Lagoon south of Del Mar, a "beach shingle railroad siding" was installed by the Santa Fe Railroad to mine the beach stones for street-paving material in the city of San Diego. During the early part of this century two other sites in northern San Diego County were "mined" to obtain abrasives.

In 1963 approximately 3000 x $10^3 m^3$ of sedimentary material (clay, sand and cobbles) was dredged from the Oceanside Harbor and placed on the beaches to the south. During the destructive winter storms of 1977-78 the buffering sand beach was rapidly removed at Oceanside as the result of wave action and high tides. Beach cobbles became exposed, were withdrawn by the waves and then thrown toward buildings located at the top of the former beach. Again in the winter of 1980 the cobbles were a destructive element, this time contributing to the undermining and collapse of the beach road, and damage to apartments and homes along the shore (Figures 10 and 11).

Another occurrence in 1978 and again in 1980 was that sea caves and weak shear zones and loosely-cemented bedding planes were seen to enlarge as cobbles abraded these weak sites during periods of high surf and high tides.

Figure 8. View in 1885 looking north along the eroding bluffs and the extremely wide sandy beach at Oceanside, California. Note: The Oceanside pier is presently located at the same location as the building seen in the photo. Photo courtesy of Scripps Institution of Oceanography; July 4, 1885.

Figure 9. 1980 aerial view looking north across the City of Oceanside toward the Camp Pendleton Boat Basin and Oceanside Harbor. Note that the width of the beaches markedly decreases to the south of the harbor. Photo courtesy of Sea Grant; February 23, 1980.

Figure 10. (Before.) 1978 photo of beach cottages located south of Wisconsin Street in Oceanside. Photo courtesy of Sea Grant; January 4, 1978.

Figure 11. (After.) Same view in 1980 showing wave and beach cobble abrasion which damaged houses along the shore during the storms in February 1980. Property damage occurred with only 4-6 ft. swell. Photo courtesy of Sea Grant; February 17, 1980.

Erosion Due to Increased Ground Water:

As an indication of what can be found by careful study of areas where casual study had suggested that there was little or no coastal erosion, we cite the results of work north of Del Mar, where many sections of the cliffs collapsed between 1973 and 1977 (Kuhn and Shepard, 1979b). Failure involved both landslides and block falls. For the most part, collapse of the cliffs along this section occurred as instantaneous failures, where masses separated from the cliff face along fractures, joint sets and contact planes. During the dry years from May 1973 to December 1977, at least 25 block falls occurred along the cliffs at Encinitas. Measuring from the cliff face landward, these blocks varied in dimensions from 1 to 12 feet across, 5 to 110 feet in length, and a maximum of 32 feet in height.

These failures are partially related to the undercutting of the sea cliff, but more importantly to a recent steady rise in ground water levels along the sea cliffs. This rise is occurring commensurate with urbanization along, and inland of the sea cliff. It is a result of excessive watering of lawns, and the attempt to grow non-native vegetation; of agricultural irrigation, septic tanks, leach lines and cess pools. Sorben and Sherrod (1977) indicate that the over-irrigation of landscaping in residential subdivisions alone is the equivalent of 50 or 60 inches of precipitation per year.

This extensive watering in the coastal area has had at least three important effects: It has caused a slow but steady rise in the water table even during the dry period preceding the heavy rains of 1978, and this has added weight to the cliff material which contributes to development of landslides; it has lubricated surfaces along which slides and block falls are initiated; and in some cases it has greatly increased solution in the underlying rocks.

Specific sites which experienced erosion during the drought years prior to 1978 have continued to collapse. Along the cliffs at the Self-Realization Fellowship property in Encinitas a large section of the cliff face separated along parallel joint sets and collapsed on April 26, 1978 (Figures 12 & 13). This recent failure measured 12 to 16 feet landward, 112 feet in width, and a maximum of 40 feet in height. Following the collapse, springs were seen to flow from the cliff face. This is the same site where the cliff collapsed during the 1940's and the temple was lost. Saint and Turner (1980) investigated conditions at this site, including the 1941 and 1978 events, and indicated that the cliffs in this area are subject to collapse as the result of a combination of factors. They stated that ". . . the large blocks, bounded by weakened joints, fractures and bedding planes move out laterally following a build up of pore pressure under the attack of wave action."

Because of the heavy rains in early 1978, cliffs that had been relatively stable were again subject to considerable retreat. A notable example is the low Quaternary alluvial cliffs south of Scripps Institution of Oceanography, which seemed to have stabilized during the dry period from 1947 to 1977. The series of houses built along the edge of these

Figure 12. (Before.) 1978 photo looking southeast along the base of the cliff at the Self Realization Fellowship at Encinitas, California. Photo courtesy of Sea Grant; February 3, 1978.

Figure 13. (After.) Same location seen following the cliff collapse of April 26, 1978. Photo courtesy of Sea Grant; May 2, 1978.

low cliffs, mostly without sea walls or any adequate protection, were subjected to erosion in January, and again in February 1978. As the rains continued and the beaches were washed out by a combination of high waves and perigean spring high tides, as well as by the erosion of rain gullying, erosion again occurred along these cliffs. The residents were threatened with loss of their property and went to desperate measures to stop the erosion. Old car bodies were placed at the slope toe, and later riprap was placed at these sites and cemented in place; temporary sea walls were installed. This was only partially successful as the houses had to be reinforced in various places to prevent their collapse. Adjacent unprotected lots were more severely eroded during the storm period. In 1979 a concrete seawall was constructed along most of this area.

Man-Induced Erosion:

Accelerated headward erosion is taking place along existing paleo-drainage avenues where culverts have been constructed under the railroad track and highways at San Onofre State Park and Camp Pendleton (Kuhn, Baker and Campen, 1980). Between 1885 and 1980, normal erosional drainage avenues have been drastically altered by man. In 1885 the Atcheson-Topeka & Santa Fe Railroad Company constructed trestles across existing gullies and canyons. These structures were designed to span channels and did not appreciably alter the existing drainage patterns. During construction of California State Highway 101 in 1912-1918, culverts were installed under the highway in order to control surface runoff under the road and down the canyons during periods of heavy rainfall. In the wet years following installation of these drains, erosion of gullies accelerated where these structures concentrated flow.

During construction of Interstate 5 in the late 1960's, large culverts and storm drains were installed directly below the San Onofre watershed and east of old Highway 101. These drains concentrated the surface runoff from numerous adjacent drainage avenues, directing it parellel to the highway and then west into culverts under Highway 101. No severe wet years occurred after completion of Interstate 5 until 1978. During January through April 1978, however, rainfall was greater than normal and headward erosion along these altered drainage avenues accelerated greatly. At Horno Canyon the southbound lanes of Highway 101 collapsed, leaving a 75-ft deep vertical scarp.

At one location no canyon existed perpendicular to the shoreline prior to the 1969 storms (Figure 14). However, between 1968 and 1976 a new canyon began to cut headward, reaching headward about 60 feet. During January and February of 1978 approximately 160 feet of headward erosion occurred. On February 14, 1980, another 5 ft of headward erosion occurred, and on the 20th of February, during one day, the canyon actually eroded approximately 235 ft landward, 50-100 ft deep and 60-90 ft in width (Figure 15). This amounted to approximately 50,000 cubic yards of sedimentary material which moved immediately out onto the beach.

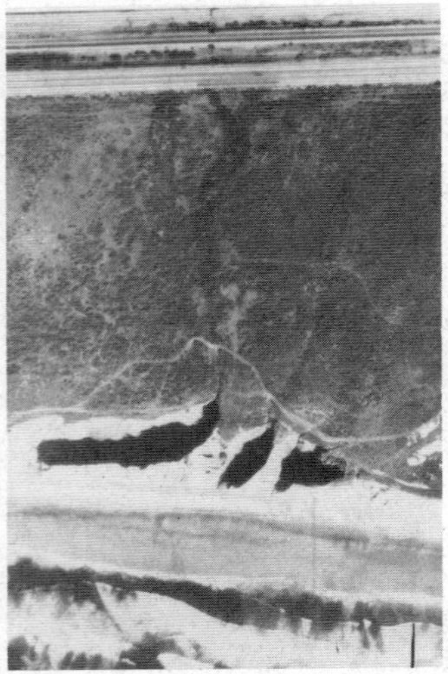

Figure 14. (Before.) 1968 vertical aerial photograph of a canyon along the bluffs at San Onofre State Park in northern San Diego County. Prior to this time only a notch existed perpendicular to the shoreline at this site. Photo courtesy of U. S. Navy; 1968.

Figure 15. (After.) March 1980 vertical aerial photograph showing development of 460 ft canyon perpendicular to the shoreline. It was actually cut 235 ft on February 20, 1980 alone. Photo courtesy of U. S. Navy; March 1, 1980.

Dead Dog Canyon, south of Horno Canyon on Camp Pendleton, experienced approximately 560 ft of headward erosion between 1932 and 1977. This same canyon eroded another 100 feet during a two-week period during January and February 1978 (Berggreen, 1979) and approximately 100 feet in February 1980.

These are but two examples of the man-induced rapid headward erosion during wet years that can occur in canyons located perpendicular to the shoreline. Rapid degradation of the bluffs also occurs wherever fences, stairways, lifeguard towers and storm drain pipes are located along former drainage avenues. Surface water tends to adhere and flow along the underside of such structures, thus eroding the bluff face. At Carlsbad State Park numerous beach stair accesses have been undermined and storm drains collapsed between 1978-1980 (Figures 16 and 17).

Bluff-Top Grading:

The grading of the bluff tops, in many cases to the edge of the slope, initiates slope failures in the sea bluff and creates drainage avenues along which further surface runoff erodes the bluff-top edge and the bluff face. It has been observed that where the existing terrace deposits and soil profile are left intact and surface drainage is diverted, the bluff edge appears to erode very slowly (Figure 18). In 1971-72 along south Solana Beach the bluff top was lowered by excavation by 15 to 18 ft and surface water was allowed to run over and down the bluff face. The upper slope continued to collapse during and following heavy rainfall. The base of the cliff along this site eroded 8-10 ft between 1972 and 1978 (Figure 19).

A condominium homeowner's association connected with the property threatened at this location has spent in excess of $1.5 million since 1978 in an attempt to forestall and retard future erosion. A concrete seawall was constructed along the slope toe and concrete "cribbing" was installed along specific sites in the unconsolidated bluff face (Figure 20). In early May 1979, the week that the concrete "crib wall" was expected to be finished, the entire structure rotated and collapsed, leaving the debris and sediment resting on the seawall below (Figure 21).

Another cliff collapse also occurred following extensive grading of the bluff top along the cliffs at Scripps Institution of Oceanography. The Quaternary cliffs in front of the Marine Biology building retreated 54 ft between 1912 and 1975 (Hannan, 1975). During extensive grading of the bluff top in 1975 the cliffs collapsed 5 to 8 ft landward.

The importance of grading in relation to construction and the removal of protective vegetation and case-hardened surface materials has also had an important effect and is clearly indicated where areas have been used for construction along bluff margins.

Conclusions

The popular concept of an average coastal-retreat rate for the evaluation of developmental projects is open to serious questioning, at

Figure 16. (Before.) 1977 view of storm drain designed to carry surface water down and away from slope face at Carlsbad State Park, California. Photo courtesy of Sea Grant; December 17, 1977.

Figure 17. (After.) 1979 view at the same location following the collapse of storm drain pipe, and severe gullying of bluff face. Photo courtesy of Sea Grant; February 11, 1979.

Figure 18. (Before.) 1954 oblique aerial view of a portion of South Solana Beach, California prior to bluff top development. Photo courtesy of Shore Processes Lab, Scripps Institution of Oceanography; 1954.

Figure 19. (After.) Same view in 1974 after development. Photo courtesy of B and A Engineering; June 1974.

Figure 20. (Before.) April 1979 view of concrete "cribwall" being installed on bluff face at Solana Beach, California. Photo courtesy of Sea Grant; April 23, 1979.

Figure 21. (After.) May 1979 view of same site following "cribwall" collapse. Photo courtesy of Sea Grant; May 4, 1979.

least in the San Diego region and probably elsewhere. Sea-cliff and bluff retreat is episodic, site-specific, and strongly related to the meteorological conditions, induration and stability of rock formations within a given area, and to a combination of erosive agents, both natural and man-induced, that may act upon them.

Both canyon head erosion and landslides are presently contributing significant quantities of sand to the beaches at San Onofre State Park and Camp Pendleton, north of Oceanside. It had previously been believed that most beach sand along the shore in San Diego County was supplied by rivers in flood, with only minor contributions from other sources.

A combination of continuing detailed observations in selected areas, detailed mapping on a large scale, careful review of historical records, and consideration of rock structures and competence along with hydrological considerations gives us the opinion that much of the coastal bluff area of San Diego County is a hazardous area for building, particularly near bluff edges.

Acknowledgments

This paper represents the results of research funded by NOAA, Office of Sea Grant under grant numbers 04-80M01-189 and project no. R/NP-1-96, and the California State Resources Agency (R/CZ-43), and the San Diego County Integrated Planning Organization contract no. 11596-0800E. We appreciate the cooperation and suggestions of David Bukry, Jeffery D. Frautschy and Robert E. Stevenson.

References

Berggreen, R. 1979. Geology of the proposed Camp Pendleton LNG Site, San Diego County, California: In Geologic Guide to San Onofre Nuclear Generating Station and Adjacent Regions of Southern California; Pacific Sections AAPG, SEPM, SEG Guidebook #461, D. L. Fife, ed., A-49-A-62.

Emery, K. O. 1955. Grain size of marine gravels, J. Geology, 63, 39-49.

Emery, K. O. 1960. The Sea Off of Southern California - A Modern Habitat of Petroleum. John Wiley and Sons, 366 pp.

Ganus, W. J. 1976. Is Southern California ready for a wet period? Environment Southwest, Winter 1976, no. 472, 9-13.

Hanna, M. L. 1926. Geology of the La Jolla Quadrangle, California. Univ. of Calif. Pub. Bull. of the Dept. of Geological Sciences, 16(7), 187-246.

Hannan, D. L. 1975. Sea Cliff Stability, West of the Marine Biology Building, Scripps Inst. of Oceanography, La Jolla, Calif: Benton Engineering, Project #75-9-18FG, 5 pp.

Inman, D. L. 1976. Man's impact on the California coastal zone, State of California, Dept. of Navigation and Ocean Development, 150 pp.

Kuhn, G. G. 1977. Coastal Zone Geology and Related Sea Cliff Erosion, San Dieguito River to San Elijo Lagoon, San Diego County, California: Integrated Planning Organization Contract #11596-0800E, County of San Diego.

Kuhn, G. G. and F. P. Shepard 1979a. Accelerated beach-cliff erosion related to unusual storms in southern California, California Geology, 32(3), 58-59.

Kuhn, G. G. and F. P. Shepard 1979b, Coastal Erosion in San Diego County, California: In Earthquakes and Other Perils, San Diego Region, ed. P. L. Abbott and W. J. Elliott, Geol. Soc. of Amer. field trip, San Diego Assn. of Geologists, November 1979, 207-217.

Kuhn, G. G., E. D. Baker and C. Campen 1980. Greatly accelerated man-induced coastal erosion and new sources of beach sand, San Onofre State Park and Camp Pendleton, San Diego County, California: Shore and Beach, in press.

Saint, P. K. and R. J. Turner 1980. Hydrogeology and groundwater conditions related to sea cliff stability on Self Realization Fellowship property (SRF), Encinitas. Unpublished report prepared for Self Realization Fellowship, 15 pp.

Shepard, F. P. and U. S. Grant, IV 1947. Wave erosion along the southern California coast, Geol. Soc. Amer. Bull., 58, 919-926.

Shepard, F. P. and H. R. Wanless 1971. Our Changing Coastlines. McGraw-Hill Co., New York, 579 pp.

Sorben, D. R. and K. L. Sherrod 1977. Groundwater occurrence in the urban environment: San Diego, California: In Geology of Southwestern San Diego County and Northwestern Baja, California, ed. G. T. Farrand. San Diego Assn. of Geologists, 67-74.

U. S. Coast and Geodetic Survey 1889. Descriptive report to accompany original field sheet entitled: Topography Pacific Coast from Horno Canyon to Las Flores and Canada Aliso, California, Topo. map #2015, U.S. Coast and Geodetic Survey Archives, Wash., D. C.

U. S. Coast and Geodetic Survey 1889. Descriptive report to accompany original field sheet entitled: Topography Pacific Coast from San Onofre Creek to Horno Canyon, California, Topo. map #2016, U. S. Coast and Geodetic Survey Archives, Wash., D. C.

Vaughan, T.W. 1932. Rate of sea cliff recession on the property of the Scripps Institution of Oceanography at La Jolla, California, Science 75(1939) 250.

SEA CLIFF EROSION IN SOUTHERN CALIFORNIA

by

Louis J. Lee[1]

INTRODUCTION

The Southern California coastline is generally classified as erosional with sand beaches at the base of sea cliffs. Gravel and bare rock are present locally in place of sand, and sea cliffs are absent where stream valleys intersect the shoreline.

The recession of coastal cliffs in Southern California is one of the large-scale coastal phenomena with which engineers and geologists must deal in near shore developments. Many of the structures that have been built above the sea cliffs are currently endangered by erosion and undermining of the cliffs, and new structures are being proposed in undeveloped areas. Consequently, it is now generally required that bluff or cliff developments be sited and designed such that the stability and structural integrity of the bluff is assured for the expected economic life span of the development. Thus, it is important to demonstrate that the cliff is stable given the current environment, and that any proposed developments (including related runoff, foot traffic, irrigation, construction, and other activities and facilities related to them) do not create or contribute significantly to the problems of erosion or geologic instability.

The erosional processes on the sea cliffs and bluffs in Southern California are a complex combination of many natural phenomena which are complicated further by the activities of man and animals. These pro-

---

1 Vice President, Woodward-Clyde Consultants, Orange, California.

cesses include: wave action at the base of the cliffs; the attack of wind and rain on the face of the bluffs; erosion of the face and the bluff tops by seepage of water, by flow of water over the bluff tops; and by human foot traffic, cave digging, and scribing graffiti on the face of the bluff; and animal burrowing.

A considerable amount of information has already been gathered pertaining to the rate of sea cliff erosion along portions of the Southern California coastline (Shepard and Grant, 1947; Nordstrom and Inman, 1973; Esgate, 1975; Pinckney and Lee, 1973; Cottonaro, 1975; Norris, 1968 and 1977; Kuhn, 1978; Griggs and Johnson, 1979). This information indicates that a variety of methods have been used for measuring rates of erosion in California, as well as a wide range in the rates of measured erosion (see Table 1 below). A study of a portion of the coastline in San Diego County, California provides more insight into the erosional process of the sea cliffs. The results of the study and a review of the literature suggests methods for monitoring erosion in the future.

TABLE 1 - SEA CLIFF EROSION RATES IN CALIFORNIA

Location	Geologic Composition	Average Annual Erosion (in/yr)	Method of Measurement
Santa Cruz	Mudstone Siltstone, Sandstone	0 to 8 20 to 45	Aerial Photographs (Griggs and Johnson, 1979)
Isla Vista	Marine Shale, Sandstone	2 to 15	Surveys & plot plans at top of bluff (Cottonaro, 1975)
Santa Barbara	Shale and Siltstone	3 to 10	Survey monuments at top of bluff (Norris, 1968)
Leucadia	Sandstone	0.1 to 0.7	Monuments at base of bluff (Lee, et al., 1976)
Sunset Cliffs San Diego	Sandstone, Shale	0.5 to 10	Photographs and topographic maps (City of San Diego, 1976)

DESCRIPTION OF STUDY AREA

Location and Coastal Setting

The study area includes approximately 9 miles of coastline between the cities of Del Mar and Leucadia (Figure 1). Approximately 7 miles of this area consist of sea cliffs that vary from approximately 25 to 100 feet in height. The inclination of the cliff typically ranges from near vertical at the base to 1 to 1 in the upper bluffs. The remaining coastline, approximately 2 miles, includes: the mouth of the San Dieguito River, which empties into the ocean south of Solana Beach; the San Elijo Lagoon, which is situated between Solana Beach and Cardiff-by-the-sea; the Agua Hedionda Lagoon located just north of Leucadia; and several other minor drainages. The tops of the bluffs are generally well developed with houses, condominiums, and apartment buildings. Approximately 5 percent of the area is undeveloped and about 20 percent is occupied by state and local parks.

The shore zone in front of the sea cliffs is generally characterized by a relatively wide, flat, thin, sandy beach which is typically over 100 feet wide and has a slope of approximately 1 to 15 (vertical to horizontal). The sand at the base of the cliff is typically at an elevation of +4 to +10 feet (MSL Datum). The beach material is composed of a medium to fine sand, which ranges from approximately 3 to 10 feet in thickness. There are local outcrops of bedrock, cobbles, and gravel. The primary sources of beach sand in the study area are the Batiquitos Lagoon, the San Elijo Lagoon, and erosion from the sea cliffs. Typically, the beaches build up in the summer time. During the winter, they lose from 3 to 7 feet of sand, and, in some areas, the ocean laps up against the base of the cliffs at high tide (Nordstrom and Inman, 1973). Portions of the beach are also severely scoured during the winter, and cobblestones are exposed on the entire beach. The sand beach laps out into an inshore zone which has a bedrock bottom with a slope inclination on the order of 50 to 1 (horizontal to vertical) extending 2,000 to 3,000 feet offshore. Typical Beach Profile-Related Terms from the U. S. Army Coastal Engineering

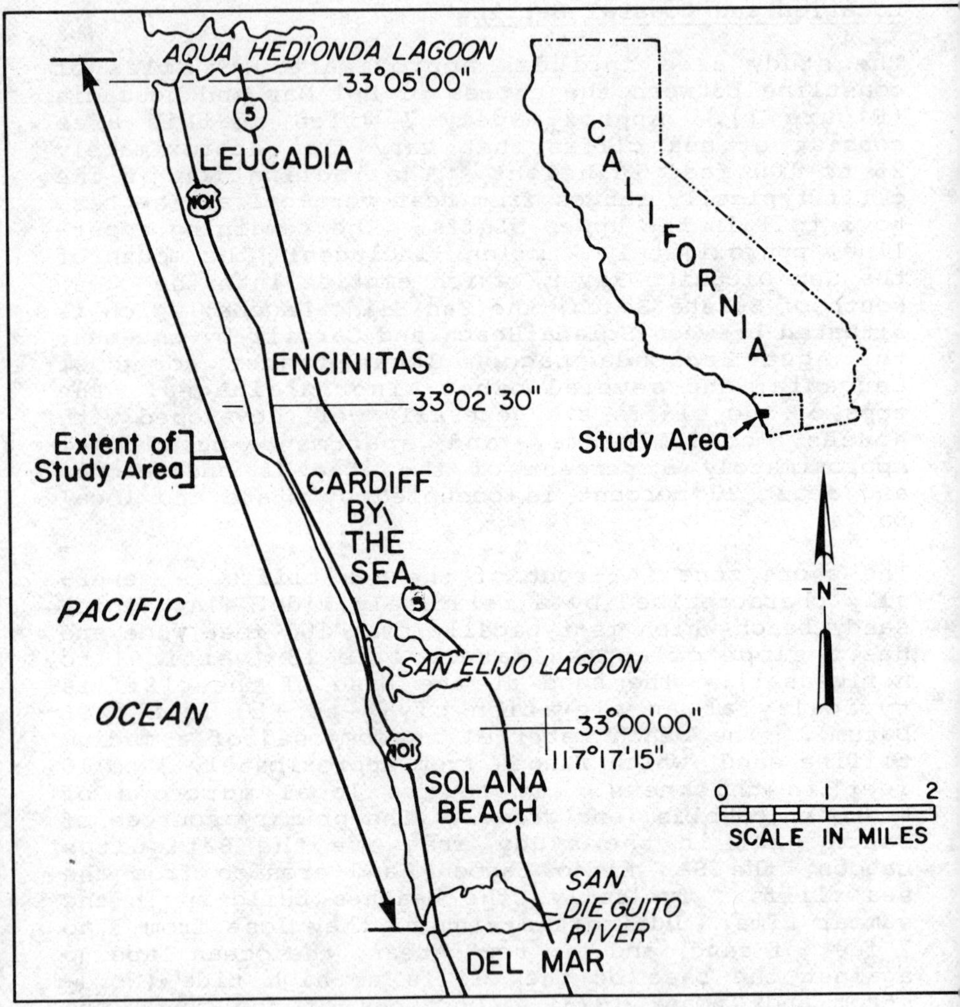

FIGURE 1 – LOCATION MAP

Research Center (1973) "Shore Protection Manual" are shown in Figure 2.

Geologic Setting

Rock types exposed in the sea cliffs between Del Mar and Leucadia have been mapped by Kennedy (1975) and Wilson (1972). These maps indicate that the cliff base is divided into a group of middle Eocene aged geologic formations with local areas of Pleistocene channel fill. The base rocks generally consist of the following units: Santiago formation (Ts), Torrey sandstone (Tt), and Delmar formation (Td). These units are predominantly composed of interbedded sandstone and siltstone with thin interbeds of clayey shale. The Torrey sandstone is composed of a massive to crossbedded, slightly cemented clayey, fine- to medium-grained sandstone with rare thin claystone interbeds. The channel fills are composed of medium-dense to dense uncemented sands and gravels with lenses of silt and clay. The base rock units exposed in the sea cliff, except for the channel fills, are generally flat to gently dipping and contain steeply dipping joints systems. Numerous small, normal faults cross the coastline within the study area. A major structure, the Rose Canyon fault zone, crosses the coast near La Jolla south of the study area and extends offshore from the study area.

The base rock is typically unconformably overlain by Quaternary terrace deposits which extend along almost the entire stretch of cliff. These sedimentary units form the upper bluffs and are generally composed of unconsolidated laminated beach sands and wind- or water-deposited sands which are locally cemented.

Tidal and Wave Climate

The maximum tidal range along the Southern California coast is approximately 9.5 feet with a mean tide level of approximately +3.0 feet and a high spring tide elevation of approximately +7.0 feet (MLLW). Typical wave heights range from 2 to 5 feet, and they do not normally break on the cliffs. Severe storm waves may reach heights of 10 to 15 feet; however, these waves generally break offshore in deeper water.

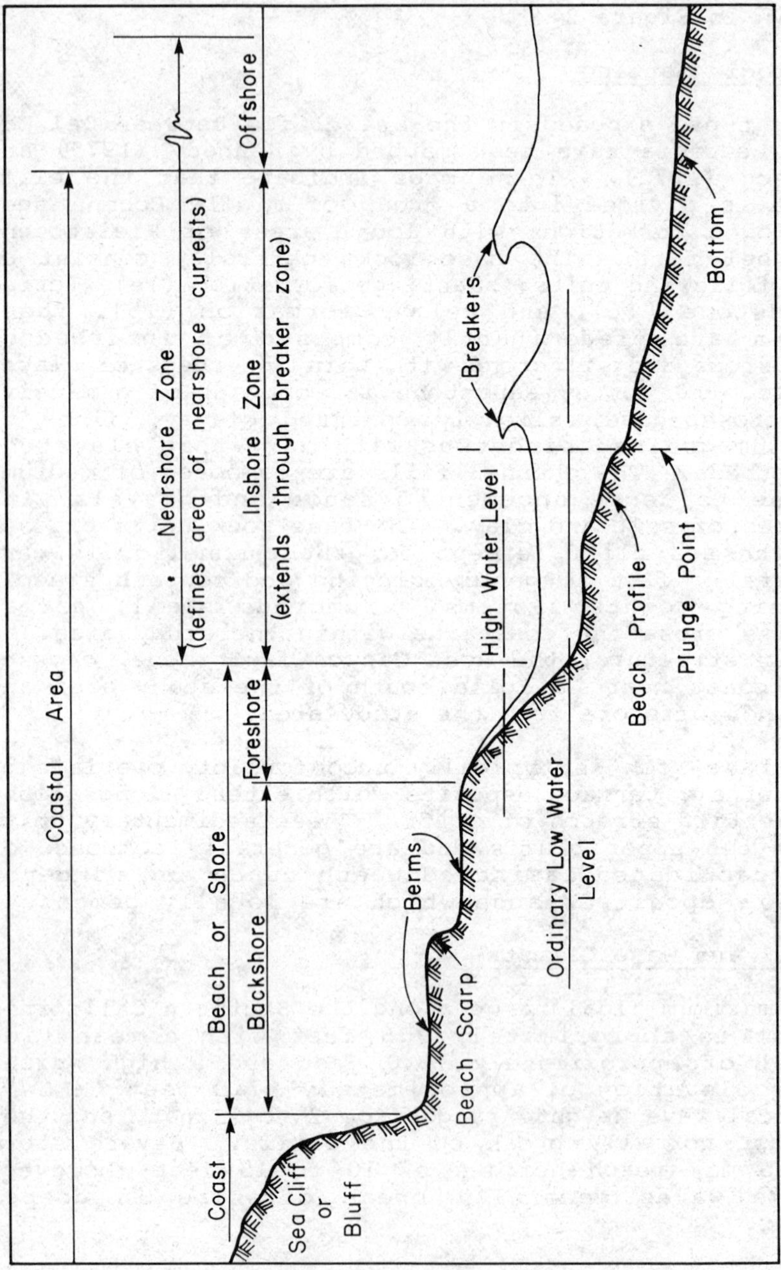

FIGURE 2 — BEACH PROFILE-RELATED TERMS

Meteorology and Rainfall

San Diego County lies on the southern edge of dry summer subtropical North America. Semi-arid conditions prevail with average annual precipitation varying from 9 to 10 inches, most of it falling in the winter. The maximum recorded rainfall is on the order of 25 to 30 inches. Occasional tropical storms reach the San Diego area in the summer, but no tropical cyclone of hurricane itensity has occurred within the San Diego area during the past 109 years (Eidemiller, 1979). Storms of sufficient intensity to produce flooding and related damage are also extremely rare during the spring and fall seasons. Erosion and related problems are most apt to occur in the San Diego coastal area during the winter occurrences of open wave, mid-latitude cyclones developing farther south than normal (Eidemiller, 1979).

EROSIONAL PROCESSES

It is important to separate and evaluate the various erosion phenomena as much as possible. This allows for a better understanding of the overall erosional process at a specific location and encourages better and more economical solutions to the problems.

Wave Action

The erosive force of waves at the base of sea cliffs depends on the wave energy developed in deep water; this energy is modified by the local submarine topography, by the width and slope of the beach, the characteristics of the beach material, and the tidal range. The waves along much of the Southern California coastline do not breach directly on the cliffs, except during storms. Thus, much of the wave erosion at the base of the sea cliffs occurs in the winter; this is partly because the weather is generally stormier then, and also because the beach is typically lower at the base of the sea cliffs during the winter.

Rain, Water, and Wind Processes

The most widespread and sometimes the most severe type of erosion is due to rain water. There are three basic types of erosion by water which are active on the sea cliffs in Southern California: raindrop or splash erosion, rill erosion, and gully erosion.

Shallow surface sloughing processes are active along the face and top of the bluffs. This results from rain or irrigation water penetrating the soils at the top or along the face of the bluffs and from ground water seeping out of the bluff face.

Wind erosion is generally related to dry periods and is most severe in arid regions. Portions of Southern California are in a semi-arid region and the sea cliffs are subject to occasional high ocean winds. Under certain conditions, soil particles are removed from the upper bluffs, transported, deposited, and mixed by these winds; however, the salt spray from the ocean develops a protective coating over much of the bluff and has limited this type of erosion.

Landslides

Landslides may result from either natural phenomena or human activities. Landslides along the Southern California coastline constitute a significant part of the erosion process. The continuing processes include surface rockfalls and major deep slides. The rock falls are generally associated with undercutting at the base of the cliff. Deep slides are generally related to the local geology and may be associated with either undercutting or ground-water seepage.

Human and Animal Activities

Numerous activities of man and animals hasten the rate of cliff erosion. Much of the man-induced erosion of sea cliffs in Southern California can be attributed to foot traffic, excavation of caves, and carving of graffiti on the cliff face. Cliff erosion can also be related to various types of development along the bluff top which encourage water runoff or which result in excessive water perculating into the ground above the cliffs. Erosion related to animals consists primarily of burrowing into the cliff face or behind protective structures.

METHODS OF OBSERVING AND MEASURING EROSION

Sea cliff recession studies have been made by various investigators throughout the world (Horikawa, 1978). Many of these studies have been limited to a comparison of measurements taken at the top of the cliff, and a few have measured the recession rates along the face or at the base of the cliff. Some of the studies also attempt to relate the rate of recession to the cliff geometry, geologic conditions, or erosional processes which may be acting on the cliff.

Investigations of the erosion of sea cliffs in Southern California have been made primarily by the comparison of old and new photographs (Shepard and Grant, 1947). This has been accomplished by establishing ranges near the left and right hand side of old photographs, checking objects in the foreground, and attempting to duplicate the scene with new photographs as closely as possible.

Studies made on the cliff retreat along the top of Sunset Cliffs in the City of San Diego (Kennedy, 1973, and City of San Diego, 1976) employed a combination of photographs, topographic maps prepared from aerial photographs, and geologic reconnaissance. The rate of face retreat of a 4-mile stretch of sea cliff in La Jolla was estimated by evaluating dated inscriptions carved in the rock (Emry, 1941). Sea cliff base erosion along approximately 21 miles of sea coast in San Diego County has been measured by nails driven into the base of the cliff (Lee, Pinckney, and Bemis, 1976).

SEA CLIFF EROSION IN THE STUDY AREA

Solana Beach

Starting at the south end of the study area near the mouth of the San Dieguito River, there are approximately 2 miles of sea cliff extending north to the San Elijo Lagoon. The base bedrock in the first 1/4 mile of this area is composed of well-consolidated, horizontally thin-bedded siltstones, mudstones, and sandstones; the base is approximately 20 to 40 feet high; and it is near vertical to vertical to overhanging. Bedrock is also exposed along portions of the beach. This section of sea cliff has numerous high angle faults and significant joint systems. Several sea caves have formed along these zones of weakness and are generally oriented in the direction of shearing or jointing. There is also evidence of past sea cave collapse and block falls through this area. The base rock exposures in approximately the next 1/2 mile are typically composed of near vertical arkosic sandstone, approximately 20 to 30 feet high. This area is heavily faulted with minor joint sets paralleling the faults. Numerous sea caves and channels have developed along these fractures and joints (over 20 caves in approximately 1,400 feet). Some of these caves extend more than 30 feet into the cliff. There is evidence of past sea cave collapse and rock falls in this area. The sandstone exposed along the base of approximately the next 3/4 mile of sea cliff is relatively unfractured, has a near vertical to vertical inclination, and is approximately 20 to 30 feet high. There are no significant faults through this area, and erosion appears to be relatively uniform. Localized sea caves have developed along open joint planes and in less resistant zones of rock. There is also local undercutting of interbedded silt and clay lenses. The base rock exposed in the last portion of this section of sea cliff includes both sandstone and some interbedded siltstone and mudstone. The base of the sea cliff ranges from approximately 15 to 30 feet high and is near vertical to vertical. There are numerous high angle faults through this area along with significant joint systems. Numerous sea caves (over 15 in less than 1,000 feet) have developed along these zones of weakness.

There are three late Pleistocene channel fills located along this stretch of cliff. The fills are generally composed of friable dense sand with gravel and shells. The channel fill near the south end is approximately 600 feet wide and is currently protected at the base by a rip rap sea wall in the northern portion and a concrete seawall in the southern portion. The next channel fill is approximately 400 feet wide and is the site of a county park and beach. The base of the cliff at this location is currently protected by concrete gunite. The last channel fill near the north end is approximately 100 feet wide and is protected at the base by concrete-filled bags.

The upper terrace deposits along the entire upper bluff are generally composed of unconsolidated friable sands with occasional gravels and shells which are locally slightly cemented. These deposits have inclinations ranging from approximately 1/2 horizontal to vertical to 1 to 1 and are near vertical locally. Rain and surface runoff has rilled, gullied, and channeled the upper bluff face. Localized sloughing and human foot traffic are evident on portions of the bluff face. Upper bluff erosion is accelerated in areas of sea cave collapse and block falls at the base.

Cardif-by-the Sea

Approximately the next 1-1/2 miles of sea cliff is indented somewhat and extends north from the mouth of the San Elijo Lagoon below a state camping ground to Sea Cliff County Park just south of Encinitas. The geologic formation at the base of the sea cliff along approximately the first 1/2 mile of this area consists of claystone with localized cemented sandstone beds exposed at the top and bottom. This material erodes fairly uniformly and has resulted in a near vertical to approximately 3/4 horizontal to 1 vertical bluff ranging from less than 5 to approximately 25 feet high. No significant rilling or gullies have been noted in this area. Erosion has been accelerated in some areas by beach access facilities, by foot traffic, and rodent activity. Portions of the base, particularly near stairways, are protected by rip rap. There are no upper terrace deposits on this section of sea cliff. Moving north the cliffs become higher (40 to 50 feet) and steeper. The base rock formation in approximately the northern 1 mile of this section of sea cliff is composed of interbedded, well-consolidated siltstones and sandstones. The rock is fractured and jointed through this area and there is some evidence of sloughing and rock falls. Significant water seepage is evident along the base of the cliff. There are occasional sea caves and channels.

The northern portion of this section of sea cliff has friable sands in the upper bluff. These sands range from several feet to over 20 feet in thickness, generally increasing in thickness moving north. The sands in the upper bluff have rilling and erosion gullies. Animal burrows are also evident. The northern portion of this section below Pacific Highway also includes an old landslide which has been repaired and replaced by an engineered fill. The toe of the fill is protected by a rip rap seawall.

Encinitas-Leucadia

North of the park the bluffs jut back out and become 60 to 100 feet high. These cliffs extend for another approximately 3-1/2 miles to Batiguitos Lagoon. The base rock in approximately the first 3/4 mile is composed of well-consolidated, horizontally bedded siltstone, mudstone, and sandstone. Bedrock is exposed along portions of the beach. The lower portions of the cliff are near vertical to vertical to slightly overhanging and are 20 to 40 feet high. The vertical portion generally decreases in height to the north. There are numerous faults and significant joint systems throughout this section. Numerous sea caves and small localized rock falls are evident through this area. There is also a major landslide (several hundred feet wide) and several smaller landslides and rockfalls near the beginning of this portion of the cliff. Significant amounts of water seep out of the joints and fractures and run down along the cliff face. At the northern end of this stretch of cliff, the bedded formation dips down and the more massive sandstone is exposed for a short section up to a small drainage which has been developed as Moonlight State Beach. The cliff to the north of the state beach continues for approximately 3/4 mile with a 20 to 25 high near vertical section of sandstone at the base overlain by the more friable sandy terrace deposits. Base erosion through this area appears to be relatively uniform. No significant faults or joint systems are exposed and no sea caves are present. A faulted area is exposed in the cliff at the end of this section, and the base rock changes in character. The upper bluff is relatively steep (3/4 to 1) and has sparse vegetation. Upper bluff erosion consists of rilling with localized gullies and sloughs.

Interbedded and more fractured material is exposed at the base through the faulted area which is approximately 300 feet wide. The base rock to north becomes light gray to yellow gray, silty to clayey sandstone. This condition extends approximately 1/2 mile to Leucadia State Park. Base rock erosion consists primarily of localized sloughs and rock falls. A major landslide is located approximately 700 to 800 feet before the park. The bedrock dips down through the park area for approximately 500 to 600 feet and the upper terrace sands extend to the bottom of the cliff. The interbedded and fractured siltstone, sandstone, and mudstone are again exposed north of the park. This condition continues for approximately the next 1-1/2 miles to the end of the study area. The vertical portion of the base rock ranges from several feet to over 15 feet in height and changes from a relatively hard material with widely spaced joints and fractures to a soft highly fractured material throughout this section. In areas where harder rock is exposed along the base, sea caves develop along zones of weakness. Where the material is weaker and more fractured, undercutting and rock falls are more prevalent. Some portions of the base rock are sandier and appear to erode in a slow, more uniform process. There are also two significant faulted areas in this section of the cliff.

The major upper bluff erosion along this section of cliff appears to be sloughing due to saturation of the sands and loss of support at the base. This portion of the bluff is also the most highly developed with stairway accesses and drainage pipes down the face. These have channelized and accelerated erosion of the friable sands in many areas.

SUGGESTED METHODS FOR MONITORING SEA CLIFF EROSION

Base Erosion

Almost all of sea cliffs in Southern California have a vertical to near vertical section at the base of the cliff. There is also either a sand (or gravel) beach or rock ledge extending out from the base of the cliff (Picture 1). The intersection of these two land forms provides a relatively distinct line which could be established and used as a basis for monitoring erosion at the base of the cliff.

Picture 1 - BEACH ALONG BASE OF CLIFF

Picture 2 - SMALL SEA CAVES ALONG BASE OF CLIFF

One of the most significant forms of base erosion is sea caves (Pictures 2 and 3). These caves can be mapped and measured (Kuhn, 1978) and can be monitored for future expansion and/or collapse.

Rock falls are also prevalent along the base of the cliff and are generally associated with highly fractured and jointed areas (Picture 4). These falls tend to be localized and leave indentations from which the depth into the cliff and the volume of rock falls can usually be measured.

All areas where shore protection measures have been constructed could be identified and their performance monitored to determine their effectiveness in preventing base erosion. In many of the areas where erosion is more uniform, graffiti has been scribed in the base rock. These areas can be photographed and monitored for changes. Large nails driven into the cliff have also proved to be an effective method of monitoring uniform erosion (Lee and others, 1976).

Upper Bluff Erosion

The top of many of the sea cliffs in Southern California are well developed and a bluff top can usually be fairly well established. The location of existing structures relative to the top of the bluff can also be established. This data could provide a good base for monitoring the erosion of the upper bluffs.

The most significant erosion of the upper bluffs appears to be water falling on, running over, or seeping through the bluff face. This results in rilling, gullying, and sloughing (Picture 5). This type of erosion is erratic and can best be monitored by photographs and by documenting the presence of sand debris at the base of the cliff.

Erosion is accelerated in many portions of the upper bluffs by construction by man (Picture 6). Erosion in the vicinity of the structures could be monitored by photographs and measurements. Many facilities, such as drainage pipes, stairways, and walls could be used as base points for monitoring erosion.

Picture 3 - SEA CAVE COLLAPSE AT BASE OF CLIFF

Picture 4 - ROCK FALL AT BASE OF CLIFF

Picture 5 - RILLS, GULLIES, AND SLOUGHS IN UPPER BLUFF

Picture 6 - ACCELERATED EROSION BELOW BLUFF TOP DEVELOPMENT

CONCLUSION

There are many forms of active erosion along the sea cliffs in Southern California. Erosion at the base is due primarily to wave action and takes the form of block falls, channels, and sea caves. Erosion of the upper bluffs is due primarily to the action of uncontrolled water and loss of support at the base and takes the form of sloughing and gullies. Erosion of both the base and upper bluff is accelerated by the activities of man. Channel fills, faults, and landslides which have made up approximately 15 to 20 percent of the study area present special erosion problems.

In any program for monitoring the rate of erosion, it is essential that the geology and soil and rock characteristics first be established. It is also necessary to gather site-specific field data in order to identify the form(s) and probable cause(s) of the erosion so that an effective method of monitoring can be established for the particular area in question. Topographic surveys, photographs, monuments, and references to local natural or man-made features can all be used effectively for monitoring long term erosion. A periodic inspection program should be established and site specific erosion information should be gathered. Each inspection should document the location, existing land forms, geologic conditions, type of erosion, type of material, extent of erosion and volume (if possible), and water conditions. General and localized erosion rates can be established from these data.

REFERENCES

City of San Diego, 1976, Sunset Cliffs - Shoreline protection study: Task Force Report, Appendix A, p. 1-5.

Cottonaro, W. F., 1975, Sea cliff erosion, Isla Vista, California: California Geology, June 1975, p. 140-142.

Eidemiller, O. I., 1979, Regional meteorology, in Abbott, P. L., and Elliott, W. J., eds: Earthquakes and other perils - San Diego Region, Prepared for Geological Society of America field trip by San Diego Association of Geologists, November 1979, p. 143-149.

Emry, K. O., 1941, Rate of surface retreat of sea cliffs based on dated inscriptions: Science, v. 93, no. 24, p. 617-618.

Esgate, Richard, 1975, San Diego County - Shore processes program proposal and background Report: San Diego County Department of Sanitation and Flood Control Report, April 1975, p. 9-17.

Griggs, G. B., and Johnson, R. E., 1979, Coastline erosion-Santa Cruz County: California Geology, April 1979, p. 67-76.

Horikawa, Kiyoshi, 1978, Coastal engineering - An introduction to ocean engineering: University of Tokyo Press, published in the U.S.A. by Halsted Press, a Division of John Wiley and Sons, Inc., p. 235-242 and 317-336.

Kennedy, M. P., 1973, Sea-cliff erosion at Sunset Cliffs, San Diego: California Geology, February 1973, v. 26, no. 2, p. 26-31.

Kennedy, M. P., 1975, Geology of the San Diego Metropolitan Area, California: California Division of Mines and Geology, Bulletin 200, Section A, p. 11-35.

Kuhn, G. G., 1978, Coastal zone geology and related sea cliff erosion, San Dieguito River to San Elijo Lagoon, San Diego County, California: Geological Research Division, Scripps Institute of Oceanography, Chapter 5, p. 1-29.

Lee, L. J., Pinckney, C. J., and Bemis, C., 1976, Sea Cliff Base Erosion: ASCE National Water Resources and Ocean Engineering Convention, San Diego, pre-print 2708.

Nordstrom, C. E., and Inman, D. L., 1973, Beach and cliff erosion in San Diego County, California," in Ross, A., and Dowlen, R., eds., Studies on the geology and geologic hazards of the greater San Diego area: San Diego, California, San Diego Association of Geologists, p. 125-131.

Norris, R. M., 1968, Sea cliff retreat near Santa Barbara, California: Mineral Information Service, June 1968, p. 87-91.

Norris, R. M., 1977, Erosion of Sea Cliffs: Geologic Hazards in San Diego, John Porter Dexter Memorial Publications, San Diego Society of Natural History, 1977, p. 36-40.

Pinckney, C. J., and Lee, L. J., 1973, Sea cliff recession study, southern one-half of San Diego County, in Elliott, W. J., ed., Field trip guide book to engineering geologic problems, Southwestern San Diego County, San Diego Association of Geologists, August 1, 1973.

Shepard, F. P., and Grant, U. S., IV, 1947, Wave erosion along the southern California coast: Bulletin of the Geological Society of America, v. 58, p. 919-926.

U. S. Army Coastal Engineering Research Center, 1973, Shore Protection Manual, v. 1, p. 1-3.

Wilson, K. L., 1972, "Geologic Map of the Encinitas Quadrangle, San Diego County, California: Unpublished, San Diego State University, Senior Report.

ELKHORN SLOUGH: THE MAKING OF AN ESTUARINE SANCTUARY

Nona B. Dennis[1/], Bruce Browning[2/], and Hal Bissell[3/]

Introduction: A Sanctuary By Any Other Name...?

The term "sanctuary" evoked an image, in its most literal sense, of a holy temple, consecrated to a religious purpose, and in its broader sense, of a place of asylum, or of a forest closed to the hunt (Oxford English Dictionary). The image further suggests a condition of refuge which is inviolable.

When the National Oceanic and Atmospheric Administration (NOAA) on March 7, 1974, proposed draft guidelines (15 CFR Part 921) and called for comment pursuant to Section 315 (then 312) of the Coastal Zone Management Act of 1972, as amended—known generally as the Estuarine Sanctuary Program—some forty-two commentators submitted suggestions concerning various sections of the proposed regulations. It is significant that no one questioned the use of *sanctuary* as the appropriate term for conferring a protected status on the lands and waters which would eventually be designated under provisions of the program. It may be that, in their attempt to avoid duplicating other broad purpose Federal preservation programs, the authors of the original Section 312 chose a classification which would not present confusion with similar institutionalized concepts such as "wildlife refuge", "research natural area", "preserve", "reserve", and so forth. In any case, the term held, and with it the implication that estuarine sanctuaries would indeed be inviolable, possibly along with lands and waters surrounding them.

It would be misleading to attribute the entire chronology of geographic factionalism, local expression of fear of land and water use restrictions implied by preservation, and accusations of "federal and state land grab" which have characterized the Elkhorn Slough case, to a specific reaction to the *sanctuary* designation *per se*. Protective status by any other name might have been equally suspect to landowners around the slough. But the term does present some ironies. If holy wars were once an integral part of consecrating and guarding the holy grounds of religious sanctuaries, then the term sanctuary is quite appropriate for Elkhorn Slough, where a certain amount of strife over the sanctuary nomination, designation, and acquisition program has proved to be a persistent feature throughout the process.

1/ President, Madrone Associates, Marin County, California.

2/ Senior Wildlife Biologist, Director of Coastal Wetlands Program, California Department of Fish and Game.

3/ Environmental Scientist, Jones and Stokes Associates, Sacramento, California.

This paper presents an informal, synoptic case history of the establishment of an estuarine sanctuary at Elkhorn Slough, Moss Landing, California (Figure 1), complete with the minor skirmishes and major battles of the "holy war" which have impeded but not blocked the making of the sanctuary.

The Estuarine Sanctuary Program

> "Estuarine Sanctuary, as defined in the Coastal Zone Management Act, means a research area which may include any part or all of an estuary, adjoining transitional areas, and adjacent uplands, constituting to the extent feasible a natural unit, set aside to provide scientists and students an opportunity to examine over a period of time the ecological relationships within the area." (15 CFR 921.2[a])

Thus the primary purpose of the program is to create natural field laboratories: for the study of natural and human processes in representative estuaries in the coastal zone; to endure into the future; and to provide through research and education some of the information essential to coastal zone management decision making. The Federal Estuarine Sanctuaries Program envisions a national network of such estuarine areas, representative of the eleven biogeographic provinces that make up the nation's coastline; two of these provinces are represented in California. At present, eight sanctuaries have been designated, including the recent designation at Padilla Bay, Washington. As many as 20 are planned.

Under this program, coastal states are eligible for Federal grants on a matching basis to cover costs of acquisition up to $2,000,000, and development and operation up to $50,000 each for three years. Subsequent operation and research and educational activities are dependent on other funding sources.

The 1979 designation of Elkhorn Slough as an estuarine sanctuary under provisions of the program represented the culmination of one of at least three parallel efforts since the late sixties and early seventies—Federal, State, and private—to bring all or part of the slough's rich natural resources under protective status. That designation, however, was but one point in a process which is not yet complete, in which acquisition of lands and waters comprising the proposed sanctuary is proceeding cautiously in the face of landowners' resistance to sell, some guarded acceptance, and pockets of local opposition. Management objectives and plans for the future are only now being formulated. The case, then, is in process. We can document what has ensued, but we can only speculate as to what will follow, and we can offer some predictions and pose the questions that will need resolution. It may be that the major battles are over.

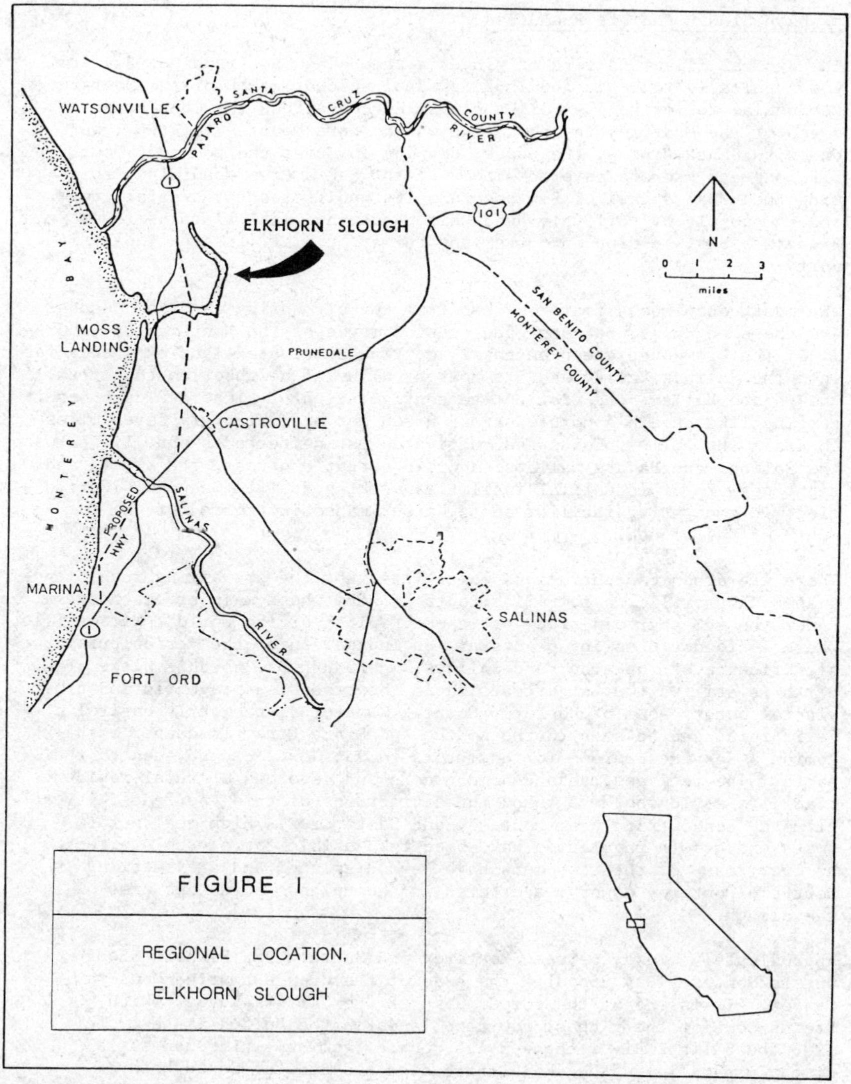

FIGURE I

REGIONAL LOCATION, ELKHORN SLOUGH

Elkhorn Slough And Its Resources

The typical flight pattern of planes approaching San Francisco Bay from Los Angeles is to begin the long, gradual descent south of the Monterey Peninsula, following a straight course angled across the long crescent sweep of the Monterey Bay shoreline which links Monterey at the lower end with Santa Cruz at the upper, dipping low over the redwood, Douglas fir, oak and madrone covered slopes of the Santa Cruz Mountains, skimming the urban sprawl of San Jose and its undifferentiated satellites and the mosaic of variously-hued salt ponds and tidal flats of lower San Francisco Bay, to drop onto the long runway of S. F. International Airport.

The most conspicuous features seen from the air during the slow descent are the sand dunes, beaches, and crested waves of the Monterey Bay shoreline; the brown and green patchwork of rich cultivated land extending far into the Salinas Valley and its nearest alluvial neighbor to the north, the Pajaro Valley; the distinctive configuration of Elkhorn Slough itself curving like an elk's horn east and north for approximately seven miles to drain the hilly uplands and cultivated marine terraces that lie between the Salinas and Pajaro Valleys; and, in abrupt contrast, the towers and tight network of industrial facilities marking Pacific Gas and Electric's electric generating facility and a Kaiser magnesium refractory at the mouth of the slough.

There are urban concentrations which interrupt the prevailing rural landscape—Castroville, a few miles to the south, whose main street passes under the 40+ year old sign "Artichoke Capital of the World"; Watsonville, where 15 food processing plants are an indication of the agricultural significance of the area; and Salinas, where John Steinbeck's place of birth is still a tourist attraction in the otherwise pragmatic and provincial county seat of Monterey County. Monterey, the early capitol of California, can be seen to the south, and Santa Cruz, in danger of becoming a seaside bedroom for commuters to San Jose, can be seen to the north. The retirees, second-home owners and devotees of rural-residential life style who have found the attractions of this "perfect all year climate" concentrate near Monterey and Santa Cruz. High real estate costs and growth regulation have been responsible for a bubble effect, however; many of the newcomers have been displaced and are settling in northern Monterey County, scattered in the uplands south and east of the slough.

Approximately midway between Monterey and Santa Cruz, Elkhorn Slough enters Monterey Bay opposite the head of the deep submarine Monterey Canyon. It is one of the former arms of the Salinas River, which drains some of the richest row-crop lands in the United States. In 1908 the Salinas River changed its course from one which had for a distance paralleled the shoreline behind a system of dunes before entering the bay 1½ miles north of the present mouth of Elkhorn Slough. The river now enters the bay 5½ miles south. Elkhorn Slough, together with three other former "arms"—McCluskey, Moro Cojo, and Tembladero Sloughs, became more obviously estuarine at that time, still draining what had been the 226 square mile northern watershed of the Salinas.

The natural diversion of the Salinas River was followed in 1946 by a Corps of Engineers dredging program and construction of a jetty to stabilize a permanent 500-foot opening for Elkhorn Slough directly into the bay. This allowed the slough to become fully tidal for at least three miles of its length.

Elkhorn Slough appears to follow the path of an ancient river, encompassing now around 2,500 acres of tidal flats, submerged and semi-submerged land, about 700 acres of which are wholly or partly reclaimed. It was undoubtedly a freshwater lagoon during its period as an arm of the Salinas River estuary, although archaeologic evidence reveals that it was a saltwater estuary in earlier periods.

The near view of Elkhorn Slough—from the ground—is best seen from State Highway 1, the scenic route along the California coast. In its varying states of improvement along Monterey Bay, State Route 1 has four lanes leaving Monterey and Santa Cruz but only two heavily used lanes in the mid-section which crosses the neck of Elkhorn Slough over a causeway-like bridge. The low bridge, which precludes all but small boats at low tide from entering the slough from the harbor, has been responsible in part for the low use of the slough. Long range plans of the California Department of Transportation call for replacing the bridge with a wider, high-level bridge and 4-lane highway across the slough to the east. Bridge abutments would be located in the sanctuary lands on the south side of the slough and national wildlife refuge lands proposed for the north side (Figure 1).

Only a limited portion of the slough can be seen from the highway and bridge. The view is brief, limited on the north by the flatness of the immediate terrain and on the south by the massive facilities of the PG&E plant (Figure 2). The major north-trending reach of the slough lies hidden from view behind low marine terraces covered with strawberries, artichokes, and other truck crops and grazing lands for dairy cows and other livestock. It is a pastoral scene, attractive to at least one major land developer who envisions a 600-acre community by the slough.

FIGURE 2
Pacific Gas and Electric
Moss Landing Facility

The tiny community of Moss Landing at the mouth of the slough consists of a small cluster of buildings across from the PG&E and Kaiser industrial complex. There is a scattering of ranches and ranch homes around the slough, many of them held for several generations by the same families.

More evident from the highway than the slough itself, is the picturesque assemblage of sailboats and fishing boats in the Moss Landing Harbor, managed by the Moss Landing Harbor District. The District, under a state land grant, holds title to all slough waters and lands to the mean high water mark, and has been an adamant opponent of the sanctuary designation. Although more than picturesque, the present harbor is less active than it was during earlier periods of whaling activity in the latter 19th century to the 1930s, and at the peak of sardine processing in the '30s and '40s. The fish processing industry dropped off as the sardines mysteriously disappeared from Monterey Bay three decades ago, leaving Monterey's Cannery Row as well as Moss Landing with idle canneries and reduction plants. The harbor is still the center of a 15-million pounds/year fishing industry, but most of the harvest is trucked out. The harbor complex also includes the former sardine cannery that houses Moss Landing Marine Laboratories, since the sixties a center of academic marine research in both the slough and Monterey Bay.

Unseen from air or ground is the pipeline which enters the shore near the slough's mouth carrying oil to the PG&E plant from a tanker off-loading platform offshore. Along with the power plant, it is a reminder of the potential of the harbor to meet future energy and navigational needs. Also largely invisible, except from a boat within the slough, is the discharge point for 713 acre-feet/day of power plant cooling water which enters the slough, several hundred yards east of the bridge.

For most travelers and recreationists the only accessible and visible part of the slough is that seen from the bridge to the east, and the mudflats and dunes of the state park and beach between the bridge and the ocean, west of the slough proper. The groups of birders and field biology students who frequent the slough know the few other access points. Local residents, until recently, regarded them with benign amusement.

Among the fish and wildlife habitat resources of the slough (Browning, *et al.*, 1972) are the vestiges of formerly extensive eelgrass beds near the mouth of the slough; 190 acres of inactive salt ponds that are summer habitat for as many as 2,000 brown pelicans, listed as endangered; populations of snowy and common egrets, and a nesting population of great blue herons. The endangered California clapper rail is an elusive resident of the approximate 800 acres of tidal *Salicornia* (pickleweed) marsh. Another endangered species, the Santa Cruz long-toed salamander, inhabits freshwater marshy areas of McCluskey Slough. MacGinitie (1935) first documented the many invertebrate species of the mud and sand flats, which include twenty-eight species of clams, *Macoma nasuta* (bent-nosed clam) being the most abundant. Sea lions and harbor seals use the slough for hauling out; 73 different species of fish have been identified in the slough, many of them using small channels and sloughs for spawning and rearing. More than 90 species of water-associated birds have been identified.

Most recreational fishing and shellfishing takes place around the harbor, jetties, and outer (seaward) slough mudflats. A launching platform at Kirby Cove provides small boat access to the upper slough. Waterfowl hunting is not a major use and is confined largely to a dozen small duck clubs. Historically, shellfish have been one of the richest resources of the slough for both recreation and commercial pursuits. Clam digging continues to be popular, and a limited operation for culturing oysters is active in spite of typically high, fecal coliform counts in the slough.

Many people come and go at Moss Landing—fishing, sightseeing, recreating in other ways, studying, or simply passing through *en route* to the tourist attractions of Monterey and Carmel. The small number of persons, such as commercial fisherman, who come from nearby towns to work, and the 600 or so permanent residents comprise a close community which displays a natural possessive sense for the slough and harbor.

Selection Process and Nomination

The nomination by the California Coastal Commission in 1978 of Elkhorn Slough as candidate to be the first estuarine sanctuary in California came after a four-year process which had begun not long after the enactment of the Coastal Zone Management Act of 1972 and adoption of Estuarine Sanctuary Guidelines in 1974.

At that time (1974), California coastal planners were grappling with the immense job of developing a coastal zone management plan. As early as 1972, California voters had passed an initiative (Proposition 20) calling for coastal protection; the interim state and regional coastal conservation zone commissions established by that initiative were working toward a 1975 deadline for legislative adoption of the plan and enabling legislation: The California Coastal Act, which was passed in 1976.

It is difficult to separate the events in the Elkhorn Slough selection chronology from the larger context of developing the California Coastal Plan. Decisions were intertwined. The evolving coastal plan had to encompass great geographic breadth and deal with major policy issues. The consideration of an estuarine sanctuary was a detail—one minute geographic segment along a 1,100 mile front where major policy battles were being fought.

It was apparent that selection would require careful study. In the interest of getting the larger job done first it was reluctantly agreed that selection of an estuarine sanctuary within the coastal planning area would be deferred until the Coastal Plan had been adopted. However, the six regional commissions were invited to submit their preliminary candidates in 1974, before the more rigorous job of evaluation was to take place. Elkhorn Slough, Tomales Bay in Marin County, augmented later by the Esteros Americano and de San Antonio in Sonoma County, and Upper Newport Bay in southern California, were proposed by the central, north central, and south coast regional commissions respectively. Of these, the first two were mentioned in the policy language of the coastal plan as possibilities for estuarine sanctuary designation, and upper Newport Bay was recommended for protection and partial acquisition.

The nomination that was finally approved by the State Coastal Commission in 1978 was the result of a "biopolitical" and philosophical decision as much as a strictly biological one, although the analysis which characterized the selection process was rigorous and designed to be objective in spite of the intense geographic partisanship which centered in the candidacy of "favorite estuaries." At the outset, the Coastal Commission was looking for an estuary that would meet the principal criteria set forth in sanctuary guidelines (15 CFR Part 921): 1) that the sanctuary fulfill the program definition of an estuary; and 2) that it be typical of the "California" biogeographic region, one of 11 such biogeographic regions or categories characterizing the nation's coastal areas:

> "...'estuary' means that part of a river or stream or other body of water having unimpaired connection with the open sea where the seawater is measureably diluted with fresh water derived from land drainage. The term includes...lagoons in more arid coastal regions."

> "*Californian*. South Pacific coast from Mexico to Cape Mendocino; shoreland influenced by coastal mountains; rocky coasts with reduced freshwater runoff; general absence of marshes and swamps; biota temperate."

The other biogeographic region represented in California is the "*Columbian*. North Pacific coast from Cape Mendocino to Canada..." At the time of the selection process this region was represented by the candidacy of South Slough in Coos Bay, Oregon.

A technical advisory panel was appointed to carry out the selection process with the help of Coastal Commission staff. They used as the basis for developing their own criteria, the list used by the Office of Coastal Zone Management for review of sanctuary applications. These criteria are:

- benefit to the coastal zone management program: including usefulness in research, and how well the proposal fits the national program of representative types
- ecological characteristics of the ecosystem: including biological productivity, diversity, and representativeness, and the extent of alteration of the natural system
- size and choice of boundaries: the proposed sanctuary area should approximate a natural ecological unit
- cost: the federal share must be less than $2 million
- proximity and access to existing research facilities

- lack of suitable alternative sites already protected which might be capable of providing the same use or benefit
- enhancement of noncompetitive uses
- lack of existing or potential competing uses
- compatibility with existing or proposed land and water use in contiguous areas

Two criteria were eliminated from the list. The availability of already protected alternative sites had been considered in the earlier selection by regional commissions of sanctuary candidates. The criterion of representativeness was particularly problematic. It was questionable whether the *Californian* biogeographic type description adequately characterized the range of California estuaries (south of Cape Mendocino). A preliminary survey by Coastal Commission staff of 125 of California's larger estuaries and coastal lakes (109 south of Cape Mendocino) distinguished five categories and three subcategories by physical form, type and periodicity of connection with the ocean, and types of habitats present (adapted from Clark, 1974):

<u>tidal river mouth</u>: narrow estuary, always open, often emptying directly from coastal mountains into the sea;

<u>lagoon</u>: confined water body with restricted inlet to the sea and little freshwater inflow;

<u>bay</u>: large, semi-enclosed waterbody, always open to the ocean, fed by one or more rivers;

<u>embayment</u>: confined water body with restricted inlet to the sea and significant freshwater inflow;

<u>slough</u>: always open, riverine or linear form, usually marshy, traversing predominantly low-lying land;

<u>periodically closed tidal river</u>, *e.g.* Estero Americano;

<u>periodically closed slough</u>, *e.g.* Salinas River;

<u>periodically closed lagoon</u> - south coast type, *e.g.* Mugu Lagoon.

A non-estuarine type was added because of its proximity to the coast:

<u>coastal pond or lake</u>: not an estuary, isolated from the ocean, for example by system of dunes.

It was apparent that the *Californian* biogeographic region covered a diversity of estuarine areas, none of which could be considered to represent a majority of those surveyed. Nor did any single estuary represent a majority of those lumped under the *Californian* biogeographic type.

The technical advisory panel turned to other criteria to augment the estuarine sanctuary guidelines, placing less emphasis on typical features and more on usefulness as a study area, with features of particular interest for estuarine research and management. The selection criteria are summarized here, because of their potential usefulness in other selection processes:

- Present research use: number of institutions and people conducting research; broadness of research; historical extent of research; existence of baseline studies; pertinence to coastal zone management decisions.
- Potential research use: access; proximity and number of research institutions; potential interest in studying estuary; suitability of estuary for answering coastal zone management questions; ability of estuary to withstand potential research use.
- Present educational use: number of students and number of general public served; broadness of educational programs and variety of educational activities and participating groups.
- Potential educational use: access; proximity of educational institutions and non-academic population to be educated; potential numbers of students and general public; broadness and variety of potential educational programs; activities and participating groups; ability of estuary to withstand potential educational use.
- Biological value: existence of properties that account for exceptional natural value of estuarine systems in general; habitat diversity; species diversity and abundance; presence of rare species; critical role in maintaining significant species population.
- Physical value: processes of geologic or hydrologic interest; presence of interesting geologic or archaeologic features.
- Naturalness: generally, freedom from human impact, minimal alteration of natural ecological processes; physically intact characteristics, such as natural flow of fresh water and tides, natural sediment transport, undiked marshlands, undredged channels, good water quality; continuation of historic populations of plants and animals.
- Manageable ecological unit: Are major influences left outside the sanctuary? does it contain enough of the watershed to be protected from degradation? can the unit be managed with respect to size, jurisdictions and powers of management agency?
- Availability of proposed acquisition area: area under private control, under public control, and controlling agency; number of owners and parcels; parcel sizes and cost; owners' willingness to sell; cost of development rights or easements, and agency to accept them; size of area to be acquired; total cost of acquisition.
- Potential for multiple compatible uses: variety of existing compatible uses; number of users; ability of estuarine environment to support additional use if declared a sanctary.
- Lack of conflicting uses: Confliciting uses could include: *resource use* - sand and gravel mining, consumptive use or alteration of water (*e.g.* for power plant cooling or other industrial

use, stock watering, irrigation), hunting, fishing, damaging recreational use (*e.g.* trampling, extensive motor-boating; dirt-biking); *development* - marinas, urbanized uplands, diked or filled marshlands, saltponds, sewage or industrial outfalls, port facilities.

- Compatibility with local plans: attitude of local citizens and officials toward sanctuary proposals; local government plans for watershed areas.

- Lack of alternative management programs, to achieve the same objectives: suitability and existing proposals for designation as a marine sanctuary, national wildlife refuge, state reserve, state wildlife area, ecological reserve, *etc.*; existing proposals for purchase by protective organization or institution and protection offered; land and water area (within proposed sanctuary) already protected by above means.

Elkhorn Slough and Tomales Bay in particular received strong political support from the sponsoring regional commissions. Each nominee had its supporters on the technical review panel. Tomales Bay is a narrow 12½ mile long ocean embayment following the San Andreas fault rift zone in west Marin County, 25 miles north and west of San Francisco Bay. It encompasses a watershed as large as that of Elkhorn Slough and its associated tributary sloughs, with seven dams intercepting freshwater inflow. The two smaller estuaries—Estero Americano and Estero de San Antonio—had been added to the Tomales Bay nomination "package" and then separated from it as unwieldy, to be considered as a distinct nomination. Upper Newport Bay is an S-shaped inland portion of Newport Bay located 24 miles southeast of the Los Angeles - Long Beach harbor.

During the staff and panel's review, Upper Newport Bay was eliminated, largely because the California Department of Fish and Game already owned and managed a major portion as an ecological reserve. Tomales Bay, which had many attributes that met the selection criteria, was eliminated because of its "unsuitability as a manageable ecological unit." Perhaps the single fact that the major part of the watershed is the source of water for 170,000 Marin County residents was a compelling basis for its elimination, but the angry Tomales Bay supporters were convinced that regional coastal politics had come into play. It remained then for a final playoff between Elkhorn Slough and the two small esteros to decide on the "winning" sanctuary candidate.

Although many arguments were presented justifying the selection of Elkhorn Slough on the basis of its usefulness as a locus for estuarine research, the ultimate deciding factor was an expedient one: available state funds. Contained in a brief memorandum from the Secretary for Resources (Johnson, January 1978), in response to the Coastal Commission's staff report, was the statement that "there is state money (already) available for the necessary match if the Coastal Commission selects Elkhorn Slough as the State's nomination. AB 2133 (Keene-Nejedly California Wetlands Preservation Act: Chapter 462, Statutes of 1976) appropriated $2,885,000 for the acquisition of five wetlands, of which Elkhorn Slough is one. There is no immediate state money available for the other areas being considered." (authors' underline)

In many respects Elkhorn Slough seemed to be an obvious candidate for sanctuary status—or preservation by any other name. The resources were there; the research and educational institutions and uses were nearby and already established; conservation interests had long recognized Elkhorn Slough's value. While endangerment was not listed as one of the criteria (an important difference from some other protective programs), it was difficult to ignore proposed plans for industrial and commercial development around the Moss Landing harbor and major proposals for land development.

There were also some disjunctive features that didn't fit the criteria: The slough was a "naturalized" product of human manipulation; there were alarmingly high coliform counts in the water, the indicator of possible management problems in the watershed; there was possible conflict between plans for industrial habor development to meet national energy needs and the maintenance of the more or less natural ecosystem; and there was that plume of thermal discharge 9°F above ambient, influencing, if not "damaging", the aquatic environment. Certainly, if the interplay between human and natural processes in an estuarine environment was to be the subject of research and a source of information for coastal management decisions, Elkhorn Slough was an ideal candidate. The question of local acceptance at that point could only be speculated.

Preacquisition Study

The selection process was successful in fulfilling its objective of naming a candidate sanctuary for California and moving Elkhorn Slough one step toward gaining protective status. It was not successful to those who felt that Tomales Bay had been eliminated for the sake of politics, or expediency, or both. Nor was it successful in arousing much local interest beyond the support of conservation groups whose home base was generally elsewhere.

The selection skirmishes had been within the ranks; the believers in estuarine protection simply differed on where the holy ground should be consecrated. Tempers flared at public meetings, leaving some breaks within the otherwise unified front of those dedicated to the larger commitment of protecting California's coast. These rifts were minor, however, compared to those separating the local, conservation, and other interests which were identified and brought to confrontation in the preacquisition study and environmental review (EIS) of the sanctuary proposal which followed (Madrone Associates, 1978; OCZM, 1979).

The practical purposes of the study were straightforward; *viz.* to provide information required under Section 921.12 of the Estuarine Sanctuary guidelines for acquisition grant applications (revised 1977, *Federal Register*, 45 [175] P. 45523) and the information base and assessment for the subsequent EIS which would be required. Beyond these it was necessary to describe the resource itself in more detail—its ecological attributes, the boundaries which would best define the ecosystem, the nature of land and water uses and potential conflicts in and adjacent to the slough, proposed research and education programs, a proposed management approach and administrative structure for the sanctuary, *etc.*

In effect, the study took the selection panel's criteria, which had been used with considerable conjecture for comparing candidates, and applied them in depth in an assessment of Elkhorn Slough. One important criterion —*viz.* availability of proposed acquisition area and associated factors such as owners, parcels, cost, willingness to sell, *etc.*—was deliberately (and in retrospect, unfortunately) omitted. As the study drew to a close, a sanctuary line began to emerge, like an image in magic ink brought out by rubbing with a pencil. The 10-foot (Mean Sea Level datum) contour was the magic line, a physically and ecologically defensible landward boundary, drawn at a small enough scale so that landowners could only speculate as to what that line might mean to them. In the absence of clarification, they feared the worst. Since the study emphasized the need for buffer lands, a concept later rejected by OCZM, they had reason for their concern.

In other respects the preacquisition study served the important purpose of identifying and sorting out the plethora of major landowners, institutions, jurisdictions, and other entities with activities or interest in the slough and its watershed. Among them:

Ownerships:

- Moss Landing Harbor District holds title to the waters of the slough, as defined above, and is authorized to regulate and monitor commerce, fisheries, and navigation in these tidal waters. (See also State Lands Commission below). The District has been a vocal opponent of the sanctuary.
- Pacific Gas and Electric Company, since 1950, has had a major electric generating facility at Moss Landing, presently fulfilling 20% of their total capacity. PG&E is the sponsor of significant applied research in connection with 316(a)(b) requirements (Clean Water Act, PL 92-500, as amended) (Smith, 1977; Nybakken, *et al.*, 1975). Their concerns are with potential restrictions on their operation and possible expansion.
- Kaiser Refractories has similar concerns. Their major operation is on adjacent Moro Cojo Slough, tributary to the Moss Landing harbor.
- Rubis Ranch includes 600+ acres on the north-west side of the slough; a large-scale residential and recreational development plan has been prepared for the ranch.
- Monterey Salt Company owns 190 acres (Browning, 1972) of inactive salt ponds used for raising brine shrimp. The Company has refused to discuss acquisition in any form.
- A dairy ranch and other grazing lands and truck farms surround most of the remainder of the slough. Runoff is a source of high coliform counts, in spite of partial control by the Regional Water Quality Control Board.
- Nature Conservancy has owned and managed up to 400 acres of wetlands as a preserve. 160 acres were sold to California Fish and Game (Wildlife Conservation Board) to start off the sanctuary acquisition program (below).

Institutions and Other Interests:

Elkhorn Slough has been a "living laboratory" for many decades, to a variety of researchers, students, and conservation organization field groups. Among the more important are:

- Moss Landing Marine Laboratories, owned and used by five California State Colleges and Universities
- U.S. Naval Postgraduate School, Monterey
- Hopkins Marine Station of Stanford University, oldest biological laboratory on the Pacific Coast, Monterey Peninsula
- University of California at Santa Cruz
- Three Community Colleges: Monterey Peninsula, Cabrillo, and Hartnell
- National Sea Grant College and Program Act is sponsoring significant research concerning the effect of various land uses and soils in the slough's watershed on sedimentation processes and biological resources.
- California Department of Fish and Game, Marine Culture Laboratory
- San Francisco Bay Area and other California chapters of National Audubon Society, and chapters of the Sierra Club
- Several areas near the mouth of the slough are leased from the Moss Landing Harbor District for oyster culture, where oyster culture is stimulated by thermal discharge from the PG&E plant. They are transferred to Tomales Bay for flushing prior to harvest.

Agencies and Jurisdictions:

- U.S. Army Corps of Engineers, has direct permit authority in the slough under Section 10 (River and Harbor Act of 1899) and Section 404 (Clean Water Act). Significantly, the San Francisco District of the Corps exercises this authority behind dikes and typically interprets 404 authority quite broadly to include historical as well as existing wetlands.
- United States Fish and Wildlife Service comments on Corps permit applications. The Service proposes a National Wildlife Refuge for the northwest side of the slough (Figure 3).
- NOAA has proposed not only an esturarine sanctuary at Elkhorn Slough but an adjacent marine sanctuary in waters of the Pacific Ocean including a major part of Monterey Bay.
- California State Lands Commission has jurisdiction over lands underlying navigable waters, which includes ownership and control of certain submerged and tidal lands in Elkhorn Slough, subject to unresolved jurisdictional determination.
- State Parks and Recreation owns a state park and beach between Highway 1 and the bay, managed for recreational use.
- State Department of Fish and Game (Wildlife Conservation Board) owns a 10-acre area and boat ramp—Kirby Cove—in the upper slough. The Department will assume primary management of the sanctuary.

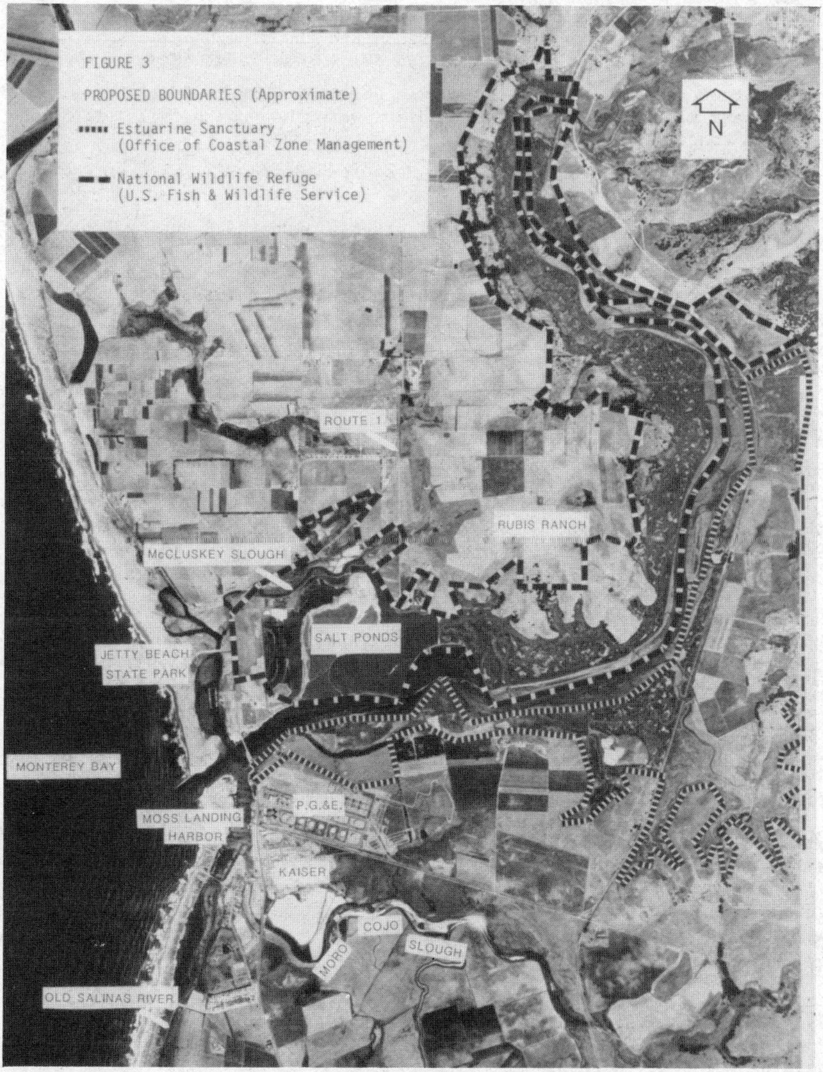
FIGURE 3
PROPOSED BOUNDARIES (Approximate)
Estuarine Sanctuary (Office of Coastal Zone Management)
National Wildlife Refuge (U.S. Fish & Wildlife Service)
N
ROUTE 1
RUBIS RANCH
McCLUSKEY SLOUGH
SALT PONDS
JETTY BEACH STATE PARK
MONTEREY BAY
P.G.&E.
MOSS LANDING HARBOR
KAISER
COJO
MORO
SLOUGH
OLD SALINAS RIVER

- State Water Resources Control Board, through Regional Boards, exercises authority over discharges into the slough and its tributaries under Sections 208 and 201 (Clean Water Act).
- State Coastal Commission exercises permit authority over the 49 square miles of the slough's immediate watershed which falls within the coastal zone.
- Others in the State include Air Resources Board and California Department of Transportation.
- Monterey County Board of Supervisors, through the Planning Department, is preparing a Local Coastal Program in compliance with the California Coastal Act, 1976, which the County will then administer; and through the Health Department, certain waste disposal systems in and near the slough are regulated.
- Other County departments include mosquito abatement and the agricultural commission.

The list appears to be a simple catalogue, but inherent in it are a multitude of overlapping but not always congruent boundaries, territories, management objectives, and programs all having an influence over land and water use in the slough.

The preacquisition study also helped to unravel the parallel pursuits toward acquisition of all or parts of the slough. Nature Conservancy's acquisition and management activity in Elkhorn Slough goes back to 1968 when the University of California's newly established Natural Land and Water Reserve System alerted the Conservancy to the slough's importance for ecological study. By 1974 the Conservancy had purchased 343 acres on both sides of the slough and was quietly pursuing additional properties in the marshes and adjacent watershed, amounting to a total of 400 acres.

Interest in Elkhorn Slough as a National Wildlife Refuge goes back almost as far. By 1978, U.S. Fish and Wildlife Service was completing an assessment and had personnel on the ground drawing alternative boundaries for future acquisition. Without fanfare and with minimal public involvement the Service was actively considering portions of large key parcels on the north and west sides of the slough. Proponents of the sanctuary who understood the differences between the refuge program and the sanctuary program (there were many who didn't) had mixed feelings about this pursuit. On the one hand it was viewed as a strong allied effort, as an opportunity to add much needed dollars to the total public investment in protecting the slough. There was also the power of eminent domain which the national wildlife refuge system would theoretically be able to exercise, in contrast to the sanctuary program, which is constrained by "voluntary sale." On the other hand, the Service had a reputation for "parochialism" and "taking over"—for wanting to manage all of its holdings. Would there be duplication of administrative effort and dollar investment? Conflicts between the sanctuary and refuge in management philosophy and techniques? It appeared that the Service was in a hurry and did not want to wait for the sanctuary program to go through its ponderous, prescribed processes. The funds were allocated and they had every intention of using them.

In 1980, U.S. Fish and Wildlife Service abruptly (it appeared to the outsider) redirected those funds to another refuge project: the Antioch Dunes at the western end of the Sacramento-San Joaquin Delta. One U.S. Fish and Wildlife Service person explained it as the relocation of a "tourniquet"—the bleeding of one natural resource, the habitat of several endangered species, was simply more profuse than the other. The Service has not dropped its intent at Elkhorn Slough; fiscal 1981 will provide funds sufficient to begin appraisals and perhaps even a token land purchase. The larger questions of boundaries, management, and coordination with the sanctuary all remain unanswered. If the Service's plans proceed as they have—low-keyed, quiet, but persistent—they will face far less opposition than has the estuarine sanctuary program. Is it the name? or are there other mysterious ingredients that transform almost identical objectives into different processes?

Even before the preacquisition study, which was to recommend boundaries, was complete, the California Wildlife Conservation Board was on the ground negotiating with certain property owners. If the sanctuary were to fail at least the state's funds, destined under the "sanctuary" scenario to provide matching funds, would be used for a State "ecological reserve."

Environmental Review And The Public

The accusation of insufficient notice is common to the public hearing process. Although most public noticing procedures are spelled out, in the interest of compliance with the law, their dry litany of minimum requirements provides little prior insight into the vested interests and emotions of "the affected public." A legal notice, compressed into a newspaper column next to the obituaries can be lost to most readers.

The accusation, when founded, usually reflects both a public preoccupied with its own pursuits until aroused by direct or imagined threats of interference with or erosion of property rights and values, and a government agency with good intentions but a naive technique in dealing with the public. The public which feels excluded, or at best pressed by a tightly orchestrated agency schedule, quickly transforms its opposition into hostility. Coupled with a traditional distrust of outside (read "state" and "federal") control, the hostility at Elkhorn Slough created its own medium for misunderstanding, irrationality, and public rebellion.

All the elements of confrontation were evident, when in the words of an articulate and rational opponent, the process of designating Elkhorn Slough as an estuarine sanctuary was "brought out into the open." In fact, the preacquisition study process appeared to have been open. There had been many personal contacts with representative interests; a large study review group was formed by the Department of Fish and Game, which met twice to hear lengthy status reports by consultants on the incipient sanctuary; these were reported in the local newspapers. Although optional under program guidelines, a public hearing on the study report was scheduled in Monterey, 20 miles distant from the proposed sanctuary, and duly noticed.

At that meeting there were grumblings and a few articulate arguments for and against the sanctuary. At this point, however, the public did not yet have a clear understanding of how or when to appropriately fight or support, as the case might be, the sanctuary designation. They were reassured that a Draft Environmental Impact Statement would be prepared by OCZM and that another (optional) hearing would be held, providing the appropriate forum for comment on both DEIS and the proposed sanctuary. Perhaps that would be the time to deploy whatever forces they had.

That hearing was scheduled, duly noticed in the legal column in the "wrong" newspapers, according to one opponent, and again accusations of "insufficient notice" and "closed procedure" were made. In response to the mounting tide of opposition, a local politician prevailed upon OCZM and the California Department of Fish and Game to conduct a pair of hearings in Castroville, the community closest to the slough and the place of residence for many of the people employed in Moss Landing. The first attracted 100, still discontent with the public process. The final public meeting was attended by over 300 people. The majority were opposed to the sanctuary although few were landowners who would be affected.

The tenor for that final meeting on the Draft EIS was presaged by the announcement which the opposition sponsored in local newspapers. By this time opponents had formed a co-ordinated group and collective voice focused principally on questions, such as restriction of property rights, public access, bureaucratic interference, and *de facto* confiscation of land. While the collective voice was typical and predictable, the individual comments were often colorful:

RESTORE OUR RIGHTS

Mass Meeting: JULY 10
Gambetta School 7 P.M.
(Cooper & Seymour Sts., Castroville)

We already have the Coastal Commission taking away our property rights . . . Now, another Agency wants to take over Elkhorn Slough and its watershed.

Sponsored by:
RESTORE OUR RIGHTS COMMITTEE
26 Empire Road
Castroville, CA. 95012
Small donations appreciated.

"I've seen the Coastal Commission grow from a cute little kiddie to one of the most awful beasts ever to walk the face of the earth. I don't want the federal government to come out and feed it more raw meat."

"This is another land grab" "...estuaries have no inherent ecological value."

"I think you (federal spokesman from OCZM) have enough to do in Washington, C.C., cleaning up your own mess. What are you doing here?"

"The sanctuary has turned into a scientific playground for students and the educational community with every one else just swept aside."

"The Sierra Club and governmental agencies are too poor to buy (land) but all powerful to control, to the point of confiscation."

It was not the right time for the many supporters of the estuarine sanctuary to rebut.

In retrospect, the meeting served its primary purpose: the voice of the people was heard. It was not an atmosphere for rational argument or decisions, however. Amidst all the rhetoric, the issues were clear and relatively straightforward:

1. Would land and water use restrictions be imposed, directly or indirectly?
2. Would landowners receive a fair value?
3. Would another level of government enter Elkhorn Slough in the guise of a sanctuary?
4. Would local interests have a voice in decisions?
5. Would public access in fact jeopardize the resources of the slough?

The Final EIS systematically and rationally sorted, interpreted and responded to the many and lengthy comments—and an estuarine sanctuary came into being at Elkhorn Slough.

The Future Of Elkhorn Slough Estuarine Sanctuary

The reader is left here, in part because of a page limitation, and in part because the case is incomplete. Of those issues raised in the environmental review process, perhaps the principal one was that concerning additional land and water use restrictions which the local residents and landowners justifiably feared.

It is necessary to return to the California Coastal Act and the Coastal Plan in order to locate the standard for regulating land and water use within the Elkhorn Slough watershed. The keystone to regulation of Elkhorn Slough is the Coastal Plan's identification of the slough as an environmentally sensitive area, and the basic policy position that requires that wetlands be protected. The legislators drew the line around 49 square miles of the 71 square mile Elkhorn Slough immediate watershed, thus firmly establishing a regulatory framework, with much more far reaching implications for future land use controls than the sanctuary program could ever imagine. Viewed in this context, the circumscribing of a specific area within the Elkhorn Slough Complex (ecosystem) as an estuarine sanctuary (or a wildlife refuge) takes on the attributes of a sanctum sanctorum—that is, a sanctuary within a "sanctuary", with little direct bearing on the governance of the outer "sanctuary" grounds. The EIS on the sanctuary said it more succintly: "There is no legal mechanism for any sanctuary to influence activities outside its boundaries."

Monterey County is preparing its required local coastal program not in response to the possible presence of an estuarine sanctuary, but rather in response to the Coastal Act. While this land use plan is likely to place tight controls on development and agriculture well beyond the sanctuary's well-defined boundaries, the determination of what will constitute acceptable use—on the land and in the slough—is ultimately a policy question. The effect of the sanctuary will be to strengthen the hand of the environmentalists in the inevitable political struggles over the final shape of the County's local coastal program.

The ending of the case at this point coincides with a positive beginning. A negotiator for the Wildlife Conservation Board is quietly pursuing acquisition or conservation easements, talking over coffee with the landowners who last year rose up in arms. 980 acres have been acquired of the 1,510 acres being sought.

An advisory committee representing most interests in the slough has had its first meeting and is beginning the process of constructive use of the slough for research and education. And the Department of Fish and Game is looking for the perfect manager—administrator, biologist, environmental educator, law enforcement officer, outdoorsperson, and public relations specialist. If communication with the Moss Landing and Northern Monterey County community remains open, the sanctuary far from being an inviolable refuge, should become an accepted and useful member of that community.

Along the Pacific coast, however, clouds are already gathering over the selection of California's second estuarine sanctuary...

REFERENCES:

Clark, John. 1974. Coastal ecosystems. The Conservation Coundation, Washington, D.C. 178 p.

Johnson, Huey D., Secretary for Resources. 1978. Letter to Executive Director, California Coastal Commission, January 1978.

MacGinitie, G. E. 1935. Ecological aspects of a California marine estuary. Amer. Midl. Natur. 16(5): 629-765.

Madrone Associates (in association with Jones and Stokes Associates). 1979. Preacquisition planning study for Elkhorn Slough Estuarine Sanctuary. Prepared for: California Department of Fish and Game. 140 p.

Nybakken, J. W., G. M. Cailliet, and W. W. Broenkow. 1975. A baseline study of the Moss Landing/Elkhorn Slough environment. Moss Landing Marine Laboratories. Progress report. 219 p.

Smith, E. H. and Associates. 1977. A preliminary report on a baseline study of Elkhorn Slough, presented to Kaiser Engineers as part of a 316(a) demonstration program for Moss Landing PG&E Power Plant, Units 1-5.

U.S. Office of Coastal Zone Management and California Management and California Department of Fish and Game. 1979. Final Environmental Impact Statement on Elkhorn Slough Estuarine Sanctuary.

MARINE SANCTUARIES, A CONTROVERSIAL CONCEPT

JOHN K. CASSELL
CHEVRON U.S.A. INC.
San Francisco, CA

ABSTRACT

The marine sanctuaries program has had a checquered history since passage of the Marine Protection, Research and Sanctuaries Act in 1972. Interested parties, and the public, have been presented with a series of elastic policy proclamations, alternately expanding and shrinking lists of proposed sanctuaries, and a number of controversial white papers, issue papers, and draft environmental impact statements. Although intended to clarify and establish the need for each proposed sanctuary, these often raised more questions than they resolved, and frequently failed to recognize important existing marine research and lengthy experience factors in various ocean regions.

The magnitude of the entire program, and of many individual proposals, has also been the object of critical discussion. In late 1977 more than 100 proposals were outstanding. These totaled about 220,000 square miles, about one-quarter of the continental shelf within the 200 meter water depth boundary. The largest were in Alaskan waters, where some exceeded 40,000 square miles in size, and where the sum total amounted to a third of Alaska's shelf area. Currently there are 70-odd areas on the "recommended" or "active" lists, but the size of each is not clearly defined.

During the past two years considerable controversy has centered around several recommended sanctuaries, the principal thrust of which has been to prohibit, or severely restrict, petroleum development. Three of these, the Georges Bank, Flower Gardens, and Santa Barbara Channel Islands proposals, have been the subject of intense critical debate. Each has been challenged on the basis of its size, the proposed regulatory regime, and the need for such a sanctuary in view of the multiplicity of existing laws and regulations protecting the marine environment.

The nature of this controversy suggests that the time has come to re-evaluate the program, with the object of determining if it is based on a coherent policy, which in turn reflects a consensus of scientific findings in fact. Since the origin of the concept there has been a dichotomy of interpretation as to the proper nature and magnitude of a marine sanctuary. On one hand sanctuaries are conceived as small discrete areas defined to protect specific resources, wherein activities would be strictly limited. The other view sees them as large areas, protecting the general marine environment, but permitting multiple-use with only such limited additional safeguards as clearly may be required.

Implementation of the program to date has not followed either course, but instead proposes to severely restrict activities, especially petroleum development, over very large ocean areas. The Georges Bank and Channel Islands proposals, 20,000 and 1,500 square miles respectively, are examples of this approach. At the Flower Garden Banks 500 square miles of sanctuary were proposed to protect one-quarter square mile of reefs.

Scientific research and lengthy experience do not support this approach. Current laws and regulations adequately protect the marine environment in all but the most unique site-specific instances. Prohibiting development of critically needed petroleum resources in large ocean areas is not necessary, nor does it serve the national interest.

INTRODUCTION

The marine sanctuaries program stems from Title III of the Marine Protection, Research and Sanctuaries Act of 1972 (U.S.C. SS 1431-1434). This brief act, Title III is only four pages in length, was but one of many environmentally oriented legislative bills, which became law in the late 1960's and early 1970's. It was accompanied in this general time-frame by the passage of laws to protect air and water quality, marine mammals, estuaries, endangered species and archaeologic and historic data, as well as the Coastal Zone Management Act of 1972.

Title III, Sec. 302(a) simply provides for the Secretary of Commerce to: ... "designate as marine sanctuaries those areas of the ocean waters .... which he determines necessary for the purpose of preserving or restoring such areas for their conservation, recreational, ecological, or esthetic values". A subsequent section [302(f)] empowers the Secretary to: ..."issue necessary and reasonable regulations to control any activities permitted within the designated marine sanctuary...." The very general and non-specific language of Title III has undoubtedly been a contributing factor to the controversy and disputation which has surrounded implementation of the sanctuary program. The Act is silent with respect to the number of sanctuaries which might be appropriate, their size, the proper magnitude of the entire program, and the application of multiple-use principles within designated sanctuaries. This is somewhat surprising in view of the Congressional debate accompanying passage of the bill. During the House discussion Congressman Aspinall moved to strike Title III from the bill because he believed it would impair the mineral leasing program. He stated: .... "The enactment of this title could result in locking up unnecessarily offshore resources valued at billions of dollars ...." Congressman Pelly, a sponsor of the bill, replied: ... "This fear is groundless. Let me assure the distinguished Chairman of the Interior Committee that it has never been our intention to stop the development of these vast resources...." Congressman Dingell, also responding to Congressman Aspinall, said: ..."This legislation is not.... going to halt oil drilling." Unfortunately none of this philosophy was subsequently included in the language of the Act, and its implementation became virtually a discretionary matter in the hands of the administrative offices of the National Oceanic and Atmospheric Administration.

Another facet of Title III apparently not considered by Congress at the time of its enactment or subsequently, is the relationship it bears to numerous other environmental laws and regulations. Various laws passed at about the same time, and several others enacted since 1972, provide environmental protection for virtually every conceivable aspect of the ocean environment. Table F-1a, taken from the Final Environmental Impact Statement on the Proposed Channel Islands Marine Sanctuary, published by NOAA in May, 1980, lists 22 federal and state authorities, implemented by 18 agencies, the thrust of which is almost entirely devoted to protecting the marine environment. To this list can be added the Coastal Zone Management Act of 1972 (amended 1976) which further empowers the states to determine that activities beyond their tidelands are consistent with their coastal zone management plans. Notably absent from the authority provided by these laws is the power to prohibit oil and gas operations on the outer continental shelf. The Outer Continental Shelf Lands Act Amendments of 1978 provide the Secretary of the Department of the Interior with authority to strictly regulate petroleum activities and to stop them altogether in the event of environmental threats or harm. But it does not provide for outright prohibition of such activities. This is the essential thrust of the marine sanctuary program as it is now evolving. In each case where a proposed sanctuary involves a part of a known or potential oil and gas province the regulations drafted for sanctuary designation contain a prohibition of oil and gas exploration and development activities. Sanctuary proposals to date have little or no regulatory impact on other activities such as commercial or sport fishing, recreation, research, shipping, kelping, etc., primarily because all these concerns are already amply controlled by existing law and regulation.

Reviewing the currently pending sanctuary proposals at the Flower Gardens, off Texas, and the Channel Islands, Point Reyes-Farallon Islands and Monterey Bay, off California, simply confirms the concern Congressman Aspinall expressed during the House debate on the bill when he said: "There is no question but that this legislation is directed at the mineral leasing program...." Having to accept this premise, one must then consider the magnitude of the sanctuaries program and its potential impact on petroleum resource development in order to evaluate the significance and import of the controversy it has engendered.

TABLE F-1a. Abbreviations of Authorities and Agencies

Abbreviations of Authorities

Federal

AA - Antiquities Act; 16 USC §§461-469i
CAA - Clean Air Act; 42 USC §§7401-7642
CWA - Clean Water Act; 33 USC §§1251-1376
ESA - Endangered Species Act; 16 USC §§1531-1543
FCMA - Fishery Conservation and Management Act; 16 USC §§1801-1882
MBTA - Migratory Bird Treaty Act; 16 USC §§703-711
MMPA - Marine Mammal Protection Act; 16 USC §§1361-1407
NHPA - National Historic Preservation Act; 16 USC §§470-470n
OCSLA - Outer Continental Shelf Lands Act; 43 USC §§1331-1343
OPA - Oil Pollution Act of 1961; 33 USC §§1001-1016
PMR - Pacific Missile Range; U.S. Navy
PWSA - Ports and Waterways Safety Act; 33 USC §§1221-1227
SSVS - Space Shuttle Vehicle System; U.S. Air Force
RHA - River and Harbor Act of 1899; 33 USC 403 Sec. 10

State

AQCA - Air Quality Control Act; California Health and Safety Code, §§39000-42708
ASBS - Areas of Special Biological Significance; California Water Code §13260
CCA - California Coastal Act; California Public Resources Code §27000
ER - Ecological Reserves; California Fish and Game Code §1580
FGC - Fish and Game Code; California Fish and Game Code, California Administrative Code, Title 14
HCRPA - Historical and Cultural Resources Protection Act; California Public Resources Code §5000
OGS - Oil and Gas Sanctuaries; California Public Resources Code §6870
WQCA - Water Quality Control Act; California Water Code §13000

Abbreviations of Agencies

Federal

BLM - Bureau of Land Management - Department of the Interior
COE - Army Corps of Engineers - Department of Defense
EPA - Environmental Protection Agency
FWS - Fish and Wildlife Service - Department of the Interior
HCRS - Heritage Conservation and Recreation Service - Department of the Interior
MMC - Marine Mammal Commission
NMFS - National Marine Fisheries Service - Department of Commerce
NPS - National Park Service - Department of the Interior
PMFC - Pacific Fisheries Management Council; Joint Federal-State
USCG - United States Coast Guard - Department of Transportation
USGS - United States Geological Survey - Department of the Interior

State

ARB - Air Resources Board
CCC - California Coastal Commission
DFG - Department of Fish and Game
HRC - Historic Resources Commission
PFMC - Pacific Fisheries Management Council; (Joint Federal-State-Private Body)
SLC - State Lands Commission
WRCB - Water Resources Control Board

MAGNITUDE OF THE MARINE SANCTUARIES PROGRAM

When the site of the wreck of the Monitor off Cape Hatteras, N.C., was designated as a mrine sanctuary in January, 1975, only one square nautical mile of ocean was included in the sanctuary boundary. The designation of the Key Largo Sanctuary off Florida in December, 1975, involved 100 square miles. These did not seem inordinately large, nor did they pose conflicts with essential resource development. Beginning in 1977, however, the program suddenly accelerated, and began to take on the aspects of a widespread effort to thwart future potential resource evaluation activities, particularly oil and gas exploration and production. Sanctuary proposals were made in great numbers and many of these involved very large areas.

In September of 1977 a listing of outstanding sanctuary proposals showed a total of 110, with the gross area involved amounting to an estimated 220,000 square miles, or approximately one-quarter of the continental shelf to the 200 meter water depth boundary. The Environmental Protection Agency had proposed 58 of these, the Department of Interior 22, and state and local agencies and environmental groups about 30. At this time there were 19 sanctuaries proposed off the California coast, 27 off Oregon and Washington, and 21 off Alaska; a total of 67 for the west coast and Alaska. Over 200,000 square miles were proposed for Alaskan waters alone, amounting to a third of the Alaskan shelf above the 200 meter water depth contour. Utilizing the information then available it was possible to superimpose a potential overlay of these proposals on the Alaskan offshore sedimentary basins as shown in Figure 1. This showed that several of the prospective offshore oil provinces would be severely impacted by the designation of large sanctuaries such as were proposed for Bristol Bay (38,500 square miles), for the Gulf of Alaska Peninsula (54,700 square miles) and for the Bering Straits (41,500 square miles). Off Southern California another 8500 square miles of sanctuary proposals included over 2000 square miles in the Santa Barbara Channel region, the most promising oil and gas region off the California coast.

The number of these proposals in late 1977, as well as the great size of many of them, caused considerable concern for those parties charged with resource development on the continental shelf. It was believed that possibly the consolidation of overlapping proposals, and the elimination of unwarranted and indefensible recommended areas, would substantially reduce the total number as well as the size of some of them. In October of 1978 another list appeared which contained 170 proposals, but which no longer provided any information with respect to the size of each or the reason for their nomination. When this list was consolidated there seemed to be about 102 separate recommended areas of which 22 were off the Atlantic coast, 12 in the Gulf of Mexico, 31 off the Pacific Coast, 13 off Alaska and 16 in the Great Lakes. Because their size and precise location were unknown it continued to be very difficult to assess the impact they might have on future offshore oil and gas operations.

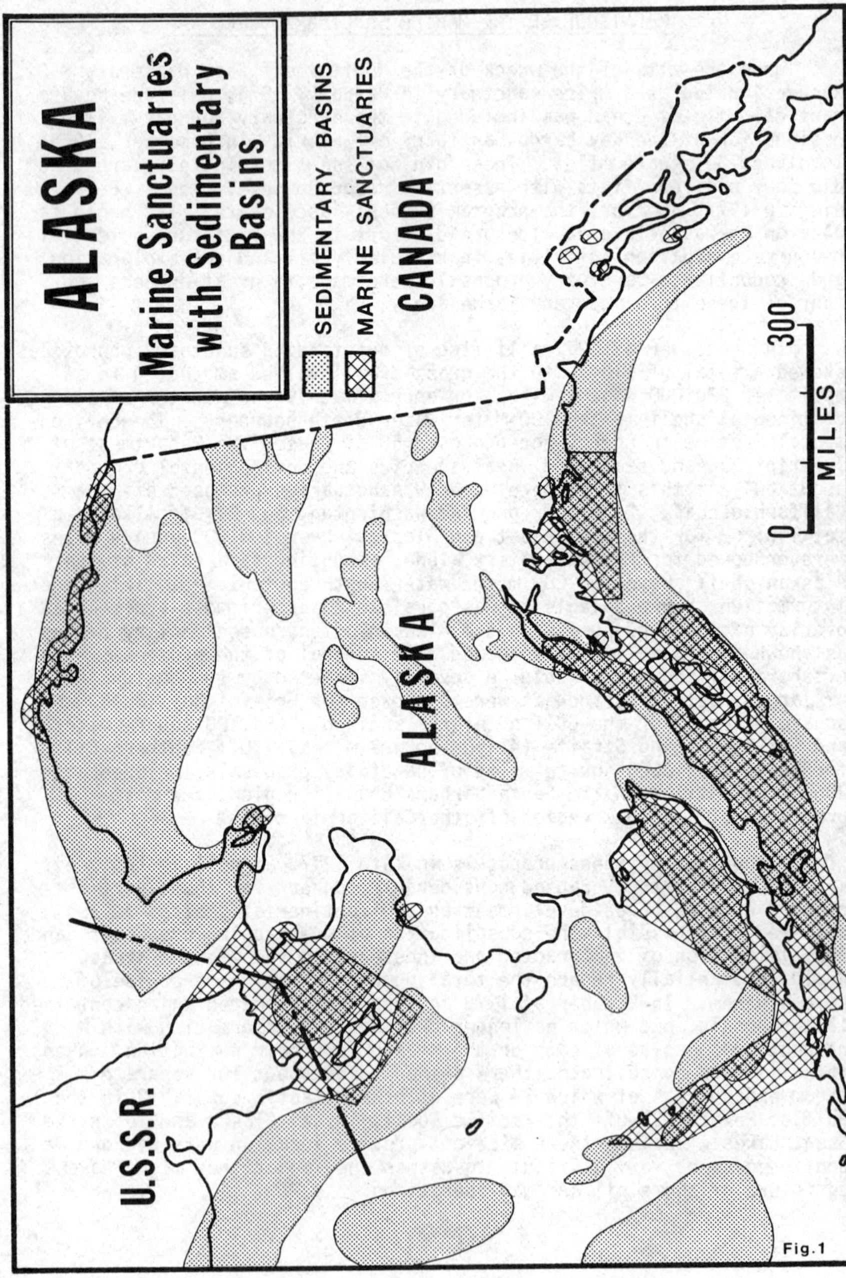
ALASKA
Marine Sanctuaries with Sedimentary Basins
SEDIMENTARY BASINS
MARINE SANCTUARIES
CANADA
ALASKA
U.S.S.R.
0
300
MILES

Fig.1

In July, 1979, NOAA published a final rule for the Designation and Management of Marine Sanctuaries. In part this attempted to set forth specific criteria for determination of the eligibility of a prospective area for consideration as a marine sanctuary candidate. However, the criteria were sufficiently broad and numerous as to permit almost any shelf area in the ocean to qualify on one basis or another. Under the new rule if an area qualified for further consideration it could then be placed on the list of Recommended Areas. Upon meeting additional specific requirements the sanctuary proposal would advance to Active Candidate status. Applying the standards of the new rule resulted in a candidate list published in October, 1979, which contained 69 Recommended Areas and 7 Active Candidates. Again, with no definition of the size or exact location of each proposal, an evaluation of their potential impact on resource development is quite uncertain. However, it is clear that, assuming prohibition of petroleum operations continues to be the principal thrust of the program, future outer continental shelf exploration and development will be significantly diminished. This appears to be especially true for California and Alaskan waters. Of the 10 sanctuaries on the Recommended or Active Lists off California it appears that 9 may adversely impact the four California OCS sales on the new 5-year schedule. Off Alaska 10 OCS sales are listed in the new 5-year schedule and 8 of these may be affected by the outstanding sanctuary proposals. Other sanctuaries recommended off the Atlantic Coast and in the Gulf of Mexico will probably also inhibit petroleum development, although perhaps not as severely as in the California and Alaskan OCS.

MARINE SANCTUARIES VERSUS PETROLEUM DEVELOPMENT

There can be little doubt that if the marine sanctuaries program proceeds as presently being implemented, and if the sanctuaries are of the dimensions and numbers proposed to date, the nation will suffer a substantial loss of petroleum resource. The latest U.S. Geological Survey estimates of undiscovered recoverable resource for the OCS to the 200 meter water depth are 25.2 billion barrels. Of this 17.6 billion are believed to lie beneath Alaskan waters, and this is the region where the greatest number of large sanctuaries have been proposed. If a quarter of this Alaskan potential is curtailed by the sanctuaries program the resource loss would be on the order of 4-5 billion barrels. If even nominal resource losses result from sanctuary designations off California, the Gulf of Mexico and the Atlantic Coast, then the total loss to the nation could approach 6-7 billion barrels. This would amount to 2.5 years of oil imports at 7 million barrels per day, and would mean an expenditure of about 200 billion dollars if the price of oil remained constant at 30 dollars per barrel. Furthermore, the Department of Energy's production goal for the OCS of 597 million barrels per year by 1990 would not be met, and other energy sources would have to be developed, unless we continued to depend on unreliable importation of foreign oil.

The present OCS 5-year leasing schedule has already been criticized in various quarters as being too slow, and offering areas too small in size, to provide sufficient opportunities for discoveries which will increase domestic production rapidly enough to displace significant amounts of imported oil. To further diminish potential productivity from the OCS via de facto withdrawals of large ocean areas, as will occur under the present marine sanctuaries program, is a proposal which must be viewed with great concern. It is on this issue that the controversy surrounding the sanctuaries program centers.

Proponents of OCS oil and gas development in general are not opposed to the concept of marine sanctuaries where they can be shown to fill a genuine need on a site-specific basis and where the regulations proposed for them are not needlessly precluding essential resource development. As noted earlier, the prohibition of oil and gas operations within the sanctuaries is the only major or significant addition to the existing authorities and agency regulations currently protecting the ocean environment. The question then arises: is this a necessary and reasonable prohibition? Long experience in the Gulf of Mexico, the Santa Barbara Channel, Cook Inlet and other offshore production areas world-wide strongly supports the view that it is not.

The record of offshore petroleum operations in domestic waters is an outstandingly fine one with respect to oil spills and other environmental considerations. About 24,000 wells have been drilled in U.S. waters with only four major spills, only one of which, the 1969 Santa Barbara spill, reached shore. Scientific studies of this spill showed rapid recovery of the marine ecology and no significant or lasting harm to the environment. Many other real-world studies in the Gulf of Mexico and other areas of long-standing marine oil and gas production have also shown no environmental damage related to this activity. Based on this lengthy favorable experience, the extensive environmental protection afforded under existing law and regulation, and especially the environmentally protective mandates of the OCS Lands Act Amendments of 1978, it is difficult to find justification for the prohibition of petroleum operations within the large sanctuary areas under consideration.

The Congressional debate of Title III of the Act prior to its passage in 1972 clearly shows that the legislators, including the sponsors of the bill, had no intention of precluding oil and gas development within marine sanctuaries. Had our lawmakers anticipated the consequences which are now unfolding, it seems highly probable they would have written into the law provisions to insure continued resource development in sanctuary areas. Unfortunately, they could not foresee the manner in which the law would be implemented and neither could they anticipate the energy crisis our nation now faces.

FUTURE OF THE MARINE SANCTUARY PROGRAM

It is difficult to predict what the ultimate outcome of the burgeoning sanctuary program, with its resource development prohibition, will be. Virtually all analysts agree that the U.S. will be facing a serious, even critical energy supply situation for many years to come. The petroleum resources of the continental shelf could greatly help in achieving the transition to new energy sources which must and will come on stream as we enter the next century. The oil and gas which the ocean region may contain is needed now, indeed we have already delayed too long in setting about finding and developing it. One might then suppose that policies such as the prohibition of petroleum development within marine sanctuaries will come under increasingly critical scrutiny in the near future.

Standing back from the issue, and trying to see it in a realistic perspective, it would seem that a few very simple and straightforward amendments to, or modifications of, Title III of the Act would remedy this conflict and provide for the application of multiple-use principles for all but the most unusual sanctuary proposals. Efforts have been made to clarify the legislation in this regard, but they have been fruitless to date. Perhaps the magnitude of the ultimate impact of the de facto withdrawal of these numerous and large ocean areas from our national resource base has not yet become sufficiently apparent. Indeed, there has been no assessment of the cumulative effects of the program based on a consolidated analysis of the outstanding sanctuary proposals. In the case of the recently adopted new 5-year OCS Leasing Schedule an environmental impact study was performed in order to facilitate just this sort of an evaluation. It would appear that a similar study of the cumulative impacts of the sanctuaries now proposed is sorely needed prior to further implementation of the program.

With such an analysis in hand it should be possible to reach a reasonable and well-founded judgement as to the appropriate manner in which to continue to implement the sanctuary program. Many marine scientists and analysts of offshore oil and gas development are convinced that this activity can proceed, even in relatively sensitive ocean areas, without causing environmental harm. Since the principal thrust of sanctuary designation is the prohibition of petroleum operations, and because there is a considerable amount of evidence based on real-world studies and past experience, indicating that this is not a necessary or reasonable proviso, the need for a long-range analysis of the program seems clear. To have abandoned opportunities for the development of critically needed energy resources without a proper weighing of the consequences and of the justification for so-doing would be contrary to the national interest. If, on the other hand, a comprehensive long-range evaluation of the sanctuary program should clearly show that all the proposed sanctuaries are necessary, and that it is essential to prohibit oil and gas development therein, then significant changes in the OCS 5-year Leasing Schedule would probably be required in order to meet production goals. If a number of the most promising frontier areas for petroleum exploration are to be withdrawn from the national resource base then it may become necessary to accelerate the opening-up of presently unavailable onshore areas, and to expedite increased use of coal and nuclear energy resources. The implications of a continuing program which prohibits offshore oil and gas development are not trifling, and are certainly deserving of a thorough study of the cumulative and related consequences.

Another consideration which has been the subject of continuing controversy has been whether it is a proper democratic procedure to withdraw numerous large areas of the public domain from multiple-use without Congressional approval. It has been noted that wilderness land withdrawals under the Federal Land Management Policy Act of 1976 are subject to Congressional approval authority, and it is argued that public lands in the OCS should receive no less consideration. Any significant reduction of the nation's energy resource base is clearly a matter of great public concern, and it seems quite reasonable to place the final decision on such actions in the hands of elected representatives of the people. Correction of this element of Title III of the Act would require only a simple amendment to the statute.

It seems highly probable that the past and current controversy over the marine sanctuary program will continue as long as the present policy of prohibiting resource development in designated sanctuary areas prevails. As on-going offshore oil and gas activities build on their already fine record, and as more real-world research continues to show that these operations have no harmful environmental consequences, perhaps the implementation of the sanctuary program will shift toward multiple-use concepts. This could occur administratively or as the result of legislative action. In this era of national energy crises it seems certain that the public interest is best served by multiple-use of the public domain to the maximum extent possible.

PROPOSED GRAY'S REEF MARINE SANCTUARY: A CASE STUDY

Carroll N. Curtis
National Oceanic and Atmospheric Administration

Jeannette H. Phillips
Georgia Department of Natural Resources

Abstract

Under the Marine Protection, Research and Sanctuaries Act of 1972, the Secretary of Commerce, with Presidential approval, may designate ocean areas as marine sanctuaries to preserve or restore their conservation, recreational, ecological or aesthetic values. The Marine Sanctuary Program is administered through the National Oceanic and Atmospheric Administration (NOAA). In June 1978, the Georgia Department of Natural Resources (DNR) nominated a "live bottom reef" on the South Atlantic Continental Shelf off Georgia as a marine sanctuary candidate. This area, Gray's Reef, is a site of limestone outcrops which support rich marine life on an otherwise flat, sandy seafloor and attract a variety of user groups from coastal and inland Georgia.

Live bottoms are known locally as fish havens, yet the role and importance of these ecosystems to fishery resources remains uncertain. Likewise, little information exists on the sensitivity of live bottom resources to human use. Unlike much-studied coral reefs of the tropics, public awareness of live bottom ecosystems is minimal. Designating Gray's Reef as a marine sanctuary would provide a mechanism for protecting the resources, promoting scientific research, enhancing public education and coordinating activities compatible with conservation.

The history of the Gray's Reef Marine Sanctuary designation process exhibits a combined effort of State and Federal governments, a high level of public involvement and perhaps a new awareness of how conservation programs can achieve a number of goals in the public interest. Recognizing the multiple values of Gray's Reef to the people of Georgia, NOAA has encouraged active public participation in predesignation planning. Proponents of the sanctuary, including scientists, resource managers, educators, fishery management council members, environmentalists, sport fishermen, divers and other concerned citizens have accepted the challenge and have assisted NOAA in the development of sanctuary boundary,

---

Ms. Curtis is a Program Analyst in the Sanctuary Programs Office, Office of Coastal Zone Management, NOAA, 3300 Whitehaven St., N.W., Washington, D.C. 20235 (202/634-4236).

Ms. Phillips is Acting Chief of the Coastal Management Services Section, Coastal Resources Division, Georgia DNR, 1200 Glynn Ave., Brunswick, Georgia 31523 (912/264-7289).

regulatory and management strategies that are sensitive to all vested interests. This combined effort has already resulted in the formation of the following proposed management goals:

- To protect and maintain physical, biological, ecological, and aesthetic resources of the live bottom ecosystem in their natural state.
- To promote scientific understanding of the ecological nature and role of the live bottom ecosystem and the functional relationships of live bottom areas throughout the South Atlantic to one another and to the overall coastal and marine ecosystems of the region.
- To promote public appreciation and wise use of regionally significant live bottom resources.

Introduction

Many persons in coastal Georgia regard Gray's Reef as a special marine area with important ecological research, recreational and aesthetic values. The live bottom is located 17.5 nautical miles east of Sapelo Island, Georgia and is one of the largest known expanses of inshore live bottom in the South Atlantic, covering an estimated 16 square nautical miles. The reef consists of a series of limerock outcrops which provide habitat for an abundance and variety of marine species at various levels in the food web. The live bottom is an important habitat for marine invertebrates and finfish of ecological, recreational and commercial value, for threatened and endangered sea turtles and for tropical reef biota which are naturally rare to the area, representing extensions of their normal geographical range.

Gray's Reef is one of the few live bottom areas in the South Atlantic which has been studied, although only to a limited extent. Hardbottom outcrops are not typically encountered in nearshore areas off Georgia; the geomorphology of the Georgia Embayment is such that flat, sand bottom covers over 80 percent of the ocean floor. Furthermore, the hardbottom structure and relief at Gray's Reef is usually only encountered in deep water locations farther offshore.

The nearshore location of the live bottom, its year round accessibility and the distinguishing nature of its resources make Gray's Reef particularly inviting for public use. It is perhaps the most highly utilized natural reef off Georgia, attracting recreational fishermen and divers, educators and scientists from coastal and inland Georgia and from neighboring southeastern states. Sport fishermen are regularly drawn to this fish haven in search of demersal reef fish and migratory pelagics; two Georgia record game fish were taken at the reef: red snapper (36 lbs.) and king mackerel (46 lbs., 4 oz.). Divers proclaim that, when visibility and currents are favorable, Gray's Reef is a good diving spot, with encrusted limestone ledges two to six feet high, hard corals and most of the tropical fish found on reefs off

Florida. Educators and researchers contend that the live bottom provides an excellent area for study and a control area for research of the biological interactions occurring at southeastern hardbottoms. Increased use of Gray's Reef is expected in the future especially in light of fuel shortages and rising costs which make travel farther offshore more difficult.

Live bottom areas have just recently been recognized as significant biotopes in the South Atlantic and very limited knowledge exists concerning their ecological nature and role. The dynamics of live bottom benthic communities and their importance to marine fishery resources of the South Atlantic have not been fully explored or interpreted. Scientific research concerning live bottom resources has been limited to qualitative biological inventories and geological characterizations.

Relatively little data exist on the impacts of human activities on live bottom systems. Preliminary studies suggest that live bottom resources are vulnerable to environmental perturbation. In combination, certain human activities (e.g., seabed alteration and construction, disposal of shipborne wastes, anchoring, bottom trawling and dredging, wire trap fishing and marine specimen collecting) could adversely impact ocean water quality, benthic habitat and living marine resources in live bottom areas.

Public knowledge of live bottoms is also limited. Most public education in marine science features sterotypical tropical coral reefs.

In light of these issues, Gray's Reef is being considered for marine sanctuary designation under Title III of the Marine Protection, Research and Sanctuaries Act of 1972. Sanctuary designation would offer multiple public benefits not currently available: long-term protection and technical monitoring of live bottom resources; coordination of activities which are compatible with conservation; promotion of the "living laboratory" for scientific research; maintenance of a control area for assessment of environmental situations in other live bottom areas; and provision for public education concerning the ecological importance and wise use of live bottom resources.

The history of the Gray's Reef Marine Sanctuary designation process exhibits a combined effort of State and Federal governments, a high level of public involvement and perhaps a new awareness of how conservation programs can be used to achieve a number of objectives in the public interest. The following is a case study which highlights predesignation management planning as a vehicle to promote public involvement and information exchange by various entities concerned with the future status and use of Gray's Reef.

Background

The Marine Sanctuary Program is administered through the Office of Coastal Zone Management of the National Oceanic and Atmospheric Administration (NOAA). It focuses on comprehensive management of marine ecosystems for the long-term preservation of natural resources and the enjoyment and benefit of society. Two national marine sanctuaries have been designated thus far: the Monitor Marine Sanctuary, surrounding the wreck of the Civil War ironclad U.S.S. MONITOR, off Cape Hatteras, North Carolina; and the Key Largo Coral Reef Marine Sanctuary just seaward of the John Pennekamp Coral Reef State Park, Florida Keys. Besides the Gray's Reef live bottom, six other marine areas are active candidates for sanctuary designation: Flower Garden Coral Banks, twin coral caps in the Gulf of Mexico; waters around Channel Island and Santa Barbara Island; Monterey Bay; and Point Reyes and the Farallon Islands, California, which support many marine mammals (seals, sea lions, sea otters, whales and dolphins) and sea birds; Florida's Looe Key coral-reef area; and lush seagrass beds and coral reefs off St. Thomas, Virgin Islands.

An extensive process of consultation and evaluation is conducted by NOAA prior to designation of any marine sanctuary. Development of various management, boundary and regulatory alternatives considered for a proposed sanctuary is coordinated with affected Federal agencies, State governments, Regional Fishery Management Councils and other interested entities. This exchange of information is incorporated into NOAA's decisionmaking process. For example, NOAA and the South Atlantic Fishery Management Council (SAFMC) signed a Memorandum of Understanding (MOU) for the exchange of advice and information concerning the management of marine resources, including the regulation of activities, in areas under their respective jurisdiction. (The SAFMC's geographical area of authority includes high sea waters between 3 and 200 nautical miles -- the Fishery Conservation Zone -- off the coasts of North Carolina, South Carolina, Georgia and eastern Florida.) As a result of the MOU, NOAA and SAFMC work closely together to coordinate activities related to fishery management plans and marine sanctuaries to ensure compatible management measures.

Similarly, NOAA views public participation as an integral part of the designation process, planning phases and day-to-day sanctuary management. Public involvement during the early stages of the designation process has proven effective in capturing various local interests, creating long-lasting avenues of communication and solving problems, if they arise. NOAA has strived to foster a highly visible and balanced public involvement program to ensure equal consideration of a wide range of public interests.

The Gray's Reef Marine Sanctuary Designation Process

In June 1978, the Coastal Resources Division of the Georgia Department of Natural Resources (DNR) nominated Gray's Reef for consideration as a marine sanctuary. NOAA preliminarily reviewed the nomination and determined, based on its distinctive marine resources and potential sensitivity to environmental perturbation, that Gray's Reef met the criteria outlined in NOAA regulations as required for placement on the sanctuary program's List of Recommended Areas (LRA).

The first step in NOAA's consultation process for Gray's Reef was to solicit information and opinions on the nomination. In July 1979, NOAA distributed the Gray's Reef nomination among Federal and State authorities, Regional Fishery Management Councils, environmental and special interest groups and interested individuals. Responses to the nomination were favorable.

Based on the information obtained through consultation and on criteria stated in the NOAA regulations, NOAA selected Gray's Reef from the LRA as an Active Candidate for sanctuary designation and announced in the Federal Register (44 Fed. Reg. 58938 10.12/79) its intent to prepare an Issue Paper and to schedule public workshops in areas affected by the proposed designation.

Preparation of the Issue Paper involved several NOAA field visits to Georgia for consultation with Georgia DNR and others familiar with the Gray's Reef area. DNR provided an initial list of key contacts who would likely be concerned with the proposed action and briefed NOAA on technical issues concerning Gray's Reef. DNR also sponsored a field trip to Gray's Reef to acquaint NOAA representatives with the affected resource.

Interests and issues which evolved early in the designation process helped identify initial field contacts. Through site visits and written and verbal correspondance, interested parties were contacted by NOAA to inform them of the proposed action and the major issues under study. Contacts were asked to provide techincal assistance and viewpoints which would help NOAA in its decisionmaking and to suggest other persons or groups who might want to take part in and be notified of the marine sanctuary proposal. All persons contacted were put on a mailing list for receipt of any Gray's Reef documents.

The Issue Paper was widely circulated in late October 1979 for public review and comment. It presented a brief summary of the important resources and uses of Gray's Reef, described the issues which prompted the proposed action and outlined a range of management, boundary and regulatory alternatives for consideration. For various reasons, gaps existed in the information presented in the Paper and NOAA solicited comment and discussion concerning the proposal.

NOAA conducted public workshops on the proposal in Brunswick and Savannah, Georgia on November 19 and 20, 1979, respectively. The purpose of these forums was to create an opportunity for public discussion and clarification of issues surrounding the proposal. Question-and-answer exchanges were a part of the format and proved successful in providing factual information, correcting rumors and identifying additional field contacts.

Written and verbal comments received on the Issue Paper aided NOAA in deciding whether to propose to Gray's Reef Marine Sanctuary and to prepare a Draft Environmental Impact Statement (DEIS) on such a proposal. Overall, the Gray's Reef proposal received considerable support, with proponents citing as beneficial impacts: (1) the coordination of uses and promotion of conservation of live bottom resources, (2) development of research and education programs; and (3) provisions for public services. Local recreational fishermen and divers took issue with the possible regulation of hook and line fishing and spearfishing and convinced NOAA that these activities were not currently threatening the live bottom resources.

Comments on the Issue Paper provided guidance on the nature of the issues addressed in the sanctuary proposal. NOAA carefully evaluated all available information and announced in the Federal Register (45 FR 2078 1/10/80) intent to go forward with the designation process and a DEIS.

During the DEIS preparation phase, NOAA conducted a scoping meeting in Washington, D.C. to solicit suggestions, comments and substantive information concerning the proposal from Federal agencies and other interested parties. Additionally, NOAA consulted extensively the scientific literature and many field contacts concerning the environment affected by the proposal and the types of activities subject to management.

The DEIS was widely circulated for public review in May 1980. To notify persons not currently on the program mailing list, announcements of its availability appeared in the Federal Register (45 FR 39597 6/11/80) and in several Georgia newspapers. Copies of the statement were also available for public review at Coastal Resources Division of Georgia DNR (Brunswick, Georgia) and in regional public libraries around the State. Additionally, a summary of the DEIS geared to local citizen's language was prepared by the Georgia DNR and distributed among fishermen, divers, and other user groups in coastal Georgia.

NOAA held public hearings on the proposal on July 7 and 8, 1980 in Brunswick and Savannah, Georgia, respectively. Announcements of the hearings appeared in the Federal Register (45 FR 41407, 6/17/80) and in several Georgia newspapers. The hearings provided local citizens with the opportunity to express their views concerning the Gray's Reef proposal. A majority of the public testimonies and written statements received on the DEIS supported the Gray's Reef proposal, recognized the various public benefits the sanctuary program would provide, and agreed with the preferred alternatives. Closing date for comments on the DEIS was August 5, 1980.

At the time of this writing, the Final Environmental Impact Statement (FEIS) on the Gray's Reef proposal is being prepared. Public opinion influended NOAA to make a few changes in the proposal. For example, in light of public request, the FEIS includes an expanded discussion on proposed sanctuary management and a clarification of the use of monitoring as a management tool. The FEIS proposes to designate 16.68 square nautical miles of live bottom habitat as a marine sanctuary and to prohibit, except by permit, seabed alteration and construction, bottom trawling and dredging, wire trap fishing, marine specimen collecting and removal of cultural and historic resources. Disposal of polluting substances, except fish bait, parts and chumming materials, marine sanitation effluents and vessel cooling waters, would be prohibited. Recreational activities, including boating, fishing, diving, spearfishing and anchoring would be allowed and sanctuary management would monitor and assess affected resources and activities.

The FEIS will be widely distributed. NOAA will receive comments for 30 days and then will consult again with Federal agencies. After review and consultation, a decision will be made whether to proceed with the designation. If so, the Secretary of Commerce will seek Presidential approval of the designation.

Predesignation Management Planning: Public Information Exchange and Education

Predesignation planning efforts involving local public information exchange and education are intertwined throughout the formal designation process described above. To enhance this effort and to utilize the experience and expertise found at the local level, NOAA entered into a Cooperative Agreement with Georgia DNR in June 1980 to intiate predesignation managment planning. As part of the joint nature of this agreement, NOAA and Georgia DNR have provided the services of staff members from each division.

One of the most important tasks involving the designation process and predesignation planning has been "getting the word out". As previously described, a first and continuing effort has been the identification of key field contatcs, or those local citizens, groups or officials who are personally involved, by virtue of their recreation, profession or interest, in the sanctuary area and proposal and whose views and technical assistance are important in planning and management. Contacts have included representatives of government, science, education, conservation, recreation and resource management. They have included both proponents and opponents of the proposal. Contact emphasis has been directed towards obtaining information on the following management concerns: (1) resources and monitoring; (2) surveillance and enforcement; (3) permit review and evaluation; (4) scientific research and education; and (5) public interest and involvement.

A variety of techniques has been used to get the word out. Georgia DNR has prepared and distributed questionnaires, document summaries and announcements to local county government seats, schools, libraries, marinas, dive shops, fishing camps and suppliers and other vehicles of local communication to help identify current and potential users of Gray's Reef and to foretell the climate of public interest. The news media has provided an indispensable avenue for contacting wider segments of the public. Newspaper and magazine articles radio and television spots have increased public awareness of the national marine sanctuary program, in general and the Gray's Reef proposal, in particular. Media attention has been stimulated by conducting interviews, releasing news bulletins, writing articles and conducting public meetings, workshops and hearings. Local newspapers in coastal Georgia have given good coverage of the Gray's Reef proposal and public involvement. Articles on the national marine sanctuary program, with reference to Gray's Reef, have appeared in recent editions of Defenders (Vol. 55, No. 3, June 1980), Skin Diver (Vol. 29, No. 7, July 1980), and Sport Diver (Vol. 4, No. 4 July/August 1980). Gray's Reef has been featured in several Georgia-wide publications, including the Atlanta Constitution, the Atlanta Journal and Coastlines Georgia (A Georgia DNR bimonthly publication). Announcements of significant developments in the sanctuary proposal have appeared in the Federal Register. Radio coverage of the proposal through the "Georgia Network" and a video segment on the live bottom habitat at "Stay and See Georgia", an annual public information exhibition in Atlanta, succeeded in reaching large segments of the Georgia public.

NOAA and Georgia DNR representatives have met with various groups and organizations in coastal Georgia, including the Coastal Audubon Society, Georgia Coastal Management Board, Georgia Conservancy, Golden Isles Divers and Sportfishing Clubs, Fishermen's Coop (McIntosh County), and the newly forming Savannah Dive Club. The purpose of these meetings has been to keep interested parties informed of the sanctuary designation process, to increase their understanding of the proposal and to solicit their input in planning and management endeavors.

In a similar vein, NOAA has briefed Federal agencies at the Washington level on the progress of the predesignation planning and solicited their input to the product. Similar meetings were held with the South Atlantic Fishery Management Council.

A record of public involvement has been kept to provide documentation of the correspondence and consultation with interested groups and individuals. Transcripts, tapes, notes and similar material from public meetings, workshops, and presentations have been filed. Similarly, all comments on documents and results of correspondence/consultation have been maintained in the record. Responses to questionnaires have been analyzed statistically, yielding valuable information on public interests, activities and demographic distribution. All interested organizations, institutions and citizens identified to date have been categorized by interest and stored on a computer source list. Interest group card files supporting the source list have been started to provide for additional information.

Future Directions

The first step in the management of the Gray's Reef Marine Sanctuary following designation will be the preparation of a formal Management Plan (MP). The MP will be a direct outgrowth of predesignation planning efforts. Final program goals and objectives will be formulated at the time the MP is prepared. They will form the heart of the Plan, providing a framework for conserving resources and allowing environmentally compatible public activities. Because a marine sanctuary concept is holistic in nature, the MP will be structured in a way that conveys the interrelationships among resources and public use and among areas of management concerns.

At the time of this writing, Georgia DNR has submitted to NOAA a proposal to develop the Gray's Reef Marine Sanctuary MP, contingent on sanctuary designation. As proposed, the MP will address specific areas of management concern: administration; surveillance/enforcement; resources management; and public education and visitor use. It will identify programs to be implemented within specific timeframes, provide the reationale behind them and relate them to other activities in the sanctuary and its vicinity. Preparation of the MP will involve considerable public input.

One mechanism being considered to facilitate public involvement in preparation of the MP is a MP Planning Workshop. The Workshop will be in the form of a panel discussion meeting and will be convened in the area affected by sanctuary designation to solicit public participation. Since planning considerations and management decisionmaking involve complicated issues in a number of areas, the panel will be composed of professionals and interest groups representing alternative viewpoints. The workshop will create an opportunity for presentation, open discussion and assessment of a wide spectrum of technical information pertinent to management decisions. It will ensure that all reasonable alternatives have been considered and that both beneficial and adverse consequences of each alternative have been identified.

Invited workshop panelists will likely be among the key contacts identified early in the designation process who have followed the development of the proposal and who have experience or interest in marine resource management. The public will be invited and will have the opportunity to present their viewpoints to NOAA, the panel and other citizens during the open session. Additional information exchange between the public audience and the panel may be facilitated through a chairman or moderator. The proceedings of the workshop will be a public document.

The MP will contain provisions for public education and public involvement in management issues. In developing the MP, preparers will examine, among other things: (1) the purpose of public involvement; (2) the methods of notifying the public about the opportunity to participate; and (3) the methods of gathering, synthesizing and utilizing public input. A desireable outcome of this evaluation will be strategies for communicating information to the public at frequent and systematic intervals to keep pace with emerging management issues. Similarly, the MP will examine a range of educational techniques for promoting public knowledge and understanding of the sanctuary and its resources.

The MP will also evaluate the desireability and feasibility of establishing an advisory committee for the purpose of advising or making recommendations to NOAA. As tentatively proposed, an advisory committee would represent a variety of interests including government, research and education, conservation and recreation, and therefore would provide broad insight into planning and management issues. A representative committee would present general public input in a timely and direct fashion which might otherwise not be available. Because committee members would include local citizens, the committee would serve to enhance public support and participation in sanctuary affairs. Regularly scheduled meetings could provide a forum for disseminating information and reviewing management proposals.

A draft MP will be distributed for review and comment, and public meetings will be convened to clarify program goals and objectives, where necessary. A final MP will be made available as a public document. Following implementation of the MP, an on-site manager will be designated by NOAA to oversee day-to-day management of the sanctuary and to carry out directives in the MP.

In conclusion, the history of the Gray's Reef Marine Sanctuary proposal represents pioneer efforts to encourage a high level of public involvement in the designation, planning and management of a marine sanctuary. Local, State, Regional and Federal interests have been consulted to foster more effective and meaningful development of the proposal. This effort has evolved as part of the required designation process and through predesignation planning in the areas affected by designation. It has proved extremely successful in formulating management goals and objectives in the best public interest.

MARINE SANCTUARY INSTITUTIONAL ANALYSIS

by

Brian J. O'Sullivan
Environmental Planning Analyst
Center for Natural Areas
Washington, D.C.

ABSTRACT

Title III of the Marine Protection, Research and Sanctuaries Act of 1972, as administered by the United States Department of Commerce Office of Coastal Zone Management's (DOC/OCZM) marine sanctuary program office, provides a means of comprehensively managing marine activities by designating and assuring the long-term protection of significant marine areas. Once nominated and having attained "active" status as a potential sanctuary site, these areas undergo further in-depth study through the preparation of an Environmental Impact Statement (EIS).

One significant element in this proposed sanctuary designation process is to assess, within an EIS framework, the adequacy of State and Federal regulatory and management arrangements for protecting sensitive marine resources and habitats against human activity threats. In operation, marine sanctuary EISs consist of a programmatic evaluation of potential marine resource protection (and conversely degradation) impacts under existing institutional arrangements (the "status quo"), as well as alternative regulatory/management mechanisms. Ideally, such an analysis helps to identify the need for a formal designation, if one in fact exists, and a "preferred" alternative for strengthening the marine environmental protection regime.

Regardless of its intent, however, the EIS is a somewhat limited vehicle for accomplishing thorough institutional evaluations. While statutory gaps and jurisdictional constraints "on paper" can be readily identified, many agency and organizational performance shortcomings (as well as strengths) relating to administrative efficiency are difficult to isolate and analyze. To date, no institutional performance evaluation criteria or methodologies to implement them have been integrated into the marine sanctuary EIS procedure. Consequently, the analytical basis for rationalizing changes in the existing regulatory/management system for a sanctuary designation often remains weak and underdeveloped. This can, for example, result in premature or unwarranted determinations that the "status quo" merits alteration based on insufficient evidence. No framework exists for reasonably determining the innovative resource management capability of existing arrangements exclusive of additional regulatory tampering.

This paper briefly outlines issues surrounding the need to incorporate institutional performance evaluations into the marine sanctuary EIS process; at the same time it highlights obstacles to their adoption.

1.0 INTRODUCTION

The establishment of a Federal marine sanctuary program (MSP) in 1972 helped place the National Oceanic and Atmospheric Administration's Office of Coastal Zone Management (NOAA/CZM) at the forefront of the nation's offshore environmental resource protection and management effort. To date, however, the MSP has moved slowly towards establishing its own identity and attaining programmatic momentum. Funding problems and administrative wrangling--both intra- and inter-agency--are in part responsible for this evolutionary lag. In addition, the MSP has become deeply embroiled in the country's more recent energy vs. environment controversy, particularly in relation to Offshore Continental Shelf (OCS) oil and gas development activities. Proposals to further restrict such environment-threatening activities in the interests of cautious, long-term marine resource species and habitat protection are increasingly falling upon deaf ears. Generally speaking, there is evidence of a backlash within both the private and public sector against the promulgation of environmental protection regulations for their sake. As is the case for numerous other environmentally-oriented Federal programs, the MSP has also had to contend with the proponents of a more conservative environmental policy and management intervention.

Largely on account of its comprehensive perspective on marine resource management, the MSP transcends the jurisdictions of many other Federal and State agencies or departments. One consequence of this role has been the MSP's subjection to interagency tensions and rivalries. It has become increasingly apparent, for example, that certain Federal-level departments strongly resent NOAA/CZM's statutory power to regulate _across_ functional agency jurisdictions. Having perceived MSP intervention as unnecessary regulatory tampering and a threat to their own missions, some departments actively seek to influence--or exert _de facto_ control over--NOAA/CZM's marine sanctuary designation decision-making process. Many State-level governmental agencies are equally as wary, although relatively less influential.

The framework utilized by NOAA/CZM to assess nominated marine sanctuary sites is the Draft Environmental Impact Statement (DEIS).

This document not only inventories in detail the significant marine resources at a specific site (as well as human activity threats), but also critically evaluates the adequacy of existing "institutional" arrangements to provide sufficient environmental protection with and without additional regulations. Wherever the status quo legal and administrative system encompassing resources and activities is non-existent or appears unsatisfactory, the MSP is authorized to promulgate reasonable regulations and management strategies to assure more thorough environmental protection.

The contention of this brief and conceptually-oriented paper is two-fold. First, it suggests that much of the interagency controversy and opposition to the MSP stems from their dislike of--or at least their suspicions about--being subjected to critical NOAA/CZM evaluations regarding their statutory authority and administrative performance. Second, it notes that many agencies probably question the analytical rigor of such evaluations and are fearful of being branded with unfavorable reputations based on insufficient or biased evidence. Until these two issue areas are clarified, and preferably integrated into the overall DEIS preparation process required to achieve marine sanctuary designation, it is doubtful that the MSP will escape the controversial atmosphere in which it now finds itself. Failure to resolve such tensions will undoubtedly weaken the program and perhaps even lead to its demise.

2.0 CONCEPTUAL AND METHODOLOGICAL ISSUES IN INSTITUTIONAL EVALUATION

Across nearly all disciplines, there is little agreement or consistency in the manner in which the term "institutions" is defined. A wide variety of disparate--and often contradictory--interpretations have emerged. To some, the concept connotes nothing more than existing laws (the MSP leans heavily toward this viewpoint); to others, institutions encompass everything from public enterprises, societal values and interest groups to communication channels and administrative frameworks--both formal and informal. A more comfortable conceptualization views institutions as complex systems of social arrangements incorporating many of the definitional permutations noted above. Most nominated marine sanctuary sites are characterized by an intricate web of such arrangements governing--either directly or indirectly --decisions about how offshore resources are utilized and protected. It seem logical, therefore, that any NOAA/CZM assessment of the adequacy of protective measures should examine the sufficiency of legal and organizational/administrative arrangements--among others. To date, however, there has been an overemphasis on the "on paper" sufficiency of statutory arrangements oriented toward marine resource protection and too little study of the performance of governmental agencies responsible for the day-to-day administration of those legal provisions. As discussed in Section 3.0 below, it remains to be seen whether the DEIS is the best forum for achieving such evaluations. Beyond the conceptual clarifications necessary to help focus MSP institutional evaluations, there are also methodological issues. What objectives and consistently applied criteria, for example, are available to assist NOAA/CZM judgements about the performance of Federal or State agencies in assuring that significant marine resources are now or will be adequately protected? At the moment, there are few. What performance level thresholds--linked presumably to the current and projected health of marine resources--can be established below which additional MSP regulations and management provisions are justified? For the time being, there is little guidance to assist NOAA/CZM in this area. Consequently, their evaluations of institutional arrangements appear incomplete and at best overly subjective. In lieu of greater conceptual clarity and a more rigorous methodology to measure performance, the agencies and departments subjected to NOAA/CZM review are bound to become more obstructive and contentious of MSP regulatory intervention.

3.0 THE DEIS FRAMEWORK

The DEIS prepared by NOAA/CZM for specific marine sanctuary sites is programmatic in nature. Its objective is to assess the natural and human environmental impacts of alternative regulatory and management strategies, i.e., a sanctuary designation package. From this analysis of alternatives--including no-action--a preferred option can be identified.

Given the conceptual and methodological uncertainties mentioned previously, however, the DEIS appears inadequate for effecting thorough institutional evaluations. To begin with, its analytical framework was not originally designed to integrate rather sophisticated legal and organizational/administrative performance studies. Also, the time period allocated for DEIS preparation, review and ultimate transformation into a Final Environmental Impact Statement (FEIS) precludes lengthy in-depth investigations of specific governmental agencies and departments.

Regardless of the limitations of the DEIS as a broadbrush institutional evaluation tool, NOAA/CZM could still isolate specific arrangements upon which to focus detailed analyses of the need for further MSP regulations or management assistance. For example, were OCS development activities deemed critically important relative to marine resource protection (as they often are), responsible Federal agency performance could be rated, e.g., the U.S. Bureau of Land Management or U.S. Geological Survey. Still, an evaluation period would have to identified as well as flexible assessment criteria. The precise arrangements chosen for in-depth study are likely to vary significantly depending on the peculiar resources and activity threats present at a given offshore site.

Supplementary DEIS institutional evaluation data--such as internal Federal agency performance assessments and annual reports or externally-authored critiques of particular agency programs and projects--could and ought to be integrated more diligently by NOAA/CZM. Consultations with agency personnel could also be arranged more systematically. Far from abandoning the programmatic DEIS framework, the MSP must work to fine-tune its own institutional evaluation needs and adapt them to achieve a truly objective decision-making document.

4.0 CONCLUSIONS

This paper has sketched a number of controversial issues--both conceptual and practical--surrounding the MSP's current approach to marine sanctuary institutional evaluations. Rather than offering any elaborate solutions, it has sought merely to bring these problem areas up for further public discussion and debate. In particular, it seems important for NOAA/CZM to take a hard look at how the MSP's DEIS preparation process can be altered to integrate more consistent and meaningful assessments of various institutional arrangements linked to marine resource protection. While certain substantive points made above may appear somewhat academic, they are nonetheless critical to maintaining the MSP on a steady, innovative course. Also, without any serious NOAA/CZM reflection on these topics, the risk remains high that the continuing alienation of other Federal-level agencies and departments will disrupt the MSP's effectiveness as a unique marine environmental protection tool.

Author's Note: The opinions expressed above are solely those of the author and do not in any way represent the sentiment of the Center for Natural Areas.

A PRACTICAL INTERGOVERMENTAL COORDINATION PROGRAM

BY MORGAN R. REES
Chief, Regulatory Branch
Army Corps of Engineers
New England Division

Introduction. With the mushrooming growth of governmental regulation, a great deal of duplication and wasted effort has developed in the regulatory process at all levels of government. Further, manpower limitations have placed pressures on existing staff which detract from the agency's ability to dispense with routine matters and respond in cases of importance with well reasoned professional judgments.

However, there are examples of successes in the field where agencies have pooled resources and shared responsibilities to achieve a high level of professionalism in the regulatory review process without needless duplication. They have in the process developed a rapport and trust among former "adversaries" which provide for speedier action without sacrificing the quality of review.

Frankly, at the outset of our consideration of joint efforts with other agencies in New England, I was a skeptic. I thought that the goals and procedures of the various agencies were too diverse to allow for much cooperation. However, our experience has been just the opposite. True, the goals and procedures are diverse. However, cooperation has been excellent.

We began by sponsoring a conference for all State and Federal agencies throughout New England. The attendance was excellent and everyone participated with a positive attitude. The Corps and the National Marine Fisheries Service provided advisers from Washington who had experience in developing joint procedures elsewhere. They were extremely helpful, particularly in setting the tone of cooperation for the conference.

After two and a half days of open, honest exchange, we all came away with a feeling of optimism and commitment which has carried through the development of various specific cooperative programs.

The agenda of the conference included discussions of problems common to each agency such as adequacy of plans and information upon which to base evaluations, importance of the principle of water dependency in projects proposed for aquatic and wetland areas, the scope of examination of alternatives and cumulative effects, and methodology for determining the extent of wetlands. We discussed general permits including the effectiveness of existing ones and possibility of expanding them, and various candidates for new general permits. We had a session on special studies to determine and evaluate the effects of permit conditions, and a full day on opportunities for joint processing activities.

To sum up in a few sentences, the conference resulted in people gaining a much better understanding of the functioning of many other agencies and in adoption of specific measures of cooperation including designation of who would do what and in what time frame.

For the very cost conscious, the conference was well worth it. The dividends received in terms of savings of staff effort, red tape, etc. will more than pay for the cost of the conference within the first year or so.

I am no longer a skeptic. I believe that with a positive attitude and a willingness to work cooperatively with agencies having differing mandates, a great savings can be realized by all and the quality of review can be improved to the benefit of the public and the individual agencies.

New England is certainly not the only area of the country to develop joint procedures. Many areas already have many different types of joint procedures. In this discussion, I have tried to cover all the bases of cooperation for processing permit applications so that other existing programs might be expanded as well as new programs begun where they do not now exist.

There are also opportunities for joint efforts in the enforcement area which we have not yet tapped in New England. The Baltimore District of the Corps and its cooperative State and regional offices of Federal agencies have developed an Enforcement Conference which meets regularly to deal with that aspect of the Corps' regulatory authority. It includes the State Attorney General's office and the U.S. Attorney as well as the State and Federal resource agencies. For further information on this program, contact the District Engineer in Baltimore. We hope to have a similar program in effect in New England in the coming months.

Various states have also taken the initiative in developing "one-stop" permit centers for the various State regulatory requirements. With proper coordination, the Federal review process can be added to the State process to result in a "one-stop" center for all regulatory requirements.

Since we deal with six states in New England, our coordination efforts are diluted six ways, but we are pursuing joint actions in all six states.

The following is a six joint program, each point of which is now operating in at least one state with the goal set as 1982 for all six points to be operational in all six states. For the purpose of organization of this discussion, the program is divided as follows:

1. Field Inspections
2. Applications
3. Public Notices
4. Public Hearings
5. Review and evaluation meetings
6. Special studies and impact evaluation

1. Joint site inspections. Prior to our coordination program, each review agency at the State and Federal levels conducted inspections of sites proposed for work subject to Corps permit requirements. Since there are too many permit applications for each agency to visit every site, each agency based on its own priorities would decide which sites to visit and which sites to pass up. This would result in some sites being visited by all agencies and other sites being visited by none. The duplication is obvious. In addition, however, there was the concern that for those sites not visited, there may be important resources not shown on the plans or the plans may not accurately represent the features of the area. Thus, proper evaluation could not be made. By jointly developing a mutually agreeable site inspection report, each agency can take its proportionate share of inspections and furnish the reports to all other interested agencies. With this process, every site is inspected, but the individual agencies reduce their total number of field visits significantly. However, if a proposal is complex and there are specific resources at risk in which any of the agencies has a special interest, each agency can still make case-by-case inspections on its own initiative.

In starting such a cooperative effort, we drafted a field report form with a checklist of items to be investigated and room for appropriate discussion. The form was reviewed by EPA, FWS and NMFS and several suggestions for improvements were incorporated. The attitude in developing the form was one of cooperation based on an understanding of other agency goals and philosophies. As a result, we reached concensus on the format and content in a very short time. In addition to the field report form which is shown in Appendix A, photographs are taken to be shown at a joint review meeting discussed below.

We divided the responsibility for site inspections geographically, based on proximity to agency field offices. Adjustments must be made occasionally to account for a balancing of the workload.

Based on the geographic distribution agreement, we have added an alpha character at the end of our numeric public notice number to indicate which agency is responsible for the field inspection. In this way, no additional communication is necessary to make the inspection assignments.

Here again, the spirit of cooperation can help overcome some potential problems relative to proportionate shares of the work. Each agency must realize that no system is perfect and that patience and compromise are keys to developing a workable program.

The field reports are used by each agency to make its own initial evaluation of a project. If the report is sufficiently detailed, no further field work is necessary. The report, however, need not have to stand alone. Sketches, photographs, and slides used by the inspector in a presentation at a joint review meeting can prompt additional questions and discussions of issues which may not be apparent from the report. We have found it helpful, almost essential, that the inspector be present at the joint review meeting (see below).

Since starting this process, we have had to make very few adjustments in its use. One major concern which remains is the training available to the field inspectors. Since the scope of review under the Corps regulatory process includes all factors which affect the public intent, field inspectors must have a basic understanding of engineering and environmental aspects of project sites. Close attention to the thoroughness of field reports, with regular feedback and instruction to the inspectors can go a long way toward achieving and maintaining high quality field reports and hence, good evaluation discussion.

Since most of the project sites involved are also subject to similar State regulatory programs, we intend to enlist the aid of State agencies which are also inspecting the same sites we are. Obviously under serious budget constraints, State agencies are not anxious to take on additional work. However, it is possible to coordinate field report forms with the State as we did among the Federal agencies, have the State agencies conduct the inspections, and transfer funds to the State for a proportionate share of the inspection costs. Most of the State's we deal with have expressed an interest in this kind of arrangement. It would address several serious governmental problems--it would add funds to lean State budgets and would allow Federal agencies to minimize their inspection staffs, thereby reducing the number of Federal employees or freeing up manpower resources to deal with other critical tasks.

Another approach to field inspection is to contract it out to private contractors. We have had an open-end contract for such services in effect for two years. We are, quite frankly, not at all satisfied with its workability. First, it requires reams of paperwork to develop and manage the contract, the minimum turn-around time from a request for inspection to filing an inspection report seems to be about a month under ideal conditions--much too long to be of much use--and the quality of the report is difficult to control. We have found that people outside the program do not readily grasp the scope of the program and, as a result, they frequently file inadequate field reports. Revision of such inadequate reports is another major delay and time consuming requirement.

In summary, field inspection burdens on individual agencies can be reduced substantially and overall costs to the government cut drastically with cooperative agreements among State and Federal agencies to share the effort.

2. Joint Applications

When there are duplicative regulatory processes at the local, State and Federal levels, and sometimes several duplicative processes at each level, the applicant is faced with a mountain of different application forms to fill in, most with identical information requirements but in different formats. With a little effort, agencies can set up a format which meets the needs of multiple agencies. Prior to 1977, Corps regulations were fairly narrow with respect to use of "the authorized" application form. With revision of the regulations effective July 19, 1977, the followng sentence appeared: "Local variations of the application form for purposes of facilitating coordination with State and local agencies may be used." (33 CFR, 325.1(b)) That was a clear signal to us in the field that the old Form 4345 had lost some of its sanctity. We went to work in those states which showed an immediate interest. The going was much slower than we had hoped. Even in the most efficient states, there are still numerous offices which may have a say in approving a new form and many with their own ideas on what the form should contain. The staff people negotiating and drafting the form must be persistent yet remain flexible in order to reach concensus on what the form should look like.

We have completed a joint application form with the State of Maine. Copies are available on request. The Maine Department of Environmental Protection will handle the distribution of the joint application forms and provide the Corps with copies of all applications. We have coordinated agency needs so that when the Maine DEP accepts an application as being complete, it can also be considered to be complete for the purposes of the Corps application process. The Corps will provide logistical support as required to print forms and may compensate the State for extra costs incurred in handling and reviewing the expanded application form. From there, each agency will follow its own processing procedures.

Maine officials were exceptionally cooperative. They are dedicated to the goals of reducing unnecesary government bureaucracy while maintaining a high level of protection of public resources.

As a side benefit, a news reporter sat through several of our coordination meetings and wrote some very favorable news articles. This not only helps the agency's "image", but more importantly, it serves to help rebuild public confidence that government officials are really concerned about how much they spend and how effective they are.

The normal progression in any development is crawl, walk, run. As significant as it is, this joint application form with Maine is like crawling. We still have a long way to go before the program runs smoothly. We have five other states, hundreds of municipalities and several Federal agencies to bring into the fold.

It may not be administratively workable to achieve joint applications with municipalities on our own initative because there are simply too many. The municipalities would most likely have to be brought together through some state mechanism such as Coastal Zone Management or the State planning office. Then the Corps and other Federal agencies might deal with one representative of all municipalities.

3. Joint Public Notice. The primary means of soliciting public comment for most regulatory processes is by public notice. When multiple permits are required, the applicant is faced with multiple notices being issued for the same proposal, the public is faced with responding to multiple agencies, and again the applicant must respond to the same questions to multiple agencies. All of this takes time, effort, reams of paper and does nothing to aid in the quality of review. As with the joint application forms, joint public notices can serve to focus in one place many of the duplicative activities now occurring at several agencies. Also as with joint applications, a commitment to cooperation and compromise can serve to overcome many obstacles of agency regulations. At

this writing, the only state with which we have joint public notices is Connecticut. The notice covers Corps application/State water quality certification. The major consideration in coordinating joint notices beyond the documents themselves, is how the administrative effort would be shared before and after the notice is issued.

In the case of Connecticut, the Corps will issue the public notices working from its own mailing list and the list provided by the State. The public will be advised that they can submit all comments to the Corps who will, in turn, furnish copies to the State. They may also submit comments directly to the State should they wish to. From there on, each agency would independently follow its own processing procedures.

This program, too, is just a beginning. It can and should be expanded to include all agencies at the State and Federal levels and possibly at the local level as well. It requires an investment of staff time at the outset, but the dividends to be gained by the public and by the agencies are well worth it.

4. Joint Public Hearings. Public hearings are time consuming, costly affairs, but in many cases, are critical to the public involvement process and must be held. Because they are such a heavy burden on staff time, we in New England are able to have formal hearings only when we absolutely cannot legally avoid it. In those cases where we don't have hearings, the public desiring the hearing is left with a reinforced feeling of an unresponsive government and we are left with a public blackeye.

However, many states and other Federal agencies face the same problems of trying to be responsive to the public, yet trying to conserve manpower and funding resources to the maximum extent.

In New England, the State agencies have many more hearings than we do. We are working on joint hearing processes with Connecticut and Maine so that we can "piggyback" on the State hearings while assisting with the staff effort and expense of conducting the hearings. At this writing, we have one hearing scheduled in Connecticut and one in Maine. As regards the details of hearing arrangements, both Maine and Connecticut would provide the meeting hall near the project site and make logistical arrangements for a PA system. The Corps will provide logistical and administrative support for any public information requirements, hand-outs and registration of attendees. Maine is providing for hearing transcripts in that State, whereas the Corps is providing for the transcript for the hearing in Connecticut. That simply points out the necessity for cooperating agencies to be flexible to achieve the most reasonably workable joint effort. For the two hearings under discussion, each agency issued its own public notice. That was only because by the time the decision was made to have a joint hearing, there was not enough time to coordinate the legalities of a joint notice. In future joint hearings, we will be prepared to issue joint hearing notices.

The procedures for the hearings are quite similar. The State hearing officer would be the lead official, but the Corps hearing officer would co-chair the hearing. Each agency would give a brief--5 minute or less--introduction covering its respective authorities, specific concerns, and regulatory procedures to occur following the hearing. From there on, the hearing would proceed according to State hearing rules. If State hearings call for cross-examination of witnesses, a procedure not allowed for in Corps hearings, the Corps hearing officer must include in his/her opening remarks a disclaimer that cross-examination would not be allowed for any matters subject solely to Corps authority. We have found, however, that witnesses are virtually always agreeable to responding to reasonable questions regarding their testimony. Hence, it is probably not necessary to have formal cross-examination. In fact, where cross-examination is allowed, the hearing frequently degenerates into legal arguments of trial procedure and the purpose of the hearing becomes secondary. Prior to the hearing, it is helpful to have a dry run of the procedures to be followed.

The project in Maine primarily involves a bridge for which a Coast Guard permit is also required. Coast Guard District One in Boston had agreed to participate jointly with us and Maine DEP. However, they were later overruled by Coast Guard headquarters in Washington and were not able to participate. I believe that action to have been an error on the Coast Guard's part as being contrary to the principles stated in the Corps/Coast Guard (DOT) Memorandum of Agreement which was reached as a result of Section 404(q) of the Clean Water Act. I would encourage the Coast Guard to reconsider its position in future similar situations and encourage Corps districts and State agencies to continue to urge the Coast Guard to participate in joint hearings.

Fortunately, we have had much better cooperation with EPA and the Federal Highway Administration. We have had one joint hearing with EPA, on a refinery proposal for Eastport, Maine, which, at least in terms of the administration of the hearing, went off without a hitch. Here again, we shared administrative and logistical burdens according to each agency's capabilities and resources.

While we have not had joint hearings with FHWA, it is only because no significant projects have developed since the MOA with DOT. FHWA officials are receptive to the idea of joint hearings. However, FHWA normally has their hearings prior to applying for Corps permits. Corps and FHWA officials must confer closely and be alert to each others programs so that joint hearings can be scheduled properly.

5. Joint Review Meetings. Before we started our regular monthly joint review meetings, each Federal agency, EPA, FWS and NMFS, would respond in writing to each public notice. When they had specific comments or questions, we would either respond directly in writing or send the comments to the applicant with a request that he/she respond to us so we could conduct further coordination with the commenting agency. This procedure takes a great deal of time to accomplish just a little coordination and in many cases, didn't really get at the heart of the issue.

We in New England had resisted monthly coordination meetings on the basis that in dealing with six states, such a meeting would be far too complicated to administer efficiently. However, based on glowing successes elsewhere and frequent urgings expressed in agency policy statements, we began joint meetings a year ago. For starters, we did not include State officials because of administrative difficulties of dealing with six states. We have found much to our pleasant surprise, that the process just among the Federal agencies works very well. While it is not conclusive that we reduce the staff time required to coordinate actions, the benefits are abundantly clear in terms of reduced response time to the applicant, reduced clerical load, and a higher quality review.

We have not yet included state agencies, and have no immediate plans to do so. However, we would welcome state participation, and have let the states know that the joint meetings are in progress. The states have shown some interest, but the scheduling of location and time each month to accommodate six states has proven to be beyond our grasp for the moment.

The agenda for each meeting consists of review of all the public notices issued subsequent to the close of the previous month's agenda and prior to a time two weeks before the meeting. This timing allows each agency to conduct the field inspections required of it under the joint inspection program discussed above, to have photos or slides developed, to prepare the field reports and to furnish copies to each participating agency at least several days in advance of the meeting.

In effect, the joint field inspection program has become an integral and indispensible part of the joint review meeting. Usually, the field inspectors particpate in the joint review meetings and make an oral presentation of their field reports. Agency representatives are then able to question the inspector to develop the best possible understanding of the project site and to develop a dialogue among the agencies which has helped to resolve concerns at the meeting which previously took several months by corresponding back and forth.

Such discussions have also served to promote trust and understanding among agency representatives where there used to be suspicion and misunderstanding. Each agency now seems to have gained a much better appreciation of the others agencies' policies and expertise. That particular benefit is not measurable in terms of dollars or time saved, but it unquestionably is a major individual contribution to improved quality of review and responsiveness to the public.

We have found that with our workload - averaging 60 to 70 public notices each month - that we can complete the agenda within one workday with a little time to spare. About 50 percent of the applications are resolved on the spot, another 30 per cent require additional discussion or information, but are resolved prior to the next meeting. The remaining 20 percent are complex enough such that they are coordinated outside the regular monthly meeting format.

Paperwork is significantly reduced because multiple exchanges of correspondence is replaced by preparing minutes of the meeting with conclusions reached and furnishing copies to each agency for their files. If any of the agencies disagree with the way the conclusions are stated in the minutes, they can notify the Corps immediately and the case would be discussed further by phone or put off for discussion at the next monthly meeting. We have also found that to date, the minutes have been accurate and no such concerns have been raised. Appendix B shows a typical portion of the monthly minutes. The Corps provides an original to each agency which in turn must furnish its own copies for its own files for each individual case listed on the page.

Several minor administrative aspects to the meeting have proven to be most helpful. A slide projector, a carousel, an overhead projector, and a screen are critical. A chalk board is also very helpful. There must be a definite agenda which is adhered to strictly and starting times must be heeded.

6. Special Studies and Impact Evaluations. Frequently, complete information on which to base a regulatory decision is lacking. Almost as often, conditions are imposed on permits to protect or enhance some resources. In both instances, individual agencies do not have the capability to obtain sufficient detailed information or to follow-up and evaluate the effectiveness of the conditions which are imposed. Through a cooperative effort, these and other matters of mutual concern can be addressed by joint studies or by jointly funding a study by one agency or a consultant. The potential of this type of cooperation is limited only by one's imagination.

We have had only minor success in this area but for reasons unrelated to the purpose of the effort. The major drawbacks to date have been to find enough staff time to develop good scopes of work and to break through the bureaucratic maze of procurement and contacting requirements.

One success we had was done informally with our own staff in consultation with National Marine Fisheries Service. It involved physically relocating habitat for mysid shrimp (*Mysidopsis* sp.) in Long Island Sound some 100 feet in order to allow a permittee to construct a boat ramp. We specified the location and method of movement of sediments, rocks and vegetation and oversaw the permittee's contractor doing the work. The habitat was successfully transplanted and field investigations a year after showed the mysid shrimp population to be thriving. The real value of the effort, however, came from an unexpected quarter. The transplant procedure was initiated because NMFS biologists had thought that mysid shrimp were uncommon in the area in question. However, after several field visits, we learned that they are quite abundant throughout that portion of Long Island Sound. That information and the knowledge that the habitat can be transplanted will be helpful in evaluating future permit applications.

It may be beneficial to report unsuccessful attempts so others may learn from our mistakes. There is great concern that construction impacts on fisheries resources may be acceptable most of the time, but would be too severe during certain stages of the life cycle of certain species. As a result, permits are frequently conditioned to prohibit construction activity during these vulnerable stages of the life cycle. Unfortunately, these times too often coincide with the prime construction season causing serious operational problems and economic loss to the construction industry. Recommendations from the fisheries agencies tend to be very conservative and include the maximum lengths of time when the fish may theoretically be susceptible to construction impacts. In an effort to better define these susceptible periods in several of the major rivers in New England, we developed with the U.S. Fish and Wildife Service a scope of work and discussed its implementation with Massachusetts Cooperative Fisheries Unit at the University of Massachusetts. We intended to transfer funds to the Fish and Wildlife Service who, in turn, can transfer them to the Coop unit. However, at the last minute, we learned that such work agreements must be reviewed and approved by FWS Washington office who imposes an additional cost to cover the expense of coordination with Washington. There may be very good reasons for such a procedure, but as for its effect on our proposed project, it killed it in two ways. It escalated the cost beyond what we had available in our budget and it delayed implementation beyond the end of the fiscal year when we had to turn back all uncommitted funds.

We did learn several things. Such plans take an extreme amount of time to bring to successful conclusion. Anyone planning cooperative studies would be well advised to allow themselves long lead times for budgeting purposes and they should dry run the administration of a study to learn what and where the pitfalls of procurement regulations and agency idiosyncracies are.

There are thousands of studies already done on as many topics relating to environmental effects of construction activity. Agencies must be thoroughly familiar with the literature to avoid repeating studies already done. In most instances, however, results of existing studies are not transferable to specific concerns at specific project sites. Existing studies can, however, be expanded upon to make them more applicable to project evaluation on a regional or site specific basis.

Some of the studies proposed for the coming years are turbidity and siltation effects, particularly as they affect anadromous fish runs; evaluation of time of year restrictions as discussed above, effects of underwater blasting, mosquito control ditching, habitat benefits of riprap slopes versus vertical bulkheads, and impact of lobster pounds on clamming.

Summary. There are many excellent opportunities for agencies to work together to reduce costs, be more responsive to the public, save time, and improve the quality of review and evaluation. All that is required is a positive, cooperative attitude, ingenuity, and an investment of time which will pay great dividends in the near future.

APPENDIX A
PERMIT APPLICATION FIELD INSPECTION REPORT

Corps File #: State or Agency File #:

Inspector/Title/Agency: Date & Time of Inspection:

Tide Level:

Applicant's Name: Agent/contractor's name:

Address: Address:

Town, State, Zip: Town, State, Zip:

Location of proposed work:

Waterway: Street, Town:

Description of proposed work:

Ecosystem Type:

Coastal________ marsh______ tidal flat____

dune______ beach_______

ledge_____ other(specify)__

Freshwater______ river_____ highgradient stream__

lowgradient stream______

lake______ pond_______

wetland (specify type)_______

other (specify type)________

Adjacent upland: Developed____ open_______

Agricultural__ wooded______

Other (specify)___________

Environmental setting (general description of the area):

PLEASE SUPPORT YOUR COMMENTS WITH PHOTOS AND SHOW THE LOCATION OF THE PHOTOS ON THE PROJECT PLANS:

1. Corroboration of application drawings (distances, sizes, wetland boundaries, etc) A site sketch should be included if preapplication or enforcement action):

2. Ecosystem types:
   - Shoreline types:
   - Waterbody types:
   - Water quality:

3. Vegetation present (D = dominant, C = common, I = infrequent)

   Species:

   Density of plant community:

4. Location and relation to nearest waterbody:
   (Note separating features, such as upland areas, berms, roads, etc., if any, and their dimensions.)

5. Source and frequency of inundation and saturation:

6. Other geologic and hydrologic characteristics (note evidence of erosion):

7. Shoreline protection facilities present on project site and adjacent properties:

8. Natural biological functions (i.e., food chain support, detrital export, specialized or general habitat, etc.):

9. Significant biological features (i.e., shellfish beds, known fisheries, etc.):

10. Overall habitat value:

11. Aquatic study areas, sanctuaries, or refugees:

12. Location and elevation of project site within wetlands:

13. Dimensions of total wetland site:
of project site:

14. Fetch:

15. Proximity to cultural development:

16. Samples taken; tests performed; (note specific location, tide, weather, etc.):

17. For dredging projects--
Grain size

Potential sediment pollutants and their source:

Description and locations of disposal area:

Potential biological availability:

18. Existing surrounding land use:

19. Potential effects of project on water-flow and flooding:

20. Anticipated long- and short-term impacts:

21. Potential cumulative effects:

22. Alternative methods or locations:

23. Mitigation and compensation options
Summary of information and views of investigator (include recommendations):

24. Names and affiliations of people met on the site:

EXPEDITING THE COASTAL APPROVAL PROCESS

By
Jeffrey A. Layton,[1] M., ASCE
and
Peter L. Buck,[2] J.D.

1. STATEMENT OF PROBLEM

1.1 Introduction. Perhaps the greatest criticism project proponents have of the coastal zone regulatory scheme is the great length of time necessary to obtain governmental permits and approvals. Many coastal projects have been delayed two to six years while working their way through the approval process.

1.2 Regulatory delay. One of the main reasons for the lengthy processing period is the requirement for multiple permits. Even for complex inland projects, local, state and federal agencies all have areas of concern which are protected by multiple regulations. In the coastal zone, additional agencies such as the Corps of Engineers become involved and other local, state and federal agencies have special sets of regulations to protect the coastal zone areas. Coastal projects require anywhere from 10 to 25 approvals.

In addition to the multiplicity of approvals, there is a problem of the lengthy process it takes to obtain each of them. For example, the Corps of Engineers' requirement of circulation and resolution of disagreements resulting from comment is inherently a lengthy process. Further, NEPA and "little NEPA" requirements add considerably to what is already a long process.

1.3 Litigation delay. Even after one has gone through the lengthy process of obtaining each of the multiple permits, a project may be further delayed through litigation by groups or agencies opposed to project approvals. Final resolution of litigation can take anywhere from six months to four years. Prior to resolution through full trial, courts have been fairly willing to provide injunctive relief against agency projects which would substantially modify the environment. Private projects may not have to be enjoined to be halted pending trial since uncertainty over final project approval by a court leads to a "constructive injunction" due to the unavailability of financing.

---

[1] Consulting Coastal Engineer, Layton and Associates, 1111 Third Avenue, Seattle, Washington 98101.

[2] Attorney, Law Offices of Peter L. Buck, P.S., 1111 Third Avenue, Seattle, Washington 98101.

1.4 Effect of delay. The effect of delay, particularly on the private sector, can be disastrous. Although a project may be feasible when starting the approval process, delay can work a significant change in a variety of manners. Market conditions may change, making the project unviable. Financing which was available at the start of the process may no longer be available. Other agencies may adopt more restrictive regulations or requirements which make a formerly feasible project infeasible. Meanwhile, various carrying costs associated with the land and with the proposed project will continue to grow.

1.5 Lack of effective response to delay problem. Unfortunately, the private sector engages in more criticism of the delay process than in effective solution of the problem. Project proponents always begin a project with much discussion about their desire to fast-track a project. In reality, however, few projects are actually put on a fast-track by their proponents.

1.6 Delays are reducible through proponent action. Other than an improvement in mental health, little is accomplished by criticizing the regulatory process or the agencies which administer it. There is little likelihood of change since the problem is largely institutional resulting from a multiplicity of permits and lengthy procedures with regard to those permits. Applicable regulations are designed to protect various concerns which society has determined are important.

The problem of delay is solvable, however, through aggressive proponent action. Effective utilization of fast-tracking procedures can cut delay at least in half.

2. CREATING A FRAMEWORK TO REDUCE DELAY

2.1 Introduction. A number of procedures exist which will allow a project proponent to reduce delay. These procedures not only allow the approval process to be accomplished much faster but improve chances of ultimately receiving approval as opposed to disapproval.

2.2 Decision to fast-track. A starting point is the making of a conscious decision that the project should be fast-tracked. This decision must be articulated and assigned a relatively high priority so that necessary resources will be available to achieve the goal.

2.3 Assigned fast-tracking responsibility. One individual must be identified as having responsibility for assuring that the project is fast-tracked. This is often the project manager, but it may be the developer. Typically, the developer is the most motivated to accomplish this task.

The designated person must then constantly push all other people working on the project. For that reason this person must have clearly understood authority, must be willing to be tough with other project team members, and must have the time to constantly follow up.

2.4 Outline approval requirements and sequence. To understand what must be accomplished in the fast-tracking procedure, it is important to develop a comprehensive list of all required approvals. Next, a sequence for obtaining those approvals should be developed. Some approvals should be obtained early to assist in obtaining others.

2.5 C.P.M. A list should be prepared for each approval including all the steps necessary to obtain that approval. Included would be all procedural steps such as notice, circulation and hearings. It would also include all required reports and information which would need to be developed. The project proponents should know the agency's procedures and requirements as well as anyone in the agency.

When all of that information is developed, a master C.P.M. chart should be prepared for the entire approval process. Additionally, a C.P.M. chart should be developed in detail for each specific approval.

2.6 Work plan for each approval. A work plan should be developed for each step of each approval process. For example, it should be determined who will initially see each relevant official and when in the process that contact will be initiated. Additionally, it should be determined what reports should be prepared to assist approval processing, what information should be gathered in that regard, and when. Such preparation can be done very early in the process and will put the project proponent in the posture of controlling the process rather than reacting to the process.

2.7 Develop and motivate a team. As early as possible, a full team should be developed. This would include the project engineers, architects, planners, environmental scientists, lawyers, etc. Consideration should be given as to how to motivate team members through high contingency fees, participation in the project, or simply paying bills on time. The project proponent should continually look for ways to keep up morale and to develop a team attitude, e.g. periodic parties or unexpected rewards for a successful team member and spouse.

2.8 Assign specific responsibility. At each point everyone should know who is assigned the responsibility to accomplish a particular task. There is a great deal of room for overlap of tasks, and it is often easy to assume that someone else had the responsibility to accomplish a particular assignment. All assignments should be put in writing, and each team member should agree on a reasonable date by which he or she can accomplish a particular task. The date for achieving a task and the acceptance of responsibility to achieve that date should always be in writing and be made known to other members of the team.

2.9 Follow up on assignments. Constant follow up is needed to check whether or not tasks are being accomplished by the assigned dates. These assignments should be checked against the master list of assignments which has previously been put in writing. The responsible member should be held accountable to the entire team through written reports reporting progress or through periodic team meetings at which time each member reports specifically on whether or not they have accomplished the task previously agreed to. Through such a procedure, the project manager can constantly create accountability for expeditious work.

2.10 Production of first-class work. The project proponent must be prepared to do first-class work if a project is to be expedited. Cutting corners on the quality of a project, the extent of mitigating measures, or the content of requested environmental reports will always achieve short-term financial savings, but will increase delay and long-term costs. A project cannot be expected to move quickly unless it is presented in a first-class manner. A "down and dirty" effort often is significantly more time consuming and expensive. For example, it is always cheaper in the short run to have a negative declaration but, if it is determined six months later that an EIS needs to be prepared, it will be significantly more costly to prepare one, and six months' delay will have occurred.

2.11 Prepare everything early. Very early in the process the project proponent should prepare all notices, applications, etc., which will be required later in the process. These should be held in a file ready to be activated when needed. Things such as legal descriptions, vicinity maps, and lists of adjacent owners will always be needed. It is much easier to obtain those things early in the process when there is a lack of pressure than it is two days prior to desired filing.

Reports should be prepared as early as possible to avoid delay and continuances. Reports which are submitted at the time of a hearing inevitably lead to a legitimate desire on the part of the agency to have more time to study them. It is usually possible to predict early in the process what reports will be necessary; their preparation should begin at that time.

Reports which help support a project should be presented at the earliest possible stage in a proceeding to have more effect. A good report or a good graphic which is to be used at the time of a hearing can be prepared earlier and can be used to help persuade staff members to make a favorable recommendation. Since the expense for the report or graphic will have to be incurred eventually, it might as well be incurred early so that the maximum benefit can be made of it.

3. WORK EARLY AND CLOSELY WITH AGENCIES

3.1 Importance. Agency personnel often hold life and death power over a particular project. Additionally, they hold the power, either intentionally or unintentionally, to speed up or delay a project indefinitely.

3.2 Get along with agency personnel. There should be a constant selling of agency personnel by designing a good project and seeing that the personnel are aware of it. Also, on a personal level there should be constant romancing of personnel at all levels, including secretarial staff who can assist in getting copies without trauma and who can improve the chance of phone messages being returned.

3.3 Use consultants recommended by the agency. If possible, consultants recommended by an agency should be used. Typically, agencies will be reluctant to give names, but, if asked in the proper manner, will always give names. Such recommendations rarely relate to personal relationships with agency personnel but instead relate to an agency's confidence in the credibility of a particular consultant. Agencies are in the best possible position to see which consultants can get a job done quickly and which cannot. Additionally, it is difficult for agencies to later criticize the quality of work of consultants they recommended.

In retaining these consultants, it is important to obtain necessary time commitments. If fast-tracking a project is a high priority, that should be explained to consultants, and the consultants should agree in writing that they will assist in that regard. Inevitably, good consultants are too busy, and it is, therefore, necessary to have a confirmation that your project will obtain the necessary priority.

3.4 Listen and respond to concerns. When dealing with agency personnel, it is important not to sell a project too hard. One must listen very carefully to concerns and find ways to solve them.

Use of an environmental design process can be of great assistance in solving agency problems. Initially, all environmental concerns should be identified through an impact statement or assessment. Aggressive use of the mitigating measures can usually lead to resolution of most problems. It is particularly important not to settle upon a final design until the end of the review process. The design should remain fluid until the concerns of relevant agencies and individuals can be met. The environmental design process is a cyclic planning process rather than a conventional, rigid, one-shot approach. (For a summary of the environmental design process, see Appendix B of "Resolution of Shoreline Disputes," Coastal Zone 80 Proceedings, and References 14 and 15.)

The vast majority of problems which agencies or individuals have with a project can be solved through the use of imagination. Mitigating measures involving change of the project plan or operating conditions should be developed through full application of the project team's imagination.

3.5 Contact all agencies. Early contact should include all agencies, not just those most heavily involved in a project. Often a seemingly innocuous requirement may be a problem. Since these agencies have to be contacted at some point, they might as well be contacted early.

3.6 Follow up. There needs to be relentless but courteous follow up with agencies. Project proponents do not hire consultants to leave phone messages. They hire consultants to talk with agency personnel and resolve problems. It is necessary for people working on a project to continually follow up with agency personnel. All of this, of course, must be done in a courteous, appropriate manner.

3.7 Insert consultants in process. The project team should work with agencies in ways even the agencies may not have contemplated. For instance, project proponents can prepare various notices for agencies and see that they get published. They can also occasionally help with such mundane things as typing. In the Corps permit process, the project team should personally circulate a copy of the public notice and explain the project to each of the consulted agencies prior to their receipt of the notice from the Corps. A time should be arranged to meet with the consulted agencies to receive their comments. Such a follow-up meeting should be assured through an appointment. The project team should get closure on the concerns of each of the agencies and continually be finding a way to modify the proposal to obtain their support.

4. CITIZEN SUPPORT OR OPPOSITION

4.1 Meet with opposition. It is usually best to meet with concerned citizens and environmental groups who may be concerned about the project. One should not do so, however, if otherwise there would be a lack of knowledge in the community about a project or if such a meeting would simply help the opposition to get organized.

4.2 Explain the project. One of the primary reasons to have such a meeting is to explain the project and to alleviate unfounded concerns. Often a group of citizens will have unfounded fears based on a misunderstanding of what is proposed.

4.3 Solve problems. Legitimate concerns of citizens groups should be identified as early as possible and solutions should be found. Just as with agencies, an environmental design process can be utilized to modify a project to make it more acceptable.

4.4 Weaken the opposition. It often will not be possible to eliminate all opposition. If even 50% of the opposition can be neutralized, it may weaken the remaining opposition to the point where it is not a serious problem. For that reason it is worthwhile to have meetings to reduce concerns through education or project modification even if only part of the opposition can be mollified.

4.5 Developing support. It is always helpful to demonstrate support for a project. Petitions and letters are helpful to neutralize the effect of opposition. A project proponent should rarely start the "petition game", however. There are usually groups such as Chambers of Commerce or trade associations which can be helpful.

5. EXPEDITING LITIGATION

Early in litigation the proponent should move for an early trial date noting the inherent problems of delay for a preliminarily enjoined project or of the "constructive injunction" effect if the project has not been enjoined but involves private funds. Most courts either have provisions for expedited trials or will accept a motion for an expedited trial. Other court appearances may also provide forums for seeking an early trial. For instance, a motion on a preliminary injunction is an excellent time to note the inherent disadvantage to all parties of a late trial date. (See Appendices)

Imagination should constantly be applied to find ways to expedite the litigation. For instance, a special master may be appropriate for complex environmental litigation.

A litigant should always be prepared to take the next step in the litigation process. The same organization and accountability discussed in section 2 above can be utilized at the litigation stage.

If one loses at any stage in the litigation, it is important to seek out appropriate remedies. For instance, if it is determined that an impact statement is required when one has previously not been prepared, defendant should attempt to continue with construction pending preparation of the impact statement. If an impact statement has been found inadequate, then the defendant should attempt to have the court retain jurisdiction. This avoids all of the delays inherent in commencing new litigation and usually will end up with a sympathetic judge if his order is followed.

6. INSTITUTIONAL SYSTEMS FOR REDUCING DELAY

Several states have instituted coordinated permit procedures to solve the problems of multiple permits. For instance, the Washington Environmental Coordinated Procedures Act allows for all permits and all appeals to be resolved in one forum with specific time limits. Few proponents have utilized those procedures, however, since it often takes control of the flow of the process out of their hands.

A number of local governments have adopted one-stop permit procedures. Generally, these have been more viable than the one-stop state procedures. See generally _The Permit Explosion: Coordination of the Proliferation_ Bosselman, Feurer and Siemon, The Urban Land Institute, Washington, D.C., 1976.

Recent attempts to improve processing pursuant to NEPA also promise some potential for reducing delay. The new CEQ regulations provide for combining NEPA and "little NEPA" processing, reduction of the length and scope of EIS's, and simplifying finalization procedures. 40 CFR 1500. The Corps of Engineers proposed NEPA regulations provide further basis for expediting the NEPA process through coordination among Corps sections. Proposed 33 CFR Sec. 230, Appendix B, Sec. 3. These provisions must be pushed by a project proponent to be effective, however.

7. CASE HISTORIES

7.1 _Typical approval process for a marina_. Since marinas typically involve dredging and protective structures, the approval process is lengthy. Typical marinas in Western Washington have taken three to six years to obtain the necessary permits and approvals.

However, by applying the expediting procedures described above, it is possible to significantly speed up the approval process. The following examples illustrate two successfully expedited marina projects in Puget Sound.

7.2 Point Roberts Marina. The Point Roberts Marina located in Washington State south of Vancouver, British Columbia, is a 1,000-berth basin dredged from former agricultural land. The full approval process took 12 months. One key to rapid permit approval was a client willing to spend money to conduct extensive environmental and engineering studies necessary to support the environmental review process. Additionally, the client continually pressed all consultants to accomplish their work quickly. Other keys to success included preparation of a detailed environmental impact statement, early contact with all applicable agencies and aggressive efforts to expedite litigation which was brought by neighboring citizens. (See References 12, 13, 14 and 15)

A representative example of an issue which was expedited during the permit review phase for the Point Roberts Marina project concerned the maintenance of water quality levels in the 40-acre backshore dredged harbor. As part of the process to obtain rapid project approval an innovative design concept was developed to solve the water quality issue. The concept involved the unique combination of curvilinear shoreline geometry and tidal flow to produce an artificial boat basin which maximized hydraulic circulation and water quality. (See References 10 and 11)

Past design practice of backshore-dredged marinas in Puget Sound concentrated primarily on optimizing dredged geometry to develop maximum maneuvering space and moorage area. Usually a rectangular basin shape was used. However, environmental studies showed that the sharp-edged corners of rectangular basins can have limited water circulation and flushing, resulting in the formation of stagnant zones and poor water quality. Typically, water temperatures rise and dissolved oxygen levels decrease in marinas with poor circulation. These changes can have dangerous consequences for the marine life inhabiting the marina. Of particular concern for the Point Roberts Marina project were newly hatched pink and chum salmon fry which migrate annually along the marina's shoreline. Because of these water quality concerns, state and federal agencies initially contacted by the project consultants felt the proposed project would not be acceptable. Their primary concern was that the potential for a fish kill due to poor water circulation in a conventional rectangular backshore boat basin was very high and therefore they could not approve a marina in this location.

The project consultants recognized this issue early in the design of the project such that sufficient time was available to thoroughly study the problem. As a result of the consultants' review, a radical design concept for the Point Roberts Marina was proposed. Conventional rectangular geometry of the boat basin was replaced with a curvilinear design to eliminate sharp-edged stagnant corners. The rounded basin geometry was then coupled with a single narrow entrance channel which was hydraulically designed to direct a flooding tidal flow into the marina. This tidal jet possesses sufficient momentum to create a large scale rotating current pattern within the basin. The resulting circulation system is capable of maintaining water quality levels at acceptable levels.

The new concept was reviewed by the appropriate agencies and their responses were encouraging. However, they felt additional verification would be required before they could approve the concept.

To provide the verification, the consultants proposed a detailed and expensive hydraulic modeling program to test the concept. The project sponsor provided the funding and the model tests were conducted. The model tests verified the concept and the regulatory agencies approved the design.

The resolution of the water quality issue was a key element in obtaining the rapid approval of the Point Roberts Marina project. By recognizing a serious problem early in the design process, sufficient time was available for the project consultants to develop an innovative solution which was acceptable to both the regulatory agencies and the project proponent. Equally important was the project proponent's willingness to commit a substantial amount of money to study the problem without any assurance that an acceptable solution could be developed.

7.3 Anacortes Marina. The Anacortes Marina is a 500-berth marina currently under construction in Anacortes, Washington. One third of the slips approved are covered. The approval process took ten months.

A key to the quick processing involved an excellent environmental impact statement. That statement and the positive comments made about it gave the decision makers a high level of confidence in making a quick decision.

During the review of this project by state and federal agencies, a potential problem arose concerning the project's impact on migratory herring which were known to periodically spawn in eel grass beds located about one-half mile from the proposed marina site. The agencies were not certain, however, if herring also spawned on eel grass beds located at the proposed marina site. Therefore, to verify whether or not herring spawned at the marina site, the agencies stated they would not approve the project until they conducted tests during the next migration season which was about six months away.

Studies by the project consultants indicated that herring did not spawn at the site and the project proponents did not want to delay the project an additional six to nine months to resolve this issue. Therefore, the proponent authorized the consultants to make a detailed search of all records in the state and federal agencies' files that concerned herring spawning at the proposed marina site. Results of the search revealed unpublished data which conclusively proved that the proposed site was not a herring spawn area. Discovery of this information allowed the agencies to go ahead and make a favorable decision without having to wait an additional six months to conduct additional studies.

Another potential problem for the Anacortes Marina project involved the destruction of the eel grass beds at the proposed site. Dredging of the boat basin would result in the elimination of about ten acres of moderately productive eel grass. The agencies were reluctant to approve the project because they felt the loss of this natural resource could not be mitigated. However, by working with all concerned agencies and being willing to offer mitigation, the project proponents were able to resolve this issue. Ultimate resolution involved a substantial donation to The Nature Conservancy to purchase other privately owned eel grass beds to be protected for the public in perpetuity.

REFERENCES

1. Buck, P.L. Environmental Litigation. "Proceedings of the Practising Law Institute's Course on Environmental Law and Practice," Litigation and Administrative Practice Course Handbook Series, Number 117, San Francisco, July 1980.

2. Buck, P.L. Challenges to EIS Adequacy. "Proceedings of Washington State Bar Association Course on Land Use and Environmental Law Practice," Seattle, June 1980.

3. Buck, P.L. A Practical Approach to Land Use Litigation and Land Use Dispute Resolution. "Proceedings of Practising Law Institute Conference on Modern Control of Land Development," Real Estate Law and Practice Course Handbook Series, Number 175, New York City, 1980.

4. Buck, P.L. A Practical Approach to Land Use Regulation. "Proceedings of Seminar on Legal Aspects of Land Use for Nonlawyers," University of Washington Institute for Environmental Studies, Seattle, 1980.

5. Buck, P.L. Citizen Litigation. "Proceedings of Litigation Workshop," sponsored by League of Women Voters, Seattle, 1979.

6. Buck, P.L. NEPA Litigation. "Proceedings of Practising Law Institute's Course on Environmental Law and Practice," Litigation and Administrative Practice Course Handbook Series, Number 110, San Francisco, 1979.

7. Buck, P.L. Successful Defense of Land Use Litigation, "Proceedings of Annual Meeting of the Washington State Association of Municipal Attorneys and Washington State Association of Prosecuting Attorneys," Olympia, Washington, 1979.

8. Buck, P.L. Project Feasibility and Analysis, Impact Statement Litigation, and Land Use Litigation and Settlement. "Proceedings of University of Puget Sound Law School Conference on Environmental Law and Land Use Planning," Kona, Hawaii, 1979.

9. Cormick, G.W. Resolving Environmental Conflicts Through Mediation: Experience, Process and Potentials. Institute for Environmental Studies, University of Washington, Seattle, 1978.

10. Layton, J.A. Hydraulic Circulation Performance of a Curvilinear Marina. "Proceedings of the 17th International Conference on Coastal Engineering," ASCE, Sydney, Australia, 23-28 March 1980.

11. Layton, J.A. Innovative Design Concept for Hydraulic Circulation of Small-Craft Harbors. Presented at the National Convention and Exposition for the American Society of Civil Engineers, Portland, Oregon, 14-18 April 1980. (Preprint paper No. 80-137)

12. Layton, J.A., J.D. Hansen and P.L. Buck. Recreational Planning Features of a Large Marina. "Proceedings of ASCE's Recreational Planning and Development Conference," Snowbird, Utah, 18-21 April 1979.

13. Layton, J.A. Design and Construction of a Curvilinear Marina. "Proceedings of ASCE's Coastal Structures 79 Conference," Washington, D.C., 14-16 March 1979.

14. Layton, J.A. Coping with Multiple Permit Projects. "Proceedings of the Practising Law Institute's Course on Modern Land Use Control," Real Estate Law and Practice Course Handbook Series, Number 143, New York City, February-March 1978.

15. Layton, J.A. and J.D. Hansen. Environmental Design Process: Small-Craft Harbor. "Proceedings of Ports 77 Conference," ASCE, Long Beach, California, 9-11 March 1977.

APPENDIX A

COURT OF APPEALS, DIVISION IV
OF THE STATE OF WASHINGTON

MARGARET ADAMS, Appellant,

v.

CITY OF SEATTLE, a municipal corporation, and WATERFRONT BUILDERS, INC., a Washington corporation, Respondents.

No. 1324

MOTION TO ADVANCE THE DATE FOR HEARING

1. Identity of Moving Party. Waterfront Builders, Inc., (herein "WBI"), respondent-cross appellant, asks for the relief specified in Part 2.

2. Statement of Relief Sought. WBI asks the Court to advance the date for hearing this case to allow the Court to enter a decision on or before July 31, 1978. WBI also requests that the Court establish a schedule for the submission of briefs by the parties that will allow for this accelerated review or that the Court hear the case based on the briefs filed in the Superior Court. WBI would suggest the following schedule to the Court:

	to be filed
Appellant's opening brief	May 5
Respondent's brief	May 15
Appellant's reply brief	May 22

3. Facts Relevant to Motion. WBI applied to the City of Seattle for a shoreline development permit on January 17, 1977, to construct a forty (40) unit condominium along the shoreline at 3100-3124 West Commodore Way in Seattle. The permit was issued by the City on April 29, 1977. Appellant Adams appealed that decision to the City of Seattle Hearing Examiner who upheld the City. She also appealed that decision to the Shorelines Hearings Board (herein the "Board"). The Board held two days of public hearings on August 10 and 11, 1977, and affirmed issuance of the permit, with slight modifications, on September 7, 1977. Appellant Adams appealed the Board's decision to the Superior Court. On January 9 and 10, 1978, the Honorable James J. Walsh heard the case on cross motions for summary judgment and granted WBI's motion, affirming the Board's decision. On February 3, 1978, appellant Adams filed this appeal from the decision of the trial court.

LAW OFFICES OF
PETER L. BUCK, P.S
13TH FLOOR SEATTLE TOWER
1218 THIRD AVENUE
SEATTLE, WASHINGTON 98101
(206) 382-9540

WBI does not presently own the property on which it plans to build its proposed condominium. It has only an option to purchase the property that expires on July 31, 1978. WBI cannot exercise its option unless it can finance the project, and it cannot finance the project while this case remains undecided. (See attached Affidavit of Rubin Miller)

4. Grounds for Relief and Argument. RAP 18.8(a) provides that the Court may, on motion of a party, shorten the time within which an act must be done under the ordinary rules of appellate procedure in order to serve the ends of justice. The Supreme Court said in State ex rel. Holme v. Fawcett, 127 Wash. 489, at 491, 221 Pac. 305 (1923) that:

> This court has power, and it is its practice, on application, to advance the hearing of causes, where it is made to appear that the judgment of the court will be without effect if the regular course of procedure is pursued.

Delay in the decision beyond July 31, 1978, will effectively deny WBI its right to seek relief from this Court because, unless this case is decided before that date, WBI will never be able to complete its proposed project. The reason for this is that WBI must exercise its option to purchase the property on which it intends to build its proposed condominiums before that option expires on July 31, 1978. In order to exercise its option WBI must obtain financing for the project. WBI cannot and will not be able to obtain such financing while this case remains undecided.

If WBI is forced to abandon this project, it will also lose the money it has spent for architectural and engineering plans, attorneys' fees, the option for the property, and the work done by its employees. These costs are in excess of One Hundred Fifteen Thousand Dollars ($115,000). (See attached Affidavit of Rubin Miller).

The effect of not deciding this case before July 31, 1978, will be that appellant will have caused this project to be abandoned even though she has failed to convince any of the three adjudicatory bodies before which she has appeared that the issuance of the permit in this case was anything but proper. The State Environmental Policy Act and the Shorelines Management Act were enacted to serve legitimate functions. None of those functions will be served by forcing this project to be abandoned because of the delay inherent in judicial proceedings. It will

LAW OFFICES OF
PETER L. BUCK, P.S
13TH FLOOR SEATTLE TOWER
1218 THIRD AVENUE
SEATTLE, WASHINGTON 98101
(206) 382-9540

hardly serve the ends of justice to strip WBI of the right that has been unanimously upheld and to award appellant the relief that has been unanimously denied all without a hearing.

Because of the nature of this case, advancement of the date for the hearing in this Court will not prejudice any of the parties to the case. This case comes to the appellate court on an appeal from the trial court's review of an administrative decision. There was no new evidence presented before the trial court. The record before the trial court for review is the same as that now before the appellate court. All the parties have had ample time to review that record. The facts and the law are unchanged, and the case has been briefed and argued once. The briefs that were filed in the trial court are part of the record before the appellate court (Clerk's Papers Nos. 15-20). This Court can review those briefs to see for itself how well the parties availed themselves of the opportunity to review the record and address the legal issues presented by it.

An additional reason why expediting this case would not work a hardship on appellant is that a substantial part of the time normally allotted to appellant after the filing of an appeal to prepare the record for review was unnecessary in this case. Specifically, the forty-five (45) days allowed to arrange for a verbatim transcription of the proceedings and the additional forty-five (45) days to complete the transcription and file the report of proceedings were not required in this case since the proceedings were transcribed and paid for before the case was heard in the Superior Court.

WBI realizes the additional effort this motion places on this Court, and would not request such extraordinary relief if the situation did not warrant it. WBI will certainly make every effort to accommodate the Court should this motion be granted. As indicated in the proposed schedule, WBI will be able to file its brief within six (6) working days from the date on which appellant's brief is filed. An attorney will be available to present oral argument on the case on short notice whenever the Court is able to schedule the argument.

It is certainly within the Court's discretion to advance the hearing of this case. Unless the Court does so, WBI's right to develop under the permit issued by the City of Seattle will be denied to it not by due process of law but rather in derogation of such due process. None of the parties to this case will be prejudiced by advancement of the date for the hearing in the

LAW OFFICES OF
PETER L. BUCK, P.S
13TH FLOOR SEATTLE TOWER
1218 THIRD AVENUE
SEATTLE, WASHINGTON 98101
(206) 382-9540

appellate court. The court should advance the date for the hearing of this case to such a date as will allow it to decide this case prior to July 31, 1978.

Respectfully submitted,

LAW OFFICES OF PETER L. BUCK, P.S.

By ______________________
Peter L. Buck
Attorneys for Respondent-Cross Appellant Waterfront Builders, Inc.

LAW OFFICES OF
PETER L. BUCK, P.S
13TH FLOOR SEATTLE TOWER
1218 THIRD AVENUE
SEATTLE, WASHINGTON 98101
(206) 382-9540

APPENDIX B

COURT OF APPEALS, DIVISION IV
OF THE STATE OF WASHINGTON

MARGARET ADAMS, Appellant, v. CITY OF SEATTLE, a municipal corporation, and WATERFRONT BUILDERS, INC., a Washington corporation, Respondents.	No. 1324 AFFIDAVIT OF RUBIN MILLER IN SUPPORT OF MOTION TO ADVANCE THE DATE FOR HEARING

STATE OF WASHINGTON)
) ss.
COUNTY OF KING)

RUBIN MILLER, being first duly sworn upon oath, deposes and says that:

I am over the age of twenty-one (21) years, competent to testify in this matter, and have personal knowledge of the matters attested to herein. I am the President and a Director of Waterfront Builders, Inc., a Washington corporation (hereinafter "WBI").

I have been involved full-time in condominium and apartment construction and real estate sales since 1969. During that time I have served as a general contractor on several different multifamily buildings which were constructed in the Puget Sound area. As such, I have had and continue to have constant and detailed involvement with all facets of the construction industry, particularly in arranging the financing for construction projects. It is the uniform and consistent practice of all lending institutions in the Puget Sound area to refuse to finance a project during the pendency of a lawsuit which challenges the basic right to construct the project.

WBI has applied for, and was granted, a substantial development permit to construct a 40-unit condominium. The issuance of that permit is the subject of this lawsuit. The issuance of that permit and the steps leading up to that issuance have been reviewed by a hearing examiner for the City of Seattle, the Shorelines Hearings Board, and the Superior Court for King County. In every instance the issuance of the permit has been

LAW OFFICES OF
PETER L. BUCK, P.S
13TH FLOOR SEATTLE TOWER
1218 THIRD AVENUE
SEATTLE, WASHINGTON 98101
(206) 382-9540

upheld. WBI has prepared final building plans for this project and is prepared to begin construction on it at any time.

WBI does not own the property on which this building is to be constructed. It has only an option to purchase the property. That option will expire on July 31, 1978. In order to exercise that option, WBI must obtain financing for the project. As I have stated, it is my experience that no lending institution in the Puget Sound area will finance this project while this lawsuit remains undecided. It is also my opinion that in this particular case the factor preventing WBI from obtaining financing for this project is the instant appeal.

Unless this case is decided before July 31, 1978, WBI's option to purchase this property will expire; and any relief the Court of Appeals could grant after that date would be of no effect. The shoreline development permit issued by the City of Seattle, which has been upheld by all three of the adjudicatory bodies that have reviewed it, will be worthless to WBI. In addition, WBI will irrevocably lose in excess of One Hundred Fifteen Thousand Dollars ($115,000), which amount WBI has already spent for attorneys' fees, engineering and architectural plans, the option to purchase the property, and the other predevelopment efforts. The only reason for such contemplated loss is the three previous appeals by appellant which have taken up over a year, although she lost at every stage, and her instant appeal.

/s/ ______________________________
Rubin Miller

SUBSCRIBED & SWORN to before me this 13th day of April, 1978.

/s/ ______________________________
Notary Public in and for the State of Washington, residing at Bainbridge Island

LAW OFFICES OF
PETER L. BUCK, P.S
13TH FLOOR SEATTLE TOWER
1218 THIRD AVENUE
SEATTLE, WASHINGTON 98101
(206) 382-9540

SCIENTIFIC COLLECTING IN SOUTH FLORIDA: PROBLEMS WITH PERMITS

Eugene A. Shinn
U.S. Geological Survey
Fisher Island Station
Miami Beach, Florida 33139

A patchwork of parks, sanctuaries, preserves, and local county regulations, combined with ever increasing numbers of visiting biologists and geologists desiring to collect marine specimens and rocks, stimulated the writing of this article. The purpose is to explain in non-legal language the various laws, regulations and permitting procedures, so that scientists may collect for educational and research purposes and avoid fines and incarceration.

At present, three State offices (Division of Recreation and Parks, Division of Marine Resources, and Bureau of Aquatic Plants, all under the Florida Department of Natural Resources) and three Federal agencies (National Park Service, National Oceanographic and Atmospheric Administration, and Fish and Wildlife Service under the Departments of Interior and Commerce) are involved in regulatory activities, while two State (Florida Marine Patrol and Recreation and Parks) and three Federal (National Park Service, U.S. Coast Guard, and Fish and Wildlife Service) agencies are involved with enforcement. Clearly, the would-be scientific collector needs to know exactly where collection is permissible and where to obtain necessary permits for areas where collection is restricted. All the agencies cited do issue permits for bona fide scientific collecting and except for occasional bureaucratic problems, these permits will be issued promptly.

It is presently illegal to collect living coral and many other organisms in all State waters adjacent to Florida (State waters extend three miles offshore on the eastern coast and three leagues, approximately nine miles, on the western coast) without a permit. In addition, collection of coral and rock from federally managed parks and sanctuaries requires a permit. Until recently, permits were required from the Bureau of Land Management for collection in all Federal waters outside existing parks and sanctuaries, but a Fifth Circuit Court ruling invalidated that Secretarial Order. Although at this time no Federal agency controls collecting in Federal waters outside the parks and sanctuaries, new regulations governing collecting will soon go into effect. As ordered by the Fishery Management Act of 1976, the Gulf of Mexico and South Atlantic Fisheries Management Councils have prepared coral and other fishery laws that will go into effect sometime in 1981. These new laws will be administered by the National Oceanographic and Atmospheric Administration (NOAA). Regulations will apply in Federal waters extending out 200 miles (where they do not infringe on sovereign waters of adjacent foreign nations). Under these regulations, certain areas termed Habitat Areas of Particular Concern (HAPCs) will receive additional protection similar to that of existing marine sanctuaries presently administered by NOAA. Looe Key reef, a five-square-mile area off the lower Florida Keys, has been designated as a HAPC. This reef has also been proposed for Marine Sanctuary status under Title III of the Marine Sanctuaries Act, but a decision has not been made at this time.

Other recent changes involve Biscayne National Monument. In June of 1980 this monument was designated as a National Park. As indicated in a previous publication (Shinn, 1979), former rules and regulations still apply, but the new status may result in future rule changes. Biscayne National Park is presently in the process of extending its boundaries northward, and it is expected that no rule changes will take place until after the planned expansion. The proposed addition to this park is indicated on Figure 1, along with the boundaries of all the previously mentioned regulated areas.

It should be pointed out that this publication concerns itself primarily with scientific collection of biological and geological specimens. It does not include all the regulations concerning fish, of which there are many aimed primarily at commercial and sports fishing. For the purposes of discussion, all the above mentioned areas and permit requirements will be described individually.

Everglades National Park
(Department of Interior's National Park Service)

As shown in Figure 1, Everglades National Park includes Florida Bay as well as the Everglades. The calcium carbonate muds of Florida Bay have long been studied by geologists because of their similarity to lithified equivalents in the geologic record. The mechanics of deposition have served, in addition, as a model for the understanding of ancient limestones, many of which are prolific oil producers.

According to Park Service regulations, nothing other than sea water and fish caught by sports fishermen with hook and line can be collected either in the Everglades or Florida Bay without a valid collecting permit. A permit is required therefore to take sediment or peat cores and surface sediment containing living or dead organisms. Park Service biologists and administrators realize, however, that lime mud is the product of living organisms (i.e., it is a renewable resource) and therefore maintain a liberal policy toward granting permits for its collection. Permit applications, however, must be justified in terms of education and research.

Nesting birds make most islands in Florida Bay off limits to everyone, as are certain parts of the Everglades. Two islands, however, Crane Key and Bob Allen Key, are open to geologists for visitation, but a permit will be required if sediment is to be collected. Both islands are excellent for observing modern supratidal sedimentation, algal stromatolites, and laminated sediments. Hundreds of geologists visit these two areas each year. They are, in a sense, like "type sections," having provided the background for understanding ancient analogues which exist throughout the geologic record. These islands and adjacent mud banks are known by geologists all over the world and are visited annually by both foreign and domestic scientists. Other islands can be visited, but because they are rookeries, landing is illegal without a special permit.

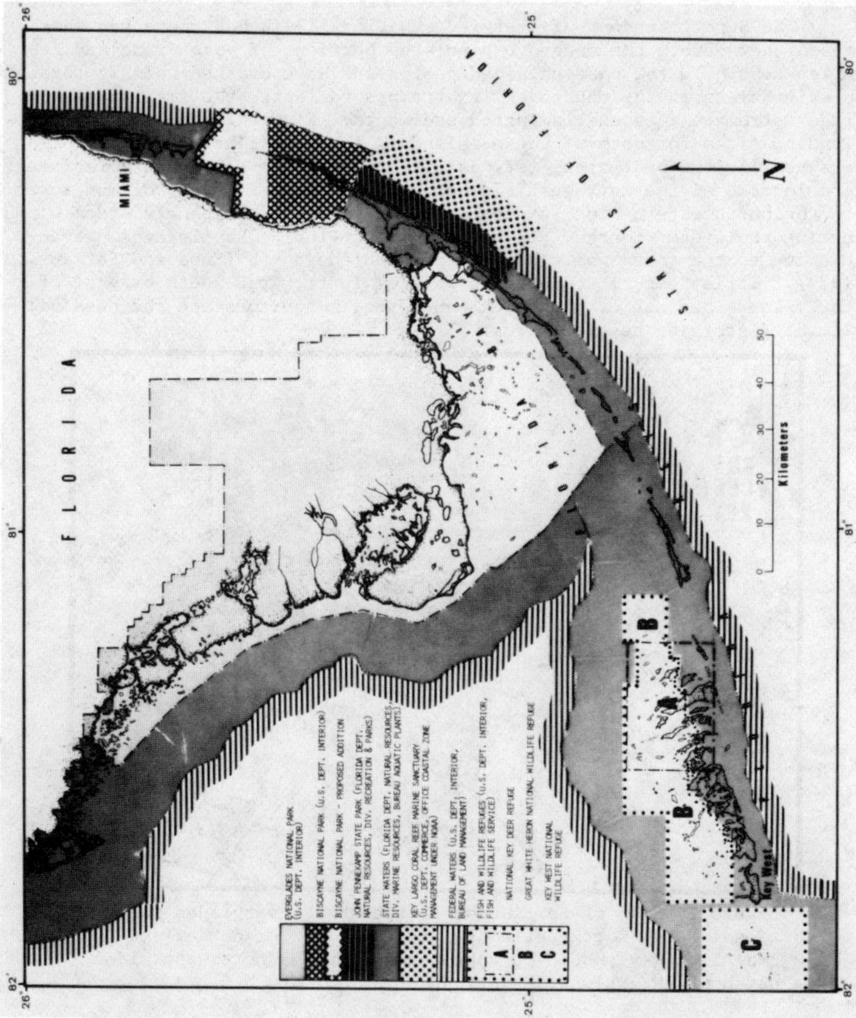
FLORIDA
MIAMI
STRAITS OF FLORIDA
N
Key West
Kilometers
EVERGLADES NATIONAL PARK (U.S. DEPT. INTERIOR)
BISCAYNE NATIONAL PARK (U.S. DEPT. INTERIOR)
BISCAYNE NATIONAL PARK - PROPOSED ADDITION
JOHN PENNEKAMP STATE PARK (FLORIDA DEPT. NATURAL RESOURCES, DIV. RECREATION & PARKS)
STATE WATERS (FLORIDA DEPT. NATURAL RESOURCES, DIV. MARINE RESOURCES, BUREAU AQUATIC PLANTS)
KEY LARGO CORAL REEF MARINE SANCTUARY (U.S. DEPT. COMMERCE, OFFICE COASTAL ZONE MANAGEMENT UNDER NOAA)
FEDERAL WATERS (U.S. DEPT. INTERIOR, BUREAU OF LAND MANAGEMENT)
FISH AND WILDLIFE REFUGES (U.S. DEPT. INTERIOR, FISH AND WILDLIFE SERVICE)
NATIONAL KEY DEER REFUGE
GREAT WHITE HERON NATIONAL WILDLIFE REFUGE
KEY WEST NATIONAL WILDLIFE REFUGE

The Superintendent of Everglades National Park has requested that researchers avoid the mangrove populated portions of both Crane and Bob Allen Keys. In the past unguided geologists have crashed noisily through nesting areas in the mangroves in attempts to locate the small areas of algal stromatolites and laminated sediments. Figure 2 shows the correct landing sites for both of these islands. The best landing areas, which are usually the best areas for observing supratidal sedimentary features, are located on the northwest side of each island. Scarcity of mangroves on the northwestern side is due to the influence of northerly storm winds during the winter months. In general, most Florida Bay islands are accreting to the south while being eroded on the north (Enos and Perkins, 1979). It is best not to attempt walking in from the south or east of the islands because of the thick mangroves, mosquitoes and the possibility of disturbing nesting birds.

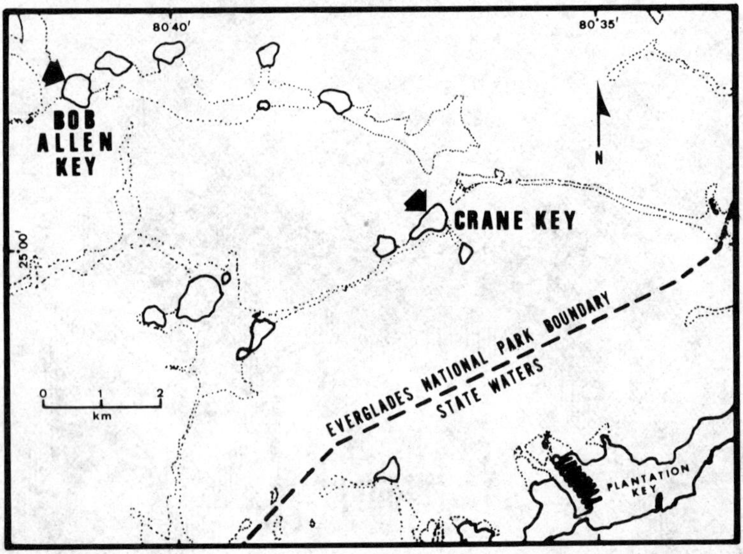

Figure 2. Map of portion of Florida Bay (Everglades National Park) northwest of Plantation Key. Crane Key and Bob Allen Key should be approached from the northwest side as shown by arrows.

For detailed park regulations or to apply for a permit, write to Research Director, South Florida Research Center, Everglades National Park, P.O. Box 279, Homestead, Florida 33030. Forty permits were granted to geologists in 1979 and at the time of this writing (August 1980), approximately thirty-one have been given. Permits are needed only by the leader of a visiting group, and since they are often reused two or three times a year, the number of visitors is therefore much higher than the number of permits issued. At least 700 geologists visited the area in 1979.

Everglades National Park also has jurisdiction over Fort Jefferson National Monument in the Dry Tortugas located 65 miles west of Key West. Permit information requests regarding Dry Tortugas therefore should be directed to the above address. Fines for collecting without a permit, whether in Everglades National Park, Fort Jefferson National Monument or Biscayne National Park, are generally $25 but can go as high as $500.

Biscayne National Park
(Department of Interior's National Park Service)

Biscayne National Park (Fig. 1), which was a National Monument until June 1980, is predominantly an underwater park but includes several Pleistocene islands. The main island is Elliott Key, site of a visitors center, and Sands Key to the north. In all there are about thirty islands or "keys" amounting to approximately 6,500 acres and a 300-foot-wide mangrove populated strip along the mainland. Biscayne National Park will eventually expand to the north (see dashed area of proposed new boundaries in Figure 1) to a total size, including water areas, of approximately 180,000 acres.

To establish the original monument, the State of Florida required that fishing be continued under existing State laws, unless it was determined to be detrimental to the purpose for which the monument was established. These stipulations are expected to apply under the new National Park status. Commercial fishing, crab and lobster trapping, and sport fishing are acceptable and desirable activities which will continue in accordance with State and special regulations. Spearfishing is currently legal but could be prohibited at some time in the future.

Collection of tropical fish, coral, rock, sand, trees, birds, land crabs and use of fish traps is illegal. A permit is necessary for scientific and educational collection of specimens.

Biscayne National Park is administered from a 60-acre administrative site located nine miles east of Homestead, Florida on North Canal Drive. To apply for a permit or more detailed information, write to: Superintendent, Biscayne National Park, P.O. Box 1369, Homestead, Florida 33030.

John Pennekamp Coral Reef State Park and Key Largo Coral Reef Marine Sanctuary
(State Department of Natural Resources, DNR, Division of Recreation and Parks and Department of Commerce's Office of Coastal Zone Management, National Oceanographic and Atmospheric Administration, NOAA)

John Pennekamp Coral Reef State Park and the Key Largo Coral Reef Marine Sanctuary are two complimentary parks in one. The State of Florida has jurisdiction over the Pennekamp State Park out to three miles, and authority of NOAA's Office of Coastal Zone Management under the Department of Commerce starts three miles from shore and extends beyond the coral reefs out to a depth of 300 feet (Fig. 1).

Named after the late John Pennekamp, the combined areas were originally a State Park. John Pennekamp was for many years the editor of the Miami Herald newspaper and a strong advocate of conservation. Mainly through his efforts, land was acquired for the Everglades National Park, but due to Federal regulations his name could not be applied to that park. Pennekamp was also very active in the establishment of the reef park, but it was given his name partially in honor of his key role in establishing Everglades National Park.

The John Pennekamp State Park was established in 1960 to protect coral and prevent spearfishing and tropical fish collection. It was a time when there were few divers and great bitterness between divers and commercial fishermen as well as local Florida Keys residents, a condition which today is reversing due to growing popularity of diving and its contribution to the local economy. New arrivals, many of whom are divers, now greatly outnumber native-born Keys residents (known locally as "conchs"). The State did not limit sports fishing, commercial hook and line fishing and lobstering with traps when the park was created, and that condition has not changed.

Because the reefs beyond the 60-foot contour were not protected under the Eisenhower proclamation and possibly because no Federal funds were available to help finance regulatory activities, the area was proposed for Sanctuary status by Dennis O'Connor of the University of Miami and Rezneat Darnell of Texas A & M University. On December 18, 1975, the area beyond the State's three mile limit was dedicated on board the glass bottom boat "Discoverer" under Title III of the Marine Protection Research and Sanctuaries Act of 1972. In essence, the existing regulations for the John Pennekamp State Park were adopted for the marine sanctuary.

Through a joint management agreement with the State of Florida, Key Largo Coral Reef Marine Sanctuary is patrolled by State Park Rangers and the U.S. Coast Guard. In actual practice the Coast Guard's involvement is minor, but State Park Rangers are efficient and diligent. Fines can be stiff. Collection or possession of coral, dead or alive, and sand or any other organism dead or alive (other than fish or lobster) can be as high as $50,000. If coral or other materials or organisms mentioned above are collected outside either John Pennekamp Coral Reef State Park or Key Largo Coral Reef Marine Sanctuary, they cannot be transported into these areas. This is similarly true of Biscayne National Park and Everglades National Park.

To apply for scientific collection permits or obtain additional information concerning the Key Largo Coral Reef Marine Sanctuary (beyond three miles from shore), write the Office of Coastal Zone Management, NOAA, Department of Commerce, 3300 Whitehaven St. NW., Washington, D.C. 20235.

To apply for a scientific collection permit or obtain additional information concerning the John Pennekamp Coral Reef State Park (within the three mile limit), write to State Department of Natural Resources, Division of Recreation and Parks, Crown Building, 202 Blount St., Tallahassee, Florida 32304. Permits for this area require special justification, because all organisms that live there can be collected in State waters outside (north or south) of the John Pennekamp Coral Reef State Park.

State Waters Outside of John Pennekamp Coral Reef State Park

(State Department of Natural Resources' Division of Marine Resources and Bureau of Aquatic Plants)

A State law protecting corals is in effect for all State waters outside John Pennekamp Coral Reef State Park (Fig. 1). Regulations are less stringent than those of John Pennekamp Coral Reef State Park in that they protect only live coral and algae. Sand, coral rubble, and other bottom dwelling organisms, except algae, may be collected without a permit. A permit is absolutely necessary, however, to collect live corals.

To apply for a permit or request other information, write Permit Secretary, Florida Department of Natural Resources, Division of Marine Resources, Crown Building, 202 Blount St., Tallahassee, Florida 32304.

The collection or importation of algae within State waters is controlled by the Bureau of Aquatic Plants. Although the law was clearly intended to control importation of noxious fresh-water plants, which thrive and choke stream flow, reduce oxygen levels and in general disrupt the ecology of lakes and streams, the law nevertheless applies to marine waters as well. Under State Law Chapter 16C-19.03, Definitions, Item 7 states that, "'Waters of the state or water' shall include but not be limited to rivers, lakes, streams, springs, impoundments, all other waters or bodies of water, including fresh, brackish, saline, tidal, surface or underground."

Although any law enforcement officer of the State of Florida may make arrests for violation of the Aquatic Plants laws, to my knowledge no arrests have been made for collection of marine plants in the Florida Keys, nor is any marine patrol office likely to enforce the law. To comply with the law, however, a permit is necessary. To obtain a permit or additional information, write the Florida Department of Natural Resources, Bureau of Aquatic Plants, Crown Building, 202 Blount St., Tallahassee, Florida 32304.

National Key Deer Refuge, Key West National Wildlife Refuge, and Great White Heron National Wildlife Refuge

(Department of Interior's Fish and Wildlife Service)

The above three refuges are administered by the National Key Deer Refuge located on Big Pine Key (Fig. 1). To collect anything on land, including rock, sediment or peat, requires a permit. Collecting from the sea floor does not require a permit from the Fish and Wildlife Service. A permit from the State of Florida is necessary, however, because these waters lie within State boundaries. Information on obtaining State permits has already been discussed.

The islands making up this complex area are patrolled by Fish and Wildlife Service Rangers. Waters surrounding the islands are patrolled by the Florida Marine Patrol.

Most permits for collecting on the islands are handled locally and therefore require less than a month to obtain. To apply, write the Manager, National Key Deer Refuge, U.S. Department of Interior, P.O. Box 510, Big Pine Key, Florida.

Fort Jefferson National Monument, Dry Tortugas
(Department of Interior's National Park Service)

Located 65 miles west of Key West, Florida are the Dry Tortugas and Fort Jefferson (not included on Fig. 1). Although the main purpose of Fort Jefferson National Monument is to protect the remains of the fort, unusually good protection is offered for the surrounding coral reefs. This area is more of a true sanctuary than all others in south Florida in that only sports fishing with hook and line is allowed. Commercial fishing, lobster trapping, tropical fish collecting, spearfishing and collecting of coral, algae, and conchs are illegal. Because of the small size and high observation posts situated atop the fort, rangers can survey almost the entire area.

At present, the boundaries are in a state of flux due to confusion between what is posted on modern marine charts of the area as "administrative boundaries" and "monument boundaries." It is likely that the administrative boundaries, which are legally and physically more defined, will be adopted.

To apply for a collecting permit, write to Research Director, South Florida Research Center, Everglades National Park, P.O. Box 279, Homestead, Florida 33030.

Discussion

Prior to 1980, the Bureau of Land Management (BLM, Department of Interior) had jurisdiction through a Secretarial Order over coral reefs in all Federal waters and issued permits that cost $25. The BLM became involved in coral reef management as a result of concern over corals near offshore drilling areas. BLM is the agency that conducts sales of offshore oil, gas and mineral areas.

As a result of a Fifth Circuit Court of Appeals decision (*United States v. Alexander*), the Secretarial Order was rescinded. As of this writing, living corals in Federal waters beyond State boundaries and outside existing parks and sanctuaries are unprotected.

Under the Fishery Management Act of 1976, generally known as the 200-mile fishing law, regional Fisheries Management Councils were set up to determine how fisheries will be managed. For various reasons corals were considered a fishery and the Gulf of Mexico Fishery Management Council contracted in cooperation with the South Atlantic Fishery Management Council for a management study by the Center for Natural Areas, a private research organization. With help from members of the Councils, the Center for Natural Areas prepared three draft fishery management plans for coral areas. The final plan is in preparation and is slated to become law in 1981, after a series of mandated public hearings. When this plan goes into effect, all coral reef areas in Federal waters will be protected from collection and harassment by persons without proper permits. At present, it is not known how and from whom the permits will be obtained.

The State of Florida is presently in the throes of streamlining regulations both for fishing and scientific collecting. The problem is compounded by numerous local County laws which often conflict with State laws. Fortunately, this is not a problem for coral areas which are beyond County jurisdiction. For the above reason, some of the information in this paper may soon become outdated.

In general, the obtaining of scientific collecting permits has been simplified in the past year. For example, prior to the (U.S. versus Alexander) court decision, permit applications for collecting in the Key Largo Coral Reef Marine Sanctuary had to be approved by the Bureau of Land Management and published in the Federal Register. Under that system, a minimum of 60 days was required to obtain a permit. It is hoped that permitting procedures will continue to improve.

The regulations of many marine plants by the Bureau of Aquatic Plants (State of Florida) stand out as unnecessary. It is clear from reading of the legislation that created this bureau that its primary purpose was to regulate importation and destruction of fresh-water aquatic plants. Such regulations are certainly necessary in light of the modifications that have occurred from the runaway growth of certain aquatic plants, such as the water hyacinth. State and local governments spend thousands of dollars each year cleaning these imported plants from choked canals and streams. There is, however, no known threat from imported marine plants, nor have there been any reports of non-native plants springing up on Florida's sea floor.

Marine waters were included in the Bureau of Aquatic Plants' jurisdiction possibly for the protection of *Thalassia* (turtle grass) and related sea grasses, such as *Halodule* and *Syringodium*, which are vital as habitats for spawning fishes and protection against sea floor erosion. These plants are vital and should be protected from dredge and fill operations, although numerous transplanting projects have shown that depopulated areas can be reseeded. There has been little legal concern for damage inflicted by motor boats, anchors and lobster traps on marine aquatic plants. Plants such as *Penicillus*, *Acetabularia* and *Goniolithon* are not known to serve vital ecological functions as do the sea grasses and appear to be in little danger of over-collection. They are rarely of interest to anyone other than researchers. *Halimeda* is important in that, since Tertiary times (the last 63 million years), it has been the major sediment producer in south Florida and most of the Caribbean. Upon death, its calcareous plates disaggregate to form sand. More than 50 percent of the sand, even on coral reefs off the Atlantic side of the Florida Keys, is composed of *Halimeda* sand. Fortunately, no one has found a commercial use for this plant and it thus has never been harvested commercially.

Since the average enforcement officer is not likely to recognize any of the marine plants, other than the common turtle and eel grasses, the aquatic plants laws probably should be revised. Plants such as *Penicillus*, *Acetabularia*, *Halimeda* and the like could easily be excluded, thus making the legal process much simpler.

The author is not a lawyer, and this article has been written in non-legal language. Proper legal code numbers and statutes have been eliminated to provide easier reading. The writer, therefore, does not assume responsibility should a person be arrested while following the advice given here. Furthermore, the laws are in a constant state of flux and some may have changed somewhat by the time this is read.

It is hoped that this paper will aid the scientist who does not have the time to search through the various agencies to learn about permit application. It is also hoped that the work of various marine researchers will have input into the development of future marine regulations.

Acknowledgments

The author gratefully acknowledges the cooperation and help of the officers and personnel of all agencies cited.

References

Enos, P. and R.D. Perkins, 1979, Evolution of Florida Bay from island stratigraphy: Geol. Soc. Amer. Bull., Part I, v. 90, p. 59-83.

Shinn, E.A., 1979, Collecting biologic and geologic specimens in south Florida: Atlantic Reef Committee Information Circular No. 1, ARC, Univ. of Miami, Fisher Island Station, Miami Beach, Fla. 4 p.

INTERGOVERNMENTAL COORDINATION IN NORTH CAROLINA

Mr. Hubert Glenn T. Dunn
Chief, Regulation Coordination and Enforcement
Office of Coastal Management
NC Department of Natural Resources and Community Development
PO Box 27687
Raleigh, NC 27611

and

Mr. Charles W. Hollis
Chief, Regulatory Functions Branch
US Army Engineer District, Wilmington
PO Box 1890
Wilmington, NC 28402

INTRODUCTION

The development of well integrated, cooperative, and complementary programs between State and Federal agencies is an often discussed subject of politicians and administrators, but committing to it by either party seems to be rare indeed. There is a natural distrust and jealousy, a perception of threat between the two levels of government, and often a very basic difference of goals: The purpose of their respective programs is seen as being incompatible with the other. It is in this framework that we see many State agencies reluctantly and stubbornly conducting programs, especially regulatory program, in accordance with insensitive Federal guidelines and under the suspicious supervision of a conceited Federal agency. As times goes on, each level of government develops very divergent goals: The State trying to outfox the watchdog Federal agency, and the Feds trying to catch them at it--all to no one's benefit and at the public's expense. In North Carolina, we've proved it can be different and better. The participants are the Wilmington, North Carolina District, US Army Corps of Engineers, and the North Carolina agencies operating within the North Carolina Coastal Management Plan. The programs are the Corps and State regulatory programs within the designated 20-county "Coastal Zone," and the benefactors are the taxpayers.

TEN YEARS OF EXPERIENCE

To understand what we are doing and why, it is necessary to look back a few years to see when and how it all started.

Prior to 1969, the Corps was conducting a regulatory program requiring permits for work of any kind in the navigable waters of the United States - those waters subject to the rise and fall of the tide or the use by vessels engaged in interstate commerce. The Corps was limited in its discretion to deny a permit such that denial could only be based on the contention that the proposal would unreasonably interfere with navigation. In North Carolina, as elsewhere at that time, there was a widespread concern for greater protection of coastal resources, especially for the productive marshlands that are so abundant along our coast. The North Carolina legislature passed in 1969 an act requiring "permits to dredge or fill in or about estuarine waters or State-owned lakes." The essence of the Act was to require State permits for dredging or filling in any coastal waters, tidelands, salt marshlands subject to even occasional flooding by lunar or wind tides, and State-owned lakes. Denials of such permits could be made based on any of the following negative findings:

1. That there will be a significant adverse effect of the proposed dredging and filling by the public; or, on the use of the water.

2. That there will be significant adverse effect on the value and enjoyment of the property of any riparian owners; or,

3. That there will be significant adverse effect on public health, safety, and welfare; or

4. That there will be significant adverse effect on the conservation of public and private water supplies; or

5. That there will be significant adverse effect on wildlife or fresh water, estuarine, or marine fisheries.

While the legislature was developing this Act and before the first permits were required on 1 January 1970, the Corps program was being changed by some significant events. First was a revision in Corps policy published in December 1968 which listed factors in addition to navigation which would be considered in the review of permit applications. Then, in 1969, the Corps in Florida, denied a permit on the grounds of its adverse impact on fish and wildlife. This denial was challenged in court but the court took note of the confirming language in the National Environmental Policy Act which Congress passed in January 1970 and pointed out that the Corps was "acting under a Congressional mandate to collaborate and consider all of these factors when it reached that decision." The Corps suddenly found itself with virtually unlimited permit denial authority based on a broad review process which began to be called a "public interest review." In North Carolina, then, we had a Corps program covering a wider range of activities than the State program and also having the discretion to deny based on essentially any factors affecting the public interest. The State program, on the other hand, although limited to dredge or fill activities which could only be denied based on the five negative findings, had a much broader geographical scope which included the vast high marshes in coastal North Carolina and the State-owned lakes. Needless to say, there were numerous projects which required both permits.

Coordinating the processing of permits was the first and easiest step taken. It had long been a requirement for the Corps to seek a "State Coordinated Viewpoint" on each case, and, with the advent of North Carolina's dredge and fill permit program, the State's permit action became the best vehicle for that viewpoint. We found that the public did not always recognize the need to apply for both permits and certainly did not understand the differences in the scope of each jurisdiction. It was, therefore, often necessary to refer permits to one another for our respective processing. In 1972, we developed a joint application form which allowed the applicant to furnish one set of information to the two regulatory offices which, in turn, would proceed with their respective administrative details. This combined form has been revised several times in a continuing effort to provide applicants with a one-stop application insofar as the required information is concerned.

Enforcement regarding both permit violations and unpermitted activities has been a very strong and continuing part of Corps-State coordination from those early years. The regulatory agency for the State dredge and fill permit program was the North Carolina Division of Marine Fisheries, who also had, as part of their regular duties, the enforcement of the various State commercial fishing regulations. To accomplish this, they were staffed with approximately 50 law enforcement personnel equipped with radios in vehicles, boats, and airplanes and assigned throughout the coastal region of the State. This staff soon became the eyes and ears of not only the State dredge and fill program, but also the Corps' regulatory program. To help coordinate this effort and also to develop interagency agreement as to the solution to various enforcement cases and problems, a monthly "Enforcement Conference" was begun in 1974. Although now held on an approximate bi-monthly frequency these conferences have remained an important element in the overall program of interagency coordination.

On the Federal side, they are attended regularly by representatives of all of the review agencies (EPA, US Fish and Wildlife, National Marine Fisheries Service) plus SCS, the Assistant US Attorney, and on occasion by the Coast Guard and others. State agencies include those responsible for any programs in the Coastal Zone such as Marine Fisheries, Coastal Management, Environmental Management, Human Resources, Wildlife Resources, Office of the Attorney General, Department of Transportation, and others. Typical attendance is about 30 people. These meetings are conducted by the Corps regulatory staff on an informal, operating staff level and now cover any and all subjects of interest such as changing policies, complex applications, updates on enforcement matters, and brainstorming solutions to interagency problems. Although there is considerable individual contact on a daily basis, the conference, followed by a seafood lunch (dutch treat), serves well to maintain the vitally important camaraderie this group has developed through the years.

In 1974, the North Carolina legislature passed the Coastal Area Management Act (CAMA), a comprehensive act which developed a regulatory/management program over areas of environmental concern which includes, but extend considerably beyond, the geographical jurisdiction of the older dredge and fill statute. Use standards for these areas were authorized by the statute and have been implemented. In addition, local governments were required to develop land use plans which include local policies for land use and development and a land classification system for all areas outside areas of environmental concern. This Act and the accomplishment of its various phases of planning and now implementation, provided the vehicle necessary to bring into play a broad matrix of local government ordinances, State approved local land use plans, AEC regulations, and systematic reviews by State agencies having an interest in the Coastal zone. This matrix forms the backbone of North Carolina's Coastal Zone Management Plan approved by the US Department of Commerce under the 1972 US Coastal Zone Management Act on 5 September 1977. Among other things, the Federal Act requires that activities proposed under a Federal permit must be consistent with the State's approved plan.

Then, in mid-1975, the Corps published revised regulations, in response to a court order, which expanded the Corps permit program, insofar as it concerned the discharge of dredged or fill material, to all waters of the United States including adjacent wetlands. This portion of the Corps responsibility, known as the Section 404 program, added a new dimension to the geographic scope of the Corps program as well as to the list of activities subject to Corps permits. The same project in North Carolina's "Coastal Zone" now may require as many as six permits:

Two from the Corps of Engineers under Section 10 of the River and Harbor Act of 1899 and one under Section 404 of the Clean Water Act;

Four from the State of North Carolina; Dredge and Fill, CAMA, an Easement to fill or otherwise occupy State-owned submerged lands, and a Water Quality Certification under Section 401 of the Clean Water Act. In addition, the non-Federal applicant must certify that his project is consistent with the North Carolina Coastal Zone Management Plan.

All of these permits, certificates, etc. are still applied for on <u>one form</u>. The applicant can obtain the forms and instructions and get help in completing them from any of the Corps or State personnel involved in the regulatory program. He then submits one copy to the Corps and one to the State who then individually process their respective actions. There remain situations where no Federal permit is required and State permits are required or vice versa. In either case the consistency provisions prevail since, by Federal law, the Federal permit must be consistent, and by State policy, the State permit must be consistent with the Management Plan. In some cases which require a Corps permit, the CAMA permit may be processed and decided by a local CAMA permit officer, usually the local building inspector. These people, some 52 of them, have had to be trained, not only in the requirements of CAMA, but to have an awareness of Corps of Engineers involvement. This has added a wider range of necesary coordination to all involved in the program, by bringing in the local perspective to State and Federal permit decisions though it admittedly called for a little extra patience as we bring these persons into a better understanding of the whole regulatory and management system.

TODAY AND TOMORROW

Those associated with the regulatory programs are familiar with the strengths and weaknesses in each level of Government (Federal, State, or local), and we are ever aware of the need to support one another or to ask for support when we need it. Enforcement action in the larger or more politically sensitive violations can often be best handled by the Corps and the US Attorney through the courts. The more numerous smaller violations are often more efficiently addressed by Civil penalties, an administrative procedure available to the State. Many cases are jointly pursued such that whichever enforcement authority can reach an acceptable settlement first has the support and added weight of pending further legal action by the other authority if the settlement fails.

Mutual support in permit processing is especially helpful in speeding the process and developing a consistent position on the application. Permit processors for the respective agencies (Corps, State, local) share the workload of site visits and field reports on small projects and jointly visit and coordinate their work on the more complex ones. This practice leads to more consistent preapplication consultation and helps the applicant avoid inconsistent and confusing advice from the different levels of Government. Our experience over the years of working this closely has led in recent weeks to the development of more innovative procedures for avoiding even more of the remaining duplicative tasks that we have. The Corps has been able to use General Permits to underwrite many of the actions under the CAMA program and thus further reduce the paperwork and delays otherwise sometimes experienced by the applicant having to respond to the Corps and the State.

Our mutual goals are not new but are continuing in that there is always more to do. Some of those goals are:

1. Develop a better recognition of the need for the autonomy of each level of regulatory authority. It is inevitable that there will continue to be occasional different decisions made on a project by different government levels. We must continue to be aware that these different levels of government are often motivated, politically or otherwise, by different interests; but more importantly, we must learn to better respect and understand those interests, however much we may disagree with them. Each level of statutory authority may be based on satisfying somewhat different needs which must, then, be reflected in the decisions made under that authority. The decision then is different but not wrong. The accomplishment of this goal lies in each participant developing an understanding of and respect for the laws, rules, regulations, and policies of each of his counterparts in other levels of government.

2. Serve the public. There are at least three subheadings under this topic which we will list, but most importantly we must not lose sight of the fact that we regulators are working for the public -- treat each public person with the respect due your boss!

2.a. Avoid duplication of effort between agencies. This was mentioned earlier, but is a continuing need. Innovation is the rule -- try at all times to build enough trust from others in what you are doing or from you in what others are doing so that it doesn't have to be repeated. We have done a reasonably good job through General Permits and CAMA exemptions to give quick approval to those cases that obviously will be approved. By issuing General Permits for certain activities and CAMA exemptions for other activities, the two agencies avoid duplication for certain classes of routine projects that can be handled adequately by a single agency. This is a particularly unique cooperative technique for saving time and resources for the agencies and especially for the applicants.

2.b. Minimize burden on the public. The one-stop application has been our giant step in this effort, but more needs to be done. What the applicant wants most is an early decision. In those cases that obviously cannot be permitted, we are still looking for ways, without avoiding the due process under the law, to furnish the applicant a quicker answer. An ever increasing amount of time and effort is spent by all personnel in pre-application consultations especially with applicants for questionable or very complex projects. Advice is given at that time as to the relative likelihood of permits being approved based on our experience with similar projects in the past. Suggestions are made for changes which would minimize environmental or other known impacts so that when the application is received, it is subjected to review by persons already familiar with it -- who can then develop their comments in a minimum of time.

2.b. Reduce paperwork. The joint application is again a big help in this effort, but much remains to be done. Expanded acceptance by the various review agencies of a single comprehensive site description and field report is an innovation that is already working well. Each agency can then draw on that report for their position unless special circumstances prevent this. A similar approach can work well in investigating violations thus reducing trip time and reports and hastening a decision on the resolution of the matter.

Tools of Interagency Coordination

In ascending order of importance, here is our attempt at listing what we feel are vital tools to accomplish effective interagency coordination:

1. Clear legislative authority with administrative flexibility. The proof of any regulatory program is the strength and clarity of the law on which it is based. Counterpart agencies must be able to understand a law in order to understand and respect the actions taken under its authority. But that law should be flexible enough to allow for reasonable administrative decisions to be made based on the circumstances found in each case. Without that flexibility, the administrator becomes a robot with no public mission to perform, and, worst of all, the public has no respect for the law or any incentive to develop projects carefully designed to comply with the spirit and intent of the law. In this program, we are regulating and/or managing, not stopping and preventing.

2. Practical (and practicable) and reasonable administrative regulations. The administrator and the public deserve well written, understandable regulations that outline the process whereby approvals may be granted or denied. Again, the counterpart agency needs to have a clear understanding of what your rules are so that he can effectively coordinate his own program with yours. With the law and the regulations clearly understood by each agency and the process carefully and consistently followed, there is assurance of understanding of not only the process, but the decision, even if other agencies disagree with it.

3. High quality people. Everything mentioned earlier is subject to this most important point -- these programs, their conduct and their coordination, are accomplished by people. Mediocre people are going to conduct only mediocre programs whereas quality programs are a product of quality people. Some of the traits of quality needed here are:

a. Sensitivity. A concerned awareness of the needs and feelings of the applicants, the larger public, of co-workers in and beyond his agency -- a need to exercise tact and diplomacy and patience. They must also have a sensitivity for the mission of the regulatory/management program, being able to balance those needs against the needs of the applicant.

b. Sense of urgency. The public's most frequent complaint is that the bureaucracy operates too slowly. A quality employee look for every opportunity to overcome that complaint. In doing so, he has a sense of urgency about getting his job done and about getting good answers where they are needed - quickly.

c. Mission oriented. Pride in a job well done. This drives a person in many directions. One direction should be to seek out and develop a clear understanding of the laws and regulations of his own as well as companion programs. He is thereby able to guide his applicant though the regulatory maze in the shortest time - and that becomes part of his mission.

d. Integrity. Consistent - dependable - honest - above reproach - all of these are vital and too often not, in the public's eye, typical of government employees. Nothing can destroy a whole program and cut into the effectiveness of even other agencies as much as some act of indiscretion on the part of even one person.

e. Candor. Tell it like it is. The public wants the truth and the facts. When a project is denied, the denying agency must tell the applicant why and be prepared to defend itself in court. Each employee must be strong enough to tell the public the truth even when it's bad news and may upset the hearer.

f. Mediator. This is a regular and necessary role of the regulator as he must often serve as the focal point for divergent views on a project and then strike a compromise that properly balances the various opinions. Here again is needed the qualities of tact, diplomacy, understanding of the various programs, and recognition of the needs of the applicant, the demands of the program, and interests of the reviewers.

Fortunately, most of these qualities can be learned - learned best by example but also through training and good supervision.

In North Carolina, we have the good fortune to have relatively good, solid coastal management laws (which have stood some early judicial tests) and their administrative regulations which have been carefully devised and published for all to use and study. The Corps, the State, and the local governments recognize their respective strengths and weaknesses and emphasize the strengths of each program. And finally, we have a pride in our overall program, its people, and its accomplishments. Everyone is proud to be on the same team and to share in its good reputation among all of its agency members, but especially the reputation within the public whom it serves. This pride and this reputation keeps driving us all to look for newer and even better ways of doing our job for the public-together.

MANAGING BARRIER ISLANDS: ROLE OF LOCAL GOVERNMENT

Kent S. Butler,[1] Kathryn N. Nichols[2] and Ray Quay[3]

ABSTRACT

Local governments on barrier islands are in a unique position to plan for and manage coastal development. While Federal and state policies, programs and regulations clearly have significant impacts on island development, local governments on these islands have the prime opportunity to execute comprehensive and integrated measures to control or manage growth. Many of the 128 islands in the United States which support communities rank among the more innovative and forward-looking of all coastal units of government.

Barrier island local governments should strive to approach the tasks of land development control aggressively and comprehensively. Responsibility for administering many Federal and state programs affecting island development is rightfully delegated to local governments, for they sense the concerns of affected citizens most keenly and shoulder the ultimate responsibility to maintain a viable economy and environment. Their challenge is to select and implement environmental management and development guidelines that are workable within the unique constraints present in their island environments. The experiences and techniques used by exemplary island communities noted in this discussion can be transferred in part to other locations. But the particular combinations of state laws, enabling legislation, development pressures, and landform and biologic characteristics on each island dictates that each be addressed differently. Federal and state governments are currently making careful assessments of their policies, programs and regulations that inadvertently affect (and sometimes promote) island development. Where island communities exist, higher units of government should not only temper their promotional influences on development, but also ensure that their programs are compatible and supportive of local governmental efforts.

---

[1]Assistant Professor, School of Architecture and Planning, Univ. of Texas at Austin, 78712, (512) 471-4911.

[2]Recreation Planner, Texas Parks and Wildlife Department, Comprehensive Planning Division, Austin, Texas.

[3]Principal Planner, City of Galveston Planning Department, P. O. Box 779, Galveston, Texas, 77550.

INTRODUCTION

The considerable attention given to barrier island development and the attendent burdens imposed on public institutions and coastal natural resources has raised many questions concerning the proper role of coastal governance at all levels. At the Federal level, the Carter Administration is currently reviewing alternative policies for barrier island protection (U.S. Dept. of the Interior, 1979), while the House of Representatives is debating alternatives for Federal acquisition and restriction of supports for undeveloped islands (H.R. 5981, 1980); and the National Park Service has made major modifications in its policies related to dune stabilization and shoreline protection on the national seashores (Dolan, 1974). State coastal governments are comprehensively addressing barrier island resources and uses with their heightened recognition of islands as "geographic areas of particular concern" (Coastal Zone Management Act Amendments of 1976). Civic and non-profit conservation organizations such as the Barrier Island Watch Group, the Nature Conservancy, and many others have taken great strides to protect island resources and promote nondegradative, recreational, scientific, or educational uses.

Local units of government, however, have probably done more to guide and manage island development than all other governmental levels combined. Their record, however, has been checkered with successes and shortcomings. To this date, there has been no comprehensive assessment *per se* of the local role in managing island resources, or of this role vis-a-vis other levels of government. The following discussion is an initial attempt to determine what their proper role should be, both in the context of previous experiences and of future actions by Federal and state government.

THE CONTEXT OF LOCAL GOVERNMENTAL INVOLVEMENT

Cities and towns, counties, and various special purpose districts have been actively involved in managing barrier island development for decades. The islands of the northern Atlantic Coast were the first to encounter significant development pressures in recent times, as upper class families acquired land for residential-recreational use. Gradually, the locus of island recreational-residential development moved southward along the Carolinas, Georgia, and into Florida as coastal population growth increased, and as the limitations of travel distance and island access were circumvented with modern modes of transportation and mainland-to-island causeways. Even today, the Gulf Coast islands are significantly less developed than their Atlantic counterparts, but localized pressures have also risen sharply there in recent years.

The urgency for greater involvement of local units of governments is most keenly illustrated by the rate of barrier island development in the past three decades. While only 12% of all Gulf Coast islands were "developed" in 1973, more than 37% of the land on the northern Atlantic islands (from New Jersey to Maine) was already developed by that time. In general, barrier islands are becoming urbanized twice as rapidly as the Nation as a whole; 3% of the Nation is considered urban, while more than 14% of the island space is urban (U.S. Dept. of the Interior, 1979). Table 1 indicates these and other trends that are causing severe problems for people and property and the natural resources on these islands.

Table 1
SELECTED INDICATORS OF BARRIER ISLAND DEVELOPMENT*

Developed Land:			
Barrier Islands:	1950	104,500 acres	6.5% of total land
	1973	242,400 acres	14% of total land
	change	6,000 acres/yr.	
Nationwide:	1973		3% of total land
Gulf States (including part of Florida):	1973		
		83,900 acres	12% of total land
South Atlantic States (Delaware to Florida):	1973		
		111,277 acres	14% of total land
North Atlantic States (Maine to New Jersey):	1973		
		47,168 acres	37.5% of total land
Maine		1,640 acres	62%
New Hampshire		700 acres	64%
New Jersey		22,700 acres	47%
Rate of Housing Development:			
Nationwide:	1960-1970		22.8%
Barrier Island Counties:	1960-1970		38.4%
Rate of Population Growth		1960-1970	1970-1975
Nationwide		14.9%	5.6%
Barrier Island Counties		30.8%	9.6%

*Taken from: U.S. Dept. of the Interior, 1979.

Traditionally, local levels of government have been vested by state legislatures with extensive land use control authorities and provisions for managing the full spectrum of urban public services. The traditional tests for appropriateness of local land use guidance or control measures are usually fourfold. They must be based on: state enabling legislation, the comprehensive plan of the community, protection of the health, safety and welfare of the people, and ultimately, sound and rational priniciples that will stand up in court. In the arena of coastal development, local controls are a part of an increasingly complex and multi-layered set of development controls involving many state and Federal agencies, and often other local governmental jurisdictions.

Most barrier island local governments have at least acknowledged that they also share the responsibility to maintain and preserve physical and biological resource assemblages that characterize island systems. There are five critical resource concerns that all barrier island local governments should address with environmental and land development controls: groundwater, beach and dune areas, wetland resources, recreational access, and human lives in the event of tropical storms or hurricanes. Many local governments have responded in good faith to these concerns, but to varying degrees. In the following section, we will examine some of the actions taken by local governments.

TECHNIQUES USED BY BARRIER ISLAND LOCAL GOVERNMENTS

There are dozens of controls and management techniques currently in effect on barrier islands. In the context of the immense development pressures faced by many local governmental jurisdictions, many of these techniques are inadequate or not ideally suited for the harsh, yet sensitive environments that characterize barrier islands. Paul Sedway's introductory comments about traditional zoning techniques in a discussion of viable coastal regulation (Sedway, 1978) were rather poignant:

> It is contended, with some evidence, that conventional zoning is: (1) fragmented; (2) exclusionary; (3) staltifying to good design; (4) inflationary; (5) unfair; (6) self-serving; (7) parochial; (8) poorly administered; and (9) unrelated to the planning process.

Perhaps an equally valid counterargument can be made for other, more forward-looking measures that have been practiced on barrier islands. Certainly, many of the 128-odd islands in the United States that support communities rank among the most innovative of all coastal units of government. A representative selection of techniques currently in operation on barrier islands is listed in Table 2. A brief portrayal of eight barrier island local governments and some of their management or control techniques will serve to illustrate the range of possibilities that are readily available at this time.

Sanibel, Florida (Sanibel Island)

The Comprehensive Land Use Plan and Land Development Code for Sanibel Island (Sanibel Planning Commission, 1976) is probably the most salient model among all barrier island development control systems. Future allowable levels of development on the island were determined on the basis of ecological carrying capacity, public service capacities, and safe population levels in the event of hurricane evacuation. Three of the more reknowned planning, legal, and conservation organizations in the Nation -- Wallace, McHarg, Roberts, and Todd; Ross, Hardies, O'Keefe, Babcock, and Parsons; and the Conservation Foundation -- were solicited to carry out basic ecological inventories and interpretations, and to develop a land allocation system and performance standards that would preserve much of the island's intrinsic values, while still accommodating a 50% growth in housing (about 2,000 additional units). The allocation system provides for phased future development, guided by maximum allowable densities that are specified for each of several ecological zones on the island. Numerous performance standards, ranging from allowable proportions of impervious cover to restrictions on the introduction of exotic plant species, were assigned to each of the ecological zones (Clark, 1976).

Table 2
REPRESENTATIVE LOCAL LAND USE CONTROLS AND MANAGEMENT TECHNIQUES

AREAWIDE TECHNIQUES
- Zoning (use, density, bulk, etc.)
- Planned Unit Developments
- Subdivision Regulations
- Flexible Districting
- Transfer of Development/ Density Rights
- Construction Moratoria
- Restrictive Covenants

PERFORMANCE AND SPECIFICATION STANDARDS
- Building Codes
- Sanitary Codes
- Special Use Permits
- Special Area Districts
- Setback Lines (dynamically determined, uniform distance, geomorphically determined, visually/aesthetically determined)
- Sand Removal Ordinances
- Vegetation Protection/Restoration Ordinances
- Utility Construction Standards (roads, drainage pipes, catch basins, etc.)
- Impervious Surface/Open Space Standards
- Waterway Engineering Standards (bulkheads, lagoons, canals, etc.)

ACQUISITION TECHNIQUES
- Eminent Domain
- Direct Purchases (mutual agreement)
- Easements (accessways, scenic, conservation)
- Open Space Dedication Requirements in Subdivision Review

FISCAL TECHNIQUES
- User Fees
- Tax Incentives
- Tax Disincentives
- Bond Programs for Shore Protection, Restoration, Beach Nourishment, etc.

MANAGEMENT CONTROL PROCEDURES
- Comprehensive Land Use Plans
- Plan-based Administrative Review
- Building Permits
- Environmental Impact Assessments and Reviews
- Public Access Plans and Provisions
- Centralized (One-Stop) Permitting
- Development Preapplication Conferences
- Developments of Regional Impact Review (Model Land Development Code)

MISCELLANEOUS TECHNIQUES
- Beach Nourishment
- Structural Shoreline Protection
- Municipal Incorporation
- Annexation of Urban Fringe Areas

Interestingly, development of the comprehensive plan was the first major item on the city government agenda after its municipal incorporation proceedings in 1974. The Lee County Planning Commission had persistently promoted high-density development over most of the island until the citizen referendum for incorporation freed the island from county jurisdiction.

Collier County, Florida (Ten Thousand Islands)

Unlike its neighbor to the north (Lee County), the Board of County Commissioners of Collier County adopted a Special Treatment ordinance in 1974 which was designed to protect environmentally sensitive areas, expecially along the Ten Thousand Island archipelago (Anonymous, Report of the Barrier Island Work Group, 1978). The islands had been owned for a long time by the Collier County Development Corporation, which had been unwilling to sell any of the islands to the state. The locally initiated ordinance attempted to circumvent this impass through the use of very restrictive site plan reviews in conjunction with a transferable development rights system. If the developer intended to build on one of the islands, a lengthy site plan and numerous performance standards had to be followed. But if the developers relinquished some of their right to build on the island (and donate the land to the county for preservation), they were compensated by the county with a proportionate increase in permitted density of housing on mainland sites.

The Special Treatment ordinance was preceded by a series of detailed environmental investigations by The Conservation Foundation (Clark, 1974). The objectives of these studies were to determine the type and extent of development that the Rookery Bay area could support without undergoing irreparable harm. The recommended standards for land development and the ensuing ordinance focus on the sensitive hydrological regime and serious water supply problems of the region. Excavation, impervious surface cover, artificial drainage facilities, and soil erosion/sedimentation are specifically controlled in the county ordinance. Either the developer meets the stringent island development controls, or he transfers development rights to a more suitable and less restrictive mainland site. Collier County was one of the pioneer government entities in promoting transferable development rights for purposes of environmental protection.

Westhampton Beach, New York (Fire Island chain)

This small community in Suffolk County, Long Island was one of the first barrier island local governments to develop a vigorous program of beach and dune protection using development controls. The village dune protection ordinance prohibits any weakening of, walking in, or building on the barrier island sand dunes except when the structures are placed on high pilings. The initial ordinance was adopted in June 1939 (Mitchell, 1974). It prescribes elevated pathways, and all buildings seaward of the dune access road must be raised proportionately higher above sea level the closer they lie to the dune crest. Dune vegetation restoration is also strongly encouraged, and neighboring residents closely monitor and report any infringements of the ordinance to the village board. Citizen groups concerned with beach and dune protection were also instrumental in adopting the ordinance and proposing amendments to strengthen its function. The Westhampton Beach experience is not the most protective of dune ordinances, but its early conception date and the close integration of residents, interest groups, and local government officials are most commendable.

Glynn County, Georgia (St. Simons Island, Sea Island)

Coastal goelogists who have sought to translate dune formation and migration processes into workable development controls have stressed the need for flexible standards, based on the dynamic factors of formation (Nordstrom, Psuty, and Fisher, 1978). In 1974, Glynn County adopted its Beach and Dune Protection District which accomplishes most of these objectives. The ordinance delineates a primary dune district with the more stringent controls, and a secondary dune field district in which conditional uses are allowed by special permit (Dudark, 1980). The primary district extends landward 40 feet from the base of the landward side of the most seaward stable dune, which is an intrinsically variable and site specific distance. Fencing and elevated boardwalks are the only permitted uses in this district. In the secondary dune-conditional use zone, buildings must be set on pilings 14 feet or higher above mean sea level, and special road access restrictions are imposed in the review and approval of subdivision site plans. The ordinance was based on a plan which was developed by an *ad hoc* working group, composed of a prominent ecologist and a planner (Eugene Odum and Anthony Catanese) and a representative of the Georgia Department of Natural Resources. The Glynn County ordinance is the most innovative and successful dune protection ordinance in Georgia. The State enacted the Shoreline Assistance Act in 1979 which regulates dune area development, but it is not as rigorous as the Glynn County ordinance.

Galveston, Texas (Galveston Island)

Unlike most other Gulf Coast barrier islands, Galveston Island's close proximity to the Houston metroplex (estimated at 1.8 million population in the 1980 census) has created tremendous recreational pressures on the waterfront area. Although the urban concentration in Galveston is clearly confined within the 12 square miles on the eastern end of the island, the city annexed almost the entire remainder of the 42 square mile island in 1975 and 1977 (Galveston City Planning Commission, 1976). The primary reason for the annexation was to institute a comprehensive growth management plan for the west end, which is characterized by moderately intensive gulf-front development and a series of resort communities. Planned unit developments are encouraged in the newly annexed portion; they must undergo an extensive site plan review involving several city agencies.

The Planning Department has initiated a beach and dune management plan and ordinance that will significantly restrict dune area development (City of Galveston Planning Department, 1980). An exemplary aspect of the proposed ordinance is the attention given to measures that will aid in the natural restoration and regeneration of the island's poorly developed and disturbed dune system. Specific standards are prescribed for heights and configurations of buildings and dune walkovers to allow for an ample amount of direct sunlight for native plant growth. Structures built in the dune area must also provide for dispersed runoff from roof drain spouts (if used) so as to collect and distribute rainwater evenly <u>under</u> the structure. Vegetative restoration of disturbed construction sites must be carried out using carefully prescribed techniques and plant species that are known to be successful on Galveston Island.

<u>Port Aransas-Corpus Christi-Nueces County, Texas (Mustang Island)</u>

The salient feature of local control of development on Mustang Island is the complex (and sometimes successful) network of multiple units of government in the review and permitting of gulf-front developments. Unlike Galveston Island, most of Mustang Island is unincorporated, but the entire island lies within the city limits or within the extraterritorial jurisdictions of Corpus Christi and Port Aransas. With the exception of its dune permit program, Nueces County (as with all Texas counties) has a very limited ordinance-making authority and no countywide planning body to respond to island development proposals.

Consequently, the various units of government on the island have been obliged to rely on each other to guide and manage growth. The County Commissioners rely heavily on the scientific and technical advice of ecologists in the Texas General Land Office in reviewing and negotiating on developers' gulf-front subdivision proposals. The Commissioners have also placed the entire City Council of Port Aransas on their dune permit review board so as to promote coordinated island-wide development policies and decisions. Although the Port Aransas building code would not normally apply to island developments beyond the city's jurisdictional boundaries, the county water district has an intergovernmental contract with Port Aransas which requires that developers seeking sewer and water connections anywhere on the island comply with the city's building code.

The County Commissioners and Port Aransas are also jointly sponsoring a comprehensive planning study to assess the island's development potential over the next few years, and to propose a more integrated development guidance system. Other more subtle intergovernmental agreements also exist among these overlapping units of government.

<u>Nantucket, Massachusetts (Nantucket Island)</u>

Nantucket Island was first settled by the colonials in the 1640's, and has much more recently encountered a rapid growth in housing development (400 lots per year). Nantucket County is the fastest growing county in New England (Nantucket Planning and Economic Development Commission, 1979). As a consequence, public and non-profit conservation organizations have attempted to protect the island's remaining natural areas, and have met with resounding success.

The subdivision regulations of the Planning Board include detailed specifications on developments affecting wetlands, floodplains, and wildlife habitats. In the relevant sections for each of these areas, implicit suggestions are made that would have the developer designate such lands and waters as common open space, to be deeded in fee simple to the Nantucket Conservation Commission, a homeowner association, or a suitable conservation organization (Nantucket Planning Board, 1976). Similarly, the non-profit Nantucket Conservation Foundation actively manages a land trust on the island which is composed of donated or acquired natural tracts. Since its inception in 1965, the Foundation has acquired more than 90 separate tracts totalling more than 5,344 acres, or roughly 17% of all the land on the island (Nantucket Conservation Foundation, Inc., 1980). As on a few other Atlantic Coast islands, the Foundation receives strong support from the local government and plays a vital role in the long-term management of critical island resources.

Chatham, Massachusetts (Cape Cod)

Although not a barrier island community, Chatham occupies portions of a barrier spit system and represents a highly useful case example of local-Federal cooperation in barrier beach protection. When plans for the acquisition and development of the Cape Cod National Seashore got underway in 1960 and 1961, the Congress authorized the establishment of land use control criteria for voluntary adoption by local towns on the Cape. The intentions of the Congress were two-fold: the criteria would ensure that land uses in the towns and on the National Seashore would be compatible, without the need for expensive acquisition of land throughout the area; and lands within the town jurisdictions could remain on the local tax rolls. If a town refused to adopt the recommended land control ordinances, then the Park Service was to condemn and acquire the needed lands (Duscik, 1974). Chatham was one of four towns to adopt the controls in 1969 (Town of Chatham, Massachusetts, 1969).

The adopted Conservancy District prohibits almost any additional residential beachfront development, but provided existing owners with a modest amount of discretion in building or repairing accessory facilities. Enforcement of the District rules has not involved the Park Service to any considerable extent, and local townspeople and officials have generally been very accepting of the controls (Hiscock, 1980). This early example of benevolent cooperation among local governments and the National Park Service may be a model for future barrier island protection strategies involving local governments.

The preceding eight case examples, summarized in Table 3, illustrate the diversity of approaches available to barrier island local governments. They also indicate the ability of governments to adjust these approaches to the particular conditions at hand. It is not possible, however, to assess the success of these cases adequately in a brief discussion. The more general issue of the positive attributes versus constraints of local governmental involvement is addressed in the following two sections.

Table 3
SELECTED CASES OF LOCAL ACTION

Sanibel, FL	Comprehensive island plan based on ecological carrying capacity and absolute growth limits Municipal incorporation
Collier County, FL	Performance standards Transferable development rights
Westhampton Beach, NY	Early establishment of dune ordinance Citizenry-interest group-governmental coordination
Glynn County, GA	Flexible dune ordinance using two-tiered district system
Galveston, TX	Island-wide municipal annexation Restoration-oriented dune ordinance with shading, runoff, and planting requirements
Port Aransas-Corpus Christi-Nueces County, TX	Intergovernmental arrangements Ongoing use of outside technical expertise Comprehensive island plan
Nantucket, MA	Encouragement of land dedication
Chatham, MA	Federal-local cooperation in protection of amenities and resources

POSITIVE ATTRIBUTES OF LOCAL GOVERNANCE

Continued Provision of Urban Public Services

There are approximately 128 islands in the United States that contain urban land, urban structures, and sizeable populations (U.S. Dept. of the Interior, 1979). Local island governments have an obligation to continue providing these populations with the kinds of public services that are generally available to all citizens: water, waste disposal, education, police, fire protection, energy, communication, recreational opportunities, and the like.

Indeed, most citizens who live on developed barrier islands are in strong support of local governance, even if it means more stringent land use controls for future development. Once residency on an island is established by people, they are often anxious to "pull the plank" behind them. The overriding vote of Sanibel Island residents in favor of incorporation and the comprehensive plan, and the strong consensus of Westhampton Beach residents to monitor closely their own dune protection ordinance are testimony to this trait.

Unique Attributes of Individual Islands

The case has been made several times that each barrier island has a unique assemblage of land and water environments and a particular response to wave climate and other outside forces. State or Federal governments would be hard pressed to adjust their management techniques to conform to conditions on each island in their domain, whereas local governments are in an ideal position to investigate thoroughly and make a proper response to particular conditions on their island.

Erosion and recession are ready illustrations of the differences among islands. The propensity of southern Atlantic Coast islands (e.g., Bogue Banks) is to be breached by inlets at random locations after great hurricanes, while western Gulf Coast islands (Mustang and Padre) typically encounter breaches through well-delineated hurricane washover channels. Central Gulf Coast islands off the Louisiana coast (e.g., Grand Isle) appear to be more prone to complete island overwash. Each of these particular attributes dictates a slightly different set of management objectives. Dune systems and groundwater flow systems are equally variable from island to island.

Tasks Not Readily Undertaken by Higher Levels of Government

There are several matters of island governance for which state or Federal agencies have rarely been able to assume responsibility. Some obvious examples include fire protection, education, and tax collection. Possibly less obvious tasks that seem more appropriate for local initiatives include comprehensive land use planning, zoning, monitoring of new developments, and local economic development planning. Local governments can review development proposals to determine whether they are consistent with local needs and with balanced land use for the area. State and Federal permit decisions often lack the comprehensive perspective of the impacts on long-term growth and development. The case for this argument is made stronger in states such as Texas, in which the State's role in managing activities on private lands outside of city limits is cut short by a lack of state coastal resource management authority. Some states have adopted programs that allow local governments to share or assume regulatory responsibilities that are otherwise vested in state government. Such cooperative state-local efforts usually ensure that the state law will be fully and fairly executed at the local level.

Inaction or Improprieties of State and Federal Government

Many coastal states have been unwilling to establish general, areawide land use controls, or even critical area programs for the protection of significant natural or cultural resources. Collier County, Glynn County, and Westhampton Beach, for example, acted on their own initiative with no state guidance; complementary state programs then ensued. Some state and Federal actions run directly counter to the goals of orderly resource protection and development on islands. Local governmental participation is needed to interpret state and Federal policies at the local level and to serve as a check against inappropriate and unwise activities.

Examples of governmental inaction or innappropriate action include the allowance of oversized waste treatment facilities based on population projections without regard for island natural resource constraints, and encouragement by banks and insurance companies to have people rebuild again on islands following major storm damages. Island developers and residents alike should be able to understand that workable, restrictive local controls which ensure safe and sane development also ensures continuing profits for developers and stable land values for occupants.

CONSTRAINTS ON LOCAL GOVERNANCE

Limited Legal Authority to Control Growth

Most barrier island development occurs outside the jurisdiction of cities and towns; and county land use control ordinances are often considerably less rigorous and sophisticated than municipal ordinances. The avoidance of municipal controls by developers, however, is not necessarily a reason for promoting widespread annexation or municipal incorporation. On the positive side, Sanibel residents were able to incorporate the island and avoid the tremendous development pressures that were being fueled by the county planning administration. Galveston was most ambitious in its annexation of the rest of the island to control sporadic growth; unfortunately, Texas annexation laws call for the provision of comparable urban services (water, wastewater, streets, etc.) to newly annexed areas within three years, or the annexed residents can request deannexation.

On the negative side, incorporation into a town or city does not necessarily strengthen local legal authorities -- it can even initiate a trend towards a laissez faire attitude about development impacts. Some communities have incorporated expressly to avoid the development controls posed by county governments. Such moves have allowed island housing developments to increase in density and have exacerbated the rate of environmental degradation. There are no assurances that municipal regulations will expand to become as strict or more strict than county authorities (Godschalk, et al., 1974).

Limited Capacity or Capability to Adopt and Administer Ordinances

Most barrier island communities are so small that their staff are comprised of one or a few persons who are usually trained as generalist planners. Certainly there are notable exceptions, but the fundamental problem remains that local governments often lack the depth and breadth to manage complex barrier island resources and respond to development pressures effectively. State and Federal governments do little to assist local planning efforts.

The mere presence of well-meaning ordinances and building codes serves little purpose if ample staff review and monitoring support do not exist. Moreover, some barrier island development standards are workable with small staff capabilities, but the scientific precepts upon which they are based might be misguided, misunderstood, inadequate, or simply wrong. Dune ordinances that restrict building construction in dunes, for example, can actually be detrimental if they fail to manage pedestrian access between the beach and the structures.

Vulnerability to Political and Vested Interest Pressures

Local governments are often guilty of allocating resources of regional significance solely on the basis of local needs, pressures, and values.

Developers can mount intensive lobbying campaigns against the enforcement of policies and fully authorized ordinances. City and county staff have often watched the pressures of developers sway city council decisions. Local officials are often torn because disallowing development often implies foregoing of future property tax. In addition to these lobbying pressures, the threat of a lawsuit creates havoc for many small coastal communities.

Burdens and Risks of Providing Public Services

Barrier island local governments have responded, as have all other local governments, to the need for public utilities, roads, schools, fire and police, and the like. Local matching funds for seawalls, bulkheads, and beach nourishment projects often amount to very sizeable investments, and yet on countless occasions these island protection measures proved futile in the long run. Recent investigations have indicated that island infrastructure costs can also be significantly higher on barrier islands than on the mainland. In an unpublished draft report prepared by the Texas Coastal and Marine Council (Davenport, 1979), the costs of providing water service to Texas barrier island residents was 2 to 3.4 times the cost to mainland residents; road maintenance costs were as much as 4 times greater than mainland road costs; and quite obviously, the unique transportation costs associated with ferries, high bridges, and long causeways were greater. The ultimate risk in providing infrastructure is the incidence of a great hurricane. Human settlements and barrier islands can suddenly become antithetical to one another. Although the risk of complete disaster for island communities is admittedly probabilistic, the inevitability of its occurrence somewhere along the Atlantic and Gulf coasts is a matter of record. (Each of the preceding eight case examples has encountered at least one great hurricane.)

CONCLUSIONS

Barrier island local governments face many constraints in their duties to guide and control development, but they appear to be the most appropriate entities. They must approach these tasks aggressively and comprehensively. Even in locations where intensive development has already occurred, restorative and regenerative measures such as revegetation, dune development, natural drainage restoration, and sometimes beach nourishment should be considered. Some islands which have recently encountered rapid growth may find their slate "wiped clean" when the next great hurricane makes a landfall; they may have the distinguished opportuinty of starting afresh to plan for and rebuild a new and different island community.

A major challenge will be to establish environmental management guidelines that are: (1) predicated on current knowledge of island dynamics and ecosystems; (2) sensitive to the unique resource constraints and development opportunities present on the particular island; and (3) workable within the capacities and capabilities of the local governmental resources. There are many exemplary island governments (some are cited in this discussion) that have translated the particular landform and biological attributes of their island into flexible development policies and guidelines. In every new case, the particular combinations of state laws, enabling legislation, development pressures, and environmental factors should dictate a unique approach. Local governments should always seek technical assistance from state and Federal agencies to strengthen their limited technical capabilities.

The Federal government is currently making careful assessments of their policies, subsidy programs, and regulations that directly or inadvertently affect island development. State governments should also take the initiative to evaluate their programs that affect islands. There is ample room for more direct coastal planning and guidance in many state governments. Every coastal state with barrier islands should endeavor to provide more technical assistance to island governments, and promote more intergovernmental cooperation on overlapping management concerns. Professional and conservation associations and foundations (American Planning Association, American Society of Civil Engineers, municipal alliances, The Conservation Foundation, and others) should continue to transfer the experiences gained by exemplary island governments to others for their consideration and emulation.

The intent of this discussion to initiate a dialogue on what the proper role of barrier island local governments is, in the context of future actions by the Federal and state governments and on the basis of their own record. Some see increased island development as a means of providing greater opportunities for recreational enjoyment, while others view development as a foreclosure on future options for public uses. Regardless of one's view, the recent trends in barrier island development command a strengthened role on the parts of local government.

BIBLIOGRAPHY

Anonymous. Report of the Barrier Island Work Group. Workshop sponsored by U.S. Dept. of Commerce, Off. of Coastal Zone Management. Washington, D.C. December 18, 1978.

City of Galveston Planning Department. Proposed Beach and Dune Ordinance. Draft, unpublished. Galveston, Tx. May 1980.

Clark, J. Rookery Bay: Ecological Constraints on Coastal Development. The Conservation Foundation. Washington, D.C. 1974.

Clark, J. The Sanibel Report. Part 3: The Sanibel Plan. The Conservation Foundation. Washington, D.C. 1976.

Coastal Zone Management Act Amendments of 1976. Public Law 94-370, 90 Stat. 1013, Sec. 3(1).

Davenport, S. S. Development on Texas Barrier Islands... Does it Cost the Taxpayer More? Unpublished preliminary manuscript. Texas Coastal and Marine Council. Austin, Tx. October 1979.

Dolan, R. and Hayden B. Adjusting to Nature in Our National Seashore. National Parks and Conservation Magazine 48(6): 9-14. June 1974.

Ducsik, D. W. Shoreline for the Public. MIT Press, p. 214. 1974. See also 16 U.S.C. 459b, et seq.

Dudark, R. Planning Director, Glynn County, Georgia. Personal communication. July 31, 1980.

Galveston City Planning Commission. West Island Zoning Plan. City of Galveston, Tx. 1976.

Godschalk, D. R., Parker, F. H., and Knoche, T. R. Carrying Capacity: A Basis for Coastal Planning? Univ. of North Carolina, Dept. of City and Regional Planning. See especially Chp. IV, Sect. B by David Owens. Chapel Hill, NC. 1974.

Hiscock, A. Chatham Conservation Commission. Personal communication. Chatham. MA. August 1, 1980.

H.R. 5981. Submitted to House of Representatives, National Parks Subcommittee, by Rep. Phillip Burton, California. 1980.

Mitchell, J. K. Community Response to Coastal Erosion. Univ. of Chicago, Department of Geography, Research Paper No. 156. Chicago, IL. 1974.

Nantucket Conservation Foundation, Inc. General Information and Properties Map 1979-80. Nantucker, MA. 1980.

Nantucket Planning and Economic Development Commission. Residential Growth Impact Study 1978. Nantucket, MA. September 1, 1979.

Nantucket Planning Board. Rules and Regulations Governing the Subdivision of Land. Nantucket, MA. Amended August 2, 1976.

Nordstrom, K. F., Psuty, N. P., and Fisher S. F. Empirical Models of Dune Formation as the Basis for Dune District Zoning. Proc. Coastal Zone '78 Conf. Vol. III, pp. 1489-1507. San Francisco, CA. March 14-16, 1978.

Sanibel Planning Commission. Comprehensive Land Use Plan, City of Sanibel, Lee County, Florida. City of Sanibel Ordinance No. 76-21. Adopted July 19, 1976.

Sedway, P. H. Regulation in the Coastal Zone. Proc. Coastal Zone '78 Conf. Vol. I, pp. 521-527. San Francisco, CA. March 14-16, 1978.

Town of Chatham, Massachusetts. Residence-Seashore Conservancy District 10-69. Protective By-laws, Sec. 3.S. Adopted 1969.

U.S. Dept. of the Interior. Alternative Policies for Protecting Barrier Islands Along the Atlantic and Gulf Coasts of the United States, and Draft Environmental Impact Statement. Prepared by Heritage Conservation and Recreation Service, et al., Washington, D.C. December 1979. See also President's Environmental Messages of 1977 and August 2, 1979.

COASTAL ZONE INTEGRATION OF STATE AND LOCAL PLANS

by

Lowdon Wingo
Professor of Urban and Regional Planning
University of Southern California

and

James Fawcett
Coastal Planning Specialist
University of Southern California

In passing the Coastal Zone Management Act of 1972 the Congress found, among other things, that "In light of competing demands and the urgent need to protect and to give high priority to natural systems in the coastal zone, present state and local institutional arrangements for planning and regulating land and water uses in such areas are inadequate."[1].

This paper constitutes an assessment of one legislature's response to the congressional finding. It is not concerned with how the new arrangements are working out in practice; that evaluation remains to be done. It is concerned with the institutional redesign of planning processes that has taken place in the turbulent political climate of California, as conflicts between environmentalists and development interests, between local government and the state, between the established state resource management agencies and a new multipurpose planning and management agency were compromised. We focus on the planning role, as distinct from regulation or the more general term management, for planning ostensibly precedes and gives coherence to management and its instruments. Furthermore, while state planning tends to be diffuse, poorly defined, and variable in its relation to policy formation, local land use and public facilities planning are conducted in a well established institutional and professional context. The crisis in the nation's coastal zones has given some urgency to the task of redefining the proper allocation of planning roles between state and local governments. In the last decade California has struggled mightily with that task.

The California Coastal Act of 1976 can be seen as an attempt by the California legislature at a substantial innovation in the relationship between the state and local government roles.[2] The complexity and multitude of public and private interests that converge on California's Pacific coast could have led the legislature to create, at one extreme, a new multipurpose coastal government, or at the other an advisory body simply to alert the governor and the legislature to emerging coastal issues. Instead, it designed an organizational strategy to preserve the relationships between affected local governments and the state interests exercised by established, single-function resource agencies of the state. It did this by setting up the coastal management program in two stages. In the first stage, the management agency, the California Coastal Commission,[3] is to oversee the translation of state policies governing the utilization of coastal zone resources into the content of local plans and policies while it controls coastal development through its power to issue coastal permits. The second phase will find primary responsibility for coastal zone management exercised by sixty-seven local jurisdictions. Each will implement the plans and policies spelled out in its local coastal program previously certified by the commission, which thereafter becomes essentially a reviewing agency. Such a strategy impairs the authority and responsibility of no state resource agency; the roles of local governments are not substantially altered, and their powers are enhanced by the authority of the Coastal Act; and, it is hoped, policy coherence is established in the conservation and efficient utilization of coastal resources. In the words of the statute:

> "....to coordinate and integrate the activities of the many agencies whose activities impact the coastal zone, and to supplement their activities in matters not properly within the jurisdiction of any existing agency, it is necessary to provide for continued state coastal planning and management through a state coastal commission."[4]

The extent to which such a strategy can be sucessful in knitting together a welter of state resource policies and local land use and housing policies into a coherent coastal management program depends on the realization of several conditions. First, the commission must have the capacity to make the policies of a number of strong state line agencies cohere among themselves in the coastal zone and with the commission's special charge with respect to recreational, scenic, and ecological resources. Second, it must be able to provide the planning base on which the mosaic of sixty-seven separate local coastal programs can be designed, amended, and implemented while preserving the policy intergrity of the coastal zone. Third, the local coastal jurisdictions must be capable of assuming -- and sufficiently receptive to -- the responsibility of acting as implementing agents for statewide objectives over and above local interests. While it would be asking much for it to be equally effective in all of these matters, no one of these conditions can completely fail without jeopardizing the outcome of this organizational strategy.

It must be admitted from the beginning that, notwithstanding the fervor of environmentalists, the values of coastal resource conservation have no inherent pre-eminence over such other societal values as those associated with low cost and reliable energy supplies, coherent urban development responsive to the demands of regional growth, and improvements in the general level of economic welfare. Clearly environmental values from time to time are made to give way before other values which circumstance has endowed with a quality of imperiousness. To argue that some values have a hierarchical superiority is to misconstrue the nature of conflict which the commission will on occassion face, for the weighting of values by society at any one moment in time depends on changing circumstances which imbue first one set of values and then another with a sense of urgency. Thus did the environmental movement surge to the forefront during the 1970's; so now have energy and productivity issues moved to the center of our social concerns. The central policy concept here is the trade-off: at any one moment how much of one value is society willing to forego to realize how much of another. Our perspective here is not so much, then, the natural superiority of environmental values over others held by society, as it is the extent to which an organizational strategy facilitates a proper assessment of the trade-offs involved in policy choices, among values.

The commission has two kinds of responsiblity for the state's interest in natural resource policy in the coastal zone. The statute charges it to "coordinate and integrate" in the coastal zone the activities of those established state line agencies who manage the traditional ntuaral resource sectors [5.] and to see to it that those policies are reflected in the local coastal programs.[6.] Then it has special responsibility for such nontraditional resources as the amenity features of the coastal landscape and ecological communities. Its primary instruments for achieving its purposes are (1) the direct regulation of private land development which threatens these resources in the coastal zone, (2) the power to deny permits for local governments public works not consistent with coastal management objectives, and (3) the exercise of a consultative and coordinative role with respect to public developments of state agencies in the coastal zone. It is in this last area that the commission's formal authority is seen at its weakest. Absent the legal authority and the political clout of established agencies, sweet reasonableness must be the commission's instrument in its dealings with state resource agencies.

In actual fact, the commission shares responsibility for the planning and management of traditional natural resources in the coastal zone with at least twenty other state agencies each of whose jurisdictions is defined by the characteristics of its resource interst rather than by a political boundary, as is the case with the coastal commission. Its domain has been imposed on a pre-existing structure of single-function state agency responsibilities. The legislature defined this relationship in the 1976 Act by declaring its intent to be "...to minimize duplication and conflict among existing state agencies carrying out their regulatory duties and responsibilities"[7.] and further specifying that the Act "...does not increase, decrease, duplicate, or supercede the authority of any existing state agency."[8.] While the statute directs "all state agencies (to) carry out their duties and responsibilities in conformity with this division,"[9.] the formal mechanism for coordination is found in the Act's direction to the commission to "...periodically ...submit recommendations designed to to encourage (other state agencies) to carry out (their) functions in a manner consistent with this division."[10] Each such agency is required to explain to the governor and the legislature its failure to implement such recommendations within six months.

Several state natural resource agencies are subjects for specific procedural guidelines as the commission pursues its coordinative task.[11] The Department of Fish and Game and the Fish and Game Commission are responsible for wildlife and fishery management programs throughout the state. Coordinating with the Department of Boating and Waterways and the coastal commission, the Department of Fish and Game is charged in the 1976 Coastal Act with "...study(ing) degraded wetlands and (identifying) those which can most feasibly be restored in conjunction with development of a boating facility...".[12]

Given the coastal commissions responsibility for coastal recreation[13] and enhancement of coastal environmental resources, especially wetlands[14] it finds itself with the role of balancing the interests of recreationists with those of the environment while its two partner agencies are each responsible for one of the competing uses.

With respect to its relations with the State Water Resources Control Board and regional water quality control boards, the 1976 Coastal Act limits the authority of the commission to that necessary to restrain development sprawl and to regulate the actual siting and visual appearance of water and sewage treatment works within the zone.[15] The statute specifies that the commission shall review and make a determination on an application for construction of a treatment work prior to the final decision of the State Water Resources Control Board. The decision of that board shall subsequently be final and binding upon the state and regional coastal commissions.[16] In a more positive sense, the statute allows the commission to "...provide or require reservations of sites for the construction of treatment works and points of discharge within the coastal zone adequate for the protection of coastal resources..."[17]

Under the Energy Resources Conservation and Development Act, the Energy Resources Conservation and Development Commission has sole authority for siting thermal power plants. Their high demand for cooling water will result in the location of many such plants, both nuclear and non-nuclear, in the coastal zone. The coastal commission has authority to specify those sites along the California coastline which would be unsuitable for the location of a thermal power generating facility[18] subject only to the constraint that it must recognize the right of existing facilities to "...reasonable expansion..."[19]

Similarly, the energy commission must file with the coastal commission any notice of intent it issues prior to issuing a construction permit for a power plant. Although the coastal commission has no statutory authority to veto such a permit, it must review the notice and file a written report with the energy commission addressing the suitability of the site for that use in terms of enumerated specific issues:[20] will the proposed project conflict with the goal of protecting coastal resources? will existing or planned nearby coastal-dependent uses be disrupted? will the proposed project conflict with certified local coastal programs in the vicinity? Again, conclusions of the coastal commission are advisory only.

The State Air Resources Board exercizes its authority through regional air quality management districts (AQMDs). The 1976 Coastal Act defines the relationship between the Air Resources Board (and the regional AQMDs) and the coastal commission in denying the commission the power to modify the standards of these agencies with respect to emissions by stationary sources within the coastal zone;[21].and second in expressly permitting them to "...recommend ways in which actions of the commission...can complement or assist in the implementation of established air quality programs."[22]. Thus, coordination between the commission and air quality management agencies is informal with the exception of the commission being prohibited from altering emission standards.

All state lands, including tidal and submerged lands, in California are managed by the State Lands Commission. The Coastal Act provides that the lands commission shall review individual local coastal plans and port plans and shall be given the opportunity to comment on provisions of these plans which may have adverse impacts on state lands prior to certification of these LCPs by the coastal commission.[23]. Aside from this provision however, the coastal act merely specifies the jurisdiction of the two agencies with respect to regulation of state lands and development thereon when those lands are found in the coastal zone.[24].

The timber industry is a major component of north coast economies in California, whose forest harvesting practices are potentially at variance with established principles of coastal resources management. Acknowledging this, the 1976 Coastal Act gave the commission the power to "...identify special treatment areas within the coastal zone in order to insure that natural and scenic resources are adequately protected.[25]. Such designations were to be accompanied by reasons for designation and "...recommendations designed to assist the State Board of Forestry in adopting rules and regulations which adequately protect the natural and scenic qualities of such special treatment areas."[26]. Beyond the designation of such special treatment areas, the commission was given no formal coordinative role with the State Board of Forestry.

Finally, the Division of Oil and Gas of the State Department of Conservation is responsible for the "...drilling, operation, maintenance and abandonment of all oil, gas, and geothermal wells in the state."[27]. Beyond those development controls granted to the coastal commission is specifically prohibited from imposing "...regulatory controls that duplicate or exceed controls established by the Division of Oil and Gas pursuant to specific statutory requirements or authorization."[28]. In making decisions over the appropriate uses of coastal lands where an application for an oil, gas, or geothermal well is at issue, the Division of Oil and Gas "...shall cooperate with the commission by providing necessary data and technical expertise regarding proposed well operations within the coastal zone."[29].

Coordination between the coastal commission and state resource agencies, thus, takes many forms. In many cases it takes the form of a "review and comment" with respect to rules and regulations. However, in a few cases notably the Department of Fish and Game and Boating and Waterways with respect to estuary restoration, the State Water Resources Control Board with repect to the siting of sewage disposal plants and the extension of service lines and the Energy Resources Conservation and Development Commission with respect to the siting of thermal power plants and transmission lines, that coordination takes the form of ongoing bilateral communications between two agencies. As might be expected, the agencies with whom closer coordinative ties are found are those which either have jurisdiction over large public works facilities or over estuaries, an important and diminishing public resource. In other words, the commission has been chartered to coordinate most with those state agencies which regulate resources or public facilities which have major spillover effects within the coastal zone.

It is, of course, possible that the commission's coordinative role be carried out with little significant conflict. This is most likely to be the case where coastal plans and policies are general or where the stakes are small. The fact is, however, that the guidelines for the local coastal programs are frequently quite specific with respect to locations and activities, a fact dictated by the regulatory mode by which local government will implement the sixty-seven LCPs which, in the aggregate, will comprise the California Coastal Plan in phase two. Furthermore, the stakes are quite likely to be high in many cases, a proposition which the recent history of the commission bears out. In the case of conflict, the established state line agencies hold all of the cards, subject only to the cumbersome and uncertain right of appeal to the governor and the legislature as specified above. Even this right is attenuated by the independent coordinative role of the governor's Office of Planning and Research assigned by the statute:

> "The Director of the Office of Planning and Research shall... recommend to the appropriate agencies actions that should be taken to minimize...conflicts...(He) shall...recommend to the Governor and the Legislature how the programs, duties responsibilities, and enabling legislation of any state agency should be changed to better achieve the goals and policies of this division."30.

However likely it may be, there is no assurance that OPR's independent assessment will always come down on the side of the commission. Another administration with less identification with the environmental values of the current administration could very well find that other values require the commission, not its opposing agency, to give way.

In general, a close reading of the statute against the background of political realities is not likely to inspire confidence in the institutional resources available to the commission to carry out its coordinative charge sufficiently well to assure coherence among state agency policies in the coastal zone. This is a conservative assessment, admittedly. The skillful conduct of interagency negotiations by the commission and its staff and the cooperative postures taken by state agencies may well make the institutional arrangements thus conservatively construed superfluous in most cases. The point is, however, that the legislation requires the commission to take account of the plans and policies of other state agencies without directly mandating that they accord similar attention to the coastal plans and policies of the commission. It is not, hence, clear that the organizational strategy laid out for California's coastal management assures even handed assessment of the trade-offs between the environmental values for which the commission is steward and other values embodied in the charges to other state agencies.

And what, then, can be said of the adequacy of the planning base of the coastal management program to integrate the sixty-seven local coastal programs into a coherent strategy for the conservation and effective deployment of coastal resources? While management and regulation are extensively discussed in both the federal and state coastal legislation, very little substantive guidance is to be found on what is to constitue acceptable "plans or planning". One gets the strong impression that the legislative sponsors were so uncertain as to follow the line that the less said the better. Indeed, the Federal Coastal Zone Management Act of 1972 (amended) requires of state management programs not "plans" but "planning processes", conditions which can likely be satisfied by pointing to the establishment of some planning body charged to "to plan..."[31]. "Plans" and "planning" consisting of whatever such a body would choose to do. The structure and substance of plans, then, is not subject to federal evaluation in terms of the intent of the Act.

In California the original definition of a state coastal plan is embodied in the public initiative known broadly as Proposition 20 and statutorially enacted as the California Coastal Zone Conservation Act of 1972, whose General Provisions and Findings include the statement that

> "...to protect the coastal zone it is necessary to prepare...a comprehensive, coordinated, enforceable plan for the orderly, long-range conservation and management of the natural resources of the coastal zone, to be known as the California Coastal Zone Conservation Plan."[32].

The Coastal Zone Conservation Act then proceeds to define five major components:

1. a precise, comprehensive definition of the public interest in the coastal zone;
2. ecological planning principles and assumptions to be used in determing the suitability of...allowed development;
3. a set of nine plan elements;
4. reservations of land and water for specific uses;
5. recommendations for government policies and powers required to implement the plan.

The nine elements specified as part of the third component are detailed as follows:[33].

land use,
transportation,
scenic and natural resources,
public access,
recreation,
public services and facilities,
ocean mineral and living resources,
population, and
educational or scientific use.

These components and elements are directed by the statute toward four objectives:[34].

"(a) the maintenance, restoration, and enhancement of the overall quality of the coastal zone environment, including, but not limited to, its amenities and aesthetic values.

(b) the continued existence of optimum populations of all species of living organisms.

(c) the orderly, balanced utilization and preservation, consistent with sound conservation principles, of all living and non-living coastal zone resources.

(d) avoidance of irreversible and irretrievable commitments of coastal zone resources."

Finally, "the commission shall...submit to the legislature for implementation the California Coastal Zone Conservation Plan," a condition amplified by the legislative requirement that "on or before December 1, 1975, the Commission shall adopt the coastal zone plan and submit it to the legislature for its adoption and implementation."35. From this mandated procedure arises some critical confusion about the adequacy of the planning base for the coherence of an ultimate coastal plan consisting of sixty-seven local government pieces.

The *California Coastal Plan* was submitted to the legislature on schedule. As a planning exercise it is an impressive document referring to the legislature 162 separate policy recommendations dealing with the management of coastal resources and 44 addressing government organization, powers, and finances. The second major section of the *Plan* applies those policies to the specific situations with which California's 1,000 miles of coastal zone are endowed. These two major parts are tied together by the following statement:

> "The maps, notes, and summary text apply the Coastal Plan policies ...but are not intended to substitute for them...; where (they) direct an action, they are, however, intended to be given the same force and effect as the policies on which they are based, and taken together with the maps, are intended to guide further planning as necessary by governmental agencies and property owners...Local governments can use (them) in their own implementation of the plan."[36].

This statement coupled with specific provision for future amendment of the *Plan* leaves little doubt that the *Plan* was intended to be the planning matrix for local government policies and plans affecting the parts of the coastal zone falling in local government jurisdictions.[37]. Indeed, recommended policy No. 161 states that

> "Local governments shall revise their general plans as necessary to comply with the Coastal Plan (and approved amplifications of it)."38.

The original California Coastal Zone Conservation Commission did, then, produce a plan to serve as the matrix for the plans and policies of both state resource and other agencies and of local governments in the coastal zone.

In late 1976 the California Coastal Conservation Act of 1972 was succeeded by the California Coastal Act of 1976, the legislative response to the original commission's submission of the coastal plan to the legislature did *not* adopt the plan, even by reference. Instead, the legislature simply noted that in addition to detailed study of the coastal zone with widespread public and private participation

> "The...Commission has prepared a plan for the orderly, long-range, conservation, use, and management of the natural, scenic, cultural, recreational, and man-made resources of the coastal zone. Such plan contains a series of recommendations which require implementation by the legislature and that some of those recommendations are appropriate for immediate implementation as provided for in this division while others require additional review."[39].

Chapter 3 of the 1976 Act entitled "Coastal Resources Planning and Management Policies" contains thirty-three sections, in contrast to the 162 policy recommendations submitted in the *Plan*. No further operative mention of the *Plan* is to be found in the new Act.

What does this action of the legislature do to the availability

the 162 policy recommendations submitted in the Plan. No further operative mention of the Plan is to be found in the new Act.

What does this action of the legislature do to the availability of a broad planning base for integration of state and local plans and policies in the coastal zone? A liberal interpretation would argue that the Legislature's recognition of the Plan gives it authority in all of those specific situations described in the Plan which are consistent with the policies in Chapter 3 of the current Act, that is to say, that in the preparation of local coastal programs the successor commission can insist on their inclusion. It would follow that in all other cases the commission may invite, but not insist on, their inclusion, provided that their implementation involves powers currently available to local governments and state agencies. This interpretation would leave intact, if somewhat attenuated, the concept of a sort of regional plan as a basis of coherence for the plans and policies of state agencies in the coastal zone and as the mechanism by which state resources planning can be translated into local plans and policies through the medium of local coastal program development and approval.

A conservative view can also be sustained. It can be argued that the legislature's refusal to "adopt" the plan with legislative amendments, as conceived in the earlier act was an express rejection, leaving it with neither force nor effect, freeing local governments and state agencies from any obligation contemplated or embodied in the Plan.

Indeed, the new Act is quite specific about where the authority rests to determine the content of local plans:

> "The precise content of each local coastal program shall be determined by the local government...in full consultation with the commission."[40].

which, when read with

> "The commission shall...assist local governments in exercising the planning...powers and responsiblities (of this Act) where the local government...requests assistance from the commission..."[41].

makes it clear that the commission, even as advocate for the Plan may statutorily intrude itself into the content of the local coastal programs (a) if invited to do so, (b) in its consultative capacity, or (c) by way of its power to deny certification of the local programs. Even that last power is seriously circumscribed:

> "...the commission shall approve and certify...a land use plan...if such commission finds that a land use plan meets the requirements of, and is in conformity with the policies of Chapter 3..."[42].

If there are several ways in which a local coastal program can be consistent with the policies set out in the new act, the commission may not insist on its prefered program as a condition of certification, even if it is embodied in a coastal plan promulgated and published by the commission. This interpretation leads to the conclusion that no coastal plan -- neither the original nor any subsequently adopted -- has other than indicative authority before the disretion explicity set forth for local governments. This, notwithstanding the provision that

> "The commission...may prepare and adopt any additional plans and maps...it deems necessary and appropriate to better accomplish the purposes, goals, and policies of this (Act)!"[43].

On the whole, the internal evidence of the 1976 Act seems to support the conservative interpretation. The current act is preoccupied with management, not with the planning role, assuming by implication that 100 local management programs for the coastal portions of the sixty-seven jurisdictions will meet the tests of the legislative objectives.[44] What is lost to some degree, if not totally, is the power of a strong, authoritative planning base to assure coherent integration of this jurisdictional mosaic into a unified, consistent program for the effective utilization of coastal resources.

To what extent can one find in the statutorily defined role of local government some promise of future coherence in the utilization and conservation of coastal resources? An effect of the passage of the 1976 Coastal Act was to abnegate the concept of a coastal plan under the stewardship of the coastal commission and imbued with the authority and broad concerns of the state expressed in legislatively enacted policies. Coastal management policies there are -- detailed in the Act's Chapter 3 -- but there is no grant of authority to the commission to embody those policies in a authoritative plan. Instead, the notion of a different kind of "plan" emerges from the legislation, one consisting of 67 separate parts, each of which falls into place upon certification by the commission, certification being the transforming step in the interaction between each local government and the coastal commission.

Even this process would carry with it the prospect of coherence, provided that the commission had adequate discretion to require in each case not only a "best" plan, but also one that took into account the obvious fact that problems and processes rarely recognize jurisdictional boundaries. As previsously noted, the only ground on which the commission may refuse to approve a local coastal program is its lack of consistency with the policies enumerated in the 1976 Act. Hence, the role of the commission is confined in the last analysis to determining within the rule of reason the degree of consistency between the subject program and the legislated policies.

This poses unique difficulties for dealing with those problems that flow over jurisdictional boundaries. Can the commission require of one jurisdiction major actions that redound to, perhaps, the trivial benefit of another? e.g. can the commission require Ventura County to restrict residential development in order to minimize congestion along the Malibu coast in Los Angeles County? It is unclear how far the commission may go in the negotiations with one local government acting as authoritative advocate for the environmental interests of other jurisdictions which may either contiguous to it or partners in the same coastal region. The answer may ultimately turn on the ability of the commission to sustain the rejection of a local coastal program that does not adequately deal with the boundary conditions, as being inconsistent with state policies for the coastal zone. The prior existence of a coastal plan might save such a determination from the appearance of arbitrariness; absent that, perhaps only the courts will be able to assert that such *ad hoc* requirements of the commission among jurisdictions are reasonably within the purposes of the legislature in passing the 1976 Act.

This is no trivial problem. Whether or not the 67 local coastal programs on certification become in reality a state management plan depends in large degree on the ability of the commission to have translated into local coastal programs perspectives that transcend the local interests of each jurisdiction. The alternative would be the emergence of 67 local plans which embrace the entire coastal zone but which would deal ineffectively with those coastal resource problems that are more regional than local in nature. It is not to be denied that even this would be a gain over the situation along the coast before the passage of Proposition 20 in 1972. Those 67 jurisdictions would have assumed responsiblilty for coastal resources previously ignored in local policy making, and this under watchful eyes of the commission with a capacity to authoritatively nudge the faint of purpose and the inconstant of will. Nor can it be denied that such an outcome would constitute a substantial retreat from the strong purposes and planning format of Proposition 20.

A recent commission document noted in a discussion of the current status and prospects for LCP certification that "...the type of coastal planning called for in the Coastal Act is still relatively novel and requires (kinds of) planning many local governments had never done before."[45]. Certainly the content of planning by local jurisdictions has been enlarged. Prior to the Coastal Act few localities had experience in dealing with such issues as marine resources, environmentally sensitive areas, visual resources and landscapes which are among the commission's prescribed issue areas. Indeed, comprehension of the expertise needed to integrate such new elements in a larger planning and policy context must strain the traditional political processes of local jurisdictions.

Further, the 1976 Act calls for a form of interactive planning between the commission and the local jurisdiction which also has its unique qualities. The joint character of the planning process warrants some examination. It begins with an "issue identification" stage [46]. where a local government and the commission work toward agreement on which of the fourteen broad issue areas reflecting the specific policy concerns of the Coastal Act should be central to the planning and management of its unique configuration of coastal resources.[47]. Concurrence on issue identification permits the local government to submit a work program [48]. detailing the need for additional resources to address the study and planning of those areas. The work program will be tailored to fit with other work program budgets within the total budget available to the commission for local assistance, always inadequate to fund planning for all identified issues in all communities. At this point the influence exercised by the commission over the local planning effort becomes one of advice and consultation at the initiative of the local government as it carries out the work program. The local government has responsiblilites for direct coordination with state agencies affected by its coastal program. Its need to coordinate horizontally is unspecified in the documentation of the process.

On completion, the local government is entitled to one informal hearing and critique before the commission, and thereafter submits its full program for hearing and certification. Thereafter it assumes the full managerial responsibility described in its program for its part of the coast, with exceptions concerning special uses which the commission will continue to handle.

The planning exercise in the preparation of a local coastal program focuses on the amendment of a local jurisdictions land use plan as it relates to the coastal zone, which

"shall incorporate a statement of applicable development and resource protection policies in the substantive text or geographic provisions of the general plan..."49.

It is in the review and certification of such plans that the new Act substitutes a test of adequacy with respect to the policies it sets out for a test of coherence implied in the planning thrust of Proposition 20.

The state-local interaction in the development of local coastal programs prescribed by the 1976 Coastal Act does not foreclose the development of a coherent fabric of coastal plans and planning, but it makes it dependent on a complicated set of negotiations between the commission and the sixty-seven coastal jurisdictions. Not all negotiations are likely to be equally sucessful. The commission has inadequate resources during this transition period to fund all of the analysis and planning to bring the fourteen issue areas into line throughout these jurisdictions. The problems of the various segments of the coast vary enormoulsy from the rural northern California coasts to the heavily developed metropolitian regions of the Bay Area, Los Angeles, and San Diego. The technical and professional planning capabilities available to local jurisdictions differ radically a matter which special significance in terms of the analytical formats required to deal with the ecological and scientific issues which use the coastal zone brings to the center of the policy stage. While the possiblilty of a coherent coastal plan is not foreclosed by the strategy embodied in the 1976 Coastal Act, it is unlikely to be achieved by the single exercise prescribed in the preparation and approval of local coastal programs.

In review, then, in the Coastal Act of 1976 the legislature rejected a concept of planning that was traditional in the field of city planning -- embodying in a map the special policy requirements of differnet parts of that complex environment known as the coastal zone. Planning through mapping is a technique for achieving coherence in a space where policy concerns vary by location, a feature which would seem richly to characterize the coastal zone. It also abandoned the concept of a strong central image or perspective of coastal zone development and protection by ultimately reducing the commission to an instrument of review. The nature of the planning process under these circumstances will necessarily evolve into something quite different from what we have called planning in the past. The "new planning" that will result is yet to be assessed, but what seems to be in danger of being lost is the quality of coherence which in the past has been touted as one of planning's significant contributions to policy making and management. Perhaps the strongest statement that can be made in this assessment is that is the legislature did not mandate planning in the coastal zone, it did, at least, allow it. Whether or not it takes place will depend on the commission's commitment to planning within the general context of coastal resource management and its success in propagating that commitment to state agencies and local jurisdictions through the mechanisms made available by the legislature in the Coastal Act of 1976.

FOOTNOTES

1. Coastal Zone Management Act of 1972, Sec. 302(g)
2. California Public Resources Code (CPRC hereafter), Chapter 20
3. CPRC, Sec. 30300 establishes the California Coastal Commission and six regional commissions. For purposes of this paper the regional commissions are simply extensions of the state commission and will be collectively referred to hereafter as "the commission."
4. CPRC, Sec. 30004(b)
5. ibid.
6. CPRC, Sec. 30522
7. CPRC, Sec. 30400
8. CPRC, Sec. 30401
9. CPRC, Sec. 30402
10. CPRC, Sec. 30404
11. CPRC, Chapter 5
12. CPRC, Sec. 30411(b)
13. CPRC, Sec. 30220
14. CPRC, Sec. 30233(a)(3)
15. CPRC, Sec. 30412(c)(1)
16. CPRC, Sec. 30412(c)(3)
17. CPRC, Sec. 30412(d)
18. CPRC, Sec. 30413(b), (c)
19. ibid.
20. CPRC, Sec. 30413(d)
21. CPRC, Sec. 30414(a)
22. CPRC, Sec. 30414(b)
23. CPRC, Sec. 30414(a)
24. CPRC, Sec. 30416(b), (c), (d)
25. CPRC, Sec. 30417(b)
26. ibid.
27. CPRC, Sec. 30418(a)
28. ibid.
29. CPRC, Sec. 30418(b)
30. CRPC, Sec. 30415
31. Federal Coastal Zone Management Act of 1972 (as amended), Sec. 305 (b)(7), (8), and (9)
32. CRPC, Sec. 27001(b)
33. CRPC, Sec. 27304
34. CRPC, Sec. 27302
35. CRPC, Sec. 27300
36. California Coastal Zone Conservation Commissions, California Coastal Plan (Sacramento, 1975), p. 199
37. ibid. p. 188, Par. 26
38. ibid. p. 176.
39. CRPC, Sec. 30002
40. CRPC, Sec. 30500(c)
41. CRPC, Sec. 30336
42. CPRC, Sec. 30512(e)
43. CRPC, Sec. 30342
44. Sec. 30511(c) of the Public Resources Code permits local coastal programs to be submitted in separate geographical

segments where "the commission finds that (such separate) areas... can be analyzed for the potential cumulative impacts of development ...independently..." The hundred segments consist of four port master plans in addition to the 96 segments approved for submission by the 67 local jurisdictions along the coastal zone.

45. California Coastal Commission, "Future Organization Task Force Report--Public Hearing (Second) Draft," April 23, 1980, p.3.

46. California Coastal Commission, Local Coastal Program Regulations, Sec. 00021 (as revised July 8, 1980).

47. The fourteen issue areas are
    1. Shoreline Access
    2. Recreation and Visitor-serving Facilities
    3. Housing
    4. Water and Marine Resources
    5. Dredging, Filling, and Shoreline Structures
    6. Commercial Fishing and Recreational Boating
    7. Environmentally Sensitive Habitat Areas
    8. Argriculture
    9. Hazards
    10. Forestry and Soils
    11. Location and Planning New Development
    12. Coastal Visual Resources and Special Communities'
    13. Public Works
    14. Industrail and Energy Development

48. California Coastal Commission, op. cit., Sec. 00023(b).

49. California Coastal Commission, Local Coastal Program Regulations (Revised July 8, 1980) Sec. 00040(d)

DEVELOPMENT OF A NORTH SEA OIL TERMINAL

Iain M. Robertson*

Introduction

The discovery of oil and gas fields in the northern North Sea in the late 1960s and early 1970s began a decade of increasing and diverse pressures on the Scottish coast to accommodate oil-related development of all sorts. Much of this pressure occurred in areas that had never before been used for anything other than subsistence agriculture and fishing and were very sparsely populated. Thus conflict between proponents of development and those of conservation of the often wild, remote and scenic Scottish coastline was inevitable. This paper describes how one such development, a major Oil Handling Terminal for the Occidental Oil Company (The Occidental North Sea Consortium) successfully overcame many of these problems and was integrated into the social, environmental and visual fabric of the Orkney Islands. Because of space limitations, the discussion will focus particularly on the resolution of the visual issues.

Types of Oil Related Development Pressure

The following description of the types of development pressure which resulted from the exploitation of the North Sea oil and gas fields indicates that at each successive stage in this process the types of coastal site required by the industry varied. Thus different parts of the coast experienced development pressures at different times for different facilities. Although presented chronologically it should not be inferred that demands for the earlier types of facility no longer exist but only that this was the general order in which the demands were first experienced. Briefly the types of facility were:

1. Oil Rig and Platform Fabrication Yards

   Major Requirements: Deep, sheltered water adjacent to flat land. Earlier rigs were usually constructed of steel and required approximately 50 feet of water depth while later concrete rigs required 200 feet approximately. Sites for these include existing shipyards on the east and west coasts and new sites located at the mouths of sheltered firths (drowned river estuaries) on the east coast and deep sea lochs on the west coast.

2. Gas and Oil Pipeline Landing and Processing Sites

   Major Requirements: Close proximity to fields to minimize expensive undersea pipelines, flat land relatively close to shore. All of these sites are either on the east coast or in the islands to the north of the mainland. The facility described in this report is located in the Orkney Islands.

*Assoc., Jones & Jones, Seattle, WA, formerly with W.J. Cairns & Partners, Edinburgh, Scotland

3. Backup Services and Industrial/Port Facilities

   Major Requirements: Existing infrastructure of roads, airports, harbours, housing, and available workforce; proximity to oil and gas fields.

   These development pressures have led to the expansion of harbour facilities and rapid growth of existing towns on the east coast, principally Aberdeen and Peterhead in the northeast, but also significantly in Wick in the far north and Dundee and Leith in the southeast. New roads, powerlines, airports, and towns and industrial estates have also been constructed.

4. Downstream Petrochemical Developments Using Oil and Gas as Feedstocks

   Major Requirements: Deep sheltered water and docking facilities; very large areas of developable land relatively close to the shore.

   Potential or possible sites for such development have been identified in all of the major east coast estuaries, the Forth, Tay, Cromarty and Dornoch Firths. The presence of an existing oil refinery in the Forth is an added inducement to siting in this area.

   In the future, as a result of the development of petrochemical complexes, a series of downstream industrial developments may be expected throughout the country.

Social Impacts of Oil Development in the Orkneys

From the above analysis it is evident that oil-related development pressures have been felt on many parts of the Scottish coastline as well as throughout the country. A northward and eastward shift in the locations of new industries is occurring, away from the old industrial Central Lowlands between Edinburgh and Glasgow, toward relatively undeveloped regions of the country. In no place has this growth caused greater changes to the existing social structure than in the remote Orkney and Shetland Islands to the north of the mainland, located between latitudes 56 and 61° north. Development pressures for oil terminals have concentrated here because of their proximity to many of the larger, more productive fields. Because of this and their traditional isolation from the rest of the country, their small populations, and often subsistence economies, the impacts of and changes resulting from development have been more widely felt here than in other parts of the country. The 1971 population of the Orkneys was approximately 17,000 while the population of the small island of Flotta, on which the terminal was located, was at that time 73.

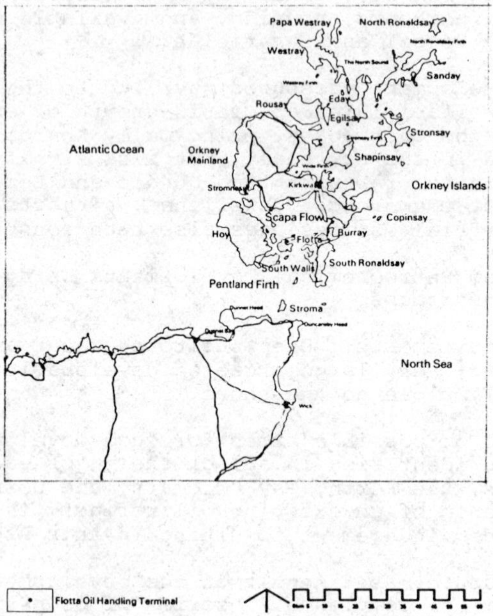

Figure 1   The Orkney Islands

To minimize disruptions to traditional life patterns of the Islands the local government agency responsible for planning and development control, the Orkney Island Council, and the developers of the terminal jointly chose Flotta as the most suitable location for the terminal. Although this caused major changes to Flotta itself, its isolation from the main islands enabled adverse social impacts such as uncontrolled residential development, and disturbances caused by the large construction workforce to be contained and minimized, while ensuring that the economic benefits of more revenues and jobs would be received by the Islands. Access to and from Flotta is controlled by the oil company which runs daily ferry services to the main islands of Orkney as well as air services to cities in the south. Throughout the construction phases all construction workers, with the exception of local Orcadians, were accommodated in a temporary work camp on site. Although the social changes on Flotta have been immense, these will, in the long run, also be beneficial as the island had been experiencing a population decline since 1881, when it stood at 425, to 1971 when it was 73. Without the terminal the island would in all probability have been abandoned by now.

Thus careful control of the workforce and buffering the local population from it reduced adverse social impacts during the most sensitive period of terminal construction. The strategy for the operating phase, projected to last at least 50 years, is to integrate operating staff as closely with Orkney society as possibly by employing native Orcadians wherever possible.

Environmental Impacts of the Oil Handling Terminal

The location of Flotta at almost 56° north might be expected to result in a very severe, cold climate; however, the proximity of the sea considerably modifies and moderates the climate. The most severe aspect of the climate is the prevalence of strong winds and gales; otherwise it is a predominantly cloudy and moist maritime climate. Environmental Analyses,[1] conducted by Planning and Landscape Architecture Consultants W.J. Cairns and Partners of Edinburgh, concluded that the part of Flotta chosen for the terminal had no ecological features of particular value that might require modifications or changes in design or location of the terminal. However, it was also concluded that the large body of partially enclosed water, Scapa Flow, in which Flotta was located was a particularly important and sensitive habitat of international significance as a wintering area for marine ducks as well as important for wading birds, cormorants and Atlantic and Common Seals. Detailed studies of the marine environment were, therefore, undertaken by the Department of Offshore Engineering of the Heriot Watt University of Edinburgh. From the consultant's studies elaborated safety procedures were developed to protect the high water quality of the Flow. These included developing strategies for the quick containment, removal or dispersion of any oil spills and recommendations that all treated ballast water from tankers should be piped out of the Flow to a point where strong tidal action would quickly mix this water with the surrounding sea. Figure Two illustrates one of the studies of tidal velocities and directions which assisted in developing the safety procedures and locating the oil loading moorings in the Flow.

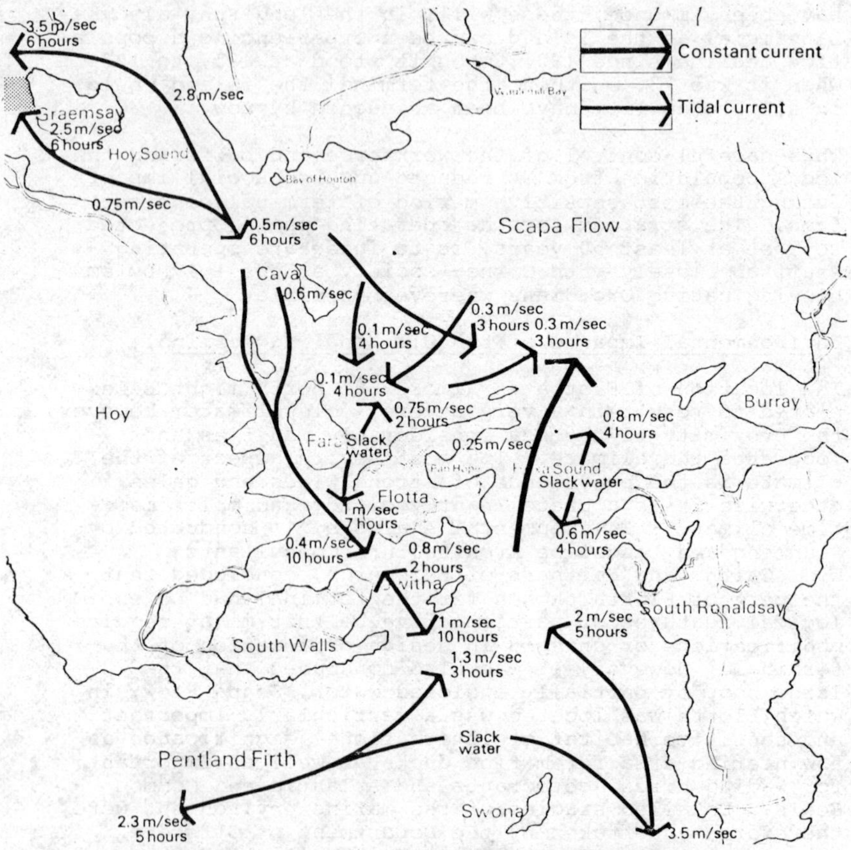

Figure 2 Tidal Patterns in Scapa Row

The other major consideration highlighted by the consultant's Environmental Survey was the importance of integrating the terminal with the visual landscape of the Orkneys. Studies to determine how this could be done were also undertaken by W.J. Cairns and Partners.(2)

Visual Integration of the Terminal with the Orkney Landscape

The visual character of the Orkney Islands with their wide, open land and seascapes composed of softly rounded, bare hills and low flat islands is illustrated in Figure Three.

Figure 3 The Landscape of the Orkney Islands

A more detailed view of Flotta showing the pastureland and heather moors which occupied the terminal site prior to its development is illustrated in Figure Four. The location of the site is on the narrow neck of land at the head of the bay in the foreground. The problem of integrating the oil terminal and the Orkney landscape was expressed in

the following way by the consultants: "... the problem that emerges is one of fitting into a predominantly flat, exposed, rural landscape, a major industrial development with component parts alien to that landscape."(2)

Integration using normal techniques such as screening development by trees or landforms was not feasible for several reasons:

1. The scale of components was very large. Oil storage tanks were 48 feet high and 280 feet in diameter and process plant and flares were considerably taller.

2. The surrounding landforms were so flat and views so open that landforms to screen development would have had to be extremely high.

3. The climate was not conducive to tree growth. The only trees on the island, a small coniferous plantation planted by the Admiralty in 1946, had reached an average height of 15 feet in sheltered areas and, where planted on more exposed locations, trees were less than knee height and showed extreme signs of dieback and wind pruning.

Thus other techniques for visually integrating the terminal and the landscape had to be sought. Figure Five illustrates the design that eventually emerged for the terminal. This layout was the result of an elaborate series of visual studies carried out by the consultants. These are described briefly below.

<u>Visual Design of the Terminal</u>

To determine the most sensitive view directions and locations of the site a map of site visibility from roads on the islands surrounding Scapa Flow and Flotta was prepared, Figure Six. Following this, and based on engineering requirements for the relative base elevations required by the component parts of the terminal, shown in Figure Five, a series of studies were undertaken to determine how the parts of the terminal should be laid out to minimize visibility. Since this could not be done entirely by screening three concepts were employed to assess designs for clarity of visual definition and therefore visibility.

1. <u>Edge</u> - Since it is the profile of new forms which makes them particularly conspicuous anything that would confuse or obscure the definition of edges would weaken the perceived image.

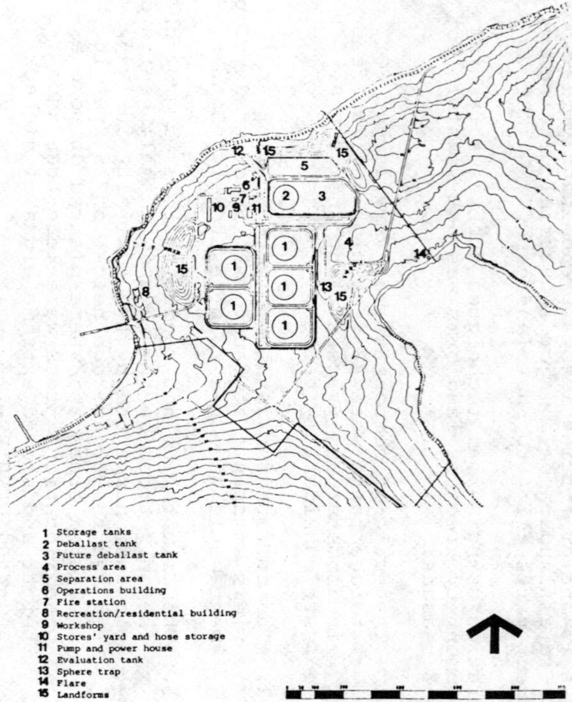

Figure 5 The Flotta Oil Handling Terminal

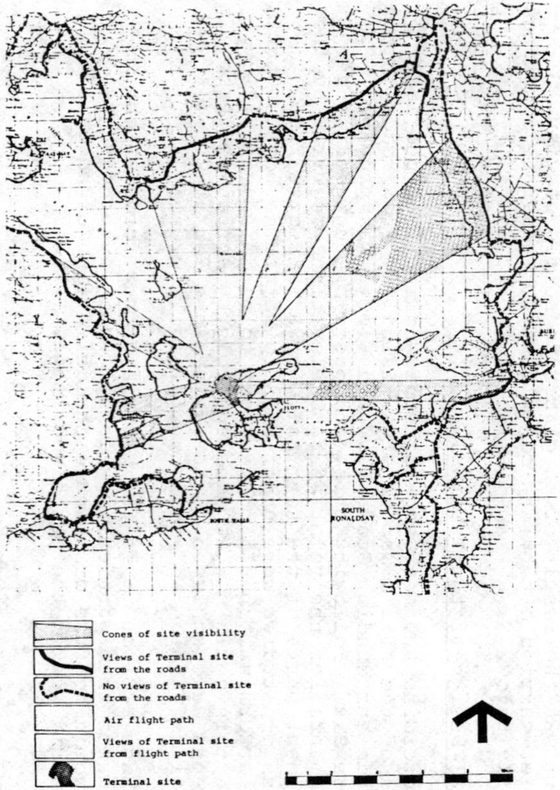

Figure 6 Views of Terminal

2. <u>Depth of Field</u> - Since objects are defined visually by seeing them in relationship to each other and to foreground and background, by making them more complex or obscuring these interrelationships their definition and hence visibility will be reduced.

3. <u>Overlap</u> - The overlap of one object with another assists in giving them good visual definition. If overlapping is simple, forms are easily read and perceived; if it is complex, then perceived relationships become diffused and break down.

These concepts were used to assess a series of layouts of the oil and ballast water storage tanks for their perceived visibility. Figure Seven illustrates three layouts with progressively lower perceived visibility, the final layout being the one proposed for the terminal and storage tanks.

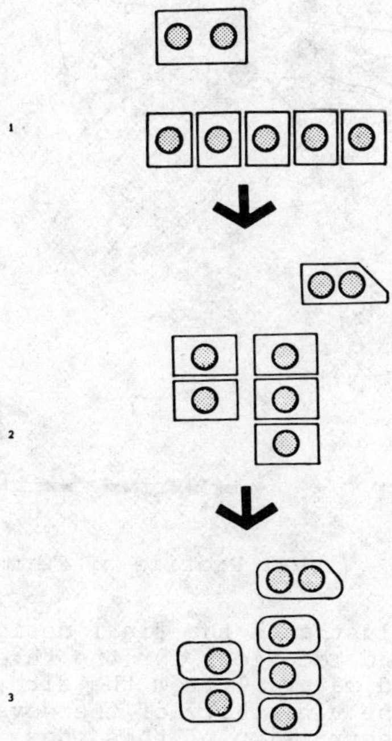

Figure 7 Storage Tank Layouts

For each of these layouts a computer check was made on the visual profile--a measure of the ease with which the tanks could be perceived. The profile of the final preferred layout is shown in Figure Eight.

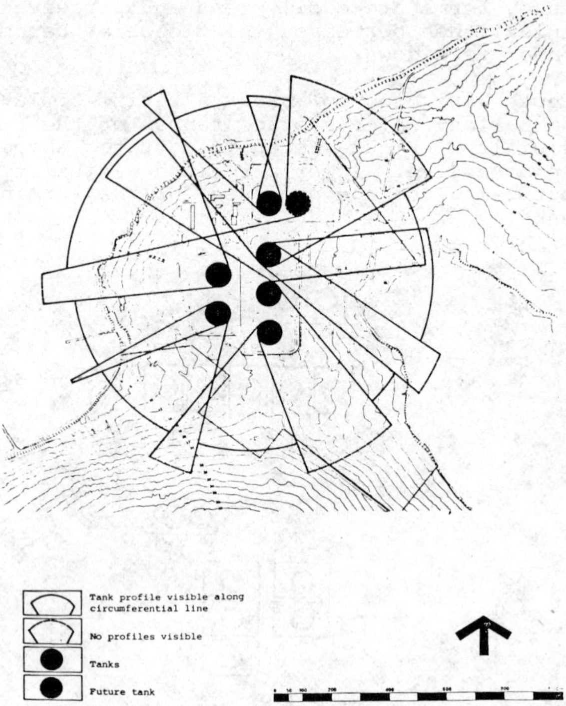

Figure 8 Visual Profile of Terminal

Figure Eight also illustrates the final design chosen for the height, slopes and locations for the three landforms, composed of excavated material from the site, which further assist in reducing the visibility of the development. Subsequent to the construction of this phase of the terminal, a further oilfield was discovered by Occidental and two more larger diameter tanks were incorporated into the design. Figure Nine illustrates a view of the completed terminal from the southeast showing how tank layout and land moulding help reduce the apparent size and prominence of the tanks.

In addition to these design studies of plant layout and landform moulding several other techniques were used to reduce the visual prominence of the terminal. These included selection of a suitable colour for plant and buildings and proposals for revegetation and future landscape management of the site.

Colour Selection

The selection of a paint colour which would help integrate the terminal with the heather moors surrounding it was done in two stages. Initially, chips of eleven different colours were laid on typical photographs of the site and a choice of the least obtrusive ones was made (Figure Ten). Subsequently

Figure 9
Terminal from
Southeast

Figure 10
Colour Selection

large panels of the two best colours were erected on site so that the final colour choice could be made under the prevailing atmospheric and lighting conditions of the Orkneys. Figure Eleven illustrates the terminal seen from the west, painted in the Walnut Brown colour finally selected as most appropriate for merging the terminal with the moorland heather of Flotta.

Figure 11 Terminal from West

Landscape Management of the Site

Because of the climate it was not feasible to grow trees on the site; however, to add visual diversity and integrate the site with its surroundings, thickets of trees and shrubs were proposed wherever landforms or terminal structures gave some shelter. On the berms surrounding the storage tanks grass was proposed and on most other disturbed areas surrounding the terminal, including the landforms, heather was to be planted. Figure Twelve shows the proposed landscape development plan.

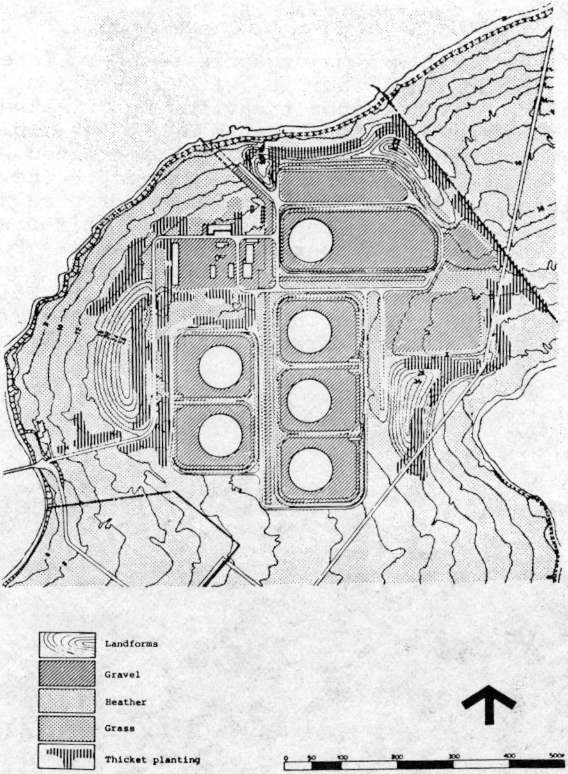

Figure 12 Landscape Development Plan

The thickets were intended to "... reinforce the landforms and to provide a pattern of separation, spatial containment and textural differentiation throughout the site."(1)

The initial proposals were subsequently modified when trial plots indicated that it would be difficult to get heather to germinate quickly and provide adequate ground cover to prevent erosion in the short term. All heather areas were, therefore, planted with grass with the unfortunate result that the carefully chosen brown colour of the structures was more often seen against a green rather than brown foreground and background.

In terms of management of the site it was proposed that grass mowing be minimized by selecting a slow growing species mix for the bund walls and by allowing local farmers to graze their sheep on the landforms and grass areas outside the security fence.

Conclusions

The terminal represents a major achievement in integrating a new, alien, industrial facility with a rural, old established society and an unspoilt environment. Although several mistakes or omissions were made in the implementation of design proposals these are only mentioned here to highlight the general success of the project. One of the major problems in implementing the design proposals was a different attitude between the landscape architects and the terminal construction and operations engineers. For example, attempts were made to tidy up a natural shore bluff with an "untidy" appearance, as shown in Figure Thirteen. However, this only resulted in increased erosion when natural shoreline processes resumed.

Figure 13 Shoreline Modification

Additionally, rejection of the "thicket" concept was made in part because operating staff did not like the windswept and often battered appearance which all trees on Orkney assumed unless well protected, preferring to have no trees or thickets if they had the scruffy, untidy appearance of those on the island (Figure Fourteen).

Figure 14 Trees on Flotta

The concept of carefully controlled sheep grazing was also rejected in favour of allowing free access to all parts of the site outside the security fence. Thus it was not possible to ensure that all areas would be grazed and it was, therefore, necessary to mow the grass on the landforms.

Despite these relatively minor problems the project can claim to be an impressive success story. Planning permission was quickly granted due to close cooperation between developer and planning authority, disruption of the traditional way of life on Orkney was minimized while economic benefits were achieved, and impressive pollution control and safety measures were adopted to protect the sensitive marine environments and prevent creeping environmental degradation.

If any general conclusions had to be drawn from the success of this project they would be (1) that cooperation between the developer and the local community and planning department reaps benefits for all parties and avoids wasteful, expensive and often bitter legal disputes, and (2) that a commitment by the developer to examine and resolve environmental problems at the planning and design stages before they occur is essential to a problem of this complexity.

* * *

Acknowledgments

The Author wishes to thank W. J. Cairns for his assistance in preparing this paper and permission to use photographs and diagrams from the published references below.

References

(1) Flotta Orkney Oil Handling Terminal Visual Impact Appraisal and Landscape Proposals by W.J. Cairns and Assocs., Edinburgh, June 1974. For Occidental of Britain Inc. and The Associated Companies.

(2) Flotta Orkney Oil Handling Terminal: An Environmental Assessment by W.J. Cairns and Assocs., Edinburgh, December 1973. For Occidental of Britain Inc. and The Associated Companies.

(3) Flotta Orkney Oil Handling Terminal: Marine Environment Protection by W.J. Cairns and Partners, Edinburgh, May 1975. For Occidental of Britain Inc. and The Associated Companies.

Modeling the Cumulative Onshore Effects of Offshore Oil and Gas Development

E. T. Smith and L. G. Hecht, Jr.

Introduction

The pattern of offshore oil and gas development over time can cause a series of related events to occur in an adjacent onshore coastal region. An onshore region, composed of one or a number of States, may receive varying levels of impact, due to these offshore events. These events include the siting of various onshore facilities, which in turn result in a time stream of primary, secondary, and higher order effects. Such effects can, in turn, impact the physical, social, and economic sturcture of the region.

Modeling Requirements

Modeling this set of events and effects, involves the use of a variety of planning and analysis techniques. Because the "OCS process" (ref. 1) generally covers long periods of time between say, an initial lease sale, discovery of oil and gas, and the placement of a related onshore facility, an investigator must conduct an analysis using the method(s) that best matches the quantity, quality, and scale of available data.

As the "OCS process" in a region moves from initial phases (e.g., initial lease sale, first discovery, initial development, etc.) to mature phases, data quantity, quality, and scale tend to move from region-specific to site-specific levels of detail. Thus, analysis techniques associated with a site-specific event may be used with increasing certainty as the "process" of mineral development in a region matures over time. The problem here is to select the best set of methods to fit a given situation. Depending on the analysis to be conducted, and data quantity, quality, and scale factors, several alternative methods can be employed. These are discussed in some detail in Williams (ref. 2). One point to note is that given the usually large areal extent of OCS oil and gas operations in a region, the process of modeling offshore events and onshore effects for purposes of planning and evaluation is probably best approached from first a regional, and then a State perspective.

For more information: Dr. E. T. Smith, U.S. Geological Survey, can be reached at (703) 860-6717 and Louis G. Hecht, Jr., U.S. Geological Survey, can be reached at (703) 860-7166.

This paper has not been edited for conformity with the publication standards of the Geological Survey.

This paper concerns onshore aspects of the OCS oil and gas development process; however, planning for onshore activities requires an understanding of the scale, magnitude, and timing of offshore activities that trigger onshore actions. There are a number of offshore factors which usually influence the numbers and kinds of onshore facilities that eventually locate in an area, given marketable discoveries of oil and gas. Generally, these include the discovered recoverable reserves of oil and gas in an offshore basin, the number and size of oil and gas fields discovered over time, production platform requirements, timing of field development and production operations, and the location and means of product tranportation and processing.

The sequence of events that occurs during the minerals leasing, exploration, development, and production process has been the subject of study by this office, with cooperation from the New England River Basins Commission for a four-year period. Much of the discussion on magnitude and timing of offshore activity and onshore facility impacts leans heavily on work done as part of the effort (ref. 3 and 4).

Magnitude and Timing of Offshore Activity

Before one can proceed with onshore facility analyses, an investigation into the functions of offshore development should be completed. Data relating to each function of the development process should then be translated into one or a series of offshore scenarios. Building a suitable and relatively certain offshore scenario requires the investigator to review historical trends about the amounts of oil and gas to be discovered over some future time period. In addition, an investigator will also need to compile data concerning the following:

- year-by-year schedules of past oil and gas finds
- the number of past sales and tracts leased at each sale
- exploratory rig activity by lease sale
- platform requirements for discovered oil and gas fields
- platform installation schedules (past and present)
- development drilling schedules for fields presently undergoing development
- production schedules for oil and gas (past, present, and projected).

Much of the data required to construct a scenario may be obtained from Summary Reports and Indexes provided by the USGS and BLM (ref. 4). For example, the production function showing magnitude and timing of offshore development for the Pacific OCS from 1966 to 1979 is shown in figure 1 (ref. 5). Supporting this graph are tables 1, 2, and 3 which show the history, current status, and projections of oil and gas development and production in the Pacific OCS (ref. 5). Note that the driving force for the production function (1968-1979) is in terms of more than one variable. Not only is the physical quantity of the resource considered, but also the number of platforms and development wells needed to handle this amount of production, as well as physical factors such as water depth and distance to shore. It is from these technological and physical variables, rather than the physical quantity of the resource itself, that estimates of onshore development and their primary and secondary impacts are derived. In the case of the Pacific OCS, production information for figure 1 was derived from those areas which are presently producing (Dos Cuadras and Carpinteria Offshore). Other areas listed on tables 1 and 2 will begin producing at a later period. Projected oil and gas production figures are found on table 3. For purposes of scenario construction covering present and projected events, present and projected amounts of production need to be graphed.

Onshore Facilities and Their Impacts

Generally, the numbers and kinds of facilities likely to occur in a region, as a result of marketable discoveries, can be estimated from several sources, such as available historical evidence, facility inventory data, analogous development occurring in similar OCS areas, or extrapolating statistical inferences from data concerning magnitude and timing of offshore production coupled with probable energy transportation strategies. Data relating to impacts of each facility type, including land, labor, economic, demographic, and environmental parameters, should also be collected from source documents such as development and production plans, environmental assessments and reports, summary reports, and indexes.

The Pacific Summary Report (ref. 5) provides one with a realistic model of the types of information necessary to undertake an analysis of onshore facilities and their effects. This report contains up-to-date planning information concerning the following:

- current resource and reserve estimates for oil and gas
- past, present, and projected lease sale activity
- past, present, and projected development activities
- past, present, and projected production data
- existing oil and gas related facilities
- new onshore oil and gas facilities.

Table 1

Status of development and production plans for the Southern California OCS

Unit/field	Operator	Date of plan submittal	Date of plan approval	Platform(s) (P=Proposed)
Santa Ynez	Exxon	11/11/71 Rev. 8/15/72	8/16/74	Hondo
Santa Rosa	Exxon	None submitted		
Dos Cuadras	Sun	late 1960's	late 1960's	Hillhouse
	Union	5/12/77	6/9/77	A, B, C
Carpinteria Offshore	Sun	6/14/77	11/11/77	Henry
	Phillips	late 1960's	late 1960's	Hogan, Houchin
Pitas Point	Texaco	9/26/79	Pending	Pitas Point (P)
Santa Clara	Chevron	12/30/76	7/14/77	Grace
	Union	12/12/79	Pending	Gilda (P)
Hueneme Offshore	Union	5/7/79	Pending	Gina (P)
Beta	Shell	11/7/79	1/3/80	Ellen, Elly
	Shell	None submitted		Eureka (P)
	Chevron	None submitted		Edith (P)

SOURCE: USGS, Development and Production Plans for Southern California OCS.

Table 2

Current and proposed oil and gas development off Southern California

Unit/Field	Platform	Lease	Operator	Number of slots	Water depth (m)	Distance to shore (km)	Status
Santa Ynez	Hondo	OCS P-0188	Exxon	28	259	8	Platform installed, drilling 12th development well, awaiting OS&T installation
Santa Rosa			No development to date				
Dos Cuadras	Hillhouse	OCS P-0240	Sun	60	58	9	Producing
	A	OCS P-0241	Union	56	56	9	Producing
	B	OCS P-0241	Union	61	57	9	Producing
	C	OCS P-0241	Union	59	59	9	Producing
Carpinteria Offshore	Henry	OCS P-0240	Sun	30	53	8	Platform installed, development drilling started 2/8/80
	Hogan	OCS P-0166	Phillips	66	46	6	Producing
	Houchin	OCS P-0166	Philips	60	49	8	Producing
Pitas Point	Pitas Point	OCS P-0234	Texaco	24	92	18	Plan approval pending
Santa Clara	Grace	OCS P-0217	Chevron	48	97	16	Platform installed, drilling started 2/20/80
	Gilda	OCS P-0216	Union	90	64	16	Plan approval pending
Hueneme Offshore	Gina	OCS P-0202	Union	15	29	6.5	Plan approval pending
Beta	Ellen	OCS P-0300	Shell	80	81	14.5	Jacket installed, platform under construction
	Elly	OCS P-0300	Shell	*	78	14.5	Approved
	Eureka	OCS P-0301	Shell	60	213	14.5	Awaits plan submission
	Edith	OCS P-0296	Chevron	60	46	11	Awaits plan submission

*Platform Elly is a production platform from which no wells will be drilled. Only initial treatment and storage of oil will be done there.

SOURCES: USGS, Conservation Division (Pacific OCS Region), 1979, unpublished documents; Development Plans for Southern California OCS; Ensele, 1980, oral commun.; Ballantyne, 1980, oral commun.

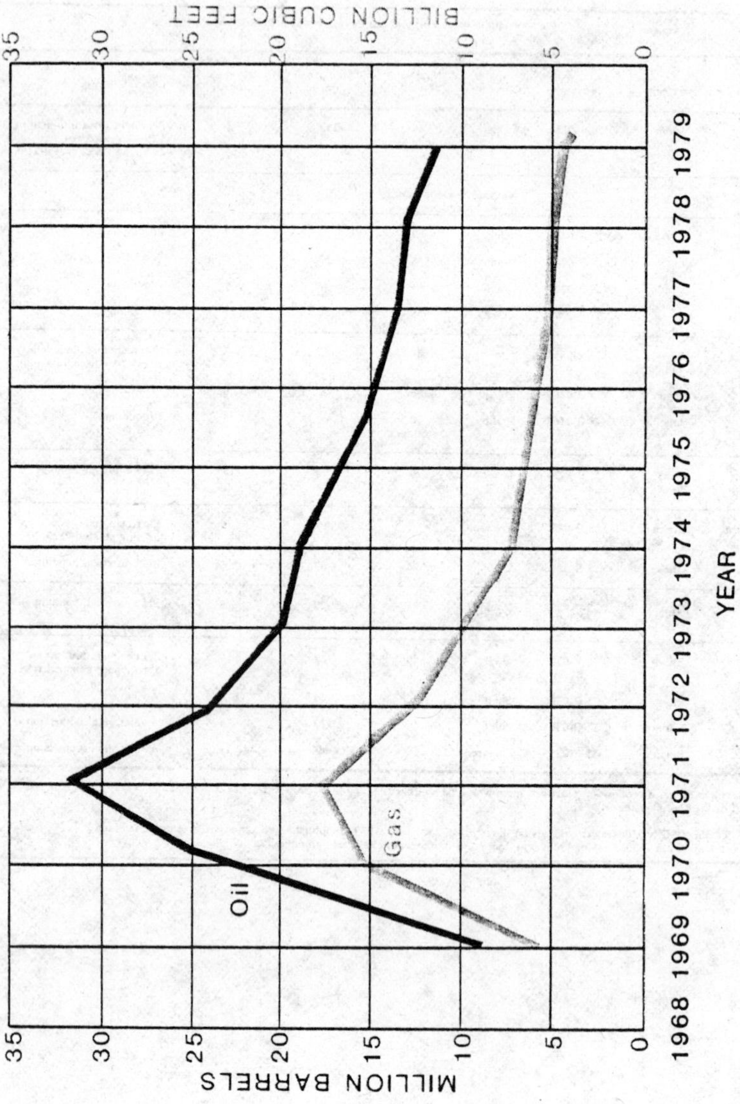

Figure 1. Annual oil and gas production from Southern California OCS, 1968-1979.

Table 3. Expected oil and gas production off Southern California

(n.a. = Information not available at time of writing. As information becomes available, it will be included in future Pacific Summary Report updates.)

Unit/Field	Platform	First year of production	Peak year	Daily peak oil production (bbl)	Ultimate oil production (million bbl)	Daily peak gas production (mmcf)	Ultimate gas production (bcf)
Santa Ynez	Hondo	1981	1983-86	30,000	94	30	30
Carpinteria Offshore	Henry	1980	1981	6,000	12.7	4	8.65
Pitas Point	Pitas Point	1982	1983	0	0	63	n.a.
Santa Clara	Gilda	1981	1983	18,000	43	19	40
	Grace	1980	1982-1983	16,000	44.6	16	44.6
Hueneme Offshore	Gina	1981	1982	6,450	9.53	1.2	1.72
Beta	Ellen	1982	1983-1985	16,000	150	0	0
	Eureka	1984	1983-1985	10,000		0	0

SOURCES: Adams, 1979; USGS, Development Plans for Southern California OCS, Ensele, 1979, oral commun.

From a planner perspective, it is important to note that the planner's job of monitoring activities on the OCS is handled by these reports due to requirements, set forth in law, that updating be undertaken on a timely basis, and when a significant event occurs to revise the report.

There exist nine major types of OCS oil and gas related onshore facilities:

- support and service bases (temporary and permanent)
- repair and maintenance yards
- platform fabrication yards
- pipe coating yards
- separation and treatment facilities
- gas processing and treatment plants
- onshore transportation facilities (pipelines, marine terminals)
- refineries
- petrochemical complexes

Depending on the circumstances surrounding actual leasing, exploration, development and production in a region some or all of the facility types listed above may become a reality. A geographic depiction of existing and proposed OCS related onshore facilities in Southern California is shown in figure 2. These facilities come about as a result of lease sales in both Federal and State waters over a 30-year period. (It should be noted that in California's case, offshore oil and gas exploration comes about as an extension of onshore oil and gas discoveries.)

For each type of onshore facility projected for an OCS region, there are at least four categories of important information that needs to be collected and kept up to date. These are (1) industry's and government's principle siting criteria, (2) land, labor, water, and capital requirements for each facility, (3) primary environmental effects, and (4) primary and secondary economic and fiscal effects.

The four information needs listed above can be translated into a set of analysis tasks for an agency to undertake when either planning for or evaluating the effects of onshore oil and gas facilities. These tasks include: regional (multi-State) analysis to determine onshore facility numbers and kinds, State analysis (siting, environmental and economic assessments), and cumulative environmental analysis.

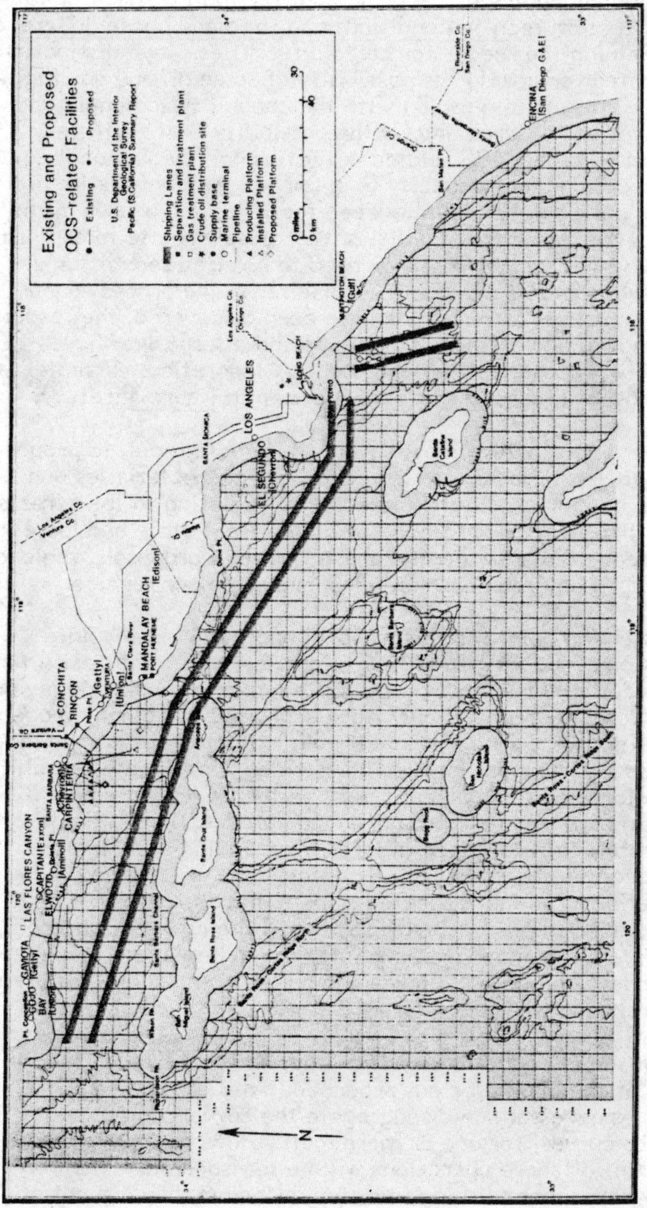

Figure 2. Existing and proposed facilities for handling OCS production, 1980.

Regional Analysis

Prior to any analysis, it is important for the investigator to account for certain facility conditions and time and phase differences when determining the "need" for certain facilities. Two distinctions to be accounted for are (1) the possibility of some oil and gas facilities already existing in a region with the capability to handle additional amounts of resource and (2) the possibility that facilities, presently not handling OCS related activities, will begin handling them with existing plant capacity (e.g, ports). Another distinction to be noted is the relationship between facilities and phases in the OCS process (ref. 1). Some facilities may closely relate to exploration phases, which means they may need to be sited soon after a lease sale, while others generally occur later in the process once commercial amounts of resources have been discovered, the resource quality is known, and offshore transportation alternatives have been formally proposed or implemented. A theoretical example of the phasing issue discussed above is shown in figure 3 (ref. 3).

As any given facility moves from the hypothetical, to proposed, and finally to approved for construction, site characteristics and impacts can be more easily tabulated. Data relating to the effects of such facilities will generally be available from State and local impact statements/assessments, industry project proposals, regional (A-95 and BLM) environmental studies, and summary reports.

The starting point for this regional analysis is to tabulate the population of existing facilities, if any, by type. The best information source for this class of data are summary reports and inventories conducted by regional and State economic and planning agencies and industry sources. The important point to note here is that the data will usually show pronounced clustering. That is, those facilities which have location tendencies to service the OCS function, will be located in near proximity of ports and harbors. This is by no means an unfailing rule, especially for those facilities which are not bound exclusively to the OCS function. The usual example is the oil refinery, which is located so as to maximize profits that may depend only, in part on the OCS (this is the case in California, for example). Other facilities, such as service bases, are much more dependent on access to ports and harbors, and hence will show a good correlation with the rule.

Given some production function that shows the magnitude and timing of the oil and/or gas produced in a source region, it next becomes necessary to disaggregate the curve into process functions, or curves, for one or more multi-State regions to which the product of offshore operations will be transported. There are several ways of accomplishing this step, although they are not of equal reliability. The best approach of course would be exact knowledge of where the product is to be shipped or use a firm plan that is in existence for transportation, such as output from the Interior Department's Intergovernmental Planning Program for OCS Oil and Gas Leasing, Transportation, and Related Facilities (ref. 6).

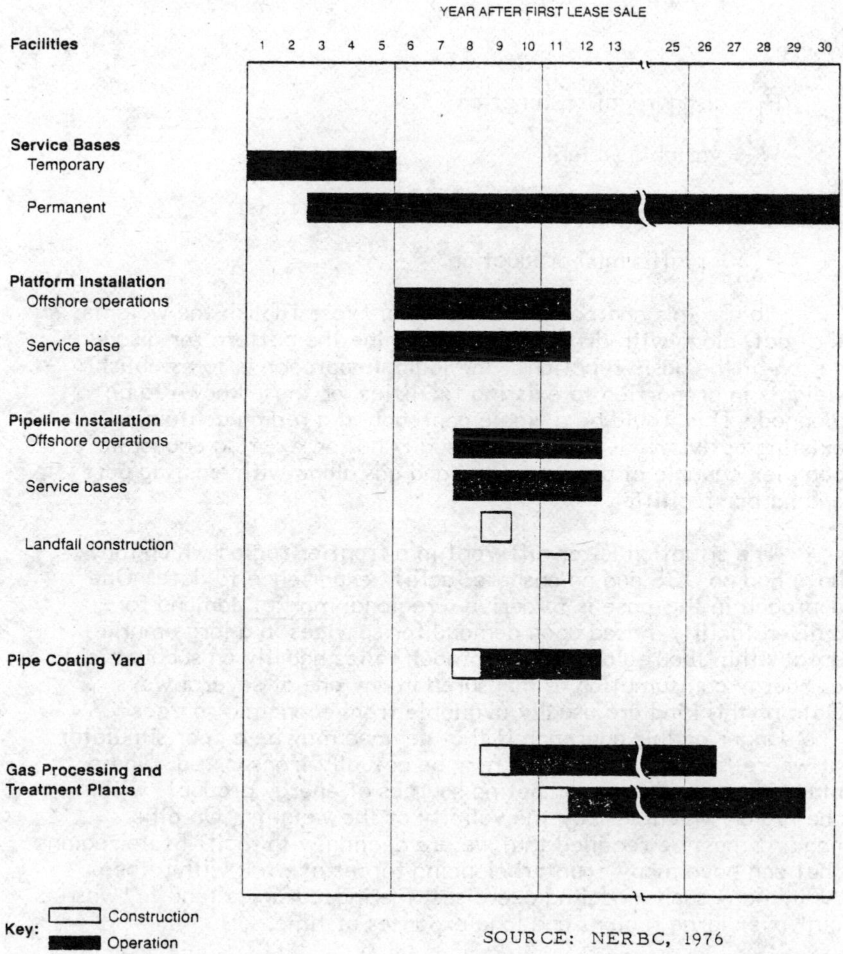

Figure 3. Timing of facilities and activities.

In the absence of real data or a plan, one needs to simulate transportation through use of a mathematical approach like the gravity model (ref. 7). A typical model formulation is shown as

$$P_i = \left(\frac{W}{d}\right)_i P_{oil}$$

Where

P_i = disaggregation function

W = variable weight

d = distance from shore to offshore platform(s)

P_{oil} = platform(s) production.

To use this approach, it is necessary to establish the weights that act, along with distance, to determine the pattern for disaggregation of the basic function. One logical approach is to establish weights in proportion to existing facilities, or those known to be planned. This would be a viable approach in a region containing existing activity. Such a region usually has an existing economic complex capable of processing oil and gas along with existing port and harbor facilities.

The situation is far different in a frontier region which may have had no OCS and no onshore facility experience to date. One approach in this case is to derive a regional market demand for a given facility, based upon demand for services in a large market area within the region. This approach relies heavily on such variables as energy comsumption as measured in any one of several ways. Data of this kind are usually available from economic sources. The danger of this approach is that demand may be a poor simulator of where the OCS production may be actually transported. There may, for example, be competing sources of energy products within the region which destroy the validity of the weights. On other hand, it must be recalled that we are dealing with multi-State regions that can have many counterbalancing forces at work within them. As in many such modeling exercises, the inaccuracies tend to "wash out" over large regions and long expanses of time.

As a next step it becomes necessary to account for the effect of the facilities already in existence. The goal is to subtract out their effect so as to isolate the effect of any new facilities that may be constructed. If some of the allocated amounts of oil and gas are handled by existing facilities, then that total amount of oil or gas handled is usefully represented by a surrogate variable, such as the number of platforms needed to produce that amount of oil or gas. This results in a measure of production in terms of amount handled and amount not handled (Ponyh). Then by employing a multiplier to express the processing capacity of a type of facility, it then becomes possible to calculate the effect of the existing facilities in reducing the demand for new facilities, and to determine the demand as

$$P_{onyh} = P_{\substack{oil \\ total}} - N_{fac} X_j$$

Where

X = oil processing capacity of facility j.

Furthermore, the number of new facilities can be computed as

$$N_{\substack{new \\ fac}} = \frac{P_{onyh}}{X_j}$$

In all cases P is measured in terms of number of platforms.

This method has, in fact, been used in the computer model DEROCS (ref. 8), which models the effects of oil and gas finds by use of a series of simple multipliers. The complexity of the situation becomes evident only when it is noted that these calculations must be repeated for each class of facility, for each region, and for each time period in an analysis.

The next step is to extrapolate the impacts for each desired impact variable using another series of multipliers for each variable, which are specific for each facility type. An equation of the following sort results

$$I_{\substack{new \\ impact \\ k}} = N_{\substack{new \\ fac}} M_{k_j}$$

Where

M_{k_j} = the multiplier for impact k of facility j all at time t in region i.

State Analysis

Up to this point, the discussion has centered on analysis concerning numbers and kinds of facilities in a multi-State region along a coastline. At this level of aggregation many of the inaccuracies of estimation are less important than at more detailed scales. The estimation issue becomes vital when any attempt is made to drive the models to the more detailed State level. It is of course possible to carry out such a formulation mathematically. The computer program DEROCS begins with source areas on the OCS and disaggregates results down to the individual port level. The question, of course, is how good are such estimates?

Assuming that the sensitivity of regional estimates are low, the simplest approach is to use the knowledge of what activity already exists. If the pattern of existing facilities within the multi-State region (by State) is known, then that pattern can be used as a guide for locating the simulated facilities and their impacts. In many cases this could be a simple proportion. Other information can enter into the estimation of how to disaggregate the overall totals. For example, any firm industry proposal to locate in an area will be just as useful at this stage of analysis as when used to simulate gravity model weights.

Just as in the case of the gravity model, the approach breaks down in new or frontier areas when there may be no activity in a State. In such cases it becomes necessary again to turn to a surrogate variable, such as market demand in a State. The uncertainties associated with this approach is far larger than in the regional case, since at this stage of analysis we are dealing with much smaller areas which vary in size, geometry, and many other characteristics.

An overall methodological approach which does not necessarily require the use of rigorous models has been developed by the New England River Basins Commission (ref. 3). The third of three planning procedures or methodologies establishes a series of seven modules that planners can use to determine site-specific suitablity for OCS facilities. The modules first attempt to identify potentially capable sites by using industry criteria. These sites are next screened by applying public policy suitability criteria, which presumably would result in the deletion of some sites. The institutional adequacy of the State's siting authority is also assessed and applied as a criterion. Candidate sites result from these steps. The legal authorities that the State has available to assure that it defines and can gain its public objectives are reviewed next. The final two components are applied only after specific proposals are made by industry: they involve a detailed impact assessment of all environmental, economic, fiscal, and social variables, and a ranking of alternative sites for suitability.

Since most of the steps in the NERBC method must be done for each facility and for each site, the process can be time-consuming. Perhaps even more important for planning, some aspects cannot be done at all until some industry proposal has been received. It seems then that many State level analyses, <u>if</u> they wish to look far enough ahead, must make use of the admittedly inaccurate, but readily available, mathematical forecasting approaches.

On the other hand, States have alternatively concentrated their resources on developing and perfecting <u>analytical tools</u> for siting and evaluating OCS related onshore facilities. State-specific studies make it possible to allocate activities at the State level more realistically. In addition, it is logical that the State would be the primary determinant of activity within its bounds, since this is where the major governmental responsibility lies. Maryland's experience with tool development is discussed below because this effort represents a comprehensive package of tools for use in planning and evaluation.

The Maryland Major Facilities Study (ref. 9) provides a transferable process for site selection by areas and sites, resolving site conflicts, and assessing the economic, fiscal, social, and environmental effects from specific facilities. All of these elements are crucial to both economic and environmental objectives in States.

The Maryland Regional Screening Process contemplates allocating activities according to a series of geographic screening criteria - threshold criteria, exclusion criteria, and potential restriction criteria. The regional screening process operates so as to identify threshold areas by eliminating areas that are clearly not suitable for facility development, then by determining the most suitable candidate areas by applying the exclusion and potential restriction criteria. By applying finer level data at the candidate area level, a number of "best" alternative sites emerge.

A major strength of this regional screening process is that it is open and invites scrutiny. The process allows anyone to review how the results were achieved, and to reapply the methods if different criteria are chosen or new data becomes available. Another strength is that the process reserves the introduction of subjective values until the last set of steps, after candidate areas have been suitably characterized by objective data. Thus, Maryland was able to select suitable candidate areas without imposing on restricted areas (e.g., unique wetlands). Having a set of alternative sites that meet minimum economic, environmental, and social criteria for industry use at some point in time provides the stimlus for both industry and the State to use this information for more detailed planning when and if a given facility becomes a necessity. Through such cooperative efforts industry can site and government can regulate in a cost effective manner while protecting valuable coastal resources.

The Economic, Fiscal, and Social Assessment Handbook contains three major elements: an economic element, a population and housing element, and a fiscal element. The economic element forecasts the levels of income and employment generated with the county, or adjacent county, as a result of a major facility being located there. It is based on a variation in the multiplier approach with the principle distinction being that a series of steps were taken to separate those economic effects that remain with the county from those that are lost to the county and go elsewhere. This is significant because a good deal of attention is given, in relevant literature, to massive multipliers for certain industries such as offshore oil and gas activities, without recognizing that most of the activity will not benefit or harm the county in which a direct effect takes place. Most of the indirect economic activity generated by such facilities deals with the provision of specialized goods and services that do not exist in local economies outside the Gulf of Mexico and highly urbanized areas. The same is true for personal consumption.

If substantial inmigration of workers occurs, rural areas generally, do not have the retail and infrastructure to absorb most of the spending of new worker households. Thus, spending goes elsewhere.

The population and housing element deals first with the movement of workers hired outside county into the county. Second, workers are converted into households and school children. Third, the households are then converted into housing demand.

The fiscal element deals with both revenues (to various jurisdictions and levels of government) generated as a result of the facility and the serivce demands and costs generated thereby. Service demands are distinquished by two types of costs: (1) demands reflected in specific increments that may be required for existing services and (2) facilities (e.g., police and sewer and water systems) and government costs, which represent the fiscal effect of increased service demands on the budget. It should be noted that service demands are very sensitive to policy choices by government. Thus, the handbook leaves the choice of inputs for service demands to the investigator. This procedure, originally developed to be used manually by local planners, is now computerized. Table 4 depicts the kinds of output an investigator may calculate with this component of the study.

The Environmental Assessment Handbook provides planners with readily useable methods for identifying, measuring, and evaluating effects of onshore facilities on the natural environment. The handbook is divided into three major parts to facilitate an orderly assessment procedure. Part one provides the investigator with information on effects of development from each major facility. These effects result from a variety of activities carried out during site preparation, facility construction, and facility operation and maintenance. Following this information is a discussion of pertinent activities and environmental factors for each facility. A matrix for each facility was developed relating activities and environmental effects.

The second part of the handbook covers measurement of effects. Twenty-six environmental factors are considered. The factors considered are, for the most part, either scarce resources that are threatened by facility development or natural hazards that affect development or are intensified by development activities.

The third part of the handbook pertains to evaluation of effects. A project evaluation worksheet was developed as a means of comparing results of environmental factor assessments and arranging them in a comprehensive format.

Table 4. Summary of major outputs from economic and fiscal assessment model

ECONOMIC EFFECTS (Chapter 1)	POPULATION EFFECTS (Chapter 2)	REVENUE EFFECTS (Chapter 3)	SERVICE & COST EFFECTS (Chapters 4 & 5)
Income from purchase of goods and services by facility (indirect income) by sector	Immigration of workers and households to county and to subareas within each county	Personal income tax from workers in added jobs	Specific service increments for the following service areas: police, fire, sewer, water, solid waste disposal, health, education (keyed to decision inputs by user) (marginal costs)
Income from personal consumption of workers at facility (induced income) by sector	Households, single individuals, and school children inmigrating to county and subareas	Sales tax from purchases of goods & services	
Employment generated by indirect and induced income by sector	Demand for housing in county by type of structure (house, apartment, mobile home) and form of tenure (own, rent)	Personal and corporate income taxes from firms providing goods & services	Fiscal impact (costs) by year for major services and for total county costs, school and nonschool (based on average costing method)
Matching of local labor supply to demand from facility	Effect of housing demand on availability of housing in county	Property taxes from house value increment (inmigrants)	
Number of outside hires resulting from facility labor demand		Recordation and transfer taxes	
Classification of labor force by wage and consumption level		Taxes paid by major facilities	
Effect of labor demand on county economic sectors		Miscellaneous taxes	

Cumulative Effects

The impacts discussed to this point are the so-called facility-specific effects; that is, effects occurring as a result of one facility locating in a region or a State. However, there also exists a series of higher order effects which when taken together constitute a set of cumulative effects in a region. In general, these cumulative effects are repercussions of primary and secondary economic and fiscal effects and primary environmental effects.

There are a number of approaches for handling this kind of effect, not all of which rely on mathematical or computer techniques (ref. 10). However, the approach this office developed for problems of this kind does utilize a relatively simple mathematical scheme based on matrix algebra.

The general impact analysis model, EIAMAT, is a matrix approach combining concepts from cross-impact analysis and related matrix algebra methods. The methodology entails combining a schedule of primary effects expected as a result of future activities with a matrix of interrelationships. Each term in the cross-impact matrix relates a change in one variable in one year to a change in another variable in the following year, by a linear multiplier. These multipliers are the most important data input necessary to utilize the methodology. The resulting environmental impact forecasts account not only for the direct effect of future activities, but also for higher order interactions that occur in later time periods. This makes it possible to provide more complete forecasts in each time interval of the effect of implementing a project, especially as it interacts with higher order effects over time.

The general model can be represented in matrix notation as follows:

$$V(t) = A\left(V(t-1)\right) + B(t)$$

Where

V(t) = The vector of the changes in primary plus secondary values of each impact variable for a given year.

A = The square matrix of multiplier coefficients that show cross impacts between pairs of variables for a one-year period.

B(t) = The vector of the changes in primary values of each impact variable for a given year (these values can be input from the DEROCS program).

It should be emphasized that this forecasting or probablistic process is far more important than the particular values used for the coefficients in the model. Even with the most exacting procedures to validate or verify the model against reality, it should be noted that no model can predict the future. There is always the possibility that the conditions under which model validation is carried out may change in some unexpected way. Instead, the strength of the forecasting technique lies in the ease with which it can be used to examine the results of a wide range of alternatives. Furthermore, this process should not be done only once, but rather could become an integral component of a continuous planning process. In this context, the need for continually analyzing large volumes of numerical information emphasizes the requirement for an automated approach.

This model was originally developed and tested on a coal strip-mining operation in Campbell County, Wyoming. As of August 1980, work is underway to expand the variable base and "fine tune" the model so that a test can be made for a series of onshore facilities. Testing with an OCS case should be completed one year from now.

References Cited

1. Benton, J. B., and others, 1979, Outer Continental Shelf Oil and Gas Information Program: Directory to Federal, State, and local OCS-related activities and contacts: U.S. Department of the Interior, Geological Survey Open-File Report 79-1481, 192 pp.

2. Williams, D. C. and Hom, K. B., 1979; Onshore impacts of offshore oil: a user's guide to assessment methods: U.S. Department of the Interior, Office of Policy Analysis, Washington, D.C., Government Printing Office, 67 pp.

3. New England River Basins Commission, 1978, Methodologies for OCS-related facilities planning: Boston, Massachusetts, 151 pp.

4. U.S. Geological Survey, 1979, CFR 30 Part 252.

5. Macpherson, G. S. and Bernstein, J. 1980, Outer Continental Shelf oil and gas activities in the Pacific (Southern California) and their onshore impacts: a summary report, May 1980. U.S. Department of the Interior, Geological Survey Open-File Report 80-645.

6. Bureau of Land Management, 1979, Intergovernmental Planning Program for OCS oil and gas leasing, tranportation, and related facilities.

7. Isard, Walter, and others, 1960, Methods of regional analysis: Cambridge, Massachusetts, MIT Press, 784 pp.

8. Marcus, P. A., Smith, E. T., Robertson, S. R., Wong, A. T., 1977, DEROCS: a computer program to simulate offshore oil and natural gas development scenarios and onshore service base requirements: Reston, Virginia, U.S. Geological Survey Open-File report 77-130, 256 pp.

9. Maryland Department of Natural Resources, 1978, Maryland major facilities study: Annapolis, Maryland, Energy and Coastal Zone Administration.

10. Schlesinger, B. and Daetz, D., 1975, Development of a procedure for forecasting long-range environmetnal impacts: Department of Industrial Engineering, Standard University, Stanford, California, Technical report 75-3, 137 pp.

CONSISTENCY REVIEW OF OCS ACTIVITIES OFF CALIFORNIA

Irwin D. Karp
Energy Counsel
California Coastal Commission

The Federal consistency provisions contained in Section 307(c) of the Coastal Zone Management Act ("CZMA") established a unique concept in Federalism, providing coastal states with substantial powers over Federal decisions and permits regarding oil and gas activities on the outer continental shelf. As has been stated by one of my colleagues, the resultant interrelationship between Federal agencies (Department of Interior, Corps of Engineers, Coast Guard and Environmental Protection Agency), the California Coastal Commission, and the oil and gas industry lies on the "frontiers of governmental processes." Most controversy and most industry anguish over coastal zone management program involvement in offshore oil and gas activities has been sparked by the consistency provisions.

Section 307(c)(3)(B) of the CZMA specifically provides for State review of activities described in detail in any OCS exploration or development plan which affect any land or water uses in the coastal zone for consistency with the State's Federally-approved coastal management program. If the state objects to any of the activities described in the OCS plan, and the objection is not overridden by the Secretary of Commerce, the Federal agency cannot issue the permit. If the State has not acted within six months, or has not notified the applicant of the status of review within three months, then concurrence with the applicant's consistency certification is conclusively presumed.

The National Oceanic and Atmospheric Administration ("NOAA") within the Department of Commerce approved the California Coastal Management Progeam ("CCMP"), as implemented by the California Coastal Commission, in November 1977. Due to litigation by the oil industry against NOAA's approval of the CCMP, however, the consistency provisions did not become effective in California until August 1978 when the Federal District Court upheld the Federal approval of the CCMP. See American Petroleum Institute v. Knecht, 456 F.Supp 889 (C.D.Cal.1978), aff'd. 609 F.2d 1306 (9th Cir.1979).

Despite the oil industry's contentions before the Courts, and later before the Congress in oversight hearings on the CZMA, that the consistency provisions would cause delay in the exploration for and development of offshore oil and gas resources, the Coastal Commission's record in processing consistency reviews indicates expeditious decisionmaking. In exercising its consistency review authority over exploration and development plans, the Coastal Commission has substantively reviewed 14 plans of exploration and 2 plans of development to date. Except for a few plans of exploration which presented particularly complex issues of protection of marine resources around the Santa Barbara Channel Islands, consistency action on plans of exploration has averaged approximately 10 weeks from receipt to Commission concurrence. Review of the first plan of development ever subject to the consistency provisions took six months, the allotted statutory period. */

Commission consistency procedures provide for consultation with applicants prior to submittal of their consistency certification to discuss the scope of consistency review and to flag potential problem issues which must be resolved. In several instances objections were avoided through applicant's agreement to revise its proposal to satisfy CCMP requirements. One such example was industry agreement to provide additional on-site oil spill containment equipment beyond that required by USGS. In another instance, a company agreed, at the last minute, to postpone exploratory drilling several months until the time when the concentrations of marine mammals in the area were at their lowest point. As oil and gas exploratory activity increases in the Santa Barbara Channel on tracts leased in Lease Sale 48, the potential for significant adverse effects on marine resources will increase and the possibility for objection exists. I remain cautiously optimistic, however, that most disagreements are capable of resolution through negotiation in such a manner as to lead to a consistency concurrence.

---

*/ As of the date of this paper (August 15, 1980), the Coastal Commission concurred in all plans of exploration and development which were substantively reviewed, and objected to only one plan of development based upon inadequate information. The plan was not acted upon by the Commission because the applicant failed to submit requested information regarding its oil spill contingency plan and air quality emissions. Because of the pendency of the six month time period after which concurrence is conclusively presumed, the Commission objected for lack of information. It was made clear that the objection was "based solely on the failure to submit the requisite information and does not presuppose the Commission's substantive decision on the proposed project...." The company is presently compiling the requested data and will resubmit the plan in the future.

By the time of the Coastal Zone 80 conference in November, Coastal Commission staff anticipates that at least 10 more plans of exploration will have been submitted for consistency review. One plan of exploration submitted by Chevron for an exploratory well in the Santa Barbara Channel, within six nautical miles of Anacapa Island and within the buffer zone of a vessel traffic lane, has received a staff recommendation for an objection based upon CCMP policies for protection of marine resources,and oil spills and navigation safety. If the Commission objects to the applicant's consistency certification, I will prepare an addendum to this paper for distribution at the conference.

The consistency provisions have not worked smoothly in relation to the Department of Interior's lease sale activities. In the course of implementation of the consistency provisions, a serious disagreement between California and the Department of Interior,concerning the issue of whether the federal activity consistency provision of the CZMA applies to Interior's pre-leasing activities, led to the first mediation under the CZMA. Section 307(c)(1) requires that:

> Each Federal agency conducting or supporting activities directly affecting the coastal zone shall conduct or support those activities in a manner which is, to the maximum extent practicable, consistent with approved state management programs.

Simply stated, the Coastal Commission believes that the final notice of lease sale, which contains the selection of tracts to be offered and the lease stipulations to be imposed, is the key Federal planning decision on the OCS which directly affects the coastal zone. The Department of Interior has historically opposed the application of the Section 307(c)(1) requirements to pre-leasing activities. Both parties were unable to resolve the disagreement during the course of the mediation proceeding over the meaning of the term "direct affect." In his final report to the Secretary of Commerce, the General Counsel of that department, who served as the mediator in the proceeding stated that:

> In my opinion, the law contemplates that pre-lease sale activities as in California be subject to a consistency determination, and moreover, that this would not impose an undue burden in a situation such as this one. The CZM program further contemplates that state plans be accorded substantial respect where consistency is at issue. DOI has, however, adopted an extremely narrow and restrictive definition of "direct affect."

The Secretary of Interior has instructed NOAA to promulgate regulations to resolve the disagreement over the "direct affect" issue and has stated that rulemaking will be guided by the General Counsel's opinion. A full discussion of the legal and policy issues presented by both Interior and California in the dispute could easily occupy a paper (possibly a volume) of its own and is beyond the scope of this paper.

This paper will explore three examples of Coastal Commission actions under the consistency provisions of the CZMA related to oil and gas activities offshore California:

-- a Chevron plan of exploration which proposed exploratory drilling within six nautical miles of one of the Santa Barbara Channel Islands; this plan violated the CCMP policy for protection of marine resources and invoked the CCMP provisions for coastal dependent industrial development;

-- a Union Oil plan of development for a platform in the Santa Barbara Channel, the first plan of development ever reviewed under the CZMA. The plan involved a consideration of the scope of consistency review as related to onshore processing facilities; and

-- the Commission procedures for multiple permit review of OCS exploratory and development activities, including EPA's NPDES discharge permits and Corps of Engineers permits for placement of structures in navigable waters.

As will be demonstrated by these three case studies, the Coastal Commission has learned a great deal during these past two years it has been on the "frontiers" of implementing consistency review over OCS activities. The level of understanding of the process and the sophistication as to policies on the part of the Coastal Commission, the Federal agencies involved, and the oil and gas industry applicants has grown and holds considerable promise to accommodate the state's very legitimate concerns over offshore drilling and its effect upon coastal resources. The case studies discussed in this paper are meant to provide the reader with examples of California's experience in implementing the consistency provisions; observations expresses herein reflect the views of the author and may not necessarily reflect the views of the California Coastal Commission. The examples provided in this paper should serve to demonstrate that, with the possible exception of the serious disagreement with Interior over the lease sale consistency issue, the consistency provisions of the CZMA work precisely as intended by Congress and provide the States with effective input into and decisionmaking authority over oil and gas activities on the OCS.

Chevron Plan of Exploration - Protection of Marine Resources Around the Santa Barbara Channel Islands

The following plan of exploration involves the proposed siting of an exploratory well in a sensitive marine resource area. The review illustrates the Coastal Commission's process for balancing of the national interest in oil and gas development with the requirement for protection of marine resources, thereby providing a classic example of the tensions inherent in energy facility siting in sensitive areas. The following discussion provides a representative, but highly complex, example of the Coastal Commission's experience in reviewing plans of exploration for consistency with the CCMP.

In early 1979, Chevron USA, Inc. submitted a plan of exploration which proposed to drill one exploratory well on the Santa Rosa-Cortez ridge approximately 4.3 nautical miles south of Santa Rosa Island, one of the Santa Barbara Channel Islands. (see map of the Santa Barbara Channel area; Figure 1). The Coastal Commission has a long-standing policy, based upon the CCMP policies for protection of marine resources, that the marine mammals and seabirds on and around the Channel Islands must be protected. This policy has been articulated in lease sale recommendations, in marine sanctuary recommendations, and in decisions on

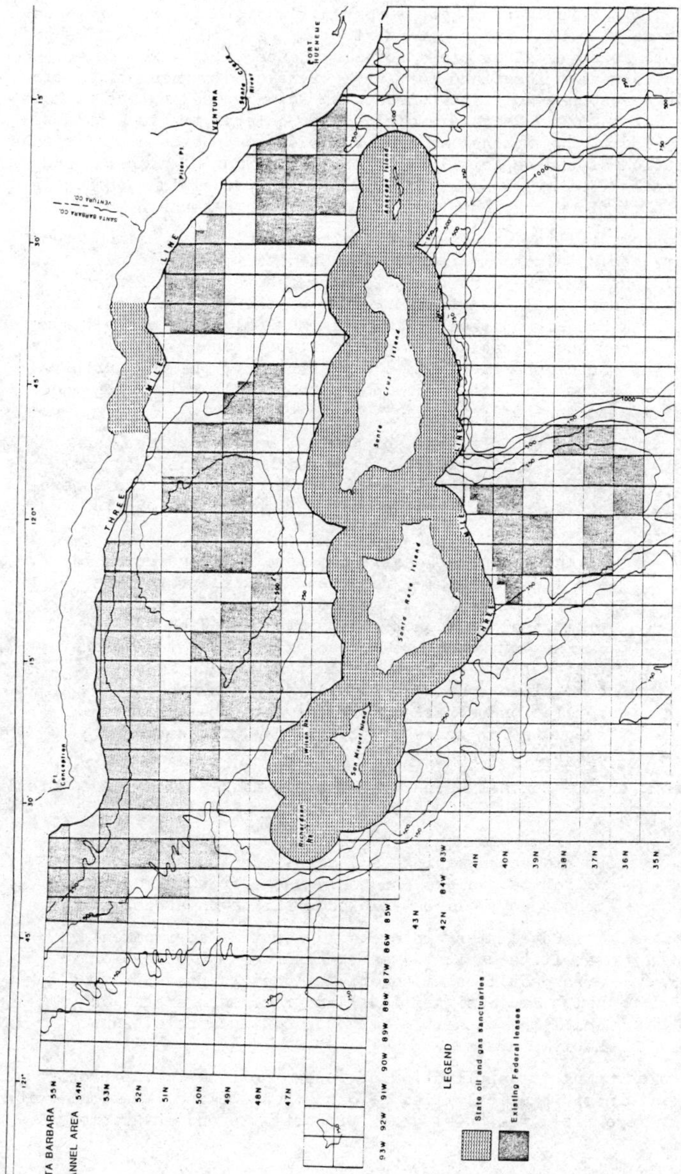
SANTA BARBARA CHANNEL AREA
LEGEND
State oil and gas sanctuaries
Existing Federal leases
THREE MILE LINE
Anacapa Island
Santa Cruz Island
Santa Rosa Island
San Miguel Island
VENTURA
PORT HUENEME
Santa Clara River
SANTA BARBARA CO.
VENTURA CO.
Pt. Conception

Figure 1

siting of a liquefied natural gas terminal. Accordingly, any proposed oil or gas development within six nautical miles of the Channel Islands and offshore rocks receives close scrutiny, especially with respect to possible alternative locations for the activity and maximum mitigation measures. In response to state comments on the proposed notice of sale for Lease Sale 48, the Secretary of Interior deleted tracts within six nautical miles of the Channel Islands in order to provide a buffer zone for protection of the especially diverse populations of seabirds and marine mammals from oil and gas activities. The parcel in question in this plan of exploration, however, was purchased in Lease Sale 35.

No prior drilling had occurred south of Santa Rosa Island, but preliminary geologic data indicated that the location proposed for the exploratory well had the most potential for recoverable reserves of natural gas. Chevron claimed that movement of the proposed well location to a site at least six nautical miles from the island in order to protect marine resources would preclude adequate exploration of the area. Since this location was deemed to be their best prospect, Chevron further claimed that if the well were a dry hole, they would relinquish their five other leases in the area.

The biological productivity and habitat values of the Channel Islands are protected by two CCMP policies with which oil and gas exploration or development activities must be consistent. The Coastal Act requires specific protection for marine resources in Section 30230 which states:

> Marine resources shall be maintained, enhanced, and where feasible, restored. Special protection shall be given to areas and species of special biological or economic significance. Uses of the marine environment shall be carried out in a manner that will sustain the biological productivity of coastal waters and that will maintain healthy populations of all species of marine organisms adequate for long-term commercial, recreational, scientific, and educational purposes.

In addition, environmentally sensitive habitat areas are afforded special protection under Section 30240 which states:

> (a) Environmentally sensitive habitat areas shall be protected against any significant disruption of habitat values, and only uses dependent on such resources shall be allowed within such areas.
>
> (b) Development in areas adjacent to environmentally sensitive habitat areas and parks and recreation areas shall be sited and designed to prevent impacts which would significantly degrade such areas, and shall be compatible with the continuance of such habitat areas.

In discussing the application of these sections to the proposed drilling within six nautical miles of Santa Rosa Island, the Commission explicitly recognized the "biological productivity and environmental sensitivity of the marine resources that thrive in the Channel Islands environment." The Commission noted that Environmental Impact Statements for Lease Sales 35 and 48 and the proposed Channel Islands Marine Sanctuary have documented the fact that marine mammals and seabirds use the Channel Islands as resting and breeding areas and feed in surrounding waters. There is ample evidence that the island related marine environment, especially within six nautical miles of the islands, qualifies as environmentally sensitive habitat area. The specific breeding and feeding areas in the southern Santa Rosa Island area were reported to the Commission in a staff summary which accompanied recommended findings.

The staff summary noted that several State and Federal agencies, in addition to the Coastal Commission, support the six nautical mile distance as a minimum buffer for protection of resources on and around the islands from oil and gas activities. The major oil spill related reasons for the buffer include provision of increased response time for oil spill cleanup efforts, and increased distance between spill points (e.g., drillships and platforms) and sensitive resource areas to allow for weathering and dilution before the spilled oil reaches concentrations of marine mammals and seabirds.

Accordingly, the Commission found that because the proposed well was to be drilled within six nautical miles of Santa Rosa Island, it did not meet the requirements of Sections 30230 and 30240 of the CCMP for protection of marine resources and environmentally sensitive habitat areas. Nonetheless, the inquiry concerning consistency with the CCMP did not end with this finding. OCS exploratory drilling qualifies for further review as a coastal dependent industrial facility under Section 30260 of the CCMP since it "requires a site on, or adjacent to, the sea to be able to function at all." Section 30260 provides that:

> Coastal-dependent industrial facilities shall be encouraged to locate or expand within existing sites and shall be permitted reasonable long-term growth where consistent with this division. However, where new or expanded coastal-dependent industrial facilities cannot feasibly be accommodated consistent with other policies of this division, they may nonetheless be permitted in accordance with this section and Sections 30261 and 30262 if (1) alternative locations are infeasible or more environmentally damaging; (2) to do otherwise would adversely affect the public welfare; and (3) adverse environmental effects are mitigated to the maximum extent feasible.

The Coastal Commission approached its analysis of the three factors to be considered under Section 30260 within the framework of a Commission recommendation to NOAA on the proposed Channel Islands Marine Sanctuary. NOAA's proposal to establish the sanctuary would preclude future leasing within six nautical miles of the Channel Islands, but would allow exploration and development on tracts within the sanctuary which had been leased under prior lease sales. Pre-existing leases on 15 tracts were located partially or entirely within six nautical miles of the

Channel Islands.

In its recommendation to NOAA, the Commission developed a policy for reviewing proposed exploration within the six nautical mile buffer zone. The policy provides that the tract must first be explored from adjacent leased areas outside of the six mile buffer zone. If such exploration indicated the likelihood of an oil or gas field extending underneath the buffer zone, exploration to delineate the size of the field would be permitted inside the zone. In the event, however, that the applicant demonstrated with geophysical data that the most favorable hydrocarbon bearing structure can only be explored from within the buffer zone, such exploratory drilling may be permitted provided maximum feasible mitigation measures were taken.

During the consistency review of the proposed exploratory well, the Commission addressed each factor under Section 30260 in turn.

(1) Alternative locations. The Commission considered two alternative locations to that proposed by Chevron: drilling of several wells outside the buffer area, or drilling of a directional exploratory well from outside the buffer area into the most-favored potential gas bearing region. Concerning the first alternative, Chevron testified that they will not drill in other areas since that would constitute an inefficient method of exploration when they believe it is absolutely necessary to drill in the proposed location. As to the second alternative, drilling contractors indicated that slant drilling from a floating drillship is more hazardous that similar drilling from a platform. Chevron presented proprietary geophysical data to the Coastal Commission staff geologist which demonstrated that the proposed well location was necessary to determine the existence of natural gas in the region south of Santa Rosa Island. The Commission found that "alternative locations to this well are infeasible at this stage of exploration and under certain circumstances have the potential for greater environmental damage." In addition, the proposed drilling location should provide better data on gas resources than the two proposed alternatives. The Commission further found that the "information will be important in determining how development/ production platforms can be located outside the 6 nautical mile buffer area consistent with CCMP policies in the event gas is discovered."

(2) Public Welfare. Through analysis of this factor in the balancing process under section 30260, the Coastal Commission considers the national interest in energy facility siting. The Commission found that "an objection to this Plan of Exploration would adversely affect the public welfare." Since there had been no previous drilling south of Santa Rosa Island and the well is proposed to be drilled on the most favorable geologic structure for a gas find, the Commission found that "to prevent this well would adversely affect the public welfare because of the resource information it will provide...."

(3) Mitigation of adverse environmental effects. This factor proved to be the most difficult point for resolution and provided some high drama at the eleventh hour during the Commission's public hearing on this plan of exploration. The primary impact from exploratory drilling activities upon marine resources identified in the Commission findings is the potential for oil spills. Although geologic information for the area indicated that any hydrocarbons present were probably gas and not

oil, the area had not been subjected to drilling before. In the event oil was discovered and a spill were to occur, Chevron agreed to the additional on-site oil spill containment equipment which had been established in previous Commission consistency determinations for plans of exploration. While the discussion of oil spill trajectories in the Lease Sale 48 EIS indicates low probability of oil movement toward the island, marine mammals and seabirds which feed in the open ocean could be impacted even if oil doesnot reach the island.

In seeking maximum feasible mitigation of the potential impacts to marine mammals and seabirds, the Coastal Commission staff consulted University of California-Santa Cruz studies prepared for BLM which define seasonal breeding, resting and feeding patterns of marine mammals and seabirds in the Southern California Bight. The studies indicated that the period from March to mid-June, coincidentally the period proposed for drilling by Chevron, contained the greatest local activity along the southern shore of Santa Rosa Island, including harbor seal breeding and pupping and seabird nesting and feeding. Although the studies showed that the lowest concentrations of marine mammals and seabirds occurred between August and December, activities south of Santa Rosa Island are significantly reduced after mid-June. Chevron claimed that an extensive drilling delay would not provide adequate time to explore other tracts outside of the buffer zone if gas was found in this well because those parcels were acquired in Lease Sale 35 and were due to expire in January 1981.

Because of the fact that marine mammal and seabird activity is significantly reduced south of Santa Rosa Island after June 15, the Commission found that:

> Commencement of drilling on or after June 15 will provide the maximum feasible protection of these marine mammals and seabirds while at the same time allowing Chevron a full six months of exploratory drilling prior to their lease termination.

You are by now probably wondering where the "high drama" mentioned above makes its appearance. The Commission staff had attempted to secure Chevron's agreement to conduct drilling after June 15 before the Commission hearing on this matter. Chevron refused, preferring to carry its case to the Commission. In its recommendation to the Commission, staff included the following information:

> If Chevron does not agree to delay drilling until on or after June 15, the Commission should object to Chevron's consistency certification for this Plan of Exploration on the grounds that potential adverse environmental effects have not been mitigated to the maximum extent feasible and therefore the requirements of Section 30260 have not been met.

When all the public testimony was completed and the Commissioners had had discussion among themselves, one of the Commissioners gave Chevron one last opportunity to agree to postpone drilling until June 15 before she made her "motion for an objection." At that time, Chevron's representative approached the podium and announced that management had authorized him to agree to postpone drilling until June 15.

Because of the Commission's long-standing interest in protection of the marine resources located on and around the Santa Barbara Channel Islands, the Commission findings on this plan of exploration included a very clear statement of its policy regarding production platforms within the buffer zone:

> Concurrence with the Exploration Plan for this one well in no way reflects a lessening in the Commission's recognition of the need for protection of the six nautical mile ocean buffer around the Channel Islands. It is the policy of this Commission that an oil and gas production platform <u>cannot be located</u> within the buffer area of six nautical miles around the Santa Barbara Channel Islands and offshore rocks consistent with the California Coastal Management Program. A production platform within this six nautical mile area would represent an unacceptable disturbance to the sensitive marine resources surrounding the Channel Islands. ... Any exploration by the oil industry within the 6 nautical mile buffer area shall be done with the knowledge of this policy. Therefore, any exploration within the buffer area shall be done to determine the extent of oil or gas resources in the area and how such resources can be produced from outside of the buffer zone. (emphasis in original).

In summary, the Commission looks very carefully at proposed exploratory activities to be conducted in proximity to sensitive habitat areas. Where substantial impacts are presented, the Commission applies a balancing analysis to weigh the national interest in oil and gas exploration and development with policies for protection of marine resources. In this instance, the balance was struck in favor or exploration where alternative locations were infeasible and the applicant ultimately agreed to mitigation measures. In other instances, the balancing may result in an objection. The Commission utilized this consistency review to provide further policy guidance on the unacceptability of production platforms within the Channel Islands buffer zone to permit oil companies to plan methods for future development.

<u>Union Plan of Development - Related Onshore Processing Facilities</u>

In late 1979, the Coastal Commission conducted the first consistency review of a plan of development under the CZMA. Review of this plan raised the CCMP policies for consolidation of facilities to avoid proliferation of facilities along the coast. The consistency review provided a forum for Commission guidance to the applicant on alternatives to their proposed onshore processing facility. The following discussion demonstrates some procedural difficulties when the agency is faced with a piecemeal rather than a comprehensive approach to planning.

The project involved a proposal by Union Oil Company to construct a 15 slot platform (Platform "Gina") in the Hueneme field 4½ miles offshore. The plan also proposed construction of pipelines from the platform to a proposed onshore processing facility. Staff was faced with the option of focusing consistency review only on the offshore activities such as platform placement and method of transportation,or of including associated onshore facilities in the Commission's review.

At the time of consistency review of Union's plan of development, an Environmental Impact Report for the proposed onshore facility was still in the preparatory stages under the lead of the City of Oxnard. Staff recommended that because a future coastal development permit would be required for both the onshore processing facility and the portion of the pipelines crossing state waters, consistency review should be limited to the Federal permit activities on the OCS but should provide policy guidance to the applicant for the onshore facilities in the coastal zone. While the Coastal Commission concurred in the consistency determination for the platform, it expressly reserved the right to consider the sizing and location of the onshore processing facility and pipeline through state waters when reviewing the coastal development permit for such facilities.

Peak production from Platform Gina was expected to be approximately 6450 barrels of oil per day ("BPD"). The plan of development included the proposal to transport the crude oil and gas from the platform to the proposed onshore processing facility via 6.5 mile long pipelines. In addition, Union proposed an onshore processing facility capable of processing 36,000 BPD of crude oil at the Mandalay Dunes near the City of Oxnard. Given the amount of oil projected from Platform Gina production, the excess capacity proposed for the processing facility raised some questions. Upon further investigation, Coastal Commission staff ascertained that the onshore facility was intended to handle production from a future platform.

In the midst of review of Platform Gina, Union submitted a second plan of development for Platform Gilda in the Santa Clara Unit, 12 miles from Platform Gina. This piecemeal approach to planning had the effect of increasing the time necessary to analyze the project and provide adequate guidance to Union to avoid conflict with the CCMP. Although the Coastal Commission utilized the full six month time period for consistency review provided in the CZMA because the plan presented new consistency issues, the consistency concurrence was the first regulatory approval provided to Platform Gina.

The proposed Mandalay Dunes onshore processing facility was being sized to handle the production from Platform Gina as well as the expected production of 18-20,000 BPD from Platform Gilda; it could also accommodate production from a lease in state waters owned by Shell Oil adjacent to the Hueneme field. The Commission's consistency concurrence in the platform was based in part upon Union's agreement to provide one slot on Platform Gina for use by Shell in its development of its adjacent state lease, thereby consolidating facilities for the offshore portion of the project.

While the Coastal Commission did not render a decision on the location of the proposed onshore facilities, it used the consistency determination to provide guidance to Union for investigation of other alternatives which might be more consistent with Coastal Act policies favoring consolidation and mitigation of adverse environmental effects. The Coastal Commission sought adequate information for resolution of the onshore planning issue of whether the proposed processing facility was necessary or whether production could feasibly be accommodated at existing processing facilities. The applicable CCMP policies are as follows:

> Section 30260 provides that:
>
> Coastal-dependent industrial facilities shall be *encouraged to locate or expand within existing sites* and shall be permitted reasonable long-term growth where consistent with this division. However, where new or expanded coastal-dependent industrial facilities cannot feasibly be accommodated consistent with other policies of this division, they may nonetheless be permitted in accordance with this section and Sections 30261 and 30262 if (1) alternative locations are infeasible or more environmentally damaging; (2) to do otherwise would adversely affect the public welfare; and (3) adverse environmental effects are mitigated to the maximum extent feasible (emphasis added).
>
> Section 30262 states that:
>
> Oil and gas development shall be permitted in accordance with Section 30260, if the following conditions are met:
>
> (b) *New or expanded facilities related to such development are consolidated, to the maximum extent feasible* and legally permissible, unless consolidation will have adverse environmental consequences and will not significantly reduce the number of producing wells, support facilities, or sites required to produce the reservoir economically and with minimal environmental impacts.... (emphasis added).

In order to have information sufficient to make findings on the consolidation policies contained in these two provisions of the CCMP, the Commission suggested two onshore alternatives for analysis in the Environmental Impact Report: pipelines to the existing Rincon processing facility partially owned by Union, or to Chevron's Platform Grace in the Santa Clara Unit which ties into existing pipelines to the Carpinteria processing facility. (See enclosed maps for location of platforms and onshore processing alternatives (Figure 2), and alternative pipeline routes (Figure 3)). Among the questions which the Coastal Commission requested Union to address prior to deliberations on the onshore facility were: whether Union could process the oil and gas at other existing

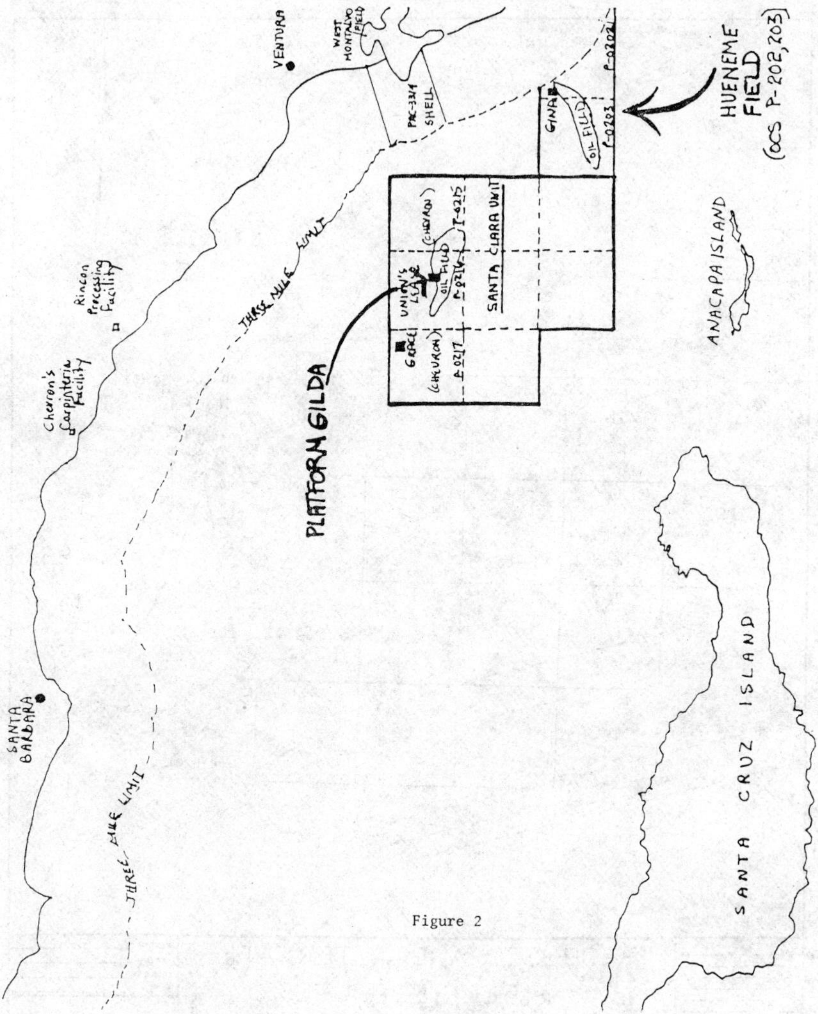
SANTA BARBARA
THREE MILE LIMIT
Chevron's Carpinteria Facility
Rincon Processing Facility
VENTURA
WEST MONTALVO FIELD
SHELL
PLATFORM GILDA
GRACE
(CHEVRON)
UNION'S LEASE
OIL FIELD
SANTA CLARA UNIT
GINA
OIL FIELD
HUENEME FIELD
(OCS P-202,203)
ANACAPA ISLAND
SANTA CRUZ ISLAND

Figure 2

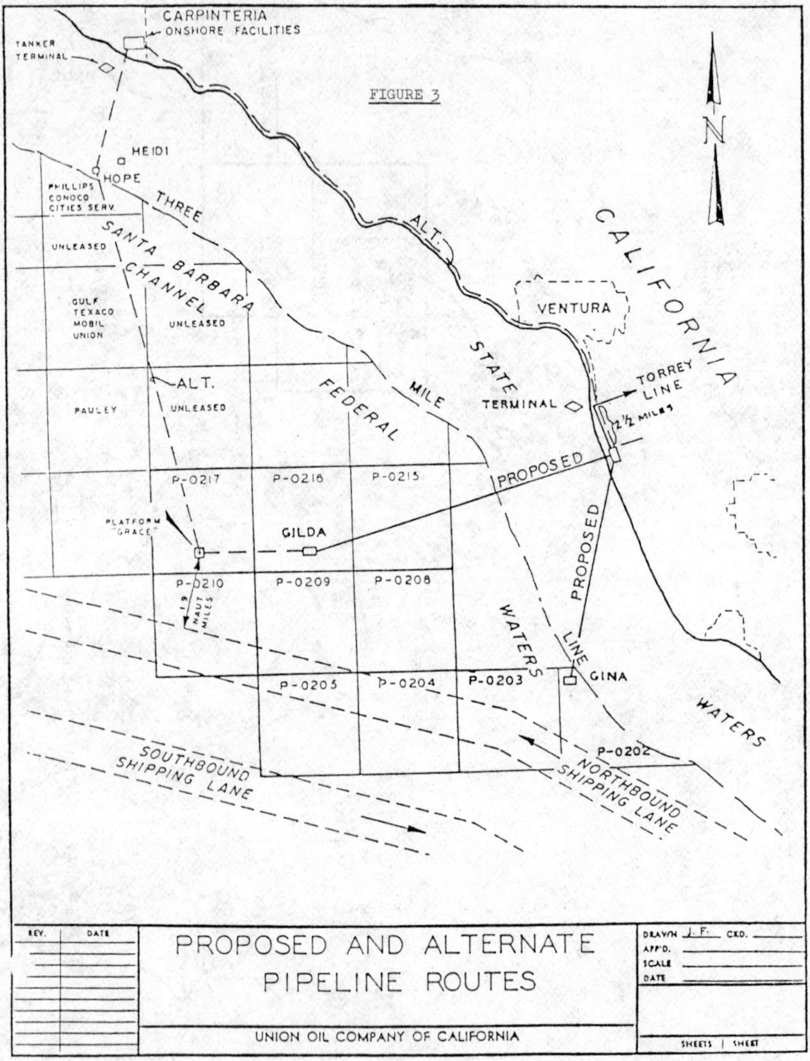
FIGURE 3
CARPINTERIA
ONSHORE FACILITIES
TANKER TERMINAL
HEIDI
HOPE
PHILLIPS
CONOCO
CITIES SERV.
UNLEASED
GULF
TEXACO
MOBIL
UNION
PAULEY
THREE
MILE
STATE
WATERS
SANTA BARBARA CHANNEL
FEDERAL
WATERS
CALIFORNIA
VENTURA
ALT.
ALT.
TERMINAL
TORREY LINE
2 1/2 MILES
PROPOSED
PROPOSED LINE
P-0217
P-0216
P-0215
PLATFORM "GRACE"
GILDA
P-0210
P-0209
P-0208
1.9 NAUT MILES
P-0205
P-0204
P-0203
GINA
P-0202
WATERS
SOUTHBOUND SHIPPING LANE
NORTHBOUND SHIPPING LANE
N
REV.
DATE
PROPOSED AND ALTERNATE PIPELINE ROUTES
UNION OIL COMPANY OF CALIFORNIA
DRAWN J. F.
CKD.
APP'D.
SCALE
DATE
SHEETS | SHEET

processing facilities such as Mobil's Rincon facility or Chevron's Carpinteria facility; whether the proposed new facility is necessary; and, if the new facility is needed, why it should be sized to accommodate production from the Santa Clara unit 12 miles away where Chevron is already installing pipelines to shore specifically sized to handle production from the entire Santa Clara Unit.

As has been mentioned above, Union also submitted a Plan of Development for proposed Platform Gilda in the Santa Clara Unit. Union proposed to run 9.9 mile long pipelines from this 90 slot platform to the Mandalay Dunes proposed onshore processing facility. Gilda is located on a parcel adjacent to Chevron's Platform Grace less than three miles away. The oil and gas pipelines from Chevron's Platform Grace tie in to Platform Hope in state waters which connects with existing lines to the Carpinteria processing facility. The Commission again subjected the Federal permit activities for installation and operation of the platform to consistency review, but reserved judgment on the onshore processing facility and portions of pipelines through state waters for analysis in the coastal development permit proceeding. The Commission concurred in the Federal permit activities for the offshore platform. Ten of the ninety slots on Platform Gilda were made available for use by Chevron in developing its adjacent lease in the Santa Clara Unit, thereby potentially eliminating the need for another Chevron platform. The Commission therefor found that "Union's provision of 10 slots in its platform for Chevron meets the consolidation policy of Section 30262(b) because it could eliminate a platform...."

Concerning the onshore processing facility planning, the City of Oxnard is the lead agency for preparation of the Environmental Impact Report and has combined the review of Platforms Gina and Gilda. Apparently because of Union's objections to alternatives, the Draft EIR included only a cursory analysis of the two processing alternatives suggested for consideration by the Coastal Commission. The Commission believed that this constituted a serious deficiency in the EIR and consulted with Union and Oxnard to secure full consideration of the alternatives. An addendum to the EIR is now under preparation to review those alternatives. The Commission does not have a preferred alternative, it merely seeks review of all available options. The Commission explicitly found that:

> it is not appropriate at this time for the Coastal Commission to select one alternative in consistency review among several in view of upcoming permit review by the Regional Commission.

The Commission did, however, request that:

> both economic and technical feasibility of these alternatives should be fully analyzed in the Final EIR to enable the Coastal Commission to have sufficient information available for its permit review on the proposed onshore processing facility.

In summary, this plan of development presented a timing problem inherent in the consistency review process. The requirement that a consistency determination be rendered by the State within six months appears to require concurrence in the offshore portion of the project before sufficient information is developed for adequate consideration of the onshore impacts. Accordingly, the review must be bifurcated. Perhaps these practical difficulties can be overcome if oil and gas companies engage in additional joint planning for onshore facilities and pipelines, and provide comprehensive information about future plans and alternatives to assist the State in planning for the siting of such facilities in the coastal zone in accord with the CCMP's policy for consolidation of facilities. The consistency review process, assuming that timing problems can be overcome, offers the potential for assuring that transportation and processing facilities for offshore oil are developed in a manner which maximizes their potential and avoids proliferation of unnecessary facilities.

Multiple Permit Review - Corps of Engineers Permits and EPA NPDES Permits

This section illustrates application of the Commission procedures for review of Federal permits for activities associated with OCS exploratory and development drilling. Included in the discussion are examples of Commission exclusion of activities from consistency review when it has been determined that the activities do not affect land or water uses in the coastal zone or will be deemed consistent if conducted according to specified guidelines.

In addition to review of the activities described in detail in plans of exploration or development for which USGS permits are issued, the CCMP listed EPA National Pollutant Discharge Elimination System ("NPDES") permits and Corps of Engineers permits for placement of structures in navigable waters as Federal permit activities which are subject to consistency review. In order to avoid separate review of the OCS plan, the Corps permit and the NPDES permit for discharges from drilling operations, the Commission's multiple review regulations "strongly encourage" applicants to include along with consistency certifications for activities described in detail in the OCS plans, consistency determinations for other OCS-related activities such as Corps and NPDES permits.

In the course of such review, the Commission addressed an interesting problem concerning the effects of discharges under NPDES permits in Federal waters upon the state's coastal zone. NPDES permits for exploratory drilling activity for tracts leased in Lease Sale 35 were grandfathered from consistency review. With the advent of Lease Sale 48 in June 1979,however, EPA proposed to modify all of the outstanding permits previously granted to dischargers from drillships for Lease Sale 35 tracts to include the Lease Sale 48 tracts. After these proposed modifications were brought to the Coastal Commission's attention by an environmental group during a public hearing on several plans of exploration, the Commission notified EPA that the modified NPDES permits could not be issued until the applicant had certified to EPA and the Commission that the activity was consistent with the CCMP and the Commission had concurred that the activity was indeed consistent.

There has been much controversy and a number of industry-sponsored studies concerning the effects of discharge of so-called "clean" (meaning oil free) drill cuttings and drilling muds from offshore drilling operations on marine life and water quality. The threshhold determination for the Coastal Commission was whether discharges of cuttings and muds in Federal waters would affect land or water uses in the coastal zone. The State's coastal zone extends seaward three miles from the mean high tide line.

Coastal Commission staff consulted with the staffs of the California Department of Fish & Game, the State and Regional Water Quality Control Boards, the State Lands Commission, and the EPA to determine the effects of discharges from exploratory drilling operations on the California coastal zone. The State Water Board staff advised that the discharge of drill cuttings and drilling muds from exploratory drillships would not have adverse consequences upon either marine life or water quality in state waters if the discharges occurred more than 1000 meters from state waters. This conclusion, concurred in by the other agencies consulted, was largely based upon indications that drill cuttings settle a comparatively short distance from the discharge point and that fine materials in the drilling muds disperse and dilute rapidly. Under the CCMP, the State and regional water boards have "primary responsibility" for water quality control (Coastal Act §30412(b)), and their technical determinations on water quality are therefore accepted by the Coastal Commission in conducting consistency reviews.

Following a public hearing on the matter, the Coastal Commission adopted the following resolution concerning consistency review of NPDES permits for exploratory drilling:

> The California Coastal Commission finds that individual activities conducted under an NPDES permit during exploratory drilling on the OCS beyond 1000 meters from the State's three nautical mile limit ("State waters") do not measurably affect the state's coastal zone although there may be long-run cumulative effects that may come under Environmental Protection Agency and Regional Water Quality Control Board regulations. Therefore, the Commission finds that no consistency review of discharge activities under the U.S. Environmental Protection Agency's NPDES permits is required when the exploratory drilling activity is beyond 1000 meters from State waters. The Commission further finds that applicants for NPDES permits which allow for discharges during exploratory drilling on the OCS need no longer submit a consistency certification when the proposed drilling will occur beyond 1000 meters from State waters. Such NPDES permits may therefore be issued by EPA without concurrence from the Coastal Commission.

The Commission made clear that any discharges from exploratory drilling activities within 1000 meters of State water would require a consistency review prior to issuance of the NPDES permit from EPA. After consultation with EPA as to implementation of the Coastal Commission policy, the NPDES permits for drillships for the tracts leased in Lease Sale 48 contain the following condition:

> This permit modification is effective 30 days from the date of issuance, provided, this permit modification shall not become effective with respect to any drilling within 1000 meters of waters of the State of California until the discharger provides EPA with a certification, concurred in by the California Coastal Commission, that the drilling is consistent with the approved State Coastal Zone Management Plan. (emphasis in permit).

The practical application of the Commission's 1000 meter NPDES determination, as well as the Commission's regulatory procedures for multiple permit review, are exemplified in a consistency certification for drilling of up to three exploratory wells by Chevron USA, Inc. approximately five nautical miles west of Point Conception in the Santa Barbara Channel. Chevron utilized the multiple permit procedure and submitted a certification that the activities described in detail in the exploratory plan for the parcel in question, the Corps of Engineers permit, and the NPDES permit comply with the CCMP and will be conducted in a manner consistent with such program. The Commission concurred in all three permit activites in one action.

Two of the three exploratory well sites proposed by Chevron were located within 1000 meters of the State's three mile jurisdictional limit and therefore came within the Commission's consistency review. Commission staff consulted with the staffs of the State and Regional Water Boards as well as the Department of Fish & Game. Since the proposed sites were not in proximity to state designated Areas of Special Biological Significance, those agencies had no objection to conduct of exploratory drilling at the proposed sites. Accordingly, the Coastal Commission concurred with Chevron's consistency certification for the NPDES-permitted discharges.

The Commission also concurred in the Corps permit to anchor the drillship in navigable waters at the specified locations as consistent with the CCMP since they were located outside the vessel traffic lanes and the 500 meter buffer zone surrounding those lanes. These lanes and buffer zones in the Santa Barbara Channel and San Pedro Bay were established by the 11th Coast Guard District for navigational safety reasons. The Commission has a long-standing policy which considers oil exploration or development activities within the vessel traffic lanes or within the 500 meter buffer zone thereof to be inconsistent with the CCMP policies and objectives for protecting against the spillage of crude oil and for navigation safety.

As of the date of this paper, the Commission staff has a recommendation before the Coastal Commission to exclude from further consistency review Corps of Engineers permits to conduct exploratory drilling on the OCS where the drilling activity would take place outside of the traffic lanes and buffer zones in the Santa Barbara Channel and San Pedro Bay. The Commission would continue to review the Corps permits, and the Coast Guard decisions rendered thereunder, for consistency with the CCMP for proposed drilling within either vessel traffic lanes or buffer zones. The Coast Guard is involved because the Corps grants the permits conditioned upon Coast Guard approval of locations within the traffic lanes or buffer zones. Accordingly, the Coast Guard decision is in effect the Federal action which would permit the placement of a drillship within a traffic lane or buffer zone and will be subject to consistency review by the Commission when the plan of exploration is reviewed.

A final example of multiple permit consistency review for a plan of development is provided by Union Oil's Platform Gilda discussed above. Development drilling of the 90 wells on Platform on Platform Gilda was estimated to occur over a four year period. Regarding consistency of the NPDES permit, the Commission found that:

> The resulting accumulation of muds and cuttings over this period could cover up to 75 acres under the platform but with a very slow rate of accretion. Some benthic communities located directly under the platform will be smothered but more mobile fauna can accommodate by recolonizing. The plumes from the discharges and traces of pollutants travel a maximum of 1000 meters in measurable quantities from the point of discharge. Because the platform will be located five miles from the coastal zone's 3-mile limit, the Commission finds that these discharges will have no effect on the coastal zone. Therefore, no consistency review of this NPDES permit is necessary.

Regarding the Corps permit for placement of the platform, the Commission found that since Platform Gilda was proposed to be located approximately 3½ miles from the vessel traffic lanes in the Santa Barbara Channel the platform would pose no significant hazard to navigation and was therefore consistent with the policies of the CCMP.

In summary, the Coastal Commission has attempted to formulate a process which permits applicants to secure consistency review of all necessary activities for OCS exploration or development at one time. Where further consistency review of a particular activity is not necessary in order to protect coastal resources, the Commission is more than willing to exclude such activities from review.

Conclusion

The California Coastal Commission's experience with consistency review of offshore oil and gas activities proposed in exploratory and development plans indicates that the consistency provisions of the CZMA provide States with the regulatory mechanism for assuring that State coastal management policies will be enforced and that coastal zone resources will be protected. The Coastal Commission procedure for balancing the national interest in development of offshore oil and gas with protection of coastal resources provides the framework for resolution of competing policies and interests. In upholding NOAA's approval of the California Coastal Management Program, the District Court noted that under the CZMA, consideration of the national interest in the siting of energy facilities must be based on a "balancing of those interests relative to the wise use, protection, and other development of the coastal zone." (456 F.Supp. 889,924).

The case studies presented in this paper demonstrate that workable solutions, mutually acceptable to industry and the State, are possible when all available alternatives are fully explored, maximum mitigation measures are adopted, and facilities are consolidated to the maximum extent practicable. We should be under no illusions, however, because as oil and gas activities increase offshore California and the potential for significant and cumulative adverse effects on coastal resources likewise increase, the balancing decisions will become more difficult and the potential for objection remains. I remain cautiously optimistic, however, that the balancing provisions of the CCMP will provide a mechanism for resolution of such difficulties.

FEDERAL CONSISTENCY - IS IT WORKING?

A PERSPECTIVE FROM THE PACIFIC NORTHWEST

BY

DONALD E. LAWYER [1/]

The Coastal Zone Management Act of 1972, as amended (CZMA), is a program where the coastal States, using Federal funds and criteria, may exercise considerable control over activities within their coastal zone, through implementation of a Coastal Zone Management Plan or Program. While the overall CZM Program is administered by the National Oceanic and Atmospheric Administration, Department of Commerce, other Federal responsibilities play an important part in a successfull CZM Program.

Section 307, Coordination and Cooperation (Federal Consistency), contains requirements for the coordination and cooperation among Federal agencies and between Federal and State agencies to achieve the purposes of the CZMA and Congressional declaration of policy. To the Federal agencies this Section is by far the most important Section of CZMA for it establishes the following requirements: (1) for coordination with them by the Secretary of Commerce, delegated to NOAA, on CZM functions and responsibilities (Section 307(a)); (2) for their views on a State's CZM Program (Section 307(b)); (3) for conducting their activities and development projects in the CZ to be consistent "...to the maximum extent practicable" to approved State programs (Section 307(c)(1) and (2)); (4) for Federal licenses or permits to be issued only after the State has concurred that the activities or project is consistent with the State's CZM Program (Section 307(c)(3)(A)); (5) for Outer Continental shelf activities to be consistent with the State's CZM Program (Section 307(c)(3)(B)); (6) for Federal assistance applications to be consistent with the State's CZM Program (Section 307(d)); (7) which do not in any way affect any requirements of the Federal Water Pollution Control Act, as amended and the Clean Air Act, as amended (Section 307 (f)); (8) for concurrence in the event a national land use program is enacted (Section 307(g)); and (9) procedures to mediate serious disagreements between any Federal agency and a Coastal State (Section 307 (h)).

1/ Donald E. Lawyer, MPA is Chief, Regulatory Functions, North Pacific Division, Corps of Engineers, P.O. Box 2870, Portland, Oregon 97208.

Direct Federal Activities (Section 307(c)(1) and (2).

While Federal lands are excluded from the State's coastal zone, Federal activities are still subject to consistency with the State's CZM Program. NOAA's regulations 15 CRF 930, in Subpart C discuss these requirements. Federal activities are defined to mean "...any function performed by or on behalf of a Federal agency in the exercise of its statutory responsibilities," this term does not include a Federal license or permit or granting of Federal assistance as these are defined separately. The term "significantly affecting the coastal zone" means a Federal action which "...causes significant (i) changes in the manner in which waters, lands or other coastal zone resources are used, (ii) limitations on the range of uses of coastal resources, or (iii) changes in the quality of coastal zone resources. The significance of the effect on the coastal zone shall be considered in terms of the primary, secondary and cumulative consequences of the activity." The term "consistent to the maximum extent practicable" means that Federal activities significantly effecting the coastal zone should be consistent with the State's CZM Program "...when such deviation is justified because some unforeseen circumstances arising after the approval of the management program present the Federal agency with a substantial obstacle that prevents complete adherence to the approved program." NOAA's 15 CFR 930.33 (b) is very specific in this regard and states in part "...Federal agencies shall consider all development projects within the coastal zone to be activities significantly affecting the coastal zone. All other types of activities within the coastal zone are subject to Federal agency review to determine whether they significantly affect the coastal zone."

Even a conservative interpretation of these consistency requirements for Federal activities indicates that the State's, once they have an approved management program, will be in a very strong position to influence the manner in which the Federal activity is accomplished or even if it its undertaken at all. Serious disagreements between a Federal agency and a State on consistency of a Federal action is resolved in one of the following methods: (1) both parties accepting the results of the Secretary of Commerce's mediation efforts; (2) use of Federal judicial action; or (3) the voluntary acceptance of either party's view by the other party.

This consistency requirement for Federal activities will increase, or initiate if none now exists, the coordination between Federal agencies and the State's on review of Federal activities.

Federal Permits and Licenses (Section 307(c)(3)(A).

Federal licenses and permits that are issued in the State's coastal zone must insure that the activity being licensed or permitted be consistent with the State's approved management program. Procedurally, this requires the applicant to provide to the Federal agency issuing the license or permit a consistency certification which describes whether or not the proposed activity will be conducted in a manner consistent to the State's management program. A copy of this consistency certification is then forwarded to the State together with other pertinent activity information for State concurrence. The State has six months to either concur or object to the consistency certification. If the State concurs in the consistency certification and so notifies the Federal agency, the Federal agency may approve the Federal license or permit. Notwithstanding State agency approval, the Federal permitting agency may deny approval of the Federal license or permit application (based on reasons other than Coastal Zone Management consistency certification). If the State objects to a consistency certification, it must notify the Federal agency and the applicant on the specific elements that are inconsistent and what alternatives might be available, also the State's objection must include information on the applicant's right of appeal to the Secretary of Commerce. In denying a Federal license or permit because of State's objection to the consistency certification, the Federal agency should hold that decision in abeyance until the applicant's rights of appeal as provided in the CZMA (Section 307(c)(3)(A)) and NOAA's regulations (15 CRF 930.120-930.134) have been exercised.

Typical types of Federal licenses and permits that would be considered under this Section 307(c)(3)(A) are:

Environmental Protection Agency.

1. Permits required under Section 402 and 405 of the Federal Water Pollution Control Act Amendment of 1972.

2. Permits and applications for reclassification of land areas under regulations for the prevention of significant deterioration (PSD) of air quality.

Corps of Engineers, US Army.

1. Permits required under Sections 10, 11, and 14 of the River and Harbor Act of 1899.

2. Permits required under Section 1 of the River and Harbor Act of 1902.

3. Permits required under Section 404 of the Clean Water Act of 1977.

4. Permits required under Section 103 of the Marine Protection, Research and Sanctuaries Act of 1972.

Bureau of Land Management, Department of the Interior.

1. Permits and licenses required for off-shore drilling and mining on public lands.

2. Plans for the exploration, development, and production from areas leased under OCS Lands Act (43 U.S.C. 1331 et seq).

U.S. Coast Guard, Department of Transportation.

1. Permits for construction of bridges and causeways under 33 U.S.C. 401, 4591-507, and 525-534.

2. Permits for deep-water ports under 33 U.S.C. 158 et seq.

Department of Energy.

1. Permits and licenses required for siting, construction and operation of nuclear power plants (formally by Nuclear Regulatory Commsission).

2. Permits and licenses required for power plant siting and transmission lines.

3. Permits and licenses required for interstate pipelines.

4. Licenses for construction and operation of hydroelectric power plants.

5. Permits for construction and operation of facilities needed to import or export natural gas. (Note: items 2-5 were formally issued by the Federal Power Commission).

This Section 307(c)(3)(A) gives the States a great deal of clout covering non-Federal license or permit activities. For without the State's approval the required activity cannot be licensed or permitted unless the State's objection is overruled by the Secretary of Commerce. This can be done if he/she finds the activity to be consistent with the objectives of the CZMA or if he/she believes that the license or permit is necessary in the interest of national security. It is conceivable that the current environmental emphasis of no or minimal development in the coastal zone could be the factor that would cause the Secretary of Commerce to override a State's objection. This could be based on the State's failure to give full consideration "to the needs for economic development". If this is true, the Secretary would have the opportunity to determine the ultimate balance between ecological, cultural, historic, and esthetic values and pressures for economic development within the State's coastal zone.

The Pacific Northwest Programs

The three coastal zone States in the Pacific Northwest all have approved CZM programs. Washington was the first State program approved by NOAA in 1976 followed by Oregon in 1977 and Alaska in 1979. Close coordination has been maintained with each of these States and their specific office responsible for the State CZM programs, so that the requirements of the CZM program and of the Corps of Engineers' permit program could be dove-tailed into an easily understood and managable program.

The following are abbreviated procedures currently in place and operating for the States of Alaska, Oregon and Washington. As Washington was the first approved State program in the North Pacific Division, it has serviced as the proving grounds for Federal Consistency and the Corps of Engineers' permit program.

Corps of Engineers:

1. Upon receipt of a DA permit application, the Seattle District will ascertain if the proposed work is located within the Washington Coastal Zone (WCZ).

2. For those projects within the WCZ, Seattle District will send a pre-printed "Consistency Certification" (Inclosure 1) to the applicant for signature prior to issuing the Public Notice on the proposed work.

3. Upon receipt of the signed "Consistency Certification", the Seattle District will: (a) Include a statement in the Public Notice that the State of Washington is reviewing the proposed work for consistency with the approved WCZ Management Program; and (b) Attach to the Corps' Public Notice a "Notice of Application for Certification of Consistency: (Inclosure 2). If the proposed work also requires a water quality certification (Section 401 of the Clean Water Act of 1977), it is included in the same State notice (Inclosure 3).

4. Seattle District, in addition to the Public Notice, will provide a copy of the "Consistency Certification" to the responsible State agency.

5. The decision on whether to issue or deny the DA permit will depend upon concurrence or nonconcurrence on the "Consistency Certification" from the State of Washington.

State of Washington:

1. Upon receipt of the Corps' Public Notice and "Consistency Certification", the State will review the proposed project for consistency with the approved WCZ Management Program, including any required coordination with other State and local agencies.

2. After review of the proposed project, provide comments to the Seattle District on concurrence or nonconcurrence with the "Consistency Certification":

a. If the Washington State decision is concurrence, the Seattle District will process the permit application to decision.

b. If the State decision is nonconcurrence, the District Engineer shall deny the permit after the applicant's rights of appeal have run their course.

Inclosures 4 and 5 are the public notices for the states of Oregon and Alaska. The procedures followed in each of these states is similar to that described for Washington.

CZM Impacts on Corps' Permit Program.

In all three states (Alaska, Washington and Oregon) full implementation of CZM programs will have positive effects upon the Corps' permit program. All three states have CZM programs that are implemented by local g overnments based upon state determined criteria or goals with the state providing a review and ajudacary role. This will change as local governments prepare better CZM programs and gain experience with the program.

At the present time, a preliminary analysis of our permit program indicates that about 60% of the permit applications in process require CZM certification. This means that of the 2600 permit applications that we will be processing this year, approximately 1560 will require CZM certification. About 450 of this 1560 will be cancelled or withdrawn due to the applicant's inability to obtain other necessary atuhorizations or resolution of objections to the proposed work. The remaining 1110 will be processed to decision.

What are the impacts on the Corps; permit program. The most obvisious is delay in reaching a decision, for the State has up to six months in which to make a decision on consistency. Also it places another public interest factor into the complex public interest review process used in reaching permit decisions. But most importantly, it places decisions for coastal zone resources in the hands of local and State governments – the ones that are closest to the resources and ones that are in effect responsible for wise and proper utilization and/or preservation of the valuable coastal zone. The greatest impact is that Federal decision on specific projects benefit from an informed, comprehensive decision on consistency that follows preparation of a balanced and approved plan and criteria. The CZM process also increases opportunity for comments from the general public.

What used to be "business as usual" now results in inquiry and consistency on processing of proposed work and activity in the coastal zone. Federal consistency in the Pacific Northwest is working well.

Certification of Consistency with
the Washington Coastal Zone Management Program

Re: Department of the Army Permit
Application No. ___________
Applicant: (NAME)

Section 307(c)(3) of the Coastal Zone Management Act of 1972 as amended by 16 USC 1456(c)(3) requires the applicant for a Federal permit to conduct an activity affecting land or water uses in the Washington coastal zone to provide certification that the activity will comply with the Washington Coastal Zone Management Program.

The proposed activity described in your permit application is within the 15 coastal counties in Washington and the following certification is required before your application can be processed to Public Notice. Upon receipt of the signed, dated certification, the Public Notice will be issued and will include the certification statement. The Public Notice will be forwarded to Washington Department of Ecology for its concurrence or objection.

CERTIFICATION STATEMENT

I certify that, to the best of my knowledge and belief, the proposed activity described in Department of the Army Application No. _________ complies with the approved Washington Coastal Zone Management Program, and will be conducted in a manner consistent with such program.

____
Signature of Applicant

____
Date

Incl 1

STATE OF WASHINGTON
Dixy Lee Ray
Governor

DEPARTMENT OF ECOLOGY
Olympia, Washington 98504 206/753-2800
Mail Stop PV-11

27 NOV. 1979

STATE OF WASHINGTON
DEPARTMENT OF ECOLOGY

Notice of Application for
Certification of Consistency with the
Washington Coastal Zone Management Program

Notice is hereby given that a request is being filed with the Department of Ecology for concurrence, as provided in Section 307(c)(3) of the Coastal Zone Management Act of 1972, as amended (P.L. 94-370; 90 Stat. 1013; 16 U.S.C. 1456(c)(3)), that the project described in the Corps of Engineers Public Notice No. 071-OYB-1-006102, will comply with the Washington Coastal Zone Management Program and that the project will be conducted in a manner consistent with that Program.

Any person desiring to present views on considerations pertaining to the project's compliance or consistency with the Washington Coastal Zone Management Program may do so by providing his views in writing to the Department of Ecology, ATTN: Interagency Operations Section, Olympia, Washington 98504, within 20 days of publication of this notice.

Incl 2

STATE OF WASHINGTON

Dixy Lee Ray
Governor

DEPARTMENT OF ECOLOGY
Olympia, Washington 98504 206/753-2800

Mail Stop PV-11

STATE OF WASHINGTON
DEPARTMENT OF ECOLOGY

Notice of Application for
Water Quality Certification
and for
Certification of Consistency with the
Washington Coastal Zone Management Program

Notice is hereby given that a request is being filed with the Department of Ecology for certification, as provided in Section 401 of the Federal Water Pollution Control Act, that a proposed discharge to navigable waters, resulting from the project described in the Corps of Engineers Public Notice No. ________________, will comply with the applicable provisions of the Federal Water Pollution Control Act.

Notice is also given that a request is being filed with the Department of Ecology for concurrence, as provided in Section 307(c)(3) of the Coastal Zone Management Act of 1972, as amended (P.L. 94-370; 90 Stat. 1013; 16 U.S.C. 1456(c)(3)), that the above described project will comply with the Washington Coastal Zone Management Program and that the project will be conducted in a manner consistent with that Program.

Any person desiring to present views on either or both (1) water pollution considerations pertaining to the project or (2) considerations pertaining to the project's compliance or consistency with the Washington Coastal Zone Management Program may do so by providing his views in writing to the Department of Ecology, ATTN: Interagency Operations Section, Olympia, Washington 98504, within 20 days of publication of this notice.

Incl 3

ROBERT W. STRAUB
GOVERNOR

Department of Land Conservation and Development

1175 COURT STREET N.E., SALEM, OREGON 97310 PHONE (503) 378-4926

Notice of
Certification of Consistency with the
Oregon Coastal Management Program

Notice is hereby given that a certification has been filed with the Department of Land Conservation and Development, as provided in Section 307(c)(3) of the Coastal Zone Management Act, that the project described in the Corps of Engineers Public Notice No. 003492, will comply with the Oregon Coastal Management Program and that the project will be conducted in a manner consistent with that Program.

Any person desiring to present views or considerations pertaining to the project's compliance or consistency with the Oregon Coastal Management Program may do so by providing his views in writing to the Division of State Lands, 1445 State Street, Salem, OR 97310; within twenty days of publication of this notice.

REVIEW CRITERIA

Comments should address consistency with the applicable elements of Oregon's Coastal Management Program. These include:

-Acknowledged Local Comprehensive Plans
-Statewide Planning Goals
-Fill and Removal Law (when a state permit is required)

ADEQUATE INFORMATION?

A copy of the applicant's consistency certification and information supporting the certification are available through the Department of Land Conservation and Development. If you believe additional information is needed to make a decision on this application, please indicate this as soon as possible. Requests for additional information must describe why more information is needed to make the consistency decision.

If following a written request the applicant fails to provide information needed to make a consistency decision, the permit may be found inconsistent.

INCONSISTENT?

If you believe this project is inconsistent with the Oregon Coastal Management Program, you should list the relevant policy (i.e., goal requirement or plan policy) and explain briefly why you believe the project is inconsistent. You should also describe how the project could be modified to make it consistent, if this is possible.

BC:ka/MC
12/26/78

Incl 4

01-A3LH

STATE OF ALASKA

OFFICE OF THE GOVERNOR

DIVISION OF POLICY DEVELOPMENT AND PLANNING

JAY S. HAMMOND, Governor

Phone 465-3512

Pouch AD - Juneau 99811

STATE OF ALASKA

DIVISION OF POLICY DEVELOPMENT AND PLANNING

Notice of Application for
Certification of Consistency with the
Alaska Coastal Management Program

Notice is hereby given that a request is being filed with the Division of Policy Development and Planning for concurrence, as provided in Section 307 (c)(3) of the Coastal Zone Management Act of 1972, as amended [P.L. 94-370; 90 Stat. 1013; 16 U.S.C. 1456 (c)(3)], that the project described in the Corps of Engineers Public Notice No. 071-OYD-4-790406, will comply with the Alaska Coastal Management Program and that the project will be conducted in a manner consistent with that program.

Any person desiring to present views on considerations pertaining to the project's compliance or consistency with the Alaska Coastal Management Program may do so by providing his views in writing to the State Clearinghouse, Division of Policy Development and Planning, Office of the Governor, Pouch AD, Juneau, Alaska 99811, within 20 days of publication of this notice.

Incl 5

STRIKING THE BALANCE IN ENERGY PROJECT SITING

By

Richard A. Du Bey,*
Cheryl R. Koshuta,**
and
Daniel I. Steinborn***

I. INTRODUCTION

The heart of the coastal zone management concept is the federally approved state coastal management program. This state program consolidates the state's laws and policies relating to coastal energy development.1/ The management program must clearly set forth the state's environmental laws and policies for the siting of energy facilities in the coastal zone. If not, the program would be unenforceable with respect to energy facility siting.

In addition to the requirement that a state management program be clear and comprehensive, it should provide for a balance between future coastal development and protection of coastal resources.2/ As a result of the 1976 Amendments to the National Coastal Zone Management Act (CZMA) the management program must include an established planning mechanism for the siting of energy facilities in the coastal zone.3/ As will be discussed below, the energy siting element of the approved Washington Coastal Zone Management Program (Management Program) gives rise to a quasi-judicial siting process conducted by a state agency empowered to preempt state environmental laws. Thus, the balance in energy project siting is struck on a case-by-case basis.

----

* Washington State Bar; Attorney, Ziontz, Pirtle, Morisset, Ernstoff & Chestnut; Seattle, Washington.

** Washington State Bar; Attorney, Office of Regional Counsel, Region X, U.S. Environmental Protection Agency (EPA).

*** Energy and Coastal Zone Specialist, Environmental Evaluation Branch, Region X, EPA.

This paper was written by the authors in their private capacity. No official support or endorsement by the EPA or Ziontz, Pirtle, Morisset, Ernstoff & Chestnut is intended or should be inferred.

This paper will examine the State of Washington energy facility planning process, developed pursuant to Section 305(b)(8) of the CZMA and approved by the Federal Office of Coastal Zone Management in February, 1979. The focus of our discussion will be upon the siting of energy facilities subject to the broad jurisdiction of the Washington Energy Facility Site Evaluation Council (Energy Council).4/ In particular, the proposed Northern Tier Pipeline Company oil terminal and pipeline project presently under review by the Energy Council will be discussed.5/

One must recognize the uncertainties associated with energy siting under state coastal zone management programs. These uncertainities were discussed by District Court Judge Kelleher who in his 1978 review of the California coastal program stated:

> [U]nder our so-called federal system, the Congress is constitutionally empowered to launch programs the scope, impact, consequences and workability of which are largely unknown, at least to the Congress, at the time of enactment; <u>the federal bureaucracy is legally permitted to execute the congressional mandate with a high degree of befuddlement as long as it acts no more befuddled than the Congress must reasonably have anticipated</u>; if ultimate execution of the congressional mandate requires interaction between federal and state bureaucracy, the resultant maze is one of the prices required under the system. (Emphasis added).6/

It should be noted at the outset that it is not our intent to single out the State of Washington and its energy facility planning process as being uniquely inadequate. However, an examination of the Washington Management Program provides a reasonable yardstick for evaluating state as well as federal schemes which provide for energy facility siting in the coastal zone.7/

II. THE NCZMA AND ENERGY PROJECTS

A. The Institutional Framework: Problems and Objectives

When Congress enacted the CZMA it did so in direct response to the apparent inability of the existing institutional arrangements to handle the problems associated with conflict resolution in the nation's coastal zone. Specifically, the Congress found:

> In light of competing demands and the urgent need to protect and to give high priority to natural systems in the coastal zone, present state and local institutional arrangements for planning and regulating land and water uses in such areas are inadequate; and the key to more effective protection and use of the land and water resources of the coastal zone is to encourage states to exercise their full authority over the lands and waters in the coastal zone. (16 U.S.C. Sec. 1451(g)-(h)) (Emphasis added).

The state management programs delimit the geographical boundaries of a state's coastal zone and must reflect national as well as state concerns in order to achieve federal approval.8/ The national interest is a particularly important element in the siting of energy facilities because the regulations promulgated under the CZMA specifically include energy facilities as within the purview of the national interest.9/ The CZMA further provides that state programs must be developed so that local zoning does not unreasonably restrict or exclude land and water uses that are of regional benefit. Thus, in developing a management program, the state must consider local, regional and national interests.

The intent of the CZMA is to improve intergovernmental relationships and to insure that state coastal zone management programs are capable of balancing coastal ecosystem preservation against necessary coastal development. The language of Section 305 of the CZMA, concerning management techniques and the essential elements of an approvable program, is directed at insuring that proposed state programs are capable of attaining these statutory objectives.10/

B. Energy Facility Planning Process

Section 305 (b)(8) of the CZMA requires that each coastal zone management program shall include:

> A planning process for energy facilities which are likely to be located in, or which may significantly affect, the coastal zone, including, but not limited to, a process for anticipating and managing the impacts from such facilities. (Emphasis added).

The language of the CZMA and the regulations promulgated thereunder make it clear that Congress intended that the energy facility planning process serve to implement substantive siting criteria. Thus, the siting process was not designed to be simply a series of procedural hoops established to test the tenacity of project proponents. Rather, the energy facility planning process is to (1) identify likely facilities, (2) contain procedures for assessing site suitability, (3) clearly articulate state policies regarding siting conditions, and (4) contain or be supported by legal authorities adequate to assure effective implementation of these state policies.[11]

The integration of state policies regarding siting conditions is an essential part of the energy facility siting process provided for by the CZMA. The objective of the siting scheme is to ensure that energy facilities are subject to the requirements of the Management Program. Thus, an energy facility siting process which provides for the waiver of environmental laws and policies established under the Management Program clearly defeats the plain meaning of the statute. Moreover, the state laws and policies relevant to energy facility siting also affect the operability and utility of the federal consistency processes mandated by Section 307(c) of the Act.

C. Section 307 Federal Consistency Determinations

Siting of energy facilities in the coastal zone necessarily involves a decision as to the effect of the facility upon coastal resources. The states under the CZMA have an important role in this decision-making process by providing, through the operation of Section 307, for an intergovernmental coordination system. Section 307(3)(1) illustrates, in general, what Section 307 is intended to accomplish. This Section provides that:

> Each Federal agency conducting or supporting activities directly affecting the coastal zone shall conduct or

> support those activities in a manner which is, to the maximum extent practicable, consistent with approved state management programs.

Regulations to implement Section 307 incorporate an extensive coordination process between the state and the federal agency proposing an action which may affect the state's coastal zone.

The consistency regulations require that the federal agency consult with the state to ensure that the proposed federal action will be undertaken in a manner which is consistent with the state's coastal program. The state is required to concur in or object to a consistency determination regarding such a federal action. If the state objects to the consistency determination, the proposed federal action cannot be approved.[12] The state thus has substantial authority to essentially veto a federal permit. Also, the state wields additional power when reviewing the consistency certification because of its authority to impose conditions on the proposed project. Such conditions would be drafted to insure that the project would be consistent with the approved management program.

The consistency provisions provide a good indicator of the relative strength of the underlying state management program. In theory, the program should provide for an objective balancing of future development needs against protection of the affected ecosystem. With a weak management program, however, the consistency determination process amounts to a case-by-case balancing without the benefit of prior established standards.

Alone, the coordinating provisions of Section 307 cannot assure that important coastal resources will be adequately protected where the energy facility siting process is out of harmony with the management program. This exaltation of process over substance is the major deficiency in the federally-approved energy facility siting portion of the Washington Coastal Zone Management Program.

III. ENERGY FACILITY SITING IN THE STATE OF WASHINGTON

A. The Washington Energy Council

The original Washington Management Program was approved by the federal Office of Coastal Zone Management in June 1976. The energy facility planning section of the Program, prepared in accordance with the 1976 Amendments to the CZMA, was federally approved in March 1980. 13/ The energy facility siting portion of the Management Program sets out the inter-relationship of the Washington Energy Facility Site Evaluation Council (Energy Council) with existing state environmental and land use authorities and with federal CZMA requirements.

Under the Washington Management Program, energy facility siting decisions are the result of contested case hearings conducted by an administrative law judge on behalf of the Energy Council. During the course of the hearing, both the applicant and intervening parties present testimony regarding the proposed project. After considering all information presented at the hearing, the Energy Council makes a siting recommendation to the Governor in the form of a draft Site Certification Agreement. Within sixty days, the Governor must approve the draft Agreement, reject the draft Agreement or direct the Energy Council to reconsider certain aspects of the draft Agreement. 14/

As a matter of state law the Energy Council has sole jurisdiction over the siting of crude or refined petroleum facilities, refineries and transmission pipelines.15/ Under the Management Program energy facilities not under the jurisdiction of the Energy Council are subject to the jurisdiction of the Washington Department of Ecology. As we are here reviewing the proposed Northern Tier oil terminal and pipeline facility, this discussion will be limited to the energy siting process administered by the Energy Council.

In carrying out its role vis-a-vis energy facility siting, the Energy Council is vested with broad preemptive authority over inconsistent state and local law. During the contested case hearing, the various state and local laws which would otherwise be binding upon an energy facility are reviewed. Where a local land use plan or zoning ordinance is to be overridden, procedures designed to protect local interests must be followed.16/ However, in all other instances the Energy Council must:

> <u>[I]nclude conditions in the draft Certification Amendment</u>...<u>to protect state or local governmental</u> or community <u>interests</u> affected by the construction

> or operation of the energy facility, and conditions designed to recognize the purpose of laws or ordinances or rules or regulations promulgated thereunder, that are preempted or superseded. (RCW 80.50.100) (Emphasis added)

The preemptive authority of the Energy Council provides that the substance of state laws which would otherwise be binding upon a proposed energy facility must be "recognized," but that full compliance is not required.

It is the basic premise of this analysis that energy facility siting decisions should be made in accordance with pre-established substantive criteria. Only where such legislative non-site specific substantive criteria are employed can a proposed project be fairly evaluated in the context of a contested case hearing conducted by the Energy Council.

It would be to the benefit of each party involved in the decision-making process for the federally approved management program to incorporate such substantive siting criteria. Energy facility proponents would thus have clear notice of the substantive requirements of state law. The Energy Council in making its quasi-judicial siting decision would also have a specific set of regulatory tools with which to measure the level of compliance of the proposed project. Finally, the public would be in a position to monitor both the agencies' compliance with its own directives as well as the actions of the project proponent.

The unhappy alternative is an imprecise, case-by-case decisional process resulting in a power struggle among the energy facility proponent, the regulating body, and interested members of the public. Unfortunately, as is often the case in such struggles, the ultimate loser is the public. For the most part, it is the public who is unable to act effectively because of lack of adequate notice, lack of adequate information, or the absence of pre-established siting criteria. Further, the public bears the brunt of the high transaction costs, including significant attorney's fees and expert witness costs, associated with mounting an attack against such an ad hoc decision making process.

The Washington Management Program fails to meet a key mandate of the CZMA which demands that states develop a mechanism for exercising their full authority in the energy facility siting process. For example, with regard to energy facility siting in Washington, State program decisions balancing coastal development and coastal preservation are not made at the time the Management Program is developed or approved but rather are made as the need for a given project arises. The Energy Council's preemptive authority may serve to further weaken the energy siting criteria incorporated within the federally approved Management Program. The basic reason for this is that although the substance of the Management Program must be "recognized," the Site Certification Agreement developed by the Energy Council binds:

> the state and each of its departments, agencies, divisions, bureaus, commissions, boards, and political subdivisions, whether a member of the council or not to the approval of the site and the construction and operation of the proposed energy facility. (RCW 80.50.120).

The existence of enforceable standards is an important element of a management program. However, the standards which exist in the Washington Management Program may not be enforceable in light of the Energy Council's preemptive authority. Although the Council is required to include conditions in the draft Site Certification Agreement to "recognize the purpose of a law or regulation which is being preempted," this provides little comfort where litigation may be the sole mechanism to review the question of whether the Energy Council has properly exercised its discretion in a particular case.

Clear guidance regarding permissable land and water uses within the coastal zone is also an essential part of the state management program. Again, the Energy Council's preemptive authority may nullify this element of the Program. Under state law the Council must comply with statutory procedures in determining whether to recommend to the Governor that local land use plans or zoning ordinances be preempted for the siting of an energy facility.17/ However, the existence of such procedures serves to further emphasize the fact that each energy facility siting decision is made on a case-by-case basis. Parties are not accorded the benefits of pre-planning even though local zoning is designed specifically to plan for a community's future based on the needs and desires of that community. The adoption of local master programs under the Washington Management Program is also intended to provide for the local interest. Furthermore, federal approval of a state coastal zone management program should be based upon a demonstration by the state that the national interest has been considered in the development of the State Program.

Accordingly, since approval of a state coastal zone management program by the federal government should indicate that the national interest has been adequately considered, the national interest need not again be reviewed in the context of each subsequent energy facility siting decision. If the Energy Council's enabling legislation is considered to be a part of the Management Program, 18/ the Energy Council is essentially provided with carte blanche in regard to the siting of an energy facility in the State of Washington since all actions taken by the Council within the scope of its discretion would be *per se* consistent with the Management Program. Furthermore, if this line of reasoning is accepted, federal agencies acting under Section 307 of the CZMA would not be acting in a manner consistent with the substance of the approved Management Program, but rather in a manner consistent with the quasi-judicial actions of the Energy Council.

This result is already evident where a consistency determination must be made by a state agency pursuant to Section 307 of the CZMA. Because the Energy Council's Site Certification Agreement binds each state agency, (RCW 80.50.120), the Department of Ecology, which is charged with the administration of the Management Program, has no independent discretion regarding the issuance of a consistency determination for an energy facility within the jurisdiction of the Energy Council. In short, the Department of Ecology may not contradict the ultimate decision of the Governor following the Energy Council's substantive review of a given energy project. 19/

Here again the primary check on whether the substance of the Washington Management Program has been adequately considered by the Energy Council is through the development of conditions in the Site Certification Agreement. Because the Governor's decision regarding the Site Certification Agreement must be in hand before the consistency determination can be made by the Department of Ecology, the Department's consistency determination serves merely to rubber stamp the Energy Council's siting decision. This emasculation of the consistency provision serves to further emphasize the basic problem of the Washington energy facility siting program. This problem is the absence of pre-established substantive criteria which must be followed in energy facility siting and the substitution of a process through which a case-by-case decision is made.

As previously noted, state and local issues and concerns of the local and state elements relating to coastal energy siting may be brought out during the Energy Council's contested case hearing. In addition, the Management Program must be addressed by the Energy Council during its environmental review of the project pursuant to the state Environmental Policy Act. Integration of local concerns is accomplished by requiring that local governments be represented on the Energy Council. Individual citizens can also alert the Council to their coastal zone concerns through participation as an intervenor in the contested case hearing. However, although the framework for local and state plan integration is established by the Washington Management Program, the preemptory authority of the Energy Council serves to chill such participation.

B. The Northern Tier Project

In general, the lack of pre-established criteria in the energy siting process results in a very resource intensive evaluation process. An example of how the process has operated with regard to a particular siting decision is the present experience with the proposed Northern Tier facility in Washington State. The Northern Tier proposal consists of a crude oil receiving terminal in Washington State and a pipeline across the northern tier states of Washington, Idaho, Montana, North Dakota and Minnesota.

The Norther Tier Pipeline Company first submitted its application to the Energy council for a Site Certification Agreement in 1976. The contested case hearings began in January of 1980 and are presently not expected to conclude until late 1980 or early 1981. Concurrent with the Council's proceeding, Northern Tier is attempting to obtain all necessary federal permits for the proposed facility.20/

Environmental concerns have traditionally been very important in the State of Washington. In 1971, the State of Washington enacted its own shoreline management program and in 1976, Washington was the first state to receive final federal approval of its Coastal Zone Management Program uner the CZMA. However, the lack of definitive criteria in the energy siting portion of the Management Program has caused considerable delay in the Northern Tier siting process. In short, the energy siting process provides no baseline against which the merits of a particular energy project can be evaluated. Furthermore, in Northern Tier's direct testimony before the Energy Council, it has stated that the proposed facility may be inconsistent with both local zoning and the Washington Management Program.

The passage of the Public Utilities Regulation Policy Act of 1978 (PURPA) and the subsequent decision of the president to give Northern Tier expedited review under PURPA, emphasizes the keen national interest in the energy siting process and may provide further justification for overriding many of the provisions designed to protect local or regional interest. 21/ Providing adequate conditions in the Site Certification Agreement to protect competing state and local interests is difficult. It may thus be easier to ignore such interests and push onward with the siting of the facility presumably in the "national interest".

The lack of substantive siting criteria places a burden upon the applicant in his or her attempt to comply with the various requirements of the Management Program. Further, the Energy Council, is given broad discretion to site according to criteria which are largely unreviewable since any deviation from the norm could be explained away as being necessary in light of the site-specific requirements of a particular project. Moreover, the public is left in the dark as to what substantive criteria will be applied in the decision-making process.

Arguably, some measure of flexibility must be retained with regard to energy siting decisions since each project is unique and the national interest is a major consideration. However, the substitution of an energy facility siting process for a substantive planning program, leaves too much to chance or political expediancy with regard to the siting of energy facilities in the coastal zone. Thus, the balance in the energy facility siting decision may be forced by the strength of a party's political power base rather than through the implementation of a carefully developed program.

IV. CONCLUSION

It has been our thesis that in order for a state to effectively exercise its full authority over energy development in its coastal zone, there must be: (1) clear state decisions regarding the balance between coastal development and coastal preservation; (2) a clear framework for integrating the state and local programs; (3) clear decisions regarding permissible land and water uses; and (4) enforceable standards that are applicable to energy development projects.

Although one may argue that the existence of pre-established enforceable standards precludes arrival at an optimal solution by reducing the range of negotiability, this is easily countered when transaction costs and power structures are considered.22/ The high transaction costs in a case-by-case analysis and the often government-applicant topheavy power structure reduce optimal decision-making regardless of the theoretical negotiating freedom of each party.

Thus, it is necessary to consider the ways in which states can exercise their full authority within the coastal management program and reach meaningful decisions regarding the balance between coastal development and coastal ecosystem preservation. There are at least three possible solutions to this problem which are available and which could serve to mitigate the problems incident to a case-by-case decision making process.

First, state coastal management agencies could provide definitive guidance to local governments, project proponents, and federal permitting agencies regarding the application of standards which exist within their respective management programs. This guidance would specifically lay out the standards limiting the range of choices available to local planners especially as they relate to the siting or nonsiting of energy facilities. If such clarification of the program were to take place, it would reduce the uncertainty associated with the matter of energy facility siting under an approved program.

Second, where no management program standards governing energy siting exist and where the energy facility impact planning process is incapable of producing decisions by a date certain, the state could amend its management program. Such management program revisions may seek to develop: (1) specific standards regarding permissable land and water uses within its coastal zone on a habitat specific or area specific basis; (2) ecosystem quality standards regarding permissible uses for the various ecosystems in the coastal region; and (3) revisions to the energy facility planning processes aimed at reducing the amount of time necessary to make well-informed decisions which are in compliance with the standards of the management program.

Such efforts may well be characterized by controversy. However, the range of energy siting problems would be significantly more limited and more manageable within a framework where generic decisions have been made. The development of such a decision baseline would not eliminate the need for case-by-case decision making associated with project specific energy facility siting. It would, rather, reduce significantly the range of issues which need to be resolved for each case and make it substantially easier for officials at all levels of government, the facility proponent, and the public to determine what would be in the "interest" of all those affected by the decisions.

If the changes recommended above are not implemented in timely manner, there may exist a third alternative which merits consideration. Assuming that federal legislation in the nature of the proposed Priority Energy Project Act of 1980 [23/] is enacted energy siting decisions may be made at the federal level without regard to approved state management programs. Congress may also, independent of any federal energy act, consider amendments to the CZMA to force these program changes. Alternatively, the Office of Coastal Zone Management could initiate appropriate state program changes though its annual program reviews under Section 312 of the CZMA. In light of increasing national attention, it may be appropriate for the federal government to consider funding *only* for those state management programs which include substantive energy facility siting provisions.

In conclusion, those charged with responsibility for the siting of an energy facility in the coastal zone must fully consider the resource base at risk and the importance of striking a proper balance between development and preservation. The apparent passion for expediency with regard to energy siting today only serves to further emphasize the pressing need for carefully legislated substantive criteria to be used in the energy siting process.

1. 16 U.S.C. Secs. 1451-1464 (1976); Pub. L. No. 92-583 as amended by Pub. L. No. 94-370 (1976). All citations in the text are to the Public Law. The term "management program" includes, but is not limited to "a comprehensive statement in words, maps, illustrations, or other media of communication, prepared and adopted by the state in accordance with the provisions of this title, setting forth objectives, policies, and standards to guide public and private uses of lands and waters in the coastal zone." 16 U.S.C. Sec. 145(11) (1976).

2. States acting under such unenforceable programs will not be following the mandate of Congress which directs states to exercise their _full_ authority with regard to coastal zone decision-making. _See_, 16 U.S.C. Sec. 1453 (1976).

3. Pub. L. No. 94-370 (1976).

4. RCW 80.50 (1978).

5. The Final Environmental Impact Statement for the proposed Northern Tier Project summarizes the project as follows:

> Northern Tier Pipeline Co. proposes port and on-shore storage facilities at Port Angeles, Washington, and a 1,491-mile pipeline system extending to Clearbrook, Minnesota. Through put capabilities would be 709,000 barrels per day initially and 933,000 per day ultimately. The system would be designed to supply an additional 350,000 barrels per day to existing refineries in the Puget Sound area. (DOI, BLM, _Final Environmental Statement Crude Oil Transportation Systems_ at iii, Vol. 1 (1979).)

6. _American Petroleum Institute, et al. v. Robert W. Knecht, et al._, 456 F.Supp. 889, 931 (C.D. Cal. 1978).

7. The Second Session of the 96th Congress has been active in the area of fast track energy legislation. S.1308, the Priority Energy Project Act of 1980, was reported out of Conference Committee in late June of this year and then in the face of stiff opposition, promptly returned to Conference. Hopes are dim that any such legislation will be forthcoming this session of Congress. See, S.1308, 96th Cong., 2d Sess. (1980) and H.R. Rep. No.96-1119, 96th Cong., 2d Sess. (1980).

8. 16 U.S.C. Sec. 1455(c)(8) (1976).

9. 15 C.F.R. Sec. 923.15 (1977).

10. Section 305 of the CZMA, 16 U.S.C. Sec. 1455, sets out several program element requirements important to the matter of energy facility siting. Such provisions are found at Sec. 305(b)(2), (4), (5) and (8).

11. See 15 C.F.R. Secs. 920.18(a)(1)-(15) and 15 C.F.F. Secs. 923.11-923.14 (1978).

12. 15 C.F.R. Secs. 930, et seq. (1978).

13. 13 U.S.C. Sec. 1455(b)(8) (1976) and 15 C.F.R. Pt. 920 (1978).

14. RCW 80.50 (1978) and W.A.C. Title 463 (1979).

15. The statute provides for specific size limitations with regard to energy facilities with the scope of the Energy Council's jurisdiction. Other sources over which the Energy council has jurisdiction include floating thermal power plants, liquified petroleum gas facilities underground reservoirs for receipt and storage of natural gas, stationary thermal power plants, liquified natural gas facilities, natural gas transmission pipelines, and synthetic fuel gas, or liquified petroleum gas transmission pipelines. RCW 80.50.020 (1978).

16. W.A.C. Sec. 463-28 (1979).

17. W.A.C. Chapter 163-28 sets forth the procedures which the Energy Council must follow. See also, Attorney General Opinion No. 1, 1977, which states that the Energy council may override local land use and zoning ordinances by virtue of its preemptive authority.

18. The status of the enabling legislation is unclear although OCZM has indicated action may be taken this fall to specifically include the Energy Council's enabling legislation as part of the approved Washington Coastal Zone Management Program.

19. Department of Ecology Washington Coastal Zone Management Program Amendment, at 24 (February 1979).

20. Three independent law suits have been filed to date in the State of Washington challenging the various federal actions taken with regard to the siting of the Northern Tier facility. Among other claims made by the plaintiffs in these actions is that the proposed Northern Tier facility would violate the Washington Management Program.

21. 43 U.S.C. Sec. 2001 et seq. (1978) - See also the decision of the President under Title V of PURPA. 45 Fed. Reg. 6480-64826 (January 28,1980).

22. See generally, the Center for Technology Assessment Research Policy Energy Facility Siting Study (April, 1979). This study was prepared for the office of Coastal Zone Management, National Oceanic and Atmospheric Administration.

23. Supra, note 7.

THE EAST EVERGLADES : PRESERVATION VS. LAND USE

Richard Collier, Jr.
Senior Environmental Planner - Wallace, Roberts and Todd
Philadelphia, PA

INTRODUCTION

The East Everglades study area is approximately 240 square miles of largely undeveloped wetland within the Southern Zone of the Coastal Lowlands of the Florida Plateau. The study area is situated in western Dade County between the City of Miami and Everglades National Park. Because of increased regional demand for undeveloped land to meet residential, agricultural, and recreational needs, the East Everglades is under increasing pressure to accommodate development. As a result of the area's unique natural attributes, land use proposals are often found to be incompatible with environmental protection objectives and land use planning policies.

Since its inception, the East Everglades Resource Planning Study has been a cooperative effort on the part of Everglades National Park, Florida Department of Environmental Regulation, South Florida Water Management District, Dade County Department of Environmental Resources Management, and the coordinating agency, Dade County Planning Department. A Steering Committee represented by each of these agencies assumed the responsibility for preparing a strategy by which future conflicts may be rationally resolved. The ultimate product of this Study is a Management Plan designed to direct the extent and type of land uses permitted in the East Everglades while generally ensuring the protection of the area's unique features. To achieve these ends, the Study was organized into three major steps: 1) data collection, 2) interpretation, and 3) plan preparation. A multi-disciplinary team of scientific and technical consultants was assembled to collect and prepare a comprehensive environmental data base. The planning firm of Wallace, Roberts and Todd (WRT) was retained as a consultant to the Planning Department to conduct the interpretation step and assist in the preparation of the Management Plan. The information developed in the interpretation step provided the basis for the recommendations of the Management Plan currently under consideration by the Dade County Commission.

The purpose of this paper is to address the conflicts which arise between land use and preservation objectives. To understand the conflicts, one must first have a fundamental understanding and appreciation for the East Everglades environment. The first part of this paper will address the most important environmental features, processes, and interrelationships which characterize the East Everglades ecosystem. This description is largely based on the data collected and presented by the project's scientific and technical consultants (see References). Due to the limited length of the paper, this section will highlight environmental characteristics in the following subject areas: hydrology, vegetation, wildlife and land use. From this description it is evident that this once pristine wetland environment is a unique natural resource, one of local, regional and national importance, and one which warrants carefully designed strategies for its protection and use.

The second part of the paper will focus on the conflicts which arise from competing objectives of preservation, resource utilization, and land use. Prospective conflicts arising from existing and future agricultural, residential, and recreational land uses in the East Everglades are discussed. Resolution of these conflicts as proposed in the Management Plan are summarized in the conclusion.

ENVIRONMENTAL CHARACTERISTICS OF THE EAST EVERGLADES

Hydrologic System

In its intermediate location between the metropolitan fringe of Miami and Everglades National Park, the East Everglades occupies a portion of a vast wetland known as the Everglades. Historically, the Everglades extended some 100 miles south from Lake Okeechobee to Florida Bay and nearly 40 miles west from the Atlantic Coastal Ridge to the Big Cypress Swamp (Figure 1). The Everglades is characterized by an almost imperceptible slope, averaging less than two inches per mile. Under historic conditions, the Everglades was seasonally inundated, and drained slowly to the south (to Florida Bay) and east to Biscayne Bay (through the transverse glades). Shark River Slough, the major Everglades subbasin draining to the southwest cross-cuts the northwest portion of the East Everglades study area. A second major subbasin, Taylor

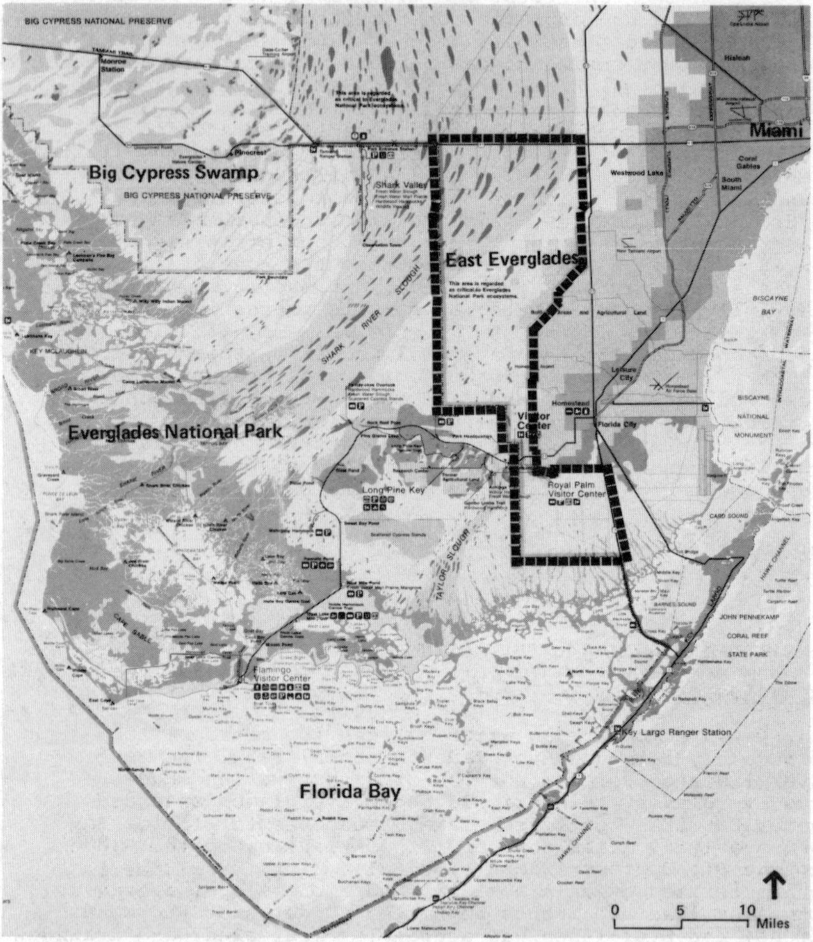

Figure 1 : Regional Context

Slough, drains to the south and occupies much of the central East Everglades study area.

Aside from the natural Everglades Basin, the major regional surface water feature is the water management system composed of a complex network of canals, flood gates, pump stations, levees, and impoundment areas. The initial purpose of the canal system which was to drain "useless" wetland, has since been replaced by the more environmentally sensitive objectives of supplying freshwater, and in some locations, regulating flooding. Except for the boundary between the Everglades National Park and the East Everglades, most of the study area is bounded by the canals of the South Dade Conveyance System. Operation of the canals and structures as determined by the South Florida Water Management District and the U.S. Army Corps of Engineers greatly affects the water regimen of the East Everglades. In the recent past, the most visible affect of the canal system has been the predominance of drier conditions (i.e., reduced hydroperiod) throughout much of the area which in turn cause changes in the biota to more mesic assemblages. The Water Management District is charged with the responsibility of maintaining flow through operation of canals to sustain the wetland character. Furthermore, the District is mandated by Congress to supply 315,000 acre feet of water to Everglades National Park through the East Everglades via the canal system and Northeast Shark River Slough.

Surface water in the East Everglades does not move in clearly defined channels but rather in a massive sheet (overland flow). The stage, rate, and volume of surface flow is greaest following storm events; its movement, however, being nearly imperceptable. It is this continuous south-southwesterly flow which inhibits saline intrusion from the Florida Bay estuary. Approximately half of the study area is inundated at the onset of the dry season. The nearly flat topography allows four diffuse subbasins to be distinguished, primarily during the dry season -- Northeast Shark River Slough, Taylor Slough, Rocky Glades, and Coastal Glades (Figure 2). Water entering the surface system comes from precipitation, canal seepage, and releases through gate structures. Aside from runoff, evapotranspiration accounts for much of the net loss of water from the system.

The groundwater and surface water systems are often indistinguishable from one another and operate as one unit throughout much of the year. In an average year, approximately half the study area has a water table within one foot of the ground surface for at least nine months of the year. Approximately 90% of the area experiences similar

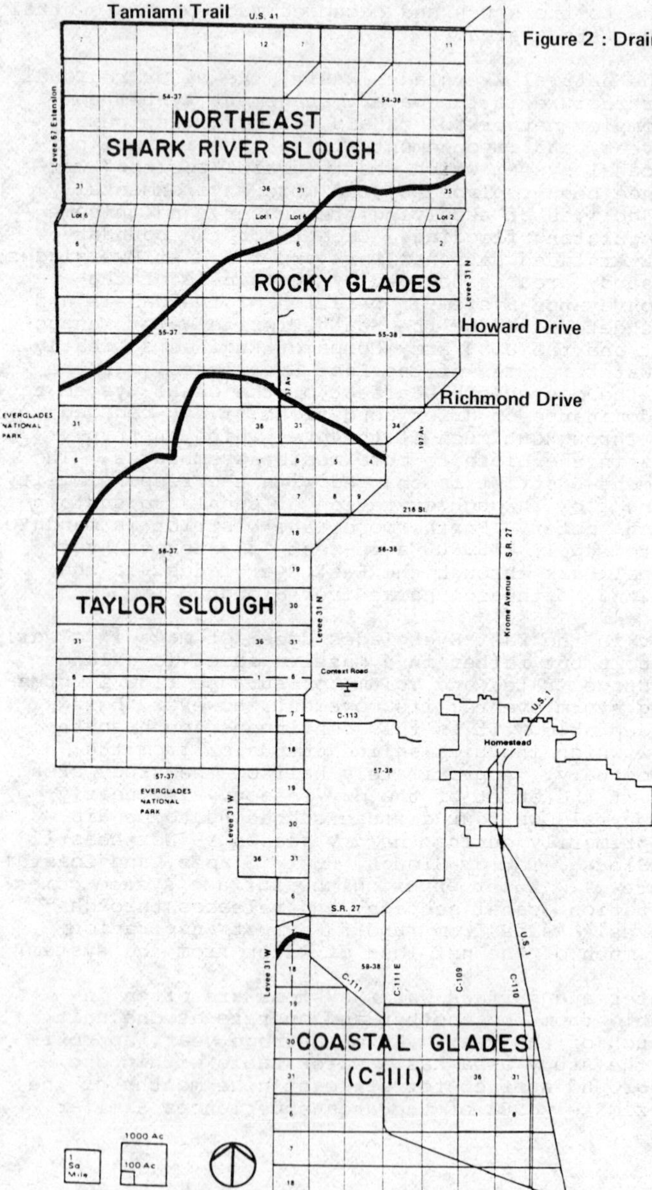

Figure 2 : Drainage Sub-basins

water levels for at least 36 days of the year. The dynamics and capacity of the groundwater system result from the characteristics of the Biscayne Aquifer. This aquifer is a highly permeable wedge-shaped formation of limestone, sand, clay, and shells ranging in thickness from 20 to 60 feet across the East Everglades (west to east). The aquifer has been designated by the EPA a sole source aquifer and is recognized as one of the most porous and productive aquifers in the world. Municipal wells in the region yield up to 7,000 gpm with minimal drawdown. Nearly all residents of south Florida rely on this aquifer for domestic and agricultural water supplies. The extensive volume of water stored in the system provides the hydraulic head which maintains the fresh/salt water contact zone and intrusion of saline water. Surface water recharges the groundwater system. When surface water quantity is reduced or the quality degraded, the ground water resources may be similarly affected. Such changes may threaten public water supplies and the health of all water-dependent features of the East Everglades.

Vegetative communities exist because of the flooding pattern and would be replaced by very different communities if the water regimen were changed to create either wetter or drier conditions. For example, tree hammock communities tolerate only occasional flooding and are not adapted to prolonged saturation associated with long-term flooding. On the other hand, wetland communities such as sawgrass marshes require standing water for at least 5-10 months each year. Organic soils such as peats exist because of the saturated conditions which persist in some places throughout much of the year. Peats which are permitted dry out are vulnerable to oxidation. Much of the seasonally flooded rockland tend to dry out as the water table drops during the dry season. Most of the diverse wildlife inhabiting the study area and adjacent areas of Everglades National Park depend on water and water-related features as its habitat. Many wading birds commonly observed in the park and the East Everglades feed on fish and other aquatic organism which inhabit areas of standing water. Many human activities depend on the amount, movement, and quality of water. Of particular concern is the availability of fresh water to meet domestic and irrigation demands in south Florida.

Vegetation

The East Everglades occupies part of three distinct physiographic provinces: Freshwater Swamps, Low Pinelands and Sloughs, and Tree Island Everglades (Craighead, 1971). The lower part of the area lies within the Saline Everglades (Egler, 1952). By definition, the East Everglades is wetland, most of which has been mapped in the Natural Wetland Inventory conducted by the U.S. Fish and Wildlife Service. All of the study area's wetlands are in the Palustrine System, with graminoid-, forested- and thicket/shrub-dominated assemblages.

Nearly fifty types of vegetation communities have been identified in the study area through extensive field investigation. Although distinct in character, similar communities can be grouped into the following broad categories:

Hammock Forest (tropical and broadleaf tree)
Thicket (cocoplum-willow, wax myrtle-salt bush, buttonwood, pond apple, scrub mangrove)
Prairie (muhly, sawgrass, beardgrass)
Marsh (sawgrass, spike rush-beak rush, maidencane, flag-pickerel weed, cattail)
Prairie - Tree Island Mosaic
Disturbed (including native and exotic species)

The natural vegetation communities of the East Everglades include the wet marsh and periphyton associations, the moderately dry prairies and thickets, and the elevated hammock forests. The marsh associations occur primarily in the northern and southern portions of the study area and the prairie associations dominate the central areas. Hammock forests or tree island association and thicket communities are scattered throughout the area as pockets of woody species among the large expanses of graminoid (grassland) communities. Disturbed vegetation communities occur predominantly along the eastern border of the study area and comprise only a small portion of it (approximately 10%). Many of the native plant communities include rare and endemic species, as well as approximately 40 species which have been placed on lists of threatened or endangered species.

Community character and changes which occur naturally are the result of various interrelated factors, the most important of which are climate, flooding, soils, fire, wildlife, and human use. The wetland associations are supported by an extended period of flooding, while upland associations can tolerate only brief incidents of inundation. Most species are intolerant of frost and are stressed by drought. The wettest associations are commonly found to occupy peat or marl soils, while the drier communities prefer the better drained rockland.

Each community has inherent value for certain wildlife species. The combination of communities provides habitat variety, and this variety is essential to maintenance of a diverse wildlife population. While plant communities play a significant role in (1) cycling water, nutrients, energy, and air,(2) in purifying contaminated water, and (3) providing recreational opportunities, human uses often disrupt or eliminate some of these critical functions.

Beside changes resulting from human use, invasion of exotic plant species and intense fires are among the most serious current threats to the perpetuation of natural vegetation. Exotic species having significant adverse affects on native communities include Australian pine (*Casuarina* spp.), Brazilian pepper (*Schinus terebinthifolius*), and cajeput (*Melaleuca quinquenervia*). Generally when natural communities are disturbed, they become highly susceptible invasion by these and other exotic species. Once established, these exotic species are nearly impossible to remove and tend to out-compete and replace indigenous species. Although natural fire is an essential element for the health and survival of many plant communities, unusually hot dry season fires that are the result of arson or carelessness can be extremely damaging to the native plant communities and their growth medium.

Wildlife

The wildlife which occurs in the East Everglades is one of the most visible and unique features in the study area and throughout south Florida. Approximately 350 animal species are known to occur in the study area, including 34 species of fish, 18 species of amphibians, 44 species of reptiles, 28 species of mammals, and 230 species of birds. These species depend on a large diversity of habitat offering opportunities to feed, reproduce, and seek protection. Valuable habitats occur throughout the East Everglades, however, two habitat distinctions can be made -- natural habitat and disturbed areas habitat. Animals which utilize natural areas frequently move freely to other natural areas within the study area and to other natural areas throughout south Florida. Some of the birds utilizing parts of the East Everglades on a seasonal basis are migrating species (e.g., woodstork, spoonbill). It is well recognized that East Everglades habitats serve an important function for local as well as regional and national wildlife populations. Moreover, species which inhabit disturbed areas of the East Everglades tend to have a more limited range than those species occupying natural areas. In general, species using disturbed habitat tend not to move into the natural, undisturbed areas.

Although all wildlife is considered a valuable resource, some species have been given special distinction. For example, some species are singled out because of their recreational value. Other species have been distinguished because they have been placed on protected lists by federal or state government. A total of 30 lists species occur in the East Everglades including 6 endangered and 2 threatened species on the federal list, 6 endangered and 4 threatened species on the state list, and 6 endangered, 7 threatened, 15 species of special concern, 1 rare, and 1 status undetermined species on the Florida Committee list. Several of the nation's rarest and unique animals survive in the East Everglades. The value of a species or subspecies is inestimable, particularly when extinction occurs as it has so frequently in recent southern history (e.g., whooping crane, ivory bill woodpecker, Carolina parakeet).

Water, vegetation, and human activities are the most critical factors affecting East Everglades wildlife. Aquatic organisms rely on flooding in both wet and dry seasons as well as the man-made canals to provide habitat. Changes in the extent of standing water have had adverse impacts on wildlife population size, distribution, and diversity. Similarly, terrestrial species depend directly on vegetation for food and shelter and indirectly on other environmental features such as flooding which maintain vegetation communities. Conversely, many plant species depend on wildlife for their continued success (e.g., seed distribution, pollination). Human activities in the East Everglades have been and continue to be instrumental in changing wildlife populations. Changes have occurred either through reducing the number of a particular species, as in the case of the Florida panther, or through the elimination of habitat. Loss of habitat usually occurs when land uses displace habitat or when characteristics of the plant community are modified. Historic reduction of flooding in most parts of the area has shifted habitats from those characterized by vegetation preferring wet conditions to those communities which require drier environments (e.g., sawgrass to muhly prairie). Therefore, current wildlife population in the area differ in many ways from those of the recent past. Besides habitat modifications, recreational activities (e.g., hunting, airboating) have also altered wildlife populations in the study area.

Land Use

Only a small portion (approximately 12%) of the land in the East Everglades is currently being used, for agriculture and residential except passive recreation which occurs generally throughout. Existing land uses occur primarily along the eastern limits of the study area between Northeast Shark River Slough and the Coastal Glades.

Existing residential development is rural in character and makes up only a small percent (0.25%) of the study area. Many of the residential parcels include ancillary uses such as grove crops and animal husbandry. Residential land is concentrated in the Howard and Richmond Drive areas, although small residential pockets also occur near Chekika State Recreational Area and along Tamiami Trail (Figure 2). Total current estimated permanent population for the East Everglades is 400 - 450 persons, with the average household size ranging from 3.5 to 4.0 persons. Only four roads near the residential area are paved. Public services in the residential areas are poor, although most homes are serviced by electric and telephone.

Agricultural development is more extensive than residential use, occupying approximately 10,000 acres or approximately fifty-five percent (55%) of the area currently in use. The area's subtropical climate makes parts of the East Everglades attrative for growing winter crops. Crops most commonly grown include winter vegetables, citrus and tropical fruits, tropical vegetables, and plant nursery stock. Continued demand for agricultural land in Dade County and decreasing availability of it elsewhere in the County has placed greater emphasis on utilizing land in the East Everglades. The most significant drawback to agriculture in the study area is higher flooding risks than would be found on agricultural land which is provided with drainage.

Recreational activities occur throughout the East Everglades on public and private land. Chekika State Recreation Area is the only public recreation facility. It offers a range of activities and facilities including camping, swimming, picnicking, and use of boardwalks and an interpretive walking trail. Much of the private land is used with or without the landowner's permission for operation of off-road vehicles (ORVs), hunting, fishing, and nature appreciation. Airboats and glade buggies are the most common type of ORVs. Airboat operation occurs primarily in the Northeast Shark River Slough area by individual users and small commercial operators. Glade buggies, large rubber tire vehicles, half and full track vehicles, jeeps, and two/three wheeled motorcycles which are commonly driven over all types of wetland marsh and rocky glades can and do have adverse impact on the environment.

Hunting and fishing occur in several locations in the study area. The most commonly hunted game includes deer, raccoon, rabbit, fox, and birds; the most commonly caught fish include panfish and black bass.

Other passive recreational activities known to occur in the study area are soaring, skydiving, and nature appreciation (e.g., birdwatching, plant identification).

Historically, some heavy commercial activities have occurred in the East Everglades. Commercial activities have included jet engine testing, fish farming, roof-truss construction, and auto testing.

CONFLICTS OF PRESERVATION AND LAND USE

Prime developable land in Dade County has continued to dwindle in supply and rise in cost as population and demands for winter crops has rapidly increased. As a result, new development is forced onto less desirable, vacant land. Whereas the prime farm and residential land exists along the elevated Atlantic Coastal Ridge, new land to accommodate future land use is available only in the interior wetlands. These wetlands, the Everglades, are unsuitable for most land uses without major land alterations. When in their natural state, these wetland serve a critical function for both the private and public benefit. Of primary concern is flood storage, water supply, and maintenance of indigenous biota. It is the demand for land and resources in competition with preservation objectives which creates serious challenge to develop optimal management policies for the East Everglades. To avoid deleterious and irreversible commitments of these wetland resources, it is considered necessary to establish a strategy whereby comprises can rationally and defensibly be made in cases where preservation and use are not mutually attainable. Conflicts arising from competing demands for land use and resource protection are discussed below for each land use. The three general categories of land use which are present in the East Everglades now are expected to continue. These are agriculture, residential, and recreation.

AGRICULTURE

The less costly East Everglades land is an attractive alternative to the limited supply of high-priced land on the Coastal Ridge for agricultural use. The climate is nearly the same throughout the region and demand for winter crops continues to accelerate.

However, much of this land in the East Everglades is flood-prone, making agriculture a high-risk venture, even though substantial economic returns can be realized in a flood-free season. The driest land lies along the eastern edge of the study area where elevations are highest and adjacent canals have an under-draining effect. While water is a major constraint for farming in most of the area, it also provides a plentiful irrigation source. Withdrawals for irrigation lowers the water table locally at the pumping site.

Agricultural interests strongly support additional drainage, whether by building new canals or modifying the operation of the existing canal network. Although advanteous to the farmer, additional drainage, disrupted flow, and reduced periods of inundation would have severe and deleterious effects on the environments of the East Everglades. Moreover, these effects will extend to Everglades National Park. Reduced hydroperiod would alter vegetation to more mesic forms and alter the extent and type of wildlife habitat. Many habitats support rare, threatened and endangered species including the American crocodile, Cape Sable sparrow, and bald eagle. Wildlife inhabiting Everglades National Park depends on these habitats in addition to those within the Park which are maintained through mandated deliveries of water to the Park through the East Everglades. Loss of habitat is likely to diminish the number of species or force them to other areas, provided that adequate habitat exists elsewhere.

Maintenance of water levels and movement in the study area is critical to public water supply and quality. Water recharged and held in the aquifer is drawn out by municipal and domestic wells to supply potable freshwater to nearly all of south Florida's population. Additionally, the water held in the system maintains the hydraulic head or pressure which retards saline intrusion along the southern reaches of the Everglades. Intrusion of salt water seriously threatens public water supply.

Because of the high water table and flooding, agrichemicals added to farmland are likely to be carried by recharge water into the aquifer. Depending on the extent of farming, nature and extent of chemical applications, and location of wells, these chemicals may enter public water supplies and aquatic habitats. While data representing the current extent of farming suggest only localized impacts, cumulative and synergistic effects cannot be accurately calculated or dismissed.

Aside from hyrologic constraints, the most significant impact of agriculture is that associated with rockplowing. The driest, best drained areas are typically those occupied by rockland. Peats are too wet and acidic, and marls soils are often too poorly drained to farm. Rockland must be broken up by heavy machinery and mixed with surface soils if they are to be successfully farmed. The process is called rockplowing. Rockland characteristically is low in organic matter, drains rapidly, and is alkaline. Once rockplowed, the substrate provides mechanical support, but extensive and frequent irrigation and fertilization are necessary. Despite these shortcomings, rockland can be productive for selected types of cash crops. However, once rockplowed, the substrate and its inherent characteristics are permanently altered, and native vegetation and natural wildlife habitat can never be restored. In fact, when the land is released from agricultural use, it will be rapidly colonized by exotic plant species which out-compete all native plants. Once established, no means are currently available for removal of exotic species. Coincidently, much of the rockland in the East Everglades is vegetated by muhly prairie, the exclusive habitat of the endangered Cape Sable sparrow. A large portion of this land is federally designated critical wildlife habitat.

RESIDENTIAL

The urban expansion factors which initiated the movement of agriculture to the East Everglades are also operative for residential relocation. Aside from land costs and availability, the East Everglades is attractive to those seeking privacy and a natural, rural setting. In general, the conflicts created by low density residential development appear to be less severe than those created by agriculture. (Existing zoning allows one units per five acres, except where grandfathered lots are a minimum of 1¼ acres.) The major concerns for residential use are those of health, safety, and welfare. Because much of the land is flood-prone, structures and septic systems must be elevated on pilings or fill. The placement of fill in wetlands is regulated by law, but where permitted, it creates associated environmental impacts. Among those impacts are disrupted surface water flow, increased runoff, reduced recharge, loss of native vegetation, and displacement of wildlife habitat. In addition to dredge and fill operations, roads require special engineering (e.g., culverting) to ensure that surface flow is not impeded and siltation is minimized. Under severe flood conditions, roads become impassable and demand costly maintenance following most flood events. Because of small scale intervention-half-acre fill pads at permitted densities and certain design requirements (culverting), most of these impacts of fill placement are localized and minimal when one considers the East Everglades as a whole. However, public safety is a matter of concern due to current county and regional policies which clearly state that additional public services are not planned for the area in the short-term future. Given that the majority of the area will experience flooding for anywhere from one to nine months of the year, the East Everglades is not the most desirable place for safe human habitation. Acceptance of this natural risk is unlikely, as other residential areas in south Florida with similar conditions have required public expenditures to ensure the health, safety and welfare of its residents usually through the provisions of positive drainage. In light of the South Florida Water Management District's adopted position that structural drainage will not be provided in these marginal areas of the East Everglades, residential development must be stringently curtailed.

Fire risk constitutes another major public safety risk for residential use in the area and conflicts with vegetation management requirements. Many of the plant communities in the area are fire climax communities, requiring natural periodic fires for their continued success. Residential use of the land necessitates fire control and management programs to ensure public safety. However, where attempts are made to suppress naturally occurring fires and where the land is drained, the inevitable fires are likely to be more intense due to an accumulated fuel loads and the abnormally dry conditions.

Residential use is also likely to have adverse effects upon many of the wildlife species in the area which require isolation from human disturbances and large expanses of unaltered habitat. Some species are capable of adapting to man, while many others cease to reproduce or flee to new habitats when forced to live in proximity to human settlement. Furthermore, it needs to be recognized that areas inhabited by potentially hazardous species such as alligators and water moccasins pose some risk to residents, especially during high water conditions when fill pads and homes offer refuge to these species.

RECREATION

Most forms of recreation occurring in the study area are passive in nature and have little effect on the natural environment. In fact, the abundant and diverse wildlife, the dense vegetative cover, the natural waterbodies, and man-made canals offer a variety of recreational opportunities. Much of the land, whether in public or private ownership is sought for some recreational purpose.

Regulations such as hunting laws and bag limits have in most instances prevented elimination or severe imbalances in animal populations. However, illegal hunting has in the past threatened to eliminate such species as the Florida panther, bobcat, and American alligator.

The impact of off-road vehicles (ORVs) on the natural environment is evident. Wheeled vehicles have marred the landscape by compacting substrate and destroying vegetation. Airboats, which are restricted to areas of standing water, have made perceptible trails through the marsh vegetation. For the most part, the current activity level of ORVs results in only a small-scale disturbance when viewed from the perspective of the East Everglades as a whole. However, it has not been clearly established how significant the impact of ORV operation (noise, intrusion) is on individual wildlife species or on populations. Aside from increased noise levels, degradation of water and air quality is a minor concern at current use levels. Future conflicts may arise with substantial increases in ORV use.

CONCLUSIONS

The resolution of land use demands and preservation objectives is sought through a series of recommendations presented in the Management Plan. The means of resolving these conflicts include strict land use controls and a series of regulations in the form of Best Management Practices (BMPs). Controls and regulations are specific to the type of land use and are tailored to meet the specific environmental conditions and projected uses within the study area. Many of these are supplemental in nature or are revisions to existing laws and regulations which are operative in the area.

The Management Plan divides the East Everglades into six management areas (Figure 3). Management recommendations which apply to the entire study area and to specific management areas within the study area are summarized below.

Areawide Management Recommendations

. Residential use should be discouraged to preserve the area's historic water storage capability.

. Areawide positive drainage is totally unacceptable for economic, environmental, and social reasons. Based on previous experience, positive drainage is inevitably provided to wetlands where residential development has taken place.

. The design and construction of residences, and residential use and activities should occur in a manner compatible with environmental characteristics -- maintain sheet flow, retain high water quality, protect plant communities and wildlife habitats, and protect life and property from flood and fire.

. Agriculture should be permitted in areas which are likely not to experience flooding during the growing season. Hammock and tree islands should not be disturbed even if they are hydrologically suited for farming.

. Agriculture should be given permanence by designating suitable areas as exclusive agricultural use zones.

. Agricultural practices should ensure minimal adverse environmental impact.

Recommendations Specific to Management Areas

Management Area 1: Modified Environment

Includes all land which has been permanently altered by human activity. Currently 10,600 acres are in agricultural use, 600 acres in residential use and 3,500 acres in vacant. The well drained rockland is relatively high, and due to its proximity to canal along the eastern border is one of the least flood-prone areas in the study area.

. Future residential use should be discouraged in favor of agriculture.

Figure 3 : Management Areas

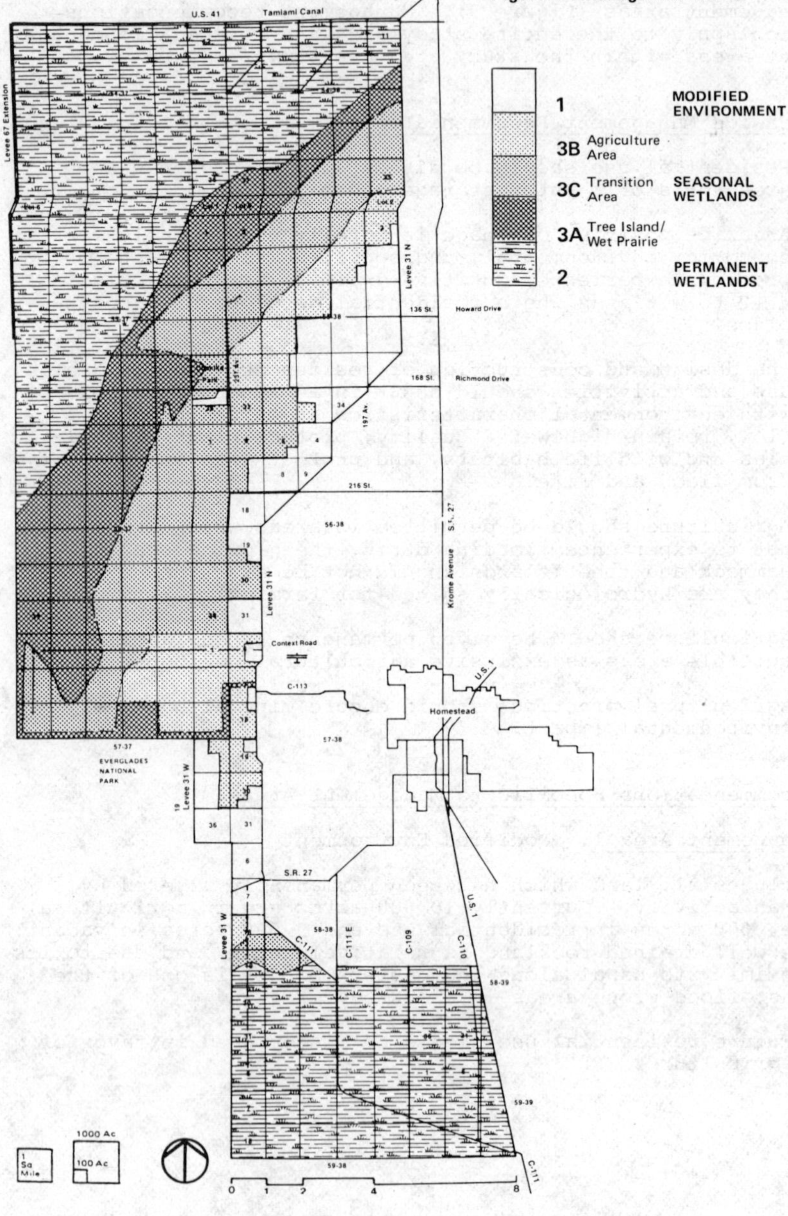

. Existing agricultural and vacant land should be reserved for agricultural use.

. Invasion of undesirable exotic plant species should be controlled wherever possible.

. Surface water flow should be preserved through regulation of access roads, raised planting beds, ditching, and ancillary structures.

. Current groundwater recharge should be maintained.

. High water quality should be maintained through regulation of agrichemicals.

Management Area 2: Permanent Wetlands

Includes all land flooded for at least 9 months of a normal water year. Most of the valuable vegetation communities of Forest/Thicket and Marsh, along with nearly half of Prairie communities, lie within this area. Also included within its boundaries are the federally designated critical habitat of American crocodile and substantial portions of valuable feeding grounds and habitats for wading birds and other wildlife species. Most of peat and marl soils also occur within this area. This management area comprises 72,000 acres or approximately 40% of the East Everglades.

. Agricultural use is not permitted.

. Residential use is discouraged but permitted at a density and character which ensures no degradation of the environment. Downzoning is recommended to meet these criteria.

. Recreational use is encouraged in a manner that private property rights and long-term environmental integrity are preserved.

. Existing limited access should remain unchanged - no elevated roads are permitted.

. Dredge and fill operation is permitted only for on-site septic systems, non-convetional waste disposal systems (e.g., chemical closet) are preferred where feasible.

. Plant species used in ornamental planting should be species native to this management area.

. Fire management should be practiced.

Management Area 3: Seasonal Wetlands

The remainder of East Everglades, in recognition of its range of hydrologic characteristics, wildlife habitats and locational attributes, is divided into the following three subareas:

Subarea 3A: Tree Island/Wet Prairie

Includes all land flooded from 3 to 9 months during an average year. Valuable vegetation communities of Tree Island and Wet Prairie are naturally related to this flood period and are included within this area. Scattered "islands" which may flood for less than 3 months, and a 300 foot wide buffer strip adjacent to the Everglades National Park boundary and the feeder canal (L 31 W), are also included.

. Agricultural use is not permitted.

. Residential development is permitted with same restrictions as specified for Permanent Wetlands.

. Access roads are permitted with design specifications for maintaing uninterrupted flow.

. Placement of fill is permitted for residential pads and septic drainfields. No dredging to obtain fill is permitted.

Subarea 3B: Agriculture Area

Includes all land, except federally designated critical habitat of the Cape Sable sparrow, flooded for less than 3 months in an average year and contiguous to the Modified Environment.

. Residential development is discouraged in favor of future agricultural use.

. Future agricultural use should be orderly; except under special circumstances, new agricultural use should be contiguous to land already in agriculture.

Subarea 3C: Transition Area

Includes the remaining land. A substantial part of the area is the federally designated critical habitat of the Cape Sable sparrow.

. Residential development is permitted as specified for Subarea 3A.

. Agriculture may be feasible and permitted in parts of this subarea when 1) no alternative location is permitted, 2) the proposed site is contiguous to previously farmed land, 3) the proposed site is hydrologically capable.

. Federal permits within designated critical wildlife habitats must be obtained.

REFERENCES

Prepared for East Everglades Resources Planning Project

Everglades National Park, 1979. "Wildlife Surveys and Freshwater Fish Study." South Florida Research Center, Everglades National Park.

Hartwell, James, 1979. "Effects of Land Use on Surface Water Quantity."

Hofstetter, R. and C. Hilsenbeck, 1979. "Description of Major Plant Communities in the East Everglades." Dept. of Biology, Univ. of Miami, Coral Gables, FL.

_______________, 1980. "Vegetational Studies of the East Everglades." Dept. of Biology, Univ. of Miami, Coral Gables, FL.

Metropolitan Dade County Planning Department, 1978. "Recreation Land Use in the East Everglades."

_______________, 1978. "Residential Land Use in the East Everglades."

_______________, 1978. "Agriculture in the East Everglades."

_______________, 1978. "Unique Natural Features in the East Everglades."

_______________, 1979. "Policy Constraints Upon Development in the East Everglades."

_______________, 1979. "Management Objectives for the East Everglades Resource Planning Project."

Parker, Garald, 1979. "Groundwater Resources of the Southeast Coastal and Everglades Area, Dade and Broward Counties, FL."

_______________, 1979. "Groundwater Levels and Directions of Flow in the Biscayne Aquifer."

_______________, 1980. "Effects of Land Uses on Ground Water Quantity."

Schneider, J. and B. Waller, 1979. "Summary of Hydrologic Data for the East Everglades." U.S. Geological Survey, Water Resources Div., Miami, FL.

South Florida Water Management District, 1979. "Water Management Requirements for Base and Alternative Scenarios." Resources Planning Dept.

_______________, 1980. "Hydrologic Scenarios for the East Everglades." Resources Planning Dept.

Waller, B., 1979. "Effect of Land Use on Ground-Water Quality in the East Everglades." U.S. Geological Survey, Water Management Div, Miami, FL.

_______________, 1979. "Effects of Land Use on Surface Water Quality in the East Everglade U.S. Geological Survey, Water Resources Div., Miami, FL.

U.S.G.S., 1979. "Baseline Water Quality Monitoring." Water Resources Div., Miami, FL.

Others

Craighead, F.C., 1971. "The Trees of South Florida", Univ. of Miami Press, Coral Gables, FL.

Edler, F.E., 1952. "Southeast Saline Everglades, Florida, and Its Management." Veg. 3:213-265.

Klein, H. and J.E. Hull, 1978. "Biscayne Aquifer, Southeast Florida." U.S.G.S. Water-Resources Investigation, 78-107.

U.S.D.A., 1958. "Soil Survey, Dade County Florida." Soil Conservation Service, Series 1947, No. 4.

California Coastal Data Collection Program - Data Collection, Analysis and Dissemination

George W. Domurat & Douglas M. Pirie*

INTRODUCTION

The purpose of the California Coastal Data Collection Program is to acquire, analyze and make avialable a data comprised of wave data and coastal processes information. This data base is to assist in the planning, design, construction and evaluation of coastal projects as well as to provide qualitative and quantitative information to better understand coastal engineering problems for efficient coastal zone management.

The data base is to be available to the Corps and other interested users through records and timely reports and also will be provided to necessary users, as near-real time wave statistics for use in Corps operations, National Weather Service transmissions, Naval operations and "customers service inquires".

To date, the California Coastal Data Collection Program has developed a wave gaging program comprised of 5 wave rider buoys and 12 pressure sensor wave gaging stations. By the end of this calendar year, 8 more buoy sites will be instrumented along the California coast with two of these being the satellite relay variety because of their distance from the coast. Also included are additional Pressur sensor gage stations. If funded until completion, the wave gaging of the California coast will consist of 65 pressure sensor stations and 34 offshore wave rider buoys.

FIELD EQUIPMENT

The success of the California Coastal Data Collection Program depends heavily on the realiability and efficient operation of all instrumentation installed in the field. Equipment used in this program has been designed to report coastal data for long periods of time virtually unattended. Emphasis is placed on the long term stability and immunity of these instruments to environmental effects such as corrosion, temperature, and vibration. High performance of all field equipment is insured by the testing of sensors and components prior to installation. Periodic calibration in the field assures the accuracy of measurements.

A complete field site consists of an onshore station containing the electronics for signal conditioning and data transmission, one or more near/offshore wave sensors, and a data link connecting the sensors to the onshore station. Additional sensors such as anemometers and current meters may also be installed.

The CCDCP uses tough weatherproof steel enclosures for shore stations, in which the electronics are contained on plug-in cards allowing direct replacement of faulty components and elimination of lenghty down-time. All other field equipment is constructed in modular form to permit easy interchange of subassemblies throughtout the system.

*U.S. Army Corps of Engineers, San Francisco, CA

INSTRUMENT TYPES-NEARSHORE WAVE MEASUREMENTS

The California Coastal Data Collection Program uses two types of commercially available pressure transducers for standard wave measurement. The first unit is manufactured by Gulton Industries. Its sensing element is inherently resistant to sea water and it can be exposed directly, eliminating the seal problems associated with other units.

This sensor was developed as an informal joint venture with Gulton Industries and Scripps Institution of Oceanography to produce a unit specifically design for nearshore measurements. The heart of the sensor is an accurate, stable and reliable pressure transducer of a design type which Gulton has used successfully for years.

Many modifications and additions were made to the original transducer to produce the sensing instrucment. The material used for the port and pressure sensing capsule were changed from stainless steel to Inconel alloys 625 and 718, which have very high resistance to salt water without seriously shortening its usuable life. The electronics housing was strengthened to protect the transducer from potting pressures and the outside forces incurred during handling and installation. The interior of this housing is filled with rigid foam material which immobilizes the components to protect them against shock and vibration. These modifications make the transducer a sturdy, reliable unit well suited to survive the hazards of the nearshore underwater environment.

In addition to these mechanical modifications, the transducer circuit was supplemented to minimize the effects of cable degradation, one of the most serious problems encountered in long term underwater installation. First, a voltage regulator was added to isolate the transducer from the effect of cable losses on supply voltage. Second, a voltage controlled oscillator (VCO) was added to convert the D.C. output of the transducer to a proportional frequency. This virtually eliminated cable degradation as a source of inaccuracy (a serious problem with D.C. signals), since the frequency of the output signal will be unaffected by the quality of the cable path.

Cable attachmet and potting complete the sensor assembly. Strain relief is provided by attaching the cable armor to a flange welded directly to the tough transducer housing. The potting operation encapsulates the transducer ina tough waterproof polyurethane material which isolates the electrical connections and housing from sea water.

The second type of wave sensor assembly consists of a Kulite model IPT-1000-50A semiconductor strain gage pressure transducer housed in a plastic pressure case together with associated electronics. The low level output signal from the pressure transducer is amplified and converted to a frequency output proportional to the applied pressure. An audio amplifier line driver provides a low impedence output to drive the interconnecting cable. Regulator circuits inside the pressure housing allow operation of the wave sensor from 15 to 35 VDC voltage sources. A Fairprene diaphram isolated the oil filled transducer from the sea water pressure medium preventing corrosion.

INSTRUMENT TYPES-DEEP WATER WAVE MEASUREMENTS

Deepwater waves are to be recorded in areas where the water depth can be as great as 200 meters. This environment precludes the use of bottom-mouted pressure sensors for their measurement. Some type of surface or near surface buoy is required to provide a platform for sensing waves.

The most successful and widely used technique for measurement of deepwater waves is that of sensing the verticl heave of a surface moored buoy by on-board accelerometers. The CCDCP utilizes the Datawell Waverider buoy which is commerically manufactured instrument proven in the field for over to years and has become the most widely used instrument of this type. It operates by mooring a small (0.7 m or 0.9 m) spherical hull containing a vertically stablized accelerometer which senses the vertical component of motion of the buoy. Outputs from the accelerometer are double intergrated on board the buoy and the resultant heave data telemetered to shore on a 23 MHz radio link for distances up to 50 km, or telemetered to the GOES satellite and recovered from NESS (Nationl Environmental Satellite Service) when distances from shore are greater than 50 km.

At the shore site, a phase-lock reciever obtains the data and may produce an output in the form of a strip chart record of sea surface motion versus time. A frequency analog of this data presented at the receiver output is adapted to the same type of digital processing and storage system as is used for botton-mouted pressure sensors. The buoy has good accuracy (±3% displacement) over a usable band width from 0.033 to 0.5 Hz.

The field endurance of this buoy is dependent upon its internal power source which provides 9 months and 16 months capacity in the 0.7 m and 0.9 m diameter hulls, respectively.

DATA COLLECTION

Raw data in the form of a frequency signal proportional to pressure is sent through a special underwater cable assembly designed to maintain water-tight integrity and electrical transmission properties. The sensor is capable of driving as much as two miles of this cable. On the receiving end of the underwater cable an input signal conditioning circuit restores any decrease of signal and presents a constant amplitude square wave to the digital counting circuit. The counting circuit multiplies the input frequency by a factor of eight to increase resolution and places the digital output value of the frequency signal on the buss of a digital memory for storage. Each value is stored in succession until the memory is full. Afterwards, the memory overflows the oldest data and adds the newest value to the memory input.

The central station consists of a data acquisition computer driving a specially designed wave height receiver and a larger data processing computer. The data acquistion computer is a Nova 1200 minicomputer with 64K bytes of core memory (maximum capacity). Two 9 track 800 BPI tape drives plus a 10 megabyte disk drive contain programs and raw data. A Monitor Labs 4100 Nova buss extender provides Input/Output connections to external devices.

The central station computer initiates a telephone call from the data recreiver to the remote site when a sample of data is required from the station. At the instant of receipt of the call, the remote site freezes the data in the digital memory (typically 1024 values), and then transmits the contents of the memory over the telephone circuit at a 1200 baud rate utilizing frequency shift keying in a synchronous mode.

At the Central Station a special data receiver locks onto the incoming data system and syncronizes the data . The data is then loaded into the bit stream Nova 1200 computer under program control via parallel word transfers. When the proper amount of data has been recorded and loaded into the computer buffer the computer terminates the telephone call. The data stored in the computer buffer is then transferred to 9-track magnetic tape along with header information supplied from the computer. Results of each station call are then printed out on a logging terminal to notify operations personnel as to the status of each call.

Periodically data tapes from the Nova computer are read into the central processor computer, a Prime 500. Raw data is placed on disc file for further processing.

DATA ANALYSIS

Data anlaysis on the Prime computer is composed of 3 seperate phases. Phase one involves the receipt of the raw data time-series as collected bu the CCDCP data acquisiton system and initiates extensive data verification and editing in preparation for phase 2. The second phase performs the Fast Fourier Transforms operations and applies the appropriate depth correction coefficients to the edited time-series. Phase 3 operates on the transformed series to produce the monthly summary reports. The operation of each of these phases is expanded as follows:

. <u>Phase One</u> - At present there are two main categories of data which are collected by the Nova and transmitted to the Prime computer. The two categories are (10 time history of pressure falctuations are sensed by a submerged pressure transducers, and (2) time history of surface elevations as sensed by a Waverider buoy. The first category further breaks down into 3 types of pressure data transmitted: (a) a standard fast sampled gauge which samples pressure flactations for a period of 17 min. at one sample per second. (b) a slow gage which samples pressure fluctuations for a period of 4.5 hrs at a sampling frequency of 0.125 Hz, and (c) a pressure transducer array comprising 4 gages, one mounted in each corner of a rigid frame measuring 6 m on a side. The array is capable of resolving directional wave spectra as well as energy spectra and normally samples each channel at one sample per second.

In the first phase data is examined for various anomalies by an automatic editing program. Data which passes as acceptable is stored on disk files while bad data is stored in a different location for later examination. The function of the editor is to examine each individual time-series for a number of preset, recognizable dysfunctions. Presently the editior's function is to check for the following conditions: spikes, mean shift exceedence, zero crossings, maximum wave height exceedence. In addition the tidal components are removed by Digital Filtering.

. <u>Phase Two</u> - The edited and corrected time series are operated upon by a fast Fourier transform routine. Pressure sensor data is corrected for pressure attenuation effects and the resultant time series of Fourier coefficients is stored in a new set of disk files.

. <u>Phase Three</u> - There are two possible paths at this stage, each corresponds to a different sensor function.

1. Pressure gage and/or Waverider data. The transformed Fourier coefficients are grouped by preselected bands, the energy content is tabulated by frequency band, the significant wave height is calculated, and a frequency distribution table of the spectra is preapred for inclusion in the monthly report. Concurrently, the data is stored in the general purpose data bank files. A maximum and minimum wave height test is held on the significant wave height to insure acceptable data (this is an extension of the editor function which can only be done with the corrected transformed coefficients).

2. Directional processing. The processing is broken down into the following sequential steps:

(a) An inner comparison is held among data from the four gages to insure that there are no unacceptable energy level differences between the gages in the array. This is an extension of the deitor function. Afterwards the standard deviations of the energy levels of the gages are normalized to remove any biases. The energy is now grouped into 9 energy bands.

(b) The array orientation is determined. This process considers the local bathymetry and azimuthal orientation to arrive at a rotational transform so that a diagonal axis of the array will coincide with the beach normal. The number of functioning gages is now ascertained, and processing may proceed only if 3 or more gages are functioning.

(c) The wave slope is determined by taking the differences of opposing gages and dividing the result by the corresponding distance. A three dimensional coordinate rotation is now applied to align a diagonal slope axis with the beach normal. A slope routine is applied to account for the linear fashion in which the slopes are computed. Finally, the cospectra of the orthogonal slopes are calculated.

(d) The spectral angles are calculated through the relationship:

```
ANGLE(I)=0.5(ARCSIN(2.*(COSSPT(I)/ENERGY(I))))
```

where I is the frequency band counter,
COSSPT is the cross spectra of the slopes, and
ENERGY is the energy content of the band.

Since this expression can resolve to no more than 45 degrees, it is ncesssary to fold the resultant angles depending on whether the directional spectra are predominantly in the longshore or offshore direction. The spectral energy components are calculated as before and tabular summary information is stored for publication in the monthly report. The summarized data is also stored in the general purpose data bank.

DATA PRODUCTS: THE GENERAL PURPOSE DATA BASE

The data base generated by the CCDCP contains all available wave data accessible to users in three distinct forms. These include published reports, direct access via remote computer terminal, and raw and processed data on magnetic tape.

Published Reports

1. Monthly reports - monthly reports are generated within 2 weeks of the end of each reporting month. These reports include both tabular and graphical output.

A. Tabular content includes:

a. Listing of percent energy contained in two-second band widths ranging from 4 to 22+ seconds. For deep water gages this range will be from 4 to 40 seconds with five-second band widths from 20 to 40 seconds.

b. Table of wave height persistence for the month including the number of consecutive days on which significant heights exceed half meter increments.

c. Table of maximum significant wave heights observed during each day of the month.

d. Listing of total longshore momentum flux (Sxy) significant angle, and indicated directions for each two-second period band from 4 to 22+ seconds. This is for each observation at all directional stations.

B. Graphical Output includes:

a. Qualitative plot of nearshore spectra for each observation once each six (6) hours. Plots of spectra for deepwater gages have an observation frequency of three (3) hours.

b. Vector plots of wave direction and energy by period bands for all directional stations for each observation.

2. Yearly Reports - Yearly reports are distributed within one month of the end of every calendar year. These reports include table, graphs, and narratives describing the wave climate for that year.

A. Tabular Reports include:

a. Table of joint distribution of significant wave height and period of peak energy.

b. Table of cumulative height distribution function.

c. Table of cumulative period distribution function.

B. Graphical Reports include:

a. Cumulative distribution function of significant wave height.

b. Cumulative distribution function of period of peak energy.

c. Cumulative qualitative three-dimensional contour plot of joint height/period distribution.

d. Quarterly probabilities of exceeding significant wave heights in half meter increments for waves having exceedence probabilities over 5 percent.

C. Narrative Reports include:

Narrative description of significant observations, analyses and findings, including extraordinary wave events.

3. Special Reports - Occassionally coastal projects require that certain special reports be published to fulfill the wave data needs of specific users. These reports include both tabular, graphical, and narrative output and consist of topics such as comparison of wave data between stations, sediment transport predictions and analysis of harbor seiche problems.

DIRECT ACCESS VIA REMOTE COMPUTER TERMINALS

As mentioned in previous sections, data collected from the field stations is edited, processed and transferred to the data base. This wave data resides on disk at the central station and is directly accessible via computer terminal with an acoustic coupler or modem. Products include:

1. System Status - a report describing the operational status of all gages within the CCDCP. This includes duration of down-time, reasons for system malfunctions, estimates as to when stations will be brought back on-line, and dates when reports are available.

2. Gage Locations - a report listing the latitude and longitude position of gages, installation depth, and station identification numbers.

3. Tabular Output - formats similar to those published in the monthly and yearly reports.

4. Graphical Output - formats similar to those found in the monthly reports with the added option of plotting a series of wave energy spectra on the same set of axes. This allows comparison of wave energy for different times at the same station, different stations all at the same time, or other combinations.

RAW AND PROCESSED DATA ON MAGNETIC TAPE

All raw and processed data is recorded on magnetic tape and sent to the National Oceanographic Data Center for archiving and general purpose use. Other archives are located at Scripps Institution of Oceanography and the U.S. Army Corps of Engineers Coastal Engineering Research Center.

All magnetic tapes used for data storage begin with a file uniquely distingishing that tape from others in the library. Also, each data file on the tape is headed with information describing the station number, sensor type, sampling time, calibration codes, etc. Tapes are standard 9-track, 800 BPI in 16-bit binary code.

ACKNOWLEDGEMENTS

The authors wish to acknowledge the Army Corps of Engineers and the state of California for access and permission to use this study material. The views of the authors do not purport to reflect the position of the Corps of Engineers, Department of Army, or the State of California.

Contents of this paper are not to be used for advertising or promotional purposes. Citation of trade names does not constitute an official endorsement or approval of the use of such commercial products.

CALIFORNIA COASTAL DATA COLLECTION PROGRAM STATUS

DOUGLAS M. PIRIE, Coastal Engineer,
U.S. Army Corps of Engineers, San Francisco

GEORGE W. DOMURAT, Oceanographer,
U.S. Army Corps of Engineers, San Francisco

INTRODUCTION

Coastal engineers and scientists recognize the need for high quality wave information. Coastal engineering studies using wave gage information include beach erosion, harbor shoaling, harbor oscillation, and jetty stabilization studies. Operational uses include dredging, fishing, general navigation, maintenance, and recreation.

To meet these needs, the Corps of Engineers began a wave-gaging program, the California Coastal Data Collection Program, in May 1978 in cooperation with the California Department of Boating and Waterways. The program utilizes buoys and pressure transducer gages for monitoring continental shelf and near-shore waves.

The wave-gaging system is developing wave statistics for coastal engineering requirements, and also is providing the National Weather Service with selected near real time wave statistics for broadcasting on the marine weather channels. Use of the data by recreational boaters, fishermen, search and rescue units and others could have enormous value in saving lives and property.

To date, the California Coastal Data Collection Program has developed a wave gaging program comprised of 5 wave rider buoys and 12 pressure sensor wave gaging stations. By the end of this calendar year, 8 more buoy sites will be instrumented along the California coast with two of these being the satellite relay variety because of their distance from the coast. Also included are additional pressure sensor gage stations. If funded until completion, the wave gaging of the California coast will consist of 65 pressure sensor stations and 34 offshore wave rider buoys. In addition to the California sites, selected sites on the Great Lakes; in Alaska, Oregon and Washington; and Pacific Ocean islands have been selected for instrumentation within the near future. These non-California sites will utilize the data processing system operating at Scripps Institution of Oceanography.

WAVE STATION TYPES AND LOCATIONS

The three types of wave gaging stations are as described below and the locations of the various sites are shown on Tables 1 and 2.

. Waverider - a buoy used to determine wave heights and periods by measuring the acceleration of the waves. Information is radioed to a land station or via satellite relay which send the data to the central data processing facility in San Diego.

. SP Gage - a single-point pressure gage used to measure wave heights and periods, connected by cable to a land station. The land station sends the data to the central data processing facility over telephone lines.

. Sxy Array - four pressure gages set in a square array and connected by cable to a land station. Data products from this type of gage are wave heights, periods, and directions.

TABLE 1

NORTHERN CALIFORNIA STATION LIST

	Location		
	N Latitude	W Longitude	1/ Status
Northern California Coastal Buoys			
Point St. George	41°51'	124°29.7	F
Crescent City	41°44'	124°18'	80
Humboldt Bay (Outer)	40°52.5'	124°24.5'	80
Humboldt Bay (Inner)	40°50.0'	124°17.0'	80
Punta Gorda	40°16.3'	124°34	F
Shelter Cove	39°58.5'	124°10.8'	F
Fort Bragg	39°26'	123°53'	80
Point Arena	38°55'	123°56.5'	F
Cordell Bank	38°04.2'	123°31.0'	F
Point Reyes	37°56.3'	123°03.8'	80
South Farallons	37°34'	122°54'	80
Pescadero Point	37°10.7'	122°34.3'	F
North Monterey Bay	36°50.4'	122°09.4'	O
Santa Cruz	36°53.4'	122°04.3'	O
South Monterey Bay	36°42'	121°58'	F
Point Sur	36°19'	122°04'	F
Northern California Coastal Nearshore Stations			
Crescent City Array	41°44.04'	124°11.53'	80
Humboldt North Spit Array	40°48.00'	124°13.0'	80
Humboldt Bay Entrance SP Gage	40°45.22'	124°13.12'	F
Shelter Cove SP Gage	40°01.05'	124°03.0'	F
Noyo Harbor* (6 SP Gages)	39°25.33'	123°48.0	80
Manchester Array	38°59.0'	123°43.0'	F
Arena Cove SP Gage	38°54.8'	123°43.0'	F
Tomales Bay Entrance SP Gage	38°14.67'	122°59.1'	F
Drakes Bay SP Gage	38°01.5'	122°57.3'	F
Stinson Beach Array	37°53.83'	122°38.92'	80
Thornton Array	37°41.8'	122°30.3'	80
Pacific SP Gage	37°38.0'	122°30'	F
El Granada SP Gage	37°29.68'	122°28.27'	F
Santa Cruz Pier SP Gage	36°57.4'	122°00.97'	O
Santa Cruz Harbor Array	36°57.4'	122°00.10'	O
Capitola Pier SP Gage	36°57.9'	121°56.92'	A
Seacliff SP Gage	36°57.13'	121°55.17'	O
Pajaro Dunes SP Gage	36°51.3'	121°49.2'	F
Moss Landing SP Gage	36°48.8'	121°47.35'	A
Marina SP Gage	36°41.3'	121°49.0'	F
Monterey Harbor SP Gage	36°36.4'	121°52.9'	O

*Special purpose harbor study using six single point stations.

TABLE 1
(Cont'd)

NORTHERN CALIFORNIA STATION LIST

	Location N Latitude	Location W Longitude	1/ Status
San Francisco Bay Area SP Gages			
Suisun Bay	38°05.53'	122°02.75'	F
Stake Point	38°03.03'	121°57.25'	F
Rodeo	38°07.50'	122°16.50'	F
Pinole Point	38°00.95'	122°22.03'	F
McNear Beach	37°59.63'	122°26.72'	F
Castro Point	37°55.12'	122°24.33'	F
Paradise Cay	37°54.75'	122°28.27'	F
Treasure Island	37°50.05'	122°23.75'	F
Fort Point	37°48.62'	122°28.12'	F
Bakers Beach	37°47.65'	122°29.17'	F
Alameda Beach	37°45.57'	122°16.67'	F
Hunters Point	37°43.17'	122°21.33'	F
San Leandro Marina	37°40.12'	122°13.57'	F
Coyote Point	37°35.75'	122°18.83'	F

1/ Status Codes: 0 = Operating
80 = To be installed in CY 1980
F = To be installed after CY 1980
A = Abandoned (past records available)

Note that locations of future stations are subject to change.

TABLE 2

SOUTHERN CALIFORNIA STATION LIST

	Location N Latitude	Location W Longitude	1/ Status
Southern California Coastal Buoys			
Cape San Martin Buoy	35°53.00'	121°30.70'	F
Morro Bay Buoy	35°21.50'	121°01.50'	F
San Luis Obispo Bay Buoy	35°03.00'	120°54.50'	F
Point Arguello Buoy	34°33.11'	120°42.38'	O
Point Arguello Buoy	34°30.90'	120°37.90'	A
Point Conception Buoy	34°22.50'	120°25.00'	F
Point Conception Buoy	34°25.39'	120°25.28'	A
Richardson Rock Buoy	34°10.00'	120°37.00'	80
Santa Barbara Channel Buoy	34°17.50'	119°43.00'	F
Santa Cruz Island Buoy	34°05.30'	119°45.20'	F
Mugu Buoy	34°03.20'	119°09.75'	F
Santa Monica Bay Buoy	33°53.00'	118°38.00'	80
San Pedro Channel Buoy	33°34.00'	118°12.00'	80
Catalina Island West End Buoy	33°29.15'	118°32.00'	F
San Clemente Island Buoy	32°48.00'	118°30.00'	80
Oceanside Buoy	33°10.50'	117°27.00'	F
Mission Bay Buoy	32°45.00'	117°18.00'	80
San Diego Buoy	32°37.00'	117°20.70'	F
Southern California Coastal Nearshore Stations			
Morro Bay Entrance Array	35°22.3'	120°52.5'	F
Point San Luis SP Gage	35°09.8'	120°45.0'	F
Pismo Beach Array	35°08.2'	120°39.2	F
Point Arguello SP Gage	34°33.3'	120°36.5'	A
Point Conception Array	34°26.75'	120°25.7'	A
Santa Barbara Point Array	34°24.1'	119°41.6'	O
Santa Barbara Array	34°24.1'	119°41.5'	O
Ventura Pier SP Gage	34°16.65'	119°17.4'	F
Ventura Marina Array	34°15.2'	119°16.56'	F
Channel Islands Array	34°10.1'	119°14.45'	F
Channel Island SP Gage 2/	34°10'	119°14.2'	O
Las Tunas Beach Array	34°02'	118°35.0'	F
Marin del Rey Array	33°58.25'	118°28.17'	F
Redondo Beach Array	33°51.5'	118°24.4'	F
Los Angeles Harbor SP Gage	33°43.0'	118°16.33'	F
Long Beach Harbor SP Gage	33°44.4'	118°11.68'	F
Seal Beach Array	33°43.87'	118°06.68	F
Sunset Array	33°42.8'	118°04.7'	80

TABLE 2
(Cont'd)

SOUTHERN CALIFORNIA STATION LIST

	Location		1/
	N Latitude	W Longitude	Status
Newport Beach Array	33°37.0'	117°57.0'	F
Balboa Beach Array	33°35.87'	117°54.1'	F
Oceanside Harbor North SP Gage	33°12.63'	117°24.0'	F
Oceanside Harbor South SP Gage	33°12.47'	117°23.6'	F
Oceanside Beach Array	33°11.42'	117°23.37'	O
Scripps Pier SP Gage	32°52.0'	117°15.4'	O
Pacific Beach Pier SP Gage	32°47.71'	117°15.64'	F
Mission Bay Entrance Array	32°45.4'	117°15.9'	O
Mission Bay System (6 SP Gages)	32°46'	117°14'	O
Ocean Beach Pier SP Gage	32°44.9'	117°15.5'	O
Silver Strand Array	32°37.79'	117°08.7'	F
Imperial Beach Pier Array	32°34.75'	117°08.42'	F

1/ Status Codes: O = Operating
80 = To be installed in CY 1980
F = To be installed after CY 1980
A = Abandoned (past records available)

2/ Formerly Port Hueneme Gage

Note that locations of future stations are subject to change.

MEASUREMENT NEEDS

U.S. Army Corps of Engineers district personnel have tentatively selected California Coastal Data Collection Program (CCDCP) measurement sites necessary to meet the wave statistic needs of present or planned projects or studies. These sites have been given priorities, and installation schedules are based upon these priorities.

The California Coastal Data Collection Program is required to provide the data needs of a large and diverse group of users. Insofar as possible, it is necessary to consolidate these needs into a compact set of requirements governing the design of the program to avoid exorbitant costs.

On 12 July 1978, a meeting of users of coastal data was hosted by the Los Angeles District. Generalizing from the results of that meeting, the following needs for wave data were expressed:

- Site specific nearshore wave climatology -- generally the extremes of height and period;

- Site specific nearshore wave climatology -- statistics in support of predictions about the general wave climate during a given season, the persistence of periods of low or high waves, and the potentials for both longshore and offshore sediment transport;

- Broad area real-time nearshore and deepwater wave conditions for operational purposes and forecasts;

- Deepwater and nearshore wave measurements at many discrete sites over a very broad region to verify hindcasting and refraction models.

The above needs are grouped into two major categories -- site specific and broad area measurements. Since some level of extrapolation is acceptable even for site specific data and since the discrete samples that provide the broad area coverage necessarily yield site specific data, it is obviousv that considerable economics may be achievved by carefully considering all the needs as a system.

The site specific, or project-oriented measurement points have been selected to meet the particular needs of that location. Among the more common specific needs are:

- A structure is under study and design data may be needed to determine maximum design wave height and period and approach directions of larger waves;

. The wave conditions affecting safe navigation are either doubtful, in dispute or unknown, and wave heights, energy spectra and approach directions may be needed to decide if a protective structure is necessary and feasible;

. Modifications to an existing structure such as a harbor entrance are planned and the wave climate data may be needed to provide criteria for modeling alternatives and to provide design data;

. A beach nourishment project is planned and the potentials for both on-offshore and longshore transport may be required to design this structure;

. A sand bypassing operation is required and estimates of maximum and net longshore transport may be required to design the plant;

. A shoreline erosion problem is being studied and the potentials for sand transport as well as the normal design wave data may be required to arrive at the optimum solution.

Since the wave dta need decision should be made long before the design decision phase of the study, these judgments are largely subjective and artistic. Only the project engineer is likely to possess the detailed knowledge of the problem, the economics and the possible alternatives necessary to determine the relative importance of a given project.

The broad area measurement sites have been evaluated upon consideration of general oceanographic features in a more straightforward way. Among the criteria considered are:

. To provide large-scale definition of deepwater wave climate at locations relatively free of the effects of refraction and shadowing to provide tests of hindcasting models;

. To provide definition of the nearshore wave climate at representative locations relatively free of severe shadowing or complex refraction to provide tests of refraction models;

. To provide real-time wave statistics in areas of abrupt changes in shoreline alignment or bathymetry.

There also exists a need for operational use of gage data. This near real-time use of the data will be realized by furnishing the National Weather Service with computer-to-computer links for their broadcasting of sea and swell data over their VHF weather channels. In addition, remote terminals at Corps and State offices allow for selecting gage data for analysis during times of storms or other times of need.

The program will make available to the general public a mass of operationally important data for commercial and recreational purposes. It must be understood that technical capabilities can grossly limit the choice of sites -- particularly in deep water -- and that final selection is necessarily a compromise.

The variety of needs of wave data, outlined in the previous paragraphs, produces a variety in the type of data required. In most cases, the type of data dictates the measurement approach.

The most commonly needed data are energy spectrum estimates for surface gravity waves in the range of periods up to about 20 seconds. These data are normally obtained by collecting relatively short (17 to 35 minute) records of sea surface elevation (or its analog, subsurface pressure). To identify this measurement form, it is convenient to refer to it as wave energy measurement.

A second commonly needed form of data is an estimate of the wave directional spectrum. This is normally limited to waves with periods less than 20 seconds and is most often obtain by considering the correlations between a number of wave energy measurement points in a properly designed array. This data type is referred to as wave direction measurement.

In certain cases, it is important to measure the energy in very long period waves — in the range from 20 to 200 seconds. In harbors or close to shore, these periods can produce resonant oscillations that can result in significant forces on structures. Much longer measurement times are required to quantify these long period waves -- on the order of 3 to 6 hours. To reduce the number of data points retained, the sampling rate is usually much lower than that used for ordinary surface gravity waves. This practice requires that the shorter period waves be filtered from the record, either mechanically, electronically, or numerically, to prevent aliasing. These data are referred to as long wave measurements.

In certain critical locations, it is necessary to know the actual maximum wave heights during a storm or the actual significant wave height at any specific time. These data can be inferred or estimated from occasionaly wave energy measurements, but with reduced reliability. In addition, the occasional measurement provides little data on the tendency for ocean waves to produce groups of larger waves. To meet the need for continuous measurement without incurring the burden of recording excessive amounts of data, some consolidation of the data is normally accomplished in the field.

One method of accomplishing this consolidation is to preserve only the maxima. By measuring zero crossing periods, an estimate of the maximum period is obtained. By measuring the elevation differences between successive crests and troughs, an estimate of the maximum wave height is obtained. By retaining only the maxima over the time period of interest, considerable data shrinkage can be accomplished. This form of data is referred to as maximum height and period measurement.

A second method of data consolidation for continuous measurement is a running estimate of the total energy, or variance, of the sea surface. For example, averaging the squared value of sea surface elevation from the mean over one minute intervals provides an estimate of the variance. This allows piecewise estimation of significant wave heights and height maxima and a means for analyzing wave groups. This data form is referred to as continuous variance measurement.

As discussed in the previous paragraphs, the best evaluation of project oriented candidate sites are made by those engineers with the greatest familiarity with project needs. It is recognized that these recommendations are, in certain cases, tentative and subject to review as study priorities change. However, these represent the best thinking at this time and are of particular value in determining the scope of the whole program.

The central processing facility is nearing completion at Scripps Institution of Oceanography with only minor hardware and software development remaining. The computer processed data products are already being produced and provided to concerned agencies, institutions, and user communities.

Although the primary reason for establishing the wave gaging system is to develop wave statistics to satisfy coastal engineering and wave hindcast modeling requirements, the spin-off of near real-time data for use by recreational boaters, fishermen, search and rescue units and other operational users should be of enormous value. The benefits of this aspect with respect to lives and property saved, should prove to be a greater value than the cost of the system.

ACKNOWLEDGMENT

The authors wish to acknowledge the Army Corps of Engineers and the State of California for access and permission to use this study material. The views of the authors do not purport to reflect the position of the Corps of Engineers, Department of Army, or the State of California.

Contents of this paper are not to be used for advertising or promotional purposes. Citation of trade names does not constitute an official endorsement or approval of the use of such commercial products.

THE USE OF SYNTHETIC FABRICS IN COASTAL DUNE STABILIZATION

Donald K. Stauble[1]

ABSTRACT

A synthetic fabric with an without a filter cloth under-liner has been used in combination with new dune vegetation to stabilize a non-vegetated dune face. The fabric known as Enkamat, is a soil matting made from nylon monofilament fused at their intersection, creating a three dimensional structure with very porous openings. This porous surface (90% by volume) is designed to hold sediment and new dune grass in place until it establishes a viable root system.

The demonstration area, a 55m (180.14 ft.) section of dune escarpment, is located along the Atlantic Ocean in Brevard County, Florida. The site has been divided into five sections, each 10m (32.8 ft.) long. To compare the effectiveness of the thickness of mat material in this high wave energy location, two different types of Enkamat have been used. The test has also included a layer of Stabilenka fiber fabric to determine if it would have any added effect on securing plant growth to the dune face under conditions of heavy wave attack. A comparison of well vegetated and non-vegetated control sections was also included.

Just three weeks after installation, Hurricane David made landfall south of Vero Beach, Florida and the eye of the storm proceeded up the coastline over the study site. Storm tides were well above normal but only affected the lower portion of the mat and dune base. From the debris line of the lower dune storm waves appeared to have risen up the dune face approximately 1 meter. The control section at the base of the natural dune eroded back, while sand was held under the Enkamat and Stabilenka.

One year after installation, the vegetation and sand has revegetated the sections with the two thicknesses of mat material to a point where the mat is almost covered. The mat/filter cloth section is still bare of all sand and vegetation. While the mat/filter fabric combination works best during storm conditions, where sand trapped under the mat and cloth is prevented from being washed out, the cloth also prevents the natural process of dune revegetation from taking place. Comparing the behavior of the mat versus controlled areas, there appears to be an advantage to having the mat stabilizing the dune base.

---

[1] Assistant Professor, Dept. of Oceanography and Ocean Engineering Florida Institute of Technology, Melbourne, Florida 32901.

INTRODUCTION

Synthetic fabrics have recently been used in solving a variety of coastal engineering problems. Enkamat is a product of American Enka Company, originally designed for streambank and lakeshore areas where erosion prevents the normal establishment of vegetation. It is a three dimensional black fabric in the form of a mat consisting of entangled nylon monofilament, forming a porous structure much thicker than the typical woven fabrics. It is designed to trap and hold sediment and roots in its tangled fibers until the vegetation establishes itself. The mat is provided in two thicknesses, the type 7010 is the thinner of the two with a thickness of 9mm, while the type 7020 is approximately 18mm thick. Both mats come in rolls of approximately 1m widths. Both thicknesses of mat are being tested in this open coast dune stabilization project. A filter fabric known as Stabilenka, has also been used under the Enkamat in some installations. The Stabilenka is a non-woven, continuous filament polyester fabric, bonded with an acrylic binder to form a permeable sheet.

LOCATION AND SITE DESCRIPTION

The demonstration area is located along the Atlantic Ocean 1.5 miles north of Indialantic, Florida in Brevard County (Figure 1). The site is a 55m (180.4 ft.) long section of dune escarpment extending from a point 16m (52.49 ft.) southward of the southermost dune crossover at Paradise Beach County Park. This point is approximately 169m (554.5 ft.) southeast of the Florida Department of Natural Resources range monument #R111, Brevard County, Florida. The site is devoid of vegetation on the dune face. The typical profile at the site has a dune face that rises on a 1. or 1.25 slope from an elevation of 2.91m (9.55 ft.) above mean sea level to around 5.47m (17.95 ft.) above sea level. The width of the slope is approximately 2.8m (9.19 ft.) along the length of the study area. While the dune face is well above the normal water elevation, the base of the dune is subject to erosion by waves generated during mild storms. The entire dune face is subject to scarping and severe erosion by wave attack during moderate to severe tropical storms (hurricanes) and extratropical storms (northeasters).

PROJECT PLAN

The demonstration site has been divided into five sections, four of which are 10m (32.8 ft.) long. Section A served as a control, and is 15m (49.2 ft.) long, with a fair amount of natural vegetation (Figure 2). To compare the effectiveness of the thickness of mat material in this high wave energy location, the two different types of Enkamat have been used. A third section also included a layer of Stabilenka fiber fabric to determine if it would have any added effect on securing plant growth to the dune face under conditions of heavy wave attack. Section B has been covered with the thinner Enkamat type 7010, Section C has been covered with the thicker Enkamat type 7020 and Section D has been covered with Enkamat type 7020 laid on top of a layer of Stabilenka type T-80. Section E is a 10m (32.8 ft.) long area with

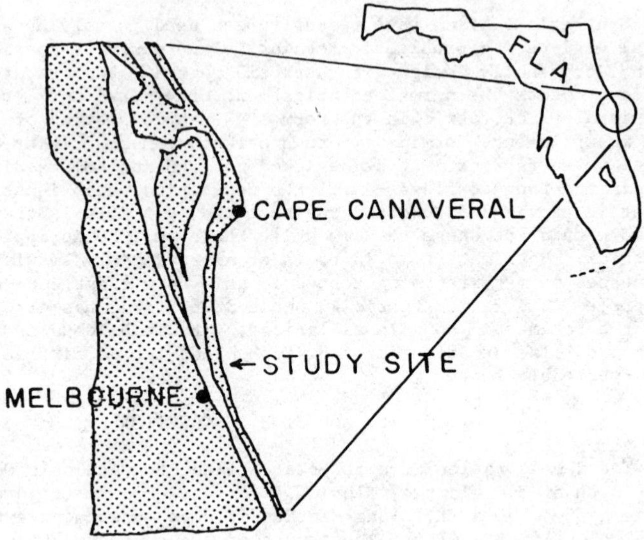

FIGURE 1. LOCATION MAP

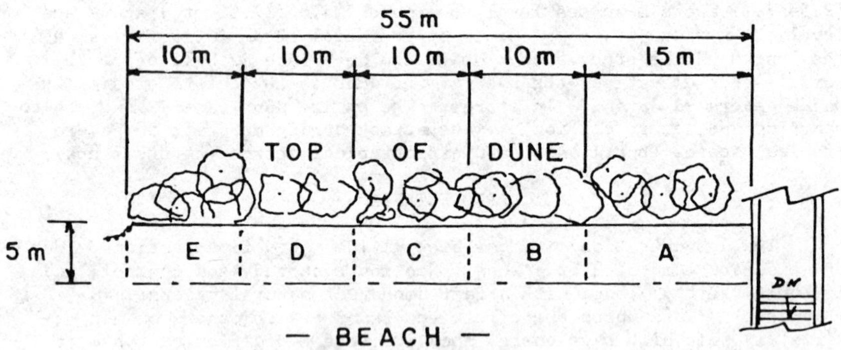

FIGURE 2. SITE LAYOUT

SECTION A - VEGETATED CONTROL
SECTION B - ENKAMAT 7010
SECTION C - ENKAMAT 7020
SECTION D - ENKAMAT 7020/STABILENKA T80
SECTION E - NON-VEGETATED CONTROL

little vegetation and was a control section also, thus allowing comparison of well vegetated with non-vegetated control sections.

INSTALLATION

The Enkamat installation was begun on 13 August 1979. A pre-installation survey was taken before any work was done. Five profiles were established at midpoint of each section A thru E. They extended from the top of the existing dune down to just seaward of daily low tide.

The mat was cut into 7m (23.1 ft.) lengths and stretched from the top of the dune to the base. A 0.5m deep trench was cut at the top of the dune and the top end of the mat was placed in this trench. The 1m (39 inch) wide mat was overlapped approximately 0.10m (3.9 inches). To secure the mat to each other and the dune staples were constructed using 5/8" diameter rebarb rod. Two designs were used, Type 1 a hook design 0.30m (1 ft.) long was used along the top and sides of the mat. Type 2 a U shaped design 0.30m (1 ft.) long and 0.10m (3.9 inches) wide was used along the base of the dune.

The mat was secured at the dune base by digging a 1m deep trench along the dune base, laying down the mat, then placing an 0.1m (4.1 inch) diameter aluminum rod along the trench. The U shaped staples were then hammered in over the rod. The excess mat was folded back over the rod and the entire unit was buried.

Zone D was installed on 14 August, which consisted of Enkamat 7020 over Stabilenka T80. The T80 was rolled down from the top of the dune along with one roll of 7020 and staked. The process was repeated until the entire zone was covered and staked. The bottom trench was then filled in after the mat was stretched and contoured to fit flush with the dune face.

The entire site was then top dressed with a layer of beach sand. Difficulties were encountered in this process due to the steep slope of the dune face and the inability to use power equipment. The project was done completely with shovels using local beach sand in front of the dune.

On 15 August the Brevard County Agricultural Extension Service supplies shore plants for the experimental planting, three different types of plants were supplied, all of which are native to the area. Sea oats (*Uniola paniculata*)were supplied for the dune base while panic grass (*Panicum amarum*)were used on the dune face. Ten sea grape (*Coccolaba unifera*) scrub plants were placed at the dune top in hopes of supplying additional holding power at the top of the mat. Knife cuts in an X pattern were required in the Enkamat and Enkamat/Stabilinka combination in order to place the individual plants into the dune. After planting, the plants and mat were then re-covered. The plants were placed in the fabric on 1m (39 inch) centers in staggered rows of alternate planting from top to base of the dune. Standard 6-6-6 fertilizer was then broadcast onto the dune and watered lightly with a garden hose.

HURRICANE DAVID

Just three weeks after installation, the first significant storm occurred. On 3-4 September 1980 Hurricane David made landfall south of Vero Beach, Florida and the eye of the storm proceeded up the coastline over the study site. Unofficial reports of winds up to 85 mph were reported in the area. The storm center passed over the site on a falling tide, which was fortunate for the coastal residences, but did not afford an acid test for the performance of the mat. Storm tides were well above normal but only affected the lower portion of the mat and dune base. All the top dressing sand was washed off from rain and some of the new vegetation was lost, having not had a chance to stabilize. From the debris line on the lower dune, storm waves appeared to have risen up the dune face approximately 1 meter. The base of the mat held in place and sand was trapped under the mat. This sand would have normally eroded during the storm judging from the dune base retreat in the control sections. Close inspection showed that sand was held exceptionally well under the Enkamat and Stabilinka Zone D. The filter fabric prevented sand from washing out and a scarp was formed by the mat base where sand from in front of the cloth was eroded. The base of dune installation technique succeeded in holding despite the scouring by wave action.

Each test section behaved slightly different during the hurricane. From profiles taken before and immediately after the storm, it was found that the thinner 7010 mat in Section B was unable to retain sediment under the mat. While the mat held in place, 51.8 cm^3/cm of sand was eroded and there was marked settlement on the dune face. The thicker 7020 mat in Section C was able to trap sand within its 3-dimensional structure for a net accretion of 59.2 cm^3/cm. The Enkamat/Stabilenka combination was completely denuded of its top dressed sand and vegetation, but trapped a significant amount of sand under the filter cloth at the base of the mat installation.

POST HURRICANE PROFILE CHANGES

In order to evaluate the performance of the various configurations of Enkamat and Stabilinka, monthly beach profiles have been measured for each of the five test sections. The profiles start at the top of the dune,(designated as the backstake) proceed down the dune face to the dune base at approximately 5m and continue out to the vicinity of the low tide swash area (45m). The area of critical interest to this study is the first 5 meters of the profile on the sloping face of the dune. Most of the discussion will center on this area of the profile. Section A is a normally vegetated control section, Section E is a non-vegetated control section, thus these areas have not been modified by the installation of the mat material. Sections B thru D, from the top of the dune (0m) to dune base (5m) have been reconstructed with the installation of the combinations of Enkamat and Stabelinka.

Comparing the year long profile changes, the Section A profile (Figure 3) of the vegetated control section shows that from the mid

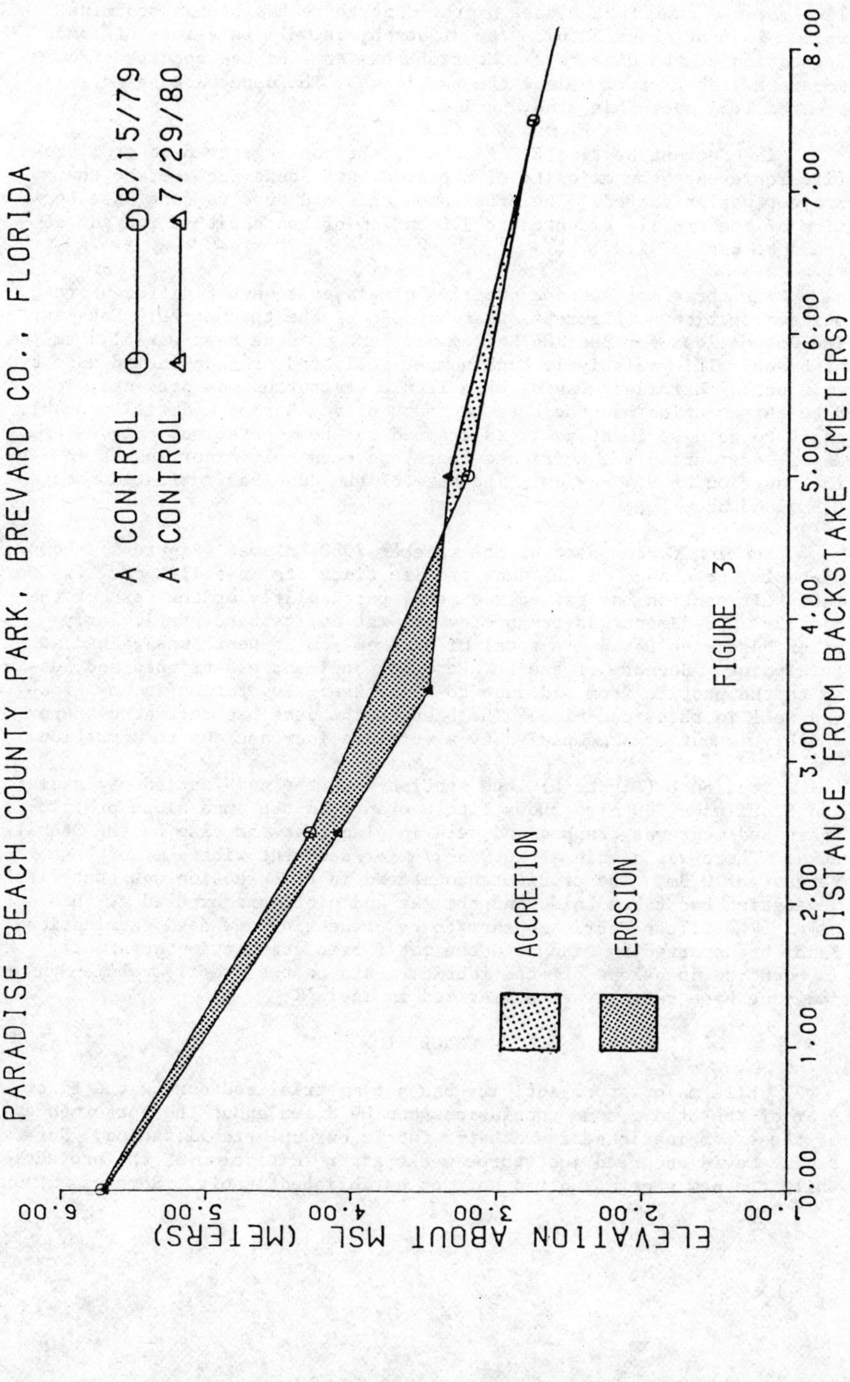
PARADISE BEACH COUNTY PARK, BREVARD CO., FLORIDA
A CONTROL 8/15/79
A CONTROL 7/29/80
ACCRETION
EROSION
ELEVATION ABOUT MSL (METERS)
DISTANCE FROM BACKSTAKE (METERS)
0.00
1.00
2.00
3.00
4.00
5.00
6.00
7.00
8.00
1.00
2.00
3.00
4.00
5.00
6.00

FIGURE 3

level of the dune to the base of the dune there has been a sediment volume loss of about 53.6 cm^3/cm of beach, as well as a loss of some vegetation at the dune base. Most of this erosion has occurred from moderate winter storms since the hurricane. The dune base has retreated 1.5m over this study period.

In contrast Section E, (Figure 7) the non-vegetated control profile represents the majority of the study area dune face before the project was initiated. The erosion on this mid dune to dune base section of the profile amounted to 3.0 cm^3/cm of beach, with a retreat of the dune base of 1.2.

The three mat section profiles displayed somewhat different results. Section B (Figure 4) was composed of the thinner 7010 Enkamat. The volume loss for Section B from mid dune to dune base was 61.2 cm^3/cm of beach. This relatively high number reflected eroison in mid dune by rain during Hurricand David, when little vegetation was present. To date this section has the best regrowth of vegetation and will probably start to gain sediment as it is trapped by the vegetation. Since the major loss during the hurricane there has been only minor losses in this section of 9.4 cm^3/cm. The base of the dune has retreated about 0.5m landward.

Section C, composed of the thicker 7020 Enkamat (Figure 5) shows virtually no change in the dune profile since its installation. In fact this section has gained sediment, particularly at the base of the mat (5m). Sediment was trapped by the mat at its base particularly after Hurricane David. A total of 31.2 cm^3/cm of beach was gained at this point. Because of the thicker mat, sediment was trapped and added to the profile from mid dune to dune base. Vegetation is now growing well in this zone also. The base of the dune has retreated approximately 1m but is accompanied by a verticle increase due to accretion.

Section D (Figure 6), the section with the combination of Enkamat and Stabilenka T80 also shows little change in the dune slope profile. Again sediment was trapped and held in place, at the base by the Stabilenka. There was a gain of 20.9 cm^3/cm of sediment with a minor base retreat of 0.3m. The problem encountered in this section was that little vegetation has taken hold, and the mat and cloth are exposed at this time. The filter fabric appears to be preventing new seed germination. Sand, transported by wind onto the mat/fabric combination apparently is bouncing downslope off the fabric. Data on the profile volume changes and dune base retreat are summarized in Table 1.

CONCLUSION

While major storm activity has not materialized during the first year of the study, some conclusions can be drawn about the performance of these combinations of synthetic fabric on dune stabilization. Hurricane David occurred just three weeks after initiation of the project, while the new vegetation had not yet established itself. Some vegetation

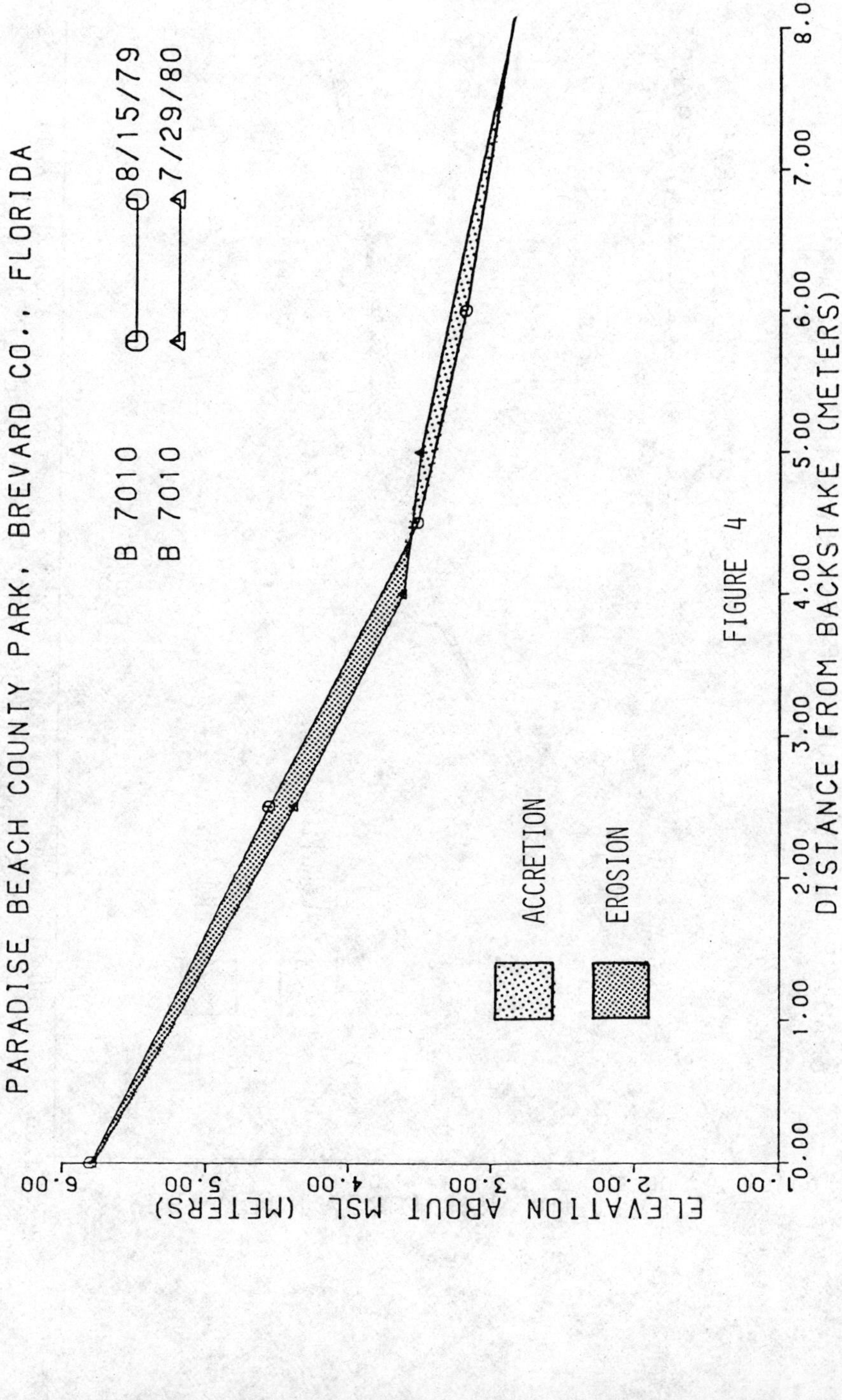
PARADISE BEACH COUNTY PARK, BREVARD CO., FLORIDA
B 7010 8/15/79
B 7010 7/29/80
ACCRETION
EROSION
ELEVATION ABOUT MSL (METERS)
6.00
5.00
4.00
3.00
2.00
1.00
DISTANCE FROM BACKSTAKE (METERS)
0.00
1.00
2.00
3.00
4.00
5.00
6.00
7.00
8.00

FIGURE 4

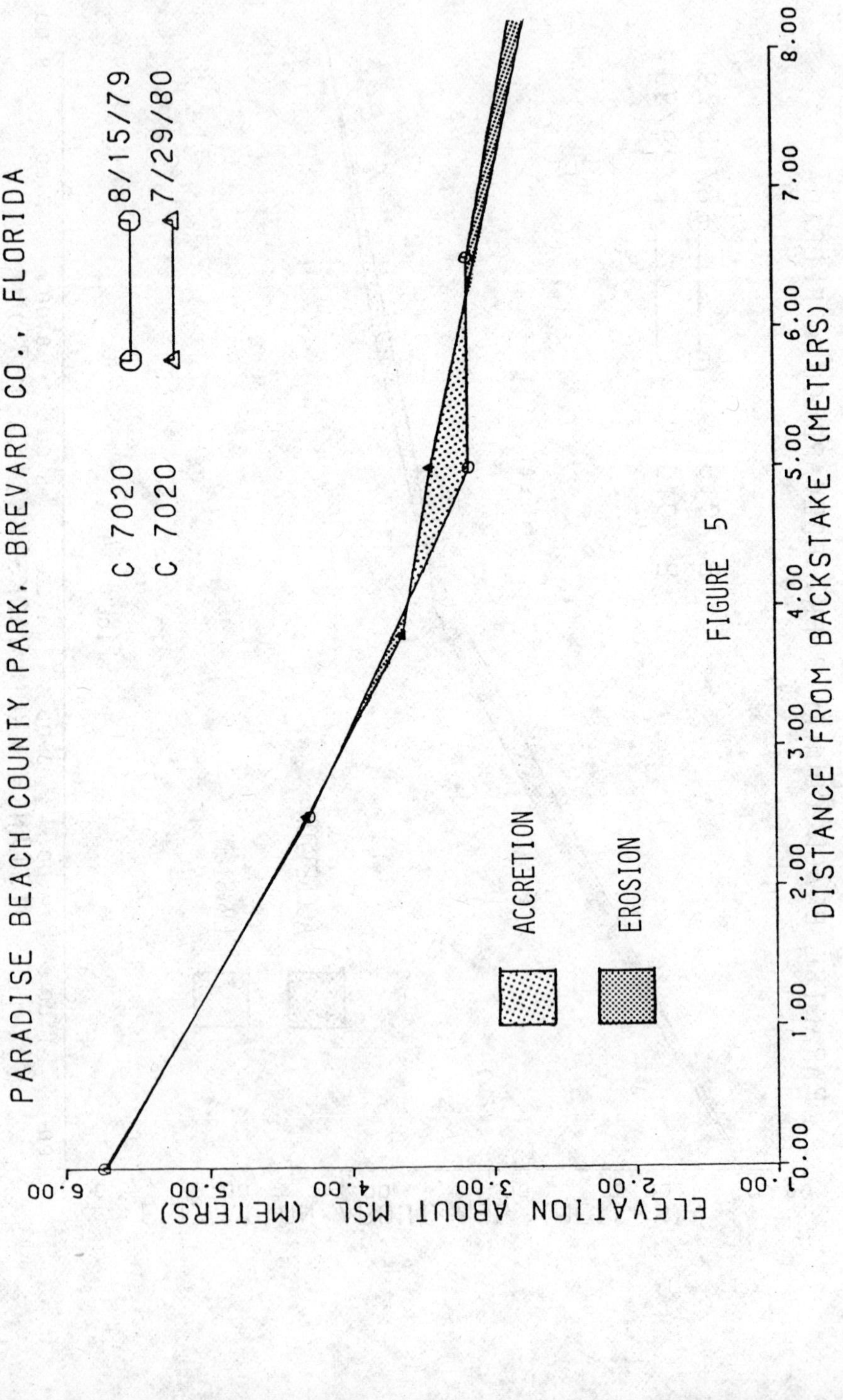
PARADISE BEACH COUNTY PARK, BREVARD CO., FLORIDA
C 7020 8/15/79
C 7020 7/29/80
ACCRETION
EROSION
ELEVATION ABOUT MSL (METERS)
DISTANCE FROM BACKSTAKE (METERS)
6.00
5.00
4.00
3.00
2.00
1.00
0.00
1.00
2.00
3.00
4.00
5.00
6.00
7.00
8.00

FIGURE 5

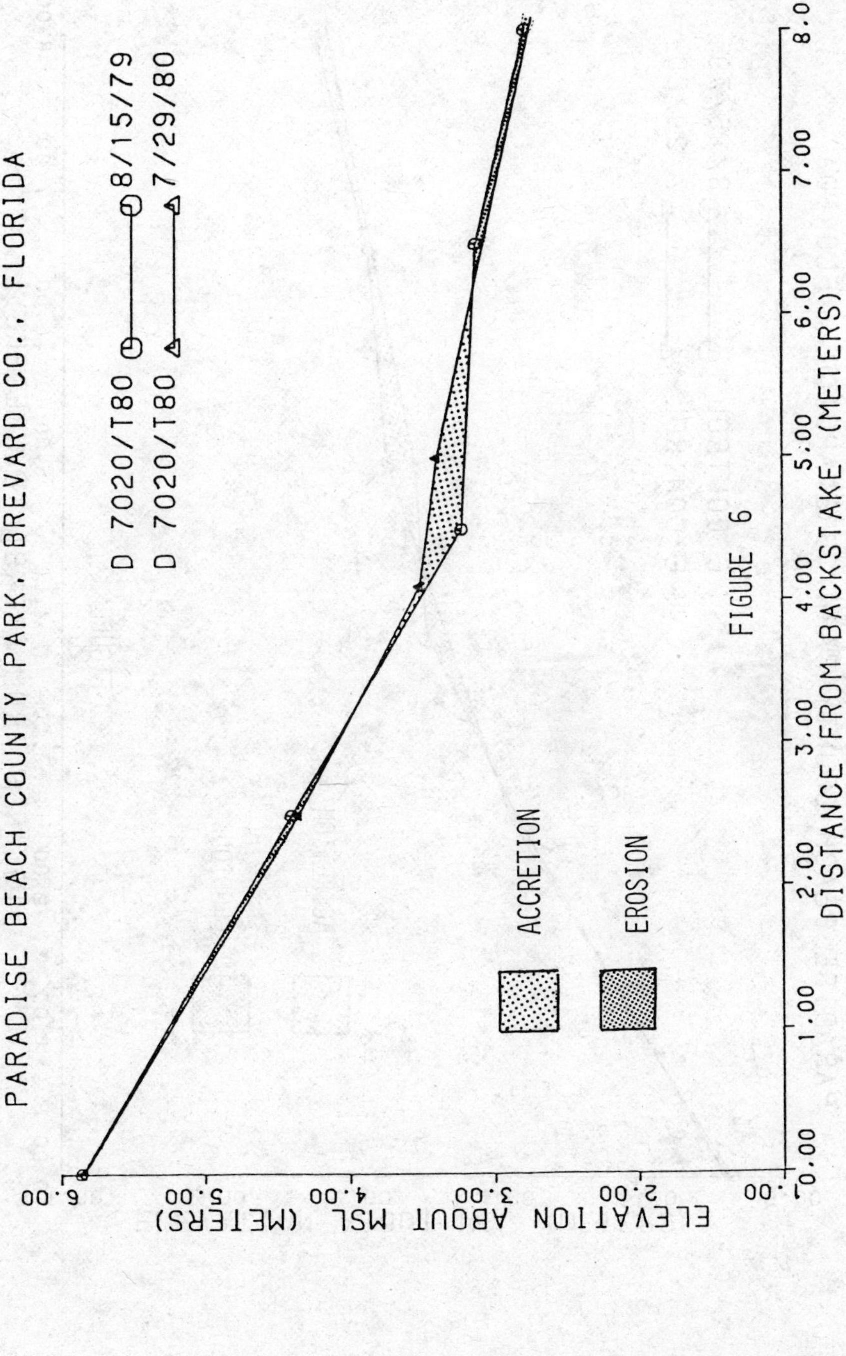
PARADISE BEACH COUNTY PARK, BREVARD CO., FLORIDA
D 7020/T80 8/15/79
D 7020/T80 7/29/80
ACCRETION
EROSION
ELEVATION ABOUT MSL (METERS)
1.00
2.00
3.00
4.00
5.00
6.00
0.00
1.00
2.00
3.00
4.00
5.00
6.00
7.00
8.00
DISTANCE FROM BACKSTAKE (METERS)

FIGURE 6

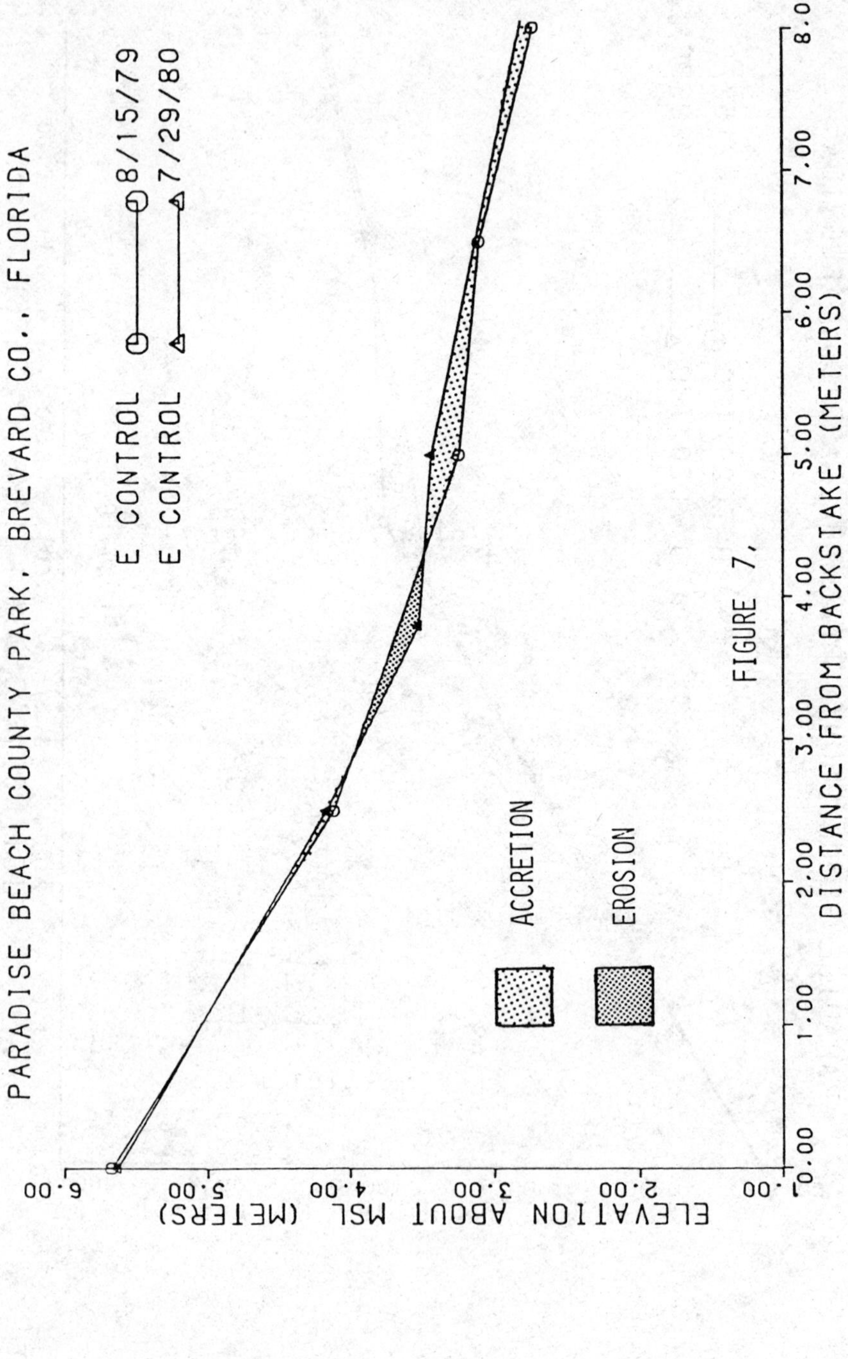
PARADISE BEACH COUNTY PARK, BREVARD CO., FLORIDA
E CONTROL 8/15/79
E CONTROL 7/29/80
ACCRETION
EROSION
ELEVATION ABOUT MSL (METERS)
DISTANCE FROM BACKSTAKE (METERS)
6.00
5.00
4.00
3.00
2.00
1.00
0.00
1.00
2.00
3.00
4.00
5.00
6.00
7.00
8.00

FIGURE 7.

TABLE 1

	Sediment Volume Changes Backstage (0m) to 8/15/79-9/7/79 (Hurricane David)	Dune Base (5m) cm^3/cm 8/5/79-7/29/80 (Total Change)
Section A Vegetated Control	+ 2.8	-53.6
Section B Enkamat 7010 (9mm) thick	-51.8	-59.2
Section C Enkamat 7020 (18mm) thick	+59.2	+31.2
Section D Enkamat 7020/ Stabilinka T80	- 7.7	+20.9
Section E Non-vegetated Control	- 7.4	- 3.0

	Dune Base Retreat 8/15/79-7/29/80
Section A Vegetated Control	-1.5 W/erosion at Base
Section B Enkamat 7010 (9mm) Thick	-0.5 W/erosion at Base
Section C Enkamat 7020 (18mm) Thick	-1.1 W/accretion at Base
Section D Enkamat 7020/ Stabilinka T80	-0.3 W/accretion at Base
Section E Non-vegetated Control	-1.2 W/erosion at Base

was lost, particularly at the dune base, but the thicker Enkamat 7020 and 7020/Stabilenka T80 combination trapped sand during the storm and maintained the dune base, compared with the natural control sections. While the thinner Enkamat 7010 did not perform well during the storm, allowing sediment to erode from under the mat, to date it is revegetated to a point where the mat is practically covered and eolian transported sand is now being trapped. The thicker, Enkamat 7020 did perform well during storm conditions, and has reestablished vegetation as well as the thinner version. Although the mat/filter cloth combination performed the best, of all test sections during storm conditions, with the cloth holding sand beneath the fabrics, the fabric also prevents natural reestablishment of the vegetation. This combination test section has not lost any sand but it is practically devoid of vegetation one year after installation. In comparing the response after one year of the natural dune sections, these areas have eroded back to a greater extent than the mat protected areas. The vegetation in these natural sections are basically unchanged within the study year. Control Section E is still basically non-vegetated while the two mat test sections are now revegetated to a point where their appearance is similar to the vegetated control section.

ACKNOWLEDGEMENTS

This paper is the partial product of a study sponsored by the American Enka Company, Enka, N. C. Plant material was provided by Richard Van Epp of the Brevard County Agricultural Extension Service. James Vencill of the Brevard County District 5 Parks and Recreation Department allowed the use of Paradise Beach Park for the study. Installation and field studies were conducted by graduate students of the Department of Oceanography and Ocean Engineering.

SOIL-CEMENT FOR PROTECTION OF SHORELINES

by

William G. Dinchak, P.E.
Director, Department of Energy and Water Resources
Portland Cement Association, Skokie, Illinois

ABSTRACT

Compacted soil-cement has a record of over 30 years' use in energy and water resources projects. Recently it was used to protect about 6,000 ft (1,800 m) of shoreline on the St. Lawrence River shore of the Gaspé Peninsula, Quebec, Canada.

Basic information on soil-cement as a construction material is presented, including embankment protection work in the United States and Canada, and considerable detail about the construction and performance of Quebec shore protection.

INTRODUCTION

During the past 30 years a new and very useful tool has been developed for designers and builders of dams and other embankments. Soil-cement, which has been used successfully as a paving material in the United States and other countries for more than 40 years, has proved, to be very effective as revetment for embankments exposed to water, particularly earth dams; highway, railroad, and other embankments; and shorelines subject to severe wave action (Fig. 1).

SOIL-CEMENT AS A MATERIAL

Soil-cement can be broadly defined as a mixture of natural soils, portland cement, and water. In its moist-compacted state it forms a strong, durable, and economical material with properties similar to

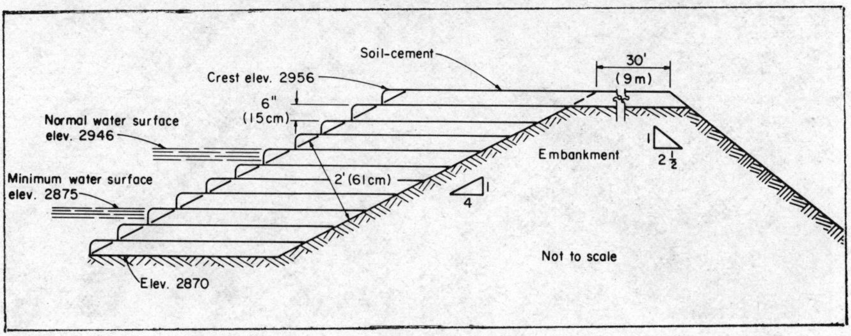

Fig. 1. Details of soil-cement embankment protection at Merritt Dam, Nebraska.

concrete and natural rock. American Society for Testing and Materials standard tests in connection with other criteria are used to determine, for any given soil, the cement content and moisture content necessary to produce the strength, durability, and other characteristics that may be desired. A typical mixture contains about 10% cement by weight of the dry soil, but actual cement contents have ranged from less than 7% to more than 14% by weight.

SOIL-CEMENT AS EMBANKMENT PROTECTION

Soil-cement was first seriously considered as embankment protection for earth dams in 1947 when it was found that rock suitable for riprap was not available at many sites where water storage was most needed. Long hauls of 200 miles (322 kilometers) or more from quarries to these sites were resulting in riprap costs of up to $18 per cubic yard ($24 per cubic meter). Such high costs threatened the economic feasibility of some of the dams. An equally durable but more economical method of embankment protection was needed. In 1980 riprap is even more costly and rock quarries are environmentally undesirable, making soil-cement even more attractive as a substitute for natural rock.

In 1951, following a great deal of laboratory research and a number of conferences between engineers of the Water and Power Resources Services (WPRS), formerly the U.S. Bureau of Reclamation, and the Portland

Fig. 2. Soil-cement test section at Bonny Reservoir, Colorado, after 29 years of exposure.

Cement Association, the WPRS built a test section of soil-cement embankment protection on the southeast shore of Bonny Reservoir in eastern Colorado. The test section site was selected for maximum exposure: it was subjected to severe wave action producing countless cycles of wetting and drying, and more than 100 cycles of freezing and thawing each year. The soil-cement test section was 24 ft (7 m) high and about 300 ft (about 100 m) long. It consisted of a series of 6-in. (15-cm)-thick by 7-ft (2-m)-wide horizontal layers of soil-cement, one on top of another, each successive layer being offset 12 in. (30 cm) from the preceding layer so that the exposed, finished embankment was 2:1 (horizontal to vertical).[1]

Very soon after completion of the test section, water stored by the dam reached the normal operating level of the reservoir, about three feet (one meter) below the top of the soil-cement. The water has been maintained at that elevation almost continuously since 1951, (Fig. 2) except that on at least one occasion the water rose higher than the top of the soil-cement. The completed test section was inspected frequently during its first year and very thoroughly in 1961, when it was 10 years old.[1] Its satisfactory performance during those 10 years led the Water and Power Resources Services to decide that the engineering feasibility of this new method of embankment protection had been proved beyond reasonable doubt. On this basis the WPRS, in 1961, prepared plans and specifications for both soil-cement and natural riprap as erosion protection for Merritt Dam, a 120-ft (37-m)-high earth dam located in the sand hills region of north-central Nebraska, and Cheney Dam, an 84-ft (26-m)-high, 5-mile (8-kilometer)-long earth dam about 30 miles (48 kilometers) west of Wichita, Kansas. Contractors were invited to bid on either or both construction materials, and many submitted bids on both. Based on the contractors' unit bid prices for soil-cement and riprap, the saving through the use of 51,000 cu yd (38,995 m^3) of soil-cement at Merritt Dam was nearly $500,000 while at Cheney Dam the use of 180,000 cu yd (137,628 m^3) of soil-cement saved about $700,000.

One of the best examples of demonstrated durability for a full-scale project is the soil-cement embankment protection for Louisiana State Highway 6 at the Toledo Bend Reservoir, near Many, Louisiana. At full

Fig. 3. Soil-cement embankment protection for Louisiana State Highway 6 at Toledo Bend Reservoir.

pool the reservoir has an area of 181,600 acres (73,492 ha) and a capacity of 4,477,000 acre feet ($5.52x10^9$ m^3). Following 14 years of service this soil-cement embankment is still in excellent condition (Fig. 3). The sharp vertical edges and the 8-ft (2.4-m)-wide, 6-in. (15-cm)-thick level layers are especially noteworthy because of the recreational and aesthetic possibilities. A motor grader with a notched blade was used to cut these edges and the wasted material was later removed from the site leaving the level layers.

TABLE 1.
SOIL-CEMENT IN ENERGY AND WATER RESOURCES

Project and Location	Quantity, yd^3	Complete soil-cement cost per yd^3
Contracts Awarded in 1979		
Gulf of St. Lawrence Shore Protection, Quebec	5,710	$52.00
Lower Running Water Draw, Site 1, Texas	6,050	26.20
Roger Road Wastewater Treatment Plant Dike, Arizona	60,000	-
Sterling Wastewater Treatment Lagoons, Colorado	13,000	20.00
Contracts Awarded in 1978		
Pawnee Plant Reservoir, Colorado	90,000	$16.86
Contracts Awarded in 1977		
Black Bayou Dam, Louisiana	20,300	$15.00
Coleto Creek Dam, Texas	38,150	18.05
Coleto Creek Power Station, Dikes and Flumes, Texas	3,540	19.60
Coleto Creek Project, Road Relocations, Texas	12,245	19.00
Coleto Creek U.S. 59 and FM2987 Highway Emb., Texas	3,950	23.52
Forest Grove, U.S. 175 Hwy. Emb., Texas	14,005	21.74
Gibbons Creek Dam, Texas	37,200	16.86
Gibbons Creek Plant, R.R. Spur Emb., Texas	51,500	15.22
Lake Fork FM17 Hwy. Emb., Texas	73,415	14.39
Lake Fork FM2225 Hwy. Emb., Texas	32,950	16.36
Lake Fork SH19S and FM 514W Hwy. Emb., Texas	18,330	14.09
Running Water Draw, Site No. 3 Dam, Texas	10,480	17.15

In 1979, construction was completed on the largest soil-cement energy and water resources contract awarded to date, the erosion protection for about 20 miles (32 kilometers) of 39- to 52-ft (12- to 16-m)-high embankment--tops of interior dikes are generally 10 ft (3 m) lower--forming the cooling water reservoir for a nuclear power plant near Bay City, Texas. Protecting these embankments required about 1.2 million cu yd (918,000 m^3) of soil-cement made from silty fine sand from borrow pits within the reservoir, mixed with cement and water in two 1,000-ton (907,200-kilogram)-per-hour mix plants, and compacted by self-propelled, dual-wheel vibratory rollers.

Table 1 provides cost information for soil-cement contracts awarded in 1977, 1978, and 1979. The contracts are for embankment projects in which the soil-cement was constructed in a stairstep configuration. The total cost includes cost of processing and cost of the cement. Cost information for contracts awarded prior to 1977 is available.(2)

From 1961 through 1979 at least 214 such projects have been built or are under construction in the United States and Canada. These are summarized in Table 2. They utilized about 9.7 million cu yd (7.4 million m^3) of soil-cement.

TABLE 2.
SOIL-CEMENT FOR ENERGY AND WATER RESOURCES
UNITED STATES AND CANADA

214 projects through 1979:		9,711,248 cu yd (7,425,240 m^3)
Alternate bids on 23 projects:		1,130,448 cu yd (864,343 m^3)
Total saving:	$4,889,227, or	$4.33 per cu yd ($5.66 per m^3)
No alternate bid submitted on 9 other projects:		743,200 cu yd (568,252 m^3)
No alternate specified for 182 projects:		7,837,600 cu yd (5,992,645 m^3)

On 23 projects for which alternates were specified and on which alternate bids were received, the total saving due to use of soil-cement was nearly $4.9 million, which is equivalent to $4.33 per cubic yard ($5.66 per m^3). Data are not available to calculate savings on the other 191 projects for which alternates were not specified or for which no contractor submitted alternate bids where they were permitted. Where adequate supplies of suitable soil aggregate are readily available, which almost always is the case, soil-cement for embankment protection usually is competitive with rock riprap that comes from other than required excavation and has to be hauled more than about 25 miles (40 kilometers).(3)

Fig. 4. Winter storm at Ste.-Anne-des-Monts on Gaspé Peninsula, Quebec, in December 1978.

In 1975 approximately 6,000 ft (1,800 m) of shoreline at three villages on the north shore of the Gaspé Peninsula, Quebec, Canada, (Figs. 5 and 6) were protected with compacted soil-cement constructed stairstep fashion to provide a solid thickness of about 2.5 ft (75 cm) measured normal to the slope. During the succeeding three winters these locations, which are in a seawater environment, were subjected to typical Atlantic Coast storms, with wind velocities up to 70 miles per hour (113 kilometers per hour), which generated waves more than 10 ft (3 m) high (Fig. 4). Some surface attrition and a few minor failures of the exposed soil-cement resulted. However, none of the structures protected by the soil-cement was damaged, and the soil-cement protection should be effective for many years to come (Fig. 7).

LABORATORY TESTS FOR SOIL-CEMENT

Laboratory testing to establish the required cement and moisture factors for these soil-cement projects requires about six weeks. Optimum moisture and maximum density are determined by ASTM test D-558. The required cement content is determined by ASTM D-559 and D-560 plus the Portland Cement Association weight-loss criteria. For sandy soils shortcut test procedures, requiring only about one week, can be used. These procedures do not always give the minimum cement factor, but they almost always provide a safe cement factor. Detailed instructions for conducting all these tests, which were developed for soil-cement road bases but have been adapted for energy and water resources structures, are contained in Portland Cement Association, IS166W.[4]

The research on soil-cement that formed the basis for these tests and the resulting design criteria was carried out by the Water and Power Resources Service,[1 5] the Portland Cement Association,[6] and other agencies.

An important part of PCA research was to establish wave-runup factors for various configurations of soil-cement facings. The wave-runup data were reported by Nussbaum and Colley.[6] Experience has shown that the stairstep effect produced by constructing soil-cement embankment protection in the manner described is effective in breaking up the waves. However, when very severe wave action is expected, as would be the case in very large reservoirs or on the ocean shores, where high wind velocities are common, the runup factor of 1.3 times the design wave height should be adequate.[7]

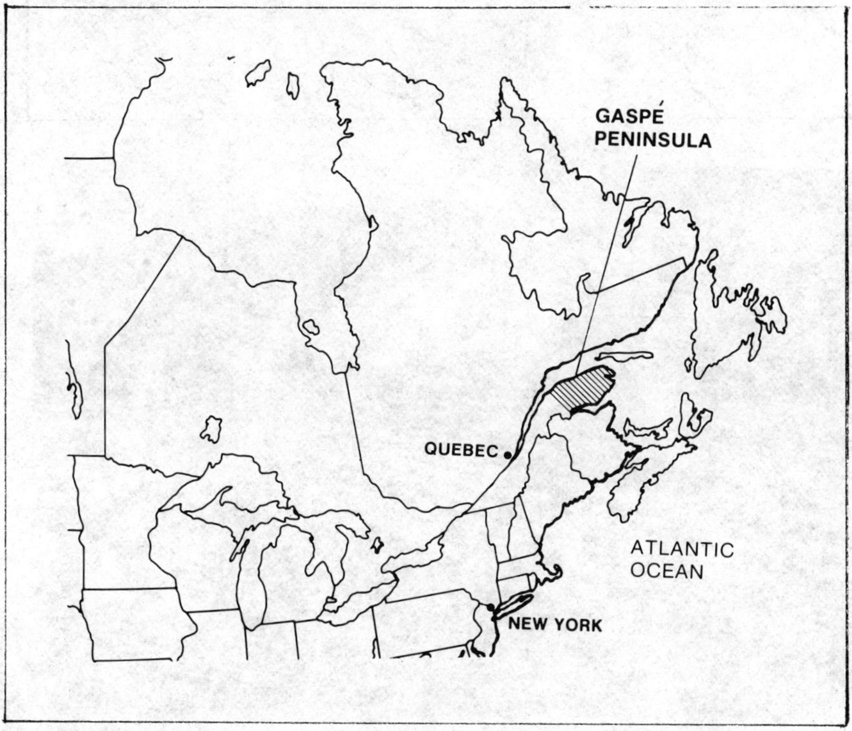

Fig. 5

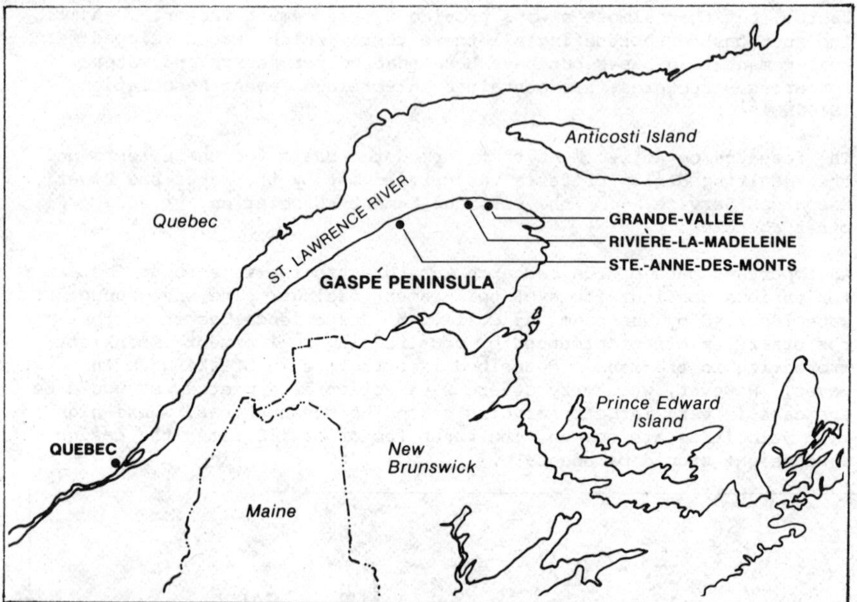

Fig. 6.

Fig. 7. Soil-cement shore protection at Grande-Vallée on Gaspé Peninsula, Quebec.

DESIGN AND CONSTRUCTION OF SOIL-CEMENT EMBANKMENT PROTECTION

Usually no special considerations are necessary in the design of an embankment that is to have a soil-cement facing.(3 7) While soil-cement is essentially impervious, some cracking is inevitable; therefore the facing cannot be considered to be totally watertight. Also, leakage may sometimes take place through the bonding surfaces between layers. Thus the possibility that water may get behind the facing should be taken into account in the design.

Three design concepts have been used to prevent damage to the embankment due to such leakage through the facing:

1. The weight and strength of the facing is considered to be sufficient to resist any uplift pressures that may develop, particularly since outward drainage through cracks in the soil-cement will serve to relieve such uplift pressures.(3)

2. The embankment should be designed so that its least permeable zone is immediately adjacent to the soil-cement facing. This method, which has been used on a number of dams, ensures that any seepage through the facing will not build up a pool of water sufficient to produce damaging uplift pressures or to erode the body of the embankment.

3. Free drainage should be provided behind and through or under the facing to prevent uplift. This has been done at several dams.

Most engineers consider that design concepts 1 and 2 will adequately minimize uplift damage from seepage through the facing. To date no distress to the soil-cement due to excessive hydrostatic uplift has been reported.(7)

The usual soil-cement facing design is for a 2-ft (60-cm) thickness measured normal to the embankment as shown in Fig. 1. This is constructed by placing approximately horizontal layers, 7 to 10 ft (2 to 3 m) wide and 6 in. (15 cm) thick, one on top of another in stairstep fashion up the embankment as illustrated in Fig. 9. Soil-cement of these dimensions can be mixed, placed, and compacted with ordinary highway construction equipment. Adequate working width for construction can be provided by building a small portion of unstabilized embankment along with the soil-cement, as was done at the Bonny test section and several other projects.

At the South Texas power plant cooling water reservoir described earlier, the soil-cement was placed by specially designed equipment in layers 6.25 ft (2 m) wide. The 12-in. (30-cm) loose thickness of soil-cement produces a compacted layer thickness of about 8 to 10 in. (20 to 25 cm). When the soil-cement is mixed in a central plant, (Fig. 8) which is recommended for most projects, it is not essential to maintain

Fig. 8. Central-mixing plant at the South Texas plant.

Fig. 9. Conveyor-belt system placing central-plant-mixed soil-cement on embankment slopes of cooling water reservoir near Sanford, Fla.

accurate and consistent grade and thickness of the successive layers. What is important is that the proper proportions of cement and water be thoroughly mixed with the soil, that the mixture be compacted to the required density throughout the layer, and that it be properly cured.(8 9)

The few minor failures that have occurred in the Gaspé Peninsula work are primarily because of construction deficiencies, the principal one being that the soil-cement was mixed in place one layer at a time. With such mixing it is practically impossible to control the thickness of a layer and to ensure thorough mixing all the way down to the top of the previously compacted layer. Occasional thin streaks of uncemented sand, which simply cannot be reached by the mixing tool, are inevitable. When they are exposed to wave action, also practically inevitable, erosion of that uncemented soil is certain. This happened not only at the Gaspé projects but also at the Bonny test section and several other small projects where the soil-cement was mixed in place.

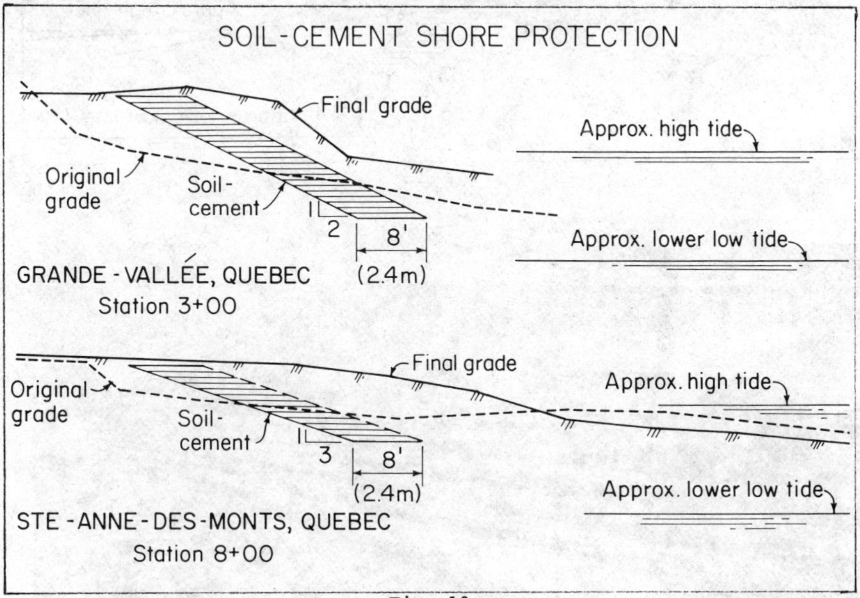

Fig. 10.

Figs. 10 and 11 show the approximate design and construction of the Quebec shore protection projects. Figs. 12 and 13 typify their condition after three winters of severe wave action. Several examples of eroded, thin, uncemented soil layers can be noted, and also that the layers are not uniform in thickness and not in alignment. However, the exposed soil-cement is hard and strong and obviously adequate to provide the protection needed by the too-close-to-the-shoreline structures--protection that large 2 to 5 ton (1,800 to 4,500 kilogram) quarrystone, probably on inadequate bedding, had not provided.

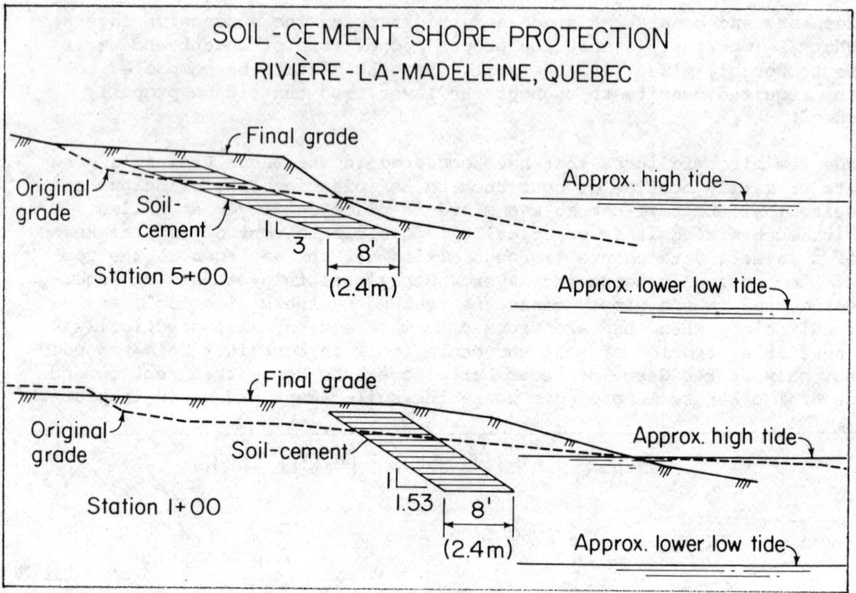

Fig. 11.

Fig. 12. Soil-cement shore protection at Riviere-la-Madeleine on Gaspé Peninsula, Quebec.

Fig. 13. Soil-cement shore protection at Ste.-Anne-des-Monts on Gaspé Peninsula, Quebec.

The deficiencies in these three Gaspé Peninsula soil-cement shore protection projects can be summarized as follows:

1. Mixing the soil-cement in place, rather than in a central mixing plant, resulted in poor control of the layer widths and thicknesses and inadequate mixing of cement to the bottom of the layer.

2. The soil was spread, not very uniformly, over the previously compacted layer. Measured thicknesses of the mixed, but as yet uncompacted, soil-cement ranged from 6 to 9 in. (15 to 23 cm) or more.

3. The cement was supplied in bags, which were dropped at the theoretical spacing for the cement content specified, opened, and the cement spread by eye over the width of the layer as specified. Observation indicated a lack of uniformity of cement from spot to spot.

4. The soil fraction of the soil-cement was beach sand and gravel, containing too many gravel particles larger than 1 in. (2.5 cm) and too much nonmineral debris, all of which made mixing and compacting difficult and adversely affected the density and overall quality of the soil-cement.

A possible danger is that the bottom of the lowest layer of soil-cement is above the daily low-water level at these locations. On inland dams and embankments it is standard practice to extend the soil-cement to at least 5 ft (1.5 m) below the lowest water level expected. Probably the soil-cement in these Quebec projects has not been undermined because of the gentle slopes of the river bottom at these locations. Quite possibly the very high waves broke and were thereby converted to simple, high-velocity surface flows before they reached the soil-cement, and thus did not undercut the soil-cement.[10]

SUMMARY

Laboratory and field research studies and project developments over the past 30 years have demonstrated that soil-cement can be effectively and economically utilized in many different kinds of energy and water resources structures. Substantial sums have been saved by using soil-cement rather than more expensive erosion protection materials. Performance in service to date has ranged from satisfactory to excellent, reflecting the conservative design criteria used and recommended and the good construction practices that have been developed.

REFERENCES

1. Coffey, C. T., and Jones, C. W., 10-year Test of Soil-Cement Slope Protection for Embankments, Earth Laboratory Report No. EM-630, Division of Engineering Laboratories, (Water and Power Resources Service, formerly U.S. Bureau of Reclamation), U.S. Department of the Interior, Denver, Colo., 1961, 22 pages.

2. Soil-Cement for Water Control, Contracts Awarded, MS261W, Portland Cement Association, Skokie, Ill., 1976, 39 pages.

3. Davis, Fred J., Gray, Edward W., Jr., and Jones, Chester W., "The Use of Soil-Cement for Slope Protection," Transactions, Eleventh International Congress on Large Dams, Volume III, 1973, pages 237-255.

4. Soil-Cement for Water Control: Laboratory Tests, IS166W, Portland Cement Association, Skokie, Ill., 1976, 26 pages.

5. Holtz, Wesley G., and Hansen, Kenneth D., "The Use of Compacted Soil-Cement in Water Control Structures," Transactions, Twelfth International Congress on Large Dams, 1976, 28 pages.

6. Nussbaum, P. J., and Colley, B. E., Dam Construction and Facing with Soil-Cement, RD010W, Portland Cement Association, Skokie, Ill., 1971, 12 pages.

7. Soil-Cement Slope Protection for Embankments: Planning and Design, IS173W, Portland Cement Association, Skokie, Ill., 1975, 10 pages.

8. Soil-Cement Slope Protection for Embankments: Construction, IS167W, Portland Cement Association, Skokie, Ill., 1975, 11 pages.

9. Soil-Cement Slope Protection for Embankments: Field Inspection and Control, IS168W, Portland Cement Association, Skokie, Ill., 1976, 16 pages.

10. Wilder, Carl R., and Dinchak, William G., "Soil-Cement for Shore Protection," Proceedings, Coastal Structures 79, Volume I, 1979, pages 301-313.

COASTAL EROSION MANAGEMENT IN AUSTRALIA

David M. Chapman,
Coastal Studies Unit,
University of Sydney,
Australia.

1. INTRODUCTION - THE PROBLEM OF COASTAL EROSION IN AUSTRALIA

The coastline of Australia is approximately 36,700 km long [1] and is backed by landscapes ranging from cliffs, dunes and dune grasslands, heath, coastal scrub and woodland to a wide variety of forest types. Within these are fresh water rivers and streams, fresh and salt water lakes and estuaries. Half of the coastline consists of sandy beach, usually backed by dunes or sand ridges, with the remainder made up of tidal flats, mangrove swamps and cliffs (Australia, Commonwealth, 1980). The coastline encloses a land area of 7.7 million km^2, and a population of merely 14 million, hence Australia would appear to be relatively well endowed with coastal resources. However, the population is highly concentrated at or near the shore, especially in the environs of the capital cities, all of which are on the coast; over a quarter of the nation's population lives within 3 km of the shore (see Table 1), and three-quarters within 40 km.

TABLE 1 - AUSTRALIA - SHORE DWELLERS

STATE	POPULATION (1976 Census)	% LIVING WITHIN 3KM OF SHORE [2]
Queensland	2037197	21.8
New South Wales (incl. A.C.T)	4974481	30.9
Victoria	3646981	16.6
Tasmania	402866	65.4
South Australia	1244756	19.8
Western Australia	1144857	38.2
Northern Territory	97090	46.6
Australia	13548228	26.4

Concentration of population within the coastal zone has created considerable pressure on coastal lands in urbanised and resort areas; favoured sites for second-home development are those within walking distance of a beach, and preferably atop a frontal dune. In these areas, coastal erosion management has emerged as a significant problem, increasingly so over the last two decades.

The problem is compounded, particularly on the east coast where coastal erosion is most severe, by the fact that most of the beaches are contained in more-or-less closed "sedimentary compartments" (Davies, 1974), each with a fixed sediment budget, apart from biogenic accessions. Unequivocal evidence of fluvial sediment supply to the east coast is hard to find, and the major longshore-transport dominated system in this region is fed from eroding Pleistocene beach-ridge complexes (Chapman, 1978).

However, about 70% of the coast remains uninhabited, especially in tropical Australia (where much of the shoreline is of mud flat or mangrove), and shorelines of the Great Australian Bight and far west coast abutting desert country. Even in the more populous south-eastern part of the country, a significant proportion of the shoreline is in National Park, Forest or Nature Reserve, or various forms of Public Open Space, due in no small part to the fact that in Australia the Crown assumed title to all land at the time of European settlement, so that it has been possible to reserve many natural areas by excluding them from the land available for alienation for private use under Crown Lands legislation.

Major erosional events on record for the state of New South Wales, which accounts for 37% of Australia's population, are summarised in the Appendix. References to beach scour and barrier island breaching (Stradbroke Island, Queensland) during storms, with occasional mention of damage to shorefront facilities, date from the mid-nineteenth century (e.g. Andrews, 1912, 1916; Sussmilch, 1912; Beasley, 1948-50; S.M.H., various). In some cases simple responses, such as timber walls or groins, were provoked.

By the 1930's many popular beaches, such as Manly and Balmoral (Sydney) or Glenelg (Adelaide) had acquired vertical sea-walls. Rather than as a defence against coastal erosion, however, these walls seem to have been constructed for the purpose of formalising the interface and providing a stable edge for a promenade, as the recreational habits of the late 19th - early 20th century beachgoer were focussed on the sand/land interface rather than the sand/water interface, as now.

Erosional events between 1929 and 1936 caused substantial cut to beaches and damage to sea-walls on southern and eastern shores, and some losses of dwellings (albeit in many cases merely shacks[3]); managerial action taken at resident or local authority level included dumping of rip-rap or other fill, repairs to sea walls, and, in at least one case, construction of grouted boulder walls and timber-piled groins (Kindler & O'Connor, 1951). Further events (especially on the east coast) in the forties and fifties and even beyond, were largely responded to at local authority level with works of dubious technical efficiency. Management by scores of local authorities meant that few persons skilled in coastal engineering were involved, no pool of collective experience developed, and similar sub-optimal or even useless adjustments were made at many points along the coast.

The problem of coastal erosion stimulated the first State legislation for coastal zone management, in Queensland (1968). Since then, two other states have passed coastal management legislation and a fourth is developing CZM within the context of environmental protection legislation passed in 1970. Characteristically (for Australia) each State has developed its individual approach to the problems of coastal management without Federal guidelines or consistency provisions, or reference to its neighbours.

2. MANAGEMENT RESPONSES TO THE PROBLEM OF COASTAL EROSION

Coastal erosion has been reported from areas representative of virtually all of the east Australian coastline south of the Great Barrier Reef and has generated public concern where it has affected the settled districts. This is not to say that erosion is unknown outside urban areas, but rather that shoreline recessions occurring in rural, state forest or national park domains cause little or no public concern. The problem has been atributed to long-term sediment starvation; the result of the attrition of a fixed sediment budget postulated for most of the coast (Thom, 1975; Roy and Crawford, 1977; Langford-Smith and Thom, 1969; Davies, 1974), and is the joint product of physical processes and human occupance.

Structural solutions, such as sea-walls, revetments, or beach nourishment have been the most popular response to the erosion problem in the past; the visible nature of such works made them attractive for their value in maintaining the credibility of the agency concerned, frequently local government. By way of contrast, land use controls are not only low in visibility, but are usually distinctly unpopular, as they are seen as an infringement on personal rights to dispose of or use property as the owner wishes.

At State level, there is considerable diversity in the procedures that have evolved to deal with the problems of coastal management. This diversity reflects the differing perception of problems, different philosophical bases of environmental management, and the level of technical knowledge at the time legislation was framed.

2.1 Queensland

2.1.1 Institutional Framework

Queensland was the first Australian State to adopt a state-wide approach to managing the coast with the establishment of the Beach Protection Authority (BPA) in 1968. The institutional framework developed largely as a response to events on the Gold Coast of Queensland, Australia's major resort. Stress generated by extreme events in the physical resource base provided the stimulus for major management decisions.

Coastal erosion did not generate stress until 1950, when property, including State Government facilities, was threatened at the Gold Coast. Official action, initiated at State Cabinet level, for an investigation, the subsequent tabling of a report, and most importantly the natural recovery of beaches, resolved stress at the time. However, investigator Kindler did not forget his observations which showed that the 1950 erosion was less severe than that of 1936.

As urban development on the Gold Coast proceeded apace in the 1950's, Kindler was conscious of the potentially devastating effects of a repetition of 1936 magnitude erosion. Isolated erosion events of the 1950's, dealt with as emergent measures by various State agencies, or by Local Authorities. No "institutional memory" of erosional events developed, but Kindler's interest in the phenomena caused him to see them, not as discrete events, but as part of an emerging problem of coastal erosion, and he advocated centralisation of coastal erosion control works for the State within the Department of the Coordinator General of Public Works. However the structure of the Department and the legislative framework within which it worked limited its effectiveness, and a consultant (Delft, 1965) recommended creation of an agency for coastal erosion management, together with an appropriate enabling Act to give it scope.

Calm weather prevailing in the mid 1960's provided little stimulus for implementing the Delft recommendations. No official action was taken until the catastrophic events of 1967 (cf. Appendix). In late 1967 the Queensland Government initiated legislation pursuant to the administrative recommendations of Delft (1965), and the Beach Protection Act was passed on 19th April, 1968.

The functions of the BPA are to investigate erosion or encroachment by the sea, to plan associated preventive and remedial measures and to record and evaluate the results of these measures. The BPA also may cause certain parts of the coast to be declared Beach Erosion Control Districts, within which the BPA is required to prepare schemes for the protection of the beaches. Beach Erosion Control Districts are usually declared over lengths of coastline having similar management problems, and extend from 1/4mi (402m) above HWM to 1mi (1609m) below LWM. There is opportunity for public participation once the scheme is prepared and input from residents and the local authority is transmitted to the Government in Council, who may approve, modify or reject the scheme. Once a scheme is approved most of the responsibility for implementing any required works rests with the local authority, which receives a 20% State subsidy on beach protection works.

The BPA also has veto power over proposals for subdivisions, structures, or road openings within a Beach Erosion Control District. However, any person whose estate is injuriously affected by the decision of the BPA is able to claim compensation.

2.1.2 Implementation

The BPA was set up as a direct result of Gold Coast erosion, and was initially concerned to a large extent with areas where erosion was threatening property and where beach replenishment projects were required. However the Authority concluded that in most areas beach replenishment was not economically feasible and that it is more realistic to concentrate effort on areas where it is possible to prevent development too close to the sea, and thus avoid the need for future beach replenishment.

The Authority has developed a philosophy that erosion should be accommodated rather than prevented, since erosion is a natural part of the coastal processes which create and maintain beaches. Implementation of this philosophy has been through the objective of preserving or restoring well managed buffer zones of sand in which coastal fluctuations can be allowed to occur without threat to property[4]. These zones are delineated by estimating the extent of foreshore which is likely to be vulnerable to erosion in the future and attempting to restore or carefully manage this land so that it remains available as a buffer between the ocean and nearby development. The extent of buffer zone is calculated by:

$$\text{BUFFER ZONE WIDTH} = ((N \times R) + C) + F \text{ (metres)}$$

where N = Planning period (usually 50 years)
F = Factor of safety (metres)
R = The rate of long term coastal erosion (metres per year)
C = The extent of the design cyclone erosion (metres)

giving rise to the structure illustrated in Fig. 1.

Based on available data on coastal fluctuations, together with equilibrium profile considerations (Swart, 1974) the Authority began designating modal buffer zones of 150m. These were mapped on large scale cadastral maps and the information forwarded to Local Authorities and relevant Government Departments for input on decision making with respect to land subdivision and land transactions. In many cases the designation of land as vulnerable to erosion placed a blight on its further development, but in some areas adverse reaction was generated by vested interests who saw the buffer zone as a threat to speculative profit-taking (cf. Fig. 2).

However, the BPA has been successful in educating many local authorities to the need for buffer zones. Local authorities not in Beach Erosion Control Districts are also informed of the advisable set-back distance and many have included appropriate recommendations in their planning schemes.

The Authority recognises that there is a conflict between the requirements of protection of coastal property and protection of beaches. Resolution of this conflict is best achieved by accommodating coastal fluctuations within a buffer zone so that beaches are retained while property is not threatened by erosion. However, where coastal land has been subdivided without leaving an adequate undeveloped buffer zone, property may be under threat of erosion by the sea, and the Authority may either:

(a) permit revetment construction and accept damage to beaches, or
(b) refuse approval to construct revetments (thus allowing property damage), or
(c) insist on the use of protective measures which would not damage beaches, but which are too costly for implementation.

As the latter alternatives are politically unacceptable the Authority has little option but to approve revetment construction where properties are under threat. Hence, the BPA has concluded that if subdivision has occurred it is inevitable that buildings will be constructed and ultimately that seawalls or revetments will be built

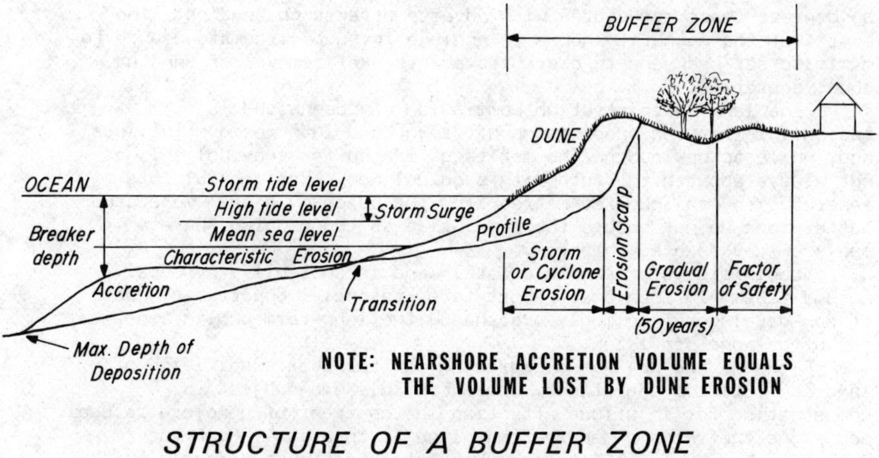

FIGURE 1. BUFFER ZONE (after BPA publications).

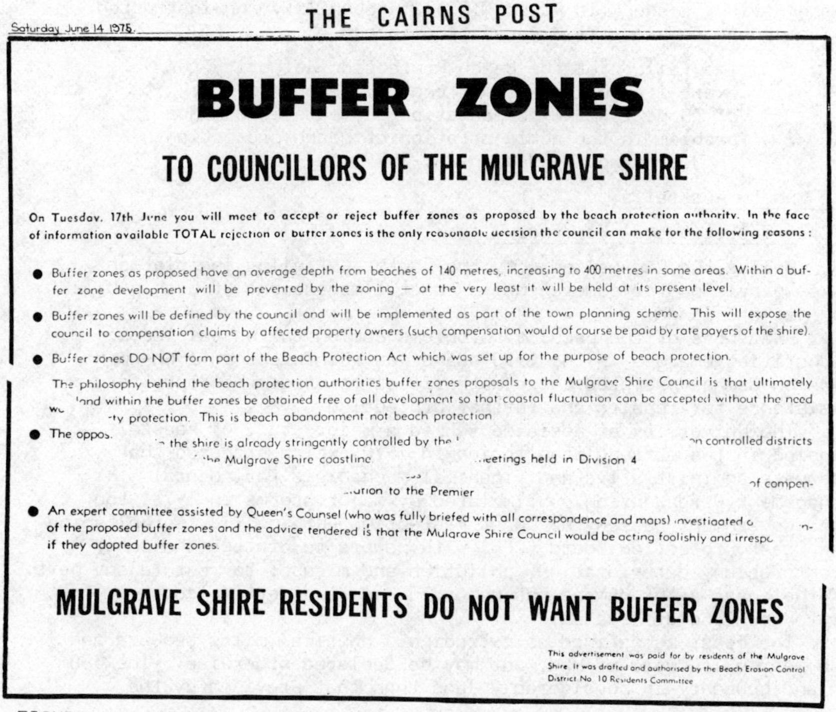

Saturday June 14 1975 — THE CAIRNS POST

BUFFER ZONES

TO COUNCILLORS OF THE MULGRAVE SHIRE

On Tuesday, 17th June you will meet to accept or reject buffer zones as proposed by the beach protection authority. In the face of information available TOTAL rejection of buffer zones is the only reasonable decision the council can make for the following reasons :

- Buffer zones as proposed have an average depth from beaches of 140 metres, increasing to 400 metres in some areas. Within a buffer zone development will be prevented by the zoning — at the very least it will be held at its present level.
- Buffer zones will be defined by the council and will be implemented as part of the town planning scheme. This will expose the council to compensation claims by affected property owners (such compensation would of course be paid by rate payers of the shire).
- Buffer zones DO NOT form part of the Beach Protection Act which was set up for the purpose of beach protection.

  The philosophy behind the beach protection authorities buffer zones proposals to the Mulgrave Shire Council is that ultimately we ... land within the buffer zones be obtained free of all development so that coastal fluctuation can be accepted without the need ...ty protection. This is beach abandonment not beach protection
- The oppos... ...n the shire is already stringently controlled by the ... ...on controlled districts ... the Mulgrave Shire coastline ... ...eetings held in Division 4 ... of compen- ...ution to the Premier
- An expert committee assisted by Queen's Counsel (who was fully briefed with all correspondence and maps) investigated ... of the proposed buffer zones and the advice tendered is that the Mulgrave Shire Council would be acting foolishly and irrespo... ...n- if they adopted buffer zones.

MULGRAVE SHIRE RESIDENTS DO NOT WANT BUFFER ZONES

This advertisement was paid for by residents of the Mulgrave Shire. It was drafted and authorised by the Beach Erosion Control District No. 10 Residents Committee

FIGURE 2. RESIDENT OPPOSITION TO BUFFER ZONE CONCEPT

to protect these structures with adverse effects on beaches. The best that the Authority hopes to achieve in such circumstances is to maximise set-back and to exert its powers over removal of sand or dune vegetation.

The land Administration Commission of Queensland is responsible for the subdivision of Crown Land into residential lots and, while acknowledging the merits of BPA buffer zone policy, has not always adopted the Authority's guidelines. Residential lots separated several hundred metres from the beach by a well vegetated buffer zone do not secure the same price as those which "have a sea view and an ocean breeze".

Subdivision of freehold coastal land is controlled by Local Authorities. In such cases short-term, potential opportunity costs of non-development commonly overshadow the long-term public benefits of buffer zone provision.

Although the BPA has a degree of control over subdivision of land and building operations, these controls are subject to compensation for injurious affection in the event of complete refusal, and since the value of shore-front land is invariably high the Authority has been inhibited and has not found it practicable to refuse approval of development except in exceptional circumstances. As the Authority is not allocated sufficient funds to pay compensation if sued, it would, in all probability, be instructed by the Court to allow development to proceed.

> The failure of the Beach Protection Authority to exert its full powers of control because of lack of funds to make compensation payments is a major problem in the administration of beach protection in Queensland. (Macdonald, 1979, pp. 41-2.)

2.2 South Australia

2.2.1 Institutional Framework

Unlike the Queensland Act, the South Australian legislation makes provision for comprehensive coastal management of both public and private lands.

Awareness of the peculiar problems common to coastal local authorities emerged in the 1950's and was reinforced in 1960 when erosion damage prompted these authorities to press the State for assistance for repairs and further protection.

The University of Adelaide tabled a major study of coastal erosion in the metropolitan region in 1970, which stressed that adequate administrative and financial mechanisms for coastal management were wanting. Stimulated by major storms in 1972, the S.A. Government passed the Coast Protection Act in 1972, establishing the Coast Protection Board (CPB) with powers to protect the coast from erosion, deterioration, pollution and misuse, to restore any part of the coast after damage has occurred and to undertake coastal improvements and research.

The coast is defined as extending 3 nautical miles seaward and 100 metres landward of HWM, but may be declared otherwise. The 100m inland boundary is considerably less than that provided by the

Queensland legislation (402m) and probably indicates the concern with erosion occurring in built-up areas at the time the Act was framed.

The Act makes provision for the declaration of Coast Protection Districts. Once a District is declared the CPB is required to investigate the area and to develop a management plan, which must also be discussed with local governments and the general public prior to its approval. The CPB is then authorised to execute all works required to implement the management plan. It can repair or restore any parts of the coast damaged by storms or pollution (and acquire land even without an approved plan). Local authorities wishing to carry out works on the coast must seek the approval of the Board, and once this is obtained the Board may grant funds to assist with the works.

The Board also has the power to control land use in a Coast Protection District.

2.2.2 Implementation

The early activities of the Board were dominated by the restoration of erosion damage and the construction of coastal protection works but since about 1975 the CPB has developed a philosophy directed to avoiding problems through land management, rather than repair.

Although it has not articulated a buffer zone policy as has the BPA in Queensland, the CPB has endeavoured to gain a buffer or set back between high water and the edge of development (esplanade or lot) sufficient to allow for likely shoreline fluctuations in the next 50 to 100 years[5]. Width of buffer proposed varies greatly; modal 200m on low to medium energy sandy shorelines, or poorly consolidated cliffs; 50m setback for aesthetic reasons on hard-rock cliffs; no development at all on transgressive dunes associated with high energy shorelines.

In order to achieve these aims, the Board has used its powers of land management and land acquisition in a variety of ways, e.g.

(a) outright purchase of subject land in whole or part,
(b) esplanade closure, with access provided from landward side for affected lots,
(c) imposition of various conditions on development (such as minimum vegetation clearance, location of house to be determined by CPB),
(d) refusal of development application and offer of alternative sites to owners,
(e) refusal of development application.[5]

In the latter case, the CPB may have, and sometimes has had, its decision reversed by the Planning Appeal Board.

Lands acquired by purchase may be leased for acceptable open space uses, handed over to the local authority for management as Public Open Space, or incorporated in National Parks. The Board considers it has already acquired the most crucial areas with respect to coastal erosion management.[5]

2.3 New South Wales

2.3.1 Institutional Framework

Although New South Wales has a long history of coastal erosion (see Appendix), specific coastal management legislation was only introduced in mid-1979 with the proclamation of the Coastal Protection Act.

However, since 1973 the State Planning and Environment Commission had been implementing a Coastal Lands Protection Scheme under which areas of significant amenity value were acquired outright by purchase, or had strict land use controls imposed. Since 1975 the State Department of Public Works had been implementing a Beach Improvement Scheme designed to improve the amenity of recreational beaches. Both of these schemes were employed at times for erosion control, to finance acquisition of endangered land, or engineering works.

The National Parks and Wildlife service has responsibility for management of 18% of the State's shoreline (a figure which could rise to 30% with execution of proposals for new parks).

Apart from agencies concerned with public engineering works and planning at State level, the agencies charged with pollution control and soil erosion control have also been involved in coastal erosion management, the latter by virtue of the fact that it is charged with the responsibility of coastal sand drift management. As early as 1968 the Soil Conservation Service published guidelines for the development of real estate on the shoreline which included recommendations for a buffer zone, and in 1977 the State Pollution Control Commission likewise published land use guidelines which stressed controls on beach access, sand removal, and intrusive engineering works.

With the passing of the Coastal Protection Act, coastal management policy shifted from mere acquisition of important areas to more aggressive control on coastal lands. The aim of the Act is to protect the New South Wales coastal zone from increasing development pressures by providing an integrated and comprehensive approach to coastal planning. The Act establishes an expert Coastal Council to advise the Minister for Planning and Environment on the protection and restoration of the coast and on the orderly and balanced use and conservation of the coastal region, having regard to the financial resources of the State and the social and economic needs of the people of the State. In particular the Council is to advise on policies, co-ordination and acquisition. It may carry out appropriate investigations and disseminate information on coastal management. The Council has no direct powers, but its advice can be implemented under the powers of the Planning and Environment Commission. The Act also requires the consent of the Public Works agency for any development works in the coastal zone. Consent can be refused if the development will adversely affect the beach or dune system or if the development is likely to be adversely affected by the sea. The consent provision is not seen as a form of planning power but as a power of veto, on specific technical grounds of coastal erosion or damage. The Works agency may also place conditions on its consent, requiring modification to the proposal.

A useful provision of the Act is its extension, for planning and investigative purposes, to a "coastal region" which is undefined in the Act save for the fact that it includes the coastal zone, thus allowing planning to be cognizant of the interrelationships between land-uses, such as catchment area land uses which affect estuaries. The coastal zone is initially defined for convenience by reference to map sheets which contain oceanic or estuarine waters, which allows more rigorous definition to be postponed until knowledge is more perfect.

2.3.2 Implementation

At the time of writing, it would be premature to attempt evaluation of the new legislation.

However, the Coastal Council has energetically set about investigating problems of coastal management, and has established a Working Group to investigate the problem of coastal erosion in some detail.

The Planning and Environment Commission has since 1970 pursued a policy of establishing land use controls on all local government areas on the coast. The main thrust of these controls has been to regulate sub-division of coastal lands. The Department of Public Works formed a Coastal Engineering Branch in 1976 which carries out investigations and advises on coastal dynamics. Curiously, the new Act brings together these two major government agencies involved in coastal management, but makes no specific provision for their interaction. It also gives certain land-use regulatory power to the works agency, and power to the Works Minister to authorise projects up to $1m. in value for coastal protection.

2.4 Western Australia

Western Australia has no specific coastal management legislation, but is evolving a coastal management policy within the framework of its Environmental Protection Act, 1971. Draft policy guidelines emphasise the dynamic nature of sandy shorelines but do not make specific provision for coastal erosion management.

Localised erosion events on built-up shorelines have resulted in substantial State expenditure and involvement in protective works, usually nourishment with or without groins, and usually with but small contribution from the local authority concerned. In at least one instance, assistance was given to property-owners to relocate from an eroding sandy spit.

The Department of Conservation and Environment (which administers the Act) espouses the philosophy that all coastal locations should have some type of buffer, with minimal reservation sufficient to provide protection against erosion. This philosophy however has not been made explicit as a policy but is used in providing advice to the Town Planning Board and to Local Authorities. The width of buffer zones is determined on a case by case basis with advice from the Department of Public Works and Department of Agriculture (sand drift control). The aim is to provide a coastal reserve most appropriate to a particular location. As a general rule the distance is determined not less than 100 metres from the permanent vegetation line or cliff edge for any shore[6].

2.5 Victoria

Coastal management in this State is still rather fragmented. In 1966 the Victorian Government established the Port Phillip Authority (PPA) to oversee and control developments within the Port Phillip Region (Melbourne and environs) because of government concern with coastal erosion, land use conflicts and lack of co-ordination between various agencies. The Authority has jurisdiction over Crown Lands and water within 200m landward and 600m seaward of HWM around the shores of Port Phillip Bay and advises on the co-ordination of development, the preservation of beaches and natural beauty, and the improvement of amenity. Because no private land is included, the Act has no influence over subdivision, which is normally controlled by local authorities.

As well as its advisory role the PPA is required to approve any structure, works or vegetation removal proposed for its area of jurisdiction, however, since it is only concerned with developments on Crown land the proposers of projects are generally government agencies or local authorities.

Following the PPA, several other regional planning bodies such as the Westernport and Geelong Regional Planning Authority have also been established to help co-ordinate local government planning in specific regions. The Department of Crown Lands and Survey is responsible for coastal reserves, which are managed by locally elected Committees of Management. The Public Works Department is responsible for most of the ports, and is also responsible for protection of the coastline from wave attack. The Soil Conservation Authority is responsible for control and repair of erosion on Crown Lands, with a strong orientation toward stabilisation of coastal sand dunes. The Town and Country Planning Board is responsible for preparation and co-ordination of planning schemes and for the preparation of policy statements. A Draft Coastal Policy has been prepared.

2.6 Tasmania

Tasmania has no specific coastal management legislation. The establishment of a Coastal Management Committee has been mooted by Government to co-ordinate the activities of various agencies and to employ staff and consultants to work on some of the problems that have been identified. Funding to service such a Committee has not been made available at the time of writing.

There is a Town and Country Planning Commission which is responsible for the formulation of land use policy and for the approval of subdivision proposals. The Lands Department is responsible for restoration of eroded sand dunes and for the management of Crown lands, whilst the National Parks and Wildlife Service manages about 20% of the Tasmanian coastline, plus about thirty offshore islands.

A problem of coastal erosion has been identified at State level, but isolated erosion events to date have been dealt with at resident or local authority level on an ad hoc basis.

2.7 Australian Commonwealth Government

The Australian Government is directly involved in the management of coastal lands used for defence installations, lighthouses, some airports, and the 7280ha Jervis Bay Commonwealth Territory (on the NSW coast), 62% of which is Nature Reserve. It has resisted involvement in coastal management elsewhere, taking the attitude that "formation of specific policy on coastal management has always been a State rather than a Commonwealth responsibility".[7]

Natural disaster insurance was agreed to in principle by the Australian Government in 1976, when a working party was formed to investigate a scheme in consultation with the insurance industry. However, in releasing the final report of the Working Group, the Treasurer announced that a number of difficulties with this type of scheme had arisen and commented on the Government's reservations about its ability to resolve the problems apparent with current natural disaster insurance arrangements. The Government felt that such a scheme would involve an excessive degree of intervention by the Government in activities traditionally the responsibility of the private sector and a likelihood that compulsion would be required on either insured and/or insurers for the scheme to be fully successful. Concern was also expressed that the selected hazards to be covered by the scheme would not be consistent with needs and risk exposures of some of the States and that major reorganisation of existing insurance arrangements would be required. Undoubtedly the possibility of potential payouts by the Commonwealth of large and unforeseeable amounts, disruptive to budgetary planning and management, was a factor in the decision.

The Treasurer also commented that improvements in risk reduction measures, such as improved policies and practices relating to building codes and land use allocation, could significantly reduce the levels of losses arising from natural hazards and the costs of natural disaster insurance cover (Australia, Treasurer, 1978, 1979).

Some State and Local governments have requested Commonwealth financial assistance for coastal erosion management, arguing that coastal erosion should be viewed as a national disaster. The Commonwealth has not shared this view, and on at least one occasion has made its attitude explicit:

> On the general question of financial assistance for restoration of beaches as a result of wave erosion, the Government attitude has been that the taxpayer should not bear the cost.[8]

A Commonwealth Government standing committee on environment and conservation recently concluded that unwise location of private and public facilities has often led to threat or damage to these facilities by wind or coastal erosion, and observed that the issues are those of equity in the use of public funds to protect people who locate buildings unwisely, the technical efficiency of some of the protective works proposed, the environmental impacts of such works on other coastal uses and the possible transfer of the erosion problem to adjacent areas (Australia, Commonwealth, 1980).

The same committee felt that there was room for greater involvement by the Australian Government, especially with respect to the development of national coastal management objectives and policies in co-operation with the States, and in the formation of a national Coastal Management Council to encourage collaboration and co-operation between agencies involved in the coastal zone, to promote and co-ordinate appropriate research, and to act as a central clearinghouse for coastal information.

3. CONCLUSION

All existing coastal management legislation in Australia has been stimulated largely or wholly by the problem of coastal erosion. The more recent legislation, in South Australia and New South Wales, exerts controls over private landholders, the latter by empowering the Minister for Public Works to prevent development on the basis of coastal engineering criteria, as well as creating an organ for the provision of expert advice on the use of existing, more general, planning powers.

In Queensland powers for the control of development were vested in the Beach Protection Authority, but it can only use them if funds are available to compensate landholders. The then Premier indicated when introducing the Beach Protection Act that this aspect of the Authority's operations would be closely controlled at Cabinet level[9]; evidently the Authority was expected to operate principally through the provision of technical expertise.

The concept of a buffer zone between development and the sea on sandy shorelines has emerged, in one form or another, as a strong thread in coastal erosion management policy in Australia. The nation is fortunate in that the system of Crown Lands legislation produced a buffer zone *de facto* over many sections of shore, in the form of reserves for esplanades, Public Open Space, and the like, which served to insulate development from the impact of many early erosional events.

Coastal management provisions in Australia to date have therefore reflected the growing realisation that erosional losses may be the result of past errors in human judgement with respect to location of land-uses or structures, rather than the caprice of Nature. Solutions are being sought in the encouragement or enforcement of wiser allocation of land-use, or location of structures, rather than simply in the control of Nature.

4. FOOTNOTES

1. The figure is scale-dependent, being derived from 0.5km chords on 1:250,000 map sheets by Aust. Division of National Mapping.

2. Data courtesy Dr. G. Yapp, C.S.I.R.O. Division of Land-Use Research.

3. Shack (colloq.) - a simple dwelling, usually without utilities, used for accommodation during beachside vacation.

4. The essentials of this concept were embodied in the primary recommendation of the Kindler & O'Connor (1951) report, which went unheeded.

5. Pers. comm., R.I. Thomas, Coastal Planner, CPB.

6. Pers. comm., Dr. B. Hamilton, Dept. of C. & E.

7. Letter, Prime Minister to House of Representatives Standing Committee on Environment and Conservation, 7 May, 1979.

8. Letter to Sutherland Shire Council, January, 1975.

9. Parliament of Queensland, Hansard, 2 April, 1968, p. 2753.

5. REFERENCES

Andrews, E.C. (1912) Beach Formations at Botany Bay. Journal of the Royal Society of N.S.W., 46: 158-185.

Andrews, E.C. (1916) Shoreline studies at Botany Bay. Journal of the Royal Society of N.S.W., 50: 165-176.

Australia, Treasurer (1978) A Natural Disaster Insurance Scheme for Australia. Report by Tech. Committee upon tech. aspects of a Nat. scheme for Natural Disaster Insurance. Canberra, The Treasury, mimeo.

Australia, Treasurer (1979) Natural Disaster Insurance. A policy information paper issued by the Treasurer, The Hon. John Howard, M.P., Canberra, A.G.P.S., 20 pp.

Australia, Commonwealth (1980) Management of the Australian Coastal Zone. Report from the House of Representatives Standing Committee on Environment and Conservation. Canberra, A.G.P.S., 86 pp.

Beasley, A.W. (1948-50) Heavy Mineral Beach Sands of Southern Queensland. Part I - The Nature, Distribution and Extent, and Manner of Formation of the Deposits. Part II - Physical and Mineralogical Composition, Mineral Descriptions, and Origin of the Heavy Minerals. Proc. Roy. Soc. Qld. 59: 109-140; 61: 59-104.

BROOKS, J.H. (1953), Stradbroke Island Erosion and Broadwater Silting Southport; Mineral Sand Deposits Tugun. Qld. Govt. Min. J. 54: 493-494, 566-569.

Chapman, D.M. (1978) Management of Sediment Budget, Lower Gold Coast, Queensland. Ph. D. Thesis, Univ. of Sydney, 279 pp.

Davies, J.L. (1974) The coastal sediment compartment. Aust. Geog. Studs. 12; 139-151.

Delft Hydraulics Laboratory (1965) Queensland Coastal Erosion. Recommendations for a Comprehensive Coastal Investigation. Report R.257, Delft Hydraulics Laboratory, Delft, The Netherlands, 1965.

Foster, D., Gordon A.D. and Lawson N.V. (1975). The Storms of May-June 1974, Sydney NSW. The Institution of Engineers, Australia Proc. Second Aust. Conf. on Coastal and Ocean Eng. pp. 1 - 11.

Foster, D. Stone D.M. (1965) Historical evidence of erosion at Cronulla, NSW. Inst. Eng. Aust. J. 37; 275-286.

Foster, D, Stone D.M., Munro, C.H. (1963), Preliminary Study of Beach Erosion on Cronulla Beach, U.N.S.W. Water Research Laboratory, Report No. 59.

Hattersley, R.T. (1968), Report to Warringah Shire Council on Erosion Control Measures at Collaroy Beach, U.N.S.W. Water Research Laboratory, Technical Report No. 68/7.

Kindler, J.E., O'Connor, C. (1951) Report on Erosion of Beaches in the Town of South Coast. Co-Ordinator General's Dept., Brisbane, 10th August, 1951, 57 pp.

Langford-Smith, T., Thom, B., (1969) New South Wales Coastal Morphology. J. Geol. Soc. Aust. 16; 572-580.

Macdonald, H.V. (1979) Beach Protection in Queensland. Thesis, M. Pub. Ad. Department of Government, University of Queensland.

McGrath, B.L. (1968). Erosion of Gold Coast Beaches, 1967. J. Inst. Engrs. Aust., 40: 7,8,155.

McGrath, B.L., ROBINSON D.A. (1972) Erosion and Accretion of Gold Coast, Qld., Beaches. Civ. Eng. Trans., Inst. Engrs. Aust. CE14: 158-163.

Queensland Co-ordinator General's Department (1975), Coastal Management: Queensland-New South Wales Border to Northern Boundary of Noosa Shire (prepared by Gutteridge, Haskins, and Davey, Brisbane).

Robinson, D.A., Patterson, D.C. (1975). The Kirra Point Groyne - A case history. The Institution of Engineers, Australia, Proc. 2nd Aust. Conf. on Coastal and Ocean Engineering, pp. 46 - 52.

Roy, P.S., Crawford E.A. (1977) Significance of Sediment Distribution in Major Coastal Rivers, Northern N.S.W. The institution of Engineers, Australia, Proc. 3rd Aust. Conf. Coastal and Ocean Eng., pp. 177 - 184.

Sussmilch, C.A. (1912) Note on some recent marine erosion at Bondi, Journal of the Royal Society of N.S.W., 46: 155-158.

Swart, D.H. (1974) Offshore sediment transport and equilibrium beach profiles. Delft Hyd. Lab. Pub. 131.

Thom, B.G. (1974), Coastal Erosion in Eastern Australia, Search, 5(5): 198-209.

Thom, B.G. (1975) Modification of coastal and deltaic terrain subsequent to deposition. Z.f. Geomorph. 22; 145-170.

6. APPENDIX

MAJOR EROSIONAL EVENTS ALONG N.S.W. COASTLINE, 1857-1980)

(Modified and expanded from Thom (1974, p. 201) and Chapman (1978, pp 14-18)

Date	Location of main recorded impacts	Notes	Source
August 1857	Sydney area	The "Dunbar" storm (referring to shipwreck off Sydney Heads) scour recorded at Lady Robinson's Beach	E.C. Andrews (1916); Sydney Morning Herald (S.M.H.) (22-8-1857, p. 4; 24-8-1857, p. 4)
September 1876	Sydney and South Coast	The "Dandenong" storm (referring to shipwreck off Jervis Bay)major dunal damage recorded at Dee Why	Andrews (1916); S.M.H. (14-9-1876, p. 5); C. Swancott (1968?), Dee Why to Barrenjoey to Pittwater (Sydney).
May 1889	Central coast	Cyclone approached from north; strong southerly gales and heavy seas; damage recorded at Manly and Kiama	Andrews (1916); S.M.H. (28-5-1889, p. 4; 29-5-1889. pp. 6,7,8) Bureau of Meteorology (1947), Results Of Rainfall Observations made in N.S.W., appendix
1896	Far North Coast	Stradbroke Is. (Qld.) breached (Jumpinpin breakthrough) and strong SE gales recorded	Beasley (1948-50) and Beach Conservation No. 1, 1970.
July 1910	Central coast	"Tremendous storm"; references to "hurricane" off South Head	E.C. Andrews (1912); S.M.H. (19-7-10, pp. 7, 8)
July 1912	Central and south coast	Two major storm periods, July 15-17 and 24-26; Southerly gales and very rough weather on coast; major erosion	Andrews (1912, 1916); C.A. Sussmilch (1912) J. Roy. Soc. N.S.W., 46, Bureau of Meteorology (1947)
May 1913	Central and south coast	Tropical depression over south Qld. moved south and developed into cyclone; very rough on coast with south east gales; damage to Bondi seawall and Manly beach facilities.	S.M.H. (17-5-13, p.5); R.B. Harris. Heroes of the Surf: Fifty years history of Manly S.L.S.C.
Easter 1927	Central and north coast	Destruction of Myall Point (Port Stephens); Newport beach breakthrough.	B.G. Thom (1960), B.A.(Hons.) Thesis U. of Sydney; Thom (1974); N.S.W. Parl. Debates, Session 1967-8, 68, p. 1025.
1927	North coast	HWM on parts lower Gold Coast farther landward than in 1950 erosion	Kindler & O'Connor (1951)

Date	Location of main recorded impacts	Notes	Source
1929	North Coast	Gold Coast erosion severe, some property destruction	S.M.H. 3-7-29; Currumbin State School Diamond Jubilee Souvenir History 1909-1969.
1933-34	North Coast	Gold Coast erosion - HWM almost at Pacific Hwy. at Kirra	S.M.H. 31-7-33; Kindler & O'Connor (1951).
1935-36	Sydney and north coast	Erosion at all Sydney beaches; Major damage reported at Coogee, Balmoral, Broken Bay, Newcastle, Ballina, Gold Coast.	S.M.H. (19-6-35, p.13; 20-6-35 p. 11); Kindler & O'Connor (1951); Robinson & Patterson (1975), McGrath & Robinson (1972).
Winter - Spring 1942	Sydney area	North Cronulla, Manly, Balmoral seawalls severely damaged	D.N. Foster, D.M. Stone and C.H. Munro (1963), U.N.S.W.; Water Res. Lab. Report 59; D.N. Foster and D.M. Stone (1965), J. I.E.A., 37(9); S.M.H. (15-10-42, p. 5).
Winter 1944	Sydney area	Property damaged at Collaroy, and also Manly and Coogee	R.T. Hattersley (1968) U.N.S.W. Water Res. Lab. Tech. Rept. 68/7; S.M.H. (26-5-44 p. 3 14-7-44 p. 4)
June 1945	Sydney area and lower north coast	Major property damage at Collaroy: 8 houses damaged (3 beyond repair), land later resumed; property threatened and road undermined at Stockton.	S.M.H. (13-6-45 pp 1,4; 14-6-45, p.4); C.E. McDonald and C.W.T. Henderson (1975), The Manly Warringah Story Hattersley (1968).
1945-46	North coast	10m cut to lower Gold Coast	Beasley (1948-50).
July - August 1946	Sydney area	South Cronulla seawall severely fractured; North Cronulla wall and beach facilities destroyed, S.L.S.C. threatened and later removed; undermining of promenade and road at Curl Curl, road later resited westward.	Foster, Stone and Munro (1963); Foster and Stone (1965); H.M. Connell (1969), B.A. (Hons) Thesis U. of Sydney S.M.H. (24-8-46, p.3; 27-8-46, p.1); T. Larsen and T. Belbin (1976), A shade of blue, a touch of gold; a fifty year history of North Cronulla S.L.S.C.
1937-50	North coast	Principally period of fine weather - little cut on Gold Coast.	Brooks (1953)
Winter 1950	Sydney and lower north coast	Major damage to Sydney beaches and facilities; seawall collapse at Manly, Balmoral, Nielson Park, and Cronulla	Foster, Stone, and Munro (1963); Foster and Stone (1965); S.M.H. (21-6-50 p. 1; 25-6-50, p. 6; 27-6-50 p. 4); Thom (1974).
1950	Far North coast	Major erosion on Gold Coast - houses relocated near Surfers Paradise, Spit breached, Pacific Hwy. threatened at Narrow Neck - rip-rap dumped.	S.M.H. 25-5-50; Kindler & O'Connor (1951); Memo 31-1-51 from Qld. Inspector of Mines to Minister; B. McGrath (personal communication to D.M.C.).
1954	Far North Coast	Tropical cyclone in S. Coral Sea 982mb central pressure - extensive cut SE Qld. (and presumably N. Coast of N.S.W.) beaches.	B. McGrath (personal communication to D.M.C.).

Date	Location of main recorded impacts	Notes	Source
May 1955	Sydney and north coast	Widespread scarping and minor property damage, esp. Kurnell, Balmoral, Manly (Fairy Bower), and Newcastle.	D.N. Foster and K.B. Higgs (1974), U.N.S.W. *Water Res. Lab. Tech. Rept.* 75/13; *S.M.H.* (21-5-55, p. 1; 23-5-55, p. 1).
1955-56	Far North Coast.	Severe cut central Gold Coast	Qld. C.O.G. Dept. (1975).
Winter-Spring 1967	Central and north coast	Cliffing and property damage; major impacts on Sydney coast at Collaroy (houses threatened) and Manly (sea-wall collapse); On north coast severe erosion at Byron Bay and Lennox Head; worst recorded erosion Gold Coast	Hattersley (1968); P.W. Stone (1967) *B.A. (Hons.)* Thesis U. of Sydney; A. Short (1967); *B.A. Thesis* U. of Sydney; *S.M.H.* (26-9-67 p. 10; 20-11-67 p. 1); McGrath (1968); Robinson & Patterson (1975)
May-June 1968	Sydney area	Severe sand depletion and seawall damage	D.N. Foster (1968), U.N.S.W. *Water Res. Lab. Tech. Rept.* 68/11; H.M. Connell (1969)
1972	Far North coast	severe erosion Gold Coast	Robinson & Patterson (1975); records of Gold Coast City Council.
February 1974	North Coast	Cyclone "Pam" beach and property damage, Gold Coast eroded for almost entire length, Southport Spit breached.	B. Caton (1975), *Search* 6(5); A.H. Wearne and J.A. Quilty (1974), *J. Soil. Cons. Ser. N.S.W.* 30(4); records of Gold Coast City Council, Qld. C.O.G. Dept. (1975).
May-June 1974	Entire coast	Severe erosional event: widespread property damage especially on central coast, additional cut to Gold Coast, some beaches North Coast N.S.W. cut by up to 33m.	J.A. Quilty and A.H. Wearne (1975), *J. Soil Cons. Serv. N.S.W.* 31(3); E.A. Bryant and R.W. Kidd (1975), *Search*, 6(11-12); D.N. Foster, A.D. Gordon, and N.V. Lawson (1975); B.C. Fleck (1975), *J. Soil Cons. Ser. N.S.W.* 31(1); *N.S.W. State Emer. Ser. Bull.* (1974), 16(3); *S.M.H.* (27-5-74, p. 1; 28-5-74, p. 1) *The Australian* (27-5-74 p. 1) D.N. Foster (1974); U.N.S.W. *Water Res. Lab. Rept.* 74/19; personal observations (D.M.C.)
1974-80	Principally Quiescent, apart from:		Personal observations
1976	Far North Coast	T. cyclone "David" - berm cut on beaches, no property damage.	Personal observations
1978	Central Coast & Metrop.	Winter storm events cut Sydney and Central Coast beaches. Some property damage on Central Coast.	Personal observations

SPOIL ISLANDS: DETERMINING HOW LOCAL GOVERNMENTS CAN MANAGE THEM

Mark L. Gumula
Treasure Coast Regional Planning Council
50 Kindred Street
Post Office Box 2395
Stuart, Florida 33494

R. Grant Gilmore
Harbor Branch Foundation, Inc.
Route 1, Box 196
Fort Pierce, Florida 33450

INTRODUCTION

Many coastal counties throughout Florida have a resource for which there has been little recognition for its value, management and administration. This particular resource, and the subject of this presentation are spoil islands. Spoil islands are sediment deposits usually produced by dredging operations. The spoil islands studied were produced when the Army Corps of Engineers dredged the Intracoastal Waterway through the Indian River lagoon in a county which bears the name. Indian River County, Florida, has had a desire to better utilize and ensure adequate protection of the 43+ spoil islands within their boundaries for some time. In March of 1974, the Environmental Study Committee of the Vero Beach-Indian River County Planning Commission completed a spoil island report which was part of an effort to make recommendations for the use and protection of land areas of critical environmental concern. In January, 1977, the County appointed a Marine Island Advisory Commission (MIAC) to provide policy guidance on spoil island management matters. In addition, a resolution was passed in April 1977, by the County Commission which declared that the spoil islands in the County were to be used and enjoyed in a manner consistent with the public's interest, emphasizing the conservation of their resources. In December of 1977, the County even went so far as to place the islands in a marine park zoning district to limit the uses of the islands to recreational, conservation and research uses.

There existed a hurdle, however, which the County would have to overcome if it was to exert additional management influence over its spoil islands. The State of Florida holds title to the majority of the islands which exist within Indian River County. The County would thus have to first gain sufficient control over the islands from the State if it was to more effectively manage them at the local level.

It was known that the state was not in the habit of transferring ownership of the islands to other jurisdictions. In addition, it was not a simple case of requesting and then receiving a lease agreement from the State for use of the islands either as strict requirements had to first be met before such a request could be granted. Indeed, Florida Administrative Code, Section 16C-12.05(1), stated that:

> "No spoil islands shall be developed except upon a clear showing that the development is in the public interest and hardship would result if the development was not authorized."

The burden was then on the county to show the State that the control over the islands should be transferred to them, as it could effectively manage them in accordance with official State mandates. The County's foresight and concern to do so led to the Florida Department of Environmental Regulation (DER), Office of Coastal Management entering into a contract agreement with the County in July, 1979, to fund the formulation of "a plan and general framework for implementing the optimal use or control of spoil islands not only in Indian River County, but in any similar coastal area with similar conditions." Monies for the study were made available by a special grant to the State from the Federal Office of Coastal Zone Management.

The resulting pilot project was primarily comprised of two major efforts which were coordinated with the County's Marine Island Advisory Commission (MIAC) and its Planning and Zoning Department. The first effort, a spoil island resource inventory (see Step 1 of the spoil island management process), was conducted by Mr. R. Grant Gilmore who was contracted out as an independent biological consultant. The Treasure Coast Regional Planning Council (TCRPC) developed the spoil island management program, which constituted the second major portion of the study and incorporated the resource inventory into the management process which was produced. The TCRPC is a voluntary association of local govenments which advises its constituent local governments on regional comprehensive planning matters.

This paper represents an account and an assessment of the pilot project which resulted in the development of a process which Indian River County and other local governments can utilize to determine how to better manage spoil islands in their area in a coordinated and consistent manner. It also discusses the spoil island management (SIM) plan developed for Indian River County through the application of the process, which reflects an attempt to employ the systematic and comprehensive aspects of the coastal management (CM) process based on sound planning principles.

Key aspects of the paper are the discussion of the resource inventory (i.e., the biophysiographic study), the cross impact analysis matrices and management program, and the critique of the SIM process and plan.

MANAGING SPOIL ISLANDS AND COASTAL MANAGEMENT

It should be stressed here that coastal management (CM) is not simply related to the proper maintenance of beaches, but rather its scope is much more encompassing. It focuses on the entire area along the coast where the sea interfaces with the land areas, an area which is referred to as the coastal zone (CZ). The fact that the CZ's most dominating force for virtually all aspects of life is the sea, water use issues are of extreme importance in CM, along with land use concerns. Traditional land use planning, on the other hand, placed little emphasis on water uses.

The chief emphasis of coastal management is on how the coastal zone's natural resources are to be managed. This is primarily due to the fact the coastal zone generally includes the most biologically productive and sensitive areas. CM is also comprehensive, in that it addresses all aspects of development which impact the natural resources of the coastal zone. This is significant as some of the most extreme development pressures are exerted in the CZ, where the majority of the population reside. Hence, in an attempt to meet the people's needs, the very resources which have attracted people to the coastal areas are many times subjected to irreversible alterations.

Spoil islands require special management considerations, like other coastal resources. While they are not generally considered to be as sensitive and biologically productive as some other coastal areas, spoil islands resources (i.e., including both the uplands and adjacent aquatic environment) can provide significant environmental values as: wildlife habitats, including diverse terrestrial island floral communities; avian habitat and nesting sites; and as rich aquatic fishery and nursery areas. With the destruction of some of the most productive coastal areas occurring as a result of man's disregard for his environment, spoil islands are likely to become even more important in providing refuge to a variety of wildlife.

While the environmental qualities of spoil islands are extremely important, their values toward meeting our navigational needs should be stressed. They serve as spoil disposition sites for navigational dredging projects to provide for our extensive waterways network. The value of spoil islands to waterbourne commerce and pleasure boaters is then significant. In addition, without having the option of placing dredged material on spoil islands, more sensitive coastal resources (e.g., adjacent marshes) may have had to been irreversibly altered, and with higher spoil removal costs.

The recreational values of spoil islands also exist because of their relative location to navigation channels and population centers. Boaters frequently use them for both picnicking and swimming activities. Aesthetic, research and educational opportunities are also prevalent on spoil islands because of their habitat values and natural beauty. In addition, spoil islands can provide a degree of storm protection to population centers on the mainland.

Spoil island resources are being adversely affected by man's carelessness. Large amounts of trash can be found on the islands, and fires are frequently started as the result of his use. In his pursuit of recreation, adjacent marine grasses have been damaged by boats; bird nesting sites violated and exotic vegetation has prevented natural species from taking hold. In addition, although significant development limitations are known to exist, people have been known to inhabit the islands, even constructing permanent residences on them.

It is then critical that the spoil islands are adequately protected if their present values are to be maintained. A comprehensive management program is necessary which provides the appropriate use and management of spoil islands through coordinated and consistent decision-making. Otherwise the damage to these valuable resources may prove to be an irretrievable loss to man and wildlife alike.

THE SPOIL ISLAND MANAGEMENT PROCESS

OVERVIEW

The spoil island management process developed for this project was adapted from a coastal planning process devised by the Treasure Coast Regional Planning Council (TCRPC) for one of its local governments in the Spring of 1979. In addition, it reflects the use of the basic precepts generally felt to be components of the comprehensive planning process: 1.) survey and analysis; 2.) goal formulation; 3.) plan making; and 4.) plan effectuation. Also, the process provides for a "plan update", in the devised management program which is often referred to as another step of the comprehensive planning process.

Key coastal management concepts and techniques were also used in this process which were developed by and for Florida's first CM agency (Florida Coastal Coordinating Council) over a period of several years. In addition, much attention was given to the federal coastal planning requirements set out in the Federal Coastal Zone Management Act Amendments of 1976. Further, the County's authorities and the project's biological consultant all participated in the selection and calibration of the SIM process. Hence, although the SIM process was primarily developed by the TCRPC over the duration of this project, its selection and composition was built upon the hard work of many interested parties over several years.

The SIM process developed during the project (see EXHIBIT 1) included five major steps:

1. Data collection;
2. Analysis of data; determination of potential land/water uses, problems, and opportunities;
3. Development of spoil island management goals and objectives;
4. Setting spoil island management strategies; and
5. Implementation technique development for spoil island management.

EXHIBIT 1

THE SPOIL ISLAND MANAGEMENT PROCESS

-1-

COLLECTION OF DATA ON SPOIL ISLANDS AND ADJACENT COASTAL RESOURCES

- Coastal resource characteristics and existing/historical conditions
- Status of present use and management

-2-

ANALYSIS OF DATA: DETERMINATION OF POTENTIAL LAND/WATER USES, PROBLEMS AND OPPORTUNITIES

- Set specific spoil island criteria
- Assess potential spoil island management problems and opportunities
- Determine potential land/water uses
- Evaluate anticipated financial resources necessary for spoil island management

-3-

DEVELOPMENT OF SPOIL ISLAND MANAGEMENT GOALS AND OBJECTIVES

- Survey of community values concerning spoil islands
- Review of spoil island management efforts by governmental agencies

-4-

SETTING SPOIL ISLAND MANAGEMENT STRATEGIES

- Develop coastal policies for spoil island management
- Define permissible land/water uses
- Set permissible use priorities
- Designate special management areas

-5-

IMPLEMENTATION TECHNIQUE DEVELOPMENT FOR SPOIL ISLAND MANAGEMENT

- Define means to control permissible land/water uses
- Describe organizational structure necessary to implement management strategies
- Recommend changes in the existing management structure to implement selected management strategies

THE PROCESS

The following is a discussion of the five major steps of the developed SIM process. Each step is reviewed as to its makeup and then an account is given on how they were applied to Indian River County's situation.

STEP 1 - DATA COLLECTION

This step would involve the collection of basic data concerning the characteristics, use and management of spoil islands and their adjacent coastal resources. Coastal resource information to be collected would include biophysical data on the spoil islands (e.g., its ecosystems, physical characteristics and wildlife) and the surrounding marine resources it interacts with, such as seagrass beds and a variety of marine life).

The use and management of those resources should also be reviewed. Data is compiled on their present use to determine the status of the area's spoil island planning and management efforts. Also, pertinent CM authorities (e.g., laws, administrative rules) would be reviewed to determine which CM agencies could or do have spoil island management responsibilities.

This phase of the process consisted of an extensive biophysiographic study, to obtain a spoil island data baseline, and the collection of other pertinent management information which was collected from a variety of sources, including the County's MIAC, a literature search of the County's spoil island administrative files, and various governmental agencies. Other information collected included specific data on each island, such as: ownership and/or lease arrangements; a determination of the governmental jurisdiction an island was within; recorded community desires for island use; and information identifying if the islands were within special designation management areas.

The biophysiographic study was an essential ingredient of the project as sufficient information on the physical nature and wildlife resources of the islands were necessary to adequately assess spoil island issues. Forty-three islands were considered by Indian River County for study. Obvious time and logistic problems limited the number of islands that could be adequately studied. Therefore, a carefully chosen sub-sample of sixteen islands were chosen from the original 43. These islands were picked as representatives of the various island types based on preliminary studies made by the county and after examination of maps and aerial photographs.

After the representative islands were chosen, they were photographed through low level aerial photography and rough field maps were drawn of each island to be visited.

The islands were then examined for physiographic and biological features. A north-south and east-west transect was made across each island. At ten meter intervals along these transects the species of vegetation at each point was noted. After the completion of these transects a table could be produced which would reveal dominant vegetation trends and accurate vegetation maps (see EXHIBIT 2) could be produced from the combination of transect data, preliminary field maps and aerial photographs.

The soil humus or organic layer (A & B horizon detritus included) was examined for relative depth over the spoil material. This value was used to give a rough estimate of the age and development (successional) state of the soil organic layer and to detemine whether fires had occurred on the island. Parent spoil material varied but was not examined or classified.

Terrestrial animal observations were limited to avian and mammal species. Estimates of the number of individuals per species were made for the time of visitation. Any nesting activity was noted.

Aquatic seagrass beds adjacent to the island perimeter (within 30 m offshore) were qualitatively examined and mapped. Aquatic faunal samples were made with a small seine net at four compass quadrates surrounding the island. Species of fishes and crustaceans taken in the seine samples were noted and given a relative abundance code. Oyster beds and clam beds were also noted and mapped.

General island physiography was noted as areas of evident erosion, sand beaches, rocky coast, height above sea level, etc. were recorded. The location of the island relative to erosional processes, i.e., prevailing wind direction, tidal water movements, etc. was also noted (see EXHIBIT 3). Proximate water depth and seasonal boat access to the island was also noted.

After these parameters were determined, the islands were qualitatively and quantitatively discussed via an annotated list and classified according to physiographic character (e.g., high, rocky, dry, without evident erosion problems) and biological resources (e.g., abundant and diverse floral and faunal assemblages were present). The islands were then discussed relative to their predicted future physiographic and biological resource values. The future state of the islands could be predicted accurately only when historical photographs could be seen. These latter determinations should be major factors in long-term management schemes.

Cursory observations were made of boat activity in the vicinity of the islands and evidence of island visitation for recreation activities, etc. (e.g., quantity of refuse, campfires). Systematic quantitative observations of human use of the islands were not made.

EXHIBIT 2
Typical spoil island resource map showing vegetation codes, seagrass and oyster beds.

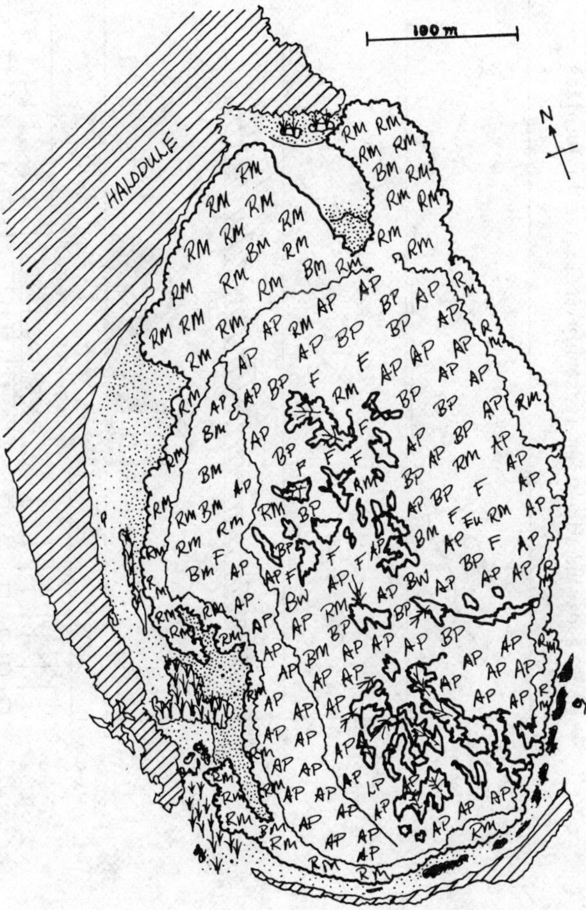

EXHIBIT 3

Diagrammatic representation of spoil island setting and pertinent data. Arrow size and direction indicate mean wind speed and wind direction (easterly) relative to location.

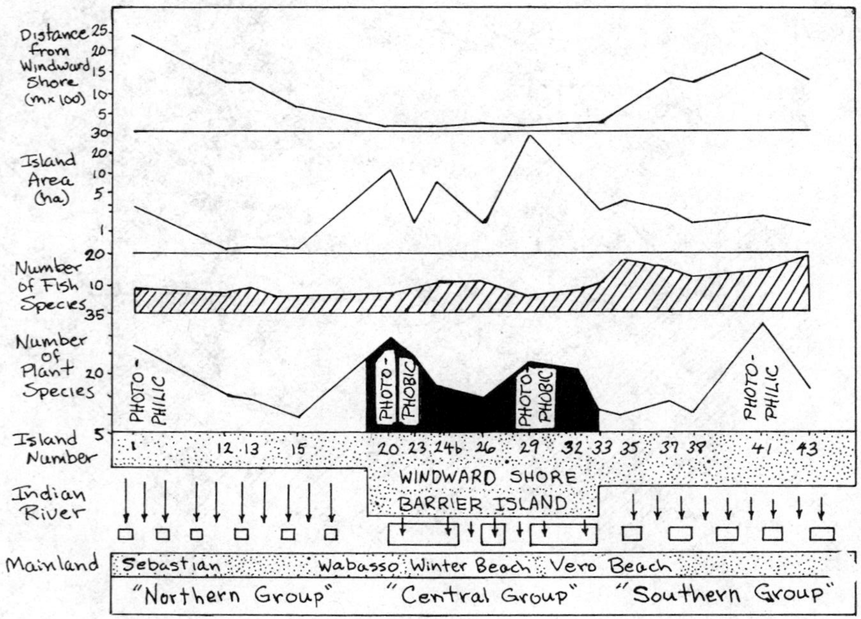

A total compilation of all spoil island resource inventory data was made available in summary form through a carefully prepared report for use in management plan development.

STEP 2 - ANALYSIS OF DATA: DETERMINATION OF POTENTIAL LAND/WATER USES AND PROBLEMS/OPPORTUNITIES.

The second phase of this management approach would primarily entail organizing the spoil island data collected in Step One of the process and determining what could be the potential land and water uses of the spoil islands and their environs. A major effort here would be to set out the data collected in a manner where spoil island characteristics are related to possible land/water uses. For instance, the availability of access is of great import in assessing the value of utilizing an island for recreational purposes. Upon setting these relationships, various characteristics of each spoil island could be assessed as to its value (positive or negative) in determining how appropriate it would be to utilize an island for a particular use. Accuracy of these assessments could be determined by seeking advice from governmental agencies and a Spoil Island Management Citizens Advisory Committee (CAC).

Spoil island characteristics which could be evaluated in this process would include:

- availability of access
- soil characteristics
- topographic characteristics
- morphological advantages
- locational advantages
- use/ownership/lease arrangement impacts
- relationship to special coastal resource designations (e.g., water classification, State aquatic preserve designation)
- institutional constraints (e.g., environmental permitting)
- community values
- special socio-economic values
- development potential
- potential for sanitary facilities
- other values/opportunities
- water quality
- avian habitat values
- other wildlife values
- marine life values
- vegetation significance
- special habitat values (e.g., endangered species, critical habitat)
- special research values
- special educational values
- fishing opportunities (including shellfishing)
- camping opportunities
- other recreational values (e.g., sandy beaches)

Spoil island uses which could be considered would be recreation, limited development, conservation, research, education and other special uses (e.g., designated for future spoil dispotion).

To accomplish this task, a cross impact analysis approach was employed which utilized a series of matrix tables to analyze the data. EXHIBITS 4 and 5 are examples of the final matrix tables used for the purposes outlined.

The approach used first entailed the setting of the relationships of various spoil island characteristics to a potential use of an island. Island characteristics are then assigned weighted values to set the relative importance of each characteristic in determining the most suitable use. In addition, the application of weighted island characteristics values is also used to determine the island's relative suitability for a potential use as compared to another island being examined. The result is that a suggested use for an island can be determined and the devised management program (see discussion on Steps 4 and 5), which was developed for that use, can then be applied. Theoretically, this approach provides for any given spoil island to be in a position of receiving the necessary management attention to set up and maintain the island in its appropriate use.

The first matrix table (EXHIBIT 4) was used to set the nature of the relationships between the selected island characteristic and the potential island use specified (i.e., favorable, unfavorable or of no apparent relationship). In addition, the matrix also fixes the relative importance of a given characteristic to another by the assigned weighted values.

The other matrix table, was employed to rate the suitability of each specific island to the potential uses indicated. If a characteristic was found to exist in a spoil island and the relationship was positive relative to the potential use, a score of "1" to "3" (i.e., depending on the assigned weighted values) was placed next to the use shown. If there was no apparent relationship, or the relationship between the characteristic present and the use was negative, a "0" score was then given. For example, since spoil island No. 1 exhibited high aquatic values, the assigned value of "3" was placed in the line which lists a conservation use, as the relationship here was deemed as being favorable. Under the two other potential uses it was felt that high aquatic value represented an unfavorable relationship; hence a "0" score was given.

The outcome of listing these values in the manner indicated is that a final score can be derived for each island, indicating which potential use would appear to be most appropriate for that island, and how it compares with other islands relative to its suitability for a given use.

It should also be noted that, the number and type of uses which were

EXHIBIT 4
SPOIL ISLAND
USE-CHARACTERISTIC RELATIONSHIPS

SELECTED CHARACTERISTICS / POTENTIAL SPOIL ISLAND USE	(1) Locational Advantages	(2) Adequate Opean Beach Access	(2) Accessibility	(1) Erosion Problems	(2) Soils Quality High (i.e., humus layer)	(2) Sufficient Developable Area	(2) Topographic Advantages	(2) Sanitary Facilities Suitability	(3) Avian Habitat Significance	(3) High Aquatic Values	(3) High Fauna Values	(3) High Flora Values	(3) Special Biotic Values	(2) Habitat Development Beneficial	(2) Receives Human Traffic	(3) Within Special Designated Area	(3) No Existing Lease Arrangements	(1) Ownership and/or Lease Limitations Exist	(1) Within Another Governmental Jurisdiction	(1) Limited Finances Exist for Management	(3) Community Value
CONSERVATION	-	-	-	-	+	-	+	o	+	+	+	+	+	+	-	+	-	+	-	+	+
PASSIVE RECREATION	+	+	+	-	+	+	+	o	-	-	-	-	-	-	+	-	-	+	-	-	+
LIMITED DEVELOPMENT RECREATION	+	+	+	-	+	+	+	+	-	-	-	-	-	-	+	-	-	-	-	-	+

Legend:

+ = Favorable relationship
- = Unfavorable relationship
o = No apparent relationship
(1) = Weighted values assigned to characteristics

EXHIBIT 5
SPOIL ISLAND
USE-CHARACTERISTIC RATINGS

SPOIL ISLAND NUMBER	POTENTIAL USE (SELECTED CHARACTERISTICS →)	Locational Advantages	Adequate Open Beach Access	Accessibility	Erosion Problems	Soils Quality High (i.e., humus layer)	Sufficient Developable Area	Topographic Advantages	Sanitary Facilities Suitability	Avian Habitat Significance	High Aquatic Values	High Fauna Values	High Flora Values	Special Biotic Values	Habitat Development Beneficial	Receives Human Traffic	Within Special Designated Area	No Existing Lease Arrangements	Ownership and/or Lease Limitations Exist	Within Another Governmental Jurisdiction	Limited Finances Exist for Management	Community Value	SCORE
1	Conservation	0	0	0	0	0	?	2	?	0	3	0	3	3	0	0	3	0	1	1	1	0	17
	Passive Recreation	1	2	2	0	0	?	2	?	3	0	3	0	0	0	2	0	0	1	1	0	3	20*
	Limited Devel. Rec.	1	2	2	0	0	?	2	?	3	0	3	0	0	0	2	0	0	0	1	0	3	19
12	Conservation	0	0	0	0	0	?	2	?	0	3	0	0	0	0	0	3	0	1	1	1	2	13
	Passive Recreation	1	2	2	0	0	?	2	?	3	0	3	3	3	0	2	0	0	1	1	0	0	23*
	Limited Devel. Rec.	1	2	2	0	0	?	2	?	3	0	3	3	3	0	2	0	0	0	1	0	0	22
13	Conservation	0	0	0	0	0	?	2	?	0	3	0	0	0	0	0	3	0	1	1	1	0	11
	Passive Recreation	1	2	2	0	0	?	2	?	3	0	3	3	3	0	2	0	0	1	1	0	2	25*
	Limited Devel. Rec.	1	2	2	0	0	?	2	?	3	0	3	3	3	0	2	0	0	0	1	0	2	24
15	Conservation	1	2	2	0	0	?	2	?	3	0	3	0	0	0	2	3	0	1	1	1	3	24*
	Passive Recreation	0	0	0	0	0	?	2	?	0	3	0	3	3	0	0	0	0	1	1	0	0	13
	Limited Devel. Rec.	0	0	0	0	0	?	2	?	0	3	0	3	3	0	2	0	0	0	1	0	0	12
20	Conservation	0	2	0	1	2	?	2	?	0	0	3	3	3	2	2	3	0	1	1	1	0	25*
	Passive Recreation	1	0	2	1	2	?	2	?	3	3	0	0	0	0	0	0	0	1	1	0	3	19
	Limited Devel. Rec.	1	0	2	1	2	?	2	?	3	3	0	0	0	0	0	0	0	0	1	0	3	18
23	Conservation	0	2	0	1	2	?	0	?	0	0	3	3	3	-	2	3	0	1	1	1	3	25*
	Passive Recreation	1	0	2	1	2	?	0	?	3	3	0	0	0	0	0	0	0	1	1	0	0	14
	Limited Devel. Rec.	1	0	2	1	2	?	0	?	3	3	0	0	0	0	0	0	0	0	1	0	0	13
24b	Conservation	1	2	0	1	2	?	2	?	0	3	3	3	3	0	2	3	3	1	0	1	3	33
	Passive recreation	0	0	2	1	2	?	2	?	3	0	0	0	0	0	0	0	3	1	0	0	0	14
	Limited Devel. Rec	0	0	2	1	2	?	2	?	3	0	0	0	0	0	0	0	3	0	0	0	0	13

selected in the case of Indian River County was limited on the basis of what seemed potentially feasible given the available information. The data was obtained by reviewing existing spoil island leases, past County ordinances, and advice from other governmental sources. Public input was also sought from the County's MIAC. In addition, the selected characteristics were chosen from the information collected and analyzed during Steps one and two of the process.

Values assigned to each spoil island characteristics were devised by a professional analysis of the data by the parties involved. Also, definitions developed for both the spoil island characteristics and potential uses were developed so as to meet Indian River County needs.

STEP 3 - DEVELOPMENT OF SPOIL ISLAND GOALS AND OBJECTIVES

Community values and governmental efforts concerning spoil island management would be assessed here for the purpose of developing a set of goals and objectives. They would serve to provide the necessary guidance as to how the area's spoil islands should be managed. Again, a spoil island CAC (composed of representative coastal interests) and the area's governmental CM agencies would serve as important sources of guidance in completing this management phase.

A viable set of overall CM goals and objectives relating to spoil islands management were devised for Indian River County to provide for the development of an effective SIM plan. They were developed to provide the "benchmark" from which the County's spoil island management strategies and implementation techniques would be based upon. They were compiled on the basis of the information collected in the early phases of the SIM process from a number of important sources, including the County's MIAC, with the intention of reflecting the County's values and needs.

STEP 4 - SETTING SPOIL ISLAND MANAGEMENT

A crucial task in the fourth phase of this management process would be to establish coastal policies for spoil island management. This is a necessary step as it provides the guidance necessary to determine the land/water uses and priority of those uses which should be permitted relative to each spoil island. Governmental CM agencies and the CAC would serve to provide guidance in this fourth phase as part of a continuing process, beginning at the early stages of this management approach.

In addition to the above, areas to receive special attention would be designated, including those areas which:

1. should be closed to public (no development);
2. are environmentally sensitive;
3. are suitable for replanting (recommending species to be removed

and those to be used for revegetation);
4. should be used for limited development for passive recreation;
5. should be developed (including specific uses for each area);
6. should be developed for picnicking, hiking, and camping;
7. should restrict boating activities.

STEP 5 - IMPLEMENTATION TECHNIQUE DEVELOPMENT FOR SPOIL ISLAND MANAGEMENT

In this fifth management phase, the necessary steps are laid out to implement the selected spoil island management program. For instance, if an area is designated to be closed to the public, certain institutional measures would have to be taken (e.g., policing the island) to bring about that desired end. In addition, other actions would have to be set in motion to ensure that the selected spoil island management would be integrated into the area's existing planning/management framework. Other spoil island implementation techniques could include transferring ownership of certain spoil islands, new and/or amended governmental ordinances, capital improvement programming, governmental agency agreements (e.g., to facilitate educational opportunities) etc.

EXHIBIT 6 provides an example of a few of the recommended list of the approximately 20 spoil island management strategies and numerous implementation techniques developed for Indian River County. The recommended management strategies define the appropriate actions the County should take to address the management problems/opportunities which exist relative to effectively managing their spoil islands. They also serve to define permissible land/water uses, set permissible use priorities, and designate special management areas.

The corresponding implementation techniques represent those methods which should be utilized to bring the County's spoil island management strategy into fruition. These include methods which involve both physical changes in various islands and changes to the existing management structure (e.g., changes in governmental procedures) relative to spoil island management.

The implications of the County's management actions are summarized in the third column of the table. The management implication section of the table also briefly discusses the management responsibilities of various governmental agencies, the general costs involved in implementing certain actions, and other pertinent information.

CONCLUSIONS

The SIM process is felt to be an extremely valuable management tool which could have wide application, if properly used. Like most management processes, however, there are a number of conditions which

EXHIBIT 6
MANAGEMENT OF SPOIL ISLAND RESOURCES
INDIAN RIVER COUNTY

MANAGEMENT STRATEGY

L. The County shall promote a comprehensive investigation into the possibility and methods of undertaking habitat development on those spoil islands identified as having a potential for such work.

MANAGEMENT IMPLICATIONS

Habitat development appears to be feasible but there are a lot of obstacles which have to be overcome prior to undertaking such efforts, including environmental permitting requirements and determining appropriate materials and techniques. Hence, it is necessary that these efforts be coordinated with such agencies as the Corps, DER and the SCS. Costs for such efforts are also a major problem. For instance, in planting mangroves in a 100 square foot area, plant material alone could easily run over $100. In addition, logistics problems (e.g., hauling the materials to the site) could cause costs to greatly increase. Major habitat development efforts would amount to at least $500-$1,000 an acre. On the other hand, significant conservation, recreational and amenity opportunities can be realized by habitat development efforts.

IMPLEMENTATION TECHNIQUES

1. Investigate, with those agencies identified in A-3, the feasibility and advisability of minor habitat development of spoil island resources, especially in regard to the environmental permitting requirements, vegetation suitability and recommended techniques. This investigation should include an examination of such habitat development effort as:
   A. avian habitat development (e.g., island #36)
   B. the removal of certain exotic plant species and planting of native vegetation (e.g., island #29)
   C. the planting of aquatic seagrasses (e.g., island #35)

2. Major habitat development efforts (i.e., complete alteration of spoil island habitat), should also be jointly investigated to determine their feasibility. Such efforts should logically build upon the U.S. Army Corps of Engineers advances in undertaking this type of work and include a consideration of recreating a spoil island into a viable marsh habitat.

3. Utilize such existing, readily available, manpower and financial resources in accomplishing habitat development as:
   A. the plant testing program of the U.S. Soil and Water Conservation Service
   B. the use of the volunteer capabilities of civic organizations (e.g., Boy Scouts, garden clubs, conservation groups, CETA personnel, prisoners, etc.)

should be met in carrying out the process if it is to be a viable tool. In addition, one must understand the limitations of the process and recognize that additional refinements to it may be necessary if the process is to meet their specific needs.

The SIM process involves the use of subjective inputs, and has specific personnel and data requirements. True, it can be argued that the process incorporates objective techniques as a systematic approach which is used to analyze island data. However, the setting of both the potential spoil island uses and pertinent characteristics is subjective as well as the treatment of those inputs (i.e., determining the data relationships and the weighted values assigned). In addition, the decided numbers and definitions for the data inputs are also subjective. On the other hand, this aspect of the process can prove to be a significant advantage if the data inputs (i.e., matrix variables) are properly developed, as it can be made to better fit individual needs.

Also, an attempt should be made to utilize all available sources of data to ensure data collection efforts are complete, including citizen groups and governmental agencies. Relative to this point one potential problem is that sufficient spoil island data may not be available to make definite decisions on spoil island management. In undertaking this pilot project, it was found that even after conducting an extensive biophysiographic study of the islands, additional data was still necessary to determine the development potential of the islands. Consequently, it is important to set the spoil island baseline data to be collected prior to the actual investigation, thereby reducing the need for additional research. However, it is likely some supplemental data may be required, even if those measures are taken.

In addition, trained and capable personnel must be used for collecting and analyzing the data if it is to be useful and accurately assessed. Trying to save finances by using untrained personnel to collect and analyze the data could end up costing the local government much more money in the long run in attempting to undo bad advice. If a local government ever finds itself in a position of having to use untrained personnel, professionals should at least be consulted at the outset of the project.

Another general concern in employing the SIM process relates to the fact that the plan can only be as effective as one uses it. In essence, the SIM plan will not perform well if not properly implemented. To be effective, the SIM plan sould be integrated into the government's comprehensive plan and adopting it, the plan can be tied directly to the decision-making and budgetary processes of the local government. Consequently, the SIM plan would then have the necessary official status to carry it out.

In properly developing and implementing the plan, it is also necessary to coordinate with all governmental quasi-public agencies and private interests which play a role in spoil island management at the outset. This would include governmental permitting agencies, property owners, local sponsors of dredging projects, governmental entities with easements for spoil disposal and various other governmental agencies which have coastal management authorities in the area (e.g., governmental management agency for a wildlife preserve). Coordination problems can be prevented by using this approach, and the plan is likely to be more complete and effective with widespread involvement of pertinent interest groups and agencies.

Possible improvements or refinements of the process which can be employed would include: 1.) the setting of priorities for management strategies; 2.) assigning negative matrix table scores when a negative relationship is felt to exist between a potential spoil island use and the selected characteristic being considered; and 3.) employing the use of graphics to portray existing conditions and the selected management strategies for those particular spoil island resources.

By setting management strategy priorities another dimension is added to the spoil island management program. It is not sufficient enough to set island use priorities, but a timetable for management strategy implementation is also needed.

To ensure that negative relationships are adequately considered, negative scores could be substracted (i.e., instead of being indicated as a "0" score) to determine the island use rating scores when negative use characteristic relationships exist. This would then provide for both negative and positive scoring in utilizing the matrix tables. A possible approach here would be to use a larger scoring continuum (e.g., instead of a possible scoring range from 0-3, a -3 to +3 range could be employed).

Also, graphics many times can be used to better portray spoil island issues and how they are to be resolved. This approach would tend to assist those involved in the SIM process such as the public citizen and elected official. In addition, by graphically indicating issues, the spoil island planner can more easily determine management strategy needs as he is developing the SIM plan. The graphics to be used need not be extensive either as sketch graphics can be very effectively used to exhibit ideas.

In summary, although the SIM process requires special considerations to be made to work properly it can be an effective management tool. By its application, a plan and a general framework can be formulated to provide for the sufficient protection and utilization of spoil island resources. It represents a technique which follows a logical progression of evaluating all pertinent inputs to the culmination of a plan of action to bring about the desired end. The end product is not solely a plan to better manage spoil island, but a vehicle to bring about coordinated and consistent management of all our coastal resources.

BIBLIOGRAPHY

1. Chadwick, George, A Systems View of Planning, Perganon Press, Oxford, England, 1971.

2. Haar, Charles M., Land Use Planning - Second Edition, Little Brown and Company, Boston, Mass., 1971.

3. Hunt, L.J., M.C. Landin, A.W. Ford, and B.R. Wells, 1978. Upland Habitat Development With Dredged Material: Engineering and Plant Propagation. Tech. Rpt. D.S.-78-17. U.S. Army Engineer Waterways Experiment Station, CE, Vicksburg, Miss., 79 p., Tables, Appendix.

4. Lewis III, R.R. and C.S. Lewis, 1978. Colonial Bird Use and Plant Succession on Dredged Material Islands in Florida, Vol. II. Patterns of Plant Succession. Tech. Rpt. p. 78-14. U.S. Army Engineer Waterways Exp. Sta., Vicksburg, 169 p.

5. Lunz, J.D., R.J. Diaz and R.A. Cole, 1978. Upland and Wetland Habitat Development with Dredged Material: Ecological Considerations. Tech. Rpt. DS-78-15, U.S. Army Engineers Waterways Exp. Sta., Vicksburg, 50 p.

6. Maxwell II, G.R. and H.W. Kale II, 1974. Population Estimate of Breeding Birds on a Spoil Island in the Indian River, Indian River County, Florida. Fla. Field Nat. Vol. 1 Fall, 1971, p. 32-39.

7. Maxwell II, G.R. and H.W. Kale II, 1977. Breeding Biology of Five Species of Herons in Coastal Florida. The Auk 94: 689-700.

8. The RMBR Planning/Design Group, Local Coastal Zone Management: A Handbook, Florida Coastal Coordinating Council, Tallahassee, Fla., undated.

9. Schreiber, R.W. and E.A. Schreiber, 1978. Colonial Bird Use and Plant Succession on Dredged Material Islands in Florida, Vol. 1: Sea and Wading Bird Colonies. Tech. Rept. D-78-14. U.S. Army Engineer Waterways Experiment Stateion, Vicksburg: 95 p + Appendix.

10. Soots, R.F., Jr. and M.C. Landin, 1978. Development and Management of Avian Habitat on Dredged Material Islands. Tech. Rpt. DS-78-18. U.S. Army Engineer Waterways Exp. Sta., Vicksburg: 96 p + Appendix.

11. Treasure Coast Regional Planning Council, Coastal Zone Management: Interim Goals, Objectives and Policies for the Treasure Coast Region, TCRPC, Stuart, Fla., 1978.

12. Treasure Coast Regional Planning Council, Coastal Zone Protection Element-City of Palm Beach Gardens, TCRPC, under contract with the City of Palm Beach Gardens, Stuart, Fla., 1979.

13. U.S. Congress, Coastal Zone Management Act Amendments of 1976, U.S. Printing Office, Washington, D.C., 1976.

14. Yorke, Thomas H., Impact Assessment of Water Resource Development Activities: A Dual Matrix Approach, U.S. Fish and Wildlife Service (FSW/OBS-78/82), Harpers Ferry, West Virginia, September 1978

INFLUENCE OF STORM TYPE AND URBANIZATION ON TIDAL FLOODING

Armando Balloffet, F·ASCE[1] and Michael L. Scheffler[2]

INTRODUCTION

The tidal hydraulics of the Hackensack Meadowlands (N. J.) was studied through the use of mathematical models under contracts between the Corps of Engineers and the Federal Emergency Management Administration (FEMA) and Tippetts-Abbett-McCarthy-Stratton. The Hackensack River is located 3 miles west of New York City, across the Hudson River (Fig. 1). Flooding of the meadowlands results primarly from tidal surges induced by northeaster (or extratropical) and hurricane storms propagating through Newark Bay (N. J.).

The proximity of the area to Metropolitan New York has over the past decade induced large scale development of the flood plain. This development has resulted in extensive landfill. The object of this study was to determine the effects of present and future landfills upon storm flood stage and discharges.

The results of model simulations show that landfill significantly increases storm tide flooding for equal storm surges. In addition, the differences between flooding due to northeasters and hurricane storm tide surges were investigated. The results show that northeasters may produce higher flooding in the Meadowlands, even with peak surges lower than those produced by hurricanes. This is due to the longer duration of the northeaster surges which results in larger volumes filling the inland area.

---

(1) Consultant, Tippetts-Abbett-McCarthy- Stratton, and Adjunct Professor, Columbia University, New York, N. Y.

(2) Oceanographer, Stone and Webster Corporation, Boston, Mass.

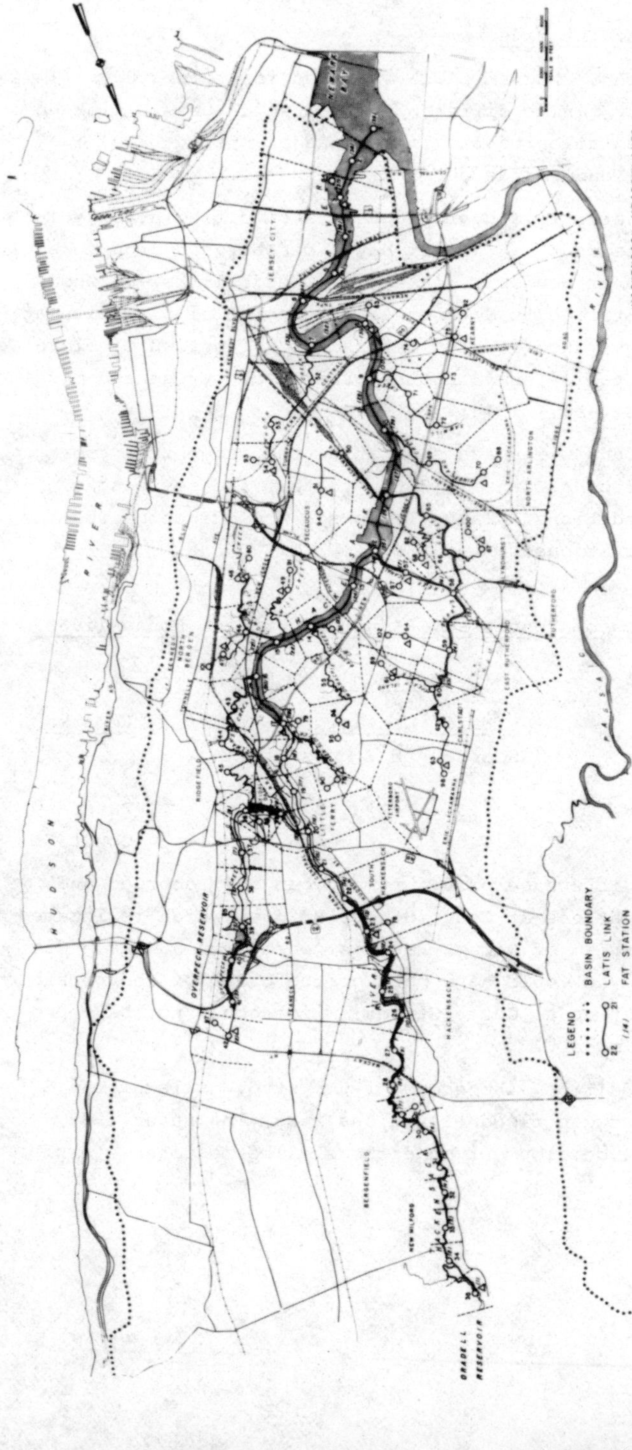

FIG I

HACKENSACK MEADOWLANDS
MATHEMATICAL MODEL
MAP OF THE MODEL

Description of the Model

The TAMS Program LATIS was used for this study. This is an explicit finite difference model which uses a link-node network presentation of the rivers and channels. Fig. 1 presents the Meadowlands network.

Boundary stage hydrographs are specified at Node 1 located at the mouth in Newark Bay. Discharge hydrographs are specified at boundary links, (or pseudo links) as shown in Fig. 1. Storage is defined as a function of elevation at nodes and conveyance is also defined as a function of elevation at links. Details regarding the program are presented elsewhere (Balloffet, at. al., 1974).

Two runs were made simulating the surge tide produced by Hurricane Donna in 1960, to test the response of the model conditions. Table 1 shows a comparison of recorded and model elevations.

Table 1

Comparison of Inputs and Results Hurricane Donna (1960)

Location	Maximum Water Elevation		Remarks
	Recorded ft MSL	Model ft MSL	
Node 1	8.3	8.3	Boundary
Node 25	6.0	5.6	Court St. Bridge

The difference of 0.4 ft between the recorded and computed elevations at Court Street Bridge, which is located 17 miles upstream of Newark Bay,was considered adequate for calibration. It would have been of the order of the probable local wind set up at Court Street, which could not be reproduced by the model.

In addition, the scope of the study called for a comparison between flood elevations resulting under 1963, 1972 and 1984 development conditions. The results of this comparison show the changes in flooding produced by landfill.

After several consultations with the Hackensack Meadowlands Development Commission, planning maps were constructed showing the areas presently developed and to be developed until 1984.

This produced two different model configurations corresponding respectively to the 1972 and 1984 levels of development (Figs.2 and 3), which were obtained by introducing suitable changes in the 1963 configuration.

The changes is storage were obtained outlining the areas subject to development within each node which have been subtracted from the volume of the tidal low lands. Similarly, link inputs were changed to take into consideration encroachment in the floodways. These changes were made with a LATIS Pre-Program, to obtain revised values of conveyances and cross sectional areas.

As a result of present development, the following Nodes were eliminated for the 1972 and 1984 configurations: 64 in the Sports Complex area and 92, 75, and 74 in the area north of Kearny which has been diked off. Therefore, the storage capacity of these nodes was completely eliminated for the new configurations. Links 92-75, 73-75, 73-72 and74-8 in the area north of Kearny and 59-64 and 64-102 in the Sports Complex were also eliminated. A new pseudo-link, 102-59, was provided to allow for the inflow hydrographs that might originate from the pumping station in the sports complex. Another pseudo-link, 103-18, was added to represent the outflow from a sewage treatment plant.

No assumption was made of possible overtopping of the levees around the sports complex and north of Kearny, as it was assumed that they should be raised in case the design criteria were upgraded as a result of these studies. Consequently, the model indicates the necessary crest elevations (without freeboard) for such levees, and results

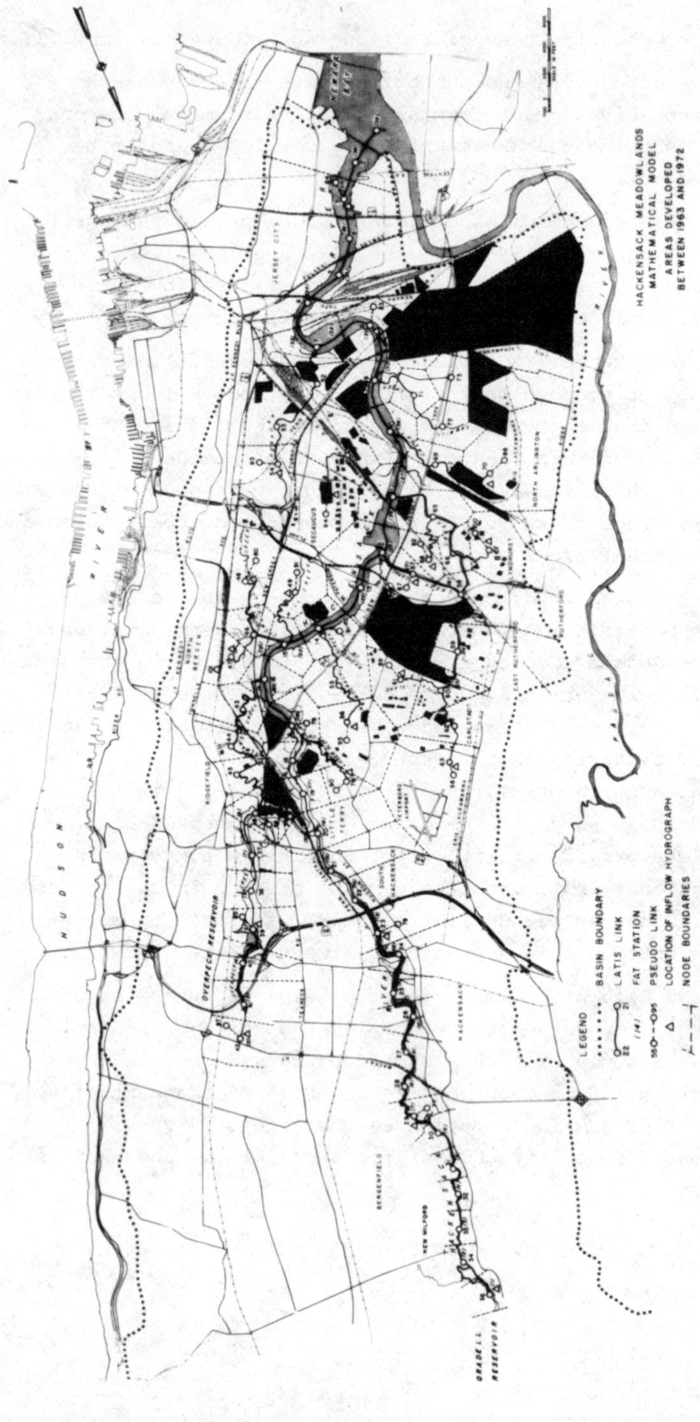
HACKENSACK MEADOWLANDS
MATHEMATICAL MODEL
AREAS DEVELOPED
BETWEEN 1963 AND 1972
LEGEND
BASIN BOUNDARY
LATIS LINK
FAT STATION
PSEUDO LINK
LOCATION OF INFLOW HYDROGRAPH
NODE BOUNDARIES
DEVELOPED AREAS
HUDSON
RIVER
JERSEY CITY
SECAUCUS
NORTH BERGEN
RIDGEFIELD
OVERPECK RESERVOIR
LITTLE FERRY
SOUTH HACKENSACK
HACKENSACK
BERGENFIELD
NEW MILFORD
ORADELL RESERVOIR
NORTH ARLINGTON
RUTHERFORD
EAST RUTHERFORD
CARLSTADT
RIVER

FIG. 2

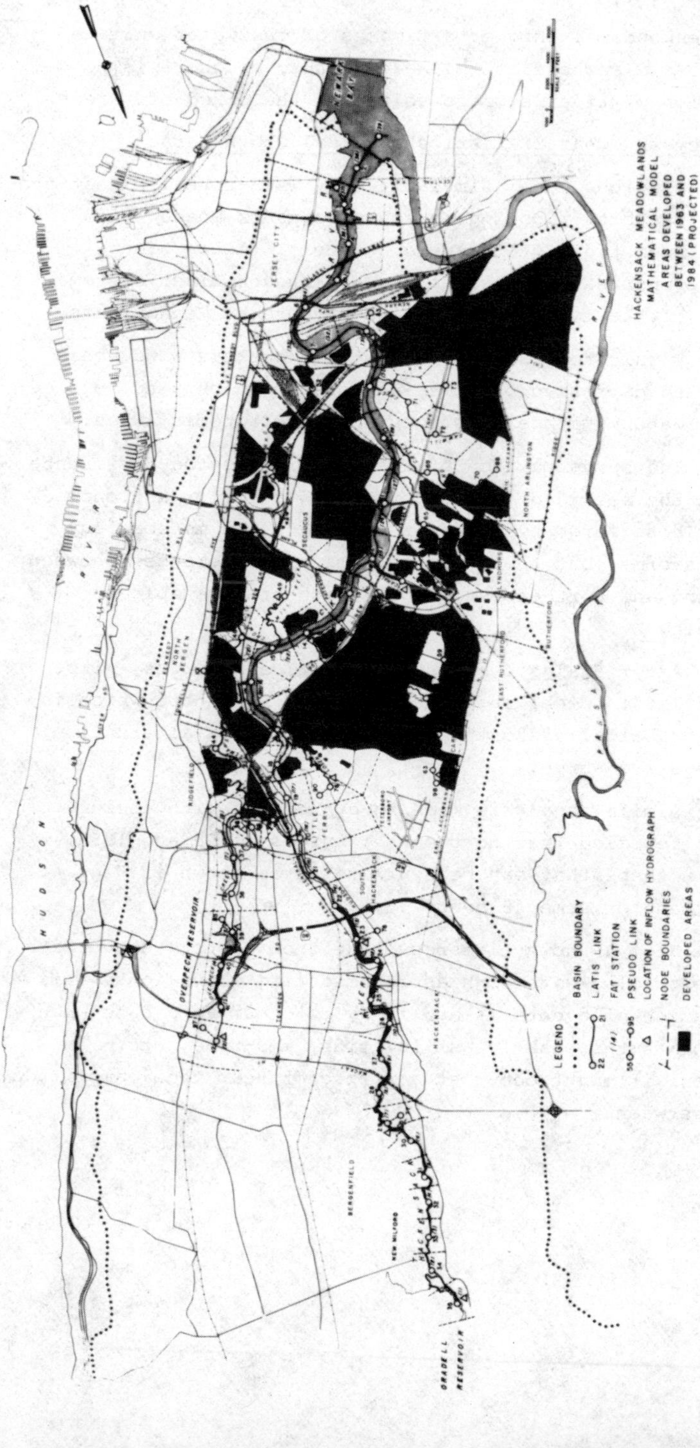

FIG. 3

HACKENSACK MEADOWLANDS MATHEMATICAL MODEL AREAS DEVELOPED BETWEEN 1963 AND 1984 (PROJECTED)

in correspondingly higher estimates of the water surface elevations elsewhere in the Meadowlands, as there is no assumption of tidal storage volume in the diked-off areas.

Frequency and Characteristics of Tidal Storm Surge

Previous model simulations by TAMS revealed that the primary cause of flooding in the Hackensack Meadowlands results from tidal storm surge. Fluvial floods affect only the upstream reaches of the Hackensack River system because of the large storage capacity of the Meadowlands.

Tidal storm surge results from onshore wind shear stress and depressed barometric pressure which tend to elevate the water surface to stages above astronomic tides.

The surge portion of a storm tide is the difference between the recorded tide and the predicted, or astronomic, tide. These surge hydrographs vary with the meteorology of the storms, but there are some major differences between hurricane and northeaster surges which are amenable to synthesis.

After analysis of several hurricane surges, Wilson (Wilson 1960) established a hurricane surge shape which is characteristically short and presents a rapid alternation to a negative surge.

Similar analysis made in connection with these studies revealed that northeaster surges (1950 and 1953) showed a shape that may be synthesized as shown in Fig. 4. with a base of some 48 hours.

Fig. 4 shows tide resulting from a storm surge with peak coinciding with high astronomic tide elevation. This combination of events is as likely to occur as a case when the storm surge peak is ast low tide, mid-tide, etc. In addition, it might occur at a spring or neap tidal amplitude, or in between.

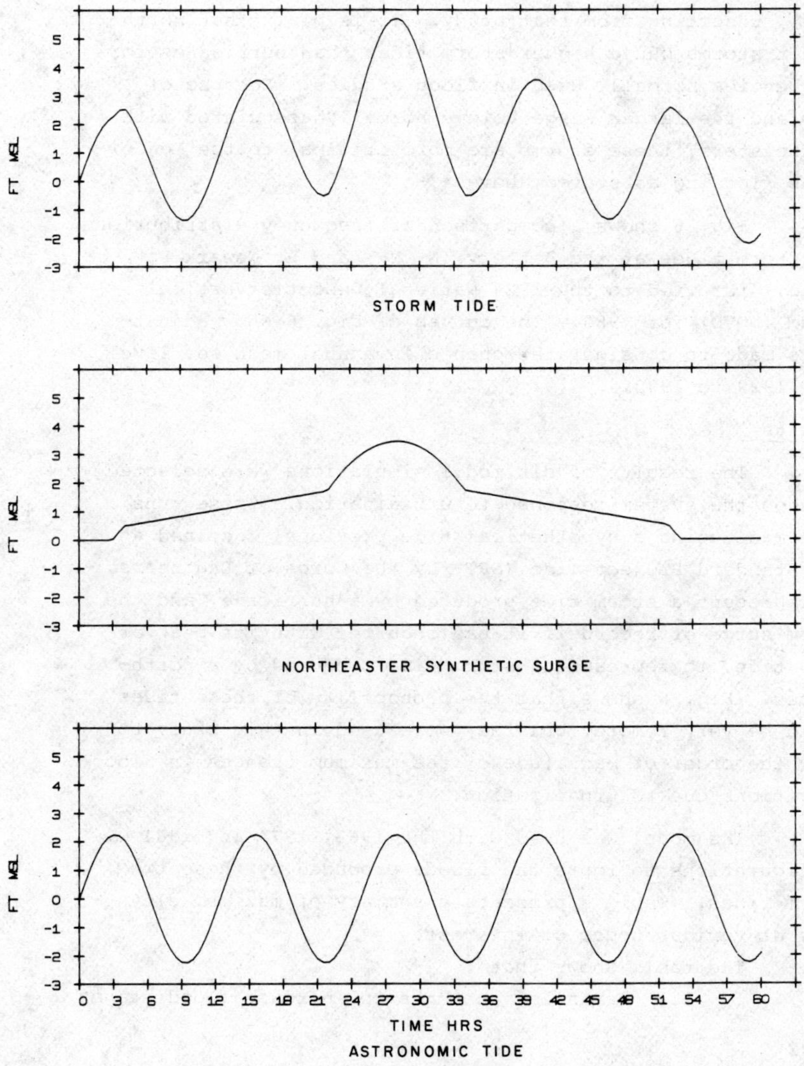

THE STORM TIDE AND ITS COMPONENTS

FIG. 4

This problem was studied under contract with FEMA using a joint probability method (Myers, 1970). Fig. 5 shows tidal elevation frequency curves for the Battery (New York), resulting from that study. It is plain that northeaster storms cause higher storm tides than hurricanes for frequencies normally used in flood studies. Because of this and the larger surge volume normally associated with northeasters, these storms are more critical to the Meadowlands flooding as stated above.

Fig. 6 shows a comparison of frequency distributions for storm tides at the Battery (N. Y.) and at Newark Bay (N. J.) referred to the 1929 National Geodetic Vertical Datum (NGVD) for 1980. The curves of Fig. 6 show adjustments made to consider the change in annual mean sea level from 1929 to 1980.

Results

The results of six model simulations were selected to show the system response to urbanization. These runs were made using a hypothetical tide previously defined as the Standard Project Tide (SPT) by the Corps of Engineers, to represent a storm tide produced by a hurricane, and the storm surge of record if it had occurred with its peak at high tide, to represent a storm tide produced by a northeaster. Fig. 6 shows that the probability of these tides would be very remote, but they were used in this study to show the order of magnitude of the maximum changes in flood elevations due to urbanization.

The model was used with the 1963, 1974 and 1984 configurations to route the floods produced by these two storm tides. Table 2 presents a summary of maximum elevations at various nodes of interest.

The table shows that :

1. Urbanization tends to increase flood stages.

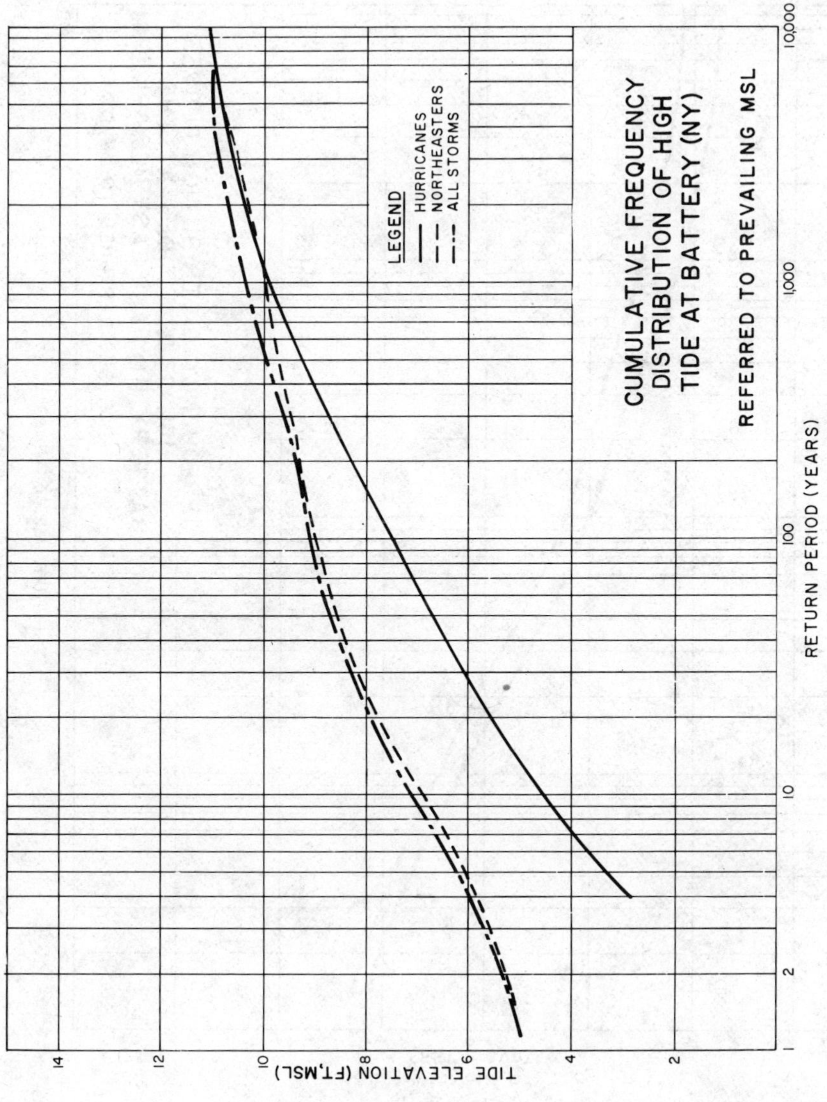
CUMULATIVE FREQUENCY
DISTRIBUTION OF HIGH
TIDE AT BATTERY (NY)
REFERRED TO PREVAILING MSL
LEGEND
HURRICANES
NORTHEASTERS
ALL STORMS
TIDE ELEVATION (FT, MSL)
2
4
6
8
10
12
14
RETURN PERIOD (YEARS)
1
2
10
100
1,000
10,000

FIG. 5

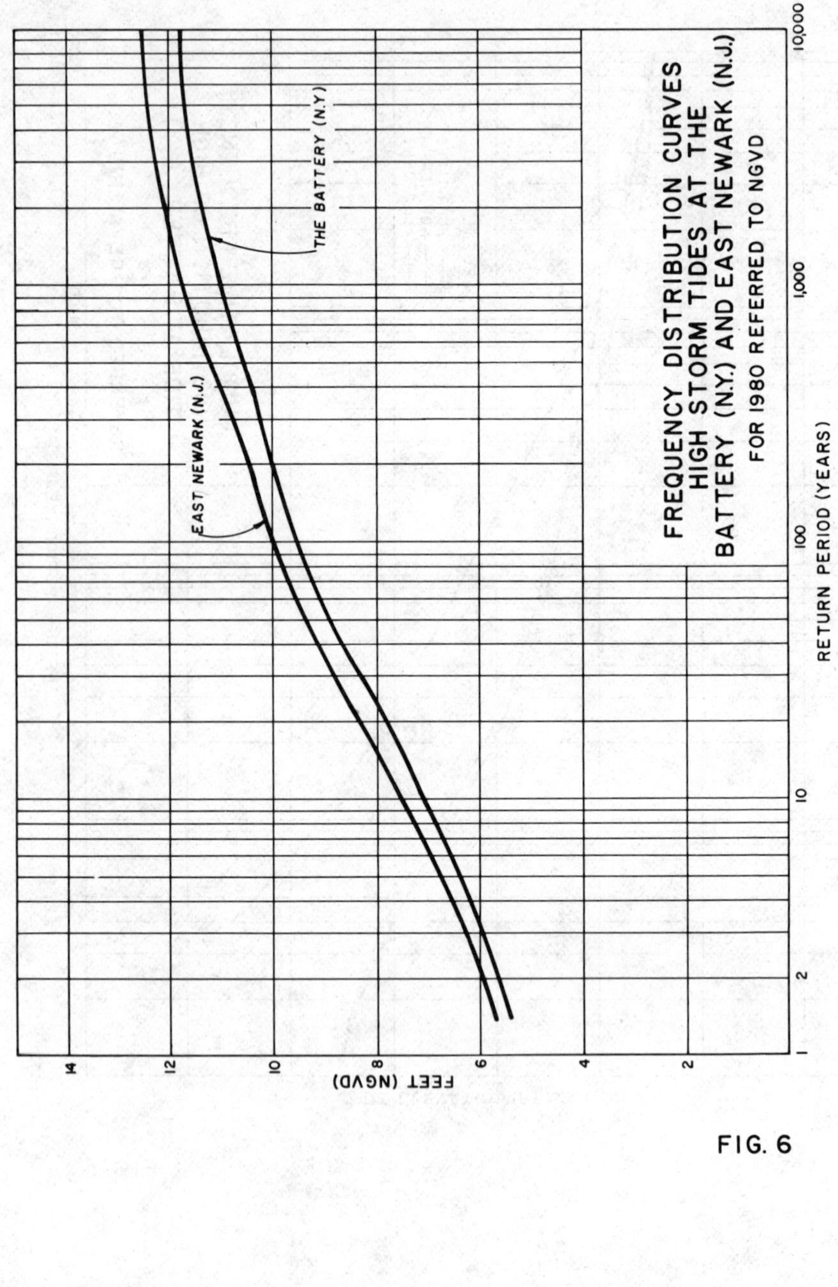
FREQUENCY DISTRIBUTION CURVES
HIGH STORM TIDES AT THE
BATTERY (N.Y.) AND EAST NEWARK (N.J.)
FOR 1980 REFERRED TO NGVD
THE BATTERY (N.Y)
EAST NEWARK (N.J.)
RETURN PERIOD (YEARS)
1
2
10
100
1,000
10,000
FEET (NGVD)
14
12
10
8
6
4
2

FIG. 6

TABLE 2

MAXIMUM ELEVATIONS AT VARIOUS KEY NODES (In Feet MSL)

Node Number	Standard Project Tide Configuration			1950 Surge on High Tide Configuration			Location
	1963	1972	1984	1963	1972	1984	
1	17.0	17.0	17.0	12.0	12.0	12.0	Mouth Boundary
8	11.8	13.0	13.3	10.2	10.7	11.0	
12	8.9	9.9	10.7	9.6	10.2	10.5	Junction of Berry's Creek Canal
15	7.9	8.6	9.7	9.6	10.2	10.6	Junction of Cromakill Creek
20	7.6	8.2	9.2	9.8	10.4	10.8	Junction of Overpeck Creek
25	7.9	8.4	9.3	10.1	10.7	11.1	Court Street Bridge
30	8.1	8.7	9.7	11.0	11.5	11.9	
58	8.0	9.0	10.2	9.4	10.1	10.4	Berry's Creek
60	6.1	7.3	8.1	9.0	10.0	10.3	
70	5.1	7.2	6.1	9.4	10.3	10.4	Kingsland Creek

2. The northeaster would have produced greater flooding within most of the Meadowlands than the SPT, despite its lower peak at the mouth of the river.

These results show that flood levels in the New York area should not be defined solely considering hurricane storms, because northeasters are more critical. This prevailance is more obvious for areas located north of New York. Conversely, for areas further to the south, such as the Atlantic City (N. J.), hurricane tides become the critical source of flooding.

Discussion

The mathematical model is able to simulate the propagation of floods and tides in the Hackensack Meadowlands for various assumed configurations.

In regard to maximum elevations for a given frequency, the tidal elevations are more critical than the river discharges in the area of the Meadowlands, even for 1984 conditions, when a large part of the available area will be filled.

The network is composed by nodes (storage areas) and links (conveyance elements). The maximum water surface elevations are influenced by the relative magnitude of the node areas and the conveyance of the links connecting them. The main channel is an example of a large conveyance with relatively small node areas connected to it. In this case, the maximum water elevations are governed by the maximum tidal discharges in the channel.

On the other hand, Node 70, for example, has a large storage area connected to the main stream by a small conveyance channel. In this case, the maximum water surace elevation depends on the duration and volume of the surge, in addition to the maximum elevation at Node 1. Runs conducted using a tide with the 10-year maximum hurricane elevation at Node 1 resulted in a maximum water surface elevation at Node 70, higher than that obtained with a 100-year peak associated with a shorter hurricane surge.

This is a counterpart of the familiar reservoir routing problem, where the critical flood may be one of large volume or large peak discharge according to the available reservoir capacity.

To have a basis of comparison of model results, tides should have the same astronomical component and surges of the same duration, with maxima according to specified maximum tide elevation, and the surge volumes should be related to the selected maximum surges. Hence, the need for some surge shape normalization as stated above.

The changes in maximum water elevations due to increased development since 1963 are very noticeable. For example, the maximum elevation at Node 25 (Court Street Bridge) would be 1.45 foot higher in 1984 than in 1963 for the standard project tide as defined for these studies.

Conclusions

The following conclusions are derived from these studies :

1. Any study of coastal flooding in the vicinity of New York should determine the flooding effects of both northeaster and hurricane storm events.
2. Since prevailing mean sea level varies seasonally, the datum should be established to account for these variations, because hurricanes and northeasters tend to occur in summer and winter respectively. This correction is not negligible and should be added to the secular variation in MSL to obtain elevations related to the 1929 National Vertical Geodetic Datum.

3. A joint probability frequency analysis of storm tides should be used to establish return periods for storm tide elevations to be used as boundary conditions for model routing.
4. Landfill in a estuarine area will effect storm tide elevations through either reducing storage or reducing channel conveyance or both.
5. The inland flooding produced by a northeaster will in general be larger than for a hurricane with the same peak elevation in the open sea.
6. Northeaster storm events occur more frequently than hurricanes in the New York Metropolitan area and consequently are critical in defining design criteria.

Balloffet, A. Cole, E. and Balloffet, A. F., "Dam Collapse Wave in a River", Journal of the Hydraulics Division, ASCE, No. 10523, May 1974.

Wilson, B. W., "The Prediction of Hurricane Storm-Tides in New York Bay", Technical Memorandum No. 120, Corps of Engineers, Beach Erosion Board, 1960.

Myers, V. A., "Joint Probability Method of Tide Frequency Analysis Applied to Atlantic City and Long Beach Island, N. J.", ESSA Technical Memorandum, WBTM HYDRO 11, Weather Bureau, 1970.

SUNSET CLIFFS STABILIZATION

SAN DIEGO, CALIFORNIA

By

Louis J. Lee[1] and Walter Crampton[2]

INTRODUCTION

The Southern California coastline is comprised of sandy beaches, rocky headlands, and coastal bluffs. The Sunset Cliffs shoreline is essentially a coastal bluff with surf often impacting directly upon the bluff. The study area extends from Newport Avenue on the north to Osprey Street at the south in the Ocean Beach area of San Diego, California. It is approximately 5,800 feet in length and is located directly down coast of the mouth of the San Diego River (Figure 1). This portion of the shoreline makes up the northern portion of the Point Loma peninsula and is very rocky with occasional narrow beaches, small pocket beaches, and rocky reefs extending from the shore with shear cliffs rising 40 to 50 feet above the coastline (Picture 1). Numerous caves, arches, and irregular stacks have been formed by wave action in this area.

Since subdivision of this area in the 1880s, there has been a gradual retreat of the cliffs as a result of the erosion processes of the sea. At present, erosion has advanced to where it is endangering both public and private improvements, and the City has had to install fences and signs to stop pedestrian access to the cliff area. One dwelling has fallen into the ocean and another four are in serious danger. A corollary problem is unsafe access. Even though

---

[1] Vice President, Woodward-Clyde Consultants, Orange, California.

[2] Senior Project Engineer, Woodward-Clyde Consultants, San Diego, California.

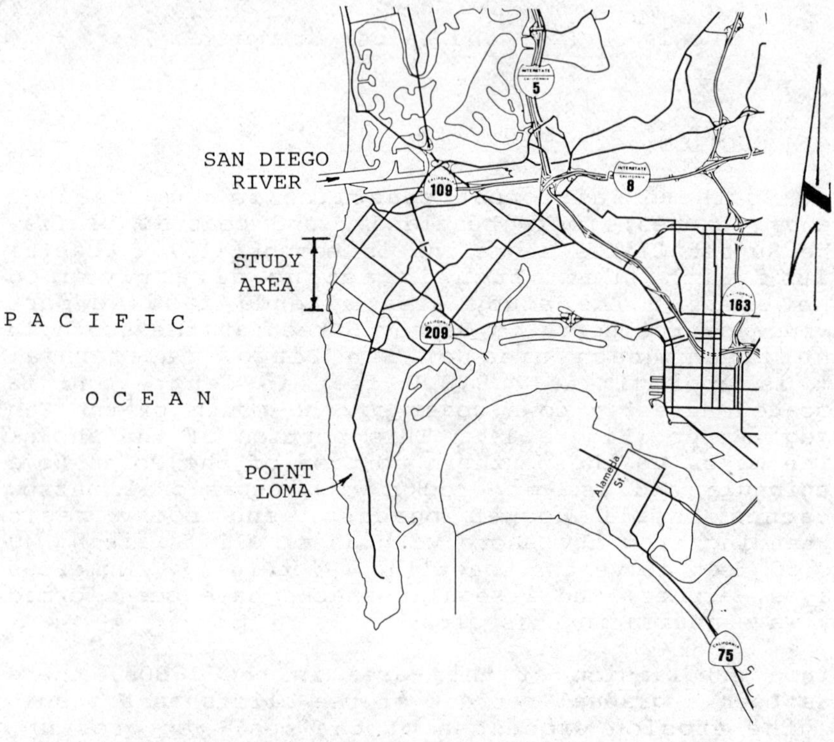

FIGURE 1 - LOCATION MAP

Picture 1 - SUNSET CLIFFS STUDY AREA

(City of San Diego Photograph)

fences have been put up to limit pedestrian traffic, people still traverse the cliffs, and several people have died or been injured in recent years. Erosion is also undermining Sunset Cliffs Boulevard in two locations. This city street is a main traffic artery providing access down the peninsula, and its loss would have a strong impact on traffic in the area.

Recognition of sea cliff erosion and the threat posed to private and public improvements dates to the 1950s. Since that time, numerous studies of cliff erosion in this area have been undertaken by the City of San Diego, the California Department of Water Resources, the U. S. Army Corps of Engineers, and others (U. S. Army Engineer District, 1965, and City of San Diego, 1976). In 1979, the City of San Diego authorized Woodward-Clyde Consultants to prepare preliminary plans and specifications for shoreline protection and upper cliff stabilization along the area, which is approximately one mile long. In order to assist in the selection of a shoreline protection program consistent with the community's needs and within a workable project budget, a Citizens' Coordinating Committee was also organized. This committee consisted of homeowners, representatives from various concerned local groups, and representatives from various government agencies.

The State of California and the City of San Diego have both maintained an aerial coastline monitoring program providing aerial photographs dating back to the 1950s. From this information, a "no project" 50-year estimated cliff retreat line was developed; problem areas were identified, and priorities were established. By using the previously developed information, Woodward-Clyde Consultants, with the help of the Citizens' Coordinating Committee and City staff, selected a preferred shoreline protection program that, when implemented, would provide a more stable coastline with improved access, additional recreational facilities, and protection for both private and public improvements.

EXISTING CONDITIONS

Topography and Land Improvements

Sunset Cliffs is characterized by an older, established, high density, residential neighborhood constructed on the Point Loma peninsula. The Point Loma peninsula is a long promontory extending about six miles southward from the low land adjacent to the San Diego River. Parts of its western shoreline are bordered by a narrow terrace with a top elevation ranging from 25 to 55 feet above sea level. The shoreline of the Point Loma peninsula is irregular in detail due to differences in the geologic structure and rock hardness. Wave erosion has etched out less resistent rock masses resulting in shallow pocket coves between rocky headlands (Picture 2). When sufficient sand is available, small sandy pocket beaches are formed.

The study area extends along approximately 5,800 feet of coastline, of which approximately 75% is privately owned. A total of 40 parcels comprise the private portion of the sea cliff. The remaining public lands consist of street ends, portions of streets paralleling the cliffs, and a public park approximately 1,000 feet in length.

In the past, erosion and the threat of property loss has prompted several of the private homeowners, as well as City staff, to place various forms of concrete rubble (usually demolition material) at the toe of cliffs which are experiencing rapid retreat. Many homeowners have constructed walls near the toe of cliffs to retard upper cliff erosion. Several homeowners have placed gunite on slopes to help protect their residences. Portions of the upper slopes have also been landscaped and are privately maintained (Pictures 3 and 4).

Geology

There are two basic geologic units exposed in Sunset Cliffs: the Point Loma formation and the Bay Point formation.

Picture 2 - WAVE EROSION AND POCKET COVES
(City of San Diego Photograph)

Picture 3 - EXISTING WALL AND LANDSCAPING
(City of San Diego Photograph)

Picture 4 - EXISTING RUBBLE, CONCRETE WALLS, AND LANDSCAPING
(City of San Diego Photograph)

The Point Loma formation is an upper Cretaceous sedimentary rock consisting of interbedded, fine-grained, gray-brown to yellow-brown sandstones and olive-gray clay shales. This near vertical rock unit is exposed along the entire base of the cliffs below an approximate elevation of +10 to +20 feet (MSL Datum) and generally dips to the northeast at 4 to 6 degrees. Isolated erosion-resistant stacks exist seaward on the coastline throughout much of the study area. Offshore, the Point Loma formation exhibits a ledgy shelf surface with seaward ledges becoming progressively deeper. The foreshore slope in this area is on the order of 1 in 25.

The Bay Point formation is a Pleistocene terrace deposit which overlies the Point Loma formation throughout the study area above an approximate elevation of +10 to +20 feet. This formational unit is composed of marine and non-marine, poorly consolidated, fine to medium grained, light brown, fossiliferous sandstone. The fossils found include mollusks, foraminifera, and ostracods indicative of a brackish water estuarine depositional environment. The Bay Point formation extends to bluff top elevations of approximately +25 to +55 feet and has inclinations ranging from near vertical to approximately 1 to 1.

The beach material, where it exists, consists predominantly of well-sorted medium to fine sand, interrupted occasionally by pebble and cobble foreshores on rocky shelves at the toe of the bluffs. Previous studies revealed that approximately 30 years ago the subject portion of Sunset Cliffs had a sand beach, typically 200 feet in width. Today only isolated beaches exist. Studies by the Corps of Engineers in the 1970s proposed to replace the beach; however, they concluded that this was not an acceptable solution.

Tides and Wave Climate

The tides along the Pacific coast have a semidiurnal inequality. The lowest tide each year is about -1.7 feet MLLW Datum; the highest tide is about +7.3 feet; and mean sea level is about +2.9 feet.

Waves that break along the San Diego County shoreline generally range in height from 2 to 5 feet; however, large waves ranging from 6 to 10 feet in height are common. Waves 12 to 15 feet in height have been observed at Ocean Beach (north of the study area) and breakers with estimated heights of 15 to 20 feet have been observed off the coastline within the study area (National Marine Consultants, 1960).

Winds and Rain

A seabreeze obtaining velocities of 10 to 20 miles an hour blows landward across the San Diego shoreline nearly every afternoon. Reciprocal land breezes at night are of much lower velocity. Storms moving in from the Pacific Ocean occasionally bring somewhat stronger winds to the San Diego area, but their duration is relatively short. Tropical cyclones from the area to the south reach San Diego on rare occasions. The maximum wind velocity of record was 51 miles per hour from the southeast in November 1944 (National Marine Consultants, 1960).

Semi-arid conditions prevail in the San Diego area with average annual rainfall ranging from approximately 9 to 10 inches. Most of the rainfall occurs between November and March. The maximum recorded rainfall was approximately 26 inches in 1883-84.

Area Use

The southern terminus of Ocean Beach Park is the beginning of the study area. This provides an access for people to walk the entire mile-long portion of coastline. Spalding Park, which is maintained by the City is situated at the southern end of the project. Four small pocket beaches exist where people may enjoy a reasonably secluded atmosphere. The remainder of the coastline provides various scenic elevated ledges and rocky headlands. Numerous tidepools also exist which attract many people. Surfing is also a popular activity in the area.

Parking is a major problem which serves to limit the number of people that visit the beach. The entire City of Ocean Beach has a maximum of approximately 500 public parking spaces servicing Ocean Beach Park and adjacent areas. Remaining visitors are forced to park on the street, which at present fails to meet even the needs of the inhabitants of the area.

SHORELINE EROSION

Base Seacliff Erosion

The Point Loma formation is exposed along the entire base of the sea cliffs in the study area; it is vertical to near vertical in most areas and is 5 to 20 feet high. Erosion at the base of the cliff is predominantly associated with direct wave impact acting upon small joints and fissures within the massive rock unit (Picture 5). Much of the Point Loma formation appears quite intact and has experienced little erosion in the last 30 years. In other areas where fractures and joints within the rock are more prevalent, the erosion is occurring at a more rapid rate. Where shear zones are encountered, surge channels and underground caves have developed.

Although no detailed measurements of base erosion have been made, it is estimated that local erosion may be several inches per year in some areas.

Upper Sea Cliff Erosion

The Bay Point sands form the upper portion of the bluffs and range from approximately 10 to 40 feet in thickness. These sands are subject to several different forms of erosion as a result of the following actions:

- direct wave impact, wave spray, and wave splash;
- undermining of the basement rock and over-steepened slopes;
- wind, rain, irrigation, and uncontrolled surface runoff; and
- human-induced erosion (cave digging, climbing on the cliffs, and channelized pedestrian traffic).

Picture 5 - BASE SEA CLIFF EROSION BY WAVES IN JOINTS AND FISSURES

(City of San Diego Photograph)

Direct wave impact rapidly erodes the friable Bay Point sands; however, this rarely occurs and is only associated with infrequent, severe storms in which waves actually impact upon the upper sands in a few locations. Along much of the study area, the Bay Point sands are exposed to the elements with little or no vegetation. Wave spray and splash often reach these unprotected sands causing saturation of the outer layer and subsequent sloughing of oversteepened slopes.

In areas where the Point Loma formation is experiencing erosion, the overlying Bay Point sands become undermined and subsequently fail through loss of vertical support. This results in oversteepened slopes which stand near vertical until the pore water tension within the soil has had a chance to dissipate and sloughing occurs. The slopes are relatively stable at inclinations of about 1 to 1.

Wind, rain, and irrigation contribute to minor erosion of the upper cliff face, especially on the more exposed, oversteepened portions of the friable sands. A considerable amount of rilling has occurred along portions of the upper cliffs as a result of this action.

Human activities probably account for a large portion of the upper cliff erosion. Because the cliffs are moderately accessible and no provision presently exists for control of traffic, people are free to traverse and dig in the cliffs, scribe graffiti on exposed areas, and generally destroy the integrity of the relatively delicate upper slope face (Picture 6). Additionally, channelized pedestrian traffic accelerates erosion in local areas causing oversteepening upslope of foot paths.

It is estimated that the 50-year retreat at the top of the bluff will be 20 to 40 feet in some areas (City of San Diego, 1976).

Picture 6 - UPPER SEA CLIFF EROSION BY HUMAN ACTIVITIES IN FRIABLE SANDS

(City of San Diego Photograph)

SHORELINE AND UPPER CLIFF STABILIZATION

Design Considerations

In order to determine an appropriate shoreline cliff stabilization program, especially in an area of extreme environmental sensitivity, it is important to address the problems of visual aesthetics, the current uses of the area, the present hazards associated with already unstable, oversteepened slopes, the potential for future erosion, and the impact of minimal or no stabilization in certain areas. Most individuals involved with the study were primarily concerned with the preservation of as much of the natural visual aesthetics of the coastline as possible. In order to satisfy this concern, the proposed plan essentially eliminates erosion control measures (1) along areas of the cliff face that aren't already experiencing cliff retreat and (2) where erosion is not expected to endanger private or public improvements within the near future. Since certain areas of the coastline are presently oversteepened and unstable, it is necessary to develop additional shoreline in these areas to provide a stable reconstructed bluff face that will resist further erosion.

Protection of the Lower Point Loma Formation

The main erosion problems at the base of the sea cliff are associated with the direct impact of waves in the areas where joints and fractures are present within the Point Loma formation. It is necessary to protect these areas. In some portions of the coastline where the upper portion of the cliff are near vertical and are marginally stable, it is also necessary to develop additional support at the base of the cliff. To accomplish these measures, approximately 1,300 feet of rock revetment is proposed to mitigate the wave impact in the required areas. In order to increase public access and use of the beach in the vicinity of the revetment, elevated sand beaches are planned in some areas (Figure 2A). Those beaches will also provide lateral access where none exists now. In localized zones of weakness, earth-toned cement grout is proposed in lieu of the rock revetment.

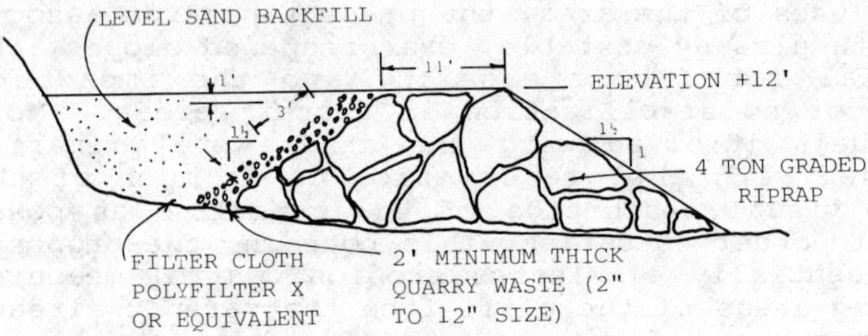

FIGURE 2A - TYPICAL RIPRAP INSTALLATION WITH SAND BACKFILL

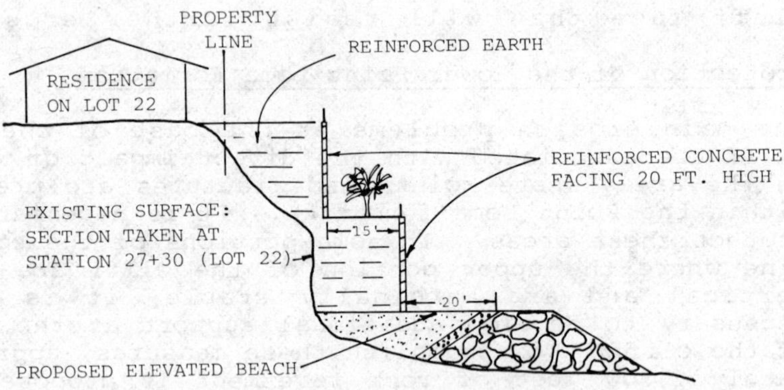

FIGURE 2B - REINFORCED EARTH WALL

In those areas where major erosion of the upper cliffs is occurring associated with undermining of the Point Loma formation, a reinforced earth wall has been proposed with the rock revetment (Figure 2B). It is felt that this type of structure more appropriately will solve the slope stability problem as well as fit the aesthetic and economic constraints of the project, than would concrete walls, a planted fill slope, or other types of shore protection facilities.

Protection of Upper Bay Point Formation

Past information and the study of existing conditions indicate that, along the upper bluffs, certain erosional processes are prevalent in different locations. It is also evident that some of the coastline already has remedial measures, such as short walls above the surf zone and planted upper slopes. Therefore the use of various erosion-control methods to stabilize the upper bluffs seems appropriate. This will best solve the local erosion problem, and the installations will blend in with existing structures.

In areas where cave digging is prevalent, low concrete walls will be utilized (over approximately 12% of the study area). These walls will also protect the Upper Bay Point Formation from erosion due to wave splash and spray. Occasionally, short fill slopes will be extended above the "splash" walls to protect oversteepened friable sands (Figure 3A). The new slopes, as well as a number of existing bare slopes will be planted with a hardy ground cover that will be drought resistant and essentially maintenance free once it is established. Approximately 145,000 square feet of landscaping is planned.

Certain areas of the coastline are subject to more pedestrian traffic than others. Approximately 700 feet of concrete walkways are incorporated into the plans to provide channelization of foot traffic and increased lateral access in precarious areas (Figure 3B). The concrete walkways are designed to blend into the topography and to have vegetation on both sides. Certain plants have also been selected for landscaping that upon maturity, will, provide an effective barrier to pedestrian traffic in non-walkway areas.

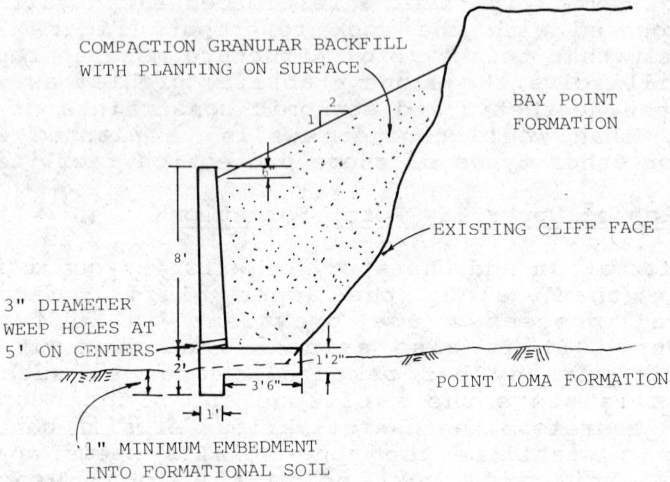

FIGURE 3A - CONCRETE SPLASH WALL

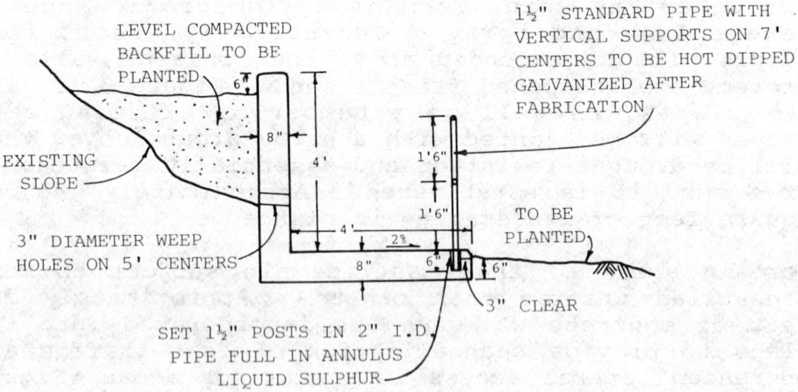

FIGURE 3B - CONCRETE WALKWAY

The project includes modification of surface drainage along the bluff tops to divert uncontrolled surface runoff away from the bluff face. Part of the Assessment District's costs included such modifications to private residences in order that they conform to the general guidelines for drainage facilities adjacent to slopes.

Four new stairways are included in the project to improve public access within the study area. An improved access roadway extending off one of the pocket beaches is also incorporated into the design to provide a permanent landscaped access to the beach. This roadway will be used by City maintenance and emergency vehicles; it will also provide a logical point of access to the beach during construction of the project.

FUNDING AND SCHEDULE

The proposed project is estimated to cost approximately 2.2 million dollars. It is proposed that the protection project be funded by the State of California, the City of San Diego, and an Assessment District made up of all the properties benefiting from the project. The State of California's proposed share will be one-half the total project cost, and the City and Assessment District will make up the rest. Assessment spreads are based on cliff frontage, total lot area and an evaluation of the relative benefit of the proposed shoreline protection measures. Appropriate formulas for calculating assessment costs have been worked out with the City Assessor's Office, and final assessment spreads will be modified after completion of design studies.

Present plans are to make final presentations to the Coastal Commission and the San Diego City Council during the summer of 1980. Final approval of the preliminary plans is expected in early fall. A contract will then be let for preparation of final plans and specifications. It is anticipated that construction will begin in late spring of 1981. Construction is estimated to take seven months, and an additional six months has been projected for establishment of the landscaping.

ACKNOWLEDGEMENTS

The authors wish to express their gratitude to the City of San Diego Engineering staff for their valuable discussions, and for contributing historic information and pictures. The authors also thank the Woodward-Clyde Consultants Professional Development Committee for their financial assistance for portions of this study.

REFERENCES

City of San Diego, 1976, Sunset Cliffs Newport Avenue to Osprey Street shoreline protection study: Task Force Report, December 1976, Appendix A, p. 1-5.

Kennedy, M. P., 1973, Sea-cliff erosion at Sunset Cliffs, San Diego, California Geology, February 1973, v. 26, no. 2, p. 26-31.

National Marine Consultants, Inc., 1960, Wave statistics for seven deep water stations along the California Coast: U. S. Army Corps of Engineers District, Los Angeles, California.

U. S. Army Engineer District, Los Angeles, Corps of Engineers, 1965, Special study of City of San Diego (Sunset Cliffs): Beach Erosion Control Report on Coast of Southern California, Appendix VII, Contract No. W-04-193-ENG-5196, p. 4-38.

Washington's Automated Coastal Data Base: The Coastal Zone Atlas

by

Sue Wenke and Marc Horton

Washington State Department of Ecology

ABSTRACT: *The Coastal Zone Atlas of Washington* was developed in response to the perceived need by many individuals for a detailed, accurate, and uniform set of environmental data in the coastal zone. A computerized spatial data information system using polygons to represent specific data variables was chosen as an efficient method for storage and update of the *Atlas*' data base, and multicolored, printed maps were chosen as the method to display this information to users. On a county-by-county basis, the *Atlas* involves twelve volumes of colored, 17" x 23.5" maps at a scale of 1:24,000. The data categories which are represented include land use/land cover, critical faunal and floral areas, sand and gravel resources, flood hazard areas, geology, slope stability, and drift sectors.

INTRODUCTION

The Shoreline Management Act of 1971 is the basis for Washington's coastal zone management program. Under this act, local governments have the primary responsibility for making land and water use decisions for the shoreline areas and marine aquatic areas under their jurisdiction. In addition, state and federal agencies, special use districts such as port districts, and private enterprises make important decisions that determine the usage of large areas of shoreline.

The effectiveness of the management effort is dependent upon three things: the quality of the applicable legislation, the data upon which decisions are based, and the skill of the user in correlating the legislation and the data into a management decision. It is the second of these, the need for a reliable data base, which prompted the Department of Ecology to develop the *Coastal Zone Atlas* and its automated data base.

ATLAS DESIGN

The design phase of any geographic information system is the most critical stage. A number of questions must be considered at the onset of such a large project to insure the utility of the system. These questions will include:

Who are the users?
What kind of information do the users need, and to what degree of accuracy?
How will the information be stored?
How will the system handle new information?
What is the available funding?

After consideration of these questions and others, with local, state, and federal officials, the designers of the information system's graphic product, The Coastal Zone Atlas, reached the following conclusions as design constraints.

Areal Coverage

The coastal zone of Washington State encompasses over 2,400 miles of marine shoreline situated in 15 counties. The critical "first tier" of the coastal zone is a strip of water, beach, and upland defined for the planning and administrative purposes of the State Shoreline Management Act and the National Coastal Zone Management Act. In general, the Coastal Zone Atlas coverage of this strip extends from a point approximately minus 30 feet below mean sea level shoreward to the mean lower low water line, across the beach to the mean higher high water line, and then inland 2,000 feet. The inland limit is extended where there are significant coastal related features or conditions which might affect the coastal zone.

Scale

It was decided that the Atlas be produced at a scale of 1:24,000. Planners and the general public are familiar with the United States Geological Survey 7.5 minute quadrangle maps published at this scale, and they are readily available to be used as base maps. A smaller scale would not provide enough specific information, and a larger scale would lead to an unwieldly document in which regional parameters may be difficult to ascertain.

Format

The *Atlas* maps are 17 inches by 23.5 inches and are printed on one side only. In such a way, they can be used on a light table as base maps for other studies or in conjunction with other data. The *Atlas* binding allows for easy extraction or addition of pages as revised or new information becomes available.

Color

Originally, the *Atlas* was planned to be produced in black and white. It was later felt that this would limit the amount of data which could be incorporated visually, and that color would be the most effective way to convey the *Atlas*' message in graphic terms. Local planners expressed the opinion that any other presentation would be nearly useless to them. In the end, the *Atlas* was produced using a combination of three colors; four tints each, plus black, allowing for a total of 120 possible colors.

Organization

The *Atlas* is organized so that each volume represents the coastal area in one county. Within each county, the shoreline is divided into map segments, each representing an area of ecological affinity such as bays, estuaries, harbors, and spits. Generally, an effort was made to maintain a northward orientation to each segment, but the association of ecological features was the major factor.

Originally, the *Coastal Zone Atlas* was planned to include all 15 coastal counties. Because the three Pacific coastal counties were so different from the inner coastal counties (Puget Sound and the Strait of Juan de Fuca), and because the project was costing more than anticipated, only the 12 inner coastal counties have been mapped.

Data Coverage

Under funding from Section 306 of the National Coastal Zone Management Act, the Department of Ecology contracted for investigations of important biological and physical parameters in the coastal zone. At a total cost of over $600,000 (1978 dollars), these investigations provided seven data categories which have been included in the *Atlas* as a set of six maps. (Figure 1.)

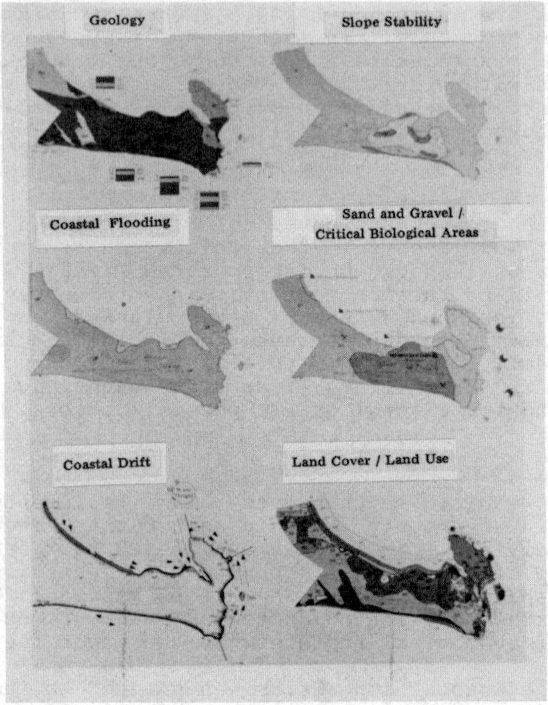

Figure 1 - Coastal Zone Atlas map

Land Use/Land Cover - The classification system being used in Washington is based on the United States Geological Survey's standardized classification system developed during the 1960s. The broad generalized classification categories of the USGS system were used in Florida's land use/land cover classification system, and Florida's system ultimately served as a model for Washington. In developing Washington's scheme, modification and improvements were made to the Florida model to reflect the unique coastal zone vegetation and land uses of our state.

Seven major categories comprise the core of this system: (1) Urban; (2) Agricultural; (3) Nonforested Vegetated Uplands; (4) Forested Uplands; (5) Water; (6) Wetlands; and (7) Exposed/Other lands.

Within each of these "first level" descriptors are further breakdowns into second, third, and fourth levels, each increasingly specific. As an example, take Forested Uplands. The first level of classification, Forested Uplands, is provided with a single-digit identifier, 4; the second level, Young Broadleaf Forest, a two-digit identifier, 42; the third level, Young Broadleaf Forest, a three-digit number, 421; and the fourth level, Young Broadleaf/Successional Shrub, has the four-digit identifier, 4211. There are 41 Level II, 132 Level III, and 59 Level IV descriptors. Each of the seven major categories is also color coded to approximate the type of land cover unit being identified and mapped.

All land use/land cover information originates from photographic interpretations of aerial photography, all of which were subjected to ground truthing to supply an accuracy of at least 90 percent. The minimum size resolution of each identified land cover type is a little less than one acre.

An important feature of the land use/land cover study is a habitat narrative description for each unit of the classification system, which is published as a supplement to the Atlas. These narratives include a discussion on the habitat itself, its associated biological community, and the impact to and with adjacent habitats. Land cover units are prioritized based on sensitivity, relationship to the coastal zone, and other criteria.

Critical Biological Areas - During 1976, critical biological areas were determined and identified for 206 significant marine and estuarine animal species. "Significant" species are considered important commercially or recreationally, or those which prey on or compete with or serve as food for commercial or recreational species.

Eight mammal; 80 bird; 81 fish, and 37 marine invertebrate species were considered in this investigation.

In determining critical biological areas, two criteria were followed:

1. The area supports a population of species that not only reproduces itself but, because of favorable

environmental conditions (currents, water temperature, salinity, etc.), provides a major source of recruitment for adjacent areas or regions whose populations do not consistently reproduce themselves.

2. The area consists of a habitat type or types that provide shelter, food, or other environmental necessities during a critical part of the species' life history. For example, nesting sites or shelter from predators during early life stages.

The information on critical faunal areas was developed by a consultant to the Department of Ecology through an analysis of available data. The consultant then drew up fact sheets on each species including its life history, its distribution in Washington, habitat requirements, critical biological areas, and known data gaps.

In the Atlas, critical biological areas are assigned a simple one-color identifier. Each faunal group has been assigned a particular identifying color and, within the group, each species is identified by its common name. Its season of criticality or importance is also indicated.

Known locations of rare, endangered, or threatened plant species in the coastal zone are also identified in the Atlas. These sites are shown by large red asterisks.

The plants selected for inclusion are those proposed by the Smithsonian Institute and published by the U.S. Department of the Interior in response to the Endangered Species Act of 1972.

Sand and Gravel Resources - Sand and gravel information contained in the Atlas is intended to help communities recognize their aggregate resources, so that rich areas are not precluded from being used through irrevocable land use decisions.

In the first year of Atlas preparation, a consultant to the department analyzed raw data from mining reports, mineral and ground water studies, geologic maps, soil surveys, well logs, and similar sources. From this information, ratings were developed which identify various grades of sand and gravel by location.

The ratings are based on relative quantity and quality gradations of the glacial deposits. More often than not, these are heterogeneous mixture of all sizes of detritus, ranging from fine clays over sand to gravel and boulders. Higher quality material is likely to occur in the outwash areas where reworking has sorted the coarse from the fine.

During the second year, some refinements were made to the methodology and the aggregate mapping was carried on concurrently with the geology and slope stability mapping, enabling a more thorough job of ground truthing.

Coastal Flood Hazard Areas - Flood damage in tidal and coastal areas is mostly the result of two factors; high still water levels and wave action. Near the mouths of coastal streams and in estuaries, high stream flows will also contribute to flooding and the resulting damages. The still water level is a result of astronomical tides and storm surge. Wave action is caused primarily by high storm winds which may accompany the high still water levels.

On December 15, 1977, extreme high tides occurred in the Puget Sound and Washington coastal areas. These were accompanied by low atmospheric pressures and high winds in some areas. The result was record highs for nearly all of Puget Sound. These tides were only a few tenths of a foot different (mostly higher) from approximate 100-year frequency still water tide levels.

Since the difference between the December 15 record high tide level and the approximate 100-year frequency tide level cannot be identified at the mapping scale, they are consequently mapped as the same. One-color shading identifies the flood hazard with an approximate one (1) percent change of flooding (100-year frequency) during any year, as well as the approximate historic record high tide elevations observed on December 15, 1977. The remainder of the areas are shaded as the approximate area subject to less than one (1) percent chance of flooding during any year which has not been studied and may include undetermined, but possible, flood hazards.

Predicted and observed tidal elevation information is also presented on the *Atlas* maps. This information for some locations was acquired from the record high day on

December 15, 1977. The difference between these revealed a rather constant value for the storm surge into Puget Sound -- being less pronounced in the Strait of Juan de Fuca and increasing slightly toward southern Puget Sound. The maps of flooding caused by the December 15, 1977 tides were prepared after visiting with the persons who observed the high tides or their immediate results along the shoreline areas.

Coastal Geology - The purpose of this data category is to provide baseline information utilized in land use planning from which useful data can be extracted for evaluation of geological hazards. A total of 75 descriptions are used to cover a range of bedrock formations from the oldest (paleozoic) exposed rock units to several quaternary and younger, mostly unconsolidated units that occur as a result of the various glaciations throughout Puget Sound.

In addition to the usual information included on a geologic map; rock types, contacts, faulting, and strike and dip of bedding planes, information on stratigraphy of the coast is provided by columnar diagrams of the bedding as observed in exposed seacliffs and bluffs.

Also included in each county volume are tables which describe general engineering considerations relative to drainage characteristics, ease of excavation, foundation, slope and seismic stabilities, and the planning significance of individual deposits.

The geolgists who mapped this information compiled their data from published reports, maps, books, theses, their own extensive files, and numerous other sources. In addition, they surveyed key locations, established standard stratigraphic sections, and ground-proofed as much of the area as possible.

Coastal Slope Stability - Slope stability maps displayed in the Atlas give a generalized picture of the stability of the earth materials in the coastal zone. They identify areas where the developer or house builder would be wise to seek professional geological and engineering advice before making plans.

Slope stability is mapped in the Atlas in relative terms under the categories of "stable," "intermediate," and "unstable." A fourth category, termed "modified," refers to slopes highly altered by man with an unpredictable response to physical process.

These terms are assigned to a particular area based on several factors. One is the Slope Angle of the land surface which is probably the most obvious element involved. Geological Materials are also a major factor in the behavior of the slope. Noncohesive materials such as gravel or sand with little silt or clay binder will flow, even when dry, to their particular angle of repose (roughly 30 to 35 degrees). Materials with high silt content such as glacial till or even some silty sands will stay near vertical for years. Rock Structure is also an element in slope stability as well as the presence of Water. Wave Erosion is the last factor which was considered, and is important because even high strength materials are made unstable by the undercutting of slopes and the waves which carry away the stabilizing debris found at the toe.

Coastal Drift Sectors

The coastal drift sector represents a segment of the shoreline along which littoral along shore movements of sediment occur at noticeable rates. It allows for uninterrupted transport or drift of beach materials from a feed source that supplies the sediment, along a driftway, to an accretion terminal where the drift is deposited. The intent is to explain the dynamics of littoral drift systems on an individual, local basis; to describe their importance as a constantly active process; and to contrast these processes with coastal land use activities, from both a historical and predictable impact analysis perspective. This will allow locating developments in areas where they will cause minimal disruption of natural systems and will reduce individual and accumulative adverse effects.

An evaluation of wind data, aerial photography, field work, and statistical analysis of some published data allowed for the mapping of the following criteria:

- Beach and intertidal substrates
- Coastal processes such as: accreting, equilibrium, and eroding shores; plus, nonfeeding or feeding uplands
- Approximate low water line

- Wave height in feet by percentage occurrence in selected locations (based on analysis of wind data, fetch, and exposure) represented as wave roses
- Net direction of littoral transport
- Natural and modified shorelines
- Principal type of man-made structures (bulkhead, breakwater, dock, jetty, launch ramp, and riprap)

AUTOMATED DATA BASE DESIGN

In many cases, the need for a computerized spatial information system has lead to the development of a graphic product such as the *Coastal Zone Atlas*. For the State of Washington, the design alternatives for the graphic product lead to the development of the system. The computerized information system can facilitate cartographic production of an atlas, and by choosing this method over manual cartographic techniques, it was felt that long-term benefits would be derived in terms of availability of the data for spatial analysis, simplicity of update, and the highest quality of production.

The disadvantages are few, if one has enough money to make the choice. The quality of computer derived line images is not as high as a manually scribed line; greater expertise is required by the owner of a computerized spatial information system; and computer hardware and software must be on-line to operate the system. But cost is the largest disadvantage, unless it is compared to the size and utility of the product. (For a tabular breakdown of *Atlas* costs, See Appendix)

Integrated Terrain Unit Map

The *Atlas* spatial information system was devised to bring together map information from seven separate sources, and to reduce that information to a compact and easily retrievable form. For this purpose, the method of Integrated Terrain Unit Mapping (ITUM) of polygon storage units was employed in assembling the data base.

The ITUM method is a photo interpretation procedure for the combination of various environmental parameters. Maps of different data categories are combined manually to create a set of polygons which are the least common denominator of all of the polygons from the data categories.

The process begins with the selection of a controlled base map; in this case, the USGS 7.5 minute topographic maps, where possible. The data collectors mapped their information on pin-registered, mylar overlays to the base maps.

These mylar overlays were then gathered to begin the process of terrain unit analysis. The first step in this method is the identification of homogeneous areas of photographic tone and texture, called photomorphic units, from aerial photography. It is assumed that boundaries between these units reflect differences in the physical and biological environment that will correspond to boundaries in the data compilations. The resulting map is used as a base for the integration of the seven separate data maps.

During the integration, data boundaries coincident with the edges of photomorphic units are identified. In places where photomorphic units are crossed with data boundaries, the units are split into two or more separate areas. Data values are then ascribed to the resulting polygons. Each data map is successively layed on the previous integration until all of the data sets have been combined into one final integrated map (Figure 2). This integrated map is a set of polygons, each of which represent as nearly as possible, a homogeneous distribution of data within each of the seven categories of aerial data.

A number of problems involved in automating large amounts of data are overcome using the ITUM method, the first of which is one of economics. The cost of data base computerization is a function of the number of maps to be automated and the complexity of those maps. By compressing all of the interrelated environmental variables as contained in the *Atlas* onto one map, the integration process effectively reduces the time required to both efficiently computerize and overlay a large number of maps. The manual integration of the data avoids expensive computer time required to automatically overlay the information and also saves on computer storage costs.

The second problem resolved through the use of the ITUM process is basically a cartographic one and is important

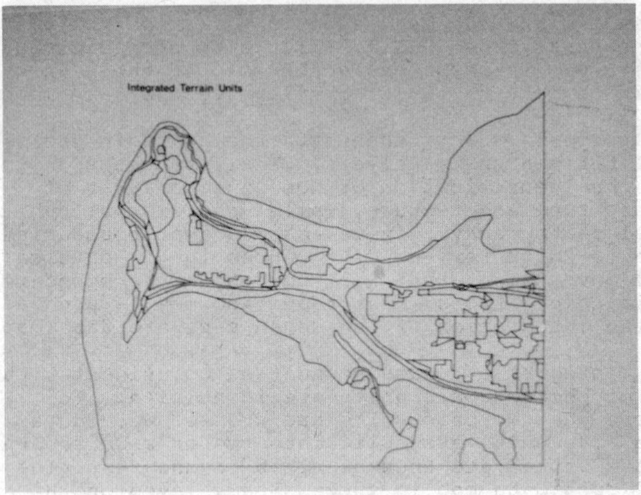

Figure 2 - Integrated Terrain Unit Map.

from the standpoint of information interpretation and display. Map variables have numerous boundaries which are naturally coincidental; e.g., slope stability areas often coincide with geological structures. Because these variables are mapped independently by different professionals with different objectives, different map accuracies, and at different times, automatically overlaying this information results in sliver errors, or small errors created by overlapping of coincidental boundaries. These errors are confusing both in cartographic display and in interpretation of the information.

The third problem resolved through the integration process is that all information is subject to a second round of verification and aerial photograph interpretation. By methodically applying routine principles of remote sensing to each data element, Atlas data are therefore consistently synchronized throughout the study area.

Interactive Cartographic Design and Automated Execution

With the ITUM for each segment in the data base, this information can be extracted to interactively edit each page before plotting. The graphical data can be automatically manipulated to maintain northern orientation; the symbolism such as labels and arrows can be rearranged; and maps can be repositioned to prevent overlap.

When the cartographer is satisfied with the design of the Atlas page, a plotting program is used to produce plates at the final printing scale to aid in manual and photomechanical cartographic procedures used in final preparation of the maps.

Spatial Analysis Software

Computer programs are available to spatially analyze the Coastal Zone Atlas data base which is structured on segment files that correspond to the map segments found in the Atlas.

The LINEAR SUMMATION program identifies the occurrence and calculates the kilometers of shoreline structures and upland processes found in the Coastal Zone Segment (CZS) file. This includes processes such as eroding bluffs or manmade structures such as riprap and bulkhead (Figure 3).

The Classification Summary program provides a listing of the total area of all classifications within a specified data category. For example, given the data category of Land Use/Land Cover, this software would compute the total area of each classification (e.g., pasture) of land use/land cover involved in that segment (Figure 4).

The AREA SUMMATION program searches the CZS file for a requested segment and calculates the aerial extent of polygons that satisy a specified classification. Upon command, the program will generate a scale plot of the resultant polygons with control points annotated. If one needs to know how much disturbed forest occurs in a specified segment, the area will be computed and a plot can be developed and then registered to the printed Atlas page to highlight these areas.

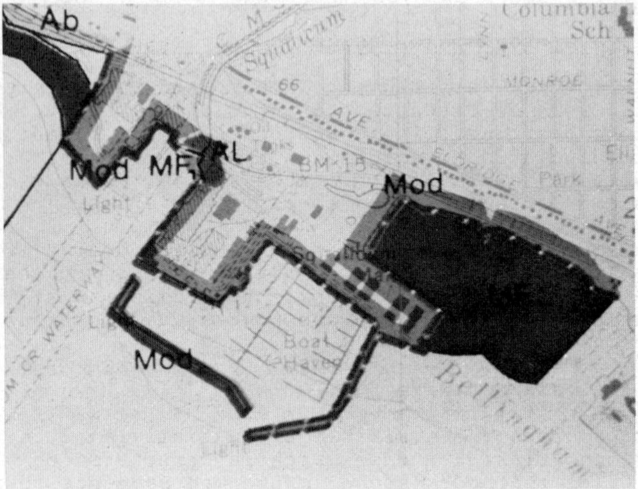

Figure 3 - Linear Program can calculate the extent of the modified (dotted line) structure.

The OVERLAY PROGRAM is used to identify areas where specified data classifications of one category (e.g., land cover-residential) intersect with specified data classifications of other categories (e.g., slope stability-unstable.) The program will provide two outputs: a summary of the search stating the number of polygons found and their total area, and a scale plot of the polygons which may be registered to the Atlas page.

PRODUCTION PROBLEMS

Only with great foresight, considerable planning, and usually someone else's previous experience, can problems be avoided in a project of this magnitude. To help others with the latter, some problems encountered by the creators and producers of the Coastal Zone Atlas will be pointed out.

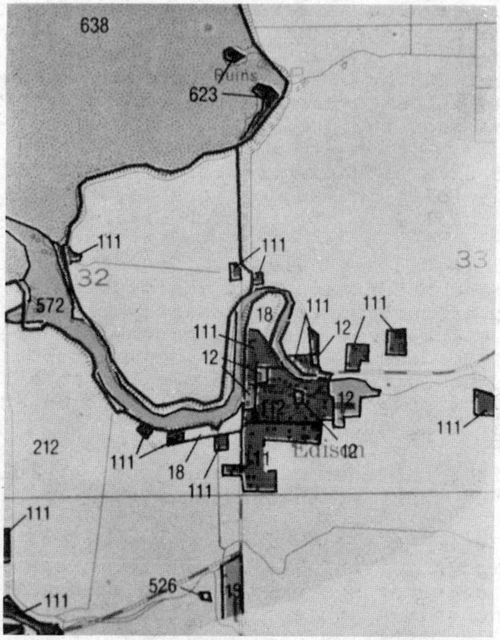

Figure 4 - Classification Summary program can calculate the area of each land use on this map.

The Department of Ecology, with little expertise in management of the magnitude of data envisioned for the Atlas, contracted for the production and storage design of this spatial information system. When an acceptable approach was agreed upon, contracts were initiated for data gathering, integration, and production, and the project commenced.

Department staff supervised the contracts and a Project Coordinator was assigned to oversee the program, but because there were no staff with enough expertise in spatial data collection or computer program development, the contractors were relied upon to manage production problems.

Issues began to develop as data collection did not meet schedules; the data integrator questioned data; and information was being reworked even during production. Each issue was handled, but the gravity of the problem became apparent a year later when cost overruns were recorded for the production process, and the production schedule had slipped.

Duplication of this situation can best be avoided by considering the following suggestions before a project the size of the Coastal Zone Atlas is undertaken:

1. A realistic schedule must be set for data gathering and production. It is important to consider that production is an assembly line process and any slippage at the beginning will affect the whole process. Therefore, all data must be thoroughly tested and judged reliable not only by the investigator but by a Project Manager in the department or company.

2. Monitor the cost of the project carefully, and set up a strict accounting system to be adhered to.

3. Assign a person with expertise in spatial data and computer system design to manage the project in-house. A project with such a large budget easily justifies hiring an employee if there is none available on staff. That person should be prepared to make decisions as the reliability of the data or changes in system design.

CONCLUSIONS

The first volume of the Atlas was available late in 1977, and the final volume was completed in 1980. Because it has been out a relatively short time, it is too early to make a proper analysis of its use at this moment. Limited feedback does show that the Atlas information is being used by local governments, libraries, consultants, and others for answering public inquiries, reference material for reports and Environmental Impact Statements, permit review, teaching, and making administrative decisions.

Future plans for the Atlas involve expansion of the data base, updating the information to show historical trends, and making the data base more accessible to others who may find it valuable.

APPENDIX

Coastal Zone Atlas Cost Summary

	1976	1977	1978	1979	1980	Total
Collateral (Base Maps and Imagery)	$ 8,056	$ 62,279	$ 48,917	$ 720	-0-	$119,972
Concept Development	-0-	$ 8,800	-0-	-0-	-0-	$ 8,800
Data Collection	$ 31,367		$195,133	$199,978	-0-	$617,351
Data Processing	-0-	$ 23,800	$ 34,650	$ 19,230	$ 34,421	$112,101
Production	-0-	$ 68,373	$199,332	$297,806	$194,645	$760,156
TOTAL	$ 39,423	$358,385	$482,877	$508,629	$229,066	

PROJECT TOTAL - $1,618,380
AS OF JULY 1, 1980

REFERENCES

Anderson, Kevin, 1979, "Coastal Zone Atlas of Washington: An Overview," a pamphlet.

Guinn, Charles and Michael Kennedy, 1975, "Avoiding System Failure: Approaches to Integrity and Utility," Urban Studies Center, University of Louisville, Louisville, Kentucky.

Northwest Cartography, Inc., 1980, "Coastal Zone Atlas of Washington: Automated Data Base," Washington State Department of Ecology, Seattle, Washington.

Youngmann, C.E., "The Coastal Zone Atlas of Washington," paper of unknown origin.

COMPUTER MODELING OF THE GREAT DISMAL SWAMP

Vincent M. Caruso
James E. Paschal, Jr.
U.S. Geological Survey
Reston, Virginia 22092

ABSTRACT

The Great Dismal Swamp, Virginia-North Carolina, is an area of slight terrain relief and dense overgrowth that preclude standard photogrammetric mapping techniques. An experimental computer simulation using control-point data was used to generate test maps of the swamp. Contour maps and related isometric plots were generated from 1,400 third-order control points randomly located by field surveys along roads. A 200-m gridding interval of approximately 18,000 nodes was generated using the Contour Plotting System, a Multics[1] gridding, contouring, and profiling software package. This simulation technique is able to compare various statistical surfaces, compute volumes, and generate isometric surfaces from extremely complex data.

INTRODUCTION AND BACKGROUND

In recent years, there has been increasing interest in mapping the topographic surface of the Great Dismal Swamp of Virginia and North Carolina. A detailed knowledge of the terrain elevations is necessary to plan for manipulating surface water and ground water in and around the swamp. Traditional techniques of obtaining adequate elevation data have been of limited usefulness in this complex and inaccessible environment so this study was undertaken to develop a representative ground surface using computer-aided graphics and simulation.

The Great Dismal Swamp is a forested wetland located on the Virginia-North Carolina border on the southern Atlantic coastal plain (Carter and others, 1977) (Figure 1). Extremely difficult field conditions within the swamp prohibit the use of conventional techniques of determining elevation. There is only limited access to the swamp by road during the dry seasons, and many roads are impassable when wet; ditches, surface flooding, and dense vegetation cover make access to certain areas by foot almost impossible. Change of surface relief is slight and difficult to measure; thick vegetation covers the swamp even during the winter, precluding the use of aerial photographic techniques. What was at one time a relatively flat surface of lagoonal and shallow marine sediments overlaid by up to 4 m (13.1 ft)

---

[1] The use of trade names in this report is for identification purposes only and does not constitute endorsement by the U.S. Geological Survey.

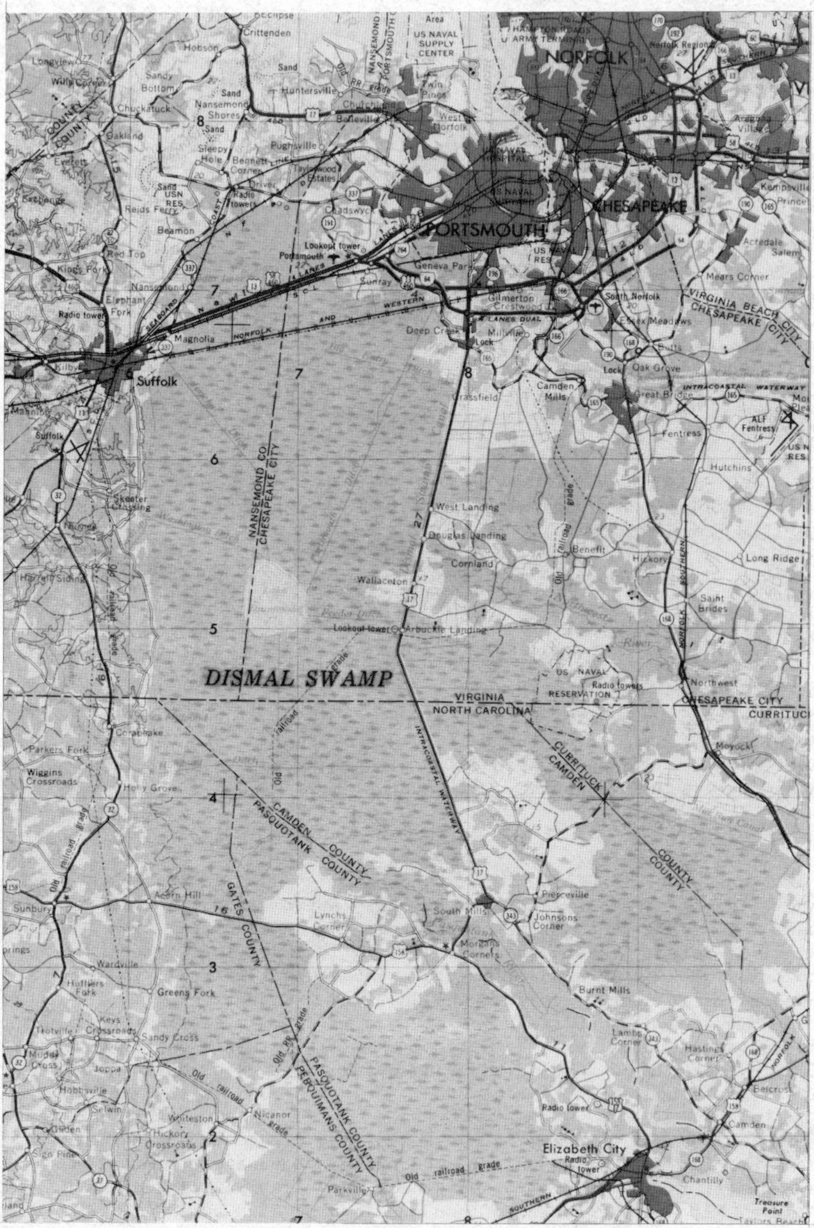

Figure 1. Map of Great Dismal Swamp area, Virginia-North Carolina. Scale: 1 inch equals approximately 7 miles.

of organic muck is now separated by roads and ditches into separate topographic units (Carter and Baker, 1978) (Figure 2). Lumbering, ditching, and fires have changed drainage patterns and in some cases have caused a loss of 0.9 to 1.2 m (3 to 4 ft) of soil surface.

The U.S. Fish and Wildlife Service is charged with preserving and managing the total wetland ecosystem of the swamp without special regard for waterfowl, other migratory birds, or specific wildlife species (U.S. Fish and Wildlife Service, 1974). The most important element affecting the swamp environment is the water regime; therefore, the refuge manager's chief priority is to properly manage the available surface and ground water (Carter and others, 1977). The topographic surface and the configuration of the roads and ditches determine the direction water will flow and the amount of water that will be available in any given area. Practically, elevation estimates within the swamp should be accurate to within 0.15 to 0.46 m (0.5-1.5 ft) to allow for correct placement of water control structures.

The objectives of this study are:

1. To obtain UTM (Universal Transverse Mercator) grid coordinates and elevations in the swamp using data from recent surveys.

2. To develop two- and three-dimensional drawings of swamp contours.

3. To provide a basis for future field data collection and field testing to improve the preliminary contour maps.

These contour maps can be used to analyze surface areas and volumes for developing alternatives for future water retention structures and drainage. The techniques developed in this study could also be applied to other wetland areas having similar problems.

EXTENT OF THE GREAT DISMAL SWAMP

The Great Dismal Swamp covers about 84,900 ha on the North Carolina and Virginia coastal plain, and of that, approximately 49,800 ha have been recommended for public ownership. The Great Dismal Swamp National Wildlife Refuge, established in 1973, is included in, but does not coincide with, this area. The U.S. Fish and Wildlife Service has the responsibility for evaluating and developing management priorities for the 49,800-ha unit (Garrett and Carter, 1977). This primary unit lies in Nansemond County and the City of Suffolk, Virginia and in Currituck, Camden, Perquimins, Gates, and Pasquotank Counties of North Carolina. Its area is contained in nine standard 7 1/2' quadrangles.

The western boundary of the swamp lies along the abrupt, linear, eastward-facing Suffolk scarp. The eastern boundary of the primary swamp unit is the Dismal Swamp Canal; the northern boundary is U.S. Highway 58; and the southern boundary is U.S. Highway 158. Lake

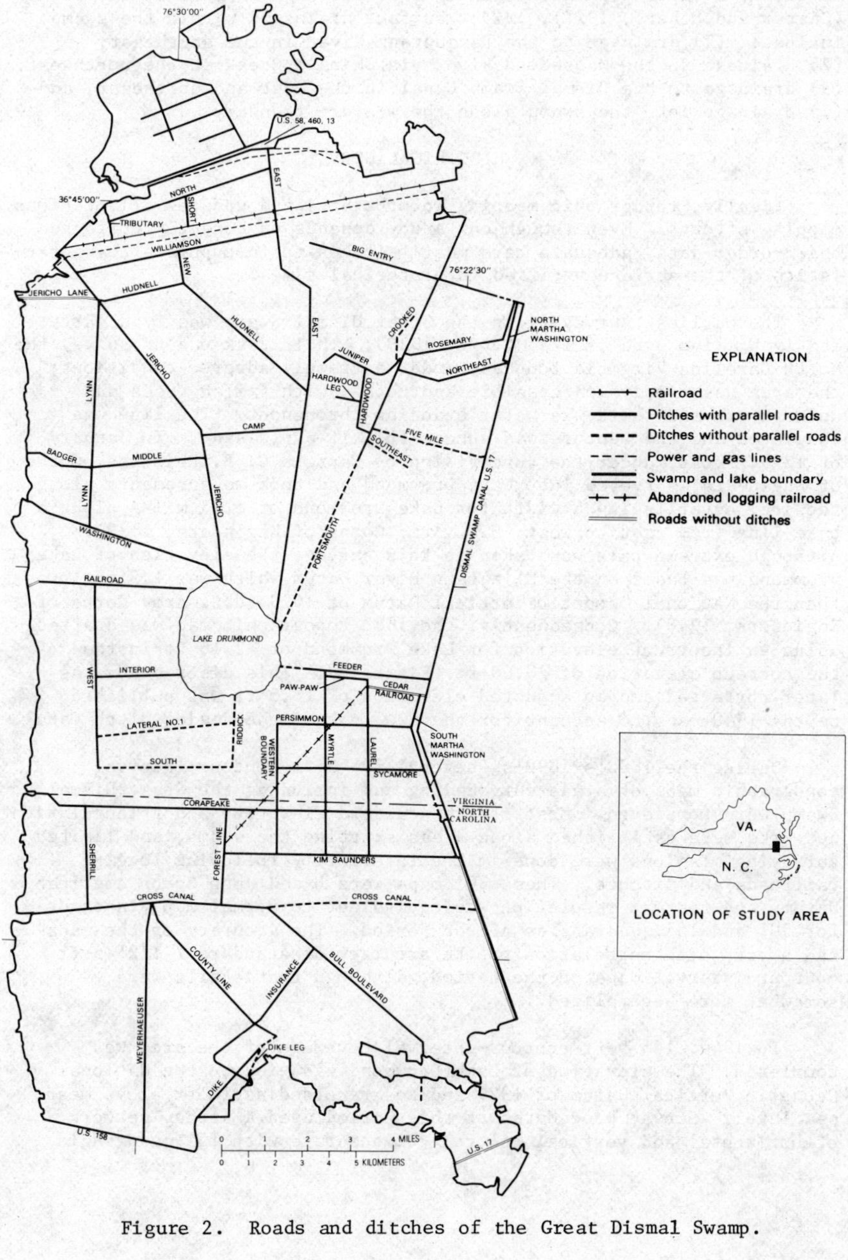

Figure 2. Roads and ditches of the Great Dismal Swamp.

Drummond is approximately 4 km (2.5 miles) in diameter and is almost centrally located within the gently east-sloping gradient of the swamp (Carter and others, 1977). Major surface drainages within the swamp include: (1) drainage to the Pasquotank River in the southeast, (2) drainage to the Nansemond River via Shingle Creek in the northwest, (3) drainage to the Dismal Swamp Canal in the east and northeast, and (4) drainage into the swamp along the western boundary.

HISTORICAL SURVEYS

Ideally, topographic mapping incorporates and updates all previous mapping efforts. Even though our study depends on recently collected third-order data, adequate care must be taken to incorporate characteristics of the surface verified in historical times.

The earliest surveying in the Great Dismal Swamp was by a party led by William Byrd in 1728 (Boyd, 1929). Their task was to survey the North Carolina-Virginia boundary under extremely adverse conditions; the area was nearly inaccessible and choked with fallen trees and undergrowth and there was water standing throughout. The line was completed and the venture was successful, if unpleasant. In January of 1879, Frost, under the supervision of Captain C. B. Phillips, Corps of Engineers, surveyed into Lake Drummond and took measurements along the ice, establishing a width for Lake Drummond of 2.27 miles along a base line from east to west (U.S. Army Corps of Engineers, 1879). Although extreme care was taken in this survey, the elevation of Lake Drummond was based on the Elizabeth River Datum which was 1.5 ft lower than the National Geodetic Vertical Datum of 1929 (U.S. Army Corps of Engineers, 1948). Consequently, the 1888 topographic map was drafted using an incorrect elevation for Lake Drummond of 22.16 ft instead of the correct elevation of 20.66 ft (Figure 3). This discrepancy was later corrected and an adjusted elevation of 18.0 ft was published on the 1940 maps to account for the new lock of the Feeder Ditch Canal.

During the 1880's-1890's, several 30' 5-ft contour-interval topographic maps of areas surrounding and including the Great Dismal Swamp were completed. Most of the standard elevation and triangulation networks were established along areas skirting the swamp, and limited surveying missions were done in the interior by following logging railroads and ditches. These 30' maps were based upon North American Datum, and the map resolution was equivalent to normal map standards for 30' and 15' quadrangles of the period. The accuracy of the maps had a very high correlation to the accuracy of standard 7 1/2' 5-ft contour-interval maps of the period, although the details were somewhat more generalized.

In 1940, 15' 5-ft contour-interval coverage of the area was completed. The elevation of Lake Drummond was tied to the National Geodetic Vertical Datum of 1929 and was recorded as 18 ft above mean sea level. Survey procedures of this period used a strong network of horizontal and vertical control commensurate with 15' quadrangle

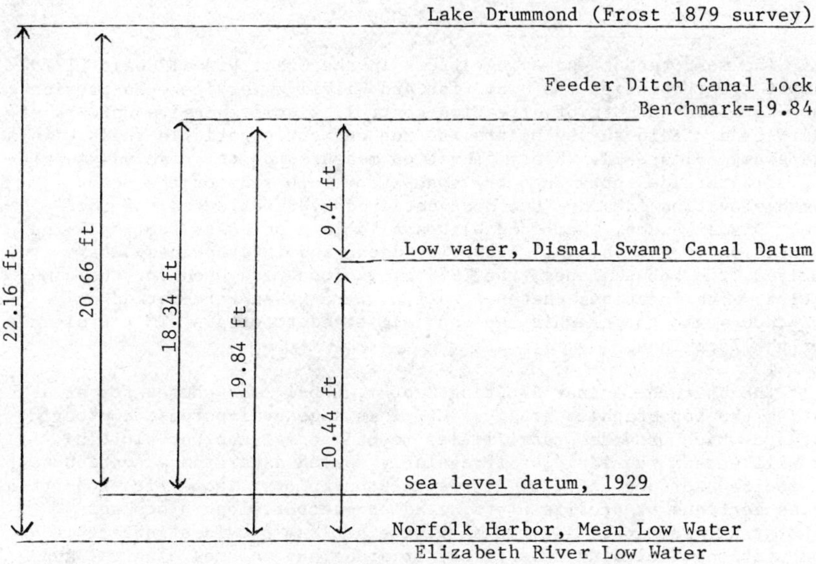

Figure 3. Comparison of Datums for Estimating the Elevation of Lake Drummond

specifications. Contour features at 5-ft intervals were generally located by planetable methods; culture and drainage information was compiled from photographs.

The 15' 1:62,500-scale maps of 1940 were revised and enlarged to 7 1/2' 1:24,000 scale and were published in 1954; no new contour information was added within the swamp. Mason (1952) documented the field conditions of the 1954 revision; during his survey, he encountered a sudden drop along Jericho Ditch which may correspond to an ancient pre-peat drainage basin in the headwaters of the Northwest River. Mason's account of his experiences shows that working in the swamp remains dangerous and difficult, but the processes of surveying had improved substantially since the days when Byrd used rifle or handgun shots from the far side of the swamp to give alinement to his crew clearing a sightline. During the 1950's, logging railways, ditches, and activities of hunters and others having interest in the swamp kept access open to survey activities. The new maps were an improvement over the previous maps; precision was higher, and more detail was included.

METHODS

The most recent surveying effort in the Great Dismal Swamp (1979) was at the request of the U.S. Fish and Wildlife Service. To provide a more-detailed network of elevation control, a comprehensive network of third-order field survey points was run on most negotiable roads within the swamp (Figure 2). Every elevation measured on the road was supplemented with side shots into the swamp from each side of the road. These elevation data are the most detailed ever collected for the Great Dismal Swamp. Combined with available maps, data from soil and vegetation transects, well elevation data, and interpolated data derived from known slopes, the 1979 elevations were used for the simulation. The locations where elevations were taken were marked on 1:24,000-scale topographic maps and digitized to obtain UTM coordinates for the locations.

The Unitech Contour Plotting System, CPS-1, was chosen for simulating the topographic surface. CPS-1 is a general-purpose contouring package which provides capabilities beyond normal contour plotting. It will accept regularly or irregularly spaced data from a continuous or faulted surface and will display contour plots, isometric projections, cross sections or profile plots, gradient-vector slope plots and annotated base maps. Additionally, the program provides surface computations including measurement of areas and volumes, logical and algebraic surface-to-surface computations, surface differentiation, and inverse sampling (Unitech, Inc., 1978).

Several basic assumptions were made prior to the processing of the elevation data with CPS-1. First, we assumed that the surface of the swamp is generally flat--Oaks and Coch (1973) stated that the surface slopes gently eastward at about "1 foot per mile." Secondly, we assume that the effect of the roads and ditches has been to divide the swamp surface into a variety of polygons, each of which is partially or completely isolated by manmade barriers. Thirdly, we assume that the elevations taken at the center of the roads are artificial; that is, they represent manmade artifacts raised above the overall surface of the swamp.

Based upon the first assumption, the best possible set of elevation points would be a uniformly spaced grid covering the entire swamp. However, the actual data were clumped or concentrated along the manmade features giving us a poor distribution from which to project generalized flat features. This distribution can lead the computer to project excessive curvature if processed incorrectly.

Based on the third assumption, elevations of road centerlines were not used because they would generally bias the swamp surface to an artifically high elevation. Paired elevations in the swamp on either side of the roads were averaged, and based on the first assumption, the averaged points were extended further from the road to assist the computer in extending a flat surface into the polygons without influence from the elevations on the opposite side of the road. The default option of CPS-1 is to follow slopes thereby building hills or valleys that do not actually exist. The options of CPS-1 were used which did not key on slopes in the final refinement of the computer-generated surface. However, in order to pick up the generally eastward slope, options were used to relate points from road edge to road edge across the polygons. Known areas of discontinuity were given special treatment using an option of CPS-1 that deals with earthquake faults.

The CPS-1 program procedures generated a contour map of the Great Dismal Swamp by performing the following basic steps:

1. Initialized the system parameters and computer storage.

2. Defined the area and grid parameters. Defined maximum and minimum coordinates for enclosing the area of the swamp and the number of x and y intervals used for producing a grid.

3. Obtained a gridded surface by interpolating the irregularly spaced x,y,z points. By using weighted least-squares computations, each control point was given a statistical weight with north-south and east-west surface slopes. Each 200-m node of the grid was examined to determine its proximity to each sample population of control points. Excessive slopes and surface curvature were removed. A surface was produced at each 200-m square of the grid which is analogous to a grid of tilted plates on the Earth's surface.

4. Printed statistical information about input and gridded data.

5. Plotted contours and isometric projections.

The actual critical procedure constants used with CPS-1 are shown in Table 1.

RESULTS AND DISCUSSION

The maps produced in this study include two-dimensional contour (1-ft intervals) and isometric maps (Figures 4-6). These maps depict the dominant features in the Great Dismal Swamp as well as subtle features that are invisible on aerial photographs. They clearly show the uniform eastward-sloping surface, and additionally, the isometric maps show the tilting to various drainage basins. The slightly elevated mesic islands show up as spikes south of Corapeake Road, east of the transition zone. Discontinuities around the roads and ditches are obvious, and a 20-ft contour is shown that corresponds closely in location to the 20-ft contour of the 1954 topographic maps. Note that because Figure 4 is a computer simulation, it does not conform to National Map Accuracy Standards.

The generally high elevations along Corapeake Ditch, which runs parallel to and south of the Virginia-North Carolina State line, show

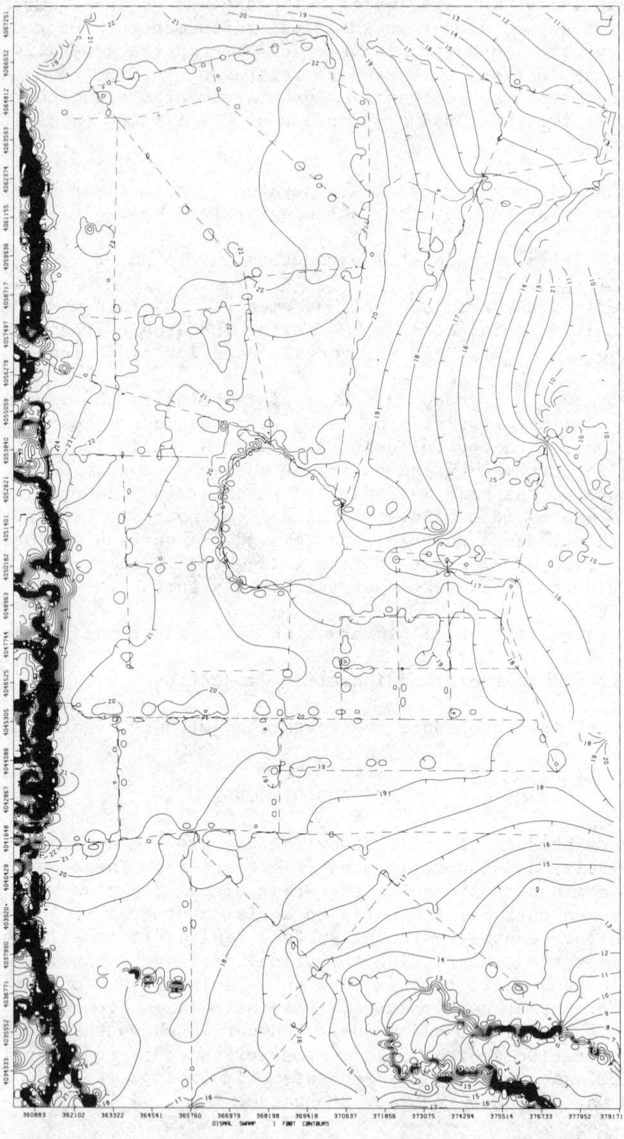

Figure 4. Computer-generated contour simulation of the Great Dismal Swamp area based on 1,400 third-order field survey control points.
(Does not conform to National Map Accuracy Standards)

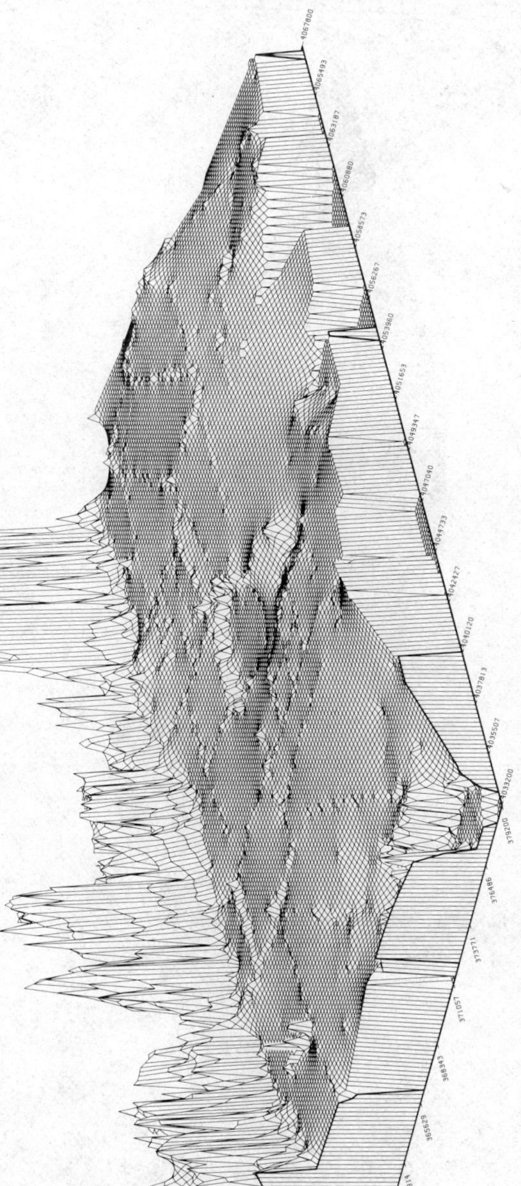

Figure 5. Computer-generated isometric plot of the Great Dismal Swamp area. (Vertical exaggeration x729; fall-off areas in right foreground indicate no data.)

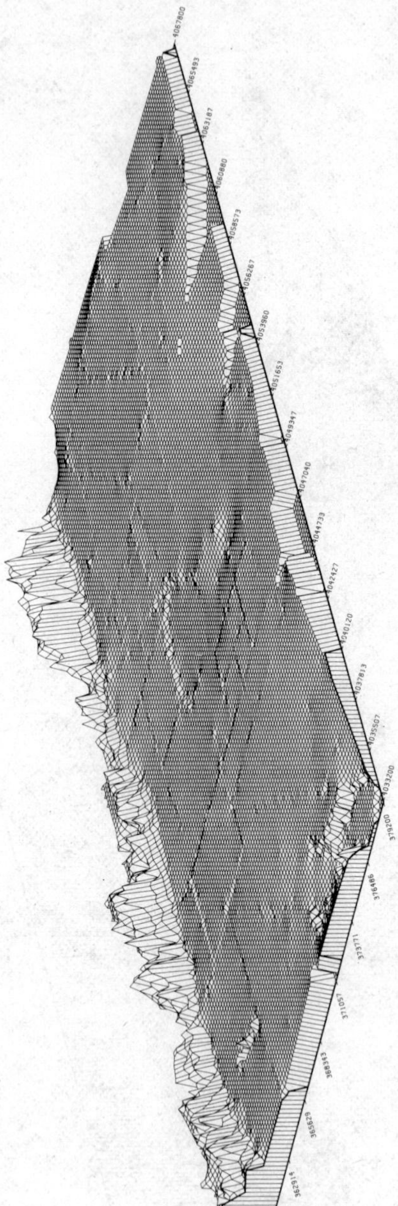

Figure 6. Computer-generated isometric plot of the Great Dismal Swamp area. (Vertical exaggeration x200; fall-off areas in right foreground indicate no data.)

Table 1. CPS-1 procedures and constants used to generate maps of the Great Dismal Swamp

(See Unitech, 1978, for program documentation)

Procedures	Critical constants									
	NDEG	DMP	ITYP	SLM	DEL2	DINC	JTYP	IGRD	NPD	NPT
TANG	1	-1	5	5000	5000	0.10	--	1	16	4
ZFIT	1	-1	5	5000	80	0.10	--	2	16	4
RFIT	--	--	3	--	--	0.10	2	--	--	--

as an east-west linear feature. Comparison of both the contour and the isometric plots is necessary to isolate this linear feature. As shown on the two-dimensional plot, the contours nose out slightly eastward along Corapeake Ditch; the isometric plot confirms the slope away from both sides of the road. Another feature shown by the plots is depicted north of Lake Drummond and south of the 22-ft contour as it crosses Jericho Ditch. This low broad west-east-trending plain on the isometric plot may show the ancient pre-peat river basin described by Oaks and Coch (1973).

The isometric map (Figure 5) would have shown nothing but a flat mesh grid if the vertical scale had not been exaggerated. Elevations in the central portion of the swamp range from 5.5 m (18 ft) to 7 m (23 ft) above sea level, a difference of only 1.5 m (5 ft) whereas each grid square is 200 m (656 ft) on a side. To show this small difference in elevation, a vertical exaggeration of 729 was used. This exaggeration gives a somewhat grotesque appearance to the transition zone; however, it vividly depicts slopes as they exist in the swamp. This plot shows the relative discontinuity between some of the polygons as an abrupt step either higher or lower than the adjoining contours, a situation that actually exists. There is some spiking shown which is only a function of the extreme exaggeration of the vertical scale. A somewhat smaller exaggeration of 200 (Figure 6), while reducing the great exaggeration of the transition zone, fails to show adequate slope detail within the interior of the swamp.

Some anomalies exist on the contour plots as circles and on the isometric plots as small localized spikes. Removing these features would require excessive smoothing, which would change the alinements of the most significant contours. For this reason, these features are retained in the final plots. Their presence indicates that the simulated surface in this vicinity is slightly higher or lower than the adjacent level.

One product of this study was a mylar sheet with a 1:24,000-scale plot of the Dismal Swamp which can be superimposed on the 7 1/2' topographic maps covering the entire area of the swamp. Other map products which could be readily obtained include strip maps and maps of individual polygons within the swamp. Strip maps with greater detail could be used to evaluate linear features such as the transition zone along the western boundary of the swamp. Maps of individual polygons would help to isolate differences of the surface circumscribed by the roads.

The highest priority of use for the maps is to aid in managing the available surface water within the swamp. The single most important element affecting the stability of the swamp is the water regime. The swamp is divided into management units consisting of one to several polygons. The amount of surface water present during each season may be determined by using staff gages within the polygons and stream and staff gages in the ditches surrounding the polygons. The maps show which points along the ditches have the lowest elevations, so the best locations for dams or drains may be selected. Additionally, the maps can be used to obtain elevation-storage capacity curves for each polygon or set of polygons. The elevation-storage capacity curve shows how much water can be stored between each increment of elevation; total storage volume may be calculated indirectly from the contours or directly using an option of CPS-1. Given depth information from gages, the height of roads (which act as dams), and surface elevations, one could estimate the volume of water in a polygon if an outlet was blocked to a certain height. Other surfaces such as the piezometric surface of ground water and depth of organic soil can be generated. Such information is useful for understanding the hydrological regime and soil-water-plant relationships. Changes in the swamp surface caused by fire, drainage, flooding, and construction can be simulated, and the impacts to the ecosystem may be projected.

CONCLUSIONS

Computer simulation is a useful method of projecting representative topographic surfaces in the Great Dismal Swamp. The simulations represent known features and can be used to evaluate surface characteristics. The computer package, CPS-1 is a versatile program for graphical simulation, and provides a wide range of graphical output. The methodology used in this study is applicable to other wetlands and may be one of the few available alternatives in areas where data are limited and where access is severely restricted.

REFERENCES

Boyd, W. K., 1929, William Byrd's histories of the dividing line betwixt Virginia and North Carolina: Raleigh Press, p. 70.

Carter, Virginia, Garrett, M. K., Shima, Lurie, and Gammon, Patricia, 1977, The Great Dismal Swamp: management of a hydrologic resource with the aid of remote sensing: Water Resources Bull., V. 13, no. 1, p. 1-12.

Carter, Virginia and Baker, C. P., 1978, Map of roads and ditches of the Great Dismal Swamp, Virginia and North Carolina: U.S. Geological Survey Open-File Map 78-86.

Garrett, M. K., and Carter, Virginia, 1977, Contribution of remote sensing to habitat evaluation and management in a highly altered ecosystem: Transactions of the Forty-Second N. Amer. Wildlife and Natural Resource Conf., Atlanta, Georgia, March 1977, p. 56-65.

Mason, C. R., 1952, Mapping in Dismal Swamp: U.S. Geol. Survey Topo. Div. Bull., p. 92-102.

Oaks, R. O., Jr. and Coch, N. K., 1973, Postmiocene stratigraphy and morphology, southeastern Virginia: Virginia Div. Min. Resour. Bull. 82., 135 p.

Unitech, Inc., 1978, CPS-1 User's Manual, training documentation, V. 1, 167 p.

Unitech, Inc., 1978, CPS-1 User's Manual, verb documentation, V. 2, 63 p.

U.S. Army Corps of Engineers, 1948, Lake Drummond spillway -- bench marks, Norfolk District, Norfolk Co., Va., Description Book 4, p. 157.

U.S. Army Corps of Engineers, 1879, Lake Drummond, Va. survey, archive map, Norfolk District, Norfolk Co., Va.

U.S. Fish and Wildlife Service, 1974, Report to the United States Congress from the Secretary of the Interior regarding actions taken and recommendations made pursuant to Public Law 92-478 (Dismal Swamp Study).

SHORE PROTECTION DECISIONS FOR SANDY HOOK

by
Donald A. Falvey, P.E.
Civil Engineer
National Park Service
Denver, Colorado

William J. Funge, P.E.
Vice President
Dravo van Houten, Inc
New York, New York

INTRODUCTION

Sandy Hook, a peninsula extending from the New Jersey coast into New York Bay (see Fig. 1) has played a role in our nation's defense since before the Revolutionary War. As early as 1689 troops were stationed at Sandy Hook to protect New York from the French Canadians. A lighthouse erected in 1764 was bombarded by American Artillery to hinder the movement of the British Fleet. In the late 1800's, Sandy Hook served as a proving grounds for artillery testing with Fort Hancock as the headquarters. World War II turned Fort Hancock into a staging area for the troops bound for Europe and for prisoners returning from Europe. More recently, Nike missles were installed on Sandy Hook to protect the New York metropolitan area. Today Sandy Hook is engaged in yet another battle - this time requiring defense against the foes of wind and waves and the resulting erosion of its beaches.

With the establishment of Gateway National Recreation Area, the Sandy Hook Unit and the National Park Service has been active in combating the severe erosion occurring on Sandy Hook's southern beaches. This erosion has forced the Park Service to pursue interim protection measures while defining long range strategies. This paper will present an examination of both issues.

CURRENT LAND USES AND ACTIVITIES

The 1665 acres and 7 1/2 miles of ocean beaches and sheltered bayside coves that comprise Sandy Hook lend themselves to a variety of uses. The U.S. Coast Guard and U.S. Corps of Engineers (see Fig. 2) maintain properties at the tip of Sandy Hook that are excluded from the park. Development is concentrated at the southern beach area, the Fort Hancock complex, and a few other areas, leaving large stretches of undisturbed land in between.

A wide variety of recreational uses occur at Sandy Hook. Both the South Beach and North Beach bathing areas are used for all types of beach-related activities. Picnicking and fishing are popular activities on the bay side. Environmental education programs are held in the marshes and grasslands near Spermaceti Cove.

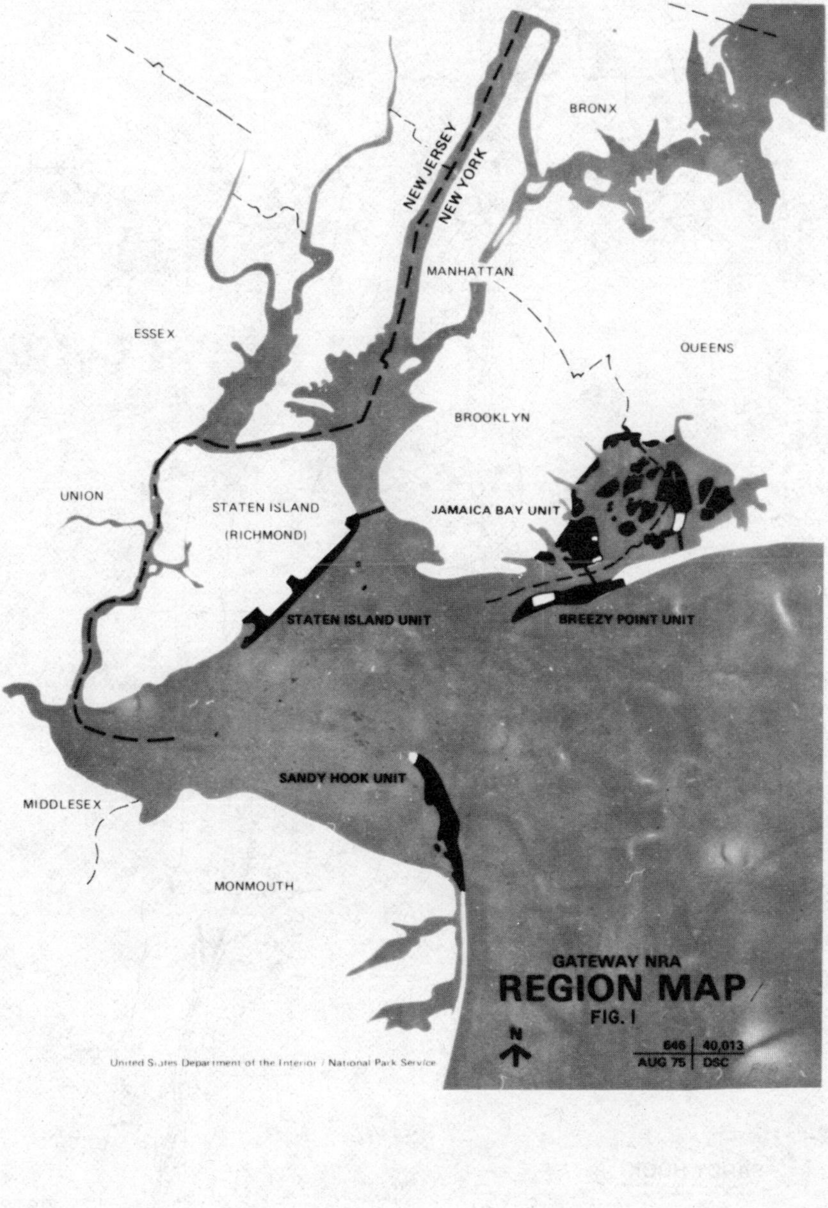
BRONX
NEW JERSEY
NEW YORK
MANHATTAN
ESSEX
QUEENS
BROOKLYN
UNION
STATEN ISLAND
(RICHMOND)
JAMAICA BAY UNIT
STATEN ISLAND UNIT
BREEZY POINT UNIT
SANDY HOOK UNIT
MIDDLESEX
MONMOUTH
GATEWAY NRA
REGION MAP
FIG. I
N
646 | 40,013
AUG 75 | DSC
United States Department of the Interior / National Park Service

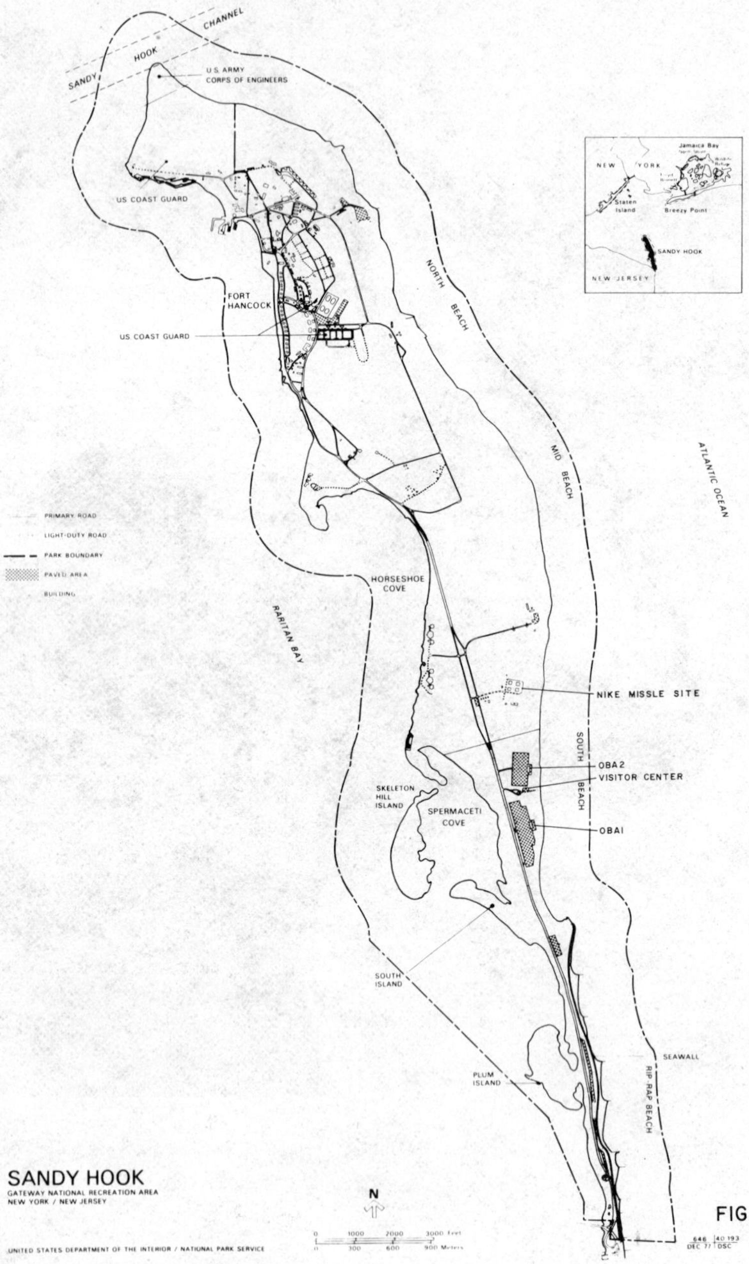
SANDY HOOK CHANNEL
U.S. ARMY CORPS OF ENGINEERS
US COAST GUARD
FORT HANCOCK
US COAST GUARD
NORTH BEACH
MID BEACH
ATLANTIC OCEAN
PRIMARY ROAD
LIGHT-DUTY ROAD
PARK BOUNDARY
PAVED AREA
BUILDING
HORSESHOE COVE
RARITAN BAY
NIKE MISSLE SITE
SOUTH BEACH
OBA 2
VISITOR CENTER
SKELETON HILL ISLAND
SPERMACETI COVE
OBAI
SOUTH ISLAND
SEAWALL
PLUM ISLAND
RIP-RAP BEACH
NEW YORK
Jamaica Bay
Staten Island
Breezy Point
SANDY HOOK
NEW JERSEY
SANDY HOOK
GATEWAY NATIONAL RECREATION AREA
NEW YORK / NEW JERSEY
UNITED STATES DEPARTMENT OF THE INTERIOR / NATIONAL PARK SERVICE
N
0 1000 2000 3000 Feet
0 300 600 900 Meters
646 | 40 193
DEC 77 | DSC

FIG. 2

Fort Hancock, a nationally recognized historic district, is the center of activity for a variety of uses. Several structures have been renovated for use by private groups and by National Park Service, National Oceanic and Atmospheric Administration, Coast Guard, Army and Navy personnel.

The prime beach activity occurs at the southern beach area at Ocean Bathing Areas (OBA) 1 and 2. Situated between OBA1 and OBA2 is the visitor center, a historic Coast Guard life saving station. Just north of OBA2 is an abandoned Nike missle battery. The bathhouses constructed by the state of New Jersey at OBA 1 and 2 when this area was a state park have been destroyed by beach erosion and the winter storms of 1977-1978. These bathhouses have now been replaced with temporary facilities to support summer beach use.

PROBLEM DEFINITION

The beach erosion that led to the destruction of the bathhouses has been occurring for some time. A team from the Center for Coastal and Environmental Studies, Rutgers University, has undertaken a comprehensive study of the erosional trends of Sandy Hook. To approach the problem, they divided the shoreline into segments to identify the interrelationship of energy and sediment supply and the combined effect on the erosional rates.

The area of most severe erosion occurs along the segment near OBA1 known as the critical zone (see Fig. 3). This zone, located just north, and downdrift, of the terminus of a seawall and extensive groin field has been experiencing a net retrogression of the shoreline. This loss is caused by a combination of factors.

In this area, the dominant waves are subjected to refraction which results in a concentration of wave energy. Erosion during storms is therefore quite extensive. Due to the high density of sand retention structures along the northern New Jersey shoreline, much of the sand normally transported by the longshore currents is not available to the system. Thus, when the shoreline erodes due to high energy conditions, the sediments that should build the beach back during periods of calm weather are no longer available.

Another factor contributing to accelerated erosion at the critical zone is related to the location of the seawall terminus which functions as a resistant headland. Removal of sediment by longshore transport initiates the beach erosion. Diffraction of waves into the shadow zone behind the headland further erodes the beach resulting in a spiral configuration.

Measurements indicate a shoreline retreat in the critical zone of approximately 125 feet between April 1976 and October 1977. These figures translate into an erosion rate of 75 to 80 feet per year (see Fig. 4). At this rate, Sandy Hook could soon be breached and become an island.

The rapid retreat of shoreline in this area has already resulted in: (1) the destruction of recreation facilities, including the bathhouses at OBA1 and OBA2; (2) removal of the dune field between the northern terminus of the seawall and OBA2; (3) closing of the road leading into the park at times of high tide and/or storm attack; and (4) considerable dune erosion between OBA2 and the South Bathing Beach.

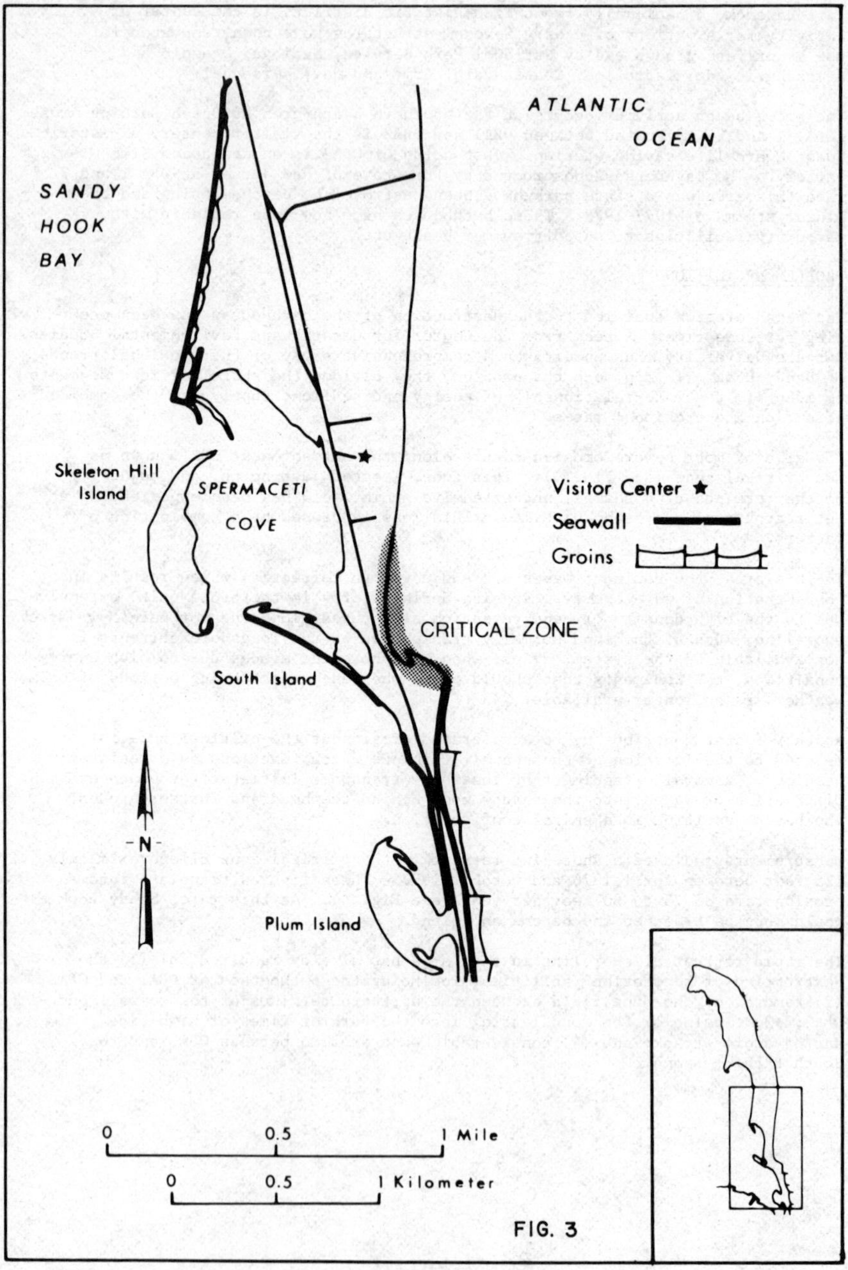
ATLANTIC
OCEAN
SANDY
HOOK
BAY
Skeleton Hill
Island
SPERMACETI
COVE
Visitor Center ★
Seawall
Groins
CRITICAL ZONE
South Island
Plum Island
N
0
0.5
1 Mile
0
0.5
1 Kilometer

FIG. 3

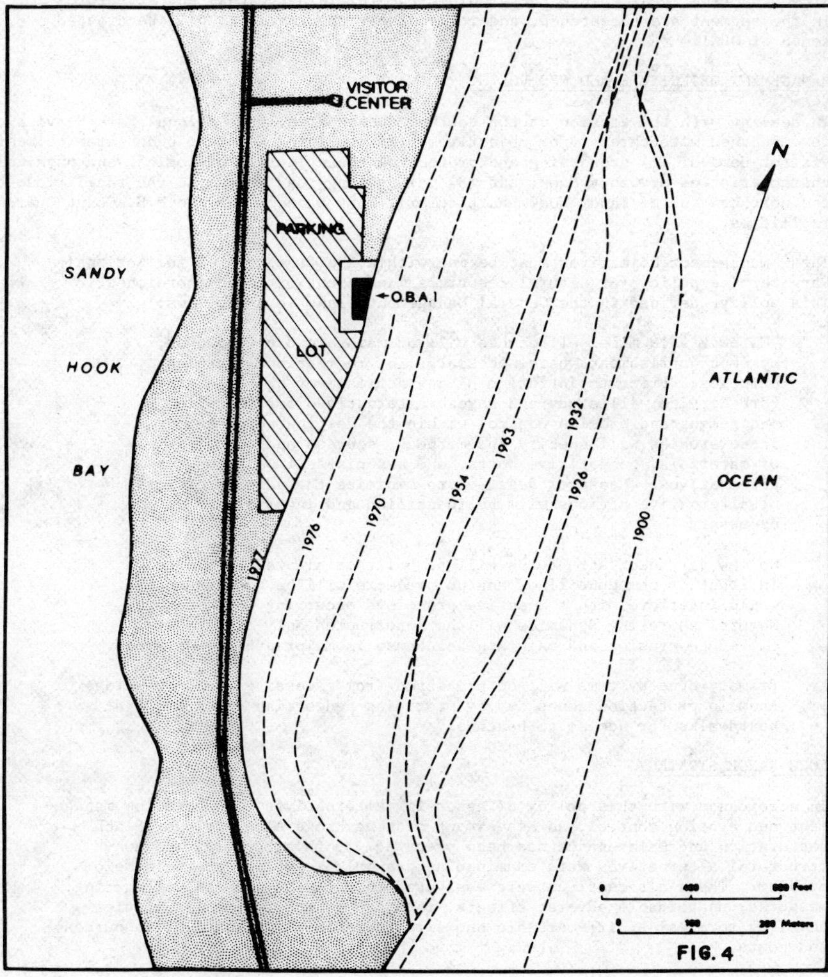
VISITOR CENTER
PARKING
LOT
O.B.A.1
SANDY
HOOK
BAY
ATLANTIC
OCEAN
N
1977
1976
1970
1954
1963
1932
1926
1900
0 400 800 Feet
0 100 200 Meters

FIG. 4

Without action to alleviate the problem, the remaining beaches and structures in the segment are threatened, and it is likely that an inlet will be created south of OBA1.

MANAGEMENT OBJECTIVES AND NPS POLICY

In dealing with the erosion at the southern beach area, the National Park Service is concerned with three major objectives: (1) providing adequate beach area to meet visitor demand; (2) preserving and protecting the cultural, biological and physical characteristics of Sandy Hook; and (3) maintaining road access to the facilities at the northern end of Sandy Hook, particularly Fort Hancock and the U.S. Coast Guard facilities.

These management objectives must be met within the framework of the National Park Service policy for natural resources management of ocean shorelines. This policy, defined in the General Management Plan, is as follows:

> "Gateway's 19 miles of beaches will be maintained or expanded, by sand nourishment where necessary, for recreational bathing purposes. Prior to intiation of any nourishment projects, the Park Service will study and assess alternatives for shoreline management and erosion control within the designated areas. These studies will specify the methods, sources and quantities of materials, comparitive costs, and associated impacts of the alternatives. Measures designed to minimize the adverse effects of alternative actions will be identified and incorporated as necessary.
>
> No new permanent structures will be built or allowed on the beach in front of the dune line, and no projects will be undertaken that would interfere with the active processes occurring in this zone. Natural shoreline dynamics will be interrupted only as necessary to reduce erosion and maintain beach use in major public-use areas.
>
> Primary dune systems will be protected from excessive use by placing them in protection zones and by providing pedestrian crossings and boardwalks for access to beaches."

LONG RANGE STRATEGY

In accordance with this policy of assessing alternatives for shoreline management and erosion control, an Assessment of Long Range Alternatives, Beach Restoration and Maintenance has been prepared. Both structural and non-structural alternatives were examined as possible solutions to the erosion problem. These alternatives were evaluated in terms of impacts, mitigating measures, unavoidable adverse effects, relation of short term liabilities and long term gains, irreversible and irretrievable commitments of resources and costs.

Non-Structural Alternative

The non-structural alternative evaluated was beach nourishment. An initial implacement of two million cubic yards at the critical zone would be followed by subsequent annual maintenance placements of 50,000 to 100,000 cubic yards, depending on the severity of winter storms. The most significant biological impact that would occur from placing sand in the critical zone is the burial of benthos. Loss of benthic animals would be an unavoidable adverse effect, however, this loss would be a short-term loss, not uncommon in this unstable environment.

More significant than the effects of placing sand are the effects of removing and transporting sand to the critical zone. An evaluation of both borrow sites and transportation schemes has been performed in a joint effort by the Center for Coastal and Environmental Studies, Rutgers University, and the consulting firm of Dravo van Houten, Inc.

Three borrow areas were identified: the Offshore Borrow Site, the Sandy Hook Channel Borrow Site, and the Shrewsbury Bay Borrow Site. All three sites contain sediment with a particle size suitable for beach nourishment.

The Offshore Borrow Site is located approximately one mile east of the critical zone and is about 1/3 mile wide and 3 miles long. If dredged to a depth of one yard, the site would yield more than 3 million cubic yards of suitable material. The site is reasonably free of debris and will not interfere with either navigation or fishing activities.

The Sandy Hook Channel Borrow Site is located at the northern tip of Sandy Hook. The Corps of Engineers performs maintenance dredging of the channel every other year, removing approximately 350,000 cubic yards of material. While insufficient for the initial emplacement, this site would yield sufficient material for maintenance nourishment.

The Shrewsbury Bay Borrow Site is located in the bay west of the critical zone. If the Shrewsbury Borrow Site were dredged to a depth of 21 feet below mean sea level, approximately four million cubic yards of sediment would be available. While this site appears to be acceptable from the gradation and quantity aspects, it is likely that concentrations of pollutants will make this site unsuitable. Further complicating the situation is the unknown effect that removal of the Shrewsbury Bar would have on the hydrodynamics on Sandy Hook and Raritan Bays. Also of concern are the primary and secondary effects of the destruction of existing benthos populations.

Transportation schemes include piplines, barges, trucking, mono-mooring and combinations of these methods. Since use of the Sandy Hook Channel Borrow Site appeared feasible for maintenance nourishment, consideration was given to stock piling sand near the tip and transporting it to the beach as required.

For emplacement of the initial two million cubic yards, only the Offshore Borrow Site contained sand of sufficient quantity and quality. The following combinations of sediment source and transportation appear feasible for the initial placement:

1. Offshore dredging - barge to oceanside - pipeline through surf zone to beach
2. Offshore dredging - barge to bayside - pipeline across spit to beach
3. Offshore dredging - barge to holding area at tip of Sandy Hook - pipeline the length of hook to beach
4. Offshore dredging - barge to holding area at tip of Sandy Hook - trucking the length of hook to beach

The following combinations of sediment source and transport technique appear to be viable for maintenance nourishment of up to 200,000 cubic yards of sand per year:

1. Offshore dredging - barge to oceanside - pipeline through surf zone to beach
2. Offshore dredging - barge to bayside - pipeline across spit to beach
3. Offshore dredging - barge to holding area at tip of Sandy Hook - pipeline the length of hook to beach
4. Offshore dredging - barge to holding area at tip of Sandy Hook - trucking the length of hook to beach
5. Dredge spoil from Sandy Hook Channel - barge to oceanside - pipeline through surf zone to beach
6. Dredge spoil from Sandy Hook Channel - barge to bayside - pipeline across spit to beach
7. Dredge spoil from Sandy Hook Channel - barge to holding area at tip of Sandy Hook - pipeline the length of hook to beach
8. Dredge spoil from Sandy Hook Channel - barge to holding area at tip of Sandy Hook - trucking the length of hook to beach

Another method of obtaining sand for maintenance placement is the installation of a mono-mooring system that would accept material dredged from the Sandy Hook Channel by the Corps of Engineers. This sediment would then be placed on the beach through a slurry pipeline.

The combinations for placing sand on the beach at the critical zone are numerous but the flexibility does exist for implementing this alternative.

Structural Alternative

The structural alternatives evaluated included a seawall, groins, a causeway and bridges. It should be noted that beach nourishment is required for each of the structural alternatives, both for initial beach reconstruction, and for protection of the structure.

The seawall alternative involves the extension of the seawall approximately 7000 feet northward, ending in the vicinity of the Nike missle site. The wall would be constructed of 12 ton stones with a finished elevation 20 feet above mean sea level. A possible source of stone is along the Hudson River near Peekskill, New York. A combination of barging and trucking would be required to transport stone to the site.

Constructing a seawall would limit visibility of the ocean from the road and would require beach users to cross the wall using steps or ramps. While protecting against breaching, the seawall would serve little advantage as beach nourishment is still required to maintain a functional beach.

Groins were also considered as an alternative. Following beach nourishment, a series of ten groins spaced at intervals of 750 feet would be constructed in the critical zone. The groins would be 500 feet long and constructed of large stone forming a section about 45 feet wide at its base and 18 to 20 feet wide at the top.

Through the use of groins, annual nourishment requirements would be reduced by half, however it is anticipated that localized erosion would occur north of the groin field. With construction of groins, the beach would be divided into compartments that would not only appear unnatural but would encourage segregation in beach use.

The causeway alternative consists of an elevated roadway one mile long with a top elevation of 12 feet above mean sea level. As with the seawall, beach nourishment is required for initial building of the beach area and for protection against wave and surge attack.

The last structural alternative considered was the construction of a bridge over a controlled inlet. Also included was the installation of jetties and placement of initial nourishment. Three types of bridges were considered: high level, low level and bascule.

The high level and bascule bridge would permit navigation under the bridge, a low level bridge would not.

The major impact associated with this alternative relates to the controlled inlet. It is likely that shoaling will occur in Sandy Hook Bay and if the inlet is to be maintained in an open state, occasional dredging will be required. Also tidal velocities could cause erosion on some bay side beaches.

In 1972, the U.S. Army Corps of Engineers conducted a model study of the New York Harbor which included Sandy Hook. As part of the study, they examined the effects of an inlet forming at the southern end of Sandy Hook. Based on these studies, it appears that in the bay salinity levels would increase, with turbidity and nutrient levels showing a decrease. There is a possiblity of bayside pollution spreading to the beaches, resulting in high coliform counts.

A cost comparison of the various alternatives (see Table I) reveals that beach nourishment has both the lowest initial cost and annual maintenance cost. Construction of the seawall or groins would cost approximately three times as much.

TABLE I. COMPARISON OF COSTS FOR EACH ALTERNATIVE

Alternative	Initial Construction Cost	Annual Maintenance Cost	Annual Cost 1/
Beach Nourishment	$ 3,750,000	$350,000	$ 350,000 2/
Seawall	11,250,000	425,000	1,150,000
Groins	12,350,000	305,000	1,100,000
Armored Causeway	7,500,000	425,000	900,000
Low Level Bridge	7,200,000	375,000	830,000
Draw Bridge	8,500,000	555,000 3/	1,100,000 4/
High Level Bridge	15,500,000	515,000	1,500,000

1/ Includes 50-year amortization at six percent interest plus annual maintenance cost.

2/ Since there is no projected life span for this alternative, the initial construction cost is not amortized.

3/ Includes annual operation cost of $100,000.

4/ Amortization over 25 years, rather than 50 years.

INTERIM SOLUTION

As shown on Fig. 3 and Fig. 4, the erosion rate in the critical zone has been increasing at an alarming rate. This pattern of erosion is threatening to undermine an access road and possibly breach the peninsula. Since the National Park Service has the obligation of maintaining the access to Sandy Hook, it has been decided to undertake emergency action to protect the access road, and implement an interim solution which would buy enough time to develop properly the long-term solution. The Park envisions implementing the long-term solution in approximately two years.

Basic Parameters of Interim Solution

To meet the objectives of the Park Service, the interim solution had to:

1. Protect the access road for a period long enough to permit the long-term solution to be implemented.

2. Be as inexpensive as possible, and fall within budgetary limitations.

3. Be quickly implemented.

4. Maintain the character of the beach with respect to the materials used; that is materials such as rock or sheet piling could not be used.

5. Provide an aesthetic blend with the natural appearance of the beach.

6. Act as a barrier to prevent entrance to the beach except through the designated access routes.

In considering these parameters, the National Park Service decided that a roadway protection dike should be built using the very large (jumbo) sandbags.

Sandbag Protection

N.P.S. selected the use of jumbo sandbags because this method best met the stated requirements, and because they had been used successfully for preventing beach erosion in New Jersey and in other parts of the United States and the world. The Park Service has used the jumbo sandbags for protecting the Cape Hatteras Lighthouse along the North Carolina Outer Banks.

These large sandbags generally weigh in the vicinity of two to three tons, and contain approximately two cubic yards of sand. They come in various sizes, depending upon the manufacturer of the bag, with lengths ranging from seven to ten feet, widths from two to four feet, and depths of approximately 1 1/2 to 2 1/2 feet. These bags are designed to have a pillow configuration when filled with sand, and the size of the bag selected is based upon the weight required for the desired use.

Due to the weight of the bags, the materials which are generally used by most manufacturers for making the bags are either nylon, polypropylene or polyethylene. These synthetic materials can also receive a coating of polyvinyl chloride to protect them from sunlight, which extends their life. The sandbags can also be made from materials, such as burlap or canvas, which would be more biodegradeable, if this characteristic were desired. However, since these materials would disintegrate quickly, and are not as strong as the synthetics mentioned, the manufacturers do not use these materials, and would have to custom fabricate burlap or canvas bags on a special order basis. Most manufacturers will guarantee the integrity of the uncoated bags for a period of years provided they are not exposed to direct sunlight. If exposed, the guaranteed life is a minimum of six months; however, tests have been run by the U.S. Corps of Engineers where the uncoated bags have survived exposure to direct sunlight of one year. When it is anticipated that the fabric will be exposed to direct sunlight, manufacturers recommend that polyvinyl chloride coating be used, and claim that it will extend the life of the bags for several years.

Because these large sandbags are usually employed for minor low cost facilities, the equipment used for filling the bags is usually simple and relatively light weight. In general, the bags are filled either hydraulically or by a dry-fill process. When the hydraulic filling process is used the bags are usually filled in place in order to eliminate the need for handling the heavy bags. This is usually done by a means of a pump of approximately five to ten horsepower which takes sand off the beach by means of a small dustpan dredge, and slurries the sand through a short hose or pipeline into the bags. A small work crew of 2 or 3 men can arrange and position the equipment to fill the sandbags hydraulically. Using the dry filled method, the bag is suspended under a small hopper which is usually fed by means of a front end loader. When filled the bags are closed and moved into position by means of a small crane, or specially equipped piece of mobile equipment.

The sandbags, when filled, are closed either by sewing the opened end or by means of a self-sealing flap which is folded into the bag. Most manufacturers provide the option of having the bags delivered with or without ties. If ties are ordered, the bags usually come with six; four are in the corners and two are at the center of the seamed edges. These ties are used primarily during the filling process to help align the bags but are also used to connect the bags together to provide a certain degree of continutiy.

Alternative Layout

In consultation with Rutgers University and Dr. Robert Dolan of Coastal Research Associates, Virginia, the extent of dike protection needed for defending the access road was determined. Considering that the erosion control dike under study is an emergency action to protect the road until permanent action can be taken, a dike of approximately 1,400 feet in length was judged to be adequate (Fig. 5).

Alternate layouts considering the possibility of a double dike in the critical zone, but shorter in length, were also considered. The advantage of this arrangement would have been that the dike adjacent to the road would receive some protection from the oceanside dike, thereby reducing the potential for undermining the dike adjacent to the roadway. The disadvantage of this double dike would have been that it would generally shorten the extent of protection along the access road because the total length of the dike which was built was determined by budgetary consideration. Since the advantage gained by the second dike did not appear to be as valuable as the added protection against flanking, which would be achieved by extending the single dike, the layout using a longer single dike was judged to be more cost effective.

The centerline of the erosion control dike was located 30 feet from the edge of the access road, except at the widened ends of the dike. This distance was selected in order to provide the maximum protection for the road without being too close and creating the possiblity of undermining the road during construction. This distance also permits a corridor to exist adjacent to the road wide enough for present and possible future utiliies. Locating the dike 30 feet from the road was judged to be approximately the limit where maximum protection would be gained from the proximity of the dike, and yet where dike undermining would not also constitute undermining of the access road.

Alternative Cross-Sections

In order to obtain the maximum erosion protection for the access road, the bottom of the dike was placed at Elevation +3 feet MSL. This elevation was used because it was considered to be as deep as possible without excavating into the water table. Installing the dike so that the bottom would be at a lower elevation would provide greater protection for the road; however, since excavation would be below the water table, the construction cost and installation time would increase considerably. Because the erosion control dike was not designed to prevent over-topping of waves during storm periods and to prevent flooding, the top of the dike was located at Elevation +8 feet MSL. This placed the top of the dike at an elevation of approximately 1 to 3 feet above the existing roadway elevation along the length of the dike. This permitted the use of a dike which was 3 bags deep and considered to be a relatively stable height of dike.

Since the National Park Service had recently completed an emergency sand filling operation at the South Beach, the elevation of the beach adjacent to the road required trenching in order to install the erosion protection dike. The variation in ground elevation required a trench of approximately 3 feet to 5 feet in depth to be cut for installing the dike. This placed

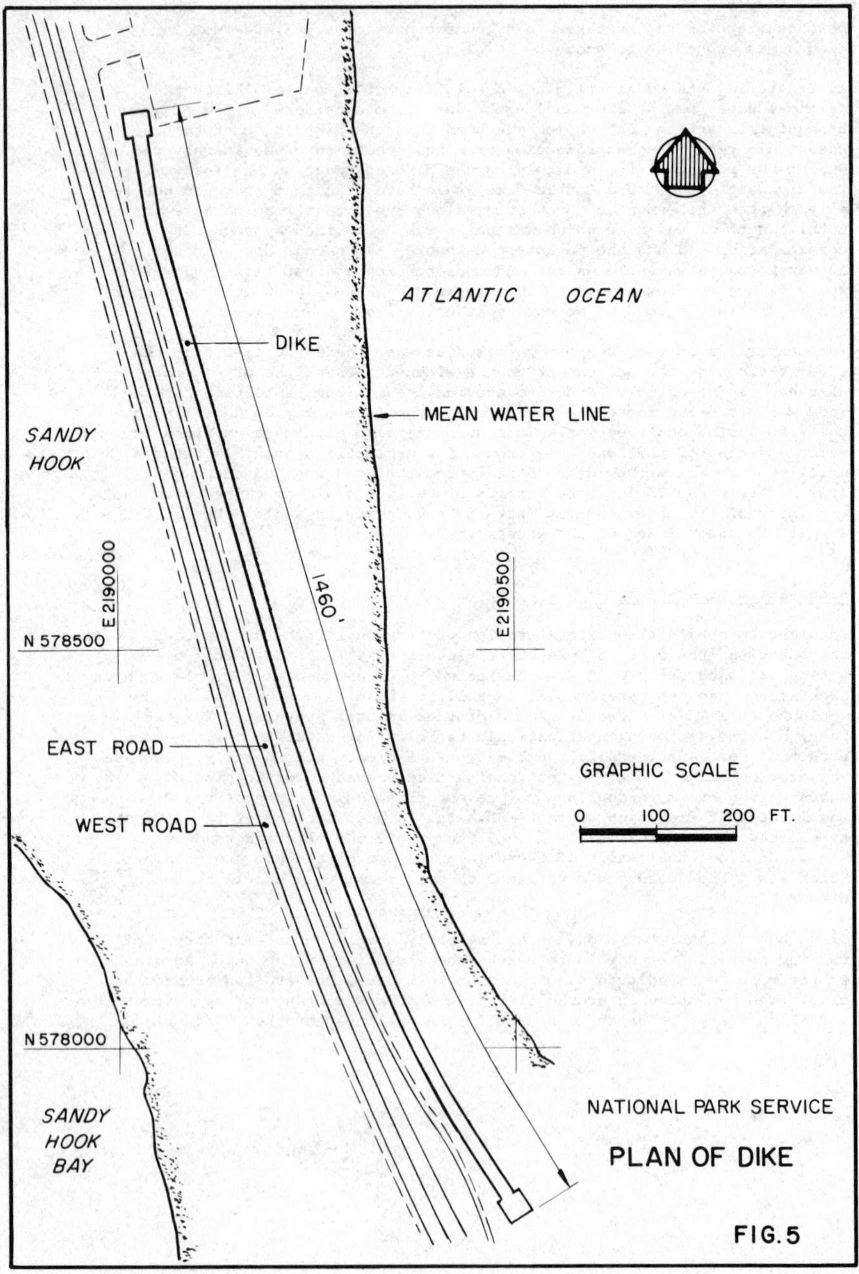
ATLANTIC OCEAN
DIKE
MEAN WATER LINE
SANDY HOOK
1460'
E 2190000
E 2190500
N 578500
EAST ROAD
WEST ROAD
GRAPHIC SCALE
0 100 200 FT.
N 578000
SANDY HOOK BAY
NATIONAL PARK SERVICE
PLAN OF DIKE

FIG. 5

the top of the dike in some areas at approximately ground elevation, and in other areas up to 2 to 3 feet above the ground.

In order to protect the sandbags in the dike from deterioration due to sunlight and damage due to vandalism, the dike received a cover of sand 2 feet deep at the centerline of the dike. This covering also had an aesthetic affect of having the appearance of a roadside dune. To prevent the loss of sand from the top of the dike and drifting onto the roadway, a sand-fence was located between the dike and the roadway. This sand-fence was also considered to be an effective way of keeping people from crossing over the top of the dike to gain access to the beach.

To analyze the stability of the erosion control dike, several cross-sections using various patterns of sandbags were considered. Two are shown on Fig. 6. The first three of these cross-sections were based upon various arrangements using the standard trapazoidal dike pattern, and the fourth scheme used a square configuration. In reviewing these alternatives, there appeared to be very little difference in the degree of erosion control, and the resultant protection, achieved if the dike were to be undermined. One of the primary considerations in the cross-sectional layout of the dike was to establish the bottom of the dike at an elevation below the natural beach line; consequently, excavation was required. This need led to the selection of the square dike since the trapazoidal shaped dikes required greater excavation for installation. Also, since additional stability was to be achieved by wrapping the dike in filter fabric material, the more compact square dike was considered to be a better shape for containment.

Dike End Details

Since the integrity of the sandbag dike is largely dependent upon the sandbags not separating along the longitudinal axis of the structure, the ends of the dike required special treatment. In order to strengthen the ends of the dike and to have the ends act as anchor points, the physical size of each end of the dike was increased. At these dike ends, the normal width of 9 feet was increased to 27 feet by tripling the cross-section over a length of 24 feet. In effect, this provides greater stability for the ends in the event that these ends receive a wave attack. In addition, to further protect the ends, the northern end of the dike was located within the southern reach of the parking lot for the South Beach, and the southern end of the dike was located beyond the southern end of the critical erosion zone in the undisturbed vegetation area.

For additional strength at the ends of the dike, the entire 24 foot lengths of both end sections were wrapped in filter fabric to have each end to act as a single unit. The sandbag stacking pattern over the 27 foot width was accomplished by repeating the pattern used for the standard 9 foot section 3 times, and by wrapping the entire 27 foot width rather than individually wrapping each 9 foot section. To maintain continuity with the rest of the sandbag dike, the bottom elevation of the end sections

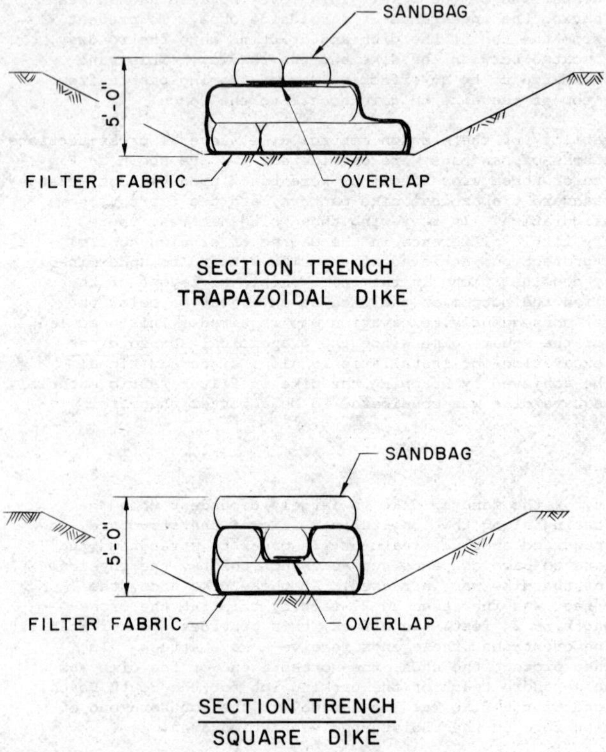

NATIONAL PARK SERVICE

CROSS SECTION

FIG. 6

was established at the same elevation as the interior of the dike, Elevation +3 feet MSL.

Improvement of Dike Stability

Previous experience with sandbag protective dikes indicated that their destruction usually resulted from the separation of bags, which in turn resulted in the loss of dike integrity, and the ultimate break-up of the entire dike.

To reduce sandbag separation and increase the dike's ability to hold together, the entire length of the dike was wrapped in a sheet of filter fabric material. Fig. 7 indicates the pattern used for wrapping the standard interior dike cross-section, and the pattern used for wrapping the end sections. The filter fabric material can be obtained in various widths and lengths; for this project, sheets 36 feet wide and 200 feet long were used. These sheets were wide enough to wrap the standard dike section. Where longitudinal sheets were mated, they were overlapped by 6 to 9 feet. The end sections of the dike were wrapped by turning the run of the fabric material so that it's length is across the dike.

Another benefit obtained from the use of the filter fabric wrapping is the reduction of bag settlement into the beach. Because the fabric material is spread out under the entire dike, it tends to provide support of local sections when they are subjected to quick conditions in the sand below the bags. The temporary quick condition occurs on beaches as the waves change the pressure of the water in the sand. Heavy objects, including individual bags, tend to settle deeper into the beach as the waves momentarily fluidize the sand beneath them. Filter fabric placed beneath the entire dike is held in position by those bags which are not affected by the waves, and prevent the sand, at points where it is fluidized, from flowing up between the bags.

Interim Solution Conclusions

The primary reasons, therefore, that the single dike of jumbo sandbags, wrapped in filter fabric, was selected for the interim protection for the critical zone of the South Beach were:

1. For the parameters given for the emergency protection of the access road, the erosion control dike was constructed using the jumbo sandbags because:

    a. They are in keeping with the temporary nature of the emergency defense required, and can be removed with less difficulty than other systems in the future, if necessary.

    b. They blend into the environment of the South Beach Area better than systems employing rock, sheet-piling, wire and stones.

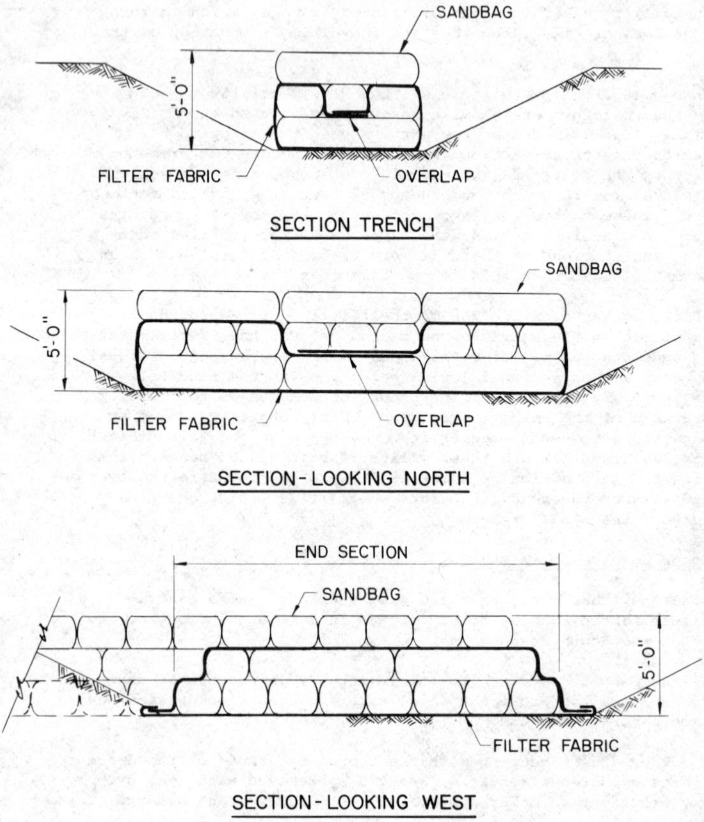

NATIONAL PARK SERVICE

WRAPPING DETAILS

FIG. 7

c. They are cost-effective for short-term use and provide a good means of stretching available emergency funds for protecting the access road.

d. They are installed quickly, thereby providing a quick solution to the road protection problem.

2. The sandbag dike layout had to be at least 1,400 feet long and positioned, as shown on Fig. 5, to cover the critical zone of beach erosion. Since flanking of the dike would seriously threaten it's integrity, a single dike was considered to be more cost-effective than a double dike arrangement of shorter length.

3. The square cross-section dike arrangement was used for placing the sandbags because:

   a. It provided a more cost-effective arrangement than the trapazoidal shaped dike since it required significantly less excavation.

   b. It was more easily wrapped with the filter fabric sheet and provided a tighter bundle.

   c. It required no more sandbags per unit length of dike than the other schemes.

4. Because of strength requirements, synthetic materials were used for the sandbags and the filter fabric sheet; and, although coating with polyvinyl chloride extends the life of this material when exposed to direct sunlight, no coating was needed for this project because the dike is covered with a two foot layer of sand.

SUMMARY

While the sand bag dike is serving its purpose, it is only a matter of time until one of the long range alternatives will be required. The National Park Service is pursuing implementation of the beach nourishment alternative, completing environment documents and attempting to obtain necessary funding. The battle of Sandy Hook goes on!

Beachfront Restoration Using Water Quality Programs

Owen W. Bludau, AICP
Director of Environmental Studies
Moore, Gardner & Associates, Inc.
Memphis, Tennessee

INTRODUCTION

Can the source of a half-billion dollar annual tourist industry be considered the proverbial sow's ear? And if it is, how do you make it into a silk purse? The unusual source of concern is the beachfront of the resort city of Myrtle Beach, South Carolina. The desired silk purse, or goal, is to protect and to significantly enhance attractiveness of the beachfront. This may seem an unusual goal for a resort which is presently the destination for over five million visitors annually. But there is a cloud over paradise. This cloud has the potential to deliver a devastating blow to Myrtle Beach's tourist industry. This particular cloud involves water quality and its potential to close the beach. However, a rainbow may exist beyond the cloud. The rainbow is the opportunity to significantly improve the beachfront area while solving the water quality problems that exist.

A litany of errors compounded of very rapid tourist-oriented development in the 1960's, lack of perceptive stormwater control planning, and willful destruction of the duneline have united to create the present situation. The situation involved violations of South Carolina State Water Quality Standards after each rainfall event. Remedying the water quality situation was the issue.

PURPOSE

This paper will examine the direct linkage between urban stormwater runoff, water quality violations in the surf, human contact with highly polluted runoff on the beach, structural and "non-structural" control recommendations, and initial efforts to restore the beachfront while achieving stormwater control goals.

The findings and actions to date of one of EPA's original National Urban Runoff Program prototype studies are summarized. Municipal, regional, state and Federal agencies, consulting engineering and landscape architectural firms and private interests have become involved in methods to eliminate water quality violations and insure the future of the city's tourist industry.

The purpose of this paper is three fold: share study findings linking urban runoff and potential health hazards; describe the developing process of integrating water quality programs with broad urban planning concepts, and to solicit information from others with similar experiences.

BACKGROUND

Myrtle Beach is a linear city fronting on 9.3 miles of the Atlantic Ocean. The Waccamaw River/Intracoastal Waterway approximately parallels the beachfront 2.5 miles to the west. A low ridgeline lying three blocks from the beachfront creates a series of small basins draining eastward directly into the surf. West of the ridgeline, drainage flows into Withers Swash and other natural drainage estuaries. The city's stormwater drainage system follows the local topography. Therein lies the problem!

Stormwater drainage east of the ridgeline discharges directly onto the beach. Many motel roof drains, swimming pool backwash drains and outside shower drains do the same. There are 289 individual pipes discharging directly onto the beach within the city limits. The pipes range in size from 1-1/2 - 72 inches in diameter.

Stormwater discharge pipes and beach conditions at Myrtle Beach.

Stormwater discharge pipes incorporated into seawall construction.

Tourists wading through stormwater discharged across the beach.

Myrtle Beach was included in a regional 208 water quality management study completed by the Waccamaw Regional Planning and Development Council in 1978. Water quality sampling conducted as part of the study by Moore, Gardner & Associates, consulting engineers, verified a 1972 EPA finding of high bacteria counts in the surf at Myrtle Beach. Some limited stormwater runoff sampling was conducted during the 208 study. Runoff samples were also found to have high bacteria counts.

N.U.R.P. PROTOTYPE STUDY

High bacteria counts identified during the 208 process were of sufficient concern for EPA to provide funding for a determination of the sources and significance of the water quality violations. Seventy-five percent study funding came from 208 Continuing Planning Program funds and a National Urban Runoff Program grant. The city of Myrtle Beach provided the local share of 25 percent.

The Waccamaw Regional Planning and Development Council was designated as the lead agency for the study. EPA's Region IV office in Atlanta and the South Carolina Department of Health and Environmental Control were the governmental agencies concerned with the study results. Their concerns were solely related to water quality issues. The Waccamaw Regional Planning and Development Council was also concerned about the water quality issue from an economic standpoint. Tourism is the region's major industry; any water quality problem creating negative publicity would adversely affect that industry and the local economy.

The City of Myrtle Beach had the greatest stake in the study. If the water quality violations were minor, it would mean "business as usual." If major problems were identified, the favorable image of the beachfront was at stake. To preserve its future, the city would have no real choice but to take whatever actions and pay whatever costs would be necessary to remedy the problem.

The consulting engineering firm of Moore, Gardner & Associates, of Asheboro, North Carolina and Surfside Beach, South Carolina, was retained for the technical portions of the study. The municipal, regional, state and Federal agencies and the engineering consultants formed the Study Coordinating Committee which reviewed and directed study progress.

INITIAL STUDY OBJECTIVES

The work program of the study was designed to identify the magnitude of water quality violations in the surf, confirm if stormwater runoff was the primary cause of the violations, and to propose methods of eliminating any significant problems. The 9.3 mile beachfront comprised the study area.

Specific study tasks included: cataloguing each discharge pipe, identifying the land uses drained, conducting a sampling program from selected and random pipes, sampling the surf and swashes, and identifying sources of pollution. Tasks for evaluating and recommending control alternatives were also specified.

INITIAL SAMPLING RESULTS

The initial sampling results were cause for concern. After rainfall, surf samples had significantly higher coliform counts than allowed by South Carolina Water Quality Standards for Class SA waters. Fecal and total coliform counts from the beach discharge pipes were orders of magnitude higher than in the surf. Fecal coliform counts exceeding 70,000/100 milliliters were frequent, and counts exceeding 800,000/100 milliliters were found on several occasions.

The Coordinating Committee realized that the work program was not going to provide the level of detail needed to address questions raised by the initial findings. Therefore, the work program was redefined and expanded in early 1979. The study area was narrowed to a 50 block section containing the major motel, commercial and amusement activities of the city. This beachfront area was designated as Section 5. This section contained 160 discharge pipes and had the highest coliform counts.

The Coordinating Committee focused all sampling and control activities into Section 5, with the exception of some regional sampling for comparative purposes. The expanded sampling program was structured to sample: pipe discharges, discharge pool and stream flows across the beach, the surf line and 50 feet into the surf. Time delay sampling procedures were provided to develop "first flush" information and data on the duration of coliform counts. Sampling from the swashes and from city streets were included to help identify possible sources and types of stormwater contaminents. An investigation of the stormwater collection system with smoke, visual inspection and television surveys was to identify possible cross-connections with the sanitary sewer system.

FINAL SAMPLING RESULTS

Twenty-one conclusions were drawn from the results of 3000 analyses conducted on 1192 water quality samples. Some of the significant conclusions presented in the "Data Interpretation & Analysis" subtask report were:

- Wet weather surf data contained significantly higher coliform counts than dry weather surf data.
- Class SA water quality standards applicable to the surf were violated for each storm event during which sampling was taken.
- A significant coliform reduction in the surf was observed between 10 and 22 hours after a storm ended.
- Street samples collected during the peak tourist season revealed extremely high coliform counts. (The geometric means of the fecal coliform data collected at 10 different street locations during the peak season was 109,207/100 milliliters.)
- The direct beach discharges, natural beach pools, and pipe streams samples had significantly higher coliform counts than the surf samples.
- Reducing the amount of impervious surfaces and pavement draining into the storm sewer system should reduce the number of coliforms in the surf.
- The stormwater runoff from the collection system poses a greater health threat potential than does violations in the surf. Stormwater runoff coming out of the pipes far exceeds public health standards for health protection.
- There is body contact between storm sewer discharges and people. The greatest potential for health impacts is in storm sewage, regardless of whether the state and/or Federal standards are applied; i.e., pathogens may exist.
- Removal of urban stormwater runoff entering the surf through drain pipes and street runoff will result in water quality improvement and reduce contact exposure to the public.

DILEMMA

These sampling conclusions created a dilemma for some of the Coordinating Committee members. The State Department of Health and Environmental Control was in an especially vulnerable position. Neither state nor Federal water quality standards apply to stormwater runoff. Stormwater is specifically exempted. Therefore, regardless of the coliform counts present in the runoff, the state could not act from a water quality standpoint. To act from a public health concern, the state would need confirmed cases of illness resulting from stormwater runoff contact, and there are none.

EPA recognized that the coliform counts were extremely high on the streets and on the beaches. But EPA's focus was on surf water quality. Neither National Urban Runoff Program nor 208 Continuing Planning Program funds could be used to construct solutions to the problems identified. EPA could recognize that a problem existed, but could do nothing until Congress authorized construction funds for urban stormwater programs.

Waccamaw Regional Planning and Development Council considered the stormwater contact on the beach as a major concern. Tourists stroll through the beach streams. Small children frequently play in the stormwater pools below the discharge pipes because they are shallower and calmer than playing in the surf. The regional agency was especially concerned about the loophole in the law exempting stormwater from water quality standards. If standards did apply, coliform counts in the runoff across the beach would violate every standard existent. However, pathogens may exist, even if laws do not apply. The potential for an outbreak of disease is evident, but legal enforcement cannot be applied.

The City of Myrtle Beach was in a particularly difficult position. The scientific and medical professions are not in agreement about the relationship and validity of coliforms as indicator organisms for pathogenic disease. Therefore, the high counts found could or could not be significant. None of the agencies involved could categorically tell the city if the high counts indicated a real threat or not.

CONTROL ALTERNATIVES

Improving surf water quality is not the only problem that exists on the beachfront. Other beachfront problems are not related to water quality, but they are also caused by stormwater runoff. The City of Myrtle Beach was also concerned about beach erosion from sheet flow and pipe discharges. The numerous protruding, rusty, often broken discharge pipes created an unsightly appearance on the beach. The city was encouraged by Moore, Gardner & Associates to use the National Urban Runoff Program to help attack these additional problems which potentially affect the health and economic vitality of the city.

Both structural and "non-structural" site runoff control alternatives were examined. Neither could solve the existing problems alone. A combined program involving structural and site runoff controls was felt necessary to resolve present and future conditions.

The existing configuration of drainage, site development and paving precluded sole reliance upon surface runoff controls. Very high land costs, lack of available useable sites, and a high water table prevented diverting flows to beachfront area detention basins. Structural control methods were determined to be necessary to prevent contamination from reaching the surf.

Subsurface injection was investigated. Methods of in-line, discharge point and volume disinfection were examined. Many variations on diversion and treatment were developed. These included: diversion to Withers Swash, collection and ocean discharge from one or more diffusers, diversion to the Intracoastal Waterway, and diversion to remote detention basins.

As uneconomical or locally impractical alternatives were eliminated, one element became common among those remaining. An interceptor system to receive and divert the flow from the 160 beach discharge pipes in Section 5 was determined to be necessary. Depending upon the final treatment alternative selected, the interceptor could be one gravity flow system taking all beach discharges to Withers Swash or a system of smaller interceptor systems feeding to individual ocean diffusers.

Two locations were proposed for the interceptor - just behind the duneline or in Ocean Boulevard (one-half block to the west). The duneline location would be cheaper to construct, at $12 million, but the Ocean Boulevard location, at $15 million, would be less susceptible to future storm damage. The city preferred the Ocean Boulevard location. Its construction would allow the placement underground of existing overhead telephone and electric lines. The city has long desired to accomplish this goal.

BEACHFRONT RESTORATION OPPORTUNITIES

While Moore, Gardner & Associates was examining the implications of the duneline location, a whole concept for beachfront revitalization developed. Backfilling the interceptor trench would allow the opportunity to restore a duneline. Landscaping the duneline and providing dune walkways would provide visual benefits. The interceptor would allow the beach discharge pipes to be eliminated, thus improving aesthetics. Even a small restored and revegetated duneline would stop sheet runoff from reaching the beach and causing erosion.

When the city indicated a preference for the Ocean Boulevard location for the interceptor, the opportunity to link dune restoration with interceptor construction was lost. But the need for the duneline became even greater. The interceptor would divert flows from west of Ocean Boulevard. Runoff eastward toward the beach would still rinse an almost solid covering of asphalt parking lots, walks, dead-end streets, and concrete pool patios.

Individual site collection and pumping facilities to get beachfront site runoff up into the interceptor were evaluated. They were eliminated for economic and maintenance considerations. Site runoff controls were recommended for containing beachfront property runoff. A combined strategy was developed using an interceptor to collect and divert all runoff reaching Ocean Boulevard and using "non-structural" site runoff controls between Ocean Boulevard and the beach.

"Non-structural" site controls that were recommended for consideration in redeveloping impervious areas were:

- use of combined broom and vacuum street cleaning,
- disconnection of drainspouts to streets and stormwater sewers,
- porous surfaces for light traffic parking areas and on pedestrian and deck areas,
- parking area detention,
- use of vegetative strips for absorption and detention,
- rooftop retention and detention,
- subsurface tanks for temporary detention,
- landscape planning to maximize control options,
- infiltration wells, drains and ditches, and
- stream channel storage.

THREE ZONE CONCEPT

To graphically describe the beneficial impacts that the recommended combination of controls would have on beachfront properties, the award-winning landscape architectural firm of Roy Mann Associates, of Cambridge, Massachusetts, was retained. Under the direction of Mr. Roy Mann, a three zone concept was further developed and graphically presented.

The concept involved three zones of human activity, stormwater controls, and aesthetic sequence. Moving from Ocean Boulevard eastward, Zone III would be the motel and parking zone. Phased redevelopment of this zone would provide for reorganized parking, introduction of more porous surfaces, landscaping, drainage control, and public access facilities. This zone would be the most active in terms of vehicular movement and land use.

Zone II would be located at the east end of the parking zone. It would consist of a grassed swale zone to catch, detain and absorb runoff escaping from Zone III. The grassed swale would be used for activities such as sunbathing and children's play areas. Landscaping Zone II with salt resistant trees and shrubs would provide islands of shade and more vegetation for stormwater absorption and improved aesthetics.

Zone I would consist of an enhanced or restored duneline. Where dunes remained, they would be extended. The duneline would be restored and revegetated where possible. Dune access walks would be provided to help dune vegetation to stabilize and extend. Sand fencing would be used initially. The duneline would provide the final filter barrier to keep urban runoff from the beach. Excess runoff from very large storms would be directed to planned overflow points in the duneline. Zone I would be the least active of the three zones. In addition to controlling stormwater runoff and erosion, it would provide more storm wave protection from the ocean than now exists. It would add a more natural image to the beachfront than now exists.

COMPLICATIONS

The combined interceptor and three-zoned site runoff control concepts have been enthusiastically received by the city. Some major hurdles remain, however, before the concept becomes reality. The cost of implementation will be very high. No construction funds are included in either the National Urban Runoff Program or 208 Continuing Planning Program funds. With costs for the basic interceptor alone estimated at $15 million, the city must find additional financial support or new taxing opportunities. Since sampling results link increased pollution with tourist numbers, passing some of the corrective costs on to the tourists appears justified. State laws now preclude this.

Preventing additional runoff from being generated by new development is also necessary. The interceptor system was sized to contain the one-year storm flow from existing development. Increasing the volume of runoff entering the storm collection system will exceed interceptor capacity and cause system overflows on a more frequent basis than planned.

Complicated legal, financial, and social issues between the city and private property owners are involved in implementing the three zone concept. Some motel parking and pool facilities must be relocated to create enough area to develop a duneline and swale. Who pays for the removal? How is parking replaced?

Parking is the critical issue involved here. Motels need adequate spaces to meet local codes and visitor requirements. Land along the beachfront is very expensive and vacant land almost non-existent. The city must become involved with this issue if it is to be resolved. Roy Mann Associates has advocated that the city acquire space across Ocean Boulevard for parking structures, by condemnation if necessary, to replace spaces that will be required for developing the duneline. Spaces would be leased to the motels. Air rights over the structures could be leased for new tourist facility development and partial cost recovery.

Swimming pools could be relocated between motels for shared use. Other facilities could also be located near the swale zone that now occupy the original duneline location.

Careful landscape and parking lot planning will enable the Zone III area to percolate and detain large amounts of stormwater. The same landscaping and planning will enhance the aesthetics of the beachfront, improve beach access, and provide a more attractive urban image for residents and tourists alike.

The structural alternatives for treatment or diversion of the stormwater from the beachfront interceptor have been narrowed significantly. The city government has eliminated ocean diffuser alternatives. They do not want any obstruction on the ocean floor that will adversely affect beach dynamics. The State of South Carolina has disapproved discharging large volumes of stormwater, even treated stormwater, into the Intracoastal Waterway. The Corps of Engineers and the South Carolina Coastal Council disapprove any alteration of Withers Swash estuary to serve as a permanent or temporary detention basin. Diverting runoff to Withers Swash for gravity flow into the ocean appears the most acceptable alternative remaining. However, the water quality effects of concentrating this large volume of urban runoff at the mouth of Withers Swash remain unknown. This may be the necessary first step, with subsequent treatment alternatives based on the impacts this flow has on surf quality.

CURRENT STATUS

The Myrtle Beach Stormwater Study entered its implementation phase in May 1980. Moore, Gardner & Associates is preparing technical drawings for construction of site runoff control measures. Technical details of soil types, groundwater table levels, and infiltration rates are being developed for use during implementation of controls.

Roy Mann Associates has prepared a classification system of the beach itself, the beach edge and land uses immediately behind the edge in the Zone I and II areas. This classification will be the basis for a Unified Concept Plan to integrate beachfront runoff control concepts, downtown redevelopment plans, park and recreation plans, and other urban improvements. Roy Mann Associates will also develop a manual for public use in understanding and applying site stormwater control concepts.

Waccamaw Regional Planning and Development Council is coordinating a search among state and Federal agencies for implementation funding. Relating aspects of the Myrtle Beach stormwater control strategy with funding objectives of grant agencies is the thrust of this effort.

CONCLUSION

It is too early to see how the Myrtle Beach stormwater control strategy will succeed. The city has indicated every intention of proceeding where implementation is now possible. A stormwater control ordinance is being developed. Four public street ends will be redesigned and reconstructed by the city to demonstrated site control methods. Four public beach access walkways will also be redesigned and reconstructed. Stabilization and revegetation funded by the South Carolina Coastal Council is proceeding on existing dune remanents. Very attractive wooden decks and beach access walkways are being constructed by city crews over existing dunes. Stormwater controls are being incorporated into redesign of a public park, the design of a new Police and Justice complex, into expanded Convention Center parking and into other budgeted city construction projects.

The city expects to provide up to $350,000 in eligible in-kind matching projects during this program year.

Civic groups and private property owners are being drawn into the process. While large motel owners appear to support the concepts, small motel owners may feel more impact and respond less favorably. A public-private partnership is developing. This partnership recognizes what is at stake on the beachfront - the economic future of each property owner, the city and the region as a whole. Maximizing water quality programs to help achieve all the program objectives is the catalyst that has started the movement forward.

TIDAL INLET EFFECTS ON BARRIER ISLAND MANAGEMENT

Duncan M. FitzGerald[1]

Miles O. Hayes[2]

ABSTRACT

Effective management of barrier islands necessitates an understanding of the processes which cause short and long-term shoreline erosional-depositional changes. Equally important is determining the locations where these changes are most likely to occur. Fieldwork at mesotidal barrier islands along the coasts of New England, South Carolina, Northwestern Germany and the Gulf of Alaska has led to the very basic conclusion that barriers are most unstable at their ends and that the shoreline changes which occur at these locations are a result of tidal inlet processes.

The ebb-tidal deltas which front inlets are huge reservoirs of sand and may be comparable in volume to that of adjacent barriers. Slight changes in the size of an ebb-tidal delta can increase or decrease the supply of sediment to the downdrift beaches, resulting in their buildup or erosion. The landward transport of sand through inlets to the flood-tidal deltas is another process which can decrease a barrier's sediment supply.

Wave refraction around the ebb-tidal delta causes a longshore transport reversal along the downdrift barrier. At this location sand is transported back toward the inlet which often results in a progradation of the updrift end of the barrier island. This sediment trapping process can also be responsible for sand starved beaches further down the barrier. During storms the ebb-tidal delta acts as an offshore breaker and protects the adjacent shoreline from storm wave attack.

Inlet migration, which occurs in areas that have high longshore sediment transport rates relative to the tidal scour ability of the inlets, results in erosion of the

---

[1]Dept. of Geology, Boston Univ., Boston, Mass. 02215

[2]Geology Dept., Univ. of S.C., Columbia, S.C. 29208

downdrift barrier and the addition of curved beach ridges to the updrift barrier. The recurved spit that builds during this process is an unstable part of a barrier island due to breachings that frequently occur along its length.

The processes of inlet sediment bypassing control the rate at which sand is transported through a tidal inlet from the updrift barrier to the downdrift barrier. The sand that is added to the downdrift barrier normally occurs in packets in the form of large landward-migrating bar complexes. The location where these bar complexes weld to the beach, substantially building up that portion of the barrier, is a function of the size of the tidal inlet, the orientation of the inner inlet channel and the nature of the channel bank sediments.

Collectively, the tidal inlet processes can account for dramatic shoreline changes along the ends of the barriers and indirectly may also be responsible for erosional-depositional patterns along the middle of the barriers. The effects of these processes can be measured over a period of several years or in as little time as a few hours. Management decisions which do not consider tidal inlet influences may suffer costly consequences.

INTRODUCTION

The development and management of a barrier island requires a knowledge of its historical shoreline erosional-depositional trends and an understanding of the processes which have produced these changes. It is clearly evident from fieldwork done by the authors along the mesotidal coasts of New England, South Carolina, Alaska, and Germany that most of the changes to barrier islands occurs in the vicinity of tidal inlets. Using Kiawah Island, South Carolina as an example, it is seen in figure 1 that the shoreline along the northeast and southwest ends of the island has experienced much greater change during a 120 year period of time than the region along the middle of the island. Beach profile data collected over a year also indicates that the ends of Kiawah Island undergo greater short-term fluctuations (Fig. 2).

This trend is a product of tidal inlet processes. Tidal inlets and their associated ebb-tidal deltas strongly influence wave energy and refraction patterns along the shoreline adjacent to an inlet. They also affect the rates of inlet sediment bypassing and the locations where sand is added to the downdrift beaches. A more obvious influence is the shoreline erosion and progradation that accompanies migrating inlets.

The degree to which inlets affect the adjacent coastline is largely dependent on their size which in turn, is a function of their tidal prisms (O'Brien, 1969; Jarrett,

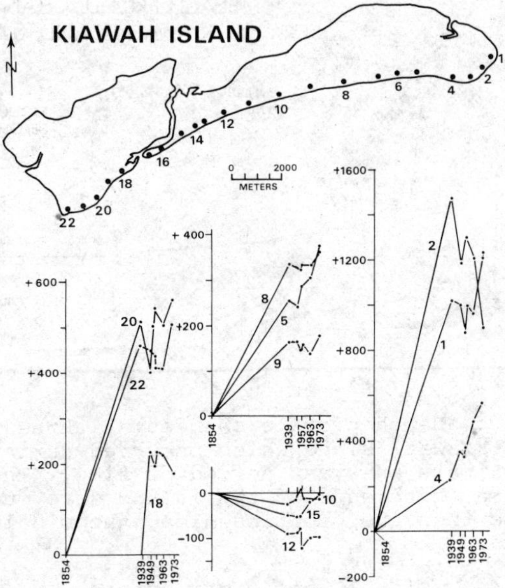

Figure 1. Historical erosional-depositional changes along Kiawah Island, South Carolina based on U.S. Coast and Geodetic Survey charts and vertical aerial photographs. The numbers on the graphs refer to locations along the island. Note that the largest scale changes occur at the ends of the barrier.

1976). An inlet's tidal prism is determined by its bay area and tidal range and less importantly on its filling characteristics. The close relationship between increasing tidal range or bay area along a coast and a corresponding increase in inlet size has been demonstrated along the South Carolina coast by Brown (1977) and along the Copper River Delta area in Alaska by Hayes (1979). Thus, the conclusion can be made that as tidal range or inlet bay area increases inlets will exert a greater influence on barrier island morphology.

It is the purpose of this paper to describe the tidal inlet processes which affect erosion and deposition along barrier island shorelines. This study focuses on mesotidal (1.0 m < tidal range < 3.5 m) barrier island coasts but many of the processes that will be discussed in this paper are applicable to microtidal coastlines as well.

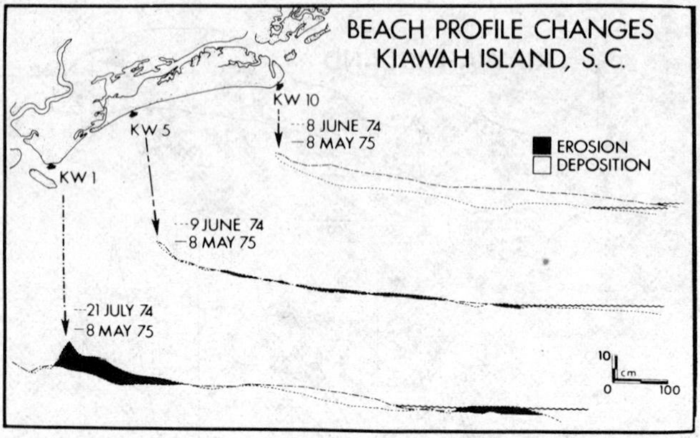

Figure 2. Beach profile changes at three locations along Kiawah Island which occurred during an approximate one year period of time. The profiles at both ends of the island were much less stable than the profiles along the middle of the island.

EBB-TIDAL DELTAS

Volume

Ebb-tidal deltas represent huge reservoirs of sand which can be comparable in volume to that of the adjacent barrier island. The sand volume of the two ebb-tidal deltas which border the Seabrook and Kiawah Island barrier complex in South Carolina (Fig. 3) has been computed by Hayes _et al_. (1978) using a method designed by Dean and Walton (1975). These calculations are given below:

Stono Island ebb-tidal delta	- 69×10^6 m^3
North Edisto Inlet ebb-tidal delta	- 128×10^6
Total	197×10^6
Seabrook-Kiawah barrier island complex	252×10^6
Difference	55×10^6

The results indicate that the two ebb-tidal deltas contain 44% of the sand of the combined inlet-barrier island system. Walton (1976) has shown that there exists a direct relationship between an inlet's tidal prism and volume of its ebb-tidal delta. Thus, as tidal range increases, and thereby inlet size and the percentage of inlet shoreline, the ratio

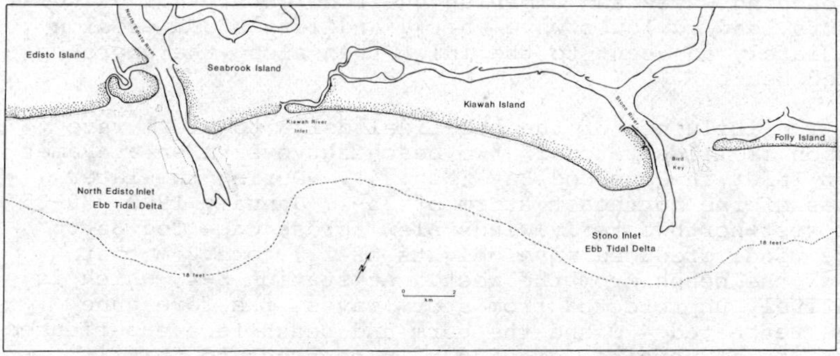

Figure 3. Ebb-tidal delta morphology of Stono and North Edisto Inlets. The sand comprised in these two deltas is comparable in volume to that of the Kiawah-Seabrook Island complex.

of the sand volume of the ebb-tidal delta to that of the barrier island complex should also increase.

Changes in the volume of the ebb-tidal delta will directly affect the nourishment of the downdrift barrier. If an ebb-tidal delta is increasing in size it means that more sand is being transported to the delta than is by-passing the inlet. This situation leads to sand starved downdrift beaches. The opposite is true when an ebb delta is decreasing in size. Historical inlet studies along the South Carolina coast by Finley (1977), FitzGerald (1976) and Hubbard (1977) have shown that ebb-tidal deltas of this region undergo periods of growth (4 to 6 years) followed by a shorter erosional period (1 to 2 years). This morphological cycle in South Carolina can be related to the landward migration of bar complexes and will be discussed in more detail in a later section of this paper. It should also be noted that some of the changes in volume of an ebb delta can be accounted for by deposition or erosion in the main inlet channel (Byrne _et al._, 1975; FitzGerald, 1977).

Breakwater Effect

The shallow nature of the ebb-tidal delta provides a partial breakwater for the shoreline behind the delta. This is especially true during lower tidal elevations when most of the wave energy is dissipated at the terminal lobe. During higher tidal stages intertidal and subtidal bars still cause waves to break some distance from the shore. Also, Hales and Herbich (1972) have demonstrated that ebb-tidal currents issuing out of the inlet interact with approaching waves and diminish their heights. All of these factors lead to lower wave energy and less storm erosion immediately adjacent to the inlet than along the unprotected beach.

The influence of the ebb-tidal delta to storm wave erosion is illustrated in two beach surveys taken at Pamet River Inlet in Cape Cod Bay (Fig. 4). During the later stages of the northeast storm of 25-27 January 1979, 20-25 knot west-northwesterly winds blew across Cape Cod Bay. These winds produced wave heights of 1.7 to 2.2 m that struck the beach from the west. At station S-4, which is relatively unprotected from storm waves, the fore dune ridge retreated 4 m and the berm and beach face experienced about 1 m of vertical erosion. In contrast to this is station S-3 where, although the beach was eroded, the dune ridge was only slightly damaged. The reason for this is that this section of the beach is fronted by several large intertidal bars which broke up much of the wave action during the storm. The degree to which regions onshore will be sheltered from storm wave attack is dependent on ebb-tidal delta morphology and the height and distribution of intertidal and subtidal bars.

Wave Refraction

Waves approaching an inlet at an oblique angle to the coast refract around the ebb-tidal delta and cause a longshore sediment transport reversal on the downdrift side of the inlet shoreline (Fig. 5). This idea was first introduced by Hayes et al. (1970) and is shown graphically in a wave refraction diagram constructed by Hubbard (1976) for the Merrimack River Inlet in figure 6. From the diagram it can be seen that the dominant northeast waves of this region bend around the ebb-tidal delta such that, the orthogonals just south of the inlet approach the shoreline from an east-southeasterly direction. A transport reversal has been documented along this section of beach by longshore current measurements (Fig. 6) and from erosional-accretionary patterns around nearby coastal structures (Hayes et al., 1970). Transport reversals have been measured at other tidal inlets including Price Inlet, South Carolina (FitzGerald, 1977); North Inlet, South Carolina (Finley, 1977; Humphries, 1977); and Norderney Seegat, Germany (Luck, 1978).

The change in the dominant direction of longshore transport near inlets has an important consequence on the morpholgy of barrier islands because it aids in trapping sand on the ebb-tidal deltas and it causes accretion along an inlet's downdrift shoreline. The latter process has been used by Hayes (1979) to explain the bulbous updrift end of barrier islands which give barriers a "drumstick" shape (Fig. 7). Drumstick barriers have been identified at many locations along the East Coast of the United States

A

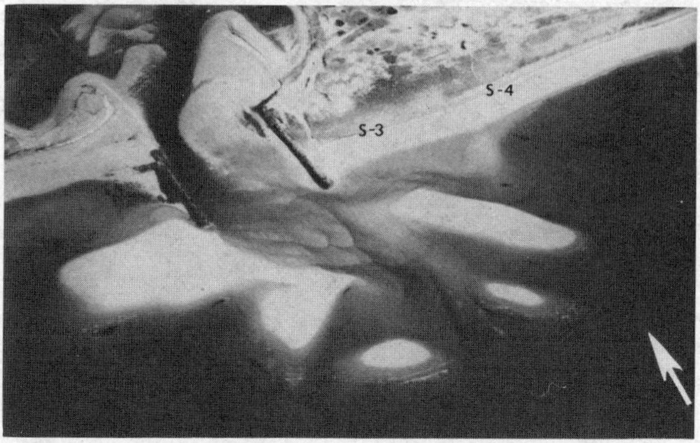

B

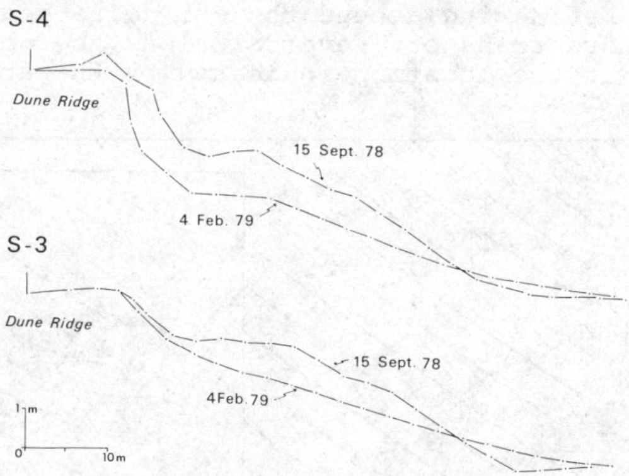

Figure 4. A. Oblique aerial photograph of Pamet River Inlet, Massachusetts. S-3 and S-4 give the locations of the beach profiles shown below. The arrow indicates the direction of wave approach during the January 1979 storm.

B. Beach profile changes attributed to the January 1979 storm. Note that more erosion occurred to the exposed beach (S-4) than the portion of beach which was protected by the nearshore intertidal bars (S-3).

Figure 5. Oblique aerial photograph of the ebb-tidal delta at Price Inlet, South Carolina. Waves refracting around the ebb delta cause a longshore transport reversal along the downdrift beach at the location indicated by the arrow.

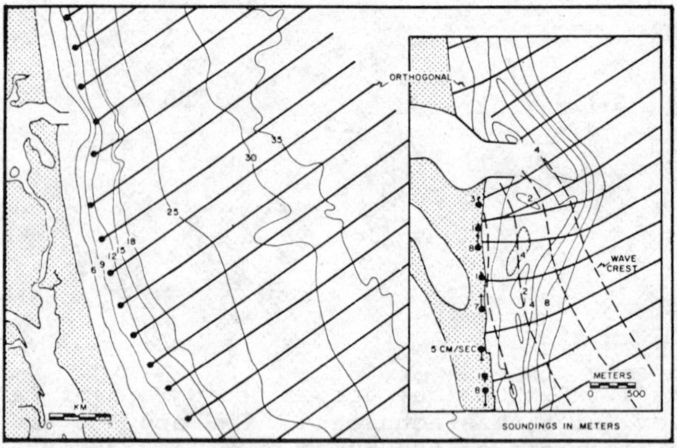

Figure 6. Wave refraction diagram for the Merrimack River Inlet region in Massachusetts (From Hubbard, 1976). The dominant wave approach for this area is from the northeast which results in a southerly longshore transport direction. Just south of the inlet the refraction of waves crest around the ebb-tidal delta cause a transport reversal which has been verified by longshore current measurements that are indicated on the diagram.

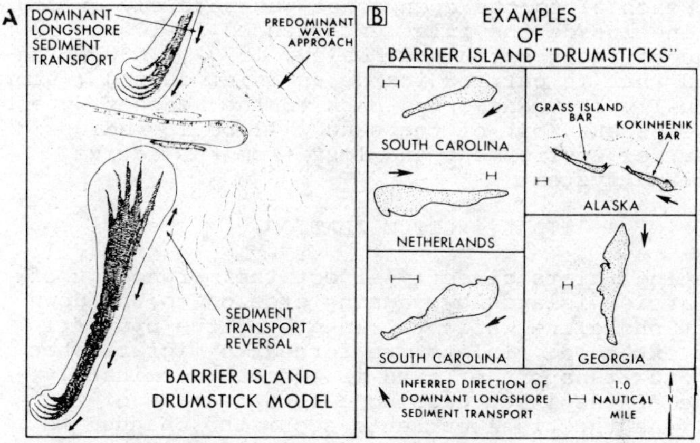

Figure 7. A. Barrier island drumstick model. Note the wave refraction processes.
B. Examples of barrier island drumstick from South Carolina, the Netherlands and Alaska (From Hayes, 1979).

and also along the coasts of Alaska, and the Netherlands.

TIDAL INLETS AS SEDIMENT SINKS

Tidal inlets not only trap sand temporarily on their ebb-tidal deltas but they also can be responsible for the more permanent loss of sand from the beach when sediment is transported through them and deposited on flood-tidal deltas, in marsh creeks, and on tidal flats. In a recent study of the transgressive Malpegue barrier system in the southern Gulf of St. Lawrence, Armon (1979) determined that 52% of the volume of sand that was transferred past the barriers from 1935 to 1968 occurred at tidal inlets and an additional 42% occurred along low areas of former inlet positions. Bartberger (1976) reported a similar trend for Assateague Island in Maryland and Virginia.

The landward transport of sand through inlets occurs as a continuous process at flood-dominated inlets, that is, inlets which have stronger flood-tidal velocities than ebb velocities, but it can also take place at ebb-dominated inlets during storms. The surge that accompanies large storms produces increased flood-tidal velocities and greater wave action at the inlet mouth which results in sediment transport into the backbarrier environment. Beach profile data collected before and after the passage of the 1979 Hurricane David along Seabrook Island, inside North Edisto Inlet and on the open coast (Fig. 3), showed that while the beach along the open coast was severely eroded the shoreline inside the inlet experienced 50-100 cm of vertical accretion (Sexton and Moslow, 1980). A portion of the sand that is carried inside an inlet during a storm will eventually be transported back to the beaches by ebb-tidal currents but most of the sand will be locked up in the backbarrier environment and lost from the active barrier island system.

TIDAL INLET MIGRATION

Migrating inlets directly affect the morphology of adjacent barrier islands by causing erosion to the downdrift inlet shoreline while the length of the updrift barrier is extended. Inlets are forced to migrate when the longshore transport of sand is added predominantly to one side of the inlet causing a constriction of the flow area. As the tidal currents scour the channel to remove this sand the downdrift side of the inlet is eroded preferentially and the inlet migrates in that direction. The rate of inlet migration is dependent on sediment supply, tidal current scour power, wave energy and channel bank composition.

Deeper inlets are less likely to migrate because they are often scoured into semi-consolidated sediments and therefore, lateral erosion is impeded. Conversely, shallower inlets, whose channels have eroded no deeper than the extent of the barrier island sands, frequently migrate. FitzGerald et al. (1978) have reported that inlets along the South Carolina coast with depths greater than 8 m have been stable during the past 100 years whereas, inlets which are shallower than 3-4 m have had a history of migration.

The spit which builds as the inlet migrates does so through the addition of curved beach ridges that weld completely or incompletely to the inlet's updrift shoreline. At Kiawah River Inlet incomplete ridge welding has resulted in a series of curved beach ridges that are separated from one another by marshy swales (Fig. 8). These low regions will eventually be filled by washovers, aeolian deposition and fine grained sediments which are deposited by tidal waters.

The history of Kiawah River Inlet, depicted in figure 9, illustrates another process which is common to migrating inlet shorelines and that is spit breaching. The relatively narrow peninsula of sand, usually less than 1 km wide, that makes up a spit is highly vulnerable to erosion, washovers and breaching during intense storm activity. This is particularly true of spits where the curved beach ridges incompletely welded to the adjacent shoreline. Marsh creeks which occupy these inter-ridge areas cut across the spit and effectively decrease its width. During the past 300 years the spit along the southwestern end of Kiawah Island

Figure 8. Mirror image of an aerial photograph of Kiawah Island spit. Note the curved beach ridges and the tidal creeks which occupy the inter-ridge areas.

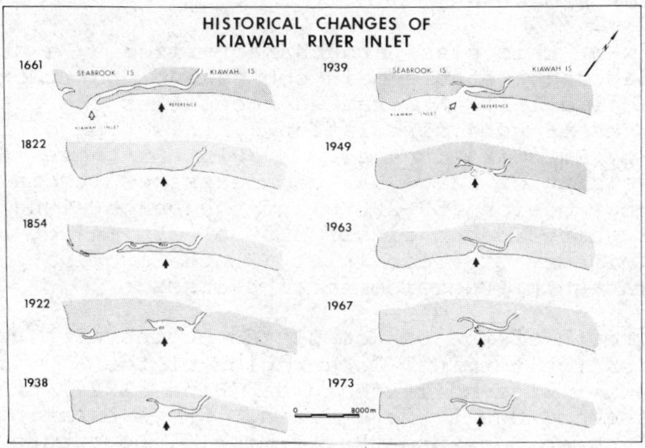

Figure 9. History of Kiawah River Inlet determined from old maps, U.S. Coast and Geodetic Surveys and vertical aerial photographs. The shallow nature of this inlet permits rapid migration.

has experienced at least three periods of breaching, occurring prior to 1822, 1922 and 1949.

INLET SEDIMENT BYPASSING

Introduction

Tidal inlet processes control the volume of sand that is transported through an inlet system and the location where the sand is added to the downdrift island. Bruun and Gerritsen (1959) first identified that sediment bypasses an inlet by wave action along outer portions of the ebb-tidal delta, by tidal currents in the channels, or by a combination of these two processes. In our study we have concluded that at most inlets wave induced sediment transport along the periphery of the ebb delta is an active process, especially at lower tidal elevations, which continually nourishes the downdrift barrier island with sand. Also, sand is intermittently added to the downdrift barrier in packets in the form of large (200-1,000 m long) landward migrating bar complexes. The rate at which this process takes place and the location where these bars weld to the beach is controlled by inlet size, orientation of the inner inlet channel and stability of the main ebb channel through the ebb-tidal delta. Bar welding events can drastically alter barrier island morphology.

Stable Inlet Processes

Inlets of this class are characterized by a stable inlet throat and a stable main ebb channel position through the ebb-tidal delta. A detailed account of the patterns and processes of sand circulation at this type of inlet is discussed by FitzGerald *et al*. (1976). At these inlets the periodic welding of large bar complexes results in a tremendous progradation (50-200 m) of the beach along a 1 km or more of shoreline on either side of the inlet. The shoreline history of Price Inlet, South Carolina in figure 11 illustrates this developmental process.

The growth of the bar complexes begins with swash bar formation on the terminal lobe region of the ebb delta. Swash bars are wave-built features (Hine, 1975) that migrate landward due to wave swash. Waves breaking over individual swash bars and the terminal lobe create bores of water which enhance the flood-tidal currents but retard the ebb currents. This causes a net landward transport of sediment on both sides of the main ebb channel and results in a landward migration of intertidal and subtidal bars. As these bars move onshore they merge together to form one or more large bar complexes. On the updrift side of the inlet a spit commonly builds out from the shore and attaches to the bar. This process is shown in several stages of formation in a photograph of Price Inlet, South

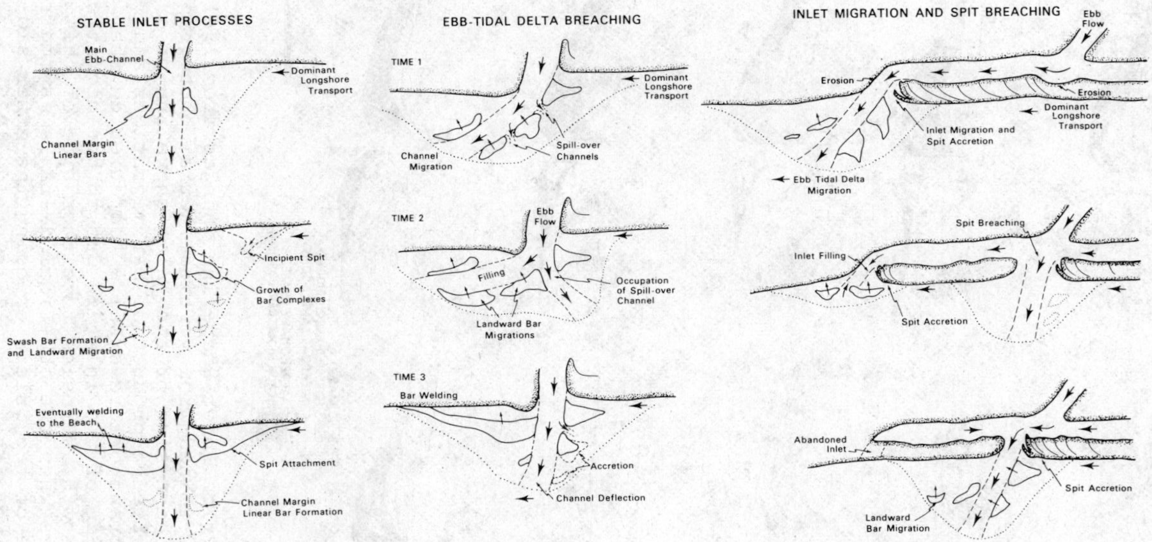

Figure 10. Models of inlet sediment bypassing for inlets along the South Carolina coast. One or more of these processes can occur at the same time at an inlet. These models are also applicable to other mesotidal, moderately sized inlets (width .8 km).

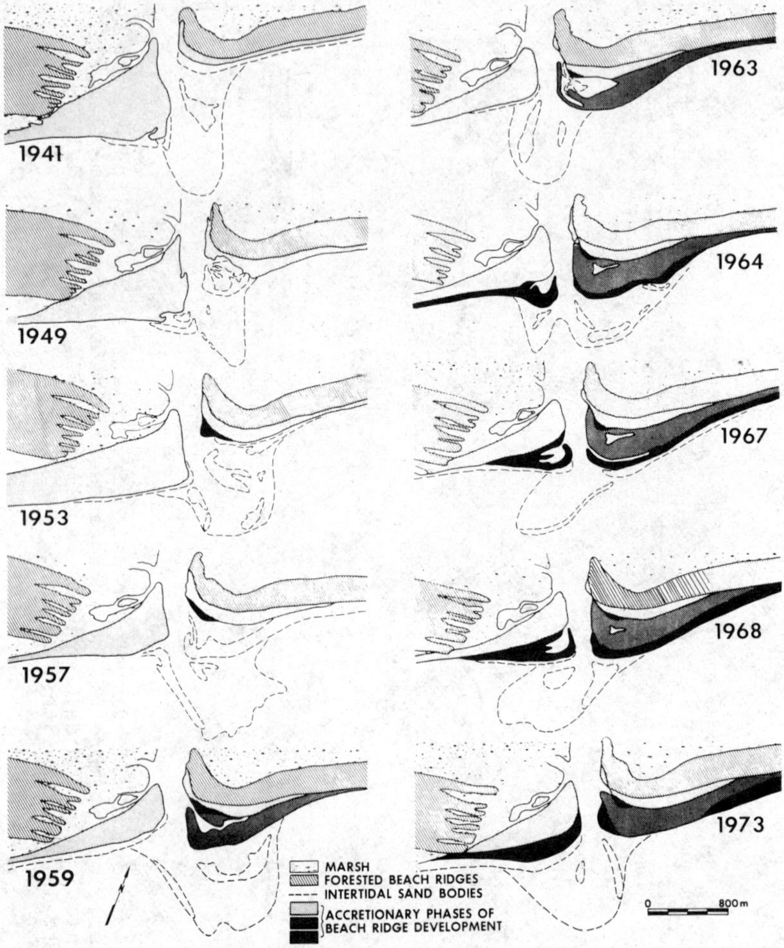

Figure 11. Price Inlet shoreline changes from 1941 to 1973 determined from vertical aerial photographs. On the southern side of the inlet shoreline progradation has been a product of bar migrations and the longshore transport reversal. Accretion on the opposite side of the inlet is due to bar migrations and spit growth. Erosion to either side of the inlet occurs when the ebb-tidal delta is asymmetric to one side of the inlet, exposing the other side to storm wave attack (From FitzGerald, 1976).

Carolina in figure 12. On the downdrift side of the inlet, once the bars weld to the shore, the sediment supply for the beaches further away from the inlet is substantially increased. Prior to the bar welding events, when the ebb-tidal delta protrudes further seaward, sediment trapping on the ebb delta is more complete due to greater refraction of incoming waves.

Figure 12. Oblique aerial photograph of Price Inlet, South Carolina. Note the landward migrating bar complexes on the northern side of the inlet and the spits which have built out from the shoreline and attached to these bars.

Ebb-Tidal Delta Breaching

The ebb-tidal delta breaching process, which is depicted in the second model in figure 10, occurs at inlets that have a stable throat position but whose main ebb channels migrate through their ebb-tidal deltas like the wag of a dog's tail. In this model the dominant longshore transport direction causes a preferential deposition of sand on the updrift side of the ebb-tidal delta which results in a deflection of the main ebb channel. The deflection process continues in some cases until the channel parallels the downdrift barrier island shoreline. Because this channel configuration is hydraulically very inefficient, the inlet will gradually divert its flow from this channel to a more direct seaward course through a spillover channel. Examples of spillover channels are shown in a photograph of Capers Inlet, South Carolina in figure 13. The diversion process can also occur more rapidly during storm conditions when tidal flow increases.

Figure 13. Oblique aerial photograph of Capers Island, South Carolina. Spillover channels, indicated by the arrows, provide the sites necessary for ebb-tidal delta breaching.

After ebb-tidal delta breaching is completed the old channel fills with sediment and the sand shoals which had previously bordered this channel migrate onshore and weld to the downdrift shoreline. Various stages of ebb-tidal delta breaching and subsequent landward bar migrational processes are shown in figures 14-16. As figure 16 illustrates, bar welding is often times incomplete causing the formation of lenticular salt water ponds. These low regions are eventually filled by washovers and aeolian sedimentation. Another accretionary process which commonly occurs as the bar complexes migrate close to shore and that is the development of tombolos. The intertidal and sometimes supratidal bar complexes act as effective offshore breakwaters which cause deposition and shoreline progradation behind the bars as they migrate onshore. Two photographs of the northern end of Kiawah Island depict this process in its early and late stages of formation (Fig. 17). During the 15 month period between the two photographs the cuspate-shaped bar migrated landward approximately 300 m on the flanks of the Stono Inlet ebb-tidal delta (Fig. 3).

Inlet Migration and Spit Breaching

The effects of inlet migration and spit breaching have already been described but it should be emphasized that spit breaching and subsequent filling and abandonment of the old inlet (Fig. 10) is an effective mechanism for bypassing large quantities of sand. This process undoubtably increases the supply of sand to the downdrift beaches which affects their erosional-depositional trends. In figure 18 a number of spit breachings and inlet abandonments can be identified at the southern opening of Santee River Inlet, South Carolina.

Figure 14. Oblique aerial photograph of Kiawah River Inlet shown in figure 3. Ebb-tidal breaching had recently occurred before this photograph was taken.

Figure 15. Vertical aerial photograph of Breach Inlet, South Carolina. At this time the flow in the main ebb channel was mostly confined to a direct seaward path through a recently breached spillover channel. The old main ebb channel was being filled by landward migrating bars and bedload sediment transport.

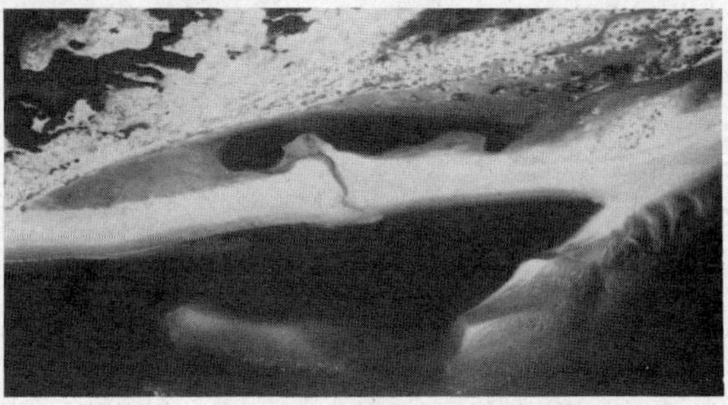

Figure 16. Oblique aerial photograph of the shoreline along Dewees Island near Capers Inlet, South Carolina. Incomplete bar welding has resulted in the formation of a lenticular salt water pond.

OCT 1978

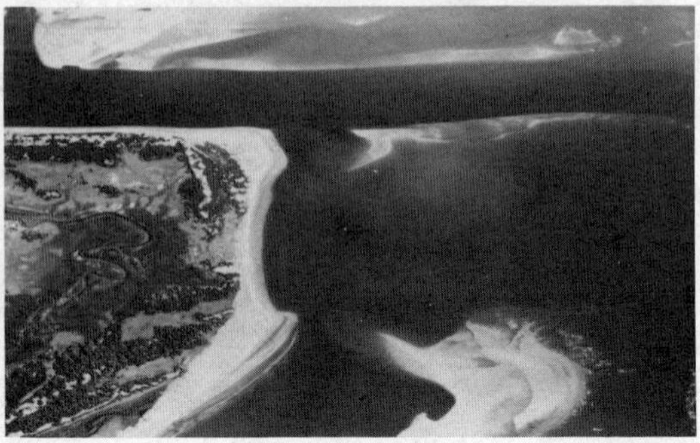

DEC 1979

Figure 17. Two oblique aerial photographs looking northeast along the northern end of Kiawah Island taken approximately 15 months apart. Note that as the cuspate bar moves onshore a tombolo-like feature builds out from the shoreline.

Location of Bar Welding

The location along a barrier island where bars migrate landward from the ebb-tidal delta and weld to the adjacent beach is dependent on the size of the inlet and the orientation of the main inlet channel. For example, the South Carolina inlets which have been discussed thus far, have moderate widths (< .8 km) and relatively straight main channels. Their ebb-tidal deltas overlap the adjacent shoreline 1-3 km on both sides of the inlet. Consequently most of the morphological changes that are directly attributed to bar migrational events occur along a 1-2 km length of shoreline downdrift and in some instances, updrift of the inlet. However, a much greater proportion of shoreline is affected by bar welding processes along the barrier

Figure 18. Oblique aerial photograph of the southern opening of Santee River Inlet depicting the spit breaching process.

island coasts of Georgia, Alaska and West Germany due to their much larger inlet widths (> 2 km) and ebb-tidal delta sizes.

At Norderney Seegat along the Friesian Island coast of West Germany the inlet is approximately 4 km wide and its ebb-tidal delta borders over one-third of the Norderney Barrier Island coastline. The sand circulation model (Fig. 19) which was developed for this inlet by Nummedal and Penland (1980) indicates that most of the bars, which migrate landward from the ebb delta, attach to the beach 3 to 5 km east of the inlet. This pattern of sedimentation is a product of: 1) the northeasterly inlet channel orientation; 2) the asymmetrical ebb-tidal delta morphology; and 3) the dominant easterly longshore sediment transport direction of this region. Luck (1978) has shown that because these bars weld to the beach at such a considerable distance from the inlet, the shoreline in the vicinity of the inlet is sediment starved (Fig. 20). The lack of a sufficient sand supply to this region has led to erosion which has necessitated the building of numerous coastal structures including groins and seawalls (Fig. 21). Conversely, the middle of the island where the bars migrate onshore is highly progradational (Fig. 22). Note that the erosional-depositional patterns along this barrier have produced a barrier island that is quite different in morphology from the drumstick barriers shown in figure 7.

The influence of the orientation of the inlet channel in controlling bar welding processes is further demonstrated at Parker River Inlet in Massachusetts (Fig. 23). The position of the inlet is anchored by two drumlins; one located on the updrift Plum Island inlet shoreline and the other on the downdrift Castle Neck shoreline. The ebb currents that issue out of the inlet are directed toward the beach along

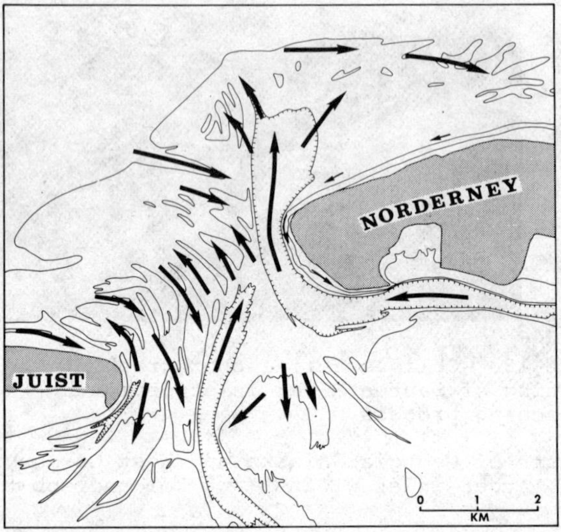

Figure 19. Sand circulation model for Norderney Seegat, West Germany (From Nummedal and Penland, 1980).

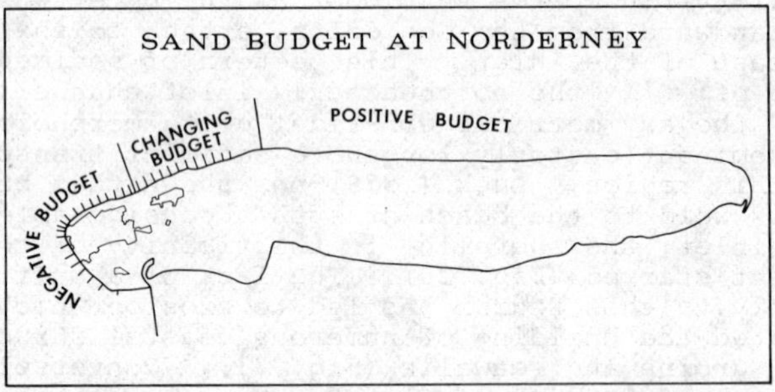

Figure 20. Sand budget for Norderney Island on the West Germany Friesian Island coast (From Luck, 1978). The positive sand budget for the middle of the island versus the negative budget at the west end of the island is a function of the location of landward migrating bars welding to the beach.

Figure 21. Oblique aerial photograph of the sediment starved west end of Norderney Island.

Figure 22. Oblique aerial photograph of the landward migrating bars nourishing the beach along the middle of Norderney Island.

Castle Neck. Ebb-tidal delta breaching at this inlet only takes place at the very southern end of the delta (Fig. 23, Location A). Consequently, bar migrations from the ebb-tidal delta nourish the middle of the island while the northern end of the island which receives little sand is highly erosional.

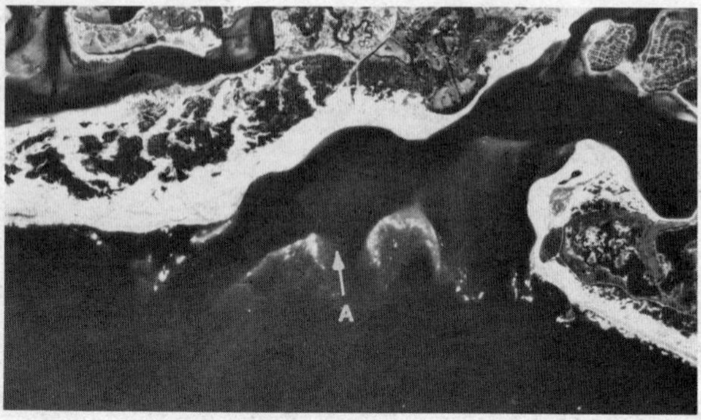

Figure 23. Vertical aerial photograph of Parker River Inlet. Note the orientation of the inner inlet channel. Ebb-tidal currents emptying out of the inlet are directed toward the downdrift Castle Neck shoreline. Location A defines the region where ebb-tidal delta breaching occurs. The bulbous middle portion of Castle Neck is due to landward bar migrations.

CONCLUSIONS

1. Ends of barrier islands experience the greatest short and long-term shoreline fluctuations and these morphological changes are caused by tidal inlet processes.

2. As tidal range or bay area increase along mesotidal ($1 \text{ m} < TR < 3.5 \text{ m}$) coasts inlet size correspondingly increases which results in a greater proportion of barrier island shoreline that is affected by tidal inlet processes.

3. The major inlet processes which are responsible for changes in barrier island morphology include: 1) the loss of sand through inlet systems to the backbarrier environment; 2) the temporary trapping of sand on the ebb-tidal delta due to a transport reversal which is set up by wave refraction around the ebb delta; 3) the breakwater effect that the inter-tidal bar complexes on the ebb-tidal delta afford the landward beaches; and 4) inlet sediment bypassing mechanisms.

4. The type of inlet sediment bypassing process and the rate and location of sand addition to the downdrift barrier island shoreline is controlled by: 1) inlet size; 2) orientation of the main inlet channel; 3) stability of the main ebb channel through the ebb-tidal delta; and 4) depth of the inlet throat and composition of the channel banks.

ACKNOWLEDGEMENTS

This investigation was supported by the U.S. Army Research Office Grant No. DAAG 29-76-6-0111 (Hayes and Nummedal, Principal Investigators, Geology Dept., Univ. of S.C.), the Office of Naval Research (Study of the German Bight, Nummedal, Principal Investigator, Geology Dept., Louisiana State Univ.; Studies of New England and Alaska, Hayes, Principal Investigator, Geology Dept, Univ. of S.C.) and a grant from the Boston University Graduate School. This manuscript was typed by Michelle Pellerin.

REFERENCES

Armon, J. W., 1979, Landward sediment transfers in a transgressive barrier island system, Canada, In: Barrier Islands (Leatherman, ed.), Academic Press, New York, p. 65-80.

Bartberger, C. E., 1976, Sediment sources and sedimentation rates, Chincoteague Bay, Maryland and Virginia, J. of Sed. Petrology, vol. 78, p. 326-336.

Brown, P. J., 1977, Variations in South Carolina coastal morphology, Southeastern Geology, vol. 18, p. 259-264.

Bruun, P., and Gerritsen, F., 1959, Natural by-passing of sand at coastal inlets, J. of the Waterways and Harbors Div., ASCE, vol. 85, p. 75-107.

Byrne, R. J., Bullock, P., and Tyler, D. G., 1975, Response characteristics of a tidal inlet: a case study, In: Estuarine Research, vol. 2 (Cronin, ed.), Academic Press, New York, p. 201-216.

Dean, R. G., and Walton, T. L., 1975, Sediment transport processes in the vicinity of tidal inlets with special reference to sand trapping, In: Estuarine Research, vol. 2 (Cronin, ed.), Academic Press, New York, p. 129-150.

Finley, R. J., 1976, Hydraulics and dynamics of North Inlet, South Carolina, 1974-75, General investigation of tidal inlets, Rpt. 10, U.S. Army Corps of Eng., Coastal Eng. Res. Center, Ft. Belvoir, Va., 126 p.

FitzGerald, D. M., 1976, Ebb-tidal delta of Price Inlet, South Carolina: geomorphology, physical processes and associated inlet shoreline changes, In: Terrigenous Clastic Depositional Environments (Hayes and Kana, eds.), Tech. Rpt. No. II-CRD, Geology Dept., Univ. of South Carolina, Columbia, S.C., p. 158-171.

FitzGerald, D. M., 1977, Hydraulics, morphology and sediment transport at Price Inlet, South Carolina, Ph. D. dissertation, Geol. Dept., Univ. of South Carolina, 84 p.

FitzGerald, D. M., Hubbard, D. K., and Nummedal, D., 1978, Shoreline changes associated with tidal inlets along the South Carolina coast, Coastal Zone '78, Sixth Symposium of the Waterway, Port, Coastal and Ocean Div. of the ASCE, San Francisco, Cal., p. 1973-1994.

FitzGerald, D. M., and Levin, D., 1980, Hydraulics, morphology and sediment transport patterns at Pamet River Inlet: Truro, Massachusetts, J. of Northeastern Geology, (in press).

FitzGerald, D. M., Nummedal, D., and Kana, T., 1976, Sand circulation patterns at Price Inlet, South Carolina, Amer. Soc. of Civil Eng., Proc. 15th Coastal Eng. Conf., p. 1868-1880.

Hales, L. Z., and Herbich, J. B., 1972, Tidal inlet current-ocean wave interaction, Amer. Soc. of Civil Eng., Proc. 13th Coastal Eng. Conf., p. 669-687.

Hayes, M. O., 1979, Barrier island morphology as a function of tidal and wave regime, In: Barrier Islands (Leatherman, ed.), Academic Press, New York, p. 1-28.

Hayes, M. O., FitzGerald, D. M., and Hubbard, D. K., 1978, Tidal inlet control of beach erosion-deposition cycles, Amer. Shore and Beach Preservation Assoc., Proc. 5th Annual Meeting, Tallahassee, Florida, p. 121-140.

Hayes, M. O., Goldsmith, V., and Hobbs, D. H., 1970, Offset coastal inlets, Amer. Soc. of Civil Eng., Proc. 12th Coastal Eng. Conf., p. 1187-2000.

Hine, A. C., 1975, Bedform distribution and migration patterns on tidal deltas in the Chatham Harbor Estuary, Cape Cod, Mass., In: Estuarine Research (Cronin, ed.), vol. 2, Academic Press, New York, p. 235-252.

Hubbard, D. K., 1976, Changes in inlet offset due to stabilization, Amer. Soc. of Civil Eng., Proc. 15th Coastal Eng. Conf., p. 1812-1823.

Hubbard, D. K., 1977, Variations in tidal inlet processes and morphology in the Georgia Embayment, Tech. Rpt. 14-CRD, Coastal Research Div., Geology Dept., Univ. of South Carolina, 79 p.

Humphries, S. M., 1977, Morphological equilibrium of a natural tidal inlet, Coastal Zone '78, Proc. 5th Symposium of the Waterways, Port, Coastal and Ocean Div., ASCE, Charleston, S.C., p. 734-753.

Jarrett, J. T., 1976, Tidal prism-inlet relationships, General Investigation of Tidal Inlets, Rpt. 3, U.S. Army Corps of Eng., Coastal Eng. Res. Center, Ft. Belvoir, Va., 76 p.

Luck, G., 1978, Islands in front of the southern North Sea Coast, In: Die Küste, Archive for Research and Technology on the North Sea and Baltic Coast, p. 84-93.

Nummedal, D., and Penland, S., 1980, Sediment dispersal in Norderney Seegat, West Germany, Proc. of National Meeting on Holocene Marine Sedimentation in the North Sea Basin, International Assoc. of Sedimentologists Spec. Pub. 5, Texel, Netherlands, (in press).

O'Brien, M. P., 1969, Equilibrium flow areas of inlets on sandy coasts, J. of Waterways and Harbors Div., ASCE, vol. 95, p. 43-52.

Sexton, W. J., and Moslow, T. F., 1980, Effects of Hurricane David, 1979, on the beaches of Seabrook Island, South Carolina, J. of Northeastern Geology, (in press).

Walton, T. L., and Adams, W. D., 1976, Capacity of inlet outer bars to store sand, Amer. Soc. of Civil Eng., Proc. 15th Coastal Conf., p. 1919-1937.

MAJOR MAPPING ACTIVITIES IN THE COASTAL ZONE

Peter F. Bermel
Assistant Division Chief for Plans and Operations
National Mapping Division
U.S. Geological Survey
Reston, Virginia 22092

ABSTRACT

The National Mapping Program is the responsibility of the U.S. Geological Survey. A major objective is to make available a variety of multi-purpose base maps and related cartographic products to the Nation's users. The planners and managers of coastal zone activities are major users of the Survey's map products and to meet their needs, planning emphasis is placed on the development of mapping programs in the coastal zone. Although 7.5-minute topographic maps and orthophotoquads are very important basic tools for planning, economic development, and land management, other mapping programs such as topographic/bathymetric mapping, land use/land cover mapping, digital mapping, and the new High-Altitude Aerial Photography Program also play an important part in coastal zone management activities. Products and data from these programs provide the most current source of accurate information on coastal resources.

NATIONAL MAPPING PROGRAM

In support of the U.S. Geological Survey's mission to provide information about the Earth and its physical resources, the National Mapping Division provides geographic and cartographic information, maps, and technical assistance and conducts related research responsive to national needs. The National Mapping Program was created to satisfy this responsibiliy, which has made the Geological Survey the Nation's largest civilian mapping organization. The goal of the Program is to provide the base map coverage and geographic data for the country, its commonwealths, and trust territories. Products include line maps at a variety of scales, topographic maps, image products, thematic maps such as land use/land cover maps, and a digital geographic/cartographic data base for multi-purpose needs.

The largest scale map series in the program is the 7.5-minute quadrangle. Products in this series include topographic line maps, orthophotoquads, and orthophotomaps. These products are published at 1:24,000 scale (1 inch = 2,000 feet) in States preferring the U.S. Customary System (USCS) units of measurement and 1:25,000 scale (1 centimeter = 250 meters) in States preferring the SI or metric units of measurement. In Alaska, the largest scale mapping is the 15-minute series published either at the 1:63,360 scale (1 inch = 1 mile) in USCS units or 1:50,000 scale (1 centimeter = 500 meters) in SI units.

These products constitute the building blocks of the National Mapping Program and each is prepared from original surveys; that is, aerial photography especially designed for each map, field-established horizontal and vertical geodetic control, photogrammetric operations, field completion, and inspection prior to publication. In many cases these sequential phases of map production produce intermediate products that are useful in their own right and which are available from the Geological Survey's National Cartographic Information Center. Please refer to Al Stevens' paper entitled NCIC: A Source for Coastal Cartographic Information, that discusses NCIC's user services with particular attention to cartographic products and related data available in the coastal zone.

The National Mapping Program also includes a wide variety of intermediate- and small-scale maps which are prepared primarily from these large-scale maps and supplemented and updated as required from other source materials and aerial photography. These products are extensively used by other Federal agencies as bases for thematic information, and by State and local organizations.

7.5-MINUTE QUADRANGLE MAPPING AND REVISION

In the coastal zone area of the conterminous United States, including the Great Lakes, about 88 percent of the 7.5-minute coverage is published and an additional 10 percent is in work. For the remaining 2 percent, published map coverage is available at 1:62,500 scale. Complete 7.5-minute coverage is expected to be available for the entire conterminous coastal zone by the late 1980's.

Maps must be kept up to date to maintain their usefulness. Using photography from the national High-Altitude Photography Program, plus other imagery, the Geological Survey has a current goal of photoinspecting about 4,000 7.5-minute quadrangles annually to determine their need for revision. Maps in urban areas and the coastal zone are photoinspected about every five years. A national average of about 40 percent of maps inspected, or 1,600 7.5-minute quadrangles, annually require revision, of which about 95 percent are updated by photorevision methods. In photorevision, the maps are revised by office procedures from the aerial photography without a field check. The revised features are printed in a distinctive purple color over the standard five-color map.

Some early 7.5-minute maps of the coastal zone were prepared before modern photogrammetric techniques were available. These maps will be replaced as soon as funds and capacity permit. The new maps will be prepared in metric units; that is, if the terrain calls for a 5-foot contour interval, the new maps will have a 1.5-meter contour interval.

Some experimental work has been done in Florida using a two-meter contour in relatively flat areas, supplemented with intermediate one-meter form lines. The one-meter contour is surveyed in and meets National Map Accuracy Standards, whereas subsequent odd-numbered contours are photogrammetrically derived. It is too early to know whether coastal zone map users will find this a satisfactory solution to the problem of providing one-meter topographic information at a cost commensurate with the resources available to map the coastal zone.

In addition to line maps available in the coastal zone, a variety of image products have been prepared as follows:

IMAGE PRODUCTS

Orthophotoquads are monocolor orthophotographs presented in a 7.5-minute quadrangle format with virtually no cartographic enhancement other than a few place names. Orthophotoquads can be prepared under cost-sharing agreements with Federal agencies or joint funding agreements with States or regional governments and organizations. As a result of a Federal program to provide a 7.5-minute map product for all areas of the conterminous United States, orthophotoquads now are available for a large percentage of the coastal zone which formerly was covered by either 15-minute maps, or not mapped at all.

Orthophotomaps are full color topographic maps, published over an image base and prepared for areas of the country where a standard line map cannot adequately portray the terrain, such as deserts, swamplands, and complex areas in the coastal zone. Orthophotomaps are available for the Florida Keys and Everglades, the Georgia coast, and portions of the Louisiana, Maryland, and South Carolina coastal zones. Orthophotomaps are currently in work for parts of the gulf coast of Louisiana and the Florida Panhandle, and the Chesapeake Bay area of Maryland.

Color Orthophotoquads. Based on favorable response to simulated natural color orthophotoquads prepared for the U.S.-Canada international border, the Survey has begun two pilot projects for similar products in the coastal zone. These simulated natural color 7.5-minute orthophotoquads at 1:24,000 scale will be prepared from high-altitude color-infrared (CIR) aerial photography at 1:80,000 scale. The CIR film will be color separated, processed photogrammetrically including differential rectification, and the three separates recombined with assigned colors. The products will be lithographed using a process color technique.

Coverage will be available next year for a 24 7.5-minute quadrangle area in North Carolina and a six 7.5-minute quadrangle area in Delaware. We are going to be interested in the user response to these quadrangles in the coastal zone. It may be that the color orthophoto base can be combined with standard cartographic data to provide a product at least as useful as the orthophotomap, for a reduced cost and with an added ability to revise periodically showing altered shorelines and other changing features.

TOPOGRAPHIC/BATHYMETRIC MAPPING IN THE COASTAL ZONE

In the early 1970's officials of the National Ocean Survey (NOS) and the U.S. Geological Survey (USGS) recognized the need for a new product in the coastal zone and outer continental shelf. USGS/NOS experiments resulted in an agreement to produce a single cartographic product, which formed the basis for the topographic/bathymetric mapping program. Under the agreement, NOS compiles the shoreline and bathymetry and provides the materials to USGS in manuscript form. USGS then incorporates the bathymetry into the 1:24,000-, 1:100,000-, or 1:250,000-scale maps and publishes the final topographic/bathymetric maps.

Bathymetry is depicted on each series for the offshore area beginning at the mean-low-water datum while topography is shown for the land area based on the national vertical geodetic datum of 1929. The mean-high-water line and mean-low-water line for all three map series are compiled by NOS from tide coordinated aerial photography. On the 1:100,000- and 1:250,000-scale series, the Bureau of Land Management (BLM) offshore protraction diagram is also shown and printed in red. The heavy red lines indicate the limits of the BLM Outer Continental Shelf Official Protracion Diagram and the thin red lines indicate individual tracts.

Each series is designed to serve the specific map need of land use planners, physical scientists, conservationists, foresters, surveyors, geologists, and offshore environmentalists. The 1:250,000- or small-scale maps show the least detail, but cover the largest area (approximately 19,400 square kilometers).

They are useful for reconnaissance or areawide administration and general planning. The 1:100,000- or intermediate-scale maps (covering about 4,700 square kilometers) provide a good base for more detailed regional studies of land use and cover, recreation, corridor planning, management, and general design. The 1:24,000- or the largest-scale maps contain the most detail, but cover the smallest area of the three series (approximately 155 square kilometers). This scale may be used for more in-depth studies such as specific management activities including appraisals and inventories, location design and boundary delineation. In addition, the 1:24,000-scale maps will continue to be a valuable tool in many recreational activities.

The initial USGS/NOS joint topographic/bathymetric effort began in 1975 with the production of two experimental maps--the Ft. Pierce, Florida, 1:24,000-scale map and the Beaufort, North Carolina, 1:250,000-scale map. In 1976, a cooperative project with the State of Georgia produced 36 maps at the 1:24,000, 1:100,000, and 1:250,000 scales covering the entire Georgia coastline. Today the joint topographic/bathymetric program has grown to the point where 256 maps are in production, 44 maps are published, and annual additions of about 100 maps are planned.

At an annual rate of 100 topographic/bathymetric maps (30 at 1:24,000 scale, 50 at 1:100,000 scale, and 20 at 1:250,000 scale), complete conterminous U.S. coverage at the 1:100,000 and 1:250,000 scales is anticipated for the mid-1980's. Complete coverage at the 1:24,000 scale is not expected for some years. Current priority requests from Federal and State agencies for the 1:24,000-scale topographic/bathymeoverage at the 1:100,000 and 1:250,000 scales is anticipated for the mid-1980's. Complete coverage at the 1:24,000 scale is not expected for some years. Current priority requests from Federal and State agencies for the 1:24,000-scale topographic/bathymetric maps include 1146 maps; 532 first priority, 408 second priority, and 206 third priority.

NATIONAL HIGH-ALTITUDE AERIAL PHOTOGRAPHY PROGRAM

Under Geological Survey leadership 12 Federal agencies have pooled their resources to begin a national High-Altitude Aerial Photography program with the goal of obtaining coverage of the conterminous United States in four to five years. The program for the first year requires approximately 600,000 square miles of coverage, and includes coastal areas in the States of Delaware, Maryland, Washington, most of Georgia, and parts of Alabama, Florida, Michigan, Mississippi, and Wisconsin.

The program will utilize two cartographic cameras--a 6" focal length camera to provide black-and-white, 1:800,000-scale quad-centered photographs and an 8 1/4" focal length camera to provide color infrared 1:58,000-scale photographs. By common agreement the original color infrared film will be stored at the Agricultural Stabilization and Conservation Service photo laboratories in Salt Lake City, and the black-and-white film and a master duplicate of the color-infrared film will be stored at the EROS Data Center, Sioux Falls, South Dakota. Reproduction from either will be available at cost to the public.

LAND USE AND LAND COVER MAPPING AND RELATED WORK

The Geography Program Office regards the completion of land use and land cover mapping at 1:250,000 scale for the coastal zones a high priority and has given special emphasis to coastal zone maps whenever adequate source materials are available. In addition, several graphic research projects involving the coastal zone have been carried out and publications resulting from these efforts are available either from the Geological Survey or the National Technical Information Service (NTIS).

Land use and land cover maps have been compiled for all 1:250,000-scale sheets along the Atlantic, Gulf, and Pacific coasts, with the exceptions of the Ukiah and Santa Rosa, California, maps, which are in compilation at present. The Louisiana coastal sheets are being recompiled in FY 1980. Some of the northern Pacific coastal areas are available at 1:100,000 scale. In most cases, associated maps (political units, hydrologic units, census subdivisions) are also available. For some areas, State and (or) Federal land ownership overlays have also been produced. Two other products are also available; (1) digitized coastal zone land use and land cover data and (2) reproducible copies of the land use map, cultural base, and any associated maps.

In 1979, the Geography Program began producing folded maps of land use and land cover overprinted on a screened culture base background (L-Series). These 1:250,000 scale maps are available for the Atlantic and Gulf coasts with the exception of Louisiana.

Several Geographic Research efforts in recent years have dealt with the coastal zone. Specific areas involved in these projects include the central Atlantic region, Atlantic and Gulf Coast barrier islands, and the Puget Sound region. The Puget Sound study involved an experimental land use and land cover classification system derived from digital data. Publications resulting from these coastal area studies are available from the Geological Survey.

CARTOGRAPHIC AND GEOGRAPHIC DIGITAL DATA

Digital mapping is a new activity initiated as an integral part of the National Mapping Program to provide cartographic and geographic data in a computer-compatible (digital) form. Following a number of pilot and demonstration projects designed to resolve technical matters such as methods, formats, and coding procedures, the Geological Survey launched the Digital Mapping Program in 1978. While the objective of the National Mapping Program is to prepare and make available multi-purpose maps and fundamental cartographic data, the purpose of the Digital Mapping Program is to make this same cartographic data available in digital form through the creation of a National Digital Cartographic Data Base and to assist users in applying these data. The principal types of data are those defined by the Geological Survey as base cartographic data. They are: reference systems; hypsography; hydrography; vegetative cover; non-vegetative features; boundaries; transportation systems; significant manmade structures; geodetic control, survey monuments, other survey markers, and landmark structures and objects; geographic names; and orthophotgraphic imagery. In addition, other categories of data may be digitized to meet major user requirements and will be referred to as non-base categories. One such non-base category is the U.S. Public Land Survey data.

Land use/land cover maps at 1:100,000 and 1:250,000 scales are also being digitized and are closely related to the base categories. The source materials for this data base are the land use and land cover maps and associated map series already published. The number of associated map overlays varies but always includes three standard products--a county boundary map, a census county subdivision map, and a drainage basin map.

Currently, most digital mapping projects have been initiated in response to requirements stated by Federal agencies responsible for managing natural and mineral resources, especially coal, forests, and wildlife. Digital cartographic projects within the Nation's coastal zone include the following at 1:24,000 scale:

- Columbia River mouth--digital line graphs for land net and boundaries between 45°30' and 46°30' north latitude and from 122° west longitude to the Pacific.

- San Francisco Bay--digital line graphs for land net and boundaries between 37°30' and 38°30' north latitude and from 121° west longitude to the Pacific.

- San Mateo County, California--digital elevation models covering the San Francisco area county.

- California shoreline--digital line graphs for the coastline along various segments of the California coast; project totals 40 7.5-minute quadrangles.

- Arcadia National Park, Maine--digital line graphs for boundaries between 68° and 68°30' west longitude.

- Cape Cod, Massachusetts--digital line graphs for boundaries east of 70°30' west longitude.

Digital data derived from the land use and land cover and associated maps have been compiled for coastal sheets in Connecticut, the Carolinas, Delaware, Florida, Georgia, New Jersey, and parts of Maryland and Texas.

Two projects at smaller scales currently in work are:

- California State Base Map--digital line graph for all planimetric features depicted on the 1:500,000-scale California State Base Map.

- National Atlas Sectional Maps--digital line graph for all features depicted on the 1:2,000,000-scale Atlas sectionals. These sectionals include all 50 States.

ADVANCE SURVEY TECHNOLOGY

The Geological Survey constantly monitors new commercial instruments, techniques, and procedures to determine their usefulness in establishing horizontal and vertical geodetic control for mapping.

One of the more promising recent developments has been the inertial survey system. In early 1979 the Geological Survey developed a program to test and evaluate such a system to determine its potential application in the National Mapping Program. The chosen test site was in Camden, Texas. Using 638 combined National Geodetic Survey horizontal and vertical control points and USGS vertical control points as the test standard, extensive control lines were run to develop appropriate statistical data. The results of the 15-day test showed that the horizontal control was within established second-order limits and the vertical control was suitable for 10-foot contour interval mapping.

A follow-on test to evaluate the application of the system was conducted later in 1979 in Trenton, New Jersey. The test project consisted of 32 7.5-minute quadrangle units. The purpose of the test was to evaluate specific maps to determine if they met National Map Accuracy Standards and to establish horizontal and vertical control for future mapping. The results were similar to those obtained in the Camden, Texas, test. During a 1-month period 1,540 kilometers of double-run survey lines were established with the system mounted in a four-wheel drive van. The standard error of 50 vertical test points was 0.12 m; the maximum error was -0.41 m; and the maximum horizontal error was 0.85 m.

The initial experience with inertial survey systems indicates that costs are somewhat higher than conventional methods. However, there are several marked advantages since (1) the system requires less man-power to establish control for an area, (2) it establishes horizontal and vertical control for every point simultaneously, (3) a project can be completed in a short timeframe, (4) it is an all-weather system, (5) the data is immediately available in digital form free of human blunders, and (6) the data is ready for use in other mapping operations.

While the Geological Survey has no immediate plans to acquire a system, it does appear to have a place in modern surveying operations. In the coastal zone it may be useful in new mapping and replacement mapping.

AERIAL PROFILING OF TERRAIN (APT) SYSTEM

The Geological Survey has been studying the concept of measuring terrain profiles from low-flying aircraft using a laser altimeter and inertial guidance technology. The system accuracy envisioned is 15 cm vertically and 61 cm horizontally, with 3-minute positional updates. Experimental work on such a system was first begun in 1974 and has progressed from a developmental phase to completion of a system design. The project is now at midpoint, and fabrication of the prototype system has commenced. A Twin Otter aircraft has been selected for the APT system, and aircraft modifications will begin in June 1981. Plans are to continue with the program through FY 1982 when it is anticipated the operational prototype will be completed and flight-tested. Further activities will be based on the results of actual field tests of the prototype system.

The APT products will include two magnetic tapes: one, a TV image tape showing the ground traversed and the second, a digital tape showing the latitude, longitude, and elevation along the profile. The output tapes will be used simultaneously in an interactive edit console display unit to be constructed where the TV tape will be used to edit the digital tape. Although the system is capable of "seeing" through trees, grasses, and other vegetation, many cultural features will show as false ground elevations on the digital tape. The data resulting from non-ground features such as houses, buildings, super-elevated roads and bridges, etc., will be smoothed manually on the digital tape.

While the APT system is so new we have not yet begun to identify all possible uses for it, we do have two immediate uses for the final digital tapes in the coastal zone and other areas of low terrain relief. In those areas where aerial photography is economically suitable for map compilation, the APT can be used to identify vertical control data in the sidelap area of parallel and overlapping flights for analytical triangulation and subsequent stereocompilation. In other areas, such as those where a 1-meter contour interval map may be appropriate in wetlands and swamplands and aerial photograph may not be economically feasible, multi-profile data can be used with computer software programs to compile the contour drawing.

While the above examples pertain to 7.5-minute maps for unmapped coastal zone areas and other areas where older maps need to be replaced or where metric conversion is in progress, State and county governments may want to become familiar with the APT system to prepare large scale flood plain maps and map other areas of local interest.

TECHNICAL ASSISTANCE

Technical assistance is provided directly to Federal, State, and county agencies in planning or in utilizing Geological Survey cartographic data and products.

One prime example is the technical assistance provided the U.S. Fish and Wildlife Service (F&WS) in support of the National Wetlands Inventory project of the Office of Biological Services, St. Petersburg, Florida. Under the terms of an interagency agreement, a Geological Survey cartographer is assigned to the office on a full-time basis. The cartographer provides technical support in coordinating the mapping and contract services of the Service as part of the overall Federal effort. The resulting products help meet F&WS requirements for maps showing physical, biotic, ecological, and socio-economic information and data to support their programs at the specific regional and national levels. The agreement also provides for acquisition and exchange of aerial photographs and map data and products.

HISTORICAL MAP FILES

Coastal zone administrators and planners should be aware that in addition to published maps, the Geological Survey makes available at cost black-and-white photographic copies of any map it has ever published. These maps can be very helpful in understanding the changes in shoreline, bays, channels, and migrating sand dunes over a substantial period of time. The maps also show the cutural development of an area, locate forgotten plant and waste sites that may be the source of current pollution, and suggest strategies for the future development of the area.

GEOGRAPHIC NAMES

The Geological Survey in cooperation with the U.S. Board on Geographic Names, has developed a program for the automatic processing of geographic name information. The Geographic Names Information System (GNIS) is a geographic names data base, which when completed in 1981, will contain about two million name entries. Based on USGS standard map series and other published and unpublished sources, and developed to meet national needs, GNIS will be capable of providing information on a number of data elements, including official name, feature class, location of names feature (State, county, and geographical coordinates), variant names, map sheet, and elevation. Preliminary information is presently available for about 20 States in the form of alphabetical finding lists, microfiche, and magnetic tape.

The Geological Survey, under the direction of the Board, is planning to publish a National Gazetteer of the United States as a series of State gazetteers and an abridged gazetteer of about 40,000 official names covering the major places and features in the

United States. These gazetteers will be published as chapters of USGS Professional Paper 1200, and changes, additions, and corrections will be published in the quarterly report of the U.S. Board on Geographic Names between updated revisions. The first State gazetteer will be published in 1981.

SPECIAL PROJECTS

One significant project undertaken by the Geological Survey was the cartographic support for the U.S. Army Corps of Engineers Institute of Water Resources. This support consisted of preparing maps at scales of 1:7,500,000 and 1:17,000,000. The maps showed the movement of waterborne domestic and international commerce on inland waterways and along the nation's coastlines. Although prepared specifically for an Institute report, the maps were reprinted in 1980 for inclusion in the U.S. National Atlas separate sales program. The maps may prove helpful.in understanding and planning for the economic needs in the coastal areas, particularly those areas where the activity is intense.

Another project of interest is the research being done in the preparation of an image product at 1:100,000 scale from Landsat RBV imagery for the Cape Cod area.

This image product is designed specifically as a base for a geologic map of an area from Boston to Martha's Vineyard. The central portion of the product will provide complete coverage for a full intermediate-scale 30- by 60-minute quadrangle unit. Plans are to reproduce this area as an experimental companion image product for this series, similar to the orthophotoquad companion to the 7.5-minute topographic quadrangle map series.

LANDSAT IMAGE MAP FOR THE BERRY ISLANDS

One demonstration product designed to give potential users an idea of the types of hydrographic products which can now be produced from Landsat digital data is the image map for the Berry Islands. The product was prepared from two digitally enhanced images of a portion of the Bahamas at a nominal scale of 1:500,000.

- One image is a false color composite made from three multispectral scanner bands. Each of the three spectral bands is assigned a different color, and since the seawater attenuation of each band is different, the resulting image provides relative depth information.

o The second image portrays the same area, but is made from one band only. This data has fewer clouds and less disturbed ocean surface, and better reveals the deeper submerged features. The image is overprinted with hydrographic data taken from Defense Mapping Agency Chart 26320. The shallow-water data is from British surveys of 1836 and 1842. The hydrographic data has been adjusted to fit the bottom topography as seen on the images.

The islands are covered by vegetation, which is highly reflective in infrared, and shows as red in the color image. Reaches and shallow sandbars saturate all bands and are white in the images, as are clouds. A smoke plume is present on the color image due to field burnoff on the southern end of Abaco Island. Dark features within the cyan zone (to 5 meters) are due to the low reflectivity of seaweed growing on the sea bottom. High winds disturbing the sea surface cause the mottled patterns seen in the color image. The monochrome image is free of this disturbance.

CONCLUSION

In this paper I have attempted to describe what we believe are our most useful cartographic and geographic products available to coastal zone map users. I use the word "available" advisedly, with the full realization that not all of the products I have described are available in all of the areas of our Nation's coastal zone. One reason this is so is that the coastal zone must compete with other areas of the country for the Survey's limited resources. There are no special funds provided in our budget exclusively for coastal zone products.

To understand how coastal zone map users may be able to influence the situation, let me briefly describe our selection process as mandated by the Federal Office of Management and Budget (OMB). OMB Circular A-16 charges us with the responsibility to annually canvass Federal map using agencies to identify their map needs and priorities, and to use this information in our selection of what maps to authorize for work. In the past few years, we have also given full status to similar information provided by duly constituted State Mapping Advisory Committees (SMAC's). Coastal Zone map users wanting to influence our selections would be well advised to work through their SMAC's , which have been created by the following coastal zone States:

Alabama
California
Delaware
Hawaii
Maine
Maryland
New Jersey
North Carolina
Oregon
South Carolina
Texas
Washington

<u>Great Lakes</u>

Illinois
Michigan
Minnesota

If your State does not have such a group, we would be pleased to send you information on how to go about having one established and who our mapping contacts have been within the State Government. Map users working with other Federal agencies should not discount the possibility of having those agencies identify their areas of need as high priority in the mapping process.

A word about funding may be useful here, too. We regularly enter into cost-share funding agreements with other Federal agencies, and into joint funding agreement with States and local governmental units (city, county, regional planning organizations, etc.). Monies available to us for mapping are merged on a 50-50 basis with monies available to those governments in order to produce cartographic products that further our basic mission and concurrently provide needed products to our cooperators. Most States have such an agreement with us, and SMAC's influence the mapping accomplished under joint funding agreements. Also, it may be of interest to note that other Federal funds (Federal Highway Administration, Office of Coastal Zone Management, Department of Housing and Urban Development, Environmental Protection Agency, etc.) and grants may be able to be used by the non-Federal governmental unit to match our funds for mapping in support of the purpose for which the other Federal funds were made available.

In conclusion, the U.S. Geological Survey is dedicated to the production of cartographic products to support the needs of coastal zone managers. We have produced innovative products in this area when we believed that our standard products could be improved, and we intend to continue this initiative and research. We would be interested in hearing from coastal zone managers about our products, their needs, and how well the two mesh. We can use such information in identifying user requirements, which in turn can influence budgets, programs, and products.

MARINE TURTLES AND PROBLEMS IN COASTAL MANAGEMENT

John G. Frazier, Research Associate
Department of Zoological Research
National Zoological Park
Smithsonian Institution
Washington, D.C. 20008

Abstract

There are seven species of marine turtles and all but one are endangered or threatened. These animals do not live only in the sea, but during their life cycles, they depend on terrestrial, estuarine, neritic, and pelagic environments. Direct exploitation has been intense and responsible for depleting many populations of turtles, but indirect contaminants, introduced by Man, include a wide varity of substances, ranging from radioactive, light and sound pollutants to thermal, chemical, and physical disturbances. Chemical pollutants are probably most insidious, and pernicious, but there is little information. This paper is an attempt to draw attention to potential problems relating to coastal management and marine turtles, and to stimulate research and remedial actions.

Introduction

Statement of purpose

Although studies of marine turtles have burgeoned in the last decade (see the number of references cited here that are in press), many basic aspects of their biology are not understood. Hence, conservation and management procedures are based on assumptions that are often unconfirmed. The situation is further confounded by the rate at which major environmental perturbations occur at the hand of Man. It is rare to find baseline data from before human disturbance, so it is difficult to establish the magnitude of environmental change that has occurred, and even more difficult to determine how this is likely to affect the turtles.

However, there is a great urgency in documenting the status of marine turtles and their requisite habitats, for populations are declining. Hence, the purpose of this paper is to outline potential problems that may be of major importance in the management of coastal areas for the conservation of marine turtles.

Problems Facing Marine Turtles

The plight of marine turtles has been well publicized by a recent World Conference on Sea Turtle Conservation, from 26 to 30 November 1980 in Washington, D.C. (Bjorndal, in press). Declines in numbers of the marine reptiles are well documented from all around the world (Parsons, 1962; Carr, 1967; King, in press; Ross, in press).

The most usual cause attributed to population declines has been hunting, and there is no denying that direct exploitation on turtles - especially breeding animals and their eggs - has been disastrous. Trade statistics from many turtle fisheries reveal an incredible quantity of marine turtles that has been killed for commerce. The usual pattern of a fishery is a rapid increase in production as the markets and suppliers develop to their fullest, followed by steady declines in production as turtles become less available (Frazier, in press a). For example, the fishery for Ridley Turtles in the East Pacific has taken as many as 300,000 animals in one year, most of which were reproductive females. Although tens of thousands of turtles are still taken annually, there have been ominous signs of depletion and extirpation of some populations (Frazier, in press b).

However, conservation and management of marine turtles is far more complicated than regulating fisheries, for direct exploitation, although most obvious, is not the only pressure on the populations. More subtle, less visible, yet more insidious and probably more important in the long run are forms of interference that are indirect and often unintentional. Elements that are foreign to the environment, either in quality or quantity, are introduced by Man's activities, and these contaminants or pollutants can disrupt ecosystems and the organisms in them. The effects may be long-lasting but not conspicuous at first examination.

The Legal Status of Marine Turtles

Marine turtles comprise seven species, and all of these - except one found only in Australia - are considered to be endangered or threatened. All marine turtles that occur in the U.S.A. are covered by U.S. legislation. The Leathery Turtle, *Dermochelys coriacea* (L.), Hawksbill Turtle, *Eretmochelys imbricata* (L.), and Kemp's Ridley, *Lepidochelys kempii* (Garman), are listed as endangered; the Green Turtle, *Chelonia mydas* (L.), Loggerhead, *Caretta caretta* (L.), and Olive Ridley, *Lepidochelys olivacea* (Eschscholtz), are regarded as threatened (U.S. Government, 1980a). Many coastal states also list sea turtles as threatened or endangered, and the international treaty to restrict trade in endangered wildlife, CITES, places all marine turtles except the Australian Flatback, *Chelonia depressa* Garman, on Appendix I. This list includes the most endangered species (U.S. Government, 1980b).

Habitats in Which Marine Turtles are Found

These reptiles are of course found in the sea, but they occur in a variety of habitats. Hatchling turtles of most species apparently go through a pelagic phase (Frick, 1976; Ireland, et al., 1978; Fletemeyer, 1978;1979; Carr and Meylan, 1980), but adults and immatures seem to live mainly in neritic environments. In some studies there is good evidence for a transition from pelagic to neritic habitats (Hughes, 1974). The leathery turtle may depart from this general scheme, spending nearly the entire life in a pelagic situation (see Pritchard, 1971).

Not only do most turtles spend a great part of their lives in coastal waters, but there is ever-increasing evidence that many populations feed and rest in sheltered bays and estuaries - even hibernating in some protected areas (Felger, et al., 1976; Ogren, in press). Recently, considerable numbers of immature Loggerheads, and Kemp's Ridleys have been found hundreds of km from the open ocean, in the Chesapeake Bay; evidently this is an important feeding area (Lutcavage, pers. comm.). There is even a record of a Loggerhead in Oklahoma, almost 1,000 km from the sea (Carr, 1952)!

Occasionally marine turtles venture onto land, but with the exception of a few remote islands, where Green Turtles crawl out onto beaches to bask or escape courting males (Bustard, 1973; Whittow and Balazs, 1979), only nesting females leave the sea. Nesting is done on sandy beaches, above the high water mark, and may be done tens of, or even one hundred, meters from the high water mark.

Marine turtles, thus, are found in a variety of habitats and ecosystems, including terrestrial, estuarine, neritic and pelagic. They are subject to interference in any of these systems, so management plans must be comprehensive.

Environmental Disturbances That May Affect Marine Turtles

Because these animals occur in a variety of environments, during the course of a normal life cycle, they are subject to a multiplicity of disturbances. However, our knowledge and understanding of these problems is minimal, and what follows is a list of topics that are known to be problematic, at least in other animals, and which need careful consideration in the management of marine turtle habitats.

Radioactive Pollution

The testing of atomic weapons on South Pacific islands, and the dumping of atomic waste at sea, introduce concentrations of radioactive substances directly into ecosystems that may affect marine turtles. It is not known to what level of pollution the animals have been subjected, or how radiation may affect them, and in this vein it would be valuable to document the marine turtle situation, past and present, in areas such as Eniwetok Atoll.

Light Pollution

A hatchling marine turtle, on leaving its nest on the beach, has an unlearned response which directs it toward the brightest part of the horizon (Mrosovsky and Shettleworth, 1974), although other cues may also be used (van Rhijn, 1979). Nesting females also use brightness cues in finding the sea (Ehrenfeld, 1968; Mrosovsky, 1978). When a nesting beach is disturbed with artificial illumination, the light characteristics of the area are altered and turtles may become disoriented. This often results in hatchlings heading inland, to be squashed by cars or to die of dessication (McFarlane, 1963;

Philibosia, 1976; Mann, 1978; Fletemeyer, 1979). It can also cause disorientation at sea, and a hatchling may actually crawl back onto land (Caldwell, 1969; Frazier, 1971; Fletemeyer, 1979).

Nesting females may be disoriented by artificial lights, but they seem to be able to overcome this problem. However, it is often claimed among turtle hunters that lights on a beach scare off nesting females. Some individuals are very sensitive to artificial lights and will abort a nesting with the slightest disturbance, but there is great variation in females' responses, and some seem unaffected by unnatural lights (Mann, 1978; pers. obs.). An indication of the effect of lights is in the density of nesting activity along a developed beach; where there are lights, nesting is less frequent (Fletemeyer, 1979; Fritts, in prep.), but this is not always the case (Mann, 1978).

The problem of light pollution is well documented and simple enough to identify, but it is just as easy to create, in all innocence. This issue has recently received attention in the southeastern U.S.A. where it is especially critical, because of the density of hotels and buildings; and attempts to control the problem are being made (Fletemeyer, 1979; Witham, in press; Fritts, in prep.).

Thermal Pollution - Terrestrial

Turtle eggs must incubate at between about 26°C and 34°C; outside this range hatching success is poor or nil, for embryos may not develop or may die (Bustard and Greenham, 1968; Yntema and Mrosovsky, 1980).

A more subtle problem is that sex differentiation of the turtle embryo seems to depend on incubation temperature (Yntema and Mrosovsky, 1980). Although certain temperature regimes may give laudable hatching success, the sex ratios could be "unhealthy" for a population if, for example, a large majority of males were produced (Mrosovsky, in press).

Another facet of the turtle's life that could be fatally disrupted by thermal perturbations is the emergence of hatchlings. "To avoid" diurnal predators and heat prostration, hatchlings emerge from the nest under cover of darkness, and this response is tied to thermal cues (Mrosovsky, 1968; Hughes, 1974). Diurnal emergences can be induced by the sudden cooling of surface sand, as after a rain (pers. obs.), and any thermal disturbances resulting in diurnal emergences could cause tremendous mortality (see Hughes, 1974).

Although the chances of disrupting emergence conditions seem remote, other than by intentionally, the possible mechanisms for thermal pollution are many. Clearing vegetation around a beach would increase the incidence of solar radiation; shading the beach would decrease it. Adding black particles to the sand, e.g. small tar balls, could increase the absorption of solar radiation. Whether these, or any of the other possibilities for altering the thermal properties of a beach, are significant threats to the turtles is unknown. This problem has not been studied.

Thermal Pollution - Marine

The rate of growth of a hatchling is related to temperature; Loggerheads kept below 17°C show little growth and become weak (Hughes, 1974). Hatchlings at sea could also succumb to heat prostration, but neither cooling nor heating of the marine habitats of turtles seem likely to be threats.

Sound Pollution

Turtles are not well known for their auditory abilities. Laboratory studies have shown a sensitivity to low frequency sounds in the Green Turtle (Ridgeway, et al., 1969), and Seychellois turtlemen claim that banging an oar in a boat can frighten these animals. If turtles are disturbed by low frequency vibrations, the chances of sound pollution off of nesting beaches or on feeding grounds will be great in areas of human activity. Low frequency sounds have the capacity to travel great distances; the calls of baleen whales, for example, may in some circumstances be able to cross oceans (Payne and Webb, 1971). As most machines, including boat motors, produce strong, low frequency sounds, they could affect turtles over a wide area.

Sound pollution may not only frighten turtles, but it could interfere with their orientation. As Mrosovsky (1972) has suggested, the sounds of waves crashing on a coastline may be important in submarine navigation. These cues could be used in beach-finding by females, either within a season or between seasons, after returning from migrations to distant feeding areas. If hatchlings "imprint" on a beach to find their way back to it on maturation (Carr, 1967), natural underwater sounds could be important cues, and sound pollution may interfere with these signals.

Sound pollution is unlikely to have much importance in terrestrial situations, but the problem, even in the marine context, has received no attention.

Air Pollution

Marine turtles respire air and would be subject to air-borne pollutants on both land and sea. The contaminants could be atomized or particulate substances such as pesticides or PCBs, and volatile toxins could also be inhaled. Thus, these problems could better be discussed under the heading Chemical Pollution (below).

Chemical Pollution - Terrestrial

Because turtles come to land to lay eggs, physical pollution in the terrestrial environment is most significant on nesting beaches. The most common chemical pollutants on a beach are likely to be petroleum derivatives, floated onto the sand during high tides. Pesticides, industrial wastes, or other toxins could leach onto beaches, and this may be common in areas of agro-business, insect control, heavy road traffic, or socio-economic development. Acid rain could also contaminate beaches.

These substances might not affect the health of an adult female on land, but they could interfere with the cues involved in selection of a nest site, which is thought to be an olfactory process. Poor nest site selection could devastate reproductive success. However, little is known about the mechanism of nest site selection, much less the effect of pollutants on the behavior.

Eggs, embryos and hatchlings are more vulnerable than the adult, for volatile and water soluble contaminants could be absorbed into the egg (see Ackerman and Prange, 1972; Prange and Ackerman, 1974; Bustard and Greenham, 1968). As sea turtle eggs incubate for nearly two months (Frazier, 1971), there is considerable time for poisoning to occur. However, studies of this topic have only begun (Fritts, pers. comm.).

Behavioral modification could also result from the disruption of cues used in orientation. Sea-finding in hatchlings and adults depends on optical cues (Mrosovsky, 1978), so may not be affected. However, it is thought that hatchlings may imprint on odors in a beach, enabling them to return to and breed in their natal area (Carr, 1967; Hughes, 1974). Pollutants could mask odors or interfere with their perception, but despite the great possibility that these problems exist, they remain unstudied.

Chemical Pollution - Marine

This may be the most critical factor in the recent decline of marine turtles; as the majority of the life cycle is spent in the sea, marine pollutants have the most time in which to affect the animals. Contaminants in the ocean can circulate widely - the occurrence of DDT from North America in Antarctic penguins is a classic example (George and Frear, 1969). Toxic pollutants can accumulate insidiously in organisms and ecosystems to produce pernicious and chronic problems in organisms and ecosystems. Examples of estuarine pollution and the interdependency of terrestrial and aquatic systems are many and have been documented for years (Bullard, 1966; Odum, 1970; Likens and Bormann, 1974). Even wildlife refuges are not immune to these problems (O'Shea, et al., 1980).

In addition to pesticides, PCBs, heavy metals and other waste products, petroleum hydrocarbons are introduced into the marine environment in phenomenal quantities. Tens of thousands of barrels of crude oil poured into the Bay of Campeche daily after the IXTOC I disaster (see Delikat, in press), and it is unusual *not* to find oil slicks or tar balls in any large body of water nowadays (NAS, 1975). There is good evidence that "... the Sargassum community is measurably contaminated by petroleum," and that "... the overall resulting petroleum load to the community is in response to the long-term general level of pollution not to local concentration of tar" (Morris, et al., 1976).

Marine turtles are likely to be exposed to high levels of contamination. Many filter-feeding and detrital feeding organisms concentrate pollutants, and these animals are important prey items for turtles in neritic habitats: Hawksbills eat sponges, ascidians and other soft-bodied reef-dwelling invertebrates (Carr and Stancyk, 1975); immature Green Turtles consume sponges, jellyfish, molluscs and other neritic invertebrates (Frazier, ms a); Loggerheads specialize on crustaceans and molluscs (Hughes, 1974); Ridleys also eat crustaceans and molluscs (Pritchard and Márquez, 1973; Márquez, et al., 1976).

Hatchling turtles are found in association with windrows of Sargassum and other floating vegetation and material (Fletemeyer, 1978; Carr and Meylan, 1980. These windrows represent the boundaries between Langmuir cells where the surface currents are colliding before pushing downward and around again to complete the circuit (see Assaf, et al., 1971). For sessile or weakly motile organisms, the Langmuir cell is virtually a closed system, and organisms in windrows will be repeatedly exposed to the same surface water, continually filtering or absorbing from it as it circulates within the two adjacent cells. This system is inherently vulnerable to concentrating pollutants "up the food web;" small predatory fishes in the Sargassum community may have levels of petroleum hydrocarbons more than 20 times that found in the Sargassum (Morris, et al., 1976).

Another distressing possibility has been suggested by Hendrickson (in press a; in press b). Leathery Turtles and immatures of other marine turtles eat pelagic jellyfish, sometimes in large quantities. These medusas, because of their place in the pelagic food web, have high concentrations of pollutants, and by specializing on this prey item, turtles are injesting high levels of contaminants.

The effect of the contaminants, however, is unknown. Aquatic turtles in various taxonomic families have been found with several pesticides and their metabolites, PCBs, heavy metals and other toxins. There are numerous cases of aquatic turtles, and box turtles, *Terrapene carolina* (L.), being killed, apparently from pesticides such as DDT, Endrin, Heptachlor, and Toxaphene (Hall, 1980). Fat in snapping turtles, *Chelydra serpentina* (L.), may have levels (ppm) of PCB's, DDE and Dieldrin of nearly 3,000, 11, and 8 respectively, and all of these are well above the allowed levels in fish for human consumption (Stone, et al., 1980). Only two studies have been done on marine turtles, involving eggs and tissues of both Loggerheads (Hillestad, et al., 1974) and Green Turtles (Thompson, et al., 1974); heavy metals, PCBs and metabolites of pesticides were found.

It is difficult to assess the significance of these findings, for there is no information on lethal levels. The problem is confounded by the great variation in sensitivity between related taxa, and the differences in effect of various contaminants and metabolites, to say nothing of problems in analysis and presentation of comparable data. Even with birds and mammals, which are better studied than turtles, there is little specific information that allows a clear interpretation of the effects of different toxins. Three generalities seem clear: reptiles have a greater sensitivity to toxic pollutants than do mammals and birds; animals higher in the trophic structure are more susceptible to concentrated pollutants; and long-lived animals are more likely to be affected. However, the available information on turtles is sparce (see Hall, 1980 for a review).

The possible effects of chemical contamination are many. Dead Green and Ridley turtles were collected in association with the IXTOC I disaster, but the causes of death have not been reported (Hall, 1980). There is evidence that crude oil can drastically alter the locomotor behavior of Green Turtles (Kleerkoper and Bennett, 1976) and Green Turtles found with tar in the mouth died or were very weak (Witham, 1978). This suggests that petroleum hydrocarbons may be relatively fast acting and fatal in certain concentrations.

Sublethal effects to the turtles could be catastrophic to populations. The olfactory sense in marine turtles appears to be highly developed (Manton, et al., 1972a; 1972b; Bennett and Kleerkoper, 1978), and it is possible that this sensory modality is essential in navigation and orientation (Koch, et al., 1969; Carr, 1972). Contaminants that mask or interfere with perception of olfactory cues may not kill the turtle directly, but if it cannot orient correctly it will be lost from the population.

In the light of these potential problems it is worth examining the history of Kemp's Ridley, most endangered of the sea turtles. Only one major nesting area is known, at Rancho Nuevo, Tamaulipas, Mexico. When first discovered in 1947, there were estimated to have been some 40,000 females nesting in a single massed nesting. By 1970, the number of females in a massed nesting was estimated to have been 2,500 and in the last few years the total numbers of females nesting annually have been several hundred (Carr, 1977; Pritchard, et al., 1979).

No other sea turtle has experienced such a catastrophic decline or exists at such a precarious state, despite the intensity and persistence with which other populations have been exploited: over the past two decades nearly two million Olive Ridleys, mostly mature females, have been slaughtered in western Mexico, and still the turtles occur in large, although reduced, numbers (Frazier, in press b). Why has Kemp's Ridley declined so rapidly while other populations have shown less vulnerability?

The distribution of major nesting areas of Ridley Turtles, of both species, show a relationship with major river mouths (Figure 1): Kemp's Ridley is concentrated around the Rio Grande and Mississippi Rivers. A large population of Olive Ridleys are in Surinam (Schulz, 1975), between the mouths of the Orinoco and the Amazon. Sizeable populations are suspected along the shores of the Gulf of Guinea (Brongersma, in press), into which the Niger and Congo Rivers flow. Olive Ridleys occur at the northern end of the Mozambique Channel (Hughes, 1972), between the Zambezi and Rufiji Rivers. There are large populations in Pakistan by the Indus River (Frazier, ms b). The largest single rookery may be in Orissa, India (Kar, 1980), not far from other large rookeries in Burma (Frazier, in press a), and all of these are near the mouth of Ganges River. There are enormous concentrations in western Mexico, Costa Rica and Ecuador (Frazier, in press b), between the Colorado and Guyas Rivers.

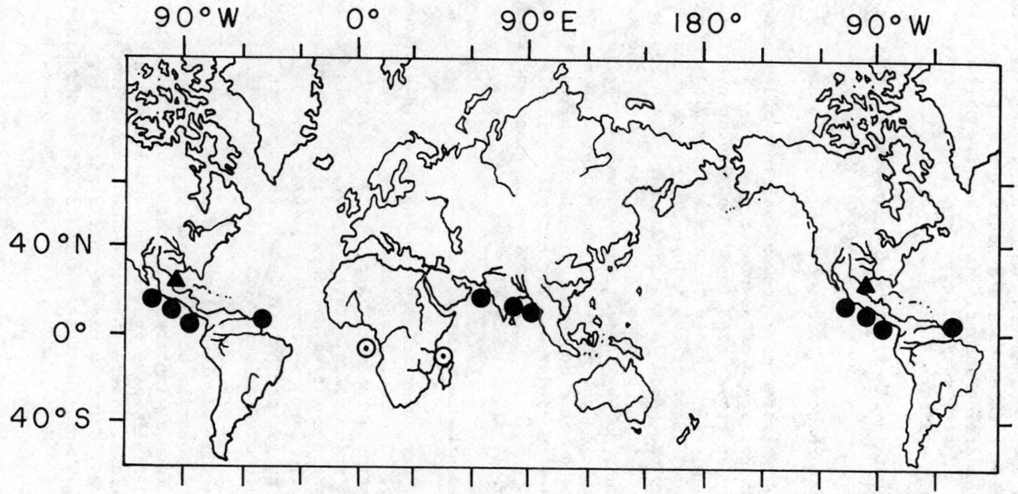

Figure 1. World distribution of major nesting areas of marine turtles of the genus *Lepidochelys*; ▲ = *L. kempii* (Rancho Nuevo, Mexico); ● = confirmed, *L. olivacea* (Surinam; Pakistan; Orissa, India; Burma; Mexico; Costa Rica); ⊙ = suspected or not major, *L. olivacea* (West Africa; Mozambique). (Note that the American continents are repeated on both sides of this figure in order to show contiguous oceans).

The relationship between major Ridley concentrations and major rivers is clear, and Hughes (1972) has pointed out that this genus is characteristically found in waters of low salinity, high turbidity, high organic content, and where prawns are commercially abundant. Is the decline of Kemp's Ridley related to its association with the Mississippi, the river that drains a continent so heavily polluted by pesticides and industrial wastes, which flows into a gulf that is perennially troubled by oil spills?

An unequivocable answer is likely to be long in coming, but the warning is clear enough.

Physical Pollution and Disturbance - Terrestrial

Obstructions to the high water mark prevent nesting, and breakwalls are the most common and drastic type of obstruction. Large pieces of debris such as logs, wood, and trash also impede or prevent nesting, and small items such as sharp pieces of glass or metal, stray wires or lines and nets are dangerous for they can injure or ensnare animals. Mangrove trees that litter the beach in French Guiana, take a considerable toll of nesting Leathery Turtles, by impeding their return to the sea, and trapping them in the sun, and by inflicting deep wounds (Fretey, 1977), although this example is not caused by human disturbance. Tar may also foul an animal, and could be devastating to hatchlings if eyes or the movements of limbs were hindered. Tar and oil could also foul eggs and sand and interfere with gas exchange (see Ackerman and Prange, 1972; Prange and Ackerman, 1974; Hall, 1980).

Physical Pollution and Disturbance - Marine

Floating plastic and tar occur universally in the world's oceans and decompose very slowly. Turtles, large and small, of several species, have been found with plastic in the gut. Apparently a plastic bag resembles a jellyfish, on which pelagic turtles prey. The plastic is indigestible and sufficient quantities may impact the gut (Brongersma, 1969; 1972; Pritchard, 1971; Carr, in Cornelius, 1975; Hays de Brown and Brown, in press). Floating tar may have a fatal effect; Green Turtles found with tar in the mouth died or were very weak (Witham, 1978).

Oil, when in a thick film, could foul the heads of turtles, especially hatchlings. If nostrils or eyes were covered, respiration and orientation could be fatally impaired. Ingestion of liquid oil, either directly, or incidentally as a coating on food items, is highly likely in areas of oil slicks. There is some evidence that ingestion of oil is fatal (Witham, 1978), but this may be more appropriately considered a chemical pollutant. However, deaths of Green and Ridley Turtles during the IXTOC I disaster (Hall, 1980), may not have been due entirely to chemical toxins.

Fishing nets, anti-shark nets and trawls greatly disturb marine turtle habitat. Turtles are caught accidentally in anti-shark nets, and many drown (Parsons, 1962; Hughes, 1974). However, this is probably of little importance, although it is poorly quantified.

Shrimp and bottemfish trawls also catch turtles accidentally, and incidental capture occurs wherever there is trawling, although the biggest problem is in the U.S.A., from Cape Hatteras to the southern tip of Florida, in water less than 80 m deep (Bullis and Drummond, 1978). In heavily fished areas there may be tremendous loss; the estimated incidental catch for just the state of Georgia in 1975 is from 1,595 to 9,417 turtles. Perhaps one third of the captured turtles die immediately (Hillestad, et al., 1977; Hillestad, in press). The total incidental capture for the U.S.A. must be considerable, and there is evidence that the size-class distribution of at least one population has been altered (Ruckdeschel and Zug, ms).

Underwater explosives can damage or kill turtles, along with other marine life. In countries, e.g. Tanzania, where dynamite is used for "fishing," this is problematic, but where explosives are used for demolition or construction the impact may be considerable. Little information is available.

The National Marine Fisheries Service is trying to resolve the problem of incidental catch by developing an excluder panel for trawls (Seidel, 1978; in press). However, there seems little hope that the plastic and tar that pollute the world's oceans will be cleaned up soon, and anti-littering campaigns on the high seas are unlikely to succeed.

Incidental Disturbances From Human Activities

Numerous animals, such as domestic and feral dogs, cats and pigs, are associated with human habitations and activities and they cause considerable problems on nesting beaches. Feral dogs and pigs are major predators in many areas, for they are efficient in finding and excavating nests, and few hatchlings may survive. Rats and other commensals, predators and scavengers, such as crabs and vultures, are attracted to human refuse, and these animals also prey on eggs and hatchlings. Predatory fishes are likewise attracted to human refuse and may prey heavily on sea-going hatchlings. The impact of these problems, however, has been rarely quantified (see Stancyk, in press).

Human activities in densely settled and developed areas cause tremendous disturbances to nesting beaches. Trampling by people, beach vehicles, and beach cleaning machines and activities can cause enormous damage to nests (Mann, 1978; Fletemeyer, 1979; Witham, in press).

Summary and Recommendations

The potential disturbances to marine turtles and their habitats seem countless, but despite the fact that pollutants and perturbations may cause damage to turtles and their populations, these problems have hardly been diagnosed, much less studied. Because of the vulnerable and endangered status of marine turtles there is considerable exigency in determining and resolving factors which threaten them. The speed and intensity at which marine environments are being polluted emphasizes this urgency.

An immediate necessity is in distinguishing major problems, and the above list can serve as an outline. Disruptions involving rapid and obvious responses, such as those involved in light pollution and physical disturbances, can be diagnosed relatively easily. However, long-term and indirect effects, such as through radioactive and chemical pollution, will not be easy to determine, and careful extended studies are needed.

Those problems that are most easily detected are likely to be quickly and directly resolved. Reducing the impact of light pollution, physical disturbances and human activities need not require complicated procedures. Resolving problems that involve indirect and cumulative effects to common resources have proven difficult and long in getting results in the past. Radioactive and chemical pollution of the sea are well known examples. The need for strictly enforced national and international regulations is undeniable. However, any effort to manage the coastal zone must also include comprehensive contingency plans (Delikat, in press), for catastrophies cannot always be prevented.

Probably the single most effective measure for managing marine turtles and their habitats is by establishing reserves in which human activities are severely restricted. Nesting and feeding reserves can at least protect turtles in these critical phases of their life history. Defending pelagic turtles from the many contaminants and perturbations that Man, the terrestrial animal, has inflicted on the world's oceans, will require immense cooperation. The establishment of the enormous Indian Ocean Sanctuary and the complimentary Indian Ocean Alliance for Conservation including some 15 nations bordering on its shores (Anon., 1980) holds great promise. Hopefully more wealthy and developed nations that border on the Atlantic and Pacific Oceans will have as much foresight.

Acknowledgements

I am grateful to Ms. Virginia Garber for typing several drafts and to Mrs. S. K. James Bruch for illustrations.

Literature Cited

Ackerman, R. A. and H. D. Prange. 1972. Oxygen Diffusion Across a Sea Turtle (*Chelonia mydas*) Egg Shell. *Comp. Biochem Physiol.* 43A:905-909.

Anon. 1980. Indian Ocean alliance gains support. *IUCN Bulletin* (New Series) 11(6):1-2.

Assaf, G., R. Gerard, and A. L. Gordon. 1971. Some Mechanisms of Oceanic Mixing Revealed in Aerial Photographs. *J. Geophys. Res.* 76(27):6550-6572.

Bennett, J. A. and H. Kleerekoper. 1978. A preliminary investigation of the effects of chemical stimulation on the locomotor behavior of hatchling green turtles (*Chelonia mydas*). *Fla. Mar. Res. Publ.* No. 33:3-7.

Bjorndal, K. (ed.). in press. *Proceedings of the World Conference on Sea Turtle Conservation*, 26 to 30 November 1979, Washington, D.C.

Brongersma, L.D. 1969. Miscellaneous Notes on Turtles. II A - B. *Proc. Kon. Ned. Akad. Wet., Amsterdam* C, 72(1):76-102.

Brongersma, L.D. 1972. European Atlantic Turtles. *Zoologische Verhandelingen* No. 121, 318 pp + 12 pls + 42 figs + 17 tables + 8 charts.

Brongersma, L. D. in press. The status of sea turtles in the East Atlantic. *In:* Bjorndal, K. (ed.), *Proceedings of the World Conference on Sea Turtle Conservation.*

Bullard, W. E. 1966. Effects of Land Use on Water Resources. *J. Water Pollut. Cont. Fed.* 38(4):645-659.

Bullis, H. R. and S. B. Drummond. 1978. Sea Turtle Captures Off the Southeastern United States By Exploratory Fishing Vessels, 1950-1976. *Fla. Mar. Res. Publ.* No. 33:45-50.

Bustard, H. R. 1973. *Kay's Turtles.* Collins, London. 128 pp.

Bustard, H. R. and Greenham, P. 1968. Physical and Chemical Factors Affecting Hatching in the Green Sea Turtle, *Chelonia mydas* (L.). *Ecology* 49(2):269-276.

Caldwell, D. K. 1969. Hatchling Green Sea Turtles, *Chelonia mydas*, at Sea in the Northeastern Pacific Ocean. *Bull. S. Calif. Acad. Sci.* 68 (pt 2):113-114.

Carr, A. 1952. *Handbook of Turtles.* Cornell Universtiy Press, Ithaca. xv + 542 pp.

Carr, A. 1967. *So Excellent A Fishe.* Natural History Press, Garden City, N.Y. x + 249 pp + 45 pls.

Carr, A. 1972. The case for long-range chemoreceptive piloting in *Chelonia.* *In:* Galler, S. R., K. Schmidt-Koenig, G. J. Jacobs and R. E. Belleville (eds.). *Animal Orientation and Navigation.* NASA, Washington, D.C. pp. 469-483.

Carr, A. 1977. Crisis for the Atlantic Ridley. *Marine Turtle Newsletter* No. 4, May 1977, pp. 2-3.

Carr, A. and A. B. Meylan. 1980. Evidence of Passive Migration of Green Turtle Hatchlings in Sargassum. *Copeia* 1980(2):366-368.

Carr, A. and S. Stancyk. 1975. Observations on the Ecology and Survival Outlook of the Hawksbill Turtle. *Biol. Cons.* 8(3):161-172.

Cornelius, S. E. 1975. Marine Turtle Mortalities Along the Pacific Coast of Costa Rica. *Copeia* 1975(1):186-187.

Delikat, D. S. in press. IXTOC I Oil Spill & Atlantic Ridley Turtle Survival. *In:* *Proceedings of the Coastal Zone 80 Conference*, 17 to 20 November 1980. Holywood, Florida.

Ehrenfeld, D. W. 1968. The Role of Vision in the Sea-Finding Orientation of the Green Turtle (*Chelonia mydas*). 2. Orientation Mechanism and Range of Spectral Sensitivity. *Anim. Behav.* 16(2-3):281-287.

Felger, R. S., K. Clifton and P. J. Regal. 1976. Winter Dormancy in Sea Turtles: Independent Discovery and Exploitation in the Gulf of California by Two Local Cultures. *Science* 191: 283-285.

Fletemeyer, R. R. 1978. Underwater tracking evidence of neonate Loggerhead sea turtles seeking shelter in drifting Sargassum. *Copeia* 1978(1):148-149.

Fletemeyer, J. 1979. 1979 Report. Sea Turtle Monitoring Project. Broward County Environmental Quality Control Board. 64 pp + 15 figs.

Frazier, J. 1971. Observations on sea turtles at Aldabra Atoll. *Phil. Trans. Roy. Soc. Lond.* B. 260:373-410.

Frazier, J. in press a. Exploitation of Marine Turtles in the Indian Ocean. *Human Ecology*.

Frazier, J. in press b. Marine Turtle Fisheries in Ecuador and Mexico: The Last of the Pacific Ridley? *Special Scientific Report*, U.S. Fish and Wildlife Service.

Frazier, J. ms a. Marine Turtles in Peru and the East Pacific. 130 pp, 26 tables, 37 figs.

Frazier, J. ms b. The Status of Marine Turtles in the Western Indian Ocean. 29 pp. 1 table, 7 figs.

Fretey, J. 1977. Causes de mortalité des tortues luths adultes (*Dermochelys coriacea*) sur le litteral guyanis. *Courrier de la Nature* No. 52:257-266.

Frick, Jane A. 1976. Orientation and Behavior of Hatchling Green Turtles (*Chelonia mydas*) in the Sea. *Anim. Behav.* 24(4):849-857.

George, J. L. and D. E. H. Frear. 1969. Pesticides in the Antarctic. *J. Appl. Ecol.* 3(Suppl.):155-167.

Hall, R. J. 1980. Effects of Environmental Contaminants on Reptiles: A Review. *Special Scientific Report - Wildlife*, U.S. Fish and Wildlife Service No. 228, 12 pp.

Hays de Brown, C. and W. M. Brown. in press. The Status of Sea Turtles in the Southeastern Pacific: Emphasis on Peru. *In:* Bjorndal, K. (ed.), *Proceedings of the World Conference on Sea Turtle Conservation*.

Hendrickson, J. in press a. The ecological strategies of sea turtles. *Amer. Zool.*

Hendrickson, J. in press b. Review of Nesting Behavior of Sea Turtles. *In:* Bjorndal, K. (ed.), *Proceedings of the World Conference on Sea Turtle Conservation.*

Hillestad, H. O. in press. Review of Incidental Catch. *In:* Bjorndal, K. (ed.), *Proceedings of the World Conference on Sea Turtle Conservation.*

Hillestad, H. O., J. I. Richardson, G. K. Williamson. 1977. Incidental Capture of Sea Turtles By Shrimp Trawlermen in Georgia. Typescript (National Marine Fisheries Service, Miami) x + 106 pp (32 figs.).

Hillestad, H. O., R. J. Reimold, R. R. Stickney, H. L. Windom, and J. H. Jenkins. 1974. Pesticides, heavy metals and radionuclide uptake in loggerhead sea turtles from South Carolina and Georgia. *Herp. Rev.* 5(3):75.

Hughes, G. R. 1972. The Olive Ridley Sea-turtle (*Lepidochelys olivacea*) in South-east Africa. *Biol. Cons.* 4(2):128-134.

Hughes, G. R. 1974. The Sea Turtles of South-East Africa. II. The biology of the Tongaland Loggerhead Turtle *Caretta caretta* L. with comments on the Leatherback Turtle *Dermochelys coriacea* L. and the Green Turtle *Chelonia mydas* L. in the study region. *Investl Rep. Oceanog Res. Inst.*, No. 36, 96 pp. figs., tables, plates.

Ireland, L. C., Jane A. Frick, and D. B. Wingate. 1978. Nighttime Orientation of Hatchling Green Turtles (*Chelonia mydas*) in Open Sea. *In:* Schmidt-Koenig, K. and W. T. Keeton (eds.), *Animal Migration, Navigation, and Homing*. Springer-Verlag, New York. pp. 420-429.

Kar, C. S. 1980. The Gahirmetha Turtle Rookery Along the Coast of Orissa, India. *Marine Turtle Newsletter*, No. 15, May 1980, pp. 2-3.

King, F. W. in press. Historic Review of the Decline of Green Turtles and Hawksbills. *In:* Bjorndal, K. (ed.), *Proceedings of the World Conference on Sea Turtle Conservation*.

Kleerekoper, H. and J. A. Bennett. 1976. Some effects of the water soluble pollution fraction of Louisiana crude on the locomotor behavior of juvenile green turtle (*Chelonia mydas*) and sea catfish (*Arius felis*). Preliminary results. Report on contract 206-76 to American Petroleum Institute. 7 pp.

Koch, A. L., A. Carr and D. W. Enrenfeld. 1969. The Proglem of Open-sea Navigation: The Migration of the Green Turtle to Ascension Island. *J. Theoret. Biol.* 22:163-179.

Likens, G. E. and F. H. Bormann. 1974. Linkages between Terrestrial and Aquatic Ecosystems. *BioScience* 24(8):447-456.

Mann, T. M. 1978. Impact of developed coastline on nesting and hatchling sea turtles in southeastern Florida. *Fla. Mar. Res. Publ.* No. 33:53-55.

Manton, M. L. A. Karr and D. W. Ehrenfeld. 1972a. An Operant Method for the Study of Chemoreception in the Green Turtle, *Chelonia mydas*. *Brain, Behav. Evol.* 5:188-201.

Manton, M., A. Karr, and D. W. Ehrenfeld. 1972b. Chemoreception in the Migratory Sea Turtle, *Chelonia mydas*. *Biol. Bull.* 143(1):184-195.

Márquez M., René, A. Villanueva O. and C. Peñaflores S. 1976. Sinopsis de Datos Biológicos Sobre la Tortuga Golfina *Lepidochelys olivacea* (Eschscholtz, 1829). Instituto Nacional de Pesca, Sinopsis Sobre la Pesca No. 2, iii + 61 pp.

McFarlane, R. W. 1963. Disorientation of loggerhead hatchlings by artificial road lighting. *Copeia* 1963(1):153.

Morris, B. F., J. Cadwallader, J. Geiselman and J. N. Butler. 1976. Transfer of Petroleum and Biogenic Hydrocarbons in the Sargassum Community. *In:* Windon, H. L. and R. A. Duce (eds.), *Marine Pollutant Transfer*. Lexington Books, Lexington, Mass. pp. 235-259.

Mrosovsky, N. 1968. Nocturnal Emergence of Hatchling Sea Turtles: Control by Thermal Inhibition of Activity. *Nature* 220(5174): 1338-1339.

Mrosovsky, N. 1972. Spectrographs of the Sounds of Leatherback Turtles. *Herpetologica* 28(3):256-258.

Mrosovsky, N. 1978. Orientation Mechanisms of Marine Turtles. *In:* Schmidt-Koenig, K. and W. J. Keeton (eds.), *Animal Migration, Navigation and Homing*. Springer-Verlag, New York. pp. 413-419.

Mrosovsky, N. in press. The Thermal Biology of Sea Turtles. *Amer. Zool.*

Mrosovsky, N. and S. J. Shettleworth. 1974. Further studies on the sea-finding mechanism in green turtle hatchlings. *Behavior* 51:195-208.

NAS (National Academy of Sciences). 1975. *Petroleum in the Marine Environment*. Washington, D.C. 107 pp.

Odum, W. E. 1970. Insidious Alteration of the Estuarine Environment. *Trans. Amer. Fish. Soc.* 99(4):836-847.

Ogren, L. in press. Hibernation in sea turtles. *In:* Bjorndal, K. (ed.), *Proceedings of the World Conference on Sea Turtle Conservation.*

O'Shea, T. J., W. J. Fleming III, E. Cromartie. 1980. DDT Contamination at Wheeler National Wildlife Refuge. *Science*, 209(4455):509-510.

Parsons, J. J. 1962. *The Green Turtle and Man.* University of Florida Press, Gainesville. viii + 126 pp.

Payne, R. and D. Webb. 1971. Orientation by means of long-range acoustic signaling in baleen whales. *Ann. N.Y. Acad. Sci.* 118:110-142.

Philibosia, R. 1976. Disorientation of Hawksbill Turtle Hatchlings, *Eretmochelys imbricata*, by Stadium Lights. *Copeia* 1976(4):824.

Prange, H. D. and R. A. Ackerman. 1974. Oxygen Consumption and Mechanisms of Gas Exchange of Green Turtle (*Chelonia mydas*) Eggs and Hatchlings. *Copeia* 1974(3):758-763.

Pritchard, P. C. H. 1971. The leatherback or Leathery Turtle. IUCN Monograph, No. 1, 39 pp.

Pritchard, P. C. H. and R. Márquez M. 1973. Kemp's Ridley Turtle or Atlantic Ridley, *Lepidochelys kempi.* IUCN Monograph No. 2, 30 pp.

Pritchard, P. C. H., D. F. Gicca and S. F. Wehrle (ed.). 1979. Report on United States/Mexico Conservation of Kemp's Ridley Sea Turtle at Rancho Nuevo, Tamaulipas, Mexico, 1978. U. S. Fish and Wildlife Service. v + 72 pp., figs. and appendicies.

Ridgeway, S. H., E. G. Wever, J. G. McCormick, J. Palin and J. H. Anderson. 1969. Hearing in the giant sea turtle, *Chelonia mydas. Proc. Nat. Acad. Sci.* 64:884-890.

Ross, J. P. in press. Historic Review of the Decline of Loggerheads, Ridleys and Leatherbacks. *In:* Bjorndal, K. (ed.), *Proceedings of the World Conference on Sea Turtle Conservation.*

Ruckdeschel, C. and G. R. Zug. ms. Mortality of Sea Turtles. (submitted to *Biol. Cons.*).

Schulz, J. P. 1975. Sea Turtles Nesting in Surinam. *Zoologische Verhandelingen* No. 143, 143 pp + 49 pls.

Seidel, W. R. 1978. Milestone Report. Sea Turtle Excluder Shrimp Trawl Development. Typescript (National Marine Fisheries Service, Pascagoula, Miss.). pag. var.

Seidel, W. R. in press. Development of a Prototype Trawl Net Which Excludes Sea Turtles from its Catch and Behavior of Sea Turtles in Trawl. *In:* Bjorndal, K. (ed.), *Proceedings of the World Conference on Sea Turtle Conservation.*

Stancyk, S. in press. Predators on sea turtles and their control. *In:* Bjorndal, K. (ed.), *Proceedings of the World Conference on Sea Turtle Conservation.*

Stone, W. B., E. Kiviat, and S. A. Butkas. 1980. Toxicants in Snapping Turtles. *N. Y. Fish and Game J.* 27(1):39-50.

Thompson, N. P., P. W. Rankin, and D. W. Johnston. 1974. Polychlorinated biphenyls and P,P'-DDE in green turtle eggs from Ascension Island, South Atlantic Ocean. *Bull. Environ. Contam. Toxicol.* 11(5):399-406.

U. S. Government. 1980a. *Federal Register* 45(99):33768-33781.

U. S. Government. 1980b. *Federal Register* 45(66):22848-22862.

van Rhijn, F. A. 1979. Optic Orientation in Hatchlings of the Sea Turtle, *Chelonia mydas.* I. Brightness: Not the Only Optic Cue in Sea-Finding Orientation. *Mar. Behav. Physiol.* 6:105-121.

Whittow, G. C. and G. H. Balazs. 1979. The Thermal Biology of Hawaiian Basking Green Turtles (*Chelonia mydas*) (Abstract). *Amer. Zool.* 19(3):981.

Witham, R. 1978. Does a Problem Exist Relative to Small Sea Turtles and Oil Spills? *In: Proceedings of the Conference on Assessment of Ecological Impacts of Oil Spills,* 14-17 June 1978, Keystone, Colorado. AIBS. pp. 629-632.

Witham, R. in press. Human Alteration of Sea Turtle Habitat. *In:* Bjorndal, K. (ed.), *Proceedings of the World Conference on Sea Turtle Conservation.*

Yntema, C. L. and N. Mrosovsky. 1980. Sexual Differentiation in Hatchling Loggerheads (*Caretta caretta*) incubated at Different Controlled Temperatures. *Herpetologica* 36(1):33-36.

Subject Index

Author Index